Dictionary of Microbiology and Molecular Biology

Third Edition, Revised

We would like to dedicate this book to the memory of Hubert Sainsbury. His lively and enquiring mind and his passion for knowledge and understanding were always an inspiration, and his enthusiasm for this Dictionary was a strong motivating force during its long gestation. The book owes more to him than he would have believed.

Dictionary of Microbiology and Molecular Biology

Third Edition, Revised

Paul **Singleton**
Diana **Sainsbury**

John Wiley & Sons, Ltd

The Atrium, Southern Gate, Chichester,
West Sussex PO19 8SQ, England

Telephone (+44) 1243 779777

Email (for orders and customer service enquiries): cs-books@wiley.co.uk
Visit our Home Page on www.wileyeurope.com or www.wiley.com

1st edition 1978 (reprinted 1980)
2nd edition 1987 (reprinted 1988, 1989, 1991, 1994)
Paperback edition 1993 (reprinted 1993, 1994, 1995, 1996 (twice), 1997 (twice), 1999, 2000)
3rd edition 2001 (reprinted 2002)
3rd edition revised 2006 (reprinted 2007)

Japanese edition (Asakura, Tokyo) 1997
Chinese edition in preparation 2006

Other Wiley Editorial Offices

John Wiley & Sons Inc., 111 River Street, Hoboken, NJ 07030, USA
Jossey-Bass, 989 Market Street, San Francisco, CA 94103-1741, USA
Wiley-VCH Verlag GmbH, Boschstr. 12, D-69469 Weinheim, Germany
John Wiley & Sons Australia Ltd, 42 McDougall Street, Milton, Queensland 4064, Australia
John Wiley & Sons (Asia) Pte Ltd, 2 Clementi Loop #02-01, Jin Xing Distripark, Singapore 129809
John Wiley & Sons Canada Ltd, 6045 Freemont Blvd, Mississauga, ONT, L5R 4J3, Canada

Wiley also publishes its books in a variety of electronic formats. Some content that appears in print may not be available in electronic books.

Library of Congress Cataloging-in-Publication Data

(to follow)

British Library Cataloguing in Publication Data

A catalogue record for this book is available from the British Library

ISBN-13 978-0-470-03545-0 (P/B)

Typeset in 10/12pt Times by Laserwords Private Limited, Chennai, India
Printed and bound in Great Britain by Antony Rowe Ltd, Chippenham, Wiltshire
This book is printed on acid-free paper responsibly manufactured from sustainable forestry,
in which at least two trees are planted for each one used for paper production.

Preface

This edition follows the style of previous editions. It has similar aims, and was written with the same enthusiasm and care.

It is vital that readers be aware of the type of alphabetization used in the Dictionary. A glance at 'Notes for the User' – particularly the first paragraph – is essential.

September 2001

Preface to the Second Edition

In writing this new edition of the Dictionary we had several aims in mind. One of these was to provide clear and up-to-date definitions of the numerous terms and phrases which form the currency of communication in modern microbiology and molecular biology. In recent years the rapid advances in these disciplines have thrown up a plethora of new terms and designations which, although widely used in the literature, are seldom defined outside the book or paper in which they first appeared; moreover, ongoing advances in knowledge have frequently demanded changes in the definitions of older terms – a fact which is not always appreciated and which can therefore lead to misunderstanding. Accordingly, we have endeavoured to define all of these terms in a way which reflects their actual usage in current journals and texts, and have also given (where appropriate) former meanings, alternative meanings, and synonyms.

A second – but no less important – aim was to encapsulate and integrate, in a single volume, a body of knowledge covering the many and varied aspects of microbiology. Such a reference work would seem to be particularly useful in these days of increasing specialization in which the reader of a paper or review is often expected to have prior knowledge of both the terminology and the overall biological context of a given topic. It was with this in mind that we aimed to assemble a detailed, comprehensive and interlinked body of information ranging from the classical descriptive aspects of microbiology to current developments in related areas of bioenergetics, biochemistry and molecular biology. By using extensive cross-referencing we have been able to indicate many of the natural links which exist between different aspects of a particular topic, and between the diverse parts of the whole subject area of microbiology and molecular biology; hence the reader can extend his knowledge of a given topic in any of various directions by following up relevant cross-references, and in the same way he can come to see the topic in its broader contexts. The dictionary format is ideal for this purpose, offering a flexible, 'modular' approach to building up knowledge and updating specific areas of interest.

There are other more obvious advantages in a reference work with such a wide coverage. Microbiological data are currently disseminated among numerous books and journals, so that it can be difficult for a reader to know where to turn for information on a term or topic which is completely unfamiliar to him. As a simple example, the name of an unfamiliar genus, if mentioned out of context, might refer to a bacterium, a fungus, an alga or a protozoon, and many books on each of these groups of organisms may have to be consulted merely to establish its identity; the problem can be even more acute if the meaning of an unfamiliar term is required. A reader may therefore be saved many hours of frustrating literature-searching by a single volume to which he can turn for information on any aspect of microbiology.

An important new feature of this edition is the inclusion of a large number of references to recent papers, reviews and monographs in microbiology and allied subjects. Some of these references fulfil the conventional role of indicating sources of information, but many of them are intended to permit access to more detailed information on particular or general aspects of a topic – often in mainstream journals, but sometimes in publications to which the average microbiologist may seldom refer. Furthermore, most of the references cited are themselves good sources of references through which the reader can establish the background of, and follow developments in, a given area.

While writing this book we were very fortunate in having exceptional and invaluable cooperation from a number of libraries in South-West England. In particular, we would like to acknowledge the generous help of Mr B. P. Jones, B.A., F.L.A., of the Medical Library, University of Bristol, Mrs Jean Mitchell of the Library at Bicton College of Agriculture, Devon, and Maureen Hammett of Exeter Central Library, Devon. Finally, we are grateful to Michael Dixon, Patricia Sharp, and Prue Theaker at John Wiley & Sons, Chichester, for their enthusiastic and efficient cooperation in the production of the book.

Paul Singleton & Diana Sainsbury
Clyst St Mary, Devon, April 1987

Notes for the User

1. *Alphabetization.* Alphabetization would need no comment if every term consisted of a single word; in practice, however, many terms consist of two or more words and often contain single letters, numbers, symbols etc. Terms consisting of two or more words can be alphabetized in either of two ways: on the basis of the first word, or on the basis that both or all of the words are run together and treated as one; thus, e.g., according to the 'first-word' ('nothing-before-something') system, *red tide* comes before *redox potential*, but according to the second system *redox potential* comes before *red tide*. Terms in this Dictionary have been alphabetized by the first-word system; in this system a single letter counts as a word (hence e.g. *R plasmid* comes before *rabies*), as does a group of letters (e.g. an abbreviation, or a gene designation). Examples:

air sacculitis	**atoxyl**	**black stem rust**	**RecA protein**
airlift fermenter	**ATP**	**black wart disease**	**recapitulation theory**
AIV process	**ATP synthase**	**black yeasts**	***recB* gene**
Ajellomyces	**ATPase**	**blackeye cowpea mosaic virus**	**RecBC pathway**

When a *hyphen* connects two complete words, or occurs between a letter (or group of letters) and a word, the hyphen is regarded as a space; however, if a hyphen is used to link a *prefix* to a word (i.e., if the letters preceding the hyphen form a part-word which cannot stand alone) the term is alphabetized as though it were a single, non-hyphenated word. (In a few cases an entry heading contains words which can be written as separate, hyphenated or non-hyphenated words, or closed up as a single word: e.g. *red water fever*, *red-water fever*, *redwater fever*; in such cases an entry or cross-reference has been included in both possible positions.) Examples:

BL-type starter	**M**	**nonsense mutation**	**preaxostyle**
***bla* gene**	**M antigen**	**nonsense suppressor**	**pre-B cell**
black beans	**M-associated protein**	**non-specific immunity**	**prebuccal area**
Black beetle virus	**M bands**	**non-specific immunization**	**precipitation**

When a *Greek letter* forms a significant part of an entry heading it is counted as a word and is alphabetized as spelt (i.e., α as alpha, β as beta, etc: see Appendix VI for the Greek alphabet). A Greek letter is ignored for the purposes of alphabetization if it is a relatively minor qualification: e.g., part of a chemical designation (which can usually be replaced by a number, as in β-hydroxybutyrate, = 3-hydroxybutyrate). Examples:

Delhi boil	**MTOC**	**pHisoHex**	**polyhedrosis**
Δ	**μ**	**ϕX phage group**	**poly-β-hydroxyalkanoate**
delta agent	**Mu**	***Phlebia***	**poly-β-hydroxybutyrate**
δ antigen	***mu* chain**	***Phlebotomus***	**Polyhymenophorea**

A *number* which forms part of an entry heading affects the position of that entry only if the number immediately follows a letter or word (but cf. chemical names, below). A number which precedes a letter or word is usually ignored, although in the few cases where a number is the first and *main* part of an entry heading it is alphabetized as spelt. Letter–number combinations come after a letter–space but before letter–letter combinations, as in the illustrative sequence A, A2, A2A, A3, A22, AA, ABA etc. Roman numerals are treated as ordinary numbers (I as 1, II as 2 etc). (The reader should bear in mind that, in an unfamiliar term, 'I' could be a letter I or a Roman one, and its location in the Dictionary will be affected accordingly; similarly, 'V' could be letter V or Roman five. O and 0 (zero) may also be confused. If in doubt check both possible positions.) Examples:

bacteriophage Pf2	**D loop**	**Fitz-Hugh–Curtis syndrome**	**T1 side-chains**
bacteriophage ϕI	**D period**	**five–five–five test**	**T-2 toxin**
bacteriophage ϕII	**12D process**	**five-kingdom classification**	**T2H test**
bacteriophage ϕ6	**D-type particles**	**five–three–two symmetry**	**T7 phage group**

Subscript/superscript numbers and letters are alphabetized as though they were ordinary numbers and letters (except in the case of ion designations: see below). Examples:

avoparcin	**B virus**	**C3 convertase**	**CO_2**
a_w	**B_{12} coenzymes**	**C_3 cycle**	**CO_2-stat**
axenic	**B663**	**C3bina**	**CoA**
axial fibrils	**Babes–Ernst granules**	**C5 convertase**	**coactin**

Primes, *apostrophes* and other *non-alphabetizable symbols* (including e.g. plus, minus and *% signs*) are ignored. Examples:

brown rust	**F antigens**	**Gautieriales**	**pluronic polyol F127**
Browne's tubes	**F^+ donor**	**Gazdar murine sarcoma virus**	**plus progamone**
Brownian movement	**F-duction**	**GC%**	**plus strand**
Brown's tubes	**F factor**	**GC type**	**Pluteaceae**

In *chemical names* qualifications such as D-, L-, *N*-, *o*-, *p*-, numbers and Greek letters, as well as hyphens between parts of chemical names, are all ignored for the purposes of alphabetization. Examples:

acetyl-CoA synthetase	**diazomycin A**	**methylmethane sulphonate**
***N*-acetyl-D-glucosamine**	**6-diazo-5-oxo-L-norleucine**	***N*-methyl-*N′*-nitro-*N*-nitrosoguanidine**
acetylmethylcarbinol	**diazotroph**	***N*-methyl-*N*-nitrosourea**
***N*-acetylmuramic acid**	**dibromoaplysiatoxin**	***Methylobacterium***

In entry headings which include an *ion designation*, the ion is treated as a word, the charge being ignored; thus, H^+ is regarded as H, Ca^{2+} as Ca, etc. Examples:

H antigens	**H^+/P ratio**	**K cells**	**Na^+-ATPase**
H^+-ATPase	**H^+-PPase**	**K^+ pump**	**Na^+-motive force**
$H^+/2e^-$ ratio	**H strand**	**K^+ transport**	**Na^+ pump**
H-lysin	**H-1 virus**	**K virus**	**nabam**

2. *Cross-references.* References from one entry to another within the Dictionary are indicated by SMALL CAPITAL letters. In order to effect maximum economy of space, information given in any particular entry is seldom repeated elsewhere, and cross-referencing has been extensively employed to ensure continuity of information. In some cases a complete understanding of an entry, or an appreciation of context, is dependent on a knowledge of information given in other entries; where it is particularly important to follow up a cross-reference, the cross-reference is followed by 'q.v.'. In other cases a cross-reference may be used to link one topic with another of related interest, or to extend the scope of a given topic in one or more directions; in such cases a cross-reference is usually preceded by 'see also' or 'cf.'. (N.B. For a variety of reasons, not every microbiological term or taxon used in the text is cross-referred – even though most of these terms and taxa are defined in the Dictionary; the reader is therefore urged to use the Dictionary for *any* unfamiliar term or taxon.)

When reading an entry for a genus, family or other taxon, it is especially important to follow up, when indicated, a cross-reference to the higher taxon to which it belongs. An entry for a given higher taxon gives the essential features applicable to all members of that taxon, and such features are usually *not* repeated in the entries for each of the constituent lower-ranked taxa; thus, in failing to follow up such cross-references, the reader will forfeit fundamental information relating to the lower taxon in question.

In some cases an entry heading is followed simply by 'See CROSS-REFERENCE'. This is *not* intended to indicate that the two terms are synonymous (usually they are not); such referral signifies only that the meaning of the term is given under the heading indicated. When the entry heading and cross-reference are synonymous, this is indicated by *Syn.*, thus: **entry heading** *Syn.* CROSS-REFERENCE.

3. *External references.* References to papers, articles etc in books or journals are given in square brackets. In order to save space, books are referred to by a 'Book ref.' number, and journal titles are abbreviated

somewhat more than is usual; keys to book reference numbers and journal title abbreviations can be found at the end of the Dictionary (after the Appendices).

A book reference is usually quoted as a source of general background information for the reader, while papers in journals are usually quoted for specific details of current information (or for reviews) and/or for their references to other literature in the field. We should emphasize that the papers we have cited are not necessarily (and are commonly not) those which were the first to report a particular fact, finding or theory; rather, we have chosen, where possible, to cite the most recent references available to us, so that the reader is referred to *current* information and can, if he wishes, trace the earlier literature via references given in the cited papers. We should also point out that the quoting of a single reference in an entry is not intended to indicate that the entry was written solely from information in that paper or book. In relatively few cases does the information in an entry derive from a single source; in the great majority of entries the information has been derived from, or checked against, a range of sources, but limitations of space have necessarily prevented us from citing all of them.

4. *Numbered definitions.* In some cases a term is used with different meanings by different authors, or it may have different meanings in different contexts; for such a term the various definitions are indicated by (1), (2), (3), etc. The order in which the numbered definitions occur is *not* intended to reflect in any way appropriateness or frequency of usage.

5. *Taxonomy.* See entries ALGAE, BACTERIA, FUNGI, PROTOZOA and VIRUS for some general comments on the taxonomy of each of these groups of microorganisms. Each of these entries (except that on bacteria) provides a starting point from which the reader can, via cross-references, follow through a hierarchical system down to the level of genus and, in many cases, species and below; similarly, the hierarchy can be ascended from genus upwards.

6. *The Greek alphabet.* See Appendix VI.

A (1) Adenine (or the corresponding nucleoside or nucleotide) in a nucleic acid. (2) Alanine (see AMINO ACIDS).

Å (Ångström unit) 10^{-10} m (= 10^{-1} nm).

2–5A See INTERFERONS.

A-DNA See DNA.

a-factor See MATING TYPE.

A layer An S LAYER associated with virulence in strains of *Aeromonas salmonicida*.

A-protein In TOBACCO MOSAIC VIRUS: a mixture of small oligomers and monomers of coat protein subunits which occur in equilibrium with the larger 'disc' aggregates under conditions of physiological pH and ionic strength; coat protein occurs mainly as A-protein under conditions of high pH and low ionic strength. (cf. PROTEIN A.)

A site (of a ribosome) See PROTEIN SYNTHESIS.

A-tubule (A-subfibre) See FLAGELLUM (b).

A-type inclusion body See POXVIRIDAE.

A-type particles Intracellular, non-infectious, retrovirus-like particles. Many embryonic and transformed mouse cells contain retrovirus-like 'intracisternal A-type particles' (IAPs) which form by budding at the endoplasmic reticulum; these particles have reverse transcriptase activity and an RNA genome coding for the structural protein of the particles. The mouse genome contains ca. 1000 copies (per haploid genome) of DNA sequences homologous to IAP-associated RNA; these sequences appear to be capable of transposition within the mouse genome – probably via an RNA intermediate [Book ref. 113, pp. 273–279], i.e., they may be RETROTRANSPOSONS. Some A-type particles are non-enveloped precursors of B-type particles (see TYPE B ONCOVIRUS GROUP).

A23187 An IONOPHORE which transports divalent cations, particularly Ca^{2+}; it can effect the transmembrane exchange of $1Ca^{2+}$ (or $1Mg^{2+}$) for $2H^{+}$ without causing perturbation in the gradients of other monovalent cations.

AAA ATPases 'ATPases associated with diverse cellular activities'. AAA ATPases occur e.g. in PEROXISOMES and as components of eukaryotic PROTEASOMES.

AAA pathway AMINOADIPIC ACID PATHWAY.

AAC Aminoglycoside acetyltransferase (see AMINOGLYCOSIDE ANTIBIOTICS).

AAD Aminoglycoside adenylyltransferase (see AMINOGLYCOSIDE ANTIBIOTICS).

AAS Aminoalkylsilane (3-aminopropyltriethoxy-silane, APES; 3(triethoxysilyl)-propylamine, TESPA): a reagent used for binding a tissue section to the surface of a glass slide (e.g. for in situ hybridization); it reacts with silica glass and provides aminoalkyl groups which bind to aldehyde or ketone groups in the tissue section.

***aat* gene** In *Escherichia coli*: a gene whose product promotes the early degradation of those proteins whose N-terminal amino acid is either arginine or lysine. *aat* encodes an 'amino acid transferase' which catalyses the addition of a leucine or phenylalanine residue to the N-terminus of the protein; this destabilizes the protein, facilitating its degradation. (See also N-END RULE.)

***Aat*II** See RESTRICTION ENDONUCLEASE (table).

Aaterra See ETRIDIAZOLE.

AAUAAA locus See MRNA (b).

AAV Adeno-associated virus: see DEPENDOVIRUS.

ab (*immunol.*) ANTIBODY.

AB-transhydrogenase See TRANSHYDROGENASE.

ABA ABSCISIC ACID.

abacavir A NUCLEOSIDE REVERSE TRANSCRIPTASE INHIBITOR.

abacterial pyuria See PYURIA.

Abbe condenser A simple two- or three-lens substage CONDENSER which is uncorrected for spherical or chromatic aberrations.

ABC (1) (*immunol.*) ANTIGEN-BINDING CELL. (2) See ABC TRANSPORTER.

ABC excinuclease See EXCISION REPAIR.

ABC exporter An ABC TRANSPORTER concerned with export/secretion. These systems are found in both prokaryotic and eukaryotic microorganisms and in higher animals, including man. (The mammalian transporters include P-glycoprotein ('multidrug-resistance protein') – a molecular pump by which some types of cancer cell can extrude anti-cancer drugs.) In *Saccharomyces cerevisiae*, an ABC exporter mediates secretion of a peptide PHEROMONE (the **a**-factor) which regulates sexual interaction.

In bacteria, ABC exporters transport various proteins (including enzymes and antibiotics) and, in some species, the polysaccharide components of the capsule; an exporter may be able to transport various related or similar molecules. [Bacterial ABC exporters: MR (1993) *57* 995–1017.]

In *Escherichia coli* the α-haemolysin is secreted via an ABC exporter – a one-step process direct from cytoplasm to environment; this exporter is in the *type I* class of protein secretory systems in Gram-negative bacteria (see PROTEIN SECRETION).

Other proteins secreted by these systems include the cyclolysin of *Bordetella pertussis* and the alkaline protease of *Pseudomonas aeruginosa*. In *Streptomyces antibioticus* an ABC exporter secretes the antibiotic OLEANDOMYCIN.

Bacterial proteins secreted by ABC exporters typically lack an N-terminal signal sequence (see SIGNAL HYPOTHESIS) but they have a C-terminal *secretion sequence* that may interact directly with the ABC protein. Exporters which transport molecules to the periplasm, or outer membrane, as the *final* destination may have fewer protein components than those exporters which *secrete* proteins.

In Gram-negative bacteria, at least some exporters appear to consist of (i) ABC proteins; (ii) a *membrane fusion protein* (MFP) (in the periplasm and cytoplasmic membrane); and (iii) an OUTER MEMBRANE component. Assembly seems to occur in a definite sequence which is promoted and/or initiated by the binding of substrate (i.e. the molecule to be secreted) to the ABC protein; in this scheme, substrate–ABC binding is followed by ABC–MFP interaction – MFP then binding to the outer membrane, presumably to complete the secretory channel [EMBO (1996) *15* 5804–5811].

ABC immunoperoxidase method An IMMUNOPEROXIDASE METHOD involving the use of a preformed avidin–biotin–peroxidase complex (ABC) which has surplus biotin-binding capacity. Initially, a ('primary') antiserum is raised against the required antigen; if the primary antiserum is derived from e.g. a rat, a 'secondary' anti-rat antiserum is prepared, and the anti-rat Ig antibodies are BIOTINylated. To locate a specific antigen, the section is treated with primary antiserum, washed, and then treated with secondary antiserum; the subsequent addition of ABC localizes peroxidase at the site of specific antigen (since the

ABC adheres non-specifically to biotin). Peroxidase (and hence antigen) is detected by incubating the section with e.g. H_2O_2 and diaminobenzidine (which results in the antigenic site being stained brown) or H_2O_2 and 4-chloro-1-naphthol (resulting in a blue stain).

The ABC method can be used for paraffin-embedded sections, frozen sections, and smears. Endogenous (tissue or cell) peroxidase may be quenched e.g. with H_2O_2 in methanol.

ABC protein See ABC TRANSPORTER.

ABC transporter (traffic ATPase) A type of TRANSPORT SYSTEM which, in bacteria, consists typically of a multiprotein complex in the cell envelope, two of the proteins having a specific ATP-binding site (termed the *ATP-binding cassette*; ABC) on their cytoplasmic surface; a (bacterial) protein with an ABC site has been called an 'ABC protein' or an 'ABC subunit'. In eukaryotes, an ABC transporter generally consists of a single polypeptide chain – which also has two ATP-binding sites. Transport mediated by an ABC transporter is energized by ATP hydrolysis at the ABC sites. [ATP-hydrolysing regions of ABC transporters: FEMS Reviews (1998) *22* 1–20.] (See also PROTEIN SECRETION.)

A given type of ABC transporter imports or exports/secretes certain type(s) of ion or molecule. Collectively, these transporters import or secrete a wide range of substances, including ions, sugars and proteins; for example, some import nutrients, or ions for OSMOREGULATION, while others secrete antibiotics or protein toxins. The LmrA transporter in *Lactococcus lactis* mediates an efflux system that extrudes amphiphilic compounds and appears to be functionally identical to the mammalian P-glycoprotein that mediates multidrug-resistance [Nature (1998) *391* 291–295]. The AtrB transporter of *Aspergillus nidulans* mediates energy-dependent efflux of a range of fungicides [Microbiology (2000) *146* 1987–1997].

ABC transporters occur e.g. in Gram-positive and Gram-negative bacteria, members of the Archaea, and in higher animals, including man. In man, certain inheritable diseases (e.g. CYSTIC FIBROSIS and adrenoleukodystrophy) result from defective ABC transporters.

The bacterial *ABC importer* is commonly called a BINDING PROTEIN-DEPENDENT TRANSPORT SYSTEM (q.v.). (See also ABC EXPORTER.)

ABE process An industrial process in which acetone, butanol and ethanol are produced by the fermentation of e.g. molasses by *Clostridium acetobutylicum.* (See also ACETONE–BUTANOL FERMENTATION.)

Abelson murine leukaemia virus (Ab-MuLV) A replication-defective, v-*onc*$^+$ MURINE LEUKAEMIA VIRUS isolated from a prednisolone-treated BALB/c mouse inoculated with Moloney murine leukaemia virus (Mo-MuLV). Ab-MuLV apparently arose by recombination between Mo-MuLV and mouse c-*abl* sequences; the v-*abl* product has tyrosine kinase activity. (See also ABL.) Ab-MuLV induces B-cell lymphoid leukaemia with a short latent period (3–4 weeks). [Abelson virus–cell interactions: Adv. Imm. (1985) *37* 73–98.]

abequose (3,6-dideoxy-D-galactose) A sugar, first isolated from *Salmonella abortusequi*, which occurs in the O-specific chains of the LIPOPOLYSACCHARIDE in certain *Salmonella* serotypes and which contributes to the specificity of O antigen 4 in group B salmonellae (see KAUFFMANN–WHITE CLASSIFICATION).

aberration (chromosomal) See CHROMOSOME ABERRATION.

abhymenial Of or pertaining to a region opposite or away from the HYMENIUM.

abiogenesis (spontaneous generation) The spontaneous formation of living organisms from non-living material; apart from its application to the evolutionary origin of life, this doctrine has long been abandoned.

abiotic Non-living; of non-biological origin.

abl An ONCOGENE originally identified as the transforming determinant of ABELSON MURINE LEUKAEMIA VIRUS (Ab-MuLV). The v-*abl* product has tyrosine kinase activity. In humans, c-*abl* normally occurs on chromosome 9, but is translocated to chromosome 22q- (the Philadelphia chromosome) in cells from patients with chronic myelogenous leukaemia (CML); in chromosome 22 it forms a chimeric fusion gene, *bcr-abl*, encoding a tumour-specific tyrosine kinase designated P210.

ablastin Antibody which specifically inhibits reproduction of epimastigote forms of *Trypanosoma lewisi* in the vertebrate host.

abomasitis Inflammation of the abomasum. (See also BRAXY; cf. RUMENITIS.)

abomasum See RUMEN.

aboral Away from, or opposite to, the mouth.

abortifacient Able to cause abortion.

abortive infection (*virol.*) A viral infection of (*non-permissive*) cells which does not result in the formation of infectious progeny virions, even though some viral genes (e.g. early genes) may be expressed. (cf. PERMISSIVE CELL.)

abortive transduction See TRANSDUCTION.

abortus Bang reaction (abortus Bang ring-probe) *Syn.* MILK RING TEST.

ABR See MILK RING TEST.

***abrB* gene** See ENDOSPORE (figure (a) legend).

abscess A localized collection of PUS surrounded by inflamed and necrotic tissue; it may subside spontaneously or may rupture and drain before healing. Abscesses may occur in any tissue and may be caused by any of a variety of organisms. Abscesses in internal organs (e.g. liver, kidney, brain) may follow bacteraemia or septicaemia and may be due to staphylococci, streptococci, coliforms, etc. A *cold* (or *chronic*) abscess is one with little inflammation, often due to tubercle bacilli. (See also DYSENTERY (b) and QUINSY.)

abscisic acid (ABA) A terpenoid PHYTOHORMONE which acts e.g. as a growth inhibitor, as an inhibitor of germination, and as an accelerator of e.g. leaf abscission. ABA is also formed (as a secondary metabolite) e.g. by the fungus *Cercospora rosicola.*

Absidia See MUCORALES.

absorption (*serol.*) The removal or effective removal of particular antibodies, antigens, or other agents from a given sample (e.g. serum) by the addition of particular antigens, antibodies, or agents to that sample; the resulting antigen–antibody (or other) complexes may or may not be physically removed from the sample. Absorption is used e.g. to remove HETEROPHIL ANTIBODIES.

absorptive pinocytosis See PINOCYTOSIS.

7-ACA 7-Aminocephalosporanic acid (see CEPHALOSPORINS).

Acanthamoeba A genus of amoebae (order AMOEBIDA) in which the pseudopodia each have a broad hyaline zone (see PSEUDOPODIUM) from which arise several to many slender, tapering, flexible, and sometimes forked projections (*acanthopodia*). Polyhedral or roughly circular cysts with cellulose-containing walls are formed. Species are widespread and common in soil and fresh water, where they prey on e.g. bacteria, yeasts etc. [Adhesion of *Acanthamoeba castellanii* to bacterial flagella: JGM (1984) *130* 1449–1458; bacterial endosymbionts of *Acanthamoeba*: J. Parasitol. (1985) *71* 89–95.] Some strains can cause e.g. eye infections, MENINGOENCEPHALITIS [pathogenicity: RMM (1994) *5* 12–20]. (cf. HARTMANNELLA.)

Acantharea A class of marine, mostly planktonic protozoa (superclass ACTINOPODA) which have elaborate 'skeletons' composed of strontium sulphate; typically, the skeleton consists of 10 spines arranged diametrically in the (more or less spherical) cell, or 20 spines which radiate from the cell centre (where they may or may not be joined at their bases, according to species). In many species the cell contains a central capsule (cf. RADIOLARIA); many species contain zooxanthellae. Five orders are recognized; genera include e.g. *Acanthochiasma*, *Acanthometra*, *Astrolophus*, *Gigartacon*.

Acanthochiasma See ACANTHAREA.

Acanthocystis See CENTROHELIDA.

Acanthoeca See CHOANOFLAGELLIDA.

Acanthometra See ACANTHAREA.

acanthopodia See ACANTHAMOEBA.

acaricide Any chemical which kills mites and ticks (order Acarina).

Acarospora A genus of LICHENS (order LECANORALES). Thallus: crustose, areolate, with prominent areolae. Apothecia are embedded in the areolae; ascospores: very small, many per ascus. All species are saxicolous, some are ENDOLITHIC; *A. smaragdula* occurs on rocks and slag rich in heavy metals.

Acarpomyxea A class of protozoa (superclass RHIZOPODA) with characteristics intermediate between those of the naked amoebae and the plasmodial slime moulds: they form small plasmodia (or large uninucleate plasmodium-like forms) which are usually branched and which sometimes anastomose to form a coarse reticulum. Spores, fruiting bodies and tests are absent; cysts are produced by some species. Orders: Leptomyxida (soil and freshwater organisms, e.g. *Leptomyxa* [Book ref. 133, pp. 143–144], *Rhizamoeba*) and Stereomyxida (marine organisms, e.g., *Corallomyxa*, *Stereomyxa*).

Acaryophrya See GYMNOSTOMATIA.

Acaulopage See e.g. NEMATOPHAGOUS FUNGI.

Acaulospora See ENDOGONALES.

acceptor site (of a ribosome) See PROTEIN SYNTHESIS.

acceptor splice site See SPLIT GENE (a).

accessory cells (*immunol.*) Those cells which, together with B LYMPHOCYTES and/or T LYMPHOCYTES, are involved in the expression of humoral and/or cell-mediated immune responses; they include e.g. MACROPHAGES, DENDRITIC CELLS, and LANGERHANS' CELLS.

accessory pigments In PHOTOSYNTHESIS: those pigments contained in LIGHT-HARVESTING COMPLEXES.

AcCoA Acetyl-COENZYME A.

Ace toxin (*Vibrio cholerae*) See BACTERIOPHAGE CTXΦ.

acellular (non-cellular) (1) Refers to an organism, usually a protozoon, which consists essentially of a single cell but in which occur functionally specialized regions sometimes regarded as analogous to the organs and tissues of a differentiated multicellular organism. (2) Refers to an organism (e.g. a VIRUS) or structure (e.g. the stalk of ACYTOSTELIUM) which is not CELLULAR in any sense. (3) Not divided into cells (as e.g. in a PLASMODIUM).

acellular slime moulds See MYXOMYCETES.

acentric (of a chromosome) Having no CENTROMERE.

acephaline gregarines See GREGARINASINA.

acer tar spot See RHYTISMA.

acervulus A flat or saucer-shaped fungal STROMA supporting a mass of typically short and densely-packed conidiophores; acervuli commonly develop subcuticularly or subepidermally in a plant host, becoming erumpent at maturity, i.e., rupturing the overlying plant tissue to allow dispersal of the conidia. Some acervuli bear setae (see SETA).

Acetabularia A genus of DASYCLADALEAN ALGAE. The vegetative thallus consists of a single cell in which the CELL WALL contains MANNAN as a major component and is generally more or less heavily calcified; the cell is differentiated into an erect stalk or axis (up to several centimetres tall) anchored to the substratum by a branching rhizoid. The single nucleus is located in one branch of the rhizoid. As the stalk grows, whorls of sterile 'hairs' develop around the tip; these hairs are eventually shed, leaving rings of scars around the stalk. When the thallus is mature, gametangia develop as an apical whorl of elongated sac-like structures which, depending on species, may or may not be joined to form a characteristic *cap* (giving rise to the popular name 'mermaid's wine-glass'). Once the gametangial sacs have developed, the primary nucleus in the rhizoid grows to ca. 20 times its original size; it then undergoes meiosis, and numerous small secondary nuclei are formed. These migrate from the rhizoid to the gametangia by cytoplasmic streaming. Within a gametangial sac, each nucleus becomes surrounded by a resistant wall, resulting in the formation of many resistant cysts; the cyst walls contain cellulose rather than mannan, and are often heavily calcified. The cysts are liberated into the sea and then undergo a period of dormancy before liberating numerous biflagellate isogametes; pairs of gametes fuse to form zygotes which then develop into new vegetative thalli.

acetate formation See e.g. ACETIFICATION and ACETOGENESIS.

acetate thiokinase See METHANOGENESIS.

acetate utilization See e.g. METHANOGENESIS and TCA CYCLE.

Acetator See VINEGAR.

acetic acid bacteria (1) *Acetobacter* spp. (2) Any bacteria capable of ACETIFICATION, including *Acetobacter* spp and *Gluconobacter* sp.

aceticlastic Able to catabolize acetate.

acetification The aerobic conversion of ethanol to acetic acid by bacteria (usually *Acetobacter* spp). Ethanol is converted to hydrated acetaldehyde ($CH_3CH(OH)_2$) which is then dehydrogenated to give acetic acid. Acetification is an exothermic process. (See also e.g. VINEGAR, BEER SPOILAGE, WINE SPOILAGE.)

Acetivibrio A genus of bacteria (family BACTEROIDACEAE) whose natural habitat is unknown. Cells: straight to slightly curved rods, $0.5–0.9 \times 1.5–10.0$ µm; in motile species the concave side of the cell has either a single flagellum or a number of flagella which arise in a line along the longitudinal axis of the cell. The cells stain Gram-negatively but the cell wall of the type species resembles those of Gram-positive bacteria. The major products of carbohydrate fermentation typically include acetic acid, ethanol, CO_2 and H_2; butyric, lactic, propionic and succinic acids are not formed. GC%: ca. 37–40. Type species: *A. cellulolyticus*.

A. cellulolyticus. Monotrichous. Substrates include cellobiose, cellulose and salicin; aesculin is not hydrolysed. The type strain was isolated from a methanogenic enrichment culture.

A. cellulosolvens. A non-motile species (isolated from sewage sludge) which can hydrolyse cellulose, cellobiose, aesculin and salicin; the cells apparently have an outer membrane. [IJSB (1984) *34* 419–422.]

A. ethanolgignens. Multitrichous. Substrates include fructose, galactose, lactose, maltose, mannitol and mannose – but not cellobiose, cellulose or aesculin. *A. ethanolgignens* is consistently present in the colons of pigs suffering from SWINE DYSENTERY.

Acetobacter A genus of Gram type-negative bacteria of the family ACETOBACTERACEAE; the organisms occur e.g. on certain fruits and flowers, are responsible for some types of BEER SPOILAGE and WINE SPOILAGE, and are used e.g. in the manufacture of VINEGAR.

Cells: typically ovoid or rod-shaped, 0.6–0.8 × 1.0–4.0 μm, non-motile or with peritrichous or lateral flagella. Most strains are catalase-positive. Typically, ethanol is oxidized to acetic acid, and acetic acid is oxidized ('overoxidation') to CO_2 (cf. GLUCONOBACTER). Principal substrates include e.g. ethanol, glycerol and lactate; most strains grow well on glucose–yeast extract–$CaCO_3$ agar (GYC agar), forming round pale colonies. (See also CARR MEDIUM.) Some strains form CELLULOSE (see PELLICLE (1)). Sugars appear to be metabolized primarily via the HEXOSE MONOPHOSPHATE PATHWAY and the TCA CYCLE; phosphofructokinase seems to be absent (cf. Appendix I(a)). The ENTNER–DOUDOROFF PATHWAY appears to occur only in cellulose-synthesizing strains. Growth on HOYER'S MEDIUM appears to involve enzymes of the glyoxylate shunt. Optimum growth temperature: 25–30°C. GC%: ca. 51–65. Type species: *A. aceti*.

A. aceti. Ketogenic with glycerol or sorbitol substrates; 5-ketogluconic acid (but not 2,5-diketogluconic acid) formed from D-glucose. No diffusible brown pigments are formed on GYC agar. Grows on sodium acetate.

A. hansenii. Ketogenic with glycerol or sorbitol substrates; 5-ketogluconic acid (but not 2,5-diketogluconic acid) is formed by some strains from D-glucose. No growth on sodium acetate. No diffusible brown pigments are formed on GYC agar. (cf. *A. xylinum*.)

A. liquefaciens. Brown diffusible pigments are formed on GYC agar. 2,5-Diketogluconic acid is formed from D-glucose. Ketogenic with glycerol as substrate.

A. pasteurianus. Ketogluconic acids are not formed from D-glucose. No brown diffusible pigments are formed on GYC agar. Some strains (formerly called *A. peroxydans*) are catalase-negative. (cf. *A. xylinum*.)

A. peroxydans. See *A. pasteurianus*.

A. suboxydans. See GLUCONOBACTER.

A. xylinum. Cellulose-producing strains formerly classified as a subspecies of *A. aceti*, then distributed between the two species *A. hansenii* and *A. pasteurianus*; *A. xylinum* has now been accepted as a revived name for cellulose-forming and cellulose-less, acetate-oxidizing strains [IJSB (1984) *34* 270–271].

[Book ref. 22, pp. 268–274.]

Acetobacteraceae A family of aerobic, oxidase-negative, chemoorganotrophic, Gram type-negative bacteria which typically oxidize ethanol to acetic acid. Metabolism: strictly respiratory (oxidative), with O_2 as terminal electron acceptor. Growth occurs optimally at ca. pH 5–6. The organisms occur e.g. in acidic, ethanol-containing habitats. GC%: ca. 51–65. Two genera: ACETOBACTER (type genus), GLUCONOBACTER [Book ref. 22, pp. 267–278].

Acetobacterium A genus of Gram-negative, obligately anaerobic bacteria which occur in marine and freshwater sediments [IJSB (1977) *27* 355–361]. Cells: polarly flagellated ovoid rods, ca. 1.0 × 2.0 μm, often in pairs. The type species, *A. woodii*, can carry out a homoacetate fermentation of e.g. fructose, glucose or lactate, or can grow chemolithoautotrophically (see ACETOGENESIS); it contains group B PEPTIDOGLYCAN. Optimum growth temperature: 30°C. GC%: ca. 39. (See also ANAEROBIC DIGESTION.)

acetogen (1) Any bacterium (e.g. *Acetobacterium woodii*, *Clostridium aceticum*, *C. thermoaceticum*) which produces acetate – as the *main* product – from certain sugars (via homoacetate fermentation and reduction of carbon dioxide) and (in some strains) from carbon dioxide and hydrogen (see ACETOGENESIS).

(2) (hydrogenogen; proton-reducing acetogen) Any bacterium which can use protons as electron acceptors for the oxidation of certain substrates (e.g. ethanol, lactate, fatty acids) to acetate with concomitant formation of hydrogen. Obligate hydrogenogens include e.g. SYNTROPHOMONAS (see also ANAEROBIC DIGESTION). Some SULPHATE-REDUCING BACTERIA appear to be facultative hydrogenogens. The synthesis of acetate by hydrogenogens is thermodynamically favourable only when the partial pressure of hydrogen is very low – e.g. in the presence of a hydrogen-utilizing methanogen.

acetogenesis Acetate formation. A variety of microorganisms can form acetate, as a major or minor product, e.g. via the MIXED ACID FERMENTATION or PROPIONIC ACID FERMENTATION. (cf. ACETIFICATION.)

The term is also used more specifically to refer to the *particular* pathways used by the ACETOGENS (sense 1). These organisms form acetate, as the *main* product, from e.g. certain hexoses in a process (*homoacetate fermentation*) in which the hexose is metabolized to pyruvate (via the EMBDEN–MEYERHOF–PARNAS PATHWAY) and thence to acetate and carbon dioxide.

Additional acetate is formed as follows. Some of the carbon dioxide is reduced to formate; this formate is bound to tetrahydrofolate (THF) and is further reduced (in an ATP-dependent reaction) to yield 5-methyl-THF. The methyl group is then transferred to coenzyme B_{12}. The remainder of the carbon dioxide is reduced to carbon monoxide (by CO dehydrogenase). Carbon monoxide reacts with methyl-coenzyme B_{12} in the presence of coenzyme A and CO dehydrogenase disulphide reductase to yield acetyl-CoA. Acetyl-CoA is converted to acetate and CoASH with concomitant substrate-level phosphorylation to yield ATP.

Some acetogens (e.g. *A. woodii*, *C. aceticum*, some strains of *C. thermoaceticum*) can form acetate from carbon dioxide and hydrogen [autotrophic pathways in acetogens: JBC (1986) *261* 1609–1615]. This process resembles the latter part of the pathway above: CO is derived from carbon dioxide, $2H^+$ and $2e^-$, and 5-methyl-THF from THF, carbon dioxide and hydrogen.

acetoin ($CH_3.CHOH.CO.CH_3$; acetylmethylcarbinol) See e.g. Appendix III(c); BUTANEDIOL FERMENTATION; VOGES–PROSKAUER TEST.

Acetomonas Former name of GLUCONOBACTER.

acetone–butanol fermentation (solvent fermentation) A FERMENTATION (sense 1), carried out by certain saccharolytic species of *Clostridium* (e.g. *C. acetobutylicum*), in which the products include acetone (or isopropanol) and *n*-butanol (collectively referred to as 'solvent'). Glucose is initially metabolized via the BUTYRIC ACID FERMENTATION, but subsequently the pH drops to ca. 4.5–5.0 and acetone and *n*-butanol are formed as major end products [Appendix III (g)]. This fermentation is carried out on an industrial scale to a limited extent. [Review: AAM (1986) *31* 24–33, 61–92.]

acetosyringone See CROWN GALL.

3-acetoxyindole See INDOXYL ACETATE.

acetylcholine (neurotransmitter) See BOTULINUM TOXIN.

acetyl-CoA synthetase See TCA CYCLE.

***N*-acetyl-L-cysteine** See MUCOLYTIC AGENT.

***N*-acetyl-D-glucosamine** (GlcNAc) *N*-Acetyl-(2-amino-2-deoxy-D-glucose): an amino sugar present in various polysaccharides – see e.g. CHITIN, HYALURONIC ACID, LIPOPOLYSACCHARIDE, PEPTIDOGLYCAN (q.v. for formula), TEICHOIC ACIDS.

acetylmethylcarbinol *Syn.* ACETOIN.

***N*-acetylmuramic acid** See PEPTIDOGLYCAN.

***N*-acetylmuramidase** *Syn.* LYSOZYME.

***N*-acetylneuraminic acid** See NEURAMINIC ACID.

A-CGT See IMMUNOSORBENT ELECTRON MICROSCOPY.

achlorophyllous *Syn.* ACHLOROTIC.

achlorotic (achlorophyllous) Lacking chlorophyll. (cf. APOCHLOROTIC.)

Achlya A genus of aquatic fungi (order SAPROLEGNIALES) in which the thallus is characteristically a branched, coenocytic mycelium; the width of the hyphae varies with species. Although *Achlya* species are typically saprotrophic some have been reported to parasitize rice plants. (See also DIPLANETISM, HETEROTHALLISM and PHEROMONE.)

Achnanthes See DIATOMS.

Acholeplasma A genus of facultatively anaerobic, urease-negative bacteria (family ACHOLEPLASMATACEAE) which are associated with various vertebrates (and possibly with invertebrates and plants), and which also occur e.g. in soil and sewage and as contaminants in TISSUE CULTURES. Cells: non-motile cocci (minimum diam. ca. 300 nm) or filaments (typically ca. 2–5 μm in length); carotenoid pigments occur in some species. The organisms resemble *Mycoplasma* spp in their general properties, but differ e.g. in that their growth is sterol-independent, and in that NADH oxidase occurs in the cytoplasmic membrane rather than in the cytoplasm. *Acholeplasma* spp are susceptible to various ACHOLEPLASMAVIRUSES. GC%: ca. 26–36. Type species: *A. laidlawii*; other species: *A. axanthum*, *A. equifetale*, *A. granularum*, *A. hippikon*, *A. modicum*, *A. morum*, *A. oculi*. [Book ref. 22, pp. 775–781.]

Acholeplasmataceae A family of bacteria of the order MYCOPLASMATALES; species of the sole genus, ACHOLEPLASMA, differ from the other members of the order e.g. in that their growth is not sterol-dependent. [Proposal for re-classifying Acholeplasmataceae as the order Acholeplasmatales: IJSB (1984) *34* 346–349.]

acholeplasmaviruses BACTERIOPHAGES which infect *Acholeplasma* species: see PLECTROVIRUS, PLASMAVIRIDAE, MV-L3 PHAGE GROUP.

achromat (achromatic objective) An objective lens (see MICROSCOPY) in which chromatic aberration has been corrected for two colours (usually red and blue), and spherical aberration has been corrected for one colour (usually yellow–green). (cf. APOCHROMAT.) A FLAT-FIELD OBJECTIVE LENS of this type is called a *planachromat*.

Achromobacter An obsolete bacterial genus.

achromogenic Refers to an organism (or e.g. reagent) which does not produce pigment (or colour); used e.g. of non-pigmented strains of normally CHROMOGENIC organisms.

achromycin See TETRACYCLINES.

aciclovir A spelling used by some authors for the drug ACYCLOVIR.

acicular Needle-shaped.

Aciculoconidium A genus of fungi (class HYPHOMYCETES) which form budding ovoid or ellipsoidal cells (occurring singly or in short chains or clusters) as well as branched septate hyphae. Conidia are formed terminally and are acicular, rounded at one end and pointed at the other. NO_3^- is not assimilated. One species: *A. aculeatum* (formerly *Trichosporon aculeatum*), isolated from *Drosophila* spp. [Book ref. 100, pp. 558–561.]

acid dye See DYE.

acid-fast organisms Organisms (e.g. *Mycobacterium* spp) which, once stained with an ACID-FAST STAIN, cannot be decolorized by mineral acids or by mixtures of acid and ethanol.

acid-fast stain Any stain used to detect or demonstrate ACID-FAST ORGANISMS – e.g. ZIEHL–NEELSEN'S STAIN, AURAMINE–RHODAMINE STAIN.

acid fuchsin See FUCHSIN.

acid phosphatase See PHOSPHATASE.

Acidaminococcus A genus of Gram-negative bacteria (family VEILLONELLACEAE) which occur e.g. in the intestine in humans and pigs. Cells: typically kidney-shaped cocci, 0.6–1.0 μm diam, occurring in pairs. Amino acids are the main sources of carbon and energy; all strains need e.g. arginine, glutamate, tryptophan and valine, and most need e.g. cysteine and histidine. In general, the organisms metabolize carbohydrates weakly or not at all. Optimum growth temperature: 30–37°C. Optimum pH: 7.0. GC%: ca. 57. Type species: *A. fermentans*.

acidophile An organism which grows optimally under acidic conditions, having an optimum growth pH below 6 (and sometimes as low as 1, or below), and which typically grows poorly, or not at all, at or above pH 7: see e.g. SULFOLOBUS, THERMOPLASMA, THIOBACILLUS. (cf. ALKALOPHILE and NEUTROPHILE; see also LEACHING.)

acidophilus milk A sour, medicinal beverage made by fermenting heat-treated, partially skimmed milk with *Lactobacillus acidophilus*. (Viable *L. acidophilus* appears to have a therapeutic effect on some intestinal disorders.) The main fermentation product is lactic acid which reaches a level of ca. 1.0%. A more palatable preparation, 'sweet acidophilus milk', is made by adding *L. acidophilus* to milk at ca. 5°C; under these conditions the cells remain viable but lactic acid is not produced. (See also DAIRY PRODUCTS.)

acidosis (1) (lactic acidosis) (*vet.*) A (sometimes fatal) condition which may occur in ruminants fed excessive amounts of readily fermentable carbohydrates (e.g. starch, sugars – found e.g. in grain and beet, respectively) or when the transfer from a roughage to a 'concentrate' diet is made too quickly. Under these conditions the rate of acid production in the RUMEN is very high; the resulting fall in pH in the rumen (due mainly to the accumulation of lactic acid) inhibits cellulolytic bacteria and protozoa, and favours the growth of certain LACTIC ACID BACTERIA – so that the pH falls still further. (See also RUMENITIS.) A gradual transition from roughage to concentrate may permit the somewhat more acid-tolerant bacterium *Megasphaera elsdenii* to metabolize the lactic acid and maintain a normal pH in the rumen. (See also THIOPEPTIN.)

(2) (*med., vet.*) A pathological condition characterized by an abnormally low pH in the blood and tissues.

Acidothermus A proposed genus of aerobic, thermophilic (growing at 37–70°C), acidophilic (growing at pH 3.5–7.0), cellulolytic, non-motile, rod-shaped to filamentous bacteria isolated from acidic hot springs; GC%: ca. 60.7. [IJSB (1986) *36* 435–443.]

aciduric Tolerant of acidic conditions. (cf. ACIDOPHILE.)

Acineria See GYMNOSTOMATIA.

Acineta See SUCTORIA.

Acinetobacter A genus of strictly aerobic, oxidase −ve, catalase +ve Gram-type-negative bacteria of the family MORAXELLACEAE (within the gamma subdivision of PROTEOBACTERIA); the organisms occur e.g. in soil and water and may act as opportunist pathogens in man. (See also MEAT SPOILAGE and SEWAGE TREATMENT.)

Cells: short rods, 0.9–1.6 × 1.5–2.5 μm, or coccobacilli (coccoid in stationary-phase cultures); cells often in pairs. Non-motile, but may exhibit TWITCHING MOTILITY. Non-pigmented.

Metabolism is respiratory (oxidative), with oxygen as terminal electron acceptor; no growth occurs anaerobically, with or without nitrate.

Most strains can grow on a mineral salts medium containing an organic carbon source such as acetate, ethanol or lactate as the sole source of carbon and energy; some can use amino acids (e.g. L-leucine, ornithine) and/or pentoses (e.g. L-arabinose, D-xylose), and some are able to degrade e.g. benzoate, *n*-hexadecane and alicyclic compounds (see HYDROCARBONS). Acinetobacters appear to contain all the enzymes of the TCA CYCLE and the glyoxylate cycle. Many carbohydrates can be used. Most strains in the *A. calcoaceticus–A. baumannii* complex (and in certain other groups) can form acid from glucose (oxidatively), but many (e.g. most strains designated *A. lwoffii*) cannot. The optimal growth temperature is typically 33–35°C. GC%: ~38–47. Type species: *A. calcoaceticus*.

The taxonomy of *Acinetobacter* is confused and unsatisfactory. Emended descriptions of the two species *A. calcoaceticus* and *A. lwoffii*, and proposals for four new species (*A. baumannii*, *A. haemolyticus*, *A. johnsonii* and *A. junii*), were published in 1986 [IJSB (1986) *36* 228–240]. Since then, a number of adjustments have been made to the taxonomic structure of the genus. [Taxonomy, and epidemiology of *Acinetobacter* infections: RMM (1995) *6* 186–195.]

Acinetobacters have been isolated in a number of hospital-associated (and other) outbreaks of disease, often as part of a mixed infection; in most cases such infections involve glucolytic strains of the *A. calcoaceticus–A. baumannii* complex – particularly *A. baumannii* (also called group 2, or genospecies 2). The most common manifestations of disease include septicaemia and infections of the urinary tract, lower respiratory tract and central nervous system. Transmission may occur by direct contact or may involve the airborne route. Acinetobacters have been reported to survive on dry surfaces for at least as long as e.g. *Staphylococcus aureus*.

One problem associated with the pathogenic role of *Acinetobacter* is that these organisms appear easily to acquire resistance to antibiotics – so that they have the potential to develop as multiresistant pathogens; currently, for example, acinetobacters are reported to be resistant to most β-lactam antibiotics, particularly penicillins and cephalosporins, and to chloramphenicol and trimethoprim–sulphamethoxazole. [Mechanisms of antimicrobial resistance in *A. baumannii*: RMM (1998) *9* 87–97.]

AcLVs AVIAN ACUTE LEUKAEMIA VIRUSES.

acne A chronic skin disorder characterized by increased sebum production and the formation of comedones ('blackheads' and 'whiteheads') which plug the hair follicles. *Propionibacterium acnes*, present in the pilosebaceous canal (see SKIN MICROFLORA), may play a causal role; it produces a lipase that hydrolyses sebum triglycerides to free fatty acids, and these can cause inflammation and comedones [JPed (1983) *103* 849–854]. *Treatment*: e.g. topical SALICYLIC ACID or benzoyl peroxide; the latter has keratinolytic activity and exerts bactericidal action on *P. acnes* by releasing free-radical oxygen.

Aconchulinida See FILOSEA.

aconitase See Appendix II(a) and NITRIC OXIDE.

Aconta Algae of the RHODOPHYTA. (cf. CONTOPHORA.)

acquired immune deficiency syndrome See AIDS.

acquired immunity (1) SPECIFIC IMMUNITY acquired through exposure to a given antigen. (2) PASSIVE IMMUNITY. (3) NON-SPECIFIC IMMUNITY acquired through exposure to certain viruses (see e.g. INTERFERONS) or by immunization with BCG.

Acrasea See ACRASIOMYCETES.

acrasids See ACRASIOMYCETES.

acrasin In cellular slime moulds: a generic term for a chemotactic substance which is produced by cells and which serves as a chemoattractant for cell aggregation. Acrasins are a diverse group of substances; they include cAMP in *Dictyostelium discoideum* (q.v.), a pterin in *Dictyostelium lacteum* [PNAS (1982) *79* 6270–6274], and a dipeptide, 'glorin', in *Polysphondylium violaceum* (q.v.).

Acrasiomycetes (acrasid cellular slime moulds; acrasids) A class of cellular SLIME MOULDS (division MYXOMYCOTA) in which the vegetative phase consists of amoeboid cells that form lobose pseudopodia; the amoebae aggregate (without streaming) to form a pseudoplasmodium which is not slug-like and does not migrate (cf. DICTYOSTELIOMYCETES). The pseudoplasmodium gives rise to multispored fruiting bodies which may have long or short stalks (but no cellulosic stalk tube) bearing e.g. simple globular sori or branched or unbranched chains of spores. Flagellated cells have been observed in only one species (*Pocheina rosea*). Sexual processes are unknown. Acrasids occur in various habitats: e.g. dung, tree-bark, dead plant materials, etc. Genera include *Acrasis*, *Copromyxa*, *Copromyxella*, *Fonticula*, *Guttulinopsis*, *Pocheina* (formerly *Guttulina*).

(Zoological taxonomic equivalents of the Acrasiomycetes include the class Acrasea of the MYCETOZOA, and the class Acrasea of the RHIZOPODA.)

Acrasis See ACRASIOMYCETES.

Acremonium A genus of fungi of the class HYPHOMYCETES; teleomorphs occur in e.g. *Emericellopsis* and *Nectria*. The genus includes organisms formerly classified as species of *Cephalosporium* [for references see MS (1986) *3* 169–170]. *Acremonium* spp form septate mycelium; conidia, often in gelatinous masses, are produced from phialides which develop from simple, single branches of the vegetative hyphae. *A. kiliense* (= *Cephalosporium acremonium*) produces cephalosporin C (see CEPHALOSPORINS). (See also MADUROMYCOSIS.)

acridine orange (basic orange, or euchrysine; 3,6-bis(dimethylamino)-acridinium chloride) A basic dye and FLUOROCHROME used e.g. in fluorescence MICROSCOPY to distinguish between dsDNA (which fluoresces green) and ss nucleic acids (which fluoresce orange-red). Sublethal concentrations of the dye are used for CURING plasmids. (See also ACRIDINES.)

acridines Heterocyclic compounds which include acridine and its derivatives. At low concentrations, aminoacridines (e.g. proflavine (3,6-diaminoacridine), QUINACRINE) appear to bind to dsDNA (or to double-stranded regions of ssDNA) primarily as INTERCALATING AGENTS. At higher concentrations there is also a weaker, secondary type of binding in which the acridine binds to the outside of dsDNA or to ssDNA or ssRNA; the two types of binding may account for the differential staining of DNA and RNA by ACRIDINE ORANGE. [Book ref. 14, pp. 274–306.] Acridines inhibit DNA and RNA synthesis and cause e.g. FRAMESHIFT MUTATIONS. They are used e.g. as antimicrobial agents (see e.g. ACRIFLAVINE), as mutagens, and as fluorescent stains for nucleic acids; they also have potential antitumour activity. (See also CURING (2).)

As *antimicrobial agents*, acridines are active against a wide range of bacteria, but they are not sporicidal; some are active against certain parasitic protozoa (see e.g. QUINACRINE and KINETOPLAST) and inhibit the replication of certain viruses. Activity is not significantly affected by proteinaceous matter. [Acridines as antibacterials (review): JAC (2001) *47* 1–13.]

As *mutagens*, acridines may be effective in replicating bacteriophages but are generally not effective in bacteria. However, compounds in which an acridine nucleus is linked to an alkylating side-chain – *ICR compounds* (ICR = Institute for Cancer Research) – can induce frameshift and other mutations in bacteria.

(6) 8, (5) 9, (4) 1, (7) 7, 2 (3), (8) 6, 3 (2), N, 5 (9), 10 (10), 4 (1)

ACRIDINE. The numbering system used in this dictionary is indicated by the numbers which are not in parentheses; an alternative numbering system (numbers in parentheses) is used by some authors.

acriflavine (acriflavin; *syn.* euflavin) 3,6-Diamino-10-methylacridinium chloride *or* (according to some authors) a mixture of this compound and 3,6-diaminoacridine (proflavine). Acriflavine is soluble in water and in ethanol, and has been used as an ANTISEPTIC. (See also ACRIDINES.)

acro- Prefix meaning tip or outermost part.

Acrocordia See PYRENULALES.

acrolein (CH_2=CH–CHO) An aldehyde used e.g. for pre-FIXATION; it penetrates tissues more rapidly than GLUTARALDEHYDE.

acronematic Refers to a eukaryotic FLAGELLUM which is smooth and tapers to a fine point.

acropetal development Development from the base, or point of attachment, towards the tip; e.g., in a chain of acropetally developing spores the first-formed spores occupy positions in the chain nearest the base of the spore-bearing structure, while spores formed later occupy positions in the distal parts of the chain. (cf. BASIPETAL DEVELOPMENT.)

acropleurogenous Located both at the tip and on the sides of an elongated structure.

Acrosiphonia A genus of branched, filamentous, siphonocladous green algae (division CHLOROPHYTA).

Acrospermum See CLAVICIPITALES.

acrylate pathway See PROPIONIC ACID FERMENTATION.

ActA protein (*Listeria monocytogenes*) See LISTERIOSIS.

actaplanin See VANCOMYCIN.

Actidione *Syn.* CYCLOHEXIMIDE.

actin (1) A protein, found in most types of eukaryotic cell, which can polymerize (reversibly) to form non-contractile filaments (*microfilaments*) that are involved e.g. in maintaining cell shape and structure (see e.g. CYTOSKELETON) and (together with MYOSIN) in CAPPING (sense 3), amoeboid movement (see PSEUDOPODIUM), CYTOPLASMIC STREAMING, PHAGOCYTOSIS, and (in higher animals) muscle contraction.

Actins from various sources are similar in structure. The monomeric form (*G-actin*) is a globular protein (MWt ca. 42000) consisting of ca. 375 amino acid residues; each molecule can bind one molecule of ATP. In most non-muscle cells, G-actin occurs in dynamic equilibrium with the polymerized (filamentous) form, *F-actin*, which consists of a helical, double-stranded chain of monomers ca. 7 nm thick. Although F-actin is itself non-contractile, its interaction with myosin can cause microfilaments to slide relative to one another – thereby bringing about movements and contractions in structures bound to the microfilaments. During the polymerization of G-actin ATP is hydrolysed; as in the assembly of MICROTUBULES, energy is not essential for – but increases the rate of – polymerization. Polymerization and depolymerization can occur at both ends of a microfilament, but one of the ends may grow (or depolymerize) at a greater rate than the other. (See also CAPPING sense 2.)

The formation and fate of microfilaments are regulated in vivo e.g. by various proteins. *Profilin* binds to G-actin, inhibiting polymerization. *Gelsolin* (in e.g. macrophages), *severin* (in *Dictyostelium*), *fragmin* (in *Physarum*), and *villin* (in microvilli) can each cleave F-actin into fragments in a Ca^{2+}-dependent reaction, thereby e.g. effecting a gel-to-sol transition. *Filamin* and *α-actinin* can cross-link microfilaments, promoting gel formation. *β-Actinin* can act as a CAPPING (sense 2) protein. *Vinculin* may help to anchor microfilaments to other cell components. [Binding of microfilaments to the cytoplasmic membrane in *Dictyostelium discoideum*: JCB (1986) *102* 2067–2075.] *Fimbrin* binds together longitudinally adjacent microfilaments to form bundles.

Actin polymerization/depolymerization is affected e.g. by agents such as CYTOCHALASINS and by phalloidin (see PHALLOTOXINS).

(2) See MACROTETRALIDES.

actin-based motility See DYSENTERY (1a) and LISTERIOSIS.

Actinichona See HYPOSTOMATIA.

α-actinin See ACTIN.

β-actinin See ACTIN.

actino- Prefix signifying a ray or rays.

actinobacillosis Any animal (or human) disease caused by a species of *Actinobacillus*. *A. lignieresii* causes granulomatous lesions in and around the mouth – particularly the tongue ('wooden tongue') – in cattle; in sheep *A. lignieresii* is associated with suppurative lesions in the skin and internal organs. *A. equuli* is pathogenic for horses (see SLEEPY FOAL DISEASE) and pigs; in pigs symptoms may include fever, haemorrhagic or necrotic skin lesions, arthritis and endocarditis. *A. suis* causes septicaemia and localized lesions in pigs. (See also PERIODONTITIS.)

Actinobacillus A genus of Gram-negative bacteria of the PASTEURELLACEAE. Cells: mostly rod-shaped (ca. 0.3–0.5 × 0.6–1.4 μm), but a coccal form often occurs at the end of a rod, giving a characteristic 'Morse code' form; filaments may occur in media containing glucose or maltose. Extracellular slime is often produced. Cells stain irregularly. Glucose, fructose, xylose, and (most strains) lactose are fermented (no gas). Growth occurs only on complex media; all species (except *A. actinomycetemcomitans*) can grow on MacConkey's agar. Most species are non-haemolytic, but *A. suis* and some strains of *A. equuli* exhibit clear haemolysis on sheep blood agar; *A. suis* causes partial haemolysis on horse blood agar. GC%: 40–43. Type species: *A. lignieresii*.

Actinobacilli occur as commensals in the alimentary, respiratory and/or genital tracts of animals: *A. lignieresii* in cattle and sheep, *A. equuli* in horses, *A. suis* in pigs(?) and horses, *A. capsulatus* in rabbits(?), *A. actinomycetemcomitans* in man. All can be opportunist pathogens (see ACTINOBACILLOSIS). (*A. muris* = *Streptobacillus moniliformis*; *A. mallei* = *Pseudomonas mallei*; *A. ureae*: see PASTEURELLA.)

[Book ref. 22, pp. 570–575; proposal to re-classify *A. actinomycetemcomitans* as *Haemophilus actinomycetemcomitans*: IJSB (1985) *35* 337–341.]

Actinobifida An obsolete genus of actinomycetes which included species with dichotomously-branching sporophores; at least some strains were transferred to THERMOMONOSPORA.

Actinobolina A genus of carnivorous ciliates (subclass GYMNOSTOMATIA). Cells: roughly ovoid, with uniform somatic ciliature, an apical cytostome, TOXICYSTS, and retractable tentacles distributed evenly over the body.

Actinocephalus See GREGARINASINA.

actinoidin See VANCOMYCIN.

Actinomadura A genus of bacteria (order ACTINOMYCETALES, wall type III; group: maduromycetes) which occur e.g. in soil; some species (*A. madurae*, *A. pelletieri*) can be pathogenic in man (see MADUROMYCOSIS). The organisms form a branching, usually stable, substrate mycelium, but (spore-forming) aerial mycelium may be common or rare according to species; some species contain only trace amounts of madurose, or none at all. GC%: reported to be within the range 65–78. Type species: *A. madurae*. [Taxonomic studies on *Actinomadura* and *Nocardiopsis*: JGM (1983) *129* 3433–3446; ecology, isolation and cultivation: Book ref. 46, pp. 2103–2117.]

Actinomucor See MUCORALES.

Actinomyces A genus of asporogenous bacteria (order ACTINOMYCETALES; wall type varies with species); species occur in warm-blooded animals e.g. as part of the microflora of the mucous membranes (particularly in the mouth) and can act as opportunist pathogens. The organisms occur as rods, branched rods or filaments, or as a rudimentary mycelium. All species can grow anaerobically, or under reduced partial pressure of oxygen; growth in vitro occurs readily on rich media at 37°C, and is typically enhanced if the partial pressure of carbon dioxide is increased. Carbohydrates are fermented anaerogenically – acetic, lactic and succinic acids being the main acidic end products of glucose fermentation in PYG MEDIUM. Most species are catalase-negative; *A. viscosus* is catalase-positive. GC%: ca. 57–73. Type species: *A. bovis*.

A. bovis (wall type VI) and *A. israelii* (wall type V) can cause chronic disease in animals and man (see ACTINOMYCOSIS); *A. naeslundii* and *A. viscosus* (both wall type V) can cause periodontitis e.g. in rodents. (See also COAGGREGATION.) *A. pyogenes* (formerly *Corynebacterium pyogenes* [JGM (1982) *128* 901–903]) is the cause of 'summer mastitis' in cattle, and is often isolated from pyogenic lesions in cattle, pigs and other animals; *A. pyogenes* typically occurs as short rods or coryneforms which secrete a soluble haemolysin. *A. hordeovulneris* [IJSB (1984) *34* 439–443] is a causal agent of actinomycosis in dogs.

Actinomycetales An order of GRAM TYPE-positive, typically aerobic bacteria; species range from those which occur as cocci and/or rods to those which form a well-developed, branching SUBSTRATE MYCELIUM and/or AERIAL MYCELIUM, and which may form sophisticated structures such as sclerotia, sporangia and synnemata. (cf. ACTINOMYCETE.) Most members of the order have a GC%>55, thus distinguishing them from species of the other major subbranch of Gram-positive bacteria: the *Clostridium–Bacillus–Thermoactinomyces* line (but cf. CORYNEBACTERIUM, RENIBACTERIUM and THERMOACTINOMYCES). Phylogenetic relationships between actinomycetes are indicated by 16S rRNA oligonucleotide cataloguing and nucleic acid hybridization; within the order, groups of genera can be distinguished on the basis of e.g. the chemical nature of the cell wall and the lipid profiles of the organisms. [The system of classification adopted in the Dictionary is based on the scheme proposed in Book ref. 73, pp. 7–164.]

Actinomycetes are widespread in nature, occurring typically in soil, composts (see COMPOSTING) and aquatic habitats; most species are free-living and saprotrophic, but some form symbiotic associations (see e.g. ACTINORRHIZA) and others are pathogenic in man, other animals, and plants (see e.g. ACTINOMYCOSIS, DERMATOPHILOSIS, JOHNE'S DISEASE, POTATO SCAB, and TUBERCULOSIS). The organisms are chemoorganotrophs; collectively they can degrade a wide range of substances which include e.g. agar, cellulose, chitin, keratin, paraffins and rubber. Some species produce important antibiotics (see e.g. STREPTOMYCES).

Ultrastructure and staining. The cell structure is that of a Gram-positive prokaryote; most species give an unequivocally positive reaction in the Gram stain (but see e.g. CELLULOMONAS), and some species are acid-fast (see e.g. MYCOBACTERIUM, NOCARDIA, RHODOCOCCUS). Cytoplasmic inclusions observed in at least some species include e.g. granules of poly-β-hydroxybutyrate, polyphosphate, and polysaccharide, and globules of lipid. The cell wall commonly appears to be either uniformly electron-dense or three-layered, the electron-density of the middle layer being somewhat less than that of the layer on either side of it. The wall contains PEPTIDOGLYCAN and other polymers, e.g. TEICHOIC ACIDS – although the latter appear not to occur in the NOCARDIOFORM ACTINOMYCETES; the cell wall is commonly surrounded by a layer of diffuse or (in sporoactinomycetes) fibrous material. Depending on the presence of certain amino acids in the peptidoglycan, and the identity of the cell wall sugars, eight wall types (chemotypes I–VIII) of actinomycetes can be distinguished [Book ref. 46, pp. 1915–1922]:

I. LL-DAP (LL-diaminopimelic acid), glycine.
II. *meso*-DAP, glycine.
III. *meso*-DAP.
IV. *meso*-DAP, arabinose, galactose.
V. Lysine, ornithine.
VI. Lysine; aspartic acid and galactose sometimes present.
VII. DAB (2,4-diaminobutyric acid), glycine; lysine sometimes present.
VIII. Ornithine.

A further wall type (IX), characterized by *meso*-DAP and numerous amino acids, was defined for species of MYCOPLANA.

In most species which form non-fragmenting mycelium (e.g. *Streptomyces* spp) the vegetative hyphae are largely aseptate, although septa (cross-walls) can be present – particularly in the older parts of the mycelium. The septa in non-fragmenting mycelium have been designated type 1 septa; each septum consists of a single layer which develops centripetally from the cell wall. Such septa may contain microplasmodesmata, each 4–10 nm in diameter.

In fragmenting mycelium each septum consists of two distinct layers, each layer eventually forming a terminal wall of one of the two neighbouring cells; such septa are designated type 2 septa.

Spore formation. Spores are formed by the septation and fragmentation of hyphae, the spore wall being formed, at least in part, from all the wall layers of the sporogenous hypha. Spore-delimiting septa are of various types, and different types may occur even within a given genus; such septa have been designated type I (two layers developing centripetally), type II (two layers which develop centripetally on a single, initially-formed annulus), and type III (a single, thick layer which develops centripetally). Spore chains are reported to develop acropetally (in e.g. *Pseudonocardia*), basipetally (in e.g. *Micropolyspora*), randomly (in e.g. *Nocardiopsis*), or more or less simultaneously (in e.g. *Streptomyces*).

In some actinomycetes the spores are formed within sporangia: see e.g. ACTINOPLANES, AMORPHOSPORANGIUM, AMPULLARIELLA, DACTYLOSPORANGIUM, FRANKIA and PILIMELIA.

Genetic aspects. Genetic exchange has been studied in various actinomycetes, particularly *Streptomyces* spp [*Streptomyces* genetics: Book ref. 73, pp. 229–286; genetics of nocardioform actinomycetes: Book ref. 73, pp. 201–228]. Actinomycetes are hosts to a number of ACTINOPHAGES, and generalized transduction with phage ϕSV1 has been recorded in strains of *Streptomyces* [JGM (1979) *110* 479–482]. Actinomycetes can contain various transmissible or non-transmissible plasmids, some of which are involved in antibiotic production. Genetic analyses have been carried out by methods involving e.g. conjugation and protoplast fusion.

Genera include: ACTINOMADURA, ACTINOMYCES, ACTINOPLANES, ACTINOPOLYSPORA, ACTINOSYNNEMA, AGROMYCES, AMORPHOSPORANGIUM, AMPULLARIELLA, ARACHNIA, ARCANOBACTERIUM, ARTHROBACTER, BREVIBACTERIUM, CASEOBACTER, CELLULOMONAS, CORYNEBACTERIUM, CURTOBACTERIUM, DACTYLOSPORANGIUM, DERMATOPHILUS, EXCELLOSPORA, FRANKIA, GEODERMATOPHILUS, INTRASPORANGIUM, KINEOSPORIA, MICROBACTERIUM, MICROBISPORA, MICROMONOSPORA, MICROPOLYSPORA, MICROTETRASPORA, MYCOBACTERIUM, NOCARDIA, NOCARDIOIDES, NOCARDIOPSIS, OERSKOVIA, PILIMELIA, PLANOBISPORA, PLANOMONOSPORA, PROMICROMONOSPORA, PSEUDONOCARDIA, RENIBACTERIUM, RHODOCOCCUS, ROTHIA, SACCHAROMONOSPORA, SACCHAROPOLYSPORA, SPIRILLOSPORA, SPORICHTHYA, STREPTOALLOTEICHUS, STREPTOMYCES, STREPTOSPORANGIUM, STREPTOVERTICILLIUM, THERMOMONOSPORA.

[Ecology, isolation, cultivation etc: Book ref. 46, pp. 1915–2123.]

actinomycete Any member of the order ACTINOMYCETALES; the name is often used to refer specifically to those species which form mycelium, i.e. excluding many members of the NOCARDIOFORM ACTINOMYCETES.

actinomycetoma See MADUROMYCOSIS.

actinomycin D (actinomycin C_1) An ANTIBIOTIC from *Streptomyces* sp; it contains a (red) substituted phenoxazone chromophore linked to two identical pentapeptide lactone rings. All cell types are potentially susceptible, any resistance being due to low permeability of cells to the drug. Actinomycin D specifically inhibits DNA-directed RNA synthesis. It binds specifically to B-DNA as an INTERCALATING AGENT (ϕ ca. 26°). The phenoxazone chromophore intercalates primarily between two adjacent (antiparallel) GC pairs, while the lactone rings fit into the minor groove [ARB (1981) *50* 171–172]. The drug dissociates from DNA only very slowly; it blocks the movement of RNA polymerase along its DNA template. (Since actinomycin D shows little binding to AT-rich PROMOTERS chain *initiation* is not inhibited.) DNA replication may be insensitive to actinomycin D because strand separation by the replicative apparatus may facilitate dissociation of the antibiotic.

actinomycosis (1) Any human or animal disease caused by a species of ACTINOMYCES: *A. israelii* in man, *A. bovis* in cattle. Infection is probably endogenous. Dense nodular lesions are formed, mainly around the jaw ('lumpy jaw'), developing into pus-discharging abscesses. Abscesses may also occur in the lungs, brain or intestine. *Lab. diagnosis*: the pathogen may be isolated from small yellow granules ('sulphur granules') present in the pus. *Chemotherapy*: e.g. penicillins.

(2) Any human or animal disease caused by an ACTINOMYCETE: e.g. actinomycosis (sense 1); MADUROMYCOSIS. (See also LACHRYMAL CANALICULITIS.)

actinophage Any BACTERIOPHAGE whose host(s) are member(s) of the ACTINOMYCETALES. Actinophages, which include both temperate and virulent types, can be isolated from e.g. soils and composts; most have a wide host range, but some (e.g. BACTERIOPHAGE ϕEC, BACTERIOPHAGE VP5) can infect only one or a few species. (See also STYLOVIRIDAE). [Soil actinophages which lyse *Streptomyces* spp: JGM (1984) *130* 2639–2649.]

Actinophryida An order of protozoa (class HELIOZOEA) in which the cells have no skeleton and no centroplast (cf. CENTROHELIDA). Some members have flagellated stages. Sexual processes have been observed in some species. Genera include e.g. ACTINOPHRYS, ACTINOSPHAERIUM, *Ciliophrys*.

Actinophrys A genus of heliozoa (order ACTINOPHRYIDA). *A. sol* is common among vegetation in freshwater ponds and lakes. The cell is ca. 40–50 μm diam., with a highly vacuolated cytoplasm; the distinction between ectoplasm and endoplasm is not clear in living cells (cf. ACTINOSPHAERIUM). The axial filaments of the axopodia (see AXOPODIUM) originate close to the single central nucleus. Reproduction occurs asexually by binary fission. Autogamy occurs when environmental conditions are unfavourable: meiosis follows encystment of the uninucleate cell, 2–4 gametes being formed; fusion of gametes results in the formation of zygotes which can remain dormant in the cyst until conditions improve.

Actinoplanes A genus of aerobic, sporogenous bacteria (order ACTINOMYCETALES, wall type II) which occur e.g. in soil, plant litter and aquatic habitats. The organisms form a branching surface mycelium, hyphal diameter ca. 0.2–1.5 μm, which may also ramify into the substratum; the mycelium later forms vertical hyphae, each developing, at its tip, a (commonly spherical) desiccation-resistant sporangium containing a number of spherical or oval spores – each bearing a polar tuft of flagella. Colonies may be e.g. yellow, orange, red, blue, brown or purple. Type species: *A. philippinensis*. [Morphology, ecology, isolation: Book ref. 46, 2004–2010; isolation: JAB (1982) *52* 209–218.]

Actinopoda A superclass of protozoa (subphylum SARCODINA) which are typically more or less spherical, typically have axopodia (filopodia in some members), and are usually planktonic. Classes: ACANTHAREA, HELIOZOEA, Phaeodarea and Polycystinea (see RADIOLARIA).

Actinopolyspora A genus of bacteria (order ACTINOMYCETALES, wall type IV); the sole species, *A. halophila*, was isolated from a salt-rich bacteriological medium. The organisms form substrate and aerial mycelium, the latter giving rise to chains of spores; at least 10% (w/v) sodium chloride is required for growth, the optimum being ca. 15–20%, and the maximum ca. 30%. GC%: ca. 64. Type species: *A. halophila*. [Book ref. 73, 122–123.]

Actinopycnidium See STREPTOMYCES.

actinorrhiza A bacterium–plant root association in which nitrogen-fixing root nodules are formed in certain non-leguminous angiosperms infected (through root hairs) by FRANKIA strains; the plants involved are typically woody pioneers of nutrient-poor soils in cold or temperate regions in the northern hemisphere. There are at least two morphological types of actinorrhizal root nodule. In the *Alnus* type, formed in *Alnus* spp (alder) and many other plants, the root nodules are coralloid (i.e., thickened and dichotomously branched). In the *Myrica* type, formed e.g. in species of *Myrica*, *Casuarina* and *Rubus*, the nodule is clothed with upward-growing (negatively geotropic) rootlets which may aid aeration in boggy habitats. In either type, the endophyte occurs within the cortical parenchyma of the nodule and does not invade vascular or meristematic tissues. In the distal part of the nodule the (young) hyphae spread from cell to cell, perforating the host cell walls. In the proximal part the hyphal tips swell to form vesicles which appear to provide a reducing environment within which NITROGEN FIXATION can occur; rates of nitrogen fixation are comparable to those in leguminous ROOT NODULES. [Book ref. 55, pp. 205–223.] (See also MYCORRHIZA.)

Actinosphaerium A genus of heliozoa (order ACTINOPHRYIDA) in which the cells are multinucleate and ca. 200 μm to 1.0 mm in diameter, according to species; the highly vacuolated ectoplasm is clearly distinct from the granular endoplasm (cf. ACTINOPHRYS). Numerous needle-like axopodia radiate from the cell, their axial filaments arising at the junction between ectoplasm and endoplasm. Asexual reproduction involves plasmotomy. Autogamy occurs when environmental conditions become unfavourable: the cell produces a gelatinous covering, and many of its nuclei degenerate; numerous uninucleate daughter cells are produced, and each encysts. Meiosis within the cyst results in two haploid gametes which fuse, and the resulting zygote remains dormant until conditions improve.

Actinosporangium See STREPTOMYCES.

Actinosporea A class of protozoa (phylum MYXOZOA) which are parasitic in invertebrates (particularly annelid worms). The spores contain 3 polar capsules (each enclosing a single polar filament) and several to many sporoplasms. The spore wall consists of 3 valves which may be smooth (e.g. in *Sphaeractinomyxon*) or drawn out into long, horn-like processes (as in *Triactinomyxon*, a parasite of tubificid and sipunculid worms). (cf. WHIRLING DISEASE.)

Actinosynnema A genus of bacteria (order ACTINOMYCETALES, wall type III) which occur e.g. on vegetable matter in aquatic habitats. The organisms form a thin, branching, yellow substrate mycelium (hyphae <1 μm diam.) on which develop synnemata (up to ca. 180 μm in height) or 'dome-like bodies'; aerial hyphae may arise from the substrate mycelium, or from the tips of the synnemata, and give rise to chains of spores which become motile (flagellated) in liquid media. GC%: ca. 71. Type species: *A. mirum*. [Book ref. 73, pp 116–117.]

activated acetic acid pathway See AUTOTROPH.

activated sludge See SEWAGE TREATMENT.

activation (*immunol.*) (1) (of lymphocytes) A process which begins with BLAST TRANSFORMATION and continues with proliferation (cell division) and differentiation; some authors use 'activation' to refer specifically to stage(s) preceding proliferation. (2) (of complement) See COMPLEMENT FIXATION. (3) See MACROPHAGE. (4) (of spores) See ENDOSPORE and SPORE.

activation-induced cytidine deaminase See RNA EDITING.

activator (1) *Syn.* COFACTOR (sense 2). (2) See SPORE. (3) See OPERON and REGULON.

active bud See LIPOMYCES.

active immunity Specific IMMUNITY (3) afforded by the body's own immunological defence mechanisms following exposure to antigen. (cf. PASSIVE IMMUNITY.)

active immunization See IMMUNIZATION.

active transport See TRANSPORT SYSTEMS.

actomyosin An actin–myosin complex (see ACTIN and MYOSIN).

aculeacin A An antibiotic which is active against yeasts, inhibiting the formation of yeast CELL WALL glucan. *Echinocandin B* is a structurally related antibiotic with an apparently similar mode of action. (cf. PAPULACANDIN B.)

aculeate Slender and sharp-pointed, or bearing narrow spines.

acute (*med.*) Refers to any disease which has a rapid onset and which persists for a relatively short period of time (e.g. days) – terminating in recovery or death. The term is also used to refer to an exceptionally severe or painful condition. (cf. CHRONIC.)

acute cardiac beriberi See CITREOVIRIDIN.

acute haemorrhagic conjunctivitis (AHC) A highly infectious form of CONJUNCTIVITIS, a worldwide pandemic of which occurred during 1969–1971; it is caused by enterovirus 70 (see ENTEROVIRUS) and is characterized by subconjunctival haemorrhages ranging from petechiae to larger areas covering the bulbar conjunctivae. Recovery is usually complete in ca. 10 days.

acute herpetic gingivostomatitis See GINGIVITIS.

acute necrotizing ulcerative gingivitis See GINGIVITIS.

acute-phase proteins Various types of protein, found in plasma, formed as a rapid response to infection; they are synthesized in the liver e.g. under stimulation from cytokines produced in a region of INFLAMMATION. These proteins include C-REACTIVE PROTEIN (CRP) and *serum amyloid A* (SAA), both of which can bind to phospholipids in the microbial cell envelope and act as OPSONINS; additionally, binding by CRP activates COMPLEMENT. CRP and SAA are so-called *pentraxin* proteins in which the molecule consists of five identical subunits.

(See also CD14.)

acute-phase serum Serum obtained from a patient during the acute phase of a disease.

acute respiratory disease See ARD.

ACV ACYCLOVIR.

ACVs See VACCINE.

acycloguanosine *Syn.* ACYCLOVIR.

acyclovir (ACV; acycloguanosine; Zovirax) An ANTIVIRAL AGENT, 9-(2-hydroxyethoxymethyl)guanine, which is active against alphaherpesviruses. It is phosphorylated by the virus-encoded thymidine kinase to the monophosphate; the monophosphate is converted by host-cell enzymes to the active triphosphate form which inhibits DNA polymerase – the viral polymerase being much more sensitive than the cellular α-polymerase. (cf. BROMOVINYLDEOXYURIDINE.) Uninfected cells do not effectively phosphorylate acyclovir, and the drug is relatively non-toxic to the host.

Acyclovir is used topically, systemically or orally in the treatment of e.g. herpes simplex keratitis, primary genital herpes, mucocutaneous herpes simplex in immunocompromised patients, progressive varicella and HERPES ZOSTER. Acyclovir is not equally active against all alphaherpesviruses – its ability to inhibit the replication of varicella-zoster virus is approximately 10-fold lower than its ability to inhibit replication of herpes simplex virus. [Use of acyclovir in the treatment of herpes zoster: RMM (1995) *6* 165–174 (167–170).]

acylalanine antifungal agents See PHENYLAMIDE ANTIFUNGAL AGENTS.

***N*-acyl-L-homoserine lactone** See QUORUM SENSING.

Acytostelium A genus of cellular slime moulds (class DICTYOSTELIOMYCETES) in which the sorocarp stalk is acellular, cellulosic, slender, and apparently tubular; no myxamoebae are sacrificed in stalk formation (cf. DICTYOSTELIUM). The stalk bears a single terminal sorus of spores. Four species are recognized [descriptions and key: Book ref. 144, pp. 393–407].

Ad Human adenovirus: see MASTADENOVIRUS.

A–D group ALKALESCENS–DISPAR GROUP.

ADA deficiency See ADENOSINE DEAMINASE DEFICIENCY.

***ada* gene** See ADAPTIVE RESPONSE.

adamantanamine See AMANTADINE.

adamantane See AMANTADINE.

Adansonian taxonomy A method of biological classification, proposed in the 18th century by Michel Adanson, in which relationships between organisms are defined by the number of characteristics which the organisms have in common; the same degree of importance ('weighting') is attached to each characteristic. (cf. NUMERICAL TAXONOMY.)

adaptation Change(s) in an organism, or population of organisms, by means of which the organism(s) become more suited to prevailing environmental conditions. *Genetic* adaptation involves e.g. mutation and selection: those (mutant) organisms in a given population which are genetically more suited to the existing environment thrive and become numerically dominant. (See also FLUCTUATION TEST.) *Non-genetic* (*phenotypic*) adaptation may involve a change in metabolic activity – e.g., by enzyme induction or repression (see OPERON). (See also CHROMATIC ADAPTATION.) *Behavioural* adaptation may involve changes in tactic responses (see TAXIS); thus, e.g., if a phototactic organism is subjected to a sudden increase in light intensity followed by steady illumination at the new intensity, the organism initially responds phototactically, but after a period of time it adapts to the new light intensity and resumes its normal pattern of motility. (See also CHEMOTAXIS.)

adaptive response A DNA REPAIR system which is induced in cells of *Escherichia coli* in response to exposure to low concentrations of certain ALKYLATING AGENTS (e.g. MNNG, MNU); the response is independent of the SOS system.

Genes involved in the adaptive response include *ada*, *aidB*, *alkA* and *alkB*.

The *alkA* gene encodes 3-methyladenine-DNA glycosylase II, an enzyme which (despite its name) cleaves various methylated bases (e.g. N^3- or N^7-methylpurines and O^2-methylpyrimidines) from alkylated DNA; studies on the crystal structure of AlkA complexed with DNA indicate that the enzyme distorts DNA considerably as it 'flips out' (i.e. exposes) the target site [EMBO (2000) *19* 758–766]. (Another enzyme of *E. coli*, DNA glycosylase I (the Tag protein; *tag* gene product), is synthesized *constitutively*; the Tag protein cleaves 3-methyladenine from DNA.)

DNA is susceptible to chemical change owing to the reactivity of its bases [Nature (1993) *362* 709–715]. As well as methylation in vitro, DNA is also subject to aberrant, non-enzymatic methylation in vivo via *S*-adenosylmethionine (normally a legitimate methyl donor) which can give rise to 3-methyladenine and/or 7-methylguanine; these aberrant methylated bases are also cleaved by the glycosylases mentioned above.

Cleavage of a chemically aberrant base, to form an AP SITE, is the first stage in the repair process; repair continues via BASE EXCISION REPAIR.

A different aspect of the adaptive response involves the *ada* gene product: a bifunctional methyltransferase which directly reverses the effects of the methylating agent. One function of Ada transfers a methyl group from a major-groove adduct – O^6-methylguanine (a potentially highly mutagenic lesion) or O^4-methylthymine – to a cysteine residue near the C-terminus of the Ada protein itself. The second function of the Ada protein transfers a methyl group from a methyl phosphotriester (formed by methylation of a phosphodiester bond in the DNA) to a cysteine residue in the N-terminal portion of the protein. Once methylated at both sites, the Ada protein is inactivated.

The Ada protein acts as a positive regulator of its own expression as well as that of *aidB*, *alkA* and *alkB*; alkylation of the Ada protein from a methyl phosphotriester is apparently the intracellular signal for induction of the adaptive response [Cell (1986) *45* 315–324].

adaptor (*mol. biol.*) A synthetic oligodeoxyribonucleotide which is similar to a *linker* (see LINKER DNA) but which contains more than one type of restriction site and may also have pre-existing STICKY ENDS.

ADCC Antibody-dependent CELL-MEDIATED CYTOTOXICITY: the killing of antibody-coated target cells by a non-phagocytic mechanism in which the effector cell (see e.g. NK CELLS) initially binds to the Fc portion of the (bound) antibodies via specific receptors. (See also CD16.)

addition mutation *Syn.* INSERTION MUTATION.

Adelea See ADELEORINA.

Adeleina A suborder of protozoa (order Eucoccida [JP (1964) *11* 7–20] or Eucoccidiida [JP (1980) *27* 37–58]) equivalent to the ADELEORINA.

Adeleorina A suborder of protozoa (order EUCOCCIDIORIDA) in which syzygy characteristically occurs (cf. EIMERIORINA, HAEMOSPORORINA). Genera include *Adelea*, *Haemogregarina*, *Klossiella*.

adenine arabinoside *Syn.* VIDARABINE.

adenitis Inflammation of gland(s).

adeno-associated viruses See DEPENDOVIRUS.

adeno-satellite viruses See DEPENDOVIRUS.

adenosine See NUCLEOSIDE and Appendix V(a).

adenosine 3′,5′-cyclic monophosphate See CYCLIC AMP.

adenosine deaminase deficiency A congenital lack of the enzyme adenosine deaminase (EC 3.5.4.4), the effects of which include a marked reduction in the numbers of functional B and T lymphocytes. (See SEVERE COMBINED IMMUNODEFICIENCY.)

The disease has been treated by GENE THERAPY.

adenosine triphosphatase See ATPASE.

adenosine 5′-triphosphate See ATP.

Adenoviridae (adenovirus family) A family of non-enveloped, icosahedral, linear dsDNA-containing viruses which infect mammals (genus *Mastadenovirus*) or birds (genus *Aviadenovirus*).

Adenoviruses are generally specific for one or a few closely related host species; infection may be asymptomatic or may result in various diseases (see AVIADENOVIRUS and MASTADENOVIRUS). Many adenoviruses can induce tumours when injected into newborn rodents, but none is known to cause tumours in natural circumstances. In cell cultures, adenoviruses cause characteristic CPE, including the rounding of cells and the formation of grape-like clusters of cells; adenovirus replication and assembly occur in the nucleus, resulting in the formation of intranuclear inclusion bodies. Virions sometimes form paracrystalline arrays. Many adenoviruses can haemagglutinate RBCs from various species.

The adenovirus virion consists of an icosahedral CAPSID (ca. 70–90 nm diam.) enclosing a core in which the DNA genome is closely associated with a basic (arginine-rich) viral polypeptide (VP), VII. The capsid is composed of 252 capsomers: 240 hexons (capsomers each surrounded by 6 other capsomers) and 12 pentons (one at each vertex, each surrounded by 5 'peripentonal' hexons). Each penton consists of a penton base (composed of viral polypeptide III) associated – apparently by hydrophobic interactions – with one (in mammalian adenoviruses) or two (in most avian adenoviruses) glycoprotein fibres (viral polypeptide IV); each fibre carries a terminal 'knob' ca. 4 nm in diam. The fibres can act as haemagglutinins and are the sites of attachment of the virion to a host cell-surface receptor. The hexons each consist of three molecules of viral polypeptide II; they make up the bulk of the icosahedron. Various other minor viral polypeptides occur in the virion.

The adenovirus dsDNA genome (MWt ca. 20–25 × 10^6 for mammalian strains, ca. 30 × 10^6 for avian strains) is covalently linked at the 5′ end of each strand to a hydrophobic 'terminal protein', TP (MWt ca. 55000); the DNA has an inverted terminal repeat (ITR) of different length in different adenoviruses. In most adenoviruses examined, the 5′-terminal residue is dCMP (dGMP in CELO virus).

Adenovirus virions are stable and are not inactivated by e.g. lipid solvents or by pancreatic proteases, low pH, or bile salts.

Replication cycle. The virion attaches via its fibres to a specific cell-surface receptor, and enters the cell by endocytosis or by direct penetration of the plasma membrane. Most of the capsid proteins are removed in the cytoplasm; the virion core enters the nucleus, where the uncoating is completed to release viral DNA almost free of virion polypeptides. Virus gene expression then begins. The viral dsDNA contains genetic information on both strands. [By convention, the AT-rich half of the DNA molecule is designated the right-hand end, and the strand transcribed from left to right is called the r-strand, the leftward-transcribed strand being called the l-strand; the DNA is divided into 100 'map units' (m.u.): Book ref. 116, pp. 40–42.] *Early genes* (regions E1a, E1b, E2a, E3, E4) are expressed before the onset of viral DNA replication. *Late genes* (regions L1, L2, L3, L4 and L5) are expressed only after the initiation of DNA synthesis. *Intermediate genes* (regions E2b and IVa_2) are expressed in the presence or absence of DNA synthesis. Region E1a encodes proteins involved in the regulation of expression of other early genes, and is also involved in transformation (see MASTADENOVIRUS). The RNA transcripts are capped (with $m^7G^5ppp^{5'}N$) and polyadenylated (see MRNA) in the nucleus before being transferred to the cytoplasm for translation.

Viral DNA replication requires the terminal protein, TP, as well as virus-encoded DNA polymerase and other viral and host proteins. TP is synthesized as an 80K (MWt 80000) precursor, pTP, which binds covalently to nascent replicating DNA strands. pTP is cleaved to the mature 55K TP late in virion assembly; possibly at this stage, pTP reacts with a dCTP molecule and becomes covalently bound to a dCMP residue, the 3′-OH of which is believed to act as a primer for the initiation of DNA synthesis (cf. BACTERIOPHAGE ϕ29). Late gene expression, resulting in the synthesis of viral structural proteins, is accompanied by the cessation of cellular protein synthesis, and virus assembly may result in the production of up to 10^5 virions per cell.

[Book ref. 116.]

adenoviruses Viruses of the ADENOVIRIDAE.

adenylate cyclase An enzyme (EC 4.6.1.1) which catalyses the synthesis of CYCLIC AMP (cAMP) from ATP. (cAMP is degraded to AMP by the enzyme cAMP phosphodiesterase.)

In *Escherichia coli* and other enterobacteria, adenylate cyclase is a single protein (product of the *cya* gene) whose activity is modified e.g. by the PTS system (see CATABOLITE REPRESSION); it may also be regulated directly by pmf, and its synthesis may be repressed by cAMP–CRP.

The EF component of ANTHRAX TOXIN and the CYCLOLYSIN of *Bordetella pertussis* both have CALMODULIN-stimulated adenylate cyclase activity; the pore-forming activity of the cyclolysin may serve to internalize its adenylate cyclase activity, resulting e.g. in raised levels of cAMP in cells of the immune system.

In higher eukaryotes, adenylate cyclase occurs as part of a plasma membrane complex which includes e.g. hormone receptors and GTP-binding regulatory components (so-called *G proteins*; = *N proteins*); G_s (= N_s) stimulates adenylate cyclase, G_i(= N_i) inhibits it. This system is the target for certain bacterial toxins: see e.g. CHOLERA TOXIN and PERTUSSIS TOXIN.

adenylate energy charge (energy charge, EC) A unitless parameter which gives a measure of the total energy associated with the adenylate system, at a given time, within a cell. It is defined as:

$$EC = \frac{[ATP] + 1/2\ [ADP]}{[ATP] + [ADP] + [AMP]}$$

Cells growing under ideal conditions have an energy charge of ca. 0.8–0.95, while senescent cells may have an EC of ca. 0.6 or less. Changes in the relative proportions of adenine nucleotides in a cell (i.e., changes in EC) have regulatory effects on various metabolic processes – the activity of certain enzymes being regulated by the actual concentration of a given adenine nucleotide or by the ratio of particular nucleotides (e.g. ATP:ADP). Thus, e.g., in certain yeasts the enzyme phosphofructokinase is inhibited by ATP (an effect which is reversed by AMP), while in *Escherichia coli* the same enzyme is stimulated by ADP; hence, glycolysis tends to be stimulated when the EC is depressed.

Energy charge is also a regulatory factor for the degree of supercoiling (superhelical density) in a cell's DNA; changes in superhelical density can, in turn, influence the activity of various gene promoters. Thus, via energy charge and superhelicity, the environment can modulate the expression of particular genes. (Interestingly, the expression of genes or operons may also be regulated by *local* changes in superhelicity due to divergent transcription from closely spaced gene promoters [Mol. Microbiol. (2001) *39* 1109–1115].)

Estimations of EC involve both rapid sampling and precautions to prevent hydrolysis or interconversion of adenine nucleotides. ATP is often measured by techniques which involve CHEMILUMINESCENCE.

adenylate kinase An enzyme (EC 2.7.4.3) which catalyses the reversible conversion of two molecules of adenosine 5′-diphosphate (ADP) to one molecule each of ATP and AMP. [Structural and catalytic properties of adenylate kinase from *Escherichia coli*: JBC (1987) *262* 622–629.]

adenylylsulphate See APS.

ADH ARGININE DIHYDROLASE.

adherent cells (*immunol.*) Cells which adhere to e.g. glass and plastics; they include e.g. MACROPHAGES and DENDRITIC CELLS.

adhesin A cell-surface component, or appendage, which mediates ADHESION to other cells or to inanimate surfaces or interfaces; there are many different types of adhesin, and a given organism may have more than one type. (The term *adhesin* is also used to include certain secreted substances which behave as adhesins – e.g. MUTAN.)

Bacterial adhesins. Many bacterial adhesins are FIMBRIAE, and in some pathogenic species fimbrial adhesins are important virulence factors which mediate the initial stage of pathogenesis (adhesion to specific site(s) in the host organism). For example, fimbria-mediated adhesion is important for virulence in ETEC – strains of *Escherichia coli* whose capacity to cause disease depends on their ability to bind to the intestinal mucosa; among strains of ETEC there are more than 10 different types of fimbrial adhesin (as well as non-fimbrial adhesins) [RMM (1996) *7* 165–177]. [Expression of fimbriae by enteric pathogens (review): TIM (1998) *6* 282–287.] Fimbrial adhesins are also important in *Haemophilus influenzae* type b for the initial binding to respiratory tract epithelium [adhesins in *Haemophilus*, *Actinobacillus* and *Pasteurella*: FEMS Reviews (1998) *22* 45–59]. (See also UPEC and P FIMBRIAE.)

Some fimbrial adhesins additionally function as an INVASIN (see e.g. UPEC) or as a phage receptor (see e.g. BACTERIOPHAGE CTXΦ).

The dimensions and charge characteristics of fimbriae are such that they experience minimal repulsion from a surface bearing charge of the same polarity; thus, fimbrial adhesins can help to bridge the gap between the charged bacterial surface (see ZETA POTENTIAL) and the surface of another cell or substratum which bears a charge similar (in polarity) to that on the bacterium.

Non-fimbrial adhesins include the filamentous haemagglutinin (FHA) of *Bordetella pertussis*, the high-molecular-weight adhesion proteins (HMW1, HMW2) of 'non-typable' strains of

Haemophilus influenzae, and the M PROTEIN of streptococci. (See also DR ADHESINS and OPA PROTEINS.)

Proteinaceous 'capsular' adhesins include K88 in certain strains of *Escherichia coli*.

Non-proteinaceous adhesins include the capsular carbohydrates of *Rhizobium trifolii* which bind to TRIFOLIIN A on the root hairs of the clover plant. (See also streptococcal lipoTEICHOIC ACIDS.)

Receptors for bacterial adhesins. Binding sites for bacterial adhesins on mammalian tissues include various cell-surface molecules, but a given adhesin typically binds only to a specific site. For example, the type I FIMBRIAE of *E. coli* bind to mannose residues, while the P fimbriae of uropathogenic *E. coli* bind to α-D-galactopyranosyl-(1–4)-β-D-galactopyranoside receptors of glycolipids on urinary tract epithelium. Some pathogens (e.g. *Bordetella pertussis*, *Borrelia burgdorferi*, *Yersinia enterocolitica*) bind to specific INTEGRINS. Protein adhesins of *Staphylococcus aureus* bind to components such as collagen and fibronectin in the mammalian extracellular matrix [TIM (1998) *6* 484–488]. Certain LECTINS (q.v.) may promote disease by enhancing attachment of ETEC to the porcine ileum.

adhesion Microorganisms often bind specifically or non-specifically to a substratum or to other cells – adhesion being mediated by specialized microbial components or structures: see e.g. ADHESIN and PROSTHECA.

For some *pathogens*, adhesion to specific host cells or tissues is a prerequisite for disease, so that, in these organisms, adhesins are important virulence factors (see ADHESIN).

The adherence of a pathogen to a host cell may be necessary simply promote the host's uptake of a secreted toxin. However, in some cases, adhesion triggers a more interactive sequence of events. For example, when the FHA adhesin of *Bordetella pertussis* binds to an INTEGRIN receptor on a monocyte, it generates signals within the monocyte that upregulate the expression of a second type of integrin – one which binds to a different site on the adhesin. Thus, this pathogen 'manipulates' the host cell's internal signalling system in order to secure for itself additional binding sites. In another example, the binding of UPEC (q.v.) to uroepithelium via type I FIMBRIAE promotes uptake (internalization) of the pathogen. In enteropathogenic *E. coli*, initial adhesion to gut epithelium is followed by complex prokaryote–eukaryote interactions that result in a unique form of colonization by the pathogen (see EPEC).

The adhesion of bacteria to inanimate surfaces can be problematical in the context of prosthetic devices. Infection associated with these devices is a serious complication which is often difficult to treat (not least because adherent bacteria are typically less susceptible to antibiotics) [prosthetic device infections: RMM (1998) *9* 163–170].

Free-living microorganisms in aquatic habitats often adhere to submerged surfaces on stones, particles of debris, other organisms or man-made structures – sometimes forming BIOFILMS. (See also EPILITHON.) Adhesion may affect the activity of such organisms because the conditions at a submerged surface differ from those in the bulk aqueous phase; for example, surfaces can adsorb nutrients and/or stimulatory or inhibitory ions, so that solid–liquid interfaces may be significantly more advantageous or disadvantageous compared with the liquid phase. Cell–cell interactions may or may not be facilitated in biofilms.

The interaction between microorganisms and a surface is governed by various physicochemical forces that may include electrostatic attraction or repulsion, hydrophobic interaction (i.e. mutual attraction between hydrophobic molecules), hydrogen bonding, and van der Waals' forces. Because some types of cell resemble colloids in their dimensions and electrical characteristics (see ZETA POTENTIAL), the interaction between a cell and a substratum (or between two cells) in an aqueous medium is sometimes considered in the context of classical colloid theory – in particular the theory of Derjaguin, Landau, Verwey and Overbeek (the 'DLVO' theory). The DLVO theory supposes that a particle which bears a distributed surface charge of given polarity is surrounded by a layer of ions of opposite charge – forming a so-called 'double layer' extending some distance from the surface of the particle; two similarly charged particles will therefore be mutually repulsive through the interaction of their double layers. An increase in the ionic concentration in the medium effectively compresses each double layer – so that the particles can then approach each other more closely; in a number of cases, cell–substratum or cell–cell contact has been shown to be facilitated by raising the concentration of electrolyte. However, a rigid application of the DLVO theory (or any other mathematically based theory) to biological systems is made difficult by a number of factors which include the susceptibility of the cell to physical deformation, the chemical complexity and non-uniformity of the cell surface, and the possibility of ionic flux across juxtaposed surfaces.

Microorganisms which are normally attached to a substratum may be dispersed by means of either non-adherent progeny or propagules – or they may be able to detach, temporarily, in order to colonize fresh surfaces. The hydrophobic, benthic cyanobacterium *Phormidium* J-1 appears to achieve dispersal by forming an emulsifying agent (EMULCYAN) which masks cell-surface hydrophobicity – permitting detachment; the emulcyan is presumed to be washed off the cells at some stage so that attachment is again possible [FEMS Ecol. (1985) *31* 3–9].

adhesion site (Bayer's junction; Bayer's patch) In Gram-negative bacteria: a localized 'fusion' between the OUTER MEMBRANE and CYTOPLASMIC MEMBRANE [Book ref. 101, pp. 167–202]. In electronmicrographs, a plasmolysed cell may show adhesion sites under some experimental conditions but not under others [JB (1984) *160* 143–152]. [Cell envelope fraction with apparent adhesion sites: JBC (1986) *261* 428–443.]

Adhesion sites appear to be osmotically sensitive, physiologically important regions of the cell envelope which serve e.g. as sites for the export of proteins (such as porins), LPS molecules [JB (1982) *149* 758–767] and filamentous phages, as sites of infection for certain phages, and as the anchorage sites of e.g. F pili. Certain proteins (e.g. penicillin binding protein 3, THIOREDOXIN [JB (1987) *169* 2659–2666]) have been associated with adhesion sites.

[Mol. Microbiol. (1994) *14* 597–607.]

(See also PERISEPTAL ANNULUS, and 'lysis protein' in LEVIVIRIDAE.)

adiaspiromycosis (adiaspirosis; haplomycosis) A non-infectious MYCOSIS which primarily affects animals, rarely affecting man. It is caused by *Chrysosporium parvum* var. *parvum* (formerly *Emmonsia parva* or *Haplosporangium parvum*) or by *C. parvum* var. *crescens* (formerly *E. crescens*). Infection occurs by inhalation of conidia (formed e.g. in soil); the conidia enlarge within the lungs to reach diameters of ca. 40 μm (var. *parvum*) or 400 μm (var. *crescens*). Granulomas may develop around the adiaspores.

adiaspirosis *Syn.* ADIASPIROMYCOSIS.

adiaspore A spore (conidium) which grows in size without dividing – see e.g. ADIASPIROMYCOSIS and CHRYSOSPORIUM.

adjuvant (1) (*immunol.*) Any substance which, when administered with or before an antigen, heightens and/or affects qualitatively the immune response in terms of antibody formation and/or the cell-mediated response. (The adjuvant L18-MDP(A) has also been reported to enhance non-specific phagocytosis by polymorphonuclear leucocytes [JGM (1982) *128* 2361–2370].) Adjuvants include e.g. BCG, aluminium hydroxide, and water-in-oil emulsions (e.g. FREUND'S ADJUVANT). (2) Any substance which is added to a drug or other chemical (e.g. a disinfectant) to enhance its activity.

Adjuvant 65 A water-in-oil emulsion ADJUVANT made by emulsifying peanut oil with mannide monooleate and stabilizing with aluminium monostearate.

***adk* gene** (*dnaW* gene) In *Escherichia coli*: the gene for adenylate kinase.

Adler test A test used for the identification of *Leishmania* spp. The organisms are cultured in an immune serum; in the presence of homologous antibodies, promastigotes develop in clusters or syncytia.

adnate (1) Of the region of lamella–stipe attachment in an agaric: extending for a length equal to much or most of the depth of the lamella.

(2) Of flagellar spines: flattened against the flagellum from which they arise.

adnexed (*mycol.*) Of the region of lamella–stipe attachment in an agaric: extending for a length equal to only a small fraction of the depth of the lamella.

adonitol *Syn.* RIBITOL.

adoptive immunity *Syn.* PASSIVE IMMUNITY.

adoral ciliary spiral See AZM.

adoral zone of membranelles See AZM.

ADP Adenosine 5′-diphosphate. (See also ATP and ADENYLATE KINASE.)

ADP-ribosylation The transfer of an ADP-ribosyl group from NAD^+ to a protein, catalysed by an ADP-ribosyl transferase. In eukaryotic cells, various proteins may be ADP-ribosylated – apparently as a normal regulatory mechanism; certain bacterial toxins act by exerting an ADP-ribosyl transferase function: see e.g. BOTULINUM C2 TOXIN, CHOLERA TOXIN, DIPHTHERIA TOXIN, EXOTOXIN A and PERTUSSIS TOXIN. ADP-ribosylation also occurs e.g. in cells of *Escherichia coli* infected with BACTERIOPHAGE T4, the host RNA polymerase undergoing phage-induced ADP-ribosylation. [Review: TIBS (1986) *11* 171–175.]

adrenalin (epinephrine) A multifunctional hormone, secreted by the adrenal gland, which affects e.g. carbohydrate metabolism and the activity of smooth muscle, particularly that of the cardiovascular and bronchial systems. Adrenalin is used e.g. for the treatment of ANAPHYLACTIC SHOCK – in which it counteracts the effects of histamine, relaxing smooth muscle and reducing vascular permeability.

adriamycin See ANTHRACYCLINE ANTIBIOTICS.

ADRY reagents Certain substituted thiophenes which can e.g. stimulate the photooxidation of cytochrome b_{559} in photosystem II (see PHOTOSYNTHESIS).

adsorption (*serol.*) Non-specific adherence of substances (in solution or in suspension) to cells or to other forms of particulate matter. (See e.g. BOYDEN PROCEDURE; cf. ABSORPTION.)

adsorption chromatography See CHROMATOGRAPHY.

adult T-cell leukaemia (ATL; adult T-cell leukaemia/lymphoma, ATLL) A T-cell LEUKAEMIA (q.v.) which affects adults; the causal agent is an exogenous retrovirus, HTLV-I (see HTLV). ATL is endemic in certain regions of Japan, the Caribbean, Africa and the Americas.

In the carrier state the incidence of ATL is low; latency can last for many decades.

The disease has been categorized into four clinical subtypes: acute, chronic, smouldering and lymphoma. Acute cases comprise the majority. Chronic and smouldering forms often progress to the acute form. In some cases disease is nodal rather than leukaemic.

Various manifestations of ATL may be seen. In many cases there are lesions in liver, spleen and/or lymph nodes, though other sites (including the central nervous system) may be affected; skin lesions may include nodules, ulcers or rashes. Hypercalcaemia may be present, and the undermined immune system may permit infection by opportunist pathogens (e.g. *Pneumocystis carinii*, cytomegalovirus).

Leukaemic cells are monoclonal, originating from a single cell infected with HTLV-I; the cells are larger than normal and may have multilobed or convoluted nuclei.

The median survival time for the acute form of ATL is reported to be ~6 months (~2 years for the chronic form).

Lab. diagnosis. Diagnosis involves e.g. clinical observation, serology (for anti-HTLV-I antibody) and detection of abnormal T cells by microscopy.

Chemotherapy. Given the absence of standard therapy, it has been recommended that treatment be limited to acute and lymphoma-type cases; for these patients combination chemotherapy may be successful, but relapses (often involving the CNS) are common.

[ATL (epidemiology, leukaemogenesis, clinical features, laboratory findings, diagnosis, treatment, prognosis and prevention): BCH (2000) *13* 231–243.]

(cf. MYCOSIS FUNGOIDES and SÉZARY SYNDROME.)

adventitious septum See SEPTUM (b).

aecial cup See UREDINIOMYCETES stage I.

aecidioid (cupulate) Refers to a peridiate, cup-shaped to cylindrical aecium of the type formed e.g. by species of *Puccinia* and *Uromyces*.

aecidiospore See UREDINIOMYCETES stage I.

aecidium See UREDINIOMYCETES stage I.

aeciospore See UREDINIOMYCETES stage I.

aecium See UREDINIOMYCETES stage I.

Aedes A genus of mosquitoes (order Diptera, family Culicidae); *Aedes* spp are vectors of certain diseases: see e.g. CHIKUNGUNYA FEVER, YELLOW FEVER.

***Aedes aegypti* EPV** See ENTOMOPOXVIRINAE.

***Aedes albopictus* cell-fusing agent** See FLAVIVIRIDAE.

Aegyptianella A genus of Gram-negative bacteria of the family ANAPLASMATACEAE. Cells: pleomorphic cocci (diam. ca. 0.3–0.8 μm) which form membrane-limited inclusion bodies (each containing up to 30 cells) within the erythrocytes of the infected host (cf. ANAPLASMA). The sole species, *A. pullorum*, is parasitic in a range of birds and can cause disease e.g. in chickens; transmission occurs mainly or exclusively via ticks. *A. pullorum* occurs in southern Europe and the Mediterranean area, in Africa south of the Sahara, and in Asia.

aequihymeniiferous See LAMELLA.

Aer See AEROTAXIS.

aerial mycelium In many actinomycetes: mycelium which projects above the level of the medium; no aerial mycelium is formed e.g. by *Arachnia* or *Intrasporangium*, while *Sporichthya* forms only aerial mycelium. Aerial mycelium and SUBSTRATE MYCELIUM differ e.g. morphologically, structurally and physiologically [Book ref. 73, pp. 169–170]; aerial mycelium is typically less branched and, in at least some species, its surface is hydrophobic. In some genera the aerial mycelium is the main, or only, part of the organism which bears spores or sporophores.

Aerobacter An obsolete bacterial genus (*nom. rejic.*). Most strains formerly regarded as *'Aerobacter aerogenes'* are referable (e.g. on the basis of motility) to *Klebsiella pneumoniae* or to *Enterobacter aerogenes*. [Book ref. 21, pp. 322, 324, 339–340.]

aerobactin See SIDEROPHORES.

aerobe An organism which has the ability to grow in the presence of oxygen, i.e., in or on media or substrates which are in contact with air – the term commonly being reserved for those organisms which, in nature, *normally* grow in aerobic habitats; some aerobes can also grow under anaerobic conditions, i.e., they are FACULTATIVE anaerobes. (cf. ANAEROBE; see also MICROAEROPHILIC and ANAEROBIC RESPIRATION.)

aerobic (1) Refers to an environment in which oxygen is present at a partial pressure similar to that in air. (cf. ANAEROBIC, MICROAEROBIC.) (2) Having the characteristic(s) of AEROBE(s).

aerobiosis (1) The state or condition in which oxygen is present. (2) Life in the presence of air.

Aerococcus A genus of Gram-positive bacteria of the family Streptococcaceae. Cells: non-motile cocci, commonly in pairs and tetrads. Growth occurs optimally under microaerobic conditions; glucose is fermented homofermentatively to L(+)-lactic acid. Some strains form a pseudocatalase. *A. viridans* (formerly *Gaffkya homari* and *Pediococcus homari*) is the causal agent of GAFFKAEMIA.

aerogenic Gas-producing.

aeromonad A species or strain of *Aeromonas*.

Aeromonadaceae See AEROMONAS.

Aeromonas A genus of Gram-negative bacteria of the VIBRIONACEAE. [Proposal to move *Aeromonas* from the Vibrionaceae to a new family, Aeromonadaceae: SAAM (1985) *6* 171–182; IJSB (1986) *36* 473–477.] Cells variable: straight, round-ended rods or coccobacilli (0.3–1.0 × 1.0–3.5 μm) – occurring singly, in pairs, or in short chains – or short filaments. The genus is divided into two groups: non-motile psychrotrophs (*A. salmonicida*) and motile mesophiles (*A. hydrophila* group). Motile cells usually have a single, polar, unsheathed flagellum (wavelength ca. 1.7 μm); some cells in young cultures on solid media may be peritrichously flagellate. *Aeromonas* spp are oxidase +ve; a range of sugars and organic acids can be used as carbon sources; acid (± gas) is formed from glucose but not e.g. from inositol; gelatinase, DNase, RNase and Tween 80 esterase (lipase) are formed; NO_3^- is reduced to NO_2^-. NaCl is not required for growth. Strains are resistant to O/129. Most strains can grow on e.g. nutrient agar or trypticase-soy agar but usually not e.g. on TCBS agar. [Media, identification etc: Book ref. 46, pp. 1288–1294.] GC%: 57–63. Type species: *A. hydrophila*.

The taxonomy of the motile strains remains unsettled. Three species have been recognized: *A. hydrophila*, *A. caviae* and *A. sobria* [Book ref. 22, pp. 545–548]. (*'A. punctata'* is regarded as a later synonym of *A. hydrophila*, with VP −ve, anaerogenic strains now included in *A. caviae*; *'A. proteolytica'* = *Vibrio proteolyticus*.) Motile strains can grow on many media used for enterobacteria (e.g. DCA, MacConkey's agar), usually forming colourless (lactose −ve) colonies. Optimum growth temperature: ca. 28°C, maximum usually 38–41°C; some strains can grow at 5°C. Species occur in fresh water, sewage, and associated with aquatic animals; strains can cause disease in fish (see e.g. RED PEST and RED MOUTH), amphibians (e.g. RED LEG), reptiles, and cattle, and in man – causing e.g. septicaemia, meningitis, gastroenteritis (enterotoxigenic strains), etc.

A. salmonicida has an optimum growth temperature of 22–25°C, maximum usually ca. 35°C; most strains can grow at 5°C. Three subspecies are recognized. Subsp *salmonicida* is indole −ve, aesculin +ve, and produces a water-soluble brown pigment when grown aerobically on media containing 0.1% tyrosine or phenylalanine. Subsp *achromogenes* is indole +ve/−ve, aesculin −ve, and does not produce brown pigment. Subsp *masoucida* is indole +ve, aesculin +ve, and does not produce brown pigment. *A. salmonicida* is haemolytic on blood agar. The species is strictly parasitic and often pathogenic in fish, causing e.g. FURUNCULOSIS and secondary infections in various other diseases.

[*Aeromonas* bacteriophages: Ann. Vir. (1985) *136E* 175–199.]

aerosol Minute (colloidal) particles of liquid and/or solid dispersed in a gas (e.g. air), formed e.g. by sneezing, coughing, liquids splashing, bubbles bursting, etc. An aerosol may contain viable microorganisms.

aerotaxis A TAXIS in which a (motile) cell migrates along an oxygen concentration gradient to a location where the concentration of oxygen is optimal for that cell.

In bacteria, aerotaxis appears to be exhibited by all aerobic and facultatively aerobic motile species. Interestingly, though, organisms which use oxygen as terminal electron acceptor in energy metabolism do not necessarily migrate to positions where oxygen is at atmospheric levels; for example, the respiratory-type bacterium *Azospirillum brasiliense* is microaerophilic, and this organism migrates towards regions where the oxygen concentration is only 3–5 μM. (This species may have a highly efficient oxidase which enables it to carry out oxygen-based respiratory metabolism in microaerobic conditions.)

For some time it has been thought that aerotaxis may be linked to changes in proton motive force (pmf). In support of this idea, it was found that different types of change in pmf (i.e. increases or decreases) are associated with different effects on motility; thus, for example, if the partial pressure of oxygen is kept constant, then an artificially induced increase in pmf encourages smooth, continual swimming, while a decrease in pmf promotes tumbling in peritrichously flagellated cells. More recently it has been shown that, in *A. brasiliense*, pmf increases when the cell swims towards the preferred concentration of oxygen and decreases when the cell swims in the opposite direction; this has suggested that the change in level of pmf acts as a signal which regulates aerotactic movements [JB (1996) *178* 5199–5204]. Aerotaxis apparently cannot occur in a bacterium in which the pmf is at a maximum; under such conditions certain other (pmf-independent) taxes – e.g. CHEMOTAXIS – may occur.

In *Escherichia coli* a sensor protein, Aer, may mediate aerotaxis by responding to redox changes in component(s) of the electron transport chain; the Tsr protein (an MCP in chemotaxis) appears to be another, independent sensor for aerotaxis and may respond to changes in pmf [PNAS (1997) *94* 10541–10546].

aerotolerant Refers to an ANAEROBE which, in the presence of air, can either survive (but not grow) or grow at sub-optimal rates.

aeruginocin *Syn.* PYOCIN.

aesculin (esculin) The 6-β-D-glucosyl derivative of 6,7-dihydroxycoumarin. Certain bacteria (e.g. most strains of the former group D streptococci, many *Bacteroides* spp – including *B. fragilis*) can hydrolyse aesculin to yield 6,7-dihydroxycoumarin which gives a brown coloration with soluble ferric salts. (Hydrolysis may also be detected by the disappearance of aesculin fluorescence under Wood's lamp.)

Various media are used for aesculin hydrolysis tests. *Aesculin broth* may contain e.g. aesculin (0.1%) and $FeCl_3$ (0.05%) in heart infusion broth. *Aesculin agar* is aesculin broth gelled with agar. *Bile–aesculin agar* contains e.g. oxgall (4%), aesculin (0.1%) and ferric citrate (0.05%) in nutrient agar. *Bacteroides bile–aesculin agar* is bile–aesculin agar with the addition of e.g. haemin and gentamicin.

The former group D streptococci can be identified by their ability to cause blackening on bile–aesculin agar slants – usually within 48 h; most non-group D streptococci do not cause blackening, although some viridans streptococci may do so.

aethalium See MYXOMYCETES.

aetiology (etiology) The study of *causation*. The aetiological agent of an infectious disease is the organism which initiates/causes the disease.

aetioporphyrins See PORPHYRINS.

AEV Avian erythroblastosis virus: see AVIAN ACUTE LEUKAEMIA VIRUSES.

AF-2 (furylfuramide) See NITROFURANS.

AFC (*immunol.*) Antibody-forming cell.

AFDW Ash-free dry weight.

affinity (*immunol.*) The strength of binding between a given antibody and a single antigenic determinant or monovalent hapten. Factors which affect affinity include the area of contact between the antibody combining site and the antigenic determinant, the closeness of fit, and the nature of the intermolecular forces involved. (cf. AVIDITY.)

affinity chromatography See CHROMATOGRAPHY.

affinity labelling A technique used to identify the COMBINING SITE of an antibody. In one method, the antibody is allowed to combine with a homologous HAPTEN (the 'affinity label') that carries an azide side-chain; on subsequent irradiation with ultraviolet light the azide forms a highly reactive nitrene group that combines with any of a range of organic groups in the combining site. The nature of the combining site may subsequently be determined e.g. by chemical analysis.

affinity maturation See ANTIBODY FORMATION.

affinity tail See FUSION PROTEIN.

aflatoxicosis A MYCOTOXICOSIS caused by ingestion of AFLATOXINS.

aflatoxins A group of MYCOTOXINS, produced by strains of *Aspergillus flavus* and *A. parasiticus*, which contain a bifuran moiety fused at the 4,5 ring positions to a substituted coumarin. (cf. STERIGMATOCYSTIN.) Aflatoxins are soluble in organic solvents, and some exhibit fluorescence on UV irradiation; they are extremely heat-stable. Aflatoxins are toxic to a wide range of eukaryotes; they can retard germination and growth in plants, inhibit the germination of several moulds (e.g. *Mucor*, *Neurospora*, *Penicillium*), and are hepatotoxic and hepatocarcinogenic in animals. Severe outbreaks of disease have occurred in domestic animals and poultry fed with aflatoxin-contaminated feedstuffs – particularly groundnut and cereal feeds. (See also HEPATITIS X; TURKEY X DISEASE; cf. RUBRATOXINS.) Aflatoxins can affect the immune system, reducing resistance to infection, and may increase the risk of hepatocellular carcinoma due to HEPATITIS B VIRUS infection. The toxins can inhibit mitosis in tissue cultures. Aflatoxins may e.g. cause errors in DNA replication by reacting with guanine bases; toxic effects may differ in different species. [Aflatoxin biosynthesis genes: AEM (2004) *70* 1253–1262; (2005) *71* 3192–3198.]

AFLP fingerprinting A PCR-based TYPING method [original description: NAR (1995) *23* 4407–4414] which requires only a small amount of purified genomic DNA. The principle of the method is outlined in the figure. (In the standard protocol, primers are end-labelled; a simplified procedure avoids the need for end-labelling of primers by using α-labelled nucleotides (α-[^{33}P]-dATP) that are incorporated into products during amplification [BioTechniques (2000) *28* 622–623].)

[Taxonomic evaluation of *Bacillus anthracis* and related species by AFLP fingerprinting: JB (1997) *179* 818–824. AFLP fingerprinting used for detecting genetic variation in *Xanthomonas*: Microbiology (1999) *145* 107–114. AFLP-based study of *Escherichia coli*: JCM (1999) *37* 1274–1279. Review of AFLP fingerprinting: JCM (1999) *37* 3083–3091.]

African farcy *Syn.* EPIZOOTIC LYMPHANGITIS.

African horse sickness An infectious HORSE DISEASE caused by an *Orbivirus* and transmitted by insects (e.g. *Culicoides* spp); it occurs in Africa, parts of the Middle East, and the Mediterranean region. The disease may be acute, with an incubation period of ca. 5–7 days, followed by fever, laboured breathing, severe paroxysms of coughing, and a profuse nasal discharge of yellowish, frothy serous fluid; death usually occurs within 4–5 days of onset. Subacute forms of the disease may occur in enzootic areas: an incubation period of up to 3 weeks is followed by oedema of the head region, spreading to the chest; cardiac and pulmonary symptoms and paralysis of the oesophagus may occur, but mortality rates are generally lower than in the acute form. A mild form involving fever and moderate dyspnoea ('horse sickness fever') may occur e.g. in partially immune animals. African horse sickness may also cause severe debility in mules and donkeys, but mortality rates are lower than in horses. Control: e.g. vaccination; control of vectors.

African swine fever A highly infectious PIG DISEASE which clinically resembles European SWINE FEVER (q.v.); it occurs e.g. in Africa, Spain and Portugal. The causal agent is a virus previously classified in the IRIDOVIRIDAE but currently considered to belong to a separate family. The virion is icosahedral, enveloped, and contains a DNA-dependent RNA polymerase and RNA-modifying enzymes; the genome is dsDNA (MWt ca. 100×10^6) in which the strands are covalently joined at each end (cf. POXVIRIDAE), and which contains terminal inverted repeats.

African trypanosomiases See TRYPANOSOMIASIS.

ag (*immunol.*) ANTIGEN.

Agamococcidiorida An order in the COCCIDIASINA.

agamont See ALTERNATION OF GENERATIONS.

agar A complex galactan which is widely used (in gel form) as a base for many kinds of solid and semi-solid microbiological MEDIUM; agar (or agarose – see below) is also used e.g. in techniques such as GEL DIFFUSION, GEL FILTRATION and ELECTROPHORESIS, and in industry as a gelling agent in foods, pharmaceuticals etc. Agar is produced by many marine rhodophycean algae and is obtained commercially from e.g. *Gelidium* and *Gracilaria* spp; in the alga it is associated with the CELL WALL and intercellular matrix. (The term 'agar' derives from the Malay *agar-agar* which refers to certain edible seaweeds.)

Agar consists of two main components: *agarose* (ca. 70%) and *agaropectin* (ca. 30%). Agarose is a non-sulphated linear polymer consisting of alternating residues of D-galactose and 3,6-anhydro-L-galactose: [-3,6-anhydro-α-L-galactopyranosyl-(1 $\rightarrow$ 3)-β-D-galactopyranosyl-(1 $\rightarrow$ 4)-]$_n$; in agarose from some species, a proportion of the D-galactose residues have 6-*O*-methyl substituents. Agaropectin is a mixture of sulphated galactans which may also contain e.g. glucuronic acid or pyruvic acid, depending on source. (Agar-like substances from non-commercial seaweeds show various structural differences from commercial agar [Book ref. 38, pp. 291–292].)

An agar gel is a translucent or transparent jelly-like substance formed when a mixture of agar and water is heated to >100°C and then cooled; gelling occurs at ca. 40–45°C.

(a) | adaptor | restriction fragment | adaptor |

(b) – – – –NNNG–3′ 5′–AATTGNNNNNNN–3′
– – – –NNNCTTAA–5′ CNNNNNNN–5′

(c) – – – –NNNGAATTGNNNNNNN–3′
– – – –NNNCTTAACNNNNNNN–5′

(d) – – – –NNNGAATTGNNNNNNN–3′
TCTTAACNNNNNNN–5′ ← primer

AFLP FINGERPRINTING (principle, diagrammatic).

Chromosomes of the test strain are initially digested with *two* types of RESTRICTION ENDONUCLEASE, commonly *Eco*RI (recognition site: G/AATTC) and *Mse*I (recognition site: T/TAA). As a result, the two sticky ends of each fragment may be created either by the same enzyme or by different enzymes.

The chromosomal fragments are then mixed with 'adaptor' molecules of two types, here designated A and B. Each type of adaptor molecule is a short DNA sequence which has *one* sticky end that corresponds to the recognition sequence of *one* of the two restriction enzymes. The reaction mixture contains a ligase, so that covalent binding between fragments and adaptor molecules gives rise to the following sequences: A-fragment-A, A-fragment-B, B-fragment-A and B-fragment-B. *Note.* In each adaptor molecule a 'mutant' nucleotide is incorporated immediately adjacent to the sticky end so that, after ligation to a fragment, the cutting site of the enzyme is *not* restored; thus, following ligation, the sequence is not susceptible to restriction.

The fragments, flanked on each side by (ligated) adaptor molecules, are now subjected to PCR under high-stringency conditions.

Each PCR primer is designed to be complementary to one or other of the adaptor molecules, including the restriction site; however, an important feature of each primer is that its 3′ end extends for one or a few nucleotides *beyond* the restriction site – i.e. into the 'unknown' fragment. The one (or few) 3′ nucleotide(s) of the primer are *selective* nucleotides, i.e. the primer will be extended only if these nucleotides are paired with *complementary* nucleotides in the fragment. Hence, while primers may bind to all fragments in the mixture, only a subset of fragments will be amplified, i.e. those fragments containing nucleotides that are complementary to the selective 3′ nucleotide(s) of the primer. A primer with one selective nucleotide has a 1-in-4 chance of binding to a complementary nucleotide in the fragment; this type of primer will amplify only about one in four of the fragments to which it binds.

The primers of one type are labelled so that, following PCR and gel electrophoresis of the products, a fingerprint of (e.g. ~50–200) detectable bands is obtained.

(a) Each restriction fragment is flanked by (ligated) adaptor molecules.

(b) *Left.* A fragment's sticky end produced by *Eco*RI (N = nucleotide). *Right.* An adaptor molecule with the complementary 5′-AATT overhang; note that, in the overhang strand, the 5′-AATT is followed by G, rather than C.

(c) Following base-pairing of the sticky ends in (b), and ligation, the resulting sequence

5′-GAATTG-3′
3′-CTTAAC-5′

differs from the cutting site of *Eco*RI and is not susceptible to cleavage by *Eco*RI.

(d) During cycling, a primer binds to one strand of the fragment–adaptor junction region. As this primer's 3′-terminal (selective) nucleotide is T, the primer will be extended only if the complementary nucleotide (A) occurs at this location in the fragment; extension will not occur on this fragment if T is mis-matched.

Reproduced from Figure 7.6, page 193, in *DNA Methods in Clinical Microbiology* (ISBN 07923-6307-8), Paul Singleton (2000), with kind permission from Kluwer Academic Publishers, Dordrecht, The Netherlands.

Media made from Japanese agars usually contain 1.5–2.0% w/v agar; however, *semi-solid* agars contain 0.5% or less, and *stiff* agars contain e.g. 8% w/v. (If New Zealand agars are used, these concentrations should be halved to give gels of similar strengths.) Sterile gels are prepared by autoclaving a suspension of agar in water (see AUTOCLAVE). Growth media are prepared by adding nutrients, selective agents etc to the agar – usually before autoclaving but sometimes after (see e.g. BLOOD AGAR). If the medium is to have a pH of 6.0 or less, the pH must be adjusted after autoclaving, since agar is hydrolysed during heating at low pH. Various refined forms of agar may be used for specific purposes: e.g. *ion-free* agar may be used in immunoelectrophoresis. (See also ELECTROENDOSMOSIS.)

Agar is a useful base for microbiological media in that it gels at moderate temperatures and, once set, the gels are stable at temperatures up to ca. 65°C or higher (although syneresis tends to occur at these temperatures); furthermore, the ability to degrade agar is confined to only a few organisms (including e.g. strains of *Streptomyces coelicolor*, certain marine pseudomonads, marine species of *Cytophaga*). (cf. GELATIN.) However, agar shortages, product variability, and rising prices have led to a search for suitable substitutes for agar; substitutes which have found some applications include CARRAGEENAN, Gelrite (see GELLAN GUM), low-methoxy PECTINS, and SILICA GELS.

agar diffusion test A DIFFUSION TEST which differs from the DISC DIFFUSION TEST in that, instead of employing antibiotic-impregnated discs, a *solution* of each antibiotic is allowed to diffuse from a separate 'well' cut into the agar.

agar dilution test See DILUTION TEST.

agar disc diffusion test *Syn.* DISC DIFFUSION TEST.

agar gel diffusion See GEL DIFFUSION.

agar plate See PLATE.

agar-slide method *Syn.* DIP-SLIDE METHOD.

agaric (1) Any fungus of the Agaricaceae. (2) Any fungus of the AGARICALES. (3) Any fungus whose hymenium is borne on lamellae (see LAMELLA).

Agaricaceae See AGARICALES.

Agaricales An order of terrestrial (typically humicolous or lignicolous), mainly saprotrophic fungi (subclass HOLOBASIDIOMYCETIDAE) most of which form mushroom-shaped, gymnocarpic or semiangiocarpic, fleshy fruiting bodies in which the hymenium is borne on radially arranged 'gills' (= lamellae, see LAMELLA) on the underside of the pileus; the pileus and (when present) stipe do not contain sphaerocysts (cf. RUSSULALES). (See also SECOTIOID FUNGI.) The order may be divided into families on the basis of e.g. basidiospore colour, the structure of the trama, and the nature of the cortical layers of the pileus [Book ref. 64, pp. 7–8]; these families include:

Agaricaceae. Basidiocarp: stipitate, typically with an annulus when mature; basidiospores: typically dark brown or colourless, but not rust- or cinnamon-coloured. Genera include AGARICUS and LEPIOTA.

Amanitaceae. Basidiocarp: stipitate, the lamellae each having a divergent BILATERAL TRAMA; basidiospores: white or pale. Some species form both a PARTIAL VEIL and a UNIVERSAL VEIL. Genera include AMANITA (volva formed), *Limacella* (volva not formed), and TERMITOMYCES.

Bolbitiaceae. Basidiocarp: stipitate; basidiospores: ochre or cinnamon to rust-brown. Genera include *Agrocybe* and *Conocybe*.

Coprinaceae. Basidiocarp: stipitate, a palisade-like layer of cells occurring in the pellis; basidiospores: dark or black, each usually containing a germ pore. Genera include COPRINUS and *Psathyrella*.

Cortinariaceae. Basidiocarp: stipitate, characteristically with a fine, cobweb-like cortina; basidiospores: rust-coloured or some shade of brown, smooth to rough. Genera include *Cortinarius*, *Galerina*, *Inocybe* (see also MUSCARINE).

Crepidotaceae. Basidiocarp: non-stipitate, or with a rudimentary lateral stipe; basidiospores: cinnamon-coloured. Genera include CREPIDOTUS.

Hygrophoraceae ('wax caps'). Basidiocarp: stipitate, often brightly coloured, the lamellae being waxy, and the basidia typically elongated; basidiospores: colourless. Genera include *Hygrocybe* and *Hygrophorus*.

Pluteaceae. Basidiocarp: stipitate with a volva, the lamellae each having a convergent BILATERAL TRAMA; basidiospores: pink. Genera include *Volvariella* (see also PADI-STRAW MUSHROOM).

Strophariaceae. Basidiocarp: stipitate (stipe often elongated), pileus e.g. buff, yellow, ochre, or (in *Stropharia aeruginosa*) greenish; basidiospores: typically brown to purplish-brown. Genera include *Hypholoma*, *Panaeolus*, *Pholiota*, *Psilocybe*, *Stropharia*. (See also HALLUCINOGENIC MUSHROOMS.)

Tricholomataceae. Basidiocarp: stipitate, lamellae with non-bilateral trama; basidiospores: white or pink, without a germ pore. Genera include ARMILLARIA, *Clitocybe*, *Collybia*, *Crinipellis* (see also WITCHES' BROOM), *Flammulina* (see also ENOKITAKE), LENTINULA, LEPISTA, *Marasmius* (see also MYCORRHIZA), *Mycena* (see also BIOLUMINESCENCE), *Omphalotus*, *Oudemansiella*, *Tricholoma*.

agaricoid Refers to the type of fruiting body which is characteristic of fungi of the AGARICALES: gymnocarpic, with the hymenium forming a layer on lamellae and giving rise to ballistospores. (cf. GASTEROID.)

Agaricus (formerly *Psalliota*) A genus of fungi (AGARICALES, Agaricaceae), most (not all) species of which are edible. Except in *A. brunnescens*, the basidia each form four basidiospores; basidiospores are dark brown and commonly ovoid. *A. arvensis* is the horse mushroom, *A. campestris* the common or field mushroom, and *A. silvicola* the wood mushroom (all edible species). *A brunnescens* (= *A. bisporus*), a species which forms two-spored basidia, is the cultivated mushroom (see MUSHROOM CULTIVATION). (See also FUNGUS GARDENS.)

agarobiose A disaccharide: 3,6-anhydro-4-*O*-(β-D-galactopyranosyl)-L-galactose, a degradation product of AGAR.

agaropectin See AGAR.

agarophyte (agarphyte) Any AGAR-producing seaweed.

agarose See AGAR.

age-dependent polioencephalomyelitis (in mice) See LACTATE DEHYDROGENASE VIRUS.

agglutinated test (*protozool.*) See e.g. FORAMINIFERIDA.

agglutination The formation of insoluble aggregates following the combination of antibodies with cells or other particulate antigens (see e.g. WEIL–FELIX TEST) or with soluble antigens bound to cells or other particles (see e.g. LATEX PARTICLE TEST), or following the combination of soluble (or particulate) antigens with cell-bound or particle-bound antibodies (see e.g. PROTEIN A); agglutination may also be mediated by e.g. LECTINS or by fibrinogen (see clumping factor in COAGULASE). Agglutination may be detected macroscopically as suspended aggregates or (subsequently) as sedimented aggregates. (See also PASSIVE AGGLUTINATION and HAEMAGGLUTINATION; cf. PRECIPITATION and FLOCCULATION sense 1.)

On sedimentation, agglutinated particles may form a mat over a relatively large area of the bottom of the test-tube; by contrast, non-agglutinated particles generally sediment to form a smaller, dense button in the control tube.

agglutination factor (*algol.*) See CHLAMYDOMONAS.
agglutination test Any test in which reactions between particulate and/or soluble entities (particularly free or particle-bound antibodies and antigens) is detected by AGGLUTINATION.
agglutinin (1) Any ANTIBODY involved in an AGGLUTINATION reaction. (2) Any substance which can agglutinate cells or inanimate particles by binding to their surface components.
agglutinogen The antigen homologous to an agglutinin.
Aggregata See EIMERIORINA.
aggregate gold standard See GOLD STANDARD.
aggregation substance See PHEROMONE.
aggressin Any product or component of a pathogenic microorganism which promotes the invasiveness of that organism – see e.g. HYALURONATE LYASE. (cf. STREPTOKINASE; see also ADHESION and IGA1 PROTEASES.)
aglycon The non-sugar portion of a glycoside.
agmatine See e.g. DECARBOXYLASE TESTS.
Agmenellum A phycological genus of 'blue-green algae' currently included in the SYNECHOCOCCUS complex.
agnogene See AGNOPROTEIN.
agnoprotein A 61-amino acid, highly basic polypeptide encoded by a gene (the 'agnogene') in the late leader region of SIMIAN VIRUS 40; it appears to play a role in viral assembly.
Agonomycetales (Mycelia Sterilia) An order of fungi (class HYPHOMYCETES) which include species that form neither conidia nor sexual structures, though some species have been found to have teleomorphs in the Ascomycotina or the Basidiomycotina. The order includes lichenized fungi (e.g. LEPRARIA) and plant pathogens (see e.g. RHIZOCTONIA and SCLEROTIUM).
agr **locus** (in *Staphylococcus aureus*) Accessory gene regulator locus: a chromosomal sequence which encodes, *inter alia*, (i) the sensor and response regulator of a TWO-COMPONENT REGULATORY SYSTEM (AgrA–AgrC), and (ii) an octapeptide 'pheromone' which is secreted by the cell and which, at appropriate concentrations (see QUORUM SENSING), activates the two-component system. When activated, the AgrA–AgrC system upregulates expression of the (*agr*-encoded) transcript RNA III which, in turn, regulates the expression of genes for exotoxins and certain cell-surface-associated virulence factors.

Another two-component system, encoded by the *srrAB* genes (*s*taphylococcal *r*espiratory *r*esponse genes), appears to regulate the expression of exotoxins and certain cell-surface virulence factors in accordance with the levels of environmental oxygen, such regulation being exerted, in part, via the *agr* system [JB (2001) *183* 1113–1123].
agranulocyte Any white blood cell which has non-granular cytoplasm, e.g., a LYMPHOCYTE.
Agrobacterium A genus of Gram-negative bacteria of the RHIZOBIACEAE. Cells: rods (0.6–1.0 × 1.5–3.0 μm), capsulated. Motile, with between one (often non-polar) and six peritrichous flagella. Optimum growth temperature: 25–28°C. Species can metabolize a wide range of mono- and disaccharides and salts of organic acids; acid (no gas) is formed from glucose (which is metabolized mainly via the ENTNER–DOUDOROFF PATHWAY and the HEXOSE MONOPHOSPHATE PATHWAY). Colonies on carbohydrate-containing media are typically mucilaginous, abundant extracellular slime (including a neutral (1 → 2)-β-glucan) being produced. (See also CURDLAN.) Some strains can use NH_4^+ and NO_3^- as nitrogen sources, while others require amino acids; some strains can carry out NITRATE RESPIRATION. Nitrogen fixation does not occur. GC%: 57–63. Type species: *A. tumefaciens*.

Agrobacterium spp occur in soil, mainly in the rhizosphere. All (except *A. radiobacter*) can infect a wide range of dicotyledonous plants (and some gymnosperms) and induce the formation of self-proliferating galls or adventitious roots. The species are defined primarily or solely on the basis of their pathogenic characteristics: *A. radiobacter* is non-pathogenic, *A. rhizogenes* causes HAIRY ROOT, *A. rubi* causes CANE GALL, and *A. tumefaciens* causes CROWN GALL. However, pathogenicity depends on the presence of a plasmid(s) and can readily be altered or lost; hence the currently recognized species do not reflect true taxonomic relationships among the agrobacteria. [Book ref. 22, pp. 244–254.]

[Media and culture: Book ref. 45, 842–855.]
agrocin 84 See AGROCINS.
agrocinopines A class of (sugar phosphodiester) opines found in CROWN GALL. (See also AGROCINS.)
agrocins Antibiotics which are produced by certain strains of *Agrobacterium* and which are active against other strains of the same genus; being non-protein in structure, agrocins are not strictly BACTERIOCINS. Agrocin 84 is produced by a non-pathogenic, nopaline-catabolizing strain of *A. radiobacter* (strain 84, NCPPB 2407) and is selectively active against agrobacteria which harbour a nopaline Ti plasmid. Strain 84 is used in the BIOLOGICAL CONTROL of CROWN GALL; a cell suspension is used to treat seeds, roots or wounded plant surfaces (e.g. graft wounds), and almost 100% control of nopaline pathogens (responsible for most of the economic damage due to crown gall) can be achieved. Agrocin 84 is an adenine nucleotide derivative containing an N^6-phosphoramidate substituent (necessary for uptake by sensitive cells) and a 5′-phosphoramidate substituent (necessary for toxicity). Agrocin 84 is taken up by sensitive strains via a high-affinity 'agrocin permease', apparently an agrocinopine transport system normally inducible by agrocinopines (sugar phosphodiesters) present in galls caused by nopaline strains. Strain 84 contains at least three plasmids: one (pAgK84, 47.7 kb) coding for agrocin 84 production, another (pAt84b, ca. 200 kb) coding for nopaline catabolism. pAgK84 is self-transmissible only at very low frequencies, but can be mobilized by the conjugative plasmid pAt84b. As transfer of pAgK84 to a crown gall pathogen could threaten the continued use of agrocin 84 in biocontrol, a transfer-deficient mutant strain was prepared. However, strain K84 apparently exerts some activity against agrocin 84-resistant pathogens independently of pAgK84 [AEM (1999) *65* 1936–1940].
Agrocybe See AGARICALES (Bolbitiaceae).
agroinfection A method for introducing viral DNA (or cDNA) into a plant. Viral DNA is initially incorporated into the T-DNA part of a Ti plasmid. The plasmid is then introduced into the bacterium *Agrobacterium tumefaciens*–which is used to infect the plant; during infection the viral DNA is transferred to plant cells within the T-DNA (see CROWN GALL). [Example of use: JV (2003) *77* 3247–3256.]
Agromyces A genus of microaerophilic to anaerobic, catalase-negative, asporogenous bacteria (order ACTINOMYCETALES, wall type VII – see also PEPTIDOGLYCAN). The organisms grow as a branched mycelium which subsequently fragments into coccoid and diphtheriod forms; metabolism: oxidative. *A. ramosus*, the type species, occurs in large numbers in certain soils; it appears to attack and destroy other species of bacteria [AEM (1983) *46* 881–888].
agropine See CROWN GALL and HAIRY ROOT.
Agropyron **mosaic virus** See POTYVIRUSES.
AHG (*serol.*) Anti-human globulin: ANTIGLOBULIN homologous to human globulins.

AHL See QUORUM SENSING.

***ahpC* gene** See ISONIAZID.

AID (activation-induced cytidine deaminase) See RNA EDITING.

AIDA-I In *Escherichia coli*: an adhesin that mediates diffuse adherence (to e.g. HeLa cells) – hence the designation adhesin involved in diffuse adherence. AIDA-I is the α domain of an autotransporter (see type IV systems in PROTEIN SECRETION); in wild-type cells the α domain is apparently cleaved (autocatalytically?), but it remains attached (non-covalently) to the cell surface.

(See also AUTODISPLAY.)

***aidB* gene** See ADAPTIVE RESPONSE.

AIDS (acquired immune deficiency syndrome) In an HIV$^+$ individual (see HIV): the stage of disease characterized by (i) counts of CD4$^+$ T LYMPHOCYTES commonly within or below the range 200–500/μl (in adults and adolescents) *and* (ii) the presence of one or more category C diseases (AIDS-defining diseases specified in the clinical staging system of the Centers for Disease Control (CDC), Atlanta, Georgia, USA); category C diseases include e.g. CANDIDIASIS of the *lower* respiratory tract; disseminated COCCIDIOIDOMYCOSIS; extrapulmonary CRYPTOCOCCOSIS; retinitis due to cytomegalovirus (BETAHERPESVIRINAE); herpes simplex oesophagitis; HIV-related encephalopathy; extrapulmonary HISTOPLASMOSIS; chronic infection with ISOSPORA; KAPOSI'S SARCOMA; primary lymphoma of the brain; extrapulmonary infection with *Mycobacterium tuberculosis* (see also MAC); pneumonia caused by PNEUMOCYSTIS CARINII; TOXOPLASMOSIS of the brain. [Infections in AIDS (MICROSPORIDIOSIS, invasive pneumococcal disease, non-typhoid salmonellae): JMM (2000) *49* 947–957.]

The normal CD4 count in adults/adolescents is ~1000/μl; in neonates it is much higher (≥2000), so that the adult-based relationship between CD4 counts and susceptibility to opportunist pathogens is not appropriate for very young children.

[AIDS Insight (various aspects): Nature (2001) *410* 961–1007.]

Transmission/containment of HIV.
HIV can be transmitted: (i) by sexual contact (male ↔ female, as well as homosexual); (ii) via transfusion of (infected) blood or blood products (e.g. contaminated Factor VIII formerly given to haemophiliacs); (iii) via the placenta; (iv) by breast-feeding; and (v) through use of contaminated needles by intravenous drug abusers.

The highest concentration of virus (free and intracellular) is found in blood; HIV also occurs e.g. in semen, milk and cerebrospinal fluid.

Efforts to prevent/limit the spread of HIV have included: (i) education (making clear the basic facts of the disease, including routes of transmission); (ii) discouraging promiscuity; (iii) encouraging the use of condoms; (iv) discouraging needle-sharing by drug addicts; (v) screening of blood donors; (vi) treatment of blood products.

Clinical manifestations of HIV infection.
Infection is commonly followed, in ~2–6 weeks, by an early 'acute' phase which is characterized by high-level viraemia; p24 antigen (the major core protein: see HIV) can often be demonstrated in serum during the viraemic phase. Levels of virus remain high for some weeks – after which there is a sharp decline (and a loss of detectable p24 antigen); this decline in viraemia appears to reflect the activity of antigen-specific CD8$^+$ cytotoxic T cells (see T LYMPHOCYTE) and may also involve non-specific killing of virus-infected cells by NK CELLS.

Antibodies (e.g. anti-gp120, anti-p24) are first detectable ~6–12 weeks after infection; their appearance may follow, or coincide with, the rapid decline in viraemia.

During seroconversion some patients experience a *seroconversion illness* which may include e.g. fever, sore throat, skin rash, generalized lymphadenopathy, pneumonitis, gastrointestinal and/or CNS involvement.

A subsequent phase of infection is characterized by *persistent generalized lymphadenopathy* (PGL) (also called *lymphadenopathy syndrome*) – swollen lymph nodes reflecting an active immune response to HIV. (In some cases, PGL is the first manifestation of disease following infection, i.e. in patients who do not exhibit an acute phase.) Infection may then become asymptomatic ('clinically latent'), and this state may continue for months or years (during which time viral replication continues).

Patients with clinically latent infection, as well as those with PGL, may pass directly to AIDS. Alternatively, both types of patient may progress to AIDS via a further stage commonly referred to as the *AIDS-related complex* (ARC) (= category B of the CDC clinical staging system). Category B diseases include bacillary angiomatosis (see BARTONELLA), oropharyngeal candidiasis, HAIRY LEUKOPLAKIA, LISTERIOSIS, PID (pelvic inflammatory disease), herpes zoster (involving at least two distinct episodes, or more than one dermatome), and peripheral neuropathy.

In AIDS, the final stage, the CD4$^+$ count is often <200 cells/μl; there is high-level viraemia, and the p24 antigen is again detectable. Note that the AIDS-defining diseases (see above) include those that result from infection by 'low-grade' pathogens, i.e. organisms which generally cause disease only in those patients who are severely immunodeficient (including HIV$^-$ individuals who may be immunodeficient for other reasons). Conversely, the 'high-grade' pathogens (which can cause disease even in the immunocompetent) may cause disease in HIV$^+$ patients whose immune system is only marginally impaired.

Immunopathogenesis.
In HIV$^+$ individuals, the reduction in numbers of CD4$^+$ T cells may arise in various ways – e.g. (i) productive infection by HIV and subsequent lysis; (ii) killing of HIV-infected cells by HIV-specific CD8$^+$ cytotoxic T cells (see T LYMPHOCYTE) or by NK CELLS; (iii) ADCC (of virus-infected cells). It also appears that *non*-infected CD4$^+$ T cells may be susceptible to attack e.g. by HIV-specific cytotoxic T cells or by NK CELLS; this may occur when free (isolated) gp120 protein (see HIV) binds to CD4 on virus-free ('bystander') cells – such cells then becoming vulnerable to the antiviral immune response. CD4$^+$ cells may also be vulnerable to APOPTOSIS if viral gp120 cross-links CD4 molecules on the cell surface, such cross-linking resulting in an upregulation of the FAS antigen.

Following depletion of CD4$^+$ T cells, the virus may persist within tissue macrophages (see under *Chemotherapy*, below).

HIV infection produces (i) a direct effect (death of infected cells), and (ii) an indirect effect (weakening of the overall immune system). The direct effect may include neurological deficits (e.g. encephalopathy) as well as immunological damage; the indirect effect is manifested in the range of ARC and AIDS-defining diseases (including both neoplastic and opportunistic diseases).

Depletion of CD4$^+$ T cells is only one of many abnormalities in the immune system of HIV$^+$/AIDS patients. Other types of cell (including B cells) are reported to display abnormalities, and DELAYED HYPERSENSITIVITY reactions may be either absent or of reduced intensity.

The role of CYTOKINES in HIV infection has yet to be clarified, but it seems that, at some stage, the cytokine milieu becomes aberrant, and that increased amounts of e.g. TNF-α and several interleukins (including INTERLEUKIN-4) are formed. It has been suggested that TNF-α may promote viral transcription through activation of the host cell's nuclear transcription factor, NF-κB (see HIV), and that increased levels of IL-4 may bring about a (deleterious) switch to the Th2 subset of T cells (see T LYMPHOCYTE).

Diagnosis of HIV infection.
In adults and adolescents, laboratory diagnosis of HIV infection is serologically based.

The viral antigen p24 (major core protein) is commonly detectable (transiently) in blood during the initial viraemic (acute) phase of infection, i.e. before any antibodies are detectable. However, not all patients exhibit an acute phase of infection; while a positive test for p24 is useful, a negative test is not a reliable indication of the absence of HIV infection.

Tests for anti-HIV antibodies (e.g. anti-gp120) may be carried out e.g. by ELISA. A positive result may be confirmed by Western blot analysis of viral proteins. Antibodies are commonly present by ~6 weeks post-infection, although positive serology is sometimes delayed. (In some cases, patients who are serologically negative have been implicated, epidemiologically, in the transmission of AIDS.)

Serology is inappropriate for neonates and very young children: if the mother is HIV$^+$ then, owing to passive transfer of maternal antibodies, the neonate will also contain anti-HIV antibodies, whether infected or not. Culture is a useful approach for diagnosis in this context.

The pre-seroconversion 'window' (the period, post-infection, before development of antibodies) presents a problem in blood transfusion because donor samples taken in this period are likely to contain virus. To minimize this risk, samples can be subjected to PCR-based tests designed to detect viral nucleic acid. Such tests may involve initial concentration of virions from plasma [Lancet (1999) *353* 359–363]. Alternatively, the procedure may test for the integrated (i.e. provirus) form of HIV; in this case, PCR is carried out on DNA extracted from blood leukocytes [PNAS (1999) *96* 6394–6399]. (See also BLOOD DNA ISOLATION KITS.)

Chemotherapy for HIV infection.
The drug zidovudine (see AZT) was the first agent used for the treatment of HIV infection. Although initially successful, the use of AZT was found to be limited by the development of drug-resistant strains of HIV. In recent years, *combinations* of ANTIRETROVIRAL AGENTS have been used in so-called 'highly active antiretroviral therapy' (see HAART). Attempts are being made to extend the range of antiretroviral agents by developing inhibitors of the viral integrase (in order to inhibit integration of HIV provirus into the host cell's genome). (See also NU-1320.)

Even with HAART, however, HIV commonly (or always) emerges after cessation of therapy, and it is generally believed that, during therapy, HIV persists in CD4$^+$ memory T cells. However, using a chimeric virus (simian immunodeficiency virus/HIV-1), it has been shown that tissue macrophages (in lymph nodes, liver, spleen etc.) form the principal reservoir of virus following depletion of CD4$^+$ T cells in rhesus macaques; this has suggested that macrophages may be a major source of HIV in the symptomatic phase of human infection [PNAS (2001) *98* 658–663].

The drug nevirapine is reported to inhibit intrauterine transmission of HIV to the fetus in a proportion of cases.

One problem associated with chemotherapy is that the combination of drugs used in HAART is often augmented by other drugs (e.g. antibiotics) used for the treatment of opportunistic infections; this provides potential for adverse drug–drug interactions. [Drug interactions in AIDS: JMM (2000) *49* 947–967 (957–962).]

One suggested approach is to combine chemotherapy (using a range of drugs to discourage the emergence of resistant strains of virus) with a programme of vaccination aimed at maintaining or increasing the population of specific cytotoxic T lymphocytes (CTLs); an appropriate level of CTLs would serve to augment the effects of chemotherapy by helping to prevent the growth of mutant strains of virus [PNAS (2000) *97* 8193–8195].

Gene therapy. See GENE THERAPY.

Anti-AIDS vaccines.
The production of an anti-AIDS vaccine is fraught with difficulties. One of the major problems is the extensive variability of the envelope protein gp120 (see HIV); a vaccine which is active against one isolate of HIV may be less active, or non-active, against many other isolates.

Early attempts at a vaccine included the use of a recombinant VACCINIA VIRUS containing the *env* sequence of HIV [Nature (1987) *326* 249–250].

More recently it has been found that a particular, hypervariable part of the gp120 glycoprotein, the V3 loop, is an immunodominant area which influences certain phenotypic features of the virus, including infectivity; within this region is a tetrapeptide subregion, GPGR (see AMINO ACID), which seems to be conserved in a large number of field isolates.

Recent studies have examined the effect of variation in the V3 loop on the immunogenic potential of gp120, and it appears that certain changes in composition have considerable influence on the immune response to this glycoprotein [Arch. Virol. (2000) *145* 2087–2103].

Considerable efforts are currently being made to find a useful anti-AIDS vaccine, emphasis often being placed on the need for combined antibody- and cell-mediated responses. In one approach, a DNA VACCINE – HIV-derived DNA linked to polylactide co-glycolide – was found significantly to improve both cell-mediated and humoral immunity in monkeys. Use is being made of various vectors in candidate vaccines – including e.g. *Salmonella*, canarypoxvirus and Semliki Forest virus. [AIDS vaccines (news focus): Science (2001) *291* 1686–1688.]

AIDS-like diseases have been recognized in certain animals – e.g. cats (see FELINE LEUKAEMIA VIRUS) and monkeys (see SIMIAN AIDS).

AIDS-related complex (ARC) See AIDS.

AIDS virus See AIDS.

Aino virus See AKABANE VIRUS DISEASE and BUNYAVIRUS.

air (microbiological aspects) Air normally contains various microorganisms (particularly spores), pollen, and other particulate matter (see also AEROSOL); the microflora varies e.g. with location, general weather conditions, and with particular factors (such as relative humidity), while the viability of the microflora depends e.g. on the extent to which the air has been exposed to ultraviolet radiation. The airborne microflora is sampled e.g. in studies on pneumonitis-type allergies [Book ref. 51, pp. 27–65], environmental microflora [AEM (1983) *45* 919–934], the stability of aerosols [AEM (1982) *44* 903–908], and organisms of potential meteorological interest [AEM (1982) *44* 1059–1063]. Instruments used to sample the airborne microflora include those of the simple gravity-type (e.g. the DURHAM SHELTER, TAUBER TRAP), and e.g. the ALL-GLASS IMPINGER, ANDERSEN SAMPLER, HIRST SPORE TRAP and ROTOROD (see also SLIT SAMPLER). (In the context of air samplers the terms 'impactor' and 'impinger' are sometimes used indiscriminately; thus, e.g. the Andersen sampler has been described as a 'sieve impinger' [Book ref. 51, pp. 59–61] and as a 'cascade impactor' [Book ref. 57, p. 163].) Individual types of sampler have particular limitations; for example, gravity-type instruments tend to collect the larger particles in preference to smaller ones, the Rotorod can be used only for short periods of time in air containing high concentrations of particulate matter, while some samplers tend to dehydrate the collecting medium and the deposited microorganisms.

The air in confined spaces can be disinfected e.g. by ULTRAVIOLET RADIATION, by sprays of e.g. propylene glycol, or by the use of hydrophobic membrane filters (see FILTRATION).

air bladder (*algol.*) *Syn.* PNEUMATOCYST.

air sacculitis A POULTRY DISEASE which affects mainly chickens and turkey poults, particularly birds reared in broiler houses. It is caused by *Mycoplasma gallisepticum*, *M. synoviae* or (in turkeys only) *M. meleagridis* – often in association with *Escherichia coli* or respiratory virus infection. Symptoms: coughing, nasal discharge, conjunctivitis etc; sinuses below the eyes are characteristically swollen. The air sacs become filled with a thick white or yellowish caseous material. Transmission may occur from parent to offspring via the egg and from bird to bird.

airlift fermenter A LOOP FERMENTER in which the circulation of the culture is typically achieved by pumping air in at the bottom of a DRAFT TUBE (or at the base of the annulus) – the air being voided via an opening in the top of the column; the air bubbles lower the hydrostatic pressure of culture in the draft tube so that culture continually flows down the annulus and up into the draft tube. (Reversed flow occurs if air is bubbled into the annulus.) The ICI *pressure-cycle fermenter* (used for Pruteen production – see SINGLE-CELL PROTEIN) is an example of a *tubular loop* airlift fermenter. It consists essentially of two tall vertical columns (of different diameters) which communicate at top and bottom, and there is no draft tube; air is pumped into the bottom of the wider column (the 'riser') and promotes circulation on the airlift principle. This fermenter achieves a high level of dissolved oxygen since oxygen solubility is increased by the increased hydrostatic pressure at the base of the (tall) riser; additionally, the physical separation of riser and downcomer facilitates the removal of heat by allowing the inclusion of a heat-exchanger in the downcomer. [Construction, behaviour and uses of airlift fermenters: Book ref. 3, pp. 67–95.]

AIV process A process for preparing SILAGE by direct acidification of vegetable matter (to ca. pH 3.5) with a mixture of dilute HCl and H_2SO_4. In the *Penthesta process* HCl and H_3PO_4 are used.

Ajellomyces A genus of fungi of the GYMNOASCALES. *A. dermatitidis* = teleomorph of *Blastomyces dermatitidis* (q.v.); it is heterothallic and produces spherical, 8-spored asci. *A. capsulatus* = teleomorph of *Histoplasma capsulatum* (q.v.).

Akabane virus disease A CATTLE DISEASE (which can also affect sheep and goats) caused by the Akabane virus (genus BUNYAVIRUS, serogroup Simbu) and transmitted by midges (*Culicoides* spp) and mosquitoes; it occurs e.g. in Africa, Australia and Japan. Infection of cows early in pregnancy causes malformation of the fetus; deformities depend on the stage of development at the time of infection, but may involve absence of cerebral hemispheres (hydranencephaly) and/or fixation of joints leading to deformities of limbs and spine (arthrogryposis). Abortion or stillbirth may occur. The cow shows no other clinical symptoms.

A similar condition is caused by the Aino virus (*Bunyavirus*, serogroup Simbu).

akinete (1) In certain CYANOBACTERIA: a specialized cell which shows some resistance to desiccation and cold and which apparently functions as an overwintering propagule. Akinetes are formed under various growth-limiting environmental conditions (e.g., nutrient limitation); in e.g. *Anabaena* spp, akinetes develop adjacent to HETEROCYSTS, while in e.g. *Nostoc* spp they develop in positions midway between two heterocysts. An akinete is typically larger than a vegetative cell; it has a thickened wall and granular cytoplasm rich in storage materials (cyanophycin, glycogen etc). Rates of photosynthesis, respiration etc are usually much lower in akinetes than in vegetative cells; in e.g. *Anabaena doliolum*, akinetes appear to be deficient in both photosynthesis and inorganic nitrogen metabolism [JGM (1984) *130* 1299–1302]. On germination of an akinete, a single cell or short filament may be released via a pore in – or by rupture of – the akinete wall, depending on species.

(2) (*algol.*) A thick-walled non-motile resting cell produced by certain algae of the CHLOROPHYTA and XANTHOPHYCEAE.

(3) (*mycol.*) A non-motile spore.

akinetoplasty Obsolete *syn.* DYSKINETOPLASTY.

AktA (of *Actinobacillus actinomycetemcomitans*) See RTX TOXINS.

AKV A replication-competent MURINE LEUKAEMIA VIRUS which occurs endogenously in various strains of mice (e.g. AKR, BALB/c). It appears to have given rise to the transforming virus 'AKR mink cell focus-forming virus' (AKR-MCF) by recombination with one or more endogenous xenotropic viruses. (See also MCF VIRUSES.)

alafosfalin (alaphosphin; L-alanyl-L-1-aminoethylphosphonic acid) A synthetic ANTIBIOTIC which is taken up by the LL-dipeptide transport system of a sensitive cell; it is subsequently hydrolysed intracellularly to release the inhibitory component, 1-aminoethylphosphonic acid (ala-P), which itself cannot cross the cytoplasmic membrane. (cf. WARHEAD DELIVERY.) Ala-P acts by competitively inhibiting alanine racemase and – at higher concentrations – by inhibiting the addition of L-alanine to UDP-MurNAc during PEPTIDOGLYCAN synthesis. Alafosfalin has a broad spectrum of activity, but is generally more effective against Gram-negative than Gram-positive bacteria.

alamethicin A water-soluble peptide antibiotic (MWt ca. 2100) which is produced – often together with a related antibiotic, *suzukacillin* – by strains of *Trichoderma viride*. It can act as an IONOPHORE [Book ref. 14, pp. 219–224].

L-alanine biosynthesis See Appendix IV(b) and AMMONIA ASSIMILATION.

ala-P (AlaP) See ALAFOSFALIN.

alaphosphin *Syn.* ALAFOSFALIN.

Alaria See PHAEOPHYTA.

alarmone A low-MWt molecule, synthesis of which serves as a trigger or signal for the redirection of cellular metabolism in response to a particular type of stress; an example is ppGpp in STRINGENT CONTROL (sense 1).

alastrim See SMALLPOX.

alazopeptin See DON.

albamycin *Syn.* NOVOBIOCIN.

Albert's stain A stain used to demonstrate METACHROMATIC GRANULES. To prepare Albert's stain: TOLUIDINE BLUE (0.15 g) and MALACHITE GREEN (0.2 g) are dissolved in 95% ethanol (2 ml) and added to 1% acetic acid (100 ml); the whole is filtered after standing for 24 h. A heat-fixed smear is stained with Albert's stain (3–5 min), washed in tap water, and blotted dry; LUGOL'S IODINE is applied for 1 min and the smear washed and blotted dry. Granules stain black, cytoplasm pale green.

albicidin An antibiotic, produced by *Xanthomonas albilineans*, which inhibits DNA synthesis in *Escherichia coli* [JGM (1985) *131* 1069–1075].

albofungin *Syn.* KANCHANOMYCIN.

albomycin See SIDEROMYCINS.

alborixin See MACROTETRALIDES.

Albugo A genus of obligately plant-parasitic fungi (order PERONOSPORALES) which are distinguished by their production of basipetally formed chains of zoosporangia. *A. candida* (economically the most important species) is the causal agent of 'white rust' of crucifers (= 'crucifer white blister' or 'white blister disease'). Within the tissues of the host plant, this fungus forms a branching, aseptate, intercellular mycelium which produces rounded haustoria. The chains of zoosporangia develop on short, club-shaped sporangiophores beneath the host's epidermis, giving rise to smooth, white, blister-like lesions within which the zoosporangia are compacted; subsequently, the epidermis ruptures, and the powdery mass of zoosporangia is dispersed by wind and rain. During sexual reproduction, oogonia and antheridia are produced within the host, and a fertilization tube is formed between them; following fertilization and meiosis, the oosphere gives rise to a thick-walled, warty oospore which later germinates to form zoospores.

albumen The white of an egg. (cf. ALBUMINS.)

albumins A class of low-MWt proteins which are soluble in dilute salt solutions and (unlike globulins) readily soluble in water.

Alcaligenes A genus (*incertae sedis*) of catalase-positive, oxidase-positive, Gram-negative bacteria which occur e.g. in soil, water, the alimentary tract in vertebrates, and in clinical specimens. Cells: non-pigmented rods (up to ca. 3.0 μm in length), coccobacilli, or cocci (ca. 0.5–1.0 μm diam.), with 1–12 flagella per cell. Metabolism is respiratory (oxidative); all strains can use O_2 as terminal electron acceptor, and some can use nitrate (anaerobic respiration). All strains can grow chemoorganotrophically on e.g. amino acids, acetate, fumarate, lactate, malate or succinate – carbohydrates being little used, although some strains can use (and form acid from) glucose and/or xylose; some (H_2-oxidizing) strains can grow chemolithotrophically. The organisms usually grow well on e.g. peptone-containing media and blood agar. Alkali is formed from the salts of certain organic acids and from some amides. GC%: ca. 56–70. Type species: *A. faecalis*.

A. denitrificans. Most strains can reduce both nitrate and nitrite to N_2. Strains of subsp. *xylosoxydans* are typically able to use glucose and xylose, those of subsp. *denitrificans* are not. The species includes strains previously named *A. ruhlandii* (H_2-oxidizing organisms which have sheathed, peritrichous flagella), and '*Achromobacter xylosoxidans*'.

A. faecalis. Most strains (including those previously named *A. odorans*) can reduce nitrite but not nitrate. No chemolithotrophic strains have been reported.

A. odorans. See *A. faecalis*.

A. ruhlandii. See *A. denitrificans*.

Species *incertae sedis* [according to Book ref. 22, pp. 370–373] include the peritrichously flagellated, aerobic, H_2-oxidizing bacteria known as *A. eutrophus*, *A. lactus* and *A. paradoxus*, and the non-fermentative, peritrichously flagellated marine bacteria known as *A. aestus*, *A. aquamarinus*, *A. cupidus*, *A. pacificus* and *A. venustus* (see DELEYA); these organisms are considered not to belong to the genus *Alcaligenes*.

[Book ref. 22, pp. 361–373.]

alcian blue A basic dye used e.g. for staining glycoproteins and polysaccharides.

alcohol oxidase (alcohol:oxygen oxidoreductase; EC 1.1.3.13) An enzyme (see ENZYMES) which oxidizes alcohols, giving the corresponding aldehydes and H_2O_2. It is obtained e.g. from *Pichia pastoris*. (See also LIGNIN.)

alcoholic beverages See e.g. BREWING, CIDER, KEFIR, KOUMISS, PULQUE, SAKE, SPIRITS, WINE-MAKING.

alcoholic fermentation (ethanol fermentation) A type of FERMENTATION (sense 1), carried out by various yeasts and other fungi (e.g. species of *Saccharomyces*, *Pichia*, *Aspergillus*, *Fusarium*, *Mucor*) and by certain bacteria (e.g. *Zymomonas*), in which ethanol is formed from D-glucose (or certain other sugars, depending e.g. on organism). In e.g. *Saccharomyces*, glucose is converted to pyruvate via the EMBDEN–MEYERHOF–PARNAS PATHWAY; pyruvate is decarboxylated to acetaldehyde by pyruvate decarboxylase and thiamine pyrophosphate, and acetaldehyde is then reduced to ethanol by NAD-dependent alcohol dehydrogenase – thus allowing reoxidation of the NAD reduced during the EMP pathway. Small amounts of side-products are usually formed, e.g., GLYCEROL (see also NEUBERG'S FERMENTATIONS), acetaldehyde, lactic acid, 2,3-butanediol, succinic and acetic acids, and FUSEL OIL; these occur in proportions which depend on organism and conditions. Alcoholic fermentation by *Saccharomyces* spp is widely exploited commercially: see e.g. BREWING, CIDER, INDUSTRIAL ALCOHOL, SPIRITS, WINE-MAKING. (cf. ZYMOMONAS.)

[Physiological function of alcohol dehydrogenases and long-chain (C_{30}) fatty acids in the alcohol tolerance (~8% ethanol) of a mutant strain of *Thermoanaerobacter ethanolicus*: AEM (2002) *68* 1914–1918.]

alcohols (as antimicrobial agents) Under appropriate conditions certain alcohols can be rapidly lethal to a range of bacteria, fungi and viruses; they have little or no effect on endospores. The mechanism of antimicrobial activity may involve the denaturation of structural proteins or enzymes and/or the solubilization of lipids (e.g. those in the bacterial cytoplasmic membrane, or in the envelope of certain viruses); methanol and ethanol can cause translational errors in protein synthesis. The antimicrobial activity of alcohols increases with molecular weight and with chain length up to ca. C_{10}; above this, insolubility becomes important. Activity decreases in the order primary, *iso*-primary, secondary, tertiary.

Methanol (methyl alcohol, CH_3OH) has poor antimicrobial activity. *Ethanol* (ethyl alcohol, C_2H_5OH) exerts maximum activity as ca. 60–90% (v/v) ethanol/water mixtures. *Isopropanol* (isopropyl alcohol, $(CH_3)_2CHOH$) is less volatile and more effective than ethanol, and is used e.g. as a skin antiseptic. *Phenylethanol* (phenylethyl alcohol, $C_6H_5(CH_2)_2OH$) is more active against Gram-negative than Gram-positive bacteria, and has been used e.g. as a selective agent in bacteriological media. *Phenoxyethanol* (phenoxetol, $C_6H_5O(CH_2)_2OH$) and *benzyl alcohol* (phenylmethanol, $C_6H_5CH_2OH$) are used e.g. as preservatives in pharmaceutical preparations; the activity of benzyl alcohol is improved by halogenation: 2,4-dichlorobenzyl alcohol is used e.g. as a skin antiseptic. *Ethylene glycol*, *propylene glycol* and *trimethylene glycol* (dihydric alcohols) have been used, in aerosol form, for the disinfection of air; a relative humidity of ca. 60% is required. *Bronopol* (2-bromo-2-nitropropan-1,3-diol) is an antibacterial and antifungal compound used e.g. as a preservative in pharmaceutical preparations. The trihydric alcohol *glycerol* is bacteriostatic at concentrations above 50%; it has been used e.g. as a preservative in vaccines. (See also DISINFECTANTS and STERILIZATION.)

aldopentose See PENTOSES.

Alectoria A genus of LICHENS (order LECANORALES); photobiont: a green alga. The thallus is fruticose, greenish-grey or fuscous black, and lacks a whitish central strand (cf. USNEA); ascospores are large, brown when mature, 2–4 per ascus (cf. BRYORIA). Species occur e.g. on trees, rocks etc.

Aleppo boil See CUTANEOUS LEISHMANIASIS.

aleukia Absence or reduced numbers of leucocytes in the blood.

Aleuria A genus of fungi (order PEZIZALES) which form sessile or stipitate, discoid or cup-shaped, minute or conspicuous apothecia in which the hymenium may be red, orange (as in *A. aurantia*, the 'orange peel fungus'), or yellow.

aleuriospore (*mycol.*) A term which has been used to refer to various types of spore – including e.g. thick-walled and thin-walled, pigmented and non-pigmented blastoconidia – and which has become meaningless owing to indiscriminate use.

Aleutian disease of mink A progressive disease of mink caused by an autonomous PARVOVIRUS; it is a virus-induced immune complex-mediated disease and is characterized by glomerulonephritis, arteritis, plasmacytosis and hypergammaglobulinaemia. Death may occur 2–24 months after infection. Virus strain and host genotype are major determinants of disease development. Mink homozygous for the (recessive) Aleutian coat colour gene are most susceptible; infected non-Aleutian mink may have a slow progressive disease, may not have disease, or may later shed the virus. [3D structure of ADM virus and implications for pathogenicity: JV (1999) *73* 6882–6891.]

Alexandrium tamarense *Syn. Gonyaulax tamarensis.*

alexin (1) (*immunol.*) Archaic *syn.* COMPLEMENT. (2) (*plant pathol.*) *Syn.* PHYTOALEXIN.

alfalfa mosaic virus (AMV) A multicomponent ssRNA-containing PLANT VIRUS which has a wide host range and is transmitted via seeds (in some plants), by aphids (non-persistently), and mechanically (under experimental conditions). The genome consists of three linear positive-sense ssRNA molecules: RNA1 (MWt ca. 1.1×10^6), RNA2 (MWt ca. 0.8×10^6) and RNA3 (MWt 0.7×10^6); coat protein mRNA ('RNA4', MWt ca. 0.3×10^6) is also encapsidated. The four RNAs are capped at the 5′ end and occur in at least four different types of virus particle, three bacilliform (B particles: 58 × 18 nm; M particles: 48 × 18 nm; Tb particles: 36 × 18 nm) and one ellipsoidal (Ta particles: ca. 28 × 18 nm). B, M and Tb particles contain one molecule of RNA1, RNA2 and RNA3, respectively; Ta particles contain two molecules of RNA4. RNAs 1, 2 and 3, together with coat protein or RNA4, are necessary for infectivity; coat protein from most ILARVIRUSES can also activate the AMV genome. AMV particles accumulate mainly in the host cell cytoplasm and may form whorled aggregates.

Alferon N™ See INTERFERONS.

ALG (*serol.*) See ANTILYMPHOCYTE SERUM.

algae A heterogeneous group of unicellular and multicellular *eukaryotic* photosynthetic organisms. (cf. CYANOBACTERIA, PROCHLOROPHYTES, RHODOSPIRILLALES; see also MICROORGANISMS.) Algae resemble higher plants in that they evolve oxygen during PHOTOSYNTHESIS, and in that their photosynthetic pigments include CHLOROPHYLL *a*; they differ from vascular plants e.g. in that they typically lack vascular conducting systems – although sieve tubes occur in some of the large brown SEAWEEDS. Algae differ from bryophytes (mosses etc) in that, in most cases, algal reproductive structures (when formed) lack a peripheral envelope of sterile cells (cf. CHAROPHYTES). Some organisms are classified in both algal and protozoal classification schemes: see PHYTOMASTIGOPHOREA.

Aquatic algae occur in fresh, brackish and marine waters (according to species) where they are often important in PRIMARY PRODUCTION. (See also PLANKTON.) Terrestrial algae occur e.g. on damp soil, on ice ('ice algae' – see DIATOMS) and snow ('snow algae' – see RED SNOW), and on tree-trunks etc. Some algae are photobionts in LICHENS or endosymbionts in various organisms (see e.g. ZOOCHLORELLAE and ZOOXANTHELLAE). [Algal symbioses: Book ref. 129.] A few algae are parasitic or pathogenic (see e.g. CHOREOCOLAX, HOLMSELLA, PROTOTHECA, RED RUST). (See also ALGAL DISEASES.)

Certain algae have domestic and/or commercial or industrial uses: see e.g. AGAR, ALGINATE, CARRAGEENAN, DIATOMACEOUS EARTH, FUNORAN, FURCELLARAN, KELP, LAVER, NUNGHAM, SINGLE-CELL PROTEIN, YAKULT.

Algae are classified on the basis of their pigments, types of storage carbohydrate, types and arrangements of flagella, CHLOROPLAST ultrastructure (including arrangement of THYLAKOIDS) and CELL WALL composition. However, there is currently no universally accepted taxonomic scheme which encompasses all the algae; moreover alternative taxonomic schemes coexist even within particular subgroups of algae: see CHLOROMONADS, CHLOROPHYTA, CHRYSOPHYTES, CRYPTOPHYTES, DIATOMS, DINOFLAGELLATES, EUGLENOID FLAGELLATES, PHAEOPHYTA, PRYMNESIOPHYCEAE, RHODOPHYTA, SILICOFLAGELLATES and XANTHOPHYCEAE.

According to species, algae range from unicellular organisms of a few micrometres to seaweeds of 50 metres or more in length. (Unicellular organisms occur in most of the main groups of algae – cf. PHAEOPHYTA.) A CELL WALL is present in most algae but is absent in a few unicellular algae (e.g. PORPHYRIDIUM). The multicellular algae exhibit a great diversity of forms which include branched and unbranched filaments or ribbons, sheets of cells etc; in the thalli of some species there is considerable differentiation – e.g. in *Laminaria* spp the thallus includes structures analogous to root, stem and leaf (*holdfast*, *stipe* and *blade*, respectively) and a system of photosynthate-conducting sieve tubes. (See also PNEUMATOCYST.) (Differentiation occurs also e.g. in the unicellular alga ACETABULARIA.) Meristematic tissue may occur in apical, intercalary and/or diffuse regions depending e.g. on species. Some unicellular algae are motile (see also MOTILITY). Colonial organization is exhibited by certain microalgae (see e.g. COENOBIUM (sense 2) and PALMELLOID PHASE).

Although normally photosynthetic, some algae (e.g. species of CHLAMYDOMONAS, CHLORELLA and SCENEDESMUS) can grow chemoorganotrophically, in the dark, on substrates such as glucose or acetate; some algae (e.g. OCHROMONAS) can ingest particulate food by phagocytosis.

Sexual reproduction (often oogamous) occurs in many algae, and a number of algae exhibit an ALTERNATION OF GENERATIONS which may be isomorphic (e.g. in *Ectocarpus*, *Ulva*) or heteromorphic (e.g. in *Laminaria*).

algal diseases ALGAE are subject to various diseases of microbial aetiology, some of which are of economic importance in seaweeds cultivated for food etc. Thus, e.g., diseases of *Laminaria japonica* ('haidai') include 'frond twist disease' caused by a mycoplasma-like organism, and various rots caused by alginate-degrading bacteria; sporelings in culture may be killed by H_2S produced e.g. by sulphate-reducing bacteria [Book ref. 130, pp. 706–708]. *Porphyra* spp are subject to 'red wasting disease' (= red rot disease, *Pythium* red rot) caused by *Pythium* spp [Experientia (1979) *35* 443–444], and to 'green spot disease' caused by localized infection with species of *Pseudomonas* or *Vibrio*. Various red algae may be attacked by *Petersenia* spp: e.g. *Petersenia palmariae* infects *Palmaria mollis* [CJB (1985) *63* 404–408, 409–418]. Other microorganisms which can infect algae, but whose pathogenicity is uncertain, include e.g. LABYRINTHULAS, PHAGOMYXA, and members of the PLASMODIOPHOROMYCETES. (See also CHOREOCOLAX, HOLMSELLA and PHYCOVIRUS.)

algal rust *Syn.* RED RUST.

algicides Chemical agents which kill algae. Algicides include e.g. copper sulphate (see also BLOOM) and TBTO. Some herbicides (e.g. Diquat, Paraquat, TERBUTRYNE) are also effective against at least some algae, as are various general disinfectants (e.g. chlorine in WATER SUPPLIES).

algin *Syn.* ALGINATE.

alginase *Syn.* ALGINATE LYASE.

alginate A salt of alginic acid: a linear polymer consisting of $(1 \rightarrow 4)$-β-linked D-mannuronic acid residues and $(1 \rightarrow 4)$-α-linked L-guluronic acid residues. (Guluronic acid is the C-5 epimer of mannuronic acid.)

Alginates occur in the CELL WALL and intercellular mucilage in phaeophycean algae; a similar polymer (differing in that at least some of the mannuronic acid residues are acetylated) occurs as capsular material in certain ('mucoid') strains of *Pseudomonas* and in the resting stage of *Azotobacter vinelandii.* (See also CAPSULE (bacterial) and CYST (bacterial).)

Algal alginic acid is insoluble in water but sodium alginate is soluble; alginate solutions form gels in the presence of Ca^{2+}.

The alginate molecule contains mannuronic acid-rich regions ('M-blocks'), guluronic acid-rich regions ('G-blocks'), and regions containing both types of residue ('MG-blocks'). The binding of calcium ions (and other divalent cations) occurs preferentially at the G-blocks; a calcium alginate gel can therefore be envisaged as a three-dimensional network of long-chain molecules cross-linked (between G-blocks) by calcium ions. In the alga, where alginate is in equilibrium with seawater, alginate is associated mainly with calcium, magnesium and sodium ions. The composition (and hence properties) of the polymer varies with species (e.g. alginate from *Laminaria* spp is rich in guluronic acid, while that from *Ascophyllum* and *Macrocystis* is rich in mannuronic acid) and also varies with environmental conditions; the composition of the polymer may even differ in different parts of the same plant.

[Biosynthesis of alginate: Microbiology (1998) *144* 1133–1143.]

Alginate in bacteria.
Genes for alginate synthesis are common in strains of *Pseudomonas aeruginosa*, but strains isolated from the general environment typically do not express these genes. In patients with CYSTIC FIBROSIS, conditions in the lung seem to select for mucoid (i.e. alginate-producing) strains; in these patients *P. aeruginosa* can form a viscous alginate slime associated with a poor prognosis. In *P. aeruginosa* the conversion of non-mucoid strains to mucoidy may occur if a specific SIGMA FACTOR – AlgU (= σ^E) – becomes available for transcription of the alginate genes. Constitutive expression of AlgU may occur as a result of mutation in the *muc* genes, the activity of AlgU being inhibited by binding of MucA. [Mucoidy of *P. aeruginosa* in cystic fibrosis: JB (1996) *178* 4997–5004. Alginate/biofilms/antibiotic resistance in *P. aeruginosa*: JB (2001) *183* 5395–5401.]

AlgU is also required for alginate production in *Azotobacter vinelandii*, and the activity of the sigma factor is similarly regulated by products of the *muc* genes; the products of both *mucA* and *mucC* have a negative role in alginate production [JB (2000) *182* 6550–6556]. [Alginate formation in *A. vinelandii* in the stationary phase: Microbiology (2001) *147* 483–490.]

Commercial applications of alginate.
Alginates have a wide range of uses. Calcium alginate fibres can be made by passing a solution of sodium alginate through a spinneret immersed in a solution of calcium chloride acidified with hydrochloric acid; calcium alginate is precipitated in the form of continuous threads which may be processed (stabilized) to form *calcium alginate wool*. This material (marketed as e.g. 'Calgitex') is used as a COTTON WOOL substitute for making SWABS or absorbent and absorbable surgical dressings. (Calcium alginate swabs have been reported to be inhibitory when used to prepare cultures of herpes simplex virus.) Calcium alginate wool can be sterilized by autoclaving or by dry heat; it can be dissolved e.g. in a 5% solution of sodium citrate or in quarter-strength Ringer's solution containing 1% sodium hexametaphosphate. Thus, a swab carrying an inoculum can be completely dissolved to release its entire complement of microorganisms.

In industry, alginates are used e.g. as emulsifiers and thickeners in foods (alginates are easily digested), cosmetics, pharmaceuticals etc., and as supports for the IMMOBILIZATION of cells or enzymes (for which purpose alginates rich in guluronic acid are preferred as they form stronger gels).

alginate lyase (alginase) Any enzyme within the categories EC 4.2.2.3 and EC 4.2.2.11 which can degrade ALGINATE; such enzymes have been isolated from many types of organism. [Alginate lyase (sources, characteristics, structure–function analysis, roles and applications): ARM (2000) *54* 289–340.]

alginic acid See ALGINATE.

algivorous Feeding on algae.

algology The study of ALGAE.

AlgU (σ^E) See ALGINATE.

alicyclic hydrocarbons See HYDROCARBONS.

alimentary toxic aleukia A severe, usually lethal MYCOTOXICOSIS caused by ingestion of mouldy grain contaminated with certain TRICHOTHECENES – usually T-2 toxin produced by *Fusarium tricinctum*. Symptoms include extreme leucopenia and multiple haemorrhages.

aliphatic hydrocarbons See HYDROCARBONS.

***alkA* gene** See ADAPTIVE RESPONSE.

Alkalescens–Dispar group Non-motile strains of *Escherichia coli* in which glucose is fermented anaerogenically and lactose fermentation is delayed or absent.

alkaline peptone water See APW.

alkaline phosphatase See PHOSPHATASE.

alkaline phosphatase test See e.g. PHOSPHATASE TEST (for milk).

alkaliphile *Syn.* ALKALOPHILE.

alkalophile (alkaliphile) An organism which grows optimally under alkaline conditions – typically exhibiting one or more growth optima within the pH range 8–11 – and which typically grows slowly, or not at all, at or below pH 7. (cf. ACIDOPHILE.) Alkalophiles include a range of bacteria – e.g. certain *Bacillus* spp (including *B. alcalophilus*, *B. firmus* and *B. pasteurii*), *Ectothiorhodospira abdelmalekii*, *Exiguobacterium aurantiacum*, species of *Natronobacterium* and *Natronococcus*, and *Thermomicrobium roseum* – and certain fungi; the organisms occur e.g. in natural alkaline lakes and in waters made alkaline by the effluents from certain industrial processes (such as rayon manufacture). (Natural alkaline environments are characterized by high concentrations of free or complexed Na_2CO_3 – and usually by high concentrations of NaCl.) A number of alkalophiles have an obligate requirement for Na^+ – an ion important e.g. in SYMPORT processes; in at least some flagellated alkalophiles flagellar rotation is driven by SODIUM MOTIVE FORCE. However, in some species capable of growth at neutral pH, Na^+ is required only under non-alkaline conditions. [Book ref. 192; the alkaline saline environment: Book ref. 191, pp. 25–54; genetic engineering of alkalophiles: Book ref. 191, pp. 297–315.]

alkane metabolism See HYDROCARBONS.

***alkB* gene** See ADAPTIVE RESPONSE.

alkene metabolism See HYDROCARBONS.

alkylating agents Agents which react with nucleophilic groups (e.g. amino, carboxyl, hydroxyl, phosphate, and/or sulphhydryl groups) in e.g. proteins and nucleic acids, substituting them with alkyl groups. (As commonly used, the term 'alkylating agent' is also applied to agents which substitute nucleophilic groups with *derivatives* of alkyl groups: e.g. hydroxyethyl groups in the case of ETHYLENE OXIDE.) *Bifunctional* alkylating agents have two reactive groups and can cause cross-linking between nucleophilic groups in proteins and/or nucleic acids (see e.g. NITROGEN MUSTARDS and MITOMYCIN C). (See also SULPHUR MUSTARDS.)

Depending e.g. on their reactivities, alkylating agents can be effective antimicrobial agents and/or MUTAGENS. Some react directly with cell components, others (e.g. alkyl-*N*-nitrosamines: see *N*-NITROSO COMPOUNDS) require prior metabolic activation (apparently to generate an alkyl carbonium cation). In general, methylating agents are more reactive with DNA than are the corresponding ethylating agents.

Alkylating agents form a range of products with DNA. However, only some of the lesions are directly mutagenic: e.g. O^6-alkylguanine can pair with thymine during subsequent DNA replication, resulting in G·C-to-A·T transitions, and alkylation of the O-4 position of thymine can cause A·T-to-G·C transitions. Most other lesions (e.g. N^7-alkylguanine, N^3-alkyladenine) are not directly mutagenic, but they may be lethal unless repaired by the cell (e.g. alkylation of the N-3 position of adenine blocks replication forks); various DNA REPAIR systems can recognize and repair alkylated bases (see e.g. ADAPTIVE RESPONSE).

The mutagenic effects of a given alkylating agent depend largely on the nature of the lesions it produces. For example, MNNG (q.v.), EMS (ethylmethane sulphonate) and MNU (*N*-methyl-*N*-nitrosourea: see *N*-NITROSO COMPOUNDS) produce relatively more directly mutagenic lesions (particularly O^6-alkylguanine), while MMS (methylmethane sulphonate) produces higher proportions of e.g. N^7-methylguanine and N^3-methyladenine but relatively little O^6-methylguanine. However, in organisms (such as *Escherichia coli*) which have an inducible error-prone repair system (see SOS SYSTEM), MMS can be mutagenic by causing lesions which induce this system; even in the case of directly mutagenic agents such as MNNG, a proportion of the mutations induced may be due to error-prone repair in *E. coli* [JB (1985) *163* 213–220].

alkyldimethylbenzylammonium chloride See QUATERNARY AMMONIUM COMPOUNDS.

***N*-alkylnitrosoureas** See *N*-NITROSO COMPOUNDS.

all-glass impinger (bubbler) An instrument used e.g. for sampling the airborne microflora. (See also AIR.) Essentially, it consists of a vertical glass cylinder, containing a volume of liquid, and a longer, narrower glass tube fitted coaxially within the cylinder and partly submerged in the liquid; when suction is applied to the annular space, air is drawn in through the narrow tube and bubbles up through the liquid – during which process particles are transferred from the air to the liquid.

allantoid Sausage-shaped; elongated and slightly curved with rounded ends.

allele (allelomorph) Any of one or more alternative forms of a given GENE; both (or all) alleles of a given gene are concerned with the same trait or characteristic, but the product or function coded for by a particular allele differs, qualitatively and/or quantitatively, from that coded for by other alleles of that gene. Three or more alleles of a given gene constitute an *allelomorphic series*. In a diploid cell or organism the members of an allelic pair (i.e., the two alleles of a given gene) occupy corresponding positions (loci) on a pair of homologous chromosomes; if these alleles are genetically identical the cell or organism is said to be *homozygous* – if genetically different, *heterozygous* – with respect to the particular gene. A *wild-type allele* is one which codes for a particular phenotypic characteristic found in the WILD TYPE strain of a given organism. (See also DOMINANCE.)

allelomorph *Syn.* ALLELE.

allergen An antigen (or autocoupling HAPTEN – e.g., certain drugs) which can initiate a state of HYPERSENSITIVITY (commonly IMMEDIATE HYPERSENSITIVITY) or which can provoke a hypersensitivity reaction in individuals already sensitized with the allergen.

allergic alveolitis See EXTRINSIC ALLERGIC ALVEOLITIS.

allergy (1) A condition in which contact with a given allergen provokes a TYPE I REACTION. (See also PRAUSNITZ–KÜSTNER TEST.) (2) A condition in which contact with a given allergen gives rise to any manifestation of HYPERSENSITIVITY (see e.g. EXTRINSIC ALLERGIC ALVEOLITIS). (3) Formerly, the condition of a PRIMED individual.

allergy of infection An early name for DELAYED HYPERSENSITIVITY.

Allerton disease An African CATTLE DISEASE which involves a mild febrile condition followed by the appearance of skin nodules; it is caused by the bovine mammillitis virus and is apparently similar or identical to pseudo-LUMPY SKIN DISEASE.

Allescheria boydii See PSEUDALLESCHERIA.

alloantigen Antigen from a genetically different individual of the same species.

allochromasy Gradual, spontaneous chemical modification which occurs in the solutions of certain dyes – a single dye becoming a mixture of dyes. (See e.g. polychrome METHYLENE BLUE and NILE BLUE A.)

allochthonous Not indigenous to a given environment. (cf. AUTOCHTHONOUS sense 1.)

alloenzyme See MULTILOCUS ENZYME ELECTROPHORESIS.

allogeneic Derived from a genetically different individual of the same species. (cf. SYNGENEIC; XENOGENEIC.)

Allogromia See FORAMINIFERIDA.

Allogromiina See FORAMINIFERIDA.

allolactose β-D-Galactopyranosyl-(1 $\rightarrow$ 6)-D-glucopyranose: a minor product of β-galactosidase action on LACTOSE; it is the natural inducer of the LAC OPERON in *Escherichia coli.*

Allomonas A genus of bacteria (family VIBRIONACEAE) which have been isolated from fresh water, sewage and faeces; GC%: ca. 57. Type species: *A. enterica* [IJSB (1984) *34* 150–154].

Allomyces A genus of fungi (order BLASTOCLADIALES) which occur in moist soils, muds, and water. The thallus is a branched, coenocytic mycelium which is attached to the substratum by branching rhizoids; in at least some species the cell wall contains CHITIN. (See also CONCENTRIC BODIES.) Some species exhibit an ALTERNATION OF GENERATIONS (q.v.). In e.g. *A. macrogynus*, the sporothallus forms both thick-walled resistant sporangia (*meiosporangia*) and thin-walled sporangia (*mitosporangia*); the meiosporangia give rise to haploid zoospores (which develop into gametothalli), while the mitosporangia form diploid zoospores (which develop into new sporothalli). Terminal branches of a gametothallus give rise to an orange-coloured distal male gametangium and a colourless subterminal female gametangium; the small male gametes fuse with the larger female gametes, and the zygote germinates to form a sporothallus. (See also PHEROMONE.)

Species which do not exhibit sexual processes are sometimes placed in the subgenus *Brachyallomyces*.

allopatric Existing in different environments or geographical regions (cf. SYMPATRIC).

allophycocyanins See PHYCOBILIPROTEINS.

allotype Any one of a range of serologically distinguishable variant forms of an Ig molecule produced as a consequence of allelic variation in the Ig-specifying genes; a given allotype is thus present only in those individuals who have the relevant allele (cf. ISOTYPE). Different allotypes usually differ in amino acid sequence in the 'constant' region of their heavy or light chains; sometimes the variable region is involved. Allotypes in man include e.g. the Gm (G1m, G2m etc) allotypes of IgG.

allulose phosphate pathway *Syn.* RMP PATHWAY.

allylamines A group of synthetic ANTIFUNGAL AGENTS which are highly active against dermatophytes and show somewhat variable activity against yeasts (e.g. *Candida* spp), apparently by inhibiting the enzyme squalene epoxidase; they include naftifine and terbinafine. Terbinafine has *in vitro* activity against *Paracoccidioides brasiliensis* [JCM (2002) *40* 2828–2831].

***Alnus* root nodule** See ACTINORRHIZA.

alopecia Loss of hair.

α (linking number) See DNA.

***alpha* chain** (*immunol.*) See HEAVY CHAIN.

α-factor See MATING TYPE.

α-granules CYANOPHYCIN granules.

alpha interferon See INTERFERONS.

α operon See RIBOSOME (biogenesis).

α peptide ('auto-α') A peptide (185 amino acids long) which is cleaved from the N-terminus of the (*lacZ*-encoded) β-galactosidase of *Escherichia coli* e.g. during autoclaving. The α peptide can restore some β-galactosidase activity to a population of cells which, owing to a deletion mutation in *lacZ*, produce an inactive enzyme lacking the N-terminal portion.

The *lacZ* sequence corresponding to the α peptide can be used as a marker in a CLONING vector. During cloning, use is made of a restriction endonuclease which cleaves at a site within this sequence; thus any insertion of exogenous DNA will usually result in loss of α peptide synthesis. The DNA is introduced into suitable *lacZ* deletion mutants, and the cells are plated on Xgal medium (see XGAL). Cells which receive the intact vector form blue colonies (due to complementation between the α peptide and the defective β-galactosidase), while those receiving recombinant DNA usually form white colonies.

α1-α2 hypothesis See MATING TYPE.

Alphaherpesvirinae (herpes simplex virus group) A subfamily of viruses of the HERPESVIRIDAE (q.v.). Alphaherpesviruses have a short replication cycle (<24 hours); they spread rapidly in cell cultures, causing mass lysis of susceptible cells.

The natural host range varies from narrow to wide, according to virus. While, in cell cultures, latent infection with (non-defective) viruses does not occur readily, latent infection often occurs in nerve ganglia within the living host.

The subfamily includes at least two genera: *Simplexvirus* (human herpesvirus 1 group) and *Poikilovirus* (proposed name) (suid herpesvirus 1 group) [Intervirol. (1986) *25* 141–143].

The type species of the genus *Simplexvirus* is human (alpha) herpesvirus 1 (HERPES SIMPLEX virus type 1, HSV-1). The linear dsDNA genome of HSV-1 (~152 kbp) contains two regions, designated L ('long') and S ('short'), each region flanked by inverted repeats; the ability of each region (L and S) to invert, independently, means that DNA isolated from virions will include four isomeric forms. The genome also includes several copies of a short sequence (*a*) – a *cis*-acting region involved in circularization of the genome. [Molecular epidemiology of herpes simplex virus type 1: RMM (1998) *9* 217–224.]

Other simplexviruses include human (alpha) herpesvirus 2 (HSV-2), bovine – or bovid – (alpha) herpesvirus 2 (bovine mammillitis virus, causal agent of e.g. BOVINE ULCERATIVE MAMMILLITIS), and probably cercopithecine (or cercopithecid) herpesviruses 1 (B VIRUS) and 2.

The type species of the genus *'Poikilovirus'* is suid (alpha) herpesvirus 1 (AUJESZKY'S DISEASE virus). Other members include human (alpha) herpesvirus 3 (varicella-zoster virus (VZV), causal agent of CHICKENPOX and HERPES ZOSTER [review: AVR (1983) *28* 285–356]) and equid (alpha) herpesvirus 1 (equine abortion virus, causal agent of e.g. abortion, respiratory disease and/or neurological disease in horses). (See also DELTA HERPESVIRUS.)

Probable members of the Alphaherpesvirinae include equid herpesvirus 3 (EQUINE COITAL EXANTHEMA virus) and felid herpesvirus 1 (FELINE RHINOTRACHEITIS virus). Possible members of the subfamily include canid herpesvirus 1 (canine herpesvirus).

Alphavirus ('arbovirus group A') A genus of viruses of the family TOGAVIRIDAE (q.v. for replication cycle etc); nearly all members are transmitted by mosquitoes, and many can cause disease (commonly encephalitis or fever with rash and arthralgia) in man and/or animals. Some alphaviruses are grouped into three serologically defined complexes – the complex-specific antigen being associated with the E1 protein (see TOGAVIRIDAE), the species-specific antigen with the E2 protein. The *Semliki Forest virus complex* includes Bebaru virus, CHIKUNGUNYA FEVER virus, GETAH VIRUS, MAYARO FEVER virus, O'NYONG-NYONG FEVER virus, ROSS RIVER VIRUS, Sagiyama virus, SEMLIKI FOREST VIRUS, and Una virus. The *Venezuelan equine encephalomyelitis* (= *Venezuelan encephalitis*) *virus complex* includes Cabassou virus, Everglades virus, Mucambo virus, Pixuna virus, and VENEZUELAN EQUINE ENCEPHALOMYELITIS virus. The *Western equine encephalomyelitis* (= *Western encephalitis*) *virus complex* includes Aura virus, Fort Morgan virus, Highlands J virus, Kyzylagach virus, SINDBIS VIRUS, WESTERN EQUINE ENCEPHALOMYELITIS virus, and What-aroa virus. Other alphaviruses include Barmah Forest virus [JGV (1986) *67* 295–299], originally thought to be a bunyavirus; EASTERN EQUINE ENCEPHALOMYELITIS (= Eastern encephalitis) virus; Middelburg virus; and Ndumu virus.

[Clinical aspects: Book ref. 148, pp. 931–953.]

ALS ANTILYMPHOCYTE SERUM.

Alsever's solution A solution containing D-glucose (20.5 g), sodium citrate dihydrate (8.0 g) and NaCl (4.2 g) dissolved in distilled water (1.0 litre); the pH is adjusted to 6.1 with citric acid. The solution is sterilized by filtration and used for the preservation of sheep blood; blood is added to Alsever's solution (1:1 by volume) and stored at 4°C.

Alternaria A genus of fungi (class HYPHOMYCETES) which include many plant-pathogenic species – e.g. *A. solani* (see e.g. EARLY BLIGHT) and *A. radicina* (causal agent of black rot of carrot seedlings and stored carrots). (See also TIMBER STAINING.) *Alternaria* spp form septate mycelium and pyriform to elongated, dark-coloured conidia which usually have both transverse and longitudinal septa; conidiogenesis is tretic (see CONIDIUM), the conidia developing singly or in chains.

alternaria rot A firm, dark rot produced in various plant hosts by species of *Alternaria*.

alternaric acid A complex compound, produced by *Alternaria solani*, which contains a diketotetrahydropyran group linked to a long-chain fatty acid; it inhibits germination of the spores of certain fungi, and causes wilting and necrosis in the tissues of higher plants.

alternate host (of heteroxenous rust fungi) (1) The *secondary* host (see UREDINIOMYCETES). (2) Sometimes, loosely: either host of a heteroxenous rust.

alternation of generations In the life cycles of some organisms: the alternating formation of one or more generations of mature haploid individuals and one or more generations of mature diploid individuals; in this context 'mature' refers to the ability of the organisms to produce reproductive cells (gametes or spores). An alternation of generations occurs e.g. in some fungi (e.g. *Allomyces* spp), in certain protozoa of the FORAMINIFERIDA, and in many algae (e.g. *Laminaria*).

An individual in the diploid phase is known variously as a *sporophyte*, *sporothallus* or *agamont*. When a sporophyte undergoes meiotic division (*sporic meiosis*) it gives rise to haploid *meiospores*. Each meiospore gives rise to a haploid individual known variously as a *gametophyte*, *gametothallus* or *gamont*; individuals in this generation produce gametes. Male and female gametes may be formed on the same gametophyte (as e.g. in *Allomyces macrogynus*) or on separate male and female gametophytes (as e.g. in *Laminaria*). The gametes fuse to form a (diploid) zygote from which a sporophyte develops. An *isomorphic* (= *homologous*) *alternation of generations* is one in which the gametophyte and sporophyte are morphologically similar. A *heteromorphic* (= *heterologous*) *alternation of generations* is one in which the gametophyte and sporophyte differ morphologically – and perhaps also in other ways: see e.g. CELL WALL (algal).

alternative splicing See SPLIT GENE.

Alteromonas A genus (*incertae sedis*) of aerobic, chemoorganotrophic, Gram-negative bacteria which occur in coastal and marine waters. [Book ref. 22, pp. 343–352.] Cells: straight or curved, round-ended rods, 0.7–1.5 × 1.8–3.0 μm, each having a single, unsheathed polar flagellum. Some species form insoluble pigments: orange and yellow non-carotenoid pigments are formed by *A. aurantia* and *A. citrea*, respectively; *A. luteoviolacea* forms violacein; *A. rubra* forms prodigiosin; some strains of *A. hanedai* form soluble brown pigments. '*A. hanedai*' (see SHEWANELLA) exhibits BIOLUMINESCENCE. Metabolism is exclusively respiratory (oxidative), with O_2 as terminal electron acceptor. All species need Na^+ for growth (optimum: 100 mM Na^+). Utilizable carbon sources vary with species; they include e.g. acetate, alcohols, amino acids, aromatic compounds and sugars. Some species can attack extracellular alginate and/or chitin; several species – including *A. communis*, *A. espejiana*, *A. haloplanktis*, *A. macleodii*, *A. undina* and *A. vaga* – can metabolize D-glucose via an inducible ENTNER–DOUDOROFF PATHWAY. (It has been proposed that *A. communis* and *A. vaga* be transferred to a new genus, *Marinomonas* [JGM (1983) *129* 3057–3074].) No species can accumulate PHB intracellularly. All species grow at 20°C. GC%: ca. 38–50. Type species: *A. macleodii*. (See also BACTERIOPHAGE PM2.)

A. nigrifaciens ([IJSB (1984) *34* 145–149], formerly '*Pseudomonas nigrifaciens*') and '*A. putrefaciens*' (formerly '*Pseudomonas putrefaciens*', now *Shewanella putrefaciens*: see SHEWANELLA) can be responsible for e.g. FISH SPOILAGE and/or MEAT SPOILAGE (see also BUTTER and DFD MEAT).

altro-heptulose *Syn.* SEDOHEPTULOSE.

***Alu* sequences** In the human genome: a family of closely related, dispersed sequences, each ca. 300 nt long, many of which contain a common cleavage site for the restriction enzyme *Alu*I; *Alu*-like sequences occur in the genomes of other mammals and of certain lower eukaryotes. (cf. REP SEQUENCE.)

***Alu*I** A RESTRICTION ENDONUCLEASE from *Arthrobacter luteus*; AG/CT.

ALV AVIAN LEUKOSIS VIRUS.

alveolar membrane (*protozool.*) See PELLICLE (sense 3).

alveolitis Inflammation of the pulmonary alveoli. (cf. PNEUMONIA; see also EXTRINSIC ALLERGIC ALVEOLITIS.)

alveolysin See THIOL-ACTIVATED CYTOLYSINS.

Alysiella A genus of GLIDING BACTERIA (see CYTOPHAGALES); *A. filiformis* occurs in the oral cavity in various vertebrates. The organisms occur as flat filaments – each composed of elongated cells (2–3 μm long) arranged side-by-side. Gliding occurs in a direction perpendicular to the long axis of the cells – i.e. along the axis of the filament. Metabolism: chemoorganotrophic.

Alzheimer's disease See CREUTZFELDT–JAKOB DISEASE.

***am* mutant** An 'amber mutant', i.e., a mutant with an amber NONSENSE MUTATION.

amaas See SMALLPOX.

amanin See AMATOXINS.

Amanita A genus of fungi (AGARICALES, Amanitaceae) which occur in deciduous and/or coniferous woodlands; some species form mycorrhizal associations – e.g. *A. muscaria* with birch (*Betula*). According to species, the colour of the pileus may be e.g. white, yellowish, red or brown. *A. muscaria* (the 'fly agaric') forms a bright red pileus to which often adhere scattered white scales (remnants of the universal veil). Some species are highly poisonous; these include e.g. *A. muscaria* (see also MUSCARINE), *A. pantherina*, *A. phalloides* (the 'death cap fungus'), *A. verna*, and *A. virosa* (the 'destroying angel'). (See also AMATOXINS.)

Amanitaceae See AGARICALES.

α-amanitin An AMATOXIN which, at low concentrations, specifically inhibits eukaryotic RNA POLYMERASE II; RNA polymerase III is inhibited at high concentrations.

amantadine (1-adamantanamine hydrochloride; 1-aminoadamantane hydrochloride) A polycyclic ANTIVIRAL AGENT (adamantane = tricyclodecane, $C_{10}H_{16}$) which inhibits the replication of certain viruses in tissue culture. It is used for the prophylaxis and early treatment of INFLUENZA caused by type A influenzaviruses; it can be administered orally or by aerosol.

At high concentrations, the action of amantadine is non-specific: it raises the pH in endosomes and thus inhibits membrane fusion following endocytosis of a virus (see ENVELOPE). At lower concentrations the drug selectively inhibits an early stage in the infection of type A influenza viruses, the primary target apparently being the M2 protein (see INFLUENZAVIRUS); the activity of the drug results in failure of the pH-dependent fusion of viral and vesicle membranes.

Rimantadine (α-methyl-1-adamantane methylamine hydrochloride) resembles amantadine in its spectrum of activity but apparently causes fewer side-effects.

amanullin See AMATOXINS.

Amapari virus See ARENAVIRIDAE.

amastigote A form assumed by the cells of many species of the TRYPANOSOMATIDAE (q.v.) during certain stages of their life cycles. (See also LEISHMAN–DONOVAN BODIES.)

amatoxins Toxic cyclic peptides which occur in some species of *Amanita*, e.g. *A. phalloides*, *A. verna*. In man, small quantities of toxin (e.g. 5 mg) may be lethal; clinical effects are produced in ca. 8–24 hours after ingestion. Initial symptoms include severe vomiting and diarrhoea; degenerative changes occur in the liver and kidneys, and death may follow within a few days.

Amatoxins include α-AMANITIN, β-, γ- and ε-amanitins, and amanin. A *non*-toxic compound of similar chemical composition, *amanullin*, also occurs in *A. phalloides*. The amatoxins are more toxic than the PHALLOTOXINS found in *Amanita* spp. (Other *Amanita* toxins include tryptamines such as *bufotenine* (see also HALLUCINOGENIC MUSHROOMS) and isoxazole alkaloids such as *ibotenic acid*.)

amber codon See GENETIC CODE.

amber mutation See NONSENSE MUTATION.

amber suppressor See SUPPRESSOR MUTATION.

ambisense RNA A viral ssRNA genome or genome segment which is positive-sense with respect to some genes but negative-sense with respect to others (see VIRUS). Ambisense genome segments have been found in the ARENAVIRIDAE and in the genus PHLEBOVIRUS. In these cases the −ve-sense sequences are transcribed directly into (viral-complementary) subgenomic mRNA; however, +ve-sense sequences are expressed only after genome replication, subgenomic mRNA being transcribed from the RNA strand complementary to the genomic strand. [Review: AVR (1986) *31* 1–51.]

Amblyospora See MICROSPOREA.

amboceptor Current usage: *syn.* HAEMOLYTIC IMMUNE BODY.

ambrosia fungi Fungi which grow in the tunnels made by wood-boring ambrosia beetles, e.g. *Xyleborus* spp (Scolytidae). The beetle larvae and adults form tunnels mainly in the sapwood of fallen timber and of dead or weakened standing trees; healthy trees are not normally attacked. The beetles derive nutrients mainly or solely from the fungal growth lining their tunnels (i.e., they are 'xylomycetophagous'); wood apparently plays little or no direct part in their nutrition. (In some cases the ambrosia fungi also appear to be necessary for reproduction or pupation of the beetles, possibly by supplying an essential sterol.) Each of the many species of ambrosia beetle is associated with one or more particular species of fungus, usually a hyphomycete (e.g. *Ambrosiella* sp, *Fusarium* sp) or an ascomycete (e.g. *Ambrosiozyma* sp, *Dipodascus* sp). The fungus grows in the tunnels as a palisade-like layer or as separate or confluent sporodochia, frequently bearing chains of conidia or terminal chlamydospores; the mycelium may penetrate the wood to a depth of a few millimetres. Ambrosia fungi appear to use only storage sugars, starch etc in the wood cells, and do not cause significant damage to the structural components of the wood. Fungal propagules (spores or yeast-like cells) are carried to new tunnels in specialized pockets (*mycetangia* or *mycangia*) in the exoskeleton of the (usually female) beetle; mycetangia contain an oily secretion and differ in structure and location in different species of beetle.

Bark-boring ('phloeophagous') beetles of the Scolytidae are also associated with various fungi, including e.g. sap-stain fungi (mostly *Ceratocystis* spp): see also DUTCH ELM DISEASE. In many cases these associations are fortuitous and non-specific, but some blue-stain fungi may be carried in mycetangium-like structures, such a structure occurring e.g. at the anterior margin of the prothorax in *Dendroctonus frontalis*.

(See also WOODWASP FUNGI.)

Ambrosiozyma A genus of fungi (family SACCHAROMYCETACEAE) which form budding yeast cells, pseudomycelium, and true mycelium with dolipore-like septa. Asci are formed on the hyphae; ascospores are bowler-hat-shaped. Species have been isolated e.g. from the tunnels of wood-boring beetles. [Book ref. 100, pp. 106–113.] (cf. AMBROSIA FUNGI.)

ambruticin An antifungal antibiotic (a cyclopropylpolyene-pyran acid) obtained from a strain of *Polyangium cellulosum*; it shows in vitro activity against e.g. *Candida* spp, dermatophytes, and other pathogenic fungi.

amdinocillin *Syn.* MECILLINAM.

ameba *Syn.* AMOEBA.

American foulbrood A BEE DISEASE which affects the larvae of *Apis mellifera* – usually after they have spun their cocoons; the causal agent is *Bacillus larvae*. Infection occurs by ingestion of food contaminated by spores of *B. larvae*. The spores germinate in the gut, and the bacteria penetrate to the haemolymph and multiply; the larvae die, turn brown, and putrefy. (cf. EUROPEAN FOULBROOD.)

amerosporae See SACCARDOAN SYSTEM.

Ames test (Mutatest; *Salmonella*/microsome assay) A test for detecting whether or not a particular agent is mutagenic (and hence possibly carcinogenic) by determining its ability to cause reversion to prototrophy in certain histidine-requiring mutants of *Salmonella typhimurium*. Various 'tester strains' of *S. typhimurium* may be used, each having a different type of MUTATION (frameshift, missense or nonsense) in the histidine operon. Many of the strains used also contain mutations in *uvrB* (preventing EXCISION REPAIR) and in *rfa* (causing LPS deficiency and hence increased permeability to certain chemicals); most contain the plasmid pKM101 which carries genes for error-prone repair (see SOS SYSTEM) and which thus enhances the mutagenic effects of DNA-damaging agents. Since certain chemicals are mutagenic/carcinogenic only after metabolic activation, the test commonly includes a preparation of microsomal enzymes from a liver homogenate (9000 *g* supernatant, fraction 'S9') obtained from rats pre-treated with a carcinogen (to induce the appropriate enzymes).

The test may be carried out e.g. as a 'plate incorporation test'. A culture of a particular tester strain of *S. typhimurium*, an S9 preparation, and the chemical under test are mixed with soft agar containing a low concentration of histidine, and this is poured onto a minimal agar plate; the whole is then incubated at 37°C for 48 h in the dark. The low level of histidine permits limited growth of the auxotrophic mutant cells, resulting in a background of confluent light growth in the upper layer of agar ('top agar'); any prototrophic revertants (whose growth is not limited) can be seen as isolated colonies. In scoring revertants, account must be taken of the (known) spontaneous reversion rate for the strain used. (Absence of background growth implies that the agent under test has general antibacterial activity, and any colonies which develop are unlikely to be revertants.) The basic test has been modified in various ways for particular purposes. [Example of use with a carbamic acid derivative: AAC (2005) *49* 1160–1168.]

(See also SOS CHROMOTEST.)

amicyanin A BLUE PROTEIN (MWt ca. 12000) present in certain methylamine-utilizing bacteria (e.g. '*Pseudomonas* AM1') – see METHYLOTROPHY.

Amies transport medium See TRANSPORT MEDIUM.

amikacin A semi-synthetic derivative of KANAMYCIN A which contains an α-aminohydroxybutyric acid residue; it is more active than kanamycin against e.g. *Pseudomonas aeruginosa* and has been used against gentamicin-resistant strains.

aminacrine *Syn.* 9-AMINOACRIDINE.

amino acids For principal biosynthetic pathways see Appendix IV; see also e.g. AROMATIC AMINO ACID BIOSYNTHESIS, GLUTAMIC ACID, OPERON (attenuator control). For standard abbreviations for amino acids see table.

AMINO ACIDS: standard abbreviations

Amino acid	Three-letter abbreviation	One-letter abbreviation
alanine	Ala	A
arginine	Arg	R
asparagine	Asn	N
aspartic acid	Asp	D
cysteine	Cys	C
glutamic acid	Glu	E
glutamine	Gln	Q
glycine	Gly	G
histidine	His	H
isoleucine	Ile	I
leucine	Leu	L
lysine	Lys	K
methionine	Met	M
phenylalanine	Phe	F
proline	Pro	P
serine	Ser	S
threonine	Thr	T
tryptophan	Trp	W
tyrosine	Tyr	Y
valine	Val	V

9-aminoacridine (aminacrine) A substituted acridine (see ACRIDINES) used e.g. as an ANTISEPTIC for the treatment of wounds, and as a laser dye.

aminoacyl-tRNA synthetase See PROTEIN SYNTHESIS.

aminoadipic acid pathway (AAA pathway) A pathway for lysine biosynthesis [see Appendix IV(e)] which occurs only in certain lower fungi (Blastocladiales, Chytridiales, Mucorales), ascomycetes (including e.g. *Saccharomyces*), basidiomycetes, and euglenoid flagellates. [Review: CRM (1985) *12* 131–151.] (cf. DIAMINOPIMELIC ACID PATHWAY.)

***p*-aminobenzoic acid** (PABA; PAB) A component of FOLIC ACID. PABA is synthesized from chorismate [see Appendix IV(f)]. Certain microorganisms – e.g. *Clostridium* and *Lactobacillus* spp – require PABA as a growth factor. (See also PAS and SULPHONAMIDES.)

7-aminocephalosporanic acid See CEPHALOSPORINS.

aminoglycoside antibiotics A class of broad-spectrum ANTIBIOTICS; a typical aminoglycoside antibiotic contains an aminosugar and either STREPTIDINE or (more commonly) 2-DEOXYSTREPTAMINE. [Book ref. 14, pp 418–442, gives detailed chemical structures.]

This group of antibiotics is generally taken to include e.g. AMIKACIN, APRAMYCIN, butirosin, FRAMYCETIN, GENTAMICIN, hygromycin B, KANAMYCIN, KASUGAMYCIN, LIVIDOMYCIN, NEAMINE, NEOMYCIN, netilmicin, PAROMOMYCIN, ribostamycin, SISOMYCIN, SPECTINOMYCIN, STREPTOMYCIN and TOBRAMYCIN.

Many aminoglycoside antibiotics are bactericidal but some (e.g. kasugamycin, spectinomycin) are bacteriostatic. Aminoglycoside antibiotics are active against both Gram-positive and Gram-negative bacteria; at least some (e.g. gentamicin, kanamycin, streptomycin) are inactive, or only weakly active, against certain archaeans (methanogens [SAAM (1985) *6* 125–131]), while others (e.g. kasugamycin, neomycin, paromomycin, STREPTOMYCIN) are active against both prokaryotic and eukaryotic microorganisms.

Aminoglycoside antibiotics are less effective under anaerobic conditions and are ineffective against obligate anaerobes.

The aminoglycoside antibiotics are widely used therapeutically, sometimes in combination with other drugs. [Clinical roles of aminoglycosides (symposium): Am. J. Med. (1985) *79* (1A) 1–76.] Side-effects (particularly with neomycin and streptomycin) include dose-dependent damage to the 8th cranial nerve (associated with ototoxicity); kidney damage and hypersensitivity reactions are also possible.

Aminoglycoside antibiotics bind to the 30S ribosomal subunit in bacteria – some (e.g. streptomycin) bind at a single site while others (e.g. gentamicin, kanamycin, neomycin) apparently bind at multiple sites. They inhibit PROTEIN SYNTHESIS. For example, low levels of streptomycin cause misreading of mRNA (i.e. incorporation of incorrect amino acids) while higher levels completely inhibit protein synthesis – apparently by blocking ribosomes specifically at the start of translation. There are other sites of activity (e.g. the outer membrane in Gram-negative bacteria), but the main effect of these antibiotics appears to result from their influence on protein synthesis.

Resistance to aminoglycoside antibiotics can arise by several distinct mechanisms: (i) mutation(s) in ribosomal proteins of the 30S subunit which can affect the binding of these antibiotics (see e.g. STREPTOMYCIN); (ii) modification (inactivation) of the antibiotics by plasmid-encoded or chromosome-encoded bacterial enzymes which carry out *O*-phosphorylation, *N*-acetylation or *O*-adenylation; (iii) reduced uptake. [TIM (1998) *6* 323–327.]

Mutations giving one-step high-level resistance to aminoglycoside antibiotics are uncommon, but resistance to some (e.g. spectinomycin, streptomycin) can occur in this way; spectinomycin-resistant mutants do not show cross-resistance to streptomycin, and streptomycin-resistant mutants do not show cross-resistance to those antibiotics (e.g. kanamycin) which appear to bind at multiple sites on the ribosome.

Aminoglycoside antibiotic-inactivating enzymes include acetyltransferases (AACs), adenylyltransferases (AADs) and phosphotransferases (APHs) [for nomenclature and sites of action see: BMB (1984) *40* 28–35]. Some of these enzymes can inactivate only a few aminoglycoside antibiotics while others can inactivate many; some are found only in Gram-negative species *or* in Gram-positive species, and some are found only in e.g. *Pseudomonas* or *Staphylococcus* spp. Some of the enzymes occur in a wide range of species. (See also e.g. Tn*5* and Tn*21*.)

Certain (mutant) bacteria exhibit drug-dependence in respect of certain aminoglycoside antibiotics: see e.g. SPECTINOMYCIN and STREPTOMYCIN.

6-aminopenicillanic acid (6-APA) A derivative of natural PENICILLINS (see β-LACTAM ANTIBIOTICS for structure): a precursor of many semi-synthetic penicillins. 6-APA itself has little or no antibacterial activity.

aminopeptidase See PROTEASES. (See also METHIONINE AMINOPEPTIDASE).

aminopterin See FOLIC ACID ANTAGONIST.

2-aminopurine See BASE ANALOGUES.

aminoquinolines A group of ANTIMALARIAL AGENTS that include 4-aminoquinolines (e.g. chloroquine, mefloquine, amodiaquine) and 8-aminoquinolines (e.g. primaquine). The quinoline-type antimalarials (including quinine) apparently kill the parasite by preventing its detoxification of the ferriprotoporphyrin IX by-product of haemoglobin metabolism (see HAEMOZOIN and QUININE).

Chloroquine is an inexpensive synthetic drug which is used orally for treating uncomplicated malaria. It has also been used parenterally to treat severe malaria; however, as resistant strains of the parasite are common, other drugs may be recommended for the treatment of severe disease.

Chloroquine is generally the drug of choice for non-falciparum malaria, and is also used (with e.g. proguanil) for prophylaxis.

Mefloquine has been used for uncomplicated falciparum malaria and (via nasogastric tube) for the treatment of severe disease. It has also been used prophylactically where, owing to resistance, chloroquine/proguanil may be ineffective. Mefloquine prophylaxis is absolutely contraindicated in the first trimester of pregnancy, and should not be used in the second or third trimesters; its therapeutic use in pregnancy, especially during the first trimester, should take into consideration the potential risk to the fetus. The drug is not recommended e.g. for treating patients with epilepsy or for prophylaxis in those with e.g. severe renal insufficiency or abnormal liver function [mefloquine (evaluation): Drugs (1993) *45* 430–475].

The mode of resistance to quinoline-type drugs is poorly understood, but a consistent finding is that resistant strains accumulate lower levels of the drug. Most studies have been carried out on chloroquine, and several mechanisms have been suggested. (i) Weakened proton pump activity in the membrane of the parasite's food vacuole may result in a higher pH within the vacuole; this would tend to decrease the level of protonation of chloroquine in the vacuole and, hence, militate against its retention within the vacuole (the non-protonated form of the drug can freely diffuse outward across the vacuolar membrane). However, the mechanism of chloroquine accumulation assumed in this model (i.e. uptake by inward diffusion into the vacuole, and retention following protonation) may not account for the high levels of drug achieved in sensitive strains of the parasite. (ii) The vacuolar membrane in *P. falciparum* includes an ABC TRANSPORTER, designated Pgh1, encoded by gene *pfmdr1*. When *pfmdr1* was expressed in Chinese hamster ovary cells, Pgh1 apparently localized within the membrane of the lysosome and mediated uptake of chloroquine. This may indicate that Pgh1 mediates the uptake of chloroquine into the food vacuole of *P. falciparum*; this is supported by the observation that 'deamplification' of *pfmdr1* occurs in mutants selected for chloroquine resistance [EMBO (1992) *11* 3067–3075]. Interestingly, when chloroquine-resistant mutants were selected for mefloquine resistance (by mefloquine-limited growth), the mefloquine-resistant mutants were found to over-express Pgh1 and to be less resistant to chloroquine (but more resistant to e.g. quinine). Hence, resistance to chloroquine and mefloquine seem to be inversely related; this is not understood. Moreover, the apparent cross-resistance between mefloquine and quinine may mean that resistance to mefloquine in geographical areas not previously exposed to mefloquine may have developed as a result of exposure to quinine [PNAS (1994) *91* 1143–1147].

[Mode of action and mechanism of resistance to antimalarial drugs: Acta Tropica (1994) *56* 157–171.]

***p*-aminosalicylic acid** See PAS.

amitosis In certain eukaryotes: an atypical form of MITOSIS (q.v.) in which the chromosomes do not condense, a spindle is not formed, the nuclear membrane persists throughout division, and the nucleus divides by constriction.

amixis In certain haploid organisms: a type of reproduction, considered to be a deviant form of the sexual process, in which karyogamy and meiosis do not occur even though morphologically differentiated structures (e.g. spore-containing asci) are formed; thus, e.g., the eight haploid ascospores formed (per ascus) by *Podospora arizonensis* are said to be derived by mitotic divisions of each nucleus of the dikaryon in the ascus mother cell. (cf. APOMIXIS.)

Ammodiscus See FORAMINIFERIDA.

Ammonia See FORAMINIFERIDA.

ammonia assimilation Ammonia can be used as the sole source of nitrogen by many types of microorganism; it may be obtained from an external source or produced intracellularly – for example, by deamination reactions or by ASSIMILATORY NITRATE REDUCTION.

Ammonia can be taken up readily through the cytoplasmic membrane; ammonium ions require a TRANSPORT SYSTEM [transport of ammonium ions in bacteria: FEMS Reviews (1985) *32* 87–100].

Within the cell, ammonia can be assimilated in different ways.

With high concentrations of ammonia, some bacteria (including *Escherichia coli, Klebsiella* spp) use ammonia for the reductive amination of 2-oxoglutarate to L-glutamate in a reaction catalysed by NADPH-dependent *glutamate dehydrogenase* (GDH; EC 1.4.1.2). L-Glutamate can act as an N-donor in transamination reactions in which amino acids are synthesized from 2-oxoacids; *E. coli* synthesizes three low-specificity transaminases: transaminase A (which preferentially catalyses the synthesis of e.g. L-alanine and L-aspartate), B (aromatic amino acids) and C (branched-chain amino acids).

With low concentrations of ammonia (e.g. <ca. 1 mM) many bacteria and other organisms assimilate ammonia via a two-step reaction which is catalysed by (i) *glutamine synthetase* (GS; EC 6.3.1.2) and (ii) *glutamine:2-oxoglutarate aminotransferase* (GOGAT; *glutamate synthase;* EC 1.4.1.13). In this pathway, glutamate is initially aminated to glutamine in an ATP-dependent, GS-catalysed reaction which consumes ammonia. Then, in an NADPH-dependent, GOGAT-catalysed amination, one molecule of glutamine and one of 2-oxoglutarate yield two molecules of glutamate (see Appendix IV(a)). Many organisms can use the initial (GS-catalysed) step for glutamine synthesis, glutamine being used as a donor of amino groups in the biosynthesis of e.g. purines, carbamoyl phosphate (a precursor of pyrimidines), histidine and tryptophan.

In *E. coli*, it appears that the GDH-catalysed pathway is used under energy-limiting conditions (the other pathway is ATP-dependent). [Pathway choice in glutamate synthesis in *E. coli*: JB (1998) *180* 4571–4575.] Mutants of *E. coli* lacking both pathways cannot use exogenous ammonia, but can grow on an external source of glutamate.

In enterobacteria, the GDH and GS/GOGAT systems are regulated at the level of transcription and (in the case of GS) enzyme activity. High concentrations of ammonia repress the synthesis of GS (and other nitrogen-regulated genes: see NTR GENES) and induce the synthesis of GDH; this situation is reversed when ammonia levels become limiting. GS activity is controlled in response to ammonia levels, the enzyme being inactivated by (reversible) adenylylation in the presence of high levels of ammonia.

In e.g. *Streptomyces venezuelae* the *anaerobic* assimilation of ammonia involves synthesis of alanine by the reductive amination of pyruvate via NADH-dependent *alanine dehydrogenase* (EC 1.4.1.1) [CJM (1985) *31* 629–634].

In *Streptomyces clavuligerus* the GS/GOGAT pathway appears to be the only one for ammonia assimilation [JGM (1986) *132* 1305–1317].

ammoniacal silver nitrate See SILVER.

ammoniated mercury See MERCURY.

ammonification The formation of free ammonia (or NH_4^+) during the microbial breakdown of nitrogenous organic matter and/or during DISSIMILATORY NITRATE REDUCTION. Ammonia does not normally accumulate under aerobic conditions (being readily assimilated by a wide range of organisms), but may do so under anaerobic conditions. (See also MINERALIZATION; NITROGEN CYCLE; SAPROBITY SYSTEM.)

ammonium salt sugars A range of media used for determining which (if any) of a series of carbohydrates are metabolized by a given organism; these media are used e.g. for those organisms which form excess alkaline products from peptone media and which, therefore, cannot be tested in PEPTONE-WATER SUGARS. The medium contains (g/l): KCl (0.2), $MgSO_4.7H_2O$(0.2), $(NH_4)_2HPO_4$ (1.0), yeast extract (0.2), agar (20) and bromcresol purple (4 ml of a 0.2% solution); the carbohydrate is added after autoclaving (to give a final concentration of 1% w/v), and the medium is poured to form slopes.

amoeba (ameba) A type of cell or organism characterized by its ability to alter its shape drastically, generally by the extrusion of one or more pseudopodia (see PSEUDOPODIUM). Such *amoeboid* cells are characteristic of the SARCODINA, but also occur in other groups – for example, in the genus HISTOMONAS (see also MACROPHAGE).

Amoeba A genus of free-living amoebae (order AMOEBIDA). The cell is differentiated into an outer hyaline layer (the *ectoplasm*) and an inner granular *endoplasm*. Each cell typically forms several lobopodia, one of which is dominant at a given time, and each of which has a hemispherical tip and (usually) a hyaline cap. Reproduction occurs asexually by binary fission.

A. proteus is common in slow-moving or still freshwater habitats – often on the undersides of floating leaves; it feeds mainly on small flagellates, ciliates and bacteria. The cell may reach ca. 600 μm diam. (See also X-BACTERIA.)

amoebiasis (1) Any disease of man or animals caused by a protozoon of the AMOEBIDA – see e.g. DYSENTERY (b) and MENINGOENCEPHALITIS. (2) *Syn.* amoebic DYSENTERY.

amoebic dysentery See DYSENTERY (b).

amoebicide Strictly: any chemical agent which kills amoebae; in practice, the term is commonly used for any drug which is active against (particularly) *Entamoeba histolytica*, even when the drug may inhibit, rather than kill, the parasite (i.e., it may act as an *amoebistat*). (See e.g. ARSENIC; DILOXANIDE; EMETINE; FUMAGILLIN; 8-HYDROXYQUINOLINE; METRONIDAZOLE.)

Amoebida An order of naked, typically uninucleate (cf. CHAOS) amoebae (class LOBOSEA) in which mitochondria are typically present (cf. PELOBIONTIDA) and flagellate stages do not occur (cf. SCHIZOPYRENIDA). Most species occur in fresh water and/or moist soil (e.g. ACANTHAMOEBA, AMOEBA, *Echinamoeba*, MAYORELLA, *Rosculus*, *Saccamoeba*, *Thecamoeba*, TRICHAMOEBA, *Vannella*, *Vexillifera*); some are marine (e.g. *Flabellula*, PARAMOEBA, most *Platyamoeba* spp), and a few are parasitic (see e.g. ENDAMOEBA, ENDOLIMAX, ENTAMOEBA, IODAMOEBA, MALPIGHAMOEBA, PARAMOEBA). (See also HYDRAMOEBA.) [Descriptions of freshwater and soil species and genera: Book ref. 133.]

Amoebidiales See TRICHOMYCETES.

Amoebidium See TRICHOMYCETES.

amoebistat See AMOEBICIDE.

Amoebobacter See CHROMATIACEAE.

amoeboflagellate An essentially amoeboid protozoon which can – at least under certain circumstances – form flagella as well as pseudopodia. (See e.g. SCHIZOPYRENIDA; HISTOMONAS.)

amoeboid AMOEBA-like.

amoeboid movement See PSEUDOPODIUM.

amoeboma See DYSENTERY (b).

amoebostome A sucker-like structure present on free-living cells of *Naegleria fowleri*; amoebostomes are apparently used to engulf food particles [JP (1985) *32* 12–19]. (cf. CYTOSTOME).

Amorphosporangium A genus of bacteria (order ACTINOMYCETALES, wall type II) which occur e.g. in leaf litter. The organisms resemble *Actinoplanes* spp but differ e.g. in that the sporangium is highly irregular in shape, and the zoospores are rod-shaped. Type species: *A. auranticolor*.

Amorphotheca resinae See HORMOCONIS.

amoxycillin A PENICILLIN; 6-(α-amino-4-hydroxybenzylamido) penicillanic acid. (See also AUGMENTIN.)

AMP Adenosine 5′-monophosphate [see Appendix V(a)]. (See also ADENYLATE KINASE; cf. CYCLIC AMP.)

Amphiacantha See RUDIMICROSPOREA.

amphibiotic (1) Amphibious. (2) Able to behave either as a commensal or as a parasite in a given host. (See also OPPORTUNIST PATHOGENS.)

amphibolic pathway A metabolic pathway which can have both catabolic and anabolic functions: e.g. the TCA CYCLE.

Amphidinium See DINOFLAGELLATES.

amphiesma The wall – including the vesicles, thecal plates (if present), etc. – of a DINOFLAGELLATE.

amphigenous Situated or growing on both sides of, or all over, a given structure.

amphipathic Having both hydrophilic and hydrophobic regions.

amphiphilic *Syn.* AMPHIPATHIC.

amphispore A type of dark, thick-walled uredospore formed e.g. by some species of *Hyalospora* and *Puccinia*.

Amphistegina See FORAMINIFERIDA.

amphithecium (*lichenol.*) *Syn.* THALLINE MARGIN.

amphitrichous Refers to the presence of a single flagellum (or e.g. fimbria) at each pole of a cell.

amphitrophic Refers to an organism which can grow photosynthetically in the light and chemotropically in the dark.

ampholyte An electrolyte whose molecule has both acidic and basic groups; the pH of a solution of a given ampholyte depends on the relative strengths of its acidic and basic groups. (See also ISOELECTRIC FOCUSING.)

Amphora A genus of pennate DIATOMS in which the cingulum is much wider on one side (the 'dorsal' side) than on the other, resulting in the two valves facing in more or less the same direction; the raphes in both valves can thus both make contact with the substratum. Species are aquatic, benthic organisms capable of GLIDING MOTILITY.

amphoteric dye See DYE.

amphotericin B A MYCOSAMINE-containing heptaene POLYENE ANTIBIOTIC produced by *Streptomyces nodosus*. It is active against a wide range of fungi and certain pathogenic protozoa; it is administered orally for the treatment of various mycoses, including systemic infections, but its clinical use is limited by its nephrotoxicity. (See also VISCERAL LEISHMANIASIS.)

amphotropic Refers to a retrovirus which can replicate both in its host of origin and in 'foreign' host cells (e.g. a murine retrovirus which can replicate in both murine and non-murine mammalian cells). Murine amphotropic viruses do not show cross-interference or cross-neutralization with the ecotropic or xenotropic murine retroviruses, i.e., their *env* glycoproteins are unrelated to those of the ecotropic or xenotropic viruses [Book ref. 114, p. 73]. (cf. DUALTROPIC.)

ampicillin See PENICILLINS.

amplicon (1) In a target molecule of nucleic acid: the specific sequence of nucleotides which is to be copied (amplified) by a nucleic-acid-amplification technique such as NASBA or PCR.

(2) A copy of the specific nucleotide sequence which has been amplified by a nucleic-acid-amplification technique such as NASBA or PCR.

(3) A defective virus vector [e.g. PNAS (1985) *82* 694–698].

amplicon containment See PCR.

amplicon inactivation See PCR.

amplification (*mol. biol.*) See GENE AMPLIFICATION.

amplification pathway (*immunol.*) See COMPLEMENT FIXATION (b).

amplified *Mycobacterium tuberculosis* direct test See TMA.

amplimer An alternative (though infrequently used) name for a primer in a PCR reaction.

AmpliTaq Gold™ DNA polymerase See HOT-START PCR.

AmpliWax™ See HOT-START PCR.

ampoule *Syn.* VIAL.

AMPPD® Trade name (Tropix, Bedford, MA, USA) of a 1,2-dioxetane substrate which gives rise to CHEMILUMINESCENCE when activated by the enzyme alkaline phosphatase.

amprenavir See ANTIRETROVIRAL AGENTS.

amprolium A coccidiostatic agent which is structurally similar to thiamine.

ampulla (1) (*protozool.*) In certain ciliates (e.g. some hymenostomes): a channel, or the wider part of a channel, leading into a contractile vacuole.

(2) (*protozool.*) In certain hypostome ciliates: an adhesive organelle.

(3) (*mycol.*) The expanded, terminal region of some types of conidiogenous cell (see CONIDIUM).

(4) (*mycol.*) The expanded, terminal region of some types of conidiophore (see e.g. ASPERGILLUS).

Ampullariella A genus of bacteria (order ACTINOMYCETALES, wall type II) which occur e.g. in soil and freshwater habitats. The organisms resemble *Actinoplanes* spp but differ e.g. in that the rod-shaped zoospores are formed in parallel rows within the cylindrical sporangium. Type species: *A. regularis*.

ampullate development See CONIDIUM.

ampulliform Flask-shaped.

***Amsacta moorei* EPV** See ENTOMOPOXVIRINAE.

AMTDT See TMA.

AMU ATOMIC MASS UNIT.

amv See MYB.

AMV Avian myeloblastosis virus: see AVIAN ACUTE LEUKAEMIA VIRUSES.

AMV reverse transcriptase See REVERSE TRANSCRIPTASE.

amygdaliform Shaped like an almond.

amygdalin A β-glycoside, present e.g. in bitter almonds, consisting of GENTIOBIOSE linked to mandelonitrile. On β-glucosidase action (or acid hydrolysis), amygdalin yields hydrogen cyanide, glucose and benzaldehyde.

amylases Enzymes which cleave glucosidic linkages in e.g. STARCH or GLYCOGEN.

α-Amylases ((1 → 4)-α-D-glucan 4-glucanhydrolases, EC 3.2.1.1) are endoenzymes which have little action on terminal (1 → 4)-α-bonds or on bonds adjacent to (1 → 6)-α branch points. They act on amylopectin and glycogen to form glucose, maltose and branched α-limit DEXTRINS, and on amylose to form first maltose and maltotriose, then slowly on maltotriose to form maltose and glucose. α-Amylases are common among microorganisms. They are obtained commercially mainly from *Bacillus* spp (see also IMMOBILIZATION (sense 1)) and are used e.g. for processing starch to form glucose syrups: insoluble starch granules are dispersed in water by heating, and the starch is partially hydrolysed with thermostable α-amylases from *B. amyloliquefaciens* and/or *B. licheniformis*. Further hydrolysis is achieved using e.g. γ-amylases (see below).

β-Amylases ((1 → 4)-α-D-glucan maltohydrolases, EC 3.2.1.2) are exoenzymes which cleave alternate bonds from the non-reducing end of a linear (1 → 4)-α-D-glucan; thus, e.g. amylose is degraded to maltose. β-Amylase action is halted at (1 → 6)-α branch points; thus maltose and β-limit dextrins are formed from glycogen and amylopectin. β-Amylases are common in plants and are produced e.g. by *Bacillus* and *Streptomyces* spp. They may be used, together with DEBRANCHING ENZYMES, in the manufacture of maltose syrups from starch. β-Amylases are inhibited by SCHARDINGER DEXTRINS.

γ-Amylases (amyloglucosidases, glucoamylases, (1 → 4)-α-D-glucan glucohydrolases, exo-(1 → 4)-α-glucosidases, EC 3.2.1.3) are exoenzymes which cleave (1 → 4)-α-bonds consecutively from the non-reducing end of a (1 → 4)-α-D-glucan to yield β-D-glucose; they can also cleave (1 → 6)-α- and (1 → 3)-α-bonds, although at a much lower rate. γ-Amylases are found mainly in fungi and are obtained commercially e.g. from *Aspergillus niger* and *Rhizopus* spp. They are used to convert malto-oligosaccharides (e.g. limit dextrins) to D-glucose (see above).

(See also BREAD-MAKING and ENZYMES.)

amyl-3-cresol See PHENOLS.

amyloglucosidases See γ-AMYLASES and DEBRANCHING ENZYMES.

amyloid (1) Starch-like; staining blue or blue-black with iodine (e.g. MELTZER'S REAGENT). (2) (*med.*) An abnormal protein deposited, intercellularly, in various tissues in certain pathological conditions.

amylolytic Capable of hydrolysing STARCH.

amylopectin See STARCH.

amyloplast A non-pigmented PLASTID which synthesizes and stores STARCH.

amylose See STARCH.

Amylostereum See APHYLLOPHORALES (Stereaceae).

amylovorin An extracellular polysaccharide which is produced by strains of *Erwinia amylovora* and which may be involved in the production of wilt symptoms in FIREBLIGHT; it consists essentially of a (1 → 3)- and (1 → 6)-β-linked galactan backbone with side-chains containing residues of galactose, glucuronic acid, glucose and pyruvate.

amylum stars Star-shaped aggregates of starch-filled cells which develop from the lower nodes in CHAROPHYTES and function in vegetative propagation.

amytal A barbiturate which acts as a mitochondrial RESPIRATORY INHIBITOR: its binding site(s) and mechanism of action appear to be similar to those of ROTENONE. Many bacterial systems are partially or completely insensitive to amytal.

Anabaena A genus of filamentous CYANOBACTERIA (section IV) in which the trichomes are made up of spherical, ovoid or cylindrical vegetative cells; heterocysts may be intercalary or terminal. GC%: 38–44. Species include e.g. *A. azollae*, *A. cylindrica*, *A. flos-aquae* and *A. variabilis*, distinguished e.g. on the basis of cell size and shape, filament form, etc; however, some of these characteristics may be variable in culture.

A. flos-aquae contains GAS VACUOLES; it can form extensive BLOOMS in (usually) freshwater habitats, and may produce various toxins (*anatoxins*) which may be lethal to animals which drink the water. Anatoxin-*a* is an alkaloid which resembles cocaine; it acts as a potent post-synaptic depolarizing neuromuscular blocking agent, causing stupor, tremors, prostration, convulsions, opisthotonos and death.

A. azollae occurs in symbiotic association with species of the small floating freshwater fern *Azolla* (Azollaceae). The nitrogen-fixing *A. azollae* occurs in a specialized cavity at the base of the upper lobe of each leaflet; it has a higher proportion of heterocysts than do free-living *Anabaena* spp, and grows poorly when isolated from the fern. *Azolla* grows e.g. in rice fields in the Far East, where it is an important source of nitrogen. The nitrogen becomes available to the rice plants only when the fern dies, sinks and decomposes; this occurs when the temperature rises and coincides with growth and tillering of the rice plants. *Azolla* has also been used as a forage crop for animals in Asia. [Expression of two kdp operons in *Anabaena* strain L-31 and relevance to desiccation: JV (2005) *71* 5297–5302.]

anabolism The metabolic reactions by which cell components, extracellular products etc are built up from organic and/or inorganic precursors. Anabolism requires an input of energy which, in organotrophs, may be derived from CATABOLISM of exogenous organic compounds (cf. LITHOTROPH). (See also ANAPLEROTIC SEQUENCES.)

Anacystis A genus of cyanobacteria; those strains designated '*A. nidulans*' are apparently strains of *Synechococcus* (restricted sense: see SYNECHOCOCCUS) [Ann. Mic. (1983) *134*B 21–36]. [Cell cycle events in '*A. nidulans*': JGM (1984) *130* 2535–2542.]

anaerobe An organism which has the ability to grow in the absence of oxygen, the term commonly being reserved for those organisms which, in nature, *normally* grow in – or can grow only in – anaerobic habitats; some anaerobes can also grow under aerobic conditions, i.e., they are FACULTATIVE aerobes. (cf. AEROBE; see also AEROTOLERANT and MICROAEROPHILIC.) Obligately anaerobic organisms occur e.g. in anaerobic river muds and in the RUMEN (see also ANAEROBIC DIGESTION and TERMITE–MICROBE ASSOCIATIONS).

The energy-yielding metabolism of an anaerobe may be primarily or solely fermentative (see FERMENTATION sense 1), respiratory (oxidative) (see ANAEROBIC RESPIRATION), or photosynthetic (see anoxygenic PHOTOSYNTHESIS).

Culture of obligate anaerobes. A strict anaerobe requires an oxygen-free gaseous phase above the surface of the medium (see ANAEROBIC JAR) and a medium free from dissolved oxygen; even under these conditions, however, many anaerobes will not grow unless the medium has been pre-reduced, i.e., poised at or below a particular REDOX POTENTIAL. The E_h of chemically pre-reduced media is usually −150 mV to −350 mV at pH 7; the precise E_h required by a given anaerobe may depend e.g. on the size of the inoculum – ongoing growth tending to lower the E_h in the surrounding medium. Apart from incubation per se, the handling of a strict anaerobe (i.e., transfer, inoculation etc) is often carried out in an oxygen-free environment (see ROLL-TUBE TECHNIQUE). However, it has been reported that certain strict anaerobes (e.g. *Fusobacterium* sp) can be handled in air, without significant loss of viability, when use is made of media (containing reducing agents) which have been stored anaerobically prior to inoculation; nevertheless, a significant loss of viability occurs when use is made of cysteine-containing media which have been stored aerobically [CJM (1984) *30* 228–235]. Reducing agents containing a sulphhydryl group (e.g., H_2S, cysteine, thioglycollate) have been found to increase the oxygen sensitivity of various sulphate-reducing bacteria [FEMS Ecol. (1985) *31* 39–45]. (See also BREWER'S THIOGLYCOLLATE MEDIUM; COOKED MEAT MEDIUM; POUR PLATE; SHAKE CULTURE (1).)

Oxygen sensitivity of anaerobes. Some anaerobes are killed by exposure to gaseous oxygen, while the growth of others may be stopped or merely retarded. In some cases oxygen sensitivity is believed to be due to the lack of enzymes such as CATALASE and SUPEROXIDE DISMUTASE – the cells thus lacking protection against the toxic metabolites H_2O_2 and O_2^-. Another suggestion is that an unacceptably high (positive or low negative) E_h may lead to a high intracellular E_h which, in turn, may e.g. prevent reduction in redox couple(s) which are normally freely reversible within the growing organism, and/or cause irreversible oxidation of vital components.

anaerobic (1) Refers to an environment in which oxygen is absent. (cf. AEROBIC.) (2) Having the characteristic(s) of ANAEROBE(s).

anaerobic digestion The anaerobic breakdown of complex animal and/or plant materials to simple substances which include a high proportion of gaseous and soluble products; it involves a diverse range of organisms whose metabolic pathways are closely interrelated, and occurs e.g. in benthic muds, in the RUMEN and in certain types of SEWAGE TREATMENT plant (see also TERMITE–MICROBE ASSOCIATIONS).

Initially, polysaccharides, lipids and proteins are cleaved, by extracellular enzymes, to (respectively) sugars, (usually) fatty acids and glycerol, and amino acids; most polysaccharides – including cellulose, pectins and starch – are readily cleaved, but LIGNIN is not readily degraded. The sugars and other small molecules are fermented by various organisms (e.g. species of *Bacteroides*, *Clostridium*, *Selenomonas*, *Succinivibrio* and *Veillonella*) to products which include acetate, butyrate, ethanol, lactate, propionate and succinate – together with carbon dioxide and hydrogen. Propionate can be metabolized to acetate and/or carbon dioxide by e.g. *Desulfobulbus*, *Desulfonema* and *Syntrophobacter*. [Propionate and succinate degradation: JGM (1985) *131* 643–650.]

Carbon dioxide and hydrogen are metabolized to acetate by some bacteria (e.g. *Acetobacterium woodii*, *Clostridium aceticum* – see ACETOGENESIS) and to methane by certain METHANOGENS. The scavenging of hydrogen by these organisms creates conditions suitable for the growth of the so-called *obligate proton reducers* or *obligate hydrogenogens* (see e.g. SYNTROPHOMONAS) – which metabolize ethanol, lactate (and other products of the initial fermentation) to acetate and hydrogen; for these organisms only protons can act as electron acceptors, and the continual scavenging of the resulting hydrogen is essential for their growth.

In some anaerobic ecosystems acetate is metabolized to methane by e.g. *Methanosarcina* and/or oxidized to carbon dioxide by SULPHATE-REDUCING BACTERIA (e.g. *Desulfobacter postgatei*) and sulphur-reducing species (e.g. *Desulfuromonas acetoxidans*); the latter organisms form hydrogen sulphide as a by-product. In the presence of high concentrations of sulphate, the sulphate-reducing bacteria can inhibit METHANOGENESIS by using both acetate and hydrogen. In the rumen, however, much of the acetate is absorbed by the ruminant itself.

The process of anaerobic digestion results in an appreciable reduction in the bulk of biodegradable solids – a particularly useful feature in SEWAGE TREATMENT. The conversion of organic carbon to methane forms an important part of the CARBON CYCLE.

In sewage treatment (and in the treatment of other organic wastes – such as agricultural and food-industry effluents) anaerobic digestion can yield a rich, relatively odourless agricultural fertilizer (rich in microbial biomass) and a useful fuel gas (*biogas*, also called *marsh gas*, *sewer gas*, *sludge gas*) which may contain 50% or more of methane; ca. 30–70% of solids may be converted to gases – the actual proportion depending e.g. on the substrate(s), the initial percentage of solids and the retention time.

anaerobic jar A container used for the incubation of materials (e.g. inoculated media) in the absence of oxygen or, in general, under gaseous conditions other than atmospheric. See e.g. BREWER JAR, GASPAK, MCINTOSH AND FILDES' ANAEROBIC JAR. (See also HOLDING–FLUSH JAR PROCEDURE.)

anaerobic respiration RESPIRATION under anaerobic conditions, the terminal electron acceptor being (according to species and/or environmental conditions): fumarate, nitrate, nitrite, nitrous oxide, elemental sulphur, sulphate etc. See e.g. DISSIMILATORY SULPHATE REDUCTION, FUMARATE RESPIRATION, NITRATE RESPIRATION, SULPHUR RESPIRATION and DENITRIFICATION. (See also dissimilatory perchlorate reduction in the entry BIOREMEDIATION.)

anaerobiosis (1) The complete absence, or (loosely) the paucity, of gaseous or dissolved elemental oxygen in a given place or environment. In nature, such conditions occur e.g. in the bottom muds of ponds and rivers and in the intestinal tracts of animals (see e.g. RUMEN). (See also ANAEROBIC JAR and ROLL-TUBE TECHNIQUE.) In aerobic habitats, anaerobic microenvironments may be created by the consumption of locally available oxygen e.g. by facultatively anaerobic organisms.

(2) Life in the absence of air.

Anaerobiospirillum A genus of Gram-negative bacteria (family BACTEROIDACEAE). Cells: round-ended helical rods, 0.6–0.8 × 3.0–8.0 μm, with bipolar tufts of flagella. Various sugars (e.g. glucose, lactose, sucrose) are fermented, the major products of glucose metabolism being succinate and acetate. GC%: ca. 44. Type species: *A. succiniciproducens*, first isolated from dogs. [In human bacteraemia: JCM (1998) *36* 1209–1213. Ileocolitis in cats: JCM (2004) *42* 2752–2758.]

anaerogenic Refers to an organism which does not produce gas from a given substrate.

Anaeroplasma A genus of obligately anaerobic, cell wall-less, sterol-requiring bacteria (class MOLLICUTES) which occur e.g. in the RUMEN. Cells: non-motile, coccoid to pleomorphic; plasmalogens are major components of polar lipids in the cytoplasmic membrane. Some strains are bacteriolytic, causing clearing in agar media containing e.g. *Escherichia coli*. Carbohydrates are fermented to e.g. acetic, formic and lactic acids (together with propionic or succinic acid in some strains), ethanol and CO_2. GC%: ca. 29–34. Type species: *A. abactoclasticum*; other species: *A. bactoclasticum*. [Nucleic acid relationships among anaerobic mycoplasmas: JGM (1985) *131* 1223–1227.]

Anaerovibrio A genus of Gram-negative bacteria (family BACTEROIDACEAE) which occur e.g. in the RUMEN. Cells: curved rods, ca. 0.5 × 1.2–3.6 μm; monotrichously (polarly) flagellated. Carbon sources include e.g. fructose, lactate and ribose; major fermentation products from these substrates include propionic and acetic acids and CO_2. Fermentation of glycerol yields mainly propionic acid. Lipolytic, producing clearing on linseed oil agar. Type species: *A. lipolytica*.

ana-holomorph (*mycol.*) A fungus for which no sexual stage or state has been identified. (See also DEUTEROMYCOTINA.)

anammox bacteria Bacteria able to oxidize ammonia to dinitrogen, *anaerobically*, using nitrite as electron acceptor [see e.g. Nature (2003) *422* 608–611].

anamnestic response In persons or animals: a heightened response to the second or subsequent administration of a particular antigen given some time after the initial administration; thus, e.g., the concentration of specific antibody increases at a rate much greater than that in response to the first dose of the antigen. (See also BOOSTER.)

anamorph (*mycol.*) The asexual (= imperfect) stage or state of a given fungus. (cf. ANA-HOLOMORPH; HOLOMORPH; SYNANAMORPH; TELEOMORPH.)

anamorph-genus *Syn.* FORM GENUS.

anamorph-species *Syn.* form species.

anaphase See MITOSIS and MEIOSIS.

anaphylactic shock A manifestation of IMMEDIATE HYPERSENSITIVITY involving a TYPE I REACTION. In man, the symptoms of anaphylactic shock include bronchospasm (contraction of bronchial smooth muscle) and cyanosis; in severe cases death may occur, in minutes or hours, from asphyxiation due to bronchiolar contraction and/or laryngeal oedema or from a major fall in blood pressure resulting from vascular permeability. First-line treatment includes administration of ADRENALIN.

anaphylatoxins The fragments C3a and C5a formed during COMPLEMENT FIXATION; they induce the release of HISTAMINE from MAST CELLS, and C5a is chemotactic for neutrophils and monocytes.

anaphylaxis *Syn.* TYPE I REACTION.

Anaplasma A genus of Gram-negative bacteria (family ANAPLASMATACEAE) – once thought to be protozoa of the PIROPLASMEA. Cells: cocci, diam. ca. 0.3 μm. They infect erythrocytes, multiplying by binary fission within membrane-limited vacuoles (formed by invagination of the erythrocyte membrane) to form inclusion bodies – each containing up to ca. 10 cells. The inclusion body typically occupies either a central or a marginal position within the erythrocyte. *Anaplasma* spp (*A. caudatum*, *A. centrale*, *A. marginale* and *A. ovis*) occur worldwide; they apparently infect only ruminants (see ANAPLASMOSIS). (cf. PARANAPLASMA.)

Anaplasmataceae A family of bacteria (order RICKETTSIALES) that included AEGYPTIANELLA, ANAPLASMA, EPERYTHROZOON and HAEMOBARTONELLA. In a proposed reorganization, changes include transfer of species of EHRLICHIEAE and WOLBACHIEAE to this family; *in this edition of the dictionary the lower taxa of this family reflect earlier classifications*. [Proposed reorganization of Anaplasmataceae and Rickettsiaceae: International Journal of Systematic and Evolutionary Microbiology (2001) *51* 2145–2165.]

anaplasmosis Any disease caused by an *Anaplasma* sp; transmission occurs mainly via ticks. In cattle, *A. marginale* can cause a severe anaemia with debility and wasting; in sheep and goats infection with *A. marginale* or *A. ovis* is usually subclinical. [Book ref. 33, pp. 875–878.]

anaplerotic sequences Sequences of reactions which serve to replenish 'pools' of intermediates which have been depleted as a consequence of biosynthesis – see e.g. TCA CYCLE.

Anaptychia A genus of foliose or fruticose LICHENS (order LECANORALES). Photobiont: a green alga. Apothecia: lecanorine; ascospores: brown, one-septate. Species may occur on e.g. rocks or bark.

anatid herpesvirus 1 See DUCK VIRUS ENTERITIS.

anatoxin (1) See ANABAENA. (2) *Syn.* TOXOID.

Ancalochloris See CHLOROBIACEAE.

Ancalomicrobium A genus of chemoorganotrophic, facultatively anaerobic PROSTHECATE BACTERIA found in aquatic habitats. In high nutrient concentrations the cells are non-motile, rounded to rod-shaped; in low nutrient concentrations they first become ovoid, knobbed, subpolarly flagellate cells, subsequently developing into non-motile forms bearing two to eight cylindrical (occasionally bifurcated) prosthecae. Cell division occurs by budding from the cell body. One species: *A. adetum*.

anchor sequence See SIGNAL HYPOTHESIS.

Ancistrocoma See HYPOSTOMATIA.

Ancistrum See SCUTICOCILIATIDA.

Ancylobacter See MICROCYCLUS (2).

Ancylonema A genus of saccoderm DESMIDS.

Andersen sampler An instrument used e.g. for sampling the airborne microflora. (See also AIR.) Essentially, the instrument permits air to be drawn, sequentially, over the surface of each of a number of vertically stacked agar plates or adhesive-coated glass or metal discs; a suction pump, connected at the base of the instrument, draws air in through an opening at the top of the stack. Before reaching each plate or disc, the air passes through a radially-arranged pattern of holes in a metal screen – the size of the holes in a given screen being smaller than those in the preceding one; hence, the size range of particles collected by a given plate depends on the position of the plate within the stack.

Andes virus See HANTAVIRUS.

Andrade's indicator A PH INDICATOR used e.g. in PEPTONE-WATER SUGARS. To 100 ml 0.5% aqueous acid FUCHSIN is added 15 ml 4% NaOH; the solution is allowed to stand at room temperature for a day, or until its colour changes to a very pale yellow. (A further 1–2 ml NaOH may be added if this has not occurred after 24h.) This solution (before or after autoclaving) is added to the medium (1 ml in 90 ml); the indicator becomes red at ca. pH 5.5.

androphage (male-specific phage) Any BACTERIOPHAGE which infects only host cells that contain a conjugative plasmid. Androphages include members of the genus INOVIRUS and of the family LEVIVIRIDAE. (cf. FEMALE-SPECIFIC PHAGE.)

anemochoric Dispersed by the wind.

anergy (*immunol.*) A state of unresponsiveness to antigenic stimulus; the term is used to refer e.g. to a B cell or T cell, or to an individual person or animal. The absence of a DELAYED HYPERSENSITIVITY reaction in a PRIMED subject on exposure to the relevant antigen would be one manifestation of anergy; for example, anergy may be exhibited when tuberculous individuals – particularly those with extensive pulmonary tuberculosis – are given a tuberculin test. 'T cell anergy' can be induced by a SUPERANTIGEN.

aneuploid (anorthoploid) Having one or more complete chromosomes in excess of, or less than, the normal haploid, diploid or polyploid number characteristic of the species. (See also HETEROPLOID.)

aneurin *Syn.* THIAMINE.

Angeiocystis See EIMERIORINA.

angiocarpic development (angiocarpous development) (*mycol.*) The mode of development of a fruiting body in which the spores reach maturity within a closed chamber or cavity in the fruiting body. (See e.g. GASTEROMYCETES.) (cf. GYMNOCARPIC DEVELOPMENT.)

Angiococcus See MYXOBACTERALES.

angiomatosis (bacillary) See BARTONELLA.

ang-kak (ang-quac, angkak; red rice) A Chinese product (used mainly as a colouring agent for foods) made by the fermentation of rice by *Monascus purpureus* strains; the fungus produces two pigments: monascorubrin (red) and monascoflavin (yellow).

angolamycin See MACROLIDE ANTIBIOTICS.

Ångström unit (Å) 10^{-10} m (= 10^{-1} nm).

anguibactin See SIDEROPHORES.

anguidine Diacetoxyscirpenol: see TRICHOTHECENES.

anicteric Without jaundice.

AN-IDENT See MICROMETHODS.

aniline dyes Dyes derived from aniline (aminobenzene, $C_6H_5NH_2$), e.g., some TRIPHENYLMETHANE DYES.

animal viruses Viruses (see VIRUS) which can infect and replicate in the cells of animals – including mammals, birds, poikilothermic vertebrates and/or invertebrates. See table for general features.

The host range for a given animal virus varies from very narrow (e.g. in nature, rubella and variola viruses apparently infect only man) to very wide (e.g. rabies virus can infect most mammals, while ARBOVIRUSES can replicate in both invertebrates and mammals). A virus may be transmitted from host to host directly (by body contact), via aerosols (DROPLET INFECTION), via food or water (see e.g. FOOD POISONING and POLYHEDRA), via a VECTOR, etc; some viruses can be transmitted transovarially or can cross the placenta and infect the fetus.

Probably, many virus infections in animals are asymptomatic, but some can result in diseases ranging from mild and self-limiting (e.g. most cases of the COMMON COLD) to severe or fatal (e.g. AIDS, RABIES, YELLOW FEVER). The virus may remain more or less localized at the site of infection (e.g. in some forms of HERPES SIMPLEX, and in PAPILLOMAS), or it may spread systemically via the bloodstream, lymphatic system or nervous system – often replicating preferentially or only in a specific tissue or organ. (See also TROPISM sense 2 and PERSISTENCE.) Disease symptoms may be due e.g. to disturbances in host cell structure and/or metabolism, host cell lysis, immune responses in the host animal, etc. Animal viruses have been implicated in certain cancers (see DNA TUMOUR VIRUS and ONCOVIRINAE) and in other diseases of unknown causation – e.g. atherosclerosis (see MAREK'S DISEASE), MULTIPLE SCLEROSIS, and even psychiatric disorders (see BORNA DISEASE and RETROVIRIDAE).

Identification of animal viruses often cannot be based entirely on clinical manifestations of infection: infection may be asymptomatic, different viruses can sometimes cause similar diseases (see e.g. ENCEPHALITIS), and similar viruses may cause widely differing diseases (see e.g. ENTEROVIRUS). Thus, identification often depends on laboratory procedures: e.g., light MICROSCOPY (for e.g. CYTOPATHIC EFFECT) or ELECTRON MICROSCOPY of clinical specimens; propagation of the virus in TISSUE CULTURE, EMBRYONATED EGGS or living animals, followed by observation of any virus-induced effects; serological tests (e.g. COMPLEMENT-FIXATION TEST, ELISA, HAEMADSORPTION-INHIBITION TEST, HAEMAGGLUTINATION-INHIBITION TEST, IMMUNOFLUORESCENCE, IMMUNOSORBENT ELECTRON MICROSCOPY, NEUTRALIZATION TEST, etc). TYPING (e.g. for epidemiological purposes) may involve e.g. serological tests and/or nucleic acid HYBRIDIZATION techniques. PCR-based examination may be used e.g. to identify drug-resistance mutations [genotyping of HIV-1: Lancet (1999) *353* 2195–2199].

In only a few cases can virus diseases in animals be controlled by chemotherapy (see ANTIVIRAL AGENTS); in most cases prevention is the only means of disease control. In humans, this may involve e.g. vaccination, isolation of infected patients, elimination of vectors, etc; SMALLPOX has been effectively eradicated by such methods. Virus diseases of other animals may be controlled by similar methods, and also e.g. by slaughter of infected animals, quarantine of animals for transportation, etc. Viruses which cause INSECT DISEASES are sometimes used in the BIOLOGICAL CONTROL of insect pests.

(cf. PLANT VIRUSES.)

animalcules (Animaliculae) An archaic term for microscopic organisms, particularly protozoa.

anisogamy The union of gametes which differ in form and/or physiology. When the gametes differ in size, the larger (*macrogamete*) is regarded as female, the smaller (*microgamete*) as male. (See e.g. OOGAMY; cf. ISOGAMY.)

ANIMAL VIRUSES: basic characteristics

Genome[a]	Presence of envelope	Nucleocapsid symmetry	Virus family[b] (host type[c] in parentheses)
dsDNA			
monopartite, ccc, sc	+	helical	Baculoviridae (I)
monopartite (partially ds)	+	isometric	Hepadnaviridae (V)
monopartite, linear	+	isometric	Herpesviridae (V)
monopartite, strands covalently joined at each end	+/−	complex	Poxviridae (V;I)
monopartite (possibly sometimes bipartite), linear	+/−	isometric	Iridoviridae (I;V)
monopartite, linear, with 5′-terminal protein	−	isometric	Adenoviridae (V)
monopartite, ccc, sc	−	isometric	Papovaviridae (V)
ssDNA			
monopartite, linear; −ve-sense, or either −ve- or +ve-sense	−	isometric	Parvoviridae (V;I)
ccc	−	isometric	porcine circovirus (V)
dsRNA (linear)			
bipartite, with terminal protein	−	isometric	Birnaviridae (V;I)
multipartite	−	isometric	Reoviridae (V;I;P)
ssRNA (linear)			
+ve-sense, monopartite	+	helical	Toroviridae[d] (V)
+ve-sense, monopartite	+	helical	Coronaviridae (V)
+ve-sense, 'diploid'	+	isometric	Retroviridae (V;I?)
+ve-sense, monopartite	+	isometric	Togaviridae (V;I;P?)
+ve-sense, monopartite	+	isometric	Flaviviridae (V;I)
+ve-sense, monopartite	−	isometric	Caliciviridae (V)
+ve-sense, monopartite	−	isometric	*Nudaurelia* β virus group (I)
+ve-sense, monopartite	−	isometric	Picornaviridae (V;I)
+ve-sense, bipartite	−	isometric	Nodaviridae (I)
−ve-sense? monopartite	+	helical	Filoviridae[d] (V)
−ve-sense, monopartite	+	helical	Paramyxoviridae (V)
−ve-sense, monopartite	+	helical	Rhabdoviridae (V;I;P)
−ve-sense[e], bipartite	+	helical	Arenaviridae (V)
−ve-sense[e], tripartite	+	helical	Bunyaviridae (V;I)
−ve-sense, multipartite	+	helical	Orthomyxoviridae (V)
ssRNA (circular)			
?-sense, monopartite	+	isometric	delta virus (V)

[a] See entry VIRUS for generalized account. ccc = covalently closed circular; ds = double-stranded; sc = supercoiled; ss = single-stranded.
[b] See separate entries for details.
[c] I = invertebrate; P = plant; V = vertebrate.
[d] Proposed family.
[e] Some segments are ambisense.

anisokont *Syn.* HETEROKONT.

Anisolpidiaceae See HYPHOCHYTRIOMYCETES.

anisomycin (2-[*p*-methoxybenzyl]-3,4-pyrrolidinediol 3-acetate) An antibiotic which inhibits protein synthesis in eukaryotes, binding to 80S ribosomes and inhibiting the peptidyltransferase reaction. Many members of the ARCHAEA (e.g. species of *Halobacterium*, *Methanobacterium* and *Thermoplasma*) are also sensitive to anisomycin, although *Methanococcus* and *Methanosarcina* spp are relatively resistant [Book ref. 157, p. 538].

anisotropic inhibitor A type of compound (e.g. monoazide ethidium) which inhibits energy transduction in OXIDATIVE PHOSPHORYLATION by binding to certain negatively charged sites on the outer surface of the mitochondrial inner membrane; the occurrence of redox reactions within the membrane stimulates the binding of anisotropic inhibitors to these sites, and this has been taken to indicate that redox reactions cause conformational changes in membrane proteins [PNAS (1985) *82* 1331–1335]. The binding sites of anisotropic inhibitors (hydrophobic proteins designated *chargerin I* and *chargerin II*) may correspond to sites in Complexes I, III and IV of the mitochondrial ELECTRON TRANSPORT CHAIN and to the F_0 moiety of the PROTON ATPASE.

anisotropic medium (optics) See BIREFRINGENCE.

Ankistrodesmus A genus of unicellular, non-motile green algae (division CHLOROPHYTA); the cells are generally long and thin and may be straight, curved, helical etc, sometimes occurring in bundles. (See also TBTO.)

anlage (*pl.* anlagen) (*ciliate protozool.*) *Syn.* INITIAL(S).

annealing (of primers) See PCR.

annelate ascus An ASCUS whose wall has a small, annular, apical thickening that is often amyloid or light-refractive; at maturity, the ascospores are released via a pore in the centre of the thickening.

annellations See CONIDIUM.

annellide See CONIDIUM.

annellidic development See CONIDIUM.

annulus (*microbiol.*) Any of various structures which are ring-shaped (either actually or in cross-section). Examples include e.g. the outermost layer (= 'sheath') of the contractile stalk in some peritrich ciliates, the PERISEPTAL ANNULUS, and the annular remnant of the PARTIAL VEIL.

Anodonta See ZOOCHLORELLAE.

Anomoeoneis See DIATOMS.

anonymous mycobacteria *Syn.* ATYPICAL MYCOBACTERIA.

Anopheles A genus of mosquitoes (order Diptera, family Culicidae); *Anopheles* spp are vectors of MALARIA.

anopheline Refers to mosquitoes of the genus *Anopheles*.

Anoplophrya See ASTOMATIDA.

anorthoploid *Syn.* ANEUPLOID.

anoxia A deficiency or lack of oxygen. (Hence *adj.* anoxic.)

anoxygenic Not producing oxygen.

anoxygenic photosynthesis See PHOTOSYNTHESIS.

Anoxyphotobacteria A proposed class of anoxygenic photosynthetic prokaryotes (division GRACILICUTES): see RHODOSPIRILLALES.

ansamycins A group of structurally related ANTIBIOTICS consisting of a macrocyclic ring which comprises an aromatic moiety spanned by an aliphatic bridge. Ansamycins (which include e.g. rifamycins, streptovaricins and tolypomycins) are produced by actinomycetes and appear to share a common mode of antibacterial action (see RIFAMYCINS); at high concentrations they also show activity against certain viruses in cell cultures.

anserid herpesvirus 1 See DUCK VIRUS ENTERITIS.

antagonism The inhibition of one entity by another of similar kind. Examples: one antibiotic may reduce or abolish the effects of another (e.g. a bacteriostatic drug will counteract the effects of a drug which acts only on growing cells); acid-producing organisms may inhibit or kill acid-intolerant organisms sharing the same environment. (cf. SYNERGISM.)

antapical organelle In the bacterium *Thiovulum majus*: the presumed source of a thin (adhesive?) thread or stalk which projects from the surface of the organism.

antenna (in PHOTOSYNTHESIS) See LIGHT-HARVESTING COMPLEX.

anterior station (*parasitol.*) The anterior part of the alimentary canal and/or the salivary glands of an arthropod vector; infective forms of parasites in the anterior station can be transmitted to the vertebrate host by the bite of the vector. (cf. INOCULATIVE INFECTION.)

antheraxanthin See CAROTENOIDS.

antheridiol See PHEROMONE.

antheridium A male gametangium.

anthracnose Any of various plant diseases – particularly those caused by fungi of the MELANCONIALES – in which discrete, dark-coloured, necrotic lesions develop on the leaves and/or fruits. For example, anthracnose of beans (*Phaseolus* spp) is caused by *Colletotrichum lindemuthianum* and is characterized by brown or black lesions (sometimes bearing masses of pink conidia) on the leaves, pods and seeds.

anthracycline antibiotics ANTIBIOTICS which have a tetrahydrotetracenequinone chromophore (containing three flat, coplanar, 6-membered rings). They function as INTERCALATING AGENTS, inhibiting DNA and RNA synthesis. *Daunomycin* (= rubidomycin, daunorubin), produced by *Streptomyces* sp, is an anticancer agent which binds to dsDNA (ϕ ca. 12°) with no apparent base preference; it contains an aminosugar (daunosamine) which projects into the major groove and stabilizes the drug–DNA complex by electrostatic interaction with phosphate in the DNA. *Adriamycin* is very similar to daunomycin in structure and function. Other anthracyclines: *nogalamycin* (antitumour agent, ϕ ca. 18°), *cinerubin*, *rhodomycin*.

anthramycin A member of the pyrrolo-(1,4)-benzodiazepine group of antibiotics; other members include sibiromycin and tomaymycin. Anthramycin (produced by *Streptomyces refuineus*) has antibacterial and antitumour activity; it binds covalently to DNA, causing e.g. inhibition of DNA and RNA synthesis. The modes of action of sibiromycin and tomaymycin are apparently similar.

anthranilate An intermediate in tryptophan biosynthesis: see Appendix IV(f).

anthrax A disease – primarily of animals, but also of man – caused by strains of *Bacillus anthracis* that are both capsulated and toxinogenic (see ANTHRAX TOXIN, BACILLUS and STERNE STRAIN).

In man the disease is usually localized, resulting from infection of skin lesions (*anthrax boil*, *cutaneous anthrax*, *malignant pustule*); sepsis may develop in untreated cases. Rarely, infection occurs via the lungs (*pulmonary anthrax*, *woolsorters' disease*) or gut (*intestinal anthrax*), leading to sepsis; germination of the endospores of *B. anthracis*, and expression of toxin genes, have been demonstrated within (murine) alveolar macrophages [Mol. Microbiol. (1999) *31* 9–17]. If treated promptly (e.g. with benzylpenicillin or other β-lactam antibiotic) mortality is low; if untreated, the septicaemic form is usually fatal.

In herbivores (e.g. cattle, sheep, horses), the ingestion of *B. anthracis* spores during grazing can lead to rapidly fatal sepsis ('splenic fever').

(See also MCFADYEAN'S TEST.)

[Anthrax (various aspects): JAM (1999) *87* 189–321.]

anthrax toxin The TOXIN, produced by *Bacillus anthracis*, which is responsible for pathogenesis in ANTHRAX. Anthrax toxin comprises three protein components: *oedema factor* (EF, factor I), *'protective antigen'* (PA, factor II) and *lethal factor* (LF, factor III); independently, each component has no biological activity. All three components are encoded by plasmid pOX1. (The capsule of *B. anthracis*, an essential anti-phagocytic attribute of the pathogen, is encoded by plasmid pOX2.)

Activity of the toxin. PA forms a β-barrel structure in the eukaryotic cell membrane; this permits translocation, into the cell, of the two catalytic components of the toxin, LF and EF.

LF is a zinc protease. Within the eukaryotic cell, LF cleaves *mitogen-activated protein* (MAP) *kinase*, inactivating it. MAP kinase is an essential component in eukaryotic signalling; inhibition of the MAP kinase pathway blocks the development of cells at a specific point in the cell cycle. (LF is also a potent inducer of cytokines, and is thought to be active at a concentration of 10^{-18} M.)

EF is an ADENYLATE CYCLASE which, within the eukaryotic cell, is stimulated by CALMODULIN, thus causing an increase in the level of cyclic AMP (cAMP). EF is responsible for the oedematous aspect of anthrax (and may also inactivate phagocytic cells).

anthroponosis (1) An infectious disease which can be contracted by humans, the pathogen being normally maintained in a reservoir consisting of human population(s). (cf. ZOONOSIS.) (2) An infectious disease which can be contracted by animals, the pathogen being normally maintained in a reservoir consisting of human population(s) [use of term (sense 2): JMM (1997) *46* 372–376].

anthropophilic Refers to a parasite or pathogen which preferentially infects man (cf. ZOOPHILIC).

antibacterial agents Agents which kill, or inhibit the growth of, bacteria – e.g. certain ANTIBIOTICS, ANTISEPTICS and DISINFECTANTS.

antibiogram The pattern of sensitivities of a given microorganism towards a range of antibiotics.

antibiosis The antagonism of one organism towards another or others.

antibiotic (1) Originally, the term used for any *microbial* product which, in low concentrations, inhibits or kills susceptible microorganisms; with this meaning, the term could distinguish between an agent such as penicillin (produced by a fungus) and synthetic (i.e. man-made) antimicrobials such as sulphonamides. This use of the term became untenable when penicillins and other natural (i.e. microbially synthesized) antimicrobial products were modified, or synthesized *de novo*, in the laboratory.

(2) Currently, the term *antibiotic* is used most commonly to refer to any natural, semi-synthetic or wholly synthetic antimicrobial agent effective at low concentrations – although products which are active against *viruses* are usually referred to as ANTIVIRAL AGENTS rather than antibiotics.

This account refers primarily to antibiotics in a medical/veterinary context, i.e. as agents used for the prevention or treatment of infectious diseases in man and other animals. However, it should be noted that natural antibiotics (e.g. BACTERIOCINS) have ecological roles. Moreover, some antibiotics (e.g. NISIN) are used as food preservatives, while others (e.g. AGROCINS) are used in BIOLOGICAL CONTROL. Certain antibiotics have been used in the control of *plant* diseases (e.g. COCONUT LETHAL YELLOWING).

For therapeutic or prophylactic use, antibiotics should, ideally, be toxic only for (the appropriate) pathogens – i.e. they should be harmless to the patient; in other words, antibiotics should exhibit an appropriate level of *selective toxicity*. Even so, many of the antibiotics currently in use are associated with important side-effects – e.g. streptomycin may cause dose-dependent damage to the 8th cranial nerve, while penicillins may provoke hypersensitivity reactions (ANAPHYLACTIC SHOCK) in certain patients. Some antibiotics are contraindicated for particular categories of patient (e.g. pregnant women). (See also THERAPEUTIC INDEX.) Of the large number of known antibiotics, relatively few are suitable for clinical use; some are too highly toxic for human or animal use, and some would be ineffective as antimicrobial agents at safe doses.

An antibiotic may be MICROBICIDAL or MICROBISTATIC; some can be microbicidal or microbistatic according e.g. to concentration.

Mixtures of antibiotics may exhibit synergism or antagonism. *Synergism* is exhibited when two antibiotics, acting simultaneously on an organism, produce an effect which is greater than the sum of their individual effects; cotrimoxazole is an example of a synergistic combination of antibiotics (trimethoprim + sulphamethoxazole) – two drugs which block different reactions in the same pathway.

Antagonism is the converse of synergism. For example, antibiotics which inhibit growth will antagonize other antibiotics (e.g. penicillins) which are active only on growing cells. In another form of antagonism, the presence of certain types of antibiotic will promote the synthesis of inducible enzymes that inactivate *other* antibiotics; thus, for example, certain β-lactam antibiotics (such as imipenem or cefoxitin) can induce the synthesis of enzymes called β-lactamases which inactivate other β-lactam antibiotics. (For this reason, those β-lactam antibiotics which are strong inducers of β-lactamases are not used in combination with other β-lactam antibiotics.)

The antimicrobial activity of an antibiotic in the human (or animal) body does not necessarily correspond to its activity in laboratory tests. For example, under in vivo conditions an antibiotic may bind to plasma proteins so that only a fraction of the administered dose is available for antimicrobial activity. Moreover, some antibiotics are administered in an inactive form that depends on in vivo metabolism for the development of its activity (see e.g. *p-N*-succinylsulphatriazole in SULPHONAMIDES).

The ability of certain antibiotics (e.g. fluoroquinolones, rifampicin) to remain in an active state *within* phagocytes is an important feature because some bacterial pathogens can survive, or even grow, following uptake by phagocytic cells [RMM (1995) *6* 228–235].

Antibiotic action. To be effective an antibiotic must (at least) enter the cell envelope; commonly, antibiotics must pass through the cell envelope in order to reach their target sites. Entry into a cell often occurs via an energy-dependent transport system (see e.g. entries CYCLOSERINE, POLYOXINS, TETRACYCLINES, WARHEAD DELIVERY).

Within the pathogen, a given antibiotic acts at specific target site(s). Depending on antibiotic, the target site may be e.g.: (i) enzymes that catalyse the synthesis of a structural polymer in the cell envelope; (ii) the cytoplasmic membrane (certain antibiotics cause leakiness incompatible with cell viability); (iii) enzymes required for the synthesis or supercoiling of nucleic acids; (iv) the ribosome (some antibiotics can inhibit protein synthesis by binding to specific sites on the ribosome); (v) enzymes required for biosynthesis of an essential coenzyme – for example, dihydrofolate reductase (required for the synthesis of tetrahydrofolate) is inhibited by trimethoprim.

Antibiotics of the same group have similar or identical target sites, and they all affect cells in the same way. However, antibiotics from different groups may have a similar target site; thus, for example, the target site of oxazolidinones appears to overlap the binding sites of both lincomycin and chloramphenicol [JB (2000) *182* 5325–5331].

Some target sites are found in only certain categories of microorganism; for example, the cell wall polymer *peptidoglycan* occurs only in bacteria – so that antibiotics directed specifically against this target are unlikely to be of use against (e.g.) fungi. Clearly, no antibiotic is effective against all pathogens, and in many cases the activity of a given antibiotic is limited to a particular subset of organisms; in the context of bacterial pathogens, a *broad-spectrum* antibiotic is one which can be effective against a wide range of species, including both Gram-positive and Gram-negative bacteria.

Some examples of antibiotics (in various target regions) are listed below; see individual entries for mode of action etc.

Bacterial cell wall (peptidoglycan). CYCLOSERINE; β-LACTAM ANTIBIOTICS; MOENOMYCIN; VANCOMYCIN.

Bacterial cytoplasmic membrane. DEPSIPEPTIDE ANTIBIOTICS; GRAMICIDINS; POLYMYXINS; TYROCIDINS.

DNA intercalating agents. ACTINOMYCIN D; ANTHRACYCLINE ANTIBIOTICS; QUINOXALINE ANTIBIOTICS.

Folic acid antagonists. CHLORGUANIDE; PYRIMETHAMINE; TRIMETHOPRIM. (See also SULPHONAMIDES.)

Fungal cell wall. ACULEACIN A; NIKKOMYCINS; PAPULACANDIN B; POLYOXINS.

Fungal cytoplasmic membrane. POLYENE ANTIBIOTICS (sense a).

Gyrase. COUMERMYCINS; NOVOBIOCIN; QUINOLONE ANTIBIOTICS.

Ribosome (protein synthesis). AMINOGLYCOSIDE ANTIBIOTICS; ANISOMYCIN; CHLORAMPHENICOL; CYCLOHEXIMIDE; FLORFENICOL; FUSIDIC ACID; LINCOSAMIDES; MACROLIDE ANTIBIOTICS; OXAZOLIDINONES; POLYENE ANTIBIOTICS (sense b); PUROMYCIN; SPARSOMYCIN; SPECTINOMYCIN; TETRACYCLINES; THIOSTREPTON; TIAMULIN; VIOMYCIN.

RNA polymerase. ANSAMYCINS (e.g. RIFAMYCINS) (see also STREPTOLYDIGIN).

Microbial resistance to antibiotics. Resistance to a given antibiotic ('drug resistance') is said to be *constitutive* (= *intrinsic*) in cells which (i) lack the specific target of that antibiotic, (ii) have an alternative form of the target that is not affected by the antibiotic, and/or (iii) are inherently impermeable to the antibiotic. For example: (i) species of *Mycoplasma*, which lack cell walls, are intrinsically resistant to penicillins (which inhibit the synthesis of a cell wall polymer); (ii) strains of *Staphylococcus aureus* that exhibit intrinsic resistance to methicillin contain an alternative form of the target (see MRSA); (iii) many Gram-negative bacteria are intrinsically resistant to e.g. benzylpenicillin because their outer membrane is impermeable to the antibiotic.

A cell may *acquire* heritable resistance to one or more antibiotics in several ways. In bacteria, gene(s) specifying resistance to one or more antibiotics may be transferred to an antibiotic-susceptible cell by CONJUGATION, TRANSDUCTION, TRANSFORMATION or CONJUGATIVE TRANSPOSITION; resistance may also develop as a consequence of MUTATION. The way in which resistance is acquired is important in that it may determine the range of antibiotics to which a cell becomes resistant. Thus, a single point mutation commonly confers resistance to only one type of antibiotic (or to those antibiotics which share the same target site). By contrast, receipt of an R PLASMID or TRANSPOSON may confer resistance to either a single antibiotic or to two or more *unrelated* antibiotics (*multiple drug resistance*).

Mechanisms of resistance to antibiotics. Six main mechanisms can be distinguished:

(i) *Mutation.* The target site of a given antibiotic may be modified as a result of a MUTATION such that the target (e.g. enzyme, ribosome) functions more or less normally in the presence of otherwise inhibitory concentrations of the antibiotic (see e.g. STREPTOMYCIN). The cell in which such a mutation has occurred can therefore grow and give rise to a population of resistant cells in the presence of the given antibiotic. Examples of mutation-mediated resistance: (i) point mutations affecting 23S rRNA which confer resistance to macrolide antibiotics [AAC (1996) *40* 477–480]; (ii) point mutations in the *rpoB* gene of *Mycobacterium tuberculosis* (encoding a subunit of RNA polymerase) conferring resistance to rifampicin. Although resistance-conferring mutations are commonly located within known sequences in the genome, resistance may also arise through mutation at novel sites; for example, in a strain of *Streptococcus pneumoniae*, mutation outside the recognized 'quinolone resistance-determining region' has been associated with enhanced resistance to certain of the newer fluoroquinolones [AAC (2001) *45* 952–995].

(ii) *Inactivation or degradation of antibiotics by enzymes.* Enzymes which inactivate or degrade antibiotics are encoded by genes that occur e.g. in plasmids or chromosomes. Such enzymes include the acetyltransferases, adenylyltransferases and phosphotransferases that are active against aminoglycoside antibiotics [TIM (1998) *6* 323–327], the enzyme CHLORAMPHENICOL acetyltransferase, and the β-LACTAMASES which inactivate penicillins and related antibiotics.

(iii) *Efflux mechanisms.* See EFFLUX MECHANISM. Some efflux mechanisms involve an ABC TRANSPORTER. [Efflux-mediated resistance to fluoroquinolones in Gram-negative bacteria: AAC (2000) *44* 2233–2241.]

(iv) *Diminished permeability.* Any change in the composition of the cell envelope which hinders the uptake of an antibiotic will either increase the MIC of that antibiotic or else give rise to high-level resistance. For example, diminished permeability in *Enterobacter aerogenes*, associated with an alteration in the outer membrane porins, has been reported to affect this organism's susceptibility to various antibiotics, including some of the fourth-generation cephalosporins (e.g. cefepime) [Microbiology (1998) *144* 3003–3009].

Interestingly, in *Escherichia coli*, phenotypic (*non*-heritable) resistance to some antibiotics (e.g. ampicillin, chloramphenicol) can be induced by the presence of certain chemorepellents – e.g. acetylsalicylate (aspirin) and benzoate – which are detected by the *tsr* gene product; this phenomenon has been attributed to inhibition of synthesis of the OmpF porin by the chemorepellent [FEMS (1987) *40* 233–237].

(v) *Increased synthesis of an affected metabolite.* Increased production of a given metabolite may overcome competitive inhibition by the antibiotic; for example, production of higher levels of PABA is one form of resistance to sulphonamides.

(vi) *Acquisition of an antibiotic-insensitive target.* One example is the acquisition of a methicillin-resistant penicillin-binding protein by strains of *Staphylococcus aureus* (see MRSA).

[Mechanisms of resistance to β-lactam, aminoglycoside, quinolone and other antibiotics in *Acinetobacter baumannii:* RMM (1998) *9* 87–97.]

Induction of resistance to antibiotics. Resistance to certain antibiotics is inducible – the presence of such antibiotics (over a certain minimum concentration) promoting resistance in the cell. For example, in some Gram-positive bacteria (e.g. staphylococci), the enzyme chloramphenicol acetyltransferase may be induced in the presence of its substrate, CHLORAMPHENICOL (see also TRANSLATIONAL ATTENUATION). Also in staphylococci, several distinct modes of resistance to β-lactam antibiotics may be induced by the presence of these antibiotics; thus, resistance may be due to the induction of β-LACTAMASES (leading to mechanism ii, above) or to induction of the specific PENICILLIN-BINDING PROTEIN (PBP) 2a (see MRSA) (mechanism vi, above). In the case of inducible resistance to β-lactam antibiotics, studies on the transmembrane signalling pathway have indicated that induced resistance – through either β-lactamases or the *mecA* product PBP 2a – may depend on a novel mechanism involving sequential cleavage of several regulatory proteins [Science (2001) *291* 1962–1965; commentary: Science (2001) *291* 1915–1916]. In this scheme, regulation of *mecA* expression involves initial binding of a β-lactam antibiotic to a cell-surface sensor–transducer: the MecR1 protein; as a consequence, MecR1 undergoes cleavage, and a (hypothetical) product of such cleavage, MecR2, directly or indirectly brings about cleavage of the transcriptional repressor protein, MecI – thus allowing expression of *mecA*. (See also β-LACTAMASES.)

Antibiotic-sensitivity (antibiotic-susceptibility) tests. Laboratory tests are often carried out to determine the susceptibility of a given pathogen to various antibiotics (the pattern of sensitivities being referred to as the pathogen's *antibiogram*); the results of such tests may enable the clinician to select optimal antibiotic(s) for chemotherapy.

A common form of test is the DISC DIFFUSION TEST (q.v.; see also NEO-SENSITABS); this kind of test is intended primarily for those bacteria which produce visible growth after overnight incubation (rather than slow-growing species such as *Mycobacterium tuberculosis*). (See also DILUTION TEST.)

The E TEST is a diffusion test which can indicate an MIC.

Other methods, such as LATEX PARTICLE TESTS, have been assessed for their ability to detect resistance to methicillin in staphylococci [rapid detection of *mecA*-positive and *mecA*-negative coagulase-negative staphylococci by an anti-penicillin binding protein 2a slide latex agglutination test: JCM (2000) *38* 2051–2054].

Molecular methods (including PCR-based tests) can also be used in certain cases for detecting antibiotic resistance. [Rapid identification of methicillin-resistant *Staphylococcus aureus* and simultaneous species confirmation using real-time fluorescence PCR: JCM (2000) *38* 2429–2433. Multiplex PCR for detection of genes for *Staphylococcus aureus* enterotoxins, exfoliative toxins, toxic shock syndrome toxin 1 and methicillin resistance: JCM (2000) *38* 1032–1035. Simultaneous identification and typing of multidrug-resistant *Mycobacterium tuberculosis* isolates by analysis of *pncA* and *rpoB*: JMM (2000) *49* 651–656.] (See also *line probe assay* in the entry PROBE.)

[DNA-based ('genotypic') antibiotic susceptibility testing: Book ref. 221, pp 203–228.]

Clinical impact of bacterial resistance to antibiotics. Although large numbers of new antibiotics have been identified in the last few decades, bacterial resistance to existing drugs has continued to emerge so that, in some cases, therapeutic options have become very limited; examples of such resistance include: vancomycin-resistant enterococci [EID (2001) *7* 183–187]; tetracycline-resistant gonococci; penicillin-resistant pneumococci [Drugs (1996) *51* (supplement 1) 1–5]; multidrug-resistant *Vibrio cholerae* O1 [Lancet (1997) *349* 924]; multidrug-resistant *Mycobacterium bovis* [Lancet (1997) *350* 1738–1742]; and low-level resistance to vancomycin in MRSA [JAC (1997) *40* 135–136]. [The crisis in antibiotic resistance: Science (1992) *257* 1064–1073.] There is thus an ongoing need to develop new antibiotics and to prevent, or delay, the emergence of resistance to those drugs that are currently useful by avoiding inappropriate usage; given the value of antibiotics in medicine (and our current dependence on them) it is essential that new drugs be used carefully and rationally [quinupristin/dalfopristin and linezolid: where, when, which and whether to use?: JAC (2000) *46* 347–350]. Already, clinical resistance to linezolid has been reported in vancomycin-resistant strains of *Enterococcus faecium* (vancomycin-resistant enterococci, VRE) [Lancet (2001) *357* 1179].

[Controlling antimicrobial resistance in hospitals (infection control and use of antibiotics): EID (2001) *7* 188–192.]

Because of ongoing development of resistance in pathogenic bacteria, new targets for antibiotics are being sought within the bacterial cell. One target in *Escherichia coli* is *de-acetylase* (a zinc-requiring enzyme, product of gene *lpxC*) which is essential for the synthesis of lipid A (an integral part of the outer membrane in Gram-negative bacteria) (see ENDOTOXIC SHOCK). New approaches to antibiotics may also involve agents such as CcdB (see F PLASMID) and certain bacteriocins. By contrast, peptide deformylase (PDF), which cleaves the N-terminal formyl group from nascent *bacterial* polypeptides (see PROTEIN SYNTHESIS), seems less than optimal as a candidate drug target; certain derivatives of hydroxamic acid are potent inhibitors of PDF, but resistance to these agents develops readily, and the drugs appear to be only bacteriostatic [AAC (2001) *45* 1058–1064].

The ongoing problem of antibiotic-resistant pathogens has prompted consideration of other anti-infection modalities, including bacteriophage therapy [AAC (2001) *45* 649–659].

Antibiotics for chemotherapy: factors affecting the choice. Antibiotics may be wasted if they are used without due regard to specificity and pharmacology; moreover, use of inappropriate drugs may forfeit the chance of helping the patient and may also exacerbate unnecessarily the general problem of antibiotic resistance. The kind of factors which affect the choice of antibiotic include:

- Effectiveness against relevant pathogens. In ideal circumstances the pathogen's identity and its sensitivity to antibiotics are known prior to treatment. However, in certain diseases (e.g. meningitis) therapy is often begun before the aetiology has been confirmed. When this happens the choice of antibiotic is based on the *probable* pathogen; probability reflects factors such as (local) prevalence of particular species/strains of pathogen, reported problems of resistance in the pathogen, and the patient's age/general condition. [Example: management of bacterial meningitis (paediatric patients): RMM (1997) *8* 171–178.]
- Prevalence of antibiotic-resistant strains in the locality. Many pathogens are now resistant to antibiotic(s) to which they were once invariably sensitive; moreover, some pathogens are resistant to a *range* of drugs so that, in some cases, therapeutic options are extremely limited.
- Antagonism towards other antibiotics – or even towards other therapeutic agents; for example, rifampicin can potentiate certain mammalian enzymes and, as a consequence, may interfere with oral contraception.
- Possible side-effects (e.g. penicillin allergy).
- In pregnant patients: possible teratogenicity.
- Pharmacokinetics. The physicochemical nature of a drug determines its ability to reach particular sites when administered via a given route. For example, if certain drugs are injected intravenously they do not reach the cerebrospinal fluid (CSF) because they cannot pass the blood–brain barrier; other types of drug can, and some can do so only if the permeability of this barrier has been increased e.g. as a result of inflammation.

(See also ANTIFUNGAL AGENTS, ANTIPROTOZOAL AGENTS, ANTISEPTICS and DISINFECTANTS.)

antibiotic-associated colitis See COLITIS.

antibody (ab) Any IMMUNOGLOBULIN molecule produced in direct response to an ANTIGEN (or to an autocoupling HAPTEN) and which can combine specifically, non-covalently, and reversibly with the antigen which elicited its formation (see also ANTIBODY FORMATION); specificity is not necessarily absolute: see CROSS-REACTING ANTIBODY. (A MYELOMA can produce Ig with antibody function in the absence of antigenic stimulation. See also PASSIVE IMMUNITY.) The term 'antibody' is also used to refer to a homogeneous or heterogeneous *population* of antibodies (see MONOCLONAL ANTIBODIES and POLYCLONAL ANTISERUM).

Antibody–antigen combination occurs by contact between specific antibody COMBINING SITES and antigenic DETERMINANTS, and may promote COMPLEMENT FIXATION. (See also VALENCY, COMPLEMENT-FIXING ANTIBODIES, and OPTIMAL PROPORTIONS.) In vivo, antibody–antigen combination may be beneficial if the antigenic stimulus derives from an invading pathogen (see e.g. OPSONIN), but can be harmful (see e.g. IMMEDIATE HYPERSENSITIVITY). (See also e.g. AGGLUTININ; BLOCKING ANTIBODIES; CYTOPHILIC ANTIBODIES; NATURAL ANTIBODIES; PRECIPITIN.)

Antibodies may be identified, assayed etc in vitro by a range of techniques – see e.g. GEL DIFFUSION; IMMUNOASSAY; IMMUNOELECTROPHORESIS; IMMUNOFLUORESCENCE; PRECIPITIN TEST. Detection and quantification of specific antibodies in samples of SERUM are important in the diagnosis of many diseases.

antibody absorption See ABSORPTION.

antibody-dependent cell-mediated cytotoxicity See ADCC.

antibody excess The presence of a high antibody-to-antigen ratio. In an agglutination test gross antibody excess may give rise to a PROZONE – possibly due to the saturation of antigenic determinants by antibodies in the absence of free antigen with which to form links.

antibody formation An IMMUNE RESPONSE in which B LYMPHOCYTES synthesize and secrete ANTIBODY of a given antigenic specificity after immunological contact with the corresponding ANTIGEN; for so-called THYMUS-DEPENDENT ANTIGENS, B cells require help from antigen-specific $CD4^+$ T LYMPHOCYTES ('T helper cells') and/or the products of such T cells (cf. THYMUS-INDEPENDENT ANTIGENS).

The response to thymus-independent antigens.

Thymus-independent (TI) antigens (such as lipopolysaccharides and bacterial cell wall polysaccharides) induce antibody formation independently of T cells. When activated by TI antigens, certain B cells – $CD5^+$ ('B1') cells (see B LYMPHOCYTE) and MARGINAL ZONE B CELLS – proliferate and secrete antibodies; however, the expanded clone of B cells does not give rise to memory cells (see later), and the antibodies formed in response to these antigens are almost exclusively of the IgM class. [T-independent immune response (new aspects of B cell biology): Science (2000) *290* 89–92.]

The response to thymus-dependent antigens.

The process of antibody formation is incompletely understood; the following is a generally accepted scenario.

To elicit antibodies, a thymus-dependent (TD) antigen must initially be taken up and *processed*, intracellularly, by an ANTIGEN-PRESENTING CELL (APC). Processing involves enzymic degradation of the antigen; a particular (peptide) fragment of the antigen is then linked to an MHC class II molecule (synthesized in the APC) and *presented*, at the surface of the APC, to an antigen-specific T cell.

B lymphocytes can act as APCs, so that antigen (plus MHC class II molecule) may be presented at the surface of a B cell to an antigen-specific T cell; T cells of the Th2 subset (see T LYMPHOCYTE) characteristically act as 'helper' cells in this context. The (antigen-specific) T cell binds to the antigen with its T CELL RECEPTOR – thus physically linking B and T cells via the antigen; the MHC class II molecule (on the B cell) binds to the CD4 antigen on the T cell.

Other 'paired interactions' also occur between binding sites on the B and T cells. Thus, certain receptors (B7.1, B7.2) on the B cell bind to antigen CD28 (q.v.) on the T cell; such binding permits the T cell to synthesize INTERLEUKIN-2 – a cytokine which promotes activation and clonal proliferation of the T cell and induces it to release cytokines, including INTERLEUKIN-4. However, once activated, T cells produce the cell-surface antigen CTLA-4 (= CD152), a molecule (similar to CD28) to which B7.1 and B7.2 can bind; interaction between B7.1/B7.2 and CTLA-4 may initiate a signal which terminates synthesis of IL-2 in the T cell. (Note that, while IL-2 is not a characteristic product of the Th2 subset of T cells, transient synthesis of this cytokine at an early stage would appear to be necessary as it seems to be an essential growth factor for T cells.)

Yet another paired interaction between the B and T cells involves CD40 (q.v.).

The mutual (physical) co-operation between the B cell and T cell (described above) is referred to as *cognate help* or *linked recognition*.

The activated, proliferating B cells form an expanded clone. Some of the cells of this clone become antibody-secreting PLASMA CELLS. Others become MEMORY CELLS; these cells do not secrete antibody but retain the potential to do so, rapidly, if subsequently challenged with the same antigen (giving a heightened *secondary response*). (See also ANAMNESTIC RESPONSE.)

Initially, antibodies secreted by plasma cells are of the IgM class. Subsequently, plasma cells form antibodies of another class – generally IgG – but of the same specificity; such *class switching* (= *isotype switching*) is promoted by several stimuli – including the binding of B cell antigen CD40 to the corresponding T cell ligand, and the activity of the T-cell-derived cytokine INTERLEUKIN-4.

Class switching occurs in so-called *germinal centres* within lymph nodes and spleen. Such sites are also involved in a process known as *affinity maturation*; in this process, mutant cells among the (proliferating) antigen-stimulated B cells are selected for high-affinity binding to the specific antigen – selective pressure presumably being a falling concentration of the given antigen. Within the B cells a process known as *somatic hypermutation* gives rise to mutations that are concentrated in the V regions of antibody genes; as a consequence, different B cells produce antibodies with different levels of affinity for the given antigen – those producing antibody with the highest affinity being selected. [Hypermutation in antibody genes: PTRSLB (2001) *356* 1–125.] (See also RNA EDITING.)

In the response to an antigen not previously encountered (a *primary* response) the numbers of specifically reactive B cells (prior to clonal expansion) are likely to be low. Under these conditions it may be that the presentation of antigen to T cells is carried out not by B cells but by other, more numerous, APCs such as macrophages or dendritic cells. In this scenario, interaction between APC and T cell activates the T cell – which then secretes various cytokines; these cytokines may induce differentiation, proliferation and antibody secretion in B cells that have bound the specific antigen and which are close to (though not necessarily in contact with) the activated T cells [Book ref. 227, pp 207–208].

When produced in response to invasion by pathogens (or their products – e.g. toxins), antibodies commonly have a protective role (see e.g. ADCC and OPSONIZATION). However, under certain conditions, antibodies can promote damage to the host's tissues (see IMMEDIATE HYPERSENSITIVITY).

anticapsin See BACILYSIN.

anticlinal Perpendicular to a given surface. (cf. PERICLINAL.)

anticoagulant Any agent which inhibits the coagulation (clotting) of blood – e.g. EDTA, HEPARIN, sodium citrate, sodium oxalate. SPS is used in BLOOD CULTURE.

anticodon See GENETIC CODE.

anticomplementary (*immunol.*) Refers to any substance (e.g. ZYMOSAN), or to any system (e.g. pre-formed antigen–antibody complexes) *other than* the specific combination of antibody with antigen, which brings about COMPLEMENT FIXATION. If present in a COMPLEMENT-FIXATION TEST, anticomplementary factors can – by fixing complement – simulate the effects of antigen–antibody combination, giving a false-positive reaction.

anti-downstream box See DOWNSTREAM BOX.

antifol *Syn.* FOLIC ACID ANTAGONIST.

antifungal agents Chemical agents (including certain ANTIBIOTICS) which either kill or inhibit the growth of fungi (fungicidal and fungistatic action, respectively). (Antifungal agents are often called 'fungicides' regardless of whether their action is fungicidal or fungistatic.) Antifungal agents are used e.g. for the treatment or prevention of fungal diseases of man, animals and plants, and as PRESERVATIVES for a wide range of materials.

(a) *Medical and veterinary* antifungal agents ('antimycotics') are used for the treatment of local and systemic mycoses (see MYCOSIS); such agents include e.g. ALLYLAMINES, AZOLE ANTIFUNGAL AGENTS, CYCLOHEXIMIDE, FLUCYTOSINE, GRISEOFULVIN, NITROFURANS, POLYENE ANTIBIOTICS, SALICYLIC ACID, TOLNAFTATE. Many general DISINFECTANTS and ANTISEPTICS also have antifungal activity.

(b) *Agricultural* antifungal agents (used against fungal diseases of crop plants) may function as ERADICANTS and/or as PROTECTANTS; they fall into two broad categories: *contact* (*surface* or *non-systemic*) antifungal agents, which adhere to plant surfaces but do not penetrate the plant tissues, and *systemic* antifungal agents, which are taken up by the plant and translocated within its tissues – usually in the transpiration stream (from roots to leaves). In general, a contact agent typically acts as a general enzyme inhibitor, while a systemic agent usually acts at a specific target site (process or structure) in the fungal cell. Fungi may rapidly acquire resistance to an agent which has a single target site, and continued use of such agents has led in many cases to the emergence and spread of resistant strains of fungal pathogens; the risk of emergence of resistant strains can be reduced by using mixtures containing two or more agents which have different modes of action. Agricultural antifungal agents include e.g. BENZIMIDAZOLES (a), BISDITHIOCARBAMATES, CARBOXIN, CHLORONEB, CHLORONITROBENZENES, copper sulphate (see COPPER), DICARBOXIMIDES, DICHLOFLUANID, DICHLONE, DINITROPHENOLS, DITHIOCARBAMATES, DODINE, ETHAZOLE, GLYODIN, HYDROXYPYRIMIDINE ANTIFUNGAL AGENTS, IMAZALIL, MORPHOLINE ANTIFUNGAL AGENTS, certain ORGANOPHOSPHORUS COMPOUNDS, PHENYLAMIDE ANTIFUNGAL AGENTS, PHTHALIMIDE ANTIFUNGAL AGENTS, PROCHLORAZ, PYRACARBOLID, QUINONE ANTIFUNGAL AGENTS, SULPHUR, THIRAM, TRIFORINE. (See also AUREOFUNGIN, GRISEOFULVIN and POLYOXINS.)

antigen (ag) Any agent which can elicit an IMMUNE RESPONSE (cf. HAPTEN); 'antigen' may refer to an individual macromolecule or to a homogeneous or heterogeneous population of antigenic macromolecules. A given antigen usually contains more than one DETERMINANT (see also POLYCLONAL ANTISERUM). An antigen may be soluble (e.g. microbial toxins, extracts) or particulate. The most effective antigens are proteins and polysaccharides (see also Z-DNA); in general, large molecules are more antigenic than are smaller ones, and antigenicity is enhanced by the presence of an ADJUVANT. Minute quantities (e.g. a few μg) of antigen may be sufficient to elicit an immune response – the actual amount required depending e.g. on the route of administration.

The surface of a microorganism typically consists of repeating patterns of antigens, and the classification of some groups of microorganisms is based on differences between the antigens of different strains (see e.g. KAUFFMANN–WHITE CLASSIFICATION). (See also e.g. ALLERGEN; ANTIBODY; ANTIGENIC VARIATION; IDIOTYPE; IMMUNIZATION; THYMUS-DEPENDENT ANTIGENS.)

antigen-binding cell (ABC) (1) Any cell which can bind only a specific antigen. (2) Any cell (including e.g. macrophages) which can bind various antigens.

antigen deletion In a microorganism: the loss of particular antigenic determinant(s) due e.g. to a mutation or to the loss of a plasmid. (cf. ANTIGEN GAIN.)

antigen excess The presence of a high antigen-to-antibody ratio. Under conditions of gross antigen excess precipitation does not occur since, according to the LATTICE HYPOTHESIS, all the combining sites of antibodies are bound to otherwise unattached antigens – leaving no combining sites free to form cross-links and hence larger (insoluble) complexes. (See also TYPE III REACTION.)

antigen gain In a microorganism: the formation of new antigenic determinant(s). Antigen gain may be due e.g. to a MUTATION, to the acquisition of a PLASMID, or to BACTERIOPHAGE CONVERSION; thus, e.g., the sex factor in a conjugative plasmid specifies certain donor-specific antigens, while certain cell-surface antigens in *Salmonella* strains are present only when the cells are lysogenized by a particular phage. (cf. ANTIGEN DELETION.)

antigen-presenting cell (APC) Any of a range of cells which bind antigen and present it to an antigen-specific T cell. APCs, which bear MHC class II cell-surface molecules (see HISTOCOMPATIBILITY RESTRICTION), include B LYMPHOCYTES, macrophages, dendritic cells, Langerhans' cells and Kupffer cells.

antigen-specific T helper cell A T LYMPHOCYTE which is specific for, i.e. activated by, a given, specific antigen; such a T cell may give COGNATE HELP to a B cell which has bound the same antigen (see ANTIBODY FORMATION). (See also CARRIER-SPECIFIC T HELPER CELL.)

antigenic (1) *Syn.* immunogenic (see IMMUNOGENICITY). (2) Of or pertaining to an ANTIGEN or antigens.

antigenic competition Inhibition of the immune response to a given antigen or DETERMINANT due to exposure to another (different) antigen or determinant; antigenic competition has been regarded as a form of natural immunosuppression.

In *intermolecular* competition, suppression of the response to a given antigen may occur as the result of the previous administration of the other antigen (*sequential competition*) or of the simultaneous administration of the other antigen. By carefully adjusting ('balancing') the relative amounts of the antigens, antigenic competition may be reduced or abolished; this is important e.g. in the preparation of MIXED VACCINES.

In *intramolecular competition* the response to a given determinant is inhibited by the presence of another determinant on the same macromolecule. For example, with IgG as antigen, good titres of anti-Fc may be produced but relatively few anti-Fab antibodies are formed – fewer than would be formed against isolated Fab fragments (i.e., relative to Fab, Fc is *immunodominant*).

antigenic determinant See DETERMINANT.

antigenic drift (immunological drift) In a given species, strain or type of microorganism: minor changes in antigenic specificity which occur over an extended period of time (e.g. years) – see e.g. INFLUENZAVIRUS. (cf. ANTIGENIC SHIFT; ANTIGENIC VARIATION.)

antigenic formula Any symbolic notation which indicates the antigenic nature of an organism – see e.g. KAUFFMANN–WHITE CLASSIFICATION.

antigenic shift In a given species, strain or type of microorganism: major change(s) in antigenic specificity which apparently occur abruptly. For example, antigenic shifts in strains of INFLUENZAVIRUS have been reported to coincide with pandemics of influenza. (cf. ANTIGENIC DRIFT; ANTIGENIC VARIATION.)

antigenic variation Successive changes which occur in the surface antigens of certain microorganisms; in some cases, each alternative antigen results from the expression of a specific allele (see PHASE VARIATION), but in other cases (e.g. the fimbriae of *Neisseria* – see below), successive new antigens are created by repeated recombinational events.

Antigenic variation may help a pathogen to evade the host's immune defence system; thus, e.g. new antigenic forms of the pathogen which arise in the host may not be susceptible to OPSONIZATION by those antibodies which have developed in response to the previous antigenic forms.

Antigenic variation occurs e.g. in *Borrelia* (see RELAPSING FEVER), INFLUENZAVIRUS, visna virus, *Leishmania*, *Plasmodium* and *Trypanosoma* spp (see VSG).

The major structural subunit in the fimbriae of *Neisseria gonorrhoeae* (which are type IV FIMBRIAE) is encoded by the chromosomal gene *pilE*. The chromosome also contains copies of a gene, *pilS*, with which *pilE* can undergo repeated RECOMBINATION; moreover, recombination can also occur between *pilE* and any homologous DNA acquired through transformation. Recombinational events continually create new versions of the subunit gene, so that the fimbriae of this organism can be synthesized from any one of a very large number of variant forms of the subunit, giving rise to extensive antigenic variation [Mol. Microbiol. (1996) *21* 433–440]. These fimbriae are also subject to phase variation (on/off switching). (The Opa adhesins of *Neisseria* undergo phase variation by DNA re-arrangements that alter reading frames in *opa* genes [TIM (1998) *6* 489–495].)

[An update of antigenic variation in African trypanosomes: TIP (2001) *17* 338–343.]

antigenicity The capacity to function as an ANTIGEN. (cf. IMMUNOGENICITY.)

antiglobulin (Coombs' reagent) Antibody homologous to the antigenic determinants of serum globulins, particularly IMMUNOGLOBULINS; antiglobulin–Ig combination can occur without involvement of the specific COMBINING SITES of the Ig molecule.

antiglobulin consumption test (*serol.*) Any CONSUMPTION TEST in which the presence of cell- or particle-bound Ig is detected by the consumption of added ANTIGLOBULIN. For example, the in vitro combination of autoantibodies with homologous cell-surface antigens can be detected by washing the cells free of uncombined antibody and incubating them with antiglobulin; if autoantibody–antigen combination has occurred, antiglobulin will be consumed and may cause agglutination by the formation of cell–autoantibody–antiglobulin complexes. The amount of antiglobulin consumed can be determined e.g. by titrating the unconsumed antiglobulin against Ig-coated erythrocytes.

antiglobulin test (*serol.*) Any test in which the presence of Ig (including BLOCKING ANTIBODIES, sense 1) is detected by exploiting the ability of ANTIGLOBULIN to agglutinate Ig-bound antigens.

anti-idiotype Antiserum to the variable-region domains of an Ig molecule (cf. IDIOTYPE); it contains antibodies homologous to each of the individual determinants (IDIOTOPES), though the different types of antibody may differ quantitatively and in their AFFINITY for their respective determinants.

antilymphocyte serum (ALS) Antiserum to the cell-surface antigens of LYMPHOCYTES; it is used e.g. for IMMUNOSUPPRESSION. Antilymphocyte globulin (ALG) is the globulin fraction of ALS.

antimalarial agents ('antimalarials') Drugs which are effective in the chemotherapy and/or chemoprophylaxis of MALARIA. Most of these drugs are effective against only one stage of the malaria parasite – e.g. chloroquine, mefloquine, quinine and QINGHAOSU are SCHIZONTICIDES. Primaquine and WR238605 are both active against hypnozoites; primaquine is also active against tissue schizonts (i.e. schizonts within liver cells), and WR238605 is also active against gametocytes.

(See AMINOQUINOLINES, QINGHAOSU, QUININE, PYRIMETHAMINE and MALARIA for details of antimalarials.)

Combinations of antimalarials are often used for treatment; this helps e.g. to limit the emergence of resistant strains of the parasite.

antimetabolite A substance which competitively inhibits the utilization, by an organism, of an exogenous substrate or endogenous metabolite. (See e.g. SULPHONAMIDES.)

antimony (as an antimicrobial agent) Antimony is a HEAVY METAL whose compounds have long been used as therapeutic agents – though their toxic effects were often greater than the benefits they bestowed. Antimonials have found use in the treatment of certain diseases of protozoal aetiology, e.g. leishmaniasis, and other diseases, e.g. schistosomiasis. It is believed that trivalent antimony is necessary for antimicrobial activity, and that pentavalent compounds owe their activity to in vivo conversion to the trivalent state. Antimonials have been shown to combine with essential sulphhydryl groups in trypanosomal enzymes involved in glucose metabolism; such inhibition may be reversed by thiol compounds, e.g. cysteine. Trivalent antimonials include e.g. potassium antimonyl tartrate (*tartar emetic*; $2[K(SbO).C_4H_4O_6].H_2O$) and *stibophen* (sodium antimony *bis*[pyrocatechol-3,5-disulphonate]). Pentavalent antimonials include e.g. *stibanilic acid* (*p*-aminobenzenestibonic acid) and its sodium salt (*stibamine*), both of which are constituents of a number of widely-used drugs – e.g. *ethyl stibamine* (*neostibosan*) which is used for the treatment of espundia and kala-azar.

antimutator gene See MUTATOR GENE.

antimycin A An antibiotic which acts as a RESPIRATORY INHIBITOR in Complex III of the mitochondrial ELECTRON TRANSPORT CHAIN, apparently blocking electron flow between *b*-type cytochromes and cyt c_1; antimycin A and HOQNO appear to act at a common site which may be associated with the *b*-type cytochromes.

antimycotic See ANTIFUNGAL AGENTS (a).

antioxidant (*food sci.*) A substance added to certain foods to inhibit oxidation of the lipids in those foods – thus enhancing stability and prolonging shelf-life. (See also OXIDATIVE RANCIDITY.) Antioxidants currently used include e.g. ascorbic acid, butylated hydroxyanisole, butylated hydroxytoluene, carotene, propylgallate and tertiary butylhydroquinone. Some antioxidants have antimicrobial properties [CRM (1985) *12* 153–183].

antiparallel (of DNA strands) See DNA.

antiplectic metachrony See METACHRONAL WAVES.

antiport (counter transport, exchange diffusion) The transmembrane transport of a solute (e.g. a given ion) coupled, directly, to the transmembrane transport of a different solute in the opposite direction. (cf. SYMPORT, UNIPORT; see also ION TRANSPORT.)

antiprotozoal agents Chemical agents which kill or inhibit protozoa; they include e.g. AMINOQUINOLINES; antimonials (see ANTIMONY); arsenicals (see ARSENIC); certain DIAMIDINES; DILOXANIDE; EMETINE; 8-HYDROXYQUINOLINE; METRONIDAZOLE; MONENSIN; NITROFURANS; PYRIMETHAMINE; SULPHONAMIDES; SURAMIN. (See also antimalarial agents under MALARIA.) [Review: Book ref. 153, pp. 95–132.]

antiretroviral agents Any of a range of compounds which can inhibit the replication of retroviruses (i.e. viruses of the

RETROVIRIDAE); these agents are used e.g. in chemotherapy against HIV infection and AIDS. [BMJ (2001) *322* 1410–1412.]

The three main groups of antiretroviral agents are: (i) NUCLEOSIDE REVERSE TRANSCRIPTASE INHIBITORS (q.v. for examples); (ii) NON-NUCLEOSIDE REVERSE TRANSCRIPTASE INHIBITORS (delavirdine, efavirenz and nevirapine) [the emerging role of NNRTIs in antiretroviral therapy: Drugs (2001) *61* 19–26]; (iii) PROTEASE INHIBITORS (amprenavir, indinavir, nelfinavir, ritonavir and saquinavir). The target for protease inhibitors is the viral protease (encoded by the *pol* region in HIV). Single-drug chemotherapy is prone to failure owing to the emergence of drug-resistant mutant viruses. (cf. HAART.)

antisense RNA Any RNA that can recognize a specific sense sequence in RNA or DNA and affect the activity/expression of that sequence: see e.g: COLE1 PLASMID, FINOP SYSTEM and R1 PLASMID.

MicroRNAs (miRNAs) are ~20–25 nt RNAs that regulate genes post-transcriptionally by binding to, and blocking, mRNAs; they are cut (by an RNase-III-like enzyme: Dicer) from longer transcripts. Human miRNAs appear to bind preferentially to AT-rich 3′-untranslated regions of mRNAs [PNAS (2005) *102* 15557–15562].

ssRNA can form triple-strand structures with dsDNA (cf. TRIPLEX DNA) and can also bind to ssDNA during transcription, in each case affecting transcription. Antisense RNA is also involved in RNA INTERFERENCE.

Note that, in contrast to the above, sequence-specific RNAs mediate RdDM (RNA-directed DNA methylation), in plant and human cells, and RNA EDITING.

antisense strand The *non*-CODING STRAND.

antisepsis ('degerming') Prophylactic or therapeutic treatment of human or animal tissues with ANTISEPTICS. (cf. SEPSIS.)

antiseptic paint ANTIBODY-containing secretions which bathe the mucous surfaces of the body.

antiseptics Chemical agents used for treating human and animal tissues (particularly skin) with the object of killing or inhibiting pathogens. Antiseptics are typically non-injurious to living tissues (cf. DISINFECTANTS). For examples of antiseptics see under ACRIFLAVINE; ALCOHOLS; AMINOACRIDINE; BISPHENOLS; CHLORHEXIDINE; CHLORINE; IODINE; PHENOLS; QUATERNARY AMMONIUM COMPOUNDS; SALICYLANILIDES; SALICYLIC ACID; TRICLOSAN.

antiserum (immune serum) SERUM containing antibodies to one or more particular antigens.

anti-sigma factor (anti-σ factor) Any molecule which can inhibit the normal activity of a given SIGMA FACTOR by binding to it.

Examples of anti-sigma factors include e.g. (i) FlgM (see SIGMA FACTOR); (ii) DnaK, which inhibits σ^{32} (see HEAT-SHOCK PROTEINS); and (iii) AsiA, a protein, encoded by bacteriophage T4, which is active during phage development in *Escherichia coli* and which can (*inter alia*) inhibit the *E. coli* σ^{70} sigma factor.

The Rsd protein ('regulator of sigma D'; *rsd* gene product) can inhibit σ^{70} in *E. coli*, and it has been proposed that Rsd is involved in the switching from σ^{70} to σ^{S} during the transition from exponential growth to the stationary phase in this organism [JB (1999) *181* 3768–3776].

[Anti-sigma factors: ARM (1998) *52* 231–286.]

antistreptolysin O test (ASLT; ASO test; ASOT) A test used for detecting/quantifying serum antibodies to STREPTOLYSIN O; it is of use e.g. in the differential diagnosis of RHEUMATIC FEVER and rheumatoid arthritis. In one method, INACTIVATED SERUM is serially diluted, and to each dilution is added a fixed volume of streptolysin O solution; after incubation at 4°C for several hours (or 37°C for ca. 15 minutes) – to allow antibody–streptolysin O combination – a fixed volume of RBC suspension is added to each dilution, and incubation carried out at 37°C for 30–45 minutes. Complete absence of haemolysis in a given tube indicates that streptolysin O in that tube has been fully neutralized by antibodies. The ASOT titre (often expressed in TODD UNITS) is given by the highest dilution of serum which completely inhibits haemolysis. ASOT can also be carried out as a LATEX PARTICLE TEST using streptolysin O-coated particles.

antitermination A process in which, during TRANSCRIPTION, the RNA polymerase fails to recognize a transcription termination signal and therefore continues transcription beyond the terminator. Antitermination may involve various mechanisms and can serve as a means for the regulation of gene expression (see e.g. BACTERIOPHAGE LAMBDA (gp*N*) and OPERON (attenuator control)) and for the prevention of erroneous transcription termination (see e.g. RRNA).

antitoxin (1) An antibody to a toxin. (2) An antiserum which contains antibodies to one or more particular toxins.

antiviral agents Agents which inhibit the replication of viruses in cells, tissues or organisms – either directly (e.g. by inhibiting a specific viral enzyme) or by inducing the synthesis of INTERFERON. (See also TILORONE.)

Owing to poor selective toxicity and/or the ready emergence of resistant strains of virus, relatively few compounds which exhibit antiviral activity in cell cultures are suitable for clinical use, and many of these are limited to topical treatment of localized infections.

ANTIRETROVIRAL AGENTS (q.v.) are antiviral agents used e.g. in chemotherapy against AIDS.

Examples of antiviral agents include: ACYCLOVIR, AMANTADINE, ANSAMYCINS, ARABINOSYL NUCLEOSIDES, ARILDONE, BROMOVINYLDEOXYURIDINE, CYCLARIDINE, DHBG, DHPA, DHPG, DIDEMNINS, ENVIROXIME, EPIPOLYTHIAPIPERAZINEDIONES, FAMCICLOVIR, GANCICLOVIR, IDOXURIDINE, LAMIVUDINE, MONENSIN, PENCICLOVIR, PHOSPHONOACETIC ACID, PHOSPHONOFORMIC ACID, RIBAVIRIN, SINEFUNGIN, THIOSEMICARBAZONES, TRIFLUOROTHYMIDINE, TUNICAMYCIN, VALACICLOVIR, ZANAMIVIR. [See also Book ref. 134.]

AOAC Association of Official Analytical Chemists (of the USA).

AOAC use-dilution test A CARRIER TEST (q.v.), devised by the AOAC, which determines the USE-DILUTION of a given disinfectant. To each of a number of dilutions of the disinfectant is added a (contaminated) small steel carrier; after immersion for 10 min, each carrier is transferred to broth to detect any surviving microorganisms. Sporicidal activity is determined by exposing the endospores of *Bacillus subtilis* ATCC 19659 or *Clostridium sporogenes* ATCC 9081 to the disinfectant and subsequently testing for viability.

AP endonuclease See BASE EXCISION REPAIR.

AP site An apyrimidinic or apurinic site in DNA (see ADAPTIVE RESPONSE and URACIL-DNA GLYCOSYLASE).

6-APA 6-AMINOPENICILLANIC ACID.

***Apa*I** See RESTRICTION ENDONUCLEASE (table).

apalcillin A penicillin (see PENICILLINS) which is active against both Gram-positive and Gram-negative bacteria (including *Pseudomonas aeruginosa*) [AAC (1982) *21* 906–911].

Apc APHIDICOLIN.

aperture diaphragm See CONDENSER.

APES See AAS.

APH Aminoglycoside phosphotransferase (see AMINOGLYCOSIDE ANTIBIOTICS).

Aphanizomenon A genus of filamentous CYANOBACTERIA; the cells are cylindrical with closely abutting ends, and the filaments aggregate to form raft-like floating colonies. GAS VACUOLES are present, and heterocysts occur but are often sparse. Trichomes commonly terminate in tapering, colourless 'hair cells'.

A. flos-aquae is an important freshwater BLOOM-forming species in temperate regions; it may also occur in brackish waters. The filaments of this species typically form wispy, flake-like colonies, the formation of which, in culture at least, appears to depend on the presence of iron. Strains of *A. flos-aquae* produce toxins ('aphantoxin') which may apparently include e.g. SAXITOXIN and neosaxitoxin; 'aphantoxin' acts as a neuromuscular blocking agent, and is toxic for amphibia, fish, etc.

Aphanoascus See GYMNOASCALES.

Aphanocapsa See SYNECHOCYSTIS.

Aphanochaete A genus of filamentous, prostrate, branched green algae (division CHLOROPHYTA) which grow as epiphytes on other freshwater algae. Some of the cells in the filaments bear unicellular 'hairs' with bulbous bases. Quadriflagellate zoospores or aplanospores are formed; sexual reproduction is oogamous.

Aphanomyces A genus of predominantly aquatic fungi of the order SAPROLEGNIALES (q.v.). A species known as the 'MG fungus' has been reported to cause a mycotic granuloma of fish [Book ref. 1, pp. 195–197], while *A. astaci* causes CRAYFISH PLAGUE. Plant-pathogenic species include *A. brassicae*, *A. camptostylus*, *A. cochlioides* (on sugar-beet), *A. euteiches* (on peas) and *A. raphani* (on radish).

aphanoplasmodium See MYXOMYCETES.

Aphanothece A phycological genus of unicellular 'blue-green algae' (Chroococcales) apparently referable to *Cyanothece* (see SYNECHOCOCCUS).

aphantoxin See APHANIZOMENON.

aphid (bacteriocytes) See BUCHNERA.

aphidicolin (Apc) A tetracyclic diterpenoid, isolated from *'Cephalosporium aphidicola'*, which – in eukaryotic cells – is a potent and specific inhibitor of DNA POLYMERASE α. Apc has no effect on bacterial DNA polymerases, but it inhibits growth and DNA synthesis in certain members of the ARCHAEA: e.g. halobacteria [FEMS (1984) *25* 187–190] and methanogens [SAAM (1985) *6* 111–118]. (cf. DDTTP.)

aphotic zone That part of an aquatic habitat below the PHOTIC ZONE.

aphthous fever *Syn.* FOOT AND MOUTH DISEASE.

Aphthovirus A genus of viruses (family PICORNAVIRIDAE) which infect mainly cloven-hooved animals, causing FOOT AND MOUTH DISEASE; there are 7 serotypes and many antigenically distinct strains. Aphthovirus virions are rapidly inactivated at pH below ca. 6–7, but are appreciably resistant to desiccation; stability of the virions is generally increased by high ionic concentrations and low temperatures.

Aphyllophorales (Polyporales) An order of terrestrial (typically humicolous or lignicolous), mainly saprotrophic fungi (subclass HOLOBASIDIOMYCETIDAE) which characteristically form tough (rather than fleshy), macroscopic, gymnocarpic fruiting bodies that may be e.g. crust-like, erect and coralloid, bracket-shaped (BRACKET FUNGI) etc; the hymenium may be borne on a smooth, toothed or irregular surface, and it may form the lining of tubular pores (in 'polypores') or the surface layer on clavate or coralloid basidiocarps or on lamellae.

Constituent families [Book ref. 64, p.20] include e.g.:

Clavariaceae. Basidiocarp: erect – simple and clavate ('club fungi'), branched, or coralloid ('coral fungi') – and often brightly coloured (e.g. pink, yellow, violet); basidiospores: white or pale cream, typically non-amyloid. Genera: e.g. *Clavaria* and *Typhula* (see also SNOW MOULD).

Coniophoraceae. Basidiocarp: characteristically resupinate; basidiospores: pigmented, with a cyanophilic spore wall. Genera: e.g. CONIOPHORA, *Gyrodontium*, and SERPULA.

Corticiaceae. Basidiocarp: characteristically resupinate or effused-reflexed, with a hymenium that may be e.g. smooth or wrinkled. Genera: e.g. *Botryobasidium*, *Christiansenia* (see also MYCOPARASITE), *Merulius*, *Mycoacia*, *Peniophora* (see also HETEROBASIDION), *Phanerochaete* (see also WHITE ROT and LIGNIN), and *Phlebia*.

Fistulinaceae. Basidiocarp: annual, pileate with a lateral stipe, fleshy and moist; the hymenophore consists of a layer of closely packed parallel tubules. *Fistulina hepatica* (the edible 'beef-steak fungus') causes BROWN OAK. It forms a fan-shaped or tongue-shaped basidiocarp whose upper surface is rusty brown, and whose lower (hymenial) surface is pale yellow when unbroken; the reddish monomitic context of the pileus exudes a red liquid when cut.

Ganodermataceae. Basidiocarp: annual or perennial, typically fan-shaped and horizontal, either non-stipitate or borne on a lateral stipe; the context of the pileus is trimitic, and the underside of the pileus bears a porous hymenophore. Basidiospores: ovoid, typically flattened at one pole; ornamentations on the inner, brown layer of the spore wall penetrate deeply into the outer, hyaline layer of the spore wall. The organisms occur e.g. on felled timber; *Ganoderma applanatum* causes heart rots in various types of tree, infection occurring via wounds.

Hydnaceae. Basidiocarp: typically mushroom-shaped, with a monomitic, non-xanthochroic context; the hymenophore consists of a layer of pendulous tooth-like processes ('spines' or 'teeth') on the underside of the pileus. Basidiospores: non-pigmented, spheroidal. *Hydnum repandum* (the 'hedgehog fungus') is an edible species which forms a thick, commonly cream-coloured pileus and a short, thick stipe which is often eccentrically located on the pileus; the teeth are often decurrent.

Hymenochaetaceae. Basidiocarp: annual or perennial, ranging from resupinate to pileate or clavate according to species; hymenophore: poroid (in e.g. INONOTUS and *Phellinus*) to smooth (in *Hymenochaete*). Basidiospores: pigmented or colourless. The organisms are commonly lignicolous.

Polyporaceae. Basidiocarp: annual or perennial, ranging (with species) from resupinate to pileate (either stipitate or sessile); the context is non-xanthochroic, and (according to species) may be mono-, di- or trimitic, and e.g. corky, leathery or woody. The hymenophore is *typically* porous, the hymenium being confined to the walls of the pores. (cf. LENTINUS, PLEUROTUS.) Basidiospores: colourless and usually smooth. The organisms grow on felled timber and on living trees. Genera: e.g. CORIOLUS, DAEDALEA, FOMES, *Grifola*, HETEROBASIDION, *Irpex*, *Laetiporus*, LENTINUS, *Lenzites*, PIPTOPORUS, PLEUROTUS, POLYPORUS, PORIA, *Rigidoporus* (see also BUTT ROT), TRAMETES and *Tyromyces*.

Schizophyllaceae. Basidiocarp: fan-shaped or lobed; basidiospores: colourless and smooth. The organisms are characteristically lignicolous. Genera: e.g. SCHIZOPHYLLUM.

Sparassidaceae. Basidiocarp: branching, flattened, monomitic lobes, the hymenium being on both sides of the erect lobes. Genera: e.g. SPARASSIS.

Stereaceae. Basidiocarp: appressed, effused-reflexed or stipitate, typically dimitic; basidiospores: colourless and smooth. The organisms are characteristically lignicolous. Genera: e.g. *Amylostereum* (see also WOODWASP FUNGI), *Chondrostereum* (see also SILVER LEAF), *Podoscypha*, STEREUM.

API system See MICROMETHODS.

apical Terminal: at the tip or distal part of a structure. *Antonym*: basal.

apical complex A collective term for the CONOID, MICRONEMES, POLAR RING and RHOPTRIES – some or all of which are found in members of the APICOMPLEXA, particularly in motile stages. (Some authors include the MICROPORES and/or the subpellicular tubules as part of the apical complex.)

apical granule *Syn.* SPITZENKÖRPER.

apical paraphysis See PARAPHYSIS.

Apicomplexa A phylum of parasitic PROTOZOA with one type of nucleus, no cilia and no flagella (some have flagellated microgametes); components of the APICAL COMPLEX are present at some stage in the life cycle. Classes: PERKINSASIDA, SPOROZOASIDA. [Cytoskeleton: Microbiology and Molecular Biology Reviews (2002) *66* 21–38.]

apiculate Having a short, pointed projection (*apiculus*) at one end.

Apiosoma See PERITRICHIA.

Aplanochytrium See THRAUSTOCHYTRIDS.

aplanogamete A non-motile GAMETE.

aplanospore A non-motile spore.

aplastic crisis (in parvovirus B19 infection) See ERYTHROVIRUS.

aplysiatoxin See LYNGBYA.

APO-1 See FAS.

apo-activator See OPERON.

apochlorotic Having lost chlorophyll – either during evolutionary development or during growth under certain conditions. Apochlorotic organisms may have lost either chloroplasts or chlorophyll; in the latter case a colourless plastid (leucoplast) may be formed. (cf. ACHLOROTIC; see also BLEACHING.)

apochromat (apochromatic objective) An objective lens (see MICROSCOPY) in which chromatic aberration has been corrected for three colours (red, green, and blue), and spherical aberration has been corrected for two colours (red and blue). (cf. ACHROMAT.) A FLAT-FIELD OBJECTIVE LENS of this type is called a *planapochromat*.

Apodachlya See LEPTOMITALES.

apoenzyme The protein component of a PROSTHETIC GROUP-requiring enzyme. (cf. HOLOENZYME.)

apokinetal stomatogenesis See STOMATOGENESIS.

apomict See APOMIXIS.

apomixis (*microbiol.*) In certain diploid organisms: a type of reproduction, considered to be a deviant form of the sexual process, in which meiosis, gamete formation and fertilization do not occur even though morphologically differentiated structures (e.g. spore-containing asci) are formed; progeny produced apomictically from a given cell (the apomict) are genetically identical to that cell. Thus, e.g., some apomictic strains of *Saccharomyces cerevisiae* form asci each containing two uninucleate diploid spores [Yeast (1985) *1* 39–47]. (cf. AMIXIS.)

apoplastidic Lacking a PLASTID.

apoptosis In eukaryotic cells: cell death which occurs in a regulated ('programmed') manner: i.e. death resulting from an organized process involving an internal mechanism that characteristically includes fragmentation of the genome.

Apoptosis is a natural process. For example, in vertebrates it occurs during embryogenesis, while in invertebrates it occurs during metamorphosis.

Apoptosis can also be induced by physical and chemical factors – e.g. heat, radiation and agents such as glucocorticoids. Moreover, apoptosis can be induced by certain pathogens. Thus, e.g. in HIV-infected individuals (see AIDS), apoptosis may occur in $CD4^+$ T cells which have bound viral protein. Again, poliovirus 2A protease is reported to induce apoptosis in human embryonic kidney epithelial cells [MCB (2000) *20* 1271–1277]. Some bacteria can induce apoptosis in host cells. Thus, for example, on phagocytosis by a macrophage, *Shigella* can escape from the vacuole and then induce apoptosis in the phagocyte by means of a plasmid-encoded protein, IpaB. IpaB, secreted by a type III secretory system (see PROTEIN SECRETION), activates the host cell's IL-1β-converting enzyme (ICE, see INTERLEUKIN-1); activated ICE (i) initiates apoptosis in the macrophage, and (ii) activates the intracellular, inactive form of IL-1β. Activated IL-1β, when extracellular, may initiate an acute inflammatory response (involving TNF-α, IL-1, IL-6 and IL-8). In one model of shigellosis, these events have a positive role in promoting a pro-inflammatory response that helps to control the infection [TIM (1997) *5* 201–204].

The *Shigella* example (above) involves an endogenous stimulus (via the IpaB protein). Exogenous stimuli include the binding of certain cytokines, or immune cells, to specific cell-surface receptors on target cells. Thus, apoptosis is one of the possible effects which may follow when tumour necrosis factor binds to its receptor (see TNF). Induction of apoptosis by $CD8^+$ T cells may be possible in at least two different ways. One way involves interaction between cell-surface molecules on the T cell and target cell (see FAS). Alternatively, following contact between T cell and target cell, the T cell may release pore-forming lethal agents (*perforins*) – and also *granzymes* which are believed to activate *caspases* (see below) in the target cell.

Whatever the stimulus that triggers apoptosis (in different types of cell under differing conditions), it appears that the mechanism of this process is dependent on certain intracellular cysteine proteases (*caspases*) that are synthesized as ZYMOGENS and activated specifically during apoptosis; these enzymes, which have a cysteine residue at the active site, cleave specific protein substrates at a site adjacent to an aspartic acid residue. The activity of a caspase on a substrate molecule may lead to either inactivation or (less commonly) activation of the given protein; for example, caspase-mediated activation of a certain nuclease (*caspase-activated DNase*, CAD) leads to fragmentation of genomic DNA and formation of the characteristic nucleosomal 'ladder'. (CAD occurs in living cells, in inactive form, complexed with an inhibitory entity referred to as ICAD.)

The activation of caspases appears to occur commonly by proteolytic cleavage of the zymogen form, and, in at least some cases, one caspase may be activated by an 'upstream' caspase in a *caspase cascade*. An understanding of the mechanism(s) involved in caspase activation is central to an appreciation of the process of apoptosis, and this area is being actively researched.

[Programmed cell death (concept, mechanism and control): Biol. Rev. (1992) *67* 287–319. Apoptosis (reviews on biochemistry and other aspects of the process): Nature (2000) *407* 770–816.]

aporepressor See OPERON.

Aporpium See TREMELLALES.

Apostomatida See HYPOSTOMATIA.

aposymbiotic Refers to an organism (aposymbiont) which has lost symbionts it normally possesses.

apothecioid Of an ASCOCARP: having, at maturity, a hymenium more or less open to the environment – as in a typical APOTHECIUM; apothecioid ascocarps may be e.g. cup-shaped, discoid (circular and flat), or lirelliform (see LIRELLA).

apothecioid pseudothecium See ASCOSTROMA.

apothecium A typically dish- or cup-shaped or discoid, sessile or stipitate ASCOCARP, the inner (or upper) surface of which is lined with a hymenial layer of asci (often interspersed with paraphyses); the main structural tissue of the apothecium is called the *excipulum*, and the thin layer of hyphae between the hymenium and the excipulum is termed the *hypothecium* or *subhymenium*. Some apothecia are not cup-like or discoid: see e.g. MORCHELLA, PODETIUM and VERPA. Apothecia range from minute to 10 cm or more in diameter; those formed by e.g. species of *Aleuria* and *Scutellinia* are brightly coloured. (See also LECANORINE APOTHECIUM.)

AP-PCR See ARBITRARILY PRIMED PCR.

appertization A method of FOOD PRESERVATION in which certain types of food are exposed to a temperature/time regime which renders them safe for the consumer and microbiologically stable for extended, or indefinite, periods of time on subsequent storage in hermetically sealed containers at temperatures below 40°C; appertized foods are not necessarily sterile, and in certain cases (see CANNING) appertized foods must be kept under refrigeration. For foods which have no intrinsic properties (e.g. acidity) inhibitory to microbial growth and/or toxin production, appertization involves treatment which, at least, destroys the spores of pathogenic bacteria (particularly *Clostridium botulinum*); acidic foods (below ca. pH 4.5), and those containing appropriate concentrations of e.g. sodium nitrite, are appertized at somewhat lower temperatures. High temperature short time (HTST) appertization involves heat processing at high temperatures for periods ranging from ca. 2 sec to several minutes (cf. HTST PASTEURIZATION). In the ultra-high temperature (UHT) appertization of milk the process involves e.g. maintenance of a minimum temperature of ca. 140°C for at least 2 or 3 sec. (cf. RADAPPERTIZATION.)

apple canker A CANKER of apple trees caused by *Nectria galligena* f. sp. *mali*; other trees – e.g. pear, poplar, hawthorn – may also be affected. Enlarged, sunken lesions develop on the branches; bark in the centre of a lesion dies and may break away. Infection occurs via wounds or leaf scars, and can be initiated by ascospores and by conidia.

apple chlorotic leafspot virus See CLOSTEROVIRUSES.

apple diseases See e.g. APPLE CANKER, APPLE SCAB, BROWN ROT (sense 2) and FIREBLIGHT.

apple mosaic virus See ILARVIRUSES.

apple scab An important disease of apple trees (*Malus*) caused by *Venturia inaequalis* (see VENTURIA). In spring, germinating ascospores infect via wounds or by means of appressoria. Small dark spots appear on leaves; on fruits, small dark velvety lesions enlarge to brown, corky scabs. Conidia, formed during late spring and summer, spread the disease. Control: antifungal sprays (see e.g. BENOMYL, DICHLONE, GLYODIN and TRIFORINE). (See also FIREBLIGHT.)

apple stem grooving virus See CLOSTEROVIRUSES.

App(NH)p (β,γ-imido)ATP: a (biologically) non-hydrolysable analogue of ATP.

appressed Lying close to, or flattened against, a surface.

appressorium A specialized, flattened region of a hypha or germ tube formed by some plant-pathogenic fungi (e.g. *Puccinia graminis*, *Venturia inaequalis*); it aids infection by adhering closely to the surface of a host cell and producing a small outgrowth of fungal tissue (*infection peg*) which penetrates the host cell wall. A plant may respond to an infection peg by depositing a small region of wall-like material (a *papilla*) near the site of attempted penetration; such a papilla can sometimes prevent infection.

apramycin An AMINOGLYCOSIDE ANTIBIOTIC which is resistant to many of the enzymes which inactivate antibiotics of this group; it is inactivated by e.g. chromosome- and plasmid-specified acetyltransferases [JGM (1984) *130* 473–482].

aprotinin A basic polypeptide which inhibits many serine PROTEASES, including e.g. trypsin, chymotrypsin, and some bacterial proteases.

APS Adenosine-5′-phosphosulphate (also called adenosine-5′-sulphatophosphate or adenylylsulphate). See ASSIMILATORY SULPHATE REDUCTION and DISSIMILATORY SULPHATE REDUCTION.

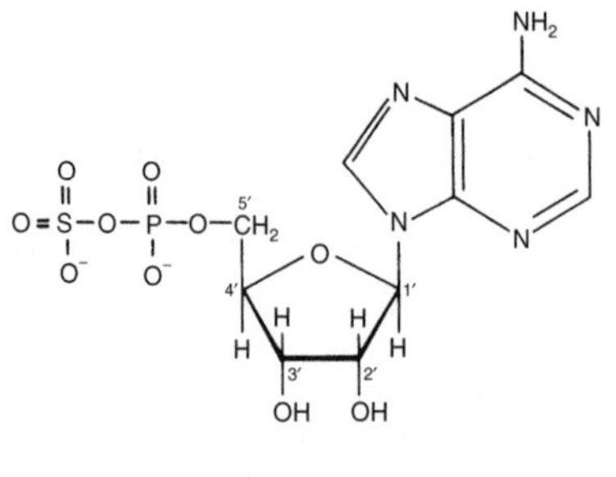

APS (adenosine 5′-phosphosulphate)

APT agar An agar-based medium containing tryptone, yeast extract, glucose, Tween 80, citrate, and various inorganic salts. [Recipe: Book ref. 46, p. 1661.] It is used for isolating organisms such as lactobacilli and *Brochothrix* from e.g. meat and meat products.

APT paper BLOTTING paper impregnated with 2-aminophenylthioether; before use, APT is chemically modified to its reactive diazo derivative, and is then able to bind (covalently) ssDNA, RNA and/or protein via the diazonium group.

APW Alkaline peptone water: 1% peptone and 1% NaCl in distilled water, pH adjusted to 8.6–9.0, used e.g. for isolating *Vibrio cholerae*.

ApxIA, ApxIIA (of *Actinobacillus pleuropneumoniae*) See RTX TOXINS.

AqpZ See MIP CHANNEL.

Aqualinderella See LEPTOMITALES.

aquaporin See MIP CHANNEL.

Aquaspirillum A phylogenetically heterogeneous genus of Gram-negative, asporogenous bacteria formerly included in the genus SPIRILLUM. Cells: fairly rigid, generally helical (0.2–1.4 × ca. 2 – >30 μm, according to species); *A delicatum* cells are mainly vibrioid, and those of *A. fasciculus* are straight rods. S LAYERS, intracellular PHB granules, and POLAR MEMBRANES are commonly present; COCCOID BODIES are commonly formed. Motile, typically with bipolar tufts of flagella (*A. polymorphum* has a single flagellum at each pole, *A. delicatum* has one or two flagella at one pole only). Cells swim in a straight line with a characteristic corkscrew-like motion; *A fasciculus* swims effectively only in media of high viscosity. Metabolism is respiratory; species are usually aerobic or microaerophilic, but some can carry out NITRATE RESPIRATION under anaerobic conditions. Carbon sources: amino acids or salts of organic acids, but usually not carbohydrates; *A. autotrophicum* can grow autolithotrophically with CO_2, H_2 and O_2. Nitrogen source: NH_4^+; *A. peregrinum*, *A. fasciculus* and strains of *A. itersonii* can fix N_2 under anaerobic conditions. Aquaspirilla are typically oxidase +ve; catalase +ve; usually phosphatase +ve; indole −ve; arylsulphatase −ve. No growth with 3% NaCl. Optimum growth temperature: usually ca. 30–32°C (ca. 20°C for *A. psychrophilum*, 41°C for *A. bengal*). GC%: 49–66. Type species: *A serpens*. About 17 species are recognized on the basis of morphology, nutrition and GC%.

Aquaspirilla occur in a wide range of freshwater habitats – e.g., stagnant waters rich in organic matter; none appears to be pathogenic.

[Book ref. 22, pp. 72–90. Media, culture etc: Book ref. 45, pp. 596–608.]

(See also MAGNETOTACTIC BACTERIA; cf. AZOSPIRILLUM; OCEANOSPIRILLUM; SPIRILLUM; SPOROSPIRILLUM.)

aquifer See WATER SUPPLIES.

***ara* operon** (arabinose operon) An OPERON concerned with the metabolism of ARABINOSE; when used without qualification, the term '*ara* operon' usually refers to the *araBAD* operon, but in e.g. *Escherichia coli* three distinct genetic loci are involved in the uptake and metabolism of L-arabinose: *araBAD* (at ca. 1 min on the chromosome map), *araFG* (at ca. 45 min) and *araE* (at ca. 61 min). The *araBAD* operon encodes enzymes for the conversion of L-arabinose to D-xylulose 5-phosphate; the *araA* gene encodes L-arabinose ketol-isomerase, *araB* encodes L-ribulokinase, *araD* encodes ribulose 5-phosphate 4-epimerase. The *araFG* operon specifies a high-affinity BINDING PROTEIN-DEPENDENT TRANSPORT SYSTEM for L-arabinose; *araF* encodes the binding protein, *araG* the membrane component. The *araE* gene specifies a protein involved in an independent low-affinity transport system.

All three operons are controlled by the product of the regulator gene *araC* which is located next to *araBAD* but transcribed in the opposite direction. (See also REGULON.) In the absence of arabinose, the AraC protein behaves as a repressor of the *araBAD* operon, binding to the OPERATOR and preventing transcription (*negative* promoter control: see OPERON). However, when arabinose is present it converts the AraC protein into an activator which binds to an initiator (*araI*) to promote *araBAD* transcription (*positive* promoter control). (The *araC* gene is itself subject to autogenous regulation, i.e., the AraC protein represses its own synthesis.) Thus, expression of the *araBAD* operon depends on the relative amounts of repressor and activator, which in turn depend on the level of arabinose (the co-activator).

The *ara* regulon is also subject to CATABOLITE REPRESSION.

ara-A *Syn.* VIDARABINE.

***araBAD* operon** See ARA OPERON.

Arabidopsis thaliana A flowering brassica widely used as a model plant for research. [Genome sequence and other data: Nature (2000) *408* 791–826.] (See also CAMALEXIN.)

arabinogalactans See PECTIC POLYSACCHARIDES. (See also WAX D.)

arabinose An aldopentose (see PENTOSES) which occurs e.g. as a component of hemicelluloses (see e.g. XYLANS) and PECTIC POLYSACCHARIDES, and in the cell walls of certain bacteria (e.g. *Nocardia* spp). Arabinose can be metabolized by various bacteria, including e.g. *Escherichia coli* and strains of *Salmonella*, *Bacillus*, and lactic acid bacteria [Appendix III(d)]. Arabinose is heat-labile; solutions should be sterilized by filtration. (See also ARA OPERON.)

arabinose operon See ARA OPERON.

arabinosyl nucleosides A group of compounds which inhibit various enzymes – including DNA polymerases and reverse transcriptase – by acting as analogues of biological nucleosides and nucleotides. Arabinosyl nucleosides inhibit DNA synthesis in a variety of eukaryotic cells, and have been used as antitumour and ANTIVIRAL AGENTS; in general, viral replication is more sensitive to these agents than are the host cells. Arabinosyl nucleosides probably must be converted to the corresponding nucleoside 5′-triphosphate for activity. (See e.g. ARA-T; CYTARABINE; FIAC; FMAU; VIDARABINE.)

arabinoxylans Heteroglycans containing both xylose and arabinose (cf. XYLANS).

arabitol A PENTITOL formed e.g. by the reduction of ARABINOSE. It occurs, often with MANNITOL, in various basidiomycetes and ascomycetes.

ara-C *Syn.* CYTARABINE.

***araC* gene** See ARA OPERON.

arachidonic acid In mammals: an essential C_{20} straight-chain fatty acid – $CH_3(CH_2)_3(CH_2CH{=}CH)_4(CH_2)_3COOH$; it occurs e.g. as a component of cell membrane phospholipids, and (bound to protein) in blood. Free arachidonic acid is formed e.g. during the activation of lymphocytes, and is a precursor of e.g. LEUKOTRIENES, PROSTACYCLIN, PROSTAGLANDINS and thromboxanes.

Arachnia A genus of anaerobic (but aerotolerant), catalase-negative, asporogenous bacteria (order ACTINOMYCETALES, wall type I). The organisms grow as a substrate mycelium which fragments into branched filaments and irregularly-shaped rods; aerial hyphae are not formed. Metabolism: fermentative, fermentation of glucose yielding primarily acetic and propionic acids; growth occurs readily in vitro on rich media at 37°C. *A. propionica*, the type species (GC%: 63–65), is pathogenic in man and other animals; it is a common causal agent of LACHRYMAL CANALICULITIS.

arachnoid Resembling a spider's web.

Arachnoidiscus See DIATOMS.

Arachnula See GRANULORETICULOSEA.

***araE* gene** See ARA OPERON.

***araFG* operon** See ARA OPERON.

aranotin See EPIPOLYTHIAPIPERAZINEDIONES.

ara-T (1-β-D-arabinofuranosylthymine) An ARABINOSYL NUCLEOSIDE isolated from a marine sponge (*Cryptoethya crypta*). It has antiviral activity against herpes simplex viruses (HSV-1 and -2), varicella-zoster virus and vaccinia virus, and is relatively non-toxic to animal cells.

arbitrarily primed PCR Certain TYPING methods in which PCR is used with primers of arbitrary (i.e. random) sequence to copy discrete sequences of chromosomal DNA. Only one type of primer is used (cf. standard PCR protocol), and during the annealing stage the primers bind to various 'best-fit' sequences on the strands of DNA. As the primer is of arbitrary sequence it is not possible to predict which chromosomal sequences will be copied; however, results are reproducible if the method is carefully standardized – including not only the primer sequence but also e.g. the use of a specific type of DNA polymerase [NAR (1993) *21* 4647–4648] and a particular procedure for preparing the sample DNA [NAR (1994) *22* 1921–1922]. Various guidelines have been recommended for promoting reliability and reproducibility in arbitrarily primed PCR [JCM (1997) *35* 339–346].

The method is outlined in the figure.

Arbitrarily primed PCR (AP-PCR) was first published as a typing procedure in 1990 [NAR (1990) *18* 7213–7218; NAR (1990) *18* 6531–6535]. Another typing method – RAPD (q.v.) – is based on the same principle; it generally employs shorter primers (~10-mer) compared with those used in AP-PCR (>20-mer), but, even so, some authors regard the two methods as identical.

Typing methods based on this principle have two main advantages. First, the entire chromosome can participate in

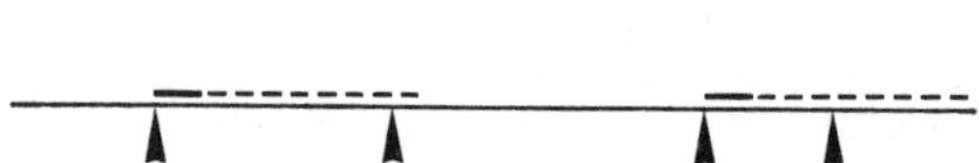

ARBITRARILY PRIMED PCR (principle, diagrammatic). In the diagram, the two strands of chromosomal DNA (which have been separated by heat) are shown as long parallel lines. During the annealing (primer-binding) stage of PCR, copies of the arbitrary primer bind at a number of 'best-fit' sequences, on both strands of DNA, under low-stringency conditions. In some cases (by chance), two primers will bind, on opposite strands, a few hundred bases apart; if strand elongation can occur efficiently from these two primers, and if elongation is time-limited, two short strands of DNA will be produced. In the diagram (*right-hand side*), two primers (short lines) have bound, close together, on opposite strands; two more primers (*left-hand side*) have bound, further apart, on opposite strands. Strand elongation from each primer (*dashed line*) has produced the four fragments shown. (Note that a fragment synthesized on one strand contains a copy of the best-fit sequence of the other strand.) Another cycle of low-stringency PCR produces more copies of the fragments. Subsequently, many cycles of PCR are carried out under higher stringency (using the same primer). Under these conditions, primers bind to best-fit sequences (rather than elsewhere) on the fragments formed by low-stringency PCR – although (due to the higher stringency), primers may not bind to best-fit sequences on *all* of the fragments, so that only a proportion of the fragments formed under low stringency may be amplified under higher stringency. The length of the fragments formed under higher stringency is shown by the distance between each pair of arrowheads.

On electrophoresis, the fragments from a given strain form a characteristic pattern of bands (the *fingerprint*).

Reproduced from *Bacteria* 5th edition, Figure 16.4(c), page 434, Paul Singleton (1999) copyright John Wiley & Sons Ltd, UK (ISBN 0471-98880-4) with permission from the publisher.

the comparison of strains. Second, no prior knowledge of the genome's sequence is required, so that, potentially, any isolate may be typed.

arboricolous Living/growing on trees.

arboviruses 'Arthropod-borne viruses': a non-taxonomic category of viruses which can replicate in both vertebrate hosts and arthropod vectors; arboviruses include e.g. members of the families ARENAVIRIDAE, BUNYAVIRIDAE, Reoviridae (ORBIVIRUS), Togaviridae (ALPHAVIRUS), and FLAVIVIRIDAE. [Vector competence of mosquitoes for arboviruses: ARE (1983) *28* 229–262.]

arbuscule See MYCORRHIZA.

ARC (AIDS-related complex) See AIDS.

Arcanobacterium A genus of asporogenous bacteria (order ACTINOMYCETALES, wall type VI). In culture the organisms initially grow as irregular rods which may later become granular and segmented or may give rise to coryneform cells; growth occurs anaerobically or aerobically and is enhanced by blood or serum or by increased partial pressures of carbon dioxide. Growth is poor on tellurite media. Type species: *A. haemolyticum* (formerly *Corynebacterium haemolyticum* [JGM (1982) *128* 1279–1281]).

Arcella A genus of testate amoebae (order ARCELLINIDA) in which the test is predominantly or entirely organic. In *A. vulgaris* the test is more or less spherical (ca. 30–100 μm diam.) and is typically somewhat translucent in young cells, often becoming darker with age. The cell contains two nuclei. Slender lobopodia are extended through a ventral aperture and are used chiefly for feeding. A gas-filled vacuole may be formed in the cytoplasm and apparently aids in the orientation of the organism. *A. dentata* has a test bearing lateral spines, giving it a stellate appearance when viewed from above. *Arcella* spp occur among vegetation in ponds, in soil, etc.

Arcellinida An order of freshwater and soil amoebae (class LOBOSEA) in which the cell is surrounded by a TEST which may be primarily organic or primarily inorganic in composition, and which is perforated by a single aperture through which pseudopodia can be extended. The cells commonly contain two or more nuclei. Reproduction occurs asexually e.g. by binary fission. Genera include e.g. ARCELLA, *Centropyxis*, COCHLIOPODIUM, DIFFLUGIA.

Archaea A DOMAIN of prokaryotic organisms which are evolutionarily distinct from members of the domain BACTERIA [see e.g. JB (1994) *176* 1–6]. These organisms were formerly classified in the kingdom Archaebacteria, being distinguished from other prokaryotes (kingdom Eubacteria) on the basis of e.g. nucleotide sequences in 16S rRNA; later, the distinction between these two groups of prokaryotes was deemed to be more fundamental than previously supposed, and each kingdom was elevated to the taxonomic rank of domain.

Many archaeans are associated with environments characterized by e.g. high temperatures or high salinity, and such organisms have been referred to as *extremophiles*. Nevertheless, some archaeans are found in 'moderate' environments such as soil [e.g. PNAS (1997) *94* 277–282]. [Archaeal dominance in the mesopelagic zone of the Pacific Ocean: Nature (2001) *409* 507–510.]

The domain Archaea has been divided into the kingdoms Euryarchaeota and Crenarchaeota. Euryarchaeota includes halophiles (e.g. HALOBACTERIUM) and METHANOGENS. Crenarchaeota includes 'sulphur-dependent' species such as DESULFUROCOCCUS and SULFOLOBUS.

Archaeans differ from bacteria in many ways. For example, gene expression (transcription and translation) appears to be closer to the eukaryotic pattern than the prokaryotic pattern [TIM (1998) *6* 222–228]. The archaeal DNA-dependent RNA POLYMERASE resembles more closely the eukaryotic (nuclear) polymerase than the bacterial enzyme. Elongation factors in protein synthesis are (unlike their bacterial counterparts) sensitive to DIPHTHERIA TOXIN. [Assembly of the archaeal signal recognition particle from recombinant components: NAR (2000) *28* 1365–1373.]

[Replication origin of archaeans: TIBS (2000) *25* 521–523.]

At least some archaeal enzymes appear to have unique structures [TIM (1998) *6* 307–314].

In some archaeans the cell wall may consist mainly or solely of an S LAYER (q.v.) closely associated with the cytoplasmic membrane. The cell wall in some species contains a peptidoglycan-like polymer, PSEUDOMUREIN (which is not cleaved by the enzyme LYSOZYME).

The cytoplasmic membrane contains lipids of a type which do not occur in species of Bacteria. In contrast to the ester-linked bacterial lipids, archaeal lipids are characteristically ether-linked molecules that contain e.g. isoprenoid or hydro-isoprenoid components. Certain archaeal and bacterial lipids are structurally analogous – e.g. the di-ether and di-ester lipids, both types of molecule having a single polar end. However, the archaeal lipids also include other types – such as tetra-*O*-di(biphytanyl) diglycerol – which contain two ether-linked glycerol residues, one at *each* end of the molecule; such molecules, which have two polar ends, may span the width of the cytoplasmic membrane.

[Protein translocation across the archaeal cytoplasmic membrane: FEMS Reviews (2004) *28* 3–24.]

The archaeal flagellum is markedly different in composition, structure and apparent mode of assembly from that found in bacteria. For example, the subunit, flagellin, is typically glycosylated, and it contains a *signal sequence* (see SIGNAL HYPOTHESIS) – suggesting passage into/through the cytoplasmic membrane (cf. bacterial flagellin, which passes through the hollow structures of the developing flagellum); this latter feature (as well as certain similarities between archaeal flagellins and bacterial type 4 fimbriae) suggests that archaeal flagella assemble from the *base* (in contrast to 'tip growth' in bacteria) [JB (1996) *178* 5057–5065]. The archaeal flagellar filament is typically much thinner than the bacterial filament.

Various similar or analogous proteins have been found in archaeans and bacteria. For example, the archaeal RadA protein is apparently analogous to the RecA protein in bacteria [GD (1998) *12* 1248–1253]. Again, an FtsZ protein occurs in members of both domains, suggesting that the cell division apparatus was similar in a common ancestor [Mol. Microbiol. (1996) *21* 313–319]. [The CELL CYCLE in archaeans: Mol. Microbiol. (2003) *48* 599–604.]

archaean A member of the domain ARCHAEA. The term is also spelt 'archaeon'.

Archaebacteria A kingdom (now obsolete) which included all those prokaryotes not classified in the kingdom EUBACTERIA. [Phylogeny of the Archaebacteria: SAAM (1985) *6* 251–256.] Some archaebacteria were placed in a separate taxon, Eocyta, on the basis of ribosomal characteristics [PNAS (1984) *81* 3786–3790]. (See also EOCYTES.) Members of the Archaebacteria are now included in the domain ARCHAEA.

Archaeobacteria A proposed class of prokaryotes (see MENDOSICUTES) corresponding to the (later) kingdom ARCHAEBACTERIA (now also obsolete).

Archangium See MYXOBACTERALES.

archicarp In ascomycetes: the cell(s) which give rise to a fruiting body or to a part of it.

archigregarines See GREGARINASINA.

Arcobacter A genus of Gram-negative, asporogenous bacteria of the family Campylobacteraceae. Cells: spiral rods with unsheathed flagella. Catalase +ve. Nitrate is reduced. Typically urease −ve. These organisms resemble *Campylobacter* spp but differ e.g. in their ability to grow in air at 15–25°C. *A. butzleri* and *A. cryophilus* have been isolated from patients with diarrhoea, including children in the developing countries; most strains do not hydrolyse hippurate but do hydrolyse indoxyl acetate.

arcuate Curved like a bow; arched.

Arcyria A genus of slime moulds (class MYXOMYCETES) which form stalked, globose to cylindrical sporangia; the peridium is evanescent, and the spores in masses may be yellow, pinkish, red, etc. *A. cinerea* is common on dead wood and humus.

ARD Acute respiratory disease: a general term for any such disease affecting closed populations of people (e.g. military recruits, school children). Major causal agents of ARDs are adenoviruses (usually Ad3, Ad4, Ad7, Ad14, or Ad21 – see MASTADENOVIRUS); an incubation period of ca. 5–7 days is followed by fever, chills, headache, malaise and coryza, but the disease is usually mild and self-limiting. A live vaccine, administered orally in enteric-coated capsules, is widely used for preventing adenoviral ARD in military recruits.

arenaceous Resembling or (referring e.g. to the test of DIFFLUGIA) incorporating or bearing grains of sand.

Arenaviridae (arenavirus group) A family of enveloped ssRNA viruses. One genus: *Arenavirus*; type species: lymphocytic choriomeningitis virus (LCMV). LCMV has a worldwide distribution; other arenaviruses are restricted to, and named after, particular geographical areas: Lassa virus, Mozambique (= Mopeia) virus (probable member), and the 'New World' arenaviruses (the Tacaribe complex) – Amapari (Brazil), Junín (Argentina), Latino and Machupo (Bolivia), Parana (Paraguay), Pichinde (Columbia), Tacaribe (Trinidad), and Tamiami (Florida, USA) viruses. In the natural host (usually a single rodent species) arenavirus infection is persistent and silent (see PERSISTENCE); transmission occurs vertically and horizontally, apparently without involving vectors. Only LCMV, Lassa, Junín and Machupo viruses cause significant human disease (see LASSA FEVER, LCM VIRUS, VIRAL HAEMORRHAGIC FEVERS).

The arenavirus virion is spherical or pleomorphic, diam. ca. 50–300 nm (average: 110–130 nm). It consists of a lipoprotein envelope enclosing a core of viral ribonucleoprotein (RNP) and several host ribosomes (each 20–25 nm diam.). The envelope is derived from the plasma membrane of the host cell and bears club-shaped, apparently hollow surface projections 5–10 nm long [JGV (1983) *64* 2157–2167]. There are two types of viral ssRNA, designated L (31–34S, MWt $2.1–3.2 \times 10^6$) and S (22–25S, MWt $1.1–1.6 \times 10^6$); the viral RNAs are complexed with the major structural protein of the virion (N protein, 'p63') to form a filamentous RNP nucleocapsid which has a beaded appearance in the electron microscope and which forms non-covalently closed, supercoiled circles of various sizes [JGV (1983) *64* 833–842]. The viral RNA is usually described as negative-sense, but certain viral proteins are apparently encoded by +ve-sense sequences in the genome – see AMBISENSE RNA; thus, e.g., the viral N protein is encoded by a sequence complementary to the 3′-half of the S RNA, while the viral glycoproteins are synthesized from a sequence corresponding to the 5′-half of the S RNA. A non-glycosylated protein with RNA-dependent RNA polymerase activity has been found in virions of Pichinde virus. Arenavirus virions can be inactivated by temperatures > 56°C, by pH <5.5 or >8.5, and by organic solvents or detergents.

Arenaviruses can replicate in a wide range of mammalian cell cultures (e.g. BHK-21, Vero and L cells). Virus replication occurs in the cell cytoplasm; the virus matures by budding through the plasma membrane, when ribosomes become incorporated in the virion.

Arenavirus See ARENAVIRIDAE.

areolate Divided up into small areas (areolae). (Used e.g. of a lichen thallus: see e.g. RHIZOCARPON.)

arg-poly(asp) *Syn.* CYANOPHYCIN.

***arg* regulon** See ARGININE BIOSYNTHESIS.

Argentinian haemorrhagic fever A VIRAL HAEMORRHAGIC FEVER caused by the Junín virus (see ARENAVIRIDAE). Mortality: usually ca. 3–15%.

argentophilic Staining well with SILVER STAINS.

L-arginine biosynthesis See Appendix IV(a). In *Escherichia coli* the genes encoding enzymes for arginine biosynthesis constitute a REGULON (the *arg* regulon); four of the genes occur in a cluster (*argECBH*), the remainder occur singly at loci scattered around the chromosome. [Biosynthesis and metabolism of arginine in bacteria: MR (1986) *50* 314–352.]

arginine decarboxylase test See DECARBOXYLASE TESTS.

arginine deiminase See ARGININE DIHYDROLASE.

arginine dihydrolase (ADH) An enzyme system which catalyses the catabolism of arginine, with a concomitant substrate-level phosphorylation; the system occurs in a range of bacteria. (i) *Arginine deiminase* hydrolyses arginine to citrulline and ammonia. (ii) *Ornithine carbamoyltransferase* catalyses a phosphorolytic cleavage of citrulline to form carbamoyl phosphate and ornithine. (iii) *Carbamate kinase* transfers the phosphate group of carbamoyl phosphate to ADP to form ATP and carbamic acid, the latter dissociating spontaneously to CO_2 and NH_3.

Tests for ADH production are widely used in the identification of certain bacteria (e.g. members of the ENTEROBACTERIACEAE). Typically, the organism is grown in a medium containing L-arginine and a pH indicator; the presence of ADH is indicated by an alkaline reaction after a few days' incubation. [Methods: Book ref. 2, pp. 411–412.]

arg-poly(asp) *Syn.* CYANOPHYCIN.

***argT* gene** See BINDING PROTEIN-DEPENDENT TRANSPORT SYSTEM.

Argyn See SILVER.

Argyrol See SILVER.

argyrome *Syn.* SILVER LINE SYSTEM.

argyrophilic *Syn.* ARGENTOPHILIC.

arildone (4-[6-(2-chloro-4-methoxyphenoxy)-hexyl]-3,5-heptanedione) An ANTIVIRAL AGENT which is active against various DNA and RNA viruses in vitro; it blocks uncoating in polioviruses and apparently also in herpes simplex viruses. Arildone may be useful clinically e.g. in the topical treatment of HERPES SIMPLEX infections.

Arizona See SALMONELLA.

arizonosis A POULTRY DISEASE, caused by *Salmonella arizonae* (= *Arizona hinshawii*), characterized by malaise, diarrhoea, and often symptoms of CNS involvement.

Arkansas bee virus See NODAVIRIDAE.

ArlS–ArlR See TWO-COMPONENT REGULATORY SYSTEM.

Armillaria (*'Armillariella'*) A genus of mainly lignicolous fungi (AGARICALES, Tricholomataceae). *A. mellea* (the 'honey fungus') grows saprotrophically on many types of wood, forms a symbiotic association with certain orchids (see MYCORRHIZA), and is parasitic on a range of deciduous and coniferous trees, on certain shrubs, and on some herbaceous plants. The fruiting body is highly variable in colour and appearance, the pileus being convex to flat (ca. 3–15 cm diam.), ochre to brown, with darker scales particularly near the centre; the upper part of the stipe, which often bears a wide annulus, is initially whitish, later reddish-brown. Basidiospores: ca. 8–9 × 5–6 μm. The fruiting bodies (which exhibit BIOLUMINESCENCE) typically occur in clusters. The organism spreads by means of tough, black RHIZOMORPHS ('boot laces') which may be found e.g. under the bark of infected trees or in soil. (See also CARBON DISULPHIDE and THREITOL.)

Armillariella See ARMILLARIA.

armoured dinoflagellates See DINOFLAGELLATES.

***aroA* gene** See STAPHYLOCOCCUS (*S. aureus*).

arogenate See AROMATIC AMINO ACID BIOSYNTHESIS.

***aroH* gene** See TRP OPERON.

aroma bacteria See DIACETYL.

aromatic amino acid biosynthesis Phenylalanine (Phe), tyrosine (Tyr) and tryptophan (Trp) are synthesized via the shikimate pathway: Appendix IV(f). Although most of the steps in this pathway are apparently more or less universal, the control mechanisms and physical organization of the enzymes (e.g. the formation of multienzyme complexes) vary between species and may be useful taxonomic criteria [CRM (1982) *9* 227–252]. The reactions by which prephenate is converted to Phe and Tyr also differ in different species. Thus, e.g., *Escherichia coli* and *Bacillus subtilis* synthesize Tyr via 4-hydroxyphenylpyruvate (HPP), and Phe via phenylpyruvate (PPy) [see Appendix IV(f)]. Certain cyanobacteria and 'coryneforms' (*Corynebacterium glutamicum*, *Brevibacterium* spp) synthesize Phe via PPy but lack prephenate dehydrogenase and synthesize Tyr via arogenate (= 'pretyrosine'). *Pseudomonas diminuta* synthesizes Phe via arogenate but Tyr via HPP, while *P. aeruginosa* can synthesize Phe and Tyr via PPy and HPP, respectively, and also via arogenate. *Euglena gracilis* synthesizes both Phe and Tyr via arogenate only.

(See also TRP OPERON.)

aromatic hydrocarbons See HYDROCARBONS.

array (DNA) See DNA CHIP.

***Arrhenatherum* blue dwarf virus** See FIJIVIRUS.

Arrhenius effect See *staphylococcal* α-HAEMOLYSIN.

Arrhenosphaera See ASCOSPHAERALES.

ARS Autonomously replicating sequence: a DNA sequence (first described from the yeast *Saccharomyces cerevisiae*) which, when linked to a non-replicative DNA fragment, promotes the capacity for autonomous intracellular replication. ARSs occur in the yeast genome with a frequency of about one ARS for every ~40 kilobases. Efficient replication of ARS-containing DNA fragments may require other factors such as the CEN (centromere) element and the minichromosome maintenance protein 1.

Many of the ARSs appear to function as origins of replication, although some are so-called *silent origins*; some of the ARSs (both active and silent) are reported to act as transcription silencers.

Most of the studies on ARSs have been carried out on eukaryotes–particularly on yeasts (e.g. *Saccharomyces, Hansenula*). However, ARSs have also been reported in prokaryotes [see e.g. JB (2003) *185* 5959–5966].

arsenate respiration See SELENATE RESPIRATION.

arsenic (a) (as an antimicrobial agent) The aromatic compounds of arsenic include some effective antimicrobial agents, some of which have found use in chemotherapy; however, some of these 'arsenicals' have been discontinued owing to toxicity.

The antimicrobial activity of arsenicals appears to involve their reaction with thiol (–SH) groups within cells – resulting e.g. in the inhibition of many enzymes; LIPOIC ACID is particularly sensitive because arsenic can bridge the two thiol groups in this coenzyme.

Apparently, pentavalent arsenic in an arsenical must be converted, in vivo, to the trivalent state before the arsenical can act as an antimicrobial agent.

The selective activity of arsenicals is believed to be due to differences in the permeability of different types of cell. Microorganisms can be protected from arsenicals by e.g. thiols, *p*-aminobenzoic acid or quinoid dyes.

Atoxyl ($NH_2.C_6H_4.As(OH)_2O$; *p*-aminophenylarsonic acid) was the first arsenical to be used against trypanosomiasis.

Salvarsan (3,3′-diamino-4,4′-dihydroxyarsenobenzene) was formerly used e.g. for the treatment of syphilis and trypanosomiasis; it is oxidized in vivo to produce the toxic agent: 3-amino-4-hydroxyphenylarsenoxide. This drug has now been superseded.

Glycobiarsol (bismuth *N*-glycolyl-*p*-arsanilate) has been used for treating amoebic dysentery.

Tryparsamide (*p*-*N*-phenylglycineamidoarsonic acid) has been used for late-stage trypanosomiasis (African trypanosomiasis: sleeping sickness). Note that arsenicals are not active against *Trypanosoma cruzi* (Chagas' disease).

Melarsoprol (a derivative of benzenearsenous acid – a trivalent arsenical) is effective against both West and East African strains of the causal agent of sleeping sickness (*Trypanosoma brucei gambiense* and *T. brucei rhodesiense*, respectively). The specific target of the drug is trypanothione (N^1,N^8-bis[glutathionyl]spermidine) [PNAS (1989) *86* 2607–2611]. Melarsoprol reacts with trypanothione to form an adduct (Mel T) which inhibits trypanothione disulphide reductase – an enzyme essential for regulating the parasite's thiol/disulphide balance. Melarsoprol crosses the blood–brain barrier less readily than eflornithine, but it is still regarded as the most effective trypanocidal drug available for treating sleeping sickness [AP (1994) *33* 1–47].

Trypanosomes exposed sub-lethally to arsenicals develop resistance quite readily, possibly owing to decreased uptake.

(b)(in energy metabolism) Some bacteria use arsenate (or selenate) as electron acceptor in anaerobic respiration, electron donors including e.g. acetate, lactate and ethanol (according to species); cell-envelope reductases have been found [FEMS Reviews (1999) *23* 615–627]. Arsenite is used as electron donor in autotrophic CARBON DIOXIDE fixation and as a respiratory electron acceptor [FEMS Ecol. (2004) *48* 15–27; JB (2004) *186* 1614–1619].

ART Automated reagin test: a qualitative or quantitative STANDARD TEST FOR SYPHILIS similar in principle to the RPR TEST.

arteannuin *Syn.* QINGHAOSU.

artefact (artifact) Any feature which does not occur in a specimen under natural conditions, but which may be seen in that specimen during experimentation. Artefacts are due to the disturbance introduced by the process of experimentation or observation; they may occur e.g. as a result of FIXATION.

artemether See QINGHAOSU.

artemisinine *Syn.* QINGHAOSU.

arteritis Inflammation of an artery or arteries.

Arterivirus A genus of viruses (family TOGAVIRIDAE); the arterivirus group currently includes LACTATE DEHYDROGENASE VIRUS, SIMIAN HAEMORRHAGIC FEVER VIRUS, the 'Lelystad' virus (see BLUE-EARED PIG DISEASE) and *equine arteritis virus*.

In equines, equine arteritis virus causes necrosis of small arteries with various clinical manifestations – for example, rhinitis, oedema, enteritis, bronchopneumonia. Infection of pregnant mares commonly results in abortion. Transmission occurs both horizontally and vertically; vectors are unknown. The virus can infect a range of vertebrate cells in vitro.

The viruses in this group are similar in their morphology and genomic organization; moreover, all exhibit a predilection for macrophages, and all tend to cause a lengthy period of viraemia.

artesunate See QINGHAOSU.

Arthonia See ARTHONIALES.

Arthoniales An order of fungi of the ASCOMYCOTINA; members include crustose LICHENS (photobiont green, often trentepohlioid) and lichenicolous and saprotrophic fungi. Ascocarps are APOTHECIOID and may be lirelliform, irregular, etc; they contain paraphysoids. Asci are bitunicate, usually clavate. Genera: e.g. *Arthonia*.

Arthopyrenia A genus of crustose LICHENS of the order DOTHIDEALES; ascospores have one to several septa. *A. halodytes* (photobiont '*Hyella*') is a marine species which grows endolithically in calcareous rocks and on the shells of limpets, barnacles etc in the intertidal zone; it occurs in Europe and N. America.

arthralgia Joint pain.

arthric conidium See CONIDIUM.

arthritis Inflammation of one or more joints. It may occur as a symptom or complication of various infectious diseases – e.g. brucellosis, gonorrhoea, Haverhill fever, Lyme disease, rickettsioses, syphilis, tuberculosis, yaws; it may also be caused by certain viruses – e.g., parvovirus [Lancet (1985) *i* 419–421, 422–425], Ross River virus, rubella virus. *Septic arthritis* is due to infection of the synovial fluid – commonly by staphylococci – e.g. secondarily to OSTEOMYELITIS in an adjacent bone, or as a complication of septicaemia; *S. epidermidis* may cause chronic septic arthritis of the hip following total hip replacement. (See also RHEUMATOID ARTHRITIS and REITER SYNDROME.)

Arthroascus A genus of fungi (family SACCHAROMYCETACEAE) which form yeast cells (which bud, often bipolarly, on a wide base) and true mycelium (which tends to break up into arthrospores). Asci are formed directly after conjugation between two cells. Non-fermentative; NO_3^- is not assimilated. Species: *A. javanensis*, isolated e.g. from soil, fruit, rotting wood. [Book ref. 100, pp. 114–116.]

Arthrobacter A genus of obligately aerobic, catalase-positive, asporogenous bacteria (order ACTINOMYCETALES, wall type VI – see also PEPTIDOGLYCAN) which occur in the soil. In culture the organisms initially grow as irregular, 'lumpy', pleomorphic rods, with or without primary branching, but stationary-phase cultures consist predominantly of spherical or ovoid cells; on subculture rod-shaped forms develop. Optimum growth temperature: ca. 25°C. *Arthrobacter* spp have an oxidative-type metabolism; they are not cellulolytic. GC%: ca. 59–66. Type species: *A. globiformis*.

Arthrobotrys A genus of fungi of the HYPHOMYCETES. Species can grow saprotrophically and can trap and digest nematodes (see NEMATOPHAGOUS FUNGI); *A. oligospora* can also attack and kill other fungi [FEMS Ecol. (1985) *31* 283–291].

arthroconidium *Syn.* ARTHROSPORE.

Arthrocystis See EIMERIORINA.

Arthroderma A genus of fungi of the GYMNOASCALES (anamorphs: CHRYSOSPORIUM; *Trichophyton*).

arthrospore (arthroconidium) (1) An *arthric* CONIDIUM. (2) Any CONIDIUM formed by *thallic* conidiogenesis.

Arthuria See UREDINIOMYCETES.

Arthus reaction A severe local inflammatory skin reaction which involves a TYPE III REACTION; the reaction becomes maximal 3–12 hours after the intradermal administration of antigen, and involves erythema, oedema, and local haemorrhage and necrosis.

Arthus-type reaction Any disorder, other than the classical ARTHUS REACTION, which involves a TYPE III REACTION – see e.g. FARMERS' LUNG and GLOMERULONEPHRITIS.

artichoke curly dwarf virus See POTEXVIRUSES.

artichoke mottled crinkle virus See TOMBUSVIRUSES.

artifact *Syn.* ARTEFACT.

***Artogeia rapae* GV** See BACULOVIRIDAE.

ARV See HIV.

arylsulphatase An enzyme which can hydrolyse aromatic sulphate esters at the O–S bond; arylsulphatases occur e.g. in some *Aspergillus* and *Mycobacterium* spp.

arylsulphatase test A test used in the identification of *Mycobacterium* spp. Essentially, the test strain is grown in a liquid medium containing tripotassium phenolphthalein disulphate (TPD); ARYLSULPHATASE-*positive* strains cleave TPD to PHENOLPHTHALEIN which is detected by a colour change on addition of alkali after 10 days (for slow-growing strains) or after 3 or 7 days (for rapidly-growing strains).

ascarylose 3,6-Dideoxy-β-L-mannopyranose: a sugar found e.g. in the LIPOPOLYSACCHARIDE of certain strains of *Yersinia pseudotuberculosis* (and in the eggs of *Ascaris* worms).

Ascetospora A phylum of PROTOZOA [JP (1980) *27* 37–58] which form spores containing one or more sporoplasms but no extrusion apparatus (no polar capsules or polar filaments). (cf. MICROSPORA; MYXOZOA.) The organisms are parasitic in invertebrates. Classes: PARAMYXEA and STELLATOSPOREA.

Aschaffenburg–Mullen phosphatase test *Syn.* PHOSPHATASE TEST (for milk).

Aschersonia A genus of fungi (order SPHAEROPSIDALES) which include parasites of scale insects and whiteflies. Strains of *Aschersonia* have been used e.g. in Florida for the biological control of scale insects.

asci See ASCUS.

ascigerous Bearing, or giving rise to, asci (see ASCUS).

ascites The condition in which fluid (ascitic fluid) accumulates in the peritoneal cavity during certain pathological conditions. (See also INFECTIOUS DROPSY (of carp).)

Ascobolus See PEZIZALES.

ascocarp (ascoma) A structure at the surface of which, or within which, asci (see ASCUS) develop; the main forms of ascocarp are the APOTHECIUM, ASCOSTROMA, CLEISTOTHECIUM and PERITHECIUM. (Ascocarps are not produced e.g. by *Saccharomyces* spp and related yeasts.)

In *ascohymenial* species ascocarp development appears to follow the sexual stimulus, i.e., plasmogamy (fertilization) precedes ascocarp development; in *ascolocular* species the initiation of ascocarp development (i.e., formation of a stroma) occurs before plasmogamy. In general, ascohymenial species form unitunicate asci, and ascolocular species form bitunicate asci.

Ascochyta A genus of fungi of the order SPHAEROPSIDALES; *A. pisi* is the causal agent of leaf spot disease of the pea plant. Conidiophores are borne in dark, thick-walled, non-setose, unilocular, ostiolate pycnidia that are immersed in the host tissue.

Ascocoryne See HELOTIALES.

ascogenous ASCUS-forming.

ascogone *Syn.* ASCOGONIUM.

ascogonium (ascogone) In ascomycetes: the female GAMETANGIUM; it may be unicellular or multicellular, simple or complex in form, and it may or may not bear a TRICHOGYNE (according to species).

ascohymenial See ASCOCARP.

Ascoideaceae See ENDOMYCETALES.

ascolichen An ascomycetous LICHEN.

Ascoli's thermoprecipitin test A serological PRECIPITIN TEST used to detect antigens of *Bacillus anthracis* (causal agent of ANTHRAX) in various animal products (e.g. hides). The material is extracted with e.g. boiling saline, and the extract is layered over a known positive antiserum in a RING TEST.

ascolocular See ASCOCARP.

ascoma *Syn.* ASCOCARP.

ascomycetes Fungi of the subdivision ASCOMYCOTINA. (In some taxonomic schemes these fungi form the class Ascomycetes.)

Ascomycotina (the 'ascomycetes') A subdivision of fungi (division EUMYCOTA) characterized by the formation of sexually derived spores (ASCOSPORES) in *asci* (see ASCUS). The ascomycetes are typically terrestrial saprotrophs or parasites; they include e.g. most YEASTS, the edible morels and TRUFFLES, the cup fungi, the POWDERY MILDEWS, BLACK MILDEWS and SOOTY MOULDS, and organisms which are better known in their asexual ('deuteromycete') states: the common 'blue moulds' and 'green moulds'. Marine ascomycetes include members of the SPATHULOSPORALES.

In most species the thallus is a well-developed, septate, branching mycelium in which the CELL WALL contains CHITIN (see also SEPTUM); however, some ascomycetes are unicellular organisms, and some are DIMORPHIC FUNGI.

In the *typical* life cycle, the germination of an ascospore leads to the development of a septate mycelium consisting of uninucleate, haploid cells. Many (but not all) ascomycetes then exhibit a conidial (= asexual, imperfect or anamorphic) phase in which conidia (see CONIDIUM) are formed; although the anamorph is part of the HOLOMORPH, anamorphs are commonly (for convenience) classified in the DEUTEROMYCOTINA. At some stage the thallus enters DIKARYOPHASE – e.g. by GAMETANGIAL CONTACT, SOMATOGAMY or SPERMATIZATION. Subsequently, karyogamy and MEIOSIS occur in the developing ASCOCARP – which commonly gives rise to asci which each contain 8 (haploid) ascospores. In the typical life cycle diplophase is thus of limited duration; however, in some ascomycetes (e.g. many yeasts) diplophase is dominant – plasmogamy and karyogamy quickly following ascospore formation, so that the vegetative cells are commonly diploid.

In some taxonomic schemes [see e.g. Book ref. 64] classes are not recognized; instead, the ascomycetes are divided into 37 orders: ARTHONIALES; ASCOSPHAERALES; CALICIALES; CLAVICIPITALES; Coryneliales; Cyttariales; DIAPORTHALES; Diatrypales; DOTHIDEALES; ELAPHOMYCETALES; ENDOMYCETALES; ERYSIPHALES; EUROTIALES; GRAPHIDALES; GYALECTALES; GYMNOASCALES; HELOTIALES; HYPOCREALES; LABOULBENIALES; LECANIDIALES; LECANORALES; MICROASCALES; OPEGRAPHALES; OPHIOSTOMATALES; OSTROPALES; PELTIGERALES; PERTUSARIALES; PEZIZALES; POLYSTIGMATALES; PYRENULALES; RHYTISMATALES; SORDARIALES; SPATHULOSPORALES; SPHAERIALES; TAPHRINALES; TELOSCHISTALES; VERRUCARIALES.

Ascophyllum See PHAEOPHYTA.

L-ascorbic acid (vitamin C; L-threo-2,3,4,5,6-pentahydroxy-2-hexenoic acid-4-lactone) A VITAMIN, essential to man but not normally to microorganisms, found e.g. in fresh fruit and vegetables; it can be synthesized e.g. by certain algae and strains of *Aspergillus niger*. It is a strong reducing agent (oxidized form = dehydroascorbic acid) used e.g. as a poising agent in media for anaerobes (see REDOX POTENTIAL). Commercial manufacture (the Reichstein process) involves the hydrogenation of glucose to D-sorbitol followed by the SORBOSE FERMENTATION; L-sorbose is converted chemically to ascorbic acid.

Ascoseira See PHAEOPHYTA.

Ascosphaera See ASCOSPHAERALES.

Ascosphaerales An order of fungi (subdivision ASCOMYCOTINA) which are associated with bees and pollen. Genera: *Ascosphaera* (see CHALKBROOD), *Arrhenosphaera* and BETTSIA.

ascospore A SPORE formed within an ASCUS. According to species, ascospores may be septate or aseptate, and may be any of a variety of shapes, sizes and colours. On germination, the ascospores of most species form germ tube(s), while those of yeasts characteristically give rise to budding cells.

Ascospore discharge. In most ascomycetes the ascospores are forcibly ejected from the ascus; forcible ejection can occur either from a UNITUNICATE ASCUS or from a BITUNICATE ASCUS. The mechanism of ascospore ejection appears to be unknown; however, ejection probably depends on the development of pressure within the ascus by the absorption of water – e.g. as a result of the enzymic breakdown of a polysaccharide with consequent increase in osmotic pressure. Different mechanisms may be involved in different ascomycetes.

In apothecial ascomycetes ('discomycetes') the ascospores are often released simultaneously by a large proportion of the asci; this results in a visible cloud of ascospores and is called *puffing*.

In cleistothecial ascomycetes the mature asci appear to swell up and rupture the cleistothecial wall. In some species the protruding (or totally freed) asci release their ascospores by violent disintegration.

In perithecial ascomycetes various strategies have evolved to ensure that the ascospores do not remain trapped within the perithecium. Thus, e.g. in *Sordaria fimicola* each of the hymenial asci, in turn, elongates, discharges its ascospores, and collapses. In ascomycetes which form *long*-necked perithecia the asci may become detached from the hymenium and are subsequently blown through the ostiole as a result of pressure in the perithecial cavity; however, in species of *Ceratocystis* the asci are evanescent, and the mature ascospores are slowly extruded in a slimy mass through the perithecial neck.

In the ascostromal ascomycete *Myriangium*, ascospore discharge is necessarily preceded by disintegration of the stroma – which occurs as a result of weathering. Once exposed to the environment the (bitunicate) asci forcibly eject their ascospores.

In e.g. *Tuber* spp (and other hypogean ascomycetes) the ascospores are not forcibly discharged. In such species ascospore dispersal is mediated e.g. by burrowing animals.

Ascospore discharge can be strongly influenced by environmental conditions (e.g. light, humidity), and a distinct periodicity of spore discharge ('endogenous rhythm') has been detected in a number of species. In ascomycetes which form deeply concave (cupulate) apothecia the asci are commonly positively phototropic; asci which line the near-vertical walls of the apothecium thus direct their spores outwards (into the environment) rather than towards the opposite wall of the apothecium. In some fungi, e.g. *Daldinia concentrica*, ascospore discharge occurs mainly or exclusively in the dark.

ascostroma An ASCOCARP consisting of a STROMA (sense 1) containing one or more cavities (*locules*), each locule containing one or more asci (see ASCUS). In each locule the CENTRUM is bounded only by the stromatic tissue; this contrasts with the ascocarp structure of those ascomycetes (e.g. *Xylaria* spp) which form perithecia immersed in a stroma: in such species a distinct perithecial wall occurs within the locule. A uniloculate ascostroma (i.e., one containing a single locule) may contain a hymenial layer of asci, and may thus resemble a perithecium; such a structure is referred to as a *perithecioid pseudothecium* or *pseudoperithecium*. (In e.g. *Rhytidhysteron* the hymenium forms an apothecium-like structure which is called an *apothecioid pseudothecium*.) In some ascomycetes (e.g. *Myriangium*) the asci develop in uniascal locules which are scattered irregularly in the stroma. An ascostroma may develop on the surface of, or within, the substratum.

Ascotricha See SPHAERIALES.

ascus (*pl.* asci) A microscopic, sac-like structure within which are formed the sexually derived (or, exceptionally, parthenogenically derived) spores (ASCOSPORES) of fungi of the subdivision ASCOMYCOTINA. Asci are commonly cylindrical or clavate, but can be e.g. cup-shaped, globose or dumb-bell-shaped, according to species; a mature ascus typically contains eight ascospores, but in some species it contains e.g. two or four ascospores, or (in e.g. *Kluyveromyces* and *Lipomyces*) more than eight ascospores. Asci may be formed within or at the surface of an ASCOCARP, and may occur singly, in groups, or in a closely packed layer (*hymenium*) which, according to species, may be interspersed with sterile structures (see e.g. PARAPHYSIS). In some species the asci are evanescent (see also PROTOTUNICATE ASCUS).

Ascus formation is usually preceded by plasmogamy – which, depending on species, occurs by GAMETANGIAL CONTACT, GAMETANGIAL COPULATION, SOMATOGAMY or SPERMATIZATION. Many ascomycetes exhibit HETEROTHALLISM. (See also MATING TYPE.) Following plasmogamy, the events leading to ascus formation vary according to species.

Ascus formation in ascogenous yeasts. After plasmogamy (which may occur e.g. by the fusion of somatic cells or ascospores), karyogamy commonly follows without delay. The zygote may undergo MEIOSIS immediately (as e.g. in *Schizosaccharomyces octosporus*) so that the ascus develops directly from the zygote; alternatively, meiosis may be delayed for one or more mitotic divisions (e.g. as in the diploid budding phase of *Saccharomyces cerevisiae*) – in which case the ascus develops following meiosis in one of the diploid descendents of the zygote. Following meiosis, the four (haploid) nuclei may develop directly into four ascospores (as in e.g. *S. cerevisiae*) or there may be subsequent mitotic division with the eventual formation of e.g. eight ascospores (as in e.g. *S. octosporus*). The process by which a nucleus gives rise to an ascospore is called *free cell formation* (see below).

Ascus formation in other ascomycetes. (The following is necessarily a *generalized* account since details of the process vary from species to species.) After plasmogamy, karyogamy is typically delayed. One or more hyphae (*ascogenous hyphae*) arise from the ASCOGONIUM, and the (haploid) male and female nuclei migrate into these hyphae. The tip of each hypha then curves to form a crook (*crozier*) such that two nuclei (one from each parent) occur in the curved upper portion of the crozier. These nuclei then undergo mitosis, simultaneously, with their mitotic spindles arranged parallel to the long axis of the ascogenous hypha; the subsequent formation of a septum creates an apical cell (the *ascus mother cell*) which contains one daughter nucleus from each of the two original nuclei. Karyogamy now occurs. The zygote undergoes meiosis followed by one or more mitotic divisions to produce the number of haploid nuclei (ascospore initials) characteristic of the species. The ascospores develop by *free cell formation*. In this process, each nucleus (with a portion of cytoplasm) becomes enclosed within an envelope (the spore-delimiting membrane, SDM) composed of two unit-type membranes. In some ascomycetes (e.g. many members of the Endomycetales) the nucleus may be enveloped directly by vesicles (derived from the GOLGI APPARATUS or Golgi-equivalent?) which develop in association with the SPINDLE POLE BODY. In other species, e.g., *Taphrina deformans* and some species of *Tuber*, each nucleus becomes enveloped by membrane derived from invaginations of the ascus plasma membrane. However, in most 'euascomycetes' the SDMs derive from a system of nuclear and/or endoplasmic reticular membranes which, initially, form a discontinuous layer around the periphery of the ascus cytoplasm (the *peripheral membrane cylinder*). That portion of the ascus cytoplasm which is *not* incorporated into ascospores is termed the *epiplasm*. Ascospore wall material is subsequently laid down between the two layers of the SDM. [Ascospore development: Book ref. 175, pp. 107–129.]

Types of ascus. There are at least nine morphologically and/or functionally distinct types of ascus [Bot. J. Lin. Soc. (1981) *82* 15–34 (29–33)]: see e.g. ANNELATE ASCUS, BITUNICATE ASCUS, OPERCULATE ASCUS, OSTROPALEAN ASCUS and PROTOTUNICATE ASCUS. (See also UNITUNICATE ASCUS and BILABIATE.)

ascus mother cell See ASCUS.

Asellariales See TRICHOMYCETES.

asepsis (1) The state in which potentially harmful microorganisms (e.g. pathogens in a medical context, spoilage organisms in an industrial context) are absent from particular tissues, materials or environments; in this sense asepsis does not necessarily involve sterility (cf. STERILE (sense 1)). (2) A state of sterility. (See also ANTISEPSIS and STERILIZATION.)

aseptate Lacking septa – see SEPTUM.

aseptic (*adj.*) Refers to ASEPSIS (sense 1 *or* 2).

aseptic meningitis See MENINGITIS (b).

aseptic technique Precautionary measures taken to prevent the contamination of cultures, sterile media etc and/or the infection of persons, animals or plants by extraneous microorganisms. Thus, e.g. all vessels for media etc must initially be STERILE (presterilized disposable petri dishes, syringes etc are often used). The working surfaces of instruments (forceps, LOOPS etc), and the rims of bottles used for dispensing sterile (non-flammable) materials, are sterilized by FLAMING. Before use, sterile material should not be exposed to non-sterile material or conditions. The risk of contamination is reduced by treating bench surfaces etc with DISINFECTANTS and/or with ultraviolet radiation (see STERILIZATION). Procedures (e.g. INOCULATION) are carried out in such a way that exposure to the atmosphere is reduced to a minimum, or is eliminated (see e.g. SAFETY CABINET).

Ashbya See METSCHNIKOWIACEAE and STIGMATOMYCOSIS; see also RIBOFLAVIN.

AsiA See ANTI-SIGMA FACTOR.

Asian flu See INFLUENZAVIRUS.

Asiatic cholera *Syn.* CHOLERA.

ASLT ANTISTREPTOLYSIN O TEST.

ASO test ANTISTREPTOLYSIN O TEST.

ASOT ANTISTREPTOLYSIN O TEST.

L-asparaginase (L-asparagine aminohydrolase; EC 3.5.1.1) An enzyme which hydrolyses L-asparagine to L-aspartate and NH_3. L-Asparaginase (e.g. from *Escherichia coli* or *Citrobacter* sp) is used as an anticancer agent – e.g. for treating acute lymphocytic leukaemia in which the cancer cells require exogenous asparagine. (See also ENZYMES.)

L-asparagine biosynthesis See Appendix IV(d).

asparenomycins CARBAPENEM antibiotics.

aspartame L-Aspartyl-L-phenylalanine methyl ester; it is used as a sweetener in certain foods and beverages.

aspartase See ASPARTATE BIOSYNTHESIS.

aspartate ammonia-lyase See ASPARTATE BIOSYNTHESIS.

L-aspartate biosynthesis For the main pathway of aspartate biosynthesis see Appendix IV(d). In many microorganisms, aspartate can also be produced from fumarate and ammonia by the enzyme *aspartase* (L-aspartate ammonia-lyase, EC 4.3.1.1), but this (reversible) reaction is probably more important in the deamination of aspartate than in its synthesis. (cf. IMMOBILIZATION sense 1.)

aspartokinase See Appendix IV(d).

aspergillic acid An antibiotic (a pyrazine derivative) formed by *Aspergillus flavus*.

aspergilloma See ASPERGILLOSIS.

aspergillosis Any disease of man or animals in which the causal agent is a species of ASPERGILLUS.

In man, the causal agent is usually *A. fumigatus*, although *A. flavus*, *A. niger* or *A. terreus* may be involved. Infection usually occurs by inhalation of air-borne conidia, but may occur via wounds or by ingestion. The disease may be non-invasive, the fungus colonizing a pre-existing lung cavity (e.g. a lung cyst or healed tuberculosis lesion) and forming compact mycelial masses called *aspergillomas* (cf. MYCETOMA sense 2); symptoms may be absent or may include chronic productive cough and haemoptysis. Less commonly, the disease may become invasive, disseminating to other organs (particularly in immunocompromised patients); this form is commonly fatal. In some (atopic) individuals inhalation of *Aspergillus* conidia can lead to allergic bronchopulmonary aspergillosis ('extrinsic bronchial asthma'); the conidia may germinate in the bronchi and sputum plugs, but the hyphae do not normally invade the tissues. Other forms of aspergillosis include e.g. infections of the ear or paranasal sinuses. *Lab. diagnosis:* cultural or microscopical (or histological) examination of e.g. biopsy material; *repeated* isolation of *Aspergillus* (e.g. from sputum) is presumptive evidence of infection. In some forms of aspergillosis specific antibodies may be demonstrated by gel diffusion techniques.

Aspergillosis can affect a wide range of animals, causing e.g. pneumonia, gastroenteritis (e.g. in calves), or placentitis leading to abortion (e.g. in cattle, sheep and horses). Birds (including poultry) are particularly susceptible; almost any organ can be affected, and the disease may be acute and (usually) fatal (as in 'brooder pneumonia' of baby chicks) or chronic (in adult birds). (See also AFLATOXINS.)

Aspergillus A genus of fungi of the class HYPHOMYCETES: some species are known to have ascomycetous teleomorphs (see later). The organisms are widespread in nature, and are characteristically saprotrophic; they can use a wide range of substrates as nutrients. Some species can be pathogenic (see e.g. ASPERGILLOSIS, MADUROMYCOSIS, STONEBROOD), and many produce toxins (see e.g. AFLATOXINS, ASTELTOXIN, CYCLOPIAZONIC ACID, FESCUE FOOT, GLIOTOXIN, OCHRATOXINS, PATULIN, STERIGMATOCYSTIN). Some species can cause deterioration in various types of material (see e.g. BREAD SPOILAGE, CHEESE SPOILAGE, COAL BIODEGRADATION, GLASS, LEATHER SPOILAGE, PAINT SPOILAGE, PAPER SPOILAGE, PETROLEUM), while certain species are used in the manufacture of particular enzymes, foods or other commodities (see e.g. AMYLASES, BRINASE, CATALASE, CITRIC ACID, COFFEE, GLUCOSE OXIDASE, HAMANATTO, KOJI, MISO, PECTIC ENZYMES, SAKE, TAKADIASTASE).

Aspergillus spp form a well-developed, septate mycelium. (See also NIGERAN and PSEUDONIGERAN.) Each conidiophore, which develops from a FOOT CELL (sense 2), consists of an erect hypha which has a more or less spherical terminal swelling (the *ampulla* or *vesicle*). In some species the ampulla is partly or completely covered by a layer of *phialides* ('primary sterigmata') – see CONIDIUM; in most species, however, the ampulla is covered by a layer of short, finger-like extensions ('primary sterigmata' or *metulae*) which give rise to the phialides ('secondary sterigmata') at their distal ends. Each phialide produces a basipetally formed chain of spherical, pigmented, aseptate conidia which, according to species, may be e.g. yellow, green or black. (Spore colour can also vary according to available trace elements in the medium; thus, e.g. *A. niger* may form yellow (instead of the normal black) conidia if levels of copper are low.) The ampulla, together with its metulae and/or phialides and associated chains of conidia, is called a *conidial head.* A PARASEXUAL PROCESS occurs e.g. in *A. nidulans*.

Species include e.g. *A. flavus*, *A. fumigatus*, *A. glaucus* (teleomorph: EUROTIUM), *A. nidulans* (teleomorph: EMERICELLA), *A. niger*, *A. oryzae*, *A. parasiticus*, *A. phialiseptus*, *A. terreus* and *A. versicolor*.

A. fumigatus (teleomorph: SARTORYA) is a major cause of aspergillosis in man. In the conidial head, the ampulla may appear domed (rather than spherical) owing to the gradual widening of the distal end of the conidiophore, and it bears a layer of phialides over the distal half to three-quarters of its surface; the proximally situated phialides are longer, and are inclined such that their free ends tend to be parallel with the more distal phialides – the chains of grey-green, spherical conidia thus forming a parallel-sided columnar spore mass which may reach ca. 1 mm in length. *A. fumigatus* has been reported to grow at 50°C. (See also FUMAGILLIN.)

A. phialiseptus is morphologically similar to *A. fumigatus*, but it forms septate phialides which are longer than those of *A. fumigatus*.

Aspidisca A genus of ciliate protozoa (order HYPOTRICHIDA) related to *Euplotes* but differing e.g. in typically having a relatively reduced oral ciliature; species occur in freshwater and marine habitats, and e.g. in some SEWAGE TREATMENT plants.

aspirin See PROSTAGLANDINS.

asporogenous Not capable of forming spores.

assay host *(plant virol.)* *Syn.* LOCAL LESION host.

assimilatory nitrate reduction The (typically aerobic) reduction of nitrate to ammonia, the ammonia being assimilated as a source of nitrogen (see AMMONIA ASSIMILATION); nitrate can be used as the sole source of nitrogen by many bacteria, various fungi, and most algae and plants. Nitrate is initially reduced to nitrite – typically by a soluble, oxygen-insensitive enzyme (*assimilatory nitrate reductase*) in a reaction in which the electron donor may be a reduced pyridine nucleotide or a reduced ferredoxin; in fungi the electron donor is often (sometimes specifically) NADPH, but in e.g. *Candida nitratophila* it can be either NADH or NADPH [JGM (1986) *132* 1997–2003]. The assimilatory reduction of nitrate to nitrite does not generate proton motive force – cf. DISSIMILATORY NITRATE REDUCTION. Nitrite is reduced to ammonia by an *assimilatory nitrite reductase*, reducing power being derived (according to organism) from NADPH, NADH or reduced ferredoxin – commonly the latter in algae and higher plants. The synthesis/activity of enzymes involved in assimilatory nitrate reduction is regulated at least partly by the extracellular concentrations of ammonia and nitrate.

assimilatory sulphate reduction Sulphate can be used as the sole source of sulphur by most microorganisms. The major assimilatory pathway in bacteria and fungi is shown in the figure. (cf. DISSIMILATORY SULPHATE REDUCTION; see also SULPHUR CYCLE.)

Astasia See EUGLENOID FLAGELLATES.

astaxanthin See PHAFFIA.

asteltoxin A polyene MYCOTOXIN isolated from toxic maize meal cultures of *Aspergillus stellatus* (= *A. variecolor*, *Emericella variecolor*). In experimental animals asteltoxin can cause e.g. paralysis of hind-limbs and respiratory impairment.

aster *(cell biol.)* See MITOSIS.

aster yellows See YELLOWS.

Asterionella A genus of freshwater and marine planktonic pennate DIATOMS. *A. formosa* occurs e.g. in lakes and reservoirs, forming blooms in spring and, to a lesser extent, in autumn. In this and other species each individual vegetative cell is long and narrow with slightly flared ends; a variable number of cells (ca. 8 under optimum conditions) form a star-shaped colony in which each corner of one end of a given cell is joined to one corner of a neighbouring cell.

Asticcacaulis A genus of chemoorganotrophic, strictly aerobic PROSTHECATE BACTERIA; habitat and life-cycle are similar to those of CAULOBACTER, but one or more prosthecae arise subpolarly and/or laterally and do not have an adhesive role – adhesive material occurring on the cell surface.

Astomatida An order of protozoa (subclass HYMENOSTOMATIA) in which the cells are typically large or long, are uniformly ciliated, contain a number of contractile vacuoles, lack a cytoproct, and are invariably mouthless; some species have an elaborate holdfast organelle. Asexual reproduction may involve budding and (in e.g. *Intoshellina*) the formation of chain-like colonies. The organisms are endoparasites of e.g. annelids, molluscs and amphibians in freshwater and marine habitats. Genera include e.g. *Anoplophrya*, *Cepedietta*, *Jirovecella*, and *Radiophrya*.

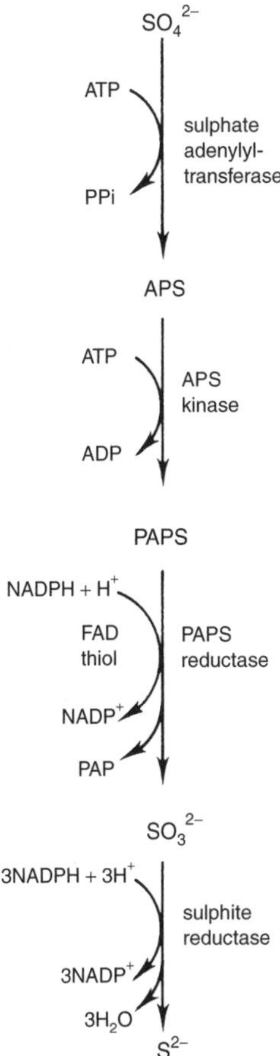

ASSIMILATORY SULPHATE REDUCTION in bacteria and fungi. APS = adenosine-5′-phosphosulphate (see APS); PAPS = 3′-phosphoadenosine-5′-phosphosulphate; PAP = 3′-phosphoadenosine-5′-phosphate. The pathway for the assimilation of sulphide depends on organism. In e.g. *Escherichia coli* sulphide is incorporated into *O*-acetylserine to form cysteine [Appendix IV(c)]. In e.g. *Yarrowia* (*Saccharomycopsis*) *lipolytica* sulphide may be incorporated into *O*-acetylserine (to form cysteine) or into *O*-acetylhomoserine to form homocysteine; homocysteine may be converted to methionine [Appendix IV(d)] or (via cystathionine) to cysteine [see e.g. MGG (1979) *174* 33–38].

astomatous Refers to any structure or cell which does not have a pore, opening, or mouth.

astome A member of the ASTOMATIDA.

Astracantha See RADIOLARIA.

Astraeus See SCLERODERMATALES.

Astrephomene A genus of VOLVOX-like green algae which form spherical coenobia of 16–128 cells.

Astrolophus See ACANTHAREA.

astropyle See RADIOLARIA.

astroviruses Spherical, ether-resistant viruses, 29–30 nm diam., which have a characteristic five- or six-pointed star-shaped surface pattern. Astroviruses have been observed in normal and diarrhoeic faeces and may cause gastroenteritis in infants, calves, lambs and piglets. Lamb astrovirus apparently contains ssRNA [JGV (1981) *53* 47–55].

ASW Artificial seawater. [Recipes: Book ref. 2, p. 436.]

asymmetric PCR A variant form of PCR in which one of the two types of primer in the reaction mixture is used at a much lower concentration (e.g. 1:50) – so that this primer will be used up quickly when cycling begins; a normal concentration of the other primer ensures that *one* strand of the amplicon will be significantly amplified. Asymmetric PCR can be used for preparing single-stranded DNA products suitable e.g. for DNA SEQUENCING.

An alternative procedure for obtaining ssDNA products from PCR involves tagging *one* of the two types of primer with BIOTIN, both primers being used at normal concentrations [NAR (1996) *24* 3645–3646]. After cycling, STREPTAVIDIN is added to the reaction mixture; streptavidin binds to the biotinylated products, i.e. it binds to one of the two types of ssDNA product (but not to the complementary product). Gel electrophoresis in a *denaturing* gel (which inhibits hybridization between the complementary ssDNA products) separates the two types of product because the electrophoretic mobility of the streptavidin-bound product is greatly reduced. If a *particular* product strand is required, the primer which forms the other strand is biotinylated.

asymmetric unit membrane See UROPLAKIN.

asymptomatic Without symptoms.

AT type See BASE RATIO.

ATCC American Type Culture Collection, 12301 Parklawn Drive, Rockville, Maryland 20852, USA.

ateline herpesvirus See e.g. GAMMAHERPESVIRINAE ('*Herpesvirus ateles*').

Athalamida See GRANULORETICULOSEA.

atherosclerosis See MAREK'S DISEASE.

Athiorhodaceae See RHODOSPIRILLACEAE.

athlete's foot (tinea pedis) RINGWORM of the foot, characterized by itching (often with cracking and scaling) of the skin – particularly that between the toes. Causal agents: usually *Epidermophyton floccosum* or *Trichophyton* spp; *Candida albicans* (see CANDIDIASIS) can cause a similar condition.

athymic Lacking a thymus.

Atkinsiella A genus of fungi (order SAPROLEGNIALES) which include organisms parasitic on e.g. the eggs of marine crustacea. Species include *A. dubia*, *A. entomophaga* and *A. hamanaensis* [Book ref. 1, pp. 201–202].

ATL ADULT T-CELL LEUKAEMIA.

Atmungsferment Cytochrome aa_3.

atomic mass unit (AMU; u) A unit of atomic mass: one-twelfth of the mass of a neutral ^{12}C atom; it is ca. $1.6605655 \times 10^{-24}$ gram. (cf. DALTON; RELATIVE MOLECULAR MASS.)

Atopobium A genus of asporogenous, obligately anaerobic, catalase-negative Gram-positive bacteria; cells: short rods or cocci which occur singly, in pairs or short chains. Non-motile. Glucose is metabolized to lactic, acetic and formic acids. *A. parvulus* was formerly classified as *Streptococcus parvulus*.

GC%: 35–46. Type species: *A. minutum*.

[Proposal for the genus *Atopobium*: FEMS (1992) *95* 235–240.]

atopy An inherited tendency to develop immunological HYPERSENSITIVITY states.

atovaquone See MALARIA.

Atoxoplasma A genus of protozoa (suborder EIMERIORINA) similar to LANKESTERELLA but parasitic in birds. [Book ref. 18, p.13.]

atoxyl See ARSENIC.

ATP Adenosine 5′-triphosphate (see figure). (See also NUCLEOTIDE.) In both prokaryotic and eukaryotic cells ATP actively participates in a range of processes and reactions which involve the conversion or expenditure of energy (see also ATPASE); energy derived from chemotrophic or phototrophic metabolism holds the mass–action ratio of the reaction (ATP ↔ ADP + Pi) at values such that ATP hydrolysis is thermodynamically highly favourable, i.e., it can yield free energy which the cell can use for specific purposes. (See also ADENYLATE ENERGY CHARGE.)

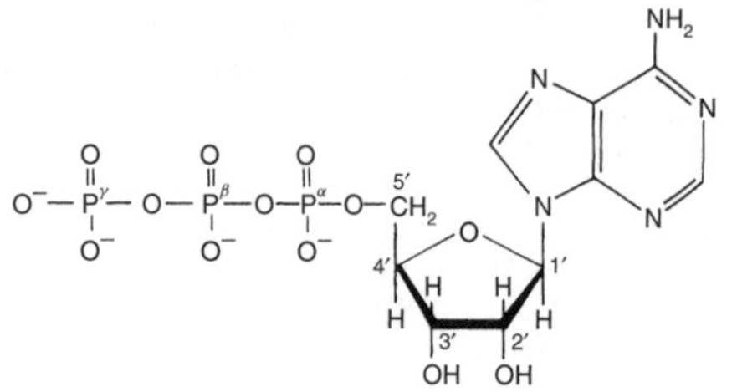

ATP (adenosine 5′-triphosphate)

The phosphorylation of ADP to ATP commonly occurs by OXIDATIVE PHOSPHORYLATION, photophosphorylation or SUBSTRATE-LEVEL PHOSPHORYLATION; in some organisms PYROPHOSPHATE hydrolysis by a PROTON PPASE can be coupled to ADP phosphorylation via proton motive force (see CHEMIOSMOSIS).

In some energy-requiring reactions nucleotides other than ATP may be used; thus, e.g., guanosine 5′-triphosphate (GTP) is involved in certain stages in PROTEIN SYNTHESIS.

Small amounts of ATP can be detected e.g. by techniques which involve CHEMILUMINESCENCE.

ATP-binding cassette See ABC TRANSPORTER.

ATP synthase See ATPASE.

ATPase Adenosine 5′-triphosphatase: any of a wide range of structurally and functionally distinct enzymes and enzyme complexes capable of hydrolysing a phosphate bond in ATP (q.v.) – cf. KINASE; many types of ATPase (sometimes called ATP synthases) can catalyse the synthesis of ATP (see e.g. OXIDATIVE PHOSPHORYLATION) as well as the hydrolysis of ATP. Most ATPases appear to be γ-phosphohydrolases, i.e., they catalyse the dephosphorylation of ATP to adenosine 5′-diphosphate (ADP) and inorganic phosphate (Pi), and/or the converse reaction.

Energy derived from the hydrolysis of ATP by a given ATPase may be used, directly, in one or other of two main types of energy-requiring activity: (a) energy conversion and/or the transmembrane translocation of ions (see e.g. PROTON ATPASE and ION TRANSPORT); (b) mechanical work (see e.g. PRIMOSOME and eukaryotic FLAGELLUM). By generating or augmenting proton motive force (see CHEMIOSMOSIS), ATP hydrolysis can supply energy for some functions (e.g. rotation of the bacterial FLAGELLUM) that cannot use ATP as a direct source of energy.

Note on nomenclature. The activity of some *proton*-translocating ATPases is enhanced by specific ions – e.g. the cytoplasmic membrane of *Escherichia coli* contains a 'Ca^{2+},

Mg^{2+}-stimulated ATPase'; confusingly, this terminology is also used for some ATPases which translocate ions other than protons: see e.g. ION TRANSPORT.

ATP-γS Adenosine-5′-(γ-thio)triphosphate: a (biologically) non-hydrolysable ATP analogue.

atractyloside A glucoside, produced by the Mediterranean thistle (*Atractylis gummitera*), which competitively inhibits the ADP/ATP exchange carrier system in the mitochondrial inner membrane. (cf. BONGKREKIC ACID; WEDELOSIDE.)

AtrB An ABC TRANSPORTER in *Aspergillus nidulans*.

atrichous (1) Hairless. (2) Lacking flagella. (3) Lacking any filamentous appendages (flagella, fimbriae etc).

atrium (*ciliate protozool.*) A shallow concavity in the cytostomal region – particularly in certain hypostomes.

***att* site** A DNA sequence at which site-specific recombination occurs during integration of the genome of a temperate bacteriophage into the chromosome of its host; the site on the phage genome is designated *attP*, that on the host chromosome *attB*. (See e.g. BACTERIOPHAGE λ.)

attaching and effacing lesions See EPEC.

***attB* site** See ATT SITE.

attenuated (1) Of e.g. a structure: becoming narrow; tapered. (2) Having a reduced virulence – see ATTENUATION (1) and (2).

attenuated vaccine A VACCINE containing live pathogens whose virulence for a given host species has been reduced or abolished by ATTENUATION.

attenuation (1) (*immunol.*) Any procedure in which the virulence of a given (live) pathogen for a particular host species is reduced or abolished without altering its immunogenicity; attenuation may be achieved by SERIAL PASSAGE and may be used to prepare VACCINES (e.g., the SABIN VACCINE). (2) A reduction in the virulence of a pathogen under natural conditions. (3) See SPIRITS. (4) (*mol. biol.*) See OPERON.

attenuator See OPERON.

***attP* site** See ATT SITE.

atypical interstitial pneumonia (of cattle) Any of a clinically distinct group of acute or chronic respiratory diseases which typically involve e.g. pulmonary emphysema and oedema, hyperplasia of alveolar epithelium and interstitial cells, and an unresponsiveness to treatment [Book ref. 33, pp. 1255–1261]. (cf. BOVINE RESPIRATORY DISEASE.) In no naturally occurring case has the causal agent been established with certainty; causes may include the effects of noxious gases, metabolic products (see FOG FEVER), dusts (see BOVINE FARMERS' LUNG), a ketone constituent of the weed *Perilla frutescens*, or IPOMEANOL. (See also PNEUMONIA.)

atypical mycobacteria (anonymous mycobacteria) Species of MYCOBACTERIUM which can cause disease in man but which (unlike *M. tuberculosis* and *M. bovis*) do not cause fatal, disseminated disease when inoculated into guinea pigs. Atypical mycobacteria often give a negative NIACIN TEST.

Au antigen See HEPATITIS B VIRUS.

Audouinella See RHODOPHYTA.

Aufwuchs (periphyton community) Organisms (including certain algae and sessilinids) which colonize and form a coating on submerged objects (stones, plants, etc) in aquatic habitats.

Augmentin A mixture of AMOXYCILLIN and CLAVULANIC ACID used e.g. in the treatment of urinary tract infections.

Augusta disease (of tulips) See TOBACCO NECROSIS VIRUS.

Aujeszky's disease (pseudorabies; mad itch; infectious bulbar paralysis) A PIG DISEASE which may also affect e.g. cattle, dogs, cats and rats; it occurs in North America, Europe, S. E. Asia, and the UK. The causal agent is a herpesvirus (see ALPHAHERPESVIRINAE). Infection probably occurs mainly via the upper respiratory tract; the virus then invades the nervous system. In piglets, symptoms may include vomiting or diarrhoea, fever, trembling, incoordination, convulsions, and prostration; mortality rates may reach 100% in neonates. In adult pigs infection may be asymptomatic or may result in e.g. fever, respiratory symptoms, anorexia, and abortion; recovery usually occurs in a few days. A carrier state is recognized. In cattle, dogs and cats, the disease may resemble the active form of RABIES, with signs of intense itching, salivation, convulsions, and death. [Book ref. 27.]

Aulacantha See RADIOLARIA.

Aulosira A phycological genus of filamentous 'blue-green algae' (Nostocales) in which both heterocysts and akinetes are formed (cf. CYANOBACTERIA section IV).

AUM See UROPLAKIN.

Aura virus See ALPHAVIRUS.

auramine O A basic, yellow, substituted diphenylmethane FLUOROCHROME; fluorescence: yellow.

auramine–rhodamine stain A fluorescent ACID-FAST STAIN. The stain is made by mixing auramine O (1.5 g) and rhodamine B (0.75 g) with distilled water (25 ml) and melted phenol (10 ml); this mixture is added to distilled water (25 ml) and glycerol (75 ml), well mixed, and filtered through glass wool. A heat-fixed smear is stained for 15–20 min at 37°C and rinsed in distilled water. Decolorization, for 2–3 min, is attempted with 70% ethanol containing 0.5% v/v conc. HCl. The smear is washed in distilled water and then flooded with 'counterstain' (0.5% aqueous potassium permanganate) for ca. 2 min to suppress non-specific fluorescence of tissue debris etc. (Prolonged counterstaining would mask the fluorescence of acid-fast organisms in the smear.) The smear is then washed in distilled water, blotted dry, and examined by fluorescence MICROSCOPY.

Aureobasidium (*Pullularia*) See HYPHOMYCETES; see also BLACK YEASTS; PAINT SPOILAGE; PULLULAN; SAP-STAIN.

aureofungin An antifungal antibiotic of the heptaene group of POLYENE ANTIBIOTICS produced by *Streptoverticillium cinnamomeum* var. *terricolum;* it is a golden-yellow powder which is insoluble in water but soluble in dilute alkali or ethanol. Aureofungin is active against a wide range of plant-pathogenic fungi, is translocated within the plant, and has been used e.g. for crop spraying and for the prevention of seed-borne disease. [Book ref. 121, pp. 137–148.]

aureomycin See TETRACYCLINES.

Auricularia A genus of fungi of the order AURICULARIALES. *A. auricula-judae* (the 'Jew's ear fungus') can be parasitic e.g. on elder (*Sambucus*). An edible species, *A. polytricha*, is cultivated on oak (*Quercus*) in China. (See also FUNGUS GARDENS.)

Auriculariales An order of fungi (subclass PHRAGMOBASIDIOMYCETIDAE) which typically form elongated, *transversely* septate basidia in gymnocarpous or semiangiocarpous fruiting bodies which may be e.g. stalked or sessile, and which are characteristically gelatinous or waxy; the basidiocarp may be e.g. cup-shaped or ear-shaped. Some species are saprotrophs; others are parasitic e.g. on mosses and on other fungi. Genera include e.g. AURICULARIA, *Helicobasidium* and *Phleogena*.

aurintricarboxylic acid (ATA) A TRIPHENYLMETHANE DYE which inhibits PROTEIN SYNTHESIS in prokaryote and eukaryote cell-free extracts; it appears to inhibit the binding of mRNA to ribosomes and, at higher concentrations, can inhibit chain elongation. ATA strongly chelates metal ions.

aurodox See POLYENE ANTIBIOTICS (b).

aurovertins A group of polyene, α-pyrone-containing MYCOTOXINS isolated from cultures of *Calcarisporium arbuscula*. [Biosynthesis: PAC (1986) *58* 239–256.] Aurovertins bind non-covalently to, and inhibit, (F_0F_1)-type PROTON ATPASES.

Australia antigen HBsAg: see HEPATITIS B VIRUS.

Australia X disease *Syn.* MURRAY VALLEY FEVER.

Australian bat lyssavirus See LYSSAVIRUS.

auto-agglutination (saline agglutination) Spontaneous agglutination which may occur when bacterial cells (or other particulate materials) are suspended in e.g. saline. (See also VW ANTIGENS.)

auto-α *Syn.* ALPHA PEPTIDE.

autoantibody An ANTIBODY produced by an individual against one of its own antigens. (See e.g. TYPE V REACTION.)

autoantigen An antigen homologous to an AUTOANTIBODY.

Autobiocounter M 4000 See HACCP.

autocatalytic splicing See SPLIT GENE (b, c and e).

autochthonous (1) Indigenous to a given environment. (cf. ALLOCHTHONOUS.) In a given environment the autochthonous microorganisms maintain more or less constant numbers or biomass – reflecting more or less constant (albeit typically low) levels of nutrients. (cf. ZYMOGENOUS.)

(2) (*immunol.*) *Syn.* AUTOLOGOUS.

autoclave An apparatus within which objects or materials can be heat-sterilized by (air-free) saturated steam, under pressure, at temperatures usually in the range 115–134°C. (The pressure itself plays little or no part in the STERILIZATION process – which depends on the combined effects of heat and water vapour.) The simplest ('bench-type') autoclaves resemble the domestic pressure cooker in both principle and appearance. Material to be autoclaved is placed on a rack in the chamber and the lid clamped securely in position with the valve open; water in the bottom of the chamber is boiled (by means of an internal electric element, or by external heating) and the steam allowed to escape, via a valve in the lid, until all the air has been displaced from the chamber. The valve is then closed; on further heating, water continues to vaporize, and the pressure and temperature of the steam rise to levels determined by the setting of the steam exhaust valve – which then opens to maintain the required pressure and temperature of steam in the chamber. For a given

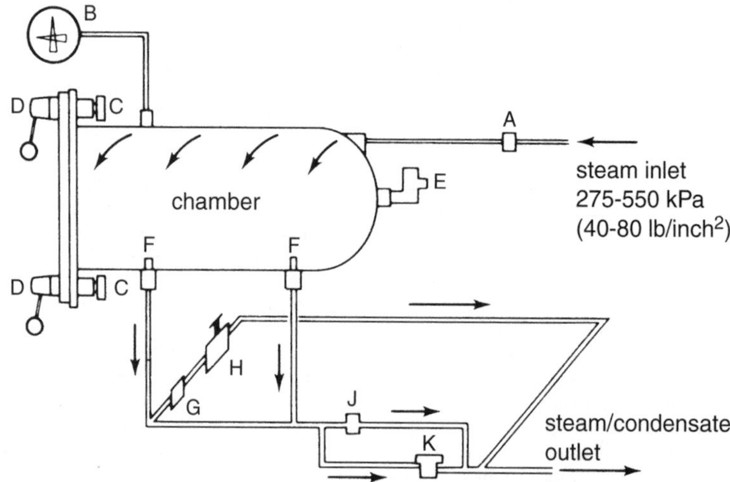

AUTOCLAVE. Simplified, diagrammatic representation of a Series 225 steam–mains autoclave (courtesy of Baird and Tatlock (London) Ltd). **A**. Steam inlet valve. The valve is opened/closed electromagnetically – its operation being governed by the pressure controller/indicator. (Control by temperature – via the thermocouple – is an available option.) **B**. Pressure controller/indicator. **C**. Door microswitch – a safety device; electrical power is connected to the autoclave only when the microswitch has been actuated, i.e., when the door has been fully closed. Additionally, this device automatically opens the steam exit valve if any attempt is made to open the door while the chamber is under pressure. **D**. Door closure bolts which operate microswitches. **E**. Safety valve set to open at 275 kPa. **F**. Filters which prevent blockage of steam discharge lines and steam trap. **G**. Thermocouple pocket: temperature control/indicator/recording option. **H**. Manually-operated by-pass valve – can be operated in case of electrical power failure. (Under such conditions valve **J** could not be opened.) The valve contains a built-in bleed which permits a small but continual flow of steam from the chamber to the exterior. **J**. Electrically-operated steam exit valve. **K**. Steam trap.

Operational sequence. 1. Insert load. 2. Secure door. 3. Set controls for required pressure/temperature and sterilizing time. 4. Switch on power. Steam enters the chamber through **A**, and the downwardly-displaced air is purged, via filters **F**, through (a) the steam trap **K** and (b) the permanent bleed **H**; **J** is closed, and when *steam* leaves the chamber, **K** closes. **A** closes when the chamber pressure reaches the selected value. Steam continues to leave the chamber via the permanent bleed **H** – permitting continual temperature monitoring by the thermocouple **G**; since **G** is located at the lowest (and hence coolest) part of the system, it measures the 'worst case' temperature – the chamber temperature always being higher than that registered by **G**. Sterilizing conditions within the chamber are maintained (against heat losses and against the continual steam bleed) by periodically admitting a pulse of steam via **A**. At the end of the sterilization cycle steam is discharged from the chamber via **J**; when atmospheric pressure has been re-established in the chamber the door can be opened and the load withdrawn.

Automatic timing of the sterilization cycle commences when the chamber pressure (or, optionally, temperature – measured by **G**) reaches the desired value.

steam temperature the time required for sterilization depends on the nature of the load – the rate of steam penetration and the heat capacity of the load being important considerations. Sterilizing temperature/time combinations used with bench-type or larger autoclaves are commonly 115°C (10lb/inch^2; ca. 69 kPa) for 30–35 min; 121°C (15lb/inch^2; ca. 103 kPa) for 15–20 min; 134°C (30 lb/inch^2; ca. 207 kPa) for 4 min. These times may be varied according to e.g. the nature of the load and the nature and degree of contamination. Large autoclaves, used e.g. in hospitals and industry, often use steam piped direct from a boiler, and frequently incorporate automatic controls for timing etc (see Figure). (See also STEAM TRAP.)

Steam quality is important for effective sterilization. If air is present in an operating autoclave, the temperature corresponding to a given pressure is lower than it would be in the complete absence of air; thus, air must be completely purged from the autoclave chamber, and all free space within the chamber must be filled with *saturated* steam at the required temperature and pressure. Saturated steam is steam which holds the maximum amount of water vapour for its temperature and pressure. (Saturated steam effects optimum heat transfer since it delivers up a large amount of latent heat when condensing on objects in the chamber during the initial (pre-sterilization) heating-up period.) In a bench-type autoclave the steam is necessarily saturated since it remains in contact with water in the chamber. In a larger autoclave, steam entering the chamber from a boiler should, for maximum efficiency, be dry as well as saturated, i.e., it should contain no water in the liquid phase. (Wet saturated steam contains droplets of water.) *Superheated* steam is an unsaturated vapour formed when the pressure of dry saturated steam (at constant temperature) is lowered, or when the temperature of dry saturated steam (at constant pressure) is raised in the absence of free water. Superheated steam tends to behave as a gas, and must be at temperatures higher than those of saturated steam to achieve sterilization; it may be formed e.g. when steam is piped from a boiler to a large autoclave – particularly at a pressure-reduction valve.

In most large autoclaves steam enters near the top of the chamber, and air is purged by downward displacement; this may permit pockets of air to become trapped at the bottom of deep, empty vessels – e.g. bottles with small openings or long, narrow necks – and such regions may not be adequately exposed to sterilizing conditions. (Air may also be trapped in items of apparatus which are totally enclosed in a wrapping of e.g. aluminium foil.) In a *vacuum autoclave* air is purged from the chamber by connecting it briefly to a vacuum line; when steam subsequently enters the chamber it can penetrate more easily into flasks and between the fibres of surgical dressings etc so that the overall sterilizing time is reduced.

At the end of the sterilization cycle in a large autoclave the steam inlet valve is closed (or closes automatically); the autoclave may be left to cool slowly, or steam within the chamber may be allowed to escape via the steam exit valve until pressure in the chamber is equal to that of the atmosphere – when the chamber can be safely opened. When bottles of liquid have been autoclaved the temperature and pressure in the chamber should be allowed to decrease slowly to prevent boiling and the consequent loss of contents and/or damage to bottles. (Screw-caps on bottles should always be loosened *prior* to autoclaving.) Sometimes a partial vacuum is allowed to form in the chamber as the steam condenses; this can be advantageous since it tends to dry materials (e.g. dressings) in the chamber. When 'breaking the vacuum', i.e. allowing ingress of air to the chamber, the entry of contaminated air is prevented by the presence of a filter in the air intake line. In vacuum autoclaves steam can be removed, and materials dried, by connecting a vacuum line to the chamber.

Regular monitoring of an autoclave's performance is important; monitoring devices should be used to check the temperatures reached at a number of different locations within the chamber. Monitoring devices may be physical (e.g. thermocouples), chemical (e.g. BROWNE'S TUBES), or biological (test envelopes containing the endospores of e.g. *Bacillus stearothermophilus*). In biological monitoring the autoclaved spores are tested for viability by attempting to culture them; although such a process is directly related to the aim of autoclaving, it involves a built-in delay (i.e. the incubation time). (See also AUTOCLAVE TAPE; BOWIE–DICK TEST.)

Some materials (e.g. petroleum jelly, liquid paraffin) cannot be sterilized by autoclaving owing to their impermeability to steam; such materials (and e.g. clean glassware) are usually sterilized in a HOT-AIR OVEN.

(See also CHEMICLAVE; LOW-TEMPERATURE STEAM DISINFECTION; STEAMER; STERILIZER.)

autoclave tape Paper tape (usually self-adhesive) which changes colour (or exhibits other visible changes) when exposed to sterilizing conditions in an AUTOCLAVE. (See also BOWIE–DICK TEST.)

autocolony A colony (or coenobium) which develops within one cell of a parent colony and which, when released, resembles the parent colony (see e.g. SCENEDESMUS). (cf. AUTOSPORE.)

autocompartmentalization See PROTEASOME.

autocoupling hapten See HAPTEN.

autodigestion (of lamellae) See BASIDIOSPORES.

autodisplay (*biotechnol.*) Autotransporter-mediated display of recombinant proteins at the cell surface of *Escherichia coli*. In this method, the α domain of the AIDA-I *autotransporter* (see type IV systems in PROTEIN SECRETION) is replaced by a recombinant protein; this is done by preparing a fusion protein consisting of the sequence encoding the β domain and that encoding the (heterologous) recombinant protein. In one study, the α domain was replaced e.g. by the cholera toxin B subunit; cells with the (normal) outer membrane protease OmpT cleaved the (recombinant) α domain, which was released from the cell, but in mutant (*ompT*$^-$) cells, which lack a functional OmpT protease, the subunit was not cleaved, but displayed at the cell surface. [Autodisplay: JB (1997) *179* 794–804.]

autoecious *Syn.* HOMOXENOUS.

autofluorescence See FLUORESCENCE.

autogamy (self fertilization) In certain protozoa: a sequence of events which culminates in the fusion of haploid nuclei or gametes derived from a single cell; autogamy occurs e.g. in certain ciliates (e.g. *Paramecium* – but apparently not in e.g. *Tetrahymena*) and in some sarcodines.

In *Paramecium aurelia* each of the two micronuclei undergoes meiosis; of the eight haploid *pronuclei* produced all except one disintegrate. The surviving pronucleus divides mitotically to form a pair of gametic nuclei; these nuclei subsequently fuse to form the (homozygous) zygotic nucleus (= *synkaryon*). Two mitotic divisions then occur; two of the resulting (diploid) nuclei become new micronuclei, the other two become macronuclei – the original macronucleus having disintegrated during the preceding events. During the next binary fission one macronucleus passes to each daughter cell; both micronuclei divide, mitotically, and a pair of micronuclei passes to each daughter cell – thus re-establishing the normal nuclear constitution.

Since only the original complement of genes is involved, the potential for genetic change in autogamy is less than that in CONJUGATION; nevertheless, new combinations of alleles may be formed during the meiotic divisions. (See also ACTINOPHRYS and ACTINOSPHAERIUM.)

In some organisms autogamy occurs spontaneously; in others it may be induced by starvation, ageing, radiation etc.

autogenous regulation (*mol. biol.*) The regulation of expression of a gene or operon by its own product(s). (See e.g. RIBOSOME (biogenesis).)

autogenous vaccine A VACCINE prepared by the culture and inactivation of pathogen(s) isolated from a patient and subsequently used to inoculate that patient.

***Autographa californica* NPV** See NUCLEAR POLYHEDROSIS VIRUSES.

autoimmune disease Any disease directly attributable to AUTOIMMUNITY. In some autoimmune diseases (e.g. acquired haemolytic anaemias) antibodies play a significant role, but many autoimmune diseases are cell-mediated. (See also e.g. MULTIPLE SCLEROSIS.)

autoimmunity An expression of IMMUNITY (1) in which the body directs immunological mechanism(s) against one or more of its own components. (See also TYPE V REACTION.)

autoinducer See AUTOINDUCIBLE ENZYME and QUORUM SENSING.

autoinducible enzyme A (repressible) enzyme whose induction in a given organism is brought about by the presence of a specific compound (*autoinducer*) that is produced by the organism itself (cf. enzyme induction in e.g. the LAC OPERON). An example is bacterial luciferase (see BIOLUMINESCENCE). In this system autoinducer is secreted by the bacteria and can accumulate to the critical, effective concentration (sufficient to bring about derepression) only when the organisms are in a confined environment (e.g. in the luminous organs of certain marine fish).

autoinfection Infection resulting from the transfer of a pathogen from one site to another in the same individual (cf. RE-INFECTION).

autologous (*immunol.*) Present in or derived from an individual's own tissues.

autologous blood transfusion See TRANSFUSION-TRANSMITTED INFECTION.

autolysate The products of AUTOLYSIS. (See e.g. YEAST EXTRACT.)

autolysin (autolytic enzyme) Any of a range of endogenous enzymes which, by degradation of certain structural cell components (e.g. PEPTIDOGLYCAN in bacteria), can bring about AUTOLYSIS or AUTOPHAGY; autolysins are apparently involved in normal processes of growth and development (e.g., spore germination; cell wall polymer extension and cell separation during normal growth). Mutant bacteria defective in autolysin activity may show e.g. abnormal growth forms (e.g. filaments) or resistance to antibiotics active against the wild-type. In prokaryotes, autolysins occur in the cell wall or (in Gram-negative species) in the periplasmic space, often as inactive precursors which are activated e.g. by proteases (see also BILE SOLUBILITY); in eukaryotes autolysins are sequestered in LYSOSOMES.

autolysis The lysis of a cell due to the action of its AUTOLYSINS – usually following the death of the cell or tissue.

autolytic enzyme *Syn.* AUTOLYSIN.

automatic volumetric spore trap *Syn.* HIRST SPORE TRAP.

autonomously replicating sequence See ARS.

autophagy The digestion, by a eukaryotic cell, of some of its own internal components – e.g., during periods of starvation. (See FOOD VACUOLE.)

autoplaque A plaque (in a bacterial lawn plate) which resembles that caused by a lytic bacteriophage, but in which no phage can be demonstrated. Autoplaques occur e.g. in cultures of certain strains of *Rhizobium leguminosarum* biotype *trifolii* and seem to result from autolysis.

autoplast A PROTOPLAST or SPHAEROPLAST which results from the action of an organism's own autolytic enzymes.

autoradiography (radioautography) The use of a photographic process to locate and/or quantify a radioactively-labelled substance previously incorporated in living cells or tissues; autoradiography can be used e.g. to study sites of biosynthesis, turnover rates, and transport. Initially, a radioactive substrate is taken up by the living cells, and the label becomes incorporated in structures, metabolites etc as would be the non-labelled analogue. (The labelled substrate may be introduced using a PULSE–CHASE TECHNIQUE; since radioactive labels can be injurious to cells, dose and duration of the pulse may have to be limited.) The cells/tissues are then prepared for microscopy (e.g., fixed, dehydrated, embedded, sectioned, stained). For ELECTRON MICROSCOPY the section (e.g. on a formvar-coated grid) is stained with e.g. lead citrate or uranyl acetate, and is coated with a thin (ca. 5 nm) layer of carbon; the carbon layer is then overlaid with a thin layer of photographic emulsion, and the whole preparation is left in the dark at e.g. 4°C for a period of the order of days or weeks. Each point source of radioactivity in the section will affect the overlying area of emulsion, giving rise (on developing and fixing the emulsion) to a local deposit of silver which, under the EM, commonly appears as a tangled filament. Under the light microscope each silver deposit appears as a dot or fleck. Since the silver deposits are superimposed on the tissue section, the cellular location of the radioactive components can be determined.

For optimum resolution, the section, carbon layer, and emulsion layer must each be very thin; given a choice of suitable radioactive sources, it is preferable to use the one with the least emission energy. Low-energy radiators, e.g. ^{3}H, tend to affect a small area of the emulsion immediately above each point source; with radiators of high emission energy, e.g. ^{32}P, each point source gives a wide-angled cone of effective radiation which affects a larger area of emulsion. Radioactive precursors used for studying particular cell components include e.g. [^{3}H]thymidine (DNA), [^{3}H]glucose (polysaccharides), and [^{3}H]proline (proteins). [Book ref. 4, pp. 235–277.]

autosome Any chromosome other than a sex chromosome.

autospore (*algol.*) A type of spore, formed asexually, which is an exact morphological replica of the parent cell. The term is usually reserved for aplanospores produced by *Chlorella* and related algae, but is sometimes extended to include e.g. the planospores of e.g. BRACHIOMONAS.

autotransporter See PROTEIN SECRETION (type IV systems).

autotroph An organism which uses CARBON DIOXIDE for most or all of its carbon requirements (cf. HETEROTROPH; LITHOTROPH); all obligate autotrophs appear to be either chemolithotrophs or photolithotrophs, and 'autotroph' is often used with the implication of lithoautotrophy (see also CHEMOTROPH). In many autotrophs CO_2 fixation occurs via the CALVIN CYCLE or via the REDUCTIVE TRICARBOXYLIC ACID CYCLE. Some ACETOGENS fix CO_2 via a different pathway (sometimes called the 'activated acetic acid pathway') in which acetyl-CoA can be synthesized by the reduction of two molecules of CO_2 (see ACETOGENESIS); a similar pathway occurs in *Desulfovibrio baarsii* in which the methyl and carboxyl groups of acetyl-CoA are derived from formate and CO_2 respectively [FEMS (1985) *28* 311–315]. An analogous pathway occurs in METHANOGENS.

autoxidation Spontaneous oxidation by atmospheric oxygen: see e.g. OXIDATIVE RANCIDITY and SULPHUR CYCLE.

auxanogram See AUXANOGRAPHIC TECHNIQUE.

auxanographic technique A procedure used e.g. for identifying the substances which a given organism can use as carbon sources, or for identifying the growth factor(s) required by an AUXOTROPH.

To identify usable carbon sources, a plate of medium which lacks a carbon source is heavily inoculated with the test organism; a small quantity of one of each of a number of different carbon sources is then placed at a separate location on the plate. Following incubation, usable carbon sources are indicated by growth at their locations on the plate. The pattern of usable carbon sources (or e.g. nitrogen sources) is termed an *auxanogram;* closely related organisms which are characterized by different auxanograms are said to be different *auxotypes* (= *auxanographic types*).

An analogous procedure is used for examining a suspected auxotroph; a solid MINIMAL MEDIUM is inoculated with the test organism, and to the plate are then added small, discrete amounts of (a) a mixture of amino acids, (b) a mixture of vitamins, and (c) one or more possible requirements for nucleic acid synthesis. Following incubation, growth at one or other location on the plate indicates the identity of the growth factor(s) required. If, for example, growth occurs at the location of the mixed amino acids, the analysis is continued by repeating the procedure with separate, discrete inoculations of individual amino acids until the specific requirement(s) is/are known.

auxanographic type See AUXANOGRAPHIC TECHNIQUE.

auxins PHYTOHORMONES which promote stem elongation and, in conjunction with CYTOKININS and/or GIBBERELLINS, play important roles in many plant processes; auxins, which are derivatives of tryptophan, are produced mainly at stem apices and in young leaves. Similar or identical compounds are also formed by certain microorganisms (including a number of plant pathogens).

In many plants the main auxin appears to be indole 3-acetic acid (IAA, also known as 'auxin' or 'heteroauxin'), a compound synthesized via the precursor indole 3-acetonitrile (IAN). In vivo, IAA can e.g. stimulate ETHYLENE production, but the mechanisms by which IAA functions in the overall regulation of growth are not well understood. Regulation of the in vivo level of IAA appears to involve an enzyme, 'IAA-oxidase', which oxidizes IAA to products which include 3-methylene-oxindole and indolealdehyde; this enzymic oxidation is stimulated e.g. by certain monophenols (e.g. *p*-coumaric acid) and by Mn^{2+}, and is inhibited by certain diphenol derivatives (e.g. CAFFEIC ACID, CHLOROGENIC ACID, QUERCETIN and SCOPOLETIN).

Certain plant diseases of microbial causation are characterized by the presence of atypical levels of auxins in the diseased plants. *Hyperauxiny* (abnormally high levels of auxins) occurs in many diseases and can arise in various ways – see e.g. CLUBROOT, CROWN GALL, OLIVE KNOT and SCOPOLETIN. The nature of the relationship between hyperauxiny and disease development is not well understood.

auxochrome In a DYE molecule: any ionizable group by means of which the CHROMOPHORE can bind to target molecule(s).

auxospore See DIATOMS.

auxotroph A strain of microorganism which lacks the ability to synthesize one or more essential growth factors; an auxotroph arises by the occurrence of one or more mutations in a PROTOTROPH. A medium used to culture a given auxotroph must contain any factors which the organism cannot synthesize: see COMPLETE MEDIUM and MINIMAL MEDIUM.

Auxotrophy results from a cell's genetically determined inability to produce (normal amounts of) functional enzyme(s) which catalyse particular stage(s) in the synthesis of essential growth factor(s); such a block may involve (a) the complete absence of enzyme; (b) the presence of normal enzyme in subnormal amounts; (c) the presence of abnormal enzyme which is devoid of or has lowered enzymic activity. (See also SYNTROPHISM.)

Apparent auxotrophy may result e.g. from a defective TRANSPORT SYSTEM; thus, e.g. an organism may appear to be an auxotroph if it is unable to take up a substrate which is necessary for the synthesis of an essential growth factor. (See also CRYPTIC MUTANT.)

Isolation of auxotrophic mutants. Since auxotrophs have nutritional requirements *in excess* of those of the corresponding (prototrophic) wild-type strains, they cannot be isolated from mixed auxotroph–prototroph populations by common selective culture methods.

In the *limited enrichment* method, a dilute suspension of mutagenized cells is inoculated onto a minimal medium which has been enriched with limiting amounts of nutrients; the dilution is chosen such that isolated colonies are obtained following incubation. Any auxotrophic cells which may be present form colonies which quickly exhaust the nutrient supply at their locations – so that colony size is restricted; a colony of prototrophic cells, which is not restricted by nutrient supply, attains a greater size. Thus, small colonies may be presumed to be those of auxotrophs.

In the *delayed enrichment* method, the mutagenized preparation of cells is first inoculated onto a minimal medium. A small quantity of molten minimal agar is then layered onto the surface of the inoculated medium and allowed to set. The plate is then incubated. Colonies are formed only by prototrophs, and the positions occupied by these colonies are recorded. Finally, complete medium is layered onto the surface of the plate and allowed to set; the plate is then re-incubated. As nutrients diffuse into the minimal agar the auxotrophic cells begin to grow and form colonies.

Auxotrophic mutants may also be isolated by a REPLICA PLATING process.

Another method is based on the differing effects of PENICILLIN on growing and non-growing cells of certain (penicillin-sensitive) organisms. If a well-washed mixture of auxotrophic and prototrophic cells is suspended in a minimal medium with an appropriate concentration of penicillin, the prototrophs are lysed while the auxotrophs, being unable to grow, remain viable. The suspension is then washed and is plated on a complete medium to recover the auxotrophs. In this method it is essential that the mutagenized cells be grown in a complete medium for several cell-division cycles *prior to penicillin treatment.* This is essential because the newly mutated cells contain the full complement of enzymes etc found in the prototroph; only after several rounds of cell division do the progeny cells exhibit a truly auxotrophic phenotype. Only low concentrations of cells should be used in this method since substances from the lysed cells may be used as nutrients by the auxotrophs – thereby rendering the latter susceptible to lysis by penicillin. (cf. STREPTOZOTOCIN.)

auxotype See AUXANOGRAPHIC TECHNIQUE.

avenacin A fluorescent polycyclic SAPONIN formed in the roots of oat plants (*Avena* spp). Avenacin confers resistance, in *Avena* spp, to many strains of *Gaeumannomyces graminis* (see TAKE-ALL); however, *G. graminis* var. *avenae* forms an extracellular glycosidase which detoxifies the compound.

avermectins Macrolide-like antihelmintic agents, obtained from *Streptomyces avermitilis*, which are active against various human and animal parasites [AEM (2003) *69* 1263–1269].

aversion zone (*mycol.*) A zone of growth inhibition separating two fungal colonies; such zones are formed e.g. when strains of *Phycomyces blakesleeanus* of similar mating type are cultured on the same plate.

Aviadenovirus (avian adenoviruses) A genus of adenoviruses (family ADENOVIRIDAE) which infect birds; type species: fowl adenovirus type 1 (= chick embryo lethal orphan (CELO) virus, = gal-1). Diseases caused by aviadenoviruses include e.g. EGG-DROP SYNDROME 1976, HAEMORRHAGIC ENTERITIS OF TURKEYS, INCLUSION BODY HEPATITIS, and MARBLE SPLEEN DISEASE. Several avian adenoviruses (including CELO virus) can induce tumours in newborn rodents (cf. MASTADENOVIRUS).

avian acute leukaemia viruses (AcLVs) A group of replication-defective, v-onc^+ type C retroviruses (subfamily ONCOVIRINAE) which cause acute erythroid and myeloid leukaemias in chickens; they may also induce carcinomas, endotheliomas and sarcomas. AcLVs include avian myeloblastosis virus, AMV (which carries v-*myb*); avian erythroblastosis virus, AEV (which carries v-*erb*); and avian myelocytomatosis virus, MC29 (which carries v-*myc*). (See also ERB, MYB and MYC.) AcLV replication requires the presence of a replication-competent helper virus (see e.g. AVIAN LEUKOSIS VIRUSES). [Molecular biology of AcLVs: Book ref. 105, pp. 38–63.]

avian encephalomyelitis (epidemic tremor) A POULTRY DISEASE caused by an ENTEROVIRUS; infection in adult birds is usually asymptomatic, but in young chicks symptoms may include ataxia, tremors and somnolence. Infection occurs e.g. by ingestion of food contaminated with faeces from infected birds; transmission can also occur via the egg.

avian erythroblastosis virus See AVIAN ACUTE LEUKAEMIA VIRUSES.

avian infectious bronchitis An acute, highly infectious POULTRY DISEASE caused by a coronavirus (IBV – see CORONAVIRIDAE). The symptoms include gasping, coughing, nasal discharge, etc; mortality rates may be high. Secondary bacterial infection may be common. [Experimentally-produced disease with mixed IBV/*Escherichia coli* infection: JGV (1985) *66* 777–786.] Live vaccines are available. A strain of the avian infectious bronchitis virus (the 'T' strain) may cause avian kidney disease (Cummings' disease).

avian infectious laryngotracheitis An acute POULTRY DISEASE, affecting mainly chickens and pheasants, caused by gallid herpesvirus 3. Mortality rates may be high.

avian leukosis complex A group of POULTRY DISEASES which includes e.g. LYMPHOID LEUKOSIS and MAREK'S DISEASE.

avian leukosis viruses (ALVs; lymphatic leukosis viruses; lymphatic leukaemia viruses) A group of avian type C retroviruses (subfamily ONCOVIRINAE) which usually induce neoplastic disease only after a long latent period (several months or more). The most common neoplasm induced is LYMPHOID LEUKOSIS. (See also OSTEOPETROSIS.) ALVs are replication-competent and v-onc^- (see RETROVIRIDAE). In some cases induction of neoplastic disease may involve the insertion of a viral LTR adjacent to and upstream of a c-*onc* (usually c-*myc*) sequence, the LTR acting as a promoter for c-*onc* expression; alternatively, or perhaps additionally, an LTR may have a promoter-unrelated transcription-enhancing function [Book ref. 105, pp. 64–68]. Neoplastic transformation may require the activation of a second cellular gene, B-*lym*, in addition to c-*myc*.

ALVs often occur in association with replication-defective v-onc^+ viruses (e.g. AVIAN ACUTE LEUKAEMIA VIRUSES) as 'helper' or 'associated' viruses; such ALVs include Rous-associated viruses (RAVs) and myeloblastosis-associated viruses (MAVs).

avian myeloblastosis virus See AVIAN ACUTE LEUKAEMIA VIRUSES and MYB.

avian myeloblastosis virus reverse transcriptase See REVERSE TRANSCRIPTASE.

avian myelocytomatosis virus See AVIAN ACUTE LEUKAEMIA VIRUSES.

avian pneumoencephalitis *Syn.* NEWCASTLE DISEASE.

avian reticuloendotheliosis viruses (REVs) A group of type C avian retroviruses of the ONCOVIRINAE. The group includes the replication-competent viruses duck infectious anaemia virus, Trager duck spleen necrosis virus and chicken syncytial virus, and the replication-defective, v-onc^+ (v-rel^+) strain T virus (REV-T) and its associated helper virus (REAV, = REV-A). REVs are pathogenic in poultry (chickens, ducks, turkeys), causing a range of (usually rapidly lethal) diseases: e.g. anaemia, visceral reticuloendotheliosis, enlargement and necrosis of the spleen, lymphomas, and infiltrative nerve lesions. REV-T and its helper (REV-A) can be lethal in chickens within ca. 7–14 days of infection; the REV-A component has been reported to induce or activate a splenic suppressor cell population which inhibits the proliferation of cytotoxic cells capable of lysing REV-T-induced tumour cells [MS (1984) *1* 107–112].

avian sarcoma viruses (ASVs) A group of v-onc^+ type C retroviruses (subfamily ONCOVIRINAE) which, after a short latent period, cause tumours in fowl and (sometimes) other animals. The group includes ROUS SARCOMA VIRUS, Fujinami sarcoma virus (which carries v-*fps*), and Yamaguchi-73 sarcoma virus (which carries v-*yes*). (cf. FES.)

avian tubercle bacillus *Mycobacterium avium*.

avianized vaccine Any vaccine containing microorganisms whose virulence for a given host has been attenuated by adaptation in live chicks and/or serial passage through chick embryos (eggs). Attenuated organisms may or may not be inactivated prior to use in a vaccine. The FLURY VIRUS is an avianized strain.

Avicel A commercial preparation consisting of ground microcrystalline (insoluble) CELLULOSE (average DP ca. 200); it is used e.g. for determining the ability of an organism or enzyme to degrade microcrystalline cellulose. (cf. CM-CELLULOSE.)

avidin A protein (MWt ca. 68000) present e.g. in the white of raw hens' eggs; the chicken avidin molecule consists of four identical subunits, and it can bind – non-covalently, but very strongly – four molecules of BIOTIN. (See also ABC IMMUNOPEROXIDASE METHOD; cf. STREPTAVIDIN.)

avidity (*immunol.*) The stability of the antibody–antigen complex formed when multivalent antigen and homologous antiserum are mixed. Avidity depends not only on the AFFINITY of the individual determinant-combining site bonds but also on the number of satisfied valencies of the antigens and antibodies since, e.g., a complex in which two antigen molecules are linked by two antibody molecules is disproportionately more stable than a complex in which the two antigen molecules are linked by a single antibody molecule.

Aviemore model See RECOMBINATION (figure 2).

Avipoxvirus (fowlpox subgroup) A genus of viruses of the CHORDOPOXVIRINAE which infect birds (see e.g. FOWL POX). Avipoxviruses are commonly transmitted (mechanically) by arthropod vectors. Infected cells form lipid-rich A-type inclusion bodies; haemagglutinin is not formed. Infectivity is ether-resistant. Members are closely related serologically. Type species: fowlpox virus; other members include canarypox, juncopox, pigeonpox, quailpox, sparrowpox, starlingpox and turkeypox viruses. (See also POXVIRIDAE.)

avirulence gene (avr gene) (*plant pathol.*) In a plant-pathogenic microorganism: a gene which interacts, functionally, with a specific 'resistance gene' in certain strain(s) of the host species, eliciting a defensive response (HYPERSENSITIVITY) which results in resistance to that particular strain of the pathogen. The products of such genes appear to be delivered to the interior of plant cells via a type III PROTEIN SECRETION system (designated Hrp: hypersensitivity response and pathogenicity) [see e.g. JB (1997) *179* 5655–5662]; under in vitro conditions, the Hrp system in *Pseudomonas syringae* pv. *tomato* has been found to include a filamentous pilus, 6–8 nm in diameter, in which HrpA is a major structural protein [PNAS (1997) *94* 3459–3464]. (See also HOP PROTEINS.)

Avirulence genes form part of the 'gene-for-gene' concept that relates to interaction between plants and their pathogens [ARPpath. (1971) *9* 275–296]. [Gene-for-gene complementarity in plant–pathogen interactions: ARG (1990) *24* 447–463.] Direct interaction between the products of a resistance gene and an avirulence gene has been reported to occur, for example, when strains of rice expressing the *Pi-ta* resistance gene are challenged with strains of the rice pathogen *Magnaporthe grisea* (= *Pyricularia oryzae*) that express *AVR-Pita*; against this challenge, the plant failed to succumb to rice BLAST DISEASE [EMBO (2000) *19* 4004–4014].

Avirulence genes from *Pseudomonas syringae* commonly have a GC% of ~40–50, i.e. markedly below the chromosomal values (59–61) typical of the species; such a difference in GC% – also commonly seen between PATHOGENICITY ISLANDS and their host chromosomes – is consistent with the concept of horizontal transfer of these genes. Also consistent with horizontal transfer is the finding of highly conserved sequences flanking avirulence genes in *P. syringae* pv. *pisi* [Microbiology (2001) *147* 1171–1182].

(See also HARPIN.)

avirulent Not exhibiting VIRULENCE.

avocado sunblotch viroid See VIROID.

avoiding reaction *Syn.* PHOBIC RESPONSE.

avoparcin A glycopeptide antibiotic, produced by *Streptomyces candidus*, which inhibits Gram-positive bacteria by interfering with peptidoglycan synthesis (see VANCOMYCIN); it is a well-established FEED ADDITIVE for pigs and poultry, and has been found to have growth-promoting properties for ruminants. In sheep, avoparcin appears to shift the balance of cellulolytic bacteria from ruminococci to *Bacteroides succinogenes* [JGM (1985) *131* 427–435].

avr gene (*plant pathol.*) See AVIRULENCE GENE.

a_w Symbol for WATER ACTIVITY.

axenic Of a culture: containing or comprising cells of a single species. Thus, e.g., an axenic bacterial culture is a pure (uncontaminated) culture of a single species or strain, while an axenic tissue culture is one which contains cells (of one or more types) from one species (or individual) with no microbial contaminants, intracellular parasites, viruses etc. Certain microorganisms – e.g. obligately intracellular bacteria, obligately predatory protozoa – cannot be grown axenically.

axial fibrils See SPIROCHAETALES.

axial filament (1) See ENDOSPORE (bacterial). (2) See SPIROCHAETALES. (3) See AXOPODIUM.

axile Situated in the centre or on the axis.

axoneme (1) See FLAGELLUM (b). (2) See AXOPODIUM.

axopodium A fine, rod-like, often tapering, relatively rigid type of PSEUDOPODIUM which has an axial core of microtubules (the *axial filament*, *axial rod* or *axoneme*). Axopodia occur e.g. in HELIOZOEA and RADIOLARIA in which they typically emanate radially from the cell body. They function mainly in feeding (but see HELIOZOEA). When an axopodium comes into contact with a prey organism, it adheres to it and may – at least in some cases – immobilize it; the axopodia of some heliozoa have extrusomes (*kinetocysts*) which may function in immobilizing prey. Ingestion may involve the retraction of the axopodium with the prey, and/or a co-operative engulfing action of a number of pseudopodia in the vicinity.

axostyle A rod-like endoskeletal structure which occurs in some protozoa (e.g. *Giardia*, *Hexamita*, *Trichomonas*); in *Trichomonas* it extends from the anterior end of the cell and projects a short distance beyond the posterior end. (cf. COSTA; see also OXYMONADIDA.)

azaserine ($CO_2H.CHNH_2.CH_2.O.CO.CH{=}NH^+{=}N^-$) An ANTIBIOTIC and antitumour agent obtained from *Streptomyces* sp. It is an analogue of glutamine and blocks a number of reactions in which glutamine acts as an NH_2 donor; in particular, it binds to and inactivates phosphoribosylformylglycinamidine synthetase, thus preventing purine (and hence nucleotide) synthesis (see Appendix V(a)). (cf. DON; HADACIDIN.)

azathioprine See IMMUNOSUPPRESSION.

azdimycin See POLYENE ANTIBIOTICS (b).

azide (N_3^-) Azide acts e.g. as a RESPIRATORY INHIBITOR by combining with, and preventing the reduction of, oxidized CYTOCHROME OXIDASES of the aa_3-type.

Sodium azide (NaN_3) is an antimicrobial agent used e.g. as a preservative in some laboratory reagents (0.1% final concentration NaN_3). It is also used e.g. in certain selective media (0.025% w/v NaN_3) for the isolation of enterococci from samples of sewage-polluted water; at this concentration coliforms and many other Gram-negative bacteria are inhibited.

3′-azido-3′-deoxythymidine See AZT.

azithromycin A MACROLIDE ANTIBIOTIC used e.g. as an alternative to clarithromycin in the prophylaxis and treatment of infections involving members of the *M. avium* complex (see MAC). The drug tends to concentrate in macrophages and tissue cells. Strains resistant to azithromycin are also resistant to clarithromycin. Side-effects are uncommon; they include nausea, diarrhoea, headaches, dizziness, deafness. (See also QUORUM SENSING.)

azlocillin See PENICILLINS.

AZM Adoral zone of membranelles: a number of MEMBRANELLES serially arranged in a definite and taxonomically important pattern along the left-hand side of the oral area in many members of the OLIGOHYMENOPHOREA and POLYHYMENOPHOREA; the corresponding ciliature in peritrichs and spirotrichs is sometimes called the *adoral ciliary spiral*. The AZM is primarily concerned with feeding (i.e., the production of water currents directed towards the cytostome) but is sometimes used also for locomotion. (cf. PARORAL MEMBRANE.)

Azoarcus See SPLIT GENE (e).

azofer See NITROGENASE.

azofermo See NITROGENASE.

azoferredoxin See NITROGENASE.

azole antifungal agents A group of synthetic, broad-spectrum ANTIFUNGAL AGENTS, many of which are useful in the treatment of mycoses in man and animals while others are effective against many fungal diseases of plants. The group contains two main categories: the *imidazole* derivatives (e.g. the agriculturally useful BENZIMIDAZOLES and the medical and veterinary drugs bifonazole, butoconazole, CLOTRIMAZOLE, econazole, fenticonazole, isoconazole, KETOCONAZOLE, MICONAZOLE, oxiconazole, sulconazole, TIOCONAZOLE, and zinconazole) and the *triazole* derivatives (e.g. the agricultural antifungals FLUTRIAFOL, PROPICONAZOLE, TRIADIMEFON, and TRIADIMENOL, and the medical drugs ITRACONAZOLE, terconazole and vibunazole). Most of the medically useful azoles are administered topically, being poorly absorbed from the gut and/or too toxic for systemic use; however, a few (e.g. ketoconazole, itraconazole and vibunazole) can be given orally.

With the exception of the BENZIMIDAZOLES, the azoles generally appear to have the same primary mode of action. At minimal fungistatic concentrations they interfere with the permeability and function of the CYTOPLASMIC MEMBRANE by inhibiting the biosynthesis of ergosterol. In the normal biosynthetic pathway ergosterol is synthesized from its precursor lanosterol by a 14α-demethylation reaction which involves a cytochrome P-450-dependent monooxygenation step; the azole compound binds (in place of oxygen) to the 6th coordination position of the P-450 haem iron, thus preventing the monooxygenation step. As a result, 14α-methylated sterols accumulate and ergosterol (and cholesterol) levels fall; this not only alters membrane permeability: the altered lipid environment of the membrane also interferes with the activity and/or control of other enzyme systems (e.g. chitin synthase). Miconazole – in addition to its effect on the cytoplasmic membrane – also exerts an apparently direct inhibitory effect on the mitochondrial ATPase in yeasts such as *Candida albicans* and *Saccharomyces cerevisiae* [Eur. J. Bioch. (1984) *143* 273–276].

Actively growing fungi exposed to minimal fungistatic concentrations of azoles (e.g. 10^{-8}–10^{-7} M for miconazole) show characteristic structural changes, including the formation of dense membrane-derived vesicles at the cell periphery. At somewhat higher fungistatic concentrations (e.g. 10^{-6} M miconazole) changes in the cell vacuole also occur, the vacuole becoming filled with vesicles and granular material. At still higher concentrations (e.g. 10^{-5} M miconazole) the azoles become fungicidal, causing degeneration of organelles (mitochondria and nuclei).

[Review of medically useful azoles: Book ref. 153, pp. 133–153.]

Azolla See ANABAENA.

Azomonas A genus of Gram-negative or (*A. macrocytogenes*) Gram-variable, motile bacteria (family AZOTOBACTERACEAE) which occur e.g. in soil and water. Cysts are not formed (cf. AZOTOBACTER). Most strains form water-soluble pigments, but none forms insoluble pigments; a fluorescent pigment is formed by some strains in iron-deficient media. Carbon sources used by all species include e.g. fructose, glucose, acetate, fumarate, lactate, and ethanol; all species can fix nitrogen and can use ammonium salts as the sole source of nitrogen. Optimum growth temperature: 30–37°C, according to species. GC%: ca. 52–59. Type species: *A. agilis*.

A. agilis. Peritrichously flagellated. Can use e.g. malonate, but not mannitol. Growth can occur at 32°C and 37°C.

A. insignis. Lophotrichously flagellated. Can use e.g. malonate, but not mannitol. Growth can occur at 32°C but not at 37°C.

A. macrocytogenes. Monotrichously flagellated (rarely biflagellated at one pole). Can use mannitol and maltose, but not malonate.

[Book ref. 22, pp. 230–234.]

azomycin See NITROIMIDAZOLES.

Azorhizobium caulinodans See NITROGEN FIXATION.

Azospirillum A genus of Gram-negative or Gram-variable, asporogenous, nitrogen-fixing bacteria. Cells: curved and straight rods, often with pointed ends, ca. 1.0 × 2.1–3.8 μm. Motile by a single polar flagellum; numerous lateral flagella (of shorter wavelength) are also formed during growth on solid media at 30°C. Enlarged, ovoid or pleomorphic, non-motile, capsulated forms ('C forms') may develop under certain conditions (e.g. in old cultures). Metabolism is mainly respiratory, with either oxygen or NO_3^- acting as terminal electron acceptor; DENITRIFICATION occurs under microaerobic conditions. Glucose or fructose may be fermented weakly. Disaccharides are not metabolized. Oxidase +ve; catalase variable; phosphatase +ve; indole −ve; weakly pectinolytic. Optimum growth temperature: 35–37°C. Colonies on potato agar are typically pink, often wrinkled, not slimy. NITROGEN FIXATION occurs under microaerobic conditions; an 'uptake hydrogenase' (see NITROGENASE) is present, and *A. lipoferum* (but not *A. brasilense*) can grow as an H_2-dependent lithoautotroph. Aerobic growth can also occur in the presence of fixed nitrogen (e.g. NH_4^+, NO_3^-). GC%: 69–71. Type species: *A. lipoferum* (formerly *Spirillum lipoferum*).

[Book ref. 22, pp. 94–104.]

Azospirilla occur in soil, both free-living and in association with the roots of grasses (including cereals) and tuberous plants; the bacteria occur on the root surface, in the mucigel, and also within the tissues of the root (outer and inner cortex and stele). The plant appears to benefit from the nitrogen fixed by the bacteria; nodule formation does not occur (cf. ROOT NODULES). There is some degree of host specificity: 'C4 plants' (e.g. maize) are infected by *A. lipoferum*, 'C3 plants' (e.g. barley, oats, rice, rye, wheat) by strains of *A. brasilense*.

Azotobacter A genus of Gram-negative, CYST-forming bacteria (family AZOTOBACTERACEAE) which occur e.g. in fertile soils of near-neutral pH. The cells are peritrichously flagellated or non-motile, and are commonly short rods or coccobacilli – though *A. paspali* regularly forms both rods and long filaments; the organisms appear to contain many copies of the chromosome per cell [JGM (1984) *130* 1603–1612]. Carbon sources used by all species include e.g. glucose, fructose, sucrose, acetate, fumarate, gluconate, and ethanol; most species can use nitrate and/or ammonium salts as a source of nitrogen. In the presence of combined nitrogen growth can occur within the pH range ca. 5–8.5; NITROGEN FIXATION occurs optimally within the pH range 7–7.5. In laboratory cultures nitrogen-fixing ability declines with age, and is very poor immediately prior to encystment. (Cyst-formation occurs maximally in old cultures on nitrogen-free media containing 0.2% butanol.) The optimum growth temperature varies with species, and is ca. 30–37°C. GC%: ca. 63–68. Type species: *A. chroococcum*.

A. armeniacus. Peritrichously flagellated. Can use e.g. caprylate and mannitol but not e.g. caproate; some strains can use propionate.

A. beijerinckii. Non-motile. Can use e.g. malonate and propionate but not caproate or rhamnose.

A. chroococcum. Peritrichously flaggellated. Can use e.g. caproate, mannitol and propionate, but not e.g. caprylate.

A. nigricans. Non-motile. Cannot use caproate, propionate or rhamnose. Some strains form e.g. a yellow or dark non-diffusible pigment or a dark diffusible pigment.

A. paspali. Peritrichously flagellated. Cannot use caproate, caprylate, malonate, mannitol, propionate or rhamnose, but can use oxaloacetate, and some strains can use propan-1-ol. Appears to occur only on (or within?) the root cortex of the grass *Paspalum notatum*.

A. vinelandii. Peritrichously flagellated; non-motile strains have been reported. Can use caproate, caprylate, malonate, mannitol, propionate and rhamnose. A yellowish-green water-soluble fluorescent pigment is formed on iron-deficient media. [Book ref. 22, pp. 220–229.]

Azotobacteraceae A family of Gram-negative (or Gram-variable), aerobic, chemoorganotrophic bacteria which occur in soil, in the rhizosphere, and in aquatic habitats; the organisms are capable of NITROGEN FIXATION – typically under free-living conditions, but sometimes in association with higher plants. Cells: ovoid, or round-ended rods (sometimes filaments), 1.5–2.0 μm or more in width, which may be motile (polarly or peritrichously flagellated) or non-motile; some species form cysts. Pigments, some fluorescent, are formed by some species. PHB can be accumulated. Catalase +ve. Oxidase +ve (most strains of most species). GC%: ca. 52–68. Two genera: AZOMONAS, AZOTOBACTER. [Book ref. 22, pp. 219–234.]

azotoflavin See FLAVODOXINS.

AZT (3′-azido-3′-deoxythymidine; zidovudine) A thymidine analogue ANTIRETROVIRAL AGENT used in the treatment of AIDS. AZT is converted by cellular enzymes to the triphosphate derivative which is then incorporated – instead of thymidine – into (provirus) DNA by the viral REVERSE TRANSCRIPTASE (see RETROVIRIDAE); the presence of the 3′-azido group inhibits further chain elongation by preventing the formation of a phosphodiester bond at the 3′ position. Cellular DNA polymerase α is much less sensitive than reverse transcriptase to AZT triphosphate.

AZT competes for intracellular phosphorylation with another NUCLEOSIDE REVERSE TRANSCRIPTASE INHIBITOR, stavudine.

As high-level resistance develops readily if AZT is used alone, it is used in combination with other drugs.

Unlike other nucleoside reverse transcriptase inhibitors, AZT undergoes significant metabolism (glucuronidation) in the liver; drugs (e.g. PROBENECID) that inhibit glucuronidation may therefore affect the plasma concentration of AZT.

The use of AZT is associated with megaloblastic changes; the drug may induce a dose-dependent macrocytic anaemia (less often a severe normocytic anaemia) and is also able to induce neutropenia [haematological aspects of HIV infection: BCH (2000) *13* 215–230].

The value of AZT in anti-AIDS combination drug therapy (against HIV-1) was indicated by an in vitro study in which those nucleoside reverse transcriptase inhibitors which lacked the 3′-azido moiety were less active against a particular mutant form of the reverse transcriptase [AAC (2005) *49* 1139–1144].

azthreonam *Syn.* AZTREONAM.

aztreonam (azthreonam) A MONOBACTAM derivative in which $R' =$ an aminothiazoleoxime group, $R'' = H, R''' = CH_3, R'''' = SO_3{}^-K^+$ (see β-LACTAM ANTIBIOTICS). It is active only against aerobic Gram-negative bacteria, having a high degree of specificity for PBP3 of Gram-negative bacteria (see PENICILLIN-BINDING PROTEINS); it is resistant to most β-lactamases. [Action, use: Drugs (1986) *31* 96–130.]

Azuki bean mosaic virus See POTYVIRUSES.

azure See *polychrome* METHYLENE BLUE.

azurin (1) A BLUE PROTEIN which occurs in certain bacteria (e.g. *Alcaligenes denitrificans*, *Bordetella pertussis*, *Paracoccus denitrificans*, *Pseudomonas aeruginosa*); depending on source, the MWt ranges from ca. 12000 to ca. 15000, and the E_m appears to be in the approximate range 230–330 mV.

(2) A solution of $CuSO_4$ and NH_4OH used as an agricultural antifungal agent.

azygospore A parthenogenically derived spore, similar to a ZYGOSPORE, formed e.g. by many species of the Mucorales.

1. Words in SMALL CAPITALS are cross-references to separate entries.
2. Keys to journal title abbreviations and Book ref. numbers are given at the end of the Dictionary.
3. The Greek alphabet is given in Appendix VI.
4. For further information see 'Notes for the User' at the front of the Dictionary.

B

B cell (*immunol.*) *Syn.* B LYMPHOCYTE.

B cell differentiation factor See LYMPHOKINES.

B cell growth factor See LYMPHOKINES.

B cell superantigen See SUPERANTIGEN.

B-DNA See DNA.

B lymphocyte (B cell) A type of LYMPHOCYTE concerned primarily with ANTIBODY FORMATION (cf. T LYMPHOCYTE). In mammals, B cells are formed initially in the fetal liver, but later they develop from haemopoietic stem cells in the bone marrow. (In birds, B cells appear to develop in the BURSA OF FABRICIUS.) B cells occur e.g. in blood, in lymph and in the spleen. In an adult, the B cell population consists of $>10^6$ clones, the cells of each clone being potential sources of antibodies with a unique and highly specific antigen-binding capacity.

In mammals, B cells mature in a series of stages. Stem cells give rise to *pro-B cells* which, in turn, develop as *pre-B cells*. Pre-B cells differ from mature B cells e.g. in that they express no surface immunoglobulin (Ig), i.e. they are sIg^-, although they contain *mu* HEAVY CHAINS (associated with so-called *surrogate light chains*) which may be located at the cell surface. The next stage, the *immature B cell*, is characterized by the presence of *monomeric* IgM at the cell surface; interaction with specific antigen at this stage may lead to inactivation of the cell. The *mature B cell* displays IgM and IgD at the cell surface (with antigen-binding sites facing outwards); the IgM and IgD have identical antigenic specificity. The B cell reaches maturity, in an antigen-independent way, within the bone marrow.

In addition to antibodies, the surface of a mature B cell displays various types of molecule involved in the recognition of exogenous factors and the initiation of intracellular signalling. These molecules include: (i) a receptor that mediates *isotype switching* (= *class switching* – e.g. the IgM → IgG switch following antigenic stimulation) (see CD40); (ii) MHC class II molecules (required for presenting antigen to a T cell: see ANTIBODY FORMATION); (iii) co-stimulatory molecules (designated B7.1 and B7.2) which, during B cell–T cell contact, bind to a specific T cell receptor and initiate signals that activate the T cell (see CD28); (iv) CD32: a low-affinity receptor for the Fc portion of IgG (the binding of antigen-bound antibody to CD32 may suppress the ongoing production of specific antibodies); (v) CD21 (involved in B cell activation, and a binding site for certain components of COMPLEMENT); (vi) receptors for CYTOKINES (including INTERLEUKIN-1; INTERLEUKIN-4, which e.g. induces isotype switching and promotes expression of MHC class II molecules on B cells; and INTERLEUKIN-6).

The body's B cell population can be divided into three subsets on the basis of the CD5 and CD23 antigens. B cells in the general circulation are mainly $CD5^-$, $CD23^+$; they have a relatively short existence and are continually renewed from the bone marrow stem cells. Two other subsets, one $CD5^+$, $CD23^-$, and the other $CD5^-$, $CD23^-$, are found mainly in mucosal sites (e.g. the peritoneal cavity) and are self-renewing (albeit with precursor cells in the bone marrow); these cells can rapidly produce IgM antibodies (of a limited range of specificities) in response to bacterial polysaccharides, and can do so without help from T cells. These two populations of B cells are viewed as a first line of defence against bacterial incursion in sites which are particularly vulnerable to infection.

No further development of the B cell occurs unless it encounters specific antigen e.g. within so-called *germinal centres* in the spleen or lymph nodes. In the event of contact with specific antigen, the outcome depends e.g. on the type of antigen involved and the contribution of T cells: see ANTIBODY FORMATION. When a B cell is appropriately stimulated it prepares for the role of antibody formation by initially undergoing BLAST TRANSFORMATION and then proliferating to form a clone of cells of identical antigenic specificity (= *clonal expansion*); some of these cells develop as PLASMA CELLS while others (commonly) become MEMORY CELLS.

B cells which do not bind specific antigen remain viable for a limited period of time and subsequently undergo APOPTOSIS.

B-tubule (B-subfibre) See FLAGELLUM (b).

B-type inclusion body See POXVIRIDAE.

B-type particles See TYPE B ONCOVIRUS GROUP.

B-type starter See LACTIC ACID STARTERS.

B virus (cercopithecine herpesvirus 1; cercopithecid herpesvirus 1; herpesvirus B; *Herpesvirus simiae*) A herpesvirus (subfamily ALPHAHERPESVIRINAE) which naturally infects monkeys of the genus *Macaca*. In rhesus monkeys (*M. mulatta*) infection may be asymptomatic, or vesicular lesions (which may ulcerate) may develop in the mouth and sometimes on the skin and conjunctivae. B virus may be transmitted to humans e.g. by monkey bites or scratches, and can cause in humans a severe (usually fatal) encephalomyelitis ('monkey-bite encephalomyelitis').

B_{12} coenzymes See VITAMIN B_{12}.

B19 parvovirus See ERYTHROVIRUS.

B663 CLOFAZIMINE.

Babes–Ernst granules METACHROMATIC GRANULES observed in bacteria.

Babesia A genus of protozoa (subclass PIROPLASMASINA) parasitic in invertebrates and in the erythrocytes (RBCs) of vertebrates (cf. THEILERIA); at least some species (including *B. equi* and *B. microti*) develop in the lymphocytes (as well as in the RBCs) of the vertebrate host, and it has been suggested that such species be transferred to other genera (e.g. *Nicollia*, *Nuttallia*). Some species (e.g. *B. bigemina*, *B. bovis*) can cause tick-borne disease in domestic animals (see REDWATER FEVER), and some (e.g. *B. divergens*, *B. microti*) can cause disease in man [ARE (1981) *26* 90–92]. The cells of *Babesia* are rounded or pyriform, ca. 1–6 μm. In the typical life cycle, sporozoites of *Babesia* are injected into the vertebrate host by the tick vector, and each sporozoite enters an RBC and undergoes schizogony to form two or four merozoites – which infect fresh RBCs when the host cell ruptures. Following the tick's meal of infected blood, gametes are formed in the tick's gut, and these fuse and give rise to a (motile) kinete which passes, via various tissues, to the salivary glands – there undergoing sporogony and forming many sporozoites. Unlike *Theileria*, *Babesia* can invade the ovary and egg of the tick, and can be transmitted transovarially to the tick's offspring. [Life cycles of *Babesia* and *Theileria*: AP (1984) *23* 37–103.]

babesiosis Any disease of man or animals caused by a species of BABESIA (e.g. REDWATER FEVER).

Bacillaria See DIATOMS.

Bacillariophyta See DIATOMS.

bacillary angiomatosis See BARTONELLA.

bacillary dysentery See DYSENTERY (a).

bacillary white diarrhoea *Syn.* PULLORUM DISEASE.

bacille Calmette–Guérin See BCG.

bacillin *Syn.* BACILYSIN.

bacillus (1) A member of the genus BACILLUS. (2) *Any* rod-shaped bacterial cell, i.e., a cell whose length is ca. two or more times greater than its width. (cf. COCCOBACILLUS and FILAMENT.) A bacillus may be straight or curved (cf. VIBRIO sense 2), with rounded, truncated or tapered ends (cf. FUSIFORM), and may occur singly, in groups, pairs, chains, etc. (See also PALISADE.)

Bacillus A genus of GRAM TYPE-positive, strictly aerobic or facultatively anaerobic, typically catalase-positive, rod-shaped, ENDOSPORE-forming bacteria. The organisms typically occur as saprotrophs in soil and water, but certain species can be pathogenic in man and other mammals (see e.g. *B. anthracis* and *B. cereus*, below) and some species are entomopathogenic (see e.g. *B. moritai*, *B. popilliae* and *B. thuringiensis*). [Insecticidal species: Book ref. 171, pp. 185–209; MR (1986) *50* 1–24.] *Bacillus* spp can cause biodeterioration (see e.g. FLAT SOUR, LEATHER SPOILAGE, SWELL), but some species are used commercially as sources of antibiotics (see e.g. BACITRACIN, POLYMYXINS) or other products (e.g. DEBRANCHING ENZYMES, GLUCOSE ISOMERASE, GLYCEROKINASE, SUBTILISINS).

Cells: typically motile rods, commonly ca. 0.5–1.5 × 2–6 μm, often in chains; some strains form a CAPSULE, and some are pigmented. (See also S LAYER and TEICHOIC ACIDS.) Metabolism may be respiratory (see RESPIRATION) – some strains being capable of NITRATE RESPIRATION – or facultatively fermentative (see FERMENTATION); a few species (e.g. *B. macerans*, *B. polymyxa*) can carry out NITROGEN FIXATION. Most species are chemoorganoheterotrophs; *B. schlegelii* can grow chemolithoautotrophically (see CARBOXYDOBACTERIA). Utilizable substrates for organoheterotrophic species range from simple sugars and other carbohydrates to e.g. uric acid (see *B. fastidiosus*) and proteinaceous materials; PENTOSES are metabolized via the HEXOSE MONOPHOSPHATE PATHWAY. Many species can grow on NUTRIENT AGAR. Storage compounds include e.g. POLY-β-HYDROXYBUTYRATE. The genus includes psychrotrophs (e.g. *B. globisporus*), thermophiles (e.g. *B. schlegelii*, *B. stearothermophilus*) and alkalophiles (e.g. *B. alcalophilus*, *B. firmus*). GC%: ca. 30–70. Type species: *B. subtilis*.

B. alcalophilus. An ALKALOPHILE (q.v.): optimum growth pH ca. 9–10; no growth below pH 7.

B. alvei. See EUROPEAN FOULBROOD.

B. amyloliquefaciens. Similar or identical to *B. subtilis*. (See also AMYLASES.)

B. amylolyticus. *Nom. rev.* [IJSB (1984) *34* 224–226].

B. aneurinolyticus. A species, related to *B. brevis*, which produces a thiaminase. (See also BACTERIOPHAGE ϕBA1.)

B. anthracis. The causal agent of ANTHRAX. Cells: non-motile, ca. 1.5 × 3–6 μm, typically square-ended and usually in chains. Virulent strains form a PLASMID-encoded toxin (see ANTHRAX TOXIN) and a plasmid-encoded poly-D-glutamic acid CAPSULE. [Capsule-encoding plasmid: see e.g. Inf. Immun. (1985) *49* 291–297.] (cf. STERNE STRAIN.) Growth occurs on nutrient agar. In biochemical tests *B. anthracis* gives results very similar to those of *B. cereus*. Lecithinase activity is weak or absent. Haemolysis on sheep-blood agar is weak or absent. *B. anthracis* is susceptible to the 'γ phage' (*B. cereus* is not). On agar containing benzylpenicillin (0.05–0.5 unit/ml) *B. anthracis* gives rise to chains of enlarged, spherical cells ('string of pearls' effect).

B. azotofixans. A proposed (nitrogen-fixing) species from Brazilian soil [IJSB (1984) *34* 451–456].

B. badius. Similar to *B. brevis*. Growth is not inhibited by 5% NaCl. The (ellipsoidal) spore does not distend the cell.

B. brevis. Typically forms no acid or gas from carbohydrates. Growth occurs at 50°C. Spore: ellipsoidal, distending the cell. (See also TYROCIDINS.)

'*B. caldolyticus* group'. Strains (including *B. caldotenax* and *B. caldovelox*) closely related to *B. stearothermophilus*.

B. cereus. A saprotroph or opportunist pathogen which can cause FOOD POISONING (q.v.) or, rarely, e.g. meningitis. [Meningitis due to *B. cereus*, and review of *Bacillus* infections other than anthrax: Israel J. Med. Sci. (1983) *19* 546–551.] Cells: typically motile rods (mean width ca. 1.5 μm), sometimes pigmented. Spore: ellipsoidal, not distending the cell. Acid is formed in anaerobic glucose broth; no acid is formed from arabinose, mannitol or xylose. No gas is formed from carbohydrates. Most strains are VP +ve. Starch and casein are hydrolysed, and lecithinases are produced. Nitrate is reduced by most strains. No growth at 50°C.

B. circulans. Similar to *B. polymyxa* but ferments carbohydrates anaerogenically; typically, a range of carbohydrates is fermented. (See also MOTILE COLONIES.)

B. coagulans. Forms lactic acid as the major product of glucose fermentation; 2,3-butanediol is also produced. No gas is formed from carbohydrates. VP +ve. (See also GLUCOSE ISOMERASE.)

B. euloomarahae. See MILKY DISEASE.

B. fastidiosus. Obligately aerobic. Uses uric acid as a source of carbon, nitrogen and energy.

B. firmus. An ALKALOPHILE. Forms acid from glucose but typically from few other carbohydrates.

B. fribourgensis. See MILKY DISEASE.

B. globisporus. A PSYCHROTROPH (maximum growth temperature ca. 25–30°C). Urease +ve. Spore: round, distending the cell.

B. larvae. See AMERICAN FOULBROOD.

B. lautus. *Nom. rev.* [IJSB (1984) *34* 224–226].

B. lentimorbus. See MILKY DISEASE.

B. licheniformis. Morphologically and biochemically similar to *B. subtilis*, but growth readily occurs anaerobically (see also DENITRIFYING BACTERIA).

B. macerans. Biochemically very similar to *B. polymyxa*. Forms SCHARDINGER DEXTRINS.

B. megaterium. Cell width commonly ca. 1.5 μm but may reach ca. 3.0 μm in carbohydrate-containing media. Spore: ellipsoidal, not distending the cell. Glucose and many other carbohydrates are utilized anaerogenically.

B. moritai. A species pathogenic e.g. in house-flies. The addition of *B. moritai* spores to faeces has been found to reduce the emergence of adult house-flies by up to ca. 90%.

B. mycoides. Morphologically and biochemically very similar to *B. cereus*. Often regarded as a non-motile variety of *B. cereus*.

B. pabuli. *Nom. rev.* [IJSB (1984) *34* 224–226].

B. pasteurii. An ALKALOPHILE (optimum growth at ca. pH 9). Cells: typically slender (ca. 0.5–1.0 μm wide). Spore: spherical, typically distending the cell. Carbohydrates are attacked either very weakly or not at all. Urease +ve. NH_3 is needed for growth.

'*B. piliformis*'. See TYZZER'S DISEASE.

B. polymyxa. Cells: typically slender (ca. 0.5–1.0 μm wide); usually motile. Spore: ellipsoidal, distending the cell. Grows well anaerobically, and produces acid and gas from carbohydrates; typically, glucose, mannitol, and many other sugars and sugar alcohols are utilized. Glucose is fermented via the BUTANEDIOL FERMENTATION. Most strains attack e.g. casein, gelatin, starch and pectins, and most can carry out NITROGEN FIXATION. (See also POLYMYXINS.)

B. popilliae. See MILKY DISEASE.

B. psychrophilus. A proposed (psychrophilic) species [IJSB (1984) *34* 121–123].

B. pulvifaciens. Proposed species (isolated from a diseased bee) [IJSB (1984) *34* 410–413].

B. pumilus. Cells: typically slender (0.5–1.0 μm wide); usually motile. Spore: elongated, not distending the cell. Metabolically similar to *B. subtilis*, but starch is not utilized. Usually VP +ve.

B. schlegelii. See CARBOXYDOBACTERIA and HYDROGEN-OXIDIZING BACTERIA.

B. sphaericus. Cells: slender (ca. 0.5–1.0 μm wide); usually motile. Spore: spherical, distending the cell. Typically, carbohydrates are not utilized. Can cause LEATHER SPOILAGE, and can be insecticidal to e.g. mosquito larvae – toxicity apparently being due to an uncharacterized toxin and not associated with sporulation. [Field evaluation of a *B. sphaericus* strain as a mosquito 'biocide': J. Inv. Path. (1986) *48* 133–138.]

B. stearothermophilus. Thermophilic; can grow at e.g. 65°C. Cells: ca. 1.0 μm wide; spore: ellipsoidal, often distending the cell. The spores are highly resistant to heat, and are sometimes used to monitor AUTOCLAVE performance. (See also FLAT SOUR.) Typically, a range of carbohydrates can be metabolized anaerobically and anaerogenically; the main product of carbohydrate fermentation commonly appears to be lactic acid. (See also GLYCEROKINASE.)

B. subtilis. Cells: slender (typically ca. 0.8 μm wide), usually motile; chains are uncommon. (See also CERULENIN and MACROFIBRE.) Spore: ellipsoidal, usually not distending the cell. Metabolism appears to be primarily respiratory. Growth does not occur in anaerobic glucose broth. Glucose, various other sugars and sugar alcohols, and starch, are metabolized anaerogenically. VP +ve. Nitrate is reduced. Casein and gelatin are hydrolysed. Phages which infect *B. subtilis* include e.g. BACTERIOPHAGE PBS1, BACTERIOPHAGE ϕ105 and BACTERIOPHAGE SPO1.

B. thuringiensis. An entomopathogenic species which is morphologically and biochemically very similar to *B. cereus*, differing primarily in its formation of DELTA ENDOTOXIN (q.v.).

B. validus. *Nom. rev.* [IJSB (1984) *34* 224–226].

Other species include e.g. *B. insolitus*, *B. laterosporus*, *B. lentus*, and *B. pantothenticus*.

Note on the identification of species. Some mutually contradictory tables of biochemical test reactions have been published. For example, the VP test reactions (obtained using traditional methods) given in Book ref. 46, p. 1731 differ significantly, in respect of a number of species, from those (obtained using the API system) given in JGM (1984) *130* 1871–1882. (Species designated as 'VP +ve' above are confirmed as such in many sources.)

bacillus Calmette–Guérin See BCG.

bacilysin ('bacillin'; 'tetaine') A dipeptide ANTIBIOTIC which is active against a wide range of Gram-positive and Gram-negative bacteria and e.g. against *Candida albicans*. It consists of a C-terminal epoxy-L-amino acid ('anticapsin') and an N-terminal L-alanine residue; it is taken up by peptide transport systems in sensitive cells, and is subsequently hydrolysed by cellular enzymes to release the anticapsin – an inhibitor of glucosamine (and hence e.g. PEPTIDOGLYCAN) synthesis. (cf. WARHEAD DELIVERY.)

bacitracin A cyclic dodecapeptide ANTIBIOTIC produced by strains of *Bacillus* spp. It is active against many Gram-positive and certain Gram-negative bacteria (e.g. *Neisseria* spp, *Haemophilus* spp). In the presence of divalent cations (particularly Zn^{2+}) bacitracin binds to bactoprenol pyrophosphate, inhibiting the regeneration of bactoprenol monophosphate during e.g. PEPTIDOGLYCAN biosynthesis; it can also inhibit other processes involving bactoprenol pyrophosphate (e.g. O-specific chain formation in LIPOPOLYSACCHARIDE biosynthesis) and can affect membrane permeability. Bacitracin is used clinically e.g. for the topical treatment of local infections, and as a FEED ADDITIVE for ruminants to decrease methane production in the RUMEN.

back focal plane Of a convex lens: the focal plane furthest from the light source.

back mutation (reverse mutation) (1) A MUTATION which reverses the effects of a FORWARD MUTATION by restoring the original nucleotide sequence. (2) Either a mutation as in sense 1 above, or an *intragenic* SUPPRESSOR MUTATION.

background mutation *Syn.* SPONTANEOUS MUTATION.

bacon spoilage See MEAT SPOILAGE.

BACTEC™ culture systems Liquid media (marketed by Becton Dickinson) which can be used e.g. for the growth of *Mycobacterium tuberculosis* – growth being detectable more rapidly (~1–2 weeks) than is usually possible on solid media (~6 weeks). These media may be used for (i) detecting *M. tuberculosis* in clinical specimens, and (ii) examining an isolate of the pathogen for susceptibility to antibiotics; for the latter purpose an isolate is tested for growth (or lack of growth) in a medium containing a known amount of the given antibiotic.

Earlier BACTEC systems were radiometric, i.e. they detected growth by detecting radioactive carbon dioxide produced from a radioactive substrate in the medium. A more recent system, the BACTEC MGIT™ 960 (MGIT = mycobacteria growth indicator tube), monitors growth by means of a fluorescent sensor system which detects the consumption of oxygen. [Evaluation of BACTEC MGIT 960: JCM (1999) *37* 748–752.]

bacteraemia (bacteremia) The condition in which viable bacteria are present in the bloodstream. (cf. PYAEMIA; SEPTICAEMIA.)

bacteremia *Syn.* BACTERAEMIA.

bacteria (*singular*: bacterium) (1) Formerly: a term which referred, collectively, to *all* prokaryotic microorganisms (see PROKARYOTE); the term was also used to refer to a population of specific prokaryote(s) of any given type.

(2) All, or particular, members of the domain Bacteria (see next entry).

Note. In literature prior to the 1980s 'bacteria' was always used with the meaning given in sense 1. With the establishment of the domain ARCHAEA (q.v.), appropriate usage is that given in sense 2.

Bacteria A taxon (DOMAIN) comprising one of the two fundamentally distinct groups of prokaryotic microorganisms (see PROKARYOTE and ARCHAEA). (See also previous entry.)

The bacteria are a diverse group of (usually) single-celled organisms. Most are free-living, occurring e.g. in soil, on plants, in various aquatic habitats, and even in antarctic snow [AEM (2000) *66* 4514–4517]; some are important in the cycles of matter (see CARBON CYCLE, NITROGEN CYCLE, SULPHUR CYCLE). Bacteria are also found as symbionts in plants, animals and certain microorganisms (see CAEDIBACTER, MYCETOCYTE, ROOT NODULES, RUMEN).

A minority of bacteria occur as intracellular or extracellular parasites or pathogens in man and/or other animals. In man, bacteria can cause a number of major and minor diseases – which include e.g. ANTHRAX, BOTULISM, BRUCELLOSIS, CHOLERA, DIPHTHERIA, DYSENTERY, ERYSIPELAS, GONORRHOEA, LEGIONNAIRE'S DISEASE, LEPROSY, LYME DISEASE, MENINGITIS, PLAGUE, PSEUDOMEMBRANOUS COLITIS, Q FEVER, SCARLET

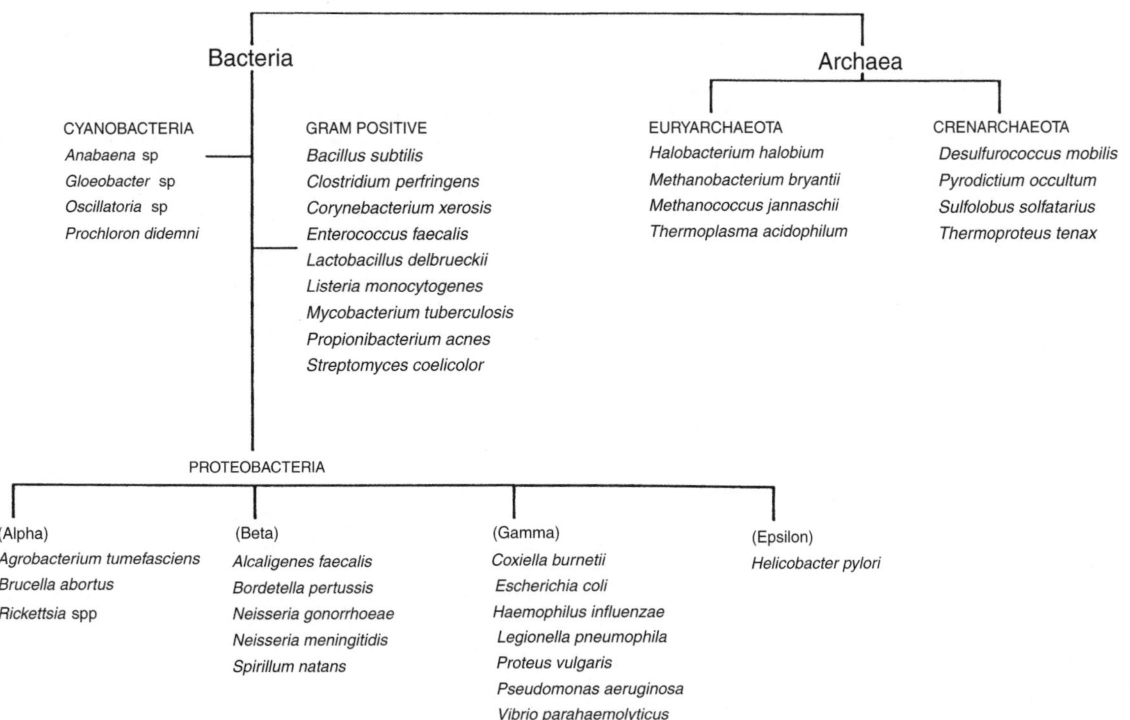

BACTERIA: examples of species in some of the groups in a current taxonomic scheme for the Bacteria (some species of the domain Archaea are also shown); the lines are not intended to reflect evolutionary distances between the organisms. A number of groups not shown in the figure are outlined in the text; note, for example, that the Gram-positive species are divided into 'high GC%' and 'low GC%'.

Reproduced from *Bacteria*, 5th edition, Figure 16.5, page 436, Paul Singleton (1999) copyright John Wiley & Sons Ltd (UK) (ISBN 0471-98880-4) with permission from the publisher.

FEVER, SYPHILIS, TETANUS, TOXIC SHOCK SYNDROME, TRACHOMA, TUBERCULOSIS, TULARAEMIA, TYPHOID, TYPHUS FEVERS, WHIPPLE'S DISEASE and WHOOPING COUGH.

Bacteria cause economically important diseases in livestock as well as infections in wild animals – see e.g. CATTLE DISEASES, FISH DISEASES, HORSE DISEASES, PIG DISEASES, SHEEP DISEASES. (See also INSECT DISEASES.)

Relatively few bacteria cause disease in plants (but see e.g. ERWINIA, SPIROPLASMA, XANTHOMONAS).

Predatory/bacteriolytic bacteria include species of ANAEROPLASMA, BDELLOVIBRIO, MYXOBACTERALES and VAMPIROVIBRIO.

Bacteria are used in a number of commercial/manufacturing processes: see e.g. BIOPOL, DAIRY PRODUCTS, GLUTAMIC ACID, LEACHING, PICKLING, RETTING, SAUERKRAUT, SOY SAUCE, SUBTILISINS, VINEGAR, VITAMIN B12. Certain antibiotics (e.g. POLYMYXINS, STREPTOMYCIN and some β-LACTAM ANTIBIOTICS) can be synthesized by bacteria. (See also BIOASSAY, BIOFUEL CELL and BIOLOGICAL CONTROL.)

In size, most bacteria are between 1 and 10 μm (maximum dimension). The smallest range from <1 μm (e.g. *Chlamydia*, *Francisella*, *Rickettsia*). One of the largest (>600 μm) is *Epulopiscium fishelsoni* [Nature (1993) *362* 239–241].

The basic shapes of bacteria are BACILLUS (sense 2) (also called 'rod'), COCCUS and SPIRILLUM (sense 1); the coccus seems likely to be a degenerate form of the rod [evolution of bacterial morphology: Microbiology (1998) *144* 2803–2808]. (See also e.g. COCCOBACILLUS, FILAMENT, L FORM, MYCELIUM, PLEOMORPHISM (sense 1), SPIROCHAETALES and VIBRIO (sense 2).) In general, the shape of a bacterium is determined primarily by its species. (cf. PROTOPLAST.)

According to species, a bacterial cell may have certain appendages: see e.g. FIMBRIAE, FLAGELLUM, PILI, PROSTHECA AND SPINA. (See also CAPSULE.)

Many types of bacteria are motile (see MOTILITY); motile species commonly exhibit CHEMOTAXIS.

Cells may occur singly or in pairs, chains, clusters, PACKETS, PALISADES etc.; some bacteria of the order ACTINOMYCETALES form a mycelium. (See also COENOCYTE and CONSORTIUM.)

Some species form EXOSPORES; some form ENDOSPORES. Members of the MYXOBACTERALES form fruiting bodies.

A bacterial cell lacks the sophisticated physical compartmentalization of eukaryotic cells. However, bacteria exhibit a 'functional compartmentalization' [FEMS Reviews (1993) *104* 327–346]; this refers to various *molecular* strategies – such as the self-assembly of protein components of composite enzyme systems (see e.g. PROTEASOME). Moreover, the localization of proteins in a bacterium now appears to be much more organized than was previously supposed [Science (1997) *276* 712–718]. [Dynamic spatial regulation in the bacterial cell: Cell (2000) *100* 89–98.]

In most species there is a characteristic type of CELL WALL in which PEPTIDOGLYCAN is a common constituent (cf. ARCHAEA); the mycoplasmas are atypical in being wall-less.

Ester-linked lipids occur in the bacterial CYTOPLASMIC MEMBRANE (cf. ARCHAEA). In common with e.g. higher animals and plants, the cytoplasmic membrane in many (not all) bacteria (and in some archaeans) contains aquaporins and/or glycerol facilitators: see MIP CHANNELS. (See also OSMOREGULATION.)

Various TRANSPORT SYSTEMS are associated with the cell envelope (see e.g. PROTEIN SECRETION and PTS). TWO-COMPONENT REGULATORY SYSTEMS regulate responses to various environmental stimuli.

Energy may be obtained by FERMENTATION (sense 1), by RESPIRATION, and/or (in e.g. CYANOBACTERIA) by PHOTOSYNTHESIS. (See also PURPLE MEMBRANE.)

According to species, bacteria may be obligate or facultative AEROBES or ANAEROBES. They may be CHEMOTROPHS and/or PHOTOTROPHS, HETEROTROPHS or AUTOTROPHS, and some are CHEMOLITHOAUTOTROPHS.

The bacterial genome commonly consists of covalently-closed circular DNA; however, in some bacteria (e.g. *Borrelia burgdorferi* and species of *Streptomyces*) the DNA is linear. The number of chromosomes per cell depends e.g. on species and on the growth rate. In *Vibrio cholerae* the genome consists of two circular chromosomes (chromosome 1 = ca. 2961 kb; chromosome 2 = ca. 1073 kb) [Nature (2000) *406* 477–483]. For some species, the complete sequence of nucleotides in the genome has been determined – e.g. *Borrelia burgdorferi* [Nature (1997) *390* 580–586], *Buchnera* [Nature (2000) *407* 81–86], *Escherichia coli* [Science (1997) *277* 1453–1474], *E. coli* O157:H7 [Nature (2001) *409* 529–533; erratum: Nature (2001) *410* 240], *Helicobacter pylori* [Nature (1997) *388* 539–547], *Mycobacterium tuberculosis* [Nature (1998) *393* 537–544] and *Neisseria meningitidis* (serogroup A strain) [Nature (2000) *404* 502–506].

In many bacteria the genome is supplemented by one or more plasmids (see PLASMID); bacterial plasmids are commonly circular, but some are linear. Certain bacteria are normally plasmid-free; they include species of *Anaplasma*, *Bartonella*, *Brucella* and *Rickettsia*. (See also BACTERIOPHAGE and LYSOGENY.)

Reproduction occurs asexually, usually by BINARY FISSION but sometimes by BUDDING or TERNARY FISSION. (See also CELL CYCLE.) A developmental cycle occurs e.g. in CAULOBACTER, CHLAMYDIA and RHODOMICROBIUM.

Despite the lack of sexual reproduction, gene transfer between bacteria can occur by CONJUGATION, CONJUGATIVE TRANSPOSITION, TRANSDUCTION or TRANSFORMATION.

Taxonomy. Until the 1980s, bacterial TAXONOMY was based primarily on criteria such as staining reaction, morphology and substrate requirements. Although some of the earlier taxa (e.g. ENTEROBACTERIACEAE) are still recognized, the modern approach is based mainly on molecular criteria, i.e. base sequences in nucleic acids. Using molecular criteria, bacteria have been divided into a number of groups which are believed to reflect evolutionary affinities; some of these groups are shown in the figure on page 71. An outline of the scheme is as follows.

- Cyanobacteria (see CYANOBACTERIA).
- *Cytophaga/Flexibacter/Bacteroides* group. As well as the named organisms, the group includes species of e.g. *Flavobacterium*, *Microscilla*, *Saprospira*, *Sphingobacterium*, *Spirosoma* and *Sporocytophaga*.
- Fibrobacteria. *Fibrobacter* spp.
- Fusobacteria. *Fusobacterium*, *Leptotrichia* spp.
- Gram-positive bacteria (high GC%). Species of *Arthrobacter*, *Bifidobacterium*, *Corynebacterium*, *Faenia*, *Frankia*, *Gardnerella*, *Mycobacterium* and *Streptomyces*.
- Gram-positive bacteria (low GC%). Species of e.g. *Bacillus*, *Clostridium*, *Desulfotomaculum*, *Enterococcus*, *Erysipelothrix*, *Gemella*, *Lactobacillus*, *Leuconostoc*, *Listeria* and *Pediococcus*.
- Green non-sulphur bacteria. Species of e.g. *Chloroflexus*.
- *Planctomyces/Chlamydia* group. Species of e.g. *Chlamydia*, *Isosphaera*, *Planctomyces*.
- Proteobacteria (purple bacteria). See PROTEOBACTERIA.
- Spirochaetes. Species of e.g. *Borrelia*, *Leptonema*, *Spirochaeta*, *Treponema*.
- Thermotogales. Species of e.g. *Fervidobacterium*, *Geotoga*, *Thermotoga*.

Bacteria from ancient sources. It has been reported that an organism resembling *Bacillus* has been cultured from a brine inclusion located *within* a salt crystal believed to be ~250 million years old [Nature (2000) *407* 897–900; discussion 844–845].

[Bacteria (general text): Book ref. 223.]

bacterial blotch of mushrooms See BROWN BLOTCH and GINGER BLOTCH.

bacterial endocarditis ENDOCARDITIS caused by bacteria.

bacterial fin rot A common FISH DISEASE in which infection e.g. by species of *Flexibacter*, *Aeromonas* or *Pseudomonas* leads to progressive necrosis of fins and tail; stress is an important factor in disease development. Secondary SAPROLEGNIASIS is common.

bacterial gill disease A FISH DISEASE affecting salmonids. Filamentous bacteria (genus uncertain) cover the gills, and death by asphyxia may result. The *primary* cause is unknown, but overcrowding is a predisposing factor.

bacterial kidney disease See KIDNEY DISEASE.

bacterial leaching (of ores) See LEACHING.

bacterial overgrowth syndrome See GASTROINTESTINAL TRACT FLORA.

bacterial vaginosis (non-specific vaginitis) A syndrome characterized by a malodorous vaginal discharge and an increase, in the vagina, in the numbers of certain bacteria – e.g. species of *Bacteroides* and GARDNERELLA (see also MOBILUNCUS); under these conditions the vaginal pH is usually >4.5, the E_h of the vaginal epithelial surface is ca. +71 mV to −257 mV (normal values ca. +322 mV to +137 mV) [JID (1985) *152* 379–382], and 'clue cells' (vaginal epithelial cells coated with small Gram-negative rods) can usually be seen in vaginal smears. In cases of bacterial vaginosis, vaginal secretions typically give a fishy, amine-like odour when treated with 10% KOH. [Diagnostic criteria: Am. J. Med. (1983) *74* 14–22; JCM (1985) *22* 686–687.]

Bacteriastrum See DIATOMS.

bactericidal (bacteriocidal) Able to kill at least some types of bacteria. (cf. BACTERIOSTATIC.)

bactericidal/permeability-increasing protein See BPI PROTEIN.

bactericidin (1) An antibody which, under appropriate conditions, can act as a bactericidal agent. (2) A non-specific bactericidal plasma factor (see e.g. COMPLEMENT FIXATION). (3) An antibacterial protein produced by an invertebrate – see e.g. SARCOTOXINS.

bacterin A VACCINE containing killed bacterial cells.

bacteriochlorophylls See CHLOROPHYLLS.

bacteriocidal *Syn.* BACTERICIDAL.

bacteriocin (bacteriocine) Any of a wide variety of (usually) protein or peptide ANTIBIOTICS, produced by certain strains of

Gram-positive and Gram-negative bacteria, which are bacteriostatic or bactericidal – often specifically to organisms that are closely related to the bacteriocin-producing strain. Bacteriocins include e.g. COLICINS, MICROCINS and LANTIBIOTICS.

Agents analogous to bacteriocins are produced by members of the Archaea; these agents (e.g. the *halocins* produced by halobacteria) are apparently not homologous to any bacteriocin in terms of amino acid sequences.

In size and structure, bacteriocins range from a simple modified amino acid (microcin A15), through short peptides and high-MWt colicins, to phage-like PYOCINS. (See also BACTERIOPHAGE ϕBA1.) Some bacteriocins (lantibiotics) undergo distinctive post-translational modification which is necessary for their activity. The lantibiotic LACTICIN 3147 (produced by *Lactobacillus lactis* subsp *lactis*) is a *two-component* bacteriocin – both peptides being necessary for activity; moreover, each of the two peptides needs modification by a separate enzyme [Microbiology (2000) *146* 2147–2154].

Evolutionary relationships are evident within some groups of bacteriocins. [Molecular mechanisms of bacteriocin evolution: ARG (1998) *32* 255–278.]

Bacteriocins are commonly encoded by PLASMIDS (see e.g. COLICIN PLASMID); bacteriocin-encoding plasmids include both conjugative and non-conjugative types. Examples of *chromosomally* encoded bacteriocins include 'bacteriocin 28b' (a colicin encoded by *Serratia marcescens*) and some of the class IIa bacteriocins produced by lactic acid bacteria (see later).

The mode of regulation of bacteriocin synthesis varies among the different groups. For example, synthesis of colicins is promoted by those conditions which trigger the SOS SYSTEM (see COLICIN PLASMID). By contrast, many bacteriocins of the lactic acid bacteria appear to be regulated by a *three-component system* which includes a histidine protein kinase, a response regulator, and an induction factor. (cf. TWO-COMPONENT REGULATORY SYSTEM). Induction factors (IFs) are small heat-stable peptides produced by the bacteriocinogenic cell. The mechanism by which IFs promote synthesis of bacteriocin is unknown, but it has been suggested that they may trigger synthesis as a result of their gradual intracellular accumulation or that their influence may reflect environmental factors. Synthesis of the bacteriocin *sakacin A* (produced by *Lactobacillus sake*) is reported to be regulated by a three-component system (which includes a 23-amino-acid cationic peptide) in a temperature-sensitive way [Microbiology (2000) *146* 2155–2160]. The bacteriocin *divercin V41*, produced by *Carnobacterium divergens* strain V41, may be regulated by a two-component system [Microbiology (1998) *144* 2837–2844].

The different types of bacteriocin are released in different ways from the cells which synthesize them. For example, release of colicins depends on a *lysis protein* which causes a (non-specific) increase in the permeability of the cell envelope – allowing dispersal of the bacteriocin (but with concomitant adverse effects on the producing cell). By contrast, certain bacteriocins (e.g. ENTEROCIN P) are secreted via a *sec*-dependent pathway, while some class IIa bacteriocins of the lactic acid bacteria are reported to be secreted by ABC TRANSPORTERS.

The import of some types of bacteriocin (e.g. colicins) into sensitive cells depends on the binding of the bacteriocin to a *specific* cell-surface receptor. In most or all cases, uptake of colicins appears to be an energy-dependent process (pmf being needed for transport across the outer membrane, and ATP hydrolysis being involved in transport across the cytoplasmic membrane); the route of translocation through the cell envelope depends on the given colicin (see groups A and B COLICINS). By contrast, studies on pediocin PA-1 (a class IIa bacteriocin produced by *Pediococcus* spp) reported binding to lipsomes in the absence of a protein receptor [AEM (1997) *63* 524–531] – suggesting that a protein receptor may not be an absolute requirement for these bacteriocins; it may be that these (cationic) bacteriocins bind to anionic phospholipid groups in the membrane.

A given bacteriocin acts on a susceptible target cell in a characteristic way. Some bacteriocins (e.g. some colicins, lantibiotics) form pores in the cytoplasmic membrane; some (e.g. microcin B17) inhibit DNA gyrase; some (e.g. LYSOSTAPHIN) disrupt PEPTIDOGLYCAN; and some (e.g. CLOACIN DF13) cleave 16S rRNA. It appears that, in some cases, a single bacteriocin molecule can be lethal for a sensitive cell. (Although bacteriocins are active primarily against bacteria, certain colicins and VIBRIOCINS have been reported to affect some types of eukaryotic cell [JAC (1980) *6* 424–427].)

Various mechanisms provide a cell with immunity to the bacteriocin(s) it encodes. For example, while cells actively secreting colicins are damaged by their lysis proteins, neighbouring cells of the *same* strain, repressed for colicin synthesis, are protected by immunity proteins (see COLICINS). Cells producing LYSOSTAPHIN achieve immunity by modifying their own peptidoglycan. Cells producing the lantibiotic EPIDERMIN achieve immunity by operating an ATP-dependent pump (transport system) which actively secretes any molecules of the agent that are taken up.

Biotechnological applications of bacteriocins. The great diversity of bacteriocins, and the ability of many of them to kill/inhibit certain pathogenic bacteria, has prompted much research into the possibility of using particular bacteriocins as antibiotics and/or as food preservatives/additives (see e.g. LACTICIN 3147).

Particularly useful bacteriocins are produced by certain LACTIC ACID BACTERIA (e.g. species of *Carnobacterium*, *Enterococcus*, *Lactobacillus* and *Lactococcus*), some of which are highly active against important food-borne pathogens (such as *Listeria monocytogenes*). According to structural and other characteristics, bacteriocins *of the lactic acid bacteria* have been classified into four classes [FEMS Reviews (1993) *12* 39–86]:

Class I. LANTIBIOTICS: small, typically pore-forming peptides containing unusual constituents (such as lanthionine) – e.g. EPIDERMIN and NISIN.

Class II. Small peptides (<10 kDa) which lack unusual constituents (such as lanthionine) and which have a specific form of processing site in the precursor molecule; they exhibit stability at e.g. 100–120°C. This category is divided into three subgroups. *Class IIa* bacteriocins contain a specific N-terminal sequence, and all are highly active against *Listeria* spp. They include PEDIOCIN PA-1 and SAKACIN A. [Biosynthesis, structure and activity of class IIa bacteriocins: FEMS Reviews (2000) *24* 85–106.] *Class IIb* bacteriocins are two-component, pore-forming agents. They include e.g. lactacin F and lactococcin G. *Class IIc* bacteriocins are thiol-activated peptides which require reduced cysteine residues for activity.

Class III. Large (>30 kDa) heat-labile proteins; they include helveticin J, lactacin A and lactacin B.

Class IV. Bacteriocins which consist of a protein moiety plus at least one non-protein (e.g. lipid) constituent which is necessary for activity. They include lactocin 27, leuconocin S and plantaricin S.

Other bacteriocins. See also PESTICIN I, PYOCINS, STAPHYLOCOCCIN and ULCERACIN 378. Bacteriocins are also produced by *Bacillus megaterium* (megacins), *Klebsiella* spp (klebicins), *Listeria monocytogenes* (monocins [Zbl. Bakt. Hyg. A (1986) *261*

12–28]) and by *Clostridium botulinum* (boticins), *C. butyricum* (butyricins) and *C. perfringens* (perfringocins). (cf. AGROCINS and KILLER FACTOR.)

bacteriocin 28b See COLICINS.

bacteriocin typing A form of TYPING in which strains of bacteria are distinguished on the basis of the BACTERIOCIN(s) they produce or the bacteriocin(s) to which they are susceptible.

In one common form of bacteriocin typing (used e.g. for colicins and pyocins) the strain under test is inoculated in a diametrical strip on a blood agar plate which is incubated for ca. 12–24 hours; growth is scraped from the plate and discarded, and any cells remaining on the plate are killed by exposure to chloroform. The chloroform is allowed to evaporate, leaving a band of bacteriocin-impregnated agar. Indicator strains are then inoculated onto the agar in lines perpendicular to, and passing through, the bacteriocin-containing band; on incubation, the growth of sensitive strains is inhibited inside the band. In this procedure the test organism is typed by the range of strains susceptible to its bacteriocin(s). [Revised method for pyocin typing: JCM (1984) *20* 47–50.]

bacteriocinogenic Able to produce a BACTERIOCIN.

bacteriocinogenic factor The earlier name for any PLASMID that encodes a BACTERIOCIN.

bacteriocuprein A bacterial CuZnSOD (see SUPEROXIDE DISMUTASE).

bacteriocyte In certain invertebrates, particularly insects: a specialized cell which contains intracellular bacterial symbionts. (cf. MYCETOCYTE.) Bacteriocytes occur e.g. in the cockroach (see BLATTABACTERIUM), in some marine sponges (containing '*Aphanocapsa*' endosymbionts) and in aphids (see BUCHNERA).

bacteriolysis The lysis (rupture) of bacterial cells. In nature, various microorganisms produce enzymes or antibiotics which lyse bacteria – either to provide nutrients or (presumably) for competitive advantage. (See e.g. ANAEROPLASMA, ENSIFER, LYSOSTAPHIN, MUSHROOM CULTIVATION, MYXOBACTERALES.) Most BACTERIOPHAGES eventually lyse their host cells to release their progeny. In the laboratory bacteria may be lysed mechanically or enzymically (see CELL DISRUPTION).

bacterio-opsin See OPSIN.

bacteriophaeophytin A PHAEOPHYTIN derivative of a bacteriochlorophyll. Bacteriophaeophytins occur e.g. in the REACTION CENTRES in a number of 'purple' photosynthetic bacteria and in the cytoplasmic membrane of some 'green' photosynthetic bacteria.

bacteriophage (phage) Any VIRUS whose host is a bacterium. (Viruses which infect cyanobacteria are conventionally called CYANOPHAGES.) Most – probably all – bacteria can be infected by particular phages; commonly, a given phage can infect only one or a few strains or species of bacteria. The consequences of phage infection depend on phage and host, and to some extent on

BACTERIOPHAGES: some representative examples[a]

Bacteriophage (or phage group)	Virion morphology	Principal host(s)
Genome: linear dsDNA[b]		
λ	isometric head + long non-contractile tail	*Escherichia coli*
Mu	isometric head + long contractile tail	enterobacteria
MV-L3	isometric head + short tail	*Acholeplasma laidlawii*
N4	isometric head + short non-contractile tail	*Escherichia coli*
P1	isometric head + long contractile tail	*Escherichia coli*
P2	isometric head + long contractile tail	*Escherichia coli*
P22	isometric head + minimal tail	*Salmonella*
φ29	elongated head + short non-contractile tail	*Bacillus subtilis*
SPO1	isometric head + long contractile tail	*Bacillus subtilis*
T4	elongated head + long contractile tail	*Escherichia coli*
T7	isometric head + short non-contractile tail	enterobacteria
Tectiviridae (e.g. PRD1, AP50)	isometric with internal lipid membrane (no tail)	various
Genome: ccc dsDNA[b]		
MV-L2	pleomorphic; envelope enclosing a DNA–protein complex	*Acholeplasma laidlawii*
PM2	icosahedral with internal lipid membrane	*Alteromonas espejiana*
Genome: ccc ssDNA[b]		
Inoviridae (e.g. f1, MV-L51)	filamentous or rod-shaped	various
Microviridae (e.g. φX174)	icosahedral	enterobacteria
Genome: dsRNA[b]		
φ6	enveloped nucleocapsid	*Pseudomonas syringae* pv. *phaseolicola*
Genome: ssRNA[b]		
Leviviridae (e.g. MS2, Qβ)	icosahedral	various

[a] See separate entries for details.
[b] In the virion.

ccc = covalently closed circular; ds = double-stranded; ss = single-stranded.

conditions. Some (*virulent*) phages always induce a LYTIC CYCLE in the host cell, while other (*temperate*) phages can establish a stable, non-lytic relationship (LYSOGENY) with the host. Some phages can replicate and produce progeny virions within the host cell without killing or lysing it (see INOVIRIDAE and INOVIRUS). Phages of the MV-L3 PHAGE GROUP apparently kill their host cells without actually lysing them.

The antibacterial activity of phages has been exploited (primarily in Eastern Europe and the former Soviet Union) for both therapeutic and prophylactic use against various bacterial pathogens (including strains of *Escherichia*, *Klebsiella*, *Proteus*, *Pseudomonas*, *Salmonella*, *Shigella*, *Staphylococcus* and *Streptococcus*) [AAC (2001) *45* 649–659].

Bacteriophages are a highly diverse group of viruses. Of the many hundreds known, relatively few have been thoroughly characterized [guidelines for phage characterization: AVR (1978) *23* 1–24]. Some phages have been included in the overall taxonomic scheme for the viruses [Book ref. 23], and these are classified in the families CORTICOVIRIDAE, CYSTOVIRIDAE, INOVIRIDAE, LEVIVIRIDAE, MICROVIRIDAE, MYOVIRIDAE, PLASMAVIRIDAE, PODOVIRIDAE, STYLOVIRIDAE and TECTIVIRIDAE; however, many phages remain unclassified.

Morphologically, phage virions may be small and icosahedral, ca. 23–32 nm diam. (e.g. Qβ and ϕX174); filamentous, ca. 760–1950 × 6 nm (INOVIRUS); pleomorphic (e.g. BACTERIOPHAGE MV-L2); or – in many phages – complex in structure with a polyhedral 'head' (the CAPSID) linked (often via a more or less complex connecting structure) to a long or short, simple or complex, contractile or non-contractile 'tail' (see e.g. PODOVIRIDAE, STYLOVIRIDAE, T-EVEN PHAGES). While in many phages the virions consist only of protein and the nucleic acid genome, some have a significant content of lipid – either as an external ENVELOPE (e.g. in ϕ6 and MV-L2) or as an internal layer (in BACTERIOPHAGE PM2 and phages of the TECTIVIRIDAE); the T-even phages contain e.g. dihydropteroyl hexaglutamate, phages of the MV-L3 group contain fucose, and MS2 contains spermidine.

The genome of a phage may be linear dsDNA, ccc dsDNA, ccc ssDNA, linear dsRNA or linear ssRNA (see Table). In some of the DNA phages the DNA contains unusual bases: e.g. PBS1 contains deoxyuracil instead of thymine, ϕW-14 contains α-putrescinylthymine, T-even phages contain hydroxymethylcytosine instead of cytosine, SP8 and SPO1 contain 5-hydroxymethyluracil instead of thymine. (See also BACTERIOPHAGE MU and RESTRICTION–MODIFICATION SYSTEM.)

The phage replication cycle begins when the virion adsorbs to the host at specific cell-surface sites: e.g., particular components of the cell wall (see e.g. BTUB PROTEIN and LAMB PROTEIN), the flagellum (see FLAGELLOTROPIC PHAGE), a sex pilus (see ANDROPHAGES), etc. Some phages adsorb only in the presence of appropriate concentrations of certain ions or of certain organic 'adsorption cofactors' (e.g. tryptophan for BACTERIOPHAGE T4). In Gram-negative bacteria penetration is believed to occur – at least in some phages – at ADHESION SITES. Either the entire virion, or the phage genome with or without certain phage proteins, enters the host cell: see e.g. INOVIRUS, LEVIVIRIDAE, and entries for bacteriophages N4, ϕ6 and T4 for examples. For at least some of the phages with contractile tails (see MYOVIRIDAE) – e.g. BACTERIOPHAGE T4 (q.v.) – the genome is understood to be injected (by a syringe-like action) into the periplasmic region of the bacterium. However, many phages have non-contractile tails, and some are entirely tail-less; hitherto, little information has been available on the way in which the genome of such phages is translocated into the host cell. Studies on phage T7 have indicated that, on attaching to the host cell, the phage ejects proteins that may assemble to form a channel across the bacterial cell envelope through which the genome is translocated; it has been speculated that two of the ejected proteins may form the components of a 'motor' that rachets phage DNA into the host cell [Mol. Microbiol. (2001) *40* 1–8].

For phage development to proceed, the phage nucleic acid must escape degradation by the host's RESTRICTION ENDONUCLEASE system. For a general account of viral replication strategies see VIRUS; for detailed accounts of particular phages see following entries. Virion assembly may occur by the spontaneous aggregation of the various phage components, but the more complex virions require the participation of non-structural phage-coded proteins (see SCAFFOLDING PROTEIN). Host cell lysis may be induced by phage-coded enzymes or by activation of host cell autolysins. (See also LYSIS FROM WITHOUT, LYSIS PROTEIN and ONE-STEP GROWTH EXPERIMENT.)

Bacteriophages can cause considerable economic problems in certain biotechnological processes such as the manufacture of dairy products, antibiotics, etc. A few phages are medically important in that they encode certain toxins (see BACTERIOPHAGE CONVERSION). In microbiology, phages are useful e.g. for PHAGE TYPING, for achieving gene transfer between bacteria (see TRANSDUCTION), for probing the molecular biology of bacteria, etc.

Phages may be cultivated and/or assayed e.g. by inoculation at low multiplicity of infection on a LAWN PLATE of susceptible bacteria (see PLAQUE and PLAQUE ASSAY). They may also be cultivated in broth cultures of host bacteria, the activity of virulent phages being indicated by the progressive decrease in turbidity of the culture as the cells lyse; a suspension of phages may be prepared from the lysate e.g. by CENTRIFUGATION or by membrane FILTRATION to remove bacterial debris.

bacteriophage 7-7-1 See FLAGELLOTROPIC PHAGE.

bacteriophage 21 See LAMBDOID PHAGES.

bacteriophage 82 See LAMBDOID PHAGES.

bacteriophage 434 See LAMBDOID PHAGES.

bacteriophage 1307 See PLASMAVIRIDAE.

bacteriophage α3 See MICROVIRIDAE and BACTERIOPHAGE G4. (Also: a dsDNA-containing phage of '*Achromobacter* sp 2' [JGV (1981) *53* 275–281].)

bacteriophage α15 See LEVIVIRIDAE.

bacteriophage AP50 See TECTIVIRIDAE.

bacteriophage Bam35 See TECTIVIRIDAE.

bacteriophage β See LEVIVIRIDAE. (cf. Corynephage β – see DIPHTHERIA TOXIN.)

bacteriophage BF23 A bacteriophage which infects *Escherichia coli* and which resembles phage T5. (See also BTUB PROTEIN.)

bacteriophage BPB1 See STYLOVIRIDAE.

bacteriophage Cf See INOVIRUS.

bacteriophage χ See STYLOVIRIDAE.

bacteriophage conversion (phage conversion) In a bacterium: the acquisition, loss or modification of one or more phenotypic characteristics as a result of infection by a BACTERIOPHAGE – typically a temperate phage (*lysogenic conversion*); it may result e.g. from the expression of phage genes or from inactivation of one or more bacterial genes due to insertion of a prophage into the bacterial chromosome. (Phenotypic alteration in a recipient cell due to TRANSDUCTION of *bacterial* DNA is not regarded as phage conversion.)

Phage conversion is responsible for toxigenicity in a number of pathogenic bacteria. For example, phages encode CHOLERA TOXIN, DIPHTHERIA TOXIN and the shiga-like toxins of enterohaemorrhagic strains of *Escherichia coli* (including O157:H7) (see SHIGA TOXIN). The enterotoxin A of *Staphylococcus aureus* (see ENTEROTOXIN) is phage-encoded in at least some strains. (See also BETACIN.)

In *Staphylococcus aureus*, lipase activity is lost on infection with phage L54a owing to inactivation of the lipase structural gene (into which the prophage inserts) [JB (1986) *166* 385–391].

Lysogenic conversion can also be manifested by a change in cell-surface antigens, i.e. a change in serotype. For example, in group E salmonellae, phage ε_{15} converts antigens 3,10 to 3,<u>15</u>, and the LPS of the latter serotype can act as a receptor for phage ε_{34} which, in turn, converts the antigen type to 3,<u>15</u>,<u>34</u> (for the meaning of underlined numbers see KAUFFMANN–WHITE CLASSIFICATION). Phage-mediated modification of serotype is important e.g. when developing vaccine strains of a pathogen [see e.g. TIM (2000) *8* 17–23].

(See also PSEUDOLYSOGENY.)

bacteriophage CTXΦ A temperate, filamentous ssDNA BACTERIOPHAGE which infects *Vibrio cholerae* and encodes CHOLERA TOXIN (as well as the Ace and Zot toxins). The double-stranded form of the genome can integrate into the bacterial chromosome in a site-specific, RecA-independent process. Isolates of lysogenic El Tor and O139 strains of *V. cholerae* typically contain tandem arrays of prophage DNA at a single locus within the large chromosome, and it appears that the presence of multiple copies of the prophage is necessary for the production of virions; it also seems that the genomes of progeny virions are formed by a process in which replicative forms of the phage develop through hybridization between adjacent prophages (or between a prophage and a (phage-related) RS1 element in the chromosome) [PNAS (2000) *97* 8572–8577]. The cell-surface receptor for CTXΦ occurs on so-called *toxin co-regulated pili* (TCP) (encoded by a chromosomal PATHOGENICITY ISLAND); genes encoding TCP and the cholera toxin are jointly, and positively, regulated by transcriptional regulator proteins (ToxR, ToxS, ToxT) which apparently become active on receipt of appropriate signals within the gut. Lysogenic conversion of (non-lysogenic) cells of *V. cholerae* can occur in the mammalian gut in the presence of phage-donating strains; this apparently reflects infection by phage following induction of TCP receptors by the gut-derived signals. [Science (1996) *272* 1910–1914; Cell (1996) *87* 795–798; Mol. Microbiol. (1997) *24* 917–926; TIM (1998) *6* 295–297.]

Infection of *V. cholerae* by CTXΦ is dependent on the *tolQRA* gene products [JB (2000) *182* 1739–1747].

In the classical biotype of *V. cholerae*, the prophages are present either singly or as two fused prophages, and integration occurs at two sites; this biotype produces cholera toxin but, apparently owing to deficiencies in the structure of arrays (and the absence of RS1), does not produce virions [JB (2000) *182* 6992–6998].

bacteriophage D108 A mutator bacteriophage which is very similar to BACTERIOPHAGE MU; the phage D108 genome shows ca. 90% homology with that of phage Mu, differing mainly in the early-gene region at the left end [EMBO (1985) *4* 3031–3037].

bacteriophages ε_{15}, ε_{34} See BACTERIOPHAGE CONVERSION.

bacteriophage f1 See INOVIRUS.

bacteriophage f2 See LEVIVIRIDAE.

bacteriophage F116 A temperate, transducing phage (family STYLOVIRIDAE) which infects *Pseudomonas aeruginosa*; in lysogenic cells the prophage appears to have an extrachromosomal location [JV (1977) *22* 844–847].

bacteriophage fd See INOVIRUS.

bacteriophage fr See LEVIVIRIDAE.

bacteriophage G4 An isometric SSDNA PHAGE of the MICROVIRIDAE. On penetrating the host cell, the ss cccDNA genome is coated with host SSB protein; a GC-rich region between genes *F* and *G* probably remains uncoated. Stage I (see SSDNA PHAGE) is independent of host *dnaB* function (as it is in microviruses St-1, α3, φK and φXtB – cf. BACTERIOPHAGE φX174). The *F–G* intergenic region seems to fold into a secondary structure [JV (1986) *58* 450–458] which is capable of direct (*dnaB*-independent) recognition by host PRIMASE (cf. PRIMOSOME and DNAB GENE). Primase synthesizes a short RNA primer at a unique site (*ori*) in the intergenic region. The *c* strand is completed as in other SSDNA PHAGES. Stage II in G4 requires *dnaB* function, presumably for initiation of *v* strand synthesis. (Phages St-1, φK and α3 do not require *dnaB* at any stage.) G4 *v* strand synthesis begins at the *v* strand origin (O_v) in gene *A*; the *c* strand origin (O_c) in the *F–G* intergenic region is on the opposite side of the molecule. *v* strand synthesis seems to proceed by the displacement of the old *v* strand as a closed loop (D LOOP) at the O_v site, and hence is independent of a gp*A*-induced nick (cf. BACTERIOPHAGE φX174). Synthesis of *v* strand by DNA polymerase III holoenzyme and Rep protein proceeds unidirectionally. When the replication fork passes O_c, primer synthesis can begin on the displaced loop, and *c* strand synthesis can proceed in the opposite direction. The parental *v* and *c* circles may separate before replication of either is complete. G4 seems to resemble φX174 (q.v.) in stage III and morphogenesis.

bacteriophage G6 See MICROVIRIDAE and BACTERIOPHAGE φX174.

bacteriophage G13 See MICROVIRIDAE and BACTERIOPHAGE φX174.

bacteriophage G14 See MICROVIRIDAE and BACTERIOPHAGE φX174.

bacteriophage HM3 See MYOVIRIDAE.

bacteriophage I3 See MYCOBACTERIOPHAGES.

bacteriophage If1 An INOVIRUS which adsorbs specifically to I-type pili of enterobacterial hosts; phage If2 is very similar.

bacteriophage If2 See BACTERIOPHAGE IF1.

bacteriophage IKe An INOVIRUS specific for enterobacteria which contain an IncN plasmid; the phage can apparently establish a pseudolysogenic infection of its host [CJM (1978) *24* 1595–1601]. [Nucleotide sequence and genetic organization of IKe genome: JMB (1985) *181* 27–39.]

bacteriophage L1 group See PLECTROVIRUS.

bacteriophage L2 group See PLASMAVIRIDAE.

bacteriophage L3 group See MV-L3 PHAGE GROUP.

bacteriophage L4 A BACTERIOPHAGE closely related to BACTERIOPHAGE P22 but defective in maintenance of lysogeny; it is commonly used as a transduction vector in *Salmonella typhimurium*.

bacteriophage L17 See TECTIVIRIDAE.

bacteriophage L34a See BACTERIOPHAGE CONVERSION.

bacteriophage λ A temperate BACTERIOPHAGE (family STYLOVIRIDAE) which infects *Escherichia coli*. Head: icosahedral, ca. 55 nm diam., composed of two major structural proteins (gp*E* and gp*D*) and several minor proteins (gp*B*, gp*C*, gp*FII*). Tail: long (ca. 150 × 10 nm), flexible, tubular, non-contractile, joined to the head via a 'neck' or 'connector' region, and terminating at the distal end in a small basal structure to which is attached a single fibre (the adsorption organelle). (See also LAMB PROTEIN.) Genome: linear dsDNA (48514 bp long [nucleotide sequence: JMB (1982) *162* 729–773]) with single-stranded 12-base 5′ STICKY ENDS (designated *cos*). On infection of a host cell, phage DNA is transferred to the host through the tail and is immediately circularized by base-pairing between the sticky ends followed by host DNA ligase action. Subsequent events may eventually lead either to host cell lysis (involving the production and release of

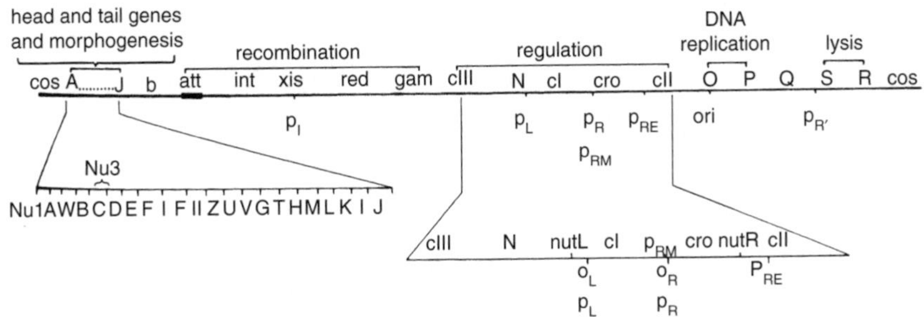

BACTERIOPHAGE λ. Simplified map of the virion genome (*not* drawn to scale). See entry for explanation.

progeny phages) or to LYSOGENY (during which λ DNA becomes integrated into the host chromosome); the initial sequence of events is common to both pathways.

Lytic cycle. Transcription of λ DNA by the host RNA polymerase proceeds leftwards from promoter p_L and rightwards from promoter p_R, stopping at rho-dependent terminators shortly beyond genes *N* and *cro* ('immediate early' genes), respectively (see figure). gp*N* functions as an antiterminator which is specific for transcription initiated at the early promoters p_L and p_R; gp*N* recognizes special sites (*nutL* and *nutR*) – regions of hyphenated dyad symmetry – downstream from p_L and p_R, and when the host RNA polymerase traverses these sites it is modified by gp*N* (in the presence of the host *nusA* gene product) such that it can read through subsequent terminators. As a result, genes to the left of *N* (*cIII . . . att*) and to the right of *cro* (*cII . . . Q*) – delayed early genes – are expressed. Expression of *cro* is necessary for the continuation of the lytic cycle (see later). gp*N* is unstable (half-life ca. 5 min), so that continued *N* expression is necessary for ongoing transcription of the delayed early genes.

λ DNA REPLICATION is initiated at a site (*ori*) in the *O* gene and requires gp*O*, gp*P*, and certain host proteins, including DnaB protein, primase (DnaG protein), RNA polymerase, and components of the DNA polymerase III holoenzyme (but not e.g. *dnaA*, *dnaC* or *dnaI* functions). λ gp*O* binds both to a 19-bp tandem repeat in the *ori* region and to λ gp*P*; gp*P* interacts with DnaB protein and appears to act as an analogue of *E. coli* DnaC protein (see PRIMOSOME). Initially, the λ cccDNA is replicated by the CAIRNS MECHANISM, but later replication proceeds by a ROLLING CIRCLE MECHANISM to yield double-stranded concatemers. Phage structural components (and e.g. lysozyme necessary for host cell lysis) are encoded by late phage genes (*S, R, A . . . J* – juxtaposed by circularization of the genome); transcription of these genes is turned on by the delayed early gene product gp*Q*, which functions as an antiterminator of late mRNA initiated at $p_{R'}$. Phage morphogenesis is complex, involving host proteins as well as many phage-coded functions. In essence, gp*B* and gp*C* appear to form a complex with gp*Nu3*, and gp*E* is then incorporated; these components undergo modification (fusion/cleavage reactions), and the scaffolding protein gp*Nu3* is cleaved during or after its elimination from the head precursor. Concatemeric λ DNA is cut (by a gp*Nu1*–gp*A* complex) at a *cos* site; a left *cos* site is inserted into the capsid precursor, and DNA insertion continues until the next *cos* site is reached, when the DNA is cleaved. (cf. COSMID.) The resulting head structure is stabilized by the addition of gp*D*, gp*W* and gp*FII*. The tail is assembled separately and interacts spontaneously with the completed head to form the mature infectious virion.

Lysogeny and the lysis/lysogeny decision. During the early period of phage transcription, the system is committed neither to lysis nor to lysogeny, expression of immediate early and delayed early genes being necessary for both pathways. Expression of late genes results in lysis, while establishment of lysogeny requires the repression of most of the λ genes by a repressor protein (gp*cI*); gp*cI* prevents both leftward and rightward transcription by binding to operator regions (o_L and o_R) that overlap p_L and p_R, respectively. When a host cell is first infected, *cI* cannot be expressed. However, transcription of *N* allows transcription of *cII* and *cIII*, and both gp*cII* and gp*cIII* are necessary for repressor synthesis. gp*cII* functions as a positive regulator for the initiation of repressor synthesis from $p_{RE}(= p_E)$, being necessary for recognition of this promoter by RNA polymerase. gp*cII* is unstable in vivo owing to proteolysis by host proteins (products of the *hflA* and *hflB* genes being involved); this proteolysis is apparently inhibited by gp*cIII* and by cAMP–CAP. The establishment of lysogeny is favoured by certain environmental conditions, e.g., by starvation of cells prior to infection, and by a high multiplicity of infection (MOI). It has been suggested that starvation may increase cAMP–CAP levels, while high MOI increases levels of gp*cIII* (a gene dosage effect); in either case the result is stabilization of gp*cII* thus favouring lysogeny.

Once synthesized, the repressor maintains its own synthesis by activating *cI* expression from an alternative promoter, p_{RM} $(= p_M)$; transcription from p_{RM} can occur only when repressor is bound to o_R. Translation of the p_{RM}-initiated mRNA is much less efficient than that of the p_{RE}-initiated mRNA. Thus, synthesis of repressor occurs at high levels by the p_{RE}-dependent 'establishment circuit' and at lower levels by the p_{RM}-dependent 'maintenance circuit'.

The repressor is antagonized by the product of the immediate early gene *cro*: gp*cro* ('antirepressor') prevents transcription of *cI* from p_{RM}, and reduces (but does not eliminate) transcription of early genes from p_L and p_R [*cro*-operator and repressor–operator interaction: ARB (1984) *53* 293–321]; thus gp*cro* prevents 'maintenance synthesis' of repressor and also (by reducing gp*cII* formation) 'establishment synthesis' from p_{RE}. Sufficient early gene transcription persists to allow expression of *Q* and hence of the late genes, so that the lytic cycle can be completed. The lysis/lysogeny decision thus depends on whether gp*cI* or gp*cro* occupies the operators o_R and o_L; this in turn depends largely on the level of gp*cII* (which also affects integration – see later) and hence is subject to influence by the host cell e.g. in response to environmental conditions.

In *Escherichia coli* lysogeny involves integration of the phage cccDNA with the host chromosome. This occurs by SITE-SPECIFIC RECOMBINATION, i.e., by reciprocal recombination between specific 'attachment sites' – one on the phage genome (*attP*) and one on the host genome (*attB*, also called att^{λ}, situated between genes *gal* and *bio*); each *att* sequence consists of a central 'core' sequence (O) which is common to both *attB* and *attP*, flanked by 'arm' sequences designated B and B′ for *attB*, P and P′ for *attP*. Integration requires the *int* gene product (integrase) and certain host proteins, including DNA gyrase and INTEGRATION HOST FACTOR, IHF. (IHF consists of two subunits, α and β: products of genes *himA* and *himD* (= *hip*), respectively.) Integrase recognizes *attB* and *attP*, while IHF recognizes only *attP*. Interaction between these recombination proteins, bound to their *att* sites, brings about recombination by the formation of a staggered break in the O region followed by strand exchange and ligation. The integrated prophage is thus flanked by hybrid *att* regions, *attL* (BOP′) and *attR* (POB′). The integration reaction is reversible, but while integration involves recognition between *attP* and *attB*, excision – the reverse reaction – requires recognition between *attL* and *attR*; excision requires an additional phage protein, the *xis* gene product (excisionase), which can bind to a specific region in the P arm of *attP* and *attR* [PNAS (1985) *82* 997–1001]. The regulation of the reaction such that integration occurs preferentially under conditions which favour lysogeny, while excision occurs only on induction of a lysogen, is achieved by control of the amounts of integrase and excisionase available.

The *int* gene can be transcribed from either of two promoters: p_L and p_I; however, integrase is produced only during establishment of lysogeny and not during the lytic cycle. Under conditions which favour lysogeny, gp*cII* not only activates *cI* expression (hence preventing transcription from p_L), it also activates *int* expression from p_I. (Since p_I overlaps *xis*, transcription from this promoter allows expression of *int* without that of *xis*, thus favouring integration.) However, under lytic conditions – during which transcription proceeds from p_L – *int* gene expression is prevented by RETROREGULATION: gp*N*-modified, p_L-initiated transcription extends beyond *int*, through *att*, and into the *b* region; the *b* region contains a sequence (*sib*) which causes the 3′ end of the corresponding mRNA to adopt a secondary structure which is recognized and cleaved by RNase III (q.v.). This cleavage is followed by degradation of the transcript in the 3′-to-5′ direction, thus preventing *int* expression. (Under conditions which favour lysogeny, the p_I-initiated transcript, which terminates before the *sib* region, is a poor substrate for RNase III, and hence *int* expression can occur.)

Induction of λ may occur spontaneously (ca. 1 per 10^2–10^5 lysogenic cells per generation); however, high levels of induction may be brought about by agents which damage DNA (e.g., UV irradiation, mitomycin C). Damage to DNA results in the 'activation' of the host RecA protein (see RECA PROTEIN), resulting in cleavage and inactivation of the repressor. (Levels of DNA damage necessary for λ induction are higher than those required to induce the SOS SYSTEM; presumably gp*cI* becomes susceptible to RecA-mediated cleavage only when host cell viability is threatened.) Since the repressor functions as a positive regulator for its own synthesis, its inactivation also prevents the synthesis of replacement repressor molecules. Inactivation of the repressor allows transcription from p_L and p_R, followed by excision of the prophage and completion of the events of the lytic cycle. In contrast to p_L-initiated transcription in the lytic cycle, p_L-initiated transcription in the prophage results in the expression of both *int* and *xis* (necessary for excision). This is possible because, owing to the permuted gene order in the prophage, the *sib*-containing *b* region is no longer immediately downstream from *int*; the p_L-initiated transcript is thus not recognized by RNase III, and both *int* and *xis* can be expressed.

[General review: Book ref. 79; λ–host interactions: MR (1984) *48* 299–325.]

bacteriophage M12 See LEVIVIRIDAE.

bacteriophage M13 See INOVIRUS.

bacteriophage MS2 See LEVIVIRIDAE.

bacteriophage Mu A temperate bacteriophage which can infect various enterobacteria (e.g. *Escherichia coli*, *Citrobacter freundii*, *Erwinia* spp, and certain mutant strains of *Salmonella typhimurium*). The phage particle has an icosahedral head (ca. 60 nm diam.) and a contractile tail (ca. 100 nm long when extended, 60 nm when contracted) to which is attached a base plate with spikes and tail fibres; the phage tail adsorbs to LPS in the outer membrane of a host cell. Phage particles contain linear dsDNA ca. 39 kb long.

An early event after phage infection – regardless of whether infection will eventually result in a lytic cycle or the lysogenic state – is the integration of Mu DNA into the host chromosome at a more or less randomly selected location. Integration often results in the inactivation of the gene in which it occurs; ca. 2–3% of bacteria in a population acquire a recognizable mutation on lysogenization by Mu, a frequency much greater than the spontaneous mutation rate in the absence of Mu – i.e., Mu acts as a mutator phage (Mu = abbr. for mutator). The initial integration event may be conservative (i.e., both strands of the Mu DNA may be inserted) and appears to be a simple (non-replicative) insertion; it requires the product of Mu gene *A* which can recognize and bind to the ends of Mu DNA [Cell (1984) *39* 387–394]. Synthesis of a repressor (the *c* gene product) results in lysogeny and immunity to superinfection by Mu.

During the lytic cycle, Mu DNA replication occurs entirely by repeated replicative transposition of the integrated DNA (which thus functions as a giant TRANSPOSON, causing chromosomal deletions, inversions, duplications etc); Mu DNA insertion results in a 5-bp duplication of the target DNA (cf. Tn*3*). Transposition occurs at very high frequency (ca. 100 transpositions per cell in ca. 30 min), copies of Mu DNA being inserted at random locations in the chromosome; this results in the death of the host cell, and ca. 100 progeny phages are released ca. 60 min after infection. Transposition is controlled by the products of phage genes *A* and *B* and requires continual synthesis of gp*A* – which apparently acts stoichiometrically; gp*A* functions as a transposase, gp*B* apparently functions in the modification of gp*A* activity. [Role of DNA topology in Mu transposition: Cell (1986) *45* 793–800.]

Phage Mu DNA contains three regions, designated α (ca. 33 kb), G (ca. 3 kb) and β (ca. 1.7 kb), flanked by random sequences of bacterial DNA (see below). The α region contains most of the phage genes, including e.g. the repressor (*c*) gene, all the early functions, all the head functions, and most of the tail functions. The G region carries the remainder of the tail functions and specifies the host range of the phage. The G segment is flanked by short inverted repeats, and can invert by reciprocal recombination between these sequences; when in one orientation, designated G(+), it specifies a host range which includes *E. coli* K12 strains, but in the alternative

G(−) orientation it specifies different tail fibres specific for a different host range. [Host cell receptors for Mu G(+) and G(−) types: FEMS (1985) *28* 307–310.] (See also G LOOP and BACTERIOPHAGE P1.) A gene designated *gin*, in the β region adjacent to the G region, is essential for G inversion; *gin* encodes a site-specific recombination enzyme which is stimulated by a small, heat-stable host protein [BBA (1986) *866* 170–177] and which apparently recognizes the inverted repeats flanking the G segment. In the lysogenic state, the G segment undergoes inversion at a slow but steady rate due to a low level of *gin* expression. The level of *gin* expression is thought to be about the same during the lytic cycle, but G inversion rarely has time to occur before the lytic cycle is complete; hence G loops are rarely observed in denatured/renatured Mu DNA in lysates from lytic infections. (See also RECOMBINATIONAL REGULATION.)

In addition to *gin*, the β region also contains a gene (*mom*) which encodes a DNA modification function. The targets for this modification system are adenine residues in pentanucleotide sequences 5′... (C/G)A(C/G)NPy ... 3′; ca. 15% of the adenine residues in the Mu DNA are modified, the modified bases being N^6-(1-acetamido)-2-deoxyadenosine. The modified DNA is resistant not only to the host restriction system in vivo but also to in vitro cleavage by a range of restriction endonucleases. The *mom* gene is apparently repressed in the prophage but is strongly expressed on induction; the host Dam methylase (product of the DAM GENE) is necessary for *mom* gene expression [mechanism: EMBO (1986) *5* 2719–2728].

During phage assembly, Mu DNA is packaged by a 'headful' mechanism: packaging begins near the 'left' (*c*) end of the Mu DNA and continues until ca. 39 kb of DNA have been incorporated into the phage head. The Mu DNA itself is ca. 37.5 kb long; thus, the packaging mechanism results in a variable length of bacterial DNA at each end of the phage genome (on average ca. 50–100 bp at the *c* end, ca. 1700 bp at the 'right' or S end). These end sequences are random because of the random insertion of Mu DNA into the host chromosome; they appear as 'split ends' in phage DNA which has been denatured and reannealed. The 'split ends' are lost when the phage DNA integrates into a new host chromosome. (See also MINI-MU.)

Mu can function as a generalized transducing phage, mediating gene transfer at frequencies of ca. 10^{-7}. Transduction by Mu can be detected only in rec^+ recipient cells. The transducing particles, which contain only host DNA, probably arise by the rare packaging into Mu heads of host DNA instead of phage DNA.

[Review: Book ref. 20, pp. 105–158; behaviour of Mu in *Salmonella typhi*: JGM (1986) *132* 83–89.]

bacteriophage μ2 See LEVIVIRIDAE.

bacteriophage MV-L1 See PLECTROVIRUS.

bacteriophage MV-L2 (MVL2, L2 strain L2) A species of the PLASMAVIRIDAE. Host: *Acholeplasma laidlawii*; plaques small, turbid. Genome: negatively supercoiled ds cccDNA (MWt ca. 7.8 × 10^6). Virion: roughly spherical, somewhat pleomorphic, (50–)80(–125) nm in diameter, sensitive to detergents, organic solvents, and heat (e.g. 60°C for 5 min). There is apparently no rigid helical or icosahedral nucleocapsid. The virion seems to consist of DNA condensed with protein and enclosed within a flexible, lipid-containing, 'unit-type' membrane apparently derived from the host cell membrane. Infection leads to lysogeny, the phage DNA becoming integrated with the host's chromosome. Progeny virions are released by budding through the host cell membrane; lysis does not occur, and the host remains viable.

bacteriophage MV-L3 See MV-L3 PHAGE GROUP.

bacteriophage MV-L51 See PLECTROVIRUS.

bacteriophage 06N 58P See CORTICOVIRIDAE.

bacteriophage N4 A virulent coliphage of the PODOVIRIDAE. Genome: linear dsDNA (ca. 71 kb) containing terminal direct repeats 400–450 bp long and a 7-base 3′ overhang at one end. The phage particle contains a rifampicin-resistant DNA-dependent RNA polymerase which is injected into the host cell with the DNA and which is necessary for transcription of N4 early genes. Transcription of intermediate (middle) genes requires the synthesis and activity of three N4 early proteins, two of which are components of a second rifampicin-resistant RNA polymerase; transcription of late genes requires *E. coli* RNA polymerase activity. Host DNA replication is blocked by an N4 early gene product, but host RNA and protein synthesis continues (except in the case of cAMP-dependent operons). The N4 genome codes for most functions required for its replication, including a DNA polymerase and DNA-binding protein; the only host functions it appears to require are gyrase, DNA polymerase 5′ → 3′ exonuclease activity, DNA ligase, and *nrdA* gene function [Book ref. 69, pp. 245–254].

bacteriophage P1 A temperate BACTERIOPHAGE (family MYOVIRIDAE) which infects *Escherichia coli* and *Shigella* spp. Virion: icosahedral head (ca. 90 nm diam.) with a long, complex contractile tail bearing a base-plate and tail fibres. Genome: dsDNA (MWt ca. 66 × 10^6), circularly permuted and terminally redundant. On infection of a host cell, the genome circularizes (with loss of redundancy) and may enter a LYTIC CYCLE or establish LYSOGENY. Relatively little is known about the lytic cycle; both θ- and σ-type modes of DNA replication occur early in infection, but later only σ-type replication can be observed. DNA packaging occurs by a 'headful' mechanism, starting at a unique site (*pac*); cleavage of DNA at a *pac* site apparently occurs early in infection, before phage heads are formed, and the enzyme responsible may remain bound to the DNA and subsequently interact with a prohead to initiate packaging. (cf. BACTERIOPHAGE P22.)

In the lysogenic state, the P1 prophage does not normally integrate into the host chromosome, but persists as an autonomous, circular PLASMID (P1 plasmid) which is maintained at one or two copies per host chromosome. The replication control system of the P1 plasmid resembles that of the F PLASMID (q.v.) e.g. in that it involves a plasmid-encoded protein (designated RepA in the P1 plasmid) which is essential for replication and which autoregulates its own synthesis. Moreover, the *repA* gene is flanked on each side by multiple 19-bp repeat sequences, and it appears that one of these multiple repeats (*incA*) may be involved in the control of copy number in that it appears to titrate the RepA protein. [P1 replication: JBC (1986) *261* 3548–3555.] The P1 plasmids are segregated (see PARTITION) with great accuracy at each host cell division, and spontaneous loss of the plasmid is extremely rare. This high degree of accuracy involves several mechanisms. For example, P1 encodes a SITE-SPECIFIC RECOMBINATION (SSR) system which facilitates accurate partition by efficiently resolving plasmid dimers into monomers. (Dimers are formed by homologous recombination between the two plasmids resulting from replication.) This system involves a site (*loxP*) on the plasmid at which SSR occurs, and a gene (*cre*) encoding the recombinase. (The *lox*-Cre system also catalyses cyclization of the phage genome on infection of the host – at least in $RecA^-$ hosts; it can also mediate the rare integration of the P1 plasmid into the host chromosome at a site designated *loxB*.) [Regulation of the *cre* gene: JMB (1986) *187* 197–212.] Partition per se involves a sequence (*par*) in the plasmid; a gene (*parA*) within this sequence encodes a protein which may promote partition by binding to the plasmid at a particular site (*incB*) and to (probably) the host cell envelope [JMB (1985) *185* 261–272].

P1 also has a second SSR system which is responsible for the inversion of a 4.2-kb segment (the C segment) of the P1 genome. The C segment was first identified as a 'C loop' analogous to the G LOOP of bacteriophage Mu. The C segment of P1 and the G segment of Mu are homologous (each specifies different host ranges in different orientations), and their controlling elements (encoded by *cin* in P1, *gin* in Mu) can complement each other (see also RECOMBINATIONAL REGULATION). However, the inverted repeats at each end of the C segment are not homologous with those of the G segment. (Efficient inversion of the C segment apparently requires an additional *cis*-acting sequence, distinct from the cross-over sites *cixL* and *cixR* [PNAS (1985) *82* 3776–3780].)

[The P1 genome: JB (2004) *186* 7032–7068.]

bacteriophage P2 A temperate BACTERIOPHAGE (family MYOVIRIDAE) which infects *Escherichia coli*. The phage virion consists of an isometric head attached to a contractile tail. Genome: linear dsDNA (MWt ca. 22×10^6) which has 19-base STICKY ENDS. On infection of a host cell, the DNA forms closed circular molecules which, in lysogenic infections, can integrate into the host chromosome at a preferred site near the *his* operon; the phage 'attachment site' (*attP*) occurs between genes *int* (encoding the P2 integrase) and *ogr* (see below). (cf. BACTERIOPHAGE LAMBDA.)

P2 DNA replication occurs unidirectionally from a single origin near the right-hand end of the genome. Replication requires P2 genes *A* and *B* and (at least) host genes *dnaB*, *polC* and *rep*; P2 gp*A* is a (preferentially *cis*-acting) DNA-binding protein. The products of DNA replication are cccDNA monomers (i.e., concatemers are not formed). Expression of P2 late genes (which encode phage structural components) depends on the expression of the *ogr* gene, which in turn depends on P2 DNA replication. (cf. BACTERIOPHAGE P4.) During phage assembly, the DNA monomer undergoes staggered cleavage at a unique site (*cos*) to form the linear molecule with sticky ends characteristic of the P2 virion.

bacteriophage P4 A satellite BACTERIOPHAGE (see SATELLITE VIRUS) which infects *Escherichia coli* and which requires all of the late genes of a helper phage such as BACTERIOPHAGE P2 to complete its replication cycle. The P4 genome is linear dsDNA (ca. 11.4 kb) with 19-nucleotide sticky ends which allow the DNA to circularize in the host cell. P4 DNA replication occurs bidirectionally from a single origin [JMB (1985) *182* 519–527]; an early P4 gene product, gpα, is required for P4 DNA replication and is apparently a (rifamycin-resistant) primase [JMB (1985) *182* 509–517]. In the absence of a helper phage, P4 can replicate its DNA and can lysogenize its host, but progeny virions can be formed only in the presence of a helper. In cells co-infected with P2 and P4, P4 capsids predominate and are assembled largely from P2 head proteins – although the P4 capsid is smaller than that of P2. P4 alters the regulation of transcription of P2 head proteins; the product of the P4 δ gene is apparently a *trans*-acting accessory transcription factor which can substitute for P2 gp*ogr* and thus circumvent the dependence of late P2 genes on P2 DNA replication [Book ref. 118, pp. 108–110].

bacteriophage P7 A BACTERIOPHAGE which is closely related to BACTERIOPHAGE P1; however, P1 and P7 are heteroimmune and differ in plasmid maintenance functions.

bacteriophage P22 A temperate BACTERIOPHAGE (family PODOVIRIDAE) which infects *Salmonella* (smooth strains). The virion consists of an icosahedral head (ca. 60 nm diam.) attached via a short tubular 'neck' to a 'tail' which is essentially no more than a hexagonal base-plate bearing short spikes. The 'tail' region has endorhamnosidase activity. Genome: dsDNA (MWt ca. 26×10^6), terminally redundant, and circularly permuted in certain regions of the genome.

On infection of a host cell, the linear DNA is circularized either by the host RecA system or by a general recombinase encoded by the P22 *erf* gene. In the lytic cycle, the DNA may undergo some replication in the circular form; subsequently, concatemers many unit genomes in length are produced. During virion assembly, a prohead consisting mainly of coat protein (gp5) and a SCAFFOLDING PROTEIN (gp8) interacts with a concatemeric DNA molecule; DNA is encapsidated, beginning at a specific site (*pac*) in the concatemer and proceeding sequentially until the head is full, and the scaffolding protein is eliminated intact (and is recycled). It appears that once the first packaging event has been completed, the remaining DNA of the concatemer is encapsidated by other proheads – i.e., the leading end of the DNA which enters the second and subsequent proheads is determined not by the *pac* site but by the length of DNA packaged by previous prohead(s). (The mechanism of this is uncertain; the initial cleavage at the *pac* site can apparently occur in the absence of encapsidation [JMB (1982) *154* 565–579].) (cf. BACTERIOPHAGE P1.) Maturation of the virion requires the addition of certain minor phage proteins. (See also TRANSDUCTION.)

The lysogenic pathway in P22 is essentially similar to that in λ (see LAMBDOID PHAGES). The prophage inserts in the *Salmonella* chromosome at a site between *proA* and *proC*; insertion is catalysed by a P22-specified integrase encoded by an *int* gene located close to the *att* site.

bacteriophage PA2 See STYLOVIRIDAE.

bacteriophage PBS1 A large, morphologically complex FLAGELLOTROPIC PHAGE (family MYOVIRIDAE) which infects *Bacillus subtilis*; its DNA contains deoxyuridine instead of thymidine. PBS1 is a pseudotemperate phage, establishing a carrier state in its host (see PSEUDOLYSOGENY); it can mediate transduction.

bacteriophage PBSX A defective bacteriophage, the prophage of which is normally present in the chromosome of its host, *Bacillus subtilis*. When induced by mitomycin C, PBSX packages 13-kb lengths of DNA which may include sequences from any region of the host chromosome, packaging probably occurring by a 'headful' mechanism [JV (1985) *54* 773–780].

bacteriophage Pf1 See INOVIRUS.

bacteriophage Pf2 See INOVIRUS.

bacteriophage ϕI See T7 PHAGE GROUP.

bacteriophage ϕII See T7 PHAGE GROUP.

bacteriophage ϕ6 The sole member of the family Cystoviridae. Host: *Pseudomonas syringae* pv. *phaseolicola*; plaques clear, diam. 1–3 mm. Some strains of *P. pseudoalcaligenes* can be infected. The virion (diam. ca. 75 nm) has a segmented genome of three pieces of dsRNA designated L, M and S (MWt ca. 5.0, 3.1 and 2.3×10^6, respectively). [Nucleotide sequence of, and translational control in, the S segment: JV (1986) *58* 142–151.] The RNA occurs in a polyhedral inner particle (composed of proteins P1, P2, P4 and P7) which is covered with a layer of protein P8 – the whole forming the nucleocapsid (diam. ca. 60 nm). The nucleocapsid is surrounded by a phospholipid- and protein-containing envelope (P3, P5, P6, P9, ?P10) which can be removed (with loss of infectivity) by treatment with detergents. An RNA polymerase (P1, P2) is associated with the capsid. The virion also contains an enzyme (P5, ?P10) active against the host cell wall; this enzyme seems necessary both for penetration of the host and for the release of phage progeny.

The virion adsorbs (P3, P6) to the sides of the subpolar host pili; the ϕ6 envelope fuses with the host outer membrane, and the nucleocapsid *minus* P8 enters the host cell. Inside the cell the phage RNA polymerase catalyses RNA synthesis; an ss mRNA, designated *l* (MWt ca. 2.2×10^6), is transcribed early and codes for P1, ?P2, P4 and P7. These four proteins form a 120S procapsid or *previrion I* which incorporates progeny dsRNA to form *previrion II*. (The synthesis of dsRNA is not understood, but it occurs within a subvirion structure. A model for the initiation of replication and transcription in bacteriophage ϕ6 has been proposed [Nature (2001) *410* 235–240].) Previrion II synthesizes ss mRNAs *m* and *s* (MWt ca. 1.4 and 1.1×10^6, respectively); *s* codes for P8 and P9, *m* for P3, P6 and P10 (and possibly also P4 and P7.) Previrion II incorporates P8 to form the complete nucleocapsid. Finally, the lipoprotein envelope is added – apparently mediated by the (non-structural) phage protein P12; although the lipid is apparently derived from host phospholipid, the membrane is formed in the cytoplasm and not in contact with the cell membrane. Phage release occurs by host cell lysis.

(Note: the name bacteriophage ϕ6 has also been used for a pilus-dependent dsDNA phage of *Caulobacter* [JV (1980) *35* 949–954].)

bacteriophage ϕ29 A BACTERIOPHAGE (family PODOVIRIDAE) which infects strains of *Bacillus subtilis* and those of certain other species of *Bacillus*. The virion has a complex morphology: an elongated head (ca. 32 × 42 nm), bearing protein fibres at each end, joined via a connector region to a short tail. Genome: linear dsDNA (ca. 18 kb) with 6-bp terminal inverted repeats. The 5′ end of each strand in the DNA genome is formed by a dAMP residue; each dAMP residue is linked covalently (via a phosphodiester bond) to a serine residue in a phage-encoded protein (gp3) – a so-called *terminal protein* (TP). The terminal proteins are essential for initiation of replication of the ϕ29 genome.

Replication of the linear ϕ29 DNA is initiated at both ends of the molecule (not simultaneously); it involves a phage-encoded DNA polymerase (gp2) which has exonuclease activity [NAR (1985) *12* 1239–1249]. The polymerase binds to a *free* molecule of the TP, and this complex localizes at the 3′ end of each strand – perhaps by interacting with TP at the adjacent 5′ end. The polymerase then mediates a dATP-dependent reaction in which dAMP is covalently linked to the complexed TP; this dAMP residue serves as a 3′-OH primer. The polymerase begins strand synthesis (5′-to-3′), and after insertion of nucleotide 10 the polymerase and TP dissociate from one another – the TP remaining covalently bound at the 5′ end of the new strand while the polymerase continues to extend the (DNA) primer. The polymerase is able to displace the (5′-to-3′) strand complementary to the template, and to achieve good processivity, without help from a helicase or from other factors [JBC (1996) *271* 8509–8512]. Replication of the ϕ29 genome is an example of so-called *protein priming* – only one example of the strategies used for replicating a linear genome. [Protein priming of DNA replication in phage ϕ29: EMBO (1997) *16* 2519–2527.] (See also ADENOVIRIDAE.)

During phage assembly, the major coat protein (gp8) is assembled with the aid of a (re-cyclable) scaffolding protein (gp7) which is displaced as the DNA is incorporated. [Morphogenesis: JV (1985) *53* 856–861.]

The packaging of DNA into the phage head requires energy from ATP hydrolysis. It also requires certain phage-encoded RNA molecules designated pRNA. [DNA packaging: Cell (1998) *94* 147–150.] In a current model, a number of pRNA molecules form a (static) ring-shaped structure around a hole in the phage head through which the genome is inserted; the connector (a dodecamer of gp10, 75 Å in length, with an axial channel) rotates, driven by ATP hydrolysis, and the 'threaded' (i.e. helical) DNA molecule enters the phage head like a bolt drawn through a nut when the nut is rotated [Nature (2000) *408* 745–750].

Host cell lysis involves a phage-encoded peptidoglycan-degrading enzyme; however, lysis also requires the product of phage gene *14* which is apparently a LYSIS PROTEIN [JB (1993) *175* 1038–1042].

bacteriophage ϕ80 See STYLOVIRIDAE and LAMBDOID PHAGES.

bacteriophage ϕ105 A temperate BACTERIOPHAGE which infects *Bacillus subtilis* (see STYLOVIRIDAE). The virion consists of an isometric head (ca. 52 nm diam.) attached to a flexible, non-contractile tail (ca. 220 nm long) which bears a baseplate with six appendages. Genome: dsDNA (MWt ca. 26×10^6), non-permuted, apparently with sticky ends. During lysogeny, the phage DNA integrates at a specific attachment site in the host chromosome but, unlike e.g. BACTERIOPHAGE λ, it appears not to circularize prior to integration, and the prophage is not circularly permuted with respect to the virion DNA; the mechanism of integration is not understood. On induction, ϕ105 DNA apparently replicates in situ, together with adjacent regions of host DNA, and the phage DNA is subsequently excised from the newly synthesized molecules. LFT (but not HFT) lysates have been described; transduction appears to require that the recipient be a ϕ105 lysogen. [Book ref. 170, pp. 251–256, 273–275.]

bacteriophage ϕA See MICROVIRIDAE.

bacteriophage ϕB See MICROVIRIDAE.

bacteriophage ϕBA1 A temperate, tailed, dsDNA-containing BACTERIOPHAGE isolated from *Bacillus aneurinolyticus*. The virions apparently have a BACTERIOCIN-like killing activity against some strains of *B. aneurinolyticus*; this activity is independent of phage gene expression (phage DNA is rapidly degraded in the sensitive cells), and may be due to a proteinaceous component of the intact virion [JV (1986) *59* 103–111]. (cf. BETACIN.)

bacteriophage ϕC (1) See MICROVIRIDAE. (2) See STYLOVIRIDAE.

bacteriophage ϕC31 A temperate bacteriophage originally isolated from *Streptomyces coelicolor*; it can infect a wide range of *Streptomyces* spp. Genome: dsDNA (ca. 41.5 kb) with sticky ends. The prophage integrates in the host chromosome at a specific attachment site.

bacteriophage ϕCbK See STYLOVIRIDAE.

bacteriophage ϕD328 See STYLOVIRIDAE.

bacteriophage ϕEC A tailed, temperate ACTINOPHAGE whose hosts are species of *Rhodococcus*; genome: linear dsDNA. ϕEC is inactivated by various lipid solvents. [Book ref. 73, pp. 216–218.]

bacteriophage ϕK See MICROVIRIDAE and BACTERIOPHAGE G4.

bacteriophage ϕNS11 See TECTIVIRIDAE.

bacteriophage ϕR See MICROVIRIDAE.

bacteriophage ϕW-14 A bacteriophage (family MYOVIRIDAE) which infects and lyses certain strains of *Pseudomonas acidovorans*. In ϕW-14 DNA ca. 50% of the thymine residues are hypermodified, occurring as α-putrescinylthymine; the hypermodified bases are apparently essential for the production of viable progeny phages – probably being required for DNA packaging [Virol. (1983) *124* 152–160].

bacteriophage ϕX174 The type species of the MICROVIRIDAE (q.v. for morphology etc). After adsorption of ϕX174 to a host cell, the ss cccDNA phage genome penetrates the cell and is coated with host SSB protein; a region between genes *F* and *G* (which can form hairpin loops) remains uncoated. Stage I (see SSDNA PHAGES) is dependent on host *dnaB* function (as it is in microviruses S13, G6, G13, G14 and ϕXahb – cf. BACTERIOPHAGE G4). A PRIMOSOME assembles at a specific n′ recognition sequence in the *F–G* intergenic region. The primosome migrates (5′ to 3′) along the *v* strand (displacing the SSB protein?), synthesizing primers at several sites. The *c* strand is completed as in other SSDNA PHAGES to form the ds RF (parental RF). It seems that most components of the primosome remain bound to the parental RF throughout this and subsequent stages, and that this obviates the need for supercoiling of the RF by gyrase – a step previously thought to be essential for initiation of stage II [PNAS (1981) *78* 1436–1440]. Stage II: ϕX174 gp*A* nicks the *v* strand of the parental RF at a specific site in gene *A* and becomes covalently bound to the 5′ end of the nicked strand. The 3′ end is elongated by host DNA polymerase III holoenzyme in a ROLLING CIRCLE MECHANISM; ahead of holoenzyme action, the two RF strands are processively unwound by a complex of gp*A* and Rep protein. (The presence of gp*A* allows Rep protein to begin unwinding from the *nick* – not normally possible for HELICASES.) As synthesis proceeds, the displaced *v* strand (attached by its 5′ end to gp*A* at the replication fork) forms a growing loop which can function as template for the discontinuous synthesis of a *c* strand (primed by the conserved parental RF primosome components). When one genome length of *v* DNA has been displaced, it is excised and circularized by gp*A* and the *c* strand is completed to form the progeny RF. Progeny RFs are supercoiled by GYRASE and seem at this stage to function primarily as templates for transcription of phage genes. Stage III is closely coupled to phage morphogenesis. RFI (see RF) is cleaved by gp*A*, as in stage II, to form an RFII–gp*A* complex which, in the presence of gp*C*, associates with a procapsid (consisting of gp*B*, gp*D*, gp*F*, gp*G* and gp*H*) to form a 50S particle. *v* strand synthesis and displacement seem to occur by a rolling circle mechanism; as the *v* strand is displaced, it is packaged within the procapsid. Packaging (but not phage DNA synthesis) requires gp*J*, a component of the phage capsid [JV (1985) *54* 345–350]. When a complete *v* strand has been displaced it is cleaved and circularized by gp*A*, releasing the RFII–gp*A* complex (for further rounds of *v* strand synthesis) and an ss cccDNA-containing immature phage particle; gp*B* and gp*D* are subsequently lost to form the mature 114S ϕX174 virion.

bacteriophage ϕXahb See BACTERIOPHAGE ϕX174.

bacteriophage ϕXtB See MICROVIRIDAE.

bacteriophage PM2 The type species of the CORTICOVIRIDAE; host: *Alteromonas espejiana* BAL-31 (= *Pseudomonas* BAL-31). The virion is icosahedral (diam. ca. 60 nm) and is sensitive to organic solvents and detergents; it comprises four major proteins, 12–14% by weight phospholipid, and supercoiled ds cccDNA (MWt ca. 6×10^6). The outer layer of the capsid is an icosahedral shell of protein PII (= sp27) with spikes of protein PI (= sp43) at the vertices. Within this is a lipid bilayer membrane whose composition reflects that of the host cell membrane at the time of phage assembly. The membrane seems to be associated with one or both of the remaining two proteins; these proteins may also be associated with the DNA, and at least one has transcriptase activity. Virions adsorb to the host cell wall. Maturation occurs at the cell periphery. The progeny are released by cell lysis.

bacteriophage PP7 See LEVIVIRIDAE.

bacteriophage PR3 See TECTIVIRIDAE.

bacteriophage PR4 See TECTIVIRIDAE.

bacteriophage PR5 See TECTIVIRIDAE.

bacteriophage PR772 See TECTIVIRIDAE.

bacteriophage PRD1 See TECTIVIRIDAE.

bacteriophage Qβ See LEVIVIRIDAE.

bacteriophage R1 See STYLOVIRIDAE.

bacteriophage R2 See STYLOVIRIDAE.

bacteriophage R17 See LEVIVIRIDAE.

bacteriophage R23 See LEVIVIRIDAE.

bacteriophage 7S See LEVIVIRIDAE.

bacteriophage S13 See MICROVIRIDAE and BACTERIOPHAGE ϕX174.

bacteriophage SP3 See MYOVIRIDAE.

bacteriophage SP6 A BACTERIOPHAGE which infects *Salmonella typhimurium*; it has a linear dsDNA genome (ca. 43.5 kb) and appears to resemble BACTERIOPHAGE T7 in its morphology and development. SP6 encodes a rifamycin-resistant RNA polymerase which is synthesized shortly after infection of a host cell. [Nucleotide sequence and expression of the cloned SP6 RNA polymerase gene: NAR (1987) *15* 2653–2664.]

bacteriophage SP8 See MYOVIRIDAE.

bacteriophage SP50 See MYOVIRIDAE.

bacteriophage SPβ See STYLOVIRIDAE and BETACIN.

bacteriophage SPO1 A virulent BACTERIOPHAGE which infects *Bacillus subtilis*. The virion consists of a large isometric head, containing >7 different polypeptides, joined via a complex connector structure to a contractile tail which terminates in a complex base-plate. The genome is a single molecule of linear dsDNA, MWt ca. 10^8, which contains hydroxymethyluracil. [Review: Book ref. 170, pp. 218–245.]

SPO1 is widely used in studies of transcriptional regulation in Gram-positive bacteria. Phage genes are expressed in three phases: early, middle and late. Early gene promoters are of the same type as those of host genes involved in vegetative functions, allowing transcription of early phage genes by the (unmodified) host RNA POLYMERASE. One early gene, gene 28, encodes a novel SIGMA FACTOR (σ^{gp28}) which replaces the host σ^{43} (formerly known as σ^{55}), resulting in a holoenzyme which can recognize only phage middle promoters. Middle genes include genes 33 and 34, the expression of which is necessary for late gene transcription; gp33 and gp34 together replace the σ^{gp28} in the RNA polymerase, resulting in an enzyme which – in the presence of the delta factor (see RNA POLYMERASE) – recognizes only late phage promoters, thus allowing the transcription of late genes (which encode phage structural components). (See also ENDOSPORE sense 1.)

bacteriophage SPO2 A temperate BACTERIOPHAGE which infects *Bacillus subtilis*. Virion: an isometric head (ca. 50 nm diam.) attached to a tail ca. 180 nm long. Genome: dsDNA, MWt ca. 26×10^6. [Book ref. 170, pp. 247–268 (256–259).]

bacteriophage SPP1 See STYLOVIRIDAE.

bacteriophage St-1 See MICROVIRIDAE and BACTERIOPHAGE G4.

bacteriophage SV-C1 See SPIROPLASMAVIRUSES.

bacteriophage SV-C2 See SPIROPLASMAVIRUSES.

bacteriophage SV-C3 See SPIROPLASMAVIRUSES.

bacteriophage T1 See STYLOVIRIDAE.

bacteriophage T2 See T-EVEN PHAGES and MYOVIRIDAE.

bacteriophage T3 A bacteriophage (family PODOVIRIDAE) which is very closely related to BACTERIOPHAGE T7.

bacteriophage T4 A virulent enterobacterial (linear dsDNA-containing) BACTERIOPHAGE – the best-known representative of the T-even phages. (See entry T-EVEN PHAGES for phage structure, nature of genome etc; see also MYOVIRIDAE.)

The T4 infection cycle is initiated when the phage attaches, via the distal tips of its long tail fibres, to specific sites on the host cell surface (see T-EVEN PHAGES). The tail fibres may occur in either of two states: extended or 'retracted' (i.e., folded back along the tail sheath and head); infection can occur only when the fibres are extended. Transition between the two states is affected by conditions: retraction (and hence loss of infectivity) is promoted by e.g. low pH, low temperature, low ionic strength, and in some strains by the absence of the adsorption cofactor tryptophan. Binding is initially reversible, but is followed by irreversible binding during which the base-plate undergoes a conformational change from a hexagonal form to an expanded, star-shaped structure with a central opening. This in turn triggers the contraction (by conformational change) of the tail sheath and the penetration of the host cell envelope by the inner tail tube. (Penetration may occur at ADHESION SITES.) The phage DNA is then transferred through the tail inner tube to the host cell. Since the inner tube tip appears to penetrate only as far as the outer surface of the host cytoplasmic membrane, DNA apparently does not enter the cytoplasm directly; uptake of the DNA by the host cell appears to require a membrane potential and can be inhibited by ionophores.

Infection of the host cell is followed by transcription of the T4 DNA. This occurs in three main phases: early (immediate early and delayed early), middle, and late; each phase occurs at a distinct time after infection, is initiated at a distinct class of promoters, and requires a different transcription apparatus. The host RNA POLYMERASE is apparently used throughout, but undergoes a series of phage-induced alterations (e.g. ADP-ribosylation of first one, then both, of its α subunits) which affect its activity and affinities (e.g. for sigma factor); T4 gp55 is a SIGMA FACTOR which is necessary for the transcription of T4 late genes. Early effects on the host cell include the cessation of host DNA, RNA and protein synthesis (within 2–5 min of infection), unfolding of the host nucleoid, and degradation of host DNA by phage-coded nucleases specific for cytosine-containing DNA. The nucleotides thus released are used for the synthesis of progeny phage DNA.

Late transcription is coupled to T4 DNA replication, which begins ca. 5 min after infection. Replication can be initiated at multiple origins [locations: JV (1985) *54* 271–277] and occurs bidirectionally. It appears to be catalysed by a complex of phage-coded proteins (a 'replisome' – see DNA REPLICATION), including gp43 (DNA polymerase), gp32 (an SSB protein), gp44/45/62 (DNA polymerase accessory proteins which e.g. increase polymerase processivity and affinity for DNA), gp*dda* (a 5′-to-3′ helicase), and gp41 and gp61 (involved in the priming of Okazaki fragments). Leading strand synthesis is probably primed initially by host RNA polymerase. Okazaki fragments in lagging strand synthesis are primed by the formation of pentaribonucleotide primers (pppApCpNpNpN, where N = any of the 4 nucleotides) by gp41 and gp61; gp41 may be analogous to the DnaB protein of *Escherichia coli* (see DNAB GENE), while gp61 may be a PRIMASE. When elongation is complete, the 3′ end of the template for each lagging strand remains single-stranded; these single-stranded regions can 'invade' homologous regions of other phage DNA molecules, forming recombinational forks (see RECOMBINATION) and resulting in the formation of complex, branching, concatemeric intermediates. Resolution of the branches appears to require gp49 (T4 endonuclease VII). Later rounds of DNA replication may be initiated at recombinational forks, allowing replication to become independent of the (altered) host RNA polymerase. (Alteration of the RNA polymerase may prevent its recognition of origin promoters.) Modification of the DNA (glucosylation and methylation) occurs on the completed polynucleotide strands; HMC substitution occurs at the level of precursor synthesis, involving hydroxymethylation of dCTP.

Phage components are encoded by late phage genes. Heads, tail fibres and base-plates are each assembled by separate pathways. Head assembly involves at least 20 genes. Initially, a prohead is assembled on the host cytoplasmic membrane; the prohead consists of a shell surrounding a central core, the latter acting as scaffolding and controlling the assembly, size and shape of the prohead. When this structure is complete, nearly all of its constituent proteins undergo proteolytic cleavage, resulting in the removal of the core and the formation of a 'cleaved prohead'. The cleaved prohead detaches from the membrane, and the main shell protein (gp23) undergoes extensive conformational changes which result in expansion of the prohead; this step is probably normally accompanied by DNA packaging. DNA is packaged by the 'headful' mechanism: one end of a linear concatemeric DNA molecule is inserted into the prohead, and the DNA continues to be incorporated until the head is full, when the DNA is cleaved (thus yielding the circularly permuted genome characteristic of T4). Base-plate assembly occurs by separate assembly of the hub and wedges (see T-EVEN PHAGES), followed by binding of the six wedges around the hub. The tail is then polymerized (inner tube first) on the base-plate. The completed tail structure associates spontaneously with the DNA-containing head, and finally the tail fibres and collar 'whiskers' are added.

Progeny phages are released by lysis of the host cell. The mechanism of lysis is still not fully understood, but requires T4-coded LYSOZYME (gp*e*), the function of gene *t* (see LYSIS PROTEIN), and possibly gp5 (the base-plate hub lysozyme – see T-EVEN PHAGES).

[Reviews on all aspects of T4: Book ref. 99.]

bacteriophage T5 See STYLOVIRIDAE.

bacteriophage T6 See T-EVEN PHAGES.

bacteriophage T7 A virulent bacteriophage of the PODOVIRIDAE which infects *Escherichia coli* and other enterobacteria. Head: isometric, ca. 60 nm diam. [capsid architecture: Virol. (1983) *124* 109–120]. Tail: ca. 17 nm long, with six short fibres; the distal end of the tail adsorbs to LPS in the outer membrane of a host cell. The T7 genome is a linear, non-permuted dsDNA molecule (ca. 40 kb) with terminally redundant ends (160 bp). T7 early genes constitute a single operon near the left end of the genome; these genes are transcribed by the host RNA polymerase and are concerned mainly with switching off host gene transcription and switching on transcription of T7 intermediate and late genes. (The large early-gene transcript is cleaved by host RNase III (q.v.) to give mRNAs.) Early gene products include a protein kinase (gp0.7), which phosphorylates and inactivates the host RNA polymerase (thus preventing further early gene transcription), and an RNA polymerase (gp1) which transcribes the T7 intermediate genes (concerned mainly with T7 DNA replication) and late genes (concerned with phage assembly and release of phage progeny by host cell lysis). Another phage product, gp2, also binds to and inactivates host RNA polymerase. The phage RNA polymerase is highly specific for T7 promoters; thus, by inhibiting host RNA polymerase, all transcriptional activity in the cell is switched to T7 DNA. [Organization and expression of T7 DNA: CSHSQB (1983) *47* 999–1007.]

Replication of T7 DNA (which remains linear throughout) is achieved mainly by proteins encoded by T7 DNA; it occurs in three stages: (i) *Initiation* occurs at a single site near the 'left' end of the genome. (Secondary sites can be used, although less efficiently, if the primary site is deleted.) The T7 gp1 RNA polymerase is necessary for initiation, apparently synthesizing a short primer from promoters in the initiation region. (ii) Bidirectional *elongation* (from the RNA primers) catalysed mainly by two phage-coded enzymes: gp5 and gp4. The protein gp5 combines with host THIOREDOXIN to form a DNA polymerase which synthesizes DNA on the RNA primers in the 5′-to-3′ direction, thus forming the leading strands. The protein gp4 is a multifunctional enzyme with primase, helicase and NTPase activity; it binds to unwound ssDNA regions of parental DNA and translocates unidirectionally in the 5′-to-3′ direction (driven by NTP hydrolysis), unwinding the parental strands ahead of the replication forks. As it encounters specific primase recognition sites in the DNA, gp4 synthesizes tetraribonucleotide primers (pppApCpCpC or pppApCpCpA) which provide the 3′-OH ends necessary for initiation of lagging strand synthesis by the T7 DNA polymerase. [Book ref. 69, pp. 135–151.] Nucleotides for T7 DNA synthesis are provided by degradation of host DNA by T7 gp3 (an endonuclease) and gp6 (an exonuclease); gp6 (and/or host DNA polymerase I) may be involved in the removal of RNA primers. (iii) The newly synthesized daughter duplexes undergo recombination, probably mainly end-to-end, to form concatemers several times the length of a T7 genome [Virol. (1982) *123* 474–479]; these concatemers undergo further rounds of replication. Each concatemer apparently contains only a single copy of the 160-bp terminal repeat sequence at the junctions between the genomes; the terminal repetition of the mature DNA is generated during maturation and packaging of the DNA into preformed prohead structures, when the concatemers are cleaved (possibly by gp3) into genome lengths. Mature progeny phages are released by host cell lysis (gp3.5 has lysozyme activity).

Studies on the rate of internalization of the phage genome during infection have led to new insight into the mechanism by which the DNA of phage T7 is transferred to the bacterial cell. It appears that, following attachment, the phage ejects certain proteins which may form a channel across the bacterial cell envelope through which DNA is transferred; it has been speculated that two of the ejected proteins may be components of a 'motor' which rachets DNA into the host cell [Mol. Microbiol. (2001) *40* 1–8].

bacteriophage Tb *Syn.* TBILISI PHAGE.

bacteriophage typing See PHAGE TYPING.

bacteriophage U3 See MICROVIRIDAE.

bacteriophage v1 See PLASMAVIRIDAE.

bacteriophage v2 See PLASMAVIRIDAE.

bacteriophage v4 See PLASMAVIRIDAE.

bacteriophage v5 See PLASMAVIRIDAE.

bacteriophage v6 See INOVIRUS.

bacteriophage v7 See PLASMAVIRIDAE.

bacteriophage VP5 An ACTINOPHAGE of the STYLOVIRIDAE which infects *Streptomyces coelicolor.*

bacteriophage W31 See T7 PHAGE GROUP.

bacteriophage Xf See INOVIRUS.

bacteriophage Xf2 See INOVIRUS.

bacteriophage Z See BETACIN.

bacteriophage ZJ/2 See INOVIRUS.

bacteriophages See BACTERIOPHAGE.

bacteriophagous Refers to organisms (e.g. certain protozoa) which consume bacteria.

bacteriorhodopsin A hydrophobic, pigment-containing protein (MWt ca. 27000) which is the major protein constituent of the PURPLE MEMBRANE in *Halobacterium salinarium*; it is involved in the (energy-generating) light-dependent transmembrane translocation of protons. The protein part of the bacteriorhodopsin molecule (apoprotein, bacterio-opsin) consists of seven membrane-spanning regions; the pigment – RETINAL – occurs within the space bounded by these seven regions, being linked to one of them via a lysine residue.

Bacteriorhodopsin occurs in hexagonal arrays in the plane of the membrane.

On illumination, bacteriorhodopsin undergoes a cyclical series of changes with a periodicity of ca. 5–10 msec, and this photocycling is associated with the pumping of protons from the inner (cytoplasmic) to the outer side of the membrane, i.e. generation of proton motive force (pmf). During photocycling at least five photointermediates appear to be formed; these are designated: K_{590}, L_{550}, M_{412}, N_{530} and O_{640} (the numbers being wavelengths, in nm, of maximum absorption). bR_{568} is the light-adapted ground state.

Recent studies have clarified the mechanism, and route, of vectorial energy-dependent transport of protons across the membrane. Light induces the isomerization of (protonated) retinal from all-*trans* to 13-*cis*, isomerization resulting in deprotonation of the retinal – the proton being transferred to the asp-85 residue of the protein. Deprotonation of the retinal causes it to change conformation, and this, in turn, modifies the protein's conformation – opening up a channel which permits entry of a proton from the cytoplasmic side of the membrane; this proton reprotonates residue asp-96 – the proton from asp-96 having been donated to reprotonate retinal. The proton on asp-85 is transferred to the exterior of the membrane; this transfer appears to coincide with the N_{530} photointermediate.

When incorporated into LIPOSOMES, bacteriorhodopsin tends (except at low pH) to adopt an orientation opposite to that in the cytoplasmic membrane – and hence to mediate inward pumping of protons.

[Bacteriorhodopsin (overview): Nature (2000) *406* 569–570. Structural changes in bacteriorhodopsin coupled to proton transport: Nature (2000) *406* 645–648. Structural alterations in the M state: Nature (2000) *406* 649–652. Molecular mechanism of vectorial proton translocation: Nature (2000) *406* 653–657.]

bacterioruberins C_{50} CAROTENOID pigments which occur in members of the HALOBACTERIACEAE.

bacteriostatic (bacteristatic) Able to inhibit the growth and reproduction of at least some types of bacteria. (cf. BACTERICIDAL.)

bacteristatic *Syn.* BACTERIOSTATIC.

bacterium See BACTERIA.

Bacterium A bacterial genus which has been obsolete for decades; various bacteria, e.g. *Escherichia coli*, *Clostridium chauvoei* etc., were once referred to as *B. coli*, *B. chauvoei* etc.

bacterium-associated haemophagocytic syndrome (BAHS) See HAEMOPHAGOCYTIC SYNDROME.

bacteriuria The presence of bacteria in the urine.

bacterivore Any organism (e.g. a ciliate or amoeba) which ingests bacteria as its main or sole source of nutrients. (cf. e.g. DETRITIVORE.)

bacterization The process of coating seeds, tubers etc. with certain bacteria (e.g. *Azotobacter*), prior to planting, with the object of improving plant growth.

bacteroid A bacterium-like cell or a modified bacterial cell – see e.g. ROOT NODULES. (In an ACTINORRHIZA, 'bacteroid' may refer e.g. to a senescent hyphal fragment.)

Bacteroidaceae A family of (typically) Gram-negative, anaerobic, chemoorganotrophic, asporogenous, motile and non-motile, typically rod-shaped or filamentous bacteria, some of which are pathogenic in man and other animals. Genera: ACETIVIBRIO, ANAEROBIOSPIRILLUM, ANAEROVIBRIO, BACTEROIDES, BUTYRIVIBRIO, FUSOBACTERIUM, LACHNOSPIRA, LEPTOTRICHIA, PECTINATUS, SELENOMONAS, SUCCINIMONAS, SUCCINIVIBRIO and WOLINELLA. In some of these organisms (e.g. *Acetivibrio*, *Butyrivibrio*, *Lachnospira*) the CELL WALL appears to resemble the Gram-positive (rather than Gram-negative) type of wall; since these organisms give a negative or variable reaction in the Gram stain (at least under some conditions) they have been grouped with the frankly Gram-negative genera [Book ref. 22, pp. 602–662].

Bacteroides A genus of Gram-negative bacteria (family BACTEROIDACEAE) which occur e.g. in the RUMEN and in the mouth and intestinal tract in man and other animals; some species are opportunist pathogens, being isolated from abscesses and other types of lesion (see e.g. FOOT-ROT; PNEUMONIA (f); SCALD). (See also ANAEROBIC DIGESTION; TERMITE–MICROBE ASSOCIATIONS; WETWOOD.) Cells: typically rods or filaments, non-motile or peritrichously flagellated, sometimes with central or terminal swellings or vacuoles. Some species form a dark or black pigment. [Black-pigmented strains from animals: JAB (1983) *55* 247–252.] Metabolism is typically fermentative, though some strains can carry out FUMARATE RESPIRATION in media containing haemin – electron donors including formate and H_2. Most species ('fermentative', 'saccharoclastic' or 'saccharolytic' species) typically attack a range of sugars, producing a mixture of products which may include acetic, formic, lactic, propionic and succinic acids; such species usually use or need CO_2 – which is assimilated by the reductive carboxylation of succinate to 2-oxoglutarate. Other species attack sugars weakly, or not at all, but utilize peptones with the formation of mixtures of either small amounts of e.g. acetic, formic, lactic and succinic acids, or mixtures of larger amounts of e.g. acetic, butyric, propionic and succinic acids – isobutyric and isovaleric acids being formed whenever butyric acid is formed. Primary isolation may be carried out on media containing e.g. peptone, yeast extract, haemin and menaquinone under at least 5% CO_2. GC%: ca. 28–61. Type species: *B. fragilis*.

B. endodontalis. A non-saccharolytic, black-pigmented species isolated from the dental root canal [IJSB (1984) *34* 118–120].

B. fragilis (formerly *Fusiformis fragilis*). Cells (in glucose broth): non-motile round-ended rods, $0.8–1.3 \times 1.6–8.0$ μm. Catalase +ve (most other species are catalase −ve). Saccharolytic; in haemin-containing media some strains form a *b*-type cytochrome and can carry out fumarate respiration. On horse- or rabbit-blood agar colonies are typically smooth, round, entire, low convex, grey and non-haemolytic; some strains produce clear ('β') haemolysis. Optimum growth temperature: 37°C. *B. fragilis* may be sensitive to e.g. CEFOXITIN and clindamycin (see LINCOSAMIDES). (The '*B. fragilis* group' commonly includes former subspecies of *B. fragilis* which are currently classified as species: *B. distasonis*, *B. ovatus*, *B. thetaiotaomicron* and *B. vulgatus*.)

B. melaninogenicus. A saccharolytic species which forms a black pigment (a haem derivative) and which has SOD activity; it occurs e.g. in the human mouth.

B. nodosus. A non-motile, non-saccharolytic species which causes foot-rot in sheep. [Surface structure and virulence of *B. nodosus*: JGM (1983) *129* 225–234.]

B. ochraceus. See CAPNOCYTOPHAGA.

B. pneumosintes (formerly *Dialister pneumosintes*). A non-saccharolytic species which occurs e.g. in the human nasopharynx and which may be a secondary invader in upper respiratory tract infections. Cells: typically $0.2–0.4 \times 0.3–0.6$ μm.

B. ruminicola. A saccharolytic species which ferments a wide range of carbohydrates, including e.g. cellobiose and starch; it occurs in the RUMEN.

B. succinogenes. A saccharolytic rumen species which attacks relatively few sugars but which can utilize cellobiose and cellulose; unlike *B. ruminicola*, this species does not digest gelatin or hydrolyse aesculin.

B. termitidis (formerly *Sphaerophorus siccus* var. *termitidis*). A saccharolytic species which occurs in the intestinal tract in termites.

Other species: *B. amylophilus*, *B. asaccharolyticus*, *B. bivius*, *B. buccae*, *B. capillosus*, *B. coagulans*, *B. corporis*, *B. denticola*, *B. disiens*, *B. distasonis*, *B. eggerthii*, *B. furcosus*, *B. gingivalis*, *B. gracilis*, *B. hypermegas*, *B. intermedius*, *B. levii*, *B. loescheii*, *B. macacae*, *B. microfusus*, *B. multiacidus*, *B. oralis*, *B. oris*, *B. ovatus*, *B. praeacutus* (cf. TISSIERELLA), *B. putredinis*, *B. splanchnicus*, *B. thetaiotaomicron*, *B. uniformis*, *B. ureolyticus*, *B. vulgatus* and *B. zoogleoformans*.

[Book ref. 22, pp. 604–631; *Bacteroides* of the human lower intestinal tract: ARM (1984) *38* 293–314.]

Bacto The designation of products of Difco Laboratories Inc., Detroit, USA.

bactoprenol (undecaprenol) In bacteria: a lipid-soluble, membrane-bound polyprenol consisting of a linear chain of 11 unsaturated isoprene units:

$$H–[CH_2.C(CH_3)=CH.CH_2]_{11}–OH$$

Bactoprenol phosphate acts as a carrier in the synthesis of a number of cell envelope and extracellular polymers: e.g., PEPTIDOGLYCAN, LIPOPOLYSACCHARIDE, TEICHOIC ACIDS and certain CAPSULE polymers; it apparently facilitates the transfer of lipophobic sugar residues across the cytoplasmic membrane. (cf. DOLICHOL; see also TUNICAMYCIN.)

Baculoviridae A family of ccc dsDNA-containing VIRUSes which infect arthropods (particularly insects of the Diptera, Hymenoptera and Lepidoptera). Baculovirus virions are structurally complex, consisting of one or more nucleocapsids within an envelope; virions containing one nucleocapsid per envelope are rod-shaped, ca. $200–400 \times 40–60$ nm. The genome is non-segmented circular supercoiled dsDNA, MWt ca. $58–110 \times 10^6$.

The family is divided (on morphological criteria) into three subgroups: *A* (the NUCLEAR POLYHEDROSIS VIRUSES, NPVs), *B* (the GRANULOSIS VIRUSES, GVs), and *C*. (cf. POLYDNAVIRIDAE.) The NPVs and GVs are characterized by the formation of intracellular crystalline protein *occlusion bodies* within which virions are embedded; the NPVs form polyhedral intranuclear occlusion bodies (called POLYHEDRA), each containing many virions, while the GVs form ellipsoidal or round-ended rod-shaped occlusion bodies (called 'granules' or 'capsules'), each containing only one (rarely two) virions. The matrix proteins of both types of occlusion body (called *polyhedrin*, or polyhedrin in NPVs and granulin in GVs) are closely related in structure and function [review: JGV (1986) *67* 1499–1513], but are apparently unrelated to the polyhedrins of the CYTOPLASMIC POLYHEDROSIS VIRUS GROUP. In contrast to the NPVs and GVs, baculoviruses of subgroup *C* do not form occlusion bodies; viruses of this subgroup have been observed in various arthropods (insects, arachnids, crustacea), the type species of the subgroup being *Oryctes rhinoceros* virus (from the coconut palm rhinoceros beetle).

An insect becomes infected with an NPV or GV when it ingests an occlusion body; the polyhedrin dissolves in the alkaline contents of the insect's midgut, releasing virions which then infect cells of the midgut. Non-occluded virions produced in these cells bud through the plasma membrane of the midgut cells and are disseminated via the haemolymph, subsequently infecting cells of a wide range of tissues. The virion-containing occlusion bodies are formed late in infection, and are eventually released by the death and dissolution of the insect host.

Baculoviruses have been extensively used for the BIOLOGICAL CONTROL of insect pests; they are particularly suitable for this purpose since they have no apparent relationship with any other known animal or plant viruses (and are therefore deemed unlikely to infect mammals or plants), and they can be lethal for – and spread rapidly among – insects in nature. Thus, e.g., NPVs have been used to control the European pine sawfly (*Neodiprion sertifer*) and European spruce sawfly (*Gilpinia hercyniae*), GVs have been used against the Small White butterfly (*Artogeia* (= *Pieris*) *rapae*, a pest of crucifers), and *Oryctes rhinoceros* virus has been used successfully to control the coconut palm rhinoceros beetle.

Since NPVs and GVs produce large quantities of polyhedrin/granulin in infected cells, the genomes of these viruses are of interest as vectors for the introduction and high-level expression of foreign genes in insects or insect cell cultures; e.g., a recombinant baculovirus containing a human α-interferon gene linked to the polyhedrin promoter yielded large quantities of α-interferon in silkworms [Nature (1985) *315* 592–594].

baculoviruses Viruses of the BACULOVIRIDAE.

Badhamia See MYXOMYCETES.

***Bae*I** See RESTRICTION ENDONUCLEASE (table).

baeocyte A small, spherical reproductive cell (diam. ca. 2–3 μm) formed by members of section II of the CYANOBACTERIA. Baeocytes (formerly termed 'endospores') are produced by multiple fission of a vegetative cell which is enclosed in a fibrous layer external to the outer membrane of the cell wall; numerous baeocytes are released on rupture of the fibrous layer. A baeocyte enlarges, without division, until it reaches the size of a vegetative cell, this growth being accompanied by a progressive thickening of the fibrous outer layer. Gliding motility may be exhibited initially by those baeocytes which lack a fibrous outer layer at the time of their release.

Baeomyces A genus of LICHENS (order LECANORALES); photobiont: e.g. *Myrmecia*. Thallus: crustose to squamulose, often sorediate. Fruiting bodies are non-lichenized, stalked, convex hymenial discs resembling miniature mushrooms (cf. PODETIUM). Species occur on soil, rocks etc.

BAF See SEWAGE TREATMENT.

bagassosis An EXTRINSIC ALLERGIC ALVEOLITIS associated with inhalation of the dust of sugar cane waste (bagasse); certain bacteria (e.g. *Thermoactinomyces sacchari*) have been implicated.

BAHS (bacterium-associated haemophagocytic syndrome) See HAEMOPHAGOCYTIC SYNDROME.

bakanae disease ('foolish seedling disease'; 'foolish rice') A disease of rice caused by *Gibberella fujikuroi* (*Fusarium moniliforme*) and characterized by conspicuous elongation of the plant stems (internodes) followed by wilting. The symptoms are believed to be due to the effects of GIBBERELLINS produced by the pathogen. (See also CEREAL DISEASES.)

baker's yeast A specialized strain of *Saccharomyces cerevisiae* which is capable of rapid fermentative activity in dough (see BREAD-MAKING), i.e., under conditions of low oxygen tension, low water activity, and high osmotic pressure. The yeast is manufactured by batch culture (in a medium containing molasses, vitamins, minerals and a nitrogen source) for 12–18 hours at ca. 30°C with full aeration (to suppress fermentation). The yeast is harvested by centrifugation, washed, and concentrated by pressing or filtration. 'Fresh yeast' is blended, extruded, cut into cakes and wrapped; it can be stored for only a few days. 'Active dry yeast' is usually dried in a stream of warm air and sold as granules (water content ca. 8.0%); it can be stored for up to ca. 18 months in sealed containers.

baking See BREAD-MAKING.

BAL British anti-Lewisite: 2,3-dimercaptopropanol; in the presence of oxygen BAL acts as a mitochondrial RESPIRATORY INHIBITOR, apparently blocking electron flow between cytochromes b and c_1 in Complex III of the ELECTRON TRANSPORT CHAIN.

Balamuthia mandrillaris A species of amoeba which can cause ENCEPHALITIS in the young, old and immunocompromised. Trophozoites, 15–60 μm diam., have a round nucleus and a large, strongly staining nucleolus; cysts are roughly spherical, ca. 6–30 μm diam. [*B. mandrillaris* infection: JMM (2001) *50* 205–207.]

balanced growth See GROWTH.

balanced salt solution (BSS) Any of several solutions used (with or without supplement) e.g. in TISSUE CULTURE to provide satisfactory ionic, pH and osmotic conditions for the maintenance and/or growth of cells (see e.g. EARLE'S BSS and HANKS' BSS). Antibiotic(s) may be added to suppress microbial growth. (See also TRANSPORT MEDIUM; DULBECCO'S PBS; EAGLE'S MEDIUM.)

Balanosporida See STELLATOSPOREA.

Balansia See CLAVICIPITALES.

balantidiasis See DYSENTERY (c).

Balantidioides See HETEROTRICHIDA.

Balantidium A genus of parasitic ciliate protozoa of the order TRICHOSTOMATIDA. *B. coli* can cause dysentery (see DYSENTERY (c)) in man, and is the only ciliate known to be pathogenic for man; it normally occurs e.g. as a harmless parasite in the intestinal tract in pigs. The trophozoite is ovoid with a tapering anterior end, ca. 40–70 μm in length, and is covered by spirally arranged longitudinal rows of cilia; the organism exhibits a rotatory movement. At the anterior end of the cell there is a deep invagination (the vestibulum) leading to the cytostome, and a smaller invagination (the cytopyge) occurs at the posterior end. The cell contains a large, kidney-shaped macronucleus (generally visible only in stained preparations) and a small micronucleus situated close to it; contractile vacuoles may be present. Reproduction occurs by binary fission; conjugation also occurs. Encystment occurs in the gut; the spherical cyst (ca. 50 μm diam.) contains a large macronucleus which can be seen in unstained preparations or in preparations stained with a solution of methyl green in dilute acetic acid. Trophozoites can be cultured in media which support the growth of parasitic intestinal amoebae.

BALB/c mice An inbred mouse strain predisposed to MYELOMA formation on intraperitoneal injection of e.g. mineral oil. [The BALB/c mouse – genetics and immunology: CTMI (1985) *122* 1–253.]

Ballerup–Bethesda group (1) Strains of *Citrobacter freundii* which ferment lactose slowly. (2) *Citrobacter freundii*.

ballistoconidium See BALLISTOSPORE.

ballistospore A spore which, at maturity, is forcibly projected from its site of attachment on the sporophore; ballistospores are formed by many basidiomycetes (cf. BASIDIOSPORE), by some imperfect fungi (e.g. *Sporobolomyces*) and by slime moulds of the PROTOSTELIOMYCETES. (cf. STATISMOSPORE and GASTEROID

BASIDIOSPORE.) An asexually-derived ballistospore is sometimes referred to as a *ballistoconidium*.

The ballistospore of a basidiomycete is formed terminally and asymmetrically on a sterigma; its surface is readily wettable. Shortly before discharge, a drop of liquid appears on the surface of the spore in a region near the attachment site (e.g. the HILAR APPENDIX); the drop grows in size (for ca. 5–40 sec) until, suddenly, both spore and drop are projected from the sterigma. [Proposed mechanisms for ballistospore discharge in *Itersonilia perplexans*: TBMS (1984) *82* 13–29.]

ballotini Small glass beads, obtainable in a range of sizes, used e.g. for the ballistic disintegration of cells; for this purpose, cells and Ballotini are shaken together in e.g. a BRAUN MSK TISSUE DISINTEGRATOR.

Baltimore classification An early system of virus classification, proposed by Baltimore [Bact. Rev. (1971) *35* 235–241], based on the nature of the genome and the strategy for viral gene expression (see VIRUS). Six classes (groups) were distinguished. *I*. dsDNA viruses. *II*. ssDNA viruses (only positive-sense ssDNA were known at the time). *III*. dsRNA viruses. *IV*. Positive-sense ssRNA viruses. *V*. Negative-sense ssRNA viruses. *VI*. RNA viruses in which DNA is synthesized (by reverse transcription) from the genome, and mRNA is transcribed from the DNA. (More classes would now be required to accommodate current knowledge of viral strategies.)

bamboo mosaic virus See POTEXVIRUSES.

***Bam*HI** A RESTRICTION ENDONUCLEASE from *Bacillus amyloliquefaciens;* G/GATCC.

banana bunchy top virus See LUTEOVIRUSES.

banana leaf spot (sigatoka disease) A disease of the banana plant (*Musa*) characterized by leaf lesions and imperfect development of the fruit; causal agent: *Mycosphaerella musicola*. The disease has been controlled by spraying with copper compounds or with mineral oil.

band centrifugation See CENTRIFUGATION.

Bangia See RHODOPHYTA.

Bang's disease See BRUCELLOSIS.

bank (gene bank) *Syn*. LIBRARY.

Banzi virus See FLAVIVIRIDAE.

bar A unit of pressure equal to 100 kPa (cf. PASCAL).

barban 4-Chlorobut-2-ynyl-3-chlorophenylcarbamate: a compound used e.g. for the control of wild oats; it also acts as a NITRIFICATION INHIBITOR.

barbital (barbitone) 5,5′-Diethylbarbituric acid. (Barbituric acid = malonylurea.)

barbone *Syn*. HAEMORRHAGIC SEPTICAEMIA (2).

Barbour Stoenner Kelly medium See BORRELIA (*B. burgdorferi*).

Barbulanympha See HYPERMASTIGIDA.

bark-boring beetles See SCOLYTIDAE.

bark necrosis See CANKER.

barley B-1 virus See POTEXVIRUSES.

barley diseases See CEREAL DISEASES.

barley stripe mosaic virus See HORDEIVIRUSES.

barley yellow dwarf virus See LUTEOVIRUSES.

barley yellow mosaic virus See POTYVIRUSES.

barley yellow striate mosaic virus See RHABDOVIRIDAE.

Barmah Forest virus See ALPHAVIRUS.

baroduric *Syn*. BAROTOLERANT.

barophile An organism which grows optimally or obligately under elevated hydrostatic pressure; under atmospheric pressure (101.325 kPa) some barophiles (e.g. *Shewanella benthica*) continue to grow (less vigorously) but others (obligate barophiles) die at a temperature-dependent rate. Barophilic bacteria in seas and oceans occur both in the nutrient-poor seawater and in the relatively nutrient-rich guts of invertebrates (e.g. holothurians – sea cucumbers); those occurring in cold regions (<4°C) are heterotrophic bacteria (cf. HYDROTHERMAL VENT). [Reviews: ARM (1984) *38* 487–514; MS (1986) *3* 205–211.] (See also BAROTOLERANT.)

barophilic Having the characteristics of a BAROPHILE.

barosensitive See BAROTOLERANT.

barotolerant (baroduric) Refers to an organism which can grow at elevated pressures (up to ca. 400–600 atmospheres; 1 atm = 101.325 kPa) but which grows optimally at atmospheric pressure (cf. BAROPHILE); many bacteria (including e.g. *Escherichia coli* and *Pseudomonas aeruginosa*) are barotolerant, but some (e.g. *Vibrio parahaemolyticus*) cannot survive for long even under relatively low pressures (e.g. ca. 200 atm) and are termed *barosensitive*.

The inhibitory effect on barotolerant organisms of pressures greater than ca. 400–600 atm appears to be due, at least partly, to inhibition of certain aspects of PROTEIN SYNTHESIS – specifically, the binding of amino acyl-tRNAs and translocation. In a number of organisms the susceptibility of protein synthesis to pressure has been linked with the 30S ribosomal subunit; interestingly, increased barotolerance occurs in those strains of *E. coli* which have become resistant to streptomycin owing to a mutation affecting the 30S subunit.

barren kinetosome See CILIFEROUS.

barrier filter See MICROSCOPY (e).

Barritt's method See VOGES–PROSKAUER TEST.

Barrouxia See EIMERIORINA.

Bartonella A genus of Gram-negative, oxidase-negative bacteria (family BARTONELLACEAE) within the alpha subgroup of Proteobacteria; the cells are small, typically capnophilic, fastidious rods which can usually be grown aerobically on blood-agar in the presence of 5% carbon dioxide [guidelines for the practical identification of isolates: JCM (1993) *31* 2381–2386].

The genus now includes several species which have been imported from the (former) genus *Rochalimaea* [IJSB (1993) *43* 777–786].

B. bacilliformis (the type species) is the causal agent of BARTONELLOSIS in man (the only natural host); the bacterium is a small bacillus or pleomorphic coccobacillus, commonly ca. 1–3 μm in length, which, in culture, bears one or more flagella at one pole and divides by binary fission; it is transmitted by sandflies, and occurs in or on erythrocytes, and in endothelial cells, in infected persons.

B. clarridgeiae has been associated with cat scratch disease [JCM (1997) *35* 1813–1818].

B. elizabethae (formerly *Rochalimaea elizabethae*) has been isolated from a case of endocarditis [JCM (1993) *31* 872–881].

B. henselae (formerly *Rochalimaea henselae*) has been associated with various types of disease, including e.g. cutaneous bacillary angiomatosis; this disease (which occurs mainly in the immunocompromised) involves reddish-coloured, typically nodular lesions which may ulcerate. (*R. quintana* can also give rise to the disease.)

B. quintana (formerly *Rochalimaea quintana*) is the causal agent of TRENCH FEVER; the cells are small rods, and the organism is aerobic and capnophilic.

B. vinsonii (formerly *Rochalimaea vinsonii*) is parasitic in voles (the 'Canadian vole agent'); it is not capnophilic.

[Diseases caused by *Bartonella* spp: RMM (1995) *6* 155–164.]

Bartonellaceae A family of bacteria which are parasitic or pathogenic in or on erythrocytes (red blood cells), and, in some cases, other cells, in various mammals, including man; the organisms can be grown in cell-free media. Genera: BARTONELLA and GRAHAMELLA.

bartonellosis (Carrión's disease) A human disease, occurring in parts of South America, caused by *Bartonella bacilliformis*; it is transmitted by the bites of sandflies of the genus *Phlebotomus*. There are two forms of the disease which usually occur in sequence: the first stage (*Oroya fever*) is characterized by fever and anaemia, the second (*verruga peruana*) by wart-like, ulcerating skin lesions.

basal body (a) (prokaryotic) See FLAGELLUM (a).

(b) (eukaryotic) (basal granule; blepharoplast; kinetosome; mastigosome) In some (ciliated or flagellated) cells: an intracellular microtubular structure which resembles a CENTRIOLE and which can serve e.g. as a MICROTUBULE-ORGANIZING CENTRE for the formation of the axoneme of a CILIUM or FLAGELLUM; unlike the prokaryotic basal body, a eukaryotic basal body plays no role in flagellar beating per se: see FLAGELLUM (b). A basal body can develop from a centriole or it can be formed de novo (e.g. during the transition of *Naegleria gruberi* from the non-flagellated to the flagellated form). In at least some cases a basal body can give rise to a centriole. A basal body differs from a centriole e.g. in that it is generally longer, and in that it has a system of microtubular and fibrillar filaments (*rootlets*) which arise from the side of the basal body and which may pass from one basal body to another.

(See also GRAIN CONVENTION; PITELKA CONVENTION.)

basal granule Obsolete term for BASAL BODY (b).

basal medium See MEDIUM.

basauxic development See CONIDIUM.

base analogues Reagents which are analogues of – and which can substitute for – normal bases in DNA; they can act as MUTAGENS since, during subsequent replication of the DNA, the base analogue may pair with the wrong normal base to generate a POINT MUTATION. For example, 2-aminopurine (AP) may be incorporated into DNA in place of adenine; during subsequent replication AP can pair with cytosine which, on further replication, pairs with guanine (an A·T-to-G·C transition). (See also 5-BROMOURACIL.) (Some base analogues in DNA may be recognized by the cell as 'damage' and may induce mutation by error-prone repair: see SOS SYSTEM.) (cf. BUDR and DDTTP.)

base composition (of DNA) See BASE RATIO and GC%.

base excision repair A form of EXCISION REPAIR that follows removal of a base from DNA (see ADAPTIVE RESPONSE and URACIL-DNA GLYCOSYLASE).

After removal of the base, the phosphodiester bond on one side of the AP SITE is cleaved by an *AP endonuclease*. The main AP endonuclease in *Escherichia coli* is *endonuclease III*; this is a multifunctional enzyme whose structure suggests that cleavage of phosphodiester bonds involves a nucleophilic attack on the P–3′O link [Nature (1995) *374* 381–386]. Once the phosphodiester bond has been cleaved, a DNA polymerase can replace the damaged nucleotide. As well as the damaged nucleotide, several adjacent nucleotides may also be replaced [NAR (1998) *26* 1282–1287]. The repair is completed by a ligase.

base pair A pair of bases, each in a separate NUCLEOTIDE, in which each base is hydrogen-bonded to the other. A 'classical' (Watson–Crick) base pair always contains one purine and one pyrimidine: adenine pairs specifically with thymine (A–T), guanine with cytosine (G–C), uracil with adenine (U–A); the two bases in a classical base pair are said to be *complementary* to each other. (Other combinations (A–A, G–T, A–C, G–G, A–G) can occur under certain circumstances [Nature (1976) *263* 285–289] but these are less stable than the classical pairs.) Classical base-pairing is responsible for holding together the strands in dsDNA (see DNA) or in dsRNA (e.g. in certain viruses); it is also involved e.g. in ordering the sequence of nucleotides in DNA REPLICATION and TRANSCRIPTION.

base ratio (dissymmetry ratio) Of microbial DNA: the ratio (A + T)/(G + C) in which A, T, G and C represent relative molar amounts of adenine, thymine, guanine and cytosine, respectively. (cf. GC%.) Species with a base ratio >1 may be called AT types; GC types have a base ratio <1.

basic dye See DYE.

basic fuchsin See FUCHSIN.

basic orange *Syn.* ACRIDINE ORANGE.

basic replicon (1) The smallest part of a replicon which encodes all the functions necessary for replication.

(2) As (1), above, with the added constraint that replication maintains the wild-type COPY NUMBER [Book ref. 161, p. 202].

Basidiobolus See ENTOMOPHTHORALES.

basidiocarp A fruiting body of a basidiomycete.

basidiole (1) An immature BASIDIUM prior to the appearance of the sterigmata. (2) An aborted basidium. (3) A sterile cell which is similar in appearance to an immature basidium.

basidiolichen A lichen (see LICHENS) in which the mycobiont is a basidiomycete.

basidiomycetes Fungi of the BASIDIOMYCOTINA.

Basidiomycotina (the 'basidiomycetes') A subdivision of fungi (division EUMYCOTA) characterized by the formation of sexually derived spores (BASIDIOSPORES) on *basidia* (see BASIDIUM). The basidiomycetes are typically terrestrial saprotrophs or parasites; they include e.g. mushrooms, puffballs, stinkhorns, and the (parasitic) rust and smut fungi. A few basidiomycetes are marine organisms (see e.g. NIA). In most species the thallus is a well-developed septate mycelium which often has CLAMP CONNECTIONS (see also DOLIPORE SEPTUM); some basidiomycetes can occur in a yeast-like (unicellular) phase (see FILOBASIDIACEAE) – as can certain fungi which appear to be the anamorphs of basidiomycetes (see e.g. SPOROBOLOMYCETACEAE).

In the *typical* life cycle, the germination of a basidiospore leads to the development of a septate mycelium consisting of uninucleate, haploid cells. At some stage the thallus enters DIKARYOPHASE (see also HYPHA), and eventually karyogamy and MEIOSIS occur in the developing basidium – which typically gives rise to four (haploid) basidiospores.

Classes: GASTEROMYCETES, HYMENOMYCETES, UREDINIOMYCETES, USTILAGINOMYCETES.

Basidiophora A genus of fungi (order PERONOSPORALES) in which the sporangiophores resemble basidia, and the sporangia are borne on short sterigmata; the organisms include parasites of the ornamental aster and of members of the Compositae.

basidiospores Sexually-derived, typically uninucleate and haploid SPORES formed on basidia (see BASIDIUM) by fungi of the BASIDIOMYCOTINA. According to species, basidiospores may be any of various shapes and sizes, and either pigmented or non-pigmented. In many basidiomycetes the basidiospores are BALLISTOSPORES, while in some (e.g. smut fungi) they are STATISMOSPORES. (cf. GASTEROID BASIDIOSPORE.) On germination, a basidiospore commonly gives rise to a germ tube, but in some cases GERMINATION BY REPETITION may occur.

Dispersal of basidiospores. In agarics with aequihymeniiferous lamellae (see LAMELLA), the ballistospores are projected more or less horizontally for ca. 0.1–0.2 mm before gravity initiates a vertical fall; the spores on a given basidium are commonly discharged in rapid succession (within seconds or minutes of one another). In these agarics the shape of the lamella facilitates spore dispersal – spores from the more proximal parts of the lamella being able to fall vertically even when the lamellae are orientated several degrees from the vertical.

In e.g. *Coprinus* spp the inaequihymeniiferous lamellae occur close together on the pileus, and the shape of the lamellae does not facilitate dispersal of the ballistospores. However, in some species cystidia (see CYSTIDIUM) may serve to hold apart adjacent hymenial surfaces. In other species the process of *autodigestion* is an aid to spore dispersal; autodigestion involves the progressive liquefaction of the lamellae, beginning at the free edge of the lamella and progressing towards the pileus as fresh crops of basidia become mature. Thus, spores discharged by a given basidium need fall for only a short distance between adjacent lamellae before becoming free of the fruiting body.

See also e.g. CYATHUS, LYCOPERDALES, SPHAEROBOLUS.

basidium (*pl.* basidia) A microscopic, unicellular or multicellular structure which gives rise to the sexually-derived spores (BASIDIOSPORES) in fungi of the BASIDIOMYCOTINA; basidiospores are formed externally, i.e., as extrusions, on the basidium (but see TETRAGONIOMYCES). In some basidiomycetes (e.g. HYMENOMYCETES) the basidia are formed in a layer on specialized fertile tissue (see HYMENIUM), while in others (e.g. UREDINIOMYCETES) they are formed individually from germinating spores.

In many species the basidium is a simple, cylindrical or clavate structure which gives rise to (usually) four apically and symmetrically situated basidiospores, each basidiospore commonly being borne on a small, narrow protrusion (STERIGMA). In other species the basidium may be apically bifurcate, each fork (sterigma) bearing a single terminal basidiospore (as in members of the DACRYMYCETALES), or it may be divided, apically, into four substantial finger-like protrusions (sterigmata), each bearing a terminal basidiospore (as in members of the TREMELLALES). A basidium which is not divided, internally, by septa is called a *holobasidium*; such basidia are formed by members of the HOLOBASIDIOMYCETIDAE. A basidium which is divided, internally, by transverse or longitudinal septa (as e.g. in *Auricularia* and *Tremella*, respectively) is called a *phragmobasidium*; such basidia are formed by members of the PHRAGMOBASIDIOMYCETIDAE and of the UREDINIOMYCETES and USTILAGINOMYCETES.

Basidium formation. A simple holobasidium typically develops from a terminal dikaryotic hyphal cell within a hymenium; the boundary between this cell and the parent hypha is often marked by a CLAMP CONNECTION. Within the developing basidium, karyogamy and MEIOSIS commonly give rise to four haploid nuclei. (See also METABASIDIUM and PROBASIDIUM.) Four small protuberances (sterigmata) then form at the apical (free) end of the developing basidium, and the terminal part of each sterigma swells to form a basidiospore initial; one nucleus migrates into each initial, and the basidiospores are subsequently delimited. In some species mitotic division occurs within the developing basidiospores; in such cases either binucleate basidiospores are formed, or uninucleate basidiospores are formed after one daughter nucleus from each pair has returned to the basidium.

In rust and smut fungi, the basidium typically develops from a short germ tube which arises from the *teliospore*. In some cases a diploid nucleus passes from the teliospore into the germ tube and undergoes meiosis; the structure which develops (the metabasidium, or 'promycelium') is septate in many species and (in rusts, and in many smut fungi) it gives rise to basidiospores both apically and laterally. Rust fungi characteristically form four basidiospores per basidium; in many smut fungi the basidium may give rise to an indefinite number of basidiospores following mitotic divisions within the metabasidium.

[Review: BBMS (1983) *17* 82–94.]

basionym (basonym) The name of the species whose specific epithet is included in a new combination (see COMB. NOV.).

basipetal development Development from the tip towards the base or point of attachment; e.g., in a chain of basipetally developing spores the first-formed spores occupy positions in the terminal or distal parts of the chain, while spores formed later occupy the more proximal positions. (cf. ACROPETAL DEVELOPMENT.)

basonym *Syn.* BASIONYM.

basophil A PMN (q.v.) which can respond to certain stimuli e.g. by rapidly secreting vasoactive products; it is primarily a circulatory cell, and is important e.g. in immediate-type hypersensitivity reactions. As in the MAST CELL, with which it shares many characteristics, the basophil surface contains e.g. many high-affinity receptor sites for the Fc portion of IgE antibodies, and its (basophilic) cytoplasmic granules also contain substances such as HISTAMINE and SEROTONIN; stimuli which cause activation and degranulation lead to the secretion of various products and to the formation of e.g. SUPEROXIDE and H_2O_2. (See also JONES–MOTE SENSITIVITY.)

batch culture (closed culture) A form of CULTURE (sense 2) in which a given volume of liquid medium is inoculated with cells (e.g. bacteria) capable of growth in that medium, and the inoculated medium is incubated for an appropriate period of time; cells growing under these conditions are exposed to a continually changing environment due e.g. to the gradual consumption of nutrients and the accumulation of metabolic wastes (cf. CONTINUOUS CULTURE and FED BATCH CULTURE).

The growth curve (see GROWTH) obtained by monitoring a batch culture commonly exhibits a sequence of four main phases of growth. In the *lag phase* the growth rate – i.e., the rate of increase in cell numbers (or biomass) – is initially minimal but subsequently rises to a value dictated e.g. by the prevailing conditions (e.g. temperature, concentration of nutrients etc). The length of the lag phase is influenced by the cultural history of the cells in the inoculum. For example, if slowly dividing cells from a nutrient-poor environment are transferred to a nutrient-rich medium which can support a higher rate of growth, there is usually a relatively long lag phase during which time the cells become adapted to the new environment; during this period of adaptation the cells exhibit *unbalanced* GROWTH. Subsequently, growth occurs at a new, higher rate permitted by the higher levels of nutrients. A long lag phase may also occur e.g. if the carbon source in the new medium differs from that previously used by the cells. (cf. DIAUXIE.) When actively dividing cells are transferred to a medium which offers conditions similar to those under which the cells were previously growing, a lag phase is not observed.

At the end of the lag phase the cells enter the *exponential* (= *logarithmic* or *log*) *phase* of growth in which, for a given organism, the growth rate is both constant and maximal for the particular growth conditions. In this phase there is an exponential increase in cell numbers (and biomass); this type of growth is referred to as *balanced* GROWTH. (See also TROPHOPHASE.)

In the *stationary phase* the growth rate declines and eventually reaches zero. (See also IDIOPHASE.)

In the *death phase* the number of viable cells in the culture (maximal in the stationary phase) declines.

batch retort In CANNING: a vessel within which filled, sealed cans (or other containers) undergo heat treatment in a batch-type process (cf. COOKER–COOLER). The cans are subjected to saturated steam under pressure or to an air–steam mixture, or (in an *overpressure retort*) are submerged in heated water under an air pressure of up to ca. 250 kPa. In some batch retorts (e.g. the *Konservomat*) the load is agitated to facilitate heat penetration.

bating (of hides) See PROTEASES.

Batrachospermum See RHODOPHYTA.

Battarrea See GASTEROMYCETES (Tulostomatales).

Battey bacillus Strain(s) of *Mycobacterium intracellulare* or, loosely, strain(s) of related species (including *M. avium*).

Bayer's junction (Bayer's patch) See ADHESION SITE.

Bayleton See TRIADIMEFON.

BB-transhydrogenase See TRANSHYDROGENASE.

BCDF (*immunol.*) See LYMPHOKINES.

BCF Bioconcentration factor: a measure of the degree to which a compound (commonly a xenobiotic), present in an aquatic environment, is accumulated in the biomass of organisms (e.g. algae) living in that environment. BCF $= C_o/C_W$ where $C_o =$ concentration of the compound in the organisms, and C_W is the concentration of the compound in the water. (See e.g. TBTO.)

BCG Bacille (bacillus) Calmette–Guérin: an attenuated strain of *Mycobacterium bovis* used e.g. as a vaccine against TUBERCULOSIS. The vaccine often gives protection against e.g. tubercular meningitis in childhood – but in a number of studies it has failed to protect against pulmonary tuberculosis [Lancet (1990) *335* 1016–1020]; moreover, it may produce disseminated infections in the immunocompromised (e.g. AIDS patients). In some instances, BCG has provided some protection against LEPROSY.

[Mycolic acids in BCG substrains: JGM (1984) *130* 2733–2736.]

(See also ESAT-6.)

BCGF (*immunol.*) See LYMPHOKINES.

Bchl Bacteriochlorophyll: see CHLOROPHYLLS.

BclI A RESTRICTION ENDONUCLEASE from *Bacillus caldolyticus*; T/GATCA.

BCNU See *N*-NITROSO COMPOUNDS.

Bd (buoyant density) See CENTRIFUGATION and GC%.

BD-type starter See LACTIC ACID STARTERS.

bdellocyst See BDELLOVIBRIO.

bdelloplast See BDELLOVIBRIO.

Bdellospora See ZOOPAGALES.

Bdellovibrio A genus of aerobic Gram-negative bacteria which are characteristically predatory on other bacteria (cf. VAMPIROVIBRIO); they occur worldwide in soil and sewage, and in freshwater and marine habitats. On primary isolation all strains are obligately parasitic on certain Gram-negative bacteria, including e.g. *Aquaspirillum serpens*, *Escherichia coli*, *Pseudomonas* spp, and *Rhodospirillum rubrum* – the range of prey species varying with the strain of *Bdellovibrio*; prey-independent, cultivable strains can develop from strictly predatory strains, and some strains are facultatively predatory. The predatory strains exhibit a biphasic life cycle: the predatory phase (in which the non-growing cells search for prey) alternating with a phase of growth and reproduction in the periplasmic space of a host cell. Predatory cells of *Bdellovibrio* are vibrioid, 0.2–0.5 × 0.5–1.4 μm, each having a single polar sheathed flagellum (the sheath being continuous with the outer membrane); cells of prey-independent strains may be larger and non-motile, and may form a yellowish pigment. All strains are chemoorganotrophs. Growth occurs optimally at 28–30°C, and metabolism is respiratory (oxidative). Three species are recognized [Book ref. 22, pp. 118–124]: *B. bacteriovorus*, *B. starrii* and *B. stolpii*; the genus is divided by phage typing into six groups: groups I–IV consist of strains of *B. bacteriovorus*, group V contains *B. stolpii* strains, and group VI consists of strains of *B. starrii* (for which no phages are currently known). GC%: ca. 33–51. Type species: *B. bacteriovorus*.

Predation. A predatory *Bdellovibrio* cell initially attaches, via a 'holdfast' region at its non-flagellar end, to a prey cell. ('Bdello' derives from the Greek for 'leech'.) The predatory cell secretes enzymes (e.g. lysozyme) and enters the periplasmic space of the prey cell via the pore produced by enzymic action. (*Bdellovibrio* can penetrate cells of *E. coli* even when they have a thick capsule [Microbiology (1997) *143* 749–754].) The prey cell quickly dies and subsequently rounds up to become a *bdelloplast*; during this process the peptidoglycan sacculus of the prey cell becomes insensitive to lysozyme, thus inhibiting predation by other *Bdellovibrio* cells. Within the periplasmic space of the bdelloplast, the predatory cell grows and develops into a long, spiral, non-motile filament; the filament subsequently fragments to form motile predatory cells which, on release, can initiate a new cycle. *Bdellovibrio* strain W can form a resting cell (*bdellocyst*) within the bdelloplast; a bdellocyst has a thickened peptidoglycan wall and has increased resistance to elevated temperatures and to desiccation.

Isolation. *Bdellovibrio* may be isolated from soil infusions etc. e.g. by membrane filtration (membrane pore size ca. 1.2 μm) – the filtrate being inoculated onto a lawn plate prepared from a sensitive strain of bacteria, or onto a soft agar which incorporates suitable prey cells; on incubation, viable cells of *Bdellovibrio* give rise to macroscopic plaques.

bDNA assay (branched DNA assay) A procedure, based on signal amplification, which is used for quantifying a specific single-stranded DNA or RNA target sequence in a sample; essentially, signal amplification involves (i) the use of a target-specific PROBE for immobilizing the target nucleic acid on a solid support, and (ii) the use of other types of probe for labelling each bound target sequence with an enzyme system that generates (from a suitable substrate) an amplified chemiluminescent signal – the intensity of which is related to the number of target molecules in the sample.

Signal amplification avoids an important problem associated with target amplification methods (such as PCR): the problem of potential contamination of apparatus and environment with copies of the (highly amplified) target nucleic acid. It has the disadvantage that, by failing to provide copies of the target, further investigation of the target molecule (e.g. by sequencing) is precluded.

bDNA-based commercial kits (Quantiplex™; Bayer Diagnostics, Leverkusen, Germany) have been used e.g. for quantifying hepatitis viruses B and C and the human immunodeficiency virus (HIV). In a given assay, a divalent probe is used initially to bind target molecules to the surface of a microtitre plate. Another divalent probe is then used to link each bound target molecule to a synthetic multi-branched DNA molecule (bDNA; the 'amplifier'); subsequently, excess enzyme-labelled probes are allowed to bind to (multiple) probe-binding sites on the bDNA. After washing, addition of the enzyme's substrate (a dioxetane) generates the chemiluminescent signal.

A bDNA assay may be used for detection, as well as quantification, of a given sequence; for example, this method has been used for detecting the *mecA* gene in blood culture bottles [JCM (1999) *37* 4192–4193].

BDProbe Tec™ ET See SDA.

beak (*mycol.*) The tubular neck of a perithecium or pycnidium.

bean common mosaic virus See POTYVIRUSES.

bean golden mosaic virus (BGMV) A member of the GEMINIVIRUSES. BGMV can cause chlorosis in beans (*Phaseolus vulgaris*); it is transmitted by the whitefly *Bemisia tabaci*, and occurs in tropical America.

bean paste *Syn.* MISO.

bean summer death virus See GEMINIVIRUSES.

bean yellow mosaic virus See POTYVIRUSES.

beard-lichens (beard-moss) Lichens of the genera ALECTORIA, BRYORIA, and/or USNEA.

Beauveria A genus of fungi of the class HYPHOMYCETES. *B. bassiana* forms a septate mycelium; spheroidal conidia are borne on the sympodially branched, narrow, apical region of each flask-shaped conidiophore. *B. bassiana* causes the SILKWORM DISEASE *muscardine*, and has been used (e.g. in the former USSR) for the BIOLOGICAL CONTROL of certain crop pests – e.g. Colorado beetle (*Leptinotarsa decemlineata*); for biological control the fungus has been used e.g. in the form of a powder, *Boverin*: a dried mixture of conidia and kaolin. *B. bassiana* forms a depsipeptide toxin, *beauvericin* (see ENNIATINS).

beauvericin See ENNIATINS.

Bebaru virus See ALPHAVIRUS.

becampicillin See PENICILLINS.

Bedsonia See CHLAMYDIA.

bee acute paralysis virus See PICORNAVIRIDAE.

bee diseases 'Domestic' hive bees – *Apis mellifera* (the European honey-bee) and *A. cerana* (the Eastern honey-bee) – are subject to a range of diseases of microbial aetiology. In general, larvae are affected more often than adults. Spread of disease is often kept under control by the activity of worker bees which remove diseased and dead larvae from the hive. *Bacterial diseases* include e.g. AMERICAN FOULBROOD and EUROPEAN FOULBROOD (see also SPIROPLASMA). *Fungal diseases* include CHALK-BROOD and STONE-BROOD. *Protozoal* parasites and pathogens of bees include e.g. *Nosema apis*, which develops in the midgut epithelial cells in adult honey-bees (infected bees may show no obvious symptoms, although they have a shortened life-span), and *Malpighamoeba mellificae* (*Vahlkampfia mellificae*), which infects the lumen of the Malpighian tubules in adult bees [SEM of developmental stages: JP (1985) *32* 139–144]. *Virus diseases* include several caused by small unclassified RNA viruses (see PICORNAVIRIDAE) and some caused by iridescent and filamentous viruses. [Book ref. 66.] (See also INSECT DISEASES.)

bee slow paralysis virus See PICORNAVIRIDAE.

bee virus X See PICORNAVIRIDAE.

beech bark disease A European and North American disease of beech trees (*Fagus* spp) caused by *Nectria coccinea*. Infection commonly follows infestation of the tree by beech scale insects (*Cryptococcus fagisuga*). Symptoms: phloem and cambium cells in the bark die in patches, allowing leakage and blackening of sap, and there may be yellowing of the leaves and die-back of the crown; infected trees may die.

Beechey ground squirrel hepatitis virus *Syn.* GROUND SQUIRREL HEPATITIS VIRUS.

beef-steak fungus *Fistulina hepatica* (see APHYLLOPHORALES).

beer-making See BREWING.

beer spoilage Off-flavours in beer may be caused by contaminating organisms – e.g. *Acetobacter* spp can produce acetic acid and cause ROPINESS; heterofermentative lactobacilli (which preferentially ferment maltose) can cause off-flavours (e.g. diacetyl), turbidity and/or ropiness. *Sarcina sickness* results from the production of diacetyl by *Pediococcus cerevisiae*. Other spoilage organisms include *Candida* and *Pichia* spp.

beet cryptic viruses See CRYPTIC VIRUSES.

beet curly top virus See GEMINIVIRUSES.

beet leaf curl virus See RHABDOVIRIDAE.

beet mild yellowing virus See LUTEOVIRUSES.

beet mosaic virus See POTYVIRUSES.

beet necrotic yellow vein virus (BNYVV) A virus which can cause severe disease ('rhizomania') in sugar beet (*Beta vulgaris*); it is transmitted by the 'fungus' *Polymyxa betae* – which is also responsible for the persistence of the virus in the soil. BNYVV may belong to the 'furovirus' group – see SOIL-BORNE WHEAT MOSAIC VIRUS. (cf. TOBAMOVIRUSES.) [BNYVV RNA: JGV (1985) *66* 345–350.]

beet western yellows virus See LUTEOVIRUSES.

beet yellow net virus See LUTEOVIRUSES.

beet yellow stunt virus See CLOSTEROVIRUSES.

beet yellows virus group *Syn.* CLOSTEROVIRUSES.

Beggiatoa A genus of GLIDING BACTERIA (see CYTOPHAGALES); species are widespread e.g. in freshwater, estuarine and marine waters and sediments, in the rice rhizosphere, and in activated sludge. The organisms are non-sheathed, non-pigmented, multicellular filaments (trichomes) which form necridia and hormogonia; according to species, the trichomes are 1 to >50 μm in diameter. The organisms appear to be chemoorganotrophic heterotrophs, although at least some strains may be able to grow mixotrophically using sulphide as an electron donor; refractile 'intracellular' granules of sulphur (located between cytoplasmic membrane and cell wall) are formed in the presence of sulphide. Metabolism is respiratory, and a complete TCA cycle is present. Most strains can use acetate and/or lactate as the sole source of carbon; none can use hexoses. The main storage product is poly-β-hydroxybutyrate; volutin is also formed. [Review: ARM (1983) *37* 341–354.]

Beggiatoaceae See CYTOPHAGALES.

Beijerinckia A genus (*incertae sedis*) of Gram-negative, aerobic, catalase-positive, chemoorganotrophic, asporogenous bacteria which occur e.g. in the phyllosphere and in soils (particularly in the tropics). Cells: typically round-ended rods, ca. 0.5–1.5 × 1.7–4.5 μm, peritrichously flagellated or non-motile; division involves constriction. Each cell characteristically contains two PHB-containing polar bodies. In some species each cell may be encysted, or several cells may occur within a single capsule. Metabolism is respiratory (oxidative), with O_2 as terminal electron acceptor. NITROGEN FIXATION is carried out under both atmospheric and microaerobic conditions; in many strains N_2 is used in preference to NO_3^-. All strains can utilize glucose, fructose and sucrose as carbon sources; peptone does not support growth. Slime (sometimes elastic) is commonly formed, particularly on agar under N_2-fixing conditions. Colonies are often some shade of pink, or orange to pale brown. No pellicle is formed on liquid media. Optimum growth temperature: 20–30°C; no growth at 37°C. Growth can occur within the pH range ca. 3–10. GC%: ca. 55–61. Type species: *B. indica*; other species: *B. derxii*, *B. fluminensis* and *B. mobilis*. [Book ref. 22, pp. 311–321.]

bell morels The fruiting bodies of VERPA spp.

belladonna mottle virus See TYMOVIRUSES.

benalaxyl See PHENYLAMIDE ANTIFUNGAL AGENTS.

Beneckea An obsolete bacterial genus, members of which are now included in the genus VIBRIO.

benign (*med.*) (1) (of disease) Mild; self-limiting; not recurrent. (2) (of tumours) Not MALIGNANT (sense 2).

benign enzootic paresis *Syn.* TALFAN DISEASE.

benign foot-rot (of sheep) See SCALD.

benign tertian malaria See MALARIA.

Benlate See BENOMYL.

benomyl (methyl-(1-*n*-butylcarbamoyl)-benzimidazole-2-carbamate; trade name: e.g. Benlate) An agricultural ANTIFUNGAL AGENT (see BENZIMIDAZOLES (a)). Benomyl is a systemic antifungal agent which has both eradicant and protectant properties against a wide range of plant diseases (e.g. many powdery mildews, apple scab, botrytis, eyespot of wheat and barley, black spot of roses); when used as a seed dressing it protects against seed-borne diseases such as smuts and bunts of cereals, and when mixed with a dithiocarbamate such as maneb or mancozeb it gives some control against glume blotch in wheat and against some cereal rusts. It is also used to prevent storage rots of fruit and vegetables (see e.g. GANGRENE sense 2). Benomyl owes its activity to its degradation, in solution, to the fungitoxic substances MBC (see BENZIMIDAZOLES) and butyl isocyanate; it also acts as a cutinase inhibitor, preventing penetration of the plant cuticle by the pathogen (see CUTIN) [Book ref. 58, pp. 94–95]. Benomyl-resistant strains of many fungal pathogens have emerged, reducing the usefulness of the fungicide in agriculture. [Bacterial degradation of benomyl: AvL (1978) *44* 283–292, 293–309.]

benquinox (1,2-benzoquinone-*N′*-benzoylhydrazone-4-oxime) A QUINONE ANTIFUNGAL AGENT which is used mainly as a seed dressing.

Benson–Calvin–Bassham cycle *Syn.* CALVIN CYCLE.

benthic Refers to the mud, sand etc, and/or to the indigenous organisms, at the bottom of a lake, sea or other body of water. (cf. PELAGIC; see also LITTORAL.)

benzalkonium chloride See QUATERNARY AMMONIUM COMPOUNDS and e.g. KLEINSCHMIDT MONOLAYER TECHNIQUE.

benzene Aerobic degradation: see HYDROCARBONS. Degradation in an anoxic environment: see BIOREMEDIATION.

benzethonium chloride See entry QUATERNARY AMMONIUM COMPOUNDS.

benzidine test A test used to detect bacterial CYTOCHROMES. Colonies of the test strain are flooded with a reagent consisting of benzidine dihydrochloride dissolved in a mixture of acetic acid and ethanol; on addition of an equal volume of H_2O_2 (5%), cells which contain cytochromes develop a green or blue–green coloration. (cf. OXIDASE TEST.)

benzimidazoles (a) (as antifungal agents) An important group of agricultural systemic ANTIFUNGAL AGENTS; they are often called 'MBC fungicides' because many of them decompose in aqueous solution (or in the plant) to form methyl benzimidazol-2-yl carbamate (MBC, CARBENDAZIM) – apparently the primary fungitoxic agent. MBC appears to act primarily in the nucleus where it disrupts the formation of spindle microtubules, inhibiting or disrupting mitosis. Single-step resistance to benzimidazoles tends to emerge readily. The group includes BENOMYL, CARBENDAZIM, FUBERIDAZOLE, THIABENDAZOLE and THIOPHANATE-METHYL.

(b) (as antiviral agents) See e.g. ENVIROXIME and HBB.

benzoic acid (C_6H_5COOH) An antibacterial and antifungal ORGANIC ACID used (as the free acid or as benzoate) as a PRESERVATIVE (e.g. in fruit juices, cordials, sauces, acidic foods, pharmaceuticals) and, with other agents, for the topical treatment of certain fungal infections; it is maximally effective below ca. pH 5, activity decreasing sharply with increase in pH above this value. Benzoic acid can be inactivated by certain colloids (e.g. kaolin). (cf. PARABENZOATES.)

benzoquinones See QUINONES.

benzoyl peroxide See ACNE.

benzyl alcohol See ALCOHOLS.

2-benzyl-4-chlorophenol See PHENOLS.

benzylpenicillin (penicillin G) One of the first of the natural PENICILLINS to be produced (from e.g. *P. chrysogenum*); $R = C_6H_5.CH_2.CO$ (see β-LACTAM ANTIBIOTICS). It is active against many Gram-positive bacteria which do not produce β-LACTAMASES, but is poorly active against Gram-negative species. It is administered parenterally.

Berenil A trypanocidal DIAMIDINE which binds to AT-rich regions of dsDNA (apparently in the same way as does NETROPSIN).

beriberi (acute cardiac) See CITREOVIRIDIN.

Berkefeld candle See FILTRATION.

Berne virus An enveloped RNA virus (proposed family: TOROVIRIDAE) isolated from a horse. The virion is pleomorphic (ca. 120–140 nm diam.) and consists of an elongated, curved nucleocapsid enclosed by a peplomer-bearing envelope. The genome is a single molecule of positive-sense ssRNA (MWt ca. 6×10^6); polyadenylated subgenomic mRNAs are apparently synthesized in infected cells. Maturation apparently involves the budding of preformed tubular nucleocapsids into Golgi vesicles and intracytoplasmic cisternae; the nucleocapsid appears to undergo a morphological change during budding [JGV (1986) *67* 1305–1314].

Berry–Dedrick phenomenon See NON-GENETIC REACTIVATION.

Besnoitia A genus of coccidian protozoa (suborder EIMERIORINA); members are classified by some authors [AP (1982) *20* 403–406] in the genus ISOSPORA. The organisms are parasitic in a range of animals – e.g. cattle, horses, sheep, cats, rodents; transmission may occur e.g. via blood-sucking flies or by ingestion of infected tissues. Disporic, tetrazoic oocysts are formed.

β (duplex winding number) See DNA.

β-barrel pore See PROTEIN SECRETION (type IV systems).

β-exotoxin *Syn.* THURINGIENSIN.

beta interferon See INTERFERONS.

***beta*-lactam antibiotics** See β-LACTAM ANTIBIOTICS.

***beta*-lactamases** See β-LACTAMASES.

β operon See RIBOSOME (biogenesis).

***beta*-rays** See IONIZING RADIATION.

Betabacterium See LACTOBACILLUS.

betacin A BACTERIOCIN-like, heat-resistant, proteinaceous substance produced by bacteriophage SPβ-containing (or bacteriophage Z-containing) lysogens of *Bacillus subtilis* when growing on agar (but apparently not in broth cultures). Betacin inhibits the growth of non-lysogens but not that of lysogens containing SPβ (or the related phage Z). [Book ref. 170, p. 280.] (cf. BACTERIOPHAGE φBA1.)

Betacoccus Obsolete name for LEUCONOSTOC.

Betadine See IODINE (a).

Betaferon™ See INTERFERONS.

Betaherpesvirinae ('cytomegalovirus group') A subfamily of viruses of the HERPESVIRIDAE (q.v.).

The Betaherpesvirinae formerly consisted of only (i) the *human cytomegalovirus group* – type species human (beta) herpesvirus 5 (= human herpesvirus 5, HHV5; human cytomegalovirus, HCMV) and (ii) *murine cytomegalovirus group* – type species murid (beta) herpesvirus 1 (= murid herpesvirus 1; mouse cytomegalovirus, mouse CMV) [Book ref. 23, pp 49–50]; possible members included suid herpesvirus 2 (pig CMV, causal agent of INCLUSION BODY RHINITIS), murid herpesvirus 2 (rat CMV) and caviid herpesvirus 1 (guinea-pig CMV). Later, certain bovine herpesviruses were classified as 'bovine cytomegaloviruses' [JGV (1984) *65* 697–706] (see also MALIGNANT CATARRHAL FEVER).

Characteristically, the replication cycle of these viruses is appreciably longer than 24 hours (in contrast to the cycles of other herpesviruses). In culture they typically give rise to enlarged, rounded cells (cytomegalic cells) and slowly spreading foci of cell lysis. DNA-containing inclusion bodies may be seen in the nuclei and sometimes in the cytoplasm of infected cells. Typically, the host range for a given virus is narrow and is often restricted to a single species. Latent infection can be established in cell cultures and in vivo.

HCMV (often truncated to CMV) is a large herpesvirus with a linear dsDNA genome of MWt ca. 1.5×10^8 containing terminal and internal repeated sequences; the genome encodes a number of structural and regulatory proteins. Each virion consists of the DNA–protein core, a capsid, and an external lipoprotein envelope bearing surface projections, the region between envelope and capsid (the 'tegument') containing several virus-encoded proteins.

In the replication cycle of HCMV, the virion initially fuses with the cell membrane of the target cell, and the (unenveloped) virus migrates towards the nucleus. Viral DNA is replicated in the nucleus, and virus-encoded proteins are synthesized in the cytoplasm; the nucleocapsid is assembled in the nucleus.

HCMV genes are transcribed in strict temporal sequence. The 'immediate–early' genes (expressed within 2–4 hours of infection) encode regulatory proteins that are needed for synthesis of 'early' and 'late' genes. The 'early' gene products include a DNA polymerase. 'Late' gene products include capsid proteins.

Culture of human CMV is carried out in human fibroblasts. (See also SHELL VIAL ASSAY.) In vivo, replication appears to occur primarily in epithelial cells (e.g. in salivary glands, kidneys); HCMV is also known to infect peripheral blood leukocytes and haematopoietic progenitor cells.

Infection with HCMV is endemic in human populations world-wide, but in most cases it is asymptomatic; primary infection may be followed by viral persistence which may take the form of low-level replication or latent infection. (The site of latency is currently unknown.) However, the virus is e.g. a common cause of congenital viral infections, and sometimes causes a disease resembling INFECTIOUS MONONUCLEOSIS. HCMV-mediated disease appears to occur most commonly among the immunocompromised (e.g. immunosuppressed transplant patients and AIDS patients) and in those with an immature immune system (e.g. neonates). Typically, HCMV-mediated disease is associated with conditions such as fever, hepatitis and leukopenia. (See also CYTOMEGALIC INCLUSION DISEASE.)

[Epidemiology and transmission of HCMV: JID (1985) *152* 243–248. Molecular biology and immunology of HCMV (review): Bioch. J. (1987) *241* 313–324. Occupational risk of HCMV: RMM (1994) *5* 33–38. HCMV in haematological disease: BCH (1995) *8* 149–163.]

The laboratory diagnosis of HCMV can be achieved in <24 hours e.g. by SHELL VIAL ASSAY or by serological or nucleic acid-based techniques [diagnosis of HCMV by detection of antigen and DNA: RMM (1994) *5* 265–276]. A NASBA-based procedure has been used for detecting the transcript of a *late* HCMV gene in order to monitor transplant patients at risk from HCMV-related disease [JCM (1998) *36* 1341–1346].

In recent years, two more herpesviruses have been added to the Betaherpesvirinae: human (beta) herpesvirus 6 (human herpesvirus 6, HHV6) and human (beta) herpesvirus 7 (human herpesvirus 7, HHV7).

HHV6 has been identified as a causal agent of EXANTHEM SUBITUM, and is able to cause severe morbidity in immunocompromised patients; it is also a possible causal agent of e.g. lymphoma and leukaemia.

HHV6 can infect different types of cell – although *replication* may not occur in all of them; for example, latent or poorly replicative viruses have been detected in macrophages and peripheral blood mononuclear cells. HHV6 has a tropism for lymphoid cells; in vitro it replicates well in $CD4^+$ cells. Saliva appears to be a good source of the virus in infected persons.

Interestingly, the HHV6 genome includes two genes with homology to chemokine genes, and a gene resembling that of a chemokine receptor. (Certain other herpesviruses, including HCMV, encode soluble chemokine receptors.)

In cultures of peripheral blood mononuclear cells, HHV6 induces the cytokines IL-1β and TNF-α.

On the basis of (i) restriction patterns (see RFLP), (ii) binding of specific monoclonal antibodies, and (iii) cultural properties, HHV6 has been divided into two variant forms (A and B). Variant A appears to be observed more frequently in HIV-infected patients.

[HHV6 (review): BCH (1995) *8* 201–223; Lancet (1997) *349* 558–563.]

HHV7 (as well as HHV6) is widespread in the human population; unlike HHV6 it has not been firmly linked to human disease. HHV7 replicates well in $CD4^+$ cells, and may (unlike HHV6) use the CD4 antigen as a receptor. In culture, HHV7 replicates more slowly than HHV6.

[HHV7 (review): Lancet (1997) *349* 558–563.]

Another recently isolated herpesvirus, HHV8, has been placed in the GAMMAHERPESVIRINAE.

betaine *N*-Trimethylglycine: $(CH_3)_3N^+CH_2COO^-$. (See also OSMOREGULATION and STICKLAND REACTION.)

Betaseron™ See INTERFERONS.

Bethell process A method of TIMBER PRESERVATION in which the timber is first subjected to a vacuum (to withdraw air) and then exposed to a PRESERVATIVE under pressure. (cf. BOUCHERIE PROCESS; SAUG–KAPPE PROCESS.)

Bethesda–Ballerup group *Syn.* BALLERUP–BETHESDA GROUP.

Bettsia A genus of fungi of the ASCOSPHAERALES. *B. alvei* ('pollen mould') apparently occurs only in stored pollen in the combs of bee-hives.

BF_0F_1 ATPase A *bacterial* F_0F_1-type PROTON ATPASE.

BFNU See *N*-NITROSO COMPOUNDS.

BFP See PATHOGENICITY ISLAND.

BFPR (BFP) BIOLOGICAL FALSE POSITIVE REACTION.

BGAV Blue-green algal virus (CYANOPHAGE).

BGG Bovine γ-globulin.

***bgl* operon** (in *Escherichia coli*) See CATABOLITE REPRESSION.

BglF (in *Escherichia coli*) See CATABOLITE REPRESSION.

BglG (in *Escherichia coli*) See CATABOLITE REPRESSION.

***Bgl*I** A RESTRICTION ENDONUCLEASE from *Bacillus globigii*; the recognition sequence is: GCCNNNN/NGGC.

***Bgl*II** A RESTRICTION ENDONUCLEASE from *Bacillus globigii*; A/GATCT.

BGMV BEAN GOLDEN MOSAIC VIRUS.

BHIA Brain–heart infusion agar.

BHIB Brain–heart infusion broth.

BHK-21 An ESTABLISHED CELL LINE derived from hamster kidney; the cells are heteroploid and fibroblast-like.

BIAcore An analytical system in which *surface plasmon resonance* (SPR) is used to study the kinetics, affinities and binding positions in antigen–antibody and other molecular interactions. Specific molecules, in solution, are allowed to bind to an immobilized ligand at a hydrophilic sensor surface; *binding affects the refractive index in the vicinity of the sensor surface*. The refractive index (and, hence, molecular interaction at the surface) can be monitored by a beam of polarized light incident on the sensor surface. [Example of use: NAR (2000) *28* 1935–1940.]

biased random walk See RANDOM WALK.

Bicine A zwitterionic pH buffer: *N*,*N*-bis(2-hydroxyethyl)glycine. Unlike HEPES, Bicine binds divalent cations. At 25°C the pK_a of Bicine is ca. 8.3.

Biddulphia See DIATOMS.

bifactorial heterothallism See HETEROTHALLISM.

bifermentolysin See THIOL-ACTIVATED CYTOLYSINS.

bifid (1) (*adj.*) Forked. (2) (*noun*) A bacterium forked at one or both ends.

Bifidobacterium A genus of Gram-positive, asporogenous, anaerobic bacteria which occur e.g. among the GASTROINTESTINAL TRACT FLORA and the VAGINA MICROFLORA, and which carry out a HETEROLACTIC FERMENTATION in which the main products of e.g. glucose fermentation are typically acetic and lactic acids in the ratio 3:2; the key enzyme in the pathway, fructose 6-phosphate phosphoketolase [see Appendix III(b)], is a diagnostic feature of the genus. Cells: non-motile (often pleomorphic) rods and bifids. GC%: ca. 57–66. Type species: *B. bifidum* (formerly e.g. *Lactobacillus bifidus*); other species include e.g. *B. adolescentis*, *B. asteroides* (from bees), *B. breve*, *B. infantis*, *B. suis*, *B. thermophilum*. (See also INDICATOR ORGANISM.) [Isolation and media: Book ref. 46, pp. 1951–1961.]

bifidus pathway See HETEROLACTIC FERMENTATION.

bifonazole See AZOLE ANTIFUNGAL AGENTS.

big bone disease See OSTEOPETROSIS.

big liver disease See LYMPHOID LEUKOSIS.

big T antigen (large T antigen) See POLYOMAVIRUS.

biguanides *Syn.* DIGUANIDES.

biguttulate (*mycol.*) Containing two GUTTULES.

bijou A glass, screw-cap, 5–7 ml bottle.

bikaverin A benzoxanthone quinone produced by *Fusarium* spp (e.g. *F. moniliforme*); it is a wine-red pigment which shows some anti-protozoal activity [Book ref. 131, pp. 223–225].

Biken test A precipitin test, based on the ELEK PLATE method, used for detecting the production of heat-labile enterotoxin by ETEC (q.v.) [JCM (1981) *13* 1–5].

bilabiate (*mycol.*) Refers to those asci in which spore discharge is preceded by the development of a vertical slit in the apex of the ascus.

bilateral trama A type of trama in which the hyphae have a bilateral arrangement. In e.g. members of the Amanitaceae, the hyphae diverge from the central plane of the lamella, passing downwards and outwards towards the lamellar surfaces (*divergent* trama). In e.g. members of the Pluteaceae, the hyphae converge towards the central plane (*convergent* trama).

bile acids Steroid carboxylic acids – typically hydroxylated derivatives of 5β-cholan-24-oic acid: *cholic acid* has α-hydroxyl groups at positions 3, 7 and 12, *deoxycholic acid* at positions 3 and 12, *chenodeoxycholic acid* at positions 3 and 7, and *lithocholic acid* at position 3. These compounds occur – usually conjugated via an amide bond with glycine (glycocholic acids) or taurine (taurocholic acids) – in bile, the relative proportions depending on species. Cholates (the salts of bile acids) are surfactants which emulsify dietary fats in the intestine. *Primary* bile acids are synthesized in the liver. *Secondary* bile acids are produced from primary acids by the action of intestinal bacteria; they are absorbed from the intestine and returned to the liver for secretion. In mammals, primary bile acids include cholic acid and small amounts of chenodeoxycholic acid. (Chenodeoxycholic acid is the main bile acid in fowl.) Lithocholic acid and deoxycholic acid are secondary bile acids; lithocholic acid, formed from chenodeoxycholic acid, is hepatotoxic and may promote liver disease associated with biliary stasis.

Bile acid salts and amides inhibit many non-enteric bacteria and are useful (e.g. in MACCONKEY'S AGAR) for selecting enteric species. Bile salts promote multidrug-efflux in *Campylobacter jejuni* [JB (2005) *187* 7417–7424].

bile–aesculin agar See AESCULIN.

bile solubility The 'solubility' (i.e., tendency to lyse) shown by most (freshly isolated) strains of *Streptococcus pneumoniae* when exposed (under alkaline conditions) to bile, sodium deoxycholate, sodium taurocholate, or e.g. lauryl sulphate; other similar streptococci do not lyse under these conditions. Lysis appears to result from the activation/stimulation by the bile of an AUTOLYSIN, *N*-acetylmuramyl-L-alanine amidase, which disrupts the cell wall peptidoglycan. Heat-killed pneumococci, and pneumococci in which choline has been replaced by ethanolamine in the wall teichoic acid (by culture in a defined medium), do not lyse with bile.

Bile solubility test (one method): 2 drops of 10% sodium deoxycholate solution are added to ca. 1 ml of a 24-hour broth culture or saline suspension of the test organism, the pH is adjusted to 7.4–7.6, and incubation is carried out (10–15 min /37°C). If the organism is soluble in bile, the turbid suspension becomes clear (positive test). [Methods: Book ref. 53, pp. 1437–1438.]

bilin See PHYCOBILIPROTEINS.

biliproteins *Syn.* PHYCOBILIPROTEINS.

binapacryl See DINITROPHENOLS.

binary fission FISSION (q.v.) in which two cells – usually of similar size and shape – are formed by the growth and division of one cell. (cf. MULTIPLE FISSION.) In binary fission cytoplasmic division is preceded by DNA REPLICATION in prokaryotes, and e.g. MITOSIS in eukaryotes. (See also CELL CYCLE.) If a cell divides across its longitudinal axis the process is termed *transverse binary fission*; if division occurs along the longitudinal axis it is termed *longitudinal binary fission*. Binary fission is a common mode of asexual reproduction in many types of microorganism.

In bacteria, binary fission is typically of the transverse type, and it usually results in the formation of two morphologically and functionally similar daughter cells; asymmetrical binary fission occurs e.g. in *Caulobacter*. Following binary fission the daughter cells may separate; alternatively, the progeny cells of sequential binary fissions may remain together to form chains, PACKETS, PALISADES etc. (See also BUDDING; GROWTH (a); SEPTUM (a); TERNARY FISSION.)

In protozoa, binary fission in ciliates is typically HOMOTHETOGENIC, and in flagellates it is typically SYMMETROGENIC. (See also STOMATOGENESIS.)

In yeasts, binary fission occurs e.g. in *Schizosaccharomyces* spp; in many yeasts (including e.g. *Saccharomyces*) cell division involves BUDDING (cf. BUD-FISSION).

binding hyphae See HYPHA.

binding protein (1) (periplasmic binding protein) In e.g. bacteria of the Enterobacteriaceae: a soluble protein, located in the periplasmic region, which binds to a specific substrate and which is involved in a BINDING PROTEIN-DEPENDENT TRANSPORT SYSTEM for that substrate; periplasmic binding proteins can also function as chemoreceptors for CHEMOTAXIS.

(2) Any protein which binds, more or less specifically, to other molecules: see e.g. DNA BINDING PROTEIN; SINGLE-STRAND BINDING PROTEIN; PENICILLIN-BINDING PROTEINS.

binding protein-dependent transport system (ABC importer; periplasmic transport system; periplasmic permease; osmotic shock-sensitive permease) In Gram-negative bacteria: an ABC TRANSPORTER in which an essential stage is the binding of (imported) substrate by a soluble *periplasmic binding protein* which transfers the substrate to a membrane complex. (Some binding proteins also function in chemosensing.) Binding proteins can be released by OSMOTIC SHOCK; as membrane-bound components cannot function without their binding proteins, this type of transport system is characterized by its sensitivity to osmotic shock.

In *Escherichia coli* and *Salmonella typhimurium* (for example) various sugars, ions, amino acids, organic acids, oligopeptides etc. (almost half of all imported material) can be transported by binding protein-dependent systems. (See also VITAMIN B_{12}.)

One example is the maltose uptake system in *E. coli* which consists of (i) membrane-associated proteins: MalF, MalG and the two (MalK) ABC proteins, and (ii) the periplasmic binding protein, MalE; uptake of maltose via the outer membrane involves a specific PORIN, the LAMB PROTEIN.

Another importer, the *E. coli* ProU system, mediates uptake of the osmoregulatory molecule glycine betaine [FEMS Reviews (1994) *14* 3–20].

A further example is the high-affinity histidine uptake system (*histidine permease*) in *Salmonella typhimurium* [e.g. JBC (1997) *272* 859–866] – one of a number of distinct histidine transporters in this organism. This importer consists of (i) HisM, HisQ and two copies of the ABC protein, HisP (all membrane-associated), and (ii) the periplasmic binding protein, HisJ. (Genes encoding the J, Q, M and P proteins constitute a single OPERON.) On binding to histidine, HisJ undergoes a conformational change and interacts with the membrane complex; histidine is translocated concomitantly with ATP hydrolysis. [HisP (crystal structure): Nature (1998) *396* 703–707.]

In some cases, binding proteins specific for different substrates can interact with the same membrane complex. For example, in *Salmonella typhimurium* the binding protein for histidine and that for lysine, arginine and ornithine (the LAO binding protein: *argT* gene product) both interact with the same membrane-bound complex. In *E. coli*, two binding proteins for branched amino acids – LIV-I (for L-leucine, L-isoleucine and L-valine) and LS (for D- and L-leucine) – also have common membrane-bound components.

binomial A designation consisting of two names. For example, a SPECIES is designated by a Latin binomial which consists of the name of the genus to which it belongs (the *generic name*) followed by the name (*specific epithet*) which distinguishes that species from all others in the genus. Thus, e.g. in *Bacillus cereus* '*Bacillus*' is the generic name and '*cereus*' is the specific epithet.

bioadsorbents See LEACHING.

bioassay Any quantitative procedure in which a given organism is used for assay purposes. Bioassay is used most often to measure trace amounts of a given substance (e.g. a vitamin) in a sample; this is achieved by using an organism for which the given substance is an essential growth factor, and determining the total amount of growth of that organism in the sample. For example, for the bioassay of PANTOTHENIC ACID, *Tetrahymena pyriformis* is inoculated into a medium containing adequate concentrations of all growth requirements except pantothenic acid – which is present at a known, suboptimal concentration; no pantothenic acid should be carried over in the inoculum. After incubation (under standard conditions) the amount of growth is measured (e.g. in mg dry weight); the amount of growth is proportional to the amount of pantothenic acid in the medium. The procedure is repeated, using different concentrations of pantothenic acid each time, and a standard curve is constructed by plotting the amount of growth against the concentration of pantothenic acid; the standard curve can then be used to quantify an unknown concentration of pantothenic acid. Clearly, a medium must not contain substance(s) which allow the test organism to bypass the restriction on growth imposed by the limiting concentration of the assayed substance; in the case of BIOTIN bioassay, for example, the presence of pimelic acid in the medium may circumvent the requirement for biotin and permit growth of the test organism. (See also FOLIC ACID; METHANOBREVIBACTER; NICOTINIC ACID; VITAMIN B_{12}; cf. BIOAUTOGRAPHY.)

Bioassay is also used e.g. for measuring concentrations of antimicrobial agents, and for assaying the pathogenicity of entomopathogenic microorganisms by determining their effect on susceptible insect populations.

bioautography A method for detecting a trace amount of a substance (e.g. a vitamin) in a complex mixture, the given substance being an essential growth requirement for the test organism used. The components of the mixture are initially separated by CHROMATOGRAPHY. The presence of a given substance, in a given chromatographic band, is then determined by extracting the band with an appropriate solvent and adding the extract to a medium which lacks a particular growth factor; the medium is then inoculated with an organism which fails to grow in the absence of the growth factor. On incubation, growth of the test organism indicates the presence of the growth requirement in the original mixture. (cf. BIOASSAY.)

Biobor An organoboron preservative used e.g. in various liquid hydrocarbon fuels.

bioburden In an industrial sterilization process: the number of contaminating organisms per product unit before sterilization.

biochemical oxygen demand See BOD.

biocide (1) *Syn.* STERILANT. (2) Sometimes used to refer to a DISINFECTANT or PRESERVATIVE.

bioconcentration factor See BCF.

biocontrol *Syn.* BIOLOGICAL CONTROL.

bioconversion (1) The conversion of a substrate to particular product(s) by cells or by isolated enzymes. (2) The conversion of substrate(s) to cell biomass.

biocytin ε-*N*-Biotinyl-lysine (see BIOTIN).

biodegradation The degradation of a substance as a result of biological (usually microbial) activity. In ecological terms, biodegradation is vital for the recycling of matter (see e.g. CARBON CYCLE). In human terms, biodegradation may be useful e.g. in the disposal of waste materials (see e.g. SEWAGE TREATMENT) and xenobiotics (see BIOREMEDIATION), and in processes such as COFFEE production and RETTING; it can also be a nuisance – see BIODETERIORATION.

biodeterioration The deterioration (spoilage) of an object or material as a result of biological – usually microbial – activity. See e.g. CATHODIC DEPOLARIZATION THEORY (of iron corrosion), COAL BIODEGRADATION, FOOD SPOILAGE, GLASS, LEATHER SPOILAGE, PAINT SPOILAGE, PAPER SPOILAGE, PETROLEUM (spoilage of fuels and oils), RUBBER SPOILAGE, TEXTILE SPOILAGE, TIMBER SPOILAGE, WOOL SPOILAGE. (See also PRESERVATIVES; cf. BIODEGRADATION.)

bioemulsifiers See BIOSURFACTANTS.

biofilm An adherent layer of microbial cells embedded in a polymer matrix secreted by the cells. Biofilms can develop on living tissues (e.g. oral cavity) on catheters, in water pipes [FEMS Ecol. (1997) *22* 265–279], and on ships' hulls; pathogens in biofilms (e.g. on prostheses) may be resistant to antibiotics (and to the immune system), but biofilms can be useful e.g. in waste-water treatment systems [TIBtech.(2000) *18* 312–320]. (See also EPILITHON.)

At least some biofilms contain fluid-filled channels which are kept open by rhamnolipid biosurfactants [JB (2003) *185* 699–700; *185* 1027–1036]. Biofilm development may be controlled by QUORUM SENSING [Microbiology (2001) *147* 2517–2528; JB (2004) *186* 1838–1850] and (in *Pseudomonas aeruginosa*) may involve three-component regulation–a sensor and two response regulators (cf. TWO-COMPONENT REGULATORY SYSTEM) [JB (2005) *187* 1441–1454].

[Symposium/review: JB (2004) *186* 4427–4440.]

biofilm fluidized-bed reactor A reactor in which the liquid to be treated is pumped upwards through a column ('bed') of small, biofilm-coated particles (e.g. sand grains of 0.5 mm diam.) at a flow rate sufficient to cause *fluidization* of the bed, i.e. a state in which the particles, though retained within the reactor, are able to move relative to one another in the liquid – rather than being sedimented and immobile.

biofuel cell A device in which certain redox reactions, catalysed by isolated enzymes or by living cells, are coupled to metal electrodes – thereby generating an electromotive force (emf) between the electrodes.

In a biofuel cell a given redox system may be coupled to an electrode via a mediator substance which facilitates electron flow between the redox system and electrode; for example, 4,4′-dipyridyl can facilitate electron flow between oxidized/reduced cytochrome *c* and a gold electrode [Nature (1980) *285* 673–674].

In one type of biofuel cell energy is derived from the metabolism of glucose by *Proteus vulgaris* [JGM (1985) *131* 1393–1401]. In another device, energy is derived from the oxidation of carbon monoxide by an isolated CO OXIDASE using e.g. ferrocene monocarboxylic acid ([carboxycyclopentadienyl]cyclopentadienyl-iron) as a mediator [MS (1986) *3* 149–153].

More recently, improved results in a microbial fuel cell have been obtained by using neutral red as an electron mediator [AEM (2000) *66* 1292–1297].

biogas See ANAEROBIC DIGESTION.

bioinsecticide See BIOLOGICAL CONTROL.

bioleaching Microbial LEACHING.

biological aerated filter See SEWAGE TREATMENT.

biological assay *Syn.* BIOASSAY.

biological clock See CIRCADIAN RHYTHMS.

biological control The exploitation, by man, of one species or strain of an organism (or a product of that organism) to control the numbers and/or activities of another organism.

Biological control is used on a commercial scale e.g. in agriculture, forestry and horticulture; it often involves introducing into the environment a microbial pathogen of insects (a microbial insecticide, *bioinsecticide* or *biopesticide*) which can infect and kill, or disable, particular insect pests of certain crop plants. Similarly, weeds may be controlled by the use of plant-pathogenic fungi (*mycoherbicides*). (See also TRICHODERMA.)

Certain strains of *Bacillus thuringiensis* can synthesize a (typically plasmid-encoded) insecticidal crystal protein (ICP); there are several types of ICP – designated Cry types I–IV (and subtypes) – and these important bioinsecticides are widely used against insect pests on a range of crops. Most of the ICPs are formed as a *parasporal crystal* (see DELTA ENDOTOXIN), i.e. they are synthesized only in sporulating cells; synthesis and sporulation are linked because transcription of the relevant ICP genes depends on sporulation-specific sigma factors such as σ^E. ICPs of the Cry III type differ in that they are formed during vegetative growth; overexpression of these ICPs occurs in mutant cells blocked in the phosphorelay (see ENDOSPORE). [ICP production: Mol. Microbiol. (1995) *18* 1–12.]

ICPs applied to crops are readily degraded – so that repeat application is necessary. To avoid this problem, *B. thuringiensis* has been genetically modified so that toxin accumulates within the cell; this (partly-protected) toxin is still highly active [Biotechnology (1995) *13* 67–71].

Biological control has also been directed against mosquitoes; strains of *B. thuringiensis*, *B. sphaericus* and *Clostridium bifermentans* produce toxins which can kill mosquitoes, and considerable effort is being made to develop a product suitable for controlling mosquito-borne diseases such as MALARIA, filariasis and YELLOW FEVER [MR (1993) *57* 838–861].

[Biological control by Gram-positive bacteria: FEMS (1999) *171* 1–19.]

In a 'competitive' form of biological control, *Pseudomonas fluorescens* reduces frost injury in certain crops by using nutrients which would otherwise be available to ICE-NUCLEATION BACTERIA.

Other biological control agents include e.g. strains of *Agrobacterium* (see AGROCINS), *Bacillus popilliae* (see MILKY DISEASE), fungi (see e.g. ASCHERSONIA, BEAUVERIA, HETEROBASIDION, METARHIZIUM, NOMURAEA), protozoa (see e.g. NOSEMA) and viruses (see e.g. BACULOVIRIDAE, CYTOPLASMIC POLYHEDROSIS VIRUS GROUP, MYXOMATOSIS).

biological false positive reaction (BFPR) In a serological test: a positive reaction obtained as a result of the occurrence, in the patient, of disease(s) or condition(s) other than those being tested for (see e.g. STANDARD TESTS FOR SYPHILIS).

biological filter See SEWAGE TREATMENT.

biological oxygen demand See BOD.

biological safety cabinet *Syn.* SAFETY CABINET.

biological transmission Transmission of a parasite by a *biological* VECTOR.

biological vector See VECTOR (1).

biological warfare Any form of warfare which exploits the harmful potential of bioactive substance(s) and/or of particular microorganism(s) – use of the latter sometimes being termed *germ warfare*. The topic is discussed in a number of books – e.g. *Superterrorism* by Schweitzer & Dorsch (1998) published by Kluwer Academic Publishers/Plenum.

[Bioterrorism as a public health threat: EID (1998) *4* 488–494. Bioterrorism: EID (1999) *5* (special issue) 491–565.]

There are plans to develop a former Russian bioweapons laboratory into an international centre for studying emerging and re-emerging diseases [see Science (2001) *291* 2288–2289].

biological washing powders See SUBTILISINS.

biologicals (*med.*) Certain biologically derived therapeutic agents which include e.g. *infliximab* (see TNF).

bioluminescence CHEMILUMINESCENCE generated by certain microorganisms and (e.g.) by fireflies and by some crustaceans and jellyfish. (cf. PHOSPHORESCENCE.)

In all cases, bioluminescence requires the presence of free oxygen – although only very low levels may be necessary. However, bioluminescence in different organisms may require different sources of energy (e.g. ATP in fireflies, NAD and FMN in bacteria); moreover, different organisms may produce light of different wavelengths (e.g. 560 nm in fireflies, ~475–505 nm in bacteria).

In some organisms the light-emitting entity has been identified as a particular compound; for example, in the firefly (*Photinus pyralis*) it is 4,5-dihydro-2-(6-hydroxy-2-benzothiazolyl)-4-thiazole carboxylic acid. The various light-emitting compounds/systems from different organisms are referred to by the generic term *luciferin*.

All bioluminescent organisms contain a species-specific thiol-containing oxidoreductase which catalyses oxidation of the light-emitting entity – giving rise to an energized form which produces light as it returns to the unexcited ground state. The generic term for this oxidoreductase is *luciferase*. In *Photobacterium fischeri*, luciferase (EC 1.14.14.3) is a protein of MWt ~80000. (The firefly luciferase (EC 1.13.12.7) is a protein of MWt ~62000.)

Bioluminescence in bacteria. The luminescent bacteria include marine and coastal species, e.g. *Alteromonas hanedai* and species of PHOTOBACTERIUM and XENORHABDUS. All the luminescent bacteria appear to generate light by similar reactions, involving the same types of component, and light is emitted continuously (i.e. not in pulses) under appropriate conditions. Bacterial luciferase is an AUTOINDUCIBLE ENZYME. The autoinducer in e.g. *P. fischeri* is *N*-(β-ketocaproyl)homoserine lactone (see QUORUM SENSING); above a certain concentration, autoinducer switches on the *lux* operon that encodes luciferase.

Bacterial bioluminescence apparently involves the initial reduction of FMN by NADH, and the formation of a luciferase–$FMNH_2$ complex. In the presence of (i) free oxygen, and (ii) an aliphatic aldehyde (containing at least seven or eight carbon atoms), a larger complex appears to develop: luciferase–$FMNH_2$–O_2–RCHO; this complex emits light and yields FMN, luciferase, water and a carboxylic acid corresponding to the aldehyde. (The aldehyde may be re-cycled after reduction of the carboxylic acid by NADH.)

Bacterial bioluminescence competes with the respiratory chain for reduced NAD; luminescence is rapidly enhanced following the inhibition of respiration by cyanide.

Bioluminescence in fungi. The fruiting bodies and/or mycelium of some agarics (e.g. *Armillaria mellea*, *Mycena* spp) produce a continuous (non-pulsing) light which, in some species, exhibits a diurnal fluctuation in intensity; it is dependent on NAD(P)H and appears to involve two types of enzyme. The emitted light may be bluish-green or green (wavelength ca. 530 nm).

Bioluminescence in dinoflagellates. On appropriate stimulation (e.g. acidification, agitation), the cells of some dinoflagellates (e.g. species of *Gonyaulax*, *Noctiluca*, *Pyrocystis*, *Pyrodinium*) emit bluish light (wavelength ca. 480 nm) in flashes of ca. 0.1 second duration. *Gonyaulax* has a system of subcellular, guanine-containing, crystal-like fluorescent structures referred to as 'scintillons'; the luciferin occurs (together with at least some luciferase) only in the scintillons, although some luciferase may also occur in other parts of the cell [JCB (1985) *100* 1435–1446]. Structures similar to scintillons have been observed in non-luminescent dinoflagellates. Bioluminescence in *G. polyedra* (and probably in many other dinoflagellates) exhibits a CIRCADIAN RHYTHM. In *Noctiluca*, luminescence is associated with subcellular organelles, the *microsources*.

Bioluminescence is also exhibited by a number of radiolarians.

The function of microbial bioluminescence, if any, is unknown. It has been suggested that it is the vestige of a system once used for removing low levels of environmental oxygen.

Applications. Bioluminescent bacteria have been used for detecting very low concentrations of dissolved oxygen.

The ATP-dependent firefly luciferase–luciferin system is used for detecting/quantifying ATP (e.g. in PYROSEQUENCING, in which pyrophosphate is first converted to ATP). A luciferase reporter system is used e.g. for studying promoter function [AEM (2005) *71* 1356–1363] and the efficacy of antisense oligomers [AAC (2005) *49* 249–255]. Three different luciferases have monitored the simultaneous expression of three genes [BioTechniques (2005) *38* 891–894].

biomass The dry weight, or other (usually quantitative) estimation of organisms (commonly microorganisms), in a given habitat (soil, water, culture medium, etc).

biomimetic technology Biologically-based systems which can achieve results similar to those hitherto dependent on abiotic electrical/mechanical/chemical methodology. One example involves the ability of a strain of *Pseudomonas stutzeri* to form (in its periplasm) crystalline particles of silver in a nanometre range of sizes; heat treatment (400°C) of a layer of particle-containing cells can form a carbonaceous matrix containing the small, homogeneously embedded particles of silver – a structure which may have applications as a coating for the absorption of solar energy. Silver–carbon films from *P. stutzeri* may also be useful (as a consequence of their high porosity) as electrodes in lithium-ion and other batteries.

In general, certain microorganisms are capable of producing intracellular or extracellular *cermet* (ceramic–metal) or *orgmet* (organic–metal) structures, or composites, which are organized on a nanometre scale and which potentially have a range of applications in materials science.

[Metal-accumulating bacteria and their potential for materials science: TIBtech. (2001) *19* 15–20.]

biomineralization The deposition of minerals due to biological activity – e.g. the formation of magnetosomes (see MAGNETOTACTIC BACTERIA) and the intracellular or extracellular deposition of sulphur by photosynthetic bacteria. (cf. MINERALIZATION.)

biomining See LEACHING.

Biomyxa See GRANULORETICULOSEA.

biont *Syn.* SYMBIONT.

biopesticide See BIOLOGICAL CONTROL.

biophore A 'host' cell in genetic engineering.

bioplastics See BIOPOL.

Biopol The trade name (Zeneca, Great Britain) for a range of biodegradable thermoplastics based on POLY-β-HYDROXYBUTYRATE (PHB). The homopolymer (PHB) is formed under suitable conditions when *Alcaligenes eutrophus* uses glucose as the sole source of carbon. When the growth medium is appropriately supplemented, *A. eutrophus* forms co-polymers of hydroxybutyrate and hydroxyvalerate; the proportion of hydroxyvalerate (controlled by adjusting the growth medium) determines the properties of these co-polymers. The intracellular granules of polymer (either PHB or co-polymer) are harvested and purified to a fine, white powder.

Biopol can be used for making containers, mouldings, fibres, films and coatings, and can be worked by blow-moulding and injection-moulding processes. While stable in normal use, Biopol is fully degradable after suitable disposal.

To reduce costs, attempts are now being made to develop transgenic plants for the commercial production of bioplastics.

[Review on bioplastics: BAB (1996) *49* 1–14.]

biopsy (1) The removal of tissue from the living body for examination, culture etc. (2) The tissue specimen referred to in (1).

biopterin 2-Amino-4-hydroxy-6-(1′,2′-dihydroxypropyl)-pteridine. Biopterin is believed to act as a coenzyme in certain hydroxylation reactions, and is required as a growth factor e.g. by *Crithidia fasciculata* and *Trypanosoma platydactyli* (formerly *Leishmania tarentolae*); the requirement for biopterin may be partially or totally overcome by high concentrations of FOLIC ACID.

bioreactor (reactor) A FERMENTER or other apparatus used for BIOCONVERSIONS.

bioremediation Biotechnological (microbe-based) clean-up of pollutants (chemical contaminants) in the environment (aqueous and terrestrial).

As a group, microorganisms are metabolically highly diverse, so that, in theory, it should be possible to degrade any of a wide range of pollutants given a suitable choice of microorganisms(s). Moreover, *microbial* degradation of pollutants may give rise to simple products – such as carbon dioxide and water – whereas alternative approaches (e.g. physical decontamination) may simply transfer the problem from one location to another. Nevertheless, while physical methods are typically rapid, often with a predictable outcome, biological methods frequently have unknown, unpredictable or unquantifiable effects in the *environment*; for example, the *bioavailability* of a pollutant (its accessibility to microorganisms) may be reduced by adsorption to soil particles, and this may limit or preclude efficient bioremediation.

To be more widely accepted and used, bioremediation must be shown to be both effective and reliable in the environment. The efficacy of a bioremediation process may be gauged e.g. by (i) assaying pollutants of all levels of biodegradability at the polluted site; (ii) assaying the long-lived breakdown products of pollutants; (iii) assessing the levels/states of *normal* constituents of the environment; and (iv) quantifying specific microbial genes involved in the catabolism of pollutants [see e.g. AEM (1996) *62* 2381–2386].

[Bioremediation: towards a credible technology (review): Microbiology (1998) *144* 599–608.]

Certain types of pollutant are susceptible to anaerobic degradation, and it has been suggested that, in the future, anaerobic microbial processes may have the potential for widespread application in bioremediation technology.

Anaerobic degradation of the perchlorate ion can be carried out by various facultatively anaerobic or microaerophilic bacteria which are capable of dissimilatory perchlorate reduction (a form of ANAEROBIC RESPIRATION); these bacteria include *Wolinella succinogenes* and (particularly) species belonging to two new genera: *Dechloromonas* and *Dechlorosoma* (both members of the beta subclass of Proteobacteria). In dissimilatory perchlorate reduction, the initial (energy-yielding) stage produces chlorite; the chlorite is split (enzyme: *chlorite dismutase*) to chloride and molecular oxygen. [Ubiquity and diversity of (per)chlorate-reducing bacteria: AEM (1999) *65* 5234–5241.] Currently (2000), biological methods for perchlorate degradation are based mainly on the use of bioreactors, and there is at present no technique for treating perchlorate contamination in the environment.

Anaerobic degradation of certain hydrocarbons (including BTEX) would be very useful when these pollutants occur e.g. in fuel-contaminated aquifers under anoxic conditions – but BENZENE (for example) typically resists anaerobic degradation. A novel approach to the anaerobic degradation of benzene involves the addition of low levels of chlorite – so that the (ubiquitous) perchlorate-reducers (see above) can convert the chlorite to chloride and (free) oxygen; the oxygen is then available for certain aerobic organisms (e.g. *Pseudomonas*) which are able to degrade cyclic aromatic hydrocarbons [see Nature (1998) *396* 730].

[Emerging techniques for anaerobic bioremediation: TIBtech. (2000) *18* 408–412.]

Degradation of biphenyls and polychlorinated biphenyls (PCBs) – an important category of environmental pollutants – can be carried out by various types of Gram-negative and Gram-positive bacteria. Genes encoding the relevant degradative activity for biphenyls/polychlorinated biphenyls form the *bph* operon. [Transcription of the *bph* operon in *Pseudomonas pseudoalcaligenes* strain KF707: JBC (2000) *275* 31016–31023.]

The pesticide pentachlorophenol (PCP) can be degraded by strains of the bacterium *Sphingomonas chlorophenolica* – the metabolic potential for such degradation probably having developed during the last few decades [TIBS (2000) *25* 261–265].

(See also POLYURETHANASE and TOL PLASMID.)

Wastewater containing terephthalic acid (used e.g. in the manufacture of plastics) can be treated anaerobically (see SEWAGE TREATMENT).

[Heavy metals: FEMS Reviews (2002) *26* 327–338.]

The development of genetically engineered bacteria for degrading specific compounds (particularly XENOBIOTICS) is likely to be assisted by information on the way in which signals from various environmental substrates are integrated at gene promoters [EMBO (2001) *20* 1–11].

Bios test A test used for assessing the integrity of the seams in a can (see CANNING). Essentially, cans are filled with a nutrient medium and are sealed and processed in the normal way – cooling being carried out with water containing a gas-producing organism; the cans are then incubated and examined for the presence of SWELLS. An analogous test is used for testing the integrity of RETORT POUCHES.

biostratigraphy The study of the age and structure of stratified rocks in terms of the fossils – e.g. FOSSIL MICROORGANISMS – they contain. [Book ref. 136.]

biosurfactants Biological molecules which function as surfactants, i.e., they lower the surface tension of water; however, as commonly used, the term includes *bioemulsifiers*: substances which act as emulsifying agents but which do not necessarily have a significant effect on surface tension. Most biosurfactants/bioemulsifiers are amphipathic cell components such as fatty acids, phospholipids, lipopolysaccharides, lipoteichoic acids, etc. Certain microorganisms produce extracellular biosurfactants or bioemulsifiers which may play a role e.g. in the adhesion of cells to – and/or their detachment from – surfaces (see e.g. EMULCYAN), or in the utilization of hydrophobic substrates such as sulphur and hydrocarbons. For example, organisms growing on HYDROCARBONS may produce extracellular fatty acids, glycolipids (e.g. RHAMNOLIPIDS, SOPHOROLIPIDS), high-MWt polymers (e.g. EMULSAN), etc (see also LIPOSAN); such substances may serve e.g. to emulsify the substrate and/or to allow cells to adhere to substrate–water interfaces. (See also SAPONINS and SURFACTIN.) [Potential use of biosurfactants in industry: MS (1986) *3* 145–149.]

biotechnology In a broad sense: collectively, the various forms of technology that exploit biological sources – usually microorganisms and/or their products and components (for examples see INDUSTRIAL MICROBIOLOGY). (See also BIOMIMETIC TECHNOLOGY and IMMOBILIZATION.) However, the term 'biotechnology' refers more commonly to those forms of technology which depend on the use of modern techniques in molecular biology/genetic engineering to construct novel organisms and/or products for industrial, medical or other purposes.

bioterrorism See BIOLOGICAL WARFARE.

biotic Living; of biological origin.

biotin (vitamin H; coenzyme R) A VITAMIN which acts as a cofactor in carboxylation and transcarboxylation reactions, acting as a carrier of CO_2; biotin is bound via its carboxyl group (see figure) to the ε-amino group of a lysine residue in the apoenzyme. (cf. BIOCYTIN.)

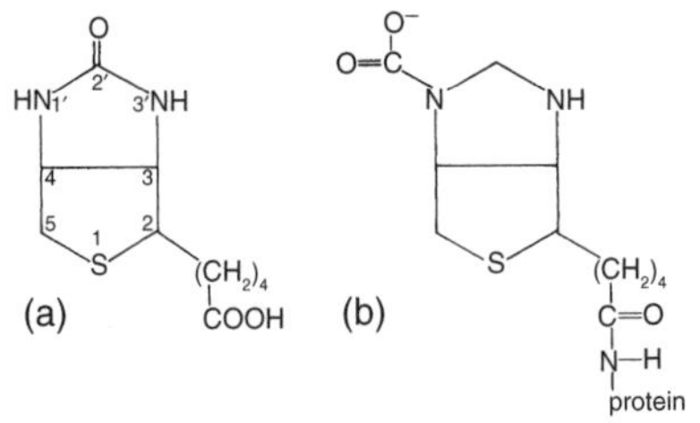

BIOTIN: (a) free, and (b) bound.

Biotin is required as a growth factor e.g. by many fungi (e.g. *Saccharomyces cerevisiae*, *Candida* spp) and bacteria (e.g. species of *Clostridium*, *Lactobacillus*, *Leuconostoc*). Some organisms can synthesize biotin from cysteine, pimelic acid and carbamoyl phosphate. (See also AVIDIN and ABC IMMUNOPEROXIDASE METHOD.)

biotin enzyme Any enzyme whose active centre is associated with a BIOTIN residue – e.g. PYRUVATE CARBOXYLASE and oxaloacetate decarboxylase.

biotope (1) The environment occupied by an organism or organisms. (2) (*med., vet.*) The location of a particular parasite or pathogen within the body. (3) The spatial distribution of the biomass in a cross-section of a river, lake, etc.

biotransformation *Syn.* BIOCONVERSION (1).

biotroph An organism which derives nutrients from the living tissues of another ('host') organism. (cf. NECROTROPH.)

biotype (biovar) Any VARIETY distinguished by metabolic and/or physiological properties.

biovar *Syn.* BIOTYPE.

biparental See PROTOPLAST FUSION.

biphenyls (degradation) See BIOREMEDIATION.

biplicity See FLAGELLUM (a).

bipolar heterothallism See HETEROTHALLISM.

Birbeck granule See LANGERHANS' CELLS.

bird flu An outbreak of disease (Hong Kong, 1997) caused by an avian strain (H5N1) of influenza virus type A (see INFLUENZAVIRUS).

bird's nest fungi See NIDULARIALES.

birefringence (double refraction) The phenomenon in which a beam of light, on entering certain (optically *anisotropic*) materials (e.g. a calcite crystal), gives rise to two beams of polarized light which travel at different velocities in mutually perpendicular planes and which often (depending on conditions) follow divergent paths.

Birnaviridae See IPN VIRUS.

birth scar (*mycol.*) See SCAR.

bisdithiocarbamates (as antimicrobial agents) Metal ethylenebisdithiocarbamates (MS.CS.NH.CH_2CH_2NH.CS.SM, in which M is a metal ion) are used as agricultural ANTIFUNGAL AGENTS – see e.g. MANCOZEB, MANEB, NABAM and ZINEB; they are believed to be converted, under field conditions, to diisothiocyanate derivatives.

Bismarck brown A brown basic diazo DYE used e.g. for VITAL STAINING.

bismuth sulphite agar See WILSON AND BLAIR'S AGAR.

bisphenols Compounds that contain two phenolic residues/molecule; in bisphenols which are effective ANTISEPTICS or DISINFECTANTS the residues are linked directly (i.e. carbon–carbon) or e.g. via $-CH_2-$ (*dihydroxydiphenylmethanes*) or –S– (*dihydroxydiphenylsulphides*). For maximum antimicrobial activity, each phenolic residue must carry an *o*-(2-)hydroxyl. Halogenation increases activity; two di- or trihalogenated phenolic groups generally confer greater activity against Gram-positive than Gram-negative species. Bacteriostatic at low concentrations, bisphenols may be bactericidal at higher concentrations; they appear to affect the cell membrane – e.g. hexachlorophane affects membrane potentials and, at higher concentrations, may (like various antibacterial agents) cause coagulation of the cytoplasm.

Bisphenols are poorly soluble or virtually insoluble in water but are soluble in dilute alkali and/or organic solvents; they are inactivated by non-ionic surfactants. Since bisphenols retain antimicrobial activity in the presence of soaps and are relatively non-toxic to man, they are widely used as antiseptics and preservatives. *Bithionol* (2,2′-dihydroxy-3,5,3′,5′-tetrachlorodiphenylsulphide) is an antibacterial and antifungal agent used e.g. in antiseptic soaps. *Bromochlorophane* (3,3′-dibromo-5,5′-dichloro-2,2′-dihydroxydiphenylmethane) is active against Gram-positive bacteria and has been used e.g. in deodorants and toothpastes. *Dichlorophane* (dichlorophene) (5,5′-dichloro-2,2′-dihydroxydiphenylmethane) is active against fungi, bacteria and algae; it is used e.g. as a preservative for paper and textiles. *Fentichlor* (2,2′-dihydroxy-5,5′-dichlorodiphenylsulphide) is active primarily against Gram-positive bacteria (e.g. staphylococci) and fungi (particularly dermatophytes); it has been used to treat infections involving dermatophytes. *Hexachlorophane* (hexachlorophene) (2,2′-dihydroxy-3,5,6,3′,5′,6′-hexachlorodiphenylmethane) is much more active against Gram-positive than Gram-negative bacteria; it has been used in antiseptic soaps, liquid soaps (e.g. *Ster-zac*) and lotions (e.g. *pHisoHex*). (cf. TRICLOSAN; see also PHENOLS.)

bisterigmate Having two sterigmata.

bis(tri-*n*-butyltin) oxide See TBTO.

bisulphite (as a MUTAGEN) Bisulphite (HSO_3^-) undergoes various reactions with nucleic acids, and can also cause cross-linking between proteins and nucleic acids. The reaction believed to be responsible for mutagenesis is the deamination of cytosine and (at a slower rate) of 5-methylcytosine to give, respectively, uracil and thymine (resulting in G·C-to-A·T transition in either case); bisulphite can apparently act on these bases only in ssDNA. (See also SITE-SPECIFIC MUTAGENESIS.)

bithionol See BISPHENOLS.

Bittner virus *Syn.* MOUSE MAMMARY TUMOUR VIRUS.

bitty cream See MILK SPOILAGE.

bitunicate ascus (fissitunicate ascus) An ASCUS whose wall consists of a thin, more or less rigid outer envelope (*ectotunica*), and a thicker, extensible inner envelope (*endotunica*); at maturity, the outer envelope ruptures and the inner envelope elongates – the ascospores being discharged via an apical pore in the inner envelope. (cf. UNITUNICATE ASCUS.) *Lecanoralean* asci release their ascospores in a similar way; they differ from other bitunicate asci e.g. in that the endotunica protrudes only to a limited extent during ascospore discharge.

biuret test A test based on the fact that peptide bond-containing substances (e.g. peptides and proteins) give a violet/pink colour when treated with a dilute, alkaline solution of copper sulphate.

bivalent (noun) See MEIOSIS.

Bivalvulida See MYXOSPOREA.

BK virus See POLYOMAVIRUS.

BL-type starter See LACTIC ACID STARTERS.

***bla* gene** A gene encoding β-LACTAMASE. (See also e.g. Tn*3*.)

black beans *Syn.* HAMANATTO.

Black beetle virus See NODAVIRIDAE.

black bread mould See BREAD SPOILAGE.

black cat/white cat principle The principle that different mechanisms for regulating a given catabolic system are equally efficient if they both (or all) ensure responsiveness to the substrate in the context of the organism's physiological state [EMBO (2001) *20* 1–11].

Black Creek Canal virus See HANTAVIRUS.

Black Death See PLAGUE.

black disease (infectious necrotic hepatitis) An acute, toxaemic disease, which affects mainly sheep and cattle (sometimes pigs), caused by infection of the liver by *Clostridium novyi*; the disease is commonly precipitated by invasion of the liver by the liver fluke (*Fasciola hepatica*). Death usually occurs rapidly.

black fluids Commercial DISINFECTANTS consisting of certain phenolic coal tar fractions solubilized with soaps; their activity is decreased by electrolytes. (cf. WHITE FLUIDS.)

black gill disease A disease of prawns and shrimps (*Pennaeus* spp) caused by *Fusarium solani* (cf. BURNED SPOT DISEASE); the gills bear many black spots and may undergo necrosis or collapse.

black light Radiation of ca. 300–420 nm.

black mat syndrome A disease of the tanner or snow crab (*Chionoecetes bairdi*) caused by the ascomycete *Trichomaris invadens* [CJB (1981) *59* 2121–2128]. The fungus forms a dense black nodular encrusting subiculum (bearing partly embedded perithecia) on the crab exoskeleton; non-pigmented hyphae penetrate the tissues beneath [AEM (1983) *46* 499–500]. (See also CRUSTACEAN DISEASES.)

black mildews (dark mildews) Plant-parasitic fungi of the family Meliolaceae (order DOTHIDEALES); the organisms form dark mycelium which, in many species, is setose (see also HYPHOPODIUM). Genera: e.g. *Meliola*. (cf. POWDERY MILDEWS and SOOTY MOULDS.)

black piedra A human MYCOSIS in which hairs (e.g. of the scalp) are infected by *Piedraia hortae*. Hard black nodules adhere firmly to infected hairs (mainly on the distal portions). The nodules are composed of masses of dark hyphae which form a stroma containing many uniascal locules; asci are spherical and contain 8 fusiform ascospores. (cf. WHITE PIEDRA.)

black pod disease A CACAO DISEASE caused by species of PHYTOPHTHORA (e.g. *P. palmivora*, *P. megakarya*) and characterized by blackening of the (normally green) pods. Pulp prepared from the beans of infected pods contains subnormal amounts of sugar so that the fermentation process (see COCOA) may be affected; such beans also contain excessive amounts of free fatty acid. The disease is usually controlled by antifungal sprays. [Black pod disease in Nigeria: Phytopathol. Paper number 25 (April, 1981).]

black pox (black spot) (*vet.*) Ulcerative lesions, each with a black centre, which occur on the teats, particularly at the orifice, in e.g. cattle. *Staphylococcus aureus* may be isolable from the lesions. *Treatment*: antimicrobial ointments or teat dips. (See also MASTITIS.)

black queen cell virus See PICORNAVIRIDAE.

black rot (*plant pathol.*) A general term for any of various unrelated plant diseases. For example, *brassica black rot*, caused by *Xanthomonas campestris* pv. *campestris*, is characterized by blackening of leaf-margins and often of leaf-veins, yellowing of leaves, stunting etc. (*Carrot black rot*: see ALTERNARIA.)

black scours *Syn.* WINTER DYSENTERY.

black smoker See HYDROTHERMAL VENT.

black spot (1) (*food microbiol.*) A type of MEAT SPOILAGE which affects red meats at sub-zero temperatures; black spots of fungal growth develop on the meat surface. Species involved include *Cladosporium cladosporioides*, *C. herbarum* and *Penicillium verrucosum* var. *corymbiferum* (*Penicillium hirsutum*). [JAB (1982) *52* 245–250.]

(2) (*vet.*) *Syn.* BLACK POX.

(3) (*plant pathol.*) Any of various plant diseases characterized by the formation of black spots e.g. on the leaves and/or fruit. Black spot of roses is caused by *Diplocarpon rosae*; irregular dark-brown or black patches appear on both surfaces of the leaves, and leaves turn yellow and fall prematurely. Bacterial black spot of mangoes, a serious necrotic disease present in most mango-growing regions, is caused by *Xanthomonas campestris* pv. *mangiferaeindicae* [IJSB (1984) *34* 77–79].

black stem rust A CEREAL DISEASE, caused by *Puccinia graminis*, which can affect wheat (causing serious economic losses e.g. in the USA) and other cereals; see UREDINIOMYCETES for details of the disease in wheat.

black wart disease (of potato) See WART DISEASE.

black yeasts Fungi which can form dark, yeast-like cells (e.g. species of *Aureobasidium* and *Cladosporium*).

blackeye cowpea mosaic virus See POTYVIRUSES.

blackfellow's bread The SCLEROTIUM of *Polyporus mylittae* which is used as food by Australian aboriginals.

blackhead (1) (*vet.*) (histomoniasis; infectious enterohepatitis) A disease which affects mainly turkeys but also e.g. chickens and pheasants; it is caused by *Histomonas meleagridis*. Transmission occurs mainly via the caecal nematode *Heterakis gallinae*, in which *H. meleagridis* is transmitted transovarially; the bird is infected on ingestion of nematode eggs passed with the faeces of an infected bird. Symptoms in turkeys: anorexia, lethargy, passage of characteristic yellow faeces; cyanosis or blackening of the head may occur. Initially, small haemorrhagic ulcers occur in the caecum; later the caecal tubes are greatly enlarged. Characteristic circular, yellowish, necrotic lesions occur in the liver, sometimes coalescing to form large areas of necrosis. In the acute form the disease is often rapidly fatal, but it may be chronic in older birds. In chickens, the disease is usually mild (although severe outbreaks may occur under intensive rearing conditions). Chickens are commonly parasitized by *H. gallinae* and can act as a reservoir of *H. meleagridis*; they should therefore not be reared alongside turkeys. *Lab. diagnosis: H. meleagridis* may be observed in scrapings from caecal lesions. Furazolidone and dimetridazole are used for prevention/treatment. (See also POULTRY DISEASES.)

(2) (*med.*) See ACNE.

blackleg (1) (blackquarter) (*vet.*) In cattle (and occasionally in other animals): an acute (usually) fatal disease, characterized by severe toxaemia and inflammation of leg (and/or other) muscles, caused by *Clostridium chauvoei*; the pathogen is presumed to reach the affected site via the intestine and bloodstream, but trauma to muscles may be a predisposing factor. *Treatment*: parenteral administration of antibiotics. (cf. MALIGNANT OEDEMA.)

(2) (*plant pathol.*) Any of various bacterial or fungal diseases of plants in which the plant stem characteristically becomes blackened and often rots.

In potatoes blackleg is caused by *Erwinia carotovora* subsp. *atroseptica*; symptoms: yellowing of foliage, rotting of stems (which become black and slimy), slimy lesions in the tubers.

In cruciferous plants 'blackleg' may refer to any of several diseases. That caused by *Xanthomonas campestris* involves leaf chlorosis and blackening of the veins; infection is seed-borne or occurs via stomata or wounds. *Leptosphaeria maculans* can be an important cause of blackleg in oil-seed rape and other brassicas.

In beet plants, blackleg may be caused by *Pleospora* (*Phoma*) *bjoerlingii*; in seedlings, the stem blackens and shrivels, while in mature plants brown leaf spots and blackening of root tissues occur. Infection is seed-borne.

Pelargonium blackleg is caused by a *Pythium* sp; stems blacken from the base upwards, foliage wilts, etc.

blackpatch disease (of clover) See SLAFRAMINE.

blackquarter *Syn.* BLACKLEG (1).

blackwater fever A complication of falciparum MALARIA characterized by profound haemoglobinuria.

bladder (*algol.*) *Syn.* PNEUMATOCYST.

bladder infection See CYSTITIS and UPEC.

bladder wrack See FUCUS.

blade (*algol.*) The flat or crinkled, essentially sheet-like part of the thallus in certain algae (e.g. *Fucus*, *Laminaria*).

Blakeslea See MUCORALES.

blanket weed See CLADOPHORA.

BLAST Basic local alignment search tool, a program for comparing sequences: http://www.ncbi.nlm.nih.gov/BLAST/

blast disease (of rice) A disease of the rice plant caused by *Pyricularia oryzae* (see PYRICULARIA). Leaf lesions are often e.g. bluish or grey, each having a dark brown margin; bluish lesions may occur in the neck of the culm (i.e., just below the flowering head) – often leading to breakage – and small brown lesions may form on the glumes. The disease is favoured e.g. by high levels of soil nitrogen. (See also HINOZAN, KASUGAMYCIN, KITAZIN and THIOPHANATE-METHYL.)

blast transformation Morphological and metabolic changes which occur in B LYMPHOCYTES and T LYMPHOCYTES on exposure to specific antigen or to a MITOGEN; in at least some cases, blast transformation appears to be triggered by the cross-linking of receptors on the surface of the lymphocyte.

In B cells, within ~12–24 hours of stimulation, the cell begins to enlarge and develops a basophilic, ribosome-rich cytoplasm, an enlarged nucleolus, and a large, pale-staining nucleus which contains an increased amount of interchromatinic protein; the transformed cell is known as a *blast cell* (= *blast*), and its formation appears to represent a shift from the G_0 to the G_1 phase of the cell cycle. For ANTIBODY FORMATION, a blast proliferates and forms PLASMA CELLS.

blastic conidiogenesis See CONIDIUM.

Blastobacter A genus of rod-shaped to spherical, Gram-negative budding bacteria which occur e.g. in ponds and lakes; some strains are flagellated, and some are pigmented.

Blastocaulis A subjective SYNONYM of PLANCTOMYCES.

Blastocladiales An order of fungi (class CHYTRIDIOMYCETES) in which the thallus consists typically of rhizoid-bearing hyphae or mycelium, and in which sexual reproduction involves the fusion of iso- or anisoplanogametes; most species live saprotrophically in water or soil, but *Coelomomyces* spp (which form wall-less, coenocytic hyphae lacking rhizoids) are obligate parasites of certain insects (e.g. mosquito larvae). [*Coelomomyces*: Book ref. 190.] Other genera include ALLOMYCES, *Blastocladiella*, and *Catenaria* (see also NEMATOPHAGOUS FUNGI).

Blastocladiella See BLASTOCLADIALES and GAMMA PARTICLE (2).

blastoconidium See CONIDIUM.

Blastocrithidia A genus of homoxenous parasitic protozoa (family TRYPANOSOMATIDAE) which occur in the gut of bugs and mosquitoes; epimastigotes, approx. 10–70 μm. *B. triatomae* forms desiccation-resistant cysts.

Blastocystis A genus of organisms currently regarded as protozoa (formerly as fungi of the Entomophthorales). *B. hominis* occurs in the intestinal tract in man and other animals; it is generally non-pathogenic, but may be associated with diarrhoea when present in large numbers [Ped. Inf. Dis. (1985) *4* 556–557]. Three morphological forms (amoebic, granular and vacuolated) have been described, but the vacuolated form is the most common in human faecal specimens; these cells are non-motile and spherical, oval or ellipsoid, (5–)8–10(–30) μm diam., and may be mistaken for cysts of certain amoebae. Staining of the vacuolated forms with e.g. iodine reveals a thin peripheral layer of cytoplasm containing one nucleus (sometimes 2–4 nuclei) and surrounding a large central or slightly eccentric body or vacuole.

Blastodinium See DINOFLAGELLATES.

Blastomyces A genus of fungi of the HYPHOMYCETES. *B. dermatitidis* (= *B. zymonema*, teleomorph *Ajellomyces dermatitidis*) is a dimorphic fungus which causes BLASTOMYCOSIS; in mammalian cells and in culture at 37°C it grows in the form of yeast-like, spherical, thick-walled cells (usually ca. 8–15 μm diam.). The cells reproduce by budding, a (usually) single bud being attached to the mother cell by a broad base. When cultured at <30°C the organism grows as a septate mycelium, producing smooth, spherical to ovoid conidia (3–5 μm diam.) on short, straight conidiophores.

Blastomycetes A class of fungi (subdivision DEUTEROMYCOTINA) which include the anamorphic yeasts; the class is sometimes divided into two families: CRYPTOCOCCACEAE (species do not form ballistospores) and SPOROBOLOMYCETACEAE (species form ballistospores). In some taxonomic schemes [Book ref. 64, p. 54] the organisms are included in the HYPHOMYCETES.

blastomycosis (North American blastomycosis; Gilchrist's disease) An acute or chronic MYCOSIS, affecting man and animals (e.g. dogs), caused by *Blastomyces dermatitidis*; it occurs e.g. in North America, Africa and Israel. Infection apparently occurs by inhalation of spores from the (presumably saprotrophic) fungus, although *B. dermatitidis* has proved difficult to isolate from environmental habitats. The disease occurs in two clinically distinct but variable forms: the cutaneous form (a chronic disease in which granulomatous, suppurative lesions are confined mainly to the skin, although the pathogen is disseminated from a primary pulmonary site), and the pulmonary form (in which lesions develop primarily in the lungs). Pulmonary blastomycosis may be subclinical, may (occasionally) manifest as an acute, self-limiting, influenza- or pneumonia-like disease, or may be chronic and progressive with dissemination to other regions of the body (e.g. bones, genitourinary tract, CNS). Mortality rates associated with untreated systemic blastomycosis may be >80%. *Chemotherapy*: e.g. amphotericin B; 2-hydroxystilbamidine. [Review of human blastomycosis: CRM (1982) *9* 139–164.] (cf. PARACOCCIDIOIDOMYCOSIS.)

blastospore See CONIDIUM.

Blattabacterium A genus of Gram negative-type bacteria which occur as intracellular endosymbionts of the cockroach (*Blatta* spp). Cells: non-motile rods, ca. 1.0 × 1.6–9.0 μm, which may give a Gram-positive or Gram-variable reaction; they occur within the bacteriocytes of the cockroach – and within the oocytes, thus ensuring their transmission to the larvae. GC%: ca. 26–28. Type species: *B. cuenoti*. [Book ref. 22, pp. 830–831.]

bleach See HYPOCHLORITES.

bleaching Loss of the chromophore from a biological pigment, or loss of pigment(s) from an organism – e.g. loss of chlorophyll from a photosynthetic organism.

bleeding canker A CANKER which affects e.g. horse chestnut, lime and apple trees; it is caused by *Phytophthora citricola*. A brownish or reddish gum oozes from the regions of dying bark and dries to form a hard black crust; dieback may occur, and the lesions may be infected with other wood-decaying fungi.

blending Vigorous agitation involving high shear forces.

bleomycin An anticancer antibiotic produced by *Streptomyces verticillus*; it is structurally complex, consisting of peptide and disaccharide moieties. Bleomycin binds to DNA and – in the presence of Fe^{2+} and O_2 – causes release of free bases and the degradation of the DNA by single- and double-stranded cleavages.

Blepharisma A genus of ciliates (order HETEROTRICHIDA). Cells: ovoid to elongate, non-contractile, with uniform somatic ciliature and an oral region extending about half the length of the cell; usually pink to red.

blepharismin See GAMONE.

blepharismone See GAMONE.

blepharmone See GAMONE.

Blepharocorythina See TRICHOSTOMATIDA.

blepharoplast (*protozool.*) *Syn.* BASAL BODY (b).

blewit The edible fruiting body of *Lepista saeva (syn. Tricholoma saevum, T. personatum)* (= field blewit), or of *Lepista nuda (syn. Tricholoma nudum)* (= wood blewit).

blight (1) (*plant pathol.*) A general term applied to any of a wide range of unrelated plant diseases: see e.g. CANE BLIGHT, CHESTNUT BLIGHT, FIREBLIGHT, FUSARIUM, HALO BLIGHT, LATE BLIGHT.
(2) (*vet.*) *Syn.* INFECTIOUS KERATITIS.

blister blight A TEA DISEASE which occurs throughout the Asian tea-growing regions; it is caused by *Exobasidium vexans*. Translucent (sometimes pinkish), circular, concave lesions commonly occur on the upper surfaces of leaves; the convex underside of each lesion (on the lower side of the leaf) becomes white and downy with spores. Only young leaves are susceptible to infection. Control: e.g. copper sprays.

blister rust (white pine blister rust) An economically important disease of the western white pine (*Pinus monticola*), and of the Weymouth pine (*P. strobus*), caused by *Cronartium ribicola* (an apparently heterothallic rust [CJB (1985) *63* 1086–1088]). The disease develops over a period of years; symptoms include the formation of large orange blisters, and of cankers which may spread until the tree dies. (See also PINE DISEASES.)

bloat (ruminal tympany) In ruminants: a (sometimes fatal) condition in which the gaseous products of fermentation become trapped (either as free gas or as a stable foam) in the RUMEN – which thus becomes distended; the cause of bloat is not always clear, but the *pasture* (or *legume*) *bloat* which may occur in cattle grazing legume-rich pasture may be due to the foam-promoting properties of soluble leaf proteins or of lipids present in the feedstuff. *Feedlot* (or *grain*) *bloat* appears to be due to certain factor(s) of microbial origin – e.g. extracellular polysaccharides or slimes. One factor which may be important in both pasture and feedlot bloats is the ability of some bacteria (e.g. *Butyrivibrio fibrisolvens*, *Megasphaera elsdenii*, *Streptococcus bovis*) to digest the salivary mucins which act as antifoaming agents.

bloater damage (in cucumbers) See PICKING.

Blochman bodies Intracellular, prokaryotic symbionts which occur in mycetomes in insects.

blocked reading frame See OPEN READING FRAME.

blocking antibodies (1) (non-agglutinating antibodies) Antibodies which fail to agglutinate homologous antigens even though antigen–antibody combination may occur; lack of agglutination may be due e.g. to an insufficiency of effective combining sites on the antibody molecules. By combining with antigen, blocking antibodies prevent ('block') agglutination by homologous agglutinating antibodies. Blocking antibodies can be detected e.g. by an ANTIGLOBULIN CONSUMPTION TEST. (2) IgG antibodies which combine with allergens and which prevent those allergens from initiating an immediate hypersensitivity reaction.

blood agar An agar-based medium containing defibrinated (or whole citrated or oxalated) blood. It is prepared by mixing 5–10% blood (drawn aseptically) with a pre-autoclaved molten agar base (e.g. TRYPTICASE–SOY AGAR) at ca. 50°C; the medium is dispensed e.g. to Petri dishes and allowed to set. (cf. CHOCOLATE AGAR.) Blood agar is used e.g. for the culture of fastidious bacterial parasites of man and animals, and for the detection of HAEMOLYSINS; since a given bacterial haemolysin may be active against the RBCs of only some animals (e.g. horse, sheep), the choice of blood is important.

blood clotting See FIBRIN; COAGULASE; ANTICOAGULANT.

blood culture A procedure for detecting the presence of viable bacteria in blood. Since blood-borne bacteria may be present only in very small numbers, isolation is attempted only after a blood sample has been incubated in a suitable medium to allow the bacteria to multiply. Blood (5–10 ml), taken aseptically, is immediately added (aseptically) to 50–100 ml of medium (e.g. trypticase soy broth) which usually contains an anticoagulant (e.g. SPS) and sometimes e.g. penicillinase (to inactivate any penicillin in the blood) and/or PABA (to counteract any sulphonamide drugs). (The medium is commonly contained in a plain bottle whose opening is sealed by a rubber disc held in place by a screw-cap with a central hole; blood is injected into the bottle through the rubber disc.) The medium is then incubated. Periodically (e.g. daily for 7 days) the medium is subcultured to a suitable solid medium which is then incubated and examined for bacterial growth. (See also CASTAÑEDA'S METHOD; CLOT CULTURE.) [ARM (1982) *36* 467–493.]

blood DNA isolation kits Commercial kits used for preparing genomic DNA from samples of blood.
In the DNA Stat™ system (marketed by Stratagene), the blood sample (25–200 µl) is added to a hypotonic solution in a spin cup resting in a 2-ml collection tube; erythrocytes (red blood cells) are lysed in the hypotonic solution. On centrifugation, plasma proteins and lysed RBCs pass through the spin cup into the collection tube; leukocytes are retained within the spin cup. A chaotropic solution is then added to the spin cup; this lyses the leukocytes and releases genomic DNA. DNA binds to the spin cup, and contaminants are removed by two washing steps. DNA can then be eluted into buffer. The resulting DNA is suitable for use as a template in PCR.

blood parasites See e.g. ANAPLASMA, BABESIA, ENDOTRYPANUM, LANKESTERELLA, PLASMODIUM and THEILERIA.

blood poisoning *Syn.* SEPTICAEMIA.

blood transfusion (infection by) See TRANSFUSION-TRANSMITTED INFECTION.

bloom (water bloom) An abundant (easily visible and often conspicuous) growth of e.g. planktonic algae (including phytoflagellates), cyanobacteria (particularly GAS VACUOLE-forming species), or occasionally protozoa (e.g. *Stentor*) at or near the surface of a body of water (reservoir, lake, fishpond, ocean, etc). Blooms are often seasonal and may be favoured by particular set(s) of environmental conditions, including e.g. thermal stratification (in lakes and reservoirs), an excess of particular nutrients (e.g. leached nitrogen-rich agricultural fertilizers), an abundance of organic matter, etc; blooms of *Chrysochromulina breviturrita* may be triggered by the deposition of selenium (apparently an essential growth factor for this species) derived from coal-fired power plants [CJFAS (1985) *42* 1783–1788]. Blooms may recur cyclically: e.g. when dinoflagellates in a bloom exhaust their nutrient supply and die, their decomposition releases nutrients which encourage bacterial growth; the bacteria in turn produce VITAMIN B_{12} (an essential growth factor for dinoflagellates), thus triggering another dinoflagellate bloom. (See also RED TIDE.)

Blooms can cause various deleterious effects on the habitat in which they occur, as well as causing problems for man. The death and sudden decomposition of the massive microbial populations can cause a severe depletion of dissolved oxygen, often resulting in the asphyxiation of fish and other aquatic animals. Many bloom-forming species produce potent PHYCOTOXINS which may cause illness or death in animals or humans drinking the water or consuming aquatic animals from affected waters. (See also RED TIDE.) Certain species can impart unpleasant tastes and odours to the water: e.g. *Chrysochromulina breviturrita* can cause a 'rotten cabbage' odour, while *Synura petersenii* can cause 'fishy' tastes and odours. (See also GEOSMIN and METHYLISOBORNEOL.) Excessive algal growth in a reservoir may, in addition to the above problems, block water filters (see WATER SUPPLIES).

The formation of blooms may be discouraged or prevented by e.g. pumping to discourage thermal stratification, adding certain algicides, preventing contamination of waters with fertilizers, etc. ALGICIDES commonly used in this context include e.g. copper sulphate (a few parts per million), although this may actually encourage some smaller, more troublesome species by decimating larger, more susceptible species. Phenanthraquinone and DICHLONE have been reported to be effective against bloom-forming cyanobacteria.

bloomed lens (coated lens) A lens coated with a thin layer of e.g. magnesium fluoride to reduce the amount of incident light reflected back from the lens surface; the thickness of the coating is calculated such that *destructive interference* occurs between rays reflected from the bloomed surface and rays reflected at the fluoride–glass interface. In practice, blooming eliminates only that reflected light (usually green or yellow) whose wavelength is about four times the thickness of the fluoride coating. Blooming of a microscope eyepiece lens reduces stray background light in the tube and improves image sharpness.

blot (*mol. biol.*) See BLOTTING.

blotch disease *Syn.* BROWN BLOTCH.

blotting Following gel ELECTROPHORESIS: the transfer of nucleic acid and/or protein molecules from a gel strip to a specialized, chemically reactive paper, or other matrix, to which the transferred molecules may bind in a pattern similar to that present in the original gel. Transfer may be effected by capillary action, in which case the matrix (e.g. nitrocellulose, DEAE paper, APT PAPER) is sandwiched between the gel and a highly absorptive pad. Alternatively, in *electroblotting*, transfer is effected by electrophoresis.

In the earliest (capillary) blotting, DNA was transferred to nitrocellulose in the eponymous *Southern blot* technique (see SOUTHERN BLOT; see also SOUTHERN HYBRIDIZATION); later, blotting of RNA (*Northern blot*) and protein (*Western blot*) was carried out.

After blotting, a particular target molecule on the matrix may be identified or assayed e.g. by labelled probes or by enzyme immunoassay techniques. In *immunoblotting*, the transfer of proteins to a matrix is followed by exposure of the matrix to labelled antibodies; the binding of specific antibodies to a particular protein on the matrix may indicate the corresponding antigen.

(See also SOUTHWESTERN BLOTTING.)

blower *Syn.* SWELL.

blown can *Syn.* SWELL.

blue bodies See EAST COAST FEVER.

blue cheese See CHEESE-MAKING.

blue-eared pig disease Porcine reproductive and respiratory syndrome (PRRS): a disease of pigs, reported first in the USA and later in Europe, associated with a variety of symptoms, including abortion/stillbirth, agalactia, respiratory distress (particularly in new-born animals) and cyanosis of the ears; some strains can cause encephalitis. The disease may be mild, or may be severe with heavy losses, depending e.g. on the virulence of the causal agent and existing health of infected pigs. Secondary infections may occur.

The causal agent is a virus (genus ARTERIVIRUS), first identified in The Netherlands (the 'Lelystad' virus); it is also referred to as the swine infertility and respiratory syndrome (SIRS) virus.

Infection may occur via the airborne route. The incubation period may be a few days to a week, or much longer. Following experimental infection via the oronasal route, the virus can be detected within 12 hours in alveolar macrophages, in lung lymph nodes within 24 hours, and in spleen within 3–4 days. A reduction in numbers of alveolar macrophages (the major target cell of the virus) may be important in predisposing to secondary infections. Virus is shed in the oronasal secretions, urine and faeces of acutely infected animals.

[PRRS: RMM (1995) *6* 119–125.]

blue-green algae CYANOBACTERIA.

blue-light fluorescence See FLUORESCENCE.

blue mould (1) (of tobacco) See PERONOSPORA.

(2) Any blue-spored species of *Penicillium*.

blue proteins Copper-containing proteins which characteristically absorb strongly at or near 600 nm; at least some blue proteins are known to have roles in electron transport, and these proteins have been referred to as 'cupredoxins'.

Blue proteins occur in bacteria, e.g. AMICYANIN, AZURIN (sense 1), RUSTICYANIN; in fungi, e.g. LACCASE; in higher plants, e.g. cucumber basic blue protein, stellacyanin, umecyanin; in higher plants, algae and cyanobacteria, e.g. PLASTOCYANIN; and in animals, e.g. ascorbate oxidase (EC 1.10.3.3) and ceruloplasmin (found in mammalian blood plasma).

blue pus Pus formed in suppurative infections involving *Pseudomonas aeruginosa*; the pus has a bluish tinge owing to the presence of pyocyanin.

blue slime disease See COSTIASIS.

blue-stain A form of SAP-STAIN in which timber becomes bluish-grey owing to the deep penetration of fungal hyphae; the bluish-grey 'stain' is due to refraction of light by the hyphae. The commonest causal agents of blue-stain in temperate zones are species of *Ceratocystis*, e.g. *C. pilifera*; *Diplodia* spp are important in the tropics. Blue-stain may reduce the value of timber, but it does not significantly affect its mechanical strength since the cellulose and lignin components of the wood are little affected.

blueberry shoestring virus See SOBEMOVIRUSES.

bluecomb disease virus See CORONAVIRIDAE.

bluetongue An acute SHEEP DISEASE caused by an ORBIVIRUS; other ruminants (e.g. cattle, deer) may be affected, but in these the disease is usually mild or subclinical. Bluetongue occurs mainly in Africa but has occurred e.g. in Europe, Asia, USA. Transmission occurs mainly via a biological vector (especially *Culicoides* spp). Incubation period: 2 days to 2 weeks. Symptoms: fever; inflammation and cyanosis of the mucous membranes of the mouth, nose, and alimentary tract. The lungs and feet may be involved. Mortality (in sheep): e.g. 5–30%.

blunt-ended DNA Linear dsDNA with flush ends, i.e., with no 3′ or 5′ single-stranded extensions. (cf. e.g. STICKY ENDS.)

blusher *Amanita rubescens.*

BLV Bovine leukosis virus: see BOVINE VIRAL LEUKOSIS.

BOD (biochemical oxygen demand; biological oxygen demand) The amount of dissolved oxygen needed for microbial oxidation of soluble, biodegradable matter in an aquatic environment. For example, when discharged to rivers etc., sewage effluents may consume large amounts of dissolved oxygen, and if dilution is inadequate they may give rise, locally, to anaerobic or microaerobic conditions (i.e. exert a high BOD); in such cases there is a risk of asphyxiation for fish and other organisms dependent on appropriate levels of dissolved oxygen.

The BOD test measures the amount of oxygen (mg) consumed per litre of (e.g.) sewage, or a known dilution of it, in 5 days at 20°C.

(See also SEWAGE TREATMENT.)

BOD test See BOD.

Bodanella See PHAEOPHYTA.

Bodo A genus of protozoa (suborder BODONINA); species occur e.g. in organically polluted waters (see SAPROBITY SYSTEM). The cells are ovoid or tapered, sometimes flattened, ca. 5–20 μm in length; two flagella arise at the anterior end (near the cytostome), one being directed posteriorly.

Bodonina A suborder of biflagellate protozoa (order KINETOPLASTIDA); some members are free-living, others are parasitic e.g. in the blood or gut of fish. Genera include BODO, *Cryptobia*, *Rhynchomonas*.

body microflora Each of the various regions of the body typically harbours a characteristic MICROFLORA (sense 1); for microflora of the human body see e.g. EAR MICROFLORA, EYE MICROFLORA, GASTROINTESTINAL TRACT FLORA, GENITOURINARY TRACT FLORA, MOUTH MICROFLORA, RESPIRATORY TRACT MICROFLORA, SKIN MICROFLORA.

body odour See SKIN MICROFLORA.

boil (furuncle) A skin lesion due to infection of a hair follicle, sebaceous gland or skin lesion by (usually) *Staphylococcus aureus*. A tender red nodule becomes pus-filled and may either subside or break at the skin surface and drain. (cf. CARBUNCLE; FURUNCULOSIS (2).)

boil disease (in fish) See MYXOBOLUS.

Boivin antigens *Syn.* O ANTIGENS.

Bolbitiaceae See AGARICALES.

Boletales An order of fungi (subclass HOLOBASIDIOMYCETIDAE) that typically form mushroom-shaped basidiocarps in which the hymenophore lines a mass of vertically orientated tubules on the underside of the pileus (porous basidiocarp) or forms a layer on lamellae; the porous basidiocarps are fleshy (cf. APHYLLOPHORALES). Genera include e.g. *Boletus* (basidiocarp porous, stipe central), *Paxillus* (lamellate), and *Phylloporus* (lamellate). Some species are edible – e.g. *Boletus edulis* (cèpe or penny bun), some form mycorrhizal associations with trees, and one species, *Boletus parasiticus*, is parasitic on *Scleroderma*. [Classification and spore structure: TBMS (1981) *76* 103–146.]

Boletus See BOLETALES.

***Boletus* virus** See POTEXVIRUSES.

Bolivian haemorrhagic fever A VIRAL HAEMORRHAGIC FEVER caused by the Machupo virus (see ARENAVIRIDAE). Mortality: usually ca. 5–30%.

***Bombyx mori* NPV** See NUCLEAR POLYHEDROSIS VIRUSES.

bongkrek See TEMPEH.

bongkrekic acid A toxic, substituted heptaenedioic acid produced by '*Pseudomonas cocovenenans*' in spoiled bongkrek (see TEMPEH); it inhibits (uncompetitively) the ADP/ATP exchange carrier system which operates across the inner mitochondrial membrane. (cf. ATRACTYLOSIDE.)

booster (*immunol.*) A second or subsequent dose of a VACCINE administered with the object of eliciting a higher rate of antibody formation (see ANAMNESTIC RESPONSE).

boot laces See ARMILLARIA.

boracic acid *Syn.* BORIC ACID.

Bordeaux mixture (*plant pathol.*) An antifungal preparation first used in Bordeaux in the late 19th century for the treatment of downy mildew of vines; it has since been widely used as an agricultural antifungal agent against various blights, mildews, rusts and leaf-spot diseases, but is now used less frequently owing to its phytotoxicity. It is prepared by mixing a solution of copper sulphate with an aqueous suspension of calcium hydroxide, and appears to contain a mixture of basic copper sulphates ($CuSO_4.xCu(OH)_2$). The freshly-prepared mixture, sprayed onto leaves, stems etc, forms an insoluble deposit which gradually releases COPPER ions as a result of the influence of e.g. atmospheric carbon dioxide and/or plant or fungal secretions.

border disease (hairy shaker disease) A congenitally acquired SHEEP DISEASE characterized by an abnormally hairy birth-coat, tremors, and poor growth; it is caused by a PESTIVIRUS. [Review: Adv. Vet. Med. (1982) *36* (90 pp.).]

Bordet–Gengou agar An agar-based medium, used for the primary isolation of *Bordetella* spp (especially *B. pertussis*), containing glycerol, soluble starch, and 20–30% fresh horse or sheep blood; peptone (1%) may be added, but certain peptones inhibit the growth of *Bordetella* spp. Penicillin may be added to suppress growth of the nasopharyngeal flora. [Recipes: e.g. Book refs. 2, p. 132; 19, p. 145.]

Bordetella A genus of aerobic, chemoorganotrophic, Gram-negative bacteria classified within the beta group of Proteobacteria; the organisms occur as parasites and pathogens in mammals. Cells: non-motile or peritrichously flagellated coccobacilli, ca. 0.2–0.5 × 0.5–2.0 μm. Metabolism: respiratory (oxidative). Carbon sources include e.g. various amino acids; carbohydrates are not used. Growth factors required include organic forms of nitrogen and sulphur. Optimum growth at 35–37°C. GC% ca. 66–70. Type species: *B. pertussis*.

B. avium. Proposed name for bacteria causing coryza (rhinotracheitis) in turkey poults [IJSB (1984) *34* 56–70].

B. bronchiseptica. Cells: motile. Growth can occur on peptone-agar, and may be visible after overnight incubation. Oxidase +ve. Catalase +ve. Urease +ve. This species occurs e.g. in dogs and other animals, and is sometimes a cause of respiratory disease in man and animals.

B. parapertussis. Cells: non-motile. Growth on peptone-agar is characterized by the formation of a brown, soluble pigment. Improved growth may be obtained on Moredun medium (see WHOOPING COUGH). Oxidase −ve. Catalase +ve. Urease +ve. *B. parapertussis* is isolated from some cases of WHOOPING COUGH (q.v.).

B. pertussis. The primary cause of WHOOPING COUGH (q.v.). Initial isolation is usually carried out by plating a pernasal swab on BORDET–GENGOU AGAR or (better) CHARCOAL BLOOD AGAR. The typical colony (ca. 1 mm after 3–6 days) is smooth and domed with a metallic ('mercury drop') appearance and a butyrous consistency. Growth is inhibited by unsaturated fatty acids or colloidal sulphur. Cells: non-motile. Oxidase +ve. Some strains are catalase +ve. Urease −ve. Citrate is not utilized. (For toxin production see WHOOPING COUGH.) (See also BVG GENES.) *B. pertussis* and *B. parapertussis* can be detected, simultaneously, by a duplex PCR [JCM (1999) *37* 606–610].

boric acid (boracic acid) (as an antimicrobial agent) Boric acid (H_3BO_3) has mild antimicrobial properties, and is widely used as an ANTISEPTIC for delicate tissues (e.g. eyes) and e.g. as a bacteriostatic agent in urine samples. It is sometimes used, in combination with borax ($Na_2B_4O_7$), as a wood preservative.

Borna disease A disease of horses (and rarely sheep) caused by the BORNA DISEASE VIRUS. Symptoms in the horse include fever, pharyngeal paralysis and hyperaesthesia; death usually occurs in 1–3 weeks.

Borna disease virus (BDV) can cause behavioural abnormalities (with aggressive and passive phases) in certain animals, and it has been suggested that BDV could be responsible for certain psychiatric disorders in humans [Arch. Gen. Psych. (1985) *42* 1093–1096]; the involvement of BDV in psychiatric disease remains controversial, but immunological and PCR-based studies for BDV in psychiatric patients and blood donors in Japan have suggested that certain individuals are infected with BDV or with a BDV-related virus [JCM (2001) *39* 419–429].

Borna disease virus (BDV) A neurotropic virus (family Bornaviridae) which is the causal agent of BORNA DISEASE; BDV has a non-segmented, negative-sense ssRNA genome and appears to be a member of the MONONEGAVIRALES. [The coding srategy of BDV: Virol. (1995) *210* 1–8.]

Bornholm disease (epidemic myalgia; devil's grip) A self-limiting, epidemic disease usually caused by a group B coxsackievirus; other enteroviruses (e.g. coxsackieviruses A4, A6, A9, A10, and echoviruses 1 and 6) have sometimes been implicated. Onset is abrupt, with spasmodic, often severe chest pain (pleurodynia) and a brief period of high fever.

Borrelia A genus of Gram-negative bacteria (family SPIROCHAETACEAE) which occur as parasites or pathogens in man and other animals (see e.g. RELAPSING FEVER). The cells are motile, helical (3 to 10 loose coils), ca. $0.2–0.5 \times 3.0–20$ μm, each containing up to 20 periplasmic flagella; they stain well with conventional dyes, and are visible by light microscopy. Some species (including *B. recurrentis*) have been grown in vitro; cultured species require complex media and are microaerophilic.

Species were differentated by their arthropod vectors: e.g. human-louse-borne borreliae *B. recurrentis*, tick-borne borreliae according to tick species. Type species: *B. anserina*.

The 19 species recognized in Book 22 (pages 57–62) included *B. anserina*, *B. brasiliensis*, *B. duttonii*, *B. graingeri*, *B. harveyi*, *B. hermsii*, *B. parkeri*, *B. persica*, *B. recurrentis*, *B. theileri* and *B. venezuelensis*.

B. burgdorferi (*sensu lato*) comprises *B. burgdorferi* (*sensu stricto*) and the species *afzelii*, *andersonii*, *bissettii*, *garinii*, *japonica*, *lusitaniae*, *tanukii*, *turdi* and *valaisiana* [Molecular typing of *B. burgdorferi sensu lato*. Clinical Microbiology Reviews (1999) *12* 633–653].

B. burgdorferi [IJSB (1984) *34* 496–497; genome: Nature (1997) *390* 580–586] causes LYME DISEASE. This organism can be grown in BSK (Barbour–Stoenner–Kelly) II medium [recipe: Book. ref. 219, pp. 570–571], although diagnosis is more often serological (confirmation of a clinical diagnosis by detection of a rising titre of specific IgM during the 3rd–6th week following first symptoms e.g. by an indirect immunofluorescent antibody test). Quantification of *B. burgdorferi* in tissue samples has been achieved with a PCR-based method [JCM (1999) *37* 1958–1963]. The outer surface protein C (OspC) of *B. burgdorferi* is possibly involved in the binding of the pathogen to host ligands, and this protein has been examined in some detail [crystal structure of OspC: EMBO (2001) *20* 971–978].

borreliosis Any disease of man or animals caused by a *Borrelia* sp: see e.g. RELAPSING FEVER and LYME DISEASE.

Bostrychia See RHODOPHYTA.

botches disease Synonym for freshwater RED PEST.

bothrosome See LABYRINTHULAS and THRAUSTOCHYTRIDS.

boticins See BACTERIOCIN.

Botrydium A genus of siphonaceous algae (class XANTHOPHYCEAE) in which the mature vegetative organism consists of a globose to elongate (typically pear-shaped) coenocytic vesicle, up to ca. 2 mm diam., anchored to the substratum (mud, damp soil, etc) by colourless branched rhizoids; inside the vesicle is a thin peripheral layer of cytoplasm (containing the chloroplasts and nuclei) surrounding a large central vacuole. Asexual reproduction occurs by the formation of zoospores or aplanospores; resistant cysts are formed under dry conditions.

Botryobasidium See APHYLLOPHORALES (Corticiaceae).

botryomycosis ('bacterial pseudomycosis') A chronic localized infection of the skin and subcutaneous tissues in man and animals. The pathogen (commonly e.g. *Pseudomonas aeruginosa*, *Staphylococcus aureus*, or *Escherichia coli*) forms one or more granules in the centre of an abscess which is surrounded by fibrous tissue. (cf. MADUROMYCOSIS.)

Botryotinia See HELOTIALES.

botrytis (*plant pathol.*) *Syn.* GREY MOULD.

Botrytis A genus of fungi of the class HYPHOMYCETES; many species are known to have teleomorphs in the genus *Botryotinia*. The organisms include many plant-pathogenic species – e.g. *B. aclada* (= *B. allii*) (see ONION ROT), *B. cinerea* (see GREY MOULD, NOBLE ROT), and *B. fabae* (see CHOCOLATE SPOT). (See also SOFT ROT sense 2.) Antifungal agents used against *Botrytis* infections include e.g. BENOMYL, DICARBOXIMIDES, DICHLOFLUANID, DICLORAN, THIRAM and ZINEB.

Botrytis spp form septate mycelium; dark, ovoid, non-septate conidia develop in a layer on the swollen, terminal region of each of the small branches of a conidiophore. Sclerotia are formed e.g. by *B. aclada*, *B. cinerea* and *B. tulipae*. [Book ref. 183.]

bottom-fermenting yeast See BREWING.

bottromycins A group of peptide antibiotics, produced by *Streptomyces* spp, which inhibit bacterial protein synthesis.

botulinolysin See THIOL-ACTIVATED CYTOLYSINS. (cf. BOTULINUM TOXIN.)

botulinum C2 toxin A powerful toxin, produced by certain strains of *Clostridium botulinum*, which apparently has no neurotoxic activity (cf. BOTULINUM TOXIN) – but which causes e.g. increased intestinal secretion. The C2 toxin consists of two components, the larger of which appears to be involved in cell-surface binding and (possibly) uptake of the smaller component. The smaller component effects ADP-RIBOSYLATION of an ATP-binding ACTIN molecule, G actin, preventing its polymerization to F actin; this disrupts the cell's microfilament cytoskeleton and leads e.g. to rounding of the cell.

botulinum cook In the food processing industry: COOKING equivalent to a 12D PROCESS in respect of the endospores of *Clostridium botulinum*. The $D_{121.1}$ value (see D VALUE) of *C. botulinum* endospores is generally taken to be ca. 0.21 min. (See also CANNING.)

botulinum toxin Any of a range of related, powerful protein toxins (MWt ca. 150000) produced mainly by strains of *Clostridium botulinum*.

Botulinum toxins are differentiated antigenically into types A–G, with type C being divided into subtypes C1 and C2. Toxins of types A–F – with the exception of BOTULINUM C2 TOXIN – are powerful neurotoxins which cause BOTULISM in man and animals. The (plasmid-encoded) botulinum toxin G has been isolated e.g. from autopsy material but has not been linked definitively with disease. The botulinum C and D toxins are encoded by genes from two different bacteriophages.

All of these toxins can be produced by strains of *C. botulinum*.

Botulinum toxin type F has also been linked with an organism resembling *Clostridium barati*, associated with infant botulism [JCM (1985) *21* 654–655]. [See also JCM (2002) *40* 2260–2262.] A toxin similar to the botulinum type E toxin is produced by strains of *C. butyricum* isolated from cases of infant botulism. The ability to form type C toxin can apparently be transferred, by phage, from *C. botulinum* type C to *C. novyi*.

All the botulism-causing toxins are similar in structure and mode of action. Proteolytic cleavage into two fragments – a heavy (H) chain and a light (L) chain (connected by a disulphide bond) – activates the toxin; cleavage is mediated by a bacterial protease or an exogenous protease. The activated toxin binds, via the H chain, to the cell membrane of motor neurones at the nerve–muscle junction; toxin is internalized within a vesicle that remains at the nerve–muscle junction. The active part of the toxin (L chain) is translocated from the vesicle to the neurone's cytoplasm. Within the cytoplasm, the toxin acts as a zinc-endopeptidase in a way similar to that of TETANOSPASMIN (q.v. for essential details); in this case, however, the toxin blocks neuroexocytosis of vesicles containing the neurotransmitter *acetylcholine* – i.e. it prevents nervous stimulation of the muscle, leading to a *flaccid* paralysis. (Compare the site and mode of action of TETANOSPASMIN.)

Botulinum toxins differ in their targets: the A and E toxins cleave (inactivate) SNAP-25 (see TETANOSPASMIN); the C toxin cleaves syntaxin; B, D, F and G toxins cleave VAMP/synaptobrevin.

[Mechanism of action of botulinum neurotoxins: Mol. Microbiol. (1994) *13* 1–8.]

Purified botulinum toxin is heat-labile, being inactivated e.g. by heating at 100°C/10 minutes.

Botulinum toxin has been used as an alternative to surgery for treating some muscular disorders of the eye (strabismus; = 'squint') [Br. J. Ophth. (1985) *69* 718–724, 891–896]. [Uses of botulinum and other microbial neurotoxins: MR (1992) *56* 80–99. Therapy with botulinum toxin: Book ref. 220. Botulinum toxin for spastic gastrointestinal disorders: BCG (1999) *13* 131–143.]

botulism A disease of man and animals caused by any strain of *Clostridium* (usually a strain of *C botulinum*) which forms a (neurotoxic) BOTULINUM TOXIN (q.v.).

(a) *Botulism in man.* The disease is commonly caused by *C. botulinum* types A, B or E, but may be caused by any neurotoxigenic strain.

'*Classical*' botulism is a food-borne intoxication which follows ingestion of foods contaminated with pre-formed toxin – although additional toxin may be formed in the gut if viable cells of the pathogen are ingested. In the mouse, the LD_{50} of botulinum toxin is reported to be of the order of 0.1–1.0 ng/kg.

Foods of a low-acid, low-salt type are often implicated. *C. botulinum* type A may be associated with e.g. improperly canned vegetables, fish and meat products (Latin *botulus* = sausage), types B and F with e.g. meat products, and type E with e.g. uncooked sea-foods.

Toxin is absorbed from the gut and disseminated via the circulation. The incubation period is typically 12–36 hours, but may be ~2–8 days. The symptoms may include nausea, blurred/double vision (due to effects of toxin on the oculomotor muscles), dysphagia, dysphonia etc.; there is a progressive flaccid paralysis, and death may follow from asphyxia or cardiac failure.

Wound botulism is a rare form of the disease resulting from contamination of a wound by a toxigenic strain of the pathogen. Incubation period ~4–14 days. Symptoms are similar to those of food-borne botulism, sometimes with the addition of fever.

Infant botulism can follow ingestion of spores of the pathogen which germinate and form toxin in the gut (= *toxicoinfection*). The disease affects mainly infants <6 months of age, and varies from subclinical to rapidly fatal; typical symptoms: e.g. constipation, lethargy, dysphagia and weakness. One source of infection was reported to be honey contaminated with *C. botulinum* spores [JPed (1979) *94* 331–338]. Infant botulism may be responsible for some cases of SUDDEN INFANT DEATH SYNDROME; sudden death may be due to weakening or flaccid paralysis of muscles of the tongue and pharynx, leading to obstruction of the airway and asphyxia.

Botulism of undetermined classification includes those cases in individuals, older than 1 year, in which 'a specific food cause cannot be implicated'. Botulism in an adult may be due to the synthesis of toxin within the intestine during a gut infection with the pathogen [NEJM (1986) *315* 239–241; Lancet (1987) *i* 357–360; (*C. barati*) JCM (2002) *40* 2260–2262].

Lab. diagnosis (all forms): assay of serum, faeces and implicated foods for botulinum toxin and organisms.

Treatment. Any unabsorbed toxin should be removed from the gut. Polyvalent antitoxin may be administered (avoiding possible hypersensitivity reactions to heterologous components). Guanidine, given orally, may counteract the effects of toxin at the nerve–muscle junction. Supportive care will be needed. (A prophylactic botulinum toxoid vaccine has been given to laboratory personnel working with *C. botulinum*, but such a vaccine is unlikely to be useful for *treatment* owing to the time required to develop an immune response.)

[Botulism (epidemiology, diagnosis, treatment): RMM (1995) *6* 58–62.]

(b) *Botulism in animals.* In cattle, the disease may be caused by *C. botulinum* types B, C or D (see e.g. LAMZIEKTE and MIDLAND CATTLE DISEASE). In horses, types B or C may be involved (see e.g. FORAGE POISONING). Birds may be affected by types A or C (see e.g. LIMBERNECK and WESTERN DUCK DISEASE). In farmed salmonid fish the disease has been caused by type E [J. Fish Dis. (1982) *5* 393–399]. In animals, the disease is characterized by progressive paralysis without loss of sensibility and without fever.

Boucherie process A method of TIMBER PRESERVATION in which the sap of newly felled trees is replaced by a preservative (under pressure) in order to prevent subsequent fungal decay of the timber. (cf. SAUG-KAPPE PROCESS; BETHELL PROCESS.)

Bouin's fluid A FIXATIVE: a mixture of picric acid (75 ml saturated, aqueous), formalin (25 ml), and glacial acetic acid (5 ml).

bound coagulase See COAGULASE.

bourbon See SPIRITS.

boutonneuse In man: a mild disease caused by *Rickettsia conorii* and transmitted by tick bite. After an incubation period of 1–2 weeks a characteristic lesion (the *tâche noir*) develops at the location of the bite; this lesion (up to 0.5 cm diam.) consists of a black necrotic centre and a reddened margin. The regional lymph nodes may be affected. Fever, headache and muscular pains are typically followed by a maculopapular rash. Diagnosis may include the WEIL–FELIX TEST.

Boveria See SCUTICOCILIATIDA.

Boverin See BEAUVERIA.

bovid herpesvirus See under *bovine* herpesvirus.

bovine cytomegaloviruses See BETAHERPESVIRINAE.

bovine diseases See CATTLE DISEASES.

bovine ephemeral fever An infectious CATTLE DISEASE caused by a virus (a probable member of the Rhabdoviridae) and transmitted by insects; it occurs e.g. in Africa and Asia. Incubation period: ca. 2–10 days. Symptoms: e.g. fever, tremors, stiffness, and enlargement of the peripheral lymph nodes; mortality rates are low, but lactation is severely depressed.

bovine farcy See FARCY (2).

bovine farmers' lung (*vet.*) A chronic ATYPICAL INTERSTITIAL PNEUMONIA of cattle; it occurs, infrequently, in housed adult cattle exposed to mouldy or dusty feedstuffs. Symptoms: coughing, dyspnoea, nasal discharge. (cf. FARMERS' LUNG.)

bovine herpesvirus 1 See HERPESVIRIDAE.

bovine herpesvirus 2 See ALPHAHERPESVIRINAE.

bovine leukosis virus See BOVINE VIRAL LEUKOSIS.

bovine mammillitis virus See BOVINE ULCERATIVE MAMMILLITIS.

bovine mastitis See MASTITIS.

bovine papilloma virus See PAPILLOMAVIRUS.

bovine parvovirus See PARVOVIRUS.

bovine pustular stomatitis virus See PARAPOXVIRUS.

bovine respiratory disease (BRD) A collective term which includes a range of acute respiratory diseases of diverse aetiology in which the symptoms may include e.g. anorexia, coughing, dyspnoea, nasal discharge, pneumonia; cattle of all ages may be affected, but young animals are generally more susceptible. Treatment generally includes broad-spectrum chemotherapy to combat primary or secondary bacterial pathogens. Specific BRDs include: *pneumonic pasteurellosis* (shipping fever), a disease which is spread by droplet infection, and which can occur with high incidence in young stressed cattle; symptoms: fever, acute bronchopneumonia and toxaemia. The causal agent is usually *Pasteurella haemolytica*, sometimes *P. multocida*. There is usually a good response to early chemotherapy, with recovery in 1–2 days. *Bovine respiratory syncytial virus infection* is an acute bronchiolitis and alveolitis which occurs in (usually) young animals and involves e.g. fever, cough and dyspnoea; the disease can be fatal. (See also ATYPICAL INTERSTITIAL PNEUMONIA; CALF PNEUMONIA; CONTAGIOUS BOVINE PLEUROPNEUMONIA; INFECTIOUS BOVINE RHINOTRACHEITIS.)

bovine rhinotracheitis See INFECTIOUS BOVINE RHINOTRACHEITIS.

bovine spongiform encephalopathy (BSE; mad cow disease) A TRANSMISSIBLE SPONGIFORM ENCEPHALOPATHY of bovines caused by a PRION; it was reported in Great Britain in 1986. Clinical signs include e.g. nervousness, kicking and locomotor difficulty [clinical picture: VR (1992) *130* 197–201]. In Great Britain there were 3.9 confirmed cases per 1000 adult animals up to the end of 1989. Epidemiological studies implicated the meat and bone meal in concentrate feeds as a source of the causal agent; dairy herds were found to be affected more than beef suckler herds, due apparently to different feeding practices. [Epidemiological features: VR (1992) *130* 90–94; TIM (1997) *5* 421–424.]

Transmission of the BSE prion to humans, via the food chain, is now believed to cause nvCJD (variant, or new variant CREUTZFELDT-JAKOB DISEASE). This aetiological link is supported by various experimental data which have accumulated over the last few years. For example: intracerebral injection of monkeys with brain homogenates from affected cows yields clinically and pathologically similar CJD [Nature (1996) *381* 743–744]; transgenic mice which express human PrP proteins react with identical clinical, biochemical and histological changes when exposed to BSE or nvCJD material [Nature (1997) *389* 448–450]; the BSE prion converts human PrP to the prion form in a cell-free system [Nature (1997) *388* 285–288]. [Review of BSE: RMM (1998) *9* 119–127.]

[Cumulative incidence of BSE in the United Kingdom during the period July 1986–June 1997, animal-associated risk factors influencing age of onset of clinical signs, and effectiveness of control measures: VR (2000) *147* 349–354.]

(See also CATTLE DISEASES.)

bovine syncytial virus See SPUMAVIRINAE.

bovine tubercle bacillus *Mycobacterium bovis*.

bovine typhus *Syn.* RINDERPEST.

bovine ulcerative mammillitis (bovine herpes mammillitis) A CATTLE DISEASE caused by bovine (alpha) herpesvirus 2 (bovine mammillitis virus – see ALPHAHERPESVIRINAE). The disease involves severe inflammation and ulceration of the skin of the udder and teats; stomatitis and facial infection may also occur, particularly in calves. (See also LUMPY SKIN DISEASE and MASTITIS.)

bovine viral leukosis (enzootic bovine leukaemia) A chronic transmissible CATTLE DISEASE caused by the bovine leukosis virus (BLV), an exogenous v-*onc*$^-$ retrovirus which is closely related to HTLV-I and HTLV-II (see HTLV). [Nucleotide sequence of the BVL genome and its relation to other retroviruses: PNAS (1985) *82* 677–681.] Natural infection occurs e.g. in cattle (*Bos taurus* and *Bos indicus*), sheep, and water buffalo. The disease involves systemic malignant neoplasia of the reticuloendothelial system; in a small proportion of infected animals, malignant lymphomas (mainly of the B cell type) may develop in almost any organ, resulting in a variety of clinical manifestations. Transmission occurs horizontally by transfer of infected lymphocytes – most commonly via surgical instruments and needles (e.g. during herd vaccinations, blood-sampling, etc); natural transmission may occur via milk or colostrum or, in tropical countries, possibly via insects or parasites. [Reviews: Book refs 105, pp. 229–260, and 110, pp. 197–216.]

bovine virus diarrhoea (BVD) A CATTLE DISEASE caused by the BVD-MD virus (see PESTIVIRUS). BVD may affect cattle (and other ruminants) at any age, and follows infection by ingestion or inhalation; the condition is usually mild and self-limiting. (cf. MUCOSAL DISEASE.)

Bovista See LYCOPERDALES.

bowel oedema *Syn.* OEDEMA DISEASE.

Bowie–Dick test A test used to monitor the performance of an AUTOCLAVE. Essentially, two pieces of AUTOCLAVE TAPE are fixed in the form of an 'X' to a piece of paper which is inserted into the middle of a stack of cotton towels; the towels are autoclaved and the tape is examined. A satisfactory sterilizing regime is deemed to have been achieved only if *all* parts of the tape indicate exposure to sterilizing conditions.

box-like bacteria Variously shaped, angular bacteria which are found in hypersaline environments and which contain bacteriorhodopsin-like pigments [JB (1982) *151* 1532–1542]. (cf. SQUARE BACTERIA.)

Boyden chamber A chamber, divided into two sections by a membrane filter, used e.g. to study macrophage migration; the filter initially separates activated T lymphocytes (or their supernatant) and macrophages – the latter subsequently migrating through the filter in response to macrophage chemotactic factor.

Boyden procedure Treatment of erythrocytes with tannic acid (see TANNINS) to enable them to adsorb soluble antigens; such tanned cells are used e.g. in PASSIVE AGGLUTINATION. (Chromium salts or e.g. *bis*-diazobenzidine have been used in place of tannic acid.)

bp BASE PAIR.

Bph BACTERIOPHAEOPHYTIN.

***bph* operon** See BIOREMEDIATION

BPI protein (bactericidal/permeability-increasing protein) A component of the azurophil granules in human and rabbit NEUTROPHILS; it is bactericidal for some Gram-negative bacteria – in which it increases the permeability of the OUTER MEMBRANE – but it is not active against e.g. POLYMYXIN-resistant enteric bacteria (such as *Proteus*) or Gram-positive bacteria [MR (1992) *56* 399]. BPI binds to LPS and can apparently neutralize endotoxicity, and it may also attenuate the inflammatory response [see e.g. RMM (1995) *6* 101–108].

BPL β-PROPIOLACTONE.

BPV Bovine papilloma virus: see PAPILLOMAVIRUS.

Brabant mastitis test See CALIFORNIA MASTITIS TEST.

Brachiomonas A genus of unicellular flagellated green algae which are closely related to CHLAMYDOMONAS. The vegetative cells have a distinctive morphology: the posterior end is pointed, and four backward-directed pointed projections extend laterally from the cell body. Motile 'autospores' are formed (the daughter cells acquiring the distinctive adult-cell morphology before being released).

Brachonella See HETEROTRICHIDA.

Brachyallomyces See ALLOMYCES.

Brachyarcus A proposed genus of gas-vacuolated, curved rod-shaped bacteria which occur, in groups (coenobia), in microaerobic or anaerobic sulphide-containing aquatic habitats. [Book ref. 22, pp. 137–138.]

Brachybasidiales An order of plant-endoparasitic fungi (subclass HOLOBASIDIOMYCETIDAE) which form bisterigmate basidia either in definite hymenia or in tuft-like fascicles. *Brachybasidium* forms non-septate basidiospores on tufts of basidia which emerge from the host plant's stomata. Members of the other genera form true hymenia – which emerge through the host's epidermis and stomata – and basidiospores which sometimes become septate. In *Dicellomyces*, the basidiocarp is gelatinous (when fresh) and the hymenium is of uniform thickness. In *Ceraceosorus* and *Proliferobasidium* the basidiocarp is waxy, and the hymenium is not of uniform thickness; *Ceraceosorus* forms intracellular hyphae, *Proliferobasidium* forms intercellular hyphae [Mycol. (1976) *68* 640–654]. (cf. EXOBASIDIALES.)

Brachybasidium See BRACHYBASIDIALES.

bracket fungi (shelf fungi) Those fungi of the APHYLLOPHORALES (e.g. *Piptoporus betulinus*) whose fruiting bodies jut out like brackets or shelves from infected trees or rotting logs etc.

bradsot *Syn.* BRAXY.

Bradyrhizobium A genus of Gram-negative bacteria of the RHIZOBIACEAE; strains were formerly included in the genus RHIZOBIUM as 'slow-growing' strains (colonies ≤1 mm diam. in 5–7 days on e.g. yeast extract–mannitol–mineral salts agar). *Bradyrhizobium* spp produce an alkaline reaction in media containing carbohydrate (e.g. mannitol) and mineral salts. Extracellular slime includes a neutral (1 → 2)-β-glucan. Some strains can grow chemolithotrophically with H_2, CO_2 and low levels of O_2; some can fix nitrogen under microaerobic conditions in media which contain e.g. a source of fixed nitrogen. Species induce ROOT NODULE formation in certain leguminous plants (mostly tropical, some temperate) – e.g. soybeans, cowpea, siratro, certain *Lotus* spp. One non-leguminous plant, *Parasponia* (Ulmaceae), is also nodulated by a *Bradyrhizobium* strain. GC%: 61–65. Type species: *B. japonicum* (formerly *Rhizobium japonicum*). [Book ref. 22, pp. 242–244.]

bradyzoite *Syn.* CYSTOZOITE.

brain fungus *Sparassis crispa*.

brain–heart infusion medium A medium containing infusions of calf-brain and beef-heart, proteose, D-glucose, NaCl, and Na_2HPO_4; it is available commercially in dehydrated form. The medium may be used as a broth or solidified with agar; it may be supplemented with e.g. haemin, menadione, yeast extract etc. It is used for the culture of a range of bacteria and medically important fungi. [Recipe: Book ref. 47, pp. 644–645.]

branch migration (*mol. biol.*) See RECOMBINATION (figures 1 and 2).

branched DNA assay See BDNA ASSAY.

branching enzyme An enzyme which introduces branches into a linear polysaccharide (see e.g. GLYCOGEN and Q ENZYME). (cf. DEBRANCHING ENZYMES.)

Branchiomyces A 'genus' of fungi of uncertain taxonomic affinity; strains of *Branchiomyces* have features in common with members of the SAPROLEGNIALES, but it is unknown whether they form heterokont zoospores or aplanospores [Book ref. 1, pp. 215–216]. The organisms have been isolated only from the gill tissues of fish (see GILL ROT).

branchiomycosis *Syn.* GILL ROT.

brand spores See USTILAGINALES.

Brandenfleckenkrankheit *Syn.* BURNED SPOT DISEASE.

brandy See SPIRITS.

Branhamella See MORAXELLA.

brassica black rot See BLACK ROT.

brassica diseases See CRUCIFER DISEASES.

Braun lipoprotein (murein lipoprotein) A small lipoprotein (MWt ca. 7500) which occurs e.g. in *Escherichia coli* (encoded by gene *lpp*) and in other enterobacteria; each cell contains ca. 10^5–10^6 copies of the lipoprotein – about one-third of which form links (in association with OmpA) between the peptidoglycan layer and the OUTER MEMBRANE (thus stabilizing the cell envelope).

Braun lipoprotein is synthesized as a precursor with a signal sequence (see SIGNAL HYPOTHESIS). The cysteine residue immediately adjacent to the signal peptide cleavage site is modified by substitution with a diacylglycerol, permitting cleavage by the signal peptidase; following cleavage, a fatty acid residue is added to the N-terminal cysteine. Such post-translational modification apparently targets the protein to the outer membrane. [Cell (1990) *61* 739–741.]

The activity of the signal peptidase is inhibited by the peptide antibiotic *globomycin* (a structural analogue of the cleavage site).

Braun MSK tissue disintegrator An apparatus for CELL DISRUPTION. The sample, together with glass or plastic beads, is placed in a cooled container which is shaken at several thousand oscillations per minute for ca. 5 minutes; cells are broken by shear forces. The apparatus can break e.g. Gram-positive bacteria and endospores.

braxy (bradsot) An acute, usually fatal disease of lambs and young sheep; symptoms: fever, anorexia, inflammation of the wall of the abomasum, and toxaemia. Braxy occurs only in winter and appears to be due to the invasion of the abomasum by *Clostridium septicum* – the ingestion of frozen grass being a possible predisposing factor. Death occurs within hours.

BRD BOVINE RESPIRATORY DISEASE.

BrdU *Syn.* BUDR.

bread-making Bread is prepared by mixing flour (usually wheat flour), water, BAKERS' YEAST and salt to form a dough; other ingredients (e.g. milk, sugar, shortening etc) may be added. After thorough mixing the dough is left to 'prove' at ca. 25°C;

during this time the yeast metabolizes sugars (released from the flour starches by AMYLASES present in the flour) and produces bubbles of CO_2 which become trapped in an elastic complex of flour proteins (gluten) – lightening the texture of the dough. (Fungal amylases may be added to increase the availability of fermentable sugars (see also PROTEASES).) In addition to its leavening (lightening) action, the yeast contributes flavour to the product and helps to modify the gluten structure. When the leavened dough is baked the yeast is usually killed and any alcohol produced by fermentation evaporates.

Sourdough breads are made by using a starter consisting of flour (usually rye) and water inoculated with dough from a previous batch. In the starter, heterofermentative lactic acid bacteria (*Lactobacillus sanfrancisco*) ferment maltose (released from the starch by amylase action) and produce lactic and acetic acids, ethanol and CO_2; the pH falls to e.g. ca. 3.8. The acid-tolerant yeast *Saccharomyces exiguus*, which does not use maltose, uses other sugars (glucose, fructose etc) to leaven the bread. The starter may then be mixed with flour, water and salt to make a dough as with ordinary bread.

[Gluten biochemistry: *Biotechnology* (1995) *13* 1185–1190.]

bread mould See BREAD SPOILAGE.

bread spoilage Bread and similar products are very susceptible to spoilage by the growth of moulds, especially *Rhizopus nigricans* (black bread mould), *Neurospora sitophila* (red bread mould), and species of *Penicillium* and *Aspergillus*. Preservatives which may be incorporated in the dough include e.g. propionates, sorbic acid, and potassium sorbate. (See also ROPINESS; ERGOTISM.)

breakpoint chlorination See WATER SUPPLIES.

breakage-and-reunion model See RECOMBINATION.

breakbone fever *Syn.* DENGUE.

breaking (*plant pathol.*) *Syn.* COLOUR-BREAKING.

Breda virus An RNA virus which causes diarrhoea in young calves; it is apparently related to the BERNE VIRUS (see TOROVIRIDAE). [Morphological study of Breda virus replication: JGV (1986) *67* 1293–1304.]

brefeldin A A fungal toxin which inhibits the activity of the Golgi apparatus and which, as a consequence, can block certain processes in various types of cell; for example, brefeldin A can block protein secretion and inhibit the development of viruses.

Breinl strain A virulent strain of *Rickettsia prowazekii*; unlike the attenuated E strain it can be cultivated in human macrophage cultures.

Bremia A genus of fungi (order PERONOSPORALES) which cause DOWNY MILDEWS on members of the Compositae (e.g. *B. lactucae* on lettuce). [*B. lactucae*–lettuce interaction: Book ref. 174, pp. 116–118.]

Bremiella See PERONOSPORALES.

Brettanomyces A genus of yeasts (class HYPHOMYCETES) which form single spheroidal, cylindrical or elongated budding cells, pseudomycelium, or branched non-septate mycelium. Growth occurs slowly e.g. on malt agar; species may be isolated on cycloheximide-containing malt agar. Most strains can ferment glucose; fermentation is usually stimulated by O_2 and acetoin ('Custers effect', 'negative Pasteur effect'). Under aerobic conditions acetic acid is produced from glucose and from ethanol. Only some species can assimilate NO_3^-. Species: *B. abstinens*, *B. anomalus*, *B. bruxellensis* (teleomorph: *Dekkera bruxellensis*), *B. claussenii*, *B. custersianus*, *B. custersii*, *B. intermedius* (teleomorph: *Dekkera intermedia*), *B. lambicus*, *B. naardenensis*. ('*B. sphaericus*' = *Candida etchellsii*; '*B. versatilis*' = *Candida versatilis*.) Species occur in beers, wines (see WINE SPOILAGE), grape must etc. [Book ref. 100, pp. 562–576.]

brevetoxins Polycyclic toxins produced by the dinoflagellate *Ptychodiscus brevis* (see RED TIDE). Brevetoxins A, B and C each consist of 11 contiguous ether rings; they are highly lipid-soluble and are potent neurotoxins, causing depolarization of nerve membranes. [PAC (1986) *58* 339–350 (346–347).]

Brevibacterium A genus of obligately aerobic, catalase-positive, asporogenous bacteria (order ACTINOMYCETALES, wall type III – see also PEPTIDOGLYCAN) which occur e.g. on certain cheeses. In culture the organisms initially grow as irregular, 'lumpy' rods, but coccoid forms predominate in the stationary phase; colonies vary from yellow to orange-red, though some strains produce pigment only in the light. Growth occurs at ca. 20–25°C e.g. on trypticase–soy agar containing 4% sodium chloride; metabolism is oxidative. GC%: ca. 60–64. Type species: *B. linens*.

Brewer jar An ANAEROBIC JAR which consists of a cylindrical glass chamber and a flat, gas-tight lid which is secured by means of a screw clamp. In use, the jar is filled with e.g. a mixture of N_2 (85%), CO_2 (5%) and H_2 (10%), and any residual O_2 is eliminated by means of an electrically-heated platinized catalyst which is fitted inside the lid and which is separated from the interior of the jar by a gauze safety screen. Alternatively, the jar may be used for the evacuation–replacement method as described for the MCINTOSH AND FILDES' ANAEROBIC JAR. Anaerobiosis within the jar is monitored e.g. with a tube containing a METHYLENE BLUE solution.

Brewer's thioglycollate medium A liquid medium (pH 7.0–7.2) used e.g. for the culture of ANAEROBES; it includes e.g. glucose, tryptone, agar (ca. 0.1%), sodium thioglycollate (a poising agent – see REDOX POTENTIAL), and a redox indicator dye. [Recipe: Book ref. 53, pp. 1409–1410.] The medium is particularly useful for testing the sterility of products which contain mercurial preservatives – the antimicrobial activity of the latter being neutralized by the thioglycollate.

brewers' yeast Any of various strains of *Saccharomyces cerevisiae* used for BREWING (q.v.); 'bottom-fermenting' strains (formerly called *Saccharomyces uvarum* or *S. carlsbergensis*: see SACCHAROMYCES) are usually used in the manufacture of lagers, while 'top-fermenting' strains are usually used for ales. (Top- and bottom-fermenting strains can generally be distinguished by the ability of bottom yeasts to produce an extracellular melibiase (α-galactosidase) and thus to hydrolyse raffinose to glucose, galactose and fructose, whereas top yeasts – which do not produce melibiase – hydrolyse raffinose to melibiose and fructose.) Particular strains of yeast may be selected for particular properties (e.g. flavour production), but all brewers' yeasts must be able to ferment the sugars (particularly maltose and maltotriose) present in wort, and must be able to tolerate the initially high levels of solutes in wort and produce – and tolerate – high levels of ethanol under brewing conditions; the yeast must also be readily separable from the fermented wort (see BREWING) and should retain high viability for re-use in subsequent fermentations. [Selection and modification of brewers' yeast strains: Food Mic. (1984) *1* 289–302.]

brewing The preparation of beer or lager by the fermentation of an aqueous extract of malted barley containing essential oils and bitter resins of the dried female flowers ('cones') of the hop (*Humulus lupulus*). During *malting* the barley is allowed to germinate, and the grain starch and protein are partially degraded by endogenous enzymes. Germination is halted by *kilning*: drying (e.g. to stop respiration, preserve enzymes and substrates etc), followed by heating (e.g. to ca. 80°C). The dried malt is ground (milled) to expose the starchy endosperm; the resulting *grist* is made into a *mash* by mixing with water. Other carbohydrate sources (*adjuncts*) may be added at this stage. The mash (pH 5.2–5.5) is subjected to controlled (variable) temperature–time cycles during which starch is broken down to fermentable sugars (glucose, maltose, maltotriose) and non-fermentable dextrins, proteins are degraded to amino acids and peptides, and phytin is hydrolysed to inorganic phosphate (which acts as a buffer) and inositol (an essential yeast growth factor). The liquor (*wort*) is separated from the spent grain and is boiled (with hops) to inactivate enzymes, coagulate proteins, destroy spoilage organisms etc. The hopped wort is cooled, aerated, and inoculated (*seeded* or *pitched*) with a pure culture of yeast: a strain of *Saccharomyces cerevisiae* (see BREWERS' YEAST). A lag period of ca. 12 h is followed by growth of the yeast. As oxygen is used up, metabolism switches from respiratory to fermentative, and sugars in the wort are converted to ethanol and CO_2 primarily via the EMBDEN–MEYERHOF–PARNAS PATHWAY. (Fermentation is exothermic, so the fermentation vessel must be cooled.) The formation of bubbles of CO_2 keeps the wort agitated and the yeast cells in suspension. At the end of the fermentation the yeast cells clump together – see FLOCCULATION (sense 2); in 'top-fermenting' strains of *S. cerevisiae* the clumps rise to the surface of the liquor, while those of 'bottom-fermenting' strains sediment. Most of the yeast is removed (e.g. by surface skimming of top yeasts or collection of sediment for bottom yeasts, and/or by centrifugation), and the beer is *matured* at low temperature (e.g. −1°C); residual yeast cells (a) carry out a 'secondary fermentation' of remaining maltotriose, and (b) reduce levels of diacetyl. H_2S and acetaldehyde may be removed by purging with CO_2. Finally, the beer is filtered, bottled or canned, and pasteurized. (See also BEER SPOILAGE.)

brick cheese See CHEESE-MAKING.

Brie cheese See CHEESE-MAKING.

bright-field microscopy See MICROSCOPY (a).

brightener See FLUORESCENT BRIGHTENER.

Brill–Zinsser disease See TYPHUS FEVERS.

brilliant green A yellowish-green TRIPHENYLMETHANE DYE which contains ethyl-substituted amino groups. It has antibacterial activity, inhibiting aerobic spore-formers more than anaerobic spore-formers.

brilliant green agar A selective medium used for salmonellae which have a broad host range (e.g. *S. typhimurium*); it contains e.g. lactose, phenol red, and brilliant green (selective agent). Lactose +ve colonies are green, lactose −ve colonies are pink.

brinase A fibrinolysin-like PROTEASE, obtained from *Aspergillus oryzae*, which can hydrolyse fibrin and fibrinogen and can lyse blood clots. (cf. STREPTOKINASE.)

British anti-Lewisite See BAL.

brittleworts See CHAROPHYTES.

broad bean mottle virus See BROMOVIRUSES.

broad bean wilt virus A virus which may be a member of the COMOVIRUSES but which has a wide host range and is transmitted by aphids.

broad-range primers Primers, used in PCR, which bind to highly conserved sequences found in the 16S rRNA gene in *all* bacteria; such primers can therefore be used to detect *any* species of bacterium whose 16S rRNA gene is accessible. Broad-range primers are also referred to as *universal primers*.

Broad-range primers are used e.g. to study bacterial diversity in environmental samples; the products are sequenced, and variable regions within the amplicon (which include species-specific sequences) are checked against the 16S rRNA database of known species in order to attempt identification of bacteria within the sample.

Some PCR-based studies use broad-range primers (simultaneously with other primers) as an amplification control [see e.g. JMM (1997) *46* 773–778; JCM (1999) *37* 2090–2092].

broad-spectrum antibiotics (1) ANTIBIOTICS active against a wide range of bacteria – including both Gram-positive *and* Gram-negative species. (2) Antibiotics active against a wide range of bacteria which may include Gram-positive species, Gram-negative species, *or* both.

broccoli necrotic yellows virus See RHABDOVIRIDAE.

Brochothrix A genus of Gram-positive, facultatively anaerobic bacteria; type species: *B. thermosphacta* (formerly *Microbacterium thermosphactum*) [IJSB (1976) *26* 102–104]. *B. thermosphacta* is common e.g. in meat and meat products (see MEAT SPOILAGE). In young cultures cells are rod-shaped (occurring in chains), in older cultures they are small coccobacilli. Growth conditions: 0–30°C (optimum ca. 23–25°C); pH 5–9 (optimum ca. 7.0); all strains can grow in the presence of 6.5% NaCl, many in 10% NaCl. Tests: MR +ve; VP +ve; indole −ve; H_2S −ve; nitrate is not reduced, catalase +ve e.g. on APT agar at 20°C, usually catalase −ve at 30°C. During aerobic growth *B. thermosphacta* forms cytochromes and can use glycerol or a range of carbohydrates as energy sources. Anaerobically, cytochromes are not formed, and glucose is fermented primarily to L-lactic acid, acetic acid, formic acid, and ethanol in proportions that vary with growth conditions [AEM (1983) *45* 84–90]. In *meat* under aerobic conditions *B. thermosphacta* uses glucose and, when glucose is depleted, glutamate; under anaerobic conditions only glucose can be used. (See also STAA MEDIUM.)

[Review: Book ref. 29, pp. 139–173.]

broken cream See MILK SPOILAGE.

bromatia See FUNGUS GARDENS.

bromcresol green A PH INDICATOR: pH 3.8–5.4 (yellow to green); pK_a 4.7.

bromcresol purple A PH INDICATOR: pH 5.2–6.8 (yellow to purple); pK_a 6.3.

brome mosaic virus See BROMOVIRUSES.

bromelain A basic glycoprotein thiol PROTEASE (EC 3.4.22.4) obtained from the pineapple plant.

bromoaplysiatoxin See LYNGBYA.

5-bromo-4-chloro-3-indolyl-β-D-galactoside See XGAL.

5-bromo-4-chloro-3-indolylphosphate-*p*-toluidine See XP.

bromochlorophane See BISPHENOLS.

bromocresol green *Syn.* BROMCRESOL GREEN.

bromocresol purple *Syn.* BROMCRESOL PURPLE.

5-bromodeoxyuridine See BUDR.

bromophenol blue *Syn.* BROMPHENOL BLUE.

bromothymol blue *Syn.* BROMTHYMOL BLUE.

5-bromouracil (BU) A BASE ANALOGUE which, in the *keto* tautomeric form (which predominates), pairs with adenine (A) but in the *enol* form pairs with cytosine (C). Thus, during DNA synthesis, keto-BU may be incorporated into DNA in place of thymine (T); if it then remains in the keto form, BU will continue to mimic T and no mutation will occur, but if it switches to the enol form just prior to further DNA replication it will pair with guanine (G) instead of A, leading to an A·T-to-G·C transition. Conversely, enol-BU may be incorporated into DNA in place of C, and a subsequent switch to the keto form will result in a G·C-to-A·T transition on subsequent DNA replication.

bromovinyldeoxyuridine (BVDU; (E)-5-(2-bromovinyl)-2′-deoxyuridine) An ANTIVIRAL AGENT which is highly active against varicella-zoster virus; it is also active against herpes simplex virus (HSV) type 1 (but less so against HSV type 2) and e.g. against pseudorabies virus, infectious bovine rhinotracheitis virus, and herpesvirus simiae. It can be administered topically or orally. BVDU is apparently converted to the 5′-monophosphate by the viral thymidine kinase; the monophosphate can be converted to the diphosphate by HSV type 1 thymidine kinase (cf. ACYCLOVIR).

bromoviruses (brome mosaic virus group) A group of tripartite ssRNA-containing PLANT VIRUSES, each of which infects a narrow range of plants; mechanical transmission can occur readily, and some members are transmitted by beetles. Type member: brome mosaic virus; other members: broad bean mottle virus; cowpea chlorotic mottle virus.

Virion: icosahedral, ca. 26 nm diam., containing a single species of coat protein. Genome: three molecules of linear, positive-sense ssRNA – RNA1 (MWt ca. 1.1×10^6), RNA2 (MWt ca. 1.0×10^6), and RNA3 (MWt ca. 0.7×10^6); 'RNA4' (MWt ca. 0.3×10^6), the coat protein mRNA, is also encapsidated. A given bromovirus particle may contain a molecule of RNA1, a molecule of RNA2, or one molecule each of RNA3 and RNA4; these different particles all have about the same S_W^{20} and all three are necessary for infectivity. The RNAs each have a capped 5′ end and a tRNA-like structure (which accepts tyrosine) at the 3′ end. Virus assembly occurs in the cytoplasm; crystalline arrays of virus particles and granular cytoplasmic inclusions may be formed.

bromphenol blue A PH INDICATOR: pH 3.0–4.6 (yellow to blue); pK_a4.0.

bromthymol blue A PH INDICATOR: pH 6.0–7.6 (yellow to blue); pK_a7.0.

bromthymol blue agar See BTB AGAR.

bronchiolitis Inflammation of the bronchioles. The disease occurs mainly in infants and is often due to human respiratory syncytial virus. Spread occurs by droplet transmission. The disease may be self-limiting or may lead to cyanosis, apnoea and death.

bronchitis Inflammation of the bronchi; the condition may be acute or chronic and often follows a (usually viral) upper respiratory tract infection. Bronchitis may be caused by bacteria (e.g. *Haemophilus influenzae*, *Mycoplasma pneumoniae*, *Staphylococcus aureus*, *Streptococcus pneumoniae*), by viruses, or by physical or chemical irritants. Symptoms: chest pain, cough with mucoid to purulent sputum, fever etc.

bronopol See ALCOHOLS.

bronze leaf wilt *Syn.* HARTROT.

brood pouch (marsupium) In certain ciliates: a cavity, formed by invagination of the pellicle, within which 'larval' cells are formed by budding; in suctorians the brood pouch is a temporary structure, but in chonotrichs (in which it is also called a *crypt*) it is permanent. In some species the larval cells are released into the brood pouch, while in others the larval cells detach from the parent cell only after the pouch has everted.

brooder pneumonia See ASPERGILLOSIS.

brosse (*ciliate protozool.*) A tuft of cilia, of unknown function, which occurs on some gymnostomes.

broth In microbiology: any of a variety of liquid media – especially NUTRIENT BROTH or any liquid medium based on nutrient broth and/or hydrolysed protein.

broth culture See CULTURE.

broth dilution test See DILUTION TEST.

broth sugars See PEPTONE-WATER SUGARS.

brown algae Algae of the PHAEOPHYTA.

brown blotch A MUSHROOM DISEASE caused by *Pseudomonas tolaasii*. Caps and stalks of infected mushrooms are pitted with wet lesions which become dark chocolate-brown. (cf. GINGER BLOTCH.) [Disease control by disinfection: JAB (1985) *58* 259–281.]

brown bodies See LEPROSY.

Brown–Boveri UV-C lamp A source of high-intensity ULTRAVIOLET RADIATION; the lamp gives a radiation intensity of up to ca. 1W/cm^2. (Conventional sources give up to ca. 10 μW/cm^2.)

Brown–Hopps modification (of the GRAM STAIN) A procedure used for staining bacteria in deparaffinized, hydrated tissue sections. Essentially, sections are stained with crystal violet; washed; treated with Gram's iodine; washed; differentiated in acetone; washed; stained with basic fuchsin; washed; treated with Gallego's solution (an aqueous solution of formalin and acetic acid); washed; dipped in acetone, picric acid–acetone, and acetone; and finally transferred, via xylene, to the mountant. Gram-positive bacteria stain blue; Gram-negative bacteria stain red. [AJCP (1973) *60* 234–240.]

brown oak Timber from an oak tree (*Quercus* sp) which has been infected by *Fistulina hepatica*; such infection causes the heartwood of the tree gradually to develop a rich brown colour. The colour appears to derive partly from pigment within the fungal hyphae and partly from extracellular fungal products. Brown oak is somewhat weaker than the normal wood, but it is valued e.g. for furniture-making.

brown rot (1) (of timber) A form of fungal TIMBER SPOILAGE in which the CELLULOSE and HEMICELLULOSES of the wood are decomposed while the LIGNIN remains virtually intact (although it may be modified), leaving the wood brown, soft and friable, and subject to cross-grain cracking. The hyphae of brown-rot fungi (e.g. *Piptoporus betulinus*, *Poria placenta*) penetrate the wood cell lumen and lie in contact with the inner surface of the cell wall; a generalized thinning of the wall occurs, suggesting a diffusion of degradative enzymes from the hyphae (cf. WHITE ROT). A novel mechanism has been proposed for the disruption of native cellulose by brown-rot fungi (which do not produce exoglucanases – see CELLULASES): H_2O_2, generated by the oxidation of hexoses by glucose oxidase, is believed to act in combination with Fe^{2+} (present in the wood) to oxidize glucopyranosyl residues, causing rupture of the cellulose chains; this could disrupt microfibrils sufficiently to allow access for endoglucanases [Book ref. 26, pp. 28–29; Book ref. 39, pp. 56–59].

(2) (of fruits) A rot which affects a wide variety of fruits – particularly apples, pears and stone fruits (cherries, plums etc) – both on the tree and in storage. Soft brown lesions develop in the fruits, initially beneath an intact epidermis; subsequently, whitish cottony pustules may develop, often in a concentric arrangement, and the fruit eventually shrivels and dries. Brown rots are caused by *Sclerotinia* spp: *S. fructigena* in e.g. apples, pears and almonds, *S. laxa* or *S. fructigena* in cherries, plums, peaches etc.

brown rust A disease of wheat and rye caused by *Puccinia recondita*; orange-brown pustules, later becoming dark or black, develop mainly on the upper (adaxial) leaf surface. On barley, a similar disease is caused by *Puccinia hordei*. (See also CEREAL DISEASES.)

Browne's tubes Sealed glass tubes containing a fluid which changes (irreversibly) from red to green when exposed to certain temperature/time combinations; they are used to monitor conditions at various positions within an operating AUTOCLAVE or hot-air oven. Type I and type II tubes are used in conventional autoclaves and high-vacuum autoclaves, respectively; type III is used in hot-air ovens, and type IV for monitoring exposures of ca. 180°C/12 min. (cf. BROWN'S TUBES.)

Brownian movement *Random* movements made by small (ca. 1 µm) particles, or organisms, freely suspended in a fluid medium; the movements are due to bombardment of the particles by molecules of the medium. Brownian movement causes *non-motile* organisms to oscillate about more or less fixed positions; by contrast, *motile* organisms change their relative positions.

Brown's tubes A set of sealed glass tubes containing aqueous suspensions of barium sulphate in increasing concentrations; the tubes thus exhibit a range of optical densities which vary from transparent to turbid. Brown's tubes are used for standardizing microbial suspensions: the opacity of a given bacterial suspension is matched (visually) with that of a particular tube, and the suspension described e.g. as a Brown's number 2 or a Brown's number 5 suspension etc. (Higher numbers correspond to greater opacities.) The unknown suspension should be examined in a tube of size and thickness equivalent to those containing the standard suspensions.

Brucella A genus of Gram-negative, aerobic, chemoorganotrophic, catalase-positive bacteria which occur, typically as intracellular parasites or pathogens, in man and other animals (see BRUCELLOSIS). The cells are non-motile bacilli, cocci or coccobacilli, ca. 0.5–0.7 × 0.6–1.5 µm, which stain well with conventional dyes (see also KOSTER'S STAIN). Metabolism is respiratory (oxidative). Growth requirements generally include certain amino acids, nicotinamide and thiamine; 5–10% CO_2 may be essential, particularly in primary isolation. Primary culture may be carried out on e.g. serum–glucose agar or chocolate agar; on primary isolation, 48-hour colonies on serum–glucose agar are typically 0.5–1.0 mm in diameter, shiny, convex and entire. Optimum growth temperature: 37°C. With the exception of *B. neotomae*, no species forms acid from carbohydrates in peptone-water sugars. MR −ve; VP −ve; most strains are oxidase +ve; most strains are urease +ve, but all strains of *B. ovis* are urease −ve (at 24 hours). The organisms are normally killed by pasteurization, are generally sensitive to tetracyclines, and are often sensitive e.g. to chloramphenicol, erythromycin and streptomycin. GC% ca. 55–58. Type species: *B. melitensis*.

Brucellae can be divided into smooth, rough and mucoid strains on the basis of their reaction in ACRIFLAVINE solution (0.1% aqueous): smooth strains are easily emulsifiable, rough strains agglutinate, and mucoid strains form stringy threads.

The brucellae are susceptible to a range of tailed, polyhedral DNA phages which have been divided into five groups that are used in phage typing (see e.g. TBILISI PHAGE, WEYBRIDGE PHAGE); group 5 phages include those which lyse non-smooth brucellae. (*Brucella* phages are sensitive to e.g. cationic detergents and oxidizing agents.)

The sensitivity of some brucellae to certain dyes, e.g. basic FUCHSIN and THIONIN, has been exploited for between-species and within-species differentiation; these dyes inhibit the growth of sensitive strains when incorporated in media at concentrations of ca. 1:50000 (w/v).

Differentiation between species and biovars is also achieved by the use of monospecific antisera prepared against each of the two main somatic antigens in the brucellae: the A and M antigens; an antiserum against rough strains (R) is used for *B. canis* and *B. ovis*.

B. abortus. Primarily a pathogen of bovines, but man and other animals can be infected; infection of pregnant animals commonly results in placentitis and abortion. All strains of *B. abortus* are lysed by Tbilisi phage at RTD – a characteristic which distinguishes this species from other brucellae. There are 9 biovars, biovars 1, 2 and 4 being sensitive to thionin, biovar 2 to fuchsin. Biovars 4, 5 and 9 do not agglutinate in A antiserum; biovars 1, 2, 3 and 6 do not agglutinate in M antiserum. *B. abortus* 'strain *19*' is similar to biovar 1 but has reduced virulence; it is used as a live vaccine for cattle.

B. canis. Primarily a pathogen of the dog, causing e.g. epididymitis, orchitis, prostatitis and the development of local granulomas; man is occasionally infected. Only rough or mucoid strains occur; these are agglutinated by R antiserum but not by A or M antisera. Insensitive to thionin; most strains sensitive to fuchsin. No biovars.

B. melitensis. Typically pathogenic in goats and sheep, but also in bovines, man and pigs. Insensitive to both fuchsin and thionin; biovar 1 is agglutinated by M antiserum, biovar 2 by A antiserum, and biovar 3 by both A and M antisera.

B. neotomae. A non-pathogenic or weakly pathogenic species. Acid (no gas) is formed from arabinose, galactose, glucose and xylose. Sensitive to both fuchsin and thionin, and agglutinated by A antiserum. No biovars.

B. ovis. A pathogen of sheep, causing epididymitis/orchitis, or abortion in pregnant ewes. Insensitive to thionin; most strains sensitive to fuchsin. No biovars. Smooth strains unknown; agglutination by R antiserum only.

B. suis. Primarily a pathogen of pigs, but most or all strains can also infect man; other hosts include e.g. dogs and hares. Insensitive to thionin; most strains sensitive to fuchsin, but biovar 3 is insensitive. Biovars 1–4 are agglutinated by A antiserum; 4 is also agglutinated by M antiserum, and 5 is agglutinated only by M antiserum.

An early proposal to include all strains of *Brucella* in a single species (*B. melitensis*) [IJSB (1985) *35* 292–295] has not been accepted.

***Brucella* ring test** *Syn.* MILK RING TEST.

brucellin An antigenic material derived from cultures of *Brucella* spp.

brucellin test A SKIN TEST, analogous to the TUBERCULIN TEST, used for detecting *Brucella*-induced hypersensitivity (cf. BRUCELLOSIS).

brucellosis Any human or animal disease caused by *Brucella* spp. In animals the reproductive organs are generally affected, often resulting in abortion (*contagious abortion*). *B. abortus* is the usual causal agent in cattle, *B. suis* in pigs, *B. melitensis* in goats, *B. ovis* in sheep, and *B. canis* in dogs, but host specificity is not absolute. Contagious abortion in cattle (*Bang's disease*) may follow infection via the mouth, vagina or wounds; the pathogen has a predilection for the udder (resulting in its presence in the milk) and for fetal and placental tissues (its growth apparently being stimulated by erythritol present in these tissues).

In man brucellosis (*undulant fever*, *Malta fever*) is an acute or chronic systemic disease contracted from animals via unpasteurized milk, infected carcases, etc; *B. abortus*, *B. melitensis* or *B. suis* is most commonly involved. Infection occurs via the mouth, conjunctivae or wounds. Incubation period: days, weeks or months. Symptoms: variable, but typically include headache, malaise, and intermittent periods of fever. Untreated cases may persist for years. (In individuals frequently exposed to *Brucella* spp disease may be due to hypersensitivity to *Brucella* antigens rather than to true infection.) *Lab. diagnosis*: (i) Blood culture (usually positive if blood is taken at the height of the fever). (ii) Culture of bone marrow or lymph node aspirates (may be useful in chronic cases). (iii) Serological assay for specific serum agglutinins. (N.B. Agglutinins against cholera vibrios cross-react with those against *Brucella* spp.) *Chemotherapy*: e.g. tetracyclines with streptomycin. (See also BRUCELLIN TEST and MILK RING TEST.)

Bryopsis A genus of siphonaceous green seaweeds (division CHLOROPHYTA) which exhibit a heteromorphic alternation of generations (cf. DERBESIA).

Bryoria A genus of fruticose LICHENS (order LECANORALES) separated from ALECTORIA e.g. on the basis of the smaller colourless ascospores (8 per ascus).

BSA Bovine serum albumin.

BS-C-1 An ESTABLISHED CELL LINE derived from the kidney of an African green monkey (*Cercopithecus aethiops*); the cells are heteroploid and epithelioid.

BSE BOVINE SPONGIFORM ENCEPHALOPATHY.

BSS BALANCED SALT SOLUTION.

***Bss*HII** See RESTRICTION ENDONUCLEASE (table).

***Bst*EII** A RESTRICTION ENDONUCLEASE from *Bacillus stearothermophilus*; G/GTNACC.

BTB agar (bromthymol blue agar) A medium (used e.g. for enterobacteria) containing e.g. peptone, lactose, glucose, yeast extract, MARANIL, thiosulphate, and bromthymol blue; pH: 7.7–7.8. [Recipe: Book ref. 22, p. 421.]

BTEX Benzene, toluene, ethyl-benzene and xylene – a group of compounds which are sometimes present as contaminants in aquifers. (See also BIOREMEDIATION.)

BtuB protein In *Escherichia coli*: an OUTER MEMBRANE protein encoded by the *btuB* gene; it acts as a receptor for bacteriophage BF23 and for various colicins, and is involved in the uptake of VITAMIN B_{12}.

BtuC protein See VITAMIN B_{12}.

BU 5-BROMOURACIL.

bubble column fermenter A FERMENTER in which mixing and circulation of the culture is induced by a stream of bubbles introduced at the base of the column (which lacks a draft tube – cf. LOOP FERMENTER); it may be used e.g. with low-viscosity cultures when efficient mixing is not essential. An increased gas–liquid interface can be achieved by placing mesh sieve plates at intervals in the column.

bubble diseases MUSHROOM DISEASES caused by *Verticillium fungicola* (dry bubble) or *Mycogone perniciosa* (wet bubble). [Control by PROCHLORAZ and other antifungals: PP (1983) *32* 123–131.]

bubble point test A test used to determine the physical integrity of a membrane filter (see FILTRATION). In a wetted membrane, water is retained in the pores by the forces of surface tension, and a considerable pressure of air is needed to dislodge this water and to force air through the pores; the minimum pressure which can do this (the bubble point pressure) increases with decrease in pore size, and is about 55 lb/inch^2 (379 kPa) for a membrane of pore size 0.22 μm. Using the relationship between bubble point pressure and pore size, the integrity of a membrane of given pore size can be tested by determining its bubble point pressure. (cf. DIFFUSION TEST.)

bubbler *Syn.* ALL-GLASS IMPINGER.

bubo An enlarged, inflamed lymph node (particularly in the axilla or groin) formed e.g. in bubonic PLAGUE.

bubonic plague See PLAGUE.

buccal cavity In some ciliates: a cavity, chamber or surface depression which is typically open to the environment and which contains, at its base, the CYTOSTOME; the ciliature of the buccal cavity is typically readily distinguishable from the somatic ciliature. (See also e.g. PARORAL MEMBRANE.)

buccal overture The entrance (outer or distal region) of the BUCCAL CAVITY.

buccokinetal stomatogenesis See STOMATOGENESIS.

Buchnera A genus of intracellular bacteria found in the BACTERIOCYTES of aphids; aphids and *Buchnera* are mutually dependent – neither can reproduce without the other.

The chromosome of *Buchnera* sp APS is one of the smallest to be sequenced so far in bacteria (a total of 640681 base pairs) [Nature (2000) *407* 81–86].

bud-fission In certain fungi (e.g. *Saccharomycodes ludwigii*): a form of BUDDING in which the septum between bud and parent cell is almost as wide as the widest part of the parent cell.

bud scar (*mycol.*) See SCAR.

budding (1) (*bacteriol., mycol.*) In certain bacteria and fungi: a process of cell division in which a daughter cell (or spore) develops from the mother cell as a localised outgrowth or protrusion (*bud*), the cell wall of the mature daughter cell apparently being composed mainly of newly synthesized material; mother and daughter cells are often physically, and sometimes functionally, distinguishable. (cf. FISSION.) In the BUDDING BACTERIA, budding often – but not always – occurs at one of the poles of the mother cell. In some fungi (e.g. *Saccharomyces cerevisiae*) buds may arise at any of a number of sites on the parent cell (*multipolar budding*); in e.g. *Kloeckera*, buds can develop at both cell poles (*bipolar budding*), while in *Malassezia* they can arise only at one pole (*monopolar budding*). (cf. TRIGONOPSIS; see also BUD-FISSION, SCAR and SPROUT MYCELIUM.)

(2) (*protozool.*) In some protozoa (e.g. members of the Astomatida, Peritrichia and Suctoria): a reproductive process in which a cell divides unequally to form a large adult and small progeny cells (external or exogenous budding), or in which progeny develop within the cytoplasm of the adult cell (interior or endogenous budding).

(3) (*virol.*) See ENVELOPE (sense 1).

budding bacteria A diverse, non-taxonomic group of bacteria all of which undergo BUDDING; they include e.g. species of BLASTOBACTER, CHAMAESIPHON, ENSIFER, GEMMIGER, GEODERMATOPHILUS, *Hyphomicrobium*, HYPHOMONAS, NITROBACTER, PASTEURIA, PLANCTOMYCES, RHODOMICROBIUM, RHODOPSEUDOMONAS and SELIBERIA.

budgerigar fledgling disease virus A virus which causes a commercially important acute, fatal disease in fledgling budgerigars (*Melopsittacus undulatus*) and may also be responsible for a milder, chronic disease ('French moult') in older birds. The virus is apparently a POLYOMAVIRUS – the first non-mammalian polyomavirus to be recorded. [Virol. (1986) *151* 362–370.]

BUdR (BrdU) 5-Bromo-2′-deoxyuridine: the deoxynucleoside corresponding to 5-BROMOURACIL. It can be phosphorylated by THYMIDINE KINASE. (cf. IDOXURIDINE.)

Buellia A genus of crustose LICHENS (order LECANORALES). Apothecia: lecideine, black; ascospores brown, one-septate. (cf. DIPLOICIA; see also ENDOLITHIC.)

buffalopox virus See ORTHOPOXVIRUS.

buffy coat The thin layer of white cells which forms at the surface of the packed red cells when unclotted blood is centrifuged.

bufotenine See AMATOXINS.

bulbil (*algol.*) See CHAROPHYTES.

Bulbochaete A genus of freshwater filamentous green algae related to OEDOGONIUM. The filaments are branched and bear long, colourless 'hairs', each with a bulbous base.

Bulgaria See HELOTIALES.

Bulgarian buttermilk *Syn.* BULGARICUS MILK.

bulgaricus milk (Bulgarian buttermilk) A beverage made by fermenting milk with *Lactobacillus bulgaricus*.

bulgecin A low-MWt glycopeptide which, in combination with a β-lactam antibiotic, causes bulge formation in susceptible bacterial cells [AAM (1986) *31* 181–205 (200–201)].

bullate With bubble- or blister-like swellings.

Buller phenomenon A form of DIKARYOTIZATION in which a dikaryotic cell or mycelium donates a nucleus to a monokaryotic (haploid) cell or mycelium.

Bullera A genus of fungi (see SPOROBOLOMYCETACEAE) which form spheroidal, ovoid or elongate, budding vegetative cells, and which may form pseudomycelium or true mycelium; the cell walls contain xylose. On malt agar, the growth is cream-coloured to slightly yellowish or brownish (cf. SPOROBOLOMYCES). Ballistospores are formed on sterigmata of varying lengths and are rotationally symmetrical, being spheroidal, obovoid, turbinate, apiculate or ampulliform. Metabolism is strictly respiratory. NO_3^- is assimilated by *B. piricola* and *B. tsugae*, but not by *B. alba*, *B. dendrophila* or *B. singularis*. Species have been isolated from insect frass and plant material. [Book ref. 100, pp. 577–584.]

bullous impetigo See IMPETIGO.

bumblefoot An arthritis of poultry due to infection via wounds – particularly in the feet – by *Staphylococcus aureus*. (See also POULTRY DISEASES.)

bumper primer See SDA.

bundle-forming pili See PATHOGENICITY ISLAND.

bundlin The major subunit of bundle-forming pili (see EPEC) – a 19.5 kDa protein encoded by the 'EPEC adherence factor plasmid' (EAF plasmid) found in strains of EPEC. (See also PATHOGENICITY ISLAND.)

bunts (stinking smuts) See e.g. COMMON BUNT and KARNAL BUNT.

Bunyamwera supergroup See BUNYAVIRUS.

Bunyaviridae A family of enveloped, ssRNA VIRUSES which infect warm- and cold-blooded vertebrates, arthropods and plants. Some members are important pathogens of man and animals, causing e.g. VIRAL HAEMORRHAGIC FEVERS or pulmonary disease; transmission may occur e.g. via ticks, mosquitoes or sandflies (*Phlebotomus* spp), while some bunyaviruses are transmitted in aerosols. (See also TOMATO SPOTTED WILT VIRUS; cf. ARBOVIRUSES.)

Virion. Roughly spherical, enveloped, ca. 90–100 nm diam., containing three major protein species (nucleocapsid protein N, and glycoproteins G1 and G2), one large but minor protein (L), and three molecules/segments of negative sense (or ambisense) ssRNA designated L, M and S (MWts ca. 2–5, 1–2.3 and $0.3–0.8 \times 10^6$ respectively); the envelope contains ca. 20–30% by weight of lipid.

Virion structure. The lipoprotein envelope, which contains G1 and G2, encloses three circular (but not covalently closed) helical strands of ribonucleoprotein (comprising RNA and N protein); the haemagglutinating activity and neutralizing antigenic determinants are associated with the glycoproteins. Virions can be inactivated by e.g. lipid solvents and detergents.

Transcriptase activity is associated with the virions, and viral replication occurs in the cytoplasm of the host cell.

In the genus *Bunyavirus*, all three genome segments are negative-sense. The S RNA segment encodes (in overlapping reading frames) the N protein and a non-structural protein (NS_s). The M RNA segment encodes G1, G2 and at least one non-structural protein (NS_m). The L RNA appears to encode the L protein. (cf. PHLEBOVIRUS.) During maturation, the virions bud into smooth-surfaced vesicles in or near the Golgi region.

In culture, some bunyaviruses can induce cell fusion.

The family is divided into genera on the basis of e.g. size of RNA segments, size and nature of viral proteins, nature of arthropod vector (if any) etc.; the genera include BUNYAVIRUS; HANTAVIRUS; NAIROVIRUS; PHLEBOVIRUS and UUKUVIRUS.

Bunyavirus ('Bunyamwera supergroup') A genus of viruses of the BUNYAVIRIDAE. Host range: various vertebrates; vectors: mainly mosquitoes. MWts of L, M and S RNAs: ca. 2.7–3.1, 1.8–2.3, and $0.28–0.50 \times 10^6$, respectively. MWts of proteins L, G1, G2 and N: ca. 145–200, 108–120, 29–41 and $19–25 \times 10^3$, respectively. The genus comprises at least 16 serological groups, including the Bunyamwera group (e.g. Bunyamwera virus (type species of the genus), Cache Valley virus), the California group (e.g. CALIFORNIA ENCEPHALITIS virus, Inkoo virus, La Crosse virus, snowshoe hare virus), and the Simbu group (e.g. Aino virus, Akabane virus – see AKABANE VIRUS DISEASE).

bunyaviruses (1) Viruses of the BUNYAVIRIDAE. (2) Viruses of the genus BUNYAVIRUS.

buoyant density See CENTRIFUGATION.

burdock stunt viroid See VIROID.

burdock yellows virus See CLOSTEROVIRUSES.

Burdon's stain A stain for bacterial intracellular lipid. A heat-fixed smear is stained with SUDAN-BLACK B for 5–10 min, drained, and blotted dry. The smear is washed in xylene, blotted dry, and counterstained for 5–10 sec in 0.5% aqueous safranin. Lipid stains black, cytoplasm red. (See also NILE BLUE A.)

Burgundy mixture (*plant pathol.*) An agricultural antifungal preparation which consists of a mixture of copper sulphate and sodium carbonate; it is used in aqueous solution as a spray. Active constituent: the COPPER ion.

Burgundy truffle See TRUFFLES.

Burkard trap See HIRST SPORE TRAP.

Burkea See MICROSPOREA.

Burkholderia A genus of Gram-negative bacteria initially proposed to accommodate seven (group II) species from the genus PSEUDOMONAS, with *Burkholderia* (formerly *Pseudomonas*) *cepacia* as the type species [Microbiology and Immunology (Tokyo) (1992) *36* 1251–1275]; the other species in this proposal were: *P. mallei*, *P. pseudomallei*, *P. caryophilli*, *P. gladioli*, *P. pickettii* and *P. solanacearum*. Two other former species of the genus *Pseudomonas* (*P. plantarii* and *P. glumae*), and a new species, *B. vandii*, have since been added to the genus [IJSB (1994) *44* 235–245].

(See also CYSTIC FIBROSIS.)

[*B. cepacia* (surface chemistry and typing methods): RMM (1995) *6* 1–9. *B. cepacia* (pathogenicity and resistance to antibiotics): RMM (1995) *6* 10–16.]

Burkitt's lymphoma (BL) A malignant monoclonal B-cell lymphoma which is endemic in parts of Africa where it is the most common tumour among children. The tumours commonly develop in the mandibular and maxillary bones, but may also develop in the ovaries, thyroid, kidneys, heart, stomach, and spinal column. BL is associated with EPSTEIN–BARR VIRUS (EBV) infection, but other factors – including the occurrence of malaria – are also important in pathogenesis. BL cells contain multiple copies of the EBV genome, and also show characteristic chromosomal rearrangements involving reciprocal translocations between chromosome 8 and chromosome 14 or, less frequently, chromosome 2 or 22. These translocations apparently result in the juxtaposition of a cellular ONCOGENE (c-*myc*, or possibly c-*mos*) and the immunoglobulin (Ig) genes on chromosomes 14, 2 or 22; c-*myc* may be activated by the Ig gene promoter, resulting in oncogenesis. These chromosomal rearrangements can occur in the absence of EBV; EBV may predispose to oncogenesis by its mitogenic effect on B cells, providing large numbers of transformed cells and thereby increasing the chances of one of the three chromosomal rearrangements occurring. Chronic malaria depresses EBV-specific cell-mediated immunity and stimulates B cell proliferation, thus potentiating the effects of EBV. 'Burkitt-like' lymphomas occur rarely and sporadically outside the endemic areas, and only ca. 20% of these are associated with EBV infection.

burned spot disease (shell disease; *Brandenfleckenkrankheit*) A marine CRUSTACEAN DISEASE which affects the exoskeletons of crabs, lobsters, crayfish etc; it is caused by chitinolytic bacteria and/or fungi which form brown erosive lesions in the calcified chitin of the exoskeleton following damage to the epicuticle (e.g. by abrasion). Once initiated, the lesions may support a varied microbial flora. The disease may be fatal, but recovery may occur if the animal survives to ecdysis. *Fusarium solani* has been associated with the disease in lobsters; the fungus may penetrate the muscles and occasionally the foregut of the lobster [TBMS (1981) *76* 25–27]. (cf. BLACK GILL DISEASE.)

Burnet's clonal selection theory See CLONAL SELECTION THEORY.

bursa of Fabricius In birds, a sac-like organ which develops from the hindgut in the early embryonic stage; within the bursa, precursors of B LYMPHOCYTES appear to undergo maturation. (See also INFECTIOUS BURSAL DISEASE VIRUS.)

Bursaria See HETEROTRICHIDA.

bursattee *Syn.* EQUINE PHYCOMYCOSIS.

burst size The average number of virus particles released per infected cell following the lytic infection of a population of sensitive cells; it may be determined e.g. by the ONE-STEP GROWTH EXPERIMENT (by comparing the latent period plaque count with the plateau plaque count) or by the SINGLE BURST EXPERIMENT.

burst test A test used for assessing the strength of the seal in a RETORT POUCH; in the test, the pouch is either inflated or is filled with water and then compressed.

Bursulla See MYXOMYCETES.

Buruli ulcer A chronic, progressive, granulomatous skin lesion caused by *Mycobacterium ulcerans*; infection occurs via wounds.

Buschke–Löwenstein tumour (giant condyloma acuminatum) A large, cauliflower-like, malignant but non-metastasizing genital tumour of humans. The genomes of HPV-6 or HPV-11 (see PAPILLOMAVIRUS) have been identified in Buschke–Löwenstein tumour cells [see e.g. JV (1986) *58* 963–966]. (cf. PAPILLOMA.)

Busse–Buschke disease *Syn.* CRYPTOCOCCOSIS.

Bussuquara virus See FLAVIVIRIDAE.

butanediol fermentation (butylene glycol fermentation) A FERMENTATION (sense 1) carried out e.g. by certain enterobacteria, including species of *Enterobacter*, *Erwinia*, *Klebsiella* and *Serratia*. The main products of glucose fermentation include ethanol, 2,3-butanediol and formic acid or CO_2 and H_2 [Appendix III(f)]; small amounts of DIACETYL ($CH_3.CO.CO.CH_3$) may also be formed from acetolactate. Organisms which form diacetyl or acetoin give a positive VOGES-PROSKAUER TEST. The amount of acid produced in the butanediol fermentation is generally insufficient to give a positive methyl red test (cf. MIXED ACID FERMENTATION).

butanol fermentation *Syn.* ACETONE–BUTANOL FERMENTATION.

butirosin An AMINOGLYCOSIDE ANTIBIOTIC.

butoconazole See AZOLE ANTIFUNGAL AGENTS.

Bütschli's granules METACHROMATIC GRANULES observed in diatoms.

butt (*microbiol.*) The thickest part of a SLOPE. The butt of a freshly-prepared slope is a microaerobic or anaerobic environment.

butt rot A TREE DISEASE in which the base (butt) of the trunk rots. In e.g. elms (*Ulmus* spp) a butt rot may be caused by *Rigidoporus ulmarius*, while in oaks (*Quercus* spp) it may be caused by *Inonotus dryadeus* (*Polyporus dryadeus*).

butter A food consisting mainly of the fatty components of milk; it contains at least 80% fat, up to 16% water, and usually ca. 2% added salt. *Sweet cream butter* is made without a bacterial starter. Cream is pasteurized, cooled, and churned at ca. 8–12°C to destabilize the fat globules in the cream. The butter is then strained from the aqueous phase (buttermilk) and salted; it may be stored for a year or more at ca. −23 to −29°C. *Ripened cream butter* (*cultured creamery butter*) is made from pasteurized cream seeded with a B- or BD-type LACTIC ACID STARTER. The main function of the bacteria is to contribute flavour (LACTIC ACID and DIACETYL). After several hours' incubation, during which the pH falls to ca. 4.6, the cream is cooled, churned etc. Cultured butters are more common in e.g. Europe than in the USA; they have better flavour than sweet cream butter, but they also undergo chemical deterioration faster – the rate of deterioration being higher if salt is added. (See also DAIRY PRODUCTS.)

Butter is generally resistant to microbial spoilage owing to its low water content. However, proteolytic and/or lipolytic psychrotrophs (e.g. *Pseudomonas fragi*, *Ps. fluorescens*) can cause putrid or rancid flavours in refrigerated butter. A cheese-like off-flavour ('surface taint') may be caused by '*Ps. putrefaciens*', and '*Ps. nigrifaciens*' can cause a black surface discoloration.

buttermilk Originally: the liquid residue formed during BUTTER-making. Currently, the term usually refers to 'cultured buttermilk': a beverage made by fermenting skim milk with lactic acid bacteria; a BD-type LACTIC ACID STARTER is usually used. [Book ref. 8, pp. 140–143.] (See also DAIRY PRODUCTS.)

button stage The (early) stage of an agaric fruiting body in which the lamellae are not yet exposed, and the pileus is roughly dome-shaped.

butyl alcohol fermentation *Syn.* ACETONE–BUTANOL FERMENTATION.

butylene glycol fermentation *Syn.* BUTANEDIOL FERMENTATION.

butyric acid fermentation A FERMENTATION (sense 1) carried out e.g. by *Butyrivibrio fibrisolvens*, certain species of CLOSTRIDIUM (e.g. *C. acetobutylicum*, *C. butyricum*, *C. pasteurianum*, *C. perfringens*) and *Fusobacterium nucleatum*. The products of glucose fermentation include acetic and butyric acids, CO_2 and H_2 [Appendix III(g)]; some species (e.g. *C. perfringens*) form lactic acid and/or ethanol in addition to the above products – lactic acid becoming a major product under conditions of iron deficiency. (See also ACETONE–BUTANOL FERMENTATION.)

butyricins See BACTERIOCIN.

Butyrivibrio A genus of bacteria (family BACTEROIDACEAE) which occur e.g. in the RUMEN and in non-ruminant faeces. Cells: motile or (infrequently) non-motile curved rods, 0.3–0.8 × 1.0–5.0 μm (occurring singly or in chains), or filaments which are sometimes helical; the cells stain Gram-negatively, although the cell wall of *B. fibrisolvens* is of the Gram-positive type (see also TEICHOIC ACIDS). [Heterogeneity in cell envelopes of *Butyrivibrio* strains: JUR (1985) *90* 286–293.] Fermentation of e.g. glucose or maltose typically yields butyric acid as the main product (although lactic acid may be the major product under certain conditions). Some strains are cellulolytic, pectinolytic and/or amylolytic. GC%: ca. 36–41. Type species: *B. fibrisolvens*.

B. crossotus. Cells are lophotrichously flagellated. Maltose is fermented, but cellobiose, sucrose and xylose are not.

B. fibrisolvens. The cells each have one polar or subpolar flagellum. Glucose, maltose, cellobiose, sucrose and xylose are fermented.

butyrolactone antifungal agents See PHENYLAMIDE ANTIFUNGAL AGENTS.

butyrous Butter-like in consistency.

BVD BOVINE VIRUS DIARRHOEA.

BVD-MD virus See PESTIVIRUS.

BVDU BROMOVINYLDEOXYURIDINE.

***bvg* genes** In *Bordetella* spp: genes of a TWO-COMPONENT REGULATORY SYSTEM controlling the synthesis of various virulence factors (adhesins, toxins); *bvgS* encodes the histidine kinase, while *bvgA* encodes the response regulator. It is not known which signal(s) activate the system in vivo; in vitro it can be regulated e.g. by temperature.

BWD See PULLORUM DISEASE.

BWYV Beet western yellows virus (see LUTEOVIRUSES).

BYDV Barley yellow dwarf virus (see LUTEOVIRUSES).

byssinosis A human pulmonary disease, associated with the handling of cotton, flax, hemp or sisal, characterized by acute symptoms of airway obstruction; aetiology may involve e.g. the endotoxins and/or allergens from contaminating Gram-negative bacteria. [AEM (1982) *44* 355–362; CIA (1984) *4* 45–47.]

Byssochlamys A genus of fungi (order EUROTIALES) found e.g. in soil; anamorphs occur in the genus PAECILOMYCES. The mature ascocarp consists of a group of asci surrounded by a few wisps of hyphae. *B. fulva* can cause spoilage of canned foods (see CANNING); the organism forms extracellular pectinases, and its ascospores can withstand ca. 85°C for 30 minutes, and higher temperatures for shorter periods. (See also PATULIN.)

byssoid Composed of slender fibrils.

bystander help (*immunol.*) Non-specific help; for example, an antigen-stimulated T cell may provide help (in the form of cytokines) for another cell which has been stimulated by a different antigen.

bystander lysis *Syn.* REACTIVE LYSIS.

1. Words in SMALL CAPITALS are cross-references to separate entries.
2. Keys to journal title abbreviations and Book ref. numbers are given at the end of the Dictionary.
3. The Greek alphabet is given in Appendix VI.
4. For further information see ‘Notes for the User’ at the front of the Dictionary.

C

C (1) (*immunol.*) COMPLEMENT. (2) (*mol. biol.*) Cytosine (or the corresponding nucleoside or nucleotide) in a nucleic acid. (3) Cysteine (see AMINO ACIDS).

c̄ See C VALUE.

C′ (*immunol.*) COMPLEMENT.

1-C compounds See C_1 COMPOUNDS.

C forms See AZOSPIRILLUM.

C loop See BACTERIOPHAGE P1.

c-onc See ONCOGENE.

C period See HELMSTETTER–COOPER MODEL.

C-reactive protein (CRP) An ACUTE-PHASE PROTEIN, increased amounts of which circulate in the body e.g. during inflammatory processes. CRP is precipitated by so-called *C-reactive substances* which occur in various microorganisms – e.g. *Streptococcus pneumoniae* (the basis of the C-reactive protein diagnostic test) and *Aspergillus fumigatus*. [Binding of CRP to C-reactive substance of *Aspergillus fumigatus*: JMM (1986) *21* 173–177.]

Changes in CRP titre may help to distinguish between bacterial and viral infections: a rise in the serum level of CRP is more likely to reflect a bacterial infection because viral infections usually do not produce such a change.

C-reactive substance See C-REACTIVE PROTEIN. (cf. C SUBSTANCES.)

C region (of Ig) See IMMUNOGLOBULINS.

C ring See FLAGELLUM (a).

c strand (C strand) Complementary strand, e.g., the strand complementary to the genomic (viral, v) strand in the RF of an ssDNA virus. (cf. CDNA.)

C substances (of streptococci) A range of serologically distinct carbohydrates – typically only one of which occurs in a given strain of *Streptococcus*; C substances are used in LANCEFIELD'S STREPTOCOCCAL GROUPING TEST. In the majority of Lancefield groups the C substance is a CELL WALL component. In Lancefield group A streptococci the C substance is a rhamnose polymer with a terminal residue of *N*-acetylglucosamine, the latter being immunodominant; in strains of group B the main determinant appears to be rhamnose, while in those of group C it is a terminal residue of *N*-acetylgalactosamine. [Structure of the C substance in group B streptococci: Biochem. (1987) *26* 476–486.] The C substance of group D strains, a glycerol TEICHOIC ACID, occurs in the periplasmic region.

C substances may be extracted from cells by various methods. In the hot-HCl method (Lancefield extraction method), packed (centrifuged) cells of the strain under test are resuspended in ca. 0.3 ml of a solution of sodium chloride (0.85%) containing HCl (at a final concentration of 0.2 N) and one drop of the pH indicator metacresol purple (0.04%); the pH should be ca. 2 (pink coloration). The tube containing the suspension is placed in boiling water for ca. 10 min with periodic shaking. The tube is then centrifuged, and the supernatant is decanted into a clean vessel; NaOH (0.2 N) is added dropwise until the pH is ca. 7.5 (*weak* purple).

In the simpler *autoclave extraction method*, packed cells are re-suspended in 0.5 ml of NaCl solution (0.85%), autoclaved (121°C/15 min), centrifuged, and the supernatant is used for the test.

[Methods of preparing antigen extracts for Lancefield grouping: Book ref. 120, pp. 170–172.]

C-type particles See TYPE C ONCOVIRUS GROUP.

C value (*C*-value; c̄) The amount of DNA in a *haploid* genome; it can be measured e.g. in picograms or in terms of the molecular weight or the number of constituent kilobase-pairs.

C_0 In the food processing industry: heating (COOKING) for 1 min at 100°C *or* an equivalent amount of heat processing.

C_1 cellulase See CELLULASES.

C_1 compounds (1-C compounds; one-carbon compounds) Carbon compounds which contain no carbon–carbon bonds and which are more reduced than is CO_2 – e.g. CO, methane, methanol, formaldehyde, formamide and formate; C_1 compounds may have more than one carbon atom: e.g. dimethyl sulphide [$(CH_3)_2S$], dimethyl sulphoxide [$(CH_3)_2SO$], dimethyl sulphone [$(CH_3)SO_2$], trimethylamine-*N*-oxide [$(CH_3)_3NO$]. (See also METHYLOTROPHY.)

C1 esterase inhibitor See COMPLEMENT FIXATION (a).

C1EI See COMPLEMENT FIXATION (a).

C1INH *Syn.* C1EI – see COMPLEMENT FIXATION (a).

C_2 carbon oxidation cycle See PHOTORESPIRATION.

C2a pathway See COMPLEMENT FIXATION (d).

C3 convertase See COMPLEMENT FIXATION.

C_3 cycle *Syn.* CALVIN CYCLE.

C3bina (C3bINA) *Syn.* FACTOR I.

C5 convertase See COMPLEMENT FIXATION.

C27 organisms See PLESIOMONAS.

C_{55} lipid carrier *Syn.* BACTOPRENOL.

Ca^{2+}-ATPase See ION TRANSPORT.

Ca^{2+}, Mg^{2+}-stimulated ATPase See ATPASE.

Ca^{2+} transport See ION TRANSPORT.

Cabassou virus See ALPHAVIRUS.

cabbage B virus *Syn.* CAULIFLOWER MOSAIC VIRUS.

cabbage diseases See CRUCIFER DISEASES.

cabinet (safety cabinet) See SAFETY CABINET.

cacao diseases For diseases of the cacao plant (*Theobroma cacao*) see e.g. BLACK POD DISEASE, FUSARIUM (*F. decemcellulare*), POD ROT, RED RUST, VASCULAR-STREAK DISEASE and WITCHES' BROOM.

cacao yellow mosaic virus See TYMOVIRUSES.

Cache Valley virus See BUNYAVIRUS.

cachectin An early name for tumour necrosis factor-α (TNF-α).

cacodylate Dimethylarsinate.

cactus virus 2 See CARLAVIRUSES.

cactus virus X See POTEXVIRUSES.

CAD See APOPTOSIS.

cAD1 See PHEROMONE.

cadang-cadang See COCONUT CADANG-CADANG VIROID.

cadaverine See e.g. DECARBOXYLASE TESTS.

cadherins A family of CELL ADHESION MOLECULES. The cadherins are large, transmembrane glycoproteins that mediate calcium-dependent *homophilic* cell–cell adhesion (i.e. cadherin binds to cadherin). The intracellular domain of the cadherin molecule binds to proteins called *catenins* which, in turn, are linked to the cell's actin cytoskeleton.

E-cadherin occurs mainly in epithelial tissue (and in the early stage of mammalian embryogenesis); E-cadherin is used as a receptor for the adhesin *internalin A* of *Listeria monocytogenes* during invasion of the intestinal epithelium. P-cadherin occurs in the placenta and epidermis, and N-cadherin is found e.g. in nerve and heart cells.

cadmium See HEAVY METALS.

caducous Tending to be shed, or to fall, early.

Caedibacter A genus of Gram-negative bacteria which are obligate endosymbionts in strains of *Paramecium aurelia*. Cells: non-motile coccobacilli or rods, ca. 0.4–1.0 μm across and 1.0–4.0 μm in length, which stain well with toluidine blue; some (not all) cells of *Caedibacter* contain an R BODY (or, rarely, two R bodies). Cell division is inhibited in bacterial cells which contain an R body. The possession of R body-containing cells of *Caedibacter* appears to confer on the host *Paramecium* the ability to behave as a killer, or as a MATE KILLER, towards sensitive paramecia (i.e. paramecia which lack endosymbiotic *Caedibacter*); in the 'killer' phenomenon, sensitive paramecia appear to ingest R body-containing bacterial cells (and/or isolated R bodies or toxins?) released by killer paramecia, and are killed as a consequence. The ability of killer paramecia to remain viable while harbouring *Caedibacter* is dependent on certain dominant allele(s). GC%: ca. 40–44. Type species: *C. taeniospiralis*.

The cells of *C. pseudomutans*, *C. taeniospiralis* and *C. varicaedens* – which occur in the cytoplasm of their hosts, and which confer the killer characteristic – are sometimes called kappa particles. *C. paraconjugatus* (see also MU PARTICLE) is associated with mate killing. Another endosymbiont, which occurs in the macronucleus of *P. caudatum* – and which confers the killer characteristic – may belong to this genus. (See also HOLOSPORA, LYTICUM, PSEUDOCAEDIBACTER, SPIN KILLING.)

[Book ref. 22, pp. 803–806; protein synthesis in kappa particles: JGM (1984) *130* 1517–1523.]

caeomatoid (caeomoid) Refers to an irregularly-shaped aecium which lacks a distinct peridium.

caesium chloride See CENTRIFUGATION.

caffeic acid 3-(3′,4′-Dihydroxyphenyl)-propenoic acid, a compound produced by higher plants (apparently as a precursor of LIGNIN). Caffeic acid may be a factor in the resistance of certain plants to certain diseases; thus, e.g. potatoes resistant to LATE BLIGHT have been found to accumulate caffeic acid on infection with *Phytophthora infestans*. (Caffeic acid inhibits *P. infestans* in vitro.) (See also AUXINS and CHLOROGENIC ACID.)

caffeine 1,3,7-Trimethylxanthine, a purine derivative present e.g. in tea leaves and coffee beans. In e.g. certain bacteria caffeine can act as an inhibitor of EXCISION REPAIR of DNA.

cag See PATHOGENICITY ISLAND.

Cairns' mechanism (θ replication) A mechanism for the replication of a ds cccDNA molecule: replication is initiated at a fixed point, progresses (unidirectionally or bidirectionally) around the circle – both strands being replicated more or less simultaneously – and ends at a defined terminus. (The partly replicated intermediate thus has a 'θ'-like structure.) Bidirectional replication occurs e.g. in the chromosome of *Escherichia coli* (see DNA REPLICATION); unidirectional replication occurs e.g. in the COLE1 PLASMID. (cf. ROLLING CIRCLE MECHANISM.)

Calberla's solution A mixture of glycerol (5 ml), 95% ethanol (10 ml) and distilled water (15 ml) used as a microscopical mounting medium for the ROTOROD.

Calcarisporium See HYPHOMYCETES; see also AUROVERTINS and CONTACT BIOTROPHIC MYCOPARASITE.

calcicolous Growing preferentially in a calcium-rich environment.

calcium alginate wool See ALGINATE.

calcium nutrient agar (CNA) NUTRIENT AGAR containing $CaCl_2$; a lawn plate prepared from CNA may be used e.g. for the cultivation of certain bacteriophages.

calcium transport See ION TRANSPORT.

calcofluor white A FLUORESCENT BRIGHTENER which binds by hydrogen-bonding to β-linked fibrillar polysaccharides such as cellulose and chitin; it can be used as a fluorescent marker to locate these polymers in cells (see MICROSCOPY (e)), and can disrupt CELLULOSE and chitin microfibril formation in growing cells (e.g. yeasts: [JGM (1983) *129* 1577–1582]).

caldoactive Refers to an extreme THERMOPHILE. The term is used more specifically by some authors to refer to an organism with a maximum growth temperature >70°C, an optimum growth temperature >65°C, and a minimum growth temperature >40°C [Sci. Prog. (1975) *62* 373–393].

caldopentamine See POLYAMINES.

calf diphtheria See NECROBACILLOSIS.

calf pneumonia (enzootic pneumonia of calves; viral pneumonia of calves) A non-specific calf disease, the causal agent(s) being one or more viruses (e.g. adenoviruses, parainfluenza virus type 3, respiratory syncytial virus) frequently in association with bacteria (e.g. *Chlamydia*, *Mycoplasma* spp); typical symptoms: fever with constipation, followed by a mucopurulent nasal discharge, PNEUMONIA and diarrhoea. Poor hygiene and housing conditions (e.g. poor ventilation) and overcrowding appear to predispose towards the disease. Mortality rates can be high e.g. in acute RSV pneumonia.

Control involves improvements in housing etc. and chemotherapy. (See also danofloxacin in entry QUINOLONE ANTIBIOTICS.)

calf scours Scouring (see SCOURS) in young calves due to any of a range of microbial agents, e.g. enterotoxigenic strains of *Escherichia coli*, coronaviruses etc; aetiology may be complex (e.g. mixed bacterial and viral causal agents) and the occurrence and severity of the disease may involve both immunological and environmental factors.

Calgitex See ALGINATE.

Caliciales A heterogeneous order of (mostly LICHEN-forming) fungi of the ASCOMYCOTINA. Three families [Book ref. 64]:

Caliciaceae. Mostly lichen-forming (photobiont: a green alga). Thallus thin crustose or immersed in the substratum (usually wood or bark). Apothecia (mazaedia) stalked or sessile. Lichen-forming genera include e.g. *Calicium*, *Chaenotheca*, *Coniocybe*, *Cyphelium*.

Mycocaliciaceae. Lichenicolous or fungicolous (possibly also lichen-forming) fungi in which the thallus is immersed in the substratum or absent. Apothecia stalked, asci thick-walled (mazaedia are not formed). Genera: e.g. *Stenocybe*.

Sphaerophoraceae. All lichen-forming (photobiont: a green alga). Thallus foliose or fruticose. Apothecia (mazaedia) often globose, marginal or terminal. Genera: e.g. *Sphaerophorus*.

Calicium See CALICIALES.

Caliciviridae A family of VIRUSES which infect vertebrates; caliciviruses were formerly classified as a genus in the family Picornaviridae. The virions are non-enveloped, roughly spherical, ca. 35–39 nm diam., and have 32 characteristic cup-shaped surface indentations – arranged with icosahedral symmetry – observable by electron microscopy of negatively stained preparations. The capsid contains a single major polypeptide species (MWt ca. 60000–71000) and a minor polypeptide (MWt ca. 15000). Genome: one molecule of positive-sense ssRNA (MWt ca. $2.6–2.8 \times 10^6$) which is apparently polyadenylated at the 3′ end but not capped at the 5′ end; the RNA is covalently linked (possibly at its 5′ end) to a protein (MWt ca. 10000–15000) which is necessary for infectivity. Replication occurs in the cytoplasm; dsRNAs (presumably replicative intermediates), genome-sized ssRNAs, and subgenomic ssRNAs can be detected in infected cells. Mature viruses are released by host cell lysis. Caliciviruses are not sensitive to lipid solvents or to mild detergents, but are inactivated at pH 3–5.

The family includes VESICULAR EXANTHEMA of swine virus (VESV), FELINE CALICIVIRUS (FCV) and SAN MIGUEL SEA LION VIRUS (SMSV); these three viruses are classified in the genus *Calicivirus* (type species: VESV serotype A). VESV, SMSV and FCV are readily propagated in cell culture and do not cause gastroenteritis (although they are pathogenic); morphologically similar viruses which have not been cultivated and which cause gastroenteritis in calves, piglets and humans (especially young children) are included as 'possible members' of the family [Book ref. 23, pp 133–134] (but see SMALL ROUND STRUCTURED VIRUSES).

Calicivirus See CALICIVIRIDAE.

California encephalitis (CE) An acute, usually mild, viral ENCEPHALITIS affecting mainly children below ca. 15 years of age; it occurs sporadically in forested areas in parts of North America. The CE virus (genus BUNYAVIRUS) occurs in small mammals and is transmitted by mosquitoes (mainly *Aedes* spp).

California mastitis test (CMT; CM test) (*vet.*) A test, used for the indirect detection of MASTITIS, in which the number of white blood cells in milk is estimated. A reagent is added to the milk, causing a degree of gelation which corresponds to the white cell count; test results designated 'negative', 'trace', 1, 2 and 3 correspond to cell counts of ca. $<10^5$, 3×10^5, 9×10^5, 27×10^5, and 81×10^5, respectively. The Brabant and Wisconsin mastitis tests also depend on a gelation principle.

callose A linear (1 → 3)-β-D-glucan which occurs e.g. associated with sieve-plates and sieve-tubes in higher plants and in algae of the Laminariales (PHAEOPHYTA). Callose may be deposited in plant cell walls e.g. in response to wounding or invasion by pathogens.

calmodulin A 16.7 kDa, heat-stable, acid-stable Ca^{2+}-binding protein found in eukaryotic cells, including those of certain protozoa and slime moulds. The complex of calmodulin with calcium ions stimulates a number of enzymes (e.g. cyclic nucleotide phosphodiesterase) and activates certain toxins (e.g. CYCLOLYSIN and the EF component of ANTHRAX TOXIN).

Calocera A genus of fungi (order DACRYMYCETALES). *C. cornea* forms an erect, columnar, pointed, usually unbranched, yellow, gelatinous basidiocarp, ca. 1 cm high, on various types of wood. *C. viscosa* forms an erect, antler-like, yellow or orange, gelatinous basidiocarp, commonly 3–5 cm high, on coniferous wood.

calomel See MERCURY.

Calonectria See HYPOCREALES.

calonectrin See TRICHOTHECENES.

Caloplaca A genus of LICHENS (order TELOSCHISTALES) in which the thallus is crustose or placodioid, usually yellow, orange or red (containing PARIETIN) – but grey in some species. Apothecia: usually yellowish or reddish; ascospores: polarilocular in most species. Species grow on rocks and/or on wood.

Caloramator See CLOSTRIDIUM.

Calostoma See GASTEROMYCETES.

Calothrix A genus of filamentous CYANOBACTERIA (section IV) in which the mature trichomes taper from base to apex and are composed of disc-shaped, isodiametric or cylindrical vegetative cells; a terminal heterocyst occurs at the base of the trichome. Hormogonia are uniform in width, and young trichomes from hormogonia each have a heterocyst at one end only. GC%: 40–44. The genus is taken to include members of the family RIVULARIACEAE. Species are common e.g. on coastal rocks in the littoral zone.

calprotectin A zinc-chelating protein released from dying NEUTROPHILS at sites of inflammation (e.g. abscesses); it inhibits bacterial growth (apparently by withholding zinc ions), and this may antagonize certain antibiotics (e.g. β-lactams) used for chemotherapy. [Zinc in antimicrobial defence: RMM (1997) *8* 217–224.]

Calvatia See LYCOPERDALES.

Calvin cycle (Calvin–Benson cycle, or Benson–Calvin–Bassham cycle; reductive pentose phosphate cycle; C_3 cycle) A cyclic pathway used for the fixation of CARBON DIOXIDE by a wide range of AUTOTROPHS (q.v.). (See also METHYLOTROPHY.) The key enzymes in the Calvin cycle are RIBULOSE 1,5-BISPHOSPHATE CARBOXYLASE–OXYGENASE (RuBisCO), responsible for the CO_2-fixing reaction, and phosphoribulokinase (ribulose 5-phosphate kinase, EC 2.7.1.19), responsible for regenerating the CO_2 acceptor: ribulose 1,5-bisphosphate (RuBP). For every three molecules of CO_2 fixed, six molecules of 3-phosphoglycerate are formed; of these, five are required for the regeneration of RuBP, while the remaining molecule is available as a source of carbon for biosynthesis (see figure on page 120). The ATP and reduced NAD(P) required to drive the cycle are supplied by the light reactions of PHOTOSYNTHESIS in phototrophs, or by the oxidation of reduced inorganic compounds in CHEMOLITHOAUTOTROPHS. [Regulation of the Calvin cycle in bacteria: AvL (1984) *50* 473–487.] (See also PHOTORESPIRATION.)

calyciform Cup-shaped.

calymma See RADIOLARIA.

Calymmatobacterium A genus (*incertae sedis*) of Gram-negative bacteria which occur as pathogens in man (see GRANULOMA INGUINALE). Cells: non-motile, pleomorphic, capsulated, round-ended rods, 0.5–1.5 × 1.0–2.0 μm. In exudates from diseased tissues the cells are usually seen in the cytoplasm of large mononuclear phagocytes. The organisms can be cultured in the chick embryo yolk sac or on specialized egg yolk-containing media. Optimum growth temperature: 37°C. Type (only) species: *C. granulomatis*. [Book ref. 22, pp. 585–587.]

Calyptralegnia See SAPROLEGNIALES.

CAM (1) Chorioallantoic membrane: see EMBRYONATED EGG.
(2) CELL ADHESION MOLECULE.

***cam* gene** See METHANOGENESIS.

CAM plasmid An IncP-2 *Pseudomonas* PLASMID (ca. 500 kb) which encodes the capacity for metabolization of camphor (compare TOL PLASMID).

cAM373 See PHEROMONE.

camalexin A PHYTOALEXIN (a substituted indole) produced by *Arabidopsis*.

Camarops See SPHAERIALES.

Camarosporium A genus of fungi of the class COELOMYCETES. (See also LEPTOSPHAERIA.)

camelpox virus See ORTHOPOXVIRUS.

Camembert cheese See CHEESE-MAKING.

cAMP CYCLIC AMP.

cAMP–CAP See CATABOLITE REPRESSION.

CAMP factor See CAMP TEST.

cAMP phosphodiesterase See ADENYLATE CYCLASE.

cAMP receptor protein See CATABOLITE REPRESSION.

CAMP test (Christie–Atkins–Munch-Petersen test) A test used for the presumptive identification of group B streptococci. (Some group A streptococci may also give a positive test, particularly under anaerobic conditions [Book ref. 46, pp. 1590–1591].) The organism under test is inoculated in a fine streak on the surface of ox- or sheep-blood agar; a second streak – perpendicular to the first but separated from it by a few millimetres – is made with a culture of a β-HAEMOLYSIN-producing strain of *Staphylococcus aureus*. The plate is then incubated (aerobically) for ca. 12 hours

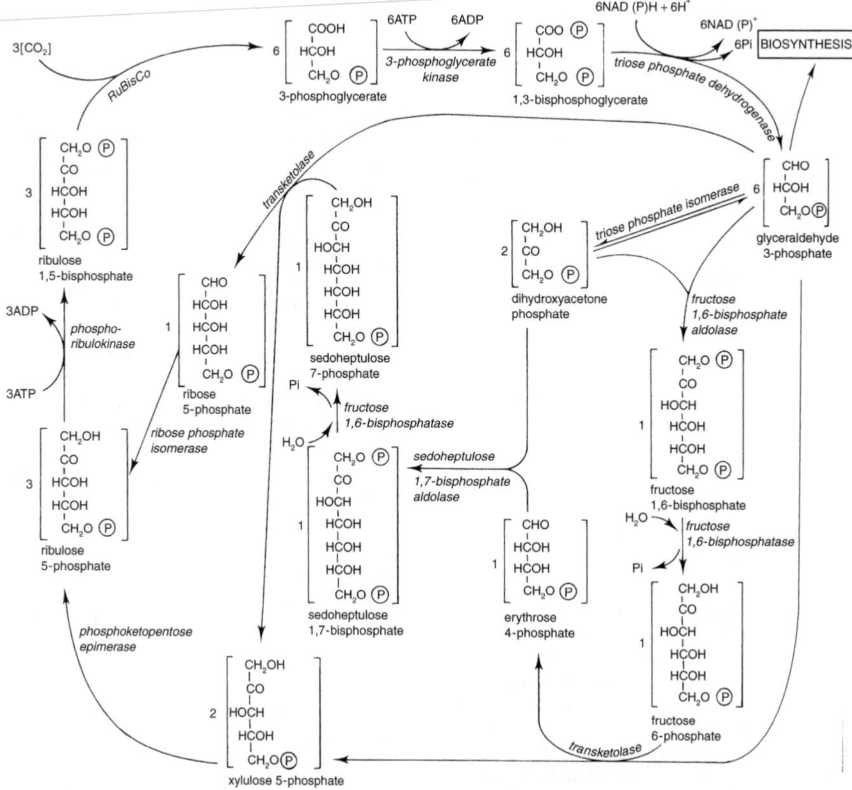

CALVIN CYCLE. For simplicity, the intermediate withdrawn for biosynthesis is shown as glyceraldehyde 3-phosphate. In reality, other intermediates may be withdrawn. For example, 3-phosphoglycerate is a precursor for the synthesis of various amino acids, fatty acids, purines and pyrimidines; withdrawal of 3-phosphoglycerate rather than glyceraldehyde 3-phosphate reduces the immediate energy cost of the cycle by 1 ATP and 1 NAD(P)H per 3 CO_2 fixed.

or more at 37°. In a positive test, a typical arrowhead- or flame-shaped area of *clear* haemolysis – due to synergy between a streptococcal extracellular polypeptide ('CAMP factor') and staphylococcal β-HAEMOLYSIN – occurs between the two lines of bacterial growth. [Conditions affecting CAMP factor-induced haemolysis: JGM (1985) *131* 817–820.]

The CAMP test is used e.g. for the detection of *Streptococcus agalactiae* (a causal agent of bovine MASTITIS) in milk; milk which is moderately or heavily contaminated with *S. agalactiae* may itself give a positive CAMP test if streaked onto the test plate in place of an isolate of the organism, as may a culture filtrate of *S. agalactiae*.

A 'reverse CAMP test', using a known group B streptococcus, can be used for the identification of β-haemolysin-producing staphylococci.

Campanella A genus of sedentary ciliate protozoa (subclass PERITRICHIA) found in freshwater and brackish habitats. *Campanella* resembles VORTICELLA (q.v.) but its zooid is larger and it is a colonial organism, one zooid occurring on each branch of a *non*-contractile stalk.

Campbell model A model, originally proposed by Campbell [Adv. Gen. (1962) *11* 101–145], in which a circular plasmid or phage genome integrates into a bacterial chromosome by a single cross-over between the two circular molecules. (See e.g. BACTERIOPHAGE λ; cf. BACTERIOPHAGE φ105.)

Campylobacter A genus of Gram-negative, asporogenous bacteria of the family CAMPYLOBACTERACEAE. Cells: slender, spirally curved rods, usually ca. 0.2–0.5 × 1.0–5.0 μm; POLAR MEMBRANES are present. (In old cultures, spherical or COCCOID BODIES may occur.) Motile; a single unsheathed flagellum occurs at one or both poles of the cell. Microaerophilic; growth is optimal in 5% oxygen, 10% carbon dioxide and 85% nitrogen. Metabolism: respiratory. Carbon sources: amino acids or TCA cycle intermediates, but not carbohydrates. Oxidase +ve. Indole −ve. Nitrate is reduced. Lipase −ve. MR −ve. VP −ve. GC%: 30–38. Type species: *C. fetus*.

Campylobacter species occur in the reproductive and intestinal tracts of man and animals. Some species are pathogenic (see CAMPYLOBACTERIOSIS.)

C. coli. Closely resembles *C. jejuni*, but e.g. does not hydrolyse hippurate.

C. concisus. Cells: small, curved. Catalase −ve. Hydrogen sulphide (TSI) +ve. No growth at 25°C. Hippurate and indoxyl acetate are not hydrolysed. Occurs in the gingival crevice flora in humans with gingivitis, periodontitis etc., but apparently non-pathogenic.

C. fetus. Cells appear comma- or S-shaped or 'gull-winged'. (Loosely wound spiral filaments (up to 8 μm in length) occur in old cultures.) Catalase +ve. Hydrogen sulphide (TSI) −ve. Growth occurs at 25°C but not at 42°C. Hippurate and indoxyl acetate are not hydrolysed. Pathogenic.

C. jejuni. Cells: small, tightly coiled spirals. On exposure of cultures to air, coccoid bodies form rapidly. SWARMING occurs on moist agar (the DIENES PHENOMENON does not occur). Catalase +ve. Hydrogen sulphide (TSI) −ve. Growth occurs at 42°C but not at 25°C. Hippurate and INDOXYL ACETATE are hydrolysed. Pathogenic. [PCR-RFLP typing of *C. jejuni*: LAM (1996) *23* 163–166; PGFE typing of Penner HS11 strains of *C. jejuni*: JMM (1998) *47* 353–357.]

C. pylori. Re-classified: see HELICOBACTER.

C. sputorum. Cells: slender, curved rods, appearing comma-shaped, 'gull-winged' or occasionally filamentous. Catalase −ve. Hydrogen sulphide (TSI) +ve. Growth occurs at 42°C (except for subsp *bubulus*) but not at 25°C (except for some strains of subsp *bubulus*). Hippurate and indoxyl acetate are not hydrolysed. Subsp *bubulus* and *sputorum* can grow anaerobically with fumarate; subsp *mucosalis* requires hydrogen or formate (as well as fumarate) for anaerobic growth. Subsp *bubulus* occurs in the genital tract in cattle and sheep, apparently as a commensal. Subsp *mucosalis* is found in pigs, apparently as a commensal in the mouth and as a pathogen, causing enteric disease. Subsp *sputorum* occurs in the gingival crevice flora of man but is apparently non-pathogenic.

(See also ARCOBACTER and HELICOBACTER.)

[Book ref. 22, pp 111–118. Culture, media etc.: Book ref. 219, pp 444–446. *Campylobacter* spp (genotyping: mini-review): AEM (2000) *66* 1–9.]

Campylobacteraceae A family comprising the genera ARCOBACTER and CAMPYLOBACTER.

campylobacteriosis (*med., vet.*) Any human or animal disease caused by a strain of CAMPYLOBACTER.

In man, disease is commonly food-borne, and is usually caused by *C. jejuni* or *C. coli*; less frequently isolated pathogens include *C. hyointestinalis* and *C. upsaliensis*. Typically, disease involves diarrhoea/enterocolitis, but disseminated infection can occur in the immunocompromised patient, and meningitis may develop in neonates. The usual sources of infection are inadequately cooked meats, especially poultry, unpasteurized milk and untreated (or contaminated) water; cross-contamination of food, particularly from raw (uncooked) poultry, is an important factor. The incubation period (~1–7 days) may be followed by several days of watery diarrhoea, often with abdominal pain. In severe cases the diarrhoea may be bloody; in some cases, symptoms resemble those of acute ulcerative colitis. In a small number of cases patients subsequently develop reactive arthritis; very rarely they develop the GUILLAIN–BARRÉ SYNDROME.

Chemotherapy. For persistent cases: erythromycin or e.g. ciprofloxacin.

[*Campylobacter* infections (reviews): RMM (1997) *8* 113–124; Microbiology (1997) *143* 5–21.]

In animals, campylobacters are commonly found as commensals in the gut. However, *C. fetus* can cause abortion in sheep (and less often in cattle). The *venerealis* subspecies of *C. fetus* causes abortion and infertility in cattle, while *C. sputorum* subsp *mucosalis* causes enteric disease in pigs. (See also WINTER DYSENTERY.)

CaMV CAULIFLOWER MOSAIC VIRUS.

Canada balsam A clear, ethanol-insoluble resin, obtained from fir trees (*Abies* sp), used (dissolved in xylene or benzene) as a MOUNTANT. It takes weeks to dry (harden) at room temperature; the refractive index of the dry resin is ca. 1.53. Canada balsam becomes yellowish with age owing to oxidation. (See also DPX.)

Canadian vole agent See BARTONELLA.

canarypox virus See AVIPOXVIRUS.

L-canavanine 2-Amino-4-guanidinohydroxybutyric acid. (See also CGB AGAR.)

cancellous (cancellate) Spongy; reticulated.

cancer Any disease of man or animals in which abnormal, unregulated proliferation of cells results in the formation of a malignant tumour (e.g. a CARCINOMA or SARCOMA); as the tumour grows, it tends to invade and destroy adjacent tissues and may shed cancer cells which can then disseminate (metastasize) to other areas of the body, resulting in the formation of secondary tumours.

The causes of cancer are complex and appear to be varied and multifactorial; in general, the detailed mechanisms are not fully understood. Oncogenesis appears commonly to involve an irreversible change in the genome of the cell, such a change ranging from a single point mutation to chromosomal rearrangements. (See also ONCOGENE.) In only relatively few cases have microorganisms been implicated as direct or indirect causes of cancer. Certain viruses are known to induce tumours and e.g. LEUKAEMIAS in animals and man (see DNA TUMOUR VIRUSES and ONCOVIRINAE); others may play an indirect role in predisposing to cancer (e.g. by suppressing the normal host immunological defence mechanisms). (cf. PARVOVIRUS.) [Viruses and cancer: Book ref. 110.]

Certain bacteria and fungi may play an indirect role in causing cancer by producing cancer-promoting substances (CARCINOGENS); e.g., certain MYCOTOXINS (e.g. AFLATOXINS) can be oncogenic, and certain bacteria of the human GASTROINTESTINAL TRACT FLORA may be responsible for the formation of substances which may play a role in the induction of bowel cancer. However, some microbial products have anticancer activity: e.g. ANTHRACYCLINE ANTIBIOTICS, ANTHRAMYCIN, BLEOMYCIN, etc.

See also e.g. IMMUNOTOXINS and NK CELLS; cf. GALLS.

Candelariella A genus of LICHENS (order LECANORALES); photobiont: a green alga. Thallus: crustose, yellow, lacking parietin. Apothecia: lecanorine.

candicidin B A heptaene POLYENE ANTIBIOTIC produced by *Streptomyces griseus*; its spectrum of activity is similar to that of AMPHOTERICIN B, and it is particularly effective against *Candida albicans*.

Candida (formerly e.g. *Monilia*) A large and heterogeneous genus of yeast-like or dimorphic imperfect fungi (class HYPHOMYCETES). (Although *Candida* has been traditionally classified in the Fungi Imperfecti, analysis of the genome of *C. albicans* has indicated a genetic repertoire that could support a sexual cycle [PNAS (2001) *98* 3249–3253].) Some species occur as commensals and/or as opportunist pathogens in man and animals (see CANDIDIASIS), and some can be isolated from plants, soil, foods etc. (See also e.g. CIDER, GARI, MEAT SPOILAGE.) The yeast-like cells are e.g. spherical, ovoid, cylindrical or elongate, and reproduce by (usually multilateral) budding. Pseudomycelium may be absent, rudimentary, or well developed, according to species and/or conditions; some species can form a true, septate mycelium. Blastospores and chlamydospores may be formed. Carotenoid pigments are not produced. Many species can ferment glucose and other sugars. Inositol and NO_3^- may or may not be assimilated; inositol-assimilating strains form pseudomycelium. The species [196 are described (with keys) in Book ref. 100, pp. 585–844] are distinguished e.g. on the basis of metabolic tests, growth requirements, etc.

C. albicans (a species which incorporates e.g. '*C. stelloidea*') occurs as a commensal e.g. in the mouth and in the genital and intestinal tracts in man; *C. albicans* is the commonest cause of CANDIDIASIS. Cells: e.g. ca. 3–6 × 5–10 μm. On e.g. cornmeal agar, pseudomycelium may be abundant, typically bearing grape-like clusters of blastospores. True mycelium may be formed, and thick-walled chlamydospores are formed e.g. on cornmeal agar at 20°C. *C. albicans* can ferment glucose and maltose, but not e.g. lactose. The ability to form GERM TUBES after incubation in serum for 2 hours at 37°C is presumptive evidence of *C. albicans*. It has been demonstrated that *C. albicans* can switch between at least seven different phenotypes (identified by colony morphology on agar media); the switching is heritable, reversible, and occurs at high frequency (increased further by low doses of UV radiation) [Science (1985) *230* 666–669]. Such switching may be important in the role of *C. albicans* as an opportunist pathogen (cf. ANTIGENIC VARIATION), but its mechanism is unknown.

C. tropicalis can also cause candidiasis. On e.g. cornmeal agar, pseudomycelium is abundant and consists of long, branched pseudohyphae bearing blastospores singly or in short chains and clusters. True mycelium and chlamydospores may also be formed.

C. utilis (= *Torulopsis utilis*; teleomorph: *Hansenula jadinii*) is an important food yeast (see SINGLE-CELL PROTEIN, TORULA YEAST, YEAST EXTRACT). On cornmeal agar an abundant pseudomycelium is formed, consisting of branched chains of short, coarse pseudohyphae and oval cells. True mycelium is not formed.

A few *Candida* spp have basidiomycetous affinities: e.g. *C. frigida*, *C. gelida*, *C. nivalis* and *C. scottii* have teleomorphs in the genus *Leucosporidium*, *C. japonica* in FILOBASIDIUM. Many species have teleomorphs among the ascomycetes: e.g., *C. ciferrii* (STEPHANOASCUS), *C. colliculosa* (TORULASPORA), *C. dattila* (KLUYVEROMYCES), *C. domercqii* (WICKERHAMIELLA), *C. globosa* (CITEROMYCES), *C. guilliermondii* (PICHIA), *C. holmii* (SACCHAROMYCES), *C. kefyr* (= *C. pseudotropicalis*) (KLUYVEROMYCES), *C. krusei* (ISSATCHENKIA), *C. lipolytica* (*Yarrowia lipolytica* – see SACCHAROMYCOPSIS), *C. lusitaniae* (CLAVISPORA), *C. pelliculosa* (HANSENULA), *C. pintolopesii* (SACCHAROMYCES), *C. pulcherrima* (METSCHNIKOWIA), *C. reukaufii* (METSCHNIKOWIA), *C. sorbosa* (ISSATCHENKIA), *C. sphaerica* (KLUYVEROMYCES), *C. utilis* (HANSENULA), *C. valida* (PICHIA), *C. variabilis* (PICHIA). (*C. hellenica*, *C. inositophila*, *C. steatolytica*: see ZYGOASCUS.)

candidate virus (1) A virus which is of uncertain taxonomic affinity but which may belong to (i.e., is a 'candidate' for) a particular taxonomic group; e.g. a 'candidate calicivirus' is a virus which may belong to the family Caliciviridae.

(2) A virus which might be the causal agent of a particular disease.

candidiasis (candidosis; moniliasis) A MYCOSIS, affecting man and animals, caused by a species of *Candida* – usually *C. albicans* (see CANDIDA); other pathogenic species include e.g. *C. parapsilosis* and *C. tropicalis*. Infection is probably endogenous in most cases; predisposing factors include e.g. suppression of the normal body microflora (e.g. by antibiotic therapy), impaired immune responses, etc. In man, candidiasis commonly occurs as a localized infection of the mouth (*oral thrush*) or vagina (*vaginal thrush*), and involves the formation of whitish mucoid plaques on the mucous membranes. (See also NON-GONOCOCCAL URETHRITIS.) Cutaneous candidiasis tends to occur on skin areas constantly exposed to moisture, regions commonly affected including e.g. the groin and axillae (see also PARONYCHIA); infected skin is swollen, red and pruritic. *Chronic mucocutaneous candidiasis* is a severe condition which occurs in immunocompromised or otherwise abnormal individuals; the skin and mucous membranes of the entire body may be affected, with chronic, granulomatous, inflammatory reactions in the underlying tissues. Other forms of candidiasis include bronchocandidiasis, pulmonary candidiasis (often secondary to other diseases), and systemic candidiasis (involving e.g. fungaemia, endocarditis, nephritis etc – cf. TORULOPSOSIS). *Lab. diagnosis*: microscopical and cultural examination of material from lesions etc (see CANDIDA).

In animals, candidiasis may involve the intestinal tract, skin, and/or various internal organs (see also MASTITIS); in calves and piglets the intestine is usually involved, and infection can cause watery diarrhoea, anorexia, dehydration, and even death. In birds, candidiasis commonly involves the digestive tract – particularly the crop.

[Book ref. 103.]

candidosis *Syn.* CANDIDIASIS.

candle (*verb*) To carry out CANDLING.

candle filter See FILTRATION.

candle jar A container within which cultures etc can be incubated under an increased partial pressure of CO_2; the CO_2 is provided by placing a lighted candle in the container prior to closure.

candle-snuff fungus See XYLARIA.

candling (*virol.*) A procedure for observing the contents of an intact egg by transmitted light. An EMBRYONATED EGG is placed over an aperture in a box containing an electric lamp; internal details of the egg – such as the positions of the embryo and blood vessels – can be seen, thus facilitating the inoculation of a particular membrane or cavity of the egg. Candling is best performed in a darkened room, and should be carried out rapidly to avoid overheating the egg.

cane blight In raspberry canes: a disease caused by *Leptosphaeria coniothyrium*. Infection occurs mainly via wounds; leaves shrivel and die, dark patches appear on the stem bases, and bark associated with these patches cracks open.

cane-cutters' disease LEPTOSPIROSIS caused by *L. interrogans* serotype *australis A*; the disease is common among cane-cutters in N. Queensland, Australia.

cane gall A disease of *Rubus* spp (raspberry, blackberry etc) in which small spherical galls or elongated ridges develop on the stems. Causal agent: *Agrobacterium rubi*.

cane spot (of raspberry) See ELSINOË.

canicola fever LEPTOSPIROSIS caused by *L. interrogans* serotype *canicola*.

canid herpesvirus 1 See ALPHAHERPESVIRINAE.

canine distemper An acute, infectious disease of dogs and other animals (e.g. foxes, mink, raccoons, wolves) caused by a virus of the genus MORBILLIVIRUS. Infection occurs mainly by droplet inhalation, also e.g. by ingestion. Incubation period: ca. 3–30 days. Symptoms initially include fever, nasal and ocular discharge, and listlessness. The fever subsides, then recurs with vomiting, diarrhoea, and often pneumonia; signs of CNS involvement (e.g. fits) may occur. (cf. HARD-PAD.) The disease affects mainly young animals. However, persistent infection may result in 'old dog encephalitis' – a condition characterized by progressive neurological degeneration. The distemper virus has also been associated (rarely) with a demyelinating disease in man.

Effective anti-distemper vaccines are available.

canine herpesvirus See ALPHAHERPESVIRINAE.

canine parvovirus A subspecies of the feline parvovirus (genus PARVOVIRUS) which appeared initially in Europe and then spread rapidly in 1978, causing worldwide outbreaks of disease in dogs. Infection may occur via the nasopharynx or via lymphoid tissues, including the tonsils. Viraemia leads to generalized infection of lymphoid tissues. Levels of erythrocytes (red blood cells) are not affected, and panleukopenia is infrequent in the dog (cf. FELINE PANLEUKOPENIA VIRUS). Infection of dividing epithelial cells in the ileum and jejunum impairs osmotic regulation and leads to diarrhoea (often containing blood/mucus); large numbers of virions are shed during the diarrhoeal phase. The disease can be associated with high mortality rates, but the severity varies in individual animals – some infections being mild/subclinical.

In neonatal puppies there is no manifestation of enteritis. In these animals infection may lead to death from myocarditis, often at 3–8 weeks of age (but sometimes up to 4 months of age). Onset of disease is rapid, typically with cardiac arrhythmia, pulmonary oedema and dyspnoea; the myocardium suffers multifocal necrosis. Infrequently, infection of the neonatal animal gives rise to lesions in a variety of tissues.

Canine parvovirus is apparently distinct from the 'minute virus of canines' (sometimes also called 'canine parvovirus') – a virus which is widespread in canine populations and which can be isolated from the faeces of asymptomatic dogs.

[Pathogenesis of canine parvovirus: BCH (1995) *8* 57–71.]

canker (*plant pathol.*) An imprecise term usually used for a plant disease which is characterized (in woody plants) by the death of cambium tissue and resulting loss and/or malformation of bark, or (in non-woody plants) by the formation of sharply delineated, dry, necrotic, localized lesions on the stem. (The term 'canker' may also be used to refer to the lesion itself, particularly in woody plants.) Cankers may be caused by bacteria (e.g. bacterial canker of cherry and plum is caused by *Pseudomonas syringae* pv. *morsprunorum*) or fungi (see e.g. APPLE CANKER, BEECH BARK DISEASE, BLEEDING CANKER, PITCH CANKER).

canning A form of APPERTIZATION in which, typically, suitably prepared foods are put into metal containers ('cans', 'tins') which are then exhausted, hermetically sealed, and heated; as an alternative, the food may be heat-treated before being placed aseptically in pre-sterilized cans which are then sealed under aseptic conditions. Most cans are made of tinplate (tin-plated steel) – see also TIN – but aluminium or other materials (e.g. lacquered tin-free steel) may be used. (See also RETORT POUCH.)

(a) *The canning process.* Conventional canning involves the following stages.

Food preparation. According to the type of food, this may involve grading, cutting, peeling, blanching (see also INDIVIDUAL QUICK BLANCH), washing etc; foods for canning must be as free as possible from microbial contamination because the efficacy of heat treatment is related to the level of contamination.

Filling the cans. This process must take into account the fact that the efficiency of subsequent processes depends on the correct volume of free space (*headspace*) between the surface of the food and the top of the can.

Exhausting the cans. The removal of air reduces the strain on the can during subsequent heating and inhibits long-term internal corrosion. Exhausting may be carried out by heating immediately prior to sealing, by sealing cold food into the can under vacuum (*vacuumizing*), or by injecting steam into the headspace during the sealing process.

Sealing. The can end is added by a double-seaming process in which the edge of the can end is wrapped around the body flange to form a seam of five thicknesses of metal – seven at the side seam, if present. (See also BIOS TEST.) (Concentric rings (*expansion rings*) on the can ends facilitate double-seaming and help to relieve strain on the can seams during the heating process.)

Heat processing. Cans may be heated in a BATCH RETORT or a COOKER–COOLER (see also STERIFLAMME PROCESS); in some cases cans are agitated to facilitate heat penetration, and in such cases an adequate headspace is essential. The temperature–time regime used in a particular case depends e.g. on the type and pH of the food. Following heat treatment, the food (i) must not contain *Clostridium botulinum* capable of growth and toxin production under the conditions of storage; (ii) should not contain organisms capable of causing spoilage; and (iii) should retain, as far as possible, its organoleptic properties. For non-acid foods (e.g. potatoes, mushrooms) – except foods containing curing salts – heat processing is usually to a *minimum* F_0 VALUE of 3.0 in order to destroy the spores of *C. botulinum* (see also BOTULINUM COOK); for foods with a pH below 4.5, an F_0 of 0.7 may be satisfactory because growth and toxin production by *C. botulinum* is generally considered not to occur at or below this pH (cf. CLOSTRIDIUM). Large cans of ham and other *cured* meats (pH above 5) are frequently processed at quite low temperatures (e.g. a centre temperature of ca. 70°C for a few minutes); such cans are necessarily stored under refrigeration – consumer safety and preservative action both relying on the combined effects of the curing salts and low temperature. Smaller cans of cured meats generally undergo a higher level of heat treatment and are shelf-stable, without refrigeration, owing to the ability of the curing salts to inhibit growth and toxin production by (heat-injured) pathogens [Book ref. 35]. Some canned foods (e.g. sweetened condensed milk) are not heated after can sealing; this product contains sufficient sugar to lower its WATER ACTIVITY to a level which inhibits most species of bacteria, and pre-PASTEURIZATION of the milk helps to prevent spoilage by fungi. Sweetened condensed milk (like many other – particularly acidic – canned foods) is usually non-sterile (cf. COMMERCIAL STERILITY) and often contains e.g. *Bacillus* spp and micrococci.

Cooling. Cans are cooled as rapidly as possible to avoid overcooking; cooling may be carried out under pressure to prevent unnecessary strain on the can seams and possible leakage. Water used for cooling should be bacteriologically clean and generally of potable standard.

(b) *Spoilage of canned foods.* Spoilage of the contents of a can may result from (i) mechanical damage to the can or leakage of imperfectly sealed cans (permitting the ingress of microorganisms), or (ii) the activities of thermoduric organisms or heat-stable enzymes in foods subjected to inadequate heat treatment. The type of spoilage depends on the type of food and the nature of the spoilage organism(s). (See also FLAT SOUR and SWELL.) Organisms which may survive the canning process include e.g. endospore-forming bacteria and certain fungi – e.g. *Byssochlamys* spp can cause pectinolytic softening in certain types of canned fruit (the ascospores of *B. fulva* can survive 87.7°C for 10 min). Canned fruit and vegetables may also be spoiled by heat-stable endopolygalacturonases (see PECTIC ENZYMES) produced by fungal contaminants prior to the canning process.

***Cantharanthus* alkaloids** *Syn.* VINCA ALKALOIDS.

Cantharellales An order of fungi (subclass HOLOBASIDIOMYCETIDAE) in which the basidiocarp may be e.g. funnel-shaped (as e.g. in *Cantharellus infundibuliformis* and *Craterellus cornucopioides*) or stalked and pileate; the hymenophore may be smooth, wrinkled, or lamella-like. *Cantharellus cibarius* (the 'chanterelle') is an edible species which forms a deep yellow, flat or centrally concave pileus on a stipe which tapers towards the base; the 'gills' are yellow, forked and decurrent, and the basidiospores are ochre, ellipsoidal, ca. 8–10 × 5 μm.

Cantharellus see CANTHARELLALES.

cap (1) (*mycol.*) The PILEUS of a mushroom-shaped fruiting body. (2) (*mol. biol.*) See MRNA (b). (3) See CAPPING sense 3.

CAP Catabolite activator protein: see CATABOLITE REPRESSION.

cap-binding protein See MRNA (b).

capacity test A type of test used to determine the capacity of a DISINFECTANT to retain activity in the presence of increasing numbers of bacteria. Essentially, aliquots of a bacterial suspension are periodically added to a fixed volume of disinfectant solution which is subcultured (to detect viable bacteria) at fixed intervals of time after the addition of each aliquot. (See e.g. KELSEY–SYKES TEST.)

capillitium A system of (non-cellular) threads or filaments which ramify through the spore mass in the fruiting bodies of certain slime moulds of the MYXOMYCETES and fungi of the GASTEROMYCETES. In myxomycetes, the capillitial threads may be solid or hollow, free or interconnected, smooth or ornamented, and in certain species they may bear nodes or encrustations of lime (see e.g. FULIGO); the nature of the capillitium is an important taxonomic feature. At least some types of capillitium may aid in spore dispersal; some are elastic, helping to expose spores when the peridium ruptures, while others are hygroscopic, their movements in response to changes in humidity helping to dislodge and release the spores.

capneic *Syn.* CAPNOPHILIC.

Capnocytophaga A genus of Gram-negative, capnophilic GLIDING BACTERIA found in the human oral cavity and associated with e.g. PERIODONTITIS, abnormalities in neutrophil function, and – in compromised patients – systemic disease. The organisms are flexible fusiform rods, ca. 0.3–0.5 × 3–8 μm. Metabolism: fermentative (glucose being fermented to succinic and acetic acids) or respiratory (with O_2 as terminal electron acceptor). Although CO_2 may be required for initial isolation, high levels of CO_2 are not necessarily required for fermentation or oxidation of glucose [IJSB (1985) *35* 369–370]. Agar is not liquefied. Optimum growth temperature: 37°C. Species: *C. gingivalis*, *C. ochracea* (formerly *Bacteroides ochraceus*), and *C. sputigena*.

Capnodium See SOOTY MOULDS.

capnophilic Refers to any organism whose growth either requires, or is stimulated by, higher-than-atmospheric levels of carbon dioxide.

capped mRNA See MRNA (b).

capping (1) See MRNA (eukaryotic).

(2) The binding of a specific protein or agent to the end of an ACTIN microfilament or a microtubule (see MICROTUBULES), resulting e.g. in cessation of polymerization/assembly at that end.

(3) In some types of eukaryotic cell: a process which is initiated when cell-surface molecules are cross-linked by multivalent ligands (e.g. antibodies, lectins). The cross-linked molecules from different regions of the cell surface initially cluster into groups or 'patches' (*patching*); subsequently, all the patches undergo an ATP-dependent translocation to one pole of the cell, forming a 'cap' (*capping*). Capping appears to occur only in cells capable of motility, and it has been suggested that it is the expression of a response similar to that given by a motile cell on contact with a surface. In *Dictyostelium discoideum*, capping appears to involve interaction between ACTIN microfilaments (which are linked, through the cytoplasmic membrane, to patches of ligand) and cortical MYOSIN [JCB (1985) *100* 1884–1893]. (See also DYSENTERY (b).)

caprine arthritis–encephalitis (CAE) A disease of goats caused by a retrovirus of the LENTIVIRINAE; symptoms include interstitial pneumonia, leucoencephalitis, inflammatory lesions in the CNS, and a progressive arthritis.

caprinized vaccine A VACCINE containing live organisms whose virulence for a given host species has been attenuated by SERIAL PASSAGE through goats.

Capripoxvirus (sheep-pox subgroup) A genus of viruses of the CHORDOPOXVIRINAE which cause disease in ungulates (see e.g. GOAT POX, LUMPY SKIN DISEASE and SHEEP POX); capripoxviruses can be transmitted mechanically by arthropod vectors. The virions are longer and narrower than those of orthopoxviruses; haemagglutinin is not produced, and infectivity can be destroyed e.g. by ether. The viruses in this group are very closely related serologically [JGV (1986) *67* 139–148].

capsduction A mode of gene transfer in strains of *Rhodopseudomonas capsulata*: DNA transfer from donor to recipient is mediated by a phage-like particle known as the *gene transfer agent* (GTA). Although the GTA morphologically resembles a tailed bacteriophage (head: apparently icosahedral, ca. 30 nm diam., with short spikes; tail: variable in length, bearing tail fibres) no phage-specific DNA has been detected; the linear dsDNA (MWt ca. 3×10^6) carried by the GTA is derived entirely from the host cell (chromosome or plasmid), and any region of the host genome can be packaged. Strains of *R. capsulata* which do not naturally produce GTA can act as recipients in capsduction but do not acquire the ability to produce GTA in consequence. Capsduction has been useful for mapping and genetic analysis in *R. capsulata*. (cf. TRANSDUCTION.)

capsid In a virion: a protein coat or shell which surrounds either the nucleic acid genome or a nucleoprotein CORE; it commonly exhibits ICOSAHEDRAL SYMMETRY. (See e.g. ADENOVIRIDAE.) In some viruses the capsid is itself surrounded by an ENVELOPE (see e.g. TOGAVIRIDAE). (cf. NUCLEOCAPSID.)

capsidiol A PHYTOALEXIN (a derivative of 5-epiaristolochene) produced e.g. by the tobacco and green pepper plants.

capsomer (capsomere) See ICOSAHEDRAL SYMMETRY.

capsule (1)(a) In prokaryotes, a layer of material external to but contiguous with the CELL WALL; the term thus includes any polysaccharide and/or protein surface layer(s) (including the bacterial S LAYER) but excludes those S layers which constitute the cell wall proper in members of the Archaea. (cf. GLYCOCALYX.) In some bacteria (e.g. *Beijerinckia*) a capsule may enclose more than one cell.

Capsules are commonly divided into three categories. (i) *Macrocapsules* or 'true' capsules are sufficiently thick to be easily visible (with negative staining – see CAPSULE STAIN) on light microscopy. (ii) *Microcapsules* cannot be observed by light microscopy, but their presence may be revealed by electron microscopy or by serological techniques (see e.g. QUELLUNG PHENOMENON); some bacteria which have lost the ability to form a macrocapsule may form a serologically identical microcapsule. (iii) *Slime layers* are diffuse secretions which may adhere loosely to the cell surface; slime layers commonly diffuse into the medium when the organism is grown in liquid culture, and may be too permeable to stains to be observable by microscopy. (Since capsules normally contain a high proportion of water, they may not be detectable in specimens dehydrated prior to microscopy.)

Capsule composition varies with species, growth conditions etc; serologically distinguishable capsules may be formed even by different strains of the same species (e.g. strains of *Streptococcus pneumoniae* can be distinguished on this basis). Most capsules consist of polysaccharide – either homopolysaccharide (see e.g. CELLULOSE; COLOMINIC ACID; DEXTRAN; LEVAN) or heteropolysaccharide (see e.g. ALGINATE; COLANIC ACID; HYALURONIC ACID). In virulent strains of *B. anthracis* the capsule consists of a homopolymer of γ-linked poly-D-glutamic acid. *Xanthobacter* spp can form both an α-polyglutamine capsule and copious polysaccharide slime. (See also M PROTEIN (sense 1).) The nature of the binding between the capsule and cell wall is largely unknown, but may involve ionic and/or covalent bonding. For example, capsular material covalently bound to peptidoglycan has been isolated from *B. anthracis*; in *Azotobacter vinelandii* and some mucoid strains of *Pseudomonas aeruginosa* retention of the alginate capsule depends on the level of Ca^{2+} in the growth medium, alginate being released into the medium at low Ca^{2+} levels.

Biosynthesis of polysaccharide capsules generally appears to involve transglycosylation from sugar–nucleotide precursors; the subunits are commonly transferred to the exterior of the cell via phosphorylated isoprenoid lipid carriers in the cell membrane.

In nature, bacterial capsules have various functions: they may act as permeability barriers (protecting e.g. against heavy metal toxicity) and/or as ion exchange systems; they may prevent desiccation, phagocytosis by protozoa, or phage infection (by covering receptor sites and/or by immobilization of the phage); they may promote bacterial adhesion to surfaces (see e.g. DENTAL PLAQUE) and/or play a role in recognition systems in interactions with other organisms (see e.g. ROOT NODULES); and they may act as a nutrient reserve (see e.g. XANTHOBACTER). Among pathogens, capsule formation often correlates with pathogenicity; in an animal host, a capsule may allow the pathogen to evade the host's defence mechanisms (see e.g. PHAGOCYTOSIS) [review: JID (1986) *153* 407–415], while in a plant host capsular material may be involved in the infection process (see e.g. CROWN GALL). (See also e.g. VI ANTIGEN.)

In vitro, colonies of capsulated bacteria on solid media are typically mucoid. Capsules may be lost on prolonged in vitro culture, but may be retained if a selective agent (e.g. sublethal levels of a surfactant or antibiotic) is incorporated in the medium.

A number of extracellular bacterial polysaccharides have important commercial applications: see e.g. CURDLAN; DEXTRAN; GELLAN GUM; XANTHAN GUM.

(b) (*mycol.*) Structures analogous to bacterial capsules occur in many fungi: gel-like or mucilaginous layers may surround hyphae or yeast cells and may sometimes diffuse into the surrounding medium. For example, in *Schizophyllum commune* a water-soluble glucan is loosely associated with the hyphal walls and may be released into the medium; the glucan consists of a $(1 \rightarrow 3)$-β-D-linked chain with single glucose residues attached by $(1 \rightarrow 6)$-β-D-linkages at (on average) every third residue [Book ref. 62, p. 139]. Cells of *Cryptococcus* spp form capsules of an amylose-like polymer e.g. when cultured in media of pH <3.0, or (*C. neoformans*) in the tissues of an infected host animal. In certain plant-pathogenic fungi, capsular material may be involved in the recognition of and/or attachment to a suitable host plant [e.g. *Drechslera (Helminthosporium) oryzae*/rice: Phytopath. (1982) *72* 285–292].

(2) (*entomopathol.*) See BACULOVIRIDAE.

(3) (*protozool.*) See e.g. RADIOLARIA and DESMOTHORACIDA.

capsule stain Bacterial CAPSULES may be demonstrated e.g. by negative STAINING: bacterial growth is mixed with a loopful of e.g. India ink or aqueous NIGROSIN on a clean slide and overlaid with a cover-glass; under the high-dry or oil-immersion objective the capsule appears as a clear zone between the cell wall and the dark background.

capsule swelling reaction *Syn.* QUELLUNG PHENOMENON.

captafol (difolatan; *N*-(1,1,2,2-tetrachloroethylthio)-cyclohex-4-ene-1,2-dicarboximide) An agricultural antifungal agent related to CAPTAN; it is used as a PROTECTANT e.g. against late blight of potato.

captan An antibacterial and ANTIFUNGAL AGENT (*N*-(trichloromethylthio)-cyclohex-4-ene-1,2-dicarboximide) widely used as an agricultural PROTECTANT; it is effective against a number of fungal plant pathogens, and is used e.g. for the prevention of apple scab and leaf-spot diseases (such as black spot of roses), and as a component of seed dressings. The activity of captan has been attributed to the $-SCCl_3$ group which may react with essential $-SH$ groups within the fungal cell. Captan is almost insoluble in water and is often formulated as a WETTABLE POWDER. (cf. CAPTAFOL.)

carate *Syn.* PINTA.

carbamate kinase See e.g. ARGININE DIHYDROLASE.

carbamic acid derivatives (as antimicrobial agents) Antimicrobial derivatives of carbamic acid (NH_2COOH) include UREA, derivatives of thiocarbamic acid, $NH_2.CS.OH$ (e.g. TOLNAFTATE), and DITHIOCARBAMATES. (See also BENOMYL.)

carbamide *Syn.* UREA.

carbanilides (as antimicrobial agents) Carbanilide (diphenylurea, $(C_6H_5NH)_2CO$) and substituted carbanilides (e.g. 3,4,4′-trichlorocarbanilide, TCC, = triclocarban) are active mainly against Gram-positive bacteria; they appear to affect the functioning of the cytoplasmic membrane. TCC is also active against certain fungi (e.g. *Trichophyton* spp).

carbapenems A class of β-LACTAM ANTIBIOTICS (q.v. for structure) produced (extracellularly) by *Streptomyces* spp; in addition to possessing antibacterial activity, many carbapenems are inhibitory to β-LACTAMASES. (cf. CLAVULANIC ACID.) Carbapenems are stable in aqueous solution near neutral pH. They include e.g. asparenomycins, epithienamycins, olivanic acids, pluracidomycins, PS-5, and THIENAMYCIN.

carbendazim (MBC) An agricultural antifungal agent – the active derivative of many of the BENZIMIDAZOLES. Carbendazim is used e.g. in the treatment of eyespot of wheat and barley and (as a seed treatment) as a protectant against damping off caused by *Rhizoctonia solani*. It is often mixed with other antifungal agents – e.g. PROPICONAZOLE or PROCHLORAZ – for use as a broad-spectrum antifungal preparation.

carbenicillin See PENICILLINS.

carbofuran A broad-spectrum carbamate insecticide used as a soil treatment for the control of insect pests on a range of crop plants. The persistence of carbofuran in the field is limited owing to the wide range of soil microorganisms which can degrade the compound [FEMS Ecol. (2000) *34* 173–180].

carbolfuchsin A red dye used e.g. in the GRAM STAIN and in ZIEHL–NEELSEN'S STAIN. Concentrated carbolfuchsin is made by adding 10% *basic* FUCHSIN in 95% ethanol (10 ml) to 5% aqueous phenol (100 ml); the solution is filtered after standing overnight.

carbolic acid *Syn.* phenol (see PHENOLS).

carbolic soap See SOAPS.

Carbomyces See PEZIZALES.

carbomycin See MACROLIDE ANTIBIOTICS.

carbon cycle In nature: the cyclical interconversion of carbon compounds (see figure); carbon in its elemental form appears to play no significant role in biological carbon metabolism (cf. NITROGEN CYCLE and SULPHUR CYCLE).

A major metabolite in the carbon cycle is carbon dioxide – see CARBON DIOXIDE (b). In AUTOTROPHS, carbon dioxide can be 'fixed' (i.e. incorporated into biomass) via several distinct pathways; the energy required for carbon dioxide fixation may be light energy (see PHOTOSYNTHESIS) or energy derived from chemical reactions. Carbon dioxide can also be used in a dissimilatory manner: see METHANOGENESIS.

The pool of carbon dioxide is replenished from organic matter by MINERALIZATION, a process in which both oxidative (respiratory) and fermentative forms of metabolism are involved. (See also SAPROTROPH.)

In *aquatic* environments, at least some dissolved organic carbon derives from phytoPLANKTON; in the oceans almost all dissolved carbon derives from this source. (See also PRIMARY PRODUCTION.) Some of the dissolved carbon is used for bacterial growth and respiration, and it appears that, at certain times and locations, the amount of carbon dioxide produced by bacterial respiration may exceed that fixed by the phytoplankton; such an imbalance in carbon cycling may mean that periods of net autotrophy (when autotrophic fixation of carbon dioxide exceeds bacterial respiratory output) may occur at different times of the year to compensate for periods of net heterotrophy [Nature (1997) *385* 148–151].

Carbon cycling is relevant to the GREENHOUSE EFFECT. Moreover, the evolution of the carbon cycle is an interesting topic in itself [FEMS Reviews (1992) *103* 347–354].

carbon dioxide (CO_2) (a) *CO_2 as a food preservative.* CO_2 (5–50%) inhibits the growth of many types of microorganism and is used in FOOD PRESERVATION – e.g. for meats (see MEAT SPOILAGE), fish, fruit and vegetables (in which CO_2 also extends storage life e.g. by delaying ripening), and for carbonated beverages. In general, CO_2 inhibits moulds more than yeasts, Gram-negative bacteria more than Gram-positive species. CO_2 dissolves in water to form carbonic acid, H_2CO_3, which dissociates to form H^+ and HCO_3^- (pK 6.37); some inhibition may thus be due to a pH effect. However, CO_2 also exerts a direct antimicrobial effect which is largely independent of either pH or oxygen concentration. The mechanism of this effect is unknown, but direct inhibitory effects of CO_2 on certain enzymes have been reported; it has also been suggested that CO_2 dissolves

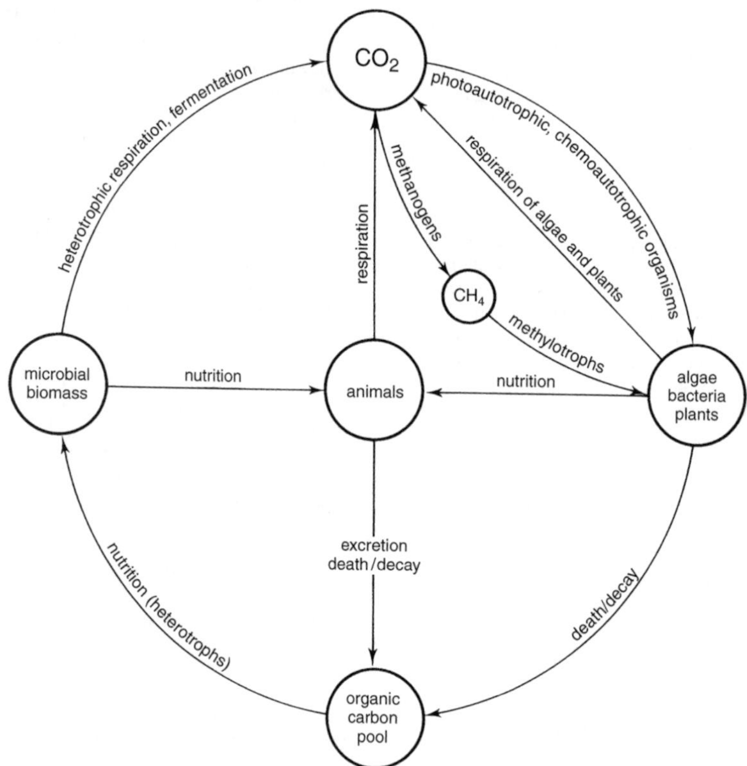

CARBON CYCLE: simplified scheme showing some of the major interconversions of carbon in nature (see entry). Bacteria have significant roles, as both autotrophs and heterotrophs, and certain members of the Archaea have unique roles as methanogens; the methylotrophs include methane-utilizing bacteria. Various microorganisms, including bacteria and fungi, are responsible for the essential conversion of 'dead' organic carbon to biomass and carbon dioxide; without this process the cycle would stop.

Reproduced from *Bacteria* 5th edition, Figure 10.1, page 256, Paul Singleton (1999) copyright John Wiley & Sons Ltd, UK (ISBN 0471-98880-4) with permission from the publisher.

in the cytoplasmic membrane and increases its fluidity, thereby impairing membrane function [Book ref. 29, pp. 335–343]. CO_2 forms complexes with amino groups in proteins, a reaction which is slowly reversible at low temperatures and under low partial pressures of CO_2; this may account for the observed residual effect in CO_2-treated meats subsequently exposed to air [Book ref. 29, pp. 369–375].

(b) *CO_2 as a carbon source.* CO_2 is assimilated via several different pathways in AUTOTROPHS (q.v.); in some algae, CO_2-concentrating mechanisms operate at low CO_2 levels, preventing (energy-wasting) PHOTORESPIRATION. In heterotrophs, CO_2 may be fixed in various carboxylation reactions (see e.g. Appendix II (b) and BACTEROIDES). (See also BIOTIN.)

carbon disulphide (CS_2) CS_2 (b.p. 46°C) is sometimes used as a soil fumigant for the control of *Armillaria mellea* root infections. It also acts as a NITRIFICATION INHIBITOR.

carbon monoxide (CO) (a) (as a metabolite) See CARBOXYDOBACTERIA, ACETOGENESIS and METHYLOTROPHY. (b) (as a RESPIRATORY INHIBITOR) Carbon monoxide can bind to the reduced a_3 component of eukaryotic cytochrome oxidase (see ELECTRON TRANSPORT CHAIN), and to many bacterial terminal oxidases, inhibiting the oxidation of the oxidase; a CO-insensitive respiratory chain occurs in the CARBOXYDOBACTERIA. At least some bacterial CO–cytochrome oxidase complexes can be dissociated by strong illumination (*photodissociation*) with light of wavelengths absorbed by the complex – with concomitant release of inhibition.

carbon monoxide dehydrogenase See ACETOGENESIS and METHANOGENESIS.

carbon monoxide oxidase. See CO OXIDASE.

carbonic anhydrase A zinc-containing ENZYME (EC 4.2.1.1) which catalyses the reaction:

$$CO_2 + H_2O \longleftrightarrow HCO_3{}^- + H^+$$

Carbonic anhydrases are found in eukaryotes and in members of the domains Bacteria and Archaea [PNAS (1999) *96* 15184–15189]. (See also acetate in the figure legend of the entry METHANOGENESIS.)

The carbonic anhydrases are categorized in classes α, β and γ; members of different classes do not share sequence homology.

Tests for the ability to produce carbonic anhydrase have been used e.g. to distinguish between species of MORAXELLA and NEISSERIA.

carbonylcyanide-*m*-chlorophenylhydrazone See PROTON TRANSLOCATORS.

carbonylcyanide-*p*-trifluoromethoxyphenylhydrazone See PROTON TRANSLOCATORS.

Carbowax See POLYETHYLENE GLYCOL.

Carboxide See ETHYLENE OXIDE.

carboxin (2,3-dihydro-6-methyl-5-phenylcarbamoyl-1,4-oxathiin) An agricultural systemic ANTIFUNGAL AGENT which is active against basidiomycetes, apparently by inhibiting mitochondrial succinate dehydrogenase. Carboxin is used e.g. for the control of *Rhizoctonia solani* and of smuts and rusts in cereals. Oxycarboxin, the corresponding sulphone, has also been used against cereal rusts.

carboxydismutase *Syn.* RIBULOSE 1,5-BISPHOSPHATE CARBOXYLASE–OXYGENASE.

carboxydobacteria A (non-taxonomic) category of bacteria characterized by the ability to use CARBON MONOXIDE, aerobically, as the sole source of carbon and energy [ARM (1983) *37* 277–310]. [Isolation and culture: Book ref. 45, pp. 899–902. Metabolism: Book ref. 132, pp. 131–151. Biotechnological aspects: MS (1986) *3* 149–153. Biochemistry and physiology: FEMS Reviews (1986) *39* 161–179.]

The oxidation of carbon monoxide to carbon dioxide is catalysed by an enzyme sometimes referred to as 'CO dehydrogenase' (a CO OXIDASE); the cytochrome oxidase of one branch of the respiratory chain is insensitive to inhibition by carbon monoxide. The carbon dioxide produced is assimilated via the CALVIN CYCLE. In at least some cases it appears that the CO oxidase is functional only when attached to a membrane-bound electron acceptor (cytochrome b_{561}) – the ability of the cells to oxidize carbon monoxide being lower during the stationary phase when some of the enzyme is cytoplasmic [FEMS Reviews (1993) *104* 332–336].

Many or all of the carboxydobacteria are able to obtain energy by the oxidation of hydrogen, and can grow more rapidly with hydrogen/carbon monoxide/oxygen mixtures than with mixtures of carbon monoxide and oxygen.

At least some species are facultatively heterotrophic.

The organisms occur e.g. in soil and polluted waters, sewage etc. Species include e.g. *Bacillus schlegelii*, *Pseudomonas carboxydohydrogena*, *P. carboxydovorans* and *P. thermocarboxydovorans*.

The carboxydobacteria were once classified as methylotrophs (users of C_1 compounds) but are now excluded from this category as they do not produce/assimilate formaldehyde.

carboxymethylcellulose See CM-CELLULOSE.

carboxypeptidase See PROTEASES.

carboxysomes Intracellular 'polyhedral bodies' (typically ca. 90–500 nm diam.) found in many autotrophic prokaryotes: e.g., in *Thiobacillus intermedius*, *T. neapolitanus*, *T. thiooxidans* and *T. thioparus* (but not in *T. novellus* or *T. versutus*); in *Nitrobacter winogradskyi* and a marine species of *Nitrosomonas* (but not in *Nitrospina gracilis* or *Nitrosococcus mobilis*); apparently in *Rhodomicrobium vannielii* (but not in green bacteria or purple non-sulphur bacteria); in most or all cyanobacteria (in vegetative cells and akinetes but not in heterocysts); in *Prochloron*; and apparently in many cyanelles. Carboxysomes from *T. neapolitanus* each consist of a number of doughnut-shaped particles (ca. 10 nm diam.) enclosed within a single-layered membrane or shell; these particles are molecules of RIBULOSE 1,5-BISPHOSPHATE CARBOXYLASE–OXYGENASE (RuBisCO). Carboxysomes from other organisms appear to resemble those of *T. neapolitanus*, and RuBisCO has been found in all carboxysomes in which it has been sought (e.g. in those from *Anabaena cylindrica*, *Chlorogloeopsis fritschii*, *Nitrobacter* sp, *Nitrosomonas* sp). The function of carboxysomes is unknown. [Biol. Rev. (1984) *59* 389–422.]

carbuncle A cluster of interconnected BOILS in which PUS is eventually discharged via several openings. It is usually caused by *Staphylococcus aureus*.

Carchesium A genus of sedentary ciliate protozoa (subclass PERITRICHIA). *Carchesium* occurs in various freshwater and marine habitats, and some species (e.g. *C. polypinum*) are common in SEWAGE TREATMENT plants. (See also SAPROBITY SYSTEM and SEWAGE FUNGUS.) In size and shape *Carchesium* resembles VORTICELLA – e.g. it has a bell-shaped zooid: however, *Carchesium* is a colonial peritrich – the mature organism consisting of a number of zooids carried on a branching stalk, each zooid and its individual branch of the stalk being capable of independent contraction.

carcinogen (1) Any agent which can cause CANCER. In general, carcinogens appear to exert their effect by interacting with the DNA of the target cell; many carcinogens are also mutagens.

(See also MUTAGENICITY TEST.) Some apparent carcinogens actually require enzymic conversion within the body to an active form, the 'ultimate carcinogen', before they can cause neoplastic transformation.

(2) *Syn.* oncogen *sensu lato* (see ONCOGENIC).

carcinoma Any malignant tumour which arises from epithelial tissue: see e.g. NASOPHARYNGEAL CARCINOMA and HEPATOCELLULAR CARCINOMA. (See also CANCER.)

cardiac beriberi See CITREOVIRIDIN.

Cardiobacterium A genus of oxidase-positive, catalase-negative, Gram-negative bacteria which occur e.g. in the upper respiratory tract (see also HACEK) and which have been isolated from the blood in cases of bacterial endocarditis. Cells: small, pleomorphic, non-motile, round-ended rods ($\sim 0.5-0.75 \times 1-3$ μm), or filaments. Chemoorganotrophic. Aerobic and facultatively anaerobic; apparently capnophilic. On anaerobic blood agar, with 5% carbon dioxide, colonies are smooth and circular, 1–2 mm diam., after 48 hours. Strains are urease −ve but may be weakly indole +ve. Nitrates are not reduced. Optimum growth temperature 30–35°C. Media which have been used for primary isolation include e.g. casein-soy broth and thioglycollate broth (each containing SPS). No growth occurs on MacConkey agar. Acid (but no gas) is formed from the fermentation of fructose, glucose, mannose, sorbitol or sucrose; lactic acid is the main product of glucose fermentation. The organism is typically sensitive to β-lactam and other antibiotics. GC%: 59–62. Type species: *C. hominis*.

cardioid condenser See MICROSCOPY (b).

cardiolipin Diphosphatidylglycerol. Such lipids occur e.g. in bacterial CYTOPLASMIC MEMBRANES. Cardiolipin prepared by alcoholic extraction of beef heart is used in STANDARD TESTS FOR SYPHILLIS.

Cardiovirus A genus of viruses (family PICORNAVIRIDAE) isolated mainly from rodents; the true natural host is uncertain. Cardiovirus virions are generally stable at pH 3–9, but are labile at pH 4.5–6.4 in the presence of 0.1 M halide ions. There is only one serotype, and the various named members of the genus – e.g. Columbia SK virus, mouse (or Maus) Elberfeld virus, mengovirus, MM virus – are regarded as strains of *murine encephalomyocarditis virus* (EMCV). EMCV causes encephalitis and myocarditis in rodents, and a variant can cause a diabetes-like syndrome in certain strains of mice [JGV (1985) *66* 727–732].

carfecillin See PENICILLINS.

caries (dental) See DENTAL CARIES.

cariogenic Refers to any factor, substance or organism which promotes DENTAL CARIES.

carlaviruses (carnation latent virus group) A group of PLANT VIRUSES in which the virions are slightly flexuous filaments (ca. 600–700 × 13 nm) containing one molecule of linear, positive-sense ssRNA (MWt ca. 2.7×10^6). Each member has a fairly narrow host range, and some can cause disease in crop plants. Transmission occurs (non-persistently) via aphids in at least some members; mechanical transmission can be achieved under experimental conditions. Type member: carnation latent virus. Other members include e.g. cactus virus 2, hop latent virus, HOP MOSAIC VIRUS, pea streak virus, potato virus M (= potato paracrinkle virus), potato virus S, red clover vein mosaic virus. Possible members include e.g. chicory blotch virus, eggplant mild mottle virus, and groundnut crinkle virus.

CARNA See CUCUMOVIRUSES.

carnation etched ring virus See CAULIMOVIRUSES.

carnation Italian ringspot virus See TOMBUSVIRUSES.

carnation latent virus See CARLAVIRUSES.

carnation mottle virus See TOMBUSVIRUSES.

carnation necrotic fleck virus See CLOSTEROVIRUSES.

carnation ringspot virus See DIANTHOVIRUSES.

carnation vein mottle virus See POTYVIRUSES.

carnidazole See NITROIMIDAZOLES.

carnivorous Refers to protozoa which feed on other protozoa, crustacea, etc. (cf. HERBIVOROUS.)

Carnobacterium A genus of heterofermentative LACTIC ACID BACTERIA found e.g. on vacuum-stored chilled meats and on meats stored in a carbon-dioxide-rich environment. *Carnobacterium* spp have been studied e.g. as sources of BACTERIOCINS. [Bacteriocin production by *C. piscicola*: JBC (1994) *269* 12204–12211. Divercin V41, a new bacteriocin from *C. divergens* V41: Microbiology (1998) *144* 2837–2844.]

Carnoy's fluid A FIXATIVE: either (a) ethanol (95% or 100%) and glacial acetic acid in the ratio 3:1, or (b) ethanol (95% or 100%), chloroform, and glacial acetic acid (6:3:1).

carotenes See CAROTENOIDS.

carotenoid band shift A light-induced change in the absorption spectrum of a photosynthetic membrane due, mainly, to a change (of a few nanometres) in the absorption spectra of the carotenoids in the membrane; this change results from the effect on the carotenoids of the light-induced transmembrane pmf (see PHOTOSYNTHESIS) and is sometimes referred to as an 'electrochromic effect'.

carotenoids A class of polyene (isoprenoid), aliphatic or alicyclic, commonly yellow or orange pigments; the class includes the *carotenes* (hydrocarbons with usually nine conjugated double bonds) and oxygen-containing derivatives of carotenes (*xanthophylls*) – see figure. Carotenoids are widely distributed in microorganisms, plants and animals; they are present in probably all photosynthetic organisms, in which they function e.g. in LIGHT-HARVESTING COMPLEXES and in protecting CHLOROPHYLLS from photodegradation. (See also SINGLET OXYGEN and ANTIOXIDANTS.) Carotenoids may also be involved in phototaxis (see e.g. EYESPOT), and a carotenoid derivative, RETINAL, occurs in certain bacterial energy-transducing systems (see PURPLE MEMBRANE). The presence of one or more particular carotenoids in a given organism may be an important taxonomic feature, especially in algae.

Carotenes include e.g. *β-carotene* ($R' = R'' = \beta$-ionone ring, see figure), widely distributed in algae and higher plants, and present in certain bacteria; *α-carotene* (an isomer of β-carotene), also widely distributed; *γ-carotene* ($R' = \beta$-ionone ring, $R'' =$ 2,6-dimethyl-1,5-heptadienyl), present e.g. in some bacteria of the Chlorobiaceae; *lycopene* ($R' = R'' =$ 2,6-dimethyl-1,5-heptadienyl), found e.g. in fruits.

Xanthophylls include e.g. *antheraxanthin* (found e.g. in green algae (Chlorophyta), in chloromonads, and in plants); *astaxanthin* (see PHAFFIA); BACTERIORUBERINS; *diatoxanthin* (in cryptophytes); *fucoxanthin* (in brown algae (Phaeophyta), chrysophytes, diatoms and prymnesiophytes – see figure); *lutein* (3,3′-dihydroxy-α-carotene, common in plants and green algae); *myxoxanthophyll* (a rhamnosyl-containing xanthophyll found in cyanobacteria); *neoxanthin* (common in green algae and higher plants); *peridinin* (in dinoflagellates); *siphonein* and *siphonoxanthin* (in siphonaceous green algae); *spirilloxanthins* (in anoxygenic photosynthetic bacteria); STAPHYLOXANTHIN; *torularhodin* (in *Rhodotorula*); *violaxanthin* (in green algae and higher plants); *zeaxanthin* (3,3′-dihydroxy-β-carotene, an isomer of lutein, found e.g. in green algae and higher plants).

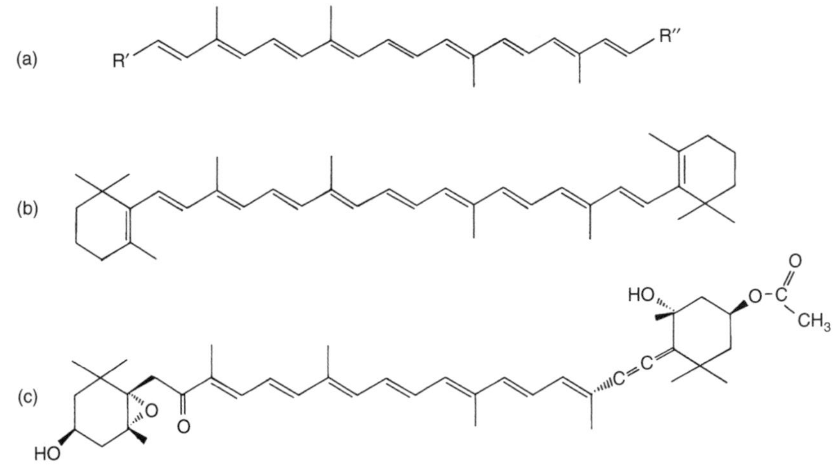

(a) (b) (c)

CAROTENOIDS. (a) General structure of a typical carotene. (b) β-Carotene (= β,β-carotene). (c) Fucoxanthin (a xanthophyll).

Some bacterial polyene pigments (e.g. FLEXIRUBINS, XANTHOMONADINS) resemble carotenoids but lack the methyl groups on the hydrocarbon chain.

(See also SPOROPOLLENIN.)

carp erythrodermatitis (Polish INFECTIOUS DROPSY) A FISH DISEASE affecting carp and certain other fish (e.g. goldfish, bream); aetiology: unknown. The disease is subacute to chronic, with variable mortality. Inflamed skin lesions become necrotic and ulcerative, and generalized oedema follows. Secondary infection with *Aeromonas* and *Pseudomonas* spp is common.

carp pox (epithelioma papulosum) A benign FISH DISEASE affecting the common carp (*Cyprinus carpio*) and other cyprinids. Multiple transient hyperplastic nodules (up to several centimetres across) develop on the skin and fins, and occasionally on the eyes and gills; the lesions do not become necrotic. Carp pox is apparently caused by a herpesvirus (see HERPESVIRIDAE).

carpomycetes Fungi which form fruiting bodies – particularly the ascomycetes and basidiomycetes.

Carr medium A medium used e.g. to distinguish between strains of ACETOBACTER and GLUCONOBACTER; it contains yeast extract (3%), agar (2%), BROMCRESOL GREEN (0.002%) and ethanol (2% v/v). *Gluconobacter* strains (which typically oxidize ethanol to acetic acid) produce colonies each surrounded by a yellow halo; *Acetobacter* strains (which typically carry out 'overoxidation') initially form similar colonies but, on further incubation, the acetic acid is metabolized and the pH indicator reverts to its original colour (i.e. the halo is lost).

carrageenan (carrageenin; carragheen) A mixture of sulphated galactans associated with the CELL WALLS of many rhodophycean algae (e.g. *Chondrus crispus*, *Gigartina* spp); they appear to prevent desiccation when the alga is exposed to the atmosphere. Carrageenans have the general structure: $-\beta\text{-}A\text{-}(1 \rightarrow 4)\text{-}\alpha\text{-}B\text{-}(1 \rightarrow 3)\text{-}\beta\text{-}A\text{-}$, where A = D-galactose 4-sulphate or D-galactose 2-sulphate, and B is any of a range of sulphated galactose or 3,6-anhydrogalactose residues. One component of the carrageenan mixture, κ-carrageenan, is precipitated by KCl; in κ-carrageenan A = D-galactose 4-sulphate, B = 3,6-anhydro-D-galactose. κ-Carrageenan forms gels with e.g. K^+, $NH_4{}^+$, Ca^{2+}, Mg^{2+}, Fe^{3+}, and with water-miscible organic solvents; it is widely used in industry e.g. as a gelling and thickening agent in foods, pharmaceuticals etc, and in supports for IMMOBILIZATION of cells and enzymes. It can also be used as an AGAR substitute in certain bacteriological media [Book ref. 2, pp. 144–145]. The KCl-soluble, non-gelling fraction of carrageenan consists of various polymers in which B = e.g. D-galactose 6-sulphate, D-galactose 2,6-disulphate, 3,6-anhydro-D-galactose 2-sulphate, etc. (The term λ-carrageenan sometimes refers to the whole KCl-soluble fraction, sometimes specifically to the component in which A = D-galactose 2-sulphate, B = D-galactose 2,6-disulphate.)

carragheen moss See CHONDRUS.

Carrel flask A squat cylindrical glass vessel which opens, at the side, via a straight, upward-sloping tubular neck.

carrier (1) (*med., vet.*) An individual who harbours a particular pathogen but shows no clinical signs of disease; such an individual can transmit the pathogen to others. Carriers are important reservoirs of infection in certain diseases – e.g. DIPHTHERIA and TYPHOID FEVER (cf. SMALLPOX). [The carrier state (involving bacteria): JAC (1986) *18* Supplement A 1–81.] (2) (*immunol.*) See HAPTEN. (3) See CARRIER TEST. (4) (*virol.*) See PSEUDOLYSOGENY and CARRIER CULTURE.

carrier culture (*virol.*) A cell culture which is persistently infected with a virus (see PERSISTENCE) but in which only a small proportion of the cells in the population is actually infected by the virus. The infected cells may release virus (e.g. by lysis) but only a small proportion of the remaining cells is then infected, and so on. The culture can be 'cured' of the virus e.g. by treatment with virus-neutralizing antiserum or by serial subculture using small inocula. In e.g. mammalian cell cultures a carrier culture may arise e.g. because INTERFERON production may render most of the cells resistant to infection. (See also PSEUDOLYSOGENY; cf. SEMIPERMISSIVE CELLS.)

carrier gene See FUSION PROTEIN.

carrier-specific T helper cell (CTh) A T LYMPHOCYTE whose TCR recognizes a specific epitope in the 'carrier' part of an antigen – another epitope (the 'hapten') on the same antigen being recognized by a B cell. (See COGNATE HELP.) If an antigen carrying a particular hapten–carrier pair is used for primary immunization, then the hapten must be paired with the same carrier for an efficient response during secondary immunization.

carrier test A type of test used to determine the ability of a given disinfectant to disinfect an object ('carrier') which has been artificially contaminated with microorganisms; the carrier may be e.g. a piece of cloth or an instrument. (See e.g. AOAC USE-DILUTION TEST.) [Evaluation of carriers used in AOAC test methods: AEM (1986) *51* 91–94.]

Carrión's disease *Syn.* BARTONELLOSIS.

carrobiose A disaccharide: 3,6-anhydro-4-*O*-(β-D-galactopyranosyl)-D-galactose, a degradation product of κ-carrageenan.

carrot black rot See ALTERNARIA.

carrot latent virus See RHABDOVIRIDAE.

carrot mosaic virus See POTYVIRUSES.

carrot mottle virus See TOGAVIRIDAE.

carrot red leaf virus See LUTEOVIRUSES.

carrot soft rot See SOFT ROT (sense 2).

carrot thin leaf virus See POTYVIRUSES.

carrot yellow leaf virus See CLOSTEROVIRUSES.

Carr's medium See CARR MEDIUM.

Carteria A genus of green algae which closely resemble CHLAMYDOMONAS but which have four flagella instead of two.

carvacrol See PHENOLS.

Cary–Blair transport medium See TRANSPORT MEDIUM.

caryonide *Syn.* KARYONIDE.

Caryophanon A genus of Gram-positive, asporogenous, obligately aerobic, chemoorganotrophic bacteria which form motile (peritrichously flagellate) TRICHOMES of diameter ca. 1.5 μm (*C. tenue*) or ca. 2.5–3.2 μm (*C. latum*). Both species have been isolated from cattle manure, but their true habitat is unknown. [Book ref. 46, pp. 1701–1707.]

Caryospora A genus of protozoa (suborder EIMERIORINA) which form oocysts each enclosing one sporocyst containing 8 sporozoites.

Caryotropha See EIMERIORINA.

casamino acids Acid-hydrolysed casein.

caseation (*med., vet.*) A type of necrosis in which diseased tissue breaks down into a dry, amorphous, cheese-like mass (cf. TUBERCULOSIS).

casein A phosphoprotein which occurs in MILK mainly in the form of micelles, each micelle consisting of aggregated molecules of the 3 main types of casein: α-, β- and κ-casein; the (Ca^{2+}-precipitable) α- and β-caseins in the micelles are stabilized by the κ-casein. κ-Casein can be cleaved by various enzymes – e.g. by pepsin (EC 3.4.23.1), or by rennin (= chymosin; EC 3.4.23.4), an enzyme derived from the abomasum (see RUMEN) of suckling ruminants; these enzymes cleave κ-casein into an insoluble N-terminal fragment, *p*-κ-casein, and a soluble C-terminal fragment, 'glycomacropeptide'. Cleavage of micellar κ-casein in milk destablizes micelles and leads to the precipitation of α- and β-caseins and *p*-κ-casein (i.e., the milk *coagulates*). (The proteolytic enzymes of some psychrotrophic bacteria can cleave κ-casein and cause 'gelation' (coagulation) in stored UHT-treated milk.)

Enzyme-independent coagulation ('clotting' – see e.g. LITMUS MILK) occurs when the pH is lowered to ca. 4.6 (i.e., at or near the isoelectric points of the caseins).

γ-Casein is a product of the enzymic cleavage of β-casein by a trypsin-like enzyme which occurs in milk in small amounts.

Caseobacter A genus of aerobic, non-motile, asporogenous bacteria (order ACTINOMYCETALES, wall type IV) which occur e.g. on certain cheeses. The organisms are pleomorphic or coryneform rods, ca. 1 × 2–4 μm, which give rise to coccoid forms in stationary-phase culture; on yeast extract–glucose agar the colonies are grey-white, pink or red. Up to 12% salt (NaCl) can be tolerated. GC%: 65–67. Type species: *C. polymorphus*.

caseous necrosis *Syn.* CASEATION.

caspases See APOPTOSIS.

cassava common mosaic virus See POTEXVIRUSES.

cassava latent virus See GEMINIVIRUSES.

cassava vein mosaic virus See CAULIMOVIRUSES.

cassette mechanism (*mol. biol.*) A system for gene regulation in which a particular gene is removed from one site ('active slot'), in which it can be expressed, to another ('inactive slot') in which it cannot be expressed – another gene being inserted into the 'active slot'. (See e.g. MATING TYPE.)

Castañeda's method A form of BLOOD CULTURE in which the culture bottle contains an agar slope which projects above the level of the liquid medium when the bottle is standing vertically. The liquid medium is inoculated in the usual way, and the bottle is incubated. After 24/48 h the bottle is tilted so that the medium washes over the exposed agar; the bottle is then incubated, vertically, and subsequently examined for bacterial growth on the agar. Tilting etc may be repeated at intervals of one or two days. By eliminating subculturing this method avoids a source of contamination.

cat (parvovirus infection) See FELINE PANLEUKOPENIA VIRUS.

cat bite fever A disease caused by *Pasteurella multocida* and transmitted by cat-bites; an abscess develops at the site of the bite. (cf. CAT SCRATCH DISEASE.)

cat diseases See under 'feline'.

cat flu virus *Syn.* FELINE CALICIVIRUS.

***cat* gene** A gene encoding chloramphenicol acetyltransferase: see CHLORAMPHENICOL and Tn*9*.

cat scratch disease (cat scratch fever) A (usually benign) human disease (more common in children) associated with the scratch or bite of a cat or, occasionally, with injuries from thorns, splinters etc. A local inflammatory skin lesion develops after 2–3 days, and this is followed, in days/weeks, by enlargement of regional lymph nodes – which suppurate in some cases. There is often a general malaise with low-grade fever. The condition usually resolves spontaneously within weeks; however, in some cases there are complications – which may involve e.g. liver, brain or eyes.

Various organisms have been proposed as the causal agent of cat scratch disease, including e.g. *Bartonella clarridgeiae* [JCM (1997) *35* 1813–1818].

(See also CAT BITE FEVER.)

catabolism The metabolic reactions by which exogenous or endogenous organic compounds are degraded to simpler organic or inorganic compounds. (cf. ANABOLISM; AMPHIBOLIC PATHWAY.)

catabolite activator protein See CATABOLITE REPRESSION.

catabolite control protein A See CATABOLITE REPRESSION.

catabolite inhibition See INDUCER EXCLUSION.

catabolite repression The repression of certain inducible enzyme systems by the presence of a 'preferred' carbon source (e.g. glucose in enterobacteria and *Saccharomyces cerevisiae*, certain TCA cycle intermediates in *Pseudomonas* spp). For example, in *Escherichia coli*, various operons concerned with the degradation of substrates other than glucose (e.g. ARA OPERON, LAC OPERON, TNA OPERON) are repressed in the presence of glucose, even when their respective inducers are present; thus, when *E. coli* is grown in a medium containing both glucose and lactose, the *lac* operon remains repressed until all the glucose has been used – at which time the *lac* operon becomes derepressed. (See also DIAUXIE, CRABTREE EFFECT.)

Among the bacteria there are several distinct mechanisms for catabolite repression. Thus, e.g. one mechanism is characteristic of *E. coli* and other enterobacteria, while others are typical of AT-rich Gram-positive species (e.g. species of *Clostridium*, *Lactobacillus* and *Staphylococcus*); each of these mechanisms is regulated by signals from components of the PTS (q.v.). Models outlining these mechanisms are described below.

(a) *Catabolite repression in enterobacteria.* In these organisms the key regulatory molecule is the IIA component of glucose permease in the PTS; note that, in this particular permease, IIA is a cytoplasmic (i.e. soluble) protein.

When glucose is *absent*, phosphorylated IIA (IIA~P) does not undergo dephosphorylation via IIB, and under these conditions IIA~P activates the enzyme ADENYLATE CYCLASE – thus stimulating the synthesis of CYCLIC AMP (cAMP). cAMP binds to, and activates, the so-called *cAMP-receptor protein* (CRP) – also called *catabolite activator protein*, CAP, or *catabolite gene activator protein*, CGA protein. (CRP is encoded by the *crp* gene.) The cAMP–CRP complex acts as a transcriptional activator by binding in the vicinity of promoters of certain operons (e.g. *lac*), allowing them to be expressed in the presence of their respective inducers. This explains why the inducibility of the repressed operons can be at least partly restored by the addition of cAMP. The precise way in which cAMP–CRP promotes transcription is still unknown. One early model for the *lac* operon supposed that the binding of cAMP–CRP displaced RNA polymerase from a weak promoter, P2, to the primary promoter, P1; that such a mechanism is unlikely to contribute significantly to the transcription of the *lac* operon has been inferred from the failure of inactivation of P2 to result in activation of P1 [Mol. Microbiol. (1992) *6* 2419–2422]. Studies on e.g. the *lac* operon suggest that the presence of cAMP–CRP is not needed after the establishment of a productive open complex, and that only transient interaction is required between cAMP–CRP and RNA polymerase [EMBO (1998) *17* 1759–1767].

When glucose (or other rapidly metabolizable substrate) is *present*, unphosphorylated IIA binds to the permeases of various sugars (e.g. lactose), inhibiting uptake of the corresponding sugars; as these sugars are involved in the induction of their respective operons, this phenomenon is called INDUCER EXCLUSION.

Note that, in some cases, cAMP–CRP may repress rather than activate gene expression. For example, in *Rhizobium meliloti* cAMP–CRP blocks transcription of the *dctA* gene whose expression is necessary for symbiotic nitrogen fixation [EMBO (1998) *17* 786–796]. (In fact, CRP – originally associated with activation of catabolic operons – is involved in a much wider range of gene-regulatory activities.)

(b) *Catabolite repression in the AT-rich Gram-positive bacteria.* In these bacteria, transcription of some catabolic operons is at least partly regulated by operon-specific regulator proteins, each regulator protein having two phosphorylation sites; each of these two sites is called a PRD (= PTS regulation domain). PRDs can be phosphorylated by certain intermediates in the PTS (IIB~P and HPr~P); one of the PRDs can be phosphorylated by IIB~P, the other by HPr~P. It seems that phosphorylation by IIB~P is inhibitory, causing the regulator protein to inhibit transcription from relevant operons, while phosphorylation by HPr~P promotes transcription from those operons. The activity of a given regulator protein thus depends on the mode of its phosphorylation; thus, when phosphorylated *only* by HPr, the regulator protein promotes transcription from all relevant operons, but phosphorylation by HPr *and* IIB, or a lack of phosphorylation in both PRDs (i.e. in the presence of both inducer and glucose), causes the regulator protein to function as a repressor of transcription. Induction of operon(s) therefore occurs only under appropriate conditions, i.e. when (i) inducer is present and (ii) glucose is absent. [PTS-dependent control of PRD-containing regulators in induction and repression of catabolic operons: Mol. Microbiol. (1998) *28* 865–874.]

A PRD-containing regulator protein in *E. coli*, BglG, is involved in the control of the *bgl* operon (a β-glucoside catabolic system concerned e.g. with the metabolism of salicin); in the *absence* of β-glucoside, BglG is phosphorylated by the IIB component of BglF (the β-glucoside permease) – thus inhibiting transcription of the *bgl* operon.

In AT-rich Gram-positive bacteria, catabolite repression involving HPr and the enzyme *HPr kinase* occurs by ATP-dependent phosphorylation of HPr at a site (Ser-46) which differs from that (His-15) phosphorylated by enzyme I in the PTS; when HPr is phosphorylated at Ser-46 it can form a complex with *catabolite control protein A* (CcpA), and this complex can bind to a regulatory site in target operons, inhibiting transcription. (*In vitro*, phosphorylation of HPr at the Ser-46 site decreases the ability of HPr to accept phosphate from the 'routine' donor, enzyme I; thus, phosphorylation of HPr at this site is likely to inhibit uptake of sugars at the PTS permeases.) HPr kinase is a bifunctional enzyme (HPr kinase/phosphatase), and its role as a kinase in phosphorylating HPr under appropriate conditions of carbon availability is not fully understood; it is possible that the kinase function of the enzyme is stimulated by fructose 1,6-bisphosphate (an intermediate in glucose catabolism). Inactivation of the HPr kinase gene (*hprK*) in *Staphylococcus xylosus* resulted in abolition of repression in the three catabolic enzyme activities examined [JB (2000) *182* 1895–1902].

catalase (H_2O_2:H_2O_2 oxidoreductase; EC 1.11.1.6) A tetrameric ENZYME, consisting of four ferriprotoporphyrin IX-containing subunits, which catalyses the reaction $2H_2O_2 \rightarrow 2H_2O + O_2$; with low concentrations of H_2O_2 the enzyme can act as a PEROXIDASE – but this ability may not be common to catalases from all sources [JGM (1983) *129* 997–1004]. Catalase plays a vital role in detoxifying HYDROGEN PEROXIDE produced during aerobic metabolism; it occurs in the majority of aerobic organisms and is absent from most obligate anaerobes (though present e.g. in *Bacteroides fragilis*). The enzyme is inhibited by e.g. H_2S, HCN and N_3^-, and is inactivated by SUPEROXIDE. Catalase is obtained commercially from e.g. *Aspergillus niger* and *Penicillium* spp; it is used e.g. in the food industry for removing excess H_2O_2 (used for cold disinfection) from milk and dairy products and from irradiated foods. (See also CATALASE TEST; METHYLOTROPHY; PEROXISOMES; cf. PSEUDOCATALASE.)

catalase test A test commonly used with the object of determining whether or not a given bacterial strain produces CATALASE. One drop of H_2O_2 is added to a bacterial colony, or to a bacterial emulsion; if the cells contain catalase, bubbles of gas (oxygen) appear immediately or within one or two seconds. Such a (positive) reaction may indicate the presence of catalase or (e.g. in some strains of *Enterococcus faecalis*) the presence of a PSEUDOCATALASE; certain bacteria can synthesize catalase only if they are provided with haem. If, for the test, bacterial growth is obtained from a blood-containing medium, care should be taken to prevent contact between H_2O_2 and the medium: erythrocytes, which contain catalase, can give rise to a false-positive reaction.

In carrying out the catalase test care should be taken to avoid exposure to the AEROSOL produced by bursting bubbles; in one method the test is carried out in a non-vented Petri dish: a drop of H_2O_2 and a speck of bacterial growth are placed at opposite sides of the dish which, with the lid in place, is tilted so that the H_2O_2 runs onto the growth.

(See also SUPEROXOL TEST.)

cataloguing (of 16S rRNA) See RRNA OLIGONUCLEOTIDE CATALOGUING.

Catapyrenium A genus of LICHENS (order VERRUCARIALES) which differ from DERMATOCARPON spp in having a squamulose thallus attached to the substratum by rhizines. The genus includes e.g. *C. lachneum* (formerly called e.g. *Dermatocarpon hepaticum*).

catarrh Inflammation of a mucous membrane, with excessive production of mucus.

catecholamides (in iron uptake) See SIDEROPHORES.

catenane A complex of two or more circular nucleic acid molecules interlocked like links in a chain. (cf. CATENATE; CONCATEMER; see also GYRASE.)

Catenaria See BLASTOCLADIALES.

catenate (1) (*verb*) To interlock circular nucleic acid molecules to form a CATENANE (hence *noun* catenation). (2) (*adj.*) Of e.g. spores: formed in chains.

catenins See CADHERINS.

catheter-associated infection INFECTION (sense 3) associated with catheterization of the blood vascular system or of the urinary tract; in general, the risk of such infection increases with the duration of catheterization. Catheter-related infections account for a significant proportion of nosocomial disease.

In catheter-associated bacteraemia, the source of infection can be investigated either with or without withdrawal of the catheter [current methods for the diagnosis of catheter-related bacteraemia: RMM (1997) *8* 189–195].

In catheter-associated infection of the urinary tract, a significant reduction in incidence during short-term (<3 weeks) catheterization may be obtained by using catheters that have anti-infection surface activity; such catheters may be either (i) impregnated with e.g. nitrofurazone, or with minocycline and rifampicin, or (ii) coated with a silver alloy–hydrogel [EID (2001) *7* 342–347].

cathodic depolarization theory A theory proposed to account for the microbial corrosion of iron. In this theory, the spontaneous solubilization of iron (by ionization: $Fe \rightarrow Fe^{2+} + 2e^-$) is promoted by (a) hydrogen formation, the electrons from iron combining with protons derived from water ionization, and (b) utilization of the hydrogen by SULPHATE-REDUCING BACTERIA. However, it has been reported that the primary cause of such corrosion may involve an extracellular (probably phosphorus-containing) product of sulphate-reducing bacteria [Book ref. 155, pp. 619–641].

cattle diseases Cattle may be affected by a wide range of diseases, some of which are specific to bovines; different species and breeds of bovines may show differences in susceptibility to a given disease (see e.g. FOOT-ROT). (a) *Bacterial diseases*: see e.g. ACTINOMYCOSIS; ANAPLASMOSIS; ANTHRAX; BLACK DISEASE; BLACKLEG; BRUCELLOSIS; CAMPYLOBACTERIOSIS; CONTAGIOUS BOVINE PLEUROPNEUMONIA; EPERYTHROZOONOSIS; FARCY (sense 2); FOOT-ROT; HAEMORRHAGIC SEPTICEMIA (sense 2); INFECTIOUS KERATITIS; JOHNE'S DISEASE; LEPTOSPIROSIS; LISTERIOSIS; MASTITIS; NECROBACILLOSIS; PARATYPHOID FEVER; RED WATER (of calves); TUBERCULOSIS; TULARAEMIA; ULCERATIVE LYMPHANGITIS; WHITE SCOURS. (See also COWDRIA; LAMZIEKTE; MIDLAND CATTLE DISEASE.) (b) *Fungal diseases*: see e.g. ASPERGILLOSIS; ZYGOMYCOSIS. (See also FESCUE FOOT; 4-IPOMEANOL; PASPALUM STAGGERS; PENITREMS; SPORIDESMINS.) (c) *Protozoal diseases*: see e.g. DALMENY DISEASE; EAST COAST FEVER; NAGANA; REDWATER FEVER. (d) *Viral diseases*: see e.g. AKABANE VIRUS DISEASE; ALLERTON DISEASE; AUJESZKY'S DISEASE; BOVINE EPHEMERAL FEVER; BOVINE ULCERATIVE MAMMILLITIS; BOVINE VIRAL LEUKOSIS; BOVINE VIRUS DIARRHOEA; CALF PNEUMONIA; FOOT AND MOUTH DISEASE; INFECTIOUS BOVINE RHINOTRACHEITIS; JEMBRANA DISEASE; LUMPY SKIN DISEASE; MALIGNANT CATARRHAL FEVER; MUCOSAL DISEASE; PSEUDOCOWPOX; RABIES; RIFT VALLEY FEVER; RINDERPEST; ROTAVIRUS; VESICULAR STOMATITIS. (e) *Diseases of unknown/variable aetiology*: ATYPICAL INTERSTITIAL PNEUMONIA; BOVINE RESPIRATORY DISEASE; CALF SCOURS; WIMMERA GRASS POISONING; WINTER DYSENTERY.

(See also BOVINE SPONGIFORM ENCEPHALOPATHY.)

cattle plague *Syn.* RINDERPEST.

cattle tick fever *Syn.* REDWATER FEVER.

Caulerpa A genus of siphonaceous, tropical and subtropical green seaweeds related to HALIMEDA (q.v.). The thallus consists of a horizontal green rhizome with colourless rhizoids and erect fronds. (See also ELYSIA.)

caulescent Becoming stalked or tailed.

cauliflower fungus *Sparassis crispa.*

cauliflower mosaic virus (CaMV; cabbage B virus) A member of the CAULIMOVIRUSES which infects various brassicas. In e.g. cauliflower (*Brassica oleracea* var. *botrytis*) symptoms of infection typically include vein-banding and/or vein-clearing; enations are occasionally formed. Transmission: chiefly stylet-borne via aphids.

A CaMV virion consists of a capsid (composed of a single type of protein) and the dsDNA genome. In the virion, the dsDNA is circular with three discontinuities at specific sites in the molecule: two in one strand (the β strand), one in the other (α strand); these discontinuities are neither nicks nor gaps, but are sites of triple-stranded DNA in which the $3'$ and $5'$ ends overlap by ca. 8–20 nucleotides [FEBS (1981) *134* 67–70]. In the host plant, CaMV DNA appears to occur in the form of covalently closed circular, supercoiled, histone-associated (chromatin-like) 'mini-chromosomes' which are apparently active in transcription but not in replication. Only the α-strand is transcribed. Viral RNAs (which are capped and polyadenylated) include a large (8.2 kb, 35S) transcript corresponding to the length of the entire genome with an additional 180-nucleotide terminal repeat; this transcript may be an intermediate in DNA replication: a molecule of plant $tRNA^{met}$ binds to the 35S transcript at a 14-bp region of homology, and this tRNA may act as a primer for cDNA synthesis by reverse transcription (see RETROID VIRUSES). [Evidence for replicative recombination in CaMV: Virol. (1986) *150* 463–468.] Viral transcription occurs in the plant cell nucleus, while virus particle assembly occurs in rounded, amorphous, electron-dense, cytoplasmic viroplasms; the site of DNA replication is unknown.

CaMV is widely studied as a model for transcription, DNA replication etc in higher plants, and as a potential (non-integrating) vector for gene transfer in plants.

[Book ref. 80, pp. 17–48.]

caulimoviruses (cauliflower mosaic virus group) A group of PLANT VIRUSES characterized by a dsDNA genome (ca. 8000 bp long). The virions are isometric, ca. 50 nm diam.; the capsid is composed of a single type of polypeptide (MWt ca. 42×10^3). Type member: CAULIFLOWER MOSAIC VIRUS; other members: carnation etched ring virus, dahlia mosaic virus, figwort mosaic virus, *Mirabilis* mosaic virus, strawberry vein-banding virus; possible members: cassava vein mosaic virus, *Petunia* vein-clearing virus, *Plantago* virus, thistle mottle virus. (See also NON-CIRCULATIVE TRANSMISSION sense 1.)

Caulobacter A genus of chemoorganotrophic, strictly aerobic, PROSTHECATE BACTERIA found e.g. in oligotrophic waters and in soils poor in organic matter. Type species: *C. vibrioides*. GC%: 64–67.

Most of the studies on this genus have been carried out on *C. crescentus* – to which the following account applies.

The non-motile, mature ('mother') cell is a straight or curved rod, ca. 1–2 μm, which typically has a single polar stalk (PROSTHECA) – the distal end of which is adhesive; cells may adhere, singly, to the substratum – or may occur in groups ('rosettes'), each group consisting of a number of cells radiating from a central mass of adhesive material. Many strains contain carotenoid pigments.

The mother cell gives rise, by asymmetric binary fission, to a daughter cell (*swarmer*) which bears a single polar flagellum. The swarmer subsequently loses its flagellum, develops a prostheca, and can then function as a mother cell. In stained mother cells the prostheca may show one or more cross-striations (*crossbands*), each of which is thought to mark one turn of the cell cycle.

The molecular mechanisms involved in the process of differentiation and asymmetric cell division (essential features of the cell cycle) are beginning to be understood through studies on the intracellular localization of proteins using techniques such as gene fusion. In some cases the targeting, or activation, of a specific protein has been linked to an essential aspect of differentiation or cell division.

Cell cycle regulation involves various TWO-COMPONENT REGULATORY SYSTEMS. One response regulator, CtrA, has an important role in the control of initiation of DNA replication and in the expression of many genes [PNAS (2002) *99* 4632–4637].

Following initiation of DNA replication in the mother cell, levels of the phosphorylated form of CtrA (CtrA~P) increase, CtrA~P (and/or CtrA) then promoting transcription of certain genes – including 'early' flagellar genes and those regulatory genes needed for expression of the later flagellar genes. The regulation of flagellar genes may be more complex than was previously supposed [PNAS (2002) *99* 4632–4637].

The FliF protein is targeted to a specific polar site in the developing daughter (swarmer) cell, i.e. the MS ring in the swarmer cell develops prior to septation; the mechanism responsible for directing FliF to the appropriate location is unknown.

After formation of the septum, the FlbD protein is activated by phosphorylation – the kinase, FlbE, occurring near the septum in the *daughter* cell; the activated FlbD protein (FlbD~P) promotes transcription of 'late' flagellar genes – resulting in the development of a polar flagellum in the daughter cell.

In the daughter cell, CtrA~P binds to the chromosomal origin, thus inhibiting replication; hence, unlike the mother (stalked) cell, the swarmer does not divide until it matures.

Under normal circumstances, the role of CtrA (i.e. the unphosphorylated form) in the mother cell includes activation of the late cell division genes *ftsQ* and *ftsA* – whose expression is necessary for cell division. If DNA replication is inhibited, then CtrA is not synthesized (thus blocking cell division) – lack of synthesis under these conditions reflecting failure of activation of *ctrA*; as activation of *ctrA* is normally dependent on CtrA~P, it has been suggested that inhibition of DNA replication may inhibit phosphorylation of CtrA – constituting a cell cycle 'checkpoint' which blocks cell division in the absence of DNA replication [EMBO (2000) *19* 4503–4512].

[Control of differentiation in *C. crescentus*: Science (1997) *276* 712–718. Signal transduction mechanisms in *C. crescentus* development and cell cycle control: FEMS Reviews (2000) *24* 177–191.]

Caulococcus See KUSNEZOVIA.

caviid herpesvirus 1 See BETAHERPESVIRINAE.

cavitation (1) (*med.*) The formation of an (abnormal) cavity or space in the body (as e.g. in certain types of pneumonia). (2) See ULTRASONICATION.

cavity slide A slide (somewhat thicker than a normal microscope SLIDE) which has a shallow, circular depression about 1 cm across in the centre of one face. (cf. RING SLIDE.)

Cavostelium See PROTOSTELIOMYCETES.

CBP Cap-binding protein: see MRNA (b).

CBS Centraalbureau voor Schimmelcultures, Oosterstraat 1, Baarn, The Netherlands; yeast division: CBS, Julianalaan 67A, Delft, The Netherlands.

***cca* gene** See TRNA.

cccDNA Covalently-closed circular DNA.

CCCP See PROTON TRANSLOCATORS.

CCCV COCONUT CADANG-CADANG VIROID.

***ccd* mechanism** See e.g. F PLASMID.

CcdB toxin See F PLASMID.

CCFAS (compact colony-forming active substance) A heat-stable, acid-labile carbohydrate, present in the cell walls of certain staphylococci, responsible for the compact morphology of colonies growing in serum soft agar. (Strains without CCFAS form diffuse colonies in such media.) CCFAS exerts its effect by interacting with fibrinogen in the serum agar, promoting the formation of fibrin. It is distinct from, but has been confused with, clumping factor (see COAGULASE). [Book ref. 44, pp. 551–552.]

CCM Czechoslovak Collection of Microorganisms, J. E. Purkyne University of Brno, Tr. Obrancu Miru 10, 66243 Brno, Czechoslovakia.

CCNU See *N*-NITROSO COMPOUNDS.

CcpA See CATABOLITE REPRESSION.

CCPs Critical control points: see HACCP.

CCY medium A medium containing casein hydrolysate and yeast extract; cultures on this medium are commonly incubated under a CO_2-enriched atmosphere.

CD (meaning of) See CELL ADHESION MOLECULE.

CD2 See IMMUNOGLOBULIN SUPERFAMILY.

CD3 A T-cell-receptor-associated antigen consisting of three transmembrane polypeptides (γ, δ and ε) which occurs in all T cells (both $\alpha\beta$ and $\gamma\delta$ types) and which is part of the so-called *T-cell receptor complex*. CD3 is not covalently linked to the T cell receptor, and it does not bind antigen; however, when antigen binds to the T cell receptor, serine and tyrosine residues on the three polypeptides of CD3 undergo phosphorylation, and this initiates part of the intracellular signalling in the T cell which follows the binding of antigen.

CD4 A CELL ADHESION MOLECULE of the IMMUNOGLOBULIN SUPERFAMILY which occurs on some $\alpha\beta$ and $\gamma\delta$ T cells (see also CD8).

CD4 binds to MHC class II ligands on antigen-presenting cells (APCs) when antigen is bound by the T cell receptor; such binding stabilizes T cell–APC contact, and causes CD4 to act as a signal-transducing molecule – this involving phosphorylation of its intracellular domain by a specific kinase.

CD4 also acts as a receptor for the human immunodeficiency virus (HIV), allowing infection of $CD4^+$ cells.

CD5 See B LYMPHOCYTE.

CD8 A CELL ADHESION MOLECULE of the IMMUNOGLOBULIN SUPERFAMILY which occurs e.g. on some $\alpha\beta$ and some $\gamma\delta$ T cells (see also CD4). CD8 binds to MHC class I ligands; like CD4, bound CD8 acts as a signal-transducing molecule.

CD11a/CD18 (*syn.* LFA-1) A CELL ADHESION MOLECULE of the INTEGRIN family ($\alpha_L\beta_2$) found on most leukocytes; when activated, it binds strongly to ICAM ligands (see e.g. INFLAMMATION). (CD11a refers to the α subunit, CD18 to the β subunit.)

CD11b/CD18 (*syn.* Mac-1) See INTEGRINS.

CD11c/CD18 The CR4 receptor on macrophages; it binds e.g. the C3bi fragment of complement.

CD14 A protein which occurs either in plasma or membrane-associated on macrophages and fibroblasts. Cell-bound CD14 binds exogenous LIPOPOLYSACCHARIDES (LPS); such binding (indirectly) stimulates the synthesis of cytokines – including pro-inflammatory (e.g. IL-1, TNF) and anti-inflammatory (e.g. IL-6, IL-10) molecules. The activity of LPS is greatly increased by the ACUTE-PHASE PROTEIN *lipopolysaccharide-binding protein* (LBP) which promotes the transfer of LPS to CD14.

CD15 (sialyl-Lewis x) A carbohydrate CELL ADHESION MOLECULE, found on leukocytes, which binds to ligands of the SELECTIN family (see e.g. INFLAMMATION).

CD16 (*syn.* FcγRIII) A CELL ADHESION MOLECULE, found on e.g. PMNs and NK cells, which binds to the Fc portion of antigen-bound antibody and is involved e.g. in ADCC.

CD18 See CELL ADHESION MOLECULE and INTEGRINS.

CD21 A CELL ADHESION MOLECULE which occurs e.g. on epithelial cells of the oropharynx; it acts as a receptor for Epstein–Barr virus. CD21 (= CR2) is also found on B lymphocytes, and is involved e.g. in antigen-dependent activation.

CD23 See EPSTEIN–BARR VIRUS and B LYMPHOCYTE.

CD28 A cell-surface molecule which occurs on T lymphocytes. During presentation of antigen by an ANTIGEN-PRESENTING CELL to a *naive* $CD4^+$ T cell, so-called *co-stimulatory signals* – which promote clonal proliferation of the T cell and secretion of cytokines – are generated when CD28 binds to certain molecules on the APC – e.g. B7.1 (also written B7-1; = CD80) or B7.2 (B7-2; CD86). Co-stimulatory signals generated by such binding (which are necessary for T cell development) lead e.g. to the synthesis of INTERLEUKIN-2 by the T cell.

During presentation of antigen to an *activated* T cell, CD80/CD86 binds preferentially to the cell-surface molecule CD152 (found only on activated T cells) – the affinity of CD80/CD86 for CD152 being greater than that for CD28. (CD152 is also called *cytokine T-lymphocyte-associated antigen-4*, or CTLA-4.) Unlike the binding between CD80/CD86 and CD28, binding between CD80/CD86 and CD152 generates a signal that is inhibitory to the T cell's response, e.g. it inhibits the synthesis of IL-2 in the lymphocyte.

(See also CD40.)

CD31 (*syn.* PECAM-1) A CELL ADHESION MOLECULE of the IMMUNOGLOBULIN SUPERFAMILY which occurs e.g. on endothelial cells, monocytes and platelets. Binding, apparently homophilic (i.e. CD31–CD31), is believed to be necessary for extravasation in INFLAMMATION.

CD32 See B LYMPHOCYTE.

CD34 A marker present on haemopoietic stem cells but not on mature haemopoietic cells.

CD35 Complement receptor 1 (CR1): a glycoprotein present e.g. in the membranes of erythrocytes (red blood cells, RBCs) and of various nucleated blood cells, including B cells and some T cells; plasma contains low levels of a soluble form of CR1 (sCR1).

CR1 belongs to a family of COMPLEMENT control proteins. The molecules of CR1 on RBCs bind immune complexes, and these complexes are removed as the RBCs circulate through the liver and spleen; CR1 promotes cleavage of the C3b and C4b components of complement, thus protecting the cell from complement-mediated lysis.

The Sl^a antigen of the CR1 glycoprotein (an antigen of the Knops blood group system) is associated with the phenomenon of 'rosetting' in falciparum MALARIA.

CD36 See MALARIA (vaccines).

CD40 A cell-surface molecule which occurs on B lymphocytes. During recognition of a thymus-dependent (TD) antigen by antigen-specific B and T cells, CD40 binds to the T cell's CD40L (gp39) antigen; such binding generates part of the stimulus which results in clonal proliferation of the B cell and class switching of antibody (IgM → IgG etc.). (The physical interaction between T and B cells also generates signals that activate clonal proliferation in the T cell: see CD28.)

(See also RNA EDITING.)

CD41b See INTEGRINS.

CD44 (*syn.* homing-associated cell adhesion molecule, HCAM) A CELL ADHESION MOLECULE found e.g. on monocytes, neutrophils, fibroblasts and memory T cells; it is apparently unrelated to other CAMs. CD44 binds to e.g. collagen and hyaluronate, and is apparently involved in 'lymphocyte homing' (i.e. translocation of lymphocytes between blood and lymph vessels).

CD46 (membrane co-factor protein, MCP) A cell-surface molecule on macrophages which acts as a receptor for COMPLEMENT component C3b. (CD46 can also act as a receptor for the measles virus; the binding of measles virus to CD46 inhibits the ability of the macrophage to form INTERLEUKIN-12.)

CD48 A macrophage cell-surface glycoprotein which can bind the type I FIMBRIAE of *Escherichia coli*. In the absence of opsonizing factors, cells of *E. coli* that have type I fimbriae can bind to CD48, initiating uptake into a phagosome which is not acidified in the normal way and in which *E. coli* can survive.

CD50 See IMMUNOGLOBULIN SUPERFAMILY.

CD54 (*syn.* ICAM-1) A CELL ADHESION MOLECULE of the IMMUNOGLOBULIN SUPERFAMILY which occurs e.g. on endothelial and epithelial cells, monocytes and fibroblasts. CD54 binds to certain integrins (e.g. LFA-1, Mac-1). (See also INFLAMMATION.)

CD55 Decay-accelerating factor (DAF): a glycoprotein present in the membranes of blood cells (red and white), endothelial cells, and epithelial cells of the urinary and gastrointestinal tracts; DAF accelerates decay of the C3 and C5 convertases in the classical and alternative pathways of COMPLEMENT FIXATION, thus helping to protect cells from complement-mediated lysis.

The DAF glycoprotein bears antigens of the Cromer blood group system, and also acts as a receptor for *Escherichia coli* and enteroviruses.

CD56 (*syn.* NCAM) See IMMUNOGLOBULIN SUPERFAMILY.

CD58 See IMMUNOGLOBULIN SUPERFAMILY.

CD61 See INTEGRINS.

CD62E (*syn.* E-selectin; also known as endothelium–leukocyte adhesion molecule-1, ELAM-1) A CELL ADHESION MOLECULE of the SELECTIN family which is inducible, on endothelial cells, by certain cytokines (e.g. IL-1β, TNF-α); it binds to sialylated carbohydrate ligands on e.g. monocytes, neutrophils and $CD4^+$ T cells. CD62E is involved in the tethering stage of INFLAMMATION.

CD62L See SELECTINS.

CD62P See SELECTINS.

CD66 A category of cell adhesion molecules (of the IMMUNOGLOBULIN SUPERFAMILY), some of which occur on polymorphonuclear leukocytes (PMNs) and/or e.g. epithelial cells. Specific molecule(s) – CD66a, c, d and/or e (not b) – can bind neisserial OPA PROTEINS and, in at least some cases, mediate uptake of bacteria by the eukaryotic cell.

CD69 An antigen detectable on immature T lymphocytes (thymocytes) during the process of maturation in vivo.

CD77 antigen A B-lymphocyte differentiation antigen which is structurally equivalent to the globotriosylceramide (Gb_3) receptor for SHIGA TOXIN and the shiga-like toxins. These toxins selectively kill B cells expressing IgG or IgA – so that the immune response to infection may be actively suppressed by toxin.

CD80 See CD28.

CD81 A protein of the tetraspanin family which is found in the cytoplasmic membrane of cells in many human tissues; CD81 acts as a receptor for the E2 envelope glycoprotein of the hepatitis C virus. [Three-dimesional structure of the extracellular domain of CD81: EMBO (2001) *20* 12–18.]

CD86 See CD28.

CD95 See FAS.

CD102 (ICAM-2) A CELL ADHESION MOLECULE of the IMMUNOGLOBULIN SUPERFAMILY found e.g. on endothelial cells, monocytes and some lymphocytes; it binds to the integrin LFA-1, and may be involved in lymphocyte homing.

CD106 (VCAM-1, vascular cell adhesion molecule-1) A CELL ADHESION MOLECULE of the IMMUNOGLOBULIN SUPERFAMILY found e.g. on endothelial cells, macrophages and fibroblasts; it binds to ligands of the integrin family, and is involved e.g. in the adhesion of leukocytes to the endothelium during the flattening stage of INFLAMMATION.

CD120a (p55) See TNF.

CD120b (p75) See TNF.

CD152 See CD28.

CDC Centers for Disease Control, a division of the US Public Health Service.

***cdc* mutant** See CELL CYCLE.

***cdl* gene** See COLICIN PLASMID.

cDNA (complementary DNA; copy DNA) A DNA molecule obtained by reverse transcription of an RNA molecule (commonly an mRNA molecule).

cDNAs prepared from *mature* mRNAs are useful e.g. for CLONING and expressing intron-containing eukaryotic genes; if the genes themselves were cloned in bacteria, the presence of intron(s) would prevent their expression. (Introns are removed during the formation of mature mRNA – see SPLIT GENE.)

Most eukaryotic mRNAs [and some bacterial mRNAs: Microbiology (1996) *142* 3125–3133] have a polyadenylate (poly(A)) tail at the 3′ end (see MRNA). To synthesize a cDNA molecule, a short oligodeoxythymidine (oligo(dT)) strand is hybridized to the poly(A) tail of the mRNA, and a REVERSE TRANSCRIPTASE, together with a mixture of dNTPs, is then added; using oligo(dT) as a primer, the reverse transcriptase synthesizes a DNA strand on the mRNA template. The mRNA strand can then be removed e.g. by alkaline hydrolysis or by RNase H. The DNA strand synthesized by reverse transcriptase often folds over at the 3′ end to form a HAIRPIN structure, thus providing a primer for the synthesis of a complementary DNA strand by an appropriate DNA polymerase; following such synthesis, the hairpin structure can be opened by endonuclease S1 to form a linear dsDNA molecule which is suitable for cloning.

CDP Cytidine 5′-diphosphate.

CDR factor See METHANOGENESIS.

CDSC Communicable Diseases Surveillance Centre, Public Health Laboratory Service, Colindale, London, UK.

CDTA A chelating agent, *trans*-1,2-diamino-cyclohexane-*N*,*N*, *N*′,*N*′-tetraacetic acid. CDTA complexes a range of cations (e.g. Ba^{2+}, Ca^{2+}, Mg^{2+}, Zn^{2+}); complexes formed with either Ca^{2+} or Mg^{2+} are more stable than those formed by EDTA or EGTA with these cations.

***cea* gene** See COLICIN PLASMID.

cecidia See GALLS.

cecidization GALL-formation.

cecropins A family of PEPTIDE ANTIBIOTICS; cecropins appear in the haemolymph of the Cecropia moth (*Hyalophora cecropia*), and certain other insects, following exposure to bacteria. (cf. SARCOTOXINS.)

cedar apples See GYMNOSPORANGIUM.

Cedecea A genus of bacteria (family ENTEROBACTERIACEAE) isolated from human clinical specimens (pathogenicity unknown). [Book ref. 22, pp. 514–515.]

cefamandole See CEPHALOSPORINS.

cefazolin See CEPHALOSPORINS.

cefepime A fourth-generation CEPHALOSPORIN. Resistance to cefepime has been associated with diminished permeability of the outer membrane in *Enterobacter aerogenes* [Microbiology (1998) *144* 3003–3009] and with hyperproduction of the β-lactamase SHV-5 in *Klebsiella pneumoniae* [JCM (1998) *36* 266–268].

cefixime A third-generation CEPHALOSPORIN. It has been used e.g. (together with potassium tellurite) as a supplement to SMAC medium (see EHEC); this supplement has been found to increase the rate of isolation of *Escherichia coli* strain O157 by suppressing the growth of other non-sorbitol-fermenters (e.g. strains of *Proteus*).

cefmenoxime See CEPHALOSPORINS.

cefonicid See CEPHALOSPORINS.

ceforanide See CEPHALOSPORINS.

cefotaxime See CEPHALOSPORINS.

cefoxitin A CEPHAMYCIN which, in e.g. *Escherichia coli*, binds preferentially to PBP-1A (see PENICILLIN-BINDING PROTEINS) and moderately well to other PBPs (but not to PBP-2).

Cefoxitin is active against a range of Gram-negative and Gram-positive bacteria, including anaerobes such as *Clostridium* and members of the *Bacteroides fragilis* group. It is resistant to many β-LACTAMASES, including extended-spectrum β-lactamases, but may be sensitive to metalloenzyme β-lactamases produced by certain strains of anaerobic bacteria.

Cefoxitin is an inducer of β-lactamases, so that it should not be used in combination with another β-lactam antibiotic.

Cefoxitin has been useful e.g. as a systemic prophylactic agent in both gastrointestinal and gynaecological surgery; when combined with doxycycline it has been used for the treatment of pelvic inflammatory disease, while a combination of cefoxitin and an aminoglycoside antibiotic has been useful for treating post-operative infections.

[Cefoxitin (review): RMM (1995) *6* 146–153.]

cefpirome See CEPHALOSPORINS.

cefsulodin See CEPHALOSPORINS.

cefsulodin–Irgasan–novobiocin agar See CIN AGAR.

ceftazidime See CEPHALOSPORINS.

ceftizoxime See CEPHALOSPORINS.

ceftriaxone See CEPHALOSPORINS.

cefuroxime See CEPHALOSPORINS.

celecoxib See PROSTAGLANDINS.

celery crown rot See CROWN ROT.

celery mosaic virus See POTYVIRUSES.

celery yellow mosaic virus See POTYVIRUSES.

celery yellow spot virus See LUTEOVIRUSES.

celery yellows A YELLOWS disease of celery which may be caused e.g. by an MLO or by the fungus *Fusarium oxysporum*; in the latter case the main symptoms (stunting and yellowing) are due to vascular occlusion by the pathogen.

celesticetin See LINCOSAMIDES.

cell adhesion molecule (CAM) In man and other animals: a membrane-associated molecule characteristically involved in cell–cell or cell–matrix adhesion and (often) in the signalling between a cell and its environment. In the latter role, incoming signals modulate various signalling pathways within the cell; signals transmitted via CAMs influence various aspects of cell physiology, including the cell cycle.

Most CAMs are proteins which consist of (i) a hydrophobic transmembrane domain; (ii) an extracellular domain – which can bind to a *counter-receptor* (a different type of molecule on another cell) or to another extracellular ligand; and (iii) an intracellular domain (often linked to the cytoskeleton of the cell). A CAM may be normally present in an inactive form; under appropriate conditions, activation may confer the ability e.g. to bind a counter-receptor. The extracellular domain of a CAM may act e.g. as a receptor for invasive bacterial pathogens (see CADHERINS, INTEGRINS), or as a means of cell migration during INFLAMMATION.

In a given cell, the type and number of CAMs depend on factors such as stage of development, function (or change in function), and the presence or absence of viral infection.

A CAM may be referred to by a 'family' name (see cross-references, below) or by a CD ('cluster of differentiation') number. A given CD number (e.g. CD54) signifies a specific antigen which has been identified by various batches of monoclonal antibodies from different laboratories. An integrin molecule may have *two* CD numbers (e.g. LFA-1 = CD11a/CD18); one number refers to the α subunit, the other to the β subunit.

See: CADHERINS, IMMUNOGLOBULIN SUPERFAMILY, INTEGRINS, SELECTINS and CD44; see also separate entries under 'CD'.

cell-bound antibodies Antibodies bound by their combining sites to homologous cell-surface antigens, or bound by other sites (see e.g. CYTOPHILIC ANTIBODIES) to cell-surface receptors.

cell culture See TISSUE CULTURE.

cell cycle In prokaryotic or eukaryotic microorganisms: the sequence of events which occur in a growing cell during the period between the cell's formation from its parent cell and the division of the cell itself into daughter cells; this period of time includes (i) initiation of DNA replication from at least one chromosomal origin, (ii) the replication of chromosome(s), (iii) the segregation of chromosomes, (iv) the formation of a septum or intercell partition, and (v) cell division. These events always occur strictly in the same order. The events which constitute the cell cycle are regulated by both extracellular (environmental) and cell-associated factors.

One event regarded as an important juncture in the cell cycle is the initiation of DNA replication; earlier workers proposed two main hypotheses to account for the regulation of this stage of the cell cycle.

In the 'G_1 event model' (devised primarily for eukaryotic cells), triggering of DNA replication, and subsequent cell division, is dependent on event(s) which occur only during a specific part of the cell cycle (the G_1 phase – see below).

In the 'continuum model' (postulated to refer to both prokaryotic and eukaryotic cells), a hypothetical initiator of DNA replication is produced by the cell during all phases of the cell cycle; initiation of DNA replication occurs when the concentration of initiator, per origin of replication in the cell, reaches a certain critical level – the concentration of initiator (per origin) then falling sharply owing to the abrupt doubling of numbers of available origins at the start of replication. (It follows from this model that a cell can govern its rate of division by regulating the rate of synthesis of the initiator.)

In both models, initiation of DNA replication is a key event which occurs late in an empirically defined phase that follows cell division and precedes the next round of DNA replication. Once DNA replication has begun the cell is normally committed to continue replication to completion; the point at which a cell becomes committed to DNA replication has been called the *restriction point* (= R point, or R) or, simply, *start*.

(a) *The eukaryotic cell cycle*. The typical cell cycle in eukaryotes consists of the following sequence of phases:

S phase. DNA replication.

G_2 phase. Gap phase 2: the period between the end of DNA replication and the start of mitosis.

M phase. MITOSIS and cell division.

G_1 phase. Gap phase 1: the period between cell division and the start of DNA replication.

The total period ($G_1 + S + G_2$) is called *interphase*. (In some lower eukaryotes, e.g. *Physarum*, the cell cycle apparently lacks a G_1 phase.)

Non-cycling ('resting') cells are said to be in the *G_0 phase*.

A given phase in the cell cycle is characterized e.g. by the cell's PLOIDY; thus, in a 'diploid' cell, the ploidy is 2n in G_1, 2n–4n in S, and 4n in G_2.

Typically, when variation occurs in the rate of cell growth and division (due e.g. to nutritional factors) it is the length of G_1 which varies, i.e. the other two phases remain relatively constant in length over a wide range of growth rates.

Regulation of the eukaryotic cell cycle. The cell cycle can be regulated by signals from outside the cell; thus, cells exposed to e.g. certain growth-inhibitory factors or sex pheromones become arrested in the G_1 phase. (Those cells which, at the time of exposure, are in the S, G_2 or M phase complete their cycling to the G_1 phase.)

To investigate regulatory mechanisms within the cell, studies have been carried out e.g. on the yeasts *Saccharomyces cerevisiae* and *Schizosaccharomyces pombe*. Particularly useful have been the cell-division-cycle mutants (*cdc* mutants) in which the cell cycle is arrested at specific stage(s). The products of some *cdc* genes have been found to influence more than one point in the cycle; for example, the *cdc2* gene product of *S. pombe* (Cdc2) affects the cycle in late G_1 and also in late G_2.

What kind of internal regulatory mechanisms are involved? At certain points in the cell cycle there are 'gates' or 'checkpoints' at which the cell makes a decision to either continue through the cycle or stop cycling; one such checkpoint is *start* (mentioned earlier). Continuation through *start* depends e.g. on the nutrient status of the cell's environment, i.e. the cycle is arrested at the *start* checkpoint if nutrient levels are inadequate.

The effector molecules involved in regulating 'continue/stop' decisions are certain types of protein. They include PROTEIN KINASES and PROTEIN PHOSPHATASES; a cyclical element is contributed by the CYCLINS (q.v.) which accumulate intracellularly, prior to checkpoints, and are cleaved on passage through the checkpoints.

(b) *The prokaryotic cell cycle*. In prokaryotes, most studies on the cell cycle have been carried out on bacteria – particularly *Escherichia coli*; the following account is based primarily on the cell cycle in *E. coli*. [Cell cycle in the Archaea: Mol. Microbiol. (1998) *29* 955–961.]

In bacteria, the cell cycle typically involves a period of growth followed by division into two similar or identical daughter cells (BINARY FISSION). (An atypical cell cycle occurs e.g. in CAULOBACTER (q.v.).) The daughter cells receive equal numbers of chromosome(s). Chromosome replication and cell division are related in a manner described by the HELMSTETTER–COOPER MODEL (q.v.).

The cell's dimensions during growth. Cells that are growing rapidly are larger (in both mass and volume) than those growing more slowly. If the doubling time is less than ~1 hour, the *C* and *D* periods (see HELMSTETTER–COOPER MODEL) are more or less constant; at the beginning of each *C* period the cell's mass (= *initiation mass*) is also virtually constant for a *given rate of growth*. Under these conditions, it follows that the faster a cell grows the greater will be the increase in cell mass and volume during the (constant) $C + D$ interval prior to cell division.

Cells of *E. coli* growing at a constant rate increase in size mainly by elongation (the diameter of the cells remaining largely unchanged). If the growth rate is increased, newly formed cells are larger than those formed at the lower rate. *During* up-regulation of the growth rate, the cell's diameter increases slowly, but the increase in cell length occurs more rapidly and, in fact, initially 'overshoots' – cells being longer than appropriate for the new, higher rate of growth; during ongoing growth at the new rate, the cell diameter increases to its new value while the cell length decreases to its final value (which is still higher than that at the lower rate of growth). If growth rate falls, the cell's diameter, as before, adjusts more slowly than the cell's length – the latter exhibiting an initial 'undershoot' [JGM (1993) *139* 2711–2714].

In *Bacillus subtilis*, determination of the rod shape appears to involve helical, actin-like filaments within the cell. Filaments composed of MreB protein may influence the cell's width. Those of Mbl protein influence longitudinal growth; they seem to guide the cell-wall-synthesizing machinery. These types of protein are found in many species, including *Escherichia coli*. Some bacteria (e.g. *Corynebacterium glutamicum*) lack these proteins; in these bacteria the rod shape may be maintained through *polar* growth. [Bacterial morphogenesis: Cell (2003) *113* 767–776.]

Synthesis of the cell envelope. During steady growth, cell envelope material is presumably being synthesized more or less continually throughout the cell cycle. Studies on the turnover of PEPTIDOGLYCAN in growing cells of the (Gram-negative) bacillus *E. coli* suggest that the lateral wall of the cell contains a monolayer of peptidoglycan and that multilayered peptidoglycan may occur at the poles [JB (1993) *175* 7–11]. In Gram-positive bacteria, wall growth is believed to follow the INSIDE-TO-OUTSIDE MODEL.

Höltje [Microbiology (1996) *142* 1911–1918] described a model for the mode of incorporation of newly synthesized peptidoglycan in the growing sacculus: see THREE-FOR-ONE MODEL.

It has been assumed that synthesis of the CYTOPLASMIC MEMBRANE proceeds passively (secondarily to the growing peptidoglycan sacculus). However, it has been reported that peptidoglycan synthesis is dependent on the synthesis of membrane phospholipids – a finding which suggests a novel mechanism for the regulation of growth of the cell envelope [Microbiology (1996) *142* 2871–2877].

DNA replication. As mentioned, co-ordination between chromosome (DNA) replication and cell division is described by the HELMSTETTER–COOPER MODEL.

Regulation of the *initiation* of a round of DNA replication is not understood – although a link between the timing of initiation and growth rate can be seen in the Helmstetter–Cooper model by comparing the start of the *C* periods in slow-growing and rapidly growing cells.

Although the precise mechanism of initiation is unknown, several factors appear to be important. [Some similarities among the initiators of DNA replication in Bacteria, Archaea and Eucarya: FEMS Reviews (2003) *26* 533–554.]

One factor is the intracellular concentration of DnaA protein (or possibly an activated form of it); thus, replication is dependent on the initial binding of a number of DnaA molecules at specific sites (*DnaA boxes*) within the *oriC* region of the chromosome – such binding being followed by localized 'melting' (i.e. strand separation) in the DNA, a prerequisite for replication. The concentration of DnaA molecules within the cell is therefore critical because replication is delayed until the DnaA boxes have been occupied. Studies with in vitro replication systems have identified mutant DnaA proteins defective in opening the duplex at *oriC* [Mol. Microbiol. (2000) *35* 454–462].

The link between initiation and growth rate, mentioned earlier, possibly involves guanosine 5′-diphosphate 3′-diphosphate (= 'guanosine tetraphosphate'; ppGpp) – a small molecule whose concentration is inversely proportional to growth rate; the high levels of ppGpp found during slow growth can e.g. inhibit synthesis of DnaA protein (at the level of transcription), and this may result in delayed initiation.

Another factor apparently involved in initiation is the Fis protein (*fis* gene product). Like DnaA, Fis has binding sites in the *oriC* region; moreover, it appears to regulate genes which include those encoding DnaA and the β subunit of DNA polymerase III [JB (1996) *178* 6006–6012]. Fis binds at *oriC* in a cell-cycle-specific way [EMBO (1995) *14* 5833–5841], forming an 'initiation-preventive' complex [NAR (1996) *24* 3527–3532]. Nevertheless, the way in which Fis influences the timing of initiation is unknown.

Initiation may also be influenced by the degree of methylation in the *oriC* region – a certain level of methylation apparently being necessary for a functional *oriC*. It has been suggested that, after a round of replication, the *oriC* region in a daughter cell may be left temporarily hemi-methylated (i.e. methylated in only the template strand), and that the time needed to achieve a functional level of methylation may affect the timing of initiation.

Decatenation and partition of chromosomes. Newly synthesized chromosomes are *catenated* (i.e. interlocked, like the links of a chain); clearly, prior to partition, the chromosomes have to be separated, and this process is carried out by TOPOISOMERASES. (If decatenation is prevented, e.g. by mutation in an essential topoisomerase gene, the result may include the formation of filamentous cells, containing interlocked chromosomes, as well as anucleate cells which have no chromosome; this can occur e.g. in cells of *E. coli* which are mutant in the *parC* gene which encodes a subunit of topoisomerase IV.)

On separation, each chromosome forms a compact nucleoid; this may involve the binding of many molecules of HU protein [Mol. Microbiol. (1993) *7* 343–350].

The way in which nucleoids are partitioned to daughter cells is unknown. Various mechanisms have been suggested: (i) attachment to the cytoplasmic membrane – the nucleoids being segregated mechanically; (ii) mutual repulsion between nucleoids; (iii) the *extrusion–capture* model, in which extrusion of DNA from the 'replication factory' (replisome) provides the energy for partition [GD (2001) *15* 2031–2041].

In *Bacillus*, the *oriC* regions may be drawn apart *actively* [JB (1998) *180* 547–555]. In *B. subtilis* the Smc protein seems to contribute to nucleoid structure – hence promoting normal segregation. [Segregation (primarily in *B. subtilis*): Microbiology (2001) *147* 519–526.]

After the report of 'mitosis-like' partitioning of plasmid R1 in *E. coli* [EMBO (2002) *21* 3119–3127] it was suggested that actin-like filaments of MreB protein (a 'cytoskeletal' element associated with the maintenance of cell shape in rod-shaped bacteria) may mediate a similar process in the segregation of nucleoids; dysfunctional MreB inhibits chromosome segregation in *E. coli* [EMBO (2003) *22* 5283–5292].

Septum formation. In *E. coli*, a cross-wall (septum) eventually divides the growing cell into two daughter cells. This process starts with the formation of a ring-shaped structure composed of FtsZ molecules (encoded by *ftsZ*) circumferentially on the inner surface of the cytoplasmic membrane – mid-way along the length of the cell; this FtsZ ring ('Z ring') marks the plane of the forthcoming septum. (Expression of FtsZ may be influenced by ppGpp [JB (1998) *180* 1053–1062].) Binding between FtsZ and ZipA (a cytoplasmic membrane protein) is reported to be necessary for cell division [Cell (1997) *88* 175–185]. [Interaction between ZipA and a C-terminal fragment of FtsZ (X-ray crystallography): EMBO (2000) *19* 3179–3191.]

Assembly of the Z ring apparently begins at a *nucleation site* and proceeds, bidirectionally, around the circumference of the cell; development of nucleation sites themselves may involve the Era protein (product of the *era* gene; a GTPase of the Ras superfamily) which seems to be a key factor in the control of cell division [Mol. Microbiol. (1998) *29* 19–26].

During cell division the septum must form mid-way along the cell – not at alternative polar sites; the Z ring is guided by the *min* gene products. In one model for this guidance system, MinE molecules form a circumferential, annular band at mid-cell. This MinE ring oscillates, i.e. it moves alternately towards one pole and then the other; this causes the MinCD proteins to oscillate between opposite poles of the cell – so that the (inhibitory) influence of MinC on FtsZ polymerization is minimal at the mid-cell location, permitting assembly of the FtsZ ring at the mid-cell position [Science (2002) *298* 1942–1946]. (During oscillation of MinCD proteins, arrival/departure at the cell's poles involves binding/hydrolysis of ATP.) One report suggested that the oscillation of MinD may involve ATP-dependent polymerization of MinD into fibres [PNAS (2002) *99* 16776–16781].

A fluorescent image of the FtsZ ring has been obtained by means of gene fusion between *ftsZ* and the gene for green fluorescent protein [PNAS (1996) *93* 12998–13003].

An FtsZ ring has been reported in anucleate cells, indicating that division sites can develop even in the absence of a chromosome [Mol. Microbiol. (1998) *29* 491–503].

The Z ring also occurs in members of the Archaea; this suggests that the apparatus for cell division was similar in an ancestral prokaryote [Mol. Microbiol. (1996) *21* 313–319].

In the plane of the Z ring, peptidoglycan grows inwards to form the septum; this begins very early in the development of the division site [JB (2003) *185* 1125–1127].

Cell division. The septum is split into two layers, each of which forms the new pole (end) of a daughter cell; this process is carried out by the EnvA protein. The OUTER MEMBRANE grows inwards to complete the cell envelope.

The *morphogenes* encode those products, such as EnvA, FtsZ and the MinCDE proteins, which ensure that, during the cell cycle, proteins are synthesized at the right times and targeted to correct locations within the cell.

Control of the cell cycle. The occurrence of a given event in the cell cycle could depend directly on the occurrence of the previous event in the sequence. An alternative view is that the cell cycle is a sequence of independent events that are co-ordinated. Co-ordination was suggested [Mol. Microbiol. (1991) *5* 769–774] for reasons that include e.g. (i) The inhibition of certain cell cycle events does not necessarily inhibit others. Thus, cell division *can* occur, for example, without proper partitioning of the nucleoids – i.e. division can proceed in the absence of a 'signal' from the event that normally precedes division. (ii) There is no evidence for *direct* coupling between chromosome replication and cell division. However, there is evidence for indirect co-ordination: damage to DNA (which may inhibit chromosome replication) triggers the so-called SOS SYSTEM which, among other effects, inhibits cell division.

More recent studies on e.g. CAULOBACTER show that the cell cycle has *checkpoints*: stages at which, normally, a given event must be completed before the next can begin [see JB (2003) *185* 1128–1146 (1135–1136)]. (See also *essential* TWO-COMPONENT REGULATORY SYSTEMS.)

cell disruption The breaking of cells – usually for analysis of their components. Methods include: (a) Exposure to physical forces (see e.g. BRAUN MSK TISSUE DISINTEGRATOR, FRENCH PRESS, HUGHES PRESS, ULTRASONICATION). (b) Enzymic digestion of specific cell wall components – e.g. the digestion of PEPTIDOGLYCAN by LYSOZYME or LYSOSTAPHIN; cells thus weakened can be lysed osmotically. The morphology and structure of a cell influence the ease with which it is broken. Cells most resistant to mechanical disruption include yeasts, bacterial endospores, and Gram-positive cocci; Gram-negative bacteria are less resistant, and the least resistant include protozoa and other animal cells. Once cells are broken, their components become susceptible to the action of e.g. AUTOLYSINS. (See also CELL FRACTIONATION.)

cell envelope (*bacteriol.*) The CYTOPLASMIC MEMBRANE together with all layers external to it – including the CELL WALL and/or (where applicable) an S LAYER.

cell fractionation The separation of cell components. Fractionation is preceded by some form of CELL DISRUPTION. Methods of fractionation include various forms of CENTRIFUGATION, CHROMATOGRAPHY and ELECTROPHORESIS.

cell-free ice nuclei See ICE NUCLEATION BACTERIA.

cell fusion See e.g. HYBRIDOMA; see also PROTOPLAST FUSION.

cell line (*tissue culture*) The heterogeneous population of cells resulting from the first subculture of a PRIMARY CULTURE, or from subsequent serial passaging of the cells. (cf. CLONED LINE and ESTABLISHED CELL LINE.)

cell-mediated cytotoxicity Lysis of specific or non-specific target cells by certain types of lymphoid cell. Lysis by $CD8^+$ *cytotoxic* T cells is triggered by *specific* antigen present at the surface of the target cell, is subject to HISTOCOMPATIBILITY RESTRICTION (involving MHC class I antigens), and is independent of both antibody and complement (see T LYMPHOCYTE). Cytotoxicity mediated by $CD8^+$ T cells is important e.g. as a defence against viral infections (antigen-specific T cells can lyse virus-infected cells which bear virus-specific cell-surface antigens). Lysis of the target cell requires a single collision with a functional cytotoxic T cell – which can kill repeatedly.

For other types of cell-mediated cytotoxicity see ADCC and NK CELLS.

All types of cell-mediated cytotoxicity appear to require divalent cations (Ca^{2+}, Mg^{2+}).

cell-mediated hypersensitivity *Syn.* DELAYED HYPERSENSITIVITY.

cell-mediated immunity (CMI) Any form of immunity which can be transferred to a non-immune individual by the transfer of cells (including e.g. LYMPHOCYTES) but not by (cell-free) serum or plasma (see DELAYED HYPERSENSITIVITY), or in which cells are *directly* involved as effectors (see CELL-MEDIATED CYTOTOXICITY). (cf. HUMORAL IMMUNITY; see also AIDS.)

cell membrane *Syn.* CYTOPLASMIC MEMBRANE.

cell sap (1) The fluid in a plant cell vacuole. (2) *Syn.* CYTOSOL.

cell sorter See e.g. FACS.

cell wall In most algae, archaeans, bacteria and fungi: the structure which forms a (usually rigid) layer external to the CYTOPLASMIC MEMBRANE and which is responsible for the shape of the organism; it protects the protoplast from mechanical damage, osmotic lysis etc., and it may also serve as a permeability barrier to ANTIBIOTICS and other substances. Microbial cell walls differ greatly in structure and composition, according to type and species.

(a) *Algal cell walls.* Although a few algae (e.g. DUNALIELLA, PORPHYRIDIUM) lack cell walls, the majority have a cell wall composed of one or more layers of microfibrillar polysaccharide embedded in a matrix of amorphous polysaccharide (cf. CHLAMYDOMONAS); in many species a layer of mucilaginous polysaccharide also occurs external to the cell wall.

The most common microfibrillar component in algal cell walls is CELLULOSE, which occurs e.g. in members of the Phaeophyta and in many members of the Chlorophyta (e.g. *Chaetomorpha, Chlorella, Scenedesmus*), Rhodophyta and Xanthophyceae (e.g. *Botrydium, Vaucheria*). The proportion of cellulose varies from species to species but is often small compared with that of the matrix polymers. In some green algae (e.g. *Bryopsis* (gametothallus), *Caulerpa, Halimeda, Penicillus, Udotea*) and red algae (e.g. *Palmaria palmata, Porphyra umbilicalis*) XYLANS replace cellulose as the main microfibrillar component of the wall. A $(1 \rightarrow 4)$-β-linked MANNAN is the main structural polysaccharide in the cell walls of e.g. *Acetabularia* and *Codium* spp; a mannan also occurs as an external layer (cuticle) in the red algae *Porphyra umbilicalis* and *Bangia fuscopurpurea*. Xylomannan is a major component of *Prasiola japonica*.

Matrix polymers in algal cell walls are many and varied; they include e.g. AGAR, ALGINATE, CARRAGEENAN and FUCOIDIN.

The walls of some algae contain inorganic materials such as silica (see e.g. DIATOMS) and calcium carbonate (e.g. as a surface encrustation in the CHAROPHYTES); calcium carbonate occurs as calcite in some species, aragonite in others (e.g. HALIMEDA). Many unicellular algae of the CHRYSOPHYTES, PRYMNESIOPHYCEAE and 'PRASINOPHYCEAE' have a cell covering composed of scales, each consisting of an organic base in which silica or calcite may be deposited. In the Prymnesiophyceae and 'Prasinophyceae' these scales are synthesized in the Golgi apparatus and are released to the exterior by fusion of the Golgi vesicles with the cytoplasmic membrane. (See also THECA.)

In those algae which have a heteromorphic alternation of generations, the different phases may have different cell wall structures/compositions; for example, in *Derbesia marina/Halicystis ovalis*, the *Derbesia* phase has cell walls containing mannan, while the *Halicystis* phase has walls containing cellulose and glucoxylan. Gametes and spores may lack cell walls.

Many algal walls may be stained for light microscopy with e.g. ALCIAN BLUE (for acidic polysaccharides), PAS, or TOLUIDINE BLUE (which gives a pink colour with carboxyl and sulphate groups).

[Review: Book ref. 38, pp 278–332.]

(b) *Bacterial cell walls.* Although some species are wall-less (see e.g. MYCOPLASMA), a cell wall is present in the majority of bacteria; moreover, there may be layer(s) external to the cell wall proper – see e.g. CAPSULE, M PROTEIN, S LAYER.

Traditionally, bacterial cell walls have been classified into two major types: the Gram-positive type and the Gram-negative type (see also GRAM TYPE). However, while generally classified as Gram-negative organisms, the cyanobacteria commonly have cell walls that exhibit features of both Gram-negative and Gram-positive species [JB (2000) *182* 1191–1199].

The bacterial cell wall is a target for certain ANTIBIOTICS (see e.g. β-LACTAM ANTIBIOTICS and VANCOMYCIN).

Gram-positive-type walls are relatively thick (ca. 30–50 nm) and appear to be largely homogeneous by electron microscopy. The wall consists mainly of PEPTIDOGLYCAN (up to ~80% dry weight), and this polymer accounts for ~20–40 nm of the wall's thickness. The backbone chains of peptidoglycan run parallel to the cell surface; they are orientated perpendicular to the long axis of the cell. During wall growth, new glycan chains appear to be intercalated between existing chains (see THREE-FOR-ONE MODEL). Various acidic and/or neutral polymers (e.g. TEICHOIC ACIDS, TEICHURONIC ACIDS) are covalently linked (via a phosphodiester bridge) to the C-6 of *N*-acetylmuramic acid residues – either directly or via 'linkage units'. In some genera the cell wall contains lipids (see e.g. MYCOBACTERIUM) or carbohydrates (e.g. C SUBSTANCE). Some compounds (e.g. lipoteichoic acids) are closely associated with, but not covalently linked to, the wall. A hydrophilic trans-wall channel occurs in STREPTOMYCES and in a species of *Corynebacterium* [JB (2003) *185* 4779–4786].

The precise composition of the Gram-positive type cell wall may vary with growth conditions and with the degree of maturity of the wall; thus, e.g. in *Bacillus* spp the presence of teichoic and/or teichuronic acids in the cell wall is influenced by environmental levels of phosphate, while (in various genera) the degree of cross-linking is greater in mature than in newly synthesized peptidoglycan. (See also INSIDE-TO-OUTSIDE MODEL.)

Gram-negative-type walls are thinner than the walls in Gram-positive species and they have a distinctly layered structure; they are ~20–25 nm in thickness, of which the peptidoglycan layer (the innermost layer of the cell wall) may contribute less than ~6 nm (being less than ~10% dry weight of the cell wall). (It has been suggested that the *lateral* wall in *Escherichia coli* may contain a monolayer of peptidoglycan [JB (1993) *175* 7–11].)

External to the peptidoglycan layer is the OUTER MEMBRANE; some of the BRAUN LIPOPROTEIN molecules of the outer membrane are covalently linked to peptidoglycan; in e.g. *E. coli* the 6-amino group of the C-terminal lysine residue of the lipoprotein is linked by a peptide bond to a DAP residue in peptidoglycan. At a number of sites, the outer membrane apparently fuses with the cytoplasmic membrane to form ADHESION SITES.

(See also CELL CYCLE; L FORMS; PERIPLASMIC REGION; SPHAEROPLAST; ZETA POTENTIAL.)

(c) *Fungal cell walls.* Cell walls are present in most fungi (absent in *Coelomomyces*) and they consist chiefly of polysaccharides which vary e.g. according to taxonomic group.

The architecture of the fungal cell wall seems to be basically similar in many cases: one or more alkali-insoluble, possibly microfibrillar polysaccharides (e.g. CHITIN, CELLULOSE, other β-glucans) form an innermost 'structural' layer, while other (commonly alkali-soluble) components occur throughout the wall – forming an outer layer and also permeating the structural layer. The various wall components may be more or less extensively bonded together by hydrogen, ionic and/or covalent bonds, and the wall may be surrounded by a gel-like or mucilaginous CAPSULE. [Survey of fungal cell walls: Book ref. 38, pp 352–394.]

In (mycelial) *basidiomycetes*, *ascomycetes* and *chytridiomycetes* the main structural components are typically chitin and β-D-glucans (usually containing (1 → 3)- and (1 → 6)-linkages in varying proportions); these polymers are closely associated (possibly covalently linked) to form an alkali-insoluble complex. (This association may prevent the formation of crystalline microfibrils of chitin in the native cell wall [Book ref. 38, pp 355–357]; microfibrils of α-chitin readily form on chemical extraction of the β-glucan.) Alkali-soluble components of these walls include glycoproteins (usually containing mannose and/or galactose in ascomycetes, xylose and mannose in basidiomycetes) and (1 → 3)-α-D-glucans.

In *zygomycetes* the cell wall typically contains CHITOSAN closely associated with poly-anionic substances such as polyphosphate and polyglucuronate (cf. MUCORAN); glucans appear to be absent.

In *oomycetes* the main structural component is cellulose, occurring with (1 → 3)-β-D-glucans and/or (1 → 6)-β-D-glucans; in some oomycetes (e.g. certain members of the Leptomitales, Peronosporales, Saprolegniales), and in some members of the Hyphochytriomycetes, the walls contain both cellulose and chitin.

In *yeasts* the cell wall tends to reflect taxonomic affinities.

In *Saccharomyces cerevisiae* (an ascomycetous yeast) the wall consists mainly of alkali-insoluble β-glucans and mannan (as phosphomannoprotein) in roughly equal amounts; chitin is only a minor component, occurring mainly in the primary septa and bud scars but also dispersed throughout the wall. The glucan forms the innermost layer which comprises two components. The major component (which apparently provides the rigid framework of the wall) is a β-glucan containing mostly (1 → 3)-linkages with some (1 → 6)-linkages, while the minor component contains mainly (1 → 6)-β-linkages with some (1 → 3)-β-linkages. The mannan component in *S. cerevisiae* is an extensively branched polymer (containing (1 → 6)-α-, (1 → 2)-α- and(1 → 3)-α-linkages) covalently bound to protein. This mannoprotein forms a continuous layer at the wall surface and penetrates the inner glucan layer to some depth; it may have a role in determining the porosity of the wall, and it can e.g. protect the glucan from the action of β-glucanases such as the Z-glucanase ZYMOLYASE [JB (1984) *159* 1018–1026]. The mannoprotein molecules are linked covalently (e.g. by disulphide bonds) and by hydrogen and ionic bonds to the glucan and to each other. (Mannoproteins with specific functions (e.g. INVERTASE, acid phosphatase, molecules involved in sexual recognition and agglutination) also occur in the cell wall and/or periplasmic space.)

The walls of other yeasts may differ from those of *S. cerevisiae*. For example, while the walls of *Candida albicans* yeast cells are essentially similar to those of *S. cerevisiae*, the alkali-insoluble β-glucans are highly branched, having a higher proportion of (1 → 6)-β-linkages [JGM (1984) *130* 3295–3301]. *Schizosaccharomyces pombe* contains no chitin, and little or no mannan; instead, it contains (1 → 3)-α-D-glucan (PSEUDONIGERAN) and (1 → 3)-β-D-glucan. The yeast *Sporobolomyces* (a basidiomycete) contains mainly chitin and galactomannan in its walls.

Different morphological forms of a given fungus may show differences in wall composition and structure; thus, for example, in *Mucor rouxii* the hyphal wall contains a high proportion of chitosan but no melanin, while spore walls of this fungus contain much less chitosan, more glucan, and up to ca. 10% by dry weight of MELANIN.

Wall structure also differs between mycelial and yeast forms of DIMORPHIC FUNGI (sense 1); for example, in *Mucor rouxii* the transition from mycelial form to yeast form is accompanied by the appearance of mannan in the walls. In *Paracoccidioides brasiliensis* the yeast-form walls contain (1 → 3)-α-D-glucan as the sole glucan, while mycelial walls contain (1 → 3)-β-D-glucan/(1 → 6)-β-D-glucan in addition to the α-glucan.

In addition to their role as structural components, some fungal wall polymers can be used as endogenous sources of nutrient (for example, during carbon starvation or fruiting body formation – see e.g. PSEUDONIGERAN).

Certain antifungal antibiotics function by interfering with cell wall biosynthesis: see e.g. ACULEACIN A; NIKKOMYCINS; PAPULACANDIN B; POLYOXINS.

(See also GROWTH (fungal).)

(d) *Archaeal cell walls*. Distinctive types of cell wall occur in members of the Archaea; there are also wall-less species (e.g. *Thermoplasma*).

Archaeal walls include those which consist mainly or solely of an S LAYER that is closely associated with the cytoplasmic membrane; this type of wall occurs e.g. in species of *Desulfurococcus*, *Halobacterium*, *Methanococcus* and *Thermoproteus*. *Heteropolysaccharides* occur in the cell walls of *Halococcus* and *Methanosarcina*, and PSEUDOMUREIN occurs in the cell wall of *Methanobacterium* and *Methanobrevibacter*.

cellar fungus *Coniophora puteana*. (cf. WET ROT.)

cellobiase See CELLULASES.

cellobiohydrolase Exoglucanase: see CELLULASES.

cellobiose A reducing disaccharide: β-D-glucopyranosyl-(1 → 4)-D-glucopyranose. (See also CELLULASES and LIGNIN.)

cellobiose:quinone oxidoreductase See LIGNIN.

cellodextrins Oligomers of (1 → 4)-β-D-linked glucopyranosyl residues formed by the action of CELLULASES on CELLULOSE.

cellular (1) Of or pertaining to a cell or cells. (2) Having the form and characteristics of a (single) cell. (3) Composed of two or more cells. (cf. ACELLULAR; see also UNICELLULAR and MULTICELLULAR.)

cellular immunity (1) CELL-MEDIATED IMMUNITY. (2) Aspects of NON-SPECIFIC IMMUNITY mediated e.g. by macrophages and other phagocytic cells.

cellular microbiology A branch of microbiology concerned with the interactions between microorganisms and eukaryotic cells at the molecular level. [Science (1996) *271* 315–316; Book ref. 218.]

cellular slime moulds SLIME MOULDS in which the vegetative phase consists of uninucleate amoeboid cells (myxamoebae) which aggregate to form a multicellular pseudoplasmodium from which the fruiting bodies arise. Classes: ACRASIOMYCETES and DICTYOSTELIOMYCETES. (cf. MYXOMYCETES.)

cellulases Enzymes which degrade at least some forms of CELLULOSE. *Endoglucanase* (endo-(1 → 4)-β-D-glucanase, endocellulase, (1 → 4)-β-D-glucan 4-glucanhydrolase, EC 3.2.1.4) attacks more or less randomly at sites within (1 → 4)-β-D-glucan chains in amorphous regions of cellulose or at the surfaces of microfibrils. *Exoglucanase* (exo-(1 → 4)-β-D-glucanase, exocellulase, (1 → 4)-β-D-glucan cellobiohydrolase, EC 3.2.1.91) releases cellobiose (and, in some cases, glucose) from non-reducing ends of β-D-glucan chains. *β-Glucosidase* (cellobiase, β-D-glucoside glucohydrolase, EC 3.2.1.21), hydrolyses cellobiose and water-soluble cellodextrins to glucose. Many cellulolytic organisms produce a number of isoenzymes.

Many fungi, several bacteria, and a few protozoa (e.g. *Trichonympha*) can degrade cellulose, but their activities depend e.g. on the degree of polymerization and crystallinity of the cellulose and its association with hemicelluloses and LIGNIN (see CELLULOSE). Crystalline cellulose is highly resistant to enzymic attack; most of the glucan chains in microfibrils are inaccessible to enzymes, and any (surface) bonds cleaved by endoglucanase action can readily be re-formed owing to the stable orientation of the (hydrogen-bonded) glucan chains. Degradation of crystalline cellulose requires the synergistic action of both endoglucanase and exoglucanase; the exoglucanase rapidly removes cellobiose units from the newly created ends formed by endoglucanase action, thus preventing the re-formation of glucosidic bonds. The two enzymes may act consecutively or in concert. Alternative pathways probably also occur – e.g. the progressive release of cellobiose and/or glucose by exoglucanase action only. [PTRSLB (1983) *300* 283–291.] Both exo- and endoglucanases are inhibited by cellobiose, and β-glucosidase action (which eliminates cellobiose) is often the rate-limiting step in cellulose degradation. (Cellobiose may also be degraded by cellobiose phosphorylase, EC 2.4.1.20, or oxidized by cellobiose oxidase or by cellobiose:quinone oxidoreductase (see LIGNIN).)

Cellulases are subject to induction (in the presence of cellulose) and to catabolite repression; SOPHOROSE (formed e.g. from cellobiose by transglycosylation reactions) is an effective inducer in many (but not all) cellulolytic organisms and may be a natural inducer – at least in some fungi. Repressors include e.g. glucose and other readily utilizable substrates. Cellobiose can act as inducer or repressor, according to concentration.

(Reference is still sometimes made to 'C_1' and 'C_x' cellulase activities. These terms derive from the early C_1–C_x concept [JB (1950) 59 485–497] but have subsequently been used in different ways by different authors [Book ref. 31, 290–292]. The terms are thus ambiguous and are best avoided.)

Cellulolytic bacteria include e.g. many actinomycetes (e.g. *Cellulomonas* spp) and RUMEN bacteria, *Clostridium thermocellum*, *Cytophaga* spp, *Polyangium cellulosum*, *Pseudomonas* spp and *Sporocytophaga myxococcoides*. In general, bacteria produce endoglucanases and β-glucosidases but not exoglucanases; they thus have little activity on highly crystalline native cellulose, but are highly active against amorphous cellulose or soluble cellulose derivatives (e.g. CM-CELLULOSE). The endoglucanases may be cell-bound, extracellular, or both. In e.g. *Cytophaga* and *Sporocytophaga*, which grow in close contact with cellulose fibres, the glucanases are associated with the outer membrane. (See also CELLULOSOME.) In rumen bacteria cellulases are retained in the GLYCOCALYX surrounding the cells, and the products of their action may also be so retained prior to their uptake by the cell [CRM (1981) *8* 303–338]. Bacterial β-glucosidases are typically cell-associated (e.g. periplasmic) or intracellular.

Cellulolytic fungi include many basidiomycetes (e.g. *Coniophora puteana*), ascomycetes (e.g. *Chaetomium* spp, *Stachybotrys atra*) and deuteromycetes (e.g. *Trichoderma* spp). (See also NEOCALLIMASTIX.) In general, the brown-rot basidiomycetes produce only endoglucanases and β-glucosidases, while ascomycetes, deuteromycetes and the white-rot basidiomycetes produce (extracellular) endoglucanases, exoglucanases, and β-glucosidases (see BROWN ROT and WHITE ROT).

Cellulases are not produced by animals; herbivores must therefore rely on symbiotic cellulolytic microorganisms to digest their dietary cellulose (see e.g. RUMEN, TERMITE–MICROBE ASSOCIATIONS).

Isolated microbial cellulases have relatively low specific activities and are costly to obtain; they currently have only limited commercial use. Pilot schemes include the use of cellulases or cellulolytic organisms for the saccharification of the cellulose in waste materials such as straw, sugarcane bagasse, sawdust, newspaper etc (see e.g. INDUSTRIAL ALCOHOL and SINGLE-CELL PROTEIN). (See also ENZYMES.)

(See also PAPER SPOILAGE.)

cellulin granules Granules, composed of CHITIN and (1 → 3)-β- and (1 → 6)-β-linked glucan, found in the cells of fungi of the LEPTOMITALES; the granules are composed of alternating concentric layers of chitin and glucan. (cf. CONCENTRIC BODIES.)

cellulitis A diffuse, spreading, inflammatory condition affecting (usually) subcutaneous tissues and characterized by oedema, redness and pain. Common causal agents are streptococci and staphylococci. *Orbital cellulitis* (inflammation of the eye-socket) may be caused by *Haemophilus influenzae*. (See also ERYSIPELAS and GAS GANGRENE.)

cellulolytic Able to degrade CELLULOSE.

Cellulomonas A genus of aerobic or facultatively anaerobic, asporogenous bacteria (order ACTINOMYCETALES, wall type VIII) which occur e.g. in soil. In young cultures the organisms are slender, irregular rods (diam. ca. 0.5–0.6 μm) and/or filaments, with or without primary branching, but shorter rods and/or coccoid forms occur in the stationary phase; the cells are Gram-positive, but easily decolorized, and cells which are apparently Gram-negative may predominate. The organisms are non-motile, or motile by one to several polar flagella [IJSB (1984) *34* 218–219]. Growth occurs e.g. on peptone meat extract at 30°C; all strains require biotin and thiamine for growth. Typically, a yellow non-diffusible pigment is formed when growth occurs on nutrient agar. Glucose is metabolized to acid both oxidatively and fermentatively. *Cellulomonas* spp are amylolytic and cellulolytic (see CELLULASES). Species include *C. biazotea*, *C. cellasea*, *C. flavigena*, *C. fimi*, *C. gelida*, and *C. uda*. GC%: 71–77. Type species: *C. flavigena*.

cellulose A linear (1 → 4)-β-D-glucan. Cellulose occurs in plants (it is the main structural component of the cell walls in most plants), in the CELL WALLS of most algae and certain fungi, in certain cellular slime moulds (e.g. in the sheath surrounding the sorocarp stalk, in the spore walls, and in the slime of migrating slugs of *Dictyostelium discoideum*), and in the cyst walls of *Acanthamoeba* spp; extracellular cellulose is produced by certain bacteria: e.g. *Acetobacter xylinum* (see PELLICLE sense 1), *Sarcina ventriculi* (most strains of which form a cellulose microcapsule apparently responsible for the formation of the characteristic packets of cells), strains of *Agrobacterium* and *Rhizobium* (in which extracellular cellulose is involved in the adhesion of bacteria to plants), and various floc-forming bacteria in activated sludge.

Native cellulose occurs largely in the form of (crystalline) *microfibrils* ('α-cellulose') each composed of stacked sheets of parallel, hydrogen-bonded glucan chains. Each chain is in extended form (held rigid by intrachain hydrogen bonds) and contains many glucosyl residues (e.g. 2000–15000, depending on source). In plant cell walls the microfibrils are embedded in a matrix of HEMICELLULOSES, LIGNIN and PECTIC POLYSACCHARIDES; regions of amorphous cellulose, infiltrated with hemicelluloses, also occur. [Plant cell wall structure: Book ref. 38, pp. 9–46.]

Cellulose may be stained for microscopy e.g. with CONGO RED or CALCOFLUOR WHITE – both of which also inhibit microfibril formation during cellulose synthesis. [Biosynthesis of cellulose in microorganisms: Book ref. 62, pp. 85–127.]

Cellulose is a vast and constantly renewable potential source of fermentable sugars which would be valuable feedstock for various biotechnological processes. However, the lignin–hemicellulose–cellulose (LHC) complex is very stable, and crystalline cellulose is very resistant to enzymic attack (see CELLULASES). Thus, before it can be used as feedstock, cellulose must be released from the LHC complex and its microfibrillar structure disrupted by various (costly) mechanical, physical or chemical procedures – e.g. grinding or treatment with steam, mineral acids or alkalis (cf. IOTECH PROCESS); this has placed a constraint on its extensive use in industry. [Lignocellulose hydrolysis: PTRSLB (1983) *300* 305–322.]

cellulosome In certain cellulolytic bacteria: a CELLULOSE-binding, multiCELLULASE-containing cell-surface organelle. The cellulosome appears to be composed of parallel rows of catalytic units; it has been suggested that these may act together, cleaving cellulose into pieces as small as cellobiose – which can be taken up by the cell [FEMS Reviews (1993) *104* 340].

cellulytic Able to lyse cells.

CELO virus See AVIADENOVIRUS.

centiMorgan (cM) A unit for indicating the distance between markers on a chromosome; the number of centiMorgans between two given markers is the (statistically corrected) RECOMBINATION FREQUENCY exhibited by the markers. Thus, e.g., a corrected RF of 10% gives a distance of 10 cM. (cf. MAP UNIT.)

Centipeda A genus of Gram-negative, anaerobic, rod-shaped bacteria; *C. periodontii*, which has HELICOTRICHOUS flagellation, occurs e.g. in human periodontal lesions.

Centraalbureau voor Schimmelcultures See CBS.

central capsule (*protozool.*) See RADIOLARIA.

central dogma (*mol. biol.*) The dogma that genetic information can be transferred only in the direction DNA → protein. Originally, the dogma specified that information could flow only in the direction DNA → RNA → protein, but with the discovery of REVERSE TRANSCRIPTASE this was modified to DNA ↔ RNA → protein.

Central European encephalitis virus See FLAVIVIRIDAE.

centric (1) (of a chromosome) Having a CENTROMERE. (cf. HOLOCENTRIC and MONOCENTRIC sense 2.) (2) Refers to that type of nuclear division in which CENTRIOLES are involved. (3) See DIATOMS.

centrifugation The use of a centrifugal field for the sedimentation of fine particulate matter – or macromolecules (see *ultracentrifugation*, below) – in a liquid medium, or for separating different types of particulate matter, or macromolecules, within a given suspension or solution. (See also CELL FRACTIONATION and MAECT.) In the most widely used form of centrifugation, the sample is placed in a suitable container at the end of one arm of a spider-shaped motor-driven metal *rotor* which has a central, vertical shaft; rotation of the shaft at high speed subjects the sample to a centrifugal field which causes particulate matter in the sample to move in an outward (radial) direction, i.e., towards the bottom of the container. (In practice one or more additional samples of equal weight must be carried by the rotor in order to achieve a balanced distribution of weight.) The apparatus described above, suitably enclosed for safety, constitutes a simple *centrifuge*. Specialized forms of centrifuge are used e.g. for refrigerated centrifugation and for continuous centrifugation; in *continuous centrifugation*, which is used for dealing with large volumes of liquid, provision is made for the continual inflow of sample (e.g. through a hollow rotor) and the continual discharge of supernatant (and of sediment in some types of centrifuge).

The strength of a centrifugal field (at a given point) is given by:

$$G = \frac{4\pi^2(\text{revs min}^{-1})^2\text{r}}{3600}$$

in which 'revs min^{-1}' refers to the number of revolutions per minute (r.p.m.) made by the rotor, and r is the distance (cm) between the given point in the sample and the axis of rotation. G is given in terms of acceleration (cm $\sec^{-2}$); since (for a given mass) acceleration is proportional to the force producing it, G gives a measure of the force acting on a body subjected to a given centrifugal field. The centrifugal field may be increased by (a) increasing the number of r.p.m. and/or (b) increasing r. Since the acceleration due to gravity (g) is ca. 980 cm $\sec^{-2}$, $G/980$ gives the number of times by which the given centrifugal field is greater than the gravitational field; $G/980$ is termed the *relative centrifugal field* (= *relative centrifugal force*), RCF. (Since, for a given value of r.p.m., the centrifugal field varies with r – i.e., may vary from one centrifuge to another – centrifugation data should be reported in the form g/time, e.g., 1000 g/10 minutes.)

G gives the acceleration which a given centrifugal field would impart to a particle in a *vacuum*. Since particles subjected to centrifugation are suspended in a liquid, factors in addition to the applied field must be taken into account when considering the effect of the field. These factors, which greatly affect the rates of movement of particles, are the size, shape and density of the particles, and the viscosity and density of the liquid in which the particles are suspended. For example, the rate of movement of a *spherical* particle in a given centrifugal field is proportional to:

$$\text{Gr}_\text{p}^2(\rho_\text{p} - \rho)/\text{viscosity of medium}$$

where r_p is the radius of the particle, and ρ_p and ρ are the densities of the particle and the suspending medium, respectively. Thus, given a mixed population of particles which differ e.g. in size and/or density, separation into homogeneous populations may be achieved on the basis of their differing sedimentation rates in a centrifugal field (*differential centrifugation*). If particle density (ρ_p) is equal to that of the suspending medium (ρ), the rate of movement in the centrifugal field is zero; if ρ_p is *less* than ρ, centrifugally accelerated flotation occurs – an effect which has been used e.g. for the purification of GAS VACUOLES. In general, the *time* taken for a spherical particle to move a given distance in a centrifugal field is given by:

$$\text{t} = \frac{9}{2}\frac{\eta}{\omega^2\text{r}_\text{p}^2(\rho_\text{p} - \rho)}\log_\text{e}\frac{\text{R}_\text{a}}{\text{R}_\text{b}}$$

in which ω is the angular velocity (radians/sec); η is the viscosity of the medium; r_p is the radius of the particle; ρ_p and ρ are the densities of the particle and medium, respectively; R_a and R_b are the distances from the centre of rotation to the positions where the particle started (a) and stopped (b); and t is the time (sec).

Density gradient centrifugation. In some types of centrifugation (e.g. isopycnic and rate-zonal centrifugation) the sample is initially layered onto a column of liquid containing a solute whose concentration increases with increasing distance down the column; solutes used to prepare such density gradients (which may be linear or stepped) include e.g. caesium chloride (CsCl), sucrose, and IODINATED DENSITY-GRADIENT MEDIA. An important practical advantage of a density gradient is that it militates against convection – which can disrupt a liquid column in the absence of a gradient.

Isopycnic centrifugation (also called *equal density*, *equilibrium density gradient*, or *equilibrium zonal centrifugation*). The sample is layered onto a density gradient column, and centrifugation is continued until equilibrium is reached, i.e., each particle has reached that part of the gradient which corresponds to its own density (its isopycnic or equal density position). Particles in a heterogeneous sample are thus separated on the basis of their individual densities: particles of different density can be separated even if they are similar in size and shape.

Rate-zonal centrifugation (also called *band, gradient differential, s zonal, zonal,* or *zone centrifugation*). The sample is layered onto a density gradient column, and centrifugation is continued until the various types of particle in the sample form layers or bands at different levels within the column – at which time centrifugation is stopped. The separation of particles into bands is thus achieved on the basis of the differing sedimentation rates of the particles.

Ultracentrifugation is used e.g. for the sedimentation of macromolecules, for the determination of molecular weights, for purifying plasmid DNA, for characterizing viruses, and as an indirect method for estimating the GC% of bacterial DNA. The principles involved are similar to those given earlier in the entry; however, while the common laboratory centrifuge can develop a maximum field of ca. 5000 to 10000 *g*, the ultracentrifuge can achieve fields of the order of 500000 *g*. In an ultracentrifuge the rotor is housed in a refrigerated and evacuated chamber, and the instrument incorporates various optical systems (e.g. the SCHLIEREN SYSTEM) and photographic systems so that the progress of sedimentation can be followed and recorded at all times. (See also SVEDBERG UNIT.)

If a species of DNA is subjected to isopycnic ultracentrifugation in a caesium chloride gradient, the DNA forms a band at a region of the gradient corresponding to its *buoyant density* (Bd); the Bd of e.g. chromosomal DNA from *Escherichia coli* is ca. 1.7 g/cm^3. Since there is a linear relationship between Bd and GC%, the value of Bd can be used for estimating GC% (q.v.).

Ultracentrifugation can also be used to separate cccDNA (e.g. plasmid DNA) from linear DNA or from circular, nicked DNA (e.g. nicked chromosomal DNA). cccDNA can incorporate smaller amounts of certain INTERCALATING AGENTS (e.g. ETHIDIUM BROMIDE) than can equivalent molecules of the other forms of DNA. The incorporation of e.g. ethidium bromide *decreases* the Bd of the DNA molecule. Thus, if e.g. a plasmid-containing bacterial lysate is centrifuged in a caesium chloride–ethidium bromide gradient, cccDNA may form a distinct band of higher Bd; such a procedure has been called *dye–buoyant density centrifugation.*

centriolar plaque *Syn.* SPINDLE POLE BODY.

centriole In many types of eukaryotic cell: an intracellular microtubular structure (see MICROTUBULES) which is involved in some types of MITOSIS and which, in some cells, can develop into a BASAL BODY; centrioles do not occur in e.g. some types of fungi. Essentially, a centriole is a hollow cylinder, ca. 300–400 nm in length, whose wall is formed of 9 longitudinally arranged triplets of microtubules; one end of the cylinder contains a dense intracentriolar material. The structure appears to lack DNA, but RNA may be present. A centriole is often surrounded by a number of dense *pericentriolar structures* which, collectively, are sometimes referred to as a *centrosome* and which can act as a MICROTUBULE-ORGANIZING CENTRE. ('Centrosome' is also used to refer to the entire structure, i.e., centriole(s) plus pericentriolar structures.) Centrioles are often formed from pre-existing centrioles, but they can also be formed de novo.

Centrohelida An order of protozoa (class HELIOZOEA) in which the cell typically has an eccentric nucleus and a central body (the *centroplast*) from which the axopodial axonemes appear to arise. Some species lack a centroplast but have a large eccentric nucleus. Typically, a centrohelid has an outer 'skeleton' of siliceous plates and/or spines embedded in a gelatinous outer covering; some species have organic spicules. Genera include e.g. *Acanthocystis, Gymnosphaera, Heterophrys,* RAPHIDIOPHRYS.

centromere The region of a replicated eukaryotic chromosome where the two CHROMATIDS are joined together; during MITOSIS a KINETOCHORE is formed on both sides of the centromere. [Molecular structure of centromeres: ARB (1984) *53* 163–194; chromatin conformation of yeast centromeres: JCB (1984) *99* 1559–1568.] (cf. HOLOCENTRIC and MONOCENTRIC (sense 2).)

centroplast See CENTROHELIDA.

Centropyxis See ARCELLINIDA.

centrosome See CENTRIOLE.

centrotype In NUMERICAL TAXONOMY: a strain regarded as that which typifies a given species cluster on the basis of its position within that cluster.

centrum (*mycol.*) The totality of fertile and sterile structures within a CLEISTOTHECIUM, PERITHECIUM, or locule in an ASCOSTROMA.

cep *Syn.* cèpe (see BOLETALES).

cèpe (cep) See BOLETALES.

Cepedea See OPALINATA.

Cepedietta See ASTOMATIDA.

Cephaleuros A genus of obligately epiphytic (and/or parasitic) algae (division CHLOROPHYTA) closely related to TRENTEPOHLIA. The heterotrichous thallus is composed of a discoid system of prostrate, compacted, branched filaments and an erect system of sterile, multicellular, uniseriate filaments and specialized zoosporangiophores; the prostrate system grows between the cuticle and epidermis in the leaves, fruits and/or young stems of a host plant, while the erect filaments and gametangia protrude through the host cuticle. Although the alga may cause considerable damage to the host plant (see RED RUST), there is apparently no evidence that it is nutritionally dependent on the plant. (*Cephaleuros* also occurs as the photobiont in certain obligately foliicolous lichens: e.g. *Strigula* spp [Book ref. 129, pp. 190–191].)

cephalexin See CEPHALOSPORINS.

cephaline gregarines See GREGARINASINA.

Cephaloascus See ENDOMYCETALES.

cephalodium (*lichenol.*) A discrete group of cyanobacteria occurring in or on the thallus in certain lichens whose main photobiont is a green alga. Cephalodia appear generally to arise by entrapment of free-living cyanobacteria by the mycobiont – usually via the lower cortex of the thallus. The cyanobacteria (usually NOSTOC) carry out NITROGEN FIXATION, fixed nitrogen being transferred to the mycobiont.

Internal cephalodia occur e.g. in species of *Lobaria, Nephroma* and *Solorina*; they typically consist of colonies of cyanobacteria which lie in the medulla and may be apparent as swellings on the surface of the thallus.

External cephalodia occur e.g. in species of *Peltigera* and *Stereocaulon*; they may occur (e.g. as warty protuberances) on the upper or lower surface of the thallus, depending on species.

Some lichens (e.g. species of *Lobaria, Peltigera* and *Sticta*) form well-differentiated, fruticose or subfoliose, cyanobacterium-containing structures which arise (usually) from the upper surface of the thallus and which are capable of independent growth. These structures were originally thought to be epiphytes and were assigned to distinct genera (e.g. *'Dendriscocaulon'*), but are now commonly believed to be elaborate cephalodia or different morphotypes ('phycosymbiodemes') of the same mycobiont. [Chimeroid associations in *Peltigera*: Lichenol. (1978) *10* 157–170.]

cephaloridine See CEPHALOSPORINS.

cephalosporin C See CEPHALOSPORINS.

cephalosporin N *Syn.* PENICILLIN N.

cephalosporin P1 A steroid antibiotic structurally related to FUSIDIC ACID.

cephalosporinase A β-LACTAMASE active mainly or solely against CEPHALOSPORINS.

cephalosporins A class of antibiotics each characterized by a molecular structure which includes a β-lactam ring fused to a dihydrothiazine ring (see β-LACTAM ANTIBIOTICS for structure and mode of action). (cf. CEPHALOSPORIN P1 and PENICILLIN N.)

Cephalosporin C (the first to be prepared) is produced e.g. by *Acremonium kiliense* (= *Cephalosporium acremonium*); this antibiotic has poor antibacterial activity.

A range of semi-synthetic cephalosporins can be prepared from the 7-aminocephalosporanic acid (7-ACA) derivative of cephalosporin C. When introduced, the earlier cephalosporins (collectively) had useful activity against a range of bacteria, and were typically (particularly cefotaxime and cefuroxime) less susceptible than PENICILLINS to β-LACTAMASES; moreover, compared to penicillins, cephalosporins were found to be less frequently allergenic.

When introduced, the first- to third-generation cephalosporins were characterized by antibacterial spectra such as: mainly Gram-positive (cephalexin, cephaloridine); good activity against Gram-positive and moderate activity against β-lactamase-negative Gram-negative species (cephalothin); Gram-positive species and e.g. *Neisseria gonorrhoeae* and *Haemophilus influenzae* (cefuroxime); limited spectrum, but some activity against *Pseudomonas aeruginosa* (cefsulodin); Gram-negative species, including *Haemophilus influenzae* (cefotaxime). (Note that cephalosporins (and e.g. the cephamycin CEFOXITIN) have little or no activity against enterococci.) The earlier cephalosporins (and some of the newer ones) are commonly susceptible to extended-spectrum β-LACTAMASES (q.v.). (See also CEPHAMYCINS.)

Some cephalosporins can be administered orally, while others (e.g. cefuroxime, cefotaxime) are administered parenterally.

First-generation cephalosporins include e.g. cefazolin, cephalexin, cephaloridine and cephalothin.

Second-generation cephalosporins include e.g. cefamandole, cefonicid [Drugs (1986) *32* 222–259], ceforanide, cefsulodin and cefuroxime.

Third-generation cephalosporins include e.g. cefixime, cefmenoxime [Am. J. Med. (1984) *77* (supplement 6A)], cefotaxime, ceftazidime [Am. J. Med. (1985) *79* (supplement 2A)], ceftizoxime and ceftriaxone.

Fourth-generation cephalosporins include e.g. CEFEPIME and cefpirome.

(*Note.* In the literature, the names of some cephalosporins are spelt with either 'f' or 'ph'.)

Cephalosporium See ACREMONIUM. (For *C. lecanii* see VERTICILLIUM.)

Cephalothamnium A genus of colonial CHRYSOPHYTES in which a cluster of biflagellate, non-pigmented cells extends from the tip of a single stalk.

cephalothin See CEPHALOSPORINS.

cephamycins Antibiotics (7-α-methoxyCEPHALOSPORINS) produced by *Streptomyces* spp or prepared from cephalosporins. Compared with those cephalosporins which lack the 7-α-methoxy group, cephamycins (e.g. CEFOXITIN) tend to have decreased antibacterial activity but increased resistance to β-LACTAMASES and improved penetrability of the Gram-negative outer membrane.

cephems Compounds which contain the cephalosporin nucleus (see β-LACTAM ANTIBIOTICS). (cf. OXACEPHEMS.)

cephradine See CEPHALOSPORINS.

Ceraceosorus See BRACHYBASIDIALES.

ceramide See SPHINGOMYELIN.

Ceramium See RHODOPHYTA

Ceratiomyxa A genus of SLIME MOULDS formerly classified with the acellular slime moulds (MYXOMYCETES); currently, the genus is placed e.g. in a separate class, Ceratiomyxomycetes, of the MYXOMYCOTA, or – in zoological schemes – is classified with the protostelids e.g. in the subclass Protosteliia (see EUMYCETOZOEA). *C. fruitculosa* is a widespread and common species found e.g. on bark and rotting wood. The vegetative PLASMODIUM is almost hyaline. Prior to fruiting, it produces copious amounts of a mucoid matrix, and plasmodium and matrix become extended to form upright, usually branched white columns (1–10 mm tall) which solidify on drying; sporocarps develop on the surface of these columns, each sporocarp consisting of a long slender stalk bearing a single round or cylindrical spore. The spores are deciduous. On germination of a spore, a quadrinucleate 'thread stage' is produced, and this eventually gives rise to 4, then 8, flagellated haploid cells which can apparently function as gametes; a diploid vegetative plasmodium develops from the zygote formed by fusion of two of these flagellated cells.

Ceratiomyxella See PROTOSTELIOMYCETES.

Ceratiomyxomycetes See CERATIOMYXA.

Ceratium A genus of DINOFLAGELLATES in which the cells typically bear 3 or 4 horn-like thecal extensions: one arising from the epicone, 2 or 3 from the hypocone.

Ceratobasidiaceae See TULASNELLALES.

Ceratobasidium See TULASNELLALES.

Ceratocystis A genus of fungi (order OPHIOSTOMATALES) which include various plant pathogens – see e.g. DUTCH ELM DISEASE and OAK WILT. (See also BLUE STAIN, PINE DISEASES, and IPOMEAMARONE.)

C. ulmi forms erumpent perithecia (see also ASCOSPORE), and in the imperfect (*Pesotum*) state can form individual conidia on simple conidiophores or chains of conidia in a drop of mucus at the tip of a dark, bristle-like synnema (ca. 1 mm high).

C. coerulescens forms superficial perithecia, and in the *Chalara* state can form cylindrical or ellipsoidal conidia.

Ceratomyces See LABOULBENIALES.

Ceratomyxa See MYXOSPOREA.

Cercophora See SORDARIALES.

cercopithecine herpesvirus 1 See B VIRUS.

Cercospora A genus of fungi (class HYPHOMYCETES) which include many plant-pathogenic species – e.g. *C. beticola* (causal agent of leaf spot of beet), and *C. apii* (e.g. on celery). The organisms form septate mycelium; conidia are characteristically thread-like and multiseptate.

cereal chlorotic mottle virus See RHABDOVIRIDAE.

cereal diseases Diseases of members of the Gramineae cultivated for their edible grain. Some of the common diseases are listed below. Although some diseases are specific to particular types of cereal, others can also affect certain wild grasses; thus, e.g. the causal agents of HALO SPOT and LEAF BLOTCH can both grow on cocksfoot (*Dactylis glomerata*) and Timothy grass (*Phleum pratense*). (See also GRASS DISEASES.)

Barley. See e.g. BROWN RUST, ERGOT, EYESPOT (sense 2), GLUME BLOTCH, HALO SPOT, LEAF BLOTCH ('rhynchosporium'), NET

BLOTCH, POWDERY MILDEW, SMUTS (sense 2), TAKE-ALL, YELLOW RUST; see also LUTEOVIRUSES (barley yellow dwarf virus) and HORDEIVIRUSES (barley stripe mosaic virus).

Maize. See e.g. CORN STUNT DISEASE, MAIZE WALLABY EAR DISEASE, SMUTS (sense 2); see also MAIZE STREAK VIRUS.

Oats. See e.g. CROWN RUST, ERGOT, SMUTS (sense 2), TAKE-ALL, VICTORIA BLIGHT.

Rice. See e.g. BAKANAE DISEASE, BLAST DISEASE, SUFFOCATION DISEASE; see also FIJIVIRUS, RICE DWARF VIRUS, RICE GALL DWARF VIRUS, RICE RAGGED STUNT VIRUS, RICE STRIPE VIRUS GROUP. (cf. YELLOW RICE.) [Engineered resistance to *Rhizoctonia solani* (sheath blight): *Biotechnology* (1995) *13* 686–691.]

Rye. See e.g. BROWN RUST, COMMON BUNT, ERGOT, HALO SPOT, LEAF BLOTCH ('rhynchosporium'), SNOW MOULD, TAKE-ALL.

Wheat. See e.g. BLACK STEM RUST, BROWN RUST, COMMON BUNT, ERGOT, EYESPOT (sense 2), GLUME BLOTCH, HALO SPOT, KARNAL BUNT, POWDERY MILDEW, SMUTS (sense 2), SNOW MOULD, TAKE-ALL, WHEAT SCAB, YELLOW RUST; see also SOIL-BORNE WHEAT MOSAIC VIRUS.

cereal striate virus See RHABDOVIRIDAE.

cereal tillering disease virus See FIJIVIRUS.

cerebriform Brain-like in appearance; convoluted.

cerebroside See SPHINGOLIPID.

cereolysin See THIOL-ACTIVATED CYTOLYSINS.

Cerinomyces See DACRYMYCETALES.

cermet See BIOMIMETIC TECHNOLOGY.

cerulenin An antibiotic – [(3R)-2,3-epoxy-4-oxo-7,10-dodecadienoyl amide] – obtained from *'Cephalosporium caerulens'*; it has antifungal and antibacterial activity, but is too unstable for clinical use. Cerulenin inhibits fatty acid and sterol biosynthesis, inhibiting e.g. the condensation reaction between acyl and malonyl thioesters and thus preventing fatty acid chain elongation. It also inhibits lipoteichoic acid synthesis in e.g. *Enterococcus faecalis*, prevents the synthesis and secretion of certain extracellular proteins in some bacteria (e.g. glucosyltransferase in oral streptococci [JGM (1983) *129* 3293–3302]), and alters morphogenesis and autolytic activity in *Bacillus subtilis* [JGM (1985) *131* 591–599].

ceruloplasmin See BLUE PROTEINS.

cervical cancer (cancer of the cervix) See PAPILLOMA.

cesspool A watertight container for the temporary storage of sewage. (cf. SEPTIC TANK.)

Cetavlon See QUATERNARY AMMONIUM COMPOUNDS.

Cetraria A genus of LICHENS (order LECANORALES); photobiont: a green alga. Thallus: either fruticose and erect or foliose with upturned margins, growing on soil, bark, wood or rocks. Apothecia have reddish-brown discs and are marginal in foliose species.

C. islandica ('Iceland moss') has an erect, branched thallus (ca. 2–7 cm in height); the strap-like branches are curved to appear more or less tubular, the inside surface being greenish-grey to dark brown, the outermost surface being paler with a scattering of whitish pseudocyphellae. The thallus margins bear short black spines. This species forms cushions on the ground in heathland, moorland, and acid woodland; it has been used as food e.g. in Scandinavian countries, and serves as fodder for e.g. reindeer and caribou.

Cetrimide See QUATERNARY AMMONIUM COMPOUNDS.

cetylpyridinium chloride See QUATERNARY AMMONIUM COMPOUNDS.

cetyltrimethylammonium bromide See QUATERNARY AMMONIUM COMPOUNDS.

CF (1) COMPLEMENT FIXATION. (2) CYSTIC FIBROSIS.

CF_0F_1 H^+-ATPase See PROTON ATPASE.

CFA/I, CFA/II See ETEC.

CFP-10 See ESAT-6.

CFT COMPLEMENT-FIXATION TEST.

cfu Colony-forming unit: see COUNTING METHODS.

CGA protein See CATABOLITE REPRESSION.

CGB agar Canavanine-glycine-bromthymol blue agar: a medium used to distinguish between certain strains of CRYPTOCOCCUS. On this medium, *C. neoformans* var. *gattii* hydrolyses glycine, raising the pH and causing the BROMTHYMOL BLUE to turn blue; the few strains of *C. neoformans* var. *neoformans* which can hydrolyse glycine are inhibited by the L-canavanine. [Recipe and method: Book ref. 100, p. 95.]

C_H region See IMMUNOGLOBULINS.

CH_{50} See HD_{50}

Chaenotheca See CALICIALES.

Chaetoceros A large genus of planktonic centric DIATOMS in which the cells are often joined together by long setae.

Chaetocladium See MUCORALES.

Chaetomella See SPHAEROPSIDALES.

Chaetomium A genus of fungi (order SORDARIALES) which form evanescent asci in perithecia which bear long, curved, coiled and/or branched hairs or bristles; the unicellular, commonly dark ascospores are discharged from the perithecium in a gelatinous mass. Most or all species are strongly cellulolytic (see CELLULASES, PAPER SPOILAGE, SOFT ROT (sense 1), TEXTILE SPOILAGE). (See also COMPOSTING and OLIVE-GREEN MOULD.)

Chaetomorpha A genus of unbranched, filamentous, siphonocladous green algae (division CHLOROPHYTA).

Chaetophora A genus of uniseriate, filamentous (heterotrichous) green algae (division CHLOROPHYTA) which occur in freshwater habitats, usually attached to rocks, plants etc. Long colourless multicellular 'hairs' extend from the ends of the (branched) filaments. Quadriflagellate zoospores and biflagellate isogametes are formed. Related genera include e.g. *Draparnaldia*, *Fritschiella*, STIGEOCLONIUM and URONEMA.

chaga fungus The sterile fruiting body of *Inonotus obliquus*.

Chagas' disease An acute or chronic, often fatal, systemic TRYPANOSOMIASIS of man which occurs in Central and South America. The causal agent, *Trypanosoma* (SCHIZOTRYPANUM) *cruzi*, is typically transmitted *contaminatively* by blood-sucking TRIATOMINE BUGS (family Reduviidae): metacyclic forms in the bug's faeces infect via wounds, abrasions and mucosal surfaces. (The pathogen can also be transmitted by blood transfusion and, via the milk, to breast-fed infants. Transplacental transmission occurs in over 1% of cases in endemic areas.) Reservoirs of bug-transmissible pathogens exist in domestic animals (e.g. dogs) and wild animals (e.g. anteaters, armadillos, opossums, certain rats and primates). [Book ref. 72, pp 161–182.]

In the patient, *T. (S.) cruzi* occurs as non-dividing trypomastigotes in the blood, and in the reproductive amastigote and epimastigote forms in pseudocysts within host cells. Symptoms may include cardiomyopathy, grossly enlarged colon ('megacolon') and oesophagus, and low levels of certain hormones resulting from damage to endocrine glands. Autoantibodies to heart and skeletal muscles, and to nerve tissue, have been detected. *Lab. diagnosis*: microscopical examination of thick blood films; XENODIAGNOSIS (using e.g. *Dipetalogaster maxima*); a CFT and/or e.g. immunofluorescence tests. *Treatment*: chemotherapeutic agents, e.g. Lampit (a nitrofuran) and benznidazole (a 2-nitroimidazole), are commonly used but are associated with significant side-effects; moreover, this treatment may or may not eliminate the intracellular forms of *T.* (*S.*) *cruzi*.

Dinitroaniline herbicides such as *trifluralin* (α,α,α-trifluoro-2,6-dinitro-*N*,*N*-dipropyltoluidine), which inhibit certain protozoan parasites by affecting tubulin polymerization, have been examined for their potential as candidate antimicrobials against *T. cruzi* [TIP (2001) *17* 136–141].

[The disease in Mexico: TIP (2001) *17* 372–376.]

chain-termination method (of DNA sequencing) See DNA SEQUENCING.

Chainia See STREPTOMYCES.

Chalara See HYPHOMYCETES and CERATOCYSTIS; see also MURAMIDASE.

chalcomycin See MACROLIDE ANTIBIOTICS.

chalk-brood A BEE DISEASE caused by *Ascosphaera apis*. Infected honey-bee larvae become fluffy and swollen, and later shrink and harden; death usually occurs after the larvae have been sealed in their cells. Some of the dead larvae remain chalky-white, others become dark-coloured. Infection occurs by ingestion of spores of *A. apis*. The spores germinate in the gut; mycelium develops in the gut lumen and penetrates the gut wall. Fruiting bodies may form on the outside of the larvae.

Ascosphaera spp can also cause chalk-brood in wild bees – e.g. *A. aggregata* in the alfalfa leaf-cutting bee *Megachile rotundata* [life cycle of *A. aggregata*: Mycol. (1984) *76* 830–842].

challenge (*immunol.*) The exposure of an animal (or other living system) to a pathogen (or other agent) – usually in order to determine the state of immunity or resistance to the pathogen or agent; for example, an animal may be challenged with a pathogen to test the efficacy of a previously administered vaccine.

challenge virus See INTERFERENCE (1).

chalybeate (of water) Containing iron.

Chamaesiphon A genus of unicellular CYANOBACTERIA (section I) in which the cells are ovoid and reproduce by repeated budding at the apical end; the buds were formerly known as 'exospores'. GC%: ca. 47.

Chamaesiphonales See CYANOBACTERIA.

Chamberland candle See FILTRATION.

chamois contagious ecthyma virus See PARAPOXVIRUS.

champagne See WINE-MAKING.

chancre A papular lesion formed at the initial site of infection in certain diseases (e.g. SYPHILIS, SLEEPING SICKNESS).

chancroid (soft chancre) A VENEREAL DISEASE, caused by *Haemophilus ducreyi*, characterized by one or more ulcerative lesions which are usually confined to the genitalia. Incubation period: 1–14 (usually 4–5) days.

Changuinola subgroup See ORBIVIRUS.

channel catfish virus disease (CCVD) A FISH DISEASE which affects cultivated channel catfish (*Ictalurus punctatus*) under 4 months of age. Symptoms include e.g. abnormal swimming movements, haemorrhages of gills, skin and internal organs, and necrotic lesions in liver, spleen etc. The causal agent is a herpesvirus (see HERPESVIRIDAE).

chanterelle See CANTHARELLALES.

Chaos A genus of large, multinucleate, freshwater amoebae of the AMOEBIDA [Book ref. 133, p. 145]. In *C. carolinense* polypodial forms may be 0.7–2.0 mm in diam., monopodial forms up to 3 mm diam.; *C. illinoisense* is generally smaller (ca. 0.5–1.5 mm).

chaperone (molecular chaperone) In most cases: a protein that is specialized for folding/stabilization or direction of a (typically newly synthesized) protein. (See also HEAT-SHOCK PROTEINS, FIMBRIAE (type I) and VIRULON.) A chaperone may bind to a nascent protein either co-translationally [JBC (1997) *272* 32715–32718] or following translation.

Although most chaperones are proteins, a membrane lipid acts as a chaperone for the LacY protein in *Escherichia coli* [EMBO (1998) *17* 5255–5264].

Chara See CHAROPHYTES.

Charales See CHAROPHYTES.

charcoal blood agar A medium used for the primary isolation of *Bordetella pertussis*; it is reported to yield better growth than that obtained with the traditional BORDET–GENGOU AGAR. Beef extract, starch, peptone, NaCl, nicotinic acid and agar are dissolved in water by heating; charcoal is added before autoclaving. Sterile, defibrinated horse blood is added to the molten agar at ~45°C. [Recipe: Book ref. 219, pp 470, 471.]

Charcot–Leyden crystals Microscopic, colourless crystals often present in the stools of patients suffering from certain intestinal diseases (e.g. amoebic dysentery) or in the sputum of patients with asthma or other chest conditions. The crystals are diamond-shaped or shaped like short, double-pointed needles; they stain with iodine.

charge-shift immunoelectrophoresis IMMUNOELECTROPHORESIS used e.g. to distinguish amphiphilic from hydrophilic proteins. Differentiation is based on the fact that when amphiphilic proteins bind non-ionic detergent at their hydrophobic sites, their electrophoretic mobility is changed by the addition of (positively or negatively charged) ionic detergent; hydrophilic proteins are unaffected.

chargerin See ANISOTROPIC INHIBITOR.

Charon A type of CLONING vector derived from the genome of bacteriophage λ; typically, a Charon is a REPLACEMENT VECTOR.

charonin An ATP-dependent protease which may also function as a molecular chaperone.

Charophyceae See CHLOROPHYTA.

Charophyta See CHAROPHYTES.

charophytes A group of macroscopic green algae variously regarded as (a) a distinct division (Charophyta) [e.g. Book ref. 123, pp. 303–330]; (b) a class (Charophyceae) within the division Chlorophyta [e.g. Book ref. 130, pp. 36–132]; or (c) an order (Charales) within a class, Charophyceae, which is broader than that in (b): see CHLOROPHYTA. Charophytes (e.g. *Chara* and *Nitella*) have several features in common with bryophytes (mosses and liverworts): e.g., the plant is structurally complex, being differentiated into root-, shoot- and leaf-like structures; the shoots are divided into nodes (each node bearing a whorl of branches) and internodes; sexual reproduction is oogamous, and an envelope of sterile cells (*shield cells*) surrounds the antheridia and oogonia; asexual reproduction may occur e.g. by the formation of bulbils ('plantlets') on the rhizoids, or of AMYLUM STARS on the lower nodes, but apparently never by the formation of zoospores. Charophytes are generally heavily calcified (hence the popular names 'stoneworts' or 'brittleworts'); they occur mainly in fresh water (lakes, ponds, etc). (See also GYROGONITES.)

chasmoendolith See ENDOLITHIC.

Chattonella See CHLOROMONADS.

chauveolysin See THIOL-ACTIVATED CYTOLYSINS.

***che* gene** In e.g. enterobacteria: a gene involved in CHEMOTAXIS.

checker colony See DRAUGHTSMAN COLONY.

Cheddar cheese See CHEESE-MAKING.

cheese-making Cheese is made by the coagulation and fermentation of MILK; differences between the various types of cheese reflect differences in e.g. mode of manufacture, microorganisms employed, and type of milk used (e.g. cows', goats' sheep's).

Initially, the milk may undergo various treatments: e.g. PASTEURIZATION, homogenization (i.e. reduction in size of fat globules), adjustment of fat content. The milk protein is then coagulated to form a solid *curd* in which milk fat and soluble components are trapped. Coagulation is usually achieved primarily by adding rennet (rennin) or microbial PROTEASES; it is aided by stirring, heating ('cooking'), and the addition of cultures of selected strain(s) of LACTIC ACID BACTERIA – which lower the pH by fermenting lactose to lactic acid. Strains of the mesophilic bacterium *Lactococcus lactis* are commonly used, but thermophilic strains (e.g. *Streptococcus thermophilus*, *Lactobacillus bulgaricus*, *L. helveticus*) are used for cheeses cooked at higher temperatures. The bacteria primarily carry out HOMOLACTIC FERMENTATION under the conditions of manufacture, but they also form small amounts of other important flavour components (acetic and propionic acids, ketones, esters etc.).

The curd is cut into pieces, and the liquid (*whey*) is removed by pressing and/or straining; the curd may then be treated to optimize pH, reduce moisture, increase salt content etc., according to the type of cheese. Finally, the cheese is ripened by a controlled storage during which flavour and texture develop due mainly to the action of microorganisms and milk enzymes.

(a) *Blue cheeses* (e.g. *Gorgonzola*, *Roquefort*, *Stilton*). As well as strains of *Lactococcus lactis*, *Penicillium roqueforti* is mixed with the curd before removal of the whey; during ripening, the mould grows through the cheese, giving the characteristic flavour and blue 'veins'.

(b) *Brick cheese* is an American soft cheese. The early stages of manufacture resemble those for Cheddar (see below); subsequently the cheese is *surface-ripened*: the relative humidity is held at 95% to allow a succession of microorganisms to develop on the surface of the cheese. First, yeasts metabolize the lactic acid, raising the pH and synthesizing vitamins etc.; this stimulates the growth of micrococci and *Brevibacterium linens*. The microbial growth is washed off before packaging and a further period of ripening. *Bel paese*, *Limburger*, *Port Salut* and *Tilsit* are made similarly.

(c) *Camembert* is a high-moisture soft, French cheese. In addition to lactic acid bacteria, the mould *Penicillium camemberti* is added either to the milk or to the separated curd. The curd is placed on racks to maximize surface exposure and facilitate growth of the mould – which forms a cottony white mat over the surface of the cheese. Yeasts (e.g. species of *Saccharomyces*, *Kluyveromyces*, *Candida*), and sometimes *Brevibacterium linens*, are also present in the surface flora. *Brie* is made in a similar manner.

(d) *Cheddar* is a hard English cheese made from raw or pasteurized whole milk. The milk is clotted enzymically, inoculated with (usually) strains of *Lactococcus lactis*, and cooked at temperatures of up to ca. 40°C. (A genetically engineered starter culture has been used to control the pathogen *Listeria monocytogenes* in Cheddar cheese [AEM (1998) *64* 4842–4845].) The curd is cut into small cubes and allowed to settle in the whey. The whey is drained off, and the curd mass is cut into blocks which are piled and turned (a process called 'cheddaring'). The curd is then cut into smaller pieces, salted, pressed into the desired shape, packaged, and ripened at 2–7°C.

Edam and *Gouda* are semi-soft Dutch cheeses made in a similar manner.

(e) *Cottage cheese* is a high-moisture, unripened soft cheese made from skim milk. The milk is clotted with a B- or BD-type LACTIC ACID STARTER. The final cooking temperature is ca. 55°C (cf. QUARG). To improve the safety of cottage cheese, use has been made of a starter culture containing organisms that produce the BACTERIOCIN lacticin 3147 – which inhibits *Listeria monocytogenes* [JAM (1999) *86* 251–256]. *Cream cheese* (for example, *Neufchatel*) is made in a similar way using cream instead of skim milk; it thus has a higher fat and lower moisture content.

(f) *Mozzarella* is a soft, usually unripened, Italian cheese formerly made from buffalo milk (but now generally made from cows' milk). Its manufacture is similar to that of Cheddar but involves higher temperatures (50–55°C) as well as thermophilic streptococci and lactobacilli.

(g) *Parmesan* is a hard, low-moisture Italian cheese; its manufacture is similar to that of Swiss cheeses (see below), but e.g. propionibacteria are not used.

(h) *Swiss cheeses* (e.g. *Emmentaler*, *Gruyère*) are manufactured in a way basically similar to that used for Cheddar; however, the process involves higher cooking temperatures (50–55°C) and uses thermophilic bacteria (*Streptococcus thermophilus*, *Lactobacillus* spp). Strains of *Propionibacterium* are used in the inoculum; during ripening, these bacteria (i) contribute flavour components (e.g. propionic acid), and (ii) produce carbon dioxide – which collects in pockets, thus forming the characteristic 'eyes' in the cheese.

(See also CHEESE SPOILAGE and DAIRY PRODUCTS.)

cheese spoilage High-moisture cheeses (e.g. cottage cheese) are very susceptible to spoilage by a wide range of bacteria and moulds, while low-moisture cheeses (e.g. parmesan) are much more resistant. Surface spoilage of cheeses is commonly due to the growth of moulds – e.g. species of *Aspergillus*, *Cladosporium* and *Penicillium*, and (in high-moisture cheeses) *Geotrichum* spp. Bacteria may produce off-flavours – e.g. H_2S (produced e.g. by clostridia), rancidity (due e.g. to lipolytic pseudomonads); undesirable holes ('eyes') may be formed during the ripening process by gas-producing bacteria (e.g. coliforms, *Bacillus* spp).

cheese-washers' lung An EXTRINSIC ALLERGIC ALVEOLITIS associated with the inhalation of allergens from *Penicillium* spp – e.g. *P. verrucosum* ('*P. casei*').

CHEF See PFGE.

chelating agents See e.g. CDTA, EDTA, EGTA, NTA.

chemical shift See NUCLEAR MAGNETIC RESONANCE.

Chemiclave An apparatus used e.g. for sterilizing surgical instruments etc; essentially, it consists of a chamber within which instruments etc are exposed to a mixture of the vapours of alcohol, formaldehyde and water at 132°C, 20 lb/inch^2 (ca. 137 kPa). (cf. AUTOCLAVE.)

chemiluminescence The production of light as a result of a chemical reaction. The term is commonly used to refer e.g. to the emission of light by phagocytes and other specialized mammalian cells under appropriate conditions, and to light emission in cell-free experimental systems, while BIOLUMINESCENCE usually refers to the emission of light from whole, living organisms; in all cases the emission of light is O_2-dependent. In those cases where the mechanism is known, light emission involves the oxidation of a particular type of organic molecule – the resulting unstable, excited molecule spontaneously returning to the (unexcited) ground state with concomitant emission of light. (cf. PHOSPHORESCENCE.) Chemiluminescence in phagocytes is associated with the respiratory burst which occurs in e.g. NEUTROPHILS during PHAGOCYTOSIS.

Experimental chemiluminescence systems employ light-emitting substances such as *lucigenin* (*bis-N*-methylacridinium nitrate), *lophine* (2,4,5-triphenylimidazole) and LUMINOL.

Luminol emits light in an alkaline solution of H_2O_2 in the presence of a suitable catalyst; catalysts include e.g. cytochromes, horseradish peroxidase, MICROPEROXIDASE, and even Mn^{2+} or Ni^{2+} ions. Other agents – e.g. lophine, lucigenin – emit light in alkaline H_2O_2 in the absence of a catalyst.

Chemiluminescence involving the (ATP-requiring) firefly luciferin–luciferase system (see BIOLUMINESCENCE) has been used to detect small amounts of ATP; e.g., very small numbers of cells can be quantified by measuring their extractable ATP [e.g. in studies on bacterial adhesion: JGM (1983) *129* 621–632]. LUMINOL-dependent chemiluminescence can be triggered in mouse spleen cells by interaction between spleen cell plasma membrane and the HN and (particularly) F glycoproteins of the Sendai virus; such chemiluminescence is apparently associated with the action of unstable oxygen species (e.g., HYDROGEN PEROXIDE, HYDROXYL RADICAL, SUPEROXIDE) on the luminol probe [Book ref. 87, pp. 451–458]. Immunoassay systems using chemiluminescent labels may offer a useful alternative to RADIOIMMUNOASSAY systems [PHLS Digest (1984) *1* 58–62]. [Reviews on applications: Book ref. 96; TIBS (1986) *11* 104–108.]

chemiluminogenic probe See e.g. LUMINOL.

chemiosmosis The phenomenon in which an energy-dependent transfer of protons or electrons across an ENERGY-TRANSDUCING MEMBRANE generates, or augments, a transmembrane proton gradient whose inherent energy (*proton motive force, pmf*) can be used for chemical, osmotic or mechanical work. A pmf can be generated e.g. by the operation of an ELECTRON TRANSPORT CHAIN; by the illumination of a PURPLE MEMBRANE; by the hydrolysis of ATP by a PROTON ATPASE; by the hydrolysis of pyrophosphate by a PROTON PPASE; by END-PRODUCT EFFLUX; or by EXTRA-CYTOPLASMIC OXIDATION. A pmf can be used e.g. for ATP (or PYROPHOSPHATE) synthesis (see OXIDATIVE PHOSPHORYLATION); for the concentration of solutes against a concentration gradient (see e.g. ION TRANSPORT); for the regulation of TRANSHYDROGENASE activity; for REVERSE ELECTRON TRANSPORT; and (in flagellated bacteria) for the rotation of the FLAGELLUM. (See also AEROTAXIS and PHOTOTAXIS; cf. SODIUM MOTIVE FORCE.)

Proton motive force is usually represented as the sum of (a) the 'electrical' energy, and (b) the 'chemical' energy associated with protons by virtue of their distribution across a transmembrane proton gradient – such a gradient being, simultaneously, both an electrical (i.e., charge) gradient and a chemical (i.e., concentration) gradient. The 'electrical' component of pmf (= *membrane potential, transmembrane electrical potential difference,* $\Delta\psi$) is generally given in millivolts (mV). The 'chemical' component of pmf corresponds to the transmembrane pH differential, ΔpH; ΔpH can be converted to millivolts by multiplying by the factor 2.3(RT/F) in which R is the gas constant, T is the absolute temperature, and F is the Faraday constant. The factor 2.3(RT/F) is commonly indicated by the letter Z. The expression for pmf (symbol: $\Delta\tilde{\mu}_{H^+}$ or Δp) is thus:

$$\Delta p = \Delta\psi - Z\Delta pH \qquad (1)$$

Z has a value of ca. 60 at 30°C. In this equation the minus sign is necessary for the correct summation of energies: ΔpH is itself commonly a negative quantity, being obtained by the subtraction of a higher pH from a lower one. (The equation above refers specifically to *proton* gradients.)

In bacteria, the outward pumping of protons during respiratory pmf generation causes the inner (cytoplasmic) side of the cytoplasmic membrane to become both alkaline and electrically negative relative to the outer (periplasmic) side; in mitochondria it is the matrix side of the inner membrane which becomes relatively alkaline and electrically negative. By contrast, in the chloroplast thylakoid membrane, protons are pumped inwards during PHOTOSYNTHESIS, so that electrochemical gradients develop in the opposite orientation.

Respiring cells of *Escherichia coli* develop a pmf of the order of 200 mV; thus, e.g. $\Delta\psi$ may be ca. 100 mV and ΔpH ca. −1.75 (Δp = 205 mV). These cells maintain an internal pH of ca. 7.4–7.8 when the extracellular pH ranges from 5.5 to 9.0, i.e., the 'bulk-phase' transmembrane ΔpH is highly dependent on the extracellular pH. (By contrast, in *Clostridium pasteurianum* a decrease in extracellular pH from 7.1 to 5.1 corresponds to a decrease in internal pH from 7.5 to 5.9 [Book ref. 82, pp. 137–138].) In respiring *E. coli*, the Δp drops sharply if the external pH rises above ca. 7.8, though Δp can be augmented under these conditions by increasing the extracellular concentration of K^+ or Na^+ [JB (1985) 163 423–429]. In chloroplasts, the contribution of ΔpH to Δp is typically much greater than it is in bacteria such as *E. coli*.

In a respiring bacterial cell there is a characteristic *proton circuit*: protons are extruded across the cytoplasmic membrane by the primary pmf generator (electron transport chain) and they re-cross the membrane via an H^+-ATPase during ATP synthesis. In such a cell the pmf acts as a thermodynamic back-pressure which opposes further extrusion of protons. However, proton extrusion can occur not only through the operation of an electron transport chain: it can also occur as a result of the tendency of the intracellular reaction (ATP ↔ ADP + Pi) to reach equilibrium; in a respiring cell, the energy derived from metabolism holds this reaction far to the left of its equilibrium position, so that the tendency to reach equilibrium (by ATP hydrolysis) is associated with a large amount of Gibbs free energy (ΔG_{ATP}; ΔG_p; *phosphate potential*; *phosphorylation potential*). ΔG_{ATP} tends to promote ATP hydrolysis (at the H^+-ATPase on the inner side of the membrane) and, hence, to promote proton extrusion in opposition to the pmf. ΔG_{ATP} and pmf may tend to maintain a dynamic balance; however, if pmf is below a certain value ATPases tend to become (reversibly) inactivated: see PROTON ATPASE. Pmf is the primary factor responsible for RESPIRATORY CONTROL.

Experimental manipulation of Δp, $\Delta\psi$ *or* ΔpH. Various IONOPHORES can be used to collapse or modify pmf or either of its components. Thus, e.g., VALINOMYCIN can alter $\Delta\psi$, without altering ΔpH, by permitting K^+ influx into the bacterial cytoplasm sufficient to reduce or eliminate $\Delta\psi$. By contrast, NIGERICIN can effect an electroneutral exchange of K^+ for H^+, thus altering ΔpH without disturbing $\Delta\psi$. UNCOUPLING AGENTS can abolish Δp (or prevent its development) and (as a consequence) can abolish respiratory control. Other agents (e.g. AUROVERTINS, DCCD, efrapeptin, OLIGOMYCIN, QUERCETIN, TENTOXIN and tributyltin chloride) can inhibit (F_0F_1)-type H^+-ATPases.

Measurement of Δp. The components of Δp ($\Delta\psi$ and ΔpH) are estimated separately and added.

$\Delta\psi$ is often estimated by allowing an indicator ion to distribute across the membrane so as to reach electrochemical equilibrium with the membrane potential; at equilibrium, a measurement is made of the concentration of the indicator ion on either side of the membrane, and $\Delta\psi$ is estimated from the Nernst equation (see equation 3 in ION TRANSPORT). The indicator ion must be permeable only by electrical uniport (so that its equilibrium distribution reflects $\Delta\psi$); it should not be bound or metabolized; it should disturb the membrane potential only minimally; and it should be measurable with ease. Ions used include the synthetic, lipophilic 'Skulachev' cations (e.g. triphenylmethylphosphonium, $TPMP^+$) and anions (e.g. tetraphenylborate, TPB^-) which disturb the gradient minimally since they can be used in very low concentrations. Methods used for measuring equilibrium concentrations of indicator ions include e.g. continual monitoring of the given ion in the incubation medium.

ΔpH is usually estimated from the equilibrium distribution of a weak acid (e.g. 5,5-dimethyl-4-oxazolidinedione, DMO) or a weak base which permeates the membrane electroneutrally.

One of the unresolved problems in chemiosmosis concerns the precise mode of transmembrane proton translocation; some workers favour the concept of VECTORIAL GROUP TRANSLOCATION, while others favour a model in which proton translocation is associated with conformational changes in the membrane proteins (see ELECTRON TRANSPORT CHAIN). Another problem concerns the extent to which the proton circuit is delocalized into the bulk phases on either side of the membrane; some evidence suggests that the proton circuit is more or less localized at the membrane surface (in accordance with the 'semi-localized' hypothesis) [Nature (1986) 322 756–758].

(See also LUNDEGARDH-TYPE RESPIRATION.)

chemoattractant See CHEMOTAXIS.

chemoeffector See CHEMOTAXIS.

chemokines A category of CYTOKINES which function primarily as chemoattractants for particular type(s) of leukocyte – and which are important in the recruitment of leukocytes to regions of infection and INFLAMMATION.

Chemokines are typically small (8–10 kDa) secreted proteins which are produced by a range of cells that include monocytes, macrophages, fibroblasts, T cells, epithelial cells and endothelial cells.

Chemokines are divided into several families on the basis of e.g. composition and function. Every cytokine molecule contains several conserved cysteine (C) residues, and the sequence of amino acids containing these residues differs in different families. Thus, in the CXC family (α-chemokines) the two initial cysteine residues are separated by one amino acid residue (X). In the CC family (β-chemokines) the two initial cysteine residues are adjacent. Members of the C family (γ-chemokines) lack some of the cysteine residues present in the α- and β-chemokines. Yet another family, CX3C, includes the membrane-bound chemokine fractalkine.

The two main families (CXC and CC) differ in the principal types of cell which they attract; thus, typically (though not exclusively), the CXC chemokines attract neutrophils while the CC chemokines attract monocytes.

The cell-surface receptors for chemokines (in most or all cases) are transmembrane molecules of the rhodopsin superfamily. The intracellular domain of such receptors is coupled to G PROTEINS which initiate intracellular signalling when chemokines bind to their cognate receptors; in at least some cases, the intracellular signalling pathway downstream of the G proteins includes protein kinase C (PKC).

Some receptors can bind only one type of chemokine (e.g. the receptor CXC-CKR2 is specific for the CXC-type chemokine IL-8) but other receptors can bind more than one type of chemokine. A receptor on erythrocytes (red blood cells) can bind both CXC and CC chemokines; such binding is non-functional, but it may serve as a 'sink' to prevent inappropriate stimulation of leukocytes.

Chemokines of the CXC family include INTERLEUKIN-8, PF4 (platelet factor 4: produced by the α granules of platelets, attractant for e.g. fibroblasts) and IP10 (produced by monocytes and endothelial cells, attractant for e.g. NK cells). (The two latter chemokines are reported to have anti-angiogenic activity, and they may inhibit tumour formation.)

Chemokines of the CC family include eotaxin (produced by macrophages and endothelial cells, attractant primarily for eosinophils) and RANTES ('regulated on activation, normal T cell expressed and secreted': produced by T cells, monocytes and epithelial cells, attractant for e.g. T cells and NK cells).

Chemokines of the C family include lymphotactin: a (murine) protein produced e.g. by activated T cells and a potent attractant for T cells.

chemoklinokinesis Klinokinesis (see KINESIS sense 2) in which the stimulus is a change in concentration of a chemical.

chemolithoautotroph A CHEMOTROPH with properties of a LITHOTROPH and an AUTOTROPH. The ability to grow chemolithoautotrophically is confined to a relatively small number of prokaryotes; these organisms make important contributions to the cycles of matter, and some are involved in LEACHING processes.

Plasmid-mediated anaerobic chemolithoautotrophic growth has been reported for a strain of *Sulfolobus ambivalens* containing the plasmid pSL10. Aerobically, this organism derives energy from the oxidation of elemental sulphur (with oxygen as the terminal electron acceptor), yielding sulphuric acid; anaerobically it obtains energy from the oxidation of hydrogen (with elemental sulphur as terminal electron acceptor), yielding hydrogen sulphide [Nature (1985) *313* 789–791].

chemolithoheterotroph A CHEMOTROPH with the properties of a LITHOTROPH and a HETEROTROPH. (See also MIXOTROPHY.)

chemolithotroph See CHEMOTROPH.

chemoorganoheterotroph A CHEMOTROPH with the properties of an ORGANOTROPH and a HETEROTROPH. The majority of bacteria and fungi are chemoorganoheterotrophs (cf. CHEMOLITHOAUTOTROPH); some such bacteria are facultative chemolithoautotrophs. Taken together, chemoorganoheterotrophs can use a very wide range of organic substrates, though particular species may be able to use only a very limited number of compounds. Because most – if not all – chemoorganotrophs are heterotrophic, the term 'heterotroph' is sometimes used as a synonym of 'chemoorganoheterotroph'.

chemoorganotroph See CHEMOTROPH.

chemoprophylaxis The use of drug(s) for PROPHYLAXIS.

chemoreceptor See CHEMOTAXIS.

chemorepellent See CHEMOTAXIS.

chemorespiration See PHOTORESPIRATION.

chemostat See CONTINUOUS CULTURE.

chemosynthetic primary production See PRIMARY PRODUCTION.

chemotaxigen Any agent which promotes the formation of a CYTOTAXIN.

chemotaxis A TAXIS in which the stimulus is a concentration gradient of a given chemical (a *chemoeffector*). A cell may make a net movement towards higher concentrations of some chemoeffectors (*chemoattractants*) but away from others (*chemorepellents*); a chemoattractant may or may not be a nutrient. Cells may exhibit chemotaxis when they occupy a physiologically suboptimal position within the concentration gradient.

A cell capable of chemotaxis must be able to (i) detect ('sense') changes in the concentration of a chemoeffector, (ii) transmit this information to locomotory organelle(s), and (iii) make the appropriate response; it must also be able to *adapt* by resuming the unbiased mode of motility when in a uniform concentration of chemoeffector, regardless of the actual concentration.

(a) In *Escherichia coli* (and other enterobacteria) FLAGELLAR MOTILITY within a *uniform* concentration of chemoeffector(s) results in a pattern of movements referred to as a RANDOM WALK. If, however, a cell be placed in a concentration gradient of e.g. chemoattractant, the cell will subsequently exhibit a modified pattern of movements; thus, if the cell were initially swimming towards the higher concentration of chemoattractant it will maintain a smooth mode of swimming (counterclockwise flagellar rotation), but if swimming away from the higher concentration of chemoattractant it will exhibit a greater tendency to tumble (clockwise flagellar rotation) – thus increasing the probability of changing to a more favourable direction.

'Sensing' the concentration gradient is achieved by means of specific receptors (in the cytoplasmic membrane) to which the chemoeffector binds; the binding of chemoeffector to a receptor results in the transmission of a signal that promotes either counterclockwise or clockwise flagellar rotation – that is, either smooth swimming or tumbling, respectively.

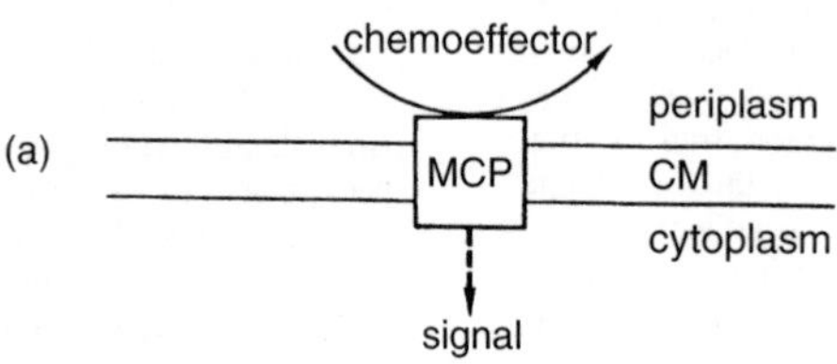

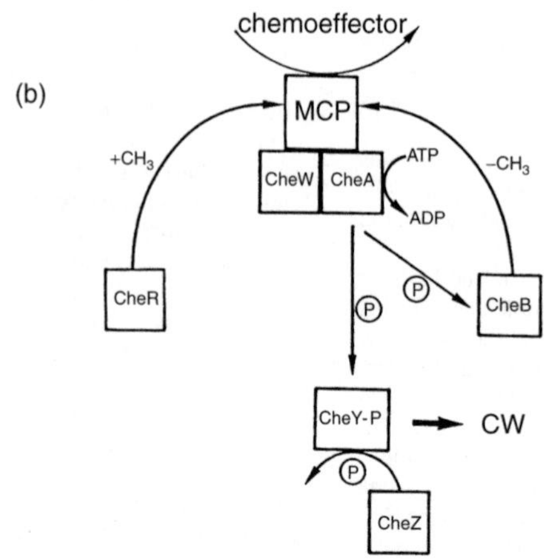

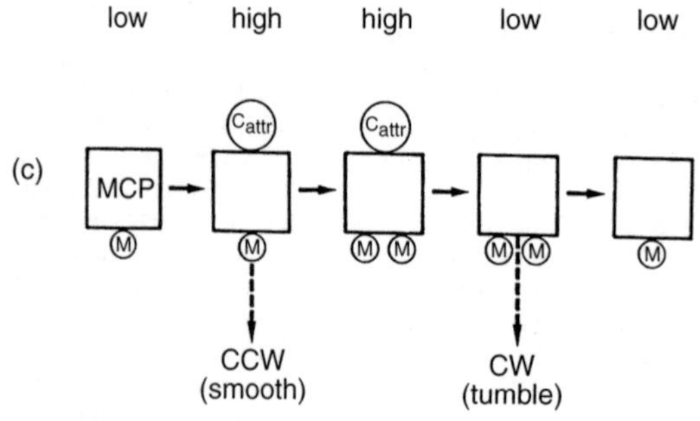

When a cell is swimming towards a higher concentration of chemoattractant, the receptors will be more likely, within a given period of time, to bind molecules of the chemoattractant than if the cell were moving in another direction; when a molecule of chemoattractant binds to a receptor it induces a signal that promotes smooth swimming. Conversely, if the cell is swimming *down* the concentration gradient, there is likely to be (over a period of time) a net loss of chemoattractant molecules bound to receptors; when a molecule of chemoattractant leaves a receptor it stimulates a signal that promotes tumbling.

The diagram (see accompanying figure) gives an outline of the mechanism of receptor-mediated chemotaxis in *Escherichia coli*. The receptors are designated *methyl-accepting chemotaxis proteins* (MCPs). In *E. coli*, MCPs include the products of genes *tap*, *tar*, *trg* and *tsr*.

Some chemoeffectors bind directly to their MCPs. Thus, e.g. serine and aspartate bind to the Tsr and Tar proteins, respectively. Other chemoeffectors, such as galactose, maltose and ribose, initially form a complex with certain periplasmic *binding proteins* before binding to their respective MCPs (see BINDING PROTEIN-DEPENDENT TRANSPORT SYSTEM); thus, e.g. the maltose complex binds to the Tar protein, and the galactose and ribose complexes bind to the Trg protein. (Uptake and metabolism of these sugars are not necessary for a chemotactic response.)

Cells can also respond chemotactically to substrates taken up by PTS (q.v.). The mechanism is not understood, but one suggestion is that particular PTS component(s) interact – via unknown intermediate(s) – with the CheY protein (see figure), thus influencing flagellar rotation. PTS-mediated chemotaxis differs from the MCP system in several ways – e.g. (i) substrates must be *transported*, not simply bound; (ii) adaptation does not involve methylation; and (iii) the system does not respond to repellents [e.g. JCB (1993) *51* 1–6].

The mechanism by which flagellar rotation is switched between counterclockwise and clockwise modes appears to be independent of pmf, although pmf is necessary for driving the flagellar motors (cf. AEROTAXIS and PHOTOTAXIS).

Interestingly, although SWARMING in *E. coli* involves components of the chemotaxis system, the substances that promote swarming appear to differ from the known chemoeffectors in chemotaxis [PNAS (1998) *95* 2568–2573].

CHEMOTAXIS in *Escherichia coli*: a model of receptor-mediated chemotaxis involving signal transduction.

(a) The receptors are called *methyl-accepting chemotaxis proteins* (MCPs), and they occur in the cytoplasmic membrane (CM). MCPs have functional regions in both the periplasm and cytoplasm. There are different types of MCP, each type recognizing its own range of chemoeffectors, and there are several hundred molecules of each type of MCP in a given cell. (In *E. coli* the MCPs are encoded by genes *tap*, *tar*, *trg* and *tsr*.) Some chemoeffectors (e.g. aspartate, serine) bind directly to their MCPs, but others (e.g. some sugars, including ribose) first form a complex with certain periplasmic proteins before binding. The binding/release of chemoeffectors by an MCP regulates an intracellular signal that influences the frequency of tumbling.

(b) The intracellular signal originates at the protein CheA (encoded by gene *cheA*). CheA is a HISTIDINE KINASE. In the diagram, CheA is bound to the MCP; CheW is a coupling protein. CheA can transfer phosphate from ATP to a regulator protein, CheY; the phosphorylated (activated) form of CheY (CheY~P) enhances clockwise (CW) flagellar rotation (and, hence, increases the frequency of tumbling) by interacting with the C ring of the flagellar motor (apparently with the FliM component).

[Role of CheY: TIM (1999) *7* 16–22.] The basic signal thus consists of the transfer of phosphate from CheA to CheY. This signal is modulated (i.e. regulated) by (i) the binding/release of chemoeffector at the MCP, and (ii) the degree of methylation of the MCP; these two factors are considered below. (Although *phosphorylation* of CheY has been regarded as the essential mode of signalling in chemotaxis, there is some evidence that, in at least some cases, *acetylation* of CheY may be involved in the response to repellents [reported in JB (2000) *182* 1459–1471 (1466)].)

The binding of a chemoattractant *inhibits* the kinase activity of CheA, thus inhibiting the transfer of phosphate to CheY. (On binding of chemoattractant, it appears that subunits of the MCP move relative to one another, and that such movement is involved in the inhibition of CheA kinase activity [see e.g. JB (2000) *182* 1459–1471 (1465–1466)].) Conversely, the release of chemoattractant from an MCP *stimulates* CheA and promotes the transfer of phosphate to CheY, thus encouraging tumbling.

The cytoplasmic side of an MCP is subject to ongoing methylation by a methyltransferase, CheR, which uses *S*-adenosylmethionine as methyl donor. Opposing this, a methylesterase, CheB, removes the methyl groups; the activity of CheB is enhanced when it is phosphorylated by CheA. The degree of methylation of an MCP thus depends on the activities of both CheR and CheB; increased methylation tends to *enhance* the activity of CheA.

(c) In a low, uniform concentration of chemoattractant (C_{attr}), tumbling occurs at a given rate. When the cell swims *up* a concentration gradient (i.e. towards higher concentrations), there is an increase in the number of molecules of C_{attr} bound to MCPs; this results in the inhibition of CheA which, in turn, inhibits phosphorylation of CheY – and thus promotes counterclockwise (CCW) flagellar rotation, i.e. smooth swimming in the same direction. Inhibition of CheA also inhibits CheB, allowing increased methylation of the MCP (as CheR methylates continually); hence, inhibition of CheA, due to binding of C_{attr}, is subsequently offset by the stimulatory effect of increased methylation of the MCP – so that the frequency of tumbling returns to its original value. The cell has thus *adapted* to the new, higher, concentration of C_{attr} (MCP in the centre).

On swimming *down* the concentration gradient (i.e. away from higher concentrations), there is a decrease in the number of molecules of C_{attr} bound to the MCPs. Release of C_{attr} stimulates CheA and promotes clockwise (CW) flagellar rotation (more frequent tumbling); at the same time, CheB is stimulated, leading to de-methylation of the MCP until CheA activity has been reduced to its original level – i.e. adaptation to the new, low, uniform concentration of C_{attr} (MCP at right-hand side). Such adaptation depends on an appropriate rate of dephosphorylation of CheY~P and CheB~P; CheB~P undergoes autodephosphorylation, while CheY~P levels are regulated by the CheZ protein. (It appears that about 30% of the intracellular pool of CheY molecules are in the phosphorylated state in fully adapted cells [EMBO (1998) *17* 4238–4248].)

Note that the degree of methylation of an MCP in an *adapted* cell reflects the extracellular concentration of the chemoattractant (compare first, third and fifth MCPs in the diagram).

Experimental modification of the concentration of various components of the chemotaxis system (e.g. CheR) indicates that, whereas variation can occur in factors such as the time taken to adapt (e.g. a >20-fold variation with variation in CheR concentration), the *precision* of adaptation (i.e. the capacity to return to the exact pre-stimulus level of tumbling) is preserved [Nature (1999) *397* 168–171].

[Control of chemotaxis: Mol. Microbiol. (1996) *20* 903–910. Aspects of chemosensing (including non-enteric bacteria): JB (1998) *180* 1009–1022. Signalling components in bacterial locomotion and sensory reception: JB (2000) *182* 1459–1471. Signalling in bacterial chemotaxis (review): JB (2000) *182* 6865–6873.]

(b) In the immune system (see e.g. INFLAMMATION), chemotaxis refers to the migration of certain types of leukocyte (e.g. MACROPHAGE, NEUTROPHIL) in the direction of higher concentrations of chemoeffectors (= *cytotaxins*); the cytotaxins include certain CYTOKINES and the ANAPHYLATOXIN C5a.

(c) In DICTYOSTELIUM (q.v.) and other cellular slime moulds, chemotaxis is involved e.g. in cell aggregation.

chemotaxonomy TAXONOMY based on chemical criteria such as DNA base composition, cell wall lipids, 16S rRNA structure etc.

chemotherapeutic agent Any drug used for chemotherapy – e.g. an ANTIBIOTIC.

chemotherapeutic index *Syn.* THERAPEUTIC INDEX.

chemotherapy The use of drug(s) (e.g. ANTIBIOTICS) for the treatment of disease.

chemotroph An organism whose energy is derived from endogenous, light-independent chemical reactions (a mode of metabolism termed *chemotrophy*). (cf. PHOTOTROPH.) A chemotroph which obtains energy by the metabolism of inorganic substrate(s) – i.e. a LITHOTROPH – is called a *chemolithotroph*; one which obtains energy by metabolizing organic substrate(s) – i.e. an ORGANOTROPH – is called a *chemoorganotroph*.

According to the type of compound from which carbon is assimilated, a chemolithotroph may be an AUTOTROPH (i.e. a CHEMOLITHOAUTOTROPH) or a HETEROTROPH (i.e. a chemolithoheterotroph). (See also MIXOTROPHY.). A heterotrophic chemoorganotroph is called a CHEMOORGANOHETEROTROPH.

The mode of metabolism may vary according to growth conditions: e.g., *Thiobacillus* strain A2 can exhibit chemolithoautotrophic, mixotrophic or chemoorganoheterotrophic growth under appropriate conditions [JGM (1980) *121* 127–138].

chemotrophy A mode of metabolism in which energy is derived from endogenous, light-independent chemical reactions (cf. PHOTOTROPHY). (See also CHEMOTROPH.)

chemotropism See TROPISM (sense 1).

chemotype Any VARIETY distinguished by structural properties – e.g. a rough mutant.

chenodeoxycholic acid See BILE ACIDS.

cherry leaf roll virus See NEPOVIRUSES.

cherry rasp leaf virus See NEPOVIRUSES.

Cheshunt compound An agricultural antifungal preparation which consists of a mixture of copper sulphate (see COPPER) and ammonium carbonate; it is used, in aqueous solution, as a protectant against DAMPING OFF.

chestnut blight (chestnut canker) A lethal disease of the American chestnut (*Castanea dentata*) caused by invasion of the vascular cambium by *Endothia parasitica*. The disease has virtually eliminated the American chestnut from parts of America; however, in Italy the disease has been ameliorated by the appearance of hypovirulent strains of the pathogen capable of transferring the HYPOVIRULENCE phenotype to virulent strains ('transmissible hypovirulence'). Hypovirulent strains contain dsRNA which can be transmitted e.g. by hyphal anastomosis; no capsid has been detected (cf. MYCOVIRUS), but the dsRNA appears to be closely associated with an RNA polymerase and may be packaged within vesicles formed by the host fungus [JGV (1985) *66* 2605–2614]. [Potential of the hypovirulence factor for biological control of *E. parasitica*: ARPpath. (1982) *20* 349–362; Science (1982) *215* 466–471.] (See also DUTCH ELM DISEASE.)

chi site (chi sequence) See RECBC PATHWAY.

chi structure See RECOMBINATION.

χ1776 A 'disabled' strain of *Escherichia coli* K12 which e.g. requires various amino acids (including diaminopimelic acid) and is highly sensitive to bile salts and other detergents. It was constructed for use in CLONING experiments to reduce the risk of propagation of potentially dangerous clones should they escape from the laboratory. However, χ1776 is difficult to grow and to transform, and is currently rarely used.

chiasma The visible manifestation of CROSSING OVER – e.g. between non-sister chromatids of homologous chromosomes during MEIOSIS.

chiasma interference See INTERFERENCE (sense 2).

chiastobasidial Refers to a *transverse* arrangement of nuclear spindles in a developing BASIDIUM. (cf. STICHOBASIDIAL.)

chick embryo lethal orphan virus See AVIADENOVIRUS.

Chick–Martin test A SUSPENSION TEST used to determine the PHENOL COEFFICIENT of a (phenolic) disinfectant; it assesses the efficacy of a disinfectant in the presence of organic matter, i.e., under conditions more closely approximating those in practical use (cf. RIDEAL–WALKER TEST). (Many disinfectants are inactivated by organic matter.) In the original test the organic matter used was 3% sterile dry faeces; a modified form of the test uses a sterilized suspension of yeast (5% dry wt/vol.). To each of several serial dilutions of the disinfectant under test are added the yeast and an inoculum of *Salmonella typhi*. After incubation for 30 min, an inoculum from each dilution is added to each of two tubes of broth which, after incubation at 37°C for 2 days, are examined for growth. The same procedure is carried out with phenol. For the *phenol* test two dilutions are identified: (i) that containing the highest concentration of phenol which permitted growth in *both* tubes inoculated from it, and (ii) that containing the lowest concentration of phenol which did not permit growth in *either* tube inoculated from it; the mean of these two concentrations is then calculated (P_{mean}). For the disinfectant under test D_{mean} is calculated in the same way. The phenol coefficient of the disinfectant is given by P_{mean}/D_{mean}.

chicken diseases See POULTRY DISEASES.

chicken syncytial virus See AVIAN RETICULOENDOTHELOSIS VIRUSES.

chickenpox (varicella) An acute, highly infectious disease which usually occurs in epidemics affecting mainly young children. The causal agent is human (alpha) herpesvirus 3 (varicella-zoster virus) – see ALPHAHERPESVIRINAE. Infection occurs by direct contact or by droplet inhalation. The incubation period is ca. 2–3 weeks. Symptoms: fever and a rash of erythematous macules which rapidly develop into thin-walled vesicles, finally becoming dry and crusty; successive crops of lesions occur. The disease is usually mild and self-limiting; complications include e.g. pneumonia (usually in adults), encephalitis, and secondary infection of skin lesions with staphylococci and streptococci. In immunosuppressed or CMI-defective patients the disease is much more severe, with extensive rash and severe systemic symptoms (*progressive varicella*) which may be fatal; vidarabine and acyclovir have been used in treatment. A live vaccine has been developed [MS (1985) *2* 249–254]. *Lab. diagnosis*: viruses can be isolated from vesicular fluid during the first 3 days of the rash; scrapings from the base of a vesicle may be stained and examined for typical varicella GIANT CELLS and eosinophilic intranuclear inclusion bodies. (See also HERPES ZOSTER.)

chiclero's ear A chronic, ulcerating lesion on the ear-lobe, caused by *Leishmania mexicana*. It occurs in Panama, Honduras, and the Amazon region.

chicory blotch virus See CARLAVIRUSES.

chicory yellow mottle virus See NEPOVIRUSES.

chikungunya fever An acute infectious human disease caused by an ALPHAVIRUS and transmitted by mosquitoes (*Aedes* spp); epidemics occur in Africa, India and S. E. Asia. The disease is characterized by severe joint and back pains with fever; rarely, a severe (often fatal) haemorrhagic form occurs. Wild primates may act as a reservoir of infection.

childbed fever *Syn.* PUERPERAL FEVER.

Chilo iridescent virus See IRIDOVIRUS.

Chilodochona A genus of marine protozoa (subclass HYPOSTOMATIA). Cells: vase-shaped, i.e., the body is narrow at the point of attachment, broadening out before constricting and finally becoming flared and convoluted at the distal end.

Chilodonella A genus of ciliates (subclass HYPOSTOMATIA). The cells are flattened both ventrally and in the dorsal anterior region, and are rounded in the dorsal posterior region; there is an anteriolateral point ('break'), and the cytostome is located subterminally in the anterioventral surface. Species occur e.g. as gill parasites of freshwater fish. and are common in stagnant ponds etc and in some SEWAGE TREATMENT plants.

Chilomastix A genus of flagellate protozoa of the order RETORTAMONADIDA. *C. mesnili* is a non-pathogen which occurs e.g. in the caecum and colon in man and other animals (e.g. pigs); it is a pear-shaped organism, ca. 10–15 μm in length, with three anteriorly directed flagella arising from the blunt anterior end, and a fourth flagellum within the conspicuous cytostome. The single nucleus is located at the anterior end of the cell. The cysts are lemon-shaped, ca. 6–10 μm; the nucleus and cytostome are usually visible in stained cysts.

Chilomonas See CRYPTOPHYTES.

chimeric plasmid (1) Any recombinant PLASMID formed from nucleic acids derived from organisms which do not normally exchange genetic information (e.g. *Escherichia coli* and *Saccharomyces cerevisiae*). (See e.g. SHUTTLE VECTOR; cf. HYBRID PLASMID sense 2.)

(2) *Syn.* HYBRID PLASMID sense 1.

Chinese letter arrangement An arrangement of the cells of certain bacteria (e.g. *Corynebacterium* spp) as seen in microscopic preparations: the cells are arranged in groups which, to the occidental eye at least, resemble Chinese characters or 'letters'.

chiniofon See 8-HYDROXYQUINOLINE.

Chionosphaera A genus of fungi (see FILOBASIDIACEAE). Basidia occur in a hymenium covering the enlarged terminal portion of a small (<1 mm), stalked fruiting body. One species: *C. apobasidiales*, isolated from dead tree-limbs. [Book ref. 100, pp. 470–471.]

chip (DNA) See DNA CHIP.

***Chironomus luridus* EPV** See ENTOMOPOXVIRINAE.

***Chironomus plumosus* iridescent virus** See IRIDOVIRUS.

chitan CHITIN in which all of the glucosamine residues are *N*-acetylated. In nature, chitan may occur only in diatom spines. (cf. CHITOSAN.)

chitin A linear polysaccharide consisting of (1 → 4)-β-linked D-glucosamine residues, most of which are *N*-acetylated (cf. CHITAN; CHITOSAN). Chitin is abundant in nature, occurring e.g. in arthropod exoskeletons, in the CELL WALLS of most fungi, in certain algae (e.g. in the spines of many planktonic centric diatoms and in the lorica of *'Poterioochromonas stipitata'* – see OCHROMONAS), and in certain protozoa (e.g. in the cyst walls of ciliates and amoebae); it is apparently never formed by bacteria.

Native chitin occurs in the form of microfibrils in which extended glucan chains are hydrogen-bonded together (antiparallel in α-chitin, parallel in β-chitin) to form sheets which, in turn, associate (by hydrogen-bonding in α-chitin but not in β-chitin) to form microfibrils.

Biosynthesis of chitin involves the enzyme chitin synthase (EC 2.4.1.16) which catalyses the polymerization of *N*-acetylglucosamine from UDP-*N*-acetylglucosamine precursors. The enzyme requires Mg^{2+} or Mn^{2+} and phospholipids for activity, and is inhibited e.g. by POLYOXINS and NIKKOMYCINS. The location of chitin synthase in fungal cells is controversial; most reports suggest that it occurs in the cytoplasmic membrane [Book ref. 38, pp. 403–405], while evidence has been presented that it is mainly cytoplasmic in *Mucor rouxii* [JGM (1984) *130* 1193–1199]. (See also CHITOSOME and GAMMA PARTICLE (sense 2).) Chitin synthase occurs in zymogen and active forms; its activity in the cell may be controlled by regulation of the proteolytic activation of the zymogen at sites of chitin synthesis. [Chitin biosynthesis in microorganisms: Book ref. 62, pp. 85–127.] (See also GROWTH (fungal).)

(See also CHITINASE.)

chitinase (poly-[1,4-β-(2-acetamido-2-deoxy-D-glucoside)] glycanohydrolase; EC 3.2.1.14) An enzyme which degrades CHITIN to chitobiose [JGM (1984) *130* 1857–1861]; chitobiose can be hydrolysed to *N*-acetylglucosamine by *chitobiase* (*N*-acetyl-β-D-glucosaminidase, EC 3.2.1.30). Chitinases are produced by a wide range of organisms: e.g., chitin-containing fungi; microorganisms parasitic on chitin-containing organisms (e.g. entomopathogens – see also BURNED SPOT DISEASE); insectivorous animals (e.g. spiders); and plant tissues (where they may play a role in resistance to fungal diseases [Nature (1986) *324* 365–367]).

chitinolytic (chitinoclastic) CHITIN-degrading: see CHITINASE.

chitobiase See CHITINASE.

chitobiose A disaccharide consisting of two (1 → 4)-β-linked *N*-acetyl-D-glucosamine residues.

chitosan (mycosin) A polymer of non-acetylated (1 → 4)-β-linked D-glucosamine residues (cf. CHITIN). It occurs in nature as a CELL WALL constituent in the hyphae of certain zygomycetes (e.g. *Mucor* spp) and as a minor component of the spores of *Fusarium solani* (in which it is a strong elicitor of PHYTOALEXIN production by pea plants [Book ref. 58, pp. 270–271]). [Coordination of chitosan and chitin synthesis: JGM (1984) *130* 2095–2102.]

chitosome A type of vesicular particle (40–70 nm diam.) comprising a membrane-like shell containing chitin synthase; chitosomes can be isolated from a wide range of fungi and can synthesize CHITIN microfibrils from UDP-*N*-acetylglucosamine in vitro. Initially, coiled microfibrils develop within the chitosome – which then apparently breaks open to release extended microfibrils. It is uncertain whether chitosomes are organelles (serving e.g. in the transport of chitin synthase from its site of synthesis to its site of action) or small vesiculated fragments of cytoplasmic membrane formed during the isolation procedure. Chitin synthase can be released from chitosomes by treatment with digitonin.

Chl Chlorophyll: see CHLOROPHYLLS.

Chlamydia (formerly e.g. *Bedsonia*, *Miyagawanella*; cf. TRIC agents) A genus of Gram-negative bacteria (order CHLAMYDIALES) which occur as obligate intracellular parasites in man and other animals. Strains of *Chlamydia* can cause various types of disease – e.g. CURTIS–FITZ-HUGH SYNDROME, INCLUSION CONJUCTVITIS, KERATOCONJUNCTIVITIS, LYMPHOGRANULOMA VENEREUM and TRACHOMA in man, PSITTACOSIS in man and animals, diarrhoea in new-born calves, lambs and piglets, and abortion (see e.g. ENZOOTIC ABORTION) and eye infections in various animals. [Human chlamydial infections: Book ref. 193. Clinical conditions associated with positive COMPLEMENT-FIXATION TESTS for *Chlamydia*: Epidem. Inf. (1987) *98* 101–108.]

The organisms undergo a development cycle, in only part of which are they infectious for fresh host cells. The cells are non-motile, coccoid and pleomorphic, 0.2–1.5 μm diam., size varying with the phase of development. The chlamydial cell envelope contains little or no muramic acid, but includes a Gram-negative-type OUTER MEMBRANE containing LPS; a heat-stable, genus-specific, KDO-containing antigen occurs in all strains throughout the development cycle and is responsible for positive FREI TEST reactions. The outer membrane contains a major protein species (MWt ca. 40000) with which is associated species-specific and type-specific antigens. The inner surface of the outer membrane consists of a monolayer of hexagonally arranged protein units [EM studies of the cell wall: JMB (1982) *161* 579–590]; in the elementary body (see below) the outer surface has a number of hemispherical projections [JB (1985) *164* 344–349]. The organisms have been cultured in laboratory animals, in the yolk sacs of chick embryos, and in cell cultures. GC%: ca. 41–44 (T_m). Type species: *C. trachomatis*.

Development cycle. The form infectious for host cells is the *elementary body* (EB): a non-dividing, osmotically-stable disseminative cell, ca. 0.2–0.4 μm diam., which has few ribosomes, and in which the condensed nuclear material occupies a major part of the cell; the cell wall proteins are extensively cross-linked by disulphide bonds [JB (1984) *157* 13–20]. EBs are metabolically limited: glucose is not metabolized, although glucose 6-phosphate and certain other substrates may yield CO_2 if ATP, NADP and other cofactors are also supplied. The development cycle begins when EBs adhere to unidentified receptor sites on a host cell and undergo cytochalasin D-sensitive endocytosis. [Mechanisms of endocytosis of *C. trachomatis*: JGM (1984) *130* 1765–1780.] Phagosomes within which EBs are internalized do not fuse with lysosomes (cf. PHAGOCYTOSIS). The EBs subsequently differentiate, becoming *reticulate bodies* (RBs): osmotically-labile, trypsin-sensitive cells, ca. 0.6–1.5 μm diam., which contain increased numbers of ribosomes and in which the nuclear material is more diffusely distributed; the disulphide inter-protein cross links in the cell wall are reduced. RBs are not infectious for fresh host cells; those of *C. trachomatis* may be capable of glycolysis and of ATP synthesis [Mol. Microbiol. (1999) *33* 177–187]. Some chlamydiae synthesize e.g. folic acid and glycogen, and all strains must produce the prokaryotic-type fatty acids and lipids which cannot be supplied by the host cell. [Protein synthesis in host-free RBs: JB (1985) *162* 938–942.] The RBs undergo binary fission ca. 8–10 hours after the initial infection of the host cell, and continue to divide for up to ca. 20 hours after infection. By about 12–15 hours after infection, the membrane-bounded population of progeny RBs (called an *inclusion* or *microcolony*) is large enough to be visible by light microscopy; the inclusions of glycogen-forming strains can be detected by staining with iodine. From ca. 20 hours after infection RBs begin to give rise to EBs, and by ca. 48–72 hours after infection the cycle is complete – each infected host cell containing ca. 10–1000 EBs in an inclusion which may occupy a major part of the cell. On release, EBs infect fresh host cells.

Antibiotic resistance. Chlamydiae are typically sensitive to e.g. tetracyclines. Penicillins may inhibit the development of EBs from RBs, although the mechanism of such inhibition is unknown. Most strains of *C. trachomatis* are sensitive to sulphonamides.

C. pecorum. A species from ruminants (including sheep and cattle) which can cause e.g. respiratory, intestinal and neurological disease. [Description of *C. pecorum*: IJSB (1992) *42* 306–308.]

C. pneumoniae. A species first isolated from a suspected case of trachoma in Taiwan (strain TW-183) and subsequently from a case of acute respiratory infection (strain AR-39). These organisms, initially thought to be strains of *C. psittaci*, were referred to as TWAR strains [see Book ref. 193, pp 321–340] but later classified in a new species, *C. pneumoniae* [description of *C. pneumoniae*: IJSB (1989) *39* 88–90]. [Phylogenetic relationship of *C. pneumoniae* to *C. psittaci* and *C. trachomatis*: IJSB (1993) *43* 610–612.] *C. pneumoniae* is associated with e.g. epidemics of subacute respiratory disease; re-infection, or chronic infection, can lead to severe disease, apparently involving immunologically-based pathogenesis.

C. psittaci. Hosts: commonly birds and mammals other than man (but see PSITTACOSIS); *C. psittaci* appears to have played a significant role in an epizootic of urogenital and respiratory-tract disease and keratoconjunctivitis in the koala (*Phascolarctos cinereus*) [AJEBMS (1985) *63* 283–286]. Folic acid and glycogen are not formed.

C. trachomatis. Natural hosts: man and mouse; no strains have been isolated from birds. Folic acid and glycogen are formed. Three biovars: trachoma (typically non-invasive, confined almost exclusively to squamocolumnar epithelial cells), lymphogranuloma venereum (LGV) (typically invasive, infects e.g. lymph nodes), and mouse (isolated only from mice; has an affinity for the lungs). Biovars trachoma and LGV can be distinguished (by means of microimmunofluorescence tests) by virtue of their type-specific antigens: A–K and L_1–L_3, respectively. [Complete genome sequence: Science (1998) *282* 754–759.]

Chlamydiaceae See CHLAMYDIALES.

Chlamydiales An order of Gram-negative bacteria which are obligate intracellular parasites in eukaryotic cells; the organisms multiply within cytoplasmic vacuoles and exhibit a developmental cycle which includes small, rigid-walled infectious forms (elementary bodies), and larger, non-infectious forms (reticulate bodies, sometimes called initial bodies) which have flexible walls and which divide by fission. One family, Chlamydiaceae, which comprises a single genus, CHLAMYDIA.

Chlamydodon See HYPOSTOMATIA.

Chlamydomonas A large genus (>500 species) of unicellular, usually flagellated green algae (division CHLOROPHYTA – cf. PHYTOMASTIGOPHOREA) which are widespread and common in a range of freshwater habitats and e.g. on damp soils. (See also RED SNOW.) The vegetative cells are usually pyriform or ovoid, ca. 5–20 μm in length, uninucleate, haploid, and usually motile by means of two similar flagella which project from the anterior end (cf. PALMELLOID PHASE). (See also FLAGELLUM (b) and FLAGELLAR MOTILITY (b).) The cell wall is thin and contains no cellulose, being composed of a complex of hydroxyproline-rich glycoproteins [wall structure: JCB (1985) *101* 1550–1568, 1599–1607]. The cell contains a single large CHLOROPLAST which may be cup-shaped, stellate, H-shaped, lobed, etc, according to species.

The chloroplast contains an EYESPOT and generally one or more PYRENOIDS. Two (or more) CONTRACTILE VACUOLES are typically present at the anterior end of the cell, and a number of elongated, branched mitochondria occur near the flagellar bases.

Asexual reproduction occurs by endodyogeny or endopolygeny: typically, a cell comes to rest, losing or withdrawing its flagella, and the protoplast undergoes one or more longitudinal cleavages within the parental wall. The 2, 4, 8 or 16 daughter protoplasts develop cell walls and flagella, and are then released by enzymic dissolution of the parent cell wall; the enzymes responsible for wall dissolution ('sporangium wall autolysins') are apparently species-specific [Book ref. 123, pp. 409–418]. [Cell cycle: New Phyt. (1985) *99* 1–40, 41–56.]

Sexual reproduction is isogamous in most species, anisogamous in some. Some species are heterothallic. Vegetative cells may develop into gametes e.g. in response to nitrogen starvation (in *C. reinhardtii*) or to general nutrient stress (in *C. eugametos* [JGM (1985) *131* 1553–1560]); gametes differ from vegetative cells in that, when cells of opposite mating types are mixed, they rapidly clump together by agglutination of their flagella. Agglutination results from the presence on the flagella of a glycoprotein (the 'agglutination factor') which is formed by the cell during gametogenesis. Cells within the clumps become sorted into pairs, their flagella separate, and the cells in each pair become joined at their anterior ends (apical papillae) to form a quadriflagellate, motile zygote; this is followed – often after a period of motility – by plasmogamy and karyogamy. The zygote then develops a thick, ornamented wall, and becomes dormant. On germination, meiosis occurs, and four or eight (occasionally 16 or 32) motile cells are released by dissolution of the zygote wall. The mating process may differ in some respects from one species to another. For example, in *C. reinhardtii* a 'fertilization tubule' extends from the apical papilla of one (mt^+) gamete, and plasmogamy is initiated when the tip of the tubule contacts a dome-shaped region near the flagellar base of the other (mt^-) gamete; by contrast, in *C. eugametos* there are no morphological differences between mt^+ and mt^- gametes, or between gametes and vegetative cells. In *C. reinhardtii* the gametes shed their cell walls following flagellar contact, while in *C. eugametos* the walls are shed after zygote formation. [Form and location of cell wall lytic enzyme in vegetative cells and gametes of *C. reinhardtii*: JCB (1987) *104* 321–329; review of cell–cell interactions in *Chlamydomonas*: ARPphys. (1985) *36* 287–315.]

Chlamydomonas spp are easily maintained in culture and have been used extensively in research e.g. on ultrastructure, physiology, etc. Mutants of *C. reinhardtii* which are defective in chlorophyll synthesis have been obtained, and these can often grow in the dark on media supplemented with acetate, forming yellow colonies. Such mutants are commonly killed by light, although a mutant has been isolated which, in the presence of acetate, can form yellow colonies both in the light and in the dark [JGM (1983) *129* 159–165].

See also e.g. BRACHIOMONAS; CARTERIA; DUNALIELLA; TETRASPORA; VOLVOX.

chlamydospore A loosely used term usually applied to a thick-walled, asexually-produced resting spore formed by certain fungi; the term is sometimes used to refer specifically to the teliospores of smut fungi: see USTILAGINALES. (cf. THALLOSPORE.)

chloramine T See CHLORINE.

Chloramoeba See XANTHOPHYCEAE.

chloramphenicol (chloromycetin; $NO_2.C_6H_4.CHOH.CH(CH_2OH).NH.CO.CHCl_2$) An ANTIBIOTIC (the D-*threo* form of four possible stereoisomers) produced by *Streptomyces venezuelae* and also made synthetically. Chloramphenicol is a broad-spectrum bacteriostatic agent active against a wide range of bacteria; it also inhibits vegetative growth and sporulation in certain fungi [JGM (1983) *129* 3401–3410].

Chloramphenicol binds to the 50S subunit of prokaryotic and mitochondrial RIBOSOMES (but apparently not to the ribosomes in at least some archaeans [SAAM (1985) *6* 125–131]); it inhibits peptidyltransferase and, hence, PROTEIN SYNTHESIS.

Resistance to chloramphenicol in bacteria is commonly due to a plasmid-specified enzyme – *chloramphenicol acetyltransferase* (CAT; also called chloramphenicol transacetylase) – which catalyses an acetyl-CoA-dependent acetylation of chloramphenicol at the C-3 hydroxy group to form a compound with little or no antibiotic activity. Plasmid-specified CAT is usually or always synthesized constitutively in Gram-negative bacteria but inducibly (see also TRANSLATIONAL ATTENUATION) in e.g. staphylococci and streptococci. (See also Tn*9*.)

CAT-producing bacteria can be detected by rosanilin dyes [JB (1982) *150* 1375–1382].

Non-enzymic resistance to chloramphenicol may be due e.g. to (mutant) ribosomes which fail to bind the antibiotic. *Pseudomonas aeruginosa* is innately resistant, being able normally to tolerate high levels of the drug.

Chloramphenicol antagonizes those antibiotics (e.g. penicillins) which inhibit only actively growing and dividing bacteria. Its clinical usage is limited by its toxicity.

chloranil (tetrachloro-*p*-benzoquinone) A QUINONE ANTIFUNGAL AGENT used e.g. as a seed dressing for the prevention of damping off, seed rots etc; it decomposes in the presence of light.

chlorazol black E A black amphoteric triazo DYE used e.g. for the VITAL STAINING of *Mycoplasma* spp and L-forms [JB (1969) *99* 1–7].

chlorbutanol (chlorbutol) Trichloro-*t*-butanol; it is used e.g. as a PRESERVATIVE (at ca. 0.5% w/v) in ophthalmic solutions. At room temperatures chlorbutanol is unstable at alkaline pH, and it can be inactivated by colloids such as bentonite and magnesium trisilicate.

chlorbutol *Syn.* CHLORBUTANOL.

Chlorella A genus of non-motile, unicellular green algae (division CHLOROPHYTA) which occur in fresh water, on soils, bark etc, and in association with various other organisms (see ZOOCHLORELLAE). (cf. VAMPIROVIBRIO.) The cells are spherical (ca. 2–15 µm diam.); each contains a single haploid nucleus and a large, usually cup-shaped chloroplast with or without a pyrenoid. The CELL WALL contains cellulose and e.g. SPOROPOLLENIN. Asexual reproduction occurs by AUTOSPORE formation: the protoplast divides to form 2, 4, 8 or 16 daughter protoplasts which develop cell walls before being released by rupture of the parent wall. Motile cells are never formed. (cf. CHLOROCOCCUM.)

The paucity of morphological features in *Chlorella* spp has led to an increased use of chemotaxonomy in this genus [Book ref. 123, pp. 391–407]. Although primarily photosynthetic, some *Chlorella* spp can ferment glucose to lactate under anaerobic conditions. Most species can reduce NO_3^-, and a few can produce extracellular gelatin- and/or starch-hydrolysing enzymes.

(cf. PROTOTHECA; see also SINGLE-CELL PROTEIN and YAKULT.)

chlorguanide (Paludrine; proguanil) N^1-*p*-chlorophenyl-N^5-isopropyldiguanide: an antimalarial drug (see MALARIA) which, itself, has little activity but which is converted in vivo to a highly active cyclic form ('cycloguanil' – a substituted dihydrotriazine). Cycloguanil binds competitively to the enzyme dihydrofolate reductase (DHFR), inhibiting the synthesis of tetrahydrofolate.

The drug has been used mainly for prophylaxis. However, it exhibits different levels of efficacy in different human populations, so that its usefulness for some groups has been questioned [BJCP (1994) *37* 67–70].

Resistance to cycloguanil develops readily. For example, a mutant form of DHFR (presumably with lower affinity for cycloguanil) arises by a serine-to-threonine change at residue 108 and an alanine-to-valine change at residue 16 of the DHFR molecule [PNAS (1990) *87* 3014–3017].

chlorhexidine (1,6-di-(4′-chlorophenyldiguanido)-hexane) An ANTISEPTIC which is bacteriostatic or bactericidal (depending on concentration) for a wide range of Gram-positive and Gram-negative bacteria; it is not active against bacterial endospores, some fungal spores, or viruses. Chlorhexidine has been reported to inhibit bacterial ATPase activity and damage the bacterial cell membrane; higher concentrations coagulate the cytoplasm. Antimicrobial activity decreases with increasing acidity, becoming very low at ca. pH 5; activity is also decreased in the presence of soaps and anions such as carbonate and chloride. An alcoholic solution of chlorhexidine is used as a skin disinfectant; water-soluble derivatives include the acetate and gluconate (e.g. *Savlon*, used as a skin antiseptic and in throat lozenges, is a mixture of chlorhexidine gluconate (Hibitane) and Cetrimide). [Review: Book ref. 65, pp. 251–270.]

chlorin A 7,8-dihydroporphyrin (see PORPHYRINS), derivatives of which occur in e.g. CHLOROPHYLLS, certain CYTOCHROMES, and sirohaem (see DESULFOVIRIDIN).

chlorinated soda solution *Syn.* DAKIN'S SOLUTION.

chlorinator See WATER SUPPLIES.

chlorine (as an antimicrobial agent) Chlorine is an effective microbicidal agent used e.g. for the disinfection of WATER SUPPLIES, swimming pools etc. Chlorine is a strong OXIDIZING AGENT; it can also react directly with certain groups in cells and viruses, and it reacts with water to form the strongly antimicrobial compound hypochlorous acid (see HYPOCHLORITES). (N.B. The word 'chlorine' is often used to refer to an aqueous solution containing Cl_2, hypochlorous acid and other active chlorine compounds – rather than to elemental chlorine; 'available chlorine' is the oxidizing capacity of a chlorine compound expressed in terms of the equivalent amount of elemental chlorine.) The antimicrobial action of chlorine is decreased by the presence of organic matter and other substances with which it can react.

Chloramines exert antimicrobial activity by decomposing slowly to release chlorine; their activity is thus much less rapid than that of hypochlorites. *Monochloramine* (NH_2Cl) has been used e.g. in the treatment of water supplies. The organic chloramines are less toxic and less irritating to animal tissues and have been used e.g. as antiseptics and/or disinfectants; they include *chloramine T* (sodium *p*-toluenesulphonchloramide) – a white, crystalline, water-soluble solid; *dichloramine T* (*p*-toluenesulphondichloramide) – water-insoluble and less rapidly effective than chloramine T; *halazone* (*p*-carboxy-*N*,*N*-dichlorobenzenesulphonamide) – used for the disinfection of small quantities of drinking water.

(See also CHLORINE DIOXIDE and IODINE.)

chlorine demand See WATER SUPPLIES.

chlorine dioxide (ClO_2) A microbicidal agent active against e.g. bacteria (including endospores), fungi and viruses; it has been used e.g. for the disinfection of swimming-pool water, and as a disinfectant in water-immersion chilling systems for poultry carcasses.

Chloriridovirus (large iridescent insect virus group) A genus of viruses (family IRIDOVIRIDAE) which infect insects; infected larvae – and purified virus pellets – exhibit a yellow-green iridescence. Virions (ca. 180 nm diam.) are ether-resistant. Type species: mosquito iridescent virus; other members include insect iridescent viruses 3–5, 7, 8 and 11–15 [Book ref. 23, p. 57].

***Chloris* striate mosaic virus** See GEMINIVIRUSES.

chlorite dismutase See BIOREMEDIATION.

β-chloroalanine A synthetic ANTIBIOTIC which acts e.g. as an inhibitor of alanine racemase.

chlorobactene See LIGHT-HARVESTING COMPLEX.

Chlorobiaceae A family of non-motile photosynthetic bacteria (suborder CHLOROBIINEAE); species occur e.g. in anaerobic, sulphide-rich freshwater and estuarine muds. The family includes cocci and rods (some species – e.g. *Prosthecochloris aestuarii* – are prosthecate); GAS VACUOLES occur in *Clathrochloris* and PELODICTYON. In green-coloured species the predominant pigments are Bchl *c* or *d* and the carotenoid chlorobactene; in brown-coloured species Bchl *e* and isorenieratene predominate. The organisms are obligate anaerobic phototrophs; they are primarily photolithotrophic autotrophs (electron donors: e.g. sulphur, sulphide – thiosulphate in some strains) which fix CO_2 by the REDUCTIVE TRICARBOXYLIC ACID CYCLE, but many strains require vitamin B_{12}, and all can use simple organic molecules (e.g. acetate) given a suitable electron donor; elemental sulphur, when produced, is deposited extracellularly.

Genera: *Ancalochloris*, CHLOROBIUM (type genus), *Clathrochloris*, PELODICTYON, *Prosthecochloris*. (cf. CHLOROHERPETON; CHLOROPSEUDOMONAS ETHYLICA; CHLOROCHROMATIUM AGGREGATUM.)

[Evidence for bacteria of the Chlorobiaceae in Palaeozoic seas: Nature (1986) *319* 763–765.]

Chlorobiineae A suborder of photosynthetic bacteria (order RHODOSPIRILLALES); the organisms contain bacterioCHLOROPHYLL *c*, *d* or *e* in CHLOROSOMES, and their REACTION CENTRES contain Bchl *a*. Families: CHLOROBIACEAE and CHLOROFLEXACEAE. The suborder appears to be taxonomically unsound; thus, on the basis of RRNA OLIGONUCLEOTIDE CATALOGUING, the species *Chlorobium limicola*, *C. vibrioforme*, *Prosthecochloris aestuarii* and *Chloroherpeton thalassium* constitute a moderately tight phylogenetic group which has no specific relationship with another, somewhat looser group comprising e.g. *Chloroflexus aurantiacus*, *Herpetosiphon aurantiacus* and *Thermomicrobium roseum* [SAAM (1985) *6* 152–156].

Chlorobium A genus of photosynthetic bacteria (family CHLOROBIACEAE). The cells are rods or vibrios, according to species, ca. 1–2.5 μm long; division occurs by binary fission. The genus includes both green species (*C. limicola*, *C. vibrioforme*) and brown species (*C. phaeobacteroides*, *C. phaeovibrioides*).

chlorobium chlorophylls See CHLOROPHYLLS.

chlorobium vesicles *Syn.* CHLOROSOMES.

Chlorochromatium aggregatum A consortium consisting of an aggregate of cells: a central, colourless, chemoorganotrophic, motile bacterium covered by green bacteria of the family Chlorobiaceae [microanatomy and ecology: JGM (1984) *130* 2717–2723]. Similar consortia, which involve species of the Chlorobiaceae, are called *Chlorochromatium glebulum*, *Cylindrogloea bacterifera*, and *Pelochromatium roseo-viride*.

Chlorochromatium glebulum See CHLOROCHROMATIUM AGGREGATUM.

Chlorociboria See HELOTIALES.

Chlorococcum A genus of unicellular green algae (division CHLOROPHYTA) which differ from CHLORELLA spp e.g. in producing biflagellate zoospores as well as autospores. Species occur e.g. on damp soil, brick-work, etc.

chlorocresol See PHENOLS.

Chlorocystis A genus of unicellular coccoid green algae (division CHLOROPHYTA).

Chloroflexaceae A family of filamentous, photosynthetic GLIDING BACTERIA (suborder CHLOROBIINEAE) which includes the genera CHLOROFLEXUS, CHLORONEMA and *Oscillochloris* [ARM (1981) *35* 339–364].

Chloroflexis See CHLOROFLEXUS.

Chloroflexus (formerly *Chloroflexis*) A genus of bacteria (family CHLOROFLEXACEAE). *C. aurantiacus* occurs in hot, neutral to alkaline (pH ca. 8) springs (often in mats in association with cyanobacteria) and grows optimally at ca. 50–55°C. Under anaerobic conditions in the light, *C. aurantiacus* can grow as a photoorganotrophic heterotroph or as a photolithotrophic autotroph (electron donor: e.g. sulphide); sulphur produced by sulphide oxidation is deposited extracellularly. The organisms contain bacterioCHLOROPHYLLS *a* and *c*. (See also CHLOROSOMES.) Aerobically, *C. aurantiacus* does not produce chlorophylls and grows chemoorganotrophically. The filaments of *C. aurantiacus* (ca. 0.45–1.0 μm in width) are bright orange under most conditions but are green when growing photolithotrophically with high sulphide levels or low light intensity. [Photochemistry in *C. aurantiacus*: PNAS (1982) *79* 6532–6536.]

chlorogenic acid 1,3,4,5-Tetrahydroxycyclohexane-1-carboxylic acid (quinic acid) esterified at the 3-hydroxy position with CAFFEIC ACID. Chlorogenic acid is an antifungal metabolite found in certain higher plants; it may be involved e.g. in the resistance of certain strains of apple to scab (*Venturia inaequalis*). Since chlorogenic acid tends to accumulate around damaged tissues at sites of infection it is sometimes regarded as a PHYTOALEXIN. (See also AUXINS.)

Chlorogloeopsis (*Chlorogloea*) A genus of filamentous CYANOBACTERIA (section V) in which the cells divide in more than one plane, producing a multicellular trichome which readily breaks up into irregular *Gloeocapsa*-like aggregates, each surrounded by a sheath (containing heteropolysaccharides and ca. 20% protein [JGM (1982) *128* 267–272]). Hormogonia are formed by rapid division of some cells in one plane, and are composed initially of small cylindrical cells which enlarge and become spherical. Heterocysts develop in terminal and intercalary positions, and akinetes are produced. GC%: 42–43. One species: *C. fritschii*.

The genus currently includes the thermophilic organism known as 'high-temperature-form (HTF) *Mastigocladus*'. This organism has an upper temperature limit for growth of 63–64°C and is apparently capable of nitrogen fixation; it occurs in hot springs, and is very sensitive to sulphide.

Chlorogonium A genus of unicellular, biflagellated green algae which are closely related to CHLAMYDOMONAS spp. (See also FLAGELLUM (b).)

Chloroherpeton A genus of green, photosynthetic, non-filamentous gliding bacteria; one species: *C. thalassium* [Arch. Micro. (1984) *138* 96–101], which is apparently closely related to *Chlorobium limicola* and *C. vibrioforme* [SAAM (1985) *6* 152–156].

Chloromonadida See CHLOROMONADS.

Chloromonadophyceae See CHLOROMONADS.

chloromonads A group of phytoflagellates (including genera such as *Chattonella*, *Gonyostomum*, *Vacuolaria*) which are variously classified as algae of the class Raphidophyceae (= Rhaphidophyceae or Chloromonadophyceae), as algae within the class XANTHOPHYCEAE, or as protozoa of the order Chloromonadida (see PHYTOMASTIGOPHOREA). The vegetative cell typically has two flagella: a tinsel flagellum directed anteriorly, and a whiplash flagellum directed posteriorly; a PALMELLOID PHASE commonly occurs e.g. in *Vacuolaria*. The numerous yellow-green chloroplasts contain e.g. chlorophylls *a* and *c*, β-carotene, and antheraxanthin. Chloromonads occur mainly in freshwater habitats.

chloromycetin *Syn.* CHLORAMPHENICOL.

Chloromyxum See MYXOSPOREA.

chloroneb (1,4-dichloro-2,5-dimethoxybenzene) An agricultural systemic ANTIFUNGAL AGENT which is taken up by plant roots and is concentrated in the roots and lower stem; it is highly fungistatic for *Rhizoctonia* spp, moderately fungistatic for *Pythium* spp, but shows little activity against *Fusarium* spp. It is used as a seed and soil treatment for the control of damping off etc.

Chloronema A genus of filamentous, mesophilic bacteria (family CHLOROFLEXACEAE [ARM (1981) *35* 339–364]) which occur e.g. in lakes. In *Chloronema* bacterioCHLOROPHYLL *d* occurs in chlorosomes.

chloronitrobenzenes (as antifungal agents) These compounds include several useful agricultural ANTIFUNGAL AGENTS: see e.g. DICLORAN, QUINTOZENE and TECNAZENE. (See also DINITROPHENOLS.)

chloropeptide (cyclochlorotine) A hepatotoxic MYCOTOXIN produced by *Penicillium islandicum* (see YELLOW RICE). It appears to bind to actin within cells, modifying the cytoskeleton.

chlorophenol red *Syn.* CHLORPHENOL RED.

Chlorophyceae See CHLOROPHYTA.

chlorophycean starch See STARCH.

chlorophylls A class of pigments involved in PHOTOSYNTHESIS; they may be regarded as derivatives of protoporphyrin IX (see PORPHYRINS) complexed with magnesium (see figure on page 158). (cf. PHAEOPHYTIN.)

Chlorophylls in algae and cyanobacteria. Chlorophyll *a* (Chl *a*) – see figure – occurs in all organisms which carry out oxygenic PHOTOSYNTHESIS. Chl *b* (see figure) occurs in euglenoid flagellates and green algae (Chlorophyta), as well as in most higher plants. Chls c_1 and c_2 occur together in brown algae (Phaeophyta), chrysophytes, diatoms, and prymnesiophytes, while Chl c_2 without Chl c_1 occurs in cryptophytes and dinoflagellates. Chl *d* occurs in some red algae (Rhodophyta).

Bacteriochlorophylls occur in anoxygenic photosynthetic bacteria (Rhodospirillales). Bchl *a* is the sole bacteriochlorophyll in most purple bacteria (Rhodospirillineae); those purple bacteria which lack Bchl *a* (*Ectothiorhodospira abdelmalekii*, *E. halochoris*, *Rhodopseudomonas sulfoviridis*, *R. viridis*, *Thiocapsa pfennigii*) have instead Bchl *b*. Probably all green bacteria (Chlorobiineae) contain Bchl *a* – usually together with Bchls *c* (= 'chlorobium chlorophyll 660 series') or Bchls *d* ('chlorobium chlorophyll 650 series'). Bchl *e* is the major bacteriochlorophyll in *Chlorobium phaeobacteroides* and *C. phaeovibrioides*, occurring together with small amounts of Bchl *a*. Bchl *g* has been identified in *Heliobacterium chlorum* (q.v.). (See also ERYTHROBACTER.)

Chlorophylls generally occur in photosynthetic cells within specialized membranes or organelles: see CHLOROPLAST, CHLOROSOME, CHROMATOPHORE (sense 2), THYLAKOID. Purified chlorophylls are unstable in the presence of light and oxygen.

Chlorophyta (green algae) A division of the ALGAE, members of which are characterized by a combination of features: e.g., they contain chlorophylls *a* and *b*, β-carotene, and xanthophylls such as antheraxanthin, lutein, neoxanthin, violaxanthin and zeaxanthin; the chloroplasts are enveloped by a double membrane but lack an additional envelope of endoplasmic reticulum; pyrenoids,

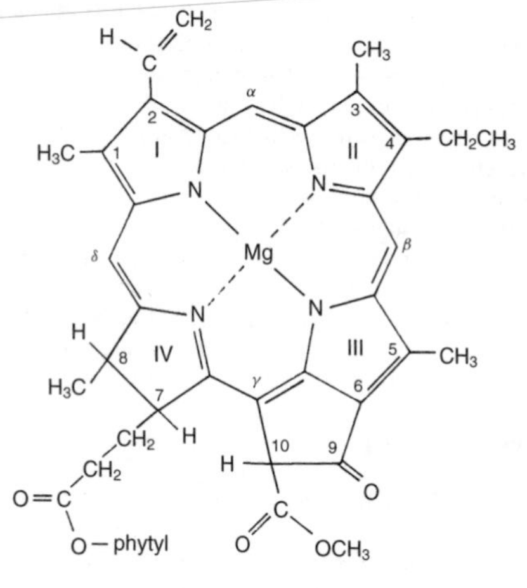

CHLOROPHYLLS: Chlorophyll *a* (phytyl = $C_{20}H_{39}$). Other chlorophylls differ from chlorophyll *a* as shown below.

when present, occur within the chloroplast; the main storage polymer is starch which is formed within the chloroplast; motile cells, when formed, bear flagella (commonly 2 or 4) which have a characteristic structure (see FLAGELLUM (b)).

The 'green algae' (which are not necessarily green in colour) are a diverse group of organisms which, in vegetative form, may be unicellular and non-motile (e.g. ANKISTRODESMUS, CHLORELLA, CHLOROCOCCUM, COCCOMYXA, TREBOUXIA; cf. ACETABULARIA and DESMIDS), unicellular and flagellated (e.g. CHLAMYDOMONAS, MICROMONADOPHYCEAE, TETRASELMIS), sarcinoid (e.g. PSEUDOTREBOUXIA), coenobial (e.g. SCENEDESMUS, VOLVOX), filamentous (e.g. APHANOCHAETE, OEDOGONIUM, SPIROGYRA, STIGEOCLONIUM, TRENTEPOHLIA, ULOTHRIX; see also SIPHONOCLADOUS), siphonaceous (e.g. CODIUM, DERBESIA, HALIMEDA, OSTREOBIUM), or thallose/parenchymatous (e.g. ENTEROMORPHA, ULVA) – reaching a relatively high degree of structural complexity and differentiation (CHAROPHYTES). *Asexual reproduction* commonly occurs by the formation of flagellated zoospores, but some species may produce e.g. aplanospores (e.g. *Aphanochaete*, *Stigeoclonium*), AUTOSPORES, or bulbils (charophytes), or may reproduce by fragmentation of the vegetative thallus (e.g. *Spirogyra*, *Stichococcus*). *Sexual reproduction* may be isogamous (e.g.

	Position	Substituent
Chlorophylls		
Chl *b*	3	–CHO
Chl c_1	7	–CH=CH.COOH
	7,8	double bond
Chl c_2	7	–CH=CH.COOH
	7,8	double bond
	4	$-CH=CH_2$
Chl *d*	2	–CHO
Bacteriochlorophylls		
Bchl *a*	2	$-CO.CH_3$
	3,4	no double bond
	7	phytyl sometimes replaced by geranylgeraniol
Bchl *b*	2	$-CO.CH_3$
	3,4	no double bond
	4	$=CH.CH_3$
Bchls *c*	2	$-CHOH.CH_3$
	4	$-C_2H_5$, $-C_3H_7$, or $-C_4H_9$
	5	$-C_2H_5$ or $-CH_3$
	7	phytyl replaced by farnesyl or stearyl
	10	–H replaces $-CO.OCH_3$
	δ-methene	$-CH_3$
Bchls *d*	2	$-CHOH.CH_3$
	4	$-C_2H_5$, $-C_3H_7$, $-C_4H_9$, or $-C_5H_{11}$
	5	$-C_2H_5$ or $-CH_3$
	7	phytyl replaced by farnesyl
	10	–H replaces $-CO.OCH_3$
Bchls *e*	As for Bchls *c*, but with –CHO at position 3, and phytyl replaced by farnesyl	
Bchl *g*	3,4	no double bond
	4	$=CH.CH_3$
	7	phytyl replaced by geranylgeraniol

in *Acetabularia*, *Chaetophora*, most *Chlamydomonas* spp, *Klebsormidium*, *Ulothrix*), anisogamous (e.g. in *Codium* and *Derbesia*), or oogamous (e.g. in *Aphanochaete*, charophytes, *Coleochaete*, *Oedogonium*); sexual reproduction is unknown in e.g. *Chlorella*. An isomorphic ALTERNATION OF GENERATIONS occurs e.g. in CLADOPHORA and ULVA, a heteromorphic alternation of generations occurs e.g. in DERBESIA and MONOSTROMA.

The majority of green algae occur in freshwater habitats, but some are found in the sea (e.g. *Acetabularia*, *Codium*, *Enteromorpha*, *Halimeda*, *Ulva*), some occur in hypersaline waters (e.g. DUNALIELLA), and some are terrestrial – growing on soil (e.g. *Chlorella* spp, *Chlorococcum*, *Friedmannia*), rocks (e.g. some desmids), snow (see e.g. RED SNOW), tree-bark (e.g. *Chlorella* spp, *'Pleurococcus'*, *Trentepohlia*), plant leaves (e.g. CEPHALEUROS), etc. Certain green algae form close stable associations with other organisms: see e.g. LICHEN and ZOOCHLORELLAE.

The classification of the green algae has been, and remains, controversial. Some schemes [e.g. Book ref. 130, pp. 86–132] divide the Chlorophyta into two classes: Chlorophyceae (divided into ca. 16 orders containing the vast majority of green algae) and Charophyceae (containing the charophytes). In another scheme [Book ref. 123, pp. 29–72] – based on comparative cytological data such as the structure of flagellated cells, the nature of cell division, etc. – the division Chlorophyta is divided into 5 classes: MICROMONADOPHYCEAE, Charophyceae (containing, in addition to the charophytes, genera such as *Coleochaete*, *Klebsormidium*, *Spirogyra* and the desmids), Ulvophyceae (containing genera such as *Monostroma*, *Ulothrix*, *Ulva*, and the siphonocladous, siphonaceous and dasycladalean algae [Book ref. 123, pp. 121–156]), Pleurastrophyceae (see TETRASELMIS and TREBOUXIA), and Chlorophyceae (containing genera such as *Chaetophora*, *Chlamydomonas*, *Oedogonium*, *Volvox*, etc).

chloropicrin Trichloronitromethane, $Cl_3C(NO_3)$; a soil fumigant with fungicidal, herbicidal and insecticidal properties.

chloroplast A semi-autonomous PLASTID, one or more of which occur in the cells of *eukaryotic* photosynthetic organisms; it is the site of PHOTOSYNTHESIS and of the biosynthesis of e.g. CHLOROPHYLLS and certain proteins and lipids. According to species, *algal* chloroplasts may be cup-shaped, discoid, lobate, spiral, stellate or irregularly shaped, and are commonly 2–20 μm in size.

Chloroplast structure. The chloroplast consists essentially of a membrane-bounded matrix (*stroma*) containing a number of flattened, membranous sacs (THYLAKOIDS) which may occur in one or more stacks (*grana* – singular: *granum*); the grana are interconnected by membranous bridges (*intergranal frets*). The stroma contains e.g. DNA (see also CYTOPLASMIC GENES), RIBOSOMES, PLASTOGLOBULI, and a range of proteins – e.g. 'fraction I protein' (= RIBULOSE 1,5-BISPHOSPHATE CARBOXYLASE–OXYGENASE) and other CALVIN CYCLE enzymes. (See also EYESPOT and PYRENOID.) The thylakoids contain the photosynthetic apparatus (e.g. chlorophylls, electron transport components), PROTON ATPASES, and (membrane-bound) ribosomes. The membrane which encloses the stroma is a *double* membrane (the *chloroplast envelope*). In some algae, one or more layers of endoplasmic reticulum (*chloroplast endoplasmic reticulum*) occur external to the envelope: one layer surrounds the chloroplast in DINOFLAGELLATES and EUGLENOID FLAGELLATES, while two layers occur in members of the PHAEOPHYTA and XANTHOPHYCEAE; none occur in members of the CHLOROPHYTA.

Chloroplast DNA (ctDNA) occurs as covalently-closed circular molecules which lack histones; typically, each chloroplast contains multiple DNA molecules. In at least some higher plants and algae, ctDNA encodes e.g. the large subunit of RuBisCO, various subunits of proton ATPase, several cytochromes, and various rRNAs and ribosomal proteins.

Chloroplast development. The following is a provisional summary. Most chloroplast polypeptides are encoded by nuclear genes and synthesized on cytoplasmic ribosomes. Such polypeptides are transported into the chloroplast by a post-translational (rather than co-translational) mechanism. Chloroplast genes, which may encode up to ca. 100 polypeptides, are essential for chloroplast development. Chloroplast-encoded polypeptides and RNAs function within the chloroplast. Light is essential for chlorophyll synthesis in angiosperms but not in most algae, mosses, ferns or gymnosperms. [Book ref. 156, pp. 1–9.]

PROTEIN SYNTHESIS in chloroplasts resembles that in bacteria: 70S-type ribosomes are used, *N*-formylmethionine is the first amino acid incorporated, and the process is inhibited e.g. by chloramphenicol.

Synthesis of galactolipids, the major polar lipids in the chloroplast, appears to occur in the chloroplast envelope [Book ref. 156, pp. 193–224].

Origin of chloroplasts. Chloroplasts appear to arise either by the division of existing chloroplasts or by the development of relatively undifferentiated plastids (*proplastids*) which lack e.g. thylakoids and specialized pigments. (See also ETIOPLAST.) The mechanism of coordination of the expression of nuclear and chloroplast genes is not fully understood.

[Chloroplasts of giant-celled and coenocytic algae: Bot. Rev. (1984) *50* 267–307.]

chloroplast endoplasmic reticulum See CHLOROPLAST.

Chloropseudomonas ethylica A species designation [Book ref. 21] which now appears to refer to a syntrophic mixture of green sulphide-oxidizing bacteria (*Chlorobium limicola* and/or *Prosthecochloris aestuarii*) and the sulphur-reducing species *Desulfuromonas acetoxidans*.

chloroquine See AMINOQUINOLINES and MALARIA.

Chloros See HYPOCHLORITES.

chlorosis (*plant pathol.*) The loss of chlorophyll from the tissues of a plant – due e.g. to microbial infection, to the action of certain phytotoxins, to lack of light, to magnesium or iron deficiency, etc. Chlorotic tissues commonly appear yellowish (cf. YELLOWS).

chlorosomes (chlorobium vesicles) Elongated, intracellular, membranous vesicles formed in all members of the CHLOROBIINEAE grown under appropriate conditions; the vesicles, which contain most of the light-harvesting ('antenna') pigments involved in PHOTOSYNTHESIS, are attached to, but differ in structure from, the cytoplasmic membrane. In members of the Chlorobiaceae the chlorosomes contain bacterioCHLOROPHYLL *c*, *d* or *e* and various CAROTENOIDS (e.g. chlorobactene in green cells, isorenieratene in brown cells), while in the Chloroflexaceae they contain bacteriochlorophyll *c* and γ- and β-carotene; a small amount of 'antenna' Bchl *a* is present in the chlorosomes of *Chloroflexus aurantiacus* and appears to occur also in the chlorosomes of members of the Chlorobiaceae. In the Chlorobiineae the cytoplasmic membrane contains the reaction centre Bchl *a* together with a small amount of antenna Bchl *a*. [Biochemistry of light-harvesting systems: ARB (1983) *52* 125–157.]

Chlorosplenium See HELOTIALES.

chloroxylenol See PHENOLS.

chlorphenol red A PH INDICATOR: pH 4.8–6.4 (yellow to red); pK_a 6.0.

chlortetracycline See TETRACYLINES.

Choanephora See MUCORALES.

Choanephoraceae See MUCORALES.

choanoflagellate A member of the CHOANOFLAGELLIDA.

Choanoflagellida An order of free-living, motile or sessile, flagellated protozoa (class ZOOMASTIGOPHOREA) characterized by a ring ('collar') of tentacles (microvilli) surrounding the proximal part of the single, anteriorly directed flagellum; the cell may have an outer membranous sheath (periplast) or a lorica composed of a basket-like arrangement of siliceous costae. Genera: e.g. *Acanthoeca*, *Codonosiga*, *Diplotheca*, *Monosiga*, *Pleurasiga*, *Savillea*, *Stephanoeca*. [Ultrastructure and deposition of silica in loricate species: Book ref. 137, pp. 295–322.]

choanomastigote A form assumed by *Crithidia* spp: see TRYPANOSOMATIDAE.

chocolate agar A medium used for the isolation and culture of certain fastidious bacteria (e.g. *Haemophilus* spp, *Neisseria gonorrhoeae*). A medium such as blood agar base or brain-heart infusion agar is autoclaved and cooled to ca. 80°C; defibrinated horse or bovine blood (10%) is added and the medium shaken gently until chocolate-brown in colour. It may be dispensed into plates, slopes etc. [Recipe: Book ref. 46, p. 1375.] Alternatively, a chocolate agar plate may be prepared by heating a set blood-agar plate to ca. 80°C.

chocolate spot A disease of beans (particularly broad beans, *Vicia faba*) caused by *Botrytis fabae*. Chocolate-brown, rounded spots develop on all parts of the plant; in severe cases entire organs or plants may be blackened and killed. Infection is usually initiated by conidia from over-wintering plant debris, but may be seed-borne. (See also WYERONE.)

choke A GRASS DISEASE (affecting e.g. *Agrostis* and *Festuca* spp) caused by *Epichloë typhina* (see EPICHLOË). Just before the host plant comes into flower, the sheath of the flag-leaf becomes encircled at its apex by a yellow to orange stroma; conidia are produced on the stromal surface, and perithecia develop within the stroma with their ostiolar pores at the surface. Flowering is partly or completely suppressed.

cholecystitis Inflammation of the gall-bladder. Inflammation may be due to bacterial infection (e.g. by bacteria from the bowel) secondary to stagnation of the bile caused e.g. by gallstones.

cholera An acute, infectious human disease involving profuse dehydrating diarrhoea; cholera is caused by certain strains of *Vibrio cholerae* (see VIBRIO) which secrete CHOLERA TOXIN. Until 1992, choleragenic strains of *V. cholerae* were associated only with serogroup O1. In 1992, non-O1 choleragenic strains were isolated in India; these strains were assigned to a new serogroup, O139, and were designated O139 Bengal [*V. cholerae* O139 Bengal: RMM (1996) 7 43–51].

Infection occurs by the faecal–oral route, usually via contaminated water. The severity (or occurrence) of disease is influenced e.g. by the number of organisms ingested; low stomach acidity is a predisposing factor. Incubation period: 1–5 (usually 2–3) days. Initially the stools contain faecal material, but they rapidly become pale-grey and watery ('rice-water stools'); 1–30 litres of fluid may be passed per day, resulting in severe dehydration and loss of electrolyte. In untreated cases mortality rates may be high (e.g. 50–75%); death may occur within hours, but usually within 1–2 days. Mild cases also occur, particularly in endemic areas.

Diarrhoeal disease may be caused by O1 strains which do not form cholera toxin; such disease may be due e.g. to the Ace and/or Zot toxins (encoded by bacteriophage CTXΦ).

V. cholerae adheres to the small intestine by 'toxin co-regulated pili' – filamentous appendages (encoded by a chromosomal PATHOGENICITY ISLAND) which form a bundle on the bacterial surface (see also BACTERIOPHAGE CTXΦ). The symptoms of cholera follow secretion and uptake of CHOLERA TOXIN (q.v.).

Lab. diagnosis. A stool specimen is incubated in alkaline peptone water for up to 6 hours and the culture used to inoculate TCBS AGAR. Colonies of interest on TCBS agar should be used to inoculate (heavily) a non-selective agar medium; growth on this medium (after ~6 hours) may be used for a slide agglutination test with appropriate antisera. For O139, a rapid screening test has been developed with monoclonal antibodies [JCM (1998) *36* 3595–3600].

Treatment. Oral or intravenous replacement of fluid and electrolyte (see e.g. ORAL REHYDRATION SOLUTION). Antibiotics (e.g. tetracyclines) may help to reduce the period of shedding and limit the risk of cross-infection (although multidrug-resistant strains of *V. cholerae* have been recorded).

Vaccines. A successful live, oral VACCINE against O1 strains was reported in 1995; however, immunity to O1 does not protect against O139. Earlier, live vaccines against O139 were being developed by deletion of certain virulence genes [JID (1994) *170* 278–283].

(See also NICED; NON-CHOLERA VIBRIOS; NON-VIBRIO CHOLERA.)

cholera toxin (choleragen) A protein ENTEROTOXIN produced by toxigenic strains of *Vibrio cholerae* and responsible for the symptoms of CHOLERA. Cholera toxin (CT) is encoded by genes *ctxA* and *ctxB* of BACTERIOPHAGE CTXΦ (toxigenic strains of *V. cholerae* are thus lysogenic for this phage).

Structurally, CT resembles PERTUSSIS TOXIN in having the AB_5 arrangement of subunits: a pentameric ring of B subunits with the central A subunit to one side of the plane of the ring. (A variant of *V. cholerae* produces B subunits ('choleragenoid') but no A subunits [PNAS (1979) *76* 2052–2056].)

The B subunits bind to glycolipid receptors (GM_1 gangliosides) in the brush border membranes of cells lining the small intestine. Internalization of the toxin occurs via a non-clathrin-coated vesicle.

Within the target cell the A subunit is proteolytically cleaved, but the two fragments remain attached, temporarily, by a disulphide bond. When the bond is reduced, the active (catalytic) fragment of the A subunit (responsible for CT activity) acts as an ADP-ribosyl transferase: it ADP-ribosylates a particular G protein, G_s, which normally stimulates the activity of ADENYLATE CYCLASE (AC). As a result, AC is converted to a permanently active form – thus raising the level of cAMP within the target cell (cf. PERTUSSIS TOXIN). CT activity thus leads to an efflux of water and electrolyte (sodium and chloride ions) into the gut lumen, and the profuse watery diarrhoea characteristic of cholera.

choleragen *Syn.* CHOLERA TOXIN.

choleragenoid See CHOLERA TOXIN.

cholesterol oxidase An enzyme (EC 1.1.3.6) which oxidizes the 3β-hydroxy group of cholesterol. The enzyme is obtained e.g. from various actinomycetes (e.g. *Nocardia* and *Streptomyces* spp) and from *Schizophyllum commune*; it is used in the detection of cholesterol in body fluids. (See also ENZYMES.)

cholic acid See BILE ACIDS.

choline See e.g. LECITHIN.

chondrioids *Syn.* MESOSOMES.

chondriosome *Syn.* MITOCHONDRION.

Chondrococcus (1) A genus of algae of the RHODOPHYTA. (2) An obsolete bacterial genus; *C. columnaris* is now *Flexibacter columnaris*.

Chondromyces See MYXOBACTERALES.

chondrosarcoma See SARCOMA.

Chondrostereum See APHYLLOPHORALES (Stereaceae).

Chondrus A genus of marine algae (division RHODOPHYTA). The thallus consists typically of dichotomously branched fronds which attach to rocks etc by means of a holdfast. *C. crispus* ('Irish moss', 'carragheen moss') is used as a source of CARRAGEENAN and, in some countries, as food. [Culture of *C. crispus*: CJFAS (1986) *43* 263–268.]

Chonotrichida See HYPOSTOMATIA.

chopped meat medium See COOKED MEAT MEDIUM.

Chorda See PHAEOPHYTA.

Chordaria See PHAEOPHYTA.

Chordopoxvirinae A subfamily of viruses, family POXVIRIDAE, which infect vertebrates; most members share a common antigen. Six genera are recognized: AVIPOXVIRUS, CAPRIPOXVIRUS, LEPORIPOXVIRUS, ORTHOPOXVIRUS, PARAPOXVIRUS, SUIPOXVIRUS. The members of a given genus are closely related serologically; genetic recombination can occur between members of a given genus, and NON-GENETIC REACTIVATION can occur between members of the same genus and of different genera.

Choreocolax A genus of algae (division RHODOPHYTA). *C. polysiphoniae* is a parasite of another red alga, *Polysiphonia lanosa*; the nuclei of *C. polysiphoniae* apparently enter the cytoplasm of the host via secondary PIT CONNECTIONS [PNAS (1984) *81* 5420–5424].

chorioallantoic membrane See EMBRYONATED EGG.

chorismate A central intermediate in the biosynthesis of aromatic compounds: see Appendix IV(f).

***Choristoneura* EPVs** See ENTOMOPOXVIRINAE.

Christensen's urea agar See UREASES.

Christie–Atkins–Munch-Petersen test CAMP TEST.

Chromatiaceae (purple sulphur bacteria; Thiorhodaceae) A family of photosynthetic bacteria (suborder RHODOSPIRILLINEAE) which occur typically in sulphide-rich aquatic and terrestrial habitats (for example, muddy soils, riverine and estuarine muds); most species contain Bchl *a* (see CHLOROPHYLLS) – the photosynthetic pigments typically occurring in vesicular intracytoplasmic membrane systems. The organisms are primarily photolithotrophic autotrophs which can use e.g. sulphur, sulphide or hydrogen as electron donors, but at least some simple organic compounds (e.g. acetate) can be used by all species under appropriate conditions; sulphur and its compounds are ultimately oxidized to sulphate, though elemental sulphur can be deposited intracellularly. Motile species are polarly flagellate (monotrichous or lophotrichous). GAS VACUOLES occur in some species. Genera: *Amoebobacter*, CHROMATIUM (type genus), LAMPROCYSTIS, THIOCAPSA, *Thiocystis*, *Thiodictyon*, *Thiopedia*, *Thiosarcina*, THIOSPIRILLUM; cf. ECTOTHIORHODOSPIRA.

chromatic adaptation In e.g. certain cyanobacteria: adaptation to changes in the wavelength of incident light by changes in the composition of the light-harvesting pigments. For example, '*Fremyella diplosiphon*' produces PHYCOBILIPROTEINS allophycocyanin and phycocyanin when grown in red light (ca. 625 nm), but when grown in green light (ca. 550 nm) the cells also produce phycoerythrin [Book ref. 75, pp. 119–126].

chromatid Either of the two longitudinally adjacent threads formed when a (eukaryotic) CHROMOSOME replicates prior to MITOSIS; the chromatids are held together at the CENTROMERE. *Sister* chromatids are derived from the same chromosome.

ChromaTide® See PROBE.

chromatin The fibrous nucleoprotein complex, containing genomic DNA, HISTONES, NON-HISTONE PROTEINS and RNA, present in the nucleus of most eukaryotic cells. (The term 'chromatin' is sometimes used in a more restrictive sense to refer specifically to the dispersed nucleoprotein material in the interphase nucleus, as distinct from the condensed, discrete CHROMOSOMES evident during mitosis and meiosis.) Two types of chromatin are distinguishable in the interphase nucleus: *heterochromatin*, in which the nucleoprotein fibres are very densely packed, and *euchromatin*, in which the fibres are less densely packed. Heterochromatin may be condensed permanently, i.e., with little variation through the cell cycle (*constitutive heterochromatin*), or temporarily (*facultative heterochromatin*). Euchromatin represents the regions within which gene expression can occur.

Chromatin released from cells appears under certain in vitro conditions as a mesh of fibres (ca. 10 nm diam.) which have a regular beaded appearance in the electron microscope; the beads are called *nucleosomes*. Treatment of these fibres with an endonuclease results in the separation of the nucleosomes, each of which then forms a structure called a *chromatosome*; a chromatosome contains ca. 165 bp (on average) of DNA, one molecule of histone H1, and a *histone octamer* comprising two molecules each of histones H2A, H2B, H3 and H4. Further endonuclease action gives rise to a *core particle*, a disc-shaped structure (ca. 11 nm in diam., 5.7 nm thick) consisting of the histone octamer around which a 146-bp sequence of B-DNA is wrapped in a left-handed superhelix. In the chromatosome, the DNA is believed to form two complete superhelical turns around the histone octamer, the turns being secured by histone H1 which binds to the sequences of DNA entering and leaving the core. In the intact 10-nm chromatin fibre the chromatosome structures are connected to one another via *linker DNA* which may vary from a few (or no) bp to >80 bp, but which is typically ca. 30–40 bp long.

In vivo, the linear chain of nucleosomes (the 10-nm fibre) is itself folded (with consequent further compaction of the DNA) into a *30-nm fibre* in which the nucleosomes are packed together – possibly in a helical arrangement, although the details of the structure are uncertain. The 30-nm fibre is apparently the basic structure both in interphase chromatin and in mitotic chromosomes. The H1 histone is essential for the formation of the 30-nm fibre.

For processes such as replication, transcription and repair, the highly compacted genome must undergo disruption (so-called 'remodelling') in order to expose the DNA to enzymes and binding proteins etc. Various mechanisms are involved. Histone properties (e.g. binding strength) may be modified by post-translational acetylation, phosphorylation, ubiquitylation etc. ATP-dependent remodelling complexes can slide nucleosomes along the DNA and replace histones [COGD (2004) *14* 165–173]. *Constitutive* variant histones (which, unlike other histones, are expressed throughout the cell cycle) can e.g. modify nucleosome properties by replacing existing histones [GD (2005) *19* 295–316].

Chromatium A genus of photosynthetic bacteria (family CHROMATIACEAE). The cells are rod-shaped and may be large (up to 15 μm long) or small (ca. 2–6 μm) according to species; they contain Bchl *a* and various carotenoids (cell suspensions are typically some shade of brown, red or purple). Gas vacuoles are absent. Motility, when it occurs, is mediated by polar flagella. Small-celled species include *C. vinosum*. The large-celled species (*C. buderi*, *C. okenii*, *C. warmingii*, *C. weissei*) are nutritionally fastidious strict anaerobes for which sulphide and vitamin B_{12} are essential; other species can grow without sulphide under anaerobic reducing conditions.

chromatography The separation of molecules (or ions) in a mixture by any technique which exploits the differing extents to which the different molecules (or ions) become distributed between a stationary phase and a contiguous mobile phase, one of the phases being either a thin film or a surface. Phases may be solid, liquid, or gaseous, according to the type of chromatography. Substances are separated primarily on the basis of their adsorption and/or solubility characteristics (cf. GEL FILTRATION); unlike ELECTROPHORESIS, chromatography does not involve the use of an electrical potential difference. Chromatography is used e.g. for the isolation, purification, and identification of molecules or ions.

In *adsorption chromatography* substances are separated on the basis of quantitative differences in their adsorption onto the surface of a stationary solid phase from a mobile liquid phase. For example, *affinity chromatography* exploits the specificity of binding which occurs between e.g. antibody and antigen, or enzyme and substrate. Essentially, one component (e.g. antigen) is immobilized on an inert support; this adsorbs the other component (antibody) from the liquid phase which flows over it. [Industrial uses: Book ref. 3, pp. 31–66.] *Ion-exchange chromatography* exploits the preferential adsorption of a particular charged species (e.g. ion, nucleic acid) to an insoluble ion-exchange material under particular conditions. (Ion-exchange chromatography is sometimes excluded from the category 'adsorption chromatography' on a mechanistic basis.)

Partition chromatography exploits the tendency of different substances to become differently distributed between two fluids in contact. In liquid–liquid chromatography one of the liquids is typically present as a coating on inert particles, or on an inert matrix, over which the mobile liquid phase is passed. For example, in *paper partition chromatography* the stationary phase is typically water which coats the fibres of a paper strip; the mobile phase, usually an organic solvent, flows along the strip by capillary action. If a spot of sample (mixture of substances) is initially applied to one end of the strip, the substances become separated – according to their different partition coefficients – as the solvent flows along the strip. In *gas–liquid chromatography* (GLC) the sample is vaporized in a carrier gas (helium, hydrogen, or nitrogen) which is pumped along a heated tube (up to ca. 350°C) containing inert particles coated with a substance that is liquid at the temperature used. Substances in the sample become partitioned between the gas and liquid; a given substance moves along the tube, in the direction of gas flow, at a characteristic rate which depends on its partition coefficient, and its exit from the tube is recorded e.g. by a thermal conductivity detector.

High-pressure liquid chromatography (high-performance liquid chromatography; HPLC) uses very high pressures to force a liquid sample through a tightly-packed column of minute particles; separations occur at the surfaces of the particles on the basis of e.g. adsorption processes. The small size of the particles increases the efficiency of interaction between solid and liquid phases; high pressures are necessary to ensure a reasonable rate of flow through the column. *Thin-layer chromatography* (TLC) employs a thin layer of e.g. silica gel on a flat glass support; the (liquid) mobile phase rises up the thin layer by capillary action.

When the substances in a sample have been separated, each may be removed (*eluted*) from the stationary phase by means of a suitable solvent or solution (the *eluent*); for example, in ion-exchange chromatography a suitable eluent may be a solution of a particular electrolyte or a buffer of particular pH. (See also GRADIENT ELUTION.) Alternatively, the separated substances may, for example, be stained in situ on the stationary phase – e.g. using specific stains which allow identification of particular types of molecule.

chromatoid bodies (*protozool.*) Ribonucleoprotein bodies (commonly rod-shaped) which occur within the young cysts of certain amoebae (see e.g. ENTAMOEBA). Chromatoid bodies stain well with iron-haematoxylin (they do not stain with iodine); one, two or more bodies per cyst may be seen. The nucleoprotein tends to disperse in the cytoplasm as the cyst ages.

chromatophore (1) One of a number of small membranous vesicles obtained by the experimental disruption of photosynthetic membranes from bacteria of the Rhodospirillineae; the vesicles are used e.g. for in vitro studies on photophosphorylation. (2) A system of vesicular, tubular or lamellar membranous structures which occurs in bacteria of the Rhodospirillineae; chromatophores, which contain bacterioCHLOROPHYLL, CAROTENOIDS, and the electron carriers involved in PHOTOSYNTHESIS, are continuous with the cytoplasmic membrane. (cf. CHLOROSOME.) (3) *Syn.* CHLOROPLAST. (4) *Syn.* CHROMOPLAST.

chromatosome See CHROMATIN.

Chromidina See HYPOSTOMATIA.

Chromobacterium A genus (*incertae sedis*) of chemoorganotrophic, Gram-negative bacteria which occur e.g. in soil and water, and as opportunist pathogens in man and other animals. Cells: round-ended, pigmented (VIOLACEIN-containing) rods, ca. 0.6–0.9 × 1.5–3.5 μm, with one polar and one or more lateral and/or subpolar flagella. Growth occurs on unenriched nutrient media. Most strains attack glucose fermentatively (forming acid but usually no gas); some strains attack glucose oxidatively. Nitrate and nitrite are usually reduced. Growth is inhibited by 6% NaCl. Catalase +ve. Usually oxidase +ve. VP −ve. GC%: ca. 50–68. Type species: *C. violaceum.*

C. fluviatile. Optimum growth temperature ca. 25°C, maximum ca. 30°C. GC%: ca. 50–52.

C. lividum. See JANTHINOBACTERIUM.

C. violaceum. Optimum growth temperature ca. 30–35°C, maximum ca. 40–44°C. GC%: ca. 65–68. (See also MONOBACTAMS and QUORUM SENSING.)

[Book ref. 22, pp. 580–582.]

chromoblastomycosis (dermatitis verrucosa; verrucose dermatitis) A chronic human MYCOSIS, occurring mainly in tropical and subtropical areas of South America and Africa, caused by any of several black moulds (e.g. *Cladosporium carrionii*, *Phialophora* spp, *Rhinocladiella* spp) which are normally saprotrophic e.g. in soil. Infection occurs via wounds, and warty, spreading, often ulcerating lesions develop in the skin and underlying tissue. In the tissues, the fungi occur as thick-walled, dark-brown *muriform cells*, 5–12 μm ('sclerotic bodies') which multiply by cell separation (not budding); in culture (e.g. on Sabouraud's dextrose agar) they form slow-growing, velvety, dark-brown to greenish-black colonies. Pathogenicity may involve the body's reaction to fungal components: e.g., lipids extracted from the fungi can induce granulomatous reactions on intravenous injection into mice [JGM (1985) *131* 187–194].

chromogenic Pigment-producing or colour-producing. (cf. ACHROMOGENIC.)

chromomycin (chromomycin A_3) An ANTIBIOTIC and antitumour agent containing an aromatic three-ring chromophore linked to five different sugar residues. *Olivomycin* and *mithramycin* are very similar to chromomycin; all three dyes bind non-covalently and non-intercalatively to dsDNA, inhibiting synthesis of RNA and DNA. Binding requires stoichiometric amounts of Mg^{2+} ions, seems to involve the 2-amino group of guanine, and results in a shift in the absorption spectrum of the dye to longer wavelengths. The dyes are used in cytochemistry as GC-specific fluorochromes, e.g., they produce specific fluorescent banding patterns in metaphase chromosomes [Science (1977) *195* 400–402].

chromomycosis May refer either to CHROMOBLASTOMYCOSIS or to PHAEOHYPHOMYCOSIS.

chromophore That part of a dye molecule which is responsible for the colour of the dye.

chromoplast (1) A PLASTID in which the main or sole pigments are CAROTENOIDS. (2) (chromatophore) Any pigmented plastid, e.g. a CHLOROPLAST or a carotenoid-rich plastid in ripening fruit.

chromosomal fingerprinting See DNA FINGERPRINTING.

chromosome In a prokaryotic cell, or in the nucleus of a eukaryotic cell: a structure consisting of or containing DNA which carries genetic information essential to the cell; the term is also commonly applied to DNA in a mitochondrion or chloroplast and to the genome of a DNA virus, but generally not to the genomic RNA of an RNA virus. (cf. GENOME sense 1 and PLASMID.)

(a) *Bacterial chromosomes.* In most bacteria the chromosome is a ccc dsDNA molecule; in some species (e.g. *Borrelia burgdorferi* [Nature (1997) *390* 580–586], species of *Streptomyces*) the DNA is linear. Bacterial chromosomes are not enclosed by a membrane (see PROKARYOTE) but they are apparently linked to the cell's cytoplasmic membrane.

Typically, a cell contains only one type of chromosome; the chromosome may be present in one, several or many copies per cell – depending e.g. on species and on growth conditions; for example, in *Escherichia coli* there may be one or more copies per cell according to the growth rate (see HELMSTETTER-COOPER MODEL). Some species of bacteria appear normally to have multiple copies of the chromosome – e.g. *Azotobacter vinelandii* may have an average of 20–40 'genome equivalents' (apparently separate chromosomes) per cell, *A. chroococcum* contains 20–25, *Deinococcus radiodurans* has four (in resting cells) to ca. 10, while *Desulfovibrio gigas* has ca. 9–17 (in growing cells) [see e.g. JGM (1984) *130* 1597–1601, 1603–1612].

The size of the chromosome (in base-pairs) varies considerably among species. One of the smallest is that of *Buchnera*, an obligate intracellular parasite of aphids (~640 kb [Nature (2000) *407* 81–86]); larger chromosomes occur e.g. in *Helicobacter pylori* ($\sim 1.7 \times 10^6$ bp) [Nature (1997) *388* 539–547] and *Mycobacterium tuberculosis* ($\sim 4.4 \times 10^6$ bp [Nature (1998) *393* 537–544]).

In *Vibrio cholerae* the genome consists of two dissimilar circular chromosomes – chromosome 1 (~2961 kb) and chromosome 2 (~1073 kb) [Nature (2000) *406* 477–483].

In *E. coli* the chromosome contains ca. 4.7×10^6 bp; if opened out, it would be ca. 1.3 mm in length. Within the cell it is extensively folded to form a compact structure variously called a *nucleoid, nucleoid body, nuclear area, 'nucleus'* etc. The nucleoid consists mainly of DNA associated with certain types of protein. The DNA is extensively supercoiled (see DNA), the degree of supercoiling being determined by the joint activities of certain enzymes (see TOPOISOMERASE). The supercoiling is segregated into many topologically distinct *domains*; a nick in one domain relaxes the DNA in that domain only. The nucleoid can be unfolded in vitro by treatment with RNase.

(See also CELL CYCLE and DNA REPLICATION.)

[The chromosome of *E. coli*: Book ref. 122, pp. 161–197, 199–227.]

(b) *Eukaryotic chromosomes.* The eukaryotic NUCLEUS typically contains one or more sets of different chromosomes (see also PLOIDY; each chromosome is believed to contain a single linear dsDNA molecule complexed with protein (see CHROMATIN). Between nuclear divisions (interphase) the chromosomes are dispersed as a mass of 30-nm chromatin fibres occupying much of the volume of the nucleus; during MITOSIS (and MEIOSIS) each chromosome becomes highly condensed to form a discrete structure with a characteristic morphology (see also KARYOTYPE) and structurally and functionally distinct regions (see e.g. CENTROMERE and TELOMERE). Chromosome replication (and transcription) occurs during interphase (see CELL CYCLE), so that at mitosis each chromosome consists of two sister CHROMATIDS.

The 30-nm chromatin fibre in a mitotic chromosome appears to be organized into a series of loops (each ca. 30–90 kb), the ends of the loops being secured by a central proteinaceous *scaffold*; when the chromosome is treated in vitro to remove most of the histones and non-histone proteins, the central scaffold – similar in shape to the sister chromatids – can be seen surrounded by a halo of naked DNA loops.

chromosome aberration An abnormality in the number or structure of chromosomes in a cell. Thus, e.g., a HETEROPLOID cell may have the diploid number of chromosomes minus one chromosome (*monosomic* cell) or plus one chromosome (*trisomic* cell). A chromosome may lose a segment (*deletion*); a segment may move from one site to another on the same or on a different chromosome (*translocation*); a segment may take up the reverse orientation in the same site (*inversion*); or a segment may be duplicated.

chromosome walking A technique with which an unknown region of a chromosome can be explored. The starting point is a LIBRARY in which the cloned DNA constitutes a series of *overlapping* fragments. A fragment containing a known gene is selected and used as a probe to identify (e.g. by COLONY HYBRIDIZATION) other, overlapping fragments which contain the same gene. The nucleotide sequences of these fragments are then characterized – thus delimiting an overall segment of the chromosome which comprises the span of the overlapping sequences. A fragment at one end of this segment is selected, and the terminal part of this fragment is cleaved and used as a probe to identify a neighbouring set of overlapping fragments, and so on.

chromotrope See METACHROMASY.

Chromulina A genus of unicellular CHRYSOPHYTES in which each cell has a long tinsel flagellum but only a very short second flagellum. Cysts resemble narrow-necked flasks.

chronic (*med.*) Refers to a disease which persists for a relatively long period of time (e.g. months, years) – terminating in recovery or death. Examples: LEPROSY; TUBERCULOSIS. (cf. ACUTE.)

chronic bacillary diarrhoea *Syn.* JOHNE'S DISEASE.

chronic mucocutaneous candidiasis See CANDIDIASIS.

chronic myelogenous leukaemia See ABL.

chronic wasting disease See TRANSMISSIBLE SPONGIFORM ENCEPHALOPATHIES.

Chroococcales See CYANOBACTERIA.

Chroococcidiopsis A genus of unicellular CYANOBACTERIA (section II) in which growth and binary fission lead to the formation of cubical aggregates of cells; baeocytes are non-motile and have a fibrous outer cell wall layer at the time of their release. GC%: 40–46.

Chroococcus A phycological genus of 'blue-green algae' included in the genus GLOEOCAPSA [JGM (1979) *111* 1–61 (p. 9)].

Chroodactylon See RHODOPHYTA.

Chroomonas See CRYPTOPHYTES.

chrysanthemum chlorotic mottle viroid See VIROID.

chrysanthemum stunt viroid See VIROID.

chrysanthemum white rust A serious disease of chrysanthemums caused by *Puccinia horiana*; typical early symptoms include pale spots on the upper surfaces of leaves, and raised, buff-coloured pustules on corresponding parts of the lower surfaces – the pustules subsequently becoming white. Common chrysanthemum rust, a less serious disease caused by *P. chrysanthemi*, is characterized by brown powdery masses on the undersides of leaves.

Chryseomonas A genus of yellow-pigmented, aerobic, catalase-positive, Gram-negative bacteria. *C. polytricha* was isolated from clinical specimens; cells are round-ended rods, motile by lophotrichous flagella. The organisms are chemoorganotrophs and are reported to be oxidatively saccharolytic but oxidase-negative. GC%: 54–56. [IJSB (1986) *36* 161–165.]

Chrysochromulina See PRYMNESIOPHYCEAE.

Chrysococcus A genus of unicellular CHRYSOPHYTES in which young cells each have a mucilaginous lorica which, as the cell ages, becomes impregnated with mineral crystals. Cytoplasmic projections extend through pores in the lorica.

chrysolaminarin (leucosin; chrysose) A reserve polysaccharide produced by CHRYSOPHYTES, DIATOMS, and members of the PRYMNESIOPHYCEAE. It is a $(1 \rightarrow 3)$-β-D-glucan with some branching via $(1 \rightarrow 6)$-β-linkages.

Chrysomonadida See PHYTOMASTIGOPHOREA.

chrysomonads *Syn.* CHRYSOPHYTES.

chrysophytes (chrysomonads) A group of eukaryotic, mainly freshwater organisms regarded as algae ('golden, golden-brown or yellow-brown algae', class Chrysophyceae or division Chrysophyta) or as protozoa (see PHYTOMASTIGOPHOREA). Species are either free-swimming (or occasionally sessile) unicellular organisms (e.g. *Chromulina*, *Mallomonas*, *Ochromonas*, *Oikomonas*), or they may form sessile colonies (e.g. CEPHALOTHAMNIUM, DINOBRYON) or motile colonies (e.g. SYNURA, UROGLENA). Most species have two flagella, one of which is directed anteriorly and bears two opposite rows of mastigonemes (see FLAGELLUM (b)), while the other is either smooth and directed posteriorly or is reduced to a short stub (e.g. CHROMULINA) or basal body. FLAGELLAR MOTILITY is achieved primarily by the tinsel flagellum.

According to species, the chrysophyte cell may be naked, may be enclosed by a LORICA (see e.g. CHRYSOCOCCUS, DINOBRYON, OCHROMONAS), or may bear scales – adherent to the outside of the plasmalemma – which may be silicified (e.g. in MALLOMONAS, PARAPHYSOMONAS, SYNURA) or organic (e.g. in *Chromulina placentula*); the scales are produced in special membranous vesicles (SILICA DEPOSITION VESICLES), not in the Golgi apparatus (cf. PRYMNESIOPHYCEAE). [Scale formation in *Synura*, *Mallomonas* and *Paraphysomonas*: JUR (1982) *18* 13–26.] Pseudopodium-like cytoplasmic projections are present in many species, and many chrysophytes (e.g. OCHROMONAS) can ingest particulate matter. The majority of chrysophytes are photosynthetic, each cell typically having a few (commonly two) golden-brown chloroplasts containing chlorophylls a, c_1 and c_2, β-carotene, and fucoxanthin. Mannitol is a major primary product of photosynthesis. Storage compounds include CHRYSOLAMINARIN.

Reproduction occurs by longitudinal binary fission. Palmelloid stages are common in some species. Spherical or bottle-shaped cysts (statospores, resting spores) are commonly formed; each cyst develops within a vegetative cell and has a silica wall which may be smooth or ornamented with e.g. spines.

[Book ref. 197. Silica in chrysophytes: Book ref. 137, pp. 201–230.]

See also SILICOFLAGELLATES.

chrysose *Syn.* CHRYSOLAMINARIN.

Chrysosporium A genus of fungi of the HYPHOMYCETES; most have teleomorphs in the Gymnoascaceae. Hyphae: hyaline, septate. Conidia: non-septate, subglobose to pyriform, 2–2.5 × 4–4.5 μm in culture (cf. ADIASPIROMYCOSIS), typically borne (singly or in pairs) on short, peg-like conidiophores. *C. pruinosum*, a saprotroph, produces adiaspores at high temperatures in vitro.

***chvE* gene** See CROWN GALL.

chymosin See CASEIN.

chymotrypsin A pancreatic serine PROTEASE.

chytrid A member of the CHYTRIDIALES.

Chytridiales (the 'chytrids') An order of fungi (class CHYTRIDIOMYCETES) which characteristically do not form true mycelium (though some species form a RHIZOMYCELIUM) and in which sexual reproduction, where known, may involve e.g. the fusion of isogametes or gametangial copulation; the organisms include parasites and pathogens of terrestrial plants, aquatic saprotrophs, parasites of algae, and MYCOPARASITES. Genera include *Chytriomyces*, NOWAKOWSKIELLA, OLPIDIUM, *Polyphagus*, RHIZOPHYDIUM, *Rozella* and SYNCHYTRIUM.

Chytridiomycetes A class of terrestrial and aquatic, saprotrophic and parasitic fungi (subdivision MASTIGOMYCOTINA) which form sporangiospores having one posteriorly-directed whiplash flagellum. (cf. NEOCALLIMASTIX.) According to species, the thallus ranges from a simple, sac-like, holocarpic structure to a well-developed, branched mycelium; in many species the cell wall contains CHITIN, and in some species the mycelium contains pseudosepta – see SEPTUM (b). Orders: BLASTOCLADIALES, CHYTRIDIALES, Harpochytriales (tropical freshwater saprotrophs), MONOBLEPHARIDALES.

Chytridiopsis See MICROSPOREA.

Chytriomyces See CHYTRIDIALES.

Ci CURIE (q.v.).

CI See COLOUR INDEX.

Ciboria See HELOTIALES.

CID CYTOMEGALIC INCLUSION DISEASE.

cider (American: hard cider) A beverage made by the ALCOHOLIC FERMENTATION of apple juice. (In the USA 'cider' refers to unfermented apple juice.) Cider apples are rich in tannin and malic acid (contributing astringent and sharp flavours, respectively). Fermentation may be carried out by the natural apple flora – including e.g. bacteria, yeasts (e.g. *Candida*, *Kloeckera*, *Pichia*, *Saccharomyces*) and other fungi (e.g. *Aspergillus*, *Penicillium*). *Kloeckera* spp begin the fermentation, suppressing both *Candida* and *Pichia*; later, *Saccharomyces* spp predominate. In modern processes the natural flora is usually suppressed (with SULPHUR DIOXIDE) or reduced (by pasteurization, filtration etc) and the juice inoculated with cider yeast – usually '*Saccharomyces uvarum*' (a strain of *S. cerevisiae*); this yeast can degrade de-esterified PECTINS to galacturonic acid. After fermentation is complete, acidity may be reduced by the MALOLACTIC FERMENTATION. The cider may then be carbonated and/or pasteurized and/or filtered.

Cider spoilage. The most common microbial spoilage is ACETIFICATION by contaminating acetic acid bacteria. *Cider sickness* is a condition caused by *Zymomonas mobilis* which produces acetaldehyde; this reacts with tannins in the cider to form a milky-white haze and a banana-like off-flavour. 'Oiliness' and ROPINESS may be caused by various lactobacilli.

cider sickness See CIDER.

Cidex An alkaline GLUTARALDEHYDE solution.

CIDS CIRCULAR INTENSITY DIFFERENTIAL SCATTERING.

CIE COUNTERCURRENT IMMUNOELECTROPHORESIS.

ciguatera A form of toxic FOOD POISONING, involving gastrointestinal and neurological symptoms, which results from the consumption of certain tropical or subtropical fish whose flesh has accumulated certain algal toxin(s); the fish may have fed directly on the algae, or may have consumed other algivorous fish. The dinoflagellate *Gambierdiscus toxicus* has been identified as the toxin producer in cases in the Gambier Islands of Polynesia; the main toxin (*ciguatoxin*) produced by this species is a highly potent neurotoxin [PAC (1986) *58* 339–350 (347–348)]. (cf. *Paralytic* SHELLFISH POISONING.)

ciguatoxin See CIGUATERA.

cilia See CILIUM.

ciliary corpuscle *Syn.* KINETID.

ciliate (1) (*adj.*) Bearing cilia (see CILIUM). (2) (*noun*) A protozoon of the CILIOPHORA.

ciliature (*ciliate protozool.*) Collectively, those cilia which occur on a ciliate or on specific region(s) of a ciliate; thus, e.g. 'somatic ciliature' refers to those cilia which occur on the body of the cell – as opposed to those which occur in the oral (buccal) cavity.

ciliferous Bearing cilia; a ciliferous kinetosome has an associated cilium, while a 'barren' kinetosome does not.

Ciliophora A phylum of PROTOZOA [JP (1980) *27* 37–58] in which the organisms are characteristically ciliated (see CILIUM) during at least some stage in their life cycles, and possess two types of nucleus (see MACRONUCLEUS and MICRONUCLEUS). (cf. PHALACROCLEPTES and STEPHANOPOGON.)

Many of the 7000 or so known species are free-living in freshwater, brackish or marine habitats; some occur in the soil (e.g. *Grossglockneria acuta* [SBB (1985) *17* 871–875]), in the RUMEN, or in SEWAGE TREATMENT plants. Some species are pathogenic (see e.g. BALANTIDIUM and ICHTHYOPHTHIRIUS) and some (e.g. members of the ASTOMATIDA) are parasitic.

In size, ciliates range from ca. 10 μm to several millimetres. The cells generally have a substantial PELLICLE, one or more CONTRACTILE VACUOLES, and most have a CYTOSTOME; some have EXTRUSOMES. CYSTS are formed by some species. Ciliates are typically colourless, but some (e.g. BLEPHARISMA, STENTOR) are pigmented. Some species have endosymbionts (see e.g. PARAMECIUM). Although typically motile, the ciliates include some sedentary species (see e.g. PERITRICHIA). Ciliates commonly feed holozoically, i.e., by PHAGOCYTOSIS.

Reproduction typically occurs by transverse, perkinetal binary fission; budding or multiple fission occurs in some species. Sexual processes (see AUTOGAMY and CONJUGATION) occur in some species.

In ciliates the GENETIC CODE differs from that previously regarded as the 'universal' code, and huge evolutionary distances are believed to separate ciliates from other organisms [MS (1986) *3* 36–40].

Currently, ciliate taxonomy involves criteria such as the nature of STOMATOGENESIS and of the PARORAL MEMBRANE. Three classes: KINETOFRAGMINOPHOREA (regarded as the most primitive); OLIGOHYMENOPHOREA; POLYHYMENOPHOREA.

(See also SILVER LINE SYSTEM.)

[Book ref. 135.]

Ciliophrys See ACTINOPHRYIDA.

cilium (*pl.* cilia) A thread-like appendage, numbers of which extend from the cells of most members of the CILIOPHORA; the coordinated movements of cilia are involved in locomotion and/or feeding. (Cilia also occur on other organisms and cells – e.g. ctenophores ('comb jellies') and certain cells in the mammalian respiratory tract.) The structure and ultrastructure of a cilium are similar to those of the eukaryotic FLAGELLUM (q.v.); however, cilia are usually shorter (often 7–10 μm) and they frequently occur in longitudinal rows (see KINETY) or in specialized groups (see COMPOUND CILIATURE). The characteristic rhythmical bending of cilia is achieved by the sliding of the ciliary microtubules relative to one another.

The basic ciliary movement begins with an *effective stroke*, i.e., a rapid swing through a wide angle in a straight, relatively rigid state due to simultaneous activity of the dynein arms on one side of the axoneme; this is followed by a *recovery stroke* in which the cilium returns to its datum position as a result of the propagation of a wave of dynein activity along the other side of the axoneme, from base to tip. Energy is therefore required for both the effective and the recovery strokes. The rhythmical beating of cilia generates currents in the ambient fluid – thus causing the cell to move or, e.g., propelling food particles towards the cytostome. (See also METACHRONAL WAVES.) Many variations of ciliary movement are found among ciliates – including e.g. the ability to change the direction of the ciliary beating. (See also CLAVATE CILIUM; cf. FLAGELLAR MOTILITY (b).) [Mechanism and controls of microtubule sliding in cilia: SEBS (1982) *35* 179–201; production of different ciliary beat patterns: SEBS (1982) *35* 139–157; regulation of ciliary motility by membrane potential via cAMP in *Paramecium*: Cell Mot. (1986) *6* 256–272.]

(2) (*lichenol.*) A vegetative appendage which structurally resembles a RHIZINE but which arises from the margin of a thallus or apothecium or, in some lichens, from the upper surface of the thallus.

CIN agar Cefsulodin–Irgasan–novobiocin agar, a medium developed for the isolation of *Yersinia enterocolitica*. At least one strain of *Y. enterocolitica* (biotype 3B, serotype O3) and strains of *Y. pseudotuberculosis* are inhibited on this medium [JCM (1986) *24* 116–120].

***cin* gene** See BACTERIOPHAGE P1.

cinchona alkaloids Alkaloids obtained from the bark of certain tropical trees (especially *Cinchona succirubra*); the alkaloids include e.g. cinchonine and QUININE.

cinerubin See ANTHRACYCLINE ANTIBIOTICS.

cingulum See e.g. DIATOMS and DINOFLAGELLATES.

cinoxacin See QUINOLONE ANTIBIOTICS.

CIP Collection de l'Institut Pasteur, Institut Pasteur, 28 Rue du Docteur Roux, 75724 Paris, Cedex 15, France.

ciprofloxacin See QUINOLONE ANTIBIOTICS.

circadian rhythms In certain parameters of a cell or organism: rhythmical changes which occur with a periodicity of ca. 24 hours and which may persist for extended periods of time (e.g. days, weeks) even when the cell or organism has been isolated from the natural diurnal fluctuations of the environment; such innate rhythms are often referred to as expressions of a 'biological clock'.

In eukaryotes, parameters that exhibit circadian rhythms include e.g. (according to organism): BIOLUMINESCENCE, cell division, chloroplast ultrastructure, mating type, motility and photosynthetic activity; thus, for example, peak photosynthetic activity in *Acetabularia* occurs with a 24-hour periodicity even when the organism is kept in continual light. (cf. INFRADIAN RHYTHMS and ULTRADIAN RHYTHMS.)

Once believed to occur only in eukaryotes, circadian rhythms have also been detected in some prokaryotes. For example, in the unicellular cyanobacterium *Synechococcus*, gene *psbAI* (which encodes a component of photosystem II) is expressed least at dawn and maximally at dusk; certain other genes are regulated in the opposite phase. Under conditions of constant light, the *timing* of cell division is influenced by the endogenous ~24-hour biological clock, even when average doubling time is only 10.5 hours [TIM (1998) *6* 407–410].

[Molecular bases for circadian clocks (review): Cell (1999) *96* 271–290. Circadian programmes in cyanobacteria: ARM (1999) *53* 389–409.]

circannual rhythms In certain parameters of a cell or organism: rhythmical changes which occur with a periodicity of ca. one year (see e.g. RED TIDE).

CIRCE See HEAT-SHOCK PROTEINS.

Circinoviridae See TT VIRUS.

circular intensity differential scattering (CIDS) A phenomenon in which left and right circularly polarized light is scattered in a characteristic way by a given type of cell, virus, or molecule. CIDS has been used for the rapid characterization of microorganisms [AEM (1982) *44* 1081–1085].

circularly permuted (*mol. biol.*) Refers to nucleic acid molecules in which the bases or genes are arranged in the same sequence but with different starting points: ABCDEFG, BCDEFGA, CDEFGAB, etc; such molecules can arise e.g. by cleavage at different positions in a given type of circular parental molecule. (See e.g. BACTERIOPHAGE T4.)

circulative transmission (of viruses) (*plant pathol.*) A mode of transmission in which viruses taken up by an insect VECTOR pass from the alimentary canal to the haemolymph and subsequently to the salivary glands and saliva of the insect; viruses which are transmitted in this way are called *circulative* or *persistent* viruses. Insects involved in circulative transmission typically carry plant-infective virions for long periods, sometimes for life. Circulative transmission characteristically involves a latent period (several hours to a number of days) during which the vector, having acquired virus from an infected plant, is unable to transmit the virus to a healthy plant. (cf. NON-CIRCULATIVE TRANSMISSION.)

Some viruses (e.g. potato yellow dwarf virus, wound tumour virus) replicate in their insect vectors; circulative transmission of such *propagative viruses* is termed *propagative transmission*. Other viruses (e.g. barley yellow dwarf virus) apparently do not replicate in their insect vectors; circulative transmission of such *non-propagative viruses* is accordingly referred to as *non-propagative transmission*.

circulative viruses See CIRCULATIVE TRANSMISSION.

circulins See POLYMYXINS.

circumsporozoite protein See PLASMODIUM.

cirramycin See MACROLIDE ANTIBIOTICS.

cirri See CIRRUS.

cirrus (*pl.* cirri) A discrete group of somatic cilia (several to over 100) which act primarily as a unified locomotive organelle; the typical cirrus is conical. Cirri are characteristic of the HYPOTRICHIDA. (See also COMPOUND CILIATURE.)

***cis*-acting** See CIS-DOMINANCE.

***cis*-dominance** The phenomenon in which a genetic element (a gene or a non-coding sequence such as a promoter) affects only the DNA molecule in which it occurs; in a COMPLEMENTATION TEST, a mutant form of such a genetic element (which is called a *cis*-acting element) cannot be complemented by (i.e., is dominant over) the wild-type form of the element present in the same cell. The converse of a *cis*-acting element is a *trans*-acting element, an element which encodes a product that can diffuse through the cytoplasm and act on other DNA molecules. For example, in the LAC OPERON, a mutation in the *lacI* gene may result in an inactive *lac* repressor and hence in the constitutive expression of the *lac* genes; the presence of a wild-type *lacI* gene (e.g. on a plasmid) in the same cell can re-establish repression of the *lac* genes, i.e., the $lacI^-$ allele is recessive to the wild-type lac^+ allele, and the *lac* repressor is said to be *trans*-acting. A mutation in the *lac* operator which renders the operator unable to bind the repressor also results in constitutive expression of the *lac* genes; however, in this case the presence of a functional *lac* operator on a plasmid in the same cell does not alter the mutant (constitutive) phenotype, i.e., the mutant operator is dominant over the wild-type form, complementation cannot occur between mutant and wild-type forms, and the operator is said to be *cis*-acting.

Although in most cases a *cis*-acting element is non-coding (e.g. a promoter, an operator, etc), certain proteins are *cis*-acting; such a protein can function only on the DNA molecule encoding it. For example, certain transposases can act efficiently only on the ends of an appropriate transposable element in the same DNA molecule as the gene which encoded the transposase (see e.g. Tn*5*). *Cis*-acting proteins are generally strongly basic; it has been suggested that, as soon as it is synthesized (perhaps even before translation is complete), a *cis*-acting protein may bind to the nearby DNA template and may subsequently move along the DNA until it reaches an appropriate site of action (e.g. the ends of a transposon in the case of a *cis*-acting transposase).

***cis–trans* test** A form of COMPLEMENTATION TEST which actually comprises two separate tests: a *cis* test and a *trans* test. The terms *cis* and *trans* refer to the arrangement of mutations in the two genomes which are brought together in the same cell to test for complementation; in the *trans* arrangement one mutation occurs in each genome, while in the *cis* arrangement both mutations are located in the same genome, the other genome being wild-type. (The doubly-mutant genome used in the *cis* test is constructed, by recombination, from the two singly-mutant genomes used in the *trans* test, so that the mutations in the *cis* arrangement occupy the same loci as those in the *trans* arrangement.) Positive results in both *cis* and *trans* tests (i.e., the wild-type phenotype is restored in both tests) indicate that the two mutations are in different genes and hence that the phenotypic characteristic under study is determined by more than one gene. A negative *trans* test with a positive *cis* test indicates that the two mutations are in the same gene (organisms in the *cis* test having one doubly-mutant copy and one wild-type copy of the gene, those in the *trans* test having only two mutant copies of the gene). However, if both *cis* and *trans* tests are negative, at least one of the mutations must be dominant (see COMPLEMENTATION TEST), i.e., the mutant phenotype is expressed in the *cis* test organisms despite the presence of a wild-type copy of the gene. The *cis* test thus provides a control to prevent the possible misinterpretation of a negative (*trans*) complementation test as indicative of monogenic control when in fact it could be the result of a dominant mutation in one gene together with a second (dominant or recessive) mutation in another gene.

cistron (1) A GENE as defined in terms of the CIS–TRANS TEST: i.e., in a diploid cell or merozygote, either of two homologous sequences in a genetic nucleic acid in which two mutations in *trans* fail to exhibit complete complementation (see COMPLEMENTATION TEST).

(2) *Syn.* GENE.

Cit plasmid An enterobacterial plasmid encoding a citrate transport system; the presence of a Cit plasmid in e.g. *Escherichia coli* confers on the (usually citrate-negative) cells the ability to utilize citrate as sole source of carbon and energy (see CITRATE TEST).

Citeromyces A genus of yeasts (family SACCHAROMYCETACEAE). Cells reproduce by multilateral budding; neither pseudomycelium nor true mycelium is formed. Asci are persistent and contain one (occasionally two) rough-surfaced spheroidal spores. Glucose is fermented; NO_3^- is assimilated. One species: *C. matritensis* (anamorph: *Candida globosa*), isolated from e.g. fruit preserved in syrup, tree exudates etc. [Book ref. 100, pp. 117–119.]

Citifluor A photofading retardant and mountant used e.g. for epifluorescence microscopy of soil [SBB (1985) *17* 739–746].

citrate metabolism See e.g. TCA CYCLE, CIT PLASMID, DIACETYL; see also CITRIC ACID.

citrate synthase See Appendix II(a).

citrate test An IMVIC TEST which determines the ability of a bacterial strain to use citrate as the sole source of carbon (see e.g. TCA CYCLE). A saline suspension of the test organism is prepared from growth on a solid medium. An inoculum from this suspension is transferred by means of a STRAIGHT WIRE to the test medium (e.g. KOSER'S CITRATE MEDIUM or SIMMONS' CITRATE AGAR); the wire is dipped (ca. 1 cm) into the suspension and then dipped into, or streaked across, the test medium. Such a small inoculum minimizes carry-over of nutrients and avoids causing turbidity in Koser's medium. The inoculated medium is incubated and examined daily for evidence of growth (turbidity in Koser's medium, blue coloration in Simmons' medium); growth is a positive reaction.

citreoviridin A yellow pigment and neurotoxic MYCOTOXIN produced by *Penicillium citreoviride* growing e.g. on polished rice. (See also YELLOW RICE.) The molecule consists of a hydrofuran ring linked, via a conjugated polyene, to an α-pyrone chromophore; it emits a brilliant yellow fluorescence on UV irradiation. Consumption, by man, of rice contaminated with citreoviridin may be responsible for *acute cardiac beriberi*: a fatal disease characterized by ascending paralysis, convulsions, and respiratory arrest; these symptoms can be reproduced in animals by feeding them with citreoviridin. In vitro, citreoviridin inhibits PROTON ATPASE isolated e.g. from mitochondrial membranes.

citric acid A tricarboxylic acid present e.g. in many fruits (especially citrus fruits) and an important metabolic intermediate in many organisms (see e.g. TCA CYCLE). Citric acid is obtained commercially mainly from strains of *Aspergillus niger* grown e.g. on molasses or starch hydrolysates; the accumulation of citric acid is apparently due to abnormal functioning of the TCA cycle. The citric acid is separated from the fermentation liquor by precipitation as the Ca^{2+} salt. Citric acid has many commercial uses – e.g. as an acidifying agent in various foods, beverages and pharmaceuticals. [Production methods etc: Book ref. 8, pp. 709–747.]

citric acid cycle *Syn.* TCA CYCLE.

citrinin A yellow pigment and MYCOTOXIN produced by *Penicillium* spp in mouldy cereals etc. It is nephrotoxic e.g. in pigs.

Citrobacter A genus of Gram-negative bacteria of the ENTEROBACTERIACEAE (q.v.). Cells: ca. 1 × 2–6 μm, usually motile. MR +ve; VP −ve citrate +ve; acid and gas from glucose; lactose +ve or −ve, or fermented slowly. Growth usually occurs on/in selective media such as DCA, SS agar, tetrathionate broth and selenite broth. GC%: 50–52. Type species: *C. freundii*.

C. amalonaticus (formerly e.g. *C. intermedius* biotype a; *Levinea amalonatica*). Indole +ve; malonate −ve; H_2S −ve; growth in KCN media.

C. diversus (= *C. koseri*; formerly e.g. *C. intermedius* biotype b; *Levinea malonatica*). Indole +ve, malonate +ve and H_2S −ve; no growth in KCN media.

C. freundii. Indole −ve; malonate (usually) −ve; H_2S + ve; growth in KCN media. Nitrogen fixation has been reported in some strains. Serotypes may be distinguished on the basis of O and H antigens, some of which are common to serotypes of *Escherichia* and *Salmonella; C. freundii* H antigens are monophasic. Certain strains are reported to have a Vi antigen serologically identical to that of *Salmonella typhi* (see VI ANTIGEN), but its presence is not related to virulence. (See also PHASE VARIATION.)

Citrobacter spp may be isolated from human or animal faeces, from various clinical specimens, and from food, water, sewage, soil etc; they may be opportunist pathogens. *C. freundii* and *C. diversus* have been associated with cases of human diarrhoea. [Book refs. 22, pp. 458–461, and 46, pp. 1140–1147.]

citrovorum factor See FOLIC ACID.

L-citrulline See e.g. Appendix IV (a).

***Citrus* exocortis viroid** See VIROID.

***Citrus* leaf rugose virus** See ILARVIRUSES.

***Citrus* leprosis virus** See RHABDOVIRIDAE.

***Citrus* tristeza virus** A member of the CLOSTEROVIRUSES which causes 'tristeza disease' of citrus trees; the virus has killed millions of orange trees in Brazil.

***Citrus* variegation virus** See ILARVIRUSES.

CJD CREUTZFELDT-JAKOB DISEASE.

C_L region See IMMUNOGLOBULINS.

Cladobotryum See HYPHOMYCETES and HYPOMYCES.

cladogram Any dendrogram which expresses phylogenetic relationships. (cf. PHENOGRAM.)

Cladonia A large genus of LICHENS (order LECANORALES); photobiont: a green alga. Species tend to be variable in appearance. There is a granular or squamulose primary thallus (often evanescent, but predominant in some species) from which may arise hollow podetia (see PODETIUM); according to species, podetia may be awl-shaped, branched and shrubby, or funnel-, cup- or wine-glass-shaped, and may bear brown or red hymenial discs ('apothecia'). Species occur on the ground (particularly on peaty soils in heaths and moors), on walls, on decaying logs, etc. Some species – e.g. *C. portentosa* (formerly *C. impexa*), *C. rangiferina*, *C. stellaris* – form cushions of extensively branched podetia ('reindeer moss') on the ground in mountainous regions of Europe and in Arctic regions, where they may serve as an important source of winter fodder for reindeer and caribou.

Cladophora A genus of filamentous, branched, siphonocladous green algae (division CHLOROPHYTA); the filaments are composed of cylindrical, multinucleate cells, the walls of which contain parallel arrays of cellulose microfibrils. An isomorphic alternation of generations occurs; sporophytes produce quadriflagellate, uninucleate zoospores, gametophytes produce biflagellate isogametes. Species occur in freshwater and marine habitats, sometimes free-floating, but usually attached to a substratum by rhizoids; some species often form extensive tangled skeins up to several metres in length ('blanket weed').

Cladosiphon See PHAEOPHYTA.

cladosporiosis Any human or animal disease caused by a species of CLADOSPORIUM.

Cladosporium A genus of fungi (class HYPHOMYCETES) which form septate mycelium and dark-coloured, aseptate or septate, ellipsoidal conidia. The organisms include saprotrophs and pathogens of plants and animals. Species include e.g. *C. carrionii* (see e.g. CHROMOBLASTOMYCOSIS), *C. cucumerinum* (see e.g. GUMMOSIS), *C. herbarum* (teleomorph: *Mycosphaerella tassiana*) (see e.g. BLACK SPOT (sense 1), SAP-STAIN, TEXTILE

SPOILAGE), and *C. oxysporum* (a pathogen of homopteran insects). (See also BLACK YEASTS, MEAT SPOILAGE, PAINT SPOILAGE, PAPER SPOILAGE.) For *C. bantianum* see XYLOHYPHA. For *C. fulvum* see FULVIA. For *C. resinae* see HORMOCONIS.

Cladostephus See PHAEOPHYTA and PHEROMONE.

clamp connection (clamp; clamp cell) In the *dikaryotic* mycelia of many (not all) basidiomycetes: a hyphal structure formed during vegetative cell division for ensuring that the daughter nuclei of a given nucleus are segregated to different cells. Prior to MITOSIS a dikaryotic cell develops a short, angled, lateral branch (the incipient clamp connection) into which *one* of the nuclei migrates. Both nuclei then undergo simultaneous mitotic division (*conjugate division*) – the spindles being orientated such that, subsequently, one end of the cell contains one daughter nucleus from each of the two parental nuclei; this pair of nuclei is then isolated from the rest of the cell by a septum (which thus delimits a heterokaryotic binucleate cell), while a second septum isolates the single daughter nucleus in the developing clamp connection. The clamp hypha now grows towards – and fuses with – the mononucleate compartment of the original (parental) cell, the clamp connection thus being completed, and a second heterokaryotic binucleate cell being concomitantly formed. The clamp connection persists as a hyphal bulge in the region of septation.

A *medallion clamp* (formed e.g. by species of *Lentinus*, *Lenzites*, *Poria* and *Polyporus*) develops when only the terminal part of the clamp hypha fuses with the mononucleate cell, a small space or eyelet remaining between the clamp hypha and the parent hypha.

Parallels have been drawn between clamp connections and the *croziers* formed during ASCUS development.

clarithromycin A MACROLIDE ANTIBIOTIC used e.g. as part of a combination of antibiotics for treating disease caused by *Helicobacter pylori*; this macrolide is more acid-stable (and is absorbed more readily) than e.g. erythromycin. In *H. pylori*, resistance to macrolide antibiotics has been associated with point mutations in 23S rRNA [AAC (1996) *40* 477–480].

class A taxonomic rank (see TAXONOMY and NOMENCLATURE).

class I/class II antigens See MAJOR HISTOCOMPATIBILITY COMPLEX.

class I/class II introns See SPLIT GENE (b) and (c).

class switching See ANTIBODY FORMATION.

clathrin See PINOCYTOSIS.

Clathrochloris See CHLOROBIACEAE.

Clathrulina A genus of freshwater protozoa (order DESMOTHORACIDA). The mature organism is attached to e.g. vegetation by a tubular stalk-like extension (ca. 200–350 μm long) of the capsule; the capsule (ca. 60–90 μm diam.) has circular to polygonal perforations through which axopodia protrude. The cell reproduces by the formation of biflagellate zoospores which lack both stalk and capsule. These cells initially form filopodia (rather than true axopodia); subsequently a stalk, a capsule, and finally axopodia, are produced.

Clathrus See PHALLALES.

clavacin *Syn.* PATULIN.

clavams A class of β-LACTAM ANTIBIOTICS which includes CLAVULANIC ACID.

Clavaria See APHYLLOPHORALES (Clavariaceae).

Clavariaceae See APHYLLOPHORALES.

clavate Club-shaped, i.e. thicker at one end.

clavate cilium (stereocilium) A short cilium whose axoneme lacks the central pair of microtubules; it does not flex. Clavate cilia occur e.g. in *Didinium* and in the scopula of some peritrichs.

clavatin *Syn.* PATULIN.

clavems A class of antibiotics; a clavem contains a clavam-like nucleus in which a double bond occurs between C-2 and C-3 (see β-LACTAM ANTIBIOTICS, Figure (f)).

Clavibacter A genus of Gram-positive, asporogenous, non-acid-fast, non-motile, obligately aerobic, phytopathogenic coryneform bacteria; *C. xyli* subsp. *xyli* causes ratoon stunting disease of sugarcane, *C. xyli* subsp. *cynodontis* causes Bermuda-grass stunting disease. [IJSB (1984) *34* 107–117.]

Claviceps A genus of fungi (order CLAVICIPITALES) which are typically parasitic on grasses (see e.g. ERGOT; see also PASPALUM STAGGERS).

C. purpurea. The cycle of infection begins when ascospores germinate and infect the ovaries of a susceptible grass. The fungus soon produces small, colourless conidia (the *Sphacelia* stage) which become suspended in a HONEYDEW secretion (possibly produced by the plant in response to infection); insects, attracted by the honeydew, disperse the conidia to the ovaries of uninfected grasses. Later, the fungus produces a SCLEROTIUM (q.v.) (the *ergot*, the overwintering stage) in the position normally occupied by the grain. In spring or early summer the sclerotia germinate, each giving rise to several erect, drum-stick-like stromata; perithecia develop in the surface layer of each stromal head, their ostiolar pores level with the stromal surface. Each ascus contains 8 thread-like ascospores.

Clavicipitales An order of fungi (subdivision ASCOMYCOTINA) which include entomogenous species, and species parasitic on other fungi or on grasses. Ascocarp: perithecioid, superficial or immersed in a stroma. Asci: unitunicate, cylindrical, with a thick apical cap. Ascospores: elongated to filiform, often multiseptate, sometimes fragmenting. Genera: e.g. *Acrospermum*, *Balansia*, CLAVICEPS, CORDYCEPS, EPICHLOË, *Hypocrella*, HYPOMYCES.

claviformin *Syn.* PATULIN.

clavine alkaloids See ERGOT ALKALOIDS.

Clavispora A genus of yeasts (family SACCHAROMYCETACEAE) which reproduce by multilateral budding; pseudomycelium may be formed. Asci are formed following conjugation between cells of different mating type. Ascospores: clavate, minutely warty, 1–4 per ascus. Glucose is fermented; NO_3- is not assimilated. One species: *C. lusitaniae* (anamorph: *Candida lusitaniae*), isolated e.g. from pigs and from human skin and sputum. [Book ref. 100, pp. 120–122.]

clavulanic acid A clavam (see β-LACTAM ANTIBIOTICS for structure). Clavulanic acid, produced e.g. by *Streptomyces clavuligerus* (ATCC 27064), binds to the PENICILLIN-BINDING PROTEINS of *Escherichia coli* (preferentially to PBP-2) but has only weak antibacterial activity; however, it is a potent inhibitor of many β-LACTAMASES from Gram-positive and Gram-negative bacteria and may thus be used in combination with β-lactamase-sensitive antibiotics (see e.g. AUGMENTIN and TIMENTIN). Clavulanic acid is unstable as the free acid and is isolated from aqueous solutions as a salt.

CLB (coccidian-like body) An organism linked with prolonged diarrhoea (duration: weeks) in humans; it appears to be water-borne. [Epidemiology: Lancet (1993) *341* 1175–1179.]

cleared lysate The supernatant obtained when bacteria in suspension are lysed and centrifuged for ca. 10–20 min at ca. 20000 *g*; it may contain plasmids (if present in the cells) but is generally free of particulate cell debris and chromosomal DNA.

clearing In preparing a specimen for microscopy: the stage in which the dehydrating agent (usually ethanol) is replaced by a solvent (*clearing agent*) such as benzene, xylene, or terpineol; if the cleared specimen is to be mounted in a MOUNTANT or impregnated with paraffin wax, the clearing agent must be miscible with both the dehydrating agent and the mountant or wax solvent. The refractive index of clearing agents is higher than that of ethanol, so that the process of clearing increases the transparency of the specimen.

CLED medium Cystine-lactose-electrolyte-deficient medium: a medium used for the culture of organisms in the absence of electrolyte (or NaCl). With added colistin, it may be used for the primary isolation of members of the PROTEEAE, since swarming is inhibited in the absence of NaCl.

cleistocarp *Syn.* CLEISTOTHECIUM.

cleistothecium (cleistocarp) A hollow, closed, typically spherical or spheroid ASCOCARP in which develop one or more asci (see ASCUS and ASCOSPORE). In some species the cleistothecial wall (*peridium*) is a more or less rigid pseudoparenchymatous structure; in others the wall may be little more than an envelope of loosely woven hyphae. In e.g. *Erysiphe* spp the cleistothecia are ca. 100–200 μm in diameter.

clindamycin See LINCOSAMIDES.

Clitocybe See AGARICALES (Tricholomataceae).

cloacin DF13 A group A colicin (see COLICINS) (MWt ca. 56000) produced by strains of *Enterobacter cloacae* and *Escherichia coli* which harbour the CloDF13 plasmid. [Proposed role for protein H (the lysis protein) in cloacin DF13 export: JGM (1986) *132* 825–834.]

In *E. coli* the receptor for cloacin DF13 is the IutA protein (which also serves as a receptor for iron–aerobactin complexes – see SIDEROPHORES); translocation across the cell envelope of the target cell is reported to require the TolAQR proteins.

Within a sensitive cell, cloacin DF13 acts as an RNase, specifically cleaving 16S rRNA – thereby inhibiting ribosome function and protein synthesis.

CloDF13 plasmid See CLOACIN DF13 and COLE1 PLASMID.

clofazimine (B663; Lamprene) A lipid-soluble substituted phenazine used e.g. as an anti-LEPROSY drug; in *Mycobacterium leprae* it binds to guanine residues in DNA and may inhibit transcription.

clonal abortion (*immunol.*) See IMMUNOLOGICAL TOLERANCE.

clonal deletion (*immunol.*) See IMMUNOLOGICAL TOLERANCE.

clonal expansion See e.g. T LYMPHOCYTE. (Clonal expansion also occurs in B cells when they react to specific antigen.)

clonal selection theory A currently accepted theory of ANTIBODY FORMATION. Essentially, it proposes that particular cells (within a cell population) respond to a specific antigen by binding the antigen on pre-existing specific cell-surface receptors (antibodies); binding promotes proliferation of these cells to form an enlarged clone of cells which are responsive to that antigen – some of the cells secreting antibodies and others acting as 'memory' cells which can produce antibodies on later exposure to the antigen. IMMUNOLOGICAL TOLERANCE in respect of certain antigens is explainable as the loss of particular clones due to exposure to self-antigens, or external antigens, at an early stage in the development of the immune system. (cf. INSTRUCTIVE THEORIES OF ANTIBODY FORMATION.)

clone (1) (*noun*) A population of cells or organisms derived asexually from a single progenitor. A clone is generally assumed to be genetically homogeneous.

(2) (*verb*) To carry out CLONING (of DNA).

cloned line (*tissue culture*) A population of cells descended directly from a single CLONE. (cf. CELL LINE.)

cloning (of DNA) A procedure (also called *gene cloning* or *molecular cloning*) used for obtaining millions of copies of a given gene (or other piece of DNA); copies of the (cloned) DNA may be used e.g. for sequence analysis (see DNA SEQUENCING) or for synthesis of the product encoded by the sequence (see EXPRESSION VECTOR).

Initially, the required gene (or other sequence) is inserted ('spliced') into a REPLICON – often a PLASMID or a phage genome. The replicon functions as a *vector* (= *cloning vector*), i.e. a molecule which, on being replicated in a cell, brings about the replication of the given gene (as the latter is an integral part of the vector).

The insertion of a gene into a plasmid vector is carried out in a procedure involving a RESTRICTION ENDONUCLEASE and a ligase (see figure).

Vector molecules, each incorporating the given gene, are introduced into cells in which they can replicate; plasmid vectors are introduced into bacteria by TRANSFORMATION or ELECTROPORATION. The cells, containing vectors, are then cultured; because a bacterial population can reach very high numbers in culture, and because each cell contains one or more copies of the vector, large numbers of vector molecules (each containing the given gene) are obtained – i.e. the gene has been *cloned*.

To harvest the cloned gene, the cells are lysed and the vector molecules are isolated e.g. by isopycnic centrifugation (cf. QIAGEN PLASMID KIT). The cloned gene is then cut from each vector molecule (by means of the original restriction enzyme), and copies of the cloned gene are separated from vector DNA e.g. by gel electrophoresis.

Cloning with plasmid vectors in bacteria.
During the initial stage (see figure), one potential problem is that some plasmid molecules may re-circularize without incorporating a copy of the gene; such plasmids may then be taken up by transformation but, lacking the gene insert, they would not contribute to the cloning process. This difficulty may be addressed by pre-treating the cleaved (linearized) plasmids with a phosphatase to remove the 5′-phosphate groups which are necessary for ligase action and re-circularization. When the gene (target) fragments are added, their 5′-phosphate groups can be ligated to 3′-OH ends in the plasmids, resulting in stable, circular hydrid plasmids which contain single-stranded nicks (at the 5′-OH ends). These hybrid plasmids can then be used for transformation – the nick sites being repaired in vivo to yield ccc dsDNA molecules.

Sometimes the target fragment lacks restriction sites, or has incompatible ones. This problem may be addressed as follows. Sticky ends, if present, are changed to blunt ends (see BLUNT-ENDED DNA); this may be achieved by either removing single-stranded overhangs (e.g with ENDONUCLEASE S1) or by synthesizing a complementary strand on each 5′ overhang (thus converting 5′ overhangs to dsDNA). To each blunt end can now be ligated a short, blunt-ended dsDNA molecule – a *linker* – which contains a suitable restriction site; one example of a linker is:

5′-GGAATTCC-3′
3′-CCTTAAGG-5′

which contains the *Eco*RI restriction site (see table in RESTRICTION ENDONUCLEASE). Because a linker is a PALINDROMIC SEQUENCE (the 5′-to-3′ sequence is the same in both strands) either end can be ligated. Blunted-ended ligation requires a higher concentration of enzyme, a lower temperature and a longer time. (See also ADAPTOR.)

Following transformation, only cells which have actually received a plasmid can contribute to the cloning process. These cells can be selected for if the plasmid vector contains an antibiotic-resistance gene; thus, if the transformed cells are

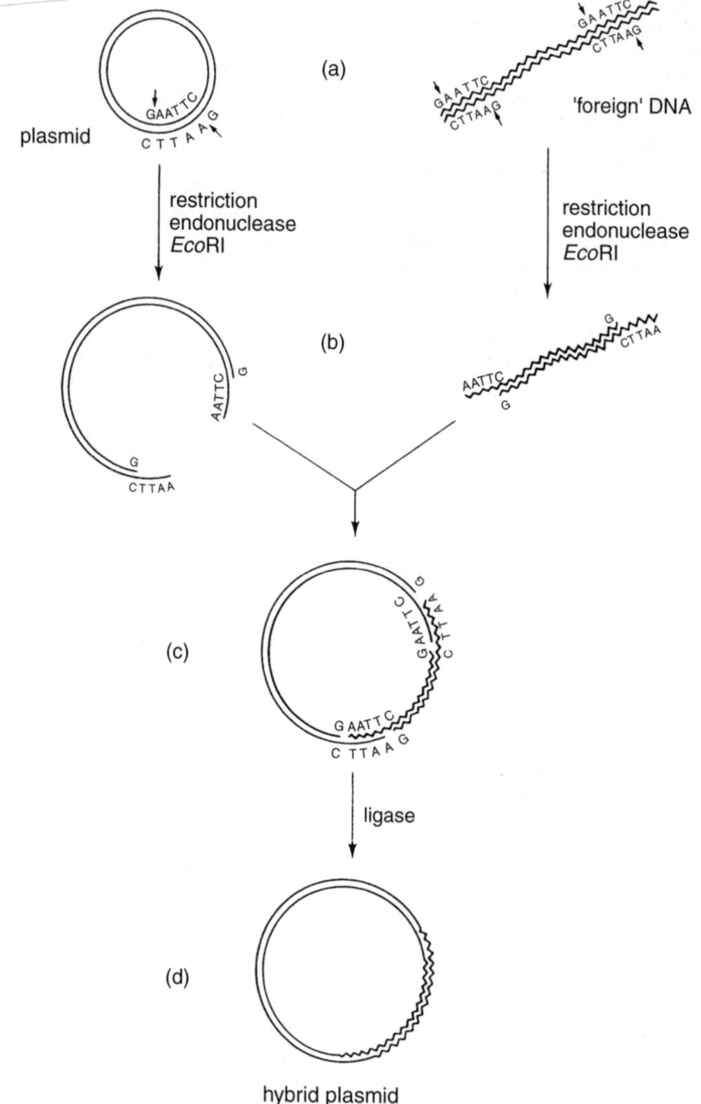

CLONING: insertion of a gene (or other 'foreign' DNA) into a plasmid vector (diagrammatic).

(a) A plasmid and a piece of 'foreign' DNA. Both molecules contain the nucleotide sequence GAATTC which is recognized by the RESTRICTION ENDONUCLEASE *Eco*RI; in the foreign DNA, the sequence to be cloned lies between the two restriction sites. *Eco*RI cuts between the guanosine (G) and adenosine (A) nucleotides, as shown by the arrows.

(b) As a result of *Eco*RI activity, the plasmid has been linearized, and both types of molecule now have 'sticky ends', i.e. terminal, single-stranded complementary sequences of nucleotides (5′-AATT).

(c) The foreign DNA has integrated with the plasmid vector molecule through base-pairing between complementary nucleotides in the sticky ends.

(d) An enzyme, DNA ligase, has catalysed a phosphodiester bond between the sugar residues of each G and A nucleotide, forming a hybrid plasmid (recombinant plasmid; also referred to as hybrid DNA, or hDNA). Note that, unless prevented, at least some of the vector molecules may simply re-circularize by base-pairing between their sticky ends; such re-circularization, which would reduce the efficiency of the procedure, may be prevented by treating the linearized plasmids with phosphatase (see entry).

A population of hybrid plasmids can be inserted into bacteria by transformation or electroporation for cloning. Note that, if required, the (cloned) fragment can be cut from the vector by using the same restriction enzyme.

Reproduced from *Bacteria* 5th edition, Figure 8.6, page 182, Paul Singleton (1999) copyright John Wiley & Sons Ltd, UK (ISBN 0471-98880-4) with permission from the publisher.

plated on a medium containing the relevant antibiotic, only the plasmid-containing cells (which are resistant to the antibiotic) will form colonies.

Some vectors are particularly useful in that, following transformation, they can signal the uptake of both vector *and* DNA insert. For example, in plasmid pBR322 (see PBR322), the tetracycline-resistance gene contains a *Bam*HI restriction site, and the ampicillin-resistance gene contains a *Pst*I restriction site; if, for example, *Pst*I is used to insert the gene (analogous to *Eco*RI in the figure), then any transformant which is (i) sensitive to ampicillin but (ii) resistant to tetracycline may be presumed to contain the complete hybrid plasmid.

A different approach to selection is provided by REPRESSOR TITRATION.

If a plasmid vector replicates under *relaxed* control (see PLASMID), the number of plasmids per cell can be greatly increased by treating the cells with chloramphenicol; in this way, the yield of cloned product can be maximized.

Cloning with phage vectors in bacteria.
In some phages (e.g. BACTERIOPHAGE LAMBDA) the genome contains a fairly high proportion of 'non-essential' DNA; thus, a sequence may be excised from the genome, using restriction endonucleases, and replaced by the fragment to be cloned. Such a vector is called a REPLACEMENT VECTOR; using the genome of phage lambda as a replacement vector it is possible to clone a fragment of about 20 kb (compared with <10 kb in plasmid pBR322). Even larger fragments, up to ~40 kb, can be cloned in a COSMID.

Phage genomes can be used as vectors only if the recombinant genome (i.e. genome plus insert or fragment) is not too large to fit into the phage head, and only if it contains, where necessary, the appropriate packaging signals.

Once incorporated into a phage particle, the recombinant genome can be injected into bacterial cells and will replicate like a plasmid. (Note that, during construction of a phage vector, any sequences in the phage genome that promote integration into the bacterial chromosome should be removed.)

ssDNA phages may also be used as vectors; filamentous phages have the advantage that their virions can accomodate a wide range of sizes of target DNA. Another option for cloning is a hybrid vector which includes sequences from a phage and a plasmid: see PHAGEMID.

Characterization of clones.
Sometimes, clones are prepared from a *mixture* of DNA fragments or from a LIBRARY, i.e. different cells within the culture will contain vectors that incorporate different fragments of DNA. In such cases one can screen for cells that contain a particular gene or sequence. Screening methods include: COMPLEMENTATION analysis (in which cells are selected when their particular fragment of cloned DNA complements a given chromosomal defect); probing for a specific gene product (e.g. by using labelled antibodies); or hybridization techniques (see e.g. COLONY HYDRIDIZATION).

Linguistic note.
The above account of cloning should make it clear that a gene (or other sequence) is *inserted into* a vector, and that a vector (and its insert) is *cloned in* a population of cells. Unfortunately, the expression '*cloned into*....' has become widely used in the literature; this piece of gibberish has been promulgated by mindless repetition. A given sequence of DNA may be cloned *in* a particular vector molecule or cloned *in* a particular type of cell.

Clonothrix A genus of IRON BACTERIA which occur as sheathed, tapering filaments. Type species: *C. fusca*.

clorobiocin An ANTIBIOTIC, obtained from *Streptomyces* spp, closely related to NOVOBIOCIN and COUMERMYCINS – which it resembles in its action.

closed complex See TRANSCRIPTION.

closed culture *Syn.* BATCH CULTURE.

closed mitosis See MITOSIS.

Closterium A large, morphologically diverse genus of placoderm DESMIDS. The cells are commonly more or less crescent-shaped; the cell is not constricted (i.e. there is no sinus), but the semicells are separated by a median suture line.

closteroviruses (beet yellows virus group) A group of PLANT VIRUSES in which the virions are highly flexuous filaments, typically 1200–2000 × 12 nm (some possible members are ca. 600–800 × 12 nm) containing one type of coat protein and one molecule of linear positive-sense ssRNA (MWt ca. $2.2–4.7 \times 10^6$). In general, the viruses have a fairly wide host range, and some are important pathogens of crop plants. At least some members are transmitted semipersistently by aphids; mechanical transmission can be achieved, often with difficulty, under experimental conditions. In the host plant the virions often aggregate in cross-banded masses in phloem cells.

Type member: SUGAR BEET YELLOWS VIRUS (SBYV). Other members include e.g. beet yellow stunt virus, burdock yellows virus, carnation necrotic fleck virus, carrot yellow leaf virus, CITRUS TRISTEZA VIRUS, clover yellows virus, *Festuca* necrosis virus, wheat yellow leaf virus. Possible members include e.g. apple chlorotic leafspot virus, apple stem grooving virus, and POTATO VIRUS T.

Clostridium A genus of ENDOSPORE-forming, chemoorganotrophic, obligately anaerobic (or, in a few cases, aerotolerant) bacteria which cannot carry out dissimilatory sulphate reduction (cf. DESULFOTOMACULUM). Species occur widely e.g. in soil and mud, and in the intestines of man and other animals; some species are saprotrophic, others are opportunistic pathogens. Cells: typically rod-shaped, ca. 0.3–1.9 × 2–10 μm, but may vary from coccoid to filamentous or (in *C. cocleatum* and *C. spiroforme*) helical; the Gram reaction is typically positive (though it may become negative in old cultures) but some species (e.g. *C. clostridiiforme*, *C. ramosum*) are commonly Gram-negative even in overnight cultures. Most species are peritrichously flagellate; some (e.g. *C. innocuum*, *C. perfringens*, *C. ramosum*) are non-motile. The endospores are usually spherical to ovoid/elongate; they typically distend the sporangium, and may be located terminally or subterminally. Except in rare cases, clostridia are catalase-negative.

Many species grow poorly on basal media and require e.g. *p*-aminobenzoic acid (PABA), biotin and/or pantothenate. Media used for the culture of clostridia include e.g. blood agar, cooked meat medium, and egg-yolk agar. (See also MCCLUNG TOABE EGG-YOLK AGAR and REINFORCED CLOSTRIDIAL MEDIUM.) SWARMING is common in some strains. Metabolism is generally described as 'fermentative'; however, some strains (e.g. of *C. perfringens*) are capable of NITRATE RESPIRATION. (See also STICKLAND REACTION.) The growth of *saccharolytic* clostridia is improved by, or is dependent on, the presence of a fermentable carbohydrate; most saccharolytic species form acetic acid and/or butyric acid as a major product (see also BUTYRIC ACID FERMENTATION), although some form, instead, products such as formic and lactic acids and/or ethanol. [Genetics and biochemistry of *Clostridium* relevant to fermentations: AAM (1986) *31* 1–60.] The *proteolytic* clostridia digest cooked meat, casein and/or other protein, and generally grow well in e.g. nutrient broth without the addition of carbohydrate; at least some of these species can derive energy by the STICKLAND REACTION, and in some species this is the preferred mode of energy metabolism. Certain species (e.g. *C. perfringens*, *C. sordellii*) are both proteolytic and saccharolytic, while others are restricted to specific substrates (see e.g. *C. cochlearium*, *C. cylindrosporum* and *C. kluyveri*). Some species can reduce nitrate. Some clostridia are thermophilic, others are psychrophilic, but most are mesophilic. GC% range for the genus: ca. 22–55. Type species: *C. butyricum*.

Currently, >80 species (some listed below) – but see end of entry.

C. aceticum. Saccharolytic; diazotrophic; acetic acid is formed from CO_2 and H_2. (See also ANAEROBIC DIGESTION.)

C. acetobutylicum. Proteolytic and saccharolytic; growth requirements appear to include PABA and biotin. (See also ACETONE–BUTANOL FERMENTATION.)

C. acidiurici. A species metabolically similar to *C. cylindrosporum* (q.v.)

C. barati. Incorporates the former species *C. perenne* and *C. paraperfringens* [IJSB (1982) *32* 77–81]. (See also BOTULINUM TOXIN.) LIPASE −ve, LECITHINASE +ve.

C. barkeri. Saccharolytic; butyric and lactic acids, CO_2 and H_2 are formed from carbohydrates.

C. beijerinckii. Saccharolytic: butyric and acetic acids and butanol are formed from glucose.

C. bifermentans. Proteolytic and saccharolytic; requires e.g. biotin, nicotinamide, pantothenate, pyridoxine and various amino acids. Spores: ovoid, subterminal. Indole-positive. The organisms form a lecithinase similar to the α-toxin of *C. perfringens*.

C. botulinum. The causal agent of BOTULISM (see also BOTULINUM TOXIN). Cells: motile, ca. $0.3–1.9 \times 3.4–9.4$ μm; spores: ovoid, subterminal. All strains can ferment glucose (none can use lactose, mannitol or sucrose), and all form metabolic products which include acetic and butyric acids. All strains can hydrolyse gelatin and produce lipases. The organisms are divided into seven types (designated A, B, C, D, E, F and G), strains of a given type producing a BOTULINUM TOXIN which is antigenically distinct from those of the other types; in some cases the strains of a given type are metabolically heterogeneous: e.g., type B strains include some which are saccharolytic as well as some which are both proteolytic and saccharolytic. Minimum growth temperatures: often ca. 10°C, but ca. 3.3° in saccharolytic type B strains and in type E strains. Minimum pH for growth is sometimes reported to be ca. 4.6 for types A and B, and ca. 5.0 for type E; under certain conditions growth and toxin formation may occur at pH values down to ca. 4.0 [Nature (1979) *281* 398–399]. The heat-resistance of endospores is strain-dependent; thus, e.g., for types A and B the $D_{112.8}$ (see D VALUE) is ca. 0.15–1.32 min, for type C the D_{100} is ca. 0.12 min (terrestrial strains) or ca. 0.4–0.9 min (marine strains), and for type E the D_{80} is ca. 0.33–3.3 min (in media) or ca. 1.6–4.3 min (in fish) [Book ref. 30, p. 327].

C. botulinum type A strains are proteolytic *and* saccharolytic; they cause botulism in man and other animals (see also LIMBERNECK).

C. botulinum type B strains are either saccharolytic or proteolytic *and* saccharolytic; they cause botulism in man and other animals.

C. botulinum type C strains are proteolytic *and* saccharolytic; they cause botulism in animals: see e.g. FORAGE POISONING, MIDLAND CATTLE DISEASE, WESTERN DUCK DISEASE.

C. botulinum type D strains are proteolytic *and* saccharolytic; they cause botulism in animals (see LAMZIEKTE).

C. botulinum type E strains are saccharolytic; they are often associated with fish and fish products, and they cause botulism in man. [Effect of oxygen and REDOX POTENTIAL on growth of *C. botulinum* type E: Food Mic. (1984) *1* 277–287.]

C. botulinum type F strains are saccharolytic, or proteolytic *and* saccharolytic. A comparatively rare cause of botulism in man.

C. botulinum type G strains are either proteolytic or neither proteolytic nor saccharolytic; they have been isolated e.g. from autopsy material.

Strains of *C. botulinum* are also divided into three metabolic groups. Group I includes proteolytic strains of types A, B, F and G. Group II includes all strains of types C and D. Group III includes non-proteolytic strains of types B, E and F.

C. butyricum. Saccharolytic, requiring biotin for growth. Ferments many carbohydrates, including starch, producing acetic and butyric acids, CO_2 and H_2. Gelatin is not hydrolysed. No extracellular lecithinases or lipases are formed. Cells: motile. Spores: ovoid, subterminal.

C. cadaveris. Saccharolytic, acetogenic; some strains are also proteolytic. Indole-positive. Spores: ovoid, terminal.

C. carnis. Saccharolytic; aerotolerant.

C. cellulolyticum. Cellulolytic; mesophilic; isolated from decayed grass. Spores: spherical, terminal. [IJSB (1984) *34* 155–159.]

C. chauvoei. Saccharolytic, forming acetic and butyric acids; some strains hydrolyse gelatin. Spores: ovoid, subterminal. Toxins: e.g. α-toxin (dermonecrotizing, haemolytic); β-toxin (DNase); γ-toxin (hyaluronidase); δ-toxin (a THIOL-ACTIVATED CYTOLYSIN). (See also BLACKLEG and MALIGNANT OEDEMA.)

C. clostridiiforme. Saccharolytic, forming acetic and lactic (but not butyric) acids, CO_2 and H_2. Some strains form indole. Cells typically stain Gram-negatively; spores (ovoid, subterminal) are seldom formed.

C. cochlearium. Non-saccharolytic; glutamate, glutamine or histidine are metabolized to acetic and butyric acids, CO_2 and H_2.

C. cocleatum. Saccharolytic; carbohydrates are fermented to acetic (but not butyric) acid. Cells: helical.

C. colinum. See QUAIL DISEASE.

C. cylindrosporum. Substrates include xanthine and other purines, but not sugars or amino acids; metabolic products include acetic and formic acids and glycine.

C. difficile. Saccharolytic; ferments glucose or mannitol, forming e.g. acetic, butyric, isobutyric and isocaproic acids. Indole-negative. No extracellular lecithinases or lipases are formed. Some strains hydrolyse gelatin. Cells: Gram-positive or Gram-variable; on cycloserine–cefoxitin–egg-yolk–fructose agar (used for primary isolation from faeces) colonies exhibit a yellow-green fluorescence under light of wavelength 360 nm. Some strains are non-motile. Ovoid spores are formed subterminally; spores are reported to occur at *both* ends of the rod to give a 'dumb-bell' appearance [Book ref. 107, pp. 327–346]. Toxinogenic strains can cause COLITIS and PSEUDOMEMBRANOUS COLITIS. [PCR ribotyping of *C. difficile* in hospitals in England and Wales: RMM (1997) *8* (supplement 1) S55–S56.]

C. histolyticum. Proteolytic; gelatin is hydrolysed, and milk is digested. Indole-negative. Some strains can grow aerobically on blood-agar plates. Spores: ovoid, subterminal. Toxins include the α-toxin (dermonecrotizing); β-toxin (collagenase); δ-toxin (a non-thiol-activated ELASTASE); and ε-toxin (a THIOL-ACTIVATED CYTOLYSIN). (See also CLOSTRIPAIN.)

C. innocuum. Saccharolytic; ferments glucose, mannitol and salicin (and sucrose in some strains), forming acetic and butyric acids. Cells non-motile. Spores: ovoid, terminal.

C. kluyveri. Non-saccharolytic, non-proteolytic; a mixture of ethanol, acetic acid and bicarbonate is metabolized to butyric and caproic acids and H_2. Biotin and PABA are required for growth.

C. limosum. Proteolytic; gelatin is hydrolysed, and milk is digested. An extracellular lecithinase is formed. Spores: ovoid, subterminal.

C. nigrificans. See DESULFOTOMACULUM.

C. novyi (= *C. oedematiens*). Proteolytic and saccharolytic. Types A, B, C and D are distinguished on the basis of the toxins they produce; types A and B are involved in human diseases (see e.g. GAS GANGRENE). (See also BLACK DISEASE and MALIGNANT OEDEMA.) Both A and B types ferment glucose, hydrolyse gelatin, and form the dermonecrotizing α-toxin. Type A also forms the γ-toxin (a phospholipase C) and ε-toxin (a lipase), and type B forms the β-toxin (a lethal dermonecrotizing phospholipase C), the ζ-toxin (a haemolysin), and the η-toxin (a tropomyosinase). Type D forms e.g. the β- and η-toxins. Some strains are non-motile. Spores: ovoid, subterminal.

C. oedematiens. See *C. novyi*.

C. paraperfringens. See *C. barati*.

C. paraputrificum. Saccharolytic; ferments glucose, lactose, maltose or salicin, forming acetic and butyric acids. Spores: ovoid, terminal.

C. pasteurianum. Saccharolytic; many carbohydrates are fermented to acetic and butyric acids, CO_2 and H_2. Requires e.g. PABA and biotin. Diazotrophic (see NITROGEN FIXATION).

C. perenne. See *C. barati*.

C. perfringens (formerly *C. welchii*). Proteolytic and saccharolytic; gelatin is hydrolysed, milk is digested, and glucose, maltose, lactose and sucrose are fermented to acetic and butyric acids. Some strains can use nitrate as electron acceptor with consequent increase in growth yield and the production of acetic rather than butyric acid. Indole-negative. Cells non-motile; on blood agar (made with human, rabbit or sheep blood) the colonies of haemolytic strains are each surrounded by a double zone of haemolysis: an inner zone of complete haemolysis, and an outer zone of incomplete haemolysis. Spores: ovoid, subterminal, but rarely formed; the spores have been reported to have a D VALUE (D_{90}) of 15–145 min for non-haemolytic strains, and 6–7 min for haemolytic strains [Book ref. 30, p. 327]. Strains of *C. perfringens* collectively form a range of toxins which include: the α-toxin (PHOSPHOLIPASE C: a LECITHINASE; see also HOT–COLD LYSIS); the β-, ε- and ι-toxins (all dermonecrotizing toxins); the δ-toxin (a lethal, haemolytic toxin); the γ- and η-toxins (both lethal, non-haemolytic toxins); the θ-toxin (a THIOL-ACTIVATED CYTOLYSIN: see also STREPTOLYSIN O); the μ-toxin (a HYALURONATE LYASE); the κ-toxin (a COLLAGENASE). On the basis of toxin production, strains of *C. perfringens* are divided into five types, designated A, B, C, D and E:

C. perfringens type A strains form the α-toxin; they are the commonest cause of clostridial FOOD POISONING, are often present in the lesions of GAS GANGRENE, and can cause haemorrhagic enteritis in young piglets.

C. perfringens type B strains form the α-, β- and ε-toxins, and cause e.g. LAMB DYSENTERY.

C. perfringens type C strains form the α- and β-toxins, and cause e.g. ENTERITIS NECROTICANS in man, STRUCK in sheep, and a fatal necrotic and haemorrhagic enteritis in young piglets.

C. perfringens type D strains form the α- and ε-toxins, and cause e.g. PULPY KIDNEY in ruminants.

C. perfringens type E strains form the α- and ι-toxins; they are associated with diseases in ruminants.

C. propionicum. See PROPIONIC ACID FERMENTATION.

C. ramosum (formerly e.g. *Ramibacterium ramosum*). Saccharolytic; ferments glucose, maltose, lactose, sucrose and salicin, forming acetic acid as a main product. Cells frequently stain Gram-negatively and are non-motile; spores: spherical to ovoid, terminal, but seldom observed.

C. septicum. Proteolytic and saccharolytic; glucose, maltose, lactose and salicin are fermented with the formation of acetic and butyric acids, and gelatin is hydrolysed. Spores: ovoid, subterminal. Toxins formed: α-toxin (dermonecrotizing, haemolytic), β-toxin (a DNase), γ-toxin (a hyaluronidase) and δ-toxin (a THIOL-ACTIVATED CYTOLYSIN). A causal agent in GAS GANGRENE and in BRAXY and MALIGNANT OEDEMA. (See also NEUTROPENIC ENTEROCOLITIS.)

C. sordellii. Proteolytic and saccharolytic; glucose and maltose are fermented to e.g. acetic acid. Gelatin is hydrolysed, and milk is digested. Indole-positive. Urease-positive. Some strains form a lecithinase (α-toxin) and e.g. a THIOL-ACTIVATED CYTOLYSIN. [Taxonomy of lecithinase-negative strains: JGM (1985) *131* 1697–1703.] Spores: ovoid, subterminal. (See also MALIGNANT OEDEMA.)

C. sphenoides. Saccharolytic; glucose, maltose, salicin and mannitol are fermented to e.g. acetic acid. Reported to be capable of growth on citrate. Some strains hydrolyse gelatin and some produce indole. Cells typically stain Gram-negatively. Spores: spherical, subterminal to terminal.

C. spiroforme. Saccharolytic; carbohydrates are fermented to acetic (not butyric) acid. Cells: helical.

C. sporogenes. Proteolytic and saccharolytic; glucose and maltose are fermented to e.g. acetic and butyric acids. Gelatin is hydrolysed, and milk is digested. Spores: ovoid, subterminal.

C. subterminale. Proteolytic; gelatin is hydrolysed, and milk is digested. Lysine is fermented to e.g. acetic and butyric acids. Spores: ovoid, subterminal.

C. tertium. Saccharolytic; glucose, maltose, lactose, sucrose, salicin and mannitol are fermented to acetic and butyric acids. Spores: ovoid, terminal. Growth occurs aerobically on blood agar.

C. tetani. The causal agent of TETANUS. (See also TETANOSPASMIN and THIOL-ACTIVATED CYTOLYSIN.) Proteolytic; gelatin is hydrolysed, and acetic and butyric acids are formed from the fermentation of peptones. No carbohydrates – not even glucose – are fermented. Some strains form indole. Spores: spherical and terminal.

C. thermoaceticum. See ACETOGENESIS.

C. thermocellum. Saccharolytic; cellobiose, CELLULOSE and hemicellulosic materials are degraded and fermented, the main products being ethanol, acetic acid, CO_2, and H_2. (See also CELLULASES and CELLULOSOME.) [Regulation of cellulase formation: JGM (1985) *131* 2303–2308.] Pentoses are apparently not utilized. Optimum growth temperature: 55–60°C.

C. thermolacticum. Saccharolytic; thermophilic. A wide range of carbohydrates (including xylan) can be fermented, the main product being L(+)lactate. Some strains are cellulolytic. [SAAM (1985) *6* 196–202.]

C. thermosaccharolyticum. Saccharolytic; many carbohydrates (including e.g. GLYCOGEN, PECTINS and STARCH) are fermented to e.g. acetic, butyric and lactic acids, CO_2, and H_2. Optimum growth temperature: 55–60°C. Spores may have a D VALUE of ca. $D_{120} = 4$ min; the organisms are sometimes responsible for hard SWELL spoilage of canned foods. (See also CANNING.)

C. thermosulfurogenes. Saccharolytic. Cells are peritrichous, Gram −ve; filaments may occur. Spores: spherical. Opt. growth at >60°C. In microbial mats from hot volcanic springs. [JGM (1983) *129* 1149.]

C. villosum. Isolated from subcutaneous abscesses in cats [IJSB (1979) *29* 241–244].

C. welchii. See *C. perfringens*.

Other species include e.g. *C. aurantibutyricum*, *C. celatum*, *C. cellobioparum*, *C. coccoides*, *C. durum*, *C. fallax*, *C. felsineum*, *C. formicoaceticum*, *C. glycolicum*, *C. haemolyticum*, *C. indolis*, *C. lituseburense*, *C. oceanicum*, *C. putrefaciens*, *C. putrificum*, and *C. sticklandii*.

The genus *Clostridium* is heterogeneous, and proposals for rationalization have included the creation of a number of new genera (*Caloramator*, *Filifactor*, *Moorella*, *Oxobacter* and *Oxalophagus*) [IJSB (1994) *44* 812–826].

clostripain A thiol PROTEASE (MWt ca. 50000; EC 3.4.22.8), obtained from *Clostridium histolyticum*, which acts specifically at basic amino acyl (arginyl, lysyl) residues. It also has amidase and esterase activity, cleaving amides and esters of amino acids.

clot culture BLOOD CULTURE in which clotted blood, minus serum, is liquefied (e.g. with STREPTOKINASE) and the resulting fluid used as the inoculum. Apparently not widely used – due e.g. to the risk of contamination prior to the inoculation stage.

clotrimazole (bis-phenyl-(2-chlorophenyl)-1-imidazole methane) One of the earliest of the clinically useful imidazole antifungal agents (see AZOLE ANTIFUNGAL AGENTS); it is used topically in the treatment of superficial mycoses (e.g. candidiasis, ringworm, pityriasis versicolor).

clotting (of blood plasma) See FIBRIN; COAGULASE; ANTICOAGULANT.

clover blackpatch disease See SLAFRAMINE.

clover rot (clover sickness; sclerotinia crown and stem rot, SCSR) A disease of clover and other forage legumes caused by *Sclerotinia* spp (particularly *S. trifoliorum*). Leaves and petioles of infected plants turn olive-brown and subsequently rot; the disease progresses from the petioles to the stem and root, leading to the death of the plant. Control by repeated applications of e.g. benomyl or quintozene is effective but economically impracticable. [Review: Bot. Rev. (1984) *50* 491–504.]

clover wound tumour virus *Syn.* WOUND TUMOUR VIRUS.

clover yellow mosaic virus See POTEXVIRUSES.

clover yellows virus See CLOSTEROVIRUSES.

cloverleaf structure (of tRNA) See TRNA.

cloxacillin See PENICILLINS.

club fungi See APHYLLOPHORALES (Clavariaceae).

clubroot A disease of cruciferous plants (cabbages, cauliflowers, turnips etc) caused by *Plasmodiophora brassicae* (see PLASMODIOPHOROMYCETES). The roots of an infected plant become swollen and distorted, developing a single large gall (the 'club' symptom) or clusters of smaller galls ('finger-and-toe disease'). The aerial parts of the plant may be stunted and may wilt in hot weather; the foliage often acquires a reddish tinge. Secondary infection of the galls by SOFT ROT bacteria commonly occurs.

P. brassicae causes both hyperplasia and hypertrophy of infected tissues, apparently by inducing hyperauxiny (see AUXINS). Healthy plant tissue normally contains, in separate compartments, an indole compound, *glucobrassicin*, and an enzyme (*glucosinolase*) which hydrolyses glucobrassicin to indoleacetonitrile (IAN), a precursor of indoleacetic acid (IAA). *P. brassicae* apparently interferes with the compartmentalization of glucobrassicin and glucosinolase, resulting in increased levels of IAN (and hence IAA) in infected tissues.

The cysts of *P. brassicae* can remain viable for years in the soil, and it is difficult to eradicate the disease from contaminated land. Soil 'sterilants' such as dazomet may reduce the numbers of cysts. The disease can be controlled to some extent by dipping the roots of plants into a suspension of calomel, thiophanate-methyl or benomyl before planting.

clue cells See BACTERIAL VAGINOSIS.

clumping factor See COAGULASE.

clumping-inducing agent See PHEROMONE.

cluster cup See UREDINIOMYCETES stage I.

cluster gene A gene encoding a multifunctional protein. (cf. GENE CLUSTER.)

cluster of differentiation See CELL ADHESION MOLECULE.

CLV Cassava latent virus (see GEMINIVIRUSES).

cM CENTIMORGAN.

CM COMPLETE MEDIUM.

CM-cellulose (carboxymethylcellulose; CMC) A soluble derivative of CELLULOSE which is readily hydrolysed by most or all cellulolytic organisms. It is used e.g. to determine the ability of an organism or enzyme to degrade non-crystalline cellulose; end-product formation or changes in the viscosity of CMC solutions may be measured. (cf. AVICEL; HE-CELLULOSE; FP-CELLULOSE.)

CM test CALIFORNIA MASTITIS TEST.

CMC (1) CM-CELLULOSE. (2) Chronic mucocutaneous CANDIDIASIS.

CMI (1) CELL-MEDIATED IMMUNITY. (2) Commonwealth Mycological Institute, Kew, Surrey, UK.

CML Chronic myelogenous leukaemia (see ABL).

CMT CALIFORNIA MASTITIS TEST.

CMV Cytomegalovirus (see BETAHERPESVIRINAE).

CNA CALCIUM NUTRIENT AGAR.

cnidosporans A group of protozoa formerly of the subphylum Cnidospora [JP (1964) *11* 7–20], currently included in the phyla MICROSPORA and MYXOZOA.

CNS (1) Central nervous system. (2) Coagulase-negative staphylococci.

CO See CARBON MONOXIDE.

^{60}Co See IONIZING RADIATION.

CO dehydrogenase A name used (confusingly) for (i) an enzyme which reduces carbon dioxide to carbon monoxide (see e.g. ACETOGENESIS), and (ii) an enzyme (cf. CO OXIDASE) which oxidizes carbon monoxide to carbon dioxide (see e.g. CARBOXYDOBACTERIA, METHANOGENESIS).

CO dehydrogenase disulphide reductase See ACETOGENESIS.

CO difference spectrum (of a cytochrome) The absorption spectrum of a given reduced, CARBON MONOXIDE-complexed CYTOCHROME *minus* the absorption spectrum of the same reduced, but non-complexed, cytochrome.

CO oxidase (CO:acceptor oxidoreductase) An inducible enzyme which occurs in CARBOXYDOBACTERIA; it catalyses the oxidation of carbon monoxide to carbon dioxide in a reaction in which oxygen is derived from water rather than from air:

$$CO + H_2O \longrightarrow CO_2 + 2H^+ + 2e^-$$

In at least some carboxydobacteria the enzyme contains a non-covalently bound molybdenopterin moiety. [Properties of CO oxidases: MS (1986) *3* 149–153.]

CO_2 See CARBON DIOXIDE.

CO_2-stat See CONTINUOUS CULTURE.

CoA COENZYME A.

***coa* gene** See STAPHYLOCOCCUS (*S. aureus*).

Coactin *Syn.* MECILLINAM.

co-activator See OPERON.

co-agglutination (1) See PROTEIN A. (2) Joint agglutination (analogous to CO-PRECIPITATION). (3) Agglutination of different strains or species of microorganism by a given antiserum (see e.g. WEIL–FELIX TEST).

coaggregation Cell-to-cell adhesion in which the cells of one species adhere more or less specifically to those of a different species; e.g. *Actinomyces viscosus* and *Streptococcus sanguis* coaggregate by LECTIN–carbohydrate interaction [JGM (1984) *130* 1351–1357].

coagulase Any bacterial component or product which causes coagulation (clotting) or PARACOAGULATION in PLASMA containing an anticoagulant such as citrate, heparin or oxalate. Coagulases are produced e.g. by certain staphylococci and by *Yersinia pestis*.

Staphylococci produce two structurally and functionally distinct types of coagulase: *free coagulase* (*staphylocoagulase*, a protein which is released into the medium) and *bound coagulase* (*clumping factor*, a protein component of the cell wall). *Staphylocoagulase* causes a true clotting of plasma (from certain species only); it interacts with a 'coagulase reacting factor', CRF, in the plasma (believed to be prothrombin or a form of it) to form a complex ('staphylothrombin') which acts on plasma fibrinogen, converting it into insoluble fibrin – apparently by limited proteolytic action similar or identical to that of normal thrombin. The resulting clot differs from a normal clot in that e.g. it is not stabilized by Factor XIII (a normal blood-clotting factor) and it is much more resistant to fibrinolysin than is a normal clot. Several antigenically distinct staphylocoagulases are produced by different staphylococcal strains. The property of staphylocoagulase production is often used to distinguish between pathogenic and non-pathogenic strains, since there is a high degree of correlation between staphylocoagulase production and virulence (although coagulase-negative staphylococci are not necessarily harmless). However, although it can apparently exert its effect freely in vivo, the role of staphylocoagulase – if any – in pathogenesis is unknown. In vitro, the activity of staphylocoagulase is inhibited by various chemicals, particularly oxidizing agents, and by certain antibiotics (e.g. some penicillins).

Staphylococcal *clumping factor* does not form a true clot in plasma; rather, it induces clumping of the cells (*paracoagulation*) in the presence of fibrinogen. When mixed with plasma, cells with clumping factor bind fibrinogen near the C-terminal ends of its γ-chains – this apparently being responsible for the clumping of the cells [Biochem. (1982) *21* 1407–1413, 1414–1420]. The role of clumping factor in pathogenesis (if any) is unknown. (cf. CCFAS.) [Isolation of clumping factor from *S. aureus*: Inf. Immun. (1985) *49* 700–708.]

Staphylocoagulase and clumping factor may be detected by different forms of the COAGULASE TEST.

[Review of staphylococcal coagulases: Book ref. 44, pp. 525–557.]

coagulase test Any procedure used to determine whether or not a given bacterial strain produces a COAGULASE. The tests described below are used for staphylococci.

The *tube test* (which detects *staphylocoagulase*) simply involves incubating the test organism with a suitable plasma sample. In one of many methods, 0.5–1.0 ml of plasma is mixed with an equal (or smaller) volume of an 18–24-hour broth culture of the organism, and the tube is incubated at 35–37°C; the tube is examined at hourly intervals, and after overnight incubation, for clot formation – a firm clot which does not move when the tube is shaken being regarded as a positive result [Book ref. 44, pp. 526–527]. Known coagulase-positive and coagulase-negative strains should be used as controls. Plasma containing anticoagulants such as citrate, oxalate, EDTA, or heparin may be used, although heparin has been reported to delay clot formation; any organism tested for coagulase formation should be unable to metabolize the anticoagulant: citrated plasma, for example, may be clotted by coagulase-negative citrate-utilizing bacteria. Preservatives (e.g. thiomersal) may inhibit staphylocoagulase. Purified fibrinogen, or freeze-dried plasma, may be used instead of fresh plasma. Staphylocoagulases from strains pathogenic for man exhibit maximum activity in human or rabbit plasma.

Strains which produce large amounts of fibrinolytic enzymes (see FIBRINOLYSIS) may not form clots, or may lyse any clot formed; such strains are more commonly found among isolates from human infections than among isolates from animals. (See also PSEUDOCOAGULASE.)

The *slide test* detects clumping factor (see COAGULASE). A thick, saline suspension is prepared (on a slide) from a colony of the test strain grown on a non-selective medium; a loopful of citrated or oxalated human or rabbit plasma (not sheep or guinea-pig plasma) is stirred into the suspension. (Fibrinogen may be used instead of plasma e.g. to avoid any antibody-mediated agglutination of the bacteria.) Clumping of cells within ca. 5 sec indicates the presence of clumping factor. (A false-negative result may be obtained with capsulated strains [Book ref. 44, p. 505].) Controls should be used (cf. AUTO-AGGLUTINATION). (See also STAPHYLOSLIDE.)

There is a high, though not total, correlation between positive results in the tube and slide tests; however, a positive slide test should not be regarded as evidence of staphylocoagulase production.

A *plate method* may be used to detect staphylocoagulase-positive organisms. The test organism is inoculated onto the surface of an agar growth medium containing citrated plasma or purified fibrinogen and prothrombin. After overnight incubation, coagulase-positive colonies are each surrounded by a dense zone of precipitated fibrin. This method is useful for estimating numbers of coagulase-positive staphylococci in e.g. foods, but is too slow and prone to misleading results for use in routine clinical diagnosis.

coal biodegradation Certain fungi are capable of growing on coal – particularly lignite ('brown coal', a form which is relatively low in carbon and rich in volatile compounds) – as the sole source of nutrients; such fungi include e.g. the basidiomycetes *Polyporus versicolor* and *Poria monticola* [AEM (1982) *44* 23–27] and various other species (e.g. *Aspergillus terreus* and species of *Candida*, *Mucor*, *Paecilomyces* and *Penicillium* [SAAM (1985) *6* 236–238]). Growth generally appears to be restricted to the surface of the coal; the nature of the substrates utilized is unknown.

Coal *products*, such as benzene and toluene (derived from coal tar), may be degraded e.g. by combinations of bacteria [Nature (1998) *396* 730]. [Aerobic and anaerobic degradation of toluene by *Thauera*: AEM (2004) *70* 1385–1392.]

(See also LEACHING.)

CoASH (CoA-SH) Uncombined COENZYME A.

coated lens See BLOOMED LENS.

coated pit See PINOCYTOSIS.

coated vesicle See PINOCYTOSIS.

cobalamin See VITAMIN B_{12}.

cobamide coenzymes See VITAMIN B_{12}.

cobinamide See VITAMIN B_{12}.

cobra venom factor (CVF) Either of the two proteins, derived from cobra venom, which affect the alternative pathway of COMPLEMENT FIXATION. CVFs mimic C3b. CVF from *Naja naja* binds Factor B and promotes its cleavage (by Factor D) to form a C3 convertase; it can also give rise to a C5 convertase. CVF from *Naja haje* can also give rise to a C3 convertase but not to a C5 convertase. Unlike C3b, CVF is not inactivated by Factors I and H; it therefore forms a C3 convertase which is not subject to regulation.

cobweb disease A MUSHROOM DISEASE caused by *Hypomyces rosellus*; it can be controlled e.g. by BENOMYL or PROCHLORAZ [PP (1983) *32* 123–131].

cocarboxylase See THIAMINE.

coccal Pertaining to a COCCUS.

cocci See COCCUS.

coccidia The common name for protozoa of the suborder EIMERIORINA.

Coccidia (1) A subclass of protozoa (class Telosporea) later incorporated, with the TOXOPLASMEA, in the subclass COCCIDIASINA. (2) A subclass of protozoa (class Sporozoea [JP (1980) *27* 37–58]) equivalent to the COCCIDIASINA.

Coccidiascus A genus of yeasts (family METSCHNIKOWIACEAE). Cells: spherical to ovoid (5–15 μm diam./length), with a large vacuole and nucleus; neither pseudomycelium nor mycelium is formed. *C. legeri* (sole species) occurs as a parasite in intestinal epithelial cells of *Drosophila* spp; it has not been cultured in vitro. [Book ref. 100, pp. 123–124.]

Coccidiasina A subclass of protozoa (class SPOROZOASIDA) parasitic mainly in vertebrates but also in invertebrates. In members of this subclass the mature gametocytes typically occur intracellularly in the host (cf. GREGARINASINA). Orders: Agamococcidiorida, EUCOCCIDIORIDA, Protococcidiorida.

coccidiasis Any subclinical or very mild infection with a coccidian parasite – as distinct from clinical COCCIDIOSIS.

Coccidioides A genus of fungi (class HYPHOMYCETES); *C. immitis* is a saprotroph in desert soils in hot, arid regions of the American continent, and is the causal agent of COCCIDIOIDOMYCOSIS. In mammalian tissues it occurs mainly as multinucleate, thick-walled, spherical (ca. 20–80 μm) cells (*spherules* or sporangia). The spherule protoplasm divides into multinucleate *protospores* and then into uninucleate sporangiospores (*endospores*) which are released on rupture of the spherule wall and develop into new spherules. *C. immitis* grows readily on various types of media, forming a mycelium which fragments into barrel-shaped arthroconidia which are highly infective and very easily dispersed by air currents.

coccidioidin Any antigenic preparation derived from *Coccidioides immitis* and used in a diagnostic SKIN TEST for COCCIDIOIDOMYCOSIS.

coccidioidomycosis (San Joaquin Valley fever; desert fever) A disease of man and animals caused by *Coccidioides immitis*; it occurs in hot, arid regions of the American continent. Infection usually occurs by inhalation of wind-borne spores (especially arthroconidia); person-to-person transmission does not occur. Infection may be asymptomatic or may result in an acute, self-limiting respiratory disease resembling a common cold or influenza – often with severe chest pain. Rarely, mycetomas ('fungus balls') may develop in the lungs. Less common symptoms include joint pains ('desert rheumatism') and/or skin lesions. The primary infection occasionally leads to progressive, often fatal, disseminated disease (coccidioidal granuloma) with lesions in e.g. skin, joints, meninges. *Lab. diagnosis*: serological tests; SKIN TESTS using COCCIDIOIDIN or SPHERULIN; demonstration of the fungus by histology, culture and/or animal inoculation. *Treatment*: ketoconazole, amphotericin B. [Book ref. 25.]

coccidiosis Any disease of man and other animals caused by protozoa of the suborder EIMERIORINA. Coccidioses are typically contracted via the oral route and may involve the intestinal tract and/or various other tissues; they may be mild to fatal. Coccidioses of domestic animals and poultry are often of economic importance.

(a) Poultry. In chickens, *Eimeria necatrix* causes severe haemorrhage in the small intestine, and *E. tenella* causes haemorrhage and necrosis in the caecum; other pathogens include e.g. *E. acervulina* and *E. brunetti*. In ducks, *Tyzzeria perniciosa* causes haemorrhagic intestinal disease. In turkeys, severe inflammation of the intestine is caused by e.g. *E. adenoeides* and *E. meleagrimitis*.

(b) Domestic animals. In cattle, haemorrhagic lesions in the intestine are caused e.g. by *E. bovis* and *E. zürnii* (*E. zuernii*); an increased incidence of disease may occur during cold weather ('winter coccidiosis'). In sheep, intestinal disease may be caused e.g. by *E. ahsata*, *E. ovina* or *E. ovinoidalis*. In pigs, enteritis, severe diarrhoea and emaciation are caused e.g. by *E. scabra*. In rabbits, intestinal disease (often fatal) is caused e.g. by *E. magna* and *E. irresidua*; *E. stiedae (E. stiedai)* causes diarrhoea, enlargement of the liver and bile ducts, and emaciation. In cats, intestinal symptoms may be due to infection by *Isospora* spp (which may also infect dogs) or *Toxoplasma* (see TOXOPLASMOSIS). (cf. SARCOSPORIDIOSIS; see also EQUINE PROTOZOAL MYELOENCEPHALITIS.)

(c) Man. Infections involving *Isospora belli* and *I. hominis* are believed to be commonly asymptomatic, but intestinal symptoms may occur. (See also CRYPTOSPORIDIOSIS, SARCOSPORIDIOSIS and TOXOPLASMOSIS.)

Diagnosis of animal coccidioses involves e.g. SPORULATION of oocysts concentrated from fresh faeces.

[Identification of *Eimeria* spp: Book ref. 7, pp. 7–30. Identification of coccidia: Book ref. 18, pp. 80–91. Coccidian pathogenicity: Book ref. 18, pp. 287–327. Pathology, lab. diagnosis, and therapy in cryptosporidiosis, isosporiasis, sarcosporidiosis and toxoplasmosis: Book ref. 118, pp. 211–236.]

coccidium An individual, or a particular strain, of an organism of the suborder EIMERIORINA.

coccobacillus A bacterial cell intermediate in morphology between a COCCUS and a BACILLUS.

Coccodiscus See RADIOLARIA.

coccoid More or less spherical: cf. COCCUS.

coccoid bodies Thin-walled coccoid forms which develop from the helical or vibrioid cells in old cultures of e.g. *Aquaspirillum*, *Campylobacter* and *Oceanospirillum* spp; their formation is enhanced by treatment with mitomycin or UV light. Coccoid bodies resemble sphaeroplasts but are resistant to osmotic lysis; the majority are apparently viable, 'germinating' to form normal cells when transferred to a fresh medium. (cf. MICROCYST.)

coccolith A calcified scale, generally measuring a few micrometres or less in diam., present on the cell surface in certain cells of the COCCOLITHOPHORIDS. Coccoliths contain crystals of calcite – either of a single type (in *holococcolithis*) or, more commonly, of different shapes and sizes (in *heterococcoliths*); coccoliths are often elliptical in shape (but may be round, polygonal, etc) and they commonly exhibit intricate patterns and ornamentations which are species-specific. The coccoliths of present-day coccolithophorids form a significant proportion of some deep-sea oozes; however, coccolithophorids were even more abundant in previous geological ages, particularly in the Cretaceous, and the coccoliths of these organisms are major constituents of Mesozoic (Jurassic and Cretaceous) and Cenozoic chalks and marls. Coccolith morphology apparently changed rapidly over the ages, making fossil coccoliths useful markers in the biostratigraphy of sedimentary rocks [Book ref. 136, pp. 329–426 (Mesozoic coccoliths), pp. 427–554 (Cenozoic coccoliths)]. (See also FORAMINIFERIDA.)

coccolithophorids Those algae of the PRYMNESIOPHYCEAE in which the cell bears a covering of one to several layers of COCCOLITHS during at least some stage of the life cycle. For example, *Emiliana huxleyi* – the commonest living coccolithophorid – has a life cycle in which two distinct types of cell are formed: the motile 'S cell', which is biflagellated and covered with one or several layers of organic scales, and the non-motile 'C cell', which lacks flagella and organic scales but is enclosed by a shell-like *coccosphere* consisting of one or several layers of coccoliths. Life cycles involving motile and non-motile phases also occur in other genera (e.g. *Coccolithus*, *Cricosphaera*, *Hymenomonas*). Coccolithophorids are planktonic marine organisms which are particularly common in tropical waters.

Coccolithus See COCCOLITHOPHORIDS.

Coccomyces See RHYTISMATALES.

Coccomyxa A genus of unicellular, non-motile green algae (division CHLOROPHYTA) which occur as photobionts in certain lichens (see e.g. NEPHROMA, PELTIGERA, SOLORINA). The cells are small, ovoid, and contain a single parietal chloroplast with no pyrenoid; flagellated stages and sexual reproduction have never been observed. (See also RIBITOL and SPOROPOLLENIN.)

Cocconeis See DIATOMS.

coccosphere See COCCOLITHOPHORIDS.

coccus (*pl.* cocci) A spherical (or near-spherical) bacterial cell. According to species, cocci may occur singly, in pairs (see DIPLOCOCCUS), in regular groups of four or more (see TETRAD and PACKET), in chains, or in irregular clusters. (cf. BACILLUS sense 2.)

Cochliobolus See DOTHIDEALES.

Cochliopodium A genus of amoebae (order ARCELLINIDA) which have a type of test known as a *tectum*: a more or less flexible, closely adherent covering of small scales (distinguishable only by electron microscopy); in some species the scales bear spines which may be visible by light microscopy, while in others sand grains and other debris may adhere to the scales. Species occur in fresh water, activated sludge, etc. [Book ref. 133, pp. 76–82.]

Cochlonema See ZOOPAGALES.

cocksfoot mottle virus See SOBEMOVIRUSES.

cocksfoot streak virus See POTYVIRUSES.

cocoa Cocoa and chocolate are made from the seeds (beans) of the cacao plant *Theobroma cacao*. The cacao fruit is a pod containing up to 50 beans covered by a white mucilage. After harvesting, the pods are opened and the beans are extracted, piled up (e.g. in perforated boxes or on plantain leaves) and covered (e.g. with leaves). A fermentation then occurs: initially, various yeasts carry out an ALCOHOLIC FERMENTATION, and this is rapidly followed by ACETIFICATION by acetic acid bacteria – the production of acetic acid and heat inhibiting further yeast action. Other bacteria (e.g. lactic acid bacteria) and moulds are also present. During the process, which lasts ca. 6 days, the beans darken in colour (owing to oxidation) and the mucilage gradually disappears. The beans are then dried. The characteristic cocoa flavour develops only after the beans are roasted. [Book ref. 5, pp. 275–292.] (cf. COFFEE; see also CACAO DISEASES.)

cocoa necrosis virus See NEPOVIRUSES.

coconut cadang-cadang viroid (CCCV) A VIROID which infects coconut palms; infected palms may be killed, and the viroid has caused severe economic losses e.g. in the Philippines (cf. TINANGAJA DISEASE). The natural mode of transmission is unknown. CCCV is unique among known viroids in that it exists as four different RNA species. Two RNAs – designated RNA-1 small or RNA-1 fast (246 nucleotides) and RNA-2 small or RNA-2 fast (492 nucleotides) – occur in infected palms during the early stages of cadang-cadang disease; later, two additional RNAs – RNA-1 large or RNA-1 slow (287 nucleotides) and RNA-2 large or RNA-2 slow (574 nucleotides) – appear and eventually predominate. The RNA-1 small is regarded as the unit CCCV RNA; the other three RNAs are apparently derived from RNA-1 small by sequence duplication. [Nature (1982) *299* 316–321.]

coconut hartrot See HARTROT.

coconut lethal yellowing A YELLOWS disease of the coconut palm caused by an MLO. The disease is controlled, commercially, by the use of oxytetracycline hydrochloride.

coconut palm rhinoceros beetle (biological control) See BACULOVIRIDAE.

co-conversion (*mol. biol.*) See RECOMBINATION.

codecarboxylase See PYRIDOXINE.

coding strand A term which (like non-coding strand) is frequently used with opposite meanings by different authors. Originally, the term was generally used to refer to that strand of a gene which acts as the template on which mRNA is synthesized during transcription – the strand whose sequence is *complementary* to that of mRNA. Currently, there appears to be a consensus for the opposite meaning, i.e. that strand of a gene which is *not* transcribed (i.e. not used as a template) and whose sequence is homologous to that of mRNA. (Even so, given the confusion present in the literature, it may be wise to define the term, in relation to mRNA, whenever it is used.) The coding strand is also called the *sense* strand or the *plus* (+) strand.

The non-coding strand of a gene (which is complementary to the coding strand) is thus the template strand; it is also called the *antisense* strand or the *minus* strand.

Interestingly, a certain gene in the fruitfly (*Drosophila*) has been shown to contain protein-encoding information in *both* of the strands [Nature (2001) *409* 1000].

Codium A genus of siphonaceous green seaweeds (division CHLOROPHYTA). The thallus is terete and dichotomously branched; it is composed of a colourless central medulla of interwoven filaments from which arises a surrounding layer of inflated green branchlets ('utricles'). Anisogametes are produced in the utricles; these are initially non-motile, later becoming biflagellate. (See also CELL WALL; ELYSIA; NITROGEN FIXATION.)

codominant gene One of two (or more) genes which, when present together, specify a phenotype unlike that specified by either (or any) of the genes individually.

codon See GENETIC CODE.

codon bias In the expression of a heterologous (i.e. 'foreign') gene (e.g. a mammalian gene in *Escherichia coli*): inefficient translation due to the presence of certain codon(s) for which the host cell has insufficient numbers of the corresponding tRNA(s); codon bias can be a problem e.g. when high-level expression in *E. coli* is required from a heterologous gene containing a high frequency of codons such as the proline codon CCC and/or the arginine codon AGG – codons which occur rarely in homologous (*E. coli*) genes. Attempts at high-level expression of genes containing relatively high frequencies of codons that are 'rare' for the host cell may lead e.g. to slowing or termination of translation and the degradation of mRNA. One solution to the problem is to insert into the host cell extra (plasmid-borne) copies of the relevant tRNA-encoding gene(s).

Codon bias may also arise as a result of a same-sense mutation which replaces a common or 'routine' codon with a synonymous but infrequent codon (see SILENT MUTATION).

Codonella See TINTINNINA.

Codonosiga See CHOANOFLAGELLIDA.

Coe virus Coxsackievirus A21: see ENTEROVIRUS.

coelichelin A tripeptide SIDEROPHORE, encoded by *Streptomyces coelicolor*, which is synthesized by a non-ribosomal synthetase [see e.g. FEMS (2000) *187* 111–114].

Coelomomyces See BLASTOCLADIALES.

Coelomycetes A class of fungi (subdivision DEUTEROMYCOTINA) which form septate mycelium and in which the conidiophores line a discrete saucer-shaped, cupulate, flask-shaped or loculate conidioma – e.g. an ACERVULUS or a PYCNIDIUM. (cf. HYPHOMYCETES.) Orders [Book ref. 64, pp. 87–88]: MELANCONIALES, PYCNOTHYRIALES and SPHAEROPSIDALES.

Coeloseira See RHODOPHYTA.

Coelosphaerium A phycological genus of unicellular 'blue-green algae' (Chroococcaceae) in which the cells occur in a single layer at the periphery of a spherical colony; the cells are ellipsoidal, divide longitudinally, and contain GAS VACUOLES. The organisms are planktonic e.g. in freshwater lakes. (See CYANOBACTERIA.)

coelozoic Refers to a parasite which lives within its host in body fluids – e.g. in the gall bladder, bloodstream (*outside* blood cells), urinary tract, etc. (cf. HISTOZOIC.)

Coemansia See KICKXELLALES.

coenobium (1) (*bacteriol.*) A MICROCOLONY (sense 1) in which, following cell division, the cells of a clone form a regular array. Thus, e.g. *Thiopedia* typically forms a coenobium which consists of a flat sheet of 16 contiguous cells in a '4×4' arrangement – but coenobia containing up to ca. 64 cells may be formed. In *Brachyarcus* the coenobia consist of symmetrically arranged cells embedded in a matrix.

(2) (*algol.*) In certain types of algae: a colonial form which consists of a number of cells arranged in a specific way; the number of cells in a coenobium is usually a stable feature of a given species or genus. Coenobia are formed e.g. by *Scenedesmus* and *Volvox*.

coenocyte (*microbiol.*) A multinucleate cell, structure, or organism, formed by the division of an existing multinucleate entity or by nuclear division without the formation of dividing walls or septa. (cf. SYNCYTIUM.) Coenocytes occur e.g. in SIPHONACEOUS and SIPHONOCLADOUS algae; the hyphae of many fungi may be regarded as coenocytic.

coenzyme (1) Any low-MWt, non-protein, freely dissociable organic molecule which is necessary for the activity of a given enzyme (cf. PROSTHETIC GROUP); e.g., dehydrogenases require electron acceptors such as NAD^+ or $NADP^+$ for catalysis of redox reactions. (2) Any low-MWt, non-protein organic molecule – whether freely dissociable or firmly bound – necessary for the activity of a given enzyme.

coenzyme F_{420} See METHANOGENESIS.

coenzyme F_{430} See METHANOGENESIS.

coenzyme I NAD (q.v.).

coenzyme II NADP: see NAD.

coenzyme A (CoA) A coenzyme, derived from PANTOTHENIC ACID (see figure), which functions as a carrier of acyl groups – with which it forms thioesters (CoA.S.CO.R). (Uncombined coenzyme A may be represented as CoA-SH or CoASH.) Acyl-coenzyme A thioesters have a high free energy of hydrolysis and are commonly involved in SUBSTRATE-LEVEL PHOSPHORYLATIONS.

coenzyme B See METHANOGENESIS.

coenzyme B_{12} See VITAMIN B_{12}

coenzyme F See FOLIC ACID.

coenzyme M See METHANOGENESIS.

coenzyme Q See QUINONES.

coenzyme R *Syn.* BIOTIN.

cofactor (1) Any low-MWt, non-protein (organic or inorganic) factor which is necessary for the activity of a given enzyme (cf. COENZYME; PROSTHETIC GROUP). (2) (*Syn.* activator) Any *inorganic* component (e.g. metal ion) necessary for the activity of a given enzyme. (3) Any organic or inorganic factor necessary for the activity of an enzyme or enzyme complex.

coffee The commercial preparation of coffee from ripe coffee fruits ('cherries') requires the removal of the sticky mucilaginous mesocarp which surrounds the two beans in each fruit. This may be achieved mechanically or chemically, but the preferred method is by fermentation – which also improves the quality and appearance of the beans. The coffee fruits are pulped to disrupt the skins and then allowed to ferment either under water or 'dry'. The mucilage is degraded both by enzymes in the fruit itself and by microbial extracellular enzymes. (Commercial preparations of mould enzymes have also been used.) Various bacteria, yeasts and moulds are associated with the fermenting coffee, the most important probably being pectinolytic species of e.g. *Bacillus*, *Erwinia*, *Aspergillus*, *Fusarium* and *Penicillium*. After the fermentation, the beans are washed, dried, blended, roasted and ground before use. (cf. COCOA.)

coffee berry disease A disease of the coffee plant characterized by sunken, anthracnose lesions on the berries; causal agent: *Colletotrichum coffeanum*. [Review: Phytopathol. Paper number 20, March 1977.]

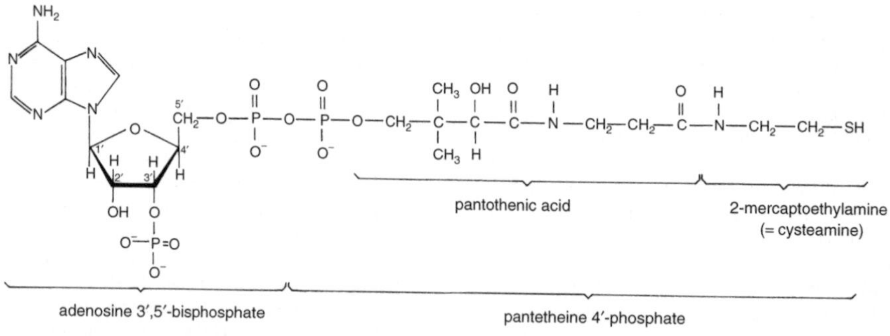

COENZYME A (CoA-SH)

coffee diseases For diseases of the coffee plant (*Coffea* spp) see e.g. COFFEE BERRY DISEASE, COFFEE RUST, FUSARIUM WILT, PHLOEM NECROSIS, RED RUST.

coffee ringspot virus See RHABDOVIRIDAE.

coffee rust A COFFEE DISEASE caused by *Hemileia vastatrix*. Yellowish spots, which become orange, appear on the undersides of the leaves, corresponding dark patches appearing on the upper surfaces; infected leaves are shed, and yields of berries are greatly reduced.

cognac See SPIRITS.

cognate In (e.g.) immunology, a word commonly used to mean 'corresponding' or 'matching'.

cognate help (linked recognition) (*immunol.*) The binding of a helper T cell to an antigen which is also bound by a B cell; the B and T cells are thus physically linked via the antigen. Such joint binding of antigen is accompanied by contact between various receptors and ligands on the two cells; this leads to activation and proliferation of both types of cell and formation of antibodies by the B cell. (See also ANTIBODY FORMATION.)

The determinant (epitope) bound by the B cell is sometimes referred to as the 'hapten', that bound by the T cell as the 'carrier' – terminology that derives from earlier hapten–carrier studies.

cohesive ends *Syn.* STICKY ENDS.

cointegrate The (circular) product of fusion between two circular replicons (e.g. two plasmids, or a plasmid and a bacterial chromosome) mediated by a TRANSPOSABLE ELEMENT (TE). Cointegration may occur as a result of a transposition event between a replicon containing a TE and one containing a target sequence for that TE. The transposition is accompanied by duplication of both the TE and a sequence at the target site: see entry TRANSPOSABLE ELEMENT. An alternative mechanism of cointegrate formation involves *recA*-dependent reciprocal recombination between homologous TEs in each of two replicons. (Some authors reserve the term 'cointegration' for the transposition event only.) (See also Tn*3*.)

(The term 'cointegrate' has also been used for any product of fusion between two replicons, regardless of the mechanism by which fusion occurred.)

coital exanthema See EQUINE COITAL EXANTHEMA.

col factor See COLICIN PLASMID.

col plasmid See COLICIN PLASMID.

Colacium A genus of EUGLENOID FLAGELLATES. The organisms differ from other euglenids in that they generally form dendroid or palmelloid colonies; flagellated cells, when formed, resemble those of EUGLENA. *C. libellae* overwinters in association with damselfly nymphs, becoming established in the rectum of the insect.

colanic acid A capsular heteropolysaccharide (the 'M antigen') produced e.g. by strains of *Escherichia coli* and *Salmonella*. K12 strains of *E. coli* produce colanic acid and form mucoid colonies at 30°C but not at 37°C; certain mutants (*lon* or *capR*) produce colanic acid and appear mucoid at 37°C.

Colanic acid consists of repeating trisaccharide units of (-glucose-fucose-fucose-), the central fucose residue of each unit carrying a trisaccharide branch (galactose-glucuronic acid-galactose-). Non-carbohydrate substituents (e.g. acetyl and pyruvyl groups) may be present, the nature and position of such groups varying e.g. with strain.

[Organization of the *E. coli* K12 gene cluster responsible for production of colanic acid: JB (1996) *178* 4885–4893.]

Colcemid (trade name) DEMECOLCINE.

colchicine An alkaloid found e.g. in the meadow saffron (*Colchicum autumnale*); the colchicine molecule consists of a tricyclic skeleton which includes an aromatic ring (substituted with three methoxy groups) and a tropolone ring. At low concentrations colchicine can e.g. bind to TUBULIN and prevent the in vitro or in vivo assembly of tubulin into MICROTUBULES; colchicine can thus e.g. inhibit MITOSIS in mammalian, plant, and other cells. (However, some complex microtubular structures – e.g. CENTRIOLES, and the ciliar and flagellar axonemes – are normally resistant to colchicine.) The in vitro effect of colchicine on microtubule assembly can be abolished e.g. by ultraviolet radiation (which converts colchicine to lumicolchicine).

CH_3O CH_3O CH_3O $NHCOCH_3$ O OCH_3

COLCHICINE

cold See COMMON COLD.

cold abscess See ABSCESS.

cold agglutinins Agglutinins which combine maximally with homologous or cross-reacting antigens at low temperatures (e.g. 4°C) but not at e.g. 37°C. Cold agglutinins may be detectable e.g. in the serum of patients suffering from PRIMARY ATYPICAL PNEUMONIA (sense 2), and they can be assayed by their ability to agglutinate human group O RBCs. Cold agglutinins occur also in certain other diseases – e.g. pneumonia caused by adenoviruses.

cold enrichment ENRICHMENT of psychrophilic or psychrotrophic organisms by incubation at low temperatures (see e.g. *Yersinia enterocolitica*).

ColD plasmid See COLICIN D and COLICIN PLASMID.

cold shock See OSMOTIC SHOCK.

cold-shock response In prokaryotes and eukaryotes: an apparently adaptive response to a down-shift in temperature (e.g. 37 → 10°C or, in general, a down-shift of at least 13°C) characterized by (i) initial cessation of growth, with resumption after a lag period; (ii) induction or increased synthesis of *cold-shock proteins;* (iii) repression of heat-shock proteins; (iv) generalized inhibition of protein synthesis, but ongoing synthesis of certain proteins involved in transcription and translation.

In *Escherichia coli*, the cold-shock proteins include CspA, RecA, IF-2 (initiation factor 2, which mediates binding of *N*-formylmethionine-charged tRNA to the 30S ribosomal subunit), NusA (involved in the last stage of transcription), and the α-subunit of DNA gyrase. So far, no cold-inducible sigma factor has been identified.

CspA is a small (70 amino-acid) protein which is inducible immediately (at the level of transcription) on temperature down-shift. It appears to interact with nucleic acids. Suggested functions of CspA include: (i) a low-temperature activator of translation; (ii) an 'RNA chaperone' that inhibits secondary structures in RNA and which may be important e.g. for efficient translation of mRNA at low temperatures [JBC (1997) *272* 196–202]; (iii) an agent which inhibits translation from specific mRNAs. (It has been reported that CspA is normally produced during early exponential growth at 37°C in *E. coli* – i.e. under non-stress conditions [EMBO (1999) *18* 1653–1659].)

The cold-shock response can also be triggered by certain inhibitors of translation – for example, the antibiotics chloramphenicol, erythromycin and tetracycline; this has suggested that a decrease in translational capacity may be an important factor in the induction of the cold-shock response [Mol. Microbiol. (1994) *11* 811–818].

cold sore (fever blister) (1) The lesion associated with labial herpes (see HERPES SIMPLEX). (2) The recurrent thin-walled vesicular mucocutaneous lesion associated with any of the various forms of HERPES SIMPLEX.

cold stability factor See MICROTUBULE-ASSOCIATED PROTEINS.

cold water disease (peduncle disease; low temperature disease) A FISH DISEASE affecting young salmonids below 10°C. The tail and peduncle undergo slow progressive necrosis. Causal agent: *Flexibacter psychrophila* ('*Cytophaga psychrophila*').

ColE1 plasmid A small (MWt ca. 4.2×10^6), non-conjugative, ccc dsDNA COLICIN PLASMID which is normally maintained in the host cell at a copy number of ca. 10–30.

In ColE1, DNA REPLICATION occurs by the CAIRNS MECHANISM. Replication is initiated from a fixed origin and (in contrast to chromosomal replication in *Escherichia coli*) proceeds unidirectionally from the origin; it is not dependent on plasmid-encoded proteins but requires the cell's RNA polymerase, DNA polymerases I and III and components of the PRIMOSOME. The first several hundred nucleotides of the leading strand are synthesized by DNA polymerase I from an RNA primer synthesized by RNA polymerase; subsequent DNA chain elongation is carried out by DNA polymerase III.

Initiation of plasmid replication is regulated at the level of primer formation. During the initiation process, *two* RNA transcripts are synthesized, one on each DNA strand, at the same region ca. 500 nt upstream from the origin (*ori*). The transcript on the H strand (RNA II) can – unless prevented (see later) – extend to the origin and give rise to a primer for synthesis of the leading strand; the transcript on the L strand (RNA I, 108 nt long) is complementary to the 5′ terminal region of RNA II and behaves as a *trans*-acting repressor of plasmid replication.

According to a current model, the 5′ end of RNA II is synthesized as a free (single-stranded) transcript, but as the RNA polymerase approaches *ori* the transcript and template form a DNA/RNA hybrid; the RNA strand in this hybrid is cleaved by RNASE H to form a free hydroxyl (–OH) terminal suitable for extension by DNA polymerase, i.e. the cleaved end of the RNA transcript, at *ori*, functions as a primer. (Although RNase H is essential for replication of ColE1 in vitro, its role in vivo has been questioned [JB (1986) *166* 143–147].)

Whether or not a functional primer is formed from the RNA II transcript, as described above, is influenced by several factors – factors which, by regulating the initiation of replication, determine the COPY NUMBER of the plasmid.

One regulatory factor is RNA I. RNA I and RNA II each form several stem-and-loop structures which, because the two transcripts are complementary, can undergo mutual base-pairing; when this happens, it inhibits development of a functional primer at the RNA/DNA hybrid in the *ori* region. (In vitro, RNA I can inhibit primer formation only if it binds to RNA II some time before the complete RNA II–DNA hybrid has been formed; once the nascent RNA II transcript has reached a length of ~360 nt it becomes resistant to RNA I.) The hybridization of RNA I with RNA II is facilitated/accelerated by a small protein (63 amino acids) encoded by the plasmid gene *rop* (= *rom*); the Rop/Rom protein (which acts as a rigid dimer exhibiting exact two-fold symmetry) thus enhances the inhibitory effect of RNA I on primer formation.

The ability of RNA I to inhibit primer formation depends on its intracellular concentration. (This, in turn, can be influenced e.g. by the host cell's RNASE E which can cleave and inactivate RNA I.) It appears that only when the copy number of ColE1 has reached its maximum is the concentration of RNA I high enough to inhibit further replication.

Interestingly, the copy number of ColE1 is affected by mutations in the *pcnB* gene (see MRNA (a)).

A similar mechanism for the regulation of plasmid replication occurs e.g. in CloDF13, p15A, pBR322, pMB1 and RSF1030 plasmids. In pBR322, RNA I is synthesized five times more often than RNA II, and it has been estimated that only 1 in 20 pre-priming event results in plasmid replication [JB (1987) *169* 1217–1222].

Coleochaete A genus of freshwater epiphytic green algae (division CHLOROPHYTA). Some species are filamentous and dichotomously branched, others are discoid; cell division is confined to apical/marginal cells. Characteristic of the genus is the formation by certain cells of long, fine, unbranched, sheathed 'hairs' (setae). Motile cells are scaly. Zoospores are biflagellate, one being formed per cell. Sexual reproduction is oogamous; an envelope of sterile cells may surround the reproductive cells. Plants may be homothallic or heterothallic.

Coleosporium See UREDINIOMYCETES.

Coleps A genus of freshwater and marine ciliates (subclass GYMNOSTOMATIA). Cells: typically barrel-shaped, ca. 50–100 μm in length, with an anterior (apical) cytostome; a semirigid endoskeleton (formed by the deposition of calcium phosphocarbonate in the pellicular alveoli) gives the appearance of a regular series of platelets forming a grid-like pattern over the entire body surface. Posteriorly directed spines may occur at the posterior edge of the cell. The organisms feed on other protozoa and algae.

colibacillosis (*vet.*) Any of certain diseases, caused by strains of *Escherichia coli*, which occur primarily in very young animals and which typically involve septicaemia and/or mild to severe diarrhoea. (See also ETEC; cf. WHITE SCOURS; OEDEMA DISEASE.)

colicin A A pore-forming colicin (see COLICINS) (MWt ca. 63000) which binds to the OmpF/BtuB protein [mechanism of action: JMB (1986) *187* 449–459].

colicin B A pore-forming colicin (see COLICINS) (MWt ca. 80000–90000) which binds to the FEPA PROTEIN.

colicin D A colicin (see COLICINS) (MWt ca. 87000) which binds to the FEPA PROTEIN and subsequently inhibits protein synthesis in sensitive cells. [Physical and genetic analysis of the ColD plasmid: JB (1986) *166* 15–19.]

colicin E A category of COLICINS (MWt ca. 64000–66000) each of which binds to the BtuB protein. E colicins are divided into nine immunity groups (E1...E9); colicins of a given group (e.g. E1) are not active against cells which contain a (repressed) plasmid encoding the *same* colicin and which are synthesizing the corresponding immunity protein. In a sensitive cell, colE1 forms pores; colE2, colE7, colE8 and colE9 are DNases; and colE3 and colE6 are RNases which cleave 16S rRNA.

colicin I A category of COLICINS (MWt ca. 80000); colIa and colIb are group B pore-forming colicins.

colicin K A group A, pore-forming colicin (see COLICINS) (MWt ca. 70000) which binds to the TSX PROTEIN; uptake across the cell envelope of the target cell is reported to require OmpFA and the TolABQR proteins.

colicin L See COLICINS.

colicin M A colicin (see COLICINS) (MWt ca. 23000) which binds to the TONA PROTEIN (= FhuA); it inhibits the synthesis of PEPTIDOGLYCAN.

colicin N A pore-forming colicin (see COLICINS) MWt ca. 39000) which binds to the OmpF protein.

colicin plasmid (col factor, or col plasmid) A PLASMID which encodes one or more COLICINS. Some col plasmids (type 1) are small, MULTICOPY PLASMIDS, while others (type 2) are large, low-copy-number plasmids. The type 1 col plasmids are non-conjugative; type 2 col plasmids are often conjugative (see CONJUGATIVE PLASMID). Conjugative col plasmids occur in various INCOMPATIBILITY (Inc) groups (e.g. ColV is an IncFI plasmid, ColB-K98 is an IncFIII plasmid, and ColIb-P9 is an Inc$I\alpha$ plasmid). Some non-conjugative plasmids have been found to exhibit incompatibility [e.g. incompatibility between ColE plasmids: JGM (1986) *132* 1859–1862].

Col plasmids contain three essential genes: (i) the colicin structural gene (= 'activity' gene); (ii) a gene encoding the immunity protein (see COLICINS); and (iii) a gene encoding the lysis protein. These genes are designated *cdl* (colicin D lysis gene), *cea* (colicin E activity gene), *cui* (colicin U immunity gene) etc.

In a population of colicinogenic bacteria only a small proportion of the cells is de-repressed for colicin synthesis; however, the immunity gene is usually expressed constitutively so that cells with repressed genes for the colicin and lysis protein are protected from the given colicin produced by their derepressed neighbours.

A col plasmid typically encodes only that type of immunity protein which protects the host cell against the particular colicin encoded by the plasmid. However, ColE9-J encodes two types of immunity protein: one protecting against colE9 and one protecting against colE5 [JGM (1986) *132* 61–71].

Transcription of col plasmids is repressed by the LexA protein so that synthesis of colicins is promoted by those agents which trigger the SOS SYSTEM.

colicin typing BACTERIOCIN TYPING with COLICINS.

colicin U A pore-forming colicin (see COLICINS) (MWt ca. 66300) produced by *Shigella boydii*; colicin U binds to sensitive cells via the OmpA and OmpF proteins, and uptake occurs via the TolABQR proteins [JB (1997) *179* 4919–4928].

colicin V The designation of a bacteriocin (MWt 6000) which acts by disrupting the membrane potential of sensitive cells [JB (1984) *158* 757–759]; it is not inducible by conditions which trigger the SOS system (cf. COLICINS), and at least some authors now classify this bacteriocin as a MICROCIN [e.g. TIM (1998) *6* 66–71].

colicin X The designation of a small bacteriocin subsequently classified as a MICROCIN.

colicin Y A recently isolated pore-forming colicin (see COLICINS) encoded by plasmid pCol-Let [complete coding sequence: Microbiology (2000) *146* 1671–1677].

colicinogenic Able to produce COLICINS.

colicinogenic plasmid *Syn.* COLICIN PLASMID.

colicins A category of high-MWt (~25–90 kDa) BACTERIOCINS produced by colicinogenic bacteria of the family Enterobacteriaceae (e.g. *Escherichia coli*, *Shigella boydii*) and active against (sensitive) strains of the same family; most colicins are encoded by plasmid genes (see COLICIN PLASMID), one exception being the chromosomally encoded 'bacteriocin 28b' – a bacteriocin (similar to colicin L) produced by strains of *Serratia marcescens*.

A given colicin is associated with a cluster of three genes: (i) the structural gene for the colicin; (ii) the immunity gene, encoding a protein which protects the cell against molecules of the *same* colicin produced by other cells in the vicinity; and (iii) a lysis gene, encoding a protein involved in the release of the colicin.

Under normal conditions the colicin and lysis genes are not expressed, but the immunity gene is expressed constitutively. The product of the immunity gene (the *immunity protein*) inhibits the activity (but not the adsorption) of any molecules of the given colicin that bind to the cell surface.

In a population of colicinogenic cells, synthesis of the colicin and lysis protein is promoted under conditions which trigger the SOS SYSTEM (see COLICIN PLASMID). However, in a colicinogenic population a small proportion of the cells (at any given time) may be derepressed for colicin synthesis, and these cells will be able to produce and externalize colicins.

Colicins synthesized in a given cell leave the cell following the activity of the lysis protein. The lysis protein localizes in the cell envelope; earlier studies indicated that this causes a non-specific increase in the permeability of the cell envelope – allowing release not only of the colicin molecules but also of various other constituent of the cell's cytoplasm [EMBO (1987) *6* 2463–2468]. During the externalization of colicin molecules, a cell may not appear to be undergoing extensive disintegration, but it becomes leaky (releasing various ions and molecules) and it fails to take up essential substrates. Synthesis and externalization of colicins is characteristically linked to the death of the colicin-producing cell.

Cells in the vicinity of a colicin-producing cell are insensitive ('immune') to the given colicin if they contain the same (repressed) colicin system and a functional immunity protein. In some cases, however, such immunity may be lost (*immunity breakdown*) in the presence of high concentrations of the given colicin; this presumably occurs because there are too many molecules of colicin for the cell's immunity system to deal with.

Externalized colicin molecules bind to specific receptors on the surface of sensitive cells, and it appears that internalization of the colicin molecule is usually or always an energy-dependent process; in at least some cases (e.g. colicin A), the colicin unfolds on binding to the receptor, such unfolding being essential for translocation through the cell envelope.

The specific receptors used by colicins include the BtuB, FepA, FhuA, IutA and Tsx proteins and the OmpF porin.

After binding, group A colicins (A, E1–E9, K, L, N, cloacin DF13 and bacteriocin 28b) are internalized via a complex of envelope proteins that include TolA, TolB, TolQ and TolR.

After binding, group B colicins (B, D, Ia, Ib, M, 5 and 10) are internalized via a complex of proteins that include the TONB PROTEIN, ExbB and ExbD.

[Colicin import into *Escherichia coli* cells: JB (1998) *180* 4993–5002.]

Within the target cell, a given colicin may act as a lethal agent by (i) forming pores in the cytoplasmic membrane, thereby collapsing proton motive force and allowing leakage of cytoplasmic constituents; (ii) functioning as a non-specific DNase; (iii) functioning as an RNase specifically against 16S rRNA; or (iv) inhibiting the synthesis of PEPTIDOGLYCAN.

A given colicin is associated with a particular mode of lethal activity. Thus, e.g. pore-formers include the group A colicins E1, A, N and K and the group B colicins B, Ia, Ib, 5 and 10; RNases include group A colicins DF13 and E6; DNases include the group A colicins E2 and E8; inhibitors of peptidoglycan synthesis include the group B colicin M.

(See also separate entries for colicin A, B, D, E, I, K, M, N, U, V, X and Y, and CLOACIN DF13.)

coliform In general: any Gram-negative, non-sporing, facultatively anaerobic bacillus which can ferment lactose, with acid and gas formation, within 48 hours at 37°C. For water bacteriologists (in the United Kingdom), a coliform is defined as any member of the family Enterobacteriaceae which grows at 37°C and which normally encodes β-galactosidase [for further information see Report 71 (1994) HMSO, London (ISBN 0 11 753010 7)].

Escherichia coli is a typical coliform.

coliform mastitis See MASTITIS (b).

coliform test A test used to detect the presence of COLIFORM organisms in e.g. water (see INDICATOR ORGANISM and WATER SUPPLIES); the test was originally designed to reveal faecal contamination (hence the designation 'faecal coliform test'). In the test, aliquots of sample are used to inoculate a number of tubes (see MULTIPLE-TUBE METHOD), each of which contains e.g. MACCONKEY'S BROTH and a DURHAM TUBE; incubation is carried out at 37°C. A 'positive' test (i.e. lactose fermentation) is shown by acidification (pH indicator) and gas production (Durham tube). The 'most probable number' of coliforms in the original sample can be calculated from the number of positive tubes and the use of statistical tables. The count thus obtained is actually a *presumptive coliform count* because, in any given tube, a 'positive' result could be due to certain endospore-forming bacteria (which also ferment lactose and form gas); this possibility is particularly important when testing *chlorinated* water samples because spore-forming bacteria are more resistant than coliforms to chlorine. Confirmation that each 'positive' result is, in fact, due to *thermotolerant* (= *'faecal'*) coliforms involves two further 24-hour tests at 44°C: the INDOLE TEST and the EIJKMAN TEST.

coli-granuloma *Syn.* HJÄRRE'S DISEASE.

coliphage A BACTERIOPHAGE that infects *Escherichia coli*. (For 'coliphage λ', 'coliphage T4' etc see BACTERIOPHAGE λ, BACTERIOPHAGE T4, etc.)

colisepticaemia A respiratory and septicaemic disease of poultry caused by strains of *Escherichia coli*. Incidence of the disease has increased greatly with the advent of modern intensive rearing methods, and often occurs secondarily to e.g. poor ventilation, vaccination of birds with live vaccines (e.g. live Newcastle disease vaccines), or various other respiratory diseases (e.g. AIR SACCULITIS, AVIAN INFECTIOUS BRONCHITIS, NEWCASTLE DISEASE). Infection occurs mainly via the respiratory tract; characteristic post-mortem findings include fibrinous pericarditis and perihepatitis.

colistins See POLYMYXINS.

colitis Inflammation of the colon. (cf. ENTERITIS.) Colitis occurs e.g. in certain infectious diseases (e.g. amoebic and bacillary DYSENTERY; see also SHIGA TOXIN). *Antibiotic-associated colitis* sometimes occurs when the normal bowel flora is suppressed by broad-spectrum antibiotic therapy; this condition ranges from mild diarrhoea to PSEUDOMEMBRANOUS COLITIS, and is commonly due to growth and toxin formation by *Clostridium difficile* (an organism normally rare in adults, but common in the bowel in infants). *C. difficile* produces at least two distinct hydrophobic protein toxins, A (an enterotoxin) and B (a cytotoxin) [JMM (1984) *18* 385–391]; toxin A causes fluid accumulation in the intestinal lumen (but *not* by activating adenylate cyclase – cf. e.g. CHOLERA TOXIN), and the B toxin disaggregates filamentous actin in tissue culture cells [GE (1984) *86* 1212 (abstr.)]. [Purification and properties of cytotoxin B: JBC (1986) *261* 1316–1321.] *Chemotherapy*: e.g. vancomycin.

colitose (3,6-dideoxy-L-galactose) A sugar, first isolated from *Escherichia coli*, which occurs in the O-specific chains of the LIPOPOLYSACCHARIDE in certain *Salmonella* serotypes (contributing to O antigen 35 in group O salmonellae – see KAUFFMANN–WHITE CLASSIFICATION) and in the LPS of *E. coli* serotypes of the O111 group.

collagen The main protein component of bone, cartilage and other connective tissues; it is a microfibrillar protein whose subunit, *tropocollagen*, consists of three polypeptide chains in which the amino acid sequences are essentially (glycine-X-Y)$_n$ where X and Y are often proline and hydroxyproline, respectively. Collagen is generally resistant to enzymic degradation (cf. COLLAGENASE). Collagenolytic organisms may cause LEATHER SPOILAGE. (See also GELATIN.)

collagenase A COLLAGEN-degrading enzyme; collagenases are produced e.g. by *Clostridium histolyticum* (the β-toxin) and *C. perfringens*. The collagenase of *C. histolyticum* appears to have maximum activity at the peptide bond preceding a glycine-proline- sequence.

collarette See CONIDIUM.

Collema A genus of LICHENS (order LECANORALES); photobiont: *Nostoc*. The thallus lacks a cortex and is homoiomerous, foliose to fruticose, brownish- to greenish-black, pulpy and gelatinous when wet, firm and cartilaginous when dry. The apothecia, when formed, generally have reddish-brown discs. According to species, globular, cylindrical, coralloid or scale-like isidia may be formed. Species occur on bark, rocks, calcareous soils, etc.

Colletotrichum A genus of fungi (order MELANCONIALES) which include some important plant pathogens (see e.g. ANTHRACNOSE, COFFEE BERRY DISEASE, RED ROT and SMUDGE). Conidiophores characteristically develop on setose acervuli (see SETA); the conidia, which lack appendages, are typically elongated, round-ended and colourless.

Collosphaera See RADIOLARIA.

Collybia See AGARICALES (Tricholomataceae).

colominic acid The capsular polysaccharide of *Escherichia coli* K1 strains; it is a linear polymer of *N*-acetylneuraminic acid residues linked by (2 → 8)-α-ketosidic bonds. [Biosynthesis: JB (1984) *159* 321–328.] An apparently identical polysaccharide occurs in strains of *Neisseria meningitidis* (groups B and C). Colominic acid is resistant to most neuraminidases; however, it can be hydrolysed by an enzyme from *Clostridium perfringens*, and a neuraminidase associated with a bacteriophage (coliphage E) *specifically* depolymerizes colominic acid [JV (1985) *55* 374–378].

colon microflora See GASTROINTESTINAL TRACT FLORA.

colonization factor antigens See ETEC.

colony (*microbiol.*) A number of cells or organisms (of a given species) which, during their growth, have developed as a discrete aggregate or group in which there is commonly direct contact or continuity between the cells.

Algal colonies. Some algal colonies are of more or less stable size and form (see COENOBIUM sense 2; cf. PALMELLOID PHASE). In some species the colony is the only form in which the organism occurs. See e.g. SCENEDESMUS and VOLVOX.

Bacterial colonies are usually formed only on, or within, a solid MEDIUM (such as an agar gel); each usually consists of a compact mass of individual cells, although certain bacteria (e.g. *Streptomyces* spp) form mycelial colonies. A discrete colony commonly comprises the progeny of a single cell (see also STREAKING). Many species of bacteria can form macroscopic colonies on appropriate media under suitable conditions; on a given medium, a colony's shape, colour, consistency, surface appearance and size (for a given incubation time) are often highly characteristic, and these features are often of use in the identification of particular bacterial species. The full description of a colony can be very detailed. Thus, e.g. the *elevation* of a (surface) colony may be flat, low convex, domed, umbonate etc; its *edge* may be e.g. entire (i.e., circular and unbroken), crenate (scalloped), lobed or fimbriate; its *texture* may be butyrous, friable or mucoid; its *surface* may be matt or glossy; it may be whitish or pigmented, or it may contain a dye taken up from the medium. When the colonies of certain species develop on blood agar, each colony is surrounded by a zone in which the blood has been lysed or greened (see HAEMOLYSIS). (See also DAISY HEAD COLONIES, DRAUGHTSMAN COLONY, fried egg colony in MYCOPLASMA, MOTILE COLONIES, SMOOTH–ROUGH VARIATION.)

Although a bacterial colony has been traditionally regarded as an aggregate of independent cells, some studies have provided evidence that the behaviour of a given cell in a colony is influenced by its position within the colony, i.e., the behaviour of a cell can be subject to 'multicellular' regulatory mechanisms [see Book ref. 198, pp. 27–69].

Other types of bacterial colony include e.g. the rosettes of CAULOBACTER spp, the arrays of *Hyphomicrobium* and RHODOMICROBIUM spp, and the networks of PELODICTYON.

Fungal colonies. Some fungi (e.g. certain yeasts) form colonies which resemble those typical of bacteria. Mycelial fungi often form circular 'fluffy' colonies, each consisting of a mass of MYCELIUM which may or may not include reproductive structures (see e.g. PENICILLIUM). (See also PELLET sense 2; PETITE MUTANT.)

Protozoal colonies. In the (few) colonial protozoa, the colony typically consists of SESSILE (sense 1) organisms which form a group of variable size (see e.g. CARCHESIUM). Chain-like colonies are formed by some members of the ASTOMATIDA.

colony-forming unit See COUNTING METHODS.

colony hybridization (modified Grunstein–Hogness procedure) A method for screening bacterial colonies for the presence of a specific sequence of DNA (e.g. among colonies containing a gene LIBRARY). As in a REPLICA PLATING procedure, cells are transferred from colonies on the master plate to a nitrocellulose (or other) filter. The cells on the filter are lysed with alkali – the alkali also serving to denature the DNA (i.e. to separate the strands of the duplex); the alkali is then neutralized, any protein is digested e.g. with proteinase K, and the filter is baked at 70–80°C under vacuum to bind the (single-stranded) DNA to the filter. The filter is then exposed to labelled probes (see PROBE) which are complementary to the sequence of interest. After washing away unhybridized probes, the given sequence, if present, is indicated by the label on bound (hybridized) probes; any probe-positive sites on the filter are then used to identify the corresponding (positive) colonies on the master plate.

A probe labelled with a radioactive isotope is detected by AUTORADIOGRAPHY.

In the original procedure [Book ref. 177, pp. 172–174], cells transferred from the master plate were allowed to grow and form colonies on the filter (which was overlaid on a plate of nutrient medium); subsequent steps were similar to those described.

This technique can also be used for yeast cells, but these cells must be converted to sphaeroplasts (with e.g. ZYMOLYASE) before lysis.

colony-stimulating factors (CSFs) CYTOKINES involved in the maturation of LEUCOCYTES. CSFs include e.g. granulocyte–macrophage colony-stimulating factor (GM-CSF), an agent which stimulates proliferation of progenitors of various blood cells (including granulocytes and macrophages), and macrophage colony-stimulating factor (M-CSF; also referred to as CSF-1), an agent formed mainly by monocytes and macrophages which, among other functions, stimulates the development of macrophages from precursor forms such as monoblasts and monocytes.

Colorado tick fever An acute human disease caused by a virus and transmitted by the tick *Dermacentor andersoni*; it occurs in the Rocky Mountain regions of the USA. Symptoms include fever, leucopenia, headache and myalgia, but no rash; CNS complications occasionally occur in young children. Wild rodents (e.g. ground squirrels) may act as a reservoir of infection. The causal agent has been regarded as an ORBIVIRUS, but has been shown to contain 12 dsRNA molecules [Virol. (1981) *112* 361–364] and may thus represent a distinct taxonomic group.

colostrum A secretion of the mammary gland produced before lactation proper; it contains e.g. immunoglobulins (mainly IgA and IgG). In man, the immunoglobulin content of colostrum is relatively low, but in some animals – e.g., ruminants (in which immunoglobulins do not cross the placenta) – large amounts are normally present. (See also PASSIVE IMMUNITY.)

colour-breaking (breaking) The development of VARIEGATION in flower petals; colour-breaking may occur as a result of virus infection (e.g., in tulips infected with tulip breaking virus, a member of the POTYVIRUSES). Flowers affected in this way are often prized for their attractive appearance.

Colour Index (CI) A comprehensive reference publication on dyes and pigments published by The Society of Dyers and Colourists, Bradford, Yorkshire BD1 2JB, England. It consists of a number of volumes which are periodically updated.

coloured field illumination *Syn.* RHEINBERG ILLUMINATION.

colourless sulphur bacteria Those non-photosynthetic SULPHUR BACTERIA which obtain energy by the oxidation of sulphur and/or reduced inorganic sulphur compounds (e.g. sulphide); although referred to as 'colourless', some strains can exhibit pigmentation owing to their content of cytochromes. The colourless sulphur bacteria include e.g. species of *Beggiatoa* and *Thiobacillus*. [Ecology of colourless sulphur bacteria: Book ref. 115, pp. 211–240.]

Colpidium A genus of ciliates (order HYMENOSTOMATIDA) which occur e.g. in freshwater habitats containing decomposing organic matter. Cells: ovoid or reniform, ca. 50–150 μm, with uniform somatic ciliature and a single macronucleus, micronucleus and contractile vacuole. In *C. colpoda* (at least) asexual reproduction occurs within a cyst.

Colpoda A genus of soil and freshwater ciliates (order COLPODIDA). Cells: typically reniform, ca. 40–120 μm in length, with uniform somatic ciliation; the cytostome is lateral. *C. steini* is a small species, ca. 20–60 μm, with a conspicuous tuft of cilia (the 'beard') projecting from the region of the vestibulum. Food consists of other protozoa, algae and bacteria. Asexual reproduction occurs within CYSTS – in which four or more individuals may be formed.

Colpodida An order of protozoa (subclass VESTIBULIFERIA) in which the vestibular ciliature tends to be highly organized. Cysts are common. The organisms are typically free-living, sometimes associated with molluscs. Genera: e.g. COLPODA, *Woodruffia*.

Colpomenia See PHAEOPHYTA.

Colsargen See SILVER.

Columbia SK virus See CARDIOVIRUS.

columella An axial or central, unicellular or multicellular structure within a fruiting body in certain fungi and in certain slime moulds of the MYXOMYCETES; in some cases it is simply an extension of the sporangiophore into the cavity of a sporangium, while in others it may be a spherical, conical or cylindrical structure extending upwards from the base of a sporangium. Although columellae are commonly described as STERILE (sense 2), the columellae of *Mucor piriformis* have been reported to germinate and give rise to secondary sporangia under certain cultural conditions [Mycol. (1985) *77* 353–357]. (cf. PSEUDOCOLUMELLA.)

column fermenter See FERMENTER.

columnaris disease (cotton wool disease; 'mouth fungus') An acute, often fatal freshwater FISH DISEASE caused by *Flexibacter columnaris*. Necrotic lesions occur on the skin and gills; cottony tufts of epithelium and bacteria appear around the mouth.

ColV plasmid Any of a heterogeneous group of IncFI COLICIN PLASMIDS which encode COLICIN V; many also encode e.g. the aerobactin iron-uptake system (see SIDEROPHORES).

ColV, I-K94 plasmid An IncFI COLICIN PLASMID which encodes COLICINS V and Ia; it occurs e.g. in *Escherichia coli* and is derepressed for conjugal transfer.

ColX plasmid See COLICIN PLASMID.

ComA See QUORUM SENSING.

Comatricha See MYXOMYCETES.

comb. nov. (*combinatio nova*; new combination) In NOMENCLATURE: the name of a species which has been moved from one genus to another but which has retained the specific epithet (suitably modified, when necessary, to agree with the form of the new genus name).

combinatio nova See COMB. NOV.

combined gold standard See GOLD STANDARD.

combined residuals See WATER SUPPLIES.

combining site (*immunol.*) That part of an ANTIBODY (or a FAB PORTION) which can combine with a determinant of the homologous antigen or hapten; it is located at the variable (N-terminal) ends of the heavy chain and light chain in an Ig molecule (see IMMUNOGLOBULINS and HYPERVARIABLE REGION). Monomeric antibodies have a VALENCY of two. (See also CYTOPHILIC ANTIBODIES and AFFINITY LABELLING.)

come-up time In the CANNING process: the time needed for the correct processing temperature to be reached in the retort.

comedone See ACNE.

co-metabolism The phenomenon in which a substrate (the *co-substrate*) which does not support the growth of a given microorganism can nevertheless be modified or degraded by that organism in the presence of a second, growth-supporting substrate. Co-metabolism is believed to result from the action of an enzyme of relatively low substrate specificity which has a different physiological function in the cell; for example, METHANE MONOOXYGENASE can oxidize certain short-chain alkanes and alkenes as well as methane, and such hydrocarbons can be oxidized by methane-grown cells (*co-oxidation*) which may be unable to use the oxidation products. In nature, co-metabolism may be important in the degradation of XENOBIOTICS.

commensal See COMMENSALISM.

commensalism SYMBIOSIS (sense 1) in which one symbiont (the *commensal*) derives benefit from the association, and the other (sometimes called the *host*) derives neither benefit nor harm. (cf. MUTUALISM.)

commercial sterility A term used when referring to the condition of a substance following APPERTIZATION.

common bunt A seed-borne wheat disease caused by *Tilletia caries* or *T. foetida* (*T. laevis*). (Rye and certain other grasses may also be affected.) Symptoms include retarded growth and the formation of long, narrow ears which have a bluish tinge; grains are filled with grey or black powdery masses of teliospores which, when released during harvesting, give off a characteristic 'rotten fish' odour (trimethylamine). *Dwarf bunt* is a similar (soil- and seed-borne) disease caused by *Tilletia contraversa* (*T. brevifaciens*); affected plants show severe stunting and chlorotic flecks on the leaves. (cf. KARNAL BUNT; see also CEREAL DISEASES.)

common cold An acute disease of the upper respiratory tract in man; the causal agent may be any of a range of viruses – usually a rhinovirus, coronavirus, influenza virus (type A, B or C), parainfluenza virus (types 1–4) or human respiratory syncytial virus. Incubation period: usually ca. 48–72 hours. Symptoms typically include coryza and nasal obstruction, sore throat, cough, sneezing, but little or no fever. Direct contact may be an important mode of transmission of rhinoviruses [AIM (1978) *88* 463]; the other viruses are probably transmitted mainly by aerosols (see DROPLET INFECTION). Transmission of rhinoviruses can be interrupted (experimentally) by the use of virucidal (citric acid-treated) paper handkerchiefs [JID (1986) *153* 352–356]. Colds are usually self-limiting, but occasional complications include e.g. OTITIS MEDIA, PNEUMONIA or SINUSITIS. (cf. INFLUENZA.) [Review: Lancet (2003) *361* 51–59.]

common fimbriae Type 1 FIMBRIAE.

common pili (1) *Syn.* FIMBRIAE. (2) *Syn.* Type 1 fimbriae.

common scab (of potato) See POTATO SCAB.

communicable disease *Syn.* INFECTIOUS DISEASE.

comoviruses (cowpea mosaic virus group) A group of PLANT VIRUSES in which the bipartite positive-sense ssRNA genome is encapsidated in (separate) icosahedral particles; both parts are necessary for infection. Most comoviruses cause systemic mosaic or mottling and stunting – occasionally with wilting, necrotic ring formation, etc. – in various plants, including e.g. cowpea (*Vigna unguiculata*), beans, red clover etc; the host range of individual members is narrow. Most comoviruses are transmitted by leaf-feeding beetles, particularly of the Chrysomelidae (e.g. *Cerotoma* and *Diabrotica* spp); seed transmission occurs in some comoviruses, and mechanical transmission can occur readily under experimental conditions. Type member: COWPEA MOSAIC VIRUS (SB isolate). Other members: e.g. cowpea severe mosaic virus, quail pea mosaic virus, red clover mottle virus, squash mosaic virus. (cf. BROAD BEAN WILT VIRUS.)

ComP See QUORUM SENSING.

compact colony-forming active substance See CCFAS.

comparative single intradermal tuberculin test (comparative test; single intradermal comparative tuberculin test) (*vet.*) A form of TUBERCULIN TEST, applied to cattle, in which two intradermal injections are given (on a single occasion – cf. STORMONT TEST) at different sites on the same side of the neck; PPD derived from *Mycobacterium bovis* is injected at one site, PPD derived from *M. avium* is injected at the other site. After 72 hours any increase in skin thickness at the injection sites is measured. If the response to PPD from *M. bovis* (increase in skin thickness) is greater than that to PPD from *M. avium* by a certain minimum amount then the animal is designated a 'reactor'. A positive reaction to PPD from *M. avium* together with a negative reaction to PPD from *M. bovis* suggests contact with *M. avium* and/or mycobacteria other than *M. bovis*.

The use of a defined antigen (ESAT-6), compared with PPD, has been associated with lower sensitivity but higher specificity [VR (2000) *146* 659–665].

comparative test (*vet.*) *Syn.* COMPARATIVE SINGLE INTRADERMAL TUBERCULIN TEST.

comparator See WATER SUPPLIES.

compatibility (1) (of plasmids) See PLASMID. (2) (in fungal sexuality) The ability, or otherwise, of a given cell or thallus to interact sexually with itself and/or with other cells or thalli of the same species. In general, the genetic constraints on mating increase, both in number and complexity, from the lower fungi, through the ascomycetes, to the basidiomycetes. HOMOTHALLISM is common in all the major groups of fungi. Bipolar HETEROTHALLISM occurs in some lower fungi (e.g. species of *Mucor* and *Rhizopus*), in many ascomycetes (including e.g. *Saccharomyces*), and in many of the rust and smut fungi. Tetrapolar heterothallism occurs in a number of basidiomycetes – including e.g. *Schizophyllum commune* and *Ustilago maydis*. (See also MATING TYPE and PARASEXUAL PROCESSES.)

compatible (*plant pathol.*) Refers to a plant–pathogen interaction which results in the development of disease in the plant. (cf. INCOMPATIBLE.)

compatible solute Any low-MWt compound which, when present intracellularly in high concentration, is compatible with metabolism and growth. Compatible solutes involved in OSMOREGULATION include: L-glutamate, GLYCEROL, ISOFLORIDOSIDE, MANNITOL, L-proline, glycine betaine [in *Escherichia coli*: FEMS Reviews (1994) *14* 3–20; in *Staphylococcus aureus*: Microbiology (1994) *140* 3131–3138], and certain sugars.

compensation level See PHOTIC ZONE.

competence See TRANSFORMATION (1).

complement (C or C′; historical *syn.* alexin or alexine) (*immunol.*) A group of functionally related proteins present in normal plasma, tissue fluids and the (freshly isolated) serum of vertebrates. (In serum, the activity of complement is abolished by exposure to room temperature for a few days, by heating to 56°C for 30 minutes, or by the action of chelating agents such as EGTA or oxalate.) When activated, the complement system is responsible for a number of important physiological activities (see e.g. COMPLEMENT FIXATION).

Before activation, many of the components exist as proenzymes and/or inactive forms (cf. FACTOR D) which, on activation, complex with, or act on, other components to form physiologically active entities.

In man, complement consists of nine numbered components (C1–C9, MWts ca. 80000–200000) together with various proteins involved e.g. in stabilization or control functions: see e.g. FACTOR B; FACTOR D; FACTOR H; FACTOR I; PROPERDIN and S PROTEIN. Component C1 is a complex macromolecule which consists of three parts: C1q, C1r and C1s; C1q itself consists of a bundle of six elongated subunits, and the macromolecule as a whole depends on Ca^{2+} for its structural integrity.

Gene-based deficiencies in the complement system are associated with various adverse effects – which depend e.g. on the particular component(s) involved. One important role of complement is the removal of immune complexes (via the liver), and deficiencies in C1, C4 and C2 may give rise to forms of disease reflecting the accumulation of such complexes. A deficiency in C3 (with reduced capacity for opsonization) may result in recurrent infections. A deficiency in MBL (lectin pathway) is associated with increased susceptibility to infection in children. Deficiencies in C5–C8 generally cause increased susceptibility to infection – particularly with *Neisseria*.

complement activation *Syn.* COMPLEMENT FIXATION.

complement fixation (complement activation) Activation of the COMPLEMENT system, i.e. promotion of the sequence ('cascade') of reactions in which various components of complement are activated in turn. In vivo, complement fixation is involved e.g. in certain immune responses and in INFLAMMATION; in vitro it provides a sensitive system for various COMPLEMENT-FIXATION TESTS. (Complement fixation is sometimes referred to as complement *inactivation* – an allusion to the fact that components of complement are 'used up' during fixation.)

Complement fixation can be triggered in various ways, the type of trigger determining which pathway is followed; four pathways are shown in the figure.

(a) *Classical pathway*. This pathway is activated e.g. when antigens bind to COMPLEMENT-FIXING ANTIBODIES in the presence of complement. When such binding occurs, sequentially activated components may bind at or near the antigen–antibody complex; according to the antigen involved, this may result in e.g. phagocytosis of the antigen–antibody complex or, if a cell-surface antigen is involved, it may lead to cell lysis (see later). The cascade of reactions which occur in the (antigen-triggered) classical pathway is as follows.

The sequence is triggered when two or more subunits of C1q (in the complement C1 molecule) bind to the FC PORTION of antigen-bound IgG or IgM antibodies. Such binding (dependent on calcium ions) activates the C1r component of C1 which, in turn, activates the C1s component. C1s has serine protease and esterase activity.

(Note that antibody-independent activation of C1 can be promoted by certain non-specific activators – e.g. C-reactive protein (an ACUTE-PHASE PROTEIN) and various proteases.)

Activated C1s cleaves C4, forming C4a and C4b; as one molecule of C1s can cleave many molecules of C4, this reaction provides initial amplification. C4b molecules can bind to bacterial or other surfaces; like C3b (see later), C4b is an OPSONIN: C4b-coated cells or particles bind (via C4b) to the CR1 receptor on the surface of phagocytes.

Surface-bound C4b may also bind C2. C2 is cleaved (in a process involving C1s) to C2a and C2b; the complex formed between the larger fragment, C2a (referred to as C2b in some sources), and C4b (C4b2a) is a *C3 convertase*: an enzyme which cleaves C3 to C3a and C3b (this being a major phase of amplification in the complement cascade). (A low level of C3 cleavage normally occurs spontaneously ('C3 tickover') but the resulting fragments are rapidly degraded.)

C3a is an ANAPHYLATOXIN.

C3b is an important OPSONIN, promoting e.g. the phagocytosis of pathogens (IMMUNE ADHERENCE); C3b-bound antigens bind (via C3b) to CR1 receptors on phagocytes.

C3b can also form a complex with C3 convertase: C4b2a3b; this complex is a *C5 convertase*, i.e. it cleaves C5 into the small fragment C5a and a larger fragment, C5b.

C5a is an ANAPHYLATOXIN and a chemoattractant for phagocytes (e.g. neutrophils).

C5b, C6 and C7 form a complex (C5b67) which binds to cell membranes – typically at the site of the initial triggering event. Component C8 binds to the complex; this promotes polymerization of a number of C9 molecules within the membrane – the C9 molecules forming a pore or channel which, in some types of cell (or even LIPOSOMES), can lead to lysis (*immune cytolysis*). C5b678 and the C9 molecules collectively form the so-called *membrane attack complex*, MAC. (Even in the absence of C9, the complex C5b678 can bring about slow lysis e.g. in erythrocytes.)

For certain pathogens (e.g. trypanosomes), MAC-inflicted damage can be lethal. In Gram-negative bacteria, pores formed by the MAC in the OUTER MEMBRANE can promote cell lysis by giving LYSOZYME access to cell-wall peptidoglycan; thus, e.g. a Gram-negative bacterium on the conjunctiva would be at risk of lysis because the tear film normally contains both complement and lysozyme.

Sometimes, C5b67 binds to a cell other than that which triggered the cascade; MAC-inflicted lysis in such a cell is called

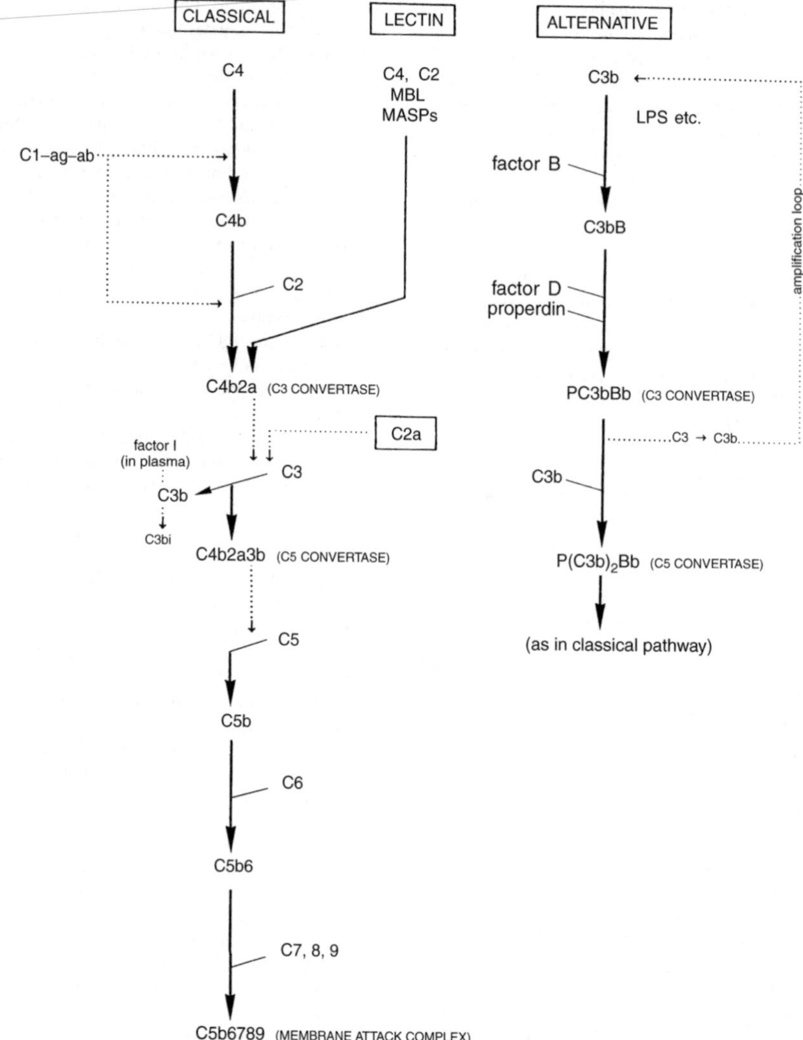

COMPLEMENT FIXATION: four pathways (simplified scheme; see text). C1, C2 etc. denote particular components of the complement system; 'a' and 'b' denote fragments of components produced by enzymic cleavage during the activation process. For clarity, the diagram does not show *all* the cleavage products; for example, C4 is cleaved to C4a and C4b, but only the C4b fragment is considered here. Dotted lines indicate those cases in which a given complex or component acts enzymically to cleave certain components of the system.

Reproduced from *Bacteria* 5th edition, Figure 11.1, page 292, Paul Singleton (1999) copyright John Wiley & Sons Ltd, UK (ISBN 0471-98880-4) with permission from the publisher.

bystander lysis or *reactive lysis*. Reactive lysis may thus damage cells adjacent to a site of infection, and to limit such damage the plasma contains factors inhibitory to the C5b67 complex (see e.g. S PROTEIN).

Because complement fixation creates highly potent physiological agents, and as it includes various stages of amplification, regulatory mechanisms are essential to prevent the system becoming an uncontrolled avalanche of physiological destruction. In fact, there are specific inhibitors at key stages in the process; these include C1EI (C1 esterase inhibitor; also called C1INH), which inhibits initial activation of the classical pathway, and FACTOR I, which cleaves C3b to an inactive form, C3bi. (C3bi, although inactive in the cascade, can nevertheless behave as an opsonin – binding e.g. to the CR4 receptor on macrophages.) Other regulatory factors are involved in the control of the alternative pathway (see later).

(b) *Alternative pathway*. In evolutionary terms, this pathway is apparently older than the classical pathway. In man, fixation via the alternative pathway is initiated by e.g. bacterial LIPOPOLYSACCHARIDES, rabbit (but not sheep) erythrocytes, aggregates of human myeloma proteins, inulin and ZYMOSAN. (In the human complement system, this pathway can be activated by sheep erythrocytes if the cell-surface sialic acids have been removed.) The sequence of reactions in this pathway is as follows.

On a suitable surface (e.g. OUTER MEMBRANE LPS), component C3b ('tickover' C3b – see above) binds FACTOR B (in the presence of Mg^{2+}) to form C3bB. The bound factor B is then cleaved by FACTOR D; this creates a C3 convertase (C3bBb) which is stabilized by binding to PROPERDIN (P). The C3 convertase (PC3bBb) cleaves C3, forming more C3b and giving rise to an amplification loop (see figure).

The C3 convertase can bind a molecule of C3b to form a C5 convertase ($P(C3b)_2Bb$); the cascade then continues as in the classical pathway.

Regulators of the alternative pathway include the plasma protein FACTOR I and the cell-bound *decay-accelerating factor* (DAF) (see CD55). Other cell-surface-associated inhibitory proteins include CR1, which can displace Bb from surface-bound C3bBb, and the *membrane co-factor protein* (MCP), which facilitates cleavage of C3b by factor I; collectively, these cell-surface factors help to protect *host* cells from the ongoing complement cascade (including the MAC) by inhibiting the cascade at the C3 convertase/C3b level.

For experimental work on the alternative pathway, COBRA VENOM FACTOR is used as an activator, while COMPLESTATIN and HEPARIN are used as inhibitors. (See also SURAMIN.)

When the *classical* pathway is stimulated, some of the C3b can bind Factor B and give rise to an alternative-pathway C3 convertase; in this way the alternative pathway can augment the classical pathway, and this sequence of reactions is sometimes called the *amplification pathway*.

The alternative pathway can be studied independently of the classical pathway by using the Ca^{2+} chelator EGTA (the latter pathway being dependent on the presence of calcium ions).

(c) *The lectin pathway*. This pathway effectively bypasses the C1 stage of the classical pathway: when the lectin pathway is triggered, a C3 convertase (C4b2a) is formed directly from C4 and C2.

The lectin pathway involves several components of normal plasma: mannose-binding lectin (MBL) and the MBL-associated *serine proteases* (MASPs) (cf. component C1s in the classical pathway); the system is triggered when MBL binds to certain groups on the microbial cell envelope. [Mannose-binding lectin: IT (1996) *17* 532–540.]

The lectin pathway may be particularly useful in the young during the 'window' period of ~12–18 months between loss of maternal antibody cover (IgG) and the development of an effective immune system. However, triggering of this pathway by a bacterial pathogen may be greatly inhibited if the pathogen has a capsule (e.g. *Neisseria meningitidis*).

(d) *The C2a pathway* (= '*salvage pathway*'). This pathway, reported in 1997, may be important in the context of pathogenic mycobacteria. In the C2a pathway, C2a acts as a C3 convertase when it binds to the mycobacterial surface. [See TIM (1998) *6* 47–49; TIM (1998) *6* 49–50.]

complement-fixation test (CFT) Any of a range of sensitive, in vitro tests in which COMPLEMENT FIXATION is used to indicate the presence or absence (or quantity) of a specific antigen or COMPLEMENT-FIXING ANTIBODY. *Principle*: antigen and antibody are allowed to interact in the presence of complement, and the amount of complement used up (a measure of the amount of antigen–antibody interaction) is estimated, indirectly, by the addition of a HAEMOLYTIC SYSTEM; the proportion of erythrocytes lysed indicates the amount of free (unfixed) complement remaining. *Practice*: to detect or quantify e.g. a given antibody in a sample of serum, SERIAL DILUTIONS of the INACTIVATED SERUM are prepared, and known, fixed quantities of COMPLEMENT and antigen are added to each dilution; the quantity of complement added (determined by titration) should be sufficient to cause lysis of ca. 90% of the erythrocytes in the haemolytic system when *no* complement-fixation has occurred. Incubation is carried out for ca. 18 hours at ca. 4°C. (Low-temperature incubation is used because some components of complement are labile at room temperature.) A fixed volume of the haemolytic system is then added to each dilution, and the whole is incubated for 15–30 min at 37°C; the test dilutions are then left at ca. 4°C to permit non-lysed erythrocytes to settle. In any given dilution, the extent of haemolysis will be inversely proportional to the extent of complement fixation in that dilution. If maximum haemolysis is observed in the lowest serum dilution (i.e. the highest serum concentration) the test is *negative* – i.e. the serum contains no detectable antibody. In a *positive* test, no haemolysis occurs in (at least) the lowest serum dilution; a test may be regarded as positive either (i) when no haemolysis has occurred in a specified number of serum dilutions (e.g. all dilutions up to 1:8 or 1:16 initial serum dilution), or (ii) when an *increase* in the titre of complement-fixing antibodies is recorded in a second serum sample taken some time after the first.

In all CFTs, controls are essential e.g. to check that the test system is functioning correctly and to preclude false-positive interpretations based on the effects of ANTICOMPLEMENTARY substances.

complement-fixing antibodies Antibodies which 'fix' or 'activate' COMPLEMENT when they combine with their homologous antigens. IgG (subclasses 1 and 3) and IgM can bring about complement fixation via the classical pathway; aggregates of IgA (subclasses 1 and 2) can activate the alternative pathway. (See also FC PORTION.)

complement receptor 1 See CD35.

complementarity-determining regions (*immunol.*) Those HYPERVARIABLE REGIONS which occur at the surface of the COMBINING SITE of an antibody.

complementary bases See BASE PAIR.

complementation The ability of a gene (or protein) to compensate for (i.e., to 'complement') a functional defect in a homologous gene (or protein) when present in the same organism or cell. For example, a wild-type gene necessary for the synthesis of a given amino acid may, when introduced into a mutant organism auxotrophic for that amino acid, complement the mutant gene (by supplying an active form of the defective gene product) and restore the wild-type (prototrophic) phenotype. (See also COMPLEMENTATION TEST; IN VITRO COMPLEMENTATION ASSAY; NON-GENETIC REACTIVATION.)

complementation test A test used to determine whether or not COMPLEMENTATION will occur in a cell with a given mutant phenotype when another mutant genome, encoding the same mutant phenotype, is introduced into that cell. The test involves bringing together, into the same cell, the two (haploid) mutant genomes (or a genome and part-genome – see MEROZYGOTE) and determining the resulting phenotype. The expression of a wild-type phenotype (a *positive* complementation test) indicates that the mutations are in different genes in the two genomes, i.e., one wild-type form of each gene can be supplied by each of the mutant genomes (*intergenic* or *intercistronic* complementation); this indicates that at least two genes are involved in the expression of that phenotype. (If the mutations were in the same gene in each genome, no wild-type form of that gene would be present and the mutant phenotype would continue to be expressed; if such a negative result is consistently obtained with a range of mutants of the same phenotype, then the phenotypic characteristic under study may be presumed to be under monogenic control.)

Interpretation of the results of a complementation test can be complicated by several factors. (a) In some cases two mutant copies of a gene – each by itself encoding an inactive polypeptide – may be able to complement one another when present together in the same organism. This may occur e.g. when the normal active gene product is a monomer of an oligomeric protein, and the two types of defective monomer encoded by the two mutant alleles can somehow interact to give an oligomer with some activity (*intragenic*, *intracistronic* or *interallelic* complementation). The resulting phenotype usually has characteristics qualitatively or quantitatively intermediate between the wild-type and mutant phenotypes. (b) Certain mutant genes are dominant over their wild-type alleles: e.g., an inactive product of a mutant gene may bind to the active product of the wild-type gene to produce an inactive hybrid oligomer (*negative complementation*: see e.g. LAC OPERON); such dominant mutations can be detected by a CIS–TRANS TEST. (c) Recombination between two mutant copies of a gene may result in the restoration of a functional gene; this possibility can be minimized by using recombination-deficient strains (e.g. *recA*$^-$ bacteria). (d) Complementation can occur only if the genes under study produce diffusible products; it cannot occur e.g. if the mutation is in a control region (e.g. a promoter) which can affect the expression only of adjacent genes in the same genomic molecule. (See also CIS-DOMINANCE.) (e) A POLAR MUTATION can give a misleading result since a single mutation can inactivate two or more adjacent genes.

(See also CISTRON.)

complementing diploids See PROTOPLAST FUSION.

complestatin A substance, produced by *Streptomyces lavendulae*, which inhibits the alternative pathway of COMPLEMENT FIXATION by combining with Factor B and preventing its binding to C3b.

complete Freund's adjuvant See FREUND'S ADJUVANT.

complete medium (CM) A type of culture medium which contains nutrients sufficient to support the growth of both PROTOTROPHS and AUXOTROPHS. (cf. MINIMAL MEDIUM.)

complete transduction See TRANSDUCTION.

complex flagellum See FLAGELLUM (a) and FLAGELLIN.

complex-mediated hypersensitivity *Syn.* TYPE III REACTION.

composting A process involving the aerobic biological degradation of waste plant matter (e.g. straw, cotton waste, corn-cobs etc) or other organic waste (e.g. paper, dewatered sewage sludge, municipal refuse); composting reduces the bulk of the waste material and converts it to an innocuous form ('compost') which can be used e.g. as a soil conditioner. (See also MUSHROOM CULTIVATION.) Composting involves the activities of a succession of microorganisms whose nature depends to some extent on the nature of the material being composted.

In the composting of e.g. wheat-straw: the damp straw is piled loosely (to allow free circulation of air); inorganic nitrogen and phosphorus – or animal or poultry manure – may be added. The activities of mesophilic bacteria generate heat, and the temperature in the centre of the heap may rise to ca. 60°C or more within a few days, creating conditions in which thermophilic actinomycetes predominate [ARM (1983) *37* 198–200]. Air circulation is maintained by convection, and the heap may be turned and mixed at intervals. The pH may rise substantially (e.g. to 8–9), due e.g. to ammonia production, but later falls to ca. 6–7. After several days the temperature may fall to ca. 30–50°C and may remain at this level for several weeks; during this stage fungi become dominant. Initially, thermophilic members of the Mucorales (e.g. *Rhizomucor pusillus*) develop; these are then replaced by thermophilic hyphomycetes (e.g. *Humicola* spp, *Thermomyces lanuginosus*) and ascomycetes (e.g. *Chaetomium thermophilum*). Finally, basidiomycetes such as *Coprinus cinereus* (often accompanied by the zygomycete *Mortierella wolfii*) become dominant [Book ref. 39, pp. 263–305]. During composting, most or all of the free soluble substrates, much of the CELLULOSE, HEMICELLULOSES and PECTIC POLYSACCHARIDES, and some of the LIGNIN, are degraded, and the loss in dry weight may reach 50% or more.

compound II See SIDEROPHORES.

compound ciliature Any form of somatic or oral ciliature which involves a discrete group of cilia (several to many) acting as a functional unit; examples: the CIRRUS, MEMBRANELLE, PARORAL MEMBRANE and SYNCILIUM.

compound trichocyst *Syn.* FIBROCYST.

Compsopogon See RHODOPHYTA.

co-mutagenesis The generation of two or more mutations at closely linked loci: see e.g. MNNG.

ComX See QUORUM SENSING.

***con* gene** See OMP.

concanavalin A (con A; jack bean lectin) A mitogenic LECTIN (tetrameric form: MWt 102000) from the jack bean, *Canavalia ensiformis*; it can stimulate T cells and, if cross-linked (e.g. by anti-con A antibody), B cells. Con A binds e.g. to terminal α-D-mannopyranosyl and α-D-glucopyranosyl residues (binding requires Ca^{2+} and Mg^{2+}), and can thus agglutinate e.g. bacteria whose cell wall TEICHOIC ACIDS contain α-linked glucosyl substituents.

concatemer Two or more identical linear nucleic acid molecules (e.g. copies of a viral genome) in tandem, i.e., covalently linked end to end in the same orientation. (cf. CATENANE.)

concatenate (1) (*verb*) To form a concatenate (sense 2). (2) (*noun*) *Syn.* CATENANE or CONCATEMER. (3) (*adj.*) Of e.g. spores: formed in chains.

concentration exponent (of a disinfectant) See DILUTION COEFFICIENT.

concentric bodies (*mycol.*) Rounded or ellipsoidal intracellular structures (ca. 300 nm diam.) which have a transparent core surrounded by concentric layers of varying optical density; they occur in almost all lichenized ascomycetes and in a few non-lichenized ascomycetes. Their origin and function are unknown. Similar structures ('concentric granules') are found in *Allomyces* and related genera; these may play a role as septal pore plugs [Mycol. (1987) *79* 44–54]. (cf. CELLULIN GRANULES.)

conceptacle (*mycol.*) A hollow structure within which spores are formed – e.g. a locule (see ASCOSTROMA).

conchate Having the shape of one-half of a bivalve shell.

***Conchocelis* stage** See e.g. PORPHYRA.

concolorous Similarly coloured.

concomitant immunity *Syn.* PREMUNITION.

condenser (substage condenser) In a microscope: a system of lenses which concentrates light and directs it onto the specimen (see MICROSCOPY, Fig. 1). Correct adjustment of the condenser is important for optimum image formation: see e.g. KÖHLER ILLUMINATION. The *numerical aperture* of a condenser (see RESOLVING POWER) should be at least as great as that of the objective lens with which it is used; an oil-immersion condenser should be used with an oil-immersion objective. Before entering the condenser, the light passes through an iris diaphragm (*aperture diaphragm, substage diaphragm*) which, by controlling the amount of light entering the condenser, controls the apical angle of the cone of light which illuminates the specimen. Hence, the aperture diaphragm should *not* be used as a means of reducing the intensity of illumination because this reduces the NA of the condenser (see RESOLVING POWER). (See also ABBE CONDENSER; CRITICAL ILLUMINATION.)

condensing enzyme *Syn.* citrate synthase (EC 4.1.3.7) [see Appendix II(a)].

conditional lethal mutant A mutant organism (e.g. a bacterium or virus) in which the mutation does not significantly affect the phenotype under one set of conditions (*permissive* conditions) but is lethal or inhibitory to the organism under another set of conditions (*restrictive* or *non-permissive* conditions). See e.g. TEMPERATURE-SENSITIVE MUTANT and HOST-DEPENDENT MUTANT.

conduction (*mol. biol.*) See CONJUGATION (1b)(i).

condyloma (*pl.* condylomas or condylomata) An elevated, wart-like lesion of the skin. *Condyloma acuminatum* = genital wart (see PAPILLOMA); *condyloma latum* = a wide, flat condyloma occurring chiefly in the anogenital region in secondary SYPHILIS.

Condylostoma See HETEROTRICHIDA.

confluent growth See CULTURE.

confocal scanning light microscopy (CSLM) A form of light MICROSCOPY in which linear resolution is superior to that obtainable with conventional light microscopes. Essentially, rays from a laser-illuminated pin-hole are focused in the plane of the specimen and collected by a lens system which rapidly scans the specimen in synchrony with the illuminating source. As the optical system scans the specimen the rays entering the collector lens are continually modulated by the specimen; the modulated rays are focused onto a detector plate – generating a signal which modulates a cathode ray tube and forms a visual display of the image.

congeneric Of the same genus.

congenital Present at birth.

conglutinated 'Glued together' – used e.g. of hyphae in certain prosenchymatous fungal or lichen tissues.

conglutination See CONGLUTININ.

conglutinin A protein, present in normal bovine serum, which can bind to a cleavage product of complement component C3b (see COMPLEMENT FIXATION); conglutinin can agglutinate particles/cells to which the cleavage product has adhered – a phenomenon termed *conglutination*.

conglutinogen Potential activity associated with complement component C3b; cleavage of C3b by FACTOR I exposes a site which can bind CONGLUTININ.

conglutinogen-activating factor *Syn.* FACTOR I.

Congo–Crimean haemorrhagic fever See VIRAL HAEMORRHAGIC FEVERS.

Congo red A red acid diazo DYE used e.g. as a PH INDICATOR (pH 3.0–4.5, blue to red) and for staining β-linked fibrillar polysaccharides (e.g. cellulose); Congo red disrupts CELLULOSE and CHITIN microfibril formation in growing cells (cf. CALCOFLUOR WHITE).

conidia Plural of CONIDIUM.

conidial head (*mycol.*) The expanded terminal portion of the conidiophore in certain fungi – see e.g. ASPERGILLUS.

conidiation The phenomenon in which conidia function as gametes.

Conidiobolus See ENTOMOPHTHORALES.

conidiogenesis See CONIDIUM.

conidiogenous cell See CONIDIUM.

conidioma Any plectenchymatous structure which bears conidia – e.g. a pycnidium, sporodochium or synnema.

conidiophore A hypha which bears one or more conidiogenous cells (see CONIDIUM). In e.g. some species of *Colletotrichum*, setae (see SETA) can function as conidiophores.

conidiospore *Syn.* CONIDIUM.

conidium (*mycol.*) An asexually-derived, non-motile SPORE formed in a blastic or thallic mode (see below) from a specialized conidiogenous (i.e., conidium-producing) cell (cf. SPORANGIOSPORE). Conidia are formed e.g. by many fungi of the DEUTEROMYCOTINA and by a few lower fungi (e.g. species of *Bremia* and *Peronospora*). The two basic modes of conidiogenesis (conidium formation) are as follows.

Blastic conidiogenesis. An apical or lateral part of a conidiogenous cell expands and develops into a mature conidium, becoming delimited from the parent cell by a septum. Such a blastic conidium (= *blastoconidium* or blastospore) secedes (i.e., breaks away) from the conidiogenous cell by centripetal splitting of the septum (*schizolysis*) – half the septum becoming the base of the conidium, the other half forming the apex of the conidiogenous cell. If the conidiogenous cell subsequently elongates from a basal growing region it is said to exhibit *basauxic* development. (Cessation of *hyphal* tip growth – see GROWTH (b) – preceding blastic conidiogenesis may depend on changes in ion gradients across the envelope of the conidiogenous cell.)

Thallic conidiogenesis. A septum-delimited part of a hypha becomes converted to a single, terminal or intercalary, conidium (a *holothallic* conidium) or to a chain of conidia (*arthric* conidia). (cf. ARTHROSPORE.) Secession of a terminal, holothallic conidium commonly occurs by *rhexolysis*: the circumferential splitting of the wall of the penultimate cell (which ruptures) a little below the septum of the conidium; arthric conidia may separate by rhexolysis (alternate cells being sacrificed) or by schizolysis. [Formation and germination of fungal arthroconidia: CRM (1986) *12* 271–292.]

There are various ways in which the conidiogenous cell can produce a succession of conidia or give rise to a crop of simultaneously formed conidia; some examples are given below.

(a) *Ampullate development.* The apex of the conidiogenous cell becomes globose ('ampullate') and gives rise, simultaneously, to a number of blastoconidia. Ampullate development occurs e.g. in *Gonatobotrys*.

(b) *Annellidic development.* After the first, terminal blastoconidium has been delimited by a septum (and either before or after its secession) a new wall is laid down on the inner surface of the old wall in the distal region of the conidiogenous cell; this new, annular wall, in conjunction with the apical septum, forms an inverted cup-like structure at the free end of the conidiogenous cell. (The formation of the new wall is referred to as *enteroblastic* proliferation of the conidiogenous cell.) Subsequently, the cup-like structure expands to form the next conidium; such a process, in which the new wall becomes continuous with the conidial wall, is referred to as *holoblastic* conidial development. Septum formation then delimits the second blastoconidium – which (sooner or later) secedes by schizolysis; the position of this septum is such that, following secession of the blastoconidium, part of the new wall remains in the conidiogenous cell, extending a little beyond the rim of the old wall. Thus, after a number of blastoconidia have been produced, the conidiogenous cell (*annellide*) exhibits a succession of bands (*annellations*) at the apical end, each annellation being the remnant of a layer of wall formed inside the previous layer.

Annellidic development occurs e.g. in *Scopulariopsis*. (See also PERCURRENT PROLIFERATION.)

(c) *Phialidic development* is very similar to annellidic development, the essential difference being that, in phialidic development, successive septa are formed at or near the level of the rim of the old wall; annellations are therefore not formed, although the edges of successive wall layers in the conidiogenous cell (*phialide*) may be visible just within the apical rim (*collarette*) formed by the original wall of the conidiogenous cell. Phialidic development occurs e.g. in *Phialophora* and *Trichoderma*.

(d) *Retrogressive development* is similar to annellidic and phialidic development but differs in that each blastoconidium formed takes with it, on secession, a part of the original cell wall of the conidiogenous cell; the conidiogenous cell thus progressively shortens. It occurs e.g. in *Cladobotryum*.

(e) *Tretic (porogenous) development*. The wall of the conidiogenous cell contains a narrow pore or channel; materials for conidial growth pass through this channel during the formation of the blastoconidia ('poroconidia' or 'porospores') – which arise either solitarily or in acropetally formed chains. Tretic development occurs e.g. in *Alternaria*.

[Reviews: MS (1984) *1* 86–89 and MR (1986) *50* 95–132.]

Conidiophores may arise individually or in discrete masses: see e.g. ACERVULUS, COREMIUM, PYCNIDIUM and SYNNEMA.

Coniocybe See CALICIALES.

Coniophora A genus of fungi of the APHYLLOPHORALES (family Coniophoraceae). *C. puteana* (= *C. cerebella*, 'cellar fungus'), a cause of WET ROT, forms a thin, resupinate fruiting body which is initially cream-coloured but later olivaceous; the basidiospores are brownish. (See also CELLULASES.)

Coniophoraceae See APHYLLOPHORALES.

coniosporiosis *Syn.* MAPLE BARK STRIPPERS' DISEASE.

Coniothyrium See SPHAEROPSIDALES.

conjugate (1) (*verb*) Carry out CONJUGATION (sense 1). (2) (*noun*) See CONJUGATION (sense 2).

conjugate division See CLAMP CONNECTION.

conjugate vaccine A VACCINE containing a *polysaccharide* antigen linked (conjugated) to a protein; such vaccines are useful for young children whose immune system does not respond adequately to (unconjugated) polysaccharide antigens. Some diseases are caused by pathogens with polysaccharide capsules, and conjugate vaccines containing such polysaccharides can elicit protective antibodies in young children. For example, a conjugate vaccine is used against diseases (epiglottitis, meningitis etc.) caused by *Haemophilus influenzae* type b [RMM (1996) *7* 231–241]. [Responsiveness of infants to capsular polysaccharides: RMM (1996) *7* 3–12.] [Vaccine against group B streptococci IV and VII: JID (2002) *186* 123–126. Meningococcal C vaccine (in teenagers): Lancet (2003) *361* 675–676.]

conjugation (1) ('mating') Any of various processes in which gene transfer (or, exceptionally, complete fusion) follows the establishment of direct contact between two or more microbial cells (which may exhibit little or no morphological differentiation in comparison to vegetative cells).

(a) (*algol.*) Conjugation occurs in several groups of algae (e.g. DESMIDS, certain DIATOMS, SPIROGYRA) and generally involves cell–cell contact followed by protoplast fusion.

(b) (*bacteriol.*) In bacterial conjugation, DNA is transferred from a 'male' (*donor*) bacterium to a 'female' (*recipient*) bacterium while the cells are in physical contact; a recipient which has received DNA from a donor is called a *transconjugant*. (cf. RETROTRANSFER.)

The donor phenotype is commonly conferred on a cell by the intracellular presence of a CONJUGATIVE PLASMID (sense 1); TRANSPOSON-mediated conjugation is considered separately in the entry CONJUGATIVE TRANSPOSITION.

Any of various plasmid-borne genes (and, in some cases, chromosomal genes) may be transferred during conjugation; such genes include those conferring resistance to antibiotic(s).

Conjugation occurs in both Gram-positive and Gram-negative bacteria; however, as the process differs in Gram-positive and Gram-negative species it is discussed under separate headings for the two categories.

(i) *Conjugation in Gram-negative bacteria*. Much of the information on this topic has derived from studies on the transfer of the F PLASMID (q.v.), and other IncF plasmids, between strains of *Escherichia coli*; the following account is based largely on this information. It is important to note that the general features of the F plasmid–*E. coli* transfer system are not common to all transfer systems found in Gram-negative bacteria.

The stages of conjugation are outlined below, in sequence, on the basis of current models and data.

Initial cell–cell contact. The presence of a (plasmid-encoded) pilus (see PILI) is believed to be essential for the donor phenotype. Moreover, different types of pilus mediate conjugation under different physical conditions, and they may function in different ways; for example, some types of pilus mediate conjugation only on moist (but not submerged) solid surfaces (e.g. agar plates), while others mediate conjugation either within the liquid phase or on a solid surface (see PILI) [see also Book ref. 177, pp 33–59 (42–45)].

In the F plasmid–*E. coli* conjugal system, initial donor–recipient contact is believed to occur when the tip of the pilus binds to a site on the surface of the recipient; the pilus is then believed to retract so that the cells achieve wall-to-wall contact. Stable cell–cell contact, involving specific plasmid-encoded proteins, is established after an initial period of unstable contact (during which donor and recipient are easily separated by mild shearing forces). (Mutant *E. coli* recipients defective in the OmpA protein (see OMP) form unstable contacts with donor cells, although this deficiency may be overcome by mating on a moist solid surface.) (See also SURFACE EXCLUSION.)

An understanding of the initial donor–recipient interaction would require a detailed knowledge of the structure and mode of action of the pilus – from initial contact to the putative retraction. This information is currently unavailable; indeed, the conjugal transfer of DNA, even in the well-studied F plasmid–*E. coli* system, and particularly in the initial stages of cell–cell contact, has been described as a 'black box' [Mol. Microbiol. (1997) *23* 423–429].

Some authors report that the region of contact between donor and recipient is characterized by a specific, electron-dense *conjugational junction* [JSB (1991) *107* 146–156; JB (2000) *182* 2709–2715]. The composition of this electron-dense layer is unknown but one suggestion is that, in F plasmid-mediated matings, it may consist of, or contain, F-pilin (the subunit of the F pilus) [Mol. Microbiol. (1997) *23* 423–429]; however, results from recent studies on RP4-mediated matings indicated that conjugational junctions are not composed of pilin [JB (2000) *182* 2709–2715].

On establishment of effective cell–cell contact, a 'mating signal' (nature unknown) promotes the *mobilization* of donor DNA.

Mobilization and transfer of DNA. Mobilization is the process in which (in at least some cases, including the F plasmid) a single, specific strand of donor DNA is prepared for transfer to the recipient.

In the F plasmid, an essential prerequisite for mobilization is the formation of a NICK at a specific site (designated *nic*), in a specific strand, within the origin of transfer (*oriT*); nicking occurs in that strand which is to be transferred to the recipient (the *T-strand*). Nicking is mediated by an endonuclease (a *relaxase*) that forms part of a so-called *relaxosome* (previously called a *relaxation complex*). The relaxosome associated with the F plasmid is a nucleoprotein complex consisting of the transfer origin (*oriT*), the relaxase, and certain other proteins. (Relaxases are encoded e.g. by the F plasmid *traI* gene, the *traI* gene of plasmid RP4, and the *nikA* gene of *Salmonella* plasmid R64.) As well as endonuclease activity, the TraI protein of the F plasmid also has (ATP-dependent) helicase activity and is referred to as *helicase I*.

Components of the relaxosome are assembled in a specific order: TraI binds after the INTEGRATION HOST FACTOR and the TraY protein [JBC (1995) *270* 28381–28386]. Recent studies confirm that TraY is required for in vivo nicking [JB (2000) *182* 4022–4027].

The relaxase is believed to mediate an ongoing cycle of nicking and ligation at the *nic* site, i.e. nicking appears not to be triggered by donor–recipient contact.

Following the mating signal in F plasmid–*E. coli* systems, it appears that the nicked strand enters the recipient in the 5′-to-3′ direction. (Single-strand transfer in the 5′-to-3′ direction is believed to occur in at least some other cases.)

Transfer of the nicked strand from donor to recipient requires that the strand be unwound from its complementary strand, and this is thought to involve helicase I (TraI); if helicase I is involved, and if it is immobilized in relation to the cell envelope, then the unwinding of the transferred strand may provide energy for strand transfer.

One early study indicated that the transferred strand could pass *through* an extended pilus to the recipient [JB (1990) *172* 7263–7264]. A later study – which used video-enhanced light-microscopy and electron microscopy – concluded that DNA is transferred while donor and recipient are in close wall-to-wall contact [JSB (1991) *107* 146–156].

The precise mode of transfer of DNA from donor to recipient is not known. For F plasmid-mediated systems, it has been postulated that the relaxosome may be coupled, via TraD, to a transport apparatus (*transferosome*) consisting of a protein complex located in the cell envelope at the base of the pilus.

In RP4-mediated mating, the transfer apparatus in the donor may include a structure that bridges the cytoplasmic and outer membranes [JB (2000) *182* 1564–1574], although this was not detected in another study on RP4-mediated mating [JB (2000) *182* 2709–2715].

DNA synthesis in the recipient cell. When a cell receives the transmitted strand of an F plasmid, a complementary strand is synthesized to form a circularized dsDNA molecule; this *recA*-independent process is called *repliconation*. Complementary strand synthesis apparently involves the recipient's DNA polymerase III and other host-encoded enzymes.

For certain IncI and IncP plasmids (e.g. ColIb, RP4), synthesis of DNA in the recipient is initiated by a donor-plasmid-encoded PRIMASE, i.e. during conjugation, primase, as well as ssDNA, is transferred from donor to recipient. In such cases, DNA synthesis is initiated from primers which are formed at various locations on the transferred strand. [Plasmid DNA primases in conjugation: Book ref. 161, pp 585–603.] Matings with ColIb-P9 involve the transfer, from donor to recipient, of molecules which include the product of the *sog* gene, which has primase activity [EMBO (1986) *5* 3007–3012].

When chromosomal genes have been conjugally transferred (following mobilization of the chromosome), recombination between donor and recipient DNA appears to involve the RECBC PATHWAY.

DNA synthesis in the donor cell. In the donor cell a complementary strand is synthesized on the non-transmitted strand in order to reconstitute the dsDNA plasmid; this synthesis, carried out by DNA polymerase III, is referred to as *donor conjugal DNA synthesis* (DCDS). DCDS is generally believed to be carried out by the ROLLING CIRCLE MECHANISM (although earlier claims that primers are needed for DCDS had made this uncertain). More recently, a study using recombinant plasmids containing two tandemly repeated R64 *oriT* sequences (one initiation-proficient but termination-deficient, the other initiation-deficient but termination-proficient) has provided evidence of rolling circle synthesis [JB (2000) *182* 3191–3196].

[DNA processing reactions in bacterial conjugation: ARB (1995) *64* 141–169.]

Mobilization of non-conjugative plasmids. The term 'mobilization' commonly includes instances in which (a) a non-conjugative plasmid (or part of the bacterial chromosome) is conjugally transferred by becoming covalently linked to a conjugative plasmid (transfer being initiated at *oriT* in the conjugative plasmid), or (b) a non-conjugative plasmid is conjugally transferred independently of (that is, without covalent linkage to) a conjugative plasmid by means of cell–cell contacts established by the conjugative plasmid; mobilization of the latter type does not necessarily involve co-transfer (i.e. concomitant transfer) of the conjugative plasmid. (Some authors [ARG (1979) *13* 99–125 (106)] have referred to (a) and (b) as *conduction* and *donation*, respectively.) (See also HFR DONOR and INTERRUPTED MATING.)

The F plasmid can mobilize certain non-conjugative plasmids, e.g. the COLE1 PLASMID, in a way that does not require covalent linkage. Plasmid ColE1 is associated with a relaxosome which can be activated to promote mobilization and transfer of plasmid DNA; the in vivo mechanism of activation is unknown. During F-mediated mobilization and transfer of ColE1, one of the proteins of the relaxosome apparently remains covalently bound to the 5′ end of the transferred strand, and it has been suggested that this protein may act as a 'pilot protein', interacting with membrane protein(s) and guiding the 5′ terminal into the recipient.

Mobilization of the non-conjugative plasmid R1162 involves at least four *trans*-acting plasmid-encoded products and a *cis*-acting site (*oriT*), *oriT* being distinct from *oriV* [JB (1986) *167* 703–710].

Effects of physicochemical parameters on conjugation. Relatively few studies have been carried out on the effects of variation in the physicochemical conditions under which cell–cell contact occurs. However, a number of early studies were conducted on the effects of variation in basic parameters such as pH, temperature, electrolyte concentration and osmotic pressure; one purpose of these studies was to characterize the response of the conjugal process to changes in 'external' conditions, thus obtaining data potentially useful e.g. for optimizing transfer in the laboratory. These studies found e.g. that transfer of the IncFII plasmid R1*drd*19 could be greatly stimulated by higher levels of electrolyte [FEMS (1983) *20* 151–153]; that the plasmid could be transferred between 37 and 17°C [AEM (1981) *42* 789–791]; that the effects of non-optimal pH and non-optimal temperature are synergistic [AEM (1983) *46* 291–292]; that transfer is subject to osmotic constraint [FEMS (1984) *25* 37–39]; and that transfer is inhibited by particles of colloidal clay [AEM (1983) *46* 756–757]. Transfer of the IncN plasmid R269N-1, which mediates surface-obligatory mating (see PILI), appears to need, or be assisted by, surface tension [FEMS (1983) *19* 179–182; JGM (1983) *129* 3697–3699].

Some early experiments [e.g. JGM (1987) *133* 3099–3107] purported to demonstrate the possibility of plasmid transfer in natural aquatic environments but were flawed by extreme artificiality.

(ii) *Conjugation in Gram-positive bacteria.* Conjugal transfer is known to occur in organisms from various genera, including *Bacillus*, *Enterococcus*, *Lactococcus*, *Staphylococcus* and *Streptomyces*. As in conjugation in Gram-negative bacteria, some plasmids can be transferred across a broad host range; for example, plasmid pAMβ_1, which encodes resistance to erythromycin, can be transferred from streptococci to e.g. *Bacillus subtilis* and *Lactobacillus casei*.

There are two main types of plasmid-mediated conjugation in Gram-positive species: PHEROMONE-mediated and (apparently) pheromone-independent mating; pili have not been demonstrated in either category.

Pheromone-mediated conjugation in e.g. strains of *Enterococcus faecalis* (see PHEROMONE) occurs in liquid (broth) cultures, and the plasmids involved commonly contain genes conferring resistance to antibiotics. Some of these plasmids also encode a peptide which antagonizes the corresponding pheromone; such a peptide may have the function of ensuring that a mating response in the donor cell is not triggered unless the relevant pheromone is in a sufficiently high concentration, i.e. conditions under which there is a good chance of a random collision between donor and recipient cells. [Review: JB (1995) *177* 871–876.]

Pheromone-independent conjugation requires that the donor and recipient be present on a solid surface (e.g. a nitrocellulose filter) – rather than in a liquid medium. Conjugation of this type occurs e.g. in strains of *Staphylococcus* and *Streptococcus*. The plasmids, which encode e.g. resistance to antibiotics, are commonly >15 kb in size.

In *Bacillus subtilis*, recombinants may be obtained by co-incubation of two apparently plasmid-less parent strains for extended periods of time (e.g. 20 hours); in such crosses most exconjugants contain the entire genome of both parent strains, and the process is believed to resemble a form of temporary cell fusion [Book ref. 199, pp 25–39]. (cf. PROTOPLAST FUSION.)

As in Gram-negative bacteria, certain plasmids may be mobilized by other plasmids, and some of the small multicopy plasmids encode function(s) which specifically promote their own mobilization. Mobilization of the streptococcal plasmid pMV158 has been studied in *Escherichia coli* using R388 or RP4 as the auxiliary plasmid [Microbiology (2000) *146* 2259–2265].

(c) (*ciliate protozool.*) Conjugation occurs in many ciliates; it involves either the temporary association of two individual cells (as e.g. in *Paramecium* or *Tetrahymena*) or the complete fusion of conjugal partners (as e.g. in peritrichs). In some ciliates (e.g. *Blepharisma*) conjugation is promoted by soluble substances (see GAMONE) which are released by the cells; in other ciliates (e.g. *Paramecium*) the chemical signals remain bound to the cell.

In *Paramecium aurelia*, two individuals of appropriate mating type (see SYNGEN) pair with their ventral surfaces in contact, and an intercellular cytoplasmic bridge forms in the cytostomal region. In each conjugant (i.e. partner) the MACRONUCLEUS disintegrates, and both of the micronuclei (see MICRONUCLEUS) undergo meiosis to form eight haploid *pronuclei;* seven pronuclei disintegrate. (In *Paramecium caudatum* the fate of a given pronucleus depends on its position within the cytoplasm; the nucleus which survives is located in the cytostomal region [JCS (1985) *79* 237–246].) The surviving pronucleus (= *gonal nucleus*) divides mitotically to form a pair of gametic nuclei, one of which passes to the conjugal partner and fuses with the stationary nucleus to form a zygotic nucleus (= *synkaryon*). Subsequently, mitotic divisions of each synkaryon lead to the formation of four diploid nuclei in each conjugant – the conjugants by now having separated; two of the nuclei give rise to two new micronuclei, while the other two form two new macronuclei. During the first binary fission after conjugation, the micronuclei (but not the macronuclei) undergo mitotic division; each daughter cell receives one macronucleus and two micronuclei – thus restoring the normal nuclear constitution of the species. In *P. aurelia* conjugation takes ca. 12–18 hours.

[DNA synthesis, methylation and degradation during conjugation in *Tetrahymena thermophila*: NAR (1985) *13* 73–87.]

In peritrichs, two morphologically dissimilar conjugants are formed: a female *macroconjugant* and a smaller, motile, ciliated *microconjugant* which swims in search of, and fuses with, a macroconjugant.

(See also AUTOGAMY and CYTOGAMY.)

(d) (*mycol.*) See e.g. MATING TYPE and PHEROMONE.

(2) (*immunol.*) The process of covalently linking two or more species of molecule to form a hybrid molecule (a *conjugate*); for example, a dye may be linked to a protein (as in fluorescein-conjugated antiglobulin), or glutaraldehyde may be used to link two different antibodies to form a hybrid antibody [SAB (1984) *22* 73–78]. (See also CONJUGATE VACCINE.)

conjugational junction See CONJUGATION (1b)(i).

conjugative pili See PILI.

conjugative plasmid (1) (self-transmissible plasmid) As commonly used: a PLASMID which encodes all the functions needed for its own intercellular transmission by CONJUGATION (sense 1b); in e.g. the F PLASMID (q.v.) these functions include pilus formation (see PILI) and the formation of proteins involved in the initial preparation of plasmid DNA for transfer. Such a plasmid may also bring about the *mobilization* of a (mobilizable) NON-CONJUGATIVE PLASMID or the mobilization of the donor's chromosome. (cf. SEX FACTOR; see also ANDROPHAGE and PROMISCUOUS PLASMIDS.)

(2) According to some authors [ARG (1979) *13* 99–125 (105–106)]: a plasmid which encodes e.g. pilus formation but which may or may not be capable of mobilization; only if such a plasmid is mobilizable is it self-transmissible.

conjugative transposition In bacteria: CONJUGATION mediated by a *conjugative transposon*, i.e. independently of a conjugative plasmid, the transposon being transferred from the *donor* cell to the *recipient* cell. First reported in Gram-positive bacteria, it is now well established in Gram-negatives (e.g. *Bacteroides*); indeed, it may permit gene transfer between Gram-positives and Gram-negatives [e.g. AEM (2001) *67* 561–568; AEM (2003) *69* 4595–4603].

A model for Gram-positive conjugative transposition was as follows. Contact between donor and recipient triggers excision of the transposon in the donor, this involving a transposon-encoded recombinase (product of gene *int*). Each end of the excised transposon carries single-stranded host DNA called a *coupling sequence*. The excised transposon then circularizes through base-pairing (albeit mis-matched) between coupling sequences. A *single* strand may be transferred to the recipient, the complementary strand being synthesized in the recipient prior to insertion. Insertion of the (double-stranded) transposon into the target site is neither site-specific nor random; each target site seems to contain an A-rich sequence of nucleotides and a T-rich sequence, the two sequences being separated by about six nucleotides.

In *Bacteroides* (Gram-negative) excision and transfer of the conjugative transposon CTnDOT (which is found in many strains of *Bacteroides*) is governed by an OPERON that may be regulated by a TRANSLATIONAL ATTENUATION mechanism [JB (2004) *186* 2548–2557].

A recipient which has received a conjugative transposon is a *transconjugant;* receipt confers the donor phenotype. (Interestingly, strains of *Lactobacillus lactis* can receive conjugative transposons but, apparently through lack of particular host function(s), transposons cannot be excised – so that *L. lactis* cannot act as a donor.)

Characteristically, conjugative transposons are resistant to restriction. It may be relevant that the *orf18* gene in Tn*916* has been found to encode a product similar to the antirestriction proteins encoded by some plasmids.

All conjugative transposons from pathogenic bacteria carry the tetracycline-resistance *tet*(M) gene [mechanism of Tet(M): JB (1993) *175* 7209–7215], and some carry other types of antibiotic-resistance gene(s). These transposons are mobile among a wide range of host species, and are often responsible for antibiotic resistance in plasmid-less Gram-positive pathogens; they have been identified in e.g. *Enterococcus faecalis*, *E. faecium* and *Streptococcus pneumoniae*.

Conjugative transposons occur in e.g. plasmids and chromosomes, and range in size from about 18 kb (Tn*916*) to >50 kb; some of the larger ones (e.g. Tn*5253*) apparently consist of a Tn*916*-like entity within another transposon (each being able to transpose independently).

Conjugative transposons include Tn*916*, Tn*925*, Tn*1545*, Tn*3710* and Tn*5253*. They differ from the 'classical' transposons in several ways: (i) they mediate conjugation; (ii) when they insert, they do not duplicate the target site; (iii) when excised, the transposon forms a cccDNA intermediate; (iv) the recipient duplex is characteristically in another cell.

conjunctivitis Inflammation of the conjunctivae. It may be due e.g. to physical or chemical irritation, allergic reaction, or to infection by bacteria (e.g. *Haemophilus influenzae*, *Moraxella lacunata*) or viruses (e.g. adenoviruses). (See e.g. ACUTE HAEMORRHAGIC CONJUNCTIVITIS; EPIDEMIC KERATOCONJUNCTIVITIS; GONORRHOEA; INCLUSION CONJUNCTIVITIS; PHARYNGOCONJUNCTIVAL FEVER; PINK-EYE; TRACHOMA.)

conk A colloquial name for the fruiting body of a wood-rotting basidiomycete (particularly a polypore).

Conocybe See AGARICALES (Bolbitiaceae).

conocyst An EXTRUSOME of the gymnostome *Loxophyllum*.

conoid A hollow cone of spirally arranged fibrils or tubules, open at the apex, found at the extreme anterior end in certain members of the APICOMPLEXA; it may assist in the penetration of the host cell.

consensus sequence (*mol. biol.*) Of the variant forms of a given type of genetic element (e.g. a PROMOTER): a theoretical 'representative' nucleotide sequence in which each nucleotide is the one which occurs most often at that site in the various forms of the genetic element which occur in nature. The phrase is also used to refer to an actual sequence which approximates the theoretical consensus.

conserved name In NOMENCLATURE: a name retained (by a recognized taxonomic body) even though it may contravene rule(s) of the relevant code.

consolidation (*med.*) See e.g. PNEUMONIA (a).

consortium A more or less stable physical association between the cells of two or more types of microorganism – advantageous to at least one of the organisms; SYNTROPHISM seems to characterize many (perhaps all) consortia. One example is CHLOROCHROMATIUM AGGREGATUM. [Review: ME (1996) *31* 225–247.]

conspecific Of the same species.

constant region (of Ig) See IMMUNOGLOBULINS.

constitutive Refers either to a gene which is expressed constantly, or to the product of such a gene. The level of expression of a constitutive gene depends largely on the efficiency of its PROMOTER.

constitutive heterochromatin See CHROMATIN.

consumption Pulmonary TUBERCULOSIS.

consumption test (*serol.*) Any test in which a measurement is made of the amount of antibody or antigen which is used up (consumed) by antigen–antibody combination in a test system; the amount consumed is determined by titrating the residual (unconsumed) antibody or antigen and comparing it with the quantity initially present. (See e.g. ANTIGLOBULIN CONSUMPTION TEST.)

contact biotrophic mycoparasite A fungus which parasitizes other fungi by means of specialized cells which make contact with – but do not penetrate or cause obvious damage to – mycelium of the (living) host; most of the known species (including e.g. *Calcarisporium parasiticum*) require biotin, thiamine, and an unidentified substance, 'mycotrophein', for growth. (See also MYCOPARASITE.)

contact-dependent secretion See type III systems in PROTEIN SECRETION.

contact dermatitis *Syn.* CONTACT SENSITIVITY.

contact inhibition See *transformation* in TISSUE CULTURE.

contact sensitivity (contact dermatitis) A form of DELAYED HYPERSENSITIVITY in which the subject is primed by cutaneous exposure to certain protein-binding substances – e.g. dinitrofluorobenzene, certain metals (e.g. nickel) and antibiotics (e.g. neomycin) – which bind skin proteins to form neoantigens; subsequent challenge, after ca. 1 week, with the sensitizing substance results in a typical lesion in 24–48 hours. (See also LANGERHANS' CELLS.)

contagious abortion See BRUCELLOSIS.

contagious acne *Syn.* CONTAGIOUS PUSTULAR DERMATITIS (2).

contagious bovine pleuropneumonia In cattle: acute lobar PNEUMONIA, with pleuritis, caused by a strain of *Mycoplasma mycoides*; infection occurs mainly by droplet inhalation. Incubation period: 3–6 weeks (sometimes longer). Onset is sudden, with fever and anorexia; subsequently there is a deep cough and dyspnoea. Death from anoxia may occur within days or weeks (mortality rate up to ca. 50%); survivors may become asymptomatic carriers. *Treatment*: antibiotic and/or other therapy is generally given only where the disease is enzootic. (See also BOVINE RESPIRATORY DISEASE.)

contagious disease (1) *Syn.* INFECTIOUS DISEASE. (2) A disease normally transmissible only by direct physical contact between infected and uninfected individuals – as e.g. in many VENEREAL DISEASES.

contagious ecthyma *Syn.* CONTAGIOUS PUSTULAR DERMATITIS (1).

contagious equine metritis (CEM) A sexually transmitted HORSE DISEASE (first recognized in Suffolk, UK, in 1977) characterized by endometritis with mucopurulent vulval discharge and temporary infertility; clinical symptoms usually resolve rapidly. A symptomless carrier state occurs in both mares and stallions. The causal agent is a previously unknown bacterial species. It is a Gram-negative coccobacillus (occasionally filamentous); catalase +ve; oxidase +ve; asaccharolytic; gives −ve reactions in most biochemical tests; grows slowly e.g. on chocolate agar at 37°C (optimum temperature); GC%: 36.1. [Book ref. 181, pp. 49–96.]

contagious hypovirulence See HYPOVIRULENCE.

contagious pustular dermatitis (1) (orf; contagious ecthyma; scabby mouth) A disease which affects mainly sheep (especially lambs) – in which pustular, scab-forming lesions form on the lips and face and in the mouth; the causal agent, a PARAPOXVIRUS, is presumed to infect via abrasions. Morbidity may be high, but mortality rates are generally very low. In cattle, lesions may be formed on the teats. In man, the disease is localized and self-limiting, involving the formation of lesions e.g. on the hands, arms or eyelids.

(2) (*syn.* contagious acne) A HORSE DISEASE characterized by the formation of pustules, particularly where the skin is in contact with harness; causal agent: *Corynebacterium pseudotuberculosis*.

contaminative infection (1) (*parasitol.*) Infection by pathogens derived from the POSTERIOR STATION of the vector. (cf. INOCULATIVE INFECTION.) (2) Infection brought about by the ingestion of material containing pathogens.

context (*mycol.*) In basidiomycetes: the sterile (i.e., non-generative) structural tissue which forms the main bulk of a fruiting body, particularly that of the pileus. (cf. TRAMA.)

continuous cell line *Syn.* ESTABLISHED CELL LINE.

continuous centrifugation See CENTRIFUGATION.

continuous cooker–cooler See COOKER–COOLER.

continuous cultivation *Syn.* CONTINUOUS CULTURE.

continuous culture (continuous-flow culture; continuous cultivation; open culture; open-system culture) The culture of microorganisms, in a liquid medium, such that the organisms can exhibit continual exponential or near-exponential growth (balanced GROWTH) for an extended period of time. (cf. BATCH CULTURE.) Essentially, the procedure involves a continual inflow of fresh medium to the culture (with which it is well mixed) and the simultaneous outflow of an equal volume of fluid consisting of a mixture of old and fresh medium and a proportion of biomass; under 'steady-state' conditions the concentration of biomass in the culture vessel remains constant. In continuous culture, growth can occur, ideally, under defined and effectively invariant conditions; for this reason continuous culture is used e.g. for studies of microbial metabolism: data from continuous culture are generally more reproducible than are those from batch culture. The commonest forms of apparatus used in continuous culture are the chemostat and the turbidostat (see also GRADOSTAT).

In the *chemostat* a culture normally grows at a *sub-maximal* growth rate, and a steady state (at a given growth rate) is achieved by controlling the concentration (always at growth-limiting levels) of an essential substrate in the incoming medium (and, hence, in the culture vessel). (By using very low concentrations of a given essential substrate it is possible to achieve very low rates of growth – similar to those believed to occur e.g. in certain aquatic habitats.) Under steady-state conditions, the growth rate and the concentration of the growth-limiting substrate often have the relationship predicted by the Monod equation (see SPECIFIC GROWTH RATE), and the specific growth rate is numerically equal to the DILUTION RATE. Under steady-state conditions the concentration of biomass in the chemostat is governed by the rate at which the growth-limiting substrate is supplied to the culture, and (for a given rate of supply of that substrate) the concentration of biomass cannot be varied by the operator. (See also YIELD COEFFICIENT.) The conditions in a chemostat tend to be unstable at or near the maximum growth rate.

In the *turbidostat* a culture normally grows at or near the *maximum* growth rate, and all substrates are usually present in excess; within certain limits, the concentration of biomass in the turbidostat can be selected by the operator, and a steady state is achieved by adjusting the dilution rate such that cell growth is matched by the rate of loss of cells from the culture vessel. Control of culture density (i.e. number of cells per unit volume) may be achieved e.g. by a photosensitive device which monitors the opacity of the culture and automatically adjusts the flow rate; however, this method may suffer from inaccuracies due e.g. to foaming and to 'wall growth' (see later). Alternatively, culture density can be controlled by monitoring other parameters which are closely linked to specific growth rate – e.g. pH (in a 'pH-stat'), CO_2 evolution (in a 'CO_2-stat'), or oxygen uptake.

In continuous culture, deviations from ideal (predictable) operation are common. Such deviations may arise e.g. as a result of the inability to achieve perfect (100%, instantaneous) mixing of the inflowing medium with the culture fluid; imperfect mixing may be indicated by a stable dilution rate which exceeds D_c. Changes in the rate of agitation can affect the mean cell volume of the cultured cells [JGM (1985) *131* 725–736]. Wall growth (adherence to, and growth of, the cultured cells on the walls of the culture vessel) may give an effect similar to (but usually greater than) that of imperfect mixing. Another type of deviation may occur when cells are cultured at low rates of growth; for example, under carbon-limited conditions a major part of the carbon source may be used to provide MAINTENANCE ENERGY (thus giving a lower-than-predicted yield of biomass), while under carbon-excess conditions the yield of biomass may be greater than predicted owing to the accumulation of intracellular reserve polymers (e.g. PHB). A different type of problem involves the emergence of mutants during the (extended) periods of culture – an effect which can change the character of the biomass if particular mutant(s) can outgrow the parent strain; nevertheless, it is possible to take advantage of this effect – e.g. when it is desired to isolate mutants which synthesize more (or more effective) enzyme(s) or which can take up the growth-limiting substrate more efficiently than can the parent strain.

continuous-flow culture *Syn.* CONTINUOUS CULTURE.

continuum model See CELL CYCLE.

Contophora Algae *except* members of the RHODOPHYTA. (cf. ACONTA.)

contour-clamped homogeneous electric field electrophoresis (CHEF) See PFGE.

contractile vacuole An osmoregulatory organelle, one or more of which may occur in the cytoplasm in most freshwater (and some saltwater) eukaryotic microorganisms (cf. PUSULE); basically, it is a membrane-limited region which, under normal environmental conditions, alternately fills with fluid (*diastole*) and then discharges the fluid to the exterior (*systole*). During diastole, the contractile vacuole may be fed by a surrounding layer of small vesicles which appear to coalesce with the main vacuole (as in amoebae), and/or – depending on species – it may be fed by a system of collecting tubules which ramify in the cytoplasm (forming the *spongiome*, = *spongioplasm*, = 'nephridial network') and apparently drain areas remote from the vacuole (a system characteristic of ciliates). In some protozoa (e.g. *Paramecium*, *Tetrahymena*) discharge occurs at a fixed site (pore) in the pellicle.

The mechanism of contractile vacuole activity is not known. Systole may involve cytoplasmic turgor pressure; at least in ciliates, the vacuole wall may be inherently contractile. In general, the length of the diastole–systole cycle (seconds, minutes or hours) appears to depend on temperature, on the size of the cell (being shorter in smaller cells), and on the osmolarity of the medium.

[Review: Biol. Rev. (1980) *55* 1–46.]

convalescent serum SERUM from a patient in the convalescent stage of a disease; in some cases it may be used to provide PASSIVE IMMUNITY.

convergent trama See BILATERAL TRAMA.

Convoluta A genus of flatworms (platyhelminths), found on sandy sea shores in the intertidal zone, which form a symbiotic association with certain unicellular green algae. On hatching, the young flatworms ingest cells of TETRASELMIS; the algal cells lose their flagella, theca and eyespot, and reside between the cells of the subepidermal tissues in the flatworm where they continue to photosynthesize. The flatworm eventually ceases to feed, its digestive organs degenerate, and it becomes completely dependent on the photosynthate produced by *Tetraselmis*. Eventually, the algal cells themselves are digested, and the animal then dies. (cf. ELYSIA; see also ZOOCHLORELLAE.)

cooked meat medium (chopped meat medium; Robertson's cooked meat medium) Any of a range of media containing boiled, de-fatted, minced lean beef, and used e.g. for the growth of ANAEROBES, for the sporulation of clostridia, and as maintenance media for e.g. clostridia and streptococci; in some formulations the medium incorporates a reducing agent (e.g. L-cysteine or thioglycollate) and e.g. haemin, vitamin K_1 and yeast extract. The medium is sterilized by autoclaving, and the final pH is about 7.4; it is stored in screw-cap bottles, and is sometimes equilibrated under O_2-free conditions at room temperature before the cap is tightened. [Recipe: (e.g.) Book Ref. 53, pp. 1423–1424.]

cooker–cooler (continuous cooker–cooler) In CANNING: a vessel within which filled, sealed cans (or other containers) undergo heat treatment in a continuous-type process (cf. BATCH RETORT). In rotary-type cooker–coolers, the cans enter and leave, through self-sealing valves, a chamber containing steam under pressure. In hydrostatic-type cooker–coolers, pressure in the steam chamber supports vertical columns of water through which the cans enter and leave the chamber.

cooking (*industrial microbiol.*) (1) In the food processing industry: heat treatment used for APPERTIZATION. (See also c_0.) (2) A stage in CHEESE-MAKING.

Coomassie brilliant blue A sensitive stain for proteins. The red (anionic) form of the dye becomes blue when bound to protein amino groups.

Coombs' reagent (*serol.*) *Syn.* ANTIGLOBULIN.

Coombs' test (*serol.*) *Syn.* ANTIGLOBULIN TEST.

Cooper–Helmstetter model *Syn.* HELMSTETTER–COOPER MODEL.

cooperative binding A mode of binding (e.g. of protein molecules to a DNA strand) in which the binding of one molecule facilitates binding of the next.

co-oxidation See CO-METABOLISM.

***cop* gene** A gene involved in the control of the COPY NUMBER of a plasmid. (See e.g. R1 PLASMID.)

cophenetic correlation coefficient In NUMERICAL TAXONOMY: a measure of the accuracy with which a phenogram represents a given S matrix.

copiotroph Any organism which grows only in the presence of high concentrations of nutrients. (cf. OLIGOTROPH.)

copper (as an antimicrobial agent) Copper is a HEAVY METAL which is essential (in trace amounts) for the activity of a number of microbial proteins (e.g. cytochrome aa_3, tyrosinase). However, in effective concentrations, copper and certain of its compounds are useful antimicrobial agents; toxicity to microorganisms appears to reside in the cupric ion.

Copper sulphate has been used as an anti-algal agent in e.g. swimming pools and as a constituent of various agricultural antifungal agents (see e.g. BORDEAUX MIXTURE, BURGUNDY MIXTURE, CHESHUNT COMPOUND). *Copper naphthenate* is a greenish, waxy solid, soluble in various organic solvents, which is used as an antifungal preservative for e.g. cellulosic textiles. *Copper soaps* are copper salts of certain fatty or oleoresinous acids (e.g. copper stearate, copper tallate) which have antifungal activity similar to that of copper naphthenate. *Cuprammonium hydroxide* is used in aqueous solution for rot-proofing fabrics. Cupric ions are complexed by 8-HYDROXYQUINOLINE (one atom of copper to two molecules of 8-hydroxyquinoline) to form *copper 8-quinolinolate* (copper-oxine; copper oxinate) which is a highly effective antifungal agent used e.g. in agriculture and as a preservative for textiles. (Copper compounds are not used in rubber products since they catalyse the oxidation of rubber.)

copper 8-quinolinolate See COPPER.

copper soaps See COPPER.

co-precipitation (*serol.*) The precipitation of otherwise non-precipitating molecules etc. as part of, or enmeshed with, an immune complex.

Coprinaceae See AGARICALES.

Coprinus A genus of humicolous, lignicolous or coprophilous fungi (AGARICALES, Coprinaceae) in which, in most species, the lamellae undergo rapid autodigestion at maturity (see LAMELLA and BASIDIOSPORE). (See also INK-CAP FUNGI.)

coproantibodies Antibodies produced by the intestinal mucosa and found in the lumen of the gut; they are mainly of the secretory IgA class.

Coprococcus A genus of Gram-positive, asporogenous, anaerobic bacteria which occur e.g. in the human gut; the organisms ferment carbohydrates with the production of butyric and acetic acids. Cells: cocci, occuring in pairs or chains. GC%: ca. 39–42.

coprogen See SIDEROPHORES.

Copromyxa See ACRASIOMYCETES.

Copromyxella See ACRASIOMYCETES.

coprophilic (coprophilous) Refers to an organism which grows preferentially or exclusively on or in animal faeces. For example, certain fungi (e.g. species of *Coprinus*, *Pilobolus* and *Sordaria*) grow more or less specifically on dung – particularly that of herbivorous animals.

co-protease (coprotease) A type of molecule which, acting non-enzymically, can promote autocatalytic cleavage of a protein; an example is RecA* in the SOS SYSTEM.

coprozoic Refers to e.g. protozoa which are COPROPHILIC.

copy-choice model See RECOMBINATION.

copy-mutant (copy-number mutant; *cop* mutant) A mutant PLASMID whose COPY NUMBER differs from that of the wild-type plasmid. In copy mutants the copy number is usually higher than that of the wild-type plasmid, but in some cases it is lower.

copy number (1) (of a bacterial plasmid) The number of copies of a given PLASMID, per chromosome, in a cell; copy number depends on the replication control system encoded by the plasmid, on the strain or species of cell in which the plasmid occurs, and on growth conditions. In plasmid-containing bacteria in the exponential phase of growth, a given plasmid occurs with a characteristic copy number; any perturbation in copy number tends to be corrected by a rise or fall in the frequency of plasmid replication. Mutations in a plasmid (and/or chromosome) may give rise to a COPY MUTANT.

Copy number is determined primarily by the regulation of initiation of plasmid replication; thus, e.g. a particular mutant of pBR322 (a plasmid related to the COLE1 PLASMID – q.v.) which encodes an altered (less effective) negative regulatory molecule (RNA I) exhibits an 8-fold increase in copy number [JGM (1986) *132* 1021–1026]. Other factors which govern copy number include the PARTITION system (if any) of the plasmid. In some plasmids the *ccd* system contributes to the stability of plasmid inheritance (see F PLASMID).

(See also INCOMPATIBILITY and MULTICOPY PLASMID.)

(2) The number of copies of e.g. a given gene product per copy of that gene (or per cell), or the number of copies of a given gene per cell etc.

CoQ Coenzyme Q: see QUINONES.

coral fungi See APHYLLOPHORALES (Clavariaceae).

coral spot See NECTRIA.

coral symbiosis See ZOOXANTHELLAE.

Corallina See RHODOPHYTA.

Corallococcus See MYXOBACTERALES.

Corallomyxa See ACARPOMYXEA.

cord factor Any of the MYCOLIC ACID diesters of TREHALOSE found in the cell walls of *Mycobacterium* spp and in other mycolic acid-containing species. The name 'cord factor' was originally given to a glycolipid (6,6′-dimycolyl-α,α'-D-trehalose) derived from tubercle bacilli characterized by growth in the form of 'cords'. Certain mycolic acid esters (e.g. 6,6′-dimycolyl-α,α'-D-trehalose, or 5-mycolyldiarabinoside) from the walls of *M. tuberculosis* can potentially act as ENDOTOXINS (sense 2) by uncoupling mitochondrial electron transport and oxidative phosphorylation [Ann. Mic. (1983) *134*B 233–239].

cordycepin (3′-deoxyadenosine) An inhibitor of RNA synthesis; it is obtained e.g. from *Cordyceps militaris*.

Cordyceps A genus of fungi (order CLAVICIPITALES) which include parasites of insects and of other fungi. At least some entomogenous species can form CHITINASES, and infection appears to be initiated via the cuticle and haemolymph rather than via the gut; the fungus subsequently forms, on the dead insect, an erect, cylindrical or clavate stroma which contains perithecia in the surface layer. The dead host is resistant to decay – apparently due to the presence of CORDYCEPIN. *C. militaris* ('scarlet caterpillar fungus') forms orange stromata (ca. 2–5 cm high) on the larvae and pupae of butterflies and moths. *C. canadensis* is parasitic on *Elaphomyces* spp.

core (1) (*virol.*) Any of various internal structures within a virion – commonly a protein–nucleic acid complex; the core may be enclosed within a CAPSID (see e.g. HERPESVIRIDAE), a membrane (see e.g. POXVIRIDAE) or an ENVELOPE (see e.g. RETROVIRIDAE). (cf. NUCLEOCAPSID.)

(2) (*bacteriol.*) See ENDOSPORE sense 1(a).

core oligosaccharide See LIPOPOLYSACCHARIDE.

core particle (of chromatin) See CHROMATIN.

coremium In some fungi (e.g. *Penicillium expansum*): an erect bundle of spore-bearing hyphae, the hyphae being associated more loosely than those in a SYNNEMA. (For some authors the terms coremium and synnema are synonymous.)

corepressor See OPERON.

coriaceous Leathery in texture.

Coriolus A genus of fungi of the APHYLLOPHORALES (family Polyporaceae) which form non-stipitate, typically semicircular, bracket-type basidiocarps on deciduous wood; the context is trimitic, usually leathery, rubbery or woody, and the hymenophore is porous. In *C. versicolor* (sometimes called *Trametes versicolor*) the upper surface of the basidiocarp is often velvety, exhibiting a number of dark bands concentric about the region of attachment, while the lower surface is pale; basidiospores: cream-coloured, ca. 6 × 2 μm. (See also WHITE ROT and LIGNIN.)

corn oil test (for lipase) See LIPASE.

corn stunt disease A leafhopper-transmitted YELLOWS disease of maize, characterized by stunting and leaf striping/discoloration, caused by the MAIZE CHLOROTIC DWARF VIRUS or by certain strains of *Spiroplasma citri*. [Proposal to regard the corn stunt spiroplasmas as a distinct species, *S. kunkelii*: IJSB (1986) *36* 170–178.]

cornmeal agar A mycological medium. Essentially, cornmeal (4% w/v) is heated in distilled water for 1 hour at 60–65°C; agar (1.5% w/v) is added to the filtrate, and the medium is autoclaved. Cornmeal agar is used e.g. to demonstrate chlamydospore formation in *Candida albicans* – which is reported to be encouraged by the addition to the medium of Tween 80.

cornute *Syn.* ROESTELIOID.

Coronaviridae A family of pleomorphic, enveloped, non-segmented ssRNA viruses which infect man, animals or birds. One genus (*Coronavirus*) is currently recognized. Members include AVIAN INFECTIOUS BRONCHITIS virus (IBV; type species), human coronavirus (a causal agent of the COMMON COLD), murine hepatitis virus (MHV), porcine haemagglutinating encephalitis virus (see VOMITING AND WASTING DISEASE), and porcine TRANSMISSIBLE GASTROENTERITIS virus; probable members include e.g. coronavirus enteritis of turkeys virus (= turkey bluecomb disease virus), and neonatal calf diarrhoea coronavirus. 'Possible members' include feline infectious peritonitis virus (= feline coronavirus) and human enteric coronavirus (a possible causal agent of human FOOD POISONING).

Virion: ca. 75–160 nm diam., consisting of a helical nucleocapsid surrounded by the envelope; characteristic club-shaped glycoprotein projections (peplomers, 'spikes') 12–24 nm long extend from the outer surface of the envelope. Genome: positive-sense ssRNA, MWt typically ca. $6–7 \times 10^6$ (8×10^6 has been reported for IBV). The RNA is polyadenylated at the 3′ end. Proteins associated with the virion include (at least) S ('spike'), M (membrane) and N (nucleocapsid). The virions are inactivated e.g. by lipid solvents and by detergents.

Virus replication occurs in the host cell cytoplasm. In IBV, 6 subgenomic mRNAs (A–F) are formed, all of which are 3′-coterminal with the genomic RNA (a so-called 'nested set' of mRNAs); each 5′ end has a common leader sequence (ca. 70 bases long) derived from the 5′ end of genomic RNA [possible mechanism: JGV (1986) *67* 221–228]. (MHV forms a 'nested set' of 7 subgenomic mRNAs, designated RNA1–RNA7.) The virions mature by budding through the endoplasmic reticulum into vesicles.

[Molecular biology of coronaviruses: AVR (1983) *28* 35–112.]

Coronavirus See CORONAVIRIDAE.

Coronie wilt *Syn.* HARTROT.

correction collar A part of a microscope objective lens which is used to adjust the lens to work with cover-glasses of various thicknesses.

corrinoids See VITAMIN B_{12}.

Corriparta subgroup See ORBIVIRUS.

corrosive sublimate See MERCURY.

cortex (1) Any of various external or outer layer(s) of a structure or organism – e.g. the outer differentiated tissue of certain LICHENS; in *ciliates* 'cortex' refers to the PELLICLE together with the INFRACILIATURE. (2) See ENDOSPORE.

corticate Having a CORTEX.

Corticiaceae See APHYLLOPHORALES.

corticicolous *Syn.* CORTICOLOUS.

corticolous (corticicolous) Growing on and/or in bark. (cf. EPIPHLOEODAL and ENDOPHLOEODAL.)

corticosteroid synthesis See STEROID BIOCONVERSIONS.

corticotype (*ciliate protozool.*) The morphological features of the CORTEX determined by staining (particularly silver staining) techniques. (See also SILVER LINE SYSTEM.)

Corticoviridae (PM2 phage group) A family of icosahedral, lipid-containing, non-enveloped BACTERIOPHAGES which contain supercoiled ds cccDNA. Type species: BACTERIOPHAGE PM2. Possible member: bacteriophage 06N 58P (host: *Vibrio*).

Corticovirus Currently the sole genus of the CORTICOVIRIDAE.

cortina A remnant of the PARTIAL VEIL (q.v.) in a mature fruiting body. ('Cortina' is also used to refer to an intact partial veil.)

Cortinariaceae See AGARICALES.

Cortinarius See AGARICALES (Cortinariaceae) and MYCETISM.

cortisol synthesis See STEROID BIOCONVERSIONS.

Corynebacterium A genus of Gram-positive, aerobic, facultatively anaerobic, chemoorganotrophic, non-acid-fast, non-motile, asporogenous bacteria (order ACTINOMYCETALES, wall type IV). The genus contains saprotrophic species found e.g. in soil and vegetable matter (including e.g. some species previously classified in the genera *Arthrobacter*, *Brevibacterium* and *Microbacterium*), and a number of species which are parasitic or pathogenic in man and other animals; all the plant-pathogenic species have been reclassified in other genera (e.g. CURTOBACTERIUM, RHODOCOCCUS). *Corynebacterium* spp exhibit certain features common to all nocardioform actinomycetes – e.g. the presence of cell wall MYCOLIC ACIDS. The organisms are straight or curved, pleomorphic, often coryneform rods. Metabolism can be oxidative or fermentative. Catalase-positive. GC%: ca. 51–59. Type species: *C. diphtheriae*.

C. autotrophicum. See XANTHOBACTER.

C. betae. See CURTOBACTERIUM.

C. bovis. Occurs e.g. associated with the cow's udder. Unlike other *Corynebacterium* spp it contains tuberculostearic acid.

C. diphtheriae. Cells rod-shaped, 0.3–0.8 × 0.8–8.0 μm. Toxinogenic strains of *C. diphtheriae* are the causal agents of DIPHTHERIA (see also DIPHTHERIA TOXIN and CROUP); the organisms may be isolated e.g. on tellurite–blood agar (see TELLURITE MEDIA) and subcultured to Loeffler's serum. (Growth from Loeffler's serum is used for the determination of cell morphology; cells grown on tellurite media have an atypical morphology.) The optimum growth temperature is 37°C. *C. diphtheriae* occurs in three main varieties, *gravis*, *intermedius* and *mitis*, which can be distinguished by cell and colonial morphology and by biochemical activity. All three varieties are urease-negative, and none can hydrolyse pyrazinamide; all produce acid (but no gas) from glucose, maltose and mannose, and some *mitis* strains can ferment sucrose [Book ref. 46, p. 1831]. *Gravis* strains typically form DAISY HEAD COLONIES (ca. 3 mm diam. at 24 hours/37°C) on blood–tellurite media; haemolysis is uncommon. Typically, starch is fermented in 24–48 hours. Very few cells contain METACHROMATIC GRANULES. *Intermedius* strains typically form small colonies (<1 mm diam. at 24 hours/37°C) on blood–tellurite media, the colonies usually being dull greyish-black with a slightly crenated edge; haemolysis does not occur. The cells rarely contain metachromic granules; stained cells exhibit a characteristic 'barred' appearance due to uptake of the stain (polychrome methylene blue) in a series of transverse bands. *Mitis* strains typically form shiny, greyish-black, entire colonies, 1–2 mm at 24 hours, on blood–tellurite media; haemolysis is very common. Each cell typically contains several metachromatic granules. The *ulcerans* variety [Book ref. 46, p. 1831] (= *C. ulcerans*) produces urease and gelatinase, and ferments starch and trehalose as well as glucose, maltose and mannose; haemolysis is common. Pyrazinamide is not hydrolysed. (See also ULCERACIN 378.)

C. equi. See RHODOCOCCUS.

C. fascians. See RHODOCOCCUS. (See also FASCIATION.)

C. flaccumfaciens. See CURTOBACTERIUM.

C. glutamicum. A species which occurs e.g. in soil and vegetable matter; certain strains are used for the industrial production of GLUTAMIC ACID. *C. glutamicum* is nutritionally fastidious; biotin is necessary for growth, and some strains are reported to need B vitamins.

C. haemolyticum. See ARCANOBACTERIUM.

C. hydrocarboclastus. See CORYNECINS.

C. minutissimum. A species [*incertae sedis*: Book ref. 21, p. 605] which is the causal agent of ERYTHRASMA; the organisms ferment glucose, maltose, mannose, and usually sucrose [Book ref. 46, p. 1831].

C. oortii. See CURTOBACTERIUM.

C. ovis. See *C. pseudotuberculosis*.

C. parvum. See PROPIONIBACTERIUM.

C. poinsettiae. See CURTOBACTERIUM.

C. pseudotuberculosis (formerly *C. ovis*). The causal agent of pyogenic infections in e.g. cattle, horses and sheep (see e.g. CONTAGIOUS PUSTULAR DERMATITIS (2), and ULCERATIVE LYMPHANGITIS). The organisms typically form yellow colonies; acid is produced from glucose, maltose and mannose, and urease is formed. Some strains are haemolytic. Pyrazinamide is not hydrolysed.

C. pyogenes. See ACTINOMYCES.

C. renale. A species isolated e.g. from urinary tract infections in cattle. Colonies are typically cream to yellow; glucose and mannose (but not maltose) are fermented, and urease is formed. Pyrazinamidase and caseinase are formed.

C. salmoninus. See RENIBACTERIUM.

C. ulcerans. See *C. diphtheriae*.

C. xerosis. A species which occurs e.g. on the conjunctivae and in the throat in man; usually non-pathogenic, but cases of endocarditis following aortic valve replacement have been reported. Colonies are greyish; typically, glucose, maltose, mannose and sucrose are fermented, and pyrazinamidase is formed. Urease-negative.

corynecins Compounds analogous to chloramphenicol in which the *N*-dichloroacetyl substituent is replaced with non-halogenated acyl groups; corynecins have been isolated from '*Corynebacterium hydrocarboclastus*'. (A type strain of *C. hydrocarboclastus* has been shown to have the properties of a species of *Rhodococcus* [JGM (1982) *128* 2503–2509].)

coryneform (1) ('club-shaped') Essentially rod-shaped with one end thickened or bulbous. (2) A name applied, loosely, to any Gram-positive, asporogenous, pleomorphic rod-shaped bacterium; as such it covers bacteria from a range of genera. (cf. DIPHTHEROID.)

Coryneliales See ASCOMYCOTINA.

corynephage A BACTERIOPHAGE which infects *Corynebacterium* spp. (See also DIPHTHERIA TOXIN.)

corynomycolic acids MYCOLIC ACIDS of *Corynebacterium* spp.

coryza Discharge from the nasal mucous membranes – a symptom of e.g. the COMMON COLD and the prodromal phase of MEASLES.

cos See BACTERIOPHAGE LAMBDA; BACTERIOPHAGE P2; COSMID.

Coscinodiscus See DIATOMS.

Cosmarium A large genus of placoderm DESMIDS. The cells are elliptical or angular in outline, with a deep sinus; the semicells may be ornamented but lack prominent projections or spines (cf. XANTHIDIUM). (See also SINGLE-CELL PROTEIN.)

cosmid A PLASMID into which has been inserted the *cos* site of BACTERIOPHAGE LAMBDA. Cosmids can be used as CLONING vectors for large fragments of DNA (e.g. ca. 25–40 kb), and have been useful e.g. in the construction of genomic libraries.

The use of a cosmid is shown diagrammatically in the figure. (See also PHASMID.)

costa (1) Any rib-like structure. (2) A deeply-staining, curved, endoskeletal rod in trichomonads; its location coincides with the region where the undulating membrane adjoins the cell. (cf. AXOSTYLE.)

Costaria See PHAEOPHYTA.

Costia A genus of flagellated protozoa. *C. necatrix* is parasitic in fish (see COSTIASIS); the cell is dorsoventrally flattened and bears two long and two short flagella. The longer flagella are used for locomotion, the shorter flagella for attachment of the organism to a host.

costiasis (blue slime disease) A freshwater FISH DISEASE caused by *Costia necatrix*. The skin and gills are affected, and large

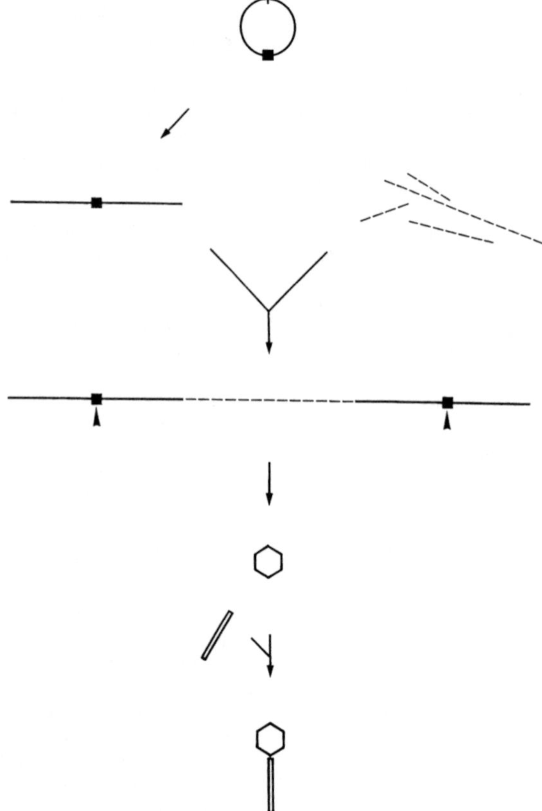

COSMID (diagrammatic). The cosmid (*top*) is a plasmid containing (i) a *cos* site (*solid black square*) from bacteriophage lambda, and (ii) the cutting site (*short vertical bar*) of a RESTRICTION ENDONUCLEASE.

Cosmids are cut at the single restriction site, yielding linearized molecules with an internal *cos* site (*top, left*). Linearized cosmids are mixed with DNA fragments (*dashed lines*) which have been cut by the same restriction enzyme.

Cosmids and fragments bind randomly via sticky ends, and in some cases a fragment will be flanked by two cosmids (*centre*). If, in such a molecule, the distance *cos*-to-*cos* is approximately 40–50 kb, then the *cos*–*cos* sequence is the right size for packaging into a phage lambda head (the phage lambda genome is ~48 kb). Enzymic cleavage at the *cos* sites (*arrowheads*) creates sticky ends, and packaging of the recombinant molecule occurs, in vitro, in the presence of phage components.

The completed virions can inject their recombinant molecules into suitable bacteria. Within a bacterium the DNA circularizes, via the *cos* sites, and replicates as a plasmid; its presence in the cell can be detected e.g. by the expression of plasmid-encoded antibiotic-resistance genes.

Screening for specific sequences in cosmids is carried out in a way similar to that used for plasmid vectors.

amounts of slime are secreted over the body. Mortality may be massive in fish hatcheries.

co-substrate See CO-METABOLISM.

cot death *Syn.* SUDDEN INFANT DEATH SYNDROME.

cotransduction The TRANSDUCTION of two or more donor genes in the same transducing virion.

co-transfer (of plasmids) See CONJUGATION (1b) (i).

co-transport *Syn.* SYMPORT.

cotrimoxazole See FOLIC ACID ANTAGONIST.

cottage cheese See CHEESE-MAKING.

cotton anthocyanosis virus See LUTEOVIRUSES.

cotton blue *Syn.* METHYL BLUE.

cotton wilt Any VASCULAR WILT of cotton plants (*Gossypium* spp). Common causal agents include *Fusarium oxysporum* f. sp. *vasinfectum* ('Frenching disease') and *Verticillium dahliae*.

cotton wool A fibrous, cellulosic material from seed coats of the cotton plant (*Gossypium*). *Non-absorbent* cotton wool contains substances (e.g. unsaturated fatty acids) inhibitory to certain organisms; *absorbent* cotton wool has been treated with solvents and bleached so that the inhibitory substances have been largely or completely removed. Plugs of non-absorbent cotton wool are often used as stoppers in tubes of bacteriological media. (See also SWAB.) (cf. cotton wool substitute in the entry ALGINATE.)

cotton wool disease See COLUMNARIS DISEASE.

cottontail rabbit papilloma virus See PAPILLOMAVIRUS.

cotype *Syn.* SYNTYPE.

***cou* gene** See GYRASE.

cough plate A method for isolating *Bordetella pertussis* from a suspected case of WHOOPING COUGH. A plate of BORDET–GENGOU AGAR is exposed to droplet inoculation from the patient's cough and is then incubated and examined for growth of *B. pertussis*. Many clinicians now use a pernasal swab (optimally, with an ALGINATE tip) in preference to the cough plate method.

Coulter counter An instrument used for counting particles (cells, spores etc) in a suspension. Essentially, it consists of two chambers separated by a partition of (electrically) insulating material which is perforated by a single hole of size similar to that of the particles to be counted. Each chamber contains an electrode. The sample is placed in one chamber and forced, under pressure, into the other; during this transfer the suspension acts as a conducting path (via the hole) between the two electrodes. As each cell passes through the hole, it momentarily decreases the electrical conductivity between the electrodes; this momentary change in conductivity is recorded as a pulse by an electronic counting circuit controlled by the two electrodes. The cell count is indicated on a digital display. (See also COUNTING METHODS.)

coumermycins A group of compounds structurally related to NOVOBIOCIN. Coumermycin A_1 inhibits ATP-dependent GYRASE activity in the same way as NOVOBIOCIN, but is much more effective both in vivo and in vitro.

Councilman body A cluster of hyaline, necrotic, eosinophilic cells in the liver of an individual suffering from YELLOW FEVER.

counter-receptor (to a CAM) See CELL ADHESION MOLECULE.

counter transport *Syn.* ANTIPORT.

countercurrent immunoelectrophoresis (CIE) IMMUNOELECTROPHORESIS (used for only certain types of antigen) in which antibodies and antigens (in separate wells in a gel strip) move towards each other when subjected to the same electric field; the antibodies move towards the cathode (see ELECTROENDOSMOSIS) while the antigens – because of their size, charge, and other physico-chemical characteristics – move towards the anode.

counterstain (1) A second, contrasting, stain used on a microscopical preparation with the object of staining those features which have not taken up the first stain. (2) In the GRAM STAIN: a stain used to stain those organisms or tissues which have been decolorized. (3) In the AURAMINE–RHODAMINE STAIN: the solution of potassium permanganate used to eliminate background fluorescence.

counting chamber (haemocytometer) An instrument used for determining the total cell count, viable cell count, or spore count etc. of a suspension of cells or spores (see COUNTING METHODS). A typical counting chamber is illustrated on p. 200. A grid, etched on the central plateau of the glass block, typically consists of a square of side 1.0 mm divided into 400 squares, each 0.0025 mm^2; the distance between grid and cover-slip may be 0.1 mm (as in the Thoma chamber) or 0.02 mm (as in the Helber chamber). When the cover-slip is correctly positioned, Newton's rings should be visible through those parts of the cover-slip in contact with the glass shoulders – indicating close contact. The suspension is introduced (with a Pasteur pipette) into the space between cover-slip and central plateau; the troughs should remain empty. The cells are allowed to settle and are counted under the microscope; since the volume between grid and cover-slip is known, the count per unit volume can be calculated. For viable cell counts, cells are subjected to VITAL STAINING.

(See also FACS.)

counting methods Methods for counting microorganisms may be divided into two categories according to the type of organism considered.

(a) *Organisms other than viruses, viroids etc.* The total number of living and dead cells in a given volume (or on a given area) of a sample is termed the *total cell count*; usually, total cell counts refer to bacteria, spores or yeasts (single-celled organisms). A total cell count may be determined by direct (visual) counting, i.e., by MICROSCOPY (see also COUNTING CHAMBER), or by the use of electronic or other instruments (see e.g. COULTER COUNTER, NEPHELOMETRY, TURBIDIMETRY); both direct and electronic methods may permit the counting of specific types of cell – e.g. by IMMUNOFLUORESCENCE microscopy and FACS (q.v.), respectively. Liquid samples containing small numbers of cells can be membrane-filtered (see FILTRATION) and the (stained) cells counted in situ. (See also DEFT.) Cell numbers in a given suspension may be estimated by comparison with standard suspensions (see BROWN'S TUBES).

Viable cell count refers to the number of living cells (per volume or area) in a given sample; viable cell counts usually refer to single-celled organisms, and they can obviously include only those organisms which are detectable by the particular method used. A viable cell count may be determined in several ways. (i) Direct examination, in a counting chamber, of a sample stained with a vital stain (see VITAL STAINING) – living and dead cells being distinguished by their different reactions to the dye. (ii) Statistical methods: see e.g. MULTIPLE-TUBE METHOD. (iii) Colony counts: the number of viable cells in a sample (or aliquot) is assessed from the number of colonies which develop on incubation of a solid medium which has been inoculated with the sample or aliquot; in such methods it is assumed that each colony developed from a single cell. The selectivity of the medium and the conditions of incubation may significantly affect the number of viable cells which give rise to colonies. For colony-counting methods see e.g. DIP-SLIDE METHOD, MILES AND MISRA'S METHOD, POUR PLATE, ROLL-TUBE TECHNIQUE and SPREAD PLATE (sense 1).

Some types of cell tend to form clumps or to grow in chains or filaments or in microcolonies. An accurate viable count of such organisms cannot be obtained by the above methods, and

(a)

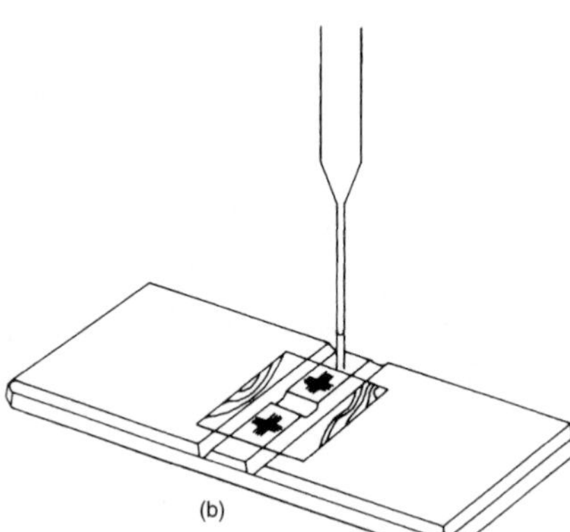

(b)

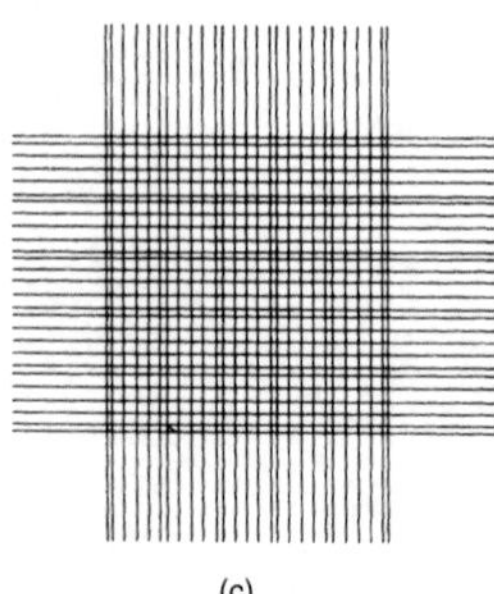

(c)

COUNTING CHAMBER (Thoma chamber). The instrument – seen from one side at (a) – consists of a rectangular glass block in which the central plateau lies precisely 0.1 mm below the level of the shoulders on either side. The central plateau is separated from each shoulder by a trough, and is itself divided into two parts by a shallow trough (seen at (b)). On the surface of each part of the central plateau is an etched grid (c) consisting of a square which is divided into 400 small squares, each 1/400 mm^2. A thin glass cover-slip is positioned as shown at (b) and pressed firmly onto the shoulders of the chamber; to achieve proper contact it is necessary, while pressing, to move the cover-slip (slightly) against the shoulders. Proper contact is indicated by the appearance of a pattern of coloured lines (Newton's rings), shown at (b).

Using the chamber. A small volume of a bacterial suspension is picked up in a Pasteur pipette by capillary attraction; the thread of liquid in the pipette should not be more than 10 mm. The pipette is then placed as shown at (b), i.e., with the opening of the pipette in contact with the central plateau, and the side of the pipette against the cover-slip. With the pipette in this position, liquid is automatically drawn by capillary attraction into the space bounded by the cover-slip and one-half of the central plateau; the liquid should not overflow into the trough. (It is sometimes necessary to tap the pipette, *lightly*, against the central plateau to encourage the liquid to enter the chamber.) A second sample can be examined, if required, in the other half of the counting chamber. The chamber is left for 30 min to allow the cells to settle, and counting is then carried out under a high power of the microscope – which is focused on the grid of the chamber. Since the volume between grid and cover-slip is accurately known, the count of cells per unit volume can be calculated.

A worked example. Each small square in the grid is 1/400 mm^2. As the distance between grid and cover-slip is 1/10 mm, the volume of liquid over each small square is 1/4000 mm^3, i.e., (for all practical purposes) 1/4,000,000 ml.

Suppose, for example, that on scanning all 400 small squares 500 cells were counted; this would give an average of $500 \div 400 (= 1.25)$ cells per small square. Thus, since 1.25 cells occur in 1/4,000,000 ml, the sample must contain $1.25 \times 4,000,000$ cells per ml, = 5,000,000 cells per ml.

If the sample was diluted before examination in the counting chamber, the count obtained must be multiplied by the dilution factor: e.g., if the sample was diluted 1 in 10, the count must be multiplied by 10.

in these cases counts may be given as the number of *colony-forming units* (cfu) per unit volume or area of the sample; a cfu can be any entity – e.g. a single cell or a chain of cells – capable of giving rise to a single colony.

The above methods are generally applicable e.g. to most bacteria and spores, and some of the methods may be used for certain algae, fungi (e.g. yeasts) and protozoa. The counting of rapidly moving microorganisms can be facilitated by the use of a viscous suspending medium – e.g. FICOLL, methylcellulose (2–5% aqueous) or poly(ethylene oxide).

(b) *Viruses*. Viruses may be counted by electron microscopy, and the largest viruses (e.g. some entomopoxviruses) may be counted by light microscopy after suitable staining; however, such methods are seldom used since e.g. they do not distinguish between infective and non-infective virions. In the plaque method of assay, serial dilutions of a viral suspension are initially prepared, and a known volume of each dilution is examined for its content of infectious, plaque-forming virions: see PLAQUE (1) and PLAQUE ASSAY. An alternative procedure is the END-POINT DILUTION ASSAY. Plant viruses may be assayed on LOCAL LESION hosts.

coupled cell division See F PLASMID.

coupling sequences See CONJUGATIVE TRANSPOSITION.

cover-glass (cover-slip) A small round or rectangular sheet of thin glass used primarily to cover a specimen on a microscope SLIDE. When a slide is viewed with a HIGH-DRY OBJECTIVE lens, the image quality is optimum only when the cover-glass is of a particular thickness (specified by the lens manufacturer); thicker or thinner cover-glasses effectively increase spherical aberration. Cover-glass thickness is relatively unimportant with low-power objectives and (usually) with oil-immersion objectives. Cover-glasses are made in thickness ranges designated No. 0 (ca. 0.1 mm), No. 1 (ca. 0.15 mm), No. 1½ (ca. 0.175 mm), No. 2 (ca. 0.2 mm), and No. 3 (ca. 0.3 mm). Most objectives are designed for use with No. 1½ cover-glasses. (See also CORRECTION COLLAR.)

cover-slip *Syn.* COVER-GLASS.

covered smut See SMUTS (sense 2).

covirus *Syn.* MULTICOMPONENT VIRUS.

Cowan I, II & III strains Strains of *Staphylococcus aureus* used e.g. for the production of standard TYPING antisera.

Cowdria A genus of Gram-negative bacteria of the tribe EHRLICHIEAE. Cells: coccoid to rod-shaped, pleomorphic, non-motile, maximum dimension ca. 0.5 μm. Growth occurs intracytoplasmically in the vascular endothelial cells of ruminants, but does not occur in cell-free media. Type species (sole species): *C. ruminantium*, the causal agent of 'heartwater' (a tick-borne, septicaemic disease of ruminants which occurs in Africa); *C. ruminantium* is transmitted trans-stadially, but not transovarially, in the tick vector. [Book ref. 22, pp. 709–710.]

cowpea aphid-borne mosaic virus See POTYVIRUSES.

cowpea chlorotic mottle virus See BROMOVIRUSES.

cowpea mosaic virus (CPMV) The type member of the COMOVIRUSES. Three types of particle can be distinguished and are designated T, M and B (S_w^{20} 58, 98 and 118, respectively); each is icosahedral, ca. 28 nm diam., the capsid consisting of 60 subunits of each of two structural proteins (MWts 22000 and 42000). T particles appear to consist only of protein. B particles contain a single molecule of RNA-1 (B-RNA, MWt ca. 2×10^6) while the M particles contain one molecule of RNA-2 (M-RNA, MWt ca. 1.2×10^6). Both RNAs are polyadenylated at the 3′ end and have a small polypeptide (VPg, MWt ca. 4000) covalently bound to the 5′ end. RNA-1 can be replicated and expressed in cowpea protoplasts in the absence of RNA-2, but requires RNA-2 for encapsidation and for cell-to-cell spread; expression of RNA-2 is completely dependent on the expression of RNA-1. Both RNAs appear to be translated (in the cytoplasm) into large POLYPROTEINS which are cleaved to form functional proteins. RNA-1 encodes at least 6 proteins, including VPg and a protease apparently involved in cleavage of the polyprotein [EMBO (1987) *6* 549–554]; RNA-2 carries genes for the capsid proteins. Cells infected with CPMV contain characteristic inclusion bodies: membranous, vesicular structures, usually adjacent to the nucleus, which contain dsRNA corresponding to RNA-1 and RNA-2, complementary (minus-strand) RNA, and replication complexes containing RNA-DEPENDENT RNA POLYMERASE molecules. (Although uninfected cowpea cells contain an RNA-dependent RNA polymerase, the activity of which increases by 10-fold or more on infection with CPMV, this enzyme appears not to be involved in CPMV replication [Book ref. 80, pp. 120–125].) Progeny CPMV particles accumulate in the host cell cytoplasm, sometimes forming crystalline arrays.

cowpea ringspot virus See CUCUMOVIRUSES.

cowpea severe mosaic virus See COMOVIRUSES.

cowpox A mild disease which primarily affects cows but which is transmissible (by direct contact) to human handlers; it is caused by the COWPOX VIRUS. In cows, lesions (which progress from papules through vesicles and pustules to crusts – cf. SMALLPOX) develop mainly on the udder and teats; other animals (e.g. cats [VR (1985) *117* 231–233]) may be affected. In man the lesions occur mainly on the hands and arms and may be accompanied by local oedema and lymphangitis.

cowpox virus A virus of the genus ORTHOPOXVIRUS; it is the causal agent of COWPOX. The virus has a relatively broad host range, including e.g. cattle, man, and many other animals; it has been recovered from wild rodents, and rodents may be the natural reservoir host. Cowpox virus forms large, haemorrhagic pocks on the CAM, and is highly lethal for chick embryos. Both A- and B-type inclusion bodies are formed in infected cells. DNA MWt: ca. 145×10^6.

coxibs See PROSTAGLANDINS.

Coxiella A genus of Gram-negative bacteria of the RICKETTSIEAE. Cells: highly pleomorphic, non-motile rods, ca. 0.2–0.4 × 0.4–1.0 μm, which stain well with the GIMÉNEZ STAIN; *C. burnetii*, the sole species, undergoes a developmental cycle in which ENDOSPORES are formed. Glucose, glutamate and pyruvate are metabolized (more readily at low pH). *C. burnetii* is not invariably killed at 63°C/30 min or 90°C for a few sec (cf. PASTEURIZATION). GC%: ca. 43.

C. burnetii is an obligate intracellular parasite in vertebrate and arthropod hosts, and is the causal agent of Q FEVER in man; in man, growth occurs e.g. in phagolysosomes (see PHAGOCYTOSIS) within mononuclear phagocytes (e.g. macrophages). Lab. culture is carried out (at ca. 35°C) e.g. in chicken embryos (by inoculating the yolk sac in hen's eggs), or in monolayer tissue cultures prepared from chicken embryos. On repeated passage, the organisms gradually change from phase I (= phase 1) – i.e., 'smooth', hydrophilic cells which are stable in suspension and are not phagocytosed in the absence of specific antibodies – to phase II (= phase 2) – i.e., 'rough', hydrophobic, autoagglutinable, readily phagocytosed cells; rapid reversion to phase I occurs when phase II cells are injected into lab. animals. [Book ref. 22, pp. 701–704.]

coxsackieviruses See ENTEROVIRUS.

cozymase NAD (q.v.).

***cpd* gene** In e.g. *Escherichia coli*: the gene encoding cAMP phosphodiesterase (see CYCLIC AMP).
CPE CYTOPATHIC EFFECT.
CPV group CYTOPLASMIC POLYHEDROSIS VIRUS GROUP.
CR1 See CD35.
CR2 (on B lymphocytes) See CD21.
Crabtree effect In certain yeasts: the occurrence of high levels of fermentative metabolism under aerobic conditions in the presence of high concentrations of a fermentable sugar (e.g. glucose in the case of *Saccharomyces cerevisiae*). When growing in batch culture with high concentrations of glucose under aerobic conditions, *S. cerevisiae* carries out a predominantly fermentative metabolism, converting glucose to ethanol (see ALCOHOLIC FERMENTATION); when the levels of glucose have been depleted, the ethanol can be used as a substrate for respiratory metabolism. Studies with continuous cultures have suggested that cells exposed to high concentrations of glucose under aerobic conditions respond by 'short-term' and 'long-term' regulatory processes. In the short term, respiratory enzymes are apparently not repressed, but ethanol is nevertheless formed; this is apparently because the cells have only a limited respiratory capacity, i.e., more pyruvate is formed than can be oxidized, and the excess pyruvate is decarboxylated to form ethanol. Long-term regulation is a process of metabolic adaptation involving repression of the synthesis of e.g. certain TCA cycle enzymes, cytochromes, and gluconeogenetic enzymes, together with proteolytic inactivation of enzymes. [JGM (1985) *131* 47–52.] (See also CATABOLITE REPRESSION.)
Craigie's tube method A method used e.g. for separating motile from non-motile organisms, and for isolating strains of *Salmonella* in the alternative flagellar phase. A piece of glass tubing (internal diam. ca. 3 mm) is placed in a test-tube of semi-solid nutrient agar such that the tubing projects above the level of the medium; the lower end of the tubing is cut obliquely, thus permitting free communication between the agar in the tubing and that in the remainder of the test-tube. If the medium inside the tubing is inoculated with motile and non-motile organisms, motile organisms may migrate into the medium in the test-tube – from which they may be subsequently isolated.

For the isolation of alternative-phase strains of *Salmonella* (see PHASE VARIATION) the medium is supplemented with anti-serum to the flagella antigens of the currently predominant phase; the antiserum inhibits motility in strains of the predominant phase, but alternative-phase strains can swim out into the test-tube from where they can be subsequently isolated.
Craterellus See CANTHARELLALES.
crayfish plague (*Krebspest*) A disease of European crayfish (e.g. *Astacus astacus*) caused by *Aphanomyces astaci*; the American crayfish (*Pacifastacus leniusculus*) is normally resistant but may be able to carry the pathogen. The disease involves mainly the exoskeleton and CNS; mortality rates are very high. (See also CRUSTACEAN DISEASES.)
***cre* gene** See BACTERIOPHAGE P1.
cream cheese See CHEESE-MAKING.
Credé procedure See GONORRHOEA.
Crenarchaeota See ARCHAEA.
crenate Scalloped.
Crenothrix A genus of IRON BACTERIA (q.v.) which occur as sheathed, sometimes terminally swollen filaments. Type species: *C. polyspora*.
creosote A highly effective antifungal wood preservative for outdoor use. It consists of a complex mixture of coal tar distillation products which include phenol, cresols, xylenols, pyridine, quinoline, anthracene and naphthalene. (See also LENTINUS.)
creosote fungus See HORMOCONIS.
Crepidotaceae See AGARICALES.
Crepidotus A genus of fungi (AGARICALES, Crepidotaceae) which grow e.g. on fallen twigs and leaves, forming typically shell-shaped or lobed fruiting bodies up to 2 or 3 cm across.
crescent (*protozool.*) The mature, crescent-shaped gametocyte formed by certain species of *Plasmodium* (e.g. *P. falciparum*).
cresol red A PH INDICATOR: pH 7.2–8.8 (yellow to red).
cresols (as disinfectants) See PHENOLS.
cresylic acid A mixture of methyl-, dimethyl- and ethyl-phenols used e.g. as an industrial biocide.
Creutzfeldt–Jakob disease (CJD) A human TRANSMISSIBLE SPONGIFORM ENCEPHALOPATHY in which the symptoms at onset typically include dementia (but which, in some patients, are limited to ataxia or to e.g. psychiatric disorders); death commonly occurs within 1 year (sometimes <6 months) of onset, but may not occur for up to 4 years in a minority of patients. Some cases involve *inherited* CJD – in which the *PRNP* gene (encoding PrP) contains particular mutations; brain extracts from such patients can transmit the disease to mice (indicating that, in these cases, the causal prion is endogenous, i.e. produced within host cells).

CJD can be transmitted e.g. via contaminated neurosurgical instruments (despite 'sterilization' with 70% ethanol and formaldehyde vapour), via transplanted corneas from infected donors, and by the administration of contaminated human growth hormone.

Several variant forms of CJD are distinguished on the basis of (i) the nucleotide sequence of the *PRNP* gene, and (ii) the particular strain of prion protein involved. Thus, cases of 'sporadic' CJD can be placed into four groups – cases in a given group differing from those in other groups e.g. by differences in initial symptoms and the rapidity with which the disease progresses.

Most or all of the strains of CJD prion can be transmitted to chimpanzees and other primates, but only some strains can be transmitted to other animals (e.g. cats, guinea pigs, mice). Goats (but not sheep) inoculated with brain tissue containing CJD prions develop a disease indistinguishable from scrapie.

A new variant form of CJD (designated nvCJD) was reported in the United Kingdom [Lancet (1996) *347* 921–925]. Cases of nvCJD have some unique features. For example, patients have died in their 2nd–5th decades, i.e. earlier than typical CJD patients. Clinical features that distinguish nvCJD from the typical CJD pattern also include psychiatric conditions, sensory abnormalities and pain preceding the more usual signs of ataxia, myoclonus and cognitive impairment; the electroencephalogram is unlike that obtained in most other cases of CJD. Moreover, histopathology includes multiple 'florid' amyloid plaques (rare in other cases of CJD) – as well as the changes that are typical of prion diseases; each of these plaques is encircled by a ring of spongiform vacuoles.

It is now believed that nvCJD is caused by a prion derived from meat or meat products from animals with BOVINE SPONGIFORM ENCEPHALOPATHY (q.v.). It has been suggested that – *before* the appearance of nvCJD – at least some cases of 'sporadic' CJD were caused by a prion derived from cattle with a rare spongiform disease, and that the BSE prion is a novel variant of this agent which, in humans, gives rise to nvCJD [RMM (1998) *9* 119–127]. Studies with transgenic mice have provided evidence in favour of the transmission of bovine prions to humans [PNAS

(1999) *96* 15137–15142] (see BOVINE SPONGIFORM ENCEPHALOPATHY for further supporting evidence).

Diagnosis may be facilitated by immunohistochemistry and/or Western blot analysis of tonsil biopsies [Lancet (1999) *353* 183–189].

At present, there is no chemotherapy for CJD. Various chemical agents have been examined for their ability e.g. to increase the incubation period of the disease in animals [therapeutic strategies (review): RMM (1998) *9* 135–151].

The causative prion of BSE has been shown to be transmissible from a sheep in the initial symptom-free phase of experimental infection to another sheep via a whole-blood transfusion; it has been suggested that symptom-free pre-nvCJD individuals may represent a risk of spread to other humans via blood transfusion [Lancet (2000) *356* 999–1000.]

[UK distribution of CJD 1994–2000: Lancet (2001) *357* 1002–1007; nvCJD (UK deaths): Lancet (2003) *361* 751–752.]

Some similarities between CJD and Alzheimer's disease have lead to attempts to demonstrate a transmissible agent in the latter disease; however, to date, there is no evidence of such an agent.

crevicular fluid A serum-like fluid which enters the mouth via the gingival crevice (= gingival sulcus) – i.e., the shallow crevice at the tooth–gum margin. It contains e.g. IMMUNOGLOBULINS (including IgA, IgG and IgM), COMPLEMENT, and various cells (e.g. LEUCOCYTES).

CRF COAGULASE reacting factor.

crib death *Syn.* SUDDEN INFANT DEATH SYNDROME.

Cribraria See MYXOMYCETES.

Cribrostomum See FORAMINIFERIDA.

cricket paralysis virus See PICORNAVIRIDAE.

Cricosphaera See COCCOLITHOPHORIDS.

Crimean–Congo haemorrhagic fever See VIRAL HAEMORRHAGIC FEVERS.

Crinipellis See AGARICALES (Tricholomataceae) and WITCHES' BROOM.

crista (1) See CRISTISPIRA. (2) See MITOCHONDRION.

Cristispira A genus of bacteria (family SPIROCHAETACEAE) which occur e.g. in the digestive tract and/or crystalline style in freshwater and marine molluscs, and in other aquatic organisms. Cells are motile, helical, ca. 0.5–3.0 × 30–180 μm, and each cell has 100 or more periplasmic flagella which may distend the outer sheath – giving rise to a longitudinal ridge, the *crista*; cells divide by transverse fission. Type species: *C. pectinis*. [Book ref. 22, pp. 46–49.]

Crithidia (*Strigomonas*) A genus of homoxenous parasitic protozoa (family TRYPANOSOMATIDAE) in which the choanomastigote form occurs; species typically occur in the gut of insects (e.g. *C. fasciculata* occurs worldwide in *Anopheles* and *Culex* mosquitoes).

crithidial form Obsolete *syn.* EPIMASTIGOTE.

critical control points See HACCP.

critical dilution rate See DILUTION RATE.

critical illumination (Nelsonian illumination) In MICROSCOPY: illumination in which the light source (e.g. lamp filament) is focused by the CONDENSER to form an image in the plane of the specimen. With electric coiled filaments critical illumination tends to give uneven illumination. (cf. KÖHLER ILLUMINATION.)

critical initiation mass See CELL CYCLE.

critical point drying A method used e.g. for dehydrating and drying specimens prior to their examination by ELECTRON MICROSCOPY; in this method the specimen is not exposed to a liquid–gas boundary and is thus not subjected to the damage caused by surface tension in air-dried specimens. In a typical schedule, the water in a fixed and washed specimen is initially replaced by an 'intermediate fluid' (e.g. ethanol or acetone), and the specimen, totally immersed in the intermediate fluid, is placed in a strong metal pressure chamber. The chamber is then closed, cooled to ca. 15–18°C, and the intermediate fluid is replaced by the 'transitional fluid' (usually liquid CO_2) by means of a valve system; the specimen remains below the liquid–gas boundary throughout. The liquid CO_2 is then partially drained and fresh liquid CO_2 is added; this is repeated several times to dilute out the intermediate fluid, the specimen always remaining fully covered by liquid CO_2. When the last traces of intermediate fluid have been replaced by liquid CO_2 the temperature of the (closed) chamber is slowly raised (e.g. 1°C/min) above 31°C: the *critical point* above which CO_2 cannot exist as a liquid. The gas is then bled off *slowly* (to avoid e.g. turbulence in the chamber).

In an alternative schedule, various halocarbons (e.g. certain Freons) may be used to displace the ethanol or acetone. Thus e.g. ethanol may be displaced by two different halocarbons, used sequentially: the first acts as an additional intermediate fluid, while the second acts as the transitional fluid. (cf. FREEZE-DRYING.)

critical temperature zone See FREEZING.

CRM Cross-reacting material: protein which cross-reacts serologically with a given biologically active protein; CRM may lack the biological activity of the given protein, e.g., it may be a product of a mutant gene or a precursor form of the given protein.

***cro* gene** See BACTERIOPHAGE λ.

Crohn's disease A (typically) chronic human disease of the intestinal tract which is characterized by granulomatous/ulcerative lesions; diarrhoea and intestinal pain are common. The acute form may resemble appendicitis.

The aetiology is not known. Possibly, mutations affecting certain intracellular molecules in the intestinal epithelium (e.g. Nod2) may affect the normal protective response to bacterial infection [JBC (2003) *278* 5509–5512, 8869–8872].

cromoglycate A therapeutic agent which stabilizes sensitized MAST CELLS, inhibiting the release of histamine; it is administered *before* the patient is exposed to antigen/allergen in order to prevent anaphylactic reactions.

Cronartium A genus of rust fungi (class UREDINIOMYCETES) which include some important parasites of pine trees (*Pinus* spp): see e.g. BLISTER RUST. (See also PERIDERMIOID.)

cross-feeding (1) *Syn.* SYNTROPHISM. (2) A particular type of SYNTROPHISM in which each organism derives essential growth factors from the other(s).

cross-linking quantitative assay A type of assay which has been used for detecting and quantifying hepatitis B virus (HBV) in serum. The assay, carried out in the solution phase, involves probes which incorporate a derivative of 7-hydroxycoumarin – a cross-linking agent that can be photo-activated by ultraviolet radiation such that each probe can be covalently bound to its target sequence in HBV DNA. Essentially, probes complementary to sequences in the major subtypes of HBV are allowed to hybridize to target sequences in the sample and are then photo-cross-linked to their targets; by means of biotin tags, the cross-linked probes are captured on streptavidin-coated DYNABEADS and subsequently labelled with alkaline phosphatase. The addition of a suitable substrate for the enzyme gives rise to a fluorescent signal. [Cross-linking assay for HBV DNA: JCM (1999) *37* 161–164.]

cross-reacting antibody Antibody which can combine with non-homologous antigen (i.e. antigen not identical to that which elicited the antibody) as well as with homologous antigen. Such cross-reaction requires that the homologous and non-homologous antigens have a certain degree of stereochemical similarity. Cross-reacting antibodies often have a much lower affinity for the non-homologous antigen (cf. HETEROCLICITY.) (See also MONOCLONAL ANTIBODIES.)

Antibodies raised against a particular microorganism may combine with related (or even unrelated) species (see e.g. WEIL–FELIX TEST); such organisms may share a common antigenic determinant or may have CROSS-REACTIVE ANTIGENS.

cross-reacting material See CRM.

cross-reactivation See REACTIVATION.

cross-reactive antigens Antigens which differ from one another but which can combine with antibodies raised against either (or any) one of them. In *type 1 cross-reactivity*, different cross-reactive ligands bind with different affinities to a given site on a given antibody molecule; in *type 2 cross-reactivity* a cross-reactive ligand reacts with *some* of the antibodies in a POLYCLONAL ANTISERUM raised against the homologous ligand [Mol. Immunol. (1981) *18* 751–763].

cross resistance The phenomenon in which the resistance of an organism to one antibiotic correlates with resistance to one or more other antibiotics.

cross-wall *Syn.* SEPTUM.

crossband (in prosthecae) See CAULOBACTER.

crossed immunoelectrophoresis See TWO-DIMENSIONAL ELECTROPHORESIS.

crossing over A RECOMBINATION process in which, e.g., exchange of double-stranded sequences occurs between two linear duplexes. In effect, crossing over involves a break in each of the two strands in both (linear) duplexes, exchange of parts of the duplexes, and ligation to form two recombinant duplexes (see RECOMBINATION). Thus, if the two duplexes carry alleles ABCD and abcd, crossing over results in recombinant molecules Abcd and aBCD, ABcd and abCD, or ABCd and abcD, depending on the site of the cross-over. In e.g. MEIOSIS, crossing over occurs between non-sister chromatids in homologous chromosomes (cf. PARASEXUAL PROCESSES).

If crossing over occurs between two *circular* molecules the result is a single larger circular molecule; conversely, crossing over between regions within a single circular molecule results in two smaller circular molecules. (See also CAMPBELL MODEL.)

crotocin See TRICHOTHECENES.

croup A type of laryngotracheitis with hoarseness, resonant cough, stridor, and respiratory obstruction. Croup occurs mainly in young children and may be of non-microbial causation (e.g. allergy) or due to infection by e.g. human respiratory syncytial virus, parainfluenza viruses, *Corynebacterium diphtheriae*.

crown gall A plant neoplasm, sometimes bearing root-like, stem-like and/or leaf-like structures, which is formed at the crown (stem–root junction) – or, less commonly, on stems or roots – in a wide range of gymnosperms and dicotyledonous angiosperms (although in few monocotyledonous angiosperms); crown gall follows infection of a wound with virulent strains of certain species of *Agrobacterium*, commonly *A. tumefaciens* (see AGROBACTERIUM). Infected plants may be stunted or even killed; the disease can cause important economic losses (for example, in stone-fruit trees and vines) e.g. in parts of Australia, Europe and the USA. Crown gall of grape is caused by strains of *Agrobacterium* which have been assigned to a new species, *A. vitis* [crown gall of grape (biology and disease management): ARPpath. (1999) *37* 53–80]. (See also GALLS.)

A. tumefaciens attaches to the host plant at the site of a wound, attachment being reported to occur via the bacterial lipopolysaccharides and via epicellular cellulose microfibrils formed by the bacteria [Book ref. 55, pp 33–54].

Virulence in *A. tumefaciens* (and in *A. vitis*) depends on the presence of a large CONJUGATIVE PLASMID called the Ti (tumour-inducing) plasmid (ca. 150–230 kb). Genes responsible for tumorigenesis are located in a small subregion of the Ti plasmid termed the T-region or T-DNA (ca. 15–24 kb); in the infection process, single-stranded T-DNA is excised from the plasmid and transferred to the plant.

Transfer of T-DNA to the plant requires certain products of the virulence (*vir*) regulon – ~25 genes in regions of the Ti plasmid outside the T-DNA; transfer is regulated by a TWO-COMPONENT REGULATORY SYSTEM consisting of the proteins VirA (sensor) and VirG (response regulator), both expressed constitutively. [Two-component regulatory system in *A. tumefaciens*: FEMS Reviews (1998) *21* 291–319 (301–303).] VirA appears to respond (directly or indirectly) to certain types of phenolic compound (e.g. *acetosyringone*) released by plant wounds; the activity of phenolics is enhanced by monosaccharides such as glucose, and it has been suggested that such enhancement results from the binding, to VirA, of a complex consisting of monosaccharide and the product of the bacterial virulence gene *chvE*. (As well as being a sensor protein, VirA may also function as a host-range determinant [EMBO (1987) *6* 849–856]; thus, different (variant) forms of VirA may differ in their response to particular plant signals.)

On activation, VirG promotes transcription in the *vir* regulon. The *virD* operon encodes a site-specific endonuclease complex which makes nicks in one strand of the plasmid at unique sites within a pair of 24-bp DIRECT REPEATS that flank the T-DNA; the (single-stranded) copy of T-DNA, with protein VirD2 covalently attached to the 5′ end, is then transferred to the plant cell nucleus. (So-called *nuclear localization signals*, associated with one of the proteins accompanying the T-DNA, appear to be required for targeting.) [Molecular signals in the interaction between plants and microbes: Cell (1992) *71* 191–199.] Pili appear to be necessary for transfer of T-DNA to plants [Science (1996) *273* 1107–1109].

Within a plant cell the processed T-DNA can apparently insert at multiple sites in the nuclear DNA, sometimes forming tandem repeats. While T-DNA is transcribed only weakly in the bacterium, in the plant cell it is transcribed into both polyadenylated and non-polyadenylated RNA [Plasmid (1981) *6* 17–29].

Genes within T-DNA encode enzymes involved in the formation of CYTOKININS [PNAS (1984) *81* 4776–4780, 5994–5998] and IAA (see AUXINS), resulting in persistently raised levels of these phytohormones which are apparently responsible for tumorigenesis.

T-DNA also carries genes which specify the production, by tumour cells, of a number of tumour-specific substances called *opines*. A given Ti plasmid may encode opines of the OCTOPINE family (e.g. octopine, histopine, lysopine, octopinic acid) or of the NOPALINE family (e.g. nopaline, nopalinic acid). Octopine Ti plasmids may also encode agropine; agropine (formerly 'null type') Ti plasmids and nopaline Ti plasmids also encode AGROCINOPINES. The different plasmids may specify tumours of different types. (cf. HAIRY ROOT; see also AGROCINS.)

A Ti plasmid carries genes for the degradation of those opine(s) synthesized by the tumour induced by its T-DNA; moreover, the pathogen – but not plant cells or other microorganisms – can use opines as the sole source of carbon and nitrogen.

The availability of nutrients (opines) promotes growth and development of a population of the pathogen. Conjugal transfer of the Ti plasmid from Ti^+ cells to Ti^- agrobacteria involves initial transcriptional activation of the regulatory gene *traR* (present in the Ti plasmid) – apparently by a gall-derived opine; a second signal, derived from a QUORUM SENSING system (based on the density of *donor* cells), is also needed to initiate the transfer process. [Tra system in *A. tumefaciens*: FEMS Reviews (1998) *21* 291–319 (299–301).]

Ti plasmids may be used as vectors for the insertion of DNA sequences into the genome of dicotyledonous plants [EMBO (1985) *4* 277–284]. (See also AGROINFECTION.) Non-oncogenic Ti plasmids and plant protoplasts may be used so that hormone levels are not disturbed, and whole plants can be regenerated from the protoplasts using standard plant cell culture techniques. Transformed protoplasts can be detected by screening for opine synthesis.

crown rot Any plant disease in which the 'crown' (i.e., the region between the root and stem) rots as a result of microbial infection; causal agents include e.g. *Erwinia rhapontici* in rhubarb, *Mycocentrospora acerina* in carrots and celery, *Phytophthora cactorum* in strawberries. (cf. FOOT-ROT (2).)

crown rust A disease of members of the Gramineae. Crown rust of oats (*Avena* spp) is an economically important disease caused by *Puccinia coronata avenae*; orange, typically elongated, uredial pustules occur on leaves and sometimes on culms. (See also CEREAL DISEASES.)

crozier See ASCUS.

CRP (1) C-REACTIVE PROTEIN. (2) cAMP receptor protein: see CATABOLITE REPRESSION.

***crp* gene** See CATABOLITE REPRESSION.

Crucibulum See NIDULARIALES.

crucifer diseases See e.g. ALBUGO (for white rusts), BLACKLEG (sense 2), CLUBROOT and FUSARIUM WILT.

crucifer white blister See ALBUGO.

cruciform (1) (*adj.*) Cross-shaped. (2) (*noun*) In a nucleic acid molecule: two HAIRPINS (or STEM-AND-LOOP STRUCTURES), opposite one another, forming a cross-shaped structure. A cruciform can be formed in a double-stranded molecule or in a double-stranded region of a single-stranded molecule. (See PALINDROMIC SEQUENCE.)

crude oil See PETROLEUM.

crustacean diseases See e.g. BLACK GILL DISEASE; BLACK MAT SYNDROME; BURNED SPOT DISEASE; CRAYFISH PLAGUE; GAFFKAEMIA; PARAMOEBA. (See also LAGENIDIALES and SAPROLEGNIALES (ATKINSIELLA and *Leptolegnia baltica*).) (See also INVERTEBRATE DISEASES.)

crustaceous *Syn.* CRUSTOSE.

crustose (crustaceous) Forming or resembling a crust. Of a lichen: having a thallus which generally lacks a lower cortex, being attached to the substratum (over the whole of its lower surface) by hyphae of the medulla. A crustose thallus may be e.g. AREOLATE, PLACODIOID, etc. (cf. FOLIOSE; FRUTICOSE.)

cryo- Prefix meaning *cold*.

cryoelectron microscopy See ELECTRON MICROSCOPY (d).

cryofixation The FREEZING of living cells preparatory to e.g. FREEZE-ETCHING; the aim is similar to that of FIXATION but cells may or may not be killed during the process. [Cryofixation techniques for soil fungi: MS (1985) *2* 225–230.]

cryogen Any agent which is used for FREEZING a specimen – e.g. liquid nitrogen. (See also FREON.)

cryopreservation See FREEZING.

cryoprotectant See FREEZING.

cryosectioning *Syn.* FREEZE-SECTIONING.

cryostat *Syn.* cryotome (q.v.).

cryosubstitution FREEZE-SUBSTITUTION.

cryotome See FREEZE-SECTIONING.

Cryoxicide A STERILANT containing ETHYLENE OXIDE (11%), trichlorofluoromethane (79%), and dichlorodifluoromethane (10%).

crypt (*ciliate protozool.*) See BROOD POUCH.

cryptic mutant A cell which lacks one or more components of a TRANSPORT SYSTEM so that a particular substrate, or substrates, cannot enter the cell – and hence cannot be utilized, even though the cell may possess all the enzymes necessary for its utilization. For example, the failure of a cell to produce enzyme I (see PTS) – as a result of a (pleiotropic) mutation – would prevent that cell (a cryptic mutant) from taking up and metabolizing a range of sugars.

cryptic plasmid A PLASMID which has no apparent effect on the phenotype of the cell in which it occurs.

cryptic viruses Virus-like isometric particles which have been detected in a wide range of symptomless plants; the particles are generally ca. 30 nm in diameter, are present in low concentrations, and appear to be seed-borne. Some have been shown to contain dsRNA [e.g. beet cryptic viruses: JGV (1986) *67* 363–366].

cryptidin See DEFENSINS.

Cryptobia See BODONINA.

cryptobiosis See DORMANCY.

Cryptocaryon A genus of cyst-forming ciliates related to ICHTHYOPHTHIRIUS; the sole species, *C. irritans*, causes WHITE SPOT in marine fish.

Cryptococcaceae A family of fungi of the class BLASTOMYCETES. The genera include CANDIDA, CRYPTOCOCCUS, KLOECKERA, RHODOTORULA and TRICHOSPORON.

cryptococcosis (torulosis; also European blastomycosis or Busse–Buschke disease) A disease of man and animals caused by *Cryptococcus neoformans* (q.v.); infection occurs on inhalation of dust contaminated with the fungus. Pulmonary infection may be mild or inapparent. However, particularly in individuals with e.g. defective CMI or certain leukaemias, the disease may become disseminated to almost any tissue (liver, bones, skin etc) but especially to the meninges (*cryptococcal* MENINGITIS – commonly fatal). *Diagnosis*: demonstration of the fungus by culture or microscopy; latex particle test for cryptococcal capsular polysaccharide. *Chemotherapy*: e.g. amphotericin B, flucytosine, possibly miconazole or ketoconazole. [Book ref. 17, pp. 180–186.]

Cryptococcus A genus of non-fermentative imperfect yeasts (class HYPHOMYCETES). The cells are spheroidal, ovoid, elongate, or polymorphic, and reproduce by multilateral budding. Pseudomycelium is rudimentary or absent. The vegetative cells usually have capsules, the composition of the capsule and the abundance of capsular material depending on growth conditions; under certain conditions the capsule contains starch-like polysaccharides which may be released into the medium. Colonies on e.g. malt agar are typically mucoid and are commonly cream- to tan-coloured, or red in some species (e.g. *C. hungaricus* and *C. macerans*). Inositol can be assimilated as sole carbon source. NO_3^- is assimilated by some species.

Most *Cryptococcus* species have basidiomycetous affinities: *C. albidus* var. *albidus* and *C. uniguttulatus* are anamorphs of FILOBASIDIUM spp, *C. neoformans* of FILOBASIDIELLA sp. Three species have ascomycetous affinities: *C. cereanus* and *C. lactativorus*, which are anamorphs of SPOROPACHYDERMIA, and *C. melibiosum* (teleomorph unknown). [Book ref. 100, pp. 845–872.]

C. neoformans is the causal agent of CRYPTOCOCCOSIS and may also be associated with e.g. bovine mastitis. Cells (e.g. in malt extract after 3 days at 25°C) are spherical, ca. 3–8 μm diam., occurring singly, in pairs, or in small groups. Pseudomycelium is not formed, and extracellular polysaccharide may be abundant. *C. neoformans* is the only *Cryptococcus* sp which can grow well at 37°C (certain other species or strains – e.g. *C. melibiosum* – may be capable of weak growth at 37°C). *C. neoformans* has two varieties, var. *neoformans* and var. *gattii* (= '*C. bacillisporus*'), distinguished e.g. on the basis of the ability of var. *gattii* (but not var. *neoformans*) to give a blue colour on CGB AGAR within 2–5 days.

Other *Cryptococcus* spp (not mentioned above) include e.g. *C. ater*, *C. dimennae*, *C. elinovii*, *C. flavus*, *C. gastricus*, *C. heveanensis*, *C. infirmo-miniatus*, *C. kuetzingii*, *C. laurentii*, *C. luteolus*, *C. magnus*, *C. skinneri*, *C. terreus*. Strains have been isolated from a wide range of habitats, including e.g. soil, plants, dew-retted flax, seawater, clinical specimens, foods and beverages, etc. Another species, *C. friedmannii*, has been isolated from rock fragments containing cryptoendolithic lichens from a dry desert area of the Antarctic [Mycol. (1985) *77* 149–153].

cryptoendolith See ENDOLITHIC.

cryptogam A plant which does not produce true flowers or seeds; cryptogams traditionally include the ALGAE, FUNGI, bryophytes (e.g. mosses) and pteridophytes (e.g. ferns).

cryptogram (virol.) A descriptive code used to summarize, in symbolic form, certain basic properties of a given virus.

Cryptomonadida See PHYTOMASTIGOPHOREA.

cryptomonads *Syn.* CRYPTOPHYTES.

Cryptomonas See CRYPTOPHYTES.

cryptophytes (cryptomonads) A category of unicellular, biflagellated, marine or freshwater organisms variously regarded as algae (class Cryptophyceae) or as protozoa (see PHYTOMASTIGOPHOREA); most cryptophytes are photosynthetic (including e.g. *Chroomonas* and *Cryptomonas*), but some (e.g. *Chilomonas*, *Cyathomonas*) are colourless and heterotrophic. The cryptophyte cell is typically flattened and dorsiventral, lacks a cell wall (having instead a plasma membrane with an internal lining of proteinaceous plates), and possesses EJECTOSOMES. In photosynthetic species each cell has a single bilobed chloroplast containing a pyrenoid, paired thylakoids, chlorophylls *a* and c_2, α-carotene, diatoxanthin, and PHYCOBILIPROTEINS. Storage products are starch and lipid. Reproduction occurs by longitudinal binary fission; sexual reproduction is unknown. [Taxonomic aspects (including genus descriptions): New Phyt. (1984) *98* 627–646.]

cryptosporidiosis A COCCIDIOSIS (q.v.) of man and other animals caused by a species of CRYPTOSPORIDIUM; the pathogen is commonly transmitted (as sporulated oocysts) via the direct or indirect faecal–oral route.

The disease in man is typically mild and self-limiting; it may involve diarrhoea, abdominal cramps, low-grade fever and headache. In the severely immunocompromised it is a life-threatening disease: usually severe, persistent watery diarrhoea; other sites (e.g. respiratory, biliary) may be affected. No effective chemotherapy is available [IJP (1995) *25* 139–195]. [The treatment dilemma: JMM (2000) *49* 207–208.]

Cryptosporidium is a cause of diarrhoea, pneumonia and disseminated infections in calves foals, lambs and piglets; it can also cause diarrhoea, inappetance and weight loss in the domestic cat [VR (1985) *116* 73–74].

[Cryptosporidiosis in animals and humans: MR (1983) *47* 84–96.]

Outbreaks of disease can be transmitted via water as thick-walled oocysts of *Cryptosporidium* are resistant to the level of chlorination normally used in WATER SUPPLIES. The viability of *C. parvum* oocysts in water has been examined by staining them with fluorescent dyes (SYTO-9, SYTO-59); oocysts which stained above a certain intensity could not establish infection in mice, but those which displayed little or no fluorescence could readily establish infection [AEM (2000) *66* 406–412].

Cryptosporidium A genus of parasitic (homoxenous) protozoa (suborder EIMERIORINA – q.v.) which are pathogenic for man and other animals (see CRYPTOSPORIDIOSIS); the organisms typically develop in the intestinal microvilli. The oocysts are very small (ca. 4–6 μm); following ENDOSPORULATION, each oocyst contains four banana-shaped sporozoites. Two types of oocyst are formed; thin-walled oocysts which release sporozoites into the host's intestine, causing re-infection ('autoinfection') of the host, and acid-fast thick-walled oocysts (ca. 80% of the total) which are voided in the faeces.

[Complete development of *Cryptosporidium* in cell cultures: Science (1984) *224* 603–605.]

The taxonomy of the genus is unsettled. It has been suggested that only 4 species should be recognized e.g. on the basis of host range: *C. crotali* (infecting reptiles), *C. meleagridis* (infecting birds), *C. muris* (infecting mammals) and *C. nasorum* (infecting tropical fish) [JP (1984) *31* 94–98]. However, '*C. parvum*', the organism which appears to be the cause of most cases of cryptosporidiosis, is apparently distinct from *C. muris* [J. Parasitol. (1985) *71* 625–629]. [*C. baileyi* from chickens: JP (1986) *33* 289–296.]

Cryptostroma See HYPHOMYCETES; see also MAPLE BARK STRIPPERS' DISEASE and SOOTY BARK.

crystal toxin (of *Bacillus thuringiensis*) See DELTA-ENDOTOXIN.

crystal violet A violet basic TRIPHENYLMETHANE DYE used e.g. in the GRAM STAIN.

CS protein See PLASMODIUM.

CS1, CS2, CS3 See ETEC.

CSF (1) Cerebrospinal fluid. (2) COLONY-STIMULATING FACTOR. (3) Cold stability factor (see MICROTUBULE-ASSOCIATED PROTEINS). (4) A pheromone of *Bacillus subtilis* (see QUORUM SENSING).

CSLM CONFOCAL SCANNING LIGHT MICROSCOPY.

CSP reaction Shedding of CS proteins by the sporozoites of *Plasmodium* spp (see PLASMODIUM) following cross-linking of the CS proteins by anti-CS antibodies. (cf. Capping phenomenon in DYSENTERY (b).)

CspA See COLD-SHOCK RESPONSE.

CSPD® Trade name (Tropix, Bedford, MA, USA) of a 1,2-dioxetane substrate which gives rise to CHEMILUMINESCENCE when activated by the enzyme alkaline phosphatase. (See also AMPPD.)

CSSD Central sterile supply department.

CTAB See QUATERNARY AMMONIUM COMPOUNDS.

ctDNA Chloroplast DNA (see CHLOROPLAST).

CTEM Conventional transmission ELECTRON MICROSCOPY.

CTh CARRIER-SPECIFIC T HELPER CELL.

CTLA-4 See CD28.

CTP Cytidine 5′-triphosphate [see Appendix V(b)].

***ctxA*, *ctxB* genes** See CHOLERA TOXIN.

cucumber 4 virus See TOBAMOVIRUSES.

cucumber green mottle mosaic virus See TOBAMOVIRUSES.

cucumber mosaic virus See CUCUMOVIRUSES.

cucumber necrosis virus See TOBACCO NECROSIS VIRUS.

cucumber pale fruit viroid See VIROID.

cucumbers (pickled) See PICKLING.

cucumoviruses (cucumber mosaic virus group) A category of ssRNA-containing PLANT VIRUSES which infect e.g. cucurbits and solanaceous plants; transmission occurs (non-persistently) via aphids and, in some hosts, via seeds, and can also occur mechanically. Type member: cucumber mosaic virus, CMV (S isolate). Other members: peanut stunt virus (PSV) and tomato aspermy virus (TAV); possible member: cowpea ringspot virus.

Virion: icosahedral, ca. 29 nm diam., containing a single type of coat protein and linear, positive-sense ssRNA. Three types of virion exist, differing in the nature of the RNA they contain; a particle may contain one molecule of RNA-1 (MWt ca. 1.27×10^6), one molecule of RNA-2 (MWt ca. 1.13×10^6), or one molecule each of RNA-3 (MWt ca. 0.82×10^6) and RNA-4 (coat protein mRNA). Some isolates of CMV and PSV also contain SATELLITE RNAS designated (respectively) CARNA (*C*MV-*a*ssociated *RNA*) and PARNA (*P*SV-*a*ssociated *RNA*). CARNA 5, the best-known example, consists of 335 nucleotides and is capped at the 5′ end; it can code for two polypeptides. The presence of a CARNA attenuates the symptoms of CMV infection in many plants, but may enhance disease severity in others. For example, in tomato plants infection with CMV (only) may cause leaf chlorosis with some leaf distortion, but infection with both CMV and CARNA 5 causes a lethal necrosis. 'White leaf' of tomatoes and 'brilliant yellowing' of tobacco are also examples of diseases caused by CARNA strains. The presence of a CARNA in a host plant also reduces the yield of CMV – apparently as a result of competition for virus replication machinery.

Cucurbitaria See DOTHIDEALES.

cud See RUMEN.

cudbear A purple dye obtained from *Ochrolechia tartarea*.

Culex A genus of mosquitoes (order Diptera, family Culicidae); *Culex* spp are vectors of certain diseases: see e.g. MURRAY VALLEY FEVER.

Culicinomyces A genus of fungi of the HYPHOMYCETES; *C. clavisporus* is pathogenic in mosquito larvae [Mycol. (1984) *76* 614–625].

Culicoides A genus of biting midges (order Diptera, family Ceratopogonidae); *Culicoides* spp are vectors of certain diseases: e.g. AFRICAN HORSE SICKNESS and BLUETONGUE.

culmination See DICTYOSTELIOMYCETES.

cultivar (cv.) A commercial or cultivated 'variety' of a given species of plant or fungus.

culture (1) (*noun*) A liquid or solid MEDIUM on or within which has grown a population of particular type(s) of microorganism or cell as a result of the prior INOCULATION and INCUBATION of that medium. The following refers primarily to microbial cultures – cf. TISSUE CULTURE. On a solid medium, microorganisms may grow as a continuous layer or film of surface growth (*confluent growth*) or as discrete (individual) colonies (see COLONY) – depending e.g. on the method of inoculation (cf. LAWN PLATE and STREAKING). Cultures in liquid media may be turbid or clear, and may or may not have a PELLICLE.

Aerobic culture. A culture prepared under AEROBIC (sense 1) conditions.

Anaerobic culture. One prepared under ANAEROBIC (sense 1) conditions. (See also ANAEROBE.)

Axenic culture. Pure culture (see below).

Batch culture. See BATCH CULTURE.

Broth culture. One prepared with NUTRIENT BROTH or with any similar liquid medium.

Closed culture. *Syn*. BATCH CULTURE.

Contaminated culture. One which has been exposed (usually unintentionally) to non-STERILE conditions and which has become contaminated with extraneous organisms.

Continuous culture. See CONTINUOUS CULTURE.

Fed batch culture. See FED BATCH CULTURE.

Mixed culture. One containing two or more species or strains of organism.

Old culture. One which has been incubated or stored for an excessive period of time and in which, as a consequence, degenerative changes may have occurred – see e.g. INVOLUTION FORMS.

Open culture. *Syn*. CONTINUOUS CULTURE.

Plate culture ('plate'). One prepared using a solid medium (e.g. nutrient agar) in a PETRI DISH.

Primary culture. See PRIMARY CULTURE.

Pure culture (AXENIC culture). One in which all the organisms are of the same species or strain.

Shake culture. See SHAKE CULTURE.

Slant culture. Slope culture (see below).

Slope culture (slant culture). One prepared using a SLOPE.

Stab culture. One produced by deep inoculation of a solid medium (e.g. the butt of an agar or gelatin slope) with a STRAIGHT WIRE; the wire is pushed vertically into the medium so that the INOCULUM (on the tip of the wire) is distributed along the length of the stab.

Subculture. See SUBCULTURE.

Synchronous culture. See SYNCHRONOUS CULTURE.

(2) (*noun*) The process of preparing a culture (sense 1).

(3) (*verb*) To encourage the growth of particular type(s) of microorganism under controlled conditions. Cultures of bacteria or fungi are usually started by seeding (inoculating) a medium with viable cells or spores from another culture, or by inoculating the medium with material expected to contain particular type(s) of viable organism; the type of MEDIUM and the incubation conditions required differ widely among the different types of microorganism. Certain organisms (e.g. *Rickettsia* spp, *Treponema pallidum*) can be cultured only in living systems. (See also ASEPTIC TECHNIQUE; ENRICHMENT; LOOP.)

culture collections Collections of characterized microorganisms. *General collections*: see e.g. ATCC, CBS, CCM, CIP, DSM, JCM, LMD, and NCTC; *agricultural* (including dairy and phytopathological organisms): see e.g. ICPB, NCDO, NCPPB, NRRL and UWO; *algae*: see e.g. SAG; *fungi* (including yeasts): see e.g. CBS and CMI; *industrial and applied*; see e.g. IAM, IFO and NCIB; *marine bacteria*; NCMB; *medical*: see e.g. CDC.

cultured (of bacteria etc.) Grown.

cultured butter See BUTTER.

cultured buttermilk See BUTTERMILK.

Cummings' disease See AVIAN INFECTIOUS BRONCHITIS.

Cunninghamella See MUCORALES.

Cunninghamellaceae See MUCORALES.

cup fungi Ascomycetes which form cup-like fruiting bodies, particularly members of the PEZIZALES.

cup lichens Species of CLADONIA whose podetia have conspicuous scyphi.

cuprammonium hydroxide See COPPER.

cupredoxin See BLUE PROTEINS.

cupulate Cup-Shaped. (See also AECIDIOID.)

curdlan A class of extracellular, water-insoluble, linear $(1 \rightarrow 3)$-β-glucans formed e.g. by *Alcaligenes faecalis* 'var. *myxogenes*' and by strains of *Agrobacterium* and *Rhizobium*. When aqueous suspensions of curdlans (e.g. 2% w/v) are heated to >90°C and then cooled slowly, they form very elastic, resilient gels.

Curdlans have potential applications in the food industry (e.g. as non-caloric gelling, thickening or stabilizing agents) and in other industries (e.g. in the formation of films and fibres for use e.g. as molecular sieves and/or as supports for immobilized enzymes). [Book ref. 62, pp. 201–229.]

curie (*abbr.* Ci) A unit of activity of a radioactive source equal to 3.7×10^{10} disintegrations per sec.

Curie point pyrolysis PYROLYSIS in which a ferromagnetic wire (coated with the sample) is heated inductively by means of a high-frequency alternating magnetic field; the temperature of the wire increases (within e.g. 0.1 sec) to the Curie point: a temperature which the inducing field maintains – but cannot increase owing to the wire's thermally promoted paramagnetic state. The Curie point depends on the chemical composition of the wire; iron-nickel wires with a Curie point of 510°C are often used.

curing (1) (of meats) A method of FOOD PRESERVATION in which meat (particularly pig meat) is permeated with a solution typically containing NaCl, $NaNO_2$ and $NaNO_3$ at a temperature of e.g. 4°C. Preservation is due mainly to the NaCl (which e.g. lowers the WATER ACTIVITY) and to the nitrite which, under appropriate conditions of e.g. a_w, pH (ca. 5.6) and temperature, inhibits the growth of vegetative bacteria and the germination and/or outgrowth of bacterial endospores. Nitrite also enhances the colour and flavour of cured meats. Brines used for curing hams contain a characteristic microflora which reduces nitrate to nitrite, lowers the pH by producing acid from carbohydrates, and (possibly) produces flavour components. The presence of these bacteria (principally halophilic, psychrotolerant *Vibrio* spp) reduces the spoilage of the cured hams; the organisms may be added to the brine as a starter culture or may derive from the meat itself.

(2) (of plasmids in bacteria) Elimination of a PLASMID without loss of bacterial viability. Sublethal doses of e.g. certain INTERCALATING AGENTS (e.g. ACRIDINES) or NOVOBIOCIN can cause preferential inhibition of plasmid replication; hence, a growing, plasmid-containing bacterial population which is exposed to these agents will give rise to an increasing number of plasmid-free progeny cells. MITOMYCIN C is often used for curing plasmids in *Pseudomonas* spp. Other curing agents include rifampicin (see RIFAMYCINS).

curling factor See GRISEOFULVIN.

Curtis–Fitz-Hugh syndrome A syndrome, observed most commonly in women, which includes acute PERIHEPATITIS in association with genital tract infection; it can be due e.g. to *Neisseria gonorrhoeae* or *Chlamydia trachomatis*. [Perisplenitis and perinephritis in the Curtis–Fitz-Hugh syndrome: Br. J. Surg. (1987) *74* 110–112.]

Curtobacterium A genus of aerobic, asporogenous bacteria (order ACTINOMYCETALES, wall type VIII – see also PEPTIDOGLYCAN) which occur in soil and plant litter; some species are phytopathogenic. In culture the organisms are short rods, coccobacilli or coccoid forms; they are non-motile, or motile with a few lateral flagella. Growth occurs optimally at ca. 25°C on e.g. media containing peptone, yeast extract, and glucose; the organisms are non-cellulolytic. GC%: ca. 66–73. Type species: *C. citreum*. [Re-classification of *Corynebacterium betae*, *C. flaccumfaciens*, *C. oortii* and *C. poinsettiae* (collectively) as *Curtobacterium flaccumfaciens*: JGM (1983) *129* 3545–3548.]

Curvularia See HYPHOMYCETES.

Custers effect (negative Pasteur effect) A phenomenon in which, apparently, the fermentative metabolism of a substrate is stimulated by the presence of O_2. (See e.g. BRETTANOMYCES; cf. PASTEUR EFFECT.)

cut-and-paste mechanism (of transposition) See e.g. Tn*10*.

cutaneous basophil hypersensitivity *Syn.* JONES–MOTE SENSITIVITY.

cutaneous herpes See HERPES SIMPLEX.

cutaneous leishmaniasis LEISHMANIASIS of man and other animals in which the pathogen primarily infects the skin and gives rise to self-healing or metastasizing lesions; the causal agents (which are transmitted by sandflies), and the patterns of disease, vary with geographical location.

Old World cutaneous leishmaniases (mainly in Africa and Asia) are typically caused by *Leishmania aethiopica*, *L. major* or *L. tropica*. In man, a papule develops (after several weeks) at the site of a sandfly bite, becoming necrotic at the centre and forming a moist, ulcerative lesion (*L. major*) or a dry lesion (*L. aethiopica*, *L. tropica*) which eventually heals. (These lesions have been called e.g. *Aleppo boil*, *Delhi boil*, *oriental sore*.) *L. aethiopica* and *L. major* infections are zoonotic, animal reservoirs including populations of e.g. the great gerbil, *Rhombomys opimus* (for *L. major*), and the rock hyrax, *Procavia capensis* (for *L. aethiopica*); *L. tropica* infections appear to be anthroponotic.

New World cutaneous leishmaniases (in South America) are caused by *L. braziliensis* or *L. mexicana*; they are zoonotic diseases in man, animal reservoirs including rodents and sloths. The clinical picture in man may resemble the Old World pattern; however, following an apparent cure the disease may re-appear at a site remote from the original lesion, and in some cases the infection may metastasize to the oral/nasal mucous membranes (see MUCOCUTANEOUS LEISHMANIASIS).

(See also CHICLERO'S EAR and PIAN BOIS.)

Diagnosis of cutaneous leishmaniasis may involve e.g. microscopical examination of exudate from lesions, culture of lesion material (in e.g. NNN MEDIUM), and the MONTENEGRO TEST; treatment may involve e.g. pentavalent ANTIMONY compounds and/or antibiotics.

(cf. VISCERAL LEISHMANIASIS.)

cutaneous T-cell lymphoma See MYCOSIS FUNGOIDES and SEZARY SYNDROME; see also ADULT T-CELL LEUKAEMIA.

cuticle (*mycol.*) *Syn.* PELLIS.

cutin An insoluble polymer which, embedded in waxes, forms the cuticle which covers the epidermal cell walls in the aerial parts of higher plants. (cf. SUBERIN.) Cutin is a polyester formed from C_{16} and C_{18} hydroxy fatty acids. It functions as a physical barrier, protecting against water loss, entry of pathogens, etc. Many plant-pathogenic fungi produce enzymes (*cutinases*) which can breach the cutin barrier prior to infection. [Book ref. 58, pp. 79–100; enzymic penetration of the plant cuticle by fungal pathogens: ARPpath. (1985) *23* 223–250.]

cutinase A CUTIN-degrading enzyme.

cutis (*mycol.*) *Syn.* PELLIS.

Cutleria See PHAEOPHYTA.

cv. CULTIVAR.

CVF COBRA VENOM FACTOR.

C_x cellulase See CELLULASES.

CY agar A medium which includes pancreatic digest of casein, yeast autolysate, and (sometimes) $CaCl_2$.

***cya* gene** See ADENYLATE CYCLASE.

***cyaA* gene** See CYCLOLYSIN.

cyanelles Endosymbiotic CYANOBACTERIA which occur in various eukaryotes: e.g. in the testate amoeba PAULINELLA, in the 'algae' *Cyanophora paradoxa* and *Glaucocystis* spp, and in certain sponges; many contain CARBOXYSOMES. A number of authors consider that cyanelles were the evolutionary precursors of CHLOROPLASTS. (See also GLAUCOPHYTA.)

cyanide (CN^-) (1) (as a RESPIRATORY INHIBITOR) Cyanide binds to and inhibits both the oxidized and reduced forms of CYTOCHROME OXIDASES of the aa_3-type. Cytochrome oxidases of the *o*-type are also generally inhibited, but those of the *d*-type are insensitive to cyanide.

(2) (as a metabolite) Cyanide is formed from glycine by certain microorganisms, e.g. a basidiomycetous fungus [JGM (1980) *116* 9–16] and various bacteria, including *Chromobacterium violaceum* [JGM (1984) *130* 521–525]; at least some of these organisms can degrade cyanide – e.g. to CO_2 or to β-cyanoalanine. A strain of *Pseudomonas fluorescens* has been reported to utilize cyanide as a source of nitrogen for growth, converting cyanide to ammonia and CO_2 in an oxygen-dependent reaction [FEMS (1983) *20* 337–341].

(3) (as an enzyme inhibitor) Cyanide inhibits e.g. CO dehydrogenase (see METHANOGENESIS).

(See also AMYGDALIN and LEACHING.)

cyanide broth See KCN BROTH.

cyanide test (KCN test) A test used to determine the ability of an organism to grow in the presence of cyanide; it is used in the identification of certain species of bacteria, particularly members of the Enterobacteriaceae. A suitable medium, e.g. KCN BROTH, is inoculated with the strain under test, and the medium is examined for growth after 24 and 48 hours; growth constitutes a positive reaction.

Cyanidium See RHODOPHYTA.

cyanobacteria ('blue-green algae') A large and varied group of prokaryotic, photosynthetic organisms which differ from other bacteria in possessing chlorophyll *a* (but not bacteriochlorophylls) and in carrying out oxygenic PHOTOSYNTHESIS. (cf. RHODOSPIRILLALES and ALGAE; see also OXYPHOTOBACTERIA and CYANOBACTERIALES.) The organisms are unicellular (cells occurring singly or in aggregates) or filamentous (usually uniseriate and unbranched, sometimes multiseriate and branched); they may form more or less well-defined and distinctive colonies in nature (see e.g. COELOSPHAERIUM, GLOEOTRICHIA, NOSTOC, TRICHODESMIUM). The essentially Gram-negative-type CELL WALL frequently has features typical of the Gram-positive-type wall [JB (2000) *182* 1191–1199]; one or more layers external to the outer membrane may be present, forming a firm or diffluent mucilaginous sheath (which may be calcified in some species) or a fibrous outer layer (see section II, below). No cyanobacterium has flagella at any stage, but many strains are capable of GLIDING MOTILITY; a novel type of motility has been reported in SYNECHOCOCCUS (q.v.). Within the cell, the photosynthetic pigments (CHLOROPHYLL *a*, CAROTENOIDS and PHYCOBILIPROTEINS) are typically attached to THYLAKOIDS (cf. GLOEOBACTER); the relative proportions of these pigments determine the colour of the organisms (blue-green, olive-green, yellow-green, red, purple, or almost black – depending on organism and environmental conditions: see also CHROMATIC ADAPTATION). Cells may also contain e.g. CARBOXYSOMES, GAS VACUOLES, and granules of CYANOPHYCEAN STARCH, CYANOPHYCIN, (occasionally) POLY-β-HYDROXYBUTYRATE, and/or POLYPHOSPHATE. In many species, cells or filaments (trichomes) may undergo differentiation in response to particular environmental conditions: see e.g. AKINETE, HETEROCYST, HORMOGONIUM.

In general, cyanobacteria are primarily aerobic, oxygenic photoautotrophs which fix CO_2 via the CALVIN CYCLE; during photosynthesis, cells accumulate storage polysaccharide which, during periods of darkness, is catabolized (via the HEXOSE MONOPHOSPHATE PATHWAY and an aerobic respiratory electron transport chain) to provide maintenance energy. *Oscillatoria limnetica* can also catabolize endogenous polysaccharide under anaerobic conditions – either by anaerobic respiration using sulphur as terminal electron acceptor (resulting in sulphide formation) or by a fermentative pathway in which lactic acid is produced. Some cyanobacteria can grow on exogenous organic compounds – either photoheterotrophically or chemoorganotrophically. [Carbon metabolism: Ann. Mic. (1983) *134*B 93–113.]

Certain species (e.g. *Cyanothece halophytica*, *Dactylococcopsis salina*, *Oscillatoria limnetica*) can, in the presence of sulphide, carry out facultative *anoxygenic* PHOTOSYNTHESIS using photosystem I with H_2S as electron donor; in *O. limnetica* the resulting sulphur accumulates extracellularly. NITROGEN FIXATION occurs in many species, usually in specialized cells called HETEROCYSTS (cf. GLOEOTHECE and TRICHODESMIUM): see also NIF GENES. In some species (e.g. *Anabaena variabilis*) diazotrophy can occur in the dark under heterotrophic conditions [energy requirement: JGM (1983) *129* 2633–2640]. [General metabolic aspects: Book ref. 76.]

Cyanobacteria occur in a wide range of habitats, from tropical to polar regions; however, they are rare or absent in habitats of pH < ca. 5. Depending on species, they may be found in fresh, brackish, marine and hypersaline waters, where they may be planktonic, benthic, epiphytic etc (see also BLOOM and OSMOREGULATION); in hot springs; on soils (including hot arid desert soils), muds, sediments and salt-marshes (where they may form extensive and conspicuous mats); on rocks (where they may be endolithic or epilithic); etc. (See also STROMATOLITE.) In many such environments (including rice fields) free-living cyanobacteria may be important providers of fixed nitrogen, and certain nitrogen-fixing cyanobacteria form symbiotic associations with eukaryotic organisms (see e.g. ANABAENA and NOSTOC). Heterocystous cyanobacteria of several genera (*Anabaena*, *Cylindrospermum*, *'Gloeotrichia'*, *Nostoc*) occur in colonies attached to the lower epidermis or in the reproductive pockets in the leaves of floating duckweed plants (Lemnaceae), possibly supplying much of their nitrogen requirement [CJM (1985) *31* 327–330]. (See also CYANELLE; RHIZOSOLENIA; RHOPALODIA.) No species appears to be directly pathogenic in any organism, but some freshwater bloom-forming strains produce potent toxins which may cause illness or death in animals drinking the water (see e.g. ANABAENA, APHANIZOMENON, MICROCYSTIS, NODULARIA; cf. LYNGBYA), and cyanobacterial blooms in reservoirs may cause tainting of the WATER SUPPLIES (see e.g. GEOSMIN). [Toxins and health: RMM (1994) *5* 256–264.] Cyanobacterial toxins (= cyanotoxins) include e.g. (i) neurotoxins such as the anatoxins (see ANABAENA) and SAXITOXIN; (ii) hepatotoxins such as microcystins (produced by strains of MICROCYSTIS and of e.g. *Anabaena*, *Nostoc* and *Oscillatoria*); (iii) dermatotoxins such as aplysiatoxin (produced by strains of LYNGBYA and of e.g. *Oscillatoria*). [Ecological and molecular investigations of cyanotoxin production (review): FEMS Ecol. (2001) *35* 1–9.]

Fossil organisms resembling cyanobacteria are among the oldest fossils known; from the fossil record and other data it is believed that cyanobacteria may have dominated the Earth's biota during the middle to late Precambrian (Proterozoic, ca. 2500–570 million years ago), and were probably responsible for the generation of oxygen in the atmosphere. [Book ref. 76, pp. 543–564.]

The taxonomy of the cyanobacteria is confused. Until quite recently they were regarded as algae and were therefore subject to the Botanical Code of nomenclature. Thus, herbarium specimens, or descriptions and illustrations, were used as type materials (living cultures not being recognized as valid types according to the Botanical Code), and emphasis was given to 'field' characteristics (e.g. colony form, pattern and frequency of false branching, presence, absence and/or consistency of sheath material) in organisms growing in nature. On this basis >150 genera and >1000 species were described. However, many of these 'field' characteristics were found to be indeterminate in culture: e.g., colony forms may be lost, mucilage production and false branching may be variable, etc. Hence, a new scheme for generic assignments, based on properties of the organisms in pure culture, was proposed [JGM (1979) *111* 1–61] and widely accepted. The genera were grouped into five sections:

Section I. Unicellular; cells occur singly or in colonies. Reproduction occurs by equal binary fission (GLOEOBACTER, GLOEOCAPSA, GLOEOTHECE, SYNECHOCOCCUS, SYNECHOCYSTIS) or by budding (CHAMAESIPHON). This section includes organisms formerly of the Chroococcales, with *Chamaesiphon* from the Chamaesiphonales.

Section II. Unicellular; cells always enclosed by a fibrous layer (F layer) external to the outer membrane. Reproduction occurs by multiple fission only (DERMOCARPA, XENOCOCCUS) or by multiple fission and binary fission (CHROOCOCCIDIOPSIS, DERMOCARPELLA, MYXOSARCINA, PLEUROCAPSA GROUP). Multiple fission occurs by rapid repeated binary fissions (unaccompanied by growth) within the F layer, resulting in the formation of BAEOCYTES which are released by rupture of the F layer. In species capable of binary fission, a series of divisions produces an aggregate of vegetative cells cemented together by their F layers; some or all of the cells in the aggregate eventually undergo multiple fission to form baeocytes. Motility, if it occurs, is restricted to baeocytes prior to the development of an F layer. [Developmental patterns: Book ref. 34, pp. 203–226.] This section includes the Chamaesiphonales (except *Chamaesiphon*) and the Pleurocapsales.

Section III. Filamentous; growth occurs by intercalary cell division in one plane only (at 90° to the long axis of the trichome), giving rise to uniseriate, unbranched trichomes (cf. FALSE BRANCHING) composed only of vegetative cells (i.e., heterocysts and akinetes are not formed). The trichomes may be helical (SPIRULINA) or straight (LPP GROUP, OSCILLATORIA, PSEUDANABAENA). This section includes some genera of the former order Nostocales.

Section IV. Filamentous; growth occurs by intercalary cell division in one plane only (at 90° to the long axis of the trichome), giving rise to uniseriate, unbranched trichomes (cf. FALSE BRANCHING); heterocysts are formed in the absence of combined nitrogen, and akinetes are produced by some members. Reproduction occurs by random breakage of trichomes, or by germination of akinetes if formed; hormogonia are produced by CALOTHRIX, NOSTOC and SCYTONEMA, but not by ANABAENA, CYLINDROSPERMUM or NODULARIA. Section IV includes those genera of the Nostocales not included in section III.

Section V. Filamentous; growth occurs by intercalary cell division which may occur in more than one plane, i.e., some cells in the mature trichome may divide in a plane parallel to the long axis of the trichome. Thus, the mature trichome may be partly multiseriate with uniseriate lateral (true) branches (FISCHERELLA), but may readily break up into cell aggregates (CHLOROGLOEOPSIS). Heterocysts are formed in the absence of combined nitrogen; akinetes are formed by some members. Reproduction occurs by random breakage of trichomes, by the formation of hormogonia, and (in some species) by the formation and germination of akinetes. Section V is taxonomically equivalent to the former order Stigonematales.

Many genera have yet to be incorporated in this scheme: see e.g. APHANIZOMENON, COELOSPHAERIUM, DACTYLOCOCCOPSIS, MICROCYSTIS, STARRIA, TRICHODESMIUM.

[Isolation, media etc: Book ref. 45, pp. 212–246.]

Cyanobacteriales An order proposed for the CYANOBACTERIA [IJSB (1978) *28* 1–6] but not accepted as legitimate owing to the lack of a genus named *Cyanobacterium*; however, the proposal of a new genus *Cyanobacterium* [Ann. Mic. (1983) *134*B 21–36] paves the way for legitimization of the order.

Cyanobacterium See SYNECHOCOCCUS and CYANOBACTERIALES.

cyanobiont A cyanobacterial symbiont: see e.g. ANABAENA and NOSTOC. (cf. PHYCOBIONT.)

Cyanobium See SYNECHOCOCCUS.

cyanocobalamin See VITAMIN B_{12}.

Cyanocyta See GLAUCOPHYTA.

cyanogen bromide (CNBr) A reagent used e.g. for the IMMOBILIZATION of a protein on the surface of a support; CNBr binds to the support (forming a 'CNBr-activated' support) and then binds spontaneously to primary amino groups on the protein. The instability of the isourea bond (between the activated support and protein) may lead to some loss of the immobilized protein; this may be avoided e.g. by cross-linking the protein with glutaraldehyde.

cyanogenic Having the capacity to produce cyanide (see also CYANIDE sense 2).

cyanoguanidine Dicyandiamide, a NITRIFICATION INHIBITOR.

cyanomycin An antibiotic produced by '*Streptomyces cyanoflavus*' and reported to be identical to PYOCYANIN [J. Antibiot. (1969) *22* 49–54 cited in JB (1980) *141* 156–163].

cyanophage Any VIRUS whose host is a cyanobacterium ('blue-green alga'). (cf. PHYCOVIRUS.) Cyanophages may be virulent or temperate, and have been isolated from filamentous and unicellular species – including members of the LPP group (cyanophages LPP-1, LPP-2), *Nostoc* (N1, N2), and *Anabaena* (A1, A2); lysis of infected intercalary cells in filamentous species results in progressive fragmentation of the filament. [Ann. Mic. (1983) *134*B 43–59; classification and nomenclature: Intervirol. (1983) *19* 61–66.]

cyanophilic Having an affinity for blue dyes such as LACTOPHENOL COTTON BLUE.

Cyanophora See GLAUCOPHYTA.

cyanophycean starch A glycogen-like storage polysaccharide found in cyanobacteria; the polysaccharide occurs as granules or rods located between the thylakoids.

cyanophycin (multi-L-arginyl-poly(L-aspartic acid); arg-poly(asp)) A high-MWt polymer consisting of equal amounts of L-aspartic acid and L-arginine; the L-aspartic acid residues occur in a linear chain, and each residue is linked via its free carboxyl group to the α-amino group of an arginine residue. Granules of cyanophycin (α granules, 'structured granules') occur in most cyanobacteria. The polymer is synthesized independently of ribosomes and serves primarily as a nitrogen reserve in both vegetative cells and heterocysts. Under CO_2-limiting conditions, cyanophycin may also serve as a source of carbon and energy by an ARGININE DIHYDROLASE pathway, at least in some species. [Review: ARM (1984) *38* 13–16.]

cyanophytes CYANOBACTERIA.

cyanosis (*med., vet.*) A bluish discoloration or darkening of the skin and mucous membranes due to inadequate oxygenation of the blood.

cyanosomes Cyanobacterial PHYCOBILISOMES.

Cyanothece See SYNECHOCOCCUS.

cyanotoxin Any toxin produced by a member of the CYANOBACTERIA (q.v. for examples).

Cyathomonas See CRYPTOPHYTES.

Cyathus A genus of fungi (order NIDULARIALES) in which the basidiocarp is funnel-shaped (ca. 1 cm across), the opening of the funnel being closed by a membrane (*epiphragm*) in the immature fruiting body. Each of the flattened, discoid peridioles is attached to the lower, inner surface of the peridium (i.e., funnel) by means of a complex structure which consists essentially of three components joined end to end. Attached directly to the peridium is the *sheath* (a short cord of aggregated hyphae) to which is joined the *middle piece* (a narrow extension of the sheath); the middle piece is attached to an elongated sac (the *purse*) to the outside of which is attached the peridiole. The purse contains a long thread of coiled hyphae (the *funiculus* or *funicular cord*), one end of which is firmly attached to the peridiole while the other (unattached) end carries a strongly adhesive body (the *hapteron*). When drops of rain fall into the funnel-shaped peridium (or 'splash cup') the peridioles are ejected by the strong upthrust of water on their undersurfaces; each peridiole flies through the air, trailing its funiculus, until the hapteron adheres to e.g. a blade of grass.

cycad symbioses See NOSTOC.

cyclaridine An ANTIVIRAL AGENT – the carbocyclic analogue of vidarabine; its activity resembles that of VIDARABINE in cell cultures, but it is resistant to deamination by adenosine deaminase.

cycles of matter See CARBON CYCLE, NITROGEN CYCLE and SULPHUR CYCLE.

cyclic AMP (cAMP) Adenosine 3′,5′-cyclic monophosphate: a cyclic nucleotide synthesized from ATP by ADENYLATE CYCLASE; cAMP is degraded to AMP by cAMP phosphodiesterase (EC 3.1.4.17).

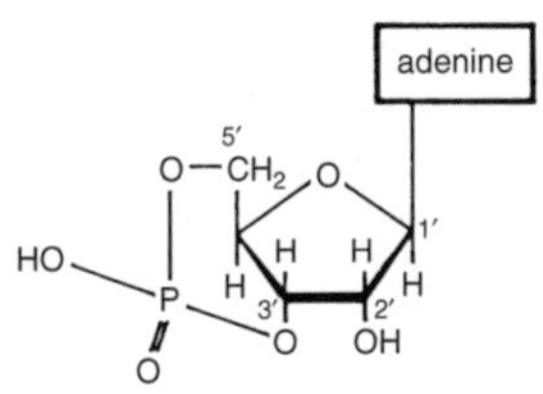

CYCLIC AMP (cAMP)

cAMP is an important regulatory molecule in various types of cell. In prokaryotes, cAMP is involved in CATABOLITE REPRESSION and e.g. in stimulating fruiting in the MYXOBACTERALES.

[cAMP in prokaryotes: MR (1992) *56* 100–122.]

In eukaryotes, cAMP is involved in various regulatory and developmental processes (see e.g. DICTYOSTELIUM); it apparently functions mainly by regulating the activity of PROTEIN KINASES. cAMP-dependent protein kinase A (PKA) consist of two types of subunit: the catalytic (C) and regulatory (R) subunits; the inactive form of PKA consists of a dimer of R subunits to which are bound two C subunits. cAMP activates PKA by binding to the R subunits; this causes the release of both C subunits – which (separately) are then able to carry out their catalytic roles. The active C subunits have a wide range of functions in both cytoplasm and nucleus – including e.g. regulation of certain transcriptional events in the nucleus.

cyclic octadepsipeptide antibiotics QUINOXALINE ANTIBIOTICS.

cyclical transmission A mode of transmission of a parasite by a VECTOR (sense 1) in which the parasite undergoes one or more essential stages of its life cycle in the vector. (See e.g. PLASMODIUM and TRYPANOSOMA.)

Cyclidium A genus of ciliates (order SCUTICOCILIATIDA) which occur e.g. in soil, fresh water (including hot springs) and marine habitats. Cells: typically ovoid and generally small, ca. 15–40 μm; the cell surface is typically not densely ciliated, but those cilia which are present tend to be long, and there is often one or more longer caudal cilia.

cyclin A type of protein involved in the regulation of the eukaryotic CELL CYCLE. During the cycle, molecules of cyclin accumulate intracellularly at specific stages and form complexes with certain PROTEIN KINASES. Phosphorylation of the kinase part of the complex – and subsequent (partial) dephosphorylation at an appropriate point in the cycle – leads to activation of the kinase. The activated kinase phosphorylates certain proteins which then promote onward cycling; it can also bring about UBIQUITIN-dependent degradation of the cyclin – so that the intracellular concentration of cyclin falls sharply on passage through the checkpoint.

There are various types of cyclin, and different types may form complexes with protein kinases at different cell-cycle checkpoints; for example, different types of cyclin form complexes with the yeast protein kinase Cdc2 in order to mediate passage through (a) the *start* checkpoint, and (b) the $G_2 \rightarrow M$ checkpoint.

cyclitol antibiotics ANTIBIOTICS which contain a cyclic alcohol – e.g. AMINOGLYCOSIDE ANTIBIOTICS (which contain a substituted INOSITOL).

cycloalkanes See HYDROCARBONS.

cycloamyloses *Syn.* SCHARDINGER DEXTRINS.

cyclochlorotine *Syn.* CHLOROPEPTIDE.

cyclodextrins *Syn.* SCHARDINGER DEXTRINS.

cycloguanil See CHLORGUANIDE.

cyclohexane metabolism See HYDROCARBONS.

cycloheximide (Actidione; β-[2-(3,5-dimethyl-2-oxocyclohexyl)-2-hydroxyethyl]-glutarimide) An ANTIBIOTIC produced by certain strains of e.g. *Streptomyces griseus*; it is a by-product of streptomycin manufacture. Cycloheximide is active against many fungi and other eukaryotes (bacteria are not affected), acting primarily by inhibiting PROTEIN SYNTHESIS; e.g., it prevents translocation by binding to the 60S subunit of 80S ribosomes. Ribosomes from different organisms may differ in sensitivity. Only the yeast-like forms of certain dimorphic fungi (e.g. *Blastomyces dermatitidis*, *Histoplasma capsulatum*) are susceptible to cycloheximide.

Certain stereoisomers of cycloheximide (e.g. naramycin B) have been isolated from *Streptomyces* cultures; these are generally less active than cycloheximide against fungi. *Streptovitacins* are monohydroxyl-substituted cycloheximides produced by *S. griseus*; streptovitacin A appears to resemble cycloheximide in its mode of action.

cyclolysin An RTX TOXIN produced by species of *Bordetella*. The ~177 kDa protein is encoded by gene *cyaA*; activity of the toxin depends on post-translational modification in which a long-chain fatty acid is linked covalently at a specific lysine residue. The toxin is secreted by an ABC EXPORTER.

Cyclolysin has ADENYLATE CYCLASE, haemolysin and pore-forming activity. A range of eukaryotic cells are susceptible; within the target cell cyclolysin is activated by CALMODULIN.

The principal role of cyclolysin in WHOOPING COUGH may involve inhibition of the activity of phagocytic cells (e.g. macrophages) by raising their intracellular levels of cAMP. Moreover, by suppressing anti-bacterial activity in phagocytes, the pathogen may be able to establish an intracellular carrier state. The raised levels of cAMP in epithelial cells may account for the secretion of fluid/mucus.

cyclo-oxygenase See PROSTAGLANDINS.

cycloparaffins See HYDROCARBONS.

cyclophosphamide See IMMUNOSUPPRESSION.

cyclopiazonic acid A MYCOTOXIN produced by species of *Aspergillus* and *Penicillium* (e.g. *A. flavus*, *A. oryzae*, *P. verrucosum* ('*P. cyclopium*') and *P. griseofulvum* ('*P. patulum*')). It is toxic e.g. for chickens [AEM (1983) *46* 698–703], rats and calves.

D-cycloserine (D-4-amino-3-isoxazolidone) A broad-spectrum ANTIBIOTIC obtained from strains of *Streptomyces* spp or synthesized chemically. It acts as an analogue of D-alanine, competitively inhibiting the two enzymes (alanine racemase and D-alanyl-D-alanine synthetase) involved in the biosynthesis of PEPTIDOGLYCAN. D-Cycloserine is actively taken up by the D-alanine/glycine transport system of the cell, and a mutation which alters this transport system can result in resistance to the antibiotic. (*Methanococcus vannielii* lacks peptidoglycan but is nevertheless susceptible to D-cycloserine [Book ref. 157, p. 529].)

cyclosis *Syn.* CYTOPLASMIC STREAMING.

Cyclospora A genus of protozoa (suborder EIMERIORINA) which form disporic, dizoic oocysts. [*C. cayetanensis* as a diarrhoeal agent: RMM (1996) *7* 143–150.]

cyclosporin A A hydrophobic peptide obtained from certain hyphomycetes (*Cylindrocarpon lucidum*, *Tolypocladium inflatum*); it is a powerful immunosuppressive drug (see IMMUNOSUPPRESSION) which appears to act mainly on T cells.

Cyclotella See DIATOMS.

***Cydia pomonella* GV** See GRANULOSIS VIRUSES.

Cylindrocarpon A genus of fungi (class HYPHOMYCETES) which include organisms previously classified as species of *Fusidium*. Some species have teleomorphs in the genus *Nectria*. (See also CYCLOSPORIN A and FUSIDIC ACID.)

Cylindrocystis A genus of saccoderm DESMIDS.

Cylindrogloea bacterifera See CHLOROCHROMATIUM AGGREGATUM.

Cylindrospermum A genus of filamentous CYANOBACTERIA (section IV) in which the heterocysts occur only at the ends of the trichomes; akinetes are always adjacent to heterocysts. Hormogonia are formed by fragmentation of whole trichomes. GC%: 42–47.

Cylindrosporium See MELANCONIALES.

Cylindrotheca See DIATOMS.

Cymbella See DIATOMS.

***Cymbidium* mosiac virus** See POTEXVIRUSES.

***Cymbidium* ringspot virus** See TOMBUSVIRUSES.

Cymbomonas See MICROMONADOPHYCEAE.

Cyniclomyces A genus of fungi (family SACCHAROMYCETACEAE) which form budding yeast cells and pseudomycelium; elevated levels of CO_2 are needed for growth. One species, *C. guttulatus*, isolated from the faeces and stomach of rabbits. [Book ref. 100, pp. 125–129.]

Cyphelium See CALICIALES.

cyphella (*lichenol.*) A round depression or pore – visible as a small white pit ca. 0.5–2.0 mm diam. – in the lower surface of the thallus in *Sticta* spp. The lower cortex forms a protruding rim around the edge of the cyphella, a medullary layer of rounded fungal cells lining the depression. Cyphellae are believed to facilitate gas exchange. [New Phyt. (1981) *88* 421–426]. (cf. PSEUDOCYPHELLA.)

Cypovirus *Syn.* CYTOPLASMIC POLYHEDROSIS VIRUS GROUP.

cyprofuram See PHENYLAMIDE ANTIFUNGAL AGENTS.

Cyrtophorida See HYPOSTOMATIA.

cyrtos (cytopharyngeal basket; nasse; pharyngeal basket) A (frequently curved) type of CYTOPHARYNGEAL APPARATUS whose walls are strengthened by longitudinal nematodesmata (which arise at apical kinetosomes) and are lined with extensions of POSTCILIARY MICROTUBULES; toxicysts are absent. The cyrtos is characteristic of the HYPOSTOMATIA. (cf. RHABDOS.)

cyst A specialized microbial cell produced either in response to adverse environmental conditions or as a normal part of the life cycle. During cyst formation (*encystment*), an organism produces a thick or thin wall within which it becomes totally enclosed. Cysts are usually resistant to desiccation, and may be resistant to e.g. ultraviolet radiation and/or heat.

(a) (*bacteriol.*) Cyst formation is uncommon among bacteria; most studies on bacterial cyst formation have been carried out on species of AZOTOBACTER, particularly *A. vinelandii*.

In *Azotobacter* spp, encystment can be promoted in vitro by the provision of substrates such as β-hydroxybutyrate, and it has been generally supposed that the intracellular accumulation of PHB is a necessary pre-requisite for cyst formation; however, it has been proposed that the stimulus for encystment may be related to the extracellular levels of both carbon and nitrogen sources [SBB (1986) *18* 23–28]. During encystment, cells lose their flagella and become spheroidal; these heavily encapsulated cells ('precysts') typically accumulate PHB. Membranous blebs develop at the cell surface, and these subsequently detach and coalesce – forming the fragmented outer layer (*exine*) of the cyst wall. Later, the cyst wall develops an electron-transparent inner layer (*intine*). The cyst wall contains ALGINATE together with protein and lipid; the exine is particularly rich in polyguluronic acid (and Ca^{2+}), while the intine is richer in polymannuronic acid. *Azotobacter* cysts are metabolically dormant, and they can remain viable in dry soil for many years; they are more resistant than the vegetative cells to desiccation, sonication and ultraviolet radiation, but they are not significantly more resistant to heat.

Azotobacter-like 'lipid cysts' are formed by some members of the Methylococcaceae.

(See also MICROCYST.)

(b) (*protozool.*) Cysts are formed e.g. by some amoebae, ciliates and phytoflagellates. During encystment, structures such as cilia or flagella are lost or resorbed, and considerable reorganization of internal structures may occur; the cell becomes enclosed by a thin or thick, commonly multilayered wall, the outer-most layer(s) being termed the *ectocyst* or *exocyst*, the innermost layer(s) the *endocyst*, and intermediate layer(s) the *mesocyst*. Cyst walls vary in composition, according to species; they commonly contain a high proportion of protein, while e.g. cellulose occurs in the endocyst of *Acanthamoeba* spp, chitin apparently occurs in *Entamoeba* cysts [BBRC (1982) *108* 815–821], and sulphated glycosaminoglycans occur in the mesocyst in *Histriculus similis* [JGM (1983) *129* 829–832]. Cysts of CHRYSOPHYTES have silica walls. (In certain loricate or testate organisms – e.g. *Euglypha* spp – a cyst is formed when the organism withdraws into its lorica/test and plugs the aperture.) *Excystment* (the release of one or more vegetative cells from the cyst) may involve rupture of the cyst wall (possibly as a result of the intake of water) and/or enzymic dissolution of the wall; in e.g. *Naegleria gruberi* the cell leaves the cyst via a pore which becomes unplugged, while in *Acanthamoeba* a lid-like operculum is removed from an exit aperture in the cyst wall.

In many protozoa the cyst serves a protective function, allowing the organism to survive e.g. the absence of food, desiccation, and/or unfavourable temperatures. Such 'resistant cysts' are generally dormant and may remain viable for years under appropriate conditions. Many such cysts (e.g. those of some euglenoid flagellates) are highly resistant to desiccation; however, those of e.g. *Didinium* and *Euplotes* cannot withstand drying, but can remain viable in water for long periods. Resistant cysts may allow dissemination of the organism e.g. by wind or by birds or animals. In parasitic protozoa – e.g. ENTAMOEBA, coccidia, GIARDIA – cysts are the form in which the parasite is transmitted from one host to another.

In some protozoa *reproductive cysts* are formed as a normal part of the life cycle; these may not be more resistant than the vegetative cell to adverse environmental conditions. For example, *Colpoda* spp form thin-walled cysts within which asexual fission occurs, and can also form thick-walled resistant cysts under adverse conditions. (See also e.g. ICHTHYOPHTHIRIUS.) Some protozoal cysts may serve both reproductive and protective (and disseminative) functions (e.g. the oocysts of the EIMERIORINA). (See also ACTINOPHRYS and ACTINOSPHAERIA.)

cystathione See e.g. Appendix IV(d).

cysteamine 2-Mercaptoethylamine (= β-aminoethanthiol): see e.g. PANTOTHENIC ACID.

L-cysteine biosynthesis See Appendix IV(c).

cysteine protease (1) See APOPTOSIS. (2) *Syn.* thiol PROTEASE.

cystibiotics Class IIa BACTERIOCINS (q.v.).

cystic fibrosis (CF) An inheritable disease [Review: Lancet (1998) *351* 277–282] usually involving defective transmembrane transport of chloride by an ABC TRANSPORTER.

Typically, the lungs are congested with a viscid dehydrated mucus (containing e.g. raised levels of calcium and magnesium ions) which may become infected with organisms such as *Burkholderia* (formerly *Pseudomonas*) *cepacia* [review: RMM (1995) *6* 1–16], mycobacteria and *Pseudomonas aeruginosa*; such infections often persist. Patients with CF are frequently given prolonged antibiotic therapy, but this may not prevent colonization of the lungs with e.g. mucoid (ALGINATE-forming) strains of *Pseudomonas aeruginosa* [mucoidy of *P. aeruginosa* in CF: JB (1996) *178* 4997–5004]; this is associated with a poor prognosis because the alginate may protect the bacteria from phagocytosis, pulmonary surfactant and antibiotics. It has been reported that high levels of salt (NaCl) on CF airway epithelia may permit bacterial colonization by inhibiting anti-bacterial activity normally associated with this habitat [Cell (1996) *85* 229–236]. (See also QUORUM SENSING.)

Results of studies on the molecular epidemiology of *P. aeruginosa* in a CF outpatient clinic indicate that cross-infection is not common within the clinic, and that acquisition of infection from a common source is unlikely [JMM (2001) *50* 261–267].

It was suggested that long-term colonization of the lungs of CF patients with *Stenotrophomonas maltophila* (formerly *Xanthomonas maltophila*) may be a significant factor in prognosis [RMM (1997) *8* 15–19]. This environmentally common, Gram-negative, non-glucose-fermenting bacillus [description: IJSB (1993) *43* 606–609] is now frequently isolated from the respiratory specimens of CF patients (and is also associated with disease in immunocompromised individuals, cancer patients and recipients of transplants). [Identification/detection of *S. maltophila* by a PCR-based approach: JCM (2000) *38* 4305–4309.]

[Genetic basis, treatment etc. of CF: Book ref. 214.]

cystidium (*mycol.*) A large, elongated, sterile cell which occurs among the basidia in the hymenia of certain basidiomycetes. Cystidia vary greatly in size and form, but they usually project some way beyond the tips of the basidia. Their function is largely unknown; however, those which arise from the surfaces of the lamellae of certain species of *Coprinus* appear to function as 'spacers' – maintaining a small but significant distance between adjacent lamellae, thus facilitating spore dispersal.

cystine–lactose–electrolyte-deficient medium See CLED MEDIUM.

cystine–tellurite–blood agar See TELLURITE MEDIA.

cystitis Inflammation of the urinary bladder. Symptoms: frequent, painful micturition, sometimes with haematuria, fever etc. Cystitis may result from the spread of an infection upwards from the urethra or downwards from the kidney. The former is more common, with *Escherichia coli* (see UPEC) and *Proteus* spp being the most common causal agents; other causal agents include enterococci and *Staphylococcus saprophyticus*. Treatment may involve e.g. copious fluid intake and therapy with broad-spectrum antibiotics.

Certain adenoviruses (e.g. type 11) have been associated with haemorrhagic cystitis (mainly in children).

(See also URINARY TRACT INFECTION.)

Cystobacter See MYXOBACTERALES.

Cystodinium A genus of DINOFLAGELLATES in which the vegetative cells are unicellular and non-motile, and reproduce by the formation of zoospores.

Cystomyces See UREDINIOMYCETES.

Cystoseira See PHAEOPHYTA.

cystosorus A cluster of cysts: see e.g. PLASMODIOPHOROMYCETES.

Cystotheca See ERYSIPHALES.

Cystoviridae (φ6 phage group) A family of BACTERIOPHAGES containing one genus (*Cystovirus*) and one member: BACTERIOPHAGE φ6.

Cystovirus See CYSTOVIRIDAE.

cystozoite (bradyzoite) In certain coccidia: a stage within host tissues (see e.g. TOXOPLASMOSIS).

cytarabine (1-β-D-arabinofuranosylcytosine; ara-C; cytosine arabinoside) An ARABINOSYL NUCLEOSIDE originally investigated as a possible systemic ANTIVIRAL AGENT for disseminated herpesvirus infections; toxicity has limited its clinical use.

cytidine See NUCLEOSIDE and Appendix V(b).

cytidine deaminase See RNA EDITING.

cytoadherence See MALARIA.

cytoadhesin subfamily A category within the INTEGRIN family which includes CD41b/CD61.

cytobiosis SYMBIOSIS in which one symbiont occurs within the cells of the other.

cytochalasins A family of fungal secondary metabolites (POLYKETIDE derivatives) – cytochalasins A, B, C etc. – which are formed e.g. by species of *Aspergillus*, *Helminthosporium* and *Phomopsis*. Cytochalasins bind to one end of an ACTIN microfilament (the fast-growing or 'barbed' end) and inhibit the addition of further monomers – thus inhibiting e.g. phagocytosis, amoeboid movement, the intracellular development of certain viruses, and other microfilament-dependent functions. (Cytochalasin B apparently slows, but does not prevent, the addition of actin monomers [JCB (1986) *102* 282–288].) Cytochalasins A and E enhance branching in certain fungi (see GROWTH (fungal)).

cytochrome See CYTOCHROMES.

cytochrome oxidase In an ELECTRON TRANSPORT CHAIN a (terminal) cytochrome which transfers electrons to molecular oxygen, reducing it to water. Cytochrome oxidases include cyts aa_3, a_1, d and o (see CYTOCHROMES). Bacteria typically have more than one type of cytochrome oxidase. The electron transport chain in the animal MITOCHONDRION appears to have only one type of cytochrome oxidase, cyt aa_3, which resembles the bacterial cyt aa_3 in containing two haem *a* molecules and two copper atoms; in at least some plant mitochondria there is an additional CYANIDE-resistant cytochrome oxidase. (See also RESPIRATORY INHIBITORS.)

cytochrome oxidase test See OXIDASE TEST.

cytochromes A class of *haemoproteins* in which the HAEM (sense 1) is usually linked to the protein via the 5th and 6th coordinate positions of the haem iron; cytochromes participate in many types of electron transfer reaction (see e.g. ELECTRON TRANSPORT CHAIN and PHOTOSYNTHESIS), such reactions involving the alternate oxidation and reduction of the haem iron. The $E_{m.7}$ (see REDOX POTENTIAL) of most cytochromes falls within the approximate range −100 to +500 mV.

Cytochrome nomenclature. Cytochromes are classified primarily by the nature of their haem group(s); individual cytochromes within a given class may be distinguished by subscripts (e.g. c_1, c_2, c_3 etc) or – see later – they may be designated by certain of their absorption characteristics. Cytochromes *a* (cyts *a*) contain haem *a* (which has a formyl group at the C-8 position, and a long hydrophobic chain at the C-2 position of the PORPHYRIN). Cytochromes *b* contain haem *b* (= *protohaem*: see HAEM). Cytochromes *c* contain haem *c* (a *mesohaem*), the haem being linked to the protein via thioether bonds (between the ethyl substituents of the haem and cysteine residues in the protein) as well as via the coordinate bonds. Some *c*-type cytochromes (e.g. cyts c_2, c_{555}) have a single haem attached near the N-terminal of the protein, while others have a single haem attached near the C-terminal; the latter, which include the c' cytochromes (= *cytochromoids*), differ chemically and spectroscopically from other *c*-type cytochromes. Cytochromes *d* contain a CHLORIN.

Certain cytochromes have more than one haem group, while others may contain e.g. a flavin group.

Some cytochromes are often referred to by their original names, even though such names do not reflect current classification; thus e.g. cyt a_2 is now cyt *d*, and cyt *o* is a *b*-type cytochrome.

Identification of cytochromes. Cytochromes may be identified/quantified by various methods, often e.g. by spectrophotometry – in which a given type of cytochrome may be detected, in situ, by its characteristic absorption pattern in the visible regions of the spectrum. Commonly a *difference spectrum* is determined: e.g., the absorption spectrum of a given cytochrome in the reduced state *minus* its absorption spectrum in the oxidized state (cf. CO DIFFERENCE SPECTRUM); such a spectrum often exhibits a maximum absorption peak in each of three bands: the α-band (ca. 540–650 nm), the β-band (ca. 510–530 nm) and the γ-band (= *Soret band*) (ca. 400–450 nm). In some cases a cytochrome may be designated by its absorption peak (in the *reduced* state) within the α-band – e.g., cyt *b*-557 (= cyt b_{557}).

Spectroscopy can sometimes lead to an incorrect *in situ* identification of a cytochrome [see e.g. JGM (1984) *130* 3055–3058]. In some cases it may be necessary to extract the haem, bind to it (coordinately) two pyridine molecules per molecule, and examine by spectroscopy the resulting *dipyridine ferrohaemochrome* (= *pyridine haemochrome*) – reduced e.g. with sodium dithionite – in an alkaline pyridine solution. For each class of cytochromes, the corresponding dipyridine ferrohaemochrome α-absorption peak falls within a narrow waveband: cyts *a* 580–590 nm; cyts *b* 556–558 nm; cyts *c* 549–551 nm; cyts *d* 600–620 nm. [Analysis of cytochromes: Book ref. 138, pp. 285–328.]

Cytochrome–ligand binding. See e.g. AZIDE, CARBON MONOXIDE and CYANIDE.

Cytochromes in eukaryotes. Cytochromes occur e.g. in the mitochondrial ELECTRON TRANSPORT CHAIN (q.v.) and in photosynthetic systems (see PHOTOSYNTHESIS). Many cells also contain e.g. certain *b*-type cytochromes – such as cyt b_5 and a cyt P-450 – associated with the endoplasmic reticulum. The mammalian cyt P-450 – the cytochrome–carbon monoxide complex absorbs at 450 nm – occurs in liver cells and mediates in hydroxylation reactions involving sterols and other lipid-soluble substances; an analogous cyt P-450 occurs e.g. in many fungi (see e.g. AZOLE ANTIFUNGAL AGENTS and HYDROCARBONS). Cytochromes appear to be absent in e.g. certain parasitic protozoa – including bloodstream forms of trypanosomes and intracellular stages of *Leishmania donovani*.

Cytochromes in prokaryotes. Cytochromes occur in both aerobic and anaerobic respiratory chains in the CYTOPLASMIC MEMBRANE; they also occur in the cytoplasm and/or periplasmic region in some cells, and appear to occur in certain photosynthetic REACTION CENTRES. In bacteria, cytochromes vary – qualitatively and quantitatively – from one species to another, and even in a given species under different growth conditions. [Cytochromes in *Escherichia coli* under different growth conditions: JGM (1982) *128* 1685–1696.] Various cytochromes may act as CYTOCHROME OXIDASES, and it is usual for a given bacterium to have more than one type of cytochrome oxidase.

Generalized patterns of cytochromes occur in the various categories of bacteria. Gram-positive aerobic chemoheterotrophs typically have a pattern of the $bcaa_3o$ type, i.e., including two types of cytochrome oxidase; exceptions include e.g. *Brochothrix thermosphacta* which has only one cytochrome oxidase (aa_3). Gram-negative chemoheterotrophs (including e.g. enterobacteria and pseudomonads) are, by comparison, more heterogeneous in their cytochrome patterns under aerobic conditions; typically, cyt aa_3 is not formed (though it is in e.g. some pseudomonads, many methylotrophs, and *Rhizobium* spp), the cytochrome pattern frequently being of the $bcoa_1d$ type, but often without cyt *c*. Cyts *a* and *d* are not formed e.g. by *Pseudomonas fluorescens*, or by certain other pseudomonads, so that these organisms may have patterns of the *bco* or *bo* type. Cytochrome oxidases are often absent or greatly decreased during anaerobic respiration. In at least some pseudomonads (and e.g. *Thiobacillus denitrificans*) which carry out DENITRIFICATION, cyt d_1c (= cd_1) acts as a nitrite reductase. Chemolithotrophs often have cytochrome patterns of the $bcaa_3o$ type. All members of the RHODOSPIRILLALES appear to contain *c*-type cytochromes, and most or all appear to have *b*-type cytochromes.

Specialized cytochromes or cytochrome patterns occur in some bacteria; thus, e.g. *Pseudomonas putida* contains a soluble cyt P-450 which, together with an iron–sulphur protein and an FAD-containing reductase, effects hydroxylation of e.g. camphor in a cyclic reaction sequence. [Structure and chemistry of cyt P-450 from *P. putida*: Book ref. 146, pp. 157–206.] (See also METHYLOTROPHY and RHP.)

Cytochromes appear to be absent from some facultative and obligate anaerobes.

[Structure, function and evolution of cytochromes (review): Prog. Biophys. Mol. Biol. (1985) *45* 1–56.]

(See also BENZIDINE TEST and OXIDASE TEST.)

cytochromoids See CYTOCHROMES.

cytocidal (*adj.*) Able to kill cells.

cytoductant See CYTODUCTION.

cytoduction A cross in which two cells undergo cytoplasmic fusion but do not undergo nuclear fusion (karyogamy) owing to a chromosomal mutation; each haploid progeny cell (*cytoductant*, *heteroplasmon*) contains the nucleus of one parent cell but may contain cytoplasmic elements from both parent cells.

cytofluorometry FLOW CYTOMETRY which involves the detection of specific fluorescence or FLUOROCHROME markers.

cytogamy (selfing) (*ciliate protozool.*) The occurrence of AUTOGAMY (instead of conjugation) in each of two ciliates which have paired.

cytohet (cytoplasmic heterozygote) A *eukaryotic* cell which is HETEROZYGOUS for one or more CYTOPLASMIC GENES.

cytokines In the human and animal body: a heterogeneous population of (glyco)proteins which form a dynamic network of intercellular messenger molecules that regulate various aspects of physiology, including the immune response to infection. (This describes, but does not define, cytokines; workers in the field have not yet indicated the essential difference(s) between those molecules which are currently regarded as cytokines and certain other regulatory molecules – thus precluding a formal definition at the present time.) Cytokines may be distinguished from the (protein) hormones in that (i) a given cytokine may be synthesized by different types of cell and/or may act on different types of cell; (ii) cytokines typically act on target cells near the source of cytokine – although if secreted into the circulatory system they may act on distant cells; (iii) cytokines may cause diverse effects (e.g. when acting on different cells under different conditions); (iv) different types of cytokine may give rise to the same physiological effect; and (v) some cytokines can act as mitogens.

Two sources [Book refs 218 and 226] recognize the following as cytokines: COLONY-STIMULATING FACTORS, INTERFERONS, INTERLEUKINS, cytotoxic agents such as tumour necrosis factor (see TNF), growth factors and CHEMOKINES.

Cytokines are synthesized mainly by leukocytes (white blood cells); some cytokines are synthesized by stationary cells (e.g. endothelial cells). Although most cytokines are soluble (secreted) products, some are, or can be, membrane-associated.

In general, transcription of cytokine-encoding genes is inducible by appropriate exogenous or endogenous stimuli. An example of an exogenous stimulus is the binding of lipopolysaccharide to CD14 receptors on macrophages (see CD14); endogenous stimuli can arise e.g. during viral infection.

On release, cytokines bind to specific receptors on target cells.

Cytokine receptors are divided into a number of families which differ e.g. in structure. (Some of the receptors found on macrophages and T lymphocytes can act as receptors for human immunodeficiency virus (HIV). Cytokine receptors have also been reported to act as binding sites for certain other viruses, including human (alpha) herpesvirus 1.) Interestingly, receptor molecules can be shed from cells; such isolated receptors can e.g. bind to (and antagonize) the corresponding cytokine, or they can bind to other cells – which may then be stimulated by the given cytokine.

The binding of a cytokine to its cognate receptor initiates an intracellular signal, the nature of which depends e.g. on the type of cell and cytokine and on the environmental and intracellular conditions under which binding takes place. Cytokine–receptor binding characteristically results (either directly or indirectly) in the activation of kinase(s) at certain stage(s) within the signalling pathway. In one intracellular signalling pathway (triggered by many of the interleukins), the binding of cytokine leads to activation of a tyrosine kinase of the JAK (Janus kinase) family which, when activated, transmits the signal by phosphorylating a molecule of the so-called 'signal transducers and activators of transcription' (STATs); STATs exist in different forms, and a given type of STAT relays the signal for only some type(s) of cytokine – thus permitting different cytokines to initiate different signals in the same cell. Phosphorylated STATs form homo- or heterodimers that enter the nucleus and promote transcription of particular gene(s). [STATs: TIBS (2000) *25* 496–502.]

As mentioned, a given cytokine may cause diverse effects. For example, the binding of tumour necrosis factor (TNF) to its receptor can result in activation of caspases (and APOPTOSIS) – or e.g. transcription of specific genes through activation of a major transcription factor: nuclear factor-κB (NF-κB). In the latter pathway, the binding of TNF to its receptor promotes phosphorylation (and consequent degradation) of a certain protein (IκBα) which, by binding to NF-κB, inactivates it; that is, phosphorylation (degradation) of IκBα results in the activation of NF-κB. NF-κB is important e.g. in the development of an inflammatory response; it promotes transcription of a range of genes – including those encoding TNF-α, IL-1β and IL-8.

In general, the binding of cytokines to their receptors may initiate cellular responses that range from secretion (of cytokines etc.), differentiation or proliferation to chemotaxis or apoptosis; currently, many of the signalling pathways in eukaryotic cells are incompletely understood. [Signalling mechanisms in prokaryotes and eukaryotes: Book ref. 218, pp 89–162; Book ref. 226, pp 111–140.]

Note regarding nomenclature of cytokines. The name of a cytokine does not necessarily correlate with its primary function(s) in vivo. For example, tumour necrosis factor was initially identified as an anti-tumour agent but is now known to be a central mediator in host defence and inflammation. Again, some interleukins, initially thought to mediate only leukocyte–leukocyte interactions, are now known to involve other types of cell; thus, e.g. interleukin-8 is secreted by endothelial cells and is classified as a chemokine.

Cytokines as factors in health and (infectious) diseases. Various roles have been attributed to cytokines in normal physiology and development. However, the contribution of individual cytokines in vivo has been difficult to establish experimentally – not least because of the complexity of the system and the existence of biochemical redundancy among cytokines; hence, in some cases, the function(s) of a cytokine have been inferred from the effects attributed to inherited deficiencies in the synthesis or activity of that cytokine (and/or its receptor). In other cases, the role(s) of cytokines have been inferred from studies on knockout mice in which genes of particular cytokine(s) have been rendered non-functional.

In diseases of microbial aetiology, cytokines typically have protective roles – e.g. in processes such as ANTIBODY FORMATION and INFLAMMATION. However, dysregulation of cytokines (enhanced production, imbalance, inhibition) may be a major factor in pathogenesis.

In some cases, the severity of, or susceptibility to, disease has been found to vary in different individuals according to the nature of the TNF-α gene promoter – particular polymorphisms in the promoter region of the gene being associated with enhanced production of the cytokine; this has been reported in cerebral MALARIA [Nature (1994) *371* 508–510] and in ENDOTOXIC SHOCK [JAMA (1999) *282* 561–568]. Again, polymorphisms in the gene encoding IL-1β have been associated with an increased risk of gastric cancer from *Helicobacter pylori* infection [Nature (2000) *404* 398–402].

In pyelonephritis caused by *Escherichia coli*, the cytokines IL-6 and IL-8 appear to be prominent – levels of IL-6 correlating with the severity of disease [e.g. PNAS (2000) *97* 8829–8835 (8833–8834)].

During infection with *Yersinia*, the pathogen's secreted YopP/YopJ (see VIRULON) down-regulates the pro-inflammatory response – e.g. inhibiting the formation of both TNF-α and IL-8; the mechanism involves inhibition of the kinase that phosphorylates (and inactivates) IκBα, thus inhibiting transcription of NF-κB-dependent genes [PNAS (2000) *97* 8778–8783 (8781–8782)].

Viruses may inhibit cytokines in various ways; for example, *soluble* forms of TNF and IL-8 receptors are encoded by Shope fibroma virus and CMV, respectively. (See also INTERLEUKIN-18.)

(See also DISSEMINATED INTRAVASCULAR COAGULATION; DNA VACCINE; HAEMOPHAGOCYTIC SYNDROME; JARISCH–HERXHEIMER REACTION; LEPROSY; MODULIN; NITRIC OXIDE; P FIMBRIAE; SUPERANTIGEN; TOXIC SHOCK SYNDROME.)

Many infections elicit an immunological response in which a particular subset of T LYMPHOCYTES – either Th1 or Th2 – is dominant. For example, Th1-type responses are typical in certain bacterial and protozoan infections, while Th2-type responses are common e.g. when infection involves helminths [IT (1996) *17* 138–146]; moreover, in at least some cases, an 'inappropriate' response (e.g. Th2 instead of Th1) is associated with increased susceptibility to the infection.

Naive CD4$^+$ T cells (i.e. those not previously exposed to antigen) may develop as either Th1 or Th2 cells on exposure to antigen; the mechanism that selects one or other subset of T cells is unknown, but the decision to develop one way or the other is influenced by the microenvironment of cytokines which are present during activation of the T cell by antigen – e.g. IL-12 promotes the Th1 response, while IL-4 promotes differentiation to Th2 cells. (A newly reported cytokine receptor, designated TCCR, appears to be necessary for the Th1-type immune response in mice [Nature (2000) *407* 916–920].) Interestingly, glucocorticoids from stress metabolism may suppress IL-12 (the main inducer of the Th1-type response); that is, stress-derived glucocorticoids may shift the balance of the immune response from Th1-type toward Th2-type [BCEM (1999) *13* 583–595].

One important difference between Th1- and Th2-type responses is that Th1 and Th2 cells secrete different types of cytokine (see table) and so have correspondingly dissimilar physiological roles. (In this context, it is interesting to note that when the molecule ICAM-1 was co-expressed with antigen (on the antigen-presenting cell) the Th2 cytokine IL-4 was found to be down-regulated in naive CD4$^+$ cells [PNAS (1999) *96* 3023–3028].)

Studies carried out in vitro and in vivo suggest that the cytokines secreted by (Th1 or Th2) cells are responsible for various aspects of the immune response observed during infections. Thus, 'inflammatory' Th1 cells secrete pro-inflammatory cytokines that are associated with important roles in cell-mediated immunity. For example, IFN-γ activates macrophages which may then (i) exhibit enhanced antimicrobial activity in phagosomes, (ii) secrete IL-8, attracting immune cells to the site of infection, and (iii) secrete IL-12, promoting further development of the Th1 subset. Moreover, IFN-γ can promote class switching to complement-fixing, opsonizing antibodies (human IgG1, IgG3; murine IgG2a). Th1 cytokines can e.g. upregulate expression of E selectins (see INFLAMMATION) and they can also upregulate MHC class II antigens. DELAYED HYPERSENSITIVITY is one manifestation of the characteristically cell-mediated Th1-type immune response.

Th2 cells are typically T-helper cells in ANTIBODY FORMATION. Th2 (anti-inflammatory) cytokines down-regulate macrophages and promote B cell activation, thus being important in the 'humoral' (antibody-mediated) immune response to infection by e.g. extracellular bacteria and helminths etc. IL-4 promotes class switching to non-complement-fixing IgG (human IgG4; murine IgG1) as well as to IgE in both man and mice. IL-5 promotes eosinophilia which may give activity against e.g. helminths and other parasites. IL-6 may contribute to antimicrobial activity by stimulating B cell proliferation and/or inducing ACUTE-PHASE PROTEINS.

Therapeutic uses of cytokines. Because cytokines regulate so many aspects of the immune defence system they are attractive as candidate therapeutic agents; for example, INTERFERONS have been useful in various contexts. However, some cytokines (e.g. TNF), though potentially useful, may be unsuitable for therapy owing e.g. to instability or toxicity in vivo; interestingly, a non-toxic TNF-mimetic peptide has been found to prevent recrudescence of *Mycobacterium bovis* (BCG) infection in CD4$^+$ T cell-depleted mice [JLB (2000) *68* 538–544]. Inhibition of TNF-α can be achieved by monoclonal antibodies or by soluble TNF-α receptor molecules. [New perspectives on the design of cytokines and growth factors: TIBtech. (2000) *18* 455–461.]

[Molecular biology of the cytokines: Book ref. 226.]

CYTOKINES: some of the cytokines secreted by Th1 and Th2 subsets of T lymphocytes

Cytokine	Th1	Th2
Interleukin-2	+	
Interleukin-3	+	+
Interleukin-4		+
Interleukin-5		+
Interleukin-6		+
Interleukin-10		+
Interleukin-13		+
Interferon-γ	+	
TNF-α	+	+[a]
TNF-β	+	

[a] Secretion from Th2 cells reported to be lower than that from Th1 cells.

cytokinesis Those events, excluding nuclear division, which occur during the division of a eukaryotic cell into progeny cells; they include the apportionment of the cytoplasm and organelles, and may include e.g. synthesis of new material for the cell wall of each progeny cell.

cytokinins (kinins, phytokinins) PHYTOHORMONES which stimulate metabolism and cell division; the cytokinins are 6-*N*-substituted adenines which are synthesized mainly at the root apex and translocated via the xylem. (Cytokinin-like activity is exhibited by certain urea derivatives, e.g. diphenylurea (found in coconut milk); it is believed that such compounds act by promoting the 6-*N*-substitution of endogenous adenine.) The mechanism by which cytokinins promote cell division is unknown; one suggestion is that cell division is encouraged by an increase in the level of endogenous cAMP brought about by the inhibition of cAMP phosphodiesterase by cytokinins.

Compounds similar or identical to cytokinins are produced by certain microorganisms; such compounds may account for the formation of ROOT NODULES and for the development of symptoms in e.g. FASCIATION.

Interestingly, cytokinins are incorporated in specificity-determining positions in a small proportion of tRNA molecules.

(See also KINETIN and ZEATIN.)

cytolysin A toxin which lyses (usually eukaryotic) cells.

cytolytic Able to lyse cells.

cytolytic T cell See T LYMPHOCYTE.

cytomegalic inclusion disease (CID) A disease caused by human cytomegalovirus (CMV) – see BETAHERPESVIRINAE. CMV can infect almost any tissue; infected cells are characteristically enlarged with large intranuclear inclusion bodies. Virus transmission occurs e.g. by direct contact, by transfusion of blood from an infected donor, or via the placenta. Congenital CID is commonly asymptomatic, but can be a severe condition with e.g. hepatosplenomegaly, encephalitis with irreversible CNS damage, and increased incidence of congenital deformities (see also TORCH DISEASES). Postnatal and adult infections are probably largely asymptomatic; in some cases the disease resembles INFECTIOUS MONONUCLEOSIS ('CMV mononucleosis', Paul–Bunnell test negative). In immunodeficient patients CMV infection may lead to e.g. pneumonia, transplant rejection, and/or death; infection may be primary or may result from reactivation of latent CMV. Patients at risk from CMV-related disease may be monitored by a NASBA-based assay of cytomegalovirus late gene *UL65* (which encodes the pp67 protein) [JCM (1998) *36* 1341–1346]; a commercial form of the assay for CMV was introduced by Organon Teknika. Known antiviral agents are not effective against CID.

cytomegaloviruses See BETAHERPESVIRINAE.

cytomegaly See CYTOPATHIC EFFECT.

cytomembranes *Syn.* INTRACYTOPLASMIC MEMBRANES.

cytomere A multinucleate structure which is formed by fragmentation of a schizont and which subsequently gives rise to merozoites.

cytopathic effect (CPE) (*virol.*) A change or abnormality in cells due to virus infection. CPEs may include e.g. INCLUSION BODY formation, cytoplasmic vacuolation (see e.g. SPUMAVIRINAE), cell enlargement (cytomegaly) (see e.g. BETAHERPESVIRINAE), SYNCYTIUM formation (see e.g. PARAMYXOVIRUS), the formation of discrete foci of cell proliferation (see e.g. FRIEND VIRUS and MCF VIRUSES), cell death (see also PLAQUE), etc. The CPEs produced by a given virus in a given cell system are often characteristic and may be useful in the identification of the virus; some viruses characteristically fail to give rise to CPE even when viral replication occurs (see e.g. RUBIVIRUS).

Cytophaga A genus of GLIDING BACTERIA of the CYTOPHAGALES; species occur in soil and in freshwater, estuarine and marine habitats. The organisms are rods or filaments, up to ca. 50 μm long, containing carotenoid pigments ranging from yellow to red; they are chemoorganotrophs and may be aerobic (respiratory metabolism using oxygen – or, in at least one strain, nitrate – as electron acceptor) or facultatively anaerobic (facultatively fermentative). Typically, *Cytophaga* spp can attack polysaccharides such as AGAR, ALGINATE, CELLULOSE and CHITIN. GC%: ca. 28–39.

Cytophagaceae See CYTOPHAGALES.

Cytophagales An order of GLIDING BACTERIA which do not form fruiting bodies (cf. MYXOBACTERALES); constituent species, all of which are Gram-negative, occur e.g. in soil and in freshwater, estuarine and marine habitats. In an early taxonomic scheme [Book ref. 21, pp. 99–119] four families were distinguished: motile rods or filaments containing carotenoid pigments (Cytophagaceae); motile filaments lacking carotenoid pigments (Beggiatoaceae); flat, non-pigmented, motile filaments found in the oral cavity in vertebrates (Simonsiellaceae); non-pigmented filaments, attached (at one end) to other filaments or to the substratum, which form gliding gonidia (Leucotrichaceae). Genera include ALYSIELLA, BEGGIATOA, CYTOPHAGA, FLEXIBACTER, HERPETOSIPHON, LEUCOTHRIX, SAPROSPIRA, SIMONSIELLA, SPOROCYTOPHAGA, THIOPLOCA, THIOTHRIX and VITREOSCILLA.

In another taxonomic scheme [ARM (1981) *35* 339–364 – cf. FLEXIBACTERIAE], of the above genera only *Cytophaga*, *Flexibacter* and *Sporocytophaga* are included in the Cytophagales; the other genera are classified as apochlorotic cyanobacteria.

cytopharyngeal apparatus (*ciliate protozool.*) Skeletal and certain other elements of the CYTOPHARYNX. Two main types of cytopharyngeal apparatus have been distinguished: the CYRTOS and the RHABDOS; this taxonomically important distinction is believed to reflect evolutionary differences in the different ciliate groups – those having a rhabdos-type structure being regarded as the more primitive.

cytopharyngeal basket *Syn.* CYRTOS.

cytopharynx (*ciliate protozool.*) An invagination of the cytoplasmic membrane which is supported, within the cytoplasm, by a tubelike system of microtubules and/or microfibrils. (See also CYTOPHARYNGEAL APPARATUS and CYTOSTOME.) Food particles which pass into the (non-ciliated) cytopharynx are taken up by PHAGOCYTOSIS through the cytoplasmic membrane at its inner end.

cytophilic antibodies Antibodies which can bind to a cell surface without involving their COMBINING SITES; the bound antibody can therefore still bind homologous antigen. (See e.g. REAGINIC ANTIBODIES.)

cytoplasmic genes (extrachromosomal genes; extranuclear genes) Genes which are not located on a bacterial CHROMOSOME or in a (eukaryotic) NUCLEUS; in eukaryotic organisms the term commonly refers to the genes in a CHLOROPLAST or a MITOCHONDRION, while in prokaryotes it has been used to refer to the genes carried by a PLASMID. (See also CYTOPLASMIC INHERITANCE.)

cytoplasmic heterozygote See CYTOHET.

cytoplasmic inheritance (extrachromosomal inheritance; non-Mendelian inheritance) Inheritance governed or influenced by CYTOPLASMIC GENES – which are not subject to the Mendelian laws of segregation and assortment; see e.g. MATERNAL INHERITANCE.

cytoplasmic membrane (CM; cell membrane; plasma membrane; plasmalemma; protoplast membrane) The lipid- and protein-containing, selectively permeable membrane which encloses the cytoplasm in prokaryotic and eukaryotic cells; in most types of microbial cell the CM is bordered externally by the CELL WALL. In microbial cells the precise composition of the CM may depend on growth conditions and on the age of the cell.

The structure of all biological membranes appears to conform to the basic *fluid mosaic model*. In this model the lipid molecules form a bilayer within which the protein molecules are partly or wholly embedded – some spanning the entire width of the bilayer; the lipid molecules are orientated such that their polar groups form the outer, hydrophilic surfaces of the bilayer while their hydrocarbon chains form the hydrophobic interior of the bilayer. (cf. UNIT MEMBRANE.) The membrane proteins are sometimes categorized as either *peripheral* (= *extrinsic*) proteins (bound to the membrane e.g. by electrostatic forces, and easily removable by electrolytes or chelating agents) or *integral* (= *intrinsic*) proteins (bound more strongly by hydrophobic bonds and extractable, with difficulty, by detergents or organic solvents). Cytoplasmic membranes are characteristically asymmetrical, i.e. the components of the outer (externally facing) side of the membrane are not identical to those of the inner (cytoplasmic) side.

Membrane 'fluidity' primarily involves lateral and rotational motion of whole lipid molecules as well as motion of the hydrocarbon chains of the lipid molecules (rather than movement of whole lipid molecules from one layer of the membrane to the other layer). The hydrocarbon chains may be disordered and flexible (the α-conformation) or ordered, rigid and perpendicular to the plane of the bilayer (the β-conformation). Fluidity greater than a certain level is essential for the normal physiological role of the CM. The degree of fluidity is governed by temperature and e.g. by the length and structure of the hydrocarbon chains; membranes containing unsaturated chains are generally more 'fluid' than are membranes containing saturated chains of the same length.

Previously, it was thought that only rarely do lipid molecules move from one side of the bilayer to the other. It is now known that such transmembrane movement (*flip–flop*) occurs with a much greater frequency than was originally supposed; in at least one case evidence has been obtained for a 'flippase' which may facilitate such translocation [Cell (1985) *42* 51–60].

Although it is believed that the membrane lipids are normally present as a bilayer, results of nuclear magnetic resonance (NMR) and other studies have indicated the presence of transient, temperature-dependent, localized non-bilayer lipid phases within biological membranes. In one of these phases, termed *hexagonal II* (or H_{II}), the lipid molecules form an array of fine (~20 Å diam.) water-filled cylinders whose walls are composed of the polar heads of lipid molecules; the array of cylinders is hexagonal in cross-section, the space between the cylinders containing the hydrocarbon chains of the lipid molecules. [Lipid structure of biological membranes: TIBS (1985) *10* 418–421.]

The CM has various functions, one of which is to regulate the cytoplasmic milieu by controlling the inward and outward passage of ions and molecules. Some uncharged and/or lipophilic molecules can pass relatively freely through the CM; these include e.g. water, carbon dioxide, oxygen, ammonia (but not ammonium ions), acetic acid (undissociated form) and ethanol. However, ions and most molecules cannot pass freely through the CM, so that their transmembrane translocation requires more or less specific TRANSPORT SYSTEMS. (Specific mechanisms also exist for OSMOREGULATION.) Transport commonly occurs at the expense of metabolic energy, and some systems depend on the presence of a transmembrane electrochemical gradient – e.g. proton motive force (pmf: see CHEMIOSMOSIS) or SODIUM MOTIVE FORCE – such gradients being a general feature of CMs.

According to species, the CM is also the site of RESPIRATION (and, in some organisms, photosynthesis).

The CM is involved in the synthesis of external structures, such as the cell wall and capsule, and also in the synthesis of components of the CM.

(a) *Bacterial cytoplasmic membranes.* In transmission electron micrographs the CM, ca. 7–8 nm thick, typically appears as a trilaminar structure: an electron-translucent layer sandwiched between two electron-dense layers. In Gram-negative bacteria, localized regions of the CM may be involved in ADHESION SITES. In many species of bacteria it has been demonstrated that the inner (cytoplasmic) face of the CM bears minute, spherical, 'stalked' particles (see PROTON ATPASE) which are involved in energy conversion. In many (not all) species, CYTOCHROMES and other components of an ELECTRON TRANSPORT CHAIN occur in the CM. (See also PURPLE MEMBRANE.)

Lipid components of bacterial CMs are mainly phospholipids; in many or all cases, these are synthesized within the membrane itself.

Phosphatidylglycerols (PG) appear to occur in all bacteria, while *phosphatidylethanolamine* (PE) is more common and more abundant in Gram-negative species; *phosphatidylcholine* is absent in Gram-positive cells and is rare in Gram-negative bacteria. *Phosphatidylinositol* occurs in some bacteria (e.g. *Mycobacterium* spp). PLASMALOGENS are found in the cytoplasmic membrane in some anaerobes. In *Escherichia coli* the main phospholipid is PE – PG and *diphosphatidylglycerol* (DPG, cardiolipin) being relatively minor components.

Glycolipids are common in small quantities; in some Gram-positive bacteria molecules of glycolipid may be covalently linked to glycerol TEICHOIC ACIDS, forming *lipoteichoic acid.*

Sphingolipids are rare in bacterial membranes.

Sterols are absent in most species (cf. MYCOPLASMATACEAE).

Hopanes (triterpene derivatives which resemble sterols in size, rigidity and amphiphilicity) are present in some bacteria [JGM (1985) *131* 1363–1367].

Lipoamino acids (*O*-aminoacylphosphatidylglycerols) occur in the membranes of some Gram-positive bacteria (e.g. *Bacillus* spp, *Clostridium* spp, *Staphylococcus aureus*). These are esters of PG and basic amino acids such as arginine, lysine or ornithine; the proportion of lipoamino acids varies according to growth phase and to the pH of the medium.

The CM *fatty acids* may be straight-chained or branched (the latter more common in Gram-positive species), saturated or unsaturated; some contain a cyclopropane ring (which is formed by methylation at a double bond). Phospholipids often contain one saturated and one unsaturated fatty acid residue per molecule.

In *E. coli* the main saturated fatty acids are hexadecanoic (palmitic) and tetradecanoic (myristic) acids, minor components including e.g. octadecanoic (stearic) and dodecanoic (lauric) acids; the main unsaturated fatty acids (all of which are *cis*-monoenes) include e.g. *cis*-Δ^9-hexadecenoic (palmitoleic) acid.

In general, the CM varies in its fatty acid composition according to growth conditions (e.g. temperature, pH). Thus, in some species the proportion of unsaturated fatty acids increases when the growth temperature decreases – e.g. in *E. coli* the proportion of unsaturated fatty acids increases from 16% to 49% when the growth temperature falls from 36°C to 25°C; such a change appears to be a compensatory response which helps to maintain optimum membrane fluidity. In contrast, almost no increase in monoenoic fatty acids occurs in the CM of *Staphylococcus aureus* when the growth temperature drops from 37°C to 25°C [Book ref. 44, p 402]. In some bacteria, adaptation to lower growth temperatures involves an increase in the *length* of fatty acid chains rather than an increase in the degree of unsaturation [JGM (1985) *131* 2293–2302].

The *lipids* of the CM clearly contribute to the essential feature of selective permeability. However, the lipids also have other functions; for example, in *E. coli* phosphatidylethanolamine can behave as a 'molecular chaperone', being required for the correct folding (maturation) of the membrane protein LacY [EMBO (1998) *17* 5255–5264].

Proteins in the bacterial CM include a variety of enzymes (involved e.g. in the synthesis of phospholipids and cell wall components); for example, penicillin-binding proteins (involved in the synthesis of PEPTIDOGLYCAN) may form part of a protein complex in the CM [FEMS Reviews (1994) *13* 1–12]. There are also components of TRANSPORT SYSTEMS, energy-converting systems (see e.g. ELECTRON TRANSPORT CHAIN and EXTRACYTOPLASMIC OXIDATION) and sensing systems (see e.g. CHEMOTAXIS).

CM proteins also include the molecular water channels called *aquaporins* (see MIP CHANNEL). (See also MECHANOSENSITIVE CHANNEL.)

[Structural dynamics of the CM of *E. coli*: Book ref. 122, pp 121–160.]

The bacterial CM is the target for a variety of ANTISEPTICS and DISINFECTANTS (see e.g. QUATERNARY AMMONIUM COMPOUNDS) and ANTIBIOTICS (see e.g. DEPSIPEPTIDE ANTIBIOTICS, GRAMICIDINS, POLYMYXINS and TYROCIDINS).

(b) *Fungal cytoplasmic membranes*. The CMs in fungi (and in other eukaryotes) differ significantly from those in bacteria. In fungal CMs the major lipids typically include phosphatidylcholine and phosphatidylethanolamine (with smaller amounts of phosphatidylinositol and phosphatidylserine) together with SPHINGOLIPIDS; as a rough generalization, the fatty acids of phospholipids in higher fungi tend to contain even numbers of carbon atoms and to be saturated or mono-unsaturated, while those of lower fungi tend to have odd numbers of carbon atoms and to be polyunsaturated. (Changes in the degree of fatty acid saturation can have interesting physiological effects; for example, when *Saccharomyces cerevisiae* strain Y185 becomes de-repressed for general amino acid permease, a higher degree of fatty acid unsaturation in the CM correlates with a more rapid expression of the permease [JGM (1985) *131* 57–65].) Most fungal membranes also contain sterols (e.g. ergosterol), so that fungi are generally susceptible to POLYENE ANTIBIOTICS (see also AZOLE ANTIFUNGAL AGENTS); vegetative cells of members of the Pythiaceae have been reported to lack sterols, although sterols are required during the reproductive phase.

Proteins in fungal CMs include those involved in transport processes, components of ATPase complexes, and various enzymes – such as those needed for the synthesis of walls and membranes.

In addition to lipids and proteins, fungal CMs contain small amounts of carbohydrate. In e.g. some slime moulds, certain CM carbohydrates are important LECTIN receptors and are involved in cell–cell recognition and/or adhesion.

(c) *Archaeal cytoplasmic membranes*. In members of the ARCHAEA the CM contains lipids of a kind which do not occur in bacteria. Unlike the ester-linked glycerol–fatty acid bacterial lipids, archaeal lipids are characteristically ether-linked molecules that contain e.g. isoprenoid or hydro-isoprenoid components. Some of the archaeal lipids are structurally analogous to those of bacteria; for example, the di-ether and di-ester lipids both have a single polar end. However, some archaeal lipids (e.g. tetra-*O*-di(biphytanyl) diglycerol) contain one ether-linked glycerol residue at *each* end of the molecule; having two polar ends, such molecules may span the width of the CM. Given the exteme habitats typically occupied by these organisms, it seems likely that some or all of the archaeans will be found to contain membrane components whose characteristics reflect an adaptation to the environment.

Some archaeans contain MIP CHANNELS.

cytoplasmic petite See PETITE MUTANT.

cytoplasmic polyhedrosis virus group (CPV group; *Cypovirus*) A genus of entomopathogenic viruses of the REOVIRIDAE. Genome: 10 dsRNA molecules which may be linked together by protein in the virion. CPVs are pathogenic in a wide range of insects of the Diptera, Hymenoptera and Lepidoptera, and can also infect certain crustaceans (*Simocephalus expinosus*); the type member, a CPV from the silkworm (*Bombyx mori*), has caused significant economic losses in the silk industry. A CPV is commercially available in Japan for the BIOLOGICAL CONTROL of the pine caterpillar *Dendrolimus spectabilis*. (cf. BACULOVIRIDAE; see also INSECT DISEASES.) CPV infection is usually chronic and is generally restricted to the columnar epithelial cells of the midgut. The mature virions occur embedded in POLYHEDRA in the cytoplasm of infected cells (cf. NUCLEAR POLYHEDROSIS VIRUS). The polyhedra may be polyhedral, cuboid, triangular etc, depending on virus and host. The infected insect shows symptoms of starvation due to reduced feeding and a reduction in the absorptive capacity of the gut cells; diarrhoea is common, with large numbers of polyhedra in the faeces. Infected larvae commonly reach adulthood, but the adults are generally small and often malformed, with greatly diminished reproductive capacity. CPVs are apparently transmitted to the larvae via the surface of the egg. [Review: Book ref. 83, pp. 425–504.]

cytoplasmic streaming (cyclosis) The intracellular flow of protoplasm which occurs in various types of eukaryotic cell; such a flow ensures e.g. that intracellular reactions are not dependent on simple diffusion. In e.g. the green alga *Chara*, cytoplasmic streaming appears to result from the interaction between peripheral ACTIN microfilaments and MYOSIN, the relative movement between actin and myosin being reflected in the flow of cytosol and organelles. In amoebae, the characteristic amoeboid movement may involve local actin–myosin-mediated contractions in the cell cortex accompanied by cytoplasmic streaming due to gel–sol interconversions in the microfibrillar networks.

cytoplast A eukaryotic cell from which the nucleus has been removed (i.e., an ENUCLEATED CELL). (cf. KARYOPLAST.)

cytoproct (cytopyge) In some protozoa: a permanent pore through which is voided particulate non-digestible material.

cytopyge *Syn.* CYTOPROCT.

cytosegresome An intracellular, membrane-limited vacuole within which a cell has enclosed some of its own constituents. Cytosegresomes are formed during AUTOPHAGY.

cytosine arabinoside *Syn.* CYTARABINE.

cytoskeleton In a eukaryotic cell: the framework, composed primarily of proteinaceous tubules and fibrils, which ramifies throughout the cytoplasm (binding e.g. to the cytoplasmic membrane and to various organelles) and which is responsible e.g. for the shape and internal organization of the cell, the intracellular transport of vesicles and organelles, and (where applicable) cell motility; in a living cell the cytoskeleton probably undergoes continual modification – being disassembled in some regions and reassembled in others, according to the needs of the cell. (cf. NUCLEOSKELETON.)

The main structural components of the cytoskeleton are microfilaments (see ACTIN), MICROTUBULES, and INTERMEDIATE FILAMENTS. Microtubules contribute e.g. to cell shape, to the correct location and orientation of intracellular structures and organelles, to certain types of motility (see e.g. axoneme in FLAGELLUM (b)), and to the formation of the spindle in MITOSIS; they can also serve as a temporary scaffolding e.g. during intracellular rearrangements. [Role of microtubules during the cell cycle in *Stephanopyxis turris*: JCB (1986) *102* 1688–1698.] Microfilaments are involved e.g. in those parts of the cytoskeleton concerned with intracellular movements (see e.g. CYTOPLASMIC STREAMING), amoeboid movement (see PSEUDOPODIUM), PHAGOCYTOSIS, and CAPPING (sense 3). Intermediate filaments may e.g. fulfil a strengthening role. There is some evidence that, in multicellular organisms and tissues, cytoskeletal elements may extend from one cell to another by penetrating cytoplasmic membranes and the intercellular matrix.

[Molecular biology of the cytoskeleton: Book ref. 166; in vitro translocation of organelles along microtubules: Cell (1985) *40* 729–730; reviews: JCS (1986) Supplement 5; the cytoskeleton in protists: Int. Rev. Cytol. (1986) *104* 153–249.]

cytosol The fluid (non-particulate) fraction of cytoplasm. (cf. CELL SAP; HYALOPLASM.)

cytostome In certain protozoa (including many ciliates and some others, e.g. *Noctiluca, Peranema*): the cell mouth, a specific region of the cell through which particulate food is ingested. (See also AMOEBOSTOME.)

In ciliates the cytostome is regarded as a two-dimensional aperture at the level of the PELLICLE; in this region the pellicle lacks both cilia and pellicular alveoli, and is typically invaginated within a supportive collar or tube-like structure of microtubules and/or microfibrils in the cytoplasm (see also CYTOPHARYNX and CYTOPHARYNGEAL APPARATUS). In advanced ciliates (e.g. *Paramecium*) that part of the pellicle which includes the cytostome typically occurs at the base of a concavity in the body (BUCCAL CAVITY); in the more primitive ciliates (e.g. the gymnostomes) the cytostome-containing region of the pellicle is typically not recessed in this way, and is characteristically situated apically or subapically in the cell. In some ciliates the ciliature surrounding the cytostome is readily distinguishable from the somatic ciliature (see e.g. AZM and PARORAL MEMBRANE).

cytotaxin Any agent chemotactic for cells.

cytotoxic hypersensitivity *Syn.* TYPE II REACTION.

cytotoxic T cell See T LYMPHOCYTE.

cytotropism See TROPISM (sense 2).

cytozoic Living (parasitic) within cells.

Cyttariales See ASCOMYCOTINA.

Czapek–Dox medium (Czapek's medium) A medium containing sucrose, $NaNO_3$, K_2HPO_4, $MgSO_4$, KCl and $FeSO_4$; it may be solidified with agar. It is used for culturing e.g. saprotrophic fungi, soil bacteria etc.

1. Words in SMALL CAPITALS are cross-references to separate entries.
2. Keys to journal title abbreviations and Book ref. numbers are given at the end of the Dictionary.
3. The Greek alphabet is given in Appendix VI.
4. For further information see 'Notes for the User' at the front of the Dictionary.

D

D (1) Dihydrouridine (see TRNA). (2) Aspartic acid (see AMINO ACIDS).

D (in continuous culture) DILUTION RATE.

d-factor See DUTCH ELM DISEASE.

D loop (*mol. biol.*) (1) (displacement loop) A single-stranded loop formed when a short ssDNA molecule pairs with a complementary region of one strand of a dsDNA molecule, displacing the corresponding region of the homologous strand (the D loop). For example, under certain (non-physiological) conditions in vitro a negatively supercoiled ccc dsDNA molecule can spontaneously take up a short complementary single (linear) strand to give rise to a D loop in a reaction driven by the energy of supercoiling; the resulting 'joint molecule' is more relaxed than the original supercoiled DNA. D-loop formation is promoted e.g. by the RECA PROTEIN; it may occur e.g. during RECOMBINATION (Fig. 2), and may be involved in the priming of certain types of DNA replication (see e.g. BACTERIOPHAGE G4). (See also SITE-SPECIFIC MUTAGENESIS; cf. R LOOP.)

(2) The loop of the 'D arm' in a tRNA molecule (see TRNA).

D period See HELMSTETTER–COOPER MODEL.

12D process In food processing: heating, at a given temperature, for a period equal to twelve times the D VALUE at that temperature. (See also BOTULINUM COOK.)

D-type particles See TYPE D RETROVIRUS GROUP.

D-type starter See LACTIC ACID STARTERS.

D value (1) (D_{10} value; decimal reduction time) The time required, at a given temperature, to reduce the number of viable cells or spores of a given microorganism to 10% of the initial number; it is usually quoted in minutes. The temperature (°C) at which the D value is determined may be indicated by a subscript, e.g., D_{112}. (In the synonym of D value, D_{10} value, the subscript refers to the 10% survival value.) (cf. F_0 VALUE; Z VALUE; see also 12D PROCESS.)

(2) (of animal feed) The percentage of digestible organic matter in the 'dry matter' (DM) – DM (in g/kg) being determined by oven-drying of feed samples and correction for e.g. loss of volatile fatty acids. (See also SILAGE.)

D_{10} value *Syn.* D VALUE (1).

Da DALTON.

***dacA* gene** See PENICILLIN-BINDING PROTEINS.

***dacB* gene** See PENICILLIN-BINDING PROTEINS.

***dacC* gene** See PENICILLIN-BINDING PROTEINS.

Dacrymyces A genus of fungi (order DACRYMYCETALES). *D. deliquescens* forms bright orange, gelatinous, globular asexual fruiting bodies (each up to ca. 5 mm diam.) on decaying wood; each fruiting body consists of a mass of radiating hyphae which, at the surface, break up into oidia. Subsequently, basidia develop at the surface of the same fruiting body, which concurrently becomes yellow.

Dacrymycetales An order of typically lignicolous fungi (subclass HOLOBASIDIOMYCETIDAE) which form the so-called 'tuning fork' type of basidium – i.e., a two-spored basidium which is apically bifurcate (forked) and Y-shaped, the two arms of the 'Y' (the sterigmata) curving inwards to become more or less parallel. The basidiocarp may be e.g. crust-like, pulvinate, cup-like and stipitate, or columnar and branched or unbranched; it may be gelatinous or waxy, and in many species it is brightly coloured (e.g. yellow, orange) due, apparently, to carotenoid pigments. Genera include CALOCERA, *Cerinomyces*, DACRYMYCES, *Dacryopinax*, *Guepiniopsis*.

Dacryopinax See DACRYMYCETALES.

dactinomycin *Syn.* ACTINOMYCIN D.

Dactylaria See HYPHOMYCETES; see also NEMATOPHAGOUS FUNGI and PHAEOHYPHOMYCOSIS.

Dactylella See HYPHOMYCETES and NEMATOPHAGOUS FUNGI.

Dactylococcopsis A genus of unicellular CYANOBACTERIA in which the cells are elongate with tapering ends, sometimes occurring in short chains. *D. salina* is a gas-vacuolate species which is blue-green when grown at low light intensities but deep orange under high light intensities; it is planktonic in hypersaline environments. Although sensitive to concentrations of $H_2S >$ ca. 10 μM, at lower concentrations *D. salina* is capable of limited anoxygenic photosynthesis using H_2S as electron donor. [Ecophysiology of *D. salina*: JGM (1983) *129* 1849–1856.]

Dactylosoma See PIROPLASMASINA.

Dactylosporangium A genus of mycelial bacteria (order ACTINOMYCETALES, wall type II) which occur e.g. in soil. The organisms form finger-like sporangia each containing a single row of three or more zoospores; non-motile spores are also formed, singly, on the substrate mycelium. Aerial mycelium is not formed. Type species: *D. aurantiacum*.

Dactylostoma See SUCTORIA.

DAEC See ENTEROADHERENT E. COLI.

Daedalea A genus of fungi of the APHYLLOPHORALES (family Polyporaceae) which occur as perthotrophs (PERTHOTROPH sense 2) and saprotrophs on various types of wood. Basidiocarp: non-stipitate, often hoof-like, commonly corky, the upper surface brown to grey-brown; hymenophore: labyrinthiform ('maze-like'), consisting of elongated pores with thick dissepiments. *D. quercina*, which grows e.g. on oak, may form imbricated basidiocarps with a trimitic context; basidiospores: ellipsoidal, hyaline, ca. 6 × 3 μm. (See also WOOD WASP FUNGI.)

DAF See CD55.

dahlia mosaic virus See CAULIMOVIRUSES.

dairy products Many dairy products are made by a LACTIC ACID FERMENTATION of milk using selected strains of lactic acid bacteria (see e.g. LACTIC ACID STARTERS): see ACIDOPHILUS MILK; BUTTER; BUTTERMILK; CHEESE-MAKING; KEFIR; KOUMISS; LEBEN; QUARG; SOUR CREAM; TAETTE; VILIA; YOGHURT. (See also MILK.)

daisy head colonies Colonies of the type commonly formed by *Corynebacterium diphtheriae* (*gravis* type) on blood–tellurite agar; after 24 hours at 37°C, each colony is 2–3 mm in diameter, dull, with a grey to black centre, translucent margin, crenated edge, and radial striations.

Dakin's solution (chlorinated soda solution) A solution of sodium hypochlorite and sodium bicarbonate used e.g. as an antiseptic for the cleansing of wounds.

Daldinia A genus of fungi (order SPHAERIALES). *D. concentrica* grows on dead or fallen branches of certain trees, particularly ash (*Fraxinus*), forming hemispherical or subglobose, superficial stromata ca. 2–10 cm across; when young, each stroma has a surface layer of branched conidiophores bearing brown conidia, but the mature stroma is a hard, brittle, shiny black structure which, in vertical cross-section, exhibits alternating light and dark bands concentric about the centre of the base of the stroma. Perithecia develop in the superficial layer of the stroma with their ostiolar pores at the surface. The bulk of the stromal tissue may act as a reservoir for water. (See also ASCOSPORE.)

Dalmau plate technique A technique used e.g. for studying the formation of pseudomycelium or true mycelium by a yeast. A streak and two point-inoculations of the test strain are made at well-separated locations on a surface-dried plate of e.g. CORNMEAL AGAR; the centre of the streak and one of the point-inoculations are then each overlaid with a sterile cover-glass. After incubation at 25°C for ca. one week the preparation is examined under the microscope.

Dalmeny disease A disease of cattle which results in e.g. abortion or death; the causal agent is believed to be a species of SARCOCYSTIS.

dalton (Da) *Syn.* ATOMIC MASS UNIT.

Dalyellia viridis See ZOOCHLORELLAE.

Dam-directed mismatch repair See MISMATCH REPAIR.

***dam* gene** In e.g. *Escherichia coli*: a gene which encodes a DNA adenine methylase (Dam methylase), i.e. an enzyme that methylates DNA at the N-6 position of adenine in the nucleotide sequence 5′-GATC-3′ ('Dam site'); a strand of DNA is methylated soon after its synthesis (see DNA METHYLATION).

Dam methylation is involved e.g. in MISMATCH REPAIR in *E. coli*. It is also involved in regulating certain genes. For example, transposition of the transposon Tn*10* (q.v.) is inhibited by Dam methylation of a sequence in the transposase gene promoter (reducing transposase synthesis) and of a sequence in IS*10* (reducing transposase activity); Tn*10* transposition is thus apparently coupled to DNA replication: it occurs just after the replication fork has passed but before Dam methylation has occurred.

Other genes which contain Dam sites in their promoters, and whose expression is sensitive to Dam methylation, include e.g. the *trpR* gene of *E. coli*, the *cre* gene of BACTERIOPHAGE P1, and the *mom* gene of BACTERIOPHAGE MU. The chromosomal origin, *oriC*, in *E. coli* contains more than 10 Dam sites which may be important in regulating the initiation of DNA replication (see CELL CYCLE (b)).

[Minireview: JB (1985) *164* 490–493.]

Dam methylation apparently occurs in various other Gram-negative bacteria (including certain cyanobacteria) and in some members of the Archaea.

Dam methylation See DAM GENE.

damping off (*plant pathol.*) A microbial disease of seedlings in which e.g. roots may rot, or the hypocotyl (lower stem) may either collapse or become wiry ('wire stem'); seedlings may die before or after they emerge from the soil (pre-emergence and post-emergence damping off, respectively). Common causal agents include species of PYTHIUM and RHIZOCTONIA. Aerial parts are often secondarily affected by GREY MOULD. Damping off may be controlled e.g. by CARBENDAZIM, CHESHUNT COMPOUND, CHLORONEB or ETRIDIAZOLE. (cf. FOOT-ROT (sense 2) and BLACK-LEG (sense 2).)

Dane particle See HEPATITIS B.

Danielli–Davson model See UNIT MEMBRANE.

Danish agar *Syn.* FURCELLARAN.

danofloxacin See QUINOLONE ANTIBIOTICS.

dansyl chloride A reagent, 5-(dimethylamino)naphthalene-l-sulphonyl chloride, which reacts with amino acids and proteins to form derivatives that exhibit an intense yellow fluorescence under UV irradiation.

Danysz' phenomenon The phenomenon in which the residual toxicity in a mixture of diphtheria toxin and antitoxin varies according to the method of admixture. Thus, if toxin is added to an equivalent amount of antitoxin in one stage (i.e., all at once) the resulting mixture is non-toxic; however, if the same quantity of toxin is added in two halves (the second ca. 30 min after the first), the resulting mixture contains free (non-neutralized) toxin. In the second method of admixture, the first portion of toxin combines with more than its equivalent of antitoxin – leaving insufficient free antitoxin to neutralize the second portion of toxin.

DAP pathway DIAMINOPIMELIC ACID PATHWAY.

DAPI 4′,6-Diamidino-2-phenylindole: a FLUOROCHROME used e.g. as a DNA-specific stain. [Limn. Ocean. (1980) *25* 948–951.] (See also DIAMIDINES.)

dapsone (DDS) A drug used e.g. in the treatment of LEPROSY and MALARIA: 4,4′-diaminodiphenylSULPHONE; DDS may inhibit FOLIC ACID metabolism. [Book ref. 54, p. 422.] (cf. CLOFAZIMINE.)

daptomycin A lipopeptide antibiotic which, in vitro, is bactericidal for a range of clinically important Gram-positive bacteria, including e.g. MRSA and vancomycin-resistant enterococci (VRE); daptomycin apparently interferes with the function of the bacterial cytoplasmic membrane.

[Resistance studies with daptomycin: AAC (2001) *45* 1799–1802.]

dark-field microscopy See MICROSCOPY (b).

dark-ground microscopy See MICROSCOPY (b).

dark mildews *Syn.* BLACK MILDEWS.

dark reaction See PHOTOSYNTHESIS.

dark repair (1) Light-independent DNA REPAIR: e.g. EXCISION REPAIR or RECOMBINATION REPAIR (as opposed to PHOTOREACTIVATION). (2) *Syn.* EXCISION REPAIR.

Darling's disease *Syn.* HISTOPLASMOSIS.

Darmbrand ENTERITIS NECROTICANS.

dasycladalean algae A small group of tropical and subtropical marine green algae (division CHLOROPHYTA) which typically grow in shallow waters. The vegetative thallus is uninucleate and is radially symmetrical, composed of an erect axis with whorls of branches and a rhizoid-like holdfast which contains the nucleus; the thallus becomes multinucleate during gametangial development. Fossil dasyclads are known from the Lower Palaeozoic onwards [Book ref. 123, pp. 297–302]; modern genera include ACETABULARIA.

Dasyscyphus See HELOTIALES.

Dasyspora See UREDINIOMYCETES.

Dasytricha A genus of ciliates (order TRICHOSTOMATIDA) which occur e.g. in the RUMEN. Cells: ovoid, with dense, uniform somatic ciliature and a cytostome located posteriorly.

DAT (*serol.*) Differential agglutination test: any in vitro diagnostic test in which the result consists of two agglutination titres – that of the test proper and that of the control; the titres may be quoted in full or expressed as a ratio.

daughter-strand gap repair See RECOMBINATION REPAIR.

daunomycin See ANTHRACYCLINE ANTIBIOTICS.

daunorubin See ANTHRACYCLINE ANTIBIOTICS.

Davson–Danielli model See UNIT MEMBRANE.

dazomet A soil fumigant (3,5-dimethyltetrahydro-2H-1,3,5-thiadiazine-2-thione) used as a fungicide and nematocide; in soil it breaks down to form e.g. formaldehyde and *N*-methylisothiocyanate. (cf. METHAM SODIUM.)

DBMIB An inhibitor of PHOTOSYNTHESIS.

D_c Critical DILUTION RATE.

DCA (deoxycholate–citrate agar) An agar medium used for the primary isolation of e.g. *Salmonella* and *Shigella*; most strains of *Escherichia* and of the Proteeae (cf. PROVIDENCIA) fail to grow on DCA. DCA generally contains meat extract, peptone, lactose, sodium citrate, ferric ammonium citrate, sodium deoxycholate, and neutral red; pH ca. 7.3. DCA should not be sterilized by autoclaving.

DCCA See DICHLOROISOCYANURATE.

DCCD *N*,*N*′-Dicyclohexylcarbodiimide; under weakly alkaline conditions DCCD binds covalently to the 'DCCD-binding protein' of F_0 in (F_0F_1)-type H^+-ATPases and inhibits both the synthesis and hydrolysis of ATP, while under weakly acidic conditions it binds to the β subunit of F_1, inhibiting the hydrolysis of ATP.

DCDS See CONJUGATION (1b) (i).

DCMU (Diuron) *N*-(3,4-dichlorophenyl)-*N*′-dimethylurea; this agent acts as a mitochondrial RESPIRATORY INHIBITOR, blocking electron flow between cytochromes *b* and c_1 in Complex III of the ELECTRON TRANSPORT CHAIN (cf. BAL), and as an inhibitor of electron flow between photosystems I and II (see PHOTOSYNTHESIS).

***dda* gene** (in bacteriophage T4) See HELICASES.

DDMR See MISMATCH REPAIR.

DDS DAPSONE.

ddTTP 2′,3′-Dideoxythymidine triphosphate: an analogue of thymidine triphosphate; it inhibits DNA synthesis since, when incorporated in a growing DNA strand in place of thymidine, it provides no 3′-OH for further polymerization. In mammalian cells it specifically inhibits DNA polymerase β action; the α enzyme is apparently unable to incorporate the analogue in a growing DNA strand. (cf. APHIDICOLIN.)

de-acetylase See ANTIBIOTIC.

dead man's fingers See XYLARIA.

Dean and Webb titration A serological titration in which a constant volume of a given antiserum is added to each of a number of serial dilutions of the homologous antigen; precipitation occurs most rapidly in that dilution in which antibody and antigen are present in OPTIMAL PROPORTIONS.

death cap fungus See AMANITA.

death phase See BATCH CULTURE.

death rate constant See STERILIZATION (b).

Debaryomyces A genus of yeasts (family SACCHAROMYCETACEAE) which reproduce by multilateral budding; pseudomycelium may be formed. Ascus formation is generally preceded by conjugation between bud and mother cell; conjugation between separate cells also occurs. Ascospores: spherical or oval, minutely warty or ridged, usually 1 or 2 (up to 4 in some species) per ascus. Fermentation occurs (weakly) in some species. NO_3^- is not assimilated. Species (*D. castellii*, *D. coudertii*, *D. hansenii*, *D. marama*, *D. melissophilus*, *D. polymorphus*, *D. pseudopolymorphus*, *D. tamarii*, *D. vanriji*) have been isolated from soil, foods etc. [Book ref. 100, pp. 130–145.]

debranching enzyme (1) (amylo-1,6-glucosidase) An enzyme which hydrolyses the $(1 \rightarrow 6)$-α branch points in e.g. amylopectin, glycogen, and/or related polysaccharides. There are two types: *pullulanase* ('limit dextrinase', pullulan 6-glucanhydrolase, EC 3.2.1.41), which can degrade PULLULAN, and *isoamylase* (glycogen 6-glucanhydrolase, EC 3.2.1.68), which has no action on pullulan. Pullulanases are obtained commercially from e.g. *Klebsiella aerogenes* ('*Aerobacter aerogenes*') and *Bacillus cereus* var. *mycoides*; the enzyme from *K. aerogenes* can debranch amylopectin and its β-limit dextrin (see AMYLASES) but has little or no activity on native glycogen, while that from *B. cereus* var. *mycoides* has little or no activity on amylopectin but can degrade its β-limit dextrin [Book ref. 31, pp. 133–137]. Isoamylases are obtained e.g. from *Pseudomonas* sp. (cf. BRANCHING ENZYME.)

(2) (*mol. biol.*) An enzyme which 'debranches' the lariat form of an excised intron (see SPLIT GENE).

débridement The removal of necrotic/infected tissue and/or foreign material from a wound.

debromoaplysiatoxin See LYNGBYA.

decarboxylase tests Tests used to determine the ability of a given bacterial strain to decarboxylate arginine, lysine and/or ornithine. Three tubes of MØLLER'S DECARBOXYLASE BROTH, each containing one of the three amino acids, are inoculated with the test organism; each broth is overlaid with a layer of sterile paraffin, incubated at 37°C, and examined daily for 4 days. Initially, in both positive and negative tests, glucose is metabolized to acidic products causing the pH indicator bromcresol purple (in the medium) to turn yellow. In a positive test the amino acid is cleaved by a specific decarboxylase in a reaction requiring pyridoxal phosphate (see PYRIDOXINE); the product is a polyamine (see POLYAMINES) which raises the pH of the medium, causing bromcresol purple to turn purple. Arginine, lysine and ornithine are decarboxylated to agmatine, cadaverine and putrescine, respectively.

decay-accelerating factor See CD55.

Dechloromonas See BIOREMEDIATION.

Dechlorosoma See BIOREMEDIATION.

decimal code (for growth stages of cereals) See ZADOKS' CODE.

decimal reduction time *Syn.* D VALUE (1).

declomycin See TETRACYCLINES.

decoy receptor See e.g. INTERLEUKIN-1.

decoyinine See PSICOFURANINE.

decurrent (*mycol.*) Refers to hymenium-bearing structures (e.g. lamellae in agarics, the layer of 'teeth' in *Hydnum*) whose region of attachment to the fruiting body extends at least partly down the stipe.

Dee disease See KIDNEY DISEASE.

deep (*noun*) (1) Any solid medium present (in a tube) at a depth sufficient to permit stab inoculation. (2) A SHAKE CULTURE (sense 1).

deep rough mutant See SMOOTH–ROUGH VARIATION.

defaunation The removal of the fauna (animal life) from a given environment – especially the removal of protozoa from a mixed microbial population (e.g. that of the RUMEN).

defective interfering particle (DI particle; defective interfering virus) A DEFECTIVE VIRUS which typically arises as a result of mutation (usually deletion mutation) in an originally non-defective virus (the *standard virus*); a DI particle can replicate only in the presence of the standard virus, and its presence reduces the yield of infectious standard virus (i.e., it exhibits INTERFERENCE). (The preferential production of DI particles at the expense of standard virus is known as *enrichment*.) The generation of DI particles and their ability to cause interference depend, at least to some extent, on the nature of the host cell. The mechanisms of DI particle generation, enrichment and interference are poorly understood. [Primary structure of poliovirus DI particle genomes and possible mechanism for their generation: JMB (1986) *192* 473–487.]

DI particles commonly accumulate in stocks of animal viruses which have been passaged at high multiplicity of infection (moi). (Passage at low moi minimizes the chances of a cell being coinfected with both DI and standard viruses, so that any DI particles which arise are unlikely to be able to replicate.) DI particles can attenuate the pathological effects of a virulent standard virus (and in some cases of other closely related viruses) both in cell cultures and in animals under experimental conditions; however, the presence of DI particles may also allow the establishment of a persistent infection of cells or animals by a virus which does not normally exhibit PERSISTENCE [e.g. Semliki Forest virus in mice: JGV (1986) *67* 1189–1194]. The role of DI particles – if any – in attenuating disease or in establishing persistent infections under natural conditions is unknown.

defective virus A virus which – inherently or e.g. as a result of mutation – lacks one or more genetic functions necessary for its replication; such a virus can produce progeny in the presence of another (non-defective) HELPER VIRUS which can provide the missing functions. (See also e.g. DEFECTIVE INTERFERING PARTICLE; RETROVIRIDAE; SATELLITE VIRUS; TRANSDUCTION.)

defensins PEPTIDE ANTIBIOTICS, produced e.g. by mucosal epithelial cells and macrophages, which are active against a range of bacteria, including both Gram-positive and Gram-negative species; examples: the *tracheal antimicrobial peptide* found in bovines, *cryptdin* in mice, and *human neutrophil peptide*. [See e.g. Nature (2003) *422* 522–526.]

defined medium See MEDIUM.

definitive host *Syn.* FINAL HOST.

DEFT Direct epifluorescent filter technique: a technique used e.g. for the rapid detection and/or quantification of microorganisms in water, milk etc. Essentially, the sample is passed through a membrane filter (see FILTRATION), and the cells retained on the filter are counted by epifluorescence MICROSCOPY – the cells having been stained with a fluorescent dye either before or after filtration. For the examination of MILK, the sample is pretreated with a protease (e.g. trypsin) and a detergent (e.g. Triton X-100) to disrupt, respectively, the casein micelles and fat globules which would otherwise block the membrane filter.

degeneracy (*mol. biol.*) See GENETIC CODE.

degerming *Syn.* ANTISEPSIS.

DegP protease See FIMBRIAE (P fimbriae).

degradosome In *Escherichia coli*: a multicomponent complex which includes RNASE E and certain other proteins involved in the degradation/processing of RNA molecules; degradosomes are associated with the cytoplasmic membrane of the cell via the N-terminal part of RNase E. Proteins which have been identified in degradosomes include polynucleotide phosphorylase, enolase, RhlB (RNA helicase), polynucleotide phosphate kinase, DnaK and GroEL. [RNA degradosomes in vivo: PNAS (2001) *98* 63–68.]

(See also PROTEASOME.)

degranulation (*immunol.*) See MAST CELL.

degron Any feature of a protein which acts as a signal for that protein's *in vivo* degradation. One example is the identity of the protein's N-terminal amino acid residue: see N-END RULE.

dehiscence Opening on maturity.

dehydration (1) (of specimens) The replacement of water, in a specimen, by a non-aqueous medium, e.g. ethanol. (See also ELECTRON MICROSCOPY; CRITICAL POINT DRYING.) (2) (of foods) See FOOD PRESERVATION (c). (3) (of microorganisms) See DESICCATION.

dehydroemetine See EMETINE.

dehydrogenase An OXIDOREDUCTASE which catalyses the removal of hydrogen atom(s) from a substrate – the hydrogen being donated to an acceptor *other than* molecular oxygen; hydrogen acceptors include pyridine nucleotides and flavin nucleotides. (cf. OXIDASE.)

Deinococcus A genus of GRAM TYPE-negative (GRAM REACTION-positive) bacteria. *D. radiodurans* (formerly *Micrococcus radiodurans*) is highly resistant to IONIZING RADIATIONS; the cells are red-pigmented cocci in which the outer layers of the cell wall (external to the peptidoglycan) include an OUTER MEMBRANE and an S LAYER. Carbohydrate chains originating at the outer membrane pass through the S layer and form the outermost surface of the cell. The S layer – also called the HPI (= hexagonally packed intermediate) layer – is composed of a single polypeptide species. [Three-dimensional structure of the HPI layer: JMB (1986) *187* 241–253.] *D. radiodurans* does not contain phosphatidylglycerol or phospholipids derived from it. [Polar lipid profiles of *Deinococcus*: IJSB (1986) *36* 202–206.] The genome is toroidal [Science (2003) *299* 254–256].

Dekkera A genus of yeasts (family SACCHAROMYCETACEAE) which reproduce by budding; pseudomycelium is usually formed. Asci are evanescent and are formed directly from (diploid) vegetative cells; ascospores are more or less bowler-hat-shaped, 1–4 per ascus. NO_3^- may or may not be assimilated. Two species: *D. bruxellensis* (anamorph: *Brettanomyces bruxellensis*) and *D. intermedia* (anamorph: *B. intermedius*), found in beers, wines etc. (See also BRETTANOMYCES.) [Book ref. 100, pp. 146–150.]

delavirdine See ANTIRETROVIRAL AGENTS.

delayed enrichment method See AUXOTROPH.

delayed hypersensitivity (DH; delayed-type hypersensitivity; cell-mediated hypersensitivity; type IV reaction) (*immunol.*) In a PRIMED individual: a HYPERSENSITIVITY (sense 1) response which (i) follows the second (or subsequent) exposure to the given antigen, (ii) is mediated by antigen-specific T LYMPHOCYTES, and (iii) involves a local inflammatory reaction that reaches maximum intensity ~24–48 hours after antigenic challenge. (cf. IMMEDIATE HYPERSENSITIVITY.)

Priming and subsequent antigenic challenge may occur in the skin or elsewhere. Initial priming (or 'sensitization'), in which the individual is initially exposed to the given antigen, involves interaction between an ANTIGEN-PRESENTING CELL (e.g. a DENDRITIC CELL) and an antigen-specific $CD4^+$ T cell. The T cell forms a Th1 clone: when activated by antigen it proliferates and secretes CYTOKINES that include INTERLEUKIN-2. Hence, a primed individual contains an expanded clone of specifically reactive T cells (referred to as T_{DH}, T_{DTH} or Tdth). Sensitization may require a period of ~1–2 weeks.

Following sensitization, a DH reaction can be elicited by further exposure to specific antigen; as before, antigen is presented by an APC, but in this instance the (primed) individual contains an expanded clone of specifically reactive T cells. The secreted cytokines include e.g. interleukin-2, INTERLEUKIN-12 and IFN-γ (see INTERFERONS); the cytokines (i) promote expansion of the Th1 subset of T cells (which secrete more pro-inflammatory cytokines) and (ii) bring about the accumulation of activated MACROPHAGES at the site of antigenic challenge. In the skin, the result is typically an indurated (hard), erythematous (red) nodule which develops in 24–48 hours of antigenic challenge.

The lesion commonly resolves slowly. However, when there is a failure to eliminate the antigen (or pathogen) the ongoing presence of a concentration of activated macrophages can result in damage to the host's tissues. In some diseases aggregations of macrophages give rise to epithelioid cells and, subsequently, GIANT CELLS within lesions called GRANULOMAS; *tubercles* are granulomas formed in tuberculosis.

DH to a given antigen can be transferred to a non-primed subject only by the transfer of *cells*, not serum or plasma alone. (cf. JONES–MOTE SENSITIVITY.)

DH reactions can be modified by various drugs; for example, glucocorticosteroids can inhibit DH reactions, and certain antitumour agents, such as mitomycin C, have been found to stimulate DH reactions in mice.

DH reactions form the basis of certain diagnostic SKIN TESTS. Thus, an individual suffering from a given microbial disease may be primed in respect of certain antigens of the pathogen; hence, if the patient is given a cutaneous injection of specific antigen, the primed state may be revealed by a DH inflammatory response (a *positive* skin test) at the site of the injection. (cf. ANERGY.)

(See also CONTACT SENSITIVITY.)

delayed-type hypersensitivity *Syn.* DELAYED HYPERSENSITIVITY.

deletion mutation A type of MUTATION in which one or more nucleotides are lost from the genome; if the number of nucleotides lost is not divisible by 3, the mutation will be a FRAMESHIFT MUTATION. (cf. INSERTION MUTATION.)

Deleya A genus which accommodates the marine species *Alcaligenes aestus*, *A. pacificus*, *A. cupidus* and *A. venustus*, and *Pseudomonas marina* [IJSB (1983) *33* 793–802]; another species, *D. halophila*, was isolated from hypersaline soils [IJSB (1984) *34* 287–292].

Delhi boil See CUTANEOUS LEISHMANIASIS.

Δ (delta plasmid; also: delta transfer factor) An Inclα, low-COPY-NUMBER enterobacterial CONJUGATIVE PLASMID. [Structure of the delta plasmid: JGM (1986) *132* 3261–3268.]

delta agent *Syn.* DELTA VIRUS.

δ antigen HDAg: see DELTA VIRUS.

***delta* chain** (*immunol.*) See HEAVY CHAIN.

δ-endotoxin A glycoprotein entomotoxin produced by *Bacillus thuringiensis*. Synthesis of the toxin is typically associated with sporulation, the toxin appearing as a (commonly bipyramidal) crystal, the *parasporal crystal*, near the spore within a sporulating cell; in *B. thuringienis* subsp. *yunnanensis* crystals are apparently produced only in asporogenous cells [J. Inv. Path. (1986) *48* 254–256]. (Genes concerned with toxin synthesis are apparently plasmid-borne.) The crystal is composed of subunits of a *protoxin*, the subunits generally being linked by disulphide bonds (which can be disrupted by reducing conditions or by alkaline pH). When crystals and spores are ingested by an insect larva, the crystals dissolve in the alkaline contents of the insect's midgut to release the protoxin subunits; the subunits are converted by midgut proteases to one or more toxic components which cause paralysis of the gut and degeneration of the midgut epithelium. This may kill the insect directly, or may allow *B. thuringiensis* to invade the insect and cause a lethal septicaemia. Larvae may be killed within hours of ingesting the toxin.

Different strains of *B. thuringiensis* are effective against different host ranges: those of 'Group A' – e.g. *B. thuringiensis* subsp. *kurstaki* – are pathogenic primarily in lepidopteran larvae (caterpillars of butterflies and moths), while those in 'Group B' – subsp. *israelensis* – are pathogenic e.g. in mosquito larvae (Diptera); subsp. *tenebrionis* is pathogenic in some coleopteran (beetle) larvae [δ-endotoxin in subsp. *tenebrionis*: FEMS (1986) *33* 261–265].

Strains of *B. thuringiensis* have been used, worldwide, as 'microbial insecticides' for the BIOLOGICAL CONTROL of various lepidopteran pests. Preparations containing spores and crystals (prepared by deep-tank culture of *B. thuringiensis*) are applied e.g. to crops; repeated applications are necessary because, e.g., both spores and (to a lesser extent) crystals tend to be inactivated by the ultraviolet component of sunlight. (To avoid the need for repeated applications, *B. thuringiensis* has been genetically modified so that toxin accumulates within the toxin-producing cells; the partly protected (intracellular) toxin is still highly active [Biotechnology (1995) *13* 67–71].) *Plodia interpunctella*, an important lepidopteran pest of stored grain, has been reported to develop stable and heritable resistance to *B. thuringiensis* crystals within a few generations [Science (1985) *229* 193–195]. (cf. THURINGIENSIN.)

[Reviews: Book ref. 171, pp. 185–209 (commercial aspects), pp. 211–249 (genetics of *B. thuringiensis*); MR (1986) *50* 1–24.]

(cf. MILKY DISEASE.)

δ-factor See RNA POLYMERASE.

delta hepatitis virus See DELTA VIRUS.

delta herpesvirus A herpesvirus which is antigenically related to human varicella-zoster virus (see ALPHAHERPESVIRINAE) and which causes a varicella-like disease in non-human primates.

delta particles See TECTIBACTER.

delta plasmid See DELTA.

δ sequence See TY ELEMENT.

delta transfer factor See DELTA.

delta virus (delta agent; hepatitis D virus; hepatitis delta virus) A defective RNA virus which can replicate only in the presence of a helper virus of the HEPADNAVIRIDAE – usually HEPATITIS B VIRUS (HBV), although under experimental conditions the delta virus can replicate in woodchucks infected with WOODCHUCK HEPATITIS VIRUS. The delta virus virion (as derived from the serum of an HBV-infected patient) is ca. 35–37 nm in diam.; it consists of a core containing the delta virus RNA genome (MWt ca. 5.5×10^5) and a delta virus-encoded protein antigen (HDAg) surrounded by an (HBV-encoded) HBsAg-containing envelope. The genome is circular ssRNA [PNAS (1986) *83* 8774–8778]. [Protein composition of the delta virus virion: JV (1986) *58* 945–950.]

In humans, the delta virus can infect an individual either simultaneously with HBV or subsequent to the establishment of an HBV infection; delta virus infection may itself be acute or chronic. The presence of the delta virus may exacerbate the symptoms of HBV infection, in some cases inducing an acute, severe or fulminant hepatitis – particularly when an acute delta virus infection is superimposed on a pre-existing chronic HBV infection. HDAg can be detected in the serum of the patient only very early in infection, but anti-HDAg antibodies apparently persist indefinitely in chronically infected patients; evidence of active delta virus replication requires immunohistological staining for HDAg in the hepatocytes.

ΔG_{ATP} See CHEMIOSMOSIS.

ΔG_p See CHEMIOSMOSIS.

$\Delta\tilde{\mu}_{H+}$ See CHEMIOSMOSIS.

Δp See CHEMIOSMOSIS.

ΔpH See CHEMIOSMOSIS.

Δψ See CHEMIOSMOSIS

dematiaceous (*mycol.*) Darkly pigmented.

demeclocycline See TETRACYCLINES.

demecolcine Deacetyl-*N*-methyl-COLCHICINE.

Demerec convention A widely-followed convention for bacterial genetic nomenclature [Genetics (1966) *54* 61–76].

demethylchlortetracycline See TETRACYCLINES.

demicyclic rusts Those rust fungi (see UREDINIOMYCETES) which form all types of spore except uredospores; they include e.g. most species of *Gymnosporangium*.

demyelination (demyelinization) The removal or destruction of the myelin sheath of one or more nerve fibres. (See also MULTIPLE SCLEROSIS; PROGRESSIVE MULTIFOCAL LEUKOENCEPHALOPATHY; VISNA.)

denaturing gradient gel electrophoresis See DGGE.

Dendriscocaulon See CEPHALODIUM.

dendritic cells Motile, non-phagocytic ADHERENT CELLS, derived from the bone marrow, which are present e.g. in spleen and lymph nodes. Dendritic cells are irregularly shaped, smooth-surfaced cells which contain pulsatile nuclei and many spherical mitochondria; they bear class II MHC antigens (cf. LANGERHANS' CELLS) and can act as ANTIGEN-PRESENTING CELLS (see e.g. DELAYED HYPERSENSITIVITY) and as accessory cells for cytotoxic T cells. [Minireviews: Cell (2001) *106* 255–274.]

dendrodochiotoxicosis (myrotheciotoxicosis) A MYCOTOXICOSIS, affecting domestic animals and man, caused by roridin and verrucarin TRICHOTHECENES produced by *Myrothecium* spp.

Dendrophoma See SPHAEROPSIDALES.

Dendrosoma See SUCTORIA.

dengue (breakbone fever) An acute, tropical and subtropical human disease caused by any of four serotypes of dengue virus (family FLAVIVIRIDAE) and transmitted by mosquitoes (usually *Aedes aegypti*). (The genetic diversity of dengue virus is reported to be increasing, and strains of the virus may now, or in the future, exhibit differences in virulence [TIM (2000) *8* 74–77].) Replication of the virus may occur chiefly in macrophages.

In most cases the disease is benign and self-limiting (*dengue fever*). After an incubation period of ~5–10 days, typical symptoms include fever, headache, joint and muscular pains, and (often) a rash; bleeding (e.g. haematuria or gingival bleeding) is present in some cases. Mortality rates for this 'classical' form of the disease are very low.

Less often, a more severe form of disease may develop: *dengue haemorrhagic fever* (DHF). The manifestations of DHF include high fever, haemorrhagic phenomena, hepatomegaly and thrombocytopenia. The severity of disease is influenced by the extent of plasma leakage (as indicated by the rise in haematocrit reading). The most severe forms of DHF (World Health Organization grades III and IV) are designated *dengue shock syndrome* (DSS). DSS develops suddenly, after several days' fever, with signs of circulatory failure/hypovolaemic shock (rapid, weak pulse, narrowed pulse pressure or hypotension, and cold clammy skin); death may occur in 12–24 hours without suitable treatment, but recovery may be rapid following volume-replacement therapy.

[Dengue haemorrhagic fever (review): RMM (1995) *6* 39–48.]

In common with many other diseases caused by arthropod-borne viruses, various manifestations of dengue infection (including those in which there is little or no haemorrhage) are characterized by a transient leukopenia [dengue-induced suppression of bone marrow: BCH (1995) *8* 249–270]. [Haematology of dengue fever and dengue haemorrhagic fever: BCH (2000) *13* 261–276.]

Detection of dengue virus in samples of serum can be achieved by rtPCR; in primary infection, detection of viral RNA seems to be optimal prior to seroconversion (soon after onset of symptoms) [JCM (1999) *37* 2543–2547].

denitrification Energy-yielding (respiratory) metabolism in which nitrate or nitrite (as terminal electron acceptor) is sequentially reduced to gaseous products, mainly dinitrogen and/or nitrous oxide (cf. DISSIMILATORY NITRATE REDUCTION). It is generally stated that denitrification occurs only under anaerobic or microaerobic conditions; however, aerobic denitrification has been reported [AvL (1984) *50* 525–544; denitrification in *Rhizobium*: SBB (1985) *17* 1–9].

In some DENITRIFYING BACTERIA (e.g. *Paracoccus denitrificans*) electrons from e.g. NADH are transferred to NO_3^- via a *b*-type cytochrome, and to NO_2^- and N_2O via *c*-type cytochromes; electron flow to NO_3^-, NO_2^- and N_2O apparently generates proton motive force. Nitrate reductase is a MOLYBDOENZYME; in some species cytochrome d_1c acts as a nitrite reductase.

(See also NITROGEN CYCLE).

denitrifying bacteria Those bacteria capable of DENITRIFICATION; they occur e.g. in soil and in marine and freshwater environments. (Some species can be useful in the elimination of nitrate from waste water.) Examples include e.g. *Bacillus licheniformis*, *Hyphomicrobium* sp, *Paracoccus denitrificans*, *Pseudomonas stutzeri* and *Thiobacillus denitrificans*.

density gradient centrifugation See CENTRIFUGATION.

densonucleosis viruses See DENSOVIRUS.

Densovirus (insect parvovirus group; also: densonucleosis viruses, DNVs) A genus of autonomous (helper-independent) viruses (family PARVOVIRIDAE) which infect insects of the Lepidoptera (and probably of the Diptera and Orthoptera). Both positive- and negative-strand ssDNA molecules are encapsidated (in separate virions). Replication occurs in most tissues of host larvae, nymphs and adults; hypertrophy of the nucleus occurs in infected cells, and virions accumulate in dense intranuclear inclusions. Infected insects may die. Type species: densovirus of *Galleria mellonella* (the greater wax moth, whose larvae – known as wax-worms – feed on pollen and wax in beehives). Other members include densovirus of *Junonia* (a genus of butterflies); similar viruses have been detected in e.g. *Aedes* (mosquitoes), *Bombyx* (silk-moth), and butterflies of the Nymphalidae.

dental caries Tooth decay: typically, a process in which acids (particularly lactic acid), formed by plaque microorganisms, cause localized breakdown (demineralization) of tooth enamel, thus exposing the dentine; an important factor in the development of caries is the formation of water-insoluble extracellular microbial glucans which promote adhesion of cariogenic bacteria to the tooth surfaces (see DENTAL PLAQUE). Various bacteria, including e.g. lactobacilli and *Streptococcus mutans*, can be cariogenic in man; the development of carries may depend on factors such as the nature of the plaque microflora, diet, and the level of immunity in the host (see also CREVICULAR FLUID).

Fluoride is a potent anti-caries agent. It converts the apatite in tooth enamel to the more stable (acid-resistant) fluorapatite, and this is believed to be the primary mode of fluoride anti-caries activity. Fluoride can also inhibit bacterial metabolism e.g. by blocking the uptake of sugars by the PEP-dependent phosphotransferase system (see FLUORIDES). However, in some oral streptococci (including *S. mutans*) there is an alternative, fluoride-insensitive, pmf-dependent uptake system for glucose; this is a low-affinity uptake system compared to the high-affinity PTS system (the latter operating as a scavenger system under conditions of substrate limitation). At low pH (e.g. 5.5) the fluoride-insensitive system appears to be more important than the PTS system for the uptake of glucose.

dental plaque On the surfaces of teeth: a thin film (up to ca. 0.5 mm thick) consisting of a mixed microbial flora embedded in a matrix composed largely of extracellular (capsular) bacterial polysaccharides and salivary polymers; it is a predisposing factor in DENTAL CARIES and PERIODONTITIS.

Early colonizers of a clean enamel surface (in man) include species of *Neisseria* and *Streptococcus* (particularly *S. sanguis*). If plaque is allowed to build up, the nature of the microflora changes with time; although streptococci tend to remain dominant, anaerobes and filamentous bacteria begin to appear in significant numbers after several days or a week. Organisms present in well-established plaque include e.g. species of ACTINOMYCES (particularly *A. viscosus*), BACTEROIDES and VEILLONELLA; there is a general correlation between the numbers of *Streptococcus mutans* (see STREPTOCOCCUS) and the dietary intake of sucrose. The nature of the plaque microflora varies e.g. from one individual to another, and from one site to another on a given tooth.

Bacterial polymers typically present in plaque include MUTAN and LEVANS.

The presence of plaque can be detected by rinsing the mouth with a solution of a dye ('disclosing agent') such as erythrosin.

Dentalina See FORAMINIFERIDA.

deodorants (skin) See SKIN MICROFLORA.

5′-deoxyadenosylcobalamin See VITAMIN B_{12}.

deoxycholate A salt of deoxycholic acid (see BILE ACIDS).

deoxycholate–citrate agar See DCA.

deoxycholic acid See BILE ACIDS.

deoxycoformycin *Syn.* pentostatin (see VIDARABINE).

2′-deoxy-2′-fluoro-5-methyl-1-β-D-arabinosyluracil See FMAU.

deoxyhemigossypol A terpenoid PHYTOALEXIN produced by the cotton plant. [Role of terpenoid phytoalexins in the resistance of a cultivar of *Gossypium barbadense* to *Verticillium dahliae*-induced wilt: PPP (1985) *26* 209–218.]

deoxynivalenol See TRICHOTHECENES.

deoxyribonuclease (DNase; DNAase) An enzyme which depolymerizes DNA. A DNase is designated DNase I (EC 3.1.21.1) or DNase II (EC 3.1.22.1) according to whether the mono- or oligonucleotide products of its action have (terminal) 5′- or 3′-phosphate groups, respectively. DNases are produced e.g. by the pancreas and by many microorganisms.

In *staphylococci*, production of a thermostable, Ca^{2+}-requiring, extracellular DNase ('thermonuclease': a 5′-phosphodiesterase with both endo- and exonuclease activity, yielding 3′-nucleotides) correlates closely with the production of COAGULASE; thermonuclease also has Ca^{2+}-dependent RNase activity. (Many coagulase-negative staphylococci produce a DNase which is generally heat-labile.) DNase activity may be detected/assayed e.g. by measuring changes in turbidity, or increasing levels of acid-soluble nucleotides (measured by spectrophotometry at 260 nm), induced in a suspension of DNA. For the detection of DNase production on solid media, an agar medium (e.g. trypticase–soy agar) containing DNA and a calcium salt is inoculated with the test strain and incubated for 18–24 hours. The plate is then flooded with 1 N HCl to precipitate DNA. A clear zone (in which the DNA has been hydrolysed) surrounds each DNase-producing colony. Certain dyes – e.g. methyl green (which combines only with highly polymerized DNA) – may be used instead of HCl, in which case cell viability may be preserved. [Book ref. 44, pp. 757–768.]

Certain *Clostridium* spp produce DNases: e.g. the β-toxins of *C. chauvoei* and *C. septicum*.

See also STREPTODORNASE.

deoxyribonucleic acid See DNA.

deoxyribonucleoside See NUCLEOSIDE.

deoxyribophage A BACTERIOPHAGE with a DNA genome.

deoxyribovirus A VIRUS with a DNA genome.

2-deoxystreptamine 2-Deoxy-1,3-diaminoinositol, a component of many AMINOGLYCOSIDE ANTIBIOTICS.

DEPC DIETHYLPYROCARBONATE.

Dependovirus (adeno-associated viruses, AAVs; adeno-satellite viruses) A genus of defective viruses (family PARVOVIRIDAE) which are totally dependent on functions of a co-infecting helper adenovirus (or herpesvirus) for their replication. Dependoviruses can infect a wide range of vertebrate hosts. Type species: AAV-1 (AAV type 1), which infects mainly monkeys; other members: AAV-2 and AAV-3 (which can infect man), AAV-4 (common e.g. in African green monkeys), avian AAV, bovine AAV and canine AAV. (Equine AAV and ovine AAV are probable members of the genus.) In vitro, a given AAV seems able to replicate in any type of cell culture which can support the replication of a suitable helper virus.

Genome: linear ssDNA, 4675 bases long in AAV-2 [sequence in AAV-2: JV (1983) *45* 555–564]. Both positive and negative strands are encapsidated (in separate virions). The DNA has terminal INVERTED REPEATS 145 bases long. Within each repeat, the first 125 bases include a PALINDROMIC SEQUENCE in which bases 1–41 are complementary to bases 125–85; between bases 41 and 85 are two shorter palindromic sequences (bases 42–62 and 64–84). Folding which permits maximum base-pairing thus results in a T-shaped structure. (cf. PARVOVIRUS.)

Initial stages of AAV infection are independent of helper virus; the virions adsorb to a host cell and subsequently enter the nucleus, where the DNA is uncoated. In the presence of a co-infecting helper virus, the DNA is replicated and transcribed, its mRNA is translated, and infectious virions are formed. DNA replication appears to be initiated at the hairpin structures and is independent of RNA primers. In the absence of a helper virus, the AAV establishes a latent state by integrating into the host DNA as a provirus. Subsequent infection of the latently infected cell by a helper virus 'rescues' the AAV DNA, allowing the formation of infectious AAV virions.

AAVs appear not to cause disease in their hosts. In vitro, they may have a strong inhibitory effect on the replication of their helper viruses, but in vivo they seem not to ameliorate symptoms of adenovirus infection to any significant extent (although AAVs have been shown to reduce the oncogenic potential of adenoviruses in hamsters and of herpes simplex virus in cell cultures). AAV-2 infecting cell cultures co-infected with human cytomegalovirus (HCMV) as helper apparently exacerbates HCMV-induced CPE [Virol. (1985) *147* 217–222].

[Reviews in Book ref. 97.]

depside An ester formed by the condensation of two or more phenolcarboxylic acids; e.g., COOH.ArO.CO.ArOH is a generalized didepside where Ar = a phenolic group. Depsides occur e.g. as 'lichen substances' (see LICHEN) and as components of TANNINS (e.g. digallic acid).

depsipeptide antibiotics ANTIBIOTICS which are derivatives of depsipeptides (molecules containing both ester and peptide bonds). They include the cyclic octadepsipeptide intercalating agents (QUINOXALINE ANTIBIOTICS) and the cyclic depsipeptide ionophores (see e.g. ENNIATINS and VALINOMYCIN). (See also DIDEMNINS.)

Derbesia A genus of siphonaceous, tropical or temperate, epilithic or epiphytic green seaweeds (division CHLOROPHYTA) which exhibit a heteromorphic alternation of generations. The sporophyte and gametophyte are so distinct morphologically that they have been described as separate plants: e.g. *Halicystis*, with a bulbous vesicular thallus, is the gametophyte of a filamentous *Derbesia* sporophyte; some species of *Bryopsis* have also been shown to have *Derbesia* sporophyte stages. *Derbesia* sporophytes produce stephanokont zoospores, while gametophytes produce biflagellate anisogametes. (See also CELL WALL.)

derepressed (*mol. biol.*) Refers to the state of a repressible gene or operon when it is not repressed, i.e., when the gene or operon is expressed. Derepression may be brought about e.g. by the presence of a specific inducer: see OPERON. (See also DRD PLASMID.)

dermatitis Inflammation of the skin. It is commonly due to mechanical or chemical irritants or to allergic reactions, but may occur in certain infectious diseases (e.g. RINGWORM.) (cf. ECZEMA.)

dermatitis verrucosa *Syn.* CHROMOBLASTOMYCOSIS.

Dermatocarpon A genus of foliose LICHENS (order VERRUCARIALES); photobiont: a green alga. Thallus: small, more or less lobed, attached to the substratum (rocks) by a central holdfast (umbilicus – cf. UMBILICARIA). Pseudothecia occur immersed in the thallus, their ostioles visible as black dots in the thallus surface. Some species – e.g. *D. rivulorum*, *D. weberi* (formerly *D. fluviatile*) – occur on rocks in or adjacent to unpolluted streams or lakes. (cf. CATAPYRENIUM.)

dermatomycosis Any MYCOSIS affecting the skin – see e.g. CANDIDIASIS, PITYRIASIS VERSICOLOR, RINGWORM.

dermatophilosis A disease of animals, transmissible (by direct contact) to man, caused by *Dermatophilus congolensis* (see e.g. LUMPY WOOL and STRAWBERRY FOOT-ROT). In man, self-limited pustular lesions develop on the skin. (cf. PITTED KERATOLYSIS.)

Dermatophilus A genus of aerobic or facultatively anaerobic, catalase-positive bacteria (order ACTINOMYCETALES, wall type III) which occur as pathogens of man and other animals (see DERMATOPHILOSIS). In culture the organisms initially form filaments (<1 μm in diameter) which become branched and undergo transverse and longitudinal septation to form long, tapering hyphae, up to 5 μm in diameter, which bear a tough gelatinous capsule. Further septation gives rise to coccoid forms which develop into zoospores (up to ca. 1 μm in diameter), each having a single tuft of flagella. Growth occurs at 37°C on enriched media (e.g. brain–heart infusion medium), and is enhanced by increased partial pressures of CO_2; the colonies are initially dry and greyish-white but later become mucoid and yellowish-orange. Clear haemolysis occurs in media containing sheep blood but not in media containing horse blood. Casein, gelatin and starch are metabolized. GC%: ca. 57–59. Type species: *D. congolensis*. [Morphology, ecology, isolation: Book ref. 46, pp. 2011–2015; differentiation between *Dermatophilus* and *Geodermatophilus* by 16S rRNA oligonucleotide cataloguing: JGM (1983) *129* 1831–1838.]

dermatophytes A group of fungi which can infect and degrade keratinized tissues (skin, nails, hair etc) in living animals (including man), causing RINGWORM. Dermatophytes belong to the HYPHOMYCETES, and many have teleomorphs in the GYMNOASCALES; the three main genera are *Epidermophyton*, *Microsporum* (teleomorph *Nannizzia*), and *Trichophyton* (teleomorph *Arthroderma*). The organisms form a hyaline, septate mycelium which in many species tends to fragment to form arthrospores. (See also RACQUET MYCELIUM.) When infecting hair, some dermatophytes (e.g. *T. tonsurans* and *T. violaceum*) form both hyphae and arthrospores largely or wholly within the hair shaft (*endothrix* infection); other species (e.g. *T. mentagrophytes*, *Microsporum* spp) may also form hyphae within the hair shaft, but arthrospores develop only on the surface of the hair (*ectothrix* infection). In culture, most dermatophytes produce both single-celled microconidia and septate macroconidia. (*Epidermophyton* spp do not form microconidia.) Microconidia are generally oval or pyriform, borne singly or in grape-like clusters along the sides of the hyphae. Characteristics of the macroconidia are used in the classification of genera and species. In *Epidermophyton* spp, macroconidia are produced abundantly and are smooth-surfaced, 2-4-septate, ca. 6–8 × 15–40 μm, and borne in groups of 2 or 3. In *Microsporum* spp, macroconidia are usually numerous (but rare in *M. audouinii*) and are rough-surfaced, usually thick-walled, 3–15-septate, ca. 3–8 × 40–120 μm, and borne singly. In *Trichophyton* spp, macroconidia are usually sparse and are smooth-surfaced, usually thin-walled, 2–8-septate, ca. 4–6 × 10–50 μm, borne singly.

Some (anthropophilic) dermatophytes occur mainly or only as human pathogens; these include e.g. *E. floccosum* (ATHLETE'S FOOT), *M. audouinii* (tinea capitis), and *T. schoenleinii* (tinea capitis, FAVUS). Zoophilic dermatophytes commonly show some degree of host specificity: e.g., *M. canis* (e.g. cats and dogs), *T. verrucosum* (e.g. cattle, horses, pigs, goats), *T. equi* (horses); most or all zoophilic species can also infect man. Certain species (e.g. *M. gypseum*, *T. ajelloi*) are said to be 'geophilic': they apparently grow as saprotrophs (e.g. in soil, birds' nests), but can also infect man and animals under certain conditions.

dermatophytosis *Syn.* RINGWORM.

Dermocarpa A genus of unicellular CYANOBACTERIA (section II) in which the vegetative cells are rounded and variable in size; baeocytes initially lack a fibrous outer cell wall layer and are motile (by gliding). GC%: ca. 38–44.

Dermocarpales *Syn.* Chamaesiphonales (see CYANOBACTERIA).

Dermocarpella A genus of unicellular CYANOBACTERIA (section II) in which growth and binary fission result in pear-shaped aggregates consisting of one or two basal cells and a larger apical cell; multiple fission of the apical cell yields baeocytes which initially lack a fibrous outer cell wall layer and are motile (by gliding). GC%: ca. 45.

Dermocystidium See PERKINSUS.

dermonecrotic Causing necrosis of the skin.

dermonecrotic toxin A TOXIN produced by species of *Bordetella*; its role (if any) in the pathogenesis of WHOOPING COUGH has not been determined, but it is known to cause e.g. re-arrangement of actin filaments within eukaryotic cells.

dermotropism See TROPISM (sense 2).

Derrick's method See IMMUNOSORBENT ELECTRON MICROSCOPY.

Derxia A genus (*incertae sedis*) of Gram-negative, aerobic, catalase-negative, chemoorganotrophic, facultatively chemolithoautotrophic bacteria which occur e.g. in acidic tropical soils (e.g. in Brazil, China, Indonesia, West Bengal). Cells: typically round-ended pleomorphic rods, ca. 1.0–1.2 × 3.0–6.0 μm, with a single flagellum at one pole (or sometimes at each pole). Metabolism is respiratory (oxidative), with O_2 as terminal electron acceptor. Carbon sources include glucose, fructose, ethanol and mannitol; little or no growth occurs on acetate, lactate, malate or succinate. NITROGEN FIXATION can occur under atmospheric conditions. Chemolithoautotrophic growth can occur with mixtures of H_2, CO_2 and O_2 (CO_2 being fixed via the Calvin cycle), N_2 or NH_4^+ serving as a source of nitrogen; growth on CH_4 or CH_3OH (see METHYLOTROPHY) has been reported. A pellicle is formed in liquid media; broth cultures become gelatinous. Colonies are slimy, becoming raised, wrinkled, and subsequently dark brown. Optimum growth temperature: 25–35°C. Growth occurs within the pH range 5.5-ca. 9.0. GC%: ca. 69–73. Type species: *D. gummosa*.

[Book ref. 22, pp. 321–325.]

desensitization (*syn.* hyposensitization) (*immunol.*) The repeated administration of known quantities of allergen to an individual in whom it usually provokes an immediate or delayed HYPERSENSITIVITY reaction; in some cases this treatment temporarily reduces or abolishes hypersensitivity to the given allergen. In one method used for treating allergy to mould(s), an aqueous extract of the allergen(s) – up to 100-fold weaker than that required to produce an intradermal hypersensitivity reaction – is administered subcutaneously; the dose, initially given weekly, is gradually increased and administered after longer intervals of time for a total period of several years [Book ref. 51, pp. 187–201].

Little is known about the mechanism(s) of desensitization. Possible mechanisms include e.g. (i) the formation of BLOCKING ANTIBODIES (sense 2), and (ii) the stimulation of suppressor T cells.

desert fever *Syn.* COCCIDIOIDOMYCOSIS.

Desert Shield virus See SMALL ROUND STRUCTURED VIRUSES.

desferrioxamine A hydroxamate SIDEROPHORE formed e.g. by *Streptomyces pilosus*. It is used e.g. in the treatment of patients with iron overload.

desiccation (of microorganisms) The removal of water: a process used e.g. for the PRESERVATION of viable populations of certain types of microorganism (including enterobacteria and spore-forming bacteria). However, the method described below is lethal for many types of microorganism, including various pathogens (e.g. *Neisseria gonorrhoeae*, *Treponema pallidum*) and many aquatic organisms; some organisms that are killed by this procedure can be preserved by FREEZE-DRYING.

Initially, the organisms are suspended in an appropriate medium (e.g. broth, serum, or 5–10% nutrient gelatin). Drops of the suspension are then placed, singly, onto a sterile filter paper (or a piece of waxed paper) which is inserted into a glass ampoule. The ampoule is placed in a desiccator which contains phosphorus pentoxide (as drying agent) and which is subsequently evacuated. The use of a vacuum is essential because it reduces the time for which cells are exposed to the deleterious effects of high concentrations of solute. When the sample is completely desiccated the ampoule is sealed. Under optimal conditions the organisms remain viable for years; spores may remain viable for decades.

In general, desiccation is tolerated to different degrees by different organisms; many do not survive for long in the air-dried state. [Desiccation tolerance of prokaryotes: MR (1994) *58* 755–805.]

desmarestene See PHEROMONE.

Desmarestia See PHAEOPHYTA.

Desmidium A genus of placoderm DESMIDS.

desmids A group of freshwater, basically unicellular green algae (division CHLOROPHYTA) which reproduce sexually by conjugation and which never form flagellated cells of any kind. In most species the cells are solitary, although in a few they are joined end-to-end to form filaments (see e.g. HYALOTHECA). The cells are generally more or less bilaterally symmetrical and may be cylindrical, fusiform, crescent-shaped, etc, sometimes bearing elaborate ornamentations. Many species are highly variable, occurring in a range of growth forms – a feature which has caused considerable confusion in speciation and nomenclature [Book ref. 123, pp. 251–269].

In *placoderm desmids* each cell is divided into two 'semicells' either by a median suture line (e.g. in *Closterium* and *Penium*) or, more commonly, by an equatorial constriction (*sinus*) – the narrow region joining the semicells being called the *isthmus*. The cell wall is perforated by pores through which mucilage may be secreted; the mucilage may allow the desmid to adhere to substrata, and is apparently necessary for the GLIDING MOTILITY often exhibited by these cells. Genera of placoderm desmids include e.g. CLOSTERIUM, COSMARIUM, *Desmidium*, *Euastrum*, HYALOTHECA, *Micrasterias*, PENIUM, STAURASTRUM, *Staurodesmus*, and XANTHIDIUM.

In *saccoderm desmids* the cells do not have a median suture line or sinus and have smooth walls without pores. They nevertheless produce mucilage, often in copious amounts, and some can exhibit gliding motility. Genera include e.g. *Ancylonema* (short cylindrical cells joined end-to-end to form short filaments), *Cylindrocystis* (cells cylindrical, round-ended, containing a stellate chloroplast at each end), and *Mesotaenium* (cells short, cylindrical, solitary, containing a single chloroplast).

Sexual reproduction in desmids involves a process of conjugation analogous to that in SPIROGYRA. Essentially, cells come together in pairs, a conjugation tube is formed between them, and the protoplasts function as amoeboid gametes, fusing to form a zygote which then develops a thick, resistant wall.

Desmids occur in various freshwater habitats, particularly in acidic, unpolluted pools, bogs etc; some (e.g. *Cylindrocystis*, *Mesotaenium*) may occur in mucilaginous, amorphous colonies on damp soil, wet rocks etc.

[Book ref. 128.]

Desmococcus See PLEUROCOCCUS.

desmodexy See RULE OF DESMODEXY.

desmosome One of a number of differentiated regions which occur e.g. where the cytoplasmic membranes of adjacent epithelial cells are closely apposed; it consists of a circular region (ca. 0.5 μm diam.) of each membrane together with associated intracellular microfilaments and an intercellular material which may include e.g. mucopolysaccharides. Desmosomes can be disrupted e.g. by hyaluronidase and Ca^{2+} chelators. (See also TRYPANOSOMA.)

Desmothoracida An order of protozoa (class HELIOZOEA) in which the cell is enclosed in a reticulate, basket-like capsule which is usually spherical and generally composed of an organic matrix often impregnated with silica; axopodia project through perforations in the capsule. In some species a stalk from the capsule attaches the organism to the substratum. There is no centroplast (cf. CENTROHELIDA). Uniflagellate or biflagellate zoospores may be formed. Genera include e.g. CLATHRULINA.

desmotubule In plant cells: an extension of the endoplasmic reticulum which is continuous from one cell to the next via the PLASMODESMA.

desoxy- A prefix equivalent to 'deoxy-'; thus, e.g., for desoxycholate see DEOXYCHOLATE.

desquamation The shedding of skin, cuticle etc in sheets or flakes.

destroying angel See AMANITA.

destruxins See METARHIZIUM.

Desulfobacter A genus of Gram-negative SULPHATE-REDUCING BACTERIA which occur e.g. in aquatic habitats (primarily brackish and marine). (See also ANAEROBIC DIGESTION.) Cells: non-motile or motile (polarly monoflagellate), rod-shaped to ellipsoidal, ca. 1.0–2.0 × 1.7–3.5 μm. Chemoorganotrophic. Acetate can be used as a source of carbon and energy, and is oxidized to CO_2 using sulphate (or another oxidized sulphur compound) as terminal electron acceptor (see DISSIMILATORY SULPHATE REDUCTION and TCA CYCLE). The organisms contain *b*- and *c*-type cytochromes; they lack desulfoviridin. Growth is inhibited by elemental sulphur. Typically, at least 0.5% NaCl and 0.1% $MgCl_2$ are required for growth. Optimum growth temperature: 28–32°C. Optimum pH: 7.3. GC%: ca. 46. Type species: *D. postgatei*. [Book ref. 22, pp. 674–676.]

Desulfobulbus A genus of Gram-negative SULPHATE-REDUCING BACTERIA which occur e.g. in freshwater, brackish and marine habitats, and in the RUMEN. Cells: ellipsoidal, ca. 1.0–1.3 × 1.5–2.0 μm; in many strains there is a single polar flagellum. Chemoorganotrophic. Lactate, propionate, pyruvate or certain alcohols are used as sources of carbon and energy, and are oxidized to e.g. acetate using sulphate (or another oxidized sulphur compound) as terminal electron acceptor (see DISSIMILATORY SULPHATE REDUCTION); the organisms contain *b*- and *c*-type cytochromes. Lactate or pyruvate can be fermented in the absence of an external electron acceptor. Optimum growth temperature: 28–39°C. GC%: ca. 60. Type species: *D. propionicus*. [Book ref. 22, pp. 676–677.]

Desulfococcus A genus of Gram-negative SULPHATE-REDUCING BACTERIA which occur e.g. in aquatic habitats (freshwater, brackish and marine) and in anaerobic sewage digesters. Cells: cocci, ca. 1.5–2.2 μm in diameter. Chemoorganotrophic. Benzoate, other aromatic compounds, acetate, lactate, pyruvate and alcohols etc are used as sources of carbon and energy; these substrates are oxidized to CO_2 using sulphate (or another oxidized sulphur compound) as terminal electron acceptor (see DISSIMILATORY SULPHATE REDUCTION). The organisms contain *b*- and *c*-type cytochromes. Lactate or pyruvate can be fermented in the absence of an external electron acceptor. GC%: ca. 57. Type species: *D. multivorans*. [Book ref. 22, pp. 673–674.]

Desulfomonas A genus of Gram-negative SULPHATE-REDUCING BACTERIA which occur e.g. in the human intestinal tract. Cells: non-motile rods. The organisms, which carry out DISSIMILATORY SULPHATE REDUCTION, contain both *b*- and *c*-type cytochromes and desulfoviridin; pyruvate is oxidized to acetate and CO_2. GC%: ca. 66–67. Type species: *D. pigra*. [Book ref. 22, pp. 672–673.]

Desulfonema A genus of SULPHATE-REDUCING BACTERIA which occur e.g. in marine sediments. Cells: gliding filaments, ca. 2.0–7.0 μm in width, several millimetres in length, each being uniseriately multicellular; the organisms stain Gram-positively. Acetate, benzoate and other substrates are oxidized to CO_2. The organisms carry out DISSIMILATORY SULPHATE REDUCTION and contain *b*- and *c*-type cytochromes; desulfoviridin occurs in strains of *D. limicola* but not in those of *D. magnum*. At least some strains have been reported to grow autotrophically. GC%: ca. 34–42.

desulforubidin A (red) sirohaem-containing sulphite reductase which occurs e.g. in *Desulfovibrio desulfuricans* strain Norway 4. (cf. DESULFOVIRIDIN.)

Desulfosarcina A genus of Gram-negative SULPHATE-REDUCING BACTERIA which occur e.g. in brackish and marine waters. Cells: irregularly shaped and sarciniform, or coccoid to ellipsoidal single cells, ca. 1.0–1.5 × 1.5–2.5 μm, occasionally motile by means of a single polar flagellum. Chemoorganotrophic, and reported to be facultatively chemolithoautotrophic. Carbon and energy sources include e.g. acetate, lactate, pyruvate, formate, propionate, and various alcohols and fatty acids; these substrates can be oxidized to CO_2 using sulphate (or another oxidized sulphur compound) as terminal electron acceptor (see DISSIMILATORY SULPHATE REDUCTION). Lactate or pyruvate can be fermented in the absence of an external electron acceptor. Growth requires at least 1% NaCl and 0.2% $MgCl_2.6H_2O$. Optimum growth temperature: 28–33°C. GC%: ca. 51. Type species: *D. variabilis*. [Book ref. 22, pp. 677–679.]

Desulfotomaculum A genus of Gram-negative, endospore-forming SULPHATE-REDUCING BACTERIA which occur e.g. in soil and in the RUMEN. Cells: motile (peritrichously flagellate) rods up to ca. 9 μm in length. Chemoorganotrophic. For most species the carbon and energy sources include lactate and pyruvate; *D. acetoxidans* can oxidize acetate. The organisms carry out DISSIMILATORY SULPHATE REDUCTION and contain *b*-type cytochromes; no species contains desulfoviridin. In addition to electron transport-coupled phosphorylation, *Desulfotomaculum* spp can carry out a substrate-level phosphorylation in which PPi (see PYROPHOSPHATE) reacts reversibly with acetate (enzyme: acetate:PPi phosphotransferase) to form acetyl phosphate and Pi; acetyl phosphate and ADP can then yield acetate and ATP in the presence of acetokinase. GC%: ca. 37–49. Type species: *D. nigrificans* (formerly *Clostridium nigrificans*); other species: *D. acetoxidans*, *D. antarcticum*, *D. orientis*, *D. ruminis*.

Desulfovibrio A genus of Gram-negative SULPHATE-REDUCING BACTERIA which occur e.g. in mud and sediments in freshwater, brackish and marine habitats. Cells: motile, typically curved (sometimes S-shaped or helical) rods, ca. 0.5–1.5 × 2.5–10.0 μm, each having one polar flagellum or a tuft of polar flagella. Metabolism is chemoorganotrophic or mixotrophic, primarily respiratory, and the cells contain *c*-type (and usually *b*-type) cytochromes; sulphate (or another oxidized sulphur compound) is used as a terminal electron acceptor and is reduced to H_2S. (See also DISSIMILATORY SULPHATE REDUCTION; cf. FERMENTATION sense 1.) Most species give a positive DESULFOVIRIDIN TEST. In the absence of oxidized sulphur compounds some strains can use fumarate and/or nitrate as terminal electron acceptor. Substrates used by most species include one or more of the following: lactate, malate, pyruvate and glycerol; these substrates are typically oxidized to acetate and CO_2, the acetate not being oxidized further. (*D. baarsii* can oxidize e.g. acetate, butyrate or formate to CO_2; it can also grow as an AUTOTROPH (q.v.).) [Use of amino acids by marine strains of *Desulfovibrio*: FEMS Ecol. (1985) *31* 11–15.] In the absence of a reducible sulphur compound some species can gain energy from e.g. pyruvate (see PHOSPHOROCLASTIC SPLIT). Most strains contain HYDROGENASE, and some can gain energy by the oxidation of H_2 (using SO_4^{2-} as terminal electron acceptor) while using acetate and CO_2 as carbon sources. (See also EXTRACYTOPLASMIC OXIDATION.) [Hydrogenase, electron-transfer proteins and energy coupling in *Desulfovibrio*: ARM (1984) *38* 551–592.] Some strains can carry out NITROGEN FIXATION. [Diazotrophy in *Desulfovibrio*: JGM (1985) *131* 2119–2122.] GC%: ca. 46–61. Type species: *D. desulfuricans*; other species: *D. africanus*, *D. baculatus*, *D. gigas*, *D. salexigens*, *D. sapovorans*, *D. thermophilus*, *D. vulgaris*. [Book ref. 22, pp. 666–672.] (For *D. sulfodismutans* see FERMENTATION sense 1.)

desulfoviridin A (green) sulphite reductase which occurs e.g. in the cytoplasm in species of *Desulfovibrio* and which is involved in DISSIMILATORY SULPHATE REDUCTION. The enzyme (MWt ca. 200000) contains several 4Fe–4S clusters and two sirohaem groups per molecule (each sirohaem consisting of an iron-containing porphyrin-like molecule, sirohydrochlorin). In vitro the products of sulphite reduction include trithionate ($S_3O_6^{2-}$) and sulphide (S^{2-}). (cf. DESULFORUBIDIN and P582.)

desulfoviridin test A test used to detect DESULFOVIRIDIN in certain SULPHATE-REDUCING BACTERIA. The pellet from a centrifuged culture (15 ml) is resuspended in growth medium, one drop of NaOH (2 M) is added, and the whole is examined under ultraviolet light (wavelength 365 nm); in a positive test the suspension gives a red fluorescence due to the presence of the free sirohydrochlorin chromophore of desulfoviridin.

Desulfurococcus A genus of chemolithoheterotrophic archaeans (order THERMOPROTEALES) which occur e.g. in Icelandic solfataras; they metabolize peptides and obtain energy by SULPHUR RESPIRATION, although at least one species can apparently survive by means of a (sulphur-independent) fermentation. The cells are cocci, ca. 1 μm diam.; *D. mobilis* is motile (flagellated), *D. mucosus* is not. GC%: 51.

Desulfuromonas A genus of Gram-negative, obligately anaerobic bacteria which occur e.g. in freshwater, brackish and marine sediments or muds. Cells: motile, round-ended rods, ca. 0.4–0.9 × 1.0–4.0 μm, which typically have a single lateral or subpolar flagellum. The organisms are chemoorganotrophs; acetate, L-malate and other substrates (e.g. alcohols in some strains) can be utilized. Metabolism is primarily respiratory, and the organisms contain *c*-type cytochromes; elemental sulphur, which acts as terminal electron acceptor, is reduced to H_2S (dissimilatory sulphur reduction), and acetate (or other substrate) is completely oxidized to CO_2. (See also ANAEROBIC DIGESTION.) (Sulphate, sulphite or thiosulphate cannot be used as terminal electron acceptor.) Fumarate or L-malate can be fermented to acetate and succinate. Optimum growth temperature: 30°C. Optimum pH: 7.2–7.5. Colonies are pink or peach-coloured, translucent to opaque. GC%: ca. 50–63. Type species: *D. acetoxidans*. Strains which are ovoid, with a polar or subpolar flagellum, and which do not use alcohols as substrates have been referred to as '*D. acetexigens*'. [Book ref. 22, pp. 664–666.]

determinant (1) (antigenic determinant; determinant group; epitope) (*immunol.*) Of an antigenic macromolecule: any region of the macromolecule with the ability or potential to elicit, and combine with, specific antibody. Determinants exposed on the surface of the macromolecule are likely to be *immunodominant*, i.e. more immunogenic than other (*immunorecessive*) determinants which are less exposed, while some (e.g. those within the molecule) are non-immunogenic (*immunosilent*). (See also ANTIGEN.)

(2) (*genetics*) A gene or functional gene group.

detritivore Any organism (e.g. earthworm, lugworm, bivalve mollusc) which feeds by ingesting detritus (such as soil particles), removing and digesting e.g. adherent microorganisms, and voiding the residue.

Detroit-6 An ESTABLISHED CELL LINE derived from human sternal bone marrow; the cells are heteroploid and epithelioid.

Dettol See PHENOLS.

Dettol chelate A disinfectant which contains chloroxylenol (see PHENOLS) and EDTA; EDTA potentiates the action of chloroxylenol on Gram-negative bacteria by increasing cell permeability. It is active against e.g. many strains of *Pseudomonas aeruginosa*.

deuteromycetes See DEUTEROMYCOTINA.

Deuteromycotina (deuteromycetes; Fungi Imperfecti) A non-phylogenetic category originally created for fungi with no known sexual stage; the category still includes fungi with no known sexual stage (and some fungi which form neither conidia nor sexual structures: see AGONOMYCETALES), but it also includes the asexual (= anamorphic, conidial or imperfect) stages of various fungi which are now known to have a sexual (= teleomorphic or perfect) stage in the Ascomycotina or the Basidiomycotina.

For convenience, 'Deuteromycotina' is generally treated as a subdivision within the EUMYCOTA. The conidium-forming deuteromycetes are arranged into form genera (see FORM GENUS) primarily on the basis of the characteristics of their conidia and their modes of conidiogenesis (see CONIDIUM). (See also SACCARDOAN SYSTEM.) The inclusion of a number of form species in a given form genus means only that those fungi have similar asexual stages; such fungi are not necessarily related (and are often unrelated) in an evolutionary sense (as determined by the sexual characteristics of the organisms, when known). Thus, a given form genus may contain e.g. anamorphs corresponding to the teleomorphs of different genera together with ANAHOLOMORPHS. (For convenience, the word 'form' is generally omitted when referring to a form genus, form species etc.) Classification of the deuteromycetes on the basis of their asexual stages facilitates the identification of those members in which the asexual stage is that which is most commonly encountered in nature (and which may form the sexual stage only rarely).

Two classes [Book ref. 64, p. 112]: COELOMYCETES and HYPHOMYCETES.

deuterosome In a eukaryotic cell: a dense region in the cytoplasm which can act as a MICROTUBULE-ORGANIZING CENTRE for the de novo assembly of a BASAL BODY.

deutomerite In a cephaline gregarine: the posterior of the two main regions of the (septate) cell (cf. PROTOMERITE); it usually contains the nucleus.

DEV Duck embryo vaccine (see RABIES).

Devarda's alloy An alloy containing aluminium (45%), copper (50%) and zinc (5%); it is used e.g. to reduce nitrite and/or nitrate to ammonia.

devil's grip *Syn.* BORNHOLM DISEASE.

dew retting See RETTING.

dexioplectic metachrony See METACHRONAL WAVES.

dextrans D-Glucans in which the glucose residues are linked mainly by (1 → 6)-α-glucosidic bonds; branches are formed by occasional (1 → 4)-α- and, less frequently, (1 → 3)-α-linkages. The size of the molecule and the nature and extent of branching depend on the source of the dextran. Extracellular dextrans are produced by a range of microorganisms, sometimes in copious amounts (see e.g. ROPINESS); they are obtained commercially from strains of *Leuconostoc mesenteroides* grown anaerobically on sucrose-containing media [Book ref. 62, pp. 1–44]. Dextrans of MWt ca. 75000 are used as plasma volume extenders for blood transfusions; they may be obtained by acid hydrolysis of higher-MWt dextrans. Artificially cross-linked dextrans (e.g. 'Sephadex') are used in GEL FILTRATION. Dextrans are relatively inert and can withstand autoclaving.

dextrins Products of the partial degradation of STARCH or GLYCOGEN by heat, acid hydrolysis, or enzyme action. *Limit dextrins* are those formed by enzymes that are unable to effect complete hydrolysis (see AMYLASES). (See also SCHARDINGER DEXTRINS.)

dextrose Dextrorotatory glucose (D-glucose).

DFD meat Dark, firm, dry meat: red meat from animals stressed before slaughter. DFD meat contains less glucose and has a higher pH (>ca. 6.0) than normal meat; during stress, muscle glycogen in the living animal is converted to lactic acid (cf. MEAT SPOILAGE) which is subsequently lost when the animal is bled. DFD meat is more susceptible than normal meat to spoilage. Increased susceptibility to aerobic spoilage appears to be due to the deficiency of glucose rather than to the high pH; since little or no glucose is available, spoilage organisms attack amino acids earlier than in normal meat and, hence, produce off-odours and off-flavours after a shorter period of storage [Book ref. 30, pp. 240–244]. Anaerobic spoilage (e.g. in vacuum packs) may be due to organisms (e.g. *Alteromonas putrefaciens*, *Serratia liquefaciens*) which are inhibited by the low pH of normal meat. *S. liquefaciens* produces off-odours. *A. putrefaciens* forms H_2S which reacts with e.g. myoglobin to produce a green compound: sulphmyoglobin ('*greening*') – cf. MEAT SPOILAGE (b); since *A putrefaciens* is inhibited at pH below 6.0, greening can be prevented by treating the meat with citrate buffer. Vacuum-packed DFD meat may also support the growth of large populations of *Yersinia enterocolitica*, although these seem not to cause significant spoilage (cf. FOOD POISONING).

DFMO (DL-α-difluoromethylornithine) See SLEEPING SICKNESS.

DFP Diisopropylfluorophosphate – see PROTEASES.

DGGE Denaturing gradient gel electrophoresis: a method for comparing samples of related *double*-stranded DNA (generated e.g. by PCR or restriction) by two-dimensional electrophoresis (cf. SSCP). In DGGE [Biotechnology (1995) *13* 137–139], fragments are separated by size in the first phase of electrophoresis. Then, in the same gel, the fragments are moved electrophoretically at right-angles to their original path. In this second phase, fragments move through an increasing concentration of DNA-denaturing agents (e.g. formamide + urea) so that, at certain levels in the gradient, localized *sequence-dependent* melting (i.e. strand separation) occurs within part(s) of the fragments (base-pairing being stronger in GC-rich sections); such localized melting affects the electrophoretic speed of those fragments in which it occurs and allows separation of fragments in the gel.

[Example of use of DGGE (differentiation of isolates of *Escherichia coli* by analysis of the 16S–23S intergenic spacer region): FEMS Ecol. (2001) *35* 313–321.]

Other uses of DGGE include e.g. characterization of organisms by comparing PCR-amplified sequences of their 16S rRNA genes [AEM (1996) *62* 340–346] and detection of *Rhizobium* spp [LAM (1999) *28* 137–141].

Another method, based on the same principle, replaces the chemical denaturing gradient of DGGE with an ongoing increase in temperature during the second phase of electrophoresis; thus, at certain levels of temperature, localized sequence-dependent melting occurs in specific part(s) of the fragments – affecting electrophoretic mobility and allowing separation within the gel. This method is referred to as *temporal temperature gradient gel electrophoresis* (TTGE).

[Comparison of DGGE with TTGE: LAM (2000) *30* 427–431.]

DHBG (9-(3,4-dihydroxybutyl)-guanine) An ANTIVIRAL AGENT which is active, both in cell cultures and in vivo, against herpes simplex viruses (HSV-1 and -2); the (*R*)-enantiomer is more effective than the (*S*)-enantiomer. The mechanism of action resembles that of ACYCLOVIR, but DHBG has higher affinity for HSV thymidine kinase.

DHBV DUCK HEPATITIS B VIRUS.

Dhori virus An unclassified virus (see ORTHOMYXOVIRIDAE) which has been isolated from ticks; antibodies to the virus have been found in man and domestic animals.

DHPA ((*S*)-DHPA; (*S*)-9-(2,3-dihydroxypropyl)-adenine) An ANTIVIRAL AGENT which acts as an analogue of adenosine; it is a potent inhibitor of 5-adenosyl homocysteine hydrolase. DHPA is active e.g. against vaccinia, herpes simplex, varicella-zoster and measles viruses; it acts synergistically with VIDARABINE. (The (*R*)-enantiomer is inactive.)

DHPG (9-(1,3-dihydroxy-2-propoxymethyl)-guanine; 2′-nor-2′-deoxyguanosine, 2′-NDG) An ANTIVIRAL AGENT which has potent activity against herpes simplex viruses (HSV-1 and -2); it is not effective against e.g. Epstein–Barr virus, varicella-zoster virus or cytomegaloviruses. The mechanism of action resembles that of ACYCLOVIR. The isomer (*S*)-9-(2,3-dihydroxy-1-propoxymethyl)-guanine is also active against HSV-1 and -2.

DI particle DEFECTIVE INTERFERING PARTICLE.

diacetoxyscirpenol See TRICHOTHECENES.

diacetyl (dimethylglyoxal; $CH_3.CO.CO.CH_3$) A water-soluble compound formed e.g. by certain lactic acid bacteria; it may also be derived by oxidation of acetoin in the BUTANEDIOL FERMENTATION. (See also VOGES–PROSKAUER TEST.)

Diacetyl produced by certain lactic acid bacteria is responsible for the characteristic 'buttery' flavour and aroma in many fermented DAIRY PRODUCTS. The diacetyl is formed from pyruvate – see Appendix III(c) – but levels of diacetyl are low if the pyruvate is derived from hexoses only. Additional pyruvate (and hence diacetyl) can be produced from citrate by certain strains of lactic acid bacteria ('aroma bacteria', 'flavour bacteria') – e.g. *Lactococcus lactis* and *Leuconostoc cremoris* (see LACTIC ACID STARTERS). (Citrate is naturally present in cows' milk at levels of ca. 0.2%, with seasonal fluctuations; it may also be added to the milk to enhance diacetyl production.) *L. lactis* produces both lactic acid and diacetyl in milk. However, *Leuconostoc cremoris* by itself grows poorly in milk, producing little lactic acid; it can take up citrate only at pH below ca. 6.0, so that diacetyl production by this organism requires the activities of a lactic acid producer such as *L. lactis*. The ability to produce diacetyl from citrate is apparently plasmid-linked in at least certain strains of *L.* (*'Streptococcus'*) *lactis* [AvL (1983) *49* 265–266]. The ratio of flavourful products (e.g. diacetyl) to flavourless products (e.g. acetoin) depends e.g. on the redox potential of the system.

Diacetyl has some antimicrobial activity [AEM (1982) *44* 525–532], being more effective against Gram-negative bacteria, yeasts and moulds than against Gram-positive bacteria.

diacridine A compound containing two ACRIDINE moieties (linked e.g. by a hydrocarbon bridge); it may act as a *bis* INTERCALATING AGENT.

diagnosis (1) In NOMENCLATURE: a list of descriptive characteristics which distinguish a proposed new taxon from other taxa. (2) (*med.*) In general, a statement of the identity of a given disease, deduced from symptoms etc., or the procedure used for arriving at such a statement.

diagnostic window See TRANSFUSION-TRANSMITTED INFECTION.

diakinesis See MEIOSIS.

Dialister pneumosintes See BACTEROIDES (*B. pneumosintes*).

dialkylnitrosamines See *N*-NITROSO COMPOUNDS.

diamidines Compounds with two amidine [$NH_2.C(=NH)-$] groups. Aromatic diamidines bind to DNA, and a number of them are trypanocidal agents which bind to kinetoplast DNA and cause structural changes in the kinetoplast; however, there is evidence that at least part of the trypanocidal effect in vivo is due to inhibition of ornithine decarboxylase, leading to depletion of putrescine. (See e.g. BERENIL; DAPI; 2-HYDROXYSTILBAMIDINE; PENTAMIDINE ISETHIONATE.)

diaminopimelic acid pathway (DAP pathway) A pathway for lysine biosynthesis [Appendix IV(e)] which occurs in bacteria, certain lower fungi (Hyphochytriales, Leptomitales, Saprolegniales) and e.g. in green plants. (cf. AMINOADIPIC ACID PATHWAY.)

diaminopyrimidine drugs See FOLIC ACID ANTAGONIST.

diamond skin disease *Syn.* SWINE ERYSIPELAS.

dianemycin See MACROTETRALIDES.

***o*-dianisidine** (fast blue B) A (carcinogenic) substance, 3,3′-dimethoxybenzidine, used e.g. in assays for HYDROGEN PEROXIDE: the peroxidase-dependent oxidation of *o*-dianisidine causes increased absorbance at 500 nm.

dianthoviruses (carnation ringspot virus group) A category of PLANT VIRUSES which have a wide host range and are transmitted mechanically and via the soil. Virion: icosahedral, ca. 31–34 nm diam., containing two molecules of positive-sense ssRNA (MWts ca. 1.5 and 0.5×10^6) and one type of coat protein (MWt 40000). Virus replication appears to occur in the cytoplasm. Type member: carnation ringspot virus; other members: red clover necrotic mosaic virus, sweet clover necrotic mosaic virus.

diapedesis See INFLAMMATION.

diaphorase Any enzyme which can catalyse the oxidation of reduced NAD or NADP by an artificial electron acceptor.

diaphoromixis Bipolar or tetrapolar *multi*allele HETEROTHALLISM.

diaplectic metachrony See METACHRONAL WAVES.

Diaporthales An order of fungi (subdivision ASCOMYCOTINA) which include saprotrophic and plant-parasitic species. Ascocarp: perithecioid, often immersed; hamathecium: absent or evanescent. Asci: unitunicate, evanescent. Genera: e.g. *Diaporthe*, ENDOTHIA, GAEUMANNOMYCES, *Gnomonia* (see also ZYTHIA).

Diaporthe See DIAPORTHALES.

diarrhoea (*American*: diarrhea) Frequent passage of fluid stools: a symptom of many types of illness. (cf. DYSENTERY; see also SCOURS.) Diarrhoea may be due to e.g. (i) the effect on the gut mucosa of particular microbial toxins which cause hypersecretion of fluid into the lumen; (ii) failure of the gut to absorb small molecules (peptides, sugars – especially lactose) due e.g. to damage to, or non-functioning of, the mucosa, leading to an osmotic effect in the gut lumen; (iii) abnormal activity of the smooth musculature of the gut wall.

Severe diarrhoea may lead to dehydration and loss of electrolyte; treatment includes rehydration with solutions containing Na^+, Cl^-, Ca^{2+}, bicarbonate ions and glucose.

Some causes of persistent diarrhoea include e.g. CLB and EAGGEC. (See also CRYPTOSPORIDIOSIS and ETEC.)

Diaspora See EIMERIORINA.

diaspore *Syn.* PROPAGULE.

diastase *Syn.* AMYLASE.

diastatic Able to metabolize starch.

diastole See CONTRACTILE VACUOLE.

Diatoma See DIATOMS.

diatomaceous earth (diatomite; *kieselguhr*) A siliceous material composed largely of fossil diatom frustules (see DIATOMS), large natural deposits of which are mined in various parts of the world (particularly Lompoc, California). Diatomaceous earth is used e.g. as a mild abrasive (in toothpastes, polishes etc), as an absorbent, as a heat-insulating material, and in various types of filter (including certain microbiological filters: see FILTRATION).

diatomite *Syn.* DIATOMACEOUS EARTH.

diatoms A large group (>10000 species) of ALGAE (often regarded as a distinct division – Bacillariophyta) which are essentially unicellular (some are colonial, some form filaments) and which have a characteristic type of CELL WALL consisting typically of a siliceous structure (the *frustule*) encased in an organic layer. Most cells have brownish chloroplasts containing chlorophylls a, c_1 and c_2 and e.g. fucoxanthin; storage products include CHRYSOLAMINARIN. Many species are facultatively heterotrophic in the dark; a few marine benthic species are obligately heterotrophic. Diatoms are unusual, if not unique, in having an absolute requirement for silicon – not only for the construction of their walls but also for general metabolism (e.g. in *Cylindrotheca fusiformis* silicon affects gene expression both directly and indirectly [JGM (1985) *131* 1735–1744]). Species occur in aquatic environments, including fresh, brackish and marine waters (where they may be benthic, planktonic, epiphytic etc. – see also AUFWUCHS), and in terrestrial environments such as soil, damp rocks, and even dry rocks and desert sands. *Achnanthes exigua* is a thermophile (optimum temperature for photosynthesis: 42°C) found in hot springs. Psychrophilic species form a major component of the 'ice algae' present on the lower surfaces of ice in polar seas [JGM (1983) *129* 1019–1023]. Diatoms occur as endosymbionts e.g. in members of the FORAMINIFERIDA. (See also RHIZOSOLENIA and RHOPALODIA.)

The diatom frustule consists of two *valves* – which, according to species, may each be saucer-shaped, bowl-shaped, boat-shaped etc. – held together by two or more siliceous bands (*girdle bands*) which, collectively, form the *cingulum*. In many species one valve (the *epivalve* or *epitheca*) is slightly larger than the other (*hypovalve* or *hypotheca*); the cingulum, which is attached to the edges of the valves, may overlap the hypovalve to a greater or lesser extent. The valves are characteristically elaborately ornamented with pores, slits, ribs, tubes, projections, etc which are usually arranged in symmetrical, species-specific patterns. The degree of cell wall silicification may vary to some extent with environmental conditions, and exceptionally the wall may be predominantly or entirely organic (e.g. in *Phaeodactylum tricornutum* – see PHAEODACTYLUM – and in *Cylindrotheca fusiformis*). [Details of wall structure: Book ref. 137, pp. 129–156.]

Two main groups of diatoms are commonly recognized: *centric diatoms*, in which the valve is radially symmetrical, and *pennate diatoms*, in which the valve is bilaterally symmetrical. (Some diatoms do not fit easily into either category: e.g. *Triceratium* spp may have 3-, 4- or 5-fold rotational symmetry; see also HEMIDISCUS.)

Centric diatoms are mostly marine and planktonic, are non-motile, and commonly exhibit oogamous sexual reproduction in which the male gametes each bear a single tinsel flagellum. Genera include e.g. *Arachnoidiscus*, *Bacteriastrum*, *Biddulphia*, CHAETOCEROS, *Coscinodiscus*, *Cyclotella*, *Hemiaulus*, MELOSIRA, *Skeletonema*, *Stephanodiscus*, *Stephanopyxis*, *Thalassiosira*, *Zygoceros*.

Pennate diatoms occur in various habitats; many are capable of gliding motility, and many exhibit isogamous sexual reproduction involving cell–cell contact (conjugation). All species which are capable of GLIDING MOTILITY (q.v.) have a *raphe*: basically, a slit which (usually) runs the length of the valve, interrupted by a *central nodule*. Such 'raphid' species are generally benthic or occur attached to a solid substratum. In a few genera (e.g. *Achnanthes*, *Cocconeis*) a raphe occurs in only one of the two valves, but in most a raphe occurs in both valves. Pennate diatoms which lack a raphe ('araphid' diatoms: e.g. *Fragilaria*, *Striatella*, *Tabellaria*) cannot glide; these organisms may have a central line, ridge, or unornamented central area called a *pseudoraphe*. (Other pennate diatoms include e.g. AMPHORA, *Anomoeoneis*, *Bacillaria*, *Cylindrotheca*, *Cymbella*, *Diatoma*, GOMPHONEMA, *Gyrosigma*, NAVICULA, *Nitzschia*, PHAEODACTYLUM, *Pinnularia*, *Pleurosigma*, *Stauroneis*, *Surirella*.)

In many centric and pennate diatoms the cells adhere to one another to form filaments or characteristic groupings (see e.g. ASTERIONELLA and MELOSIRA); the cells may be held together by mucilage secreted via tubular projections in the frustule and/or by the interlocking of spines present on the margins of the valves. Secreted mucilage may also serve to attach some forms to the substratum, and apparently serves an essential role in GLIDING MOTILITY.

Asexual reproduction occurs by binary fission. The nucleus (which is diploid in vegetative cells) moves from its position near the epivalve to the cell centre, and the protoplast expands, pushing the two valves apart. New girdle bands are formed, and mitosis occurs. The original parent hypovalve becomes the epivalve of one daughter cell, and both daughter cells synthesize a new hypovalve. Thus, the epivalve is always the older of the two valves in a cell, and – in species in which the epivalve is larger than the hypovalve – one of the daughter cells must be smaller than the parent cell. When division is complete, the nucleus in each cell returns to its position near the epivalve face. The new siliceous structures (valves and cingulum) are synthesized in SILICA DEPOSITION VESICLES beneath the plasmalemma, reaching the exterior by fusion between the silicalemma and plasmalemma. [Diatom wall formation: Book ref. 137, pp. 157–200; silicon 'biomineral' synthesis in diatoms: TIBS (1987) *12* 151–154.]

Since, in many species, the average size of cells in a population progressively decreases with increasing numbers of divisions, maximum cell size must eventually be re-established; this may be achieved by the formation of resting spores or auxospores. *Resting spores* generally have thick ornamented walls; on germination, the protoplast expands to maximum cell size prior to wall formation. *Auxospores* are usually or always formed as a result of fusion of gametes, after which the zygote protoplast escapes from the parent wall, expands, and then synthesizes a wall which is initially organic; a siliceous wall is then synthesized.

(See also DIATOMACEOUS EARTH.)

diatoxanthin See CAROTENOIDS.

diatretyne nitrile See POLYACETYLENES.

Diatrypales See ASCOMYCOTINA.

diauxie The phenomenon in which, when provided with two sources of carbon, an organism preferentially metabolizes one source (completely) before starting to metabolize the other; the two phases of growth are commonly separated by a lag phase in which the organism produces enzyme(s) necessary for the utilization of the second source of carbon. (See also CATABOLITE REPRESSION.)

diauxy *Syn.* DIAUXIE.

diazaborines A range of antibacterial agents [JB (1989) *171* 6555–6565] which apparently inhibit fatty acid biosynthesis by inhibiting the enzyme enoyl-acyl carrier protein reductase (ENR).

diazomycin A See DON.

6-diazo-5-oxo-L-norleucine See DON.

diazotroph Any organism capable of NITROGEN FIXATION.

dibromoaplysiatoxin See LYNGBYA.

dibromomethylisopropyl-*p*-benzoquinone An inhibitor of PHOTOSYNTHESIS.

dibromopropamidine An aromatic diamidine used as an antiseptic; it is active mainly against asporogenous Gram-positive bacteria.

DIC DISSEMINATED INTRAVASCULAR COAGULATION.

dicarboximides A group of agricultural ANTIFUNGAL AGENTS which are effective against *Botrytis cinerea*; they are used on a wide range of crops, functioning primarily as surface protectants. Dicarboximides include e.g. iprodione and vinclozolin.

Dice–Leraas diagrams Diagrams that are used for identifying and describing trypanosomes on the basis of their measurements (e.g. length, distances between organelles).

Dicellomyces See BRACHYBASIDIALES.

dicentric (of a eukaryotic chromosome) Having two CENTROMERES.

dichlofluanid A sulphur-containing ANTIFUNGAL AGENT used e.g. for the control of *Botrytis cinerea* in plants, and in antifungal washes for the prevention of mould in buildings.

dichlone (2,3-dichloro-1,4-naphthoquinone) A QUINONE ANTIFUNGAL AGENT used as a seed dressing (e.g., for the seeds of legumes or cotton) and as a foliar spray; it is effective against a range of diseases, including apple scab, damping off, etc. Dichlone is toxic to certain plants, and is also effective in controlling BLOOMS of certain cyanobacteria.

dichloramine T See CHLORINE.

dichloroisocyanurate (DCCA) (as an antimicrobial agent) Sodium or potassium dichloroisocyanurate, available in powder or tablet form, hydrolyses in water to form hypochlorous acid (see HYPOCHLORITES). When dry, DCCA salts are stable at room temperature.

dichlorophane (dichlorophene) See BISPHENOLS.

3-(3,4-dichlorophenyl)-1,1-dimethylurea See DCMU.

Dichothrix See RIVULARIACEAE.

Dick test A SKIN TEST used to determine whether or not an individual is susceptible to SCARLET FEVER. The test procedure (similar to that used in the SCHICK TEST) involves an intradermal injection of the erythrogenic toxin of *Streptococcus pyogenes*. A *positive* Dick test (indicating susceptibility) consists of an inflammatory response (at the site of the injection) which becomes evident within ca. 12 hours and reaches peak intensity within ca. 24 hours of the injection. In immune individuals the toxin is neutralized by antitoxin, thus giving a *negative* test.

dicloran An agricultural antifungal agent (2,6-dichloro-4-nitroaniline) which is used against various *Botrytis* infections. It is water-insoluble – hence very persistent – and is usually formulated as a dust; it has very low phytotoxicity. (cf. CHLORONITROBENZENES.)

dicloxacillin See PENICILLINS.

Dictyocha See SILICOFLAGELLATES.

Dictyoglomus A genus of bacteria. One species: *D. thermophilum*, an anaerobic, caldoactive, asporogenous chemoorganotroph; cells: rod-shaped, occurring in large spherical bodies consisting of a few to many separate cells. [IJSB (1985) *35* 253–259.]

Dictyosiphon See PHAEOPHYTA.

dictyosome (1) One of the stacks of membranous vesicles which form a GOLGI APPARATUS. (2) *Syn.* Golgi apparatus.

dictyosporae See SACCARDOAN SYSTEM.

Dictyostelia See EUMYCETOZOEA.

dictyostelids See DICTYOSTELIOMYCETES.

Dictyosteliia See EUMYCETOZOEA.

Dictyosteliomycetes (dictyostelid cellular slime moulds; dictyostelids) A class of cellular SLIME MOULDS (division MYXOMYCOTA). (cf. EUMYCETOZOEA.) The organisms occur mainly in soil, humus and dung, particularly in tropical regions. The vegetative phase consists of myxamoebae which form filose subpseudopodia; flagellated cells are not formed. The life cycle involves a feeding stage, during which myxamoebae ingest food (mainly bacteria) and divide mitotically. When the food supply is exhausted, the amoebae stop growing and, after a period of starvation (*pre-aggregation phase* or 'interphase'), the *aggregation phase* begins: numerous myxamoebae converge (chemotactically – see ACRASIN) and join together to form one to many multicellular, macroscopic pseudoplasmodia. In most species, the pseudoplasmodium can migrate over the surface of the substratum and can show tactic responses. Eventually, the pseudoplasmodium stops moving and differentiates to form a multicellular, multispored, stalked fruiting body, the *sorocarp* (a process known as *culmination*); the sorocarp stalk consists of a cellulosic 'stalk tube' which, when mature, is either hollow or filled with dead cells. The spores are dispersed by wind, rain, etc; on germination, each spore releases a myxamoeba, thus initiating a new cycle. (In some species myxamoebae may not always pass through this life cycle, forming instead MICROCYSTS and/or MACROCYSTS.) Genera: e.g. ACYTOSTELIUM, DICTYOSTELIUM, POLYSPHONDYLIUM.

[Dictyostelids – natural history, life cycles, cultivation: Book ref. 144.]

Dictyostelium A genus of slime moulds (class DICTYOSTELIOMYCETES) in which the sorocarp stalk tube is filled with a mesh of empty cell walls (cf. ACYTOSTELIUM) and bears a spherical sorus of spores at its apex. In e.g. *D. discoideum* the sorocarp consists of a single unbranched stalk that tapers from base to tip; the proximal end is flared into a basal attachment disc, and the distal end bears a single, more or less spherical, white to yellowish mass (sorus) of ellipsoidal to reniform spores (each typically ca. 2.5–3.5 × 6.0–9.0 μm). In some species the sorocarp may be sparingly and irregularly branched (cf. POLYSPHONDYLIUM); in *D. polycephalum* several sorocarps occur in a coremium-like cluster in which the stalks are fused over much of their length, becoming free at their distal, spore-bearing ends. Sorocarps of *Dictyostelium* spp vary in size from e.g. ca. 0.2–0.6 mm in *D. deminutivum* to 30 mm or more in *D. giganteum*; those of *D. discoideum* may reach a maximum height of ca. 4.0–4.5 mm, but are usually smaller. [Species descriptions and key: Book ref. 144, pp. 246–367.]

During the feeding stage of the life cycle, *Dictyostelium* myxamoebae feed mainly on bacteria, but sometimes ingest other myxamoebae of the same species ('cannibalism'); *D. caveatum* is unusual in that it feeds extensively on the myxamoebae of other slime moulds as well as on bacteria. [Self/non-self recognition in *D. caveatum*: JCB (1986) *102* 298–305.]

D. discoideum is the best-known species, being widely used in studies on differentiation and cell-cell interactions. It is easily cultured in the laboratory (e.g. on hay infusion agar or SM MEDIUM, with *Escherichia coli* or *Klebsiella pneumoniae* as food); some strains can be grown in cell-free media. (The main natural habitat of *D. discoideum* is apparently deciduous forest soils and leaf-litter.) The vegetative amoebae are commonly ca. $13-16 \times 9-11$ μm; they are haploid and uninucleate, and contain one or more contractile vacuoles, food vacuoles, etc. They are attracted to their prey by CHEMOTAXIS, chemoattractants apparently including e.g. folic acid released by the bacteria. When food is exhausted, cells enter the *pre-aggregation phase* (see DICTYOSTELIOMYCETES) during which the myxamoebae discharge their food vacuoles, become smaller, show altered staining properties, and lose the ability to respond chemotactically to folic acid but acquire the ability to be attracted by cyclic AMP (cAMP). Each cell contains six chromosomes, the largest being chromosome 2 (about 25% of the genome). [Sequence and analysis of chromosome 2: Nature (2002) *418* 79–85.]

In *D. discoideum* the aggregation phase begins when a subset of myxamoebae begins to secrete cAMP (an ACRASIN) in slow, rhythmic pulses. The cAMP diffuses out from these cells and binds to cell-surface cAMP receptors on nearby myxamoebae. Binding of cAMP to the receptors triggers several responses in the cell: e.g., it (transiently) stimulates adenylate cyclase activity within the cell, resulting in greatly increased production and release of cAMP from the cell, and it triggers migration of the cell towards the source of the cAMP. Thus, the pulses of cAMP are progressively amplified and relayed from the initial subset of cAMP-producing cells throughout the population, resulting in the formation of converging streams of cells which move in regular steps with intervening 'rest' periods. In order to maintain the cAMP gradient and to prevent swamping of the receptors by excess cAMP, the concentration of extracellular cAMP is strictly controlled by regulation of the relative proportions of (at least) two proteins: a cAMP phosphodiesterase (which occurs in both soluble extracellular and membrane-bound forms, and which hydrolyses cAMP), and a glycoprotein phosphodiesterase inhibitor (which binds to and inactivates the soluble – but not the bound – form of the enzyme). The aggregated cells adhere to one another, eventually forming a pseudoplasmodium.

The multicellular, elongated pseudoplasmodium (called a 'slug' or 'grex') consists of a mass of cells enclosed within a slime sheath which is composed of cellulose microfibrils embedded in a protein- and carbohydrate-containing matrix. The slug varies in size (e.g. ca. 0.5–2.0 mm long) depending on the number of cells it contains. It migrates over the surface of the substratum, exhibiting e.g. aerotaxis, phototaxis and thermotaxis. The slime sheath does not move relative to the substratum, and may provide traction against which the cells within can move; thus, as the slug moves forwards, a slime trail of collapsed sheath material remains behind. New sheath material is synthesized along the length of the slug, so that the sheath is thinnest at the anterior end, becoming progressively thicker towards the posterior. A migrating slug can split into two smaller slugs, or two slugs can merge to form a larger slug; the size (but not the proportions) of the sorocarp which eventually develops depends on the size of the slug at culmination.

The fates of the cells during culmination are apparently determined at the slug migration stage (i.e., before culmination begins): the cells making up the anterior third of the slug are *prestalk cells*, destined to form the sorocarp stalk, while the cells of the posterior two-thirds are *prespore cells*. [Patterns of cell differentiation within the slug: Book ref. 67, pp. 255–274.] The culmination process begins when the slug stops moving and reorientates such that the leading (anterior) tip is raised vertically above the rest of the cell mass, forming a nipple-like apical projection. A short cellulosic 'stalk tube initial' is secreted by cells near the apex; the entire cell mass then flattens, bringing the tube down through the cell mass to the substratum. Prestalk cells within the lower end of the tube enlarge, become vacuolated and compacted, and eventually die; the stalk lengthens upwards – the remaining prestalk cells progressively undergoing vacuolation and death to result eventually in the characteristic mesh of cellulose cell walls which fills the cellulose stalk tube. Meanwhile, the mass of prespore cells is gradually elevated on the growing stalk, becoming progressively differentiated into spores from the periphery to the centre of the mass. Differentiation is complete when all the cells have become either stalk cells or spores.

The sorocarp is usually orientated such that the sorus is at the maximum distance from the substratum or from adjacent objects (including other sorocarps); this orientation is believed to be due to a tropism regulated by an (unidentified) gas or vapour produced by the developing sorocarp.

The spores have thick cellulosic walls and are resistant to e.g. desiccation. Germination generally occurs only in the presence of an adequate supply of amino acids. Spores in masses fail to germinate owing to the presence of an autoinhibitor apparently produced during culmination.

In *D. discoideum* macrocysts may be formed, but microcysts have not been observed.

Dictyota See PHAEOPHYTA.

Dictyuchus See SAPROLEGNIALES.

dicyandiamide (DCD; Didin) Cyanoguanidine, a NITRIFICATION INHIBITOR. [Mineralization of DCD in acid soils: SBB (1985) *17* 253–254.]

***N,N′*-dicyclohexylcarbodiimide** See DCCD.

didanosine See NUCLEOSIDE REVERSE TRANSCRIPTASE INHIBITORS.

didemnins Cyclic depsipeptide ANTIVIRAL AGENTS isolated from Caribbean tunicates (sea-squirts); they are probably too cytotoxic for therapeutic use.

dideoxy fingerprinting See TUBERCULOSIS (antibiotic resistance testing).

dideoxy sequencing See DNA SEQUENCING.

dideoxyribonucleotide See NUCLEOTIDE (figure legend).

dideoxythymidine triphosphate See DDTTP.

Didesmis See GYMNOSTOMATIA.

Didin See DICYANDIAMIDE.

Didinium A genus of freshwater ciliates (subclass GYMNOSTOMATIA). Cells: radially symmetrical, ovoid or barrel-shaped, ca.

60–200 μm in length; the flattened apical region has a prominent cone-shaped 'proboscis' at the apex of which is the cytostome. The cytopharyngeal region is reinforced with nematodesmata. The cell may be encircled by two pectinellae: one somewhat posterior of the midline, the other surrounding the base of the proboscis. The macronucleus is U-shaped. *Didinium* feeds primarily or exclusively on *Paramecium*; in the absence of food *Didinium* encysts. (See also TOXICYST.)

Didymella A genus of saprotrophic and plant-parasitic fungi of the order DOTHIDEALES; anamorphs occur in the genera ASCOCHYTA and PHOMA. [Taxonomy: CJB (1981) *59* 2016–2042.]

Didymium See MYXOMYCETES.

didymosporae See SACCARDOAN SYSTEM.

Dienes phenomenon (Dienes reaction) The failure of two swarms of different strains of *Proteus* growing on the same nutrient agar plate to penetrate each other, with the result that a sharp line of demarcation always occurs between them; the phenomenon does not occur between two swarms of the same strain. The Dienes phenomenon is used to distinguish between strains of *Proteus* in epidemiological studies – the presence of a demarcation line (a positive reaction) being regarded as evidence for strain difference. (The results of a *negative* Dienes reaction may not be entirely reliable [Book ref. 22, p. 491].) The Dienes phenomenon appears to result from the production of BACTERIOCINS.

Dientamoeba A genus of protozoa of the TRICHOMONADIDA. *D. fragilis* occurs in the large intestine in man; it is non-pathogenic. The trophozoites (5–20 μm diam.) contain numerous vacuoles and (usually) two nuclei, each nucleus lacking peripheral chromatin and containing (usually) four closely grouped karyosomes. Pseudopodia are blunt and are formed slowly. Cysts are not formed.

Dieterle silver stain (Dieterle silver impregnation stain) A complex, non-specific, silver-impregnation staining procedure used e.g. for staining spirochaetes, cells of *Legionella* in tissue sections, etc.; stained bacteria appear brown or black. [Method: Book ref. 53, p. 1388.]

***N,N*-diethyl-*p*-phenylenediamine** (DPD) See WATER SUPPLIES.

diethylpyrocarbonate (DEPC; $C_2H_5.O.CO.O.CO.C_2H_5$) DEPC is an effective inhibitor of yeasts and moulds once used as a preservative in wines etc.; owing to toxicity (and suspected carcinogenicity) it is no longer a permitted preservative in the USA and UK.

In molecular biology, DEPC is useful as a reagent for the carbethoxylation of exposed N-7 groups of (unpaired) adenine residues in DNA or RNA. It also (covalently) inactivates RNases, and is used for treating water, solutions, glassware etc. to prevent degradation of sample RNA by extraneous RNases; whenever possible, water or solutions should be treated with DEPC (0.1%) for a minimum of 1 hour at 37°C and subsequently autoclaved to eliminate any remaining DEPC.

Dieudonné alkaline blood agar A medium, containing defibrinated ox blood and sodium hydroxide, formerly used for the primary isolation of *Vibrio cholerae*. (cf. MONSUR MEDIUM.)

difference spectrum (of cytochromes) See CYTOCHROMES and CO DIFFERENCE SPECTRUM.

differential centrifugation See CENTRIFUGATION.

differential display A technique for detecting those (primarily eukaryotic) genes which are expressed only under specific conditions; it involves isolation, and comparison, of mRNAs from two or more populations, each population having been exposed to different conditions. Initially, mRNAs from each population are converted to cDNAs; this is done by using, as a primer-binding site, the 3′-AAA... tail on polyadenylated mRNA molecules. The task is made manageable by converting only *some* of the (many) mRNAs from each population; selectivity is achieved with a primer such as 5′-TT...TTGG-3′ which permits conversion of only those mRNAs in which a cytosine (C) residue occurs in the appropriate position after the 3′-AAA... tail.

cDNAs from each of the populations are then fingerprinted by subjecting them to ARBITRARILY PRIMED PCR and separating the products by gel electrophoresis. When fingerprints from different populations are compared, any band(s) of interest – such as band(s) present in one fingerprint but not in other(s) – can be further examined by removal from the gel and amplification by the same arbitrary primer; the amplified fragments may be e.g. sequenced and/or used to probe a library.

[Differential display methodology: NAR (1998) *26* 5537–5543; review: Biotechniques (2002) *33* 338–346.]

differential host (*plant pathol.*) A plant host which, on the basis of disease symptoms, serves to distinguish between various strains or races of a given plant pathogen.

differential interference-contrast microscopy See MICROSCOPY (d).

differential medium See MEDIUM.

diffluent Readily dissolving or breaking up in water.

Difflugia A genus of testate amoebae (order ARCELLINIDA) in which the test is reinforced with sand grains, diatom frustules, sponge spicules etc which are initially ingested by the cell. *D. urceolata* is multinucleate and has a rounded test (ca. 200–230 × 150–200 μm) with a pointed top and a rim around the ventral aperture through which several narrow, round-ended pseudopodia emerge. *Difflugia* spp occur in ponds, swamps, bogs, soil etc.

diffusely adherent *E. coli* See ENTEROADHERENT E. COLI.

diffusion-coupled gradostat See GRADOSTAT.

diffusion test (1) In ANTIBIOTIC-sensitivity testing: any test which involves the diffusion of antibiotic(s) through agar. See e.g. DISC DIFFUSION TEST and AGAR DIFFUSION TEST.

(2) (for membrane filters) A test used to determine the physical integrity of a membrane filter (see FILTRATION). If a wetted membrane is subjected to a pressure lower than its bubble point pressure (see BUBBLE POINT TEST), some air can pass through the pores by simple diffusion; the amount of air transmitted in this way can be significant in filters of large surface area. In the diffusion test a wetted filter is subjected to pressure at ca. 80% of its bubble point pressure, and the volume of air transmitted is determined; the integrity of the filter is assessed by comparing this volume with that expected from an intact filter.

DL-α-difluoromethylornithine See SLEEPING SICKNESS.

difolatan *Syn.* CAPTAFOL.

DIG Abbreviation for DIGOXIGENIN.

digenetic Refers to a parasite which carries out part of its life cycle in each of two different host species. (cf. HETEROXENOUS.)

di George syndrome (congenital thymic aplasia) The condition, resulting from an undeveloped thymus gland, which is characterized by a lack of T LYMPHOCYTES; those with this syndrome are susceptible to a range of infections, especially by intracellular pathogens. The condition has been treated by transplantation of fetal thymus gland (using tissue from a fetus at <14 weeks' gestation).

digitonin A mixture of steroid SAPONINS found in the seeds of the purple foxglove (*Digitalis purpurea*).

digoxigenin (DIG) A (poisonous) steroid obtained from species of the plant *Digitalis*. Digoxigenin is used e.g. as a 'tag' for non-radioactive labelling of oligonucleotide probes (both RNA and DNA) – see e.g. DOT-BLOT.

Incorporation of digoxigenin into RNA probes can be carried out by in vitro transcription on a DNA template of the target sequence using DIG-UTP in the reaction mixture; the DNA template itself may be prepared e.g. as a cloned (then linearized) target sequence carrying a promoter at the 3′ end.

DIG-dUTP is incorporated into DNA probes during synthesis; this tagged nucleotide can be used as a substrate by a range of polymerases, including e.g. the KLENOW FRAGMENT, *Taq* DNA polymerase, phage T4 DNA polymerase, and at least some reverse transcriptases.

DIG-dUTP can also be covalently bound to the 3′ end of DNA strands using the enzyme terminal nucleotidyltransferase.

diguanides (as antimicrobial agents) See e.g. CHLORGUANIDE and CHLORHEXIDINE.

dihydrofolate reductase See FOLIC ACID.

dihydrofolate reductase inhibitor See FOLIC ACID ANTAGONIST.

dihydromocimycin See POLYENE ANTIBIOTICS (b).

dihydrowyerol See WYERONE.

dihydroxyacetone fermentation A commercial aerobic FERMENTATION (sense 2) in which GLYCEROL is oxidized to dihydroxyacetone by strains of *Gluconobacter oxydans* (*'Acetobacter suboxydans'*). The product is used in the pharmaceutical industry and e.g. as a sun-tanning agent.

dihydroxyacetone pathway *Syn.* XMP PATHWAY.

9-(3,4-dihydroxybutyl)-guanine See DHBG.

9-(1,3-dihydroxy-2-propoxymethyl)-guanine See DHPG.

(*S*)-9-(2,3-dihydroxy-1-propoxymethyl)-guanine See DHPG.

(*S*)-9-(2,3-dihydroxypropyl)-adenine See DHPA.

diiodohydroxyquin See 8-HYDROXYQUINOLINE.

diisopropylfluorophosphate See PROTEASES.

dikaryon (dicaryon) (1) A binucleate cell, or a mycelium composed of binucleate cells, in which the two nuclei are haploid and are genetically dissimilar. (See also *secondary* HYPHA.)

(2) A pair of (usually genetically dissimilar) nuclei.

dikaryophase In the life cycles of some fungi: a phase characterized by dikaryotic mycelium (see DIKARYON sense 1).

dikaryotization Any process which leads to the formation of a DIKARYON (sense 1); it typically involves SOMATOGAMY or SPERMATIZATION in which one mononucleate cell donates its nucleus to another. (See also BULLER PHENOMENON.)

Dileptus A genus of carnivorous, cyst-forming ciliates (subclass GYMNOSTOMATIA). Cells: elongate, up to ca. 500 μm in length, the anterior end having a long TOXICYST-bearing 'neck' region at the base of which is the cytostome and its associated nematodesmata; the posterior end is pointed.

diloxanide A synthetic amoebicide which, as e.g. diloxanide furoate ('furamide'), is effective in the treatment of some forms of amoebiasis.

dilution coefficient (concentration exponent; η) Of a given DISINFECTANT: a number which indicates the effect of dilution on the rate of disinfection under standardized conditions. The concentration and killing rate of a disinfectant are related by the equation:

$$tc^{\eta} = k$$

where t is the time required for 100% kill of cells in a test suspension, c is the concentration of the disinfectant, η is the dilution coefficient, and k is a constant. For example, if a given undiluted disinfectant with an η-value of 4 is diluted by a factor of 2, the expression c^{η} becomes $(1/2)^4$, i.e. 1/16; thus t must increase 16-fold in order for tc^{η} to remain numerically constant. Disinfectants with high dilution coefficients – e.g. ethanol ($\eta = 10$) and phenol ($\eta = 6$) – lose activity with dilution more than those with lower coefficients – e.g. chlorhexidine ($\eta = 2$), QACs ($\eta = 1$).

dilution end-point assay See END-POINT DILUTION ASSAY.

dilution rate (*D*) In CONTINUOUS CULTURE: the parameter *F/V* in which *F* is the rate at which medium enters (and leaves) the culture vessel, and *V* is the volume of liquid in the vessel. Dilution rate has the same dimensions as SPECIFIC GROWTH RATE (i.e., 1/time) and its unit is $hour^{-1}$. Under steady-state conditions dilution rate is numerically equal to specific growth rate. The *critical dilution rate* (D_c) is the dilution rate which corresponds to μ_{max}; if the dilution rate exceeds D_c the culture is eventually diluted to extinction and is said to have undergone 'wash-out'.

dilution test In ANTIBIOTIC-sensitivity testing: a procedure in which an organism is tested for its ability to grow in the presence of certain fixed concentrations of a given antibiotic. (cf. DIFFUSION TEST.)

In the *agar dilution test* (= *plate dilution test*) a series of agar plates containing progressively lower concentrations of a given antibiotic (and an antibiotic-free control plate) are each surface-inoculated with the test organism and incubated; the MIC (q.v.) is indicated by the lowest concentration of antibiotic at which growth does not occur.

The *broth dilution test* (= *tube dilution test*) is essentially similar (in principle and interpretation) to the agar dilution test except that growth of the test organism is attempted in each of a series of broths containing progressively lower concentrations of antibiotic. If each broth is ca. 0.2 ml or less the process may be called a *microdilution broth test*; otherwise it is a *macrodilution broth test*.

Dimargaritales See ZYGOMYCETES.

Dimastigamoeba gruberi A former name of *Naegleria gruberi*.

dimension (1) (*serol.*) In GEL DIFFUSION procedures: the mode in which antibody and antigen come together in order to form precipitate. In *single dimension* tests, antigen and/or antibody diffuse through the gel in a single direction – i.e., parallel to the long axis of a gel column. In *double dimension* tests, a given reactant (antigen and/or antibody) diffuses radially outwards from a hole or 'well' cut into the gel. (See also SINGLE DIFFUSION and DOUBLE DIFFUSION.)

(2) Either stage of TWO-DIMENSIONAL ELECTROPHORESIS.

Dimerella See GYALECTALES.

dimethirimol (5-*n*-butyl-2-dimethylamino-4-hydroxy-6-methylpyrimidine) A HYDROXYPYRIMIDINE ANTIFUNGAL AGENT which is active against powdery mildews of chrysanthemums, cucurbits and sugar beet.

dimethyldithiocarbamate See DMDC.

dimethylglyoxal *Syn.* DIACETYL.

dimethyloxazolidinedione (DMO) See CHEMIOSMOSIS.

dimethylsulphoxide (DMSO; $(CH_3)_2SO$) A non-ionized polar solvent used e.g. as a cryoprotectant in FREEZING. Both hydrophilic and lipophilic substances are soluble in DMSO.

dimetridazole (Emtryl) 1,2-Dimethyl-5-nitro-imidazole: a NITROIMIDAZOLE antimicrobial agent used in veterinary medicine e.g. for the prevention and treatment of BLACKHEAD in turkeys and of SWINE DYSENTERY.

dimictic Of, or pertaining to, DIMIXIS.

dimidiate (1) Split into two. (2) Lacking – or appearing to lack – one half (applied e.g. to a semicircular, non-stipitate fungal fruiting body).

dimitic See HYPHA.

dimixis (1) *Morphological* HETEROTHALLISM. (2) One-locus two-allele physiological heterothallism.

dimorphic fungi (1) Those fungi (e.g. species of *Blastomyces* and *Candida*) which – according to environmental conditions – can grow in either of two distinct vegetative states: a unicellular ('yeast-phase') state or a mycelial state. (See GROWTH (fungal); MORPH.) (See also QUORUM SENSING.)

(2) Fungi which exhibit DIPLANETISM.

(3) *Syn.* DIOECIOUS fungi.

***din* genes** In *Escherichia coli*, DNA *d*amage-*in*ducible genes: genes inducible in the SOS SYSTEM. For *dinB* see DNA POLYMERASE IV.

dinactin See MACROTETRALIDES.

dinitrogen fixation *Syn.* NITROGEN FIXATION.

dinitrogenase *Syn.* NITROGENASE.

2,4-dinitrophenol See PROTON TRANSLOCATORS.

dinitrophenols Dinitrophenols exhibit antifungal, insecticidal and herbicidal properties. As ANTIFUNGAL AGENTS they are used mainly for the treatment of powdery mildews; they include *binapacryl* (2,4-dinitro-6-*sec*-butyl-phenyl-3-methylcrotonate), which is also an acaricide, *dinobuton* (2,4-dinitro-6-*sec*-butylphenylisopropylcarbonate), and *dinocap* (a mixture of three isomers of each of the compounds 2,4-dinitro-6-*sec*-octylphenol and 2,6-dinitro-4-*sec*-octylphenol in which the octyl group may be 1-methylheptyl, 1-ethylhexyl or 1-propylpentyl). Dinocap is generally used as the crotonate; it functions as an antifungal ERADICANT and PROTECTANT. (See also PROTON TRANSLOCATORS.)

Dinobryon A genus of loricate, generally colonial CHRYSOPHYTES. Each cell has two flagella of unequal length, and sits in a stalked, vase-shaped LORICA composed of helical bands of cellulose microfibrils which are laid down by the slowly rotating cell within. Following cell division, one daughter cell becomes attached to the rim of the parent lorica and secretes a new lorica around itself, so that eventually a branching, fan-shaped colony is formed.

dinobuton See DINITROPHENOLS.

dinocap See DINITROPHENOLS.

dinoflagellates A large group of photosynthetic and/or heterotrophic organisms regarded either as algae (e.g. class Dinophyceae, division Dinophyta) or as protozoa (see PHYTOMASTIGOPHOREA). Most dinoflagellates are unicellular and biflagellated (but cf. e.g. CYSTODINIUM, DINOTHRIX and GLOEODINIUM). The cell typically has a *theca* composed essentially of a layer of flattened membranous vesicles beneath the plasmalemma; these vesicles may be empty (in 'unarmoured' dinoflagellates such as *Gymnodinium*, *Gyrodinium*, and *Oxyrrhis*) or may each contain a cellulosic plate varying from very thin (e.g. in *Katodinium* and *Woloszynskia*) to thick (in 'armoured' dinoflagellates such as *Ceratium* and *Peridinium*). The edges of thick plates are commonly bevelled, allowing the plates to move freely relative to each other. The number of plates per cell ranges from two (e.g. in PROROCENTRUM) to several hundred, the number, form and arrangement of plates being characteristic for a given dinoflagellate. (See also e.g. CERATIUM; ORNITHOCERCUS.) On cell division, the thecal plates may be shared between the daughter cells, each cell then synthesizing new plates to replace those taken by the other; in e.g. *Gonyaulax* shedding of the theca ('ecdysis') by the parent cell occurs prior to cell division.

In the majority of dinoflagellates (including e.g. *Amphidinium*, *Ceratium*, *Glenodinium*, *Gonyaulax*, *Gymnodinium*, *Oxyrrhis*, *Peridinium* and *Woloszynskia*) the vegetative cell is divided by a transverse groove (the *girdle*, *cingulum* or *annulus*) into two parts: the *epicone* and the *hypocone*. (The corresponding thecal regions in armoured dinoflagellates are termed the *epitheca* and *hypotheca*.) A longitudinal groove, the *sulcus*, extends perpendicularly from the girdle into the hypocone. One of the two flagella (the *transverse* flagellum) lies in the girdle and encircles the cell, while the other (*longitudinal*) flagellum arises from the sulcus and may project beyond the cell. The transverse flagellum [form and function: JP (1985) *32* 290–296] is longer than the longitudinal one, is helical in shape, contains a PARAXIAL ROD, and bears a single row of fine hairs; the longitudinal flagellum may be smooth or may bear two rows of stiffer hairs. (cf. e.g. PROROCENTRUM; see also PUSULE.) Dinoflagellate cells contain contractile, non-actin filaments which are apparently responsible for cell contraction and shape changes in at least some species (*Kofoidinium* and related genera [Cell Mot. (1985) *5* 1–15]).

The dinoflagellate nucleus contains chromosomes which are unique among eukaryote chromosomes in lacking centromeres and containing little or no protein; dinoflagellate nuclear organization has been considered to be intermediate between prokaryotic and eukaryotic, and has been termed 'mesokaryotic' or 'dinokaryotic'. During MITOSIS the nuclear membrane remains intact; the spindle is typically extranuclear – spindle MTs passing through the nucleus via channels lined with nuclear membrane – but in *Oxyrrhis marina* the spindle is wholly intranuclear [JCS (1986) *85* 161–175].

The majority of dinoflagellates are photosynthetic, typically possessing several chloroplasts per cell and containing chlorophylls *a* and c_2, β-carotene, and peridinin (replaced by fucoxanthin in *Glenodinium foliaceum*). Other dinoflagellates (e.g. NOCTILUCA) are colourless and heterotrophic. Both heterotrophic and photosynthetic species can apparently ingest particulate matter in the region of the sulcus.

Resistant 'resting spores' (cysts, hypnospores) are produced in some species. [Cyst formation in *Gonyaulax tamarensis*: JEMBE (1985) *86* 1–13.] (cf. HYSTRICHOSPHAERIDS.) Sexual reproduction has been observed in some species.

The majority of dinoflagellates are free-living in marine, brackish or freshwater habitats, sometimes forming extensive blooms (see e.g. RED TIDE). Some dinoflagellates (e.g. species of *Gonyaulax*, *Noctiluca*, *Pyrocystis*, *Pyrodinium*) exhibit BIOLUMINESCENCE, and some produce potent toxins (see e.g. SAXITOXIN). Dinoflagellates also occur in association with other organisms: e.g. *Blastodinium* is parasitic in the intestines of copepods, *Oodinium* is a parasite of fish (see VELVET DISEASE), and '*Symbiodinium*' is an endosymbiont in various invertebrates (see ZOOXANTHELLAE).

Dinoflagellida See PHYTOMASTIGOPHOREA.

dinokaryotic See DINOFLAGELLATES.

Dinophyceae See DINOFLAGELLATES.

Dinothrix A genus of marine DINOFLAGELLATES in which the cells are non-motile and are joined together to form branching filaments.

Diodoquin See 8-HYDROXYQUINOLINE.

dioecious Refers to an organism in which the male and female reproductive structures occur in different individuals. (cf. MONOECIOUS.)

dioecism Morphological HETEROTHALLISM.

1,2-dioxetane A type of substrate which gives rise to CHEMILUMINESCENCE when activated by certain enzymes (examples: AMPPD and CSPD).

dioxygenase See OXYGENASE.

dip-slide method (agar-slide method) A method for detecting or estimating microbial populations on solid surfaces and in various fluids (e.g. machine-tool cooling fluids, urine samples). The dip-slide, a plastic slide coated with a sterile nutrient agar, is attached to the inside of the cap in a screwcap cylindrical tube; for use, the cap is unscrewed and the slide is dipped into, or pressed against, the sample and it is subsequently incubated. Microbial numbers can be estimated from the number of colonies which develop on the slide. (See also COUNTING METHODS.)

diphasic medium Any MEDIUM, e.g. NNN MEDIUM, in which a solid (usually agar-based) component is at least partly covered by a liquid 'overlay'.

diphasic serotypes See H ANTIGEN (*Salmonella*).

diphenyliodonium chloride See IODONIUM COMPOUNDS.

diphenylthiourea See THIOUREA DERIVATIVES.

diphenylurea See CARBANILIDES.

diphosphopyridine nucleotide See NAD.

diphthamide See DIPHTHERIA TOXIN.

diphtheria (membranous croup) An acute infectious human disease, usually affecting children, caused by toxinogenic strains of *Corynebacterium diphtheriae*. Infection occurs by droplet inhalation or ingestion of contaminated food, milk etc. Asymptomatic carriers are often sources of infection. Incubation period: 1–10 (usually 2–5) days. Symptoms: fever, sore throat, headache, and the development of a characteristic whitish or grey membrane at the site of infection – usually the tonsils but sometimes e.g. the nasopharynx, larynx, or nose; the membrane and swelling of adjacent tissues can cause respiratory obstruction and difficulty in swallowing. Other sites occasionally affected include skin lesions or wounds (*cutaneous diphtheria*) and the genitals. The membrane results from the local effects of the DIPHTHERIA TOXIN which causes cell death and subsequent deposition of fibrin, blood cells, cellular debris, and *C. diphtheriae* cells. Local secondary infection with e.g. *Streptococcus pyogenes* (*septic diphtheria*) may occur. *C. diphtheriae* normally remains localized at the membrane, but the toxin is disseminated via the blood and lymph and can cause severe, often fatal, systemic effects – including myocarditis and nerve demyelination. *Treatment*: primarily with antitoxin. Penicillin G or erythromycin may be used in treatment and for eradication of *C. diphtheriae* from carriers. *Lab. diagnosis*: culture of *C. diphtheriae* (see CORYNEBACTERIUM) from swabs, and/or identification of the toxin (e.g. by gel diffusion tests).

[Review: RMM (1996) *7* 31–42.]

diphtheria toxin A heat-labile protein exotoxin (MWt ca. 60000) produced by certain lysogenic strains of *Corynebacterium diphtheriae* and responsible for the symptoms of DIPHTHERIA; it is lethal for many animal species and is also cytotoxic in some types of cell culture – e.g. Vero cells are highly susceptible, though murine cells are almost totally resistant. (cf. EXOTOXIN A.)

Diphtheria toxin (DT) is encoded by *tox*, a gene carried by various temperate CORYNEPHAGES (e.g. β, ω, γ) [tox^+ corynephages (DNA relationships): Inf. Immun. (1985) *49* 679–684].

The toxin is synthesized as a single polypeptide chain. The Y-shaped molecule consists of three domains: translocation domain, receptor-binding domain (RBD) and catalytic domain (CD).

Uptake of toxin involves initial binding of the RBD to a receptor protein, and internalization of the toxin within a vesicle via a clathrin-coated pit (PINOCYTOSIS). In the vesicle, CD is proteolytically cleaved but remains attached via a disulphide bond; the translocation domain undergoes an acid-induced conformational change in which it inserts into the membrane of the vesicle – apparently forming a pore through which CD is translocated into the cytoplasm of the target cell. The disulphide link undergoes reduction (scission) – releasing CD into the target cell's cytoplasm.

CD is a heat-stable fragment with resistance to cellular proteolysis. It has ADP-ribosyltransferase activity, and, in the cell, inhibits PROTEIN SYNTHESIS by catalysing the ADP-RIBOSYLATION (inactivation) of elongation factor EF-2. (One molecule of CD can kill a sensitive cell.) In EF-2, the target for ADP-ribosylation is a rare amino acid, 'diphthamide', (2-[3-carboxyamido-3-(trimethylammonio)propyl]-histidine); diphthamide appears to occur only in EF-2 – a single residue lying within a sequence which is highly conserved in eukaryotes. [Review: TIBS (1987) *12* 28–31.]

Elongation factors in members of the Archaea (formerly Archaebacteria) appear also to contain a single diphthamide residue, and are characteristically sensitive to diphtheria toxin [Book ref. 157, pp. 379–410]. Bacteria are not susceptible to the toxin.

Synthesis of DT is repressed while the intracellular concentration of iron remains above a certain level. Repression of *tox* is mediated by a repressor protein, DtxR, which, when activated by transition metal ions, binds to the operator of the *tox* gene and blocks transcription [see Nature (1998) *394* 502–506].

diphtheroid (1) Any non-pathogenic strain of *Corynebacterium*, particularly one isolated from the skin or mucous membranes. (2) Any strain of *Corynebacterium* other than those of *C. diphtheriae*. (3) *Syn.* CORYNEFORM (2).

diphtheroid stomatitis A disease of adult chickens, apparently caused by '*Spirillum pulli*' (see SPIRILLUM). Yellowish-white, firm lesions (ca. 2–20 mm), adherent to the underlying tissues, occur in the mouth, larynx and pharynx. (See also POULTRY DISEASES.)

dipicolinic acid (DPA) Pyridine-2,6-dicarboxylic acid; DPA occurs in bacterial ENDOSPORES, and has been reported e.g. in *Penicillium citreoviride*. In bacteria, dipicolinic acid is synthesized via a branch of the diaminopimelic acid pathway [see Appendix IV(e)]. [Characterization of spores of (mutant) *Bacillus subtilis* which lack dipicolinic acid: JB (2000) *182* 5505–5512.]

diplanetism In certain OOMYCETES: the phenomenon in which two types of morphologically distinct zoospore are formed at separate stages (*swarm periods*) in the asexual life cycle. *Primary zoospores* are formed in the first swarm period; in e.g. *Achlya* and *Saprolegnia*, primary zoospores are pyriform with two flagella arising at the apex. After release from the sporangium, primary zoospores may encyst directly (e.g. in *Achlya*) or may swim for some time before encysting (e.g. in *Saprolegnia*). A cyst subsequently gives rise to a *secondary zoospore*; in e.g. *Achlya* and *Saprolegnia*, secondary zoospores are reniform with two flagella arising from the concavity. After a period of swarming, secondary zoospores encyst and subsequently germinate to form a vegetative thallus. (cf. MONOPLANETISM and POLYPLANETISM.)

Diplocalyx See SPIROCHAETALES.

Diplocarpon A genus of fungi of the order HELOTIALES; anamorphs occur in the genera *Marssonina* and *Entomosporium*. *D. earlianum* causes strawberry leaf scorch. *D. rosae* causes BLACK SPOT (sense 3) of roses; the blackish acervuli of the *Marssonina* stage, bearing uniseptate conidia (the main disseminative spore), develop on leaves and/or on overwintering lesions on the shoots. The ascigerous stage can occur in overwintering leaves.

***Diplocarpon rosae* virus** See MYCOVIRUS.

diplococcin An antibiotic, structurally related to NISIN, produced by *Lactococcus* sp.

diplococcus A pair of cocci (see COCCUS) – the form commonly assumed by e.g. *Enterococcus faecalis* and *Streptococcus pneumoniae* and by species of *Neisseria*.

Diplococcus pneumoniae Incorrect name for *Streptococcus pneumoniae*.

Diplodia See SPHAEROPSIDALES.

Diplodinium A genus of ciliates (order ENTODINIOMORPHIDA) which occur e.g. in the RUMEN. The cytostome is close to the anterior pole of the cell; somatic ciliature is limited, cilia occurring in tufts in the oral and anteriodorsal regions. (See also DZM.) The pellicle may be drawn out posteriorly into one or more spines.

Diploicia A genus of placodioid LICHENS (order LECANORALES). Apothecia: lecideine, black; ascospores: brown, one-septate. *D. canescens* (formerly *Buellia canescens*) has a whitish thallus bearing whitish-grey farinose soralia; this species is common in Atlantic regions of Europe, growing on rocks, wood, bark etc. – particularly in light shade.

diploid Having a PLOIDY of *two*. (cf. HAPLOID and POLYPLOID.) In a prokaryotic cell, *partial diploid* or *merodiploid* refers to the possession of extra copies of only *some* of the chromosomal genes – see MEROZYGOTE.

Diplomastigomycotina See OOMYCETES.

Diplomonadida An order of protozoa (class ZOOMASTIGOPHOREA); cells contain either one (suborder Enteromonadina) or two (suborder Diplomonadina) KARYOMASTIGONTS, those containing two exhibiting twofold rotational symmetry or (in *Giardia*) bilateral symmetry. Each karyomastigont has 1–4 flagella. Mitochondria and Golgi apparatus are absent.

Members of the Enteromonadina are parasitic; cysts are formed in at least one genus. Genera: e.g. *Enteromonas*, *Trimitus*.

In the Diplomonadina each karyomastigont has four flagella, one of which is recurrent. Cysts are formed by at least some species. The suborder includes free-living and parasitic species. Genera: e.g. GIARDIA, HEXAMITA, *Trepomonas*.

Diplomonadina See DIPLOMONADIDA.

diplont (1) (*noun*) An organism in whose life cycle only the gametes are haploid. (cf. HAPLONT.) (2) The diploid form of an organism, e.g. a sporophyte (see ALTERNATION OF GENERATIONS). (3) (diplontic) (*adj.*) Refers to either (1) or (2) above.

diplophage A bacteriophage which infects *Streptococcus* ('*Diplococcus*') *pneumoniae*.

diplophase In organisms which reproduce sexually: the diploid phase of the life cycle (between syngamy and MEIOSIS).

Diplornaviridae A family proposed for all dsRNA-containing viruses; the family has not been accepted since differences in genome structure, replication and expression among dsRNA viruses are too great to be encompassed in a single family.

Diploschistes See GRAPHIDALES.

diplostichomonad membrane See PARORAL MEMBRANE.

diplotene stage See MEIOSIS.

Diplotheca See CHOANOFLAGELLIDA.

Dipodascaceae See ENDOMYCETALES.

Dipodascus See ENDOMYCETALES and GEOTRICHUM; see also AMBROSIA FUNGI.

dipyridine ferrohaemochrome See CYTOCHROMES.

direct epifluorescent filter technique See DEFT.

direct immunofluorescence See IMMUNOFLUORESCENCE.

direct repeat (DR) Either of two (or more) regions of a nucleic acid molecule in which the sequences of bases or base pairs are similar or identical and have the same polarity and orientation, e.g.:

$$5' \ldots GGCT \ldots GGCT \ldots 3'$$
$$3' \ldots CCGA \ldots CCGA \ldots 5'$$

The repeated sequences may or may not be contiguous. (cf. INVERTED REPEAT.)

direct repeat (DR) locus (in *Mycobacterium tuberculosis*) See SPOLIGOTYPING.

directly observed therapy, short-course (DOTS) See TUBERCULOSIS.

disc (lichenol.) The hymenial layer of an apothecium; it may be circular (flat or convex), elongated (lirelliform), or irregular in shape.

disc diffusion test (also: agar disc diffusion test) In ANTIBIOTIC-sensitivity testing: a DIFFUSION TEST in which the entire surface of an agar plate is inoculated with the organism under test, and several small absorbant paper discs – each impregnated with a different antibiotic – are placed at different locations on the agar surface; on subsequent incubation, antibiotics diffuse from the discs, and a zone of growth inhibition develops around each disc which contains an antibiotic to which the organism is susceptible. Precise details of the test (e.g. inoculum size, medium) commonly conform to the so-called 'Kirby–Bauer technique' [AJCP (1966) *45* 493–496].

Strains of *Staphylococcus aureus* containing an *inducible* β-LACTAMASE commonly form zones of inhibition around discs containing an antibiotic sensitive to the enzyme; these zones are typically bordered by relatively heavy growth containing discrete colonies. (Zones formed by sensitive bacteria generally exhibit a gradual or abrupt transition between regions of growth and non-growth.) Cells close to the disc are killed before adequate amounts of β-lactamase can be synthesized. Cells at the periphery of the zone are exposed to gradually increasing concentrations of the (outwardly diffusing) antibiotic; such cells can synthesize sufficient β-lactamase to permit survival, and they give rise to relatively large colonies by using nutrients forfeited by inactivated cells in their immediate vicinity.

(cf. AGAR DIFFUSION TEST.)

disclosing agent See DENTAL PLAQUE.

discocarp *Syn.* APOTHECIUM.

discoidins See LECTINS.

Discomycetes A class of ascomycetes in which the ascocarp is typically an apothecium (though some members, e.g. *Tuber* spp, form closed, hypogean ascocarps). The class is not recognized in most modern taxonomic schemes.

Discorbis See FORAMINIFERIDA.

disease A term used loosely for any state of an individual (human, animal or plant) characterized by a deviation from the condition regarded as normal or average for members of the species or type, and which is usually wholly detrimental; exceptionally, a disease may be associated with some benefit: see e.g. sickle-cell anaemia in the entry MALARIA. A disease may be due e.g. to a metabolic or genetic defect, to infection by a PATHOGEN, or to the ingestion of a TOXIN or other poison – but not (in conventional use) to physical wounds. By custom, the term 'disease' is generally applied only to multicellular organisms – including certain multicellular microorganisms (such as mushrooms); however, analogous states occur in unicellular organisms: e.g., a bacterium may suffer metabolic or genetic lesion(s), it may be infected e.g. by a virulent bacteriophage, or it may be affected adversely by an antibiotic, a bacteriocin, or some other toxic substance.

disinfectant Any chemical agent used for DISINFECTION; disinfectants which are non-injurious to human or animal tissues may also be used as ANTISEPTICS. (cf. PRESERVATIVES.) Disinfectants for general use should be active against a wide range of common pathogens and should be microbicidal rather than microbistatic at the concentrations used; microbicidal disinfectants may become microbistatic when diluted, and at low concentrations some disinfectants can be metabolized by certain organisms – e.g. *Pseudomonas* spp can metabolize phenol. A given disinfectant may be effective against only one particular category of organisms (e.g. Gram-negative bacteria), but is not necessarily effective against all species in that category. Most disinfectants are primarily antibacterial, but some are also (or primarily) antifungal or antiviral; most disinfectants have little or no activity against bacterial ENDOSPORES. (See also ACRIDINES; ALCOHOLS; BISPHENOLS; CHLORINE; FORMALDEHYDE; HEAVY METALS; HYPOCHLORITES; IODINE; OXIDIZING AGENTS; PHENOLS; PINE DISINFECTANTS; QUATERNARY AMMONIUM COMPOUNDS; SALICYLIC ACID.)

Different disinfectants may require different conditions for maximum effectiveness. Important factors include e.g. pH (cf. CHLORHEXIDINE and HYPOCHLORITES), the presence or absence of organic matter (cf. CHLORINE and ACRIDINES), and humidity (see e.g. FORMALDEHYDE). The detergent properties of QUATERNARY AMMONIUM COMPOUNDS permit them to penetrate grease barriers which may protect organisms from non-detergent disinfectants. In general, the activity of a disinfectant rises with temperature (see TEMPERATURE COEFFICIENT) and concentration (see DILUTION COEFFICIENT). The action of some disinfectants is reversible; thus e.g. endospores inactivated by mercuric ions may germinate after exposure to hydrogen sulphide (which combines with mercuric ions). The efficacy of a disinfectant may be estimated by certain tests – see e.g. CAPACITY TEST; CARRIER TEST; CHICK–MARTIN TEST; KELSEY–SYKES TEST; PHENOL COEFFICIENT; RIDEAL–WALKER TEST; SUSPENSION TEST; USE-DILUTION TEST.

disinfection The destruction, inactivation or removal of those microorganisms likely to cause infection or other undesirable effects (e.g. spoilage); disinfection does not normally involve STERILIZATION. The term 'disinfection' commonly refers to the use of chemical agents (DISINFECTANTS) for the treatment of inanimate objects and surfaces; however, it is sometimes also used to refer to ANTISEPSIS.

Disinfection can also be achieved by physical proceduces such as PASTEURIZATION and TYNDALLIZATION. (See also ULTRAVIOLET RADIATION and LOW-TEMPERATURE STEAM DISINFECTION.) A PULSED ELECTRIC FIELD technique may have applications e.g. in the food industry; this approach has been used e.g. to reduce significantly the numbers of viable cells of *Mycobacterium paratuberculosis* in milk [AEM (2001) *67* 2833–2836].

[Disinfection kinetics: JAM (2001) *91* 351–363.]

disjunctor cells Cells which occur between spores in a chain of spores; their disintegration may aid spore release. (cf. NECRIDIUM.)

disk See DISC.

dislodgement (*mol. biol.*) The phenomenon (mechanism unknown) in which the introduction of a second, compatible, plasmid into a cell sometimes causes the elimination of the resident plasmid.

disodium phosphonoacetate See PHOSPHONOACETIC ACID.

displacement loop See e.g. D LOOP and R LOOP.

displacement vector *Syn.* REPLACEMENT VECTOR.

disporic Of a coccidian oocyst: containing two sporocysts.

disseminated intravascular coagulation (DIC) A condition characterized by an imbalance between pro- and anticoagulant activities within the blood vascular system, resulting in the formation of disseminated clots; DIC is usually a result of sepsis, but other causes include tissue necrosis, hypothermia and some types of snake bite.

DIC is incited when bacterial components (such as LIPOPOLYSACCHARIDE) bind to specific receptors on endothelium and leukocytes; this stimulates the release of a number of CYTOKINES – including the pro-inflammatory TNF-α, IL-1 and IL-6. Cytokine activity appears to promote coagulation and also to inhibit anticoagulant activity. Additionally, cytokines upregulate CELL ADHESION MOLECULES on the endothelial membrane; adhesion molecules contribute to coagulopathy by binding monocytes and platelets. Activated platelets are degranulated, releasing various products (including fibrinogen); moreover, the platelet membrane gives rise to the potent vasoconstrictor thromboxane A_2.

Initiation of coagulation apparently involves cytokine-mediated upregulation of an endothelial membrane protein, *tissue factor* (TF), which is believed to play a key role in triggering the coagulation cascade. In an animal model, anti-TF antibodies were able to limit coagulopathy in the presence of *Escherichia coli* sepsis.

In vivo, the inhibitor of TF is a protein synthesized in the endothelium: *tissue factor pathway inhibitor* (TFPI); in at least some studies, the level of TFPI appears to be reduced in cases of DIC.

[Disseminated intravascular coagulation: BCH (2000) *13* 179–197.]

dissepiment In a polypore: the tissue which lies between, and supports, the pores. (cf. TRAMA.)

dissimilatory nitrate reduction Energy-yielding (respiratory) metabolism in which nitrate is reduced to e.g. nitrite, nitrous oxide, dinitrogen or ammonia, the product(s) typically being excreted: see DENITRIFICATION and NITRATE RESPIRATION. (cf. ASSIMILATORY NITRATE REDUCTION.)

dissimilatory perchlorate reduction See BIOREMEDIATION.

dissimilatory reduction of nitrate to ammonia A respiratory process carried out by certain bacteria (e.g. species of *Enterobacter* and *Vibrio*) and reported to occur in habitats such as marine sediments. The process appears to occur more readily when the concentration of nitrate is low – higher concentrations of nitrate being reduced to nitrite and correspondingly smaller amounts of ammonia [FEMS Ecol. (1996) *19* 27–38]. (cf. DISSIMILATORY NITRATE REDUCTION.)

dissimilatory sulphate reduction (sulphate respiration) The use of sulphate (or other oxidized compound of sulphur) as terminal electron acceptor in the anaerobic respiratory metabolism of the SULPHATE-REDUCING BACTERIA. Initially, sulphate is 'activated', i.e., converted to adenosine-5′-phosphosulphate (APS) in an energy-requiring reaction (enzyme: sulphate adenylyl transferase):

$$SO_4^{2-} + ATP \longrightarrow APS + PPi$$

This reaction is encouraged by the hydrolysis of PPi; the pyrophosphatase involved is inactivated under aerobic conditions, so that ATP is conserved when growth cannot occur.

The reduction of APS is catalysed by adenylyl sulphate reductase in the reaction:

$$APS \longrightarrow AMP + SO_3^{2-}$$

Sulphite is reduced to sulphide by an enzyme complex which, in many species of sulphate-reducing bacteria, includes e.g. DESULFOVIRIDIN. Different intermediates (e.g. thiosulphate, trithionate) have been reported from in vitro studies on sulphite reduction; these intermediates may or may not occur in vivo – or may occur in only some species. A small amount of reduced sulphur is assimilated, but nearly all is eliminated as H_2S (cf. ASSIMILATORY SULPHATE REDUCTION).

The reduction of APS and sulphite involves electron transport via e.g. *c*- and/or *b*-type cytochromes, and is accompanied by OXIDATIVE PHOSPHORYLATION. In many species electrons are derived from the oxidation of certain organic substrates to e.g. acetate; in e.g. *Desulfobacter postgatei* and *Desulfovibrio baarsii* electrons can be derived from the oxidation of acetate to CO_2. Some sulphate-reducing bacteria can grow mixotrophically, gaining energy from the oxidation of H_2 (at the expense of sulphate) and using acetate and CO_2 as sources of carbon.

In vitro inhibition of sulphate reduction can be achieved e.g. with chromate, molybdate, monofluorophosphate or selenate ions; molybdate appears to form an analogue of APS, thus depleting the cell of ATP.

(See also SULPHUR CYCLE.)

dissimilatory sulphur metabolism See DISSIMILATORY SULPHATE REDUCTION, SULPHUR-OXIDIZING BACTERIA, SULPHUR-REDUCING BACTERIA and THIOBACILLUS.

dissociation (*bacteriol.*) A phenotypic change in a given strain (see e.g. SMOOTH–ROUGH VARIATION). (cf. SALTATION.)

dissymmetry ratio See BASE RATIO.

distamycin A See NETROPSIN.

distemper (*vet.*) (1) *Syn.* CANINE DISTEMPER. (2) *Syn.* STRANGLES.

Distigma See EUGLENOID FLAGELLATES.

distoseptum See SEPTUM (b).

Distrene-80 A high-MWt polystyrene (see DPX).

distromatic Refers to a structure which is two cells thick (see e.g. ULVA). (cf. POLYSTROMATIC.)

ditalimfos See ORGANOPHOSPHORUS COMPOUNDS.

dithanes See MANEB (dithane M-22), NABAM (dithane D-14) and ZINEB (dithane Z-78).

dithianon (2,3-dicyano-1,4-dihydro-1,4-dithia-anthraquinone) A QUINONE ANTIFUNGAL AGENT used e.g. for the control of apple scab and other fungal diseases of apples and related fruits; it is not effective against apple mildew.

dithiocarbamates Dithiocarbamates (salts of dithiocarbamic acid derivatives: $R_2N.CS.SH$) include several agricultural ANTIFUNGAL AGENTS (e.g. METHAM SODIUM) and preservatives for paper and board; the antifungal activity of at least some dithiocarbamates appears to depend on their conversion to substances which are more highly fungitoxic – methylisothiocyanate in the case of metham sodium. (See also BISDITHIOCARBAMATES; DMDC; THIURAM DISULPHIDES.)

dithiothreitol See MUCOLYTIC AGENT.

Diuron *Syn.* DCMU.

divercin V41 A 43-amino-acid BACTERIOCIN produced by *Carnobacterium divergens* strain V41 [Microbiology (1998) *144* 2837–2844].

divergent trama See BILATERAL TRAMA.

divergent transcription Transcription of different genes (on the same DNA molecule) in opposite directions. When divergent transcription occurs from closely spaced promoters, the (transcription-dependent) generation of negative supercoiling behind each advancing polymerase causes an increase in negative superhelicity in the region between the active polymerases; this increase in superhelicity may affect the expression of genes or operons [Mol. Microbiol. (2001) *39* 1109–1115].

division A taxonomic rank (see TAXONOMY and NOMENCLATURE).

dizoic Of a coccidian oocyst: containing two sporozoites per sporocyst.

DLVO theory See ADHESION.

DMDC DimethylDITHIOCARBAMATE; DMDCs have antifungal (and antibacterial) activity and are used in agriculture as broad-spectrum antifungal agents – e.g. for the treatment of fruit and vegetables and as a constituent of seed and turf dressings. DMDCs include *ferbam*, $[(CH_3)_2.N.CS.S]_3Fe$, and *ziram*, $[(CH_3)_2.N.CS.S]_2Zn$. Oxidation of dimethyldithiocarbamic acid gives THIRAM.

DMO 5,5-Dimethyl-4-oxazolidinedione: see CHEMIOSMOSIS.

DMSO DIMETHYLSULPHOXIDE.

DNA Deoxyribonucleic acid: a NUCLEIC ACID consisting of deoxyriboNUCLEOTIDES, each of which *typically* contains one of the bases adenine, guanine, cytosine and thymine (cf. e.g. BACTERIOPHAGE and DNA METHYLATION). (See also RNA.) DNA is the repository of genetic information in all cells and in many VIRUSES (see GENETIC CODE). In eukaryotes DNA occurs chiefly in CHROMOSOMES but also in organelles: see MITOCHONDRION (mtDNA), CHLOROPLAST (ctDNA) and KINETOPLAST; in prokaryotes DNA constitutes the CHROMOSOME. (See also PLASMID.) DNA can be detected in situ by specific STAINING (see e.g. ACRIDINE ORANGE, DAPI, FEULGEN REACTION).

A single chain of deoxyribonucleotides (*single-stranded DNA*, = ssDNA) occurs in some viruses, usually as a covalently closed 'circle' (ss cccDNA) (see e.g. SSDNA PHAGE; cf. PARVOVIRIDAE).

In cells and in many viruses DNA is normally *double-stranded* (dsDNA, duplex DNA): the two strands are held together by Watson–Crick BASE PAIRing between bases on opposite strands; thus, the base sequence of each strand is *complementary* to that of the other. The two strands in dsDNA are *antiparallel*, i.e., orientated 'head-to-tail' ($5' \ldots 3'$ adjacent to $3' \ldots 5'$). Under physiological conditions the two strands are usually twisted around a common axis to form a *double helix*. In its commonest form (the Watson–Crick double helix, or B-DNA) the helix is right-handed (i.e., when viewed from one end it winds away from the viewer in a clockwise direction) with on average ca. 10.4 base pairs (bp) per turn, each turn being ca. 3.4 nm long and ca. 2 nm in diameter. The (planar) bases are stacked roughly perpendicular to the helix axis but with a propeller twist of ca. 11.5–17.3° between the bases in each pair. Two spiral grooves are discernible in B-DNA: the *minor groove*, located between the two strands, and the *major groove*, which runs between the turns of the helix. Substituents on the bases protrude into these grooves and are thus accessible e.g. for recognition by enzymes. B-DNA is now thought to have local variations in conformation which seem to reflect the nature of the local base sequences; this may further assist specific recognition of particular base sequences by proteins. In vivo the double helix is stabilized by hydrophobic interactions between the stacked bases, and by neutralization of the charges on the phosphate groups by e.g. divalent metal ions or (possibly) POLYAMINES in bacteria, or by HISTONES in eukaryotes.

dsDNA may adopt different helical conformations under different conditions [review: ARB (1982) *51* 395–427]. For example, A-DNA is a right-handed helix with ca. 11 bp/turn and bases tilted at an angle to the helix axis; the bases are relatively inaccessible, the sugar–phosphate backbone being the most prominent part of the molecule. A-DNA occurs in vitro e.g. at low (75%) relative humidity in the presence of Na^+, K^+ or Cs^+ as counterion; in vivo a similar conformation occurs e.g. in DNA–RNA hybrid duplexes and probably also in dsRNA. (cf. Z-DNA; see also HAIRPIN; CRUCIFORM.)

In most bacterial chromosomes and plasmids, certain viruses, mitochondria and chloroplasts, the DNA is double-stranded and covalently closed 'circular' (ds cccDNA), i.e., it has no free $3'$ or $5'$ ends. Such DNA can occur in distinct topoisomeric forms (see TOPOISOMER). *Relaxed* cccDNA can be regarded as linear B-DNA with covalently joined ends, i.e., the molecule can theoretically lie in a plane, without strain, with a pitch (p) of ca. 10.4 bp/turn. The number of times one strand winds around the other in any ds cccDNA molecule *lying in a plane* is called the *linking number* (= *topological winding number*) of that molecule and is designated *Lk* (or α); by convention, linking numbers are designated positive in right-handed helices. In a relaxed molecule, the linking number is the same as that of an equivalent B-DNA molecule (ca. $N/10.4$ where $N =$ the number of base pairs in the molecule) and is called the *duplex winding number*, Lk_0 (or β). The linking number of any cccDNA molecule is a topological property of that molecule; it cannot change without strand breakage. If a relaxed molecule is broken and the broken ends are twisted (in either direction) before rejoining, the linking number of the resulting molecule will be greater or less than Lk_0. Such a molecule is under strain (envisaged as a tendency to restore the stable pitch of ca. 10.4 bp/turn); a nick in one strand would allow rotation of the intact strand at the site of the nick and, hence, relaxation of the molecule. Unlike a relaxed molecule, the strained molecule can lie in a plane only if forced to do so, such constraint necessarily causing a deviation in pitch from 10.4 bp/turn. When such a molecule is *unconstrained*, the pitch of the helix is kept as nearly as possible to 10.4 bp/turn by contortion of the molecule such that the axis of the double helix itself becomes helical; this contortion is termed *writhe*, and the molecule is said to be *supercoiled* or *superhelical*. (See also DNA I, DNA II, etc.) Supercoiled DNA is much more compact than equivalent relaxed or linear DNA and thus has a much higher sedimentation coefficient and lower viscosity.

When all the strain of supercoiling is accommodated by writhe, the amount of writhe (*writhing number*, *Wr* – i.e., the number of superhelical turns) in a given cccDNA molecule is indicated by the *superhelix winding number*, τ, given by the difference between the linking number of the supercoiled molecule and that of the equivalent relaxed molecule: $\tau = Lk - Lk_0$. If the molecule has *fewer* turns than the equivalent relaxed molecule (i.e., the molecule is *underwound*), $Lk < Lk_0$, and τ is negative (i.e., the molecule is negatively supercoiled). If the molecule is *overwound*, $Lk > Lk_0, \tau$ is positive. (When $Lk = Lk_0, \tau = 0$, and the molecule is relaxed.) In general, naturally occurring supercoiled DNA is negatively supercoiled (cf. REVERSE GYRASE). The *superhelical density*, σ, of a given molecule is the number of superhelical turns per turn of the helix in relaxed DNA, i.e., $\sigma = 10.4\tau/N$.

The equation $\tau = Lk - Lk_0$ assumes that the pitch of the double helix, and hence the number of turns in the helix, remains constant (at the value for the relaxed molecule, Lk_0), so that any supercoils are accommodated entirely by writhe. However, the strain of supercoiling may force a change in pitch, and hence a change in the number of turns in the double helix – the actual number of turns, the *twist* (*Tw*), being related to the pitch (p) of the helix by the expression $Tw = N/p$; by convention, *Tw* is positive for a right-handed helix. Pitch, and hence twist, can also be altered by the insertion of an INTERCALATING AGENT. Twist differs from the linking number in that, in a given cccDNA molecule, the amount of twist can vary depending on the amount of writhe; the linking number is invariable and is the sum of twist and writhe. Thus, $Wr = Lk - Tw$, and, since *Lk* is constant for a given molecule, any change in writhe must be accompanied by an equal and opposite change in twist, and vice versa; when $Tw = Lk_0, Wr = \tau$, and when $Wr = 0, Tw = Lk$. [PNAS (1976) *73* 2639–2643.]

The positive free energy associated with negative supercoiling has a number of important consequences for the biological behaviour of DNA. It facilitates the formation of certain secondary structures (e.g. HAIRPINS, CRUCIFORMS, localized regions of Z-DNA) and also the unwinding of DNA strands e.g. during DNA REPLICATION, RECOMBINATION, TRANSCRIPTION, etc. The degree of supercoiling of DNA in vivo is controlled by enzymes called TOPOISOMERASES.

Normal cellular function requires that the degree of supercoiling be maintained within certain limits. In some circumstances – e.g. during the transition from aerobic growth to anaerobic growth – there is a fall in the cell's energy charge and a corresponding, transient fall in 'global' (i.e. overall) superhelical density; such changes in superhelicity alter the expression of a range of genes (e.g. by affecting the activity of their promoters). *Local increases* in superhelicity (i.e. in specific regions of the DNA molecule) can be brought about by divergent transcription from closely spaced promoters; such increased negative superhelicity may affect the expression of particular genes/operons [Mol. Microbiol. (2001) *39* 1109–1115].

Pure linear dsDNA clearly cannot be supercoiled. However, linear dsDNA (and discrete regions or *domains* within linear or ccc dsDNA) can maintain a superhelical conformation if there is a constraint on the rotation of the double helix to prevent unwinding of the supercoils (see CHROMATIN and CHROMOSOME (bacterial)).

Certain drugs exert their toxic effects by binding to DNA; the drug may bind e.g. to the major or minor grooves of dsDNA (see e.g. NETROPSIN) or may intercalate between the base-pairs (see INTERCALATING AGENTS). [DNA structure and its perturbation by drug binding (review): Bioch. J. (1987) *243* 1–13.]

See also GC%; DNA HOMOLOGY; THERMAL MELTING PROFILE.

DNA I Supercoiled ds cccDNA (see DNA).

DNA I′ *Syn.* DNA IV (q.v.).

DNA II Nicked relaxed circular dsDNA formed by an ss 'nick' in DNA I.

DNA III Linear dsDNA derived from DNA I or DNA II by strand breakage.

DNA IV (DNA I′) Relaxed (intact) ds cccDNA (see DNA).

DNA base composition See e.g. BASE RATIO and GC%. (See also DNA HOMOLOGY.)

DNA binding protein (1) *Syn.* SINGLE-STRAND BINDING PROTEIN. (2) In general: *any* non-enzymic protein which binds to ssDNA *or* dsDNA – e.g. a repressor protein, or a single-strand binding protein.

DNA blotting See BLOTTING.

DNA chip A microarray of short, single-stranded DNA molecules – having diverse (but known) sequences – immobilized on a small piece ('chip') of glass, silicon or plastic. The immobilized DNA molecules commonly function as PROBES. A chip can be used e.g. to determine the sequence of an unknown, labelled, single-stranded fragment of nucleic acid by testing the ability of the fragment to hybridize with a *particular* probe in the array, hybridization being detected by the fragment's label – using e.g. confocal epifluorescence microscopy.

A large microarray of probes can cover all possible combinations of nucleotides for a strand of given length. For example, a complete set (4^8) of nucleotide combinations for an 8-nucleotide probe – 65536 different probes – can be synthesized on a 1 cm^2 glass chip.

Chips are useful e.g. for detecting mutations, including those that confer resistant to antibiotics. Thus, that section of a genome in which mutations confer resistance to particular antibiotic(s) can be amplified (e.g. by PCR using fluorophore-labelled primers), and the amplicons tested against a microarray which includes all possible mutations (as well as the wild-type sequence). For example, the protease gene in clade B isolates of the human immunodeficiency virus (HIV-1) has been investigated in this way with the object of linking particular mutations with resistance to protease inhibitors [Nature Medicine (1996) *2* 753–759].

Chip technology can also be used for examining the differential expression of genes: with an array of immobilized, single-stranded cDNAs (which bind mRNAs of expressed genes), fluorophore-labelled probes (with binding sites elsewhere on the mRNA molecules) can be used to detect those genes which are active under specific conditions.

Chips can be made by (i) immobilizing pre-existing molecules, or (ii) 'on-chip synthesis' – in which high-speed robotic devices can synthesize a vast range of nucleotide combinations on a single chip.

One problem with microarrays is that (single-stranded) DNA probes have a tendency for intra-strand base-pairing – which may obscure target sequences. A possible solution may be the use of microarrays containing PNA probes because PNA–DNA hybridization can occur in the absence of salts (that is, under conditions which inhibit intra-strand base-pairing in ssDNA).

Among the growing number of published uses, high-density arrays have been used to identify species of *Mycobacterium* and (*simultaneously*) to test for resistance to rifampicin [JCM (1999) *37* 49–55].

The need to include a sufficiently comprehensive range of probes for a given investigation has been demonstrated in another study in which non-clade-B strains of HIV-1 were examined for resistance to antiretroviral agents using an assay system designed for clade B isolates of HIV-1; the assay was unsatisfactory for non-clade-B isolates – indicating the need to ensure that future assays can accomodate newly emergent strains of HIV-1 [JCM (1999) *37* 2533–2537].

[DNA chips (reviews): Nature Biotech. (1998) *16* 27–31 & 40–44; TIBtech. (1999) *17* 127–134. Biomedical discovery with DNA arrays (review): Cell (2000) *102* 9–15.]

DNA cloning See CLONING.

DNA delay mutant A mutant of BACTERIOPHAGE T4 which, on infection of a host cell, shows delayed synthesis of T4 DNA (particularly in *Escherichia coli* B strains) and a reduced burst size. The mutation occurs in genes 39, 52 or 60 – genes encoding subunits of the T4 TOPOISOMERASE; a DNA delay mutant requires a functional host GYRASE for replication, the gyrase presumably replacing T4 topoisomerase in an essential function in the initiation of early DNA replication.

DNA-dependent RNA polymerase See RNA POLYMERASE.

DNA–DNA hybridization See DNA HOMOLOGY.

DNA fingerprinting (chromosomal fingerprinting; also: restriction enzyme analysis, restriction endonuclease analysis, REA) A method used for TYPING. Essentially, chromosomal DNA, isolated from a culture of the given strain, is first cleaved by a RESTRICTION ENDONUCLEASE; the fragments (of many different lengths) are then separated by gel electrophoresis into a series of bands – which are stained. The pattern of bands, reflecting the cutting sites of the given RE in the chromosome, is the *fingerprint* of the strain. Strains which differ e.g. through loss or gain of a restriction site (perhaps through a point mutation) will give different fingerprints; similarly, insertion or deletion mutations (or e.g. insertion/excision of phage DNA) will alter the length of particular fragments and will also give rise to changes in the fingerprint. Hence, strains can be typed on the basis of their fingerprints.

One problem with this method is that it may yield too many fragments – giving a complex fingerprint which is difficult to interpret. One solution to the problem is to use a 'rare-cutting' RE (see table in RESTRICTION ENDONUCLEASE); this gives fewer, larger fragments that can be analysed e.g. by PFGE.

Alternatively, the original procedure can be carried out with additional use of a labelled probe that reveals only those (few) fragments to which the probe binds; an example of this approach is RIBOTYPING (q.v.).

[Evidence from DNA fingerprinting of endoscopic cross-infection in post-endoscopic acute gastritis (*Helicobacter pylori*): JCM (2000) *38* 2381–2382.]

(See also FINGERPRINTING.)

***dna* genes** In *Escherichia coli*: those genes whose products are involved in at least some types of DNA REPLICATION. See separate entries for genes *dnaA*, *dnaB*, *dnaC* (= *dnaD*), *nrdA* (= *dnaF*), *dnaI*, *dnaJ*, *dnaK*, *dnaP* and *dnaT*. For *dnaE*, *dnaN*, *dnaQ*, *dnaX* and *dnaZ* (= *dnaH*) see DNA POLYMERASE. For *dnaG* see PRIMASE. For *dnaW* see *adk* gene.

DNA glycosylase See ADAPTIVE RESPONSE and URACIL-DNA GLYCOSYLASE.

DNA gyrase *Syn.* GYRASE.

DNA helicases *Syn.* HELICASES.

DNA homology The degree of similarity ('relatedness') between base sequences in different DNA molecules (or in different parts of the same DNA molecule); two DNA molecules which are 100% homologous have identical sequences of nucleotides.

In microbial TAXONOMY the degree of homology between samples of chromosomal DNA from different organisms can be used to indicate the taxonomic relationship of the organisms. The homology between two samples of dsDNA can be assessed by various indirect methods (as well as by sequencing each sample and making a direct comparison of the nucleotides); two approaches are outlined below.

In *DNA–DNA hybridization*, single-stranded (heat-denatured) chromosomal DNA from one test strain (strain A) is bound to a membrane (e.g. nitrocellulose). Chromosomal DNA from another test strain (strain B) is fragmented (for example, by ULTRASONICATION) and the fragments denatured by heat to single-stranded pieces. Similar, single-stranded fragments are also made from *labelled* DNA of strain A. Under appropriate conditions, the unlabelled, single-stranded DNA from strain A (which is bound to the membrane) is then exposed to the single-stranded fragments from strains A and B; the concentration of B fragments is very much higher than that of the A fragments. This allows the fragments to hybridize, by base-pairing, with complementary sequences in the bound DNA.

The fragments from strains A and B compete for sequences on the bound DNA. However, the concentration of B fragments is very much higher than that of A fragments; consequently, if strains A and B are very similar, few (if any) of the (*labelled*) A fragments will hybridize. On the other hand, if strains A and B are very different many of the A fragments will hybridize. Thus, the similarity between strains A and B is indicated by the number of A fragments which hybridize; this can be determined by measuring the label on the A fragments after all unbound fragments have been washed away. Such an assay requires controls; in one control, the assay is carried out with only labelled A fragments – this indicating the maximum amount of binding by A fragments. Results are meaningful only if the A fragments hybridize stably; the stability of binding can be assessed by monitoring the dissociation of A fragments from the bound DNA as the temperature is gradually raised.

Homology may also be assessed by comparing the *thermal stability* of HETERODUPLEXes formed from two samples of DNA with the thermal stability of the corresponding homoduplexes. This involves determining the mid-point temperature [T_m] of the thermal melting curve of both the homoduplexes and heteroduplexes (see THERMAL MELTING PROFILE). The homology of DNAs from strains A and B are then assessed by comparing the T_m of the A homoduplexes with that of A–B heteroduplexes – the difference between the two values of $T_m [= \Delta T_m]$ indicating the degree of base pair mis-matching in the heteroduplex.

[Methods for determining the homology of nucleic acids: Book ref. 138, pp 33–74.]

DNA ligase A LIGASE which can make a phosphodiester bond between the 3′-OH end of one ssDNA strand and the (physically juxtaposed) 5′-phosphate end of the same or another ssDNA strand; thus, e.g., a DNA ligase can repair a NICK in one strand of a dsDNA molecule, the ends being held in position by base-pairing with the intact strand. (See also e.g. DNA REPLICATION.)

DNA looping (transcriptional control) See OPERATOR.

DNA methylation Methylation of certain bases in DNA: a normal process which occurs in many or most organisms. In prokaryotes, methylation typically occurs at the N-6 position of adenine and/or at the C-5 position of cytosine, while in most eukaryotes methylation occurs mainly at the C-5 position of cytosine; the bases which are methylated generally occur in a specific sequence of nucleotides – e.g. the second cytosine residue in the sequence CC(A/T)GG is methylated in *Escherichia coli* strain C. (See also DAM GENE and FROG VIRUS 3.)

Methylation serves various functions: see e.g. RESTRICTION–MODIFICATION SYSTEM. It also promotes a certain type of SPONTANEOUS MUTATION. (See also MISMATCH REPAIR.)

[DNA methylation: molecular biology and biological significance: Book ref. 222.]

DNA modification A physiological process in which bases in DNA are substituted with methyl or other groups – commonly shortly after synthesis of the DNA (see e.g. DNA METHYLATION; cf. BACTERIOPHAGE MU and BACTERIOPHAGE T4). Modification may occur as part of a RESTRICTION–MODIFICATION SYSTEM or may be involved e.g. in the regulation of gene expression (cf. DAM GENE).

DNA photolyase See PHOTOREACTIVATION.

DNA polymerase A type of enzyme which forms a polymer of deoxyriboNUCLEOTIDES by condensing deoxyribonucleoside triphosphates (dNTPs) with the elimination of pyrophosphate. A DNA polymerase adds nucleotides to the 3′-OH end of a pre-existing strand of DNA or an RNA primer; the order in which the nucleotides are added is dictated by the nucleotide sequence of a template strand of DNA (see DNA REPLICATION) or, in some cases, RNA (see REVERSE TRANSCRIPTASE). DNA polymerases function processively (see PROCESSIVE ENZYME).

Prokaryotic DNA polymerases. The bacterium *Escherichia coli* encodes three main DNA polymerases which are designated I, II and III (see also DNA POLYMERASE IV and DNA POLYMERASE V). All three main polymerases have exonuclease activity as well as polymerase activity; all can remove nucleotides sequentially from the 3′ end of a strand, and polymerases I and III can also cleave nucleotides from the 5′ end. (See also KLENOW FRAGMENT.) The 3′-to-5′ exonuclease activity of a DNA polymerase is required for proof-reading during DNA REPLICATION.

Polymerase I (Kornberg enzyme, product of gene *polA*) is active in DNA repair (see EXCISION REPAIR); it is also involved in the removal of RNA primers during DNA replication, and is needed for the replication of certain plasmids (e.g. ColE1). Polymerase I can act as an RNA-dependent DNA polymerase (i.e. a reverse transcriptase) but it exhibits poor processivity in this capacity and requires longer incubation times compared with a reverse transcriptase [EMBO (1993) *12* 387–396].

Polymerase II (*polB* product) is induced in the SOS SYSTEM (q.v.).

Polymerase III is the main replicative enzyme in *E. coli* and is responsible for replication of the chromosome and of phage DNA etc. 'DNA polymerase III' refers to a *core enzyme* consisting of three different subunits: (i) subunit α (product of *dnaE*/*polC*), which is involved in polymerization; (ii) subunit ε (product of *dnaQ*/*mutD*), which has 3′-to-5′ exonuclease activity and is apparently involved in proof-reading [PNAS (1983) *80* 7085–7089] (see also MUTATOR GENE); and (iii) θ (of unknown function).

The core enzyme of polymerase III can catalyse limited DNA synthesis on single-stranded gaps in dsDNA, but for full efficiency, accuracy and processivity on longer templates it requires a number of additional subunits which include: (i) the β subunit (*dnaN* product; ?'Factor I'), which apparently contributes to processivity; (ii) the γ subunit (*dnaZ* product), which apparently also contributes to processivity; (iii) the δ subunit (*dnaX* product; ?'Factor III'), which apparently enhances processivity; and (iv) the τ subunit (product of the *dnaX-Z* region), which acts as a link between two core enzymes. The core enzyme together with the *full range* of subunits is called *DNA polymerase III holoenzyme*. (DNA polymerase III* refers to the holoenzyme minus the β subunit. DNA polymerase III′ refers to a complex of core enzyme plus the τ subunit.)

In a current model of DNA replication, the holoenzyme functions as a dimer – the enzymes being physically linked via their τ subunits; during replication, one enzyme forms the leading strand while the other forms the lagging strand, i.e. they work in opposite directions.

In other bacteria, DNA polymerases appear to be essentially similar to those in *E. coli*. In the Archaea, DNA polymerases may be significantly different (see e.g. APHIDICOLIN).

In some prokaryotes (those that inhabit high-temperature environments), the DNA polymerases are thermostable, and this feature is exploited in various techniques used for the amplification of nucleic acids in vitro (e.g. PCR); these enzymes include:

(a) *Taq* DNA polymerase. An enzyme (from the bacterium *Thermus aquaticus*) which has been widely used in PCR (particularly during the early years); this enzyme lacks 3′-to-5′ exonuclease activity and (therefore) has no proof-reading capacity.

(b) Stoffel fragment. A modified, recombinant form of the *Taq* polymerase (the C-terminal region of the protein) which is more thermostable than the parent enzyme; it lacks 5′-to-3′ (in addition to 3′-to-5′) exonuclease activity. This enzyme is active in a wide range of concentrations of magnesium ions, and has been found useful e.g. in those forms of PCR which use arbitrary primers.

(c) UlTma™ DNA polymerase (Perkin-Elmer). This recombinant form of the DNA polymerase of *Thermotoga maritima* has 3′-to-5′ exonuclease activity and can repair 3′ mismatches during PCR – such proof-reading ability making the enzyme useful for high-fidelity copying.

(d) *Pfu* DNA polymerase™ (Stratagene). This highly thermostable enzyme from *Pyrococcus furiosus* retains 95% activity after 1 hour at 98°C; it has excellent proof-reading ability.

Eukaryotic DNA polymerases. Eukaryotes also have a number of DNA polymerases, including (in most eukaryotic cells) polymerases α, β and γ. Each type of polymerase appears to be associated with a particular function – for example, the γ polymerase carries out DNA replication in mitochondria.

Some *viruses* (e.g. bacteriophage T4) encode their own DNA polymerases; others (e.g. phages fd, ϕX174, G4) use host enzymes.

DNA polymerase I (Kornberg enzyme) In *Escherichia coli*: an enzyme associated e.g. with EXCISION REPAIR and with the removal of RNA primers during DNA replication. (See also DNA POLYMERASE.)

DNA polymerase II In *Escherichia coli*: a polymerase induced in the SOS SYSTEM. (See also DNA POLYMERASE.)

DNA polymerase III In *Escherichia coli*: the major DNA polymerase; it is involved in chromosomal replication and e.g. synthesis of phage DNA. For further details see entry DNA POLYMERASE.

DNA polymerase IV In *Escherichia coli*: the product of gene *dinB* (induced in the SOS SYSTEM), a protein which has DNA polymerase activity and which appears to be involved in mutagenesis; DinB has no proof-reading activity.

DNA polymerase V In *Escherichia coli*: a protein complex, consisting of two UmuD′ proteins and one UmuC protein, which has DNA polymerase activity and which may carry out some instances of DNA repair (translesion synthesis) during induction of the SOS SYSTEM.

DNA polymorphism See POLYMORPHISM (sense 3).

DNA primase *Syn.* PRIMASE.

DNA repair Any physiological process in which damaged DNA within a cell is recognized and repaired; *damage* in this context generally means any alteration or distortion of the normal double-helical structure of dsDNA resulting e.g. from the presence of abnormal bases or base pairs, nicks or gaps in one strand, covalent linkages between bases etc. (but *not* from replacement of one normal base pair with another: cf. MUTATION). *Repair* may involve e.g. direct reversal of the damage (e.g. Ada protein in ADAPTIVE RESPONSE) or removal of the damaged region followed by its replacement with normal DNA (e.g. EXCISION REPAIR).

Repair systems may be constitutive or inducible.

See e.g. ADAPTIVE RESPONSE, BASE EXCISION REPAIR, EXCISION REPAIR, MISMATCH REPAIR, PHOTOREACTIVATION, RECOMBINATION REPAIR and SOS SYSTEM.

DNA replication The process in which a copy (i.e. a *replica*) of a double-stranded DNA molecule is synthesized.

In vivo, circular, dsDNA molecules may be replicated by the CAIRNS' MECHANISMS or the ROLLING CIRCLE MECHANISM; the following refers mainly to the replication of circular dsDNA molecules by the Cairns' mechanism (replication of linear dsDNA is also considered briefly). (For replication of DNA in vitro see e.g. PCR and SDA.)

During replication, a new strand of DNA is synthesized on each of the two pre-existing strands, each pre-existing strand being used as a pattern or *template;* a template dictates the sequence of nucleotides in a new strand, thus conserving the genetic information encoded in the template strand. Essentially, molecules of deoxyribonucleoside triphosphate (dNTPs) base-pair with complementary nucleotides in each template strand and are sequentially polymerized (with elimination of pyrophosphate) by a DNA POLYMERASE. Note that each newly synthesized strand is complementary to – not a copy of – the template strand on which it is synthesized; thus, each of the two new daughter duplexes produced from a given parent duplex consists of one parent strand and one newly synthesized strand. (This type of replication is called *semiconservative replication* because only *one* strand in each daughter duplex has been newly synthesized.)

Initiation of DNA replication occurs at one or more special sites, called *origins* (designated *ori*); any DNA molecule which lacks an origin cannot be replicated in vivo. (See also REPLICON.)

Initiation of replication involves the recognition of an *ori* site by various initiation factors and enzymes. An essential prerequisite for replication is separation of the two strands of dsDNA (so-called 'melting') in the *ori* region (see CELL CYCLE (b)).

As no known DNA polymerase can *initiate* synthesis of a new strand of DNA on a template, initiation of DNA replication (by the Cairns' mechanism) requires the provision of a 'starter' sequence of nucleotides (a *primer*) which the DNA polymerase can elongate (in the 5′-to-3′ direction) to form a new strand of DNA; DNA synthesis is usually primed by a short strand of RNA which is transcribed on the DNA template strand in the *ori* region.

Chain elongation (i.e. ongoing strand synthesis) involves ongoing separation (unwinding) of the two strands of the parental duplex.

Termination of DNA synthesis usually occurs when the entire duplex has been replicated (cf. ROLLING CIRCLE MECHANISM); it involves dissocation of the replicating machinery, joining of the two ends of each daughter strand (in the case of circular dsDNA molecules), and separation of the two circular daughter molecules (*decatenation*).

Newly synthesized strands usually undergo modification before the next round of replication can begin (see RESTRICTION–MODIFICATION SYSTEM).

Replication of the chromosome in Escherichia coli. The way in which a round of replication is *initiated* (i.e. triggered) is not known, but various factors appear to be relevant; regulation of the initiation of DNA replication in *E. coli* is discussed in the entry CELL CYCLE.

Once triggered, replication of the ccc dsDNA CHROMOSOME of *E. coli* occurs by the Cairns' mechanism: replication begins at the origin – *oriC* (see ORIC) – and proceeds bidirectionally, both template strands being replicated more or less simultaneously. Thus, two migrating, Y-shaped *replication forks* (in which the stem of the Y is the parental duplex, and the arms the daughter duplexes) move in opposite directions around the chromosome (see figure).

Because the strands of the parent duplex are antiparallel, and because both new strands are synthesized more or less simultaneously in the direction of movement of a replication fork, one daughter strand is synthesized in the 5′-to-3′ direction while the other is synthesized from the 3′ direction. The strand synthesized in the 5′-to-3′ direction (the *leading strand* – see figure) is synthesized continually, but the other strand (*lagging* strand) cannot be synthesized continually because no known DNA polymerase can synthesize DNA in the 3′-to-5′ direction. The lagging strand is actually synthesized as a series of fragments (*Okazaki fragments* – see figure), each fragment being synthesized in the 5′-to-3′ direction from a separate RNA primer. Primers are subsequently excised and replaced with DNA (by extension from the preceding fragment) and the ends are joined together by a DNA LIGASE.

Chromosomal replication is a complex process requiring a number of different proteins. These proteins include the *dnaA* gene product (see DNAA GENE); a dimer of DNA polymerase III (see DNA POLYMERASE); GYRASE; topoisomerase I (see TOPOISOMERASE); HU PROTEIN; the DNAB GENE product (a helicase); PRIMASE and RNA POLYMERASE. At least some of these proteins are components of an initiation complex (earlier name:

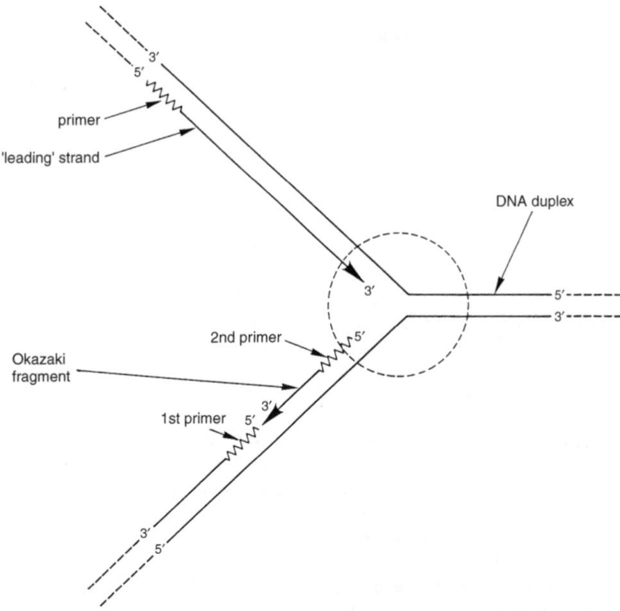

DNA REPLICATION (Cairns-type): disposition of template and newly synthesized strands at a replication fork (diagrammatic; see entry). In a circular DNA duplex the strands have separated, locally, in the *ori* region. Only one half of this region is shown; in this half, the strands of the duplex will continue to separate in a left → right direction (i.e. the replication fork is moving to the right). On one template strand (top) an RNA *primer* (zigzag line) has been synthesized (see text); ongoing addition of deoxyribonucleotides to the 3′ end of this primer will form the 'leading' strand of DNA. Synthesis of the leading strand is continuous in the 5′-to-3′ direction.

On the other template strand (below), the 1st primer has been synthesized (by a primase), and this (RNA) primer has been extended as a strand of DNA by sequential addition of deoxyribonucleotides from the 3′ end. This new strand of DNA may form the first Okazaki fragment of the lagging strand; alternatively, it may act as a primer for leading strand synthesis in the *other* replication fork (not shown). Following synthesis of the 1st primer, the replication fork has moved further to the right – opening up the duplex to the extent shown in the diagram. The 2nd primer has been synthesized and has been extended as a strand of DNA (Okazaki fragment) as far as the 5′ end of the first primer. As the replication fork moves even further to the right additional template will become available, and the 3rd, 4th, 5th ... etc. primers synthesized on this template will (like the 2nd primer) be extended as Okazaki fragments; the sequence of Okazaki fragments will be joined together (forming the lagging strand) when the (RNA) primers are replaced by DNA.

Modified from *Bacteria*, 5th edition, Figure 7.8(a), page 116, Paul Singleton (1999) copyright John Wiley & Sons Ltd, Chichester, UK (ISBN 0471-98880-4) with permission from the publisher.

'orisome'). (During strand synthesis, parts of the parent DNA molecule – around the replication forks – necessarily exist in a single-stranded state, and these ssDNA sequences are stabilized by SINGLE-STRAND BINDING PROTEINS.)

A short RNA primer, synthesized on one of the template strands, is elongated by DNA polymerase III in the 5′-to-3′ direction to form the leading strand (see figure); synthesis of the primer may be mediated by RNA polymerase (RNA polymerase is needed for initiation of DNA replication).

Exposure of the other template strand permits synthesis of the lagging strand. Primers for the lagging strand are synthesized by a PRIMASE (*dnaG* gene product) that forms part of a PRIMOSOME. Each primer is elongated in the 5′-to-3′ direction to form an Okazaki fragment (~1000–2000 nt in length) that extends to the 5′ end of the previous primer. (Primers are subsequently removed and replaced by DNA; this involves the enzyme DNA polymerase I.) As the first replication fork moves away, the first fragment of the lagging strand may act as a primer for leading strand synthesis in the opposite direction. At 37°C a replication fork moves at about 1000 nt per second.

In a current model for DNA replication, a processive complex of proteins ('replisome') is associated with each replication fork. The complex includes a DNA polymerase III *dimer* (the two enzymes physically linked via their τ subunits and orientated in opposite directions) – one enzyme synthesizing the leading strand while the other synthesizes the lagging strand. The complex also contains a PRIMOSOME which, in turn, includes a helicase and the DnaG protein (a primase); the helicase is described as a hexamer (a ring-shaped structure of six DnaB proteins) which unwinds the duplex [Cell (1996) *86* 177–180]. In this model, a section of lagging-strand template (liberated by ongoing leading-strand synthesis) forms a loop at the replication fork. This allows primer synthesis for the lagging strand (by primase) followed by extension of the primer by DNA polymerase III; when a newly synthesized fragment approaches the 5′-end of the previous fragment, the new (double-stranded) section is released by the complex, and the process is repeated on the next region of exposed lagging-strand template. Thus, the complex synthesizes both leading and lagging strands concurrently as it moves along with the replication fork.

The physical link between the two enzymes in the DNA polymerase III dimer is apparently needed for co-ordination of synthesis of the leading and lagging strands. It appears that coupling is also needed between the helicase and the polymerases – loss of such coupling being associated with a drastic reduction in efficiency of the helicase [Cell (1996) *84* 643–650].

Moving in opposite directions, the two replications forks meet at the *ter* site on the chromosome, 180° from *oriC*.

During replication, the DNA polymerase appears to recognize only the overall shape of a base-pair, i.e. the same active site of the enzyme must serve for all four possible base-pair combinations. The accuracy of DNA synthesis therefore depends initially on base-pairing specificity between incoming nucleotides and template DNA. This might be expected to result in a relatively high frequency of errors because tautomerism in the bases can interfere with the accuracy of base-pairing. However, the overall error rate in *E. coli* DNA replication has been estimated to be ca. one mistake in 10^8–10^{10} nucleotides polymerized – a much higher level of accuracy than can be accounted for by base-pairing and polymerase specificities. This level of accuracy is achieved by 'proof-reading' and 'editing' systems which recognize and correct errors in a newly synthesized strand. The 3′-to-5′ nuclease activity of DNA polymerase III holoenzyme serves a proof-reading function, removing mispaired nucleotides from the growing (3′) end of the chain before continuing chain elongation. Errors which escape this system may be detected and corrected subsequently by the post-replicative MISMATCH REPAIR system.

In growing cells, chromosomal replication must be co-ordinated with cell division in order that daughter cells each receive the full genetic complement; the rate of DNA replication in *E. coli* is increased during faster growth by the initiation of additional rounds of synthesis (see HELMSTETTER–COOPER MODEL). Conditions which inhibit DNA replication may cause inhibition of cell division (see SOS SYSTEM).

Replication of linear genomes. In some bacteria a linear chromosome is replicated bidirectionally from an internal origin of replication. This raises the question of how the replication machinery avoids leaving the 3′ end of the template strand (in each daughter duplex) as single-stranded DNA. In one of several models, DNA synthesized *on the new strand* (at the fully duplexed end) displaces the 5′ terminal nucleotide sequence of the template strand – which is then transferred to (complementary) single-stranded DNA on the other daughter duplex (thus filling the single-stranded gap). [Replication of linear DNA molecules: TIG (1996) *12* 192–196.]

In some viruses, replication of the linear dsDNA genome involves so-called *protein priming* – in which a nucleotide, bound covalently to a (virus-encoded) 'terminal protein', forms the 5′ terminus of a new strand; this system is found e.g. in BACTERIOPHAGE φ29 [protein priming in phage φ29: EMBO (1997) *16* 2519–2527] and in members of the ADENOVIRIDAE.

Chromosomal replication in eukaryotes. Replication of eukaryotic CHROMOSOMES occurs during the S phase of the cell cycle. DNA synthesis is initiated at many separate sites along the chromosome, and replication forks proceed bidirectionally until they meet and fuse. Each DNA molecule is replicated only once in a cell cycle. [Eukaryotic DNA replication (minireview): Cell (1987) *48* 7–8.]

Plasmid and phage replication. Mechanisms of replication in PLASMIDS and (DNA) phages may differ significantly from chromosome replication in the corresponding host cell, but in most or all cases at least some of the components involved in host chromosomal replication are required.

In plasmid ColE1, replication is of the Cairns' type – but it occurs *uni*directionally from the origin; in the F PLASMID replication is bidirectional (as in the *E. coli* chromosome).

Many of the small circular plasmids in Gram-positive bacteria replicate by means of a ROLLING CIRCLE MECHANISM [replication control in rolling circle plasmids: TIM (1997) *5* 440–446].

Plasmid replication, and the partitioning to daughter cells, seems likely to involve some kind of association between the plasmid and the cytoplasmic membrane of the host cell [Mol. Microbiol. (1997) *23* 1–10].

Replication of the (linear) genome of bacteriophage φ29 was mentioned above.

In certain phages the rolling circle mechanism is involved in replication (see e.g. SSDNA PHAGE, BACTERIOPHAGE λ, BACTERIOPHAGE φX174).

[The enzymology of DNA replication (historical perspective): JB (2000) *182* 3613–3618.]

DNA restriction See RESTRICTION–MODIFICATION SYSTEM.

DNA sequencing Any procedure for determining the sequence of nucleotides in a sample of DNA. Sequencing is important in taxonomy, identification and characterization.

In the 'chemical' (Maxam–Gilbert) method, end-labelled copies of an unknown sequence are cleaved by base-specific reagents; analysis follows electrophoresis of the fragments. The dideoxy (Sanger's) method is described in the figure on page 249. (See also PYROSEQUENCING.)

Automated DNA sequencing is often based on the dideoxy method. Primers are labelled with fluorescent dyes, a different dye being used for each of the four reactions. The products from all four reactions are combined and then subjected to electrophoresis from a single well in a polyacrylamide gel. On excitation by an argon laser, fragments from each reaction mixture can be distinguished from those of other reaction mixtures by their dye-specific emission characteristics; emissions from the fragments of all four reaction mixtures are detected automatically and stored e.g. in digital form.

Sequencing provides definitive information on a given sample of nucleic acid. For duplex DNA, greater accuracy is obtained if both strands of the duplex are sequenced.

RNA can be retrotranscribed into cDNA; the (single-stranded) cDNA is then amplified, and the amplified product is sequenced. The sequence of nucleotides in the RNA sample is then deduced. Amplified DNA copies of an RNA sample may be obtained by reverse-transcriptase PCR (rtPCR).

DNA splicing See CLONING (of DNA).

DNA Stat™ See BLOOD DNA ISOLATION KITS.

DNA synthesis See DNA REPLICATION.

DNA topoisomerase *Syn.* TOPOISOMERASE.

DNA toroid A highly compacted genomic DNA, found e.g. in *Deinococcus radiodurans* and ENDOSPORES, which may account at least partly for resistance to DNA-damaging agents (e.g. radiation). Toroidal structure may facilitate repair of double-stranded breaks in a RecA- and template-independent way [JB (2004) *186* 5973–5977].

DNA tumour viruses Oncogenic DNA-containing animal viruses; they include members of the ADENOVIRIDAE, Hepadnaviridae (see e.g. HEPATITIS B VIRUS), Herpesviridae (subfamily GAMMAHERPESVIRINAE), and PAPOVAVIRIDAE.

DNA unwinding protein Any protein capable of unwinding and separating the strands of double-helical DNA. (See HELICASES; cf. SINGLE-STRAND BINDING PROTEIN.)

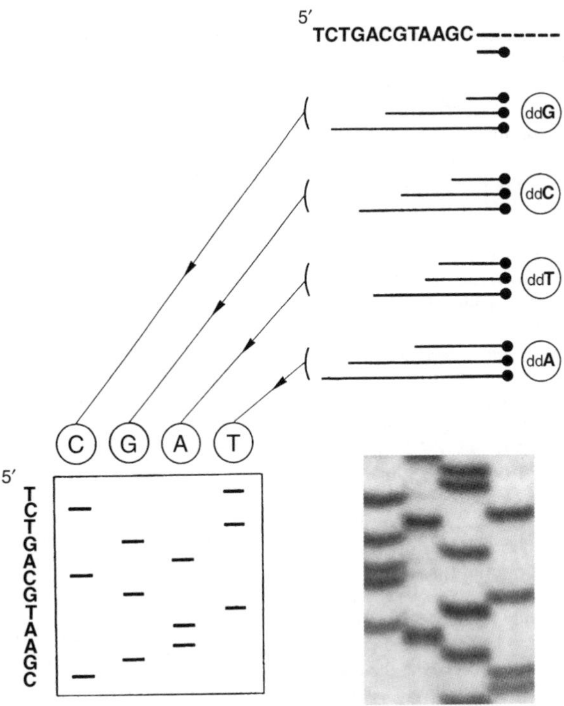

DNA SEQUENCING: determining the nucleotide sequence of a cloned DNA fragment by the dideoxy method (Sanger's chain-termination method) (diagrammatic). Initially, the fragment to be sequenced is obtained in single-stranded form (e.g. by cloning in a phage M13 vector). Regardless of the cloning method used, the unknown sequence will be flanked on its 3′ side by single-stranded DNA of *known* sequence derived from the vector molecule; this permits the design of a labelled primer which can bind at a site next to the unknown sequence such that the first nucleotide to be added to the primer will pair with the first 3′ nucleotide of the unknown sequence. In the diagram, the 'unknown sequence' is TCTGACGTAAGC; a primer (short line), which carries a label (black disc), has bound at a site flanking the 3′ end of the unknown sequence.

DNA synthesis in vitro is normally carried out with a reaction mixture that includes (i) templates (in this case, single-stranded fragments containing the unknown sequence); (ii) primers; (iii) the four types of deoxyribonucleoside triphosphate (dATP, dCTP, dGTP and dTTP); and (iv) DNA polymerase. When base-paired to the template strand, the primer is extended (5′-to-3′) by the sequential addition of nucleotides as dictated by the template.

In sequencing, there are four separate reaction mixtures (G, C, T, A), each containing all the constituents mentioned above (including millions of copies of the unknown sequence, and of the primer). In addition, each mixture contains a given *di*deoxyribonucleoside triphosphate (ddNTP); thus, the G mixture contains dideoxyguanosine triphosphate (ddG), the C mixture ddC, the T mixture ddT, and the A mixture ddA.

When a dideoxyribonucleotide is added to a growing strand of DNA it prevents addition of the *subsequent* nucleotide because dideoxyribonucleotides lack the 3′-OH group necessary for making the next phosphodiester bond. Thus, primer extension will stop (= chain termination) at any location where a dideoxyribonucleotide has been incorporated. In each reaction mixture the concentration of ddNTP is such that, in most growing strands, synthesis will be stopped, at some stage, by the incorporation of a dideoxyribonucleotide; because a ddNTP may pair with *any complementary base* in the template, chain termination can occur at different sites on different copies of the template strand – so that product strands of different lengths will be formed. For example, with ddG (see diagram) the three products are of different lengths because, during extension of the primers, ddG has paired with cytosine residues at three different locations in the template; note that, in this case, *the length of a given product strand is related to the location of a particular cytosine residue in the unknown sequence*. Analogous comments apply to reaction mixtures containing the other ddNTPs.

At the end of the reaction, product strands are separated from templates by formamide. Each reaction mixture is then subjected to electrophoresis in a separate lane of a polyacrylamide gel. Short product strands move further than longer ones, in a given time; products that differ in length by only one nucleotide can be distinguished, the shorter product moving just a little further.

The gel contains urea; this inhibits base-pairing between product strands and templates. It also inhibits *intra*-strand base-pairing in the product strands. This is essential: to deduce the sites of a particular base in the template it is necessary to compare the *lengths* of all product strands in the given reaction, and this requires proportionality between strand length and electrophoretic mobility; were intra-strand base-pairing to occur it could alter electrophoretic mobility – in which case a product's length would not necessarily be indicated by the position of its band in the gel. (*Continued on page 250*.)

DNA uptake site (in *Haemophilus*) See TRANSFORMATION (1).

DNA vaccine Any parenterally administered VACCINE consisting of DNA. A DNA vaccine consists e.g. of copies of a PLASMID that encodes specific antigen(s); the encoded antigen(s) are synthesized in vivo (i.e. within the recipient of the vaccine) and may then induce both humoral (antibody) and cell-mediated responses. A DNA vaccine may have prophylactic and/or therapeutic uses. The possibility of using DNA vaccines against the major diseases (malaria, tuberculosis etc.) has attracted much interest.

It is known that a DNA vaccine encoding e.g. the 65 kDa heat-shock protein (Hsp65) of *Mycobacterium tuberculosis* can protect mice against infection by this pathogen; the vaccine appears to stimulate an antigen-specific subset of Th1 T cells which (i) have cytotoxic activity against infected cells, and (ii) secrete interferon-γ (IFN-γ).

More recently, it has been shown that a DNA vaccine encoding Hsp65 can act therapeutically in mice during an experimentally established infection with *Mycobacterium tuberculosis* [Nature (1999) *400* 269–271]. In this study, virulent cells of *M. tuberculosis* H37Rv were injected intravenously and the infection allowed to develop for 8 weeks. DNA vaccine was then administered in four intramuscular doses at 2-week intervals; each dose of vaccine consisted of 50 μg plasmid DNA in 50 μl saline. Different types of vaccine were used in different groups of animals; the vaccine encoding Hsp65 was found to be significantly more effective than other vaccines encoding different mycobacterial antigens (Hsp70 and ESAT-6). The therapeutic effect of the Hsp65 vaccine (manifested by a marked decrease in the numbers of live tubercle bacilli in lung and spleen) was associated with a switch from a Th-2-dominated to a Th-1-dominated response in the mice.

Interestingly, non-methylated CpG regions in the DNA itself can induce secretion of INTERLEUKIN-12 (IL-12) from antigen-presenting cells (such as macrophages). IL-12 is known to promote the development of the Th-1 phenotype in T cells exposed to antigen. Because IL-12 is one of the factors essential for effective immunity to *M. tuberculosis* in mice, a plasmid encoding IL-12 (rather than a mycobacterial antigen) was tested in the same study [Nature (1999) *400* 269–271]; this vaccine was found to cause the maximum reduction in tubercle bacilli by 11 weeks, i.e. its performance exceeded that of the mycobacterial antigens.

A DNA vaccine has also been used against Ebola virus: see FILOVIRIDAE.

Any future large-scale development/use of DNA vaccines is dependent on appropriate technology for preparing pure plasmid DNA (e.g. free of contaminating chromosomal DNA) in sufficient quantity. [Biochemical engineering approaches to the challenges of producing pure plasmid DNA: TIBtech. (2000) *18* 296–305.]

[Plasmids for Therapy and Vaccination: Book ref. 225.]

(See also GENE GUN.)

DnaA boxes See CELL CYCLE (b).

***dnaA* gene** A DNA GENE whose product is necessary for the initiation of DNA REPLICATION in *bacteria*; in archaeans the Orc1 and CdC6 proteins carry out the same function. DnaA proteins bind to 9-bp sequences ('DnaA boxes') in the *oriC* region during initiation (see CELL CYCLE (b)) and this is followed by localized strand separation. Defective (mutant) DnaA proteins have been reported [Mol. Microbiol. (2000) *35* 454–462].

DnaA was earlier reported to be needed for the replication of certain plasmids (e.g. pSC101) but is now seen to be specific to the replication of *chromosomal* DNA. In eukaryotes, the DnaA function is carried out by six proteins (Orc1–6). [Initiators of DNA replication: FEMS Reviews (2003) *26* 533–554.]

DNAase DEOXYRIBONUCLEASE.

***dnaB* gene** A DNA GENE whose product is necessary for many cases of primer formation; in a current model of DNA REPLICATION in *Escherichia coli*, a hexamer of DnaB proteins functions as a HELICASE – forming part of the PRIMOSOME and unwinding the parent duplex as the primosome migrates with the replication fork.

It appears that ATP *binding* is necessary for the binding of DnaB to ssDNA, while ATP *hydrolysis* permits its release. In the absence of single-strand binding protein (SSBP), DnaB–ATP can bind to many sites on ssDNA, and interaction with primase leads to the formation of a short RNA primer at each site (*general priming*); general priming is inhibited by SSBP. *Specific priming* occurs in the presence of SSBP and depends on the presence of other proteins of the PRIMOSOME.

***dnaC* gene** (*dnaD* gene) A DNA GENE whose product is required for PRIMOSOME formation.

***dnaD* gene** *Syn. dnaC* gene (q.v.).

***dnaE* gene** *Syn. polC* gene (see DNA POLYMERASE).

***dnaF* gene** *Syn. nrdA* gene (q.v.).

***dnaG* gene** See PRIMASE.

***dnaH* gene** *Syn. dnaZ* gene (see DNA POLYMERASE).

***dnaI* gene** A DNA GENE whose function is required for initiation of DNA REPLICATION at the replication origin.

***dnaJ* gene** A DNA GENE which is closely linked to the *dnaK* gene (q.v.) and which, like *dnaK*, is necessary for viability of *Escherichia coli* at high temperatures and apparently for phage λ DNA replication.

***dnaK* gene** A DNA GENE whose function is essential for the viability of *Escherichia coli* at high temperatures, and which is also reported to be required for phage λ DNA replication. The product of *dnaK* was identified as protein B66.0, a HEAT-SHOCK PROTEIN. [Properties of DnaK: JBC (1984) *259* 8820–8825.]

DNA SEQUENCING (*continued*)

Bands in the gel are revealed e.g. by staining or chemiluminescence; if the primers carry a radioactive label, the gel is examined by autoradiography following electrophoresis.

The locations of bands indicate the relative lengths of product strands – shorter products having moved further down the gel (from top to bottom in the diagram). Note that the first unknown 3′ nucleotide (C) is identified by (i) the shortest product strand (which has moved the furthest), and (ii) the fact that this product is from the ddG mixture – indicating a base that pairs with G, i.e. C. Similarly, the next unknown (G) is the next shortest product strand which came from the ddC mixture, thus indicating G. The entire unknown sequence can be deduced in this way.

The figure also shows part of an autoradiograph of a sequencing gel (courtesy of Joop Gaken, Molecular Medicine Unit, King's College, London).

Figure reproduced from *Bacteria*, 5th edition, Figure 8.23, pages 222–223, Paul Singleton (1999) copyright John Wiley & Sons Ltd, UK (ISBN 0471-98880-4) with permission from the publisher.

***dnaL* gene** An uncharacterized DNA GENE.

***dnaM* gene** An uncharacterized DNA GENE.

***dnaN* gene** See DNA POLYMERASE.

***dnaP* gene** A DNA GENE whose function is required for initiation of DNA REPLICATION at the replication origin.

***dnaQ* gene** See DNA POLYMERASE.

DNase DEOXYRIBONUCLEASE.

***dnaT* gene** A DNA GENE whose product may interact with DnaC protein in vivo; *dnaT* mutants cannot destabilize the replication complex at chain termination.

***dnaW* gene** *Syn. adk* gene (q.v.).

***dnaX* gene** See DNA POLYMERASE.

***dnaZ* gene** See DNA POLYMERASE.

dNTP DeoxyriboNUCLEOSIDE-5′-triphosphate.

Dobiella See EIMERIORINA.

Dobrava-Belgrade virus See HANTAVIRUS.

DOC Dissolved organic carbon. (cf. POC.)

docking protein See SIGNAL HYPOTHESIS.

Döderlein's bacilli Aciduric, Gram-positive bacilli, probably *Lactobacillus* sp(p), observed by Döderlein (1892) in human vaginal secretions.

dodine (dodine acetate; *n*-dodecylguanidine acetate) An agricultural (non-systemic) ANTIFUNGAL AGENT used e.g. for the control of apple scab – against which it has eradicant as well as protective action. Dodine has the properties of a cationic surfactant.

dog (parvovirus infection) See CANINE PARVOVIRUS.

dog lichen *Peltigera canina.*

dolichol A polyprenol consisting of 14–24 α-saturated isoprene units; phosphorylated dolichols function in eukaryotes as membrane-bound sugar carriers, assisting the transfer of hydrophilic sugar residues from soluble nucleotide precursors through membranes (e.g. of the rough endoplasmic reticulum and Golgi apparatus) to polymers such as polysaccharides, glycoproteins etc. (see e.g. PALADE PATHWAY). (cf. BACTOPRENOL.)

Dolichomastix See MICROMONADOPHYCEAE.

doliform (doliiform) Barrel-shaped.

dolipore septum A type of SEPTUM present in the hyphae of most basidiomycetes (cf. AMBROSIOZYMA). The central portion of the septum is thickened and barrel-shaped, the septal pore forming a small tunnel (ca. 0.1 μm diam.) through this thickened region. A curved double membrane (the *septal pore cap* or *parenthesome*) typically occurs on each side of the septum, these structures appearing (in cross section) as brackets (parentheses) around the septa. The parenthesomes appear to be formed from the endoplasmic reticulum and may be perforate or imperforate; they may serve as screens, regulating the passage of cellular structures between adjacent cells. Parenthesomes are lacking e.g. in *Filobasidiella.*

dollar spot A disease of turf grasses (particularly red fescue, *Festuca rubra*) caused by *Sclerotinia homeocarpa*; yellowish-brown patches, up to ca. 8 cm across, develop on the turf in mild, wet weather.

domain (*immunol.*) Any one of the several compact, globular sections of a heavy chain or a light chain in an IMMUNOGLOBULIN molecule; each domain consists of a part of the polypeptide chain which forms a loop (closed by an intra-chain disulphide bond) and which is extensively folded. In IgG the heavy chain domains, in sequence, are designated V_H (variable region), and C_H1, C_H2 and C_H3 (constant region); the light chain domains are designated V_L and C_L. (Domains are also written as e.g. CH1, CH2 etc, or as CH_1, CH_2 etc.)

domain (*taxon.*) A category ranked above kingdom [origin of term: PNAS (1990) *87* 4576–4579] – see e.g. ARCHAEA and BACTERIA.

Domestos See HYPOCHLORITES.

dominance (genetics) In a diploid (or merodiploid) heterozygous cell or organism: the tendency of certain (*dominant*) alleles to be expressed in preference to their corresponding (*recessive*) alleles. (cf. EPISTASIS.) The complete dominance of one allele over another occurs in only some cases; in such cases the phenotype of the heterozygote Aa is identical to that of the homozygote, AA (A being dominant). When partial dominance is exhibited the Aa phenotype is intermediate between the AA and aa phenotypes.

dominant allele See DOMINANCE.

Domiphen bromide See QUATERNARY AMMONIUM COMPOUNDS.

DON (1) (6-diazo-5-oxo-L-norleucine; $CO_2H.CHNH_2.(CH_2)_2.CO.CH{=}N^+{=}N^-$) An ANTIBIOTIC and antitumour agent obtained from *Streptomyces* sp; its mode of action resembles that of AZASERINE. Other antitumour DON derivatives are produced by *Streptomyces* spp: e.g. diazomycin A (*N*-acetyl-DON) and alazopeptin (apparently comprising one alanine and two DON residues). (cf. HADACIDIN.)

(2) Deoxynivalenol – see TRICHOTHECENES.

donation (*mol. biol.*) See CONJUGATION (1b) (i).

donor Any cell or organism which donates genetic information to another cell or organism (the *recipient*): see CONJUGATION, TRANSDUCTION and TRANSFORMATION.

donor conjugal DNA synthesis See CONJUGATION (1b) (i).

donor site (of a ribosome) See PROTEIN SYNTHESIS.

donor splice site See SPLIT GENE (a).

donor-suicide model (for transposition) See e.g. Tn*10*.

Donovan bodies See GRANULOMA INGUINALE.

donovanosis *Syn.* GRANULOMA INGUINALE.

L-dopa See MELANIN.

Dorisa See EIMERIORINA.

Dorisella See EIMERIORINA.

dormancy (hypobiosis) The state of an organism or spore which exhibits minimal physical and chemical change over an extended period of time; when *no* physical or chemical change can be detected the state is sometimes referred to as *cryptobiosis.* (See also ENDOSPORE (sense 1(a)) and SPORE.)

dorsal zone of membranelles See DZM.

Dorset's egg A MEDIUM used e.g. for the maintenance of *Mycobacterium* spp; it consists of a saline suspension of homogenized whole hens' eggs which has been inspissated in sloped bijou or universal bottles.

dorsiventral Having two unlike (upper and lower) surfaces.

dosa A traditional fermented food in India; it is prepared from milled rice and black gram (cf. IDLI). [Microbiology of dosa: Food Mic. (1986) *3* 45–53.]

dosage (of genes) See GENE DOSAGE.

dot-blot A PROBE-based method for detecting or quantifying a specific sequence of nucleotides in a given sample. A drop of the sample is added to a suitable membrane and treated so that any nucleic acids present in the sample bind to the membrane in single-stranded form. The membrane is then exposed to probes (which are complementary to the target sequence) under conditions permitting probe–target hybridization; when unbound probes are removed by washing, the presence of bound probes indicates the presence of the required target sequence. If, for example, the probe had been tagged with DIGOXIGENIN, the bound probe is detected by treating the membrane with a *conjugate* consisting of anti-digoxigenin (antibody to digoxigenin) covalently linked to an enzyme (such as alkaline phosphatase); the binding of conjugate to digoxigenin is determined by adding a colour-generating substrate for the enzyme after all unbound conjugate has been removed by washing.

Alternatively, a radioactive probe can be used.

For quantification, serial dilutions of the sample are examined as described above; the quantity of target present in the sample can then be assessed by noting the strength of label (i.e. colour, radioactivity etc.) in comparison with that of controls of known concentration.

***dot* genes** (of *Legionella pneumophila*) See end of section (a) in PHAGOCYTOSIS.

Dothidea See DOTHIDEALES.

Dothideales The largest order of fungi in the subdivision ASCOMYCOTINA; the organisms include saprotrophs, plant parasites and lichenized fungi. Ascocarp: ascolocular (see ASCOCARP and ASCOSTROMA), ostiolate or closed. Asci: bitunicate, cylindrical to spheroidal. Some species exhibit one or more asexual (conidial) stages. Genera: e.g. ARTHOPYRENIA, *Cochliobolus* (see also DRECHSLERA), *Cucurbitaria* (syn. *Phialospora*), DIDYMELLA, *Dothidea*, ELSINOË, LEPTOSPHAERIA, MYCOSPHAERELLA, *Myriangium* (see also ASCOSPORE and ASCOSTROMA), *Ophiobolus*, PIEDRAIA, *Pleospora* (see also BLACKLEG sense 2), *Pyrenophora* (see also DRECHSLERA and NET BLOTCH), *Sphaerulina* (see also NORMANDINA), VENTURIA. (See also BLACK MILDEWS and SOOTY MOULDS.)

DOTS Directly observed therapy, short-course: see TUBERCULOSIS.

double diffusion (*serol.*) GEL DIFFUSION in which antigen and antibody *both* diffuse through the gel (cf. SINGLE DIFFUSION). In one procedure, this is achieved by placing antibody and antigen at opposite ends of a gel column; in the *Oakley–Fulthorpe test* (double diffusion, single DIMENSION) antibody is incorporated in a layer of gel, and this is separated by a layer of plain gel from the aqueous solution of antigen which is placed on top. (cf. OUCHTERLONY TEST.)

double dimension (*serol.*) See DIMENSION.

double fixation See ELECTRON MICROSCOPY (a).

double helix See DNA.

double lysogen See e.g. TRANSDUCTION.

double-negative T cells See T LYMPHOCYTE.

double recessive HOMOZYGOUS for a given recessive allele, or alleles.

double refraction *Syn.* BIREFRINGENCE.

double-strand break–repair model See RECOMBINATION Fig. 2. (See also RECOMBINATION REPAIR.)

double thymidine blockade See SYNCHRONOUS CULTURE.

double-vial See FREEZE-DRYING.

doubling dilutions See SERIAL DILUTIONS.

doubling time See GROWTH (a).

Doulton filter See FILTRATION.

dourine (mal du coit) A chronic or subacute, often fatal, sexually transmitted disease of equines caused by *Trypanosoma equiperdum*; it occurs in parts of Africa, America, Asia and Europe. Initial symptoms include oedematous, inflamed lesions of the genitalia, and transient skin lesions (*plaques*) up to 5 cm in diameter; later symptoms include anaemia and neurological involvement (e.g. paralysis). Quinapyramine and suramin have been used therapeutically. A carrier state occurs.

down mutation (down-promoter mutation) A MUTATION in a PROMOTER which results in a decreased level of transcription from that promoter. (cf. UP MUTATION.)

downcomer In a LOOP FERMENTER: the descending column of liquid *or* that part of the fermenter which contains it. (cf. RISER.)

downstream (*mol. biol.*) In, or in relation to, the direction in which a nucleic acid strand or a polypeptide chain is synthesized; the converse of 'downstream' is referred to as 'upstream'. (See e.g. PROMOTER.)

downstream box In certain prokaryotic and phage genes: a sequence of nucleotides, downstream from the start site, which has been regarded as a translation enhancer. The ribosomal 16S rRNA may contain a complementary region (called the *anti-downstream box*), although suggestions that this may bind to the downstream box have been disputed [JB (2001) *183* 3499–3505].

downstream promoter element In some RNA polymerase II PROMOTERS: a region downstream of the start site with a role apparently similar to that of the TATA box.

downy mildews Plant diseases caused by certain members of the PERONOSPORALES (e.g. BREMIA, PERONOSPORA, PLASMOPARA); downy mildews are characterized by the formation of superficial hyphal growth in which, typically, individual spore-bearing structures can be distinguished under low magnification. (cf. POWDERY MILDEWS.) Downy mildews can be controlled e.g. with copper-based antifungals or ZINEB.

doxycycline See TETRACYCLINES.

D_p Pattern difference. In the comparison of two strains by NUMERICAL TAXONOMY: a coefficient which indicates the degree of *dissimilarity* corrected for any difference(s) due solely to inter-strain differences in metabolic vigour. (cf. entry S_p.)

DP Degree of polymerization: the number of monomeric units per molecule of a polymer.

DPD *N*,*N*-diethyl-*p*-phenylenediamine (see WATER SUPPLIES).

DPN Diphosphopyridine nucleotide: see NAD.

DPT A MIXED VACCINE containing diphtheria toxoid, pertussis vaccine and tetanus toxoid.

DPX A neutral, synthetic MOUNTANT of refractive index ca. 1.53 when dry; it consists of *D*istrene and a *p*lasticizer (tritolyl phosphate) dissolved in *x*ylol.

DR DIRECT REPEAT.

Dr adhesins A category that includes both fimbrial and non-fimbrial adhesins, members of which are found e.g. on UPEC strains of *Escherichia coli*; the Dr adhesins bind to CD55 (at a site corresponding to the Dr^a blood group antigen).

draft tube In a LOOP FERMENTER: a wide, vertical tube typically situated coaxially within the column so as to leave an annular space between itself and the wall of the fermenter; when the fermenter is operating, the draft tube is completely submerged in the culture. Culture is made to flow up (or down) the draft tube in order to generate a circulatory flow in the column – culture flowing down (or up) the annulus.

***Dra*I** See RESTRICTION ENDONUCLEASE (table).

Draparnaldia See CHAETOPHORA.

draughtsman colony (checker colony) The form of COLONY typically (though not invariably) produced by *Streptococcus pneumoniae* growing on blood agar: round and raised, with steep sides and a flat top, i.e., resembling one of the pieces used in a game of draughts (= checkers); after overnight growth under a raised partial pressure of CO_2 the colonies are each ca. 1 mm in diameter and surrounded by a zone of α-haemolysis. Very young colonies of *S. pneumoniae* are typically convex (domed); the flattened colony form apparently results from autolysis of the constituent cells on continued incubation, and further incubation may result in the development of a concavity in the upper surface of the colony.

draw tube In some microscopes: a tube coaxial with the body tube (see MICROSCOPY) used to adjust the tube length.

***drd* plasmid** A mutant CONJUGATIVE PLASMID in which the TRANSFER OPERON is permanently DEREPRESSED.

Drechslera A genus of fungi of the HYPHOMYCETES; teleomorphs occur in the genera *Cochliobolus* and *Pyrenophora*.

The organisms form septate mycelium and pigmented, septate, commonly cylindrical conidia; conidiogenesis is tretic (see CONIDIUM). *Drechslera* spp are typically graminicolous (cf. e.g. PHAEOHYPHOMYCOSIS). Species which have been transferred from HELMINTHOSPORIUM to this genus include e.g. *D. gramineae, D. maydis, D. oryzae* (causal agent of brown spot disease of rice), and *D. victoriae* (see also VICTORIN).

Dreyer's tube A small, conically-based glass test-tube with a flared rim and an internal diameter of ca. 5 mm; it is used in serology for the clear observation of reactions involving agglutination or precipitation.

drift See GENETIC DRIFT and ANTIGENIC DRIFT.

DRNA DISSIMILATORY REDUCTION OF NITRATE TO AMMONIA.

drop plate method *Syn.* MILES AND MISRA'S METHOD.

droplet infection INFECTION (sense 1) by inhalation of an AEROSOL of saliva, mucus etc. (contaminated with pathogens) from an infected individual.

dropsy *Syn.* OEDEMA. (See also INFECTIOUS DROPSY.)

***Drosophila* A virus** See PICORNAVIRIDAE.

***Drosophila* C virus** See PICORNAVIRIDAE.

***Drosophila* P virus** See PICORNAVIRIDAE.

***Drosophila* X virus** (DXV) A virus which causes anoxia sensitivity and death in *Drosophila melanogaster* and which can be cultivated in *Drosophila* cell lines [JGV (1979) *42* 241–254]. DXV closely resembles IPN VIRUS in morphology, genome etc. (See also SIGMA VIRUS.)

drug resistance *Syn.* ANTIBIOTIC resistance.

dry bubble See BUBBLE DISEASES.

dry rot (1) (of timber) A BROWN ROT of structural (and other) timbers caused by the cellulolytic fungus *Serpula lacrymans*; only wood having a moisture content greater than about 20% (see TIMBER SPOILAGE) can be attacked initially, but water produced during metabolism can be transported via moisture-conducting rhizomorphs, enabling the fungus to spread to drier regions of wood (or across brickwork etc). Infected timbers often exhibit longitudinal and cross-grain cracking and a surface growth of whitish mycelium containing yellow and/or lilac patches; the leathery fruiting bodies bear a mass of rust-coloured spores. Control of dry rot involves removal of decayed wood and disinfection of remaining timbers by heat and/or the application of fungicides (see TIMBER PRESERVATION). Infections similar to dry rot may be caused e.g. by *Poria* spp. (cf. WET ROT.)

(2) A storage rot of potato tubers caused e.g. by *Fusarium coeruleum* (= *F. solani* var. *coeruleum*) and *F. sulphureum*. The tubers develop dark, sunken lesions which develop into mycelium-filled cavities; affected tubers lose water and shrink, becoming wrinkled and eventually drying into a hard mass. Under humid conditions, however, infection by SOFT ROT organisms may result in rapid decomposition of the tubers. (See also TECNAZENE.)

dryad's saddle *Polyporus squamosus* (q.v.).

drying (1) (of foods) See FOOD PRESERVATION (c). (2) (of plates) See PLATE. (3) (of microorganisms) See DESICCATION.

ds (of DNA or RNA) Double-stranded.

***dsb* genes** See PROTEIN SYNTHESIS (protein folding).

ds(c)DNA Double-stranded circular DNA.

DSI stain See DIETERLE SILVER STAIN.

DSM Deutsche Sammlung von Mikroorganismen (culture collection of microorganisms), Grisebachstr. 8, D-3400 Göttingen, Germany.

DTH Delayed-type hypersensitivity: see DELAYED HYPERSENSITIVITY.

dTMP Deoxythymidine 5′-monophosphate [see Appendix V(b)].

DTP *Syn.* DPT.

dualtropic (polytropic) Refers to a recombinant retrovirus derived from an ecotropic virus and a xenotropic virus; the recombinant virus has an extended host range and can, like AMPHOTROPIC viruses, replicate in both homologous and heterologous cells. However, a dualtropic murine virus differs from an amphotropic murine virus in that it possesses *env* glycoprotein antigenic determinants in common with those of the parental ecotropic and xenotropic murine viruses, and may thus be neutralized by antiserum to the *env* glycoprotein of either or both of these viruses. [Book ref. 114, p. 73.] (See also MCF VIRUSES.)

duck diseases See POULTRY DISEASES.

duck embryo vaccine (DEV) See RABIES.

duck hepatitis B virus (DHBV) A virus of the HEPADNAVIRIDAE which infects Pekin ducks (*Anas domesticus*) and other domestic ducks; strains of DHBV have also been detected in wild mallard [JGV (1986) *67* 537–547]. DHBV can be transmitted vertically via the egg. Infected ducks commonly exhibit persistent viraemia and may develop chronic liver disease; integrated DHBV DNA has been detected in a hepatocellular carcinoma in a domestic duck [PNAS (1985) *82* 5180–5184] (cf. HEPATITIS B VIRUS).

duck infectious anaemia virus See AVIAN RETICULOENDOTHELIOSIS VIRUSES.

duck plague *Syn.* DUCK VIRUS ENTERITIS.

duck virus enteritis (duck plague) An acute, highly infectious disease of ducks, geese and swans; it is caused by anatid (or anserid) herpesvirus 1 (see HERPESVIRIDAE). Symptoms: loss of appetite, thirst, nasal and ocular discharge, neurological signs; multiple haemorrhages occur in internal organs. Transmission is believed to occur mainly via pond-water.

duck virus hepatitis (DVH) An acute disease of young ducklings, caused by an ENTEROVIRUS; older ducks and other birds are not susceptible. Infection occurs by ingestion of food contaminated with faeces of infected birds. Incubation period: 24–48 hours. The disease progresses rapidly and death may occur within one or a few hours. On postmortem examination the liver is found to be enlarged and haemorrhagic. Vaccines are available for prevention; immune serum has been used in treatment.

Dulbecco's PBS Phosphate-buffered saline originally formulated to contain (g/l): NaCl (8.0), KCl (0.2), Na_2HPO_4 (anhydrous) (1.15), $CaCl_2.2H_2O$ (0.132), KH_2PO_4 (0.2), and $MgCl_2.6H_2O$ (0.1).

dulcitol *Syn.* GALACTITOL.

dulse See PALMARIA.

Dunaliella A genus of unicellular, naked (wall-less), halotolerant, biflagellated green algae (division CHLOROPHYTA) which closely resemble CHLAMYDOMONAS spp except for their lack of cell wall. *Dunaliella* spp can grow in aquatic environments of a wide range of salinities, up to and including saturated brines, and are common e.g. in hypersaline environments (salt lakes etc) throughout the world – often forming extensive red or green blooms. The organisms have an exceptional osmoregulatory capacity, using glycerol as an osmoregulator (see OSMOREGULATION); the intracellular concentration of glycerol can be varied rapidly to compensate for changes in the osmotic strength of the surrounding medium. The glycerol can be formed either from stored starch or as a direct product of photosynthetic carbon fixation; a reduction in glycerol concentration apparently occurs by glycerol degradation.

When certain species are subjected to hypo-osmotic shock, the cells swell, change shape (from elongated to spherical), and lose motility. (In *D. salina* expansion of the cytoplasmic membrane during swelling is apparently achieved by fusion of the membrane with numerous small membranous vesicles present in the cytoplasm [JCB (1986) *102* 289–297].) Some of the shocked cells lose buoyancy and sink; they may then continue to reproduce slowly or remain dormant, but they recover their motility when transferred to a medium of more favourable salinity. It has been suggested that this is a survival strategy whereby the organism can escape from the upper layers of (potentially damaging) low salinity resulting e.g. from rain or land drainage [JEMBE (1985) *91* 183–197].

(See also SINGLE-CELL PROTEIN.)

Duncan disease *Syn.* XLP SYNDROME.

Duovirus *Syn.* ROTAVIRUS.

duplex DNA Double-stranded DNA.

duplex PCR See MULTIPLEX PCR.

duplex winding number See DNA.

Durham shelter (gravity slide sampler) An instrument used for sampling airborne particles, particularly pollen. (See also AIR). Essentially it is an adhesive-coated microscope slide mounted horizontally inside a 'shelter' consisting of two parallel metal discs joined together by three equidistant 10-cm struts; the upper disc acts as a rain shield. Particle deposition is greatly affected by e.g. the speed and direction of the wind and by turbulence, and the volume of air sampled is unknown.

Durham tube A transparent glass tube (2–4 cm long, internal diam. ca. 3–4 mm), closed at one end, used to detect gas production by microorganisms during growth in a liquid medium. An *inverted* Durham tube is placed in a test-tube of liquid medium before sterilization; when the medium is autoclaved all the air is driven from the Durham tube which thus sinks to the bottom of the test-tube. On inoculation and incubation, any gas formed is trapped in the Durham tube.

Durvillaea See PHAEOPHYTA.

Dutch elm disease A disease of elm trees (*Ulmus* spp) caused by *Ceratocystis ulmi*. In the 1970s/1980s a new aggressive strain of *C. ulmi* caused massive epiphytotics, destroying millions of elms throughout Europe, SW Asia and North America. The disease is spread by bark beetles of the Scolytidae (commonly *Scolytus* spp) in a complex series of interactions [review: Book ref. 77, pp. 271–306]. Essentially, adult beetles, carrying *C. ulmi* propagules, emerge during the spring and summer and feed in the tops of healthy trees; the beetles feed in the crotches of young twigs, and some of the resulting wounds become contaminated with *C. ulmi*. The pathogen develops and spreads within the xylem vessels ('pathogenic phase') and infected twigs wilt and show characteristic streaks or spots; TYLOSES are formed, and the whole tree may die within months. The inner bark (phloem) of dying and dead trees becomes a suitable breeding ground for the scolytid beetles. Female beetles excavate extensive breeding galleries in the bark. *C. ulmi* colonizes these galleries ('saprotrophic phase') and sporulates abundantly (forming conidia, including synnematospores, and ascospores) during the autumn and winter when the larvae are developing. The new generation of adult beetles, carrying spores of *C. ulmi*, emerges the following year to feed in the twigs of healthy trees, thus completing the cycle.

Various 'diseased' strains of *C. ulmi*, characterized by abnormal or reduced growth and impaired reproductive fitness, have been isolated. The disease is apparently caused by an infectious agent, called a 'd factor' (for disease factor), which is spread between mycelia by hyphal fusion during the bark stage of the fungus; it has been found that certain diseased isolates contain 10 distinct dsRNA segments whereas healthy isolates contain 0–4 dsRNA segments, suggesting that specific dsRNA segments may be associated with the development of disease [PP (1986) *35* 277–287]. (See also CHESTNUT BLIGHT and MYCOVIRUS.)

Duttonella A subgenus of TRYPANOSOMA within the SALIVARIA; species include parasites and pathogens of a range of wild and domestic animals (see e.g. NAGANA). The trypomastigote form typically has a large, terminal kinetoplast and a free flagellum.

T. (D.) vivax occurs in the vertebrate in the trypomastigote form (ca. 20–30 μm in length) which divides by longitudinal binary fission; it is transmitted cyclically by *Glossina* spp – in which epimastigotes and metacyclic forms develop solely in the proboscis. Mechanical transmission by other biting flies (e.g. *Tabanus*) occurs outside the tsetse belt.

T. (D.) uniforme is similar to *T. (D.) vivax*, though somewhat smaller.

Duval's bacillus *Shigella sonnei*.

Duvenhage virus See LYSSAVIRUS.

DVH DUCK VIRUS HEPATITIS.

dw Dry weight.

dwarf bunt See COMMON BUNT.

DWELL See ETRIDIAZOLE.

dyad symmetry See e.g. PALINDROMIC SEQUENCE.

dye A water-soluble, aromatic compound which has coloured anions and/or cations that can bind to particular substance(s); binding may be primarily ionic but may involve e.g. hydrogen-bonding. (See also AUXOCHROME; CHROMOPHORE; LEUCO COMPOUND.) (In microbiology 'dye' is often used interchangeably with 'stain' to refer not only to dyes, as defined, but also to other substances – e.g. LYSOCHROMES – used for STAINING; 'stain' may also refer to a *mixture* of dyes and/or to the process of using a dye or stain.) An *acid dye* has a coloured anion which combines with cationic groups. A *basic dye* has a coloured cation which combines with anionic groups. A *neutral dye* is a compound of acid and basic dyes in which each ion contains a chromophore. (The pH of a solution of such dyes does not necessarily indicate whether the dye is acid, basic or neutral.) An *amphoteric dye* is either acidic or basic according to pH. Although dyes are commonly used for STAINING, some have additional uses: see e.g. ACRIDINES, ETHIDIUM BROMIDE, METHYLENE BLUE, TRIPHENYLMETHANE DYES; see also PH INDICATORS, RESAZURIN TEST.

Dyes are classified according to the nature of their chromophores. Those based on a (*para-* or *ortho-*) quinonoid ring include the TRIPHENYLMETHANE DYES, xanthene dyes (e.g. EOSIN, pyronin), azine dyes (e.g. JANUS GREEN, NEUTRAL RED, NIGROSIN), azine-related dyes (oxazines, e.g. NILE BLUE A; thiazines, e.g. METHYLENE BLUE, thionin, toluidine blue), and HAEMATEIN. Dyes with a chromophore of one or more azo groups (–N=N–) include BISMARCK BROWN, CHLORAZOL BLACK E, CONGO RED and TRYPAN BLUE; Janus green also contains an azo group. Dyes with a nitro group as chromophore include 2,4,6-trinitrophenol (picric acid).

dye-buoyant density centrifugation See *ultracentrifugation* in the entry CENTRIFUGATION.

dye test *Syn.* TOXOPLASMA DYE TEST.

Dynabeads® Microscopic beads used in *magnetic separation*: a technique in which a particular type of cell or molecule is separated from a mixture (in the liquid phase) by allowing the required cells or molecules to bind to beads coated with a specific ligand – the beads then being segregated by use of a magnetic field.

Dynabeads® (Dynal, Skøyen, Oslo, Norway) are microscopic spheres that contain a mixture of iron oxides; they are *superparamagnetic*, i.e. they do not exhibit magnetic properties until placed in a magnetic field. The beads can be coated with any of a variety of ligands – the particular type used depending on the required cell or molecule. For example, if beads are coated with the oligonucleotide T-T-T-T-T-T-T they will bind the poly-A tails of mRNA molecules; this approach can be used to separate mRNAs from total RNA.

Essentially, the beads are well dispersed within the sample by thorough mixing; the required cells or molecules bind to the beads – which are then drawn to one side of the vessel by a magnetic field. Unwanted material is discarded, and the required cells/molecules can be washed etc. prior to use. (Note that permanently magnetic beads would not be suitable: they would clump together rather than disperse within the sample.)

Beads coated with specific antibodies are useful for isolating particular pathogens – *E. coli* O157, *Salmonella* spp, *Listeria monocytogenes* etc. – from food or other specimens; the bead–pathogen complex can be plated on a suitable medium. This approach can also be used to isolate a specific type of eukaryotic cell from a mixture of cells. This method is called *immunomagnetic separation* (IMS).

Examples of use include: enrichment of *Mycobacterium paratuberculosis* in milk [AEM (1998) *64* 3153–3158]; detection of *E. coli* O157 in meat [LAM (1998) *26* 199–204]; preparation of human glomerular endothelial cells for use in tests with *E. coli* verocytotoxin [Kidney International (1997) *51* 1245–1256]; detection of specific mycobacterial DNA prior to PCR ('sequence capture' PCR) [JCM (1996) *34* 1209–1256].

dynein See FLAGELLUM (b).

dysentery A disease characterized by inflammation of the intestine (particularly the colon), abdominal pain, and frequent passage of fluid stools that may contain blood, mucus and/or pus. 'Dysentery' (rather than DIARRHOEA) is the term generally used to refer to a diarrhoeal condition in which (i) blood/mucus is typically present in the stool, *and* (ii) the intestinal epithelium is actively *invaded* by the causative organism (as is the case with e.g. *Shigella dysenteriae*, enteroinvasive *E. coli* and *Entamoeba histolytica*).

(1) (*med.*) (a) *Bacillary dysentery*. The classical (severe) form of dysentery is caused by *Shigella dysenteriae* serotype 1 (found mainly in tropical countries); somewhat less severe disease is caused by *S. sonnei* (a common causal agent in temperate zones), *S. boydii* (mostly tropical) and *S. flexneri*. As well as the *shigellosis*, caused by *Shigella* spp, bacillary dysentery can be caused e.g. by EIEC.

Dysentery caused by *S. dysenteriae* type 1 is transmitted via the faecal–oral route – commonly via food or water contaminated with faeces from patients with dysentery (or from a carrier of the pathogen). Incubation period: 1–6 days. Onset is abrupt, with fever followed by diarrhoea and abdominal cramps. *S. dysenteriae* proliferates in the gut lumen and invades the epithelium of the terminal ileum and colon. While dysentery due to *S. sonnei* and *S. boydii* is typically self-limiting (lasting days), that caused by *S. dysenteriae* may cause ulceration of the intestinal mucosa and may last for weeks, leading to exhaustion and anaemia; mortality may be high in untreated cases.

In animal studies, *Shigella* is taken up via the so-called 'M cells' in Peyer's patches [M cells in infection (review): TIM (1998) *6* 359–365]; from this location the bacteria can spread from cell to cell within the epithelial layer. Such spreading (an essential feature of dysentery) depends on the presence of a (plasmid-encoded) outer membrane protein: IcsA (= VirG). (IcsA is the α-domain of an autotransporter: see type IV secretory systems in PROTEIN SECRETION. An analogous protein occurs in EIEC.) IcsA promotes *actin-based motility*: i.e. it induces polymerization of actin filaments at one pole of the cell – forming a growing 'tail' of actin which propels the bacterium (through the host cell cytoplasm) in a direction opposite to that of the developing tail. The tail remains stationary within the host cell: ongoing deposition of actin causes bacterial motility. The motile bacterium pushes into the wall of the adjacent cell, either forming a pocket or becoming enclosed within a vacuole (bounded by a double membrane); *Shigella* lyses the membranes, thus entering the adjacent cell. Mutants lacking a functional IcsA do not spread, and do not cause dysentery.

Shigella taken up by M cells may be engulfed by macrophages which occur beneath the M cells. *Shigella* can kill macrophages (see APOPTOSIS), and in the resulting inflammatory response PMNs may migrate *between* epithelial cells to reach the gut lumen; such migration is likely to expose basolateral surfaces of epithelial cells to invasion by *Shigella* [TIM (1997) *5* 201–204].

Shigella can induce uptake by the so-called 'trigger mechanism', invasion apparently occurring via the basolateral surfaces of epithelial cells; the bacterial adhesin(s) are unknown, but epithelial cell receptors may include integrins (which occur on basolateral surfaces). Invasion involves a type III secretory system (see PROTEIN SECRETION) encoded by the plasmid-borne *mxi-spa* genes. The same plasmid also encodes IpaA; this protein is translocated to the epithelial cell where it induces certain changes (e.g. actin polymerization) needed for uptake. During uptake (*macropinocytosis*), the bacterium is engulfed by extrusions of host cell membrane which are supported by actin microfilaments. The bacterium thus induces its own uptake; moreover, the host cell's GTPase 'Rho' also appears to be required. Following uptake (into a vacuole), *Shigella* lyses the membrane and escapes into the host cell's cytoplasm; such lysis may involve the (plasmid-encoded) IpaB protein. Cell-to-cell translocation within the epithelial layer can then occur as described above.

[Invasion of host cells by *Shigella* and subsequent inflammation: FEMS Reviews (2001) *25* 3–14.]

Shigella dysenteriae serotype 1 produces SHIGA TOXIN (q.v.); the role of this toxin in dysentery in not fully understood, but it may be at least partly responsible for the vascular damage and bloody stools.

Lab. diagnosis. Culture from stool, or rectal swab, followed by serological typing; colicin and/or phage typing may also be useful.

Treatment. Antibiotics (when indicated, and in accordance with results from susceptibility tests); fluid and electrolyte replacement. (See also ORAL REHYDRATION SOLUTION.)

(b) *Amoebic dysentery* (amoebiasis) is caused by *Entamoeba histolytica* (see ENTAMOEBA). Infection follows ingestion of cysts from the faeces of patients or carriers. Following excystment, the organisms localize in the large intestine and attack the mucosa; establishment of infection apparently requires the presence of normal gut bacteria (see GASTOINTESTINAL TRACT FLORA). Symptoms range from mild, intermittent diarrhoea to severe, and occasionally fatal, dysentery with blood and mucus in the stools. (Trophozoites produce cyto/enterotoxin(s) [Inf. Immun. (1985) *48* 211–218].) Chronic relapsing cases, particularly in endemic regions, may be associated with the development of tumour-like granulomatous masses (called *amoebomas*) in the large intestine. Severe amoebic dysentery may be associated with *amoebic hepatitis* (apparently *not* due to amoebic invasion of the liver). The parasite may also cause abscesses, usually in the liver but also in the lungs, brain etc.; the presence of viable amoebae in patients with high titres of antibodies may be at least partly explained by the ability of amoebae to shed aggregates of absorbed antibodies ('caps' – see CAPPING sense 3) [JID (1986) *153* 927–932].

Lab. diagnosis involves e.g. demonstration of trophozoites and/or cysts in stools or lesions. (See also FLOTATION.)

Chemotherapy: see AMOEBICIDE.

(c) *Balantidial dysentery* (balantidiasis) is a severe, sometimes fatal, dysentery caused by *Balantidium coli* (see BALANTIDIUM). Infection follows ingestion of cysts in food or water contaminated with infected swine faeces. Excystment occurs in the gut; trophozoites cause extensive ulceration of the colon mucosa.

Lab. diagnosis: identification of trophozoites or cysts in faeces.

Chemotherapy: e.g. diiodoHYDROXYQUINOLINE.

(2) (*vet.*) See e.g. LAMB DYSENTERY, SWINE DYSENTERY, WINTER DYSENTERY. (cf. SCOURS.)

dysgonic See EUGONIC.

dyskinetoplasty In e.g. certain naturally-occurring strains of *Trypanosoma evansi*: the absence of a normal KINETOPLAST – the mitochondrion containing, instead, a small, spherical, electron-dense body, the 'kinetoplast remnant', which is not visible in Giemsa-stained preparations. Dyskinetoplastic strains of *T. evansi* and other trypanosomes can be obtained by exposure to e.g. Berenil.

dysphotic zone See PHOTIC ZONE.

dyspnoea (dyspnea) Difficult or laboured breathing.

Dysteria See HYPOSTOMATIA.

DZM Dorsal zone of membranelles: a SYNCILIUM (or syncilia) occurring in the anteriodorsal region of some ciliates, e.g. *Diplodinium*, *Epidinium*.

1. Words in SMALL CAPITALS are cross-references to separate entries.
2. Keys to journal title abbreviations and Book ref. numbers are given at the end of the Dictionary.
3. The Greek alphabet is given in Appendix VI.
4. For further information see 'Notes for the User' at the front of the Dictionary.

E

E (1) See REDOX POTENTIAL. (2) Glutamic acid (see AMINO ACIDS).

E-cadherin See CADHERINS.

E. coli An abbreviation which usually refers to the bacterium *Escherichia coli* (see entry ESCHERICHIA) but which is also used e.g. for the protozoan *Entamoeba coli*. (This illustrates the need to ensure that the name of a genus is spelt out in full – on first usage – before using an abbreviated form.)

E* precursor cells (*immunol.*) Cells on whose surface COMPLEMENT FIXATION has occurred, at 4°C, but which do not lyse unless the temperature is raised; lysis can be further inhibited, even at higher temperatures, by EDTA or zinc or uranyl salts.

E protein See F PLASMID.

E-selectin *Syn.* CD62E.

E site (of a ribosome) See PROTEIN SYNTHESIS.

E test A diffusion test for determining the MIC of a given bacterial strain with respect to particular antibiotic(s). One side of a plastic strip (placed in contact with the inoculated plate) carries a given antibiotic, the concentration of which decreases uniformly from one end of the strip; the other side of the strip is graduated with the concentration of antibiotic. Following incubation, the MIC is read by noting the lowest concentration of antibiotic (on the scale) which corresponds to inhibition of growth. Several strips, each with a different antibiotic, can be used simultaneously on a standard-sized plate.

The E test has been used for various bacteria, including *Pseudomonas aeruginosa* [JCM (1991) *29* 533–538] and [*Helicobacter pylori* [JCM (1997) *35* 1842–1846]. [*Mycobacterium tuberculosis*: JCM (2002) *40* 2282–2284.]

E_0 (E^0) See REDOX POTENTIAL.

E'_0 See REDOX POTENTIAL.

(E_1E_2)-type H^+-ATPase See PROTON ATPASE.

e14 element See RECOMBINATIONAL REGULATION.

E5531 See ENDOTOXIC SHOCK.

EA Erythrocyte–antibody (see e.g. HAEMOLYTIC SYSTEM). (cf. EAC.)

EAC Erythrocyte–antibody–complement. EAC 142... indicates that COMPLEMENT components 142... have been fixed. (cf. EA.)

***eae* gene** See PATHOGENICITY ISLAND.

EAEC (1) ENTEROADHERENT E. COLI [RMM (1995) *6* 196–206]. (2) Enteroaggregative *E. coli* (see EAGGEC) [MR (1996) *60* 167–215 (p. 186)].

EAF plasmid See PATHOGENICITY ISLAND and BUNDLIN.

EAggEC Enteroaggregative *Escherichia coli*: strains of *E. coli* associated with persistent diarrhoea – usually (though not exclusively) in children in the developing countries. (See also ENTEROADHERENT E. COLI.) The organisms form toxins and adhesins, but the mechanism of pathogenesis is unknown. EAggEC may be particularly important as pathogens for undernourished/immunosuppressed patients. [Enteroaggregative and diffusely adherent *E. coli*: RMM (1995) *6* 196–206.]

Strains of O111:H12 in Brazil were reported to have the properties of EAggEC [FEMS (1997) *146* 123–128].

(See also EAEC and EAST 1.)

Eagle's medium Any of a number of growth or maintenance media (used in TISSUE CULTURE) consisting basically of EARLE'S BSS or HANKS' BSS supplemented with e.g. amino acid(s), vitamin(s), antibiotic(s), serum etc.

ear microflora The microflora of the human ear (external auditory canal) commonly includes e.g. strains of *Staphylococcus*, *Corynebacterium*, *Mycobacterium* and various yeasts. (See also BODY MICROFLORA.)

Earle's BSS Earle's BALANCED SALT SOLUTION; it contains (g/l): NaCl (6.8), KCl (0.4), $MgSO_4.7H_2O$ (0.2), $NaH_2PO_4.2H_2O$ (0.158), $CaCl_2.2H_2O$ (0.264), $NaHCO_3$ (2.2), glucose (1.0), and phenol red (0.01); pH: 7.6–7.8.

early blight (of potato) A POTATO DISEASE caused by *Alternaria solani*. Small dark spots appear on leaves, each spot having concentric rings of necrotic tissue which give a characteristic 'target' appearance; tubers may exhibit a superficial brown, dry, corky rot.

early genes See VIRUS.

earth balls See SCLERODERMATALES.

earth stars See LYCOPERDALES.

earth tongues Fruiting bodies of species of e.g. *Geoglossum* or *Trichoglossum*.

East Coast fever An East African tick-borne disease of cattle caused by species of THEILERIA; *T. parva* and *T. lawrencei* cause a febrile disease which is usually fatal, while *T. mutans* typically causes non-febrile anaemia which may be fatal. After the tick vector has ingested blood from an infected bovine host, sexual stages of *Theileria* appear in the tick's gut; zygotes develop in the gut, and each forms a motile *kinete* which migrates to the tick's salivary glands and undergoes schizogony to form infective sporozoites. The tick can then infect a fresh bovine host. In the new (vertebrate) host the parasite grows in fixed and/or circulating lymphoid cells and forms *macroschizonts* ('Koch's blue bodies'; 'blue bodies'): schizonts, 10–20 μm in size, whose cytoplasm stains blue with Giemsa's stain. Subsequently, intra-erythrocytic forms ('piroplasms') of *T. parva* (or *T. mutans*) become common, but may be scanty/undetectable in *T. lawrencei*. *T. parva* and *T. lawrencei* infections usually kill after 15 days. [Life cycles of *Babesia* and *Theileria*: AP (1984) *23* 37–103. Anti-*T. parva* vaccine: Annals New York Acad Sci (2000) *916* 464–473.]

EAST 1 A heat-stable toxin secreted e.g. by EAGGEC, many strains of EHEC O157:H7, and by *Yersinia enterocolitica*; it is an analogue of the endocrine hormone *guanylin* (which stimulates fluid outflow from the intestinal mucosa by activating guanylyl cyclase).

eastern equine encephalomyelitis (EEE; eastern equine encephalitis; eastern encephalitis) An acute ENCEPHALITIS (or encephalomyelitis) of man and horses, caused by an ALPHAVIRUS; it occurs in eastern North America, the Caribbean, and parts of Central and South America. EEE virus can also cause disease in domestic birds (pigeons, pheasants). The virus occurs in wild birds, in which transmission occurs mainly via *Culiseta melanura*; other mosquitoes (e.g. *Aedes* spp) may be responsible for the sporadic transmission of EEE to animals and man. In man, the mortality rate may be ca. 25–70% or higher. Epizootics in horses may be controlled by vaccination.

Eaton's agent *Mycoplasma pneumoniae*.

EB ELEMENTARY BODY.

EB virus EPSTEIN–BARR VIRUS.

EBERs (EBV) See EPSTEIN–BARR VIRUS.

***ebna* genes** (EBV) See EPSTEIN–BARR VIRUS.

Ebola virus See FILOVIRIDAE.

EBV EPSTEIN–BARR VIRUS.

EC (1) Energy charge (See ADENYLATE ENERGY CHARGE). (2) Enzyme Commission (see ENZYME).

ECA ENTEROBACTERIAL COMMON ANTIGEN.

ecad A distinct population, within a given species, which has adapted phenotypically to its environment. (cf. ECOTYPE.)

ecchymosis See PURPURA.

Eccrinales See TRICHOMYCETES.

ecdysis The shedding of an outer layer: e.g., the exoskeleton of an invertebrate, or the theca in certain DINOFLAGELLATES.

Echinamoeba See AMOEBIDA.

echinocandin B See ACULEACIN A.

echinomycin See QUINOXALINE ANTIBIOTICS.

Echinosporangium See MUCORALES.

Echinosteliales See MYXOMYCETES.

Echinosteliopsidales See MYXOMYCETES.

Echinosteliopsis See MYXOMYCETES.

Echinostelium A genus of slime moulds (class MYXOMYCETES) which form protoplasmodia and small, stalked sporangia (<0.5 mm tall). *E. lunatum* is the smallest known myxomycete, forming stalked sporangia <50 μm tall and bearing only 4–8(–14) spores. [Encystment of myxamoebae and protoplasmodia in *E. minutum*: Mycol. (1985) *77* 253–258.]

echinulate Covered with small spines.

echoviruses See ENTEROVIRUS.

eclipse complex See TRANSFORMATION (1).

eclipse period See ONE-STEP GROWTH EXPERIMENT.

***Eco*B** A type I RESTRICTION ENDONUCLEASE.

***Eco*K** A type I RESTRICTION ENDONUCLEASE.

econazole See AZOLE ANTIFUNGAL AGENTS.

Economo's encephalitis *Syn.* VON ECONOMO'S ENCEPHALITIS.

***Eco*PI** A type III RESTRICTION ENDONUCLEASE.

***Eco*RI** A RESTRICTION ENDONUCLEASE from *Escherichia coli*; G/AATTC.

***Eco*RV** See RESTRICTION ENDONUCLEASE (table).

ecotropic Refers to a (endogenous or exogenous) retrovirus which can replicate only in hosts of the species in which it originated. (cf. XENOTROPIC, AMPHOTROPIC.)

ecotype A distinct population, within a given species, which has adapted genetically to its environment. (cf. ECAD.)

ectendomycorrhiza See MYCORRHIZA.

ectocarpene See PHEROMONE.

Ectocarpus See PHAEOPHYTA.

ectocyst See CYST (b).

ectomycorrhiza See MYCORRHIZA.

ectoparasite See PARASITE.

ectoplasm See e.g. SARCODINA.

ectoplasmic net See LABYRINTHULAS and THRAUSTOCHYTRIDS.

ectosymbiont An organism which lives on another organism in a symbiotic association (SYMBIOSIS sense 1); an ectosymbiont remains external to the cells and tissues of a host, although it may live e.g. within the gut cavity of the host. (cf. ENDOSYMBIONT.)

Ectothiorhodospira A genus of photosynthetic bacteria originally placed in the family CHROMATIACEAE but later proposed as type genus of a new family, Ectothiorhodospiraceae [IJSB (1984) *34* 338–339]. Cells: spiral, vibrioid or rod-shaped; they contain Bchl *a* or Bchl *b* (according to species – see CHLOROPHYLLS) and various CAROTENOIDS (cell suspensions are typically brown to brown-red). The pigments occur in lamellar intracytoplasmic membrane systems which are continuous with the cytoplasmic membrane. Motility is mediated by a single polar flagellum or by a polar tuft of flagella. Gas vacuoles occur in some strains. The organisms occur typically in alkaline sulphide-containing saline habitats; they grow optimally at pH 7.5–9.1 with a salt concentration of ca. 3–30%, sulphide being oxidized to elemental sulphur which is deposited extracellularly. Species include e.g. *E. abdelmalekii*, *E. halochloris*, *E. halophila* and *E. vacuolata*.

ectothrix infection See DERMATOPHYTES.

ectotrophic mycorrhiza See MYCORRHIZA.

ectotunica See BITUNICATE ASCUS.

Ectrogella See SAPROLEGNIALES.

ectromelia virus (mousepox virus) A virus (genus ORTHOPOXVIRUS) which infects mice, causing a disease (mousepox) in which internal organs (mainly liver and spleen) are involved, and characteristic lesions develop on the skin. Ectromelia virus replicates slowly on the CAM, forming small, white, irregularly shaped pocks.

eczema An inflammatory condition of the skin involving redness, itching, development of papules and vesicles etc; it is usually an allergic reaction. *Eczema herpeticum* and *eczema vaccinatum* are severe, sometimes fatal, generalized conditions resulting from disseminated infection with herpes simplex virus and vaccinia virus, respectively; they occur most commonly in patients with atopic dermatitis or certain immunodeficiency syndromes.

(See also PROBIOTIC)

ED pathway ENTNER–DOUDOROFF PATHWAY.

ED_{50} Effective dose (50%): that dose of a given agent which, when given or applied to each of a number of experimental test animals or systems, affects 50% of those animals or systems under given conditions.

Edam cheese See CHEESE-MAKING.

edaphic Of or pertaining to terrestrial habitats – particularly soil, but also e.g. leaf litter and the surfaces of living plants.

EDDA An iron chelator: ethylenediamine-di(*o*-hydroxyphenyl-acetic acid).

edeines A group of basic linear peptide antibiotics produced e.g. by *Bacillus brevis*; they are effective against a broad spectrum of organisms, including Gram-positive and Gram-negative bacteria, fungi, and some mammalian cells. At low concentrations edeines are bacteriostatic, selectively and reversibly inhibiting DNA synthesis; at higher concentrations they may be bactericidal, inhibiting the synthesis of proteins but apparently not of RNA.

Edelfäule (German) *syn.* NOBLE ROT.

edema *Syn.* OEDEMA.

edifenphos See HINOZAN.

editing (RNA) See RNA EDITING.

EDTA Ethylenediaminetetraacetic acid: $(CH_2.N(CH_2COOH)_2)_2$. EDTA strongly chelates divalent metal ions (e.g. Zn^{2+}, Ca^{2+}, Mg^{2+}); it is used e.g. as an ANTICOAGULANT (preventing clotting by complexing plasma Ca^{2+}) and as a potentiating agent in certain antibacterial preparations (see e.g. DETTOL CHELATE). EDTA causes a non-specific increase in the permeability of Gram-negative bacteria, and under certain conditions it can inhibit or lyse many Gram-negative species – *Pseudomonas aeruginosa* being particularly susceptible; the antibacterial activity of EDTA may involve one or more of several phenomena: e.g., loss of structural integrity of the OUTER MEMBRANE due to chelation of membrane-stabilizing divalent cations, and potentiation of peptidoglycan-degrading autolysins. EDTA has been used in the preparation of SPHAEROPLASTS of Gram-negative bacteria (see also OSMOTIC SHOCK). (The susceptibility of *P. aeruginosa* to EDTA can be decreased by growth in Mg^{2+}-limited media owing to changes in the outer membrane proteins [JGM (1983) *129* 509–517].) (cf. CDTA; EGTA; NTA.)

Edwardsiella A genus of Gram-negative bacteria of the ENTEROBACTERIACEAE (q.v.). Cells: ca. 1 × 2–3 μm, motile (except

E. ictaluri). VP −ve; acid (commonly with gas) from glucose; lactose −ve; phenylalanine deaminase −ve; arginine dihydrolase −ve; lysine decarboxylase +ve. Vitamins and amino acids are required for growth. Optimum temperature: 37°C (ca. 25°C for *E. ictaluri*). The organisms are usually resistant to colistin but sensitive to other antibiotics (including penicillins); R plasmids are rare. GC%: 53–59. Type species: *E. tarda*.

E. anguillimortifera. *Nomen dubium* [Book ref. 22, pp. 488–489].

E. hoshinae. Indole usually −ve; MR +ve; H_2S −ve (on TSI); acid from e.g. sucrose, D-mannitol and trehalose. *E. hoshinae* occurs in the intestines of animals (including mammals, fish and reptiles).

E. ictaluri. Non-motile. Indole −ve; MR −ve; H_2S −ve (on TSI); no acid from sucrose, D-mannitol or trehalose. Causal agent of ENTERIC SEPTICAEMIA in channel catfish.

E. tarda. Indole +ve; MR +ve; H_2S +ve (on TSI); usually no acid from sucrose, D-mannitol or trehalose. A few strains ('biogroup 1') produce acid from e.g. sucrose and D-mannitol but not from trehalose. *E. tarda* occurs in the intestines of animals (including mammals, fish and reptiles). It can be an opportunist pathogen in man, and may be a cause of diarrhoea; it can also cause localized and generalized sepsis in fish.

[Book ref. 22, pp. 486–491.]

EEE EASTERN EQUINE ENCEPHALOMYELITIS.

Eel Virus European See IPN VIRUS.

EES Ethylethane sulphonate: a mutagenic ALKYLATING AGENT.

EF (1) Elongation factor: see PROTEIN SYNTHESIS. (2) Oedema (edema) factor: see ANTHRAX TOXIN.

efavirenz See ANTIRETROVIRAL AGENTS.

EF_0F_1 H^+-ATPase See PROTON ATPASE.

effective publication See NOMENCLATURE.

efficiency of plating See EOP.

efflux mechanism (of antibiotic resistance) In some microorganisms: any of various systems (usually or always energy-dependent) which can mediate resistance to an ANTIBIOTIC (or to more than one antibiotic) by extruding such agent(s) through the cytoplasmic membrane or the cell envelope; efflux involves a TRANSPORT SYSTEM that often consists of a complex of proteins but which, in some cases, comprises a single protein. Efflux transporters are classified according to their overall structure and e.g. the homology of amino acid sequences in component proteins; some efflux systems are ABC TRANSPORTERS, while others are found e.g. in the major facilitator superfamily (MFS) or resistance–nodulation–division (RND) superfamily. (See also MATE.)

For examples of efflux-dependent antibiotic resistance (and/or references) see e.g. entries ABC TRANSPORTER, MACROLIDE ANTIBIOTICS, QUINOLONE ANTIBIOTICS, TETRACYCLINES.

effused-reflexed (*mycol.*) Refers to a sheet-like fruiting body which lies flat against the substratum except at the edge(s), the latter growing outwards from the substratum.

EF-G See PROTEIN SYNTHESIS (elongation).

eflornithine See SLEEPING SICKNESS.

efrapeptin A lipophilic polypeptide which binds non-covalently to and inhibits (F_0F_1)-type PROTON ATPASES.

efrotomycin See POLYENE ANTIBIOTICS (b).

EF-Ts, EF-Tu See PROTEIN SYNTHESIS (elongation).

EGF Epidermal growth factor. (See e.g. ONCOGENE.)

egg (for cultivating viruses) See EMBRYONATED EGG.

egg-drop syndrome In general: any fall in egg production in poultry, due e.g. to poor housing, incorrect nutrition, various POULTRY DISEASES, etc. 'Egg-drop syndrome 1976' is a specific syndrome which came into prominence in the UK and Ireland in 1976; it is caused by an adenovirus which can agglutinate fowl RBCs, and which may be indigenous in duck populations [Book ref. 116, pp. 535–536]. In this syndrome egg-drop is preceded by egg abnormalities – e.g. eggs may be soft-shelled or shell-less.

egg wrack *Ascophyllum nodosum*.

egg-yolk agar (EYA; lecithovitellin (LV) agar) An agar-based medium incorporating the yolks of hens' eggs (which contain the lipoprotein lecithovitellin). It is used e.g. to test for bacterial extracellular LECITHINASE activity; on EYA the colony of a lecithinase-producing strain is surrounded by a zone of opacity due to the deposition of insoluble diglycerides resulting from lecithin cleavage. (See also NAGLER'S REACTION.) On EYA, strains which produce LIPASES form colonies which are each surrounded by a pearly or iridescent zone due to a thin film of free fatty acids.

eggplant mild mottle virus See CARLAVIRUSES.

eggplant mosaic virus See TYMOVIRUSES.

eggplant mottled crinkle virus See TOMBUSVIRUSES.

eggplant mottled dwarf virus See RHABDOVIRIDAE.

EGME Ethylene glycol monomethyl ether; it is used e.g. as a BIOCIDE (and antifreeze) in aircraft fuel.

EGS (*mol. biol.*) External guide sequence: see SPLIT GENE (a).

EGTA A chelating agent, ethyleneglycol-bis(β-aminoethylether)-N,N,N',N'-tetraacetic acid. EGTA complexes a range of cations (e.g. Ca^{2+}, Mg^{2+}, Zn^{2+}); Mg^{2+} is complexed much less strongly than is Ca^{2+}.

Egtved disease (viral haemorrhagic septicaemia; haemorrhagic septicaemia of trout) A FISH DISEASE affecting mainly rainbow trout (*Salmo gairdneri*); the causal agent is a virus (a probable member of the RHABDOVIRIDAE). Symptoms may include skin darkening, bulging of the eyes, and abdominal swelling, or there may be few external symptoms; mortality may be high. The disease was first observed in Egtved, Denmark.

eguttulate Lacking GUTTULES.

E_h See REDOX POTENTIAL.

EHC See OPTOCHIN.

EHEC Enterohaemorrhagic *Escherichia coli*: pathogenic strains of *E. coli* which secrete (at least) one or both shiga-like toxins (see SHIGA TOXIN). Shiga-like toxins are also called *verocytotoxins* (VT1, VT2) because they are toxic for Vero cells (kidney cells of the African green monkey); hence, EHEC is sometimes referred to as verocytotoxic *E. coli* (VTEC). A minority of authors refer to EHEC as shiga-toxin-producing *E. coli* (STEC).

EHEC is not an invasive, intracellular pathogen (cf. EIEC; see also EAGGEC, EPEC, ETEC).

EHEC is typically a food-borne pathogen which produces symptoms ranging from mild diarrhoea to severe bloody diarrhoea (*haemorrhagic colitis*); particularly (though not exclusively) in children, infection can give rise to HAEMOLYTIC URAEMIC SYNDROME. Vascular damage has been associated with the shiga-like toxins.

Strains of EHEC typically contain the LEE PATHOGENICITY ISLAND, and many secrete EAST 1; these factors may be relevant to the inflammatory and secretory symptoms of EHEC-mediated disease [MR (1996) *60* 167–215 (pp 186–188)]. It has been reported that the expression of operons on the LEE pathogenicity island is regulated by QUORUM SENSING (q.v.).

Any of >25 serotypes of EHEC may cause disease, but, world-wide, O157:H7 is the serotype isolated most often in haemorrhagic colitis; O157:H$^-$ is also quite common. [Genome sequence of O157:H7: Nature (2001) *409* 529–533; erratum:

Nature (2001) *410* 240.] EHEC occurs e.g. in undercooked meat (particularly minced beef), milk and dairy products, and water (both drinking and recreational); the minimum infective dose is apparently <100 cells. In the UK (at least), farm animals (cattle, sheep) appear to constitute the main reservoir of infection. [EHEC (VTEC) O157 and other strains in England and Wales, 1995–1998 (epidemiology): JMM (2001) *50* 135–142.]

Under starvation conditions, strains of *E. coli* O157:H7 have been found to lose antigenicity corresponding to the O157 antigen – but to retain the ability to produce shiga-like toxins; this has suggested that immunological methods based on O157 antigenicity may be unsuitable for detecting toxin-producing strains which have been under starvation conditions for a long time [AEM (2000) *66* 5540–5543].

Lab. diagnosis. Culture on sorbitol–MacConkey agar (SMAC medium) is based on the inability of (most) O157 strains to ferment sorbitol; these strains form pale colonies, while most strains of *E. coli* (which ferment sorbitol) form pink colonies. Increased sensitivity of detection may be achieved by pre-culture enrichment of samples with DYNABEADS [detection of *E. coli* O157 in food [LAM (1998) *26* 199–204]. [Sorbitol-fermenting EHEC: JMM (2002) *51* 713–714.]

[Lab diagnosis: JCM (2002) *40* 2711–2715.]

Molecular methods include a PCR-based assay for the *rfb* gene whose product is essential for synthesis of the O157 lipopolysaccharide [JCM (1998) *36* 1801–1804]. This method would not distinguish non-toxigenic strains of O157, but it could form part of a multiplex PCR which also assayed for the toxin gene.

A LATEX PARTICLE TEST for the O157 serotype is available from Oxoid (Basingstoke, UK; product number DR620).

Detection of 'free' toxin in patients' faeces may be achieved by a 'reversed passive latex agglutination test' (see LATEX PARTICLE TEST); this test involves observation of latex particles (conjugated with anti-toxin antibodies) in the presence of various dilutions of sample (faecal extracts) [LAM (2001) *32* 370–374].

Serotypes other than O157 may be identified by serotyping; probe-based tests can be used to detect toxigenicity.

(See also STARFISH and SYNSORB PK.)

Ehrlichia A genus of Gram-negative bacteria of the tribe EHRLICHIEAE. Cells: coccoid, often pleomorphic, non-motile, ca. 0.5–1.0 µm. Growth occurs intracellularly in circulating leucocytes; the bacteria are commonly seen as a group of cells within a membrane-limited inclusion body ('morula'). Growth does not occur in cell-free media. Type species: *E. canis*.

E. canis is the causal agent of canine ehrlichiosis: a disease which occurs worldwide and which is characterized by fever, nasal discharge and anorexia; a severe haemorrhagic form of the disease ('tropical canine pancytopenia') occurs in some dogs – particular breeds (e.g. German shepherd) being more susceptible than others. *E. phagocytophila* causes ehrlichiosis in cattle and sheep; *E. equi* causes equine ehrlichiosis; *E. risticii* causes POTOMAC HORSE FEVER (and can be transmitted experimentally to cattle [VR (2001) *148* 86–87]); *E. sennetsu* (formerly *Rickettsia sennetsu*) causes 'sennetsu rickettsiosis' in man: a disease which, in the severe form, involves fever, anorexia and lymphadenopathy; the disease, which occurs in Japan, has been called 'glandular fever', 'Hyuga fever', 'infectious mononucleosis' and 'Kagami fever'. (cf. INFECTIOUS MONONUCLEOSIS.)

Ehrlichieae A tribe of bacteria of the family RICKETTSIACEAE; species are primarily pathogens of animals (including cattle, dogs, goats and sheep), though *Ehrlichia sennetsu* is pathogenic in man. The genera: COWDRIA, EHRLICHIA, NEORICKETTSIA.

ehrlichiosis Any human or animal disease caused by a species of EHRLICHIA.

Ehrlich's reagent An INDOLE TEST reagent: *p*-dimethylaminobenzaldehyde (1 g) is dissolved in 95–100% ethanol (95 ml), and conc. HCl (20 ml) is added; stored at 4°C in the dark.

EIA (*serol.*) (1) ENZYME IMMUNOASSAY. (2) ERYTHROIMMUNOASSAY.

EIAV Equine infectious anaemia virus: see SWAMP FEVER and LENTIVIRINAE.

eicosanoids Polyunsaturated fatty acids which contain 20 carbon atoms; examples include ARACHIDONIC ACID and LEUKOTRIENES.

EIEC Enteroinvasive *Escherichia coli*: strains of *E. coli* which can invade and destroy epithelial cells in the ileum/colon, causing watery diarrhoea or (bacillary) DYSENTERY. Invasiveness is dependent on virulence genes carried by a plasmid (p*Inv*). EIEC encodes a cell-surface protein analogous to the IcsA protein of *Shigella* (which promotes actin-based motility within epithelial cells: see DYSENTERY); the presence of this protein (in EIEC and *Shigella*) is associated with the *absence* of a specific cell-surface protease (OmpT in EIEC and SopA in *Shigella*) [Mol. Microbiol. (1993) *9* 459–468]. EIEC produces at least one enterotoxin: ShET2 (named after *Shigella* enterotoxin 2: a similar toxin produced by *S. flexneri*); this toxin is encoded by the *sen* gene – a (plasmid) gene formerly referred to as *set2*. [Toxins produced by EIEC: MR (1996) *60* 167–215 (p. 188).]

At least part of the secretory (diarrhoeal) effect brought about by invasive organisms (such as EIEC and *Shigella dysenteriae*) is likely to be due to their ability to upregulate cyclooxygenase activity in the intestinal mucosa – this leading to increased expression of PROSTAGLANDINS (such as PGE_2) with resulting increase in fluid secretion.

EIEC causes disease in both children and adults – outbreaks occurring e.g. in schools, hospitals and institutions. EIEC typically resembles *Shigella* – often being non-motile, and giving negative results in tests for lactose utilization and lysine decarboxylase.

Eijkman test An IMVEC TEST which determines the ability of an organism to produce gas from lactose at $44°C \pm 0.2°C$; the test is often used to confirm the presence of coliforms in a water sample (see COLIFORM TEST) – for which purpose use is made of a medium such as lauryl tryptose lactose broth that includes an agent which inhibits endospore-forming bacteria.

Eikenella A genus (*incertae sedis*) of anaerobic, facultatively aerobic, oxidase-positive, catalase-negative, chemoorganotrophic, Gram-negative bacteria which occur e.g. in the human mouth and intestine, and which can be opportunist pathogens. Cells: non-motile, round-ended rods (0.3–0.4 × 1.5–4.0 µm) or short filaments; rods may exhibit TWITCHING MOTILITY. Growth occurs e.g. on blood agar; haemin is required for aerobic growth of freshly-isolated strains. Plate cultures have a 'bleach-like' odour. Optimum growth temperature: 35–37°C. No acid is formed from carbohydrates. Nitrate is reduced to nitrite. Urease −ve. GC%: ca. 56–58. Type species: *E. corrodens*. [Book ref. 22, pp. 591–597.]

Eimeria A genus of parasitic, homoxenous protozoa of the suborder EIMERIORINA (q.v. for life cycle). *Eimeria* spp form tetrasporic, dizoic oocysts (cf. e.g. ISOSPORA). No species is known to be pathogenic in man, but many cause disease e.g. in domestic animals (see COCCIDIOSIS).

eimeriid Of the suborder EIMERIORINA.

Eimeriina (1) A suborder of protozoa (order Eucoccida) later incorporated – together with organisms of the TOXOPLASMEA – in the suborder EIMERIORINA. (2) A suborder of protozoa (order

Eucoccidiida [JP (1980) *27* 37–58]) equivalent to the EIMERIORINA.

Eimeriorina ('the coccidia') A suborder of parasitic protozoa (order EUCOCCIDIORIDA) in which syzygy generally does not occur (cf. ADELEORINA) and in which motile zygotes are not formed (cf. HAEMOSPORORINA); sporozoites are typically enclosed within SPOROCYST(s). Genera include e.g. BESNOITIA, CRYPTOSPORIDIUM, EIMERIA, FRENKELIA, ISOSPORA, LANKESTERELLA, SARCOCYSTIS, TOXOPLASMA and TYZZERIA. (Other genera include *Aggregata, Angeiocystis, Arthrocystis, Atoxoplasma, Barrouxia, Caryospora, Caryotropha, Cyclospora, Diaspora, Dobiella, Dorisa, Dorisella, Grasseella, Hoarella, Mantonella, Merocystis, Octosporella, Ovivora, Pfeifferinella, Pseudoklossia, Pythonella, Schellackia, Selenococcidium, Selysina, Sivatoshella, Skrjabinella, Wenyonella.*) *Generalized* life cycles are given below.

(a) *Homoxenous species* (e.g. *Eimeria* spp). The disseminative forms (*oocysts*) are ingested by a new host, and *excystation* occurs in the intestine: the oocyst breaks down and liberates a number of banana-shaped, non-flagellated (but motile by flexion) *sporozoites* ca. 5–20 μm in length. Sporozoites penetrate and usually grow within cells of the intestinal mucosa, each sporozoite rounding up and becoming a feeding cell (*trophozoite*). Each trophozoite undergoes SCHIZOGONY and subsequently ruptures its host cell to release many *merozoites* (small cells resembling sporozoites). Merozoites penetrate fresh host cells, become trophozoites, and schizogony is repeated; several schizogonous cycles may occur. Later, intracellular merozoites develop into *gametocytes*. Some gametocytes become large, uninucleate female *macrogametes* which remain in situ; others undergo repeated division and produce small, uninucleate, flagellated cells, the male *microgametes*, which escape from the (ruptured) host cell. Each macrogamete is fertilized in situ, and the zygote encysts to form an *oocyst*. Oocysts are often spherical or oval, ca. 10–40 μm according to species; a MICROPYLE may be present. The oocyst is liberated from the host cell and voided in the faeces. Outside the host the oocyst undergoes SPORULATION (cf. ENDOSPORULATION) in which the zygote gives rise to a number of (haploid) sporozoites. In many species the sporozoites are formed within one or more sacs (*sporocysts*) within the oocyst; the number of sporozoites per sporocyst, and the number of sporocysts per oocyst, are taxonomically important features. The sporulated oocysts are able to infect fresh hosts and complete the life cycle.

(b) *Heteroxenous species*. In these species the asexual and sexual phases of the life cycle typically occur in different host species – referred to as the *intermediate host* and *final* (*definitive*) *host*, respectively; some species (e.g. *Toxoplasma gondii*) are facultatively heteroxenous, others (e.g. most or all species of *Sarcocystis*) are obligately heteroxenous. These coccidia often parasitize extra-intestinal tissues – particularly in the intermediate host. In *Sarcocystis* spp the intermediate host becomes infected on ingestion of oocysts and/or sporocysts, and the parasite typically undergoes extra-intestinal schizogony, forming cysts in the muscles (see SARCOCYSTIS); when sarcocysts (or sporocysts) are ingested by the final host, sexual reproduction (only) occurs in the intestinal tissues and sporulated oocysts, or sporocysts, are shed in the faeces. (See also TOXOPLASMOSIS.)

[See also Book ref. 18, pp. 102–165 (ultrastructure), pp. 229–285 (growth in tissue culture and avian embryos).] (See also COCCIDIOSIS.)

ejectosome ('protrichocyst') A type of EXTRUSOME present in CRYPTOPHYTES. Ejectosomes occur just below the cell surface in the region of the anterior depression (when present) and around the cell periphery; each consists of two interconnected, tightly coiled ribbon-like structures enclosed by a single membrane. On appropriate stimulation of the organism ejectosomes are discharged, the coiled structures of each being ejected to form a long tube. Ejectosomes appear not be toxic; their function is unknown.

El Tor vibrio See VIBRIO (*V. cholerae*).

EL1–EL34 proteins See RIBOSOME.

elaioplast A lipid-storing PLASTID.

ELAM-1 *Syn.* CD62E.

Elaphomyces See ELAPHOMYCETALES.

Elaphomycetales An order of ectomycorrhizal, hypogean fungi of the ASCOMYCOTINA. The dark, globose ascocarp lacks a hamathecium. Asci: globose and unitunicate; evanescent. The ascocarp of *Elaphomyces* can be parasitized e.g. by *Cordyceps* spp.

Elasmomycetaceae See RUSSULALES.

elastase A protease which cleaves the protein *elastin* (the main component of elastic fibres in e.g. arterial walls and ligaments). An elastase is formed e.g. by *Clostridium histolyticum* (the δ-toxin) and by *Pseudomonas aeruginosa*; the enzyme formed by *P. aeruginosa* (a zinc metalloenzyme) can also degrade human IgG in vitro (i.e., it can act as an 'IgG protease'), and it appears capable of degrading mouse IgG in vivo [CJM (1984) *30* 1118–1124].

elasticotaxis The tendency, observed in many species of GLIDING BACTERIA, to become orientated parallel to stress lines induced in an agar medium.

electrical transport A term used by some authors to include both ELECTROGENIC TRANSPORT and ELECTROPHORETIC TRANSPORT.

electricity generation See BIOFUEL CELL.

electroblotting See BLOTTING.

electrochromic effect See CAROTENOID BAND SHIFT.

electroendosmosis (electro-osmosis) The movement of a charged fluid, relative to a fixed medium carrying the opposite charge, under the influence of an electrical gradient. Electroendosmosis occurs e.g. during the IMMUNOELECTROPHORESIS of serum in agar gels. In this procedure (often carried out at pH 8.6) some negatively charged proteins (e.g. γ-globulins) migrate towards the negative pole (cathode). This is due to the ionization of the AGAR; the agar carries a net negative charge, and the fluid contains an excess of positive ions. The fluid thus moves towards the cathode – the flow being sufficiently strong to overcome the attraction of the anode for some negatively charged proteins and to carry them towards the cathode.

electrofocusing *Syn.* ISOELECTRIC FOCUSING.

electrofusion A wholly electrical method for bringing about the fusion of microbial or plant protoplasts, or animal cells (cf. HYBRIDOMA); essentially, protoplasts or cells are aggregated by a high-frequency alternating electric field and then fused by a short pulse (e.g. 1 ms) of direct current. (cf. ELECTRO-TRANSFECTION.)

electrogenic transport Transport across an energy-transducing membrane resulting in the generation of, or change in, a transmembrane electrical potential difference. (cf. ELECTRICAL TRANSPORT and ELECTRONEUTRAL TRANSPORT.)

electroimmunoassay (1) *Syn.* ROCKET IMMUNOELECTROPHORESIS. (2) Loosely, any form of IMMUNOELECTROPHORESIS.

electrokinetic potential *Syn.* ZETA POTENTIAL.

electron microscopy MICROSCOPY in which an electron beam interacts with a specimen and subsequently contributes, directly or indirectly, to the formation of an image of the specimen. (The specimen observed may be the original biological specimen

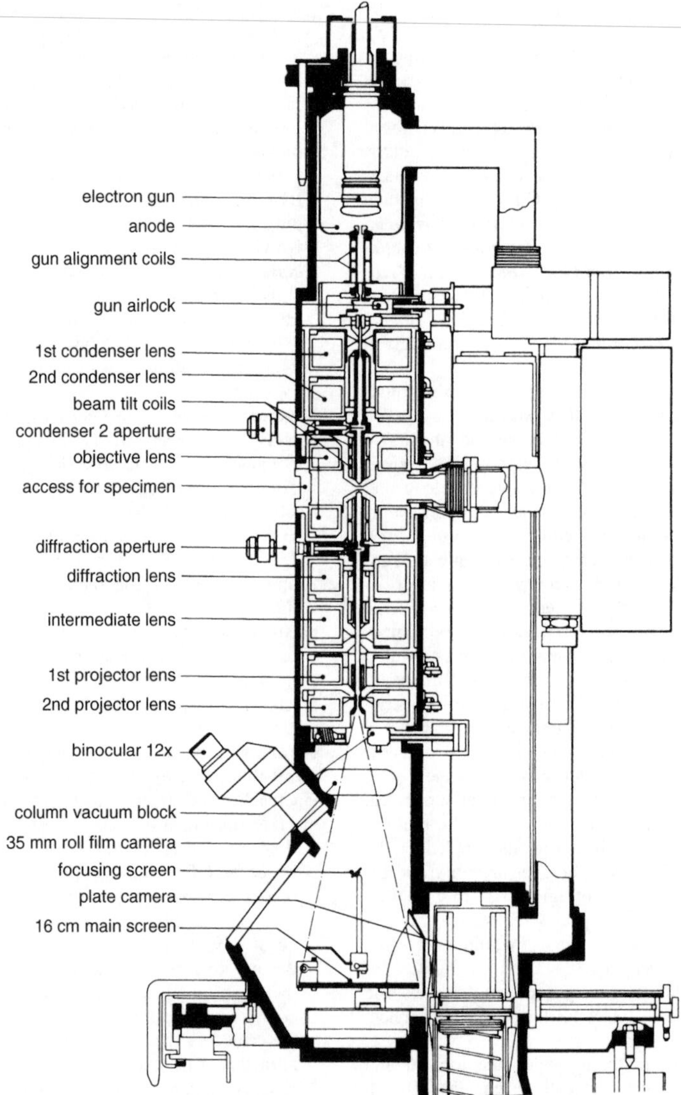

ELECTRON MICROSCOPY. Cross section of a transmission electron microscope (the Philips EM400). (Courtesy of N. V. Philips' Gloeilampenfabrieken, Eindhoven, Holland.)

or a replica of it.) Electron microscopy can give magnifications within the range ca. 10× to 1000000×. Resolving power depends e.g. on the nature of the specimen and on the type of instrument used. The best resolving power is obtained with the transmission electron microscope (see (a) below) which has resolved a structure of ca. 1 nm periodicity in the purple membrane of *Halobacterium salinarium*; for non-periodic biological specimens the best resolution may be ca. 3–5 nm.

Electron microscopy is used e.g. for examining viruses, macromolecules, and the ultrastructure of cells, and for determining the intracellular locations of e.g. specific enzymes and ions (see also IMMUNOELECTRON MICROSCOPY). Although living material has been examined in the electron microscope (EM), it is usual to examine non-living specimens which have undergone various pre-treatments; treatments vary according to e.g. the nature of the specimen and the purpose of the investigation.

(a) *Transmission electron microscopy.* In the transmission electron microscope (TEM) an electron beam passes *through* the specimen which deflects or scatters some of the electrons; electrons whose paths are slightly deviated by the specimen are focused onto a fluorescent screen to form an image of the specimen. In general, electrons undergo little or no deflection when they pass through thin layers of biological material; hence, unless specially treated, such materials do not yield images in the electron microscope. However, biological materials can be 'stained' with certain heavy metal compounds which become localized in specific regions of the specimen; this enhances the electron-deflecting properties of those regions and/or increases contrast between the specimen and its background.

Preparation of thin sections for the TEM. Typically, a tissue or cell initially undergoes *double fixation*, e.g. pre-FIXATION with GLUTARALDEHYDE and post-fixation with OSMIUM TETROXIDE. The fixed specimen is *dehydrated*, e.g. by equilibration in increasing concentrations of ethanol (or acetone) in water. From 100% ethanol the specimen is commonly transferred to, and equilibrated with, propylene oxide before being *embedded*, i.e., impregnated with an epoxy resin (e.g. SPURR'S MEDIUM) or a methacrylate – substances which polymerize and harden without significant change in volume. (Equilibration of the specimen with propylene oxide between dehydration and embedding is advantageous since propylene oxide is more miscible than ethanol with epoxy resins.) The embedded specimen is then *sectioned* (see MICROTOME) to obtain ultrathin sections of maximum thickness ca. 100 nm. A section is placed on a specimen holder (*grid*): a metal (usually copper) disc, 2–3 mm in diameter, perforated by rows of square holes each ca. 0.1 × 0.1 mm. Before receiving the section the surface of the grid may be coated with a *support film*: a sheet of e.g. FORMVAR or nitrocellulose 20–50 nm thick; such materials have a low, uniform electron-scattering power. (A support film is sometimes omitted when the section is resin-embedded; however, it is essential for a liquid specimen – see e.g. negative staining below.) The support film itself may be coated with a film of carbon (ca. 20 nm thick) which stabilizes it and improves resolution at high magnifications. The section, on the prepared grid, is now *stained*, i.e., immersed in a solution of a heavy metal compound (e.g. uranyl acetate, lead acetate, or a salt of phosphotungstic acid) which combines or complexes with particular components (e.g. lipids, proteins) – a type of staining called *positive staining*; images subsequently formed in the EM thus reflect the distribution of heavy metal atoms and the groups with which they combine. (Staining is sometimes carried out *before* the embedding stage by subjecting the specimen to prolonged exposure to e.g. osmium tetroxide or uranyl acetate; this is called 'in-block' staining.) The section is allowed to dry and is then ready for examination in the TEM.

Negative staining (*negative contrast*) is used to reveal the outlines and surface contours of particulate specimens (e.g. macromolecules, viruses, bacteria). Essentially, the particles are immersed in a thin film of electron-opaque stain on a prepared grid; the stain penetrates the interstices of each particle and forms a thin background layer but usually does not permeate the particles themselves. Electrons are more readily transmitted through those regions occupied by the (relatively) electron-transparent particles, so that in the image these regions appear bright against a dark background; the presence of stain in surface hollows and grooves indicates surface topography. For example, the TOBACCO MOSAIC VIRUS virion appears as a light rod in which a dark, central, axial line indicates the presence of stain in the hollow core. To examine bacterial morphology, one drop of (fixed or unfixed) bacterial suspension is left on a prepared grid for one or more minutes (determined by trial and error) so that cells adhere to the surface; most of the drop is then drawn off with the edge of a filter paper. One drop of stain (e.g. 2% aqueous phosphotungstic acid or 2% aqueous ammonium molybdate) is then placed on the grid; after 30–60 seconds this is drawn off, as before. The fine film on the grid is left to dry and is examined in the TEM. (Alternatively, stain and suspension can be mixed, and the stained suspension applied to the grid.)

Shadowing (*shadow-casting*) is used e.g. to study the morphology and/or dimensions of e.g. viruses and bacteria, and is also involved in the KLEINSCHMIDT MONOLAYER TECHNIQUE for examining nucleic acids. Essentially, the particles (e.g. viruses) on a prepared grid are exposed to a point source of the vapour of a heavy metal; the point source is so placed that metal is deposited on only one side of each particle – leaving a non-metallized (electron-transparent) 'shadow' on the remainder of the particle and on that part of the support film shielded (by the particle) from the vapour. The image of a shadowed particle can give information on the shape and size of the particle; its size can be calculated from the length of the shadow and the angle at which the metal vapour was incident on the particle.

Replicas of e.g. a freeze-etched specimen (see FREEZE-ETCHING) may be examined in the TEM. Essentially, this involves shadowing the surface of the frozen specimen with a heavy metal under vacuum; commonly, shadowing is effected at an angle of ca. 30° with a mixture of platinum and carbon vapours. Shadowing throws into relief the features of an irregular surface. Areas protected from the vapour remain non-metallized and are therefore structurally weak; the *primary* replica is therefore strengthened by deposition of a layer of carbon from a vapour source placed directly above the specimen. The replica is cleaned by digesting away the specimen with strong oxidizing and hydrolysing agents (e.g. bleach, sulphuric acid, chromic acid). The replica is mounted on a grid and examined in the TEM.

The transmission electron microscope (*TEM*). A typical TEM is shown in the figure. The electron beam originates at a heated metal filament in the electron gun. The electrons are accelerated through a hole in the anode and pass down through the axis of the instrument; acceleration is caused by a large potential difference ('accelerating voltage') maintained between the anode (at 'earth' or zero potential) and the filament (commonly −100 kV to −500 kV). The accelerating voltage determines the wavelength of the electron beam which, in turn, determines the resolving power; in general, higher accelerating voltages give better resolving power. (Some specimens can be damaged more than others by high accelerating voltages; resolving power is therefore partly determined by the nature of the specimen.) The various magnetic lenses each consist basically of a soft-iron reel on which are wound many turns of copper wire; current passing through the wire generates a magnetic field which controls the electron beam passing axially through the lens. The specimen, on a grid, is placed in the path of the electron beam; electrons which pass through the specimen are focused to form an image on a screen consisting of a metal plate coated with a fluorescent material. During operation, the interior of the tube-like body ('column') of the microscope must be evacuated to prevent e.g. deflection and scattering of the electrons by molecules of gas; pressure in the column is usually less than ca. 6.5×10^{-5} Pa (5×10^{-7} torr; 0.5 nm Hg).

(b) *Scanning electron microscopy* is used to examine the *surface* of a specimen. The specimen is scanned by an electron beam which moves in a zig-zag raster; a second electron beam synchronously scans a cathode ray tube (CRT) with a similar raster. Electrons which are secondarily emitted from the specimen are collected and used to control the intensity of the beam in the CRT. Since the secondarily emitted electrons vary in quantity from point to point on the specimen's surface (according to surface topography) the electron beam in the CRT will be continually modulated as the specimen is scanned; hence, the CRT screen will display a pattern of bright and less-bright spots corresponding to the surface topography of the specimen.

The scanning electron microscope (SEM) can give magnifications of ca. 10× to 100000×; for biological specimens it is commonly used at magnifications below ca. 10000×. Resolution is usually ca. 10–20 nm. The SEM provides a great depth of field, thus giving valuable three-dimensional images.

Preparation of specimens for the SEM. Completely untreated specimens have been examined in the SEM. However, specimens (particularly plant material) are often given the following sequence of treatments. (i) Fixation and dehydration – as for TEM thin sections described above. (ii) Drying (removal of all liquid); this is sometimes achieved by FREEZE-DRYING (simultaneous dehydration and drying) but more usually by CRITICAL POINT DRYING. (iii) Attachment of the specimen to a metal (often aluminium) specimen holder (*stub*) by means of a 'glue' (e.g. SILVER DAG) which conducts both heat and electricity. (iv) Coating of the specimen, under vacuum, with carbon and/or metal – usually by SPUTTER-COATING; this prevents thermal and electrical build-up on the specimen when it is placed in the electron beam (thus avoiding image distortion and damage to the specimen) and enhances secondary electron emission. (As an alternative to coating, the specimen may be thoroughly impregnated, before dehydration, with e.g. osmium tetroxide.) The specimen is then ready for the SEM.

Replicas of a specimen may be made e.g. if repeated observations are to be made of the same specimen. Essentially, fluid plastic or latex is poured over the specimen and allowed to harden. This gives a negative replica which is peeled off and lined with a suitable material that hardens to form the positive replica; the latter is sputter-coated and examined in the SEM.

(c) *Scanning transmission electron microscopy.* In the scanning transmission electron microscope (STEM) electrons pass through the specimen, and the energies of the emergent electrons are measured in an analyser. The pattern of energies of the emergent electrons is used to construct an image of the specimen. The resolving power of the STEM can be similar to that of the TEM. [Introduction to STEM: JUR (1984) *88* 94–104; imaging biological structures: JUR (1984) *88* 105–120, 177–206.]

(d) *Low temperature electron microscopy* (*cryoelectron microscopy*). A type of electron microscopy in which very small specimens are cooled rapidly to very low temperatures, without prior dehydration, such that the water in the specimen forms a 'glass-like' solid without undergoing crystallization – a phenomenon called *vitrification* [review: TIBS (1985) *10* 143–146].

(e) *Three-dimensional electron microscopy.* A procedure intended for three-dimensional examination of biological structures at atomic resolution. [ARBB (1981) *10* 563–592.]

electron paramagnetic resonance *Syn.* ELECTRON SPIN RESONANCE.

electron spin resonance (ESR; electron paramagnetic resonance, EPR) The absorption and consequent emission of electromagnetic radiation by unpaired electrons in certain types of atom in the presence of a strong magnetic field; in general principle ESR resembles NUCLEAR MAGNETIC RESONANCE (q.v.).

Molecules in which atom(s) contain unpaired, ESR-detectable electrons are paramagnetic and are commonly called 'free radicals'. In biological materials free radicals are produced during many types of metabolic process but they are typically short-lived; however, a stable, paramagnetic probe (*spin label*) can be introduced into an experimental system – e.g. by biosynthetic incorporation or by covalent linking to specific compounds. Most spin labels contain the nitroxide (NO) group; they include derivatives of fatty acids, of the steroids androstane and cholestane, and of 2,2,6,6-tetramethylpiperidine. (In some studies on membrane systems the presence of the spin label at high concentration can cause significant perturbation of the membrane characteristics.) In ESR the exciting electromagnetic radiation may involve frequencies in the *X-band* (ca. 9.3 GHz) or the *Q-band* (ca. 35 GHz). (1 GHz = 10^9 cycles per second.)

[Book ref. 151.]

electron transport chain (ETC; electron transfer chain) A sequence of redox agents along which electrons can be translocated with concomitant conversion of energy; an ETC occurs within an ENERGY-TRANSDUCING MEMBRANE, the redox agents forming a series of redox couples which, collectively, span a REDOX POTENTIAL gradient. In many cases an ETC operates as a *proton pump*: energy yielded during the flow of electrons towards a less negative (or more positive) redox potential is used to extrude protons across the membrane; the transmembrane proton gradient thus formed can be used as a source of energy (see CHEMIOSMOSIS). (In some organisms an ETC acts as an Na^+-pump: see SODIUM MOTIVE FORCE.) An ETC which is used in the respiratory metabolism of a substrate (see RESPIRATION) is often called a *respiratory chain*; ETCs are also involved in PHOTOSYNTHESIS.

Components of ETCs typically include flavoproteins (see RIBOFLAVIN), CYTOCHROMES, IRON–SULPHUR PROTEINS, QUINONES, and copper-containing proteins (e.g. PLASTOCYANIN); atypical ETCs include e.g. those of *Mycoplasma* spp which lack both cytochromes and quinones. The best-studied ETCs include those which occur in the inner membrane of the MITOCHONDRION – particularly the mitochondria of certain types of mammalian cell (e.g. bovine heart, rat liver). Although the mitochondrial ETCs in many other types of cell are often presumed to be essentially similar to those in mammalian cells, those in e.g. plant mitochondria characteristically exhibit some important differences (see later).

Mitochondrial ETCs. The mammalian mitochondrial ETC consists of four groups of functionally interrelated components (Complexes I–IV) arranged in a sequence (see figure). *Complex I* (NADH–UQ oxidoreductase) contains a flavoprotein which is believed to transfer electrons from NADH (formed e.g. in the TCA CYCLE) to the Fe–S centres (see IRON–SULPHUR PROTEINS) and thence to UQ; electron flow between NADH and UQ can be inhibited by e.g. AMYTAL, PIERICIDIN and ROTENONE. *Complex II* (succinate–UQ oxidoreductase) contains succinate dehydrogenase (an FAD-containing enzyme), Fe–S centres, and a *b*-type cytochrome (e.g. b_{560} in bovine heart, b_{559} in *Neurospora* sp) of unknown function. In reconstituted ETCs, electron flow between Complex II and UQ is inhibited by TTFA. *Complex III* (UQ–cytochrome *c* oxidoreductase) contains (at least) two *b*-type haems (which appear to be bound to a single polypeptide): b_{562} (formerly b_K) and b_{566} (formerly b_T); an Fe–S protein (a RIESKE PROTEIN); cytochrome c_1; and, in at least some species, semiquinones (believed to be ubisemiquinones). Electron flow through Complex III is inhibited by e.g. ANTIMYCIN A, BAL, DCMU and HYDROXYQUINOLINE-N-OXIDES (HOQNOs). (See also XANTHOMEGNIN.) *Complex IV* (cytochrome oxidase; cytochrome aa_3; ferrocytochrome c–O_2 oxidoreductase) contains two haems and two protein-bound copper atoms per functional unit; the haems (a, a_3) are chemically identical but have different spectroscopical and other properties owing to differences in the way in which they are bound to protein. Complex IV catalyses the irreversible reduction of oxygen (the terminal electron acceptor) to H_2O; it is inhibited by e.g. AZIDE, CARBON MONOXIDE, sulphide and CYANIDE.

Transmembrane translocation of protons can occur at three 'sites' which correspond to Complexes I, III and IV. The

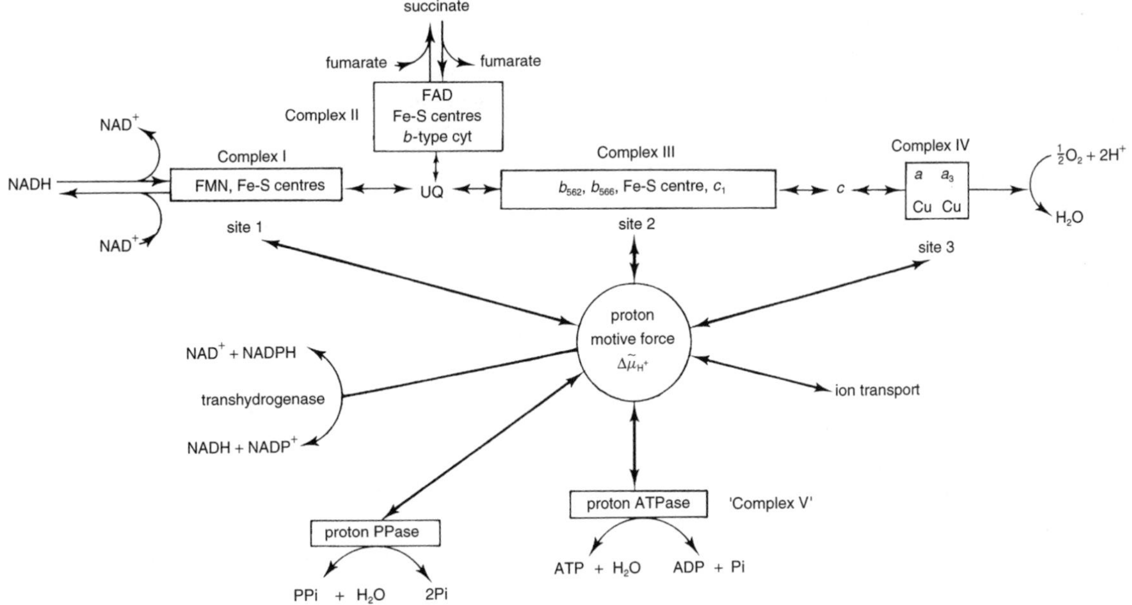

ELECTRON TRANSPORT CHAIN. A scheme for electron transport in an animal-type mitochondrion, showing electron transfer from NADH and/or succinate to the terminal electron acceptor, oxygen. (Electron transport is believed to be similar in the mitochondria of many types of microorganism.) Sites 1, 2 and 3 indicate sites of transmembrane proton translocation during electron flow. Thick lines indicate energy conversion (or regulation in the case of transhydrogenation). FMN = flavin mononucleotide; Fe–S centre = iron–sulphur centre; FAD = flavin adenine dinucleotide; b_{562}, c_1 etc = cytochromes; UQ = ubiquinone; proton PPase = proton pyrophosphatase.

mechanism of proton translocation is not understood; moreover, even the stoichiometry of proton translocation has not yet been settled: in Complex I the $H^+/2e^-$ ratio (q.v.) is believed, by various authors, to be 2, 3 or 4. Proton translocation is the subject of two main types of model: the LOOP MODEL and the conformational pump model; in the latter model, proton translocation involves changes in the pK of ionizable groups on certain proton-binding ETC components brought about by electron-dependent redox and conformation changes in those components. (See also ANISOTROPIC INHIBITOR and Q-CYCLE.) [Mitochondrial electron transport: ARB (1985) *54* 1015–1069.]

Plant mitochondrial ETCs are characteristically branched at the ubiquinone, one branch terminating with a conventional cytochrome oxidase, the other terminating with a cyanide-resistant oxidase; the latter branch is apparently non-proton-translocating, so that respiration via this branch is less efficient in terms of energy transduction.

Bacterial ETCs are generally located in the CYTOPLASMIC MEMBRANE; in at least some cyanobacteria, however, respiratory electron transport pathways occur in both the cytoplasmic membrane and the thylakoid membranes [Book ref. 75, pp. 199–218]. The composition of bacterial ETCs exhibits wide variability among the various species – and even in a given species under different growth conditions or in different physiological states (e.g. vegetative and sporulating cells of *Bacillus cereus* [JB (1984) *160* 473–477]). Moreover, many bacteria have branched ETCs and may use different pathways for different terminal electron acceptors. Although the components of ETCs in aerobically grown bacteria may resemble those in mitochondrial ETCs, the order in which they occur is typically unlike that in mitochondria.

In a given species of bacterium more than one type of cytochrome may be used as a terminal oxidase. For example, in *Escherichia coli* the major terminal oxidase under aerobic conditions is a *b*-type cytochrome, cyt *o*, while under oxygen-limited conditions it is cyt *d*. During NITRATE RESPIRATION, *E. coli* uses an inducible cytochrome which functions specifically as part of the nitrate reductase enzyme system; this cytochrome is designated ^{nr}b or $b_{556}^{NO_3^-}$. [Respiratory chains in *E. coli*: MR (1984) *48* 222–271.] Cytochromes specific for different terminal electron acceptors may coexist in the same cell [JB (1983) *154* 708–727] and, under certain conditions, two terminal electron acceptors may be used simultaneously [JGM (1979) *112* 379–383].

electron-transport particle See ETP.

electron transport phosphorylation The phosphorylation of ADP to ATP, or the formation of PYROPHOSPHATE from inorganic phosphate, in a reaction driven by the energy of a transmembrane gradient of ions generated by means of an ELECTRON TRANSPORT CHAIN. (In all known cases of electron transport phosphorylation, the ETC generates a transmembrane gradient of protons (proton motive force – see CHEMIOSMOSIS); although some respiratory chains generate a SODIUM MOTIVE FORCE, as yet there have been no reports of an Na^+-motive respiratory chain being used for the synthesis of ATP.) Electron transport phosphorylation can occur in respiratory metabolism (see OXIDATIVE PHOSPHORYLATION), in PHOTOSYNTHESIS, and in certain types of FERMENTATION (sense 1) (q.v.).

'Electron transport phosphorylation' excludes not only SUBSTRATE-LEVEL PHOSPHORYLATION but also those modes of phosphorylation dependent on a transmembrane ion gradient that is not produced by electron transport; such ion gradients can be derived e.g. by direct proton pumping (as in the PURPLE MEMBRANE) or by END-PRODUCT EFFLUX. With the exception of substrate-level phosphorylation, these forms of phosphorylation resemble electron transport phosphorylation in that they are dependent on ion-motive force and involve a membrane-bound ATP- or PPi-synthesizing enzyme system; all forms of phosphorylation which have these two characteristics may be referred to as *ion-motive force phosphorylation* (imf phosphorylation).

electroneutral transport Transport across an energy-transducing membrane of charged or uncharged species resulting in no *net* transfer of charge across the membrane; it occurs when an uncharged species is carried by a UNIPORTer, during the SYMPORT of two ions of opposite charge, or during the ANTIPORT of two ions of the same charge. (cf. ELECTROGENIC TRANSPORT.)

electro-osmosis *Syn.* ELECTROENDOSMOSIS.

electropherotype (*virol.*) A virus type (see TYPING) recognized on the basis of ELECTROPHORESIS of e.g. its genome. (See also OLIGONUCLEOTIDE FINGERPRINTING.)

electrophoresis Any procedure by means of which the members of a heterogeneous population of charged particles can be translocated or separated (or, exceptionally, brought together: see COUNTERCURRENT IMMUNOELECTROPHORESIS) by virtue of their mobility in an electric field.

Electrophoresis is used e.g. for identifying or quantifying proteins, nucleic acids etc., for molecular weight determinations (e.g. SDS-PAGE), for preparing fingerprints in various forms of TYPING, and for identifying/confirming products in nucleic-acid-amplification methods. (See also CHROMATOGRAPHY and ISOELECTRIC FOCUSING.)

In *free electrophoresis* the molecules to be separated are present in a liquid medium (usually water). In one form (*moving-boundary electrophoresis*) a mixed population of molecules is subjected to electrophoresis in a U-tube, each limb of which contains a column of buffer above the test suspension; the electric field is applied by placing a positive electrode in one limb and a negative electrode in the other. During electrophoresis the different types of molecule begin to separate into distinct zones, boundaries between the homogeneous and heterogeneous populations moving as the process continues.

In *surface electrophoresis* the molecules in a sample move through a thin film of buffer on the surface of a strip of paper or e.g. cellulose acetate; the mobility of a given molecule may therefore by influenced by interaction with the solid phase. Essentially, a droplet of sample is applied near the centre of a strip which has equilibrated with the vapour of an appropriate buffer solution in a closed chamber; an electrical potential difference is then applied between the two ends of the strip. After electrophoresis, the strip may be stained to reveal discrete zones, bands or spots corresponding to the various components in the sample.

In *gel electrophoresis* the molecules in a sample move through a gel composed, usually, of agarose (see AGAR) or polyacrylamide. Essentially, the sample is placed in a well cut into one end of a gel strip, and an electrical potential difference is applied between the two ends of the strip; after electrophoresis, the separated components may be e.g. stained in situ, extracted for further study, or transferred to another medium (see BLOTTING and SOUTHERN HYBRIDIZATION).

(See also ELECTROENDOSMOSIS, IMMUNOELECTROPHORESIS, PFGE, SDS-PAGE, TWO-DIMENSIONAL ELECTROPHORESIS, ZYMOGRAM.)

electrophoretic transport Transport of a charged species across an energy-transducing membrane in response to a pre-existing transmembrane electrical potential difference. (cf. ELECTRICAL TRANSPORT.)

electroporation An *in vitro* method for increasing the efficiency of TRANSFORMATION. Essentially, a mixture of cells and e.g. plasmids is exposed to an electric field (up to ~16 kV/cm) for a fraction of a second; the mechanism of uptake is unknown, but the frequency of transformation varies linearly with the concentration of DNA over a wide range of values, and the efficiency of transformation varies with cell concentration.

[Electroporation in *Helicobacter pylori*: AEM (1997) *63* 4866–4871; in *Escherichia coli*: NAR (1999) *27* 910–911.]

'Electrocompetent' (electroporation-competent) cells with high-level transformation efficiency are available commercially. The electrocompetent cells marketed by Stratagene (ElectroTen-Blue™) are electroporated at 1.7 kV and 600 ohms resistance.

ElectroTen-Blue™ See ELECTROPORATION.

electro-transfection Electrically-assisted TRANSFECTION; thus, e.g. tobacco protoplasts have been transfected with RNA from the tobacco mosaic virus under the influence of 50-µsec pulses of DC current at 550–800 V/cm [JGV (1986) *67* 2037–2042]. (cf. ELECTROFUSION.)

Elek plate An in vitro method for detecting the formation of a diffusible toxin by a given strain of microorganism, e.g. *Corynebacterium diphtheriae*. A plate of suitable medium is inoculated, in a straight line, with the strain under test; the streak is then overlaid, perpendicularly, with a strip of paper which has been impregnated with relevant antitoxin. After incubation (24–48 hours), toxin production is indicated by the development of lines of whitish precipitate which approximately bisect each of the four right angles formed by the paper strip and the line of microbial growth; the precipitate results from the combination of toxin and antitoxin following their diffusion from the line of growth and paper strip, respectively. Non-specific lines of precipitate may be formed, so that known negative and positive strains must be used as controls.

Several unknown strains, and the controls, may be examined on a standard-size Petri dish.

In a modified form of the test, an antitoxin-impregnated disc is placed at the centre of the plate, and the test strain(s), together with positive and negative controls, are inoculated in a circular pattern at a fixed radius; this form of test can be read after 16 hours [JCM (1997) *35* 495–498].

Diphtheria toxin can also be detected – directly from clinical specimens – by a PCR-based method which detects the A and B subunits of the toxin; a result can be available within hours of collecting the specimen. This assay is used in the Diphtheria Laboratory at the Centers for Disease Control in the USA [JCM (1997) *35* 1651–1655].

elementary body (EB) (1) See CHLAMYDIA. (2) An archaic name for a virion.

ELISA (enzyme-linked immunosorbent assay) A highly sensitive IMMUNOASSAY for specific antibodies or antigens. For the assay of specific antibodies, the antiserum (or other test material) is allowed to react with homologous antigens which have been adsorbed e.g. to the inner surface of a plastic tube; uncombined antibodies are removed by washing. Antibodies which have combined with the immobilized antigens are then detected by treating the test system with a *conjugate* (see CONJUGATION sense 2) consisting of anti-immunoglobulin antibodies each covalently linked to an enzyme of a particular type; the test system is washed free of any uncombined conjugate and examined for bound enzyme by incubating with an appropriate substrate and assaying for the products of enzymic cleavage.

Antigen may be detected or quantified e.g. by a procedure similar to that used for antibody except that the *antibody* is adsorbed, and enzyme is conjugated to anti-antigen antibody. In an alternative procedure, antigen is detected/quantified by its ability to sequester added antibody – the amount of antibody sequestered (an indirect measure of antigen) being assessed by measuring the amount of free antibody remaining in the test mixture; free antibody is measured by allowing it to bind to immobilized antigen, and detecting the bound antibody with enzyme-linked anti-Ig antibodies.

[Developments in ELISA and other solid-phase immunoassay systems: JIM (1986) *87* 1–125.]

ELISPOT assay (enzyme-linked immunospot assay) A type of assay, based on the ELISA principle, for detecting and quantifying specific molecules or cells. For example, blood has been assayed to quantify those T cells which respond to the tuberculosis-associated antigen ESAT-6 (q.v.) by secreting the cytokine interferon-gamma (IFN-γ; see INTERFERONS). Any ESAT-6-specific T cells (present in a blood sample in the well of an assay dish) react to (added) ESAT-6 antigen by secreting IFN-γ. Such IFN-γ (which is present in high concentration in the immediate vicinity of a secreting T cell) is detected by adding a conjugate consisting of an enzyme linked to a molecule that binds to IFN-γ; enzymic activity (on an appropriate substrate) then leads to the formation of a visible 'spot' on the floor of the well in the location of a secreting T cell. Hence, each spot corresponds to an ESAT-6-specific T cell. Enumeration of ESAT-6-specific T cells has been used in epidemiological studies (contact tracing) in the context of tuberculosis [Lancet (2001) *357* 2017–2021].

Ellobiophrya See PERITRICHIA.

elm butt rot See BUTT ROT.

Elokomin fluke fever See NEORICKETTSIA.

Elphidium See FORAMINIFERIDA.

elsinan A water-soluble, essentially linear glucan, consisting of blocks of $(1 \rightarrow 4)$-α-linked maltotriose units connected by $(1 \rightarrow 3)$-α-bonds, isolated from culture fluids of *Elsinoë leucospila*.

Elsinoë A genus of fungi (order DOTHIDEALES) which include some important plant pathogens – e.g. *E. veneta*, which causes purple-bordered, sunken grey lesions ('cane spot') on the canes of e.g. raspberry (*Rubus* sp). *Elsinoë* spp form uniascal locules at various levels in a pseudoparenchymatous ASCOSTROMA (the overwintering stage); the ascospores are septate, colourless or pale yellow. Anamorphs occur in the genus *Sphaceloma*.

***eltor* vibrio** See VIBRIO (*V. cholerae*).

elution (*serol., virol.*) The dissociation of an adsorbed particle.

Elysia A genus of marine molluscs which feed on siphonaceous green algae (e.g. CAULERPA, CODIUM) by piercing the cells and sucking out the cytoplasm; the algal chloroplasts are retained intact within the cells of the digestive gland (hepatopancreas) where they continue to photosynthesize for a time. The chloroplasts do not divide in the animal, and are sooner or later digested. [Book ref. 129, pp. 97–107.] (cf. CONVOLUTA.)

Elytrosporangium See STREPTOMYCES.

EM Electron microscope.

$E_{m,7}$ See REDOX POTENTIAL.

emarginate (*mycol.*) *Syn.* SINUATE (sense 2).

EMB agar (eosin–methylene blue agar) An agar-based medium used e.g. for the primary isolation of – and for differentiating between – enterobacteria. The medium contains (w/v): peptone (1%); lactose and sucrose (each at 0.5%) or lactose only (at 1%); K_2HPO_4 (0.2%); eosin Y (0.04%); methylene blue (0.0065%); agar (ca. 1.5%). Final pH: ca. 7.2. Colonies of *Escherichia coli* usually have dark centres and a greenish metallic sheen, those of *Enterobacter aerogenes* are usually mucoid, brownish in the centre, and only occasionally show a metallic sheen, while those of *Salmonella* and *Shigella* are usually colourless and translucent.

Embden–Meyerhof–Parnas pathway (EMP pathway; Embden-Meyerhof pathway; hexose bisphosphate pathway; glycolysis) A sequence of reactions in which glucose is broken down to pyruvate*. The EMP pathway occurs in a wide range of organisms, both prokaryotic and eukaryotic, and can operate under both aerobic and anaerobic conditions. By this pathway one molecule of glucose yields two molecules each of pyruvate and NADH and two (net) of ATP (from SUBSTRATE-LEVEL PHOSPHORYLATION); if glucose 6-phosphate is derived from GLYCOGEN, the net yield of ATP is 3 molecules. [See Appendix I(a).] In respiratory modes of metabolism (see RESPIRATION) the pyruvate is commonly converted to acetyl-CoA which enters the TCA CYCLE; NADH is oxidized via a respiratory chain (see ELECTRON TRANSPORT CHAIN). In fermentative modes (see FERMENTATION) the fate of pyruvate depends on species and/or conditions, but must always involve regeneration of NAD^+ from NADH by reduction of an endogenous substrate (see e.g. ALCOHOLIC FERMENTATION, HOMOLACTIC FERMENTATION, and Appendix III).

The initial step of the EMP pathway, the formation of glucose 6-phosphate from glucose, is catalysed by hexokinase in e.g. yeasts and some bacteria, while in other bacteria (e.g. enterobacteria, streptococci) glucose is phosphorylated during its uptake by a PTS (q.v.). *Phosphofructokinase* is a key enzyme in the EMP pathway, and the presence of this enzyme in an organism is generally taken as evidence that the EMP pathway occurs in that organism. In yeasts (and e.g. mammals) phosphofructokinase is inhibited by high levels of ATP and citrate, and is stimulated by high levels of AMP; in e.g. *Escherichia coli* phosphofructokinase is inhibited by phosphoenolpyruvate (but not by citrate), and is stimulated by ADP and GDP (but not by AMP). (See also ADENYLATE ENERGY CHARGE.) (In certain organisms fructose 6-phosphate is phosphorylated by a pyrophosphate:phosphofructose dikinase – see PYROPHOSPHATE.)

The EMP pathway functions not only in energy-yielding metabolism but also in supplying intermediates for biosynthesis: e.g., dihydroxyacetone phosphate can be reduced to glycerol phosphate for use in the biosynthesis of lipids; phosphoenolpyruvate may be drawn off for aromatic amino acid biosynthesis [Appendix IV(f)], 3-phosphoglycerate for cysteine, glycine and serine biosynthesis [Appendix IV(c)], and pyruvate for alanine, leucine and valine biosynthesis [Appendix IV(b)]. (See also GLUCONEOGENESIS.)

(See also ENTNER–DOUDOROFF PATHWAY; GLYCOSOMES; HEXOSE MONOPHOSPHATE PATHWAY; METHYLGLYOXAL BYPASS.)

*The terms 'EMP pathway' and 'glycolysis' are often used for the complete pathway for glucose degradation: e.g. glucose $\rightarrow$ lactate (i.e., HOMOLACTIC FERMENTATION) or glucose $\rightarrow$ ethanol (i.e., ALCOHOLIC FERMENTATION).

embedding See MICROTOME and ELECTRON MICROSCOPY.

embryonated egg (embryonating egg) An egg (usually a hen's or duck's egg) which contains a live, developing embryo. Such eggs are used in virology e.g. for the identification, culture, and/or assay of certain viruses (see e.g. POCK sense 2), and as a source of cells for TISSUE CULTURE.

An embryonated egg is prepared by incubating a fertile egg in a humid atmosphere for ca. 5–14 days at e.g. 37–39°C – the period and temperature of incubation depending e.g. on the region of the egg to be inoculated, the virus to be used, etc. Essentially, an egg is candled (see CANDLING) and inoculated (aseptically) with the virus via a small hole drilled in the shell; inoculations may be made into the yolk sac (egg pre-incubated ca. 5–8 days), the chorioallantoic membrane (CAM) (pre-incubation 11–13 days), the allantoic cavity (pre-incubation 9–12 days), or the amniotic cavity (pre-incubation 10–14 days). Following inoculation, the hole is sealed and the egg is incubated (at an appropriate temperature and humidity) for e.g. 1–7 days; the egg can then be opened and material removed for examination etc.

EMCV See CARDIOVIRUS.

Emericella A genus of fungi (order EUROTIALES) which include the teleomorph of *Aspergillus nidulans*. The ascocarp is a globose, red or violet cleistothecium containing spherical or ovoid evanescent asci; it is surrounded by a layer of *hülle cells*: thick-walled cells which develop from the hyphae that envelope the cleistothecium. The ascospores are typically reddish or violet.

Emericellopsis See EUROTIALES.

Emerson enhancement effect See ENHANCEMENT EFFECT.

emetine A toxic alkaloid, obtained from ipecacuanha root, used in the treatment of amoebic DYSENTERY. Emetine inhibits eukaryotic protein synthesis [Book ref. 14, pp. 453–454]; at clinical doses it is effective against trophozoites but not cysts. 2-Dehydroemetine is said to cause fewer clinical side-effects.

Emiliana See COCCOLITHOPHORIDS.

Emmentaler cheese See CHEESE-MAKING.

Emmonsia See ADIASPIROMYCOSIS.

Emmonsiella *Syn.* AJELLOMYCES.

EMP pathway EMBDEN–MEYERHOF–PARNAS PATHWAY.

empty magnification See MAGNIFICATION.

empyema Accumulation of pus in a body cavity (usually the lungs).

EMRSA Epidemic MRSA clone. Each distinct EMRSA is given a number (e.g. EMRSA-1, EMRSA-2).

EMS Ethylmethane sulphonate: see ALKYLATING AGENTS.

EMTA A pH buffer based on 3,6-endomethylene-1,2,3,6-tetrahydrophthalic acid.

Emtryl See DIMETRIDAZOLE.

emulcyan An extracellular amphipathic polymer produced by stationary-phase filaments of a strain of *Phormidium* – a benthic organism which normally adheres to surfaces by virtue of hydrophobic interactions, and which does not produce hormogonia. In the presence of cations, emulcyan acts as an emulsifying agent; it is believed to mask cell-surface hydrophobicity, thus serving to detach filaments and allow their dispersal. [FEMS Ecol. (1985) *31* 3–9.] (See also BIOSURFACTANTS.)

emulsan An extracellular, macromolecular bioemulsifier (see BIOSURFACTANTS) produced by *Acinetobacter calcoaceticus* RAG-1 during growth on e.g. hexadecane, ethanol or acetate. Emulsan (MWt ca. 10^6) consists of a polysaccharide (containing *N*-acetyl-D-galactosamine and an *N*-acetylhexosaminuronic acid) esterified with fatty acids [FEBS (1979) *101* 175–178]. Emulsan initially accumulates on the surface of the growing cells, forming a 'minicapsule', and is released into the medium e.g. as the cells approach the stationary phase; only the free extracellular form has emulsifying activity. An esterase which interacts with emulsan has been detected on the cell surface and free in the medium; this enzyme may play a role in the release of emulsan from the cell surface [JB (1985) *161* 1176–1181].

emulsification (1) (*bacteriol.*) The process of preparing a suspension of bacteria in water, saline etc; using a loop, bacterial growth is transferred to, and mixed with, a drop of the liquid on a slide.

(2) The formation of an emulsion, i.e., a suspension of droplets of one liquid in another with which it is non-miscible (e.g. oil in water). Substances which stabilize emulsions are called emulsifying agents. (See also BIOSURFACTANTS.)

enanthem A rash on a mucous membrane (cf. EXANTHEM).

enation (*plant pathol.*) A localized proliferation of leaf tissue caused e.g. by viral infection.

encapsidation During virion assembly: the process in which nucleic acid becomes incorporated in the viral capsid or – e.g. in certain bacteriophages – in a head/capsid precursor.

encephalitis Inflammation of the brain; it is often accompanied by inflammation of the brain meninges (*encephalomeningitis*, MENINGOENCEPHALITIS) or the spinal cord (*encephalomyelitis*) or both (*meningoencephalomyelitis*). Symptoms range from mild (fever, headache, malaise, myalgia, stiffness of the neck and back) to severe (irritability or somnolence, coma, death). Causes of encephalitis are many and varied. *Toxic encephalitis* results from severe toxaemia early in the course of an acute infectious disease. *Post-infective encephalitis* occurs after the onset of certain diseases (e.g. CHICKENPOX, MEASLES, MUMPS, WHOOPING COUGH). *Post-vaccinal encephalitis* may follow active immunization against e.g. rabies, smallpox, whooping cough, yellow fever. *Viral encephalitis* may be caused by any of a range of viruses: see e.g. B VIRUS; CALIFORNIA ENCEPHALITIS; EASTERN EQUINE ENCEPHALOMYELITIS; HERPES SIMPLEX; JAPANESE B ENCEPHALITIS; POWASSAN ENCEPHALITIS; RUSSIAN SPRING–SUMMER ENCEPHALITIS; ST LOUIS ENCEPHALITIS; SUBACUTE SCLEROSING PANENCEPHALITIS; VENEZUELAN EQUINE ENCEPHALOMYELITIS; WESTERN EQUINE ENCEPHALOMYELITIS. *Amoebic encephalitis* may involve e.g. certain species of ACANTHAMOEBA or NAEGLERIA – or BALAMUTHIA MANDRILLARIS.

(See also MENINGITIS, TALFAN DISEASE, TESCHEN DISEASE; TRANSMISSIBLE SPONGIFORM ENCEPHALOPATHIES.)

Encephalitozoon A genus of protozoa (class MICROSPOREA). *E. cuniculi* (= *Nosema cuniculi*) is parasitic (but generally non-pathogenic) in rodents and other animals, and has been recorded as the cause of a case of human encephalitis; it has also been isolated from eye, intestine, kidney, liver and peritoneum. [Mechanism of resistance to *E. cuniculi* in mice: J. Imm. (1984) *133* 2712–2719.] The spores of *E. cuniculi* (ca. 2.5 × 1.5 μm) can be isolated from the urine of infected animals.

E. intestinalis can cause travellers' diarrhoea [PCR-based identification of *E. intestinalis*: JCM (1998) *36* 37–40].

encephalomeningitis See ENCEPHALITIS.

encephalomyelitis See ENCEPHALITIS.

encephalomyocarditis virus (murine) See CARDIOVIRUS.

encystment The formation of a CYST.

end-point dilution assay (*virol.*) A method of quantifying a given type of virus in a given sample, i.e., measuring the 'infectivity' of that virus in the sample. In this method, infectivity is measured by determining the extent to which the sample must be diluted in order that a given volume of the dilution corresponds to the LD_{50}, $TCID_{50}$ or (more generally) the ED_{50} of the virus. Initially, log (or semi-log) dilutions of the sample are prepared. Each dilution is then examined, individually, by inoculating each of a number of similar test units with a fixed volume (dose) of that dilution; test units may be e.g. susceptible laboratory animals or tissue cultures which undergo detectable forms of degeneration when infected with the virus. After incubation, each test unit is examined. Animals or cultures inoculated with a dose from the lower dilutions (i.e. higher concentrations of virus) may react positively in every case; of those inoculated with a dose from the higher dilutions some may react positively while others are unaffected. The important dilution is that which corresponds to the ED_{50}; such a '50% end point' is not always obtained experimentally, but it may be calculated from the experimental data – provided that some test units have given a negative reaction. The 50% end-point dilution may be calculated by the Reed–Muench method or the Kärber method; however, these methods can be used only if the dilutions are in a regular logarithmic sequence, and only if a similar number of test units is used for each dilution. Examples of the use of these methods are given below with reference to the specimen data in the accompanying table.

END-POINT DILUTION ASSAY: specimen data

Column A: dilution of virus	Column B: mortalities in 5 test animals	Column C: cumulative mortalities	Column D: cumulative survivors	Column E: $C/(C+D)\%$
10^{-6}	0/5	0	9	0
10^{-5}	2/5	2	4	33.3
10^{-4}	4/5	6	1	85.7
10^{-3}	5/5	11	0	100

Reed–Muench method. The logarithm (to base 10) of the 50% end-point dilution lies between −5 and −4, and is obtained by linear interpolation – appropriate values from column E being substituted in the following equation:

$$\log_{10} 50\% \text{ end point} = -5 + \frac{50 - 33.3}{85.7 - 33.3}$$

The logarithm of the 50% end point is thus $10^{-4.68}$; since the antilog of 4.68 is ca. 48000 a dose from the 1/48000 dilution of the sample will contain the LD_{50}.

Kärber method. Substitutions are made in the following equation:

$$\log_{10} 50\% \text{ end point} = L_l - L(S - 0.5)$$

in which L_l = logarithm of lowest dilution tested; L = log interval between dilutions; S = sum of the proportion of positive reactions at each dilution; 0.5 is a constant. In the accompanying table, $S = 0 + 0.4 + 0.8 + 1.0 = 2.2$; hence, the required value is $10^{-4.7}$.

An *estimate* of the number of infectious virions in the original (undiluted) sample can be derived from the expression:

$$n = -\log_e(1 - p)$$

in which n = average number of infectious virions *per dose*; p = proportion of positive reactions in the test units inoculated from a given dilution; e = 2.7183. The value of n (calculated for one of the dilutions) multiplied by the dilution factor gives an estimate of the viral titre of the sample; such an estimate is less reliable than that made from a PLAQUE ASSAY.

end-point titration (*serol.*) A method of assaying the concentration of a specific antibody or antigen, in a given sample, by determining the highest dilution of the sample which gives a reaction regarded as positive under the conditions of the test. For example, in an agglutination test the end point may be regarded as the highest dilution which exhibits agglutination in the test system. (See also TITRE.)

end-product efflux The outflow, from a growing cell, of the end products of metabolism; in some cases such efflux involves end-product–proton SYMPORT (e.g. lactate–proton symport) capable of generating, or augmenting, a proton motive force (see CHEMIOSMOSIS). Such a mechanism for energy conversion may be particularly important e.g. in at least some obligately fermentative bacteria. In these organisms (which lack a respiratory chain) the pmf required e.g. for certain TRANSPORT SYSTEMS can be generated by ATP hydrolysis at a PROTON ATPASE – though this process is likely to consume much of the ATP generated by substrate-level phosphorylation; pmf generation by end-product efflux has been reported e.g. in *Lactococcus lactis* (formerly *Streptococcus cremoris*) [AvL (1984) *50* 545–555; JB (1985) *162* 383–390].

Endamoeba A genus of cyst-forming amoebae (order AMOEBIDA) parasitic in the gut of invertebrates (e.g. *E. blattae* in cockroaches). The nucleus lacks a karyosome.

endemic (1) Of a disease: commonly or constantly present in a given location or geographical region; typically, a relatively low proportion of the local population exhibits clinical signs of the disease. (2) Constantly present in a given location or geographical region.

Endemosarca A genus of endoparasitic, plasmodium-forming eukaryotic organisms which are parasitic in ciliates (*Colpoda* spp); they appear to be related to members of the PLASMODIOPHOROMYCETES. [Book ref. 147, pp. 203–206.]

endergonic reaction Any reaction requiring an uptake of energy. (cf. EXERGONIC REACTION.)

endobiotic Living *within* a host organism.

endocarditis Inflammation of the endocardium, i.e. the tissue lining the cavities of the heart.

Infective (*infectious*) *endocarditis* may be due to infection by any of a range of microorganisms – e.g. *Haemophilus* sp, *Staphylococcus aureus*, *S. epidermidis* (especially in patients with prosthetic heart valves), *Enterococcus faecalis*, viridans streptococci (see also NUTRITIONALLY VARIANT STREPTOCOCCI), *Neisseria gonorrhoeae*, *Candida*, *Aspergillus*, *Histoplasma*. Infrequently, the causal agent is one of the more recently described species of coagulase-negative staphylococci: *S. lugdunensis* [RMM (1995) *6* 94–100]. (See also CARDIOBACTERIUM.)

Infective endocarditis may be acute or subacute; *S. aureus* may cause either acute or subacute endocarditis. Infection occurs via the bloodstream. Organisms may gain access to the bloodstream e.g. during dental treatment (e.g. oral viridans streptococci), catheterization or insertion of an intrauterine contraceptive device (e.g. faecal streptococci), intravenous injection (e.g. staphylococci) etc.

Symptoms: e.g. fever, malaise, heart murmurs, weight loss, clubbing of fingertips, embolism; late symptoms of subacute bacterial endocarditis (SBE) include vasculitis, petechial rash, and (almost pathognomonic) small, tender, often bluish swellings on the skin (Osler's nodes). Damage to heart valves may lead to heart failure.

Endocarditis may occur as a complication of other infectious diseases – e.g. Q FEVER, SYPHILIS, TUBERCULOSIS. (See also RHEUMATIC FEVER.)

Endocarpon See VERRUCARIALES.

endocellulase See CELLULASES.

endocyst See CYST (b).

endocytobiosis *Syn.* CYTOBIOSIS.

endocytosis The ingestion of materials, by a cell, by PHAGOCYTOSIS and/or PINOCYTOSIS.

endodyogeny Asexual reproduction (found e.g. in *Chlamydomonas, Sarcocystis, Toxoplasma*) in which two daughter cells form within the parent cell (which is destroyed during the process).

endoenzyme An ENZYME which cleaves bonds within a polymer chain: see e.g. α-AMYLASES. (cf. EXOENZYME.)

endoflagellum See SPIROCHAETALES.

endogenote See MEROZYGOTE.

endogenous infection Infection by an organism which is part of the normal body microflora. (See e.g. CANDIDIASIS; cf. OPPORTUNIST PATHOGENS.)

endogenous retrovirus A retrovirus (see RETROVIRIDAE) which occurs, in provirus form, in the DNA of all normal cells of the host species or strain, and which is transmitted genetically via the germ line from one generation to the next. Endogenous retroviruses have been detected in a very wide range of vertebrates (including man [see e.g. JV (1986) *58* 955–959]); they are believed to have arisen by exogenous infection of germ line tissue in the (recent or possibly ancient) ancestors of the host. Many endogenous retroviruses are related to horizontally transmissible EXOGENOUS RETROVIRUSES. Most endogenous proviruses are transcriptionally silent. Many appear to be defective, but some can be activated under certain conditions (e.g. by host hormones, or by exposure to radiation or certain chemicals – e.g. BUdR), when expression may range from transcription of selected virus genes to production of complete virus particles which may be infectious or non-infectious, ECOTROPIC or XENOTROPIC, etc. (See also MOUSE MAMMARY TUMOUR VIRUS.)

endoglucanase See e.g. CELLULASES.

Endogonales An order of fungi (class ZYGOMYCETES) which form zygospores and in which asexual reproduction involves the formation of chlamydospores, azygospores or sporangiospores. Many species are involved in MYCORRHIZA formation (see also RHIZOSPHERE). Genera include *Acaulospora, Endogone, Gigaspora, Glomus* and *Sclerocystis*.

Endogone See ENDOGONALES.

Endolimax A genus of protozoa (order AMOEBIDA) parasitic in vertebrates and invertebrates. *E. nana* occurs in the human large intestine; it is non-pathogenic. Trophozoites of *E. nana* (ca. 6–12 μm diam.) contain numerous vacuoles and form blunt pseudopodia slowly; the single nucleus lacks peripheral chromatin but contains a large, eccentrically located karyosome. Cysts are usually ovoid, 4–8 × 8–16 μm, and contain four nuclei when mature; chromatoid bodies are usually absent.

endolithic Growing within rock, stone etc (cf. SAXICOLOUS). For example, certain cyanobacteria, lichens and microfungi can grow endolithically in environments too extreme (e.g. too hot or too cold) to allow normal EPILITHIC growth. Endolithic organisms (endoliths) have been categorized into three principal groups: *chasmoendoliths*, which colonize fissures and cracks in the rock; *cryptoendoliths*, which colonize structural cavities within porous rocks; and *euendoliths*, which actively penetrate the rock (e.g. by releasing substances which dissolve the rock). [J. Sed. Pet. (1981) *51* 475–478.] In e.g. hot desert regions, cyanobacteria and other bacteria are generally the dominant cryptoendoliths, while in the dry desert valleys of the Antarctic chasmoendolithic and cryptoendolithic lichens (mainly *Acarospora, Buellia, Lecanora* and *Lecidea* spp) are dominant [Science (1982) *215* 1045–1053]. These organisms generally inhabit a narrow zone below the rock surface, any fruiting structures being formed at the surface. (See also e.g. ARTHOPYRENIA; LECANORA; OSTREOBIUM.)

endolysin A PEPTIDOGLYCAN-degrading enzyme (e.g. LYSOZYME) encoded by many types of bacteriophage; the function of an endolysin is to lyse the host cell's envelope to release phage progeny. In many or all cases, endolysins (which lack a 'signal sequence') are enabled to cross the cytoplasmic membrane (to reach peptidoglycan) by means of a LYSIS PROTEIN.

endometritis Inflammation of the endometrium. In humans and animals it may result from infection of the endometrium by various opportunist pathogens (e.g. *Klebsiella pneumoniae, Pseudomonas aeruginosa*, group A streptococci). (See also CONTAGIOUS EQUINE METRITIS.)

endomitosis Chromosome replication without nuclear division. (cf. MITOSIS.)

Endomyces See ENDOMYCETALES.

Endomycetaceae See ENDOMYCETALES.

Endomycetales An order of mainly saprotrophic fungi of the ASCOMYCOTINA; the organisms include both unicellular and dimorphic species. Ascocarps are not formed. Asci (which develop from individual cells, or from cells within hyphae): prototunicate, cylindrical to spheroidal. Families: Dipodascaceae (genera: e.g. *Dipodascus, Phialoascus*), Endomycetaceae (formerly Ascoideaceae; genera: e.g. *Cephaloascus, Endomyces*), METSCHNIKOWIACEAE, SACCHAROMYCETACEAE.

Endomycopsis An obsolete fungal genus; species are now included e.g. in SACCHAROMYCOPSIS.

endomycorrhiza See MYCORRHIZA.

endonuclease A NUCLEASE which cleaves phosphodiester bonds *within* a nucleic acid strand. See also RESTRICTION ENDONUCLEASE; cf. EXONUCLEASE.

endonuclease III See BASE EXCISION REPAIR.

endonuclease S_1 (nuclease S_1, endonuclease S1, etc) A zinc-containing ENDONUCLEASE, first obtained from *Aspergillus oryzae*, which specifically cleaves ssDNA and ssRNA to yield mainly 5′-nucleoside monophosphates. It can cleave local regions of ssDNA in a dsDNA molecule.

endonucleobiosis SYMBIOSIS in which one symbiont occurs within the nucleus of the other (see e.g. HOLOSPORA). [Endonucleobiosis in ciliates: Int. Rev. Cytol. (1986) *102* 169–213.]

endoparasite See PARASITE.

endoparasitic slime moulds See PLASMODIOPHOROMYCETES.

endopectate lyase See PECTIC ENZYMES.

endopectin lyase See PECTIC ENZYMES.

endopeptidase See PROTEASES.

endoperidium See LYCOPERDALES.

endoperoxide synthase *Syn.* cyclo-oxygenase (see PROSTAGLANDINS).

endophloeodal (endophloedal; hypophloeodal) Growing *within* bark (cf. CORTICOLOUS).

Endophyllum See UREDINIOMYCETES.

endophyte (1) An organism parasitic partly or wholly within a plant.

(2) Any of certain fungi parasitic in grasses; such fungi include e.g. *Epichloë typhina*. Infection with certain endophytes can have beneficial and harmful effects on the growth and utilization of pasture grasses. [Artificial infection of grasses with endophytes: Ann. Appl. Biol. (1985) *107* 17–24.]

endoplasm See e.g. SARCODINA.

endoplasmic reticulum (ER) Within a eukaryotic cell: a system of membranes which ramifies through the cytoplasmic region and forms the limiting boundaries of compartments and channels whose lumina are completely isolated from the cytoplasm; the ER membrane is a protein-containing lipid bilayer. In some regions the outer surface of the ER membrane (i.e., that surface in contact with the cytoplasm) bears many ribosomes; such regions are called *rough* ER (RER), while those regions which lack ribosomes are called *smooth* ER (SER). A given compartment or channel may be bounded partly by rough ER and partly by smooth ER. Within the lumina of the ER occur e.g. the early stage(s) of glycosylation of proteins in the PALADE PATHWAY – typically the addition of mannose and/or glucosamine residues. (See also GOLGI APPARATUS; MICROSOME.)

endopolygalacturonase See PECTIC ENZYMES.

endopolygalacturonate lyase *Syn.* endopectate lyase (see PECTIC ENZYMES).

endopolygeny Asexual reproduction (found e.g. in *Chlamydomonas* and *Toxoplasma*) in which many daughter cells form within the parent cell (which is destroyed during the process).

endoral membrane (in *Paramecium*) See PARORAL MEMBRANE.

endoribonuclease See RNASE.

endosome (1) A phagosome or pinosome. (2) (*ciliate protozool.*) A Feulgen-positive body which occurs within the paramere of a heteromerous MACRONUCLEUS in some hypostomes. (3) In euglenoid flagellates: a type of NUCLEOLUS which persists, and divides, during mitosis.

endospore (1) A type of SPORE formed *intracellularly* by the parent cell or hypha. (cf. EXOSPORE.)

(a) (*bacteriol.*) Endospores are formed under conditions of nutrient limitation by species of e.g. *Bacillus, Clostridium, Coxiella, Desulfotomaculum, Sporolactobacillus, Sporomusa* and *Thermoactinomyces* (see also SPOROSPIRILLUM). In general, endospores are considerably more resistant than vegetative cells to heat, desiccation, antimicrobial agents and radiation; their irreversible inactivation can be ensured only by the procedures used for STERILIZATION (sense 1). Endospores can remain dormant for long periods of time: the age of some specimens may be measurable on a geological scale [Microbiology (1994) *140* 2513–2529]. Studies on the formation and germination of endospores have been carried out mainly with *Bacillus subtilis.*

Sporulation in Bacillus subtilis. When growth becomes limited through depletion of nutrients, the cell re-organizes its metabolism. This involves the induction and repression of certain genes – although there is no abrupt switch from active growth to active sporulation; during this *transition state* the decision is made to either (i) maintain the vegetative condition, with low-level metabolism but without cell division, or (ii) initiate sporulation. [Transition state in *B. subtilis*: PNARMB (1993) *46* 121–153.]

The cell's decision whether to sporulate or not is influenced by various environmental and intracellular signals which must be integrated and interpreted. For example, sporulation is inhibited if calcium levels drop to 2 μM; this reflects the need for calcium during development of the endospore, and may be seen as a mechanism that prevents sporulation from aborting at an intermediate stage.

The (unknown) signals that initiate sporulation appear to stimulate at least two types of sensor molecule which are designated kinase A and kinase B (KinA and KinB). On receipt of appropriate signals, these kinases undergo ATP-dependent autophosphorylation and initiate a so-called *phosphorelay* (see figure, part (a)); Spo0F appears to act as the main junction through which are channelled various external and internal signals. The key factor which determines the decision between ongoing vegetative growth and sporulation appears to be the intracellular level of Spo0A~P.

Spo0A~P is de-phosphorylated (i.e. negatively regulated) by the phosphatase Spo0E (see figure). Indirect regulation of Spo0A~P involves other phosphatases, the RapA and RapB proteins, which de-phosphorylate Spo0F~P; RapA and RapB are themselves regulated at the level of transcription by factors dependent on the cell's physiological state. Thus, whether or not sporulation occurs appears to depend on the opposing influences of kinases (promoting sporulation) and phosphatases (inhibiting sporulation) [TIM (1998) *6* 366–370].

During the process of sporulation, gene expression is regulated, at least partly, at the level of transcription – relevant genes being transcribed in a specific temporal sequence; such regulation involves certain SIGMA FACTORS (see figure).

Elucidation of the mechanism of sporulation has been assisted by the isolation of various mutants, each blocked at a specific stage of the process. Such *spo* mutants are designated according to the stage beyond which endospore development does not progress in the mutant cell; mutations which affect the same stage but which occur at different loci are designated A, B etc. Various mutations at *spo0* loci prevent the development of the asymmetrical septum. No stage I mutants have been detected. In general, *spo* mutations have little or no effect on vegetative growth.

The figure (part b) outlines the stages of development of an endospore.

Dormancy. Freshly formed endospores usually enter a period of DORMANCY in which little or no metabolic activity can be detected.

Activation. This is a commonly reversible process in which a dormant endospore prepares for subsequent germination and differentiation into a vegetative cell; it appears to involve changes in the configuration of the endospore's macromolecules – presumably early steps in the mobilization of metabolic potential. Activation may be brought about e.g. by a short period of sublethal heating, by 'ageing', or by subjecting the endospores to low pH or to certain chemicals. The mode of activation in nature is apparently unknown.

Germination is an irreversible process in which an endospore becomes metabolically active. The process is largely degradative; it involves e.g. hydrolysis and de-polymerization of certain constituents of the endospore, and is characterized by the release of dipicolinic acid, calcium ions, and the breakdown products of cortical peptidoglycan. Concurrently there is a loss of resistance to heat, and a fall in optical density.

Germination requires, or is promoted by, certain substances (*germinants*) such as L-alanine, L-proline, certain sugars (e.g. glucose), ions (e.g. potassium, calcium, manganese, strontium), and surfactants; a given germinant may be effective for the endospores of some species but not for those of others. Germination is also affected by factors such as pH, and may be promoted by mechanical damage to the spore wall.

Certain substances (e.g. D-alanine, sodium bicarbonate) inhibit germination of the endospores of certain species.

Outgrowth is the process in which a vegetative cell develops from a germinated endospore. (See also MICROCYCLE SPORULATION.)

(b) (*cyanobacteriol.*) See BAEOCYTE.

(c) (*mycol.*) See e.g. COCCIDIOIDES, OOSPORIDIUM, RHINOSPORIDIUM, SARCINOSPORON, TRICHOSPORON.

(2) (*mycol.*) (*syn.* endosporium) The innermost layer of a fungal spore wall.

endospore stains Bacterial endospores stain poorly (or not at all) with simple staining procedures in the cold. They can be stained e.g. by a modified form of ZIEHL–NEELSEN'S STAIN in which ethanol (only) is used for decolorization. In another method (Schaeffer–Fulton stain) an air-dried, heat-fixed smear is flooded with 0.5% (w/v) aqueous malachite green and heated to near-boiling for 5 min; the smear is then washed in running water,

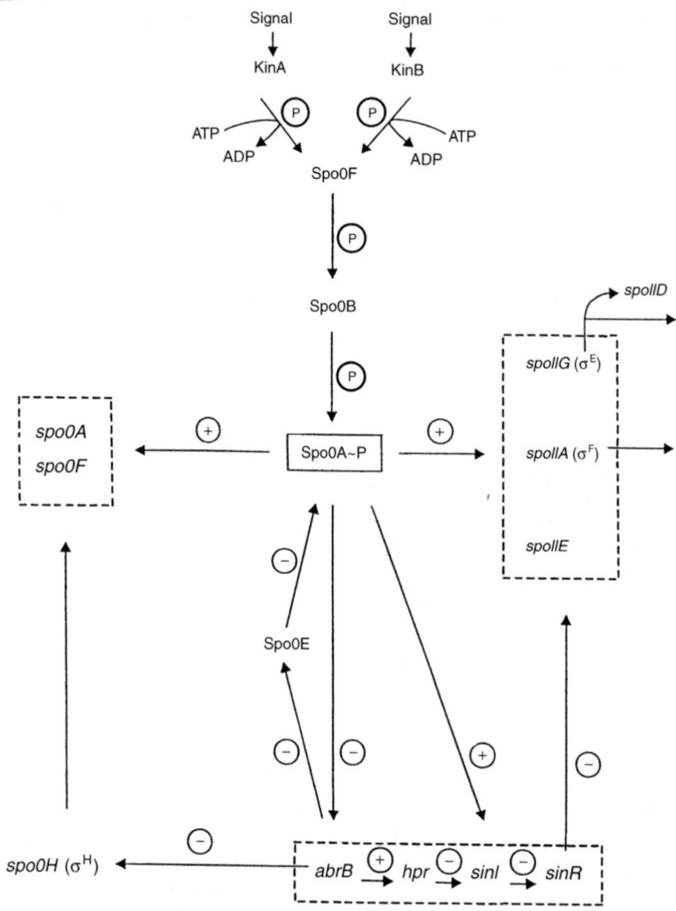

ENDOSPORE (a): simplified scheme for the regulation of gene expression during the initiation of sporulation in *Bacillus subtilis*.

The mechanism for initiating sporulation appears to recognize both external (environmental) and internal (intracellular) signals. These signals are believed to promote ATP-dependent autophosphorylation of certain kinases – shown in the diagram as KinA and KinB. The kinases transfer phosphate (symbol (P)) to a sequence of proteins: Spo0F → Spo0B → Spo0A (the so-called *phosphorelay*) [COGD (1993) *3* 203–212; JCBiochem. (1993) *51* 55–61]. The intracellular concentration of the phosphorylated form of Spo0A (Spo0A~P) appears to be the key factor in initiating sporulation; this protein is a transcription factor which promotes the expression of certain sporulation-specific genes (dashed box, centre right) and also regulates genes shown in the two other dashed boxes.

In the diagram, inhibitory influences are indicated by the symbol (−) ; (+) indicates that the expression of a given gene is promoted. Control is commonly exercised at the level of transcription.

Spo0A~P promotes the expression of genes *spo0F* and *spo0A* (dashed box, left) by a positive feedback loop [Mol. Microbiol. (1993) *7* 967–974]. Also, by repressing *abrB* (lower dashed box), Spo0A~P promotes the expression of *spo0H* – whose product (the SIGMA FACTOR σ^H) is required e.g. for the transcription of *spo0F* and *spo0A*. σ^H is produced at low level during vegetative growth, but its production is greatly enhanced following the initiation of sporulation. During sporulation, σ^H is needed for transcription of the *ftsZ* gene (see FtsZ in CELL CYCLE) from a special promoter, p2, which is distinct from the *ftsZ* promoter used during exponential growth; FtsZ is required for the formation of the asymmetrical septum.

Spo0E is a negative regulator of the phosphorelay [PNAS (1994) *91* 1756–1760]; it is an enzyme which de-phosphorylates (and thus inactivates) Spo0A~P and which may therefore help to prevent the initiation of sporulation until the cumulative effect of the various (extracellular and intracellular) signals dictates that sporulation is necessary.

Development of the asymmetrical septum (i.e. at a polar rather than a mid-cell location) initially involves the formation of a mid-cell Z ring (see CELL CYCLE) – i.e. as occurs during vegetative cell division. Later, this Z ring develops into a helical structure that extends into both poles of the cell and eventually forms two separate polar Z rings; these developments require the SpoIIE protein. Only one polar site is used – initiation of septation at one site blocks the other site. [Science (2002) *298* 1942–1946.]

During vegetative growth, SinR (encoded by *sinR* in the lower dashed box) represses expression of the *spoII* genes (dashed box, centre right). During sporulation, Spo0A~P represses *abrB* (lower dashed box) – thus (indirectly) inhibiting expression of SinR and helping to promote expression of the *spoII* genes; SinR is repressed at the protein–protein level, i.e. inhibited by the SinI protein. Note that Spo0A~P also directly promotes the transcription of *sinI*. (*Continued on page 273*.)

ENDOSPORE (a) (*continued*)

The product of *spoIIG* (dashed box, centre right) is a sigma factor, $\sigma^E (= \sigma^{29})$, which is needed for transcription of (i) the *spoIID* gene, and (ii) certain genes in the mother cell. SpoIID is one of several products (others are SpoIIM, SpoIIP and SpoIIB) which are involved in the degradation of peptidoglycan in the asymmetric septum – a necessary preliminary to the engulfment of the prespore by the mother cell (early stage II).

The product of *spoIIE* localizes on the prespore side of the asymmetrical septum where, through its phosphatase activity, it can activate σ^F [GD (1998) *12* 1371–1380]. The product of *spoIIA* (σ^F) is needed for transcription in the forespore.

Figure reproduced from *Bacteria*, 5th edition, Figure 7.14, page 154, Paul Singleton (1999) copyright John Wiley & Sons Ltd, UK (ISBN 0471-98880-4) with permission from the publisher.

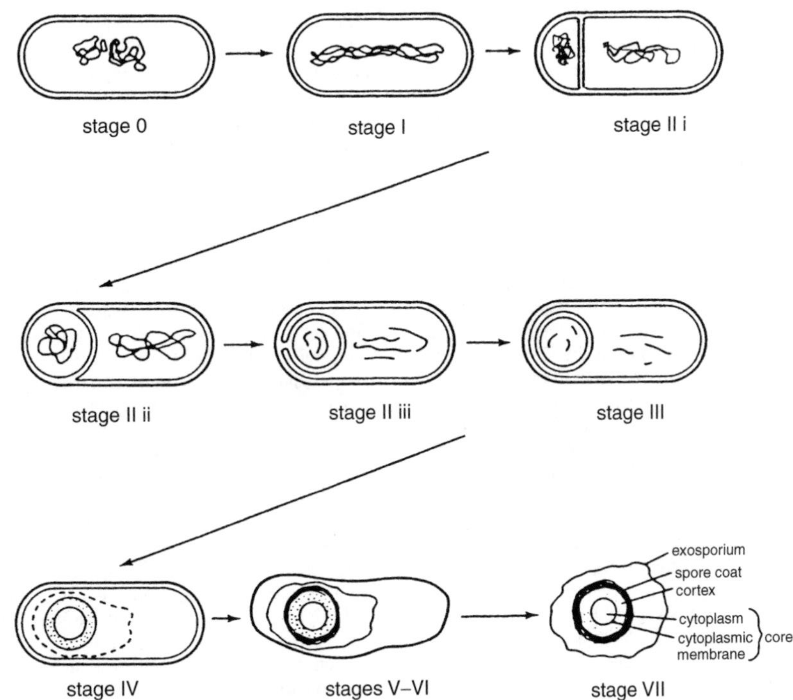

ENDOSPORE (b): stages in the development of an endospore (diagrammatic).

Stage 0. Prior to sporulation, a (vegetative) cell is said to be in stage 0.

Stage I. The cell contains an *axial filament* composed of two chromosomes. (Some authors consider that sporulation proper begins at stage II.)

Stage II. Initially, an asymmetric (polar) septum divides the cell into two protoplasts. The smaller protoplast (= *prespore*, *forespore*), which contains a single chromosome, is the precursor of the endospore; the other protoplast is referred to as the *mother cell*. Before further development, peptidoglycan in the asymmetric septum must be degraded; this involves SpoIID and other proteins (see legend of figure (a)). The cytoplasmic membrane of the mother cell then begins to engulf the smaller protoplast so that the latter subsequently becomes bounded by a double membrane; the final stages of this process (in *Bacillus subtilis*), in which the mother cell membrane fuses to complete the outermost membrane, may involve the SpoIIIE protein [PNAS (1999) *96* 14553–14558].

Stage III. Completion of development of the double membrane.

Stage IV. Modified PEPTIDOGLYCAN is laid down between the two membranes to form a rigid layer called the *cortex*. The peptidoglycan differs from that in vegetative cell walls in that e.g. it contains fewer cross-links, and in that the lactyl groups of some muramic acid residues form lactam rings by displacing the *N*-acetyl group. A loose protein envelope called the *exosporium* may begin to develop at about this time.

Stages V–VI. A multilayered protein *spore coat* is deposited outside the outermost membrane (stage V). The spore coat is rich in e.g. cystine residues, and it is strongly hydrophobic. The spore matures (stages V–VI) to develop its characteristic resistance to heat and its bright, refractile appearance under the light microscope; during this period, calcium dipicolinate (see DIPICOLINIC ACID) accumulates in the protoplast ('core'). The endospore reaches maturity.

Stage VII. The endospore is released by lysis of the mother cell. (In some species the mature endospore carries a loose outer membrane (*exosporium*) external to the spore coat; the origin and composition of the exosporium appear to be unknown.)

Figure reproduced from *Bacteria*, 5th edition, Figure 4.2, page 61, Paul Singleton (1999) copyright John Wiley & Sons Ltd, UK (ISBN 0471-98880-4) with permission from the publisher.

counterstained with safranin (0.5% aqueous) for ca. 30 sec, washed again with water, and blotted dry. Endospores stain green, cells red-brown.

endosporium *Syn.* ENDOSPORE (sense 2).

endosporulation In coccidia: SPORULATION which occurs *within* the host. Endosporulation occurs e.g. in all species of *Sarcocystis* and in *Cryptosporidium*, and it has been reported to occur in certain coccidia parasitic in fish (e.g. *Eimeria carpelli*).

ENDO-STAPH See TEICHOIC ACIDS.

Endostelium See PROTOSTELIOMYCETES.

endosymbiont An organism which lives within the cells or tissues of another organism in a symbiotic association (SYMBIOSIS sense 1). (cf. ECTOSYMBIONT.)

endothelial cells Cells which form a lining (one-cell thick) in the blood vessels, heart and lymphatic system.

endothelium–leukocyte adhesion molecule *Syn.* CD62E.

Endothia A genus of fungi of the order DIAPORTHALES. *E. parasitica* [ultrastructure: CJB (1982) *61* 389–399] causes CHESTNUT BLIGHT (q.v.). (See also PROTEASES.)

endothrix infection See DERMATOPHYTES.

endotoxic shock (septic shock) An often-fatal condition (mortality rate ≥50%) associated with blood-borne ENDOTOXINS of Gram-negative bacteria (Gram-positive bacteria are reported to provoke a similar or identical condition); in the USA alone, the number of deaths from endotoxic shock is over 100000 per year. Microbial components act on e.g. monocytes/macrophages and promote the secretion of CYTOKINES such as IL-1β and (particularly) TNF-α; these agents can recruit others, leading to symptoms such as fever, a marked fall in blood pressure and occlusion of blood vessels by white cells. Death may result from a progressive failure of organs.

Individuals with the *TNF2* allele appear to be particularly susceptible to endotoxic shock; this allele, which contains a variant form of the TNF-α promoter, correlates with enhanced production of TNF-α, spontaneously or under stimulation, both in vitro and in vivo. [Association of *TNF2*, a TNF-α promoter polymorphism, with septic shock susceptibility and mortality: JAMA (1999) *282* 561–568.]

Experimental vaccination against endotoxic shock with anti-endotoxin monoclonal antibodies has been tried unsuccessfully [RMM (1994) *5* 183–190].

Some recent candidate antibiotics – for example compound L-573,655 [Science (1996) *274* 980–982] – are directed against lipid A biosynthesis, and it seems possible that such antibiotics may reduce the risk of endotoxic shock because their activity decreases the levels of lipid A. A possible treatment for endotoxic shock follows the discovery of various molecules with endotoxin antagonist activity; one of these, E5531 (a synthetic analogue of lipid A from *Rhodobacter capsulatus*), has been found to block endotoxin challenge in human trials. [Antibacterial and anti-inflammatory agents that target endotoxin: TIM (1998) *6* 154–159.]

endotoxin (1) Traditionally, the LIPOPOLYSACCHARIDES (LPS) of Gram-negative bacteria; the toxic component of LPS is lipid A – although the polysaccharide moiety may contribute to toxicity by conferring water-solubility.

(2) Although many authors [e.g. RMM (1994) *5* 183–190] equate endotoxin with LPS, others [Book ref. 218, p 26] have used the term to refer to a complex that consists of LPS and outer-membrane protein(s); using this nomenclature, the protein associated with LPS is designated either lipid A-associated protein (LAP) or endotoxin-associated protein (EAP).

(3) Any microbial toxin which is released only on cell lysis. (See also DELTA ENDOTOXIN.)

(See also ENDOTOXIC SHOCK.)

endotrophic mycorrhiza See MYCORRHIZA.

Endotrypanum A genus of protozoa (family TRYPANOSOMATIDAE) parasitic in the erythrocytes of sloths (in Central and South America) and in sandflies (*Lutzomyia* spp) – which transmit the protozoon to sloths. One species (*E. schaudinni*) forms intraerythrocytic epimastigotes, the other (*E. monterogei*) forms intraerythrocytic trypomastigotes; in culture both species form promastigotes.

endotunica See BITUNICATE ASCUS.

endoxylanase See XYLANASES.

endozoite (tachyzoite) In certain coccidia: a stage within host tissues (see e.g. TOXOPLASMOSIS).

energy charge (EC) See ADENYLATE ENERGY CHARGE.

energy-transducing membrane Any biological membrane which contains components that are involved in the conversion of one form of energy (e.g. light, ATP) to another (metabolically usable) form of energy. Such membranes are involved e.g. in RESPIRATION and PHOTOSYNTHESIS; they include the bacterial CYTOPLASMIC MEMBRANE (and the CHLOROSOME and CHROMATOPHORE membranes), the inner membrane of the MITOCHONDRION, the THYLAKOID membranes of cyanobacteria and CHLOROPLASTS, and the PURPLE MEMBRANE in certain halobacteria. (In cyanobacteria, respiratory electron transport occurs in the cytoplasmic membrane and, in at least some species, may occur also (in addition to photosynthetic electron transport) in the thylakoids [Book ref. 75, pp. 199–218].) (See also CHEMIOSMOSIS.)

enFeLV viruses See FELINE LEUKAEMIA VIRUS.

enhancement effect (Emerson enhancement effect) In algae and green plants illuminated with light of wavelength above ca. 680 nm: an increase in PHOTOSYNTHESIS which occurs when the illumination is augmented with an additional beam of light of shorter wavelength. (cf. RED DROP.)

enhancer A *cis*-acting DNA sequence, present in the genomes of higher eukaryotes and of various animal viruses, which can greatly increase transcription; an enhancer can generally function in either orientation and at various distances (upstream or downstream) from a given promoter. (cf. UAS.) The enhancer of e.g. simian virus 40 (SV40) can function in a wide variety of animal cells, but enhancers of certain other viruses and of cellular genes show more or less strict specificity for particular cell or tissue types. Enhancer function requires the participation of *trans*-acting protein factors; for example, the SV40 enhancer apparently consists of multiple elements, each of which is recognized and bound by specific transcription factors [Book ref. 189, pp. 19–26]. Interactions between protein factors bound to enhancers and to other genetic elements (e.g. PROMOTER) may be important; in yeast, DNA looping between enhancer and promoter suggests that apposition of these elements is involved in enhancer action [NAR (2005) *33* 3743–3750].

enhancer (translational) See DOWNSTREAM BOX.

enhancer-dependent sigma factor (σ^{54}) See SIGMA FACTOR.

enniatins Cyclic DEPSIPEPTIDE ANTIBIOTICS, obtained e.g. from *Fusarium* spp, which behave as IONOPHORES (for monovalent cations) and which are active against certain Gram-positive bacteria, e.g. *Mycobacterium tuberculosis*. The enniatin molecule includes three residues of D-α-hydroxyisovaleric acid, each occurring between two residues of *N*-methyl-L-isoleucine (*enniatin A*), *N*-methyl-L-valine (*enniatin B*), *N*-methylleucine (*enniatin C*), or *N*-methylphenylalanine (*beauvericin*, produced by *Beauveria bassiana*); the residues are linked by alternating peptide and ester bonds.

enokitake (winter mushroom) *Flammulina velutipes*, a mushroom cultivated for food e.g. in Japan. Enokitake is grown on sawdust supplemented with rice bran; the mycelium is grown at 20–25°C, and fruiting is induced by lowering the temperature to 10–12°C. The cultivated fruiting bodies differ from those which develop in nature, the mushroom being whitish or cream-coloured with a long slender stipe. [Mycol. (1983) *75* 351–360.]

enoxacin See QUINOLONE ANTIBIOTICS.

enoyl-acyl carrier protein reductase (ENR) See DIAZABORINES and TRICLOSAN.

enriched medium Any (liquid or solid) basal MEDIUM which has been supplemented (enriched) with serum or blood etc – or with any of a range of nutrients or growth factors – in order to enable it to support or enhance the growth of particular type(s) of organism. Examples: BLOOD AGAR, serum sugars (see PEPTONE-WATER SUGARS). (cf. ENRICHMENT MEDIUM.)

enrichment Any process which increases the proportion of a given microorganism in a mixed population. For example, faeces from a typhoid case can be cultured in an enrichment medium (e.g. SELENITE BROTH) which suppresses growth of the normal faecal flora while permitting growth of the pathogen, *Salmonella typhi*; *S. typhi* is therefore 'enriched' so that its detection is facilitated. (See also DEFECTIVE INTERFERING PARTICLE.)

enrichment medium Any medium used for ENRICHMENT; it may contain substance(s) which encourage growth of the required organism and/or inhibit the growth of other type(s) of organism. (See e.g. SELENITE BROTH and TETRATHIONATE BROTH; cf. ENRICHED MEDIUM.)

Ensifer A genus of Gram-negative, aerobic, chemoorganotrophic and facultatively predatory, soil-inhabiting BUDDING BACTERIA; cells: rod-shaped (ca. 0.7–1.1 × 1.0–1.9 μm), motile by means of a subterminal tuft of 3–5 flagella. The cells of *E. adhaerens* (the type species) attach, end on, to a suitable prey bacterium (e.g. *Micrococcus luteus*), forming a palisade arrangement; under certain conditions the prey cell may be lysed. [IJSB (1982) *32* 339–345.]

ensiform Sword-like; narrow and pointed.

ensilage (1) The process of making SILAGE. (2) The silage itself.

Ent plasmid Any plasmid carrying genes for the ST and/or LT enterotoxins of ETEC (q.v.).

***entA* gene** (*Staphylococcus aureus*) See ENTEROTOXIN.

***entA–entG* genes** (in *Escherichia coli*) See SIDEROPHORES.

Entamoeba A genus of protozoa (order AMOEBIDA) characterized by the presence of beaded chromatin on the inner surface of the nuclear membrane. Most species (not *E. gingivalis*) form cysts which may contain CHROMATOID BODIES and glycogen-containing vacuoles. All species except *E. moshkovskii* occur as parasites/commensals in the alimentary tract in animals.

E. coli. A non-pathogen found in human faeces. Trophozoites are ca. 10–50 μm diam., contain numerous vacuoles, and slowly form blunt, granular pseudopodia; the single karyosome is located eccentrically, and the peripheral chromatin is coarsely and irregularly beaded. Mature cysts contain 8 nuclei.

E. gingivalis. A non-pathogenic, non-cyst-forming species found in the mouth in man and other animals. Trophozoites: ca. 5–35 μm diam., morphologically similar to those of *E. histolytica*.

E. hartmanni. A non-pathogenic species found in human faeces. It resembles *E. histolytica*, but both trophozoites (5–10 μm diam.) and cysts (<10 μm diam.) are smaller.

E. histolytica. A parasite and potential pathogen in the human intestine; it can cause amoebic DYSENTERY. Non-invasive trophozoites living in the gut lumen are ca. 20 μm in diam.; invasive forms which penetrate and develop in the gut mucosa are usually 20–50 μm diam. The cells contain few vacuoles, but may contain ingested red blood cells (an almost pathognomonic feature); finger-shaped, hyaline pseudopodia are produced rapidly. The (single) nucleus contains a small central karyosome, and the peripheral chromatin is finely and regularly beaded. In the gut lumen, some trophozoites expel their food particles and shrink to become rounded, non-motile *precysts*; a precyst secretes a resistant wall and becomes a spherical CYST (usually >10 μm and <20 μm diam.). A cyst initially contains a single nucleus, but two nuclear divisions result in four nuclei in the mature cyst; each nucleus contains a central karyosome. Young cysts may contain round-ended chromatoid bodies in the cytoplasm. Trophozoites, precysts and cysts may all be found in the faeces of the host, but only the cysts are infective for another host. Cysts are resistant to various environmental conditions – and to normal chlorination levels in water supplies – but are sensitive to temperatures >40°C and < −5°C and to desiccation. Cysts may be stained with iodine (cytoplasm yellow, glycogen vacuoles reddish-brown); trophozoites stain well with e.g. iron–haematoxylin after fixation in PVA (cytoplasm grey or pale violet, nuclear structures black). *E. histolytica* may be cultured anaerobically on media containing e.g. inspissated horse serum, rice starch, or egg albumen, together with bacteria or flagellate protozoa; it has also been cultured in sterile media. Strains have been classified on the basis of growth characteristics ('classical' strains grow best at 37°C, while atypical, less pathogenic 'Laredo-type' strains grow well at 37°C and at 25°C), or by the presence or absence of the enzyme phosphoglucomutase (present in organisms from patients with clinical amoebiasis, but not in those from asymptomatic carriers). (See also PYROPHOSPHATE.)

E. moshkovskii. A non-pathogen found in sewage; it resembles *E. histolytica* morphologically, but apparently occurs only in the free-living state.

E. paulista. A hyperparasite found in the cytoplasm of *Opalina ranarum* in frogs.

Other species include e.g. *E. bovis* (from cattle), *E. invadens* (pathogenic in reptiles), *E. muris* (from rats and mice), and *E. polecki* (from pigs).

[Biochemistry and functional morphology of *Entamoeba*: JP (1985) *32* 221–240.]

enteric Of or relating to the intestine (particularly the small intestine). 'Enteric bacteria' may refer to any or all bacteria normally found in the (small or large) intestine, or may refer specifically to members of the ENTEROBACTERIACEAE.

enteric adenoviruses See MASTADENOVIRUS.

enteric fever Any typhoid-like human salmonellosis – caused e.g. by *Salmonella paratyphi* (cf. PARATYPHOID FEVER), *S. typhimurium* etc. – as distinct from *Salmonella* FOOD POISONING. Some authors include TYPHOID FEVER among the enteric fevers.

enteric redmouth (red vent disease) A disease of rainbow trout (*Salmo gairdneri*) caused by *Yersinia ruckeri*. Symptoms include e.g. haemorrhagic lesions in and around the mouth and vent. (cf. RED MOUTH.)

enteric septicaemia (of fish) A disease of the channel catfish (*Ictalurus punctatus*) caused by *Edwardsiella ictaluri*; the disease has become an important problem in the intensive cultivation of catfish. Naturally infected fish often have a characteristic erosive lesion on the head ('hole-in-the-head'); chronic cases may involve enteritis, hepatitis, meningoencephalitis, etc. Primary infection can apparently occur via the gut or via the nares. [Pathogenesis: CJFAS (1986) *43* 36–42.] (See also FISH DISEASES.)

Enteridium See MYXOMYCETES.

enteritis Inflammation of the lining of any part of the intestine – particularly the small intestine; it may result from the same causes as GASTROENTERITIS. (cf. COLITIS.)

enteritis necroticans (necrotizing enteritis; 'pig bel'; *Darmbrand*) A rare, severe (often fatal), haemorrhagic enteric disease caused by ingestion of food (commonly pig meat) contaminated with *Clostridium perfringens* type C. (cf. FOOD POISONING.) The organisms adhere to the intestinal mucosa and produce the necrotizing β-toxin, causing necrosis of the mucosa. Diet influences disease development: the β-toxin is normally readily inactivated by digestive proteases; however, high-carbohydrate, low-protein diets reduce the secretion of proteases, and e.g. sweet potatoes contain a trypsin inhibitor which further lowers proteolytic activity – factors which allow the β-toxin to persist.

enteroadherent *E. coli* Diarrhoeagenic strains of *Escherichia coli* characterized by their ability to adhere to (mammalian) HEp-2 cells; strains can be differentiated on the basis of their patterns of adherence: 'diffusely adherent *E. coli*' (DAEC) adheres primarily as isolated, individual cells in a diffuse manner, while enteroaggregative *E. coli* (EAggEC: see EAGGEC) forms a pattern of scattered aggregates in which most of the cells are obviously grouped with other cells. (EPEC adheres to HEp-2 in a localized manner, forming microcolonies.) The category 'enteroadherent' has been discontinued: strains of HEp-2-adherent *E. coli* (excluding EPEC) are now divided into the two main groups EAggEC and DAEC (each of which, though still heterogeneous, is associated with distinct virulence factors and epidemiology) [RMM (1995) *6* 196–206].

(See also EAEC.)

enteroaggregative *E. coli* See EAGGEC.

Enterobacter A genus of Gram-negative bacteria of the ENTEROBACTERIACEAE (q.v.). Cells: ca. 0.6–1.0 × 1.2–3.0 μm, motile (generally 4–6 peritrichous flagella). Most strains possess fimbrial haemagglutinins [JGM (1983) *129* 2175–2180]. Typical reactions: indole −ve; MR −ve; VP +ve; citrate +ve; H_2S −ve; acid and gas from glucose at 37°C (no gas at 44.5°C); lactose +ve. Optimum growth temperature: ca. 30°C. GC%: 52–60. Type species: *E. cloacae*.

Enterobacter spp occur e.g. in the intestines of man and animals, in soil, water and sewage, in foods (e.g. dairy products), on plants etc. Most species can be opportunist pathogens, causing nosocomial infections (e.g. meningitis, pneumonia, septicaemia, UTI).

E. aerogenes. Lysine and ornithine decarboxylases +ve; arginine dihydrolase −ve; D-sorbitol +ve; non-pigmented. (It has been proposed that *E. aerogenes* be transferred to the genus *Klebsiella* as '*K. mobilis*' [see Book ref. 22, p. 463, 466 and 469].) (cf. AEROBACTER.)

E. agglomerans. A heterogeneous species comprising organisms alternatively included in *Erwinia herbicola* (see ERWINIA). Lysine and ornithine decarboxylases −ve; arginine dihydrolase −ve; lactose −ve. Some strains are yellow-pigmented.

E. cloacae. Lysine decarboxylase −ve; ornithine decarboxylase +ve; arginine dihydrolase +ve; D-sorbitol +ve; DNase −ve at 25°C. Non-pigmented. Naturally resistant to ampicillin.

E. sakazakii includes yellow-pigmented strains previously included in *E. cloacae*. D-Sorbitol −ve; DNase delayed +ve at 25°C.

Other species: *E. amnigenus*, *E. gergoviae*, *E. intermedium*. (*E. hafniae* and *E. alvei*: see HAFNIA; *E. liquefaciens*: see SERRATIA.)

[Book refs. 46, pp. 1173–1180; and 22, pp. 465–469.]

enterobacteria Bacteria of the ENTEROBACTERIACEAE.

Enterobacteriaceae A family of Gram-negative, asporogenous, facultatively anaerobic bacteria. Cells: rod-shaped, typically ca. 0.3–1.0 × 1–6 μm (cf. OBESUMBACTERIUM), motile (peritrichously flagellate – cf. TATUMELLA) or non-motile. Metabolism is chemoorganotrophic and may be respiratory or fermentative, depending e.g. on conditions. Most species can reduce nitrate to nitrite. All species are oxidase −ve and (except for *Shigella dysenteriae* serotype 1 and *Xenorhabdus nematophilus*) catalase +ve. Reproduction occurs by binary fission. Type genus: *Escherichia*.

Enterobacteria occur e.g. as parasites, pathogens or commensals in man and/or animals, as saprotrophs or pathogens in plants (see ERWINIA), or as saprotrophs (or faecal contaminants) in soil and water. Many are important opportunist causes of nosocomial infections.

Enterobacteria typically grow readily on a range of basal media (e.g. NUTRIENT AGAR). Media selective for enterobacteria commonly contain a sugar (often lactose), bile salts (as selective agents), and a pH indicator (see e.g. MACCONKEY'S AGAR).

Genera and species are traditionally differentiated on the basis of biochemical tests – e.g. the IMVIC TESTS, the ability to produce certain enzymes (e.g. UREASE, ARGININE DIHYDROLASE), to utilize particular sugars (mono- and disaccharides, polyols etc), to decarboxylate or deaminate certain amino acids (see e.g. DECARBOXYLASE TEST), to produce HYDROGEN SULPHIDE from thiosulphate (see e.g. TSI AGAR), etc. [Identification tests: Book ref. 184.] Reactions to such tests may be affected e.g. by mutation or by the presence of a metabolic PLASMID, and in some cases are strongly dependent on the reaction conditions (see e.g. YERSINIA). DNA HOMOLOGY ('DNA relatedness') studies have been widely used to indicate degrees of relatedness between strains, species and genera. However, the genera *Escherichia* and *Shigella*, for example, continue to be recognized even though they have been shown to be sufficiently closely related genetically to be regarded as a single genus (or even as a single species). [Book ref. 46, pp. 1108–1112.]

Many enterobacteria can be serotyped (see TYPING) on the basis of O ANTIGENS, H ANTIGENS, K ANTIGENS and/or VI ANTIGENS; cross-reactions between members from different genera are quite common (see e.g. CITROBACTER). All enterobacteria (except *Erwinia chrysanthemi*) possess the ENTEROBACTERIAL COMMON ANTIGEN. Colicin, bacteriophage and antibiotic TYPING may be used for particular genera. [Serology/serotyping in enterobacteria: Book ref. 68.]

Genera: CEDECEA; CITROBACTER; EDWARDSIELLA; ENTEROBACTER; ERWINIA; ESCHERICHIA; EWINGELLA; HAFNIA; KLEBSIELLA; KLUYVERA; KOSERELLA; LEMINORELLA; MOELLERELLA; MORGANELLA; OBESUMBACTERIUM; PROTEUS; PROVIDENCIA; RAHNELLA; SALMONELLA; SERRATIA; SHIGELLA; TATUMELLA; XENORHABDUS; YERSINIA. (See also SODALIS.)

enterobacterial common antigen (ECA; Kunin antigen) A heteropolysaccharide antigen found in the outer membrane in all wild-type strains of the Enterobacteriaceae (excluding *Erwinia chrysanthemi*) and e.g. in strains of PLESIOMONAS. It consists of alternating residues of (1 → 4)-linked *N*-acetyl-D-glucosamine and *N*-acetyl-D-mannosaminuronic acid which may be esterified with small amounts of e.g. palmitic and acetic acids. ECA may be linked to phospholipid (when it is poorly immunogenic), or it may be bound e.g. to the core region of LPS (when it is highly immunogenic); it is readily accessible to homologous antibodies in non-capsulated rough strains, but may be inaccessible e.g. in smooth, capsulated strains [JGM (1982) *128* 1577–1583].

enterobacterial repetitive intergenic consensus sequence See ERIC SEQUENCE.

enterobactin See SIDEROPHORES.

enteroblastic proliferation (*mycol.*) See CONIDIUM.

enterochelin See SIDEROPHORES.

enterochromaffin cells Cells in the mammalian gut epithelium which secrete certain hormones – e.g. SEROTONIN (a stimulator of intestinal secretion).

enterocin P A broad-spectrum BACTERIOCIN produced by *Enterococcus faecium* strain P13 and secreted by a *sec*-dependent pathway; it inhibits strains of e.g. *Listeria monocytogenes*, *Staphylococcus aureus*, *Clostridium perfringens* and *C. botulinum*. [Biochemical and genetic characterization of enterocin P: AEM (1997) *63* 4321–4330.]

enterococci (1) Members of the genus ENTEROCOCCUS. (2) A term which has been used very loosely to refer to some or all of the organisms referable to the genus ENTEROCOCCUS.

Enterococcus A genus of Gram-positive, asporogenous, chemoorganotrophic, facultatively anaerobic, coccoid bacteria; the genus was originally proposed to accomodate organisms previously classified as *Streptococcus faecalis* and *Streptococcus faecium* [proposal to transfer *S. faecalis* and *S. faecium* to the genus *Enterococcus* as *E. faecalis* and *E. faecium* respectively: IJSB (1984) *34* 31–34]. [Proposal to transfer other streptococci to this genus: *S. avium* (as *Enterococcus avium*), *S. casseliflavus* (*E. casseliflavus*), *S. durans* (*E. durans*), *S. faecalis* subsp *malodoratus* (*E. malodoratus*) and *S. gallinarum* (*E. gallinarum*): IJSB (1984) *34* 220–223.]

Members of the genus occur in the animal and human intestine; they can usually grow at 10°C and 45°C, and can grow in 6.5% sodium chloride and at pH 9.6. Metabolism: typically fermentative, but at least some strains can carry out respiratory metabolism when provided with haemin under aerobic conditions. (See also PYR.) GC% 37–45. Type species: *E. faecalis*.

E. faecalis. Organisms formerly classified as *Streptococcus faecalis* (Lancefield group D). Cells: ovoid, occurring singly or in pairs or short chains; motile strains are rare. Non-haemolytic or, rarely, β-haemolytic. *E. faecalis* can survive 60°C/30 min. Various sugars are fermented (e.g. D-tagatose, but not e.g. D- or L-arabinose), and acid is formed from glycerol both aerobically and anaerobically. Energy can be obtained from pyruvate, citrate and malate. Growth can occur in the presence of 0.02% sodium AZIDE, 0.04% tellurite, or 0.01% TETRAZOLIUM. Some strains form a PSEUDOCATALASE.

Some strains of *E. faecalis* can transmit/receive plasmids that confer resistance to antibiotic(s) (see CONJUGATION). A current problem is the acquisition of (plasmid-borne) resistance to VANCOMYCIN; the appearance of vancomycin-resistant enterococci (VRE) is important because enterococci are frequent causal agents of nosocomial bacteraemia. It has been estimated that ~15% of enterococci are resistant to vancomycin [EID (1998) *4* 239–249]. Vancomycin resistance may be conferred by one or more plasmid-borne *van* genes; *vanA* and *vanB* both confer resistance to vancomycin, while *vanA* also confers resistance to teichoplanin. (These two genes have been reported in both *E. faecalis* and *E. faecium*.) Multiplex PCR has been used for rapidly detecting *vanA* and *vanB* in presumptive colonies of VRE [JCM (1999) *37* 2090–2092].

[Vancomycin-resistant *E. faecium* with a VanB phenotype in a Warsaw hospital: JCM (2001) *39* 1781–1787.]

E. faecium. Organisms formerly classified as *Streptococcus faecium* (Lancefield group D). The organisms resemble *E. faecalis* in morphology and general characteristics, including resistance to heat, high pH and salinity; some strains are α-haemolytic. Acid is formed from arabinose but not from tagatose.

Unlike *E. faecalis*, *E. faecium* cannot use citrate, malate or pyruvate as sources of energy. Hydrogen peroxide may accumulate under aerobic conditions.

Enterocytozoon A genus of protozoa (phylum MICROSPORA); *E. bieneusi* has been reported from the gut of a human AIDS patient [JP (1985) *32* 250–254].

enterodiol See LIGNANS.

enterohaemorrhagic *E. coli* See EHEC.

enteroinvasive Refers to pathogenic organisms which invade the intestinal mucosa (see e.g. EIEC).

enteroinvasive *E. coli* See EIEC.

enterolactone See LIGNANS.

Enteromonadina See DIPLOMONADIDA.

Enteromonas See DIPLOMONADIDA.

Enteromorpha A genus of macroscopic green algae, closely related to ULVA spp, in which the thallus consists of hollow tubes one cell thick; species occur in coastal waters and on estuarine mudflats, often flourishing in conditions which also favour *Ulva lactuca*.

enteropathogenic Pathogenic for the intestine. (See also FOOD POISONING; for enteropathogenic *Escherichia coli* see EPEC.)

Enteropogon See TRICHOMYCETES.

enterotoxigenic (enterotoxinogenic) Refers to an organism that produces one or more ENTEROTOXINS. (See also ETEC; cf. ENTEROINVASIVE and ENTEROPATHOGENIC.)

enterotoxigenic *E. coli* See ETEC.

enterotoxin A TOXIN which acts on the intestine, either when ingested or when produced by an organism within the intestine (see e.g. CHOLERA TOXIN and ETEC). [Enteric bacterial toxins (review): MR (1996) *60* 167–215.]

The enterotoxins of *Staphylococcus aureus* (A, B, C1–C3, D and E) are SUPERANTIGENS, and their ability to stimulate the production of cytokines may account for at least some of the observed symptoms of staphylococcal FOOD POISONING; gut epithelial cells express MHC class II molecules and may present enterotoxin to submucosal T cells [MR (1996) *60* 195]. [See also Clinical Microbiology Reviews (2000) *13* 16–34.]

Staphylococcal enterotoxins are small proteins (~25–30 kDa) with similar biological activities; they may be encoded by chromosomal, plasmid-borne or phage-borne genes – in at least some strains, the gene encoding enterotoxin A (*entA*) is phage-borne. Each toxin is a single polypeptide chain which is resistant to many proteolytic enzymes, and can generally withstand boiling for up to 30 minutes. The toxins are produced during active growth.

(See also TSST-1 under TOXIC SHOCK SYNDROME.)

Enterotube II See MICROMETHODS.

Enterovioform See 8-HYDROXYQUINOLINE.

Enterovirus A genus of viruses (family PICORNAVIRIDAE), most of which replicate primarily in the mammalian intestinal tract – although subsequently other tissues may become infected. The virions are stable at pH 3–9 and at or below room temperature, and are resistant to e.g. ethanol (70%), Lysol (5%), quaternary ammonium compounds, detergents and bile salts; they are generally inactivated by heat (e.g. 50°C), formaldehyde (3%), HCl (0.1 N), or chlorine (ca. 0.3–0.5 ppm free residual Cl_2). They can be protected against heat inactivation by 1 M $MgCl_2$. Many enteroviruses can be propagated in primary cultures of human or monkey kidney cells and in some cell lines (e.g. HeLa, Vero, WI-38).

Most human enteroviruses are classified into one of three groups (see below) on the basis of e.g. pathogenicity and host range; enterovirus infections are commonly or usually subclinical, but may result in various diseases ranging from mild to severe or fatal [review: Book ref. 148, pp. 739–794].

Coxsackieviruses (first isolated in Coxsackie, New York) are generally pathogenic for newborn mice: *group A* coxsackieviruses cause a widespread inflammation of skeletal muscles (myositis) resulting in flaccid paralysis, *group B* coxsackieviruses cause a focal myositis with necrotizing inflammation of fatty tissues and e.g. encephalitis, pancreatitis, etc. In humans, coxsackieviruses can cause e.g. aseptic MENINGITIS; BORNHOLM DISEASE; HAND, FOOT AND MOUTH DISEASE; HERPANGINA; paralysis (very rare); pericarditis and myocarditis (mainly coxsackieviruses B1–B5) which may result in chronic cardiovascular disease; PHARYNGITIS; PNEUMONIA; URTIs (e.g. A21, = 'Coe virus', can cause a disease resembling the COMMON COLD). (The causal agent of SWINE VESICULAR DISEASE is apparently closely related to human coxsackievirus B5.)

Echoviruses ('enteric cytopathogenic human orphan viruses') are generally non-pathogenic in newborn mice; in humans they can cause e.g. aseptic MENINGITIS, encephalitis, paralysis, URTIs, etc.

Polioviruses are the causal agents of POLIOMYELITIS in man. There are three serotypes: Brunhilde (type 1), Lansing (type 2), and Leon (type 3). [Molecular biology of polioviruses: Book ref. 162; antigenic structure of polioviruses: JGV (1986) *67* 1283–1291.]

Unclassified *human enteroviruses* may cause e.g. bronchiolitis (type 68); ACUTE HAEMORRHAGIC CONJUNCTIVITIS (type 70); MENINGITIS and paralysis (types 70 and 71), and hepatitis (type 72, = HEPATITIS A virus).

Porcine enteroviruses cause e.g. SWINE VESICULAR DISEASE and TESCHEN DISEASE. In breeding animals porcine enteroviruses ('SMEDI viruses') can cause *s*tillbirth, *m*ummification of the fetus, *e*mbryonic *d*eath, and *i*nfertility, depending e.g. on the period of gestation at which infection occurs.

Other animal enteroviruses include bovine enteroviruses, simian enteroviruses, THEILER'S MURINE ENCEPHALOMYELITIS VIRUS, and the causal agents of AVIAN ENCEPHALOMYELITIS and DUCK VIRUS HEPATITIS.

Entner–Doudoroff pathway (ED pathway) A metabolic pathway [Appendix I(c)] for the degradation of glucose in a wide range of bacteria (e.g. species of *Acetobacter*, *Agrobacterium*, *Pseudomonas*, *Rhizobium*, *Serratia*, *Xanthobacter*, *Xanthomonas* and *Zymomonas*); some bacteria (e.g. *Escherichia coli*) which metabolize glucose mainly via the EMBDEN–MEYERHOF–PARNAS PATHWAY can metabolize gluconate and other aldonates via reactions of the ED pathway.

The fates of pyruvate and glyceraldehyde 3-phosphate (GAP) produced in the ED pathway depend on organism and conditions. In some bacteria GAP may be converted to pyruvate via phosphoenolpyruvate, as in the latter part of the EMP pathway [Appendix I(a)]; in pseudomonads, GAP is recycled (via fructose 1,6-bisphosphate, fructose 6-phosphate and glucose 6-phosphate – see GLUCONEOGENESIS) to 6-phosphogluconate (which can enter the ED pathway). Some of the 6-phosphogluconate may be diverted into the HEXOSE MONOPHOSPHATE PATHWAY. In aerobic bacteria, pyruvate can generally be metabolized via acetyl-CoA and the TCA CYCLE. In (fermentative) ZYMOMONAS species, pyruvate is largely decarboxylated to acetaldehyde – which is then reduced to ethanol.

In addition to the (phosphorylative) pathway for glucose metabolism, many pseudomonads can oxidize glucose directly (i.e. without phosphorylation) to gluconate – see EXTRACYTOPLASMIC OXIDATION. A proportion of the resulting gluconate is taken into the cell by specific transport systems and then phosphorylated to form 6-phosphogluconate (which can then be metabolized via the ED pathway and/or the HMP pathway).

[ED pathway (history, physiology, molecular biology): FEMS Reviews (1992) *103* 1–28. ED pathway in *E. coli*: JB (1998) *180* 3495–3502.]

Entodiniomorphida An order of protozoa (subclass VESTIBULIFERIA) in which the cells are naked except for tufts or bands of oral and (often) somatic syncilia. STOMATOGENESIS is apparently apokinetal. In some species the pellicle is drawn out into terminal spikes or spines. The organisms occur e.g. in the alimentary canal in herbivorous mammals. Genera: e.g. DIPLODINIUM, ENTODINIUM, *Epidinium*, *Metadinium*.

Entodinium A genus of ciliates (order ENTODINIOMORPHIDA) related to DIPLODINIUM. Typically, the anterior pole of the cell is flattened and bears a conspicuous tuft of cilia; there is no DZM, and the posterior end of the cell may or may not be drawn out into spines.

entomo- Prefix denoting insect–e.g. entomopathology: the study of INSECT DISEASES; entomopathogenic: pathogenic for insects.

entomogenous Growing in or on insects.

entomopathogenic Pathogenic for insects (see INSECT DISEASES).

Entomophaga See ENTOMOPHTHORALES.

Entomophthora A genus of fungi (order ENTOMOPHTHORALES) which include saprotrophs and parasites of insects. The vegetative thallus is commonly an aseptate or septate mycelium with a tendency to fragment; the 'conidia' are forcibly discharged. AZYGOSPORES are formed by some species. [Key to *Entomophthora* (sens. lat.): BBMS (1982) *16* 113–143; infection of houseflies by *E. muscae*: TBMS (1983) *80* 1–8; ultrastructural studies of primary spore formation and discharge in *Entomophthora*: J. Inv. Path. (1986) *48* 318–324.]

Entomophthorales An order of fungi (class ZYGOMYCETES) which include a number of parasites of insects and mammals. Asexually derived spores – often called 'conidia' – are forcibly discharged in members of the families Basidiobolaceae and Entomophthoraceae. (Members of the ZOOPAGALES are sometimes included as a third family in this order.) Genera include *Basidiobolus* (see also ZYGOMYCOSIS and EQUINE PHYCOMYCOSIS) and (in the family Entomophthoraceae) *Conidiobolus* (see also ZYGOMYCOSIS), *Entomophaga*, ENTOMOPHTHORA, ERYNIA, *Gonimochaete* (see NEMATOPHAGOUS FUNGI), and *Zoophthora*. (See also LOBOA and HYPHAL BODY; cf. BLASTOCYSTIS.)

entomophthoromycosis See ZYGOMYCOSIS.

Entomopoxvirinae (entomopoxviruses, EPVs) A subfamily of viruses, family POXVIRIDAE, which infect insects – including members of the Coleoptera (beetles etc), Diptera (flies etc), Lepidoptera (butterflies and moths), and Orthoptera (locusts, crickets, grasshoppers etc). EPV virions are brick-shaped or ovoid, ca. 300–450 × 170–250 nm, with a characteristic 'mulberry-like' appearance in negatively stained preparations – the globular surface 'units' apparently being folds of the outer membrane. The viruses multiply in the cytoplasm of (mainly) leucocytes and adipose cells of the host. Late in the replication cycle infected cells may contain one or two types of inclusion body: *spheroids* (= 'spherules', each consisting of mature virions occluded within a matrix of protein, *spheroidin*) and *spindles* (virus-free, spindle-shaped structures composed of a protein which is apparently distinct from spheroidin). A new host is infected when it ingests a spheroid and the virions are released in the alkaline contents of the gut. EPVs apparently vary in their pathogenicity, but at least some can cause diseases ('spheroidoses') in their hosts; they are currently little used for biological control of insect pests.

Three genera of EPVs have been proposed on the basis of virion morphology, host range, and genome MWt [Book ref. 23, pp. 44–46]; members of different genera show no serological relationship either to each other or to other poxviruses.

Genus A (coleopteran EPVs). Type species: *Melolontha melolontha* (cockchafer) EPV. Virion: ovoid, ca. 450 × 250 nm, containing a kidney-shaped (unilaterally concave) core with one lateral body (cf. POXVIRIDAE) and bearing surface globular units ca. 22 nm diam.; DNA MWt: ca. 170–240 × 10^6. Similar viruses have been isolated from other coleopteran insects – e.g. *Phyllopertha horticola* (the garden chafer). Spindles are formed by most of these viruses.

Genus B (lepidopteran and orthopteran EPVs). Type species: *Amsacta moorei* EPV. Virion: ovoid, ca. 350 × 250 nm, containing a cylindrical core and a 'sleeve-shaped' lateral body, and bearing surface globular units ca. 40 nm diam.; DNA MWt: ca. 132–142 × 10^6. Similar viruses have been isolated from various lepidopteran insects – e.g. *Choristoneura biennis* (the eastern spruce budworm) and other *Choristoneura* spp – and from the orthopteran *Melanoplus sanguinipes* (a short-horned grasshopper). Spindles are formed by most of these viruses.

Genus C (dipteran EPVs). Type species: *Chironomus luridus* EPV. Virion: brick-shaped, ca. 320 × 230 × 110 nm, with a biconcave core and two lateral bodies; DNA MWt: ca. 165–250 × 10^6. Similar viruses have been isolated from other dipteran insects, including other *Chironomus* spp (non-biting midges) and *Aedes aegypti* (see AEDES). Spindles are not formed in cells infected with any of the viruses in this genus.

[Review: AVR (1984) *29* 195–213.]

entomopoxviruses See ENTOMOPOXVIRINAE.

Entomosporium See MELANCONIALES.

entomotoxin A TOXIN effective against insects.

Entorrhiza A genus of smut fungi (order USTILAGINALES) which are parasitic on the roots of rushes and sedges.

Entosiphon A genus of EUGLENOID FLAGELLATES closely related to PERANEMA.

entry exclusion *Syn.* SURFACE EXCLUSION.

Entyloma A genus of smut fungi (order USTILAGINALES) which form pale-coloured spores; the host plants range from rice to spinach.

enucleated cell A CYTOPLAST, or a (eukaryotic) cell in which the nucleus has been inactivated by a 'chemical enucleator'.

enumeration of microorganisms See COUNTING METHODS.

***env* gene** See RETROVIRIDAE.

EnvA See CELL CYCLE (b).

envelope (1) (peplos) (*virol.*) A lipoprotein membrane which forms the outermost layer of the virion in certain ANIMAL VIRUSES, PLANT VIRUSES and BACTERIOPHAGES. Animal virus envelopes have been the most extensively studied, and the account below refers to them.

In general, the lipid components of the viral envelope are derived from host cell membranes, while the protein components (typically glycoproteins) are virus-encoded; many of the proteins project from the surface as 'spikes' or *peplomers*. Envelope proteins play important roles in the infection process; they are responsible for the attachment of the virion to receptor sites on the host cell surface, and they bring about the release of the nucleocapsid into the host cell cytoplasm by triggering fusion between the envelope and host membranes. In most enveloped animal viruses (but not in e.g. PARAMYXOVIRUSes) the activity of the membrane-fusing envelope proteins requires a low pH (ca. 5–6), and hence in these viruses fusion generally cannot occur between the viral envelope and the host plasmalemma in vivo. Such viruses are taken up (intact) by the host cell by receptor-mediated endocytosis, and only when the contents of the endosome become sufficiently acidic does fusion occur between the viral envelope and the endosomal membrane – resulting in the release of the nucleocapsid into the cytoplasm. [Review of endocytosis of enveloped viruses: Bioch. J. (1984) *218* 1–10.]

In many enveloped viruses (e.g. influenzaviruses) the individual virions do not share a common structure or composition; for example, the envelope glycoproteins in influenzaviruses do not occur in a fixed number in all the virions. By contrast, a defined structure is found in the virions of alphaviruses and flaviviruses [see e.g. Cell (2001) *105* 5–8].

In most cases a virus acquires its envelope (during virion assembly) by a process of *budding* (cf. FLAVIVIRIDAE). The virus-encoded envelope proteins each contain an anchor peptide (see SIGNAL HYPOTHESIS) and become anchored in the appropriate cell membrane (e.g. endoplasmic reticulum, plasmalemma, nuclear membrane – according to virus); these proteins then interact with proteins in the preformed nucleocapsid, causing the associated region of membrane to wrap around the nucleocapsid. This region of membrane eventually becomes detached from the host membrane, completing the assembly process.

Enveloped virions are typically readily inactivated by lipid-disrupting reagents and procedures: e.g. organic solvents such as ether and chloroform, detergents, heat, etc.

[Membranes in animal viruses: Book ref. 148, pp. 45–67.]

(2) (*bacteriol.*) *Syn.* CELL ENVELOPE.

environmental clean-up See BIOREMEDIATION.

environmental pollution See e.g. BIOREMEDIATION, GREENHOUSE EFFECT, SEWAGE TREATMENT, WATER SUPPLIES and XENOBIOTIC. (See also AEROSOL and AIR.)

enviroxime (2-amino-1-(isopropylsulphonyl)-6-benzimidazolyl-phenyl ketone oxide) An ANTIVIRAL AGENT which specifically inhibits rhinovirus replication in cell cultures; it does not appear to be effective in preventing or treating human rhinovirus infections in vivo.

EnvZ See OSMOREGULATION.

enzootic Refers to an ENDEMIC animal disease.

enzootic abortion (enzootic ovine abortion) Abortion in sheep caused by *Chlamydia psittaci*. Laboratory diagnosis may involve e.g. detection by microscopy of elementary bodies in stained placental smears, or culture of placental material [VR (1983) *113* 413–414]. (cf. INFECTIOUS ABORTION.)

enzootic bovine leukaemia *Syn.* BOVINE VIRAL LEUKOSIS.

enzootic ovine abortion See ENZOOTIC ABORTION.

enzootic pneumonia of calves See CALF PNEUMONIA.

enzyme A protein which acts as a highly efficient and specific biological catalyst. (Certain RNA molecules are capable of catalytic activity in the absence of protein, and these are sometimes called enzymes (cf. RIBOZYME); the account below refers specifically to *protein* catalysts.) An enzyme increases the *rate* of a (thermodynamically feasible) reaction by decreasing the activation energy; it cannot alter the equilibrium constant of a reaction. Many enzymes require non-protein COENZYMES, PROSTHETIC GROUPS, or metal ions/atoms for activity. (See also ZYMOGEN.)

Most enzymes have trivial names which usually indicate the nature of the substrate(s) and/or the reaction catalysed (e.g. *urease* splits urea, *alcohol dehydrogenase* dehydrogenates alcohol). However, a systematic scheme for the classification and precise nomenclature of enzymes has been established by the IUB Commission on Enzymes. The *classification* involves the division of all known enzymes into six general classes (numbered 1–6): 1 = OXIDOREDUCTASES; 2 = TRANSFERASES; 3 = HYDROLASES; 4 = LYASES; 5 = ISOMERASES; 6 = LIGASES. According to the nature of the reaction catalysed, each class is divided into (numbered) subclasses and sub-subclasses; within a given sub-subclass, each individual enzyme is given a specific (arbitrarily assigned) serial number. Thus, any enzyme has a unique classification number (EC number) consisting of four figures. For example, pyruvate kinase has the number EC 2.7.1.40, where 'EC' = Enzyme Commission, '2' means that the enzyme is a transferase, '7' that a phosphate group is transferred, and '1' that an alcohol group accepts the phosphate; '40' is the arbitrarily assigned serial number for pyruvate kinase. The systematic *nomenclature* is intended to reflect the nature of the substrate(s), the reaction catalysed, and any coenzyme involved. Thus, e.g., 'alcohol dehydrogenase' (EC 1.1.1.1) is *alcohol:NAD oxidoreductase*, 'pyruvate kinase' is *ATP: pyruvate 2-O-phosphotransferase*.

Microbial enzymes have a number of commercial applications, although their use is still restricted in many cases by e.g. economic factors, stringent safety regulations, and problems of finding suitable enzymes which are stable under the required conditions. Enzymes have a number of advantages over purely chemical processes: they can, under mild conditions, efficiently catalyse reactions which might otherwise require extreme conditions of e.g. temperature, pressure, pH etc; they are usually highly specific (and stereospecific); and they can (at least in theory) be extensively re-used. Either whole microbial cells or purified enzymes may be used. Purified enzymes may be costly, but they generally convert a higher proportion of the substrate (living cells may use some for maintenance energy and biomass production), and fewer side-reactions generally occur; however, isolated enzymes are often unstable. Enzymes from thermophiles are considerably more stable than those from mesophiles and are preferred when available. (See also IMMOBILIZATION sense 1.) For examples of commercial applications of microbial enzymes see AMYLASES, ASPARAGINASE, CATALASE, CELLULASES, β-GALACTOSIDASE, GLUCOSE ISOMERASE, GLUCOSE OXIDASE, GLYCEROKINASE, INVERTASE, LIPASE, PECTIC ENZYMES, PROTEASES, STEROID BIOCONVERSIONS.

[Industrial and diagnostic enzymes: PTRSLB (1983) *300* 237–434.]

Enzymes from 'extremophiles' are reported to have unique structures and/or to be stabilized by factors such as thermoprotectants; these enzymes are valuable for studies on enzyme stability, and they have potential applications in biotechnology [TIM (1998) *6* 307–314].

In certain pathogenic microorganisms enzymes may play an important role in pathogenesis – see e.g. HYALURONATE LYASE, LECITHINASE, PECTIC ENZYMES, SPHINGOMYELINASE, STREPTOKINASE.

enzyme I, II and III See PTS.

enzyme-catalysed interesterification See INTERESTERIFICATION.

enzyme-coupled probes Small, biotinylated probes labelled with a STREPTAVIDIN–alkaline phosphatase conjugate *prior to* hydridization [NAR (1999) *27* 703–705].

enzyme immunoassay (EIA) Any IMMUNOASSAY in which an enzyme (e.g. a peroxidase) is used (as a marker) to indicate the presence of specific antigens, antibodies etc: see e.g. ELISA and IMMUNOPEROXIDASE METHOD.

enzyme-linked immunosorbent assay See ELISA.

Eocyta See ARCHAEBACTERIA.

Eocytes A sub-category of PROKARYOTES which share with the eukaryotes a similar sequence of 11 amino acids in the (protein synthesis) elongation factor EF-1α, suggesting that the eocytes and eukaryotes are sister taxa with a common ancestor [Science (1992) *257* 74–76].

EOP Efficiency of plating: the PLAQUE TITRE of a given type of virus expressed as a proportion of the total number of viruses of that type, per unit volume, in a given suspension. Under ideal conditions the EOP of some bacteriophages may exceed 90%; the EOP of animal viruses is often less than 10%.

eosin (eosin Y) TetrabromoFLUORESCEIN: a water-soluble red acidic FLUOROCHROME used e.g. as a cytoplasmic stain; it fluoresces greenish-yellow. (See also PHOTODYNAMIC EFFECT.)

eosin–methylene blue agar See EMB AGAR.

eosinophil A PMN (q.v.) in which cytoplasmic granules contain basic polypeptides that stain strongly with acidic dyes (e.g. eosin).

Eosinophilia (an increase in the number of eosinophils in the blood) may be seen e.g. in coccidioidomycosis and allergic pulmonary aspergillosis (sometimes tuberculosis), but, in general, *acute* infections of bacterial, fungal or viral aetiology tend to cause a fall in the number of blood eosinophils (i.e. *eosinopenia*). Protozoan pathogens (e.g. *Entamoeba histolytica*) generally do not give rise to eosinophilia, but eosinophilia is commonly associated with metazoan parasites (such as flatworms and flukes) [eosinophilia and helminthic infections: BCH (2000) *13* 301–317].

Eosinophils can participate in ADCC (see also HEPARIN); they can also be phagocytic, but this is not their main function. Eosinophils have cell-surface receptor sites for e.g. IgG, IgE and certain components of COMPLEMENT (e.g. C3b) – sites which may be involved in cell stimulation and release of (cytotoxic) granule contents. Eosinophils are stimulated e.g. by interleukins 3 and (particularly) 5.

eotaxin (chemokine) See CHEMOKINES.

Epalxella See ODONTOSTOMATIDA.

Epalxis See ODONTOSTOMATIDA.

EPEC Enteropathogenic *Escherichia coli*: a food-borne and water-borne pathogen which causes diarrhoea primarily in infants (sometimes severe outbreaks: 'infantile gastroenteritis') – particularly under conditions of poor hygiene. EPEC is a major cause of infant mortality in developing countries.

Strains of EPEC characteristically do not invade the gut epithelium (cf. EIEC), do not form ST or LT toxins (cf. ETEC), and do not form shiga-like toxins (cf. EHEC). They attach to the intestinal mucosa and produce characteristic lesions termed *attaching and effacing* (A/E) lesions. Each lesion involves an initial loss of local microvilli and the subsequent development of a 'pedestal': an actin-based, columnar protrusion from the eukaryotic host cell which extends into the gut lumen for several micrometres; the distal end of a pedestal may carry one EPEC cell or a small microcolony of such cells. (An early account noted that strains of EPEC destroy the brush border microvilli "... fashioning for themselves in the process a cup-like pedestal of bare plasma membrane in which individual bacteria sit like eggs in egg-cups" [ADC (1984) *59* 395–396].)

The ability of EPEC to produce A/E lesions is encoded by genes on the 35 kb chromosomal LEE PATHOGENICITY ISLAND (q.v.). (*Note.* Many strains of EHEC can produce EPEC-like lesions in vitro; at least some strains of *E. coli* O157:H7, and other EHEC, are known to contain the LEE pathogenicity island. A/E lesions are also produced by certain other Gram-negative bacteria, including *Hafnia alvei*.) The expression of operons on the LEE pathogenicity island is reported to be regulated by QUORUM SENSING (q.v.).

Initial binding of EPEC to the intestinal epithelium was thought, by some authors [TIM (1997) *5* 109–114; Book ref. 218], to be mediated by plasmid-encoded bacterial *bundle-forming pili* (BFP; type IV fimbriae). However, in one study, BFP did not adhere to cultures of (paediatric) small intestine – even though they adhered to (mammalian) HeLa cells. In another study, adherence to HeLa cells by (LEE-containing) EHEC apparently depended on filamentous bacterial appendages containing the (LEE-encoded) EspA protein [Mol. Microbiol. (1998) *30* 147–161].

One model [TIM (1998) *6* 169–172] of EPEC pathogenesis is outlined below. When EPEC comes into contact with an intestinal cell, products of the (LEE-encoded) *esc* genes, and of the *espA* gene, together form a type III secretory system (see PROTEIN SECRETION) – the EspA proteins forming short, filamentous links between the two cells. The Tir and EspB proteins (at least) are secreted directly into the eukaryotic cytoplasm. (Energy requirements may derive from ATP hydrolysis at a putative ATPase, EscN, associated with the type III assembly.) Within the eukaryotic cell, Tir is inserted into the cell membrane – where it acts as a receptor for the bacterial cell-surface adhesin *intimin*. (Intimin may also bind to other host receptors, e.g. β_1 integrins.) Phosphorylation of Tir (in the eukaryotic cell) stimulates the polymerization of cytoskeletal elements (e.g. actin) in preparation for pedestal development. (EPEC cells subsequently adherent at the free end of each pedestal may bind to other EPEC, via BFP, to form microcolonies.)

EspB also has regulatory roles in the epithelial cell; it activates various enzymes, resulting e.g. in increased secretion of chloride ions into the gut lumen. The activation of myosin light-chain kinase leads to an opening of 'tight junctions' between epithelial cells (increasing the permeability of the intestinal mucosa).

Diarrhoea may result from e.g. (i) stimulation of $C1^-$ secretion (through several mechanisms); (ii) increased permeability of the epithelium (due to opening of tight junctions); and/or (iii) reduced absorptive capacity due to loss of microvilli.

EPEC strains commonly belong to serogroups O26, O55, O86, O114, O119, O125, O126, O127, O128 and O142. O111 is also often listed in this context; however, in Brazil, strains of O111:H12 were reported to have the properties of EAggEC [FEMS (1997) *146* 123–128].

EPEC adherence factor plasmid See PATHOGENICITY ISLAND.

eperezolid See OXAZOLIDINONES.

Eperythrozoon A genus of Gram-negative bacteria within the family ANAPLASMATACEAE. Cells: pleomorphic cocci (diam. ca. 0.5–1.0 μm) which occur in the blood of the infected host – either free or loosely attached to the outer surface of erythrocytes. *Eperythrozoon* spp (*E. coccoides*, *E. ovis*, *E. parvum*, *E. suis* and *E. wenyonii*) have a wide host range (including ruminants and non-ruminants) and occur worldwide. (See also EPERYTHROZOONOSIS.)

eperythrozoonosis Any disease caused by an *Eperythrozoon* sp; the pathogen is transmitted e.g. by biting insects. In stressed feeder pigs, *E. suis* can cause acute icteroanaemia which involves e.g. mild fever, paleness of the mucosae, wasting, and (often) jaundice; *E. suis* can also cause anaemia in newborn pigs. Similar types of disease can occur in cattle and sheep (caused by *E. wenyonii* and *E. ovis*, respectively). [Book ref. 33, pp. 878–879.]

ephemeral fever See BOVINE EPHEMERAL FEVER.

epicellular At or on the surface of a cell.

Epichloë A genus of fungi (order CLAVICIPITALES) which include parasites and pathogens of grasses. Perithecia are formed within a stroma, the asci each containing eight filiform ascospores which often fragment in the ascus. *E. typhina* causes CHOKE. (See also FESCUE FOOT and ENDOPHYTE.)

epicone See DINOFLAGELLATES.

epidemic In a human population within a given geographical region: an outbreak of an infectious disease in which, for a limited period of time, a high proportion of individuals in the population exhibits overt symptoms of the disease (cf. ENDEMIC); epidemics are characterized by a sudden onset. (An analogous outbreak of disease among animals or plants is called an *epizootic* or *epiphytotic*, respectively.) (See also PANDEMIC and EPIDEMIOLOGY.)

Currently, the term 'epidemic' is often used in a broader sense to include the occurrence of any disease – either infectious or non-infectious – whose incidence in a given population clearly exceeds that common in other populations.

epidemic encephalitis *Syn.* VON ECONOMO'S ENCEPHALITIS.

epidemic haemorrhagic fever *Syn.* KOREAN HAEMORRHAGIC FEVER.

epidemic hepatitis *Syn.* HEPATITIS A.

epidemic keratoconjunctivitis ('shipyard eye') A highly infectious KERATOCONJUNCTIVITIS caused by an adenovirus (commonly Ad8: see MASTADENOVIRUS). Infection appears to require direct inoculation of the eye and is typically acquired from inadequately disinfected ophthalmological instruments, contaminated eye ointments etc. (The virus is resistant to organic solvents such as alcohol.) Conjunctivitis with oedema, pain, photophobia and a watery discharge may be followed by erosion of the cornea – sometimes resulting in permanent visual impairment.

epidemic meningitis Meningococcal MENINGITIS.

epidemic myalgia *Syn.* BORNHOLM DISEASE.

epidemic spread (infectious spread) (of plasmids) An effect which occurs when a population of conjugal donor cells (see CONJUGATION sense 1b), carrying a plasmid that is repressed for transfer, is mixed with a population of recipient cells. Initially, a small proportion of the donor cells – becoming spontaneously DEREPRESSED for transfer – donate their plasmids to recipient cells (*low frequency transfer*; LFT). In each of these recipient cells, the newly acquired plasmid remains derepressed owing to the absence of repressor molecule(s); such a cell, and/or progeny derived from it, can transmit the plasmid to a fresh recipient. This effect is cumulative, resulting in the rapid ('epidemic') spread of the plasmid throughout the population (*high frequency transfer*; HFT). Cells in the HFT state gradually (over several generations) become repressed for transfer as the intracellular concentration of repressor molecule(s) builds up.

epidemic tremor *Syn.* AVIAN ENCEPHALOMYELITIS.

epidemic vomiting disease See FOOD POISONING (i).

epidemiology The study of the interrelationships between a given pathogen, the environment, and *groups* or *populations* of the relevant hosts; the object is to investigate the factors and mechanisms which govern the spread of disease within a community or population. Three main factors are involved. (a) The virulence of the pathogen. Since the observed virulence of a pathogen is related to host susceptibility (a variable) it is difficult to define virulence in absolute terms. (b) Herd immunity. This may be regarded as the *collective* immunity or resistance to a given disease exhibited by a community or population (human or animal) in the setting of its own environment. (Immunity in individual members of a population may result from e.g. VACCINATION or recovery from the disease.) An important feature of herd immunity is the ratio of *immunes* (individuals resistant to the disease) to *susceptibles* (those able to contract the disease); a high proportion of immunes reduces the probability of contact between an infectious case and a susceptible, thus impeding the spread of disease. The proportion of susceptibles may often be drastically reduced by vaccination, and the risk to remaining susceptibles may be decreased by placing in quarantine any individuals who have contracted the disease; under such conditions CARRIERS (sense 1) may be important factors in the persistence of disease within the community. (c) Environmental conditions (extrinsic factors) play an important part in the containment or spread of disease. Particularly important are those features of society which favour or impede the transmissibility of the pathogen – e.g., standards of personal hygiene, overcrowding, the quality of communal WATER SUPPLIES, and the general state of sanitation (see e.g. SEWAGE TREATMENT); for a disease transmitted by a VECTOR, the prevalence of the vector and the measures taken to reduce its numbers are important factors. Because, in the present context, a population and its environment are inseparable, these extrinsic factors determine, in part, the observed herd immunity of a given population. The totality of factors which determines the transmissibility of disease and the distribution of immunes and susceptibles within a population has been termed the *herd structure* of that population.

An increase in the proportion of susceptibles in a population may lead from an ENDEMIC to an EPIDEMIC situation; an epidemic may also be initiated when a population immune to certain strains of a given pathogen is exposed to a different strain of that pathogen (see also ANTIGENIC DRIFT).

In a given outbreak of disease, the causal organism isolated from each patient may be examined by TYPING procedures in order to detect any variation which may help to establish the source(s) of infection and route(s) of transmission.

epidermin A 21-amino-acid, pore-forming LANTIBIOTIC secreted by *Staphylococcus epidermidis* and active against a range of Gram-positive bacteria.

epidermodysplasia verruciformis (EV) A rare, chronic, human skin disease involving the formation of flat, wart-like lesions on the face, neck, hands and feet, and sometimes also on the body. The lesions have a tendency to become malignant, ca. 30–50% of EV patients developing malignant squamous cell carcinoma; exposure to sunlight is believed to be a potentiating factor in the conversion to malignancy. The disease depends on genetic and immunological factors together with infection by certain human papilloma viruses (HPVs – see PAPILLOMAVIRUS). Many HPVs have been detected in EV lesions, only some of which are specific to EV. HPV-5 DNA has been detected in EV carcinoma cells, and HPV-5 is believed to be involved in carcinogenesis; HPV-17 has also been implicated in EV carcinogenesis [Virol. (1985) *144* 295–298].

epidermolytic toxin (ET; exfoliative toxin; exfoliatin) A protein toxin produced by strains of *Staphylococcus aureus* responsible for staphylococcal SCALDED SKIN SYNDROME (SSSS), and also e.g. by many of the strains which cause IMPETIGO. There are (at least) two serologically distinct ETs, variously designated ETA and ETB, TA and DI, type i and type ii, etc. ETA appears to be specified chromosomally, ETB is apparently plasmid-specified. ET-producing staphylococci may produce either or both types of toxin. Both toxins have the same histological effect: neither is cytolytic, but both cause separation of cells of the stratum granulosum from underlying tissue in the skin of only certain species (including humans, mice and monkeys, but not e.g. rats), resulting in a positive NIKOLSKY SIGN. ET apparently plays a role in the pathogenesis of SSSS, but its mechanism of action is unknown; neither is it known why toxin-producing strains produce localized lesions (impetigo) in some individuals and extensive, spreading 'scalding' lesions in others. [Book ref. 44, pp. 599–617.]

Epidermophyton See DERMATOPHYTES.

Epidinium A genus of ciliates related to DIPLODINIUM.

epifluorescence microscopy See MICROSCOPY (e).

epigean Occurring on the surface of the ground (cf. HYPOGEAN).

epiglottitis Inflammation of the epiglottis. Acute epiglottitis is a life-threatening condition caused primarily by *Haemophilus influenzae*; it occurs mainly in small children. Fever and sore throat are followed rapidly by inspiratory stridor and dysphagia; the epiglottis becomes oedematous and cherry-red, and may cause complete respiratory obstruction.

epilimnion The surface layer of a lake which, due to thermal effects and wind, is well-mixed and aerobic and approximately uniform in temperature. (cf. METALIMNION.)

epilithic Growing on the surface of rock or stone etc. (cf. SAXICOLOUS.)

epilithon The community of sessile organisms which grow on the surfaces of underwater stones. [Composition of epilithon: Oikos (1984) *42* 10–22.]

epimastigote A form assumed by the cells of species of *Blastocrithidia*, *Endotrypanum* and *Trypanosoma* during at least certain stages of their life cycles: see TRYPANOSOMATIDAE.

epimerite In cephaline gregarines: a differentiated region of the PROTOMERITE, delineated by a septum, by means of which the developing parasite remains attached to the host cell (cf. MUCRON).

epimicroscopy See MICROSCOPY (g).

epinasty (*plant pathol.*) The downward curvature of a plant structure (particularly a leaf petiole) due to the faster growth of the dorsal side of the structure relative to the ventral side. Epinasty can be induced e.g. by ETHYLENE, and is an early symptom in certain plant diseases.

epinephrine *Syn.* ADRENALIN (q.v.).

epineuston See NEUSTON.

epipelic Growing on mud.

epiphloeodal (epiphloedal) Growing on the surface of bark (cf. CORTICOLOUS).

epiphragm See CYATHUS.

epiphyte A plant or 'plant-like' organism (e.g. a lichen) which grows upon another plant but does not derive nutrients from it – i.e., an epiphyte is non-parasitic. (cf. ENDOPHYTE.)

epiphytotic See EPIDEMIC.

epiplan objective A FLAT-FIELD OBJECTIVE LENS used for bright-field epimicroscopy (see MICROSCOPY (g)).

epiplasm (1) (*protozool.*) See PELLICLE. (2) (*mycol.*) That cytoplasm in an ASCUS which is not incorporated into ascospores.

epipodophyllotoxins See PODOPHYLLOTOXIN.

epipolythiapiperazinediones A group of fungal ANTIBIOTICS. Some (e.g. aranotin, GLIOTOXIN, hyalodendrins, sirodesmins) inhibit the replication of certain RNA viruses – apparently by binding to and inhibiting viral RNA-dependent RNA polymerase; however, they are too toxic for therapeutic use.

episome (1) A PLASMID which can exist either in an autonomous (extrachromosomal) state or integrated in the host cell's chromosome.

(2) *Syn.* PLASMID.

episporium See EXOSPORIUM (sense 2).

epistasis The dominance of one (*epistatic*) allele over a *non*-allelic (*hypostatic*) allele. (cf. DOMINANCE.)

Epistylis See PERITRICHIA.

epitheca (1) See DINOFLAGELLATES. (2) See DIATOMS.

epithecium In the hymenium of some ascomycetes: a layer of prosenchymatous tissue which covers the surface of the layer of asci; it consists of the interwoven, branched tips of paraphyses. (cf. PSEUDOEPITHECIUM.)

epithelioma papulosum *Syn.* CARP POX.

epitheliotropism See TROPISM (sense 2).

epithienamycins CARBAPENEM antibiotics.

epitope *Syn.* antigenic DETERMINANT.

epitunica *Syn.* EXOSPORIUM (sense 2).

epivalve See DIATOMS.

epixylous Growing on wood. (cf. CORTICOLOUS.)

epizootic See EPIDEMIC.

epizootic haemorrhagic disease subgroup See ORBIVIRUS.

epizootic lymphangitis (pseudoglanders; equine histoplasmosis; equine blastomycosis; African farcy) A chronic cutaneous and subcutaneous disease of horses and other equines, caused by *Histoplasma farciminosum* (cf. HISTOPLASMOSIS); it occurs in Asia, Africa and Mediterranean countries. Nodules and suppurating ulcers develop on the skin – commonly on the shoulders, neck and limbs – and spread along lymph vessels; associated lymph nodes become inflamed. Small yeast-like cells can be detected – either free or within macrophages – in the pus.

EPR See ELECTRON SPIN RESONANCE.

EPS Extracellular polysaccharide.

***epsilon* chain** (*immunol.*) See HEAVY CHAIN.

Epstein–Barr virus (EBV) Human (gamma) herpesvirus 4, the prototype of the subfamily GAMMAHERPESVIRINAE (q.v.). [DNA sequence and expression of the type strain (B95-8) of Epstein–Barr virus: Nature (1984) *310* 207–211.]

EBV is ubiquitous in human populations; it is transmitted horizontally, mainly via saliva (e.g. by kissing). (Large numbers of infectious virions can occur in saliva.) In developing countries, primary (i.e. initial) infection generally occurs in early childhood, and in almost all cases these infections are subclinical. In developed countries, primary infection generally does not occur until adolescence, and a significant proportion of infected individuals develop INFECTIOUS MONONUCLEOSIS. The reasons for the difference in impact of EBV in developing and developed countries are not known.

EBV infection results in a life-long carrier state in which the virus occurs in latent form within circulating B cells. In some individuals a productive infection results in virus shedding, thus permitting horizontal transmission of EBV.

Although the target cells of EBV were believed to be limited to B lymphocytes and epithelial cells, they are now known to include e.g. T lymphocytes, monocytes, macrophages and endothelial cells. This extended range of host cells may help to explain the fact that EBV is associated with a wide spectrum of diseases.

Infection of epithelial cells characteristically leads to viral replication and the release of infectious virions. By contrast, infection of B cells is usually followed by viral latency and the development of a permanent proliferative state in the B cells.

The B cell receptor for EBV is CD21. In cultured cells, a protein, EBV nuclear antigen-1 (EBNA-1), is detectable within several hours of viral infection. Subsequently, DNA synthesis occurs in the B cells – which then proliferate continually (i.e. so-called 'immortalization'). In proliferating B cells, the viral genome exists within the cell nucleus in a circular, 'plasmid' form; EBNA-1 binds to oriP (in the viral origin of replication), such binding apparently being necessary to keep the plasmid within the nucleus. EBNA-1 also appears to be required for replication of the plasmid form of EBV. (See also ZEBRA.)

The *ebna-1* gene is only one of nine or ten viral genes (out of ~80) which may be expressed during viral latency. The *ebna-2* gene is reported to be necessary for transactivation of *lmp-1* and *lmp-2* whose products (the latency membrane proteins 1 and 2) are incorporated into the B cell membrane.

The products of *ebna-2, ebna-3A* and *ebna-3C* may also be expressed at the cell surface.

In addition to proteins, latent EBV gives rise to two small polyadenylated RNA molecules called EBV-encoded RNAs (EBER-1 and EBER-2); these molecules are expressed at high levels, but their function is unknown. The EBERs have been used as targets for detection of EBV.

Certain genes in latent EBV are also involved in the transactivation of some of the B cell genes; these B cell genes encode e.g. ICAM-1 (see CD54), LFA-1 (see INTEGRINS) and LFA-3 (see IMMUNOGLOBULIN SUPERFAMILY). CD23, a B-cell-encoded activation-related marker, is also produced; this molecule is reported to act as an autocrine B-cell growth factor when shed from the cell surface.

In vivo, proliferating B cells (with cell-surface molecules of EBNA-2, EBNA-3A, EBNA-3C and LMP-1) are normally targeted by a subset of EBV-specific cytotoxic T cells; moreover, these virus-infected immortalized B cells are targets for NK (natural killer) cells, and are also susceptible to interferons. Thus, the number of proliferating B cells is normally controlled in immunocompetent individuals.

Cells containing latent EBV do not necessarily express all of the proteins referred to above; in some cases, however, all of these proteins (and others) may be expressed. In fact, several different types of EBV latency have been distinguished. In *type I latency* only EBNA-1 is produced. Cells which express only EBNA-1 are found e.g. in some gastric carcinomas; as these cells lack the targets for EBV-specific cytotoxic T cells (see above) they escape this form of immune control. In *type II latency* the cell expresses EBNA-1 and LMPs 1 and 2; this pattern is found e.g. in some T-cell lymphomas and in nasopharyngeal carcinoma. In *type III latency* the cell expresses the full complement of these proteins; this pattern is found e.g. in B cells from the acute phase of infectious mononucleosis. (EBERs appear to be expressed in all forms of latency.)

EBV-associated diseases include: BURKITT'S LYMPHOMA; HAIRY LEUKOPLAKIA; INFECTIOUS MONONUCLEOSIS; NASOPHARYNGEAL CARCINOMA; XLP SYNDROME, gastric carcinoma and diffuse polyclonal B-cell lymphomas. In both Burkitt's lymphoma and nasopharyngeal carcinoma, pathogenesis seems to depend on proliferation of host cells latently infected with EBV, whereas in hairy leukoplakia the main pathogenic mechanism appears to involve viral replication; in infectious mononucleosis, the immune response of the host appears to be a major factor in pathogenesis.

Diagnosis of EBV-associated diseases involves laboratory tests (see e.g. INFECTIOUS MONONUCLEOSIS) and clinical observation. Levels of positivity of anti-EBV antibodies have been well documented, so that serology may help to distinguish e.g. between a primary (initial) infection (or a previous infection) and reactivation – in the latter case higher levels of antibodies are formed against viral capsid antigen (VCA), early antigen (EA) and EBNA. The EBERs may be detected e.g. by in situ hybridization or by RNA-amplification techniques.

Treatments for various EBV-associated diseases include antivirals (such as ACYCLOVIR), immunosuppressants (e.g.

cyclophosphamide), anti-B cell monoclonal antibodies and inferon-gamma (IFN-γ).

[Haematological associations of EBV: BCH (2000) *13* 199–214.]

Epulopiscium fishelsoni See BACTERIA.

EPV Entomopoxvirus: see ENTOMOPOXVIRINAE.

equal density centrifugation See CENTRIFUGATION.

equid (alpha) herpesvirus 1 See ALPHAHERPESVIRINAE.

equid herpesvirus 3 *Syn.* EQUINE COITAL EXANTHEMA virus.

equilibrium density gradient centrifugation See CENTRIFUGATION.

equilibrium zonal centrifugation See CENTRIFUGATION.

equine abortion virus See ALPHAHERPESVIRINAE.

equine arteritis virus See ARTERIVIRUS.

equine blastomycosis *Syn.* EPIZOOTIC LYMPHANGITIS.

equine coital exanthema An acute, usually mild, usually venereally transmitted HORSE DISEASE caused by equid herpesvirus 3 (see ALPHAHERPESVIRINAE). Typically, vesicular or papular lesions develop on the external genitalia (sometimes also on the conjunctivae, lips or nasal mucosa); the lesions progress rapidly to pustules (which may ulcerate), and generally heal in ca. 2 weeks.

equine diseases See HORSE DISEASES.

equine encephalosis subgroup See ORBIVIRUS.

equine herpesvirus 2 See GAMMAHERPESVIRINAE.

equine histoplasmosis *Syn.* EPIZOOTIC LYMPHANGITIS.

equine infectious anaemia *Syn.* SWAMP FEVER.

equine monocytic ehrlichiosis *Syn.* POTOMAC HORSE FEVER.

equine phycomycosis (hyphomycosis destruens; swamp cancer; bursattee; Florida horse leech) A tropical and subtropical HORSE DISEASE characterized by ulcerating, spreading, granulomatous lesions on the skin and mucous membranes; the lesions contain masses of yellow-grey necrotic tissue ('grains', 'leeches') which may calcify. The causal agent is '*Hyphomyces destruens*', a *Pythium* sp. In Australia, outbreaks are associated with grazing in areas of permanent standing water [TBMS (1983) *80* 13–18]. Clinically similar conditions are caused e.g. by *Basidiobolus haptosporus* and by certain helminthic parasites. (See also ZYGOMYCOSIS.)

equine protozoal myeloencephalitis (equine protozoal myelitis) A HORSE DISEASE which affects young and old horses of both sexes; it occurs in certain areas of the USA. The symptoms are those typical of CNS involvement; they include stumbling and falling, sometimes with partial paralysis of the face and/or tongue, muscle atrophy, and lameness. The causal agent is believed to be a member of the coccidia (suborder EIMERIORINA) [IJP (1987) *17* 615–620].

equine rhinoviruses See PICORNAVIRIDAE.

equivalence (*serol.*) *Syn.* OPTIMAL PROPORTIONS.

ER ENDOPLASMIC RETICULUM.

Era protein See CELL CYCLE (b).

eradicant (*plant pathol.*) Any chemical agent which eliminates particular pathogen(s) from diseased plants treated with that agent. (cf. PROTECTANT.)

erb An ONCOGENE originally identified as the transforming determinant in avian erythroblastosis virus (AEV: see AVIAN ACUTE LEUKAEMIA VIRUSES). In AEV, v-*erb* consists of two contiguous genes, v-*erb*-A and v-*erb*-B; the cellular homologues (c-*erb*-A and c-*erb*-B) are not linked in the chicken genome.

The cellular homologue of v-*erb*-A (c-*erb*-A) encodes the *thyroid hormone receptor*. This protein lacks tyrosine kinase activity and is apparently not tumorigenic on its own.

The cellular homologue of v-*erb*-B (c-*erb*-B) encodes a membrane protein that functions as the *epidermal growth factor receptor* (EGF receptor); unlike the v-*erb*-A product, this protein does have tyrosine kinase activity.

The oncogenic activity of the virus appears to depend on more than one gene. Thus, the product with which the avian erythroblastosis virus transforms cells is apparently a fusion protein that includes the *erb*-B product.

Eremothecium See METSCHNIKOWIACEAE; see also RIBOFLAVIN.

***erf* gene** See BACTERIOPHAGE P22.

ergine See ERGOT ALKALOIDS.

ergobasine See ERGOT ALKALOIDS.

ergochromes (secalonic acids) Pigments present in ERGOT sclerotia and also produced e.g. by *Penicillium oxalicum*; they are mostly weakly acidic, yellow xanthone derivatives, and at least some (e.g. secalonic acid D) are MYCOTOXINS.

ergocristine See ERGOT ALKALOIDS.

ergocryptine See ERGOT ALKALOIDS.

ergolines *Syn.* ERGOT ALKALOIDS.

ergometrine See ERGOT ALKALOIDS.

ergonovine See ERGOT ALKALOIDS.

ergopeptines See ERGOT ALKALOIDS.

ergosome *Syn.* polyribosome: see PROTEIN SYNTHESIS.

ergot A CEREAL DISEASE, caused by *Claviceps purpurea* (see CLAVICEPS), which can affect e.g. rye, wheat, barley and oats; it affects only the flowering parts. The earliest infections in the year are initiated by ascospores; conidia initiate later infections. Elongated, hard, black sclerotia ('ergots') become conspicuous in the ripening ears. The sclerotia overwinter and give rise in the following year to stromata and ascospores. (See also ERGOT ALKALOIDS and ERGOTISM.)

ergot alkaloids (ergolines) Alkaloids obtained primarily from *Claviceps* spp; they are also produced by other fungi (e.g. *Aspergillus*, *Penicillium*, *Rhizopus* spp) and by certain plants. Natural ergot alkaloids stimulate smooth muscle, especially that of peripheral blood vessels (causing vasoconstriction) and of the pregnant uterus at term; certain hydrogenated derivatives have the reverse effect, causing vasodilation. Ergot alkaloids have several important clinical uses (cf. ERGOTISM).

There are two classes of ergot alkaloids: the clavine alkaloids and the lysergic acid derivatives; the clinically useful alkaloids belong to the latter class, and may be simple amides or cyclic tripeptide derivatives. The amides include e.g. *ergine* (lysergic acid amide) and *ergometrine* (also called *ergonovine* or *ergobasine*: lysergic acid L-2-propanolamide); ergometrine is used to control postpartum bleeding. Cyclic tripeptide derivatives (*ergopeptines*) include *ergotamine*, used (as ergotamine tartrate) in the treatment of migraine, and *ergotoxine* (a mixture containing *ergocristine*, *α-ergocryptine* and *ergocornine*); hydrogenated derivatives of ergotoxine are used for the treatment of peripheral and cerebral vascular disorders and of essential hypertension. Synthetic derivatives of ergot alkaloids include e.g. the hallucinogen *lysergic acid diethylamide* (LSD).

Commercial production of ergot alkaloids originally involved extraction from sclerotia of *C. purpurea* in naturally infected rye; later, fields of rye were inoculated with selected strains of *C. purpurea*. Currently, many lysergic acid derivatives are obtained from *C. paspali* grown in submerged culture.

ergotamine See ERGOT ALKALOIDS.

ergotism (St Anthony's fire) A MYCOTOXICOSIS caused by the ingestion of ERGOT ALKALOIDS present e.g. in bread made from ergotized grain; outbreaks of ergotism have been responsible for numerous deaths in Europe – especially during the Middle Ages and the 18th century. Mild cases may involve a prickling sensation in the skin, with numbness, fatigue, and e.g. diarrhoea and vomiting. Severe ergotism may take one of two forms, depending on the relative proportions of the ergot alkaloids ingested. The convulsive type is characterized by neurological symptoms, with spasms, epilepsy-like convulsions, and severe pain. In the gangrenous type the fingers and toes – or whole limbs – become blackened and gangrenous and are eventually lost. Severe cases are usually fatal, and in non-fatal cases complete recovery is rare. In animals, ergotism may result e.g. from grazing any of a wide range of fodder grasses infected with *Claviceps* spp. (See also PASPALUM STAGGERS.)

ergotoxine See ERGOT ALKALOIDS.

ERIC-PCR See REP-PCR.

ERIC sequence Enterobacterial repetitive intergenic consensus sequence (also called intergenic repeat unit): one of the repeated intergenic sequences of nucleotides in the genomic DNA of enteric and other bacteria; an ERIC sequence contains a highly conserved PALINDROMIC SEQUENCE. ERIC sequences are distinct from REP SEQUENCES (they do not share significant homology). [ERIC sequences: Mol. Microbiol. (1991) *5* 825–834.] ERIC sequences have been exploited in typing [see e.g. JCM (1999) *37* 103–109; JCM (1999) *37* 2473–2478; JCM (1999) *37* 2772–2776].

ericoid mycorrhiza See MYCORRHIZA.

Erlenmeyer flask A conical flask.

error-prone repair See SOS SYSTEM.

ertapenem A broad-spectrum 1-β-methylCARBAPENEM antibiotic with a long serum half-life (up to 4.5 hours). [In vitro activity of ertapenem against a range of bacteria from clinical specimens: AAC (2001) *45* 1860–1867.]

erumpent Bursting *through*.

Erwinia A genus of Gram-negative bacteria of the ENTEROBACTERIACEAE (q.v.). Cells: 0.5–1.0 × 1–3 μm, motile (except *E. stewartii*). Growth occurs e.g. on nutrient agar or YGC AGAR; optimum temperature: 27–30°C. Acid (little or no gas) is produced from glucose, fructose, galactose, sucrose, and β-methylglucoside. GC%: 50–58. Type species: *E. amylovora*.

Erwinia spp typically occur as saprotrophs or pathogens in or on plants; some may also occur in insects, and certain strains are opportunist pathogens in man and animals.

The genus is commonly divided into three groups. The Amylovora group includes species (e.g. *E. amylovora*) which require organic nitrogen for growth and cause dry-necrotic or VASCULAR WILT diseases in plants. The Carotovora group (sometimes regarded as a separate genus, *Pectobacterium*) includes strongly pectinolytic species (e.g. *E. carotovora*) which reduce nitrate to nitrite and cause SOFT ROT (sense 2) in a variety of plants. The Herbicola group includes species (e.g. *E. herbicola*) which typically form yellow (carotenoid) pigments and are not normally primary pathogens.

E. amylovora. Non-pectinolytic. Causes FIREBLIGHT. Colonies on sucrose nutrient agar are mucoid (due to levan production). Anaerobic growth is weak. *E. amylovora* transfers virulence proteins into target cells by means of a type III PROTEIN SECRETION system; it has been shown that secreted proteins are guided along the Hrp pilus assembly that forms between the pathogen and the eukaryotic target cell [Mol. Microbiol. (2001) *40* 1129–1139].

E. carotovora. Pectinolytic. Subspecies *atroseptica* causes BLACKLEG (sense 2) of potato plants and storage rot of potato tubers. Subspecies *carotovora* causes SOFT ROT (sense 2) in a wide range of plants. Subspecies *betavasculorum* causes soft rot of sugar-beet.

E. chrysanthemi. Pectinolytic. Causes vascular wilts in a wide range of plants. Some strains produce a blue bipyridyl pigment, indigoidine, on YGC agar.

E. herbicola. Common as saprotrophs on the surfaces of land plants; some strains can fix nitrogen. May act as secondary invaders in plant lesions. (See also HERBICOLIN A.) Some strains can cause GALL formation. Certain strains are opportunist parasites or pathogens in man and animals; such strains are regarded by clinical microbiologists as *Enterobacter agglomerans*. [Attempted classification of strains of the Herbicola–Agglomerans group: IJSB (1984) *34* 45–55.]

E. nigrifluens. Causes bark necrosis of Persian walnut (*Juglans regia*).

E. rhapontici. Causes e.g. crown rot of rhubarb (*Rheum rhaponticum*) and pink grain of wheat; it can also grow saprotrophically in plant lesions. A pink diffusible pigment is formed on sucrose–peptone agar.

E. rubrifaciens. Pectinolytic. Causes phloem necrosis (bark canker) of Persian walnut (*Juglans regia*). A pink pigment is produced on YGC agar.

E. salicis. Pectinolytic. Causes WATERMARK DISEASE in willows (*Salix* spp). Produces yellow pigment on autoclaved potato tissue.

E. stewartii. Non-motile. Causes vascular wilt of e.g. maize (*Zea mays*); overwinters primarily in the flea beetle *Chaetocnema pulicaria*.

E. tracheiphila. Causes vascular wilt in *Cucurbita* spp; overwinters in cucumber beetles (*Diabrotica* spp).

E. uredovora. A parasite of the uredia and uredospores of *Puccinia graminis*. Produces yellow pigment on nutrient agar.

Other recognized species: *E. ananas* (causes pineapple rot), *E. cypripedii*, *E. mallotivora*, *E. quercina*.

[Taxonomy: Book ref. 22, pp. 469–476. Habitats, culture etc: Book ref. 46, pp. 1260–1271.]

See also COFFEE; PECTIC ENZYMES.

Erynia A genus of fungi (order ENTOMOPHTHORALES) which include parasites of insects; *E. neophidis* is a pathogen of aphids. [Persistence of infectious spores of *E. neophidis*: Ann. Appl. Biol. (1985) *107* 365–376.]

erysipelas Typically, well-demarcated lesions of erythema – often affecting the face, although other areas may be involved; fever, prostration and sepsis may occur. The causal agent is usually a strain of *Streptococcus pyogenes*, and infection generally occurs via wounds or abrasions.

(cf. ERYSIPELOID; SWINE ERYSIPELAS.)

erysipeloid A disease involving localized inflammation or CELLULITIS of skin (usually on the fingers and hands); it is caused by *Erysipelothrix rhusiopathiae* and is contracted mainly by handlers of fish and meat products. Infection occurs via wounds etc. Rarely, septicaemia or endocarditis may occur.

Erysipelothrix A genus of Gram-positive, asporogenous, non-motile bacteria; sole species (type species): *E. rhusiopathiae* (formerly *E. insidiosa*). Cells may be slightly curved, round-ended rods (ca. 0.2–0.4 × 0.8–2.0 μm) which form smooth, (initially) transparent colonies (S-form), or occur mainly as filaments (up to 60 μm long) which form rough, opaque colonies (R-form). S-forms show a strong tendency to become R-forms. (See also PEPTIDOGLYCAN.) The organism is microaerophilic (facultatively aerobic or anaerobic). Growth occurs at 15–42°C (optimum 37°C). Indole −ve; catalase −ve; urease −ve; VP −ve; MR −ve; H_2S +ve; aesculin not hydrolysed. In gelatin stab cultures S-forms produce a typical 'pipe-cleaner' (becoming a 'test-tube brush') form of growth, with little or no gelatin liquefaction. GC%: 38–40.

E. rhusiopathiae is parasitic in man, animals, birds and fish, and can be found e.g. in animal faeces, abattoir effluents etc. Some strains are pathogenic (see e.g. ERYSIPELOID, JOINT-ILL, SWINE ERYSIPELAS). The organism is resistant e.g. to salting, pickling, smoking, and low concentrations of phenol; in dried form (e.g. in fish-meal, soil) it can remain viable for years.

[Book ref. 46, pp. 1688–1700.]

Erysiphaceae See ERYSIPHALES.

Erysiphales An order of homothallic and heterothallic plant-parasitic fungi of the subdivision ASCOMYCOTINA (see also POWDERY MILDEWS); all species are placed in the family Erysiphaceae. The organisms form at least partially superficial, septate mycelium which is typically colourless and which commonly exhibits numerous haustoria. (*Dark* mycelium is formed e.g. by *Sphaerotheca lanestris*; cf. BLACK MILDEWS.) Ascocarp (ca. 50–350 μm diam.): cleistothecioid with hamathecium lacking, often dark and rounded, and often with surface appendages; uniascal or containing a small number of asci. Asci: bitunicate, clavate or pyriform. Ascospores: aseptate, ellipsoidal, colourless. Anamorphs occur e.g. in the genera *Oidiopsis*, *Oidium* and *Ovulariopsis*; conidia are often formed in abundance, basipetally, on straight, unbranched conidiophores. Genera: e.g. *Cystotheca*, ERYSIPHE, *Leveillula*, *Microsphaera*, *Phyllactinia*, *Podosphaera*, *Sphaerotheca*, *Uncinula*.

[Book ref. 187.]

Erysiphe A genus of homothallic and heterothallic fungi (order ERYSIPHALES) which include parasites and pathogens of grasses and other plants (e.g. *E. betae* on beet crops, *E. graminis* on cereals). *Erysiphe* spp form superficial mycelium on host plants; most species form globose haustoria, but those of *E. graminis* are digitate (see also TRIFORINE). The organisms give rise to dark-coloured cleistothecia. The cleistothecial surface appendages (when present) are hypha-like, not dichotomously branched or terminally coiled; each cleistothecium contains more than one ascus. The ovoid to elongated conidia of *E. graminis* and *E. polygoni* are highly vacuolated; the conidiophores of *E. graminis* are distinctive in that in each the basal cell is slightly swollen. Conidia germinate by a germ tube which forms an appressorium.

erythema Redness of the skin due e.g. to INFLAMMATION or infection.

erythema infectiosum (fifth disease; slapped cheek syndrome) A benign, infectious disease, affecting mainly children, characterized by exanthem which begins on the face and gives a 'slapped cheek' appearance. The causal agent is parvovirus B19. (cf. EXANTHEM SUBITUM.)

erythema nodosum leprosum See LEPROSY.

erythrasma A chronic skin infection caused by *Corynebacterium minutissimum*. The reddish-brown lesions occur in the groin and armpits and fluoresce coral-pink under Wood's lamp.

erythritol The POLYOL corresponding to erythrose. It occurs e.g. in certain fungi, particularly in the Ustilaginales and Agaricales (where it apparently functions as a storage carbohydrate), and as a major photosynthetic product in certain unicellular and filamentous green algae (e.g. *Phycopeltis*, *Trentepohlia*). It can be metabolized e.g. by many yeasts.

Erythrobacter A genus of Gram-negative, aerobic, halophilic, ovoid to rod-shaped, orange- or pink-pigmented bacteria which are motile by subpolar flagella. The cells contain bacteriochlorophyll *a* which is synthesized only under aerobic conditions (cf. RHODOSPIRILLALES). Metabolism appears to be chemoorganotrophic (mainly respiratory); autotrophic growth does not occur. Strains have been isolated from seaweeds growing near high-tide levels (e.g. *Enteromorpha*, *Porphyra*). Type species: *E. longus*. [IJSB (1982) *32* 211–217.]

erythroblastosis The presence of erythroblasts (nucleated precursors of erythrocytes) in the blood.

erythrocyte A red blood cell.

erythroimmunoassay (EIA) A sensitive IMMUNOASSAY for detecting and quantifying specific antigens or antibodies. For the detection of antibodies, the inner surface of a well (or tube) is coated with homologous antigen, and the patient's serum – pre-absorbed with sheep erythrocytes (see ABSORPTION) – is added to the well; a hybrid antibody conjugate (see CONJUGATION sense 2) is then added – each hybrid antibody having combining sites for both human immunoglobulins and sheep erythrocytes. Hybrid antibodies bind to any antigen–antibody complexes on the surface of the well, and the presence of bound hybrid antibodies is detected by the addition of sheep erythrocytes. [SAB (1984) *22* 73–78.]

erythromycin A MACROLIDE ANTIBIOTIC (q.v. for mode of action etc.); it is produced by *Streptomyces erythreus*. (In some strains of group A streptococci, resistance to erythromycin has been associated with the ability to invade, and survive within, epithelial cells of the human respiratory tract [Lancet (2001) *358* 30–33].) (See also POLYKETIDE.)

erythrose An aldotetrose; erythrose 4-phosphate is an important intermediate e.g. in the CALVIN CYCLE and HEXOSE MONOPHOSPHATE PATHWAY, and is a precursor for the synthesis of aromatic amino acids [Appendix IV(f)].

Erythrovirus A genus of autonomous viruses (family PARVOVIRIDAE) which infect primates. The genus was originally created for the only parvovirus known to be pathogenic for man, i.e. parvovirus B19; some authors also include the *simian parvovirus* [Virol. (1995) *210* 314–322]. The following details refer to B19 parvovirus. (B19 was formerly referred to as 'human parvovirus' (abbreviated to HPV), but this terminology has been abandoned owing to confusion with human papillomavirus. The designation 'B19' derives from the laboratory number of the sample in which the virus was first detected.)

The B19 virion is 22–26 nm, icosahedral, and contains a single copy of the linear, ssDNA genome (~5.5 kb); identical PALINDROMIC SEQUENCES (383 nt) occur at the two ends of the genome. In a given population of B19 virions, some capsids enclose a negative-sense strand while others enclose a positive-sense strand. B19 virions are stable at 56°C/60 minutes, and (in high concentration) are reported to be stable at 80°C/72 hours); lacking an envelope, they are insensitive to lipid solvents. The virions can be inactivated by e.g. oxidizing agents, formalin and γ-radiation.

For viral DNA replication see entry PARVOVIRUS.

Transcription of the genome yields a number of overlapping RNA transcripts, all of which originate from a single, strong promoter, P6; the function of some of the encoded products is unknown.

Proteins encoded by the genome include capsid proteins VP1 and (the smaller) VP2; VP2 is the major capsid protein, accounting for >95% of the capsid. (Recombinant VP2 proteins self-assemble into capsids within various mammalian and insect cell lines.)

The major non-structural protein, NS1, can be detected in the nucleus of infected cells; it may be associated with the cytotoxicity of B19. One hypothesis supposes that NS1 acts as a positive regulator of viral DNA replication during the early stage of infection and that, subsequently, high concentrations of NS1 bring about cell death and lysis.

B19 has been cultured in the erythroid cells from human bone marrow and fetal liver; *in vitro* propagation of the virus depends on the presence of the hormone erythropoietin (which is specific to erythroid cells). The cell-surface receptor for B19 is the red cell P antigen (*globoside* – a tetrahexose ceramide), the virus binding with high affinity to the carbohydrate moiety. P antigen is expressed e.g. on erythroid progenitors, mature red cells, megakaryocytes, endothelial cells and heart cells. (Individuals with the rare *p phenotype*, who do not express P antigen, are not susceptible to B19.)

Viral replication is dependent on mitotically active cells, the virus being highly tropic for human erythroid cells in the late S phase of the cell cycle.

B19 DNA can be demonstrated in the nasal secretions of individuals with B19 viraemia, and the respiratory route is believed to be important in virus transmission. Infection can also be transmitted via blood and blood products.

Clinical manifestations of infection appear to depend primarily on host factors such as immunocompetence. In immunocompetent persons with a normal erythroid turnover, infection may lead to a short-term (~4–8 days) halt in red cell production without development of anaemia. However, given a high turnover of red cells, even a temporary loss of red cell production may bring about an *aplastic crisis*. In immunocompromised patients infection may lead to chromic anaemia. In the fetus, infection may result e.g. in HYDROPS FETALIS. Other manifestations of B19 infection include arthralgia and rash. (See also ERYTHEMA INFECTIOSUM.)

Among adults, seroprevalence of B19 IgG antibody is high, and the proportion of seropositives increases with age; IgG antibody develops within ~2 weeks of infection and persists for life.

[Persistent B19 infection: RMM (1995) *6* 246–256. Molecular epidemiology of parvovirus B19: RMM (1997) *8* 21–31. Haematological consequences of parvovirus B19 infection: BCH (2000) *13* 245–259.]

ES1–ES21 proteins See RIBOSOME.

ESAT-6 A 6 kDa secreted protein (*e*arly *s*ecreted *a*ntigenic *t*arget) encoded by *Mycobacterium tuberculosis* and by *M. bovis* and *M. africanum* (members of the '*M. tuberculosis* complex') but apparently not encoded by any of the (vaccine) strains of *M. bovis* BCG. (The operon encoding ESAT-6 is also reported to encode a novel, low-molecular-weight protein which has been designated CFP-10 [Microbiology (1998) *144* 3195–3203].)

Yet another secreted protein, MPT64, appears to be encoded by all strains of *M. tuberculosis* and BCG Tokyo but not by strains of BCG Danish 1331 [BCID (1997) *4* 157–172 (167–168)].

Antigens such as ESAT-6 may be useful e.g. in SKIN TESTS for distinguishing between sensitization due to (i) tubercle bacilli, (ii) vaccination strains of BCG, or (iii) 'environmental' mycobacteria (cf. TUBERCULIN TEST).

A test using ESAT-6 has been compared with a PPD-based test for the diagnosis of bovine tuberculosis; ESAT-6 was associated with a lower sensitivity but with increased specificity – useful e.g. for testing in areas which have a low incidence of tuberculosis [VR (2000) *146* 659–665].

The specificity of ESAT-6 for the *M. tuberculosis* complex has been exploited (by means of an ex-vivo ELISPOT ASSAY) for the enhanced tracing of contacts in epidemiological studies on human tuberculosis [Lancet (2001) *357* 2017–2021].

ESBLs Extended-spectrum β-LACTAMASES.

***esc* genes** See PATHOGENICITY ISLAND.

eschar Dead tissue which separates (sloughs) from the underlying living tissue. An eschar is formed e.g. in SCRUB TYPHUS.

Escherichia A genus of Gram-negative bacteria of the family ENTEROBACTERIACEAE (q.v. for general features); in molecular taxonomy, *Escherichia* is placed in the gamma group of the Proteobacteria. GC% ~50. Type species: *E. coli*.

E. coli has long been the principal 'guinea pig' of the microbiologist, and has been studied more extensively than any other species. (Nevertheless, *E. coli* should not be regarded as 'the typical bacterium'; the great diversity of bacteria makes such a concept meaningless.) The following refers specifically to *E. coli* (other species are mentioned at the end of the entry).

Cells are typically straight, round-ended rods, ca. 0.5–0.8 × 1–4 μm, normally occurring singly or in pairs; they stain well with ANILINE DYES.

The cell has a typical Gram-negative-type CELL WALL consisting of an OUTER MEMBRANE and a layer of PEPTIDOGLYCAN. The CYTOPLASMIC MEMBRANE contains e.g. various TRANSPORT SYSTEMS, energy-converting systems and chemosensors as well as components of osmoregulatory systems (see OSMOREGULATION); the latter include MECHANOSENSITIVE CHANNELS [mechanosensitive channels in *E. coli*: ARP (1997) *59* 633–657], *aquaporins* and *glycerol facilitators* (see MIP CHANNELS).

Cells are typically motile (peritichously flagellate) [second flagellar system in *E. coli*. JB (2005) *187* 1430–1440]. FIMBRIAE are usually present. Type I fimbriae (the most common type) may be suppressed by P FIMBRIAE (q.v.).

Some cells have a CAPSULE or microcapsule (see e.g. COLANIC ACID, COLOMINIC ACID; cf. TEICHOIC ACIDS).

Numerous strains of *E. coli* are distinguished serologically (on the basis of their O ANTIGENS, H ANTIGENS and K ANTIGENS), or by PHAGE TYPING, colicin typing, antibiotic-resistance patterns etc.

E. coli exhibits CHEMOTAXIS; the Tsr protein (an MCP in chemotaxis) may also function as a pmf-dependent sensor in aerotaxis [PNAS (1997) *94* 10541–10546]. [Ecological role of energy taxis: FEMS Reviews (2004) *28* 113–126.]

The chromosome (cccDNA) is about 4.7 million base-pairs in length. [Complete sequence: Science (1997) *277* 1453–1474.] Replication occurs by the Cairns-type mechanism (bidirectionally from the origin, *oriC*). [Localization of chromosomal segments during the cell cycle: GD (2000) *14* 212–223.] There are several systems for DNA REPAIR. (See also SOS SYSTEM.)

Many of the genes occur within OPERONS which may encode e.g. metabolic enzymes and/or components of TRANSPORT SYSTEMS (e.g. LAC OPERON).

TRANSCRIPTION of most genes involves the SIGMA FACTOR σ^{70}; other sigma factors include σ^{32} (see HEAT-SHOCK PROTEINS). A proportion of mRNA transcripts are polyadenylated at the 3′ end [characterization of the *E. coli* poly(A)polymerase: NAR (2000) *28* 1139–1144].

Gene transfer can occur *in vivo* by CONJUGATION or TRANSDUCTION. Competence in TRANSFORMATION can be induced in the laboratory in ice-cold solutions of calcium chloride [possible mechanism: JB (1995) *177* 486–490]. High-efficiency transformation can be achieved by ELECTROPORATION [NAR (1999) *27* 910–911].

Metabolism is respiratory under aerobic conditions (see RESPIRATION). FERMENTATION (sense 1) or ANAEROBIC RESPIRATION occurs anaerobically. Glucose, mannitol and various other substrates are taken up by the PTS (q.v.). Lactose is taken up by proton–lactose symport, and melibiose by Na^+/melibiose symport (see SODIUM MOTIVE FORCE). Glucose is fermented (usually aerogenically) by a MIXED ACID FERMENTATION (cf. ALKALESCENS-DISPAR GROUP).

Sulphur requirements may be obtained from sulphate; sulphate is reduced, intracellularly, to sulphide – which is incorporated into *O*-acetylserine, forming cysteine. Nitrogen may be taken up as ammonia (see AMMONIA ASSIMILATION).

Typical results of 'biochemical' tests: MR +ve; VP −ve; indole +ve; citrate −ve; lactose +ve; malonate −ve; urease −ve; LDC +ve; hydrogen sulphide (on TSI) −ve; no growth with KCN; gelatin hydrolysis −ve.

Some strains give atypical reactions which, in at least some cases, are plasmid-mediated; for example: citrate utilization (Cit plasmid), hydrogen sulphide production (Hys plasmid [FEMS (1983) *20* 7–11]), HAEMOLYSIN production (Hly plasmid, often associated with Ent and K88 plasmids in ETEC strains), urease production etc.

Growth occurs on a range of media (e.g. NUTRIENT AGAR, EMB AGAR, MACCONKEY'S AGAR) but (usually) not on e.g. DCA. The optimum growth temperature is 37°C, and the minimum doubling time is ~20 minutes. Colonies of *E. coli* on nutrient agar (37°C/24 hours) are 1–3 mm in diameter and may be smooth, entire, low-convex, greyish-translucent, butyrous and easily emulsifiable – or rough, dry, and not easily emulsifiable (see SMOOTH–ROUGH VARIATION). Mucoid and slime-forming strains also occur. Identification can be aided by the MUG TEST.

Strains of *E. coli* have various modes of PROTEIN SECRETION (q.v.).

E. coli and its components (e.g. RESTRICTION ENDONUCLEASES) are widely used in biotechnology (see e.g. OVERPRODUCTION).

E. coli is a member of the GASTROINTESTINAL TRACT FLORA in man and animals, and may occur in soil and water as a result of faecal contamination. (It is often regarded as an INDICATOR ORGANISM for faecal pollution.)

Pathogenic strains of *E. coli* include EAGGEC, EIEC, EPEC, ETEC and UPEC (for *E. coli* O157:H7 see EHEC). (See also COLIBACILLOSIS; COLISEPTICAEMIA; CYSTITIS; HAEMOLYTIC URAEMIC SYNDROME; HJÄRRE'S DISEASE; MENINGITIS; OEDEMA DISEASE; OSTEOMYELITIS; PNEUMONIA; URINARY TRACT INFECTION.)

[PCR-based detection of *E. coli* O157: JCM (1998) *36* 1801–1804.]

[Target genes for virulence assessment of *Escherichia coli* isolates from food, water and the environment: FEMS Reviews (2000) *24* 107–117.]

Other species within the genus. E. blattae is an apparently harmless commensal in the hindgut of the cockroach (*Blatta orientalis*). Test reactions: indole −ve; malonate +ve; lactose −ve.

Other proposed species include *E. fergusonii* [JCM (1985) *21* 77–81], *E. hermanii* [JCM (1982) *15* 703–713] and *E. vulneris* [JCM (1982) *15* 1133–1140].

***escN* gene** See PATHOGENICITY ISLAND.

esculin *Syn.* AESCULIN.

Eσ An RNA POLYMERASE holoenzyme.

***espA* gene** See PATHOGENICITY ISLAND.

***espB* gene** See PATHOGENICITY ISLAND.

EspB protein See PROTEIN SECRETION (type III systems).

***espE* gene** See PATHOGENICITY ISLAND.

EspE protein See PROTEIN SECRETION (type III systems).

espundia See MUCOCUTANEOUS LEISHMANIASIS.

ESR ELECTRON SPIN RESONANCE.

***estA* gene** See ETEC.

established cell line (continuous cell line) A population of cells – derived either from a tumour, or from a TISSUE CULTURE following *transformation* – which appears to be capable of unlimited in vitro propagation. See e.g. BHK-21, BS-C-1, DETROIT-6, HELA, HEP-2, VERO and WISH.

***estB* gene** See ETEC.

ET EPIDERMOLYTIC TOXIN.

η-value (of a disinfectant) See DILUTION COEFFICIENT.

etamycin *Syn.* VIRIDOGRISEIN.

etanercept See TNF.

ETC ELECTRON TRANSPORT CHAIN.

ETEC Enterotoxigenic *Escherichia coli*: strains of *E. coli* which adhere to the small intestine in man and/or animals and form 'heat-labile' ENTEROTOXIN (LT-I/LT-II: inactivated at ~60°C/30 min) and/or 'heat-stable' enterotoxin (ST-I/ST-II: not inactivated at 100°C/30 min). Strains of ETEC cause mild to severe (cholera-like) diarrhoea in humans of all ages and/or similar disease in animals. (See also COLIBACILLOSIS and FOOD POISONING; cf. EAGGEC, EIEC, EHEC and EPEC.)

The ability to adhere to (i.e. colonize) the small intestine – crucial for virulence – involves various plasmid-encoded FIMBRIAE. Strains from humans have non-fimbrial and fimbrial *colonization factors*, e.g. CS6 and 'colonization factor antigens' CFAI (F2) and CFAII (F3); CFAII has three components: CS1, CS2 and CS3 (CS = 'coli surface'). Strains of ETEC from animals have e.g. K88 or K99 FIMBRIAE (see also F6 FIMBRIAE). [Fimbriae of human strains of ETEC and their possible use in vaccines: RMM (1996) *7* 165–177.]

LT-I closely resembles the CHOLERA TOXIN in structure: both toxins have the AB_5 arrangement of subunits (found also e.g. in SHIGA TOXIN), and there is an ~80% homology in amino acids. LT-I and CT cross-react antigenically, and they also have similar binding sites and a common mode of action; however, for unknown reasons, disease produced by LT-I is generally less severe, and of shorter duration, than that due to CT.

LT-II occurs e.g. in pigs, rarely in humans. The A subunit is <60% homologous to that of LT-I, and the B subunits are essentially non-homologous (so that the binding site differs); the basic mode of action is similar to that of LT-I [MR (1996) *60* 167–215 (pp 191, 192)].

Genes encoding the LT toxins are carried on 'Ent plasmids'.

ST-I (= STa) is a (methanol-soluble) peptide containing 18 or 19 amino acids produced from a 72-amino-acid precursor; following cleavage, three disulphide bonds are catalysed in the peptide by the protein DsbA prior to secretion. ST-I binds to receptors in the brush border membrane; the receptors have guanylate cyclase activity, and binding by ST-I activates the cyclase – leading to increased levels of intracellular cyclic GMP that apparently stimulate secretion of chloride and/or inhibit the absorption of NaCl. The net result is believed to be secretion of fluid into the gut. An alternative view of the mode of action of this toxin involves interruption of the luminal acidification mechanism (apparently through inhibition of the Na^+/H^+ exchanger in the brush border membrane) leading to a reduction in the absorption of fluid [JAM (2001) *90* 7–26]. Interestingly, the receptor sites for ST-I are the same as those used by *guanylin* – an endocrine hormone (a peptide of 15 amino acid residues) whose normal function apparently involves regulation of NaCl and water homeostasis in the gut. (The number of receptor sites has been reported to decrease with age – infants having a large number of receptors; this may give a reason for the increased severity of diarrhoea in infants and young children with ST-I infection.) [MR (1996) *60* 167–215 (pp 190, 191); MP (1998) *24* 123–131.]

ST-II (STb) is a (methanol-insoluble, trypsin-sensitive) polypeptide containing 48 amino acid residues (derived from a precursor of 71 amino acid residues). The toxin does not cause secretion in the small intestine of mice or rats owing to its inactivation by trypsin. It is apparently without effect on human intestine under in vitro (USSING CHAMBER) conditions; strains secreting ST-II are infrequently involved in human disease. The mode of action is unknown but apparently does not involve cyclic nucleotides (i.e. the mode of action differs from that of ST-I). [ST-II: Microbiology (1997) *143* 1783–1795.]

ST-I is encoded by *estA* – a plasmid-borne gene located within a transposon (Tn*1681*). ST-II is encoded by the plasmid-borne gene *estB*.

[Heat-stable enterotoxin associated with resistance to colon cancer: PNAS (2003) *100* 2695–2699.]

ETEC strains commonly belong to serogroups O6, O8, O15, O25, O27, O63, O78, O115, O148, O153 and O159.

ethambutol (Myambutol) A synthetic drug, *d*-(2,2′-(ethylene diimino)-di-1-butanol), used (e.g. in combination with ISONIAZID) in the treatment of TUBERCULOSIS; it inhibits growing cells of *Mycobacterium tuberculosis* and *M. bovis*. Major adverse effects include optic neuritis and a loss of the ability to discriminate red/green (particularly in patients on a high-dose regimen) [BCID (1997) *4* 53–54].

ethanol (as an antimicrobial agent) See ALCOHOLS.

ethanol fermentation See ALCOHOLIC FERMENTATION and INDUSTRIAL ALCOHOL.

ethazole (5-ethoxy-3-trichloromethyl-1,2,4-thiadiazole) An agricultural systemic ANTIFUNGAL AGENT used e.g. as a seed dressing for protection against *Pythium*, *Phytophthora* and *Rhizoctonia*.

ethidium bromide (2,7-diamino-10-ethyl-9-phenylphenanthridinium bromide) A trypanocidal, bacteriostatic dye. It is an INTERCALATING AGENT ($\phi = 26°$) which inhibits transcription and DNA replication; it binds specifically to dsDNA (or to base-paired regions of ssDNA) with no apparent preference for particular base sequences. In eukaryotes it is highly selective for mtDNA and kinetoplast DNA. Ethidium bromide can induce nuclease attack and (non-enzymic) photochemical nicking of DNA.

Owing to their limited capacity to unwind, small cccDNA molecules can bind less ethidium bromide than can equivalent linear or nicked DNAs, and hence show a smaller reduction in density in the presence of ethidium bromide; this is exploited in methods for separating ccc from linear/nicked DNA by isopycnic ultraCENTRIFUGATION.

ethirimol (5-*n*-butyl-2-ethylamino-4-hydroxy-6-methylpyrimidine; trade names: e.g. Milstem, Milgo) A HYDROXYPYRIMIDINE ANTIFUNGAL AGENT which is used as a seed dressing and as a foliar spray for the control of powdery mildews in cereals; it also controls powdery mildew in e.g. sugar beet.

ethyl stibamine See ANTIMONY.

ethylene (*plant pathol.*) A PHYTOHORMONE which, in higher plants, can induce e.g. germination, flowering, fruit ripening, senescence – depending on the plant and its stage of development. Ethylene is also produced by plant tissues in response to stress, damage, or infection by certain pathogens, and large quantities are produced by plant tissues exhibiting HYPERSENSITIVITY; it may give rise to certain symptoms of disease–e.g. EPINASTY, premature ripening of fruit or leaf-fall.

ethylene glycol See ALCOHOLS.

ethylene oxide A colourless, highly reactive, water-soluble cyclic ether (b.p. 10.8°C) used (as a vapour) for the DISINFECTION or STERILIZATION of e.g. medical equipment and certain heat-labile materials. The vapour forms explosive mixtures with air in a wide range of proportions and is commonly diluted with carbon dioxide, nitrogen or fluorohydrocarbons: e.g. *Carboxide* contains 10% ethylene oxide, 90% CO_2 (see also e.g. CRYOXICIDE, PENNOXIDE). Ethylene oxide is an ALKYLATING AGENT which reacts with e.g. proteins and nucleic acids. Antimicrobial activity depends e.g. on temperature, concentration, duration of action, water content of the target organisms (dried spores etc can be highly resistant), and relative humidity. The use of ethylene oxide as a STERILANT thus requires careful control of conditions. For reliable sterilization the optimum relative humidity is variously reported as 35–60% [Book ref. 2, pp. 481–483] and ca. 100% [Book ref. 12, pp. 548–568]. The vapour has moderate powers of penetration, and sterilization usually necessitates exposure for at least several hours. Ethylene oxide is often used in an apparatus resembling a vacuum autoclave to allow conditions to be controlled and to reduce sterilization time. Ethylene oxide is absorbed by, or reacts with, substances such as hessian, rubber, and certain plastics; this may significantly reduce the effective concentration of the vapour. Ethylene oxide is readily hydrolysed to ethylene glycol; the toxic substance ethylene chlorohydrin (b.p. 129°C) is formed by hydrolysis in the presence of chloride ions. (N.B. Ethylene oxide is irritant and may be carcinogenic.)

The sterilizing activity of ethylene oxide at a given location can be monitored by a *Royce indicator sachet*: a small plastic bag containing $MgCl_2$, HCl and the pH indicator BROMPHENOL BLUE; under sterilizing conditions ethylene oxide penetrates the sachet and is hydrolysed to ethylene chlorohydrin, causing a rise in pH. Bacterial endospores may be used as biological monitors [see e.g. JAB (1983) *55* 39–48].

ethylenediaminetetraacetic acid See EDTA.

ethylethane sulphonate A mutagenic ALKYLATING AGENT.

ethylhydrocupreine hydrochloride See OPTOCHIN.

ethylmethane sulphonate See ALKYLATING AGENTS.

etiology See AETIOLOGY.

etioplast An achlorophyllous PLASTID formed e.g. in dark-grown (etiolated) seedlings of most types of angiosperm; most algae, mosses and ferns can synthesize chlorophylls in the dark. (See also PROLAMELLAR BODY.)

etioporphyrins See PORPHYRINS.

ETP (electron-transport particle; sub-mitochondrial particle, SMP) A type of vesicle formed by the ultrasonic disruption of mitochondria (see MITOCHONDRION and ULTRASONICATION). An ETP consists of *inverted* mitochondrial membranes containing electron-transport and H^+-ATPase functions (although the actual composition varies considerably depending e.g. on the mode of preparation); ATP synthesis in an ETP thus correlates with the *outward* passage of protons. (cf. KABACKOSOME.)

etridiazole (trade names: Aaterra, DWELL, Pansoil, Terrazole) 5-Ethoxy-3-trichloromethyl-1,2,4-thiadiazole, a NITRIFICATION INHIBITOR which is also used as a seed and soil fungicide – e.g. for the control of damping off caused by *Phytophthora* and *Pythium*. [Review: Book ref. 121, pp. 45–62.]

etruscomycin A MYCOSAMINE-containing tetraene POLYENE ANTIBIOTIC structurally very similar to PIMARICIN.

Euascomycetes A class of fungi comprising most or all ascomycetes other than the HEMIASCOMYCETES. The class is not recognized in modern taxonomic schemes.

Euastrum A genus of placoderm DESMIDS.

Eubacteria (*taxon.*) A kingdom (now obsolete) which included all those prokaryotes which were not classified in the kingdom ARCHAEBACTERIA. [The major eubacterial taxa defined: SAAM (1985) *6* 143–151.] Members of the Eubacteria are now included in the domain BACTERIA (q.v.).

Eubacterium An ill-defined genus of Gram-positive, asporogenous, anaerobic, rod-shaped bacteria consisting of those species which are excluded from the genera ACTINOMYCES, ARACHNIA, BIFIDOBACTERIUM, LACTOBACILLUS and PROPIONIBACTERIUM on the basis of the nature of the products of fermentation. *Eubacterium* spp typically form butyric and other fatty acids; some species produce acetic and formic acids (and ethanol), and some do not form acids. The organisms occur e.g. in the human gut and in the RUMEN. [Book ref. 46, pp. 1903–1911.]

Eubenangee subgroup See ORBIVIRUS.

Eucapsis A phycological genus of unicellular 'blue-green algae' (Chroococcales) in which the cells occur (in nature) in cubical packets. See SYNECHOCYSTIS.

eucarpic Refers to those organisms in which only part of the thallus is involved in the formation of reproductive organs. (cf. HOLOCARPIC.)

eucaryote *Syn.* EUKARYOTE.

Eucheuma A genus of tropical marine algae (division RHODOPHYTA). The organisms are used as a source of CARRAGEENAN.

euchromatin See CHROMATIN.

euchrysine *Syn.* ACRIDINE ORANGE.

Eucoccida An order of protozoa (subclass Coccidia) later incorporated – together with organisms of the TOXOPLASMEA – in the order EUCOCCIDIORIDA.

Eucoccidiida An order of protozoa [JP (1980) *27* 37–58] equivalent to the EUCOCCIDIORIDA.

Eucoccidiorida An order of protozoa (subclass COCCIDIASINA), parasitic in vertebrates and invertebrates, in which both schizogony and sexual reproduction occur. Suborders: ADELEORINA, EIMERIORINA, HAEMOSPORORINA.

Eucoronis See RADIOLARIA.

euendolith See ENDOLITHIC.

euflavin *Syn.* ACRIFLAVINE.

eu-form rusts *Syn.* MACROCYCLIC RUSTS.

Euglena A genus of photosynthetic EUGLENOID FLAGELLATES (q.v. for general features). Most species are green, but some (e.g. *E. rubra*) contain a bright red pigment which may mask the green colour under certain conditions. The vegetative cell is typically elongated and fusiform (ca. 15–400 μm long, according to species) but can vary its shape extensively. There is only one emergent flagellum; this bears a single helical row of fine hairs along its length, as well as a coating of shorter hairs. A lateral swelling (the *paraflagellar body*) is present on the long flagellum in the region within the reservoir, and this swelling lies opposite an EYESPOT present in the cytoplasm. A contractile vacuole empties into the reservoir. Other cell contents include the chloroplasts, a single large nucleus, paramylon granules etc.

Euglena spp exhibit two distinct types of motility: a spiral swimming motion (see FLAGELLAR MOTILITY (b)) and 'euglenoid movement' (*metaboly*) – a type of crawling motility involving progressive waves of swelling and constriction of the cell body; metaboly is the main type of motility in *E. deses*. [Regulation of cell shape in *E. gracilis* – localization of actin and myosin in the region of the pellicle: JCS (1985) *77* 197–208.]

Most *Euglena* spp (e.g. *E. gracilis*, *E. rubra*, *E. spirogyra*) occur in freshwater habitats, particularly those rich in organic matter (e.g. farm ponds) where they may form dense green or red blooms. (See also SAPROBITY SYSTEM.) *E. cyclopicola* frequently occurs attached by its anterior end to microscopic crustacea (*Cyclops*, *Daphnia* etc), when it resorbs its flagellum. Some species occur in brackish waters (e.g. in estuarine mudflats), a few occur in the sea. *E. deses* is common in soil.

Euglenamorpha See EUGLENOID FLAGELLATES.

Euglenida See PHYTOMASTIGOPHOREA.

euglenids *Syn.* EUGLENOID FLAGELLATES.

euglenoid flagellates (euglenids; euglenoids) A group of unicellular organisms regarded either as algae (class Euglenophyceae, division Euglenophyta) or as protozoa (see PHYTOMASTIGOPHOREA). The vegetative cell has no cell wall, but has a *pellicle* consisting of spirally arranged, interlocking, proteinaceous strips beneath the plasmalemma. The pellicle may be flexible (in e.g. *Astasia*, *Distigma* and EUGLENA) or rigid (in e.g. *Menoidium*, PHACUS and *Rhabdomonas*). Mucilage-secreting *muciferous bodies* occur in helical rows beneath the pellicle. In some genera (e.g. *Klebsiella*, TRACHELOMONAS) the cell is partly or totally enclosed in a mineralized LORICA. At the anterior end of the cell is an invagination, the *reservoir* (also called the 'gullet' although it does not function in feeding), which opens to the exterior via a narrow *canal*. A contractile vacuole commonly empties into the reservoir. In most genera two flagella arise from basal bodies at the base of the reservoir, but in some genera (e.g. *Astasia*, *Colacium*, *Euglena*, *Phacus*) only one flagellum (containing a PARAXIAL ROD) actually emerges from the cell via the canal; in e.g. *Distigma*, *Eutreptia*, PERANEMA and HETERONEMA two flagella are emergent, one of which is directed anteriorly, the other laterally or posteriorly. (Parasitic euglenids may have 3 or more emergent flagella.)

Many euglenids are photosynthetic, possessing green chloroplasts containing chlorophylls *a* and *b* and various carotenoids. However, photosynthetic species can apparently also grow heterotrophically on e.g. acetate or ethanol in the dark, and many genera (e.g. *Astasia*, *Khawkinea*, *Rhabdomonas*) are colourless and obligately heterotrophic; PERANEMA and related genera are holozoic. A storage compound characteristic of both photosynthetic and heterotrophic euglenids is PARAMYLON.

Euglenids reproduce chiefly by longitudinal binary fission; MITOSIS occurs within the nuclear membrane and does not involve a spindle. Palmelloid phases are common in some species. (cf. COLACIUM.) Desiccation-resistant cysts are formed in e.g. *Distigma* and *Euglena* under adverse conditions. Sexual reproduction is virtually unknown.

Euglenoid flagellates are mainly free-living aquatic organisms, occurring in fresh water (e.g. most *Euglena* spp, *Phacus*) or in brackish or sea water (e.g. some *Euglena* spp, *Eutreptia*, *Klebsiella*). Some species are parasitic (e.g. *Euglenamorpha* in the gut of tadpoles).

euglenoid movement See EUGLENA.

Euglenophyceae See EUGLENOID FLAGELLATES.

Euglypha A genus of testate amoebae (order Gromiida, class FILOSEA) in which the test (ca. 30–160 μm diam.) is composed of siliceous scales which are synthesized within the cytoplasm before incorporation into the test. Pseudopodia are generally narrow, filamentous, and sometimes anastomosing. Species occur in fresh water, activated sludge, anaerobic regions in acidic bogs, etc. (cf. PAULINELLA.)

eugonic Refers to a strain of e.g. *Mycobacterium* which, on culture, produces growth which is more luxuriant than that produced by other (*dysgonic*) strains; the growth itself may also be described as eugonic (or dysgonic).

eugregarines See GREGARINASINA.

eukaryote (eucaryote) A type of CELLULAR (sense 2 or 3) organism in which the CHROMOSOMES are separated from the cytoplasm by a membrane (see NUCLEUS) and usually contain HISTONES (cf. DINOFLAGELLATES); the CYTOPLASMIC MEMBRANE contains sterols; mitochondria are commonly present; chloroplasts occur in photosynthetic species; RIBOSOMES in the *cytoplasm* are of the 80S type; the CELL WALL (when present) contains e.g. CELLULOSE or CHITIN; storage compounds apparently never include poly-β-hydroxybutyrate; flagella or cilia (when present) are structurally relatively complex (see FLAGELLUM (b) and CILIUM). Eukaryotic MICROORGANISMS include ALGAE, FUNGI, LICHENS and PROTOZOA.

Eukaryotes may have arisen some 2–3.5 billion years ago [Science (1996) *271* 470–477] from an energy-based symbiotic relationship between cells of the ARCHAEA and BACTERIA [the hydrogen hypothesis: Nature (1998) *392* 37–41].

(cf. PROKARYOTE.)

eumycetoma See MADUROMYCOSIS.

Eumycetozoa See EUMYCETOZOEA.

Eumycetozoea According to one protozoal classification scheme [JP (1980) *27* 37–58]: a class of SLIME MOULDS of the RHIZOPODA. Members of the class typically form myxamoebae with filose pseudopodia or subpseudopodia; flagella, when formed, lack mastigonemes. Aerial fruiting bodies are produced. Subclasses: Protosteliia (equivalent to the PROTOSTELIOMYCETES together with CERATIOMYXA); Dictyosteliia (equivalent to the DICTYOSTELIOMYCETES); Myxogastria (equivalent to the MYXOMYCETES). According to another classification scheme [Book ref. 147, pp. 4–5], equivalent taxa are the class Eumycetozoa (subphylum MYCETOZOA) and its constituent subclasses Protostelia, Dictyostelia and Myxogastria.

Eumycota In some mycological classification schemes [e.g. Book ref. 64]: a division which includes the 'true fungi' (subdivisions ASCOMYCOTINA, BASIDIOMYCOTINA, DEUTEROMYCOTINA, MASTIGOMYCOTINA and ZYGOMYCOTINA). (cf. MYXOMYCOTA.)

***Euphorbia* mosaic virus** See GEMINIVIRUSES.

euphotic zone See PHOTIC ZONE.

euploid (1) Having the basic haploid number of chromosomes or any *exact* multiple of the haploid number. (cf. ANEUPLOID.) (2) Having a chromosome complement characteristic of the species. (cf. HETEROPLOID.)

Euplotes A genus of highly evolved ciliate protozoa (order HYPOTRICHIDA); species occur in freshwater, brackish and marine habitats, and some (e.g. *E. moebiusi*) are common in some SEWAGE TREATMENT plants. Cells: roughly ovoid, ca. 50–150 μm in length, with a flattened ventral side and a dorsal side which is often markedly ridged; the cytostome is anterior and ventral, the distal part of the AZM forming a conspicuous, protruding fringe. The ventral surface carries a number of cirri which occur in localized groups and on which the organism is capable of 'creeping' or 'walking' movements; other forms of somatic ciliature are absent. The macronucleus is typically band-like and curved, the micronucleus small and spherical. [Macronuclear development in *E. crassus*: JCB (1985) *101* 79–84.] Some strains contain endosymbionts (see OMICRON). A contractile vacuole is present. Asexual reproduction occurs by binary fission. CONJUGATION occurs (see SYNGEN). TRANSLATIONAL FRAME-SHIFTING seems common [Cell (2002) *111* 763–766]. Food: algae, bacteria, other protozoa.

European bat lyssavirus See LYSSAVIRUS.

European blastomycosis *Syn.* CRYPTOCOCCOSIS.

European foulbrood A BEE DISEASE which affects larvae (ca. 4–5 days old) of *Apis mellifera* and *A. cerana*; infected larvae die, turn brown and putrefy. The causal agent is *Melissococcus pluton* (formerly called *Streptococcus pluton*) [JAB (1983) *55* 65–69]; infection occurs by ingestion, and the bacteria multiply within – and sometimes fill – the midgut. Secondary invaders may include e.g. *Enterococcus faecalis* and *Bacillus alvei*. (cf. AMERICAN FOULBROOD.)

European swine fever *Syn.* SWINE FEVER.

Eurotiales An order of mainly saprotrophic fungi of the subdivision ASCOMYCOTINA; anamorphs occur e.g. in the genera ASPERGILLUS and PENICILLIUM. Ascocarp: cleistothecioid, sometimes (e.g. in *Talaromyces*) very loosely constructed of interwoven hyphae, with no appendages; the development of the ascocarp may involve intertwining of the ascogonium and antheridium. Asci: unitunicate, thin-walled, formed at various levels in the ascocarp. Ascospores: unicellular, dark or colourless. Genera: e.g. BYSSOCHLAMYS, EMERICELLA, *Emericellopsis* (see also ACREMONIUM), EUROTIUM, *Pleuroascus*, SARTORYA, TALAROMYCES, THERMOASCUS.

Eurotium A genus of fungi (order EUROTIALES) which include the teleomorph of *Aspergillus glaucus*. The ascocarp is a globose, commonly yellow, cleistothecium containing globose to pyriform, evanescent asci.

eury- A prefix signifying 'wide' or 'extensive' – e.g. a *euryhaline* organism is one able to tolerate a wide range of salt concentrations. (cf. STENO-.)

Euryarchaeota See ARCHAEA.

euryxenous Having a broad host range.

euseptum See SEPTUM (b).

Eutreptia See EUGLENOID FLAGELLATES.

eutrophic Of lakes rivers etc: rich in nutrients, particularly those which support the growth of aerobic photosynthetic organisms. (cf. OLIGOTROPHIC; see also EUTROPHICATION.)

eutrophication The enrichment of a habitat or environment with inorganic materials – in particular those substances (e.g. nitrogen and phosphorus compounds) which encourage the growth of plants and algae. In rivers and lakes, eutrophication may occur as a result of SELF PURIFICATION.

evanescent Having a transient existence.

EVE See IPN VIRUS.

Everglades virus See ALPHAVIRUS.

Evernia A genus of LICHENS (order LECANORALES); photobiont: a green alga. The thallus is strap-like, dichotomously branched, typically hanging in festoons from trees; that of *E. prunastri* (oak-moss), a species used commercially in the perfume industry, is greenish-grey above, whitish below.

evernic acid A depside LICHEN substance which has antibacterial properties; it occurs e.g. in *Evernia prunastri*.

evernimicin (SCH 27899) An antibiotic with good in vitro activity against various Gram-positive bacteria [JAC (2001) *47* 15–25].

Ewingella A genus of bacteria of the ENTEROBACTERIACEAE. *E. americana* (formerly 'enteric group 40') produces acid but no gas from glucose, is VP +ve, DNase −ve, lipase −ve, LDC and ODC −ve. [Ann. Mic. (1983) *134A* 39–52.]

exanthem A skin rash, or a disease characterized by a skin rash (cf. ENANTHEM).

exanthem subitum (roseola infantum; sixth disease; pseudorubella) A benign disease, of infants and young children, in which high fever (for 3–4 days) is followed by a rubelliform rash on the trunk, spreading to the limbs and face. Human (beta) herpesvirus 6 has been recognized as a causal agent. (cf. ERYTHEMA INFECTIOSUM.)

***exbB* gene** See TONB PROTEIN.

***exbD* gene** See TONB PROTEIN.

Excellospora A genus of thermophilic soil bacteria (order ACTINOMYCETALES, wall type III; group: maduromycetes). The organisms form both substrate and aerial mycelium, both of which give rise to spores. Type species: *E. viridilutea*. [Ecology, isolation, cultivation: Book ref. 46, pp. 2103–2117.]

exchange diffusion *Syn.* ANTIPORT.

exciple *Syn.* EXCIPULUM.

excipulum (1) See APOTHECIUM. (2) The peridium (wall) of a PERITHECIUM.

excipulum proprium See PROPER MARGIN.

excipulum thallinum See THALLINE MARGIN.

excision repair Any DNA REPAIR process which can repair lesions affecting only a single strand of dsDNA (not, for example, interstrand cross-linking) and which involves removal and replacement of one or more nucleotides. Excision repair in *Escherichia coli* can be considered under three main headings:

(i) BASE EXCISION REPAIR, i.e. repair of an AP SITE after preliminary cleavage of a base (see ADAPTIVE RESPONSE and URACIL-DNA GLYCOSYLASE).

(ii) MISMATCH REPAIR, used mainly for errors that escape proofreading during DNA replication.

(iii) UvrABC-mediated repair (nucleotide excision repair; *short patch repair*). This common excision repair mechanism recognizes and repairs DNA which has been damaged/distorted by ultraviolet radiation or by certain other causes; damage may consist e.g. of a thymine dimer or a Pyr(6-4)Pyo photoproduct (see ULTRAVIOLET RADIATION). The ATP-dependent enzyme system UvrABC (also called 'ABC excinuclease') consists of three proteins encoded by genes *uvrA*, *uvrB* and *uvrC*. Initially a UvrA dimer binds to the damaged DNA. UvrB then binds to the dimer, and the complex $UvrA_2B$ translocates to the precise site of damage; such translocation is apparently ATP-dependent (the zinc-binding protein UvrA being an ATPase). UvrC displaces UvrA, and the UvrBC complex nicks the damaged strand on either side of the lesion, the 3′ nick being made first. UvrD (= helicase II) then displaces UvrC together with the short sequence of nucleotides containing the site of damage. DNA polymerase I fills the single-stranded gap, displacing UvrB, and the repair is completed by a ligase. In *E. coli* only about 10 nucleotides in and around the damaged site are excised (thus accounting for the name 'short patch repair'). (See also PHOTOLYASE.)

The UvrABC system is enhanced during the SOS RESPONSE. Another repair system, the so-called 'error-prone repair' system (involving *translesion synthesis* of DNA), appears to operate only during the SOS RESPONSE (q.v.).

excisionase See e.g. BACTERIOPHAGE λ.

exciter filter See MICROSCOPY (e).

exclusion limit See PORIN.

exclusion zone See MICROTUBULES.

exconjugant A cell which has separated from another following CONJUGATION.

excystation *Syn.* EXCYSTMENT.

excystment (excystation) The release of one or more vegetative (or other) cells from a CYST.

exergonic reaction Any reaction which is accompanied by the liberation of energy. (cf. ENDERGONIC REACTION.)

exflagellation See PLASMODIUM.

exfoliatin *Syn.* EPIDERMOLYTIC TOXIN.

Exidia See TREMELLALES.

Exiguobacterium A genus of Gram-positive, asporogenous, facultatively anaerobic, rod-shaped or coccoid, motile bacteria [JGM (1983) *129* 2037–2042]. *E. aurantiacum* is an ALKALOPHILE which has been isolated from potato processing effluent.

exine See CYST (a).

Exo Abbr. for exonuclease; for example, Exo III is EXONUCLEASE III.

Exobasidiales An order of plant-endoparasitic, characteristically GALL-inciting fungi (subclass HOLOBASIDIOMYCETIDAE) which form hymenia that emerge through the epidermis in parasitized plants. Genera include *Exobasidium* (see also BLISTER BLIGHT) and *Kordyana*. (cf. BRACHYBASIDIALES.)

Exobasidium See EXOBASIDIALES and BLISTER BLIGHT.

exobiology Extraterrestrial biology.

exocellular *Syn.* EXTRACELLULAR (sense 1 or 2).

exocellulase See CELLULASES.

exochelins Low MWt (<ca. 1000) iron-solubilizing peptides secreted by *Mycobacterium* spp – particularly in iron-deficient media. 'MB-type' exochelins (formed e.g. by *M. bovis* BCG and by *M. tuberculosis*) are compounds which can be extracted by chloroform after they have chelated ferric iron. 'MS-type' exochelins (formed e.g. by *M. smegmatis*) are not extractable by organic solvents.

exocyst See CYST (b).

exocytosis The secretion of material by a cell in a process which resembles ENDOCYTOSIS in reverse. (See also TETANOSPASMIN.)

exoenzyme (1) An ENZYME which sequentially removes units from one end of a polymer. (See e.g. β-AMYLASES; cf. ENDOENZYME.)

(2) Sometimes used as a synonym of 'extracellular enzyme'.

exoenzyme S An enzyme produced by strains of *Pseudomonas aeruginosa*; it appears to be capable of ADP-ribosylation of eukaryotic proteins (cf. EXOTOXIN A), but its role in virulence – if any – is unknown.

exoerythrocytic schizogony (in *Plasmodium*) See PLASMODIUM.

exogenote See MEROZYGOTE.

exogenous retrovirus A retrovirus (see RETROVIRIDAE) which infects a cell from a source outside that cell – e.g. a retrovirus transmitted horizontally from another animal, or one which has spread from another site in the same animal. Such a virus is absent from normal host cells (cf. ENDOGENOUS RETROVIRUS).

exoglucanase See CELLULASES.

exon See SPLIT GENE.

exonuclease A NUCLEASE which sequentially removes nucleotides from one end of a strand of NUCLEIC ACID. (cf. ENDONUCLEASE.)

exonuclease I See RECF PATHWAY.

exonuclease III An enzyme of *Escherichia coli* (product of the *xthA* gene) which degrades either strand of a dsDNA molecule from the 3′ end, leaving single-stranded 5′ ends.

exonuclease V See RECBC PATHWAY.

exonuclease VII An enzyme of *Escherichia coli* (product of the *xseA* gene) which degrades ssDNA processively from either end but leaves dsDNA intact.

exonuclease VIII See RECF PATHWAY.

exopectate lyase See PECTIC ENZYMES.

exopeptidase See PROTEASES.

exoperidium See LYCOPERDALES.

Exophiala See HYPHOMYCETES.

exopolygalacturonase See PECTIC ENZYMES.

exoribonuclease See RNASE.

exospore (1) A type of SPORE formed from the parent organism by budding or by septum formation and fission (cf. ENDOSPORE sense 1); see e.g. ACTINOMYCETALES and RHODOMICROBIUM. (See also CHAMAESIPHON and METHYLOCOCCACEAE.)

(2) *Syn.* EXOSPORIUM (sense 2).

exosporium (1) (*bacteriol.*) See ENDOSPORE (sense 1(a)). (2) (*mycol.*) (exospore; epitunica; trachytectum) The layer external to the two innermost layers (the endosporium – see ENDOSPORE sense 2 – and episporium) of a fungal spore wall.

exotoxin An extracellular TOXIN, i.e., a toxin which is secreted by a living cell. (cf. ENDOTOXIN sense 2.)

exotoxin A A protein exotoxin (MWt ca. 66000) produced by strains of *Pseudomonas aeruginosa*; it is believed to be an important virulence factor in *P. aeruginosa* infections in patients suffering from extensive burns, immunosuppressive therapy, etc.

The mode of action of exotoxin A appears to be identical with that of DIPHTHERIA TOXIN, although the two proteins differ in other respects: they show no immunological cross-reactivity; they are toxic for different cell lines (e.g. Vero cells are resistant to exotoxin A, murine cells are sensitive); they bind to different cell-surface receptors; they appear to be taken up by sensitive cells via different mechanisms; fragment A of exotoxin A is heat-labile. [Review: MR (1984) *48* 199–221; structure: PNAS (1986) *83* 1320–1324.] (See also IMMUNOTOXIN.)

exotoxin A (streptococcal) See SUPERANTIGEN.

exoxylanase See XYLANASES.

exp-1 protein See PLASMODIUM.

expanded gold standard See GOLD STANDARD.

exponential growth rate constant See GROWTH (a).

exponential phase (of growth) See BATCH CULTURE.

exponential silencing Downregulation of activity at a given gene promoter as a consequence of exponential growth in a rich medium [see e.g. EMBO (2001) *20* 1–11 (5)].

export (of a protein) The transmembrane translocation of a protein, towards the cell's exterior, without eventual release to the external environment. (cf. SECRETION.)

export apparatus See FLAGELLUM (a).

expression vector (*mol. biol.*) A CLONING vector that encodes functions for the transcription/translation of an inserted fragment of DNA. In addition to an origin of replication and a marker gene (e.g. an antibiotic-resistance gene), an expression vector includes e.g. (i) a promoter, commonly a strong, hybrid promoter (e.g. the TAC PROMOTER); (ii) a ribosome-binding site to ensure that, following transcription of the insert, the mRNA contains a Shine–Dalgarno sequence (needed for binding to the ribosome); (iii) a multiple cloning site (MCS) – a POLYLINKER with a range of restriction sites that permit flexibility in the preparation of the insert; (iv) a transcription terminator to inhibit unwanted readthrough.

Using an expression vector, a given insert can be initially cloned and then expressed. The insert is expressed by initiating transcription, and this can be achieved in various ways – for example, by a specific change in temperature. In an alternative mode of control, the vector incorporates a *lac* operator (see LAC OPERON) which regulates transcription of the insert and which is itself regulated by the host cell's LacI protein; activity of the promoter can be switched on e.g. by adding IPTG to the medium. (For cells which do not synthesize LacI, use can be made of a vector which incorporates a *lacI* gene.)

If required, a gene can be cloned in bacteria and then expressed in eukaryotic cells, a single vector carrying a prokaryotic origin of replication and control functions suitable for expression in eukaryotes.

extein See INTEIN.

extended batch culture See FED BATCH CULTURE.

extended-spectrum β-lactamases See β-LACTAMASES.

external guide sequence See SPLIT GENE (a).

extracellular (1) Refers to enzymes or structures etc which are external to the cells that produced them – whether or not they remain attached to, or form part of, the surface of the cell.

(2) Refers to enzymes or structures etc which are released by the cell into the surrounding medium. (cf. EPICELLULAR.)

extrachromosomal genes *Syn.* CYTOPLASMIC GENES.

extrachromosomal inheritance *Syn.* CYTOPLASMIC INHERITANCE.

extract broth See NUTRIENT BROTH.

extracytoplasmic Of a bacterial enzyme or redox carrier: located in the periplasmic region, either free or attached to the outer face of the cytoplasmic membrane [MR (1985) *49* 140–157 (141)].

extracytoplasmic oxidation In certain bacteria: the oxidation of a substrate at the external face of the cytoplasmic membrane, or within the periplasmic region, giving rise to (or augmenting) a proton motive force (pmf: see CHEMIOSMOSIS). During such oxidations, protons formed at the extracytoplasmic site contribute to pmf, while electrons (derived from the substrate) pass through the cytoplasmic membrane (a process sometimes called 'vectorial electron transfer') and are involved in electron- and proton-consuming reactions. [Early review: MR (1985) *49* 140–157.]

Substrates oxidized extracytoplasmically include e.g. (depending on species): hydrogen, formate, methanol, glucose and various ions – particularly ferrous ions. Intracellular terminal electron acceptors include e.g. oxygen, fumarate, sulphate ions and nitrate ions.

Oxidations carried out extracytoplasmically include e.g. the oxidation of hydrogen by *Desulfovibrio* spp, *Escherichia coli* and by those HYDROGEN-OXIDIZING BACTERIA containing an NAD-independent membrane-bound HYDROGENASE. In *E. coli* the hydrogenase is located on the external surface of the cytoplasmic membrane, while in *Desulfovibrio* spp a soluble hydrogenase is located in the periplasmic region.

In *Beggiatoa*, sulphide appears to be oxidized extracytoplasmically to elemental sulphur. It has been proposed that *Thiobacillus thiooxidans* can oxidize insoluble (colloidal) sulphur extracytoplasmically.

In *Thiobacillus ferrooxidans*, the oxidation of ferrous to ferric ions occurs extracytoplasmically; it appears to involve e.g. an oxidase, *c*-type and *a*-type cytochromes and RUSTICYANIN [see e.g. MR (1994) *58* 39–55 (47–48)].

In some bacteria, ions such as Mn^{2+} or U^{4+} may be oxidized extracytoplasmically.

Some bacteria (e.g. *Pseudomonas aeruginosa*) can generate pmf by extracytoplasmic oxidation of glucose to gluconate by glucose dehydrogenase (a QUINOPROTEIN) [JB (1985) *163* 493–499]; in these organisms the enzyme glucose dehydrogenase, with its cofactor pyrroloquinoline quinone (PQQ), occurs on the outer surface of the cytoplasmic membrane. The gluconate formed can then be transported across the membrane, phosphorylated to 6-phosphogluconate, and fed into the ENTNER–DOUDOROFF PATHWAY.

E. coli (and some related bacteria) normally form a membrane-bound glucose dehydrogenase which lacks the cofactor (PQQ); such organisms can carry out extracytoplasmic oxidation only if they are supplied with exogenous PQQ [IJSB (1989) *39* 61–67]. However, a mutant strain of *E. coli* with a non-functional PTS system is able to synthesize PQQ and to carry out extracytoplasmic oxidation; this has suggested that PQQ-encoding genes are normally present but not expressed [JGM (1991) *137* 1775–1782].

extranuclear genes *Syn.* CYTOPLASMIC GENES.

extravasation See INFLAMMATION.

extremophile See ARCHAEA.

extrinsic allergic alveolitis (EAA; hypersensitivity pneumonitis) A form of pneumonitis, which appears to involve an ARTHUS-TYPE REACTION and a DELAYED HYPERSENSITIVITY (type IV) reaction, caused by the inhalation of certain allergen(s). Examples of EAA include e.g. BAGASSOSIS, CHEESE-WASHERS' LUNG, FARMERS' LUNG, MAPLE BARK STRIPPERS' DISEASE, SUBEROSIS. [CIA (1984) *4* 173–192.]

extrusome A generic term for an extrusile organelle (function often unknown), various types of which occur near the cell surface in many ciliates and in some other organisms. Examples:

conocyst, EJECTOSOME, FIBROCYST, haptocyst, kinetocyst (see AXOPODIUM), PEXICYST, rhabdocyst, TOXICYST, TRICHOCYST (cf. *polar filament* in MICROSPOREA and MYXOSPOREA).

exudative dermatitis *Syn.* GREASY PIG DISEASE.

EYA See EGG-YOLK AGAR.

eye microflora The surface of the eye and the conjunctival membranes normally bear only a sparse microflora owing to the constant flushing action of the secretions of the lacrimal glands; these secretions contain certain antimicrobial agents – e.g. LACTOFERRIN, LYSOZYME, and immunoglobulins (see IGA). Organisms which may be isolated from the eye surfaces include certain members of the SKIN MICROFLORA, e.g. staphylococci. (See also BODY MICROFLORA and CONJUNCTIVITIS.)

eyepiece micrometer See MICROMETER.

eyespot (1) (stigma) In certain algae: a body consisting of a group, layer or layers of closely-packed lipid globules which, according to species, occurs within the CHLOROPLAST (e.g. in members of the Chlorophyta) or outside it (e.g. in euglenoid flagellates). In most or all cases an eyespot contains orange or red CAROTENOID pigments. In certain dinoflagellates (e.g. *Nematodinium armatum*) there is a large complex eyespot (termed an *ocellus*) which incorporates a lens and cup-shaped structure containing one or more rows of pigment granules. In at least some algae the eyespot is believed to be involved in PHOTOTAXIS.

[The eyespot in green algae: Book ref. 167, pp. 193–268.]

(2) A CEREAL DISEASE which affects mainly winter wheat and barley; causal agent: *Pseudocercosporella herpotrichoides* (which can overwinter in stubble). (cf. HALO SPOT.) Pale oval spots, each with a diffuse dark margin, develop in the spring on leaf sheaths and stalks of the tillers – usually close to soil level; each eyespot lesion has a greyish centre and may contain one or more black dots. A grey mycelium develops within the stem; lodging may occur. Control: e.g. use of resistant varieties, or early foliar spray with e.g. BENOMYL or THIOPHANATE-METHYL. In *sharp eyespot* (caused by *Rhizoctonia solani*) the eyespot lesions have sharper margins, and they are more numerous – and develop further up the stem – than those of 'true' eyespot; mycelium does not develop within the stem.

1. Words in SMALL CAPITALS are cross-references to separate entries.
2. Keys to journal title abbreviations and Book ref. numbers are given at the end of the Dictionary.
3. The Greek alphabet is given in Appendix VI.
4. For further information see 'Notes for the User' at the front of the Dictionary.

F

f. See FORMA SPECIALIS.

F Phenylalanine (see AMINO ACIDS).

F-actin See ACTIN.

F antigens Fimbrial antigens – see FIMBRIAE. (cf. K ANTIGENS.)

F^+ donor See F PLASMID.

F-duction See SEXDUCTION.

F factor See F PLASMID.

F layer See CYANOBACTERIA (section II).

F-like pili See PILI.

F pili See PILI.

F plasmid (formerly F factor, 'fertility factor') An IncFI PLASMID which occurs e.g. in *Escherichia coli* with a COPY NUMBER of usually 1 or 2; it consists of a ccc dsDNA molecule (ca. 95 kilobases) which includes two copies of the INSERTION SEQUENCE IS*3*, one each of IS*2* and IS*1000*, and sequences relating to replication, incompatibility and CONJUGATION (sense 1b). An F plasmid confers on the host cell the characteristics of a conjugal donor, and it can also affect the cell's phenotype in other ways: see e.g. FEMALE-SPECIFIC PHAGE. (See also PIF.)

Within the host cell, an F plasmid exists either in the autonomous (extrachromosomal) state, i.e. as a free, circular plasmid, or integrated in the host's chromosome. (cf. EPISOME.) Integration occurs by homologous RECOMBINATION between an insertion sequence in the plasmid and one in the chromosome (see CAMPBELL MODEL); copies of insertion sequences IS*2* and IS*3* are common in the chromosome of *E. coli*. An integrated F plasmid may undergo aberrant excision from the chromosome: see PRIME PLASMID.

Conjugation mediated by the F plasmid. The F plasmid can mediate conjugation either in the autonomous state or in the integrated state.

In the autonomous (extrachromosomal) state an F plasmid can efficiently promote its own intercellular transfer by CONJUGATION (sense 1b). A cell containing an autonomous F plasmid is called an F^+ donor, while a cell lacking an F plasmid (but able to receive one by conjugation) is called an F^- recipient. (Such cells are sometimes referred to as 'male' and 'female' cells, respectively.) In a mixed population of F^+ and F^- cells, a copy of the F plasmid is transferred at high frequency, i.e. all, or nearly all, of the F^- cells receive a copy (and hence become F^+ donors themselves); in such $F^+ \times F^-$ crosses, chromosomal genes are usually not transferred.

Integration of an F plasmid into the chromosome gives rise to an HFR DONOR (q.v.) – in which the chromosome is *mobilized* for transfer by the integrated plasmid (see CONJUGATION).

Whether conjugation is mediated by an autonomous or integrated F plasmid, the mechanism of transfer depends on an extensive (ca. 33 kilobase) region of the plasmid – the transfer region – containing over 30 genes. This region is commonly referred to as 'the transfer operon', although it contains more than one promoter; however, most of the *tra* genes are found in a single large operon (*traY–I*). The *traJ* and *traM* genes (both located immediately downstream of *oriT*) are outside the *traY–I* operon; *traJ* has one promoter, *traM* has two. *traY*, the first gene of the *traY–I* operon, is adjacent to *traJ*. The *traY–I* operon is transcribed by the $E\sigma^{70}$ holoenzyme; activation of the *traY* promoter is dependent on TraJ, but other factors appear to be needed for full activation. Apparently some genes within the *traY–I* operon (e.g. *traS*, *traT*) have their own promoters, and their expression seems to be independent of TraJ.

About 15 genes are needed for synthesis and assembly of the F pilus. The *traA* gene encodes the precursor molecule that is processed to form the subunit, pilin, from which PILI are assembled; the initial product of *traA* (MWt ca. 13000) is processed to a final size of MWt ca. 7000. Genes *traQ* and *traX* are reported to be involved in processing the *traA* gene product.

Products of the *traG* and *traN* genes have been associated with stabilization of donor–recipient mating contacts.

The *traI* gene product is a HELICASE which is needed for site- and strand-specific nicking at *oriT* (the origin of transfer) and for unwinding DNA during conjugal transfer [JBC (1991) *266* 16232–16237].

The *traM* gene product has binding sites in the *oriT* region (one of which, sbmC, is associated with DNA transfer); it is also linked with the cytoplasmic membrane, possibly via TraD. The role of TraM in conjugation has not been clarified, but it is apparently essential for transfer of DNA. In a related IncF plasmid (R1), TraM stimulated nicking at *oriT* by TraI [JMB (1998) *275* 81–94]; in the F plasmid, TraY and the INTEGRATION HOST FACTOR were reported to stimulate nicking at *oriT* in vitro [JBC (1995) *270* 28374–28380]. TraM may have a role in relaying the (currently unknown) 'mating signal' on establishment of stable cell–cell contact.

The products of *traS* and *traT* are involved in SURFACE EXCLUSION.

traM is autoregulated but needs TraY for full expression; expression is affected by external factors, and is negligible in stationary phase cells. Most genes in the *traY–I* operon are positively regulated by TraJ, and in most IncF plasmids, expression of *traJ* is under negative control of the FINOP SYSTEM (q.v.) so that the transfer genes are normally repressed. The F plasmid lacks a functional *finO* gene, i.e. in this particular plasmid, *traJ* is not under negative control; consequently, the transfer genes are constitutively *derepressed* (that is, permanently 'switched on'). (See also FERTILITY INHIBITION.)

[Control of transfer genes in F-like transfer systems: FEMS Reviews (1998) *21* 291–319 (291–295).]

Replication of the F plasmid. In wild-type F plasmids replication occurs bidirectionally from an origin and is under *stringent control* (see PLASMID). The F plasmid has two origins of replication: *oriV* (= *ori*1) and *oriS* (= *ori*2); in wild-type F plasmids, a PRIMOSOME assembly site is present near *ori*2 [PNAS (1983) *80* 7132–7136]. (In MINI-F PLASMIDS containing both origins, replication occurs preferentially, and bidirectionally, from *ori*1.) Host factors needed for F plasmid replication include the products of genes *dnaB*, *dnaC* and *dnaE*; replication of the plasmid is independent of the *dnaA* product. (See also INTEGRATIVE SUPPRESSION.)

Control of replication involves the (plasmid-encoded) RepE protein (= E protein). Molecules of the E protein (MWt ca. 29000) initiate replication by binding to two sequences of 19-bp DIRECT REPEATS (ITERONS), designated *incB* and *incC*, located either side of the E-encoding gene (close to the origin of replication); the bound molecules of E protein contribute to a nucleoprotein complex which is a pre-requisite for replication. Control of replication (and, hence, of COPY NUMBER) is achieved by controlling the intracellular level of E protein to that which will permit only one (sometimes two) rounds of plasmid replication prior to division of the bacterial cell. When a plasmid enters a daughter cell, subsequent binding of E protein molecules by the iterons 'titrates' the E protein, tending to lower its intracellular concentration; this promotes synthesis of E protein and restoration of an appropriate intracellular level. Lowering the concentration of E protein stimulates its synthesis because the gene encoding E protein is autorepressible at the level of transcription, i.e. the E protein represses its own transcription by binding to a sequence in, or overlapping, the gene's promoter.

Interestingly, *monomers* of E protein bind to iterons whereas *dimers* are involved in the autorepression of its own gene.

Incompatibility. The *incB* and *incC* sequences are determinants of INCOMPATIBILITY in IncFI plasmids.

Partition. As in other low-copy-number plasmids, an efficient system for PARTITION must be present during cell division to ensure stable maintenance of the F plasmid.

To ensure stability of inheritance in bacterial populations, the F plasmid also encodes another system (*ccd* – coupled cell division) involving the genes *ccdA* and *ccdB* (formerly *letA* and *letB*, respectively); this system results in the death of any cell which does *not* contain a plasmid following cell division. CcdB, a lethal toxin for the host cell, is a stable molecule. CcdA neutralizes CcdB, but this antidote is gradually degraded by the host's Lon protease – so that its protective role is available only while it is being synthesized from the (plasmid-encoded) *ccdA* gene, i.e. only in those cells containing the plasmid. Following cell division, a plasmid-less cell will be killed by CcdB as there is no *ccdA* gene from which CcdA can be synthesized.

CcdB induces the SOS SYSTEM, and it kills cells by inhibiting DNA gyrase. Interestingly, CcdB has been considered for development as a possible therapeutic agent [TIM (1998) *6* 269–275].

F′ plasmid See PRIME PLASMID.

F-prime plasmid See PRIME PLASMID.

F^- recipient See F PLASMID.

f. sp. See FORMA SPECIALIS.

F_0 component (of ATPase) See PROTON ATPASE.

F_0 value The time (in minutes) required, at 121.1°C, to inactivate a population of cells or spores of a given species *calculated from* the time required for inactivation at another temperature, assuming a Z VALUE of 10°C. The F_0 value, sometimes written $F^{10}_{121.1}$, is one particular value in a range of F values; thus, e.g., if the inactivation time at 121.1°C is calculated assuming a z value of 9°C, the relevant F value is indicated by $F^{9}_{121.1}$. (See also D VALUE.)

(F_0F_1)-type H^+-ATPase See PROTON ATPASE.

F1 antigen See YERSINIA (*Y. pestis*).

F_1 component (of ATPase) See PROTON ATPASE.

F1 fimbriae Type 1 FIMBRIAE.

F2 fimbriae *Syn.* CFA I (see ETEC).

F-2 toxin *Syn.* ZEARALENONE.

F3 fimbriae *Syn.* CFA II (see ETEC).

F_3T TRIFLUOROTHYMIDINE.

F4 fimbriae K88 FIMBRIAE.

F5 fimbriae K99 FIMBRIAE.

F6 fimbriae (987P fimbriae) FIMBRIAE which occur on certain strains of ETEC which are pathogenic for newborn piglets. [Organization and expression of genes involved in biosynthesis of 987P fimbriae: MGG (1986) *204* 75–81.]

F41 fimbriae FIMBRIAE which occur on some strains of ETEC which are pathogenic for piglets; they promote MRHA.

F_{420} See METHANOGENESIS.

F_{430} See METHANOGENESIS.

Fab portion (Fab fragment) Either of two identical portions of an IMMUNOGLOBULIN monomer which can be released by certain enzymes (e.g. PAPAIN); the remainder of the molecule is termed the Fc portion (q.v.). Each Fab portion consists of a light chain linked via a disulphide bond to the N-terminal part of a heavy chain, i.e., it is one of the two limbs of the Y-shaped Ig molecule. Each Fab portion of an antibody contains a single COMBINING SITE and behaves as a non-precipitating univalent antibody. Under acid conditions, MERCAPTOETHANOL splits the Fab portion into the heavy chain segment (the *Fd portion*) and the light chain.

$F(ab')_2$ portion Part of an IMMUNOGLOBULIN monomer released by the proteolytic action of PEPSIN (which also degrades at least part of the remainder of the Ig molecule); it includes both FAB PORTIONS connected by the HINGE REGION. The $F(ab')_2$ portion of an antibody behaves as a bivalent, precipitating, non-complement-fixing antibody.

facial eczema (*vet.*) See SPORIDESMINS.

facilitated diffusion An energy-independent TRANSPORT SYSTEM in which the diffusion of a substance across a membrane is facilitated by a membrane-bound protein carrier system; in facilitated diffusion a substance necessarily passes *down* a concentration gradient. Unlike simple diffusion, facilitated diffusion exhibits Michaelis–Menten saturation kinetics, it is relatively substrate-specific, and it is subject to competitive inhibition – the uptake of one substance being inhibited by the presence of another, structurally similar, substance. In e.g. *Saccharomyces cerevisiae* various sugars, amino acids and vitamins can be taken up by facilitated diffusion – although, depending on conditions, the same substances may also (or alternatively) be taken up by other transport mechanisms. In *Escherichia coli*, glycerol can cross the cytoplasmic membrane by facilitated diffusion, the membrane containing a specific *facilitator protein* (see also MIP CHANNEL); once inside the cell the glycerol is phosphorylated.

FACS Fluorescence-activated cell sorter: an instrument used for analysing and/or sorting cell populations by FLOW MICROFLUOROMETRY. Essentially, the cells (in suspension) move down a thin, open-ended tube and, in doing so, each cell passes momentarily through a laser beam which crosses the liquid column at right angles. As each cell passes through the beam it gives rise to a scatter signal (scattered laser light), and also – if labelled with a FLUOROCHROME (e.g. TEXAS RED) – a FLUORESCENCE signal. By using antibody-conjugated fluorochromes it is possible e.g. to determine the presence or number of homologous antigens or receptor sites on the cell surface, since the strength of the fluorescence signal is proportional to the amount of fluorochrome bound to the cell surface. In the 'dual parameter' mode, cells may be pre-treated, simultaneously, with antibodies of two different specificities – each being conjugated with a different fluorochrome; each cell passes through two separate laser beams (of wavelengths dependent on the fluorochromes chosen) and the two fluorescence signals are interpreted separately.

The FACS can also be used e.g. to monitor the proportion of proliferating lymphocytes in a population, measurement being based on the increased content of DNA in proliferating cells; cells are initially permeated with an agent (e.g. mithramycin) which can interact with intracellular DNA to form a complex which fluoresces on irradiation.

Cell *sorting* involves the physical separation of cells into subpopulations. Essentially, ultrasonic energy is applied to the nozzle through which the stream of liquid (i.e. cell suspension) passes, thus breaking up the stream into a series of very small droplets – each droplet (ideally) containing a single cell. Immediately before each droplet is discharged from the nozzle, the fluorescence (or scatter) signal received from the cell is interpreted (electronically) and the droplet is charged positively or negatively according to the nature of the signal received. As each (charged) droplet falls under gravity it passes between a pair of deflector plates – one positively charged, the other negatively charged; each droplet is therefore deflected to the right or left and subsequently falls into one of two collecting vessels.

factor 420 See METHANOGENESIS.

factor 430 See METHANOGENESIS.

Factor B (*immunol.*) A β-globulin (MWt ca. 93000) found in normal plasma; during COMPLEMENT FIXATION by the alternative pathway, Factor B is cleaved to fragments Ba and Bb.

Factor D (*immunol.*) A glycoprotein (MWt ca. 23000) found in very low concentrations in normal plasma; it circulates as an active enzyme whose substrate is C3b-bound FACTOR B.

Factor H (*immunol.*) A protein (MWt ca. 150000) present in normal serum; it regulates (inhibits) the C3 convertase in the alternative pathway of COMPLEMENT FIXATION, and also acts as an essential cofactor for FACTOR I.

Factor I (KAF; C3bina; C3bINA; conglutinogen-activating factor; C3b/C4b inactivator) A protein (a β-globulin, MWt 90000), present in normal plasma, involved in regulating the sequence of reactions in COMPLEMENT FIXATION. Together with an essential co-factor, FACTOR H, Factor I cleaves (inactivates) C3b and C4b. A cleavage product of C3b is involved in *conglutination*: see CONGLUTININ.

factory area (*virol.*) *Syn.* VIROPLASM.

factumycin See POLYENE ANTIBIOTICS (b).

facultative 'Optional', referring to the ability of an organism to adopt an alternative life style, mode of nutrition etc; thus, e.g. an AEROBE may be a facultative ANAEROBE. The word 'facultative' is often followed by the mode *not* normally adopted; thus, e.g. 'facultative anaerobe' often refers to an organism which normally grows aerobically but which *can* grow anaerobically. (cf. OBLIGATE.)

facultative heterochromatin See CHROMATIN.

FAD See RIBOFLAVIN.

faecal coliform test See COLIFORM TEST.

faecal streptococci A general term for certain Gram-positive cocci which occur primarily in human and animal intestines and faeces; they include *Enterococcus faecalis* and *E. faecium* (both formerly classified in the genus *Streptococcus*) and species such as *Streptococcus bovis* and *S. equinus*.

faecapentaenes *Syn.* FECAPENTAENES.

faeces See e.g. GASTROINTESTINAL TRACT FLORA; INDICATOR ORGANISMS; SEWAGE TREATMENT.

Faenia See MICROPOLYSPORA.

FAIDS See FELINE LEUKAEMIA VIRUS.

falcarindiol A polyacetylenic PHYTOALEXIN produced by tomato plants.

false branching In certain sheathed filamentous bacteria (particularly CYANOBACTERIA, sections III–V): apparent branching which results from the protrusion through the sheath of the broken end(s) of a trichome, or of a loop of trichome which subsequently breaks to generate a pair of false branches.

false neisseriae Bacteria regarded either as species *incertae sedis* in the genus *Neisseria* (*N. caviae*, *N. cuniculi*, *N. ovis*) or as members of the subgenus *Branhamella* (see MORAXELLA).

false tinder fungus *Fomes igniarius*.

false truffle See TRUFFLES.

falx A specialized pellicular region in opalinids.

famciclovir An oral ANTIVIRAL AGENT that resembles ACYCLOVIR in mode of action and antiviral activity; famciclovir is the 6-deoxy diacetyl ester prodrug of the guanine nucleoside analogue *penciclovir*. After absorption in the intestine, famciclovir is rapidly converted to penciclovir; within cells, the half-life of the active form (penciclovir triphosphate) is much longer than that of acyclovir triphosphate. Famciclovir is used e.g. for the treatment of herpes zoster.

FAME Fatty acid methyl ester: a methylated fatty acid derivative prepared for use in a microbial identification scheme; in such a scheme, identification is based on differences in the fatty acid content of *whole cells* of different species and strains of microorganism. Cells (grown under standard conditions) are harvested, and their fatty acids are initially saponified e.g. by heating in a solution of NaOH containing methanol. The product is then methylated (to increase volatility) e.g. by heating to ca. 80°C with an HCl–methanol mixture. FAMEs are then extracted with organic solvent(s); the resulting solution is washed by agitation with aqueous NaOH, and its FAME content is then analysed by gas chromatography.

family A taxonomic rank (see TAXONOMY and NOMENCLATURE).

Fansidar See MALARIA (*chemotherapy*).

farcy (*vet.*) (1) (in equines) See GLANDERS. (2) (in cattle) *Bovine farcy* is a chronic condition involving purulent lymphangitis and lymphadenitis, and the development of hard, subcutaneous lesions with thickening of the associated lymph nodes and ducts; the causal agent can be either *Mycobacterium farcinogenes* or *M. senegalense* [Book ref. 54, pp. 508–509]. The disease occurs mainly in the tropics. (cf. EPIZOOTIC LYMPHANGITIS.)

farcy buds, farcy pipes See GLANDERS.

farmers' lung An EXTRINSIC ALLERGIC ALVEOLITIS associated with inhalation of the spores of certain actinomycetes (e.g. *Micropolyspora faeni*) and/or other bacteria (e.g. *Thermoactinomyces vulgaris*) present e.g. in mouldy hay; 4–8 hours after inhalation of these spores there is severe pulmonary dysfunction due to an intrapulmonary TYPE III REACTION. [CIA (1984) *4* 175–184.]

Farr technique A RADIOIMMUNOASSAY used for quantifying antibody. Essentially, the antiserum under test is added to excess radiolabelled antigen, and the antigen–antibody complex is separated from uncombined antigen by precipitation with ca. 50% ammonium sulphate; the antibody is measured (indirectly) by determining the amount of radiolabelled antigen in the precipitate. (The method can be used only for antigens which remain soluble in the ammonium sulphate.)

Fas (CD95, APO-1) On mammalian cells: the cell-surface receptor for Fas ligand (Fas L). Fas ligand is a membrane-anchored cytokine which occurs on $CD8^+$ T cells. The effect of binding between Fas ligand and its receptor depends on the specific signals triggered; in some cases, binding activates caspases and leads to APOPTOSIS.

fasciation An abnormal type of plant growth resembling a flattened bundle of coalesced shoots – a symptom of certain plant diseases. Fasciation may be caused in certain plants (e.g. sweet peas) by *Rhodococcus fascians*, and may result from cytokinin production by the pathogen.

fast blue B *Syn.* DIANISIDINE.

fast death factor See MICROCYSTIS.

FAT FLUORESCENT ANTIBODY TECHNIQUE.

fat spoilage See e.g. RANCIDITY.

fat stain See e.g. BURDON'S STAIN.

fatal familial insomnia See TRANSMISSIBLE SPONGIFORM ENCEPHALOPATHIES.

fatty acid methyl ester See FAME.

Faulschlamm *Syn.* SAPROPEL.

favus (tinea favosa) A severe form of RINGWORM (usually tinea capitis) caused by *Trichophyton schoenleinii*. It is characterized by a spreading alopecia with pathognomonic lesions called *scutula* (sing. *scutulum*): round or oval, cup-shaped, sulphur-yellow crusts which develop from infected hair follicles. Hyphae, but few arthrospores, occur within the hair shaft, as may fine tubular canals left by the disintegration of hyphae. A bluish-white fluorescence may be observed in infected hairs under Wood's lamp.

The term favus also refers to a POULTRY DISEASE caused by *Trichophyton gallinae*. White powdery spots appear on the unfeathered parts of the head and develop to form wrinkled crusts or scabs; in severe cases feathered parts of the body are affected, with resulting loss of feathers.

FBJ osteosarcoma virus Finkel–Biskis–Jinkins murine sarcoma virus: a retrovirus complex consisting of a replication-competent murine leukaemia virus (FBJ-MuLV) and a replication-defective transforming murine sarcoma virus (FBJ-MSV); FBJ-MSV carries the oncogene v-*fos* and induces osteosarcomas in mice after a latent period of ca. 3 weeks. The v-*fos* product does not have tyrosine kinase activity. Homologues of c-*fos* have been identified in the genomes of various vertebrates as well as in *Drosophila*. [Organization and expression of *fos*: Book ref. 113, pp. 309–321.]

FBP Ferrous sulphate–sodium metabisulphite–sodium pyruvate; FBP destroys H_2O_2 and superoxide anions and, when added to culture media (0.025%), can increase aerotolerance in certain microaerophilic and anaerobic organisms.

Fc portion (Fc fragment) A portion of an IMMUNOGLOBULIN monomer which can be released by certain enzymes (e.g. PAPAIN). (cf. FAB PORTION.) The Fc portion corresponds to the stem of the Y-shaped Ig molecule and consists of the C-terminal sections of the two heavy chains linked by one or more disulphide bonds; it is e.g. the site of complement fixation in COMPLEMENT-FIXING ANTIBODIES. (See also PROTEIN A.)

FCA Freund's complete adjuvant.

FCCP See PROTON TRANSLOCATORS.

FcγRIII *Syn.* CD16.

Fd FERREDOXIN.

Fd portion (Fd fragment) See FAB PORTION.

Fe protein See NITROGENASE.

fecapentaenes A group of mutagens, each having a highly unsaturated conjugated enol ether structure, produced by certain bacteria. It has been proposed that they act as alkylating agents as a result of the formation of carbocations [Science (1984) *225* 521–523].

feces See FAECES.

fed batch culture A modified form of BATCH CULTURE in which a solution of nutrient(s) is added at specific time(s) in order e.g. to optimize the production of particular metabolite(s). If volumes of culture are periodically removed to permit repeated feeding, the process is called *extended batch culture*.

feed additives Substances, such as certain antibiotics, that are added to the feedstuffs of domestic animals (including poultry, pigs and/or ruminants) in order to increase growth rates and/or increase feed conversion efficiency. Ideally, feed additives should not be absorbed from the alimentary tract at the dosages used (e.g. ca. 10–60 mg/kg dry-weight feedingstuff, according to additive); absorption may necessitate a 'withdrawal period' prior to slaughter in order to allow elimination of the additive. Feed additives include e.g. AVOPARCIN, LASALOCID, MONENSIN, NARASIN, SALINOMYCIN and VIRGINIAMYCINS. [Antibiotics as feed additives for ruminants: Book ref. 121, pp. 331–347.]

feedlot bloat See BLOAT.

Feekes' scale (*plant pathol.*) A scale of numbers which indicate the various growth stages of wheat and barley: 1–5, shoot emergence and tillering; 6–10, stem elongation; 10.1–10.5, heading (ear development); 11.1–11.4, grain ripening. Feekes' scale does not allow as detailed a designation of plant development as does ZADOKS' CODE, and is limited to wheat and barley.

felid herpesvirus 1 *Syn.* FELINE RHINOTRACHEITIS virus.

feline AIDS See FELINE LEUKAEMIA VIRUS.

feline calicivirus (FCV; cat flu virus; 'feline rhinotracheitis virus'; 'feline picornavirus') A virus (family CALICIVIRIDAE) which infects the respiratory tract (occasionally also the gastrointestinal tract) of cats, causing e.g. rhinitis, conjunctivitis, oral ulceration, and pneumonia; the disease ranges from mild to fatal. An attenuated strain of FCV is used as a vaccine. FCV can be propagated in feline cell cultures. (cf. FELINE RHINOTRACHEITIS.)

feline coronavirus See CORONAVIRIDAE.

feline immunodeficiency virus (FIV) A retrovirus of the subfamily LENTIVIRINAE, originally isolated as a 'T-lymphotropic virus' from domestic cats with an immunodeficiency-like syndrome [Science (1987) *235* 790–793]; the virus can give rise to illness resembling AIDS in humans and FAIDS (see FELINE LEUKAEMIA VIRUS) in cats.

The FIV virion, ~120–150 nm in size, contains a characteristic lentivirus-type nucleocapsid. The genome – ~9.5 kb – contains some open reading frames (as well as *gag*, *pol* and *env* regions) within terminal LTRs; the ORFs apparently encode regulatory function(s).

Horizontal transmission is believed to occur typically as a result of fighting and biting (cf. FeLV). Vertical transmission by asymptomatic, chronically infected mothers has not been reported, but the virus can be transmitted via milk following experimental infection.

[Haematological disorders associated with feline retrovirus infections (FeLV and FIV): BCH (1995) *8* 73–112.]

feline infectious peritonitis virus See CORONAVIRIDAE.

feline leukaemia virus (FeLV) A retrovirus (subfamily ONCOVIRINAE, type C), strains of which cause important diseases in cats; there are many exogenous and some endogenous strains of the virus.

The FeLV virion is roughly spherical and is ~110–120 nm in diameter. The proviral DNA (of replication-competent strains) – approximately 8.4 kb in length – includes the usual retroviral *gag*, *pol* and *env* regions between 5′ and 3′ LTRs; the expression of proviral genes is under the regulation of *cis*- acting promoter and enhancer regions located in the U3 sequence of the LTR.

FeLV is transmitted horizontally, apparently via the oronasal route, by repeated social contact; transmission also occurs vertically, by intrauterine infection of the developing embryo. Animals which develop chronic viraemia are sources of high-titre infectious FeLV virions (which occur in plasma and in nasal secretions and saliva).

In over one half of the cats exposed to FeLV, the humoral and cell-mediated defence systems succeed in abolishing detectable virus in plasma. Nevertheless, in at least a proportion of animals the virus can remain latent for a period of time in myelomonocytic precursors in the bone marrow and in stromal fibroblasts; however, it is thought that spontaneous reactivation of latent virus in nature occurs only rarely.

FeLV causes neoplastic, degenerative and immunosuppressive forms of disease, the latter including a syndrome resembling human HIV-AIDS (*feline AIDS* or *FAIDS*); other conditions include myeloid leukaemias, anaemia, marrow aplasia, and lymphosarcomas. (A FAIDS-like syndrome can also be caused by the FELINE IMMUNODEFICIENCY VIRUS.)

Typically, *chronic* infection with FeLV involves a gradual decrease in the numbers of lymphocytes and a reduction in the function of both T and B cells, thus leading to clinical immunodeficiency. Certain strains of FeLV, however, cause an *acute* immunodeficiency syndrome (FAIDS).

Isolates of FeLV are often replication-competent and v-onc^-, but ONCOGENE-containing sarcoma-inducing recombinant viruses ('feline sarcoma viruses') have been isolated from FeLV-associated tumours. Several oncogenes (e.g. FES, *abl*, *sis*) have been reported to confer on FeLV the ability to induce sarcomas.

Insertional mutagenesis by FeLV of the host cell's c-*myc* oncogene is found in a significant proportion of spontaneous and experimentally induced lymphomas.

In some experimentally induced lymphomas, the location at which the provirus integrates in the cell's genome suggests that enhancer activity associated with the proviral LTR region may stimulate expression of c-*myc* and, in this way, contribute to pathogenesis.

The exogenous, replication-competent FeLVs are divided into subgroups A, B and C on the basis of (i) differences in the viral envelope glycoprotein, (ii) the interference pattern in vitro, and (iii) susceptibility of the viruses to neutralization by specific antibodies; differences in the envelope glycoprotein are associated with differences in the host cell range of the various strains of FeLV.

Subgroup A strains of FeLV are widespread and readily transmissible – although apparently not highly pathogenic. Subgroup B strains may arise through in vivo recombination and may be more pathogenic than the A strains; in infected animals, B strains are reported to be invariably accompanied by A strains. C strains are said to be comparatively rare, and are associated with pure red cell aplasia.

Various germ-line-transmissible endogenous retroviruses may be detectable in the genome of uninfected animals. In some cases the endogenous virus contains sequences homologous to FeLV (enFeLV strains); in other cases (RD-114 viruses) there is no apparent genetic relationship with FeLV. Although enFeLV proviral DNA may not give rise to infectious virions, it may nevertheless contribute to pathogenicity; thus, e.g. expression of enFeLV genes has been detected in both FeLV-positive and FeLV-negative lymphomas. Moreover, enFeLV-encoded product(s) may represent the entity referred to as FOCMA.

Diagnosis of persistent infection by FeLV. In cells that harbour a replication-competent strain of FeLV, viral proteins and virions are commonly produced at a high rate. One particular protein, the Pr65 *gag* precursor protein, is produced in excess and can be easily detected in the blood of infected cats by immunoassay techniques.

[Haematological disorders associated with feline retrovirus infections (FeLV and feline immunodeficiency virus): BCH (1995) *8* 73–112.]

feline panleukopenia virus (FPV) A subspecies of feline parvovirus (genus PARVOVIRUS). Like CANINE PARVOVIRUS (CPV), the ssDNA genome (~5100 nt) encodes two structural proteins (VP1, VP2: the mRNAs transcribed from overlapping sequences) and two non-structural proteins. DNA from FPV and CPV isolates is almost identical; the host range of these viruses depends on small differences in the composition of the capsid: less than 10 amino acids determine whether a given virus can relicate in dogs *or* cats. The virion retains infectivity for days/weeks in the environment.

Feline panleukopenia virus causes disease in cats, raccoons and some of their relatives. The main targets for viral replication include dividing cells in lymphoid tissues and in the intestinal epithelium.

Infection probably occurs via the nasopharyngeal route and/or via lymphoid tissues.

In the adult animal, generalized infection of lymphoid tissues follows viraemia. Levels of erythrocytes (red blood cells) are not affected. Often, infection leads to PANLEUKOPENIA, the total leukocyte count sometimes falling to ~1000 per cubic millimetre, with neutrophils <200 per cubic millimetre; the fall in lymphocyte numbers is less marked, and there is minimal change in numbers of eosinophils, basophils and monocytes.

Intestinal infection (in adult animals) involves viral replication in the rapidly dividing epithelial cells of the ileum and jejunum – with consequent diarrhoea (often with blood/mucus).

In neonatal kittens there is no manifestation of enteritis. Infection characteristically affects the cerebellum, leading to cerebellar hypoplasia and, typically, to ataxia in surviving animals.

[Pathogenesis of FPV: BCH (1995) *8* 57–71.]

feline parvovirus See PARVOVIRUS.

feline picornavirus *Syn.* FELINE CALICIVIRUS.

feline rhinotracheitis (feline viral rhinotracheitis) A severe upper respiratory tract disease of cats caused e.g. by felid herpesvirus 1 (see ALPHAHERPESVIRINAE). This virus replicates mainly in the oral and nasal mucous membranes; symptoms include fever and discharge from the eyes and nose. Latent infection generally follows recovery from the acute disease; latently infected animals may periodically shed virus either spontaneously or after stress. The virus has been isolated from the trigeminal ganglia of latently infected cats [JGV (1985) *66* 391–394]. (cf. FELINE CALICIVIRUS.)

feline rhinotracheitis virus FELINE CALICIVIRUS or felid herpesvirus 1 (see FELINE RHINOTRACHEITIS).

feline sarcoma viruses See FELINE LEUKAEMIA VIRUS.

feline syncytial virus See SPUMAVIRINAE.

feline T-lymphotropic lentivirus See FELINE IMMUNODEFICIENCY VIRUS.

felon *Syn.* WHITLOW.

FeLV FELINE LEUKAEMIA VIRUS.

***fem* genes** See MRSA.

female-specific phage Any phage whose EOP on bacteria containing certain plasmid(s) is significantly lower than that on plasmid-free strains of the same bacteria. For example, phages of the T7 group (e.g. T7, ϕI, W31) are inhibited in cells of *Escherichia coli* K12 which contain the F plasmid; in the case of T7, the presence of the plasmid inhibits translation of the intermediate and late phage proteins. In some cases phage DNA is the target of plasmid-encoded restriction endonucleases; e.g., phages F116 and G101 are restricted in cells of *Pseudomonas aeruginosa* containing the plasmid pMG7 [Plasmid (1977) *1* 115–116]. (cf. ANDROPHAGE.)

FeMo protein See NITROGENASE.

FeMoco The FeMo cofactor of NITROGENASE.

fenpropimorph A MORPHOLINE ANTIFUNGAL AGENT used e.g. to control yellow rust, brown rust, and powdery mildew on wheat and barley.

Fentichlor See BISPHENOLS.

fenticonazole See AZOLE ANTIFUNGAL AGENTS.

fentin Triphenyltin; two derivatives, fentin acetate ($(C_6H_5)_3SnO.CO.CH_3$) and fentin hydroxide ($(C_6H_5)_3SnOH$), are used (sometimes in conjunction with MANEB) as antifungal agents e.g. for the control of late blight of potato and powdery mildew of sugar-beet. (See also TIN.)

FepA protein In *Escherichia coli*: an OUTER MEMBRANE protein encoded by the *fepA* gene; it acts as a receptor for B and D colicins, and is involved in the uptake of Fe^{3+}-enterobactin (see SIDEROPHORES).

ferbam See DMDC.

FERM Fermentation Research Institute of the Agency of Industrial Science and Technology, Tsukuba, Japan.

ferment (1) (*noun*) An archaic name for an enzyme. (2) (*verb*) To carry out FERMENTATION.

fermentation (1) Energy-yielding metabolism in which an energy substrate is metabolized without the involvement of an *exogenous* electron acceptor; fermentation characteristically occurs under anaerobic or microaerobic conditions (cf. e.g. CRABTREE EFFECT). The absence of an exogenous electron acceptor (= oxidizing agent, 'electron sink') necessarily means that the products of fermentation – collectively – have the same oxidation state as that of the substrate, the oxidation of any intermediate in the fermentation pathway being balanced by equivalent reduction of other intermediate(s) in the pathway; since the substrate undergoes no *net* oxidation, energy derived from the fermentation of a given substrate is less than that obtainable by the RESPIRATION of that substrate. Until the 1980s it was believed that all fermentable substrates were organic compounds; it is now known that inorganic sulphur compounds (such as sulphite) can be fermented e.g. by *Desulfovibrio sulfodismutans* and by some other SULPHATE-REDUCING BACTERIA [Nature (1987) *326* 891–892]. Although the synthesis of ATP during fermentative metabolism commonly involves SUBSTRATE-LEVEL PHOSPHORYLATION, ATP synthesis by ELECTRON TRANSPORT PHOSPHORYLATION can occur (e.g. when lactate is fermented via the succinate–propionate pathway of the PROPIONIC ACID FERMENTATION); in some organisms, proton motive force can also be generated by excreting the end products of fermentation: see END-PRODUCT EFFLUX. (See also e.g. ACETONE–BUTANOL FERMENTATION; ALCOHOLIC FERMENTATION; BUTANEDIOL FERMENTATION; BUTYRIC ACID FERMENTATION; homoacetate fermentation (in ACETOGENESIS); LACTIC ACID FERMENTATION; MIXED ACID FERMENTATION; PROPIONIC ACID FERMENTATION.)

(2) As commonly used in industrial microbiology: *any* of a wide range of processes carried out by microorganisms, regardless of whether fermentative or respiratory metabolism is involved. (See e.g. DIHYDROXYACETONE FERMENTATION; flor fermentation in WINE-MAKING; MALOLACTIC FERMENTATION; SORBOSE FERMENTATION. See also FERMENTER.)

fermentative nitrate reduction See NITRATE RESPIRATION.

fermented foods See e.g. BREAD-MAKING, DAIRY PRODUCTS, DOSA, GARI, IDLI, MISO, NATTO, OGI, ONCOM, PICKLING, SALAMI, SAUERKRAUT, SOY SAUCE, TEMPEH, TOFU. (cf. ANG-KAK, COFFEE, COCOA, FOOD PRESERVATION (e).) (See also SILAGE.)

fermenter (fermentor) A vessel in which an aerobic or anaerobic FERMENTATION (sense 2) can be carried out either in BATCH CULTURE or in CONTINUOUS CULTURE. (cf. BIOREACTOR.) Fermenters are typically vertical, essentially closed, cylindrical steel vessels. The traditional STIRRED TANK REACTOR is a squat vessel, but other types are 'column' (= 'tower') fermenters in which the vessel (the 'column') has a height-to-diameter ratio of up to ca. 10 (e.g. the typical AIRLIFT FERMENTER, BUBBLE COLUMN FERMENTER, JET LOOP FERMENTER, and PROPELLER FERMENTER). Important features of fermenter design include provision for: (i) Adequate heat transfer from the culture. (The heat produced from a given substrate can be estimated from the value of 110 kcal/mole O_2 taken up – a value obtained empirically using various substrates and organisms [Book ref. 11, p. 32].) (ii) Efficient mixing of the culture. (iii) Efficient oxygen-to-liquid transfer. (iv) Adaptability to a range of operating conditions. (v) Ease of scale-up from the laboratory or pilot stage to industrial use. These factors largely determine the economics of a fermentation and the type of fermenter required for a particular purpose; thus, e.g., an STR – but not a bubble column or airlift fermenter – would generally be considered suitable for the fermentation of a (viscous) mycelial culture since only the STR (which has a greater input energy) would be able to ensure adequate mixing.

Fernández–Morán particles The stalked particles (the F_1 parts of (F_0F_1)-type PROTON ATPASES) on the inner surface of the mitochondrial inner membrane.

Fernandez reaction See LEPROMIN TEST.

ferredoxins A category of simple IRON–SULPHUR PROTEINS which are involved only in electron transfer processes; a given ferredoxin may contain one or more iron–sulphur centres of the type [2Fe–2S], [3Fe–3S] and/or [4Fe–4S]. Ferredoxins typically have low-potential (i.e., highly negative) E_m values; thus, e.g. a [4Fe–4S] ferredoxin (Fd I) in *Desulfovibrio gigas* has an E_m of −455 mV, and a (2[4Fe–4S]) ferredoxin in *Clostridium pasteurianum* has an E_m of −400 mV. However, a [2Fe–2S] ferredoxin in *Pseudomonas putida* (putidaredoxin) has an E_m of −240 mV. (See also HIPIP; cf. FLAVODOXINS.)

ferrichrome See SIDEROPHORES.

ferrimycins See SIDEROMYCINS.

ferrioxamines See SIDEROPHORES.

ferritin An iron-storage protein which occurs e.g. within various mammalian, plant and fungal cells; iron is stored within the large, hollow ferritin molecule. (cf. SIDEROPHILINS; see also IMMUNOELECTRON MICROSCOPY.)

ferrocene monocarboxylic acid See BIOFUEL CELL.

ferrochelatase See HAEM.

ferruginous Rust-coloured.

fertility factor *Syn.* F PLASMID.

fertility inhibition (of the F plasmid) Inhibition of expression of the *traJ* gene (and, hence, inhibition of the TRANSFER OPERON)

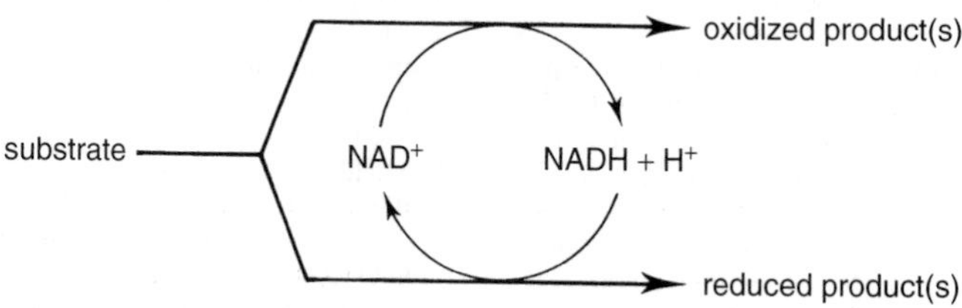

FERMENTATION (diagrammatic): the fermentation of an organic substrate showing intermediates undergoing mutual oxidation and reduction.

by the intracellular presence of an IncF plasmid which encodes a functional *finO* product (see FINOP SYSTEM); a plasmid which encodes such a *finO* product is designated fi^+ (fertility inhibition positive) or fin^+. Compatible plasmids which do not encode a functional *finO* product are designated fi^-.

ferulate 3-Methoxy-4-hydroxycinnamate.

Fervidobacterium See BACTERIA (taxonomy).

fes An ONCOGENE present in the Snyder–Theilen and Gardner–Arnstein strains of feline sarcoma virus (see FELINE LEUKAEMIA VIRUS); v-*fes* is closely related to v-*fps* (present in Fujinami AVIAN SARCOMA VIRUS) and may be derived from a related c-*onc* sequence. The products of v-*fes* and v-*fps* have tyrosine-specific protein kinase activity.

Fe–S protein (Fe/S protein) See IRON–SULPHUR PROTEINS.

fescue foot (fescue toxicity syndrome) A condition which affects animals (mainly cattle) grazing pastures dominated by tall fescue grass (*Festuca arundinacea*); symptoms: lameness, followed by dry gangrene of the extremities (tail tip, hooves, ears). It appears to be caused by a vasoconstrictive mycotoxin produced by fungi parasitic or pathogenic in the grasses; fungi which have been implicated include *Aspergillus terreus* and biotypes of *Epichloë typhina* [AEM (1977) *34* 576–581].

***Festuca* necrosis virus** See CLOSTEROVIRUSES.

Feulgen reaction A specific staining reaction for DNA in situ. Mild acid hydrolysis of DNA (e.g. with 1 N HCl) removes purine bases and makes available the aldehyde group of the deoxyribose; aldehyde groups react with SCHIFF'S REAGENT to give a purple coloration.

fever blister *Syn.* COLD SORE.

Ff phages F-specific filamentous phages (e.g. fd, f1, M13, ZJ/2): see INOVIRUS.

ff. sp. See FORMA SPECIALIS.

Ffh See PROTEIN SECRETION.

FH_4 TetrahydroFOLIC ACID.

FHA See WHOOPING COUGH.

FhuA protein *Syn.* TONA PROTEIN.

***fhuB–fhuD* genes** See SIDEROPHORES.

FhuE See IRON.

fi^- plasmid See FERTILITY INHIBITION.

fi^+ plasmid See FERTILITY INHIBITION.

FIAC (2′-deoxy-2′-fluoro-5-iodo-1-β-D-arabinosylcytosine) An ARABINOSYL NUCLEOSIDE which has antiviral activity against herpes simplex viruses 1 and 2, varicella-zoster virus and cytomegaloviruses in cell cultures; its mode of action resembles that of ACYCLOVIR. In trials, FIAC gave better control of progressive herpes zoster than did vidarabine; it can be administered orally.

fibrillae See FIMBRIAE.

fibrillum See FIMBRIAE.

fibrin A fibrous, insoluble protein present e.g. in blood clots; during normal blood-clotting, fibrin is formed from the plasma protein FIBRINOGEN by the action of thrombin in the presence of Ca^{2+}. (See also COAGULASE.)

fibrinogen The soluble glycoprotein precursor of FIBRIN. It consists of two identical monomers – each comprising three chains designated α, β and γ; the chains and monomers are held together by disulphide bridges. [Review of fibrin and fibrinogen: ARB (1984) *53* 195–229.]

fibrinolysin (plasmin) An enzyme which, in mammals, is responsible for dissolving blood clots by the proteolytic degradation of fibrin. Fibrinolysin is formed from an inactive precursor (profibrinolysin, plasminogen) normally present in the blood. The term fibrinolysin has also been applied to various microbial enzymes capable of direct or indirect FIBRINOLYSIS.

fibrinolysis The degradation of FIBRIN, and, hence, the lysis of blood clots (see FIBRINOLYSIN). Various microbial enzymes have fibrinolytic activity; such activity may be (i) direct, due to proteolytic activity on fibrin itself (see e.g. BRINASE), or (ii) indirect, due to the activation of plasminogen (profibrinolysin) (see e.g. STREPTOKINASE). The production of plasminogen activators by certain bacteria has been linked to invasiveness [TIM (1997) *5* 466–467].

fibrinolytic Capable of FIBRINOLYSIS.

Fibrobacter See BACTERIA (taxonomy).

fibrocyst (compound trichocyst) (*protozool.*) A type of TRICHOCYST formed by certain hypostome ciliates; the filament carries an umbrella-like tip.

fibroma A benign tumour of connective tissue. (cf. SARCOMA.)

fibronectin A heterodimeric glycoprotein, produced e.g. by fibroblasts and macrophages, which is found e.g. in extracellular matrix material and in plasma; it binds e.g. to various INTEGRINS, and its major roles include that of an adhesive molecule.

fibrosarcoma A SARCOMA arising from collagen-producing fibroblasts.

Ficoll A synthetic, water-soluble, non-ionic co-polymer of sucrose and epichlorhydrin, used e.g. in the preparation of density gradients for CENTRIFUGATION. It is also used for increasing the viscosity of a medium in order to slow down rapidly-motile organisms – a technique useful in certain COUNTING METHODS and in studies of ciliar and flagellar motility; in this context Ficoll has been reported to be superior to e.g. methylcellulose in that Ficoll causes less perturbation of hydrodynamic characteristics [Nature (1979) *278* 349–351].

FID Free induction decay: see NUCLEAR MAGNETIC RESONANCE.

fide In a literature citation: an indication that a given reference has not been read by the author citing that reference.

field blewit See BLEWIT.

field diaphragm (in MICROSCOPY) (1) See KÖHLER ILLUMINATION. (2) (*Syn.* field stop) A metal annulus attached to the inside of an eyepiece at the focal plane of the eyepiece lens.

field mushroom See AGARICUS.

Field's stain A water-based staining procedure for detecting e.g. *Plasmodium* spp and trypanosomes in thick blood smears. The smear is immersed in a solution of azure, rinsed, immersed in a solution of eosin, rinsed, and allowed to dry in air; the water used for staining and rinsing should have a pH of 7.0–7.2. Any parasites present will not be obscured by erythrocytes since the latter are lysed during the procedure; lysis occurs because a methanol fixation stage is omitted.

fièvre boutonneuse *Syn.* BOUTONNEUSE.

fifth disease *Syn.* ERYTHEMA INFECTIOSUM.

figwort mosaic virus See CAULIMOVIRUSES.

Fiji disease A SUGARCANE DISEASE, caused by a FIJIVIRUS and transmitted by planthoppers (*Perkinsiella* spp), which is characterized by stunting and the formation of GALLS on the underside of leaves. *Pseudo*-Fiji diseases of sugarcane include one similar to MAIZE WALLABY EAR DISEASE.

Fijivirus (plant reovirus subgroup 2) A genus of plant viruses of the REOVIRIDAE. Genome: 10 dsRNA molecules. Members infect plants of the Gramineae, and some can cause important diseases of crop plants; symptoms typically include stunting, excessive tillering, enations, and unnaturally dark-green leaves. Transmission occurs propagatively via plant-hoppers (*Perkinsiella* spp) in which transovarial transmission may occur. Type member: Fiji disease virus. Three serotypes of fijiviruses are recognized (host and geographical range in parentheses):

Group I: cereal tillering disease virus (CTDV) (barley, oats, maize, wheat; Sweden); maize rough dwarf virus (MRDV) (maize, oats, wheat and other grasses; Europe); pangola stunt virus (PSV) ('pangola grass' – *Digitaria decumbens* – and other *Digitaria* spp; S. America, Taiwan etc); rice black-streaked dwarf virus (RBSDV) (rice, oats, maize, wheat; Japan, China).

Group II: Fiji disease virus (FDV) (sugarcane – *Saccharum officinarum*; Australia, Fiji, etc).

Group III: *Arrhenatherum* blue dwarf virus (ABDV) (*Arrhenatherum elatius*; Germany); *Lolium* enation virus (LEV) (ryegrass – *Lolium* spp; Germany); oat sterile dwarf virus (OSDV) (oats, *Lolium* spp; *Arrhenatherum elatius* etc; Europe).

MRDV, CTDV, RBSDV and PSV may be strains of the same virus or closely related viruses, as may OSDV, ABDV and LEV. (cf. RICE RAGGED STUNT VIRUS; LEAFHOPPER A VIRUS.)

[Review: Book ref. 83, pp. 505–563.]

filament (1) (*bacteriol.*) An elongated bacterial cell, i.e., one in which the length exceeds the width by ca. 10 times or more. (2) (*bacteriol.*) A (sheathed or unsheathed) chain of cells. (3) *Syn.* TRICHOME.

filament pyrolyser See PYROLYSIS.

filamentous haemagglutinin See WHOOPING COUGH.

filamentous phages *Syn.* INOVIRUS.

filamin See ACTIN.

filar micrometer See MICROMETER.

Fildes' enrichment agar (Fildes' digest agar) A medium which contains the X FACTOR and the V FACTOR; it is used for isolating *Haemophilus* spp. A peptic digest of blood is made by mixing 1 volume of defibrinated sheep blood with 3 volumes of physiological saline (0.87% NaCl) and adding conc. HCl (3 ml/100 ml) and granular pepsin (0.5 g/100 ml); after incubation (56°C/ca. 4 hours, with occasional shaking), the pH is adjusted to 7 with conc. NaOH. The digest is stored at 4°C; chloroform (0.25% v/v) may be added as preservative. Before use, any chloroform present should be driven off by heating (56°C/30 min); 2–5% (v/v) of the digest is then added to e.g. molten nutrient agar which has been autoclaved and cooled to 56°C.

Fildes' peptic digest of blood See FILDES' ENRICHMENT AGAR.

filia *Syn.* VILIA.

Filibacter A genus of Gram-negative, strictly aerobic, filamentous gliding bacteria; *F. limicola* has been isolated from the sediment of a eutrophic lake [JGM (1984) *130* 2943–2959].

Filifactor See CLOSTRIDIUM.

filiform Thread-like.

filipin A pentaene POLYENE ANTIBIOTIC produced by *Streptomyces filipinensis*. It has a higher affinity for cholesterol than for ergosterol, and is more toxic to mammalian cells than to fungi; it appears not to form pores in cytoplasmic membranes but can cause disruption of membranes in sensitive cells.

filli *Syn.* VILIA.

film See SMEAR.

film and spots A test used for differentiating species of *Mycoplasma*. The organisms are grown on a medium containing egg yolk or ca. 20% horse serum. Species which give a positive reaction (e.g. *M. gallinarum*, *M. synoviae*) give rise to a wrinkled, pearly surface film, and to minute dark or black spots in the medium under and around the colonies; these cultural features indicate lipolytic activity: the film contains cholesterol and phospholipids, and the spots are small deposits of Ca^{2+} and Mg^{2+} salts of fatty acids liberated by lipolysis. A negative test is given e.g. by *M. meleagridis*, *M. mycoides* and *M. pneumoniae*. (See also LIPASE.)

Filobasidiaceae A family of basidiomycetous fungi characterized by the formation of long, slender, non-septate basidia with terminal sessile basidiospores; blastospores may be formed, teliospores are not. Members have a yeast-like vegetative phase, with globose, oval, apiculate or elongate cells which reproduce by budding. Conjugation between two haploid cells is followed by the formation of a dikaryotic mycelium on which the basidia develop. Most strains are heterothallic. The family includes the genera CHIONOSPHAERA, FILOBASIDIELLA, and FILOBASIDIUM. (See also WALLEMIA.)

The taxonomic position of the family is unsettled. It has been included e.g. in the Ustilaginales, and in a separate order (Filobasidiales) [discussion and refs: Book ref. 100, pp. 468–469]; some authors [Book ref. 64] include the above three genera in the SPORIDIALES.

Filobasidiella A genus of fungi (see FILOBASIDIACEAE). The yeast-like vegetative cells have CAPSULES, the size of the capsule depending on environmental conditions. The dikaryotic hyphae have clamp connections and dolipore septa which lack parenthesomes. Basidiospores are formed in long chains by basipetal budding from 4 sites on the apex of the basidium. The sole species, *F. neoformans*, has two varieties – *F. neoformans* var. *neoformans* (anamorph: *Cryptococcus neoformans* var. *neoformans*) and *F. neoformans* var. *bacillispora* (anamorph: *Cryptococcus neoformans* var. *gattii*) – and four serotypes (A–D). Strains have been isolated from pigeon excreta, soil, and humans (see CRYPTOCOCCOSIS). [Book ref. 100, pp. 472–482.]

Filobasidium A genus of fungi (see FILOBASIDIACEAE). The dikaryotic hyphae have clamp connections and dolipore septa with or without parenthesomes. Basidia are produced laterally and terminally on the hyphae, and sessile basidiospores are produced on a whorl at the apex of the metabasidium. Species: *F. capsuligenum* (anamorph: *Candida japonica*), isolated e.g. from cider and saké; *F. floriforme* (anamorph: possibly *Cryptococcus albidus* var. *albidus* on the basis of physiological characteristics), isolated from plants; and *F. uniguttulatum* (anamorph: *Cryptococcus uniguttulatus*), isolated from clinical specimens. [Book ref. 100, pp. 483–491.]

filoplasmodium See LABYRINTHULAS.

filopodium See PSEUDOPODIUM.

filose Thread-like.

Filosea A class of amoebae (superclass RHIZOPODA) which form long, tapering, hyaline filopodia that often branch and sometimes anastomose, and that are used for trapping prey. Spores and flagellated stages are not formed. Orders: Aconchulinida (cell naked, i.e., non-testate: e.g. NUCLEARIA, *Vampyrella*) and Gromiida (cell enclosed by a siliceous or proteinaceous and membrane-like test: e.g. EUGLYPHA, *Gromia*).

Filoviridae A family of viruses originally proposed to accomodate Marburg virus and Ebola virus [Intervirol. (1982) *18* 24–32], causal agents of clinically similar VIRAL HAEMORRHAGIC FEVERS in man (see MARBURG FEVER). The viruses occur in various parts of Africa; man is presumed to be infected only incidentally. Monkeys and rodents can be infected experimentally, infection usually resulting in fatal disease.

Marburg and Ebola viruses are morphologically similar and genetically related but serologically distinct. The virions are pleomorphic filaments of uniform width (~80 nm) but variable in length (up to ~14000 nm); the unit length (i.e. the shortest length with maximum infectivity) is ~790 nm for Marburg virus, ~970 nm for Ebola virus. Ring-shaped or branched structures are sometimes seen.

The virion consists of a helical nucleocapsid surrounded by an envelope bearing surface spikes. [Ebola virus proteins: Virol. (1985) *147* 169–176]. The genome is one molecule of negative-sense ssRNA. Virions are stable at room temperature but can be inactivated by heat (e.g. 60°C/30 min), ultraviolet radiation and gamma radiation, organic solvents, phenolic disinfectants, hypochlorite and β-propiolactone.

Virus replication occurs in the host cell's cytoplasm; maturation involves budding of preformed nucleocapsids through the plasma membrane.

Marburg and Ebola viruses can be grown in various types of cell culture (e.g. Vero cells). Infected cells exhibit CPEs such as intracytoplasmic vesiculation, swelling of mitochondria and degeneration of organelles.

[Book ref. 148, pp 1111–1118.]

Attempts to elicit protective immunity to Ebola virus in the traditional way have not been successful. However, an approach termed *genetic vaccination* (= *genetic immunization*), in which plasmid DNA (encoding viral proteins) is injected into the host, has achieved some success in an animal model: when vaccinated in this way, guinea pigs were protected against a lethal challenge of Ebola virus, protection correlating with both antibody titre and the antibody-specific T cell response [Nature Medicine (1998) *4* 37–42]. (See also DNA VACCINES.)

The Ebola virus glycoprotein is reported to be the main viral determinant of vascular cell cytotoxicity and damage [Nature Medicine (2000) *6* 886–889]. The synthesis of increased amounts of glycoprotein by a mutant virus was associated with a higher level of cytotoxicity [Science (2001) *291* 1965–1969].

A combination of 'DNA immunization' and boosting with an adenoviral vector (encoding relevant antigenic proteins) has been successful in generating protective immunity against the Ebola virus in non-human primates (cynomolgus macaques) [Nature (2000) *408* 605–609].

[Asymptomatic Ebola infection (in contacts of symptomatic patients): Lancet (2000) *355* 2210–2215. On the trail of Ebola and Marburg viruses: Science (2000) *290* 923–925.]

filter-paper cellulose See FP-CELLULOSE.

filterable virus An obsolete term originally applied to any infectious agent which could pass through the early microbiological filters (e.g. the Berkefeld candle). Such agents included viruses but also e.g. *Chlamydia* and mycoplasmas.

filtration Filtration can be used to separate microorganisms from a liquid or gas in which they are dispersed; it is used e.g. for the following purposes. (a) The sterilization of e.g. heat-labile liquids (known to be free of small viruses or viroids) by passing them through a filter capable of retaining all other microorganisms. (b) The preparation of cell-free culture filtrates for the recovery of extracellular enzymes, antibiotics etc. (c) The counting of small numbers of bacteria in a large volume of liquid – used e.g. in water bacteriology. A measured volume of the sample is passed through a membrane filter (pore size e.g. 0.45 μm – see below) and the membrane is subsequently incubated, face upward, on a pad of absorbent material which has been saturated with a suitable liquid culture medium. The number of bacteria per unit volume is calculated from the number of colonies which develop on the membrane. Membranes overprinted with a grid facilitate colony counting. (See also DEFT and COUNTING METHODS.) (d) Estimation of particle size in a monodisperse system. The suspension is filtered through a series of membranes of progressively smaller pore size; assuming minimal adsorption, particle size corresponds approximately to the pore size of the first membrane to effect significant retention of the particles. (See also *sand filter* in WATER SUPPLIES.)

Many different types of filtering apparatus have been devised. In the majority of these, liquid is drawn through the filter as a result of reduced pressure in the receiving vessel. The *candle filter* type consists essentially of a thick-walled test tube-shaped structure made either of unglazed earthenware (e.g. the *Chamberland candle* or *Doulton filter*) or wholly or partly of DIATOMACEOUS EARTH (e.g. the *Berkefeld candle* or *Mandler filter*); when assembled, the cavity of a candle filter communicates with the receiving vessel, and the liquid sample is contained in a jacket which surrounds the exterior of the candle. Such filters may be used repeatedly, but should be cleaned and sterilized after each use. The *Seitz filter* consists of a flat pad of asbestos (or asbestos and cellulose) suitably mounted between the sample container and the receiving vessel; the filter is used once and then discarded. In *sintered glass filters* the filter proper consists of a layer of minute glass particles which have been fused together ('sintered', 'fritted') to form a porous mass; such filters may be re-used after having been cleaned (e.g. in acid and/or hypochlorite solutions), washed, dried, and sterilized e.g. in a HOT-AIR OVEN.

Membrane filters are thin films of e.g. cellulose nitrate (*collodion*), cellulose acetate, nylon, polycarbonate, polytetrafluoroethylene (PTFE, *Teflon*), or polyvinylidene; they are manufactured in a range of pore sizes – commonly used membrane filters having a mean pore diameter of 0.45 μm or 0.22 μm. (Filtration through membranes with such small pores is often referred to as *ultrafiltration*.) Pores may be tortuous (as in Gelman- or Millipore-type membranes) or cylindrical (as in Nuclepore membranes). In order to avoid blocking the fine pores of a membrane filter with coarse particulate matter it may be necessary to use one or more stages of *prefiltration*; this may involve e.g. passing the sample through a Seitz filter or through pads of matted glass-fibre before filtering it through the membrane. Hydrophobic membranes (made e.g. of PTFE) are used e.g. for the filtration of air in the brewing industry. Membranes with hydrophobic rims can filter aqueous fluids in the central region and air in the annular region; these membranes can prevent air-locks forming upstream of the filter. The physical integrity of a membrane filter can be monitored e.g. by the BUBBLE POINT TEST or the DIFFUSION TEST; the presence of surfactants in a membrane can affect the results of the former test.

Although filtration depends largely on a mechanical sieving action, suspended particles, organisms and macromolecules may also be adsorbed to the material of the filter as a result of electrostatic or other forces. [Adsorption of poliovirus to membrane filters: AEM (1983) *45* 526–531.] Some filters (e.g. the Seitz filter) are highly adsorptive, while others (such as the modern porcelain filters) are relatively non-adsorptive. The charge-modified filters such as Zetapor and Zeta Plus, which carry a net positive charge, combine a sieving action with electrostatic adsorption; such filters are used e.g. for the concentration of viruses from water samples [AEM (1983) *45* 232–237]. Some membranes are markedly anisotropic in that the two faces of the membrane have different pore characteristics; an effect of this is that flow rate is higher when filtration is carried out in one particular direction.

In some cases substances can be leached from a filter during the filtration process: e.g., Seitz filters tend to enrich the filtrate with magnesium ions. In the manufacture of some types of membrane, additives such as glycerol and surfactants are incorporated to improve flexibility and wettability, respectively; unless such substances are leached out prior to use (e.g. during sterilization) they can subsequently contaminate the filtrate.

(See also HEMMING FILTER and HEPA FILTER.)

***fim* operon** See FIMBRIAE.

fimbriae (*singular*: fimbria) Thin, proteinaceous filaments which extend from the surface of some types of microbial cell; fimbriae may be sparse, or they may occur in large numbers on a given cell. More than one type of fimbria may be present on a given cell. Fimbriae are functionally distinct from both flagella (see FLAGELLUM) and PILI.

Note. A number of authors use the two terms *fimbriae* and *pili*, interchangeably, to refer to the appendages described in this entry. By contrast, appendages described in the entry PILI (which, unlike fimbriae, are concerned with the transfer of DNA between cells) are referred to consistently by all authors as 'pili'. A logical, scientific use of terminology would require that different entities be referred to by different terms; for this reason, the terms 'fimbriae' and 'pili' are used here with distinct and mutually exclusive meanings.

Fimbriae are quite common on Gram-negative bacteria (including cyanobacteria) but less common on Gram-positive species; they occur on various fungi – including the yeasts *Candida albicans* and *Saccharomyces cerevisiae* [fimbriae of yeasts and yeast-like organisms: Bot. Gaz. (1982) *143* 534–541] and some SMUTS.

The account below refers specifically to fimbriae of Gram-negative bacteria.

Fimbriae may be distributed over the entire surface of a given cell or they may occur only in particular region(s) of the cell surface.

Fimbriae may promote cell-to-cell adhesion, or adhesion between the cell and substratum. The adhesive fimbriae of some pathogenic bacteria are important virulence factors which play a critical role in the process of infection.

Some polar fimbriae – i.e. fimbriae located at the pole (end) of a rod-shaped bacterium – are associated with TWITCHING MOTILITY.

A fimbria is essentially a rod-like structure with a uniform width (typically 2–8 nm) and a length between ~100 nm and several micrometres; although it may contain more than one type of component, the main structure is composed of a single type of protein subunit. In at least some cases (type I fimbriae of *Escherichia coli* – see below), the individual fimbria has been seen as a tightly coiled *helical* filament under the electron microscope, the filament itself comprising a linear sequence of protein subunits.

Short fimbriae (e.g. K88) may be seen as a tangled mass on the cell surface, appearing as an amorphous layer under low-power magnification in the electron microscope; such fimbriae have been referred to as 'capsular material' or *fibrillae* (but see later for an alternative use of 'fibrillae'). It has been suggested that such short fimbriae may in fact represent broken fimbriae [FEMS Reviews (1996) *19* 25–52 (35)].

A fimbria is constructed essentially of linearly repeating molecules of the main protein subunit. Protein subunits have been termed *fimbrillins* or *pilins*, the latter term being used more commonly. (The subunits have also been called *fimbrins*, but this may cause confusion because 'fimbrins' are involved in the formation of actin bundles in eukaryotic cells.) (See also PSEUDOPILINS.) Pilins frequently contain a high proportion of non-polar amino acids, so that fimbriate cells are commonly (not always) more hydrophobic than afimbriate cells. (In this context hydrophobicity may be indicated by the tendency of cells to congregate at an air–water or oil–water interface.)

The occurrence, number and composition of fimbriae on a cell depend on various factors which include e.g. genetic switching mechanisms (see later) and temperature; for example, K88 and K99 fimbriae are not formed below ~18°C.

Classification of fimbriae. The different types of fimbria on Gram-negative bacteria can be grouped into categories on the basis of e.g. (i) homology in the amino acid sequences in major subunit proteins; (ii) physical, antigenic and adhesive characteristics; and (iii) the mechanism of secretion and assembly.

One group comprises the type I fimbriae (also called type 1 fimbriae); these include: (i) the 'common' type I fimbriae of *Escherichia coli*, (ii) the P FIMBRIAE and (iii) the K88 and K99 'fibrillae'.

Another well-studied category of fimbriae, the type IV (= type 4) fimbriae, are found on a range of pathogenic bacteria; these fimbriae occur e.g. on strains of EPEC and ETEC, on *Neisseria* spp and *Pseudomonas aeruginosa*, and on *Vibrio cholerae* ('toxin co-regulated pili' (TCP) which constitute an important virulence factor in CHOLERA).

Type I fimbriae (type 1 fimbriae; F1 fimbriae; 'common pili'). These fimbriae are chromosomally encoded and are common e.g. on *Escherichia coli* and other enterobacteria. Each fimbria consists primarily of the main rod-shaped region, about 7 nm in diam., which comprises several thousand copies of the major pilin subunit, FimA (encoded by the *fimA* gene). The distal end of the fimbria consists of a short (~16 nm) and narrow (2–3 nm diam.) flexible structure – the *fibrillum* (plural: *fibrillae*) – which is composed of the minor pilin subunits FimF, FimG and FimH; FimH is a mannose-binding ADHESIN which appears to be the most distal pilin in the fibrillum.

Type I fimbriae can adhere to various cells and surfaces, including guinea-pig and horse erythrocytes – causing HAEMAGGLUTINATION; such adhesion is inhibited by D-mannose, or by methyl-α-D-mannoside, i.e. binding is 'mannose-sensitive', and type I fimbriae are said to cause 'mannose-sensitive haemagglutination' (MSHA). Fimbriae whose adhesion is not inhibited in this way are said to cause 'mannose-resistant haemagglutination' (MRHA).

Type I fimbriae can also bind to UROPLAKINS (see also UPEC), such adhesion being a preliminary step in the development of many URINARY TRACT INFECTIONS (such as CYSTITIS). Moreover, type I fimbriae can promote the *invasion* of human urinary tract epithelium [EMBO (2000) *19* 2803–2812].

Genetics of type I fimbriae. In *Escherichia coli*, type I fimbriae are encoded by the *fimA–fimH* operon; *fimA* and *fimFGH* are structural genes (see above), while the *fimCD* products are involved in assembly (see later).

The *fimBE* genes have a regulatory role, being involved in PHASE VARIATION: the on/off switching of fimbrial synthesis. Switching involves inversion of a 314-bp sequence that contains the promoter of the first structural gene in the operon, *fimA*. When the control sequence is in one orientation the promoter is located correctly for transcription of *fimA* and of the other genes (so that fimbriae are synthesized); when the control sequence is in the opposite orientation, *fimA* is promoter-less (so that fimbriae are not synthesized) (cf. flagellar PHASE VARIATION in *Salmonella*). The products of the *fimBE* genes are recombinases which mediate inversion of the control sequence; FimB can bring about inversion in either direction (i.e. it can switch synthesis from on to off *and* from off to on), whereas FimE can mediate only the on-to-off switch. (Under in vivo conditions, switching also requires other (intracellular) factors, including INTEGRATION HOST FACTOR and the H-NS PROTEIN.)

Interestingly, there is evidence of regulatory cross-talk between the *fim* operon and the *pap* operon (which encodes P FIMBRIAE) – this resulting in inhibition of expression of type I fimbriae. In this cross-talk, PapB, a *pap* regulatory protein, is reported to (i) increase the expression of FimE (thus promoting the on-to-off switch), and (ii) inhibit the activity of FimB (thus blocking an off-to-on reversal of FimE activity) [EMBO (2000) *19* 1450–1457].

Secretion and assembly of type I fimbriae. The individual protein subunits are synthesized in the cytoplasm; each protein has a signal sequence (see SIGNAL HYPOTHESIS) and is transported across the cytoplasmic membrane in a *sec*-dependent manner (see type II PROTEIN SECRETION). In the periplasm, each protein is bound by a 'chaperone', FimC. FimC folds the protein and prevents it from forming abortive contacts with other subunits; the chaperone also appears to target its subunit to the outer membrane.

In the OUTER MEMBRANE, the *usher protein*, FimD, forms an oligomeric ring structure with a central pore (of ~2 nm diam.) through which the oligomerized subunits of the nascent fimbria are translocated as a linear (not helical) filament. The first subunits to be externalized are those of the fibrillum; FimG is thought to have a critical role in 'nucleating' the fibrillum [PNAS (2000) *97* 9240–9245]. Subsequently, the FimA subunits, bound together in linear sequence, are translocated through the usher pore and added to the base of the growing fimbria; only when the FimA subunits have reached the extracellular side of the usher pore do they adopt a helical form.

The order in which the various types of subunit are externalized via the usher pore may depend on the affinity with which each type of subunit binds to the pore structure; thus, e.g., under in vitro conditions, the FimH adhesin has the highest affinity for the usher.

[Secretion and assembly of type I fimbriae: FEMS Reviews (2000) *24* 21–44 (27–31).]

P fimbriae. The 11 genes encoding P FIMBRIAE occur in the (chromosomal) *pap* operon, which is quite similar to the *fim* operon. PapA is the major subunit; PapC is the usher protein; PapD is the chaperone; PapKEFG form the fibrillum, PapG being the adhesin. PapH binds at the base of the fimbria, forming an anchor and signalling the end of fimbrial growth. (In mutant cells lacking the PapD chaperone, the subunits are degraded by a periplasmic protease, DegP; in *papD degP* mutants, subunits accumulate in the periplasm.)

K88 (F4) fimbriae. These (plasmid-encoded) fimbriae occur on strains of *E. coli* pathogenic for piglets. The major fimbrial subunit, FaeG, also functions as an adhesin; the binding sites for K88 fimbriae contain Gal-α(1–3)Gal.

K99 (F5) fimbriae. These (plasmid-encoded) fimbriae occur on strains of *E. coli* pathogenic for calves and lambs. The major subunit, FanC, has adhesive properties.

[Chromatographic purification of K99 fimbriae: JGM (1983) *129* 1975–1982. Production of K88, K99 and F41 fimbriae in relation to growth phase: FEMS (1985) *26* 15–19. K88, K99 and 987P fimbrial antigens used in a vaccine against porcine colibacillosis: VR (1985) *117* 408–413.]

Type II fimbriae (type 2 fimbriae). These fimbriae occur on some strains of *Salmonella* and other enterobacteria; they lack the adhesive and haemagglutinating properties of type I fimbriae but are otherwise similar.

Type III fimbriae (type 3 fimbriae) occur on various members of the Enterobacteriaceae [JMM (1985) *20* 113–121]; their adhesive properties are not inhibited by mannose. These fimbriae promote haemagglutination of ox erythrocytes only when the latter have been treated with tannic acid (the so-called 'tanned ox' haemagglutination). [Comparative study of type 3 fimbriae in *Klebsiella* spp: JMM (1985) *20* 203–214.]

Type IV fimbriae (type 4 fimbriae). All the fimbriae placed in this category are secreted and assembled in a similar way. These fimbriae include important (adhesive) virulence factors in a range of Gram-negative pathogens such as *Neisseria gonorrhoeae* and *N. meningitidis*, *Pseudomonas aeruginosa*, *Vibrio cholerae* and certain strains of *Escherichia coli* (e.g. the 'bundle-forming pili' of EPEC). (See also LONGUS.)

The type IV fimbriae of e.g. *N. gonorrhoeae* are subject to PHASE VARIATION (on/off switching) as well as to extensive ANTIGENIC VARIATION.

Most type IV fimbriae are encoded chromosomally (see also PIL GENES). The 'bundle-forming pili' of EPEC are an exception, being encoded by a virulence plasmid (see also PATHOGENICITY ISLAND).

Type IV fimbriae are typically flexible, rod-like appendages, ~5–6 nm in diam. and 1–2 μm in length; constituent pilins are apparently organized in a helical fashion.

In a recent model for the assembly of type IV fimbriae in *N. gonorrhoeae* [FEMS Reviews (2000) *24* 21–44 (31–35)], the assembly of these appendages (unlike the assembly of type I fimbriae) begins at a protein complex in the cytoplasmic membrane; that is, the base of the fimbria is linked to the cytoplasmic membrane – a column of *assembled* subunits passing from this origin, through the periplasm, through the outer membrane (see later), and into the extracellular environment.

The adhesive properties of type IV fimbriae are due, at least partly, to the major (structural) pilins – in contrast to type I fimbriae, in which there are minor pilins specialized for adhesion in the distal fibrillum. Exceptionally, tip adhesins (encoded by the *pilC* gene) have been indentified in the two major pathogenic species of *Neisseria* [Nature (1995) *373* 357–359; Mol. Microbiol. (1997) *23* 879–892].

Secretion and assembly of type IV fimbriae. In the model for assembly of type IV fimbriae in *N. gonorrhoeae*, the major pilin subunit, PilE, is translocated across the cytoplasmic membrane but initially remains attached to the membrane by an N-terminal hydrophobic sequence; this sequence is subsequently cleaved, so that the PilE subunits can be incorporated, sequentially, into the nascent fimbria – incorporation being mediated by a complex of proteins (including PilFGT) located within the cytoplasmic membrane. A SECRETIN, PilQ, forms a pore (~5.5 nm diam.) in the outer membrane through which the developing fimbria extends to the exterior; the pore is stabilized by PilP. In *N. gonorrhoeae* (and *N. meningitidis*), the distal end of the fimbria has an adhesive subunit, PilC (which is synthesized with a *sec*-dependent signal sequence); the incorporation of PilC is one of the initial steps in the assembly process.

[Type IV fimbriae (review): ARM (1993) *47* 565–596.]

PSA fimbriae ('PSA pili') occur as a polar tuft in many strains of *Pseudomonas aeruginosa*; they are apparently chromosomally encoded, retractable filaments, ca. 6 nm in diam. and <1 to several micrometres in length. PSA fimbriae act as receptors for a range of bacteriophages.

[Secretion and assembly of regular surface structures (fimbriae, flagella and S layers) in Gram-negative bacteria: FEMS Reviews (2000) *24* 21–44.]

fimbriate (1) Having FIMBRIAE. (2) Having a fringed edge.

fimbrillin See FIMBRIAE.

fimbrin See ACTIN.

***fin*$^+$ plasmid** See FERTILITY INHIBITION.

fin rot See BACTERIAL FIN ROT.

final host (definitive host) In heteroxenous coccidia: the host in which the sexual phase occurs (see EIMERIORINA).

finger-and-toe disease See CLUBROOT.

finger millet mosaic virus See RHABDOVIRIDAE.

fingerprinting Any procedure which gives a strain-specific pattern of products when a given organism is examined; the products commonly consist of fragments of DNA (or RNA) distributed electrophoretically in a gel. A fingerprint may be generated e.g. by restriction of chromosomes followed by ELECTROPHORESIS of the products (see e.g. DNA FINGERPRINTING) or by electrophoresis of the products formed in TYPING procedures that involve nucleic acid amplification (e.g. AFLP).

Finkel–Biskis–Jinkins murine sarcoma virus See FBJ OSTEOSARCOMA VIRUS.

FinOP system In most IncF plasmids: a system in which the plasmid's *finO* and *finP* products act, jointly, to inhibit translation of the *traJ* gene (the gene which exerts a positive regulatory influence on the transfer genes); as a consequence of this activity, the transfer function of such plasmids is repressed.

The FinO products of most IncF plasmids are interchangeable but the FinP products of different plasmids are commonly not interchangeable. [Nucleotide sequences of *finP* alleles of five IncF plasmids: JB (1986) *167* 754–757.]

The *finP* product is an antisense RNA molecule which binds to the 5′ end of *traJ* mRNA (including the ribosome-binding sequence), thus inhibiting translation; the *finO* product (a polypeptide) binds to, and promotes the stability of, the FinP–mRNA duplex – resulting in repression of the *traJ* gene and (therefore) repression of transfer (so-called fertility inhibition). (The FinOP binding site was formerly designated *fisO*, O_J or *traO*.)

Mutant (defective) *finO* and/or *finP* products may fail to inhibit the expression of *traJ* – in which case the negative regulation of the transfer genes is abolished, and the transfer function is expressed constitutively. Plasmids containing mutant *finO* genes include R1*drd*19 and R100-1.

It was originally thought that the (constitutively derepressed) F PLASMID lacked a *finO* locus [Book ref. 161, p 605]. *finO* is now known to be present but is inactivated by an INSERTION SEQUENCE (IS*3*) located within the coding region of the gene.

(See also FERTILITY INHIBITION.)

fireblight A disease which can affect many members of the Rosaceae, e.g. apple and pear trees, cotoneaster and hawthorn; it is characterized by wilting and by browning and necrosis of blossoms, fruit, leaves and twigs. The causal agent, *Erwinia amylovora*, infects e.g. via blossoms and/or wounds; virulent strains of *E. amylovora* form AMYLOVORIN. [Review: Book ref. 58, pp. 45–63.]

firefly luciferin See BIOLUMINESCENCE. (See also CHEMILUMINESCENCE.)

Firmibacteria A proposed class of bacteria comprising all organisms of the FIRMICUTES other than the THALLOBACTERIA.

Firmicutes A proposed division of the PROCARYOTAE comprising bacteria in which the CELL WALL is of the Gram-positive type; classes: FIRMIBACTERIA; THALLOBACTERIA.

Fis protein See CELL CYCLE (b).

Fischerella A genus of filamentous, thermophilic CYANOBACTERIA (section V) in which some of the cells in a mature trichome divide in more than one plane, resulting in a partly multiseriate trichome with lateral uniseriate branches. Hormogonia are formed from the ends of the trichomes or from lateral branches, and are composed of small cylindrical cells which enlarge and become rounded; heterocysts initially develop mainly in intercalary positions, and are mainly intercalary or lateral in mature trichomes. Akinetes may be formed. GC%: 42–49.

The genus includes '*Mastigocladus laminosus*', an organism found in hot springs; it has an upper temperature limit for growth of ca. 58°C (cf. CHLOROGLOEOPSIS) and is capable of nitrogen fixation. [Cell division and branching in *M. laminosus*: JGM (1984) *130* 2079–2088.]

FISH See PROBE.

fish diseases Marine and freshwater fish are susceptible to a wide range of infectious diseases, the severity and spread of which tend to be enhanced under high-density fish farming conditions. (a) *Bacterial diseases* include e.g. BACTERIAL FIN ROT, BACTERIAL GILL DISEASE, COLD WATER DISEASE, COLUMNARIS DISEASE, ENTERIC REDMOUTH, ENTERIC SEPTICAEMIA, FURUNCULOSIS, KIDNEY DISEASE, RED MOUTH, RED PEST, tuberculosis, ULCER DISEASE, vibriosis (see VIBRIO). [Book ref. 9; control of bacterial diseases by antimicrobial compounds: Book ref. 121, pp. 255–268.] (b) *Fungal diseases* include e.g. GILL ROT, ICHTHYOPHONOSIS, SAPROLEGNIASIS. (See also APHANOMYCES.) [Book ref. 10.] (c) *Protozoal diseases* include e.g. COSTIASIS, ICHTHYOPHTHIRIASIS, PROLIFERATIVE KIDNEY DISEASE, VELVET DISEASE, WHIRLING DISEASE; see also GLUGEA, HENNEGUYA and MYXOBOLUS. (d) *Viral diseases* include e.g. CARP POX, CHANNEL CATFISH VIRUS DISEASE, EGTVED DISEASE, INFECTIOUS HAEMATOPOIETIC NECROSIS, INFECTIOUS PANCREATIC NECROSIS, LYMPHOCYSTIS, SPINNING DISEASE, SPRING VIRAEMIA OF CARP.

Fish may also be affected or killed by algal or cyanobacterial BLOOMS due to the production of toxins (see e.g. RED TIDE) and/or to oxygen depletion resulting from the sudden decomposition of vast numbers of algae or cyanobacteria.

fish spoilage Freshly caught, gutted, iced fish and crustacea (shrimps etc) are susceptible to spoilage by a mixed psychrotrophic flora derived from the fish itself and/or from equipment, handlers etc [Book ref. 29, pp. 283–294]. Common spoilage organisms include *Photobacterium* spp (from the fish's intestine) and *Alteromonas putrefaciens*. These organisms are facultative anaerobes which can reduce trimethylamine-*N*-oxide (TMAO) – present in the flesh of the fish – to the strongly malodorous substance trimethylamine (TMA); TMAO may function as the terminal electron acceptor for anaerobic electron transport [JGM (1983) *129* 3689–3696]. Other malodorous products of bacterial spoilage include ammonia and hypoxanthine. In certain types of fish (e.g. mackerel) dangerously high – but often organoleptically undetectable – levels of HISTAMINE may accumulate, during storage, as a result of the decarboxylation of histidine by contaminating bacteria (e.g. *Clostridium* spp, *Escherichia coli*, Proteeae); consumption of such fish causes 'scombroid poisoning' [Food Mic. (1984) *1* 263–267]. Fish spoilage may be delayed by the addition of preservatives (usually benzoic acid) and/or by irradiation.

Off-flavours and taints in fish may be due to substances produced by certain actinomycetes and cyanobacteria present in the aquatic environment (see e.g. GEOSMIN and 2-METHYLISOBORNEOL).

fisheye spoilage (of olives) See PICKLING.

***fisO* locus** See FINOP SYSTEM.

fission Cell division in which overall (i.e., not localized) cell growth is followed by septum formation which typically divides the fully grown cell into two similar or identical cells. (See also BINARY FISSION; cf. BUDDING.)

fissitunicate ascus *Syn.* BITUNICATE ASCUS.

Fistulina See APHYLLOPHORALES (Fistulinaceae).

Fistulinaceae See APHYLLOPHORALES.

FITC Fluorescein isothiocyanate: a reactive derivative of FLUORESCEIN used e.g. for labelling proteins in IMMUNOFLUORESCENCE techniques. FITC fluoresces greenish-yellow.

Fite–Faraco stain An acid-fast stain used for detecting *Mycobacterium leprae* in tissue sections; the procedure resembles that for the ZIEHL–NEELSEN STAIN but haematoxylin is used as counterstain.

Fitz-Hugh–Curtis syndrome See CURTIS–FITZ-HUGH SYNDROME.

FIV FELINE IMMUNODEFICIENCY VIRUS.

five-five-five test (5-5-5 test) A quantitative SUSPENSION TEST in which a test organism is exposed for 5 min to each of a range of dilutions of the test disinfectant; after neutralization of the disinfectant, surviving bacteria are counted by subculturing to pour plates. To be considered effective, a disinfectant must give a minimum 5-log (10^5-fold) reduction in cell numbers. Sporicidal activity is assessed by using a suspension of endospores of *Bacillus cereus* ATCC 9139 preheated at 80°C for 60 sec; an 'effective' sporicidal disinfectant must produce a minimum 1-log decrease in viable spores.

five-kingdom classification See KINGDOM.

five-three-two symmetry (5-3-2 symmetry) See ICOSAHEDRAL SYMMETRY.

fixation The process of killing cells while preserving their structure and organization in as life-like a condition as possible (cf. CRYOFIXATION). Fixation involves the inactivation of enzymes, particularly AUTOLYSINS; ideally, it also hardens cellular structures and renders cell components compatible with particular stains. Most or all methods of fixation create ARTEFACTS. (a) *Heat fixation* can be used for the fixation of smears of bacteria prior to staining. The SMEAR is passed *rapidly* through a flame two or three times; such *heat-fixed smears* are suitable e.g. for the GRAM STAIN but are not suitable for observations on cell structure or morphology. (b) *Chemical fixation.* A chemical *fixative* permeates cells and stabilizes their components by binding to them or denaturing them. Some fixatives (e.g. ethanol, mercuric chloride, picric acid) precipitate proteins and thus considerably disturb cellular organization. Others (e.g. FORMALDEHYDE, GLUTARALDEHYDE, OSMIUM TETROXIDE, potassium dichromate) tend to preserve proteins in situ. Osmium tetroxide and potassium dichromate are also important lipid fixatives. For ELECTRON MICROSCOPY osmium tetroxide and glutaraldehyde are often used in e.g. alkaline phosphate buffer or cacodylate buffer. Glutaraldehyde may be used first to fix proteins (*pre-fixation*), and osmium tetroxide used subsequently for the *post-fixation* of lipids. (See also BOUIN'S FLUID, CARNOY'S FLUID, POLYVINYL ALCOHOL FIXATIVE, SCHAUDINN'S FLUID, ZENKER'S FLUID.)

fixative A substance used for FIXATION.

fixed virus A strain of rabies virus attenuated by serial passage e.g. through rabbits. (cf. FLURY VIRUS; STREET VIRUS; SEMPLE VACCINE.)

FkpA See PROTEIN SYNTHESIS (protein folding).

***fla* genes** In e.g. enterobacteria: genes involved in the synthesis and function of bacterial flagella. [ARBB (1984) *13* 51–83.]

flabelliform (flabellate) Fan-shaped.

Flabellula See AMOEBIDA.

flacherie An economically important disease of silkworms caused by a small RNA virus resembling a picornavirus (see PICORNAVIRIDAE).

flagella Plural of FLAGELLUM.

flagellar motility (a) (bacterial) MOTILITY produced by the rotation of a FLAGELLUM, or the rotation of a number of flagella.

In peritrichously flagellated bacteria (such as *Escherichia coli* and *Salmonella typhimurium*), the flagella rotate independently of one another [Cell (1983) *32* 109–117]. For most (95%) of the time, each flagellum rotates counterclockwise (CCW). When the majority of flagella are rotating CCW they bunch together at one pole of the cell – this propelling the cell forward with the opposite pole leading. Within a uniform medium, such 'smooth swimming' is interrupted at intervals of ca. 1 second by *tumbling*: a random, chaotic movement (lasting ca. 0.1 second) resulting from a switch from CCW to clockwise (CW) in some of the bunched flagella; when this happens the flagellar bundle is disrupted, causing the random movement. (The timing of the switch from CCW to CW in a given flagellum is apparently not related to the timing of the switch in the other flagella.) Smooth swimming is resumed when the majority of the flagella are again rotating CCW.

The pattern of alternate smooth swimming and tumbling results in a three-dimensional RANDOM WALK. Certain environmental stimuli can influence the pattern of motility – leading to a TAXIS (see e.g. CHEMOTAXIS).

Peritrichously flagellated bacteria generally reach speeds of ca. 10–30 μm/second – although peritrichously flagellated cells of *Thiovulum majus* have been reported to swim at ~600 μm/second [Microbiology (1994) *140* 3109–3116].

Speeds reported in monotrichously flagellated bacteria include 70 μm/second (*Pseudomonas aeruginosa*) and 100 μm/second (*Bdellovibrio*). In these organisms, flagellar rotation may occur for approximately equal periods of time in the CCW and CW modes; during rotational reversal, conformational changes in the filament and/or hook may bring about randomization of direction (equivalent to tumbling in peritrichous cells) – thus permitting responses to environmental stimuli.

In amphitrichously flagellate bacteria the two flagella rotate in opposite directions.

In bacteria of the order SPIROCHAETALES, motility is associated with rotation of the periplasmic flagella – this generating a helical waveform in the outer membrane which enables the cells to 'screw' their way through the medium. Thus, CCW flagellar rotation in *Leptonema illini* causes the anterior (i.e. forwardly directed) end of the cell to assume a helical shape and the cell to rotate clockwise about its longitudinal axis – moving forward with a screw-like action. Spirochaetes can reverse direction and flex – the latter movement perhaps being a means of randomizing direction. The cells can swim through media which are too viscous to permit motility of cells which rely on external flagella. ('Creeping' or 'crawling' movements have also been recorded for spirochaetes in contact with solid surfaces.)

[Structure/motility of spirochaetes: JB (1996) *178* 6539–6545.]

(b) (eukaryotic) The eukaryotic FLAGELLUM commonly beats in an undulatory manner (usually in one plane, but sometimes – e.g. in *Euglena* – in a spiral) or in an oar-like manner (as in *Chlamydomonas*). Undulations commonly originate at the base of the flagellum and travel towards the tip.

The beating of a smooth flagellum generates forces which act on the surrounding medium in the direction of the propagated waves, so that the cell moves in the opposite direction (i.e. the flagellum 'pushes' the cell from behind). Conversely, if waves travel from the tip to the base of the flagellum (as e.g. in *Trypanosoma*), the cell swims with the flagellum leading.

In *tinsel* flagella, the mastigonemes appear to act as rigid, passive structures which lie in the plane of the beat and generate forces which act on the surrounding medium in a direction *opposite* to that of the propagated waves; thus, a tinsel flagellum undulating from base to tip will 'pull' the cell from the front (as e.g. in *Ochromonas*, and in zoospores of *Phytophthora*).

In some species the pattern of flagellar beating can be changed e.g. in response to environmental conditions. For example, in *Trypanosoma* the flagellum normally propagates waves from tip to base, but during an 'avoiding reaction' (PHOBIC RESPONSE) the direction of wave propagation can be reversed. *Chlamydomonas* normally swims with an oar-like 'breast stroke' ('ciliary-type' action), but in an avoiding reaction the two flagella extend forwards and propagate waves from base to tip ('flagellar-type' action).

In many species the flagellum is apparently capable of only one type of movement; in these organisms changes in direction may be achieved by a temporary cessation of beating, the flagellum remaining bent during this period. [Dynamics of eukaryotic flagellar movement: SEBS (1982) *35* 289–312.]

Flagellar movements appear to be regulated by the levels of Ca^{2+} in the flagellum; some environmental stimuli appear to cause changes in the permeability of the flagellar membrane to calcium ions, and consequent changes in motility patterns. In at least some species, levels of calcium ions within the flagellum may be regulated by a Ca^{2+}-ATPase located in the flagellar membrane. In *Chlamydomonas reinhardtii* the photophobic response is apparently initiated by Ca^{2+}-stimulated phosphorylation of certain axonemal proteins [JCB (1985) *101* 1702–1712].

flagellar motor See FLAGELLUM (a).

flagellar phase variation See PHASE VARIATION.

flagellar stain See FLAGELLUM (a).

flagellate (1) (*adj.*) Bearing one or more flagella (see FLAGELLUM). (2) (*noun*) A protozoon of the MASTIGOPHORA.

flagellin The protein subunit of the filament of a bacterial FLAGELLUM. The composition of flagellin (MWt ca. 30000–55000) varies with species, and determines e.g. the shape of the flagellum, its antigenic specificity (see H ANTIGEN and PHASE VARIATION), and susceptibility to FLAGELLOTROPIC PHAGES. Typically, flagellins are relatively rich in glutamate and aspartate but poor in aromatic amino acids; cysteine is usually absent. Flagellins from the alkalophile *Bacillus firmus* have a lower content of basic amino acids [JGM (1983) *129* 3239–3242]; the 'complex' flagella of *Rhizobium* spp have a higher content of hydrophobic amino acids. A flagellar filament may be dissociated into its flagellin subunits e.g. by treatment with detergents or acid; under appropriate conditions, and given a primer (e.g. a small fragment of flagellum), the flagellin subunits can re-assemble spontaneously.

flagellotropic phage Any BACTERIOPHAGE which adsorbs to a host cell flagellum and hence is specific for flagellated host cells (see also FLAGELLIN). Flagellotropic phages include e.g. bacteriophage χ (see STYLOVIRIDAE) in enterobacteria, BACTERIOPHAGE PBS1 (q.v.) in *Bacillus subtilis*, and bacteriophage 7-7-1 in '*Rhizobium lupini*'. In e.g. phage χ, the tip of the phage tail bears a fibre which appears to wrap around the flagellum, and the phage then appears to move down to the base of the flagellum to infect the cell.

flagellum (*plural*: flagella) (a) (bacterial) A thread-like appendage which (commonly) projects from the surface of the cell and which is responsible for MOTILITY in most types of motile bacteria; such flagella may occur e.g. singly, in groups or tufts, or distributed over the surface of the cell (see AMPHITRICHOUS, HELICOTRICHOUS, LOPHOTRICHOUS, MONOTRICHOUS and PERITRICHOUS).

By contrast, in bacteria of the SPIROCHAETALES the flagella occur *between* layers of the cell envelope. This entry refers primarily to flagella which project from the cell surface – although the structural and mechanistic features of spirochaete flagella are apparently similar.

Flagella can be detected by dark-field MICROSCOPY – but not by bright-field microscopy unless suitably stained (see e.g. LEIFSON'S FLAGELLA STAIN). [Simple method for staining bacterial flagella: JCM (1989) *27* 2612–2615.]

The bacterial flagellum consists essentially of three parts: (i) a protein *filament* which projects from the cell surface and which, by *rotating*, provides the thrust for FLAGELLAR MOTILITY; (ii) a curved protein structure, the *hook*, contiguous with the proximal end of the filament, and (iii) the *flagellar motor* (*basal body*) – a structure which anchors the hook in the cell envelope and incorporates the energy-converting apparatus.

The *filament* (ca. 5–10 μm × 20 nm) is a semi-rigid, left-handed *helix* (of periodicity ~2 μm) composed of eleven longitudinally arranged fibrils, each fibril being a chain of FLAGELLIN subunits; the fibrils are closely associated (like the strands of a rope) to form a structure with an axial channel. (In some species the cells may have two types of flagellum which have different periodicities, a phenomenon known as *biplicity* – see e.g. VIBRIO.) In some Gram-negative species the filament is *sheathed*, i.e. covered by an extension of the OUTER MEMBRANE. Yet further variation in the filament occurs e.g. in strains of *Rhizobium* which have 'complex' flagella that differ e.g. in their fine structure [JMB (1986) *190* 569–576] and in their greater fragility.

The *hook* is a tightly coiled right-handed helix, with an axial channel, which is curved through ~90°; its protein subunits are distinct from flagellin.

The *flagellar motor* (= *basal body*) is a cylindrical structure which spans the width of the cell envelope; its composition varies with species. In *Escherichia coli* the components include a stack of three protein rings and a hollow, axial protein *rod* (*shaft*) – together with a fourth protein ring which projects into the cytoplasm (see figure). The L (for 'lipopolysaccharide') ring is coplanar with the OUTER MEMBRANE, the P (for 'peptidoglycan') ring is coplanar with the PEPTIDOGLYCAN layer of the cell envelope, and the MS ring (once believed to be two distinct rings – the M ring and the S ring) lies within the cytoplasmic membrane; the C (for 'cytoplasm') ring, which is associated with the MS ring, carries the components of a switching mechanism which allows the direction of rotation to be reversed.

A different number of rings has been reported in other bacteria – e.g. two in *Bacillus subtilis* and five in *Caulobacter crescentus*. In *Campylobacter* a 'cartwheel structure' with eleven 'spokes' appears to be associated with the basal body [JGM (1984) *130* 1307–1310].

In a current model of flagellar rotation (in e.g. *Escherichia coli*, *Salmonella typhimurium* and related species) the MS ring and the C ring jointly rotate relative to the stationary L and P rings; this rotation causes the axial shaft (and hence the filament) to rotate, the filament converting torque into thrust.

Flagellar rotation needs energy in the form of an ion gradient across the cytoplasmic membrane. In *E. coli* and *S. typhimurium* energy is supplied by proton motive force (pmf) (see CHEMIOSMOSIS), and rotation is driven by an inward flow of protons through the flagellar motor (perhaps via a channel in the MotA/MotB protein assembly); torque generation appears to involve interaction between the rotor (MS ring) and the stator (MotA/MotB assembly) – specifically, between the FliG protein, associated with the MS ring, and MotA. [Models for the flagellar motor (theoretical considerations): PTRSLB (2000) *355* 491–501.]

The C ring (apparently the FliM component of the C ring – see later) plays a pivotal role in the switching mechanism in CHEMOTAXIS.

The speed of rotation of a flagellum varies linearly with the level of the cell's pmf. The minimum (threshold) value of pmf

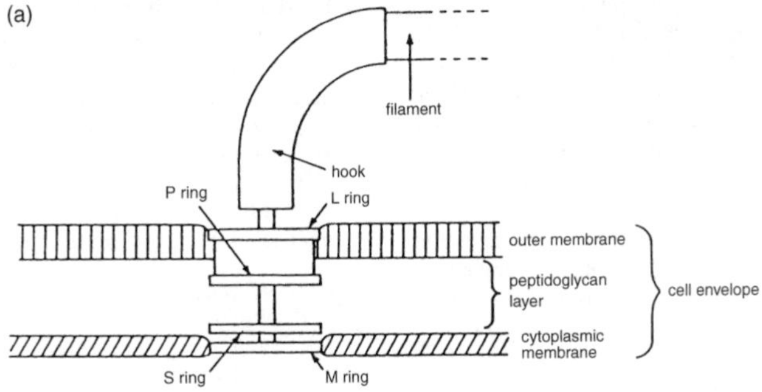

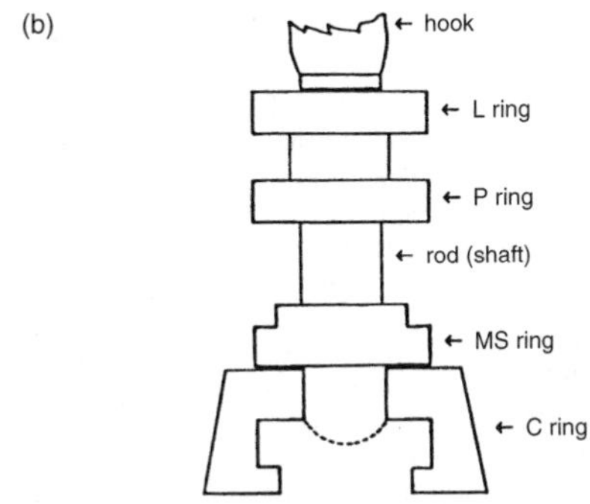

FLAGELLUM: the origin of a flagellum in the cell envelope of *Escherichia coli*, *Salmonella typhimurium* and related Gram-negative bacteria (diagrammatic, not drawn to scale; see entry).

(a) An earlier model.

(b) A current view based on several sources [e.g. JB (1996) *178* 4582–4589; JMB (1996) *255* 458–475]. As in the earlier model, the L and P rings are associated with the outer membrane and peptidoglycan layer, respectively. The MS ring – previously regarded as two distinct structures (the M and S rings) – lies in the cytoplasmic membrane. The C ring, which includes proteins FliM and FliN (see table, text), is in the cytoplasm; it is associated with the MS ring and is believed to rotate with the MS ring when the flagellar motor is running. The C ring is involved in switching between counterclockwise and clockwise rotation of the flagellum in chemotaxis.

The annular region of the cytoplasmic membrane which encircles the MS ring contains the MotA and MotB proteins; this region acts as the stator of the flagellar motor. It appears that MotA and MotB form a complex, with MotB bound to peptidoglycan, and MotA interacting with molecules of the FliG protein (which are bound to the periphery of the MS ring); the latter interaction (details currently unknown) generates torque and, hence, flagellar rotation.

The *flagellar motor* (= *basal body*) includes the L, P and MS rings, the axial rod (shaft) and the MotA/MotB complex. The C ring is included by some authors; other authors include the C ring as part of an 'extended' basal body.

The dotted line enclosed by the C ring represents a structure which has been termed the *export apparatus* (see entry) – and which now appears to be a form of type III PROTEIN SECRETION system.

Reproduced from *Bacteria* 5th edition, Figure 2.8, page 28, Paul Singleton (1999) copyright John Wiley & Sons Ltd, UK (ISBN 0471-98880-4) with permission from the publisher.

(below which rotation does not occur) appears to vary with species; it has been reported to be ~30 mV for *E. coli* and *Bacillus subtilis* but less for e.g. streptococci.

Although flagellar rotation is commonly driven by pmf, SODIUM MOTIVE FORCE is used in some species – e.g. smf is used to drive the polar flagellum in *Vibrio cholerae* [JB (1999) *181* 1927–1930].

Assembly of the flagellum. Assembly is a highly organized process in which the components are added in strict sequence. The following refers to flagellar assembly in *E. coli*, *S. typhimurium* and related organisms.

First, the MS ring – consisting of about 25 molecules of the *fliF* gene product (FliF protein) – is inserted into the cytoplasmic membrane. Then the FliG, FliM and FliN proteins are incorporated; it appears that molecules of FliG bind around the periphery of the MS ring (on the cytoplasmic side), while FliM and FliN jointly form the C ring, with FliN being more distal to the membrane [JB (1997) *179* 813–817].

At this stage an *export apparatus* assembles within the C ring (see figure). The purpose of the export apparatus is to permit the operation of a functional channel through the MS ring without compromising the cell's structural integrity; this channel allows specific components subsequently to pass through the MS ring when required. The export apparatus is a multiprotein structure that includes FlhA, FlhB, FliH, FliI, FliO, FliP, FliQ and FliR.

Interestingly, some of the proteins of the basal body exhibit homology with proteins of type III secretion systems (see PROTEIN SECRETION), and the group of proteins involved in transport of the flagellar components has been referred to as a type III secretion system.

In addition to these proteins, FliJ is needed for the export of components – and may be part of the transport apparatus; this protein also appears to act as a 'chaperone', preventing intracellular aggregation of various export substrates [JB (2000) *182* 4207–4215].

FlgB, FlgC and FlgF are translocated via the export apparatus and form the proximal part of the hollow rod (shaft); FlgG then forms the distal part of the rod.

The L and P rings are formed by molecules of FlgH and FlgI, respectively; these proteins, which have a signal sequence (see SIGNAL HYPOTHESIS), are secreted through the cytoplasmic membrane into the periplasmic space, i.e. unlike rod components, they do not pass through the MS ring.

Molecules of FlgE exit via the MS ring and give rise to the hook. When the hook has reached an appropriate length, annuli of hook-associated protiens FlgK and FlgL, and a terminal 'cap' of FliD protein, are added, in that order, at the distal end. (The mechanism which regulates the length of the hook is unknown, but it appears to involve FliK.) FliD forms a permanent distal *cap* while the flagellin subunits (FliC protein) pass through the axial channel of the rod and hook and are polymerized at the FlgL–FliD junction to form the growing filament. Incorporation of FliC may be promoted by *rotation* of the flagellar cap [Science (2000) *290* 2148–2152]. Growth of the filament thus occurs at the *distal* end ('tip growth' – compare archaeal flagella in (c), below); during growth, the FliC subunits are incorporated by self-assembly.

Genetic regulation of flagellar assembly. The first genes to be expressed – class I genes – are those in the *flhCD* operon; these genes encode transcriptional activators, and their expression is essential for the expression of all other flagellar genes. (The *flhCD* genes are expressed at peak levels in synchrony with the cell cycle [JB (1997) *179* 5602–5604].)

FlhC and FlhD promote the expression of class II genes, including: *fliF* (MS ring), *fliG* (switch), *fliMN* (switch/C ring), *flgE* (hook) and genes from two operons, *flhAB* and *fliHIOPQR*, whose products are required for the export apparatus.

Another class II gene, *fliA*, encodes a SIGMA FACTOR (σ^{28}) which activates class III genes, but FliA is temporarily inactivated by the binding of FlgM (FlgM is an ANTI-SIGMA FACTOR); however, completion of the hook serves as a signal for the release of FlgM (which is externalized via the axial canal) – thus allowing FliA to mediate expression of the class III genes.

The class III genes include *fliC* (the flagellin subunit of the filament) and the *motAB* genes (which encode components of the stator).

Some of the genes and gene products involved in the assembly of the flagellum are listed in the table.

[Genetics and assembly of flagella in Gram-negative bacteria: FEMS Reviews (2000) *24* 21–44 (22–27).]

(b) (eukaryotic) A thread-like appendage (ca. 0.15–0.3 μm in diam., up to ~40 μm in length), one or more of which extend from vegetative and/or reproductive cells of certain algae, fungi and protozoa. Eukaryotic flagella typically act as locomotory organelles but may be involved e.g. in mating (see e.g. CHLAMYDOMONAS) or attachment (see e.g. TRYPANOSOMA) (cf. HAPTONEMA).

The eukaryotic flagellum is completely different from the bacterial flagellum in both structure and mechanism of action. The main structural component of the eukaryotic flagellum is the *axoneme*: a system of longitudinally arranged MICROTUBULES (each ca. 25 nm diam.) which extend the length of the flagellum. These microtubules are arranged in a characteristic '9 + 2' pattern: two adjacent tubules (enclosed by a membranous sheath in some species) run through the centre of the axoneme and are surrounded by a circular arrangement of nine *pairs* of tubules (*doublets*). Each doublet consists of an A-tubule (= 'A-subfibre') and a B-tubule ('B-subfibre'); the A-tubules are complete microtubules, but the B-tubules are incomplete, sharing part of the wall of the A-tubule (giving the B-tubules a somewhat D-shaped appearance in cross-section). The A-tubule in each doublet bears a series of paired (inner and outer) 'arms' composed of high-MWt, Ca^{2+}- and Mg^{2+}-activated ATPase called *dynein*. The A-tubules are linked to the central

FLAGELLUM (bacterial): some of the genes and gene products involved in the assembly of the flagellum in *Escherichia coli*, *Salmonella typhimurium* and related species

Gene (product)	Function of gene product
Class I genes	
flhC (FlhC)	Transcriptional activation of class II genes
flhD (FlhD)	Transcriptional activation of class II genes
Class II genes	
flgB (FlgB)	Rod (proximal)
flgC (FlgC)	Rod (proximal)
flgE (FlgE)	Hook
flgF (FlgF)	Rod (proximal)
flgG (FlgG)	Rod (distal)
flgH (FlgH)	L ring
flgI (FlgI)	P ring
flgK (FlgK)	Hook (distal end)
flgL (FlgL)	Hook (distal end)
flgM (FlgM)	Anti-sigma factor (delays FliA activity)
flhA (FlhA)	Export apparatus
flhB (FlhB)	Export apparatus
fliA (FliA)	Sigma factor (σ^{28}) for class III genes
fliD (FliD)	Filament cap
fliF (FliF)	MS ring
fliG (FliG)	Switch/C ring?/torque generation
fliH (FliH)	Export apparatus
fliI (FliI)	Export apparatus
fliJ (FliJ)	Export apparatus/chaperone
fliK (FliK)	Regulation of hook length
fliM (FliM)	C ring/switch
fliN (FliN)	C ring/switch
fliO (FliO)	Export apparatus
fliP (FliP)	Export apparatus
fliQ (FliQ)	Export apparatus
fliR (FliR)	Export apparatus
Class III genes	
fliC (FliC)	Flagellin (protein subunit of filament)
motA (MotA)	Torque generation
motB (MotB)	Torque generation

tubules (or to the sheath surrounding them) by radial 'spokes'; protein (*nexin*) links also occur between the peripheral doublets. [Axoneme substructure: JBC (1985) *100* 2008–2018.]

The axoneme is embedded in a flagellar *matrix*, the whole being enclosed by a sheath-like *flagellar membrane* which is apparently continuous with the cytoplasmic membrane. The proximal end of the flagellum is connected to a BASAL BODY via a more or less distinct region, the *transition region* (cf. CILIUM).

Although most eukaryotic flagella conform to the basic structure described above, there are differences in detail among the species. For example, many eukaryotic flagella bear characteristic surface structures attached to the flagellar membrane – structures such as minute scales (see PRASINOPHYCEAE); a fine *tomentum* of minute hairs; or larger, more distinct hairs called *mastigonemes* (or *flimmer*) which are commonly arranged in two longitudinal rows, one on each side of the flagellum. (See also TRENTEPOHLIA.) (A distinction is sometimes made between the terms 'mastigoneme' and 'flimmer': mastigonemes are shorter and more rigid, while flimmer are longer, finer, and more flexible.) A flagellum which lacks obvious scales or mastigonemes is said to be of the *smooth*, *simple*, *whiplash* or *peitschengeissel* type, while a flagellum bearing mastigonemes is said to be of the *tinsel*, *flimmergeissel* or *hispid* type.

The nature and number of a cell's flagella, and the sites and orientation of the flagella, are important taxonomic features. (See also HETEROKONT, ISOKONT and STEPHANOKONT.) Ultrastructural details of flagella may also be of taxonomic value: e.g. in flagellated cells of the green algae (CHLOROPHYTA) the A-tubule of *one* of the nine peripheral doublets in the axoneme lacks an outer dynein arm – a characteristic apparently universal in, and unique to, green algae. The flagellum of green algae also has a microfibrillar structure in the transition region which gives a characteristic stellate pattern in cross-section. [Review of flagellar ultrastructure in green algal classification: Book ref. 123, pp 73–120.] (See also PARAXIAL ROD.)

The eukaryotic flagellum can move independently of the basal body: flagella which have been severed from the cell are capable of independent movement, and in e.g. *Chlorogonium* the cells remain motile when the basal bodies have detached from the flagella prior to cell division [JCB (1985) *100* 297–309]. Movement is thus a property of the flagellar shaft itself – cf. *bacterial* flagellum (a), above. The movement involves sliding of axonemal microtubules relative to one another, the energy being supplied by dynein-catalysed hydrolysis of ATP; movement is apparently regulated by Ca^{2+}: see FLAGELLAR MOTILITY (b).

(c) (archaeal) The archaeal flagellum differs from the bacterial flagellum in composition, structure and apparent mode of assembly. Thus, the subunits (flagellin) of the archaeal filament are typically glycosylated proteins, and the filament itself is typically much thinner than the bacterial filament. Moreover, archaeal flagellins have a signal sequence – suggesting passage through the cytoplasmic membrane rather than through an export apparatus; this feature, together with certain similarities between archaeal flagellins and bacterial type 4 fimbriae, have suggested that the archaeal flagellum is assembled from the *base* – in contrast to the 'tip growth' of the bacterial flagellum [JB (1996) *178* 5057–5064].

Interestingly, genes homologous to those which encode bacterial flagellar structures have not been detected in the genomes of archaeans, even in those whose complete genome sequence is known; this has suggested that the archaeal flagellum is a unique form of organelle [FEMS Reviews (2001) *25* 147–174].

Flagyl See METRONIDAZOLE.

flaming Brief exposure of an object (or part of an object) to the hottest part of a flame – usually for the STERILIZATION of the surface thus treated. Flaming is part of the ASEPTIC TECHNIQUE. Objects such the LOOP and STRAIGHT WIRE are flamed for ca. 3–5 sec; somewhat shorter exposures are used for the rims of e.g. bottles from which sterile (non-flammable!) substances are to be poured. Flaming for very short times may be used for the heat FIXATION of bacterial smears.

Flammulina See AGARICALES (Tricholomataceae) and ENOKITAKE.

flash-freezing *Syn.* QUENCH-FREEZING.

flash pasteurization See PASTEURIZATION.

flask fungi PERITHECIUM-forming fungi.

flat-field objective lens An objective lens (see MICROSCOPY) which can give an image in which all parts of the field are simultaneously in focus. Flat-field objectives are denoted by the prefix *plan-* or *plano-* (see e.g. ACHROMAT).

flat sour A can (see CANNING) which appears normal (i.e. the ends are flat or slightly concave – cf. SWELL) but whose contents have been soured as a result of microbial growth; *Bacillus stearothermophilus* is one of several common causal agents in low-acid foods. (See also SOFT CORE HAM.)

flavacol A compound produced by *Aspergillus flavus*: 3-hydroxy-2,5-diisobutylpyrazine.

flavin adenine dinucleotide See RIBOFLAVIN.

flavin mononucleotide See RIBOFLAVIN.

Flaviviridae ('arbovirus group B') A family of enveloped ssRNA-containing VIRUSES [Intervirol. (1985) *24* 183–192] formerly regarded as a genus within the family TOGAVIRIDAE. Flaviviruses infect a wide range of vertebrates, and many also infect invertebrates (mosquitoes or ticks) which act as VECTORS. Some flaviviruses are important pathogens of man or animals. The family contains a single genus, *Flavivirus*; type species: yellow fever virus.

The flavivirus virion is more or less spherical (ca. 40–50 nm diam.); it consists of a core (comprising a single type of core protein, C, associated with the genome), a 'membrane-like' protein, M, which surrounds the C–RNA complex, and a lipoprotein envelope containing one type of virus-specific glycoprotein, E, which acts as a haemagglutinin (e.g. with chick RBCs) and is the target of neutralizing antibodies. The genome is a single linear positive-sense ssRNA molecule (MWt ca. 4×10^6) which is capped at the 5′ end but is apparently not polyadenylated at the 3′ end. Subgenomic mRNA is not produced; the genome-length RNA functions as the sole messenger and contains only one long open reading frame – at least in West Nile virus [Virol. (1986) *149* 10–26] and in yellow fever virus [Science (1985) *229* 726–733] – suggesting that viral proteins are formed by post-translational cleavage of a polyprotein. The genes encoding structural proteins are located at the 5′ end (5′-C–M–E...); several non-structural proteins are also produced. A region at the 3′ end is not translated and has the potential for forming a stem-and-loop secondary structure [JGV (1986) *67* 1183–1188]. Viral RNA synthesis (which involves the formation of both replicative intermediates and replicative forms) apparently occurs in the perinuclear region; maturation may involve a process of 'condensation' rather than budding, and appears to occur within cisternae of the endoplasmic reticulum.

Flaviviruses can replicate in various types of vertebrate cells in culture, often causing CPE and plaque-formation; infected cells typically exhibit extensive proliferation of the endoplasmic reticulum. Host protein synthesis is not significantly inhibited. Invertebrate cells may also be infected, with or without CPE.

Most flaviviruses can be classified in 7 serologically defined complexes (subgroups); most of those of greatest medical and veterinary importance belong to one of three such subgroups. (YELLOW FEVER virus is serologically distinct from other flaviviruses.) The *dengue virus subgroup* includes DENGUE viruses types 1–4. The *West Nile virus subgroup* includes e.g. Kunjin virus, JAPANESE B ENCEPHALITIS virus, MURRAY VALLEY FEVER virus, Rocio virus, ST LOUIS ENCEPHALITIS virus, and West Nile virus. The *tick-borne encephalitis virus subgroup* includes e.g. KYASANUR FOREST DISEASE virus, LOUPING-ILL virus, Negishi virus, Omsk haemorrhagic fever virus, POWASSAN ENCEPHALITIS virus, and tick-borne encephalitis (TBE) virus – including the eastern subtype (RUSSIAN SPRING–SUMMER ENCEPHALITIS virus) and the European subtype (Central European encephalitis virus). Other flaviviruses which are occasionally responsible for human disease include e.g. Banzi virus, Bussuquara virus, Ilheus virus, Rio Bravo virus, Sepik virus, Spondweni virus, Wesselsbron virus and Zika virus.

Possible members of the Flaviviridae include *Aedes albopictus* cell-fusing agent and SIMIAN HAEMORRHAGIC FEVER VIRUS.

[Medical aspects: Book ref. 148, pp. 955–1004.]

Flavivirus See FLAVIVIRIDAE.

Flavobacterium A genus (*incertae sedis*) of aerobic, chemoorganotrophic, Gram-negative bacteria which occur e.g. in soil, water, raw meats, milk and in some clinical specimens (*F. meningosepticum* is a causal agent of neonatal MENINGITIS). Cells: non-motile, round-ended rods, ca. 0.5 × 1.0–3.0 μm, which typically contain orange or yellow non-fluorescent, water-insoluble pigments; in *F. breve* and *F. odoratum* (at least) the pigments may be of the flexirubin type. *F. meningosepticum* is typically weakly pigmented or non-pigmented. Metabolism is respiratory (oxidative). Acid (but no gas) is formed from carbohydrates in media which have a low content of peptone. Most species (not *F. aquatile*) grow well in nutrient broth and on nutrient agar; *F. aquatile* requires media enriched with e.g. casein digest, glucose, and yeast extract. *F. meningosepticum* is non-haemolytic on blood agar. Catalase +ve. Oxidase +ve. GC%: ca. 31–42. Type species: *F. aquatile*; other species: *F. breve*, *F. balustinum*, *F. meningosepticum*, *F. multivorum*, *F. odoratum*, *F. spiritivorum* [Book ref. 22, pp. 353–361], and *F. gleum* [IJSB (1984) *34* 21–25]. (See also SPHINGOBACTERIUM.)

flavodoxins A category of flavoproteins which have low (i.e. highly negative) E_m values which can replace FERREDOXINS as electron carriers in various reactions – though their efficiency as electron carriers is less than that of ferredoxins. Flavodoxins contain 1 FMN per molecule but do not contain metal or labile sulphur. In many organisms (e.g. *Clostridium* spp, certain algae) flavodoxins are synthesized in place of ferredoxins in response to iron deficiency. Algal flavodoxin is also known as *phytoflavin*. In *Azotobacter vinelandii* a flavodoxin, *azotoflavin*, acts as electron carrier in NITROGEN FIXATION.

Flavomycin *Syn.* MOENOMYCIN.

flavoproteins See RIBOFLAVIN.

flavour microorganisms Microorganisms which form metabolic products ('flavour compounds') that contribute to the flavour and/or aroma of certain foods: see e.g. DIACETYL. (See also GLUTAMIC ACID and YEAST EXTRACT.) [Review: AAM (1983) *29* 29–51.]

flax retting See RETTING.

flax rust (linseed rust) A disease of the flax plant (*Linum usitatissimum*) caused by the macrocyclic, homoxenous rust *Melampsora lini*. Symptoms include the presence of orange uredia on both sides of the leaf and, later, dark-coloured telia on the stems.

flax wilt See FUSARIUM WILT.

Flectobacillus A genus of pink-pigmented bacteria (family SPIROSOMACEAE). The cells are typically C-shaped, 0.3–1.0 × 1.5–5.0 μm, but spiral forms occur – as do filaments up to 50 μm; 'ring-shaped' structures may be apparent due to overlap of the ends of curved cells. Acid is formed from many carbohydrates. GC%: 34–40. Type species: *F. major*.

Flexibacter A genus of GLIDING BACTERIA (see CYTOPHAGALES) which occur in aquatic habitats; some species are pathogenic for fish (see e.g. COLD WATER DISEASE and COLUMNARIS DISEASE). The cells are short, non-motile rods in old cultures but are thin, thread-like and motile in young cultures; typically, they contain pigments ranging from yellow to red. Metabolism: chemoorganotrophic.

Flexibacteriae A class of non-photosynthetic GLIDING BACTERIA comprising the orders CYTOPHAGALES and MYXOBACTERALES [ARM (1981) *35* 339–364].

flexible pouch *Syn.* RETORT POUCH.

flexirubins Polyene pigments found e.g. in members of the CYTOPHAGALES.

flexuous hyphae (receptive hyphae) Hyphae which project from a pycnium and which function as female (receptive) structures. (cf. TRICHOGYNE.)

***flgM* gene** See SIGMA FACTOR.

***fliA* gene** See SIGMA FACTOR.

flimmer See FLAGELLUM (b).

flimmergeissel flagellum See FLAGELLUM (b).

flip–flop (of membrane lipids) See CYTOPLASMIC MEMBRANE.

flippase See CYTOPLASMIC MEMBRANE.

flipper A can (see CANNING) which appears normal but in which one end 'flips out' (i.e. becomes convex) if the can is struck (cf. SWELL).

***flo_{St}* gene** See FLORFENICOL.

floc blanket clarifier See WATER SUPPLIES.

flocculation (1) (*serol.*) A term used with different meanings by different authors and for different phenomena: e.g., (i) the development of a flaky or fluffy precipitate in certain precipitin tests; (ii) any form of precipitate; (iii) any form of agglutination, but particularly the agglutination of bacteria by antiflagellar antibodies.

(2) The aggregation of (initially separate) cells to form flocs. Spontaneous flocculation occurs e.g. in certain sewage bacteria (see e.g. ZOOGLOEA and CELLULOSE) and in strains of BREWERS' YEAST during the stationary phase of batch culture (e.g. at the end of the fermentation stage of BREWING). Flocculation of brewers' yeast aids in the separation of the yeast cells from the fermented wort. (Yeast cells which remain dispersed in all phases of batch culture are called 'powdery yeasts'.) Prior to flocculation, the cells are presumed to remain dispersed because of their mutually repulsive (usually negative) surface charge (see ZETA POTENTIAL); flocculation is thought to be initiated by a change in the surface structure of the cells, and Ca^{2+} appears to play a role by bridging the negative charges on adjacent cells [review: Book ref. 108, pp. 323–335]. (cf. COAGGREGATION.)

flood plate A PLATE in which the surface has been flooded with liquid inoculum and the excess drawn off with a sterile Pasteur pipette. (See also LAWN PLATE.)

flor sherry, flor yeast See WINE-MAKING.

florfenicol A synthetic, broad-spectrum antibiotic which has been used e.g. against *Mycoplasma bovis* in the treatment of calf pneumonia. In vitro tests on field isolates of *M. bovis* indicate that significant levels of resistance have developed against florfenicol [VR (2000) *146* 745–747]. In *Salmonella typhimurium* DT104 (an emergent multidrug-resistant pathogen), mutation in gene flo_{St} can result in resistance to both florfenicol and chloramphenicol [JCM (1999) *37* 1348–1351].

Florida horse leech *Syn.* EQUINE PHYCOMYCOSIS.

floridean starch An amylopectin-like D-glucan: the main storage carbohydrate in members of the Rhodophyta. [Book ref. 37, pp. 475–478.]

floridoside 2-*O*-Glycerol-α-D-galactopyranoside: a glycoside found in members of the Rhodophyta; it serves as a reserve carbohydrate and as an osmoregulatory compound. (cf. ISOFLORIDOSIDE.)

flotation A method for concentrating e.g. protozoan cysts in a sample of faeces; the sample is dispersed in a concentrated solution of e.g. $ZnSO_4$ or NaCl such that any cysts present in the sample float to the surface. (cf. FORMALIN–ETHER METHOD.)

Zinc sulphate flotation. The faecal sample (ca. 1 g) is mixed with about ten times its volume of water, centrifuged (ca. 350 *g*/3 min), and all the supernatant discarded. The sediment is re-suspended in ca. 10 ml of a solution of $ZnSO_4$ of S.G. 1.18 (made by adding water to 331 g $ZnSO_4.7H_2O$ to give a final volume of 1 litre) and filtered through a double layer of wet gauze; the filtrate is centrifuged (ca. 350 *g*/3 min) – the centrifuge being allowed to stop naturally to avoid turbulence within the liquid. A loopful of liquid from the surface is transferred to a drop of iodine solution on a slide; a cover-glass is added, and the preparation is examined under the microscope.

Salt flotation. A solution of NaCl (70–100% saturated) is often used for the concentration of coccidian cysts in faecal samples. The sample is subjected to several cycles of centrifugation in water (each ca. 350 *g*/2–3 min), the supernatant being discarded each time. The sediment is then dispersed in salt solution and centrifuged at ca. 200 *g*/5 min. The surface layer of supernatant is transferred, by pipette, to ten times its volume of water; this suspension is centrifuged and the sediment collected for study. Prolonged contact with saturated salt solution causes plasmolysis; however, oocysts generally remain viable, and they regain their original appearance on transfer to water.

flotation form (of an amoeba) See PSEUDOPODIUM.

flow cytometry Any procedure in which individual cells in a suspension are counted, sorted, or otherwise analysed by the use of an apparatus in which the cells pass, individually, through a small hole or tube; the term is commonly used to refer specifically to those techniques in which cell populations are analysed by detecting fluorescence or FLUOROCHROME-labelled antigens or other structures on/in the cell (see e.g. FACS and FLOW MICROFLUOROMETRY; cf. COULTER COUNTER).

flow microfluorometry Any procedure in which populations of cells, or of other particles, are counted, sorted, or otherwise analysed by passing them individually through a small hole or tube and detecting specific fluorescence or FLUOROCHROME labels. (See e.g. FACS; cf. FLOW CYTOMETRY.)

flowers of tan See FULIGO.

***FLP* gene** See TWO-MICRON DNA PLASMID.

flu *Syn.* INFLUENZA.

flucloxacillin See PENICILLINS.

fluctuation test A test devised by Luria and Delbrück in 1943 for investigating the mode in which bacterial populations respond to changes in the environment. At that time two hypotheses co-existed. (i) Genetic changes occur *adaptively*, i.e., as a result of environmental influences on the cells; those cells which adapt to the environment survive and proliferate. (ii) Genetic changes occur *spontaneously*, i.e., independently of the environment; those cells which have become particularly well suited to certain environmental conditions proliferate (preferentially) if such conditions arise.

If bacteria adapted to a new environment, a similar proportion of adapted cells should arise in each of a number of independent cultures. If, however, mutants arose spontaneously and randomly, the number of mutant cells should vary considerably from culture to culture.

In the original test, a measurement was made of the statistical variance among the numbers of bacteriophage-resistant cells in samples from each of a number of separate cultures of phage-sensitive bacteria; the very high variance could not be attributed to sampling error since a very much lower variance was obtained for the numbers of resistant cells in each of a number of samples taken from a single (bulk) culture. (Phage-resistant cells were counted by plating each sample on a separate plate, subjecting the plate to virulent phage, and incubating; resistant cells formed colonies.) The wide fluctuation (high variance) in mutant numbers among the separate cultures indicated that mutant clones had been initiated in some cultures much earlier than in others – i.e., resistant cells had appeared at different times; this indicated that mutant (resistant) cells had arisen *before* the cells had been exposed to phage – thus supporting hypothesis (ii).

flucytosine (5-fluorocytosine) An ANTIFUNGAL AGENT which is active mainly against yeasts and yeast-like fungi; it is used (usually with AMPHOTERICIN B) in the treatment of e.g. candidiasis and cryptococcosis. It functions as a pyrimidine analogue. After uptake by the yeast cell, it is deaminated to 5-fluorouracil by cytosine deaminase; 5-fluorouracil may then be phosphorylated and incorporated into RNA, or converted to 5-fluorodeoxyuridine monophosphate which inhibits thymidylate synthase (see Appendix V(b)), thereby inhibiting DNA synthesis.

fluid mosaic model See CYTOPLASMIC MEMBRANE.

fluid-phase pinocytosis See PINOCYTOSIS.

fluidized-bed fermenter A FERMENTER in which a nutrient solution is pumped upwards through a vertical column of small particles (e.g. stainless steel balls) coated with cells. The upward flow of liquid causes the particles to exhibit small, rapid movements which (a) facilitate interaction between the adherent cells and the medium, and (b) result in the removal (by friction) of a proportion of the adherent cells; a steady state is achieved when the loss of cells is balanced by cell growth.

flunidazole See NITROIMIDAZOLES.

fluor *Syn.* FLUOROCHROME.

FluoReporter® See PROBE.

fluorescein A pale-yellow FLUOROCHROME which fluoresces greenish-yellow; it is a substituted xanthene used e.g. in IMMUNOFLUORESCENCE techniques. (See also FITC.) Solubility in water is very low; in ethanol it is ca. 2% w/v. (cf. PYOVERDIN.)

fluorescence The emission of light which occurs when certain substances absorb radiation; emission occurs as electrons excited to the singlet state (S_1) are returning to the (unexcited) ground state (S_0). The light emitted is of wavelength longer than that of the exciting radiation, and emission typically lasts for ca. 10^{-9} sec (cf. PHOSPHORESCENCE). In fluorescence microscopy (see MICROSCOPY (e)) the exciting radiation is usually UV light, but blue light is used in some cases ('blue-light fluorescence'). *Primary* fluorescence (= autofluorescence) is that produced by a substance which is inherently fluorescent – e.g. chlorophyll fluoresces red under UV radiation. *Secondary* fluorescence is that produced by a non-fluorescent substance which has been dyed with a FLUOROCHROME.

fluorescence-activated cell sorter See FACS.

fluorescence *in situ* hybridization (FISH) See PROBE.

fluorescence microscopy See MICROSCOPY (e).

fluorescence resonance energy transfer See PCR.

fluorescent antibody technique (FAT) Any procedure in which IMMUNOFLUORESCENCE is used (e.g. for clinical lab. diagnosis, for counting or detecting specific microorganisms in water samples, etc).

fluorescent brightener (brightener; optical bleach; optical brightener) A compound which exhibits FLUORESCENCE under ultraviolet radiation, and which may be used e.g. to treat (white) paper or textiles etc in order to enhance their 'whiteness' in normal daylight (by effectively enhancing the blue end of the visible spectrum); fluorescent brighteners which are absorbed by – and are non-toxic (or selectively toxic) for – microorganisms have also been used e.g. for STAINING microorganisms, the presence of the brightener being demonstrated by fluorescence MICROSCOPY.

Type A brighteners (e.g. CALCOFLUOR WHITE) are (anionic) derivatives of 4,4′-diamino-2,2′-stilbene disulphonic acid. Type N brighteners (e.g. TINOPAL AN) are (cationic) oxacyanine compounds which tend to be more inhibitory than type A brighteners for microorganisms.

fluorescent dye Any dye capable of FLUORESCENCE.

Fluoribacter See LEGIONELLACEAE.

fluorides The fluoride ion is inhibitory to a wide range of organisms, inhibiting many enzymes; it inhibits e.g. the fermentation of sugars by streptococci – apparently chiefly by (competitively) inhibiting enolase and thereby preventing the formation of phosphoenolpyruvate (necessary for uptake of PTS sugars) from 2-phosphoglycerate. (See also PEP POTENTIAL.) The inhibitory effect of fluoride is greater at lower pH; the cytoplasmic membrane may be impermeable to F^-, and fluoride may enter the cell as HF.

Fluorides are used e.g. in the prevention of DENTAL CARIES; they act primarily by becoming incorporated into the tooth enamel, thereby decreasing its susceptibility to acid demineralization; oral bacteria are apparently relatively insensitive to fluoride in the presence of saliva [Inf. Immun. (1986) *51* 119–124]. Sodium and ammonium fluorides have also been used e.g. as antifungal wood PRESERVATIVES; their efficacy is said to be greater than that of zinc or copper compounds.

fluorite objective (semiapochromatic objective) An objective lens (see MICROSCOPY) in which chromatic and spherical aberrations are each corrected for two colours (cf. ACHROMAT, APOCHROMAT). A FLAT-FIELD OBJECTIVE LENS of this type is called a *planofluorite*.

fluorochrome Any fluorescent dye – e.g. ACRIDINE ORANGE, EOSIN, FITC, FLUORESCEIN.

5-fluorocytosine *Syn.* FLUCYTOSINE.

fluoroquinolone antibiotics See QUINOLONE ANTIBIOTICS.

5-fluorouracil See e.g. PHENOTYPIC SUPPRESSION.

Flury virus An avianized strain of the rabies virus used e.g. for anti-rabies vaccination in animals. The Flury LEP strain has been passaged (see SERIAL PASSAGE) in chick embryos ca. 40–60 times; the Flury HEP strain has been passaged >180 times.

flutriafol An agricultural AZOLE ANTIFUNGAL AGENT which is active against certain cereal diseases: e.g. powdery mildew, leaf blotch, septoria, net blotch, rusts, and MBC-sensitive and -resistant eyespot.

fly agaric See AMANITA and MUSCARINE.

FMAU (2′-deoxy-2′fluoro-5-methyl-1-β-D-arabinosyluracil) An ARABINOSYL NUCLEOSIDE which has antiherpesvirus and antileukaemia activity; its mode of antiviral action resembles that of ACYCLOVIR.

FMDV FOOT AND MOUTH DISEASE virus.

FMN See RIBOFLAVIN.

fms An ONCOGENE present in the McDonough strain of feline sarcoma virus (see FELINE LEUKAEMIA VIRUS); the v-*fms* product appears not to have tyrosine kinase activity. The c-*fms* gene probably encodes the receptor for the macrophage-specific colony-stimulating factor 1 (CSF-1).

foamy viruses See SPUMAVIRINAE.

FOCMA Feline oncornavirus-associated cell-membrane antigen, an antigen on the surface of feline lymphosarcoma cells which reacts with sera from cats exposed to FELINE LEUKAEMIA VIRUS (or from cats having recovered from feline sarcoma virus infection). Anti-FOCMA antibodies do not neutralize FeLV, but may protect against the development of lymphosarcoma by exerting a cytotoxic effect on feline leukaemia cells in the presence of complement.

focus-forming viruses Viruses which form foci of morphologically transformed cells in cell cultures or in tissues – see e.g. FRIEND VIRUS and MCF VIRUSES.

Foettingeria See HYPOSTOMATIA.

fog fever (acute bovine pulmonary emphysema and oedema) An acute ATYPICAL INTERSTITIAL PNEUMONIA in adult cattle, characterized by a sudden onset within ca. 1 week of being moved to new, lush green pasture; death may occur quickly, without warning, or animals may exhibit laboured breathing and fail to graze. Movement or exertion may cause death within minutes. The cause of fog fever is unknown. One hypothesis is that fog fever may result from an increased intake of tryptophan; in the rumen, tryptophan is metabolized to 3-methylindole – a substance which, when administered experimentally, produces a syndrome clinically indistinguishable from fog fever. The conversion of tryptophan to 3-methylindole in the rumen is inhibited e.g. by MONENSIN.

foliaceous *Syn.* FOLIOSE.

folic acid (pteroylglutamic acid) A water-soluble, photolabile VITAMIN. (See figure for structure.) Some microorganisms form pteroylpolyglutamic acid derivatives (typically containing 3–7 glutamic acid residues), certain of which may be specifically required for particular reactions.

The coenzyme form of folic acid (coenzyme F) is 5,6,7,8-tetrahydrofolic acid (THF, FH_4), formed by the reduction of folic acid (via 7,8-dihydrofolic acid) by dihydrofolate reductase and NADPH. THF functions as a carrier of one-carbon units (e.g. formyl, methyl, methylene or hydroxymethyl groups); such groups are carried at the N^5 or N^{10} positions (e.g. N^{10}-formyl-THF) or bridge these two positions (e.g. N^5,N^{10}-methylene-THF). THF is involved e.g. in glycine–serine interconversion [see Appendix IV (c)], in methionine biosynthesis [Appendix IV (d)], in the biosynthesis of purines and deoxythymidine [Appendix V], in the catabolism of e.g. histidine, in ACETOGENESIS, etc.

An exogenous source of folic acid is necessary for the growth of certain microorganisms: e.g. *Enterococcus faecalis* and *Lactobacillus casei* (both used for folic acid BIOASSAY), *Crithidia fasciculata*, *Tetrahymena* spp, and *Trypanosoma platydactyli* (formerly *Leishmania tarentolae*); most fungi can synthesize folic acid, although a few require the precursor *p*-AMINOBENZOIC ACID. *Leuconostoc cremoris* (formerly *L. citrovorum*) requires N^5-formyl-THF (= citrovorum factor, folinic acid, leucovorin).

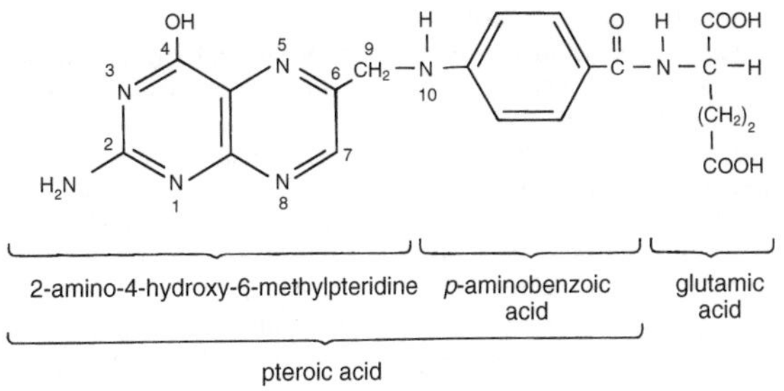

OH, H, O H COOH, 2-amino-4-hydroxy-6-methylpteridine, p-aminobenzoic acid, glutamic acid, pteroic acid

FOLIC ACID (pteroylglutamic acid)

Drugs which interfere with folic acid metabolism include e.g. FOLIC ACID ANTAGONISTS, PAS and SULPHONAMIDES.

(cf. BIOPTERIN and *tetrahydromethanopterin* under METHANOGENESIS.)

folic acid antagonist (antifol) A synthetic drug (typically a diaminopyrimidine derivative) which acts as a structural analogue of FOLIC ACID and which binds to and inhibits the enzyme dihydrofolate reductase (DHFR) – preventing the reduction of dihydrofolate (or folic acid) to tetrahydrofolate (THF) and thereby inhibiting THF-dependent reactions (e.g. synthesis of deoxythymidine and hence of DNA). The folic acid antagonists *aminopterin* and *methotrexate* have high affinities for mammalian DHFRs, and are used e.g. as anti-cancer and anti-leukaemia drugs. (See also HYBRIDOMA and IMMUNOSUPPRESSION.) PYRIMETHAMINE and TRIMETHOPRIM are more effective against microbial DHFRs, and are used in the treatment of certain infectious diseases. A microorganism may become resistant to a folic acid antagonist e.g. by acquiring a plasmid encoding a drug-resistant DHFR (see e.g. O/129), or by a mutation resulting in the overproduction of (normal) DHFR.

The action of a folic acid antagonist is often synergistic – at least in laboratory tests – with that of a SULPHONAMIDE, each type of drug inhibiting folic acid-dependent reactions but by different mechanisms; an antifol–sulphonamide mixture (e.g. *cotrimoxazole*: a mixture of trimethoprim and sulphamethoxazole) is often microbicidal, even though each drug is microbistatic when acting independently. However, sulphonamide–diaminopyrimidine combinations have a wide variety of toxic effects in humans, and their continued clinical use is open to question [JAC (1986) *17* 694–696].

foliicolous Living or growing on leaves.

folinic acid See FOLIC ACID.

foliose (foliaceous) Leaf-like or leafy. Of a LICHEN: having a thallus which is 'leaf-like' (i.e., flattened and dorsiventral) and either lobed, with part or most of its lower surface being more or less firmly attached to the substratum (e.g. by RHIZINES), or disc-like, attached to the substratum by a holdfast in the centre of the lower surface (see e.g. UMBILICARIA). (cf. CRUSTOSE and FRUTICOSE.)

folpet (*N*-(trichloromethylthio)phthalimide) An agricultural ANTIFUNGAL AGENT closely related to CAPTAN – with which it shares a similar antifungal spectrum.

Fomes (1) A genus of lignicolous fungi of the APHYLLOPHORALES (family Polyporaceae) which form woody, hoof-like, perennial fruiting bodies (with porous, stratified hymenophores) on both wood and living trees; the context is trimitic and has clamp connections. Some species have been transferred to other genera – e.g. *F. annosus* to HETEROBASIDION.

(2) *Fomes* is sometimes used as a form genus to include all polypores which form hard (corky or woody) perennial fruiting bodies. (cf. POLYPORUS sense 2.)

fomites Objects or materials which have been associated with infected persons or animals and which potentially harbour pathogenic microorganisms.

Fonsecaea See HYPHOMYCETES.

Fontana's stain (for spirochaetes) An air-dried smear on a clean, grease-free slide is fixed for 2 min in 2% aqueous formalin containing 1% acetic acid, and washed for 3 min in absolute ethanol. The smear is then flooded with 5% aqueous tannic acid containing 1% phenol, and heated so that the solution steams (not boils) for 30 sec. The smear is washed with distilled water and flooded with a solution prepared by adding 10% w/v aqueous ammonium hydroxide, drop by drop, to 0.5% w/v aqueous silver nitrate until the precipitate which forms just dissolves; the solution is made to steam (not boil) until the smear becomes brown. The slide is washed in distilled water and dried. Spirochaetes appear dark brown.

Fonticula See ACRASIOMYCETES.

food cup (in *Amoeba*) See PSEUDOPODIUM.

food microbiology For examples of the use of microorganisms as food or in the manufacture of foods see ALCOHOLIC BEVERAGES; COCOA; COFFEE; DAIRY PRODUCTS; FERMENTED FOODS; LAVER; MUSHROOM (senses 3 and 4); SINGLE-CELL PROTEIN; UNDARIA; VINEGAR; YAKULT. Microbial products used as food additives include e.g. ALGINATE; CARRAGEENAN; CITRIC ACID; GLUTAMIC ACID; NISIN; XANTHAN GUM; YEAST EXTRACT.

(See also FOOD PRESERVATION and FOOD SPOILAGE.)

food poisoning 'Food poisoning' generally refers to acute GASTROENTERITIS caused by ingestion of food contaminated with certain pathogenic bacteria and/or their toxins, or contaminated with certain viruses; clinically distinct conditions that may be food-borne (e.g. CHOLERA) are generally excluded, although, traditionally, BOTULISM has been included. (cf. e.g. GIARDIASIS, MYCOTOXICOSIS.)

Bacterial food poisoning may involve infection of the intestine ('food-borne infection') and/or the action of bacterial toxin(s) produced either in the food, prior to ingestion, or in the intestine ('food-borne intoxication', 'toxic food poisoning'). Food that is contaminated with pathogens often shows no organoleptic signs of deterioration. (See also FOOD SPOILAGE and FISH SPOILAGE.)

Some of the common types of food poisoning are described below (under causal agent).

(a) *Bacillus cereus* can cause two distinct syndromes. (i) *Diarrhoeal form.* After an incubation period of ~8–16 hours: diarrhoea, abdominal pain and nausea, but little or no vomiting. The (heat-labile) diarrhoeagenic enterotoxin causes a watery diarrhoea by stimulating adenylate cyclase (cf. CHOLERA TOXIN). Foods implicated: unrefrigerated meat and vegetable dishes, contaminated spices. (ii) *Vomiting form.* The incubation period (~1–7 hours) is followed by nausea, vomiting and abdominal pain, but little or no diarrhoea. The pathogen may produce a (heat-stable) emetic toxin in association with sporulation. Foods commonly implicated in this type of illness include cooked rice which has been stored unrefrigerated. (Other species of *Bacillus*, e.g. *B. subtilis* and *B. megaterium*, have been associated with food poisoning.)

(b) *Campylobacter.* Strains of *C. jejuni* or *C. coli* are usually involved: see CAMPYLOBACTERIOSIS; these organisms do not grow at 25°C but survive refrigeration and freezing – although they are killed by moderate heating (e.g. PASTEURIZATION).

(c) *Clostridium.* Strains of *C. perfringens* type A are the commonest causal agents of clostridial food poisoning. Most frequently isolated are type A2 strains (non-haemolytic/weakly haemolytic on horse-blood agar; endospores especially resistant to heat); isolated less often are type A1 strains (β-haemolytic; endospores relatively sensitive to heat). After an incubation period of ~8–24 hours: abdominal pain and diarrhoea, usually with little vomiting and no fever; symptoms last ~12–24 hours. Food (especially meat and poultry) may be contaminated with spores present in dust or in human or animal faeces etc.; the spores may survive cooking and, if the food is allowed to cool slowly or is stored without refrigeration, may germinate and grow rapidly. A large number of cells must be ingested in order to initiate infection as many are killed by stomach acidity. In the gut, the surviving organisms multiply and sporulate, releasing an enterotoxin which is distinct from the four major clostridial toxins (α, β, ε, ι) formed by *C. perfringens*; this toxin damages the tips of villi of the epithelial cells, inhibiting absorption of glucose and causing an efflux of sodium ions, chloride ions and water.

(See also ENTERITIS NECROTICANS.)

(d) *Escherichia.* Certain strains of *E. coli* can cause gastroenteritis after ingestion of e.g. raw or improperly cooked foods, unpasteurized milk or water which has been contaminated from a faecal source. See e.g. ETEC (travellers' diarrhoea) and EPEC.

(e) *Salmonella.* Food-poisoning strains include a range of serotypes – e.g. *S. typhimurium*, *S. hadar*, *S. virchow*, *S. agona.*

Infection commonly occurs by consumption of contaminated foods of animal origin: undercooked meat/poultry, meat products, eggs, raw milk etc. Foods may become faecally contaminated during preparation – e.g. via the hands of the cook – but in some cases (e.g. eggs) contamination may be present at source. Infection from pets has been reported. In meat, salmonellae compete poorly with the normal meat flora (cf. MEAT SPOILAGE); the meat flora is destroyed by cooking, so that salmonellae subsequently introduced are likely to thrive. Salmonellae can survive deep freezing but are killed by adequate cooking.

After an incubation period of ~12–48 hours there is a sudden onset with diarrhoea, abdominal pain, vomiting and fever; the stools may contain blood/mucus.

The pathogen multiplies in the gut, adheres to microvilli, and then invades the epithelial cells. Invasiveness in salmonellae depends on a type III secretory system (see PROTEIN SECRETION) encoded by the *inv-spa* complex of genes located on a chromosomal PATHOGENICITY ISLAND. The initial attachment of a bacterial cell may be mediated by type I fimbriae; attachment appears to promote the formation of *invasomes*: bacterial appendages which resemble very short, thick flagella and which are believed to be essential for invasion. *Salmonella* induces its own uptake – which occurs by the 'trigger mechanism'. Effector molecules (secreted via the type III system) are translocated into the epithelial cell where they bring about certain changes (e.g. actin polymerization). During uptake, extrusions of the host cell membrane (supported e.g. by actin elements) engulf the bacterium – which is thus internalized in a vacuole containing extracellular fluid. Unlike *Shigella* (see DYSENTERY), *Salmonella* invades via the *apical* (lumen-facing) surface of gut epithelial cells.

Salmonella multiplies within the vacuole.

Although much is now known about the way in which salmonellae invade the epithelium, less is known about the mechanism(s) by which infection induces intestinal secretion. Two mechanisms have been considered: (i) the action of toxin(s), and (ii) the effect of invasion *per se*. *S. typhimurium* is known to produce at least one toxin: Stn, a 25 kDa protein, encoded by gene *stn*, which is predicted to function as an ADP-ribosylating agent; however, only low levels of this toxin are formed, and its significance in pathogenesis is uncertain. It has been suggested that invasion may stimulate certain host cell secretory pathways, although invasion and secretion are not always correlated. (In a veterinary context, the cattle pathogen *Salmonella dublin* has been reported to secrete an effector protein, SopB, via a type III secretory system encoded by the *inv-spa* genes; SopB is believed to mediate enteropathogenicity by being translocated into epithelial cells and inducing both inflammation and fluid secretion [Mol. Microbiol. (1997) *25* 903–912].)

Recovery from *Salmonella* food poisoning commonly begins within a few days, but a carrier state may persist. Complications include sepsis and abscess formation in the meninges (in small children), arteries (leading to arteritis and e.g. rupture of the artery), bones and joints (cf. OSTEOMYELITIS).

(f) *Shigella.* See DYSENTERY.

(g) *Staphylococcus.* Staphylococcal food poisoning usually follows ingestion of food containing one or more ENTEROTOXINS (q.v.) formed by toxigenic, coagulase-positive strains of *S. aureus*. Commonly implicated are cooked foods eaten cold (e.g. hams, poultry), and prepared foods such as sandwiches, custards etc. The source of staphylococcal contamination is man (typically a food-handler); the organisms derive from the nares or skin, or from skin lesions (e.g. boils). [Detection of staphylococcal enterotoxins: RMM (1994) *5* 56–64.]

Incubation period: 1–6 hours. Onset is abrupt, with nausea, vomiting, abdominal pain and prostration, sometimes with diarrhoea. Symptoms may last for about 12 hours, but weakness and malaise may persist.

Vomiting may result from stimulation, by the toxin, of neural receptors in the upper intestinal tract – this being suggested by abolition of the emetic reflex (in monkeys) following vagotomy and sympathectomy. The superantigenic nature of the enterotoxins suggests that cytokines may play an important role in generating at least some of the symptoms of staphylococcal food poisoning.

(h) *Vibrio parahaemolyticus.* Gastroenteritis due to *V. parahaemolyticus* (see VIBRIO) is associated with the consumption of raw or inadequately cooked seafoods (fish, shellfish etc.). Most outbreaks occur during warm months. The pathogen may be concentrated in the gut of filter-feeding shellfish (cf. SHELLFISH POISONING) and may proliferate in contaminated seafoods stored without refrigeration.

Incubation period: often 12–24 hours, but may be less than 2 hours (a shorter period typically correlating with more severe symptoms). Gastroenteritis: diarrhoea commonly watery, sometimes explosive and dysenteric; other symptoms include nausea, abdominal cramps, vomiting, fever and chills. The illness may last ~3–6 days.

The main diarrhoeagenic factor of the pathogen is a 23 kDa protein, TDH; TDH is the haemolysin responsible for the KANAGAWA PHENOMENON. The binding site for this haemolysin (in the rabbit) is probably a trisialoganglioside designated $GT1_b$; in vitro studies indicate that the toxin promotes the secretion of chloride.

V. parahaemolyticus is reported to be usually sensitive to tetracycline, chloramphenicol, nalidixic acid and gentamicin but resistant to a number of β-lactam antibiotics, including ampicillin.

[*V. parahaemolyticus* in humans (review): RMM (1995) *6* 137–145.]

(i) *Viruses*. Strains of ROTAVIRUS are common causal agents of viral gastroenteritis in infants; they infect the mucosa of the jejunum and ileum, causing desquamation and villous atrophy. Infection occurs by ingestion of faecally contaminated food or water. Incubation period: ~2–3 days. Onset is abrupt, with vomiting and diarrhoea (often associated with an upper respiratory tract infection). Recovery may occur in ~2–3 days, or the disease may be fatal. Infection in adults seems to be largely asymptomatic.

The SMALL ROUND STRUCTURED VIRUSES (also called e.g. 'parvo-like viruses' or 'parvovirus-like agents') are major causes of gastroenteritis in both children and adults; these infections have been called e.g. 'epidemic vomiting disease', while outbreaks during the winter months have been designated 'winter vomiting disease'. Immunity following infection may be short-lived. Infection is thought to occur by consumption of faecally contaminated food (see e.g. SHELLFISH POISONING) or water. Incubation period: ~1–2 days. Onset is sudden, with nausea and vomiting, sometimes with mild diarrhoea, fever, myalgia, malaise etc.

(j) *Yersinia enterocolitica* is associated with gastroenteritis mainly in children. The condition is often mild, with fever, a watery diarrhoea, vomiting and abdominal pain, usually resolving in 1–2 days. However, severe cases may involve ileitis – often in association with mesenteric lymphadenitis; in these cases the severe abdominal pain may mimic acute appendicitis ('pseudoappendicitis').

Various foods have been implicated as vehicles for the pathogen (e.g. meat, especially pork, and milk); large numbers of cells of *Y. enterocolitica* have been reported in certain vacuum-packed meats (see DFD MEAT). The pathogen can grow at, and below, 4°C (i.e. even under refrigeration).

Strains of the pathogen secrete a heat-stable enterotoxin (designated Yst) apparently similar to the ST-I of ETEC in both structure and activity; this toxin may be responsible for the diarrhoeal symptoms.

Y. enterocolitica invades the intestinal mucosa – studies in animals indicating that initial entry occurs (as in *Shigella*) via the M cells of Peyer's patches. [Invasion by *Y. enterocolitica*: e.g. Book ref. 218, pp 225–228.] In one model, cells of the pathogen pass *through* the M cells – so that they gain access to the basolateral surfaces of adjacent epithelial cells on which are found the receptors (*integrins*) of several bacterial adhesins. Adhesins of *Y. enterocolitica* include the outer membrane protein *invasin* (chromosomally encoded by the *inv* gene), and the YadA protein (encoded by plasmid pYV) which forms a cell-surface fibrillar matrix. Invasion via the basolateral surfaces of epithelial cells is likely to resemble the process seen in vitro. In vitro, the bacterial adhesin invasin binds to $\alpha_5\beta_1$ integrins, such binding inducing the development of further integrin molecules at the binding site – leading to multiple points of attachment between bacterium and epithelial cell (*zippering*); this induces uptake in a close-fitting vacuole (within which the pathogen apparently survives). As well as consolidating bacterial attachment (zippering), the greater concentration of integrins also induces tyrosine kinase activity within the epithelial cell; this is necessary for the invasion process, but the precise details are not known.

Various invasive pathogens – including e.g. *Shigella dysenteriae* and EIEC – induce secretion of certain CYTOKINES by the gut epithelial cells; one of these, IL-8 (a chemoattractant for inflammatory leukocytes), probably represents a host response aimed at overcoming invasion. However, virulent strains of *Y. enterocolitica* have been found to induce significantly less IL-8 compared with non-virulent strains.

Virulent strains of *Y. enterocolitica* also encode various effector molecules ('Yops') and an associated type III secretory system (see PROTEIN SECRETION) [the *Yersinia* Yop virulon: Mol. Microbiol. (1997) *23* 861–867]; the genes for this system are located on plasmid pYV. An important function of the Yop system is to avoid phagocytosis by disrupting the phagocyte's uptake mechanism.

[Infections with enteropathogenic *Yersinia* species (review): RMM (1998) *9* 191–205.]

(k) Various other organisms have been linked with gastroenteritis/diarrhoea – e.g. species of *Aeromonas* [RMM (1997) *8* 61–72], *Citrobacter*, *Enterobacter*, *Klebsiella*, *Proteus*, *Pseudomonas* and *Streptococcus*.

food preservation The various methods of food PRESERVATION aim to prevent or delay microbial and other forms of FOOD SPOILAGE, and to guard against FOOD POISONING; such methods therefore help to retain the nutritive value of the product, extend its shelf-life, and keep it safe for consumption. It is generally considered preferable to use physical methods of preservation (e.g. refrigeration) when possible.

(a) *Heat treatment*. See e.g. APPERTIZATION; CANNING; PASTEURIZATION.

(b) *Low-temperature preservation*. Low temperatures delay or prevent spoilage e.g. by reducing the metabolic activities of contaminating organisms and/or the activity of endogenous enzymes in the food; spoilage by PSYCHROTROPHS may nevertheless still occur. (i) *Above 0°C*. Refrigeration between 0°C and 5°C is often used for the short-term storage of foods; however, spoilage may occur at these temperatures (see e.g. MEAT SPOILAGE). (ii) *Below 0°C*. FREEZING may kill a proportion of contaminating organisms, and it also lowers the food's WATER ACTIVITY – thus making it a less hospitable environment for surviving contaminants. Nevertheless, spoilage may still occur (see e.g. BLACK SPOT (1)); care must therefore be taken to exclude or reduce contamination before freezing.

(c) *Dehydration*. If the WATER ACTIVITY of a food is progressively decreased, a point is reached at which the food will not support the growth of contaminating organisms. The a_w of foods may be reduced by evaporation through heating (e.g. dried milk) or e.g. by the addition of sodium chloride (salting – e.g. fish products) or sugar syrups (e.g. fruits). FREEZE-DRYING has been used e.g. for coffee. (See also CURING (1) and SMOKING.)

(d) *Acidification* – see PICKLING.

(e) *Fermentation*. Fermentation of certain types of raw food not only enhances the keeping qualities of the food, it may also improve the flavour and (often) the nutritive value of the food, and avoids the need for expensive heat treatment, freezing etc. The preservative effect of fermentation may be due e.g. to (i) the lowering of pH, usually by the formation of lactic or acetic acids (see e.g. PICKLING); (ii) production of antimicrobial substances – e.g. ethanol (as in ALCOHOLIC BEVERAGES), CARBON DIOXIDE, antibiotics (e.g. NISIN, DIPLOCOCCIN in DAIRY PRODUCTS), H_2O_2 (see e.g. LACTIC ACID BACTERIA; cf. LACTOPEROXIDASE–THIOCYANATE–HYDROGEN PEROXIDE SYSTEM); (iii) the metabolism of readily utilizable substrates (e.g. simple sugars) which would otherwise be available to potential spoilage organisms or pathogens. (See also FERMENTED FOODS.)

(f) *Preservatives*. The addition of PRESERVATIVES to particular foods is generally subject to governmental control; a preservative permissible in one country may be banned in another. A preservative should ideally be innocuous to the consumer and should not be an allergen; it should be microbicidal, rather than microbistatic, to a range of possible contaminants; it should not be inactivated by constituents of the food or by the products of microbial contaminants. Food preservatives include e.g. BENZOIC ACID; nitrites (see e.g. CURING (1)); PARABENZOATES; SORBIC ACID; SULPHUR DIOXIDE (including e.g. metabisulphite e.g. in SAUSAGES). Antibiotics, particularly those which have a medical use, have limited application in food preservation (cf. NISIN and PIMARICIN).

(g) IONIZING RADIATION has been used in some countries e.g. for chicken and for fish fillets, strawberries, onions and spices. Toxicological testing of foods irradiated with doses <10 kGy is regarded as unnecessary, and this may lead to an increase in the use of this form of preservation [Book ref. 50, pp. 145–171].

(h) A PULSED ELECTRIC FIELD technique has been able e.g. to reduce significantly the concentration of viable cells of *Mycobacterium paratuberculosis* in milk [AEM (2001) *67* 2833–2836].

food spoilage Any change in the condition of food which causes it to become less palatable (and sometimes toxic); such changes may involve alterations in taste, smell, appearance or texture. (Foods contaminated with pathogens and/or microbial toxins (see e.g. FOOD POISONING and MYCOTOXINS) do not necessarily exhibit obvious deterioration.) Spoilage organisms include a variety of bacteria and fungi – although particular foods may be susceptible to only a limited range of species; the nature of the spoilage flora is determined e.g. by the types of nutrient present in the food, its WATER ACTIVITY and pH, the conditions of storage (particularly temperature and accessibility of air), and the presence or absence of PRESERVATIVES (see also FOOD PRESERVATION). For examples see BEER SPOILAGE; BREAD SPOILAGE; BUTTER; CANNING; CHEESE SPOILAGE; CIDER; FISH SPOILAGE; MEAT SPOILAGE; MILK SPOILAGE; PICKLING; RANCIDITY; ROPINESS; WINE SPOILAGE. (See also BIODETERIORATION.)

food vacuole (gastriole) A closed, membrane-limited intracellular sac which contains particle(s) of food; food vacuoles are formed e.g. by certain protozoa as a consequence of PHAGOCYTOSIS. One or more food vacuoles may be present in a given cell at any given time. Soon after its formation, a food vacuole coalesces with a LYSOSOME, and its contents become acidic and enzyme-rich; the food is digested by lysosomal enzymes (which include e.g. acid phosphatases, nucleases, amylases and esterases) and the digested food appears to enter the cytoplasm by pinocytosis – pinocytotic vesicles developing in the vacuolar membrane. Undigested food is voided when the vacuole subsequently coalesces with the cytoplasmic membrane at the cytoproct.

Under certain conditions, e.g. starvation, a cell may digest some of its own internal structures (a process called autophagy); such structure(s) are initially enclosed within membranous vesicles to form food vacuoles.

food yeast See SINGLE-CELL PROTEIN.

foolish seedling disease *Syn.* BAKANAE DISEASE.

foot (*mycol.*) In fungi of the Laboulbeniales: a dark, basal cell which anchors the thallus to the host organism. (cf. FOOT CELL.)

foot and mouth disease (aphthous fever) An acute, highly infectious disease which can affect all cloven-hoofed animals – e.g. cattle, goats, pigs, sheep – and e.g. hedgehogs and rodents; horses are immune, and man is affected only rarely. The causal agent is an APHTHOVIRUS. Infection may occur by ingestion of contaminated food or inhalation of virus-bearing aerosols; the incubation period is usually several days to a week, but may be up to ca. 20 days. Symptoms include fever and the formation of vesicles (fluid-filled blister-like lesions) on the mucous membranes of the mouth and on the feet – particularly where the horny material and skin are contiguous; the copious saliva contains viruses derived from the ruptured vesicles. Vesicles may also appear on the teats, and viruses may also be found in the milk, blood, urine and faeces. In adult animals the disease is usually not fatal; however, it causes serious economic losses in terms of meat and milk production. *Control*: in countries where outbreaks occur only sporadically, control may be effected by slaughter of infected and suspect animals; vaccination has been used in some cases. [Protection of cattle using a synthetic peptide vaccine: Science (1986) *232* 639–641.] A carrier state occurs in cattle, goats and sheep (but not pigs). *Disinfection*. The virus may persist for months in infected premises, stability being promoted e.g. by low temperatures. The virus may be destroyed e.g. by extremes of pH – citric acid solution or a 1% solution of sodium hydroxide (caustic soda) being effective disinfectants. Other disinfectants which have been used include iodophors (see IODINE) and hypochlorites; phenolics are apparently effective only if used in concentrated solutions (e.g. 1:6, 1:8). *Laboratory diagnosis*: e.g. culture of the virus (from pharyngeal swabs etc) and/or a CFT.

In February, 2001, the first cases of a major outbreak of foot and mouth disease were confirmed in the United Kingdom; during the period February–May the number of reported cases was particularly high in Cumbria, Dumfries and Galloway, County Durham and Devon. (A flare-up in Northumberland occurred in late August.) [Epidemiology (diagram of links between confirmed cases as at March 14th, 2001): VR (2001) *148* 322–325.] Government policy involved timely slaughter of both infected and suspect animals, including those on adjacent farms (the 'contiguous cull'). Large numbers of carcases were burnt or buried, but burial was considered inappropriate for cattle over a certain age owing to the presumed risk of environmental contamination with prions of BOVINE SPONGIFORM ENCEPHALOPATHY.

Despite the availability of vaccines, vaccination was resisted; an important factor in this decision was the desire to maintain the UK's 'disease-free status' (thus protecting the export trade).

In this outbreak, some veterinary inspections revealed lesions of unknown aetiology which could complicate the diagnosis of foot and mouth disease [photographs of oral lesions in sheep and cattle in Dumfries and Galloway that could complicate the diagnosis of foot and mouth disease: VR (2001) *148* 720–723].

Currently, work is ongoing to improve the performance of ELISA-based serological tests that distinguish between (i) antibodies to whole virus particles (such as those in inactivated vaccines), and (ii) antibodies to viral proteins exposed only during replication of the virus; tests that could (reliably) distinguish between infected, vaccinated and 'carrier' animals could lead to the possibility of vaccination compatible with 'disease-free status' [see Nature (2001) *410* 1012].

In *man*, the disease involves fever, malaise, and the formation of vesicles in the mouth and on the lips, hands and feet; infection may occur via wounds or by ingestion of contaminated dairy products. Complete recovery usually occurs. (cf. HAND, FOOT AND MOUTH DISEASE.)

(See also SWINE VESICULAR DISEASE; VESICULAR EXANTHEMA; VESICULAR STOMATITIS.)

foot cell (1) (*mycol.*) A foot-shaped terminal cell in a macroconidium of FUSARIUM. (2) (*mycol.*) A hyphal cell which gives rise to a conidiophore. (cf. FOOT.)

foot-rot (foot rot; footrot) (1) (*vet.*) In cattle and sheep: a disease involving infection and inflammation of the skin–horn junction and adjacent sensitive areas of the foot, frequently resulting in severe lameness; in sheep the causal agent is *Bacteroides nodosus* (cf. SCALD), while in cattle the condition may be caused by *Fusobacterium necrophorum* and/or *Bacteroides* (or other) species. Foot-rot occurs typically in warm, wet weather. (Some breeds are more susceptible than others. *Bos taurus* cattle appear to be much more susceptible than *Bos indicus* (Zebu-type, Brahman) breeds, and the Australasian Merino sheep tend to be more susceptible than British breeds.) *Symptoms*: the foot is tender and inflamed and has a characteristic necrotic odour, but pus is generally not formed in large amounts. Fever may be present. *Treatment*: parenteral administration of antibiotics and/or topical application (to the cleaned, prepared foot) of e.g. a 5% w/v $CuSO_4$ solution; foot-baths of e.g. 5% $CuSO_4$, 5% formalin, or 10% $ZnSO_4$ are useful for control.

In pigs, foot-rot is not unlike that found in cattle and sheep, but the condition appears to be more obviously linked to mechanical damage to the feet, and the causal agent(s) appear to include a wider range of species (e.g. *F. necrophorum*, staphylococci); a dietary deficiency of biotin predisposes towards foot lesions in the pig. Topical treatment and foot-baths can be effective.

(2) (*plant pathol.*) Any of various plant diseases in which the base of the stem rots as a result of microbial infection; causal agents include species of e.g. FUSARIUM, *Pythium*, *Rhizoctonia*, *Sclerotium*, *Thielaviopsis*, etc. These fungi commonly soften the stem tissue by producing pectinolytic – and possibly cellulolytic – enzymes, causing the plant to collapse. (cf. DAMPING OFF.)

footprinting In the context of protein–DNA interactions: any of various methods used to determine the particular sequence of DNA to which a given protein binds; essentially, the binding site of the protein is determined by identifying that region of the DNA molecule in which the presence of bound protein protects ('shields') the DNA from cleavage by a nuclease (e.g. DNase I) or by certain chemical agents.

In one approach, two sets of homologous, end-labelled DNA fragments – one set carrying the bound protein – are exposed to a particular nuclease, or a chemical agent, under conditions in which, ideally, each fragment is cut at only one of a number of potential cleavage sites. The result is a number of labelled sub-fragments in a range of different sizes – reflecting cleavage at different sites in different fragments. If, in *protein-bound* fragments, the presence of protein obscures one or more cleavage sites, then no sub-fragment will terminate at such protected site(s). Consequently, if the two sets of sub-fragments are compared by gel electrophoresis, the *absence* of certain size(s) of sub-fragment from the protein-bound set will be seen as a gap in the gel; this gap, resulting from shielding of corresponding cleavage site(s) by the protein, is the protein's *footprint*.

The use of DNase I as the DNA-cleavage agent commonly gives a large, clear footprint because it tends not to cut at sites close to the bound protein. The binding site of the protein may be determined with greater resolution by employing chemically mediated cleavage e.g. with in vitro-generated hydroxyl radical; this agent can cut at all unprotected phosphodiester bonds, so that the electrophoretic pattern in this case includes a band for each unshielded position in the sequence – a relatively narrow range of missing bands in the gel indicating a protein-binding site.

[Example of DNase I-mediated footprinting to investigate protein binding to the regulatory region of the *gal* operon in *Escherichia coli*: NAR (1994) *22* 4375–4380.]

An entirely *unrelated* procedure has been termed GENETIC FOOTPRINTING (q.v.).

forage poisoning A form of BOTULISM in horses caused by *C. botulinum* type C_β (cf. MIDLAND CATTLE DISEASE); it may occur in epizootics.

foramen (*pl.* foramina) A perforation or aperture in a structure: see e.g. FORAMINIFERIDA.

foramina See FORAMEN.

Foraminiferida An order of free-living (mainly marine), testate, amoeboid protozoa (class GRANULORETICULOSEA). The nature of the test (<0.2 mm to >5 mm diam., depending on species) is an important taxonomic criterion. In the suborder Allogromiina (genera: e.g. *Allogromia*, *Iridia*, *Myxotheca*) the test is predominantly organic. In the suborder Fusulinina (e.g. *Fusulina*, *Schwagerina*) the test may be composed of microgranular calcite or may consist of two or more differentiated layers. Members of the suborder Miliolina (e.g. *Quinqueloculina*, *Triloculina*) form smooth, shiny, porcellanous (porcelain-like) tests which may or may not be perforated. In the suborder Rotaliina (e.g. *Ammonia*, *Elphidium*, *Rosalina*) the test is hyaline and calcareous. Members of the suborder Textularina (e.g. *Saccammina*, *Textularia*) form 'agglutinated' tests, i.e., tests composed of foreign matter (sand, small shells etc) incorporated into a (commonly calcareous) matrix. [JP (1980) *27* 37–58.]

In a few foraminifera (e.g. *Allogromia*) the test is *unilocular*, i.e., it consists of a single chamber (*locule*), but in the majority the test is *multiloculate*, new locules being added sequentially to the first-formed chamber (the *proloculus*) as the organism grows. The arrangement of locules in multiloculate species may be e.g. rectilinear and uniserial (i.e., arranged in a single row, as e.g. in *Dentalina*, *Nodosaria* and *Orthocerina*), rectilinear and biserial (i.e., in two adjacent rows, as in e.g. *Cribrostomum* and *Textularia*), planispiral (i.e., in a flat coil, as in e.g. *Ammodiscus* and *Discorbis*) or conical (i.e., in a cone-shaped spiral, as in e.g. *Elphidium* and *Glomospira*). Adjacent locules communicate via an opening (the *foramen*) between them, and the reticulopodium extends from the distal opening (*aperture*) of the most recently formed locule. In some genera reticulopodia also emerge from perforations in the test, and in some a thin layer of cytoplasm covers the outer surface of the test.

The life cycle in foraminifera is unique among protozoa in that it generally involves an ALTERNATION OF GENERATIONS. In e.g. *Elphidium* spp, meiosis occurs in the mature, multiloculate, diploid organism (*agamont*) and haploid, amoeboid cells (young gamonts) are liberated; each cell secretes a test and grows to form a mature, multiloculate, haploid *gamont* which eventually releases numerous biflagellate isogametes. Following fertilization, each zygote (young agamont) secretes a test and grows to form a mature agamont. In e.g. *Allogromia* the mature gamonts and agamonts are morphologically indistinguishable, while in e.g. *Elphidium crispum* the proloculus in the agamont test is smaller than that in the gamont test. Variations may occur in this generalized life cycle: e.g., in *Glabratella sulcata* the gametes are exchanged directly between two gamonts in contact, rather than being released into the sea.

Most foraminifera are marine benthic organisms, although some (e.g. GLOBIGERINA, GLOBIGERINOIDES) are pelagic (planktonic), and a few (e.g. *Nonion germanicum*) occur in brackish waters. The organisms are typically holozoic and omnivorous, feeding on bacteria, small protozoa, microalgae, etc. [Invasive activity of *Allogromia* pseudopodial networks: JP (1985) *32* 9–12.] Photosynthetic endosymbionts occur in many foraminifera, including most or all of the large forms (e.g. *Amphistegina*) which occur – often abundantly – in shallow tropical and subtropical seas; depending on species, these endosymbionts may be dinoflagellates (e.g. *Amphidinium*, '*Symbiodinium*'), green algae (e.g. *Chlamydomonas*, *Chlorella*), frustule-less diatoms (which can reconstitute their frustules in culture), or a unicellular red alga (*Porphyridium*?). The endosymbionts seem able to satisfy at least a significant proportion of the carbon and energy requirements of the protozoon, but ingestion of prey is apparently necessary to supply adequate levels of e.g. nitrogen and phosphorus [e.g. in *Globigerinoides sacculifer*: Limn. Ocean. (1985) *30* 1253–1267]. In *Nonion germanicum* chloroplasts from ingested algae appear to survive digestion for long enough to supply the protozoon with some photosynthate. [Symbiosis: Book ref. 129, pp. 37–68; potential symbionts in bathyal species: Science (2003) *299* 861.]

Fossil foraminiferan tests occur abundantly in certain oceanic deposits and sedimentary rocks, sometimes being a major component of thick deposits of chalk (such as the white cliffs of Dover and Normandy). (See also COCCOLITH.) Numerous genera and species of fossil foraminifera have been documented, the earliest dating from the late Precambrian to early Cambrian (primitive globular tests with a single aperture). Fossil tests are commonly ca. 0.2–2.0 mm in diam., but some 'fusulinids' (Carboniferous to Permian) and 'nummilitids' (Eocene to Oligocene) are much larger: e.g. the discoid multiloculate tests of *Nummilites gizehensis* may be >120 mm. A rapid succession of distinctive species of fossil foraminifera has occurred through the geological ages, making them ideal markers in the biostratigraphy of sedimentary rocks e.g. for oil prospecting. [Stratigraphy of planktonic foraminifera: Book ref. 136, pp. 11–328.]

foreign body giant cell See GIANT CELL.

forespore See ENDOSPORE (sense 1 (a)).

forest yaws *Syn.* PIAN BOIS.

form genus (*mycol.*) A non-phylogenetic category, equivalent to 'genus', which is distinguished on the basis of one or more (typically) morphological features. Form species within a form genus are referred to by Latin binomials (see NOMENCLATURE). (The prefix 'form' – although taxonomically correct – is frequently omitted.)

In the DEUTEROMYCOTINA, form genera are used to classify e.g. those fungi which have no known sexual stage; such form genera are based primarily on the characteristics (including mode of development) of the conidia. If the sexual stage of a form species is discovered, the name of the form species – based on the anamorph – can continue to be used, but the taxonomically valid name is that of the teleomorph.

Form genera are also used for certain members of the Aphyllophorales: see e.g. FOMES, POLYPORUS and PORIA.

form species See FORM GENUS.

forma specialis (special form; f. sp., sometimes written f.; *plural*: *formae speciales*, ff. sp.) An intraspecific taxonomic rank in which the taxa are distinguished on a physiological basis – particularly on the basis of adaptation to (or pathogenicity for) one or more specific hosts. In mycology, *forma specialis* is a taxonomic rank lower than (in ascending order) form, subvariety, VARIETY and SUBSPECIES, and higher than PHYSIOLOGICAL RACE [Book ref. 64, p. 83].

formaldehyde (HCHO; methanal) (a) (as an antimicrobial agent) A colourless, water-soluble and ethanol-soluble, pungent gaseous aldehyde which functions as an ALKYLATING AGENT, substituting amino and other groups (in proteins and nucleic acids) with hydroxymethyl groups (cf. GLUTARALDEHYDE); it can be mutagenic and is a presumed carcinogen. Formaldehyde is used e.g. for FIXATION, as a DISINFECTANT, as a biocide in a low-temperature STERILIZATION process, and as an inactivating agent in certain INACTIVATED VACCINES. It is active against a wide range of bacteria and fungi and some viruses; activity against bacterial endospores is very slow and can involve reversible inhibition of germination. For maximum antimicrobial activity, gaseous formaldehyde requires a high relative humidity (>60%); the gas has poor powers of penetration and is used e.g. as an indoor surface disinfectant or sterilant (cf. FUMIGATION). An aqueous solution of formaldehyde (*formalin*) is used e.g. as a preservative for biological specimens and as a general disinfectant; alcoholic solutions have been used e.g. for disinfecting surgical instruments. Antimicrobial activity is inhibited by organic matter. *Formaldehyde-releasing agents* include e.g. HEXAMINE, NOXYTHIOLIN, PARAFORMALDEHYDE, TAUROLIN. (See also LOW-TEMPERATURE STEAM–FORMALDEHYDE STERILIZATION.)

(b) (as a metabolite) See METHYLOTROPHY.

formalin See FORMALDEHYDE.

formalin–ether method (formol–ether method) A method for concentrating protozoan cysts (and e.g. parasite larvae) in a sample of faeces. The sample (e.g. 1–2 g faeces) is emulsified in ca. 10 ml physiological saline containing 10% formalin, and left for 30 min; the suspension is then filtered through a double layer of wet gauze and the filtrate is centrifuged (ca. 350 *g*/2–3 min). The supernatant is discarded, and the sediment is shaken in a stoppered tube with 10 ml formalin–saline and 3 ml diethyl ether. The tube is then centrifuged (ca. 350 *g*/2–3 min). Cysts, if present, are found in the sediment. This method is unsuitable for the concentration of coccidian cysts if the latter are subsequently to be sporulated. (cf. FLOTATION.)

formate hydrogen lyase An enzyme system which splits formic acid into carbon dioxide and hydrogen (see e.g. MIXED ACID FERMENTATION and BUTANEDIOL FERMENTATION). The system includes a *formate dehydrogenase* (which eliminates CO_2 from formic acid) and a HYDROGENASE; electrons derived from the first reaction appear to be transferred to hydrogenase via one or more electron carriers. Anaerobically, *Escherichia coli* (and certain other bacteria) can oxidize formate using nitrate or fumarate as terminal electron acceptor; a different formate dehydrogenase is involved, and hydrogen is not evolved.

***N*-formimidoylthienamycin** See THIENAMYCIN.

formol–ether method See FORMALIN–ETHER METHOD.

Formvar Polyvinyl formal: a substance, soluble in e.g. chloroform, used in the preparation of support films for ELECTRON MICROSCOPY.

Forssman antigen (1) Any HETEROPHIL ANTIGEN. (2) Any heterophil antigen which can elicit antibodies capable of combining with determinant(s) on sheep RBCs. (The first-described Forssman antigen is an antigen found e.g. on sheep RBCs and on certain guinea-pig cells.)

Fort Bragg fever (pretibial fever) Human LEPTOSPIROSIS, caused by *Leptospira interrogans* serotype *autumnalis*, which is characterized by a pretibial rash, malaise, coryza and fever; it occurs in SE Asia, Japan, and the USA.

Fort Morgan virus See ALPHAVIRUS.

fortimicin An AMINOGLYCOSIDE ANTIBIOTIC which contains neither streptidine nor deoxystreptamine.

forward mutation A MUTATION which results in a *mutant* phenotype, i.e., a phenotype which differs from that of the WILD-TYPE organism; a forward mutation typically involves the inactivation of e.g. a structural gene. (cf. BACK MUTATION.)

forward primer In the context of a given gene whose nucleotides are numbered: a primer whose sequence of nucleotides corresponds to a sequence within the gene which reads in a numerically ascending order *in the 5′-to-3′ direction*; thus, a forward primer may have a sequence which corresponds to e.g. nucleotides $10 \rightarrow 25$ (5′-to-3′) of a given gene. The nucleotides of a *reverse primer* read in a numerically descending order in the 5′-to-3′ direction – e.g. nucleotides $350 \rightarrow 335$ in the other strand.

fos See FBJ OSTEOSARCOMA VIRUS.

Foscarnet sodium See PHOSPHONOFORMIC ACID.

fosfomycin (phosphonomycin; *cis*-1,2-epoxypropyl-1-phosphonic acid) A broad-spectrum ANTIBIOTIC (obtained from *Streptomyces* strains) which blocks PEPTIDOGLYCAN biosynthesis by binding (irreversibly) and inhibiting phosphoenolpyruvate: UDP-GlcNAc enolpyruvyl transferase – thus preventing the synthesis of *N*-acetylmuramic acid; other reactions involving PEP are apparently not inhibited. Fosfomycin is taken up by the glycerolphosphate transport system in susceptible cells, and resistance to the drug develops readily by alteration or loss of this transport system; plasmid-borne resistance involving intracellular modification of the drug also occurs [JGM (1985) *131* 1649–1655].

Fosfonet sodium See PHOSPHONOACETIC ACID.

Foshay test A SKIN TEST in which the antigen is a suspension of (or a protein extract from) killed cells of *Francisella tularensis*.

Foshay's vaccine An anti-TULARAEMIA vaccine prepared from killed cells of *Francisella tularensis*.

fossil microorganisms See e.g. COCCOLITH; CYANOBACTERIA; DASYCLADALEAN ALGAE; FORAMINIFERIDA; GYROGONITES; RADIOLARIA; STROMATOLITES. (See also BIOSTRATIGRAPHY.) [Microfossils of the Proterozoic/Phanerozoic periods in China: Nature (1985) *315* 655–658.]

foulbrood See AMERICAN FOULBROOD and EUROPEAN FOULBROOD.

Fourier transform See NUCLEAR MAGNETIC RESONANCE.

foveate Pitted with small holes or depressions.

foveolate Minutely pitted.

fowl adenovirus See AVIADENOVIRUS.

fowl cholera (avian pasteurellosis) A POULTRY DISEASE (affecting e.g. chickens, turkeys, ducks, game birds) caused by *Pasteurella multocida*. Symptoms of the acute form of the disease may include anorexia, bloody diarrhoea, discharge from nostrils, eyes and mouth, and haemorrhages of internal organs and mucous membranes; death may occur within hours or days. In the chronic form the disease mainly affects the respiratory system.

fowl paralysis *Syn.* MAREK'S DISEASE.

fowl pest (1) *Syn.* FOWL PLAGUE. (2) *Syn.* NEWCASTLE DISEASE.

fowl plague A POULTRY DISEASE caused by strains of influenzavirus A. Symptoms resemble those of NEWCASTLE DISEASE; mortality rates may be high. Wild birds may be an important source of infection.

fowl pox A POULTRY DISEASE caused by an AVIPOXVIRUS. Infection occurs via wounds (particularly in unfeathered regions such as the comb, wattles etc) or mucous membranes. Greyish-white or yellow eruptions form and may develop into large wart-like growths, and/or yellowish, caseating vesicles develop in the throat, forming a characteristic diphtheritic membrane. Live attenuated vaccines are available.

fowl typhoid An acute infectious POULTRY DISEASE caused by *Salmonella gallinarum*; it affects mainly adult chickens and turkeys. Symptoms: e.g. lethargy, anorexia, yellow or greenish diarrhoea; the liver is enlarged and has a typical bronzed appearance post-mortem. Death may occur within days, or the disease may become chronic with severe, progressive emaciation and anaemia. Survivors become chronic carriers. Transmission occurs via the faecal–oral route or transovarially. Wild birds (e.g. rooks, pigeons) may provide a reservoir of infection. *Lab. diagnosis* and *treatment*: as for PULLORUM DISEASE; antisera which are positive for *S. pullorum* cross-react with *S. gallinarum*.

fowlpox subgroup See AVIPOXVIRUS.

foxfire BIOLUMINESCENCE in decaying wood.

FP-cellulose 'Filter-paper cellulose'; filter paper may be used as a substrate for determining the ability of an organism or enzyme to degrade crystalline CELLULOSE.

FP2 plasmid See INCOMPATIBILITY.

fps See FES.

fractalkine See CHEMOKINES.

Fraction 1 antigen See YERSINIA (*Y. pestis*).

fraction I protein See CHLOROPLAST.

fractional sterilization *Syn.* TYNDALLIZATION.

fractionation See CELL FRACTIONATION.

Fragilaria See DIATOMS.

fragilis yeast See SINGLE-CELL PROTEIN.

fragmin See ACTIN.

framboesia (frambesia) *Syn.* YAWS.

frameshift mutation (phase-shift mutation) A type of MUTATION in which a number of nucleotides *not* divisible by three is inserted into or deleted from a coding sequence, causing an alteration in the reading frame of the entire sequence downstream of the mutation. For example, if a wild-type gene containing a sequence UUG-GAG-UGU-AGU (encoding Leu-Glu-Cys-Ser: see GENETIC CODE) suffers an insertion mutation in the UUG codon which generates the sequence UAUG (a +1 frameshift), the reading frame will be shifted backwards by one nucleotide, and the sequence will read UAU-GGA-GUG-UAG-U (encoding Tyr-Gly-Val-*stop*, UAG being a nonsense codon). Similarly, deletion of one nucleotide (a −1 frameshift) will shift the reading frame forwards: e.g., if the second U is deleted, the above sequence becomes UGG-AGU-GUA-GU, encoding Trp–Ser–Val. A −1 frameshift may suppress the effects of a +1 frameshift: see SUPPRESSOR MUTATION.

Frameshift mutations may occur spontaneously (see SPONTANEOUS MUTATION) and are induced by certain types of mutagen (e.g. certain INTERCALATING AGENTS).

framycetin An AMINOGLYCOSIDE ANTIBIOTIC produced by a strain of *Streptomyces lavendulae*; it may be identical to one form of neomycin.

Francisella A genus (*incertae sedis*) of aerobic, chemoorganotrophic, Gram-negative bacteria which occur e.g. as parasites and pathogens in man and other animals. Cells (when actively growing): non-motile cocci, coccobacilli or rods, according to species and growth conditions; subsequently filamentous and pleomorphic. The cells stain poorly and often bipolarly; a bipolarly stained rod may give the false impression of two cocci. Metabolism: respiratory (oxidative). One species (*F. tularensis*) is nutritionally fastidious (requiring e.g. cysteine for growth), while the other (*F. novicida*) can grow e.g. on nutrient gelatin. Carbohydrates are metabolized slowly and anaerogenically. Optimum growth temperature: ca. 37°C. Oxidase −ve. Catalase test: weakly positive. GC%: ca. 33–36. Type species: *F. tularensis*.

F. novicida. Cells: non-capsulated, ca. 0.2–0.3 × 0.3 μm in tissues, 0.5 × 0.5–0.9 μm on solid media, 0.7 × 1.7 μm in liquid media. Unlike *F. tularensis*, *F. novicida* does not require cysteine for growth. *F. novicida* is apparently not pathogenic for man but can produce tularaemia-like lesions, experimentally, in e.g. guinea pigs and white mice.

F. tularensis (formerly *Pasteurella tularensis*). The causal agent of TULARAEMIA. Cells: ca. 0.2 × 0.2–0.7 μm; virulent strains have a non-immunogenic capsule, ca. 0.02–0.04 μm in thickness. *F. tularensis* can be grown e.g. on glucose–cysteine–blood agar; colonies (2–4 days, 37°C) are 1–4 mm diam., typically smooth, grey, easily emulsifiable, each being surrounded by a zone of greenish discoloration. *F. tularensis* is generally sensitive to e.g. kanamycin, streptomycin and tetracyclines.

[Book ref. 22, pp. 394–399.]

frange A band of perioral ciliature characteristic of some members of the HYPOSTOMATIA.

Frankia A genus of bacteria (order ACTINOMYCETALES, wall type III) which form ACTINORRHIZAE but which can also occur as free-living soil saprotrophs that appear not to fix nitrogen. (At least some of the *nif* genes have been found to be located on a large plasmid in one strain of *Frankia* [MGG (1986) *204* 492–495].) The organisms form branching, septate hyphae, and give rise to non-motile spores in sporangia which develop on the hyphae. GC%: ca. 68–72. Type species: *F. alni*. [Media and methods: Book ref. 46, pp. 1991–2003.]

Frateuria A genus of Gram-negative, obligately aerobic, chemoorganotrophic bacteria (family PSEUDOMONADACEAE) which occur e.g. on the fruit of the raspberry (*Rubus parvifolius*). Cells: polarly flagellate or non-motile rods, 0.5–0.7 × 0.7–3.5 μm. Round, entire, dark brown colonies are formed on glucose–yeast extract–$CaCO_3$ agar; optimum growth temperature: 25–30°C. Growth can occur at pH 3.6. Acid is formed from various carbon sources, including ethanol. Catalase +ve. Oxidase −ve. GC%: 62–64. Type species: *F. aurantia*. [Book ref. 22, pp. 210–213.]

Frazier's medium An agar medium containing 0.4% GELATIN; it is used to detect gelatin hydrolysis. Following culture of the test organism on the medium, the plate is flooded with a solution of mercuric chloride in dilute HCl; unhydrolysed gelatin rapidly becomes opaque – clear zones indicating gelatin hydrolysis.

***frd* genes** See FUMARATE RESPIRATION.

free cell formation See ASCUS.

free electrophoresis See ELECTROPHORESIS.

free induction decay (FID) See NUCLEAR MAGNETIC RESONANCE.

free-living Living without being directly dependent on other organisms. (cf. SYMBIOSIS.)

free residuals See WATER SUPPLIES.

freeze-cleaving *Syn.* FREEZE-FRACTURING.

freeze-drying (lyophilization) A process used e.g. for the long-term preservation of certain microorganisms, certain foodstuffs etc; in this process volatile substances (mainly water) are removed from the deep-frozen specimen or material by sublimation under high vacuum (cf. CRITICAL POINT DRYING).

Freeze-drying of bacterial cultures. A fresh broth culture of e.g. 10^8 cells/ml is gently centrifuged and the cells resuspended in a cryoprotectant (e.g. 20% skim milk) to a cell density of ca. 10^8 cells/ml. An aliquot of the suspension (ca. 0.1 ml) is put into a sterile VIAL. (This account assumes the use of a vial 10 cm long, 6 mm internal diameter.) The COTTON WOOL plug is replaced and pressed into the vial to a position ca. 2–3 cm above the suspension. With precautions to avoid heating the suspension, a region of the vial 2–3 cm from its open end is heated and drawn out into a capillary neck. The suspension is then frozen (see FREEZING) at −60 to −80°C, e.g. by placing the closed end of the vial in a mixture of solid CO_2 ('dry ice') and ethanol. The vial is transferred to a freeze-drying machine (where the low temperature is maintained) and subjected to a high vacuum for a period of 18 hours to several days, depending on temperature. The vial is removed, fitted to a vacuum line, and sealed by melting the capillary neck. Vials are best stored at 5°C or below. (For many species, viability is retained for 10–30 years or longer.) When a culture is required, a vial is opened and the contents reconstituted with broth or distilled water. A sealed vial can be opened (preferably in a SAFETY CABINET) by 'nicking' the glass over the centre of the cotton wool plug and momentarily applying a red-hot glass rod to the nick; the vial (usually) cracks around its circumference, permitting the slow ingress of air – which is filtered through the cotton wool.

In the double-vial method, the material is freeze-dried in a vial which is sealed only with a cotton wool plug; this vial is placed in an outer vial, containing a desiccant, which is evacuated and sealed by fusion of the glass.

freeze-etching A technique used for examining the topography of surface(s) exposed by the fracture (cleavage) of e.g. a deep-frozen cell. The specimen is frozen rapidly (see FREEZING). While at ca. −100 to −196°C (depending on apparatus), the specimen is fractured: it may be split with a knife-blade in a microtome-based apparatus, or, in one form of the 'snapped tube' apparatus, a capillary tube containing a homogenate is deep-frozen and snapped at a pre-scored position. During fracture, part of the cleavage generally follows the boundaries of internal structures (e.g. organelles), and membranes may be split to expose their hydrophobic faces. Surfaces exposed by cleavage are then 'etched' by sublimation of volatiles (mainly water) from the surfaces at a temperature of ca. −100°C under high vacuum; organelles and non-volatiles are thus thrown into relief. A thin replica of the etched surface is then prepared (see ELECTRON MICROSCOPY (a)) and examined under the TEM. [Book ref. 4, pp. 279–341. For nomenclature of freeze-etched surfaces see Science (1975) *190* 54–56. For potential artefacts see JUR (1983) *82* 123–133.]

freeze-fracturing The process of FREEZE-ETCHING omitting the etching step.

freeze pressing The operation of a HUGHES PRESS.

freeze-sectioning (cryosectioning) The preparation of sections, for MICROSCOPY or ELECTRON MICROSCOPY, by cutting a FROZEN-HYDRATED SPECIMEN with a *cryotome* (a modified MICROTOME which maintains the specimen in a frozen state).

freeze-substitution A procedure for preparing a specimen e.g. for X-RAY SPECTROCHEMICAL ANALYSIS. Essentially, the specimen is rapidly frozen and then immersed for e.g. 14 days in anhydrous ethanol or acetone at e.g. −70°C; during this time the ice is gradually replaced by ethanol or acetone. This method helps to maintain the positions of e.g. water-soluble ions within the specimen – thus facilitating subsequent determination of the in vivo locations of these ions.

freezing Sub-zero (°C) freezing is used for the preservation (*cryopreservation*) of certain microorganisms, foods etc, and e.g. in FREEZE-DRYING and FREEZE-ETCHING techniques (see also CRYOFIXATION).

(a) *Cryopreservation of living cells*. Cryopreservation may be used to preserve viability in many species of bacteria, fungi and protozoa, many viruses, and tissue cultures; this method of preservation (or freeze-drying) is generally preferable, when practicable, to preservation by repeated subculture (see STABILATE). However, unless precautions are taken, cells may be killed or (reversibly) damaged when they are frozen. Damage or death may be due to one (or both) of two main causes. First, the formation of intracellular and extracellular ice crystals may result in e.g. mechanical or osmotic damage. (Even after water freezes, ice crystals continue to grow at temperatures above a certain *critical temperature*; the temperature range between the freezing point and the critical temperature is called the *critical temperature zone*.) Secondly, as the temperature continues to fall below 0°C, the ongoing formation of crystals of pure ice results in the development of concentrated intracellular solutions of salts as the eutectic temperature is approached; these concentrated solutions are believed to damage certain cell components (e.g. enzymes). The continued presence of these solutions at temperatures found in certain types of commercial freezer (e.g. −10 to −20°C) precludes the use of such equipment for effective microbial preservation.

Cell damage can be minimized if, prior to freezing, the cells are equilibrated with a *cryoprotectant*: an antifreeze which, by depressing the freezing point, reduces the time that cells are subjected to temperatures in the critical temperature zone. Cryoprotectants may be used in both rapid and slow freezing (see below); they include e.g. glycerol (widely used e.g. for bacteria and yeasts) and DIMETHYLSULPHOXIDE (DMSO) (used e.g. for protozoa and tissue cultures). These agents are used e.g. at 10% v/v in growth medium; they can penetrate cell membranes and permeate the cytoplasm. Other cryoprotectants (e.g. skim milk, dextran, polyvinylpyrrolidone) appear to function extracellularly.

Rapid freezing (e.g. ca. 100 to 1000 centigrade degrees per second) tends to prevent ice crystal formation by encouraging *vitrification*, i.e. the non-crystalline (amorphous) solidification of water; it is commonly used to preserve viability in eukaryotes (e.g. *Saccharomyces cerevisiae*). The use of a cryoprotectant encourages vitrification by compressing the critical temperature zone. The specimen (e.g. one drop of suspension) may be initially quench-frozen in e.g. FREON 22 at ca. −150°C and then transferred to liquid nitrogen (−196°C). Liquid nitrogen may be used as the sole cryogen (despite the LEIDENFROST PHENOMENON) if the specimen is frozen in subcooled nitrogen (at ca. −200 to −210°C); heat transferred from the specimen raises the temperature of the nitrogen – but not as far as its boiling point. [Instruments which facilitate rapid freezing: JM (1982) *126* 221–229; relative efficiency of cryogens used for plunge-cooling biological specimens: JM (1987) *145* 89–96.]

Slow freezing is often used e.g. for bacteria; it may permit extracellular ice formation and intracellular dehydration by osmosis. A cell suspension is equilibrated with a cryoprotectant, and aliquots are placed in vials; vials containing non-pathogens are usually sealed. Freezing may be carried out at a rate of ca. one degree/min down to ca. −30°C, then at 30 degrees/min down to −150°C; vials are then stored in liquid nitrogen or in the vapour phase above liquid nitrogen (ca. −150°C).

Tissue culture cells may be initially equilibrated for 1 hour at 4°C in a mixture of growth medium, serum and ca. 10% v/v DMSO to permit penetration of DMSO. The cells, in sealed vials, are then transferred to the 'freezing chamber' of a liquid nitrogen apparatus for 20–30 min, and subsequently placed in liquid nitrogen.

Thawing should be rapid to preserve viability [Nature (1980) *286* 511–514]. Vials in liquid nitrogen may be equilibrated in solid CO_2 (−78°C) before being transferred to a 37°C water-bath.

When working with liquid nitrogen, thawing vials etc, it is usual to wear a protective face-shield and gloves.

(b) See FOOD PRESERVATION (b).

Frei test A SKIN TEST, analogous to the TUBERCULIN TEST, used in the diagnosis of LYMPHOGRANULOMA VENEREUM; it involves the intradermal injection of antigen derived from a chick embryo culture of *Chlamydia trachomatis*.

Fremyella See RIVULARIACEAE.

French moult See BUDGERIGAR FLEDGLING DISEASE.

French press (French pressure cell) An apparatus for CELL DISRUPTION. A suspension or paste of cells is put into a cooled steel cylinder, and a piston is forced into the cylinder by hydraulic pressure (ca. 10 tons/inch^2, i.e., ca. 160 MPa). Cells emerge from a small hole in the cylinder and are broken by the high shear forces during their exit; they tend to explode owing to the sudden drop in pressure. The process readily breaks most Gram-negative bacteria but not Gram-positive cocci or endospores.

Frenching disease See COTTON WILT.

Frenkelia A genus of coccidian protozoa (suborder EIMERIORINA) parasitic in rodents (intermediate hosts) and birds (final hosts); the organisms closely resemble species of SARCOCYSTIS.

Freon (trade name) Any of a range of halogenated hydrocarbons. For example, Freon 22 (= halocarbon 22; $CHClF_2$) melts at −160°C and boils at −41°C; it is used e.g. as a CRYOGEN. Freon 113 (= halocarbon 113; $CCl_2F.CClF_2$) melts at −35°C and boils at +48°C; it is used e.g. as an intermediate fluid in CRITICAL POINT DRYING.

FRET See PCR.

Freund's adjuvant Freund's *incomplete* ADJUVANT is a water-in-oil emulsion made with a mineral oil. Freund's *complete* adjuvant consists of the incomplete adjuvant together with dead mycobacteria dispersed in the oil phase; in a modified form, MDP (q.v.) is used instead of the mycobacteria. In both types of adjuvant the antigen is dispersed in the water phase. (cf. ADJUVANT 65.)

fried egg colony See MYCOPLASMA.

Friedländer's bacillus A species of *Klebsiella* isolated by Friedländer (in 1883) from the lungs of postmortem cases of PNEUMONIA.

Friedländer's pneumonia See PNEUMONIA (d).

Friedmannia A genus of sarcinoid, soil-inhabiting green algae related to TREBOUXIA.

Friend virus (FV) A virus complex consisting of a mixture of a replication-competent MURINE LEUKAEMIA VIRUS (F-MuLV or Fr-MuLV) and a replication-defective component designated 'spleen focus-forming virus' (SFFV) because of its ability to induce macroscopic foci of erythroid cell proliferation in the spleen of an infected mouse. (FV does not transform cells in vitro.) The F-MuLV component alone can induce erythroleukaemia only in newborn mice of certain strains, or in adult mice after a long latent period; the F-MuLV-SFFV complex can cause disease in both adult and neonatal mice of most inbred strains, causing anaemia (variant FV-A, containing $SFFV_A$) or polycythaemia (variant FV-P, containing $SFFV_P$) after a short latent period. Both SFFV variants are believed to have arisen by recombination between F-MuLV and an endogenous xenotropic virus sequence; the glycoprotein product of the recombinant *env* gene has been implicated in the enhanced oncogenicity of FV as compared with that of F-MuLV alone. (cf. MCF VIRUSES; RAUSCHER VIRUS.)

Frings Acetator See VINEGAR.

Frings trickling generator See VINEGAR.

Fritschiella See CHAETOPHORA.

frog virus 3 (FV3) A virus (genus RANAVIRUS) originally isolated from a leopard frog (*Rana pipiens*); it is apparently non-pathogenic in adult leopard frogs, but is lethal for tadpoles and for Fowler toads. FV3 can replicate in a wide range of cells in culture, including amphibian, avian, mammalian and piscine cells. Growth temperature: ca. 12–32°C.

The virion (ca. 160–200 nm diam.) apparently contains ca. 29 structural proteins and at least six enzymic activities. An envelope may or may not be present; enveloped virions, which are more infective than naked virions, enter the host cell by endocytosis (see ENVELOPE), while the naked virions usually enter by fusion between the virion shell and the host cell plasmalemma [JGV (1985) *66* 283–293]. Initial rounds of viral DNA replication occur in the host cell nucleus, resulting in DNA molecules up to twice the length of the viral genome. The DNA is then transferred to the cytoplasm where subsequent rounds of replication occur; concatemers >10 times the genome length are formed – apparently by recombination (as in e.g. BACTERIOPHAGE T4). The virion DNA and cytoplasmic viral DNA – but not the viral DNA synthesized in the nucleus – is highly methylated, >20% of the cytosine residues being methylated at the C-5 position; methylation is carried out in the cytoplasm by a viral early-gene product (a methyltransferase), and may protect the DNA from virus-encoded endonucleases. Virus assembly occurs in morphologically distinct regions of the cytoplasm delimited by aggregated host-cell INTERMEDIATE FILAMENTS [JGV (1986) *67* 915–922].

[Review: AVR (1985) *30* 1–19.]

frond twist disease See ALGAL DISEASES.

front focal plane Of a convex lens: the focal plane nearest the light source.

Frontonia See HYMENOSTOMATIDA.

frost damage (in plants) See e.g. ICE NUCLEATION BACTERIA.

frosty pod (of cacao) See POD ROT.

frozen-hydrated specimen A specimen which has been frozen hard without prior dehydration – as e.g. in FREEZE-SECTIONING.

fructan (fructosan) A polysaccharide composed mainly or solely of fructosyl residues (see e.g. INULIN and LEVAN). Fructans are synthesized by the addition of fructose residues to SUCROSE, and hence each fructan chain ends in a (non-reducing) sucrose residue. Fructans can be hydrolysed by a wide range of microorganisms.

fructification (1) *Syn.* FRUITING BODY. (2) The formation or development of a fruiting body.

fructosan *Syn.* FRUCTAN.

fructose ('fruit sugar') A ketohexose found e.g. in plants (e.g. in sweet fruits), in honey (with glucose and sucrose), and as a component of many oligo- and polysaccharides (e.g. MELEZITOSE, RAFFINOSE, SUCROSE; see also FRUCTAN). D-Fructose ('laevulose') can be metabolized by a wide range of microorganisms.

Fructose is sweeter than sucrose and is being used on an increasing scale as a sweetener in foods (see e.g. HIGH FRUCTOSE CORN SYRUP). (See also INULINASE.)

fructose bisphosphatase See GLUCONEOGENESIS.

β-fructosidase See INVERTASE and INULINASE.

fruit (*verb*) To produce a FRUITING BODY.

fruit spoilage See e.g. CANNING; GREY MOULD; PECTIC ENZYMES; PICKLING; SOFT ROT (2).

fruiting body (fructification; fruit body; sporocarp) Any specialized structure (plectenchymatous in fungi) which bears or contains sexually- or asexually-derived spores. Examples include e.g. APOTHECIUM, PYCNIDIUM, pycnium and the resting structures of the MYXOBACTERALES. Some fruiting bodies (e.g. those formed by species of *Heterobasidion* and *Poria*) are perennial.

frustule See DIATOMS.

fruticose Shrub-like. Of a LICHEN: having a thallus which is thread-like (terete) or strap-like (more or less flattened) and either erect and shrubby or pendulous; the thallus may be attached to the substratum by a holdfast or may be unattached. The mechanical strength of the thallus may derive from a dense axial strand (in USNEA) or from the cortex; thalli of the latter type may or may not be hollow. (cf. CRUSTOSE; FOLIOSE.)

FTA-ABS test (fluorescent treponemal antibody-absorption test) A TREPONEMAL TEST used in the diagnosis of SYPHILIS. Initially, the patient's serum is absorbed with an extract of Reiter treponemes to remove non-specific anti-treponemal antibodies, and the absorbed serum is applied to a slide on which there is a film of killed, fixed cells of *Treponema pallidum*; the slide is then briefly incubated to allow any specific antibodies in the serum to combine with the surface antigens of *T. pallidum*. The slide is rinsed free of serum and briefly re-incubated with fluorescent conjugate (anti-human globulin conjugated with a fluorescent dye); uncombined conjugate is rinsed off and the slide is examined by fluorescence MICROSCOPY. Antibodies which combine with *T. pallidum* combine also with the conjugate – their presence being indicated by fluorescent treponemes.

FTO agar A selective medium used for the culture of *Micrococcus* spp. It consists of trypticase–soy agar containing 0.1% yeast extract, 0.5% Tween 80, and small amounts of furoxone and Oil Red O. [Preparation and use: Book ref. 46, p. 1542.]

***ftsA* gene** (*Caulobacter*) See CAULOBACTER.

FtsH An ATP- and zinc-dependent protease within the cytoplasmic membrane of *Escherichia coli* and many other bacteria; it is encoded by gene *ftsH* (also called *hflB* and *tolZ*). In *E. coli*, FtsH is involved e.g. in the degradation of certain membrane and regulatory proteins (including the HEAT-SHOCK PROTEIN σ^{32}), and it may also act as a chaperone. [FtsH (review): FEMS Reviews (1999) *23* 1–11.]

***ftsI* gene** See PENICILLIN-BINDING PROTEINS.

***ftsQ* gene** (*Caulobacter*) See CAULOBACTER.

FtsY protein See PROTEIN SECRETION (type II systems).

***ftsZ* gene** See SOS SYSTEM.

FtsZ protein See CELL CYCLE (b) and SOS SYSTEM.

5-FU 5-Fluorouracil (see e.g. PHENOTYPIC SUPPRESSION).

fuberidazole (2-(2′-furyl)-benzimidazole) An agricultural antifungal agent (see BENZIMIDAZOLES); it is particularly useful as a seed treatment for the control of *Fusarium* infections (e.g. *F. nivale* in cereals).

Fucales See FUCUS and PHAEOPHYTA.

fucan A polymer of FUCOSE. (See also FUCOIDIN.)

fuchsin (fuchsine) *Basic* fuchsin is a red-purple DYE consisting mainly of pararosanilin and rosanilin (see TRIPHENYLMETHANE DYES). (See also CARBOLFUCHSIN and BRUCELLA.) *Acid* fuchsin, derived from basic fuchsin by sulphonation, is used e.g. in ANDRADE'S INDICATOR.

fucidin See FUSIDIC ACID.

fucoidin (fucoidan; fucan) A group of complex, branched, sulphated heteropolysaccharides present in phaeophycean algae; fucoidins contain fucose (usually sulphated), xylose, glucuronic acid, and – in some species – galactose and/or mannose, and they may be linked to protein. Fucoidins occur in the algal CELL WALLS and intercellular matrix and in the mucilage exuded from the surfaces of the fronds. They appear to play a role in preventing desiccation of the alga when it is exposed by tides: algae which are permanently submerged contain much less fucoidin than those growing in the intertidal zone.

fucosans See PHAEOPHYTA.

L-fucose 6-Deoxy-L-galactose.

fucosterol See PHAEOPHYTA.

fucoxanthin See CAROTENOIDS.

Fucus A genus of algae (division PHAEOPHYTA) which occur primarily in marine and intertidal zones. The vegetative organism consists typically of dichotomously branched, flat leathery blades, each having a pronounced central midrib; according to species, the blade may have entire or serrated edges. The stipe is attached to rocks etc by a holdfast. Some species, e.g. *F. vesiculosus* (bladder wrack), have PNEUMATOCYSTS. As in other members of the Fucales, the haploid phase of the life cycle is represented only by the gametes of the sexual organs in diploid individuals.

fuel biodeterioration See PETROLEUM.

fuel cell See BIOFUEL CELL.

fuel ethanol See INDUSTRIAL ALCOHOL.

Fujinami sarcoma virus See AVIAN SARCOMA VIRUSES.

Fuligo A genus of slime moulds (class MYXOMYCETES) which form aethalia containing both capillitium and pseudocapillitium; the capillitial threads bear nodes of lime. *F. septica* – a common species on rotting wood, humus, manure, etc. – forms aethalia measuring up to 70 cm across (the largest sporocarp of any myxomycete); the aethalia (sometimes known as 'flowers of tan') are initially brownish-yellow, and have a thick calcareous covering which eventually breaks open to reveal the black spores.

Fulvia A genus of fungi of the HYPHOMYCETES. *F. fulva*, formerly *Cladosporium fulvum*, is the causal agent of leaf mould of the tomato plant.

fumagillin An ANTIBIOTIC produced by *Aspergillus fumigatus*; it has a complex structure which includes two epoxide rings and is an ester of decatetraenedioic acid. Fumagillin apparently has no activity against bacteria or fungi but it is a potent AMOEBICIDE; it also inhibits the development of certain bacteriophages, and may show some antiviral activity in tissue cultures.

fumarase See Appendix II(a).

fumarate reductase See e.g. FUMARATE RESPIRATION and PROPIONIC ACID FERMENTATION [Appendix III(h)].

fumarate respiration A type of ANAEROBIC RESPIRATION, carried out by a range of bacteria and by certain eukaryotic microorganisms, in which exogenous fumarate acts as the terminal electron acceptor for re-oxidation of reduced pyridine nucleotides, for oxidative phosphorylation, etc; fumarate is reduced to succinate, the E_0' for the succinate/fumarate redox couple being +30 mV. (cf. e.g. PROPIONIC ACID FERMENTATION.)

In *Escherichia coli*, fumarate respiration occurs during anaerobic growth in the presence of a non-fermentable carbon source (e.g. glycerol) and fumarate; the respiratory chain involved comprises fumarate reductase (EC 1.3.99.1), menaquinone, dehydrogenases (e.g. formate, NADH and lactate dehydrogenases), and possibly a *b*-type cytochrome. Fumarate reductase is distinct from succinate dehydrogenase (which catalyses the same reaction): fumarate reductase is repressed in the presence of O_2 and NO_3^- but is expressed under anaerobic conditions in the presence of fumarate, whereas succinate dehydrogenase is induced under aerobic conditions and repressed under anaerobic conditions. Fumarate reductase is bound to the cytoplasmic face of the cytoplasmic membrane [JGM (1984) *130* 2851–2855]; fumarate respiration thus requires a fumarate transport system. Fumarate reductase in *E. coli* consists of an FAD-containing protein subunit (*frdA* gene product), an iron–sulphur protein subunit (*frdB* gene product), and two small, hydrophobic 'anchor proteins' (*frdC* and *frdD* gene products) which anchor the enzyme in the membrane; these four cistrons occur in an operon with a promoter–operator region and a transcriptional terminator. The energy yield of fumarate respiration in *E. coli* is uncertain, but has been calculated to be ca. 0.66 ATP/fumarate [MR (1984) *48* 222–271].

In *Wolinella succinogenes*, fumarate is used as an electron acceptor for the oxidation of formate ($HCOO^- + H^+ +$ fumarate $\rightarrow$ succinate $+ CO_2$) or H_2, yielding, by oxidative phosphorylation, 1–2 ATP per fumarate reduced.

fumigant See FUMIGATION.

fumigation The exposure of an enclosed space (e.g. a horticultural glasshouse, a hospital room) to a gaseous or vapour-phase DISINFECTANT or STERILANT. Control of humidity is commonly important for effective fumigation. Agents used for fumigation (*fumigants*) include e.g. DAZOMET, ETHYLENE OXIDE, FORMALDEHYDE, SULPHUR DIOXIDE.

fumitremorgin A tremorgenic MYCOTOXIN produced e.g. by *Aspergillus fumigatus* and *Penicillium piscarium*.

functional immunity *Syn.* IMMUNITY (3).

fungaemia (fungemia) The presence of fungi in the bloodstream.

fungemia See FUNGAEMIA.

fungi (*sing.* fungus) Unicellular, multicellular or coenocytic, heterotrophic, eukaryotic MICROORGANISMS which do not contain chlorophyll (cf. ALGAE) and which characteristically form a rigid CELL WALL containing CHITIN and/or CELLULOSE (cf. PROTOZOA). In some taxonomic schemes [see e.g. Book ref. 64] the fungi are divided into two divisions: EUMYCOTA ('true fungi') and MYXOMYCOTA; since members of the Myxomycota are commonly regarded as protozoa, the following account refers specifically to the 'true fungi'.

In most fungi the vegetative form (thallus) consists of hyphae (see HYPHA), while in e.g. many YEASTS the thallus is predominantly or exclusively unicellular (cf. DIMORPHIC FUNGI). (See also GROWTH (fungal).) The majority of fungi are non-motile; however, flagellated reproductive/disseminative forms are produced by many of the LOWER FUNGI. (Phylogenetically, motility is regarded as a 'primitive' feature among the fungi.)

Most fungi can reproduce asexually – often by the formation of conidia (see CONIDIUM). In sexual reproduction, interaction between individuals may involve e.g. anisogamy, GAMETANGIAL CONTACT, isogamy, oogamy (see MONOBLEPHARIDALES), or SOMATOGAMY. (See also PARASEXUAL PROCESSES.) Specialized asexual and/or sexual SPORE-bearing structures (see e.g. FRUITING BODY) are formed by most fungi; many are macroscopic, and some (e.g. the mushrooms and puffballs) may be several centimetres or more in size.

Fungi are osmotrophic chemoheterotrophs which, collectively, utilize substrates ranging from simple sugars to e.g. cellulose (see CELLULASES), HYDROCARBONS, LIGNIN, pectins (see PECTIC ENZYMES) and xylans (see XYLANASES). Energy-yielding metabolism may involve RESPIRATION and/or FERMENTATION. Although fungi are typically aerobic organisms, some are facultative or obligate anaerobes (see e.g. RUMEN).

The fungi are widespread in nature; most species are terrestrial, but some (see e.g. NIA, OOMYCETES, SPATHULOSPORALES) are primarily or exclusively aquatic. [Review of marine fungi: Biol. Rev. (1983) *58* 423–459.] The terrestrial species include free-living saprotrophs which occur e.g. in soil and dung; parasites and pathogens of man and other animals – see e.g. INSECT DISEASES, MYCOSIS, NEMATOPHAGOUS FUNGI (see also MYCOTOXINS); plant-pathogenic species – see e.g. DOWNY MILDEWS, POWDERY MILDEWS, RUSTS (sense 2) and SMUTS (sense 2); MYCOPARASITES; and fungi involved in other associations – see e.g. AMBROSIA FUNGI, FUNGUS GARDENS, LICHENS, MYCETOCYTE, MYCORRHIZA, TERMITE–MICROBE ASSOCIATIONS. Some of the aquatic fungi are pathogenic – see e.g. BLACK MAT SYNDROME, BURNED SPOT DISEASE, FISH DISEASES and OSTRACOBLABE IMPLEXA. (See also ANTIFUNGAL AGENTS.)

Certain fungi are used commercially e.g. in the production of antibiotics (e.g. some PENICILLINS), enzymes (see e.g. PROTEASES), FERMENTED FOODS, INDUSTRIAL ALCOHOL and SINGLE-CELL PROTEIN. Fungi may also be important agents in the BIODETERIORATION of certain materials.

Fungi Imperfecti See DEUTEROMYCOTINA.

fungicidal Able to kill at least some types of fungus. (cf. FUNGISTATIC.)

fungicide See ANTIFUNGAL AGENTS.

fungicidin *Syn.* NYSTATIN.

fungicole Any organism which grows on or in fungi; hence *adj.* fungicolous.

fungimycin *Syn.* PERIMYCIN.

fungistatic Able to inhibit the growth and reproduction of at least some types of fungus. (cf. FUNGICIDAL.)

fungitoxic Toxic to fungi.

fungus See FUNGI.

fungus ball *Syn.* MYCETOMA (sense 2).

fungus combs See TERMITE–MICROBE ASSOCIATIONS.

fungus gardens Subterranean 'gardens' of fungi cultivated by parasol ants (Formicidae, subfamily Myrmicinae, tribe Attini). The ants cut leaf fragments or collect other organic particles and carry them to the nest, where the fragments are pulped, treated with faecal material, inserted into the garden, and inoculated with fungal mycelium acquired from established parts of the garden or – if a new garden is being started – from a small pellet of the fungus carried in an infrabuccal pocket at the back of the mouth. Hyphae growing in the garden commonly develop terminal swellings (variously called *bromatia*, *gongylidia* or *staphyla*), and these are eaten by adult ants and fed to the larvae; the fungi are the sole source of food for the larvae, but the adults may also feed on plant juices. The fungi appear to occur only in fungus gardens; they generally appear to be basidiomycetes of the genera *Auricularia*, *Agaricus* or *Lepiota*. The fungal hyphae contain proteases (as well as cellulases, pectic enzymes etc), but proteases are not secreted to any significant extent, and the fungi grow slowly if supplied with polypeptides as sole source of nitrogen. When the fungus is ingested by the ants, the enzymes pass through the ants' gut and are still active in the faecal material. Thus, treatment of plant fragments with faecal material increases availability of amino acids by proteolysis, augmenting growth of the fresh inoculum of fungi, and probably also facilitates fungal penetration of the plant material by pectinolysis. [Book ref. 77, pp. 163–166.] The fungus is maintained in almost pure culture by the activities of the ants, although it is not clear how this is achieved. (See also INSECT–MICROBE ASSOCIATIONS.)

fungus gnats Small flies (family Mycetophilidae) whose larvae feed on fungal mycelium and fruiting bodies; fungus gnats (particularly *Sciara* spp) can cause severe damage in mushroom beds.

funicular cord See CYATHUS.

funiculus See CYATHUS.

funoran A mucilaginous sulphated galactan produced by *Gloiopeltis furcata*; it consists mainly of D-galactose and 3,6-anhydro-L-galactose residues, with some L-galactose residues. In Japan, funoran is used as an adhesive and sizing agent.

Fur protein A multifunctional regulatory protein encoded by the *fur* gene. In many bacteria, Fur is involved e.g. in the control of ferric iron uptake (hence ferric uptake regulator); for example, after uptake of adequate ferric iron in *Escherichia coli*, accumulated intracellular ferrous ions cause the Fur protein to repress transcription from various genes (e.g. *fep* genes) which encode cell envelope proteins involved in ferric iron uptake. Repression involves the binding of Fe^{2+}–Fur to an operator sequence (the 'iron box') upstream of the coding region in the affected genes. (See also OPERATOR.) Fur has many functions other than iron regulation – e.g. in *Salmonella typhimurium* it is involved in the acid tolerance response [JB (1996) *178* 5683–5691].

furacin Nitrofurazone: see NITROFURANS.

furadantin Nitrofurantoin: see NITROFURANS.

furalaxyl See PHENYLAMIDE ANTIFUNGAL AGENTS.

furaltadone See NITROFURANS.

furamide See DILOXANIDE.

furazlocillin A β-LACTAM ANTIBIOTIC which binds exclusively to PBP-3 (see PENICILLIN-BINDING PROTEINS).

furazolidone See NITROFURANS.

furcellaran (Danish agar) A sulphated galactan, similar to κ-CARRAGEENAN, isolated from *Furcellaria fastigiata*; it has been used as a gelling agent in the food industry.

Furcellaria See RHODOPHYTA.

furfuraceous Scurfy; covered with small scales.

furocoumarin *Syn.* PSORALEN.

furoviruses See SOIL-BORNE WHEAT MOSAIC VIRUS.

furoxone Furazolidone: see NITROFURANS.

furuncle *Syn.* BOIL.

furunculosis (1) A FISH DISEASE affecting mainly freshwater fish, particularly salmonids. Pathogen: *Aeromonas salmonicida*. The typical disease occurs in two forms. In the acute form external symptoms are usually absent; internal organs become inflamed and haemorrhagic, and death is usually rapid. In the subacute form furuncle-like swellings (containing necrotic tissue, blood and bacteria) appear on the body. A chronic, subclinical carrier state is also known. Infection may occur via gills or wounds; high temperature (12–19°C) is a predisposing factor. Atypical forms of the disease ('ulcerative furunculosis'), caused by atypical strains of *A. salmonicida*, occur in salmonids and non-salmonids (e.g. goldfish) and involve progressive ulceration of skin and underlying muscle. (cf. ULCER DISEASE.)

A. salmonicida produces several extracellular toxins, including a leucocytolytic factor, proteases ('caseinase' and 'gelatinase'), and two haemolysins: H-lysin and T-lysin; H-lysin is a broad-spectrum haemolysin [purification and properties: JGM (1985) *131* 1603–1609], while T-lysin appears to be active specifically against trout RBCs. These products may, together, play a role in the pathogenicity of *A. salmonicida*.

(2) (*med.*) The (simultaneous or sequential) formation of a number of BOILS.

furylfuramide (AF-2) See NITROFURANS.

fusarenon-X See TRICHOTHECENES.

fusaric acid (2-carboxyl-5-*n*-butylpyridine) A TOXIN produced – sometimes together with dehydrofusaric acid and 10-hydroxyfusaric acid – by several species of *Fusarium*. It apparently damages plant cell membranes and may play a role in the pathogenesis of FUSARIUM WILT; the toxin can also inhibit plant polyphenol oxidases, thus possibly interfering with the host defence mechanisms. Some plants can convert fusaric acid to the less toxic *N*-methyl derivative.

fusaritoxicosis A MYCOTOXICOSIS due to toxins produced by FUSARIUM spp.

Fusarium A genus of economically important fungi (class HYPHOMYCETES) which form a septate mycelium and produce macroconidia and, in some species, microconidia and/or chlamydospores. The macroconidia are produced from phialides which are generally borne on sporodochia; they are hyaline, usually crescent-shaped and multiseptate, and each has a characteristically shaped *foot cell* (forming one pole of the macroconidium) in which a 'heel' is present. Microconidia are also formed from phialides and are usually single-celled and spherical or ovoid. Conidia intermediate in form between macro- and microconidia may be produced. Fusaria are notoriously variable in culture, often making identification difficult. Many species have teleomorphs in the Hypocreales (e.g. GIBBERELLA, NECTRIA), but some – e.g. *F. culmorum*, *F. oxysporum* – have no known teleomorph.

Fusarium spp have a world-wide distribution and occur e.g. in soil, decaying organic materials, etc. Some strains may be wholly saprotrophic, but the majority can also be weakly to highly pathogenic – mainly in plants and invertebrates; no strain appears to be obligately parasitic.

Examples of plant-pathogenic species: *F. avenacearum* (*Gibberella avenaceae*) and *F. culmorum* cause seedling blight, foot and root rot, and head blight in cereals; *F. coeruleum* causes DRY ROT (sense 2) in potatoes; *F. decemcellulare* causes gall formation on cacao pods; *F. graminearum* (*Gibberella zeae*) causes seedling blight and foot and root rots in small-grain cereals (cf. WHEAT SCAB), and stalk and ear rot of maize; *F. lateritium* (*Gibberella baccata*) causes wilt, dieback and canker in woody plants, bud rot in apples, etc; *F. moniliforme* (*Gibberella fujikuroi*) causes ear and stalk rot of maize, BAKANAE DISEASE of rice, PITCH CANKER of pines, etc; *F. nivale* causes pre-emergence blight, root rot and head blight of cereals (see also SNOW MOULD); *F. oxysporum* causes e.g. FUSARIUM WILT, a form of CELERY YELLOWS, and basal rots of bulbs and corms (e.g. *Narcissus*, tulips, gladioli); *F. poae* causes SILVER TOP in grasses; *F. solani* (*Nectria haematococca*) causes root rots in a wide range of plants; *F. sulphureum* (*Gibberella cyanogena*) can cause dry rot of stored potatoes. Many plant-pathogenic fusaria produce non-host-specific TOXINS (see e.g. FUSARIC ACID, LYCOMARASMIN, NAPHTHAZARINS), although the significance of these in pathogenesis is uncertain (cf. GIBBERELLINS). In chlamydospore-forming species the chlamydospores are believed to be the main agents of infection [Book ref. 131, pp. 71–93].

Fusarium spp can be important pathogens in e.g. crustacea (see e.g. BLACK GILL DISEASE and BURNED SPOT DISEASE) and insects – e.g. strains of *F. solani* can be pathogenic in larvae and adults of the Southern pine beetle *Dendroctonus frontalis*, and in the larvae of the elm beetle *Scolytus scolytus*, while *F. larvarum* and *F. lateritium* have been reported to be pathogenic in scale insects. Toxins such as the naphthazarin pigments may account for at least some of the insecticidal activity of these fungi. (See also AMBROSIA FUNGI.)

In man and other mammals, *Fusarium* spp are rarely pathogenic, although they may indirectly cause disease by producing mycotoxins in foodstuffs (see e.g. TRICHOTHECENES and ZEARALENONE). Certain strains have been associated with e.g. eye infections and skin lesions in man and animals; such infections are usually or always secondary to trauma or debilitation in the host.

In immunocompromised patients, invasive infection by *Fusarium* may mimic aspergillosis; a PCR-based examination may be useful for detecting this organism in clinical specimens [JCM (1999) *37* 2434–2438].

Fusarium spp have been used commercially as sources of ENNIATINS, GIBBERELLINS for horticultural use, ZEARALENONE derivatives for medical and veterinary use, and in the production of single-cell protein. (See also PAPER SPOILAGE and TIMBER STAINING.)

[Applied aspects: Book ref. 131.]

fusarium wilt A VASCULAR WILT caused by a *Fusarium* sp – usually *F. oxysporum* ('oxysporum wilt'). Strains of *F. oxysporum* may have a narrow, intermediate or wide host range: e.g. *F. oxysporum* f. sp. *cubense* infects banana plants ('Panama disease'); f. sp. *lini* infects flax; f. sp. *lycopersici* infects tomatoes (cf. LYCOMARASMIN); f. sp. *conglutinans* infects cruciferous plants; f. sp. *vasinfectum* infects e.g. coffee, cotton, cowpea, rubber (*Hevea*), soybean, and many other plants. Typical symptoms of fusarium wilts include brown or black vascular discoloration and wilting, sometimes preceded by e.g. epinasty, vein-clearing and leaf chlorosis; affected plants generally die.

fusarubin See NAPHTHAZARINS.

Fuscidea See LECIDEA.

fusel oil A mixture of higher alcohols (mainly amyl, isoamyl, isobutyl, and propyl alcohols) formed in small quantities as by-products of ALCOHOLIC FERMENTATION; the alcohols are formed by deamination, decarboxylation and reduction of amino acids, and from keto acid precursors of amino acids.

fusicoccin A toxin, produced by *Fusicoccum amygdali*, which can affect a range of plants. Fusicoccin can increase the rate at which a plant loses water, e.g. by promoting the opening of stomata. It can also stimulate H^+ efflux across the plasmalemma, thus facilitating uptake of e.g. K^+ and certain energy-yielding substrates; this may at least partly account for the observed growth-promoting activity of fusicoccin.

Fusicoccum A genus of fungi of the COELOMYCETES.

fusidic acid An ANTIBIOTIC synthesized e.g. by *Cylindrocarpon* spp (= *Fusidium coccineum*) and *Acremonium* (*Cephalosporium*) spp; it is a steroid structurally related to cephalosporin P1 – with which it exhibits cross-resistance. Sodium fusidate (fucidin) is active (bacteriostatic) against Gram-positive bacteria, particularly staphylococci; it inhibits PROTEIN SYNTHESIS by binding to EF-G, preventing the dissociation of EF-G and GDP from the ribosome following translocation. The EF-Gs (= EF-2s) of certain archaeans (e.g. methanogens, halobacteria) are sensitive to fusidic acid, whereas those from e.g. *Desulfurococcus mobilis*, *Sulfolobus solfataricus*, *Thermococcus celer* and *Thermoplasma acidophilum* are insensitive [JB (1986) *167* 265–271].

Fusidium See CYLINDROCARPON.

fusiform Spindle-shaped: tapered at both ends.

Fusiformis An obsolete bacteria genus. (See BACTEROIDES and FUSOBACTERIUM.)

fusigen A SIDEROPHORE produced by e.g. *Aspergillus* spp.

fusion protein (1) In e.g. viruses of the genus PARAMYXOVIRUS (e.g. Sendai virus): a protein which promotes fusion between host cells, as well as between virus envelope and cell membrane.

(2) (*biotechnol.*) A protein containing amino acid sequences from each of two distinct proteins; it is formed by the expression of a recombinant genetic construct prepared by joining together

two coding sequences with their reading frames in phase. (See also GENE FUSION.) Fusion proteins are constructed for various purposes.

A fusion protein may be constructed e.g. in order to facilitate identification/detection/assay of a given gene product which (for any reason) may be difficult to identify, detect or assay on its own. In such cases, the gene in question may be fused with a (*partner* or *carrier*) gene such as the *lacZ* gene of *Escherichia coli*; this results in a fusion protein with the properties of β-galactosidase – which can be readily detected/assayed e.g. on media containing XGAL. Another type of partner gene is that encoding glutathione *S*-transferase (GST); the resulting fusion protein can be isolated e.g. by affinity chromatography (using glutathione in the stationary matrix). In this latter example GST is the *affinity tail* on the protein of interest.

If a given gene encodes an intracellular protein, the product may be made secretable by fusing the gene to a partner gene whose product is secreted; this may facilitate isolation of the required gene product.

Fusion may also be used to determine the intracellular location of a given gene product. For example, the location of the FtsZ protein during cell division in *Escherichia coli* has been studied by fusing *ftsZ* with the gene encoding GREEN FLUORESCENT PROTEIN – the (fluorescent) fusion protein being detected by fluorescence microscopy [PNAS (1996) *93* 12998–13003]. Again, the gene for GFP has been fused with *smc* in a study on chromosome partitioning [GD (1998) *12* 1254–1259]. [Other GFP fusions: TIM (1998) *6* 234–238.]

Another use of fusion is to obtain the product of a gene which is normally not highly expressed – the gene of interest being fused downstream of a highly expressed partner gene in an expression vector. A well-chosen partner gene may also confer the additional benefit of greater solubility and/or stability.

If a fusion protein has been isolated it may be cleaved to yield separate products for each of the two genes. One method is to use a site-specific protease. An alternative approach is to use a chemical agent such as cyanogen bromide (which cleaves at methionine residues) or hydroxylamine (which cleaves between asparagine and glycine residues); potential problems with chemical agents include (i) possible cleavage at sites *within* the required protein, and (ii) unwanted chemical modification of the required protein.

Fusion is also useful for studying the *regulation* of gene expression. If the expression of a gene is difficult to monitor (owing, e.g. to difficulty in assaying the gene product) the activity of the gene's *promoter* may be studied separately by fusing it upstream of a promoter-less *reporter gene* – i.e. a gene whose product can be readily monitored and whose expression reflects the activity of the promoter. Reporter genes include those encoding e.g. galactokinase, chloramphenicol acetyltransferase (CAT) and green fluorescent protein. While this is generally a useful experimental approach, certain reporter genes have been found to influence the activity of some promoters [JB (1994) *176* 2128–2132].

('Fusion protein' is also used to refer to the products of certain viral ONCOGENES – e.g. MYC.)

Fusobacterium A genus of Gram-negative bacteria (family BACTEROIDACEAE) which occur e.g. in the human mouth and intestine, and in the RUMEN, and which include human and animal pathogens (see NECROBACILLOSIS and FOOT-ROT; see also SWINE DYSENTERY). Cells: typically fusiform or non-fusiform rods or filaments; all species are non-motile. Fermentation of peptone or carbohydrates yields butyric acid as a major product; isobutyric and isovaleric acids are not formed, but small amounts of acetic, formic, lactic or propionic acids may be formed. Optimum growth temperature: ca. 37°C. GC%: ca. 26–34. Type species: *F. nucleatum*.

F. necrophorum (formerly *Fusiformis necrophorus*, *Sphaerophorus necrophorus*). Fusiform or round-ended cells. Propionate is formed from both threonine and lactate; aesculin is not hydrolysed; indole +ve; copious gas is usually formed from glucose.

F. nucleatum (formerly *Fusiformis fusiformis*, *Fusobacterium polymorphum*, *Sphaerophorus fusiformis*). Cells: typically fusiform rods or filaments. Propionate is formed from threonine but not from lactate; aesculin is not hydrolysed; indole +ve; typically, little gas is formed from glucose.

Other species: *F. gonidiaformans*, *F. mortiferum*, *F. naviforme*, *F. necrogenes*, *F. perfoetens*, *F. prausnitzii*, *F. russii* and *F. varium*.

Fusulina See FORAMINIFERIDA.

fusulinids See FORAMINIFERIDA.

Fusulinina See FORAMINIFERIDA.

futile cycle The cyclic interconversion of two compounds by irreversible reactions catalysed by two or more enzymes which are active at the same time within a given cell, the result being an apparently useless dissipation of energy. For example, in e.g. *Escherichia coli* phosphofructokinase catalyses ATP-dependent phosphorylation of fructose 6-phosphate to fructose 1,6-bisphosphate [see Appendix I(a)], and fructose bisphosphatase hydrolyses fructose 1,6-bisphosphate to fructose 6-phosphate and inorganic phosphate (Pi) (see GLUCONEOGENESIS); if both enzymes were to be active simultaneously, the net result would be the hydrolysis of ATP to ADP + Pi. Presumably, futile cycles are normally prevented by control of the presence/activity of enzymes with opposing functions; however, they may have a role under certain conditions: e.g., it has been suggested that a futile cycle that results in ATP hydrolysis could provide Pi from endogenous ATP when the exogenous supply of Pi is interrupted [ARM (1984) *38* 459–486 (471–473)].

FV3 FROG VIRUS 3.

G

G (1) Guanine (or the corresponding nucleoside or nucleotide) in a nucleic acid. (2) Glycine (see AMINO ACIDS).

G-actin See ACTIN.

(G + C)% value See GC%.

G loop If the DNA of BACTERIOPHAGE MU is extracted from virions obtained by induction of a lysogenic population of bacteria and is then denatured and re-annealed, a proportion of the resulting dsDNA molecules show a 'bubble' or loop of unpaired strands in the G region of the DNA; this loop – the *G loop* – reflects localized non-homology generated by inversion of the G segment during lysogeny. (G loops are rarely observed in phage DNA derived from lytic infections.)

G phases (of cell cycle) See CELL CYCLE.

G proteins (GTP-binding proteins) In eukaryotic cells: GTP-binding proteins involved mainly in intracellular signalling. There are two main types of G protein: (i) the heterotrimeric type (each G protein consisting of α, β and γ subunits), and (ii) the (smaller) proteins of the Ras superfamily (= p21 family).

Heterotrimeric G proteins are characteristically associated with the cytoplasmic domain of certain transmembrane receptors (e.g. the receptors for CHEMOKINES). (See also ADENYLATE CYCLASE.) In the inactive state, a GDP residue binds to the α subunit of the protein. During activation of the G protein (through receptor–ligand binding), the α subunit separates from the other two subunits (which stay together), and GDP on the α subunit is exchanged for GTP; when activated in this way (i.e. when binding GTP), the α subunit can behave as an effector molecule in the signalling pathway – as can the $\beta + \gamma$ entity. During de-activation of the G protein, GTPase activity (inherent in the α subunit) cleaves the terminal phosphate from GTP – so that the α subunit is then bound to GDP; the α subunit (with bound GDP) then re-associates with $\beta + \gamma$ to form the inactive G protein.

G proteins of the Ras superfamily interact with several factors during their cycle of activation and de-activation. During activation, a guanosine nucleotide exchange factor promotes the exchange of (bound) GDP for GTP; the activated G protein (i.e. with bound GTP) can then act as an effector molecule in the signalling pathway. During de-activation, a GTPase-activating factor stimulates the latent GTPase activity of the G protein, thus giving rise to the inactive (GDP-bound) state. Another factor – guanosine nucleotide dissociation inhibitor – inhibits the exchange of GDP for GTP.

G_1-event model See CELL CYCLE.

Gaeumannomyces A genus of fungi of the order DIAPORTHALES. *G. graminis* (formerly *Ophiobolus graminis*), the causal agent of TAKE-ALL, forms dark, thick-walled perithecia which occur immersed in leaf sheaths and stem bases with their beaks (i.e., necks) protruding; the ascospores are septate, filiform or fusiform. (See also MYCOVIRUS.)

***Gaeumannomyces graminis* virus** See MYCOVIRUS.

gaffkaemia (gaffkemia) A fatal septicaemic disease of lobsters (*Homarus* spp) caused by *Aerococcus viridans* (formerly *Gaffkya homari*); infection occurs via breaks in the integument.

Gaffkya An obsolete bacterial genus [*nom. rejic.* IJSB (1971) *21* 104–105]. (See also AEROCOCCUS and PEPTOSTREPTOCOCCUS.)

***gag* gene** See RETROVIRIDAE.

***gal* operon** See OPERON and OPERATOR.

galactan A polymer of GALACTOSE: see e.g. AGAR; AMYLOVORIN; CARRAGEENAN; FUNORAN; FURCELLARAN; GALACTOCAROLOSE; PECTIC POLYSACCHARIDES.

galactitol (dulcitol) The POLYOL corresponding to galactose. It occurs in certain HONEYDEWS, in two red algae (*Iridaea laminarioides* and *Bostrychia scorpioides*), in certain yeasts and other fungi, and in certain higher plants. Galactitol can be metabolized e.g. by a number of fungi.

galactocarolose An extracellular polysaccharide from *Penicillium charlesii*; it consists of $(1 \rightarrow 5)$-linked D-galactofuranosyl residues.

galactose An aldohexose; both D- and L-forms occur naturally. Galactose is found e.g. in GALACTANS, LACTOSE, MELIBIOSE, RAFFINOSE and STACHYOSE. Many bacteria can metabolize galactose; uptake may involve a PEP-PTS or permease transport system, and subsequent metabolism may involve the tagatose 6-phosphate or Leloir pathways, respectively – see Appendix III(a).

galactose 1-phosphate pathway *Syn.* Leloir pathway: see LACTOSE and Appendix III(a).

β-galactosidase (lactase; β-D-galactoside galactohydrolase; EC 3.2.1.23) An ENZYME which hydrolyses LACTOSE to glucose and galactose. (cf. ALLOLACTOSE.) β-Galactosidases are produced by a range of microorganisms and are obtained commercially mainly from yeasts (e.g. *Kluyveromyces marxianus*) or moulds (e.g. *Aspergillus* spp). The commercial enzyme is used mainly to hydrolyse the lactose in milk and milk products (to make them acceptable to lactose-intolerant consumers, to increase their sweetness, and to avoid crystallization of lactose in frozen or condensed milk products) and in WHEY (to make it more amenable as a microbial substrate). (See also ONPG TEST and XGAL.)

galacturonan A polysaccharide composed of galacturonic acid residues – see PECTINS.

galacturonic acid The URONIC ACID corresponding to galactose. (See also PECTINS.)

Galerina See AGARICALES (Cortinariaceae).

gall See GALLS.

Gallego's solution See BROWN–HOPPS MODIFICATION.

gallid herpesviruses 1 and 2 See GAMMAHERPESVIRINAE.

gallid herpesvirus 3 See HERPESVIRIDAE.

Gallionella A genus of Gram-negative, microaerophilic IRON BACTERIA. The cells are stalked and are coccoid to kidney shaped; they divide by binary fission, the non-stalked daughter cell being flagellated. The stalk consists of a bundle of (apparently inanimate) fibrils. Metabolism is apparently chemolithoautotrophic; energy appears to be obtained by the oxidation of Fe^{2+} (thus accounting for the deposition of ferric hydroxide during growth). (See also EXTRACYTOPLASMIC OXIDATION.) Type species: *G. ferruginea*.

galls (cecidia) (*plant pathol.*) Abnormal plant structures formed e.g. in response to parasitic attack by certain insects or microorganisms. Galls may develop either by localized cell proliferation or increase in cell size. In at least some cases gall-formation involves a PHYTOHORMONE imbalance. Gall-inducing organisms often exhibit considerable host specificity; the host plant is rarely killed.

Bacterium-induced galls include e.g. CROWN GALL, FASCIATION (some instances), HAIRY ROOT, and OLIVE KNOT; galls can also be induced by *Erwinia herbicola*. (See also ROOT NODULES.)

Fungus-induced galls (*mycocecidia*) include e.g. CLUBROOT, WART DISEASE and WITCHES' BROOM; galls are also induced by e.g. *Albugo candida* on certain crucifers, by some smut fungi (see SMUTS sense 2), and by members of the EXOBASIDIALES. (See also 'cedar apples' under GYMNOSPORANGIUM.)

Virus-induced galls include those caused by the RICE GALL DWARF VIRUS and the WOUND TUMOUR VIRUS. (See also ENATIONS and FIJI DISEASE.)

galvanotaxis A TAXIS in which the stimulus is an electrical potential gradient. An organism may move (with its anterior end pointing in the direction of motion as in normal motility) towards the cathode (negative galvanotaxis) or anode (positive galvanotaxis). *Paramecium* is said to be negatively galvanotactic, but with a higher potential difference it may move towards the anode with its (normally) posterior end leading. Some protozoa are positively galvanotactic.

Gambierdiscus See CIGUATERA.

gametangial contact (gametangy) A process in which one or more male nuclei pass from a male gametangium to a female gametangium either (according to species) via a pore which develops at the point of contact of the two gametangia, or via a TRICHOGYNE which grows between the gametangia.

gametangial copulation The occurrence of plasmogamy and karyogamy in a process which may involve either the complete fusion of two gametangia ('gametangial fusion') or the transfer of the contents of one gametangium into the other.

gametangial fusion See GAMETANGIAL COPULATION.

gametangium A structure which gives rise to gametes or which, in its entirety, functions as a gamete.

gametangy *Syn.* GAMETANGIAL CONTACT.

gamete A haploid cell or nucleus involved in sexual reproduction – during which two gametes fuse to form a zygote. (See also ANISOGAMY; ISOGAMY; OOGAMY.)

gametic meiosis MEIOSIS, in a meiocyte other than a zygote, preceding gamete formation in a life cycle in which DIPLOPHASE predominates. (cf. ZYGOTIC MEIOSIS.)

gametocytaemia The presence in the bloodstream of gametocyte stages of a parasite.

gametocyte Any cell which gives rise to gamete(s); a gametocyte may be haploid (see ALTERNATION OF GENERATIONS) or diploid (see GAMETIC MEIOSIS).

gametogony (gamogony) (1) The formation of gametocytes and gametes. (2) The formation of gametes from gametocytes.

gametophyte See ALTERNATION OF GENERATIONS.

gametothallus See ALTERNATION OF GENERATIONS.

***gamma* chain** *(immunol.)* See HEAVY CHAIN.

gamma-delta (γδ) See entry Tn*3*.

***gamma* globulins** A subclass of serum globulins (which includes the IMMUNOGLOBULINS) characterized by low electrophoretic mobility towards the anode – or, under certain conditions, movement towards the cathode (see ELECTROENDOSMOSIS).

gamma interferon See INTERFERONS.

gamma particle (1) See PSEUDOCAEDIBACTER. (2) (*mycol.*) In the zoospores of some members of the Chytridiomycetes (e.g. *Blastocladiella emersonii*): an ovoid particle, one or more of which appear to play a role in cyst wall formation; in *B. emersonii* the gamma particles (each ca. 500 nm diam.) exhibit chitin synthase activity. [Book ref. 175, pp. 544–545.]

***gamma*-rays** (γ-rays) See IONIZING RADIATION.

Gammacell A laboratory apparatus used as a source of *gamma*-radiation. (See also IONIZING RADIATION.)

γδ See entry Tn*3*.

Gammerferon™ See INTERFERONS.

Gammaherpesvirinae ('lymphotropic or lymphoproliferative group') A subfamily of viruses of the HERPESVIRIDAE (q.v.).

The prototype virus of the subfamily is the EPSTEIN–BARR VIRUS (human (gamma) herpesvirus 4; human herpesvirus 4, HHV4; type species of the genus *Lymphocryptovirus* [Intervirol. (1986) *25* 141–143]). Other herpesviruses which have been either accepted or proposed as members of the subfamily include *Herpesvirus saimiri* (main natural host: squirrel monkeys, *Saimiri sciureus*), equine herpesvirus 2 (EHV2); *Herpesvirus ateles* (main natural host: spider monkeys, *Ateles* spp), gallid herpesvirus 1 (MAREK'S DISEASE virus), gallid herpesvirus 2 (turkey herpesvirus) and leporid herpesvirus 1 (rabbit herpesvirus).

These gammaherpesviruses vary in the duration of their replication cycles and in the type of CPEs which they produce. They are generally able to replicate in lymphoblastoid cells, and some can also lyse epithelioid and fibroblastoid cells. In either B or T lymphocytes (according to virus), infection often does not result in the production of progeny virions; the viral genome may persist in the cells with minimal viral gene expression and no cell lysis, or cells may lyse without formation of complete virions. Latent infection in vivo appears to occur commonly in lymphoid cells.

Gammaherpesviruses are associated with the formation of tumours either in their natural hosts (see e.g. EPSTEIN–BARR VIRUS and MAREK'S DISEASE) or in experimental animals; for example, neither *Herpesvirus saimiri* nor *Herpesvirus ateles* are pathogenic in their main natural hosts but are highly oncogenic in marmosets (*Saguinus* spp) and in other New World primates, causing malignant tumours of the lymphatic system [review: Book ref. 139, pp 253–332].

The most recent addition to the subfamily is human (gamma) herpesvirus 8 (human herpesvirus 8, HHV8). The genome is reported to be ~170 kb but to vary in size in different isolates (owing to duplication events); it includes sequences which resemble those of various human regulatory genes (e.g. those encoding a CYCLIN and a CYTOKINE). HHV8 has a tropism for B lymphocytes (and also infects endothelial cells) and it has been associated with B-cell lymphomas and with KAPOSI'S SARCOMA.

[HHV8 (review): Lancet (1997) *349* 558–563. PCR-based study on specimens of biopsy-proven Kaposi's sarcoma: Am. J. Path. (1997) *150* 147–153.]

Activation of latent HHV8 in vitro has been achieved by demethylation of the promoter region of a transactivator [PNAS (2001) *98* 4119–4124].

gamogony *Syn.* GAMETOGONY.

gamone A name used for various types of sex PHEROMONE. For example, the two mating types of *Blepharisma japonicum* (types I and II) form, respectively, gamone 1 (= blepharmone; a glycoprotein) and gamone 2 (= blepharismone or blepharismin: calcium 3-(2′-formylamino-5′-hydroxybenzoyl)-lactate) [Book ref. 28, pp. 339–340].

gamont (1) *Syn.* GAMETOCYTE. (2) See ALTERNATION OF GENERATIONS.

ganciclovir A nucleoside analogue used as an ANTIVIRAL AGENT; it is activated by phosphorylation, intracellularly, and inhibits viral DNA synthesis. Ganciclovir is used mainly for the treatment of cytomegalovirus infections; it has a number of side-effects and is contraindicated in pregnancy.

gangliosides See SPHINGOLIPID.

gangrene (1) (*med., vet.*) Death of tissues or organs in the body owing to a deficiency in or loss of local blood supply. *Dry* *gangrene* results from a gradual reduction in local blood supply, with darkening, drying and shrivelling of the tissues. *Moist gangrene* is sudden in onset, resulting e.g. from injury followed by bacterial infection and putrefaction; it may spread very rapidly. (cf. GAS GANGRENE; see also e.g. ERGOTISM.)

(2) (*plant pathol.*) A disease of stored potato tubers, caused by *Phoma exigua* (e.g. var. *foveata* in the UK), in which the tubers become decayed and hollow. Surface contamination of the tubers occurs during harvesting, an important source of the pathogen being pycnidia from senescent potato stems; infection occurs via wounds made during or after harvesting. The disease may be controlled with e.g. benomyl or thiabendazole. (See also POTATO DISEASES.)

gangrenous coryza *Syn.* MALIGNANT CATARRHAL FEVER.

Ganoderma See APHYLLOPHORALES (Ganodermataceae).

Ganodermataceae See APHYLLOPHORALES.

gap (in dsDNA) See NICK.

GAR-936 See TETRACYCLINES.

Gardner–Arnstein feline sarcoma virus See FES.

Gardnerella A genus (*incertae sedis*) of catalase-negative, oxidase-negative, chemoorganotrophic, apparently Gram type-negative bacteria which occur in the human genital and urinary tracts and which are commonly associated with BACTERIAL VAGINOSIS. Cells: pleomorphic rods, ca. 0.5 × 1.5–2.5 μm. The genus includes both obligately anaerobic and facultatively anaerobic strains. The organisms are nutritionally fastidious: little or no growth occurs on nutrient agar. Neither X FACTOR nor V FACTOR is required. Growth occurs e.g. on certain blood-containing media and on peptone–starch–D-glucose medium; most strains can grow aerobically on blood agar at 35°C. Clear ('β') haemolysis occurs on human (but not sheep) blood plates. Most strains ferment maltose (and some ferment lactose) forming acid but not gas. Nitrate is not reduced. Optimum growth temperature: 35–37°C. GC%: ca. 42–44. Type species: *G. vaginalis* (formerly *Haemophilus vaginalis*). [Book ref. 22, pp. 587–591.]

Gardneriella See RHODOPHYTA.

gari A West African food made by the fermentation of cassava pulp. The essential organisms appear to be *Leuconostoc* and *Candida* spp.

gas bladder *(algol.)* *Syn.* PNEUMATOCYST.

gas gangrene A type of acute, rapidly spreading GANGRENE (sense 1) which typically results from the infection of a wound with certain anaerobic bacteria, particularly *Clostridium* spp (e.g. *C. perfringens* type A, *C. septicum*, *C. novyi*); these organisms produce a variety of extracellular enzymes (see CLOSTRIDIUM) which promote tissue destruction. The α-toxin of *Clostridium perfringens* (which hydrolyses lecithin and also has sphingomyelinase activity) appears to be the major lethal toxin [Mol. Microbiol (1995) *15* 191–202]. Gas gangrene may occur in e.g. subcutaneous tissues (clostridial CELLULITIS), muscle (clostridial myonecrosis), or internal organs. Infected tissues become necrotic, discoloured, and swollen with serosanguinous fluid and pockets of gas (hydrogen, carbon dioxide) produced by the clostridia; the gas exerts pressure on adjacent tissues, further disrupting the blood supply to these tissues and promoting spread of the gangrene. *Treatment*: e.g. surgery; antibiotic therapy; treatment with hyperbaric oxygen. (Other organisms occasionally implicated in gas gangrene include anaerobic streptococci and *Escherichia coli*; mixed infections also occur.)

gas–liquid chromatography See CHROMATOGRAPHY.

gas vacuole A subcellular organelle, found only in prokaryotes, which consists of clusters of hollow, cylindrical, gas-filled vesicles (*gas vesicles*). Gas vacuoles appear bright and refractile under the light microscope. They occur in planktonic cyanobacteria, particularly bloom-forming species (e.g. *Anabaena flos-aquae*, *Trichodesmium*), and in certain archaeans and bacteria (e.g. strains of *Halobacterium*, *Methanosarcina*, *Pelodictyon* and *Rhodopseudomonas*).

In cyanobacteria, gas vacuole formation is usually constitutive, but it may be inducible in hormogonia (see HORMOGONIUM).

The gas vesicles of cyanobacteria are hollow cylinders with conical ends; they are ca. 60–250 nm in diameter and of variable length (e.g. 300 nm); they have a ribbed construction. Vesicles from other prokaryotes (including *Methanosarcina*) are generally similar in both morphology and construction (although they may differ in size), but those of *Halobacterium* are usually spindle-shaped.

Vesicles are composed of a rigid monolayer of a single type of protein (*gas vesicle protein* – GVP); they are permeable to gases but not to water. GVPs from e.g. cyanobacteria and halobacteria show some homology [JGM (1984) *130* 2709–2715].

The primary function of a gas vacuole is to give buoyancy to the (free-floating) organism. Gas vesicles collapse when subjected to pressure (ca. 5 atmospheres for many freshwater cyanobacteria, ca. 50 atmospheres for *Trichodesmium*). The vesicles in some freshwater species can be collapsed by high turgor pressures within the cells, thus providing a mechanism whereby these cells can regulate their buoyancy and, hence, their position in the water: an increase in light intensity causes an increase in turgor pressure in the cell (due to light-stimulated accumulation of potassium ions and the accumulation of photosynthate), resulting in the collapse of part of the gas vacuole and hence loss of buoyancy. New gas vesicles form when the light intensity is low, thus increasing buoyancy. (There is some evidence that the GVP of collapsed vesicles disaggregates and may be re-used in the assembly of new vesicles [JGM (1984) *130* 1591–1596].)

Another mechanism for the regulation of buoyancy occurs e.g. in *Oscillatoria agardhii* and in a strain of *Microcystis* [JGM (1985) *131* 799–809]: when the intensity of light is high, gas vesicles do not collapse – but existing vesicles are 'diluted out' by cell growth and division.

gas vesicle (1) See GAS VACUOLE. (2) *Syn.* PNEUMATOCYST.

gasohol A mixture of gasoline and (up to ca 20% v/v) ethanol which is used as a fuel. (See also INDUSTRIAL ALCOHOL.)

GasPak An ANAEROBIC JAR which consists of a strong, cylindrical, transparent plastic (polycarbonate) chamber with a flat, gas-tight lid which is secured by means of a screw clamp. The jar is loaded (with e.g. plates), and an indicator of anaerobiosis (e.g. a pad soaked with a METHYLENE BLUE solution) is placed inside the jar. Water is then added to a small foil sachet containing (i) sodium borohydride and (ii) a mixture of powdered sodium bicarbonate and citric acid – which (with water) yield H_2 and CO_2, respectively; the sachet is quickly placed in the jar in an upright position and the lid is clamped in place. Attached to the inside of the lid is a 'cold' catalyst (see MCINTOSH AND FILDES' ANAEROBIC JAR) which promotes the elimination of O_2. Anaerobiosis within the jar develops within a few hours. Since this type of jar does not require evacuation, plates can be incubated upside down (i.e. lid-side down); this avoids the problem (which can arise with the McIntosh and Fildes' jar) in which water of condensation drops from the inside of the Petri dish lid onto the agar surface.

gasteroid Refers to the type of fruiting body which is characteristic of fungi of the GASTEROMYCETES: angiocarpic, forming STATISMOSPORES. (cf. AGARICOID.)

gasteroid basidiospore A term proposed for a BASIDIOSPORE which is *not* forcibly discharged from the basidium [BBMS (1983) *17* 82–94 (86)]. (cf. STATISMOSPORE.)

Gasteromycetes A class of fungi (subdivision BASIDIOMYCOTINA) characterized by the formation of epigean or hypogean, sessile or stipitate basidiocarps which undergo ANGIOCARPIC DEVELOPMENT and give rise to STATISMOSPORES. The typical basidiocarp consists of e.g. fleshy or cartilaginous fertile tissue (the GLEBA) enclosed by a limiting wall (PERIDIUM); in some species the basidiocarp contains a definite hymenium, while in others the fertile cells are scattered, singly or in groups, throughout the spore-bearing tissue. (See also CAPILLITIUM.) The dispersal of mature basidiospores depends e.g. on the disintegration or perforation of the peridium. Gasteromycetes occur typically as saprotrophs on soil, dead wood and dung; some species form mycorrhizal associations.

The orders [Book ref. 64, p. 160] are:

Gautieriales. Hymenium-forming, hypogean saprotrophs or mycorrhizal fungi.

Hymenogastrales. Hymenium-forming, typically hypogean saprotrophs, mycorrhizal fungi and parasites. Some authors include gasteroid members of the Russulales in this order. Genera include e.g. *Hymenogaster* and *Rhizopogon* (which forms mycorrhizal associations with conifers).

LYCOPERDALES (q.v.).

Melanogastrales. Hypogean or marine; mature gleba mucilaginous, lacking a true hymenium. Genera include *Melanogaster* and NIA.

NIDULARIALES (q.v.).

PHALLALES (q.v.).

Podaxales. An order which includes some of the SECOTIOID FUNGI.

SCLERODERMATALES (q.v.).

Tulostomatales. An order which includes the stipitate puffballs (e.g. *Battarrea*, *Calostoma*).

gastric flu A term sometimes used to refer to the syndrome caused by infection with *Influenzavirus* type B – in which some studies have recorded involvement of the gastrointestinal tract.

gastriole *Syn.* FOOD VACUOLE.

gastroenteritis Inflammation of the lining of the stomach and intestine (usually the small intestine), usually resulting in DIARRHOEA and vomiting. (cf. ENTERITIS.) It may be due e.g. to allergic reactions, chemical poisoning (e.g. by excessive alcohol consumption), or to microbial infection or toxins (see e.g. FOOD POISONING).

gastrointestinal tract flora The mammalian gastrointestinal tract is normally sterile at birth, but thereafter becomes rapidly colonized by a range of bacteria. The gut flora of the human infant initially includes *Lactobacillus* and *Bifidobacterium* spp, coliforms and clostridia, but the flora depends to some extent on diet: the gut of breast-fed infants contains a higher proportion of bifidobacteria and lactobacilli – and fewer coliforms – than that of infants bottle-fed on cows' milk [Book ref. 107, pp 1–26]. Acid production by the lactic acid bacteria is believed to discourage the growth of enterobacteria, including enteric pathogens. On weaning, the gut flora becomes more complex and soon resembles that of the adult.

In adult humans, many of the bacteria ingested with food (and oral bacteria swallowed with saliva) are killed by the acidity in the stomach. However, despite the normal acidity of the stomach, it appears that this region of the gut is colonized by *Helicobacter pylori* (see HELICOBACTER) in a high proportion of the human population. (A greater number and variety of organisms may be present in the stomach in cases of achlorhydria.)

Bacterial numbers (in colony-forming units/gram) increase from about 10^3 in the duodenum to about 10^{11} in the colon. Anaerobes such as *Bacteroides* and *Fusobacterium* generally predominate but many other organisms are usually present–e.g. species of *Clostridium* and *Peptostreptococcus*, enterobacteria, various yeasts (e.g. *Blastocystis hominis, Candida* spp), protozoa (e.g. *Endolimax nana*, *Entamoeba* spp, *Trichomonas hominis*) and viruses such as enteroviruses.

The numbers and nature of the gut microorganisms may be affected by e.g. the nature of the diet, malnutrition, local immune responses, antibiotic therapy etc. Under certain conditions – e.g. abnormal peristalsis, trauma, diminished gut secretions – abnormally high numbers of bacteria may develop in the small intestine; symptoms of such 'bacterial overgrowth' include steatorrhoea, malabsorption of carbohydrates, and symptoms of iron and vitamin B_{12} deficiencies. TRANSLOCATION (sense 2) of gut bacteria may occur normally but may sometimes lead to extraintestinal disease; however, the importance of translocation in pathogenesis has been questioned [Critical Care Medicine (2003) *31* 598–607].

The microflora has various roles: (i) production of short-chain fatty acids – e.g. butyric acid (a nutrient for epithelial cells); (ii) recycling of BILE ACIDS. (iii) production of vitamin K (see QUINONES); (iv) stimulation of the gut immune system – helping to maintain a normal immune response; (v) inhibition of pathogens by competing for nutrients and space, and by producing antimicrobial agents (e.g. BACTERIOCINS) [Lancet (2003) *360* 512–519]. (The gut epithelium also secretes antimicrobial peptides (e.g. *defensins*) [Nature (2002) *415* 389–395, (2003) *422* 478–479].)

However, a possible role for gut bacteria in the aetiology of carcinoma of the colon has been postulated: certain substances (particularly bile acids) in the gut may be metabolized by gut bacteria to potential carcinogens or co-carcinogens [Book ref. 107, pp 347–363]; the bacteria implicated include the so-called 'nuclear dehydrogenating clostridia' (NDC) which can desaturate the A-ring in various bile acid derivatives to form 4-ene-3-one structures [JMM (1985) *20* 233–238]. (See also LIGNANS.) Certain gut parasites (e.g. *Entamoeba histolytica*) appear to require the presence of gut bacteria (as food) in order to establish infection.

GATC Dam methylation site: see DAM GENE.

gatifloxacin A QUINOLONE ANTIBIOTIC (q.v.).

gating The switching on or off of a function or facility as a consequence of the value of a given parameter reaching a critical level. Thus, e.g., certain transport systems and membrane PROTON ATPASE activity are inhibited when proton motive force falls below a certain level.

Gäumannomyces *Syn.* GAEUMANNOMYCES.

Gautieriales See GASTEROMYCETES.

Gazdar murine sarcoma virus (Gz-MSV) A MURINE SARCOMA VIRUS originally isolated from a spontaneous tumour in a (NZB × NZW)F1 mouse. Gz-MSV resembles MOLONEY MURINE SARCOMA VIRUS (Mo-MSV) in containing V-*mos*, but the *mos* sequence is inserted at a different location in the genome. The pathologies of Gz-MSV and Mo-MSV infections are similar.

Gb_3 See SHIGA TOXIN and STARFISH.

GC% (GC value; %GC value; %G + C value; mol% G + C; etc) Of (usually chromosomal) DNA: the ratio $(G + C)/(A + T + G + C)$ in which G, C, A and T represent the relative molar amounts of guanine, cytosine, adenine and thymine, respectively, in a given organism; the ratio is expressed as a percentage. (cf. BASE RATIO.) GC% values vary according to genus and species, and provide a useful criterion in microbial TAXONOMY; however, a similarity in GC% values, by itself, does not necessarily indicate a close taxonomic relationship. The possession of a PLASMID (particularly one with a high copy number) may affect the observed GC% value of a cell unless the method used for extracting chromosomal DNA discriminates against extrachromosomal nucleic acids.

The GC% of a DNA sample may be obtained, indirectly, by determining its 'melting point' (T_m: see THERMAL MELTING PROFILE), since T_m is linearly related to GC% (i.e., a high T_m correlates with a high GC%). Alternatively, an indirect determination of GC% may be achieved by buoyant density (Bd) measurement (e.g. in a CsCl density gradient – see CENTRIFUGATION) since Bd is also linearly related to GC%.

GC% values in bacteria fall within the approximate range 25–75; as a guideline, intra-species variation should be not more than 4–5% and intra-genus variation not more than 10% [Book ref. 59, p. 14].

[Methods for determining DNA base composition: Book ref. 138, pp. 1–31.]

GC type See BASE RATIO.

GDP Guanosine 5′-diphosphate.

Geastrum See LYCOPERDALES.

gel diffusion In serology: gel diffusion (= immunodiffusion) refers to any procedure in which antibodies (usually in an antiserum) and/or antigens diffuse through a gel medium (e.g. AGAR or agarose) – forming a precipitate in the gel where homologous (or cross-reacting) antigens and antibodies meet in OPTIMAL PROPORTIONS. (Agar/agarose at concentrations of ca. 0.5% to 1.5% allows the diffusion of antibody-sized molecules but hinders random mechanical and thermal movements in the medium.) Lines, bands or haloes of whitish precipitate are obtained (within hours or a day or two) in number(s) and position(s) which depend on the number and nature of the homologous (or cross-reacting) pairs of antigens and antibodies within the diffusing system. The precipitate may or may not be visible to the naked eye; visibility may be enhanced by washing the gel free of uncombined antigen and antibody, and staining the precipitate in situ with a protein stain (e.g. COOMASSIE BRILLIANT BLUE).

Gel diffusion may involve SINGLE DIFFUSION or DOUBLE DIFFUSION of the single- or double-DIMENSION type. (See also IMMUNOELECTROPHORESIS and ELEK PLATE.)

gel electrophoresis See ELECTROPHORESIS.

gel filtration (molecular sieving) A method for separating molecules from a mixture by passing the mixture through a column of e.g. Sephadex, polyacrylamide, or agarose gel; molecules are separated primarily according to their size and shape.

gelatin The product obtained by boiling COLLAGEN. Gelatin is soluble in water at temperatures above ca. 40°C, and a solution of concentration >ca. 1–2% (w/v) forms a gel on cooling below ca. 28–35°C; gels of ca. 4–12% are used e.g. to test the ability of certain microorganisms to liquefy (i.e., hydrolyse) gelatin (see GELATINASE). (See also FRAZIER'S MEDIUM.)

gelatinase A GELATIN-hydrolysing enzyme. Gelatin can be hydrolysed (to soluble oligopeptides) by clostridial COLLAGENASES and by any of a range of extracellular proteolytic enzymes produced e.g. by various bacteria and fungi. The activity of many gelatinases requires, or is stimulated by, calcium ions.

Gelbstoff Collectively, the various dissolved and colloidal organic substances which occur e.g. in lakes and coastal waters and which may absorb significant amounts of light – particularly at the blue end of the spectrum.

Geleia See KARYORELICTID GYMNOSTOMES.

Gelidium See RHODOPHYTA.

gellan gum An extracellular polysaccharide produced by '*Pseudomonas elodea*'. Gellan gum is a linear polymer containing residues of glucose, rhamnose and glucuronic acid which may be mainly (1 → 4)-β-linked; the polymer may be *O*-acetylated. In water containing monovalent or divalent cations, gellan gums form gels on heating and cooling; the nature of the gel depends on the degree of acetylation – the more de-acetylated the polymer the firmer and more brittle the gel.

A clarified, de-acetylated form of gellan gum (*Gelrite*; Keloc, San Diego, USA) can be used as a substitute for AGAR in bacteriological media; the setting temperature varies from 35°C to >50°C, depending on the concentrations of Gelrite and cations. The gels can be re-melted under standard autoclaving conditions. Gelrite gels containing 0.1% calcium chloride are not prone to SYNERESIS at high temperatures (unlike agar gels) and may thus be suitable for use in media for thermophiles. Gelrite may also be used e.g. as a substitute for agarose in gel electrophoresis (Gelrite gels are clearer than agar gels).

Gelrite has been investigated for use as a gelling agent, in place of agar, in transport media used for PCR-based studies. This investigation was prompted by the finding that agar inhibits PCR in a concentration-dependent manner [JCM (1998) *36* 275–276], an observation suggesting that detection of pathogens by PCR can be inhibited by the use of an agar-containing transport medium (such as STUART'S TRANSPORT MEDIUM). These studies yielded superior results from gellan gum, inhibition of PCR occurring with agar [JMM (2001) *50* 108–109].

In industry, the polymer may be useful e.g. as a substitute for pectins or gelatin in jams, jellies etc. or as a matrix for the IMMOBILIZATION of cells or enzymes. [Book ref. 62, pp 231–253.]

Gelman filter See FILTRATION.

Gelrite See GELLAN GUM.

gelsolin See ACTIN.

Gemella A genus of Gram-type-positive bacteria of the family Streptococcaceae; the organisms occur e.g. in the mammalian respiratory tract. Cells: cocci (diam. <1 μm), which occur singly or in pairs. Colonies on rabbit- or horse-blood agar are each surrounded by a zone of clear haemolysis. GC%: ca. 33. Type species: *G. haemolysans*.

gemifloxacin A QUINOLONE ANTIBIOTIC (q.v.).

geminiviruses A group of small, ssDNA-containing PLANT VIRUSES in which the virions have a unique morphology: each consists of a pair of isometric particles (incomplete icosahedra, each ca. 18 × 20 nm), composed of a single type of protein (MWt ca. $2.7–3.4 \times 10^4$). The group includes: MAIZE STREAK VIRUS (MSV; type member), BEAN GOLDEN MOSAIC VIRUS (BGMV), cassava latent virus (CLV), *Chloris* striate mosaic virus (CSMV), tomato golden mosaic virus (TGMV; possibly = tomato yellow mosaic virus). Probable members include: beet curly top virus (BCTV), *Euphorbia* mosaic virus, mungbean yellow mosaic virus, *Paspalum* striate mosaic virus, tobacco leafcurl (= tomato yellow dwarf) virus, tobacco yellow dwarf virus (TYDV; = bean summer death virus), tomato yellow leafcurl virus, wheat dwarf virus. Each virus has a narrow host range; some cause economically important diseases in cultivated plants, symptoms including e.g. curling and distortion of the leaves (e.g. BCTV) or chlorosis (e.g. BGMV). Depending on virus, transmission occurs via leafhoppers or whitefly – in which the viruses are persistent (circulative).

Each geminivirus virion contains one molecule of circular, positive-sense ssDNA, MWt ca. $7–8 \times 10^5$ (<2.6 kb). In some (whitefly-transmitted) members (e.g., CLV, BGMV, TGMV) the genome appears to be bipartite, i.e., two ssDNA molecules – similar in size but differing in nucleotide sequence – are present in a given population of virions [nucleotide sequence of CLV genome: Nature (1983) *301* 260–262, and of BGMV: PNAS (1985) *82* 3572–3576]; however, the (leafhopper-transmitted) members CSMV and (possibly) TYDV seem to have only one type of ssDNA [Book ref. 80, pp. 59–62]. Virus replication occurs in the plant cell nucleus, where large aggregates of virus particles accumulate. There is evidence (at least in the case of BGMV and TGMV) that replication may involve a supercoiled ccc dsDNA intermediate (which may also act as a template for transcription), and that a rolling circle mechanism may be involved [Book ref. 80, pp. 49–58]. [Model for gene regulation in geminiviruses: PNAS (1985) *82* 3572–3576.]

[Review: ARPpath. (1985) *23* 55–82; AVR (1985) *30* 139–177.]

gemma A thick-walled, irregularly-shaped or spheroidal, asexually derived spore formed by certain fungi (e.g. *Saprolegnia* spp) from a portion of a somatic hypha; gemmae may be formed singly or in chains.

Gemmata obscuriglobus A species of freshwater BUDDING BACTERIA; the life cycle includes a multitrichous swarmer stage. GC%: ca. 64. [AvL (1984) *50* 261–268.]

Gemmiger A genus of Gram-variable, anaerobic, budding bacteria which can ferment carbohydrates to butyric acid. Cells: cocci, occurring in pairs or chains. GC%: ca. 59. [*G. formicilis* in the avian caecum: JAB (1983) *54* 7–22.]

gen. nov. See GENUS NOVUM.

gene A sequence of nucleotides in a genetic nucleic acid (chromosome, plasmid etc) which encodes a functional polypeptide chain or RNA molecule (rRNA, tRNA, etc). In certain cases a gene may share a sequence with another gene (see OVERLAPPING GENES); some genes contain non-coding sequences (see e.g. MRNA and SPLIT GENE). (See also ALLELE; CISTRON; GENETIC CODE; PSEUDOGENE.)

gene amplification The generation of multiple copies of one or more genes in a cell or nucleus – e.g. as an arrangement of tandemly repeating genes within a chromosome, or in the form of extrachromosomal elements (see e.g. RRNA). (See also GENE DOSAGE.) Amplification occurs in both eukaryotic and prokaryotic cells. For example, in some mammalian cells in culture (and in bacteria) resistance to certain drugs involves the amplification of genes specifying enzymes which either detoxify or bypass the effects of the drug. [Gene amplification in eukaryotes: ARB (1984) *53* 447–491.]

In bacteria, amplification occurs spontaneously at frequencies which depend e.g. on the location of the amplified sequences, and is generally readily reversible. Amplification of particular genes may help the organism to adapt to particular environmental conditions: e.g., a substrate normally used inefficiently by a given strain may be used more effectively following amplification of genes specifying the appropriate catabolic enzymes. Amplified regions are generally found to be flanked by directly repeated sequences, and amplification commonly appears to involve RecA-dependent homologous RECOMBINATION (e.g. between daughter molecules). [Possible mechanism: CSH-SQB (1984) *49* 443–451.]

gene bank *Syn.* LIBRARY.

gene cloning See CLONING.

gene cluster A cluster of functionally related genes, each of which encodes a separate protein. (cf. CLUSTER GENE.)

gene conversion See RECOMBINATION.

gene dosage The number of copies of a given gene per cell or nucleus. Increased gene dosage may cause higher levels of gene product if the gene is not autogenously regulated.

gene-for-gene concept (*plant. pathol.*) See AVIRULENCE GENE.

gene fusion In genetic engineering: the fusion of two DNA sequences from different genes to form a 'hybrid' gene (see FUSION PROTEIN, sense 2). The term is sometimes used synonymously with OPERON FUSION. (See also MUDLAC SYSTEM.)

[Gene fusions (commentary): JB (2000) *182* 5935–5938.]

gene gun A device for inoculating a DNA VACCINE directly into the epidermis; it bombards the surface with gold particles (~ 1 μm) that are coated with the target DNA. [Example of use: Inf. Immun. (2005) *73* 2974–2985.]

gene library See LIBRARY.

gene therapy Genetic modification of certain cells specifically to correct an inherited disorder or to treat some infectious diseases; it has been successful e.g. with adenosine deaminase deficiency but has apparently caused leukaemia-like illness in one patient [Nature (2002) *420* 116–118].

In the *ex vivo* mode, cells from the patient are modified *in vitro* and then re-introduced. For example, in ADENOSINE DEAMINASE DEFICIENCY the patient's *stem cells* are supplemented with the normal, wild-type gene before re-introduction; lymphocytes – derived from the stem cells – can then develop normally. (In some cases stem cells from a normal (histocompatible) donor are introduced into the patient.) For selecting stem cells the CD34 marker can be used in immunomagnetic enrichment techniques (see DYNABEADS). In the *in vivo* mode, modification takes place *within* the patient.

Viruses used as *vectors* (for carrying genetic material into cells) include adenoviruses (see ADENOVIRIDAE), adeno-associated viruses (see DEPENDOVIRUS) and certain retroviruses (RETROVIRIDAE), e.g. lentiviruses. Viral vectors are genetically modified to be replication-deficient and non-pathogenic; the gene (etc.) to be introduced into cells is inserted into the viral genome and enters cells during infection.

Sequences/transgenes can be inserted into various types of dividing or non-dividing cell e.g. by AAVs (small sequences), adenoviruses (larger sequences) and lentiviruses.

With some viruses the broad range of host cells is problematic when *specific* type(s) of cell are to be targeted. In such cases a virus can be modified by PSEUDOTYPING – e.g. adenovirus type 5 was modified to produce a more specifically targeted vector [JV (2001) *75* 2972–2981].

In CYSTIC FIBROSIS, respiratory epithelial cells have been targeted (via an inhaler) by LIPOSOMES containing the normal gene. [Bioch. J. (2005) *387* 1–15.]

Gene therapy against infectious diseases

For this type of disease a wider range of modalities may be employed. For example, RIBOZYMES can be targeted to specific pathogen-associated RNA; useful activity was reported for the anti-*pol* ribozyme in infection by simian immunodeficiency virus (SIV) [JGV (2004) *85* 1489–1496].

Intrabodies (= single-chain antibodies) are engineered antibodies that remain intracellular [see NAR (2003) *31* e23].

DNA VACCINES are being developed against e.g. *Mycobacterium tuberculosis* [Inf. Immun. (2005) *73* 5666–5674] and *Leishmania donovani* [Inf. Immun. (2005) *73* 812–819].

RNA INTERFERENCE (RNAi) has shown promise *in vitro* but pathogens may escape RNAi by mutation [e.g. JV (2004) *78* 2601–2605; Blood (2005) *106* 818–826].

Transdominant negative proteins (TNPs) – encoded e.g. in the genome of a viral vector – can be expressed in infected or pre-infected cells; TNPs may be engineered anti-function versions of particular pathogen-associated proteins which are necessary e.g. for the completion of a given phase of replication. A TNP may be inhibitory e.g. by forming an inactive complex with its wild-type counterpart or by competing with it for specific binding site(s).

One anti-AIDS concept is to modify $CD34^+$ bone marrow stem cells or $CD4^+$ T cells so that they inhibit HIV replication. In one *in vitro* study the TNP consisted of a modified form of the Rev protein (see HIV) whose functions include facilitating export of viral transcripts from the nucleus; HIV-1 replication was strongly inhibited [JV (2001) *75* 3590–3599].

(See also TRIPLEX DNA.)

gene transfer agent See CAPSDUCTION.

GeneAmp® *In situ* PCR System 1000 See IN SITU PCR.

genera Plural of *genus* (see TAXONOMY).

general export pathway See PROTEIN SECRETION.

general recombination See RECOMBINATION.

general secretory pathway See PROTEIN SECRETION.

generalized transduction See TRANSDUCTION.

generation time (of bacteria) *Syn.* doubling time – see GROWTH (b).

generative hyphae See HYPHA.

generic Of a genus: e.g. *generic name* (see BINOMIAL).

genetic code The 'code' in which information for the synthesis of proteins is contained in the nucleotide sequence of a DNA molecule (or, in certain viruses, of an RNA molecule). During PROTEIN SYNTHESIS, the coded information in DNA is initially transmitted to mRNA (q.v.), i.e., the deoxyribonucleotide sequence is transcribed into a complementary sequence of ribonucleotides. (In RNA viruses the viral RNA may itself function as mRNA, or may need to be transcribed into complementary RNA: see VIRUS.) The mRNA is then translated into one or more polypeptides, each amino acid of a polypeptide being encoded by a particular sequence of three nucleotides (called a *codon*) in the mRNA; an mRNA molecule contains a series of non-overlapping nucleotide triplets (codons) which, together, specify the amino acid sequence of a polypeptide. Polypeptide synthesis begins at a specific *initiator codon* in the mRNA (see PROTEIN SYNTHESIS), and the translating machinery reads the triplets sequentially from this point; thus, although (theoretically) triplets in a sequence of nucleotides could be read in three different ways (*reading frames*) – e.g. AUC-AUC-AUC-AUC *or* A-UCA-UCA-UCA-UC *or* AU-CAU-CAU-CAU-C – in practice the correct reading frame is determined by the fixed starting point. (cf. FRAMESHIFT MUTATION.) (N.B. The term *codon* is also applied to the corresponding triplet in the coding or non-coding strand of DNA.)

A codon does not interact directly with its corresponding amino acid; the amino acid must be linked to an adaptor molecule, tRNA (q.v.), which contains a triplet of nucleotides (the *anticodon*) complementary to the codon for that amino acid. For example, a codon ACG is recognized by a tRNA containing the anticodon CGU. (Codons and anticodons, like other nucleotide sequences, are conventionally written in the 5′-to-3′ direction: hence, the *first* base in a codon pairs with the *third* base in the anticodon.)

The four bases (A, C, G and U) in mRNA can generate 64 possible triplet combinations; 61 of these encode 20 amino acids: i.e., in most cases an amino acid is encoded by two or more (*synonymous*) codons, a phenomenon called *degeneracy* (see table). The remaining three codons – UAA (*ochre codon*), UAG (*amber codon*) and UGA (*opal* or *umber codon*) – are *nonsense codons* ('stop codons', termination codons) which specify the termination of polypeptide chain synthesis. (cf READTHROUGH.)

GENETIC CODE: the 'universal' code

First base (5′ end)	Second base U	C	A	G	Third base (3′ end)
U	Phe	Ser	Tyr	Cys	U
	Phe	Ser	Tyr	Cys	C
	Leu	Ser	*ochre*	*opal*	A
	Leu	Ser	*amber*	Trp	G
C	Leu	Pro	His	Arg	U
	Leu	Pro	His	Arg	C
	Leu	Pro	Gln	Arg	A
	Leu	Pro	Gln	Arg	G
A	Ile	Thr	Asn	Ser	U
	Ile	Thr	Asn	Ser	C
	Ile	Thr	Lys	Arg	A
	Met	Thr	Lys	Arg	G
G	Val	Ala	Asp	Gly	U
	Val	Ala	Asp	Gly	C
	Val	Ala	Glu	Gly	A
	Val	Ala	Glu	Gly	G

The three bases of a codon do not contribute equally to the specificity of the codon: the 2nd base contributes maximally, the 3rd minimally; thus, certain codons which specify the same amino acid differ from each other only in the 3rd base (see table). (See also WOBBLE HYPOTHESIS.) There is some relationship between codons which specify related amino acids: e.g. all codons in which U is the 2nd base specify hydrophobic amino acids. [Hypothesis on the physicochemical rationale of the genetic code: FEBS (1985) *189* 159–162.]

The frequency with which synonymous codons occur in mRNAs is believed to parallel the population sizes of the corresponding tRNAs in the cell, i.e., the most commonly used codons correspond to the most abundant of the isoaccepting tRNA molecules. The codons in a given gene may thus affect the rate of translation of that gene; e.g., it has been suggested that the presence of rare codons in sequences encoding signal peptides (see SIGNAL HYPOTHESIS) may serve – or may have served at some evolutionary stage – to delay translation, allowing time for the ribosome–polypeptide complex to become associated with the cell membrane [FEBS (1985) *189* 318–324]. (See also CODON BIAS.)

The genetic code as shown in the table is often said to be 'universal', i.e., applicable to all living systems. However, there are exceptions. For example in at least some plant, animal and fungal mitochondria – and apparently also in *Mycoplasma capricolum* [PNAS (1985) *82* 2306–2309] – UGA functions not as a stop codon but as a sense codon specifying tryptophan; in mammalian mitochondria AUA specifies methionine, while AGA and AGG function as stop codons. In the ciliate MACRONUCLEUS, UAA and UAG specify glutamine, UGA apparently acting as the sole stop codon. It is believed that these distinct codes must have evolved from a common, more primitive code rather than one from another. [Evolutionary aspects: MR (1992) *56* 229–264.]

genetic drift Changes in the genotype of a species or strain which are due not to selection pressures but to the persistence of neutral mutations, i.e., mutations which appear to be neither advantageous nor disadvantageous to the organism. (cf. ANTIGENIC DRIFT.)

genetic engineering The in vitro manipulation of nucleic acid molecules e.g. to generate new combinations of genes or sequences (see e.g. FUSION PROTEIN), to place a given gene or genes under the control of a different regulatory system (see e.g. OPERON FUSION), to introduce specific mutations into a molecule (see SITE-SPECIFIC MUTAGENESIS), etc. The 'engineered' DNA may be produced in quantity by any of various CLONING techniques. (See also BIOTECHNOLOGY.)

genetic footprinting A procedure which may be used to determine the need for expression of each of a range of specific genes during growth under particular, chosen conditions. (cf. FOOTPRINTING.) The procedure involves three stages. Initially, a population of cells is subjected to transposon mutagenesis, transposons inserting randomly and extensively in the genome; from this mutagenized population, a sample (labelled T0, for time zero) is taken and set aside for subsequent analysis. The remainder of the mutagenized cells are then grown for many population doublings under conditions that repress further transposition. Finally, genomic DNA is isolated from (i) cells of the T0 population and (ii) cells of the cultured mutagenized population, and PCR is used to amplify sequences from specific genes from each of these two groups of cells. If a given gene is necessary for viability under particular growth conditions, then transposon insertions may be detected in that gene from the T0 population, but fewer (or none) will be found in that gene from cultured populations of the cells. [Genetic footprinting in bacteria (*Escherichia coli*): JB (2001) *183* 1694–1706.]

(See also IVET.)

genetic immunization The use of a DNA VACCINE. (See also FILOVIRIDAE.)

genetic reactivation See REACTIVATION.

genetic vaccination The use of a DNA VACCINE. (See ALSO FILOVIRIDAE.)

genital herpes See HERPES SIMPLEX.

genital wart See PAPILLOMA.

genitourinary tract flora The distal part of the human urethra (in either sex) may harbour any of a range of organisms which include various Gram-positive cocci (e.g. *Staphylococcus* spp), Gram-negative bacteria (e.g. species of *Acinetobacter*, *Escherichia*), and yeasts (e.g. *Candida albicans*); in the female, the urethral microflora is generally influenced by the resident vaginolabial flora (see VAGINA MICROFLORA). The microflora of the external genitalia in both sexes may include those organisms mentioned above together with e.g. various coliform bacteria and *Mycobacterium smegmatis*. (See also UREAPLASMA.) Other regions of the genitourinary tract (e.g. urinary bladder, ureters) do not normally have a resident microflora. (See also BODY MICROFLORA and URINARY TRACT INFECTION.)

genome A term which is used with various meanings, including the following:

(1) The genetic material of an organism. For example: (i) the chromosome in a bacterial cell (or one of each type of chromosome if more than one type is present – e.g. both the small and large chromosomes in *Vibrio cholerae* [Nature (2000) *406* 477–483]); (ii) the DNA or RNA in a virion; (iii) the chromosome(s) together with any associated plasmids – for example, the chromosome and the two small plasmids in the bacterium *Buchnera* [Nature (2000) *407* 81–86].

(2) All the (different) genes in a cell or virion.

(3) The *haploid* set of chromosomes or genes in a cell.

The 'complete genome sequence' (i.e. the complete sequence of nucleotides in the genome) has been published for a number of bacterial species – references for some of them can be found in the entry BACTERIA. Sequences have also been published e.g. for the genome of the plant *Arabidopsis thaliana* [Nature (2000) *408* 791–826] and the human genome [Nature (2001) *409* 813–958].

genomic library See LIBRARY.

genomic masking See PHENOTYPIC MIXING.

genophore A nucleic acid which carries genetic information; the term is usually applied to a prokaryotic chromosome, viral genome, ctDNA or mtDNA.

genospecies Any given group of strains among which genetic exchange can occur.

genotoxin Any DNA-damaging agent (e.g. a MUTAGEN).

genotype The genetic constitution of an organism: i.e., the organism's content of genetic information, either in total or with respect to one or more particular named alleles, regardless of whether or not that information is being – or can be – expressed under a given set of conditions. (cf. GENOME; PHENOTYPE.)

gentamicin Any of several (related) AMINOGLYCOSIDE ANTIBIOTICS (gentamicins A, B, C_1, C_{1a}, C_2 etc) produced by *Micromonospora* spp. The drug 'gentamicin' used clinically contains gentamicins C_1, C_{1a}, and C_2; it is active against e.g. some strains of *Pseudomonas aeruginosa*.

gentian violet A mixture of dyes which includes CRYSTAL VIOLET and METHYL VIOLET.

gentiobiose A reducing disaccharide: β-D-glucopyranosyl-(1 → 6)-D-glucose. It occurs naturally in several plant glycosides (e.g. AMYGDALIN) and as a degradation product of glucans containing (1 → 6)-β-D-glucosyl linkages.

genus A taxonomic rank (see TAXONOMY and NOMENCLATURE).

genus novum (gen. nov.; gen. n.; n. gen.) A designation used to indicate a newly proposed genus at the time of its initial publication.

Geodermatophilus A genus of aerobic, catalase-positive bacteria (order ACTINOMYCETALES, wall type III) which occur in soil. The organisms form a simple substrate mycelium of thick strands, each a mass of cuboid cells formed by repeated division in both transverse and longitudinal modes; the mycelium gives rise to zoospores and to non-motile cuboid or coccoid cells. Budding occurs in zoospores and in young mycelia. GC%: ca. 73–75. Type species: *G. obscurus*. (cf. DERMATOPHILUS.)

Geoglossum See HELOTIALES.

geophilic Living or growing in or on the earth. (cf. DERMATOPHYTES.)

geosmin (*trans*-1,10-dimethyl-*trans*-9-decalol) A substance which is produced by certain aquatic actinomycetes and cyanobacteria (e.g. *Anabaena*, *Lyngbya* and *Oscillatoria* spp) and which imparts an 'earthy' or 'musty' taste and odour to the water and to fish living in the water. (cf. METHYLISOBORNEOL.)

geotaxis A TAXIS in which the stimulus is gravity; a negatively geotactic organism moves upwards.

Geotoga See BACTERIA (taxonomy).

Geotrichum A genus of fungi of the class HYPHOMYCETES; teleomorphs occur in the genus *Dipodascus*. [DNA relatedness between *Geotrichum* and *Dipodascus*: CJB (1985) *63* 961–966.] The mycelium is septate, each septum having a single, central pore; pores as small as 50 nm (*micropores*) have been reported e.g. in *G. candidum*. Conidiogenesis is thallic-arthric (see CONIDIUM). In e.g. *G. candidum* the septa which separate adjacent conidia are multiperforate. (See also CHEESE SPOILAGE, PAPER SPOILAGE, SEWAGE FUNGUS, VILIA.)

geotropism See TROPISM (sense 1).

GEP See PROTEIN SECRETION.

GERL Part of the GOLGI APPARATUS from which LYSOSOMES are derived.

germ sporangium *(mycol.)* A SPORANGIUM produced at the distal end of a germ tube formed by a germinating zygospore.

germ tube *(mycol.)* A short, hypha-like structure which develops from certain types of spore on germination, or – in e.g. *Candida albicans* – from yeast cells undergoing yeast-to-mycelium transition. A germ tube usually develops into a hypha.

germ warfare See BIOLOGICAL WARFARE.

Germall 115 A water-soluble imidazolidinyl urea compound with antibacterial and antifungal properties; it is used as a PRESERVATIVE in e.g. cutting oil emulsions.

German measles *Syn.* RUBELLA.

germicide (1) A general term for any antimicrobial chemical agent used for disinfection, antisepsis or sterilization – regardless of whether its action is microbicidal or microbistatic. (2) Any microbicidal disinfectant, antiseptic or sterilant.

germinal centres See ANTIBODY FORMATION.

germinant See ENDOSPORE (sense 1(a)).

germination (1) The process in which a spore gives rise to a vegetative cell or hypha. (2) (of bacterial endospores) See ENDOSPORE (sense 1(a)).

germination by repetition (mycol.) On the germination of a spore: the formation of a secondary spore (rather than a germ tube).

Gerstmann–Sträussler–Scheinker syndrome A human, inherited, autosomal-dominant TRANSMISSIBLE SPONGIFORM ENCEPHALOPATHY (one of a number of inherited prion diseases). Characteristically, PrP–amyloid plaques in the cerebellar cortex are associated with pyramidal tract degeneration. Onset occurs in the third–seventh decade, and the mean duration of the illness is 5 years; clinical features (including the presence/absence of dementia) appear to depend on genotype.

Getah virus An ALPHAVIRUS which is closely related to ROSS RIVER VIRUS; it is apparently non-pathogenic in man, but in horses it can cause a febrile disease with e.g. oedema of the limbs.

GF stain GRIDLEY FUNGUS STAIN.

GFP GREEN FLUORESCENT PROTEIN.

ghost (1) An empty phage capsid (i.e., a capsid devoid of its genome).

(2) An empty vesicle, composed of cytoplasmic membrane, obtained e.g. from a bacterial cell by removal of its peptidoglycan followed by cell lysis.

G_i protein See ADENYLATE CYCLASE.

giant cells Large multinucleate cells found in the tissues in certain diseases – e.g. CHICKENPOX, MEASLES, TUBERCULOSIS (cf. GRANULOMA); they are formed by the fusion of MACROPHAGES. *Langhans* giant cells contain relatively few nuclei which are situated at the periphery of the cytoplasm; *foreign body* giant cells contain many nuclei distributed throughout the cytoplasm.

giant condyloma acuminatum *Syn.* BUSCHKE–LÖWENSTEIN TUMOUR.

Giardia A genus of parasitic flagellate protozoa of the DIPLOMONADIDA. Strains occur in the intestinal tract in a wide range of vertebrates; the organisms are transmitted as cysts.

G. lamblia (= e.g. *G. intestinalis*, *Lamblia intestinalis*) occurs in the small intestine of man (causing GIARDIASIS), other primates, and pigs. Morphologically, the trophozoite resembles one half of a longitudinally-bisected pear, the flat side being ventral; it is commonly ca. 10–20 × 5–10 μm. The two nuclei occur symmetrically in the broad anterior region. The 8 flagella arise ventrally between, and slightly anterior to, the nuclei; three pairs emerge laterally, and the distal ends of the fourth pair become free at the extreme posterior end of the cell. The anterior ventral surface of the cell body bears a large 'sucking disc' or 'sucker' by means of which the parasite is believed to adhere to the intestinal wall of its host (see also GIARDIASIS). The cell contains a pair of AXOSTYLES, and one or two deeply staining, rod- or comma-shaped bodies (median or 'parabasal' bodies) which occur in the centre of the cell and are orientated more or less perpendicularly to the long axis of the cell. The trophozoites reproduce by binary fission; they stain well with e.g. iron–haematoxylin. [Culture of *G. lamblia* in serum-free medium: J. Parasitol. (1983) *69* 1181–1182. Lipid metabolism in *Giardia*: TIP (2001) *17* 316–319.]

G. lamblia forms ovoid cysts (ca. 11–15 × 7–10 μm) in the gut lumen, and these are expelled with the faeces. Mature cysts generally contain 4 nuclei which are commonly clustered at one pole. The nuclei and portions of the median bodies and flagella can be seen in cysts stained with e.g. iodine.

The classification of *Giardia* strains into species remains controversial. Strains are often named according to the host in which they are found – e.g. *G. canis* (from dogs) – but such speciation may not be justified. It has been proposed that only three species be recognized on the basis of morphology and host range: *G. agilis* (from frogs and tadpoles), *G. muris* (from rodents and birds), and *G. duodenalis* (from mammals); strains currently known as '*G. lamblia*' would thus be included in the species *G. duodenalis*.

giardiasis (lambliasis) A human disease caused by *Giardia lamblia*. Infection occurs by ingestion of cysts; transmission may occur by person-to-person contact of via contaminated food or drinking water (cysts can survive levels of chlorination used for treating WATER SUPPLIES).

Incubation period: ~4–7 days. Infection may be asymptomatic, or there may be acute or chronic diarrhoea, often with nausea, weight loss, fatty stools and flatulence. Symptoms are believed to be due to mechanical obstruction of the intestinal mucosa; tissue invasion does not occur. Although adhesion of *G. lamblia* to the mucosa of the small intestine is generally believed to depend on the organism's 'sucker' (see GIARDIA), adhesion may involve a specific giardial LECTIN which is activated by gut proteases [Science (1986) *232* 71–73].

Diagnosis. Trophozoites may be detected by microscopy in (fresh) saline-mounted specimens of diarrhoea; cysts, rather than trophozoites, may be found in specimens of formed stool. An ELISA-based method, using polyclonal antibodies, may be used to detect *Giardia* antigens in faeces.

Chemotherapy: e.g. METRONIDAZOLE, QUINACRINE.

[*Giardia* in drinking water supplies (cyst levels, parasite viability, health impact): AEM (1996) *62* 47–54.]

Gibberella A genus of fungi (order HYPOCREALES) which include plant-pathogenic species (see e.g. BAKANAE DISEASE); anamorphs occur in the genus FUSARIUM. *Gibberella* spp form blue- or violet-coloured superficial perithecia, and ovoid to fusiform ascospores. (See also GIBBERELLINS and ZEARALENONE.)

gibberellic acid GIBBERELLIN A_3.

gibberellins A class of PHYTOHORMONES which are synthesized e.g. in the leaves of plants and which regulate stem elongation, seed germination, etc; gibberellins were originally discovered as products of the fungus *Gibberella fujikuroi* (*Fusarium moniliforme*). Many gibberellins have now been identified and designated gibberellins A_1, A_2 etc (or GA_1, GA_2 etc); all share a common tetracyclic structure, but differ in the position and nature of their substituents.

The production of gibberellins by *G. fujikuroi* is believed to play an important role in the pathogenesis of BAKANAE DISEASE; another example of a plant pathogen which produces a gibberellin (A_4) is *Sphaceloma manihoticola*, causal agent of 'superelongation disease' in cassava [BBRC (1979) *91* 35–40]. Stunting of plants infected with certain viruses may be reversed by the application of gibberellins; it has been suggested that such viruses may affect the functioning of endogenous plant gibberellins.

Gibberellins obtained commercially from *G. fujikuroi* have been used e.g. for controlling premature fruit-drop, improving fruit setting, improving the quality and yield of grapes, and inducing rapid and uniform germination in malting barley (e.g. for BREWING).

gibbon ape leukaemia virus (GALV) A C-type retrovirus (subfamily ONCOVIRINAE) first isolated from a female gibbon with disseminated lymphosarcoma; GALV can be transmitted horizontally and can cause a substantial incidence of leukaemia in gibbon colonies. GALV is infectious for a wide range of host cells, including e.g. human, monkey, cat, mink, rabbit, rat (but not mouse), and various avian (but not chicken) cells. [Book ref. 114, pp. 130–137.] (cf. SIMIAN SARCOMA VIRUS.)

Giemsa's stain See ROMANOWSKY STAINS.

Gigartacon See ACANTHAREA.

Gigartina A genus of marine algae (division RHODOPHYTA). The organisms, which attach to rocks etc by means of a holdfast, vary widely in form according to species – often being flat (ribbon-like) and more or less branched; the surface frequently bears numerous papillate outgrowths. *Gigartina* spp are used e.g. as a source of CARRAGEENAN.

Gigaspora See ENDOGONALES.

Gilbertella See MUCORALES.

Gilchrist's disease *Syn.* BLASTOMYCOSIS.

gill *(mycol.)* *Syn.* LAMELLA.

gill disease (in fish) See BACTERIAL GILL DISEASE.

gill fungi Fungi which form lamellae: see LAMELLA.

gill rot A FISH DISEASE caused by *Branchiomyces sanguinis* or *B. demigrans*. The fungus grows on and into the gills, obstructing capillary circulation and leading to necrosis; the fish soon dies from suffocation. Gill rot is encouraged by high temperatures and the presence of decaying organic matter.

***Gilpinia hercyniae* NPV** See BACULOVIRIDAE.

Giménez stain A procedure for staining rickettsiae. Add 4% aqueous phenol (250 ml) and distilled water (650 ml) to 10% basic fuchsin in 95% ethanol (100 ml); allow to stand for 2 days at 37°C, and then add 2 volumes to 5 volumes of phosphate buffer (0.2 M, pH 7.45), mix, and filter immediately before use. A heat-fixed smear is stained for 1–2 min and washed with tap water; the smear is then stained for 5–10 sec with 0.8% w/v malachite green oxalate, washed in tap water, stained again for 5–10 sec with malachite green oxalate, washed thoroughly in tap water, and blotted dry. Rickettsiae stain red, background green.

gin See SPIRITS.

***gin* gene** See BACTERIOPHAGE MU.

ginger blotch A MUSHROOM DISEASE caused by a distinct strain of the *Pseudomonas fluorescens* complex; superficial pale-brown to ginger-coloured lesions develop on the caps [JAB (1982) *52* 43–48]. (cf. BROWN BLOTCH.)

gingivitis An acute or chronic inflammation of the gingiva (i.e., the gums) affecting areas of gum tissue (the 'attached gingiva') other than those immediately adjacent to the tooth–gum junction (which are referred to as the 'free gingiva'); unlike PERIODONTITIS, gingivitis does not bring about a permanent loss of periodontal structures.

Acute herpetic gingivostomatitis occurs most commonly in children and generally lasts about two weeks. It involves reddening of the gingiva and the formation of vesicular lesions on various tissues within the mouth; systemic symptoms may include anorexia, fever, and the involvement of the submandibular and cervical lymph nodes. The causal agent is the type 1 HERPES SIMPLEX virus.

Acute necrotizing ulcerative gingivitis occurs principally in young adults and is apparently stress-linked. It involves necrosis of the gingiva and the development of ulcers which usually become secondarily infected by spirochaetes. The primary aetiological agents are believed to include *Bacteroides intermedius*.

In *chronic (non-specific) gingivitis* the gingiva may show signs of oedema, hyperplasia or atrophy.

The severity of gingivitis often appears to correlate with stress and/or with hormonal influences; thus, e.g. the severity of the disease may increase during puberty or pregnancy.

Giovannolaia A subgenus of PLASMODIUM.

girdle See e.g. DINOFLAGELLATES.

girdle bands See DIATOMS.

Glabratella See FORAMINIFERIDA.

glabrous Hairless; smooth.

glanders An acute or chronic, often fatal disease which affects mainly equines (horse, mule, donkey), although some other animals – including man – are susceptible; the causal agent, *Pseudomonas mallei*, appears to enter the body mainly via the mouth or, less commonly, via the nose or skin lesions. Among equines, the acute form of the disease (more common in mules and donkeys) typically involves fever, coughing, a highly infectious nasal discharge, ulceration of the nasal mucosa, and nodular skin lesions on the limbs or abdomen; death from septicaemia may occur within days. Chronic glanders (more common in horses) may be largely pulmonary, with coughing, dyspnoea and epistaxis; alternatively, the prominent features may include ulcerating nasal lesions and/or subcutaneous ulcerating nodules (often in the hock region) which discharge a dark-honey-coloured pus. In *farcy*, a cutaneous manifestation of glanders, ulcerating lesions occur in cutaneous and subcutaneous tissues, and the regional lymph nodes and ducts become swollen and hard (the so-called *farcy buds* and *farcy pipes*, respectively). *Lab. diagnosis*: e.g. a CFT and/or the MALLEIN TEST. *Treatment*: e.g. sulphonamides together with a formalinized vaccine prepared from *P. mallei*. (cf. EPIZOOTIC LYMPHANGITIS.)

In man, the disease (which can be fatal) affects mainly the respiratory tract and/or cutaneous tissues, but the viscera, muscles or bones may also be affected.

glandular fever *Syn.* INFECTIOUS MONONUCLEOSIS.

Glanzmann's thrombasthenia See INTEGRINS.

glass (biodeterioration of) In warm, humid conditions certain fungi (e.g. *Aspergillus glaucus*, *A. fumigatus*, *Cladosporium* spp, *Penicillium* spp) can grow on glass surfaces (e.g. microscope lenses), deriving nutrients either from adjacent substrates (e.g. lens sealing compounds) or from the film of material which tends to condense and collect on the glass; the surface of the glass may be etched by the extracellular products of such fungi. Certain lichens can grow on and etch stained glass windows etc. Damage to glass may be avoided by frequent cleaning, storage under dry conditions (where possible), use of fungicides, etc.

Glasser's disease (infectious polyarthritis and polyserositis) An acute disease of young pigs, often fatal within days. Onset is sudden, with fever, anorexia, dyspnoea, and lameness with swollen joints; the causal agent is a species of *Haemophilus* – often *H. parasuis* – although *Mycoplasma* spp can also be isolated in some cases. *Treatment*: e.g. tetracyclines, given early.

glaucescent Becoming GLAUCOUS.

Glaucocystis See GLAUCOPHYTA.

Glaucoma A genus of ciliates (order HYMENOSTOMATIDA). *G. scintillans* is ovoid, ca. 40–80 μm, with uniform somatic ciliature; it is common in freshwater habitats containing decomposing organic matter (see also SAPROBITY SYSTEM). *G. dragescui* also is ovoid, but *G. frontata* is elongate with a pointed posterior end.

Glaucophyta A phylum of 'algae' which includes those species that contain CYANELLES instead of CHLOROPLASTS. The phylum contains wall-less organisms (e.g. the flagellated, cryptomonad-like *Cyanophora*) and species which have a cell wall (e.g. *Glaucocystis*). In *Cyanophora* the (cyanobacterial) endosymbiont is a species of *Cyanocyta*; synthesis of phycocyanin by the endosymbiont is apparently partially dependent on the host's ribosomes.

glaucous Bluish-green, or covered with a bluish-green powdery or waxy bloom.

GLC Gas–liquid CHROMATOGRAPHY.

GlcNAc *N*-ACETYL-D-GLUCOSAMINE.

gleba In the fruiting bodies of e.g. fungi of the GASTEROMYCETES: the spore-bearing tissue which is (at least initially) enclosed by the PERIDIUM (or by the wall of the peridiole in members of the NIDULARIALES).

Glenodinium See DINOFLAGELLATES.

gliding bacteria A non-taxonomic group of bacteria and cyanobacteria which exhibit GLIDING MOTILITY during at least some stage in their life cycles; gliding bacteria typically occur in soil and in freshwater and marine habitats. The group comprises the genera listed under CYTOPHAGALES, the MYXOBACTERALES, the CHLOROFLEXACEAE, members of the cyanobacteria – both unicellular (e.g. *Synechococcus*) and filamentous (e.g. *Anabaena, Oscillatoria, Phormidium, Spirulina*) – and e.g. *Capnocytophaga* and *Desulfonema*.

Although not formally included among the gliding bacteria, some species of *Mycoplasma* exhibit a form of surface-associated locomotion referred to as gliding.

gliding motility A continuous-type (i.e. smooth, non-jerky) MOTILITY which occurs in some prokaryotes (GLIDING BACTERIA), algae (e.g. some diatoms and desmids) and protozoa (e.g. *Toxoplasma gondii*) which have no obvious locomotory organelles; gliding motility appears to occur only on a solid surface.

In the diatom *Amphora coffeaeformis*, some studies have suggested a possible mechanism that involves actin- and tubulin-based structures; this is compatible with the theory that gliding results from the translocation of membrane proteins along the raphe canal – the outermost part of such proteins being attached to the immobile slime secretion which has adhered to the substratum [Cell Mot. (1985) *5* 103–122].

In the gregarine *Gregarina garnhami* (which apparently does not contain actin) it has been suggested that gliding may involve two sets of filaments which run longitudinally in the pellicle [JUR (1984) *88* 66–76].

Gliding motility has been extensively investigated in prokaryotes [early review: ARM (1981) *35* 497–529]. In some species, individual cells glide quite slowly (a few micrometres per minute) while in other species the cells may achieve speeds of up to ca. 150 micrometres per minute; some gliding bacteria can also perform other types of movement (e.g. flexing, twitching – see e.g. SPIRULINA). The path of a gliding cell is marked by a slime trail, and on an agar surface the path may be 'etched' into the agar; at least some gliding bacteria can penetrate agar gels. Gliding is also exhibited by trichomes and by aggregates of cells ('swarms').

Typically, gliding bacteria glide when nutrient levels are low; high levels can suppress gliding. Gliding appears to require the presence of calcium ions and is inhibited by EGTA in a number of bacteria and cyanobacteria. In bacteria of the *Cytophaga–Flexibacter* group, sulphonolipids in the cell envelope appear to be necessary for gliding motility [Nature (1986) *324* 367–369].

In some gliding bacteria certain stages in the life cycle (e.g. the myxospore in myxobacteria) are non-motile. In *Flexibacter elegans* (cells ca. 1–50 μm) gliding occurs only in those cells which are longer than ca. 5 μm.

In some cyanobacteria, motility has been associated with the *junctional pore complex* (JPC): a channel in the cell envelope through which mucilage (carbohydrate) is secreted; each organism has a number of JPCs, and secretion via these channels is believed to provide the thrust that is necessary for locomotion [JB (2000) *182* 1191–1199].

Results of studies on *Myxococcus xanthus* (involving shock-freezing and freeze-drying) have indicated that the motility apparatus in this organism resembles a helical 'continuum' or band (~170–380 nm wide) that is wrapped around the protoplast (i.e. within the periplasmic space) to form a closed loop; 'nodes', associated with the continuum, may travel along the trichomes in a wave-like fashion (but with variable internode distance), each node possibly corresponding to a region of close contact, or adhesion, between the trichome and substratum [Microbiology (2001) *147* 939–947].

Gliocladium See HYPHOMYCETES and GLIOTOXIN.

gliotoxin An epidithiapiperazinedione MYCOTOXIN (see EPIPOLYTHIAPIPERAZINEDIONES) obtained from *Trichoderma viride* and from species of *Aspergillus, Gliocladium* and *Penicillium*; it inhibits replication of polioviruses, echoviruses and measles virus, and also has antibacterial, antifungal and antitumour activity. In eukaryotic cells an early effect of gliotoxin is apparently disruption of microfilaments [JCS (1986) *85* 33–46].

***glmM* gene** In *Helicobacter pylori*: a gene (formerly *ureC*) which encodes phosphoglucosamine mutase.

***gln* genes** See e.g. NTR GENES.

global regulatory system *Syn.* REGULON.

global warming See GREENHOUSE EFFECT.

Globigerina A genus of pelagic foraminifera (order FORAMINIFERIDA). The test is calcareous and multiloculate, the rounded locules being spirally arranged. *G. bulloides* may reach ca. 800 μm diam., and numerous fine calcareous spines radiate from the test; *G. bulloides* and e.g. *G. inflata, G. pachyderma* and *G. quinqueloba* occur in subpolar and cold temperate seas.

Globigerinoides A genus of pelagic foraminifera (order FORAMINIFERIDA). *G. sacculifer* is often abundant in tropical and subtropical seas, occuring in the upper euphotic zone; it is ca. 0.5–1.0 mm diam., bears radial spines 1–2 mm long, and contains numerous zooxanthellae.

globomycin See e.g. BRAUN LIPOPROTEIN.

globose Spherical.

globoside (parvovirus receptor) See ERYTHROVIRUS.

Gloeobacter A genus of unicellular CYANOBACTERIA (section I) in which the rod-shaped cells divide by equal binary fission in one plane and occur in colonial aggregates surrounded by a sheath; the cells are characterized by their lack of THYLAKOIDS – photosynthesis and respiration apparently being associated with the cytoplasmic membrane. Phycobiliproteins occur as an electron-dense layer of characteristically shaped phycobilisomes (apparently composed of bundles of rods) beneath the cytoplasmic membrane [Arch. Micro. (1981) *129* 181–189]. GC% (one strain): 64.

Gloeocapsa A genus of unicellular CYANOBACTERIA (section I) in which the cells are coccoid and divide by equal binary fission in two or three planes, giving rise to clusters of cells enclosed in a sheath. GC%: 40–46. The organisms grow – often in masses – on wet rocks e.g. in the supralittoral zones of rocky coasts. (cf. GLOEOTHECE.)

Gloeochloris A genus of algae of the XANTHOPHYCEAE. The vegetative cells occur in a spherical palmelloid colony; asexual reproduction occurs by the formation and release of biflagellate zoospores.

Gloeodinium A genus of DINOFLAGELLATES which form palmelloid colonies and reproduce by zoospore formation. One species (*G. montanum*), found in freshwater peaty marshes.

Gloeothece A genus of unicellular CYANOBACTERIA (section I); the cells are rod-shaped (ca. 5–6 μm wide), possess thylakoids (cf. GLOEOBACTER), divide by equal binary fission in one plane (cf. GLOEOCAPSA), and occur in colonial aggregates surrounded by a well-defined sheath (cf. SYNECHOCOCCUS). The function of the sheath seems not to include the protection of nitrogenase from oxygen [oxygen relationships in nitrogen fixation: MR (1992) *56* 346–347]. Strains of *Gloeothece* can carry out NITROGEN FIXATION aerobically, despite the absence of heterocysts, and were formerly regarded as nitrogen-fixing strains of GLOEOCAPSA. GC%: 40–43.

Gloeotrichia A phycological genus of the RIVULARIACEAE (now included in the genus CALOTHRIX). '*G. echinulata*' is a gas-vacuolate, freshwater planktonic species which may form blooms; in nature, it forms characteristic spherical urchin-like colonies in which the terminal heterocysts occur in the centre of the colony and the trichomes radiate outwards. Hair cells may be formed e.g. under phosphorus-deficient conditions.

Gloiopeltis See RHODOPHYTA.

glomerulonephritis Inflammation of the kidney glomeruli. Acute glomerulonephritis is characterized by blood and red cell casts in the urine, and a reduced glomerular filtration rate leading to oliguria, oedema and hypertension. It may be due to deposition in the glomeruli of antigen-antibody complexes formed during certain infectious diseases, e.g., SYPHILIS (see also POST-STREPTOCOCCAL GLOMERULONEPHRITIS); in mice it may be induced e.g. by the LCM VIRUS. (See also PYELONEPHRITIS.)

Glomospira See FORAMINIFERIDA.

Glomus See ENDOGONALES.

glorin See POLYSPHONDYLIUM.

Glossina (tsetse fly) A genus (22 species) of haematophagous (blood-sucking) flies (order Diptera) which include vectors of salivarian trypanosomes; they occur in sub-Saharan Africa in a belt approximately 15°N–15°S, and in SE Africa, including Mozambique. *Glossina* spp are distinguished from other dipterans e.g. by the chopper-shaped 'hatchet cell' in the wing venation. Both the male and female flies suck blood.

glove box A cabinet, the interior of which is accessible via gloves attached to openings in the front panel. (See also SAFETY CABINET.)

glove juice test A test used (in the USA) to determine the efficacy of a given antimicrobial agent used as a surgical hand scrub [Book ref. 65, pp. 958–961].

GlpF See MIP CHANNEL.

glucan A homoglycan (see POLYSACCHARIDE) consisting of glucose residues – see e.g. CELLULOSE, GLYCOGEN, LUTEOSE, NIGERAN, PARAMYLON, STARCH. (See also CELL WALL (fungal).)

glucitol *Syn.* SORBITOL.

glucoamylases *Syn.* γ-AMYLASES.

glucobrassicin See CLUBROOT.

glucocorticoids Steroid hormones, produced in the adrenal cortex, whose activity includes regulation of carbohydrate and protein metabolism. Glucocorticoids also have anti-inflammatory activity; thus, e.g. they inhibit the ability of IFN-γ and lipopolysaccharides to induce the synthesis of CYTOKINES in macrophages.

glucogenesis *Syn.* GLUCONEOGENESIS.

glucomannans See MANNANS.

gluconate See GLUCONIC ACID.

gluconeogenesis (glucogenesis) The synthesis of glucose (and hence other hexoses) from non-carbohydrate substrates such as acetate, glycerol, pyruvate, succinate etc. Essentially, this may be achieved by conversion (where necessary) of the substrate to an intermediate of the EMBDEN–MEYERHOF–PARNAS PATHWAY [Appendix I(a)], followed by reversal of reactions of the EMP pathway – the three irreversible reactions in this pathway being bypassed by other enzymes. Thus, e.g., in *Escherichia coli* phosphoenolpyruvate can be synthesized from pyruvate or from the TCA CYCLE intermediates oxaloacetate and malate [see Appendix II(b)]; phosphoenolpyruvate is then converted to fructose 1,6-bisphosphate by reversal of reactions of the EMP pathway. Fructose 1,6-bisphosphate is hydrolysed by *fructose bisphosphatase* to fructose 6-phosphate which can in turn be converted to glucose 6-phosphate by phosphohexose isomerase.

gluconic acid Gluconic acid and its salts are metabolized by various microorganisms (see e.g. ENTNER–DOUDOROFF PATHWAY), and are synthesized (by the oxidation of glucose) e.g. by *Aspergillus niger* and other hyphomycetes and by *Gluconobacter oxydans*. *A. niger* and *G. oxydans* are used as commercial sources of these compounds which have a range of uses in the food and pharmaceutical industries; for example, gluconic acid (a chelating agent) is used for sequestering metal ions; sodium gluconate is used for cleaning glassware and for removing rust from metals; calcium and iron gluconates are used therapeutically for the treatment of calcium and iron deficiencies.

Gluconobacter A genus of Gram-type-negative bacteria of the ACETOBACTERACEAE; the organisms occur e.g. on certain fruits and flowers, and are used e.g. for the production of MONOBACTAMS and in the SORBOSE FERMENTATION. Cells: typically ovoid to rod-shaped, 0.5–0.8 × 0.9–4.2 μm, non-motile or lophotrichously flagellate. Catalase-positive. Typically, ethanol is oxidized to acetic acid on e.g. CARR MEDIUM; neither acetate nor lactate is oxidized to CO_2. Ketogenic growth occurs on polyalcohol substrates, and all strains form 2-ketogluconic acid (and most form 5-ketogluconic acid) from D-glucose. Substrates utilized include e.g. D-mannitol, sorbitol, glycerol (see DIHYDROXYACETONE FERMENTATION), D-fructose and D-glucose; all strains require growth factors (which may include e.g. pantothenate, niacin, thiamine, PABA) when D-mannitol is used as the sole carbon source. The organisms lack a complete TCA cycle; the breakdown of sugars and polyols to CO_2 occurs mainly via the HEXOSE MONOPHOSPHATE PATHWAY. Optimum growth temperature: 25–30°C. GC%: ca. 56–64. Type species: *G. oxydans* – which incorporates strains previously referred to as *Acetobacter suboxydans*.

glucose dehydrogenase See QUINOPROTEIN and EXTRACYTOPLASMIC OXIDATION.

glucose effect (glucose repression) (1) *Syn.* CATABOLITE REPRESSION. (2) *Syn.* CRABTREE EFFECT.

glucose isomerase (xylose isomerase; D-xylose ketolisomerase; EC 5.3.1.5) An intracellular enzyme which converts D-glucose to D-fructose, although its main function in vivo is probably the conversion of D-xylose to D-xylulose; the enzyme is typically induced in the presence of xylose. Glucose isomerase, obtained e.g. from *Bacillus coagulans* (see IMMOBILIZATION (1)) or *Streptomyces* spp, is used commercially on a large scale for the conversion of D-glucose to D-fructose e.g. in the manufacture of HIGH-FRUCTOSE CORN SYRUP. (See also ENZYMES.)

glucose oxidase (β-D-glucose:oxygen 1-oxidoreductase, EC 1.1.3.4) An enzyme (produced e.g. by *Aspergillus* and *Penicillium* spp) which oxidizes glucopyranose in the presence of oxygen to D-glucono-δ-lactone and H_2O_2 – the former rapidly hydrolysing to D-gluconic acid. Glucose oxidase from e.g. *Aspergillus niger* (a glycoprotein containing 2FAD/molecule) is widely used in clinical tests for the detection of glucose in urine, blood etc; in general, such tests rely on a colour change resulting from the oxidation of a chromogen by H_2O_2 released by glucose oxidase in the presence of glucose and oxygen. Glucose oxidase is also used in the food industry e.g. for removing glucose from foods (e.g. eggs preparatory to drying) to prevent browning, and for removing oxygen from e.g. beer, wine, mayonnaise etc. (See also ENZYMES.)

β-glucosidase See CELLULASES.

β-glucoside permease (in *Escherichia coli*) See CATABOLITE REPRESSION.

glucosinolase See CLUBROOT.

Glugea A genus of protozoa (class MICROSPOREA) which are parasitic in fish; tumour-like cysts (XENOMAS, 'glugea cysts') ca. 2–4 mm diam. develop in the gut mucosa and liver of the infected fish.

glume blotch A CEREAL DISEASE, caused by *Leptosphaeria nodorum* (*Septoria nodorum*), which can affect e.g. wheat and barley. Leaf lesions are initially yellow, later becoming golden-brown and sometimes coalescing. Lesions on glumes (usually spreading from the tip) are a dark purplish-brown, and they later exhibit dark or black pycnidia. Control: e.g. CARBENDAZIM or PROCHLORAZ with a DITHIOCARBAMATE.

L-glutamate See GLUTAMIC ACID.

glutamate dehydrogenase See AMMONIA ASSIMILATION.

glutamate synthase See AMMONIA ASSIMILATION.

L-glutamic acid An amino acid synthesized by the reductive amination of 2-oxoglutarate – see Appendix IV(a). It is produced commercially on a large scale for use in the food industry – mainly as a flavour enhancer (monosodium glutamate) – and is obtained from strains of bacteria (e.g. *Corynebacterium glutamicum*, 'glutamic acid bacteria') grown aerobically on e.g. molasses or starch hydrolysates. (Some strains can be grown e.g. on acetic acid.) The strains used require biotin for growth and have little or no 2-oxoglutarate dehydrogenase activity (so that metabolism of 2-oxoglutarate via the TCA cycle is limited or cannot occur). Production of glutamic acid is maximal under suboptimal growth conditions (achieved e.g. by limiting biotin). The glutamic acid must be secreted by the cells to avoid feedback inhibition of its synthesis; permeability of the cells may be increased e.g. by biotin deficiency, by treatment with fatty acid derivatives or antibiotics (e.g. penicillin), or by using strains auxotrophic for glycerol or oleic acid grown on low levels of glycerol and oleic acid, respectively.

Some L-glutamine is also formed during glutamic acid production; the proportion of glutamine formed can be increased by adjusting conditions of pH, nutrient availability etc.

L-glutamine An amino acid synthesized from GLUTAMIC ACID: see Appendix IV(a). It is an important intermediate in various metabolic pathways, being involved e.g. in some types of AMMONIA ASSIMILATION and in various transamination reactions – e.g. in the biosynthesis of tryptophan and of purine nucleotides [see Appendixes IV(f) and V(a)].

glutamine synthetase See AMMONIA ASSIMILATION.

glutaraldehyde ($CHO(CH_2)_3CHO$) A dialdehyde whose two aldehyde groups can substitute amino and other groups in proteins and nucleic acids and which can thus form cross-links in and between molecules (cf. FORMALDEHYDE); it is used e.g. for pre-FIXATION (e.g. 1–10% in phosphate or cacodylate buffer) and as a DISINFECTANT or STERILANT (e.g. for medical and other equipment). Glutaraldehyde (e.g. 2% aqueous) is rapidly lethal for a wide range of vegetative microorganisms and viruses, but activity may be slow against e.g. *Mycobacterium tuberculosis* and bacterial ENDOSPORES: the dried spores of *Bacillus subtilis*, for example, have been reported to need up to 10 hours in 2% glutaraldehyde for complete inactivation (cf. STERILANT). The antimicrobial activity of glutaraldehyde is much greater (but stability is reduced) under alkaline or neutral conditions than under acidic conditions. [For antimicrobial activity and uses of glutaraldehyde see JAB (1980) *48* 161–190.]

glutarimide antibiotics A group of ANTIBIOTICS characterized by two structural components: a β-(2-hydroxyethyl)-glutarimide and a cyclic or non-cyclic ketone. They include e.g. CYCLOHEXIMIDE and *inactone*.

glutathione peroxidase (glutathione:hydrogen peroxide oxidoreductase; EC 1.11.1.9) An enzyme which catalyses the oxidation of reduced glutathione by hydrogen peroxide. (See also CATALASE.)

glycan *Syn.* POLYSACCHARIDE.

glyceollin Any of four isoflavonoid PHYTOALEXINS (glyceollins I–IV) formed by the soybean plant.

glycerate pathway See TCA CYCLE.

glycerokinase (ATP:glycerol 3-phosphotransferase; EC 2.7.1.30) An enzyme which catalyses the ATP-dependent phosphorylation of glycerol. It is obtained commercially from e.g. '*Candida mycoderma*' and *Escherichia coli*, but a more stable enzyme is obtained from *Bacillus stearothermophilus*. It is used in diagnostic kits for the determination of serum triglyceride levels [PTRSLB (1983) *300* 399–410]. (See also ENZYMES.)

glycerol ($CH_2OH.CHOH.CH_2OH$) Glycerol is produced by various organisms as a (usually minor) product of e.g. ALCOHOLIC FERMENTATION and other fermentations; it is formed by the reduction of dihydroxyacetone phosphate (DHAP, an intermediate of the EMBDEN–MEYERHOF–PARNAS PATHWAY) to glycerol 3-phosphate, followed by hydrolysis to yield free glycerol. (See also NEUBERG'S FERMENTATIONS.)

Some bacteria can use glycerol as a growth substrate: e.g. *Escherichia coli* can use glycerol (taken up by FACILITATED DIFFUSION) as sole source of carbon and energy in the presence of an external electron acceptor such as O_2, NO_3^-, or fumarate. Glycerol is phosphorylated by an ATP-dependent kinase, the resulting glycerol 3-phosphate being oxidized by an NAD-dependent dehydrogenase to yield DHAP which can be further metabolized via the EMP pathway. (See also DIHYDROXYACETONE FERMENTATION.) *Klebsiella pneumoniae* and certain lactobacilli can ferment glycerol: some glycerol is converted to DHAP as above, and the NADH thus formed is reoxidized by a parallel pathway in which glycerol is converted to β-hydroxypropionaldehyde (by a coenzyme B_{12}-dependent glycerol dehydrase) which is in turn reduced to 1,3-propanediol by an NADH-dependent reductase.

(See also ALCOHOLS (as antimicrobial agents) and COMPATIBLE SOLUTE.)

glycerol facilitator See MIP CHANNEL.

glycerol fermentation See NEUBERG'S FERMENTATIONS and GLYCEROL.

glycine betaine See OSMOREGULATION.

glycine biosynthesis See e.g. Appendix IV(c).

glycobiarsol See ARSENIC.

glycocalyx In eukaryotic cells: a layer of polysaccharide and/or glycoprotein that surrounds many types of cell. In bacterial cells: a CAPSULE – of either a polysaccharide or S LAYER type [ARM (1981) *35* 299–324] – or, specifically, a *polysaccharide* capsule.

glycocholic acids See BILE ACIDS.

glycogen A highly branched D-glucan formed as a storage polysaccharide in animal cells, in many bacteria, fungi (excluding oomycetes) and protozoa, and in certain plants. Glycogen resembles amylopectin (see STARCH) but the average chain length is generally shorter (average 10–12 glucose residues) and the molecule is more extensively branched. It forms reddish-brown complexes with iodine and may be stained (non-specifically) e.g. by the PAS REACTION.

Biosynthesis of glycogen involves the successive transfer of glucose residues from UDP-glucose (in eukaryotes) or ADP-glucose (in bacteria) to a $(1 \rightarrow 4)$-α-D-glucan acceptor in the presence of glycogen synthase; the branches are then introduced by a branching enzyme: an enzyme that transfers a segment of a $(1 \rightarrow 4)$-α-glucan chain from a 4-OH to a 6-OH position. (ADP/UDP-glucose is formed from ATP/UTP and glucose 1-phosphate, the latter being formed from glucose 6-phosphate by phosphoglucomutase.) In bacteria, at least, initiation of glycogen synthesis may involve the addition of glucosyl residues to a protein acceptor [BBA (1978) *540* 190–196]; different enzymes may be necessary for the synthesis of the glucoprotein primer and the glycogen molecule. [Book ref. 62, pp. 171–178; ARM (1984) *38* 419–458.]

Degradation of glycogen typically involves the transfer of glucose residues from the non-reducing end of a chain to inorganic phosphate, a reaction catalysed by glycogen phosphorylase. (See also amylases and DEBRANCHING ENZYMES.)

glycolate metabolism (glycollate metabolism) See e.g. entry TCA CYCLE.

glycolate pathway See PHOTORESPIRATION.

glycols (as antimicrobial agents) See ALCOHOLS.

glycolysis *Syn.* EMBDEN–MEYERHOF–PARNAS PATHWAY.

glycomacropeptide See CASEIN.

glycosaminoglycan *Syn.* MUCOPOLYSACCHARIDE.

glycosaminopeptide *Syn.* MUCOPEPTIDE.

glycosome A MICROBODY which contains glycolytic enzymes and which occurs in members of the KINETOPLASTIDA; in e.g. *Trypanosoma brucei* most of the enzymes involved in the glycolytic pathway from glucose to 3-phosphoglycerate occur exclusively in glycosomes. [BBA (1986) *866* 179–203 (190–193).]

glycosylase (DNA glycosylase) See ADAPTIVE RESPONSE and URACIL-DNA GLYCOSYLASE.

glyodin (2-heptadecyl-2-imidazoline acetate) An ANTIFUNGAL AGENT used e.g. for the control of apple scab and various fungal diseases of cherries and ornamental plants.

glyoxalase See METHYLGLYOXAL BYPASS.

glyoxylate cycle A pathway which occurs in a variety of organisms, including some bacteria and fungi: see entry TCA CYCLE and Appendix II(b); a key enzyme in the cycle is isocitrate lyase. The glyoxylate cycle permits the use of 2-carbon compounds and e.g. fatty acids as sources of carbon. (See also PHOTORESPIRATION.)

Following phagocytosis, the persistence of *Mycobacterium tuberculosis* within macrophages is reported to require the activity of the (bacterial) isocitrate lyase [Nature (2000) *406* 735–738].

The glyoxylate cycle is reported to be required for the expression of virulence in certain fungi [Nature (2001) *412* 83–86].

glyoxylate metabolism See TCA CYCLE.

glyoxylate shunt See TCA CYCLE.

glyoxysome A type of MICROBODY which occurs in plant cells, and in some eukaryotic microorganisms, and which contains enzymes of the glyoxylate cycle.

glypiated protein A protein which is anchored to a component of the eukaryotic cell membrane (glycosylphosphatidylinositol) via a carbohydrate chain (to which the protein is attached).

Gm allotype See ALLOTYPE.

GM-CSF See COLONY-STIMULATING FACTORS.

GM gangliosides See SPHINGOLIPID.

GMP Guanosine 5′-monophosphate: see Appendix V(a).

GMS Gomori METHENAMINE–SILVER STAIN.

GN broth 'Gram-negative enrichment broth' (used e.g. for the enrichment of shigellae from food or faeces); it contains enzymic digest of casein and animal tissue, D-glucose, citrate, deoxycholate, NaCl, phosphate buffer (pH 7.0). [Recipe: Book ref. 47, p. 652.]

Gnomonia See DIAPORTHALES.

gnotobiotic Refers to a microbiologically monitored environment or animal, i.e., one in which the identities of all microorganisms present are known. (cf. SPECIFIC PATHOGEN FREE.)

goat pox A goat disease, clinically similar to SHEEP POX, caused by a *Capripoxvirus* which is antigenically distinguishable from the sheep pox virus.

GOGAT See AMMONIA ASSIMILATION.

gold leaching See LEACHING.

gold standard (*lab. microbiol.*) Any well-established procedure (e.g. a diagnostic test) which is generally regarded as the superior and definitive method, and which is used as a reference against which other, proposed or alternative methods may be compared (often in terms of sensitivity and/or specificity).

In many cases *culture* is regarded as the gold standard for diagnostic purposes. However, with the advent of nucleic-acid-based methods, some authors refer to an *expanded* or *combined* gold standard; using such a standard, a true-positive result is one that is positive by the gold standard method (usually culture) *and* by one nucleic-acid-amplification test [example of use of the term *expanded/combined* gold standard: JCM (1997) *35* 957–959].

For diagnostic tests on *Chlamydia trachomatis*, some authors have used an *aggregate* gold standard in which a sample is deemed to give a true-positive result if it is either culture-positive *or* positive in two non-culture assays.

Goldberg–Hogness box See PROMOTER.

golden algae CHRYSOPHYTES.

Golgi apparatus An intracellular organelle consisting typically of a number of stacks of flattened membranous vesicles (*cisternae*); at least one Golgi apparatus occurs in most types of eukaryotic cell (cf. e.g. DIPLOMONADIDA). Within the cell the Golgi apparatus is characteristically orientated such that one end of the stacked cisternae (the proximal or 'forming' end) is nearer the nucleus; the opposite end is called the distal or 'maturing' end. Functions of the Golgi apparatus include the formation of LYSOSOMES and the formation of secretory vesicles within which e.g. macromolecules or cell envelope substructures can be transported to the cell surface (see e.g. PALADE PATHWAY and CELL WALL (a)). Secretory molecules reach the Golgi apparatus primarily from the ENDOPLASMIC RETICULUM; thus, e.g., a portion of smooth ER forms a vesicle which encloses molecules from the lumen of the ER, and the vesicle passes through the cytoplasm and fuses with a similar vesicle or with a cisterna at the 'forming' end of the Golgi apparatus. Presumably as a result of repeated budding and fusion, the contents of a vesicle eventually reach the 'maturing' end of the Golgi apparatus and are finally enclosed within a secretory vesicle which fuses with the cytoplasmic membrane. During transit through the Golgi apparatus a molecule characteristically increases its degree of glycosylation – fucose, galactose and/or neuraminic acid residues typically being added.

(See also DICTYOSOME.)

Gomori stain *Syn.* METHENAMINE-SILVER STAIN.

Gomphonema A genus of pennate DIATOMS. In e.g. *G. olivaceum* the cell adheres to the substratum by a 'stalk' of secreted mucilage.

gonal nucleus See CONJUGATION (1)(c).

Gonapodya See MONOBLEPHARIDALES.

Gonatobotrys See HYPHOMYCETES and CONIDIUM.

Gonatobotryum See HYPHOMYCETES.

gongylidia (*sing.* gongylidius) Nodular fungal structures which develop in the FUNGUS GARDENS of parasol ants or in the fungus combs of termites (see TERMITE-MICROBE ASSOCIATIONS).

gonidium A cell which is involved in asexual reproduction: see e.g. LEUCOTHRIX and VOLVOX. (The term was once applied to a photobiont cell in a lichen in the mistaken belief that such cells served a reproductive function.)

Gonimochaete See ENTOMOPHTHORALES.

gonococcal ophthalmia See GONORRHOEA.

gonococcus *Neisseria gonorrhoeae.*

***Gonometa* virus** See PICORNAVIRIDAE.

gonorrhoea (gonorrhea) An acute or chronic VENEREAL DISEASE which in nature affects only man; it is caused by *Neisseria gonorrhoeae* (the 'gonococcus' – see NEISSERIA). Incubation period: 2–5 days or longer. Gonococci adhere to the urethral epithelium by means of fimbriae; they penetrate the mucosa, causing an acute inflammatory response, and the infection spreads readily to adjacent tissues. In males, infection may be asymptomatic or may involve URETHRITIS, with painful urination and a yellowish mucopurulent discharge; prostatitis and epididymitis may occur. In females, infection may be asymptomatic or there may be a purulent vaginal discharge; infection may spread via the endometrium, fallopian tubes and ovaries to the pelvic peritoneum (*pelvic inflammatory disease*, PID). (See also CURTIS-FITZ-HUGH SYNDROME.) In either sex, sterility may result. Some strains of gonococci may become disseminated: bacteraemia, associated with e.g. fever, may be followed by skin lesions, arthritis, and (sometimes fatal) endocarditis and/or meningitis. Gonococcal contamination of the eyes may lead to conjunctivitis (*gonococcal ophthalmia*) which can cause blindness if untreated; infants may be infected during birth (*ophthalmia neonatorum*), and neonates may therefore be prophylactically treated with eye-drops containing e.g. 1% w/v silver nitrate (*Credé procedure*) or penicillin. *Lab. diagnosis*: smears of the discharge are stained (e.g. by the Gram stain or fluorescent antibody techniques) to reveal large numbers of diplococci in PMNs. Culture (e.g. on THAYER-MARTIN AGAR) may be necessary, especially for diagnosing females. A PROBE-based test (the PACE 2C test) has been used for detecting *N. gonorrhoeae* and *Chlamydia trachomatis* in a single assay (particularly useful because co-infection with these two pathogens is quite common) [evaluation of the PACE 2C test with endocervical specimens: JCM (1995) *33* 2587–2591]. A test for detecting *N. gonorrhoeae* based on the ligase chain reaction (see LCR) reported results superior to those obtained by culture [JCM (1997) *35* 239–242].

Chemotherapy depends on (changing) patterns of antibiotic resistance in the pathogen; penicillins (with probenecid), tetracycline, erythromycin and spectinomycin have been used.

Gonyaulax See DINOFLAGELLATES; BIOLUMINESCENCE; RED TIDE.

gonyautoxins Toxins produced by *Gonyaulax* spp and other RED TIDE dinoflagellates. The toxins are sulphonated derivatives of SAXITOXIN and neosaxitoxin; they have a low toxicity, but can be converted to saxitoxin/neosaxitoxin by low pH (e.g. in the stomach).

Gonyostomum See CHLOROMONADS.

Good buffers Zwitterionic pH buffers, e.g. BICINE and HEPES.

goose parvovirus See PARVOVIRUS.

Gordona An obsolete genus of bacteria currently included in the genus RHODOCOCCUS.

Gorgonzola cheese See CHEESE-MAKING.

Gouda cheese See CHEESE-MAKING.

Gower coefficient See entry S_G.

gp (1) Gene product. For example, gp6 = the polypeptide product of gene 6. (2) In retroviruses, gp designates a *glycoprotein* gene product (see RETROVIRIDAE).

Gracilaria See RHODOPHYTA. (See also HOLMSELLA.)

Gracilicutes A proposed division of the PROCARYOTAE comprising bacteria in which the CELL WALL is typically of the Gram-negative type; classes: ANOXYPHOTOBACTERIA; OXYPHOTOBACTERIA; SCOTOBACTERIA.

gradient differential centrifugation See CENTRIFUGATION.

gradient elution Elution during which the eluent, *A*, is progressively diluted with another solvent or solution, *B*, until it is finally pure *B*.

gradient plate An agar plate within which a given antibiotic is distributed such that its concentration varies, in gradient fashion, from one side of the plate to the other; the plate is used to isolate antibiotic-resistant mutants. To prepare a gradient plate, a volume of antibiotic-free agar is allowed to set in a Petri dish – one side of which is raised ca. 5 mm above the horizontal during setting; with the Petri dish level, an equal volume of antibiotic-agar is then added and allowed to set. Plates are left for ca. 1 day before inoculation; this permits equilibration of antibiotic concentrations in a vertical (but not lateral) direction.

A two-dimensional gradient plate may be prepared in a square dish; two wedge-shaped layers are prepared as described above, and a further two layers are added perpendicular to the first two. A given layer may incorporate a specific electrolyte, acid etc [JGM (1985) *131* 2865–2869].

Gradocol filter A type of membrane filter. (See also FILTRATION.)

gradostat An apparatus used for a specialized form of CONTINUOUS CULTURE in which two different solutions flow, simultaneously but in opposite directions, through each of a number of serially-linked chemostats; under steady-state conditions the concentration of each of the given solutes exhibits a stepped gradient along the line of chemostats. In one type of gradostat five chemostats are arranged at progressively higher levels so that one of the solutions can flow through the system under gravity – i.e., overflowing a weir in each chemostat and passing down to the chemostat below it; the other solution is pumped into the lowest chemostat, and pumped from each chemostat to the one above it. In this type of gradostat organisms as well as solutes are transferred between chemostats. By contrast, in the *diffusion-coupled gradostat* (Herbert model) several chemostats (each with its own independent inflow and outflow of medium) are connected in a horizontal row such that the medium in one chemostat is separated from that in the next by a membrane filter; thus, solutes can diffuse between chemostats but the organisms in a given chemostat remain in that chemostat. Such a system can be used e.g. to investigate metabolic interactions between different species [FEMS Ecol. (1985) *31* 239–247].

[Gradostats and other novel growth systems: MS (1985) *2* 53–60.]

Graff–Reinet disease *Syn.* MAEDI.

Grahamella A genus of Gram-negative bacteria (family BARTONELLACEAE); the organisms occur worldwide and are parasitic (though rarely, if ever, pathogenic) in various small mammals (e.g. rodents) but not in man. *Grahamella* spp (*G. talpae*, the type species; *G. peromysci*) are non-flagellated bacilli or coccobacilli, commonly ca. 1 μm long; the organisms are transmitted e.g. by fleas and occur in the erythrocytes of the infected vertebrate host.

grain bloat See BLOAT.

Grain convention A system for numbering the microtubular triplets in a kinetosome. Looking at a cross-section of a ciliary kinetosome from the direction of the *interior* of the cell, the number 1 triplet is that which is aligned with the axis of the KINETY and which is near the posterior side of the kinetosome (relative to the cell's anterior–posterior axis); triplets 2–9 follow in a clockwise sequence from number 1. (cf. PITELKA CONVENTION.)

Gram-negative See GRAM STAIN.

Gram-negative enrichment broth See GN BROTH.

Gram-positive See GRAM STAIN.

Gram reaction The result (positive, negative, or variable) of the GRAM STAIN; the Gram reaction does not *necessarily* indicate the chemical nature of the cell envelope (cf. GRAM TYPE). In a non-staining method for determining Gram reaction, bacterial growth is emulsified with one drop of aqueous KOH (3%); Gram-negative bacterial growth becomes viscid, or gels, in 5–60 sec [AEM (1982) *44* 992–993].

Gram stain An important bacteriological STAINING procedure discovered empirically in 1884 by the Danish scientist Christian Gram. When bacteria are stained with certain basic dyes the cells of some species (*Gram-negative* species) can be easily decolorized with organic solvents such as ethanol or acetone; cells of *Gram-positive* species resist decolorization. The ability of bacteria to either retain or lose the stain generally reflects fundamental structural differences in the CELL WALL, and is an important taxonomic feature; the Gram stain is therefore used as an initial step in the identification of bacteria.

Procedure (one of many variations). A heat-fixed smear of bacteria on a slide is stained for ca. 1 min with one of certain basic TRIPHENYLMETHANE DYES (commonly crystal violet); most bacteria take up the dye. The smear is then rinsed briefly under running water, treated for 1 min with LUGOL'S IODINE, and again rinsed with water. Decolorization is then attempted by treating the smear with ethanol, acetone, or iodine–acetone. This is the critical step: solvent is allowed to run over the tilted slide until dye no longer runs *freely* from the smear (ca. 1–3 sec), and the smear is *immediately* rinsed under running water; at this stage any Gram-negative bacteria in the smear will be colourless, while any Gram-positive cells will still be violet. The smear is then counterstained for ca. 30 sec (e.g. with dilute carbolfuchsin) to stain any Gram-negative bacteria present in the smear. Finally, the slide is rinsed briefly under the tap, blotted dry, and examined under the microscope. (In *Hucker's method* the dye solution is made by adding 20 ml crystal violet (10% w/v in 95% ethanol) to 80 ml of 1% w/v aqueous ammonium oxalate, and filtering after 24 h. The decolorizing agent is 95% ethanol. The counterstain is safranin O.) (See also BROWN–HOPPS MODIFICATION and KOPELOFF MODIFICATION.)

The mechanism of Gram-staining is still not understood. The integrity of the CELL WALL is necessary for Gram-positivity: disrupted Gram-positive cells do not retain the dye. It may be that Gram-positive cells retain the dye–iodine complex owing to the lower permeability of their cell walls: the PEPTIDOGLYCAN of Gram-positive cell walls is more extensively cross-linked than that in Gram-negative species. (N.B. *Mammalian* cells are uniformly Gram-negative.)

The cells of some bacteria (e.g. *Bacillus* spp) are strongly Gram-positive when young, but tend to become Gram-negative in ageing cultures; this may reflect degenerative changes in the cell wall. Some bacteria give a *Gram-variable* reaction, i.e., they are sometimes Gram-positive, sometimes Gram-negative; this could reflect e.g. minor variations in staining technique. To avoid the taxonomic problems of Gram-variable bacteria it has been suggested that classification be based on GRAM TYPE rather than on GRAM REACTION.

Gram type A taxonomically useful designation (either *Gram type positive* or *Gram type negative*) which is based on biochemical and anatomical properties of a bacterial cell (see e.g. CELL WALL) rather than on GRAM REACTION; designation on this basis avoids the taxonomic problems caused by a Gram-variable reaction in the GRAM STAIN. [JGM (1982) *128* 2261–2270.]

Gram-variable See GRAM STAIN.

gramicidins A group of linear peptide ANTIBIOTICS, each of which has a formyl group at the N-terminal and an ethanolamine residue at the C-terminal. ('Gramicidin J', also called 'gramicidin S', is misnamed: it is a TYROCIDIN.) Gramicidins act as channel-forming IONOPHORES, forming transient ion-conducting dimers in lipid bilayers; such channels permit the passage of a range of hydrated monovalent cations as well as protons. Against Gram-positive bacteria gramicidins may be bacteriostatic or (at higher concentrations) bactericidal; most Gram-negative bacteria tend to be resistant.

grana See CHLOROPLAST.

granulin See BACULOVIRIDAE.

granulocyte See PMN.

granuloma A type of lesion resulting from a local inflammatory response to a persistent antigen or toxin at a given site in the body; the lesion consists of a hard mass composed mainly of macrophages and strands of fibrous connective tissue – although other types of cell (e.g. lymphocytes, plasma cells, polymorphs etc) may also be present, and the lesion may also contain GIANT CELLS. Granuloma formation involves a T cell-dependent immune response. (See also DELAYED HYPERSENSITIVITY.)

granuloma inguinale (donovanosis) A chronic human disease, occuring mainly in the tropics and subtropics, in which granulomatous ulcers develop chiefly on the genitals; the causal agent, *Calymmatobacterium granulomatis*, is transmitted by intimate contact. In smears (stained e.g. with Wright's stain) the pathogen appears as clusters of blue or black 'Donovan bodies' in the cytoplasm of large mononuclear cells.

granuloplasm See e.g. SARCODINA.

Granuloreticulosea A class of protozoa (superclass RHIZOPODA) which form delicate, finely granular or hyaline reticulopodia (or, rarely, finely pointed, granular, non-anastomosing pseudopodia). Orders: Athalamida (amoebae naked: e.g., *Arachnula*, *Biomyxa*); Monothalamida (amoebae enclosed in a single-chambered organic or calcareous test, not exhibiting an alternation of generations: e.g., *Lieberkuehnia*); FORAMINIFERIDA.

granulose (1) (*noun*) A $(1 \rightarrow 4)$-α-glucan which accumulates in the cells of *Clostridium* spp prior to sporulation; it is rapidly degraded during sporulation, and may thus function as an endogenous reserve of carbon and energy for sporulation.

(2) (*adj.*) Having a granular appearance.

granulosis An INSECT DISEASE caused by a GRANULOSIS VIRUS.

granulosis viruses (GVs) Viruses belonging to subgroup *B* of the BACULOVIRIDAE (q.v.). The viruses replicate mainly in the nucleus, but some replication can also occur in the cytoplasm. The virion contains a single nucleocapsid within the envelope. Type species: *Trichoplusia ni* (cabbage looper) GV (TnGV); other members: *Plodia interpunctella* (Indian meal moth) GV, *Cydia pomonella* (codling moth) GV, and viruses isolated from a wide range of other lepidopteran insects.

granum (*pl.* grana) See CHLOROPLAST.

Granville wilt (Southern bacterial wilt) A soil-borne TOBACCO DISEASE, involving wilting and yellowing of leaves, caused by *Pseudomonas solanacearum*. (See also SCOPOLETIN.)

granzymes See APOPTOSIS.

Graphidales An order of fungi of the ASCOMYCOTINA; members include crustose LICHENS (photobiont generally trentepohlioid). Ascocarp: APOTHECIOID, or immersed and ostiolate, with paraphyses. Asci: unitunicate and cylindrical with a thickened apex. Genera: e.g. *Diploschistes*, GRAPHIS, *Phaeographis*, *Thelotrema*.

Graphiolaceae See USTILAGINALES.

Graphis A genus of crustose LICHENS (order GRAPHIDALES); photobiont: a green alga. Thallus: thin, smooth or wrinkled. Apothecia: black LIRELLAE; ascospores: colourless, multiseptate. Species occur mainly on bark.

graphitization An early term for the anaerobic corrosion of iron by sulphate-reducing bacteria.

Graphium A synnema-forming genus of the HYPHOMYCETES. (See also SAP-STAIN.)

GRAS 'Generally recognized as safe' – a category of food additives (including some food PRESERVATIVES) recognized in the USA.

grass diseases Diseases which can affect non-cereal grasses include e.g. CHOKE, DOLLAR SPOT, SILVER TOP, SNOW MOULD, and ophiobolus patch disease (see TAKE-ALL). (See also FIJIVIRUS, CEREAL DISEASES and SUGARCANE DISEASES.)

Grasseella See EIMERIORINA.

graticule See MICROMETER.

gratuitous inducer See OPERON.

gravity slide sampler *Syn.* DURHAM SHELTER.

gray (Gy) A unit of absorbed radiation equal to 1 joule of energy absorbed by 1 kg of material. (1 Gy = 100 RAD; 10^3 Gy = 1 kGy.)

gray mold See GREY MOULD.

greasy pig disease (exudative dermatitis) An acute, often fatal disease of piglets characterized by a non-pruritic dermatitis, the presence of a greasy exudate on the skin, dehydration and emaciation; aetiology: uncertain, but a Gram-positive coccus (*Staphylococcus hyicus*) and/or a virus may be involved.

green algae Algae of the CHLOROPHYTA.

green bacteria (green sulphur bacteria) (1) Bacteria of the suborder CHLOROBIINEAE. (2) Bacteria of the family CHLOROBIACEAE.

green ear disease See PHYLLODY.

green fluorescent protein (GFP) A fluorescent protein, found in jellyfish of the genus *Aequorea*, which emits green light ($\lambda = 508$ nm) when irradiated with blue light ($\lambda = 395$ nm). The gene for GPF has been widely used as a partner in gene fusion experiments (see FUSION PROTEIN sense 2).

[Green fluorescent protein (review): ARB (1998) *67* 509–544.]

green laver See LAVER and ULVA.

green manure See *Azolla* in NITROGEN FIXATION.

green monkey fever *Syn.* MARBURG FEVER.

green oak Oak wood which has been stained green as a result of infection by *Chlorociboria aeruginascens* (= *Chlorosplenium aeruginascens*, often called *Chlorosplenium aeruginosum*). Green oak is valued as a medium for inlay work etc. (cf. BROWN OAK.)

green spot disease See ALGAL DISEASES.

green sulphur See SULPHUR.

green sulphur bacteria See GREEN BACTERIA.

Greenberg–Smith sampler An ALL-GLASS IMPINGER.

greenhouse effect The warming of the planet due to rising levels of CO_2, methane and certain other gases (which effectively 'trap' some of the radiant energy reaching the Earth's surface).

In the CARBON CYCLE, large amounts of CO_2 are produced by animals, plants and microbes during respiration or fermentation, but large amounts are used ('fixed') during PHOTOSYNTHESIS; globally, the important balance between biological production and uptake of CO_2 is being upset e.g. by the burning of vast amounts of fossil fuel (gas, petroleum etc.). (See also METHANOTROPHY.)

It might be thought that raised levels of CO_2 would stimulate growth in plants, and, thereby – due to greater sequestration of CO_2 by photosynthesis – tend to offset the rise in CO_2. Some agricultural, crop-type plants do respond to increased CO_2 by improved growth. However, experimental tropical ecosystems under increased CO_2 have shown, in addition to increased root growth, an increased efflux of CO_2 from soil to atmosphere; this efflux was apparently due to stimulated metabolic activity of microorganisms in the RHIZOSPHERE. Thus, an increased uptake of CO_2 by plants may itself be offset by increased efflux of CO_2 from the soil [Science (1992) *257* 1672–1675].

More recently it was suggested that certain pasture grasses in the South American savannas may offer a useful sink for carbon in their particularly massive and deep root systems [Nature (1994) *371* 236–238].

It has been pointed out that global warming may itself result in increased levels of greenhouse gases – i.e. a potential positive feedback effect [BC (1996) *5* 1069–1083].

It has been assumed that the avoidance of deforestation is of major importance in the context of global warming because such a policy preserves a valuable means for controlling levels of atmospheric carbon dioxide [Nature (2001) *410* 429]. However, recent experiments have indicated that the ability of forest systems to sequester carbon dioxide may be limited by shortages of nutrients and water and/or by the relatively rapid turnover of organic carbon in the litter layer [Nature (2001) *411* 469–472; Nature (2001) *411* 466–469; commentary: Nature (2001) *411* 431–433].

greening (1) See HAEMOLYSIS. (2) See MEAT SPOILAGE and DFD MEAT.

gregarin The trophozoite form of a gregarine.

Gregarina A genus of the GREGARINASINA.

Gregarinasina (Gregarinidia) A subclass of protozoa (class SPOROZOASIDA) parasitic in invertebrates (e.g. annelids, insects). In members of this subclass the mature gametocytes occur extracellularly in the host (cf. COCCIDIASINA); the mature organism generally has a MUCRON or an EPIMERITE. In the *cephaline* gregarines (e.g. *Gregarina* spp) the mature organism is septate – the nucleus usually being in the DEUTOMERITE; in *acephaline* gregarines (e.g. MONOCYSTIS) the mature organism is aseptate. Species whose life cycles do not include schizogony (the *eugregarines*) are commonly non-pathogenic; species which undergo schizogony as well as sexual reproduction (the *schizogregarines*) are commonly pathogenic. Some authors divide the schizogregarines into the *archigregarines* (parasitic in annelids) and the *neogregarines* (parasitic in insects). Genera include *Actinocephalus*, *Stenophora*, *Urospora*.

gregarine A member of the GREGARINASINA.

gregarine movement See MOTILITY.

Gregarinia (1) A subclass of protozoa (class Telosporea) later classified as the subclass GREGARINASINA. (2) A subclass of protozoa (class Sporozoea [JP (1980) *27* 37–58]) equivalent to the GREGARINASINA.

Gregarinidia *Syn.* GREGARINASINA.

grepafloxacin See QUINOLONE ANTIBIOTICS.

grex See DICTYOSTELIUM.

grey mould ('botrytis') A disease, caused by *Botrytis cinerea*, which can affect a wide range of plants as well as stored fruits and vegetables (e.g. beans, peas, raspberries and strawberries); the disease is encouraged by cool, damp conditions, and is characterized by a grey, fluffy surface mould overlying a soft, commonly brown, rot. (See also DAMPING OFF.)

grid See ELECTRON MICROSCOPY (a).

Gridley fungus stain A staining procedure that depends on the oxidation (by chromic acid) of adjacent hydroxyl groups in fungal cell wall polysaccharides, followed by the reaction of the resulting aldehyde groups with e.g. SCHIFF'S REAGENT. [Method: Book ref. 53, p. 2207.]

Griffith's tube An apparatus used for grinding tissues. It consists of a stout, large-diameter glass tube which is closed at the bottom; the lower, inner part of the tube has a roughened surface – against which the tissue is ground by means of a close-fitting pestle-like glass rod.

Griffith's typing (of group A streptococci) A serological (slide agglutination) TYPING method based on strain differences in the M and/or T antigens.

Griffithsia See RHODOPHYTA.

Grifola See APHYLLOPHORALES (Polyporaceae).

grippe *Syn.* INFLUENZA.

griseofulvin (7-chloro-4,6-dimethoxycoumaran-3-one-2-spiro-1′-(2′-methoxy-6′-methylcyclohex-2′-en-4′-one)) An antifungal ANTIBIOTIC produced by *Penicillium griseofulvum*. It is inhibitory for actively growing dermatophytes and certain other filamentous fungi but is inactive against e.g. yeasts and bacteria. Griseofulvin acts primarily as a MICROTUBULE inhibitor, preventing the assembly of tubulin dimers into microtubules and thereby inhibiting e.g. mitosis. Additionally, the antibiotic interferes with the pattern of glucan and glycoprotein deposition in the fungal cell wall; as a result the tips of the hyphae become characteristically curled (hence the early name for the antibiotic: 'curling factor').

Griseofulvin is widely used in the treatment of dermatophyte infections of the skin, hair and nails. It is administered orally (it is not effective topically) and accumulates in keratinized tissues; it apparently forms a complex with keratin which may decrease the susceptibility of the keratin to fungal keratinases. Griseofulvin-resistant dermatophytes occur [Arch. Derm. (1981) *117* 16–19].

Griseofulvin is active against certain plant-pathogenic fungi (e.g. *Botrytis* spp, *Alternaria solani*). It can be taken up by plant roots and has low phytotoxicity; it has limited use as a systemic ANTIFUNGAL AGENT in plants.

griseoviridin An antibiotic structurally related to streptogramin A (see STREPTOGRAMINS).

gRNAs See RNA EDITING.

Grocott–Gomori stain *Syn.* METHENAMINE–SILVER STAIN.

***groEL* gene** See HEAT-SHOCK PROTEINS.

***groES* gene** See HEAT-SHOCK PROTEINS.

Gromia See FILOSEA.

Gromiida See FILOSEA.

groove (in DNA) See DNA.

Grossglockneria See CILIOPHORA.

ground squirrel hepatitis virus (GSHV; ground squirrel hepatitis B virus, GSHBV) A virus of the HEPADNAVIRIDAE which infects Beechey ground squirrels (*Spermophilus beecheyi*); infected animals show little or no evidence of hepatitis and appear not to develop hepatocellular carcinoma. (cf. WOODCHUCK HEPATITIS VIRUS.)

groundnut crinkle virus See CARLAVIRUSES.

groundnut eyespot virus See POTYVIRUSES.

groundwater See WATER SUPPLIES.

group I/II introns See SPLIT GENE (b) and (c).

group-phase antigens See PHASE VARIATION.

group stage See TOXOPLASMOSIS.

group translocation (1) See TRANSPORT SYSTEMS. (2) *Syn.* VECTORIAL GROUP TRANSLOCATION.

group treponemal antigens *Syn.* REITER ANTIGENS.

grouping antiserum Any antiserum which can agglutinate all the members of a particular group of serotypes within a given genus, i.e., serotypes which have in common one or more antigens which are not found in serotypes outside the group (see e.g. KAUFFMANN–WHITE CLASSIFICATION).

growth In a single living cell: a *coordinated* increase in the mass of essential cell components leading, typically, to progress through the CELL CYCLE; in a population of cells or a mycelium: a coordinated increase in biomass, or an increase in the number of cells in the population. (An increase in mass does not necessarily signify growth since it may be due e.g. to the accumulation of an intracellular storage compound.)

Growth is said to be *balanced* when the increase in mass of all essential cell components occurs exponentially and is coincident with the rate of increase in total biomass or in cell numbers. In *unbalanced* growth there is a change in the rate of synthesis of some cell component(s) relative to that of other components; under these conditions (which occur e.g. during the adaptive changes in the lag phase of BATCH CULTURE) changes in biomass are not directly proportional to changes in cell numbers. (See also CONTINUOUS CULTURE.)

Growth may be measured e.g. by monitoring cell numbers (see COUNTING METHODS), by measuring the increase in dry weight of biomass formed in a given time interval, by monitoring the uptake and metabolism (or release) of particular substances, or by measuring the amount of a radioactive metabolite incorporated in biomass in a given time. Growth may be expressed in the form of a *growth curve*: a graph in which the number of cells, or the biomass, is plotted against time.

(a) *Bacterial growth.* Growth in individual bacterial cells commonly leads to (asexual) reproduction (i.e., cell division) – usually by BINARY FISSION, but sometimes by BUDDING or by MULTIPLE FISSION; the events which occur during growth in a single cell are described under CELL CYCLE. In some species, growth may be followed by differentiation, e.g. the formation of a dormant cell (see e.g. ENDOSPORE).

The growth of a bacterial population. When cells divide by *binary* fission, each individual cell gives rise to 2, 4, 8, 16, 32. ... cells after 1, 2, 3, 4, 5. ... generations (i.e. rounds of division). Hence, the number of cells (N) derived from a single cell after n generations is given by:

$$N = 2^n$$

Similarly, after n generations, the number of cells derived from an initial cell population of N_0 is given by:

$$N = 2^n N_0$$

For bacteria dividing by binary fission, a growth curve can be prepared by plotting cell numbers on a $\log_2$ scale; this permits the number of generations per unit time to be read off directly (since each unit on the $\log_2$ scale is equal to one generation), and the graph also indicates clearly any changes in the growth rate. (The $\log_2$ and $\log_{10}$ of any number can be interconverted by use of the formula: $\log_{10} N = 0.301 \log_2 N$.)

The rate of growth may be expressed in terms of the SPECIFIC GROWTH RATE, μ. When μ is constant (e.g. during exponential growth in batch culture), the relationship between μ and the *doubling time* (i.e. the time taken for the biomass – or population – to double) is given by:

$$t_d = \frac{0.693}{\mu}$$

in which t_d is the doubling time. For some cells (including those of e.g. *Escherichia coli*) the *minimum* doubling time (= minimum generation time) is of the order of minutes, while for others (e.g. those of certain *Mycobacterium* spp) it may be many hours. (Variation may occur among the doubling times of individual cells in a single-species population – see also SYNCHRONOUS CULTURE.)

Provided that μ is constant, the number of generations (= number of doublings of biomass, or of cell numbers) per unit time is given by k:

$$k = \frac{(\log_2 N_t - \log_2 N_0)}{t}$$

– k being the *exponential growth rate constant* (sometimes, confusingly, symbolized by μ instead of k), and N_t and N_0 being the populations at time t and time zero, respectively. The quantity 1/k is the *mean generation time* (= *mean doubling time*).

(b) *Fungal growth.* According to species and conditions, growth may involve or lead to (i) an increase in cell size or (in mycelial fungi) an increase in hyphal length and/or hyphal branching; (ii) vegetative differentiation – e.g. a phase change in DIMORPHIC FUNGI; (iii) asexual reproduction; (iv) sexual reproduction.

Hyphal growth. Linear hyphal growth occurs by the incorporation of new materials at the extreme distal end of the hypha (i.e., 'tip growth'); precursors of the CELL WALL and CYTOPLASMIC MEMBRANE are assembled in the GOLGI APPARATUS and then transported, in vesicles, to the hyphal apex where fusion occurs between the vesicles and the cytoplasmic membrane. The unidirectional flow of vesicles to the hyphal apex appears to be related to a phenomenon observed in growing (but not in non-growing) hyphae: the continual circulation of positive charges through the terminal region of the hypha; thus, positive charges from the environment pass into the hypha at, and close to, the hyphal apex, and pass out into the environment a short distance proximal to the apical region. Such transhyphal currents of positive charge have been observed in several fungi [JGM (1984) *130* 3313–3318] and may be common to all growing fungal hyphae. (Axially aligned currents of positive charge also occur e.g. in zygotes of the brown algae *Fucus* and *Pelvetia* and in growing plant roots and pollen tubes.) In the water mould *Achlya bisexualis* the current consists of protons; the proton current is apparently due to the preferential distribution of proton–amino acid symporters in the apical region of the hypha, and PROTON ATPASES in the adjacent region. (In e.g. *Pelvetia* the circulating current appears to consist primarily of calcium ions.) The way in which these circulating currents contribute to the unidirectional flow of intracellular cell-building vesicles is not clear; it has been suggested e.g. that currents of charge may promote the 'electrophoretic' translocation of vesicles, or (in *Achlya bisexualis*) that the localized decrease in pH in the hyphal tip may affect vesicle–cytoskeleton interaction, thereby regulating the translocation and localization of vesicles. Interestingly, the polarity of growth of various filamentous fungi can be manipulated experimentally by applied electrical fields [JGM (1986) *132* 2515–2525].

The development of a hyphal *branch* is apparently preceded by the commencement of charge influx at the given location. Branching can be enhanced by CYTOCHALASINS A and E (which may affect microfilament systems involved in vesicle transport) and by IONOPHORES such as A23187 and CCCP (thus implicating cytoplasmic ions in the control of branch initiation) [JGM (1986) *132* 213–219].

Hyphal growth (both linear and branching) requires the coordinated action of autolytic and synthetic enzymes in the region of growth – together with an intracellular turgor pressure to act as driving force for surface expansion. Antibiotics which inhibit cell wall formation (e.g. POLYOXINS) cause the development of bulbous hyphae and distorted germ tubes in isotonic media; hyphal tips (and budding yeasts) burst in hypotonic media.

[Cell wall synthesis in apical hyphal growth: Int. Rev. Cytol. (1986) *104* 37–79.]

Growth in unicellular fungi. Growth may involve BUDDING or BINARY FISSION. BATCH CULTURES of unicellular fungi exhibit lag, exponential, stationary and death phases of growth.

Growth in dimorphic fungi. Whether growth occurs in the unicellular or mycelial phase may be determined by temperature (e.g. in *Paracoccidioides brasiliensis*), by the availability of particular nutrients (e.g. in *Candida albicans, Mucor rouxii*), or by a combination of factors which include temperature and redox potential (e.g. in *Histoplasma capsulatum*). In a given organism the composition of the CELL WALL (q.v.) may differ in the unicellular and mycelial phases. In *C. albicans* the distribution of ACTIN in a growing bud differs from that in a growing hypha [JGM (1986) *132* 2035–2047].

(See also SECONDARY METABOLISM.)

growth curve See GROWTH and BATCH CULTURE.

growth cycle (1) *Syn.* CELL CYCLE. (2) A term used by some authors to refer to the (non-cyclical) sequence of phases in BATCH CULTURE (i.e. lag phase to death phase).

growth precursor cell (shut-down cell) A cell which differs from its parent cell in that it proceeds through the CELL CYCLE only under (presumably) favourable environmental conditions. Cells conforming to this description include e.g. the swarmer cells of *Caulobacter crescentus, Hyphomicrobium* sp, and *Rhodomicrobium vannielii* – cells which are characterized by low levels of endogenous metabolism, no synthesis of DNA or rRNA, and the apparent ability to respond to external growth-triggering stimuli. It has been suggested that a growth precursor cell may function as a survival cell and/or as a stage from which one of various cell types may develop according to particular environmental stimuli. [Review: Book ref. 28, pp. 187–247.]

growth rate constant (1) *Syn.* SPECIFIC GROWTH RATE. (2) The exponential growth rate constant – see GROWTH (a).

growth-specific yield coefficient See YIELD COEFFICIENT.

growth yield coefficient *Syn.* YIELD COEFFICIENT.

Grunstein–Hogness procedure *Syn.* COLONY HYBRIDIZATION.

Gruyère cheese See CHEESE-MAKING.

G_s protein See ADENYLATE CYCLASE.

GS/GOGAT See AMMONIA ASSIMILATION.

GSHBV See GROUND SQUIRREL HEPATITIS VIRUS.

GSHV GROUND SQUIRREL HEPATITIS VIRUS.

GSP See PROTEIN SECRETION.
GSR Generalized SHWARTZMAN REACTION.
GT ... AG rule See SPLIT GENE (a).
GT1$_b$ See FOOD POISONING (*Vibrio*).
GTA Gene transfer agent: see CAPSDUCTION.
GTP Guanosine 5′-triphosphate. (See also ATP and NUCLEOTIDE.)
GTP-binding proteins *Syn.* G PROTEINS.
GU ... AG rule See SPLIT GENE (a).
guanidine ($(H_2N)_2C{=}NH$) Guanidine salts (e.g. guanidine hydrochloride: $(H_2N)_2C{=}NH_2{}^+Cl^-$) act as ANTIVIRAL AGENTS against a number of picornaviruses (e.g. polioviruses) by inhibiting viral RNA synthesis. Although cytotoxicity is low, guanidine is not used clinically since resistant mutants arise rapidly; guanidine-dependent mutants have also been isolated.
guanidine antifungal agents See e.g. DODINE and GUAZATINE.
guanosine See NUCLEOSIDE and Appendix V(a).
guanosine tetraphosphate (guanosine 5′-diphosphate 3′-diphosphate) See STRINGENT CONTROL (sense 1).
guanylate cyclase See ETEC (ST-I).
guanylin See EAST1 and ETEC (ST-I).
guanylyl transferase See MRNA (b).
Guarnieri bodies (B-type inclusion bodies) Eosinophilic INCLUSION BODIES, each surrounded by a clear non-staining 'halo', characteristically formed in the cytoplasm of cells infected with vaccinia, variola or cowpox viruses (see POXVIRIDAE).
guazatine A guanidine ANTIFUNGAL AGENT used e.g. as a seed treatment for the control of bunt of wheat, and – mixed with e.g. imazalil – as a spray against cereal smuts, net blotch, etc.
Guepiniopsis See DACRYMYCETALES.
guide RNAs (gRNAs) See RNA EDITING.
guide sequence See SPLIT GENE.
Guillain–Barré syndrome (GBS) An immune-mediated disorder of the peripheral nerves involving e.g. a progressive motor weakness in at least two limbs. In the majority of patients GBS is preceded by an infectious disease – e.g. a (very) small proportion of patients with a *Campylobacter* infection may develop GBS.
Guilliermondella A genus of fungi (family SACCHAROMYCETACEAE) which form budding yeast cells as well as a true septate mycelium. One species: *G. selenospora*. [Book ref. 100, pp. 151–153.]
guinea-pig agent *Legionella micdadei*.
guinea-pig cytomegalovirus See BETAHERPESVIRINAE.
gullet (1) The BUCCAL CAVITY in certain ciliates. (2) See EUGLENOID FLAGELLATES.
guluronic acid See ALGINATE.
gum xanthan *Syn.* XANTHAN GUM.
Gumboro disease *Syn.* infectious bursal disease (see INFECTIOUS BURSAL DISEASE VIRUS).
gumma See e.g. SYPHILIS.
gummosis Any plant disease in which the lesions exude a sticky liquid; examples: gummosis of cucumber (caused by *Cladosporium cucumerinum*), of *Axonopus* spp (Gramineae) (caused by *Xanthomonas axonopodis*), and of sweet orange (*Citrus sinensis*) (caused by *Phytophthora parasitica*).
Gunflintia See STROMATOLITES.
Gunnera See NOSTOC.
gut flora (human) See GASTROINTESTINAL TRACT FLORA.
guttulate Containing one or more GUTTULES.
guttule *(mycol.)* A globule or droplet (of oil?) within a spore.
Guttulina See ACRASIOMYCETES.
Guttulinopsis See ACRASIOMYCETES.
GV GRANULOSIS VIRUS.
Gy See GRAY.
Gyalecta See GYALECTALES.
Gyalectales An order of mainly lichenized fungi of the ASCOMYCOTINA. Ascocarp: APOTHECIOID; hamathecium: paraphyses. Asci: clavate or cylindrical. Genera: e.g. *Dimerella*, *Gyalecta*, *Petractis*.
GYC agar See ACETOBACTER.
Gymnamoebia A subclass of naked (non-testate) amoebae (AMOEBIDA, PELOBIONTIDA and SCHIZOPYRENIDA). (cf. TESTACEALOBOSIA.)
Gymnoascales An order of mainly keratinophilic fungi of the subdivision ASCOMYCOTINA; many species have anamorphs among the DERMATOPHYTES. Ascocarp: cleistothecioid, with or without appendages, sometimes insubstantial (consisting e.g. of a loose hyphal network); sessile or stipitate. Asci: unitunicate. Ascospores: unicellular, spherical, discoid or fusiform, sometimes forming a mazaedial mass at maturity. Genera: e.g. AJELLOMYCES, *Aphanoascus*, ARTHRODERMA, *Gymnoascus*, *Myxotrichum*, NANNIZZIA and *Onygena* (see also MAZAEDIUM); of these, *Aphanoascus* and *Onygena* belong in the family Onygenaceae, and the others in the family Gymnoascaceae.
Gymnoascus See GYMNOASCALES.
gymnocarpic development (also: gymnocarpous development) (*mycol.*) The mode of development of a fruiting body in which the spores are exposed to the environment throughout their differentiation and maturation. (See e.g. APHYLLOPHORALES; cf. ANGIOCARPIC DEVELOPMENT.)
Gymnodinioides See HYPOSTOMATIA.
Gymnodinium See DINOFLAGELLATES; RED TIDE; ZOOXANTHELLAE.
Gymnomyxa A protistan phylum (equivalent to the mycological division MYXOMYCOTA) comprising the subphyla MYCETOZOA, Plasmodiophorina (equivalent to PLASMODIOPHOROMYCETES), and Labyrinthulina (equivalent to LABYRINTHULOMYCETES). [Book ref. 147, pp. 4–5.]
Gymnosphaera See CENTROHELIDA.
Gymnosporangium A genus of typically heteroxenous rust fungi (class UREDINIOMYCETES) whose primary and secondary hosts are commonly members of the families Juniperaceae and Rosaceae, respectively. Most species are DEMICYCLIC RUSTS. The aecial and telial stages often form on GALLS which develop on the host plant as a response to infection by the rust; thus, e.g. *G. juniperi-virginianae*, which parasitizes the juniper (*Juniperus*) and apple (*Malus*), produces telia which develop in (and later project from) spherical galls ('cedar apples') which form on the juniper. (The cedar (*Cedrus*) is not susceptible to this rust.)
Gymnostomatia A subclass of ciliates (class KINETOFRAGMINOPHOREA) in which the cytostome is superficial (i.e., it does not occur at the base of an organized buccal cavity or vestibulum), and the cytopharyngeal apparatus is of the RHABDOS type. Somatic ciliature is commonly uniform. The organisms are typically large and carnivorous. Genera include e.g. *Acaryophrya*, *Acineria*, ACTINOBOLINA, COLEPS, *Didesmis*, DIDINIUM, DILEPTUS, *Holophrya*, *Lacrymaria*, KENTROPHOROS, LOXODES, *Loxophyllum*, *Mesodinium*, *Metacystis*, *Nanophrya*, PRORODON, STEPHANOPOGON, TRACHELOPHYLLUM and *Urotricha*.
(See also KARYORELICTID GYMNOSTOMES; PRIMOCILIATID GYMNOSTOMES.)
gymnostome A ciliate of the GYMNOSTOMATIA.
***gyrA*, *gyrB* genes** See GYRASE.
gyrase (DNA gyrase) A type II TOPOISOMERASE which catalyses the negative supercoiling of relaxed or positively supercoiled

ds cccDNA (see DNA) in an ATP-dependent process. (In the *absence* of ATP, the gyrase from *Escherichia coli* can relax negative, but not positive, supercoils.) Gyrase functions as a tetramer consisting of two of each of two types of subunit, A and B (i.e. A_2B_2); subunits A and B are the products of genes *gyrA* (= *nalA*) and *gyrB* (= *cou*), respectively. The enzyme binds to dsDNA at specific (but apparently non-homologous) sites. The DNA becomes wrapped around the surface of the enzyme in a positive supercoil; this necessarily introduces a compensating negative supercoil elsewhere in the molecule. The remaining steps effectively remove the positive supercoil. The B subunits bind ATP, a step which alters the functional (conformational?) state of the enzyme, and the A subunits bring about a staggered break in the dsDNA, the resulting 5′ ends each projecting beyond the 3′ ends by four bases. [Determinants of site-specific DNA breakage by gyrase: EMBO (1986) *5* 1411–1418.] Each 5′ end is covalently bound to one of the A subunits via a phosphotyrosine bond; this allows conservation of the energy of the phosphodiester bond for later use in re-sealing. A length of DNA from elsewhere in the molecule (possibly from the region wrapped around the enzyme) is passed through the (double-stranded) break while both ends of the break are held so that they cannot rotate; the break is then re-sealed. This decreases the linking number of the molecule by two. (If the translocated length of DNA belongs to a second molecule, rather than to another part of the same molecule, the result is a CATENANE.) Finally, the ATP bound to the B subunits is hydrolysed and the original conformation of the enzyme is restored; the whole cycle can then begin again. (Note: ATP *binding* allows a single round of supercoiling, but ATP *hydrolysis* is necessary for subsequent rounds.) The ability of gyrase to introduce negative supercoiling appears to decrease with increasing levels of negative supercoiling. In *E. coli*, the appropriate level of supercoiling in vivo is apparently achieved/maintained by the opposing activities of gyrase, on the one hand, and topoisomerases I and IV (see TOPOISOMERASE) on the other hand.

Gyrase is apparently essential in vivo for e.g. DNA REPLICATION, recombination, transposition (see TRANSPOSABLE ELEMENT), and some cases of TRANSCRIPTION; this requirement seems to be due, at least in part, to the need for supercoiled DNA in these processes, and implies that gyrase has an important role in generating and maintaining the correct level of supercoiling of DNA in the cell. (cf. ILLEGITIMATE RECOMBINATION.) There is also evidence for a more direct role for gyrase at the replication forks during DNA replication. (cf. REVERSE GYRASE; see also COUMERMYCINS; NALIDIXIC ACID; NOVOBIOCIN; OXOLINIC ACID.)

[QUINOLONE ANTIBIOTICS and other gyrase-targeted antibiotics: TIM (1997) *5* 102–109.]

Some candidate antibiotics affect gyrase at sites distinct from those targeted by quinolone antibiotics (and do not exhibit cross-resistance with the quinolones) [TIM (1998) *6* 269–275].

Gyrodinium See DINOFLAGELLATES.

Gyrodontium See APHYLLOPHORALES (Coniophoraceae).

gyrogonites Fossilized charophyte oogonia.

Gyromitra See PEZIZALES.

gyrose Sinuous; curving; marked with undulating lines. In lichenology: refers to a type of apothecium (e.g. in some *Umbilicaria* spp) in which there is a spiral or maze-like pattern of fertile ridges and sterile furrows.

Gyrosigma See DIATOMS.

1. Words in SMALL CAPITALS are cross-references to separate entries.
2. Keys to journal title abbreviations and Book ref. numbers are given at the end of the Dictionary.
3. The Greek alphabet is given in Appendix VI.
4. For further information see 'Notes for the User' at the front of the Dictionary.

H

H Histidine (see AMINO ACIDS).

H antigens Bacterial flagellar antigens. H antigens are detected in the serological characterization of certain enterobacteria (see e.g. KAUFFMANN–WHITE CLASSIFICATION). Agglutination of cells with homologous H antiserum occurs rapidly, is characteristically floccular, and the agglutinated cells can be readily dispersed by shaking (since flagella are easily broken). H antigens are heat-labile and are destroyed by alcohol but not by dilute formalin. (See also IMMOBILIZATION TEST.)

In a few *Salmonella* serotypes (e.g. *S. typhi*) and other enterobacteria the flagella exhibit only one type of antigenic pattern, i.e. the H antigens are *monophasic*. However, most salmonellae can produce either of two different patterns of H antigens, i.e. in these organisms the H antigens are *diphasic* (see PHASE VARIATION).

H^+-ATPase See PROTON ATPASE.

$H^+/2e^-$ ratio During the operation of an ELECTRON TRANSPORT CHAIN: the number of protons translocated across the membrane during the passage of two electrons along the chain.

H-lysin See FURUNCULOSIS (sense 1).

***h* mutant** HOST-RANGE MUTANT.

H^+/O ratio During the operation of an ELECTRON TRANSPORT CHAIN with oxygen as terminal electron acceptor: the number of protons translocated across the membrane per oxygen atom reduced. (cf. P/O RATIO.)

H^+/P ratio The number of protons passing through a PROTON ATPASE per molecule of ATP released at the catalytic site.

H^+-PPase PROTON PPASE.

H strand 'Heavy strand': that strand of a dsDNA molecule which exhibits the higher buoyant density (owing to a higher content of thymine and guanine) when the dsDNA is denatured and subjected to CsCl density gradient CENTRIFUGATION; the other ('light') strand is designated the *L strand.* The designation 'heavy' or 'H' is also used to refer to a strand labelled with a heavy isotope such as ^{15}N, the unlabelled strand being designated 'light' or 'L'.

H1 protein *Syn.* H-NS PROTEIN.

H-1 virus See PARVOVIRUS.

H_{II} phase See CYTOPLASMIC MEMBRANE.

HAA See HEPATITIS B VIRUS.

HAART Highly active antiretroviral therapy: a form of chemotherapy, used for HIV-positive and AIDS patients, involving a combination of different kinds of ANTIRETROVIRAL AGENT; such treatment should, theoretically, inhibit completely the replication of HIV, and in some studies HIV is reported to be undetectable in plasma.

One example of HAART is the combination of (i) efavirenz (a non-nucleoside reverse transcriptase inhibitor), (ii) nelfinavir (a protease inhibitor) and (iii) a nucleoside reverse transcriptase inhibitor [NEJM (1999) *341* 1874–1881].

Even with effective HAART, low-level replication can occur. This may involve (i) long-lived HIV-infected cells and (ii) selection of drug-resistant mutants [JV (2005) *79* 9625–9634].

Haber–Weiss reaction See HYDROXYL RADICAL.

HACCP Hazard analysis critical control point system: detailed appraisal of a production process and control of known/potential hazards at certain critical points in the process. ('Hazard' refers to any influence or factor which would, unless avoided, affect adversely the integrity of the process and/or the quality of the final product.)

Initially, an objective assessment of known and potential hazards is carried out using a flow-chart which shows all aspects of the process – details that may include e.g. source(s) of raw materials, a step-wise coverage of all stages of processing, route(s) of distribution of the product, and the movement of product to the point of sale (if applicable).

Central to the HACCP system is the identification of a number of *critical control points* (CCPs) – specific stages in the process at which effective action can be taken to prevent or counteract the effects of particular hazard(s). Monitoring is carried out at CCPs to ensure that processing continues within pre-determined tolerance limits (in terms of quality, contamination etc.).

Microbiological monitoring at CCPs may aim to detect specific organism(s) or it may assess the level of general microbial contamination. Monitoring tests should give rapid results so that corrective action, if necessary, can be taken swiftly; the traditional approach to monitoring, based on colony counting, may be too slow to permit timely intervention in a modern production process.

In the food industry, many of the microbiological tests are designed to detect contamination by non-specific organisms; these tests monitor factors such as carbon dioxide (produced by contaminating organisms) and/or the presence of ATP (released, in vitro, from contaminating organisms). An example of such tests is the screening of UHT milk for microbial contamination by the Lumac® Autobiocounter M 4000 (manufactured by Perstorp Analytical LUMAC, Landgraaf, The Netherlands); in this system, samples of UHT milk are examined for their content of (intracellular) ATP – the presence of which indicates contamination with metabolizing (living) organisms. The Autobiocounter M 4000 is used e.g. for testing samples derived from a continuous production process that operates on a batch system (see below). Before testing, each sample is pre-incubated (at e.g. 30°C for 48 hours) to allow the growth of any contaminants that may be present. After incubation, an aliquot (e.g. 50 μl) of sample is transferred to an ATP-free cuvette (container) and is treated first with ATPase to hydrolyse (destroy) any background ATP in the sample. Subsequently, the aliquot is treated with a reagent that degrades (permeabilizes) the cytoplasmic membrane of any contaminating cells – allowing leakage of ATP from such cells. ATP, if present, is then detected by the luciferin–luciferase system (see BIOLUMINESCENCE) which produces light in the presence of even minute quantities of ATP; emitted light is measured by a photomultiplier (light meter), the result being shown digitally. Measurement of ATP is a fully automated procedure that is completed in about 15–20 minutes. Each batch of product is held back until the test results on samples from that batch become available – being released for distribution when tested samples are found to be satisfactory; despite the delay for sample preparation (incubation), the manufacturing process itself can be continuous, each batch of product being held only for the period required to examine samples.

The LIMULUS AMOEBOCYTE LYSATE TEST, which detects and/or quantifies LIPOPOLYSACCHARIDE (LPS), can be used to monitor contamination by Gram-negative bacteria; any LPS-containing cell, and even isolated LPS, can give a positive test, so that this approach can detect both living and dead cells in the sample.

A particular type of microbial contaminant (e.g. a specific pathogen) may be detectable by using a PROBE-based approach. Moreover, PCR may provide a useful test system given organism-specific primers.

HACEK Acronym for a group of bacteria which occur in the mouth and upper respiratory tract and which are sometimes responsible for lesions on heart valves: *Haemophilus* spp, *Actinobacillus actinomycetemcomitans*, *Cardiobacterium hominis*, *Eikenella corrodens* and *Kingella* spp.

hadacidin ($CHO.NOH.CH_2.CO_2H$) An ANTIBIOTIC and antitumour agent. It is an analogue of L-aspartic acid and blocks adenylosuccinate synthetase (and hence adenine nucleotide synthesis: see Appendix V(a)) but not other aspartic acid-requiring reactions. (cf. AZASERINE.)

***Hae*II** A RESTRICTION ENDONUCLEASE from *Haemophilus aegyptius*; PuGCGC/Py.

***Hae*III** A RESTRICTION ENDONUCLEASE from *Haemophilus aegyptius*; GG/CC.

Haeckel's theory See ONTOGENY.

haem (heme) (1) A generic term for any compound in which iron is coordinately bound to a PORPHYRIN or related compound (e.g. a CHLORIN); the four pyrrole nitrogen atoms each contribute an electron pair to form coordinate bonds with the iron atom – which may be in the ferrous or ferric state. The 5th and 6th coordination positions of the iron may be filled by electron pairs donated e.g. from water, chloride, or – in a *haemoprotein* – basic amino acid residues in the protein. Iron–protoporphyrin IX (= *protohaem IX*, 'protohaem') is the prosthetic group in haemoproteins such as CATALASE and certain CYTOCHROMES. Biosynthesis of protohaem IX involves the insertion of an iron atom into the porphyrin, catalysed by *ferrochelatase*. (See also X FACTOR.)

(2) The term 'haem' is often used to refer specifically to protoporphyrin IX–ferrous iron – cf. HAEMIN.

haemadsorption The adherence of RBCs to cells which are infected with certain types of virus (e.g. paramyxoviruses); thus, if a tissue culture is infected by such viruses infection may be detected by the cells' ability to adsorb the RBCs of certain species (e.g. guinea pig) under appropriate conditions.

haemadsorption-inhibition test Any test used to detect serum antibodies homologous to HAEMADSORPTION-promoting viruses. Essentially, the patient's INACTIVATED SERUM (in various dilutions) is incubated (e.g. for 1 hour) with a suspension of such a virus – the whole then being used to inoculate a tissue culture; after ca. 5 days' incubation the (washed) tissue culture is tested for haemadsorption by the addition of RBCs. In a positive test at least some dilutions of the patient's serum will have neutralized the virus – thus preventing viral infection of the cells and (hence) preventing subsequent haemadsorption.

haemagglutination The AGGLUTINATION of red blood cells (RBCs). Haemagglutination may result from any of the following. (a) Combination between RBCs and antibodies homologous to their surface antigens. (b) Combination between certain types of virus (e.g. orthomyxoviruses, paramyxoviruses) and specific cell membrane receptor sites on the RBCs of certain species – the viruses acting as links between the RBCs. After a period of time (which depends e.g. on temperature) some viruses may subsequently elute, spontaneously, owing to their NEURAMINIDASE activity which destroys the neuraminic acid-containing receptor sites. RBCs from which such viruses have eluted cannot again be agglutinated by viruses of the same strain, and are said to be 'stabilized'. (Sometimes RBCs stabilized by one strain of virus can be agglutinated by a different strain.) Spontaneous elution (e.g. of influenza viruses) can be inhibited by carrying out haemagglutination tests at 4°C. (c) Passive haemagglutination: see PASSIVE AGGLUTINATION. (d) The action of certain LECTINS. (e) Combination between RBCs and bacterial fimbrial haemagglutinins (see FIMBRIAE).

haemagglutination-inhibition test (HAI or HI test) Any test in which specific antibody or antigen is detected or quantified by its ability to inhibit HAEMAGGLUTINATION. HI tests are used e.g. as clinical diagnostic tests to detect serum antibodies to haemagglutinating viruses (e.g. influenza and rubella viruses); essentially, such tests determine the ability of a patient's serum to neutralize the haemagglutinating properties of a suspension of the given virus – neutralization (a positive test) being indicated by the failure of the serum–virus mixture to agglutinate added RBCs. In many cases the serum used in such tests must be pre-treated to (i) abolish its ability to inhibit haemagglutination *non*-specifically, and/or (ii) to abolish its ability to *promote* haemagglutination. For example, serum may be pre-absorbed with kaolin (to sequester certain non-specific inhibitors of haemagglutination) and/or pre-treated with RBCs (of the type used in the test) in order to absorb HETEROPHIL ANTIBODIES; serum used in HI tests for e.g. influenza viruses is pre-treated with e.g. the enzyme RDE, and is then heated (prior to the test) in order to destroy RDE activity. [Book ref. 53, pp. 2034–2048.]

HI tests for soluble antigens may involve inhibition of *passive haemagglutination* (see PASSIVE AGGLUTINATION) – antigen being preabsorbed to RBCs. Thus, the amount of antigen in a sample can be measured by the amount of homologous antibody it can complex (sequester) when mixed with a fixed volume of antibody; the amount of free (uncombined) antibody remaining in the mixture (an indirect measure of antigen) is determined by the addition of antigen-coated RBCs (see BOYDEN PROCEDURE.) To ensure antigen-specificity in the test, the antiserum should be preabsorbed with (non-coated) RBCs of the type used in the test.

haemagglutinin Any agent or substance which can bring about HAEMAGGLUTINATION.

Haemamoeba A subgenus of PLASMODIUM.

haematein A red amphoteric DYE prepared by atmospheric oxidation of an aqueous solution of the colourless compound haematoxylin; haematein is usually referred to as 'haematoxylin'.

haematin Protoporphyrin IX–Fe(III) hydroxide. (cf. HAEMIN.)

haematogenous (hematogenous) Produced by or derived from the blood, or disseminated via the bloodstream.

haematophagous Feeding on blood.

haematotropic *Syn.* HAEMOTROPIC.

haematoxylin See HAEMATEIN.

haematuria The presence of blood in the urine.

haemin (1) Any protoporphyin IX–ferric iron complex: cf. HAEM sense 2. (2) Specifically: protoporphyrin IX–Fe(III) chloride. (cf. HAEMATIN.)

Haemobartonella A genus of Gram-negative bacteria of the family ANAPLASMATACEAE. Cells: cocci, coccobacilli or rods (range 0.1–1.5 μm) which occur tightly bound to, or within, erythrocytes in the infected host. *Haemobartonella* spp (*H. canis*, *H. felis* and *H. muris*) have a wide host range and occur worldwide; they are transmitted e.g. by biting insects.

haemocytometer See COUNTING CHAMBER.

haemoglobinuria The presence of free haemoglobin in the urine.

Haemogregarina See ADELEORINA.

haemogregarines An imprecise name for those protozoa of the suborders ADELEORINA and EIMERIORINA which infect red and/or white blood cells in the vertebrate host.

haemolysin (1) Any antibody homologous to the surface antigens of red blood cells (RBCs). (See also IMMUNE HAEMOLYSIS.)

(2) (*Syn.* haemolytic toxin) Any of various microbial products which cause HAEMOLYSIS by damaging the cytoplasmic membrane of RBCs.

(a) *Aeromonas salmonicida* haemolysins. See e.g. FURUNCULOSIS (H-lysin and T-lysin).

(b) *Bacillus* haemolysins. See e.g. SURFACTIN.

(c) *Clostridium* haemolysins. See CLOSTRIDIUM.

(d) *Escherichia coli* haemolysins.

α-Haemolysin is an extracellular haemolysin (MWt ca. 107000) specified by either chromosomal or plasmid-borne genes (*hly* genes); chromosomal and plasmid-borne *hly* genes are closely homologous, and identical *hly* genes can occur on plasmids of different incompatibility groups. In cultures of *E. coli*, production of α-haemolysin occurs under aerobic and anerobic conditions, but is enhanced by aerobiosis and inhibited by >100 μM iron. α-Haemolysin can lyse the RBCs of e.g. sheep, rabbits and man – causing clear haemolysis on blood agar plates; it is cytotoxic for human leucocytes and is lethal when injected into mice. α-Haemolysin appears to insert into lipid bilayers, generating hydrophilic transmembrane pores of effective diameter ca. 3 nm [Inf. Immun. (1986) *52* 63–69]. [Review: MR (1984) *48* 326–343; role of α-haemolysin in the pathogenesis of experimental *E. coli* infection in mice: JGM (1985) *131* 395–403.]

β-Haemolysin is a cell-bound haemolysin active against e.g. rabbit and human RBCs; it causes clear haemolysis on blood agar plates.

γ-Haemolysin has been reported in nalidixic acid-resistant *E. coli* mutants, and its production is stimulated by nalidixic acid; it is not lytic for human or rabbit RBCs.

(e) *Staphylococcus* haemolysins are produced by both coagulase-positive and -negative strains; the best known are the protein EXOTOXINS α-, β-, γ- and δ-haemolysins, although others also occur. No particular pattern of haemolysin production has been associated with virulence.

α-Haemolysin (α-toxin) is a hydrophobic, surface-active protein with (apparently) no enzymic activity. It shows the Arrhenius effect: it is inactivated on heating to 60°C but is partly reactivated on further heating to ca. 100°C. α-Haemolysin is >100 times more effective against rabbit RBCs than human RBCs. It appears to bind to the cell membrane, causing release of K^+ – possibly due to the formation of transmembrane pores by amphiphilic hexamers of the haemolysin. α-Haemolysin can be lethal for man and animals; depending on animal species, it can be leucocidal (= the Neisser–Wechsberg LEUCOCIDIN), can aggregate and lyse blood platelets, and has a range of toxic effects e.g. on the nervous and vascular systems (e.g. it increases vascular permeability). When injected subcutaneously it can have dermonecrotic effects which may be due to direct necrotizing activity or to prolonged vasospasm. α-Haemolysin has been shown to play a major role in the pathogenesis of gangrenous mastitis in sheep and cattle.

β-Haemolysin (β-toxin) is a SPHINGOMYELINASE C specific for sphingomyelin (or lysolecithin); it is inactivated at 60°C, requires Mg^{2+} for activity, and is inhibited e.g. by Zn^{2+} and chelating agents. It can hydrolyse the sphingomyelin in sphingomyelin-rich RBCs (e.g. those from sheep or ox) without causing haemolysis; however, lysis can subsequently be induced by chilling (see HOT–COLD LYSIS) or by the addition of the CAMP factor (see CAMP TEST). β-Haemolysin is most frequently produced by staphylococci isolated from animals other than man. Reports that it can be lethal are controversial. It is not dermonecrotic (although it may cause necrosis in lactating mammary gland), but it can show toxic effects on the cardiovascular system and can lyse platelets and guinea-pig macrophages (but not human PMNs). Its role in pathogenesis, if any, is unknown.

γ-Haemolysin (γ-toxin) consists of two protein components which appear to act synergistically in bringing about haemolysis. Rabbit RBCs are more susceptible than human or sheep RBCs; horse RBCs are insensitive. γ-Haemolysin can also increase membrane permeability in cultured fibroblasts, and may have some leucocidal activity. It is inactivated at 60°C and by certain acidic polymers – including agar, heparin, and dextran sulphate (but not e.g. agarose, hyaluronic acid or chondroitin sulphate), by various lipids (e.g. cholesterol, fatty acids), and by EDTA and citrate.

δ-Haemolysin (δ-toxin) is a heat-stable, hydrophobic, strongly surface active peptide containing 26 amino acid residues. It can lyse RBCs from many (possibly all) species; it is also active on a wide range of natural and synthetic membranes, lysing e.g. leucocytes, all kinds of mammalian cells in culture, certain bacteria, mitochondria, protoplasts, liposomes, etc. At relatively high concentrations, it probably exerts its effect by non-specific solubilization of membrane lipids and/or proteins; at low concentrations it may act by forming transmembrane pores and/or by activating endogenous phospholipase A_2. It is inhibited by various phospholipids and serum lipoproteins, and is poorly antigenic. Different δ-haemolysins are produced by different biotypes of *S. aureus*.

[Reviews: Book ref. 44, pp. 619–744.]

(f) *Streptococcus* haemolysins. See e.g. STREPTOLYSIN O and STREPTOLYSIN S.

(g) *Vibrio parahaemolyticus* haemolysin. See KANAGAWA PHENOMENON.

haemolysis The lysis of erythrocytes (red blood cells). Haemolysis may occur in any of the following ways. (a) By the action of a COMPLEMENT-dependent or -independent HAEMOLYSIN. (b) By a complement-fixing union between an antibody and an erythrocyte-bound particulate antigen (*passive haemolysis*). (c) By REACTIVE LYSIS. (d) By osmotic lysis in a hypotonic medium. (e) By viral activity: certain viruses (e.g. the mumps virus) can lyse the erythrocytes of certain species.

The ability of certain microorganisms to cause haemolysis can be detected e.g. by growing the organisms on a suitable BLOOD AGAR; an extracellular haemolysin causes the development of a differentiated zone around each colony. Some types of haemolysin bring about a glass-clear, colourless zone. Such *clear haemolysis* is often called 'β-haemolysis'; however, 'β-haemolysis' can be a misleading term since clear haemolysis can be caused e.g. by the α-haemolysin of staphylococci or by the α- or β-haemolysins of *Escherichia coli*. A zone in which the blood is greenish, but still opaque, is often called 'α-haemolysis'; such *greening* can be caused by various types of haemolysin. The *absence* of haemolysis is sometimes called (confusingly) 'γ-haemolysis'.

haemolytic immune body (HIB; amboceptor) Antibody to the surface antigens of erythrocytes. (See also HAEMOLYTIC SYSTEM.)

haemolytic system In a COMPLEMENT-FIXATION TEST: a suspension of *sensitized* erythrocytes, i.e. erythrocytes (commonly sheep erythrocytes) which have been exposed to a subagglutinating dose of antibodies homologous to their surface antigens. The system exhibits no HAEMOLYSIS in the absence of COMPLEMENT, and almost complete haemolysis (ca. 90% of the erythrocytes lysed) when exposed to the maximum concentration of complement used in a CFT; with amounts of complement between these extremes, the proportion of erythrocytes which lyse is dependent on the amount of free (i.e. unfixed) complement. Thus, the haemolytic system indicates the amount of complement (if any) remaining unfixed in each dilution at the end of the first stage in a CFT. (See also EAC.)

haemolytic uraemic syndrome (HUS) A potentially fatal disease (of children and adults) involving renal impairment/acute renal failure (see also TNF). An early stage of (bloody) diarrhoea is common, though not invariably present. Most cases with prodromal diarrhoea appear to be caused by O157 strains of *Escherichia coli* (see EHEC) or by *Shigella dysenteriae*.

The major symptoms of HUS are toxin-dependent; attempts are being made to sequester toxins in the gut (see SYNSORB P[k]). Complications may include e.g. CNS involvement. Antibiotics have not been found useful, and may be actually harmful.

Non-diarrhoeal HUS is sometimes associated with EHEC but may be induced by certain drugs (e.g. mitomycin C, quinine), and there are several other distinct causes.

haemophagocytic syndrome (HS) A syndrome which is characterized by fever and e.g. (commonly) cytopenia, coagulopathy, hepatosplenomegaly and *haemophagocytosis* – i.e. phagocytosis of haemopoietic cells by HISTIOCYTES. Primary HS is associated with inherited conditions. Secondary HS may be associated with certain infections, malignancy or non-malignant conditions (e.g. autoimmune disease), or may be drug-associated (phenytoin).

Infection-associated HS (IAHS) includes virus-associated HS (VAHS) and bacterium-associated HS (BAHS).

The pathogenesis of HS is not understood but it appears to involve unregulated T lymphocytes, monocytes, macrophages and cytokines; elevated levels of e.g. IFN-γ (interferon-γ) and IL-18 (interleukin-18) have been found in serum, and an imbalance in Th1/Th2 cells has also been reported. Some 60% of cases have been reported to occur in patients with pre-existing immunodeficiency, suggesting that immune status may be important as a predisposing factor.

Diagnosis includes demonstrating phagocytosis of erythrocytes (red blood cells), leukocytes and platelets by examination of bone marrow.

Microorganisms associated with HS include: Epstein–Barr virus (apparently responsible for most cases of VAHS) and other herpesviruses (e.g. cytomegalovirus, HHV6 and HHV8); species of *Borrelia*, *Brucella*, *Mycobacterium*, *Rickettsia*, *Staphylococcus* and *Streptococcus*; *Cryptococcus*; and *Leishmania donovani*.

[Haemophagocytic syndrome associated with infections: BCH (2000) *13* 163–178.]

Haemophilus A genus of Gram-negative bacteria of the PASTEURELLACEAE. Cells: pleomorphic coccobacilli, rods (often ca. 0.4 × 1–2 μm) or – especially under suboptimal growth conditions – filaments. Some strains have capsules. Fimbriae occur in haemagglutinating strains (*H. parasuis*, *H. paragallinarum*, strains of *H. aegyptius*). Optimum growth temperature: 35–37°C. Growth may require the X FACTOR and/or the V FACTOR (see PORPHYRIN TEST and SATELLITE PHENOMENON). Media for primary isolation include e.g. CHOCOLATE AGAR, FILDES' ENRICHMENT AGAR, or LEVINTHAL'S MEDIUM; some species (e.g. *H. ducreyi*, *H. parasuis*) require serum. Growth in some species (e.g. *H. aphrophilus*, *H. paragallinarum*) requires, or is enhanced by, incubation with 5–10% CO_2. Colonies on rich media are usually non-pigmented or slightly yellowish, 0.5–2.0 mm diameter (48 hours/37°C); those of capsulated strains are more mucoid and may appear iridescent in obliquely transmitted light. Most species produce acetic, lactic and succinic acids (usually without gas) from glucose. Most strains do not ferment lactose. NO_3^- is reduced to NO_2^-; NO_2^- may also be reduced. GC%: 37–44. Type species: *H. influenzae*.

Haemophilus spp are parasitic on the mucous membranes (particularly of the respiratory tract) in man and other animals. Some species are pathogenic; others can be opportunist pathogens, causing e.g. dental abscesses, ENDOCARDITIS etc. Some strains form IgA1 proteases (q.v.). Plasmid-mediated antibiotic resistance has become common in some species (e.g. *H. influenzae*).

H. actinomycetemcomitans. See ACTINOBACILLUS.

H. aegyptius. Causes e.g. PINK-EYE, mainly in hot climates. Requires X FACTOR (q.v.) and V factor. Indole −ve; urease +ve; non-haemolytic; usually forms acid (no gas) from glucose but not from xylose; catalase +ve.

H. ducreyi. Causes CHANCROID. Requires X factor but not V factor. Indole −ve; urease −ve; non-capsulated; some strains weakly haemolytic; sugars are generally not metabolized (acid from glucose may be delayed +ve); peptidases are produced; catalase −ve.

H. influenzae. Parasitic and pathogenic in man. Requires V and X factors; non-haemolytic; acid (no gas) from glucose and xylose; catalase +ve. Eight biotypes (I–VIII) are distinguished e.g. on indole, urease and ODC reactions. Some strains have capsules, and six serotypes are distinguished on the basis of capsular antigens (polysaccharides, at least some of which contain ribosyl groups). Strains of serotype b (usually biotype I) may cause acute diseases such as CELLULITIS, EPIGLOTITIS, MENINGITIS (in children), and PNEUMONIA (caused also by other serotypes). Antisera to serotype b cross-react with antigens from a wide range of bacteria – e.g. staphylococci, streptococci, *Bacillus* spp. Non-capsulated strains (usually biotypes II or III) occur as commensals (e.g. in the nasopharynx) but can occasionally cause chronic conditions such as BRONCHITIS, CONJUNCTIVITIS, OTITIS MEDIA and SINUSITIS. [Conjugate vaccine against *H. influenzae* type b (for protection against e.g. meningitis and epiglottitis); RMM (1996) 7 231–241.]

Haemophilus influenzae: differentiation of biotypes

Biotype	Indole	Urease	ODC[a]
I	+	+	+
II	+	+	−
III	−	+	−
IV	−	+	+
V	+	−	+
VI	−	−	+
VII	+	−	−
VIII	−	−	−

[a] Ornithine decarboxylase test (see DECARBOXYLASE TESTS).

H. paragallinarum (including strains formerly called *H. gallinarum*). Causes coryza in fowl. Requires V factor but not X factor. Indole −ve; urease −ve; non-haemolytic; acid (no gas) from glucose and (some strains) xylose; catalase −ve.

H. parainfluenzae. Commensal in man, occasionally implicated in e.g. endocarditis. Requires V factor but not X factor. Indole −ve; urease −ve (biotype I) or +ve (biotypes II and III); non-haemolytic; acid (±gas; biotype III: no gas) from glucose but not from xylose; catalase +ve/−ve.

H. parasuis (including most strains formerly called *H. suis* or *H. influenzae-suis*). Causes GLASSER'S DISEASE in pigs. Requires

V factor but not X factor. Indole −ve; urease −ve; non-haemolytic; acid (no gas) from glucose but not from xylose; catalase +ve.

H. piscium. A fish pathogen (see ULCER DISEASE) now regarded as an atypical strain of *Aeromonas salmonicida*.

H. vaginalis: see GARDNERELLA.

Other species: *H. aphrophilus*, *H. avium*, *H. haemoglobinophilus*, *H. haemolyticus*, *H. paracuniculus*, *H. parahaemolyticus*, *H. paraphrohaemolyticus*, *H. paraphrophilus*, *H. pleuropneumoniae*, *H. segnis*. Species *incertae sedis*: '*H. somnus*', '*H. agni*' and '*H. equigenitalis*', pathogenic in cattle, sheep and horses, respectively.

[Book ref. 22, pp. 558–569.]

haemoproteins See HAEM.

Haemoproteus A genus of protozoa of the HAEMOSPORORINA. Species are parasitic in e.g. birds; the vectors are hippoboscid flies.

haemorrhagic colitis COLITIS associated with bloody diarrhoea; when caused by EHEC it usually occurs without fever.

haemorrhagic enteritis of turkeys An acute disease of turkeys (particularly turkey poults 6–12 weeks of age) caused by an AVIADENOVIRUS. Symptoms: depression, loss of appetite, bloody faeces; haemorrhages also occur in the muscles and internal organs. The disease may be fatal. [Book ref. 116, pp. 537–539.] (cf. MARBLE SPLEEN DISEASE.)

haemorrhagic fever with renal syndrome See HANTAVIRUS.

haemorrhagic fevers See VIRAL HAEMORRHAGIC FEVERS.

haemorrhagic nephrosonephritis *Syn.* KOREAN HAEMORRHAGIC FEVER.

haemorrhagic septicaemia (1) (in fish) See INFECTIOUS DROPSY and EGTVED DISEASE. (2) (barbone) An acute, often fatal, septicaemic CATTLE DISEASE which appears to be a primary PASTEURELLOSIS caused by strains of *Pasteurella multocida*. Onset is sudden, with fever, profuse salivation, and petechial haemorrhages in submucosal tissues; death may occur within ca. 1 day. A carrier state occurs. A vaccine is available.

haemorrhagic syndrome See TRICHOTHECENES.

Haemosporina See HAEMOSPORORINA.

Haemospororina A suborder of protozoa (order EUCOCCIDIORIDA) previously classified as the suborder Haemosporina (*sic*) of the order Eucoccida [JP (1964) *11* 7–20] or Eucoccidiida [JP (1980) *27* 37–58]. In this suborder the organisms lack a conoid, syzygy does not occur, and motile zygotes are formed (cf. ADELEORINA, EIMERIORINA); some species form pigment from host cell haemoglobin (cf. PIROPLASMASINA). The organisms are heteroxenous, the asexual phase taking place in a vertebrate host and the sexual phase occurring in a blood-sucking insect. Genera: e.g. HAEMOPROTEUS, LEUCOCYTOZOON, PLASMODIUM.

haemotropic (haematotropic; hemotropic; hematotropic) Having an affinity for the blood. (Used e.g. of blood parasites such as *Plasmodium*.)

haemozoin (malarial pigment) A brown or black pigment which occurs in the intraerythrocytic schizont of *Plasmodium* spp (see PLASMODIUM and MALARIA). Haemozoin derives from the erythrocyte's haemoglobin; the parasite ingests haemoglobin, uses the protein part, and (normally) detoxifies the remaining ferriprotoporphyrin IX (FP) by polymerizing it to haemozoin. (Unless detoxified by polymerization, FP can lyse the parasite's membranes.)

Haemozoin remains behind in the parasitized erythrocyte following release of merozoites.

Quinoline-type antimalarial agents (see AMINOQUINOLINES, QUININE) appear to act by blocking the parasite's 'haem polymerase' activity, thus killing the parasite by inhibiting detoxification of FP [Acta Tropica (1994) *56* 157–171].

In order to facilitate studies on (i) the mechanism by which *Plasmodium* detoxifies FP, and (ii) the mode of action of quinoline-type antimalarials, the crystal structure has been determined for a synthetic compound, β-haematin, which is chemically identical to haemozoin [Nature (2000) *404* 307–310].

Hafnia A genus of (usually non-capsulated) Gram-negative bacteria of the ENTEROBACTERIACEAE (q.v.). Most strains are motile at 25–30°C, but many are non-motile at 35°C. MR and VP reactions are variable at 35–37°C but MR −ve, VP +ve at 22–25°C. Lactose −ve (but Lac plasmids may occur); lysine and ornithine decarboxylases +ve; arginine dihydrolase −ve. Growth occurs e.g. on DCA and in KCN media.

One species: *H. alvei* (formerly e.g. *Enterobacter alvei* or *E. hafniae*); many biotypes. *H. alvei* occurs in man, animals and birds, and in soil, sewage and water. *H. alvei* has been associated with diarrhoeal disease [JCM (1994) *32* 2335–2337]; strains of *H. alvei* can produce A/E lesions (see entry EPEC). *H. alvei* has also been associated with liver abscesses, septicaemia, peritonitis and pneumonia [characterization of *H. alvei* isolates from human clinical extra-intestinal specimens: JMM (2001) *50* 208–214].

(cf. KOSERELLA and OBESUMBACTERIUM.)

hag gene In enterobacteria: the structural gene for FLAGELLIN.

HAI test HAEMAGGLUTINATION-INHIBITION TEST.

haidai See LAMINARIA.

hair cell A narrow, elongated, apparently non-pigmented cell present at the ends of trichomes in certain filamentous CYANOBACTERIA. In 'rivularian' species (e.g. GLOEOTRICHIA), at least, hair cells seem to be formed under conditions of phosphorus or fixed nitrogen limitation; their function, if any, is unknown.

hairpin In a single strand of DNA or RNA: a double-stranded region formed by base-pairing between sequences in the strand which are complementary and opposite in polarity (e.g., 5′... GGCATG. ... CATGCC ...3′). (cf. CRUCIFORM; see also STEM-AND-LOOP STRUCTURE.)

hairy-cell leukaemia (1) A T cell LEUKAEMIA (see HTLV-II in HTLV). (2) A B cell leukaemia. Both diseases are rare, and in each the cells have fine, hair-like projections.

hairy leukoplakia (oral hairy leukoplakia) In e.g. HIV-infected individuals: whitish areas, typically found on the sides of the tongue, associated with replication of the EPSTEIN–BARR VIRUS; it may resemble oral candidiasis but, unlike candidiasis, it cannot be removed. A high proportion of patients with hairy leukoplakia subsequently develop AIDS. Hairy leukoplakia has been treated with e.g. high-dose ACYCLOVIR but it tends to recur on cessation of treatment.

[Opportunistic oral infections in patients infected with HIV-1: RMM (1996) *7* 151–163 (hairy leukoplakia: 157–158).]

hairy root A disease which can affect various dicotyledonous plants; it is characterized by the proliferation of adventitious roots from the site of a wound infected with *Agrobacterium rhizogenes* (see AGROBACTERIUM). Pathogenesis appears to be analogous to that in CROWN GALL, involving the transfer, integration, and expression in plant host cells of a segment of an Ri (root-inducing) plasmid. (Ri and Ti plasmids belong to different incompatibility groups.) Conjugal transfer of Ri plasmids appears to occur constitutively at rather high frequencies, and induction of conjugal transfer by opines or other compounds has not been observed. Two types of Ri plasmid are known: the agropine type and the mannopine type; some of the catabolic functions of the agropine strains are carried not by the Ri plasmid but by another plasmid which can cointegrate with it [MGG (1983) *190* 204–214].

[Book ref. 55, pp. 271–286.]

hairy shaker disease *Syn.* BORDER DISEASE.

halazone See CHLORINE.

half-a-Gram stain A non-specific staining procedure used e.g. for detecting cells of *Legionella* in smears of sputum or aspirates etc. The slide is flooded for 0.5–1 min with a solution of crystal violet, drained, and flooded with Gram's iodine solution for 1–2 min. The slide is then rinsed well in tap-water and dried in air. (cf. GRAM STAIN.)

half-chiasma See RECOMBINATION.

half-reduction potential See REDOX POTENTIAL.

Halicystis See DERBESIA.

Halidrys See PHAEOPHYTA.

Halimeda A genus of siphonaceous, mainly tropical or subtropical green seaweeds (division CHLOROPHYTA). The thallus is segmented and calcified, and contains both chloroplasts and amyloplasts. The cell wall contains a xylan (see XYLANS) as the main structural component, and is more or less heavily calcified with aragonite. When mature, the entire thallus is converted to gametangia, after which it dies. Related holocarpic, usually calcified algae include species of CAULERPA, PENICILLUS and UDOTEA.

Haliscomenobacter A genus of Gram-negative, rod-shaped bacteria which occur e.g. in activated sludge. Individual cells are apparently non-motile; they commonly occur in sheathed trichomes. The organisms can utilize a range of carbon and nitrogen sources; PHB is not formed. *H. hydrossis* has a GC% of ca. 49. [Book ref. 45, pp. 425–440 (436–439).]

hallucinogenic mushrooms Certain agarics whose fruiting bodies contain hallucinogenic compounds; they include certain species of *Conocybe*, *Psilocybe* (e.g. *P. mexicana*, *P. semilanceata*) and *Stropharia*. At least some of these fungi contain *psilocybin* and/or *psilocin*, hallucinogenic tryptamine derivatives related to serotonin. Some species of *Lycoperdon* also contain hallucinogenic substance(s).

halo blight A typically seed-borne disease of beans (*Phaseolus*) caused by *Pseudomonas phaseolicola*. Small, translucent lesions on leaves later darken, each becoming surrounded by a yellowish 'halo'; leaves later exhibit interveinal yellowing, and brownish spots may develop on pods.

halo spot A CEREAL DISEASE, caused by *Selenophoma donacis* (*Septoria oxyspora*) which can affect e.g. barley, wheat and rye. Small, pale lesions, each with a purplish-brown margin, occur mainly on the leaves, and rows of dark-coloured pycnidia develop in line with the veins. (Halo spot is called 'eyespot' in some parts of the world: cf. EYESPOT sense 2.)

halobacteria (1) Strains or species of *Halobacterium*. (2) Members of the family Halobacteriaceae.

Halobacteriaceae A family of extremely halophilic archaeans which occur e.g. in salt lakes, evaporated brines, and salted fish (in which they can cause spoilage). Cells: rods, cocci or discs, containing orange, red or mauve carotenoid pigments (predominantly bacterioruberins). Division occurs by binary fission. The Gram reaction is negative. (See also CELL WALL, GAS VACUOLE and S LAYER.)

Growth requires a minimum concentration of 1.5 M NaCl, good growth typically needing 3–4 M NaCl; cells accumulate electrolyte (mainly KCl) to an intracellular concentration at least equivalent to that in the environment. High concentrations of electrolyte are required to maintain the structural integrity of e.g. the cytoplasmic membrane and the ribosomes; in dilute solutions some species lyse (owing to weakening of the cell envelope rather than to an osmotic effect per se).

The organisms are chemoorganotrophs which use amino acids and/or carbohydrates as principal sources of carbon, and which typically grow aerobically; metabolism is respiratory (oxidative). The organisms form menaquinones rather than ubiquinones. Under microaerobic or anaerobic conditions some strains can use light energy (see PURPLE MEMBRANE; see also PHOTOTAXIS).

Two genera: HALOBACTERIUM (the type genus) and HALOCOCCUS.

Halobacterium A genus of catalase-positive, oxidase-positive archaeans of the HALOBACTERIACEAE. The cells are lophotrichously flagellated (or non-motile) frequently pleomorphic rods or filaments, or discs; they often contain gas vacuoles (which are plasmid-encoded in at least some strains), and they lyse in dilute solutions (e.g. NaCl ~ca. 0.8–1.5 M). Binary fission occurs by constriction.

Most strains can be cultured in e.g. well-aerated tryptone – yeast extract – salt media. Optimum growth temperature: 40–50°C.

Some strains form a PURPLE MEMBRANE under microaerobic or anaerobic conditions.

The DNA in *Halobacterium* spp is often composed of a major component (ca. 60–90% of the total DNA) with a GC% of ca. 66–68, and a minor component ('satellite DNA' or plasmid DNA) with a GC% of ca. 57–60. Type species: *H. salinarium*.

H. halobium. See *H. salinarium*.

H. pharaonis. Motile, alkalophilic (optimum pH: 8.5) rods which do not use sugars; substrates include e.g. formate, fumarate, pyruvate. Growth occurs optimally in 3.5 M NaCl.

H. saccharovorum. Motile rods which form acid (mainly acetic acid) from sugars. Growth occurs optimally in 3.5–4.5 M NaCl.

H. salinarium (= *H. halobium*). Motile, typically rod-shaped cells which grow aerobically or (strains with a purple membrane) anaerobically in the light. Amino acids are used for carbon and energy, although growth may be stimulated by carbohydrates (without acid formation). Growth occurs optimally in 4–5 M NaCl.

H. vallismortis. Motile, very pleomorphic rods which use sugars, often with the formation of acid. Facultatively anaerobic. Growth occurs optimally in 4.3 M NaCl.

H. volcanii. Mainly discoid or cup-shaped cells which are (apparently) non-flagellated but which are sometimes capable of a rotatory movement. Sugars are used for carbon and energy. Growth occurs optimally in 1.5–2.5 M NaCl.

halocarbon See FREON.

halocins See BACTERIOCINS.

Halococcus A genus of catalase-positive and oxidase-positive archaeans of the HALOBACTERIACEAE. The cells are non-motile, orange- or red-pigmented cocci (diameter ca. 0.8–1.5 μm) which occur in pairs or in regular or irregular groups. At least 2.5 M NaCl is needed for growth; growth occurs optimally in 3.5–4.5 M NaCl. The organisms are much more resistant than are *Halobacterium* spp to hypotonic solutions. Amino acids are used for carbon and energy. Optimum growth temperature: 30–37°C. GC%: ca. 61–66. Type species: *H. morrhuae* [Book ref. 22, pp. 266–267].

halofantrine See MALARIA.

halogens (as antimicrobial compounds) See entries CHLORINE, FLUORIDES, IODINE.

Halomonas A genus (*incertae sedis*) of aerobic, facultatively anaerobic, chemoorganotrophic, halotolerant, Gram-negative bacteria which occur e.g. in salterns and (presumably) in salt lakes etc. Cells: non-pigmented or yellow-pigmented, polarly or laterally flagellated rods, 0.6–0.8 × 1.6–1.9 μm, sometimes pleomorphic or filamentous. Metabolism may be respiratory (oxidative), with O_2 or nitrate (anaerobic respiration) as terminal electron acceptor, or fermentative; anaerobic growth, without nitrate, can occur with glucose but not with other carbohydrates. Carbon sources include a wide range of alcohols, amino acids, organic acids and sugars. Salt (NaCl) tolerance: 0.1–32.5% w/v. Catalase +ve. Oxidase +ve. GC%: ca. 60. Type species: *H. elongata*. [Book ref. 22, pp. 340–343.]

halonitrosoureas See *N*-NITROSO COMPOUNDS.

halo-opsin See OPSIN.

halophile An organism which grows optimally only in the presence of electrolyte (commonly NaCl) at concentrations above ca. 0.2 M, and which typically grows poorly, or not at all, in low concentrations of electrolyte; examples include e.g. species of ACTINOPOLYSPORA, DACTYLOCOCCOPSIS, HALOBACTERIUM, HALOCOCCUS, PARACOCCUS and VIBRIO. (cf. HALOTOLERANT.)

A number of authors have accepted the following categories: slightly halophilic (optimum growth in the presence of 0.2–0.5 M electrolyte); moderately halophilic (optimum growth in 0.5–2.5 M electrolyte); extremely halophilic (optimum growth in >2.5 M electrolyte).

[Halophilic and halotolerant organisms: Book ref. 157, pp. 171–214, and Book ref. 191, pp. 55–81.]

Halopteris See PHAEOPHYTA.

halorhodopsin A RETINAL-containing protein pigment (MWt ca. 25000) which occurs in the PURPLE MEMBRANE of *Halobacterium salinarium* and which is involved in transmembrane ion translocation. Maximum absorption appears to occur at ca. 580 nm, the absorption characteristics being stabilized by anions (primarily Cl^-). On illumination, halorhodopsin undergoes cyclical changes in its absorption characteristics with a periodicity of ca. 5–10 msec; several photointermediates appear to be involved. Halorhodopsin undergoes bleaching when illuminated in the presence of hydroxylamine. Some strains of *Halobacterium salinarium* (e.g. JW-5, W-296), which cannot synthesize retinal, produce bacterio-opsin but are repressed in the synthesis of halo-opsin.

Halothrix See PHAEOPHYTA.

halotolerant Refers to the ability of a *non*-HALOPHILE to grow in the presence of high concentrations of electrolyte (e.g. up to ca. 2.5 M); examples of halotolerant organisms include e.g. HALOMONAS and strains of *Staphylococcus*.

Halteria A genus of freshwater ciliates (order OLIGOTRICHIDA). Cells: roughly spherical, with very conspicuous oral ciliature and a number of groups of 'cirri' or 'bristles' forming an equatorial band. *H. grandinella* is ca. 25–50 μm in diam., has a 'bouncing' type of movement, and occurs e.g. in ponds and organically polluted waters (see SAPROBITY SYSTEM) where it feeds on bacteria; the organism is fairly tolerant of low O_2 levels, but is sensitive to NH_3 and H_2S.

ham See CURING (1); MEAT SPOILAGE; SOFT CORE HAM.

hamanatto (black beans) An oriental food made by fermenting whole soybeans with strains of *Aspergillus oryzae*. Beans are steamed, coated with wheat flour, inoculated with *A. oryzae*, and incubated 1–2 days. Brine and spices are added and incubation is continued 6–12 months; the beans are then dried.

hamathecium A term which refers, collectively, to the various types of structure which may occur between and around the asci in an ascocarp: see e.g. PARAPHYSIS, PARAPHYSOID, PSEUDOPARAPHYSIS and PERIPHYSES.

Hamigera See TALAROMYCES.

Hammondia hammondi *Syn. Isospora datusi.*

hamycin A heptaene POLYENE ANTIBIOTIC produced by *Streptomyces pimprina*; it is effective against only a limited range of fungi, but is particularly effective against *Candida albicans*.

hand, foot and mouth disease An acute, highly infectious, self-limiting disease (mainly of children) characterized by the development of vesicular-ulcerative lesions in the mouth with similar lesions scattered on the soles of the feet and on the palms of the hands. The causal agent is usually coxsackievirus A16, but A4, A5, A9, A10, B2 and B5 have also been implicated. (cf. FOOT AND MOUTH DISEASE.)

hanging drop A direct, visual procedure used for determining the MOTILITY of microorganisms in a fluid medium. A COVER-GLASS is placed flat on the bench, and a drop of fluid (e.g. from a broth CULTURE) is transferred (e.g. by a LOOP) from the culture to the centre of the cover-glass. A plasticine ring (diam. ca. 1.5 cm, ca. 4 mm thick) is gently pressed onto one face of a SLIDE; the ring (attached to the slide) is then gently pressed onto the cover-glass so that the ring becomes sandwiched between the (parallel) faces of slide and cover-glass. The whole is then inverted, so that the drop hangs *freely* beneath the cover-glass, and can be examined under the microscope. This method avoids microcurrents which could give a false impression of motility.

Hanks' BSS Hanks' BALANCED SALT SOLUTION; it contains (g/l): NaCl (8.0), KCl (0.4), $CaCl_2.2H_2O$ (0.185), $MgSO_4.7H_2O$ (0.1), $MgCl_2.6H_2O$ (0.1), $Na_2HPO_4.H_2O$ (0.06), KH_2PO_4 (0.06), $NaHCO_3$ (0.35), glucose (1.0), and phenol red (0.01); pH: 7.0–7.2.

Hanseniaspora A genus of fungi (family SACCHAROMYCETACEAE). The genus contains the teleomorphs of KLOECKERA spp. Ascospores may be bowler-hat-shaped, 1–4 per ascus, and released at maturity, or they may be spherical, with or without an equatorial ridge, smooth or warty, and not released at maturity. [Book ref. 100, pp. 154–164.]

Hansen's bacillus *Mycobacterium leprae*.

Hansen's disease *Syn.* LEPROSY.

Hansenula A genus of yeasts (family SACCHAROMYCETACEAE) in which the cells are generally spheroidal, ellipsoidal or elongate; vegetative reproduction occurs by multilateral budding. Pseudomycelium or true mycelium may be formed. Asci are generally dehiscent, occasionally persistent. Ascospores: hemispheroidal, Saturn-shaped, or bowler-hat-shaped. Species may be homothallic or heterothallic. Most species can ferment glucose, some can also ferment other sugars, and some (e.g. *H. canadensis*) are non-fermentative. NO_3^- is assimilated. Thirty species are recognized, including e.g. *H. anomala* (anamorph: *Candida pelliculosa*), *H. canadensis* (conspecific with *H. wingei*), *H. capsulata*, *H. jadinii* (anamorph: *Candida utilis*), *H. polymorpha* (see also METHYLOTROPHY), *H. wickerhamii*. Species have been isolated from tree exudates, soil, insect larvae and frass, etc. [Book ref. 100, pp. 165–213.]

Hantaan virus See HANTAVIRUS.

Hantavirus A genus of viruses of the family BUNYAVIRIDAE [taxonomic proposal: Virol. (1983) *131* 482–491]. The enveloped, spherical virions each contain a negative-sense ssRNA genome consisting of three fragments.

Approximately half of the >20 known species are associated with human disease – either (i) the hantavirus pulmonary syndrome (HPS), or (ii) haemorrhagic fever with renal syndrome (HFRS).

Unlike other members of the Bunyaviridae (which are maintained in arthropods) hantaviruses are found in populations of (persistently infected) rodents – including rats (Seoul virus), mice (Sin Nombre virus, Dobrava-Belgrade virus, Hantaan virus) and voles (Puumala virus); human infection appears to occur mainly by inhalation of aerosols of the infected urine, faeces etc. of rodents.

Species of *Hantavirus* include: Hantaan virus (HFRS; see also KOREAN HAEMORRHAGIC FEVER), Seoul virus (HFRS), Puumala virus (HFRS), Dobrava-Belgrade virus (HFRS), Sin Nombre virus (HPS), Andes virus (HPS), Black Creek Canal Virus (HPS) and Laguna Negra virus (HPS).

[Hantaviruses (persistence in rodent reservoirs): TIM (2000) *8* 61–67. Hantaviruses (review): JMM (2000) *49* 587–599.]

Haploangium See MYXOBACTERALES.

haploid Having a PLOIDY of *one*. (cf. DIPLOID, POLYPLOID.)

haploidization (*mycol.*) See PARASEXUAL PROCESSES.

haplokinety (1) The infraciliary base (only) of a *stichodyad* PARORAL MEMBRANE. (2) A stichodyad paroral membrane.

haplomycosis *Syn.* ADIASPIROMYCOSIS.

haplont (1) (*noun*) An organism in whose life cycle only the zygote is diploid. (cf. DIPLONT.) (2) The haploid form of an organism, e.g. a gametophyte (see ALTERNATION OF GENERATIONS). (3) (haplontic) (*adj.*) Refers to either (1) or (2) above.

haplophase In organisms which reproduce sexually: the haploid phase of the life cycle. In many higher fungi haplophase and DIPLOPHASE are separated by DIKARYOPHASE.

Haplospora See PHAEOPHYTA.

Haplosporangium See ADIASPIROMYCOSIS.

Haplosporidium A genus of protozoa (class STELLATOSPOREA) which are parasitic in various aquatic invertebrates (including annelids and molluscs). The spores are operculate and characteristically bear filamentous extensions which arise from the outer spore coat and are typically coiled around the spore. [Spore ultrastructure in *H. lusitanicum*: J. Parasitol. (1984) *70* 358–371.] The vegetative stage is plasmodial.

haplosporosomes See STELLATOSPOREA.

hapten A substance which can elicit an immune response only when combined with another molecule or particle (*carrier*); if administered on its own it fails e.g. to stimulate antibody production (but cf. autocoupling hapten, below). The free hapten can, however, combine with antibodies raised against the hapten–carrier complex; such combination may or may not result in the precipitation of the hapten–antibody complex. A substance which acts as a hapten in one species may act as a complete antigen in another – e.g. the pneumococcal capsular polysaccharides are haptenic in the rabbit but antigenic in man. A hapten may be artificially bound to a protein in order to increase the immunological specificity of that protein. *Autocoupling* haptens bind spontaneously to tissue carriers in vivo, thus becoming antigenic. (See also ALLERGEN.)

hapteron See CYATHUS.

haptocyst An EXTRUSOME in the tentacles of suctorian ciliates.

haptonema A thread-like appendage present on unicellular algae of the PRYMNESIOPHYCEAE. Structurally, a haptonema bears some resemblance to a eukaryotic FLAGELLUM but contains fewer microtubules (which are arranged differently from those in a flagellum) and has a sheath composed of several concentric membranes, the innermost enclosing the microtubules. Haptonemata range from relatively short (e.g. in *Prymnesium parvum*) to very long (>80 μm in *Chrysochromulina* spp); most haptonemata (but not that of *Prymnesium*) can coil and uncoil: e.g., in *Chrysochromulina* spp the haptonema remains coiled while the organism is swimming and becomes extended when the cell comes to rest. The haptonema is believed to function as an attachment organelle.

Haptophyceae *Syn.* PRYMNESIOPHYCEAE.

hard cider (American) *Syn.* CIDER.

hard gill See WATERY STIPE.

hard-pad A form of CANINE DISTEMPER which begins with tenderness and keratinization of the feet, usually followed by CNS involvement and death.

hard swell See SWELL.

hare fibroma virus See LEPORIPOXVIRUS.

Harpellales See TRICHOMYCETES.

harpin (*plant. pathol.*) Any of certain types of glycine-rich protein, encoded by plant-pathogenic bacteria, which (at least in culture) can be secreted via a type III PROTEIN SECRETION system and which are able to elicit a heat-stable HYPERSENSITIVITY reaction in the leaves of tobacco and certain other plants [see e.g. JB (1997) *179* 5655–5662].

(See also AVIRULENCE GENE.)

Harpochytriales See CHYTRIDIOMYCETES.

Harposporium See HYPHOMYCETES and NEMATOPHAGOUS FUNGI.

Hartig net See MYCORRHIZA.

Hartmannella A genus of amoebae which are similar to ACANTHAMOEBA spp except that they do not produce acanthopodia from the (usually well-defined) hyaline cap of the pseudopodium.

Hartmannula See HYPOSTOMATIA.

hartrot (bronze leaf wilt; Coronie wilt) A disease of the coconut palm (*Cocos nucifera*) caused by a species of PHYTOMONAS which invades the phloem. Symptoms: yellowing and falling of leaves and blackening of inflorescences; unripe, internally blackened coconuts fall prematurely. Trees may die within three months of the initial signs.

hart's truffle See TRUFFLES.

Harvey murine sarcoma virus (Ha-MSV) A replication-defective, v-onc^+, fibroblast-transforming MURINE SARCOMA VIRUS which was isolated from BALB/c mice inoculated neonatally with rat-passaged MOLONEY MURINE LEUKAEMIA VIRUS (Mo-MuLV). Ha-MSV carries the oncogene v-*ras* (v-Ha-*ras*) and apparently arose by recombination between Mo-MuLV and a rat cellular *ras* sequence (Ha-*ras*: see RAS). Infection of newborn rodents with Ha-MSV results in anaemia, splenomegaly (resulting from proliferation of erythroblasts) and erythroleukaemia, as well as sarcomas. (cf. KIRSTEN MURINE SARCOMA VIRUS.)

HAT medium See HYBRIDOMA.

haustorium A specialized hyphal structure formed e.g. by certain plant-parasitic fungi (e.g. *Erysiphe*, *Peronospora*) in order to obtain nutrients from the host plant; it develops *within* a host cell after a small hyphal branch has penetrated the host cell wall, but it remains external to the host cell's cytoplasmic membrane. A haustorium may be spherical, club-shaped or elongated (branched or unbranched) according to the species of fungus which forms it. (Haustoria are also formed by the mycobionts of certain lichens.)

Haverhill fever See RAT-BITE FEVER.

Hawii virus See SMALL ROUND STRUCTURED VIRUSES.

hay fever See TYPE I REACTION.

Hayflick medium A medium used for the culture of *Mycoplasma*. spp. It consists of heart infusion broth supplemented with membrane-filtered horse serum (which provides sterols), fresh yeast extract solution, calf thymus DNA solution, thallium acetate and benzylpenicillin; the pH is adjusted to 6.8–7.8 according to strain. The solid medium is made by using a purified agar (e.g. Noble agar) as gelling agent.

HBB (2-(α-hydroxybenzyl)-benzimidazole) An ANTIVIRAL AGENT which selectively inhibits the replication of a number of picornaviruses in cell cultures; it inhibits the initiation of viral

RNA synthesis. HBB-resistant mutants emerge rapidly; HBB-dependent mutants have also been isolated.

HBcAg See HEPATITIS B VIRUS and HEPATITIS B.

HBeAg See HEPATITIS B VIRUS and HEPATITIS B.

HBsAg See HEPATITIS B VIRUS and HEPATITIS B.

HBV HEPATITIS B VIRUS.

HCAM *Syn.* CD44.

HCC HEPATOCELLULAR CARCINOMA.

HCMV Human cytomegalovirus: see BETAHERPESVIRINAE.

HCV HEPATITIS C VIRUS.

HCVs (human caliciviruses) See SMALL ROUND STRUCTURED VIRUSES.

HD protein Helix-destabilizing protein (see SINGLE-STRAND BINDING PROTEIN).

HD_{50} (CH_{50}; $C'H_{50}$) Haemolytic dose (50%): the quantity of COMPLEMENT needed to lyse 50% of a standardized suspension of *sensitized* erythrocytes. In a COMPLEMENT-FIXATION TEST the HD_{50} may be used as the unit in place of MHD (q.v.); this gives a more accurate end-point (since the graph '%lysis of cells versus quantity of complement' is sigmoidal).

HDAg See DELTA VIRUS.

HDCV Human diploid cell vaccine (see RABIES).

hDNA Hybrid DNA. (See e.g. CLONING.)

H-DNA See TRIPLEX DNA.

HE-cellulose (HEC) Hydroxyethylcellulose: a neutral cellulose derivative sometimes used in preference to CM-CELLULOSE in CELLULASE assays.

HE markers See TRANSFORMATION (1).

headful mechanism A mechanism of genome packaging during virion assembly: packaging is initiated by insertion into a prohead (or provirion) of one end of a concatemeric form of the genome (which, at or before insertion, may be cut at a particular sequence); incorporation of the genome then continues until the head is full. Thus, termination of packaging is determined only by the length of the nucleic acid packaged, and not by a specific sequence in the nucleic acid. (See e.g. BACTERIOPHAGE MU and BACTERIOPHAGE P22; cf. e.g. BACTERIOPHAGE λ.)

Heaf test A form of TUBERCULIN TEST. PPD is spread over a small area of skin and is then pressed into the skin with a 'Heaf gun': an instrument which punctures the skin with several small needles.

heart rot See TREE DISEASES.

heartwater See COWDRIA.

heat-fixed smear See FIXATION.

heat-shock proteins (HSPs) Proteins which are synthesized at greatly (but transiently) increased levels in an organism in response to a sudden rise in temperature or to certain other types of stress (e.g. ultraviolet radiation, virus infection, ethanol, or the intracellular accumulation of abnormal proteins); HSPs may be necessary for survival of the organism at higher temperatures, and they are produced at the expense of 'normal' proteins. Such a *heat-shock response* has been observed in a wide range of species – from bacteria and yeasts to plants and animals; at least some HSPs have been highly conserved through evolution, but the response is regulated differently in different organisms.

In *Escherichia coli* there are at least 17 HSPs which include the products of genes *dnaJ*, *dnaK*, *groEL*, *groES*, *rpoD*, *lysU* and *lon*. Synthesis of these proteins reaches a peak ca. 5–10 min after a rise in temperature (for example, 30 → 42°C), and subsequently declines to a new steady state which is greater than that at the lower temperature. Some HSPs apparently cope with stress-induced damage – for example, the molecular *chaperones* GroES, GroEL and DnaK seem to prevent misfolding or aggregation of unfolded proteins, while the Lon protease degrades abnormal proteins.

The heat-shock genes constitute a REGULON composed of several unlinked operons (for example: promoter–*dnaK*–*dnaJ* and promoter–*groES*–*groEL*). The heat-shock regulon is under the control of a positive transcriptional activator encoded by gene *rpoH* (previously called *htpR*). RpoH (MWt ca. 32000) is a SIGMA FACTOR (σ^{32}) which interacts with RNA POLYMERASE, producing a holoenzyme ($E\sigma^{32}$) that promotes transcription from heat-shock promoters more efficiently than does $E\sigma^{70}$. (Transcription of the *rpoH* gene itself requires $E\sigma^{70}$ [JB (1986) *166* 380–384].)

Following heat shock, the increased levels of RpoH are due mainly to increased translation of the protein – rather to increased transcription of *rpoH*. The heat-induced translation from *rpoH* mRNA involves an effect of temperature on a secondary structure formed in the mRNA by base-pairing between a site immediately downstream of the start codon (a *downstream box*) and another region in the coding sequence. This secondary structure normally inhibits translation, but on temperature upshift it is destabilized so that translation can occur; thus, the mRNA acts as a thermosensor (an *RNA thermometer*) for expression of σ^{32} [GD (1999) *13* 633–636].

Although synthesized under normal conditions, σ^{32} has a short half-life (about 1 min); it appears that, within the cell, σ^{32} forms complexes with DnaK, DnaJ and GrpE and is subsequently degraded by the FtsH protease. During heat shock DnaK binds preferentially to denatured proteins – leaving RpoH free to function as a sigma factor [EMBO (1996) *15* 607–617].

The heat-shock response in other bacteria. In some bacteria the mechanism for regulating the heat-shock response differs from that in *E. coli*. In some cases there is a regulatory sequence upstream of heat-shock genes; this inverted repeat sequence, termed CIRCE (controlling inverted repeat of chaperone expression), occurs e.g. in strains of *Bacillus* and *Clostridium* and is also found in some Gram-negative bacteria. In *Bradyrhizobium japonicum* the control mechanism includes at least two distinct regulatory systems involving CIRCE and a σ^{32}-like sigma factor [JB (1996) *178* 5337–5346].

More recently it has been reported that control of the heat-shock response involves negative regulation by repressor proteins (in conjunction with *cis*-acting DNA sequences) which inhibit transcription of the HSP genes under normal physiological conditions [Mol. Microbiol. (1999) *31* 1–8].

(See also SOS SYSTEM.)

Strains of *E. coli* mutant in the *rpoH* gene have been useful in the OVERPRODUCTION of recombinant proteins.

heated lid cycler See THERMOCYCLER.

heavy chain A class-specific polypeptide IMMUNOGLOBULIN component (MWt ca. 50000–70000, depending on Ig class). The various types of heavy chain are designated *alpha*, *gamma*, *delta*, *epsilon* and *mu* – corresponding to Ig classes IgA, IgG, IgD, IgE and IgM, respectively. In IgM, different C-terminal sequences occur in the heavy chains of the membrane-associated and pentameric forms. Certain Ig classes can be divided into subclasses on the basis of minor differences in their heavy chains.

heavy metals (and their compounds) (as antimicrobial agents) Although trace amounts of certain heavy metal ions are essential for the growth of microorganisms, higher concentrations may exhibit antimicrobial activity. Heavy metal ions bind to certain groups (particularly thiol groups) and appear to exert antimicrobial activity largely by inactivating proteins, nucleic acids and other microbial constituents; their activity is often microbistatic, and they typically have little or no activity against bacterial endospores. The antimicrobial activity of e.g. mercury ions may be reversed by the presence of substances such as cysteine, glutathione etc which presumably complex free ions. In some cases the antimicrobial activity of one metal can be antagonized by another; thus, e.g., the inhibition of *Lactobacillus* spp by zinc can be reversed by the addition of manganese. Some ions appear to have antimicrobial activity only when they are in a particular redox state (see e.g. ANTIMONY and ARSENIC). (See also OLIGODYNAMIC EFFECT.)

Among bacteria, resistance to heavy metals and their compounds can be PLASMID-borne. Thus, e.g., some bacteria can synthesize an inducible, plasmid-encoded, NADPH-linked mercuric ion reductase which can reduce Hg^{2+} to (relatively non-toxic) metallic mercury which is readily lost from the cell; such cells may also synthesize an organomercurial lyase which cleaves the carbon–mercury bond in organomercurials such as phenylmercuric acetate. [Microbial resistance to mercury and organomercurials: MR (1984) *48* 95–124.] In *Staphylococcus aureus*, plasmid-encoded resistance to cadmium may involve a pmf-dependent efflux of cadmium ions (via a cadmium–proton antiporter) which prevents intracellular accumulation of Cd^{2+}. [Reviews on bacterial resistance to metal ions: Book ref. 160, pp. 345–367; FEMS Reviews (1985) *32* 39–54.] (See also METALLOTHIONEIN; TN501.)

Heavy metals and their compounds have been used e.g. as agricultural antifungal agents, as antiseptics, as chemotherapeutic agents for the treatment of diseases of bacterial, fungal and protozoal aetiology, as preservatives for e.g. textiles and wood, and in the disinfection of water.

(See also ANTIMONY; ARSENIC; COPPER; MERCURY; SILVER; TIN; ZINC; and TELLURITE MEDIA.)

heavy strand (*mol. biol.*) See H STRAND.

HEC See HE-CELLULOSE.

hedgehog fungus *Hydnum repandum* (see APHYLLOPHORALES).

Hektoen medium A medium selective for certain enterobacteria; it contains lactose, sucrose, salicin, bromthymol blue, Andrade's indicator, ferric citrate, and deoxycholate. Strains (e.g. most salmonellae) which do not ferment sucrose, lactose or salicin form blue-green colonies; those which can ferment one or more of these substrates form salmon-pink colonies. H_2S-producing colonies have black centres.

HeLa An ESTABLISHED CELL LINE derived, in 1952, from a human cervical carcinoma; the cells are heteroploid and epithelioid. HeLa cultures support the replication of a wide range of viruses.

Helber chamber See COUNTING CHAMBER.

helicases (DNA unwinding proteins) Proteins with ssDNA-dependent ATPase activity which can unwind and separate the strands of dsDNA given a single-stranded 'leader sequence' or gap.

Escherichia coli has several types of helicase: the Rep protein (encoded by the *rep* gene), the DnaB protein (see DNAB GENE), helicases II and III, and the F PLASMID-specified helicase I.

The Rep protein binds to a region of ssDNA in a dsDNA molecule, moving along (3′-to-5′) into the double-stranded region and unwinding it processively; the separated strands may be stabilized by SINGLE-STRAND BINDING PROTEIN. Two molecules of ATP are hydrolysed for each base pair separated. The Rep protein is required for RF replication in certain phages (see SSDNA PHAGES), but its role in the host cell is unknown.

Helicase II (*uvrD* gene product) is involved in EXCISION REPAIR. This enzyme may act processively in the 3′-to-5′ direction [see e.g. JBC (1987) *262* 2066–2076].

Helicases I and III both act processively and move in the 5′-to-3′ direction along the strand to which they are bound. Helicase I is encoded by the *traI* gene of the F plasmid [PNAS (1983) *80* 4659–4663]; it seems to be responsible for nicking at *oriT* in the F plasmid [JBC (1991) *266* 16232–16237], and may be involved in unwinding the plasmid DNA for transfer during CONJUGATION (sense 1b). Helicase III cannot replace either the Rep protein or helicase II.

Some bacteriophages encode their own helicases; for example, the *dda* gene product of phage T4 is a helicase, while the gp4 of phage T7 is a protein with both helicase and PRIMASE activities.

The DnaB helicase, which is active e.g. in DNA replication, is a ring-shaped hexamer of DnaB proteins [see e.g. Cell (1996) *86* 177–180].

Helicobacter A genus of asporogenous Gram-negative bacteria classified within the ε (epsilon) subdivision (= rRNA superfamily VI) of the taxon Proteobacteria (which also includes the family CAMPYLOBACTERACEAE). The cells are typically curved/helical rods with one or more sheathed flagella (unsheathed e.g. in *H. pullorum*). Currently there are ~20 named species, many of which have been isolated from the intestines (and often liver) of animals (e.g. dogs, sheep, pigs, rodents, chickens). However, attention has been focused primarily on the species *H. pylori* owing to its association with e.g. gastritis and peptic-ulcer disease in humans.

H. pylori was formerly classified within the genus *Campylobacter*, but was re-classified [IJSB (1989) *39* 397–405]. Cells: curved/helical, ca. 0.5–0.9 μm wide and up to ~3 μm in length; motile (several flagella). Genome: circular, approx. 1700 kb; GC%: average 39, but significant differences occur in specific regions – one such region including the *cag* PATHOGENICITY ISLAND [*H. pylori* strain 26695 (complete genome): Nature (1997) *388* 539–547].

Chemoorganoheterotrophic. Microaerophilic. Culture on enriched media (e.g. blood, chocolate agar) needs high humidity (freshly poured, undried plates) and a gaseous phase of e.g. oxygen/carbon dioxide/nitrogen (5%/10%/85%); grey, translucent colonies (<2 mm in diam.) are obtained after 3–7 days at 35–37°C. The medium may be made selective for *H. pylori* by adding e.g. amphotericin, nalidixic acid and vancomycin (to which the organism is resistant).

Oxidase +ve, catalase +ve, urease +ve.

The pathogenesis of *H. pylori*-associated gastric disease is not understood. One suggestion was based on the presence of lipopolysaccharide antigens identical to the Lewis x and Lewis y antigens found e.g. in human gastric mucosa; it was postulated that these antigens might provoke an antibody response and give rise to autoimmune inflammation [molecular mimicry of *H. pylori*: TIM (1997) *5* 70–73]. However, this idea was questioned because the predominant anti-LPS antibodies found in a number of infected patients were unrelated to the given LPS antigens [Inf. Immun. (1998) *66* 3006–3011].

Pathogenesis seems likely to involve the *cag* pathogenicity island and an unlinked gene, *vacA*, which is often jointly expressed. The joint expression of *cag* and *vacA* (in 'type I' clinical strains) appears to correlate with severe gastrointestinal disease, while type II strains (which lack *cag*, and do not express *vacA*) are generally not associated with severe disease.

The reason why infection with *H. pylori* gives rise to gastric cancer in only some individuals may depend on host genetic factors such as polymorphisms which affect levels of the pro-inflammatory cytokine interleukin-1β (IL-1β) [Nature (2000) *404* 398–402].

Species of *Helicobacter* have also been detected in liver tissue from patients with primary sclerosing cholangitis and primary biliary cirrhosis [JCM (2000) *38* 1072–1076].

[*H. pylori* infection (diagnosis, treatment): Lancet (1997) *349* 265–269. Vaccines against *H. pylori*: Drugs (1996) *52* 799–804; BCID (1997) *4* 413–433. E test (susceptibility to antibiotics): JCM (1997) *35* 1842–1846. Electroporation of *H. pylori*: AEM (1997) *63* 4866–4871. *H. pylori* epidemiology, metabolism, genetics, pathogenesis, diagnosis, therapy etc.: BCID (1997) *4* 239–484. *H. pylori* (detection by PCR): JCM (1999) *37* 772–774.]

Helicobasidium See AURICULARIALES.

helicosporae See SACCARDOAN SYSTEM.

helicotrichous Refers to flagella inserted in a helical zone on the cell surface of *Centipeda periodontii* [JGM (1984) *130* 185–191].

Heliobacterium A genus of photosynthetic bacteria; *H. chlorum* contains a novel bacteriochlorophyll (Bchl *g*: see CHLOROPHYLLS) which, when in situ, shows a long-wavelength absorption peak at 788 nm. From 16S rRNA sequence analysis, *H. chlorum* is apparently related more closely to the Gram-positive bacteria than to other photosynthetic bacteria [Science (1985) *229* 762–765].

Heliozoea A class of protozoa (superclass ACTINOPODA) in which numerous separate axopodia or filopodia radiate from the (more or less spherical) cell body, giving rise to the popular name of 'sun animalcules'. The cells lack a central capsule (cf. RADIOLARIA); some species have a 'skeleton' of siliceous scales or spines. The cytoplasm is more or less clearly differentiated into a highly vacuolated ('frothy') ectoplasm and a less vacuolated endoplasm which contains e.g. one or more nuclei, contractile vacuole(s), food vacuoles, etc. The ectoplasmic vacuoles appear to aid flotation. In general, the axopodia are not effective locomotory organelles, serving mainly to capture prey; however, in some members a type of rolling motility can be achieved by the sequential retraction of the axopodia. (See also TAXOPODIDA.)

Reproduction occurs by binary fission or (in some uninucleate species) by budding. Sexual processes have been recorded in some species.

Most heliozoa are free-floating freshwater organisms; however, some occur attached to a substratum by a stalk, and some are marine. They feed on other protozoa, microalgae, etc.

Orders: ACTINOPHRYIDA, CENTROHELIDA, DESMOTHORACIDA, TAXOPODIDA.

helix–coil transition In proteins or nucleic acids: the transition from an ordered, helical conformation to a disordered 'random coil' conformation. When used of double-helical DNA the term usually means simply *strand separation* ('melting').

helix destabilizing protein *Syn.* SINGLE-STRAND BINDING PROTEIN.

Helminthosporium A genus of fungi (class HYPHOMYCETES) which form septate mycelium and pigmented, septate conidia; conidiogenesis is tretic. The organisms occur e.g. on woody plants. (See also SILVER SCURF.) Some species have been transferred from *Helminthosporium* to the genus DRECHSLERA.

***Helminthosporium maydis* virus** See MYCOVIRUS.

Helmstetter–Cooper model ($I + C + D$ model) A model which relates chromosome replication (in bacteria) to the (prokaryotic) CELL CYCLE. Essentially, the cell cycle is divided into a linear sequence of three main periods designated I, C and D. Chromosome replication starts at the beginning of the C period and finishes at the end of the C period. Septum formation begins, and is completed, during the D period – at the end of which the cell divides. The I period is the inter-initiation period, that is, the period between successive initiations of DNA replication; it is equal to the doubling time. As each I period ends, another I period begins.

The principle of the Helmstetter–Cooper model is illustrated in the figure. Some of the factors that may influence or determine the length of the I period are indicated in the entry CELL CYCLE.

Helotiales An order of fungi (subdivision ASCOMYCOTINA) which are typically saprotrophs or plant parasites. Ascocarp: APOTHECIOID with paraphyses, sessile or stipitate, sometimes setose. Asci: unitunicate, typically cylindrical; inoperculate, but with an apical pore. Ascospores: typically *not* elongated. Genera: e.g. *Ascocoryne*, *Botryotinia*, *Bulgaria*, *Chlorociboria* (see also GREEN OAK), *Chlorosplenium*, *Ciboria*, *Dasyscyphus*, DIPLOCARPON, *Geoglossum*, *Korfia*, *Microglossum*, *Mitrula* (see also MYCOPARASITE), *Monilinia*, *Pezizella* (see also MYCORRHIZA), *Phacidium* (see also SNOW BLIGHT), SCLEROTINIA, *Spathularia*, *Trichoglossum*.

helotism (*mycol.*) Literally: *serfdom*. The term has been used to refer to the (supposedly) subservient role of the photobiont in a LICHEN.

helper component (*plant virol.*) See NON-CIRCULATIVE TRANSMISSION.

helper T cell See T LYMPHOCYTE.

helper virus Any virus which can provide functions necessary for the replication of a DEFECTIVE VIRUS or e.g. of a SATELLITE RNA.

Helvella A genus of fungi (order PEZIZALES) in which the ascocarp is a stipitate APOTHECIUM which is typically saddle-shaped ('saddle fungi'); the smooth to warty, guttulate ascospores are commonly colourless or pale brown.

helveticin J See BACTERIOCINS.

heme See HAEM.

Hemiascomycetes A class of ascomycetes characterized by the formation of naked asci (i.e., asci not formed on or in an organized ascocarp) which do not develop from a system of ascogenous hyphae; the hemiascomycetes include most of the yeasts (e.g. *Saccharomyces* spp) and e.g. *Taphrina* spp. The class is not recognized in most modern taxonomic schemes; most of the hemiascomycetes are now placed in the ENDOMYCETALES.

Hemiaulus See DIATOMS.

Hemibasidiomycetes See TELIOMYCETES.

hemicelluloses The non-cellulosic polysaccharide fraction obtained from isolated plant cell walls by extraction with alkali after the PECTIC POLYSACCHARIDES have been extracted. Hemicelluloses are closely (but non-covalently) associated with CELLULOSE and occur in the matrix of the plant cell wall; they include e.g. galactomannans, glucomannans (see MANNANS), mixed β-glucans, XYLANS, xyloglucans etc.

hemidesmosome See TRYPANOSOMA.

Hemidiscus A genus of DIATOMS which are generally regarded as centric, although the valves are often markedly asymmetric. Blooms of *H. hardmannianus* in estuarine and neritic waters in India have been associated with high mortalities of fish and invertebrates [Limn. Ocean. (1985) *30* 910–911].

Hemileia See UREDINIOMYCETES and COFFEE RUST.

hemimethylated DNA dsDNA in which only one of the strands is methylated (see DNA METHYLATION).

Hemitrichia A genus of slime moulds (class MYXOMYCETES). *H. serpula* forms conspicuous, yellowish, sessile, reticulate plasmodiocarps (up to several centimetres across) on humus or on rotting wood.

Hemming filter An apparatus used for FILTRATION. Two bijou bottles are clamped together, mouth-to-mouth, with a filter

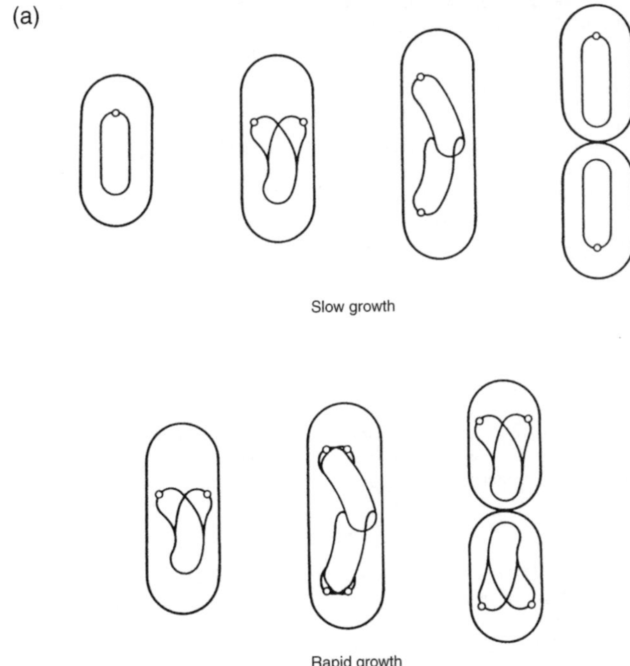

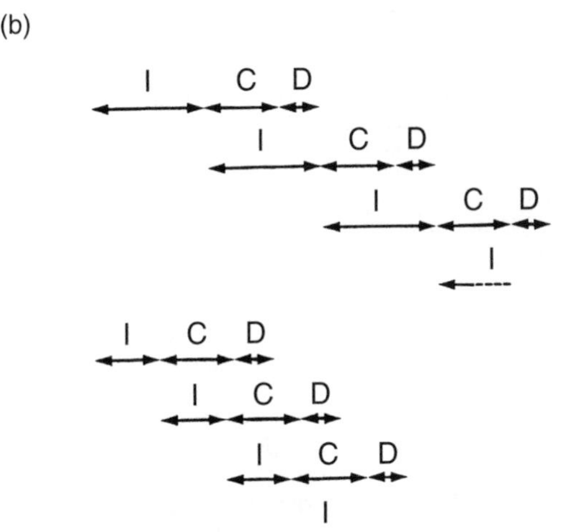

HELMSTETTER–COOPER MODEL (see entry). (a) Chromosome replication during growth at different rates (cells and chromosomes are represented diagrammatically). Replication begins at a specific location in the chromosome, the *origin*, which is shown here as a small circle. During slow growth (*upper sequence*) each new daughter cell has exactly one chromosome because, following duplication of the parental chromosome, a new round of replication does not begin before the parent cell has divided fully.

During faster growth (*lower sequence*), a new round of replication has begun before the previous round has been completed – well before cell division; as a result, each new daughter cell has fractionally more than one chromosome.

(b) The cell division cycle may be represented as a linear sequence of three periods: *I*, *C* and *D*. *C* is the period during which chromosome replication occurs. *D* is the period in which the septum forms – cell division occurring at the end of the *D* period. *I* is the period between each successive *initiation* of chromosomal replication. Each *I* period starts when the previous one ends. (*Continued on page 362*.)

between them; one bijou contains the sample, the other acts as the receiving vessel. On centrifugation, liquid passes through the filter into the receiver.

hemp retting See RETTING.

Henneguya A large genus of protozoa (class MYXOSPOREA) which are important histozoic parasites in various fish. Species are commonly specific for a given host and for a particular tissue within that host: e.g. the Channel catfish (*Ictalurus punctatus*) may be parasitized by *H. adiposa* (adipose fin), *H. diversis* (liver and kidney), or *H. exilis* (gills), while the Pacific salmon (*Oncorhynchus* spp) can be infected by *H. salminicola* (muscles). Typically, conspicuous spore-filled cysts (up to 15 mm across in e.g. salmon infected with *H. salminicola*) are formed. Heavy infestations may kill the fish (particularly when the gills are affected).

HEP High egg passage: refers to the number of times that a given strain of virus has undergone SERIAL PASSAGE in eggs. (See e.g. FLURY VIRUS; cf. LEP.)

HEp-2 (H.Ep2, Hep-2, hep-2 etc) An ESTABLISHED CELL LINE derived from a human laryngeal carcinoma; the cells are heteroploid and epithelioid. HEp-2 cultures support the replication of a wide range of viruses.

HEPA filter High-efficiency particulate air filter, used e.g. in a SAFETY CABINET.

Hepadnaviridae A family of enveloped DNA-containing animal viruses which can cause hepatitis in man, animals or birds; the family includes HEPATITIS B VIRUS (HBV) of man, GROUND SQUIRREL HEPATITIS VIRUS (GSHV), DUCK HEPATITIS B VIRUS (DHBV), and WOODCHUCK HEPATITIS VIRUS (WHV) [Intervirol. (1986) *25* 14–29], and probably tree squirrel (*Sciurus carolinensis pennsylvanicus*) hepatitis B virus (THBV) [PNAS (1986) *83* 2994–2997]. These viruses closely resemble one another in structure and genetic organization, but differ e.g. in their interactions with their hosts. The host's immune response to hepadnavirus infection has been implicated in the pathology of the virus-induced liver disease. Although all of the hepadnaviruses are highly hepatotropic, HBV DNA has also been detected in e.g. pancreatic, spleen and kidney cells, and DHBV can apparently infect duck pancreatic and kidney cells.

The infectious hepadnavirus virion (ca. 42 nm diam.) consists of a nucleocapsid (an internal core containing the viral genome) surrounded by a lipoprotein envelope. The core contains a major protein (designated $P22^C$ in HBV) and is associated with a DNA polymerase and a protein kinase which can phosphorylate the major core protein. In HBV, at least, the envelope contains three proteins: a 'major protein' (glycosylated form: $GP27^S$, non-glycosylated form: $P24^S$), the 'middle protein' (two glycosylated forms: $GP33^S$ and $GP36^S$), and the 'large protein' (glycosylated form: $GP42^S$; non-glycosylated form: $P39^S$).

The genome, as present in hepadnavirus virions isolated from the blood of an infected host, has a characteristic structure: it consists of a circular (but not covalently closed), partially double-stranded DNA molecule in which one strand (the L, 'long', or (−) strand) is of fixed length (ca. 3000–3300 nucleotides long) and has unique 3′ and 5′ ends and a protein (the 'DNA-linked protein') covalently attached to the 5′ end. The other strand (the S, 'short', or (+) strand) has a fixed 5′ end but a variable 3′ end, and varies in length from 50 to 100% of that of the L(−) strand. (The 3′ end of the S(+) strand can be extended by the endogenous viral polymerase, the L(−) strand acting as template.) The circularity of the molecule is maintained by base-pairing between the 5′ ends of the two strands (the 'cohesive overlap'); in all hepadnaviruses there is an 11-bp direct repeat sequence on either side of the cohesive overlap.

In mammalian hepadnaviruses, the L(−) strand has four open reading frames, orientated in the same direction, designated S, C, P and X regions. The S region comprises the S gene, the pre-S1 region and the pre-S2 region; it encodes the 'major protein' (S gene), 'middle protein' (pre-S2 + S gene) and 'large protein' (pre-S1 + pre-S2 + S gene) of the viral envelope. The C region comprises the C gene (which encodes the major core protein) and the pre-C region. The product of the P region (which overlaps the whole of the S region and parts of the C and X regions) is a large histidine-rich protein (MWt ca. 90000) which is a DNA polymerase with REVERSE TRANSCRIPTASE activity. The X region includes the cohesive overlap sequence and encodes a polypeptide of unknown function. (DHBV lacks an X region.) The viral genes are expressed via at least two unspliced primary transcripts. No known viral protein is encoded by the S(+) strand.

Hepadnavirus DNA replication occurs by a characteristic mechanism involving reverse transcription of an RNA intermediate (see RETROID VIRUSES). When a virion infects a host cell, the viral DNA enters the nucleus and is converted to a supercoiled ccc dsDNA molecule. The L(−) strand is transcribed to form multiple copies of a 3.5-kb RNA called the *pre-genome*; this RNA is then encapsidated with the viral DNA polymerase and 'DNA-linked protein', and an L(−) strand of DNA is synthesized by reverse transcription of the pre-genome. The pre-genome is then apparently degraded by an RNase H activity, and an S(+) strand is synthesized on the L(−) strand template. The 'DNA-linked protein' may act as primer for L(−) strand DNA synthesis, while a small fragment of RNA may be the primer for S(+) strand synthesis. (cf. RETROVIRIDAE.) The assembly of the virions and their release from the cell can occur before S(+) strand synthesis is complete, accounting for the partially double-stranded nature of the genome in virions liberated into the bloodstream.

[Review (mainly of HBV): Nature (1985) *317* 489–495; Book ref. 110, pp. 23–41.]

heparin A highly sulphated proteoglycan present e.g. in MAST CELLS and other cells. Heparin varies in MWt according e.g.

HELMSTETTER–COOPER MODEL (*continued*)

In a given set of $I + C + D$ sequences, the relationship between a particular sequence and the one immediately above it is that of a daughter cell to its parent cell. The upper set of $I + C + D$ sequences represents successive cell cycles during slow, steady growth. Notice that, when $I = C + D$, a new round of chromosome replication does not start until after the D period, i.e. after completion of cell division.

The lower set of $I + C + D$ sequences represents successive cell cycles during faster steady growth. Here, with I shorter than C, chromosome replication is initiated before the previous round has been completed. Hence, at the end of D the new round of replication is already well advanced. Thus, in cells growing rapidly, the chromosome in a new daughter cell will have reached a stage of replication which is determined by the timing of initiation (i.e. start of the C period) in the parent cell.

In *Escherichia coli*, C and D increase linearly in generation times >60–70 minutes [Microbiology (2003) *149* 1001–1010].

to the methods used for its isolation; that used as an ANTICOAGULANT (heparin inhibits thrombin activity) has a MWt of ca. 25000. In addition to anticoagulant activity, heparin can e.g. modify (limit) delayed hypersensitivity reactions in mice and guinea pigs (mechanism unknown), stimulate phagocytosis, inhibit the alternative pathway of complement fixation (by enhancing the activity of Factor H), and inhibit eosinophil-mediated ADCC. Heparin is cleaved by heparinases, and its activity can be neutralized by e.g. protamines.

hepatitis Inflammation of the liver. Hepatitis is often caused by viruses (e.g. HEPATITIS A, HEPATITIS B, HEPATITIS C); hepatitis D and hepatitis E viruses are less-common causal agents. The significance of infection with hepatitis G virus is currently uncertain [transfusion-associated hepatitis G virus infection: NEJM (1997) *336* 747–754; RMM (1998) *9* 207–215]. Hepatitis may also occur as a symptom of non-viral infectious diseases (see e.g. *amoebic* DYSENTERY). [Diagnostic perspective of viral hepatitis: RMM (1997) *8* 197–207.]

hepatitis A (infectious hepatitis; epidemic hepatitis; short-incubation hepatitis) A viral HEPATITIS caused by the hepatitis A virus (HAV, = ENTEROVIRUS type 72); it tends to occur in epidemics among children and young adults, particularly in closed institutions (e.g. mental hospitals). Man is probably the only natural host, but certain other primates (e.g. chimpanzees) can be infected. A carrier state is not known. Infection occurs mainly by consumption of water, milk or food contaminated with human faeces – e.g. shellfish harvested from sewage-polluted waters. HAV replicates in hepatocytes and is presumed to reach the intestine (and hence faeces) via the bile duct. The peak excretion of virus occurs during the latter half of the incubation period (2–6 weeks). Symptoms: fatigue, anorexia, low-grade fever, some abdominal pain, and sometimes jaundice. The disease is usually self-limiting within a few weeks; rarely, a severe, fulminant, often fatal hepatitis develops. *Lab. diagnosis*: demonstration of HAV in faeces; serological tests for anti-HAV IgM antibodies in serum.

hepatitis B (serum hepatitis; homologous serum jaundice; long-incubation hepatitis) A viral HEPATITIS caused by the HEPATITIS B VIRUS (HBV). Incubation period: 2–6 months (but occasionally as short as 2 weeks). Infection occurs by direct inoculation or by contamination of mucous membranes. The virus is present in the blood and body fluids (saliva, semen, mucus, wound exudates etc) of an infected person. Transmission may thus occur by the transfusion of blood or blood products from infected donors; by tattooing, ear-piercing, vaccination, renal dialysis, etc, using contaminated needles; by sexual contact; or (in infants) during birth. (Intrauterine infection is uncommon since HBV does not normally cross the placenta.) HBV reaches the liver via the blood and replicates in the hepatocytes. The disease may be mild or subclinical, and a carrier state is recognized (see HEPATITIS B VIRUS). In clinical cases, onset is gradual with jaundice and e.g. nausea, malaise, pains in muscles and joints, with or without fever; occasionally acute hepatic failure occurs. (See also DELTA VIRUS.) Pathogenesis appears to involve the host's immune response to viral antigens in the plasma membranes of infected hepatocytes. The condition may be self-limiting, or chronic liver disease may develop; chronic infection is associated with an increased risk of hepatocellular carcinoma (HCC: see HEPATITIS B VIRUS) and possibly other tumours, including KAPOSI'S SARCOMA [JAMA (1984) *251* 1007–1008]. HCC is a major cause of death in regions where HBV is endemic and neonatal infection is common (e.g. in regions of China and tropical Africa).

Lab. diagnosis involves detection of HBV antigens in the serum. HBsAg (see HEPATITIS B VIRUS) appears in the serum during the incubation period and persists through the symptomatic phase of the disease. It usually begins to disappear ca. 2–3 months after the onset of symptoms; if it persists for >6 months, the patient has become a carrier and may develop chronic liver disease. Anti-HBsAg antibodies appear only late in convalescence; the high levels of HBsAg present in the blood may induce immune tolerance to this antigen. HBeAg appears in the serum just before symptoms begin; its appearance is associated with high infectivity and the presence of whole (infectious) virions ('Dané particles') in the blood. HBcAg does not appear in the serum, but anti-HBcAg antibodies can be detected as symptoms begin. Anti-HBcAg IgM antibodies disappear late in the convalescence phase, but anti-HBcAg IgG antibodies may persist for years (possibly for life), and detection of the latter is a useful indicator of previous infection. [Screening blood samples for HBV by PCR: Lancet (1999) *353* 359–363.]

Treatment. There is no specific treatment for the clinical disease. Antiserum may be used for pre- or post-exposure prophylaxis. Vaccines containing HBsAg are available [PM (1984) *75* 199–211].

hepatitis B virus (HBV) A virus of the HEPADNAVIRIDAE (q.v.) which causes HEPATITIS B in man, and which is also apparently a causal agent of human hepatocellular carcinoma (HCC). (HBV can also infect chimpanzees, but does not cause chronic hepatitis or HCC in these animals.) (cf. WOODCHUCK HEPATITIS VIRUS.)

On infection of a new host (see HEPATITIS B), the HBV virion attaches to the surface of a hepatocyte and penetrates the cell by endocytosis or by fusion of the viral envelope with the host cell plasma membrane. HBV surface (envelope) antigen (HBsAg) and core antigen (HBcAg) become associated with the plasma membrane of the infected cell, and subsequently large amounts of HBsAg are released into the circulation in the form of non-infectious (DNA-free), spherical or tubular lipoprotein particles (ca. 22 nm diam.). HBsAg (formerly known e.g. as 'Australia antigen', Au antigen, or hepatitis-associated antigen, HAA) is a dimer of two molecules, linked by disulphide bonds, of the envelope 'major protein' (see HEPADNAVIRIDAE). HBcAg is the major capsid polypeptide ($P22^C$). The soluble e antigen (HBeAg) is a product of proteolytic cleavage of HBcAg. The appearance in the serum of infected individuals of antibodies to these antigens provides a useful means of monitoring disease development (see HEPATITIS B).

HBV strains can be serotyped on the basis of their HBsAg. HBsAg contains a group-specific determinant (designated *a*), common to all subtypes, together with subtype-specific determinants (designated *d*, *y*, *w* and *r*); strains typically contain either *d* or *y* with either *w* or *r*. The four principal serotypes, which differ in geographical distribution, are *adw*, *adr*, *ayw* and *ayr*. The subtype determinants are useful epidemiological markers.

In ca. 5–10% of individuals infected with HBV, HBsAg persists in the circulation for >6 months, and a carrier state is established which may persist for life. The carrier state may be asymptomatic (latent infection) or symptomatic (with e.g. chronic active hepatitis, cirrhosis, and/or HCC), and is most likely to develop in e.g. perinatally-infected or immunodeficient individuals. In HBV carriers, viral DNA may integrate at one or more (apparently random) sites in the host cell genome; the integrated sequences may be complete genomes or subgenomic fragments, and in HCC cells they are often extensively rearranged, with e.g. multiple deletions, inversions, duplications, etc. Rearrangements also occur in the flanking cellular sequences.

[Duplications of flanking cellular sequences and model for HBV DNA integration: PNAS (1985) *82* 4458–4462.] Integrated viral DNA may continue to be expressed, resulting in the continued synthesis of e.g. HBsAg; however, in HCC cells virus replication and gene expression appear generally to be absent. Integrated HBV DNA has been found in the majority of HCC cells examined, and integration is generally assumed to precede carcinogenesis; however, the relationship between viral DNA integration and the induction of HCC is unknown. HCC generally does not develop for many years after the establishment of the carrier state; in at least some cases additional factors – such as genetic predisposition, immunological factors, or exposure to chemical co-carcinogens (e.g. AFLATOXINS) – may be involved.

[Biology and epidemiology of HBV: Book ref. 110, pp. 43–65; molecular biology of HBV: Nature (1985) *317* 489–495.]

hepatitis C A viral HEPATITIS caused by the HEPATITIS C VIRUS (HCV); most cases of hepatitis formerly referred to as NANB (non-A-non-B) hepatitis are now believed to have been hepatitis C. In nature, hepatitis C is limited to humans (chimpanzees can be infected experimentally). The route of transmission of HCV is commonly parenteral (including blood transfusion); sexual and perinatal transmission also occur.

The incubation period (4–12 weeks) may be followed by non-specific symptoms that resolve within weeks or months. Infection is often asymptomatic; jaundice is uncommon. HCV RNA is detectable in serum (e.g. by rtPCR) before seroconversion. In a high proportion of patients infection becomes chronic (indicated e.g. by the continuing presence of HCV RNA in serum); usually asymptomatic, the chronic disease may nevertheless give rise to e.g. hepatocellular carcinoma or cirrhosis in a proportion of cases after 20–30 years. (Preliminary data have suggested that there is a cerebral effect (altered cerebral metabolism) in patients chronically infected with HCV [Lancet (2001) *358* 38–39].)

Lab. diagnosis: e.g. screening (with a third-generation ELISA) for specific antibodies; antigens used in the assay consist of synthetic peptides/recombinant proteins representing products which are encoded by particular regions of the HCV genome. Any specimen positive in this assay is re-tested by a supplemental assay such as the *recombinant immunoblot assay* (RIBA). In a RIBA test, a nitrocellulose strip is coated with peptides/recombinant proteins similar to those used in the ELISA, each type of peptide and protein forming a separate band on the strip. The strip is exposed to a sample of serum (permitting antibody–antigen binding) and then to an anti-human IgG–enzyme conjugate. After washing, bound antibodies are detected by adding the enzyme's substrate and examining the strip for the (coloured) product of hydrolysis; the presence of particular antibodies in the serum is deduced from the development of colour at particular band(s) on the strip. Nucleic-acid-based tests are useful for the early detection of HCV – and also for following the efficacy of antiviral therapy. [Quantification of hepatitis C virus by bDNA assay: JCM (1997) *35* 187–192; rtPCR-based screening for HCV in a blood bank: Lancet (1999) *353* 359–363.]

hepatitis C virus (HCV) A small, enveloped, positive-sense ssRNA virus which causes HEPATITIS C in man. A population or sample of HCV from a given source exists as a QUASISPECIES. HCV has not been cultured.

The 5′ non-coding end of the HCV genome is followed (5′ → 3′) by sequences which encode: core protein, envelope proteins, membrane-binding function, helicase and protease, membrane-binding function, RNA polymerase and a 3′ non-coding region. One of the envelope protein genes includes a hypervariable region designated HVR-1.

Six major genotypes and >60 subtypes of HCV have been distinguished. Treatment of hepatitis C is apparently influenced by the specific genotype involved and the level of viraemia; disease caused by genotype 2 or 3, in conjunction with low-level viraemia, is reportedly more likely to respond in a sustained fashion to treatment with IFN-α, while disease caused by genotype 1 or 4 is apparently more refractory.

[Quantification of HCV in plasma samples: JCM (1997) *35* 187–192.]

In many cases HCV causes persistent infection. Persistence may be facilitated by the relatively poor immunogenicity of the virus and the low-level viraemia. Changes in viral antigens, due e.g. to mutations in HVR-1, may also contribute to persistence; however, genetic drift in HCV may be independent of the host's (weak) immune pressure [JV (2000) *74* 2541–2549].

(See also CD81.)

hepatitis D, E See HEPATITIS.

hepatitis D virus *Syn.* DELTA VIRUS.

hepatitis delta virus *Syn.* DELTA VIRUS.

hepatitis G virus See HEPATITIS.

hepatitis X Hepatitis in dogs caused by ingestion of food contaminated with AFLATOXINS; the liver becomes necrotic and congested, with haemorrhagic oedema of the gall bladder mucosa.

hepatocellular carcinoma (HCC) A CARCINOMA of liver cells. Primary HCC can be caused e.g. by HEPATITIS B VIRUS and WOODCHUCK HEPATITIS VIRUS. (See also HEPATITIS C.)

hepatotoxin A TOXIN which acts on the liver.

hepatotropism See TROPISM (sense 2).

HEPES A zwitterionic pH buffer: *N*-2-hydroxyethylpiperazine-*N′*-2-ethanesulphonic acid. HEPES is widely used e.g. in TISSUE CULTURES, and has also been used in bacterial cultures; it is readily soluble in water, and does not bind divalent cations such as Ca^{2+}, Mg^{2+} or Cu^{2+}. At 37°C the pK_a of HEPES is ca. 7.3; at 25°C it is ca. 7.5. (cf. BICINE.)

Herbert's pits See TRACHOMA.

herbicolin A An acyl peptide antibiotic produced by *Erwinia herbicola*. It can inhibit the growth of yeasts and filamentous fungi, and can kill sterol-requiring mollicutes [JAC (1985) *16* 449–455]; it is inactive against other bacteria.

herbivorous Refers to protozoa which feed on bacteria and/or algae. (cf. CARNIVOROUS.)

herd immunity See EPIDEMIOLOGY.

herd structure See EPIDEMIOLOGY.

Heribaudiella See PHAEOPHYTA.

Hericium A genus of fungi of the APHYLLOPHORALES.

hermaphroditism The phenomenon in which an individual has both male and female sexual organs; such an individual may exhibit HOMOTHALLISM or HETEROTHALLISM.

herpangina An acute (usually self-limiting) infectious PHARYNGITIS, occurring mainly in children and young adults, caused by coxsackieviruses A2–A6, A8 and A10. Small, vesicular, ulcerating lesions develop in the pharyngeal region, and there may be varying degrees of fever and/or prostration.

herpes (1) Any disease caused by a herpesvirus (especially HERPES SIMPLEX but also e.g. HERPES ZOSTER) and characterized by the formation of small vesicular lesions on skin and mucous membranes. (2) Used increasingly as a synonym of genital herpes (see HERPES SIMPLEX).

herpes genitalis See HERPES SIMPLEX.

herpes gladiatorum See HERPES SIMPLEX.

herpes labialis See HERPES SIMPLEX.

herpes simplex An infectious disease, caused by herpes simplex virus (HSV) type 1 or 2 (see ALPHAHERPESVIRINAE), which (typically) involves the formation of thin-walled vesicles that ulcerate, crust and heal; vesicles occur, often in clusters, on skin and/or mucous membranes. Transmission occurs mainly by person-to-person contact – e.g. sexual contact, kissing, contact sports such as wrestling ('herpes gladiatorum').

Infection may occur at any of a range of body sites; in general, HSV type 2 is associated with genital (and hence neonatal) infections; HSV type 1 accounts for most of the other forms.

Incubation period: 2–12 (average 6) days. The disease varies from subclinical to severe, and is occasionally fatal. HSV can remain latent in nerve cells near the site of infection; reactivation may occur spontaneously or in response to other infections, stress, immunosuppression etc. In general, recurrent episodes become progressively less severe and less frequent. In neonates and immunodeficient individuals HSV may become disseminated and may involve e.g. the liver, adrenal glands, brain etc.

Cutaneous herpes. Skin lesions may occur on any part of the body, often in clusters which may coalesce. Other symptoms of primary infection may include pain, fever, oedema etc. HSV type 1 tends to be associated with skin lesions above the waist, type 2 with those below the waist. (See also ECZEMA and PARONYCHIA.)

Encephalitis. HSV type 1 is a common cause of non-epidemic viral ENCEPHALITIS. Onset is acute, with headache, nausea, vomiting, fever, myalgia, mental deterioration, convulsions, coma etc.; the fatality rate is typically >50%. Neurological sequelae are common in survivors. (See also aseptic MENINGITIS.)

Genital herpes (herpes genitalis) is a SEXUALLY TRANSMITTED DISEASE which usually involves HSV type 2. It may be mild and self-limiting. Lesions occur in the genital region and may spread to adjacent areas of skin (e.g. buttocks, thighs). Other symptoms may include e.g. fever, dysuria, pain, malaise. In women the cervix may be the main site of genital infection (*herpetic cervicitis*); this may be asymptomatic, or there may be ulceration with hyperplasia. (HSV type 2 infection in women is associated with an increased risk of abortion and of cervical cancer.) Genital infections may give rise to subclinical shedding of virus, with associated risk of transmission.

Labial herpes (herpes labialis) is typically a recurrent lip lesion (cf. COLD SORE) which usually crusts and heals in 3–10 days. Labial herpes may be associated with lip cancer.

Neonatal herpes is usually acquired (during birth) from a mother with genital herpes – hence HSV type 2 is commonly responsible. Herpetic vesicles or ulcers, or keratoconjunctivitis, may or may not be present. Fatality rates may be 50% or more in untreated cases; surviving infants commonly exhibit neurological and/or ocular sequelae. (See also TORCH DISEASES.)

Ocular herpes. HSV infection of the eye is a recurrent, usually unilateral, keratitis, sometimes with conjunctivitis; corneal scarring can lead to blindness. The characteristic lesion is a dendritic (branching) corneal ulcer with can be readily stained with either FLUORESCEIN or ROSE BENGAL [photograph of stained lesion: Book ref. 224, pp 67–68].

Oral herpes (herpetic stomatitis) occurs as a manifestation of initial HSV infection in adults and children. Primary herpetic gingivostomatitis involves shallow vesicular and ulcerative lesions on the labial and buccal mucosa, and tongue, with GINGIVITIS, fever and regional lymphadenopathy. The lesions usually heal in 7–10 days. Reactivation rarely involves the oral mucosa (except in immunocompromised individuals, in whom the oral lesions may be atypical in appearance). (Oral lesions may also be due to infection with the varicella zoster virus (VZV); in chickenpox, such lesions may appear before development of the characteristic skin lesions.) [Virology of the mouth (including HSV infections): RMM (1994) *5* 209–216.]

Lab. diagnosis. For infections other than encephalitis: scrapings from the edges of lesions, or biopsies from skin or liver, are examined microscopically for multinucleate giant cells with eosinophilic intranuclear inclusion bodies; HSV can be demonstrated e.g. by immunofluorescence techniques, ELISA etc.

For encephalitis, laboratory diagnosis may be made e.g. by PCR-based examination of cerebrospinal fluid (CSF), thus avoiding the risks of brain biopsy [see e.g. JCM (1997) *35* 691–696]. The PCR-based detection of HSV (types 1 and 2) may be facilitated by a commercial system involving the use of MOLECULAR BEACON PROBES that hybridize to amplicons of both HSV type 1 and HSV type 2 (HSVision™; Stratagene). (HSV type 2 is not a common cause of viral encephalitis; it may be a causal agent e.g. in immunocompromised individuals.)

Chemotherapy. A number of antiviral agents (e.g. ACYCLOVIR, IDOXURIDINE, PENCICLOVIR, trifluorothymidine, VIDARABINE) are active against HSV and are useful in some cases; for example, vidarabine has been used in cases of HSV encephalitis, while acyclovir, or penciclovir, is used in creams for the topical treatment of lesions on skin and mucous membranes. In general, such agents are not effective in preventing recurrence or transmission.

herpes simplex virus group See ALPHAHERPESVIRINAE.

herpes zoster A human disease resulting from the reactivation of a latent human (alpha) herpesvirus 3, present in a dorsal nerve ganglion, subsequent to an attack of CHICKENPOX that may have occurred (many) years earlier. The virus replicates in the ganglion, and in adjacent nervous tissue, and is transported, within the axon, to nerve endings in the corresponding region of skin; viral replication is believed to account for pain experienced during the acute phase of the disease – the pain being accompanied by an eruption of characteristic clusters of vesicular lesions. (Lesions may develop in areas not served by the affected nerves; this is thought to be due to viral dissemination in the bloodstream.) (See also RAMSAY HUNT SYNDROME.) Reactivation of the virus may occur spontaneously or may be provoked e.g. by immunosuppression, certain drugs, or other diseases.

Post-herpetic neuralgia (PHN) is prolonged pain which appears to be due to damage sustained by the peripheral nervous system as a result of viral replication. Some regard PHN as pain which persists following the disappearance of the rash, while others have defined PHN as e.g. pain which persists for at least 30 days from the onset of the rash. In an attempt to provide a usable definition (for clinical trials of antiviral agents), the concept of *zoster-associated pain* (ZAP) has been introduced; ZAP includes the continuum of acute *and* chronic pain.

Chemotherapy. The use of antiviral agents in the acute phase of herpes zoster is generally beneficial, and is particularly important in immunocompromised patients in whom herpes zoster may become a life-threatening condition; in the latter patients, intravenous ACYCLOVIR is the first-choice therapy (which may decrease the risk of dissemination).

For immunocompetent patients with moderate to severe pain during the acute phase, systemic antiviral therapy should reduce the duration of the rash and pain if given immediately after the onset of the rash (within a maximum of 72 hours after onset); useful drugs: ACYCLOVIR, FAMCICLOVIR, VALACICLOVIR (each of which can be administered orally). Acyclovir should be given intravenously if severe complication (e.g. encephalitis) develop.

[Treatment of zoster: RMM (1995) *6* 165–174.]

Herpesviridae A large family of enveloped VIRUSes containing a linear dsDNA genome. All herpesviruses are structurally similar: the virion (ca. 120–200 nm diam.) contains a core, i.e. a DNA–protein complex, within an icosahedral capsid (ca. 100–110 nm diam.; 12 pentameric and 150 hexameric capsomers) [capsid protein (structure): EMBO (2003) *22* 757–765]; the nucleocapsid is enclosed by a lipoprotein envelope (bearing glycoprotein projections at the surface) – the (apparently) amorphous region (*tegument*) between capsid and envelope containing virus-encoded proteins. Virions are readily inactivated by lipid solvents, lipases or heat.

Herpesviruses have been isolated from a wide range of animals, including mammals, birds, reptiles, amphibians and fish; some of the viruses have a narrow host range.

Many herpesviruses can cause disease in their primary host(s). Many (possibly all) can remain latent in the host's tissues, often for the lifetime of the host.

The transmission of herpesviruses commonly occurs by direct contact between mucosal surfaces – but some can be transmitted via body fluids, such as milk, or via the placenta etc. (See also DNA TUMOUR VIRUSES.)

The herpesvirus genome ranges from ~125 to ~240 kb in length and it characteristically contains repeated terminal and/or internal sequences which can take part in recombinational events (and which may therefore allow the development of variant forms of the genome). In the large genome (e.g. ~80 genes in EPSTEIN–BARR VIRUS), most of the genes encode structural or regulatory functions; relatively few genes are expressed during viral latency.

Interestingly, in some human herpesviruses the genome contains sequences that are partially homologous to certain CYTOKINES and/or cytokine receptors.

Herpesviruses may be typed (fingerprinted) by techniques such as RFLP for epidemiological and other purposes [e.g. molecular epidemiology of herpes simplex virus type 1: RMM (1998) *9* 217–224].

Replication cycle. Initially, the virion attaches to a host cell and the viral envelope then fuses with the cell membrane (see ENVELOPE). The nucleocapsid is transported across the cytoplasm to the nuclear membrane where the DNA is released into the nucleus via a nuclear pore. Within the nucleus, the viral immediate–early genes are transcribed, and the transcripts enter the cytoplasm for synthesis of viral polypeptides. Products of the immediate–early genes include proteins that serve regulatory functions and which are necessary for the expression of early genes (encoding e.g. a DNA polymerase) and late genes (encoding e.g. components of the capsid). Viral DNA is replicated, and the nucleocapsids are assembled, in the nucleus, with various structural proteins that enter the nucleus from the cytoplasm. Nucleocapsids can leave the nucleus in two ways. They can bud through the inner lamella of the nuclear membrane, thus acquiring an envelope; in this case, the mature virions are transported to the cell surface within membrane-bounded vacuoles and are then released to the exterior. Alternatively, nucleocapsids may enter the cytoplasm in an immature form which, only later, acquires an envelope by budding into a cytoplasmic vacuole.

Classification and nomenclature. Some of the (many) known herpesviruses have been allocated to one of three subfamilies: ALPHAHERPESVIRINAE, BETAHERPESVIRINAE and GAMMAHERPESVIRINAE [Book ref. 23, pp 47–51] – mainly on the basis of biological characteristics.

The naming of individual herpesviruses has been complicated by the widespread use of common names which are based on a variety of criteria – such as the disease caused by the virus (e.g. Marek's disease virus, herpes simplex virus); CPEs induced by the virus (e.g. cytomegalovirus); host species (e.g. *Herpesvirus saimiri*); names of pioneer workers (e.g. Epstein–Barr virus, Lucké virus).

It has been recommended [Book ref. 139, pp 1–23] that the name of a herpesvirus should usually be based on the family or subfamily to which its primary host belongs (subfamily for bovine and primate viruses, family in other cases) – viruses isolated from humans being prefixed by 'human'; the name may, where applicable, indicate the subfamily of herpesviruses to which the virus belongs (alpha, beta or gamma), and is followed by a number – numbers being allocated sequentially as new viruses are discovered. Thus, e.g. Marek's disease virus becomes gallid herpesvirus 1; herpes simplex type 1 virus becomes human (alpha) herpesvirus 1; Epstein–Barr virus becomes human (gamma) herpesvirus 4.

Many herpesviruses, including some important pathogens of animals, were not listed in an earlier (1982) classification scheme [Book ref. 23, pp 47–51]. These viruses include e.g. bovine herpesvirus 1 (causal agent of various cattle diseases, such as INFECTIOUS BOVINE RHINOTRACHEITIS, encephalitis in newborn calves, infectious pustular vulvovaginitis and abortion in cows etc.); anatid herpesvirus 1 (= anserid herpesvirus1, causal agent of DUCK VIRUS ENTERITIS); gallid herpesvirus 3 (causal agent of AVIAN INFECTIOUS LARYNGOTRACHEITIS); herpesviruses of amphibians (e.g. Lucké virus, causal agent of renal adenocarcinoma in frogs, *Rana pipiens*) [review: Book ref. 140, pp 367–384]; and herpesviruses of reptiles and fish (see e.g. CARP POX and CHANNEL CATFISH VIRUS DISEASE) [review: Book ref. 140, pp 319–366].

Herpesvirus ateles See GAMMAHERPESVIRINAE.

herpesvirus B *Syn.* B VIRUS.

Herpesvirus saimiri See GAMMAHERPESVIRINAE.

Herpesvirus simiae *Syn.* B VIRUS.

herpetic Pertaining to or caused by a herpesvirus – see e.g. HERPES SIMPLEX.

herpetic cervicitis See HERPES SIMPLEX.

herpetic stomatitis See HERPES SIMPLEX.

Herpetomonas A genus of homoxenous parasitic protozoa (family TRYPANOSOMATIDAE) which occur (typically in the gut) in flies, bugs and hymenopterous insects. Promastigote, paramastigote and opisthomastigote forms occur during the life cycle; biflagellate organisms may result from rapid division of promastigotes.

Herpetosiphon A genus of GLIDING BACTERIA (see CYTOPHAGALES) which occur in aquatic (freshwater to marine) habitats. The organisms are gliding filaments (up to ca. 2 μm wide) which may be sheathed; they typically contain carotenoid glycoside pigments ranging from yellow or orange. Metabolism appears to be chemoorganotrophic. GC%: ca. 44–53.

Herpetosoma A subgenus of TRYPANOSOMA within the STERCORARIA; species include parasites of rodents and man. The trypomastigote form typically has a subterminal kinetoplast, a pointed posterior end, and a free flagellum.

T. (*H.*) *lewisi*, a parasite in rats, is transmitted cyclically and contaminatively by fleas, e.g. *Xenopsylla cheopis*; it is cultivable e.g. in NNN medium at 25–27°C. In the vertebrate host, the reproductive forms (epimastigotes) undergo multiple fission or unequal binary fission, resulting eventually in the production of non-dividing trypomastigotes.

T. (*H.*) *rangeli* is parasitic in man (and in e.g. cats, dogs etc) in Central and South America, and is transmitted cyclically – mainly by bugs of the genus *Rhodnius*. Although classified in the Stercoraria, this species is transmitted by bite: the organism develops both in the gut and (as epimastigotes and trypomastigotes) in the salivary glands of the vector, but forms which develop in the gut are not infective for mammals. In man, bloodstream forms are trypomastigotes, ca. 25–35 μm in length. *T.* (*H.*) *rangeli* appears to be non-pathogenic. It can be cultivated e.g. in NNN medium.

Other species include *T. microti*, *T. nabiasi* and *T. zapi*.

Herpomyces See LABOULBENIALES.

Herxheimer reaction *Syn.* JARISCH–HERXHEIMER REACTION.

Hessea See MICROSPOREA.

Hesseltinella See MUCORALES.

heteroantibody Any antibody formed in one species against an antigen derived from another species. (cf. ISOANTIBODY.)

heteroauxin See AUXINS.

Heterobasidion A genus of fungi of the APHYLLOPHORALES (family Polyporaceae). *H. annosum* (formerly *Fomes annosus*) forms perennial, bracket-shaped or resupinate, typically corky or woody fruiting bodies in which the context is dimitic and lacks clamp connections; the bracket-shaped fruiting body typically has a rust-coloured upper surface and a pale, porous, lower surface. Fruiting bodies of several years' standing consist of vertically stacked ('stratified') layers of vertically orientated tubules. Basidiospores: spheroidal, colourless, ca. 5×4 μm. *H. annosum* grows on felled timbers, and is parasitic on conifers and e.g. red oak (*Quercus borealis*); in e.g. parts of the United Kingdom, the stumps of freshly-felled trees are inoculated with *Peniophora gigantea* which colonizes the cut surface, inhibiting infection by *H. annosum*. (See also TIMBER SPOILAGE.)

heterocaryon *Syn.* HETEROKARYON.

Heterochlorida See PHYTOMASTIGOPHOREA.

Heterochloris See XANTHOPHYCEAE.

heterochromatin See CHROMATIN.

heteroclicity The phenomenon in which a cross-reactive antigen binds an antibody with an AFFINITY greater than that shown by the homologous antigen; the cross-reactive antigen, and the antibody, are described as *heteroclitic*.

heteroclitic See HETEROCLICITY.

heterococcolith See COCCOLITH.

heterocyst A type of cell, morphologically and functionally distinct from a vegetative cell, which occurs under certain conditions in some filamentous CYANOBACTERIA (members of sections IV and V); heterocysts function as specialized anaerobic compartments within which NITROGEN FIXATION can occur in aerobic environments. (cf. AKINETE.) Heterocysts differentiate from vegetative cells (intercalary or terminal, depending on species) in response to limiting levels of combined nitrogen; differentiation is inhibited in the presence of e.g. nitrate and ammonia.

Mature heterocysts apparently cannot divide.

Differentiation from a vegetative cell involves e.g. (i) the formation of a thick, multilayered envelope – whose thickness (in *Anabaena flos-aquae*) has been found to increase with increased partial pressure of oxygen [JGM 1992) *138* 2673–2678]; (ii) rearrangement of thylakoid membranes (heterocysts are filled with contorted membranes); (iii) inactivation of photosystem II, resulting in the cessation of oxygen evolution – cyclic photophosphorylation (which involves photosystem I) apparently continues to occur; (iv) degradation of phycocyanin (see PHYCOBILIPROTEINS); (v) cessation of carbon dioxide fixation (e.g. by inactivation of RuBP carboxylase); (vi) synthesis of NITROGENASE (see NIF GENES). [Role of transcription in differentiation: JGM (1984) *130* 789–796]. Respiratory activity apparently helps to maintain the anaerobic state. [Protection of nitrogenase from oxygen in heterocystous cyanobacteria: MR (1992) *56* 352–362.]

In *Anabaena* strain PCC 7120, which forms heterocysts in both terminal and intercalary locations, differentiation of heterocysts depends on a functional *hetR* gene. In genetically engineered cells of this strain, controlled expression of *hetR* can induce differentiation of heterocysts even under conditions that would normally repress their formation. However, in cells with a mutant *patA* gene – which form heterocysts almost exclusively at *terminal* locations – the expression of *hetR* did not induce the formation of intercalary heterocysts; this was interpreted to mean that, while raised levels of HetR are necessary to promote the formation of heterocysts, PatA may play a role in regulating the expression of *hetR* [PNAS (2001) *98* 2729–2734].

During nitrogen fixation, the heterocysts and vegetative cells are metabolically interdependent. The vegetative cells supply heterocysts with e.g. fixed carbon, reduced sulphur, and glutamate; in the heterocyst, fixed nitrogen is transferred to glutamate by glutamine synthetase (see AMMONIA ASSIMILATION), and the resulting glutamine is passed back to vegetative cells to supply their nitrogen requirements. Communication between a heterocyst and adjacent vegetative cell(s) occurs via fine pores (*microplasmodesmata*) in their juxtaposed cytoplasmic membranes; juxtaposition of these membranes occurs at the vegetative-cell end of a channel (the *pore channel*) which passes through the thick heterocyst envelope. (Microplasmodesmata are also found between vegetative cells.) [Heterocyst differentiation and function: Book ref. 75, pp. 219–242, 265–280.]

heterocytotropic antibodies CYTOPHILIC ANTIBODIES (particularly REAGINIC ANTIBODIES) which can attach to the cells of two or more different host species. (cf. HOMOCYTOTROPIC ANTIBODIES.)

heteroduplex Any double-stranded nucleic acid in which some, or many, of the bases in one strand are not complementary to bases in the corresponding positions in the other strand; such mismatching of base pairs is reflected in the thermal stability of the duplex. (See also DNA HOMOLOGY and HOMODUPLEX.) The term 'heteroduplex' is also used for any duplex in which each strand originates from a different parent duplex (see e.g. RECOMBINATION), even though such strands may in some cases be strictly complementary.

heteroecious *Syn.* HETEROXENOUS.

heterofermentation (1) Any fermentation in which there is more than one major end-product. (2) *Syn.* HETEROLACTIC FERMENTATION.

heterogeneous nuclear RNA See HNRNA.

heterogenote See MEROZYGOTE.

heteroglycan See POLYSACCHARIDE.

heteroimmune phages See SUPERINFECTION IMMUNITY.

heteroimmunization The stimulation of an immune response to antigens derived from another species. (cf. ISOIMMUNIZATION.)

heterokaryon (heterocaryon) (*mycol.*) A hyphal cell, mycelium, organism, or spore which contains genetically different nuclei. (cf. HOMOKARYON.)

heterokaryosis (*mycol.*) The condition in which genetically different nuclei exist in the same cell, mycelium or spore; it may arise e.g. by the entry of a genetically different nucleus into a HOMOKARYON (e.g. by hyphal fusion) or by mutation in a binucleate or multinucleate homokaryotic cell or mycelium. Heterokaryosis is important e.g. in PARASEXUAL PROCESSES.

heterokont Refers to a pair of flagella (on a biflagellate cell) in which one flagellum differs from the other in length and often also in type: see e.g. XANTHOPHYCEAE. (cf. ISOKONT.)

Heterokontae Former taxon for the xanthophytes.

heterolactic fermentation A type of LACTIC ACID FERMENTATION in which sugars (e.g. lactose, glucose) are fermented to a range of products. There are two distinct pathways, *phosphoketolase* being a key enzyme in each. The 'classical' (6-phosphogluconate) pathway occurs in certain LACTIC ACID BACTERIA (e.g. *Leuconostoc* spp, betabacteria – e.g. *Lactobacillus brevis*): glucose is fermented to lactic acid, CO_2, and acetic acid and/or ethanol (the ratio of acetic acid to ethanol depending e.g. on the redox potential of the system); PENTOSES are fermented to lactic and acetic acids [Appendix III(b)]. In the *Bifidobacterium* ('bifidus') pathway the products of glucose fermentation are lactic and acetic acids in the molar ratio 2:3 [Appendix III(b)].

heterologous (1) Derived from or associated with a species different from that being referred to (cf. HOMOLOGOUS (2)). (2) See ALTERNATION OF GENERATIONS. (3) (*immunol.*) Of an antibody (or antigen): one which is not HOMOLOGOUS (sense 4) to a given antigen (or antibody). A heterologous antibody, for example, may be a CROSS-REACTING ANTIBODY or may not combine at all with the given antigen. A heterologous vaccine is a vaccine prepared against one organism but capable of giving protective immunity against another.

heterologous interference See INTERFERENCE (1).

Heteromastix See MICROMONADOPHYCEAE.

heteromerous (1) (*lichenol.*) Refers to a lichen thallus in which the photobiont is confined to a distinct layer within the thallus (see LICHEN). (2) (*mycol.*) In an agaric: refers to a pileus and stipe composed of two distinct types of cell: SPHAEROCYSTS and hyphae (see RUSSULALES). (3) (*ciliate protozool.*) See MACRONUCLEUS.
(cf. HOMOIOMEROUS.)

heteromixis Collectively: DIMIXIS, DIAPHOROMIXIS and *secondary* HOMOTHALLISM.

heteromorphic Morphologically dissimilar.

heteromorphic alternation of generations See ALTERNATION OF GENERATIONS.

Heteronema A genus of EUGLENOID FLAGELLATES, closely related to PERANEMA, in which the second emergent flagellum is not attached to the pellicle. *H. acus* occurs e.g. in activated sludge.

heterophil antibodies (heterophile antibodies) Antibodies which can combine with specific HETEROPHIL ANTIGENS. Such antibodies may occur in a host which has not had immunological contact with the corresponding antigen; e.g., patients with INFECTIOUS MONONUCLEOSIS usually develop heterophil antibodies which agglutinate the red blood cells of the sheep, horse and ox – species whose RBCs share a common heterophil antigen (see PAUL–BUNNELL TEST).

heterophil antigen (heterophile antigen) Any antigen which occurs (in an identical or closely related form) in several or many widely differing species (e.g. sheep, horse). If introduced into a species in which it is normally absent, a heterophil antigen elicits the synthesis of HETEROPHIL ANTIBODIES. (See also FORSSMAN ANTIGEN.)

heterophilic binding (of cell adhesion molecules) See IMMUNOGLOBULIN SUPERFAMILY.

Heterophrys See CENTROHELIDA.

heteroplasmon See CYTODUCTION.

heteroploid Having a complement of chromosomes differing from that characteristic of the species. (cf. EUPLOID sense 2.) A heteroploid cell may be EUPLOID (sense 1) – e.g. a polyploid cell arising in a haploid or diploid population – or it may be ANEUPLOID.

heteropolysaccharide See POLYSACCHARIDE.

heteroresistance (1) The resistance (of an organism) to two or more *related* antibiotics – e.g. two or more β-lactam antibiotics. (2) See MRSA.

heterothallism The phenomenon in which sexual reproduction requires the involvement of two different thalli – an individual thallus being self-sterile even when (as is common) it is hermaphroditic; gametes etc which can unite sexually are said to be *compatible*. (cf. HOMOTHALLISM; see also COMPATIBILITY sense 2.)

In *morphological heterothallism* (= *dioecism*) compatibility is determined solely on a sexual basis: there are separate male and female thalli, and sexual reproduction can occur between any male thallus and any female thallus of the same species. Dioecism is uncommon among fungi; it occurs e.g. in some species of *Achlya* and of *Laboulbenia*.

In *physiological heterothallism* male and female organs generally occur on the same thallus, and in these cases compatibility is determined not only on a sexual (male–female) basis: sexual union can occur between male and female gametes etc only if each has appropriate *mating type allele(s)* (see MATING TYPE); gametes etc which have the same mating type allele(s) cannot unite sexually and are said to be *incompatible*. In e.g. many zygomycetes, male and female organs are not distinguishable; in such cases sexually compatible thalli are often designated 'plus' and 'minus'.

In *bipolar physiological heterothallism* compatibility is determined at *one* mating type locus which may contain either of 2 alleles; if these alleles be designated *A* and *a*, then mating can occur between one individual (mating type) which has the *A* allele and another which has the *a* allele. Bipolar heterothallism is also called one-locus two-allele heterothallism, *unifactorial* heterothallism (since only one locus is involved), or DIMIXIS (since two alleles are involved).

In *tetrapolar physiological heterothallism* compatibility is determined at *two* mating type loci, each locus containing one of two alleles. If these loci be designated *A*, *B* (with alleles *a*, *b*, respectively), four mating types can be distinguished: *AB*, *Ab*, *aB* and *ab*. Mating can occur only between individuals which contain complementary alleles at *both* loci; thus, e.g. mating can occur between *AB* and *ab* individuals, and between *Ab* and *aB* individuals. Tetrapolar heterothallism is also called *bifactorial* heterothallism.

In some cases of one- and two-locus heterothallism, a locus may be occupied by any allele of an allelomorphic series; in such cases there can be many mating types. (See also DIAPHOROMIXIS.)

Heterothallism provides a constraint on inbreeding, and consequently increases the potential for genetic reassortment within a species.

Heterotrichida An order of free-living and parasitic ciliate protozoa (class POLYHYMENOPHOREA) in which the organisms are typically large, often contractile (see MYONEME), and sometimes pigmented, and in which somatic ciliature is commonly abundant and oral ciliature is well developed. Genera include e.g. *Balantidioides*, BLEPHARISMA, *Brachonella*, *Bursaria*, *Condylostoma*, *Metopus*, SPIROSTOMUM and STENTOR.

heterotrichous (1) Having different types of cilia or flagella. (2) Of certain filamentous algae: having a vegetative thallus consisting of both prostrate and erect systems of filaments (as e.g. in CHAETOPHORA).

heterotroph An organism which uses organic compounds for most or all of its carbon requirements. (cf. AUTOTROPH; ORGANOTROPH.) The term 'heterotroph' is often used to refer specifically to CHEMOORGANOHETEROTROPHS – although chemolithotrophs and phototrophs may also be heterotrophic.

heteroxenous (heteroecious) Refers to a parasite which carries out part of its life cycle in each of two or more different host species. (cf. HOMOXENOUS.)

heterozygote A HETEROZYGOUS cell or organism.

heterozygous Refers to a cell or organism (e.g. a diploid eukaryote or a bacterial MEROZYGOTE) in which one or more specified genes, or all of the genes, are present in the form of different ALLELES. (cf. HOMOZYGOUS.)

***hex* genes** See TRANSFORMATION (1).

Hexacapsula See MYXOSPOREA.

hexachlorophane (hexachlorophene) See BISPHENOLS.

hexagonal II phase See CYTOPLASMIC MEMBRANE.

hexamer See ICOSAHEDRAL SYMMETRY.

hexamethylenetetramine See HEXAMINE.

hexamine (hexamethylenetetramine; methenamine) A condensation product of ammonia and formaldehyde used as a urinary antiseptic. It has no intrinsic antimicrobial activity, but under acid conditions it slowly releases FORMALDEHYDE. It is not effective against urea-splitting bacteria since these produce ammonia which raises the pH of the urine. Hexamine is usually used with MANDELIC ACID (hexamine mandelate, Mandelamine) or with hippuric acid (Hiprex).

Hexamita A genus of protozoa of the DIPLOMONADIDA. *H. meleagridis* is an important pathogen of turkeys (see HEXAMITIASIS); cells are pear-shaped, ca. 10 μm in length, with two anteriorly located nuclei, two clearly separate, parallel, longitudinal axostyles, and 8 flagella – two pairs orientated anteriorly, one pair laterally, and the fourth pair projecting posteriorly. *H. intestinalis* is a cyst-forming species which occurs e.g. in the cloaca of the frog.

hexamitiasis An acute infectious disease of e.g. turkeys and pheasants (but not affecting chickens, ducks or geese); it is caused by *Hexamita meleagridis*. Infection occurs by ingestion of food, water etc contaminated with faeces from infected or carrier birds. Symptoms: listlessness; anorexia; passage of watery, frothy, foul-smelling stools; coma. Mortality rates are often very high in young birds. Survivors commonly become carriers. Treatment: e.g. furazolidone. (See also POULTRY DISEASES.)

hexokinase An enzyme which phosphorylates various hexoses in the 6-position (see e.g. EMBDEN–MEYERHOF–PARNAS PATHWAY).

hexon See ADENOVIRIDAE.

hexose bisphosphate pathway *Syn.* EMBDEN–MEYERHOF–PARNAS PATHWAY.

hexose monophosphate pathway (HMP pathway; HMP shunt; oxidative pentose phosphate pathway; pentose phosphate pathway/cycle; phosphogluconate pathway; Warburg–Dickens pathway) A metabolic pathway present in a wide range of prokaryotic and eukaryotic microorganisms as well as in plants and animals; it involves the oxidative decarboxylation of glucose 6-phosphate, via 6-phosphogluconate, to ribulose 5-phosphate, followed by a series of reversible, non-oxidative interconversions whereby hexose and triose phosphates are formed from pentose phosphates. The generally accepted scheme for the HMP pathway is shown in Appendix I(b).

The HMP pathway can serve various functions, the major ones probably being to provide NADPH (2 molecules per molecule of glucose converted to ribulose 5-phosphate) necessary for biosyntheses (e.g. of fatty acids), and to provide precursors for various biosynthetic pathways (e.g. pentoses for histidine and nucleotide biosynthesis [Appendices IV(g) and V], erythrose 4-phosphate for aromatic amino acid biosynthesis [Appendix IV(f)]). Fructose 6-phosphate may be converted to glucose 6-phosphate and re-enter the pathway, or may be converted to pyruvate via reactions of the EMBDEN–MEYERHOF–PARNAS PATHWAY; similarly, glyceraldehyde 3-phosphate may be converted to pyruvate via the latter part of the EMP pathway. In organisms with a functional TCA CYCLE, pyruvate can be oxidized to yield energy via the TCA cycle and a respiratory chain. In organisms which lack a complete TCA cycle, pyruvate may be converted to acetyl-CoA and thence to acetic acid (as in some acetic acid bacteria). Alternatively, under certain conditions, glyceraldehyde 3-phosphate can be converted to glucose 6-phosphate (by reactions of GLUCONEOGENESIS) which can then re-enter the HMP pathway; in this case, for every six molecules of glucose entering the pathway, one molecule is effectively completely oxidized. If reducing equivalents from NADPH can be transferred to NAD^+ (see TRANSHYDROGENASE) and thence to an electron acceptor via a respiratory chain, the pathway can be used to generate energy even in the absence of a TCA cycle. Other functions of the HMP pathway include the metabolism of those pentoses which can be converted to intermediates of the pathway. (See also RMP PATHWAY.)

hexulose phosphate pathway *Syn.* RMP PATHWAY.

hexuronic acids URONIC ACIDS corresponding to hexoses.

hexylresorcinol See PHENOLS.

***hfl* genes** (in *E. coli*) See BACTERIOPHAGE λ.

***hflB* gene** See FTSH.

Hfr donor High frequency recombination donor: a conjugal donor (see CONJUGATION sense 1b) created by the insertion of a CONJUGATIVE PLASMID (sense 1) into a bacterial chromosome. The chromosome of an Hfr donor can be mobilized by the plasmid so that chromosomal genes can be transmitted to a recipient during conjugation; thus, in a mixed population of recipients and Hfr donors, recombination between recipient and (transferred) donor DNA will occur at high frequency in the recipient population. The (relatively few) conjugative plasmids which can give rise to Hfr donors include the F PLASMID.

A conjugative plasmid which cannot normally integrate into the bacterial chromosome may be able to do so if it first integrates e.g. with the genome of an A^+B^- strain of BACTERIOPHAGE MU, integration of the plasmid–Mu complex into the chromosome then being mediated by Mu; a recipient mated with such an Hfr donor must be lysogenic for a c^+ strain of Mu to avoid ZYGOTIC INDUCTION when the Mu prophage is transferred to the recipient.

(See also PRIME PLASMID and INTERRUPTED MATING.)

HFRS Haemorrhagic fever with renal syndrome (see HANTAVIRUS).

HFT (of plasmids) See EPIDEMIC SPREAD.

HFT lysate See TRANSDUCTION.

HGV Hepatitis G virus (see HEPATITIS).

HHV1 Human (alpha) herpesvirus 1 (see ALPHAHERPESVIRINAE).

HHV2 Human (alpha) herpesvirus 2 (see ALPHAHERPESVIRINAE).

HHV3 Human (alpha) herpesvirus 3 (see ALPHAHERPESVIRINAE).

HHV4 Human (gamma) herpesvirus 4 (see GAMMAHERPESVIRINAE).

HHV5 Human (beta) herpesvirus 5 (see BETAHERPESVIRINAE).

HHV6 Human (beta) herpesvirus 6 (see BETAHERPESVIRINAE).

HHV7 Human (beta) herpesvirus 7 (see BETAHERPESVIRINAE).

HHV8 Human (gamma) herpesvirus 8 (see GAMMAHERPESVIRINAE).

HI test HAEMAGGLUTINATION-INHIBITION TEST.

HIB HAEMOLYTIC IMMUNE BODY.

Hibitane See CHLORHEXIDINE.

high-dry objective Any objective lens which has a numerical aperture between 0.65 and 0.95, and which is *not* used as an immersion lens.

high-fructose corn syrup A sweetening agent produced industrially from glucose by GLUCOSE ISOMERASE (see also IMMOBILIZATION).

high-mobility group proteins See NON-HISTONE PROTEINS.

high mutability gene *Syn.* MUTATOR GENE.

high-performance liquid chromatography See CHROMATOGRAPHY.

high-potential iron–sulphur protein See HIPIP.

high-pressure liquid chromatography See CHROMATOGRAPHY.

high zone tolerance See IMMUNOLOGICAL TOLERANCE.

higher fungi Fungi of the subdivisions ASCOMYCOTINA, BASIDIOMYCOTINA and DEUTEROMYCOTINA. (cf. LOWER FUNGI.)

Highlands J virus See ALPHAVIRUS.

highly active antiretroviral therapy See HAART.

HIGM See RNA EDITING.

Hikojima variant See VIBRIO (*V. cholerae*).

hilar appendix A small protuberance on a basidiospore near its region of attachment to the sterigma.

Hildenbrandia See RHODOPHYTA.

Hill reaction In PHOTOSYNTHESIS: light-dependent evolution of oxygen by isolated CHLOROPLASTS in the absence of CO_2 but in the presence of e.g. ferric oxalate which acts as an electron acceptor for PS-II.

hilum The mark, scar or small projection on a spore corresponding to the region at which it was formerly attached to the spore-bearing structure.

***him* genes** (in *E. coli*) See BACTERIOPHAGE λ.

Himanthalia See PHAEOPHYTA.

Himmelweit pipette A PASTEUR PIPETTE with a hole in the side of the capillary end; it is used e.g. in the harvesting of allantoic fluid from an EMBRYONATED EGG.

***hin* gene** See PHASE VARIATION.

***Hind*II** A RESTRICTION ENDONUCLEASE from *Haemophilus influenzae*; GTPy/PuAC.

***Hind*III** A RESTRICTION ENDONUCLEASE from *Haemophilus influenzae*; A/AGCTT.

***Hin*fI** See RESTRICTION ENDONUCLEASE (table).

hinge region In the heavy chain of an IMMUNOGLOBULIN: the amino acid sequence at the junction of the Fd portion and the Fc portion; it appears to confer conformational flexibility on the molecule. Structurally, the hinge region is the most variable part of the 'constant' region of the heavy chain. In *alpha* and *gamma* chains (but not in *mu* chains) the hinge region is rich in proline.

Hinozan (edifenphos; *O*-ethyl-*S,S*-diphenyl phosphorodithioate) An antifungal ORGANOPHOSPHORUS COMPOUND used as a protectant against rice blast disease; it can act e.g. as a powerful cutinase inhibitor (see CUTIN).

***hip* gene** (in *E. coli*) See BACTERIOPHAGE λ.

HiPIP High-potential iron–sulphur protein: a type of IRON–SULPHUR PROTEIN which has a single [4Fe–4S] centre and a highly positive E_m; the HiPIP in *Chromatium vinosum* has an E_m of +350 mV.

hippurate hydrolysis Certain bacteria (e.g. *Streptococcus* spp, *Campylobacter* spp) produce a hippuricase which hydrolyses hippurate ($C_6H_5.CO.NH.CH_2.CO_2^-$) to benzoate and glycine. Hippuricase activity may be detected by adding a calculated amount of acid $FeCl_3$ to a 4-day culture of the organism in a medium containing 1.0% sodium hippurate; the $FeCl_3$ forms a persistent precipitate with benzoate (positive test). (A precipitate is also formed with hippurate, but this is more soluble with excess $FeCl_3$; the amount of $FeCl_3$ used in the test is that which just dissolves the hippurate precipitate initially formed in an uninoculated control tube on dropwise addition of the reagent.) In a quicker method, a solution of sodium hippurate is heavily inoculated with the test organism; after incubation (37°C/2hours), ninhydrin reagent is added. A deep purple colour indicates the presence of glycine (positive test). [Methods: Book ref. 2, p. 416.]

Hiprex See HEXAMINE.

hircinol A dihydroxyphenanthrene PHYTOALEXIN produced by the orchid *Loroglossum hircinum* in response to infection with e.g. *Rhizoctonia* spp. (cf. ORCHINOL.)

Hirst spore trap (automatic volumetric spore trap) An instrument used e.g. for sampling the airborne microflora. (See also AIR.) Air, drawn in through a vertical slit in an upright cylindrical housing, strikes a slowly-moving greased microscope slide or, in the modern version (*Burkard trap*), a slowly-rotating drum coated with transparent adhesive tape; the slit is kept facing into the wind by a vane attached to the housing.

Hirsutella A genus of fungi of the HYPHOMYCETES. Most species are entomopathogenic; *H. thompsonii* infects mites.

***his* operon** (histidine operon) An OPERON (located at ca. 44 min on the chromosome of *Escherichia coli*) which contains 9 structural genes (*hisGDCBHAFIE*) encoding enzymes necessary for the biosynthesis of histidine [Appendix IV(g)]. The transcription of the *his* structural genes is inversely related to the level of histidyl-tRNA in the cell. Regulation is apparently achieved solely by attenuation (see OPERON, cf. TRP OPERON); the sequence encoding the leader peptide contains seven successive histidine codons.

HisJ, HisM, HisP, HisQ See BINDING PROTEIN-DEPENDENT TRANSPORT SYSTEM.

hispid Bearing hairs, bristles or spines. *Hispid flagellum*: see FLAGELLUM (b).

Hiss serum water A medium made by mixing 1 volume of serum (horse, ox or sheep) with 3 volumes of water, and adding phenol red and a sugar (final concentrations ca. 0.0025% and 1.0%, respectively).

histamine An amine formed by the decarboxylation of histidine; it occurs e.g. in MAST CELLS and is involved in ANAPHYLACTIC SHOCK and INFLAMMATION. (See also ANAPHYLATOXINS.) Histamine causes contraction of e.g. tracheobronchial and intestinal smooth muscle, and increases vascular permeability in the skin and other regions by acting as a vasodilator of terminal arterioles and as a vasoconstrictor of postcapillary venules. [Agents that release histamine from mast cells: ARPT (1983) *23* 331–351.]

The microbial production of toxic amounts of histamine in certain foods has been implicated as a cause of food poisoning: for example, *Lactobacillus buchneri* can produce histamine in Swiss cheese [AEM (1985) *50* 1094–1096] (see also FISH SPOILAGE).

histamine-sensitizing factor *Syn.* PERTUSSIS TOXIN.

histidase See HISTIDINE DEGRADATION.

L-histidine biosynthesis See Appendix IV(g); see also HIS OPERON.

histidine degradation Histidine can be used as a substrate by various bacteria; in e.g. *Pseudomonas aeruginosa* histidine is initially deaminated by histidase to urocanate which is eventually degraded to glutamate and formate. Some bacteria can decarboxylate histidine to histamine: see e.g. FISH SPOILAGE. (See also BINDING PROTEIN-DEPENDENT TRANSPORT SYSTEM.)

histidine kinase A KINASE whose active site contains a histidine residue. (See also CHEMOTAXIS and TWO-COMPONENT REGULATORY SYSTEM.) [Histidine kinases and signal transduction (review): TIG (1994) *10* 133–138.]

histidine operon *Syn.* HIS OPERON.

histidine permease See BINDING PROTEIN-DEPENDENT TRANSPORT SYSTEM.

histiocyte A MACROPHAGE which occurs in extravascular tissues.

histocompatibility restriction (MHC restriction) The restriction on (physical) cell–cell interaction, involving cells of the immune system, imposed by the need for such cells to have, or to recognize, particular MHC-specified cell-surface antigens as a prerequisite for interaction.

histolyticolysin See THIOL-ACTIVATED CYTOLYSINS.

Histomonas A genus of protozoa of the TRICHOMONADIDA. *H. meleagridis* is the causal agent of BLACKHEAD in turkeys. Intracellular cells of *H. meleagridis* are spherical to ovoid, non-flagellate (though they contain kinetosomes), ca. 10–15 μm; they exhibit amoeboid motility. Within the caecal lumen of the turkey, cells of *H. meleagridis* (ca. 10–25 μm) may exhibit both amoeboid and flagellar motility, each cell having 1–4 flagella (commonly one).

histomoniasis *Syn.* BLACKHEAD.

histone-like proteins Bacterial DNA-binding proteins believed to resemble the HISTONES associated with eukaryotic DNA. (Compare HU PROTEIN.)

histones Basic proteins, rich in arginine and/or lysine, which are major components of CHROMATIN in most eukaryotes. (cf. DINOFLAGELLATES.) The main classes of histones are designated H1 ('lysine-rich'), H2A and H2B ('slightly lysine-rich'), and H3 and H4 ('arginine-rich'). Histones interact with DNA primarily via salt bridges between the positively-charged arginine/lysine residues and the negatively-charged phosphate groups of the DNA.

Constitutive histone *variants* can effect changes in chromatin structure and dynamics that affect e.g. DNA replication, repair and transcription [GD (2005) *19* 295–316].

Histoplasma A genus of fungi of the HYPHOMYCETES; teleomorph: AJELLOMYCES. *H. capsulatum*, the causal agent of HISTOPLASMOSIS, is a dimorphic fungus which occurs e.g. in soil that is contaminated with the droppings of birds, bats and other animals. At room temperature or e.g. 25°C, *H. capsulatum* forms white to golden, cottony, septate mycelium, but at 37°C it forms ovoid, budding, yeast-like cells 2–4 μm (var. *capsulatum*) or 8–15 μm (var. *duboisii*); in vivo, grape-like clusters of cells may occur in macrophages or in regions of caseation. *H. capsulatum* gives rise to microconidia (2–5 μm) and macroconidia (ca. 8–15 μm), the latter (also called 'chlamydospores') typically bearing short, finger-like projections.

H. farciminosum. See EPIZOOTIC LYMPHANGITIS.

(See also ENDOCARDITIS and PAS REACTION.)

histoplasmin test A SKIN TEST, analogous to the TUBERCULIN TEST, used in the diagnosis of HISTOPLASMOSIS; it involves the intradermal injection of *histoplasmin* (a filtrate of a mycelial culture of *Histoplasma capsulatum*). The skin usually becomes reactive to histoplasmin within one to two months of the onset of disease, and reactivity may persist for many years; cross-reactions with other fungal antigens (e.g. coccidioidin) occur.

histoplasmosis (Darling's disease) A disease of man and animals, found worldwide, caused by *Histoplasma capsulatum* (cf. EPIZOOTIC LYMPHANGITIS). Infection occurs by inhalation of spores (e.g. from disturbed soil). The disease may be subclinical, acute or chronic. The acute condition may resemble a common cold or influenza. In the chronic condition, granulomatous, inflamed lesions occur in the lungs; the primary lesion may become inactive and calcify on healing. The primary infection is usually self-limiting, but occasionally progressive lung disease occurs and may eventually be fatal due e.g. to loss of lung function or to secondary bacterial infection. Infrequently, histoplasmosis becomes disseminated (usually in young children or in the elderly, debilitated or immunocompromised); lesions may occur in lymph nodes, liver, spleen, intestine, skin, heart, kidneys, CNS etc. Disseminated histoplasmosis is often rapidly fatal. *Lab. diagnosis*: serological tests (e.g. a CFT); microscopic and cultural examination of clinical specimens. *Chemotherapy*: e.g. amphotericin B, ketoconazole.

histotope A site on an MHC class I or II antigen recognized by a T lymphocyte.

histozoic Living (parasitic) within tissues. (cf. COELOZOIC.)

Histriculus See HYPOTRICHIDA.

hit-and-run oncogenesis Oncogenesis which is initiated by a particular agent (e.g. a virus) but, once initiated, can proceed in the absence of the initiating agent.

HIV Human immunodeficiency virus, infection with which can give rise to AIDS (q.v.). HIV is a retrovirus (family RETROVIRIDAE) classified within the subfamily LENTIVIRINAE. HIV was previously referred to as AIDS-associated retrovirus (ARV), 'HTLV-III' (cf. HTLV), immunodeficiency-associated virus (IDAV) and lymphadenopathy-associated virus (LAV).

Evidence for the involvement of HIV in AIDS was obtained in 1983 [Science (1983) *220* 868–871]. Subsequently a distinct, but related, retrovirus was detected in AIDS patients in West Africa [Nature (1987) *326* 662–669]. These two viruses are now designated HIV-1 and HIV-2 respectively. HIV-1 and HIV-2 are similar in both structure and genome.

The origin of HIV is unknown. One hypothesis supposes that the major (M) strain of HIV arose through the use of an anti-poliomyelitis vaccine that had been prepared by culturing polio virus in non-human primate cells; it is proposed that a simian strain of immunodeficiency virus (initially present in the culture cells) contaminated the vaccine and (thus) entered the vaccinated human population (located mainly in the Congo) in the late 1950s. Subsequent studies on viruses from the Congo area do not support this hypothesis [Nature (2001) *410* 1047–1048].

The HIV virion is ~100 μm in diameter. Its innermost region consists of a cone-shaped core that includes (i) two copies of the (positive-sense) ssRNA genome, (ii) the enzymes REVERSE TRANSCRIPTASE, integrase and protease, (iii) some minor proteins, and (iv) the major core protein. The core is surrounded by a protein matrix, and the whole is enclosed within an envelope (a lipid bilayer); the envelope is penetrated by a number of 'spikes' (gp41), each spike bearing a distal (outermost) trimeric glycoprotein (gp120). [Fine structure of HIV: Virol. (1987) *156* 171–176.]

The genome (~9.2 kb) includes sequences corresponding to the *gag*, *pol* and *env* regions of other exogenous retroviruses (see RETROVIRIDAE).

The *gag* region encodes structural proteins, including the major core protein (p24) and the matrix protein (p17).

The *pol* region encodes regulatory proteins: REVERSE TRANSCRIPTASE, protease (involved in the maturation of virions), RNASE H and integrase. (The enzymes reverse transcriptase and protease are major targets for antiretroviral drugs.)

The *env* region encodes the envelope-associated proteins, gp120 (required for initial binding of virus to target cell) and gp41 (involved in fusion of the viral envelope with the membrane of the target cell).

Other HIV genes include: *tat* and *rev* (whose products are needed for viral replication); *vif* ('viral infectivity factor'); and *vpu* (whose product promotes efficient budding of the (HIV-1) virion during productive infection of a cell). (See also Rev in GENE THERAPY.)

Attachment of the virion to a target cell.
The principal high-affinity binding site for viral gp120 glycoprotein is the CD4 antigen (see CD4) (formerly referred to as OKT4); target cells for HIV therefore include the $CD4^+$ ('T helper') subset of T LYMPHOCYTES, monocytes, macrophages and dendritic cells. $CD4^+$ cells also occur in the central nervous system (microglial cells), and infection of these cells by HIV may account for neurological symptoms in HIV-positive patients; multinucleate (syncytial) cells have been observed in brain tissue from such patients.

For *infection* of a cell by HIV, the cell must express not only CD4 antigen but also receptors for particular CHEMOKINES – e.g. those for certain CXC-type and CC-type chemokines (the CXCR4 and CCR5 receptors) [JLB (1999) *65* 552–565]. Some (T-cell-tropic) strains of HIV require the co-receptor CXCR4 while other (M-tropic) strains require CCR5; M-tropic strains can infect macrophages and monocytes as well as T cells. Yet other (dual-tropic) strains of HIV can use both co-receptors. On the basis of their use of particular co-receptors, strains of HIV have been categorized as X4, R5 and R5X4 respectively.

Some $CD34^+$ haemopoietic stem cells co-express the CD4 antigen (and co-receptors) but these cells are apparently not susceptible to infection by pathogenic strains of HIV; however, such cells are susceptible to infection by genetically engineered viral 'constructs' (based on HIV) which may be of use in GENE THERAPY. Although stem cells may not be susceptible to natural strains of HIV, inhibition of haemopoiesis may nevertheless occur in HIV^+ patients owing to perturbation of the cytokine milieu in bone marrow. [Haematological aspects of HIV infection: BCH (2000) *13* 215–230.]

Internalization and integration of HIV.
Fusion between viral envelope and cell membrane is achieved in a gp41-dependent manner; the viral core then enters the target cell's cytoplasm. On (intracellular) release of viral RNA and enzymes, a dsDNA form of the genome is synthesized using viral reverse transcriptase and RNase H (see: RETROVIRIDAE). [HIV *entry inhibitors*: PNAS (2002) *99* 16249–16254.]

The 3′ and 5′ LTRs of the dsDNA are involved in circularization of the molecule (forming a plasmid-like structure); this structure enters the nucleus and is integrated into the host's genome in a process involving the viral integrase. In some cases the integrated (provirus) form of HIV remains latent (inactive), or weakly active, for some time (e.g. ~10 years), the patient remaining essentially asymptomatic; in other cases disease progresses rapidly.

Replication of HIV.
Replication is apparently an inefficient process in which a high proportion of progeny virions are defective.

Transcription of the integrated viral genome appears to depend on various (viral and host) regulatory proteins which have binding sites on the LTR. For example, the viral *tat* gene product, with a binding site designated Tat-responsive region (TAR), has an important transactivation role; the activity of this protein is essential in the upregulation of replication.

The host's nuclear transcription factor, NF-κB (see CYTOKINES), is also important as a (positive) regulator of viral gene expression and it, too, has a binding site on the LTR; it is therefore possible that expression of viral genes may be promoted by superinfection with certain opportunist pathogens that activate NF-κB by inducing synthesis of appropriate cytokines. (This suggests a possible mechanism for the exacerbation of HIV-mediated disease that may be seen during opportunist infections.)

The LTR also includes a regulatory region for polyadenylation of transcripts.

Synthesis of viral mRNA, and of genomic RNA, is carried out by the host's RNA polymerase. Viral envelope proteins are translocated to, and insert into, the cell membrane (with gp120 located at the outer surface). The assembled core acquires an envelope by budding through the cell membrane.

Culture.
Culture may be useful e.g. for diagnosis of HIV infection in neonates; in these young patients the presence of anti-HIV antibody may simply reflect acquisition from an HIV-positive mother.

Cultures of HIV characteristically exhibit CPE (cytopathic effects), including the formation of giant multinucleated cells and cell lysis. Culture can be carried out in $CD4^+$ T cells; the virus can also infect other types of cell such as monocytes and promyelocytes [Virol. (1985) *147* 441–448]. HIV may give rise to a persistent, non-cytopathic and productive (virion-producing) infection of human $CD4^+$ T cells in culture [Science (1985) *229* 1400–1402].

Inactivation of HIV by disinfection.
Hypochlorite (effective conc. 10000 parts per million available chlorine) is a useful disinfectant. Glutaraldehyde (dangerous chemical!!!) has been used for disinfecting surfaces in a well-ventilated environment; safety regulations must be observed.

Phenolics and ethanol are unsuitable.

In serum, inactivation of HIV may require more than 30 minutes at 56°C [e.g. Book ref. 219, p 814].

Genetic variability of HIV isolates.
Isolates of HIV from different patients tend to be genetically heterogeneous, particularly in the *env* sequence [minireview: Cell (1986) *46* 1–4], and variation may also be observed in isolates obtained sequentially from the same patient over a period of time [Science (1986) *232* 1548–1553]. As *env* encodes the (antigenic) viral surface protein gp120, such ANTIGENIC VARIATION presents a problem for the development of an effective anti-HIV vaccine; in this context, studies have been carried out to determine the effect of genetic variation on the immunogenetic potential of gp120 [Arch. Virol. (2000) *145* 2087–2103].

HIV-positive Refers to an individual who is infected with the human immunodeficiency virus (see HIV). (See also AIDS.)

hives *Syn.* URTICARIA.

Hjärre's disease (coli-granuloma) A POULTRY DISEASE in which granulomatous lesions are formed in the wall of the intestine (particularly in the caecum); the causal agent is *Escherichia coli*.

HLA-A, HLA-B etc. See MAJOR HISTOCOMPATIBILITY COMPLEX.

Hly plasmid See ESCHERICHIA.

HlyA (of *Escherichia coli*) See RTX TOXINS.

HMA Hydroxymethionine analogue: DL-α-hydroxy-γ-methiolbutyrate; HMA is used in animal feedstuffs as a sulphur amino acid supplement [ARB (1983) *52* 215–216]. (See also SINGLE-CELL PROTEIN.)

HMG box That region, in certain types of protein, which can e.g. interact with sharply angular or kinked DNA (in eukaryotic cells); it consists of a sequence of about 80 amino acid residues (including a high proportion of aromatic amino acids) with a net positive charge. HMG boxes are found in certain NON-HISTONE PROTEINS. (See also HU PROTEIN.)

HMG proteins See NON-HISTONE PROTEINS.

HMP pathway HEXOSE MONOPHOSPHATE PATHWAY.

HMS-1 β-lactamase See β-LACTAMASES.

HMT toxin A long-chain polyketide host-specific toxin produced by *Drechslera* (*Helminthosporium*) *maydis* race T; it has been suggested that HMT toxin acts as an ionophore.

hnRNA Heterogeneous nuclear RNA: the high-MWt RNA synthesized by RNA polymerase II in the nucleus of a eukaryotic cell; hnRNA occurs in the form of ribonucleoprotein (hnRNP) and includes pre-mRNA (i.e., MRNA before the generation of 3′ ends, polyadenylation and splicing).

H-NS protein (*syn.* H1 protein) In bacteria (e.g. *Escherichia coli*), a small protein (136 amino acids, 15.6 kDa) encoded by the *hns* gene. H-NS protein is able to (i) regulate the expression of various unrelated genes (e.g. *bgl*, *proU*), (ii) constrain negative supercoiling, and (iii) contribute to the physical organization of the bacterial nucleoid.

In uropathogenic strains of *E. coli* (UPEC) the H-NS protein has been implicated in the temperature-dependent expression of P FIMBRIAE: at the non-permissive temperature of 25°C, the H-NS protein is reported to act as a methylation-blocking factor, repressing transcription [Mol. Microbiol. (1998) *28* 1121–1137]. In *Proteus mirabilis* it is reported to act as a repressor of the transcriptional activator gene, *ureR*, in the absence of urea induction [JB (2000) *182* 2649–2653].

H-NS has also been implicated in the regulation of DNA repair in *Shigella* [JB (1998) *180* 5260–5262].

H-NS protein is able to condense supercoiled plasmid DNA, and appears to play a major part in condensing DNA in the nucleoid. H-NS-mediated condensation of DNA has been investigated by studying the structure of H-NS–DNA complexes by atomic force microscopy [NAR (2000) *28* 3504–3510].

Hoarella A genus of protozoa (suborder EIMERIORINA) which form dizoic oocysts each containing 16 sporocysts.

Hoechst 33258 A *bis*-benzimidazole derivative: a FLUOROCHROME used e.g. for staining chromosomes.

hog cholera *Syn.* SWINE FEVER.

hog diseases See PIG DISEASES.

Hogness box See PROMOTER.

holdfast An organ or organelle by means of which an organism or cell attaches to the substratum or to another organism. Examples include the root-like region of LAMINARIA, and the specialized regions in SPORICHTHYA and members of the ASTOMATIDA.

holding–flush jar procedure In the culture of anaerobes: a procedure which minimizes the exposure of plates to O_2; essentially, uninoculated plates, or inoculated plates awaiting incubation, are held in an anaerobic jar (with unclamped lid) while a stream of O_2-free gas (e.g. N_2 or CO_2) is continually directed through a tube to the bottom of the jar.

holin See LYSIS PROTEIN.

Hollandina See SPIROCHAETALES.

Holliday junction See RECOMBINATION.

Holliday model See RECOMBINATION (Figure 1).

Holliday structure See RECOMBINATION.

Holmsella A genus of algae (division RHODOPHYTA). *H. pachyderma* is a colourless organism which is parasitic on the red alga *Gracilaria*; PIT CONNECTIONS have been observed between *H. pachyderma* and *Gracilaria*.

Holobasidiomycetidae A subclass of fungi (class HYMENOMYCETES) characterized by the formation of *holobasidia* (see BASIDIUM). Orders [Book ref. 64, p. 189]: AGARICALES, APHYLLOPHORALES, BOLETALES, BRACHYBASIDIALES, CANTHARELLALES, DACRYMYCETALES, EXOBASIDIALES, RUSSULALES and TULASNELLALES.

holobasidium See BASIDIUM.

holoblastic development See CONIDIUM.

holocarpic Refers to those organisms in which the entire thallus takes on a reproductive function. (cf. EUCARPIC.)

holocellulose A complex of cellulose and hemicelluloses (formed e.g by the removal of lignin from lignocellulosic plant material).

holocentric (holokinetic) Refers to a chromosome in which the CENTROMERE is diffuse, being distributed along the length of the chromosome; the kinetochores are correspondingly long, and microtubules are attached along the whole length of the chromosome.

holococcolith See COCCOLITH.

holoenzyme A 'whole' enzyme – i.e., an enzyme in its complete, active form – consisting either of an apoenzyme plus PROSTHETIC GROUP or of two or more distinct protein subunits (see e.g. DNA POLYMERASE).

holokinetic *Syn.* HOLOCENTRIC.

holomictic lake A lake in which the entire water mass undergoes a complete turnover. (cf. MEROMICTIC LAKE.)

holomorph (*mycol.*) Any fungus considered in its entirety, i.e., including all latent or expressed (sexual and/or asexual) forms and potentialities. (cf. ANAMORPH; ANA-HOLOMORPH; TELEOMORPH.)

Holophrya See GYMNOSTOMATIA.

holophytic nutrition Plant-type nutrition, i.e., the endogenous formation of nutrients by photosynthesis. (cf. HOLOZOIC NUTRITION.)

Holospora A genus of Gram-negative bacteria which are obligate endosymbionts in the micronucleus (*H. elegans*, *H. undulata*) or the macronucleus (*H. caryophila*, *H. obtusa*) of *Paramecium* spp; *Holospora* spp do not confer a killer characteristic (cf. CAEDIBACTER). Cells: non-motile rods or filaments. Type species: *H. undulata*. [Book ref. 22, pp. 802–803; *H. obtusa* in *P. caudatum* (infection and maintenance): JCS (1985) *76* 179–187.]

Holosticha See HYPOTRICHIDA.

holothallic conidium See CONIDIUM.

holotoxin The form in which an exotoxin is secreted by a pathogen. After uptake by a eukaryotic cell, the holotoxin may be processed (e.g. by proteolytic cleavage) to release the active component (see e.g. DIPHTHERIA TOXIN and SHIGA TOXIN).

holotrich Any of the relatively primitive ciliate protozoa previously classified in the subclass Holotrichia [JP (1964) *11* 7–20] but now classified in the classes KINETOFRAGMINOPHOREA and OLIGOHYMENOPHOREA.

holotype The actual specimen which an author has designated as the NOMENCLATURAL TYPE of a new taxon.

holozoic nutrition Animal-type nutrition, i.e., nutrition involving the ingestion of other organisms or components of other organisms. (cf. HOLOPHYTIC NUTRITION.)

homing (intron, bacterial) See INTRON HOMING.

homing-associated cell adhesion molecule *Syn.* CD44.

homoacetate fermentation See ACETOGENESIS.

homocaryon *Syn.* HOMOKARYON.

homocysteine See e.g. Appendix IV(d).

homocytotropic antibodies (1) *Syn.* REAGINIC ANTIBODIES. (2) Antibodies which bind only to cells of the species in which they were formed. (cf. HETEROCYTOTROPIC ANTIBODIES.)

homodiaphoromixis See HOMOTHALLISM.

homodimixis See HOMOTHALLISM.

homoduplex Any double-stranded nucleic acid in which each strand is completely complementary to the other. (cf. HETERODUPLEX.)

homoeomerous *Syn.* HOMOIOMEROUS.

homofermentation (1) Any fermentation in which there is only one major end-product. (2) *Syn.* HOMOLACTIC FERMENTATION.

homogeneous immersion See IMMERSION OIL.

homogenization (of milk) See MILK.

homogenote See MEROZYGOTE.

homoglycan See POLYSACCHARIDE.

homoheteromixis See HOMOTHALLISM.

homoimmune phages See SUPERINFECTION IMMUNITY.

homoiomerous (homoeomerous) (1) (*lichenol.*) Refers to a lichen thallus in which the photobiont is more or less evenly distributed throughout the thallus (see LICHEN). (2) (*mycol.*) In an agaric: refers to a pileus or stipe composed of hyphae only. (3) (*ciliate protozool.*) See MACRONUCLEUS. (cf. HETEROMEROUS.)

homoisocitrate pathway See icl$^-$ SERINE PATHWAY.

homokaryon (homocaryon) (*mycol.*) A hyphal cell, mycelium, organism or spore in which all the nuclei are genetically identical. (cf. HETEROKARYON.)

homokaryotic (1) (*mycol.*) Refers to a HOMOKARYON. (2) (*ciliate protozool.*) Having only one type of nucleus: see PRIMOCILIATID GYMNOSTOMES.

homolactic fermentation A type of LACTIC ACID FERMENTATION in which e.g. glucose is converted entirely, or almost entirely, to lactic acid. Many LACTIC ACID BACTERIA (e.g. *Lactococcus lactis*, thermobacteria and streptobacteria) carry out a homolactic fermentation of glucose (or lactose) under conditions of glucose (or lactose) excess (cf. LACTIC ACID STARTERS); glucose is metabolized by the EMBDEN–MEYERHOF–PARNAS PATHWAY, and most or all of the pyruvate thus formed is converted to lactic acid. (See also LACTOSE.) However, under glucose or lactose limitation, or with different substrates (e.g. PENTOSES), the same organisms may form other products, sometimes with little or no lactic acid; for example, many strains of *L. lactis* produce mainly formic and acetic acids and ethanol when grown in glucose- or lactose-limited continuous culture (see Appendix III(c)). (See also DIACETYL.)

homologous (1) Similar in form or structure, but not necessarily in function; homology suggests evolutionary relatedness. (2) Derived from or associated with the same species as that being referred to. (cf. HETEROLOGOUS (1).) (3) See ALTERNATION OF GENERATIONS. (4) (*immunol.*) The antibody elicited by a given antigen is said to be *homologous* to that antigen; similarly, the antigen is said to be homologous to the antibody. (cf. HETEROLOGOUS (3).) (5) (*genetics*) Of chromosomes or chromatids: containing the same sequence of genes – but not necessarily identical alleles. (6) (*genetics*) Of nucleic acids: see e.g. DNA HOMOLOGY.

homologous interference See INTERFERENCE (1).

homologous recombination See RECOMBINATION.

homologue (of chromosomes) In a diploid eukaryotic nucleus: each of a pair of CHROMOSOMES which encode the same types of characteristics and which are usually morphologically similar; in general, homologous chromosomes carry the same sequences of genes, though not necessarily the same alleles.

homology (of DNA) See DNA HOMOLOGY.

homomixis See HOMOTHALLISM.

homonym Any name used to refer to each of two (or more) different microbial taxa; a name of a higher animal or plant which is identical to that of a microorganism is not considered to be a homonym. The name published first is the senior homonym, while any name published later is a junior homonym; junior homonyms are usually suppressed.

homophilic binding (of cell adhesion molecules) See IMMUNOGLOBULIN SUPERFAMILY.

homoplasmid segregant See INCOMPATIBILITY.

homopolymer tailing See TAILING.

homopolysaccharide See POLYSACCHARIDE.

homoserine See e.g. Appendix IV(d).

homothallism (homomixis) Self-fertility: the phenomenon in which sexual reproduction can involve the fusion of e.g. gametes derived from the same individual (as well as the fusion of those derived from different individuals); no 'mating types' are involved (cf. HETEROTHALLISM).

Secondary homothallism (= *homoheteromixis*) occurs in some fungi which are basically heterothallic; it can arise e.g. when nuclei of compatible mating types are incorporated in the same spore – the thallus developing from such a spore being self-fertile. (This occurs regularly in e.g. *Neurospora tetrasperma*.) If secondary homothallism involves nuclei whose compatibility is determined on a one-locus two-allele basis the phenomenon is called *homodimixis*; if it is determined on a one- or two-locus *multi*-allele basis the phenomenon is called *homodiaphoromixis*.

homothetogenic Refers to the type of cell division typical of ciliate protozoa: transverse (PERKINETAL) BINARY FISSION in which the daughter cells (the PROTER and OPISTHE) are not mirror images of one another – although they may be similar or identical. For example, when *Tetrahymena* divides, the anterior end of the opisthe is near the plane of division. (cf. SYMMETROGENIC; see also INTERKINETAL.)

homoxenous (monoxenous, autoecious) Refers to a parasite which completes its life cycle in a single host species. (cf. HETEROXENOUS.)

homozygote A HOMOZYGOUS cell or organism.

homozygous Refers to a cell or organism (e.g. a diploid eukaryote or a bacterial MEROZYGOTE) in which one or more specified genes, or all of the genes, are present in the form of identical ALLELES. (cf. HETEROZYGOUS.)

honey-bee diseases See BEE DISEASES.

honey fungus See ARMILLARIA.

honeydew A sugary secretion produced e.g. by sap-sucking insects such as aphids. Sap-sucking insects feed on phloem sap which is rich in sugars but low in nitrogen; thus, large quantities of sap must be ingested to satisfy the insect's nitrogen requirements, the excess sugars being excreted as honeydew. Insect honeydews vary in composition (depending e.g. on insect); some contain e.g. melezitose (rare in nature), and some are rich in polyols (galactitol, ribitol, etc). Certain fungi – particularly SOOTY MOULDS – may grow abundantly on insect honeydews, and it has been suggested that honeydews falling on soil below insect-infested plants may enhance nitrogen-fixation by free-living bacteria in the soil [SBB (1984) *16* 203–206]. (See also CLAVICEPS.)

Hong Kong flu See INFLUENZAVIRUS.

Hoogsteen hydrogen bonding See TRIPLEX DNA.

hook (flagellar) See FLAGELLUM (a).

hop latent virus See CARLAVIRUSES.

hop mosaic virus A member of the CARLAVIRUSES which can cause severe to lethal disease in certain (Golding-type) cultivars of hops (*Humulus lupulus*); symptoms include chlorotic vein-banding, curling of leaf-margins, stunting etc. Transmission occurs via aphids. Weeds (e.g. *Urticaria urens*, *Chenopodium album*, *Plantago major*) may be important reservoirs of the virus. Sensitive hop cultivars have been replaced by tolerant cultivars in endemic areas.

Hop protein (*plant pathol.*) Hrp-dependent outer protein: a proposed designation for proteins secreted via the Hrp system (see AVIRULENCE GENE) – some of which may lack an avirulence function (and are therefore inappropriately referred to as avr proteins) [JB (1997) *179* 5655–5662 (5659–5660)].

hop stunt viroid See VIROID.

hopanes See *bacterial* CYTOPLASMIC MEMBRANE.

HOQNO HYDROXYQUINOLINE-N-OXIDE.

hordeiviruses (barley stripe mosaic virus group) A group of multicomponent PLANT VIRUSES in which the virions are elongated, rigid particles (ca. 100–150 × 20 nm) composed of ssRNA and helically-arranged subunits of a single type of coat protein (MWt ca. 21000). The genome appears to consist of 2–4 molecules (depending on strain) of linear positive-sense ssRNA, two or three of which are necessary for infection. The 3′ end of the RNA has an internal poly(A) sequence and a 3′-terminal tRNA-like structure which can be amino-acylated with tyrosine in vitro [Virol. (1982) *119* 51–58]. Viruses accumulate mainly in the cytoplasm but also in the nuclei of infected cells. Type member: barley stripe mosaic virus (BSMV), an important pathogen of barley in many regions of the world; some strains of BSMV can also infect wheat, oats etc. Transmission is mechanical and seed-borne. Seeds from BSMV-infected plants may show increased frequencies of triploidy and aneuploidy, and some BSMV strains induce a heritable genetic anomaly ('aberrant ratio phenomenon' [ARPpath (1984) *22* 77–94]) characterized by abnormal segregation ratios for one or more genetic markers.

hordeolum *Syn.* STYE.

horizontal resistance (*plant pathol.*) In a given cultivar: the existence of similar levels of resistance to each of the races of a given pathogen. Unlike VERTICAL RESISTANCE, horizontal resistance tends to be a permanent (though not necessarily complete) form of resistance which does not readily break down when genetic changes occur in the pathogen; it is commonly inherited polygenically.

horizontal transmission (lateral transmission) Transmission of a disease or parasite from one individual to another *contemporary* individual. (cf. VERTICAL TRANSMISSION.)

Hormidium *Syn.* KLEBSORMIDIUM.

Hormiscium dermatitidis See WANGIELLA.

Hormoconis A genus of fungi of the HYPHOMYCETES. *H. resinae* (formerly *Cladosporium resinae*, the 'creosote fungus' or 'kerosene fungus') can utilize e.g. creosote and various hydrocarbons as sources of carbon (see also PETROLEUM). [Mechanism of dodecane uptake by *H. resinae*: JGM (1986) *132* 751–756.] The teleomorph of *H. resinae* is the ascomycete *Amorphotheca resinae*. [Discussion and references on re-naming of *C. resinae*: MS (1986) *3* 169.]

hormocyst In certain cyanobacteria: short filaments – composed of granular cells surrounded by a common, condensed sheath – formed in intercalary or terminal positions in the vegetative trichome; hormocysts become detached from the parent trichome and appear to function as propagules (cf. AKINETE and HORMOGONIUM). (See also LEMPHOLEMMA.)

Hormodendrum dermatitidis See WANGIELLA.

Hormodendrum resinae *Syn. Hormoconis resinae*.

hormogonium A short filament of undifferentiated cells (i.e., lacking heterocysts or akinetes) formed by filamentous CYANOBACTERIA (members of sections III, IV and V) and by e.g. *Beggiatoa*; hormogonia may be formed by the fragmentation of whole filaments (e.g. in *Oscillatoria* and *Cylindrospermum*) or may develop from the tips of trichomes (e.g. in *Scytonema*). (See also NECRIDIUM.) Hormogonia may differ from the parent trichome e.g. in having smaller cells; they are typically motile (by gliding) and usually move some distance before developing into mature filaments, thus functioning as propagules. In some cyanobacteria (e.g. *Nostoc muscorum* [JGM (1983) *129* 263–270], *Calothrix* spp), GAS VACUOLES are formed only in the hormogonia, the increased buoyancy they confer presumably aiding in the dispersal of the hormogonia.

hormone A substance which is released from specialized cells in specific part(s) of a multicellular organism and which, in small quantities, can bring about one or more specific responses when translocated to other part(s) of the same individual.

hormone A See PHEROMONE.

hormone B See PHEROMONE.

horse diseases Horses may be affected by a wide range of diseases, some of which are specific to equines. (a) *Bacterial diseases*: see e.g. CONTAGIOUS EQUINE METRITIS; CONTAGIOUS PUSTULAR DERMATITIS (sense 2); GLANDERS; PARATYPHOID FEVER; POTOMAC HORSE FEVER; SLEEPY FOAL DISEASE; STRANGLES; TULARAEMIA; TYZZER'S DISEASE; ULCERATIVE LYMPHANGITIS. (See also RHODOCOCCUS (*R. equi*).) (b) *Fungal diseases*: see e.g. ASPERGILLOSIS; EPIZOOTIC LYMPHANGITIS; EQUINE PHYCOMYCOSIS; SPOROTRICHOSIS. (See also PASPALUM STAGGERS and PENITREMS.) (c) *Protozoal diseases*: see e.g. DOURINE; EQUINE PROTOZOAL MYELOENCEPHALITIS; SURRA. (d) *Viral diseases*: see e.g. AFRICAN HORSE SICKNESS; BORNA DISEASE; EASTERN EQUINE ENCEPHALOMYELITIS; EQUINE COITAL EXANTHEMA; SWAMP FEVER (equine infectious anaemia); VENEZUELAN EQUINE ENCEPHALOMYELITIS; VESICULAR STOMATITIS; WESTERN EQUINE ENCEPHALOMYELITIS.

horse mushroom See AGARICUS.

horse sickness fever See AFRICAN HORSE SICKNESS.

horseradish peroxidase A haemin-containing PEROXIDASE, used e.g. in the IMMUNOPEROXIDASE METHODS.

host-controlled modification The modification of bacteriophage DNA by its host's modification enzymes (see RESTRICTION–MODIFICATION SYSTEM). If a given phage replicates in a particular bacterial strain, the DNA of the progeny phages will carry the modification pattern characteristic of that strain; such phages will be able to infect other cells of the same strain with high efficiency. If these phages infect another strain of bacteria in which there is a *different* R-M system, a low efficiency of plating is generally observed since, in most of the cells, infection is followed by restriction of the phage DNA; however, in a few cells restriction may fail to occur, and the phage DNA then acquires the modification pattern characteristic of the new strain. The modified DNA can replicate, the daughter strands being modified by the host enzymes; the progeny phages can efficiently infect cultures of the new strain, but will infect cultures of the original strain with low efficiency.

host-dependent mutant A CONDITIONAL LETHAL MUTANT bacteriophage which contains e.g. an amber mutation in an essential gene and can thus replicate only in a (permissive) host which contains an intergenic amber SUPPRESSOR MUTATION.

host-range mutant (*h* mutant; *hr* mutant) A mutant virus in which the mutation alters the host range of the virus (i.e. the type(s) of cell which it can infect).

hot-air oven An apparatus used for the dry-heat STERILIZATION of e.g. clean glassware and those materials (such as mineral oils) which cannot be sterilized in an AUTOCLAVE owing to their impermeability to steam. It consists of an electrically heated, thermostatically controlled, heat-insulated cabinet which, ideally, incorporates an internal fan to prevent temperature stratification. Typical sterilizing conditions are 160–170°C/1 hour.

hot–cold lysis A phenomenon in which incubation of susceptible red blood cells (RBCs) with e.g. staphylococcal β-HAEMOLYSIN at 37°C causes little or no haemolysis, but subsequent chilling to 0–4°C causes rapid and extensive haemolysis. The sensitivity of RBCs from different species depends on the SPHINGOMYELIN content of their membranes, those from sheep, goat and ox being the most sensitive. Hot–cold lysis is still not fully understood; it may reflect the combined effects of degradation of sphingomyelin in the outer leaflet of the cell membrane (with consequent condensation of the resulting ceramide into droplets in the outer layer) and the weakening of hydrophobic forces on cooling [Book ref. 44, pp. 721–725]. Hot–cold lysis with staphylococcal β-toxin is most easily demonstrated in the test-tube; blood agar plates may give less consistent results.

Hot–cold lysis can also occur with the α-toxin of *Clostridium perfringens*.

hot-spot (*mol. biol.*) A region of DNA which is particularly prone to e.g. transposition (see TRANSPOSABLE ELEMENT) or mutation (see SPONTANEOUS MUTATION).

hot-start PCR A form of PCR in which an essential component of the reaction mixture is withheld, or blocked, until the temperature of the mixture has, for the first time, risen above the primer-binding temperature; the object of this procedure is to avoid mis-priming (i.e. binding of primers to inappropriate sequences) – which, in the standard form of PCR, tends to occur primarily in the initial pre-cycling stage, i.e. when all components are present but the mixture is still at room temperature. By avoidance (or minimization) of mis-priming, the hot-start technique promotes the specificity of a PCR assay; it can also promote the sensitivity of an assay by concentrating the full potential of the system on amplification of the required target sequence. Avoidance or minimization of non-specific products also serves to reduce the background against which the legitimate target is to be detected.

In one commercial procedure, a chemically modified form of the polymerase (AmpliTaq Gold™ DNA polymerase; Perkin-Elmer) is initially inactive (when added to the reaction mixture) but is thermally activated at 95°C when the temperature first rises to the denaturation level.

In another commercial system, the reaction mixture is prepared with all components except the polymerase; a layer of molten wax (AmpliWax™; Perkin-Elmer) is allowed to set on the surface of the mixture, and the polymerase is then added above the wax barrier. Once cycling begins, the wax melts when the temperature first exceeds ~75–80°C; the reactants then mix, and PCR proceeds normally.

[See also Nature (1996) *381* 445–446.]

housefly virus (HFV) A virus pathogenic in the housefly (*Musca domestica*); it resembles viruses of the REOVIRIDAE in having a dsRNA genome (10 genes), but is apparently not serologically related to reoviruses. [Book ref. 83, p 5.]

Howell–Jolly body In some erythrocytes in a stained blood smear: a small, dark, rounded body generally considered to be a fragment of the precursor cell's disintegrating nucleus.

Hoyer's medium (Frateur's modification) A medium in which ethanol is the sole source of carbon, and $(NH_4)_2SO_4$ the sole source of nitrogen; it includes K_2HPO_4, KH_2PO_4, $MgSO_4$, $FeCl_3$, biotin, calcium pantothenate, thiamine, folic acid, PABA, vitamin B_{12}, pyridoxal-HCl, niacin and riboflavin. (See e.g. ACETOBACTER.)

***Hpa*I** A RESTRICTION ENDONUCLEASE from *Haemophilus parainfluenzae*; GTT/AAC.

***Hpa*II** A RESTRICTION ENDONUCLEASE from *Haemophilus parainfluenzae*; C/CGG.

HPI layer See DEINOCOCCUS.

HPLC High-pressure liquid CHROMATOGRAPHY.

HPr See PTS.

HPr kinase See CATABOLITE REPRESSION.

***hprK* gene** (in *Staphylococcus xylosus*) See CATABOLITE REPRESSION.

HPUra (Hpura) 6-(*p*-Hydroxyphenylazo)uracil: an inhibitor of DNA polymerase III in e.g. Gram-positive bacteria.

HPV Human papillomavirus – see PAPILLOMA and PAPILLOMAVIRUS. (*Note*. 'Human parvovirus' was formerly referred to as HPV but is now referred to as 'parvovirus B19' or 'B19 parvovirus' (see ERYTHROVIRUS).)

HQNO HYDROXYQUINOLINE-N-OXIDE.

***hr* mutant** HOST-RANGE MUTANT.

HR756 *Syn.* cefotaxime (see CEPHALOSPORINS).

Hrp secretion pathway (*plant pathol.*) See AVIRULENCE GENE.

HRP-2 protein See MALARIA.

HS HAEMOPHAGOCYTIC SYNDROME.

HSA (*serol.*) Human serum albumin.

***hsp* genes** Genes encoding HEAT-SHOCK PROTEINS.

HSV HERPES SIMPLEX virus.

HT-2 toxin See TRICHOTHECENES.

HTF *Mastigocladus* See CHLOROGLOEOPSIS.

HTLV Human T-lymphotropic virus (= human T-cell leukaemia virus, human T-cell leukaemia/lymphoma virus): a generic term for exogenous, replication-competent human retroviruses (see RETROVIRIDAE) which (at least HTLV-I and HTLV-II) can infect and transform $CD4^+$ T LYMPHOCYTES; other types of cell (including macrophages) can also be infected.

HTLV-I (formerly 'HTLV') was first identified in the neoplastic T cells of patients with ADULT T CELL LEUKAEMIA (ATL) and is now regarded as a causal agent of that disease; this virus has also been associated with some non-malignant diseases which include spastic paraparesis and uveitis.

Infection with HTLV-I is permanent (i.e. life-long). A carrier state is recognized. Transmission of the virus appears to occur mainly via breast-feeding, sexual contact and blood transfusion. Endemic areas include Japan, the Caribbean, parts of Africa and Latin America.

The virion of HTLV-I is ~75 μm in diameter.

The genome of HTLV-I integrates at random sites in the host's DNA; transformed T cells from ATL patients contain only one, or a few, copies of the viral genome per cell.

The genome of HTLV-I (~9–10 kb) includes the genes *gag*, *pol* and *env* (analogous to those of other retroviruses). A further gene, *pro*, encodes a protease. Near the 3′ end of the genome (the pX region) are several genes that include *rex* and *tax*, both of which are involved in viral replication. The genome lacks a viral oncogene.

In addition to its role in viral replication, the Tax protein induces the expression of a number of host genes – including that encoding INTERLEUKIN-2 and the ONCOGENE c-*fos*. Expression of such host proteins may promote leukaemogenesis. (In vitro, HTLV-I is reported to transform normal human peripheral blood T cells, the transformed cells continuing to proliferate in the absence of exogenous interleukin-2.)

HTLV-I is closely related to SIMIAN T-CELL LEUKAEMIA VIRUS (strains of which are >85% homologous with HTLV-I in terms of nucleotide sequence); it is also related to bovine leukosis virus (BLV).

HTLV-II has been rarely isolated – initially from a T cell line derived from a case of (T cell) HAIRY CELL LEUKAEMIA [Science (1982) *218* 571–573]. The nucleotide sequence of HTLV-II is 70% homologous with that of HTLV-I. The pathogenic potential of HTLV-II is uncertain, although one report described a patient with disease closely resembling HTLV-I-associated myelopathy in a patient infected with HTLV-II.

'HTLV-III' was a former name for the human immunodeficiency virus (HIV).

Another human T-cell lymphotropic retrovirus – related to but distinct from other HTLVs – was detected in some cases of MULTIPLE SCLEROSIS (MS), but an aetiological role (if any) for such a virus in MS has yet to be determined [Nature (1985) *318* 154–160].

[Early review (HTLV-I -II -III): Nature (1985) *317* 395–403. Human T-cell leukaemia virus: BCH (1995) *8* 131–148. Detection of HTLV-I and HTLV-II by multiplex PCR: PNAS (1999) *96* 6394–6399. Human T-lymphotropic virus type I infection: BCH (2000) *13* 231–243.]

***htpR* gene** See HEAT-SHOCK PROTEINS.

HTST See APPERTIZATION and PASTEURIZATION.

HU HYDROXYUREA.

HU protein A small, basic protein, abundant in *Escherichia coli*, which binds strongly to angular/kinked dsDNA and which can inhibit cruciform extrusion from PALINDROMIC SEQUENCES in vitro. It functions e.g. in replication from *oriC* in *E. coli*, in some SITE-SPECIFIC RECOMBINATION systems, and in Tn*10* transposition. It resembles mammalian HMG1 protein (whose DNA-binding is more structure- than sequence-dependent) [Mol. Microbiol. (1993) *7* 343–350].

The HU protein is reported to be involved in regulating the expression of gene *rpoS* (encoding SIGMA FACTOR σ^S) [Mol. Microbiol. (2001) *39* 1069–1079].

Hucker's method See GRAM STAIN.

Huffia A subgenus of PLASMODIUM.

Hugh and Leifson's test OXIDATION–FERMENTATION TEST.

Hughes press An apparatus for CELL DISRUPTION. A *frozen* suspension or paste of cells in a cylinder is forced by hydraulic pressure (up to ca. 30 tons/inch^2, i.e., ca. 480 MPa) through a small hole in the containing vessel; as the cells emerge they are broken by the solid shear forces due to the presence of intracellular ice crystals. The frozen sample may incorporate an abrasive (e.g. finely powdered glass) for more efficient cell breakage.

Huilia See LECIDEA.

hülle cell See EMERICELLA.

human (alpha) herpesviruses See ALPHAHERPESVIRINAE.

human (beta) herpesviruses See BETAHERPESVIRINAE.

human caliciviruses See SMALL ROUND STRUCTURED VIRUSES.

human coronavirus See CORONAVIRIDAE.

human cytomegalovirus See BETAHERPESVIRINAE.

human diploid cell vaccine (HDCV) See RABIES.

human enteric coronavirus See CORONAVIRIDAE.

human foamy viruses See SPUMAVIRINAE.

human (gamma) herpesviruses See GAMMAHERPESVIRINAE.

human genome (sequence) See GENOME.

human herpesvirus (HHV) Any virus of the family HERPESVIRIDAE which can infect humans; HHVs occur in each of the three subfamilies ALPHAHERPESVIRINAE, BETAHERPESVIRINAE and GAMMAHERPESVIRINAE. (See also HHV.)

human immunodeficiency virus See HIV.

human papilloma virus See PAPILLOMAVIRUS.

human parvovirus See PARVOVIRUS.

human T-cell leukaemia/lymphoma virus See HTLV.

human T-lymphotropic virus See HTLV.

Humicola See HYPHOMYCETES; see also COMPOSTING and SOFT ROT (sense 1).

humicolous Growing in or on soil or humus.

humoral antibodies Antibodies present (free) in plasma and in other body fluids.

humoral immunity (1) ANTIBODY-dependent IMMUNITY; humoral immunity can be transferred to a non-immune individual by the transfer of cell-free plasma or serum. (cf. CELL-MEDIATED IMMUNITY; see also IMMUNIZATION (sense 1); ADCC; JONES–MOTE SENSITIVITY.) (2) Immunity derived from any factor(s) in the body fluids – e.g. antibodies, LYSOZYME, COMPLEMENT.

humus tank See SEWAGE TREATMENT.

Hungate technique See ROLL-TUBE TECHNIQUE.

Hungate tube A gas-tight, tube-shaped glass vessel with a rubber stopper which is held in place by a screw cap; it is used in the ROLL-TUBE TECHNIQUE.

***hupA, hupB* genes** In *Escherichia coli*, the genes which encode the two subunits of the HU PROTEIN.

Huroniospora See STROMATOLITES.

HUS See HAEMOLYTIC URAEMIC SYNDROME.

***hut* genes** Genes involved in histidine utilization (see HISTIDINE DEGRADATION).

Hutchinson's teeth See SYPHILIS.

HUVEC Human umbilical vascular endothelial cell.

HVR-1 (in hepatitis C virus) See HEPATITIS C VIRUS.

hyaline Transparent, translucent, or colourless.

hyaline cap See PSEUDOPODIUM.

hyalodendrins See EPIPOLYTHIAPIPERAZINEDIONES.

hyalodidymae See SACCARDOAN SYSTEM.

hyaloplasm (1) *Syn.* CYTOSOL. (2) See SARCODINA.

Hyalospora See UREDINIOMYCETES and AMPHISPORE.

Hyalotheca A genus of filamentous placoderm DESMIDS. The filaments are composed of chains of cylindrical, flat-ended cells; each cell has only a slight median constriction.

hyaluronan *Syn.* HYALURONIC ACID.

hyaluronate lyase ('spreading factor'; EC 4.2.2.1) A LYASE which cleaves HYALURONIC ACID at the (1 → 4)-β-linkages; the end product is a disaccharide containing a 4,5-unsaturated uronic acid residue. (cf. HYALURONIDASE). It is produced e.g. by most coagulase +ve staphylococci, *Streptococcus pneumoniae*, *S. pyogenes* and *Clostridium perfringens* (μ toxin); it may promote invasiveness. It can be assayed by its ability to reduce viscosity in a solution of hyalauronic acid. [Action of enzyme from *S. pneumoniae*: EMBO (2000) *19* 1228–1240.]

hyaluronic acid A linear polysaccharide in which repeating disaccharide units of β-D-glucuronopyranosyl-(1 → 3)-*N*-acetyl-D-glucosamine are connected by (1 → 4)-β-linkages. Hyaluronic acid occurs e.g. as an intercellular constituent of various animal tissues, in synovial fluid, and in the capsules of certain group A streptococci. Hyaluronic acid forms highly viscous solutions in water and readily forms gels.

hyaluronidase Any enzyme which cleaves HYALURONIC ACID; they include HYDROLASES and LYASES. The hydrolases include hyaluronate 4-glucanhydrolase (EC 3.2.1.35) – present e.g. in testicular extracts and produced e.g. by certain streptomycetes. The name 'hyaluronidase' is often used (erroneously) as a synonym for the enzyme HYALURONATE LYASE.

hybrid antibody See CONJUGATION (2).

hybrid-arrested translation A method for identifying a protein encoded by a given cDNA. mRNAs from the cell are translated in the presence of cloned, single-stranded copies of the cDNA. mRNAs which do *not* bind cDNA yield proteins incorporating a *labelled* amino acid. Translation is repeated *without* cDNAs, allowing translation of the mRNA previously blocked by hybridization to cDNA. The protein of interest is the *extra* protein formed in the second translation.

hybrid plasmid (1) Any recombinant PLASMID (see e.g. CLONING). (2) A recombinant plasmid containing sequences derived from organisms which can normally exchange genetic information. (cf. CHIMERIC PLASMID.)

hybridization (1) The formation of a double-stranded nucleic acid by base-pairing between single-stranded nucleic acids derived (usually) from different sources; some authors use the term specifically for the association of ssRNA with ssDNA. (See also DNA HOMOLOGY and SOUTHERN HYBRIDIZATION.)

(2) The formation of e.g. a HYBRID PLASMID.

(3) The formation of a hybrid cell (see e.g. HYBRIDOMA), or of a hybrid organism (e.g. by a cross between genetically dissimilar organisms).

hybridization protection assay See TMA.

hybridoma The product and/or progeny of cell fusion ('somatic cell hybridization') between a tumour cell and a non-tumour cell; the in vitro production of hybridomas is carried out to provide continuously replicating (hybrid) cells which exhibit some or all of the characteristics of the non-tumour cell. Hybridomas have been formed, for example, between normal B LYMPHOCYTES and MYELOMA cells; such hybridomas are used e.g. as sources of MONOCLONAL ANTIBODIES.

In one method of *B cell hybridoma* formation, spleen (or lymph node) cells are mixed with similar numbers of myeloma cells and centrifuged. The cell pellet is briefly exposed to polyethylene glycol (PEG) which promotes cell fusion. (Lysolecithin or inactivated Sendai virus has been used in place of PEG.) (cf. ELECTROFUSION.) The PEG is diluted out, and the fusion mixture is then centrifuged; the pellet is re-suspended in a growth medium, divided into aliquots, and incubated. Small cluster(s) of hybridoma cells appear in the aliquots after ca. 1 week. (The number of hybridomas depends e.g. on the initial cell density: the fusion rate may be e.g. 1 in 10^4.)

In order to prevent the culture from being overgrown by the myeloma cells, use is made of (pre-selected) mutant myeloma cells which are blocked in one of their two (normal) pathways for nucleotide synthesis; such mutants lack either thymidine kinase (TK^- mutants) or hypoxanthine-guanine-phosphoribosyltransferase ($HGPRT^-$ mutants) so that they cannot use the ancillary ('salvage') pathway for nucleotide synthesis and (hence) cannot use exogenous thymidine or hypoxanthine (or guanine). Accordingly, to suppress the growth of non-fused myeloma cells, the fusion mixture is incubated in a medium (HAT medium) which contains aminopterin (a folic acid analogue which blocks the *main* pathway of nucleotide synthesis) together with hypoxanthine and thymidine; thus, the main pathway for nucleotide synthesis is blocked in both myeloma and non-myeloma cells, while *fused* (hybridoma-forming) myeloma cells can continue to grow by using TK or HGPRT supplied by their non-mutant partner cells.

In order to prevent the formation of hybrid Ig (molecules which contain heavy and/or light chains from each of both partner cells), use is made of mutant myeloma cells which do not secrete Ig components.

Activated or proliferating B cells fuse to myeloma cells more readily than do quiescent B cells; hence, prior to hybridoma formation, B cells of the required antibody specificity should be activated e.g. by antigenic challenge. (The consequent expansion of particular B cell clone(s) will also increase the incidence with which the cells of such clones participate in hybridoma formation.)

T cell hybridomas have been used e.g. in the study of cytokines, and for characterization of the different T LYMPHOCYTE subsets.

Hydnaceae See APHYLLOPHORALES.

Hydnum See APHYLLOPHORALES (Hydnaceae).

Hydra viridis See ZOOCHLORELLAE.

Hydramoeba A genus of amoebae (order AMOEBIDA) which occur both free-living and as parasites in the gastrodermis of *Hydra* sp.

hydrocarbon inclusions See HYDROCARBONS.

hydrocarbons Compounds which contain only carbon and hydrogen; they occur e.g. in PETROLEUM. Many types of hydrocarbon can be used as substrates for growth by various microorganisms. Such organisms are ecologically important e.g. in the degradation of petroleum pollutants, and some may be commercially useful e.g. in the production of SINGLE-CELL PROTEIN from hydrocarbons; however, certain hydrocarbon-utilizing organisms can cause spoilage of hydrocarbon products such as fuels (see PETROLEUM). Hydrocarbon metabolism is strictly aerobic and appears always to involve the introduction of oxygen into the molecule in a process requiring a monooxygenase (= hydroxylase) or dioxygenase (see OXYGENASE).

Aliphatic hydrocarbons. Straight-chain paraffins (*n*-alkanes) can be utilized by bacteria (e.g. strains of *Acinetobacter*, *Corynebacterium*, *Mycobacterium*, *Nocardia*, *Pseudomonas*), by yeasts (e.g. species of *Candida*), and by mycelial fungi (e.g. species of *Aspergillus*, *Botrytis*, *Fusarium*, *Helminthosporium*, *Hormoconis*, *Penicillium*). Some organisms can use only short-chain alkanes, some can use only long-chain alkanes. In most cases *n*-alkane metabolism appears to occur by *ω-oxidation*, i.e., a terminal methyl group of the alkane is oxidized by a monooxygenase to form a primary alcohol which is in turn (apparently) oxidized, via the aldehyde, to the corresponding fatty acid by alcohol dehydrogenase and aldehyde dehydrogenase activities; the fatty acid can then be degraded by conventional β-oxidation. In e.g. *Pseudomonas* strains the alkane-oxidizing enzyme system is complex, involving an ω-monooxygenase, a rubredoxin (see IRON–SULPHUR PROTEINS), and an NADH-rubredoxin oxidoreductase; in eukaryotes, and possibly in certain bacteria (e.g. *Acinetobacter* strains [FEMS (1981) *11* 309–312]), the alkane monooxygenase is linked to the cytochrome P-450 electron-carrier system (see CYTOCHROMES). (See also METHANOTROPHY.)

In some organisms an alkane may be oxidized at *both* ends, resulting in the formation of a dicarboxylic acid. Subterminal oxidation may also occur.

Branched-chain alkanes (alkylalkanes) and unsaturated hydrocarbons (alkenes, olefins) are generally somewhat less susceptible to microbial degradation than are *n*-alkanes; they may be oxidized in the same way as *n*-alkanes, but alkenes may also be oxidized at the double bond, resulting in the formation of a diol.

Organisms growing on alkanes have certain characteristic structural features. For example, yeasts growing on hydrocarbons generally contain numerous PEROXISOMES which are apparently involved in fatty acid metabolism rather than in alkane oxidation per se. Alkane-utilizing bacteria generally contain intracytoplasmic membranes together with 'hydrocarbon inclusions': electron-translucent, spherical bodies (ca. 0.2 μm diam.) limited by a non-unit-type (monolayer) membrane; the inclusions occur at the cell periphery or in close association with the intracytoplasmic membranes, and apparently contain unmodified alkane, protein, phospholipid and neutral lipid.

Since alkane metabolism occurs intracellularly, the hydrocarbon must be taken into the cell. Uptake may occur by different mechanisms in different organisms, and in at least some cases may require prior emulsification of the hydrocarbon by an extracellular BIOSURFACTANT or bioemulsifier.

Alicyclic hydrocarbons (cycloparaffins, cycloalkanes) are cyclic, non-aromatic hydrocarbons; they are generally less susceptible to microbial attack than either aliphatic or aromatic compounds. A suggested pathway for the degradation of e.g. cyclohexane by strains of *Nocardia* or *Pseudomonas* involves oxidation of the cyclohexane to cyclohexanol by a cyclohexane monooxygenase; cyclohexanol is oxidized to cyclohexanone, and then an oxygen atom is introduced into the ring (forming a lactone) by a cyclohexanone monooxygenase. The lactone can then be hydrolysed to form a (non-cyclic) dicarboxylic acid.

Aromatic hydrocarbons (benzene, naphthalene, anthracene etc) are present in e.g. petroleum and are formed by the incomplete combustion of almost any organic material; they are therefore common pollutants, and many are recognized carcinogens. Bacteria (e.g. *Pseudomonas* spp) metabolize aromatic hydrocarbons by initially incorporating two atoms of oxygen into the substrate to form a *cis*-dihydrodiol; the reaction is catalysed by a multicomponent enzyme system comprising a dioxygenase, a flavoprotein, and IRON-SULPHUR PROTEINS. The *cis*-dihydrodiol is oxidized to a catechol which is in turn a substrate for another dioxygenase system which breaks open the aromatic ring. In contrast, fungi oxidize aromatic hydrocarbons using a cytochrome P-450-dependent monooxygenase to form a reactive arene oxide; this can either undergo isomerization to form a monohydric phenol, or enzymic hydrolysis to form a *trans*-dihydrodiol.

[Reviews on microbial hydrocarbon metabolism: Book ref. 155.]

Hydroclathrus See PHAEOPHYTA.

Hydrodictyon A genus of freshwater green algae (division CHLOROPHYTA) which form free-floating net-like colonies ('water nets') composed of large, coenocytic cells.

hydrogen (as a metabolite) Gaseous (molecular) hydrogen is produced e.g. in various types of FERMENTATION (sense 1), and is formed during NITROGENASE activity. Gaseous hydrogen may be consumed e.g. during METHANOGENESIS and may be used for lithotrophic metabolism by e.g. HYDROGEN-OXIDIZING BACTERIA and some SULPHATE-REDUCING BACTERIA. [Production and uptake of molecular hydrogen by unicellular cyanobacteria: JGM (1985) *131* 1561–1569.] (See also FORMATE HYDROGEN LYASE and INTERSPECIES HYDROGEN TRANSFER.)

hydrogen bacteria *Syn.* HYDROGEN-OXIDIZING BACTERIA.

hydrogen dehydrogenase *Syn.* HYDROGENASE.

hydrogen hypothesis See EUKARYOTE.

hydrogen-oxidizing bacteria (the 'hydrogen bacteria'; knallgas bacteria) A phrase sometimes used to refer specifically to a (non-taxonomic) category of aerobic bacteria which can grow chemolithoautotrophically by obtaining energy from the oxidation of gaseous hydrogen by oxygen via an electron transport chain (see KNALLGAS REACTION); most species can also grow chemoorganoheterotrophically and/or mixotrophically. [Details of the obligately chemolithoautotrophic hydrogen-oxidizing species *Hydrogenobacter thermophilus* (gen. nov., sp. nov.): IJSB (1984) *34* 5–10.] The hydrogen-oxidizing bacteria, *sensu stricto*, are distinct from those bacteria which can oxidize H_2 without autotrophic CO_2 fixation and/or which can carry out anaerobic oxidation of H_2 coupled to the reduction of e.g. SO_4^{2-}, CO_2 or fumarate. The hydrogen-oxidizing bacteria (which include most or all CARBOXYDOBACTERIA) include e.g. those strains of *Alcaligenes denitrificans* formerly called *A. ruhlandii*; *Aquaspirillum autotrophicum*; *Bacillus schlegelii* [JGM (1979) *115* 333–341]; *Paracoccus denitrificans*; *Pseudomonas facilis* and *P. saccharophila*; and bacteria referred to as *Alcaligenes eutrophus*, *Nocardia autotrophica*, and *N. opaca*.

Most hydrogen-oxidizing bacteria (including e.g. *Paracoccus denitrificans*) have only a membrane-bound, NAD-independent HYDROGENASE (see also EXTRACYTOPLASMIC OXIDATION); some strains have only a soluble (cytoplasmic), NAD-reducing hydrogenase ('hydrogen dehydrogenase'), while others (e.g. strains of '*Alcaligenes eutrophus*') have both forms of the enzyme.

[Habitats, culture, and descriptions of individual hydrogen-oxidizing bacteria: Book ref. 45, pp. 865–893.]

hydrogen peroxide (a) (as an antimicrobial agent) Hydrogen peroxide (H_2O_2) is an OXIDIZING AGENT which can be an effective disinfectant at low concentrations (e.g. 0.1% or less) and can act as a sterilant (for inanimate objects) at high concentrations (e.g. 10–25%); sporicidal activity at high concentrations is enhanced by ultrasonic energy and by certain metal ions, particularly Cu^{2+}. Dilute aqueous solutions of H_2O_2 are used for the treatment of wounds; here, antimicrobial activity may be largely mechanical – tissue CATALASE causing a rapid evolution of oxygen which can dislodge contaminating foreign matter, thus facilitating its removal. However, the strongly oxygenated environment thus created in the wound can inhibit the development of anaerobic pathogens.

The antimicrobial action of H_2O_2 may be due to its reduction to the highly reactive HYDROXYL RADICAL which can react e.g. with membrane lipids and nucleic acids. [Book ref. 65, pp. 240–244.]

Escherichia coli and *Salmonella typhimurium*, when exposed to sublethal concentrations of H_2O_2, acquire resistance to higher doses of H_2O_2 and other oxidants; in *S. typhimurium* this 'adaptation' to oxidative stress involves the induction of various proteins, some of which are apparently encoded by a regulon under the positive control of a locus designated *oxyR* [Cell (1985) *41* 753–762]. [Toxicity, mutagenesis and stress responses induced in *E. coli* by H_2O_2: JCS (1987) Supplement 6 289–301.]

Peracids (acids containing the peroxy group, –O–O–) – which can be regarded as derivatives of H_2O_2 – are oxidizing agents whose activity is greater than that of H_2O_2. *Peracetic acid* ($CH_3.COO.OH$) is the most active antimicrobial agent of the organic peracids, and can act as a sterilant at quite low concentrations (e.g. 1% or less). An aqueous solution of *sodium perborate* ($NaBO_3$) acts as a mixture of borate and hydrogen peroxide; a paste of sodium perborate in water or glycerol has been used for the treatment of oral infections involving anaerobes (e.g. Vincent's angina).

[H_2O_2 and peracetic acid as antimicrobial agents: JAB (1983) *54* 417–423.]

(b) (as a metabolite) H_2O_2 is produced e.g. by the spontaneous or enzymatic dismutation of SUPEROXIDE (see also SUPEROXIDE DISMUTASE). Endogenously formed H_2O_2 may be inactivated, intracellularly, by e.g. CATALASE or GLUTATHIONE PEROXIDASE, thus preventing the toxic effects of H_2O_2 (H_2O_2 can inactivate some types of superoxide dismutase, and can give rise to HYDROXYL RADICAL and SINGLET OXYGEN). However, extracellular H_2O_2, produced by the activities of certain organisms, may be important e.g. in LIGNIN degradation and in BROWN ROT (sense 1). (See also DIANISIDINE and LACTOPEROXIDASE–THIOCYANATE–HYDROGEN PEROXIDE SYSTEM.)

hydrogen sulphide (H_2S) Certain bacteria can produce H_2S e.g. by the reductive degradation of sulphur-containing amino acids (e.g. cysteine) or by DISSIMILATORY SULPHATE REDUCTION; H_2S production can be plasmid mediated – see e.g. Hys plasmid in ESCHERICHIA. (See also KLIGLER'S IRON AGAR and TSI AGAR.) H_2S can be formed from sulphite by *Saccharomyces cerevisiae* [JGM (1985) *131* 1417–1424].

H_2S can be used by some bacteria as an electron donor in lithotrophic metabolism; these bacteria include e.g. species of *Beggiatoa* and *Thiobacillus*, members of the Rhodospirillaceae and other photosynthetic bacteria, and the cyanobacterium *Oscillatoria*.

(See also SULPHUR CYCLE.)

hydrogen swell A SWELL caused by hydrogen production due to internal corrosion of the can (see also TIN). Hydrogen swells, characterized by a high proportion of hydrogen in the headspace, may take months to develop; they are commonly associated with certain acidic fruits and with foods containing curing salts.

hydrogenase An OXIDOREDUCTASE (EC 1.12.–. –) which catalyses the reaction: $H_2 \leftrightarrow 2H^+ + 2e^-$. Hydrogenases occur in various algae (see e.g. PHOTOREDUCTION) and e.g. in HYDROGEN-OXIDIZING BACTERIA, some SULPHATE-REDUCING BACTERIA, some nitrogen-fixing bacteria (see NITROGENASE), *Chloroflexus aurantiacus* [FEMS (1985) *28* 231–235], and *Escherichia coli* (e.g. as part of the FORMATE HYDROGEN LYASE system). Hydrogenases are IRON–SULPHUR PROTEINS, and in many species of bacteria they have been shown to contain nickel. Hydrogen-oxidizing bacteria may contain a membrane-bound NAD-independent hydrogenase (see also EXTRACYTOPLASMIC OXIDATION) and/or a soluble NAD^+-reducing hydrogenase ('hydrogen dehydrogenase'). [Hydrogenases – their structure and applications in hydrogen production: Book ref. 132, pp. 75–102.] (See also KNALLGAS REACTION.)

Hydrogenobacter See HYDROGEN-OXIDIZING BACTERIA.

hydrogenogen See ACETOGEN.

Hydrogenomonas An obsolete bacterial genus which included certain species of HYDROGEN-OXIDIZING BACTERIA (e.g. *H. facilis*, now *Pseudomonas facilis*).

hydrogenosome A membrane-limited intracellular organelle present in certain eukaryotic microorganisms – e.g. trichomonads (but not e.g. *Entamoeba histolytica*); in trichomonads it contains various IRON–SULPHUR PROTEINS and flavoproteins as components of an electron transport chain that is used in the oxidation of pyruvate to acetate, CO_2 and H_2 (protons being used as electron acceptors). In isolated hydrogenosomes of *Trichomonas vaginalis*, METRONIDAZOLE is reduced, anaerobically, to an active (cytotoxic) derivative [JGM (1985) *131* 2141–2144]; in vivo such active product(s) may damage the hydrogenosome itself and/or may leave the hydrogenosome and affect other target(s). (cf. MICROBODY.)

hydrolases ENZYMES (EC class 3) which catalyse the hydrolytic cleavage of bonds, including C–O, C–N and C–C bonds. The names of such enzymes are commonly formed by adding '-ase' to the name of the substrate (e.g. esterase, glycosidase, peptidase); the systematic name has the form *substrate hydrolase*.

hydrophobia *Syn.* RABIES.

hydrops fetalis (non-immune) A condition in which the fetus develops e.g. anaemia and oedema; death may result from severe anaemia. In a proportion of cases hydrops fetalis is due to infection by parvovirus B19 (see ERYTHROVIRUS).

hydroresorufin See RESAZURIN TEST.

hydrothermal vent In certain geographical locations: a region of the ocean floor where seawater, which has permeated the earth's crust for several kilometres, emerges as a warm (5–25°C) or hot (270–380°C) hydrothermal fluid which is particularly rich in sulphide; vents which discharge hot fluids ≤300°C are called 'white smokers', while those discharging fluids at ca. 350°C are called 'black smokers' – an allusion to the clouds of black metal sulphides formed when these fluids meet the cold (2°C) ambient seawater. One of the best-studied vents is situated at the Galápagos Rift (near the equator at 86°W).

The sulphide (and e.g. hydrogen, methane and ferrous iron) in hydrothermal fluids supports the chemolithotrophic growth of various bacteria (e.g. species of THIOBACILLUS and THIOMICROSPIRA) – thus permitting chemosynthetic PRIMARY PRODUCTION; these bacteria form the basis of a food chain which supports communities of e.g. giant clams, limpets, mussels and tube worms in and around the vents. Some of the lithotrophic prokaryotes form symbiotic associations with some of the invertebrates; thus, e.g. prokaryotes occur in the gill cells of the giant clam (*Calyptogena magnifica*) and in the TROPHOSOME tissue of the giant tube worm, *Riftia pachyptila*. A living community at a given vent appears to last for several years to several decades.

[Biology and microbiology of hydrothermal vents: Science (1985) *229* 713–725; evidence for chemosynthetic primary production in hydrothermal vents: Book ref. 202, pp. 319–360.]

hydroxamates (in iron uptake) See SIDEROPHORES.

hydroxyapatite (hydroxylapatite; $Ca_{10}(PO_4)_6(OH)_2$) A mineral which is used as a stationary phase in adsorption CHROMATOGRAPHY e.g. for separating ssDNA from dsDNA or RNA. Under certain ionic conditions the mineral adsorbs dsDNA; the dsDNA is desorbed at higher ionic concentrations.

hydroxycobalamin See VITAMIN B_{12}.

hydroxyethylcellulose See HE-CELLULOSE.

hydroxyl radical (OH· or ·OH) A hyper-reactive radical formed e.g. during a reaction involving SUPEROXIDE and HYDROGEN PEROXIDE (the 'Haber–Weiss reaction'); this reaction can be catalysed by iron complexes (e.g. Fe–EDTA, Fe–transferrin) and may be represented in two phases: (i) the reduction of ferric complexes by O_2^- (with the formation of ferrous complexes and SINGLET OXYGEN), and (ii) the oxidation of ferrous complexes by H_2O_2 (with the formation of ferric complexes, OH^- and OH·) [FEBS (1978) *86* 139–142]. (Hydroxyl radical is also reported to be formed e.g. in a reaction involving superoxide and hypochlorite, and by the action of ultraviolet radiation on hydrogen peroxide.) During PHAGOCYTOSIS, LACTOFERRIN from activated NEUTROPHILS may catalyse the formation of OH·, thus enhancing antimicrobial activity. Scavengers of OH· used in experimental systems include e.g. benzoate, dimethyl sulphoxide, mannitol and salicylate.

hydroxylamine (as a MUTAGEN) Hydroxylamine (NH_2OH) reacts mainly with cytosine residues (and, much more slowly, with adenine residues) in DNA (particularly ssDNA), replacing the amino group with a hydroxyamino group (–NHOH); this favours tautomerization of the base such that, during subsequent DNA replication, mispairing occurs (resulting predominantly in G·C to A·T transitions). Although hydroxylamine is an effective mutagen when used to treat certain viruses or e.g. transforming DNA, any mutagenic effect it may have in living cells is generally masked by its toxic or lethal effects. The related compound *methoxyamine* (NH_2OCH_3) acts in a similar way.

hydroxylapatite *Syn.* HYDROXYAPATITE.

hydroxylase *Syn.* MONOOXYGENASE.

hydroxymethionine analogue See HMA.

6-(*p*-hydroxyphenylazo)uracil See HPURA.

hydroxypyrimidine antifungal agents A group of agricultural systemic ANTIFUNGAL AGENTS which have a highly selective action against powdery mildews; resistance to these fungicides tends to develop rapidly. Hydroxypyrimidines include e.g. DIMETHIRIMOL and ETHIRIMOL.

8-hydroxyquinoline (oxine; 8-quinolinol) (as antimicrobial agent) A chelating agent which, in the presence of Cu^{2+} or Fe^{2+}, has antifungal and antibacterial activity; Gram-positive bacteria are much more susceptible than Gram-negative bacteria. The potentiating effects of Cu^{2+} and Fe^{2+} are antagonized by trace amounts of Co^{2+}, while Ni^{2+} acts as a competitive inhibitor. COPPER–oxine has been used in agricultural antifungal sprays and as a rot-proofing agent for e.g. textiles.

Some halogenated derivatives of oxine are used e.g. for the treatment or prophylaxis of amoebic dysentery. They include

7-iodo-8-hydroxyquinoline-5-sulphonate (chiniofon); 5,7-diiodo-8-hydroxyquinoline (diiodohydroxyquin; Diodoquin); 5-chloro-7-iodo-8-hydroxyquinoline (iodochlorohydroxyquin; Enterovioform; Vioform). (Some of these derivatives also have useful antibacterial and antifungal properties.)

hydroxyquinoline-*N*-oxide (HOQNO; HQNO) 2-Alkyl-4-hydroxyquinoline-*N*-oxides (e.g. 2-*n*-heptyl-4-HOQNO) act as RESPIRATORY INHIBITORS in mitochondria and in certain bacteria – some bacteria are apparently impermeable to HOQNOs; HOQNOs appear to resemble ANTIMYCIN A (q.v.) in their mode of action. HOQNOs also inhibit the Na^+ pump in *Vibrio alginolyticus* (see SODIUM MOTIVE FORCE).

2-hydroxystilbamidine An aromatic DIAMIDINE which has trypanocidal and antifungal activity; it has been used e.g. in the treatment of blastomycosis and African trypanosomiasis.

5-hydroxytryptamine *Syn.* SEROTONIN.

hydroxyurea (HU; $HONH.CO.NH_2$) A reagent which prevents the synthesis of deoxyriboNUCLEOTIDES by specifically inhibiting ribonucleoside diphosphate reductase.

Hyella See PLEUROCAPSA GROUP.

Hygrocybe See AGARICALES (Hygrophoraceae).

hygromycin B An AMINOGLYCOSIDE ANTIBIOTIC which, like STREPTOMYCIN, appears to act at a single ribosomal site.

hygrophilic (hygrophilous) Requiring a moist habitat for optimum growth.

Hygrophoraceae See AGARICALES.

Hygrophorus See AGARICALES (Hygrophoraceae).

hymenium (*mycol.*) A layer of ascospore- or basidiospore-forming tissue – e.g. that lining an APOTHECIUM or that forming the surface of a LAMELLA; a hymenium may also contain sterile structures – see e.g. PARAPHYSIS and PSEUDOPARAPHYSIS.

Hymenochaetaceae See APHYLLOPHORALES.

Hymenochaete See APHYLLOPHORALES (Hymenochaetaceae).

Hymenogaster See GASTEROMYCETES (Hymenogastrales).

Hymenogastrales See GASTEROMYCETES.

Hymenomonas See COCCOLITHOPHORIDS.

Hymenomycetes A class of fungi (subdivision BASIDIOMYCOTINA) in which the typical fruiting body is a well-developed, macroscopic, gymnocarpous or semiangiocarpous structure containing BALLISTOSPORE-bearing basidia in a HYMENIUM. Subclasses: HOLOBASIDIOMYCETIDAE and PHRAGMOBASIDIOMYCETIDAE.

hymenophore That part of a fruiting body which bears a HYMENIUM or SUBHYMENIUM, or the entire (hymenium-bearing) fruiting body. (cf. SPOROPHORE.)

Hymenostomatia A subclass of (mostly) freshwater ciliates (class OLIGOHYMENOPHOREA) in which somatic ciliature is often uniform, the buccal cavity (when present) is ventral, and kinetodesmata are usually conspicuous; sedentary and colonial types are not common. Orders: ASTOMATIDA, HYMENOSTOMATIDA, SCUTICOCILIATIDA.

Hymenostomatida An order of protozoa (subclass HYMENOSTOMATIA) in which the buccal cavity is well defined and contains characteristic membranelles. A SCUTICA does not appear during stomatogenesis. Genera (which may be distinguished e.g. by the number of post-oral kineties) include e.g. COLPIDIUM, *Frontonia*, GLAUCOMA, ICHTHYOPHTHIRIUS, LAMBORNELLA, *Ophryoglena*, PARAMECIUM and TETRAHYMENA.

hyperauxinic Containing or producing abnormally high levels of AUXINS.

hyperchromic shift A change in the amount of ultraviolet radiation absorbed by a nucleic acid during changes from the double-stranded to the single-stranded condition, or vice versa (see e.g. THERMAL MELTING PROFILE).

hypericin See STENTORIN.

hyper-IgM syndrome See RNA EDITING.

hyperimmune Refers to the condition of an individual whose plasma contains a high titre of a particular antibody following repeated exposure of the individual to the homologous antigen. *Hyperimmune serum* is serum derived from such an individual.

Hypermastigida An order of protozoa (class ZOOMASTIGOPHOREA) which have numerous flagella, the kinetosomes being arranged in complete or incomplete circles; the organisms are associated with insects. Genera: e.g. *Barbulanympha*, *Lophomonas*, *Microjoenia*, *Spirotrichonympha*, TRICHONYMPHA.

hyperparasite A parasite of a parasite.

hyperplasia The enlargement of an organ or tissue owing to an increase in the number of cells. (Hence *adj.* hyperplastic.) (cf. HYPERTROPHY and NEOPLASIA.)

hypersensitivity (1) (*immunol.*) The state of a PRIMED individual who, on further exposure to the relevant antigen, gives an exaggerated immune response that causes varying degrees of harm – ranging from a mild local inflammatory reaction to death. (See IMMEDIATE HYPERSENSITIVITY and DELAYED HYPERSENSITIVITY; see also BRUCELLOSIS.)

(2) (*plant pathol.*) The expression of extreme reactivity by a plant in response to a potential parasite or pathogen, the plant's response commonly serving to limit or prevent parasitization/disease. Hypersensitivity typically involves rapid cell death at the (localized) site of infection and the concomitant induction of e.g. PHYTOALEXIN(s) by microbial elicitors; the induction of phytoalexins by microbial elicitors involves the expression of new host genes and the formation of new species of mRNA and new enzymes. (In addition to phytoalexin production, microbial elicitors can also apparently e.g. stimulate the formation of ETHYLENE and bring about the accumulation of LIGNIN.) (See also PATHOGENESIS-RELATED PROTEINS.) In many cases the combination of cell death and accumulation of phytoalexin(s) at the site of infection appears to provide the level of resistance needed to prevent disease development. When the challenging microorganism is an obligate parasite (i.e., a BIOTROPH), the death of host cells may, in itself, be sufficient to prevent the spread of the parasite to uninfected tissues.

hypersensitivity pneumonitis *Syn.* EXTRINSIC ALLERGIC ALVEOLITIS.

hypertrophy Enlargement of an organ or tissue owing to an increase in the size of pre-existing cells. (cf. HYPERPLASIA.)

hypervariable region (*immunol.*) Any of several amino acid sequences, located in the VARIABLE REGIONS of the heavy chain and light chain of IMMUNOGLOBULINS, which exhibits a degree of variability (in composition) greater than that found in other parts of the variable region. (See also COMPLEMENTARITY-DETERMINING REGIONS.)

hypha (*pl.* hyphae) (1) (*bacteriol.*, *mycol.*) In many (mycelial) fungi and in some bacteria (see ACTINOMYCETALES): a branched or unbranched filament, many of which together constitute the vegetative form of the organism and (in some species) form the sterile portion of a fruiting body; a mass of vegetative hyphae is referred to as a MYCELIUM.

Fungal hyphae. Each hypha has a tubular CELL WALL which, in many species, is divided into compartments or cells by cross-walls (see SEPTUM (b)); in both septate and aseptate species, the cell wall is external to the CYTOPLASMIC MEMBRANE which encloses the cytoplasm, NUCLEUS or nuclei, and other components typical of eukaryotic organization. (See also VACUOLE.) The diameter of a hypha ranges from about 1 μm to macroscopic dimensions, depending e.g. on species. Hyphae grow mainly by apical extension – see GROWTH (fungal).

In basidiomycetes the *primary* hyphae are uninucleate haploid hyphae which develop when basidiospores germinate; the *secondary* (dikaryotic) hyphae develop following DIKARYOTIZATION.

A *basidiocarp* may, according to species, consist of one, two or three distinct types of hypha – corresponding, respectively, to a *monomitic*, *dimitic* or *trimitic* sporocarp. A monomitic sporocarp contains only *generative hyphae*: typically thin-walled, branched, septate hyphae which give rise to basidia; generative hyphae may contain CLAMP CONNECTIONS if the organism normally forms clamps on its vegetative mycelium. A dimitic sporocarp contains generative hyphae together with either skeletal hyphae or binding hyphae (both of which originate from generative hyphae). *Skeletal hyphae* are typically thick-walled and unbranched; they are sterile (i.e., do not form basidia) and do not form clamps. *Binding* (= *ligative*) *hyphae* are sterile, highly branched, and do not form clamps; in trimitic sporocarps they bind together the generative and skeletal hyphae.

(2) (*bacteriol.*) See PROSTHECA.

hyphal body In some species of the ENTOMOPHTHORALES: a piece of fragmented mycelium which can germinate to form a spore-bearing structure.

hyphenated dyad symmetry See e.g. PALINDROMIC SEQUENCE.

Hyphochytriaceae See HYPHOCHYTRIOMYCETES.

Hyphochytriales See HYPHOCHYTRIOMYCETES.

Hyphochytridiomycetes See HYPHOCHYTRIOMYCETES.

Hyphochytriomycetes A class of fungi (subdivision MASTIGOMYCOTINA) which form zoospores having one anteriorly-directed tinsel flagellum. The organisms include terrestrial species as well as aquatic saprotrophs and parasites of other fungi and of freshwater and marine algae; they closely resemble members of the CHYTRIDIALES in morphology and life cycle, but they apparently do not carry out sexual processes. One order: Hyphochytriales; three families: Anisolpidiaceae, Hyphochytriaceae and Rhizidiomycetaceae (e.g. RHIZIDIOMYCES).

This class has been called 'Hyphochytridiomycetes' by many authors. [Validation of the name 'Hyphochytriomycetes': Book ref. 174, p. 285.]

Hypholoma See AGARICALES (Strophariaceae).

Hyphomicrobium A genus of PROSTHECATE BACTERIA found in soils and in freshwater, estuarine and marine habitats. Morphologically, *Hyphomicrobium* resembles RHODOMICROBIUM, and its cell cycle is very similar to the 'simplified' cell cycle of *R. vannielii*; it differs from *R. vannielii* e.g. in the absence of pigments and in that its swarm cells bear a single subpolar flagellum. (Phages specific for developing daughter cells have been isolated [JGV (1979) *43* 29–38].) The organisms are chemoorganotrophs which can use one-carbon compounds (e.g. methanol, methylamine) as sole sources of carbon and energy (see METHYLOTROPHY); one-carbon compounds are metabolized via the icl$^-$ serine pathway. Growth occurs either aerobically or anaerobically with nitrate (which is reduced to nitrogen via nitrite); optimum growth temperature: 25–30°C. GC%: ca. 59–67. [Review: ARM (1981) *35* 567–594.]

Hyphomonas A genus of Gram-negative, aerobic, asporogenous, non-photosynthetic, heterotrophic, PROSTHECATE BACTERIA which reproduce by budding; daughter cells are motile (swarm cells), having one or more flagella according to species. The cell body may vary in size and shape during the cell cycle, and usually has a single polar prostheca ('hypha'). Iron and manganese salts are not deposited on the cell surface (cf. PEDOMICROBIUM). Amino acids are the preferred substrates for growth; one-carbon compounds are not used (cf. HYPHOMICROBIUM). Species: *H. neptunium* (isolated from seawater) and *H. polymorpha* (type species, isolated from nasal mucus from a patient with sinusitis) [IJSB (1984) *34* 71–73]. Proposed species: *H. hirschiana*, *H. jannaschiana* and *H. oceanitis* [IJSB (1985) *35* 237–243].

Hyphomycetales See HYPHOMYCETES.

Hyphomycetes A class of fungi of the subdivision DEUTEROMYCOTINA; most of the conidium-forming species do not form conidiomata (order Hyphomycetales), while the others form sporodochia (order Tuberculariales) or synnemata (order Stilbellales) – and some species do not form conidia at all (order AGONOMYCETALES). (cf. COELOMYCETES.) The vegetative form of the organisms is commonly a well-developed mycelium, but the class includes some fungi which can grow as unicellular organisms and which are sometimes classified in the class BLASTOMYCETES.

The class Hyphomycetes (ca. 1000 genera) includes e.g. ACICULOCONIDIUM, ACREMONIUM, ALTERNARIA, ARTHROBOTRYS, ASPERGILLUS, AUREOBASIDIUM, BEAUVERIA, BLASTOMYCES, BOTRYTIS, BRETTANOMYCES, CALCARISPORIUM, CANDIDA, CERCOSPORA, *Chalara*, CHRYSOSPORIUM, *Cladobotryum*, CLADOSPORIUM, COCCIDIOIDES, CRYPTOCOCCUS, CRYPTOSTROMA, CULICINOMYCES, *Curvularia*, CYLINDROCARPON, DACTYLARIA, DACTYLELLA, DRECHSLERA, *Exophiala*, *Fonsecaea*, FULVIA, FUSARIUM, GEOTRICHUM, *Gliocladium*, *Gonatobotrys*, *Gonatobotryum*, GRAPHIUM, HARPOSPORIUM, HELMINTHOSPORIUM, HIRSUTELLA, HISTOPLASMA, HORMOCONIS, HUMICOLA, HYPHOZYMA, KLOECKERA, MADURELLA, MALASSEZIA, MERIA, METARHIZIUM, MONACROSPORIUM, *Monodictys*, *Mycocentrospora*, MYCOGONE, MYROTHECIUM, NOMURAEA, *Oidiopsis*, *Oidium*, OOSPORIDIUM, *Ovulariopsis*, PAECILOMYCES, PARACOCCIDIOIDES, PENICILLIUM, *Periconia*, *Pesotum*, PHAFFIA, PHIALOPHORA, PITHOMYCES, PSEUDOCERCOSPORELLA, PYRICULARIA, *Ramularia*, RHINOCLADIELLA, RHODOTORULA, RHYNCHOSPORIUM, *Scedosporium*, SCHIZOBLASTOSPORION, *Scopulariopsis*, *Sphacelia*, SPOROTHRIX, SPOROTRICHUM, STACHYBOTRYS, STENELLA, STERIGMATOMYCES, THERMOMYCES, THIELAVIOPSIS, *Tolypocladium*, TORULA, TRICHODERMA, TRICHOSPORON, TRICHOTHECIUM, TRIGONOPSIS, *Triposporina*, *Tritirachium*, VERTICILLIUM, WANGIELLA, XYLOHYPHA, ZALERION.

hyphomycosis destruens *Syn.* EQUINE PHYCOMYCOSIS.

hyphopodium In the mycelium of the BLACK MILDEWS: a short, one- or two-celled branch on a somatic hypha. In a *capitate hyphopodium* the distal cell is enlarged; this type of hyphopodium may act as an appressorium and give rise to a haustorium.

Hyphozyma A genus of yeast-like fungi of the HYPHOMYCETES. [Key to species: AvL (1986) *52* 39–44.]

hypnospore A thick-walled resting cell.

hypnozoite A dormant form of a parasite in a living host. (See e.g. PLASMODIUM.)

hypobiosis See DORMANCY.

Hypocenomyce See LECIDEA.

hypochlorites (as antimicrobial agents) Hypochlorites (and hypochlorous acid) are strong OXIDIZING AGENTS which are microbicidal e.g. for many bacteria (including endospores) and viruses; their antimicrobial activity is not inhibited by most anionic and non-ionic detergents, but they are readily inactivated by organic matter and they have a tendency to decompose, forming e.g. chlorate (ClO_3^-) and HCl. (Decomposition is enhanced by e.g. heavy metal ions, acidity, light, and heat.) In solution, hypochlorites occur in equilibrium with hypochlorous acid, HOCl. The highly microbicidal *undissociated* form of HOCl is favoured by low pH; however, low pH reduces the stability of hypochlorites, so that commercial sodium hypochlorite solutions are often stabilized with sodium hydroxide. Sodium hypochlorite is used in a range of DISINFECTANTS and bleaches (e.g. Chloros, Domestos, Milton). Aerosols of hypochlorite solutions have been used for the disinfection of air. (N.B. Hypochlorite solutions react with formaldehyde to form the carcinogen *bis*-chloromethylether.) (See also DICHLOROISOCYANURATE.)

hypocone See DINOFLAGELLATES.

Hypocrea See HYPOCREALES.

Hypocreales An order of mainly saprotrophic, plant-parasitic or fungicolous fungi of the subdivision ASCOMYCOTINA; anamorphs occur e.g. in the genera ACREMONIUM, CYLINDROCARPON and FUSARIUM. Ascocarp: perithecioid or closed, frequently on or within a stroma and often brightly coloured. Asci: cylindrical to clavate. Ascospores: usually colourless, aseptate to multiseptate or muriform. Some genera in the order have traditionally been distinguished on the basis of the number of septa in their ascospores; spore septation has not been recognized as a valid taxonomic criterion [Mycol. Pap. (1983) Nr *150*]. Genera: e.g. *Calonectria*, GIBBERELLA, *Hypocrea*, NECTRIA.

Hypocrella See CLAVICIPITALES.

Hypoderma See RHYTISMATALES.

hypogean Occuring underground (cf. EPIGEAN).

Hypogymnia A genus of LICHENS (order LECANORALES). The thallus is foliose, with often hollow, inflated lobes, grey to greyish-brown above, black beneath; rhizines are absent. *H. physodes* is very common e.g. in the UK; it is tolerant of air pollution and is often the only macrolichen to be found in and around towns and cities.

hypolimnion In a lake: the layer of water between the METALIMNION and the lake bottom; characteristically, it is anaerobic and rich in sulphide.

Hypomyces A genus of mainly fungicolous fungi of the order CLAVICIPITALES; anamorph: *Cladobotryum*. (See also COBWEB DISEASE.)

hyponeuston See NEUSTON.

hypophloeodal *Syn.* ENDOPHLOEODAL.

Hypopylaria A category ('section') of LEISHMANIA species which develop in the midgut and hindgut of the arthropod vector; it includes '*L. agamae*' and '*L. ceramodactyli*'. (cf. PERIPYLARIA, SUPRAPYLARIA.)

hyposensitization (immunol.) *Syn.* DESENSITIZATION.

***Hyposoter exiguae* virus** See POLYDNAVIRIDAE.

hypostatic allele See EPISTASIS.

Hypostomatia A subclass of ciliates (class KINETOFRAGMINOPHOREA) in which the cytostome (when present) generally occurs on the ventral surface, and the cytopharyngeal apparatus is of the CYRTOS type; the body is typically cylindrical or dorsoventrally flattened, and the somatic ciliature is often reduced. The orders are:

Apostomatida. Cytostome inconspicuous or absent; rosette (secretory organelle?) commonly present in the oral area; somatic ciliature spirally arranged in mature organisms; cysts common (see also TOMITE); often associated with marine crustacea. Genera: e.g. *Chromidina*, *Foettingeria*, *Gymnodinioides*, *Polyspira*.

Chonotrichida. Typically vase-shaped and sedentary with a non-contractile stalk, attached to crustacea; reproduction by budding, producing motile ciliated forms (see also BROOD POUCH); ciliature in adult cell restricted to atrial region. Genera: e.g. *Actinichona*, CHILODOCHONA, *Spirochona*, *Stylochona*.

Cyrtophorida. Body flattened, FRANGE absent or vestigial. Genera: e.g. CHILODONELLA, *Chlamydodon*, *Dysteria*, *Hartmannula*.

Nassulida. Frange typically limited and restricted to left side of ventral surface. Genera: e.g. *Microthorax*, NASSULA, *Pseudomicrothorax*.

Rhynchodida. Cells with a single, TOXICYST-bearing anterior tentacle, typically associated with the gills of marine bivalve molluscs. Genera: e.g. *Ancistrocoma*.

Synhymeniida. Body commonly cylindrical and completely ciliated; frange extensive. Genera: e.g. *Orthodonella*.

hypothallus (1) (prothallus) (*lichenol.*) A pale or black layer of non-lichenized (i.e. photobiont-free) hyphae which extends from the periphery of the thallus in certain crustose (or squamulose) lichens; a hypothallus may also be evident between the areolae of certain areolate species (see e.g. RHIZOCARPON).

(2) (of myxomycetes) A thin, often transparent, sometimes calcified deposit which is secreted by the plasmodium during fruiting, and which forms a base for the fruiting bodies, remaining on the substratum following fruiting body formation. In some cases the hypothallus may be simply a remnant of the plasmodial slime sheath.

hypotheca (1) See DINOFLAGELLATES. (2) See DIATOMS.

hypothecium See APOTHECIUM.

Hypotrichida An order of typically free-living ciliate protozoa (class POLYHYMENOPHOREA). Cells: typically ovoid to elongate and dorsoventrally flattened, with a conspicuous oral ciliature and a number of groups of cirri on the ventral side; the dorsal side often carries pairs of short cilia ('sensory bristles'). Genera include e.g. ASPIDISCA, EUPLOTES, *Histriculus*, *Holosticha*, *Isosticha*, *Klonostricha*, *Oxytricha*, *Psammomitra*, *Strongylidium*, *Stylonychia*, *Trachelostyla*, *Uronychia*, *Urosoma*, *Urostyla*.

hypovalve See DIATOMS.

hypovirulence A reduced level of virulence in a strain of pathogen due e.g. to genetic changes in the pathogen, or to the effects on the pathogen of an infectious agent (*contagious* or *transmissible hypovirulence*: see e.g. CHESTNUT BLIGHT and DUTCH ELM DISEASE).

hypoxanthine The base in the nucleoside *inosine* [see Appendix V(a) for structure]. (See also NITROUS ACID and WOBBLE HYPOTHESIS.)

hypoxanthine-DNA glycosylase See NITROUS ACID.

Hypoxylon See SPHAERIALES.

Hys plasmid See ESCHERICHIA.

hystrichosphaerids (hystrichospheres; hystricospores) Fossil resting spores of DINOFLAGELLATES; they date from the late Triassic onwards. [Book ref. 136, pp. 847–964.]

Hyuga fever See EHRLICHIA.

I

I Isoleucine (see AMINO ACIDS).

(I + C + D) model *Syn.* HELMSTETTER–COOPER MODEL.

I-like pili See PILI.

I period See HELMSTETTER–COOPER MODEL.

Ia antigens Class II antigens of the murine MAJOR HISTOCOMPATIBILITY COMPLEX. (See also IR GENES.)

Ia protein The *Escherichia coli* OmpF PORIN.

IAA Indole 3-acetic acid (see AUXINS).

IAEA International Atomic Energy Agency.

IAHS Infection-associated haemophagocytic syndrome: see HAEMOPHAGOCYTIC SYNDROME.

IAM Institute of Applied Microbiology, University of Tokyo, Yayoi, Bunko-Ku, Tokyo, Japan.

IAP Intracisternal A-TYPE PARTICLE.

iatrogenic Refers to any disease or infection which is caused or exacerbated, unintentionally, as a result of medical intervention – e.g. examination or treatment.

IAVI International AIDS Vaccine Initiative.

Ib protein The *Escherichia coli* OmpC PORIN.

ibotenic acid See AMATOXINS.

IBV Infectious bronchitis virus (CORONAVIRIDAE).

ICAD See APOPTOSIS.

ICAM-1 *Syn.* CD54.

ICAM-2 See CD102.

ICAM-3 See IMMUNOGLOBULIN SUPERFAMILY.

ICDH Isocitrate dehydrogenase: see TCA CYCLE.

ICE (*syn.* caspase-1) See APOPTOSIS.

ice algae See DIATOMS.

ice nucleation bacteria Those bacteria which, at temperatures just below 0°C, promote water-to-ice transition by acting as nuclei around which ice crystals can form. Such bacteria (e.g. strains of *Erwinia herbicola*, *Pseudomonas fluorescens*, *P. syringae* and *Xanthomonas campestris*) – which include some common epiphytes – have been implicated as contributory factors in frost damage in various agricultural crops.

A DNA fragment from *P. fluorescens*, when cloned in *Escherichia coli*, can confer the ice-nucleation phenotype; the fragment may encode an OUTER MEMBRANE protein of MWt ca. 180000 [EMBO (1986) *5* 231–236]. When grown at 15°C, most ice-nucleation-positive strains of *Erwinia herbicola* shed into the medium vesicles composed of outer membrane; these vesicles ('cell-free ice nuclei') can behave as ice-nucleation centres which are active at −2°C to −10°C [JB (1986) *167* 496–502].

Some plant tissues survive, without damage, when 'supercooled' to several degrees below 0°C – but may suffer frost damage if ice-nucleation bacteria are present. Above (approx.) −5°C, the incidence of frost injury can be decreased by reducing the size of the populations of plant-contaminating ice-nucleation bacteria. This can be done e.g. by treatment of plant surfaces with antibiotics such as streptomycin or oxytetracycline; an alternative method is to treat plant surfaces with non-ice-nucleating bacteria (a form of BIOLOGICAL CONTROL). A combination of antibiotic treatment and biological control has been found to act additively in the control of frost injury in pear trees [Phytopathol. (1996) *86* 841–848].

Iceland moss See CETRARIA.

ich See ICHTHYOPHTHIRIASIS.

ichthyo- Prefix denoting fish – e.g. ichthyopathology: the study of FISH DISEASES; ichthyotoxic: toxic to fish.

ichthyophonosis A systemic, granulomatous FISH DISEASE caused by *Ichthyophonus hoferi* (see ICHTHYOPHONUS); external symptoms vary in nature and severity (cf. TAUMELKRANKHEIT). *I. hoferi* occurs in various host tissues, mainly as spherical, thick-walled, multinucleate cells ('cysts', 'resting spores') up to 200 μm or more in diameter; immediately after the death of the host the 'cysts' germinate to form branching hyphae. The disease can affect e.g. mackerel and salmonids, and has caused mass fatalities in Atlantic herring; infected fish are, or rapidly become, unsuitable for human consumption as the muscles rapidly degenerate after the death of the fish. [Review: Book ref. 1, pp. 243–269.]

Ichthyophonus A genus of fungi (or protozoa? – taxonomic affinity unknown) parasitic in fish. Species: *I. hoferi* (see ICHTHYOPHONOSIS) and *I. gasterophilum*. (cf. ICHTHYOSPORIDIUM.)

ichthyophthiriasis (white spot, ich, ick) A freshwater FISH DISEASE caused by the ciliate *Ichthyophthirius multifiliis*. White spots 1 mm or so across, each containing one or more ciliates, appear on skin and fins; affected fish are weakened and may die. Secondary SAPROLEGNIASIS is common. *I. multifiliis* needs well-oxygenated waters and may be controlled e.g. by raising the temperature to diminish dissolved oxygen levels.

Ichthyophthirius A genus of ciliates (order HYMENOSTOMATIDA); *I. multifiliis* is the causal agent of ICHTHYOPHTHIRIASIS in freshwater teleost fish. (cf. CRYPTOCARYON.) The life cycle of *I. multifiliis* involves three stages. The stage infective for fish is an ovoid or fusiform cell (called a *tomite* or *theront*), ca. 15 × 40 μm, which is covered with cilia, and which has a filamentous projection at the anterior end. Within the fish's epidermis the tomite becomes spherical and develops an oral apparatus – becoming a *trophozoite* (= *trophont*) which grows to ca. 100–1000 μm diam. When free of the host, the trophozoite forms a gelatinous exocyst, loses its oral apparatus, and undergoes schizogony. Cysts, each several hundred micrometres in size, may contain up to ca. 1000 tomites which are released to complete the life cycle.

[SEM of stages in the life cycle of *I. multifiliis*: J. Parasitol. (1985) *71* 218–226; development of *I. multifiliis* in gill epithelium: JP (1986) *33* 369–374.]

Ichthyosporidium A genus used by some to include members of the genus ICHTHYOPHONUS, and by others to include microsporidean protozoa parasitic in fish.

ichthyotoxin A TOXIN which is active against fish.

ick See ICHTHYOPHTHIRIASIS.

icl See SERINE PATHWAY.

ICL Isocitrate lyase: see TCA CYCLE and SERINE PATHWAY.

***icm* genes** (of *Legionella pneumophila*) See end of section (a) in PHAGOCYTOSIS.

ICMSF International Commission on Microbiological Specifications for Foods, an organization of the International Union of Microbiological Societies.

ICNV International Committee on Nomenclature of Viruses, superseded by the ICTV.

icosahedral symmetry (5-3-2 symmetry) The symmetry exhibited by an ICOSAHEDRON and by the capsid of certain types of virus: 5-fold rotational symmetry through each of the 12 apexes; 3-fold rotational symmetry about an axis through the centre of each of the 20 triangular faces; and 2-fold rotational symmetry about an axis through the centre of each of the 30 edges.

A viral capsid having *strict* icosahedral symmetry would be constructed from exactly 60 identical building units arranged in positions of exact equivalence on the surface of a hypothetical sphere, the units being held together via identical (strictly equivalent) interunit contacts throughout. In most viruses, however, the genome is too large to fit into a capsid composed of 60 reasonably sized units, and in many 'icosahedral' viruses the number of units which form the capsid is a *certain* multiple of 60 (n60); in some viruses (e.g. polioviruses) the n60 units include units of more than one type, while in other viruses all the units are identical. The construction of a symmetrical capsid with n60 units requires that the units be arranged in 60-unit sets, the members of each set being distributed throughout the capsid. In such an arrangement of (n60) units, the interunit contacts are not precisely identical throughout the capsid; however, since all interunit bonding involves the same general type of contact, the interunit bonds in such a capsid are described as 'quasi-equivalent'.

In the smaller icosahedral capsids consisting of n60 units, the units are (of necessity) arranged into 5- and 6-membered rings, each ring forming a tight cluster – referred to as a PENTAMER or *hexamer*, respectively – that may be distinguishable in electron micrographs; the clustering of units in this way maximizes contact between the units. In such capsids a pentamer occupies each of the 12 apexes. (The existence of 5- or 6-membered clusters does not necessarily mean that members of a given cluster are bound more strongly to each other than they are to adjacent units; thus, e.g. many viruses which have 180-unit capsids yield dimers (2-unit particles) when gently disrupted.) In some cases a pentamer or hexamer may appear, under the electron microscope, as a single entity – which may be referred to as a *morphological unit* or *capsomer* (= *capsomere*).

In those icosahedral viruses whose capsids comprise up to 240 building units, the value of *n* (in n60) can be 1, 3 or 4 (depending on virus); in these viruses, n (i.e., the number of 60-unit sets in the capsid) corresponds to the *triangulation number* (T number, or T). The large icosahedral viruses (e.g. adenoviruses) deviate from the geometrical and structural criteria obeyed by the smaller icosahedral viruses.

[Principles of virus structure: Book ref. 148, pp. 27–44.]

icosahedron A solid figure contained by 20 plane faces, all the faces being equilateral triangles of the same size. Many viruses have ICOSAHEDRAL SYMMETRY.

ICP (insecticidal crystal protein) See BIOLOGICAL CONTROL.

ICPB International Collection of Phytopathogenic Bacteria, Department of Bacteriology, University of California, Davis, California 95616, USA.

ICR compounds See ACRIDINES.

IcsA protein See DYSENTERY and PROTEIN SECRETION (type IV systems).

icteroanaemia (porcine) See EPERYTHROZOONOSIS.

icterus Jaundice.

ICTV International Committee on Taxonomy of Viruses. (cf. ICNV.)

id reaction The formation of skin lesions at sites remote from a focus of infection; such lesions are commonly sterile, and they may represent a local HYPERSENSITIVITY response to the infection. If the infecting organism is e.g. *Trichophyton* the reaction is termed trichophytid.

ID_{50} Infectious dose (50%): that dose of a given infectious agent which, when given to each of a number of experimental test systems or animals, brings about the infection of 50% of the systems/animals under given conditions.

IDAV See HIV.

idiogram See KARYOTYPE.

idiolite A product of SECONDARY METABOLISM.

idiopathic (of a disease) Of unknown cause.

idiophase In BATCH CULTURE, that phase in which SECONDARY METABOLISM is dominant (cf. TROPHOPHASE); it corresponds to the latter part of the log phase together with the stationary phase.

idiosome (*protozool.*) See XENOSOME (2).

idiotope See IDIOTYPE.

idiotype An Ig molecule defined by the sum total of antigenic determinants (*idiotopes*) on its variable (V_H and/or V_L) domains.

idli A type of steamed bread from India. It is made from a batter of milled rice and black gram (*Phaseolus mungo*) which is left to ferment overnight. The fermentation is carried out mainly by *Leuconostoc mesenteroides* which produces acid and gas; *Enterococcus faecalis*, and sometimes *Pediococcus cerevisiae*, contribute acidity, and yeasts may also be involved. The leavened batter is then steamed.

idling During DNA replication: repeated cycles of addition and excision of nuleotides at a 3′ terminal with no net synthesis or degradation of DNA. [Idling as a mechanism in lagging strand synthesis: GD (2004) *18* 2764–2773.]

idoxuridine (IDU; 5-iodo-2′-deoxyuridine; IUdR) An ANTIVIRAL AGENT which acts as an analogue of thymidine (5-methyl-2′-deoxyuridine). Selective toxicity is poor, and the drug is too toxic for systemic use; it may be used topically for the treatment of e.g. herpes simplex keratitis, but is being superseded by more recent antiviral drugs.

IDU IDOXURIDINE.

IEF ISOELECTRIC FOCUSING.

IEM IMMUNOELECTRON MICROSCOPY.

IEP (1) ISOELECTRIC POINT. (2) IMMUNOELECTROPHORESIS.

IF INTERMEDIATE FILAMENT.

IF-1, IF-2, IF-3 See PROTEIN SYNTHESIS.

IF_1 protein See PROTON ATPASE.

IFA test *Syn.* IFAT.

IFAT See indirect IMMUNOFLUORESCENCE.

IFN INTERFERON.

IFO Institute for Fermentation, 17–85 Juso-Honmachi, 2-Chome, Yodogawa-Ku, Osaka, Japan.

IFT *Syn.* IFAT.

Ig IMMUNOGLOBULIN.

IgA Immunoglobulin A, the predominant IMMUNOGLOBULIN in saliva, tears, colostrum and other body fluids, but comprising only ca. 10% of the total plasma Igs. The IgA monomer (MWt ca. 160000; S^{20}_{W} 7) contains *alpha* HEAVY CHAINS, *kappa* or *lambda* LIGHT CHAINS, and ca. 8% carbohydrate. Most IgA at mucosal surfaces (secretory IgA, sIgA, SIgA) is in dimeric form: two monomers are linked via a J CHAIN, and the whole is linked to the SECRETORY COMPONENT. Polymers larger than dimers are also formed. IgA can fix complement via the alternative pathway. There are two subclasses: IgA1 and IgA2. (See also entry IgA1 proteases.)

IgA1 proteases Various extracellular bacterial enzymes which specifically cleave the human IgA1 antibody at the heavy-chain hinge region – but which do not affect IgA2; some are sensitive to EDTA. These enzymes are synthesized by e.g. *Haemophilus influenzae*, *Neisseria gonorrhoeae*, *N. meningitidis* and *Streptococcus pneumoniae* – bacteria which are typically pathogenic at or via the mucosal surfaces; presumably, IgA1 proteases counteract the host's IgA1 defences, but the role of these enzymes in pathogenesis has not been established.

At least some IgA1 proteases are secreted by a type IV system (see PROTEIN SECRETION).

IgD Immunoglobulin D, an IMMUNOGLOBULIN present in very low concentrations in plasma; it occurs mainly as a surface receptor for antigen on the membranes of B LYMPHOCYTES. The IgD molecule (MWt ca. 175000; S_W^{20} 7) contains *delta* HEAVY CHAINS, *kappa* or *lambda* LIGHT CHAINS, and ca. 12–13% carbohydrate.

IgE Immunoglobulin E, an IMMUNOGLOBULIN found in minute quantities in plasma. The IgE molecule (MWt ca. 190000; S_W^{20} 8) contains *epsilon* HEAVY CHAINS, *kappa* or *lambda* LIGHT CHAINS, and ca. 12% carbohydrate. IgE binds to basophils and mast cells via its Fc portion, causing degranulation in the presence of specific allergen and thus effecting a TYPE I REACTION; cytophilic activity is abolished e.g. by heating (56°C/3–4 hours).

IgG Immunoglobulin G, the predominant IMMUNOGLOBULIN in plasma (ca. 9–16 mg/ml, 75% of total plasma Ig). The IgG molecule (MWt ca. 150000; S_W^{20} 7) contains *gamma* HEAVY CHAINS, *kappa* or *lambda* LIGHT CHAINS, and ca. 3% carbohydrate. There are four subclasses – IgG1, IgG2, IgG3 and IgG4 – distinguished by differences in amino acid sequences and by serology; human IgG subclasses are numbered in order of their concentrations in plasma: IgG1 comprises ca. 65% of plasma IgG. IgG1 and IgG3 are bivalent COMPLEMENT-FIXING ANTIBODIES (half-life ca. 21 and 7 days, respectively) which are important e.g. as opsonins and antitoxins in extravascular regions as well as in the bloodstream; in humans, they can cross the placenta and are important in the protection of the fetus and neonate. (Efficient binding of complement component C1 appears to require an IgG doublet, i.e. two molecules side by side; a single IgG molecule binds C1 only weakly.) IgG antibodies generally appear to predominate in the secondary (anamnestic) humoral reponse to antigen. (See also ALLOTYPE.)

IgG protease See e.g. ELASTASE.

IgM (macroglobulin) Immunoglobulin M, a minor plasma IMMUNOGLOBULIN (comprising ca. 5–10% of total plasma Ig); phylogenetically, IgM is apparently the most primitive class of Igs. The IgM monomer (MWt ca. 180000) contains *mu* HEAVY CHAINS, *kappa* or *lambda* LIGHT CHAINS, and ca. 12% carbohydrate; monomeric IgM occurs as a surface receptor for antigen on the B LYMPHOCYTE membrane. The typical form of IgM is a pentamer (S_W^{20} 19) in which the five monomers are arranged radially (Fc portions directed towards the centre) and connected via disulphide bonds, with a J CHAIN at the centre of the polymer; in the free (uncombined) form, an IgM antibody is therefore stellate, but pentameric IgM is sufficiently flexible to assume a spider-like conformation in order to combine with multiple-repeating antigens on a surface. (IgM antibodies are thus particularly good agglutinators of those antigens, e.g. bacteria, which display a pattern of repeated antigenic determinants.) IgM antibodies are COMPLEMENT-FIXING ANTIBODIES with a theoretical valency of 10. Maximum valency is generally exerted only with small haptens; with larger antigens the effective valency is often only 5. (It appears that an IgM antibody must bind antigen at more than one of its combining sites in order to be able to fix complement [Mol.Immunol. (1981) *18* 863–868, cited in Book ref. 42, p. 647].) In man, IgM does not pass through blood vessel walls into the tissues, and does not cross the placenta; its half-life is ca. 5 days. IgM antibodies are often the first to be formed in the humoral response to antigen; the humoral response to THYMUS-INDEPENDENT ANTIGENS (such as purified pneumococcal polysaccharides) generally involves, almost exclusively, the formation of IgM antibodies. The Wassermann antibody (see WASSERMANN REACTION) and the rheumatoid factor (see RHEUMATOID ARTHRITIS) are largely or exclusively IgM antibodies.

IGS (*mol. biol.*) Internal guide sequence: see SPLIT GENE (b).

IHF INTEGRATION HOST FACTOR.

IκBα See CYTOKINES.

IL Abbreviation for interleukin. See separate entries for INTERLEUKIN-1 (IL-1), INTERLEUKIN-2 (IL-2) etc.

IL-1 INTERLEUKIN-1.

IL-1γ *Syn.* INTERLEUKIN-18.

IL-2 INTERLEUKIN-2.

ilarviruses ('isometric labile ringspot viruses'; tobacco streak virus group) A group of tripartite ssRNA-containing PLANT VIRUSES which have a wide host range; transmission occurs via seeds and pollen and can readily occur mechanically. Type member: tobacco streak virus; other members include e.g. apple mosaic virus, *Citrus* leaf rugose virus, *Citrus* variegation virus, *Prunus* necrotic ringspot virus, spinach latent virus, Tulare apple mosaic virus.

Virions: quasi-isometric, occasionally bacilliform; several types occur, differing in size (ca. 26–35 nm diam.) and in S_W^{20}. Genome: three molecules of linear positive-sense ssRNA: RNA1 (MWt ca. 1.1×10^6), RNA2 (MWt ca. 0.9×10^6) and RNA3 (MWt ca. 0.7×10^6); the coat protein mRNA ('RNA4') is also encapsidated. RNAs 1, 2 and 3, together with coat protein or RNA4, are required for infectivity (cf. ALFALFA MOSAIC VIRUS).

ileocaecal syndrome *Syn.* NEUTROPENIC ENTEROCOLITIS.

Ilheus virus See FLAVIVIRIDAE.

illegitimate name In NOMENCLATURE: a validly published name which contravenes any rule(s) of the relevant nomenclatural code.

illegitimate recombination Fortuitous RECOMBINATION which occurs between DNA sequences which are non-homologous or which have very short regions of homology. In *Escherichia coli* illegitimate recombination is *recA*-independent and may be mediated by GYRASE: strands cleaved during gyrase action at different sites may undergo crossing over by the exchange of gyrase subunits covalently bound to the cut ends [model: PNAS (1983) *80* 2452–2456].

imazalil An agricultural systemic ANTIFUNGAL AGENT, an imidazolyl derivative which acts as an inhibitor of sterol biosynthesis, thereby disrupting the cytoplasmic membrane in susceptible fungi. Imazalil is used – mixed with e.g. guazatine or thiophanate-methyl – as a seed treatment against many seed-borne pathogens of cereals (e.g. *Ustilago* and *Pyrenophora* spp). It is also effective against *Penicillium* diseases of citrus fruits.

imbricated Overlapping, like tiles on a roof.

ImD unit See MICROCOMPLEMENT FIXATION.

IMF INTERMEDIATE FILAMENT.

imf phosphorylation See ELECTRON TRANSPORT PHOSPHORYLATION.

Imhoff tank A large tank formerly widely used for SEWAGE TREATMENT. Essentially, crude sewage flows into a sludge settlement chamber located within a larger tank; sludge falls from a slot in the settlement chamber into the main digester compartment where it undergoes ANAEROBIC DIGESTION. The tank is vented for the escape of gases.

IMI Imperial Mycological Institute, the former name of the Commonwealth Mycological Institute, Kew, UK.

imidazole antifungal agents See AZOLE ANTIFUNGAL AGENTS.

imipenem See THIENAMYCIN.

immediate hypersensitivity (*immunol.*) HYPERSENSITIVITY mediated by humoral antibodies: see TYPE I REACTION, TYPE II REACTION, TYPE III REACTION, TYPE V REACTION (cf. DELAYED HYPERSENSITIVITY). Immediate hypersensitivity reactions typically occur within minutes or hours of contact with specific antigen; they are also referred to as *immediate-type* hypersensitivity reactions to include those which require hours or days to develop.

immediate-type hypersensitivity See IMMEDIATE HYPERSENSITIVITY.

Immedium filter See SEWAGE TREATMENT.

immersion lens See RESOLVING POWER.

immersion oil Cedarwood or synthetic oil (refractive index ca. 1.5) used with oil-immersion objectives (see RESOLVING POWER). In *homogeneous immersion* the immersion oil, objective lens, and cover-glass all have the same refractive index. Oil should be removed from the objective with a lens tissue immediately after use; unless otherwise recommended, a suitable solvent is 1,1,1-trichloroethane, but benzene or xylene can be used. Prolonged contact with solvents may damage lens mountings.

immobilization (1) (*biotechnol.*) Microbial, plant or animal cells, or macromolecules, may be 'immobilized' by attachment to solid structures, incorporation in gels etc for use in e.g. BIOCONVERSIONS (sense 1) – in some cases on an industrial scale. Various immobilized macromolecules are used e.g. in procedures involving affinity CHROMATOGRAPHY.

Enzymes are immobilized because soluble enzymes are difficult to separate from substrates or products (and, hence, cannot be re-used in industrial processes), and most are unstable under conditions of use. The main methods of enzyme immobilization are: (a) Binding to a solid carrier or support. Supports used for covalent binding include e.g. cellulose, ceramic, glass, steel and synthetic polymers; usually, the support is 'activated' (see e.g. CYANOGEN BROMIDE), and the enzyme is then allowed to bind (commonly via its amino or carboxyl groups) to the activated support. The active site of the enzyme can be protected by allowing binding to occur in the presence of the enzyme's substrate. Supports used for *ionic* binding include ion exchangers such as DEAE-cellulose. (b) Cross-linking with bifunctional reagents to form insoluble aggregates. Reagents used include GLUTARALDEHYDE (which binds enzymes via their amino groups), or diamines (e.g. hexamethylenediamine) which bind enzymes via their carboxyl groups after these groups have been 'activated' with carbodiimides; e.g., immobilized glucose isomerase is manufactured by treating pellets of homogenized cells of *Bacillus coagulans* with glutaraldehyde, yielding water-insoluble aggregates containing glucose isomerase and other proteins. (c) Encapsulation. Enzymes are enclosed within LIPOSOMES or hollow fibres which are permeable to low-MWt substrates and products. (d) Entrapment within polymeric gels such as calcium ALGINATE, κ-CARRAGEENAN and polyacrylamide. Enzymes are added to a solution of the polymer which is then gelled e.g. by the addition of a gelling agent. Leakage of enzymes from the gel may be counteracted by cross-linking them e.g. with glutaraldehyde. Entrapment is suitable primarily for bioconversion of low-MWt substrates which can diffuse through the gel.

Immobilized enzymes are used for simple one-step or two-step bioconversions in which there is no need for regeneration of coenzymes. [Potential for immobilization of coenzyme-dependent enzymes: PTRSLB (1983) *300* 335–367.] In some cases immobilization can improve the stability of enzymes which may otherwise be inadequate under operational conditions; e.g. immobilization of proteases inhibits their mutual degradation, and multipoint binding of an enzyme to its support helps to retain the conformational integrity of the enzyme – protecting it from inactivation by various denaturing agents (e.g. heat).

Immobilized cells can be used for bioconversions which are not possible with isolated enzymes; the use of cells saves the cost and labour of preparing purified immobilized enzymes. Cells can be immobilized by the methods used for enzymes (see above), but entrapment is the most commonly used method; gels used include e.g. agar, alginate, κ-carrageenan, polyacrylamide, and polyurethane. Dead cells may be used for simple one-step or two-step reactions (provided they retain specific enzymic activity), but living cells are necessary for multistep transformations in which several enzymes act sequentially and/or in which there is a need for the regeneration of cofactors. Examples of the industrial use of dead immobilized cells include e.g. *Escherichia coli* in a κ-carrageenan gel for the production of L-aspartic acid from ammonium fumarate, and '*Brevibacterium flavum*' in a κ-carrageenan gel for the production of L-malic acid from fumaric acid. High cell densities can be achieved with immobilized living cells. Thus, e.g., in one process used for the production of ethanol, *Saccharomyces cerevisiae* ('*S. carlsbergensis*') is immobilized in carrageenan, and 'beads' of the gel (containing 3.5×10^6 cells/ml) are incubated in a shaken nutrient medium for 60 hours: cell density rises to over 5×10^9 cells/ml – ca. 10 times higher than that possible in a conventional broth culture – with the yeast cells densely packed within the surface layer of each bead. When such beads are packed into a column and fed with a glucose-containing nutrient medium they produce ethanol efficiently for long periods of time. Other uses of immobilized living cells include the production of hydrogen from cells of *Anabaena cylindrica* bound to glass beads, and the production of an extracellular α-amylase by *Bacillus subtilis* immobilized in a polyacrylamide gel.

[Book ref. 3, pp. 203–222, and PTRSLB (1983) *300* 369–389 (cells); Book ref. 31, pp. 331–367 (enzymes); Science (1983) *219* 722–727.]

(2) The inhibition of MOTILITY in a microorganism – e.g. by a specific antiserum (see IMMOBILIZATION TEST).

immobilization test Any test which involves the inhibition of MOTILITY in a microorganism. Immobilization tests can be used e.g. to detect specific antibodies (see e.g. TPI TEST) or organisms (e.g. a given strain of *Vibrio cholerae* may be identified by determining which types of specific antibody inhibit its motility). Antibodies may immobilize a motile organism e.g. by mediating the adhesion of flagella, or by promoting IMMUNE CYTOLYSIS.

immune (1) (*adj.*) Refers to the state of IMMUNITY (sense 2 or 3). (2) (*noun*) See EPIDEMIOLOGY.

immune adherence A phenomenon in which antigen–antibody complexes that have triggered COMPLEMENT FIXATION adhere firmly to receptors on phagocytes; the enhanced adhesion (which adds to that due to direct antigen–antibody–phagocyte interaction) is due mainly to component C3b.

The phrase 'immune adherence' is sometimes used also to refer to the totality of pathogen–phagocyte adherence, i.e. including both antibody-mediated and complement-mediated adhesion.

immune complex An antigen–antibody complex. Small, soluble (i.e. non-precipitating) immune complexes are formed e.g. under ANTIGEN EXCESS conditions. (See also TYPE III REACTION.)

immune cytolysis The lysis of a cell which has been sensitized (SENSITIZATION sense 3) and which has triggered COMPLEMENT FIXATION; lysis may result directly or indirectly from the formation of a pore, in the membrane, created by the membrane attack complex. (cf. REACTIVE LYSIS.)

immune deficiency-associated virus See AIDS.

immune electron microscopy IMMUNOELECTRON MICROSCOPY.

immune globulin Antiserum (or its immunoglobulin fraction) containing antibodies to a particular antigen or antigens.

immune haemolysis IMMUNE CYTOLYSIS of erythrocytes.
immune lysis *Syn.* IMMUNE CYTOLYSIS.
immune response The response(s) of a person or animal to immunological contact with an antigen; such responses may involve e.g. priming (see PRIMED), ANTIBODY FORMATION, an ANAMNESTIC RESPONSE, a HYPERSENSITIVITY reaction, or IMMUNOLOGICAL TOLERANCE. (See also ANTIGENIC COMPETITION.)
immune serum *Syn.* ANTISERUM.
immune surveillance *Syn.* IMMUNOSURVEILLANCE.
immune tolerance See IMMUNOLOGICAL TOLERANCE.
immunity (1) A state characterized by the tendency of the body to reject, eliminate or otherwise counteract foreign ('non-self') or seemingly foreign materials, or organisms, on or within its tissues; in this (broadest) sense immunity includes AUTOIMMUNITY, HYPERSENSITIVITY and IMMUNOLOGICAL TOLERANCE as well as other forms of SPECIFIC IMMUNITY, and NON-SPECIFIC IMMUNITY. (2) The state of an individual who, having had prior contact with a given antigen, can react to further contact with that antigen more vigorously than can a non-immune individual; in this sense immunity includes all the (beneficial and harmful) manifestations of HUMORAL IMMUNITY and CELL-MEDIATED IMMUNITY, including HYPERSENSITIVITY. (3) (protective immunity; functional immunity) Relative insusceptibility to specific harmful agents (e.g. pathogens) involving only the beneficial effects of humoral immunity and/or cell-mediated immunity.
immunity breakdown See COLICINS.
immunity groups See e.g. COLICIN E.
immunity protein See e.g. COLICIN PLASMID; COLICINS; KILLER FACTOR.
immunization (1) Any procedure in which specific microorganisms and/or antigenic materials (see VACCINE), or pre-formed antibodies, are introduced into the body in order to bring about specific *protective* IMMUNITY. (cf. INOCULATION (2) and VACCINATION.) Stimulation of the body's own immune response (by administration of a vaccine) is referred to as *active immunization*; the (parenteral) administration of pre-formed antibodies is called *passive immunization* (see also PASSIVE IMMUNITY). (2) The administration of antigen to elicit any form of SPECIFIC IMMUNITY. (See also ISOIMMUNIZATION; HETEROIMMUNIZATION; NON-SPECIFIC IMMUNIZATION.)
immunoadjuvant See ADJUVANT (1).
immunoassay Any procedure in which the specificity of the antigen–antibody reaction is used for detecting and quantifying antigens, antibodies, or substance(s). (See e.g. ELISA; ERYTHROIMMUNOASSAY; ENZYME IMMUNOASSAY; IMMUNORADIOMETRIC ASSAY; PREGNANCY TEST; RADIOIMMUNOASSAY.)
immunoblotting See BLOTTING.
immunocompromised Unable to exhibit a normal immune response.
immunoconglutinins Autoantibodies homologous to certain bound components of COMPLEMENT, particularly C3b; titres of serum immunoconglutinins are raised in long-term autoallergic conditions and in some diseases of microbial aetiology. Immunoconglutinins possibly play a role in the agglutination and phagocytosis of small, C3b-containing complexes. (cf. CONGLUTININ.)
immunocyte Any functional cell of the immune system.
immunocyte adherence technique (rosette technique) A method used e.g. for detecting cells which form specific antibodies. When such cells are incubated with homologous particulate antigen (or with erythrocytes coated with homologous soluble antigens) the antigenic particles adhere to specific receptor sites on the surfaces of the antibody-forming cells, forming rosettes. (See also PROTEIN A.)
immunodeficiency The inability to respond with a normal immune response to antigenic stimulation. (See also e.g. AIDS.)
immunodepression *Syn.* IMMUNOSUPPRESSION.
immunodiffusion See GEL DIFFUSION.
immunodominant See DETERMINANT and ANTIGENIC COMPETITION.
immunoelectron microscopy A procedure used e.g. for locating particular antigens in cells or tissues. Essentially, the given antigen is isolated and an antiserum is raised against it. The homologous antibodies are labelled with e.g. FERRITIN and then allowed to combine with the antigen in the cells or tissue; on examination by transmission ELECTRON MICROSCOPY the location of the given antigen is indicated by the ferritin label. (The use of ferritin in this way is called the *immunoferritin technique*.) (cf. IMMUNOSORBENT ELECTRON MICROSCOPY; see also IMMUNOGOLD TECHNIQUE.)
immunoelectrophoresis (IEP) Any procedure in which antibodies (or other serum proteins) and/or antigens are subjected to ELECTROPHORESIS prior to their detection, characterization, or quantification (e.g. by means of specific antigen–antibody precipitation or staining). In one form of IEP the serum (or other) sample is first subjected to electrophoresis in a strip of agarose (or other) gel. This separates components into discrete zones – each zone containing one or more proteins characterized by a specific electrophoretic mobility. (Some proteins migrate in an unexpected direction – see ELECTROENDOSMOSIS.) A rectangular slot ('trough'), cut into the gel strip parallel to the line of electrophoretic migration, is filled with antiserum containing antibodies to some or all of the proteins in the sample; the proteins and their homologous (and/or cross-reacting) antibodies diffuse through the agar and form lines or arcs of precipitate where they meet in OPTIMAL PROPORTIONS (see GEL DIFFUSION). (See also CHARGE-SHIFT IMMUNOELECTROPHORESIS; COUNTERCURRENT IMMUNOELECTROPHORESIS; ROCKET IMMUNOELECTROPHORESIS; TWO-DIMENSIONAL ELECTROPHORESIS.)
immunoferritin technique See IMMUNOELECTRON MICROSCOPY.
immunofluorescence The use of a fluorescent antibody (i.e. antibody conjugated with a FLUOROCHROME such as FITC, FLUORESCEIN, or RHODAMINE isothiocyanate) for the detection and/or quantification of a specific antigen (or antibody). In the simplest (*direct*) immunofluorescence techniques, used to detect antigen, the specimen (tissue section, smear etc) is exposed to the conjugate (i.e. dye-linked antibody) for an appropriate time, washed free of conjugate, and examined by fluorescence MICROSCOPY; antigens homologous to the fluorescent antibodies are readily identified by regions of fluorescence in the specimen. In the *indirect* immunofluorescence technique (e.g. the indirect fluorescent antibody test, IFAT) the presence of a given antigen can be determined by first exposing the specimen to *unstained* antibodies (e.g. a serum containing antibodies homologous to the given antigen); the specimen is then washed free of uncombined antibodies, exposed to fluorescent anti-Ig antibodies, washed, and examined by fluorescence microscopy. Any antigen–antibody combination on the specimen can thus be detected by the presence of fluorescent antibodies (which bind to the unstained antibodies); if particular antigens are known to be present in the specimen, the test can be used to detect homologous antibodies in the serum. Another type of indirect immunofluorescence technique is referred to as the SANDWICH TECHNIQUE.

In all immunofluorescence techniques adequate controls are necessary to preclude an interpretation based on the non-specific localization of conjugate, and to take into account any *primary* FLUORESCENCE.

(See also FLOW MICROFLUOROMETRY.)

immunogen (1) *Syn.* ANTIGEN. (2) An antigen which can stimulate *protective* immunity.

immunogenicity The capacity to function as an IMMUNOGEN. (Hence *adj.* immunogenic.)

immunoglobulin superfamily A category of CELL ADHESION MOLECULES characterized by a molecular structure homologous to that of immunoglobulins – i.e. a number of immunoglobulin domains, each containing an intra-domain disulphide bond. These molecules are involved e.g. in calcium-independent cell–cell adhesion and antigen-specific interaction.

The first member of the family to be isolated was the neural cell adhesion molecule (NCAM) which appears early in ontogeny and is important e.g. in the development of the central nervous system; binding is *homophilic*, i.e. NCAM binds to NCAM – in contrast to most members of the superfamily which bind to dissimilar molecules (i.e. *heterophilic* binding).

Other members are important in the immune system – e.g. the T CELL RECEPTOR and the CD4 and CD8 molecules. Interactions between T cells and other cells also involve the *leukocyte function-associated antigens* LFA-2 (= CD2) and LFA-3 (= CD58); CD2 occurs on T cells, while CD58 occurs on many types of cell. CD2 and CD58 bind to one another. (N.B. LFA-1 is a member of the integrin family: see CD11a/CD18.)

The *intercellular adhesion molecules* (ICAMs) have important roles in cell–cell adhesion. ICAM-1 (= CD54), ICAM-2 (CD102) and ICAM-3 (CD50) – widely distributed (e.g. on endothelium, monocytes, lymphocytes) – bind to integrin ligands (see CD11a/CD18). In at least some circumstances, cell-surface expression of ICAMs can be upregulated or down-regulated.

(See also CD31 and CD106.)

immunoglobulins (Igs) A class of proteins present e.g. in plasma, colostrum, tears and other body fluids; all antibodies are immunoglobulins (see ANTIBODY and ANTIBODY FORMATION). The (monomeric) form of an immunoglobulin consists of four polypeptide chains: two identical HEAVY CHAINS and two identical LIGHT CHAINS which are linked by interchain disulphide bonds to form a Y-shaped macromolecule. The two heavy chains are adjacent for part of their length, their N-terminal ends diverging (at the HINGE REGION) to form the two limbs of the Y; two or more disulphide bonds connect the heavy chains at the hinge region. One light chain runs alongside each of the two limbs of the Y, and is attached to it by a disulphide bond. The distal (N-terminal) end of each limb of the Y is, compositionally, a highly *variable region* – V_H and V_L referring to the so-called variable (distal) DOMAINS (*V regions*) of the heavy and light chains, respectively. The composition of the rest of the molecule is relatively constant, and is therefore termed the constant region (*C region*). The C region of each heavy chain comprises several DOMAINS which are designated C_H1, C_H2 etc (numbering from the V region end), while the constant region of the light chain consists of a single domain designated C_L.

Certain enzymes, e.g. PAPAIN, cleave the Ig monomer at the hinge region, producing two identical FAB PORTIONS and an FC PORTION. Pepsin cleaves monomeric Ig to form an $F(ab')_2$ portion (q.v.).

Five Ig classes are distinguished: see entries for IgA, IgD, IgE, IgG and IgM.

[Book ref. 42, pp. 131–219.]

immunogold technique A form of IMMUNOELECTRON MICROSCOPY in which the specific antibodies (or PROTEIN A molecules) are complexed with gold prior to use – the gold serving as an electron-dense marker.

immunoliposome See LIPOSOME.

immunological drift *Syn.* ANTIGENIC DRIFT.

immunological paralysis IMMUNOLOGICAL TOLERANCE – particularly that induced by pneumococcal polysaccharides.

immunological surveillance *Syn.* IMMUNOSURVEILLANCE.

immunological tolerance Inability to respond normally to a given antigen (by humoral and/or by cell-mediated mechanisms) as a consequence of prior exposure to that antigen; unrelated antigens may elicit normal responses in the same individual. (cf. SELF TOLERANCE.) Tolerance may be total (i.e. no detectable response to antigen) or may involve e.g. failure to produce certain class(es) of antibody or some type(s) of immune response. Failure to react normally to some antigens may be beneficial in that harmful HYPERSENSITIVITY reactions may be avoided.

The extent, duration, and type of acquired tolerance is influenced e.g. by the size of dose of the inducing antigen (the *toleragen*), the nature of the antigen, the route of administration, and the state of immunological maturity of the individual when tolerance is induced; the induction of tolerance is inhibited e.g. by adjuvants and is favoured e.g. by ANTIGENIC COMPETITION and by non-specific immunosuppression.

Typically, thymus-dependent antigens (e.g. proteins) can induce tolerance when administered in doses higher than the normal immunizing dose (*high-zone tolerance*) or lower than the normal immunizing dose (*low-zone tolerance*); in high-zone tolerance both specific B cell and T cell clones are unresponsive (*clonal deletion*), while in low-zone tolerance only the helper T cell clone is unresponsive. Thymus-independent antigens usually induce only high-zone tolerance. Direct access of antigens to the gastrointestinal tract favours the induction of tolerance (see also SULZBERGER–CHASE PHENOMENON). In general, tolerance to a given antigen is more readily induced in the fetus or neonate than in the adult, and in a naive subject than in a primed one. The duration of tolerance may be extended by repeated administration of antigen.

Diverse mechanisms of tolerance induction have been suggested (including e.g. enhancement of suppressor T cell activity), and different types of tolerance may involve different mechanisms. Exposure of *immature* B lymphocytes to specific antigen or to anti-Ig makes them subsequently unresponsive (i.e., unable to secrete antibody) when exposed to specific antigen (*clonal abortion*); this suggests a reason why tolerance can be induced more easily in the fetus or neonate than in adults.

immunological unresponsiveness (1) *Syn.* IMMUNOLOGICAL TOLERANCE. (2) Inability, or reduced ability, to respond to antigens in general, due e.g. to IMMUNOSUPPRESSION.

immunomagnetic separation See DYNABEADS.

immunomodulation Specific or generalized alteration of the immune response (see IMMUNOSTIMULATION; IMMUNOSUPPRESSION; IMMUNOLOGICAL TOLERANCE).

immunopathogenesis The development of pathological effects as a direct result of the immune response.

immunoperoxidase method Any method in which antibody-conjugated, or otherwise complexed, peroxidases (e.g. horseradish peroxidase) are used to locate specific antigens on cells or tissues (see e.g. ABC IMMUNOPEROXIDASE METHOD and PAP TECHNIQUE).

immunopotentiation (1) *Syn.* IMMUNOSTIMULATION. (2) Non-specific immunostimulation.

immunoprecipitable Capable of being precipitated by homologous antibodies.

immunoradiometric assay A highly sensitive IMMUNOASSAY by which specific antigens are quantified using radioactive antibodies – which may be prepared as follows. Antigen is adsorbed to an inert carrier (e.g. cellulose) and exposed to purified immunoglobulins obtained from an antiserum containing the homologous antibody; uncombined antibodies are removed by washing, and the combined antibodies are radiolabelled (e.g. with ^{125}I). The radioactive antibodies are then eluted for use in the assay. In one form of assay, excess radioactive antibody is added to the sample containing antigen at an unknown concentration; the reaction mixture is then exposed to fresh, cellulose-bound antigen which adsorbs the uncombined antibodies and permits their separation from the reaction mixture. The amount of combined antibody can then be measured by determining the level of radioactivity remaining in the reaction mixture; this allows the concentration of antigen in the sample to be determined from a previously prepared standard curve (radioactivity versus antigen concentration). (cf. RADIOIMMUNOASSAY.)

immunorecessive See DETERMINANT.

immunosilent See DETERMINANT.

immunosorbent assay Any IMMUNOASSAY in which antigen or antibody is immobilized by adsorption to a solid surface – see e.g. ELISA.

immunosorbent electron microscopy (ISEM) ELECTRON MICROSCOPY in which particulate antigens (e.g. specific viruses) are detected or examined by first complexing them with homologous antibodies; prior to antigen–antibody interaction the antibodies may be e.g. fixed to an electron microscope grid (Derrick's method, antibody-coated grid technique, A-CGT), adsorbed to a PROTEIN A-coated grid (Shukla's method, protein A-coated grid technique, PA-CGT), or adsorbed directly to the protein A of cells of *Staphylococcus aureus* (protein A-coated bacteria technique, PA-CBT). [Review: AVR (1984) *29* 169–194.] (cf. IMMUNOELECTRON MICROSCOPY.)

immunostaining The staining of a specific antigen or structure by any method in which the stain (or stain-generating system) is complexed with specific antibody (e.g. ABC IMMUNOPEROXIDASE METHOD).

immunostimulation Specific or non-specific potentiation of the immune response – e.g. by the administration of a VACCINE or by the use of an ADJUVANT. (cf. IMMUNOPOTENTIATION.)

immunosuppression (immune suppression) Complete or partial suppression of normal IMMUNE RESPONSES – either in respect of specific antigen(s) (see also IMMUNOLOGICAL TOLERANCE) or in respect of all antigens (generalized suppression). Immunosuppression can be induced e.g. by ionizing radiation; specific antimetabolites (e.g. the folic acid antagonist methotrexate); a variety of antimitotic/antitumour agents (e.g. 6-mercaptopurine, prednisone, CYCLOSPORIN A, cyclophosphamide, azathioprine); antilymphocyte serum. Immunosuppression induced by radiation or by drugs tends to be generalized.

immunosurveillance The continual monitoring of the tissues for abnormal antigens (e.g. on tumour cells) by cells of the immune system.

immunotoxin An antibody–toxin conjugate intended to destroy specific target cells (e.g. tumour cells) which bear antigens homologous to the antibody. [Anti-CD22 immunotoxin BL22, which uses a fragment of EXOTOXIN A, in chemotherapy-resistant HAIRY-CELL LEUKAEMIA: NEJM (2001) *345* 241–247.]

Imotest A form of TUBERCULIN TEST in which a disposable plastic unit is used. The tuberculin is contained in a sealed plastic tube which is first broken and then squeezed to inoculate the nine points used to penetrate the skin.

IMP Inosine 5′-monophosphate: see Appendix V(a).

impactor See AIR.

impedimetry A technique in which microbial growth and activity are measured in terms of changes in the electrical impedence of the growth medium due to the effects of metabolism. [JAB (1982) *53* 423–426.]

imperfect stage (*mycol.*) See ANAMORPH.

impetigo A superficial, highly infectious skin disease, common in children, caused by *Streptococcus pyogenes* and/or *Staphylococcus aureus*. In the streptococcal form, spreading, inflamed pustules develop, rupture, and form thick brownish-yellow crusts; lesions may be secondarily infected by staphylococci. In the staphylococcal form (*bullous impetigo*) the lesions contain watery fluid rather than pus, and a thin crust forms over the centre of the lesion. Bullous impetigo occurs most commonly in very young infants; superinfection with streptococci is rare. *Chemotherapy*: e.g. oral penicillin (for streptococcal impetigo) or erythromycin. [Book ref. 17, pp. 341–346.]

impinger See AIR.

IMS See DYNABEADS.

IMVEC tests IMVIC TESTS plus the EIJKMAN TEST.

IMViC tests Tests used for the identification of bacteria e.g. of the family Enterobacteriaceae: INDOLE TEST, METHYL RED TEST, VOGES–PROSKAUER TEST, CITRATE TEST. (cf. EIJKMAN TEST.)

in-block staining See ELECTRON MICROSCOPY (a).

in situ gene amplification See IN SITU PCR.

in situ hybridization See PROBE.

in situ PCR PCR-based amplification of a target inside intact cells. The cells may be initially attached to glass slides (e.g. with AAS in acetone) and are fixed (e.g. with a formalin-based solution). The cells are then treated (*permeabilized*) so as to allow entry of PCR reagents without encouraging the exit of PCR products. The reaction mixture is added to the cell layer (or tissue section), and is held in place with a cover slip. Temperature cycling may be carried out in a specialized form of THERMOCYCLER in which each slide is processed in a close-fitting slot that ensures good thermal contact (e.g. the GeneAmp® *In situ* PCR System 1000; Perkin-Elmer). After amplification, another phase of fixation promotes intracellular retention of the products.

Detection of products may be direct or indirect. Direct detection involves the use of labelled nucleotides or primers in the reaction mixture, and detection of the label histochemically. One disadvantage of this approach is that *all* amplified products will be labelled – even those resulting from mis-priming; as a consequence, the specificity of the assay is reduced. Improved specificity is obtained by the indirect mode of detection which involves the use of a labelled probe that is complementary to an internal sequence in the amplicon; the use of such a probe is equivalent to adding a stage of in situ hybridization (ISH), and procedures which involve indirect detection of products in this way have been referred to as PCR-ISH ('in situ PCR' being reserved for those procedures involving direct detection of products). Collectively, in situ PCR and PCR-ISH have been referred to as *in situ gene amplification* (IS-GA) [RMM (1997) *8* 157–169].

This approach has been widely used for studying virus-infected cells, and may be particularly useful e.g. for those viruses which remain latent in tissues in small numbers. One early study detected intracellular lentiviral DNA [PNAS (1990) *87* 4971–4975]. PCR-ISH has been used e.g. in studies on γ-herpesvirus type 8 [AJP (1997) *150* 147–153].

in vitro complementation assay An assay based on the COMPLEMENTATION of a defective (mutant) protein by an active (wild-type) form of that protein; such assays are used e.g. to determine which of several proteins is/are involved in a particular process. For example, to identify a protein involved in DNA synthesis, a cell-free DNA-synthesizing system is initially prepared from cells which have a temperature-sensitive mutation in a gene essential for DNA synthesis; such a system can synthesize DNA at permissive temperatures but not at restrictive temperatures. Proteins purified from a wild-type strain of the same species can then be added, separately, to the system to determine which of them can restore DNA-synthesizing activity at restrictive temperatures.

in vivo expression technology See IVET.

Inaba variant See VIBRIO (*V. cholerae*).

inactivated serum Serum which has been heated to 56°C for 30 min; heating inactivates COMPLEMENT in the serum.

inactivated vaccine (killed vaccine) Any VACCINE consisting of microorganisms which have been treated (e.g. with formalin) so that they are no longer capable of multiplying, although their PROTECTIVE ANTIGENS remain effective. (See e.g. SALK VACCINE; SEMPLE VACCINE.)

inactivation (of a microorganism) (1) The killing of a microorganism. (2) The temporary (reversible) inhibition of growth (or other activity) of a microorganism.

inactone See GLUTARIMIDE ANTIBIOTICS.

inaequihymeniiferous See LAMELLA.

Inc group See INCOMPATIBILITY.

***incB* locus** See F PLASMID.

IncC group See INCOMPATIBILITY.

***incC* locus** See F PLASMID.

incertae sedis Of uncertain taxonomic position.

IncF groups See INCOMPATIBILITY.

IncI groups See INCOMPATIBILITY.

incident light microscopy See MICROSCOPY (e) and (g).

inclusion blennorrhoea See OPHTHALMIA NEONATORUM.

inclusion bodies ('inclusions') Discrete structures of various types present (normally or abnormally) within cells.

In *virology*, the term generally refers to those structures which develop in certain virus-infected cells; such structures consist of virions and/or viral components and/or cellular material, and may be characteristic in form and location for a given type of virus. (See also CYTOPATHIC EFFECT; GUARNIERI BODIES; NEGRI BODIES; POLYHEDRA; VIROPLASM.)

In *bacteriology*, the term may refer to structures within a bacterial cell (e.g. granules of GLYCOGEN, POLY-β-HYDROXYBUTYRATE or POLYSULPHATE; 'hydrocarbon inclusions' (see HYDROCARBONS); CARBOXYSOMES; GAS VACUOLES; R BODY) or to bacterial cells within a (parasitized) eukaryotic host cell (see e.g. ANAPLASMA and CHLAMYDIA).

(See also OVERPRODUCTION.)

inclusion body hepatitis A POULTRY DISEASE apparently caused by certain strains of AVIADENOVIRUS (e.g. the 'Tipton' strain). The liver of an infected bird is enlarged, yellowish, with haemorrhagic patches; haemorrhages may also occur in the muscles. The disease may be associated with INFECTIOUS BURSAL DISEASE VIRUS infection.

inclusion body rhinitis A common, generally mild disease of young pigs, caused by suid herpesvirus 2 (see BETAHERPESVIRINAE). Infection occurs by inhalation; symptoms: e.g. sneezing, nasal discharge etc. The disease may become generalized, resulting in anaemia and sudden death.

inclusion conjunctivitis In adults: CONJUNCTIVITIS caused by *Chlamydia trachomatis*, commonly serotype D, E, F, G, H, I, J or K; if untreated, the disease (which is also called *paratrachoma*) typically resolves spontaneously after some months. (cf. TRACHOMA.) In newborn infants: see OPHTHALMIA NEONATORUM.

IncN group See INCOMPATIBILITY.

incompatibility (in plasmids) The inability of PLASMIDS to co-exist, stably, within the same cell when they have similar or identical systems for replication (see PLASMID) and/or PARTITION. Two incompatible plasmids which occupy the same cell will (in the absence of a selective pressure for both plasmids) tend to segregate to different cells during cell division. The stable intracellular co-existence of one plasmid with another requires that each plasmid be able to control, independently of the other, its own replication/partition such that it can establish and maintain a stable COPY NUMBER; the inability of a given plasmid to maintain a stable copy number in the presence of another plasmid is the characteristic feature of incompatibility.

In many cases, incompatibility arises from the synthesis of plasmid-encoded *trans*-acting elements involved in the negative control of plasmid replication. Since plasmids with identical replication systems form identical *trans*-acting elements, the elements synthesized by one plasmid will affect the replication of other plasmids (in the same cell) which have an identical replication system. In this case, plasmid replication initially occurs randomly – e.g., two (incompatible) plasmids will each have an equal chance of being replicated. Once replicated, however, a plasmid of one type will be present at a higher copy number and will then be more likely to be involved in subsequent rounds of replication; this effect tends to be cumulative and to give rise to progeny cells that contain only one type of plasmid ('homoplasmid segregants').

In plasmids which have similar but not identical replication systems, incompatibility may result in the preferential exclusion of one particular plasmid; for example, in a cell containing certain derivatives of the COLE1 PLASMID and of plasmid pMB1, the ColE1 derivative is rapidly excluded – possibly owing to differential sensitivity of the two plasmids to RNA I. (See also DISLODGEMENT and ONE-WAY INCOMPATIBILITY.)

In a number of plasmids the negative control elements which govern incompatibility are small RNA molecules: see e.g. RNA I under COLE1 PLASMID, and CopA under R1 PLASMID. In the pT181 family of *Staphylococcus aureus* plasmids, too, replication is regulated by negative control elements (countertranscripts – see ANTISENSE RNA); however, in the pT181 family of plasmids incompatibility is not governed by such elements because the negative control elements of different plasmids in this group are not interchangeable. Among pT181 plasmids incompatibility is due to the existence of a common (interchangeable) Rep protein (an initiator protein which promotes replication by binding to the origin) [MGG (1986) *204* 341–348].

Incompatibility and the classification of plasmids. The expression of incompatibility or compatibility between plasmids provides a useful criterion for classifying them. Thus, plasmids are classified into so-called *incompatibility groups* (Inc groups) in such a way that all the plasmids in a given Inc group express mutual incompatibility. Inc groups exist e.g. for plasmids which occur in enterobacteria, and separate Inc groups exist for plasmids which occur in other bacteria; some ('promiscuous') plasmids can occur e.g. in both enterobacteria and *Pseudomonas* spp, and some of these plasmids have been allocated to an Inc group in both the enterobacterial and pseudomonad grouping schemes.

Enterobacterial Inc groups are designated by the prefix 'Inc' followed by a letter and, sometimes, by a Roman numeral or a Greek letter – e.g. IncFII, IncIα. The letter (e.g. F, I) indicates the type of *conjugation* system specified by the TRANSFER OPERON; thus, e.g. an IncF plasmid specifies a conjugative system like that of the F PLASMID or of an F-like plasmid, and encodes either F pili or F-like pili (see PILI). The Roman numeral or Greek letter indicates a particular type of plasmid *replication* system; thus e.g. IncFI and IncFII plasmids have different replication systems (i.e., they are compatible) though they specify similar conjugation systems. Some examples of enterobacterial Inc groups are given below.

IncFI includes the F PLASMID, the COLICIN PLASMID ColV-K94, and the R PLASMID R386.

IncFII includes the R1 PLASMID, R6 and R100.

Inc$I\alpha$ (sometimes written IncI1 or IncI_1) includes ColIb-P9, the delta plasmid (see DELTA), and R64.

Inc$I\gamma$ includes R621a.

IncN includes N3, R46, and R269N-1.

IncX includes the R6K PLASMID.

Pseudomonas Inc groups are designated IncP-1 to IncP-13, the numbers referring to each of the 13 different types of *replication* system. The conjugation systems of *Pseudomonas* plasmids are less well known than those of enterobacterial plasmids, but the pili encoded by *Pseudomonas* plasmids have been characterized. [Pili encoded by *Pseudomonas* plasmids: JGM (1983) *129* 2545–2556; *Pseudomonas* R plasmids: Book ref. 198, pp. 265–293.] Some of the *Pseudomonas* Inc groups are given below.

IncP-2 includes the CAM PLASMID, the OCT PLASMID, and the R plasmid pMG1.

IncP-6 includes the R plasmid Rms149.

IncP-7 includes Rms148.

IncP-8 includes plasmid FP2 (which encodes resistance to mercury).

IncP-9 includes R2, the SAL PLASMID, and the TOL PLASMID.

IncP-10 includes R91.

IncP-11 includes RP8 and R151.

IncP-12 includes R716.

IncP-13 includes pMG25.

Shared enterobacterial and pseudomonad Inc groups. Some of the shared groups are given below, the enterobacterial Inc group being given first.

IncC ($\equiv$ IncP-3) includes R55.

IncP ($\equiv$ IncP-1) includes R68, R751, RK2, RP1 and RP4.

IncQ ($\equiv$ IncP-4) includes RSF1010. (A plasmid apparently related to this group, pFM739, has been isolated from *Neisseria sicca* [JGM (1986) *132* 2491–2496].)

incompatible (non-compatible) (*plant pathol.*) Refers to an interaction between a plant and a microorganism (e.g. an avirulent strain of a pathogen) which does not result in the development of disease in the plant. (cf. COMPATIBLE.)

incomplete antibodies (1) *Syn.* BLOCKING ANTIBODIES (sense 1). (2) FAB PORTIONS.

incomplete Freund's adjuvant See FREUND'S ADJUVANT.

IncP groups See INCOMPATIBILITY.

IncQ group See INCOMPATIBILITY.

incubation The maintenance of e.g. inoculated media or other types of material at a particular ambient temperature over a period of hours, days, weeks, or longer; the purpose of incubation is to provide conditions suitable for growth etc. Incubation is commonly carried out within closed, thermally insulated, thermostatically controlled chambers (*incubators*) or within vessels suspended in a WATER BATH; refrigerated incubators maintain the lower range of temperatures. Specialized incubation permits control of humidity, light, radiation etc as well as temperature. (cf. PHYTOTRON.)

Cultures or specimens are sometimes left outside the incubator, i.e., they are exposed to the (often variable) conditions within the laboratory – a procedure known as 'incubation at room temperature'.

incubation period (*med., vet.*) The time interval between INFECTION (sense 1), or e.g. exposure to a TOXIN, and the appearance of the first symptoms of disease. (cf. PREPATENT PERIOD.)

incubator See INCUBATION.

IncX group See INCOMPATIBILITY.

independent assortment (*genetics*) During e.g. MEIOSIS: the *chance* distribution of unlinked genes (see LINKAGE) among progeny cells. Thus, e.g., if the (diploid) parent cell contained the unlinked allelic pairs *A/a* and *Z/z*, any one of the four possible allele combinations (*A* and *Z*, *A* and *z*, *a* and *Z*, *a* and *z*) may occur, with equal probability, in a given (haploid) progeny cell.

indicator (pH) See PH INDICATOR.

indicator organism Any organism whose presence and/or numbers serve to indicate the condition or quality of a material or environment. For example, in studies on faecal pollution in rivers, counts of the faecal bacteria *Escherichia coli* and *Enterococcus faecalis* are often used to indicate the degree of sewage pollution at given sites. The same organisms are also used for testing the efficiency of disinfection in treated drinking water, their presence in treated water supplies indicating a failure of the treatment process or contamination by sewage effluent subsequent to treatment. These organisms are used as indicators because they are common intestinal bacteria: their presence in water signals the *potential* presence of enteric pathogens. Water samples are not routinely tested for each of the (many) different types of enteric pathogen because, even in sewage-contaminated water, a given pathogen may occur only intermittently. Moreover, in sewage-contaminated water the indicator bacteria greatly outnumber pathogens so that they permit detection of very low levels of faecal pollution. Endospores of *Clostridium perfringens*, which remain viable for long periods of time, have been used to indicate *previous* contamination. *Bifidobacterium* spp may specifically indicate *human* faecal pollution [JAB (1983) *55* 349–357], while faecal pollution from human and animal sources may be differentiated by differences in the antibiotic resistance patterns of 'faecal streptococci' [AEM (1996) *62* 3997–4002].

Traditional tests for faecal indicator bacteria include the MULTIPLE-TUBE METHOD (for *E. coli*) and membrane FILTRATION for FAECAL STREPTOCOCCI.

Indicator organisms are also monitored in the assessment of food-handling conditions/hygiene [indicator organisms in meat: J. Food Protect. (1984) *47* 672–677].

(See also SAPROBITY SYSTEM and *lichens and pollution* in the entry LICHEN.)

indigoidine See ERWINIA (*E. chrysanthemi*).

indinavir See ANTIRETROVIRAL AGENTS.

indirect agglutination *Syn.* PASSIVE AGGLUTINATION.

indirect fluorescent antibody test (IFAT) See indirect IMMUNOFLUORESCENCE.

indirect immunofluorescence See IMMUNOFLUORESCENCE.

individual quick blanch (IQB) In the preparation of certain foods for CANNING: blanching by exposure of individual items of food to steam for a short period of time.

indole 3-acetic acid See AUXINS.

indole test An IMVIC TEST used to determine the ability of an organism to produce indole from tryptophan [see Appendix IV(f)]. The organism is grown in PEPTONE WATER or TRYPTONE WATER at 37°C. (N.B. The presence of carbohydrate (which may repress TRYPTOPHANASE) or nitrite in the medium can give false-negative results.) Test procedures vary. (a) KOVÁCS' INDOLE REAGENT (0.5 ml) is added to ca. 5 ml of e.g. a 48-hour culture with gentle shaking; the reagent forms a floating layer which, in a positive test (i.e. indole produced), becomes pink or red. (b) Xylene (ca. 1 ml) is added to ca. 5 ml of e.g. a 48-hour culture with vigorous shaking (to extract the indole). EHRLICH'S REAGENT (ca. 0.5 ml) is then allowed to run down the side of the slanted tube and forms a layer beneath the (floating) xylene; in a positive test a red colour develops in the reagent layer. (c) An oxalic acid-impregnated paper strip is inserted between the tube and stopper *before* incubation; during incubation (up to 7 days) indole (volatile at 37°C) reacts with the test strip to give a pink or red colour.

indophenol oxidase test See OXIDASE TEST.

indoxyl acetate (3-acetoxyindole) A reagent used in identification tests for e.g. species of *Arcobacter* and *Campylobacter*. Growth (from a non-selective medium) is suspended in distilled water, and a paper disc, impregnated with indoxyl acetate, is added; on incubation at room temperature for 20 minutes, a blue coloration indicates hydrolysis of indoxyl acetate.

inducer exclusion (catabolite inhibition) The inhibition by glucose (or certain other readily metabolizable, 'preferred' substrates) of the uptake of certain substrates by bacteria; for example, if the bacteria are growing in a medium containing two substrates, one of the substrates is preferentially taken up and metabolized, and this substrate prevents the uptake of the other.

There are several mechanisms for inducer exclusion; for exclusion of e.g. lactose in enterobacteria see CATABOLITE REPRESSION. Uptake of other (structurally unrelated) non-PTS sugars – e.g. maltose, melibiose – is inhibited in a similar way.

In other cases, inducer exclusion may be due to competition between structurally related substrates for the same transport system or for shared components of the PTS; for example, in enterobacteria glucose can directly inhibit the uptake of galactose via a transport system common to both sugars.

In certain Gram-positive bacteria (e.g. *Streptococcus* spp, *Lactobacillus casei*), accumulated sugar phosphates are actually expelled from the cell when e.g. glucose is added (*inducer expulsion*); this involves a two-step mechanism in which intracellular dephosphorylation of the sugar is followed by efflux of the free sugar. In these organisms regulation of sugar accumulation has been associated with the CcpA system (see CATABOLITE REPRESSION); phosphorylation of HPr at the Ser-46 site reduces its ability to be phosphorylated at the (routine) His-15 site by enzyme I, and this should reduce phosphate transfer to the PTS permeases and (hence) inhibit the uptake of sugars.

inducer expulsion See INDUCER EXCLUSION.

inducer protein *Syn.* activator: see OPERON.

induction (1) (of enzymes, gene expression) See e.g. OPERON. (2) (of bacteriophage) See LYSOGENY.

industrial alcohol (microbiological aspects) Large amounts of ethanol are produced commercially by the microbial fermentation of carbohydrates; ethanol is used primarily as a fuel (e.g. in GASOHOL), as a solvent, and as a feedstock in the chemical industry. 'Fermentation ethanol' (= 'fuel ethanol') may be produced by the ALCOHOLIC FERMENTATION of substrates such as MOLASSES and STARCH hydrolysates by strains of the yeast *Saccharomyces*; laboratory- and pilot-scale evaluations have been made of the economics of ethanol production from materials as diverse as cassava, Jerusalem artichoke, LIGNOCELLULOSE, sweet sorghum and WHEY – using various yeasts, certain bacteria, and mixtures of organisms.

Since many organisms can utilize only mono- or disaccharides, polymers such as cellulose and starch must be subjected to acid or enzyme-catalysed hydrolysis before they can be fermented; strains of e.g. *Clostridium thermocellum* (which form extracellular CELLULASES) can use lignocellulose without prior hydrolysis, but the potential of this organism for ethanol production is limited by its poor tolerance of ethanol. On hydrolysis, most usable polysaccharides yield mainly hexoses – from which ethanol can be formed as a major product by various yeasts (including e.g. species of *Candida*, *Kluyveromyces*, *Saccharomyces* and *Schizosaccharomyces*) and by some bacteria (e.g. *Zymomonas mobilis*). The hydrolysis of HEMICELLULOSES yields significant quantities of PENTOSES, e.g. xylose; some organisms (e.g. *Fusarium oxysporum*) can convert xylose to xylulose, and can ferment the latter to ethanol via reactions of the HEXOSE MONOPHOSPHATE PATHWAY and the EMBDEN–MEYERHOF–PARNAS PATHWAY. Certain other organisms (e.g. *Candida tropicalis*, *Pachysolen tannophilus*) can also form ethanol from xylose, while a number of organisms (e.g. *Saccharomyces*, *Schizosaccharomyces* spp) can form ethanol from xylulose. A strain of *Paecilomyces* can give good yields of ethanol from a range of hexoses and pentoses [Nature (1986) *321* 887–888].

industrial microbiology For examples of microorganisms and/or microbial products used on an industrial or commercial scale see e.g. AGAR, ALCOHOLIC BEVERAGES, ALGINATE, ASCORBIC ACID, BIOLOGICAL CONTROL, BIOMIMETIC TECHNOLOGY, BIOPOL, CARRAGEENAN, CITRIC ACID, COCOA, COFFEE, DAIRY PRODUCTS, DEXTRANS, DIATOMACEOUS EARTH, ENZYMES, ERGOT ALKALOIDS, FERMENTED FOODS, FUNORAN, FURCELLARAN, GELLAN GUM, GLUCONIC ACID, GLUTAMIC ACID, INDUSTRIAL ALCOHOL, INTERESTERIFICATION, KELP, LEACHING, MUSHROOM (senses 3 and 4), OVERPRODUCTION, RETTING, RIBOFLAVIN, SILAGE, SINGLE-CELL PROTEIN, STEROID BIOCONVERSIONS, VINEGAR, VITAMIN B_{12}, XANTHAN GUM and ZEARALENONE.

infant botulism See BOTULISM (a).

infantile gastroenteritis See EPEC.

infantile paralysis *Syn.* POLIOMYELITIS.

infection (1) The initial entry of a pathogen into a host. (2) The condition in which a pathogen has become established in or on the cells or tissues of a host; such a condition does not necessarily constitute or lead to disease. (3) Synonymous with disease of microbial aetiology. (See also CATHETER-ASSOCIATED INFECTION and TRANSFUSION-TRANSMITTED INFECTION.) (4) The establishment of a microbial symbiont in a host organism (see e.g. ROOT NODULES).

infection-associated haemophagocytic syndrome (IAHS) See HAEMOPHAGOCYTIC SYNDROME.

infection court (*plant pathol.*) The site at which a pathogen or parasite initiates infection of a plant (e.g. a wound).

infection immunity *Syn.* PREMUNITION.

infection peg See APPRESSORIUM.

infection thread See ROOT NODULES.

infectious abortion (*vet.*) See e.g. BRUCELLOSIS, CAMPYLOBACTERIOSIS, ENZOOTIC ABORTION.

infectious ascites (of fish) See INFECTIOUS DROPSY.

infectious bovine keratoconjunctivitis See INFECTIOUS KERATITIS.

infectious bovine rhinotracheitis (red nose) A highly infectious CATTLE DISEASE caused by strains of bovine herpesvirus 1 (see HERPESVIRIDAE). (Rarely, pigs may be affected.) Infection occurs mainly by droplet inhalation. Incubation period: 3–7 days or longer. Onset is sudden, with anorexia, fever, necrotic lesions on the mucous membranes of the nasal septum, serous discharge from the eyes and nose, increased salivation etc; sudden death may result from obstructive bronchiolitis. Abortion is a common sequel. [Book ref. 33, pp. 798–804.]

infectious bronchitis See AVIAN INFECTIOUS BRONCHITIS.

infectious bulbar paralysis *Syn.* AUJESZKY'S DISEASE.

infectious bursal disease virus (IBDV) A virus which causes severe inflammation of the BURSA OF FABRICIUS in chickens. IBDV closely resembles IPN VIRUS in morphology, genome etc.

infectious centre See ONE-STEP GROWTH EXPERIMENT.

infectious disease (communicable disease) Any disease which, under natural conditions, can be transmitted from one individual to another by a causal agent which passes either directly (by physical contact) or indirectly (e.g. by DROPLET INFECTION or via FOMITE or VECTOR) from the infected to the non-infected individual (i.e. HORIZONTAL TRANSMISSION); diseases passed on by VERTICAL TRANSMISSION include some which are normally 'infectious' (e.g. AIDS) as well as others (e.g. certain TRANSMISSIBLE SPONGIFORM ENCEPHALOPATHIES) which are hereditary, i.e. not considered 'infectious'.

A *transmissible* disease may be transmitted from the affected individual to another individual, or to a fetus, by any means, including those processes that are involved in horizontal and vertical transmission as well as those involved in experimental infection (the latter form of transmission may or may not occur naturally in any given instance).

TETANUS is an example of a non-infectious disease.

(cf. CONTAGIOUS DISEASE; see also IATROGENIC.)

[The transmission of infection (review): RMM (1995) *6* 217–227.]

infectious dropsy (of carp) (*syn.* rubella; infectious ascites; also: haemorrhagic septicaemia) An 'umbrella term' for at least two distinct FISH DISEASES affecting carp (*Cyprinus* spp): see SPRING VIRAEMIA OF CARP and CARP ERYTHRODERMATITIS; abdominal distension with generalized oedema are common to both diseases.

infectious endocarditis See ENDOCARDITIS.

infectious enterohepatitis *Syn.* BLACKHEAD.

infectious equine arteritis virus *Syn.* equine arteritis virus: see ARTERIVIRUS.

infectious foot-rot See FOOT-ROT.

infectious haematopoietic necrosis (IHN) A FISH DISEASE characterized by massive destruction of the blood-forming tissues; the causal agent is a virus (see RHABDOVIRIDAE).

infectious hepatitis *Syn.* HEPATITIS A.

infectious jaundice See LEPTOSPIROSIS.

infectious keratitis (infectious bovine keratoconjunctivitis; pink-eye) In cattle: KERATITIS or KERATOCONJUNCTIVITIS which occurs world-wide, particularly in young animals, usually in summer and autumn. The common causal agent is apparently *Moraxella bovis*. [Microbial flora of the bovine eye: VR (1986) *118* 204–206.] (See also CATTLE DISEASES.)

infectious mononucleosis (*syn.* glandular fever) An acute, self-limiting infectious disease of the lymphatic system which primarily affects children and young adults; it is caused by the EPSTEIN–BARR VIRUS (EBV). Transmission occurs by direct oral contact ('kissing disease') and may also occur by droplet infection.

The incubation period is 1–7 weeks. In the typical 'glandular fever' syndrome (the anginose form), common in young adults, there is fever, pharyngitis (often with exudative tonsilitis) and swollen tender lymph nodes (with or without rash); jaundice and/or hepatosplenomegaly occur less commonly. Infrequently, the disease develops with fulminant hepatitis and VAHS (see HAEMOPHAGOCYTIC SYNDROME). Therapeutic approaches to the severe form of the disease may involve the use of agents such as ACYCLOVIR, CYCLOSPORIN A and corticosteroids.

Lab. diagnosis may include demonstration of (a) increased levels of heterophil antibodies (see PAUL–BUNNELL TEST); (b) mononucleosis and lymphocytosis; (c) the presence of atypical lymphocytes (enlarged, with foamy, vacuolated cytoplasm). (The atypical lymphocytes are not virus-infected B cells; they are T cells expressing the CD3 molecule.)

[Haematology and cell-mediated immunity in infectious mononucleosis: BCH (1995) *8* 165–199 (170–175).]

(See also XLP SYNDROME; cf. CYTOMEGALIC INCLUSION DISEASE and EHRLICHIA.)

infectious necrotic hepatitis (*vet.*) *Syn.* BLACK DISEASE.

infectious nucleic acid (*virol.*) A nucleic acid which, when extracted from virions, can infect a host cell and give rise to progeny virions. Such a nucleic acid must be capable of being expressed and replicated in the cell in the absence of *virion* enzymes. Examples include DNA genomes which can gain access to the cell nucleus and which can be recognized and transcribed by host enzymes, and positive-sense ssRNA genomes which can be recognized and translated by host protein-synthesizing machinery. Negative-sense ssRNA genomes are not infectious since an RNA-dependent RNA polymerase (present in the virion) is necessary for the synthesis of positive-sense mRNA. An infectious nucleic acid (e.g. that from polioviruses) may be able to infect cells not normally susceptible to the intact virion since the nucleic acid does not depend on the presence of a specific cell-surface receptor to initiate the infection process. (See also TRANSFECTION.)

infectious pancreatic necrosis (IPN) A FISH DISEASE affecting hatchery-reared salmonid fry during the onset and early stages of feeding. Pancreatic tissue degenerates and becomes necrotic; mortality is often high. The causal agent is the IPN VIRUS.

infectious polyarthritis (of pigs) *Syn.* GLASSER'S DISEASE.

infectious spread (of plasmids) See EPIDEMIC SPREAD.

infectious stunting syndrome (of chickens) See STUNTING SYNDROME.

infective endocarditis See ENDOCARDITIS.

infectivity (1) (of pathogenic microorganisms) The ability of a pathogen to become established on or within the tissues of a host. (See also VIRULENCE.)

(2) (of a disease) The ease with which an infectious disease may be contracted under given conditions.

(3) (viral) See END-POINT DILUTION ASSAY.

inflammation In man and other animals: an acute or chronic response to tissue damage or to the presence of an allergen, or certain types of microorganism, within the tissues; inflammation may be initiated and sustained by microbial components and/or products. Inflammation is a protective response which commonly enables the body to overcome infection and return to normal function.

Acute inflammation is characterized by redness, swelling, warmth (in the affected region), pain and loss of function. This manifestation of the body's innate response to infection involves a diverse range of intercellular and intracellular signalling events – illustrated in the following generalized account of inflammation triggered by a Gram-negative bacterial pathogen which has breached the epithelial barrier. For convenience only, events mediated by complement and those mediated by cytokines are considered below under separate headings; the events described are representative of a process which is much more complex and which is still not understood fully.

Complement-mediated events in inflammation. Activation of COMPLEMENT by LIPOPOLYSACCHARIDES (LPS) gives rise to (among other products) the fragments C3a and C5a; these fragments are anaphylatoxins which elicit HISTAMINE from e.g. MAST CELLS; C5a also acts as a chemotactic factor which attracts MONOCYTES and NEUTROPHILS to the infected site.

Histamine increases the permeability of small blood vessels, allowing efflux of plasma (hence the swelling) and exposing the pathogen to (i) increased amounts of antimicrobial agents (e.g. complement and antibodies) and (ii) cells of the immune system attracted by C5a. Histamine is also reported to stimulate the early expression of SELECTIN molecules (specifically P-selectins) on local vascular endothelium (apparently within minutes of stimulation).

Cytokine-mediated events in inflammation. CYTOKINES (e.g. IL-1, IL-8, TNF-α) are produced by local cells on stimulation by the pathogen. IL-1 and TNF-α promote the expression of E-selectins on local vascular endothelium.

Leukocytes in local blood vessels bind weakly to endothelial selectins; binding involves certain carbohydrate ligands in the leukocyte membrane – e.g. sialyl-Lewis x (= CD15). This (labile) selectin–CD15 binding causes leukocytes to slow down and *roll* along the vascular endothelium in the direction of blood flow, bonds being continually made and broken; this is referred to as *tethering*. As well as CD15, the leukocyte cell membrane also contains adhesion molecules of the INTEGRIN family – although, in the absence of infection, these molecules are normally inactive; an example is the *leukocyte function-associated antigen 1* (LFA-1; CD11a/CD18). Chemokines, e.g. IL-8 (see above), promote the activation of LFA-1 (via phosphorylation), so that LFA-1 can then bind (strongly) to counter-receptors (e.g. ICAM-1, ICAM-2) on vascular endothelium. [Leukocyte rolling and firm adhesion (biophysical view): PNAS (2000) *97* 11262–11267.] The captured leukocytes flatten, and then 'crawl' over the endothelium to a cell–cell junction – subsequently migrating *between* endothelial cells (a process called *diapedesis* or *extravasation*) into the affected area (apparently guided by the chemokine gradient). The mechanism of extravasation is not known, but *platelet–endothelial cell adhesion molecule-1* (PECAM-1) appears to be required.

As indicated, cytokines are important for the recruitment of immune cells to the local site of infection. (Neutrophils are commonly the dominant type of cell in the early stages of acute inflammation; macrophages often become prominent after ~8 hours, and their ability to secrete FIBRONECTIN may be an important factor in tissue repair [wound healing (review): Science (1997) *276* 75–81].) The importance of leukocyte recruitment to infected tissues is emphasized by those genetic disorders in which this process is inhibited (e.g. LEUKOCYTE ADHESION DEFICIENCY).

Cytokines are also involved in the production of ACUTE-PHASE PROTEINS (q.v.).

Antigen-stimulated T lymphocytes release γ-interferon (IFN-γ) which activates MACROPHAGES.

The pain associated with inflammation may be due e.g. to pressure on nerves and/or to PROSTACYCLIN.

Chronic inflammation, which occurs e.g. in some persistent intracellular infections, is characterized by the formation of new connective tissue and by the presence of concentrations of macrophages; it often leads to the formation of a GRANULOMA.

infliximab See TNF.

influenza (flu; grippe) An acute, highly infectious human disease caused by influenza virus A (often widespread epidemics, occasionally pandemics) or influenza viruses B or C (typically sporadic outbreaks, often among children and young adults). (See INFLUENZAVIRUS for details of strains, including those in the 1997 bird flu in Hong Kong, the 1999 outbreak in Guangdong (China), and the 2004–2006 problem with strain H5N1; see also INFLUENZA VIRUS C, GASTRIC FLU and COMMON COLD.)

Infection generally occurs by inhalation of virus-containing aerosols, virions being deposited onto alveolar membrane or (larger droplets) onto mucous membrane lining the respiratory tract; in the latter case, virions may bind to sialic acid residues in mucoproteins (rather than to receptors on respiratory epithelium) – but such (abortive) binding can be broken by the viral neuraminidase.

Incubation period: typically 1–3 days. Infection may be asymptomatic, or there may be a sudden onset with chills, fever, headache, muscular aches, anorexia, malaise etc.; fever subsides after 1–5 days, and respiratory symptoms (coryza, cough etc.) then become prominent. Recovery is usually rapid, although cough and malaise may persist for several weeks.

Viral infection damages ciliated epithelium in the trachea and bronchi; loss of cilia predisposes to secondary bacterial infection and development of e.g. bronchial pneumonia involving organisms such as *Streptococcus pneumoniae* and/or *Haemophilus influenzae*. Secondary infections are particularly problematic e.g. in the elderly, debilitated and immunocompromised; apart from pneumonia, secondary infections may lead to complications such as tracheo-bronchitis, sinusitis, otitis media etc. (One early report indicated that, in *mice*, the severity of disease can be reduced by ozone [AEM (1982) *44* 723–731].)

Chemotherapy. Drugs such as AMANTADINE and rimantadine may afford some protection against influenza virus type A if administered before or soon after infection. (See also ZANAMIVIR.)

Vaccines. The original, formalin-inactivated, whole-virus vaccines produce local and systemic side-effects. *Split-product* (*split-virus*) vaccines, containing H and N components of the virus (as opposed to entire virions), are tolerated better.

Various alternative approaches to vaccines have been tried. For example, attempts have been made to prepare attenuated 'live' vaccines by using REASSORTANT VIRUSes containing e.g. surface glycoproteins from a currently epidemic strain and internal components from an attenuated strain. In some cases the attenuated strain was a temperature-sensitive (cold-adapted) virus, while in other cases it was a strain from another species with low virulence for man [e.g. avian strains: Book ref. 86, pp 395–405]. Vaccines have been prepared from synthetic oligopeptides that mimic glycoprotein epitopes [oligopeptide vaccines: MS (1984) *1* 55–58] and with virosomes [Lancet (1994) *344* 160–163].

Owing to the effects of antigenic variation, the components of influenza vaccines must be regularly reviewed to ensure that, when used, a given vaccine will generate protection relevant to the existing strain(s) of virus.

[Efficacy of influenza vaccines: RMM (1996) *7* 23–30.]

influenza virus A See INFLUENZAVIRUS.

influenza virus B See INFLUENZAVIRUS.

influenza virus C A virus of the ORTHOMYXOVIRIDAE. It is physicochemically and morphologically similar to members of the genus INFLUENZAVIRUS, but differs in containing only 7 RNA segments (total MWt ca. $4–5 \times 10^6$) and in having only one type of surface glycoprotein (HA) which has both haemagglutinating and receptor-destroying activities, but apparently no neuraminidase activity. It has been shown that, contrary to previous assumptions, sialic acid may nevertheless be an essential component of the host cell-surface receptor [Virol. (1985) *141* 144–147]. Influenza virus C primarily infects man, although it has been isolated from pigs in China; it generally causes only mild or subclinical disease of the upper respiratory tract. Antigenic drifting occurs slowly, but antigenic shift has not been observed (cf. INFLUENZAVIRUS).

Influenzavirus A genus of viruses (family ORTHOMYXOVIRIDAE) which includes influenza virus type A and influenza virus type B; both viruses cause sporadic or epidemic INFLUENZA in man, but type A strains also cause epizootics in pigs, horses or birds (see e.g. FOWL PLAGUE and SWINE INFLUENZA) and (occasionally) pandemics in humans. (See also INFLUENZA VIRUS C.)

Influenzaviruses are transmitted via aerosols or water, or by direct contact.

Virions. The enveloped virions are pleomorphic, ca. 80–120 nm across; 'filamentous' forms of up to several micrometres in length are common. The genome consists of eight molecules of linear, negative-sense ssRNA (total MWt ca. 5×10^6); these RNA segments form a helical ribonucleoprotein complex with the (virus-encoded) NP protein. Associated with the RNA–NP complex are the viral proteins P1, P2 and P3 (also called PA, PB1 and PB2) which, together, constitute a *transcriptase* (i.e. an RNA-dependent RNA polymerase) which is required for synthesis of viral mRNA and for replication of the RNA genome. [Function of transcriptase (model): Book ref. 86, pp 73–84.]

The ribonucleoprotein is surrounded by a layer of (virus-encoded) *matrix protein* M1; this layer of M1 protein is wholly enclosed by the viral envelope.

The viral M2 protein forms part of the envelope. It is encoded by genome segment 7 (as is M1); M2 (MWt ca. 13×10^3) is translated from a spliced transcript.

The viral envelope (containing lipids derived from host cell membrane) forms the outermost layer of the virion.

Two types of structure arise from the surface of the envelope. These structures are usually referred to as *spikes* and *mushroom-shaped projections* (names which reflect their appearance under the electron microscope).

Spikes. Each spike (~10–15 nm long) consists of three identical glycoproteins. Because these glycoproteins can bind to red blood cells (and cause HAEMAGGLUTINATION) they are referred to as haemagglutinins (HA). However, the glycoproteins can bind to any type of cell which has a suitable sialic acid (= *N*-acetylneuraminic acid, NANA) receptor – e.g. cells of the respiratory epithelium.

Each molecule of HA glycoprotein consists of two dissimilar polypeptide subunits which are linked by a disulphide bond; these subunits are termed HA1 (the larger subunit) and HA2. (The HA protein is encoded by segment 4 of the viral genome. It is synthesized as a single polypeptide chain which undergoes post-translational cleavage: an N-terminal signal sequence is removed, and the molecule is commonly cleaved to form HA1 (328 amino acid residues) and HA2 (221 amino acid residues) – which remain linked by a disulphide bond.) In each spike, the proximal HA molecule is attached to the envelope via a site in the HA2 subunit.

Mushroom-shaped projections. Each individual mushroom-shaped projection is a tetramer of another glycoprotein, NEURAMINIDASE (NA) (q.v.); each protein in the tetramer has a functional enzymic site.

Propagation of influenzaviruses in vitro. The influenzaviruses can be propagated in embryonated hens' eggs and e.g. in chick embryo cell cultures. Some authors prefer monkey kidney or human embryo kidney cells for primary isolation. (Eggs are used for vaccine virus production.)

Inactivation of virions. Virions are rapidly inactivated by low pH (<5), by mild heat (e.g. 60°C/30 min), by lipid solvents, detergents and soaps, by formaldehyde and ultraviolet radiation, and by phenolic disinfectants and other agents; they are apparently resistant to drying at room temperature, and survive at low temperatures in water. The virions can be preserved for long periods at −70°C.

Infection of host cells; replication. The virion binds, via the HA protein, to specific cell surface receptors containing sialic acid residues; it is then internalized in a vesicle (endocytosis). Low pH within the vesicle promotes cleavage of the HA protein (by a host protease), and this leads to fusion of the viral envelope with the vesicle's membrane; as a result, viral ribonucleoprotein and transcriptase are released into the cytoplasm of the host cell. (See also ENVELOPE.)

The infectivity of a virion is abolished if antibodies have bound to the globular HA1 portion of the HA glycoprotein.

Transcription of the (negative-sense) viral RNA occurs in the nucleus; it involves viral transcriptase but depends on the priming activity of the host's (α-AMANITIN-sensitive) RNA polymerase II. (See also RIBAVIRIN.) Transcription gives rise to ten molecules of mRNA (two from each of genome segments 7 and 8 – each of which have two reading frames). Viral mRNAs are polyadenylated in the nucleus.

Replication of the viral genome also occurs in the nucleus and involves viral transcriptase and the host's RNA polymerase II. The progeny ribonucleoprotein structures are assembled in the nucleus, but details of the mode of assembly are not available.

HA and NA proteins are synthesized on rough endoplasmic reticulum and subsequently migrate to localized sites in the plasma membrane.

The type B influenzaviruses (but apparently not type A viruses) encode another glycoprotein, NB, whose reading frame overlaps that of NA; NB does not occur in the virion. Other non-structural (NS) proteins are produced in cells infected with A- or B-type viruses; their role(s) are unknown.

Maturation of the virion occurs by budding through the cell membrane at sites containing HA and NA.

Classification/nomenclature of influenzaviruses. Influenzaviruses are classified as subtypes on the basis of their HA antigens (15 different subtypes) and NA antigens (9 different subtypes). A given strain of virus is designated by a formula which indicates the following data: (i) type (i.e. A, B or C); (ii) the animal from which the strain was first isolated (omitted if the host was human); (iii) the place where the strain was first isolated; (iv) the strain number; (v) the year of isolation; and (vi) the particular HA (= H) and NA (= N) antigens. Example:

A/duck/Ukraine/1/63(H3N8)

The pandemic of 1957 – so-called Asian flu – was caused by a strain formerly referred to as 'strain A2' but now designated:

A/Singapore/1/57(H2N2)

The 1968 pandemic – so-called Hong Kong flu – was due to:

A/Hong Kong/1/68(H3N2)

It appears that these two strains of influenzavirus were circulating human strains which had acquired novel antigens through genetic reassortment with avian strains (see REASSORTANT VIRUS). The sudden acquisition of new antigens ('antigenic shift' – see below) meant that radically new strains of virus had appeared in a human population which, immunologically, was completely naive.

It was previously assumed that if a 'new' subtype appears then the 'old' strain ceases to circulate. However, subtype H1N1, which was active early in the 20th century, re-appeared in 1977 and was involved in the pandemic of so-called Russian flu.

The surface glycoproteins of influenzaviruses undergo ANTIGENIC VARIATION; by this means, the virus may avoid the host's immune response and give rise to new outbreaks of influenza. New strains of virus are generally not inactivated by antibodies induced by a previous strain – i.e. antibody-based protection tends to be strain-specific.

Variation in surface antigens may involve ANTIGENIC DRIFT or ANTIGENIC SHIFT.

Antigenic drift involves relatively small changes in the surface antigens, and this is associated with the regular occurrence of epidemic influenza.

Antigenic shift involves major (and abrupt) changes in the surface antigens, and it happens infrequently. It has been recorded only in type A influenzaviruses; this apparently reflects the fact that type A strains can infect animals as well as man, and that, in mixed infections (involving viruses from different species), segments of the genome may undergo reassortment – resulting in the formation of novel hybrid subtypes containing genome segments from different co-infecting parent virions.

In recent times, human strains typically displayed H1, H2 or H3, and N1, N2 or N3 antigens. Many other types of H and N antigens occur in animal strains, suggesting scope for the formation of new pandemic strains by reassortment. Fortunately, influenzaviruses are generally highly species-specific – so that transfer across the species barrier is assumed to occur only rarely. Transfer occurred when American harbour seals were infected by an avian H7N7 virus in 1980 [Virology (1981) *113* 712–724] and pigs were infected by avian strains [JV (1996) *70* 8041–8046].

Transfer across the species barrier may occur in a two-step process; for example, transmission of the human H3N2 subtype to avian hosts can be achieved after reassortment in pigs.

The 1997 outbreak of *bird flu* in Hong Kong, which caused six deaths, resulted from avian-to-human transmission of a fowl plague strain (H5N1). Subsequently, no deaths from this strain were reported following large-scale slaughter of poultry.

In 1999, two children in Hong Kong were infected with influenza A subtype H9N2 (strains A/Hong Kong/1073/99; A/Hong Kong/1074/99); these strains closely resemble a strain isolated from quail in 1997 (A/quail/Hong Kong/G1/97). Also in 1999, H9N2 viruses were isolated from patients with influenza-like illness in Guangdong province (China).

The H5N1 and (Hong Kong) H9N2 human isolates are similar in the six genes that encode *internal* components of the virion, suggesting that H9N2 arose from reassortment of H5N1. Avian strains with these six genes might have a tendency to infect humans and may therefore have the potential to give rise to a novel human pathogen. [Relationship between H9N2 and H5N1 human isolates: PNAS (2000) *97* 9654–9658].

H5N1 caused over 50 human deaths in East Asia in 2004–2005. Isolates from humans and those from birds consisted of two distinct clades. In all isolates the genes were of avian influenzaviruses, indicating that they were not derived through reassortment with human strains. Human-to-human transmission was described. [Evolution of H5N1 avian influenza viruses in Asia: EID (2005) 11(10). Influenza (special issue): EID (2006) 12(1) (January).]

informational suppressor See SUPPRESSOR MUTATION.

informosome See MRNA (b).

infraciliature (*ciliate protozool.*) The subpellicular system which includes the kinetosomes and all the various associated microfibrillar and microtubular structures (see e.g. KINETODESMA; POSTCILIARY MICROTUBULES; TRANSVERSE MICROTUBULES).

infradian rhythms In certain parameters of a eukaryotic cell or organism: innate rhythmical changes which occur with a periodicity >24 hours. (cf. CIRCADIAN RHYTHMS.)

infrared radiation See dry-heat STERILIZATION.

infrasubspecific rank Any rank below subspecies (see e.g. VARIETY).

infundibulum (1) In some ciliates: the lower or inner (often tubular) region of the BUCCAL CAVITY; in peritrichs it corresponds to that part of the buccal cavity which excludes the PERISTOME (sense 1). (2) Any funnel-shaped structure.

infusion agar Infusion broth (see NUTRIENT BROTH) gelled with 1.5–2.0% w/v AGAR.

infusion broth See NUTRIENT BROTH.

infusoria An archaic term for the various microscopic organisms – particularly ciliates – found in aqueous infusions of vegetable matter.

INH Isonicotinic acid hydrazide: ISONIAZID.

inhibitory medium See MEDIUM.

initial(s) (*microbiol.*) Any part of a cell or organism which subsequently develops and/or differentiates to become a distinct and characteristic structure; the earliest or early stage(s) of a given structure.

initial body See CHLAMYDIALES.

initiation complex See PROMOTER.

initiation mass See CELL CYCLE.

initiator (of an operon) See OPERON.

ink-cap fungi Species of COPRINUS whose autodigesting lamellae form a black, basidiospore-containing, ink-like liquid.

Inkoo virus See BUNYAVIRUS.

***inlA* gene** See LISTERIOSIS.

innate (*mycol.*) Refers to any fungal structure which is embedded in a substratum, or in the fungal thallus, such that its outer surface is more or less level with the surface of the substratum or thallus.

inner primers See NESTED PCR.

inner veil (*mycol.*) *Syn.* PARTIAL VEIL.

inoculation (1) The placing of material (the INOCULUM) on or into a MEDIUM, TISSUE CULTURE, animal etc using an ASEPTIC TECHNIQUE. If the purpose of inoculation is to initiate a CULTURE, the inoculated medium must be incubated (see INCUBATION). (2) (*immunol.*) *Syn.* VACCINATION (1) or (2).

inoculative infection (*parasitol.*) Infection in which a pathogen is introduced into the vertebrate host from the ANTERIOR STATION. (cf. CONTAMINATIVE INFECTION.)

inoculum (1) Material used for the INOCULATION of a MEDIUM, TISSUE CULTURE, animal etc; it typically comprises or contains viable microorganisms. (2) (*plant pathol.*) Pathogenic microorganisms which have become established within a plant and which may give rise to disease.

inoculum effect In the in vitro testing of bacteria for resistance to antibiotic(s): the phenomenon in which the MIC increases when the size of the test inoculum (cells/ml) is increased; for a given strain of bacteria the inoculum effect may differ greatly with different antibiotics.

Inocybe See AGARICALES (Cortinariaceae).

Inonotus A genus of fungi of the APHYLLOPHORALES (family Hymenochaetaceae) which form basidiocarps in which the context is reddish-brown and fibrous. *I. dryadeus* (formerly *Polyporus dryadeus*) forms bracket-type basidiocarps (context: monomitic) with a pale grey (later darkening) upper surface and a porous underside; basidiospores: whitish, spheroidal, ca. 7 μm diameter. It is parasitic on various types of oak (*Quercus*). (See also BUTT ROT and POCKET ROT.)

inoperculate Lacking an OPERCULUM.

inorganic pyrophosphate See PYROPHOSPHATE.

iNOS See NITRIC OXIDE.

inosine See NUCLEOSIDE and Appendix V(a).

inositol Hexahydroxycyclohexane. Theoretically there are nine stereoisomers; of these, *meso*-inositol (= *myo*-inositol), scyllitol, *d*-inositol and *l*-inositol occur naturally, and only *meso*-inositol is common. *Meso*-inositol occurs widely e.g. in the phospholipids of microorganisms; it is required, in small amounts, for the growth of a number of microorganisms – including many yeasts (see e.g. BREWING). Inositol is used as a substrate in biochemical characterization tests for bacteria; it is attacked e.g. by *Klebsiella pneumoniae*. (See also CYCLITOL ANTIBIOTICS.)

Inoue–Melnick virus A virus which has been isolated from patients with chronic CNS disease (including MULTIPLE SCLEROSIS) [JCM (1985) *21* 698–701].

Inoviridae A family of SSDNA PHAGES in which the virions are rod-shaped or filamentous, the capsid subunits being arranged helically around one piece of ss cccDNA; the virions are chloroform-sensitive. On entry into a host cell the ssDNA is converted into a double-stranded replicative form from which progeny ssDNA can be synthesized. Mature phages are extruded through the host cell envelope. The host remains viable but grows more slowly than an uninfected cell. There are two genera: INOVIRUS and PLECTROVIRUS. [Book ref. 23, pp. 78–79.]

Inovirus (filamentous phages) A genus of non-lytic SSDNA PHAGES of the INOVIRIDAE. Hosts: enterobacteria (for e.g. phages fd, f1, M13, If1, If2, IKe, ZJ/2), *Pseudomonas* (phages Pf1, Pf2), *Xanthomonas* (phages Xf, Xf2, Cf), and *Vibrio* (phage v6); most or all inoviruses can infect only 'male' cells (i.e. those containing CONJUGATIVE PLASMIDS). Proposed type member: bacteriophage fd; f1 and M13 are very similar to fd. The inovirus virion is a flexible filament ($760–1950 \times 6$ nm, according to phage) comprising ss cccDNA (MWt ca. $1.9–2.7 \times 10^6$), one major coat protein (gp8, B-protein), and minor coat proteins A (gp3), C (gp9) and D (gp6) at 5, 10 and 5 copies/virion, respectively. Proteins A and D occur at one end of the virion; C seems to occur at the opposite end. The A-protein (gp3) is necessary for adsorption; the roles of C and D are unknown. The DNA loop fits inside the capsid (internal diam. ca. 2 nm) such that the two adjacent (antiparallel) strands of the loop are wound around each other in a type of double helix (pitch ca. 2.7 nm) despite the lack of homology between them; the helix seems to be stabilized by the coat protein [JMB (1982) *157* 321–330].

The following refers to fd, f1 and M13; other inoviruses seem to be generally similar. Phages fd, f1 and M13 adsorb specifically to the *tips* of F and F-like pili (cf. BACTERIOPHAGE IKE, BACTERIOPHAGE IF1, LEVIVIRIDAE). During penetration of the host (which may involve pilus retraction) the capsid proteins (at least gp3 and gp8) become localized asymmetrically in the cell membrane. At it penetrates the cell, the phage DNA becomes coated with host SSB protein, initiating stage I (see SSDNA PHAGES). A unique 508-nucleotide intergenic (IG) region between genes 2 and 4 may remain uncoated; this region binds host RNA POLYMERASE holoenzyme which may recognize the secondary or tertiary structure of DNA in the IG region (which can form several hairpin loops). The RNA polymerase synthesizes an RNA primer at a specific site (*ori*) in the IG region. The complementary (*c*) strand, and hence RF, is completed as in other ssDNA phages (q.v.). RF is supercoiled by GYRASE to become RFI (see RF). The *c* strand of RFI functions as template for transcription of phage genes. An early product is an endonuclease (gp2) which initiates stage II by cleavage of the RFI *v* strand at a specific site in the IG region; the free 3′ end is elongated as in other ssDNA phages (q.v.). Genome lengths of displaced *v* ssDNA are cleaved by gp2, and each is circularized to form *v* ss cccDNA. At first, new *v* strands are converted to RFs. However, a phage-coded SSB protein (gp5) gradually accumulates, eventually coating the newly formed *v* strands and blocking their conversion to RFs (stage III). Phage coat proteins (at least gp8 and gp3) are synthesized as soluble cytoplasmic precursors with a 'signal peptide' at the amino end (see SIGNAL HYPOTHESIS); these proteins become inserted asymmetrically into the cytoplasmic membrane, the signal peptides are removed, and the proteins are 'anchored' within the membrane [PNAS (1982) *79* 5200–5204]. Phage assembly – which requires THIOREDOXIN [PNAS (1985) *82* 29–33] – occurs (in f1) at zones of adhesion between the cytoplasmic and outer membranes, f1 gp1 possibly being involved in the formation of these adhesion zones [JB (1985) *163* 1270–1274] (cf. ADHESION SITE); gp8 displaces gp5, and the mature virion is finally extruded through the host cell envelope.

insect diseases *Bacterial* pathogens of insects include e.g. *Bacillus* spp (see e.g. AMERICAN FOULBROOD, DELTA-ENDOTOXIN, MILKY DISEASE), *Melissococcus pluton* (see EUROPEAN FOULBROOD), and species of RICKETTSIELLA, SERRATIA and SPIROPLASMA. (See also CECROPINS and SARCOTOXINS.)

Fungal entomopathogens include species of e.g. *Ascosphaera* (see CHALK-BROOD); ASCHERSONIA; *Aspergillus* (see e.g. STONEBROOD); BEAUVERIA; *Coelomomyces* (see BLASTOCLADIALES); CORDYCEPS; CULICINOMYCES; ENTOMOPHTHORA, ERYNIA and other fungi of the ENTOMOPHTHORALES; FUSARIUM; HIRSUTELLA; METARHIZIUM; NOMURAEA; SEPTOBASIDIALES; TERMITARIA; VERTICILLIUM.

Protozoal entomopathogens include e.g. species of MALPIGHAMOEBA, NOSEMA and THELOHANIA. (See also CRITHIDIA, HERPETOMONAS and LEPTOMONAS.)

For *viral* entomopathogens see e.g. BACULOVIRIDAE; CYTOPLASMIC POLYHEDROSIS VIRUS GROUP; DENSOVIRUS; DROSOPHILA X VIRUS; ENTOMOPOXVIRINAE; FLACHERIE; HOUSEFLY VIRUS; NODAVIRIDAE; NUDAURELIA β VIRUS GROUP; PICORNAVIRIDAE; POLYDNAVIRIDAE; SIGMA VIRUS.

(See also INVERTEBRATE DISEASES and BIOLOGICAL CONTROL.)

insect iridescent viruses See CHLORIRIDOVIRUS and IRIDOVIRUS.

insect–microbe associations See e.g. AMBROSIA FUNGI; DUTCH ELM DISEASE; FUNGUS GARDENS; FUNGUS GNATS; INSECT DISEASES; MYCETOCYTE; TERMITE–MICROBE ASSOCIATIONS; WOODWASP FUNGI.

insecticidal crystal protein (ICP) See BIOLOGICAL CONTROL.

insertion mutation (addition mutation) A type of MUTATION in which one or more nucleotides are inserted into the genome; if the number of nucleotides inserted is not divisible by 3, the mutation will be a FRAMESHIFT MUTATION. Insertion mutations commonly result from the insertion of a TRANSPOSABLE ELEMENT or prophage (see also BACTERIOPHAGE CONVERSION). (cf. DELETION MUTATION.)

insertion sequence (IS element) A TRANSPOSABLE ELEMENT which contains no genetic information other than that necessary for its transposition. (cf. TRANSPOSON.) An IS element may transpose independently, when it may be detectable by the effect(s) of its insertion at a new site, or it may occur as a terminal part of a composite (class I) TRANSPOSON. An IS element is defined by its ends: an INVERTED REPEAT sequence of ca. 10–40 bp has been found to occur at both ends in every IS element so far analysed; these inverted repeats are apparently necessary for transposition and are probably the recognition sites for the corresponding transposition enzymes. Most IS elements also have the potential for encoding 1–3 polypeptides, presumed to include TRANSPOSASE component(s). In many IS elements a small open reading frame occurs within the larger open reading frame but in the opposite orientation; IS*50* has overlapping genes (see Tn*5*).

IS elements (designated IS*1*, IS*2* etc) may be classified according to their terminal inverted repeats and the number of base pairs duplicated at the target site on insertion. Different IS elements transpose with different frequencies depending e.g. on the structure of the donor and target sequences, on the physiological state of the host cell, etc; frequencies of transposition range from ca. 10^{-9} to 10^{-5} per IS element per cell division. (See also e.g. IS*1*, IS*5*, IS*101*.)

[Review: Book ref. 20, pp. 159–221.]

insertion vector A CLONING vector which has a single site at which a sequence of exogenous DNA can be inserted (cf. REPLACEMENT VECTOR).

inside-to-outside model A model which describes the mode of incorporation of PEPTIDOGLYCAN into the CELL WALL of a growing Gram-positive bacillus [FEMS Reviews (1986) *32* 247–254; J. Theor. Biol. (1985) *117* 137–157]. Essentially, layers of newly formed peptidoglycan are laid down just external to the cytoplasmic membrane – a region in which they are subject to minimal hydrostatic stress; as growth continues, the layers move outwards towards the cell surface (being replaced by new layers) and accordingly bear an increasing proportion of the hydrostatic pressure. Finally, near the region of maximum stress, the peptidoglycan becomes increasingly susceptible to autolysins, the (older) layers of peptidoglycan commonly being shed in fragments.

inspissation The process of thickening. In microbiology: a process in which certain heat-coagulable substances (e.g. serum, homogenized hens' eggs) – or media containing such substances – are solidified by heating to ca. 85°C for ca. 1 hour. (See e.g. LÖWENSTEIN–JENSEN MEDIUM.)

instructive theories of antibody-formation Theories (now untenable) which hold that the ability to produce specific antibodies develops in antibody-producing cells only *after* their initial exposure to the antigen, i.e., it is assumed that pre-existing specificity does not occur. In such theories antigen is supposed to act as a template for the production of specific antibodies or for the formation of an antibody-specifying nucleic acid. (cf. CLONAL SELECTION THEORY.)

int (1) See e.g. BACTERIOPHAGE LAMBDA and BACTERIOPHAGE P22. (2) See MOUSE MAMMARY TUMOUR VIRUS.

integral membrane protein See CYTOPLASMIC MEMBRANE.

integrase See e.g. BACTERIOPHAGE λ.

integrase-type recombinase See SITE-SPECIFIC RECOMBINATION.

integration host factor (IHF) In e.g. *Escherichia coli*: a heterodimeric protein consisting of polypeptide chains of apparent MWt 11000 and 9500 [JBC (1981) *256* 9246–9253]; IHF, which has the ability to induce bends in DNA, has been associated with various roles that include e.g. stimulating the transcription of nitrogen-fixing operons [Cell (1990) *63* 11–22], promoting expression of the *tra* operon in the F plasmid [JB (1990) *172* 4603–4609], stimulating helicase-catalysed nicking of the F plasmid at *oriT* [JBC (1995) *270* 28374–28380] and promoting gene specificity for a given sigma factor [EMBO (2000) *19* 3028–3037].

IHF is also one of the factors needed for the on–off switching of type I fimbriae in *Escherichia coli*.

(See also SIDD.)

integrative suppression The phenomenon in which the integration of a plasmid into a bacterial chromosome suppresses the effect of a chromosomal mutation which has inhibited the initiation of chromosome replication; replication of the joint plasmid–chromosome replicon is apparently initiated at a site within the plasmid. Thus, e.g., as a result of integrative suppression, a temperature-sensitive *dnaA* mutant of *Escherichia coli* can carry out limited chromosome replication at a non-permissive temperature.

integrins A family of CELL ADHESION MOLECULES found on various types of cell and involved in many aspects of cell physiology; the molecule is a heterodimer of proteins consisting of an α subunit non-covalently linked to a β subunit – the particular combination of α subunit (many types) and β subunit (few types) determining the biological activity of each integrin.

Integrins of the β_1 subfamily are found e.g. on various types of epithelial cell – either on the basolateral surface or (e.g. $\alpha_6\beta_1$) on the basal surface (i.e. adjacent to the basement membrane) [integrins (review): HJ (1993) *25* 469–477]. These integrins have a major role in cell–cell and cell–matrix interaction – binding to ligands that include collagen, fibronectin and laminin; however, such interaction involves more than cell position and maintenance of tissue structure: integrin–ligand binding can also trigger signals that modulate a wide range of cell functions, including aspects of the cell cycle.

The β_1 integrins are also implicated as receptors for the adhesins of invasive pathogenic bacteria; for example, adhesins of *Yersinia enterocolitica* have been found to bind (in vitro) to the $\alpha_5\beta_1$ integrin of epithelial cells.

The β_2 integrins are found primarily on leukocytes; their expression on a given cell is apparently dependent on activation – e.g. cytokine-stimulated cells may exhibit the *leukocyte function-associated antigen-1* (LFA-1; = CD11a/CD18; = $\alpha_L\beta_2$) and the Mac-1 integrin (= CD11b/CD18; = $\alpha_M\beta_2$); both of these integrins can bind to ICAM ligands (see IMMUNOGLOBULIN SUPERFAMILY and INFLAMMATION).

The β_2 integrins can also serve as receptors for bacterial adhesins. Thus, many integrins bind (via their α subunit) to ligands that contain the amino acid sequence Arg-Gly-Asp (RGD), and this sequence is mimicked by the adhesin *filamentous haemagglutinin* (FHA) of *Bordetella pertussis*; the binding between FHA (RGD sequence) and Mac-1 ($\alpha_M\beta_2$) of an activated macrophage assists bacterial invasion of the phagocyte.

The β_3 integrin $\alpha_{IIb}\beta_3$ (gpIIbIIIa; = CD41b/CD61) is found on platelets and megakaryocytes; its ligands include fibrinogen, VON WILLEBRAND FACTOR and fibronectin. Failure to synthesize adequate amounts of this integrin results in deficient platelet aggregation (*Glanzmann's thrombasthenia*).

(See also LEUKOCYTE ADHESION DEFICIENCY.)

[Defects in human integrins: Lancet (1999) *353* 341–343.]

integron A region of DNA which includes (i) a gene encoding a site-specific recombinase (see SITE-SPECIFIC RECOMBINATION) and (ii) a corresponding site for recombination into which one or more genes may be inserted [see e.g. JAC (1999) *43* 1–4]. Integrons may have a role as a general gene-capture mechanism in bacterial evolution [PNAS (2001) *98* 652–657].

intein In certain types of protein: an internal sequence of amino acids apparently capable of catalysing a self-splicing reaction in which the intein is excised (forming a separate protein) and the terminal parts of the original polypeptide (the two *exteins*) are joined to form a functional protein. [Review: TIBS (1995) *20* 351–356. Intein sequences (review): NAR (1997) *25* 1087–1093.]

interallelic complementation See COMPLEMENTATION TEST.

intercalary Not apical or terminal: e.g. formed or situated between apex and base of a given structure or between two cells in a chain of cells.

intercalating agent Any dye possessing a planar chromophore which can insert (intercalate) *between* adjacent base pairs in dsDNA or in base-paired regions in ssDNA (e.g. HAIRPIN loops); *bifunctional* (*bis*) intercalating agents have two such chromophores per molecule (see e.g. QUINOXALINE ANTIBIOTICS). Intercalation forces the base pairs apart and causes (a) local unwinding of the helix, the degree of unwinding (*unwinding angle*, ϕ) depending on the nature of the dye, and (b) an increase in the length of the DNA. In linear dsDNA these effects lead to an increase in viscosity and a decrease in sedimentation coefficient (s) of the DNA in proportion to the amount of dye intercalated. Intercalative binding is limited to a maximum of one dye molecule per 2–3 bp; the binding of a dye molecule at one site precludes binding of others at the adjacent sites (the *neighbour exclusion principle*).

In circular dsDNA intercalation causes a local increase in pitch and thus a decrease in local (and hence average overall) twist. It follows from $Wr = Lk - Tw$ (see DNA) that, as increasing amounts of dye intercalate in a *negatively* supercoiled molecule, the amount of writhe progressively decreases to zero (molecule behaves as relaxed), and then increases again as the molecule becomes *positively* supercoiled. The changes in writhe are accompanied by changes in viscosity of the DNA solution (increase to a maximum, then decrease) and in sedimentation coefficient of the DNA (decrease, then increase). Since intercalation relieves the strain of negative supercoiling, a negatively supercoiled cccDNA molecule has a higher affinity for intercalating agents than does the equivalent nicked or linear DNA, while a relaxed or positively supercoiled cccDNA has a lower affinity than the equivalent nicked or linear DNA. However, a cccDNA molecule has a limited capacity to unwind, and can therefore bind fewer molecules of intercalating agent than can equivalent linear or nicked circular DNA molecules (see e.g. ETHIDIUM BROMIDE).

Many intercalating agents have antimicrobial and antitumour activity, inhibiting e.g. transcription, DNA replication etc, and inducing FRAMESHIFT MUTATIONS. In vivo, intercalation is often followed by DNA strand breakage, possibly due to the action of nucleases which recognize distortion of the helix at the intercalation sites. At lower concentrations some intercalators (e.g. aminoACRIDINES, ETHIDIUM BROMIDE) can cause the selective loss of small cccDNAs such as plasmids (see CURING (2)), mtDNA (see PETITE MUTANT), ctDNA, and kinetoplast DNA. (See also ACTINOMYCIN D; ANTHRACYCLINE ANTIBIOTICS; QUINOXALINE ANTIBIOTICS; TILORONE. cf. MITOMYCIN C and PSORALENS.)

[Review of DNA-binding drugs: Bioch. J. (1987) *243* 1–13.]

intercistronic complementation See COMPLEMENTATION TEST.

intercrines *Syn.* CHEMOKINES; thus, α-intercrines = α-chemokines, β-intercrines = β-chemokines.

interesterification An industrial process in which fatty acyl residues are interchanged among the various triglycerides in a mixture of lipids; the process, which is catalysed e.g. by sodium methoxide, is carried out in order to modify the composition (and hence properties) of oils and fats. Interesterification can be carried out using certain microbial LIPASES for the production of useful mixtures of triglycerides which cannot be made by conventional chemical processes. [Enzyme-catalysed modification of oils and fats: PTRSLB (1985) *310* 227–233.]

interference (1) (*virol.*) The phenomenon in which the replication of one virus (the *challenge* virus) is partially or completely inhibited by the presence in the host cells of another (*interfering*) virus. In *homologous interference* the challenge and interfering viruses are of similar or identical types, while in *heterologous interference* they are unrelated (see e.g. RUBIVIRUS). Interference may result e.g. from competition between the viruses for components of the replication apparatus, etc. (See also DEFECTIVE INTERFERING PARTICLE.)

(2) (genetics) The phenomenon in which the occurrence of one recombinational event affects the chances of another occurring between adjacent regions of the same molecules of nucleic acid. In *chiasma interference* (= *positive interference*) the occurrence of a cross-over tends to prevent the occurrence of a second cross-over between nearby regions in the same duplexes; this effect decreases with increasing distance from the cross-over. In markers which are closely linked, a localized *negative interference* is sometimes observed: the occurrence of one recombinational event promotes recombination in an adjacent region; this can be due to the correction of adjacent sequences in hybrid DNA during gene conversion (co-conversion – see RECOMBINATION).

interference-contrast microscopy See MICROSCOPY (d).

interfering virus See INTERFERENCE (1).

interferons (IFNs) A category of CYTOKINES which are able to inhibit the replication of many types of virus in (i) the IFN-producing cell, and (ii) cells exposed to exogenous IFNs; IFNs also have other important roles in host defence mechanisms (see later).

Human interferons are divided into two main groups that are designated types I and II.

The type I interferons (stable at pH 2) include IFN-α ($\sim$17–26 kDa), IFN-β ($\sim$21 kDa) and IFN-ω; these interferons appear to be related – their genes (all of which are intron-less) may have developed from a common ancestral gene. The IFN-α gene exists in many allelic forms (subtypes).

The type II (acid-labile) interferon IFN-γ ($\sim$25 kDa) exhibits little homology with type I interferons in terms of amino acid sequence; even so, the three-dimensional structure of the molecule has features in common with IFN-β. The gene encoding IFN-γ contains three introns.

Distinct types of interferon are found in other animals; for example, IFN-τ (structurally related to IFN-ω) occurs in cows and sheep. These animals also form e.g. interferons related to IFN-β. (A given type of interferon from one species may or may not be active in another species.)

Human interferons of both types have antiviral activity, although this function appears to be of major importance primarily in type I interferons. The type II interferon (IFN-γ) plays a major role in various immune defence mechanisms.

Alpha interferon (IFN-α). IFN-α is produced by leukocytes. At least 14 subtypes of IFN-α occur in humans; these are designated IFN-α1, IFN-α2 etc.

Beta interferon (IFN-β). IFN-β is produced by fibroblasts. Compared with IFN-α, the amino acid sequence of IFN-β exhibits only 30% homology; the two IFNs are antigenically distinct.

Both IFN-α and IFN-β can induce intracellular antiviral effects that include degradation of viral RNA and inhibition of protein synthesis (described later). These IFNs also upregulate MHC class I molecules at the surface of virus-infected cells; this is of value in that class I molecules act as binding sites for $CD8^+$ cytotoxic T cells (which kill virus-infected cells).

Gamma interferon (IFN-γ). IFN-γ is produced by antigenically and/or mitogenically stimulated T lymphocytes. (See also

INTERLEUKIN-18.) The active form of IFN-γ is a non-covalently-bound dimer.

Like type I interferons, IFN-γ can upregulate class I MHC molecules. It can also upregulate class II MHC molecules on certain types of cell (including fibroblasts and endothelial cells); class II molecules promote the ability of cells to present antigen to $CD4^+$ T cells (for which MHC class II molecules are binding sites).

IFN-γ is categorized as a pro-inflammatory cytokine (see CYTOKINES). In addition to its ability to upregulate MHC molecules, IFN-γ can also e.g. (i) activate macrophages, (ii) stimulate the synthesis of NITRIC OXIDE, and (iii) promote class switching in B cells – upregulating e.g. IgG1 (and murine IgG2a) antibodies.

Induction of interferons. Agents which induce the synthesis of interferons include many types of virus, certain bacteria and protozoa, mitogens, lipopolysaccharides, and double-stranded nucleic acids (e.g. poly(I:C) – a polymer consisting of one strand each of polyinosinic acid and polycytidylic acid). (See also TILORONE.)

Many of the studies on induction of interferons have investigated transcriptional regulation of the gene encoding IFN-β – apparently because this gene lacks the many subtypes of IFN-α and because, unlike IFN-γ, it lacks introns.

Induction of the gene encoding IFN-β during viral infection appears to involve activation of the transcription factor NF-κB; this seems to depend on phosphorylation of IκB by a dsRNA-dependent kinase designated PKR [PNAS (1994) *91* 6288–6292]. (IκB normally inhibits NF-κB by binding to it; phosphorylation (and hence degradation) of IκB therefore results in the activation of NF-κB.) Viral infection also appears to promote synthesis of 'interferon regulatory factor-1' (IRF-1) which, together with activated NF-κB, may promote transcription from the regulatory region of the IFN-β gene.

Antiviral activity of interferons. Exogenous interferons initially bind to specific cell-surface receptors. There are various kinds of receptor for type I interferons; some (not all) are common to IFN-α, IFN-β and IFN-ω. The receptor for IFN-γ is distinct from those that bind type I interferons. In all cases, however, the receptor for an interferon consists of a heterodimeric structure associated with tyrosine kinases on the cytoplasmic side; the kinases are involved in initiating the intracellular signal that leads to transcriptional upregulation of certain genes.

A generalized account of antiviral activity is as follows. The binding of interferon to a specific receptor activates a JAK–STAT signalling pathway (see JAK and STATs in CYTOKINES); this promotes transcription of certain genes encoding antiviral proteins. The type of interferon-inducible protein produced in any given case will depend on the type of cell and the type of interferon. Proteins induced by interferons include e.g. receptors for TNF; class I and II MHC antigens; nitric oxide synthetase; CD54 (inducible only by IFN-γ); Mx protein (an inhibitor of influenzavirus replication inducible only by type I interferons); PKR (a protein kinase); and 2–5A synthetase. The two latter proteins are involved in generalized antiviral activity, as follows.

When activated by dsRNA, the PKR kinase phosphorylates (and therefore inactivates) the α subunit of the protein synthesis initiation factor 2; this inhibits formation of initiation complexes and, hence, inhibits the synthesis of both viral and cellular proteins.

The 2–5A synthetase (activated by dsRNA) catalyses the synthesis of 2–5A (= 2,5-oligoadenylate). The 2–5A activates an enzyme, RNase L (also called RNase F), which can cleave viral and cellular RNA; however, 2–5A is unstable, being enzymically degraded by 2′,5′-phosphodiesterase.

The antiviral activity of interferons has been assayed in vitro by a plaque-reduction method in which cultured cells are treated with IFN and subsequently challenged with virus; the titre of IFN may be taken as the reciprocal of the highest dilution of IFN that produces a 50% decrease in the number of plaques.

Therapeutic uses of interferons. Both natural and recombinant forms of interferon (types I and II) have been evaluated for clinical use in a range of diseases, including various types of malignancy, Kaposi's sarcoma (IFN-α), multiple sclerosis (IFN-β), diseases of viral aetiology, and disease caused by *Mycobacterium tuberculosis* (IFN-γ). Commercial products include Alferon N (multicomponent IFN-α; Interferon Sciences, USA); Betaferon and Betaseron (recombinant IFN-β; Schering AG); Gammaferon (recombinant IFN-γ; Rentschler, Germany).

[Clinical uses of interferons: Book ref. 226, pp 237–271.]

The enhanced capacity of lymphocytes to secrete IFN-γ when thalidomide is used for adjuvant therapy may offer an explanation for the ability of thalidomide to bring about clinical improvement in patients infected with e.g. *Mycobacterium tuberculosis* and HIV [AAC (2000) *44* 2286–2290].

(See also ELISPOT ASSAY (for TB).)

intergenic complementation See COMPLEMENTATION TEST.

intergenic repeat unit *Syn.* ERIC SEQUENCE.

intergenic spacer region See RIBOTYPING.

intergenic suppression See SUPPRESSOR MUTATION.

intergenote A partial zygote formed by heterologous TRANSFORMATION (i.e., transformation in which the donor DNA is derived from a species different from that of the recipient).

intergranal frets See CHLOROPLAST.

interkinesis See MEIOSIS.

interkinetal Refers e.g. to the plane of cell division typical of members of the OPALINATA: longitudinal, between the rows of kineties. (cf. HOMOTHETOGENIC.)

interleukin-1 (IL-1) A cytokine (see INTERLEUKINS) secreted by various types of cell, including monocytes, fibroblasts, dendritic cells and endothelial cells. There are several forms of IL-1: IL-1α (previously called lymphocyte activating factor, LAF) and IL-1β (both pro-inflammatory cytokines with similar or identical functions) and IL-1ra; IL-1ra binds to the cognate receptor of IL-1α and IL-1β but, having no agonist activity, it functions as an antagonist of the α and β forms of IL-1. (There are two types of receptor for IL-1 but only one is functional; the other ('decoy') receptor does not initiate intracellular signals when IL-1α or IL-1β binds to it.)

For IL-1γ see INTERLEUKIN-18.

The mature forms of IL-1α and IL-1β are produced by intracellular cleavage of precursor pro-proteins. IL-1β is cleaved (activated) by a cysteine protease referred to as IL-1β-converting enzyme (ICE). (ICE also has a role in APOPTOSIS.)

Cells stimulated to produce IL-1 generally form both the α and β types; however, in at least some cases (e.g. a monocyte responding to LPS), the amount of IL-β produced is much greater than that of IL-1α.

The known or presumed functions of IL-1α and IL-1β in vivo include:

Activation, and induction of cytokine synthesis in macrophages.

Induction of ACUTE-PHASE PROTEINS in the liver.

Stimulation of activated B cells to proliferate and form antibody (and possibly stimulation of T cells).

Induction of synthesis of endothelial adhesion molecules during INFLAMMATION; induction of e.g. IL-1 and colony-stimulating factors in endothelial cells.

Induction of fever (apparently by stimulation of prostaglandins in the temperature regulatory region of the brain).

The biological activities of IL-1 overlap with those of tumour necrosis factor (TNF); IL-1β and TNF-α have been found to be synergistic when jointly administered to animals. Moreover, IL-1β and TNF-α stimulate the production of each other (as well as their own synthesis). However, despite the overlap in activities, it appears that, unlike TNF-α, IL-1 cannot promote cytotoxic activity via apoptosis.

interleukin-2 (IL-2; killer helper factor, KHF; T cell growth factor, TCGF) A cytokine (see INTERLEUKINS) produced by T lymphocytes. In vitro, IL-2 can stimulate proliferation of antigenically or mitogenically activated T cells; in vivo, it appears to promote the growth/differentiation of T cells and to enhance the cytotoxicity of $CD8^+$ T cells and NK cells. (See also CD28 and ANTIBODY FORMATION.)

IL-2 promotes development of the Th1 subset of T lymphocytes, and is one of the cytokines secreted by these pro-inflammatory cells.

The gene encoding IL-2 (like that encoding IFN-γ) contains three introns.

(See also SEVERE COMBINED IMMUNODEFICIENCY.)

interleukin-4 (IL-4) A cytokine (see INTERLEUKINS) produced by lymphocytes of the Th2 subset. IL-4 promotes differentiation of Th0 to Th2 cells, and in antigenically stimulated B cells it seems to e.g. induce class switching to IgG4 and IgE (IgG1 and IgE in mice). IL-4 is also involved in the formation of eosinophils – probably through its ability to promote the Th2 subset and (consequent) secretion of interleukin-5 (a cytokine required for the normal development of eosinophils).

The ability of naive $CD4^+$ T cells to secrete IL-4 is inhibited when ICAM-1 is co-expressed with antigen on the antigen-presenting cell [PNAS (1999) *96* 3023–3028].

interleukin-5 (IL-5) A cytokine (see INTERLEUKINS) produced by the Th2 subset of T lymphocytes (see also INTERLEUKIN-4). IL-5 appears to be the main stimulus for eosinophilia in certain parasitic infections. [Eosinophilia and helminthic infections: BCH (2000) *13* 301–317.]

interleukin-6 (IL-6) A cytokine (see INTERLEUKINS) produced by a variety of cells, including monocytes/macrophages, fibroblasts, endothelial cells and the Th2 subset of T lymphocytes. Synthesis of IL-6 in monocytes/macrophages can be induced by LPS (lipopolysaccharides) or by cytokines (TNF-α or IL-1); the P FIMBRIAE of uropathogenic *Escherichia coli* are reported to be potent inducers of IL-6.

IL-6 appears to have many roles in vivo. These roles include stimulation of growth, differentiation and antibody production in B cells, and induction of ACUTE-PHASE PROTEINS; in at least some cases IL-6 may contribute to the development of fever.

interleukin-8 (IL-8) A cytokine (see INTERLEUKINS) produced by many types of cell, including monocytes/macrophages, neutrophils, fibroblasts, epithelial and endothelial cells, keratinocytes, NK cells and T cells. Agents that stimulate production of IL-8 include LPS and the cytokines IL-1 and TNF; production of IL-8 is inhibited by certain cytokines (e.g. IFN-γ and IL-4) and by glucocorticoids.

IL-8 acts as a chemoattractant and activator primarily for neutrophils (see CHEMOKINES), and its main role in vivo is believed to be the recruitment of leukocytes to sites of infection/INFLAMMATION. IL-8 is found in many inflammatory diseases but, unlike e.g. TNF, it does not induce shock when administered experimentally.

interleukin-10 (IL-10) A cytokine (see INTERLEUKINS) produced e.g. by the Th2 subset of T cells and by monocytes stimulated by LPS. IL-10 has anti-inflammatory activity; for example, it inhibits proliferation and release of cytokines in antigen-stimulated Th1 cells, it inhibits release of pro-inflammatory cytokines from monocytes and macrophages, and it promotes proliferation and synthesis of antibodies in activated B cells.

interleukin-12 (IL-12; natural killer stimulatory factor) A cytokine (see INTERLEUKINS) produced by B cells and also by macrophages acting as antigen-presenting cells. IL-12 appears to be essential for development of the Th1 subset of T cells from naive T cells; it also stimulates cytotoxicity in NK cells and enhances the release of cytokines from NK cells, macrophages and T cells. IL-12 also has anti-angiogenic activity (it inhibits development of new blood vessels).

The ability of macrophages to synthesize IL-12 is inhibited by the binding of measles virus to the complement receptor CD46.

interleukin-18 (*syn.* interleukin-1γ) An INTERLEUKIN produced e.g. by macrophages, keratinocytes and (murine) Kupffer cells. IL-18 is a pro-inflammatory cytokine which induces IFN-γ in T cells (and murine spleen cells), enhances cytotoxicity in NK cells, and promotes the Th1 (rather than Th2) immune response. IL-18 also triggers activation of NF-κB and synthesis of e.g. IL-1β, TNF-α and Fas ligand.

IL-18 is synthesized in a biologically inactive form; it is activated by IL-1β converting enzyme (ICE; also referred to as caspase-1).

IL-18-mediated induction of IFN-γ can be inhibited by the binding of poxvirus-encoded proteins to IL-18 [PNAS (1999) *96* 11537–11542 (correction: PNAS (2000) *97* 11673)].

interleukin-21 An INTERLEUKIN which, in vitro, influences the proliferation/maturation of NK cells from bone marrow, the proliferation of B cells co-stimulated with anti-CD40, and the proliferation of T cells co-stimulated with anti-CD3. IL-21 is synthesized by activated $CD4^+$ T cells. [Nature (2000) *408* 57–63.]

interleukins Certain CYTOKINES which are secreted by, and effective on, *leukocytes* (white blood cells) – but which, in some cases, can also be secreted by other types of cell (e.g. endothelial cells). It should be noted that the nomenclature is inconsistent in that some cytokines which conform to this description (e.g. tumour necrosis factor) are not formally referred to as interleukins; moreover, some interleukins are also classified in other named categories (e.g. interleukin-8 is a chemokine).

Each interleukin (IL) is designated by a number – e.g. IL-4, IL-6, IL-21.

(See also separate entries for individual interleukins.)

intermediate filaments (IFs; IMFs) In most types of eukaryotic cell: protein filaments, ca. 7–11 nm thick, which form part of the CYTOSKELETON. IMFs appear to be much more stable than either microfilaments or microtubules, and their roles are believed to include the strengthening of the cytoskeleton. IMFs differ e.g. according to cell type and species, and the monomers of different IMFs vary widely in size; the monomer *vimentin* (MWt ca. 55000) occurs in many types of cell.

intermediate host In heteroxenous coccidia: the host in which the asexual phase occurs (see EIMERIORINA).

internal guide sequence See SPLIT GENE (b).

internalin A See LISTERIOSIS.

interphase (1) See MITOSIS. (2) See DICTYOSTELIOMYCETES.

interrupted gene *Syn.* SPLIT GENE.

interrupted mating A technique used for studying gene transfer during bacterial CONJUGATION. For example, a population of cells of an Hfr strain of *Escherichia coli* is mixed with an F^- (recipient) population (time zero). At regular intervals of time, a sample is taken from the mating mixture, agitated to separate mating cells, diluted to reduce the probability of further mating contacts, and plated on a medium which will support the growth of particular recombinant(s) but not of the parent Hfr or recipient cells. Samples taken within ca. 5 min of time zero generally do not contain recombinants; samples taken after increasing periods of time will contain recombinants with an increasing number of donor genes. Since donor genes are transferred sequentially, a given donor gene will appear in a recombinant after a characteristic time interval which depends on the distance of the gene from the origin of transfer and on the efficiency with which the gene recombines with the recipient's chromosome. Thus, the time of transfer of a gene gives an indication of the position of that gene in the donor chromosome, and interrupted mating can be a useful means of genetic mapping. (In practice, the further a gene is from the origin of transfer, the less frequently will it be transferred owing to the increased chances of strand breakage during the longer conjugation times; thus, the *yield* of recombinants expressing a given donor gene is also indicative of the position of that gene.) In *E. coli* the time taken to transfer the entire chromosome is typically ca. 100 min at 37°C (estimated using several Hfr strains differing in the position of the origin of transfer). The position of any gene can be given in terms of the time (in *minutes*) of transfer of that gene relative to a given origin of transfer – arbitrarily standardized at the *thr* (threonine synthesis) locus (zero minutes).

interspecies hydrogen transfer The mutually beneficial, unidirectional transfer of H_2 from H_2-producing to H_2-utilizing organisms in a given ecosystem (see e.g. ANAEROBIC DIGESTION).

interstitial pneumonia See PNEUMONIA.

intervening sequence See SPLIT GENE.

intestinal tract flora See GASTROINTESTINAL TRACT FLORA.

intimin See EPEC and PATHOGENICITY ISLAND.

intine See CYST (a).

Intoshellina See ASTOMATIDA.

intoxication Poisoning. See e.g. FOOD POISONING and MYCOTOXICOSIS.

intra-epithelial lymphocytes (IELs) See T LYMPHOCYTE.

intra vitam staining See VITAL STAINING.

intracisternal A-type particles See A-TYPE PARTICLES.

intracistronic complementation See COMPLEMENTATION TEST.

intracytoplasmic membranes (also: cytomembranes) (*bacteriol.*) Membranous systems or structures present within the cytoplasm in certain bacteria; according to type, the membranes may or may not be continuous with the CYTOPLASMIC MEMBRANE. Intracytoplasmic membranes occur e.g. in members of the METHYLOCOCCACEAE and NITROBACTERACEAE, in HYDROCARBON-utilizing bacteria, in photosynthetic bacteria (see CHLOROSOMES, CHROMATOPHORE (sense 2), THYLAKOIDS), etc. (See also MESOSOME and POLAR MEMBRANE.)

intragenic complementation See COMPLEMENTATION TEST.

intragenic suppression See SUPPRESSOR MUTATION.

Intrapes See UREDINIOMYCETES.

Intrasporangium A genus of aerobic bacteria (order ACTINOMYCETALES, wall type I). The organisms form branching substrate mycelium (hyphae: 0.4–1.2 μm in diam.) but no aerial hyphae; intercalary or subterminal swellings ('sporangia') occur in old cultures. Type species: *I. calvum*.

intravital staining See VITAL STAINING.

intrinsic resistance (to antibiotics) See ANTIBIOTIC.

intron See SPLIT GENE.

intron homing (in bacteria) The process in which introns (of either group I or group II) spread, in replicative fashion, to *allelic* intron-less genes. The mechanism of intron homing differs in the two types of intron.

Intron homing in group I introns is initiated when a DNA endonuclease (*homing endonuclease*), encoded within the intron sequence, acts at the potential insertion site in an intron-less allele, typically making a double-stranded cut. The lesion is repaired by enzymes which use the intron-containing gene as a template; in this way, a copy of the intron is synthesized in the insertion site of the intron-less allele.

In group II introns, the process begins with an intron-encoded protein which forms a complex with an RNA copy of the intron. This RNA–protein (RNP) complex is able to recognize the insertion site in an intron-less allele. The RNP complex can then (i) cut both strands at the target site, covalently inserting the RNA into one strand, and (ii) form a cDNA copy of the RNA (by virtue of the protein's reverse transcriptase function); the RNA is then replaced by DNA. The involvement of a reverse-transcribed copy of the intron (cf. group I introns) is reflected in the term *retrohoming* for the process in bacterial group II introns. Intron homing (in both group I and II bacterial introns) is outlined in a recent minireview [JB (2000) *182* 5281–5289 (5281–5282)].

inulin A linear FRUCTAN consisting of $(2 \rightarrow 1)$-β-linked fructofuranose residues. Inulins occur e.g. as storage polysaccharides in certain plants (e.g. in dahlia tubers).

inulinase An enzyme (β-fructosidase, 2,1-β-D-fructan-fructanohydrolase, EC 3.2.1.7) which hydrolyses INULIN to FRUCTOSE. Many microbial inulinases also show INVERTASE activity, and some can hydrolyse bacterial LEVAN. Microbial inulinases may prove useful in the commercial production of fructose from plant inulins [AAM (1983) *29* 139–176].

***inv* gene** (*Yersinia*) See FOOD POISONING (*Yersinia*).

***inv–spa* genes** See PATHOGENICITY ISLAND and FOOD POISONING (*Salmonella*).

invasin In certain pathogenic bacteria: a type of cell-surface molecule which promotes uptake of the pathogen by (eukaryotic) host cells. Examples of invasins include 'invasin' (see FOOD POISONING (*Yersinia*)) and 'internalin A' (see LISTERIOSIS). In strains of UPEC (q.v.), the FimH adhesin appears to function as an invasin.

invasion (of phagocytes by pathogens) See end of section (a) in PHAGOCYTOSIS.

invasiveness (of a pathogenic microorganism) The ability of a pathogen to spread through a host's tissues; the degree of invasiveness reflects the relative insusceptibility of the pathogen to the host's defense mechanisms. (See also AGGRESSIN.)

invasomes See FOOD POISONING (*Salmonella*).

inversion (*mol. biol.*) See e.g. CHROMOSOME ABERRATION and RECOMBINATIONAL REGULATION.

invert sugar See INVERTASE.

invertase (1) (saccharase; sucrase; β-fructosidase; β-D-fructofuranoside fructohydrolase; EC 3.2.1.26) An enzyme which hydrolyses SUCROSE to glucose and fructose; it can also hydrolyse raffinose to fructose and melibiose. Invertase occurs in many yeasts and other fungi and in higher plants; it is obtained commercially mainly from BAKERS' YEAST in which it occurs in the cell wall as a mannoprotein containing ca. 50% mannose. Invertase is used chiefly in the conversion of sucrose to 'invert sugar' (a mixture of glucose and fructose) used as a sweetening agent. (See also ENZYMES.)

(2) (*mol. biol.*) A recombinase responsible for the inversion of a segment of DNA: see RECOMBINATIONAL REGULATION.

invertebrate diseases See e.g. CRUSTACEAN DISEASES; INSECT DISEASES; NEMATOPHAGOUS FUNGI; OYSTER DISEASES.

inverted repeat (IR) In a double-stranded nucleic acid molecule: either of two regions in which the sequences of base pairs are similar or identical and have the same polarity but are opposite in orientation, e.g.:

$$5' \ldots GGCT \ldots AGCC \ldots 3'$$
$$3' \ldots CCGA \ldots TCGG \ldots 5'$$

(The repeated sequences may or may not be contiguous.)

In a single-stranded nucleic acid, IRs are sequences which are complementary and opposite in polarity, i.e., equivalent to one of the strands of a double-stranded IR.

(cf. DIRECT REPEAT; see also HAIRPIN and PALINDROMIC SEQUENCE.)

invertible DNA See RECOMBINATIONAL REGULATION.

***invJ* gene** (*Salmonella typhimurium*) See NEEDLE COMPLEX.

involution forms (of bacteria) Morphologically or otherwise aberrant cells often seen in old cultures or among organisms cultured under harsh or hostile conditions (e.g. in the presence of sub-lethal concentrations of antibiotics). Involution forms are generally regarded as degenerate cells, degeneracy commonly being ascribed to the action of autolysins, failure of cell wall synthesis, or failure of cell division without cessation of macromolecule synthesis. (cf. L-FORM.)

Iodamoeba A genus of protozoa of the AMOEBIDA. *I. bütschlii* occurs in the human intestine; it is non-pathogenic. Trophozoites (ca. 9–20 μm diam.) contain numerous vacuoles and form blunt pseudopodia slowly; the single nucleus lacks peripheral chromatin and contains a large karyosome which is variable in shape and location and which is surrounded by clusters of granules. Cysts are irregular in shape, ca. 5–16 μm in diam., containing (usually) a single nucleus and a large persistent glycogen granule which stains a characteristic dark-brown with iodine; chromatoid bodies are usually absent.

iodinated density-gradient media Media used e.g. for the separation of macromolecules and sub-cellular fractions by isopycnic ultraCENTRIFUGATION. High-density, non-ionic media in this category include e.g. a tri-iodinated benzamido derivative of glucose (*Metrizamide*), and a tri-iodinated derivative of benzoic acid (substituted with three highly hydrophilic aliphatic side chains) having a MWt of 821 and a density of 2.1 g/ml (*Nycodenz* – trade mark of Nyegaard & Co., Oslo, Norway).

iodine (a) (as an antimicrobial agent) Iodine has microbicidal activity against a wide range of bacteria (including endospores), fungi, and viruses; in many cases its antimicrobial activity appears to be due to the combination of molecular iodine with proteins (e.g. tyrosine is irreversibly iodinated). Iodine is readily soluble in organic solvents; *tincture of iodine* (used as an ANTISEPTIC) contains ca. 2.5% iodine in an ethanol–water solution of potassium iodide. The (low) solubility of iodine in water is increased by the presence of iodide, when the tri-iodide (I_3^-) is formed; I_3^- itself is not antimicrobial, but it readily decomposes to release free iodine. For skin antisepsis aqueous solutions are generally preferred since they are less irritant than alcoholic solutions; the latter have been used e.g. for the disinfection of thermometers. The antimicrobial activity of aqueous iodine solutions is sometimes said to be maximal under acid conditions (pH 6 or below) – when dissolved iodine is present largely in the molecular form (I_2); however, an iodine solution was reported to inactivate poliovirus (at 25°C) more efficiently at pH 10 than at pH 6 (at pH 10 iodine occurs mainly as hypoiodous acid, HIO) [AEM (1982) *44* 1064–1071]. Both I_2 and HIO are strongly microbicidal; the ratio I_2:HIO in a solution of iodine is affected not only by pH but also by the total concentration of dissolved iodine and by the presence of organic matter [Book ref. 65, pp. 183–196]. The staining, corrosive and irritant properties of iodine can be overcome by complexing the element with certain high-MWt surface-active compounds – e.g. polyvinyl pyrrolidone, certain quaternary ammonium compounds; such complexes are referred to as *iodophores* (or iodophors). Iodophores have the antimicrobial properties of iodine together with the detergent properties of the surfactant, but only ca. 80% of the bound iodine is released; their response to pH is similar to that of iodine. Iodophores have been used e.g. for the disinfection of swimming-pool water and dairy equipment; polyvinyl pyrrolidone–iodine (*povidone–iodine, PVP–iodine*) has been used for treating skin and mucous membrane infections involving species of *Streptococcus*, *Staphylococcus* and *Candida*. *Betadine* and *Wescodyne* are iodophores used as antiseptics/disinfectants. (cf. IODOFORM; IODONIUM COMPOUNDS; see also CHLORINE.)

(b) (as a stain) See e.g. LUGOL'S IODINE.

iodinin A red, water-soluble pigment: 1,6-phenazinediol-5,10-dioxide; crystals of iodinin are formed in the culture medium e.g. by *Microbispora parva*.

iodochlorohydroxyquin See 8-HYDROXYQUINOLINE.

5-iodo-2′-deoxyuridine *Syn.* IDOXURIDINE.

iodoform (CHI_3) A compound once widely used as an antiseptic; its antimicrobial activity is poor compared with that of IODINE.

iodonium compounds Compounds of the form R_2IX, in which R is an organic group and X is an inorganic ion; such compounds (e.g. diphenyliodonium chloride) contain trivalent iodine, are strongly basic, and possess antimicrobial properties, but their activity against spores is doubtful.

iodophores (iodophors) See IODINE (a).

ion-exchange chromatography See CHROMATOGRAPHY.

ion-motive force phosphorylation See ELECTRON TRANSPORT PHOSPHORYLATION.

ion transport Ion fluxes across ENERGY-TRANSDUCING MEMBRANES are involved e.g. in certain types of energy conversion (see e.g. PROTON ATPASE, PROTON PPASE, END-PRODUCT EFFLUX), in the ion-linked uptake of certain substrates, in the regulation of intracellular osmotic pressure and pH, in the generation of electrochemical gradients of specific ions, and (apparently) in morphogenetic processes (see GROWTH (fungal)). Since ions cannot pass freely across such membranes (cf. OUTER MEMBRANE) their transmembrane translocation requires a driving force and more or less specific TRANSPORT SYSTEMS which may involve UNIPORT, ANTIPORT or SYMPORT processes. The energy needed for transmembrane ion translocation may be provided, directly, by (a) a concentration gradient of the ion itself; (b) a gradient of another ion – e.g. H^+ (see pmf in CHEMIOSMOSIS) or Na^+ (see SODIUM MOTIVE FORCE); (c) ATP hydrolysis (see e.g. PROTON ATPASE); (d) the direct coupling of metabolism to ion transport (see ELECTRON TRANSPORT CHAIN and SODIUM MOTIVE FORCE); or (e) light (see PHOTOSYNTHESIS and PURPLE MEMBRANE).

Ion electrochemical gradients as driving forces in ion transport. A difference in the concentrations of a given ion on either side of a membrane constitutes an electrochemical gradient which may be able to provide a driving force for the transmembrane transfer of the ion itself or of other types of ion. However, the presence of such a gradient, and the existence of transmembrane route(s) for a given ion, does not necessarily mean that the ion will be translocated: ion transport is subject to various intracellular and extracellular regulatory factors; moreover, the transmembrane route taken by a given ion may vary according e.g. to the extracellular concentration of that ion.

The potential energy of a transmembrane gradient of ions can be considered in terms of the change in Gibbs free energy which occurs when a charged species is transferred down an electrochemical gradient. Thus, e.g., if one mole of an ion, X^{m+}, is transferred down an electrical potential gradient of $\Delta\psi$ millivolts, from concentration $[X^{m+}]'$ to $[X^{m+}]''$, the net change in Gibbs free energy (ΔG) can be obtained from the general electrochemical equation:

$$\Delta G = -mF\Delta\psi + 2.3RT\log_{10}\frac{[X^{m+}]''}{[X^{m+}]'} \qquad (1)$$

in which R is the gas constant, T is the absolute temperature, F is the Faraday constant, and m is the charge on the ion. This expression can be readily converted to the ion electrochemical potential gradient, $\Delta\tilde{\mu}_{X^{m+}}$:

$$\Delta\tilde{\mu}_{X^{m+}} = m\Delta\psi - Z\log_{10}\frac{[X^{m+}]''}{[X^{m+}]'} \qquad (2)$$

where Z is the constant 2.3RT/F. When applied to the gradient of an ion which crosses the membrane by electrogenic or electrophoretic uniport, equation 2 gives a measure of the driving force (due to the gradient itself) tending to cause movement of that ion across the membrane. (When an ion is translocated across a membrane by antiport or symport mechanisms equation 2 must be modified so as to accommodate the additional forces acting on the ion.)

If a proton gradient is maintained (imposed) across an energy-transducing membrane (e.g. by the use of metabolic energy) it will influence the electrochemical equilibria of each of the various types of ion distributed across the membrane. A given cation which crosses the membrane by electrical uniport will tend to redistribute across the membrane so as to reach an electrochemical equilibrium with the membrane potential of the pmf. For example, in the case of the bacterial cytoplasmic membrane (interior negative) a given cation which, initially, is largely extracellular will tend to be accumulated within the cytoplasm; if the cation reaches electrochemical equilibrium, the Gibbs free energy associated with its gradient becomes zero (i.e., $\Delta\tilde{\mu}_{X^{m+}} = 0$), and the relationship of the imposed membrane potential to the intracellular and extracellular concentrations of the given cation is then given by the Nernst equation:

$$\Delta\psi = \frac{Z}{m}\log_{10}\frac{[X^{m+}]_{int}}{[X^{m+}]_{ext}} \qquad (3)$$

Equation 3 is obtained by re-arranging equation 2 in the special case when $\Delta\tilde{\mu}_{X^{m+}} = 0$.

In living cells, transmembrane ion translocation may affect either or both components of pmf. When ions are translocated by antiport or symport processes, the effect of such translocation on pmf depends e.g. on the stoichiometry of the process. For example, under some conditions the Na^+/H^+ antiporter in *Escherichia coli* consumes pmf, H^+ influx exceeding Na^+ efflux. (Since the H^+/Na^+ ratio is >1, operation of the antiporter in this mode consumes $\Delta\psi$ as well as ΔpH.) However, with high extracellular pH and Na^+ the antiporter operates in such a way that H^+ efflux is greater than Na^+ influx – with consequent augmentation of pmf.

Ion uptake or extrusion commonly leads to compensatory ion fluxes. Thus, e.g. in mitochondria the effects of pmf-dependent Ca^{2+} uptake are modified by the operation of Ca^{2+} efflux systems, and a steady-state level of intramitochondrial Ca^{2+} is maintained by the continual cycling of Ca^{2+} across the membrane.

Modes of ion transport include:

(a) *Uniport driven by pmf.* Examples include pmf-mediated ATP synthesis (see PROTON ATPASE) and the uptake of Ca^{2+} by mitochondria [Book ref. 126, pp. 251–255].

(b) *Uniport coupled to ATP hydrolysis.* One example is the so-called Kdp transport system for K^+ uptake in *Escherichia coli*. In this system, ATP hydrolysis (rather than pmf) at a membrane-bound K^+-ATPase ('K^+-dependent ATPase', 'K^+-stimulated ATPase', 'K^+-pump') brings about the transmembrane translocation of K^+ into the cell against a concentration gradient; the Kdp system becomes derepressed under conditions of K^+ starvation. [K^+ pathway in *E. coli*: Book ref. 126, pp. 653–666.] *Enterococcus faecalis* has been reported to use an Na^+-ATPase for Na^+ extrusion [JB (1984) *158* 844–848], and appears to use a Ca^{2+}-ATPase for Ca^{2+} extrusion.

(c) *Antiport driven by pmf or smf.* A Ca^{2+}/H^+ antiporter has been reported to operate in membrane vesicles of *Bacillus subtilis* for Ca^{2+} extrusion [JB (1985) *164* 1294–1300]. Similar systems appear to function in *E. coli* and *Azotobacter vinelandii* for Ca^{2+} extrusion, while a Ca^{2+}/Na^+ antiporter may occur in halophilic organisms (e.g. *Halobacterium* spp). Under some conditions the Na^+/H^+ antiporter in *E. coli* can be driven by pmf, extruding Na^+ and thus generating a SODIUM MOTIVE FORCE which can be used e.g. for melibiose uptake (Na^+/melibiose symport) and/or for the regulation of intracellular pH. However, in media of high pH, containing high concentrations of Na^+, the Na^+/H^+ antiporter can generate pmf by proton extrusion [JB (1985) *163* 423–429]. The K^+/H^+ antiporter in *E. coli* is believed to function e.g. for the prevention of over-alkalinization of the cytoplasm in respiring cells.

(d) *Antiport coupled to ATP hydrolysis.* An Na^+/K^+-ATPase ('sodium pump') occurs in eukaryotic cells but not in prokaryotic cells or in mitochondria; energy derived from ATP hydrolysis at this ATPase is used to pump Na^+ out and K^+ in – the Na^+/K^+ exchange ratio probably being ca. 3:2. The sodium pump is used e.g. to maintain intracellular osmotic stability, and to generate the smf needed for the uptake of certain substrates; it is inhibited by ouabain.

(e) *Symport systems.* In *E. coli*, LACTOSE is generally taken up by an H^+/lactose symport, while melibiose uptake involves an Na^+/melibiose symport. *Bacillus subtilis*, grown aerobically on citrate-containing media, can accumulate Ca^{2+} by a Ca^{2+}/citrate symport. An electrogenic H^+/lactate symport is used by some fermentative bacteria for energy conversion (see END-PRODUCT EFFLUX). In mitochondria, phosphate (Pi) uptake occurs by electroneutral symport with H^+.

(f) *Other ion transport systems.* The so-called Trk system is responsible for much of the uptake of K^+ in *E. coli*, and its activity is influenced by extracellular osmotic pressure. Trk is apparently not an ATPase, but it may be regulated by ATP or by a related compound [JGM (1985) *131* 77–85].

ionizing radiation (in sterilization) Ionizing radiations (e.g. *gamma*-rays, *beta*-rays (electrons), X-rays) are used e.g. for the STERILIZATION of pre-packed medical and biological equipment such as surgical sutures, syringes, plastic Petri dishes, etc; currently they are not used to any great extent for the treatment of foodstuffs because they can cause toxicity and a deterioration in organoleptic properties in foods (but see FOOD PRESERVATION (g)). (See also SEWAGE TREATMENT.)

Ionizing radiations sterilize by supplying energy which permits a great variety of lethal chemical reactions to occur in contaminating organisms. In general, resistance to ionizing radiation increases in the order: multicellular organisms; Gram-negative bacteria; Gram-positive bacteria and moulds; bacterial endospores, viruses and viroids. Exceptions include e.g. *Deinococcus radiodurans*, which is more resistant than most or all bacterial endospores. Low (sublethal) doses of radiation can be mutagenic: e.g., in *Escherichia coli* the SOS SYSTEM – and hence error-prone (mutagenic) repair – may be induced. (In some cases, susceptibility to ionizing radiation is enhanced by the presence of water and/or oxygen.) Typically, toxins and other microbial products are more resistant than cells to ionizing radiations. Ionizing radiation may affect the materials being sterilized – e.g., a sterilizing dose of radiation can damage certain plastics.

Large-scale industrial sterilization is generally carried out with either a source of *gamma*-rays or a source of high-energy electrons. In a *gamma*-radiation plant the source of radiation is ^{60}Co (cobalt-60): a radioactive isotope (half-life ca. 5.3 years) which provides both *gamma*-rays and (relatively) low-energy electrons. High-energy electrons (ca. 5–10 MeV) are produced in electron accelerators. The high-energy electron radiation used for sterilization carries more energy than does the *gamma*-radiation, although it has poorer powers of penetration; sterilization processes using high-energy electrons require exposures of the order of seconds, while those using *gamma*-rays require minutes or hours. The actual sterilizing dose used should, ideally, take into account e.g. the numbers and types of contaminants. In many countries the recommended minimum sterilizing dose is ca. 2.5 Mrad (25 kGy, see GRAY); a higher minimum is recommended in Scandinavia.

X-rays (a form of electromagnetic radiation) have been used e.g. for sterilizing items of high density which are impermeable to a beam of high-energy electrons.

[Quality assurance in radiation sterilization (theoretical treatment): JAB (1985) *58* 303–313.]

(See also ULTRAVIOLET RADIATION.)

β-ionone ring See CAROTENOIDS.

ionophore Any substance which actually or effectively increases the permeability of biological membranes (or of artificial lipid bilayers) to one or more types of ion. *Channel-forming* ionophores (e.g. GRAMICIDINS) form pores in the membrane; such ionophores typically exhibit little discrimination between different ions, and may allow the passage of up to ca. 10^7 ions per pore per second. *Mobile carrier* ionophores (e.g. PROTON TRANSLOCATORS) diffuse within the membrane; they can exhibit a high degree of specificity for ions. Mobile carriers may effect ion UNIPORT (see e.g. VALINOMYCIN) or ion ANTIPORT (see e.g. A23187 and NIGERICIN). (See also ENNIATINS, LASALOCID, MACROTETRALIDES, SALINOMYCIN and ION TRANSPORT.)

Iotech process A process, involving explosive depressurization, in which the lignin–hemicellulose–cellulose complex from plant material is disrupted to release CELLULOSE for use as a feedstock in certain biotechnological processes (e.g. SINGLE-CELL PROTEIN or INDUSTRIAL ALCOHOL production).

IP10 (chemokine) See CHEMOKINES.

IpaA protein See DYSENTERY.

IpaB protein See APOPTOSIS, DYSENTERY and PROTEIN SECRETION (type III systems).

IPN virus (IPNV) The causal agent of INFECTIOUS PANCREATIC NECROSIS disease of salmonid fish. The virion has an unenveloped, icosahedral capsid (ca. 59 nm diam.) containing a genome of two linear pieces of dsRNA (MWts ca. 2.3×10^6 and 2.5×10^6). Each strand of both RNA molecules is covalently linked by its 5′ end to a protein – the first known case of genome-linked protein in a dsRNA virus. A minor protein component (VP105) is reputed to have dsRNA-dependent RNA transcriptase activity. [Review: CJM (1983) *29* 377–384.] Virus transmission occurs vertically and horizontally.

IPNV is representative of a family of bisegmented dsRNA animal viruses, Birnaviridae [approved: Intervirol. (1986) *25* 141–143], which includes the INFECTIOUS BURSAL DISEASE VIRUS of chickens, DROSOPHILA X VIRUS, and various 'IPNV-like' viruses isolated from freshwater and marine fish (e.g. Eel Virus European, EVE, isolated from Japanese eels: *Anguilla japonica*; see also SPINNING DISEASE) and from marine bivalve molluscs (e.g. oysters – *Ostrea* and *Crassostrea* spp – and *Tellina tenuis*).

ipomeamarone The (+)-enantiomer of 2-methyl-2-(4-methyl-2-oxopentyl)-5-(3-furyl) tetrahydrofuran: a PHYTOALEXIN produced by the sweet potato (*Ipomoea batatas*) in response to infection by certain fungal pathogens (e.g. the 'black rot' fungus *Ceratocystis fimbriata*). Ipomeamarone and its (−)-enantiomer ngaione (a normal constituent of certain trees and shrubs) are both hepatotoxic and can cause liver disease in livestock. (See also 4-IPOMEANOL.)

4-ipomeanol A substance, (1-(3-furyl)-4-hydroxy-1-pentanone), produced by the sweet potato (*Ipomoea batatas*) in response to infection by certain fungal pathogens. It is toxic to animals, causing pulmonary oedema and massive pleural effusion. Together with e.g. ipomeanine (a dihydro derivative of 4-ipomeanol) it may be the 'lung oedema factor' responsible for the fatal respiratory disease in livestock (especially cattle) which follows ingestion of mould-damaged sweet potatoes. (cf. IPOMEAMARONE.)

iprodione See DICARBOXIMIDES.

IPTG Isopropyl-β-D-thiogalactoside: a gratuitous inducer of the LAC OPERON.

IQB INDIVIDUAL QUICK BLANCH.

IR INVERTED REPEAT.

Ir genes Immune response genes: genes, associated with the MHC, which determine the ability of a mouse to make humoral and cell-mediated immune responses to a wide range of antigens; certain of these genes encode the IA ANTIGENS.

Irgasan DP 33 *Syn.* TRICLOSAN.

Iridaea See RHODOPHYTA.

iridescence Coloured light formed by interference when white light is reflected from both surfaces of a thin film.

iridescent insect viruses See CHLORIRIDOVIRUS and IRIDOVIRUS.

Iridia See FORAMINIFERIDA.

Iridoviridae A family of icosahedral dsDNA-containing VIRUSES which infect invertebrates (mainly insects) and poikilothermic vertebrates. The family contains four genera: CHLORIRIDOVIRUS, IRIDOVIRUS, LYMPHOCYSTIVIRUS and RANAVIRUS. (cf. AFRICAN SWINE FEVER.)

Virion: icosahedral (ca. 120–300 nm diam.) and complex in structure, consisting of a nucleoprotein core surrounded by a layer composed of a lipoprotein membrane beneath an icosahedral protein lattice. Some virions may have an additional lipoprotein ENVELOPE derived from the host cell plasmalemma or endoplasmic reticulum; the envelope is apparently not essential for infectivity. Virions are generally stable at pH 3–10 and at 4°C; they can be inactivated e.g. at 55°C for 15–30 min. Members of the genera *Ranavirus* and *Lymphocystivirus* are sensitive to ether and non-ionic detergents.

Genome: one molecule (possibly two in some cases) of linear dsDNA (MWt ca. $100–250 \times 10^6$); in at least some iridoviruses

(Chilo iridescent virus, frog virus 3, lymphocystis virus) the DNA is circularly permuted and terminally redundant – features unique among animal viruses (but quite common among bacteriophages). The molecular biology of iridoviruses has been studied mainly in FROG VIRUS 3.

Iridovirus (small iridescent insect virus group) A genus of viruses (family IRIDOVIRIDAE) which infect insects; infected larvae – and purified virus pellets – exhibit a blue iridescence. Virions (ca. 120 nm diam.) are ether-resistant. Type species: Chilo iridescent virus (insect iridescent virus type 6) [Intervirol. (1986) *25* 141–143]. Other members include e.g. *Tipula* iridescent virus, which infects e.g. craneflies (*Tipula* spp) and many other insects, and insect iridescent viruses 1, 2, 9, 10 and 16–29. Probable member: *Chironomus plumosus* iridescent virus (*C. plumosus* is a type of non-biting midge, with haemoglobin-containing larvae known as 'blood-worms'). Possible member: *Octopus vulgaris* disease virus.

Irish moss See CHONDRUS.

iron Most microorganisms need iron to form e.g. CATALASE, various oxidases, IRON–SULPHUR PROTEINS, ribonucleotide reductase, certain SUPEROXIDE DISMUTASES and/or CYTOCHROMES. Rarely – as e.g. in the type strain of *Lactobacillus plantarum* [FEMS (1983) *19* 29–32] – iron is not an absolute requirement for growth. In many cases microorganisms synthesize and export SIDEROPHORES (q.v.) specifically for the sequestration and uptake of iron, particularly when iron is scarce.

Theoretically, siderophores could behave simply as shuttles – remaining extracellular and delivering iron from the environment to the cell surface for internalization by a membrane TRANSPORT SYSTEM. However, in *Escherichia coli*, the siderophore *enterochelin* (= enterobactin) carries (ferric) iron into the *cytoplasm* in an energy-dependent uptake process – binding initially at the cell-surface FEPA PROTEIN; within the cytoplasm, cleavage of the siderophore (by an esterase) releases the iron. Fe^{3+} is reduced to Fe^{2+}.

Some strains of *E. coli* produce *aerobactin*, a siderophore whose outer membrane receptor is the Iut protein.

Interestingly, *E. coli* can also use iron–coprogen and iron–ferrichrome complexes (receptors: FhuE and FhuA (= TonA) proteins, respectively) – even though *E. coli* produces neither coprogen nor ferrichrome. *E. coli* can also take up iron–citrate complexes.

Iron uptake: regulation. In e.g. *E. coli*, accumulation of intracellular Fe^{2+} results in binding of ferrous ions to the regulator FUR PROTEIN which then represses the expression of a number of genes (including *fep*).

Iron in pathogenicity. Pathogens in animals must compete successfully with the host for available iron. [Iron in pathogenicity: RMM (1998) *9* 171–178.] Thus, many pathogenic strains of *E. coli* secrete aerobactin; this promotes bacterial growth in extracellular concentrations of iron ca. 500-fold lower than those needed by the enterochelin system [Inf. Immun. (1986) *51* 942–947]. Aerobactin seems able to compete successfully with the SIDEROPHILIN *transferrin* [Mol. Microbiol. (1994) *14* 843–850].

For extracellular growth, *Mycobacterium tuberculosis* uses EXOCHELINS (which may obtain iron from siderophilins) and cell-surface ferric iron chelators: MYCOBACTINS (which internalize iron transferred from the exochelin).

Some pathogens obtain iron *directly* from siderophilins; thus, e.g. *Neisseria gonorrhoeae* and *N. meningitidis* have receptors for human TRANSFERRIN.

Listeria monocytogenes uses siderophores from other bacteria and various iron-chelating agents in the environment and mammalian hosts; a cell-surface ferric reductase may recognize all of these agents.

A pathogen's absolute need for iron has prompted the search for new vaccines that stimulate protective antibodies against the cell-surface receptors for iron-chelating agents.

Pseudomonas aeruginosa can reduce PYOCYANIN to leucopyocyanin which, in turn, can reduce the (ferric) iron in iron–transferrin complexes [Inf. Immun. (1986) *52* 263–270]; iron thus released from transferrin may be available to iron-limited bacteria.

Virulence in the fish pathogen *Vibrio anguillarum* (see VIBRIO) depends partly on a plasmid-encoded siderophore, *anguibactin*. [Plasmid-mediated iron transport and bacterial virulence: ARM (1984) *38* 69–89.]

In plant-pathogenic *Erwinia chrysanthemi*, mutants defective in iron transport were non-pathogenic [JB (1985) *163* 221–227].

[Iron metabolism in pathogenic bacteria: ARM (2000) *54* 881–941.]

Applied aspects. Siderophore-producing strains of *Pseudomonas putida* used for the BACTERIZATION of seed potatoes apparently increase the yield in short-rotation crops [NJPP (1986) *92* 249–256]; thus, siderophores may modify the RHIZOSPHERE microflora in favour of the potato plant. Interestingly, some microbial ferrisiderophores are taken up by plants [see for example FEBS (1986) *209* 147–151].

iron bacteria Bacteria whose growth is associated with the extracellular deposition of oxides or hydroxides of iron and/or manganese; such deposits usually cover or impregnate the bacterial capsule, sheath or stalk. Iron bacteria occur e.g. in various iron- and manganese-containing soils and waters, and in water distribution systems (see also TUBERCLE (sense 2)). Evidence that the deposition of metal oxides/hydroxides is a *direct* result of microbial metabolism has not been obtained for the majority of the iron bacteria (cf. GALLIONELLA) – many of which have not even been grown in pure culture; the opportunity for biological oxidation of Fe^{2+} is relatively poor under aerobic conditions (in which Fe^{2+} undergoes rapid autoxidation) but becomes significant under microaerobic conditions, particularly at pH 5–6, in which autoxidation is slow. (Autoxidation of Mn^{2+} is slow below pH 9.) The iron bacteria include e.g. CLONOTHRIX, CRENOTHRIX, GALLIONELLA, LEPTOTHRIX, METALLOGENIUM, OCHROBIUM and SIDEROCAPSA; some authors include the acidophilic iron oxidizers, e.g. *Thiobacillus ferrooxidans*.

[Review: ARM (1984) *38* 515–550.]

iron box See OPERATOR.

iron corrosion See CATHODIC DEPOLARIZATION THEORY.

iron–sulphur centre See IRON–SULPHUR PROTEINS.

iron–sulphur proteins (Fe–S proteins) A category of iron- and sulphur-containing proteins which are widely distributed in cells of all types and which participate e.g. in electron transfer processes (see e.g. ELECTRON TRANSPORT CHAIN and NITROGENASE); except in rubredoxins (see later) each iron–sulphur protein contains two or more iron atoms linked to one another by an equal number of (acid-labile) sulphur atoms, the entire complement of iron and acid-labile sulphur forming one or more discrete *iron–sulphur centres* (= *iron–sulphur clusters*) linked to the protein typically by thiolate side-chains of cysteine residues. (See also RHODANESE.)

Rubredoxins are traditionally included with the Fe–S proteins, although they contain only one iron atom and no acid-labile sulphur; the iron is linked to the protein via four thiolate ligands. A rubredoxin with an E_m of ca. −60 mV occurs in *Clostridium pasteurianum*. (See also HYDROCARBONS.)

Other iron–sulphur proteins contain, per molecule, one or more of the following types of centre: [2Fe–2S] (see also RIESKE PROTEIN), [3Fe–3S] (found e.g. in *Azotobacter vinelandii*), and/or [4Fe–4S]; 2S, 3S and 4S refer specifically to *acid-labile* sulphur.

Iron–sulphur proteins can be classified as either *simple* (e.g. rubredoxins, FERREDOXINS) or *complex* (= *conjugated*), the latter having additional prosthetic groups such as a flavin or haem or a non-iron metal atom (e.g. molybdenum). Complex iron–sulphur proteins include e.g. the sulphite reductase of *Escherichia coli* (EC 1.8.7.1), which includes four [4Fe–4S] centres together with molecules of FAD, FMN and sirohaem, and a chloroplast nitrate reductase (EC 1.7.7.1) containing one [4Fe–4S] centre and sirohaem.

[Review: Book ref. 146, pp. 79–120.]

ironophores *Syn.* SIDEROPHORES.

Irpex See APHYLLOPHORALES (Polyporaceae).

IRS Internal resolution site: see Tn*3*.

IS element See INSERTION SEQUENCE.

IS*1* A 768-bp INSERTION SEQUENCE present e.g. in R plasmids (see also Tn*9*), the genome of BACTERIOPHAGE P1, the chromosome of *Escherichia coli* K12 (in multiple copies), etc. IS*1* seems to transpose preferentially to target sites which are rich in AT, frequently cleaving at a GC pair within such a region. Transposition can apparently generate either 9- or 8-bp duplications of target DNA [PNAS (1985) *82* 839–843]. IS*1* encodes two genes, *insA* and *insB*, which are required for IS*1*-mediated transposition and cointegration.

IS*5* A 1195-bp INSERTION SEQUENCE present in the chromosome of *Escherichia coli* K12 strains; it has a high target specificity (see TRANSPOSABLE ELEMENT).

IS*10* An INSERTION SEQUENCE that has been found only in association with Tn*10* (q.v.).

IS*50* See Tn*5*.

IS*101* A 209-bp INSERTION SEQUENCE which is defective and carries no genes; its transposition depends entirely on γδ-encoded transposase and resolvase functions (see Tn*3*). (cf. Tn*951*.) Its inverted repeat sequences are closely related to, but not identical with, those of γδ.

IS*900* An INSERTION SEQUENCE reported to be specific to *Mycobacterium paratuberculosis*. IS*900* has been used e.g. as a target for PCR-based detection of *M. paratuberculosis* in milk [AEM (1998) *64* 3153–3158].

IS*1000* *Syn.* γδ (see Tn*3*).

IS*6110* An INSERTION SEQUENCE regarded as specific to members of the *Mycobacterium tuberculosis* complex (i.e. *M. tuberculosis*, *M. bovis*, BCG, *M. africanum* and *M. microti*). The number of copies of IS*6110* per genome varies with strain; a few strains of *M. tuberculosis* apparently lack the sequence, while most strains of *M. bovis* appear to contain only a single copy. Some 16 copies of IS*6110* occur in the genome of the H37rv reference strain of *M. tuberculosis* [Nature (1998) *393* 537–544].

Some authors have questioned the specificity of IS*6110*, indicating that a central region of the sequence is similar to a sequence in non-tuberculous mycobacteria [JCM (1997) *35* 799–800]; others have affirmed the specificity of this sequence for members of the *M. tuberculosis* complex [JCM (1997) *35* 800–801].

IS*6110* has been widely used for TYPING isolates of *M. tuberculosis*. However, using SPOLIGOTYPING, some authors have found that, in a particular subset of multidrug-resistant strains, the rate of transposition of IS*6110* within the genome may be too high for reliable RFLP-based typing even over a period of a few years [JCM (1999) *37* 788–791]. By contrast, others have reported that IS*6110*-based typing gives insufficient discrimination [JCM (1999) *37* 3022–3024].

isatin-β-thiosemicarbazone (IBT) An ANTIVIRAL AGENT. IBT and some of its derivatives (e.g. methisazone: *N*-methyl-IBT) inhibit the replication of poxviruses, apparently acting at a late stage in the replication cycle. Methisazone has been used prophylactically in cases of exposure to smallpox, and in the treatment of complications of smallpox vaccination. (See also THIOSEMICARBAZONES.)

ISEM IMMUNOSORBENT ELECTRON MICROSCOPY.

IS-GA See IN SITU PCR.

ISH See PROBE.

isidiate (*lichenol.*) Bearing or consisting of isidia (see ISIDIUM).

isidium (*lichenol.*) A small protuberance (ca. 0.01–0.3 × 0.5–3.0 mm) which arises from the surface of the thallus in certain lichens; it consists of photobiont cells and medullary tissue, typically covered by a cortex, and may be e.g. spherical, flattened, cylindrical, branched ('coralloid'), according to species. In many isidiate lichens the isidia are easily broken off and probably serve as propagules; in other cases they may serve to increase the surface area – and hence the assimilative capacity – of the thallus.

islandic acid A water-soluble, acidic, extracellular, (1 → 6)-β-linked glucan containing malonic acid residues; it is produced by *Penicillium islandicum*.

islet-activating protein *Syn.* PERTUSSIS TOXIN.

isoaccepting tRNA See TRNA.

isoamylase See DEBRANCHING ENZYMES.

isoantibody Any antibody formed against an antigen derived from a genetically different individual of the same species.

isocitrate dehydrogenase See TCA CYCLE.

isocitrate dehydrogenase kinase–phosphatase See TCA CYCLE.

isocitrate lyase See TCA CYCLE and SERINE PATHWAY.

isoconazole See AZOLE ANTIFUNGAL AGENTS.

isodiametric Rounded or polyhedral: having roughly the same diameter in all directions.

isoelectric focusing (IEF) A type of ELECTROPHORESIS in which amphoteric molecules (e.g. proteins) are separated from one another on the basis of their differing ISOELECTRIC POINTS (IEPs). In principle, a column of gel (e.g. polyacrylamide) containing a mixture of AMPHOLYTES (each with a different IEP) is subjected to an electrical potential difference until the ampholytes are distributed (according to their IEPs) to form a stable pH gradient. The sample (e.g. mixture of proteins) is then added and subjected to an electrical potential difference so that each protein migrates to the region corresponding to its IEP; proteins may then e.g. be stained in situ or removed from the gel.

isoelectric point (IEP; pI) In substances (e.g. proteins) whose charge depends on the pH of the medium: the pH at which the molecule has no *net* charge and hence zero electrophoretic mobility.

isoelectrofocusing *Syn.* ISOELECTRIC FOCUSING.

isoenzyme (isozyme) One of a number of ENZYMES which catalyse the same reaction(s) but differ from each other e.g. in primary structure and/or electrophoretic mobility. (See also ZYMODEME.)

isofloridoside 1-*O*-Glycerol-α-D-galactopyranoside: a storage and osmoregulatory glycoside in e.g. *Poterioochromonas malhamensis* and also (often with FLORIDOSIDE) in members of the Rhodophyta.

isogamy The union of gametes which are alike in form and physiology. (cf. ANISOGAMY.)

isogenic Refers to two or more organisms or cells which have identical genotypes.

isoimmunization The stimulation of an immune response to antigens derived from another individual of the same species. (cf. HETEROIMMUNIZATION.)

isokont Refers to a pair of flagella (on a biflagellate cell) which are similar in length and type. (cf. HETEROKONT.)

isolation (1) (*microbiol.*) Any procedure in which a given species of organism, present in a particular sample or environment, is obtained in pure CULTURE. According to organism, isolation may involve e.g. culture on a selective MEDIUM, ENRICHMENT, and/or the use of techniques such as FILTRATION or MICROMANIPULATION.

(2) (*med., vet.*) The separation of patient(s) or animals(s) from others in order e.g. to prevent the spread of disease. (See also QUARANTINE.)

L-isoleucine biosynthesis Many microorganisms synthesize isoleucine by the pathway shown in Appendix IV(d). Certain bacteria generate α-oxobutyrate by alternative pathways: e.g., from L-glutamate in *Escherichia coli*, from L-methionine in obligately anaerobic bacteria, from pyruvate in *Leptospira* sp, and from propionate in *Clostridium sporogenes* [JGM (1984) *130* 309–318, q.v. for refs].

isolichenin A linear α-D-glucan containing $(1 \rightarrow 3)$ and $(1 \rightarrow 4)$ linkages; it occurs in certain lichens (e.g. *Cetraria islandica*). (cf. LICHENIN.)

isomaltose A reducing disaccharide: α-D-glucopyranosyl-$(1 \rightarrow 6)$-D-glucopyranose. (cf. GENTIOBIOSE.)

isomarticin See NAPHTHAZARINS.

isomerases ENZYMES (EC class 5) which catalyse isomerizations (i.e., intramolecular rearrangements). They include e.g. *epimerases* (e.g. that catalysing the interconversion of D-ribulose 5-phosphate and D-xylulose 5-phosphate), *racemases* (e.g. those catalysing L- and D-amino acid interconversions), etc.

isometric labile ringspot viruses *Syn.* ILARVIRUSES.

isometric ssDNA phages *Syn.* MICROVIRIDAE.

isometric virus A virus in which the virion is more or less spherical, usually exhibiting ICOSAHEDRAL SYMMETRY.

isomorphic Morphologically similar or identical.

isomorphic alternation of generations See ALTERNATION OF GENERATIONS.

isoniazid A synthetic ANTIBIOTIC, isonicotinic acid hydrazide (INH), bactericidal for actively growing cells of *Mycobacterium tuberculosis*, *M. bovis* and, to a lesser extent, other *Mycobacterium* species. INH is an important anti-TUBERCULOSIS drug whose function depends on prior activation by a catalase (KatG; *katG* gene product) within the bacterium; the activated drug apparently inhibits enzymic step(s) in the synthesis of essential cell-wall MYCOLIC ACID (Microbiology (2000) *146* 289–296]).

Resistance to isoniazid can follow mutation in any of several genes, including *katG* and *ahpC*; isoniazid is used in combination therapy for the treatment of tuberculosis. (See also RISE-RESISTANT TUBERCULOSIS.)

In *M. tuberculosis*, susceptibility to isoniazid, oxidative stress response and iron regulation are related features, but precise details of relationships remain to be elucidated [see TIM (1998) *6* 354–358].

isopanose α-Maltosyl-$(1 \rightarrow 6)$-D-glucose.

isopenicillin N See PENICILLIN N.

isopropanol See ALCOHOLS.

isopropyl-β-D-thiogalactoside A gratuitous inducer of the LAC OPERON.

isopsoralen See PCR.

isopycnic centrifugation See CENTRIFUGATION.

isorenieratene See LIGHT-HARVESTING COMPLEX.

isoschizomers RESTRICTION ENDONUCLEASES which recognize the same sequence in DNA but which are derived from different species. [Restriction endonucleases and their isoschizomers: NAR (1991) *19* 2077–2109.]

Isosphaera See BACTERIA (taxonomy).

Isospora A genus of parasitic protozoa (suborder EIMERIORINA) which form disporic, tetrazoic oocysts (cf. e.g. EIMERIA); the genus, as originally defined, includes only homoxenous species, but it has been expanded by some authors [see AP (1982) *20* 403–406] to include heteroxenous species which form disporic, tetrazoic oocysts (typically shed unsporulated) and which may form tissue cysts in the intermediate host. The expanded genus thus includes e.g. organisms otherwise placed in the genera BESNOITIA and TOXOPLASMA, as well as the obligately heteroxenous *Isospora datusi* (= *Hammondia hammondi*).

For generalized homoxenous and heteroxenous life cycles see EIMERIORINA; see also COCCIDIOSIS (q.v. for refs).

Isosticha See HYPOTRICHIDA.

isosulfazecin See MONOBACTAMS.

isotopic labelling (probe) See PROBE.

Isotricha A genus of ciliates related to DASYTRICHA.

isotype (class) Each of the five main variant forms (classes) of IMMUNOGLOBULIN (IgA, IgD, IgE, IgG and IgM); isotypes differ in the amino acid sequence in the constant region of their heavy chains. Immunoglobulins of all five isotypes occur in every normal individual (cf. ALLOTYPE).

A given isotype may be divided into subclasses (distinguishable serologically). For example, there are several subclasses of IgG (designated IgG1, IgG2 etc.).

isotype switching See ANTIBODY FORMATION.

isovelleral See LACTARIUS.

IsoVitaleX A commercial preparation – containing e.g. vitamin B_{12}, amino acids, purines, D-glucose, thiamine and TPP, NAD^+, $Fe(NO_3)_3$, PABA – used for the enrichment of media for isolating and/or cultivating fastidious bacteria (e.g. *Haemophilus* spp).

isozyme *Syn.* ISOENZYME.

Issatchenkia A genus of yeasts (family SACCHAROMYCETACEAE) which reproduce by multilateral budding; pseudomycelium is formed. Asci are persistent. Ascospores: spheroidal, rough-surfaced. Cells contain ubiquinone-7 (Q-7). Sugars may be fermented; NO_3^- is not assimilated. A pellicle develops on liquid cultures. Species – *I. occidentalis* (anamorph: *Candida sorbosa*), *I. orientalis* (anamorph: *Candida krusei*), *I. scutulata*, *I. terricola* – have been isolated from soil, fruit, etc. [Book ref. 100, pp. 214–223.]

isthmus (*biol.*) Any narrow region connecting two structures in an organism or cell: see e.g. *placoderm* DESMIDS.

iteron One of a number of repeated ('reiterated') nucleotide sequences which occur in and/or near the replication origin(s) in certain plasmids – e.g. the F PLASMID and the R6K PLASMID.

Itersonilia A genus of fungi *incertae sedis*; the organisms form mycelium containing CLAMP CONNECTIONS, and it has been suggested that *I. perplexans* forms a one-spored basidium [TBMS (1983) *80* 365–368].

itraconazole A clinically useful AZOLE ANTIFUNGAL AGENT (a triazole); it has a high MWt and is highly lipophilic. Itraconazole has a broad spectrum of activity, inhibiting yeasts, dimorphic fungi and filamentous fungi; it appears to be less toxic to mammalian tissues than e.g. miconazole or ketoconazole owing to its much higher selectivity for the fungal rather than the mammalian cytochrome P-450. [Review: Arch. Derm. (1986) *122* 399–402.]

[Pharmocology of itraconazole (review): Drugs (2001) *61* (supplement 1) 27–37.]

IUB International Union of Biochemistry.

***iucA–iucD* genes** See SIDEROPHORES.

IUdR See IDOXURIDINE.

IUPAC International Union of Pure and Applied Chemistry.

***iutA* gene** See SIDEROPHORES.

ivanolysin See THIOL-ACTIVATED CYTOLYSINS.

IVET (in vivo expression technology) A technique used for detecting those genes (of a pathogen) which are transcribed only during infection of the host; such genes *may* be virulence genes, and this can be studied further e.g. by animal tests involving strains of the pathogen which are mutant for the given gene. IVET is outlined in the figure.

[Some examples of IVET: TIM (1997) *5* 509–513. Behaviour of bacteria in the host, and the methodology for studying it: TIM (1998) *6* 239–243.]

IVS (*mol. biol.*) Intervening sequence (= *intron*): see SPLIT GENE.

iwatake The edible lichen *Umbilicaria esculenta*.

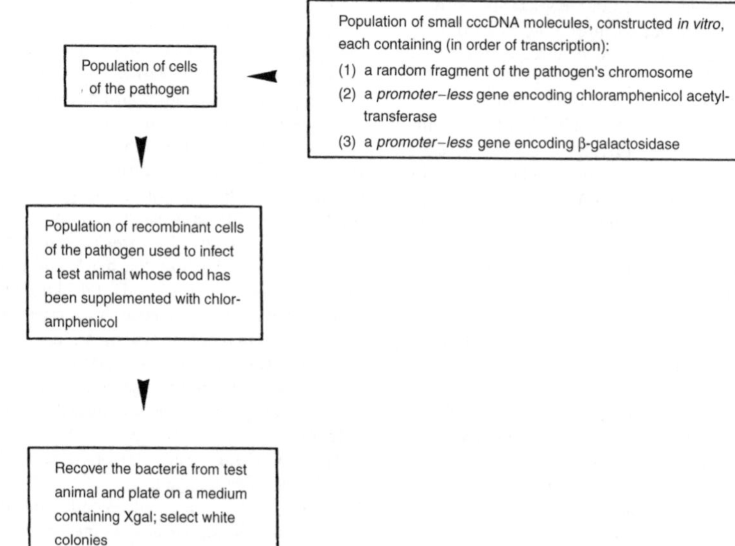

IVET (in vivo expression technology): the principle (diagrammatic). The figure shows one of several forms of IVET.

IVET detects those genes of a pathogen whose expression is induced *during* (and *only* during) infection of the host animal.

Vector molecules (see *top*, *right*) are inserted, by transformation, into a population of cells of the pathogen. In each transformed cell, the (random) fragment of chromosome in the vector inserts into the corresponding part of the pathogen's chromosome (by an 'insertion–duplication' mechanism); because the vector molecules contain different fragments of the chromosome they will insert into different chromosomal sites in different cells – forming a heterogeneous population of recombinant cells.

In some recombinant cells, the vector's two promoter-less genes will have been inserted 'in frame' with an upstream promoter. In such cells, *both* of these genes will be transcribed *if the promoter is active*.

The recombinant cells of the pathogen are used to infect a test animal whose food contains the antibiotic chloramphenicol (to which the pathogen is normally susceptible). Under these conditions, a recombinant cell can grow only if it produces chloramphenicol acetyltransferase (CAT), i.e. only if the CAT gene (in the vector molecule) is controlled by an *active* promoter; thus, the fact that a given recombinant cell *grows* within the animal indicates that its CAT gene is controlled by a promoter *which is active within the test animal*. (Because synthesis of CAT points to an active promoter, the CAT gene is sometimes called a 'reporter' gene.) Cells producing CAT can form large populations which greatly outnumber cells that do not form CAT.

We need to know whether the promoter controlling an expressed CAT gene is active *only* when the pathogen is in the test animal – or whether it is also active when the pathogen is cultured (e.g. on agar media). If active only in the test animal, this indicates that the gene *normally* controlled by that promoter is induced during infection; such a gene is of interest because of its possible association with virulence. To study promoter activity further, the recovered bacteria are plated on a medium which lacks chloramphenicol but which contains the agent XGAL (q.v.); all the cells will grow. If a given promoter is active in culture then β-galactosidase will be formed and will give rise to a *blue-green* colony; such a colony tells us that activation of the given promoter does not occur *only* within the test animal. A *white* (*lac*$^-$) colony indicates that β-galactosidase (and CAT) can be formed *only* within the test animal, i.e. the relevant promoter is active only when the pathogen is actually infecting the animal. The gene which is normally controlled by this promoter can be isolated, cloned and sequenced and examined for its role in virulence.

J

J chain A cystine-rich polypeptide (MWt ca. 15000) involved in the formation of dimeric and polymeric forms of both IgA and IgM (q.v.). In an IgA dimer the J chain forms a link between cystine residues situated near the C-terminal ends of the heavy chains in the two monomers.

jaagsiekte (pulmonary adenomatosis) A SHEEP DISEASE characterized by chronic progressive pneumonia in which the alveolar spaces become occluded by adenomatous ingrowths of the epithelium; the causal agent(s) appear to be viral (a herpesvirus and/or a retrovirus), but the causal agent of MAEDI is not involved. Incubation period: 1–3 years. Symptoms: dyspnoea, emaciation, and a profuse watery nasal discharge which may contain hyperplastic adenomatous epithelial cells; fever and inflammation are absent. Death occurs within ca. 4 months. [Book ref. 33, pp. 807–808.]

Jaccard coefficient See entry S_J.

jack bean lectin CONCANAVALIN A.

jack-in-the-box ascus *Syn.* BITUNICATE ASCUS.

Jacquard coefficient See entry S_J.

JAK See CYTOKINES.

janiemycin A polypeptide ANTIBIOTIC which inhibits transglycosylation in PEPTIDOGLYCAN synthesis.

Janthinobacterium A genus of Gram-negative, strictly aerobic, catalase-positive, chemoorganotrophic bacteria which occur e.g. in soil and water; the organisms are motile (flagellated) rods, 0.8–1.2 × 2.5–6.0 μm. Metabolism is respiratory (oxidative); acid (no gas) is formed from e.g. glucose. Growth occurs in/on simple media (e.g. peptone water); violet, often gelatinous, colonies are formed on nutrient agar. Optimum growth temperature: 25°C, maximum 32°C. MR −ve; VP −ve; usually oxidase +ve. GC%: ca. 61–67. Type species: *J. lividum* (formerly *Chromobacterium lividum*). [Book ref. 22, pp. 376–377.]

Janus green A blue basic DYE which has both azine and azo chromophores; it is used e.g. in VITAL STAINING.

Janus kinase See CYTOKINES.

Japanese B encephalitis (Japanese encephalitis; Russian autumn encephalitis) A viral ENCEPHALITIS which affects man, pigs and horses; it occurs in epidemics in e.g. Japan, China, Korea and parts of India. The disease is usually acute, but may run a protracted course with acute exacerbations. The causal agent is a flavivirus (see FLAVIVIRIDAE) which occurs e.g. in wild birds and is transmitted by mosquitoes (usually *Culex* sp). An inactivated vaccine has been used in Japan. [Review: ARM (1986) *40* 395–414.]

Japanese river fever *Syn.* SCRUB TYPHUS.

Japonochytrium See THRAUSTOCHYTRIDS.

Jarisch–Herxheimer reaction (Herxheimer reaction) A potentially fatal reaction which may follow the first effective chemotherapy against e.g. brucellosis or syphilis – or, in general, diseases caused by certain bacteria (particularly spirochaetes) and protozoa. Symptoms (e.g. an initial rise in temperature) are associated with a cascade of CYTOKINES which seem to be responsible for at least some of the pathophysiology. [Review: BCID (1994) *1* 65–74.] The reaction has been prevented by treatment with antibodies against tumour necrosis factor (TNF) [NEJM (1996) *335* 311–315].

jarrah dieback A disease of the jarrah (*Eucalyptus marginata*) – an important timber tree e.g. in Western Australia; the causal agent is generally believed to be *Phytophthora cinnamomi*, but waterlogging of the soil may be an important contributory factor [New Phyt. (1985) *101* 743–753]. (See also TREE DISEASES.)

javanicin See NAPHTHAZARINS.

JC virus See POLYOMAVIRUS.

JCM Japanese Collection of Microorganisms, Riken, Wako-shi, Saitama 351, Japan.

jelly fungi Those fungi of the TREMELLALES (e.g. *Exidia glandulosa*, *Tremella mesenterica*) which form gelatinous fruiting bodies – or all fungi (including e.g. *Auricularia* spp) which do so.

jelly lichens Gelatinous lichens e.g. of the genera COLLEMA, LEMPHOLEMMA, LEPTOGIUM.

Jembrana disease A CATTLE DISEASE, involving e.g. fever, generalized lymphadenopathy, nasal discharge, haemorrhage and splenomegaly; it was formerly assumed to be a tick-borne rickettsial disease but is now known to be caused by the bovine immunodeficiency virus.

jet loop fermenter A LOOP FERMENTER in which culture is continually withdrawn from the column and pumped back (together with air/gas) via a nozzle at the base of the DRAFT TUBE; this promotes liquid circulation and dispersion of the injected gas. In the *plunging jet* design of the Vogelbusch IZ fermenter, re-cycled culture (and air) is introduced under pressure at the *top* of the column (which lacks a draft tube); this design achieves a high level of dissolved oxygen.

Jew's ear fungus See AURICULARIA.

Jeyes fluid A household and horticultural DISINFECTANT consisting of coal tar acids solubilized with a soap prepared from pine resin and alkali.

Jirovecella See ASTOMATIDA.

JK coryneforms Coryneform bacteria which are resistant to (usually many) antibiotics; they are being increasingly reported as causes of serious nosocomial infections in immunosuppressed patients and patients with implants (pacemakers, prosthetic heart valves, etc). [Biochemical and cultural characteristics: JCP (1986) *39* 654–660.]

Johne's bacillus *Mycobacterium paratuberculosis*.

Johne's disease (chronic bacillary diarrhoea; paratuberculosis) A chronic intestinal CATTLE DISEASE which may also affect other ruminants (e.g. sheep, goats, deer); it occurs worldwide and is caused by *Mycobacterium paratuberculosis*. Infection occurs on ingestion of contaminated water, grass etc. In cattle, the incubation period is long (usually more than 12 months); symptoms are uncommon in animals under 2 years old. The pathogen multiplies in the intestinal wall which becomes thickened and poorly absorptive; scouring is at first intermittent but becomes persistent and severe. The animal becomes emaciated and anaemic. There is no effective treatment. In goats, the mesenteric lymph nodes become oedematous and there is emaciation, but scouring is not a feature [VR (1983) *113* 464–466]. In sheep, the disease involves emaciation but not scouring.

joint-breaker fever *Syn.* O'NYONG-NYONG FEVER.

joint-ill (*vet.*) In e.g. calves, foals and lambs: lameness involving inflammation of the articular surfaces of the joints, caused by any of a variety of organisms – e.g. *Actinobacillus equuli*, *Actinomyces pyogenes*, *Erysipelothrix rhusiopathiae* (particularly in piglets), salmonellae, *Staphylococcus aureus*.

joint molecule (*mol. biol.*) A term sometimes used to refer to a structure composed of two dsDNA molecules held together only by hydrogen bonding – as e.g. in early intermediates formed during homologous RECOMBINATION.

Jones–Mote sensitivity (cutaneous basophil hypersensitivity) A form of hypersensitivity which may develop e.g. when a human subject is primed with a soluble protein antigen, or a guinea pig is primed with a protein in incomplete Freund's adjuvant; subsequent challenge with the relevant allergen (within a few days of priming) leads to a weak, non-indurated, erythematous skin reaction in which the lesion contains a high proportion of BASOPHILS. (These reponses do not occur in mice.) Jones–Mote sensitivity can be transferred to a non-primed animal by serum (cf. DELAYED HYPERSENSITIVITY).

Jonesia A genus which contains the former species *Listeria denitrificans* [IJSB (1987) *37* 266–270].

Jopling reactions See LEPROSY.

JPC See JUNCTIONAL PORE COMPLEX.

Jud. Comm. JUDICIAL COMMISSION.

Judicial Commission A taxonomic body concerned with the interpretation and/or amendment of the rules of microbial nomenclature.

juglone (5-hydroxy-1,4-naphthoquinone) A compound produced in the leaves and roots of walnut trees (*Juglans* spp); it has antifungal and antibacterial activity, and its presence in the tree may play a role in resistance to disease.

[Juglone as a transcription-blocking agent: NAR (2001) *29* 767–773.]

jumping gene *Syn.* TRANSPOSABLE ELEMENT.

juncopox virus See AVIPOXVIRUS.

junctional pore complex (JPC) A carbohydrate-export apparatus, reported to occur in certain motile cyanobacteria, which (in at least some cases) includes a tubular structure (minimum diameter 13 nm) that is capable of spanning the cell envelope. It has been suggested that JPCs form the mechanistic basis of GLIDING MOTILITY in cyanobacteria [JB (2000) *182* 1191–1199].

jungle yellow fever See YELLOW FEVER.

Junín virus See ARENAVIRIDAE and ARGENTINIAN HAEMORRHAGIC FEVER.

junior synonym See SYNONYM.

jute retting See RETTING.

1. Words in SMALL CAPITALS are cross-references to separate entries.
2. Keys to journal title abbreviations and Book ref. numbers are given at the end of the Dictionary.
3. The Greek alphabet is given in Appendix VI.
4. For further information see 'Notes for the User' at the front of the Dictionary.

K

K Lysine (see AMINO ACIDS).

K antigens Capsular antigens – usually capsular polysaccharides (see CAPSULE). Examples include the capsular antigens of *Streptococcus pneumoniae*, COLOMINIC ACID, and VI ANTIGENS. In Gram-negative bacteria, K antigens can mask O ANTIGENS; in some bacteria the K antigens can be removed by heating (see e.g. VI ANTIGEN), but in others (e.g. *Klebsiella* spp) they are heat-stable. In *Escherichia coli*, several surface antigens originally designated K – e.g. K88, K99 – are actually (proteinaceous) fimbrial antigens, and it has been proposed that they be renamed F antigens (K88 = F4, K99 = F5) [Book ref. 68, pp. 61–64]; polysaccharide K antigens may occur together with fimbrial antigens in certain strains of *E. coli*. (See also ETEC and TEICHOIC ACIDS.)

K^+-ATPase See ION TRANSPORT.

K cells Killer cells: lymphoid cells which have cytotoxic/cytolytic activity against target cells; *syn.* NK CELLS.

K^+ pump See ION TRANSPORT.

K^+ transport See ION TRANSPORT.

K virus See POLYOMAVIRUS.

K vitamins See QUINONES.

K1 killer strain (of *Saccharomyces cerevisiae*) See KILLER FACTOR.

K1 RNA, K2 RNA See RNASE P.

K88, K99 In *Escherichia coli*: fimbrial antigens of certain strains pathogenic in animals – see ETEC, FIMBRIAE and K ANTIGENS.

Kabackosome A type of vesicle formed by the hypotonic lysis of a SPHAEROPLAST. A Kabackosome (which contains little or no cytoplasm) is composed of CYTOPLASMIC MEMBRANE; the inner and outer faces of the membrane correspond to those in the original cell. (cf. ETP.)

KAF *Syn.* FACTOR I.

Kaffir pox See SMALLPOX.

Kagami fever See EHRLICHIA.

Kahn test A STANDARD TEST FOR SYPHILIS.

kala-azar See VISCERAL LEISHMANIASIS.

Kanagawa phenomenon The phenomenon in which those strains of *Vibrio parahaemolyticus* isolated from human patients exhibit clear (β) haemolysis when grown on WAGATSUMA AGAR containing human RBCs but not on that containing horse RBCs (a Kanagawa +ve reaction), while almost all strains isolated from other sources, including food suspected of causing *V. parahaemolyticus* food poisoning, do not (i.e. are Kanagawa −ve). (Discoloration (α-haemolysis) and clear haemolysis on both human *and* horse RBC-containing media are both regarded as Kanagawa −ve results.) The Kanagawa haemolysin is heat-stable, extracellular, cytotoxic and cardiotoxic, and is haemolytic for human, dog and rat RBCs, weakly so for rabbit and sheep RBCs, and inactive against horse RBCs. In feeding experiments (in man), only Kanagawa +ve strains were capable of causing gastroenteritis, but the role of the haemolysin in pathogenesis is unknown; Kanagawa +ve strains appear to be better able to multiply in the intestine than are Kanagawa −ve strains. (See also FOOD POISONING (h).)

kanamycin Any of several related AMINOGLYCOSIDE ANTIBIOTICS (kanamycins A, B, C) produced by *Streptomyces kanamyceticus*; the drug used clinically is composed mainly of kanamycin A.

kanchanomycin (albofungin) A complex polycyclic ANTIBIOTIC, produced by *Streptomyces* sp, which has both antibacterial and antitumour activity. In the presence of divalent cations, kanchanomycin binds to DNA and inhibits DNA and RNA synthesis.

Kaposi's sarcoma (KS) A rare multifocal neoplastic disease which occurs in two forms: (i) slow and indolent (limited mainly to the skin), and (ii) rapid and fulminant (involving skin and gastrointestinal tract). The milder form occurs in certain ethnic groups (e.g. Ashkenazi Jews). The aggressive form occurs in children in tropical Africa and is also a common feature in HIV-infected patients. (Note that Kaposi's sarcoma occurs also in some immunosuppressed transplant patients.)

Kaposi's sarcoma appears to be associated with human (gamma) herpesvirus 8 (HHV8) [see e.g. Lancet (1997) *349* 558–563] in conjunction with the immunosuppressive effects of HIV. [PCR-based investigation of HHV8 in KS biopsies: Am. J. Path. (1997) *150* 147–153.] Activation of latent HHV8 in vitro has been achieved by demethylation of the promoter of a transactivator region by means of the reagent tetradecanoylphorbol acetate (TPA), and studies on the level of methylation of the transactivator region in biopsies have suggested a relationship between methylation status and the development of HHV8-associated disease [PNAS (2001) *98* 4119–4124].

Kaposi's varicelliform eruption May refer either to eczema herpeticum or eczema vaccinatum (see ECZEMA).

***kappa* chain** See LIGHT CHAIN.

kappa particles See CAEDIBACTER.

Karatomorpha See PROTEROMONADIDA.

Kärber method See END-POINT DILUTION ASSAY.

Karelian fever See SINDBIS VIRUS.

Karnal bunt (partial bunt; new bunt) A wheat disease caused by *Neovossia indica* (formerly e.g. *Tilletia indica*); originally a minor disease confined to NW India, it has recently spread through northern India and has become established in Afghanistan, Iraq, Pakistan, and Mexico, apparently transmitted on and in wheat seed. Usually only some of the grains in an ear are attacked; infected parts of the grain are initially grey but gradually turn black and emit a foul odour (trimethylamine). [Bot. Rev. (1983) *49* 309–330.] (See also COMMON BUNT.)

karyogamy The coalescence of nuclei (cf. PLASMOGAMY).

karyogram See KARYOTYPE.

karyokinesis *Syn.* MITOSIS.

karyological relict (*ciliate protozool.*) Any present-day ciliate whose characteristics (particularly nuclear constitution) resemble those of (presumably) phylogenetically ancient ciliates. (See also KARYORELICTID GYMNOSTOMES and PRIMOCILIATID GYMNOSTOMES.)

karyolysis (*histopathol.*) The dissolution of a cell's nucleus with consequent loss of affinity for basic dyes. (cf. KARYORRHEXIS.)

karyomastigont A nucleus together with its associated flagellum (or flagella) and, when present, axostyle. (See also DIPLOMONADIDA.)

karyomere See MACRONUCLEUS.

karyonide (caryonide) (*ciliate protozool.*) A clone of cells in which all the macronuclei have been derived from the same macronucleus.

karyoplast A nucleus which has been isolated from a (eukaryotic) cell and which is enclosed within a sac of cytoplasmic membrane containing a small amount of cytoplasm. (cf. CYTOPLAST.)

karyorelictid gymnostomes Presumptively primitive ciliates (subclass GYMNOSTOMATIA, order Karyorelictida) which contain diploid, non-dividing macronuclei; genera include e.g. *Geleia*, KENTROPHOROS, LOXODES, *Remanella* and *Tracheloraphis*.

karyorrhexis (*histopathol.*) The fragmentation of a cell's nucleus (cf. KARYOLYSIS; PYKNOSIS).

karyoskeleton *Syn.* NUCLEOSKELETON.

karyosome A NUCLEOLUS or nucleolus-like body.

karyotic (of cells) Nucleated.

karyotype The chromosomal constitution of a (eukaryotic) cell in terms of the number, size and morphology of the chromosomes at metaphase. A systematized diagrammatic representation of a karyotype is called an *idiogram*; a systematized photographic representation may be referred to as an idiogram or as a *karyogram*.

kasugamycin An atypical AMINOGLYCOSIDE ANTIBIOTIC whose molecule contains neither streptidine nor deoxystreptamine. Kasugamycin is bacteriostatic and is also active against certain fungi (e.g. *Pyricularia oryzae* and *Rhizoctonia solani*).

kat See KATAL.

Kata virus See PESTE DES PETITS RUMINANTS.

Katadyn silver Metallic SILVER – containing traces of impurities (gold, palladium etc) – deposited on sand (or other filtering medium) used for the filtration and disinfection of water; the impurities facilitate ionization of the silver.

katal (*abbrv.* kat) A unit of enzyme activity: that which increases the rate of conversion of a given chemical reaction by one mole per second under defined conditions.

***katG* gene** See ISONIAZID.

Katodinium See DINOFLAGELLATES.

Kauffmann–White classification A scheme for the classification and identification of the numerous serotypes of SALMONELLA (e.g. for epidemiological purposes). Each serotype is defined by its O ANTIGENS and, where applicable, its H ANTIGENS and VI ANTIGENS, and is given a specific ANTIGENIC FORMULA which indicates the nature of these antigens in the order O, Vi (if present) : H phase 1 : H phase 2. For example, *S. typhi* is designated by the formula 9, 12, [Vi]:d:–. This means that the organism has O antigens 9 and 12, may have Vi antigen (variable presence indicated by [ ]), and has phase 1 flagellar antigen d and no phase 2 flagellar antigen (indicated by a dash) – i.e. PHASE VARIATION does not occur in this serotype. The antigenic formula for *S. panama* is $\underline{1}$, 9, 12:1, v:1, 5 where $\underline{1}$, 9 and 12 are O antigens, and either phase 1 flagellar antigens 1 and v or phase 2 flagellar antigens 1 and 5 may be present; this serotype has no Vi antigen. Underlining (as in O antigen $\underline{1}$ in *S. panama*) indicates that the antigen is present as a result of BACTERIOPHAGE CONVERSION. (Phage conversion may lead to a change in serotype: thus, e.g., *S. anatum*, 3, 10:e, h:1, 6, may be converted to *S. newington*, 3, $\underline{15}$:e, h:1, 6. Similar effects can occur in bacteria of other genera, including e.g. *Shigella* [serotpe-converting bacteriophages and O-antigen modification in *Shigella flexneri*: TIM (2000) *8* 17–23].) For further examples of antigenic formulae see SALMONELLA. (See also SMOOTH–ROUGH VARIATION.)

Serotypes thus identified are placed into groups (O groups), the serotypes in each group having in common at least one major O antigen (the *group antigen*) which is not found in members of any other group. For example, the group antigen of group A is O antigen 2; that of group B is O antigen 4; group C, O antigen 6; group D, O antigen 9; group E, O antigen 3. Thus, both *S. typhi* and *S. panama* are group D strains. (Minor O antigens may occur in more than one group: e.g., O antigen 12 occurs in strains from groups A, B and D.)

Kawasaki disease An acute, occasionally fatal, febrile disease of unknown cause which affects infants and young children. Kawasaki disease may be a manifestation of an immune-complex-mediated systemic vasculitis [ADC (1984) *59* 405–409]; more recently it has been suggested that at least some of the symptoms may be due to a SUPERANTIGEN.

kb (kilobase) Of a DNA or RNA strand: a unit of length equal to 10^3 bases; the unit may also be used for 10^3 base pairs (bp) in dsDNA or dsRNA – although kbp (kilobase-pairs) is also used in this context.

kbp See previous entry.

Kcat mechanism See PHASEOLOTOXIN.

KCN broth (cyanide broth; Møller's [= Moeller's] cyanide medium; potassium cyanide broth) A MEDIUM used in the CYANIDE TEST; it consists of an aqueous solution of peptone (1.0%), NaCl (0.5%), KH_2PO_4 (0.0225%), Na_2HPO_4 (0.56%), and KCN (0.0075%). The medium may be stored at 4°C for up to ca. 2 weeks.

KCN test See CYANIDE TEST.

kDa Kilodalton: 10^3 DALTONS.

kDNA KINETOPLAST DNA.

KDO See LIPOPOLYSACCHARIDE.

Kdp system See ION TRANSPORT.

kefir A fermented milk beverage made in parts of Russia, Bulgaria and the former Yugoslavia. Milk is fermented by a mixed and varied population of organisms which usually include lactobacilli [SAAM (1983) *4* 286–294; JAB (1984) *56* 503–505], *Lactococcus lactis*, and yeasts (for example, *Saccharomyces* spp); lactic acid, ethanol and carbon dioxide (which causes foaming) are the main products. The organisms become embedded in an extracellular polysaccharide (kefiran) to form whitish, gelatinous granules [scanning electron microscopy: JAB (1980) *48* 265–268] which are carried to the surface by bubbles of carbon dioxide. The granules are collected and used as an inoculum for subsequent fermentations; they can be stored for several days in cold milk or water. (See also DAIRY PRODUCTS.)

kefiran See KEFIR.

Kellogg classification (of *Neisseria gonorrhoeae*) An early classification based on e.g.: (i) appearance of colony; (ii) autoagglutinability; and (iii) virulence. Types T1 and T2 correspond to fimbriate, virulent strains, while types T3 and T4 correspond to afimbriate, non-virulent strains.

keloidal blastomycosis *Syn.* LOBOMYCOSIS.

kelp (1) Any of various seaweeds that are used to obtain kelp (sense 2). (2) The ashes obtained by burning various large brown seaweeds such as *Ascophyllum*, *Fucus*, *Laminaria*, *Macrocystis*; such ashes have been used as an agricultural fertilizer and as a source of iodine, potash and soda (sodium carbonate).

Kelsey–Sykes test A CAPACITY TEST used to determine the efficacy of each of several dilutions of a given disinfectant under simulated practical conditions. (cf. USE-DILUTION TEST.) To 3 ml of a given dilution of disinfectant is added a 1-ml aliquot of bacterial suspension at times 0, 10 and 20 min; at 8, 18 and 28 min the disinfectant is subcultured to each of five tubes of nutrient broth which are then incubated to detect the presence or absence of viable bacteria. A given dilution passes the test if no growth is obtained in at least two of the five tubes inoculated from it at 8 and 18 min. The test organism used is *Pseudomonas aeruginosa* NCTC 6749, *Proteus vulgaris* NCTC 4635, *Escherichia coli* NCTC 8196 or *Staphylococcus aureus* NCTC 4163; in a given test the organism used is that which is the most resistant to the test disinfectant. To examine the efficacy of the disinfectant under 'dirty' conditions a yeast suspension is incorporated in each dilution of the disinfectant. The test conditions are standardized – e.g. the bacteria, yeast and disinfectant are each diluted in standard hard water. [PJ (1974) *213* 528–530.]

Kemerovo subgroup See ORBIVIRUS.

kennel cough See MASTADENOVIRUS.

Kentrophoros A genus of marine ciliates (subclass GYMNOSTOMATIA) related to LOXODES. Cells: very elongated and flattened (ribbon-like), ciliated only on one side, and associated with ectosymbiotic bacteria; there is no cytostome.

keratin A highly insoluble protein found e.g. in hair, wool, horn, skin, feathers. Keratin is degraded by relatively few organisms – e.g. *Candida albicans*, DERMATOPHYTES. (See also PILIMELIA and THERMOMONOSPORA.)

keratitis Inflammation of the cornea. (cf. KERATOCONJUNCTIVITIS; see also INFECTIOUS KERATITIS.)

keratoconjunctivitis Inflammation of the cornea and conjunctiva. It may be caused e.g. by *Chlamydia trachomatis* (see e.g. TRACHOMA), *Staphylococcus aureus*, HERPES SIMPLEX virus, or adenoviruses (see e.g. EPIDEMIC KERATOCONJUNCTIVITIS).

keratomycosis KERATITIS due to a fungus; it is usually due to an opportunist pathogen (e.g. *Aspergillus* spp, *Candida* spp).

kerion (exudative ringworm; tinea kerion) A severe form of RINGWORM (usually tinea capitis) in which the dry, scaling skin lesions become inflamed and suppurative.

kerogen Solvent-insoluble, condensed (i.e., covalently cross-linked and/or aromatized) organic matter which forms the major part of OIL SHALE; it is formed by the gradual transformation of sedimented biomolecules in the absence of high temperatures and pressures. (*Protokerogen* is a less mature form of kerogen from more recent sediments.) Kerogen is believed to be the precursor material of PETROLEUM and natural gas; on heating it yields HYDROCARBONS. (See also MICROBIAL MAT.)

kerosene See PETROLEUM.

kerosene fungus See HORMOCONIS.

kethoxal (2-keto-3-ethoxyl-*n*-butyraldehyde) A reagent which reacts with and modifies only unpaired guanine residues in DNA or RNA.

ketoconazole (1,3-dioxolanylmethylimidazole) An AZOLE ANTIFUNGAL AGENT which is effective against various mucocutaneous and cutaneous mycoses (including those caused by fungi resistant to GRISEOFULVIN) as well as systemic mycoses such as aspergillosis, blastomycosis, coccidioidomycosis, histoplasmosis, etc; unlike most of the imidazole antifungal agents, ketoconazole can be administered orally.

ketogenic fermentations Commercial fermentation processes in which polyhydric alcohols are converted to ketoses: see e.g. DIHYDROXYACETONE FERMENTATION and SORBOSE FERMENTATION.

α-ketoglutarate dehydrogenase See TCA CYCLE.

ketopentose See PENTOSES.

kGy See GRAY.

Khawkinea See EUGLENOID FLAGELLATES.

KIA KLIGLER'S IRON AGAR.

Kickxellales An order of fungi (class ZYGOMYCETES) which are typically saprotrophic in soil and dung; all species form one-spored sporangiola – numbers of which are borne on a short hyphal branch (*sporocladium*). Genera: e.g. *Coemansia*, *Linderina*, *Martensella* (species can be parasitic on other fungi), and *Spiromyces*.

kidney disease (Dee disease) A systemic, usually chronic FISH DISEASE affecting salmonids. The kidneys become pale and swollen with characteristic greyish necrotic lesions; mortality may be high. Pathogen: *Renibacterium salmoninarum*.

kidney stones (in humans) See UROLITHIASIS

kieselguhr *Syn.* DIATOMACEOUS EARTH.

kievitone An isoflavonoid PHYTOALEXIN produced by the French bean (*Phaseolus vulgaris*). Cell-free extracts of *Rhizoctonia solani* elicit high levels of kievitone from excised bean hypocotyls, but those of *Fusarium solani* elicit only trace amounts. Kievitone can be converted to the less fungitoxic kievitone hydrate by an extracellular enzyme produced by *Fusarium solani* f.sp *solani*.

Kilham rat virus See PARVOVIRUS.

killed vaccine *Syn.* INACTIVATED VACCINE.

killer cells (*immunol.*) Commonly refers to NK CELLS (q.v.) but may also refer e.g. to cytotoxic T cells.

killer factor Any of several protein toxins secreted by 'killer' strains of *Saccharomyces cerevisiae* and encoded by one of two cytoplasmically co-inherited dsRNA elements; the K1 toxin (formed by K1 or type 1 killer strains) binds initially to a cell wall D-glucan of a sensitive cell and then transfers to the cytoplasmic membrane where it disrupts the membrane potential. One of the dsRNA elements (designated M) encodes both the toxin and an 'immunity protein' which confers on the producing cell immunity to the toxin; the toxin contains two subunits, α and β, both derived by proteolytic processing of a precursor polypeptide during toxin secretion. The precursor polypeptide apparently functions as the immunity protein – possibly by competing with the mature toxin for binding sites on the membrane [Cell (1986) *46* 105–113]. The other dsRNA element (designated L) encodes a protein required for the (separate) encapsidation of M and L. The encapsidated particles (termed VLPs: virus-like particles) are variously regarded as plasmids or as viruses (*Saccharomyces cerevisiae* viruses, ScVs – see MYCOVIRUS). [Review: MR (1984) *48* 125–156.] (cf. KILLER PLASMIDS; see also BACTERIOCIN.)

killer helper factor *Syn.* INTERLEUKIN-2.

killer paramecia Strains of *Paramecium* which are able to kill other ('sensitive') paramecia. See: CAEDIBACTER, LYTICUM, MATE KILLER, PSEUDOCAEDIBACTER, R BODY, SPIN KILLING. (cf. TECTIBACTER.)

killer plasmids (in yeasts) In some strains of *Kluyveromyces marxianus* var. *lactis* the cells contain multiple copies of each of two linear, cytoplasmically inherited dsDNA plasmids, pG*l*1 and pG*l*2; cells containing these plasmids secrete a glycoprotein toxin capable of killing other (sensitive) strains of e.g. *Candida*, *Kluyveromyces* and *Saccharomyces* by inhibiting adenylate cyclase activity. These plasmids can be transferred to *Saccharomyces cerevisiae* (e.g. by protoplast fusion) on which they confer the killer phenotype. [ARM (1983) *37* 253–276.] (cf. KILLER FACTOR.)

killer yeasts See KILLER FACTOR and KILLER PLASMIDS.

kilobase See entry kb.

kimchi A Chinese and Korean food made by the lactic acid fermentation of vegetables (usually cabbage or radishes). Preparation resembles that of SAUERKRAUT.

KinA, KinB See ENDOSPORE (sense 1).

kinase An ENZYME which catalyses the transfer of a phosphate group from one substrate (commonly ATP) to another; examples include hexokinase and pyruvate kinase (both involved in the Embden–Meyerhof–Parnas pathway: see Appendix I(a)), ADENYLATE KINASE, and PPi:phosphofructose dikinase (see PYROPHOSPHATE).

Kineosporia A genus of bacteria (order ACTINOMYCETALES, wall type I). The organisms form a substrate mycelium but no aerial hyphae; sporangia, each containing one zoospore, develop on the hyphae. Growth occurs at/below 30°C, but not at 37°C. Type species: *K. aurantiaca*.

kinesis (1) A behavioural response in which the speed (but not the direction) of locomotion of a motile organism depends on the intensity of an external stimulus; e.g., when light intensity increases, an organism may swim faster (positive photokinesis) or slower (negative photokinesis). (See also PHOTOTAXIS.) Kinesis, unlike TAXIS, is a continuous response and is not cancelled by ADAPTATION.

(2) A behavioural response in which the rate of activity of an organism depends on the intensity of an external stimulus. When the activity concerned is speed of swimming, the phenomenon is known as *orthokinesis* (*syn.* kinesis sense 1). When the activity concerned is frequency of directional change, the phenomenon is termed *klinokinesis*. [Photochem. Photobiol. (1977) *26* 559–560.]

kinete A motile zygote, or a motile form derived from a zygote.

kinetic response *Syn.* KINESIS.

kinetid (ciliary corpuscle; kinetosomal territory) In ciliates: a unit of the KINETY; it typically includes a CILIUM with its associated kinetosome, KINETODESMA, cytoplasmic membrane and alveolus (see PELLICLE), and may also include e.g. a MUCOCYST, PARASOMAL SAC and TRICHOCYST together with various microfibrils and/or microtubules (e.g. nematodesmata).

kinetin 6-Furfurylaminopurine, a compound formed during the hydrolysis of DNA; it has CYTOKININ activity but it apparently does not occur in plants.

kinetochore An electron-dense MICROTUBULE-ORGANIZING CENTRE, one of which develops on each side of the CENTROMERE during MITOSIS.

kinetochore fibre See MITOSIS.

kinetocyst See AXOPODIUM.

kinetodesma (kinetodesmos) In some ciliates: one of a series of overlapping endoplasmic fibrils which, by light microscopy, may appear as a single fibre running parallel with, and to the right of, a row of basal bodies in a somatic KINETY; each kinetodesma arises from a kinetosome (near triplets 5–8, GRAIN CONVENTION), passes a little to the right, and then runs anteriorly to terminate in a position where it may overlap the kinetodesma(ta) of the more anterior kinetosome(s). Kinetodesmata have cross-striations of periodicity ca. 30 nm. The role of the kinetodesmata is unknown; they are commonly believed *not* to be involved in ciliary coordination. (See also RULE OF DESMODEXY; SILVER LINE SYSTEM.)

kinetofragments In many members of the KINETOFRAGMINOPHOREA: patches or short rows of kinetids (sometimes with non-ciliferous kinetosomes) in the vicinity of the oral area; their evolutionary origin is presumed to have been the anterior ends of somatic kineties.

Kinetofragminophorea A class of protozoa (phylum CILIOPHORA) which characteristically have an apical or subapical cytostome and a CYTOPHARYNGEAL APPARATUS which is usually conspicuous; COMPOUND CILIATURE is typically absent, and ciliature in the oral region is not obviously differentiated from that in other region(s) of the body. STOMATOGENESIS is typically telokinetal (apparently apokinetal in the ENTODINIOMORPHIDA). This class includes many of the ciliates previously referred to as the 'lower holotrichs' (e.g. the apostomes, chonotrichs, gymnostomes and trichostomes) together with e.g. the suctorians. Subclasses: GYMNOSTOMATIA, HYPOSTOMATIA, SUCTORIA, VESTIBULIFERIA.

kinetoplast In protozoa of the KINETOPLASTIDA: a unique structure which forms a distinct region of the mitochondrion; it comprises a complex network of numerous catenated circular DNA molecules (kDNA), including several thousand small circles (*minicircles*) and fewer (ca. 20–50) identical larger circles (*maxicircles*). The maxicircles are the functional counterpart of the mitochondrial DNA of other eukaryotes, encoding e.g. rRNA and proteins. Minicircles encode most of the gRNAs for RNA EDITING (their only known function). [kDNA: Eukaryotic Cell (2002) *1* 495–502].

During cell division, the kinetoplast (and mitochondrion) divides before nuclear division, Kinetoplast division is inhibited by certain drugs, including e.g. ACRIDINES, DIAMIDINES and PHENANTHRIDINE derivatives. (See also DYSKINETOPLASTY.) The kinetoplast stains red with ROMANOWSKY STAINS.

Kinetoplastida An order of protozoa (class ZOOMASTIGOPHOREA); each organism has a single nucleus, one or two flagella (which arise from flagellar pockets and which often contain a PARAXIAL ROD) and a single, simple or branched, typically elongated mitochondrion which usually contains a KINETOPLAST located close to the flagellar basal body. (See also GLYCOSOME.) Two suborders: BODONINA and TRYPANOSOMATINA.

kinetosomal territory *Syn.* KINETID.

kinetosome (*protozool.*) *Syn.* BASAL BODY (b).

kinety In ciliates: typically, a longitudinal ('meridianal') row of somatic KINETIDS; a kinety corresponds to the 'primary meridian' in the SILVER LINE SYSTEM.

kingdom A major taxonomic category (see TAXONOMY) ranking above phylum. Organisms are grouped into kingdoms in various ways according to different schemes; e.g., according to the 'five-kingdoms' classification scheme the kingdoms are Animalia, FUNGI, MONERA, Plantae, and PROTISTA (sense 2).

Kingella A genus of catalase-negative, oxidase-positive bacteria (family NEISSERIACEAE) which occur e.g. in the upper respiratory tract. (See also HACEK.) Cells: rods, $1 \times 2–3$ μm, in pairs or chains. Growth occurs aerobically, and has been reported to occur anaerobically on blood agar with a gaseous phase of 95% hydrogen and 5% carbon dioxide. The organisms are nutritionally fastidious. Optimal growth temperature: 33–37°C. GC%: ~47–55. Type species: *K. kingae*.

kinins See CYTOKININS.

Kinyoun stain An ACID-FAST STAIN which resembles ZIEHL–NEELSEN'S STAIN but differs in that the stain contains higher concentrations of phenol and basic fuchsin and is used cold; methylene blue may be used as counterstain. [Methods: Book ref. 53, pp. 1381, 1384–1385.]

Kirby–Bauer technique See DISC DIFFUSION TEST.

kirromycin See POLYENE ANTIBIOTICS (b).

kirrothricin See POLYENE ANTIBIOTICS (b).

Kirsten murine sarcoma virus (Ki-MSV) A replication-defective MURINE SARCOMA VIRUS isolated from rats inoculated neonatally with a cell-free extract from thymic lymphomas of C3H mice. Ki-MSV resembles HARVEY MURINE SARCOMA VIRUS in carrying the oncogene v-*ras* (v-Ki-*ras*: see RAS) and in causing erythroleukaemia as well as sarcomas in newborn mice.

kissing disease See INFECTIOUS MONONUCLEOSIS.

Kitasatoa A genus of bacteria which resemble *Streptomyces* spp in e.g. wall type and phage sensitivity; species have been regarded as members of the genus *Streptomyces* [JGM (1983) *129* 1743–1813] and as species *incertae sedis* [Book ref. 73, pp. 97–98].

kitazin (*O*,*O*-diethyl-*S*-benzyl phosphorothioate) An ORGANOPHOSPHORUS COMPOUND used as an antifungal agent against rice blast disease; kitazin and its isopropyl analogue (kitazin P) is readily absorbed by plant roots and is rapidly translocated in the transpiration stream. It appears to function by inhibiting CHITIN synthesis.

klebicin (klebecin, klebocin) See BACTERIOCIN.

Klebs–Löffler bacillus *Corynebacterium diphtheriae.*

Klebsiella (1) A genus of Gram-negative bacteria of the ENTEROBACTERIACEAE (q.v.). Cells: straight rods, ca. 0.3–1.0 × 0.6–6.0 μm, occurring singly, in pairs, or in short chains; non-motile. The cells are capsulated and usually form convex, glistening, viscid (mucoid) colonies on carbohydrate-rich media. Serotyping [Book ref. 68, pp. 143–164] is based on capsular (K) rather than on O antigens. Some strains may possess mannose-sensitive (type 1) and/or mannose-resistant (type 3) FIMBRIAE. Typical reactions: acid (sometimes with gas) from glucose; H_2S −ve; phenylalanine deaminase −ve; ornithine decarboxylase −ve; myo-inositol +ve; growth in KCN media. Many strains are lactose +ve. GC%: 53–58. Type species: *K. pneumoniae*.

K. oxytoca. Indole +ve; MR −ve; VP +ve; no gas from lactose at 44.5°C; growth occurs at 10°C; melezitose sometimes +ve; pectate +ve. Occurs in the intestines of man and animals and in plant and aquatic environments.

K. planticola (= *K. trevisanii* [IJSB (1986) *36* 486–488]). Indole and MR reactions variable; VP +ve; no gas from lactose at 44.5°C; growth occurs at 10°C; melezitose −ve; pectate −ve. Occurs mainly in plant, soil and aquatic environments.

K. pneumoniae (including strains sometimes called *K. aerogenes* and *K. edwardsii* – cf. AEROBACTER). Indole −ve; no growth at 10°C; melezitose −ve; pectate −ve. Some strains carry out NITROGEN FIXATION. Strains occur in the intestines and respiratory tract of man and animals and may be pathogenic, causing e.g. bovine MASTITIS, equine metritis, OZAENA, PNEUMONIA, and nosocomial URINARY TRACT INFECTION. Subspecies *pneumoniae*: gas from lactose at 44.5°C; MR −ve, VP +ve (but see METHYL RED TEST); urease +ve. Subspecies *ozaenae* (formerly *K. ozaenae*): lactose fermentation variable; MR (usually) +ve; VP −ve; urease variable. Subspecies *rhinoscleromatis* (formerly *K. rhinoscleromatis*): lactose −ve; MR +ve; VP −ve; urease −ve.

K. terrigena. Indole −ve; no gas from lactose at 44.5°C; growth occurs at 10°C; melezitose +ve; pectate −ve. Occurs mainly in soil and aquatic environments.

K. trevisanii. See *K. planticola*.

[Book ref. 22, pp. 461–465.]

(2) A genus of EUGLENOID FLAGELLATES found in marine or brackish waters; each cell is partly enclosed in a yellow to brown LORICA.

Klebsormidium (*Hormidium*) A genus of filamentous, unbranched, freshwater or terrestrial green algae (division CHLOROPHYTA). The filaments lack a holdfast and are composed of uninucleate cells each containing a single plate-like chloroplast. Zoospores (rarely formed) are asymmetric, biflagellate and naked, and are released via a pore in the mother cell wall. Sexual reproduction is isogamous; gametes are biflagellate. (cf. STICHOCOCCUS.)

Kleinschmidt monolayer technique A method for preparing a nucleic acid for examination by ELECTRON MICROSCOPY. Essentially, a solution of e.g. DNA is mixed with a solution of the globular protein cytochrome *c* (which adheres to DNA), and one drop of this mixture is allowed to run down a sloping surface onto an air–liquid interface. This gives rise to a monolayer of denatured cytochrome *c* containing extended molecules of DNA which are thickened by a coating of cytochrome *c*; dsDNA appears thicker than ssDNA. A small area of the monolayer is transferred to a prepared grid, and the preparation is either shadowed on a rotating platform or stained e.g. with uranyl acetate; the thread-like molecules of DNA thus become more electron-dense than the background. In a modification of this technique, benzalkonium chloride is used instead of cytochrome *c*; since, in this case, the DNA is not thickened with adherent protein, its relationship to other macromolecules (e.g. polymerases) can be determined more easily.

Klenow fragment (Klenow enzyme) The larger of two fragments obtained by proteolytic cleavage (using e.g. subtilisin) of the DNA POLYMERASE I of *Escherichia coli*. The Klenow fragment retains 5′-to-3′ polymerase and 3′-to-5′ exonuclease activities but lacks 5′-to-3′ exonuclease activity; it has various uses in genetic engineering techniques and in DNA sequencing methods.

Klett–Summerson colorimeter See TURBIDIMETRY.

Klett unit A unit of turbidity as measured with the Klett–Summerson colorimeter using monochromatic light of a specified colour.

Kligler's iron agar (KIA) A double-sugar–iron agar (cf. TSI AGAR) used e.g. for distinguishing between members of the Enterobacteriaceae. It is prepared as a slope with a deep butt, and consists of nutrient agar (pH 7.4) supplemented with glucose (ca. 0.1%), lactose (1%), sodium thiosulphate (0.05%), ferric ammonium citrate (0.05%), and phenol red (ca. 0.0025%). In general, bacterial reactions with KIA are similar to those with TSI agar; however, the absence of sucrose in KIA means that the lactose-negative reaction given by some organisms is not masked by their ability to attack sucrose.

klinokinesis See KINESIS (sense 2).

Kloeckera A genus of yeasts (class HYPHOMYCETES) which have teleomorphs in the genus HANSENIASPORA. Vegetative cells are more or less lemon-shaped (apiculate) and reproduce by bipolar budding in basipetal succession; pseudomycelium may be formed. All species ferment glucose; some ferment sucrose, none ferments maltose (although some species can assimilate maltose). NO_3^- is not assimilated. All species require inositol and pantothenate for growth. Species (*K. africana*, *K. apiculata*, *K. apis*, *K. corticis*, *K. japonica*, *K. javanica*) have been isolated from fruit, soil, bark etc.

[Book ref. 100, pp. 873–881.]

Klonostricha See HYPOTRICHIDA.

Klossiella See ADELEORINA.

Kluyver effect The phenomenon in which certain yeasts can utilize particular disaccharides aerobically, but not anaerobically. The mechanism underlying the effect is unknown [JGM (1982) *128* 2303–2312].

Kluyvera A genus of motile bacteria of the ENTEROBACTERIACEAE. MR +ve; VP −ve; usually indole +ve and citrate +ve. 2-Oxoglutarate is formed from glucose. The organisms occur in food, soil and sewage, and may be opportunist pathogens in man. [Book ref. 22, pp. 511–513.]

Kluyveromyces A genus of yeasts (family SACCHAROMYCETACEAE) in which the cells are spheroidal, ovoid, ellipsoidal, or cylindrical to elongate; vegetative reproduction occurs by multilateral budding. Pseudomycelium may be formed. Asci are evanescent; they contain one to many spores which may be crescent-shaped, reniform, spheroidal etc, and which tend to agglutinate after liberation. Cells contain ubiquinone-6 (Q-6). All species can ferment glucose; *K. marxianus* var. *lactis* and some strains of *K. marxianus* var. *marxianus* can ferment LACTOSE. NO_3^- is not assimilated. Eleven species have been recognized primarily on the basis of hybridization data; on this basis some former species (e.g. *K. bulgaricus*, *K. fragilis*, *K. lactis*) have been relegated to varieties of *K. marxianus* – e.g., '*K. fragilis*' is *K. marxianus* var. *marxianus* (anamorph: *Candida kefyr*, = '*C. pseudotropicalis*'), and

'*K. lactis*' is *K. marxianus* var. *lactis* (anamorph: *Candida sphaerica*). Other species include e.g. *K. aestuarii*, *K. africanus*, *K. blattae*, *K. polysporus*, *K. thermotolerans* (anamorph: *Candida dattila*). [Book ref. 100, pp. 224–251.] *Kluyveromyces* spp occur in a wide range of habitats: e.g. sea-water (*K. aestuarii*), soil, insects, fruit and other plant material, foods and beverages, etc. (See also CHEESE-MAKING (c), SINGLE-CELL PROTEIN, YEAST EXTRACT, and KILLER PLASMIDS.)

k-MTs MICROTUBULES attached to a KINETOCHORE.

knallgas bacteria *Syn.* HYDROGEN-OXIDIZING BACTERIA.

knallgas reaction (oxyhydrogen reaction) The oxidation of gaseous hydrogen by oxygen. In the HYDROGEN-OXIDIZING BACTERIA this reaction, which provides energy for growth, involves one or more types of HYDROGENASE and a respiratory chain which may contain e.g. *a*-, *b*-, *c*- and *o*-type cytochromes, iron–sulphur proteins and ubiquinones. [Ubiquinones in hydrogen-oxidizing bacteria: SAAM (1983) *4* 181–183.] In e.g. chemolithoautotrophically grown *Paracoccus denitrificans*, electrons from hydrogen appear to be transferred to oxygen via hydrogenase, UQ-10 and cytochromes b_{562} and *o*. In e.g. '*Alcaligenes eutrophus*' (which has both NAD-independent and NAD-reducing hydrogenases) it has been suggested that electrons from hydrogen may be donated to the respiratory chain and/or to NAD – the NADH being oxidized via the respiratory chain.

knockout mice Mice in which specific gene(s) have been disrupted (by genetic manipulation at the embryo stage) such that the corresponding gene product(s) are not synthesized in active form.

knopvelsiekte The severe form of LUMPY SKIN DISEASE.

knot (in DNA) An entanglement within a single, circularly closed DNA molecule, resolution of which (to form a knot-free cccDNA molecule) requires the activity of a TOPOISOMERASE (q.v.).

koala (chlamydiosis in) See CHLAMYDIA.

Koch–Weeks bacillus *Haemophilus aegyptius*.

Koch's blue bodies See EAST COAST FEVER.

Koch's phenomenon The phenomenon in which different responses are given by healthy and tuberculous guinea pigs to a subcutaneous injection of virulent tubercle bacilli; reactions in the *tuberculous* animals are manifestations of DELAYED HYPERSENSITIVITY (q.v.).

Koch's postulates According to Robert Koch (1843–1910): a set of conditions which should be fulfilled in order to establish that a given organism is the causal agent of a particular disease. (i) The organism must be present in every case of the disease. (ii) It should be isolable in pure culture. (iii) Inoculation of susceptible animals with the isolated organism must produce the disease. (iv) The organism must be observable in, and/or isolable from, the (experimental) diseased animals.

Koehler illumination *Syn.* KÖHLER ILLUMINATION.

Kofoidinium A genus of DINOFLAGELLATES.

Köhler illumination (Koehler illumination) In MICROSCOPY: illumination in which the entire field is lit with uniform intensity, regardless of any non-uniformity in the light source (cf. CRITICAL ILLUMINATION). To obtain Köhler illumination with an external lamp having a condensing lens with an iris diaphragm (*field diaphragm*) outside it: (i) Close the substage CONDENSER diaphragm (*aperture diaphragm*). (ii) Open the field diaphragm. (iii) Using the microscope's *plane* mirror, image the lamp's filament on the underside of the aperture diaphragm; thereafter, keep lamp and microscope in the same relative positions. (iv) Open the aperture diaphragm and almost close the field diaphragm. (v) Focus on the specimen with a low-power (e.g. $\times 10$) objective, and adjust the substage condenser *slightly* so that the edge of the field diaphragm is in sharp focus in the plane of the specimen. (vi) Open the field diaphragm until the small disc of light in the centre of the field of view has expanded to cover the entire field. Under these conditions an image of the lamp's filament is formed in the lower focal plane of the substage condenser, and divergent rays from *each* point of the filament's image pass through the condenser and emerge as parallel rays which illuminate the specimen.

koji A preparation consisting of mould (usually *Aspergillus oryzae*) growing on cooked cereal and/or soybeans; the mould produces enzymes (including a range of proteases, amylases, pectinases, glutaminase, etc) and is used in the production of e.g. SOY SAUCE and MISO.

The inoculum for making koji is a *koji starter* or *tane koji*: a powder consisting of mould spores. Tane koji is typically made by inoculating steamed polished rice with spores of selected fungal strains; the rice is spread in shallow trays, incubated at 30°C for ca. 5 days, and the spores are then harvested and dried.

kojibiose A reducing disaccharide: α-D-glucopyranosyl-(1 $\rightarrow$ 2)-D-glucopyranose. It can occur e.g. as a product of enzymic degradation of certain plant carbohydrates, and is present in glycerol TEICHOIC ACIDS of group D streptococci (in which it is linked to the C-2 position of glycerol and may be esterified with D-alanine).

kojic acid (5-hydroxy-2-hydroxymethyl-4-pyrone) A secondary metabolite produced e.g. by *Aspergillus* spp (particularly the *flavus–oryzae* group) when grown on glucose, xylose or certain other sugars. It is a chelating agent which gives a strong blood-red colour with Fe^{3+}; it also has weak antibiotic activity (enhanced by certain metal ions), being effective mainly against Gram-negative bacteria.

Kolmer CFT See STANDARD TESTS FOR SYPHILIS.

Kolpoda *Syn.* COLPODA.

kombu See LAMINARIA.

Konservomat See BATCH RETORT.

Kopeloff modification A modification of the GRAM STAIN used for staining anaerobic bacteria. The technique involves the addition of ca. 5 drops of $NaHCO_3$ solution (5%) to the smear after the latter has been flooded with crystal violet (1%); the iodine solution (iodine 2%, KI 0.1%) includes NaOH (0.4%), decolorization is carried out with acetone–alcohol, and safranin (2%) is used for counterstaining.

Koplik's spots Small, white, necrotic lesions formed in the mouth in the early stage of MEASLES.

Kordyana See EXOBASIDIALES.

Korean haemorrhagic fever (epidemic haemorrhagic fever) A VIRAL HAEMORRHAGIC FEVER involving renal dysfunction, proteinuria and oliguria; the mortality rate may be up to 30%. The causal agent is the Hantaan virus (see HANTAVIRUS); reservoirs of infection are found in rodent populations. Infection typically occurs by inhalation of aerosols of urine, faeces etc. from infected rodents. The disease occurs in Asia and Europe.

Korfia See HELOTIALES.

Kornberg enzyme See DNA POLYMERASES.

Koserella A genus of Gram-negative bacteria of the ENTEROBACTERIACEAE. *K. trabulsii* (formerly known as 'enteric group 45' and originally identified as an atypical strain of *Hafnia alvei*) has been isolated from clinical specimens; it is MR +ve; VP −ve; indole −ve; citrate +ve; H_2S − ve; urease −ve; LDC

+ve; ODC +ve; and it produces acid from cellobiose and melibiose but not from glycerol, lactose or sucrose. [JCM (1985) *21* 39–42.]

Koser's citrate medium A medium, used for the CITRATE TEST, containing (per 100 ml of water): NaCl (0.5 g), $MgSO_4 \cdot 7H_2O$ (0.02 g), $(NH_4)H_2PO_4$ (0.1 g), K_2HPO_4 (0.1 g), citric acid (0.2 g); pH 6.8.

Koster's stain A stain for detecting *Brucella* spp in mammalian tissues. The smear or section is stained for 1 min in a solution made by mixing saturated aqueous safranin and normal KOH (2:5 v/v); the slide is then washed in tap water and differentiated in 0.1% sulphuric acid (two changes for a total period of 10–20 sec). The slide is washed in tap water and counterstained with 1% carbol methylene blue. *Brucella* stains orange, tissues blue.

Kotonkan virus See LYSSAVIRUS.

koumiss (kumiss) An acidic, mildly alcoholic, fermented milk beverage made in the former USSR. Traditionally made from mares' milk, it is now also made from cows' milk. Organisms involved in the fermentation include *Lactobacillus bulgaricus* (the main acid-producer) and lactose-fermenting yeasts; the main products of fermentation are lactic acid, ethanol and CO_2. (See also DAIRY PRODUCTS.)

Kovács' indole reagent An INDOLE TEST reagent: *p*-dimethylaminobenzaldehyde (5 g) is dissolved in 75 ml amyl (or isoamyl) alcohol, and conc. HCl (25 ml) is slowly added; it is stored at 4°C in the dark.

Kovács' oxidase reagent A 1% aqueous solution of tetramethyl-*p*-phenylenediamine dihydrochloride. (See also OXIDASE TEST.)

kPa 10^3 Pa (see PASCAL).

***Kpn*I** See RESTRICTION ENDONUCLEASE (table).

krad 10^3 rad (see RAD).

kraft pulps See PAPER SPOILAGE.

Krebs cycle *Syn.* TCA CYCLE.

Krebspest *Syn.* CRAYFISH PLAGUE.

Kruse's bacillus Archaic name for *Shigella sonnei* (Kruse's '*Bacillus pseudodysenteriae* type E').

Kudoa See MYXOSPOREA.

kumiss *Syn.* KOUMISS.

Kunin antigen *Syn.* ENTEROBACTERIAL COMMON ANTIGEN.

Kunjin virus See FLAVIVIRIDAE.

Kupffer cells Actively phagocytic cells which line the liver sinusoids.

Kurthia A genus of Gram-positive, asporogenous, catalase-positive, obligately aerobic bacteria which occur e.g. in meat and meat products. In the exponential growth phase the organisms occur in long chains of rods (or of short filaments) but in the stationary phase coccoid forms or short rods predominate; individual rods are peritrichously flagellated. *Kurthia* spp are chemoorganotrophs which can utilize certain alcohols (e.g. ethanol, ethanediol), amino acids (e.g. L-alanine) and fatty acids (e.g. butyric acid). VP −ve. Indole-negative. The type species, *K. zopfii*, grows in the temperature range 5–35°C and is killed by heating to 55°C/20 minutes. *K. gibsonii* grows in the range 5–45°C and survives heating to 55°C/20 minutes [SAAM (1983) *4* 253–276]. GC%: ca. 36–38.

kuru A human TRANSMISSIBLE SPONGIFORM ENCEPHALOPATHY [early description: NEJM (1957) *257* 974–978]. The incubation period may range from a few years to some 30 years or more; characteristically there is ataxia, tremor, dysarthria and sometimes late dementia, and death occurs within a year of the onset of symptoms. Kuru is found in certain regions of Papua New Guinea and is believed to result from the ingestion of prion-containing flesh during ritual cannibalism; as this practice has declined, the incidence of kuru has decreased. Kuru can be transmitted experimentally to chimpanzees and to other primate and non-primate animals; goats (but not sheep) inoculated with tissue from kuru victims develop a disease that closely resembles SCRAPIE. [Review of kuru: Book ref. 159, pp. 483–544 (497–510).]

Kusnezovia A genus of poorly-characterized, manganese-depositing bacteria found in mud; the organisms have not been obtained in pure culture, and the validity of the genus (and of an apparently similar genus, *Caulococcus*) has been questioned [Book ref. 45, pp. 529–530].

Kyasanur Forest disease An acute, tick-borne human disease (vector: mainly *Haemaphysalis* sp) caused by a flavivirus (see FLAVIVIRIDAE); it occurs in Mysore State, India. Onset is sudden, with e.g. fever, headache, severe myalgia, haemorrhages, etc. Mortality rates are generally low (<5%). Monkeys may act as a reservoir of infection.

Kyzylagach virus See ALPHAVIRUS.

1. Words in SMALL CAPITALS are cross-references to separate entries.
2. Keys to journal title abbreviations and Book ref. numbers are given at the end of the Dictionary.
3. The Greek alphabet is given in Appendix VI.
4. For further information see 'Notes for the User' at the front of the Dictionary.

L

L Leucine (see AMINO ACIDS).

L-flagella Lateral flagella – see VIBRIO.

L-form (L-phase variant) A defective bacterial cell of spherical or irregular shape, formed either spontaneously (e.g. by *Streptobacillus moniliformis*) or as a result of various stimuli (e.g. temperature shock, osmotic shock, or antibiotics which inhibit cell wall biosynthesis); in an L-form the cell wall is either partly or totally absent. (cf. PROTOPLAST and SPHAEROPLAST.) On removal of the stimulus, an L-form may resume cell wall synthesis and revert to the condition of the original strain, or it may continue to reproduce as an L-form (*stable L-form*). L-forms have been observed in various bacteria – including species of *Bacillus*, *Proteus*, *Streptococcus* and *Vibrio*. L-form colonies often resemble those of *Mycoplasma* spp. (L-forms were named after the Lister Institute of Preventive Medicine, London.) (See also CHLORAZOL BLACK E.)

L-phase variant *Syn.* L-FORM.

L protein See LEVIVIRIDAE.

L ring See FLAGELLUM (a).

L-selectin See SELECTINS.

l-strand (in adenoviruses) See ADENOVIRIDAE.

L strand See H STRAND.

L-type starter See LACTIC ACID STARTERS.

L1 acholeplasmavirus group *Syn.* PLECTROVIRUS.

L1–L34 proteins See RIBOSOME.

L2 acholeplasmavirus group *Syn.* PLASMAVIRIDAE.

L3 acholeplasmavirus group *Syn.* MV-L3 PHAGE GROUP.

L18-MDP(A) See ADJUVANT.

L-573,655 See ENDOTOXIC SHOCK.

L-749,345 ERTAPENEM.

La Crosse virus A virus of the genus BUNYAVIRUS; this virus is an occasional causal agent of human meningoencephalitis.

La France disease See WATERY STIPE.

labial herpes See HERPES SIMPLEX.

label (probe) See PROBE.

Laboulbenia See LABOULBENIALES.

Laboulbeniales An order of fungi (subdivision ASCOMYCOTINA) which include species that are obligately parasitic on insects. The minute, non-mycelial but multicellular thallus attaches to the host via a specialized organelle (see FOOT). Some species are monoecious, others dioecious (see also HETEROTHALLISM). The ascocarp is a perithecium. Conidial stages are apparently unknown. Genera: e.g. *Ceratomyces*, *Herpomyces*, *Laboulbenia*, *Rickia*. (cf. SPATHULOSPORALES.)

Labyrinthomorpha A phylum of the PROTOZOA which includes the LABYRINTHULAS and the THRAUSTOCHYTRIDS.

Labyrinthula See LABYRINTHULAS.

labyrinthulas (labyrinthulids; 'net slime moulds') A group of eukaryotic organisms of uncertain taxonomic position; they are classified together with the THRAUSTOCHYTRIDS, being variously regarded as fungi e.g. of the LABYRINTHULOMYCETES, as protozoa of a distinct phylum, Labyrinthomorpha, or as protists of the subphylum Labyrinthulina, phylum GYMNOMYXA. There is probably only one genus, *Labyrinthula*, species of which generally occur in marine habitats on plants and algae; they appear to be at least partially parasitic, and are believed to play a role in the ecologically important 'wasting disease' of eelgrass (*Zostera marina*). One *Labyrinthula* species occurs in fresh water and is apparently parasitic on *Vaucheria*.

The vegetative stage of the organism consists of spindle-shaped cells which produce a branching, anastomosing network of membrane-enclosed 'slime-ways' (variously called the ectoplasmic net, filoplasmodium, net plasmodium, rhizoplasmodium, etc); the cells move (by gliding) within these slime-ways. The vegetative cells contain a characteristic organelle, the *sagenogenetosome* or *sagenogen* (formerly 'bothrosome'), which is apparently involved in the production of the ectoplasmic net and possibly also in the movement of the cells through the net. The cells feed osmotrophically; enzymes are secreted from the ectoplasmic net, and the soluble products of digestion diffuse back to the cells within the net.

In at least some labyrinthulas biflagellate zoospores are formed, each bearing one anteriorly directed tinsel flagellum and one posteriorly directed whiplash flagellum; an orange eyespot occurs at the base of the flagellar apparatus. The zoospores are haploid, their formation involving meiosis. They apparently develop directly into spindle-shaped cells; copulation is presumed to occur between two of these cells.

labyrinthulids See LABYRINTHULAS.

Labyrinthulina See GYMNOMYXA.

Labyrinthuloides See THRAUSTOCHYTRIDS.

Labyrinthulomycetes A class of organisms (division MYXOMYCOTA) which includes the LABYRINTHULAS and the THRAUSTOCHYTRIDS; these organisms are apparently unrelated to other members of the Myxomycota.

***lac* operon** (lactose operon) An OPERON containing genes which encode proteins involved in the uptake and utilization of β-galactosides such as LACTOSE. The *lac* operon occurs e.g. in *Escherichia coli* – in which it is located at ca. 8 minutes on the chromosome map. A simplified scheme for the *lac* operon is shown in the figure. The *lacZ* gene encodes β-galactosidase, *lacY* encodes a transport protein, 'β-galactoside permease', and *lacA* encodes thiogalactoside transacetylase. (LacA catalyses the transfer of an acetyl group from acetyl-CoA to a β-galactoside. Its function *in vivo* is unknown; it has been suggested that it may be involved in the detoxification of non-metabolizable sugars by bringing about their acetylation and, hence, excretion.)

According to an earlier view, control of the *lac* operon (illustrated in the figure) is as follows. In the absence of an inducer, the regulator protein (repressor), which is active in tetramer form, binds tightly to the OPERATOR. The operator (O) is a region of ca. 16 bp which occurs immediately downstream of the promoter (P); the operator is an imperfect PALINDROMIC SEQUENCE centred at about nucleotide +10 in relation to the transcription start site (see PROMOTER). The binding of repressor to the operator reduces transcription to a minimal level (only a few molecules of product being produced per cell).

If an inducer (ALLOLACTOSE or e.g. IPTG) is present, it binds to the operator-bound repressor and causes the repressor to undergo a conformational change that reduces its affinity for the operator; the repressor then detaches from the operator, and the level of *lacZYA* transcription increases rapidly; up to ca. 10000 molecules of β-galactosidase may appear in the cell within minutes of induction. (The presence of inducer does not *necessarily* lead to expression of the *lac* operon: see CATABOLITE REPRESSION.)

While the above description gives the essential mode of control of the *lac* operon, the actual process is rather more complex [Mol. Microbiol. (1992) *6* 2419–2422]. Thus, in

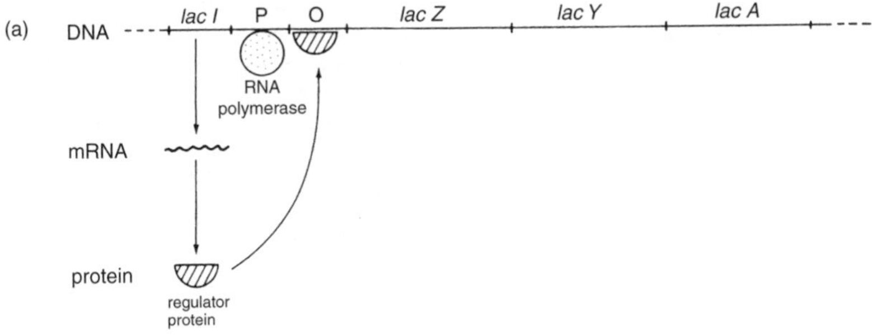

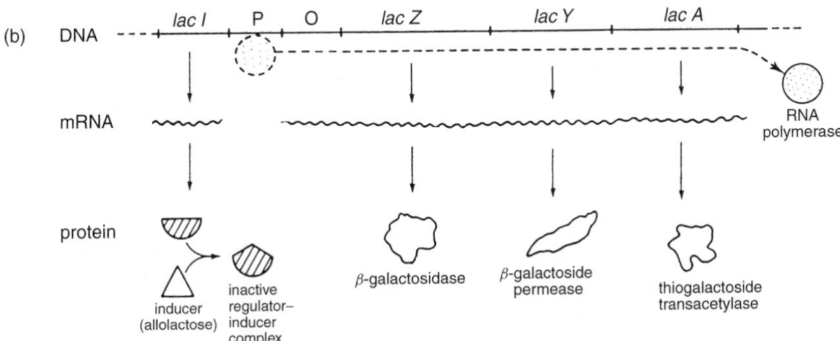

THE *lac* OPERON in *Escherichia coli* (a simplified scheme). The *lac* operon consists of (i) three regulated genes: *lacZ*, *lacY* and *lacA* (see entry); (ii) a promoter (P); (iii) an operator (O); and (iv) the regulator gene, *lacI*, encoding a repressor. *lacI* is transcribed constitutively, at a low rate, from its own independent promoter.

(a) In the absence of an inducer, the regulator protein (LacI) binds to the operator, O, and minimizes transcription of genes *lacZ*, *lacY* and *lacA* by the RNA polymerase (seen here at the promoter); there is some 'leakage': a few molecules of product are formed under these conditions in each cell.

(b) In the presence of lactose (taken up by proton–lactose symport), β-galactosidase converts some of the lactose to *allolactose*, the inducer of the *lac* operon. Allolactose binds to, and inactivates, the regulator protein – which dissociates from the operator, thus permitting transcription and translation of the regulated genes. (The single mRNA transcript contains an initiator codon and a stop codon for each of the three genes.)

When all the lactose has been metabolized, transcription of the *lac* operon is no longer needed; the inducer (allolactose) is no longer formed, and the (now active) regulator protein represses the operon.

The *lac* operon is subject to CATABOLITE REPRESSION (q.v.).

See entry for further details.

Reproduced from *Bacteria*, 5th edition, Figure 7.11, page 138, Paul Singleton (1999) copyright John Wiley & Sons Ltd, UK (ISBN 0471-98880-4) with permission from the publisher.

addition to the primary operator (O), two other specific sites have been found to bind repressor (albeit with lower affinity): O-2 (located within the *lacZ* gene) and O-3 (found upstream of the promoter); a tetramer of the repressor protein appears to bind to the primary operator (O) and, simultaneously, to either O-2 or O-3, thus forming a *DNA loop* that may stabilize repressor–operator binding.

Moreover, in addition to the primary promoter (P1), there are at least five other overlapping promoter-like sequences – two upstream and three downstream of P1 – but their role(s) are unknown; some of these promoter-like sequences can be activated by a single base-pair substitution to give a CRP-independent Lac$^+$ phenotype (but the significance of this is unknown).

As well as the main CRP-binding site (centred ~62 bp upstream from the start site), there is a secondary CRP-binding site which coincides with the (primary) operator sequence. CRP has less affinity for this site, and its role is unknown.

Isopropyl-β-D-thiogalactoside (IPTG) is a *gratuitous* inducer of the *lac* operon, i.e. it acts as an inducer of the operon but is not metabolized by the cell. IPTG is useful e.g. in recombinant DNA technology.

Mutants defective in regulation of the operon. *lacI*$^-$ mutants either fail to produce a repressor or produce an inactive one; transcription of *lacZYA* in these mutants occurs constitutively, but it can be repressed by supplying a functional repressor, i.e. in this case the *lac*$^-$ phenotype is recessive. However, in so-called *lac*$^{-d}$ mutants the mutant phenotype is dominant over the wild-type; in these cells, a defective (mutant) repressor subunit can interact with wild-type subunits to form a defective tetramer (an example of *negative complementation*).

Mutants designated *lacI*^{s} produce a 'super-repressor' which retains its affinity for the operator but has negligible affinity for the inducer – the repressor remains tightly bound to the operator even in the presence of inducer (i.e. the operon is non-inducible), and the defect cannot be overcome by the presence of normal repressor (i.e. the mutant phenotype is dominant).

A mutation in the operator may reduce the affinity of the operator for the repressor; such mutants (designated *lacO*^{c} or *o*^{c}) express the *lac* genes constitutively, and the *o*^{c} allele exhibits CIS-DOMINANCE.

Lac plasmid *Syn.* LACTOSE PLASMID.

***lacA* gene** See LAC OPERON.

laccase A monophenol oxidase (EC 1.14.18.1). Many basidiomycetes (and some ascomycetes and deuteromycetes) produce extracellular laccase which appears to be involved in the degradation of LIGNIN and e.g. in the induction of basidiocarp development [JGM (1982) *128* 2763–2770]. Most or all fungal laccases are copper-containing 'blue proteins' which can oxidize *o*- and *p*-phenols and aromatic amines.

Lachnospira A genus of bacteria (family BACTEROIDACEAE) which occur e.g. in the bovine RUMEN. Cells: curved rods, 0.4–0.6 × 2.0–4.0 μm, with one lateral or subpolar flagellum per cell; the cells from young cultures may give a weak Gram-positive reaction, while those from older cultures may stain Gram-negatively. Glucose fermentation yields acetate, ethanol, formate, lactate, CO_2 and H_2; these products, together with methanol, are also formed from pectin fermentation. Butyrate, propionate and succinate are not formed. Growth is flocculent in glucose–rumen fluid media. Type species: *L. multiparus*. [Book ref. 22, pp. 661–662.]

lachrymal canaliculitis Inflammation of the lachrymal ducts and adjacent tissues, together with the formation of yellowish concretions within the ducts; the disease is non-invasive, usually mild, and the causal agent is commonly *Arachnia propionica* or *Actinomyces israelii*.

laciniate Lobed or notched.

lacrimoid (lacrioid) Shaped like a tear-drop.

Lacrymaria See GYMNOSTOMATIA.

lactacin A See BACTERIOCINS.

lactacin B See BACTERIOCINS.

lactacin F See BACTERIOCINS.

β-lactam antibiotics A family of ANTIBIOTICS, each of which contains the β-lactam ring (see figure). CEPHALOSPORINS and PENICILLINS are produced by fungi, while the more recently discovered members (see CARBAPENEMS, CLAVAMS, MONOBACTAMS, NOCARDICINS) are produced by bacteria [carbapenems and clavams: TIM (1998) *6* 203–208]; synthetic and semi-synthetic members have been manufactured. Most β-lactam antibiotics are antibacterial, but some of the newer ones are antifungal. As well as possessing weak or moderate antibacterial activity, CLAVULANIC ACID and many CARBAPENEMS also inhibit β-LACTAMASES (see also β-LACTAMASE INHIBITOR).

The antibacterial action of penicillins and cephalosporins has been widely studied in *Escherichia coli*. One of the earliest proposals was that penicillin acted as an analogue of D-alanine–D-alanine, thus inhibiting the cross-linking of PEPTIDOGLYCAN with resultant lysis of the structurally weakened cells. The β-lactam antibiotics are now known to have multiple killing sites, and various PENICILLIN-BINDING PROTEINS (PBPs) have been identified as lethal targets. In many cases a given β-lactam antibiotic has a roughly similar degree of affinity for more than one essential PBP – e.g., amoxycillin (PBPs 1A and 2), ampicillin (2, 3, 1A), benzylpenicillin (2, 3), carbenicillin (1A, 3), cefotaxime (1A, 3), cefuroxime (3, 1A), and cloxacillin (1A, 3). However, some β-lactam antibiotics bind preferentially or exclusively to one particular PBP – e.g., cefsulodin, cephalexin, cephaloridine and cephalothin bind preferentially to PBP-1A, mezlocillin binds preferentially to PBP-3, furazlocillin binds exclusively to PBP-3, and mecillinam binds exclusively to PBP-2 (cf. CEFOXITIN, CLAVULANIC ACID and THIENAMYCIN).

Resistance to β-lactam antibiotics may be intrinsic or acquired. In *Pseudomonas aeruginosa* resistance to many β-lactam antibiotics is due to poor penetration of these antibiotics through the outer membrane. In *Escherichia coli* resistance can be due e.g. to a mutation resulting in the loss of the *ompF* PORIN. (In Gram-positive species the CELL WALL offers little hindrance to penetration by molecules the size of β-lactam antibiotics.) Another form of resistance involves mutation resulting in PBPs with lower affinities for β-lactam antibiotics; such PBPs have been detected in e.g. methicillin-resistant strains of *Staphylococcus aureus*, and in benzylpenicillin-resistant strains of *Streptococcus pneumoniae* [JGM (1983) *129* 1247–1260]. Resistance to β-lactam antibiotics may also be due to the action of chromosomal or plasmid-specified β-LACTAMASES.

An important source of resistance to β-lactam antibiotics is the *mecA* gene which encodes a novel PBP that has low affinity for β-lactams; this gene can confer high-level resistance to methicillin (and other β-lactams) e.g. in *Staphylococcus aureus* – see MRSA. [Evidence for *in vivo* transfer of *mecA* between staphylococci: Lancet (2001) *357* 1674–1675.]

Cephalosporins and penicillins are decomposed by ions of the transition metals (e.g. copper, zinc).

[Nomenclature: JAC (1982) *10* 365–372; data review of clinically useful β-lactam antibiotics and those under evaluation: JAC (1986) *17* 5–36. Clinical pharmacology and therapeutic uses of penicillins (review): Drugs (1993) *45* 866–894.]

β-lactamase inhibitor A compound which inhibits the inactivation of β-LACTAM ANTIBIOTICS by β-LACTAMASES; β-lactamase inhibitors include certain β-lactam compounds (e.g. CLAVULANIC ACID) which themselves have little or no antimicrobial activity but which can be combined with β-lactamase-sensitive antibiotics to form clinically useful drugs: see e.g. AUGMENTIN and TIMENTIN.

β-lactamases A class of enzymes (EC 3.5.2.6) which hydrolyse the β-lactam (C–N) bond in susceptible β-LACTAM ANTIBIOTICS, such hydrolysis inactivating the antibiotics. (Certain molecular features of a given β-lactam antibiotic may confer resistance to specific β-lactamases; such features include the 7-α-methoxy group in CEPHAMYCINS and the cysteamine side-chain in THIENAMYCIN.)

Bacteria generally exhibit a substantial level of resistance to a given β-lactam antibiotic if they encode a β-lactamase which is active against that antibiotic (but see β-LACTAMASE INHIBITOR). (See also *Microbial resistance to antibiotics* in the entry ANTIBIOTIC.)

β-Lactamases are produced by many species of bacteria, both Gram-positive and Gram-negative. In some cases the enzymes are inducible and liberated into the medium, while in other cases they are synthesized constitutively and retained within the cell envelope. (Some β-lactam antibiotics – including e.g. cefoxitin and imipenem – are particularly good inducers of β-lactamases.) Studies on the transmembrane signalling pathway in staphylococci indicate that the induction of β-lactamases may depend on a novel mechanism that involves sequential β-LACTAM ANTIBIOTICS. (a) *Penicillin nucleus*. In 6-aminopenicillanic acid (6-APA): R = H. (b) *Cephalosporin nucleus*. In 7-aminocephalosporanic acid (7-ACA): R′ = H, R″ = $O.CO.CH_3$; in cephalosporin C: R′ = $CO.(CH_2)_3.CHNH_2.COOH$, R″ = $O.CO.CH_3$. (c) *Nocardicin nucleus*. (d) *Monobactam nucleus*. (e) *Carbapenem nucleus*. (f) *Clavulanic acid* (a clavam).

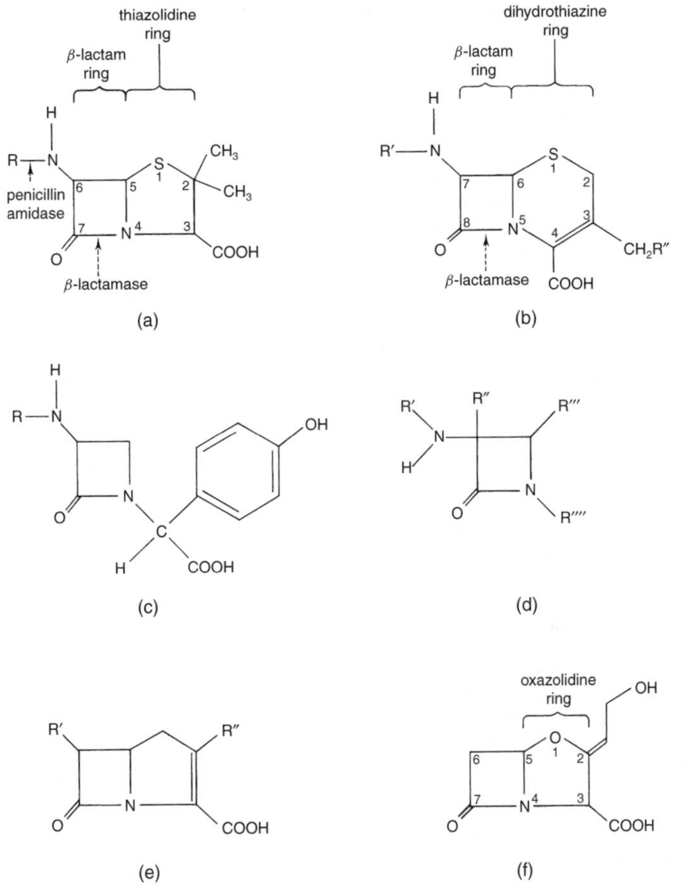

cleavage of regulatory proteins [Science (2001) *291* 1962–1965; commentary: Science (2001) *291* 1915–1916]. In this scheme, the regulation of β-lactamase genes involves initial binding of a β-lactam antibiotic to a cell-surface sensor–transducer protein, BlaR1; the resulting cleavage of BlaR1 gives rise to a (hypothetical) product BlaR2 which, directly or indirectly, brings about cleavage of the transcriptional repressor of β-lactamase genes, allowing expression of these genes.

β-Lactamases may be encoded by chromosomal genes and/or by genes in a PLASMID or transposon (see e.g. entry for Tn*3*). Genes which encode β-lactamases are designated *bla* genes.

The detection of β-lactamases in vitro can be facilitated e.g. by the use of NITROCEFIN.

Classification of β-lactamases. An early scheme classified the β-lactamases according e.g. to their substrate specificities and the location of genes encoding them. The type I β-lactamases included those chromosome-encoded enzymes of enterobacteria and pseudomonads that are active mainly against cephalosporins (i.e. cephalosporinases); type II β-lactamases included chromosome-encoded enzymes of *Proteus* spp that are active primarily against penicillins (i.e. penicillinases); type III enzymes included plasmid-encoded β-lactamases active against both cephalosporins and penicillins, but not against oxacillin or carbenicillin; type IV β-lactamases included the chromosome-encoded enzymes of *Klebsiella* spp that are active primarily against penicillins; type V β-lactamases included plasmid-encoded enzymes active against oxacillin and/or carbenicillin.

Subsequently, classification of plasmid-encoded β-lactamases of Gram-negative bacteria was based on ISOELECTRIC FOCUSING, and more than 11 types were distinguished [BMB (1984) *40* 18–27]. This classification included β-lactamases designated e.g. TEM-1, TEM-2, SHV-1 and HMS-1; these particular enzymes are active against many of (the early) cephalosporins and penicillins, although poorly active against carbenicillin, cloxacillin, methicillin and oxacillin. The OXA-1, -2 and -3 enzymes are active against oxacillin (and e.g. ampicillin and cloxacillin) but poorly active against carbenicillin. The PSE-1, -2, -3 and -4 enzymes of *Pseudomonas* spp are active against carbenicillin (and e.g. ampicillin), while PSE-2 is also active against e.g. oxacillin, methicillin and moxalactam. Subsequent to the designation of these (early) β-lactamases, many new β-lactam antibiotics have been introduced, but β-lactamases have tended to evolve in parallel – see later under *Evolution of β-lactamases*.

The β-lactamases have also been classified into four categories (A–D) on the basis of sequence homology (apparent evolutionary relationships). For example, class A β-lactamases (which include e.g. TEM-1, encoded by gene bla_{TEM-1}) contain a serine residue at the active site of the enzyme; they are found in both Gram-positive and Gram-negative bacteria and are mainly plasmid-encoded enzymes. The class B β-lactamases include zinc-requiring cephalosporinases from *Bacillus cereus*, while class C enzymes include chromosome-encoded cephalosporinases from Gram-negative bacteria.

Evolution of β-lactamases; extended-spectrum β-lactamases (ESBLs). The use of successive 'generations' of β-lactam antibiotics, sometimes with β-lactamase inhibitors, created a selective pressure for the development of new β-lactamases that are able to protect bacteria against some of the most recently introduced drugs. Initially, a new generation of so-called 'extended-spectrum' β-lactam antibiotics were designed to be effective in the presence of the (then) existing β-lactamases, but, following the use of these newer drugs, new variant (mutant) forms of β-lactamase have evolved; these can hydrolyse at least some of the newer antibiotics and are becoming widely disseminated. These so-called *extended-spectrum β-lactamases* (ESBLs) are able to inactivate oxyimino-β-lactam antibiotics (such as the third-generation cephalosporin *cefotaxime*) and the monobactam AZTREONAM. The ESBLs appear to have evolved mainly from TEM and SHV β-lactamases (both class A enzymes) [β-lactamases (evolution): TIM (1998) *6* 323–327]. (Interestingly, evolutionary relationships may exist between β-lactamases and PENICILLIN-BINDING PROTEINS [ARB (1983) *52* 853–856].) ESBLs include e.g. TEM-50 (which incorporates resistance to inhibitors) [AAC (1997) *41* 1322–1325] and SHV-18 (a plasmid-encoded enzyme from *Klebsiella pneumoniae* K6) [AAC (2000) *44* 2382–2388]. Resistance to the fourth-generation cephalosporin *cefepime* has been associated with hyperproduction of SHV-5 [JCM (1998) *36* 266–268].

ESBLs are frequently encoded by plasmids that also carry other antibiotic-resistance genes, so that the choice of drugs for chemotherapy against ESBL-producing strains may be limited. It has been suggested that the choice of β-lactam drugs suitable for treating ESBL-positive infections may (at present) be limited to CARBAPENEMS or CEPHAMYCINS [treatment options for extended-spectrum β-lactamase producers: FEMS (2000) *190* 181–184].

Infections with strains of *Acinetobacter* and *Pseudomonas aeruginosa* that encode ESBLs of the PER-1 type have been reported to be associated with an increased level of fatalities [JMM (2001) *50* 642–645].

Lactarius A genus of fungi (family RUSSULACEAE) which form mushroom-shaped, lamellate fruiting bodies that exude a milk-like fluid ('latex') when cut; *Lactarius* spp are sometimes called 'milk caps'. The lactiferous (latex-containing) hyphae appear to store precursors of the pungent dialdehydes velleral and isovelleral; these dialdehydes may protect the fungus from mycovores [Mycol. (1984) *76* 355–358].

lactase *Syn.* β-GALACTOSIDASE.

lactate dehydrogenase virus (LDV; lactic dehydrogenase virus; lactate dehydrogenase-elevating virus; Riley virus) An enveloped, roughly spherical RNA virus which causes a persistent, life-long infection in mice of the species *Mus musculus* and *M. caroli*; apparently, other animals, including other rodents, are not susceptible.

The LDV genome is a single molecule of positive-sense ssRNA, and the virus has been placed in the arterivirus group (see ARTERIVIRUS).

LDV replicates only in a particular subset of MACROPHAGES, the macrophages being destroyed in the process. [Determinants of macrophage susceptibility: JGV (1985) *66* 1469–1477.] Infection results in a persistent viraemia (maintained by the ongoing production of new susceptible cells) and gives rise to elevated levels of certain plasma enzymes – particularly certain (positively charged) lactate dehydrogenase isoenzymes. Elevated levels of these enzymes result from their reduced rates of clearance from the plasma; such enzymes may be cleared in the normal mouse by the particular subset of macrophages destroyed by LDV.

Infected mice generally show little sign of disease, but they exhibit slight splenomegaly and disturbances in the immune system (altering the growth of tumours, response to infection etc.). In the acute phase of infection there is transient polyclonal activation of antibody-producing cells – resulting in increased immunoglobulin levels – and transient depression of cell-mediated immunity; in the chronic phase, the humoral response is depressed, while cell-mediated immunity returns to normal. Antibodies to LDV are produced, and virus–antibody complexes circulate in the blood; the immune complexes cause little tissue reaction – possibly because they do not bind complement. LDV can apparently cause age-dependent polioencephalomyelitis (resulting in paralysis) in certain strains of mouse.

LDV is apparently transmitted by blood-sucking parasites and by fighting; transplacental transmission can occur if the mother becomes infected during pregnancy or lactation, but generally does not occur in chronically infected mothers.

[LDV (review): JGV (1985) *66* 2297–2312.]

lactic acid (CH_3.CHOH.COOH) A compound present e.g. in many fermented DAIRY PRODUCTS, and also widely used as a food additive. Only the L(+)-isomer can be readily assimilated by man and animals. L(+)-Lactic acid is obtained commercially by the LACTIC ACID FERMENTATION of sucrose by LACTIC ACID BACTERIA; it is used primarily in the food industry, being added e.g. to confectionary, pickles, beverages, etc [AvL (1983) *49* 86].

lactic acid bacteria A (non-taxonomic) group of Gram-positive, non-sporing bacteria which carry out a LACTIC ACID FERMENTATION of sugars; it includes species of *Lactobacillus*, *Lactococcus*, *Leuconostoc* and *Pediococcus*. (*Bifidobacterium* is sometimes included.)

Lactic acid bacteria are used extensively in the food industry (e.g. BREAD-MAKING (sourdough), DAIRY PRODUCTS, SALAMI); however, they can also cause food spoilage (see e.g. MEAT SPOILAGE; cf MALOLACTIC FERMENTATION). (See also SILAGE.)

Lactic acid bacteria are nutritionally fastidious and weakly proteolytic [for proteolytic systems see AvL (1983) *49* 225–245]. Many activities of the lactic acid bacteria (e.g. LACTOSE metabolism) appear to be plasmid-linked [AvL (1983) *49* 259–274].

Although lactic acid bacteria are primarily fermentative, many strains can grow in the presence of air and can carry out certain oxidation reactions using flavin-containing oxidases, peroxidases, or NAD-independent dehydrogenases – oxidizing at least some of the pyruvic and lactic acids (formed by fermentative pathways) to acetic acid and carbon dioxide [AvL (1983) *49* 209–224]. (See also LEUCONOSTOC.) The hydrogen peroxide formed during these oxidations has some preservative action in fermented foods (see FOOD PRESERVATION (e)).

[Genetics, metabolism and applications of lactic acid bacteria (symposium); FEMS Reviews (1993) *12* 1–272.]

lactic acid fermentation A type of FERMENTATION (1), carried out e.g. by LACTIC ACID BACTERIA, in which sugars (e.g. LACTOSE, glucose, PENTOSES) are converted either entirely (or almost entirely) to lactic acid (HOMOLACTIC FERMENTATION) or to a mixture of lactic acid and other products (HETEROLACTIC FERMENTATION) [AvL (1983) *49* 209–224]. Lactic acid bacteria produce either L(+)- or D(−)-lactic acid, or both – the nature of the isomer(s) formed being an important taxonomic characteristic (see e.g. LACTOBACILLUS). Which isomer is produced depends on the specificity of the NAD-dependent lactate dehydrogenase (LDH) present; species which produce racemic mixtures usually possess both D- and L-specific LDHs, but in certain lactobacilli the mixture results from the combined action of L-specific LDH and lactic acid racemase. (See also LACTIC ACID and DAIRY PRODUCTS.)

lactic acid starters STARTER cultures used to initiate a LACTIC ACID FERMENTATION in the commercial production of e.g. DAIRY PRODUCTS; they may be supplied in freeze-dried or frozen concentrated form.

The starters used for dairy products usually contain 'homolactic' lactobacilli or lactococci as the main lactic acid producers, with leuconostocs and/or strains of *Lactococcus lactis* as 'aroma bacteria' (see DIACETYL). The lactic acid-producing strains carry out HOMOLACTIC FERMENTATION of the lactose in milk – conditions of lactose excess for these bacteria; cows' milk contains ca. 4–7% w/v lactose, much of which remains in the product after fermentation. Only strains which use the PTS/P-β-gal system of LACTOSE (q.v.) metabolism achieve sufficiently rapid growth and produce enough lactic acid to be useful commercially.

Mesophilic starters (used e.g. for making BUTTER) usually contain strains of *Lactococcus lactis* as the main acid-producers; starters which contain only acid-producers (no aroma bacteria) are called N-type (= O-type) starters. B-type (= L-type) starters contain, in addition to N-type organisms, a *Leuconostoc* aroma species – usually *Leuconostoc cremoris* (= '*Betacoccus cremoris*', hence 'B' starter). D-type starters contain N-type organisms and aroma strains of *Lactococcus lactis*. BD-type (= BL-type) starters contain both B- and D-type organisms.

Thermophilic starters (used e.g. for YOGHURT and some CHEESE-MAKING processes) include e.g. *Lactobacillus bulgaricus*, *L. helveticus* and/or *Streptococcus thermophilus*.

A genetically engineered starter culture has been used to control the (food-borne) pathogen *Listeria monocytogenes* in cheddar cheese [AEM (1998) *64* 4842–4845]. For cottage cheese, use has been made of a starter containing organisms that produce lacticin 3147 [JAM (1999) *86* 251–256].

Starter strains are susceptible to infection by certain phages, and this can lead to problems such as slow (or absent) fermentation associated with economic losses. Efforts have therefore been made to produce 'engineered' starter strains that incorporate resistance to specific phages. For example, some engineered strains exhibit the Per (phage-encoded resistance) phenotype which is conferred by the presence of a high-copy-number plasmid that contains a sequence equivalent to the origin of replication of a particular infecting phage; it appears that the plasmid-borne phage origin (in multiple copies) inhibits phage replication by competing with replication function(s) of the infecting phage. In a different approach, transcription of a plasmid-borne sequence (within the engineered bacterium) gives rise to an antisense RNA molecule that can bind to a specific phage mRNA and prevent its translation. [Improvement and optimization of two engineered phage-resistance mechanisms in *Lactococcus lactis*: AEM (2001) *67* 608–616.]

lactic acidosis (*vet.*) See ACIDOSIS (1).

lactic dehydrogenase virus *Syn.* LACTATE DEHYDROGENASE VIRUS.

lacticin 3147 A pore-forming LANTIBIOTIC produced by *Lactobacillus lactis* subsp *lactis*. It consists of two peptide components, both of which are needed for activity; each component requires modification by a separate enzyme [Microbiology (2000) *146* 2147–2154]. In sensitive cells, the lesions formed by lacticin 3147 are ion-specific pores.

A lacticin 3147-producing starter culture has been tested for its ability to inhibit *Listeria monocytogenes* in the manufacture of cottage cheese [JAM (1999) *86* 251–256].

lactiferous hyphae See LACTARIUS.

Lactobacillaceae A family of bacteria (sole genus: LACTOBACILLUS). [Book ref. 21, p. 576.]

Lactobacillus A genus of Gram-positive, asporogenous bacteria which are anaerobic, microaerophilic, or facultatively aerobic. Cells are rods or coccobacilli, occurring singly or in chains. Most strains are non-motile. Catalase −ve, but may be catalase +ve in haem-containing media. Metabolism is predominantly fermentative, lactic acid being formed from glucose either homofermentatively or heterofermentatively. (See LACTIC ACID BACTERIA.) Species are commonly aciduric or acidophilic. Nutritional requirements are complex, generally including carbohydrates, peptones, vitamins etc. Lactobacilli may be grown e.g. on APT agar; growth is stimulated by Tween 80.

The genus (GC%: ca. 32–53) is divided into three subgenera: *Betabacterium*, *Streptobacterium* and *Thermobacterium*. Betabacteria carry out HETEROLACTIC FERMENTATION of glucose (and hence require thiamine and produce CO_2 from glucose); fructose is fermented and can be reduced to mannitol (see e.g. SILAGE), and PENTOSES may be fermented. Streptobacteria and thermobacteria carry out HOMOLACTIC FERMENTATION of glucose (CO_2 not produced, thiamine not required); fructose is usually fermented but it cannot be reduced to mannitol. Streptobacteria can grow at 15°C (and may or may not grow at 45°C), can ferment pentoses, and produce CO_2 from gluconate. Thermobacteria can generally grow at or above 45°C but not at 15°C, and cannot ferment pentoses or produce CO_2 from gluconate. Species are distinguished by biochemical tests. The nature of the lactic acid produced can be a useful criterion: betabacteria produce DL-lactic acid, streptobacteria and thermobacteria produce (mainly) L(+)-, D(−)- or DL-lactic acid, according to species. Betabacteria include e.g. *L. brevis*, *L. buchneri* (possibly a subspecies of *L. brevis*), *L. cellobiosus*, *L. confusus*, *L. desidiosus*, *L. fermentum*, *L. hilgardii*, *L. viridescens*; streptobacteria include e.g. *L. alimentarius*, *L. casei*, *L. coryniformis*, *L. curvatus*, *L. farciminis*, *L. plantarum*, *L. xylosus* (cf. LACTOCOCCUS); thermobacteria include e.g. *L. acidophilus*, *L. bulgaricus*, *L. delbrueckii* (type species of the genus), *L. helveticus*, *L. lactis*, *L. leichmannii*, *L. ruminis*, *L. salivarius*. [Book ref. 46, pp. 1653–1679.] (*L. bifidus*: see BIFIDOBACTERIUM.) [Classification of *L. bulgaricus*, *L. lactis* and *L. leichmannii* as subspecies of *L. delbrueckii*: SAAM (1983) *4* 552–557.]

Lactobacilli are found e.g. in man and animals (see e.g. DENTAL CARIES, GASTROINTESTINAL TRACT FLORA, VAGINA MICROFLORA), in MILK, DAIRY PRODUCTS and other fermented foods and beverages (see e.g. sourdough BREAD-MAKING, PULQUE, SOY SAUCE), and – usually in very small numbers – on plant material (see e.g. SILAGE, PICKLING). (See also BEER SPOILAGE, MALOLACTIC FERMENTATION, MEAT SPOILAGE, MILK SPOILAGE, WINE

SPOILAGE.) Lactobacilli are generally non-pathogenic, although some strains (particularly of *L. casei* subsp. *rhamnosus*) have been associated with endocarditis, septicaemia and abscesses, and another species (proposed name: *L. piscicola* [IJSB (1984) *34* 393–400]) is apparently an opportunist pathogen in salmonid fish.

[*Lactobacillus casei* strain Shirota has been reported to have potent PROBIOTIC activity when administered intraurethrally in an *Escherichia coli* urinary tract infection in mice: AAC (2001) *45* 1751–1760.]

lactocin 27 See BACTERIOCINS.

lactococcin G See BACTERIOCINS.

Lactococcus A genus of LACTIC ACID BACTERIA which occur e.g. in DAIRY PRODUCTS; the organisms were previously classified in the genus *Streptococcus* (most strains react positively with streptococcal Lancefield group N antiserum). Cells: non-motile cocci, or coccoid forms, which occur singly or in pairs or chains. Metabolism: fermentative, L(+)-lactic acid being the main product of glucose fermentation. Growth can occur at 10°C but not at 45°C. GC% 34–43. Type species: *L. lactis*.

L. lactis includes organisms previously known as *Streptococcus lactis*, including the subspecies *lactis* and *diacetylactis* (= *diacetilactis*), *Streptococcus cremoris* (including *S. lactis* subspecies *cremoris*), and *Lactobacillus xylosus*. It was suggested that *Lactococcus* should also include the species *L. plantarum* (= *Streptococcus plantarum*) and *L. raffinolactis* (= *Streptococcus raffinolactis*). [Validation of the names *Lactococcus*, *L. lactis*, *L. lactis* subspecies *cremoris* and *lactis*, and *L. plantarum*: IJSB (1986) *36* 354–356.]

lactoferrin A SIDEROPHILIN (MWt ca. 90000) which has two iron-binding sites per molecule; it occurs e.g. in milk and tears, and in granules within NEUTROPHILS. Lactoferrin in milk may serve an antimicrobial function by sequestering iron (IRON being an essential requirement for most microorganisms). The lactoferrin in neutrophils may serve an antimicrobial function (during PHAGOCYTOSIS and INFLAMMATION) by catalysing the formation of HYDROXYL RADICAL; levels of lactoferrin in plasma are known to rise during inflammatory processes. Lactoferrin released by neutrophils may enhance the adhesion of neutrophils to tissue cells, and to each other, and may promote chemotaxis in other neutrophils.

lactoflavin *Syn.* RIBOFLAVIN.

lactoperoxidase–thiocyanate–hydrogen peroxide system (LP system; LPS) An antimicrobial system, present in various body fluids, in which lactoperoxidase (an enzyme present e.g. in saliva, milk etc) catalyses the oxidation, by hydrogen peroxide (H_2O_2), of thiocyanate (SCN^-). SCN^- is a product of the metabolism of e.g. sulphur-containing amino acids and certain glycosides; it is present at relatively high concentrations e.g. in milk from cattle fed on clover or cruciferous plants such as kale or rape. The H_2O_2 may be derived from indigenous microorganisms (as a metabolic by-product – see HYDROGEN PEROXIDE (b)) or from PMNs. Although the final products of the oxidation (SO_4^{2-}, CO_2 and NH_3) are much less toxic than H_2O_2, various short-lived antibacterial intermediates – including e.g. hypothiocyanite, $OSCN^-$ – are generated during the reaction. The LP system inhibits various bacterial enzymes (including e.g. the EMP enzyme glyceraldehyde 3-phosphate dehydrogenase) and damages the bacterial cytoplasmic membrane, increasing the permeability of the membrane to ions and small molecules. The system can be bactericidal for Gram-negative, catalase-positive bacteria (e.g. enterobacteria, pseudomonads), but is usually bacteriostatic against Gram-positive, catalase-negative bacteria (e.g. streptococci and lactobacilli).

The LP system has a potential application in the preservation of (refrigerated or uncooled) milk; for this purpose, levels of H_2O_2 in the milk could be boosted e.g. by the addition of a glucose/glucose oxidase H_2O_2-generating system and (if necessary) SCN^-. Components which activate the LP system have been incorporated into an 'anti-plaque' toothpaste to inhibit oral streptococci (see DENTAL CARIES).

[Review: J. Food Protect. (1984) *47* 724–732.]

lactophenol cotton blue A general mycological MOUNTANT and DYE. Phenol (20 g), water (20 ml), lactic acid (20 ml), and glycerol (30 ml) are warmed together, and cotton blue (0.05 g) is added.

lactose A disaccharide (β-D-galactopyranosyl-(1 $\rightarrow$ 4)-D-glucopyranose) present in MILK. Lactose is the major source of carbon and energy for LACTIC ACID BACTERIA growing in milk (see also LACTIC ACID FERMENTATION). The ability to metabolize lactose is an important characteristic in the identification of members of the family Enterobacteriaceae.

In *Escherichia coli* and other enterobacteria, and in most lactobacilli (though not *Lactobacillus casei*), lactose is generally taken up by a lactose-specific proton symport TRANSPORT SYSTEM; the lactose is then split into glucose and galactose by β-galactosidase (see also LAC OPERON), the glucose and galactose being metabolized via the EMP and Leloir pathways, respectively: see Appendix III(a).

In e.g. *Lactobacillus casei*, *Lactococcus lactis* and *Staphylococcus aureus*, lactose is taken up by a PEP-dependent phosphotransferase system (see PTS) – entering the cell as lactose phosphate (apparently as lactose bisphosphate in *L. lactis* [Book ref. 11, pp 196–203]) which is then hydrolysed by phospho-β-D-galactosidase (P-β-gal); the glucose (glucose 6-phosphate in *L. lactis*) and galactose 6-phosphate moieties are metabolized via the EMP and D-tagatose 6-phosphate pathways, respectively – see Appendix III(a). In at least some strains of *L. lactis*, all the enzymes for lactose metabolism – from transport to the formation of triose phosphates via the D-tagatose 6-phosphate pathway – appear to be plasmid-encoded [JB (1983) *153* 76–83]; strains lacking the lactose plasmid can metabolize lactose via the permease/β-galactosidase mechanism and Leloir pathway, but their rates of lactose metabolism and growth are much reduced (cf. LACTIC ACID STARTERS).

In yeasts, *Kluyveromyces marxianus* var. *lactis* and var. *marxianus* (some strains) are among the few capable of fermenting lactose (see SINGLE-CELL PROTEIN); however, other yeasts may be able to metabolize lactose oxidatively (cf. KLUYVER EFFECT).

lactose killing Lactose-mediated SUBSTRATE-ACCELERATED DEATH.

lactose operon *Syn.* LAC OPERON.

lactose plasmid (Lac plasmid) A PLASMID which specifies the uptake and metabolism of LACTOSE. Conjugative lactose plasmids occur e.g. in some lactococci.

lacunose Covered with pits or indentations.

***lacY* gene** See LAC OPERON.

***lacZ* gene** See LAC OPERON.

LAD LEUKOCYTE ADHESION DEFICIENCY.

Laemmli electrophoresis *Syn.* SDS-PAGE.

laeoplectic metachrony See METACHRONAL WAVES.

Laetiporus See APHYLLOPHORALES (Polyporaceae).

laevulose D-Fructose.

LAF See INTERLEUKIN-1.

lag phase (of growth) See BATCH CULTURE.

Lagarde retort An air–steam BATCH RETORT.

Lagenidiales An order of aquatic fungi (class OOMYCETES) which are parasitic on e.g. other fungi, oyster larvae and crabs' eggs. The thallus is either a single cell or a short filament, according to organism. Genera include *Lagenidium*, *Olpidiopsis*, *Petersenia* (see also ALGAL DISEASES) and *Sirolpidium*.

Lagenidium See LAGENIDIALES.

Lagenophrys See PERITRICHIA.

lager-making See BREWING.

lagging-strand See DNA REPLICATION.

Lagos bat virus See LYSSAVIRUS.

Laguna Negra virus See HANTAVIRUS.

laking (of erythrocytes) Lysis, with release of haemoglobin.

LAL test LIMULUS AMOEBOCYTE LYSATE TEST.

lamb dysentery An acute, fatal disease of young lambs (<10 days old) involving diarrhoea, dysentery and toxaemia; the causal agent is *Clostridium perfringens* type B. Overcrowding is a predisposing factor.

LamB protein ('maltoporin') In *Escherichia coli*: an OUTER MEMBRANE protein encoded by the *lamB* gene; it acts as a receptor for bacteriophage lambda, and as a channel-forming protein involved in the uptake of maltose and maltodextrins – and certain other solutes (e.g. some unrelated sugars, and amino acids). (cf. PORIN.) LamB-deficient mutants exhibit impaired active transport of maltose when the external maltose concentration is <10 μM. The channel-forming role of LamB in the uptake of maltodextrins (and possibly also maltose) may depend on its interaction with the periplasmic maltose-binding protein (see BINDING PROTEIN-DEPENDENT TRANSPORT SYSTEM). Synthesis of the LamB protein is inducible by maltose and is subject to CATABOLITE REPRESSION.

***lambda* chain** See LIGHT CHAIN.

lambda particles See LYTICUM.

lambdoid phages A group of temperate phages whose genomes can undergo recombination with each other; the group includes e.g. phages λ, 21, 82, 434, ϕ80, and P22 (see BACTERIOPHAGE λ and BACTERIOPHAGE P22). The genomes of these phages have a more or less common gene sequence, but many of the gene products are genome-specific: e.g., the gp*N* antiterminators of phages λ, 21 and P22 show little homology and interact only with their own genomes. Phages λ, ϕ80, 21 and 434 each has a unique immunity region (genes *cI*, *cro*, o_L and o_R) and hence a lysogen containing one of these phages does not show SUPERINFECTION IMMUNITY to the others.

Lamblia intestinalis See GIARDIA.

lambliasis *Syn.* GIARDIASIS.

Lambornella A genus of ciliates of the HYMENOSTOMATIDA; *L. clarki* is endoparasitic in certain mosquitoes.

lamella (gill) (*mycol.*) Blade-like sheets of tissue which hang vertically from the underside of the pileus in fruiting bodies of fungi of the AGARICALES and of e.g. PLEUROTUS spp; each lamella consists of a TRAMA (see also SPHAEROCYST) bearing a HYMENIUM which gives rise to basidiospores. In some species (e.g. *Agaricus*) the lamellae are arranged radially around a central stipe; in others (e.g. *Pleurotus*) they radiate fan-wise from a lateral stipe. Lamellae may branch and/or anastomose; their colour at maturity usually reflects the colour of the basidiospores. (See also SPORE PRINT.) The shape of lamellae can be taxonomically important.

In most agarics (including *Agaricus*) the lamellae are *aequihymeniiferous*, i.e., they are wedge-shaped in transverse cross-section (tapering from the attached to the free end), and they form basidia which mature at approximately the same time at all regions in the hymenium. *Inaequihymeniiferous* lamellae occur e.g. in *Coprinus* spp; they are rectangular in transverse cross-section (opposite faces of the lamella being parallel), and the basidia at the lower (free) end reach maturity before those higher up the lamella. (See also BASIDIOSPORES.)

lamellate (*mycol.*) Having lamellae (see LAMELLA).

lamina (*algol.*) *Syn.* BLADE.

laminaran *Syn.* LAMINARIN.

Laminaria A genus of large marine algae (division PHAEOPHYTA) which exhibit a heteromorphic ALTERNATION OF GENERATIONS. The sporophyte is a large blade with a stipe and holdfast; growth occurs at the region between stipe and blade, and at the MERISTODERM. The gametophyte is a microscopic filamentous structure. *L. japonica* is used as a food in China (where it is called 'haidai') and e.g. in Japan ('kombu') [Book ref. 130, pp. 688–690]. (See also ALGINATE.)

laminaribiose A reducing disaccharide: β-D-glucopyranosyl-(1 → 3)-D-glucopyranose.

laminarin (laminaran) The storage polysaccharide characteristic of members of the Phaeophyta. Laminarins are essentially linear (1 → 3)-β-D-glucans in which the chains may terminate with D-mannitol (M-chains, non-reducing) or with D-glucose (G-chains, reducing); low levels of branching via (1 → 6)-β-glucosidic linkages may occur. Laminarins from different sources vary in their mannitol content (usually ca. 2–4%), degree of branching, and average chain length. [Book ref. 37, pp. 481–486.] Laminarins may be attacked e.g. by certain fungi and by *Nocardia* spp. (cf. CHRYSOLAMINARIN.)

lamivudine (−)2′-deoxy-3′-thiacytidine: an ANTIVIRAL AGENT with activity against the hepatitis B virus; this nucleoside analogue inhibits the viral DNA polymerase, thus inhibiting viral replication. As a candidate therapeutic agent, lamivudine was able significantly to reduce serum levels of viral DNA in phase III clinical trials. However, resistance to the drug may appear during long-term treatment programmes; such resistance has been found to be associated with mutant forms of the polymerase. [PCR-based detection of mutant strains of HBV with reduced susceptibility to lamivudine: JCM (1999) *37* 3338–3347.]

Lamivudine is also an ANTIRETROVIRAL AGENT which acts as a NUCLEOSIDE REVERSE TRANSCRIPTASE INHIBITOR; it is one of the drugs used against AIDS.

Lampit See CHAGAS' DISEASE.

Lamprene CLOFAZIMINE.

Lamprocystis A genus of photosynthetic bacteria (family CHROMATIACEAE). Cells of the sole species, *L. roseopersicina*, are motile cocci, 3–3.5 μm, which contain Bchl *a*, various carotenoids (cell suspensions are typically purple), and gas vacuoles. *L. roseopersicina* is a strict anaerobe which requires sulphide and vitamin B_{12}.

Lamproderma See MYXOMYCETES.

Lampropedia A genus (*incertae sedis*) of aerobic, oxidase-positive, catalase-positive, chemoorganotrophic, Gram-negative bacteria which have been isolated e.g. from stagnant and polluted waters. Cells: non-flagellated, non-pigmented, rounded to cuboid, ca. 1.0–1.5 × 1.0–2.5 μm, which occur in thin, hydrophobic sheets (on liquid and solid media) – each sheet consisting of a number of roughly square 'tablets' of 16–64 cells arranged, with sides adjacent, in the same plane; cell division occurs synchronously. Metabolism: respiratory (oxidative). Carbon and energy sources include TCA cycle intermediates; carbohydrates are not used. Optimum growth temperature: 30°C. GC%: ca. 61. Type species: *L. hyalina*.

[Book ref. 22, pp. 402–406.]

LAMPs See PHAGOCYTOSIS (a).

lamziekte A form of BOTULISM in cattle. It occurs in S. Africa and is contracted by the ingestion of carrion contaminated with *C. botulinum* (usually) type D.

Lancefield's streptococcal grouping test A procedure for identifying and classifying streptococci; it involves the extraction and identification, by serology, of cell components known as C SUBSTANCES. On the basis of their particular C substances, most strains of STREPTOCOCCUS can be placed into one or other of a number of *Lancefield groups* which are designated A, B, C. ... etc; all the strains in a particular Lancefield group possess a C substance which is serologically distinct from those C substances which characterize the other Lancefield groups. (Some streptococci have more than one type of C substance; thus, e.g., some strains of enterococci contain the C substances of Lancefield groups D and Q.)

The test is often carried out as a RING TEST in a capillary tube: an extract of the strain under test is layered onto an antiserum to a given C substance, and the test is repeated with antisera to different C substances until a positive result (a precipitate at the extract–antiserum interface) is obtained.

Langerhans' cells Non-phagocytic cells, derived from bone marrow, found in the skin. Morphologically they resemble DENDRITIC CELLS but, in addition to class II MHC antigens, they bear cell-surface receptors for complement components and for IgG Fc portions; they also contain racket-shaped cytoplasmic organelles (*Birbeck granules*). Langerhans' cells are involved in CONTACT SENSITIVITY.

Langhans giant cell See GIANT CELL.

lankamycin See MACROLIDE ANTIBIOTICS.

Lankesterella A genus of coccidian protozoa (suborder EIMERIORINA) parasitic in amphibia and transmitted by invertebrate vectors. Asexual and sexual development occur in the vertebrate host, sporozoites occurring in red and/or white blood cells.

lantibiotics Polycyclic peptide BACTERIOCINS whose molecules include some uncommon residues – e.g. the sulphur-containing amino acid lanthionine (hence *lantibiotic*) and D-amino acids. Lantibiotics are produced by strains of e.g. *Bacillus* and *Lactococcus*, and are typically active against Gram-positive bacteria; they form pores in the cytoplasmic membrane of a target cell, thereby causing ionic leakage and loss of pmf. Examples include NISIN and SUBTILIN.

The presence of lanthionine (and of e.g. D-amino acids) in these peptides results from post-translational modification [TIBS (1992) *17* 481–485].

[Biosynthesis and biological activities of lantibiotics: ARM (1998) *52* 41–79.]

[Lantibiotics: structure, biosynthesis and mode of action: FEMS Reviews (2001) *25* 285–308.]

LAO binding protein See BINDING PROTEIN-DEPENDENT TRANSPORT SYSTEM.

LAP See ENDOTOXIN.

lapinized vaccine A VACCINE containing live organisms whose virulence for a given host species has been attenuated by SERIAL PASSAGE through rabbits.

Laredo-type strains See ENTAMOEBA (*E. histolytica*).

large granular lymphocytes See NK CELLS.

large iridescent insect viruses See CHLORIRIDOVIRUS.

large T antigen See POLYOMAVIRUS.

lariat (lariat RNA) See SPLIT GENE (a) and (c).

***las* genes** See QUORUM SENSING.

Lasallia A genus of LICHENS (order LECANORALES). The thallus resembles that of UMBILICARIA spp but is pustulate in appearance (convexities in the upper surface corresponding to hollows in the lower surface); asci contain one or two large spores (those of *Umbilicaria* contain 8 smaller spores). (*L. pustulata*: formerly *U. pustulata*.)

lasalocid A MACROTETRALIDE antibiotic which is used as an anticoccidial agent in poultry, and which has been used as a FEED ADDITIVE for ruminants (its mode of action apparently being similar to that of MONENSIN).

Lassa fever A VIRAL HAEMORRHAGIC FEVER caused by the Lassa virus (see ARENAVIRIDAE); it occurs in Africa, especially Nigeria, Liberia and Sierra Leone. Early symptoms resemble those of other VHFs; petechiae may be present, and there is marked pharyngitis with white patchy exudate adhering to the soft palate and pharyngeal region. In fatal cases death usually occurs during the second week of infection; mortality rates may be 30–66% among hospital cases, but subclinical infections may occur. The natural reservoir of the virus is the multimammate rat *Mastomys natalensis*. Human infection may occur e.g. by contamination of wounds with urine from infected rats; person-to-person transmission also occurs.

Lassa virus See ARENAVIRIDAE and LASSA FEVER.

latamoxef *Syn.* MOXALACTAM.

late blight (of potato) A POTATO DISEASE caused by *Phytophthora infestans* (see PHYTOPHTHORA). Dark patches appear on the upper surfaces of leaves (and sometimes on stems), and patches of white 'mildew' (sporangium-bearing sporangiophores) develop on corresponding regions on the undersides of leaves. Disease may spread rapidly from plant to plant. Tubers develop dark surface patches and succumb to a dry brown rot – and sometimes to a secondary soft rot. Protectants used against late blight include e.g. MANEB and ZINEB. Late blight was responsible for the Irish potato famine in the 1840s.

[Review: Book ref. 58, pp. 3–17.]

late genes See VIRUS.

latent infection (1) (*animal virol.*) An asymptomatic infection in which infectious virions are not produced. (See PERSISTENCE.)

(2) (*plant virol.*) Any viral infection in which symptoms do not occur – whether or not virus replication occurs.

latent period See ONE-STEP GROWTH EXPERIMENT.

latent virus A virus which can establish a LATENT INFECTION (sense 1 or 2).

lateral body (*virol.*) See POXVIRIDAE.

lateral transmission See HORIZONTAL TRANSMISSION.

laterosporolysin See THIOL-ACTIVATED CYTOLYSINS.

latex (1) A product of the rubber tree, *Hevea brasiliensis*. Natural latex is a colloidal suspension of polyisoprenoid particles in an aqueous phase containing e.g. carbohydrates and proteins; it is readily attacked by a variety of microorganisms. (See also RUBBER SPOILAGE.)

(2) See LATEX PARTICLE TEST.

(3) See LACTARIUS.

latex particle test (latex test) (*serol.*) A PASSIVE AGGLUTINATION test in which antibody or soluble antigen is adsorbed to particles of a synthetic latex: e.g. a suspension of particles of polystyrene (commonly ca. 0.81 μm diam.), a substance to which proteins tend to adsorb.

For example, to detect antibodies to a given antigen, antigen-coated particles of latex are exposed to the patient's serum; the presence of homologous antibodies in the serum is indicated by agglutination of the latex particles (which are bound together by antigen–antibody bridges). Uncoated particles of latex, and known negative serum, are used for controls.

Latex particle tests have a wide range of uses, being able not only to detect serum antibodies and soluble antigens but also to distinguish between different strains of a given pathogen [e.g. rapid detection of *mecA*-positive and *mecA*-negative strains of coagulase-negative staphylococci by an anti-PBP 2a slide latex agglutination test: JCM (2000) *38* 2051–2054.]

When specific *antibodies* are bound to the latex particles, the test is often described as a *reverse* (or *reversed*) *passive latex agglutination test*. Thus, e.g. staphylococcal enterotoxins can be tested for (in e.g. food samples) by using a reverse passive latex agglutination test in which the particles are coated with antibodies to the various enterotoxins; agglutination of such particles may occur in the presence of very low concentrations of enterotoxins in the samples.

[Reversed passive latex agglutination test for detecting 'free' verocytotoxin produced by strains of *Escherichia coli* O157 (EHEC): LAM (2001) *32* 370–374.]

Latin binomial See BINOMIAL.

Latino virus See ARENAVIRIDAE.

lattice hypothesis (*serol.*) A hypothesis advanced by Marrack in 1934 to account for the observation that antibody and antigen can combine in varying proportions; it postulates that each molecule of antibody has a valency of at least two, and that antigens are multivalent. Thus, when antigens and antibodies combine in suitable proportions (see OPTIMAL PROPORTIONS) large insoluble aggregates (lattices) are formed; under ANTIGEN EXCESS conditions small, soluble complexes are formed. The precipitability of a complex may depend e.g. on its hydrophobicity as well as on its size.

Laurell's rocket immunoelectrophoresis *Syn.* ROCKET IMMUNOELECTROPHORESIS.

lauryl tryptose lactose broth A medium, containing sodium lauryl sulphate (sodium dodecyl sulphate), used e.g. in the EIJKMAN TEST.

Lauth's violet *Syn.* THIONIN.

LAV (1) See HIV. (2) LEAFHOPPER A VIRUS.

laver (laver bread) Edible algae, particularly certain species of *Porphyra* (known as 'zicai' in China, 'nori' in Japan, and 'purple laver' in some other countries). (See also ULVA.)

Laverania A subgenus of PLASMODIUM.

law of recapitulation See ONTOGENY.

lawn plate A PLATE whose entire surface bears confluent growth of microorganisms of one species or strain; it may be prepared e.g. by incubating a FLOOD PLATE.

Lb LEGHAEMOGLOBIN.

LB broth See LURIA–BERTANI BROTH.

LBP See CD14.

Lc protein A PORIN of *Escherichia coli*.

LCM virus (lymphocytic choriomeningitis virus, LCMV) The type species of the genus *Arenavirus* (see ARENAVIRIDAE). The main natural host is the house mouse (*Mus musculus*), but the syrian or golden hamster (*Mesocricetus auratus*) can also be an epidemiologically important host. LCM virus can replicate in a wide range of cultured cells in which infection is usually persistent and without CPE. Embryo or neonatal (immunologically immature) mice inoculated with LCMV survive the infection to become persistent carriers, shedding virus in their urine, faeces, saliva etc throughout their lives (see PERSISTENCE). However, adult mice inoculated intracerebrally with LCMV rapidly succumb to the infection by a mechanism which seems to involve interferon: a 'docile' strain of LCMV which does not induce interferon does not cause a lethal infection [JGV (1983) *64* 1827–1830]. (See also GLOMERULONEPHRITIS.) In man, LCMV is a rare cause of disease – which may be an influenza-like illness, an acute, benign aseptic MENINGITIS, or a severe meningoencephalomyelitis.

LCR Ligase chain reaction: a PROBE-based method for copying ('amplifying') a given sequence of nucleotides in DNA; like PCR (but unlike NASBA or SDA), LCR involves thermal cycling. Essentially, sample DNA is heat-denatured in order to expose the target sequence (*amplicon*) in one strand – and its complementary sequence in the other strand; on lowering the temperature, two oligonucleotide probes bind (*anneal*) in *adjacent* positions to cover the entire amplicon – while another two probes bind to the amplicon's complementary sequence. If a given pair of probes binds correctly to the target, the 5′ terminus of one probe will abut the 3′ terminus of the other; when this occurs, a covalent link can be made between the 5′ and 3′ termini by a heat-stable DNA LIGASE. A pair of ligated probes will thus form a copy of the target sequence and can be used as a target (by other probes) in the subsequent rounds of amplification. Ligation and amplification of the sequence-specific probes argues for the presence of the given target sequence in the sample DNA.

Note that, for ligation to occur, both juxtaposed bases in a given pair of probes must base-pair correctly with the corresponding nucleotides in the template strand (even if mismatches occur elsewhere in the probe–target duplex); moreover, the 5′ end of one probe must be phosphorylated in order for ligation to occur.

Commercial LCR-based assays (LCx® assays; Abbott Diagnostics) are used to detect specific sequences of DNA from certain pathogens in clinical specimens; such assays include tests for *Chlamydia trachomatis*, *Neisseria gonorrhoeae* and *Mycobacterium tuberculosis*. These assays differ from the (basic) format described above. Thus, the two probes in a bound pair are separated by a gap of one to several nucleotides. Consequently, part of each cycle includes a period of extension, from the 3′ end of one probe, by a DNA polymerase present in the reaction mixture; closure of the gap is followed by ligation. Cycling thus involves three phases of temperature; in the Abbott LCx assay for *Chlamydia trachomatis*, these phases are: (i) 93°C for denaturing the dsDNA sample, (ii) 59°C for annealing, and (iii) 62°C for extension/ligation. Corresponding values in the LCx assay for *M. tuberculosis* are (i) 94°C, (ii) 64°C and (iii) 69°C.

An amplification of at least 10 million-fold is commonly achieved in an assay of 30–40 cycles.

Products of the LCx® assays are detected by a system called *microparticle enzyme immunoassay* (MEIA). Each of a pair of probes carries – at its *non*-ligating end – a molecule (*hapten*) that is able to bind a specific antibody; the two probes carry different haptens – i.e. they bind different types of antibody. Hence, after cycling, each ligated probe pair consists of a copy of the target sequence flanked by two different haptens (haptens 1 and 2). Microparticles (<1 μm diam.), coated with hapten 1-specific antibodies, are added to the system; the microparticles bind hapten 1 on both ligated and non-ligated probes. The microparticles are then immobilized on a matrix. The *non*-ligated probes bearing hapten 2 are removed by washing. A conjugate, consisting of alkaline phosphatase covalently linked to a hapten 2-specific antibody, is then added to the bound microparticles; the conjugate binds to hapten 2 (via the antibody), thus labelling ligated probe pairs (on the microparticle) with alkaline phosphatase. A substrate for the enzyme, 4-methylumbelliferyl phosphate, is then added and is cleaved to the fluorescent product 4-methylumbelliferone; under ultraviolet radiation, the rate at which fluorescence is generated is proportional to the number of (enzyme-labelled) ligated probe pairs. Readings above a certain cut-off value indicate a positive result, i.e. evidence of a given sequence in the sample; readings in an 'equivocal zone' (defined in terms of the ratio: sample reading/cut-off value) indicate repeat testing.

Examples of assays with LCx® systems include studies on *Chlamydia trachomatis* [JCM (1998) *36* 94–99], *Neisseria gonorrhoeae* [JCM (1997) *35* 239–242], and *Mycobacterium tuberculosis* [JCM (2002) *40* 2305–2307].

Improved reproducibility with LCx® assays for *Chlamydia trachomatis* and *Neisseria gonorrhoeae* has been reported following modification of the recommended equivocal zones: from 0.8–0.99 to 0.8–1.5 for *C. trachomatis*, and from 0.8–1.2 to 0.8–2.0 for *N. gonorrhoeae* [JCM (2000) *38* 2416–2418].

[LCR in clinical microbiology: Book ref. 221, pp 152–167.]

LD_{50} Lethal dose (50%): that dose of a given lethal agent which, when given to each of a number of test animals, kills 50% of those animals under given conditions.

LDC Lysine decarboxylase: see DECARBOXYLASE TESTS.

LE markers See TRANSFORMATION (1).

leaching (of ores) Bacteria can promote the leaching (solubilization) of certain metals from their ores. Bacterial leaching (*biomining*) is carried out e.g. for the commercial recovery of copper from low-grade ores containing chalcopyrite ($CuFeS_2$) and for the recovery of gold from recalcitrant arsenopyrite ores.

In one process for recovering copper, a liquid containing sulphuric acid and chemolithotrophic bacteria is continually recycled through a mound of crushed ore; in *dump leaching* the mound consists of rough, ungraded ores and waste, while in *heap leaching* it consists of more concentrated, finer ore. Ferrous ions, which leach spontaneously, are oxidized by the bacteria to ferric ions – resulting in ongoing solubilization of iron; the solubilization of iron releases sulphur, which is oxidized to sulphate. These processes may be carried out by e.g. '*Leptospirillum ferrooxidans*' (an iron oxidizer), *Thiobacillus ferrooxidans* (an iron and sulphur oxidizer) and *Thiobacillus thiooxidans* (a sulphur oxidizer). The bacteria obtain nutrients from the percolating liquor and from the ore itself – for example, carbon dioxide may be derived from acidification of carbonates within the ore. Copper in the leachate (up to 5 grams per litre) is periodically removed (e.g. by electrophoresis); the liquor is recycled for months, being supplemented with water to replace losses due to evaporation. Heat generated by bacterial metabolism maintains a temperature of about 40–50°C in the leachate, and the convective upflow of air helps to provide an aerobic environment.

The development of an efficient leaching flora may arise via a succession of species, each being particularly suited e.g. to a given pH. Very low pH is advantageous for leaching copper ores because it inhibits the precipitation of ferric ions; leaching under these conditions has been achieved with strains identified as *Thiobacillus thiooxidans* and '*Leptospirillum ferrooxidans*' – the organisms being identified by characterization of their rRNA genes in DNA extracted from the leachate [AEM (1997) *63* 332–334].

Studies on the sulphur chemistry of bacterial leaching suggest that the leaching of pyrite may involve a cyclical degradative process in which thiosulphate and polythionates are among the intermediate products; ferric ions, produced by the ongoing bacterial oxidation of ferrous ions, may have an active role in the oxidation of both pyrite and thiosulphate [AEM (1996) *62* 3424–3431].

Bioleaching is also used in e.g. South Africa, Australia and Brazil as a preliminary process in the extraction of gold from certain recalcitrant (arsenopyrite) ores. In these ores, the gold is embedded in a matrix of arsenopyrite so that it cannot be extracted by cyanidation in the usual way. The ore is crushed and subjected to bacterial action in highly aerated bio-oxidation tanks; arsenopyrite (FeAsS) is solubilized by oxidation (to $FeAsO_4$ and sulphuric acid), thus exposing the gold for cyanidation. After cyanidation, gold is separated from the gold–cyanide complex e.g. by adsorption to charcoal, and the spent, cyanide-containing liquor is detoxified by bacteria; detoxification involves the oxidation of cyanide to carbon dioxide and urea, and the oxidation of urea to nitrate.

In addition to their use in leaching, bacteria have also been used as 'bioadsorbents' for the recovery of low levels of uranium in certain liquid wastes.

(See also THIOBACILLUS.)

[Microbiology of metal leaching (symposium report): FEMS Reviews (1993) *11* 1–267. Mining with microbes (review): Biotechnology (1995) *13* 773–778. Predominant types of bacteria used in commercial bioleaching processes: Microbiology (1999) *145* 5–13.]

leader (leader sequence) A region of an RNA molecule (e.g. certain mRNA molecules, pre-rRNA) between the 5′ (promoter) end and the sequence corresponding to the first (or only) structural gene. The leader sequence is commonly non-coding, but may encode a small regulatory peptide (a *leader peptide*): see OPERON (attenuator control) and TNA OPERON.

leader peptidase See SIGNAL HYPOTHESIS.

leader peptide (1) See LEADER. (2) See SIGNAL HYPOTHESIS.

leading strand See DNA REPLICATION.

leaf blotch (scald) A CEREAL DISEASE, caused by *Rhynchosporium secalis*, which can affect e.g. barley and rye; large, pale brown lesions with dark brown or purple margins appear mainly on leaves, and crop losses may be severe. (See also FLUTRIAFOL and PROPICONAZOLE.)

leaf curl (of peach) See TAPHRINA.

leaf nodules See PHYLLOBACTERIUM.

leafhopper A virus (LAV) An unclassified reovirus-like virus which was isolated from the leafhopper *Cicadulina bimaculata* during the search for the causal agent of MAIZE WALLABY EAR DISEASE; attempts to infect maize plants with LAV have been unsuccessful, and any connection between LAV and MWED has yet to be proved. Although LAV resembles viruses of the genus FIJIVIRUS, it cannot be included in the genus since it is not known to infect any plant species.

leaky mutation A MUTATION which results in only partial inactivation of the sequence in which it occurs. (cf. NULL MUTATION.)

leather spoilage Leather has a relatively low pH (usually ca. 3–4) and even at high humidities is fairly resistant to bacterial attack; however, at relative humidities of >80% leather is highly susceptible to attack by moulds which may cause significant loss of tensile strength. *Aspergillus* spp are commonly implicated, but other moulds (e.g. species of *Mucor*, *Paecilomyces* and *Rhizopus*) may also be involved. Heavy fungal contamination may raise the pH sufficiently to allow some bacterial growth; e.g. *Bacillus sphaericus* has been isolated from mouldy leather. Preservatives which have been used for treating leather include phenylmercuric salts of fatty acids, *p*-nitrophenol, and β-naphthol. (For the bating of hides by microbial enzymes see PROTEASES.)

leben A type of concentrated YOGHURT consumed in e.g. Egypt. Milk is fermented with organisms which include *Lactococcus lactis*, *Streptococcus thermophilus*, *Lactobacillus bulgaricus* and certain lactose-fermenting yeasts. The curd is concentrated by draining off the whey and may then be dried. (See also DAIRY PRODUCTS.)

LECAM-1 See SELECTINS.

Lecanactis See OPEGRAPHALES.

Lecanidiales An order of fungi of the ASCOMYCOTINA; members include crustose LICHENS (photobiont commonly green), and saprotrophic and lichenicolous fungi. Ascocarp: APOTHECIOID, sometimes lirelliform. Asci: bitunicate. Genera: e.g. RHIZOCARPON.

Lecanora A large genus of LICHENS (order LECANORALES); photobiont: a green alga. Thallus: crustose; apothecia: lecanorine, sessile; spores: aseptate. The genus includes several species which are highly tolerant of air pollution. *L. conizaeoides* is common (on trees, non-basic rocks, brick etc) in industrial and densely populated regions of Europe, but is rare elsewhere; it forms a grey-green, granular-sorediate, sometimes areolate thallus, and apothecia with yellowish- or greenish-brown discs are usually present. *L. dispersa* may occur on concrete, asbestos etc; the thallus may be superficial and whitish, or may grow endolithically, appearing only as a dark stain on rocks, walls etc. The apothecia have yellowish- to dark-brown discs which are often pruinose when young. *L. muralis* has become increasingly common on pavements, walls etc in towns; it forms a more or less circular thallus with radiating lobes and apothecia crowded in the centre.

lecanoralean ascus See BITUNICATE ASCUS.

Lecanorales An order of fungi of the ASCOMYCOTINA; members include many LICHENS (photobiont usually green), and saprotrophic and lichenicolous fungi. Ascocarp: APOTHECIOID, typically discoid (sometimes stipitate), but podetia (see PODETIUM) are formed by *Cladonia* spp; paraphyses are present. Asci: bitunicate (see BITUNICATE ASCUS), commonly elongated with a thickened apex. Genera: e.g. ACAROSPORA, ALECTORIA, ANAPTYCHIA, BAEOMYCES, BRYORIA, BUELLIA, CANDELARIELLA, CETRARIA, CLADONIA, COLLEMA, DIPLOICIA, EVERNIA, HYPOGYMNIA, LASALLIA, LECANORA, LECIDEA, LEMPHOLEMMA, LEPTOGIUM, LICHINA, PANNARIA, PARMELIA, PHYSCIA, PILOPHORUS, RAMALINA, STEREOCAULON, UMBILICARIA, USNEA.

lecanorine apothecium (*lichenol.*) An APOTHECIUM which has a THALLINE MARGIN; such apothecia occur e.g. in lichens of the genera *Lecanora* and *Xanthoria*. (cf. LECIDEINE APOTHECIUM.)

Lecidea A genus of crustose LICHENS (order LECANORALES); photobiont: a green alga (e.g. *Trebouxia*). Apothecia: lecideine. Many species of the (originally very large) genus have been transferred to other genera, including e.g. *Fuscidea*, *Huilia*, *Hypocenomyce*, *Lecidella*, *Psilolechia*, *Psora*, *Trapelia*. Thus, e.g., *L. limitata* (thallus: crustose, greenish- to yellowish-grey, sometimes with a dark hypothallus; apothecia: grey or black) is now *Lecidella elaeochroma*; *L. lucida* (thallus: leprose, yellow to yellow-green; apothecia: yellow to pale-brown) is now *Psilolechia lucida*; *L. macrocarpa* (thallus: crustose, grey, with a black hypothallus; apothecia: black) is now *Huilia macrocarpa*; *L. scalaris* (thallus: squamulose, with imbricated squamules) is now *Hypocenomyce scalaris*.

lecideine apothecium (*lichenol.*) An APOTHECIUM which lacks a THALLINE MARGIN, i.e., it has only a PROPER MARGIN. Such apothecia occur e.g. in lichens of the genera *Buellia*, *Lecidea* and *Rhizocarpon*. (cf. LECANORINE APOTHECIUM.)

Lecidella See LECIDEA.

lecithin Phosphatidylcholine, a type of phospholipid found e.g. in egg-yolk and in the CYTOPLASMIC MEMBRANES of mammalian cells and of some Gram-negative bacteria. Lecithins are derivatives of *phosphatidic acid* (itself a derivative of glycerol 3-phosphate in which the hydroxyl groups at positions 1 and 2 are each esterified with a long-chain fatty acid); in a lecithin, the phosphate group of phosphatidic acid is esterified with choline: $(CH_3)_3N^+(CH_2)_2OH$.

lecithinase A PHOSPHOLIPASE (q.v.) which cleaves LECITHIN. Extracellular lecithinases are formed e.g. by *Bacillus cereus*, *Clostridium bifermentans*, *C. novyi* (γ-toxin), *C. perfringens* (α-toxin), and strains of *Staphylococcus aureus*; the lecithinase of *C. perfringens* apparently requires Zn^{2+} for activity. Those organisms (e.g. *C. novyi*, *C. perfringens*) which form lecithinases that are C-type phospholipases produce zones of opacity around colonies when grown on e.g. EGG-YOLK AGAR due to the formation of insoluble, phosphate-free diacylglycerides ('diglycerides'). (See also MILK SPOILAGE.)

lecithovitellin agar *Syn.* EGG-YOLK AGAR.

lectin pathway See COMPLEMENT FIXATION (c).

lectins Various non-enzymic proteins (usually glycoproteins) which are not products of the immune system (thus excluding antibodies) and which can each bind to specific carbohydrate group(s) and can agglutinate (or precipitate) cells or other entities having these particular (and accessible) group(s). The term 'lectin' originally referred to certain plant products which bind to and agglutinate particular types of erythrocyte (cf. PHYTOHAEMAGGLUTININ sense 2); it is now used for materials of plant, animal or microbial origin which have specific carbohydrate-binding ability – though not necessarily in relation to erythrocytes. Lectins have been obtained e.g. from plants [Book ref. 37, pp. 537–547; ARPphys. (1985) *36* 209–234], lichens [Lichenol. (1983) *15* 303–305], the snail (*Helix pomatia*), the horseshoe crab (*Limulus polyphemus*), slime moulds (e.g. discoidins from *Dictyostelium discoideum*, pallidins from *Polysphondylium pallidum*), fungi and bacteria (e.g. mannose-sensitive FIMBRIAE).

Lectins have been used e.g. for haemagglutination in blood grouping studies. Many lectins behave as MITOGENS and are used e.g. for lymphocyte activation in immunological studies. In nature, lectins may be involved in plant–microbe symbioses [Book ref. 38, pp. 658–677]; they may play a role in the pathogenesis of some plant diseases [ARPpath. (1983) *21* 300–302] and possibly in certain animal diseases (e.g. lectins in raw beans may promote disease in pigs by enhancing attachment of ETEC to the ileal epithelium [CRM (1981) *8* 303–338 (324)]). (See also GIARDIASIS.) [Interaction of bacteria and fungi with lectins: ARM (1981) *35* 85–112.] (See also CONCANAVALIN A; LIMULIN; PHYTOHAEMAGGLUTININ; POKEWEED MITOGEN; TRIFOLIIN A; WHEAT GERM AGGLUTININ.)

lectotype A specimen, chosen to serve as NOMENCLATURAL TYPE, derived from the original material studied by an author who did not specify a HOLOTYPE; it has nomenclatural precedence over a NEOTYPE.

LEE See PATHOGENICITY ISLAND.

leek yellow stripe virus See POTYVIRUSES.

left-handed DNA See e.g. Z-DNA.

left splice site See SPLIT GENE.

leghaemoglobin (leghemoglobin) A red haem-containing pigment present (in multiple forms) in the host cell cytoplasm in the ROOT NODULES of leguminous plants. The protein and haem moieties are specified by plant genes and bacteroid genes, respectively. Leghaemoglobins are structurally and functionally similar to animal myoglobins and haemoglobins but have a much higher affinity for oxygen; they release bound oxygen only at low levels of oxygen and thus permit aerobic respiration in the bacteroids while protecting NITROGENASE from oxygen inactivation. [Book ref. 55, pp. 73–79; ARPphys. (1984) *35* 443–478.]

Legionella See LEGIONELLACEAE.

Legionellaceae A family of Gram-negative, asporogenous bacteria. Cells: rod-shaped or filamentous, 2.0 to >20.0 ×

0.3–0.9 μm; motile, with one, two or more polar or lateral flagella. Cellular fatty acids are predominantly branched. Aerobic; oxidase −ve or weakly +ve. Chemoorganotrophic, using amino acids (non-fermentatively) as carbon and energy sources; carbohydrates are generally not metabolized. Growth requires L-cysteine and iron (Fe^{3+}) salts, and can occur e.g. on blood agar supplemented with L-cysteine and iron, on Mueller–Hinton agar supplemented with haemoglobin and IsoVitaleX, or on buffered charcoal–yeast extract agar (BCYE), but not e.g. on unsupplemented blood agar or nutrient agar. Optimum growth temperature: 35–37°C. Gelatinase +ve; catalase +ve; urease −ve; NO_3^- not reduced. Several species exhibit a bluish-white autofluorescence under WOOD'S LAMP, and most species produce a diffusible brown pigment on tyrosine-containing media. Species occur in various aquatic habitats: e.g. surface waters, moist soils, thermally polluted streams, domestic water systems, etc; most or all species can be pathogenic for man (see LEGIONELLOSIS). GC%: ca. 39–43; type genus: *Legionella*.

According to some authors [Book ref. 22, pp. 279–288], *Legionella* (type species *L. pneumophila*) is the sole genus of the Legionellaceae, but other authors have divided the family into three genera: *Fluoribacter*, *Legionella* and *Tatlockia* [IJSB (1980) *30* 609–614; IJSB (1981) *31* 111–115]. Species include e.g. *L. bozemanii* (*Fluoribacter bozemanae*; 'WIGA bacterium'), *L. dumoffii* (*F. dumoffii*; 'Tex-KL bacterium'), *L. feeleii* [AIM (1984) *100* 333–338], *L. gormanii* (*F. gormanii*), *L. jordanis*, *L. longbeachae*, *L. micdadei* (*Tatlockia micdadei*; 'TATLOCK agent'; 'guinea-pig agent'; Pittsburg pneumonia agent), *L. oakridgensis*, *L. pneumophila*, *L. sainthelensi* (AEM (1984) *47* 369–373], *L. wadsworthii*).

[Taxonomy and typing of legionellae: RMM (1994) *5* 79–90.]

Sequencing of the *mip* gene has been reported to provide a basis for unambiguous identification of 39 species of the genus *Legionella* [Mol. Microbiol. (1997) *25* 1149–1158; JCM (1998) *36* 1560–1567.]

legionellosis Any (human) disease caused by a species of *Legionella* (see LEGIONELLACEAE). Most species appear to be capable of causing PNEUMONIA in man. *Legionnaires' disease* is a pneumonia caused by *L. pneumophila*. Infection is believed to occur by inhalation of contaminated aerosols (common sources of infection are air-conditioning cooling towers, showers, nebulizers, etc). [Survival of *L. pneumophila* in a model hot water distribution system: JGM (1984) *130* 1751–1756; distribution and persistence of *Legionella* in water systems: MS (1985) *2* 40–43.] Incubation period: 2–10 days. Symptoms: malaise, myalgia, headache, followed by fever, chills, and a consolidating pneumonia which primarily involves the alveoli and terminal bronchioles; an intra-alveolar exudate is characteristic of the disease. Other common symptoms include chest and abdominal pains, vomiting, diarrhoea, and mental confusion. Mortality rates may be high, particularly in compromised individuals. [Review: Chest (1984) *85* 114–120.] An important feature of pathogenesis is the ability of *L. pneumophila* to invade alveolar macrophages, to inhibit phagosome–lysosome fusion, and to multiply within these cells; such activity may be facilitated by expression of the pathogen's *icm* (= *dot*) genes (see PHAGOCYTOSIS (a)) because mutations in these genes can give rise to modified virulence [TIM (1998) *6* 253–255]. Moreover, the results of in vitro studies have suggested that the pathogen's zinc metalloprotease (a 38 kDa secreted protein; = *tissue-destructive protease*) may promote pathogenesis by inhibiting both the oxidative burst and chemotaxis in phagocytic cells [JMM (2001) *50* 517–525].

Pontiac fever is an acute, febrile, non-pneumonic, self-limiting illness caused by *L. pneumophila* (and by *L. feeleii* [AIM (1984) *100* 333–338]). Incubation period: ca. 36 hours. Symptoms: fever, chills, headache, myalgia, arthralgia, diarrhoea, and sometimes neurological symptoms. *Pittsburg pneumonia* is a pneumonia caused by *L. micdadei*.

Lab. diagnosis: direct or indirect immunofluorescence tests on e.g. sputum samples; microscopic examination of specimens stained with e.g. DIETERLE SILVER STAIN; analysis of the fatty acids of the isolated organism by GLC.

[Treatment: JAC (1999) *43* 747–752.]

Legionnaires' disease See LEGIONELLOSIS.

legume bloat See BLOAT.

legume yellows virus See LUTEOVIRUSES.

leguminous root nodules See ROOT NODULES.

***Leidenfrost* phenomenon** The formation of a heat-insulating layer of gaseous nitrogen around a specimen when it is plunged into liquid nitrogen. (See also FREEZING.)

Leifson's flagella stain (for bacterial flagella) Mix equal volumes of 1.5% w/v aqueous NaCl and 3% w/v aqueous tannic acid; to this mixture add half its volume of a solution of pararosanilin acetate (0.9% w/v) and pararosanilin hydrochloride (0.3% w/v) in 95% ethanol. Prepare a medium-free, slightly cloudy suspension of bacteria in distilled water. Using a wax pencil, draw a ring on a chemically clean glass slide enclosing ca. 2 cm^2. Spread one loopful of the bacterial suspension within the ringed area; dry at room temperature. Do not fix. Add 1 ml stain; allow to act for ca. 10 min and rinse off gently under the tap. Flagella appear to be coated with a colloidal tannic acid–dye complex, rendering them visible by light microscopy.

Leighton tube A test-tube in which part of the wall (near the closed end) forms a flat, rectangular region which can accommodate a cover-glass on which a cell monolayer can be formed.

Leiotrocha See PERITRICHIA.

Leishman–Donovan bodies Amastigote forms of *Leishmania* in the vertebrate host.

Leishmania A genus of parasitic protozoa (family TRYPANOSOMATIDAE) which resemble TRYPANOSOMA species in their ultrastructure. The organisms are the causal agents of LEISHMANIASIS in mammals, and are transmitted by sandflies of several genera (e.g. *Lutzomyia*, *Phlebotomus*, *Sergentomyia*); species of *Leishmania* which are normally dermatotropic sometimes give rise to visceral leishmaniasis, while typically visceralizing strains sometimes cause cutaneous infections. Most leishmaniases are zoonotic.

In the vertebrate host *Leishmania* grows in the intracellular, amastigote form (ca. 2–10 μm in diam.) – primarily in mononuclear phagocytes (e.g. macrophages). (*Leishmania* species can grow within phagolysosomes: see PHAGOCYTOSIS; amastigote forms of *L. donovani* have been found to contain CATALASE and glutathione peroxidase in amounts greater than those found in the promastigote stage. The ability of *Leishmania* spp to undergo ANTIGENIC VARIATION may contribute to their pathogenicity.) In the sandfly the pathogen occurs in the paramastigote and promastigote forms, the latter being the form infective for man. In cell-free media (e.g. in NNN MEDIUM), and in tissue culture, the organisms grow in the promastigote form.

Leishmania spp include *L. aethiopica*, *L. braziliensis*, *L. donovani*, *L. major*, *L. mexicana* and *L. tropica*. Some species are divided into subspecies; e.g., *L. braziliensis* includes e.g. *L. braziliensis braziliensis* (= *L. b. braziliensis*), *L. b. guyanensis* and *L. b. peruviana*. Organisms which cause visceral leishmaniasis have been given separate names in particular geographical regions: e.g. *L. infantum* (Mediterranean), *L. chagasi* (S. America), and *L. archibaldi* (Sudan); some authors regard all of these as one species, *L. donovani* [Book ref. 72, p. 215].

Leishmania spp have been divided into groups ('sections') on the basis of their location within the arthropod vector: see HYPOPYLARIA, PERIPYLARIA, SUPRAPYLARIA; at least some of the reptile-infecting species in the Hypopylaria and Peripylaria appear to be trypanosomes ('*L. tarentolae*' has been renamed *Trypanosoma platydactyli*).

leishmanial form Obsolete *Syn.* AMASTIGOTE.

leishmaniasis Any disease caused by a species of LEISHMANIA; see CUTANEOUS LEISHMANIASIS, MUCOCUTANEOUS LEISHMANIASIS, POST-KALA-AZAR DERMAL LEISHMANIASIS and VISCERAL LEISHMANIASIS. [Book ref. 72, pp. 183–249.]

leishmaniasis recidivans An intractable form of CUTANEOUS LEISHMANIASIS which may develop after an apparent cure of a previous episode of cutaneous leishmaniasis; lesions, typically non-ulcerating, form around the rim of scar tissue.

leishmanin test *Syn.* MONTENEGRO TEST.

Leishman's stain See ROMANOWSKY STAINS.

Leloir pathway See LACTOSE and Appendix III(a).

Lelystad virus See BLUE-EARED PIG DISEASE.

Lemanea See RHODOPHYTA.

Leminorella A new genus of non-motile Gram-negative bacteria of the ENTEROBACTERIACEAE; two species have been described: *L. grimontii* (the proposed type species) and *L. richardii* [JCM (1985) *21* 234–239]. Reactions: e.g. VP −ve; indole −ve; H_2S +ve; LDC −ve; ODC −ve; ADH −ve; no growth with KCN. Species are relatively inactive biochemically; acid is produced from L-arabinose and D-xylose, but not from a wide range of other sugars (including lactose). Strains have been isolated from human faeces and urine.

Lempholemma A genus of gelatinous LICHENS (order LECANORALES); photobiont: *Nostoc*. Thallus: homoiomerous, lacking a cortex. HORMOCYSTS are formed in specialized swollen regions of the thallus (termed *hormocystangia*), and hyphae of the mycobiont are – or may be – associated with the gelatinous sheath of the hormocyst.

lens-shaped A phrase sometimes used to describe the shape of a structure or colony etc; as a descriptive phrase it is of doubtful value since a lens may be double-convex, plano-convex, meniscus etc. (cf. LENTICULAR.)

lentic Pertaining to aquatic habitats with standing (still) water (e.g. ponds, swamps). (cf. LOTIC.)

lenticular (lentiform) Shaped like a lentil or a double-convex lens.

lentiform *Syn.* LENTICULAR.

Lentinula A genus of fungi of the family Tricholomataceae (order AGARICALES). *L. edodes* (formerly *Lentinus edodes*) forms edible fruiting bodies: see SHII-TAKE. [Chemically defined medium for the fruiting of *L. edodes*: Mycol. (1983) *75* 905–908; note on the reclassification of *Lentinus edodes* to *Lentinula edodes*: MS (1986) *3* 171; effect of light and aeration on fruiting of *L. edodes*: TBMS (1987) *88* 9–20.] (See also MYCOVIRUS.)

Lentinus A genus of lignicolous fungi of the APHYLLOPHORALES (family Polyporaceae) which form tough but pliant basidiocarps in which the hymenium is borne on lamellae (see LAMELLA). *L. lepideus* forms a mushroom-shaped basidiocarp in which the fawn-coloured pileus bears dark scales and is supported by a central or eccentric stipe; the lamellae are whitish and decurrent, and the basidiospores are ellipsoidal, hyaline, and ca. 10×5 μm. *L. lepideus* causes TIMBER SPOILAGE (see also STAG'S HORN FUNGUS); it can tolerate quite high concentrations of CREOSOTE. (For *L. edodes* see LENTINULA.)

Lentivirinae A subfamily of viruses of the RETROVIRIDAE which includes HIV and the FELINE IMMUNODEFICIENCY VIRUS; other viruses which have been classified in this subfamily (on the basis of e.g. nucleotide sequences) include a number of viruses isolated from diseased sheep and goats – e.g. CAPRINE ARTHRITIS-ENCEPHALITIS virus, VISNA virus and zwoegerziekte virus – and the causal agent of SWAMP FEVER (equine infectious anaemia virus).

At least some lentiviruses (including HIV and visna virus) can be cultured; the high rate of transcription of the visna virus in infected cells has been attributed to the presence of both *cis*-acting enhancer sequences and *trans*-acting transcriptional activating factors [Science (1985) *229* 482–485].

As well as replicating within (and killing) host cells, at least some lentiviruses (HIV and feline immunodeficiency virus) have been associated with instances of neoplastic disease (e.g. lymphomas); however, in these cases the mechanism of pathogenesis is unclear, and it may be that lentivirus infection permits neoplastic development by (e.g.) undermining immune surveillance.

Lenzites See APHYLLOPHORALES (Polyporaceae).

LEP Low egg passage: refers to the number of times that a given strain of virus has undergone SERIAL PASSAGE in eggs. (See e.g. FLURY VIRUS; cf. HEP.)

Lepiota A genus of fungi (AGARICALES, Agaricaceae) which typically form colourless spores. *L. procera* (the edible 'parasol mushroom') forms a flattened, umbonate, brownish pileus whose upper surface bears loose, darker brown scales. (See also FUNGUS GARDENS.)

Lepista A genus of fungi (AGARICALES, Tricholomataceae) which form pink basidiospores. Some species are edible: see BLEWIT.

leporid herpesvirus 1 See GAMMAHERPESVIRINAE.

Leporipoxvirus (myxoma subgroup) A genus of viruses of the CHORDOPOXVIRINAE which infect leporids (hares and rabbits) and squirrels. The viruses commonly cause tumours (fibromas, myxomas) in their hosts, and are commonly transmitted (mechanically) by arthropod vectors. Haemagglutinin is not formed; infectivity of the virion is destroyed e.g. by ether. Type species: myxoma virus (see MYXOMATOSIS); other members: hare fibroma virus, rabbit (Shope) fibroma virus, and squirrel fibroma virus. (See also NON-GENETIC REACTIVATION.)

lepra cells See LEPROSY.

lepra reaction See LEPROSY.

Lepraria A genus of LICHENS (of the AGONOMYCETALES) in which the (leprose) thallus forms a soft, powdery crust consisting of loosely associated fungal hyphae and green algal cells; fruiting bodies are unknown. In e.g. *L. incana* the thallus is thick, whitish to greenish grey, and parts may consist only of the (white) mycobiont; *L. incana* is fairly tolerant of air pollution and occurs e.g. on walls, bark etc in damp situations.

leproma See LEPROSY.

lepromin A preparation obtained by grinding, autoclaving and filtering lepromatous tissue (see LEPROSY); it thus contains heat-stable fractions and includes human antigens unrelated to leprosy.

lepromin test A SKIN TEST involving the intra-dermal administration of LEPROMIN. There are three types of reaction: an immediate (*Medina*) reaction; the *Fernandez reaction*, an inflammatory response 24–48 hours after the injection; and the *Mitsuda reaction* which occurs after ca. 3–4 weeks.

In a positive Mitsuda reaction a hard granulomatous nodule occurs at the site of injection. Since normal adults can give a positive Mitsuda reaction it is not diagnostic for leprosy, but a negative Mitsuda reaction is considered to indicate susceptibility to leprosy. The lepromin test can help to classify leprosy: the Mitsuda reaction is negative in lepromatous cases, positive in tuberculoid cases.

leprose (*lichenol.*) Refers to a thallus which lacks a cortex and is covered with or consists entirely of soredia (see e.g. LEPRARIA).

leprosins Ultrasonicates of tissue-free cells of *Mycobacterium leprae* (derived from lesions).

leprosy (Hansen's disease) A chronic human disease, caused by *Mycobacterium leprae*, which is currently found mainly in the tropics and subtropics. It is not readily transmissible. Infection may occur by direct contact, by droplet infection, and/or by contact with fomites. Incubation period: ca. 1–10 years.

M. leprae has a predilection for the peripheral nervous system; the CNS is apparently not affected. The disease characteristically involves destruction of nerves – leading to loss of sensitivity (allowing damage to fingers etc. by trauma) as well as weakness/paralysis of muscles.

Leprosy occurs in different forms; the specific type of T cell response to *M. leprae* in a given host correlates with the form of disease.

In *lepromatous* leprosy the $CD4^+$:$CD8^+$ T cell ratio in lesions is ca. 1:2; major T-cell-derived cytokines include interleukins 4 and 10. Bacilli are abundant (e.g. 10^{10} bacilli/gram tissue). The multiple skin lesions enlarge/coalesce to form nodules (*lepromas*) – commonly on ear lobes, nose, brows and forehead (but possible elsewhere). Affected skin is typically dry and hairless. Lepromas develop through infiltration of the dermis by *lepra cells* (= *Virchow cells*): histiocytes (macrophages) containing large numbers of bacilli; the infiltrate also contains compact masses of (intra- and extracellular) bacilli associated with cellular debris – 'globi' or 'brown bodies' – which may reach 200 μm in diameter. Nerve damage is associated with an inflammatory process. Various organs may become involved – e.g. eyes (sometimes leading to blindness), oesophagus, lymphatic system, liver.

In a diffuse form of lepromatous leprosy the infiltrate occurs as a continuous thin sheet in the dermis; an associated necrotizing vasculitis leads to ulceration (the *Lucio phenomenon*).

In *tuberculoid* leprosy the $CD4^+$:$CD8^+$ T cell ratio in lesions is ca. 2:1; major T-cell-derived cytokines include INF-γ and interleukin 2. Skin lesions (epithelioid-cell granulomas infiltrated with lymphocytes) are few, dry, anaesthetic, hypo- or hyperpigmented (or erythematous); Virchow cells are absent, and bacilli are rare/absent. The granulomas, which apparently result from delayed-type hypersensitivity, often damage nerve trunks.

Borderline (= *dimorphous*) leprosy is characterized by histopathological features of both lepromatous and tuberculoid forms; the features of one form may predominate. Most cases of leprosy are borderline.

Some patients treated for leprosy may develop one of two types of *reaction state* (acute, immunologically-based episodes). The *lepra reaction* (= Jopling type 1 reaction; reversal reaction) occurs in some borderline cases and may involve e.g. necrosis/ulceration of bacilli-containing tissues and the formation of abscesses in nerve trunks. *Erythema nodosum leprosum* (ENL; Jopling type 2 reaction) occurs in some lepromatous and borderline cases (not in tuberculoid cases); it may result from vasculitis caused by antibody–antigen complexes, and inflammation may affect e.g. eyes, joints, nose and peripheral nerves. ENL may recur repeatedly.

Laboratory diagnosis. Biopsies from skin lesions, or e.g. nasal secretions, are stained by acid-fast stains; in lepromatous leprosy the foamy Virchow cells, packed with acid-fast bacilli, are diagnostic. Tuberculoid leprosy may be difficult to differentiate from sarcoidosis or non-leprosy mycobacterial granulomas. (See also LEPROMIN TEST.) The carbohydrate moiety of PGL-I (an *M. leprae* cell-surface glycolipid) may be useful in a serodiagnostic test.

Chemotherapy. Currently, antimicrobial treatment involves a combination of drugs: DAPSONE and rifampicin with or without CLOFAZIMINE. Other drugs (e.g. ofloxacin, minocycline) are also effective for treating leprosy. Thalidomide is useful in some cases of ENL.

[Various aspects: TRSTMH (1993) *87* 499–517; epidemiology: FEMS (1996) *136* 221–230; review: RMM (1998) *9* 39–48.]

lepto- Prefix meaning fine, narrow, thin.

Leptogium A genus of gelatinous LICHENS (order LECANORALES); photobiont: *Nostoc*. The thallus has a cortex, but otherwise resembles that of COLLEMA spp (although generally less gelatinous and often felt-like or tomentose). [Studies on some species: Lichenol. (1983) *15* 109–125.]

Leptolegnia See SAPROLEGNIALES.

Leptomitales An order of aquatic fungi (class OOMYCETES) which are typically saprotrophic – though one species, *Leptomitus lacteus*, may be an opportunist invader of epidermal lesions in fish. The thallus consists of hyphae which bear characteristic constrictions along their lengths, the constrictions often being plugged with CELLULIN GRANULES; in some genera (*Apodachlya*, *Leptomitus*) there is evidence that the cell wall contains both CHITIN and CELLULOSE [Book ref. 174, pp. 62–63]. Other genera include *Aqualinderella* (an anaerobe found in polluted waters) and *Rhipidium*.

Leptomitus See LEPTOMITALES.

leptomonad form Obsolete *syn.* PROMASTIGOTE.

Leptomonas A genus of homoxenous parasitic protozoa (family TRYPANOSOMATIDAE) which typically occur in the gut (sometimes in the haemocoele or in the salivary glands) of insects. The organisms occur in amastigote and promastigote forms, the latter usually being 10–40 μm in length (excluding flagellum), or about twice as long including the flagellum; in some species the cell may be helical, up to 200 μm in length.

leptomycin B A product of SECONDARY METABOLISM (in *Streptomyces* spp) which blocks progression through the cell cycle in eukaryotic cells.

Leptomyxa See ACARPOMYXEA.

Leptonema A genus of bacteria of the order SPIROCHAETALES. [Motility in *L. illini*: see e.g. JB (1996) *178* 6539–6545.]

Leptosphaeria A genus of fungi (order DOTHIDEALES) which include saprotrophs and plant parasites (see e.g. BLACKLEG (sense 2), CANE BLIGHT and GLUME BLOTCH); anamorphs occur in e.g. *Camarosporium* and *Septoria*.

Leptospira A genus of obligately aerobic bacteria (family LEPTOSPIRACEAE) which occur as free-living organisms in soil and aquatic (freshwater and marine) habitats, and as parasites or pathogens in man and other animals (see LEPTOSPIROSIS). The cells are motile, helical ($\geq$18 coils/cell), 0.1 μm in diameter and from 6 to over 12 μm in length; a single periplasmic flagellum arises at each end of the cell. In liquid media, one or both ends of the cell are typically bent or hooked. Growth occurs on e.g. serum-containing media; optimum growth temperature: 28–30°C. Metabolism is respiratory (oxidative); fatty acids are metabolized by β-oxidation to acetate and CO_2. Catalase and/or peroxidase +ve; oxidase +ve. GC%: 35–41, but one strain 53. Type species: *L. interrogans*.

Two species are recognized [Book ref. 22, pp. 62–67]: *L. biflexa* (free living, capable of growth at 13°C, usually resistant to 225 μg/ml 8-azaguanine) and *L. interrogans* (parasitic/pathogenic, usually sensitive to 225 μg/ml 8-azaguanine, and growth inhibited at 13°C). Each species consists of a large number of named serotypes (= serovars) which are classified into named serogroups; thus, e.g. the Icterohaemorrhagiae serogroup of *L. interrogans* contains nearly twenty serotypes, including e.g. *dakota* and *icterohaemorrhagiae*.

Leptospiraceae A family of aerobic bacteria of the order SPIROCHAETALES; the cells typically have hooked ends, their cell wall PEPTIDOGLYCAN contains the diamino acid diaminopimelic acid, and their carbon and energy sources are long-chain fatty acids or long-chain fatty alcohols (cf. SPIROCHAETACEAE). One genus: LEPTOSPIRA.

Leptospirillum ferrooxidans See LEACHING.

leptospirosis A disease of man and animals, caused by any of various serotypes of *Leptospira interrogans*. A wide range of animals can act as carriers, the leptospires normally occurring in the kidney tubules; e.g., serotypes *icterohaemorrhagiae* and *australis A* are carried by rats and other rodents, *canicola* by dogs and pigs, *pomona* by cattle and pigs, *hardjo* and *grippotyphosa* by cattle. In domestic animals the disease ranges from mild to fatal, depending on the animal and serotype involved; symptoms commonly include fever, jaundice, haematuria (cf. RED WATER), and abortion. (Rodents appear not to be significantly affected by leptospire infection.) Animals may continue to excrete leptospires in their urine for several months after recovery.

Man acquires infection either by direct contact with animals or animal carcasses (cf. ZOONOSIS) or via water contaminated with urine from infected animals. Infection occurs via wounds or mucous membranes. Incubation period averages ca. 10 days. In the mild form of leptospirosis, onset is abrupt with shivering, fever, prostration, headache, anorexia and myalgia; nausea and vomiting may occur. Renal function is typically impaired, with proteinuria, haematuria etc. Uveitis is a late complication. A severe form of leptospirosis (*Weil's disease*, *infectious jaundice*) is caused by serotype *icterohaemorrhagiae*; symptoms are more intense, with e.g. marked prostration, severe nausea, vomiting and diarrhoea, enlargement of the liver, jaundice, renal dysfunction, haemorrhages, MENINGITIS. Mortality rates: ca. 15–40%. *Lab. diagnosis*: demonstration of rising titres of antibodies (which appear 7–10 days after onset); identification/culture of the pathogen from blood during the first week, later from urine. [ELISA for detection of specific IgM and IgG in human leptospirosis: JGM (1985) *131* 377–385.]

leptotene stage See MEIOSIS.

Leptotheca See MYXOSPOREA.

Leptothrix A genus of Gram-negative IRON BACTERIA. Cells: polarly flagellated single rods, or chains of rods within a sheath. Type species: *L. ochracea*.

Leptotrichia A genus of Gram type-negative bacteria (family BACTEROIDACEAE) which occur e.g. in the human mouth. Cells: non-motile, straight or slightly curved rods, 0.8–1.5 × 5.0–15.0 μm, often occurring in pairs, chains or septate filaments; the cells of young cultures may stain Gram-positively, but the cell wall resembles those of Gram-negative bacteria. Acid (no gas) is formed from carbohydrates; glucose fermentation yields lactic acid with small amounts of acetic and succinic acids. GC%: ca. 25. Type species: *L. buccalis*.

lesion A localized region of tissue which has been damaged physically or altered by any pathological process.

***letA* gene** See F PLASMID.

***letD* gene** See F PLASMID.

lethal mutation A mutation which kills the organism in which it occurs (cf. CONDITIONAL LETHAL MUTANT).

lethal yellowing disease (of coconut) See COCONUT LETHAL YELLOWING.

lethal zygosis During CONJUGATION (sense 1b) between HFR DONORS and F^- recipients: the phenomenon in which recipients are killed if the ratio of donor cells to recipient cells is high (e.g. ca. 10:1) and the culture is aerated during mating. The effect is not exhibited in $F^+ \times F^-$ crosses.

lettuce big vein agent A virus-like agent which causes big vein disease in lettuces (*Lactuca sativa*) and is pathogenic in other plants of the Compositae. In lettuces, symptoms include vein-clearing and vein-banding, and significant economic losses may result – particularly in hydroponically cultured plants. The agent is transmitted by *Olpidium brassicae*.

lettuce downy mildew See BREMIA.

lettuce mosaic virus See POTYVIRUSES.

lettuce necrotic yellows virus See RHABDOVIRIDAE.

L-leucine biosynthesis See Appendix IV(b).

leucine-responsive regulator protein See LRP.

leuco compound A colourless derivative of a dye. (See e.g. leuco METHYLENE BLUE.)

leucocidin Any microbial toxin which can damage or kill one or more types of LEUCOCYTE. Leucocidal activity may be shown by e.g. certain staphylococcal HAEMOLYSINS and STREPTOLYSIN S; the PANTON–VALENTINE LEUCOCIDIN has *specific* leucocidal activity.

leucocyte (leukocyte) Any white cell of the blood and lymphoid system, including granulocytes (= PMNS), MACROPHAGES, LYMPHOCYTES, and the precursor cells of each of these types. (The term 'leucocyte' is often used to refer solely to the PMNs.) (See also LEUCOCIDIN.)

leucocytosis An increase in the number of leucocytes in the blood.

leucocytosis-promoting factor *Syn.* PERTUSSIS TOXIN.

Leucocytozoon A genus of protozoa of the HAEMOSPORORINA. Species are parasitic in birds; the vectors are commonly blood-sucking flies of the Simuliidae.

leucolysin A toxin which lyses leucocytes.

leucomycin See MACROLIDE ANTIBIOTICS.

leuconocin S See BACTERIOCINS.

Leuconostoc A genus of Gram-positive bacteria of the family Streptococcaceae; the organisms occur e.g. in various dairy products and in fermented beverages and fermenting vegetables (see e.g. BUTTERMILK, CHEESE-MAKING, IDLI, LACTIC ACID STARTERS, PULQUE, ROPINESS, SAUERKRAUT and WINE SPOILAGE). Cells: non-motile, spherical or lenticular (maximum dimension ca. 1 μm), typically in pairs or chains. Growth requirements include a suitable carbohydrate and a range of amino acids and vitamins, e.g. BIOTIN, NICOTINIC ACID and THIAMINE. Glucose is metabolized anaerobically by a HETEROLACTIC FERMENTATION in which D(−)-lactic acid, CO_2 and ethanol are the main products; in the presence of O_2, glucose is used more efficiently than it is under anaerobic conditions, and the end-products include acetic acid rather than ethanol – NADH being reoxidized by NAD(P)H oxidase activity [JGM (1986) *132* 1789–1796]. (See also LACTIC ACID BACTERIA.) Optimum growth temperature: 20–30°C. GC%: ca. 38–44. Type species: *L. mesenteroides*.

L. cremoris (formerly *L. citrovorum*) can ferment very few carbohydrates (e.g. glucose, galactose and lactose). In the presence of a suitable carbohydrate (as energy source), *L. cremoris* can break down citrate in a non-energy-yielding manner with the formation of acetoin, DIACETYL, acetate and CO_2. Dextrans are not formed.

L. dextranicum. Many carbohydrates (although not e.g. arabinose) are fermented. DEXTRANS are formed.

L. lactis. Many carbohydrates (though not e.g. dextrin) are fermented. Dextrans are not formed.

L. mesenteroides. Many carbohydrates (including e.g. arabinose) are fermented. DEXTRANS are formed.

L. oenos. Occurs in wine (see also MALOLACTIC FERMENTATION). Fermentable carbohydrates include e.g. fructose and trehalose. Dextrans are not formed. Growth can occur in the presence of 10% ethanol, and can occur at pH 3.7.

L. paramesenteroides. Many carbohydrates (including dextrin) are fermented. Dextrans are not formed.

It has been proposed that *L. cremoris* and *L. dextranicum* be re-classified as subspecies of *L. mesenteroides* [IJSB (1983) *33* 118–119].

[Characterization of *Leuconostoc* spp: Book ref. 143, pp. 147–178.]

Leucopaxillus A genus of fungi of the Tricholomataceae (AGARICALES).

leucopenia The presence of reduced numbers of leucocytes in the blood.

leucoplast Any colourless PLASTID.

leucosin *Syn.* CHRYSOLAMINARIN.

leucosis *Syn.* LEUKOSIS.

Leucosporidium See SPORIDIALES and CANDIDA.

Leucothrix A genus of GLIDING BACTERIA (see CYTOPHAGALES) which occur e.g. as epiphytes on marine algae. The organisms are non-sheathed, sessile (attached), colourless filaments whose constituent cells (including non-terminal cells) can round up and become individual gliding gonidia; sulphur is not deposited intracellularly. Metabolism appears to be chemoorganotrophic. GC%: ca. 46–51.

Leucotrichaceae See CYTOPHAGALES.

leucovorin See FOLIC ACID.

leukaemia Any of various progressive, malignant diseases in which increased numbers of (usually abnormal or immature) leucocytes occur in the blood; numbers of circulating erythrocytes and platelets are reduced. Symptoms characteristic of leukaemia include anaemia, a tendency to haemorrhage, and increased susceptibility to infection. Leukaemias are classified on the basis of the time course of the disease (acute or chronic) and the types of cell involved (e.g., granular PMNs and their precursors, B or T lymphocytes or lymphoblasts, or monocytes). Certain retroviruses can cause leukaemias in man, animals or birds: see e.g. ADULT T-CELL LEUKAEMIA; AVIAN ACUTE LEUKAEMIA VIRUSES; BOVINE VIRAL LEUKOSIS; FELINE LEUKAEMIA VIRUS; MURINE LEUKAEMIA VIRUSES.

leukocyte See LEUCOCYTE.

leukocyte adhesion deficiency (LAD) A genetic disorder (autosomal, recessive) characterized by the failure of leukocytes adequately to express β_2 integrin adhesion molecules (see INTEGRINS); such cells are unable to give a normal response during INFLAMMATION, and individuals with LAD have correspondingly lower resistance to infection. [See also Lancet (1999) *353* 341–343.]

leukocyte function-associated antigen See INTEGRINS.

leukosis (leucosis) The (abnormal) proliferation of leucocyte-forming tissue. (See e.g. BOVINE VIRAL LEUKOSIS; LYMPHOID LEUKOSIS.)

leukotrienes Derivatives of ARACHIDONIC ACID, and of certain other fatty acids, formed e.g. by leucocytes and macrophages as a result of immunological or non-immunological stimuli; leukotrienes can e.g. cause contraction of smooth muscle, stimulate vascular permeability, and act as attractants for leucocytes. [Review: ARB (1983) *52* 355–377.]

leukoviruses Former name for retroviruses of the ONCOVIRINAE, particularly the TYPE C ONCOVIRUS GROUP.

levamisole A drug used to stimulate the immune response; inhibit tumour formation; stimulate the phagocytic activity of macrophages; inhibit alkaline phosphatases; treat intestinal disease caused by certain helminths – e.g. that caused by *Ascaris lumbricoides*.

levan A linear FRUCTAN consisting mainly of $(2 \rightarrow 6)$-β-linked fructofuranose residues. Levans are widespread as storage polysaccharides in higher plants (especially grasses), and are produced as extracellular slime by certain bacteria (e.g. species of *Bacillus* and *Streptococcus*). (See also ROPINESS and DENTAL PLAQUE.)

Leveillula See ERYSIPHALES.

Levinea See CITROBACTER.

Levinthal's medium A medium used for the isolation of *Haemophilus* spp; it contains both V FACTOR and X FACTOR. Levinthal stock is made by adding defibrinated horse blood (10%) to a boiling brain–heart infusion broth and filtering through e.g. glass wool and then through a membrane filter. Autoclaved peptone broth is then added, and the whole may be mixed with sterile molten agar and dispensed into Petri dishes. [Recipe: Book ref. 46, pp. 1375–1376.]

Leviviridae A family of BACTERIOPHAGES [Book ref. 23, p. 136]; the genome is linear, positive-sense ssRNA (MWt ca. $1.0–1.3 \times 10^6$), and the virion is icosahedral (ca. 22–26 nm diam.). One genus: *Levivirus*.

Leviviruses occur e.g. in enterobacteria (e.g. phages α15, β, f2, fr, M12, MS2, µ2, Qβ, R17 and R23), in *Pseudomonas* (e.g. PP7, 7S) and in *Caulobacter*.

Enterobacterial leviviruses (common in sewage) have been classified into four groups (I–IV) on the basis of serological and physiological properties. Phages of group I (e.g. f2, fr, MS2, R17) and group III (e.g. Qβ) have been the most intensively studied. (MS2, f2 and R17 apparently differ only in a number of point mutations.)

The group I virion consists of (i) 180 copies of a single type of coat protein, (ii) one molecule of A protein (= 'maturation protein' – necessary for infectivity and phage maturation), (iii) the RNA genome, and (iv) at least in MS2, spermidine (presumably to neutralize the negative charge on RNA). (The Qβ virion is similar but contains one molecule of an A2 protein (equivalent to A protein in group I phages) and about three molecules of an A1 protein (= IIb protein); the A1 protein is formed by READTHROUGH of the coat protein gene.) [Capsomeric structure of Qβ: Nature (1982) *298* 819–824.]

The group I phage genome encodes: (i) the A protein; (ii) the coat protein; (iii) a replicase subunit; and (iv) a 'lysis protein' (the gene for which overlaps the coat and replicase genes in the +1 reading frame [Nature (1982) *295* 35–41]). The lysis protein (L protein) of MS2 forms specific adhesion sites (junctions of the cytoplasmic and outer membranes); these sites may facilitate lysis of the cell envelope [JB (1989) *171* 3331–3336].

In Qβ and other leviviruses the replicase subunit combines with three host proteins – the S1 protein of the 30S ribosomal subunit, and the elongation factors EF-Tu and EF-Ts (see PROTEIN SYNTHESIS) – to form the active *replicase* (i.e. RNA-dependent RNA polymerase) necessary for phage genome replication. Expression of viral genes is controlled in both quantity and time; thus, e.g. coat protein is synthesized greatly in excess of A protein, and it acts as a repressor which, later in infection, terminates translation of the replicase subunit gene. Regulation of gene expression may involve changes in the secondary structure of the RNA which alter the availability of ribosome binding sites [e.g. Nature (1982) *295* 35–41]. Virion assembly apparently occurs largely or exclusively by spontaneous aggregation.

Most or all enterobacterial leviviruses are 'male-specific' (ANDROPHAGES); f2, MS2, R17, fr, M12 and Qβ all adsorb specifically to the *sides* of F pili (cf. INOVIRUS). The *Pseudomonas* phages PP7 and 7S adsorb to the chromosomally specified polar (PSA) fimbriae. *Caulobacter* phages also adsorb to polar fimbriae; only the pre-divisional and swarmer stages are susceptible to infection. Infection of the host cell may, at least in some cases, involve pilus (or fimbria) retraction. The genome enters the host cell together with – and probably mediated by – the A/A2 protein (which is cleaved during penetration).

Levivirus See LEVIVIRIDAE.

levofloxacin A QUINOLONE ANTIBIOTIC (q.v.).

levulose D-Fructose.

***lexA* gene** See SOS SYSTEM.

LFA-1 See INTEGRINS.

LFA-2 See IMMUNOGLOBULIN SUPERFAMILY.

LFA-3 See IMMUNOGLOBULIN SUPERFAMILY.

LFT (of plasmids) See EPIDEMIC SPREAD.

LFT lysate See TRANSDUCTION.

LGLs Large granular lymphocytes: see NK CELLS.

LHC LIGHT-HARVESTING COMPLEX.

LHC complex See CELLULOSE.

library (gene bank; gene library) (*mol. biol.*) A collection of recombinant DNA molecules which, together, commonly represents the entire genome of an organism (a *genomic library*) (see figure).

Another type of library consists of a collection of CDNA molecules derived from a mixture of polyadenylated mRNAs isolated from a population of cells of a given strain; this kind of library does not represent the entire genome of the strain.

Licea See MYXOMYCETES.

Liceales See MYXOMYCETES.

lichen A 'composite organism' consisting of a fungus (the mycobiont) and an alga or cyanobacterium (the photobiont – cf. PHYCOBIONT) in symbiotic association; a lichen thallus is morphologically and physiologically quite unlike that of either symbiont growing separately. Lichens are widely distributed, being found from tropical to polar regions in a very wide range of habitats, both terrestrial (e.g. on trees, rocks and soil) and aquatic (freshwater and marine).

Taxonomically, lichens are generally regarded as 'lichenized fungi', classification being based mainly on the nature and origin of their fruiting structures. Most mycobionts appear not to occur in free-living form (but cf. XANTHORIA). The majority are ascomycetes ('ascolichens') belonging mainly to the orders ARTHONIALES, CALICIALES, DOTHIDEALES, GRAPHIDALES, GYALECTALES, LECANIDIALES, LECANORALES, OPEGRAPHALES, PELTIGERALES, PERTUSARIALES, PYRENULALES, TELOSCHISTALES, and VERRUCARIALES. Some lichens belong to the Deuteromycotina, and a few to the Hymenomycetes ('basidiolichens').

The photobiont is usually either a green alga (e.g. *Coccomyxa*, *Myrmecia*, *Phycopeltis*, *Pseudotrebouxia*, *Trebouxia* or *Trentepohlia*) or a cyanobacterium (usually *Nostoc* or *Calothrix*). Some lichens may have more than one photobiont: see e.g. SOLORINA and CEPHALODIUM. Free-living counterparts of lichen photobionts occur in nature, but in the lichen the cells

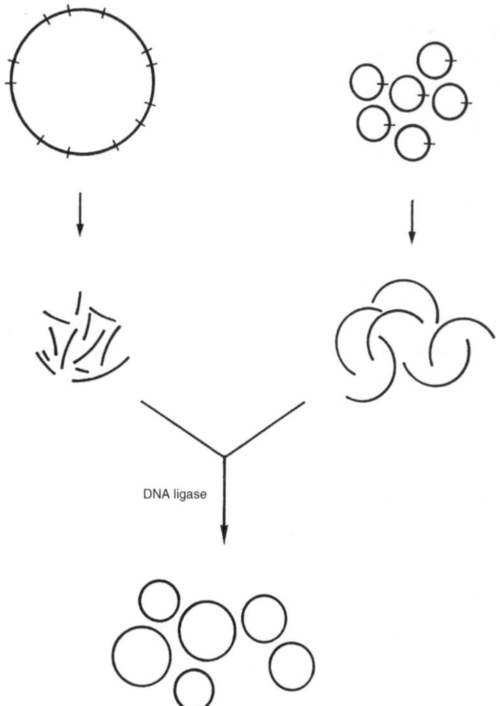

LIBRARY: making a (bacterial) genomic library (diagrammatic). Initially, chromosomes are isolated from a population of bacteria of a given strain. The chromosomes are exposed to a RESTRICTION ENDONUCLEASE which cuts at specific sites in each chromosome (*top, left*), forming many fragments of different sizes (*centre, left*).

As in CLONING (q.v.), copies of a given plasmid (*top, right*) provide vector molecules. The plasmid is one which can be cut by the same enzyme as that used to cut the chromosomes; note, however, that the plasmid has only *one* cutting site for the given enzyme – so that cutting simply linearizes the plasmid molecule (*centre, right*).

Chromosomal fragments and linearized plasmids are then mixed in the presence of ATP and DNA ligase, an enzyme that catalyses the formation of phosphodiester bonds. Fragments and plasmids join together, randomly, via their sticky ends to form a large number of recombinant molecules, many consisting of a plasmid circularized with chromosomal DNA. (Plasmid–fragment binding may be promoted by the phosphatase approach described in CLONING.) Because the fragments are of different sizes, the recombinant plasmids will also be of different sizes (*bottom of diagram*). As many millions of fragments and plasmids are used, the recombinant plasmids will, collectively, carry all the DNA of the chromosome, and this collection of recombinant molecules is therefore called a *genomic library*.

By using TRANSFORMATION or ELECTROPORATION, the library can be inserted into a population of bacteria – different cells receiving different fragments of the chromosome; these cells can then be allowed to form individual colonies, and colonies containing a particular fragment of the chromosome can be detected by various screening processes (see e.g. CLONING).

Reproduced from *Bacteria* 5th edition, Figure 8.7, page 184, Paul Singleton (1999) copyright John Wiley & Sons Ltd, UK (ISBN 0471-98880-4) with permission from the publisher.

may be modified (e.g., they may be smaller, may lack motile stages, and filamentous forms may become unicellular).

The bionts of a lichen can be separated (e.g. by micromanipulation) and can be maintained in culture, but it is notoriously difficult to induce them to re-synthesize a lichen thallus; re-synthesis has been achieved e.g. for *Endocarpon pusillum*. Culture of the photobiont may be necessary for its identification.

Morphology and anatomy. The lichen thallus may be e.g. CRUSTOSE, FOLIOSE, FRUTICOSE, LEPROSE or SQUAMULOSE, or it may have a composite form, with e.g. a squamulose *primary thallus* from which arises a fruticose *secondary thallus* (see PSEUDOPODETIUM; cf. PODETIUM). The simplest type of thallus consists of a loose association of fungal hyphae and photobiont cells, with little or no differentiation (see e.g. LEPRARIA). In gelatinous lichens such as COLLEMA and LEMPHOLEMMA, both bionts are more or less evenly distributed throughout the thallus (i.e. the thallus is *homoiomerous*), with little differentiation into tissues (cf. LEPTOGIUM). In most crustose, foliose and fruticose lichens the thallus is *heteromerous* (stratified): the mycobiont is differentiated to form a surface layer, the *cortex* (a layer of compressed, conglutinated cells or hyphae), which covers the *medulla* (a fibrous tissue of loosely interwoven hyphae), and the photobiont occurs in a layer between the upper cortex and the medulla. One or more differentiated somatic structures may also be present, depending on species: see e.g. CEPHALODIUM; CILIUM (sense 2); CYPHELLA; ISIDIUM; PSEUDOCYPHELLA; RHIZINE; TOMENTUM. (See also PORED EPICORTEX.)

Within the thallus, the hyphae of the mycobiont may be closely appressed to the cells of the photobiont, the walls of both bionts being thinner at the point of contact, or the mycobiont may penetrate the photobiont cells by means of haustoria; the haustoria do not penetrate the plasma membrane in living cells, but may do so in dead cells. [Cytology of mycobiont–photobiont interaction: Lichenol. (1984) *16* 111–127.]

Physiology. Although traditionally regarded as a mutualistic association, a lichen is now thought to be more a case of 'controlled parasitization' of the photobiont by the mycobiont. It has long been assumed that all the carbon requirements of the mycobiont are fulfilled by photosynthate released by the photobiont; cyanobionts release glucose, while phycobionts release a sugar alcohol (*Trebouxia*, *Myrmecia* and *Coccomyxa* release ribitol, *Trentepohlia* releases erythritol, *Stichococcus* releases sorbitol), and these compounds are converted by the mycobiont to e.g. mannitol and/or arabitol. However, the carbon requirement for lichen growth is small, and could – at least theoretically – be fulfilled e.g. by absorption of organic carbon from the environment; it has also been shown that, in the dark, the mycobiont in *Peltigera* sp can fix substantial amounts of CO_2 mainly via PEP carboxylase activity [Physiol. Plant. (1983) *57* 285–290]. The photosynthate received by the mycobiont from the photobiont may thus be necessary not so much for growth as for the maintenance of high levels of soluble carbohydrate (particularly polyols) which appear to function as a reserve of respirable carbon (necessary e.g. when the thallus is hydrated, and hence metabolically active, during periods of darkness) and to protect cell components during periods of desiccation.

Nitrogen requirements may be fulfilled by absorption of e.g. nitrate from the environment. Some lichens contain nitrogen-fixing cyanobacteria – either as the primary photobiont or as an additional biont in specialized cephalodia (see CEPHALODIUM); nitrogen fixed by the cyanobacteria is transferred to the mycobiont.

Vitamins found to be necessary for the growth of the mycobiont in culture are assumed to be supplied by the photobiont in nature.

Lichens accumulate high concentrations of certain metal ions (including heavy metals), but lichen mineral metabolism is not well understood.

Water (in liquid or vapour form) is readily absorbed by lichen thalli; the absorbed water occurs intercellularly and is readily lost by evaporation. The ability to absorb and retain water may be influenced by thallus morphology – e.g., surface area and hence absorptive capacity will be greater in a thallus which is extensively lobed or perforated than in one which is not. (See also RHIZINE.)

'Lichen substances' ('lichen acids'). Lichen mycobionts produce a range of water-insoluble secondary metabolites (e.g. higher aliphatic acids and esters, depsides, depsidones, dibenzofurans, anthraquinones) which may be deposited on the medullary hyphae, on the surface of the cortex (often giving a pruinose appearance), in the hymenial discs (see e.g. RHODOCLADONIC ACID), etc. Many of these substances are unique to lichens; they are generally not formed by the isolated mycobionts. Their functions are largely unknown, but may include e.g. protection of the photobiont from high light intensities (see e.g. PARIETIN); inhibition of potentially competing plants (particularly bryophytes) and/or soil microflora in the vicinity of the thallus (e.g. EVERNIC ACID and USNIC ACIDS have antibiotic activity); prevention of grazing of the thallus by invertebrates; etc.

The content of lichen substances in a given lichen is a useful feature in taxonomy and identification. The compounds may be identified e.g. by chromatography, mass spectrometry, microcrystallography etc. Some lichen substances give characteristic colour reactions with certain reagents, and this forms the basis of simple colour tests widely used for identification of lichens in the field. The reagents used are: saturated aqueous calcium hypochlorite ('C'), 10–25% aqueous KOH ('K'), and 1–5% ethanolic *p*-phenylenediamine ('P' or 'PD'). For example, parietin and other anthraquinones give a purple colour with K.

Growth of a lichen thallus generally appears to occur at or near the periphery (or apex in fruticose thalli). [Growth in circular thalli: Lichenol. (1981) *13* 265–287.] Lichen growth is characteristically slow; the radius of a circular thallus may increase by (e.g.) 0.01–2.50 mm per year, depending on species, rainfall, light intensity, temperature, etc. (See also LICHENOMETRY.) In certain crustose and squamulose lichens (e.g. *Pannaria*, *Pertusaria*, *Rhizocarpon*) a photobiont-free margin (the hypothallus) may be formed around the thallus as the hyphae of the mycobiont grow outwards; the photobiont reproduces by cell division or e.g. by autospore formation, and the daughter cells spread into the region of new fungal growth.

Reproduction. Asexual (vegetative) reproduction in lichens may be achieved e.g. by SOREDIUM, ISIDIUM or HORMOCYST formation, or by fragmentation of the thallus. Some lichens may be able to generate new thalli from conidia (which may or may not be produced in pycnidia, according to species).

Sexual reproduction commonly involves ascocarp formation. Lichen ascocarps tend to be longer-lived than those of non-lichenized ascomycetes, and may continue to produce ascospores for a period of years. Since ascospores do not contain the photobiont, the germinating ascospores must acquire the photobiont from the environment. In a few cases (e.g. *Endocarpon* sp, *Pertusaria pertusa*) cells of the photobiont may adhere to the ascospores as they are released, thereby allowing their dissemination with the ascospores. In most cases, however, the fungus must incorporate free-living cells, or may acquire the appropriate photobiont from propagules (e.g., soredia: probably an important source for TREBOUXIA-containing lichens) present on the substratum. In some lichen species the ascospores may germinate only in the presence of cells of the photobiont; in others, germination gives rise to a mat of hyphae which subsequently incorporates algal or cyanobacterial cells in its immediate vicinity. The mechanism by which the 'correct' photobiont is selected is unknown; some evidence implies a role for LECTINS in the initial recognition step [Lichenol. (1984) *16* 103–110]. However, in at least some cases the initial incorporation of algal or cyanobacterial cells seems to be non-selective, all cells except those of the prospective photobiont being killed by the fungus [Nature (1981) *289* 169–172].

Habitat. Lichens occur in a wide range of habitats – usually attached to substrata such as rocks, soil, bark, leaves, mosses etc, but sometimes unattached. Saxicolous species may be ENDOLITHIC (q.v.) or EPILITHIC, and are often specific for particular types of rock (e.g. acidic or basic); they may facilitate the weathering of rocks both by physical mechanisms (e.g. rhizine penetration, thallus expansion and contraction on wetting and drying) and by chemical mechanisms (e.g. metal chelation by lichen substances, corrosion by CO_2 and/or oxalic acid). Some species can grow on and etch the surface of glass, marble etc, and have caused significant deterioration of stone monuments, stained glass windows etc.

Many lichens can grow in situations where they are exposed to extremes of temperature and desiccation which can be tolerated by few other forms of life – e.g. in hot arid deserts, in polar regions, and at high altitudes. In some harsh environments lichens are important primary producers, and terricolous species may play an important role in stabilization of the soil in such environments.

A few lichens (e.g. *Dermatocarpon weberi*, *Verrucaria aquatilis*) are aquatic, occurring e.g. on rocks submerged in streams. Some lichens occur on rocks by the sea – at or above the intertidal zone, depending on species (see e.g. LICHINA, VERRUCARIA).

Lichens have been found in some unusual locations: e.g. on the carapaces of giant tortoises, and on the backs of (living) flightless weevils of the genus *Gymnopholus*.

Lichens and pollution. Many lichens are highly sensitive to air pollution, probably as a result of their ability to accumulate high concentrations of pollutants. They are killed by low concentrations of e.g. SO_2, fluoride, fertilizer dust etc; the pollutant primarily responsible for lichen death in industrial regions is generally SO_2, one effect of which is to degrade chlorophyll to phaeophytin. The various sensitivities of different lichen species to pollution makes them useful biological indicators of pollution levels in a given area (see also INDICATOR ORGANISM); however, sensitivities may be influenced by moisture, temperature and substratum (e.g. corticolous lichens are more sensitive than saxicolous ones). Some lichens are inherently resistant to pollution (see e.g. LECANORA).

Lichen thalli can accumulate high concentrations of heavy metal ions [Lichenol. (1984) *16* 173–188] – apparently with little adverse effect on themselves; lichens have thus been used to monitor environmental levels of heavy metals. (See also ACAROSPORA.)

Since lichens are used as food (by man and animals) in some regions, their ability to accumulate toxic substances can be a potential health hazard. For example, lichens accumulated radioactive isotopes (e.g. strontium-90, caesium-137) in arctic regions which were subjected to fall-out from atmospheric nuclear weapons tests in the 1950s/1960s; lichens in this region form the basis of the food chain: lichen → reindeer (caribou) → man, and correspondingly high levels of e.g. ^{137}Cs were recorded in Lapps who consumed reindeer meat at that time (although the levels were not considered to be dangerous).

[Lists of references on lichens and air pollution are published regularly in *The Lichenologist*.]

Uses of lichens. Some lichens are used as food by people in certain countries: e.g. *Cetraria islandica* ('Iceland moss') is consumed e.g. in Iceland, *Umbilicaria esculenta* ('iwatake') in Japan. Species of *Cladonia* and *Cetraria* form a major part of the winter diet of reindeer (caribou) and other deer in arctic regions. Various lichens have been used since ancient times as sources of dyes: e.g. *Ochrolechia tartarea* yields the purple dye cudbear, *Parmelia* spp a brown dye, and *Roccella* spp LITMUS and the purple dye orchil. Certain lichens – particularly *Evernia prunastri* ('oakmoss') – are used in the manufacture of perfumes, both for their own scent and for their ability to stabilize (fix) other scents. Lichens were formerly widely used in folk medicine; currently, USNIC ACIDS have some medical uses. In Scandinavian countries, *Cladonia stellaris* is harvested on a commercial scale for ornamental uses in wreaths, floral decorations etc [Lichenol. (1979) *11* 85–89].

lichen starch *Syn.* LICHENIN.

lichenan *Syn.* LICHENIN.

lichenicolous Growing on lichens.

lichenin (lichenan; lichen starch; 'moss starch') A linear β-D-glucan consisting mainly of (1 → 3)-β-linked cellotriose units (i.e. units of three (1 → 4)-β-linked glucose residues); it occurs in certain lichens (e.g. *Cetraria islandica*). (cf. ISOLICHENIN.)

lichenometry A method for dating glacial deposits, stone monuments etc, based on a knowledge of the growth rates of particular (usually crustose) lichens. The assumption is made that the largest crustose thalli present on a given exposed surface started to grow when the surface first became exposed; the thalli are measured, and their ages estimated from a standard growth curve for the species – thus giving an approximate age for the substratum. Certain thalli of *Rhizocarpon geographicum* have been estimated to be ca. 1000–4500 years old. [Lichenometry with *R. geographicum*: Lichenol. (1983) *15* 249–261.]

Lichina A genus of small, fruticose LICHENS (order LECANORALES); photobiont: a cyanobacterium. The thallus is more or less gelatinous, brown to black, erect, branching, terete (*L. confinis*) or flattened (*L. pygmaea*), forming tufts or mats on rocks on the seashore at or just above the high-tide level. Apothecia occur immersed in the swollen ends of the branches.

Lieberkuehnia See GRANULORETICULOSEA.

Lieberkühn's organelle (watchglass organelle) A refractive, lenticular body which occurs, subpellicularly, in or near the buccal cavity in certain hymenostomes.

***lif* gene** See LYSOSTAPHIN.

***lig* gene** Gene for DNA LIGASE.

ligase chain reaction See LCR.

ligases (synthetases) ENZYMES (EC class 6) which catalyse covalent bond formation using energy obtained from the cleavage of a pyrophosphate bond (e.g. in ATP); C–O, C–S, C–N and C–C bonds are formed by different subclasses of ligases. (See e.g. DNA LIGASE; cf. SYNTHASE.)

ligative hyphae See HYPHA.

light chain (*immunol.*) The smaller (MWt ca. 23000) of the two types of polypeptide chain in an IMMUNOGLOBULIN monomer. About half the (ca. 215) amino acid residues in a light chain constitute the variable, N-terminal region (V_L), while the remainder form the constant, C-terminal region (C_L). There are two structurally distinct types of light chain: *kappa* and *lambda*; both types occur in all classes of immunoglobulin, but a given Ig molecule contains only *kappa* or only *lambda* chains. In human Igs, the ratio *kappa:lambda* is ca. 70:30, while in mouse Igs it is ca. 95:5.

light green A green acid TRIPHENYLMETHANE DYE.

light-harvesting complex (LHC) In or on membranes involved in PHOTOSYNTHESIS: a pigment-containing complex which receives radiant energy and transfers energy to REACTION CENTRES; an *antenna* is an array of LHCs. Protein–pigment complexes in LHCs are sometimes indicated by the letter 'B' followed by the absorption maximum or maxima – e.g. B800 indicates absorption at 800 nm.

LHCs in oxygenic organisms. In e.g. green algae (CHLOROPHYTA), DIATOMS and DINOFLAGELLATES, LHCs occur within the photosynthetic membranes; in addition to CHLOROPHYLL *a*, LHCs in these organisms include Chl *b* (green algae) or Chls c_2 and/or c_1 (diatoms, dinoflagellates). In CRYPTOPHYTES, RHODOPHYTA and CYANOBACTERIA the LHCs include PHYCOBILIPROTEINS which, in cyanobacteria and Rhodophyta, occur in PHYCOBILISOMES attached to thylakoid membranes.

Factors such as light quality and the availability of nitrogen and sulphur can influence the composition of LHCs. [Environmental effects on the LHC of cyanobacteria: JB (1993) *175* 575–582.]

LHCs in anoxygenic organisms. In bacteria of the RHODOSPIRILLALES the LHCs include bacteriochlorophylls and CAROTENOIDS. In the 'purple' photosynthetic bacteria (RHODOSPIRILLINEAE) LHCs occur in 'intracytoplasmic membranes' (CHROMATOPHORE sense 2) and include bacteriochlorophyll *a* or *b* as well as various carotenoids. In the 'green' photosynthetic bacteria (CHLOROBIINEAE) the LHCs characteristically occur in CHLOROSOMES; in green-coloured species of the Chlorobiaceae, the LHCs include Bchls *c* or *d* and the carotenoid chlorobactene, while those of brown-coloured species include Bchl *e* and isorenieratene. In members of the Chloroflexaceae the LHCs include Bchl *c* and β-carotene.

light reaction See PHOTOSYNTHESIS.

light strand (*mol. biol.*) See H STRAND.

light trap (1) A zone of focused light in a liquid medium. Organisms capable of behavioural responses to light may accumulate in, or be excluded from, such a zone of light. Accumulation may occur e.g. when the light–dark boundary offers an intensity gradient sufficient to trigger a PHOBIC RESPONSE such that any attempt by the organisms to leave the light zone incites a change in direction.

(2) The reaction centre in a photosynthetic pigment system.

LightCycler® See PCR.

LightCycler Red 640 See PCR.

LightCycler Red 705 See PCR.

lignans A class of dibenzylbutane derivatives which occur in higher plants and in fluids (bile, serum, urine etc) in man and other animals. Mammalian lignans include *enterodiol* (2,3-*bis*(3-hydroxybenzyl)butane-1,4-diol), and *enterolactone* (*trans*-2,3-*bis*(3-hydroxybenzyl)-γ-butyrolactone); these compounds, which have a potential anti-cancer role, can be synthesized in vitro by human faecal flora [JAB (1985) *58* 37–43].

lignicolous Living or growing in or on wood.

Ligniera See PLASMODIOPHOROMYCETES.

lignin A complex, variable, hydrophobic, cross-linked, three-dimensional aromatic polymer of *p*-hydroxyphenylpropanoid units connected by C–C and C–O–C links; it is formed by the peroxidase-initiated dehydrogenative polymerization of substituted *p*-hydroxycinnamyl alcohols (*p*-coumaryl, coniferyl and sinapyl alcohols). Lignin is a major component in the vascular tissues of terrestrial plants, and constitutes ca. 20–35% of the weight of wood; it occurs with hemicelluloses in the matrix of the cell walls, 'waterproofing' the xylem vessels and providing mechanical strength and protection from enzymic attack. [Lignification in disease resistance in plants: ARPpath. (1980) *18* 259–288.] Lignin is tightly bonded to CELLULOSE and HEMICELLULOSES – mainly by hydrogen bonds but also by some covalent bonds.

Lignins are extremely resistant to chemical and enzymic degradation; biological degradation is achieved mainly by fungi – most efficiently by white-rot basidiomycetes (see WHITE ROT) – but also by certain actinomycetes. In the white-rot fungi *Phanerochaete chrysosporium* and *Coriolus versicolor*, lignin degradation is a secondary metabolic process, occurring under low levels of nutrient nitrogen and requiring the presence of a carbon source such as glucose or cellulose [Book ref. 39, pp. 67–90]; it appears not to yield significant carbon or energy, but allows access to the polysaccharide components of the cell walls (see CELLULASES). *P. chrysosporium* produces an extracellular haemoprotein 'ligninase' which, in the presence of H_2O_2, can degrade lignin by bringing about oxidative cleavage of the backbone between Cα and Cβ, oxidation and hydroxylation of benzylic methylene groups, oxidation of phenols and benzyl alcohols, etc. The haemoprotein functions as a peroxidase. Reaction of the enzyme with H_2O_2 generates a high-redox-potential porphyrin cation radical (oxy-ferryl complex) which can extract a single electron from an aromatic ring in the lignin substrate to generate an aromatic cationic radical; this is followed by a variety of spontaneous degradative reactions via radical and cation intermediates [FEBS (1985) *183* 7–12, 13–16]. *P. chrysosporium* produces multiple forms of the haemoprotein lignin peroxidase [homology among multiple extracellular peroxidases from *P. chrysosporium*: JBC (1987) *262* 419–424]. Other enzymes implicated in aerobic lignin degradation include phenol oxidases (e.g. LACCASE) and alcohol oxidase. Quinones, which can be toxic to organisms, are common products of phenol oxidase action; several white-rot fungi produce a cellobiose:quinone oxidoreductase which couples the oxidation of cellobiose with the reduction of quinones [Book ref. 26, pp. 25–26], thus linking cellulose and lignin degradation. [*Anaerobic* lignin degradation: AEM (1984) *47* 998–1004.]

ligninolytic Capable of degrading LIGNIN. (cf. LIGNOLYTIC.)

lignite biodegradation See COAL BIODEGRADATION.

lignocellulose A complex containing LIGNIN, CELLULOSE, and (usually) HEMICELLULOSES.

lignolytic Capable of degrading wood. (cf. LIGNINOLYTIC.)

Limacella See AGARICALES (Amanitaceae).

limax amoebae 'Slug-like' amoebae, i.e., cylindrical (not flattened) MONOPODIAL AMOEBAE, or cylindrical amoebae which 'flow' along without forming obvious pseudopodia.

limberneck A form of BOTULISM in birds, especially poultry, caused by *C. botulinum* type A. (cf. WESTERN DUCK DISEASE.)

Limburger cheese See CHEESE-MAKING.

lime sulphur See SULPHUR.

limit dextrin See AMYLASES.

limit dextrinase See DEBRANCHING ENZYMES.

limit of resolution See RESOLVING POWER.

limited enrichment method See AUXOTROPH.

limulin A LECTIN from the horseshoe crab, *Limulus polyphemus*; it is presumed to bind at sites in LIPOPOLYSACCHARIDES and can agglutinate *Escherichia coli* and many other Gram-negative bacteria.

***Limulus* amoebocyte lysate test** (LAL test) A test used for the detection or quantification of LIPOPOLYSACCHARIDE. The test depends on the ability of LPS to cause gelation of a lysate of the blood cells (amoebocytes) of the horseshoe crab, *Limulus polyphemus*. Quantities of LPS of the order of 10^{-9} g/ml can be detected; different types of LPS differ in their ability to cause gelation. Substances other than LPS (e.g. lipoteichoic acids, polynucleotides) can also cause gelation [FEMS (1983) *20* 343–346].

lincomycin See LINCOSAMIDES.

lincosamides A group of MLS ANTIBIOTICS which contain lincosamine (6-amino-6,8-dideoxyoctose); the group includes e.g. lincomycin (consisting of a substituted pyrrolidine linked to methyl-α-thiolincosamine, produced by *Streptomyces lincolnensis*), clindamycin (7-chloro-7-deoxylincomycin, a semisynthetic derivative of lincomycin), and celesticetin. Lincomycin is active mainly against Gram-positive bacteria (e.g. staphylococci, streptococci, clostridia, *Corynebacterium diphtheriae*) and may be useful e.g. against penicillin-resistant strains. Clindamycin is more active than lincomycin against Gram-negative bacteria, and is effective e.g. against most Gram-positive and Gram-negative anaerobic pathogens (including *Bacteroides fragilis*); it is also active against *Plasmodium* spp, and has been used (with QUININE) in the treatment of malaria. Lincosamides inhibit PROTEIN SYNTHESIS; they bind to the 50S subunit of 70S ribosomes and appear to block the peptidyltransferase reaction. Antiplasmodial action is presumably due to inhibition of mitochondrial protein synthesis.

lincosamine See LINCOSAMIDES.

Linderina See KICKXELLALES.

line probe assay See PROBE.

linezolid See OXAZOLIDINONES.

linkage (genetics) The degree to which the alleles of two or more given genes are inherited together. Alleles of *unlinked* genes exhibit independent assortment and occur e.g. on different chromosomes. *Linked* genes occur on the same chromosome (or other genetic molecule) and are said to constitute a *linkage group*; the closer together two genes are on a chromosome, the more tightly linked they are – i.e., the lower is their RECOMBINATION FREQUENCY. (Genes far apart on a chromosome may behave as unlinked genes; in bacteriology, genes on the chromosome are sometimes said to be unlinked if they are merely non-contiguous, regardless of the distance between them.)

linked recognition (*immunol.*) *Syn.* COGNATE HELP.

linker DNA (1) See CHROMATIN. (2) A synthetic oligodeoxyribonucleotide which contains the cleavage site of a particular restriction endonuclease, and which can be ligated to a vector or exogenous DNA fragment prior to CLONING; a linker may serve e.g. to facilitate retrieval of the cloned fragment. (See also POLYLINKER.)

linker polypeptide See PHYCOBILISOMES.

linking number (α) See DNA.

linseed rust *Syn.* FLAX RUST.

lipase (triacylglycerol acylhydrolase; EC 3.1.1.3) A type of enzyme (see ENZYMES) which hydrolyses glycerides to free fatty acids and (ultimately) glycerol. Non-specific lipases release fatty acids from all three positions of the glycerol moiety of a triglyceride – 1,2(2,3)-diglycerides, 1,3-diglycerides and monoglycerides being formed as intermediates. Other lipases release fatty acids only from the outer (1 and 3) positions of the glycerol moiety. Lipases function well at the interface between their hydrophobic substrates and the aqueous phase, and can act on triglycerides in micelles or emulsions.

Microbial lipases have some commercial applications. They are used e.g. in digestive aids, in the manufacture of certain cheeses (to improve or accelerate the development of flavour and aroma) and in INTERESTERIFICATION.

Tests for bacterial lipase production: (a) growth on EGG-YOLK AGAR; (b) TWEEN HYDROLYSIS; (c) inoculation onto a nutrient agar medium containing a fat (or e.g. corn oil) and NILE BLUE (lipase-producing colonies have a blue halo). A given lipase may not react positively in all tests. (See also FILM AND SPOTS.)

lipid A See LIPOPOLYSACCHARIDE.

lipid A-associated protein See ENDOTOXIN.

lipid cyst See METHYLOCOCCACEAE.

lipid stain See e.g. BURDON'S STAIN.

lipid X See LIPOPOLYSACCHARIDE.

lipoamino acids See *bacterial* CYTOPLASMIC MEMBRANE.

lipofection Lipid-aided insertion of nucleic acid into mammalian (and other) cells; uptake is promoted when (e.g.) DNA forms LIPOSOMES with mixed cationic and neutral lipids.

lipoic acid (α-lipoic acid; thioctic acid) A coenzyme which is involved in the oxidative decarboxylation of keto-acids: e.g. pyruvate and 2-oxoglutarate are decarboxylated to acetyl-CoA and succinyl-CoA, respectively [see Appendix II(a)]. (See also THIAMINE.) The oxidized form of the coenzyme contains a disulphide bond (–S–S–) which can be reduced to form two –SH groups, the reduced coenzyme (dihydrolipoic acid) being 6,8-dithiooctanoic acid: $CH_2SH.CH_2.CHSH.(CH_2)_4.COOH$.

Lipoic acid is required as a growth factor e.g. by *Tetrahymena* spp; it can replace acetate as a growth factor for e.g. *Lactobacillus casei*.

(See also ARSENIC.)

Lipomyces A genus of yeasts (family SACCHAROMYCETACEAE). Cells are spherical or ovoid; vegetative reproduction occurs by multilateral budding or by bud-fission. In most species the cells are each surrounded by a thick mucoid capsule which often contains a starch-like polysaccharide. Cells accumulate lipid, and those from older cultures usually contain one or more lipid globules. Asci typically develop from 'active buds'; an active bud resembles a vegetative bud but develops into an ascus – either following conjugation between the active bud and a protuberance on another cell (or the same cell), or in the absence of conjugation. The mother cell of the active bud may also become an ascus. Ascospores are ellipsoidal or oblong-ellipsoidal, light amber to brown, smooth, or e.g. rough with longitudinal ridges, 2–20 or more per ascus (depending on species). [Ascospore morphology and ultrastructure: IJSB (1984) *34* 80–86.] The organisms are non-fermentative and do not assimilate NO_3^-. Species (*L. anomalus*, *L. kononenkoae*, *L. lipofer*, *L. starkeyi*, *L. tetrasporus*) have been isolated from soil. [Book ref. 100, pp. 252–262.]

lipo-oligosaccharide (LOS) In certain Gram-negative bacteria (e.g. *Neisseria meningitidis*): a (normal) component of the OUTER MEMBRANE which is anatomically equivalent to LIPOPOLYSACCHARIDE (q.v.) but in which the O-specific chains are shorter than those (for example) in the enterobacteria [CRM (1998) *24* 281–334].

lipopolysaccharide (LPS) Strictly: any lipid-containing polysaccharide; however, the term 'lipopolysaccharide' is commonly used to refer specifically to the endotoxic component of the OUTER MEMBRANE in Gram-negative bacteria. Outer membrane LPS is a complex molecule consisting of three covalently linked

```
                                                                  (KDO-PEtN)
                                                                      ↓
                     GlcNAc          (Gal)    (Hep)    (PPEtN)     (KDO)
                       ↓               ↓        ↓         |           ↓
[O-specific chain] → Glc → Gal  →    Glc  →   Hep   →   Hep    →    KDO → [Lipid A]
                                                |
                                                P
```

LIPOPOLYSACCHARIDE. General structure of the core oligosaccharide in *Salmonella typhimurium*; the structure is somewhat variable, and possible sites of variation are indicated by parenthesis. Glc = D-glucose; Gal = D-galactose; Hep = 'heptose' (L-glycero-D-mannoheptose); KDO = 3-deoxy-D-manno-octulosonate; GlcNAc = *N*-acetyl-D-glucosamine; PEtN = phosphorylethanolamine; PPEtN = pyrophosphorylethanolamine; P = phosphate. [Data from e.g. Book ref. 138, pp. 157–207.]

regions: lipid A–core oligosaccharide–O-specific chain. (cf. SMOOTH–ROUGH VARIATION.)

Lipid A is a glycolipid which is largely responsible for the endotoxic activity of LPS (see ENDOTOXIN sense 1) and for the ability of LPS to act as a POLYCLONAL ACTIVATOR. In enterobacteria, it consists of a substituted β-linked glucosamine disaccharide – usually (1 → 6)-linked, but apparently (1 → 4)-linked e.g. in strains of *Escherichia coli*; the disaccharide is fully substituted with a number of long-chain (e.g. C_{12-16}) saturated fatty acids (including 3-hydroxyacids, usually 3-hydroxymyristic acid), phosphate, and the core oligosaccharide (linked to the C-6 position of the non-reducing glucosamine residue). In many enterobacteria the hydroxy fatty acids are 3-*O*-acylated to form characteristic 3-myristoxymyristic acid residues. Variations in the structure of lipid A occur in other bacteria: e.g., in certain photosynthetic bacteria the glucosamine disaccharide is replaced by a diaminohexose monomer (e.g. 2,3-diamino-2,3-dideoxyglucose in *Rhodopseudomonas viridis*) [LPS in photosynthetic bacteria: ARM (1979) *33* 215–239]; 3-hydroxymyristic acid is absent e.g. in *Brucella* strains; 2-hydroxy fatty acids are present e.g. in *Pseudomonas aeruginosa* strains.

Core oligosaccharide is a complex oligosaccharide usually linked to lipid A via '2-keto-3-deoxyoctonate', KDO (= 3-deoxy-D-manno-octulosonate). In enterobacteria (e.g. *S. typhimurium* – see figure) cores appear to contain a KDO dimer. *E. coli* K12 strains appear to have cores similar to that of *S. typhimurium*, while B strains have cores containing only glucose, heptose, and KDO. In *P. aeruginosa* the core contains glucose, heptose, KDO, rhamnose, galactosamine and alanine. Certain bacteria (e.g. strains of *Bacteroides*) appear to lack both KDO and heptose.

O-specific chain (O side chain) is the immunodominant part of the LPS molecule in the intact bacterial cell. It consists of a variable-length chain of identical (linear or branched) oligosaccharide subunits; the number of subunits may vary even in the same cell. In *Salmonella* serotypes each oligosaccharide subunit typically contains 3, 4 or 5 monosaccharide residues (including hexoses, pentoses, 6-deoxyhexoses, and 2,6-dideoxyhexoses). The O-specific chains determine the specificity of the O ANTIGENS of a given serotype, antigenic specificity being determined by the presence of a particular sugar (see e.g. ABEQUOSE, COLITOSE, PARATOSE, or TYVELOSE), by the positions of glycosidic bonds (e.g. 1 → 4 or 1 → 6 linkages), by the anomeric configuration (α or β) of the glycosidic linkages, by substituents (e.g. acetyl groups) on sugar residues, etc.

LPS appears to be bound in the outer membrane by hydrophobic bonds and by divalent cations. (Early reports of pyrophosphate bridges between lipid A moieties appear to have received little experimental support.) A proportion of the LPS can be released from enterobacteria by treatment of cells with EDTA. LPS can be extracted from cells with a phenol–water mixture (ca. 45% w/v phenol in water); an LPS–protein complex (the Boivin antigen) can be extracted by treatment of cells with trichloroacetic acid (TCA). (See also LIMULUS AMOEBOCYTE LYSATE TEST.)

Biosynthesis of LPS. (Most of the studies on LPS biosynthesis have been carried out with *S. typhimurium*.) LPS is assembled at the cytoplasmic membrane, and completed LPS molecules are subsequently transferred to the outer membrane – apparently at a limited number of sites which may correspond to ADHESION SITES. Lipid A is built up from glucose 1-phosphate via monosaccharide intermediates such as 2,3-diacylglucosamine 1-phosphate ('lipid X') [JBC (1985) *260* 15536–15541] and variously substituted glucosamine disaccharide derivatives. [For lipid A biosynthesis see e.g. TIM (1998) *6* 154–159.] In *S. typhimurium*, the core oligosaccharide is synthesized on the lipid A precursor by the sequential addition of KDO (from a CMP precursor) and other sugars (mostly from UDP precursors, although the nature of the heptose precursor is unknown). The oligosaccharide subunits of the O-specific chains are synthesized on BACTOPRENOL carrier molecules by the sequential addition of sugars from nucleotide precursors (presumably) at the inner face of the cytoplasmic membrane; the bactoprenol–oligosaccharide complex is presumed to be translocated across the cytoplasmic membrane, where the subunits are polymerized on the core oligosaccharide of an incomplete LPS molecule. (Bactoprenol pyrophosphate is released and is dephosphorylated by a specific phosphatase, the monophosphate derivative returning to the inner face of the cytoplasmic membrane for another round of synthesis.) Branches on the O-specific chains may be added before or after polymerization: e.g., in *S. typhimurium* the abequose residue is added to the bactoprenol-linked oligosaccharide, whereas in *S. minneapolis* and other serogroup E_3 (O3, 15, 34) salmonellae various residues in the O-specific chains are glucosylated after polymerization.

Inhibition of lipid A biosynthesis is a property of the compound L-573,655; this agent inhibits the enzyme deacetylase (product of the *lpxC* gene) that plays an essential role in the biosynthesis of lipid A [see e.g. Science (1996) *274* 980–982].

The antibiotic BACITRACIN inhibits polymerization of the O-specific chain.

[Reviews: CTMT (1982) *17* 79–151; Book ref. 92; LPS nomenclature: JB (1986) *166* 699–705.]

lipopolysaccharide-binding protein See CD14.

lipoprotein signal peptidase See SIGNAL HYPOTHESIS.

liposan A bioemulsifier (see BIOSURFACTANTS) produced by *Candida lipolytica* when growing e.g. on hexadecane [AEM (1985) *50* 965–970].

liposarcoma See SARCOMA.

liposome An artificial phospholipid vesicle. Liposomes have many uses in biology. For example, various components of biological membranes (e.g. PORINS, viral ENVELOPE proteins) can be incorporated into liposomes, allowing investigation of their properties under controlled conditions. Enzymes can be immobilized (see IMMOBILIZATION) by encapsulation within liposomes. Certain drugs may be encapsulated within liposomes e.g. to protect them from metabolic degradation and/or to prevent them from exerting general toxicity when delivered into the bloodstream; the liposomes fuse with or are endocytosed by cells of the reticuloendothelial system, where they release their contents. [Potential of liposome-mediated antiviral therapy: Antiviral Res. (1985) *5* 179–190.] Drug-loaded liposomes can be 'targeted' to specific cells by the incorporation of antibodies to antigens in the target cell surface; thus, e.g., 'immunoliposomes' incorporating a monoclonal antibody to herpes simplex virus glycoprotein D were found to bind specifically to HSV-infected rabbit corneal cells in vitro [J. Imm. (1986) *136* 681–685].

liposome swelling assay A method for determining the rate of diffusion of ions or molecules through PORIN channels; purified porins are incorporated into LIPOSOMES, and the rate of influx of specified ions or molecules is estimated from the effects of the (osmotically-induced) influx of water. [Method: JB (1983) *153* 241–252.]

lipoteichoic acid See TEICHOIC ACIDS.

Lipschütz body An intranuclear inclusion body formed (late in infection) in cells infected with e.g. herpes simplex virus type 1 or 2.

lirella (lirelliform apothecium) (*lichenol.*) A slit-like (long, narrow) apothecium; lirellae may appear as black branched or unbranched lines with (in *Graphis* spp) or without (in *Opegrapha* spp) raised margins.

lirelliform apothecium *Syn.* LIRELLA.

lissamine rhodamine (acid rhodamine B; also: sulphorhodamine B) A red FLUOROCHROME (fluorescence: orange-red) used for labelling proteins; it is initially treated with phosphorus pentachloride and acetone to form the highly reactive sulphonyl chloride derivative, and protein–dye conjugation is then effected by a sulphamido condensation.

Listeria A genus of Gram-positive, asporogenous, aerobic/facultatively anaerobic bacteria. Cells: rods or coccobacilli (ca. 0.5 × 0.5–2 μm), occurring singly or in small groups ('Chinese letter' arrangements), palisades or filaments. Optimum temperature range: 25–37°C; growth occurs slowly at 4°C. Most strains catalase +ve. Oxidase −ve. Indole −ve. Aesculin is hydrolysed. Sugars are fermented to acid (no gas). Generally motile. Colonies of *L. monocytogenes*: grey, translucent, about 1 mm in diameter after 24–48 hours; a characteristic bluish colour in obliquely transmitted light. *L. monocytogenes* may give clear haemolysis on e.g. horse- and/or sheep-blood agar.

Listeria spp occur (apparently as saprotrophs) e.g. in soil, decaying plant matter, silage, etc., and have been isolated from e.g. faeces of healthy humans and animals; some haemolytic strains of *Listeria* can be pathogenic in man and also in animals (see LISTERIOSIS and MENINGITIS). [Wild birds and silage as reservoirs of *Listeria* in the agricultural environment: JAB (1985) *59* 537–543.] [The pathogenicity of *L. monocytogenes* (a public health perspective): RMM (1997) *8* 1–14.]

Some (human) infections with *L. monocytogenes* have been linked to the ingestion of certain types of cheese. One approach to the control of *L. monocytogenes* in cheese (during manufacture) involves the use of specific types of starter culture [see e.g. AEM (1998) *64* 4842–4845; JAM (1999) *86* 251–256].

Studies of e.g. DNA relatedness among strains of *L. monocytogenes* sensu lato have established five distinct 'genomic groups', each of which has been given a species designation. Group 1 contains the type strain of *L. monocytogenes* and related strains, and corresponds to *L. monocytogenes* sensu stricto. Group 2 (*L. monocytogenes* serovar 5) has been named *L. ivanovii* [IJSB (1984) *34* 336–337]. Group 3 (now *L. innocua*) includes non-haemolytic, non-pathogenic strains. Groups 4 and 5 (now *L. welshimeri* and *L. seeligeri*, respectively [IJSB (1983) *33* 866–869]) also include strains which are (apparently) non-pathogenic.

L. grayi and *L. murrayi* have been retained in the genus [IJSB (1987) *37* 298–300]; *L. denitrificans* is now in *Jonesia* [IJSB (1987) *37* 266–270]. [Taxonomy of *Listeria*: IJSB (1991) *41* 59–64, (1993) *43* 26–31.]

[Typing of *L. monocytogenes* by repetitive element sequence-based PCR: JCM (1999) *37* 103–109.]

[Culture etc: Book ref. 46, pp. 1680–1687.]

listeriolysin See THIOL-ACTIVATED CYTOLYSINS.

listeriosis Any disease of man or animals caused by *Listeria monocytogenes*.

In man, the groups commonly affected are pregnant women (particularly during the third trimester), infants, immunosuppressed adults and the elderly. In pregnant women, the disease may involve e.g. fever, chills, backache and flu-like symptoms; while the mother herself is usually not severely affected, intrauterine infection may lead to premature birth, stillbirth or abortion. (Note that *L. monocytogenes* is one of the few bacterial pathogens which can cross the blood–brain barrier and the placenta.) Early-onset neonatal disease (i.e. within ~2 days) may be largely 'septicaemic', with disseminated infection, cardiopulmonary and CNS problems; mortality may be ~50–60%. Late-onset neonatal disease (starting after ~5–6 days) typically involves meningitis; compared with early-onset disease, mortality is somewhat lower. In adults, the common manifestations are CNS involvement, septicaemia and endocarditis.

L. monocytogenes can invade intestinal epithelium, invasion being promoted by the bacterial cell-surface INVASIN *internalin A* (encoded by gene *inlA*); the eukaryotic cell-surface receptor for internalin A is a CADHERIN (E-cadherin) which occurs on basolateral surfaces of epithelial cells. Interaction between internalin A and E-cadherin apparently initiates uptake of the bacterium by the host cell. (Interestingly, even latex beads coated with internalin A are internalized by some types of mammalian cell [Inf. Immun. (1997) *65* 5309–5319].) Invasion (uptake of the pathogen) resembles the 'zipper' mechanism involved in the uptake of *Yersinia enterocolitica* (see *Y. enterocolitica* under FOOD POISONING). *L. monocytogenes* escapes from the phagosome by secreting a pore-forming toxin, *listeriolysin O* (a THIOL-ACTIVATED CYTOLYSIN). Once free in the host cell's cytoplasm, the pathogen grows and spreads to adjacent cells by *actin-based motility* (as described for *Shigella dysenteriae* in DYSENTERY); in *L. monocytogenes*, the ActA protein plays the same role as IcsA in *Shigella*. When propelled into an adjacent cell, the pathogen becomes enclosed by a double membrane (one membrane from each cell); escape (into the cytoplasm of the adjacent cell) involves secretion of the enzyme *lecithinase* (encoded by gene *plcB*) – which hydrolyses the lecithin content of the double membrane. [Invasion and actin-based motility: EMBO (1998) *17* 3797–3806. Internalin–E-cadherin complex (structure): Cell (2002) *111* 825–836. Pathogenicity of *L. monocytogenes* (review): RMM (1997) *8* 1–14.]

Infection appears to occur primarily via food; foods which have been implicated include e.g. Mexican-style cheese, certain other cheeses, coleslaw and milk. However, risk assessment is problematic as isolates of *L. monocytogenes* vary widely in their pathogenic potential, and there is no simple test for pathogenicity. A DNA-based method has been used to type 64 isolates of *L. monocytogenes* from human, animal and food sources; the strains clustered primarily according to origin, and, on the basis of the results, it was suggested that only a small proportion of food-borne strains may be pathogenic for humans [JCM (1999) *37* 103–109].

In animals, outbreaks may be associated with the consumption of incompletely fermented SILAGE (pH > 5.5); *L. monocytogenes* cannot multiply in properly made silage (pH 4–4.5). The disease may affect e.g. cattle, sheep and goats, and may lead to manifestations such as abortion. The disease is more common in temperate regions.

Listonella A genus of Gram-negative bacteria that includes the species *L. anguillarum*, *L. damsela* and *L. pelagia* (formerly *Vibrio anguillarum*, *V. damsela* and *V. pelagia*, respectively) [SAAM (1985) *6* 171–182; validation of names: IJSB (1986) *36* 354–356]. (See also PHOTOBACTERIUM.)

L. anguillarum is an important pathogen of freshwater and marine fish (causing e.g. disease in salmonids, and RED PEST in eels); it is also pathogenic for oyster larvae.

The organism is facultatively halophilic, capable of growing on nutrient agar with 0.5–6.0% NaCl, but not with 8% NaCl. Oxidase +ve. Sucrose (usually) +ve. [Agglutination typing of *Vibrio* (*Listonella*) *anguillarum*: AEM (1984) *47* 1261–1265.]

lithobiont A SAXICOLOUS organism.

lithocholic acid See BILE ACIDS.

Lithophyllum See RHODOPHYTA.

lithophyte Refers to a SAXICOLOUS plant.

lithotroph An organism which uses an inorganic substance as substrate in energy metabolism. (cf. ORGANOTROPH; AUTOTROPH.) Among *chemolithotrophs* (see CHEMOTROPH) energy may be obtained by the oxidation of e.g. NH_3 (see NITRIFICATION), H_2 (see e.g. HYDROGEN-OXIDIZING BACTERIA, DISSIMILATORY SULPHATE REDUCTION), sulphur, sulphide and/or thiosulphate (see e.g. THIOBACILLUS), or Fe^{2+} (e.g. in *Thiobacillus ferrooxidans*). *Photolithotrophs* use inorganic substrates (e.g. water, sulphur, sulphide, H_2, or thiosulphate) as electron donors in PHOTOSYNTHESIS.

litmus A PH INDICATOR: pH 4.5–8.3 (red to blue); it is obtained from lichens (*Roccella* spp) and is a complex mixture of compounds.

litmus milk A bacteriological medium which consists of skim milk (whole MILK without cream) containing sufficient litmus solution to impart a blue-purple colour. Prior to use, the medium is either steamed (20/30 min on each of 3 successive days) or autoclaved (115°C/10 min), and the pH is adjusted to 7.

A given bacterial strain growing in litmus milk may give one of the following reactions. (a) No visible change in the medium. (b) Acid production from lactose (indicated by the LITMUS). (c) Alkali production – commonly due to hydrolysis of casein. (d) Reduction (decolorization) of the litmus. (e) The production of an *acid clot*; the clot, which is soluble in alkali, forms at about pH5 (at or near the IEPs of the various types of CASEIN). The production of acid *and* gas may give rise to a STORMY CLOT. (f) The formation of a clot at or near pH7. Such clots are formed by certain bacteria (e.g. *Bacillus* spp) which produce rennin-like proteases; the litmus may be reduced before or during clot formation, and peptonization (digestion of the clot) may subsequently occur. (See also MILK SPOILAGE.)

little t antigen (small t antigen) See POLYOMAVIRUS.

littoral A term used with different meanings by different authors; thus, the littoral region of the *marine* benthic zone is variously regarded as e.g. (a) the 'intertidal region', i.e., the zone which is uncovered at low tide and under water at high tide; (b) a zone below the average low-tide mark which is uncovered only occasionally; (c) that region between the high-tide mark and a depth of 200 metres. In *freshwater* environments the littoral region may refer to those parts of the benthic zone which occur in shallower waters of unspecified depth.

LIV-I binding protein See BINDING PROTEIN-DEPENDENT TRANSPORT SYSTEM.

live vaccine Any VACCINE containing viable, usually attenuated, microorganisms (see e.g. BCG and STERNE STRAIN).

liver of sulphur See SULPHUR.

liver-stage-antigen 1 See MALARIA.

liverwort symbioses See e.g. NOSTOC and MYCORRHIZA.

lividomycin Either of two AMINOGLYCOSIDE ANTIBIOTICS (lividomycins A, B), both of which may be regarded as derivatives of NEOMYCIN B.

Lk Linking number: see DNA.

LktA (of *Pasteurella haemolytica*) See RTX TOXINS.

LMD Culture Collection, Laboratory of Microbiology, Julianalaan 67A, 2628 BC Delft, The Netherlands.

lmp **genes** (EBV) See EPSTEIN–BARR VIRUS.

LmrA (*Lactococcus lactis*) See ABC TRANSPORTER.

lobar pneumonia See PNEUMONIA.

Lobaria A genus of foliose LICHENS (order PELTIGERALES). Photobiont: *Myrmecia* (in e.g. *L. pulmonaria*) or *Nostoc* (in e.g. *L. scrobiculata*); *Myrmecia*-containing species may have *Nostoc*-containing cephalodia – cephalodia being fruticose in *L. amplissima* (see CEPHALODIUM). Thallus: large (e.g. up to ca. 20 cm across), lobed, both surfaces corticate; the underside is usually more or less tomentose, often with glabrous patches, and in *L. linita* and *L. pulmonaria* the upper surface is coarsely reticulate. Soredia may be present in e.g. *L. pulmonaria* and *L. scrobiculata*. Apothecia: lecanorine, usually with reddish or brown discs, abundant (in e.g. *L. laetevirens*) or scarce. Found on rocks, bark etc, and generally very sensitive to air pollution.

Loboa A genus of fungi (? ENTOMOPHTHORALES). *L. loboi* causes LOBOMYCOSIS and has not been cultured in vitro. In tissues, it occurs as spherical, thick-walled, hyaline, yeast-like cells (ca. 5–12 µm diam.) which reproduce by budding; sequential budding may lead to the formation of chains of cells.

lobomycosis (keloidal blastomycosis) A MYCOSIS, which apparently affects only man and dolphins, caused by *Loboa loboi*; it occurs in Central and South America. Infection is presumed to occur via wounds; chronic, nodular, keloidal (elevated and scar-like) skin lesions develop chiefly (in man) on exposed areas of the body.

lobopodium See PSEUDOPODIUM.

lobose pseudopodium See PSEUDOPODIUM.

Lobosea A class of naked and testate amoebae (superclass RHIZOPODA) which form either lobopodia or filiform subpseudopodia from a broader hyaline lobe. The cells are typically uninucleate; the few multinucleate members do not form plasmodia. Fruiting structures do not occur, but many members can form resistant cysts. Orders: AMOEBIDA, ARCELLINIDA, PELOBIONTIDA, SCHIZOPYRENIDA, TRICHOSIDA.

lobster diseases See e.g. BURNED SPOT DISEASE and GAFFKAEMIA.

local lesion (*plant virol.*) A discrete lesion (e.g. a patch of chlorosis or necrosis, or a RINGSPOT) which develops as a result of localized viral infection in a leaf, stem etc. A plant which readily develops local lesions when infected with a given virus may be used for assaying that virus, and is termed a *local lesion host* (= *assay host*); the number of infective virus particles present in a preparation can be assessed from the number of lesions produced when the leaves of a suitable local lesion host are inoculated with a known volume of the preparation. (See also PLANT VIRUSES.)

Locke's solution A solution of inorganic salts used e.g. as a liquid overlay for certain solid media. [Recipe: Book ref. 53, p. 2175.]

lockjaw *Syn.* TETANUS.

locomotor fringe *Syn.* TROCHAL BAND.

locule (loculus) A small chamber or cavity: see e.g. ASCOSTROMA and FORAMINIFERIDA.

Loculoascomycetes A class of ascomycetes characterized by the formation of bitunicate asci in ascostromata; the genera include e.g. *Elsinoë, Mycosphaerella, Myriangium* and *Venturia*. The class is not recognized in most modern taxonomic schemes.

locus (*genetics*) The position of a given gene, operon, mutation etc on a chromosome, plasmid, or other genetic molecule.

locus of enterocyte effacement See PATHOGENICITY ISLAND.

Lodderomyces A genus of yeasts (family SACCHAROMYCETACEAE). Cells are spheroidal, ellipsoidal, or cylindrical, and reproduce by multilateral budding; pseudomycelium may be formed. Asci are persistent. Ascospores: smooth, oblong to fusiform, 1 or 2 per ascus. One species, *L. elongisporus*, isolated e.g. from fruit juice and soil. [Book ref. 100, pp. 263–265.]

lodging (*plant pathol.*) The collapse of cereal plants due to bending or breaking of the stems; lodging may occur in certain diseases.

Loeffler's methylene blue See METHYLENE BLUE.

Loeffler's serum A solid medium, usually prepared as a slope, used e.g. for the culture of *Corynebacterium diphtheriae*. The medium contains sterile serum and nutrient broth (3:1 by volume) and glucose (ca. 0.25–0.5% w/v); this mixture may be coagulated by TYNDALLIZATION or, if precautions are taken to prevent foaming, by autoclaving.

log dilutions See SERIAL DILUTIONS.

log phase (of growth) See BATCH CULTURE.

logarithmic phase (of growth) See BATCH CULTURE.

***Lolium* enation virus** See FIJIVIRUS.

lomasomes Membranous tubular, vesicular or sac-like structures which occur, generally between the cell wall and cytoplasmic membrane, in many fungi; they may be formed by invagination of the cytoplasmic membrane, and often contain particulate matter. The function of lomasomes is unknown; they have been implicated e.g. in cell wall synthesis.

***lon* gene** In *Escherichia coli*, a gene which encodes an ATP-dependent protease that e.g. inactivates the SulA (= SfiA) protein (see SOS SYTEM); the Lon protein is a HEAT-SHOCK PROTEIN.

long-incubation hepatitis *Syn.* HEPATITIS B.

longitudinal binary fission See BINARY FISSION.

longus A large plasmid-encoded fimbria* (up to 20 μm in length) formed by strains of ETEC. The appendage consists of a repeated 22 kDa subunit, the N-terminal sequence of which shows some homology with that of the subunits of various type IV FIMBRIAE (including the TCP of *Vibrio cholerae* and 'bundle-forming pili' of EPEC). [Mol. Microbiol. (1994) *12* 71–82. *In this paper the appendage is referred to as a 'pilus'.]

lonomycin See MACROTETRALIDES.

loop An instrument widely used in bacteriology and mycology for the manipulation of small quantities of solid or liquid material – e.g. during INOCULATION; the rod-shaped metal handle holds a piece of platinum or nickel–steel wire (3–10 cm long) formed into a closed loop at the free end. Before *and* after *each* use the loop and adjacent wire are sterilized by FLAMING and allowed to cool. In use, the sterile loop is brought into contact with e.g. a colony so that a small quantity of material adheres to it; this can then be used as an INOCULUM. If a sterile loop is dipped into e.g. a broth CULTURE and withdrawn, the loop retains a circular film of liquid (containing microorganisms) which can be used as an inoculum; the volume of liquid may be varied by using loops of different sizes. In use, no part of the loop *handle* should enter the culture vessel; loops with longer wires may be used e.g. for inoculating media in bottles. (See also STRAIGHT WIRE; STREAKING.)

loop fermenter A FERMENTER in which the culture is continually and efficiently circulated from the bottom of the column to the top – 'loop' referring to the circulatory nature of the flow. Most loop fermenters consist of a vertical cylindrical column containing a coaxial DRAFT TUBE (see e.g. AIRLIFT FERMENTER; JET LOOP FERMENTER; PROPELLER FERMENTER).

loop model (of proton translocation) A model, based on the concept of VECTORIAL GROUP TRANSLOCATION, which proposes that electrons travel in a loop within an energy-transducing membrane, starting and finishing at the same surface of the membrane, while protons pass to the opposite side of the membrane. In this model, the binding and release of protons at certain sites in an ELECTRON TRANSPORT CHAIN is seen as a direct consequence of the variation in pK of those sites due to their reduction or oxidation during electron flow.

looping (DNA) (transcriptional control) See OPERATOR.

loose smut See SMUTS (sense 2).

LOPAT Levan formation, oxidase reaction, potato rotting capacity, arginine dihydrolase production, and tobacco sensitivity: criteria which have been used for classifying plant-pathogenic fluorescent pseudomonads.

lophine See CHEMILUMINESCENCE.

Lophomonas See HYPERMASTIGIDA.

lophotrichous Of e.g. flagella: arranged as a tuft at one or both poles of a cell.

Lordsdale virus A SMALL ROUND STRUCTURED VIRUS.

lorica In certain eukaryotic microorganisms – including certain ciliates (e.g. TINTINNINA), choanoflagellates (see CHOANOFLAGELLIDA), CHRYSOPHYTES, and EUGLENOID FLAGELLATES: a vase-like or bottle-like, commonly mineralized, loose-fitting shell which partially or completely surrounds the cell. (cf. TEST.) Loricate organisms may be free-swimming (e.g. TRACHELOMONAS) or sedentary (e.g. DINOBRYON).

loricate Possessing a LORICA.

LOS LIPO-OLIGOSACCHARIDE.

loss mutant (loss variant) A variety of an organism which has lost one or more phenotypic traits as a result of mutation(s).

lotic Pertaining to aquatic habitats with running water (e.g. streams, rivers). (cf. LENTIC.)

louping-ill (ovine encephalomyelitis) An acute, sometimes fatal, usually tick-borne encephalomyelitis which affects mainly sheep; the causal agent is a flavivirus (see FLAVIVIRIDAE), and the tick vector is probably always *Ixodes ricinus*. Incubation period: ca. 1–2 weeks; viraemia and fever are followed by e.g. leaping ('louping') movements and paralysis. Death may occur in ca. 1 week. The louping-ill virus can also cause a fatal infection in Red grouse (*Lagopus lagopus scoticus*). Infection of cattle may result in a disease similar to that in sheep. In man, the virus can cause a mild, febrile, influenza-like illness.

low temperature disease See COLD WATER DISEASE.

low-temperature steam disinfection A process in which e.g. blankets, endoscopes and other types of hospital equipment can be disinfected by exposure to saturated steam below 100°C at subatmospheric pressure in an apparatus resembling a vacuum AUTOCLAVE. Essentially, the chamber – containing the load – is first evacuated to an absolute pressure of ca. 2 kPa or less; steam is then admitted to the chamber and held at a temperature of ca. 73–80°C for 10–15 min – the absolute pressure in the chamber being ca. 39–47 kPa; saturated steam at these temperatures disinfects more efficiently than does water at corresponding temperatures. The load is then dried under vacuum.

low-temperature steam–formaldehyde sterilization (LTSFS) A process which resembles the LOW-TEMPERATURE STEAM DISINFECTION process but which incorporates the use of FORMALDEHYDE. One of several possible schedules is as follows. The chamber is evacuated and formaldehyde is admitted; this step is carried out three times. Subsequently, formaldehyde, and then steam, are admitted to the chamber, and a temperature of ca. 73°C is maintained for two hours. The chamber is then evacuated, steam is admitted, and the chamber is again evacuated; this last evacuation removes traces of formaldehyde and steam and gives rise to a dry, odourless load. [Biological indicators for LTSFS: JAB (1985) *58* 207–214.]

low zone tolerance See IMMUNOLOGICAL TOLERANCE.

Löwenstein–Jensen medium A complex solid medium used for isolating *Mycobacterium* spp from e.g. sputum. The medium is prepared by the INSPISSATION of a mixture of homogenized whole hens' eggs and a solution containing asparagine, KH_2PO_4, $MgSO_4$, citrate, glycerol, potato starch, and malachite green; the pH is ca. 7.0. The medium is prepared (as slopes) in screw-cap bijou or universal bottles. [Recipe: Book ref. 53, p. 1415.]

lower fungi Fungi of the MASTIGOMYCOTINA and the ZYGOMYCOTINA (collectively: the phycomycetes). For some authors 'lower fungi' includes the MYXOMYCOTA as well as the phycomycetes.

***lox* site** See BACTERIOPHAGE P1.

Loxocephalus See SCUTICOCILIATIDA and PARATENY.

Loxodes A genus of freshwater ciliates (subclass GYMNOSTOMATIA) related to KENTROPHOROS. Cells: somewhat flattened, elliptical in outline, with a slit-like cytostome located in a subapical concavity (giving the anterior end of the cell a hook-like appearance). The cytoplasm contains MÜLLER'S VESICLES. [Photobehaviour of *Loxodes*: JP (1986) *33* 139–145.]

Loxophyllum See GYMNOSTOMATIA.

LP system LACTOPEROXIDASE–THIOCYANATE–HYDROGEN PEROXIDE SYSTEM.

***lpp* gene** In *Escherichia coli*: the gene which encodes the BRAUN LIPOPROTEIN. (See also MRNA (a).)

LPP group A category of filamentous CYANOBACTERIA (section III) which includes the poorly defined genera LYNGBYA, PHORMIDIUM and PLECTONEMA, as well as MICROCOLEUS and *Schizothrix*. In LPP group A the trichomes are straight, sheathed and non-motile, and are composed of disc-shaped cells; motile hormogonia are produced. In LPP group B the trichomes are straight, sheathed or non-sheathed, motile or non-motile, and are composed of isodiametric or cylindrical cells. GC% for the group: 42–67. Species occur in freshwater and marine habitats. (See also STROMATOLITE and EMULCYAN.)

LPS (1) LIPOPOLYSACCHARIDE. (2) LACTOPEROXIDASE–THIOCYANATE–HYDROGEN PEROXIDE SYSTEM.

***lpxC* gene** See ZINC and LIPOPOLYSACCHARIDE.

Lrp In *Escherichia coli*: leucine-responsive regulator protein, a homodimeric protein involved in the regulation of a range of genes/operons – apparently by acting as an activator or (in some cases) repressor of transcription; the activity of Lrp is inhibited by exogenous leucine. Lrp is reported to be one of several factors which regulate expression of $E\sigma^{70}$ (see SIGMA FACTOR).
[Lrp: JBC (2002) *277* 40309–40323.]

LS binding protein See BINDING PROTEIN-DEPENDENT TRANSPORT SYSTEM.

LSD See ERGOT ALKALOIDS.

LT See ETEC.

LTA LipoTEICHOIC ACID.

LTH See PASTEURIZATION.

LTLT See PASTEURIZATION.

LTR Long terminal repeat: see RETROVIRIDAE.

LuIII virus See PARVOVIRUS.

lubimin A sesquiterpenoid PHYTOALEXIN produced by the potato plant (*Solanum tuberosum*).

lucerne enation virus See RHABDOVIRIDAE.

lucerne transient streak virus (LTSV) See VELVET TOBACCO MOTTLE VIRUS and VIRUSOID.

Lucibacterium An obsolete bacterial genus; *L. harveyi* is now *Vibrio harveyi* (see VIBRIO).

luciferase See BIOLUMINESCENCE.

luciferin See BIOLUMINESCENCE.

lucigenin See CHEMILUMINESCENCE.

Lucio phenomenon See LEPROSY.

Lucké virus See HERPESVIRIDAE.

Ludwig's angina An inflammatory, purulent condition of tissues around the submaxillary gland due to infection by oral bacteria – usually streptococci.

luer fitting A type of joint between tubular parts of two items of apparatus: one part is tapered and fits coaxially into the other (correspondingly flared) part, forming a rigid, airtight joint.

Lugol's iodine Iodine (5% w/v) and potassium iodide (10% w/v) in distilled water. Diluted Lugol's iodine is used e.g. in the GRAM STAIN, and as a stain for starch granules, cysts of *Entamoeba histolytica*, etc.

lumber spoilage See TIMBER SPOILAGE.

lumicolchicine See COLCHICINE.

luminescence See CHEMILUMINESCENCE and BIOLUMINESCENCE.

luminol The compound 5-amino-2,3-dihydro-1,4-phthalazinedione which emits light when exposed to unstable oxygen species (e.g. $O_2^{\cdot-}$, H_2O_2, $OH\cdot$); it is used e.g. as a probe ('chemiluminogenic probe') in studies on CHEMILUMINESCENCE – luminol either acting as the sole light emitter or supplementing natural light emitter(s).

Lumi-Phos 530 A dioxetane substrate (Lumigen, Detroit, USA) which yields chemiluminescence when cleaved by alkaline phosphatase.

lumper See TAXONOMY.

lumpy jaw See ACTINOMYCOSIS (1).

lumpy skin disease Either of two infectious diseases of cattle characterized by the sudden development of flattened, firm, intradermal nodules (up to ca. 4 cm in diameter), regional or generalized lymphadenitis, oedema and dysgalactia; lesions may also occur on the nasal and buccal mucosae. The more severe of the two diseases (*knopvelsiekte*, found in various parts of Africa) is caused by the 'Neethling virus' – a *Capripoxvirus* [genome: JV (2001) *75* 7122–7130], while the milder disease (*pseudolumpyskin disease*) is caused by a member of the Alphaherpesvirinae apparently identical to bovine mammilitis virus; these diseases may be transmitted by insect vectors. (cf. ALLERTON DISEASE.)

lumpy wool ('mycotic dermatitis') A SHEEP DISEASE, caused by *Dermatophilus congolensis*, characterized by skin lesions and matting of the wool. Infection appears to occur via minor abrasions or other lesions – particularly when the skin is wet or damp. (cf. DERMATOPHILOSIS.)

lunar caustic See SILVER.

lunate Crescent-shaped.

Lundegårdh-type respiration A type of respiration in which electrons, derived from the substrate, are translocated across a membrane [Arkiv för Botanik (Stockholm) (1945) *32A* (No. 12) 1–139 (23–49)]. (cf. EXTRACYTOPLASMIC OXIDATION.)

lung oedema factor (*vet.*) See 4-IPOMEANOL.

lupinosis (*vet.*) A MYCOTOXICOSIS, involving liver damage with jaundice, caused by ingestion of lupin plants (*Lupinus* spp) infected with (or supporting the saprotrophic growth of) *Phomopsis leptostromiformis*. The disease occurs mainly in sheep; other animals (e.g. cattle) may be affected but are less commonly fed lupins. Lupinosis may follow consumption of living plants or lupin hay, and either sweet or bitter lupin cultivars may be involved; the condition is distinct from 'lupin poisoning' caused by alkaloids present in bitter cultivars. In lupins (especially *L. luteus*), *P. leptostromiformis* causes 'stem blight' ('stem spot'), a disease characterized by brownish, sunken, linear lesions on the stem, circular lesions on the pods, and brownish discolorations on the seeds. The fungus can grow saprotrophically on dead lupin plant material and on mature living plants.

Luria–Bertani broth A medium containing (per litre): 10 g tryptone, 5 g yeast extract and 10 g (or 5 g) of sodium chloride (NaCl); the pH is adjusted to 7.5 (or 7) with sodium hydroxide (NaOH). (The alternative values indicate variation in different sources of information.)

Luria–Delbrück fluctuation test See FLUCTUATION TEST.

luteic acid A water-soluble, acidic, extracellular, $(1 \rightarrow 6)$-β-linked glucan containing malonic acid residues; it is produced by '*Penicillium luteum*'.

lutein See CAROTENOIDS.

luteose A β-$(1 \rightarrow 6)$-linked glucan produced by '*Penicillium luteum*'.

luteoskyrin See YELLOW RICE.

luteoviruses (barley yellow dwarf virus group) A group of PLANT VIRUSES in which the virions are icosahedral (ca. 25–30 nm diam.), each containing a single molecule of linear, positive-sense ssRNA (MWt. ca. 2×10^6). Luteoviruses can infect various monocotyledonous and/or dicotyledonous plants, and many can cause important economic losses in crop plants; transmission occurs persistently (circulatively) via aphids, and weeds may provide important reservoirs of infection. Virus replication is confined to the phloem tissues of the host plant. Symptoms of infection typically include severe stunting with yellowing or reddening of the leaves (which may also become brittle); the severity of the symptoms depends on the virus, species and cultivar of host plant, and environmental conditions. Type member: barley yellow dwarf virus (BYDV), MAV isolate. Other members include e.g. beet mild yellowing virus, beet western yellows virus (BWYV), carrot red leaf virus, legume yellows virus, pea leaf roll virus, potato leaf roll virus, soybean dwarf virus, subterranean clover red leaf virus, tobacco necrotic dwarf virus, turnip yellows virus. Probable members: milk-vetch dwarf virus, subterranean clover stunt virus, and tomato yellow top virus. Possible members include e.g. banana bunchy top virus, beet yellow net virus, celery yellow spot virus, cotton anthocyanosis virus, millet red leaf virus, raspberry leaf curl virus, tobacco vein distorting virus, tobacco yellow net virus, tomato yellow net virus.

***lux* operon** See BIOLUMINESCENCE.

***luxS* gene** See QUORUM SENSING.

LV agar See EGG-YOLK AGAR.

lyases ENZYMES (EC class 4) which catalyse either the non-hydrolytic removal of a group from a substrate with the resulting (sometimes transient) formation of a double bond, or the reverse reaction; when the reverse reaction is the more significant (physiologically) the enzyme is often called a SYNTHASE. Subclasses are recognized on the basis of the nature of the bond split: e.g., C–C, C–O or C–N. Lyases include e.g. decarboxylases, aldolases and dehydratases. (See e.g. HYALURONATE LYASE and PECTIC ENZYMES.)

Lyb antigens In mice: antigens which occur on certain mature B lymphocytes. Lyb 5^+ cells can respond to polyclonal B cell activators (e.g. LPS) and also to soluble polysaccharide antigens (e.g. certain bacterial capsular antigens); Lyb 5^- cells can respond to polyclonal B cell activators but not to soluble polysaccharides.

Lycogala A genus of slime moulds (class MYXOMYCETES) in which the fruiting bodies are aethalia. *L. epidendrum* – a common species on rotting logs – forms clusters of hemispherical, puff-ball-like, pink to yellowish-brown aethalia, each ca. 0.3–1.5 cm diam.

lycomarasmin A TOXIN produced by certain strains of *Fusarium oxysporum* (e.g. some strains of *F. oxysporum* f. sp. *lycopersici*, the causal agent of FUSARIUM WILT in tomatoes). It chelates metal ions, and at high concentrations can cause wilting in plants; however, its role in pathogenesis – if any – is unknown.

lycopene See CAROTENOIDS.

Lycoperdales An order of fungi (class GASTEROMYCETES) in which the basidiocarp consists of highly convoluted, hymenium-bearing fertile tissue enclosed by a peridium composed of two or more layers; the organisms are characteristically humicolous or lignicolous saprotrophs.

The so-called 'earth stars' (e.g. *Geastrum*) form globose or onion-shaped basidiocarps ca. 2–10 cm in diameter. When the basidiocarp is mature, the *exoperidium* (the outer layer(s) of the peridium) ruptures and peels back to form a stellate fringe; in *Geastrum*, an apical pore develops in the remaining peridium (*endoperidium*) to permit spore release. (In *Myriostoma coliformis* many pores develop in the endoperidium.)

The (non-stipitate) 'puffballs' include species of *Bovista*, *Calvatia* and *Lycoperdon*. In *Calvatia* spore release depends on peridial rupture and disintegration; *C. gigantea* (the giant puffball) may reach ca. 1 metre across. In *Lycoperdon* the spores are released via a pore which develops in the peridium.

Lycoperdon See LYCOPERDALES.

lycoperdon nut A hypogean fruiting body of a member of the Elaphomycetales.

Lyell's syndrome *Syn.* SCALDED SKIN SYNDROME.

Lyme disease (Lyme borreliosis) A disease of humans (and e.g. canines [VR (1995) *136* 244–247]) caused by *Borrelia burgdorferi* (see BORRELIA) and transmitted by the bite of (mainly ixodid) ticks.

Symptoms (in humans): typically, an initial rash (*erythema migrans*, EM) spreading from the tick bite, followed (after days/weeks) by neurological/musculoskeletal/cardiac signs and, later, joint/neurological/skin symptoms; symptoms vary in both nature and timing.

Diagnosis: mainly serological (ELISA) (seroconversion after ~4 weeks); detection of spirochaetes e.g. by culture or PCR examination of blood.

Chemotherapy: e.g. ampicillin (during EM), ceftriaxone (for arthritis, cardiac infection). Failure of chemotherapy may be due to inaccessibility to antibiotics of intracellular spirochaetes [see e.g. AAC (1996) *40* 1552–1554].

[Review: RMM (1998) *9* 99–107; quantitative assay of *B. burgdorferi* in tissues: JCM (1999) *37* 1958–1963; treatment of early Lyme disease (review): Drugs (1999) *57* 157–173.]

lymphadenitis Inflammation of lymph node(s).

lymphadenopathy A pathological condition of lymph node(s).

lymphadenopathy-associated virus See AIDS.

lymphadenopathy syndrome See AIDS.

lymphangitis Inflammation of lymph vessel(s).

lymphatic leukosis viruses See AVIAN LEUKOSIS VIRUSES.

Lymphocryptovirus See GAMMAHERPESVIRINAE.

lymphocystis A FISH DISEASE affecting higher freshwater and marine teleosts. Small, benign, whitish, tumour-like nodules or clusters of nodules develop on skin and fins; the nodules contain greatly hypertrophied cells (to 1 mm diam.) rich in DNA. The fish seems to be little harmed. The causal agent is a virus of the genus *Lymphocystivirus* (family IRIDOVIRIDAE).

Lymphocystivirus A genus of viruses (family IRIDOVIRIDAE) which infect fish; the genus includes the causal agent of LYMPHOCYSTIS.

lymphocyte A type of LEUCOCYTE found e.g. in blood, lymph and lymph nodes; lymphocytes are essential for humoral immunity (see ANTIBODY FORMATION) and for CELL-MEDIATED IMMUNITY. (See B LYMPHOCYTE and T LYMPHOCYTE.) Lymphocytes are roughly spherical (ca. 8–15 μm in diameter) with non-granular cytoplasm and a large, non-lobed, rounded nucleus which occupies much or most of the cell. Morphologically, B and T lymphocytes are identical; they can be distinguished e.g. by their cell-surface antigens (e.g. using immunofluorescence techniques). B and T lymphocytes can be separated e.g. by passing a mixed suspension through a column of glass beads coated with antiglobulin; the B cells are retained. B cells are less numerous than T cells in the bloodstream. (See also NK CELLS and NULL CELLS.)

lymphocyte activating factor See INTERLEUKIN-1.

lymphocyte homing See CD44 and CD102.

lymphocytic choriomeningitis virus See LCM VIRUS.

lymphocytosis The condition in which increased numbers of lymphocytes are present in the blood.

lymphogranuloma venereum A VENEREAL DISEASE caused by strains of *Chlamydia trachomatis*; symptoms typically include a lesion on the genitals, followed by inflammation and swelling of lymph nodes in the groin. *Chemotherapy*: e.g. tetracyclines or sulphonamides. (See also FREI TEST.)

lymphoid leukosis A POULTRY DISEASE caused by an avian type C oncovirus (see AVIAN LEUKOSIS VIRUSES); B cell lymphomas originate in the bursa of Fabricius and metastasize to other organs, including the liver ('big liver disease'), spleen, kidney, etc. Transmission occurs chiefly via the egg, although horizontal transmission can occur. Only a small proportion of the birds infected actually develop the disease. (cf. MAREK'S DISEASE.)

lymphokines The former name for a wide range of biologically active glyco(proteins), produced and secreted by lymphocytes and certain other cells, which are now called CYTOKINES. Among the agents regarded as lymphokines were: (i) B cell growth factors (BCGFs), a term which may refer e.g. to the cytokine now known as interleukin-14; (ii) B cell differentiation factors (BCDFs), which may have included e.g. interleukin-6; (iii) INTERFERONS; (iv) chemotactic factors (see also CHEMOKINES and MIF); (v) various factors that regulate haemopoiesis. [Early account of B cell growth and differentiation factors: Imm. Rev. (1984) *78*.]

lymphoma Any tumour of lymphoid tissues, or one derived from lymphoid cells. Lymphomas may be classified e.g. on the basis of the predominant cell type involved and on the degree of differentiation of the neoplastic tissue. (See also BURKITT'S LYMPHOMA, BOVINE VIRAL LEUKOSIS, MAREK'S DISEASE.)

lymphomatosis A condition in which multiple lymphomas are formed.

lymphon An individual's entire immune system, including e.g. lymphoid tissues, LYMPHOCYTES, MACROPHAGES, and the COMPLEMENT system.

lymphopoiesis The development of LYMPHOCYTES.

lymphoproliferative virus group See GAMMAHERPESVIRINAE.

lymphosarcoma See SARCOMA.

lymphotactin (chemokine) See CHEMOKINES.

lymphotoxin α *Syn.* Tumour necrosis factor-β (TNF-β).

lymphotoxin β A membrane-anchored glycoprotein related to, but distinct from, TNF-α and TNF-β.

lymphotropic virus group See GAMMAHERPESVIRINAE.

lymphotropism See TROPISM (sense 2).

Lyngbya A genus of freshwater and marine cyanobacteria of the LPP GROUP. *L. majuscula* is responsible for causing dermatitis in swimmers (*swimmers' itch*) e.g. in waters off the coasts of Hawaii and Japan; this species can produce several toxins, including the brominated phenolic *aplysiatoxin*, various derivatives of aplysiatoxin (e.g. *debromo-*, *bromo* and *dibromoaplysiatoxin* [structures: PAC (1986) *58* 339–350 (345)]), the indole alkaloid *lyngbyatoxin A*, etc. (See also GEOSMIN and PHORMIDIUM.)

lyngbyatoxin See LYNGBYA.

lyophilization *Syn.* FREEZE-DRYING.

lysate The product of cell lysis.

lysergic acid diethylamide See ERGOT ALKALOIDS.

lysigenous (lysigenic) *Syn.* LYSOGENOUS.

lysin An antibody or other entity which, under appropriate conditions, can cause the lysis (rupture) of cells. See e.g. HAEMOLYSIN.

lysine biosynthesis See AMINOADIPIC ACID PATHWAY and DIAMINOPIMELIC ACID PATHWAY.

lysine decarboxylase test See DECARBOXYLASE TESTS.

lysis from without Lysis of a host cell when large numbers of virions adsorb to the host cell surface. (See e.g. T-EVEN PHAGES.)

lysis inhibition In certain bacteriophages (e.g. T-even phages, RNA phages): a delay in the lysis of an infected host cell on superinfection of the cell with another phage of the same strain. Mechanism: unknown. Lysis inhibition results in the formation of plaques which each have a turbid halo. (cf. RAPID LYSIS MUTANT.)

lysis protein (1) (*syn.* holin) A bacteriophage-encoded protein that forms a pore in the cytoplasmic membrane of an infected bacterium, thus enabling an ENDOLYSIN to gain access to the peptidoglycan. In bacteriophage λ, for example, a holin is encoded by gene *S*; this protein oligomerizes in the cytoplasmic membrane, forming lesions which apparently allow the *R* gene product to reach peptidoglycan [JB (1990) *172* 912–921]. The requirement for the *S* gene product in bacteriophage λ apparently reflects the absence of a signal sequence (see SIGNAL HYPOTHESIS) on the peptidoglycan-degrading *R* gene product, i.e. the latter, by itself, could not cross the cytoplasmic membrane and therefore could not effect lysis. In phage T4 the product of the *t* gene appears to have the same role. In BACTERIOPHAGE ϕ29 (which infects Gram-positive bacteria), an analogous product is encoded by gene *14* [JB (1993) *175* 1038–1042].

In a phage-infected cell, a holin must be active only after normal phage assembly has been completed; if lysis occurred too early then fewer (or no) mature phages would be released when the cell burst (i.e. the BURST SIZE would be affected). Such early lysis was reported with a strain of phage λ containing a mutant *S* gene [Mol. Microbiol. (1994) *13* 495–504].

In some holin-encoding genes there are two start codons (separated by one or two codons), i.e. the gene encodes two polypeptides of slightly different length. The shorter product is the holin, while the longer polypeptide seems to be an inhibitor of holin activity; this arrangement may be a mechanism for controlling the timing of cell lysis [Mol. Microbiol. (1996) *21* 675–682].

(2) The product of the 'lysis gene' in a colicin plasmid (see COLICINS).

(3) (*syn.* holin) A protein, product of the *tcdE* gene in *Clostridium difficile*, which appears to promote the release (across the cytoplasmic membrane) of toxin molecules (encoded by genes *tcdA* and *tcdB*) that lack a signal sequence [JMM (2001) *50* 613–619].

Lysobacter A genus of unicellular GLIDING BACTERIA; GC%: 65–68. [IJSB (1978) *28* 367–393.]

lysochrome Any non-ionic coloured substance which *dissolves in* and thus imparts colour to the lipids of cells or tissues. (See e.g. SUDAN BLACK B and NILE BLUE A; cf. DYE.)

lysogen A lysogenic cell, or a population of lysogenic cells: see LYSOGENY.

lysogenic (1) See LYSOGENY. (2) *Syn.* LYSOGENOUS.

lysogenic conversion See BACTERIOPHAGE CONVERSION.

lysogenic immunity See SUPERINFECTION IMMUNITY.

lysogenous (lysogenic; lysigenous; lysigenic) Refers to a structure, cavity or aperture (e.g. in a fungal fruiting body) which is formed as a result of the breakdown of cells.

lysogeny The phenomenon in which the genome of a (temperate) BACTERIOPHAGE establishes a stable, non-lytic presence in its living host without producing progeny virions; the host cell (which is said to be *lysogenic*) can continue to grow and divide, and replication of the phage genome (the *prophage*) is coordinated with that of the host chromosome such that, at each cell division, the prophage is inherited by both daughter cells. The prophage is maintained either by integration into the host chromosome (as in the majority of temperate phages: see e.g. BACTERIOPHAGE λ, BACTERIOPHAGE MU and BACTERIOPHAGE φ105) or as an extrachromosomal plasmid (see e.g. BACTERIOPHAGE P1 and BACTERIOPHAGE F116). (cf. PROVIRUS.) The host cell may or may not exhibit an altered phenotype (see BACTERIOPHAGE CONVERSION).

A prophage normally retains the potential to initiate a lytic cycle in which progeny virions are produced and released by cell lysis; maintenance of the lysogenic state necessitates the repression of lytic functions in the prophage by means of phage-directed synthesis of a repressor protein (see e.g. BACTERIOPHAGE λ). (See also SUPERINFECTION IMMUNITY.) Derepression – e.g. by inactivation of the repressor protein – results in *induction* of the lytic cycle; induction may occur spontaneously in a small proportion of cells in a lysogenic population. (See also ZYGOTIC INDUCTION.)

(See also PSEUDOLYSOGENY and TRANSDUCTION.)

Lysol See PHENOLS.

lysopeptide *Syn.* signal peptide: see SIGNAL HYPOTHESIS.

lysophagosome See PHAGOCYTOSIS.

lysopine See CROWN GALL.

lysosome In eukaryotic cells (only): a closed, intracellular membranous sac (often ca. 0.5 μm) containing a wide range of enzymes which, collectively, can degrade various classes of biological macromolecules; the enzymes typically have a pH optimum of ca. 5.0. (The acidity within a lysosome is maintained by a PROTON ATPASE in the lysosome membrane.) [The lysosome membrane: TIBS (1986) *11* 365–368.] Lysosomes develop from the distal end of the GOLGI APPARATUS; they are involved primarily in intracellular digestion: see AUTOLYSIN, FOOD VACUOLE, PHAGOCYTOSIS. A *primary* lysosome is one which has not yet coalesced with an autophagic or food vacuole or a phagosome; following such coalescence the whole is termed a *secondary* lysosome. When digestion is complete the entity remaining is termed a *residual body*.

Certain viruses (e.g. *Reovirus*) depend on lysosomal enzymes for uncoating.

Lysosomes may be damaged or disrupted e.g. by streptolysin O or phalloidin.

lysostaphin A 22-kDa BACTERIOCIN produced by *Staphylococcus simulans* (bv *staphylolyticus*) which is active against strains of *Staphylococcus aureus*. Lysostaphin is an endopeptidase which cleaves the glycyl–glycyl peptide bridges in staphylococcal PEPTIDOGLYCAN, causing lysis. Immunity to lysostaphin in the secreting organism is apparently due to modification of its own peptidoglycan by an enzyme encoded by the *lif* gene. Lysostaphin has been investigated for use in reducing numbers of staphylococci in the nose and throat of carriers and for the treatment of staphylococcal infections resistant to β-lactam antibiotics.

lysozyme (*N*-acetylmuramidase; endolysin; endo-β-*N*-acetylmuramide glycanhydrolase; EC 3.2.1.17) An enzyme which hydrolyses the *N*-acetylmuramyl-(1 → 4)-β-linkages in PEPTIDOGLYCAN; it can thus cause BACTERIOLYSIS in species whose peptidoglycan is accessible to enzyme action. (*Micrococcus luteus* – formerly *M. lysodeikticus* – is particularly susceptible and is used in lysozyme assays.) Lysozyme is widely distributed in nature. Animal lysozyme (MWt ca. 14500; IEP ca. 11) occurs e.g. in egg-white, milk, saliva, tears, mucosal secretions, macrophages etc (cf. NON-SPECIFIC IMMUNITY). Structurally distinct enzymes are produced by some bacteria and are coded for by certain bacteriophages (e.g. T2, T4, λ). An extracellular lysozyme-like enzyme (endo-β-*N*-acetylglucosaminidase, EC 3.4.99.17, which cleaves the glucosaminidyl-(1 → 4)-β-linkages in peptidoglycan) is produced by most coagulase-positive and some coagulase-negative staphylococci [Book ref. 44, pp. 746–748].

lyssa *Syn.* RABIES.

Lyssavirus (rabies virus group) A genus of viruses within the RHABDOVIRIDAE; type species: RABIES virus. The rabies virus is neurotropic; the amino acid sequence of the virion glycoprotein has been shown to resemble that of certain snake venom curaremimetic neurotoxins which bind to acetylcholine receptors on nerve cells [Science (1984) *226* 847–848]. The rabies virus can multiply in nearly all types of mammalian and avian cells in culture; it apparently replicates in the host cell cytoplasm, budding predominantly from intracytoplasmic membranes. Unlike VSV (see VESICULOVIRUS), it does not inhibit cellular macromolecule synthesis. (See also FIXED VIRUS; FLURY VIRUS; STREET VIRUS.)

The genus currently contains seven genotypes – all of which, except Lagos bat virus, are able to cause death in humans and/or animals in nature; the genotypes are:

1. Rabies virus
2. Lagos bat virus
3. Mokola virus
4. Duvenhage virus
5. European bat lyssavirus type 1

6. European bat lyssavirus type 2
7. Australian bat lyssavirus

The first four genotypes correspond to the four earlier serotypes. Lagos bat virus was isolated from a fruit bat. Duvenhage virus can cause a rabies-like disease in man. Mokola virus (isolated e.g. from shrews and cats) can cause CNS disease in man. Australian bat lyssavirus (reported in 1997) is infective for humans.

Kotonkan virus, isolated from *Culicoides* sp, may cause ephemeral febrile disease in animals and man. This virus, and the Obodhiang virus, isolated from *Mansonia uniformis*, have been associated with the genus *Lyssavirus* but have not been assigned to a genotype.

Phylogenetic analysis of the genus *Lyssavirus* has been able to distinguish two phylogroups: phylogroup I (rabies virus, European bat lyssavirus types 1 and 2, Duvenhage virus and the Australian bat lyssavirus), and phylogroup II (Lagos bat virus and Mokola virus); it is believed that this classification reflects more closely the biological characteristics of lyssaviruses than do the serotypes and genotypes [JV (2001) *75* 3268–3276].

***lysU* gene** See HEAT-SHOCK PROTEINS.

Lyt antigens Cell-surface antigens of murine T cells.

lytic bacteriophage *Syn.* VIRULENT BACTERIOPHAGE.

lytic cycle (*virol.*) The sequence of events which begins with adsorption of a virus to a susceptible cell and terminates with cell lysis and release of progeny virions.

lytic infection (*virol.*) The infection and subsequent lysis of susceptible cells by a VIRUS. (cf. e.g. LYSOGENY.)

Lyticum A genus of Gram-negative bacteria which occur as endosymbionts in the cytoplasm of certain strains of *Paramecium aurelia*. Cells: peritrichously flagellated rods, ca. 0.6–0.9 × 2.0–10 μm, which stain well with toluidine blue; in vitro cultivation has been reported. *Lyticum* spp do not contain R bodies (cf. CAEDIBACTER) but they produce labile toxin(s) which lyse sensitive paramecia. Type species: *L. flagellatum*.

L. flagellatum (earlier name: lambda particles) includes typically straight rods, while *L. sinuosum* (earlier name: sigma particles) includes curved and spiral rods.

[Book ref. 22, pp. 808–811.]

lyxose An aldopentose. (cf. STREPTOSE.)

1. Words in SMALL CAPITALS are cross-references to separate entries.
2. Keys to journal title abbreviations and Book ref. numbers are given at the end of the Dictionary.
3. The Greek alphabet is given in Appendix VI.
4. For further information see 'Notes for the User' at the front of the Dictionary.

M

m Maintenance coefficient: see YIELD COEFFICIENT.

M Methionine (see AMINO ACIDS).

M antigen (1) Streptococcal M PROTEIN. (2) A protein–LPS surface antigen in *Brucella* spp. (3) The COLANIC ACID capsular antigen in certain enterobacteria.

M-associated protein See MAP.

M bands (of *Stentor*) See MYONEME.

M cells See DYSENTERY.

M fibres (of *Stentor*) See MYONEME.

M fimbriae FIMBRIAE which occur on some pyelonephritic strains of *Escherichia coli*; they bind to the amino-terminal part of glycophorin A^M.

M phase (of cell cycle) See CELL CYCLE.

M phenotype In strains of *Streptococcus pyogenes* and *S. pneumoniae*: an efflux-based resistance to MACROLIDE ANTIBIOTICS (q.v.).

M plasmid See MICROCINS.

M protein (1) (M antigen) (of streptococci) A heat-stable, trypsin-sensitive protein antigen associated with virulence in certain (M^+) strains of group A streptococci. The M protein occurs as a fibrillar 'fuzz' covering the entire cell surface: the protein molecules apparently occur in extended form as double-stranded, α-helical coiled coils (ca. 50 nm long) anchored in the cell wall by their C-terminal ends [PNAS (1981) *78* 4689–4693]. The M protein protects the cell from phagocytosis, and may play a role in adhesion. M^+ strains often form matt or mucoid colonies on solid media, and may under certain cultural conditions produce an extracellular protease which degrades the M protein. M protein occurs in different antigenic forms, providing a basis for the serological typing of M^+ strains. (See also MAP; R PROTEIN; T PROTEIN.)

(2) (in e.g. *Escherichia coli*) The product of the *lacY* gene of the LAC OPERON.

M ring See FLAGELLUM (a).

M1 RNA See RNASE P.

mAb Monoclonal antibody.

MAC (1) (*immunol.*) Membrane attack complex: see COMPLEMENT FIXATION.

(2) (*bacteriol.*) (*syn.* MAIS complex) *Mycobacterium avium* complex: a category of slow-growing mycobacteria which include *M. avium*, *M. intracellulare* and *M. scrofulaceum*. These organisms are causal agents in three main types of infection: (i) cervical adenitis in children, (ii) pulmonary disease in adults, and (iii) disseminated disease in patients with advanced AIDS; treatment for cervical adenitis involves resection of affected node(s), while treatment for pulmonary and disseminated infections involves extended chemotherapy with a multiple drug regimen that includes clarithromycin or azithromycin. [Infections due to MAC and their treatment: BCID (1997) *4* 25–61. PCR-based detection of MAC bacteraemia in AIDS patients: JCM (1999) *37* 90–94. Cross-reactivity between *M. avium* (and *M. kansasii*) and *M. tuberculosis* in AMTDT: JCM (1999) *37* 175–178.]

Mac-1 See INTEGRINS.

MacConkey's agar An agar-based medium used e.g. for the primary isolation of *Salmonella* and *Shigella* from faeces, for the differentiation of lactose-fermenting enterobacteria from NLFs, etc. The medium contains (w/v): peptone (2%); lactose (1%); sodium taurocholate or glycocholate (0.5%); NaCl (0.5%); NEUTRAL RED (0.003%); agar (ca. 1.5%). Final pH: ca. 7.4. Colonies of lactose fermenters (e.g. *Escherichia coli*) are pink or red, those of NLFs (e.g. most strains of *Proteus*, *Salmonella* and *Shigella*) are colourless. Most non-enteric bacteria are inhibited by the bile salts. Some commercial preparations also contain crystal violet (1 mg/l) to inhibit Gram-positive bacteria.

MacConkey's broth A broth with the same formulation as MACCONKEY'S AGAR but without the agar. Bromcresol purple may be used instead of neutral red as pH indicator.

Machupo virus See ARENAVIRIDAE and BOLIVIAN HAEMORRHAGIC FEVER.

macrocapsule See CAPSULE.

macroconidium The larger of two types of CONIDIUM formed by certain fungi (see e.g. FUSARIUM).

macroconjugant See CONJUGATION (1) (c).

macrocyclic rusts (eu-form rusts) (1) Those rust fungi (e.g. *Puccinia graminis*) which form five clearly distinct types of spore: pycniospores, aeciospores, uredospores, teliospores and basidiospores (see UREDINIOMYCETES).

(2) Those rust fungi which form more than one type of *binucleate* spore – i.e., aeciospores and/or uredospores in addition to teliospores. (cf. DEMICYCLIC RUSTS.)

macrocyst A type of resting stage formed under certain conditions by (at least) some dictyostelid slime moulds (e.g. strains of *Dictyostelium discoideum*, *D. mucoroides*, *Polysphondylium violaceum*). It is formed when myxamoebae aggregate, and the aggregate becomes enclosed in a fibrillar sheath; one of the cells (the 'phagocytic' or 'cytophagic' cell) enlarges and begins to engulf neighbouring amoebae. When all the peripheral amoebae have been ingested, the remaining giant cell develops a cellulose wall, its single large nucleus divides to form numerous smaller nuclei, and the resulting multinucleate macrocyst enters a period of dormancy. On germination, the macrocyst protoplast divides into large, uninucleate 'proamoebae' which in turn divide to form normal-sized vegetative myxamoebae; these are released when the macrocyst wall breaks down.

Macrocysts apparently represent a sexual stage in the dictyostelid life cycle. The large nucleus of the cytophagic cell is believed to be that of a zygote; meiosis is presumed to occur when this nucleus divides to form the multinucleate cyst.

[Review: Book ref. 144, pp. 178–221.]

Macrocystis See PHAEOPHYTA.

macrodilution broth test See DILUTION TEST.

macrofibre In a culture of *Bacillus subtilis* grown under certain conditions: a bundle of aligned chains of cells which can exhibit a right-handed or left-handed helical conformation; the cell wall in the cells of a helically twisted macrofibre is believed to be under torsional stress. The helical conformation appears to depend on the presence of intact glycan backbone chains (rather than intact cross-links) in the cell wall PEPTIDOGLYCAN [JGM (1986) *132* 2377–2385].

macrogamete See ANISOGAMY.

macroglobulin (1) *Syn.* IgM (q.v.). (2) α_2-Macroglobulin is a fibrinolysin inhibitor secreted by macrophages.

macrolide antibiotics A large group of MLS ANTIBIOTICS which includes e.g. angolamycin, carbomycin, chalcomycin, cirramycin, erythromycin, lankamycin, leucomycin (= turimycin), megalomycin, methymycin, narbomycin, niddamycin, oleandomycin, relomycin, roxithromycin, spiramycins, troleandomycin and tylosin; they are produced e.g. by species of *Streptomyces*.

The molecule of a macrolide antibiotic contains a large lactone ring (12–14 carbon atoms) substituted with various functional groups and linked glycosidically to one or more sugars (including an aminosugar and, frequently, an unusual deoxysugar); thus, e.g. *erythromycin* (see also POLYKETIDE) consists of a 14-member ring linked to the aminosugar desosamine and the deoxysugar cladinose. (Macrolides which contain a conjugated polyene system form a separate group: see POLYENE ANTIBIOTICS (a).)

The macrolides are typically bacteriostatic and are active mainly against Gram-positive bacteria (including *Streptococcus* spp and *Clostridium* spp), although certain Gram-negative bacteria (e.g. *Haemophilus* spp, *Neisseria* spp, some species of *Bacteroides*) may be susceptible.

Macrolides inhibit PROTEIN SYNTHESIS by binding to the 50S subunit of 70S ribosomes, apparently inhibiting the peptidyltransferase reaction and/or affecting translocation. The macrolide AZITHROMYCIN inhibits the ability of *Pseudomonas aeruginosa* strain PAO1 to produce virulence factors by inhibiting the pathogen's QUORUM SENSING circuitry [AAC (2001) *45* 1930–1933].

In e.g. *Escherichia coli*, resistance to macrolides may be due to a mutation which results in the modification of a protein (L4, or possibly L22) in the 50S ribosomal subunit.

In staphylococci, streptococci and streptomycetes resistance to macrolides (and other MLS antibiotics) may be due to methylation of the ribosomal 23S rRNA (see STREPTOGRAMINS).

In *Streptococcus pyogenes* and *S. pneumoniae*, strains exhibiting the 'M phenotype' contain an efflux-based mechanism which confers resistance specifically to macrolides – i.e. the mechanism does not affect susceptibility to lincosamides or B streptogramins. The efflux mechanism is encoded by a sequence (*mefA*) within a 7244-bp chromosomal element designated Tn*1207.1* [AAC (2000) *44* 2585–2587].

macronucleus (*ciliate protozool.*) The larger of the two types of nucleus characteristically present in ciliates. (cf. MICRONUCLEUS.) Many ciliates contain a single macronucleus, but some contain more than one; according to species, the macronucleus may be e.g. ovoid, elongated, moniliform, C-shaped etc. Macronuclei contain nucleoli and are typically polyploid (they are typically diploid in the karyorelictid gymnostomes – e.g. *Kentrophoros*, *Loxodes*); macronuclear DNA, which is actively transcribed, directs the 'somatic' activities of the cell. During CONJUGATION the macronucleus usually disintegrates – a new one subsequently developing from a micronucleus. In *Oxytricha nova* macronuclear DNA is in the form of small, gene-sized molecules; during its development (from micronuclear DNA) the DNA undergoes a complex series of changes which include polytenization (with the loss of certain sequences), fragmentation, modification (the addition of inverted terminal repeats), and encapsulation within vesicles followed by release and replication [PNAS (1982) *79* 3255–3259].

A *heteromerous macronucleus* is one which is subdivided into two parts (*karyomeres*) – the *orthomere* and *paramere* – which differ in their content of DNA and RNA; such nuclei occur in many hypostomes. (See also ENDOSOME (2).) A *homoiomerous* (= 'homomerous') macronucleus is one which is not subdivided; such nuclei occur in the majority of ciliates.

macrophage A large (10–20 μm), relatively long-lived, amoeboid, phagocytic and pinocytotic LEUCOCYTE present in blood, lymph and other tissues; macrophages develop via the myeloid pathway from haemopoietic stem cells in the bone marrow, early stages including the monoblast and MONOCYTE. [Macrophages: JCS (1986) supplement 4, 267–286.] Macrophages have important roles in host resistance to pathogenic microorganisms (see PHAGOCYTOSIS, INFLAMMATION, ANTIBODY FORMATION and DELAYED HYPERSENSITIVITY). (See also GIANT CELLS, HISTIOCYTE, KUPFFER CELLS and PMN.)

Macrophages can express a wide range of cell-surface receptors, some of which are expressed constitutively while others can be induced by appropriate stimuli. They include: (i) receptors for the FC PORTION of IgG antibodies; (ii) receptors for various components of COMPLEMENT (e.g. CD46 and CD11c/CD18); (iii) binding sites for lipopolysaccharides (see CD14); (iv) binding sites for type I FIMBRIAE (see CD48); and (v) receptors for various CYTOKINES.

Macrophages also display MHC class II molecules and function as ANTIGEN-PRESENTING CELLS.

Under appropriate conditions (and with suitable stimulation), macrophages secrete a wide range of products, including: (i) certain components of COMPLEMENT (e.g. FACTOR B, PROPERDIN, C1–C5); (ii) FIBRONECTIN; (iii) certain PROSTAGLANDINS; and (iv) various CYTOKINES (e.g. INTERLEUKIN-1, INTERLEUKIN-6, INTERLEUKIN-8, INTERLEUKIN-12, INTERLEUKIN-18, TNF).

Activation of macrophages. On activation, a macrophage e.g. (i) becomes more aggressive towards pathogens which have been engulfed by phagocytosis, and (ii) secretes various cytokines. Thus, an activated macrophage is associated with a strong *respiratory burst* (see NEUTROPHIL) in which it produces significant amounts of e.g. SUPEROXIDE, hydrogen peroxide and NITRIC OXIDE; in the activated state a macrophage is able to kill certain ingested pathogens which may not be killed in non-activated macrophages.

Activation may occur e.g. when a macrophage presents antigen to an antigen-specific T LYMPHOCYTE of the Th1 subset (this is an MHC class II-restricted interaction – see HISTOCOMPATIBILITY RESTRICTION). Following such interaction with the macrophage, the T cell secretes an important macrophage-activating factor: interferon-gamma (IFN-γ) (see INTERFERONS). IFN-γ causes the macrophage to mount a respiratory burst (see above) and to secrete cytokines such as IL-1, IL-6, IL-12, IL-18 (see individual entries) and TNF-α (see TNF). Apparently, IFN-γ and lipopolysaccharides (LPS) can act synergistically to enhance the respiratory burst [JBC (1990) *265* 20241–20246] and to stimulate the synthesis of cytokines by macrophages. (The ability of IFN-γ and LPS to stimulate cytokine production in macrophages may be inhibited by glucocorticoids.)

Cytokines secreted by a T cell may provide BYSTANDER HELP (q.v.).

Following uptake, some pathogenic bacteria can kill macrophages [see e.g. EMBO (1996) *15* 3853–3860; TIM (1998) *6* 253–255]; *Yersinia* down-regulates production of the pro-inflammatory cytokine TNF-α in macrophages [Mol. Microbiol. (1998) *27* 953–965]. [Interactions between *Mycobacterium tuberculosis* and phagocytes: TIM (1998) *6* 328–335.]

Macrophages may be maintained for some time in culture. Viable macrophages adhere to glass and plastics and accumulate dyes such as TRYPAN BLUE.

macrophage-activating factor See MAF.

macropinocytosis See DYSENTERY.

macroscopic Visible to the unaided eye.

macrostome See TETRAHYMENA.

macrotetralides A group of ANTIBIOTICS which function as IONOPHORES; they inhibit Gram-positive bacteria. The *cyclic* macrotetralides (called 'actins' or 'nactins') each contain four similar tetrahydrofuran rings linked via ester bonds; *nonactin* is substituted with four methyl groups, *monactin* with three methyl and one ethyl, *dinactin* with two methyl and two ethyl, and *trinactin* with one methyl and three ethyl groups. *Open-chain* macrotetralides (polyether antibiotics) include alborixin, dianemycin, LASALOCID, lonomycin, MONENSIN, NIGERICIN and tetronomycin. [Structure and biosynthesis of polyether antibiotics: J. Nat. Prod. (1986) *49* 35–47.]

macule (macula) On e.g. skin: an unelevated spot of discoloration. (cf. PAPULE.)

maculopapular Refers to a rash in which both MACULES and PAPULES occur.

mad cow disease (BSE) See BOVINE SPONGIFORM ENCEPHALOPATHY.

mad itch *Syn.* AUJESZKY'S DISEASE.

Madin–Darby canine kidney cells See MDCK CELLS.

Madura foot *Syn.* MADUROMYCOSIS.

Madurella See HYPHOMYCETES and MADUROMYCOSIS.

maduromycetes Certain sporoactinomycetes with have a type III cell wall (see ACTINOMYCETALES) and which typically contain madurose; the group includes *Actinomadura*, *Excellospora*, *Microbispora*, *Microtetraspora*, *Planobispora*, *Planomonospora*, *Spirillospora* and *Streptosporangium*, but excludes the (morphologically distinguishable) genera *Dermatophilus* and *Frankia* (both wall type III, and both containing madurose) [Book ref. 73, pp. 105–115].

maduromycosis (Madura foot) A non-infectious human disease characterized by the formation – usually on feet or hands – of granulomatous, suppurative tumours (mycetomas) of the subcutaneous tissues; eventually the bone may be affected. The causal agent may be any of various fungi (e.g. *Acremonium* spp, *Aspergillus nidulans*, *Madurella* spp, *Pseudallescheria boydii*) or actinomycetes (e.g. *Actinomadura* spp, *Nocardia* spp, *Streptomyces somaliensis*) – the tumours being termed, respectively, eumycetomas (= eumycotic mycetomas) and actinomycetomas (= actinomycotic mycetomas). Infection occurs via wounds. Within the tissues, the pathogen forms characteristic aggregates (called granules or grains); depending on pathogen, the granules may be white, red, yellow, brown or black, soft or hard, and range in size from <15 µm (*Nocardia* spp) to >4 mm (e.g. *Actinomadura madurae*). (cf. ACTINOMYCOSIS.) Treatment of eumycetomas is usually surgical; some actinomycetomas can be controlled with antibiotics (e.g. sulphamethoxazole-trimethoprim + streptomycin).

madurose A sugar (3-*O*-methyl-D-galactose) which occurs e.g. in MADUROMYCETES.

MAECT Miniature anion exchange centrifugation technique: a method for detecting the presence of e.g. trypanosomes in the blood of patients suspected of suffering from SLEEPING SICKNESS. Blood, diluted with buffer, is passed through a DEAE (ion-exchange) column; erythrocytes are retained in the column, but any trypanosomes which may be present pass through the column and can be concentrated by centrifugation and examined by dark-field MICROSCOPY.

maedi (maedi-visna; ovine progressive interstitial pneumonia; Graff–Reinet disease) A chronic, progressive, fatal pneumonia which occurs mainly in sheep (also in goats); the causal agent, a virus of the LENTIVIRINAE, is apparently identical to the VISNA virus, but maedi and visna rarely occur together in one animal. Incubation period: at least two years. Symptoms: progressive emaciation and dyspnoea, with or without coughing and nasal discharge, typically lasting up to six months. (cf. JAAGSIEKTE.) [Book ref. 33, pp. 805–807.]

maedi-visna See MAEDI.

MAF Macrophage-activating factor: a former name for the CYTOKINE interferon-γ (IFN-γ).

magenta *Syn.* basic FUCHSIN.

Magnaporthe See POLYSTIGMATALES and AVIRULENCE GENE.

magnetic separation See DYNABEADS.

magnetosome See MAGNETOTACTIC BACTERIA.

magnetotactic bacteria A non-taxonomic grouping of aquatic, microaerophilic, flagellate, Gram-negative bacteria, each of which forms intracellular, enveloped magnetic bodies (*magnetosomes*); the presence of magnetosomes causes the cells to align with lines of magnetic force so that cell motility is given direction (but not assistance) by a magnetic field (*magnetotaxis*).

These bacteria are widespread geographically, occurring in sediments in freshwater and marine habitats. *Aquaspirillum magnetotacticum* has been obtained in pure culture; each cell contains a chain of ca. 20 cuboidal or polyhedral magnetosomes (of side ca. 40 nm) which contain magnetite (Fe_3O_4).

[Precipitation of magnetite in magnetotactic bacteria: PTRSLB (1984) *304* 529–536. Magnetosomes (review): JGM (1993) *139* 1663–1670.]

magnetotaxis See MAGNETOTACTIC BACTERIA.

magnification The extent to which the image of an object is larger than the object itself. With a compound (light) microscope the total magnification is the magnification of the objective lens multiplied by that of the eyepiece lens; it is determined by the focal lengths of the objective and eyepiece lenses. RESOLVING POWER involves other factors and is distinct from magnification. Different objective lenses having the same magnification may have different resolving powers; resolving power (but not magnification) may be affected by sub-optimal illumination of the specimen. Any increase in magnification which is not accompanied by an increase in resolution is said to be *empty magnification*. The maximum *useful* magnification obtainable with a given objective lens is approximately 1000 times its numerical aperture. Good-quality light microscopes can give a maximum magnification of about 1200×. Higher magnifications are obtainable by ultraviolet microscopy (see MICROSCOPY (f)). In ELECTRON MICROSCOPY magnification can be e.g. 1000000×.

Maillard reaction A reaction between a reducing sugar (e.g. lactose) and the free amino groups in proteins, the product being a brown-coloured compound; the extent of the reaction depends on the concentrations of the sugar and protein and on temperature. (See also MILK (treatment processes).)

maintenance coefficient (*m*) See YIELD COEFFICIENT.

maintenance energy The energy used by a cell for purposes other than growth; maintenance energy is used e.g. for motility, for the maintenance of osmotic pressure in the cell, for maintaining the energized state of the cytoplasmic membrane etc. At high rates of growth the maintenance energy requirement is generally relatively minimal, but at low rates of growth maintenance energy may account for an appreciable proportion of the overall energy requirement. (See also YIELD COEFFICIENT.)

maintenance medium See MEDIUM and TISSUE CULTURE.

MAIS complex *Mycobacterium avium–intracellulare–scrofulaceum* complex: a non-taxonomic grouping of species which is convenient for laboratory diagnostic purposes. (See MAC.)

maize chlorotic dwarf virus (MCDV; Ohio corn stunt agent) A PLANT VIRUS in which the virions are polyhedral (ca. 30 nm diam.) and contain one molecule of positive-sense ssRNA (MWt ca. 3.2×10^6). Host range is limited to members of the Gramineae; symptoms (in maize) include stunting and chlorotic striping of tertiary leaf veins (cf. CORN STUNT DISEASE). Transmission occurs via leaf-hoppers (e.g. *Graminella nigrifrons*) and is semipersistent. The virus overwinters in 'Johnson grass' (*Sorghum halepense*) and other weed grasses. (cf. RICE TUNGRO VIRUS.)

maize diseases See CEREAL DISEASES.

maize dwarf mosaic virus See POTYVIRUSES.

maize mosaic virus See RHABDOVIRIDAE.

maize rayado fino virus group A group of isometric ssRNA-containing PLANT VIRUSES which are obligately transmitted by leafhoppers.

maize rough dwarf virus See FIJIVIRUS.

maize streak virus (MSV) The type member of the GEMINIVIRUSES. MSV is an important cause of disease in maize (*Zea mays*) in many parts of Africa; leaves of infected plants show e.g. broken, narrow streaks of chlorosis along the veins. MSV is transmitted by leafhoppers of the genus *Cicadulina* [epidemiology: ARE (1978) *23* 259–282].

maize stripe virus See RICE STRIPE VIRUS GROUP.

maize wallaby ear disease (MWED) A disease of maize (*Zea mays*) first described in Australia; symptoms include small swellings on secondary veins on the lower surfaces of leaves, stunting etc. MWED is associated with infestation by leafhoppers (*Cicadulina*, *Nesoclutha*); the causal agent was thought to be a virus (cf. LEAFHOPPER A VIRUS), but there is evidence to suggest that the disease is due to an insect toxin [Ann. Appl. Biol. (1983) *103* 185–189].

major facilitator superfamily (MFS) See MFS.

major groove (in DNA) See DNA.

major histocompatibility complex (MHC) In all mammalian species studied: linked genetic loci that dominate the immune response; these genes, and their products, are designated by the prefixes HLA (*human leukocyte antigens*) in man, H-2 in mice, RT1 in rats, ChLA in chimpanzees etc.; the human (HLA) genes occur on chromosome 6.

MHC gene products occur on cell surfaces and serve as markers which e.g. distinguish 'self' (or syngeneic) tissue from 'non-self' (foreign) tissue; thus, MHC gene products determine the outcome of tissue transplants (see also MLR), and they are also involved e.g. in ANTIBODY FORMATION.

Genes in the MHC complex occur in three main groups (classes I, II and III). Most of the genes belong to classes I and II, and the products of these genes are designated *class I* and *class II* antigens, respectively. (Class III genes encode the COMPLEMENT components C2, C4 and factor B.)

Class I and class II antigens are distinguishable on a structural and functional basis.

Class I antigens (encoded by HLA-A, HLA-B and HLA-C genes) each consist of a single transmembrane polypeptide. These antigens are expressed on most nucleated cells in the body, and they are involved e.g. in the recognition of target cells by cytotoxic $CD8^+$ T cells.

Class II antigens (encoded by the HLA-DP, HLA-DQ and HLA-DR genes) each consist of a two-polypeptide transmembrane structure. These antigens have a limited distribution, being found e.g. on B cells and other ANTIGEN-PRESENTING CELLS; they are involved e.g. in the recognition of APCs by $CD4^+$ T cells.

For some cells the expression of MHC antigens may be induced, or upregulated, by certain CYTOKINES. For example (at least in vitro), IFN-γ (see INTERFERONS) can induce expression/upregulation of class II antigens on e.g. endothelial cells and macrophages (and can also stimulate expression of class I antigens).

mal de caderas In South America: (1) *Syn.* SURRA. (2) A fatal paralytic disease of cattle caused by the rabies virus and transmitted by blood-sucking vampire bats.

mal del pinto *Syn.* PINTA.

mal du coit *Syn.* DOURINE.

malachite green A green basic TRIPHENYLMETHANE DYE used e.g. as an ENDOSPORE STAIN.

malaria An acute or chronic, sometimes recurrent, infectious disease caused by species of PLASMODIUM and responsible, each year, for some 1–2 million deaths (including many children); the disease is prevalent in tropical and subtropical regions. Malaria may be caused by *Plasmodium falciparum*, *P. malariae*, *P. ovale* or *P. vivax*. (Mixed plasmodial infections can also occur; moreover, *P. knowlesi* and *P. simium*, normally parasites of non-human primates, can infect man under natural conditions, causing 'simian malaria'.)

Natural transmission of the malaria parasite – from an infected to an uninfected individual – occurs via an insect vector: a female mosquito of the genus *Anopheles*. [Systematics of malaria vectors (in particular, the *Anopheles punctulatus* group): IJP (2000) *30* 1–17.] If, during a blood meal, the mosquito ingests viable plasmodial gametocytes, male and female gametes can develop and fertilization can take place in the insect's gut; eventually, *sporozoites* – forms of the parasite infective for a human host – appear in the salivary glands of the mosquito, and these can be passed to an uninfected individual during the insect's next blood meal (see figure). [The malaria parasite in the mosquito: PT (2000) *16* 196–201.]

Transmission may also occur e.g. by transfusion of infected blood [NEJM (2001) *344* 1973–1978].

Incubation period: ca. 7–28 days, or longer. The typical illness is characterized by a series of paroxysms, each involving an attack of rigors followed by high fever and then profuse sweating. These paroxysms occur every ~36–48 hours in falciparum malaria ('malignant tertian malaria'), ~42–48 hours in vivax malaria ('benign tertian malaria'), ~50 hours in ovale malaria ('tertian malaria'), and ~72 hours in malariae malaria ('quartan malaria'). Other symptoms include e.g. headache, malaise, nausea and vomiting, anaemia, and pains in the muscles, joints and abdomen; enlargement of the liver and/or spleen may occur.

Falciparum malaria exhibits the most severe symptoms, and causes the majority of deaths from malaria. In certain blood vessels (the post-capillary venules), parasitized erythrocytes adhere to endothelial cells – although the significance of this in pathogenesis is not fully understood. Such adherence may lead to mechanical obstruction of the vessels; it may, perhaps additionally, activate or damage endothelial cells and/or e.g. stimulate the release of cytokines. Some authors indicate that disease manifestations result almost entirely from cytoadherence (adherence of red blood cells to the endothelium) together with the induction of certain cytokines (TNF-α, IL-1 and IL-6) [BCID (1995) *2* 227–247]. In some cases, severe disease has been correlated with a variant form of the TNF-α gene promoter [Nature (1994) *371* 508–510]. Another factor in falciparum malaria is 'rosetting': the phenomenon in which uninfected erythrocytes bind to infected red cells; again, the significance of this in pathogenesis is not clear, although some studies suggest an association between rosetting and cerebral malaria. Cytoadherence, rosetting and impaired erythrocyte deformability have been discussed in the context of abnormal microcirculatory blood flow [PT (2000) *16* 228–232]. Complications that may be associated with severe malaria include anaemia, DISSEMINATED INTRAVASCULAR COAGULATION, hypoglycaemia and renal failure. (See also BLACKWATER FEVER.) [Pathogenesis of severe falciparum malaria: BCID (1995) *2* 249–270.]

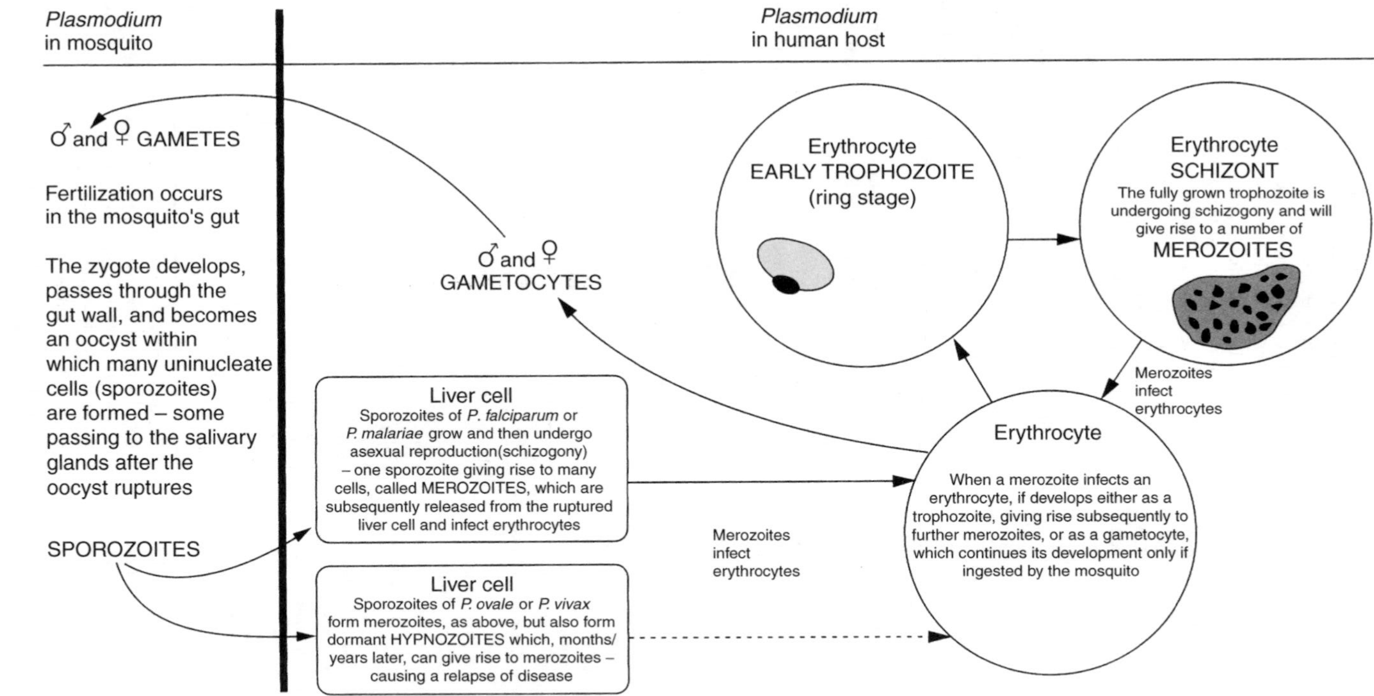
Plasmodium
in mosquito
Plasmodium
in human host
♂ and ♀ GAMETES
Fertilization occurs
in the mosquito's gut
The zygote develops,
passes through the
gut wall, and becomes
an oocyst within
which many uninucleate
cells (sporozoites)
are formed – some
passing to the salivary
glands after the
oocyst ruptures
SPOROZOITES
♂ and ♀
GAMETOCYTES
Liver cell
Sporozoites of P. falciparum or
P. malariae grow and then undergo
asexual reproduction(schizogony)
– one sporozoite giving rise to many
cells, called MEROZOITES, which are
subsequently released from the ruptured
liver cell and infect erythrocytes
Liver cell
Sporozoites of P. ovale or P. vivax
form merozoites, as above, but also form
dormant HYPNOZOITES which, months/
years later, can give rise to merozoites –
causing a relapse of disease
Merozoites
infect
erythrocytes
Erythrocyte
EARLY TROPHOZOITE
(ring stage)
Erythrocyte
SCHIZONT
The fully grown trophozoite is
undergoing schizogony and will
give rise to a number of
MEROZOITES
Merozoites
infect
erythrocytes
Erythrocyte
When a merozoite infects an
erythrocyte, if develops either as a
trophozoite, giving rise subsequently to
further merozoites, or as a gametocyte,
which continues its development only if
ingested by the mosquito

MALARIA. The life cycle of the malaria parasite (*Plasmodium*) (diagrammatic) showing the stages which are targets for antimalarial drugs (see table).

Sporozoites, injected into the bloodstream via mosquito bite, penetrate liver cells. A sporozoite grows and develops into a *tissue schizont* which undergoes asexual reproduction (*schizogony*) to produce a number of cells called *merozoites*. On rupture of the liver cell, merozoites enter the bloodstream and infect erythrocytes (red blood cells). Within an erythrocyte, a merozoite develops as a *trophozoite* (growing stage); the young trophozoite is known as the *ring stage* owing to its appearance in stained preparations under the microscope.

The mature trophozoite (*blood schizont*) undergoes schizogony, and many merozoites are released [in a two-step process: PNAS (2001) *98* 271–276] when the erythrocyte ruptures. These merozoites infect fresh erythrocytes. Some merozoites become *gametocytes* (i.e. cells with the potential to develop into male and female gametes). If ingested by a mosquito, gametocytes give rise to gametes – and the sexual phase of the life cycle follows.

In two species of *Plasmodium* the sporozoite can form a dormant stage (*hypnozoite*) in liver cells; hypnozoites may remain dormant for months/years. For this form of malaria a 'radical cure' involves treatment with an agent that kills hypnozoites (e.g. primaquine) as well as other drug(s) to deal with the intra-erythrocytic parasites.

Reproduced from *Antimicrobial Drug Action*, Figure 7.1, pages 120–121, R. A. D. Williams, P. A. Lambert & P. Singleton (1996) copyright Bios Scientific Publishers Ltd, UK (ISBN 1-872748-81-3) with permission from the publisher.

Natural resistance to falciparum malaria occurs in people heterozygous for the sickle-cell anaemia gene. Resistance to falciparum malaria may also be associated with an absence of the Sl[a] antigen, or low levels of CR1 (CD35), on erythrocytes, as Sl[a] is involved as a binding site in the rosetting phenomenon [Nature (1997) *388* 292–295].

Vivax, ovale and malariae malarias may be self-limiting, even without treatment; however, in subsequent years, species which form *hypnozoites* can give rise to relapses – at increasing intervals of time – as merozoites periodically emerge from the liver.

The receptor for *P. vivax* merozoites on the red blood cell is the Duffy glycoprotein (a marker in the Duffy blood group system), and this glycoprotein is essential for infection [see e.g. Blood (1997) *89* 3077–3091]. Many black Africans lack this receptor [a mutation inhibits expression of the Duffy gene: Nature Genetics (1995) *10* 224–228] and are accordingly resistant to vivax malaria.

The immune response to plasmodial infection. Interactions between plasmodia and the immune system are complex. Immunity develops slowly and tends to be transient; the majority of antigens do not induce immunological memory. Nevertheless, there is an antibody response to sporozoites, asexual blood-stage parasites and gametocytes, and a cell-mediated response which involves e.g. cytotoxic T cells and NK (natural killer) cells. [Clinical immunology of malaria: BCID (1995) *2* 227–247.]

Vaccines against malaria. Recognition of the need to stimulate immunity by using an 'antigenic cocktail' lead to the development of the Colombian synthetic subunit vaccine (SPf66). The future development of vaccines is aimed at improving the protective immune response. [Strategies towards the design of a synthetic subunit malaria vaccine: BCID (1995) *2* 389–403.]

The plasmodial antigen PfEMP1 (*Plasmodium falciparum* erythrocyte membrane protein 1) is displayed at the surface of those red blood cells (RBCs; erythrocytes) which are parasitized by *P. falciparum*. PfEMP1 promotes cytoadherence, i.e. the adherence of parasitized RBCs to sites such as the CD36 receptor on endothelial cells; PfEMP1 is therefore a potential target for candidate antimalarial vaccines. In the context of vaccines, however, one problem is that PfEMP1 occurs in variant forms in different strains of the pathogen, and an immune response to a given variant is predominantly variant-specific; nevertheless, studies using monoclonal antibodies have detected cross-reacting epitopes on a number of strains of *P. falciparum*, suggesting the possibility of developing an effective vaccine directed against PfEMP1 [PNAS (2001) *98* 2664–2669].

The case has been argued for developing a vaccine against the liver (pre-erythrocytic) stage of the pathogen, thus attacking the disease in the initial stage of infection; a candidate antigen for such a vaccine is liver-stage-antigen-1 (LSA-1), a 200 kDa protein that is expressed specifically by the parasite within liver cells and that appears to have a generally conserved sequence among different isolates. [Pre-erythrocytic immunity: the case for an LSA-1 vaccine: TIP (2001) *17* 219–223.]

Lab. diagnosis. (i) Detection of parasitaemia by staining blood films (Giemsa). (ii) Detection of *P. falciparum* protein-2 (HRP-2) in haemolysed blood by means of anti-HRP-2 MONOCLONAL ANTIBODIES [PT (1994) *10* 494–495]. This principle is used e.g. in the *Para*Sight-F™ test (Becton Dickinson, Maryland, USA). (iii) PCR-based testing with species-specific primers [TRSTMH (1993) *87* 649–653]. (iv) A genus-specific and (multiplex) species-specific approach by PCR [JCM (1999) *37* 1269–1273].

[Rapid diagnostic techniques: TIP (2001) *17* 307–309.]

Chemotherapy. Most antimalarials kill intraerythrocytic stages of the parasite. The choice of drug(s) depends e.g. on the local prevalence of parasites resistant to given antimalarials. (See QUININE for the formulae of some antimalarials.)

Fansidar (= PYRIMETHAMINE + sulphadoxine) is an oral treatment for uncomplicated falciparum malaria.

Quinoline-type antimalarials include QUININE, quinidine (used e.g. in the USA), chloroquine and mefloquine (see AMINOQUINOLINES). *Parenteral* quinine is considered to be the drug of choice for severe falciparum malaria because few strains of the parasite are resistant to this form of treatment.

Chloroquine is used orally for uncomplicated falciparum malaria and for most cases of vivax and ovale malaria.

Mefloquine is used orally for uncomplicated malaria and, via nasogastric tube, for severe malaria; neuro-psychiatric side-effects of this drug may be important [BMJ (1995) *310* 709–714].

Halofantrine is used orally for uncomplicated malaria. This drug (9-phenanthrene methanol), unlike e.g. quinine, seems to be active against the ring-stage trophozoite. The mechanism of action is unknown. Halofantrine is not used for prophylaxis owing e.g. to variable absorption. High levels of the drug have been (rarely) linked with cardiac arrhythmias [Lancet (1993) *341* 1054–1056].

Atovaquone (hydroxynaphthoquinone) is a candidate oral antimalarial. It appears to inhibit electron transport in the parasite.

Primaquine (an 8-aminoquinoline) is active against hypnozoites (but not against blood forms) and is used to prevent relapse in vivax and ovale malaria; for a 'radical cure', primaquine is used in conjunction with e.g. chloroquine.

WR238605 (a primaquine analogue) is being developed by the US army (Walter Reed Army Institute of Research). It has shown success in animal systems.

ANTIMALARIAL DRUGS[a]

Drug	Chemical description	Active against[b]
Artemisinin	Sesquiterpene lactone endoperoxide	BS
Atovaquone[c]	Hydroxynaphthoquinone	BS
Chloroquine	4-Aminoquinoline	BS
Halofantrine	9-Phenanthrene methanol	RS?
Mefloquine	4-Aminoquinoline	BS
Primaquine	8-Aminoquinoline	TS, HYP
Proguanil	Diguanide	TS, BS
Pyrimethamine	Diaminopyrimidine	BS
Quinidine	Quinoline-type (alkaloid)	BS
Quinine	Quinoline-type (alkaloid)	BS
Sulphadoxine	Sulphonamide	BS
WR238605[c]	8-Aminoquinoline	HYP, GAM

[a] See entry QUININE for the formulae of some antimalarials.
[b] See the figure for stages in the plasmodial life cycle. BS = blood schizont; TS = tissue schizont; HYP = hypnozoite; GAM = gametocyte; RS = ring stage.
[c] Candidate antimalarial.

Reproduced from *Antimicrobial Drug Action*, Table 7.1, page 121, R. A. D. Williams, P. A. Lambert & P. Singleton (1996) copyright Bios Scientific Publishers Ltd, UK (ISBN 1-872748-81-3) with permission from the publisher.

Proguanil (= Paludrine, CHLORGUANIDE) is a FOLIC ACID ANTAGONIST; with e.g. chloroquine it is used for prophylaxis. Of alternative prophylactic agents, mefloquine is absolutely contraindicated in early pregnancy.

See also QINGHAOSU.

In general, while antimalarials are used prophylactically by travellers, such use is considered undesirable for residents in endemic areas because it lowers the level of acquired resistance, resulting in high mortality when the drug is withdrawn.

[Chemotherapy: Book ref. 213, pp 119–128.]

[Malaria feature (vaccines, drugs, mosquito engineering etc.): Science (2000) *290* 428–441.]

malarial pigment *Syn.* HAEMOZOIN.

Malassezia (*syn. Pityrosporum*) A genus of non-fermentative imperfect yeasts (class HYPHOMYCETES) which occur on the skin in man and animals. The cells are spherical, oblong-ellipsoidal, or cylindrical, and reproduce by monopolar bud-fission. True hyphae are formed infrequently in culture, but are common in skin scales. Growth is commonly stimulated by certain oils (e.g. olive oil); optimum growth temperature: 35–37°C. The organisms can grow on glucose-peptone-yeast extract agar (overlaid with olive oil for *M. furfur*); they do not grow in the liquid media used in standard physiological tests for yeasts. Two species are recognized [Book ref. 100, pp. 882–885]: *M. furfur* (including '*Pityrosporum ovale*' and '*P. orbiculare*', which are currently considered to be variant cultural forms of *M. furfur*), and *M. pachydermatis* (= *P. pachydermatis*, *P. canis*). *M. furfur* occurs e.g. on human skin (see also PITYRIASIS VERSICOLOR); *M. pachydermatis* has been isolated from the ears of dogs (in which it may be associated with otitis externa) and from human skin.

malate synthase See TCA CYCLE.

male-specific phage *Syn.* ANDROPHAGE.

malic enzyme (1) See TCA CYCLE. (2) *Syn.* malolactic enzyme (see MALOLACTIC FERMENTATION).

malignant (*med.*) (1) (of disease) Becoming progressively worse (when untreated), usually resulting in death. (2) (of tumours) Refers to a neoplasm which invades adjacent tissues and which may metastasize (see METASTASIS). (cf. CANCER.)

malignant catarrhal fever (gangrenous coryza) An acute CATTLE DISEASE caused by a cytomegalovirus-like herpesvirus (see BETAHERPESVIRINAE). The disease is characterized by high fever with severe inflammatory and degenerative lesions in the mucous membranes of the upper respiratory tract and intestinal tract; the eyes may be affected, and blindness may result. Mortality rates are generally high (usually >95%). The disease may also affect certain other ungulates (e.g. sheep, wildebeest, deer). [Review of the disease: VR (1984) *114* 581–583.]

malignant oedema (*vet.*) Tissue oedema and necrosis, often involving frank GAS GANGRENE, due to wound infection by one or more species of *Clostridium* – e.g. *C. chauvoei*, *C. novyi*, *C. septicum*, *C. sordellii*. Symptoms: fever, with a soft, red, local swelling which becomes dark; in some cases gas production gives rise to a frothy exudate from the wound. Death usually occurs within 1–2 days. *Treatment*: parenteral administration of antibiotics, antitoxin therapy (appropriate only in early stages of the disease), irrigation of the wound with hydrogen peroxide. (See also SWELLED HEAD.)

malignant pustule Cutaneous ANTHRAX.

malignant tertian malaria See MALARIA.

mallein test (*vet.*) A test for GLANDERS (sense 1) involving the intradermal injection of mallein (containing antigens of the pathogen) e.g. into the lower eyelid; a positive reaction is indicated, after 48 hours, by local oedema and a severe purulent conjunctivitis.

Mallomonas A genus of unicellular CHRYSOPHYTES which resemble *Chromulina* spp but have a coating of silicified scales; in some species the scales bear long siliceous spines.

malolactic fermentation A type of FERMENTATION (sense 2), carried out e.g. by species of *Lactobacillus*, *Leuconostoc* and *Pediococcus*, in which L-malic acid is converted to L-lactic acid and CO_2. The reaction is apparently catalysed by a single enzyme (the *malolactic enzyme*) which has no L-lactate dehydrogenase activity; the malolactic enzyme from e.g. *Lactobacillus plantarum* has an absolute requirement for Mn(II) and NAD^+, but the NAD^+ is not reduced. There is no evidence that the reaction yields ATP.

In certain fermented products (e.g. wines, pickled cucumbers, soy sauce, silage) the malolactic fermentation has the effect of reducing the acidity of the product (lactic acid being a weaker acid than malic acid); this may or may not be desirable, depending on the degree and nature of acidity required in the product (see e.g. WINE-MAKING). Malolactic fermentation in wine tends to be slow and may be unpredictable; where required, it may be speeded up by the addition of a pure culture of suitable bacteria (commonly *Leuconostoc oenos*). The gene for malolactic fermentation has been transferred from *Lactobacillus delbrueckii* to a wine yeast in an attempt to produce a yeast capable of simultaneous malolactic and alcoholic fermentations [AEM (1984) *47* 288–293]; however, the resulting yeast exhibited only low levels of malolactic fermentation.

Malpighamoeba A genus of amoebae (order AMOEBIDA). *M. mellificae* is parasitic in bees (see BEE DISEASES), *M. locustae* is parasitic in locusts [Parasitol. (1976) *72* 127–135].

malt agar (malt extract agar) A medium made by diluting MALT EXTRACT to a specific gravity of 10° Balling and solidifying it with 2% agar.

malt extract An extract of malt, used e.g. as a medium for the culture of a wide range of yeasts and moulds. Malt (1 kg) is mixed with water (2.6 litres). The mixture is held at 45°C for 3 h (with stirring), and then at 63°C for 1 h; it is then filtered and autoclaved. The specific gravity of the filtrate is adjusted to 15° Balling using a Balling saccharometer. Malt extract is available commercially in dehydrated form. (See also BREWING.)

Malta fever See BRUCELLOSIS.

Maltaner antigen A cardiolipin–lecithin antigen used in certain STANDARD TESTS FOR SYPHILIS.

maltase An α-glucosidase (EC 3.2.1.20) which hydrolysis (1 → 4)-α-D-glucosidic bonds; thus, e.g. maltose is converted to free glucose.

maltobiose *Syn.* MALTOSE.

maltoporin See LAMB PROTEIN.

maltose (maltobiose) A reducing disaccharide: α-D-glucopyranosyl-(1 → 4)-D-glucopyranose. Maltose is an intermediate in the metabolic mobilization of STARCH in plants (e.g. during the germination of cereal grains – see e.g. BREWING); it also occurs e.g. as a major product of photosynthesis in *Chlorella* symbionts of *Hydra viridis* and of *Paramecium bursaria* (although it is not produced by the non-symbiotic alga). Maltose can be metabolized by a wide range of microorganisms. (For maltose transport in e.g. *Escherichia coli* see BINDING PROTEIN-DEPENDENT TRANSPORT SYSTEM.) (cf. ISOMALTOSE; see also AMYLASES and MALTASE.)

maltotriose A trisaccharide: α-D-glucopyranosyl-(1 → 4)-α-D-glucopyranosyl-(1 → 4)-D-glucopyranose. (See also AMYLASES.)

Mamiella See MICROMONADOPHYCEAE.

Mancini test See SINGLE DIFFUSION.

mancozeb An agricultural antifungal agent consisting of a complex of zinc and MANEB; it is used, for example, against late blight of potato.

Mandelamine See HEXAMINE.

mandelic acid Hydroxyphenylacetic acid: C_6H_5.CHOH.COOH. Mandelic acid has bacteriostatic properties and is used as a urinary antiseptic – being excreted unchanged if the urine is sufficiently acidic; it may be used in combination with HEXAMINE.

Mandler filter See FILTRATION.

maneb (dithane M-22) Manganese ethylene-BISDITHIOCARBAMATE, an agricultural antifungal agent with a spectrum of activity similar to that of ZINEB; it is used e.g. to control diseases caused by *Alternaria solani* and *Phytophthora infestans*.

manganese (in microbiology) See e.g. IRON BACTERIA, PSEUDOCATALASE and SUPEROXIDE DISMUTASE.

mango black spot See BLACK SPOT (3).

mannans Polysaccharides composed of manosyl residues. Linear (1 → 4)-β-linked mannans occur in some algae of the Chlorophyta (e.g. *Acetabularia*, *Codium*) – in which they may replace cellulose as the main structural component of the CELL WALL – and in certain plants (e.g. in many palm seeds); in many yeasts the CELL WALL has a high mannan content. Heteroglycans consisting of (1 → 4)-β-linked D-glucosyl and D-mannosyl residues (*glucomannans*) are major HEMICELLULOSES in gymnosperms.

mannitol The POLYOL corresponding to mannose; it occurs e.g. in many plants, algae, fungi and lichens, and is formed by the reduction of fructose by certain bacteria (see e.g. SILAGE). Mannitol is a major carbohydrate in members of the Phaeophyta and in chrysophytes (in which it is the primary photosynthetic carbohydrate) and in ascomycetous and basidiomycetous fungi; earlier reports that it occurs in the Rhodophyta appear to be incorrect [Book ref. 37, pp. 162–165]. Mannitol can have an osmoregulatory role in certain algae and fungi exposed to fluctuating salinities.

mannopine An opine of the agropine type: see e.g. HAIRY ROOT.

mannose An aldohexose found e.g. in MANNANS and in animal mucolipids and mucoproteins. (See also NEURAMINIC ACIDS; cf. MANNITOL.)

mannose-binding lectin See COMPLEMENT FIXATION (c).

mannose-resistant haemagglutination See FIMBRIAE.

mannose-sensitive haemagglutination See FIMBRIAE.

mantle (*mycol.*) See MYCORRHIZA.

Mantonella See EIMERIORINA.

Mantoniella See MICROMONADOPHYCEAE.

Mantoux test See TUBERCULIN TEST.

manure gas poisoning Poisoning of man or animals by gases (e.g. H_2S, NH_3, CO_2) produced by anaerobic bacteria in manure stored in large slurry pits; exposure to high concentrations of these gases – e.g. during mixing of slurries prior to emptying pits – can be rapidly fatal.

MAP (1) (M-associated protein) A protein often associated with the streptococcal M PROTEIN; it also occurs in streptococci of groups C and G that have M-like antigens.

(2) MICROTUBULE-ASSOCIATED PROTEIN.

(3) METHIONINE AMINOPEPTIDASE.

MAP kinase See ANTHRAX TOXIN.

map lichens *Rhizocarpon* spp, particularly *R. geographicum*.

map unit (map distance) (1) *Syn.* CENTIMORGAN. (2) Any unit used for indicating the position of markers in a genome: e.g. the chromosome of *Escherichia coli* is divided into 100 *minutes* (see INTERRUPTED MATING).

maple bark strippers' disease An EXTRINSIC ALLERGIC ALVEOLITIS associated with inhalation of the spores of *Cryptostroma corticale*.

Maranil Dodecylbenzolsulphonate: an agent used e.g. in BTB AGAR (0.005%) to inhibit swarming by *Proteus* spp.

Marasmius See AGARICALES (Tricholomataceae); see also MYCORRHIZA.

marble bone *Syn.* OSTEOPETROSIS.

marble spleen disease A disease of pheasants caused by an adenovirus very similar to the causal agent of HAEMORRHAGIC ENTERITIS OF TURKEYS. The disease is acute and rapidly fatal, and involves severe pulmonary oedema with enlargement and mottling of the spleen. [Book ref. 116, pp. 537–539.]

Marburg fever (green monkey fever) A VIRAL HAEMORRHAGIC FEVER, caused by the Marburg virus (see FILOVIRIDAE), first reported in 1967 when an outbreak occurred in Marburg (W. Germany) among laboratory personnel working with infected tissues of the African green monkey (*Cercopithecus aethiops*). Incubation period: ca. 6–10 days. Onset is sudden, with e.g. fever, mental confusion, diarrhoea, haemorrhages from mucous membranes, and often a maculopapular rash. The disease is often fatal. (Ebola virus causes a similar disease.)

Marburg virus See FILOVIRIDAE.

Marek's disease (fowl paralysis) A POULTRY DISEASE caused by gallid herpesvirus 1 (see GAMMAHERPESVIRINAE); chickens, turkeys, pheasants, ducks, pigeons and other birds are susceptible, but the principal natural hosts appear to be domestic chickens and their close relatives (*Gallus* spp). Infection seems to occur primarily by inhalation of 'feather dust' released from the feather follicles of infected birds. The classical (chronic) form of the disease involves the development of T-cell lymphomas in the peripheral nerves, causing progressive paralysis. A visceral (acute) form also occurs in which lymphomas develop in internal organs; mortality rates for the acute disease are commonly high. Infected birds remain carriers – probably for life. Live vaccines are available: e.g., vaccination with gallid herpesvirus 2 ('turkey herpesvirus') protects chickens and turkeys against Marek's disease. [Review of Marek's disease and its virus: Book ref. 105, pp. 155–183.]

Marek's disease virus can also cause chronic atherosclerosis in SPF chickens [AJP (1986) *122* 62–70]; MDV infection apparently prevents activation of cytoplasmic cholesteryl esterase in arterial smooth muscle cells [JBC (1986) *261* 7611–7614]. These observations lend support to the hypothesis that herpesvirus infection may play a role in the pathogenesis of atherosclerosis ('hardening of the arteries') in man [review: MS (1986) *3* 50–52].

Marfanil (sulphamylon; *p*-aminomethylbenzenesulphonamide: $NH_2CH_2.C_6H_4.SO_2.NH_2$) A drug which has been used e.g. in the topical treatment of wounds, burns, etc. It differs from other SULPHONAMIDES in that it is not a sulphanilamide derivative, the *p*-aminomethyl group replacing the *p*-amino group of sulphanilamide.

marginal zone B cells (MZ B cells) A subset of B LYMPHOCYTES found within the marginal zone of the spleen (i.e. the junction of red and white pulp) – together with DENDRITIC CELLS and macrophages; cells in the marginal zone are considered to represent an early form of defence against blood-borne pathogens. MZ B cells are important in responses to T-independent antigens, and they are particularly sensitive to lipopolysaccharides. [See e.g. Science (2000) *290* 89–92.]

Marinomonas See ALTEROMONAS.

marker (*genetics*) A genetic locus which is associated with a particular (usually readily detectable) phenotypic characteristic: e.g., an antibiotic resistance gene.

marker rescue Conferment on a gene, or genes, of the ability to replicate as a result of the integration of the gene(s) with a replicon. (See also REACTIVATION sense 1.)

Marseilles fever *Syn.* BOUTONNEUSE.

marsh gas See ANAEROBIC DIGESTION.

Marssonina See MELANCONIALES.

marsupium *Syn.* BROOD POUCH.

Marteilia See STELLATOSPOREA.

Martensella See KICKXELLALES.

marticin See NAPHTHAZARINS.

Mason–Pfizer monkey virus (MPMV) An exogenous, xenotropic retrovirus of the TYPE D RETROVIRUS GROUP. MPMV was isolated from a carcinoma of a rhesus monkey; it has not been shown to be oncogenic either in its natural host or in other primates. (cf. SIMIAN AIDS.)

MASP MBL-associated serine protease: see COMPLEMENT FIXATION (c).

mast cell A cell which can respond to certain stimuli by rapidly secreting e.g. vasoactive products; it is important e.g. in immediate-type hypersensitivity reactions and in acute INFLAMMATION. Unlike the BASOPHIL, the mast cell is primarily a non-circulating cell, occurring e.g. in blood vessels and in the lymphatic system. The (basophilic) cytoplasmic granules in the mast cell contain a variety of biologically active products – e.g. HISTAMINE, HEPARIN, SEROTONIN (significant amounts in some species), and proteases – which are released, within 15–20 sec, on activation and degranulation of the cell. Mast cells can be activated and degranulated e.g. by certain CYTOKINES, ANAPHYLATOXINS, IgE (and certain IgG) antibodies together with their homologous antigens, IgE aggregates (not monomeric IgE), bivalent anti-IgE receptor antibodies, and some LECTINS (e.g. PHA). The mast cell surface contains many high-affinity receptor sites for the Fc portion of IgE, and the cross-linking of surface-bound IgE antibodies by antigen (allergen) appears to be a trigger for degranulation (see also TYPE I REACTION). (Degranulated mast cells gradually replace their histamine.) Activated mast cells also produce e.g. SUPEROXIDE, H_2O_2, and factors which are chemotactic for NEUTROPHILS and EOSINOPHILS.

Mastadenovirus (mammalian adenoviruses) A genus of adenoviruses (family ADENOVIRIDAE) which infect mammals (including man); type species: human adenovirus type 2 (= h2, = Ad2). Almost all mammalian adenoviruses share a common (group-specific) antigen (a hexon surface antigen); subgenus-specific and species-specific antigens can also be detected.

Human adenovirus species are designated by the prefix Ad (or h) followed by the species (serotype) number: Ad1, Ad2, etc (or h1, h2, etc). Human adenoviruses can cause a wide range of diseases (as well as subclinical infections) in humans. Several can induce tumours (mainly sarcomas) when injected into newborn rodents, but none is known to cause tumours in man under natural conditions; however, most human adenoviruses can cause morphological transformation in certain types of cell cultures, causing the formation of characteristic foci of small, often polygonal cells containing scanty cytoplasm and capable of dividing indefinitely. Transformation of cells by adenoviruses involves the integration of viral DNA sequences into host chromosomal DNA. Depending on the host cell, the whole viral genome or (more commonly) a fragment of it may be integrated; integration apparently occurs at random sites in the host DNA, and the viral DNA – together with flanking host sequences – often undergoes amplification. Transformation requires the integration of at least the E1a region, but the mechanism of transformation is unknown. [Adenoviral genes and transformation: Book ref. 110, pp. 125–172.]

Human adenoviruses have been classified into subgenera (subgroups) on the basis of e.g. DNA characteristics, oncogenicity, subgenus-specific antigens, morphology, haemagglutinating properties, etc:

Subgenus A (Ad12, 18 and 31): GC% of the DNA: 47–49; length of virion fibres: 28–31 nm. Viruses in this subgenus are highly oncogenic when injected into newborn hamsters or rats, causing tumours after ca. 2–4 months; in man, they are associated with upper respiratory tract infections and diarrhoea, or infection may be subclinical.

Subgenus B (Ad3, 7, 11, 14, 16, 21, 34 and 35): GC% of the DNA: 49–52; length of the virion fibres: 9–11 nm. Viruses in this subgenus are weakly oncogenic when injected into newborn hamsters, causing tumours at low incidence after ca. one year; in man, they cause e.g. ARD (Ad3, 7, 14, 21), acute haemorrhagic CYSTITIS (Ad11), PHARYNGOCONJUNCTIVAL FEVER, pneumonia, etc.

Subgenus C (Ad1, 2, 5 and 6): GC% of the DNA: 57–59; length of the virion fibres: 23–31 nm. These viruses are non-oncogenic in newborn rodents, but can cause transformation of rodent cells in vitro; they can cause mild to severe upper respiratory tract disease in young children.

Subgenus D (Ad8–10, 13, 15, 17, 19, 20, 22–30, 32, 33, 36, 37–39): GC% of the DNA: 57–60; length of the virion fibres: 12–13 nm. The viruses are non-oncogenic in newborn hamsters, but may cause mammary adenomas in newborn female rats and can transform rodent cells in vitro. Ad8 and Ad19 are causal agents of EPIDEMIC KERATOCONJUNCTIVITIS.

Subgenus E (Ad4): GC% of the DNA: 57; length of virion fibres: 17 nm. Non-oncogenic in newborn hamsters. Ad4 can cause EPIDEMIC KERATOCONJUNCTIVITIS, acute respiratory disease (see ARD), and PHARYNGOCONJUNCTIVAL FEVER.

Two further subgenera, F (Ad40) and G (Ad41), have been recognized but are not yet well characterized; these include the 'enteric' or 'fastidious' adenoviruses [review: Arch. Virol. (1986) *88* 1–17] which cause diarrhoea in infants and young children, and which cannot be propagated in conventional cell cultures. They can be grown in an Ad5-transformed human embryonic kidney cell line.

Most human adenoviruses will not replicate readily in simian cell lines, but may do so in the presence of simian virus 40 (SV40); SV40 can interact extensively with human adenoviruses, e.g. forming SV40–adenovirus hybrids [Book ref. 116, pp. 399–449].

Animal adenovirus nomenclature follows at least two different conventions; a species may be designated by a three-letter prefix derived either from the name of the animal plus 'AV' (e.g. BAV for bovine adenovirus, SAV for simian adenovirus [Book ref. 116, p. 500]) or from the genus name of the host animal (e.g. bos for cattle, sus for pigs, equ for horses, mus for mice [Book ref. 23, p. 60]), followed in either case by the serotype (species) number – e.g. BAV-3 or bos-3, MAV-1 or mus-1, etc. Infection of animals by adenoviruses may result in (usually mild) respiratory or diarrhoeal illnesses, conjunctivitis, etc. Canine adenovirus type 1 (CAV-1, can-1) can cause encephalitis in foxes and e.g. hepatitis in dogs; CAV-2 is thought to be a causal agent of laryngotracheitis ('kennel cough') in dogs. Several animal adenoviruses (11 simian, 2 bovine and 2 canine species) can induce tumours when injected into newborn hamsters. [Animal adenoviruses: Book ref. 116, pp. 497–562.]

Mastigocladus See FISCHERELLA and CHLOROGLOEOPSIS.

Mastigomycota In some taxonomic schemes: a division equivalent to the subdivision MASTIGOMYCOTINA.

Mastigomycotina A subdivision of unicellular and (more commonly) mycelial fungi (division EUMYCOTA); most species (though not e.g. *Peronospora* spp) form flagellated spores or gametes. Classes: CHYTRIDIOMYCETES, HYPHOCHYTRIOMYCETES, OOMYCETES.

mastigoneme See FLAGELLUM (b).

Mastigophora A subphylum of protozoa (phylum SARCOMASTIGOPHORA) which have one or more flagella; cell division is typically symmetrogenic. Classes: PHYTOMASTIGOPHOREA and ZOOMASTIGOPHOREA.

mastigosome *Syn.* BASAL BODY (b).

mastitis Inflammation of the mammary gland; it may or may not be of microbial causation.

(a) (*med.*) *Puerperal mastitis* may occur in lactating women, commonly ca. 1 month after childbirth, and may be caused by staphylococci or streptococci.

(b) (*vet.*) Mastitis is quite common among farm animals, but is uncommon in mares. In cows, common causal agents include e.g. *Actinomyces pyogenes* ('summer mastitis' in dry cows and pre-calving heifers), *Escherichia coli* ('coliform mastitis' is caused by *E. coli* or other coliforms), *Staphylococcus aureus* (see also BLACK POX), and *Streptococcus agalactiae*, but any of a wide range of organisms may be involved, including various other bacteria (e.g. species of *Pasteurella*, *Pseudomonas*), fungi (e.g. *Aspergillus*, *Candida*) and algae (*Prototheca* spp). In ewes, mastitis is commonly caused by *E. coli*, *S. aureus* or *S. agalactiae* but may be due e.g. to *Mycoplasma agalactiae*, while in mares the causal agent may be e.g. *Streptococcus equi*. In cows, predisposing factors include age and the stage of lactation; thus, e.g. cows which have had ca. four or more lactations, and those within the first two months of lactation, are more susceptible than others. Symptoms may range from mild to an acute inflammatory condition with fever and other systemic effects; the milk contains an increased number of white blood cells, and may contain a significant number of pathogenic microorganisms. (See also CALIFORNIA MASTITIS TEST and CAMP TEST.) *Treatment*: e.g. parenteral administration or intramammary infusion of antibiotics [Book ref. 33, pp. 451–500]. [Chemoprophylaxis in bovine mastitis: Book ref. 121, pp. 193–204.]

[DNA fingerprinting and ribotyping in epidemiological investigations of outbreaks of *Pseudomonas aeruginosa* mastitis among Irish dairy herds: AEM (1999) *65* 2723–2729.]

mat (microbial) See MICROBIAL MAT.

***MAT* locus** See MATING TYPE.

matching (in taxometrics) See entry S_{SM}.

MATE Multidrug and toxic compound extrusion: an EFFLUX MECHANISM which e.g. mediates resistance to antibiotic(s) in certain bacteria (including e.g. *Escherichia coli* and *Vibrio parahaemolyticus*). [Mol. Microbiol. (1999) *31* 393–395.]

mate killer The term for certain (endosymbiont-containing) strains of ciliate protozoa which, on conjugation (mating), cause the death of the conjugal partner; mate killers include strains of *Paramecium aurelia* (endosymbiont: *Caedibacter paraconjugatus* or *Pseudocaedibacter conjugatus*) and of *Euplotes crassus* and *E. patella* (endosymbionts unnamed). In each case mating is essential for killing; thus, e.g. the *ingestion* of *C. paraconjugatus* by sensitive paramecia does not cause harmful effects.

maternal immunity PASSIVE IMMUNITY acquired by a fetus or neonate from its mother either via the placenta or from COLOSTRUM.

maternal inheritance A form of CYTOPLASMIC INHERITANCE in which certain characteristics are transmitted to sexually derived progeny only from the female parent cell. Various hypotheses have been proposed to account for maternal inheritance, and different mechanisms may be applicable in different cases. Thus, in certain cases where maternally inherited genes occur in mitochondria, mitochondria from the male parent are presumed to be excluded from the zygote (see e.g. POKY MUTANT). To account for the maternal inheritance of genes in chloroplast DNA (a phenomenon recorded in some isogamous green algae) it has been proposed that chloroplast DNA from the male parent is preferentially digested [MS (1985) *2* 267–270].

mating *Syn.* CONJUGATION (sense 1).

mating type A strain of an organism which can interact sexually only with other, genetically distinct, strains of the same species. (See also HETEROTHALLISM and SYNGEN.)

In *Saccharomyces cerevisiae* mating type is determined at a single genetic locus, designated *MAT*, on chromosome III. In haploid cells, *MAT* is occupied by either the *MAT***a** allele or the *MAT*α allele – which occur in **a** and α mating types respectively. According to the widely-accepted 'α1-α2 hypothesis', *MAT***a** and *MAT*α regulate the same groups of structural genes, but their different modes of regulation result in the development of phenotypically different mating types; thus, groups of genes known as '**a**-specific genes' and 'α-specific genes' (which specify the **a** and α phenotypes, respectively) occur in the cells of both **a** and α mating types, and the way in which these genes are controlled by the *MAT* locus determines the cell's mating type.

MATα is transcribed in two parts from a central promoter region, the two RNA transcripts being copied from different DNA strands; the parts are called *MAT*α1 and *MAT*α2. *MAT*α1 expression is necessary for the transcription of the α-specific genes – whose products include the *α-factor* (a PHEROMONE).

MATα2 is a negative regulator of **a**-specific genes.

The (single) *MAT***a** product has no effect in the **a**-type haploid vegetative cell; by default, **a**-type genes (including that encoding the **a**-type pheromone **a**-*factor*) are expressed in cells of this mating type.

In most natural isolates of *S. cerevisiae*, spontaneous conversion of the *MAT***a** allele to the *MAT*α allele, and vice versa, occurs quite frequently – with concomitant switching of mating type; a given population will therefore consist of a mixture of compatible mating types (**a** and α) and will give the appearance of being homothallic. (Stability at the *MAT* locus involves another locus, designated *HO* for homothallism, located on chromosome IV.) Switching involves a cassette mechanism, *unexpressed* copies of *MAT***a** and *MAT*α being stored on chromosome III at sites designated, respectively, *HMR* and *HML*; these genes are silenced by the transcription-resistant form of their chromatin. Switching requires a double-stranded cut at the *MAT* locus, made by the HO ('YZ') endonuclease, followed by excision of the resident *MAT* allele and its replacement with a *copy* of the allele of the other mating-type (*gene conversion*).

When **a**-type and α-type cells are mixed, the pheromone from a given mating type causes cells of the other type to stop in the G_1 phase of the CELL CYCLE, and to exhibit a mating type-specific glycoprotein component of the cell wall which promotes fusion between cells of opposite mating types. The zygote, which lacks mating potential and mating type characteristics, is designated **a**/α.

In the **a**/α zygote, the α2 and **a**1 proteins jointly switch off various genes, including **a**-specific and α-specific genes and the 'haploid-specific' genes active in both mating types.

matrix protein (*bacteriol.*) A former term for PORIN.

maturase See SPLIT GENE (b) and (c).

Maurer's clefts (Maurer's dots) Large dots, streaks, or 'commas' sometimes observed, on suitable staining, in erythrocytes infected with certain stages of *Plasmodium falciparum*.

Maus Elberfeld virus See CARDIOVIRUS.

maxicell A bacterial 'cell' obtained by UV irradiation of *recA*, *uvrA* mutant strains of e.g. *Escherichia coli*. Irradiation of such mutants results in damage to and degradation of chromosomal DNA (cf. RECKLESS DNA DEGRADATION); however, if small, multicopy plasmids are present in such cells, some of the plasmids may be undamaged by the UV radiation and can subsequently replicate. These cells therefore synthesize plasmid-encoded proteins almost exclusively; the proteins can be labelled by adding e.g. [^{35}S]methionine to the medium. [JB (1979) *137* 692–693.] (cf. MINICELL.)

maxicircle DNA See KINETOPLAST.

maximum specific growth rate (μ_{max}) See SPECIFIC GROWTH RATE.

Mayaro fever An epidemic febrile human disease caused by an ALPHAVIRUS and transmitted by mosquitoes (e.g. *Haemagogus janthinomys*); it occurs in the Caribbean and parts of S. America. Symptoms: e.g. headache, myalgia, arthralgia, and a maculopapular rash. Monkeys are probably the main natural hosts.

Mayorella A large genus of amoebae (order AMOEBIDA) which characteristically form several bluntly conical, hyaline pseudopodia of similar lengths. (See also UROID.) Some species have a surface covering of complex scales (not discernible by light microscopy). Species occur in freshwater habitats and e.g. sewage treatment plants. *M. viridans* contains ZOOCHLORELLAE.

maytansine See MICROTUBULES.

mazaedium A type of ascocarp in which the asci disintegrate to release a powdery mass of ascospores (with sterile elements); mazaedia are formed in members of the Caliciales and in *Onygena*. (The term 'mazaedium' may also refer to the spore mass per se.)

MBC fungicides See BENZIMIDAZOLES (a).

MBL Mannose-binding lectin: see COMPLEMENT FIXATION (c).

Mbl protein See CELL CYCLE (b) (determination of shape).

MBP Mannose-binding protein, referred to more commonly as MBL (mannose-binding lectin) – see COMPLEMENT FIXATION (c).

MBSA Methylated bovine serum albumin.

MC29 See AVIAN ACUTE LEUKAEMIA VIRUSES.

Mcc plasmid See MICROCINS.

McCartney bottle A bottle similar to a UNIVERSAL BOTTLE but which has a narrower neck.

McClung Toabe egg-yolk agar (modified) A medium used for the isolation and identification of clostridia and other anaerobes; it permits detection of lecithinase, lipase and proteolytic activity. The medium contains tryptone, yeast extract, glucose, egg yolk, Na_2HPO_4, NaCl, $MgSO_4$ and agar.

McDonough feline sarcoma virus See FMS.

MCF viruses Mink cell focus-forming viruses: variant MURINE LEUKAEMIA VIRUSES (MuLVs) which can be detected and assayed by their ability to induce foci of transformed cells in monolayer cultures of a mink lung cell line. MCF viruses were originally isolated from mouse strains (e.g. AKR) which show a very high incidence of spontaneous leukaemia. The *env* gene of an MCF virus resembles a mixture of those of murine ecotropic and xenotropic viruses, and confers on it a DUALTROPIC host range. MCF viruses are believed to have arisen by recombination between an ecotropic MuLV and an endogenous xenotropic retrovirus. Since many MCF viruses have a leukaemogenic potential greater than that of the parent MuLV, it has been suggested that an MCF component may be responsible for 'MuLV-induced' leukaemogenesis. (See also AKV; cf. FRIEND VIRUS.)

McFadyean's test A confirmatory test for ANTHRAX carried out, post-mortem, on an infected guinea pig. Blood (from the heart), or a spleen imprint, is stained with polychrome methylene blue; in a positive test, the stained smear reveals large, blue, square-ended bacilli surrounded by a reddish granular capsule.

McIntosh and Fildes' anaerobic jar (modified) An ANAEROBIC JAR which consists of a strong cylindrical metal chamber with a flat, circular, gas-tight lid; such jars are referred to by various trade names (e.g. Torbal jar). Media etc are placed in the jar, and the lid is secured by means of a screw clamp. The jar is then evacuated, by a suction pump, via one of two screw-controlled needle valves in the lid; this valve is then closed. (To prevent the vacuum sucking the agar from Petri dishes, plates are incubated lid-side up – cf. GASPAK.) The jar is filled, via the other valve, with gas(es) – e.g. H_2, or CO_2 and H_2 – from a compressed-gas cylinder or from a gas-filled rubber bladder; this valve is closed, and the cycle of evacuation and re-filling is carried out several times. Attached to the inside of the lid is a gauze envelope containing a catalyst (e.g. palladium-coated pellets of alumina) which promotes chemical combination between hydrogen and the last traces of oxygen in the jar; such catalysts work at room temperatures and are called 'cold' catalysts. The jar also has a side-arm to which is attached a small, thick-walled, closed glass vessel which communicates with the interior of the jar and which contains an indicator of anaerobiosis (e.g. METHYLENE BLUE solution).

MCP (1) Methyl-accepting chemotaxis protein (see CHEMOTAXIS).

(2) Membrane co-factor protein (see COMPLEMENT FIXATION (b).)

MCS (multiple cloning site) See POLYLINKER.

M-CSF See COLONY-STIMULATING FACTORS.

MDa Megadalton: 10^6 DALTONS.

MDCK cells Madin–Darby canine kidney cells: a heteroploid cell line derived from the kidney of an apparently normal dog. MDCK cell cultures have been used e.g. for the primary isolation and passage of influenzavirus type A [PNAS (2000) *97* 9654–9658].

MDP Muramyl dipeptide: *N*-acetylmuramyl-L-alanyl-D-isoglutamine; a water-soluble component of a modified form of complete FREUND'S ADJUVANT.

mean doubling time See GROWTH (a).

mean generation time See GROWTH (a).

measles (morbilli; cf. RUBEOLA) An acute and highly infectious human disease which affects mainly children. (Other primates can be affected.) The causal agent is a MORBILLIVIRUS. Transmission occurs by droplet infection or via fomites; the conjunctivae may be important sites of infection. Incubation period: 8–12 days. In the prodromal stage (2–4 days) there is coryza, cough, fever, and the appearance of KOPLIK'S SPOTS. Subsequently, a maculopapular rash develops on the head, then on the limbs and trunk. The spots are initially bright pink but become deeper red and then brownish – a phenomenon known as 'staining'. Viruses are present in the urine, nasopharyngeal secretions etc during the prodromal period and for several days after the appearance of the rash. Complications may include ENCEPHALITIS [NEJM (1984) *310* 137–141]; secondary bacterial infection may result in e.g.

OTITIS MEDIA or PNEUMONIA. Measles is usually self-limiting, but may be fatal in young infants or in malnourished children; death is usually due to respiratory complications. Life-long immunity usually follows recovery (cf. SUBACUTE SCLEROSING PANENCEPHALITIS). *Lab. diagnosis*: microscopic examination of smears of the nasal mucosa for epithelial GIANT CELLS; detection of the measles virus in blood, urine or nasal secretions. *Chemotherapy*: none. Measles immune globulin may be used prophylactically in high-risk patients exposed to measles. Live attenuated vaccines (cf. MR and MMR) may be given to infants and young children.

meat spoilage The occurrence, nature and extent of meat spoilage depends on e.g. the initial numbers and types of contaminating microorganisms, the conditions of storage (particularly temperature and oxygen tension), the length of storage, and the nature of the meat itself (e.g. its WATER ACTIVITY, glucose content [see e.g. AEM (1982) *44* 521–524], pH). Spoilage may involve slime formation, discoloration, souring, and off-odours (due e.g. to the volatile products of microbial amino acid metabolism).

(a) *Carcasses*. Carcass meat is normally surface-contaminated with bacteria and fungi derived from the animal's skin, intestine, faeces etc., from soil, and e.g. from the hands and tools of workers. Even under good hygienic conditions, carcasses are likely to carry 10^2-10^5 bacteria/cm^2, but spoilage commonly becomes apparent only after bacterial numbers reach ca. 10^7-10^9 per cm^2. After slaughter, the pH of meat gradually falls as the muscle glycogen is converted to lactic acid – the final pH depending on the amount of glycogen present at the time of death. The final pH of red meats is commonly 5.4–5.8; chicken leg meat normally has a pH of ca. 6.5, although the breast meat usually has a pH of ca. 5.7. Meat from stressed animals has a higher pH and is more susceptible to spoilage – see DFD MEAT.

For *short-term storage*, carcasses may be chilled to e.g. 0–10°C. Below ca. 10°C psychrotrophic spoilage organisms increase greatly in numbers; by contrast, *pathogenic* organisms (cf. FOOD POISONING) grow very slowly, and their presence is usually not obvious. In most meats, including uncured pig meat and poultry, the commonest spoilage bacteria under aerobic conditions are strains of *Pseudomonas* [JAB (1982) *52* 219–228 (taxonomic study)], often occurring together with smaller numbers of bacteria such as *Acinetobacter*, *Alteromonas*, *Brochothrix* and *Lactobacillus* spp. Of the spoilage organisms, *Pseudomonas* spp have the highest growth rate at low temperatures, and this advantage tends to become more marked with decreasing temperature. (In poultry, *Acinetobacter* replaces *Pseudomonas* as the commonest spoilage organism at temperatures *above* ca. 10°C.) However, *Brochothrix thermosphacta* and *Lactobacillus* spp, for example, are more tolerant than pseudomonads of low a_w, and these may become dominant on drying carcasses. On *cured* meats (see CURING), e.g. bacon, the commonest spoilage bacteria under aerobic conditions are halophilic, psychrophilic vibrios (e.g. *Vibrio costicola*) and species of *Acinetobacter* and *Micrococcus*; the micrococci are lipolytic and proteolytic, and some strains can reduce nitrite. Fungi found on chilled carcasses include *Cladosporium*, *Penicillium*, *Mucor*, *Rhizopus* and *Thamnidium*, and the yeasts *Candida*, *Cryptococcus* and *Rhodotorula*; these may cause e.g. surface discolorations.

For *long-term storage*, temperatures of ca. −12°C to −18°C are often used; the process of freezing kills or damages a proportion of contaminants, although lower temperatures (e.g. −30°C) are less lethal (see FREEZING). At temperatures between ca. −5°C and −10°C fungal contaminants may become important spoilage agents (see e.g. BLACK SPOT sense 1).

(b) *Packaged meats*. As well as the factors cited above, the nature of spoilage of packaged meats depends on the type of packaging – particularly on the permeability of the packing material to air. Spoilage of chilled red meats typically occurs more rapidly in the presence of oxygen than under vacuum. Under aerobic conditions (e.g. with oxygen-permeable packing film) *Pseudomonas* spp are the commonest spoilage organisms; *Acinetobacter* may be present in significant numbers if the pH is above ca. 6.0. Under anaerobic conditions (e.g. in oxygen-impermeable vacuum packs) these organisms are inhibited, permitting the growth of e.g. *Lactobacillus* spp and *Brochothrix thermosphacta*; CARBON DIOXIDE per se (ca. 10–20% v/v) inhibits the growth of e.g. *Pseudomonas* spp, *Alteromonas putrefaciens*, *Acinetobacter* sp, and *Yersinia enterocolitica*, but not of e.g. *B. thermosphacta* or lactobacilli. (cf. DFD MEAT.)

At low temperatures (e.g. 5°C) vacuum-packed cured meats (e.g. bacon) are susceptible to spoilage by e.g. *Lactobacillus* spp (which may cause e.g. souring) [AvL (1983) *49* 327–336]; lactobacilli are resistant to nitrite, smoke (see SMOKING), and low concentrations of salt. At higher salt concentrations, *Micrococcus* spp may become dominant. Cooked hams and cooked sausage (in which CATALASE has been destroyed) may undergo a type of *greening* caused by e.g. lactobacilli (especially *L. viridescens*); these organisms produce H_2O_2 which oxidizes the porphyrin ring in meat haem pigments to form a green compound (cf. DFD MEAT). (Higher concentrations of H_2O_2 can produce yellow or colourless compounds.)

[Book ref. 30.]

(See also CANNING; FOOD SPOILAGE; SALAMI; SOFT CORE HAM.)

***mecA* gene** See MRSA.

MECAM A (synthetic) functional analogue of enterobactin used e.g. in studies on IRON uptake [see e.g. JB (1986) *167* 666–680].

mechanical transmission Transmission of a parasite by passive transfer from one host to another, e.g. via a mechanical VECTOR, via contaminated tools or instruments etc.

mechanical vector See VECTOR (1).

mechanosensitive channel (stretch-activated channel) A channel through the CYTOPLASMIC MEMBRANE which responds to increases in the cell's turgor pressure by increasing in pore size, thus facilitating efflux of water and certain solutes [JBC (1997) *272* 32150–32157]; one example is the MscL channel in *Escherichia coli* [mechanosensitive channels in *E. coli*: ARP (1997) *59* 633–657]. Such channels occur in many types of bacteria and appear to have an important role in OSMOREGULATION; they open when the cell's turgor pressure is just below the level at which fatal disruption would occur [EMBO (1999) *18* 1730–1737].

[The gating mechanism of the large mechanosensitive channel in *Mycobacterium tuberculosis*: Nature (2001) *409* 720–724.]

***mecI*, *mecR1* genes** See MRSA.

mecillinam (Coactin; amdinocillin) A derivative of 6-β-amidinopenicillanic acid (cf. PENICILLINS) which, in e.g. *Escherichia coli*, binds exclusively to PBP-2 (see PENICILLIN-BINDING PROTEINS).

medallion clamp See CLAMP CONNECTION.

median lethal dose *Syn.* LD_{50} (q.v.).

medical flat A flat-sided glass screw-cap bottle obtainable in various sizes.

medicarpin An isoflavonoid PHYTOALEXIN produced e.g. by the broad bean (*Vicia faba*). (cf. WYERONE.)

Medina reaction See LEPROMIN TEST.

medium In microbiology: any liquid or solid preparation made specifically for the growth, storage or transport of microorganisms or other types of cell. Growth media for microorganisms may be used e.g. for the initiation of a CULTURE (or a SUBCULTURE), for ENRICHMENT, or for diagnostic (identification) tests – i.e. tests in which the identity of a given organism may be deduced from the characteristics of its growth in or on particular media. (See also TRANSPORT MEDIUM.) Specialized media are used for the growth/maintenance of mammalian (and other) cells in TISSUE CULTURE (q.v.). Before use, a medium is normally STERILE (sense 1).

Many chemolithautotrophic bacteria can be cultured in simple aqueous media containing mainly, or only, mixtures of inorganic salts. Nutritionally undemanding heterotrophs can be grown on/in a range of culture media (see e.g. NUTRIENT BROTH and PETONE WATER). Nutritionally 'fastidious' heterotrophs may be grown on an ENRICHED MEDIUM. Strict anaerobes are often cultured on 'prereduced' media (see ANAEROBE).

Solid media usually consist of liquid media which have been solidified ('gelled') with an agent such as AGAR or GELATIN; other gelling agents include e.g. ALGINATE, Gelrite (see GELLAN GUM), and PLURONIC POLYOL F127. (See also SILICA GEL.) Solid media (such as e.g. NUTRIENT AGAR) may be used in order to obtain colonies (see COLONY) of a given organism; colonies may be required for identification purposes or e.g. for the determination of viable cell counts (see COUNTING METHODS). Media containing an unusually high concentration of solidifying agent (e.g. *stiff* AGAR) are used e.g. to inhibit SWARMING; those containing less than normal concentrations of solidifying agent (e.g. *semi-solid* or 'sloppy' agar) may be used e.g. for CRAIGIE'S TUBE METHOD.

Basal medium. One which, without supplement, can support the growth of various nutritionally undemanding species; see e.g. NUTRIENT BROTH.

Defined medium. One in which all the constituents (including trace substances) are quantitatively known.

Differential medium. A solid medium on/in which different types of organism may be distinguished by their different forms of growth; see e.g. MACCONKEY'S AGAR; SS AGAR.

Diphasic medium. See DIPHASIC MEDIUM.

Enriched medium. See ENRICHED MEDIUM.

Enrichment medium. See ENRICHMENT MEDIUM.

Inhibitory medium. One containing certain constituent(s) which suppress the growth of certain type(s) of microorganism.

Maintenance medium. One used for the initial growth and subsequent storage (under conditions of minimal growth) of microorganisms or other cells. For microorganisms, a maintenance medium is used to prepare a CULTURE of the given organism which is then stored either at room temperature or under refrigeration; subculture is required at intervals ranging from 1 to 12 months. Ideally, the constituents of a maintenance medium are the minimum consistent with the need to maintain viability of the organisms. Examples: COOKED MEAT MEDIUM and DORSET'S EGG. (See also TISSUE CULTURE.)

Pre-reduced medium. See ANAEROBE.

Selective medium. One which allows or encourages the growth of some type(s) of organism in preference to others. The term usually refers to media such as DCA, MacConkey's agar (which generally supports the growth of enteric – but not non-enteric – bacteria), and all the various types of enrichment media; however, all media are selective to some extent.

Test medium. One used for diagnostic (identification) tests. Test media are often made by incorporating additional substance(s) into standard or modified liquid or solid media; often a PH INDICATOR is included to monitor certain aspects of metabolic activity. Examples: KCN BROTH, LITMUS MILK, TSI AGAR.

Transport medium. See TRANSPORT MEDIUM.

MEE MULTILOCUS ENZYME ELECTROPHORESIS.

***mefA* gene** See MACROLIDE ANTIBIOTICS.

mefloquine See AMINOQUINOLINES and MALARIA.

megacins See BACTERIOCIN.

megacolon See CHAGAS' DISEASE.

megalomycin See MACROLIDE ANTIBIOTICS.

Megasphaera A genus of Gram-negative bacteria (family VEILLONELLACEAE) which occur e.g. in the RUMEN and in the human intestine. Cells: cocci, ca. 2.0–2.5 μm diam., commonly occurring in pairs. The fermentation of glucose yields mainly caproate, while that of lactate yields acetate, propionate, valerate, C_4 fatty acids, much CO_2 and a little H_2 – but little or no caproate. GC%: ca. 54. Type species: *M. elsdenii* (formerly *Peptostreptococcus elsdenii*).

Megatrypanum A subgenus of TRYPANOSOMA within the STERCORARIA; species are parasitic in e.g. cattle and sheep. The organisms are large, sometimes greater than 100 μm in length; the kinetoplast is situated approximately mid-way between the nucleus and the posterior end of the cell, and there is a free flagellum. *T.* (*M.*) *theileri* occurs in cattle, and is probably transmitted by contaminative infection of mucous membranes etc with metacyclic forms via tabanid vectors; it is generally non-pathogenic. *T.* (*M.*) *melophagium* occurs in sheep, and *T.* (*M.*) *theodori* occurs in goats.

megrims *Syn.* STAGGERS.

MEIA (microparticle enzyme immunoassay) See LCR.

meiocyte A cell whose nucleus undergoes MEIOSIS.

meiosis (reduction division) The process in which a eukaryotic nucleus divides into nuclei whose ploidy is lower than that of the parent nucleus (typically, haploid nuclei being formed from diploid nuclei) and in which RECOMBINATION usually occurs. (cf. MITOSIS.) Meiosis occurs e.g. during the formation of gametes from diploid cells, and at the inception of haplophase in those organisms which exhibit an ALTERNATION OF GENERATIONS. (See also GAMETIC MEIOSIS; SPORIC MEIOSIS; ZYGOTIC MEIOSIS.)

The best-known form of meiosis is that which occurs in higher animals; other forms of meiosis (which differ in one or more aspects) occur in various microorganisms. Meiosis in higher animals involves two distinct nuclear divisions: the first meiotic division (meiosis I) and the second meiotic division (meiosis II) – each division exhibiting a prophase, metaphase, anaphase and telophase. (cf. MITOSIS.)

Prophase I. Individual chromosomes, each having previously replicated to form two chromatids, become visible as long, single threads (*leptotene* stage). In the *zygotene* stage each chromosome pairs, lengthwise, with its HOMOLOGUE, the pair of closely adjacent chromosomes being called a *bivalent*; the pairing of chromosomes (*synapsis*) involves the SYNAPTONEMAL COMPLEX. In this stage and/or the next (*pachytene*) stage CROSSING OVER occurs between pairs of non-sister chromatids. In the *diplotene* stage each chromosome begins to separate from its homologue – but the chromosomes of each homologous pair remain in contact at sites of crossing over (see CHIASMA). In the final stage of prophase I (*diakinesis*) the chromosomes condense, and each homologous pair is seen to consist of four chromatids (i.e., a *tetrad*). By the end of diakinesis the nucleolus and nuclear membrane have disintegrated.

Metaphase I. Homologous pairs of chromosomes form a layer at the equator of a microtubular *spindle* (as in MITOSIS).

Anaphase I. The two chromosomes in each homologous pair move to opposite poles of the spindle.

Telophase I. Each of the two groups of chromosomes becomes enclosed by a nuclear membrane.

The interval between the first and second meiotic divisions is called *interkinesis* or *interphase II.*

The second meiotic division is analogous to mitosis, and consists of prophase II, metaphase II, anaphase II and telophase II.

[Control of meiosis: SEBS (1984) *38*.]

meiosporangium A SPORANGIUM within which MEIOSIS occurs (see e.g. ALLOMYCES).

meiospore A haploid spore formed by meiotic cell division. (See also ALTERNATION OF GENERATIONS.)

meiotic Of or pertaining to MEIOSIS.

Mel B *Syn.* melarsoprol (see ARSENIC).

Mel T See ARSENIC (a).

Melampsora A genus of homoxenous and heteroxenous rust fungi (class UREDINIOMYCETES) which typically form caeomatoid aecia, peridiate uredia, and sessile teliospores. (See also FLAX RUST.)

Melanconiales An order of fungi (class COELOMYCETES) in which the vegetative mycelium occurs *within* the substrate or host, and the conidiomata are formed in the superficial layers of the host, becoming erumpent at maturity; mature conidiomata are either acervuli or sporodochia. Some species are important plant pathogens (see e.g. ANTHRACNOSE). Genera include COLLETOTRICHUM, *Cylindrosporium, Entomosporium, Marssonina* and *Pestalotia.* (See also DIPLOCARPON.)

melanin A high-MWt, amorphous polymer of indole quinone; melanins are pigments found in plants, animals (including the skin and hair in man), insects etc, and also in certain microorganisms (e.g. in actinomycetes, in hyphal walls of dematiaceous fungi, in the spore walls of e.g. *Mucor* spp). Melanins are relatively non-specific enzyme inhibitors; they may e.g. protect fungal cell walls from enzymic attack [Book ref. 38, pp. 373–374] and prevent cell lysis in old cultures. Melanins in plants may play a role in resistance to fungal infection.

The first step in melanin biosynthesis involves the *o*-hydroxylation of L-tyrosine (by tyrosinase) to form L-3,4-dihydroxyphenylalanine (L-dopa); L-dopa is used e.g. in the treatment of Parkinson's disease, and attempts have been made to adapt melanin-producing microorganisms for the commercial production of L-dopa [Book ref. 31, p. 567].

Melanogaster See GASTEROMYCETES (Melanogastrales).

Melanogastrales See GASTEROMYCETES.

Melanoplus sanguinipes **EPV** See ENTOMOPOXVIRINAE.

Melanospora See SORDARIALES.

Melanotaenium A genus of smut fungi (order USTILAGINALES) which are parasitic on the pteridophyte *Selaginella.*

melarsoprol See ARSENIC.

Melasmia A genus of fungi of the COELOMYCETES. (See also RHYTISMA.)

melezitose A trisaccharide – α-D-glucopyranosyl-(1 $\rightarrow$ 3)-β-D-fructofuranosyl-(2 $\leftrightarrow$ 1)-α-D-glucopyranoside – found in nature e.g. in certain HONEYDEWS. It can be metabolized e.g. by certain *Lactobacillus* spp. Hydrolysis of melezitose by α-glucosidase yields glucose and turanose (α-D-glucopyranosyl-(1 $\rightarrow$ 3)-D-fructose).

melibiose A reducing disaccharide: α-D-galactopyranosyl-(1 $\rightarrow$ 6)-D-glucopyranose, formed from RAFFINOSE by the action of β-fructosidase (invertase). In *Escherichia coli* melibiose uptake occurs by Na^+/melibiose symport (see ION TRANSPORT).

melioidosis A disease of man and animals (particularly rodents) caused by *Burkholderia* (*Pseudomonas*) *pseudomallei* – commonly by strains of the biotype that are unable to assimilate L-arabinose. The disease occurs primarily in tropical regions, particularly in S. E. Asia. Infection usually occurs via wounds or (infrequently) by inhalation of contaminated dust or by ingestion.

The causal organism occurs as a free-living bacterium in both soil and water; infection by *B.* (*P.*) *pseudomallei* (through cuts, abrasions) has been reported to occur more often during the wet season.

Person-to-person transmission of the disease has been reported [Lancet (1991) *337* 1290–1291].

In man, the disease is very variable, ranging from a local suppurative lesion at the site of infection to an acute, fulminating septicaemia. The disease may flare up after long periods of time (as found e.g. in American veterans of the Vietnam war).

In a recent study of melioidosis in Taiwan, all the isolates of *B.* (*P.*) *pseudomallei* were reported to be susceptible to the following antibiotics: amoxycillin-clavulanate, piperacillin-tazobactam (tazobactam is a β-lactamase inhibitor), imipenem and meropenem [EID (2001) *7* 428–433].

Meliola See BLACK MILDEWS.

Melissococcus See EUROPEAN FOULBROOD.

Melittangium See MYXOBACTERALES.

Melksham virus See SMALL ROUND STRUCTURED VIRUSES.

Melolontha melolontha EPV See ENTOMOPOXVIRINAE.

Melosira A genus of freshwater and marine centric DIATOMS in which the cylindrical cells (diam. ca. 5–100 μm) are typically joined valve-to-valve to form long chains. In many species the valves are highly pleomorphic; in e.g. *M. islandica* valve morphology apparently depends on nutrient (phosphorus and silicon) levels [Limn. Ocean. (1985) *30* 414–418]. Sexual reproduction is oogamous.

melting (of DNA) See HELIX-COIL TRANSITION.

Meltzer's reagent A mycological stain used e.g. for studying spore septation and ascus structure; it consists of iodine (0.5 g), potassium iodide (1–1.5 g), and chloral hydrate (20 g) in distilled water (20 ml).

membrane anchor sequence See SIGNAL HYPOTHESIS.

membrane attack complex See COMPLEMENT FIXATION (a).

membrane co-factor protein See COMPLEMENT FIXATION (b).

membrane filter See FILTRATION.

membrane fusion protein (MFP) In certain transport systems of Gram-negative bacteria: a protein which appears to act as a link between inner and outer membranes (see e.g. ABC EXPORTER and RND).

membrane-inlet mass spectrometry A method used for the direct and continuous monitoring of gas(es) dissolved in a liquid medium, e.g. a bacterial culture. Essentially, a hydrophobic membrane (e.g. PTFE – see FILTRATION) separates the liquid from a vacuum; dissolved gas(es) pass through the membrane into the vacuum and thence into the mass spectrometer. [MS (1984) *1* 200–203; Book ref. 132, pp. 239–262; some applications: Book ref. 132, pp. 271–294.]

membrane potential See CHEMIOSMOSIS.

membrane trigger hypothesis See SIGNAL HYPOTHESIS.

membranelle (*protozool.*) A discrete tuft or band of closely packed, apparently coherent cilia which behave as a unified organelle. (cf. SYNCILIUM.) The term is sometimes used to refer specifically to any one of the serially arranged membranelles of the AZM (q.v.). (See also COMPOUND CILIATURE.)

membranous croup *Syn.* DIPHTHERIA.

memory cells (*immunol.*) Lymphocytes which have had initial contact with specific antigen and which, on subsequent challenge with antigen, can give a rapid and heightened response (compared with the response that followed initial challenge).

Unlike other lymphocytes in the circulation, memory cells persist for long periods of time; they are also distinguishable by certain cell-surface antigens, e.g. CD44.

(See also ANTIBODY FORMATION.)

menadione See QUINONES.

menaquinones See QUINONES.

Mendosicutes A proposed division of the PROCARYOTAE comprising the single class ARCHAEOBACTERIA.

mengovirus See CARDIOVIRUS.

meningitis Inflammation of the meninges (membranes covering the brain and spinal cord). Meningitis may be of non-microbial causation, or it may result from primary or secondary infection by any of a range of microorganisms which gain access to the meninges via the blood or lymph systems, head wounds, or paranasal sinuses.

(a) *Bacterial meningitis*. Symptoms typically include a severe, throbbing headache, stiff neck, fever, delirium and coma; bacterial meningitis may be rapidly fatal. *Lab. diagnosis*: the pathogen can often be detected by microscopic examination of a Gram-stained smear of CSF sediment.

Meningococcal meningitis ('epidemic meningitis') – caused by *Neisseria meningitidis* (the meningococcus) – occurs sporadically or in epidemics. Transmission occurs by direct contact or by droplet infection; asymptomatic carriers harbour meningococci in the nasopharynx. Incubation period: ca. 2–10 days. Onset is usually abrupt, with rapid progression to confusion and coma; meningitic symptoms are often accompanied by petechial or purpuric skin lesions which may become gangrenous. Mortality rates: ca. 25–75% in untreated cases, ca. 7–10% in treated cases. A severe, fulminating form involving cyanosis, coma, and skin and adrenal haemorrhages (*Waterhouse–Friderichsen syndrome*) may be fatal within a few hours. *Chemotherapy*: benzylpenicillin, chloramphenicol. Rifampicin can eliminate meningococci from carriers. [A meningococcal C vaccine in teenagers: Lancet (2003) *361* 675–676.]

Haemophilus influenzae type b is a common cause of bacterial meningitis in infants and young children, often secondary to otitis media, pneumonia or viral respiratory disease. Mortality rates (treated cases): ca. 3–10%. *Chemotherapy*: e.g. chloramphenicol.

Pneumococcal meningitis (caused by *Streptococcus pneumoniae*) occurs mostly in infants and the elderly or debilitated. It is commonly associated with pneumococcal pneumonia, but may follow otitis media, head wounds etc. Mortality rates: ca. 20–70%. *Chemotherapy*: e.g. benzylpenicillin, chloramphenicol.

Various other bacteria can cause meningitis in adults – e.g. other streptococci, staphylococci, *Acinetobacter calcoaceticus*, *Listeria monocytogenes*, enterobacteria. *Neonates* are particularly vulnerable to meningitis caused by *Escherichia coli* and other enterobacteria, group B streptococci, *Staphylococcus aureus*, *L. monocytogenes*, *Flavobacterium meningosepticum* etc; mortality rates may be high, and motor or intellectual impairment often persists in survivors.

(b) *Aseptic meningitis* is generally a benign, self-limiting condition characterized by a lymphocytic infiltration of the meninges and raised protein levels in the CSF. It is usually caused by a virus – commonly mumps virus or enteroviruses (e.g. echoviruses or coxsackieviruses), also e.g. ENCEPHALITIS viruses, herpes simplex virus and LCM VIRUS. A lymphocytic meningitis may also occur in e.g. CRYPTOCOCCOSIS, LEPTOSPIROSIS, SYPHILIS and TUBERCULOSIS; following administration of nerve-tissue rabies vaccines (e.g. Semple vaccine); and in certain non-microbial diseases (e.g. cerebral tumours). (See also MENINGOENCEPHALITIS.)

meningococcus *Neisseria meningitidis*.

meningoencephalitis Inflammation of the brain and its meninges (cf. ENCEPHALITIS and MENINGITIS). *Primary amoebic meningoencephalitis* is a human disease which affects mainly children and young adults and is usually fatal. It is usually caused by *Naegleria fowleri*. Infection occurs via the nasal mucosa and is usually associated with swimming in e.g. freshwater lakes rich in organic matter. Symptoms may include headache, fever, vomiting, disturbances to taste, smell and vision, and coma; death may occur within ca. 72 h. Amoebae of the *Acanthamoeba-Hartmannella* group may cause a subacute or chronic granulomatous meningoencephalitis. *Lab. diagnosis*: identification of amoebae in the CSF. *Chemotherapy*: e.g. amphotericin B (see also MICONAZOLE).

meningoencephalomyelitis See ENCEPHALITIS.

meningomyelitis Inflammation of the spinal cord and its meninges (cf. MENINGITIS).

Meniscus A genus of Gram-negative, anaerobic (but aerotolerant) bacteria which occur e.g. in anaerobic digester sludge. The cells are capsulated, straight or curved, non-motile rods, 0.7–1.0 × 2.0–3.0 μm, which contain gas vacuoles. The organisms are chemoorganotrophs with a strictly anaerogenic fermentative metabolism; growth requirements include vitamin B_{12}, thiamine, and a raised partial pressure of CO_2. Optimum growth temperature: ca. 30°C. GC%: ca. 45. Type species: *M. glaucopsis*. [Book ref. 22, pp. 135–137.]

Menoidium See EUGLENOID FLAGELLATES.

menthol See PHENOLS.

mepacrine *Syn.* QUINACRINE.

merbromin See MERCURY.

2-mercaptoethanesulphonic acid Coenzyme M: see METHANOGENESIS.

mercaptoethanol ($HSCH_2CH_2OH$) A reagent used e.g. to reduce disulphide bonds in proteins (e.g. immunoglobulins – see FAB PORTION).

Mercurochrome See MERCURY.

mercury (as an antimicrobial agent) Mercury is a HEAVY METAL whose compounds include many effective antimicrobial agents; such agents are typically reversibly microbistatic, and they have little or no effect on the viability of bacterial endospores – though they can delay germination. The functional part of an antimicrobial mercurial is the mercury cation or the mercury-containing moiety; antimicrobial activity may involve the binding of these entities to certain groups (particularly thiols – but also e.g. amides and amines, purines and pyrimidines) in e.g. enzymes, cell walls, cytoplasmic membranes and nucleic acids. Microbistatic activity can be reversed with e.g. thiols – which probably compete with the mercury-binding sites of the organisms. In bacteria, resistance to mercurials can be plasmid-mediated.

Inorganic mercurials are toxic and irritant to tissues; their activity is readily inhibited or reversed by organic matter (e.g. blood or serum). *Mercuric chloride* (corrosive sublimate) has been used for the disinfection of inanimate objects when other methods are unsuitable, and as a preservative for e.g. paper and timber. *Mercurous chloride* (calomel) has restricted use as a fungicide in horticulture – e.g. as a control against clubroot. *Yellow mercuric oxide* has been used in ointments for the treatment of conjunctivitis and blepharitis. *Ammoniated mercury* (aminomercuric chloride; white precipitate; $HgNH_2Cl$) has been used in ophthalmic preparations and for the treatment of superficial mycoses.

Organic mercurials are generally less toxic (to man), are less irritant, and are typically more effective antimicrobial agents. *Mercurochrome* (merbromin; disodium 2′,7′-dibromo-4′-hydroxymercurifluorescein) was one of the first organic mercurial antiseptics, but it is among the least effective and is no longer widely used. *Nitromersol* (Metaphen; 4-nitro-3-hydroxymercuri-*o*-cresol anhydride) is used e.g. in antimicrobial solutions and ointments for the treatment of e.g. conjunctivitis and superficial mycoses. (See also THIOMERSAL.) *Phenylmercuric borate*, *nitrate* and *acetate* (the most soluble) all have similar antimicrobial properties; they have been used as antiseptics, as preservatives for pharmaceutical preparations, paper, wood etc, and as seed dressings.

Microbial resistance to mercury and organomercurials: see HEAVY METALS.

meri- Combining form signifying 'part'.

Meria See HYPHOMYCETES and NEMATOPHAGOUS FUNGI.

Merismopedia A phycological genus of unicellular 'blue-green algae' (Chroococcales) in which the cells occur (in nature) in rectangular 'plates'. See SYNECHOCYSTIS.

meristem A region of an organism at which new, permanent tissue is formed, i.e., a 'growth region' of a plant or multicellular microorganism.

meristematic nodules See ROOT NODULES.

meristoderm In certain algae (e.g. *Laminaria*): a surface layer of cells which, on division, gives rise to an increase in the thickness of the thallus.

meristogenous development (*mycol.*) The development of certain fungal structures (e.g. pycnidia, stromata) by the proliferation and differentiation of a small number of adjacent cells in a single hypha. (cf. SYMPHOGENOUS DEVELOPMENT.)

mermaid's wine-glass See ACETABULARIA.

mero- Combining form signifying 'part'.

Merocystis See EIMERIORINA.

merodiploid See DIPLOID.

merogony *Syn.* SCHIZOGONY.

meromictic lake A permanently stratified lake in which only the upper layers undergo turnover, the lower layer (*hypolimnion*) being typically stagnant, anaerobic, and rich in sulphide. (cf. HOLOMICTIC LAKE.)

meropenem A broad-spectrum 1-β-methylCARBAPENEM antibiotic. (See also ERTAPENEM.)

merosporangium An elongated ('cylindrical') SPORANGIUM containing either a single spore or a small number of spores in a row; merosporangia are formed e.g. by *Piptocephalis* and *Syncephalastrum*.

merozoite See SCHIZOGONY.

merozygote A bacterium which is part DIPLOID, part haploid; it may be formed in certain processes in which chromosomal genes pass from one bacterium (the donor) to another (the recipient) – see CONJUGATION, TRANSDUCTION and TRANSFORMATION. The genetic complement of a merozygote comprises the *endogenote* (the recipient's own chromosome) and the fragment of genetic material (the *exogenote*) received from the donor. If corresponding ALLELES in the endogenote and exogenote are identical the merozygote is said to be a *homogenote*; if not, it is said to be a *heterogenote*.

merthiolate *Syn.* THIOMERSAL.

Merulius See APHYLLOPHORALES (Corticiaceae). (cf. SERPULA.)

MES A pH buffer based on 2-(*N*-morpholino)-ethanesulphonic acid.

Meselson–Radding model See RECOMBINATION (Figure 2).

mesocyst See CYST (b).

Mesodinium See GYMNOSTOMATIA.

***meso*-inositol** See INOSITOL.

mesokaryotic See DINOFLAGELLATES.

mesophile An organism whose optimum growth temperature lies within a range generally accepted as ca. 20–45°C; bearing in mind the definition of PSYCHROPHILE, it would be reasonable to extend this range to 15–45°C.

mesoplankton See PLANKTON.

mesoporphyrins See PORPHYRINS.

mesosaprobic zone See SAPROBITY SYSTEM.

mesosomes (chondrioids) Intracellular membranous structures, apparently continuous with the CYTOPLASMIC MEMBRANE, which have been observed in electron micrographs of many bacteria. In Gram-positive species they may appear e.g. as parallel or concentric lamellae or as vesicles (perhaps tubules) which may be interconnected; in Gram-negative bacteria mesosomes appear to be smaller and less intricate, and are often seen as simple invaginations of the cytoplasmic membrane. Mesosomes are commonly associated with the developing septum in a dividing cell, but their function – if any – is unknown; they are widely believed to be artefacts derived from the cytoplasmic membrane during preparation of specimens for electron microscopy [see e.g. AvL (1984) *50* 433–460 (439–443)]. (cf. INTRACYTOPLASMIC MEMBRANES.)

Mesostigma See MICROMONADOPHYCEAE.

Mesotaenium A genus of saccoderm DESMIDS.

messenger RNA See MRNA.

messenger RNA-interfering complementary RNA See ANTISENSE RNA.

metabasidium A developing BASIDIUM at the stage at which MEIOSIS occurs, or that part of a developing basidium in which meiosis occurs. According to species, the metabasidium may be a later stage of (but morphologically identical to) the PROBASIDIUM, or it may arise – as a separate structure – from the probasidium. According to species, a metabasidium gives rise to sterigmata (each STERIGMA bearing one terminal basidiospore) or to basidiospores borne directly on its surface (i.e., without sterigmata). A metabasidium may be septate (as e.g. in *Tremella*) or aseptate (as e.g. in *Agaricus*). In e.g. rust fungi the metabasidium is a septate structure (= 'promycelium') which grows out from the teliospore.

metabiosis The phenomenon in which one organism alters environmental conditions in such a way as to allow the growth of another or others; e.g., utilization of oxygen by aerobic organisms may create anaerobic microenvironments in which strict anaerobes can grow.

metabisulphites See SULPHUR DIOXIDE.

metabolic inhibition test Any test in which an agent is detected or quantified by its ability to inhibit metabolic activity. For example, a toxin, virus, or other cytocidal agent can be assayed by determining its ability to inhibit the normal metabolic processes in a tissue culture; normally, the medium in a tissue culture becomes acidic with time, and the titre of the cytocidal agent can be estimated by determining the highest dilution of the agent which delays or inhibits acidification of the medium.

metaboly See EUGLENA.

metachromasy The phenomenon in which certain substances (termed *chromotropes*) develop a colour different to that of the dye used to stain them. See e.g. METACHROMATIC GRANULES.

metachromatic dye Any DYE which gives rise to METACHROMASY when used to stain *chromotropic* substances.

metachromatic granules Intracellular POLYPHOSPHATE-containing granules which, on staining with certain basic dyes (e.g. polychrome methylene blue or toluidine blue – see ALBERT'S STAIN), exhibit METACHROMASY. [Composition of metachromatic granules: JB (1984) *158* 441–446.]

metachronal waves (*protozool.*) Waves which may be seen on the surface of a ciliate due to the beating of its cilia – an effect similar to that produced by an intermittent breeze on a field of barley or tall grass. While the cilia of a given ciliate beat with similar or identical frequencies, they do not, in general, beat synchronously – i.e., all the cilia do not pass simultaneously through the same phase of the beat cycle (see CILIUM). Instead, ciliary beating is usually *metachronous*: adjacent cilia pass through slightly different phases of the beat cycle at any given instant – the phase difference between a given cilium and other cilia in a KINETY increasing and decreasing cyclically with distance from the given cilium. (This results in the smooth motion characteristic of ciliates.) In one type of metachrony (*symplectic* metachrony) a sequence of cilia which, at a given instant, are all passing through (different stages of) the effective stroke of the beat cycle, tend to bunch together to form the crest of a metachronal wave; the crests of such waves move in the direction of the effective stroke. In *antiplectic* metachrony, cilia passing through the effective stroke do not bunch together, and the metachronal crests move in a direction opposite to that of the effective stroke. In both symplectic and antiplectic metachrony the effective and recovery strokes occur in the plane of the kinety; these types of metachrony are termed *orthoplectic* metachrony. In *diaplectic* metachrony the effective and recovery strokes occur in a plane perpendicular to the kinety; the effective stroke may be to the right (*dexioplectic* metachrony) or to the left (*laeoplectic* metachrony) when looking in the direction of the metachronal wave.

Symplectic metachrony is quite rare in ciliates, but antiplectic and diaplectic (dexioplectic) types of metachrony are quite common.

metacyclic forms (metatrypanosomes) Trypanosomes, infective for the vertebrate host, which develop at the end of the period of cyclical development in the invertebrate vector; they are trypomastigote in form, with or without a short free flagellum. In *Trypanosoma vivax* a 'surface coat' (see VSG) has been reported to occur on metacyclic forms prior to contact with the mammalian host [Parasitol. (1986) *92* 75–82].

Metacystis See GYMNOSTOMATIA.

Metadinium See ENTODINIOMORPHIDA.

metal leaching See LEACHING.

metalaxyl See PHENYLAMIDE ANTIFUNGAL AGENTS.

metalimnion (thermocline) In a lake: the layer of water between the EPILIMNION and the HYPOLIMNION; this layer is characterized by the maximum rate of decrease in temperature with depth.

Metallogenium A genus of IRON BACTERIA. The organisms have been reported to be capable of passing through a filter of pore size 0.2 μm and of inhibiting various prokaryotic and eukaryotic microorganisms when cultured with them [Curr. Micro. (1984) *11* 349–356].

metalloporphyrins See PORPHYRINS.

metallothionein A cysteine-rich, low-MWt, inducible protein which binds (and may detoxify) HEAVY METAL ions; metallothioneins from a range of organisms (e.g. *Synechococcus*, *Neurospora*, *Saccharomyces*, crab, horse, man) are structurally similar.

metaphase See MITOSIS and MEIOSIS.

metaphase plate See MITOSIS.

Metaphen See MERCURY.

Metarhizium A genus of entomopathogenic fungi (class HYPHOMYCETES). *M. anisopliae* can infect a wide variety of insects; it produces depsipeptide entomotoxins (*destruxins*) which apparently play an important role in pathogenesis. *M. anisopliae* has been used for the BIOLOGICAL CONTROL of e.g. froghoppers (*Mahanarva posticata*) in Brazilian sugar-cane plantations.

metastasis (*med.*) The dissemination of disease from a localized site in the body, with the formation of new foci at distant sites. (Usually used in the context of CANCER, but also of certain infectious diseases – e.g. TUBERCULOSIS.) (Hence *verb* metastasize, *adj.* metastatic.)

metatrypanosomes *Syn.* METACYCLIC FORMS.

Metchnikovella See RUDIMICROSPOREA.

methacycline See TETRACYCLINES.

metham sodium (vapam) Sodium *N*-methyl-DITHIOCARBAMATE (CH_3.NH.CS.SNa), a soil fumigant which, in soil, breaks down to form the volatile methyl-isothiocyanate. (cf. DAZOMET, QUINTOZENE.)

methanal *Syn.* FORMALDEHYDE.

methane monooxygenase (MMO) A MONOOXYGENASE which occurs in methanotrophs (see METHYLOCOCCACEAE) and which catalyses the oxidation of methane to methanol by molecular oxygen; MMO can also catalyse the oxidation of substrates such as *n*-alkanes and *n*-alkenes: e.g., propene (propylene) is oxidized to 1,2-epoxypropane. (See also CO-METABOLISM.)

MMO may be 'particulate' (i.e., membrane bound) or soluble, depending on growth conditions. Thus, e.g., in cells of *Methylococcus capsulatus* and '*Methylosinus trichosporium*' grown on methane, the MMO is soluble or particulate according to whether the culture medium contains low (e.g. 1 μM) or high (e.g. 5 μM) levels of Cu^{2+} respectively. However, in *M. capsulatus* only particulate MMO is formed when growth occurs on methanol – in this case an increase in the level of Cu^{2+} enhancing MMO activity [JGM (1985) *131* 155–163]. The soluble MMO of *M. capsulatus* is a three-component enzyme comprising proteins A (believed to be the oxygenase [JBC (1984) *259* 53–59]), B and C, and that of '*Methylosinus trichosporium*' is apparently similar.

The substrate specificities of particulate and soluble MMOs may differ appreciably. Thus, e.g. the soluble MMO of '*M. trichosporium*' can oxidize *n*-alkanes, *n*-alkenes, aromatic and alicyclic compounds, while the particulate MMO of the same organism cannot oxidize aromatic or alicyclic substrates [JGM (1984) *130* 3327–3333].

In various types of methanotroph the reducing power requirement for MMO activity in vitro (i.e. in cell extracts) appears to be satisfied by NADH for both the soluble and particulate forms of the enzyme. However, it has been proposed that the reducing power needed for in vivo particulate MMO activity in '*M. trichosporium*' may be derived from a reduced *c*-type cytochrome. It has been suggested that reducing power for in vivo particulate MMO activity in *M. capsulatus* may be supplied by an electron transfer protein (e.g. an Fe–S protein) the reduction of which depends on reversed electron transport [JGM (1983) *129* 3487–3497].

methane production See METHANOGENESIS.

methane utilization See METHANOTROPHY.

Methanobacillus omelianskii A syntrophic association of two species of prokaryote: *Methanobacterium bryantii* and an obligate proton reducer (the 'S' organism) which oxidizes ethanol to acetate and hydrogen gas.

Methanobacteriaceae A family of METHANOGENS (order METHANOBACTERIALES); two genera: METHANOBACTERIUM and METHANOBREVIBACTER. (This family formerly included *all* the methane-producing species.)

Methanobacteriales An order of METHANOGENS comprising those species whose cell wall contains PSEUDOMUREIN. Two families: METHANOBACTERIACEAE, METHANOTHERMACEAE.

Methanobacterium A genus of facultatively autotrophic archaeans (family METHANOBACTERIACEAE). Cells: non-motile rods. Methane may be produced from carbon dioxide and hydrogen and/or from formate. GC% ca. 33–50.

M. bryantii. Mesophilic. Cells: morphologically indistinguishable from *M. formicicum* (see below), but the organisms cannot use formate. Growth is stimulated by acetate and B vitamins. (See also METHANOBACILLUS OMELIANSKII.)

M. formicicum. Cells: round-ended rods, ca. 0.6 × 2–15 μm, often in chains. Formate can be used as an electron donor, and acetate stimulates growth.

M. thermoautotrophicum. Thermophilic (opt. temperature: 65–70°C). Formate is not metabolized.

Methanobrevibacter A genus of archaeans of the family METHANOBACTERIACEAE. Cells: cocci or short rods, often in pairs or chains. Mesophilic. Methane may be produced from carbon dioxide and hydrogen and/or from formate. GC% ca. 27–32.

M. arboriphilus. A facultative autotroph which occurs e.g. in soil (see also WETWOOD). Formate is not used.

M. ruminantium. Cells: rods, ca. 0.7 × 2.0 μm, often in chains. Habitat: the RUMEN. Methane can be formed from carbon dioxide and hydrogen or from formate. Strain M-1 has been used for the BIOASSAY of coenzyme M.

M. smithii. Morphologically similar to *M. ruminantium*. Coenzyme M is not required for growth.

methanochondrion A term proposed for the apparent organelle formed (in a number of methane-producing archaeans) from invaginated cytoplasmic membrane and assumed to be the site of methanogenesis. (Such a structure is not invariably present in methanogenic species [CJM (1984) *30* 594–604].)

Methanococcaceae See METHANOCOCCALES.

Methanococcales An order of METHANOGENS comprising one family (Methanococcaceae) and one genus, METHANOCOCCUS.

Methanococcoides A genus of METHANOGENS (see METHANOSARCINACEAE). The cells of *M. methylutens* are non-flagellated cocci (ca. 1 μm diam.); the cell wall consists of an S LAYER only. Methanol and methylamine are methanogenic substrates. GC%: ca. 42.

Methanococcus A genus of mesophilic and thermophilic archaeans (order METHANOCOCCALES) which occur e.g. in saltmarsh sediments and marine muds. Cells: motile cocci in which the cell wall consists solely of an S LAYER; the GRAM REACTION is negative. Methane is typically produced from carbon dioxide and hydrogen and/or from formate. Most species (but not *M. voltae*) can grow autotrophically. (The original type species, *M. mazei*, differs markedly from the other species, and has been transferred to METHANOSARCINA (*Methanosarcina mazei*), *Methanococcus vannielii* being designated the new type species of the (conserved) genus *Methanococcus* [IJSB (1986) *36* 491].) GC%: ca. 30–33.

M. jannaschii. Thermophilic (optimum: 86°C). Methane is produced from carbon dioxide and hydrogen, but not from formate. The growth rate is the highest of all known methanogens (doubling time: 30 min). Isolated from a HYDROTHERMAL VENT.

M. maripaludis. Cells 1.2–1.6 μm diam. Mesophilic.

M. mazei. See METHANOSARCINA.

M. thermolithotrophicus. Cells ca. 1.5 μm diam. Thermophilic (optimum: 65°C).

M. vannielii. Cells ca. 3–4 μm diam. Mesophilic. Methane is produced from carbon dioxide and hydrogen and from formate.

M. voltae. Cells ca. 0.5–3.0 μm diam. Mesophilic. Growth is optimal in 0.4 M NaCl. Methane is produced from carbon dioxide and hydrogen and from formate.

methanofuran See METHANOGENESIS.

methanogen Any member of the domain ARCHAEA which can carry out METHANOGENESIS; all known methanogens are obligate anaerobes which occur e.g. in muds and in the RUMEN (some forming close physical associations with protozoa [MS (1986) *3* 100–105]). (See also ANAEROBIC DIGESTION and WETWOOD.)

Methanogens have been classified in three orders (METHANOBACTERIALES, METHANOCOCCALES, METHANOMICROBIALES) on the basis of 16S rRNA sequence analysis.

Many methanogens can grow chemolithoautotrophically in mineral salts solution with carbon dioxide and hydrogen (although some need e.g. acetate and amino acids). Methanogens are not necessarily confined to simple substrates such as acetate, methanol and methylamines; thus, pure cultures of e.g. *Methanospirillum* can be grown on 2-propanol and certain other alcohols [AEM (1986) *51* 1056–1062].

The methanogens have been divided into two groups: group I (carbon dioxide and hydrogen, or formate, typically used), and group II (carbon dioxide and hydrogen, or formate, typically not used); group II includes the family METHANOSARCINACEAE [AvL (1984) *50* 557–567].

Autotrophic carbon dioxide fixation apparently occurs via a pathway resembling that used for the synthesis of acetyl-CoA from carbon dioxide in ACETOGENESIS, although formyl tetrahydrofolate synthase does not seem to be involved. Acetyl-CoA appears to be reductively carboxylated to pyruvate – which is converted to phosphoenolpyruvate (PEP); carboxylation of PEP gives rise to oxaloacetate. Oxaloacetate appears to be metabolized to 2-oxoglutarate as in the TCA cycle [see Appendix II(a)] – via citrate in *Methanosarcina barkeri*, via malate in *Methanobacterium thermoautotrophicum*.

Some species (e.g. *Methanococcus thermolithotrophicus*, *Methanosarcina barkeri*) can carry out NITROGEN FIXATION [Nature (1984) *312* 284–288; FEMS Ecol. (1985) *31* 47–55].

methanogenesis The biosynthesis of methane (CH_4): an energy-yielding process carried out by certain members of the domain Archaea (METHANOGENS) in anaerobic environments that are characterized by an E_h below about −330 mV (see e.g. ANAEROBIC DIGESTION, RUMEN and WETWOOD). Methane can be formed from several substrates: carbon dioxide and hydrogen (or carbon dioxide and formate), acetate, methanol, or methylamines. Methanogenesis involves a number of novel enzymes and coenzymes/cofactors [FEMS Reviews (1999) *23* 13–38].

Methane from carbon dioxide and hydrogen. Essentially, the carbon atom of carbon dioxide is bound, sequentially, to each of several C_1-carrier molecules and undergoes progressive reduction (via formyl and methyl stages) to methane (see figure). The final stage (reduction of CoM-S-CH_3) is thermodynamically favourable ($\Delta G^{0'}$ greater than −100 kJ).

Methane from acetate. Growth on acetate is slower than that on carbon dioxide and hydrogen; the free energy of the reaction

$$CH_3CO_2^- + H_2O \longrightarrow CH_4 + HCO_3^-$$

is about −37 kJ. Note that this pathway differs from the carbon dioxide/hydrogen pathway e.g. in the requirement for an initial stage of ATP-dependent activation.

Methanogenesis from acetate is inhibited by cyanide (which inhibits CO dehydrogenase).

Methane from methanol. In this pathway, the methyl group from one molecule of methanol is oxidized to provide electrons

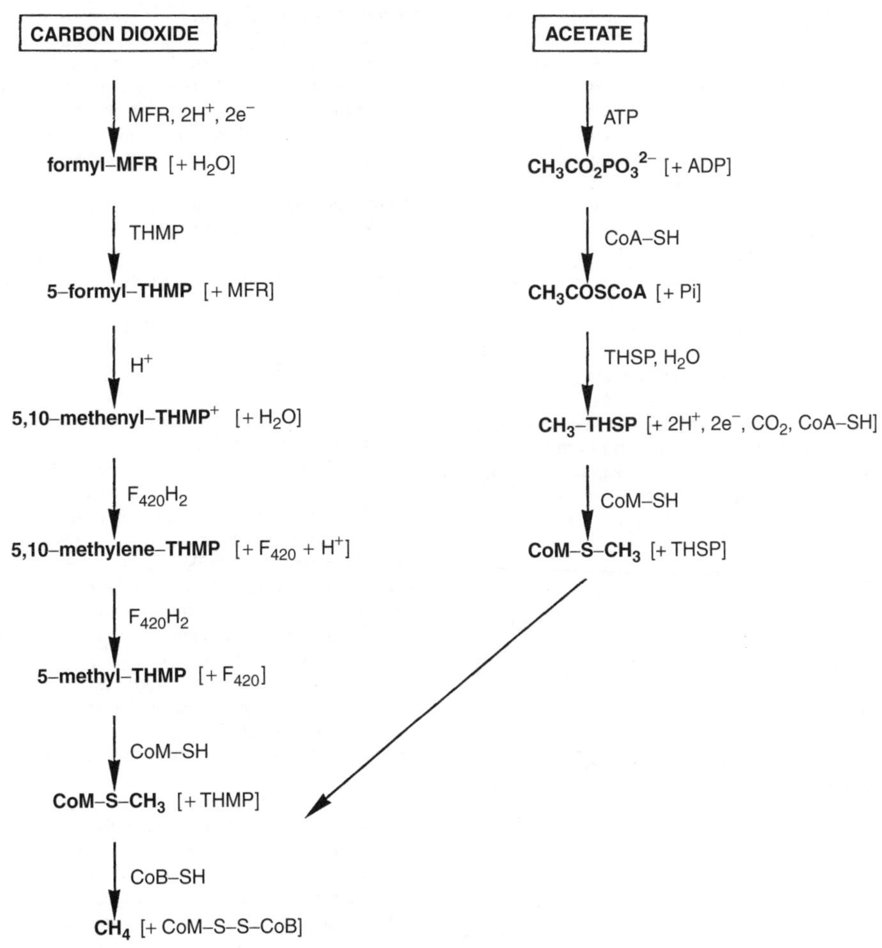

METHANOGENESIS. Two major pathways in the biosynthesis of methane: (i) synthesis from carbon dioxide, and (ii) synthesis from acetate. (Methanol and methylamines are alternative substrates.) The final stage of biosynthesis, i.e. reduction of methyl-coenzyme M (see later), is common to all the methane-forming pathways.

(i) *Methane from carbon dioxide*. In this pathway, the reduction of carbon dioxide can be linked to the oxidation of either hydrogen or formate; the reactions shown occur in both hydrogen- and formate-oxidizing pathways.

Initially, carbon dioxide is reduced to a formyl residue on the coenzyme *methanofuran* (MFR; earlier name: carbon dioxide reduction factor, CDR factor). Methanofuran is a long-chain molecule containing an aromatic ring and a terminal furan ring. This initial reduction is catalysed by formylmethanofuran dehydrogenase. The formyl group is then transferred to *tetrahydromethanopterin* (THMP). Part of the THMP molecule resembles FOLIC ACID, but the molecule is much larger than that of folic acid and it includes a sugar residue. The conversion of 5-formyl-THMP to 5,10-methenyl-THMP$^+$ is catalysed by the enzyme N^5, N^{10}-methenyltetrahydromethanopterin cyclohydrolase.

Subsequent reductions in some methanogens (e.g. strains of *Methanosarcina barkeri*) involve an enzyme acting in association with *coenzyme F_{420}* (factor 420), a 5-deazaflavin derivative that serves as an electron carrier. Strains of e.g. *Methanococcus thermolithotrophicus* use an F_{420}-independent dehydrogenase to convert 5,10-methenyl-THMP$^+$ to 5,10-methylene-THMP.

Finally, the methyl group is transferred from THMP to *coenzyme M* (CoM-SH; 2-mercaptoethanesulphonic acid; $HS(CH_2)_2SO_3^-$) in preparation for its reduction to methane by the enzyme *methyl-coenzyme M reductase*; in all cases, the electron donor for this reduction is *coenzyme B* (CoB; 7-mercaptoheptanoylthreonine phosphate). In the reductases from *Methanobacterium thermoautotrophicum* the active site of the enzyme contains two molecules of *coenzyme F_{430}* (factor 430), a nickel-containing porphinoid.

Reduction of CoM-S-S-CoB regenerates the two coenzymes.

(ii) *Methane from acetate*. The sequence of reactions shown are those for *Methanosarcina thermophila*; the pathway differs slightly in some methanogens. (*Continued on page 470.*)

for the reduction of methyl groups from (three) other molecules of methanol, the products being methane, carbon dioxide and water. For methane synthesis, methyl groups must be bound to coenzyme M in preparation for methanogenic reduction involving CoB-SH (see figure). Contributing to this pathway in *Methanosarcina barkeri* is an enzyme system that includes methanol:corrinoid methyltransferase which catalyses the initial transfer of methyl groups to the corrinoid moiety; further enzymic activity transfers the methyl groups to CoM-SH.

In nature, methanogens are believed to produce most of the methane from acetate; in vitro studies on the degradation of rice straw in anoxic paddy soil have indeed shown that a high proportion of methane is synthesized from acetate under those conditions [FEMS Ecol. (2000) *31* 153–161]. Earlier work had suggested that acetate may be the preferred substrate at 10°C, with hydrogen being used at higher temperatures [FEMS Ecol. (1997) *22* 145–153]. However, despite the overall picture, in certain locations the contribution of hydrogen to methanogenesis may be ~70–100% [FEMS Ecol. (1999) *28* 193–202].

The production of methane in anoxic soils can be suppressed e.g. by nitrate; such inhibition appears to be due to the effects of intermediates of the denitrification process [FEMS Ecol. (1999) *28* 49–61].

Methanogenium A genus of archaeans of the family METHANOMICROBIACEAE. Cells: cocci in which the cell wall consists of an S LAYER only; the GRAM REACTION is negative. Species include *M. cariaci* and *M. marisnigri*, both peritrichously flagellated organisms with an optimum growth temperature of ca. 20°C.

methanol utilization See METHYLOTROPHY.

Methanolobus A genus of METHANOGENS (see METHANOSARCINACEAE). The cells of *M. tindarius* are monoflagellated cocci (ca. 1 μm diam.); the cell wall consists of an S LAYER only. Methanol and methylamine are methanogenic substrates. GC%: ca. 46.

Methanomicrobiaceae A family of METHANOGENS (order METHANOMICROBIALES); genera: METHANOGENIUM, METHANOMICROBIUM, METHANOSPIRILLUM.

Methanomicrobiales An order of METHANOGENS comprising the families METHANOMICROBIACEAE, METHANOPLANACEAE and METHANOSARCINACEAE.

Methanomicrobium A genus of archaeans of the family METHANOMICROBIACEAE. Cells: typically short rods in which the cell wall consists of an S LAYER only; the GRAM REACTION is negative.

M. mobile. Polarly flagellated rods which need e.g. acetate, THIAMINE, PYRIDOXINE and PABA for growth. Habitat: RUMEN.

M. paynteri. Non-motile coccobacilli which occur in marine sediments.

Methanoplanaceae A family of METHANOGENS (order METHANOMICROBIALES) [validation: IJSB (1984) *34* 270]. One genus: METHANOPLANUS.

Methanoplanus A genus of archaeans of the family METHANOPLANACEAE. The cells of *M. limicola*, the sole species, are plate-like and are bounded by an S LAYER; this organism has strong affinities with *Methanogenium* spp [Book ref. 157, p 17].

Methanosaeta A genus of obligately acetotrophic METHANOGENS (methanogens which synthesize methane only from acetate). The organisms, which include *M. concilii* and *M. thermoacetophila*, are found e.g. in rice fields.

Methanosarcina A genus of archaeans of the family METHANOSARCINACEAE. Cells: non-motile, coccoid to irregularly shaped forms, ca. 1–3 μm; the heteropolysaccharide-containing cell wall is a thick (up to ca. 200 nm) layer which may be laminated at the periphery. GAS VACUOLES are found in some strains. The original type species, *M. methanica*, has been rejected and replaced by *M. barkeri* [IJSB (1986) *36* 492].

M. acetivorans. An acetotrophic marine species [AEM (1984) *47* 971–978].

M. barkeri. Cells grow as aggregates. Most strains can produce methane from carbon dioxide and hydrogen, acetate, methanol or methylamine. Strain FR-19 lyses spontaneously in substrate-depleted media [JGM (1985) *131* 1481–1486].

M. mazei (cf. METHANOCOCCUS) is similar to *M. barkeri* but differs e.g. in that *M. mazei* metabolizes acetate and methanol readily but carbon dioxide and hydrogen poorly or not at all.

Methanosarcinaceae A family of METHANOGENS (order METHANOMICROBIALES). Depending on species or strain, some or all of the following substrates are used for METHANOGENESIS: acetate, $CO_2 + H_2$, methanol, methylamines. Genera: METHANOSARCINA, METHANOTHRIX; it has been proposed that the genera METHANOCOCCOIDES and METHANOLOBUS be incorporated in the family [IJSB (1984) *34* 444–450].

Methanospirillum A genus of archaeans of the family METHANOMICROBIACEAE. *M. hungatei* is a facultative autotroph which occurs in sewage sludge; the cells are curved, sheathed rods or filaments that are lophotrichously flagellated.

Methanothermaceae A family of archaeans of the order METHANOBACTERIALES. One genus: METHANOTHERMUS.

Methanothermus A genus of archaeans (family METHANOTHERMACEAE) in which the cell wall consists of two layers: a PSEUDOMUREIN-containing layer and an (outermost) S LAYER. Thermophilic (optimum: 83°C).

M. fervidus. Cells: non-motile rods, ca. 0.4 × 1–3 μm; the GRAM REACTION is positive. Methane is formed from carbon dioxide and hydrogen but not from formate; yeast extract is needed for growth. GC%: ca. 33.

Methanothrix A genus of archaeans of the family METHANOSARCINACEAE. *M. soehngenii* is common in anaerobic digestors; cells: rods, ca. 0.8 × 2 μm, or long, sheathed filaments – each filament often consisting of several hundred rods. The GRAM REACTION is negative. Methane is formed from acetate but not from carbon dioxide and hydrogen or from formate, methanol or methylamine.

METHANOGENESIS (*continued*)

Initially, acetate undergoes ATP-dependent activation to acetyl-CoA – catalysed by the two enzymes *acetate kinase* and *phosphotransacetylase*. (In *Methanosaeta*, this section of the pathway differs in that the formation of acetyl-CoA is catalysed by *acetate thiokinase*.) The methyl group is then transferred to *tetrahydrosarcinapterin* (THSP; analogous to THMP, above) by an enzyme complex that includes *CO dehydrogenase* and *CoA synthase*; the activity of these enzymes results in the cleavage of carbon–carbon and carbon–sulphur bonds – and also oxidation of the carbonyl group (from acetyl-CoA) to carbon dioxide. Electrons derived from oxidation of the carbonyl group are involved in the final methanogenic reduction of the methyl group after the latter has been transferred to coenzyme M.

In *Methanosarcina thermophila* the *cam* gene encodes a CARBONIC ANHYDRASE which apparently includes a signal peptide (see SIGNAL HYPOTHESIS); it has been suggested that, by hydrating carbon dioxide in the periplasm, this enzyme may facilitate removal of carbon dioxide from the cytoplasm – and may thus promote conditions thermodynamically more favourable for methanogenesis.

M. concilii is an acetotrophic species isolated from sewage sludge [CJM (1984) *30* 1383–1396]. [Acetate and carbon dioxide assimilation by *M. concilii*: JB (1985) *162* 905–908.]

methanotroph Any organism capable of METHANOTROPHY.

methanotrophy The use of methane as the sole source of carbon and energy: a mode of metabolism which occurs under aerobic and microaerobic conditions in certain bacteria (see e.g. METHYLOCOCCACEAE). (cf. METHYLOTROPHY.) Most methanotrophic bacteria are obligately methanotrophic; these organisms include species of *Methylococcus* and *Methylomonas*.

The methanotrophic bacteria occur e.g. in soil, water and sediments. The oxidation of methane by methanotrophs significantly affects the global methane budget in that much less methane reaches the atmosphere; this reduces the contribution of methane to global warming.

In members of the Methylococcaceae, methane is initially oxidized to methanol by METHANE MONOOXYGENASE (q.v.); the subsequent oxidations are: methanol → formaldehyde → formate → carbon dioxide. When methane is used as the sole source of carbon and energy, some formaldehyde is assimilated into cell biomass while the remainder is oxidized to yield energy.

[Methane oxidation in natural waters: AEM (1982) *44* 435–446. Methane oxidation by microorganisms: Book ref. 132, pp 173–200. Methanotrophic bacteria (review): MR (1996) *60* 439–471. Ecology of methanotrophic species studied by molecular methods: FEMS Ecol. (1998) *27* 103–114. Detection and classification of atmospheric methane-oxidizing bacteria in soil: Nature (2000) *405* 175–178.]

methazotrophic Refers to any organism capable of using methylamines as the sole source of nitrogen (but not as sole source of carbon and energy); methazotrophs include various yeasts, e.g. *Candida utilis*. [Oxidation of methylamines by yeasts: Book ref. 169, pp. 155–164.]

methenamine See HEXAMINE.

methenamine–silver stain (Grocott–Gomori stain; Gomori stain) A stain used e.g. for detecting actinomycetes or fungi in tissue sections. The section is oxidized in 5% chromic acid for 1 h, washed in tap water, washed for 1 min in 1% sodium bisulphite to eliminate chromic acid, and washed for 10 min in tap water. After 3–4 changes of distilled water the section is immersed in a solution made by adding 0.4% aqueous borax (108 ml) to a mixture of 10% silver nitrate (7 ml) and 3% methenamine (100 ml); the temperature is kept at 58–60°C until the section becomes yellow-brown (30–60 min). The slide is rinsed repeatedly in distilled water, washed for 2–5 min in 2% sodium thiosulphate to remove unreduced silver, washed in tap water, and counterstained for ca. 1 min in a solution of light green (0.04% w/v) containing 0.04% v/v glacial acetic acid. Mycelium stains brown-black, tissues green.

methicillin See PENICILLINS.

methicillin-resistant *Staphylococcus aureus* See MRSA.

methicillinase See MRSA.

methionine aminopeptidase (MAP) In bacteria: an enzyme required for excision of the N-terminal methionine of a newly synthesized protein. In e.g. *Escherichia coli*, the N-terminal methionine is cleaved from about 50% of proteins. Whether or not such cleavage occurs in a given protein is influenced by the identity of the *penultimate* N-terminal amino acid; cleavage, or otherwise, of the N-terminal methionine thus determines the identity of the N-terminal amino acid in the *mature* protein, and, in this way, influences the protein's half-life (according to the N-END RULE).

MAP belongs to a diverse family of physiologically important aminopeptidases [bacterial aminopeptidases: FEMS Reviews (1996) *18* 319–344].

L-methionine biosynthesis See Appendix IV(d). (See also MICROCINS.)

methionine deformylase In bacteria: an enzyme which cleaves the formyl group from methionine at the N-terminal of a newly synthesized protein.

L-methionine-DL-sulphoximine See MSX.

methisazone See ISATIN-β-THIOSEMICARBAZONE.

Methocel See METHYLCELLULOSE.

methotrexate See FOLIC ACID ANTAGONIST.

methoxatin See QUINOPROTEIN.

methoxsalen 8-MethylPSORALEN.

methoxyamine (as a mutagen) See HYDROXYLAMINE.

6-methoxymellein A PHYTOALEXIN elicited from the carrot plant e.g. by cupric or mercuric chloride.

methyl-accepting chemotaxis protein See CHEMOTAXIS.

methyl blue (cotton blue) A blue acid TRIPHENYLMETHANE DYE; cf. LACTOPHENOL COTTON BLUE.

methyl bromide (CH_3Br) Methyl bromide vapour is fungicidal, insecticidal and herbicidal; it has good powers of penetration and is used as a fumigant.

methyl green A green basic TRIPHENYLMETHANE DYE. (See also METHYL GREEN–PYRONIN STAIN.)

methyl green–pyronin stain (Unna–Pappenheim stain) A stain which can distinguish DNA (which stains green) from RNA (which stains red); it can be used e.g. to distinguish lymphocytes from (ribosome-rich) plasma cells.

methyl orange A PH INDICATOR: pH 3.0–4.4 (red to orange-yellow); pK_a 3.6.

methyl red A PH INDICATOR: pH 4.4–6.2 (red to yellow); pK_a 5.1.

methyl red test (MR test) An IMVIC TEST which determines the ability of an organism to acidify a phosphate-buffered glucose–peptone medium to pH 4.4 or below. The medium is inoculated and incubated for 48 h at 37°C or 5 days at 30°C. Two drops of METHYL RED solution (0.02% w/v in ca. 50% ethanol) are added to the culture; a red coloration indicates a positive reaction. (N.B. Too short an incubation time may give a false-positive reaction: some organisms (e.g. *Klebsiella pneumoniae* subsp. *pneumoniae*) initially form acidic metabolic products which are later metabolized further.) The VOGES–PROSKAUER TEST can be performed after the MR test using the same culture.

methyl violet A composite TRIPHENYLMETHANE DYE which includes crystal violet and 4,4′-*bis*(dimethylamino)-4″-(methylamino) triphenylmethyl chloride.

methyl viologen (Paraquat) 1,1′-Dimethyl-4,4′-dipyridinium, a widely used herbicide. (See also PHOTOSYNTHESIS.) The reduced form of methyl viologen is used e.g. as a reducing agent in laboratory studies.

3-methyladenine-DNA glycosylase I See EXCISION REPAIR.

3-methyladenine-DNA glycosylase II See ADAPTIVE RESPONSE.

methylation (of DNA) See DNA METHYLATION.

methylcellulose (Methocel) An unbranched polymer (MWt ca. 143000) used e.g. to increase the viscosity of a medium in order to slow down rapidly motile organisms – see COUNTING METHODS. (cf. FICOLL.)

methylcobalamin See VITAMIN B_{12}.

methylene blue A blue basic thiazine DYE used e.g. in STAINING and VITAL STAINING and as a redox indicator (E_0 = +11 mV at pH 7). (See also e.g. METHYLENE BLUE TEST; TOXOPLASMA DYE TEST.) *Leuco methylene blue* is the reduced (colourless) form of methylene blue. *Loeffler's methylene blue* is prepared by adding 0.01% KOH (100 ml) to 1% methylene blue in 95% ethanol (30 ml). If allowed to 'ripen' (i.e. oxidize slowly) over several months this stain becomes *polychrome methylene blue*: a mixture of methylene blue and certain products of its spontaneous oxidation (e.g. the metachromatic dyes azure A and azure B).

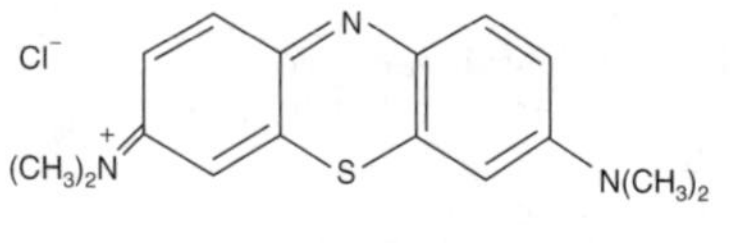

Cl^- $(CH_3)_2N^+$ N S $N(CH_3)_2$

METHYLENE BLUE

As a redox indicator (see e.g. GASPAK and REDOX POTENTIAL), methylene blue is often used in a TRIS- or phosphate-buffered solution at pH 9; at pH 9 the dye becomes colourless at a lower E_h than it does at pH 7.

methylene blue test (for milk) A test used to assess the numbers of (certain types of) microorganisms in untreated or pasteurized MILK. To a standard volume of milk is added a fixed volume of a solution of METHYLENE BLUE (or methylene blue thiocyanate); the whole is incubated at 37°C and examined periodically. If the milk contains large numbers of organisms that are metabolically active at 37°C, the dye will be quickly reduced to the colourless leuco form; decolorization is delayed if the milk contains relatively few organisms. The time taken for dye decolorization is thus a *guide* to the microbial load. (cf. RESAZURIN TEST.)

methylglyoxal bypass A sequence of reactions which can bypass reactions of the EMBDEN–MEYERHOF–PARNAS PATHWAY between triose phosphates and pyruvate. Methylglyoxal synthase converts dihydroxyacetone phosphate to methylglyoxal (2-oxopropanal: CH_3.CO.CHO) and inorganic phosphate; methylglyoxal can then be converted to D-lactate (by two glutathione-dependent enzymes, glyoxalase I and glyoxalase II), or to pyruvate (by methylglyoxal dehydrogenase). Although these enzymes have been identified in various bacteria and yeasts, their significance in vivo remains unclear; they could allow the formation of pyruvate (and hence acetyl-CoA) under conditions of phosphate deficiency which would otherwise limit the formation of 1,3-bisphosphoglycerate from glyceraldehyde 3-phosphate.

Methylglyoxal is cytotoxic, apparently reacting with 7-methylguanosine in rRNA and thereby inhibiting ribosomal function; the glyoxalase system probably plays an important role in detoxifying methylglyoxal formed from dihydroxyacetone phosphate or during other metabolic processes (e.g. the catabolism of certain amino acids).

[Methylglyoxal metabolism: ARM (1984) *38* 49–68.]

2-methylisoborneol A substance which is produced by certain aquatic actinomycetes and cyanobacteria (e.g. *Lyngbya*); in concentrated form, 2-methylisoborneol has a camphor-like smell, but when diluted it has a 'musty' or 'earthy' odour which can taint WATER SUPPLIES and fish living in the water. (cf. GEOSMIN.)

methylmethane sulphonate See ALKYLATING AGENTS.

***N*-methyl-*N*′-nitro-*N*-nitrosoguanidine** See MNNG.

***N*-methyl-*N*-nitrosourea** See ALKYLATING AGENTS.

Methylobacterium A genus of facultatively methylotrophic and methanotrophic, rod-shaped bacteria which can alternatively use glucose and other complex substrates as sole sources of carbon and energy. Type species: *M. organophilum*. (See also METHYLOCOCCACEAE and PROTOMONAS.)

Methylocaldum A genus of methylotrophic bacteria classified within the (gamma) PROTEOBACTERIA.

Methylococcaceae A family of Gram-negative, obligately methylotrophic bacteria characterized by the ability to use methane as the sole source of carbon and energy under aerobic or microaerobic conditions (see METHANOTROPHY); bacteria which can use other C_1 compounds, but not methane, are excluded. Two genera: METHYLOCOCCUS and METHYLOMONAS. [Book ref. 22, pp. 256–261.] (cf. METHYLOMONADACEAE.)

Members of the Methylococcaceae occur e.g. in soil and in water overlaying anaerobic mud. The cells range from cocci to rods, and all contain a complex arrangement of intracytoplasmic membranes when grown on methane; some strains are motile (polarly flagellated). Pigments, some water-soluble, are formed by some strains and may be e.g. blue, brown, red or yellow. Certain strains form desiccation-resistant resting stages; these stages may be cysts (similar to those of *Azotobacter* spp), 'lipid cysts' (which contain PHB inclusions, lack intracellular membranes, and have a thick pericellular coating), or 'exospores' (which develop by polar budding, and which are resistant to 85°C/15 min). Metabolism is respiratory (oxidative) with oxygen as the terminal electron acceptor; cytochromes of the *a*, *b*, *c* and *o* types occur in those strains examined. According to strain, the TCA CYCLE may be complete or e.g. may lack 2-oxoglutarate dehydrogenase (in which case the pathway is primarily biosynthetic). All strains can use methanol as a sole source of carbon and energy, and many or all can use formaldehyde. Methane oxidizers can use other carbon compounds, simultaneously with methane, as co-oxidizable substrates to provide carbon and energy; such supplementary carbon and energy sources include e.g. acetate, alkanes, formate, primary alcohols, and various alicyclic, aromatic and heterocyclic compounds. At least one species, *Methylococcus capsulatus*, can fix CO_2 via the Calvin cycle, and some methanotrophs can carry out NITROGEN FIXATION [JGM (1983) *129* 3481–3486]. All strains which have been examined are catalase +ve and oxidase +ve. Aerobic methane-oxidizing bacteria were divided into two major physiological groups; these groups included species which were not formally included in the Methylococcaceae. Type I organisms (e.g. *Methylococcus capsulatus*, *Methylomonas methanica*) form intracytoplasmic membranes comprising bundles of vesicular discs, may form cysts, have an incomplete TCA cycle, and assimilate carbon by the RMP PATHWAY. Type II organisms (e.g. METHYLOBACTERIUM, METHYLOSINUS TRICHOSPORIUM) form paired intracytoplasmic membranes aligned with the cell periphery, may form 'lipid cysts' or 'exospores', have a complete TCA cycle, and assimilate carbon by the SERINE PATHWAY.

[Ecology, isolation and culture: Book ref. 45, pp. 894–902.]

Methylococcus A genus of Gram-negative bacteria (family METHYLOCOCCACEAE) which occur e.g. in mud, soil and water. Cells: non-pigmented, non-motile cocci, ca. 1 μm in diameter. The organisms may be cultured on e.g. nitrate–mineral salts media with a gaseous phase consisting of a 30:70 mixture of methane and air; growth occurs between 30 and 50°C (optimally at 37°C), enrichment and primary isolation being carried out at 45°C. GC%: ca. 62.5. Type species: *M. capsulatus*.

Methylomonadaceae An obsolete bacterial family which consisted of methylotrophic organisms – the use of methane not being an obligatory requirement. (cf. METHYLOCOCCACEAE.)

Methylomonas A genus of Gram-negative bacteria within the family METHYLOCOCCACEAE. Cells: straight or curved, sometimes branched, motile (polarly flagellated) rods, 0.5–1.0 × 1.0–4.0 μm. Growth occurs optimally at 30°C, no growth occurring at 37°C; enrichment and primary isolation are otherwise as described under METHYLOCOCCUS. A pink pellicle is formed at 30°C; colonies are pink or yellow, consistently pink on subculture. A blue diffusible pigment may be formed in iron-deficient media. GC%: ca. 52. Type species: *M. methanica*. ('*M. methanitrificans*' and '*M. methanooxidans*' may be synonyms of *Methylosinus trichosporium*.)

Methylophaga A proposed genus of marine methylotrophic bacteria. The organisms are Gram-negative, strictly aerobic rods which grow on methanol, methylamine(s) or fructose (but not methane) and which require vitamin B_{12}; those strains tested use the RMP pathway for carbon assimilation. Two species: *M. marina*, *M. thalassia*. [IJSB (1985) *35* 131–139.]

Methylophilus methylotrophus A species of obligately methylotrophic, non-methanotrophic bacteria (see METHYLOTROPHY) which assimilate carbon via the RMP pathway and which have been used in the production of SINGLE-CELL PROTEIN.

Methylosinus trichosporium A species of methanotrophic, obligately methylotrophic, typically rod-shaped or vibrioid bacteria (see METHANOTROPHY, METHANE MONOOXYGENASE, METHYLOCOCCACEAE). [Effect of growth conditions on intracytoplasmic membranes and MMO activity: JGM (1981) *125* 63–84.]

methylotroph Any organism capable of METHYLOTROPHY.

methylotrophy Oxidative metabolism characterized by (i) obligate or facultative use of C_1 COMPOUNDS as the sole source of carbon and energy, *and* (ii) assimilation of formaldehyde (produced during methylotrophic metabolism) as a major source of carbon. Methylotrophy includes METHANOTROPHY.

Methylotrophic organisms include certain bacteria (e.g. *Hyphomicrobium*, *Methylobacterium* spp, members of the family METHYLOCOCCACEAE, *Methylophaga* spp, *Methylophilus methylotrophus*, *Microcyclus* and *Paracoccus denitrificans*) and some fungi (e.g. species of *Candida*, *Hansenula* and *Pichia*). (cf. CARBOXYDOBACTERIA.)

(a) *Methylotrophy in bacteria*. Most of the C_1 compounds can be oxidized via a sequence of reactions which end in $HCHO \rightarrow HCOOH \rightarrow CO_2$. Formaldehyde may be assimilated by either the RMP PATHWAY or the SERINE PATHWAY, according to species. (The serine pathway also assimilates some carbon dioxide.) Energy may be derived from direct (non-cyclical) oxidative pathways (e.g. methanol → formaldehyde → formate → carbon dioxide) or from the (cyclical) RMP PATHWAY operating in a dissimilatory mode. The TCA cycle appears to play little or no part in the generation of energy from C_1 substrates.

Methanol is used by most methylotrophs. It is oxidized to formaldehyde by a QUINOPROTEIN: *methanol dehydrogenase* (MDH; EC 1.1.99.8), whose electron acceptor is a soluble form of cytochrome c_L. [Roles for *c* cytochromes in methylotrophs: JGM (1984) *130* 3319–3325.]

Methane is oxidized to formaldehyde via methanol (see METHANOTROPHY).

Methylamines are oxidized directly to formaldehyde in reactions that usually involve dehydrogenases or monooxygenases; electrons from these substrates may be transferred to cyt *b* (via a flavoprotein), or may be passed to cyt *c* via the 'blue protein' *amicyanin*.

The oxidation of formaldehyde to formate may be catalysed by various enzymes – e.g. (i) MDH; (ii) NAD- and reduced-glutathione-dependent formaldehyde dehydrogenase (EC 1.2.1.1); or (iii) NAD(P)-independent non-specific aldehyde dehydrogenase.

Formate is oxidized by NAD-dependent formate dehydrogenase (EC 1.2.1.2). (Some methylotrophs can use exogenous formate as a source of energy and carbon.)

[Formaldehyde metabolism: Book ref. 169, pp 315–323. Physiology and biochemistry of methylotrophic bacteria: AvL (1987) *53* 23–28. Energetics of bacterial metabolism of C_1 compounds: AvL (1987) *53* 37–45.]

(b) *Methylotrophy in fungi*. Of C_1 compounds, only methanol has been studied extensively (as a substrate for certain yeasts); however, other C_1 compounds have been reported to be used as substrates by fungi.

The assimilatory and energy-yielding pathways of methanol metabolism in yeasts differ from those in bacteria.

Methanol can be oxidized via the sequence: methanol → formaldehyde → formate → carbon dioxide. The oxidation of methanol to formaldehyde takes place within PEROXISOMES and is catalysed by 'methanol oxidase' (EC 1.1.3.13), a non-specific alcohol oxidase which can oxidize e.g. several alkanols and alkenols, and which consists of eight FAD-containing subunits. The oxidation of methanol yields hydrogen peroxide as well as formaldehyde; CATALASE, within the peroxisome, eliminates the peroxide (as water and oxygen).

Formaldehyde may be assimilated, via the XMP PATHWAY, or may be oxidized to yield energy. The oxidation of formaldehyde occurs within the cytoplasm. In *Candida boidinii* this reaction appears to occur in three stages, the first stage involving NAD^+- and GSH-linked (reduced-glutathione-linked) formaldehyde dehydrogenase (EC 1.2.1.1):

$$GSH + HCHO + NAD^+ \longrightarrow HCO\text{-}SG + NADH + H^+$$

HCO-SG (*S*-formylglutathione) is hydrolysed by *S*-formylglutathione hydrolase (EC 3.1.2.12):

$$HCO\text{-}SG + H_2O \longrightarrow GSH + HCOOH$$

Finally, HCOOH may be oxidized to carbon dioxide by NAD-linked formate dehydrogenase – although the low substrate affinity of this enzyme in cell-free systems has raised doubts as to its function in vivo.

Methylotrophy has also been recorded in the yeast *Hansenula polymorpha*.

NADH generated in methanol-grown yeasts is oxidized in a (ROTENONE-insensitive but cyanide-sensitive) reaction involving an NADH dehydrogenase located at the outer surface of the inner mitochondrial membrane – the maximum energy yield being 2 ATP/NADPH. (In e.g. glucose-grown cells, most NADH is generated intra-mitochondrially and is oxidized by an NADH dehydrogenase located at the inner surface of the inner mitochondrial membrane, the maximum yield of energy being 3 ATP/NADH.)

[Methylotrophic yeasts: AvL (1987) *53* 29–36.]

methymycin See MACROLIDE ANTIBIOTICS.

metiram An agricultural antifungal agent used e.g. for the control of late blight of potato; it is a complex of ZINEB and polyethylene THIURAM DISULPHIDE.

Metopus See HETEROTRICHIDA.

metritis Inflammation of the uterus. (cf. ENDOMETRITIS; see also CONTAGIOUS EQUINE METRITIS.)

Metrizamide See IODINATED DENSITY-GRADIENT MEDIA.

metrocyte See SARCOCYSTIS.

metronidazole (Flagyl) 1-(2-Hydroxyethyl)-2-methyl-5-nitroimidazole: a NITROIMIDAZOLE antimicrobial agent used to treat e.g. amoebic and balantidial DYSENTERY, GIARDIASIS, genitourinary TRICHOMONIASIS and infections caused by anaerobic bacteria, e.g. PSEUDOMEMBRANOUS COLITIS and VINCENT'S ANGINA. The bactericidal effect of metronidazole on *Helicobacter pylori* appears to depend on the reduced (active) form of the drug rather than on toxic oxygen radicals formed during its reduction [JAC (1998) *41* 67–75].

Metschnikowia A genus of unicellular and pseudomycelial fungi (family METSCHNIKOWIACEAE) in which multilateral budding occurs. Ascospores: acicular, 1–2 per ascus. Six species are recognized; they occur in aquatic and terrestrial habitats and as parasites in invertebrates. *M. pulcherrima* and *M. reukaufii* are teleomorphs of *Candida pulcherrima* and *Candida reukaufii*, respectively. [Book ref. 100, pp. 266–278.]

Metschnikowiaceae (Nematosporaceae; Spermophthoraceae) A family of fungi (order ENDOMYCETALES) which includes unicellular, pseudomycelial and mycelial forms. Ascospores: fusiform to acicular (cf. SACCHAROMYCETACEAE). Genera include e.g. *Ashbya*, COCCIDIASCUS, *Eremothecium*, METSCHNIKOWIA, NEMATOSPORA, *Spermophthora*. Members are generally parasitic in plants and/or invertebrates (see e.g. STIGMATOMYCOSIS). Species of *Ashbya* and *Eremothecium* are important commercial sources of riboflavin.

metula In the conidiophores of certain fungi: a cell, or a short extension or branch (or part of a branch), which bears one or more *phialides* (see CONIDIUM) at its apex – see e.g. ASPERGILLUS and PENICILLIUM.

Mexico virus A SMALL ROUND STRUCTURED VIRUS.

mezlocillin See PENICILLINS.

MF_0F_1 H^+-ATPase See PROTON ATPASE.

MFP See MEMBRANE FUSION PROTEIN.

MFS Major facilitator superfamily: a category of TRANSPORT SYSTEMS which occur in prokaryotes and eukaryotes; they include pmf-dependent efflux pumps which mediate extrusion of antibiotics – and which therefore confer resistance to particular antibiotic(s). Examples of MFS systems are the various efflux pumps in Gram-negative bacteria which confer resistance to tetracyclines. Each pump encoded by a *tetA* gene is an integral inner membrane protein containing multiple transmembrane helices; *tetA* genes have been found on transposons and plasmids.

MG fungus See APHANOMYCES.

MGIT™ 960 A fully automated BACTEC system (Becton Dickinson) (MGIT = mycobacteria growth indicator tube) used e.g. for detecting *Mycobacterium tuberculosis* [evaluation: JCM (1999) *37* 748–752].

MHA (*immunol.*) Major histocompatibility antigen.

MHA-TP test Microhaemagglutination assay for *Treponema pallidum* antibodies, a miniaturized form of the TPHA TEST.

MHC (*immunol.*) MAJOR HISTOCOMPATIBILITY COMPLEX.

MHC restriction *Syn.* HISTOCOMPATIBILITY RESTRICTION.

MHD (1) (*serol.*) Minimum haemolytic dose. (a) The smallest quantity of COMPLEMENT needed to lyse completely a fixed quantity of a standardized suspension of *sensitized* erythrocytes. (See also HAEMOLYTIC SYSTEM.) (b) The smallest quantity of a complement-dependent haemolysin needed to lyse completely a fixed quantity of erythrocytes in the presence of excess complement. (c) The smallest quantity of streptolysin O needed to lyse completely a fixed quantity of erythrocytes in the ANTISTREPTOLYSIN O TEST. (The MHD may be referred to as '1 unit'.) (See also HD_{50}.)

(2) (*virol.*) Minimum haemagglutinating dose: the smallest quantity of a haemagglutinating virus capable of bringing about maximum agglutination of the erythrocytes in a given volume of a standardized suspension.

***mi-1* gene** See POKY MUTANT.

MIC (of an antibiotic) Minimum inhibitory concentration: the lowest concentration of a given antibiotic which inhibits a given type of microorganism under given conditions. (See also E TEST.)

Mickle shaker An apparatus used e.g. for the ballistic disintegration of tissue or cells; in principle it resembles the BRAUN MSK TISSUE DISINTEGRATOR. (See also BALLOTINI, CELL DISRUPTION and ULTRASONICATION.)

miconazole (1-[2,4-dichloro-β-(2,4-dichlorobenzyloxyl)phenethylimidazole]) An AZOLE ANTIFUNGAL AGENT with a broad spectrum of antifungal activity; it also has activity against certain bacteria (particularly Gram-positive species) and e.g. against *Naegleria fowleri*. Miconazole is used mainly in the treatment of superficial mycoses; it is usually administered topically, but can be given intravenously (in cases of invasive or systemic mycosis) or orally (e.g. to reduce intestinal populations of *Candida* spp).

Micrasterias A genus of placoderm DESMIDS.

microaerobic Refers to an environment in which oxygen is present (cf. ANAEROBIC) but at a partial pressure which is (usually significantly) lower than that in air.

microaerophilic Refers to any organism which grows optimally in a MICROAEROBIC environment. (Frequently the term is also used as a synonym of microaerobic.)

microalgae Microscopic algae – particularly unicellular algae such as CHLAMYDOMONAS, CHLORELLA, DIATOMS, etc.

microarray (DNA) See DNA CHIP.

Microascales An order of fungi of the ASCOMYCOTINA; the organisms are primarily saprotrophs in soil and dung, but some can be pathogenic in man and other animals. Ascocarp: perithecioid or cleistothecioid, dark and solitary, sometimes setose; hamathecium: absent. Asci: rounded, thin-walled, evanescent. Ascospores: aseptate, pigmented. Genera: e.g. *Microascus*, PSEUDALLESCHERIA.

Microascus See MICROASCALES.

Microbacterium A genus of catalase-positive, asporogenous bacteria (order ACTINOMYCETALES, wall type VI – see also PEPTIDOGLYCAN) which frequently occur in dairy products (e.g. spray-dried milk, some cheeses) – presumably as a result of improper cleansing of dairy equipment. In culture the organisms are slender, pleomorphic, non-motile rods. Growth occurs optimally at ca. 30°C on e.g. media containing yeast extract, peptone and milk; L-(+)-lactic acid is formed from glucose. Some strains produce a yellowish pigment. All strains grow aerobically. The organisms are typically thermoduric, surviving e.g. 63°C/30 min and 72°C/15 min. GC%: ca. 69–70. Type species: *M. lacticum*. (For *M. thermosphactum* see BROCHOTHRIX.)

microbe *Syn.* MICROORGANISM.

microbial insecticides See BIOLOGICAL CONTROL.

microbial leaching (of ores) See LEACHING.

microbial mat A complex, benthic microbial community which forms a cohesive layer and which is typically dominated by CYANOBACTERIA or other photosynthetic prokaryotes – and occasionally e.g. by microalgae; microbial mats occur in those regions where environmental stress tends to exclude, or reduce the numbers of, metazoans – e.g. in intertidal and hypersaline regions, in hot springs, and around HYDROTHERMAL VENTS. Microorganisms which occur in microbial mats include e.g. species of *Lyngbya* and *Microcoleus*, various purple photosynthetic bacteria, *Chloroflexus*, and certain aerobic and anaerobic chemotrophs, including SULPHATE-REDUCING BACTERIA.

Microbial mats are believed to have given rise to STROMATOLITES, and, under certain conditions, may have been precursor material for KEROGEN and hydrocarbon deposits; in the latter context it has been suggested that oil is more likely to have been derived from marine organic deposits than from terrestrial plant material which, under appropriate conditions, gives rise to coal. [Book ref. 154.]

microbicidal Able to kill at least some types of microorganism. (cf. MICROBISTATIC.)

microbiocoenosis A natural community of microorganisms (including algae, bacteria and protozoa) characterized by phases of short-lived stability.

microbiological safety cabinet *Syn.* SAFETY CABINET.

microbiology The study of MICROORGANISMS and their interactions with other organisms and the environment (cf. CELLULAR MICROBIOLOGY). There is considerable overlap between microbiology and certain other disciplines: e.g. biochemistry, immunology, molecular biology and PARASITOLOGY. Microbiology is also directly relevant to certain aspects of medicine, veterinary science, agriculture, the food industry (see FOOD MICROBIOLOGY) and various commercial processes (see INDUSTRIAL MICROBIOLOGY) as well as ecology and pollution studies (see e.g. CARBON CYCLE, NITROGEN CYCLE, SULPHUR CYCLE, SEWAGE TREATMENT and WATER SUPPLIES).

microbiota Microscopic organisms, particularly those in soil.

Microbispora A genus of bacteria (order ACTINOMYCETALES, wall type III; group: maduromycetes) which occur e.g. in soil. The organisms form branched aerial and substrate mycelium, the former (commonly pink) giving rise to pairs of elongated spores; when grown on certain solid media some species (e.g. *M. parva*) deposit crystals of IODININ in the medium. The GC% of the type species (*M. rosea*) has been reported to be ca. 74. [Ecology, isolation and cultivation: Book ref. 46, pp. 2103–2117.]

microbistatic Able to inhibit the growth and reproduction of at least some types of microorganism. (cf. MICROBICIDAL.)

microbody A particle, typically less than 1 μm diam., consisting of a collection of functionally related enzymes enclosed in a membranous sac; microbodies occur in eukaryotic cells: see e.g. GLYCOSOME, GLYOXYSOME, PEROXISOME. [Microbodies in urate-utilizing yeasts: AvL (1985) *51* 33–43. 'How proteins get into microbodies' (review): BBA (1986) *866* 179–203.] (cf. HYDROGENOSOME; LYSOSOME.)

microcapsule See CAPSULE.

microcarrier cell culture A system for culturing eukaryotic cells in which the cells grow to form a confluent layer on the surface of small solid particles suspended in a slowly agitated medium. [Review: AAM (1986) *31* 139–179.] (See also TISSUE CULTURE.)

microcins A category of low-MWt BACTERIOCINS produced by members of the Enterobacteriaceae and effective e.g. against bacteria of the same family. The microcins differ from COLICINS in their size, and also differ in that their synthesis is not induced by conditions which trigger the SOS SYSTEM. Typically, microcins are synthesized in the stationary phase, and synthesis tends to be repressed in rich media.

The typical microcin ranges in size from a single modified amino acid (microcin A15 appears to be a derivative of methionine) to short or medium-sized peptides (up to ~50 amino acid residues in length). Microcins are characteristically thermostable (e.g. 100°C/30 min), soluble in methanol–water (5:1), and resistant to extremes of pH and to certain proteases.

The microcins can be grouped according to their modes of action: type A microcins inhibit certain metabolic pathways, type B microcins inhibit DNA replication etc. For example, the bacteriostatic microcin A15 inhibits homoserine succinyltransferase in the methionine biosynthetic pathway (see Appendix IV(d)). The bactericidal microcin B17 inhibits DNA gyrase and induces the SOS SYSTEM in *Escherichia coli* K12 [JGM (1986) *132* 393–402]. Microcin C7 is a heptapeptide which blocks protein synthesis. (See also COLICIN V.)

Plasmids which specify microcins (M plasmids or, preferably, Mcc plasmids) appear not to encode resistance to conventional antibiotics. Some of these plasmids are conjugative – e.g. that encoding B17 (pMccB17, an IncFII plasmid) and that encoding C7 (pMccC7, an IncX plasmid).

Micrococcaceae A family of asporogenous, Gram-positive cocci which includes the genera MICROCOCCUS (the type genus), STAPHYLOCOCCUS and STOMATOCOCCUS.

Micrococcus A genus of Gram-positive, aerobic, chemoorganotrophic, asporogenous, catalase-positive, generally non-motile bacteria which have a GC value (see GC%) of ca. 65–75 and which are typically resistant to lysostaphin; the micrococcal cell wall typically contains little or no teichoic acid (cf. TEICHURONIC ACIDS), and the ratio of glycine to lysine in the PEPTIDOGLYCAN is characteristically <2. (cf. STAPHYLOCOCCUS.) *Micrococcus* spp typically do not utilize glucose anaerobically, and typically do not form acid, aerobically, in a glycerol–erythromycin medium. Metabolism is respiratory (see RESPIRATION). The organisms occur e.g. on mammalian skin and in raw MILK. They can be isolated e.g. on FTO AGAR. Cells: cocci (ca. 1–2 μm diam.) which occur in tetrads and clusters. Most species form a water-insoluble pigment. Type species: *M. luteus*.

M. agilis. A red-pigmented, flagellated, psychrophilic species which does not form acid from glucose or mannose, is resistant to lysozyme, and is VP −ve.

M. denitrificans. See PARACOCCUS.

M. kristinae. An orange-pigmented species which forms acid from glucose and mannose, is resistant to lysozyme, and is VP +ve.

M. luteus (formerly *M. lysodeikticus, Sarcina pelagia*). A yellow-pigmented species. Acid is generally not formed from glucose or mannose. Most strains are sensitive to LYSOZYME. VP −ve.

M. radiodurans. See DEINOCOCCUS.

M. roseus (formerly *M. tetragenus*, *M. roseofulvus*). A red- or pink-pigmented species which often forms acid from glucose (but not from mannose), which may be weakly or strongly resistant to lysozyme, and which can hydrolyse gelatin.

Other species: *M. halobius*, *M. lylae*, *M. nishinomiyaensis*, *M. sedentarius*, *M. varians*.

[Book ref. 46, pp. 1539–1547.]

Microcoleus A phycological genus of filamentous 'blue-green algae' (Nostocales) which has been included in the LPP GROUP [JGM (1979) *111* 1–61 (p. 23)]. The organisms may form extensive mats on sedimented silt in shallow water [JGM (1984) *130* 983–990], in salt-marshes, on mudflats etc, and may also grow on damp soil.

microcolony (1) A small clone of bacteria in which the cells may be e.g. in direct physical contact or embedded in a common polysaccharide matrix; such colonies occur e.g. in aquatic habitats. (cf. COENOBIUM, CONSORTIUM.) [Microcolony formation and consortia: Book ref. 108, pp. 373–393.] (2) See CHLAMYDIA. (3) Any very small colony.

microcomplement fixation A method for comparing the amino acid sequence in a given reference protein (e.g. an enzyme) with that in the same protein derived from a different strain or organism; the method is used e.g. in microbial TAXONOMY. Essentially, the protein under test is examined for its ability to fix COMPLEMENT when mixed with an antiserum raised against the purified reference protein; the (quantitative) difference in the complement-fixing ability of the test and reference proteins is taken as a measure of the difference between the amino acid sequences of the two proteins. The results are usually given in *ImD units*; 5 ImD units represent ca. 1% difference between the amino acid sequences.

microconidium (1) A SPERMATIUM. (2) A small CONIDIUM (cf. MACROCONIDIUM).

microconjugant See CONJUGATION (1)(c).

microcycle sporulation Sporulation immediately following outgrowth of a spore. [Microcycle sporulation in *Bacillus subtilis*: Book ref. 164, pp. 163–167.]

microcyclic rusts Those rust fungi (see UREDINIOMYCETES) which form only one type of *binucleate* spore, the teliospore – i.e., aeciospores and uredospores are not formed; they include e.g. some species of *Puccinia*.

Microcyclus (1) A genus of fungi of the Dothideales. (2) A genus of Gram-negative, obligately aerobic bacteria which occur e.g. in soil and freshwater habitats. The cells are capsulated curved rods, 0.3–1.0 × 1.0–3.0 μm; one strain has a single polar flagellum, but most strains are non-motile. The cells of some strains contain GAS VACUOLES. The organisms grow chemoorganotrophically, but chemolithoautotrophic growth, with H_2 as electron donor, has been reported for some strains; all strains tested are facultative methylotrophs (see METHYLOTROPHY) which can utilize e.g. formate and methanol. Catalase and oxidase +ve. GC%: 66–69. Type species: *M. aquaticus*. (Since the fungal genus *Microcyclus* (Saccardo 1904) takes precedence, a new name has been proposed for the bacterial genus: *Ancylobacter* [IJSB (1983) *33* 397–398].) [Book ref. 22, pp. 133–135.]

microcyst (1) (*bacteriol.*) A type of resting cell – which may or may not be resistant to heat, desiccation etc – formed by certain bacteria: e.g. *Sporocytophaga* and certain *Nocardia* spp. (2) (*bacteriol.*) In the MYXOBACTERALES: *syn.* myxospore. (3) (*bacteriol.*) *Syn.* coccoid body (see COCCOID BODIES). (4) In some SLIME MOULDS: a resting structure formed by the rounding up and encystment of an individual myxamoeba (cf. MACROCYST).

microcystin See MICROCYSTIS.

Microcystis A phycological genus of unicellular 'blue-green algae' (CYANOBACTERIA) which may be referrable to SYNECHOCYSTIS. The cells are spherical, occur in mucilaginous colonies, and contain GAS VACUOLES. Species are traditionally distinguished on the basis of cell size and colony form: *M. aeruginosa* and *M. flos-aquae* form colonies in which numerous cells are embedded in a diffluent mucilage, while colonies of *M. marginata* are surrounded by a distinct envelope. (Colony form may be lost in culture.) The organisms do not fix nitrogen.

Microcystis spp are planktonic in fresh water, often forming BLOOMS e.g. in reservoirs. Some strains of *M. aeruginosa* produce a cyclic peptide toxin, microcystin, which has hepatotoxic effects in animals drinking the contaminated water; the toxin appears to associate with cytoskeletal actin in a manner similar to phalloidin (see PHALLOTOXINS). Another (peptide?) toxin ('fast death factor', microcystin-*c*) is a potent neurotoxin responsible for convulsions, respiratory failure, and rapid death in animals.

microdilution broth test See DILUTION TEST.

Microellobosporia See STREPTOMYCES.

microfauna See MICROFLORA (2).

microfilament Any very thin filament – particularly one of polymerized ACTIN.

microflora (1) In a microbiological context: the totality of microorganisms normally associated with a given environment or location (see e.g. BODY MICROFLORA); the term 'flora' is often used as an abbreviation in this context. (2) In a broader biological context: the microscopic plants and 'plant-like' organisms (e.g. bacteria, fungi, algae etc) normally present in a given environment or location, the term *microfauna* being used to refer to the microscopic animals and 'animal-like' organisms (e.g. protozoa, nematodes) of that environment/location.

microforge See MICROMANIPULATION.

microfungi Those fungi in which the individual fructifications are too small to be seen in detail with the unaided eye.

microgamete See ANISOGAMY.

microgamont A cell which gives rise to microgametes (i.e., male gametes).

Microglossum See HELOTIALES.

Microjoenia See HYPERMASTIGIDA.

micromanipulation (micrurgy) Any procedure involving the physical manipulation of individual cells (or spores etc) or their components. Examples include: the isolation and culture of single cells for genetic studies; dissection of the ciliate infraciliature in cytophysiological studies; intercellular transplantation of nuclei, mitochondria etc. Dissection and transplantation procedures are sometimes referred to as *microsurgery*.

Micromanipulation may be carried out with any conventional microscope that is fitted with a good-quality mechanical stage and a high-power dry (i.e. non-immersion) objective lens; phase contrast microscopy is preferable for certain types of operation. A vibration-free workplace is essential. The material under study may be arranged e.g. on the surface of an agar block, or as a hanging drop preparation, and is commonly contained within a partially enclosed moist chamber on the microscope stage. (The use of a heated objective lens mounting prevents condensation of water vapour on the lower lens.)

The actual manipulation is carried out with glass *microtools* which are prepared in a *microforge*: essentially, an apparatus used to control the movement of a thin glass tube or rod relative to an electrically heated filament – the operation being observed under the low power of a microscope. *Needles* are fine glass filaments which, in the region of the point, may be ca. 0.2 μm in diameter. *Loops* are made in various sizes. *Pipettes*, which may be <1 μm diameter, are made by drawing out glass tubing which has been heated to plasticity; they have been used e.g. for holding cells during microsurgery: the (heat-polished) tip of the pipette is held against a cell and a gentle suction is applied. Each microtool is fixed in a metal holder which is clamped to a *micropositioner*: essentially, a means whereby the coarse movements of the hand are scaled down to give the finer movements necessary for the control of the microtools. By such means, microtools can make controlled movements, with speed and precision, along all three axes in space. In a simpler method (used for uncomplicated operations) the microtool holder is clamped to the nosepiece of a second microscope; rotation of the fine focusing control of this microscope moves the microtool in a vertical plane, while movements of the microtool relative to the specimen in a horizontal plane can be made by moving the *specimen* with the mechanical stage of the first microscope.

micrometer In light MICROSCOPY: a device used for the measurement of a specimen or parts of a specimen. The simplest micrometer is the *ocular micrometer* (= *eyepiece micrometer*, *graticule*, *reticle*, *reticule*): a glass or plastic disc, marked with a series of regular (though arbitrary) graduations, which is placed in the front focal plane of the eyepiece lens, i.e., resting on the FIELD DIAPHRAGM (sense 2); when the micrometer is in place its graduations can be seen superimposed on the image of the specimen. Calibration of the micrometer may be carried out by focusing the microscope on a *slide micrometer*: a glass slide etched with graduations precisely 5 or 10 μm apart; if, say, 10 graduations of the ocular micrometer correspond to 5 μm on the slide, then 1 graduation will measure 0.5 μm in a specimen.

The *filar micrometer* consists of a thin moveable wire in the front focal plane of the eyepiece, the wire being moveable by means of a knob that controls a fine-threaded screw; calibration is carried out e.g. with a slide micrometer, the distance (in μm) travelled by the wire being read off on a scale around the control knob.

micromethods (miniaturized methods) In clinical microbiology: various procedures used for routine testing of specimens or microbial isolates in a manner economical of time and/or space and material. Micromethods are commonly used e.g. to carry out, simultaneously, a range of biochemical identification tests on a given (usually bacterial) isolate; these procedures generally make use of commercially available kits, some of which are briefly described below.

The *API* system consists of a plastic strip holding a number of microtubes, each containing a different dehydrated medium; each microtube is inoculated with a suspension of the test organism, mineral oil is added to certain microtubes (to exclude air), and the strip is incubated. Subsequently, reagents are added to detect particular metabolic products, and the results are read and analysed.

The *Enterotube II* system consists of a tube divided into a sequence of 12 compartments, each compartment containing a different agar-based medium; the media are inoculated by passing an inoculum-bearing needle axially through the tube.

The *Minitek* system consists of a plastic plate containing a number of wells, each well containing a paper disc impregnated with a dehydrated medium; each disc is inoculated with a suspension of the test organism, some are then overlaid with mineral oil, and the plate is incubated. Reagents are then added (to certain wells) and the results are read and analysed.

The *PathoTec* system consists of a number of test strips, each impregnated with the dehydrated medium appropriate to a given test; each strip is incubated in a suspension of the test organism (or inoculated directly from a colony) and the test is read after a specified time.

Some systems are designed for anaerobic incubation. The *AN-IDENT* system permits the testing of anaerobes in a strip incubated aerobically.

[Descriptions of some identification kits: Book ref. 120, pp. 52–65.]

(See also BACTEC.)

micrometre (μm) 10^{-6} m, i.e., 1/1000th mm.

Micromonadophyceae A proposed class of unicellular, scaly or naked, flagellated green algae (division CHLOROPHYTA) in which the interzonal mitotic spindle persists during cytokinesis [Book ref. 123, pp. 29–72]. Examples of constituent genera are *Cymbomonas*, *Dolichomastix*, *Mamiella* (scaly), *Mantoniella* (scaly), *Mesostigma*, *Micromonas* (naked), *Monomastix*, *Nephroselmis* (= *Heteromastix*) (scaly), *Pedinomonas* (naked), *Pterosperma*, *Pyramimonas* (scaly), *Scourfieldia* (naked), *Trichloris*; many of these genera were formerly included in the PRASINOPHYCEAE. (cf. TETRASELMIS.)

Micromonas See MICROMONADOPHYCEAE.

Micromonospora A genus of bacteria (order ACTINOMYCETALES, wall type II) which occur e.g. in soil, decaying vegetation and aquatic habitats. The organisms form a non-fragmenting, branching substrate mycelium, but usually do not form an aerial mycelium; most species form a pigment (e.g. yellow, blue, purple). Non-motile, darkly-pigmented spores are borne singly at the tips of sporophores which arise from the mycelium; the spores have some resistance to heat (e.g. 70°C/30 min) and to desiccation. Most species are aerobic; some obligately anaerobic species are currently included in the genus, but the GC% of one ('*M. ruminantium*') has been reported to be much lower (ca. 53–56) than that characteristic of the genus (71–73). Many species are cellulolytic. Some species produce useful antibiotics (see e.g. GENTAMICIN). Type species: *M. chalcea*. [Book ref. 73, pp. 67–69.]

micron (μ) 10^{-6} m (= MICROMETRE).

micronemes Small, convoluted, thread-like structures, oval or circular in cross-section, which occur in members of the APICOMPLEXA; function unknown.

micronucleus (*ciliate protozool.*) The smaller (commonly <5 μm) of the two types of nucleus characteristically present in ciliates (cf. MACRONUCLEUS); according to species a cell may contain one, several or many micronuclei, although micronuclei are absent in some viable strains of species which are normally micronucleate. Micronuclei lack nucleoli, are characteristically diploid, and their DNA (which occurs in chromosome form) is transcriptionally inactive; they are involved in genetic recombination (see AUTOGAMY and CONJUGATION) and in the regeneration of macronuclei.

microorganisms According to common usage: microscopic organisms (and taxonomically related macroscopic organisms) within the categories ALGAE, ARCHAEA, BACTERIA, FUNGI (including LICHENS), PROTOZOA, VIRUSES and SUBVIRAL AGENTS.

Some authors do not regard viruses and subviral agents as microorganisms – reserving the term 'organism' for an entity which consists of one or more cells [Book ref. 159, pp. 11–13]; other authors include PLASMIDS (as well as viruses) within the category 'organisms' [Book ref. 161, pp. 3–16].

micropalaeontology The study of microscopic fossils. (See also FOSSIL MICROORGANISMS.)

microparticle enzyme immunoassay See LCR.

microperoxidase Part of a cytochrome *c* molecule (the haem moiety together with part of the protein) which has peroxidase activity and which is used e.g. in CHEMILUMINESCENCE studies.

microphotography Photography in which very small photographs are made of macroscopic or large objects. (cf. PHOTOMICROGRAPHY.)

microplasmodesmata Minute pores which occur in the septa in certain filamentous prokaryotes: e.g. in hyphae of certain actinomycetes (see ACTINOMYCETALES) and in cyanobacterial TRICHOMES (see also HETEROCYST).

Micropolyspora A (proposed) genus of bacteria (order ACTINOMYCETALES, wall type IV) which are found e.g. in mouldy hay, compost and soil. The organisms form substrate and aerial mycelium, both of which give rise to chains of spores; they do not form mycolic acids. *M. brevicatena*, the former type species of *Micropolyspora*, forms mycolic acids and has been transferred to the genus *Nocardia* [JGM (1982) *128* 503–527], leaving *Micropolyspora* a proposed genus *nomen conservandum* [Book ref. 73, pp. 123–124]. [Proposal to conserve the name *Micropolyspora* (with *M. faeni* as type species) and new genus description: IJSB (1984) *34* 505–507.] *M. faeni*, a common causal agent of FARMERS' LUNG, forms substrate mycelium of some shade of yellow, orange or brown, and a white aerial mycelium. (*M. faeni* and *M. rectivirgula* appear to be synonyms [JGM (1979) *115* 343–354].) *M. viridinigra* and *M. rubrobrunea* have been transferred to the genus *Excellospora*.

Some authors use the name *Faenia rectivirgula* as a synonym of *M. faeni*.

micropore (1) (formerly: micropyle, ultracytostome) In members of the APICOMPLEXA: a small invagination of the pellicle (usually) coincident with a discontinuity of the inner layer of the pellicle. One or more micropores may be present in a given cell; they may function as sites for the ingestion of food. (2) See GEOTRICHUM.

micropositioner See MICROMANIPULATION.

micropyle (1) In some species of coccidia (e.g. *Eimeria stiedae*): a thin, polar region in the oocyst wall. (2) See MICROPORE (1).

microRNA (miRNA) See ANTISENSE RNA.

Microscilla See BACTERIA (taxonomy).

microscope slide *Syn.* SLIDE.

microscopy The use of an instrument (*microscope*) for examining objects or details which are too small to be seen (or seen clearly) by the unaided eye; the microscope forms an enlarged *image* of the specimen. In *light microscopy* images are formed by light or ultraviolet radiation (cf. ELECTRON MICROSCOPY).

(a) *Bright-field microscopy* is the commonest form of light microscopy: the form usually meant by the term 'microscopy'. Essentially, light from a lamp is concentrated by the CONDENSER

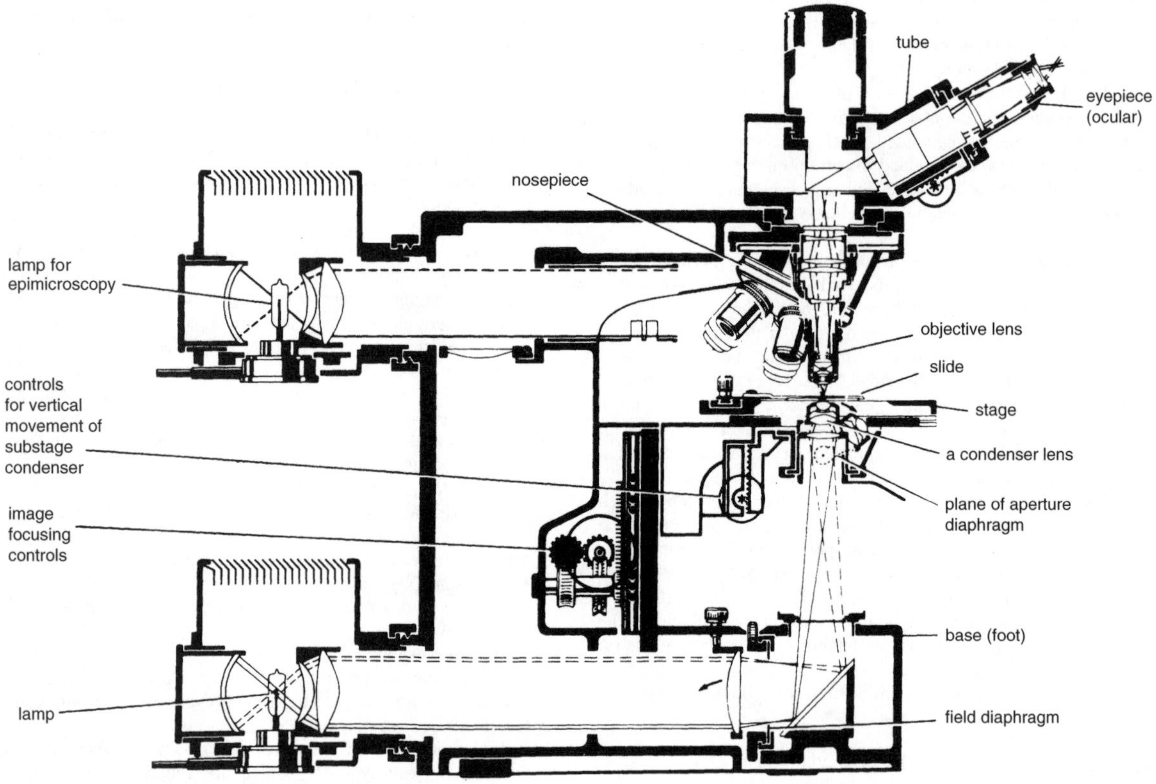

MICROSCOPY. Figure 1. A research microscope (Leitz Orthoplan) set up for Köhler illumination. Reproduced by permission of Leitz (Instruments) Ltd, Luton, United Kingdom.

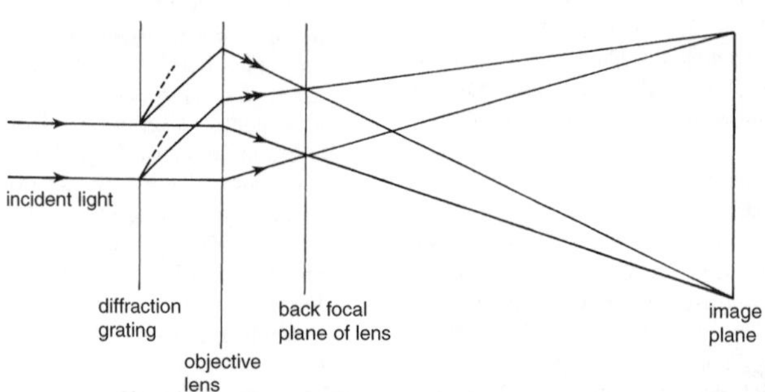

MICROSCOPY. Figure 2. Image formation in bright-field microscopy, using a diffraction grating as a specimen. Light entering one slit of the grating gives rise to zero-order rays (single arrowhead), first-order rays (double arrowhead) and second-order rays (terminally dashed lines). An image is formed as a result of interference between diffracted and non-diffracted rays.

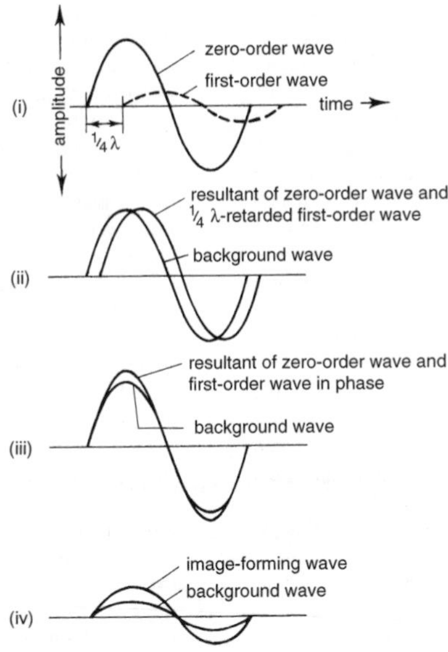

MICROSCOPY. Figure 3. Amplitude and phase relationships in phase-contrast microscopy. (See text for explanation.)

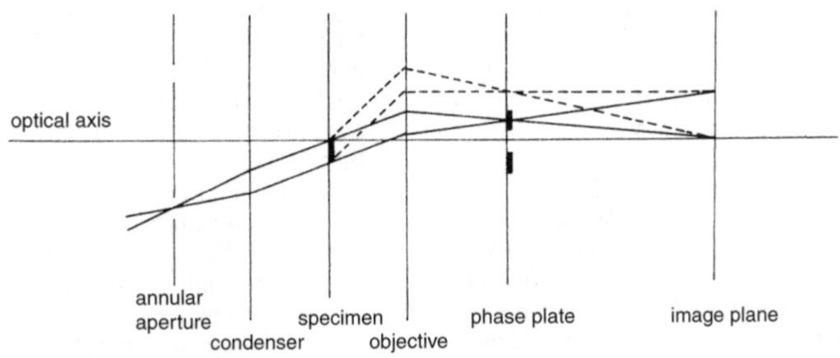

MICROSCOPY. Figure 4. Image formation in phase-contrast microscopy: simplified diagrammatic scheme showing spatial relationships between zero-order rays (solid lines from specimen to image plane) and first-order diffracted rays (dashed lines). (See text for explanation.)

and directed onto the specimen (Fig. 1). The objective lens ('objective') forms a magnified image of the specimen, and this image is further magnified by the eyepiece lens (see MAGNIFICATION). The specimen is usually examined on a SLIDE which rests on the stage. (See also HANGING DROP; MICROTOME; SMEAR; WET MOUNT.)

For an image to be formed, light which has passed through the specimen must differ from the background light in amplitude (intensity) and/or in composition (e.g. colour); thus, an image of the specimen is usually seen against a lighter (bright-field) background. Some specimens are translucent and are therefore poorly visible; such specimens can be made light-absorbing by suitable STAINING.

Image formation is often depicted by ray diagrams. However, for objects below a certain size, image formation can be described satisfactorily only in terms of wave optics; this is because such objects cause significant diffraction of light, and the formation of an image depends on interference between diffracted and non-diffracted light waves. Image formation is shown in Fig. 2; the object shown is a diffraction grating, but biological specimens form images in a similar way. Fig. 2 shows that, for image formation, the objective must collect at least the undeviated (zero-order) light waves *and* the diffracted (first-order) light waves. For a perfect image all the diffracted light should be collected, but this is impracticable. To collect the maximum amount of diffracted light an objective should have a high *numerical aperture*: see RESOLVING POWER. The specimen should be uniformly illuminated, and the (divergent) beam of light from the specimen should fill the entire aperture of the objective (see CRITICAL ILLUMINATION; KÖHLER ILLUMINATION). When using a HIGH-DRY OBJECTIVE, good image formation also requires the use of a COVER-GLASS of the correct thickness. (See also *oil immersion objective* under RESOLVING POWER; cf. CONFOCAL SCANNING LIGHT MICROSCOPY.)

(b) *Dark-field microscopy* (dark-ground microscopy) is used e.g. for examining objects too small to be seen by bright-field microscopy. For low- or medium-power work, the underside of the substage condenser is fitted with a central opaque disc (the *stop* or *patch stop*) which allows only the peripheral rays to pass through the condenser – thus forming a hollow cone of light whose apex is focused in the plane of the specimen; in the absence of a specimen these rays diverge at such an angle that none enters the objective so that the field of view appears dark or black. (The NA of the condenser must be greater than that of the objective.) A specimen scatters the rays, some of which enter the objective and form a bright image of the specimen against a dark background. For high-power work use is made of a special oil-immersion condenser (e.g. a *cardioid condenser*) within which the incident rays are twice reflected so that the emergent rays form a hollow cone of large apical angle. (See also RHEINBERG ILLUMINATION.)

(c) *Phase-contrast microscopy*. A colourless specimen which absorbs little light (e.g. a non-pigmented living cell) is not clearly visible by bright-field microscopy but may be seen by phase-contrast microscopy. When such a specimen diffracts light, the first-order waves are retarded by approximately 1/4-wavelength ($1/4\lambda$) and are of amplitude smaller than that of the zero-order waves (Fig. 3(i)). If these zero-order and first-order waves are allowed to interfere they will produce resultant waves of amplitude (intensity) similar to that of the background waves but differing from them in phase (Fig. 3(ii)); the phase difference cannot be detected by the eye so that no image is seen. (Background waves are waves which do not pass through the specimen before entering the objective lens of the microscope.) If, however, the zero-order and background waves are first retarded by $1/4\lambda$ (to eliminate the phase difference between the first-order waves and the zero-order and background waves), interference between zero-order and first-order waves will produce a resultant wave (the image-forming wave) of amplitude greater than that of the background wave (Fig. 3(iii)), i.e., a visible image will be formed.

In a phase-contrast microscope the condenser has a ring-shaped aperture in its front focal plane so that a *hollow* cone of light can be focused (as a small bright ring) onto a *phase plate* located in the back focal plane of the objective (Fig. 4). The phase plate is a glass disc on which has been deposited a ring of material (e.g. magnesium fluoride) of such thickness that it retards by $1/4\lambda$ the light it transmits. In the absence of a specimen all the incident light is focused on and transmitted via the ring. When a specimen is examined, the zero-order and background light is transmitted via the ring, but the first- and higher-order beams pass through the phase plate via regions *not* located in the ring (Fig. 4). In the image plane, interference occurs between zero-order and diffracted light; the resultant waves form a visible image because their amplitude is greater than that of the background waves (Fig. 3(iii)). To increase image contrast further the phase plate ring has deposited on it a thin layer of light-absorbing material which decreases the amplitude of background and zero-order waves; thus the amplitude of the image-forming wave becomes greater *relative to* that of the background wave (Fig. 3(iv)). (The clarity of the image can be improved by using a green filter in the microscope condenser.)

In the above scheme (*negative* or 'bright' phase-contrast microscopy) the zero-order beam is *retarded* by $1/4\lambda$, and the specimen appears lighter than the background. In *positive* or 'dark' phase-contrast microscopy the zero-order beam is *advanced* $1/4\lambda$; the specimen appears darker than the background because destructive interference occurs between the zero-order and first-order beams.

[Some improvements in phase-contrast microscopy: JM (1983) *129* 49–62.]

(d) *Interference-contrast microscopy* (differential interference-contrast microscopy). This form of microscopy avoids the 'haloes' which surround images formed by phase-contrast microscopy, and gives clearer images of fine detail. It can detect minute differences in specimen thickness and/or refractivity – such differences being indicated by colour differences in the image. In microscopes based on the *shear* system (e.g. the Nomarski microscope) light from the source is initially split into two parallel beams. The *object beam* passes through the specimen, while the *reference beam* passes through a clear or relatively clear area of the slide; subsequently the beams interfere to form an image. In the *double-focus* system both beams are co-axial – the object beam being focused in the plane of the specimen, and the reference beam being focused either above or below this plane; subsequently, the beams interfere to form an image.

(e) *Fluorescence microscopy*. In fluorescence microscopy a (usually) FLUOROCHROME-treated specimen is irradiated with ultraviolet radiation (or, in some cases, blue light) and the light emitted (see FLUORESCENCE) forms the image of the specimen in a manner similar in principle to that in bright-field microscopy. (Some types of specimen exhibit autoFLUORESCENCE.) In a fluorescence microscope, the beam from a mercury-arc lamp, or a tungsten lamp, is initially passed through an *exciter filter* which transmits only the exciting radiation. The exciting radiation is then focused onto the specimen either from below (*transmitted-light fluorescence microscopy*) or from above (*incident-light fluorescence microscopy*, or *epifluorescence microscopy*). In epifluorescence microscopy exciting radiation enters the microscope tube from the side and is reflected downwards, by a dichroic mirror, through the objective onto the specimen; the image-forming light emitted from the specimen passes through the objective, through the dichroic mirror (which does not reflect light of visible wavelength), and through the eyepiece. In both transmitted-light and epifluorescence microscopes, a *barrier filter* (= *stopping filter*) is placed below the eyepiece lens; this transmits emitted light of longer wavelength but stops the short wavelength exciting radiation – thus protecting the eyes and, in photomicrography, protecting the film against fogging. One advantage of epifluorescence microscopy, compared with transmitted-light fluorescence microscopy, is that the emitted light is not partially absorbed within the thickness of the specimen; thus a brighter image can be obtained. (See also CITIFLUOR.) (N.B. In all types of fluorescence microscopy it is necessary to use non-fluorescent lenses, slides, cover-glasses, immersion oil, etc.) (See also IMMUNOFLUORESCENCE; ACRIDINE ORANGE; AURAMINE–RHODAMINE STAIN; CALCOFLUOR WHITE.)

(f) *Ultraviolet microscopy* is a form of microscopy in which ultraviolet radiation is used both for illumination of the specimen and for image formation; the image is recorded on a fluorescent screen or on a photographic plate. Ultraviolet microscopy can resolve finer detail than can bright-field microscopy because the wavelength of the image-forming radiation is shorter than that of visible light (see RESOLVING POWER).

(g) *Epimicroscopy* is any form of microscopy in which the illuminating beam is focused on the specimen by the objective; in this type of illumination (= *vertical* or *incident-light* illumination) the objective acts simultaneously as the condenser and objective lens. An example is epifluorescence microscopy: see (e) above.

microsome One of many small heterogeneous membranous vesicles obtained e.g. when eukaryotic cells are homogenized and centrifuged; microsomes include e.g. vesicles formed from the rough and smooth ENDOPLASMIC RETICULUM and from the CYTOPLASMIC MEMBRANE.

microsource See BIOLUMINESCENCE.

Microsphaera See ERYSIPHALES.

Microspora A genus of filamentous, unbranched green algae (division CHLOROPHYTA).

Microspora A phylum of PROTOZOA [JP (1980) *27* 37–58] which are obligate intracellular parasites in a wide range of animals, including vertebrates and invertebrates. Mitochondria are absent. The organisms form spores which are small (commonly <10 μm diam.), unicellular in origin, and have walls which lack pores and are not composed of valves (cf. MYXOZOA); each spore contains a single uninucleate or binucleate *sporoplasm* (an amoeboid infective cell which is released on germination of the spore), and a single tubular, coiled, extrusile *polar filament* which is not enclosed within a polar capsule. A *polar cap* covers the attached end of the tubular filament. In many species there is a membranous (vacuolar) region beneath the polar cap (the *polaroplast* or polar sac). Classes: MICROSPOREA; RUDIMICROSPOREA. (See also ENTEROCYTOZOON; cf. ASCETOSPORA.)

Microsporea A class of protozoa (phylum MICROSPORA) which are parasitic in insects and other invertebrates, in poikilothermic vertebrates, and (rarely) in mammals. The microsporean spore has (typically) a three-layered, dense, refractile wall, and contains a single sporoplasm and a complex extrusion apparatus comprising the polar filament (which extends back from the polar cap and coils around the inside of the spore wall) and (typically) a polaroplast. When a spore is ingested by a host animal, the polar filament is discharged explosively, everting to form a long tube; the explosive force is such that the filament penetrates any host cell in its path, and the infective sporoplasm can then pass through the filament and into the host cell. [Role of pH-triggered Ca^{2+} influx in the discharge mechanism: JCB (1985) *100* 1834–1838.] The class contains two orders: Minisporida (e.g. *Burkea*, *Chytridiopsis*, *Hessea*) and Microsporida (e.g. *Amblyospora*, ENCEPHALITOZOON, GLUGEA, NOSEMA, *Pleistophora*, THELOHANIA).

Microsporida See MICROSPOREA.

microsporidean (Pertaining to) a protozoon of the MICROSPORA (formerly of the class Microsporidea, subphylum Cnidospora).

microsporidiosis Any disease caused by protozoa of the phylum MICROSPORA. In man, infections occur in both immunocompetent and immunocompromised individuals, although most isolations appear to come from the latter group (particularly AIDS patients).

Infections may occur in any of various sites. For example, travellers' diarrhoea in immunocompetent hosts can be due to intestinal infection by e.g. *Encephalitozoon intestinalis* or *Enterocytozoon bieneusi*, while *Encephalitozoon hellem*, *Nosema ocularum* and *Vittaforma corneae* have been isolated from the eye. Species of *Pleistophora* infect skeletal muscle, while *Trachipleistophora hominis* infects skeletal muscle, eye, kidney and nasopharynx. [Lab identification: JCM (2002) *40* 1892–1901.] In immunocompromised individuals, infection by microsporidia is reported to be more common in patients with $CD4^+$ counts below 100/μl.

[Microsporidiosis: JMM (2000) *49* 947–967 (948–952).]

Microsporum See DERMATOPHYTES.

microstome See TETRAHYMENA.

microstrainer See SEWAGE TREATMENT.

microsurgery See MICROMANIPULATION.

Microtetraspora A genus of bacteria (order ACTINOMYCETALES, wall type III; group: maduromycetes) which occur e.g. in soil. The organisms form a substrate mycelium (usually colourless, grey, or yellowish) and aerial hyphae which give rise to spores (typically) in chains of four. Type species: *M. glauca*. [Ecology, isolation, cultivation: Book ref. 46, pp. 2103–2117.]

Microthamnion A genus of filamentous green algae related to TREBOUXIA; the filaments are branched and possess holdfasts.

Microthorax See HYPOSTOMATIA.

Microthrix parvicella A species of filamentous bacteria isolated from activated sludge [see e.g. JGM (1984) *130* 2035–2042].

microtome An instrument for cutting thin sheets (*sections*) of tissue (or of individual cells) for examination by MICROSCOPY. Essentially, the instrument consists of a fixed knife and a mechanism which moves the specimen relative to the knife; after each section is cut the specimen is moved towards the knife by a distance equal to the thickness of the section. Sections for ELECTRON MICROSCOPY are cut with an *ultramicrotome*: a more sophisticated instrument having a glass or diamond knife. To ensure rigidity during cutting, the fixed, dehydrated specimen is embedded in paraffin wax (for light microscopy) or e.g. in an epoxy resin (for electron microscopy). Sections for light microscopy are usually about 5–10 μm thick. Sections for electron microscopy are 100 nm or less; these are cut and selected with the aid of a dissecting microscope.

microtubule-associated proteins (MAPs) Proteins associated stoichiometrically with the surfaces of MICROTUBULES; certain MAPs are present in only some types of cell. MAPs include e.g. Mg^{2+}-activated GTPase; the ATPase dynein; the various *tau* proteins (apparently involved in the assembly and/or stability of MTs); the 'high-MWt MAPs' designated MAP1 and MAP2 (which occur mainly in the cells of higher animals); and the cold stability factor (CSF) which occurs e.g. in mammalian brain cells and which increases the resistance of MTs to cold. (CSF–MT binding is Ca^{2+}-sensitive in brain cells.) [Review: ARCB (1986) *2* 421–457.]

microtubule-organizing centre (MTOC) In eukaryotic cells: any intracellular body or region involved in the initiation and/or assembly of MICROTUBULES: see e.g. BASAL BODY (b), *centrosome* in CENTRIOLE, DEUTEROSOME, KINETOCHORE and SPINDLE POLE BODY.

microtubules (MTs) Non-contractile protein tubules, each having a mean external diameter of ca. 24 nm, which occur singly, in pairs or triplets, or in bundles, in most types of *eukaryotic* cell – including the cells of microorganisms, higher plants and animals. (MTs do not occur in *disc*-shaped mammalian erythrocytes, although they are present in the elliptical erythrocytes of the Camelidae, and even in blood platelets.) MTs are involved e.g. in the maintenance of cell shape and structure (see e.g. CYTOSKELETON), in some types of MOTILITY (see e.g. CILIUM and FLAGELLUM (b)), in spindle formation during MITOSIS, and (apparently) in the intracellular translocation of vesicles, LYSOSOMES and other organelles.

The wall of each MT consists of several (typically 13) longitudinally adjacent rows of TUBULIN molecules, each row being called a *protofilament*; MTs can change their length (and, apparently, their intracellular location) by the assembly/disassembly of tubulin units at one or both ends of the tubule. The outer surface of the MT wall bears a number of MICROTUBULE-ASSOCIATED PROTEINS (MAPs). Even when MTs occur in bundles there is seldom direct contact between one MT and another; they are usually separated from one another by an 'exclusion zone' of ca. 10 nm width – a zone into which the MAPs project.

The assembly of tubulin into MTs is apparently associated with the binding of guanine nucleotides to the tubulin molecule: GTP binds at the 'E' site and (more firmly) at the 'N' site; at some stage (during or after incorporation of the tubulin molecule) GTP at the E site is hydrolysed by an Mg^{2+}-activated GTPase (an MAP) and inorganic phosphate is released (GDP remaining bound to the tubulin). However, the in vitro assembly of MTs can occur (albeit more slowly) with GDP–tubulin (or with non-hydrolysable GTP analogues) – suggesting that energy is not required for MT assembly. In contrast, MT disassembly seems to need GTP and to be energy-dependent. Although assembly and disassembly can occur at both ends of an MT, there is evidence that, at least under certain in vitro conditions, 'treadmilling' can occur; treadmilling means the incorporation of tubulin at one end of a microtubule with concomitant disassembly at the other end.

The initiation of MT assembly in vivo occurs usually, but apparently not always, at a MICROTUBULE-ORGANIZING CENTRE.

Various factors and agents can promote or inhibit MT assembly or disassembly. Thus, e.g. the in vitro assembly of tubulin into MTs occurs only above a certain minimal temperature, and is promoted by GTP, Mg^{2+}, and a low concentration of Ca^{2+}. Disassembly is promoted e.g. by cold, by high hydrostatic pressures, and by dilution (MTs disassemble below a certain critical concentration); in some cases MT disassembly can be brought about by Ca^{2+} (above a certain concentration), while in other cases Ca^{2+} is effective only in combination with CALMODULIN. Other agents which can inhibit the assembly of MTs include e.g. BENZIMIDAZOLES, COLCHICINE, GRISEOFULVIN, the macrolide *maytansine*, PODOPHYLLOTOXIN, *steganacin* (a polycyclic lactone with a trimethoxy-substituted aromatic ring), and certain VINCA ALKALOIDS. MT assembly is promoted e.g. by TAXOL; high concentrations of glycerol stabilize MTs by lowering the rates of assembly and disassembly.

[Reviews: Book ref. 165; MT structure studied by cryoelectron microscopy: JM (1986) *141* 361–373.]

Microviridae (φX phage group; isometric ssDNA phages) A large, diverse family of icosahedral, lytic, ssDNA-containing BACTERIOPHAGES; hosts: enterobacteria. Members are abundant e.g. in sewage. One genus: *Microvirus*; type species: φX174. Microviruses φX174, S13, G4, G6, G13, G14, α3, φA, φB, φC, and φR infect C strains of *Escherichia coli*; St-1, φK, φXtB, and U3 infect *E. coli* K12. Growth temperature range, antiserum cross-reactions, host receptor site, and host *dna* function requirements suggest close relationships between φX174, S13, G6, φA and φB, probably between St-1 and φK, and possibly between G4 and φC; G14 and U3 differ from one another and from other microviruses.

The microvirus has an icosahedral capsid (28–32 nm diam.) whose vertices each bear a spike; spikes seem to function in adsorption of phage to host LIPOPOLYSACCHARIDE. The ss cccDNA genome (MWt ca. 1.7×10^6) comprises 11 genes (*A–H*, *J*, *K* and *A**), four of which are OVERLAPPING GENES: gene *B* occurs within *A*, *E* within *D*, *K* overlaps *A* and *C*; *B*, *E* and *K* are translated in different reading frames from *A*, *D* and *C*. gp*A** is the C-terminal portion of gp*A*. Gene products from different microviruses differ slightly in composition. gp*F* (60 molecules per virion) is the major component of the capsid; each spike comprises gp*G* (5 molecules) and gp*H* (1 molecule at the tip). Several copies of gp*J* and possibly one of gp*A** are associated with the capsid. gp*A* functions in genome replication; gp*B*, gp*C*, gp*D*, and possibly gp*J* function in phage morphogenesis. gp*E* is necessary for host cell lysis; it may interfere with the cell membrane and/or may activate host autolytic enzymes. (For microvirus replicative cycles see SSDNA PHAGE, BACTERIOPHAGE φX174 and BACTERIOPHAGE G4.)

Microvirus See MICROVIRIDAE.

microwave radiation (in sterilization) Microwaves (2.45 gHz) have been used e.g. for the STERILIZATION of plastic tissue-culture vessels [AEM (1982) *44* 960–964].

micrurgy *Syn.* MICROMANIPULATION.

Middelburg virus See ALPHAVIRUS.

middle T antigen See POLYOMAVIRUS.

Middlebrook 7H-9 broth A growth medium used for *Mycobacterium* spp; it contains mineral salts, pyridoxine hydrochloride, biotin, sodium glutamate, sodium citrate, and *either* Tween 80 *or* glycerol. [Recipe: Book ref. 53, p. 1415.]

midland cattle disease A form of BOTULISM in cattle; the disease is due to ingestion of forage or carrion contaminated with *C. botulinum* type C_β (cf. FORAGE POISONING).

mid-point potential See REDOX POTENTIAL.

Miescher's tubes Sarcocysts in tissues (see SARCOCYSTIS).

MIF Migration inhibition factor: a CYTOKINE originally obtained from specifically reactive T lymphocytes exposed to the relevant antigen; in vitro, this factor is characterized by it ability to inhibit the movement of macrophages away from the source of MIF. (Other in vitro activities include the activation of NO synthase (see NITRIC OXIDE) in macrophages.) MIF is associated with DELAYED HYPERSENSITIVITY reactions and may account for the accumulation and activity of macrophages in DH lesions.

The human MIF is a 12 kDa protein; its sources include the pituitary gland. The precise in vivo role of MIF has not been established; it may e.g. exert pro-inflammatory activity as a counter-balance to raised levels of (anti-inflammatory) glucocorticoids resulting from stimulation of the hypothalamic–pituitary–adrenal axis.

migration inhibition factor See MIF.

mikamycins See STREPTOGRAMINS.

mildews (*plant pathol.*) See e.g. DOWNY MILDEWS and POWDERY MILDEWS.

Miles and Misra's method (drop method; drop plate method) A COUNTING METHOD for determining the viable cell count of a bacterial suspension. Serial dilutions (often log dilutions) of the suspension are prepared. One drop (of known volume) from each dilution is placed at a separate, recorded position on the surface of a pre-dried agar PLATE; the drops are allowed to dry. The plate is then incubated until visible colonies develop in the small, circular areas corresponding to the drops. If a sufficient number of dilutions has been prepared, a drop from one of the dilutions will give rise to a countable number of discrete colonies. Assuming that each viable cell in the drop gave rise to a separate colony (and knowing the volume of the drop) the viable count can be calculated. In practice, a count is calculated from each of several drops on the plate, and the average is taken.

milfuram See PHENYLAMIDE ANTIFUNGAL AGENTS.

Milgo See ETHIRIMOL.

miliary Resembling millet seed. Miliary TUBERCULOSIS: a form in which numerous small tubercles develop.

Miliolina See FORAMINIFERIDA.

milk (microbiological aspects) Milk consists primarily of water containing proteins, soluble carbohydrates, electrolytes, lipids and vitamins.

The major protein constituents of milk are CASEINS (which occur in the form of complex micelles); there are also the so-called WHEY PROTEINS such as lactalbumin and lactoglobulin. IMMUNOGLOBULINS occur in high concentrations in COLOSTRUM and in relatively lower concentrations in milk. Various enzymes, e.g., alkaline phosphatase (see PHOSPHATASE TEST), are normally present in raw (i.e., untreated) milk. (See also LACTOFERRIN and LACTOPEROXIDASE–THIOCYANATE–HYDROGEN PEROXIDE SYSTEM.)

The major carbohydrate, LACTOSE (ca. 50 g/l in cow's milk), contributes ca. 50% of the osmotic pressure of milk (which is isotonic with blood plasma); small amounts of e.g. glucose and galactose are also present.

Electrolytes include Ca^{2+}, Mg^{2+}, Na^{+} and K^{+} together with phosphate, chloride and citrate ions.

Milk lipids are mainly triacylglycerides ('triglycerides') containing the residues of various saturated and unsaturated, branched and unbranched fatty acids; there are also some free fatty acids and phospholipids. Lipids occur as globules (up to ca. 6 μm diam. in freshly secreted milk), each globule being surrounded by a phospholipid-rich 'milk fat globule membrane' derived from the cytoplasmic membrane of the secretory cell. (*Homogenization*, which involves pumping milk under high pressure, reduces the size of the globules (which then do not rise to the surface as 'cream'); the resulting partial loss of phospholipid membrane, and the increased total surface area of the globules, renders the lipid more susceptible to enzymic lipolysis.)

The microflora of raw (untreated) milk. Milk drawn *aseptically* from a healthy animal usually contains few, if any, microorganisms. However, milk may contain large numbers of bacteria if e.g. it is drawn from an animal suffering from asymptomatic or clinical MASTITIS. Freshly drawn milk may be subsequently contaminated e.g. by microorganisms on the exterior of the udder, on milking equipment, or in manure or SILAGE; the temporary storage of raw milk in refrigerated vessels provides an opportunity for the growth of PSYCHROTROPHIC contaminants (e.g. some species of *Pseudomonas*). Organisms frequently found in raw milk include certain species of e.g. ESCHERICHIA (and other enterobacteria), LACTOBACILLUS, LEUCONOSTOC, MICROBACTERIUM, MICROCOCCUS, PSEUDOMONAS, STAPHYLOCOCCUS and *Streptococcus* (see also LACTOCOCCUS), as well as certain yeasts and moulds; various actinomycetes, other organisms, and endospores of e.g. *Bacillus* spp may also be present. The nature and extent of the microbial contamination of raw milk can affect the keeping qualities of the milk following processing; thus, e.g., certain contaminating organisms form heat-stable extracellular enzymes (including lipases) that remain active after the cells that produced them have been killed during processing.

Microbiological examination of milk. An assessment of microbial contamination may be made by a total bacterial count (= standard plate count); this can be carried out e.g. by plating an aliquot of milk on yeast extract–milk agar (or similar medium) and incubating aerobically at ca. 30–32°C for several days; low-temperature incubation (for a longer period) may be used to detect psychrotrophs. Colony counts may range from <1000 to $>10^6$ colonies ml^{-1} milk according to the sample's history. (See also DEFT, METHYLENE BLUE TEST and SPIRAL PLATE COUNT METHOD.) [Assessing the bacterial content of milk (review): JAB (1983) *55* 187–201.]

Milk treatment processes. See e.g. PASTEURIZATION and APPERTIZATION (for UHT treatment). (Certain processes involving heat treatment are carefully designed to avoid or minimize the occurrence of the MAILLARD REACTION.) When milk is derived from animals that have been treated with penicillin, the milk may contain sufficient penicillin to inhibit STARTER strains; in e.g. some CHEESE-MAKING plants such milk is routinely treated with PENICILLINASE.

Pathogenic microorganisms in milk can be derived from a site of infection within the milk-producing animal itself; the consumption of such contaminated milk can give rise to disease such as e.g. BRUCELLOSIS, Q FEVER (caused by the ENDOSPORE-forming bacterium *Coxiella burnetii*) and TUBERCULOSIS. (See also LISTERIOSIS and ZOONOSIS.) Any of a variety of pathogenic microorganisms may contaminate (and survive within) milk following collection.

(See also DAIRY PRODUCTS and MILK SPOILAGE.)

milk caps See LACTARIUS.

milk ring test (abortus Bang reaction/ringprobe; ABR) A test for detecting BRUCELLOSIS in e.g. cattle. Milk from cows infected with *Brucella abortus* contains antibodies to *B. abortus* adsorbed to the fat globules. In the test, a small sample of milk is mixed with a suspension of killed, stained cells of *B. abortus*, shaken, and incubated for up to one hour. In a *positive* test (antibodies present) the stained cells react with the antibodies and are carried to the surface with the fat globules to form a coloured layer ('ring') of cream. In a *negative* test the cells remain dispersed in the milk.

milk spoilage Souring (acidification) of milk (which often results in clotting – see CASEIN) can be caused e.g. by the growth of *Lactococcus lactis* or other LACTIC ACID BACTERIA.

ROPINESS can be caused by slime-forming strains of *Lactobacillus* spp (e.g. *L. brevis*, *L. casei*).

'Broken cream' ('bitty cream'), usually due to *Bacillus cereus* or *B. mycoides*, is characterized by the development (in the cream layer) of small particles which fail to emulsify on shaking; it is thought to involve the activity of microbial LECITHINASES on the milk fat globule membrane (see MILK) with consequent coalescence of globules. Usually, the milk beneath a layer of broken cream subsequently exhibits a non-acid clot ('sweet curdling') due to the effects of extracellular rennin-like microbial enzymes on the milk CASEIN.

Malty off-flavours can be due e.g. to certain strains of *Lactococcus lactis* which convert amino acids to aldehydes (particularly leucine to 3-methylbutanal).

(See also LITMUS MILK and FOOD SPOILAGE.)

milk-vetch dwarf virus See LUTEOVIRUSES.

milkers' nodule (milkers' node) In man: non-ulcerating, red, nodular (papular) lesions on the hands of milkers caused by localized infection (via abrasions, cuts etc) with the PSEUDOCOWPOX virus.

milky disease An INSECT DISEASE affecting larvae of beetles of the family Scarabaeidae. Originally, two forms of the disease were recognized: type A, caused by *Bacillus popilliae*, and type B, caused by *B. lentimorbus* (= *B. popilliae* var. *lentimorbus*); other species associated with milky disease – e.g. *B. fribourgensis* and *B. euloomarahae* – may be varieties of *B. popilliae*. Beetle larvae become infected when they ingest spores of the pathogen; the spores germinate in the gut, and the vegetative cells proliferate, subsequently invading the haemolymph. Sporulation then occurs, resulting in ca. 5×10^9 spores/ml and giving the haemolymph the characteristic milky appearance. *B. popilliae*, but not *B. lentimorbus*, forms a parasporal crystal similar to that of *B. thuringiensis* (see DELTA-ENDOTOXIN), but this apparently plays no role in milky disease.

The milky disease pathogens have been used successfully in the USA for the BIOLOGICAL CONTROL of the soil-dwelling larvae of the Japanese beetle (*Popillia japonica*); the spores are produced commercially by the artificial inoculation of living insect larvae. The pathogen may persist in suitable environments, precluding the need for repeated application. *B. popilliae* cannot grow at 37°C and is therefore unlikely to infect mammals.

millet red leaf virus See LUTEOVIRUSES.

millimicron (mμ) 10^{-3} μ (= NANOMETRE).

Millipore filter See FILTRATION.

Milstem See ETHIRIMOL.

Milton See HYPOCHLORITES.

***min* genes** See CELL CYCLE (b).

MinCDE See CELL CYCLE (b).

Minchinia A genus of protozoa (class STELLATOSPOREA) which are parasitic in aquatic invertebrates, including oysters; *M. costalis* and *M. nelsoni* are economically important pathogens of the American oyster *Crassostrea virginica*, sometimes causing massive mortalities in oyster-beds. *Minchinia* spp have a plasmodial vegetative phase.

mineralization In nature: the process in which organic materials are broken down into inorganic materials; mineralization is effected mainly by saprotrophic bacteria and fungi. (cf. BIOMINERALIZATION.) Mineralization in organically polluted rivers and lakes may lead to SELF PURIFICATION (cf. EUTROPHICATION). Mineralization plays an essential part in the cycles of matter: see e.g. CARBON CYCLE, NITROGEN CYCLE, SULPHUR CYCLE.

miniature anion exchange centrifugation technique See MAECT.

miniaturized methods See MICROMETHODS.

minicell In bacteria: an abnormal, small product of cell division which lacks a chromosome and which is formed from a mutant parent cell by septation at a polar (rather than mid-cell) location; although lacking genomic DNA, a minicell may contain one or more plasmids derived from the parent cell. Minicells do not grow, but they contain the components necessary for transcription and translation; they can be useful in genetic studies in that, following infection or transfection with e.g. phage DNA, they will synthesize only phage-encoded proteins.

Minicells can be formed in e.g. *Escherichia coli* and *Bacillus subtilis* by mutation in the *minCD* genes (see prokaryotic CELL CYCLE); MinCD are septation-inhibitory proteins, and their absence permits septation at polar sites as well as at the mid-cell site. Minicells can also be formed from cells in which the FtsZ protein (see CELL CYCLE) is overproduced, i.e. raised levels of FtsZ can override the septation-inhibitory activity of the *min* proteins at polar sites.

(See also MAXICELL.)

minichromosome (1) A form in which the genome of certain DNA viruses occurs in the host (animal or plant) cell; the circular viral DNA complexes with histones to form a structure similar or identical to that of the host's CHROMATIN. Minichromosomes are formed e.g. by CAULIFLOWER MOSAIC VIRUS, EPSTEIN–BARR VIRUS, and viruses of the PAPOVAVIRIDAE. Those of SV40 are widely used in studies on chromatin structure; a mature SV40 minichromosome contains ca. 27 nucleosomes, or ca. 25 nucleosomes with a 'nucleosome-free gap'. [Structure of replicating SV40 minichromosomes: JMB (1986) *189* 189–204.]

(2) A plasmid which contains a chromosomal replication origin (usually the *oriC* of *Escherichia coli*) as sole origin of replication.

minicircle DNA See KINETOPLAST.

mini-F plasmid Any small, self-replicating plasmid constructed from a fragment of the F PLASMID – see e.g. PML31.

minimal medium (MM) A type of culture medium which lacks certain growth factors – i.e., it does not support the growth of some or all auxotrophic strains of a given organism, but permits the growth of prototrophic strains. (cf. COMPLETE MEDIUM.)

mini-Mu A defective form of BACTERIOPHAGE MU in which the genome lacks certain sequences but has intact ends; owing to the 'headful' mechanism by which Mu DNA is packaged into virions, the shortened genome of a mini-Mu phage is associated with a long sequence of bacterial DNA at the S end. Although mini-Mu phages are defective, they can be replicated and matured in the presence of helper Mu.

A mixture of mini-Mu and helper Mu phages can function more effectively in generalized TRANSDUCTION to rec^+ recipients than can Mu alone (which transduces only at very low frequencies); this is probably due to the higher proportion of phage particles containing significant amounts of bacterial DNA which can undergo *rec*-dependent recombination with homologous regions of the recipient's chromosome. However, if the mini-Mu DNA has an intact *A* gene, $recA^-$ recipients can be used; in this type of transduction (termed *mini-muduction*) the transduced DNA is integrated at random locations in the recipient's chromosome, the inserted DNA being flanked by two mini-Mu genomes in the same orientation.

(See also MUDLAC SYSTEM.)

mini-muduction See MINI-MU.

minimum haemagglutinating dose See MHD.

minimum haemolytic dose See MHD.

minimum inhibitory concentration (of an antibiotic) See MIC.

minimum lethal dose (MLD) (1) The minimum dose of a given lethal agent sufficient to cause 100% mortality in a population of test animals under given conditions.

(2) The minimum dose of a given lethal agent sufficient to kill an individual of a given species of specified body weight under given conditions.

Minisporida See MICROSPOREA.

Minitek system See MICROMETHODS.

mink Aleutian disease See ALEUTIAN DISEASE OF MINK.

mink cell focus-forming viruses See MCF VIRUSES.

mink enteritis virus See PARVOVIRUS.

mink leukaemia virus (MiLV) An endogenous type C retrovirus (subfamily ONCOVIRINAE) isolated from the Mv-1-Lu mink lung cell line. It is highly infectious for dog and donkey cells, weakly infectious for mink cells, and non-infectious for murine, goat, human and monkey cells. [Book ref. 114, pp. 118–120.]

minocycline See TETRACYCLINES.

minor groove (in DNA) See DNA.

minus-progamone See PHEROMONE.

minus strand (of a gene) (*mol. biol.*) See CODING STRAND.

minus strand (*virol.*) See VIRUS.

minus 10 sequence See PROMOTER.

minus 35 sequence See PROMOTER.

minute virus of canines See CANINE PARVOVIRUS.

minute virus of mice See PARVOVIRUS.

minutes (in relation to genetic loci) See INTERRUPTED MATING and MAP UNIT.

MIP channel A molecular pathway, consisting of a single protein in the cytoplasmic membrane (CM), which permits transmembrane movement of water and/or certain other uncharged molecules. Such a channel (designated Aqp1) was first described in the erythrocyte (red blood cell) membrane; it was later shown to be analogous to the major intrinsic protein (MIP) in mammalian lens fibre, to the tonoplast intrinsic proteins in plants, and to the GlpF protein in *Escherichia coli*. MIP proteins (found in all classes of living organisms) share highly conserved regions.

A MIP channel consists of a single polypeptide which apparently passes through the CM a number of times in a series of loops; this structure is believed to form a pathway for either water (channel = *aquaporin*) or glycerol (and/or other uncharged molecules) (channel = *glycerol facilitator*). (In *Lactococcus lactis* the MIP channel transports both water and glycerol.)

The aquaporin in *E. coli* (AqpZ) is encoded by gene *aqpZ*, and the glycerol facilitator (GlpF) is encoded by gene *glpF*. Mutants in *aqpZ* appear to grow poorly under low osmotic pressure – conditions which, in wild-type strains, appear to stimulate expression of *aqpZ*; this suggests that AqpZ has an important physiological role.

MIP channels occur in some members of the Archaea (e.g. *Archaeoglobus fulgidus*, *Methanobacterium thermoautotrophicum*) but apparently not in others (e.g. *Methanococcus jannaschii*). They also seem to be absent in e.g. *Chlamydia trachomatis*, *Helicobacter pylori*, *Mycobacterium tuberculosis* and *Treponema pallidum*.

[The importance of aquaporin water channel protein structures: EMBO (2000) *19* 800–806. Microbial MIP channels: TIM (2000) *8* 33–38.]

***Mirabilis* mosaic virus** See CAULIMOVIRUSES.

mirror yeasts (shadow yeasts) BALLISTOSPORE-forming yeasts (see SPOROBOLOMYCETACEAE); when these yeasts are grown in a Petri dish, the pattern of discharged ballistospores adhering to the lid is a mirror image (or shadow) of the pattern of growth beneath.

Mischococcus A genus of freshwater epiphytic algae (class XANTHOPHYCEAE) in which the coccoid, non-motile vegetative cells occur in colonies composed of dichotomously branching gelatinous tubes.

mismatch repair A DNA REPAIR system which can correct mismatched (i.e. non-Watson–Crick) base pairs – e.g. those formed during DNA replication when misincorporated nucleotides have escaped proof-reading. Mismatch repair effectively enhances the accuracy of replication. (cf. MUTATOR GENE.)

In *Escherichia coli*, mismatch repair (Dam-directed mismatch repair; DDMR) in the newly replicated daughter strand occurs soon after the failure of proof-reading and before Dam methylation (see DAM GENE) has occurred, i.e. while the new daughter strand is still (transiently) undermethylated. The mismatched base-pair is detected by MutS (*mutS* gene product), and this activates an endonuclease, MutH. [Mutational analysis of the MutH protein from *Escherichia coli*: JBC (2001) *276* 12113–12119.] (MutH exhibits sequence homology with a restriction endonuclease, *Sau*3AI [EMBO (1998) *17* 1526–1534].) MutH cleaves the new strand at a site 5′ to an *unmethylated* GATC (Dam) sequence – which may be up to ~1000 nucleotides from the actual site of mismatch; co-ordination between the mismatch location and the GATC cleavage site may be mediated by the MutL protein. When nicked by MutH, the sequence of daughter strand between the nick site and a site on the other side of the mismatch is removed. It has been assumed that removal of this sequence depends on helicase II (= MutU; = UvrD) and a DNA exonuclease; however, mutants lacking the 3′-to-5′ exonuclease ExoI *and* the 5′-to-3′ exonucleases ExoVII and RecJ appear to be capable of normal mismatch repair [JB (1998) *180* 989–993]. On removal of the sequence, the single-stranded gap is filled by a DNA polymerase and sealed by a ligase.

The mismatch repair system in *E. coli* is apparently not equally effective against all possible mismatched base pairs [PNAS (1985) *82* 503–505].

Mismatched nucleotides can arise not only by misincorporation of bases during replication but also by the formation of heteroduplex DNA during RECOMBINATION. In *E. coli*, if one strand of a heteroduplex is unmethylated, DDMR can apparently correct that strand; if both strands are fully methylated, *either* strand can apparently be corrected by a mismatch repair system involving MutS and UvrD [JMB (1986) *188* 147–157]. (See also TRANSFORMATION.)

Strains of *E. coli* lacking *mut* products have defective DNA repair systems, and they exhibit a mutator phenotype. (In humans, some types of cancer have been linked to mutations in genes homologous to *mutS* and *mutL*.)

Recent work suggests that asymmetry of ATPases on the MutS dimer is crucial for mismatch repair and controls timing of the repair cascade [EMBO (2003) *22* 746–756].

miso (soy paste; bean paste) A condiment made by fermenting soybeans, usually mixed with cereal. A KOJI is prepared by inoculating steamed rice with *Aspergillus oryzae*. Salt and steamed soybeans are added, and the mixture is inoculated with a starter culture of yeasts and bacteria (e.g. *Zygosaccharomyces rouxii*, *Pediococcus halophilus*, *Enterococcus faecalis*), blended, packed tightly into vats, and fermented at 25–30°C for ca. 2–3 months. After ageing for ca. 2 weeks, the product is blended, pasteurized and packaged, or freeze-dried and sold as a powder. [Book ref. 5, pp. 39–86.]

mis-sense mutation A MUTATION in which a codon specifying one amino acid is altered so as to specify a different amino acid. The effects of such a mutation range from none (a SILENT MUTATION) to total inactivation of the gene product, depending e.g. on the location of the amino acid in the product and on the nature of the substituted amino acid relative to that of the original. (cf. NONSENSE MUTATION; see also POINT MUTATION.)

MISTRESS Milk into steam treatment research equipment small scale: equipment used for research into high-temperature (e.g. UHT) methods of preservation in the food industry. Essentially, milk is sprayed into a steam-containing vessel and, at the end of the holding time, it is drawn through a hole into a vacuum chamber where it cools by expansion.

mithramycin See CHROMOMYCIN.

mitochondrion A semi-autonomous intracellular organelle, one or more (usually many) of which occur in most types of eukaryotic cell (being absent e.g. in protozoa of the orders Diplomonadida and Pelobiontida); RESPIRATION and the reactions of the TCA CYCLE occur within the mitochondrion. Mitochondria differ in size (<1 μm to >10 μm), shape, and number, according to the type of cell in which they occur and to the physiological state of that cell; a mitochondrion may be spherical, ovoid, elongated, calyciform or irregularly shaped, and some are branched.

Structure and composition. A mitochondrion consists of two closed membranous sacs, one fitting closely within the other, and an amorphous *matrix* (enclosed by the inner membrane) which contains the mitochondrial DNA and enzymes of the TCA cycle. The mitochondrial inner membrane is extensively corrugated and forms plate-like, tubular, or finger-like structures (*cristae*) which project into the matrix – often more or less perpendicular to the longitudinal axis of the mitochondrion; the inner membrane contains components of the ELECTRON TRANSPORT CHAIN, PROTON ATPASES and TRANSPORT SYSTEMS for e.g. nucleotides, inorganic phosphate, and Ca^{2+}. (See also ION TRANSPORT.) The mitochondrial outer membrane contains PORINS (see also VDAC).

Mitochondrial DNA (mtDNA) is typically in the form of covalently closed circular, double-stranded molecules (differing from the nuclear DNA e.g. in base composition and buoyant density); mtDNA is linear in e.g. certain ciliates (including *Paramecium*), *Physarum polycephalum*, and *Hansenula mrakii*. Mitochondrial genes employ a GENETIC CODE (q.v.) which differs in some respects from the 'universal' code, and some fungal mitochondrial genes contain introns (see SPLIT GENE). Genetic recombination can occur between mtDNAs. (See also PETITE MUTANT.)

Origin and semi-autonomy of mitochondria. It is generally believed that mitochondria are formed by the division or fragmentation of pre-existing mitochondria – or (see later) by the development of promitochondria – and that these organelles incorporate new material (i.e., grow) during interdivision periods. The components of mitochondria are encoded partly by the cell's nuclear DNA and partly by the mtDNA. [Nuclear genes encoding mitochondrial proteins in yeast: TIBS (1985) *10* 192–194.] Thus, in *Saccharomyces cerevisiae*, mtDNA encodes e.g. tRNAs, two types of rRNA, one ribosomal protein, the apoprotein of cytochrome *b* in Complex III, and a protein designated 'var1'. Certain mitochondrial components are synthesized in the mitochondrion itself. However, while a mitochondrion can synthesize DNA and RNA, and can carry out protein synthesis, many of the proteins needed for these processes are encoded by nuclear genes, synthesized on cytoplasmic ribosomes, and incorporated into the mitochondrion; control of the synthesis of these proteins may be largely at the transcriptional level.

Mitochondrial protein synthesis differs from cytoplasmic protein synthesis e.g. in that it is sensitive to those agents (e.g. CHLORAMPHENICOL, ERYTHROMYCIN) which inhibit bacterial PROTEIN SYNTHESIS; it is not sensitive to CYCLOHEXIMIDE; and it is characterized by the incorporation of *N*-formylmethionine as the first amino acid in a polypeptide chain. These features have lent support to a popular hypothesis which supposes that mitochondria have their evolutionary origins in endosymbiotic prokaryotes. [Mitochondrial origins: PNAS (1985) *82* 4443–4447.] (An alternative hypothesis supposes that mitochondria evolved from plasmids.)

In at least some facultatively fermentative organisms (including e.g. *Saccharomyces cerevisiae*) cells growing under anaerobic conditions do not contain functional mitochondria; such cells contain *promitochondria*: organelles which resemble mitochondria but which lack components of the electron transport chain. On exposure to aerobic conditions promitochondria apparently develop into functional mitochondria; reversion to promitochondria occurs if anaerobic growth is resumed.

In e.g. certain fungi mitochondrial defects can alter the nature of the cell's metabolism (see e.g. PETITE MUTANT and POKY MUTANT), and can affect the expression of nuclear genes which are involved e.g. in determining the nature of cell-surface antigens [TIBS (1982) *7* 147–151]; inhibition of mitochondrial protein synthesis can affect both meiotic and apomictic sporulation in *Saccharomyces cerevisiae* [Yeast (1985) *1* 39–47].

mitogen Any agent which, under appropriate conditions, can promote MITOSIS, non-specifically, and can bring about BLAST TRANSFORMATION. (cf. POLYCLONAL ACTIVATORS.) Mitogens include e.g. certain LECTINS (such as PHYTOHAEMAGGLUTININ). (Substances which can promote early events in the activation process – including DNA synthesis – but which cannot promote mitosis are not, strictly speaking, mitogens.)

mitogen-activated protein kinase See ANTHRAX TOXIN.

mitomycin C ('mitomycin') An ANTIBIOTIC (from *Streptomyces* spp) with general cytotoxic activity; it contains a quinone group and an aziridine (ethyleneimine) ring. It binds to DNA and causes inhibition of DNA replication followed by degradation of DNA; synthesis of proteins and RNA may continue for a time. Mitomycin activity depends on its reduction in vivo (by NADPH-dependent enzymes) to the highly reactive hydroquinone derivative which behaves as a bifunctional ALKYLATING AGENT and forms e.g. covalent cross-links between complementary strands of dsDNA; the semiquinone seems to be formed as an intermediate and may bind DNA non-covalently and intercalatively (see INTERCALATING AGENT). Preferred sites for alkylation are the O-6 groups of guanine residues. (Reduced mitomycin will also react with RNA and, to a lesser extent, protein.) Alkylation of B-DNA by mitomycin can cause conformational changes that may involve the transition to Z-DNA. Mitomycin is a potent inducer of lysogenic bacteriophages; viral DNA synthesis is relatively resistant to mitomycin.

mitoplast The inner membrane of a MITOCHONDRION together with the mitochondrial matrix.

mitosis (karyokinesis) A sequence of events which culminates in the division of a eukaryotic nucleus into two genetically similar or identical nuclei whose ploidy is the same as that of the parent nucleus; mitosis occurs during asexual (vegetative) cell division. (cf. MEIOSIS.)

Different types of mitosis occur among the various types of eukaryotic cell. The best-known ('typical') form of mitosis is that which occurs in the cells of higher animals and e.g. in some protozoa; it involves the following stages (in chronological order).

Prophase. Individual CHROMOSOMES, each having previously divided into two chromatids, condense and become visible by light microscopy. The NUCLEOLUS becomes indistinct and then invisible. MICROTUBULES of the CYTOSKELETON disassemble, forming a pool of tubulin subunits. From each of two pairs of CENTRIOLES (lying close to each other in the cytoplasm) arises a radial array of microtubules (referred to as an *aster*). Those microtubules which extend from each centriole pair towards the other elongate – the centriole pairs concomitantly moving apart until each eventually forms one pole of a bipolar microtubular *mitotic spindle*; microtubules from each of the centriole pairs (= *polar microtubules* or *polar fibres*) overlap in the mid-region of the spindle.

Prometaphase. The nuclear membrane breaks into pieces. KINETOCHORES develop on the chromosome centromeres, and from each kinetochore *kinetochore microtubules* (= k-MTs; *kinetochore fibres*) grow out more or less perpendicularly to the long axis of the chromosome. The k-MTs apparently begin to interact with the spindle (in some as yet unknown way) causing the chromosomes to execute rapid movements.

Metaphase. The chromosomes become arranged in a layer (the *metaphase plate*) midway between the poles of the spindle and perpendicular to its axis. The metaphase plate may be maintained for some time, each chromosome possibly being held in dynamic equilibrium by equal and opposite forces generated within the spindle.

Anaphase. The two chromatids of each chromosome appear to move along stationary k-MTs to opposite poles of the spindle; de-polymerization of the k-MTs apparently occurs at the kinetochores – the kinetochores possibly being involved in generating the force needed to propel the chromatids towards their respective poles [JCB (1987) *104* 9–18]. As the k-MTs shorten, the polar microtubules elongate and the spindle poles move further apart.

Telophase. The chromatids become less clearly visible, and a nuclear membrane assembles around each group of chromatids to form two new nuclei – in each of which a new nucleolus appears. [Cell-free system for studying reassembly of the nuclear envelope: Cell (1986) *44* 639–652.]

Telophase is followed by CYTOKINESIS. The non-dividing nucleus is called the *interphase* nucleus; DNA replication occurs during interphase.

In many microorganisms mitosis differs significantly from that described above; thus, e.g. there are at least four distinct types of mitosis–cytokinesis even within the green algae [Book ref. 130, pp. 95–99]. In many algae (e.g. *Dunaliella*, DINOFLAGELLATES) and fungi (e.g. *Albugo* spp, *Mucor hiemalis*, *Saprolegnia ferax*, many yeasts) the nuclear envelope persists throughout mitosis ('closed mitosis') – although in some of these organisms the nuclear envelope develops a number of holes (*fenestrae*). In some organisms the spindle may be wholly intranuclear; this occurs e.g. in many yeasts, including *Saccharomyces*, in which the spindle microtubules arise, at each pole, from a SPINDLE POLE BODY. In EUGLENOID FLAGELLATES the nuclear envelope remains intact, the nucleolus is persistent, and microtubules develop within the nucleus; however, the microtubules apparently do not function as a conventional spindle. In some algae part of the spindle is used as a section of the newly forming cell wall (see PHRAGMOPLAST). (cf. AMITOSIS; see also CELL CYCLE and PARASEXUAL PROCESSES.)

[Mitosis in fungi: Book ref. 168, pp. 12–26, 85–112; sites of microtubule assembly and disassembly in the spindle: Cell (1986) *45* 515–527.]

mitosporangium A SPORANGIUM within which MITOSIS occurs (see e.g. ALLOMYCES).

mitotic crossing over See PARASEXUAL PROCESSES.

mitotic spindle See MITOSIS.

Mitrula See HELOTIALES.

Mitsuda reaction See LEPROMIN TEST.

mixed acid fermentation A FERMENTATION (sense 1) carried out e.g. by certain enterobacteria, including *Escherichia coli* and species of *Proteus*, *Salmonella* and *Shigella*. In *E. coli* the products of glucose fermentation may include e.g. acetic and lactic acids, smaller amounts of formic acid (or $CO_2 + H_2$) and succinic acid, and ethanol [see Appendix III(e)]; products are formed in relative proportions which depend on organism and growth conditions. Those organisms (e.g. *E. coli*) which have a FORMATE HYDROGEN LYASE system split formic acid (under acid conditions) to CO_2 and H_2; some of the CO_2 may be assimilated during the formation of succinate. In *E. coli* formate accumulates under alkaline conditions. Species of e.g. *Shigella*, which lack the formate hydrogen lyase system, carry out the fermentation anaerogenically. Enterobacteria which carry out the mixed acid fermentation give a positive METHYL RED TEST. (See also PHOSPHOROCLASTIC SPLIT.)

mixed bed exchangers Ion-exchange material containing both cation and anion exchangers.

mixed culture See CULTURE.

mixed-function oxidase *Syn.* MONOOXYGENASE.

mixed infection The concurrent infection of a given individual with more than one pathogen.

mixed leucocyte reaction See MLR.

mixed lymphocyte reaction See MLR.

mixed vaccine A single VACCINE designed to give protection against more than one pathogen; such a vaccine contains PROTECTIVE ANTIGENS and/or toxoids from each of the pathogens. (See e.g. TRIPLE VACCINE; cf. POLYVALENT VACCINE.)

mixis The fusion of gametes (fertilization).

mixoploid Refers to a population of cells in which the chromosome complement per cell varies among the cells of the population.

mixotrophy A mode of metabolism in which energy is obtained by the oxidation of an inorganic substrate, and carbon is obtained from an organic substrate (and sometimes also from the fixation of CO_2 via an autotrophic pathway). In some cases of mixotrophy, energy appears to be derived from the metabolism of an organic substrate as well as an inorganic substrate; this has been reported e.g. for an iron-oxidizing bacterium [JGM (1984) *130* 1337–1349] and for '*Alcaligenes eutrophus*' [JGM (1984) *130* 1987–1994]. (See also CHEMOLITHOHETEROTROPH.)

Miyagawanella See CHLAMYDIA.

MK Menaquinone: see QUINONES.

MK broth MÜLLER–KAUFFMANN BROTH.

MK-0787 See THIENAMYCIN.

MLD MINIMUM LETHAL DOSE.

MLEE MULTILOCUS ENZYME ELECTROPHORESIS.

MLOs *Mycoplasma*-like organisms: wall-less, non-helical, prokaryotic organisms which are associated with invertebrates and plants and which are the causal agents of a number of plant YELLOWS diseases. MLOs have not been cultured in vitro and have not been formally classified, although they are usually associated, taxonomically, with the MOLLICUTES. Various MLOs have been shown to lack a *Spiroplasma*-specific antigen [JGM (1983) *129* 1959–1964]. [Book ref. 22, pp. 792–793; wall-less prokaryotes and plants: ARPpath. (1984) *22* 361–396.]

From the mid-1990s, MLOs were referred to by the trivial name *phytoplasma*. Individual MLOs are now being called by names such as *Candidatus* Phytoplasma *australiense*.

[Phytoplasma: ARM (2000) *54* 221–255.]

MLR (*immunol.*) Mixed lymphocyte reaction (= mixed leucocyte reaction or mixed leukocyte reaction): mutual interaction, with blast transformation and proliferation of T cells, which occurs when leukocytes from two allogeneic individuals are mixed and cultured; the greater the disparity between antigens (e.g. MHC class II antigens) on donor and recipient cells the greater the degree of proliferation. (The degree of proliferation may be assessed e.g. by measuring the amount of radioactive thymidine (a DNA precursor) taken up by the cells.)

MLR (which takes several days to perform) is used e.g. to assess the compatibility of transplant donors and recipients. A negative reaction (no proliferation of T cells) is given when cells are derived from identical twins.

MLS antibiotics Macrolide, lincosamide and streptogramin B-type antibiotics: a group of structurally distinct antibiotics all of which act on the 50S subunit of 70S ribosomes (see MACROLIDE ANTIBIOTICS, LINCOSAMIDES and STREPTOGRAMINS). [Symposium on MLS antibiotics: JAC (1985) *16* (supplement A).]

Co-resistance to MLS antibiotics in staphylococci, streptococci and streptomycetes can be due to N^6-methylation of adenine residues in 23S rRNA (see STREPTOGRAMINS).

Strains of *Streptococcus pyogenes* and *S. pneumoniae* that express the 'M phenotype' contain a macrolide-efflux mechanism which confers resistance to macrolides concomitant with sensitivity to lincosamides and B streptogramins (see MACROLIDE ANTIBIOTICS).

$MLS_{B/c}$ strains See STREPTOGRAMINS.

MLST (multilocus sequence typing) A method for TYPING pathogenic bacteria which permits long-term, global-scale tracking of virulent and antibiotic-resistant strains, with information made available on the Internet. In MLST, strains are characterized and classified on the basis of nucleotide sequences in *specific* alleles, this approach giving rise to categorization which is apparently stable over long periods of time.

[MLST (review): TIM (1999) *12* 482–487. MLST of methicillin-resistant and methicillin-sensitive clones of *Staphylococcus aureus*: JCM (2000) *38* 1008–1015.]

***Mlu*I** See RESTRICTION ENDONUCLEASE (table).

MLV MURINE LEUKAEMIA VIRUS.

MM MINIMAL MEDIUM.

MM virus See CARDIOVIRUS.

MMLV reverse transcriptase See REVERSE TRANSCRIPTASE.

MMO METHANE MONOOXYGENASE.

MMR Measles–mumps–rubella vaccine.

MMS Methylmethane sulphonate: see ALKYLATING AGENTS.

MMTV MOUSE MAMMARY TUMOUR VIRUS.

MNNG (*N*-methyl-*N*′-nitro-*N*-nitrosoguanidine; also: 'nitrosoguanidine', NTG) A mutagenic *N*-NITROSO COMPOUND, the mutagenic activity of which appears to depend primarily on methylation of the O-6 position of guanine (see ALKYLATING AGENTS). MNNG acts preferentially at the replication fork in replicating DNA, producing clusters of closely linked mutations (*co-mutagenesis*); it also methylates proteins.

***M*NPV** See NUCLEAR POLYHEDROSIS VIRUSES.

MNU *N*-Methyl-*N*-nitrosourea: see ALKYLATING AGENTS.

mobile genetic element See TRANSPOSABLE ELEMENT.

Mobilina See PERITRICHIA.

mobilization (*mol. biol.*) See CONJUGATION (1b) (i).

Mobiluncus A genus of anaerobic, asporogenous bacteria which occur in the human vagina and which are associated with BACTERIAL VAGINOSIS. Cells: Gram-negative and Gram-variable curved rods (ca. 1.5–3.0 μm in length) which are motile by multiple subpolar flagella; the CELL WALL is of the Gram-positive type [IJSB (1986) *36* 288–296]. Metabolic products include acetate and succinate. GC%: 49–52 (T_m). Type species: *M. curtisii*; other species: *M. mulieris*. [IJSB (1984) *34* 177–184.]

mocimycin See POLYENE ANTIBIOTICS (b).

modification (of DNA) See DNA MODIFICATION.

modulator protein See Tn*21*.

modulin Any virulence factor which causes damage by inducing inappropriate activity of CYTOKINES. The modulins include e.g. endotoxins, exotoxins and SUPERANTIGENS. [MR (1996) *60* 316–341.]

Moellerella A genus of bacteria of the ENTEROBACTERIACEAE. *M. wisconensis* (formerly 'enteric group 46') has been isolated from human faeces and is apparently associated with diarrhoea in humans. It is non-motile; MR +ve; citrate +ve (Simmons); VP −ve; indole −ve; LDC, ODC, ADH and phenylalanine deaminase −ve; anaerogenic. On MacConkey's agar colonies are bright red and surrounded by precipitated bile. [JCM (1984) *19* 460–463.]

Moeller's cyanide medium See KCN BROTH.

Moeller's decarboxylase broth See MØLLER'S DECARBOXYLASE BROTH.

moenomycin (Flavomycin) A phosphorus-containing glycolipid ANTIBIOTIC in which the lipid moiety (moenocinol) is a C-25 compound structurally analogous to undecaprenylphosphate; it inhibits transglycosylation in PEPTIDOGLYCAN synthesis in Gram-positive bacteria and, at higher concentrations, in Gram-negative bacteria.

MoFe protein See NITROGENASE.

moi MULTIPLICITY OF INFECTION.

moiré image In microscopy: a misleading image due to overlap among multiple structures.

moist heat Refers to the use of steam for STERILIZATION. (See also AUTOCLAVE.)

Mokola virus See LYSSAVIRUS.

mol% G + C See GC%.

molar growth yield See YIELD COEFFICIENT.

molasses A by-product of the sugar industry: the syrupy liquid which remains after the crystallization of sucrose from sugar-cane or sugar-beet syrup. Molasses contains ca. 50–60% sucrose (which cannot be removed by crystallization), 16–20% water, and various non-sugar and inorganic ('ash') components; it is widely used as a relatively cheap substrate for the commercial production of various microorganisms and microbial products (see e.g. BAKERS' YEAST; CITRIC ACID; INDUSTRIAL ALCOHOL; SINGLE-CELL PROTEIN).

mold See MOULD.

molecular beacon probes Oligonucleotide probes used for (i) monitoring progress in a nucleic-acid-amplification procedure, and (ii) (in PCR), estimating the initial (pre-amplification) number of target sequences in the sample. The general structure and mode of action of a molecular beacon probe are described in the figure.

Like TAQMAN PROBES (q.v.), molecular beacon probes are added (in large numbers) to the reaction mixture prior to amplification.

In NASBA, molecular beacon probes can be used for real-time monitoring of amplification [NAR (1998) *26* 2150–2155].

In PCR, the probes are useful for estimating the initial number of target sequences in a sample (see TAQMAN PROBES for rationale).

Because these probes can be labelled with different kinds of fluorophore (having different emission characteristics), it is possible to carry out a multiplex PCR in which the amplification of each type of target can be followed separately by monitoring the distinct emission from target-specific probes. [Multiplex detection of four pathogenic retroviruses using molecular beacons: PNAS (1999) *96* 6394–6399.]

Compared with linear probes (such as TaqMan), molecular beacon probes exhibit greater specificity in binding; for example, a beacon probe of equivalent length can more readily distinguish a one-nucleotide difference between two sequences. [Thermodynamic basis of the enhanced specificity of structured DNA probes: PNAS (1999) *96* 6171–6176.]

molecular chaperone See FIMBRIAE.

molecular cloning See CLONING.

molecular sieving *Syn.* GEL FILTRATION.

Møller's cyanide medium See KCN BROTH.

Møller's decarboxylase broth A medium used for DECARBOXYLASE TESTS. It contains (g/l): peptic digest of animal tissue (5.0), beef extract (5.0), glucose (0.5), BROMCRESOL PURPLE (0.1), cresol red (0.005), pyridoxal (0.005) and L-arginine *or* L-lysine *or* L-ornithine (1% final concentration). The pH of the broth is adjusted to 6.0.

mollicute A proposed trivial name for any member of the class MOLLICUTES; use of this name could permit 'mycoplasmas' to be used specifically for members of the genus *Mycoplasma*.

Mollicutes A class of small, cell-wall-less bacteria which are typically parasitic (sometimes pathogenic) in animals or plants. (cf. TENERICUTES; see also MOLLICUTE.) The class comprises a single order, MYCOPLASMATALES, and two genera *incertae sedis* (ANAEROPLASMA and THERMOPLASMA) [Book ref. 22, pp. 740–793]. (See also MLO.)

molluscum contagiosum A human skin disease caused by a virus of the POXVIRIDAE; the virus is apparently unrelated to other poxviruses, and has not been grown in cell cultures or in the CAM. Molluscum contagiosum is characterized by the formation of multiple firm, rounded, whitish, translucent nodules which become crater-like, revealing a white core; the lesions may persist for months or even years. The virus is spread by body contact (e.g. sexual contact) or via fomites.

Moloney murine leukaemia virus (Mo-MuLV) A replication-competent V-*onc*$^-$ MURINE LEUKAEMIA VIRUS originally isolated following passage of a cell-free extract of a transplantable tumour ('Sarcoma 37') in neonatal mice. Mo-MuLV can induce

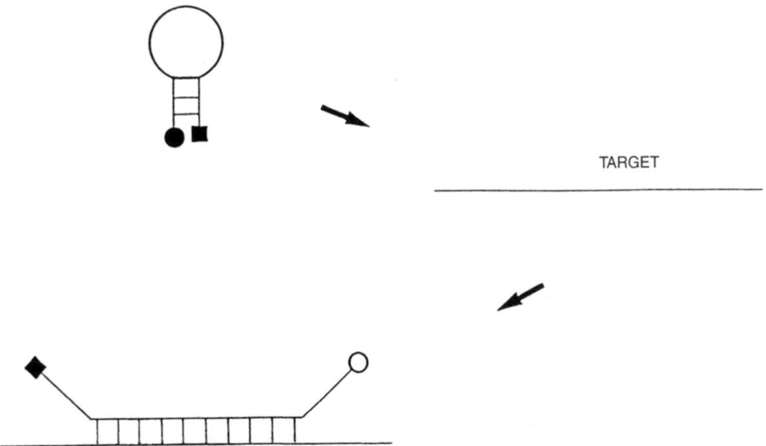

MOLECULAR BEACON PROBES: structure and mode of action (diagrammatic).

A molecular beacon probe is an oligonucleotide probe sequence flanked by two, short mutually complementary oligonucleotides (side-arms), the molecule being capable of adopting a stem-and-loop conformation (*top*); the terminals of the beacon probe are labelled with a fluorophore (●) and a quencher (■) such that fluorescence is quenched (by juxtaposition of fluorophore and quencher) when the molecule adopts the stem-and-loop conformation.

When a probe (the loop region) binds to the target sequence in an amplicon, the probe–target duplex is both longer and more stable than the stem duplex; hence, on probe binding, the two side-arms of the stem dissociate – thus separating the fluorophore from the quencher and giving rise to a functionally fluorescent probe.

Like TAQMAN PROBES, molecular beacon probes are added to the reaction mixture prior to amplification; they are added in large numbers so that a probe will be available for each amplicon at the end of amplification.

In NASBA, an isothermal process, fluorescence from the bound probes is monitored continually. In PCR, fluorescence is monitored during the primer annealing stage (because only at this stage are fluorescent probe–amplicon duplexes formed).

Reproduced from Figure 4.13, page 102, in *DNA Methods in Clinical Microbiology* (ISBN 07923-6307-8), Paul Singleton (2000), with kind permission from Kluwer Academic Publishers, Dordrecht, The Netherlands.

thymic leukaemia, disseminated lymphosarcoma, or lymphatic leukaemia in mice after a latent period of ca. 2–3 months. (See also MOLONEY MURINE SARCOMA VIRUS.)

Moloney murine leukaemia virus reverse transcriptase See REVERSE TRANSCRIPTASE.

Moloney murine sarcoma virus (Mo-MSV) A replication-defective v-*onc*$^+$ MURINE SARCOMA VIRUS isolated from a BALB/c mouse infected with MOLONEY MURINE LEUKAEMIA VIRUS (Mo-MuLV) and suffering from a rhabdosarcoma; several variants are known, each containing Mo-MuLV-derived sequences and the oncogene v-*mos* (see MOS). Infection of newborn mice with Mo-MSV can result in the appearance of sarcomas within a few days; sarcomas induced in older mice commonly undergo spontaneous regression and recurrence. (cf. GAZDAR MURINE SARCOMA VIRUS.)

molybdoenzymes The molybdenum-containing enzymes; they include e.g. aldehyde oxidase, CO OXIDASE, dissimilatory nitrate reductase, NITROGENASE, sulphite oxidase and xanthine oxidase.

molybdoferredoxin See NITROGENASE.

***mom* gene** See BACTERIOPHAGE MU.

momilactone Either of two diterpene PHYTOALEXINS (momilactones A and B) formed by the rice plant (*Oryza sativa*). (See also WL 28325.)

Monacrosporium See HYPHOMYCETES and NEMATOPHAGOUS FUNGI.

monactin See MACROTETRALIDES.

-monas A suffix signifying 'a unit'.

monascoflavin See ANG-KAK.

monascorubrin See ANG-KAK.

Monascus See PEZIZALES.

monensin (monensin sodium; *syn.* rumensin) A MACROTETRALIDE antibiotic synthesized by *Streptomyces cinnamonensis*; it is used e.g. as an anti-coccidial agent in poultry, and as a FEED ADDITIVE for ruminants (e.g. cattle, sheep). (See also FOG FEVER.)

As a feed additive, monensin appears to act by inhibiting those RUMEN microorganisms which produce acetic acid and hydrogen, thereby favouring those which produce propionic acid; thus, it appears that energetically wasteful methanogenesis is suppressed (by as much as 30–40%) as a result of the lack of hydrogen (rather than by direct inhibition of methanogens), and that it is the corresponding increase in propionic acid which leads to increased feed conversion efficiency.

In pure cultures of methanogens both monensin and LASALOCID directly affect membrane function and proton motive force, and in fermenting organic wastes monensin inhibits methanogens and causes an accumulation of acetate. The different susceptibilities of rumen and sewage methanogens may be due to the high Na^+ concentration in rumen fluid; the inhibition of methanogenesis by monensin is known to be prevented by high concentrations of Na^+ [Book ref. 157, pp. 530–531].

Monensin can inhibit the replication of certain viruses whose maturation involves the PALADE PATHWAY.

Monera The *prokaryotic* protists – see PROTISTA.

Monilia *Syn.* CANDIDA.

Moniliales An order of fungi; depending on taxonomic scheme, the Moniliales includes either most or all of the organisms which constitute the class HYPHOMYCETES.

moniliasis *Syn.* CANDIDIASIS.

monilicolin A See PHASEOLLIN.

moniliform Resembling a string of beads.

Monilinia See HELOTIALES.

Moniliophthora A genus of fungi of the APHYLLOPHORALES; *M. roreri* (anamorph: '*Monilia roreri*') causes POD ROT of cacao.

monkey-bite encephalomyelitis See B VIRUS.

monkeypox A rare disease of man and other primates; it is caused by the monkeypox virus (see ORTHOMYXOVIRUS). In man, the disease involves generalized exanthema and closely resembles SMALLPOX; it is not readily transmissible from person to person. Monkeypox generally affects people who have not been vaccinated against smallpox and who live in villages in the tropical rain forests of West and Central Africa; the natural reservoir of the virus is unknown. Monkeypox also occurs e.g. in captive monkeys; the symptoms are similar to those in man.

The monkeypox virus is highly lethal in mice and typically forms small, opaque, haemorrhagic pocks on CAM; only B-type inclusion bodies are seen in infected cells. Genome MWt: ca. 128×10^6.

monoazide ethidium See ANISOTROPIC INHIBITOR.

monobactams A class of monocyclic β-LACTAM ANTIBIOTICS (q.v. for formula). Natural monobactams include e.g. *sulfazecin* (= 'monobactam I', $R' =$ D-glu-D-ala, $R'' = OCH_3$, $R''' = H$, $R'''' = SO_3H$), produced e.g. by *Gluconobacter* and *Acetobacter* spp, and *isosulfazecin*, produced by *Agrobacterium radiobacter* and *Pseudomonas* sp. Synthetic derivatives include e.g. AZTREONAM and the *monophosphams* (2-oxoazetidine-1-phosphonates, where $R'''' = PO_3X$), *monocarbams* ($R'''' = CONHSO_2X$), and *monosulfactams* ($R'''' = OSO_3^-$).

[Review: AAM (1986) *31* 181–205.]

Monoblepharella See MONOBLEPHARIDALES.

Monoblepharidales An order of typically aquatic, saprotrophic fungi (class CHYTRIDIOMYCETES) in which sexual reproduction is *oogamous*, involving a small, motile male gamete and a large non-motile female gamete (a type of sexual reproduction not found in other fungi) (cf. OOMYCETES). The thallus is a well-developed branching mycelium in which the cytoplasm is highly vacuolated ('foamy'); in at least some species the cell wall contains CHITIN. Asexual reproduction involves the formation of elongated, typically cylindrical sporangia containing large numbers of zoospores; each zoospore can give rise to a new thallus. In sexual reproduction, distinct male and female gametangia (antheridia and oogonia, respectively) are formed; the male gametes ('antherozoids') swim to the non-motile oospheres. (In some species the oogonium forms more than one oosphere, and oospheres may be fertilized within or outside the oogonium, according e.g. to species.) After fertilization, the oosphere develops into a thick-walled resting *oospore* which later germinates to form a new thallus. Genera include *Gonapodya*, *Monoblepharella* and *Monoblepharis*.

Monoblepharis See MONOBLEPHARIDALES.

monocarbams See MONOBACTAMS.

monocentric (1) Refers to a thallus in which only one region has a reproductive function. (cf. POLYCENTRIC.)

(2) Refers to a chromosome which has a single, discrete CENTROMERE. (cf. HOLOCENTRIC.)

Monocercomonas See TRICHOMONADIDA.

Monocercomonoides See OXYMONADIDA.

monochloramine See CHLORINE.

monocins See BACTERIOCIN.

monocistronic mRNA See MRNA.

monoclonal antibodies A population of identical antibodies (see ANTIBODY), all of which recognize the same specific DETERMINANT on a simple or complex antigen or hapten. (cf. POLYCLONAL ANTISERUM.) A given population of monoclonal antibodies may be able to react with each of two apparently unrelated antigens if, by chance, those antigens happen to share a common determinant; thus, e.g., monoclonal antibodies which combine with a particular mouse myeloma protein can combine with human C-reactive protein. (cf. CROSS-REACTING ANTIBODY.) Monoclonal antibodies are obtained from B-cell HYBRIDOMAS.

Monoclonal antibodies have an enormous range of actual or potential uses. These include e.g. characterization and detection of viruses [e.g. plant viruses: MS (1984) *1* 73–78] and other microorganisms; tissue typing; the study of changes in cell-surface antigens on lymphoid cells during maturation; and the preparation of IMMUNOTOXINS.

monocystid gregarine *Syn.* acephaline gregarine.

Monocystis A genus of protozoa (subclass GREGARINASINA) many species of which are parasitic in the earthworm (*Lumbricus* spp). Sporozoites enter the seminal vesicles. The developing parasites (*trophozoites*) occur intracellularly in the sperm morulae; the fully-grown, extracellular acephaline trophozoites (*gametocytes*) associate in pairs (syzygy) and each pair forms a *gametocyst* within which many gametes develop. The gametes fuse in pairs, and each zygote forms a cyst (*sporocyst*, *pseudonavicella*) within which 8 sporozoites develop.

monocyte A large (diam. 10–20 μm) phagocytic LEUCOCYTE which contains a single, large, spherical or indented nucleus, and azurophilic peroxidase-containing granules; ca. 200–800 monocytes/ml occur in normal adult human blood. Monocytes differentiate to become MACROPHAGES. Suspensions of monocytes may be prepared e.g. by density flotation [JIM (1984) *69* 71–77] or by the use of Sephadex [JIM (1984) *71* 25–36].

monocytosis *Syn.* MONONUCLEOSIS.

Monod equation See SPECIFIC GROWTH RATE.

Monodictys See HYPHOMYCETES; see also SOFT ROT (sense 1).

monoecious Refers to an organism in which the male and female reproductive structures occur in the same individual. (cf. DIOECIOUS.)

monogeneric Refers to a taxon which has only one genus.

monogenic mRNA See MRNA.

monokaryotic Having one nucleus per cell.

monokines Certain soluble, biologically active products of MACROPHAGES – e.g. INTERLEUKIN-1.

monolayer culture See TISSUE CULTURE.

Monomastix See MICROMONADOPHYCEAE.

monomitic See HYPHA.

monomorphic Refers to an organism which occurs in only one form or MORPHOTYPE. (cf. POLYMORPHISM.)

Mononegavirales A taxon containing viruses which have non-segmented, negative-sense ssRNA genomes, including members of the FILOVIRIDAE, PARAMYXOVIRIDAE and RHABDOVIRIDAE – and (probably) the BORNA DISEASE VIRUS.

mononuclear leucocyte (mononuclear phagocyte) (1) *Syn.* MACROPHAGE. (2) *Syn.* MONOCYTE. (3) A macrophage or monocyte.

mononuclear phagocyte See MONONUCLEAR LEUCOCYTE.

mononucleosis (monocytosis) The condition in which increased numbers of monocytes are present in the blood. (cf. INFECTIOUS MONONUCLEOSIS.)

monooxygenase (hydroxylase; mixed-function oxidase; mixed-function oxygenase) An OXYGENASE (EC 1.13.12) which catalyses the incorporation of one atom of molecular oxygen into a substrate molecule, the other oxygen atom being reduced to water; the reducing power needed for monooxygenase activity may be supplied e.g. by NADH. (See also METHANE MONOOXYGENASE.)

monophasic serotypes (in *Salmonella*) See H ANTIGEN.

monophosphams See MONOBACTAMS.

monophyletic Refers to taxa which have arisen from a common evolutionary line. (cf. POLYPHYLETIC.)

monoplanetism In certain OOMYCETES: the phenomenon in which only one type of zoospore is formed and only one swarm period occurs. (cf. DIPLANETISM.)

monopodial (*adj.*) Refers to a 'tree-like' form or mode of development of a branching structure – i.e., one with a main, central stem from which lateral branches arise. (cf. SYMPODIAL.)

monopodial amoebae Amoebae which normally produce only one lobose PSEUDOPODIUM at a time (although a second, lateral pseudopodium may be produced when the organism changes direction). (See also LIMAX AMOEBAE; cf. POLYPODIAL AMOEBAE.)

Monosiga See CHOANOFLAGELLIDA.

monosodium glutamate See GLUTAMIC ACID.

monosome See PROTEIN SYNTHESIS.

monosomic See CHROMOSOME ABERRATION.

monospecific Refers to a genus which has only one species.

monospecific antiserum An antiserum (e.g. a GROUPING ANTISERUM) which is capable of reacting specifically with one particular antigen or antigenic determinant – rather than with several (or many) of the antigens present on cells etc in a given suspension. (cf. POLYVALENT ANTISERUM.)

monostichomonad membrane See PARORAL MEMBRANE.

Monostroma A genus of green seaweeds (division CHLOROPHYTA) which show a heteromorphic alternation of generations; the gametophyte is a monostromatic sheet or blade (cf. ULVA), the sporophyte is unicellular.

monostromatic Refers to a structure which is only one cell thick (see e.g. MONOSTROMA). (cf. DISTROMATIC.)

monosulfactams See MONOBACTAMS.

Monothalamida See GRANULORETICULOSEA.

monothetic classification Any form of classification based on a small number of carefully selected criteria – which are usually interpreted unequivocally (e.g. 'present' or 'absent'); such classification is necessarily biased by the choice of criteria. (cf. POLYTHETIC CLASSIFICATION.)

monotrichous Refers to the presence of only one flagellum (or e.g. pilus) per cell.

monotype (1) The sole species of a newly proposed genus. (2) Any taxon based on a single TYPE (sense 1).

monoxenous *Syn.* HOMOXENOUS.

Monsur medium (Monsur's tellurite–bile salt medium) A medium for the primary isolation of *Vibrio cholerae*. It contains e.g. peptone, NaCl (1%), Na_2CO_3, sodium taurocholate, and (added after autoclaving) potassium tellurite; pH 9.0–9.2. The liquid medium may be used as a TRANSPORT MEDIUM for specimens containing *V. cholerae*, or may be solidified with agar. [Recipe: Book ref. 46, pp. 1276–1277.] (See also TELLURITE MEDIA.)

Montenegro test A SKIN TEST sometimes used in the diagnosis of cutaneous leishmaniasis; the antigen is a suspension of dead *Leishmania* promastigotes.

Moore swab (sewer swab) A SWAB used for detecting certain types of pathogen (e.g. species of *Salmonella* and *Vibrio*) in a sewer; it consists of a pad of cotton gauze, or of a length of such gauze wrapped around a wire. The swab, attached by a cord, is suspended in flowing sewage for several days; when retrieved, the swab (or liquid expressed from it) is transferred to an appropriate liquid medium and incubated. The Moore-swab technique appears to be more effective in small sewers [AEM (1986) *51* 425–426].

Moorella See CLOSTRIDIUM.

Mopeia virus See ARENAVIRIDAE.

Morax–Axenfeld bacillus See MORAXELLA.

Moraxella A genus of oxidase +ve, typically catalase +ve Gram type-negative bacteria of the family MORAXELLACEAE which occur on mucous membranes (e.g. respiratory tract) in man and other mammals; strains of *Moraxella* can act as opportunist pathogens.

Cells: short rods/cocci which may occur singly or in pairs; non-motile, but may exhibit TWITCHING MOTILITY. Non-pigmented.

Some moraxellae are nutritionally fastidious; all strains can grow on blood agar (some better on chocolate agar); many strains are able to grow on rich, non-blood-containing media or even nutrient agar. Acid is not produced from carbohydrates. Nitrate is reduced by some strains. Optimum growth temperature: 33–36°C. Unlike species of *Neisseria*, no strain produces CARBONIC ANHYDRASE. GC%: ~40–47.5. Type species: *M. lacunata*.

Species include:

M. atlantae. Non-haemolytic. Serum is needed for growth; coagulated serum is not liquefied. Growth is stimulated by bile salts. Nitrate is not reduced.

M. bovis. A causal agent of e.g. INFECTIOUS KERATITIS. Some strains are catalase −ve. Usually haemolytic (on human blood). Coagulated serum is usually liquefied. Nitrate is usually not reduced. Acetate, butyrate, ethanol and lactate can be metabolized in complex media.

M. lacunata (the Morax–Axenfeld bacillus). A causal agent of CONJUNCTIVITIS. Non-haemolytic on human blood. Serum is required for growth; coagulated serum is liquefied. Nitrate is reduced. Phenylalanine deaminase and urease are not formed. Acetate, butyrate and lactate can be metabolized in complex media.

M. catarrhalis (formerly *Branhamella catarrhalis*). A causal agent of e.g. respiratory tract infections, acute OTITIS MEDIA and SINUSITIS. Non-haemolytic. Many strains can grow on nutrient agar. On blood agar, the colonies are whitish, convex, and 1–2 mm in diam. after 24 hours. Most strains reduce nitrate and nitrite, and most or all strains are reported to produce a DNase. [Review: Drugs (1986) *31* (supplement 3) 1–142.]

M. caviae, *M. cuniculi* and *M. ovis* have been regarded as members of a subgenus, *Branhamella*, within the genus *Moraxella* or as species *incertae sedis* within the genus *Neisseria* (so-called 'false neisseriae') [Book ref. 22, pp 290–296].

Moraxellaceae A family of Gram type-negative bacteria comprising the genera ACINETOBACTER, MORAXELLA and PSYCHROBACTER [IJSB (1991) *41* 310–319].

morbidity (1) The state or condition of being diseased. (2) (morbidity rate) The number of cases of a given disease in relation to the overall population (quoted e.g. as cases per 100000 or per million per annum). (cf. MORTALITY RATE.)

morbilli *Syn.* MEASLES.

morbilliform Resembling measles.

Morbillivirus (measles–rinderpest–distemper (MRD) virus group) A genus of viruses of the PARAMYXOVIRIDAE in which the virions have haemagglutinin but no neuraminidase activity. Virions contain proteins N (nucleocapsid), P (nucleocapsid-associated phosphorylated), H (haemagglutinin), F (fusion) and M (membrane or matrix). (cf. PARAMYXOVIRUS.) All members share a common antigen. Type species: MEASLES virus (MV); other members: CANINE DISTEMPER virus (CDV), PESTE DES PETITS RUMINANTS virus (= Kata virus), and RINDERPEST virus (RV). [Antigenic relationship between MV, CDV and RV: JGV (1986) *67* 1381–1392.]

In addition to causing acute disease in their respective hosts, MV and CDV are also associated with persistent infection which may lead to progressive, degenerative neurological disease (see CANINE DISTEMPER and SUBACUTE SCLEROSING PANENCEPHALITIS). It has also been suggested that MV may play a role in the pathogenesis of MULTIPLE SCLEROSIS – possibly by triggering the formation of autoantibodies to neural antigens in certain genetically predisposed individuals [Book ref. 88, pp. 155–165].

Morchella A genus of fungi of the order PEZIZALES. The fruiting body, a yellowish to brown APOTHECIUM, consists of a hollow, ovoid or elongated, waxy to brittle sponge-like pileus borne on a stout, hollow stipe – the whole being ca. 5–12 cm in height and ca. 5 cm at the broadest part; the outer surface of the pileus is conspicuously ridged and pitted, the hymenium-lined pits being separated by sterile ridges. Asci: cylindrical, operculate. *M. esculenta* and *M. vulgaris* form edible fruiting bodies ('morels').

mordant See STAINING.

Moredun medium See WHOOPING COUGH.

morel See MORCHELLA.

Morganella A genus of Gram-negative bacteria of the tribe PROTEEAE. Cells: ca. 0.6–0.7 × 1.0–1.7 μm. Swarming does not occur. Reactions: urease +ve; lipase −ve; gelatin not liquefied at 22°C; H_2S −ve; citrate −ve; acid from glucose and mannose. Both nicotinic and pantothenic acids are required for growth. Species: *M. morganii* (formerly *Proteus morganii*), found (rarely) in the intestines of animals; in man, it can be an opportunist pathogen and may cause diarrhoea. GC%: 50. [Book ref. 22, pp. 497–498.]

morganocin Any BACTERIOCIN from *Morganella*.

Moro test See TUBERCULIN TEST.

moromi See SOY SAUCE.

morph (*mycol.*) Any (sexual or asexual) stage or state of a given fungus. (See e.g. ANAMORPH; TELEOMORPH; and DIMORPHIC FUNGI (sense 1).)

morphogene See CELL CYCLE.

morpholine antifungal agents A group of agricultural systemic ANTIFUNGAL AGENTS which have eradicant properties. They function by inhibiting ergosterol biosynthesis, thereby disrupting the CYTOPLASMIC MEMBRANE in susceptible fungi; the mode of action differs from that of e.g. PROPICONAZOLE. Morpholines include e.g. FENPROPIMORPH and TRIDEMORPH.

morphological heterothallism See HETEROTHALLISM.

morphological unit (*virol.*) See ICOSAHEDRAL SYMMETRY.

morphology agar A complex agar-based medium used for determining morphological characteristics of yeasts; it contains amino acids, D-glucose, vitamins, salts and trace elements [composition: Book ref. 100, pp. 100–101]. It is available commercially in dehydrated form.

morphotype Any distinct, more or less stable morphological form of a given organism. (See also VARIETY and POLYMORPHISM sense 1.)

morphovar See VARIETY.

mortality rate Of a given disease: the number of deaths expressed as a proportion of the number of individuals who have contracted the disease in a particular outbreak, or over a specified period of time. (cf. MORBIDITY.)

Mortierella See MUCORALES.

Mortierellaceae See MUCORALES.

mos An ONCOGENE first identified as the transforming determinant of MOLONEY MURINE SARCOMA VIRUS (Mo-MSV). The c-*mos* sequences are hypermethylated, apparently transcriptionally silent (or expressed at very low levels) in normal and transformed murine cells, and lack introns. Wild-type Mo-MSV expresses v-*mos* only at very low levels in cell culture, but a temperature-sensitive mutant (*ts*110) produces higher levels of an unstable *gag-mos* fusion protein (MWt ca. 85000) which has serine or threonine protein kinase activity [review: JGV (1985) *66* 1845–1853]. [Properties of mouse and human c-*mos* loci: Book ref. 113, pp. 281–289.]

mosaic (*plant pathol.*) A form of leaf VARIEGATION in which two or more colours, or shades of colour, develop – the differently coloured areas being somewhat angular with relatively sharp, clear-cut boundaries between them. (cf. MOTTLE.) Mosaics occur in plants infected systemically with certain PLANT VIRUSES; however, similar leaf patterns may be an inherited characteristic or may result from the effects of toxins produced by leaf-feeding arthropods.

mosquito iridescent virus See CHLORIRIDOVIRUS.

moss starch *Syn.* LICHENIN.

most probable number method See MULTIPLE-TUBE METHOD.

***motB* gene** See FLAGELLUM (a).

motile (*microbiol.*) Capable of MOTILITY.

motile colonies (1) (*bacteriol.*) The colonies of certain bacteria, e.g. *Bacillus circulans*, can migrate across the surface of a culture plate; the track of such movement is often marked by lines of bacterial growth which arise from cells left behind by the migrating colony.

(2) (*algol.*) See e.g. SYNURA and VOLVOX.

motility (*microbiol.*) Locomotion, i.e. an active process in which a cell or organism moves from place to place by expending energy to exert force(s) against the surrounding medium or the contiguous substratum; passive movements due to changes in buoyancy (see e.g. GAS VACUOLE) are not regarded as true motility (see also TWITCHING MOTILITY).

Not all microorganisms are *motile*, i.e. capable of exhibiting motility; however, those which are motile occur among the algae, archaeans, bacteria, fungi and protozoa. Some non-motile organisms have motile stages in their life cycles – e.g. propagules, gametes.

Motility is an advantage because it enables an organism to respond e.g. to certain environmental stimuli. (See also CHEMOTAXIS.)

Locomotion may be achieved in various ways, depending on organism.

(a) *Amoeboid movement* is characteristic e.g. of some sarcodines (see PSEUDOPODIUM).

(b) CILIUM-mediated locomotion occurs in members of the CILIOPHORA (see also METACHRONAL WAVES).

(c) *Euglenoid movement*: see EUGLENA.

(d) FLAGELLAR MOTILITY involves the activity of one or more flagella (or periplasmic flagella in the SPIROCHAETALES) and is found in many bacteria, protozoa (see MASTIGOPHORA) and algae, and in some fungi (limited to certain lower fungi – e.g. the zoospores of chytrids).

(e) GLIDING MOTILITY occurs in some prokaryotes (e.g. *Beggiatoa*, *Oscillatoria*), and e.g. in some diatoms and desmids. [The motility apparatus in shock-frozen gliding cells of *Myxococcus xanthus* (electron micrographs): Microbiology (2001) *147* 939–947.]

(f) *Gregarine movement*, which occurs in some gregarines, is apparently similar or identical to GLIDING MOTILITY (q.v.).

(g) Motility (mechanism unknown) reported in certain strains of SYNECHOCOCCUS (q.v.) [see also JB (1997) *179* 2524–2528].

(h) In *Serratia marcescens*: flagellum-independent *spreading* (which may involve surface tension) [JB (1995) *177* 987–991].

(i) *Actin-based motility*: motility exhibited by certain pathogenic bacteria *within* eukaryotic cells – see DYSENTERY (1a) and LISTERIOSIS.

Direct (microscopical) determination of motility can usually be achieved easily. However, in some organisms – particularly archaeans and bacteria – BROWNIAN MOVEMENT may be mistaken for motility. (See also HANGING DROP and COUNTING METHODS.) Motility in certain organisms may be determined macroscopically: see e.g. SWARMING.

Motile and non-motile bacteria may be separated e.g. by CRAIGIE'S TUBE METHOD.

MOTT Mycobacteria other than tuberculosis: refers to certain species of *Mycobacterium* (excluding *M. leprae*) which may be associated with human disease but which are not members of the *Mycobacterium tuberculosis* complex [Book ref. 219, p 329]. The phrase *non-tuberculous mycobacteria* (NTM) is synonymous, and is preferred by some authors [e.g. RMM (1997) *8* 125–135].

mottle (*plant pathol.*) A form of leaf VARIEGATION in which two or more colours, or shades of colour, develop – the differently coloured areas being somewhat ill-defined with indistinct boundaries between them. (cf. MOSAIC.) Mottling occurs in plants infected systemically with certain PLANT VIRUSES.

Mougeotia A genus of filamentous green algae related to SPIROGYRA. Each cell contains a single plate-like chloroplast in the centre of the cell; the chloroplast can rotate within the cell in response to changes in light intensity.

mould (mold) A general, imprecise term for any fungus which forms a visible layer of mycelium and/or spores on the surface of foods, walls etc, but which does not form macroscopic fruiting bodies. (See e.g. BREAD SPOILAGE.)

mountant (mounting medium) Any substance used to immerse or impregnate a specimen or smear for microscopical examination. For temporary preparations (see e.g. WET MOUNT) the mountant may be e.g. water, saline, or glycerol. Mountants for permanent preparations include e.g. DPX and CANADA BALSAM; before being mounted in these substances, specimens must be fixed (see FIXATION), dehydrated, and cleared (see CLEARING). For fungal material LACTOPHENOL COTTON BLUE is a useful mountant (and stain) for temporary or permanent preparations.

mouse Elberfeld virus See CARDIOVIRUS.

mouse mammary tumour virus (MMTV; Bittner virus) A type B retrovirus (see TYPE B ONCOVIRUS GROUP) which induces carcinomas in the mammary glands of several inbred strains of mice. The virus replicates in the alveolar cells of the mammary gland in lactating mice, and is transmitted to the suckling offspring via the milk. MMTV also occurs as an ENDOGENOUS RETROVIRUS in most inbred mouse strains, the provirus (usually present in multiple copies) being transmitted genetically (see RETROVIRIDAE).

MMTV is replication-competent, v-*onc*$^-$, does not transform cells in culture, and induces carcinomas after a long latent period. The MMTV genome contains – in addition to *gag*, *pol* and *env* sequences – a potential coding sequence, designated *orf* (for open reading frame), near the 3′ end distal to and partly overlapping *env*; the function of this sequence, if any, in carcinogenesis is unknown. Tumours induced by MMTV are clonal (arising mainly from a single infected cell); several factors may influence the development of tumours: e.g. genetic predisposition, hormonal status, number of pregnancies, etc. A critical factor in tumorigenesis is believed to involve the insertion of the provirus at particular sites in the mouse genome; two such sites have been implicated (*int*-1 and *int*-2), neither of which shows homology with any known cellular or viral ONCOGENE. Integration of an MMTV provirus near one of these loci appears to activate the expression of a host gene which is normally silent, possibly by exerting an enhancer function. [Reviews of oncogenesis by MMTV: Book ref. 110, pp. 175–195; JGV (1985) *66* 931–943.]

mouse poliomyelitis See THEILER'S MURINE ENCEPHALOMYELITIS VIRUS.

mousepox See ECTROMELIA VIRUS.

mouth fungus (in fish) See COLUMNARIS DISEASE.

mouth microflora The microflora of the human mouth may include e.g. species of ACTINOMYCES, BACTEROIDES, CAMPYLOBACTER, CAPNOCYTOPHAGA, EIKENELLA, FUSOBACTERIUM, LEPTOTRICHIA, ROTHIA, SELENOMONAS, STAPHYLOCOCCUS, STREPTOCOCCUS, TREPONEMA, WOLINELLA, VEILLONELLA, and various yeasts (e.g. *Candida albicans*) and protozoa (e.g. *Entamoeba gingivalis* and *Trichomonas tenax*). Some of these organisms occur primarily in microaerobic or anaerobic microenvironments such as the gingival sulcus. (See also BODY MICROFLORA; DENTAL PLAQUE; GASTROINTESTINAL TRACT FLORA; RESPIRATORY TRACT MICROFLORA; STOMATITIS.)

moving-boundary electrophoresis See ELECTROPHORESIS.

moxalactam (latamoxef) An OXACEPHEM antibiotic which is active against a wide range of Gram-negative bacteria.

moxifloxacin A QUINOLONE ANTIBIOTIC (q.v.).

Mozambique virus See ARENAVIRIDAE.

mozzarella cheese See CHEESE-MAKING.

MPa 10^6 PASCAL.

MPED$_n$ In a heat STERILIZATION process: the most probable effective dose of thermal energy required to achieve an n-$\log_{10}$ reduction in the numbers of a given microbial contaminant.

***mpl* gene** See PEPTIDOGLYCAN.

MPN method See MULTIPLE-TUBE METHOD.

MppA See PEPTIDOGLYCAN.

MPT64 See ESAT-6.

M_r RELATIVE MOLECULAR MASS.

MR test See METHYL RED TEST.

MR vaccine Measles–rubella vaccine.

Mrad 10^6 rad (see RAD).

***mrcA* gene** See PENICILLIN-BINDING PROTEINS.

***mrcB* gene** See PENICILLIN-BINDING PROTEINS.

MRD virus group See MORBILLIVIRUS.

MreB protein See CELL CYCLE (b) (determination of shape).

MRHA Mannose-resistant haemagglutination (see FIMBRIAE).

mRNA Messenger RNA: any RNA molecule which functions as a template for the assembly of amino acids during PROTEIN SYNTHESIS (see GENETIC CODE). In all cells, and in DNA viruses, the mRNA is transcribed from DNA (see TRANSCRIPTION); in RNA viruses, the genome may or may not be able to function directly as mRNA (see VIRUS).

(a) *Prokaryotic mRNAs.* Many bacterial mRNAs are functional as primary transcripts, i.e. they do not require post-transcriptional maturation. An mRNA molecule may contain the transcript of a single gene (*monocistronic* or *monogenic* mRNA) or, more commonly, transcripts of two or more genes (*polycistronic* or *polygenic* mRNA). (See also OPERON.)

A polycistronic mRNA contains several distinct regions: a 5′ LEADER; *coding regions* (structural gene transcripts), each beginning with a translation initiation codon and consisting of a linear sequence of codons; and sometimes a 3′-terminal *trailer*. The coding regions may or may not be separated by non-coding *intercistronic regions* (see also OVERLAPPING GENES).

The synthesis, translation and degradation of bacterial mRNA are closely linked phenomena; *typically*, bacterial mRNAs have a very short half-life, and in some cases degradation may begin even before transcription has been completed. mRNA is particularly prone to degradation in the absence of translation: see POLAR MUTATION. (cf. RRNA.) (See also S LAYER.)

The secondary structure of bacterial mRNA may be important in determining its stability. Stable mRNAs typically contain multiple secondary structures, particularly in the 3′-terminal region, and these may serve to protect the molecule from 3′-to-5′ processive exonucleases present in the cell. (One function of the GC-rich stem-and-loop structure in a rho-independent terminator – see TRANSCRIPTION – may be to protect the 3′ end of the transcript from degradation by exonucleases.) (See also RETROREGULATION.)

Some bacterial mRNAs – e.g. that transcribed from the *Escherichia coli* gene *lpp* (which encodes an outer membrane lipoprotein) – have atypically long half-lives. Moreover, the rate of degradation of mRNAs can be affected by growth conditions. For example, in *E. coli* the half-life of at least some mRNAs is greatly extended during slow anaerobic growth (doubling time: 700 minutes); in these conditions the rate of *synthesis* of mRNA is much lower, so that extended half-life may be a mechanism for maintaining the expression of such genes during anaerobiosis [Mol. Microbiol. (1993) *9* 375–381].

In *Escherichia coli*, no known exonuclease can degrade RNA in the 5′-to-3′ direction (i.e. in the direction of transcription and translation). mRNA degradation appears to involve the activity of endonuclease(s) and (3′-to-5′) exonuclease(s) (e.g. RNASE II); small pieces of RNA (~10 nucleotides in length) are apparently degraded to single nucleotides by an enzyme designated RNase* [JB (1991) *173* 4653–4659].

Polyadenylation. Many prokaryotic mRNAs have a short (e.g. 10–50 nt) polyadenylate 'tail' at the 3′ end. This poly(A) (A-A-A...) tail, whose synthesis involves a *poly(A)polymerase*, may influence the half-life (stability) of those mRNAs whose 3′ terminus contains a stem–loop structure encoded by a rho-independent transcription terminator; such 3′ stem–loop structures tend to resist enzymic degradation, and polyadenylation may facilitate degradation by providing a suitable binding site for enzymes such as 3′-exonuclease polynucleotide phosphorylase.

Interestingly, mutations in gene *pcnB* – which encodes the *Escherichia coli* poly(A)polymerase I (PAP I) – can reduce the copy number of plasmid ColE1. Apparently, the plasmid-encoded antisense molecule RNA I (which inhibits RNA II) is normally polyadenylated, polyadenylation assisting degradation of the molecule; the absence of polyadenylation in *pcnB* mutants tends to stabilize RNA I, i.e. to enhance its inhibitory action on RNA II – causing inhibition of plasmid replication and, hence, a reduction in copy number.

A poly(A) tail of 10 nucleotides was reported earlier in the archaean *Methanococcus vannielii* [JB (1985) *162* 909–917.]

The poly(A) tail facilitates isolation of mRNAs by affinity chromatography; thus, mRNAs can be isolated by exploiting base-pairing specificity between the A-A-A... of mRNAs and the T-T-T... of oligo(dT)-cellulose in the matrix. Isolation of mRNAs can also be achieved e.g. by the use of DYNABEADS coated with oligo-dT.

[Polyadenylation of mRNA in bacteria (review): Microbiology (1996) *142* 3125–3133. Characterization of *Escherichia coli* poly(A)polymerase: NAR (2000) *28* 1139–1144. Polyadenylation in mycobacteria: Microbiology (2000) *146* 633–638.]

Introns. As in typical eukaryotic genes, some bacterial, archaean and phage genes include introns (see SPLIT GENE). In the transcripts of such genes any part corresponding to an intron must be removed in order to produce a 'mature' mRNA that correctly encodes the gene product. Bacterial introns are *autocatalytic*, i.e. enzymic activity that develops within a correctly folded mRNA molecule automatically 'splices out' the relevant sequence of nucleotides. In at least some cases, splicing may occur even while the nascent transcript is bound to the ribosome, the ribosome pausing during excision of the intron sequence. [Splicing and translation in bacteria: GD (1998) *12* 1243–1247.]

(b) *Eukaryotic mRNAs.* In eukaryotes mRNA is synthesized (by RNA POLYMERASE II) in the nucleus and must be exported to the cytoplasm for translation; typically, mRNA is synthesized as a large precursor molecule (pre-mRNA) which undergoes an extensive maturation process involving *capping* at the 5′ end (see below), polyadenylation of the 3′ end, and (commonly) splicing out of introns (see SPLIT GENE). The mature mRNA molecule is typically functionally monocistronic. (cf. POLYPROTEIN; see also SUBGENOMIC MRNA.)

Some eukaryotic genes are atypical in that they do not contain introns – thus obviating the need for splicing of mRNA; such genes include most or all of those encoding histones as well as e.g. the gene encoding interferon-β.

In the cytoplasm, mRNAs are relatively stable, and they occur complexed with proteins to form *messenger ribonucleoproteins*, mRNPs (= 'informosomes'). [Review: TIBS (1985) *10* 162–165.]

Polyadenylation. Most eukaryotic mRNAs have a poly(A) (polyadenylate) sequence – about 200 adenylate residues – at the 3′ end. This poly(A) tail is not added to the *complete* transcript: at the end of transcription an endonuclease cuts the transcript at a specific site ~10–30 nucleotides downstream of an AAUAAA (or similar) sequence – thus creating an appropriate 3′ terminus from which another enzyme, poly(A)polymerase, can synthesize the polyadenylate tail. The function of the poly(A) tail is unknown.

Some eukaryotic mRNAs do not have a poly(A) tail; for example most or all mRNAs that encode histones are apparently poly(A)$^-$.

Capping. Almost all eukaryotic mRNAs are modified *(capped)* at the 5′ end. Capping occurs shortly after the beginning of transcription: an enzyme *(guanylyl transferase)* adds to the newly formed 5′ end of the molecule an 'inverted' guanosine residue, i.e. a guanosine residue linked 5′-to-5′ to the 5′-triphosphate end of the growing pre-mRNA molecule. A methyl group is then transferred (by a specific 7-methyltransferase) to the N^7 position of the G cap to form the characteristic cap structure:

$$m^7G^{5'}ppp^{5'}N^{3'}p^{5'}N^{3'}p\ldots$$

A cap methylated only at this position is said to be of *type 0*. *Type 1* caps have an additional methyl group at the $O^{2'}$-position of the next (2nd) nucleotide, and sometimes also at the N^6-position of a penultimate adenine residue; *type 2* caps contain, in addition to the type 1 methyl group(s), a methyl group at the $O^{2'}$-position of the 3rd nucleotide. Apparently, unicellular eukaryotes contain only type 0 caps; type 1 caps predominate in other eukaryotes.

The 5′ cap enhances translation by promoting the formation of the translation initiation complex; the recognition of the cap by specific *cap-binding protein(s)* (CBP) is an important factor in the regulation of gene expression. (The cap may also be involved e.g. in pre-mRNA processing.)

[Cap-binding proteins: Cell (1985) *40* 223–224. Cap structure and interaction with CBP: Prog. Mol. Subcell. Biol. (1985) *9* 104–155.]

(See also SNRNA.)

Among plant and animal viruses, viral RNA may be capped but not polyadenylated (e.g. TOBACCO MOSAIC VIRUS), polyadenylated but not capped (e.g. poliovirus), or polyadenylated *and* capped (e.g. ADENOVIRIDAE, POXVIRIDAE). Polioviruses apparently bring about the switch from translation of host (capped) mRNA to translation of viral (uncapped) mRNA at least partly by inactivating the host CBP. (See also e.g. REOVIRUS.)

mRNP See MRNA (b).

MRSA Methicillin-resistant *Staphylococcus aureus*. Resistance to methicillin is generally accompanied by resistance to other β-lactam antibiotics, and is often accompanied by resistance to e.g. aminoglycosides and macrolides. MRSA has been commonly susceptible to VANCOMYCIN and/or TEICOPLANIN (see also QUINUPRISTIN/DALFOPRISTIN) [but see e.g. Lancet (1999) *353* 1587–1588; JCM (2001) *39* 591–595]. High-level resistance seems to have been acquired from *Enterococcus faecalis* via transposon Tn*1546* [Science (2003) *302* 1569–1571].

One report has suggested that strains of MRSA may arise, in vivo, through transfer of the *mecA* gene (see below) between strains of staphylococci [Lancet (2001) *357* 1674–1675].

High-level ('intrinsic') resistance to methicillin is (chromosomally) encoded by the *mecA* gene whose product is a novel 78 kDa PENICILLIN-BINDING PROTEIN, PBP 2a (= 2′), with a low affinity for β-LACTAM ANTIBIOTICS.

Expression of *mecA* is controlled by various regulatory genes. For example, some strains of MRSA contain a *mecI* gene which encodes a strong repressor; thus, resistance to methicillin may arise e.g. through mutation which inactivates the MecI product or which affects transcription from *mecI*. MecR1 is a sensor–transducer protein in the cell envelope. The binding of a β-lactam antibiotic to MecR1 brings about, indirectly, transcription of the *mecA* gene; transcription involves prior cleavage (i.e. inactivation) of MecI, the repressor of the *mecA* gene. Such induction of *mecA*-based resistance may parallel the induction of β-LACTAMASES (q.v.) [see e.g. Science (2001) *291* 1962–1965; commentary: Science *291* 1915–1916]. [PCR-based detection of *mecA*, *mecR1* and *mecI* genes in clinical isolates: JAC (2001) *47* 297–304.]

(See also 'Induction of resistance to antibiotics' within the entry ANTIBIOTIC.)

mecA-mediated resistance can be lost e.g. by mutations in the *fem* genes (*fem* = *f*actor *e*ssential for *m*ethicillin resistance). The *femAB* operon encodes factors needed for synthesis of normal PEPTIDOGLYCAN cross-links, and in some *femAB* mutants the (abnormal) cross-links are poor substrates for the (transpeptidase) activity of PBP 2a; as the 'routine' (methicillin-*sensitive*) PBPs are enzymically more effective than PBP 2a on the abnormal cross-links, the (mutant) cell is susceptible to methicillin.

Non-*mecA*-based methicillin resistance can be due to overproduction of β-lactamases (in which case it may be overcome by inhibitors such as CLAVULANIC ACID), to mutant PBPs which have lower affinity for β-lactam antibiotics, or to a reported methicillinase.

[Genetics of MRSA (review): RMM (1998) *9* 153–162.]

The phenotypic expression of methicillin resistance is governed by factors that include temperature and salt concentration; culture media typically contain 4% NaCl and they are incubated at e.g. 30–35°C. MRSA grows in 24–48 hours when inoculated (10^4 cfu) onto Mueller–Hinton agar containing NaCl (4%) and oxacillin (6 μg/ml) or methicillin (10 μg/ml) [Book ref. 120, p. 151].

Heteroresistance refers to the expression of resistance in only a minority of cells in a genetically resistant population; most strains of MRSA are heteroresistant. [Mechanisms of heteroresistance: AAC (1994) *38* 724–728.]

Nasal MRSA in carriers may be eliminated by MUPIROCIN; hair and skin may be disinfected by povidone–iodine (see IODINE) or by 4% CHLORHEXIDINE. Some strains of MRSA exhibit plasmid-encoded resistance to mupirocin and/or to chlorhexidine.

MRSA is a serious problem in many areas. Levels of MRSA are low e.g. in Holland – where control includes strict isolation of infected/colonized patients [e.g. RMM (1998) *9* 109–116].

[Rapid discrimination between methicillin-sensitive and methicillin-resistant *S. aureus* by intact cell mass spectroscopy: JMM (2000) *49* 295–300.]

[Multilocus sequence typing of methicillin-resistant and methicillin-sensitive *S. aureus*: JCM (2000) *38* 1008–1015.]

[Rapid (~1 hour) detection of MRSA without prior culture: JCM (2004) *42* 1875–1884.]

[Rapid detection of *mecA*-positive and *mecA*-negative coagulase-negative staphylococci by an anti-PBP 2a slide latex agglutination test: JCM (2000) *38* 2051–2054.]

[Genotyping of European isolates of MRSA by fluorescent amplified fragment length polymorphism analysis and PFGE: JMM (2001) *50* 588–593.]

MS ring See FLAGELLUM (a).

MscL channel See MECHANOSENSITIVE CHANNEL.

msDNA See RETRON.

MseI A RESTRICTION ENDONUCLEASE; T/TAA.

MSHA Mannose-sensitive haemagglutination (see FIMBRIAE).

***Msp*I** See RESTRICTION ENDONUCLEASE (table).

MSU Mid-stream urine (see URINARY TRACT INFECTION).

MSV MAIZE STREAK VIRUS.

MSX L-Methionine-DL-sulphoximine: an inhibitor of glutamine synthetase – and hence of AMMONIA ASSIMILATION via the GS/GOGAT pathway.

MT Microtubule (see MICROTUBULES).

mtDNA Mitochondrial DNA (see MITOCHONDRION).

MTOC MICROTUBULE-ORGANIZING CENTRE.

μ (mu) (1) An abbreviation for 'micro-' (i.e. one-millionth). (2) MICRON. (3) SPECIFIC GROWTH RATE. (4) One of the symbols used to designate the exponential growth rate constant – see GROWTH (a). (cf. BACTERIOPHAGE MU.)

Mu See BACTERIOPHAGE MU.

***mu* chain** (*immunol.*) See HEAVY CHAIN.

2μ circle *Syn.* TWO-MICRON DNA PLASMID.

mu particle The trivial name for certain bacteria (see CAEDIBACTER and PSEUDOCAEDIBACTER) which are endosymbionts in strains of *Paramecium* and which confer the MATE KILLER characteristic.

2μ plasmid See TWO-MICRON DNA PLASMID.

3μ plasmid See THREE-MICRON DNA PLASMID.

***mucA* gene** (1) See SOS SYSTEM. (2) See ALGINATE.

Mucambo virus See ALPHAVIRUS.

mucate A salt of MUCIC ACID.

***mucB* gene** See SOS SYSTEM.

mucic acid $HOOC(CHOH)_4COOH$. It is obtained e.g. by the oxidation of galactose.

muciferous body See MUCOCYST.

mucigel A mucilaginous layer covering the root tips and root hairs of plants; it is composed mainly of pectin-like polysaccharides and is produced by the epidermal cells of the root tips. The mucigel may protect against desiccation, aid in the uptake of plant nutrients, and help to bind soil particles; it often contains large numbers of bacteria. [Book ref. 38, pp. 565–568.]

mucigenic body See MUCOCYST.

mucocutaneous leishmaniasis LEISHMANIASIS in man and other animals in which the pathogen gives rise (in man) to invasive granulomatous lesions primarily in the oral/nasal mucous membranes; in man the disease may follow an earlier, apparently cured, episode of CUTANEOUS LEISHMANIASIS with a latent period of up to many years. The disease occurs mainly in South America where it is called *espundia*, the causal agent being *Leishmania braziliensis*; a mucocutaneous leishmaniasis caused by *L. donovani* occurs in parts of Africa.

Diagnosis may involve e.g. the MONTENEGRO TEST; pentavalent ANTIMONY compounds and/or antibiotics (e.g. amphotericin B) may be used for treatment.

mucocyst (muciferous body; mucigenic body) In many ciliates and some flagellates: a small, sac-like organelle which, when appropriately stimulated, discharges a mucoid material through a pore at the cell surface. In ciliates the mucocyst is an invagination of the outermost membrane, the closed, inner part of the sac being located in the cytoplasm; mucocysts often occur in longitudinal rows parallel to the rows of kinetosomes. In some species mucocysts may be involved e.g. in cyst formation.

mucoid Viscous; slimy; mucus-like. The term is often applied to the colonies of certain capsulated bacteria (see CAPSULE; see also CYSTIC FIBROSIS).

mucolytic agent A chemical agent used e.g. on a sample of sputum prior to use of the sample as an inoculum on a culture plate; the purpose of the agent is to liquefy the sputum so that (i) the sputum can be manipulated, and (ii) any target cells within the sputum will be exposed to reagents.

Mucolytic agents include dithiothreitol [example of use: JCM (1997) *35* 193–196] and *N*-acetyl-L-cysteine [examples of use: JCM (1999) *37* 175–178; JCM (1999) *37* 137–140].

mucopeptide (glycosaminopeptide) (1) *Syn.* PEPTIDOGLYCAN. (2) Any peptide-containing mucopolysaccharide.

mucopolysaccharide (glycosaminoglycan) A polysaccharide that contains a high proportion of amino sugars and (commonly) uronic acids: e.g. HYALURONIC ACID.

mucoprotein A protein covalently linked to amino sugars.

mucopurulent Containing mucus and pus.

Mucor A genus of fungi (order MUCORALES) which typically occur as saprotrophs on soil, vegetable matter and dung. The vegetative thallus is usually a branched, aseptate mycelium, though some species are dimorphic. The organisms form ZYGOSPORES (see also SPOROPOLLENIN); many species exhibit HETEROTHALLISM (see also COMPATIBILITY sense 2 and PHEROMONE). (For *M. miehei* and *M. pusillus* see RHIZOMUCOR.)

Mucoraceae See MUCORALES.

Mucorales An order of fungi (class ZYGOMYCETES) which occur as saprotrophs on decaying vegetable matter, soil or dung, or as parasites or pathogens of plants or animals (see e.g. ZYGOMYCOSIS). The vegetative thallus is typically a well-developed aseptate mycelium; some species (e.g. *Mucor rouxii*, other species of *Mucor*, *Mycotypha* spp) are DIMORPHIC FUNGI (sense 1). The organisms commonly form asexually derived spores in sporangia and/or sporangiola. Most species form ZYGOSPORES (q.v.) – cf. Saksenaeaceae (below).

Nine families are recognized [Book ref. 64, pp. 247–248] on the basis of reproductive structures:

Choanephoraceae. Sporangia and sporangiola are borne on different types of sporangiophore. Genera: *Blakeslea* (see also PHEROMONE), *Choanephora*. [The family has been redefined to contain *Blakeslea* (sporangiola dehiscent), *Choanephora* (sporangiola non-dehiscent), and *Poitrasia* (sporangiola not formed; sporangia distinctive): Mycol. Pap. (1984) Nr *152*.]

Cunninghamellaceae. Sporangiola: one-spored; sporangia absent. Genera: e.g. *Cunninghamella* (see also ZYGOMYCOSIS), *Thamnocephalis*.

Mortierellaceae. Columellae vestigial or absent. Genera: e.g. *Mortierella* (see also COMPOSTING and ZYGOMYCOSIS).

Mucoraceae. Sporangia are columellate. Genera: e.g. *Absidia* (see also ZYGOMYCOSIS), *Actinomucor*, *Gilbertella*, MUCOR, *Phycomyces* (see also THIAMINE), RHIZOMUCOR, RHIZOPUS and *Zygorhynchus*.

Pilobolaceae. Sporangia are columellate; in *some* species the sporangia are forcibly discharged. Genera: e.g. PILAIRA, PILOBOLUS.

Radiomycetaceae. Sporangia are absent; sporangiola are often formed on stoloniferous sporangiophores. Genera: e.g. *Hesseltinella*.

Saksenaeaceae. Zygospores are apparently not formed. Sporangia are columellate or acolumellate. Genera: *Echinosporangium*, *Saksenaea* (see also ZYGOMYCOSIS).

Syncephalastraceae. Merosporangia are formed. Genera: *Syncephalastrum*.

Thamnidiaceae. Sporangiola (each containing one to several spores) are formed – sometimes with columellate sporangia on the same or on separate (morphologically similar) sporangiophores. Genera: e.g. *Chaetocladium*, *Mycotypha*, *Thamnidium*.

mucoran A water-soluble, alkali-soluble heteroglucan (containing fucose, mannose and glucuronic acid) present e.g. in the cell walls of *Mucor rouxii*; it appears to be maintained in an insoluble form in the cell wall as a result of binding to insoluble glucosamine-containing polymers.

mucormycosis See ZYGOMYCOSIS.

mucosal disease (MD) A severe CATTLE DISEASE caused by the BVD-MD virus (see PESTIVIRUS). It apparently results from persistent BVD-MD virus infection acquired congenitally (with the consequent development of specific immune tolerance to the virus). MD is characterized by fever, anorexia, nasal discharge, profuse watery diarrhoea, buccal ulceration, etc; dehydration and emaciation are usually followed by death. [VR (1985) *117* 240–245.] (cf. BOVINE VIRUS DIARRHOEA.)

mucron In acephaline gregarines: a differentiated region of the anterior end of the cell by means of which the developing parasite remains attached to the host cell (cf. EPIMERITE).

Mu*dlac* system A system for bringing about GENE FUSIONS or OPERON FUSIONS using a defective form of BACTERIOPHAGE MU ('Mud phage') which carries the *lacZ* gene of *Escherichia coli* (and e.g. an antibiotic resistance marker); when the genome of such a phage transposes in the chromosome of a host cell, *lacZ* is inserted at more or less random sites and hence can be fused to almost any gene or operon in the chromosome. Two basic types of Mud phage can be used. Mud I phages contain an intact *lacZ* gene which can be expressed only if it can be transcribed from the promoter of the gene into which it has inserted; *lacZ* expression can be detected e.g. by plating the cells on Xgal plates (see XGAL). Mud II phages contain *lacZ* minus its translation initiation signals; in this case *lacZ* expression depends on both the promoter and the translation initiation signals of the gene into which it has inserted, the product being a FUSION PROTEIN. (See also MINI-MU.)

muduction See MINI-MU.

Mueller–Hinton medium A general-purpose medium containing beef infusion, casamino acids (or peptone), and starch. It can be used as a broth or gelled with agar, and may be used e.g. as a base for blood-containing media. [Recipe: Book ref. 53, p. 1403.]

MUG test A test which helps to detect *Escherichia coli* in a mixed microflora. An agar-based medium which includes the reagent 4-methylumbelliferyl-β-D-glucuronide (MUG) is inoculated with the sample; after incubation, colonies are examined under ultraviolet radiation (wavelength 366 nm). Most strains of *E. coli* form β-glucuronidase – an enzyme which cleaves MUG to a compound that fluoresces greenish-blue under radiation of 366 nm; the presence of fluorescent colonies allows a presumptive identification of *E. coli*. (N.B. Other sources of β-glucuronidase – found e.g. in shellfish samples – can give rise to a false-positive indication.)

In an alternative approach, a colony is smeared on filter paper which has been impregnated with MUG; the filter paper is then examined under ultraviolet radiation.

***mukB* gene** See CELL CYCLE (b).

Müller–Kauffmann broth A tetrathionate-containing ENRICHMENT MEDIUM used e.g. in the isolation of salmonellae from pre-enrichment cultures of meat products in buffered peptone media.

Müller's vesicle In certain gymnostome ciliates (e.g. *Loxodes*): a composite organelle consisting of a membrane-enclosed mineral body (the *statolith*), a vacuole, and structures derived from the overlying kinety; it appears to act as a gravity sensor. [JP (1986) *33* 69–76.]

multi-L-arginyl-poly(L-aspartic acid) *Syn.* CYANOPHYCIN.

multicellular Consisting of more than one (usually many) cells.

multicistronic mRNA *Syn.* polycistronic mRNA (see MRNA).

multicomponent virus (multipartite virus; multiparticle virus; covirus) A virus which exists as two or more separate particles, each containing only part of the genome; all the particles must be present in a given host cell for the complete replication cycle to occur. (cf. SATELLITE VIRUS.) Many PLANT VIRUSES are multicomponent viruses.

multicopy inhibition The phenomenon in which transposition of a single-copy Tn*10* element, resident in a bacterial chromosome, is inhibited 3- to 10-fold in the presence of a multicopy plasmid carrying IS*10*-R (see Tn*10*). The phenomenon is due to inhibition of translation (not transcription) of the transposase gene – apparently as a result of pairing between the start region of the transposase gene transcript and a short complementary RNA molecule (see ANTISENSE RNA) transcribed from the opposite IS*10* strand. [Book ref. 20, pp. 275–277.] (cf. TRANSPOSITION IMMUNITY.)

multicopy plasmid A PLASMID whose COPY NUMBER (in a given cell) is high, e.g., 10–30. Multicopy plasmids include e.g. the COLE1 PLASMID and plasmids pBR322, pMB1 and R6K. In a cell containing a multicopy plasmid, the replication of individual plasmids occurs randomly (as indicated by Meselson–Stahl density-shift experiments), and there is no active mechanism for PARTITION.

multicopy single-stranded DNA See RETRON.

multidrug and toxic compound extrusion (MATE) See MATE.

multifidene See PHEROMONE.

multiline cultivar (*plant pathol.*) A plant cultivar which is genetically non-uniform, comprising a mixture of plant lines which exhibit different degrees of resistance to a given pathogen or to particular race(s) of a given pathogen. The spread of a particular pathogen through such a cultivar is inhibited, and total crop loss during an epiphytotic is less likely than it would be for a genetically uniform susceptible cultivar. [Review: ARPpath. (1985) *23* 251–273.]

multiloculate (multilocular) Consisting of or having many LOCULES: see e.g. FORAMINIFERIDA.

multilocus enzyme electrophoresis (MEE) A method in which the electrophoretic mobilities of a range of enzymes from one organism are compared with those of equivalent enzymes from one or more closely related organisms; MEE is used e.g. for TYPING microorganisms. In this method, enzymes are isolated and tested under conditions in which enzymic activity is retained; enzymes can be identified in gels by specific colour-generating substrates.

Variation in the electrophoretic mobility of a given enzyme can arise e.g. when a variant form of the enzyme is specified by a different allele of the enzyme-encoding gene; such an enzyme – which may exhibit a different mobility when only a single amino acid differs – is called an *alloenzyme*.

multilocus sequence typing See MLST.

multipartite virus *Syn.* MULTICOMPONENT VIRUS.

multiple cloning site *Syn.* POLYLINKER.

multiple drug resistance See ANTIBIOTIC.

multiple fission A type of fission in which one cell gives rise to many cells. Multiple fission may involve repeated BINARY FISSION within a common envelope (as e.g. in CYANOBACTERIA of section II), or it may involve a series of nuclear divisions followed by the incorporation of each nucleus in a separate daughter cell (see e.g. SCHIZOGONY).

multiple sclerosis A chronic demyelinating disease of the CNS in humans. It is generally believed to be an autoimmune disease, possibly triggered by viral infection; it has been suggested that any of various enveloped viruses may be able to trigger the autoimmune response – which could be mounted against host lipids incorporated (with viral proteins) in the viral envelope [Nature (1986) *321* 386]. (See also HTLV; INOUE–MELNICK VIRUS; MORBILLIVIRUS; cf. THEILER'S MURINE ENCEPHALOMYELITIS VIRUS.)

multiple-tube method (most probable number (MPN) method) A method for estimating the number of viable organisms (usually bacteria) suspended in a liquid. The method is used e.g. for the bacteriological examination of natural and treated waters; for this purpose, one of several possible procedures is as follows. Five tubes, each containing a suitable liquid growth medium, are each inoculated with a fixed volume of the sample water (or a dilution of it). Another five tubes, containing the same medium, are each inoculated with a smaller fixed volume of the sample, and further sets of tubes are similarly inoculated with progressively smaller volumes. Following incubation, each tube is examined for the presence or absence of growth. (It is assumed that growth will occur in any tube which receives at least one viable organism.) Growth usually occurs in those tubes inoculated with the larger volumes of the sample since these tubes are more likely to have received at least one viable organism in the inoculum. The number of cells in the original sample can be calculated from the pattern of positive (growth) and negative (no growth) tubes by the use of statistical (probability) tables; if a diluted sample were used, the count indicated by the tables is multiplied by the dilution factor. By using a selective medium in the tubes it is possible to estimate the numbers of a particular type of organism in the sample. The multiple-tube method is characterized by a large sampling error. (See also COUNTING METHODS.) [Improved method of computing MPNs: AEM (1983) *45* 1646–1650.]

multiplex PCR A form of PCR in which two* or more different targets are amplified, simultaneously, in a single assay. The reaction mixture usually contains two types of primer for each target;

in some cases, however, one primer can be shared by two targets – e.g. 16S rRNA sequences from two species, *Bacteroides forsythus* and *Prevotella intermedia*, were simultaneously amplified with one forward primer (a BROAD-RANGE PRIMER common to both species) and two species-specific reverse primers [JCM (1999) *37* 1621–1624].

Owing to the use of multiple primers, it is particularly important to ensure the absence of complementarity between the 3′ ends of the primers in order to avoid the formation of PRIMER-DIMERS.

The possibility of competition between different targets in a multiplex PCR may be investigated by adding a few copies of another target which is amplifiable but distinct from the other targets. Failure of *this* target to amplify (when other(s) are positive) would suggest that competition/inhibition has affected the efficiency of amplification; in these circumstances, the failure of amplification of other target(s) in the same assay would not be considered meaningful [JCM (1998) *36* 191–197].

Examples of the use of multiplex PCR include: detection of chlamydiae [JCM (1999) *37* 575–580]; detection of vancomycin-resistance genes in enterococci [JCM (1999) *37* 2090–2092]; detection of *mecA* and *coa* genes in *Staphylococcus aureus* [JMM (1997) *46* 773–778]; detection of different types of virus in cerebrospinal fluid (CSF) [JCM (1997) *35* 691–696]; and detection of HIV-1, HIV-2, HTLV-I and HTLV-II [PNAS (1999) *96* 6394–6399].

*The amplification of *two* targets is also referred to as *duplex PCR*.

multiplicity of infection (moi) The ratio of virions to susceptible cells in a given system; e.g., a high moi means a large number of virions per cell.

multiplicity reactivation See REACTIVATION.

multiseriate Arranged in or forming several rows; more than one cell wide.

multisite mutation A MUTATION which involves the alteration of two or more contiguous nucleotides. (cf. POINT MUTATION.)

multivalent vaccine *Syn.* POLYVALENT VACCINE.

Multivalvulida See MYXOSPOREA.

MuLV MURINE LEUKAEMIA VIRUS.

μm MICROMETRE.

2μm circle (2μm plasmid) *Syn.* TWO-MICRON DNA PLASMID.

mumps (contagious, infectious or epidemic parotitis) An acute infectious human disease which affects mainly children; it is caused by a PARAMYXOVIRUS. Transmission occurs by droplet infection. Incubation period: 2–3 weeks. The disease ranges from subclinical to severe. Typical symptoms include fever with pain and swelling of one or more salivary glands (usually the parotid). Common complications include MENINGITIS, often with mild encephalitis, and orchitis; oophoritis, mastitis, pancreatitis etc may also occur. Infection seems to confer life-long immunity. *Lab. diagnosis*: serological tests (CFT, HAI test) for mumps virus antibodies; isolation of mumps virus e.g. from pharynx or urine.

mundticin A 43-amino-acid class IIa BACTERIOCIN produced by *Enterococcus mundtii*.

mungbean yellow mosaic virus See GEMINIVIRUSES.

mupirocin (pseudomonic acid A) An antibiotic produced by strains of *Pseudomonas fluorescens*. It is active against staphylococci and streptococci associated with skin infections, and the topically-applied antibiotic has been shown to be effective against experimental infection of surgical wounds by *Staphylococcus aureus* [JAC (1985) *16* 519–526]. Mupirocin appears to act by inhibiting isoleucyl-tRNA synthetase. [Review: JAC (1987) *19* 1–5.]

muramic acid 3-*O*-Lactylglucosamine: see PEPTIDOGLYCAN.

muramidase An enzyme which cleaves glycosidic bonds in PEPTIDOGLYCAN: see e.g. LYSOZYME. A muramidase obtained from the bacteriolytic hyphomycete *Chalara* (= '*Chalaropsis*') sp is commonly used in studies on peptidoglycan; this enzyme hydrolyses the same glycosidic bond as lysozyme but differs e.g. in substrate specificity [JBC (1967) *242* 5586–5590].

muramyl dipeptide See MDP.

murein *Syn.* PEPTIDOGLYCAN.

murein lipoprotein *Syn.* BRAUN LIPOPROTEIN.

mureinoplast Any Gram-negative bacterial cell from which the outer membrane, but not the peptidoglycan, has been removed.

murid (beta) herpesvirus 1 See BETAHERPESVIRINAE.

murid herpesvirus 2 See BETAHERPESVIRINAE.

muriform Refers to a spore which has both transverse and longitudinal septa.

murine Of or pertaining to rats or mice (Muridae).

murine cytomegalovirus group See BETAHERPESVIRINAE.

murine encephalomyocarditis virus See CARDIOVIRUS.

murine hepatitis virus See CORONAVIRIDAE.

murine leukaemia viruses (MLVs; MuLVs) A group of type C retroviruses (subfamily ONCOVIRINAE), nearly all of which are replication-competent and v-*onc*$^-$ (but cf. ABELSON MURINE LEUKAEMIA VIRUS), and some of which can induce e.g. lymphatic leukaemia in mice after a long latent period (see e.g. MOLONEY MURINE LEUKAEMIA VIRUS). (All replication-competent murine type C retroviruses are termed 'murine leukaemia viruses', although most are usually non-leukaemogenic [Book ref. 114, p. 70].)

MuLV-induced leukaemogenesis requires viraemia (i.e., it requires expression and replication of the virus) and an active cellular immune response to the MuLV *env* gene product; it has been suggested that chronic immune stimulation may be a factor in leukaemogenesis. Although the *env* gene product is thought to play a direct role in leukaemogenesis, it has been shown that the primary oncogenic determinant in one MuLV lies outside the *env* region – possibly in the LTR [JV (1983) *47* 317–328]. (See also MCF VIRUSES.)

MuLVs have given rise to a range of defective, often acutely oncogenic variants: see e.g. FRIEND VIRUS, HARVEY MURINE SARCOMA VIRUS, MOLONEY MURINE SARCOMA VIRUS, RAUSCHER VIRUS.

murine poliovirus *Syn.* THEILER'S MURINE ENCEPHALOMYELITIS VIRUS.

murine sarcoma viruses (MSV) A group of highly oncogenic, replication-defective, C-type retroviruses (subfamily ONCOVIRINAE): see e.g. FBS OSTEOSARCOMA VIRUS, GAZDAR MURINE SARCOMA VIRUS, HARVEY MURINE SARCOMA VIRUS, KIRSTEN MURINE SARCOMA VIRUS, and MOLONEY MURINE SARCOMA VIRUS. MSV replication requires the presence of e.g. a MURINE LEUKAEMIA VIRUS as helper.

murine typhus See TYPHUS FEVERS.

MurNAc *N*-Acetylmuramic acid: see PEPTIDOGLYCAN.

muropeptides Products of muramidase (e.g. LYSOZYME) digestion of PEPTIDOGLYCAN. In *Escherichia coli* they are designated C1–C8, according to their migration rates during chromatography; e.g., C3 is a dimeric subunit consisting of two GlcNAc–MurNAc disaccharides bound by a cross-link; C4 consists of a tetrasaccharide – $(GlcNAc–MurNAc)_2$ – in which there is a cross-link between the two MurNAc residues; C5 and C6 are monomeric subunits (GlcNAc–MurNAc) linked to short peptides. [Book ref. 122, pp. 79–81.]

murrain *Syn.* REDWATER FEVER.

Murray Valley fever (Australia X disease) An acute human disease caused by a flavivirus (see FLAVIVIRIDAE) and transmitted by mosquitoes (e.g. *Culex annulirostris*); it occurs in parts of Australasia. The disease is usually a mild, febrile condition, but encephalitis may occur and may be fatal in young children. Birds may act as a reservoir of infection.

Murraya See LISTERIA.

muscarine (muscarin) A quaternary ammonium derivative of tetrahydrofuran produced e.g. by *Amanita muscaria* (the fly agaric) and *Inocybe* spp; it is toxic to flies, and infrequently causes a fatal intoxication in humans. (See also MYCETISM.)

muscicolous Growing on mosses.

mushroom (1) Loosely: any umbrella-shaped fungal fruiting body; such a fruiting body consists of a horizontally orientated, circular PILEUS supported, at its centre, by a vertical, columnar STIPE. 'Mushroom' is often reserved for those fruiting bodies in which the pileus has radially arranged lamellae (see LAMELLA) or lamella-like structures on its underside.

(2) Any AGARIC.

(3) The (edible) fruiting body of *Agaricus brunnescens*, the common cultivated mushroom (see AGARICUS and MUSHROOM CULTIVATION). (See also MUSHROOM DISEASES.)

(4) Any edible mushroom – including e.g. ENOKITAKE, OYSTER FUNGUS and SHII-TAKE. (cf. TOADSTOOL.)

mushroom cultivation The mycelium of the edible mushroom *Agaricus brunnescens* (= *A. bisporus*) is cultivated on composted material – traditionally a mixture of straw and horse manure (see COMPOSTING; [SEM of flora: JAB (1983) *55* 293–304]). After an initial composting period, the compost undergoes 'peak-heating' ('pasteurization') to ca. 50–60°C; during this stage thermophilic actinomycetes and fungi continue to grow, but the number of mesophiles – including many mushroom parasites and pathogens – is reduced. (See also OLIVE-GREEN MOULD.) The compost is then allowed to cool to ca. 25°C and inoculated with SPAWN of a selected strain of *A. brunnescens*. After ca. 14 days, during which the mycelium of *A. brunnescens* grows through the compost, a 'casing layer' of soil or a mixture of peat and chalk or limestone is spread over the compost. The temperature is maintained at ca. 25°C (RH >90%) for ca. 1 week. When the mycelium reaches the surface of the casing layer, the temperature and RH are allowed to fall (e.g. to ca. 16–18°C or less, RH 80–90%) to induce initiation of fruiting. Mature fruiting bodies begin to appear ca. 3 weeks after casing, and then appear in flushes at intervals of ca. 7–10 days. During the growth of *A. brunnescens* the previous (thermophilic) flora declines, and progressive colonization by mesophilic fungi (e.g. species of *Aspergillus* and *Trichoderma*) occurs [TBMS (1980) *74* 465–470].

The composting and pasteurization processes appear to be necessary to provide low C:N ratios with high levels of organic nitrogen (ammonia being inhibitory to *A. brunnescens*). The microbial population established during these stages may act as a direct source of nutrients for *A. brunnescens*, which produces a variety of extracellular enzymes (including a β-*N*-acetylmuramidase) that can lyse and degrade bacteria [JGM (1984) *130* 761–769] and fungal mycelium [JGM (1985) *131* 1735–1744]. The casing is necessary to switch growth of *A. brunnescens* from vegetative to reproductive. This switch appears to be associated with the activities of bacteria (especially pseudomonads) in the casing layer, possibly involving the removal by the bacteria of certain 'self-inhibitory' fungal metabolites; the casing may also serve to trap some volatile product(s) of *A. brunnescens* or other fungi in the compost [Book ref. 39, p. 291].

Certain other edible agarics are cultivated in some countries; some (e.g. PADI-STRAW MUSHROOM) are grown on waste plant materials, others (e.g. ENOKITAKE, OYSTER FUNGUS, SHII-TAKE) are grown on wood, sawdust, woodchips etc.

(See also MUSHROOM DISEASES.)

mushroom diseases Diseases of the cultivated mushroom *Agaricus brunnescens* (= *A. bisporus*) can be caused by bacteria (see e.g. BROWN BLOTCH, GINGER BLOTCH), fungi (see e.g. BUBBLE DISEASES, COBWEB DISEASE, OLIVE-GREEN MOULD) or viruses (see e.g. WATERY STIPE). (See also FUNGUS GNATS.)

mushroom virus 3 See MYCOVIRUS.

mussel poisoning See SHELLFISH POISONING.

must See WINE-MAKING.

mustard gas See SULPHUR MUSTARDS.

***mut* genes** See MUTATOR GENE.

mutagen An agent (chemical or physical) which promotes MUTAGENESIS. Most mutagens interact directly with (and damage) DNA, but mutagenesis commonly requires the active participation of the cell itself; e.g., a mutagen may alter a base in DNA such that, although not itself a mutation, it may pair with the wrong base during a subsequent round of DNA replication, resulting in a mutation. The damaging effects of some potential mutagens can be counteracted, to varying extents, by particular DNA REPAIR systems, one of which (error-prone repair: see SOS SYSTEM) plays an important role in mutagenesis induced by certain mutagens.

A given mutagen may be mutagenic in one organism but not in another, or may act differently in different organisms. Physical agents which promote mutagenesis in at least some organisms include e.g. IONIZING RADIATION, ULTRAVIOLET RADIATION and heat (heat apparently increases the rates of spontaneous reactions in DNA, thus increasing the SPONTANEOUS MUTATION rate). Chemical mutagens include e.g. ALKYLATING AGENTS, BASE ANALOGUES, BISULPHITE, HYDROXYLAMINE, certain INTERCALATING AGENTS, and NITROUS ACID; many of these are more reactive with ssDNA than with dsDNA and thus probably act preferentially in single-stranded regions generated e.g. during replication. [Chemical mutagenesis: ARB (1982) *52* 655–693.]

mutagenesis The generation of one or more MUTATIONS. Mutagenesis may involve SPONTANEOUS MUTATION (see also MUTATOR GENE) or it can be induced in vitro by the use of a MUTAGEN or by recombinant nucleic acid technology (see SITE-SPECIFIC MUTAGENESIS, TRANSPOSON MUTAGENESIS and SIGNATURE-TAGGED MUTAGENESIS).

mutagenic Capable of causing MUTAGENESIS.

mutagenic repair See SOS SYSTEM.

mutagenicity test A test used to detect the mutagenic potential of a particular (usually chemical) agent. Mutagenicity tests are widely used for screening chemicals for potential carcinogenic activity since most direct CARCINOGENS are mutagenic (although not all mutagens are carcinogenic). See e.g. AMES TEST and SOS CHROMOTEST.

mutan A water-insoluble glucan, containing mostly $(1 \rightarrow 3)$-α- (with some $(1 \rightarrow 6)$-α-) glycosidic bonds, produced by certain strains of *Streptococcus mutans*; mutan is a component of DENTAL PLAQUE. [Mutan synthesis and cariogenicity in *S. mutans*: JGM (1981) *127* 407–415.]

mutanolysin Endo-*N*-acetylmuramidase.

mutant (*adj.* or *noun*) (Refers to) an organism, population, gene, chromosome etc which differs from the corresponding WILD TYPE by one or more MUTATIONS.

mutasynthesis The synthesis of novel end-products by a mutant organism which is blocked in the given biosynthetic pathway and which is provided with unnatural substrates with which to complete that pathway. Mutasynthesis has been used e.g. to synthesize new antibiotics.

Mutatest *Syn.* AMES TEST.

mutation A stable, heritable change in the nucleotide sequence of a genetic nucleic acid (DNA or – in RNA viruses, viroids etc – RNA) which typically results in the generation of a new allele and a new phenotype (cf. SILENT MUTATION and CONDITIONAL LETHAL MUTANT). (In a diploid cell, a mutant allele may be recessive to the corresponding WILD TYPE allele present in the same cell, in which case it will not affect the phenotype of the cell.) The generation of a mutation (*mutagenesis*) may occur spontaneously (see SPONTANEOUS MUTATION) or may result from the activity of a MUTAGEN. See also e.g. BACK MUTATION; DELETION MUTATION; FORWARD MUTATION; FRAMESHIFT MUTATION; INSERTION MUTATION; LEAKY MUTATION, MIS-SENSE MUTATION; MULTISITE MUTATION; NONSENSE MUTATION; NULL MUTATION; PLEIOTROPIC mutation; POINT MUTATION; POLAR MUTATION; SAME-SENSE MUTATION; SUPPRESSOR MUTATION.

mutation rate (*bacteriol.*) The probability that any given cell in a growing population will undergo, spontaneously, a specific MUTATION during its division cycle. The concept of mutation rate can be expressed:

$$a = \frac{m}{d}$$

where a = mutation rate, m = the number of specific mutations which have occurred within the population, and d = the number of individual division cycles within the population. In a SYNCHRONOUS CULTURE, d may be regarded as $(n - n_0)$ where n is the final number of cells, and n_0 is the initial number; thus $(n - n_0)$ indicates the number of newly formed chromosomes (assuming one per cell) – i.e., the number of genomes in which a mutation could occur. However, in a normal, non-synchronous culture $(n - n_0)$ will not indicate the number of newly formed chromosomes because chromosome replication will have occurred in some cells which have not divided at the time of counting; $(n - n_0)$ therefore gives an underestimate of the new genomes, and a statistical correction is applied to the expression for mutation rate, which becomes:

$$a = \frac{m \times 0.69}{n - n_0}$$

Examples of some measured mutation rates: resistance to phage T1 in *Escherichia coli* 3×10^{-8}; loss of ability to ferment galactose (*E. coli*) 10^{-10}. (A value such as 5×10^{-8} means that, on average, one mutant is formed for every 2×10^{7} division cycles within the population.) For two independent mutations having rates of, say, 10^{-6} and 10^{-8}, the mutation rate for a double mutant (i.e., a cell having both mutations) would be 10^{-14}.

The actual (observed) mutation rate in respect of a given phenotypic characteristic may be influenced e.g. by the presence of a TRANSPOSABLE ELEMENT. (See also MUTATOR GENE.)

mutator gene (mutator) Any gene (designated *mut*) within which certain mutations cause an increase in the rate of SPONTANEOUS MUTATION in other genes. For example, in *Escherichia coli*, mutations in the gene encoding the ε subunit of DNA POLYMERASE III (*dnaQ* and *mutD* alleles) can result in extremely high levels of spontaneous mutation; mutant alleles may differ from the wild-type allele e.g. by only one or two single base substitutions, leading to one or two amino acid changes in the ε subunit [MGG (1986) *205* 9–13]. The mutation(s) may lead to reduced accuracy in the polymerizing (nucleotide selection) activity and/or in the proof-reading activity of the enzyme.

Other mutator genes in *E. coli* include genes involved in the MISMATCH REPAIR system.

Another mutator, designated *mutT*, specifically increases the level of A·T → C·G transversions.

[*Escherichia coli* mutator genes: TIM (1999) *7* 29–36.]

Some DNA polymerase mutants of *E. coli* exhibit mutation rates which are *lower* than those in wild-type cells; the corresponding gene is then called an *antimutator gene*. An antimutator gene product presumably permits fewer errors than occur normally during replication – possibly by increasing the efficiency of the proof-reading/editing function of the enzyme.

(See also BACTERIOPHAGE MU.)

mutator phage Any phage which, on infection of a host bacterium, causes an increase in the rate of mutation in the host cell (see BACTERIOPHAGE MU and BACTERIOPHAGE D108).

***mutD* gene** See DNA POLYMERASE.

***mutH* gene** See MISMATCH REPAIR.

Mutinus See PHALLALES.

***mutL* gene** See MISMATCH REPAIR.

***mutS* gene** See MISMATCH REPAIR.

mutualism SYMBIOSIS (sense 1) in which both symbionts derive benefit from the association. (cf. COMMENSALISM.) Some authors reserve the term for an obligatory association, i.e., an association of organisms which, in nature, cannot live independently. Optional ('facultative') mutualism is sometimes called PROTOCOOPERATION.

MV-L3 phage group (L3 acholeplasmavirus group) A group of MYCOPLASMAVIRUSES probably belonging to the PODOVIRIDAE. Type member: bacteriophage MV-L3 (= MVL3; = L3 strain L3). The MV-L3 virion has a polyhedral head (diam. ca. 60 nm) with a short tail (ca. 20 × 9 nm) attached at one vertex via a collar; it includes five or more proteins and ca. 2% by weight of fucose. Genome: linear dsDNA (MWt 2.6×10^{7}). Host: *Acholeplasma laidlawii*; plaques clear, minute. Infected cells are killed but not lysed; progeny virions seem to be released in membrane vesicles which subsequently rupture.

Mx protein See INTERFERONS.

***mxi-spa* genes** See DYSENTERY.

myalgia Muscle pain.

Myambutol *Syn.* ETHAMBUTOL.

myb An ONCOGENE originally identified as the transforming determinant in avian myeloblastosis virus (AMV: see AVIAN ACUTE LEUKAEMIA VIRUSES); v-*myb* is an altered form of a cellular sequence *amv*, differing from *amv* in gene structure, transcript structure, gene product structure, and in the intracellular location (nucleus) of its product [Book ref. 113, pp. 143–151]. v-myb^{+} AMV can transform chicken haematopoietic cells in culture, but differs from other acutely transforming retroviruses in that it does not transform fibroblasts in culture; it causes a rapidly fatal leukaemia only in chickens.

myc An ONCOGENE originally identified as the transforming determinant of avian myelocytomatosis virus (MC29: see AVIAN ACUTE LEUKAEMIA VIRUSES). The MC29 v-*myc* product is a *gag-myc* fusion protein ($P110^{gag-myc}$) which has no protein kinase activity; it binds to dsDNA and occurs – possibly as a chromatin component – in the nucleus. In humans, c-*myc* is located on chromosome 8 and is involved in the pathogenesis of BURKITT'S LYMPHOMA. In chickens, c-*myc* activation by AVIAN LEUKOSIS VIRUSES appears to result in the development of lymphoid leukosis.

mycangium *Syn.* MYCETANGIUM.

Mycelia Sterilia See AGONOMYCETALES.

mycelium A group or mass of discrete hyphae (see HYPHA): the form of the vegetative thallus in many types of fungi and in certain bacteria (see ACTINOMYCETALES). (See also AERIAL MYCELIUM, SPROUT MYCELIUM; SUBSTRATE MYCELIUM; cf. PLECTENCHYMA.)

Mycena See AGARICALES (Tricholomataceae) and BIOLUMINESCENCE.

mycetangium (mycangium) In certain insects: a specialized region within which symbiotic fungi are carried; see e.g. AMBROSIA FUNGI and WOODWASP FUNGI.

mycetism (mycetismus) Poisoning due to the ingestion of certain mushrooms (MUSHROOM sense 1) – e.g. the poisonous species of *Amanita* or *Cortinarius*. (cf. MYCOSIS, MYCOTOXICOSIS; see also e.g. AMATOXINS, MUSCARINE and PHALLOTOXINS.) Mycetism can occur in e.g. sheep and cattle as well as in humans; thus, e.g. the toxins in *Cortinarius speciosissimus* are known to cause renal failure in both humans and sheep. (See also ORELLANIN POISONING.)

mycetismus *Syn.* MYCETISM.

mycetocyte In certain invertebrates, particularly insects: a specialized cell which contains intracellular bacterial or fungal symbionts; if the endosymbiont is a bacterium the term BACTERIOCYTE may be used – although 'mycetocyte' is often used regardless of the nature of the endosymbiont. Mycetocytes may be irregularly distributed in certain tissues (e.g. the gut lining) or they may be aggregated into specialized organelles (*mycetomes*) which are usually associated with the gut. In at least some cases the microflora of the mycetome supplies essential nutrients to the insect host. (See also MYCETANGIUM and TROPHOSOME.) [Molecular biology of symbiotic bacteria in aphid mycetocytes: MS (1986) *3* 117–120.]

mycetoma (1) *Syn.* MADUROMYCOSIS. (2) (fungus ball) A tumour-like mycelial mass formed in the tissues in certain mycoses (see e.g. ASPERGILLOSIS and COCCIDIOIDOMYCOSIS).

mycetome See MYCETOCYTE.

mycetophagous *Syn.* MYCOPHAGOUS.

Mycetozoa A subphylum (phylum GYMNOMYXA) comprising two classes: Eumycetozoa (see EUMYCETOZOEA) and Acrasea (see ACRASIOMYCETES).

Mycoacia See APHYLLOPHORALES (Corticiaceae).

mycobacteriophage Any BACTERIOPHAGE which can infect one or more *Mycobacterium* spp. Most mycobacteriophages have a hexagonal head and a non-contractile tail (contractile in I3), and many are readily inactivated by organic solvents. Mycobacteriophages include both temperate and virulent types; in certain cases phage progeny may be released from the living host cell. [Book ref. 54, pp. 326–342.]

Mycobacterium A genus of Gram-positive, aerobic to microaerophilic, non-motile, asporogenous bacteria (order ACTINOMYCETALES, wall type IV) that are acid-fast during at least some stage of growth.

Cells: typically straight or curved rods, ca. 0.2–0.8 × 1–10 μm, but may occur as coccoid forms, branched rods or fragile filaments; some strains are capsulated (see also MYCOSIDE C). Individual cells may stain uniformly or may exhibit banding or beading. The CELL WALL contains MYCOLIC ACIDS (see also CORD FACTOR, WAX D and PEPTIDOGLYCAN).

Some strains form CAROTENOID pigments: see PHOTOCHROMOGEN and SCOTOCHROMOGEN.

Species of *Mycobacterium* occur in soil as free-living saprotrophs, in water [review: JAB (1984) *57* 193–211], on plants, and as parasites and pathogens of man and other animals (including fish) (see e.g. BURULI ULCER, JOHNE'S DISEASE, LEPROSY, SCROFULA and TUBERCULOSIS).

Metabolism is respiratory and, typically, chemoorganotrophic – although chemolithotrophic strains of e.g. *M. marinum* and *M. smegmatis* have been reported. Nutritionally, mycobacteria are generally not fastidious; sources of carbon and nitrogen include e.g. sugars, hydrocarbons and amino acids. In a number of species glycerol and asparagine are preferred sources of C and N, respectively. Growth may be stimulated e.g. by serum or egg-yolk (see also LÖWENSTEIN–JENSEN MEDIUM) or by an increase in the partial pressure of carbon dioxide; in members of the *Mycobacterium tuberculosis* complex growth is enhanced by pyruvate or (in some species) by glycerol. [Nutrition/metabolism in mycobacteria: Book ref. 54, pp 185–271.]

In 'slow-growing' (SG) strains, visible growth on solid media may not appear for ~4–6 weeks (up to ~12 weeks in *M. malmoense*). In 'rapidly growing' (RG) strains, visible growth on solid media appears within 1 week, often 4–6 days. Note that in BACTEC (liquid) media growth can be detected more rapidly than it can on solid media.

Tests used in the identification of mycobacteria include e.g. the ARYLSULPHATASE TEST, catalase test (that is, persistence of CATALASE activity after incubation at 68°C/20 minutes in neutral phosphate buffer [Book ref. 53, p 1707]), NIACIN TEST, NITRATE REDUCTION TEST, T2H TEST and TWEEN HYDROLYSIS. There are also various molecular methods that are used for detecting, identifying, characterizing and typing certain species: see e.g. PROBE, SPOLIGOTYPING and TMA.

The genus comprises ~50 species. Medically important species are found in the following groups: (i) the *M. tuberculosis* complex (*M. tuberculosis*, *M. bovis*, BCG, *M. africanum* and *M. microti*) – all species associated with tuberculosis; (ii) the *M. avium* complex (*M. avium*, *M. intracellulare* and *M. scrofulaceum*) – see MAC; (iii) *M. leprae*; (iv) 'non-tuberculous mycobacteria', or 'mycobacteria other than tuberculosis': see MOTT. There is a growing number of infrequently isolated mycobacteria – some with actual or potential ability to cause human disease [RMM (1997) *8* 125–135].

GC%: ca. 62–70. Type species: *M. tuberculosis*.

The genus *Mycobacterium* includes the following species:

M. africanum. SG; microaerophilic. Similar to *M. bovis*, but results are variable in the niacin and nitrate reduction tests. Sensitive to PYRAZINAMIDE. Can cause human tuberculosis; found primarily in Africa.

M. avium. SG; non-pigmented. Typical tests: arylsulphatase −ve; catalase (68°C) variable; growth at 25–42°C +ve; niacin −ve; nitrate reduction −ve; T2H test +ve; Tween hydrolysis −ve. Pathogenic for birds. *M. avium*, *M. intracellulare* and *M. scrofulaceum* form the '*M. avium* complex'; these species are associated with various human disease syndromes: see MAC.

M. bovis ('*M. tuberculosis* var. *bovis*'). SG; microaerophilic. Non-pigmented. Typical reactions: catalase (68°C) −ve; growth at 42°C −ve; growth enhancement by pyruvate +ve; niacin −ve; reduction of nitrate −ve; T2H test −ve; Tween hydrolysis variable. Resistant to pyrazinamide. Most strains of *M. bovis* are reported to contain only a single copy of the insertion sequence IS6110.

M. bovis is a causal agent of TUBERCULOSIS in both animals and man. Differentiation between *M. bovis* and *M. tuberculosis* may be facilitated by SPOLIGOTYPING.

M. bovis BCG is a strain of *M. bovis* which differs e.g. in that growth occurs aerobically. (See also BCG.)

M. branderi. SG; non-pigmented/scotochromogenic. Typical tests: arylsulphatase +ve; niacin −ve; reduction of nitrate weak; Tween hydrolysis −ve; growth 25–45°C +ve. [Original description of species: IJSB (1995) *45* 549–553.] *M. branderi* has been isolated from the respiratory tract on a number of occasions and may be a causal agent of disease.

M. celatum. SG; non-pigmented. Typical tests: catalase (68°C) +ve; arylsulphatase variable; niacin −ve; reduction of nitrate −ve; Tween hydrolysis −ve; T2H test +ve. [Original description of species: IJSB (1993) *43* 539–548.]

Strains of *M. celatum* have been classified into three types (I–III) on the basis of RFLP analysis. Type I strains may give false-positive results with a commercial probe for *M. tuberculosis* [JCM (1994) *32* 536–538].

M. celatum has been isolated from a range of infections, some in patients with AIDS [e.g. Lancet (1994) *344* 332; Lancet (1994) *344* 1020–1021].

M. chelonei (= *M. chelonae*). RG; non-pigmented. Similar to *M. fortuitum* but e.g. does not grow at 42°C (optimum growth temperature ~25°C) and does not reduce nitrate. *M. chelonei* has been isolated e.g. from infected wounds.

M. farcinogenes. RG. Resembles *M. fortuitum* (on the basis of DNA homology studies). A causal agent of bovine FARCY.

M. flavum. See XANTHOBACTER.

M. fortuitum. RG; non-pigmented. Typical tests: arylsulphatase +ve; growth at 42°C +ve; Tween hydrolysis variable; reduction of nitrate +ve. An opportunist pathogen. [DNA relatedness study of the *M. fortuitum–M. chelonae* complex: IJSB (1986) *36* 458–460.]

M. gordonae (previously called *M. aquae*). SG; scotochromogen. The organism occurs in water, and is an occasional cause of pulmonary disease; it is not susceptible to isoniazid or pyrazinamide but is sensitive to some other antimycobacterial drugs. A DNA probe (for identifying isolates) is available commercially.

M. haemophilum. SG; non-pigmented. Requires haemin for growth. Growth occurs at 30°C, not at 37°C. Causes disease in the immunosuppressed and in healthy children. [Clinical and laboratory aspects of *M. haemophilum* infections: RMM (1998) *9* 49–54.]

M. interjectum. SG; scotochromogen. Typical tests: catalase (68°C) +ve; arylsulphatase variable; niacin −ve; reduction of nitrate −ve; Tween hydrolysis variable; growth at 31–37°C +ve; T2H test +ve. [Original description of species: JCM (1993) *31* 3083–3089.]

M. interjectum was first isolated from a child with chronic submandibular lymphadenitis (and a positive TUBERCULIN TEST). Partial resection of the lymph node and treatment with antituberculosis drugs failed to resolve the infection; subsequent total resection of the node and treatment with isoniazid, clarithromycin and protionamide was successful. *M. interjectum* has been associated with cervical lymphadenitis in a child [JCM (2001) *39* 725–727].

M. intracellulare. SG; non-pigmented. Similar to *M. avium* but typically arylsulphatase +ve, catalase (68°C) +ve. Tween hydrolysis −ve. Growth occurs at 25–40°C (some strains grow at 40–45°C). (See also MAC.)

M. kansasii. SG; usually photochromogenic. Typical tests: arylsulphatase +ve; catalase (68°C) +ve; growth at 25–40°C +ve (some strains grow at 42°C); niacin −ve; reduction of nitrate +ve; T2H test +ve; Tween hydrolysis +ve.

M. kansasii can cause human disease similar to pulmonary tuberculosis; moreover, *M. kansasii* can cross-react with *M. tuberculosis* in the AMTDT [JCM (1999) *37* 175–178]. However, chemotherapy against *M. kansasii* differs from that used against *M. tuberculosis* (e.g. *M. kansasii* is not susceptible to pyrazinamide); consequently, early identification of the pathogen is essential to ensure correct treatment. [Identification of *M. kansasii* isolates with a DNA probe: JCM (1999) *37* 964–970.]

M. leprae. SG. The causal agent of LEPROSY. This species can be cultured in the footpads of mice but has not been cultured in cell-free laboratory media. [Review: Book ref. 54, pp 273–307. Various aspects: Ann. Mic. (1982) *133B* 5–171. Carbon metabolism in *M. leprae*: JGM (1983) *129* 1481–1495.] An examination of the genome sequence of *M. leprae* has revealed massive gene decay, with less than 50% of the genome containing protein-encoding genes [Nature (2001) *409* 1007–1011].

M. lepraemurium. SG; non-pigmented. The causal agent of murine leprosy. Limited growth has been reported to occur on egg-yolk media when very large inocula are used.

M. malmoense. SG; growth on egg media at 37°C may take up to 12 weeks. This organism can cause disease similar to that associated with members of the *M. avium* complex (see MAC). To facilitate detection/identification of *M. malmoense* a PCR-based method has been devised and tested successfully [JCM (1999) *37* 1454–1458].

M. marinum. SG; photochromogenic. Typical tests: arylsulphatase +ve; catalase (68°C) +ve; growth at 30°C +ve; growth at 37°C −ve on primary isolation but +ve after serial subculture; niacin −ve; reduction of nitrate −ve; Tween hydrolysis +ve. This species causes disease in fish and skin granulomas/ulcers in humans [Arch. Derm. (1986) *122* 698–703]; it is more common in temperate regions than in tropical regions (cf. *M. ulcerans*).

M. microti. SG; aerobic. Non-pigmented. Similar to *M. tuberculosis* but typically gives a variable nitrate reduction test and a negative T2H test. Causal agent of tuberculosis in voles.

M. paratuberculosis ('*M. johnei*'). SG; non-pigmented. Typically (results from few strains): catalase (68°C) +ve; niacin −ve; reduction of nitrate −ve; T2H test +ve; Tween hydrolysis variable. (See also IS900 and MYCOBACTINS.)

M. paratuberculosis is the causal agent of JOHNE'S DISEASE. (See also CROHN'S DISEASE.) (See also PULSED ELECTRIC FIELD.)

M. phlei. RG; scotochromogenic. Typical tests: arylsulphatase −ve (at 3 days) or variable (at 1 week); growth at 52°C +ve; Tween hydrolysis +ve. *M. phlei* is found in soil and on vegetation; it has also been isolated from (i) synovial fluid in a case of Reiter's syndrome, and (ii) a foot wound, and is regarded as an opportunist pathogen.

M. scrofulaceum. SG; scotochromogenic. Typical test results: arylsulphatase −ve; catalase (68°C) +ve; growth at 25–42°C +ve; niacin −ve; reduction of nitrate −ve; T2H test +ve; Tween hydrolysis −ve. A causal agent of SCROFULA. (See also MAC.)

M. senegalense. RG; scotochromogen. A causal agent of bovine FARCY.

M. simiae. SG; some strains photochromogenic, others non-chromogenic. Generally similar to *M. scrofulaceum* but the niacin test is +ve in some strains.

M. smegmatis. RG; non-pigmented. Typical test results: arylsulphatase +ve (at 1 week); growth at 45°C +ve; Tween hydrolysis +ve. Found in smegma; believed to be non-pathogenic.

M. terrae. SG; non-chromogenic. With *M. nonchromogenicum* and *M. triviale*, forms the 'Terrae group' or *M. terrae* complex – species which are pathogenic but apparently rare.

M. thermoresistibile. RG; scotochromogen. Typical test results: arylsulphatase −ve; niacin −ve; reduction of nitrate

−ve; growth at 52°C +ve; Tween hydrolysis +ve; T2H test +ve. Found e.g. in soil. Reported to be pathogenic in lung (several cases); skin and lymph nodes; and breast tissue [JCM (1992) *30* 1036–1038].

M. triplex. SG; non-pigmented. Isolated from various clinical specimens. [Description of species: JCM (1996) *34* 2963–2967.] [*M. triplex* associated with infection in a liver transplant patient: JCM (2001) *39* 2033–2034.]

M. tuberculosis. SG; aerobic. Non-pigmented. Typical test results: catalase (68°C) −ve; growth at 42°C −ve; growth at 37°C +ve; growth enhanced by glycerol and by pyruvate; niacin +ve; reduction of nitrate +ve; Tween hydrolysis variable; sensitive to pyrazinamide. Usually forms rough, raised, whitish/pale buff colonies ('eugonic' growth).

A casual agent of TUBERCULOSIS.

[Genome sequence of *M. tuberculosis*, strain H37rv: Nature (1998) *393* 537–544.] The insertion sequence IS6110 (q.v.) is frequently used for TYPING isolates of *M. tuberculosis*.

M. ulcerans. SG; pigmentation variable. Typical test results: catalase (68°C) +ve; growth at 30°C +ve; growth at 37°C −ve; niacin variable; reduction of nitrate −ve; Tween hydrolysis −ve. This species is found in tropical regions e.g. on vegetation; it causes BURULI ULCER.

M. xenopi. SG; pigmentation variable. Found in water, and able to cause lesions in the (human) lung. Does not grow at 25°C, grows poorly at 37°C, and has an optimum growth temperature of ca. 42–45°C.

mycobactins A family of complex, lipophilic compounds which occur in the cell envelope in most species of *Mycobacterium* (not in *M. paratuberculosis* or in some strains of *M. avium*); they chelate trivalent metal ions, particularly solubilized ferric ions, and are believed to function in iron transport – iron being released after enzymic reduction to the ferrous form. For in vitro growth *M. paratuberculosis* needs mycobactin *or* e.g. ferric ammonium citrate. [Structure of mycobactins: Book ref. 54, pp. 242–245.] (See also EXOCHELINS and SIDEROPHORES.)

Related compounds occur in *Nocardia*.

mycobiont A fungal symbiont – e.g. in a LICHEN or MYCORRHIZA.

Mycobionta *Syn.* EUMYCOTA.

Mycocalia See NIDULARIALES.

Mycocaliciaceae See CALICIALES.

mycocecidia GALLS induced by fungi.

Mycocentrospora See HYPHOMYCETES; see also CROWN ROT.

mycochrome See PHOTOINDUCTION and PHOTOINHIBITION.

mycodextran *Syn.* NIGERAN.

Mycogone See HYPHOMYCETES; see also BUBBLE DISEASES.

mycoherbicide See BIOLOGICAL CONTROL.

mycolic acids α-Substituted, β-hydroxylated fatty acids (having the general formula R′CHOH.CHR″.COOH), esters of which are found in the cell walls of e.g. species of *Corynebacterium*, *Mycobacterium*, *Nocardia*, and *Rhodococcus*; in *Mycobacterium* spp the mycolic acids fall within the approximate range C_{60}–C_{90}, in *Nocardia* C_{40}–C_{60}, and in *Corynebacterium* C_{20}–C_{40}.

In mycobacterial mycolic acids R′ is usually a C_{50}–C_{60} chain which often includes double bonds, cyclopropane rings etc, while R″ is a C_{22}–C_{24} chain. [Structure and biosynthesis: Book ref. 54, pp. 113–128. Mycolic acid patterns in various strains of *Mycobacterium*: JGM (1983) *129* 889–891; (1984) *130* 363–367, 2733–2736.] (See also WAX D.)

[Mycolic acids in *Corynebacterium* spp: JGM (1984) *130* 513–519.]

mycology The study of FUNGI.

mycoparasite A fungus which is parasitic on other fungi. Mycoparasites include e.g. *Christiansenia pallida* [life history: Mycol. (1984) *76* 9–22]; PIPTOCEPHALIS; and *Rozella* spp: endobiotic and holocarpic organisms which parasitize e.g. *Polyphagus euglenae* (itself a parasite of *Euglena* spp) [Mycol. (1984) *76* 1039–1048] and other fungi and algae. Certain mycoparasites can apparently exert some control on pathogens of higher plants – e.g., in cases of CLOVER ROT, *Trichoderma viride* (or e.g. *Mitrula sclerotiorum*) can bring about a reduction of the numbers of sclerotia in the soil [Bot. Rev. (1984) *50* 491–504]. [Susceptibility of e.g. *Pythium* spp to the mycoparasite *Pythium oligandrum*: SBB (1986) *18* 91–96.] (See also CONTACT BIOTROPHIC MYCOPARASITE.)

mycophagous (mycetophagous) Fungus-eating.

mycophenolic acid An ANTIBIOTIC produced e.g. by *Penicillium brevicompactum* in aerial hyphae formed on solid media [AEM (1981) *41* 729–736]. It has antimicrobial and antitumour activity, blocking GMP synthesis by inhibiting the formation of XMP from IMP (see Appendix V(a)).

Mycoplana A genus of Gram-negative, aerobic bacteria of uncertain taxonomic affinity; species occur e.g. in soil. The organisms form branching filaments which fragment into irregular, flagellated rods. Some strains can fix nitrogen under microaerobic conditions [JGM (1982) *128* 2073–2080]. GC%: ca. 64–69. [Book ref. 46, pp. 2118–2119.]

mycoplasma (1) A member of the class MOLLICUTES. (cf. MOLLICUTE.) (2) A member of the genus MYCOPLASMA.

Mycoplasma A genus of cell wall-less, sterol-requiring, catalase-negative bacteria (family MYCOPLASMATACEAE) which occur as parasites and pathogens e.g. in the respiratory and urogenital tracts in man and other animals; diseases caused by, or associated with, *Mycoplasma* spp include e.g. AIR SACCULITIS, BRONCHITIS, CONTAGIOUS BOVINE PLEUROPNEUMONIA, GLASSER'S DISEASE, NON-GONOCOCCAL URETHRITIS, ovine MASTITIS, and PRIMARY ATYPICAL PNEUMONIA (sense 2). (*Mycoplasma* spp are also common contaminants in TISSUE CULTURES.) Cells: typically non-motile (but see below) and pleomorphic, ranging from spherical, ovoid or pear-shaped (ca. 0.3–0.8 μm diam.) to branched filamentous forms of near-uniform diameter, several μm to ca. 150 μm in length; filaments, the typical forms in young cultures under optimum conditions, subsequently transform into chains of coccoid cells which later break up into individual cells that are capable of passing through membrane filters of pore size 0.22 μm or 0.45 μm. The cells of some species have a 'tip' structure (possibly part of a microfibrillar 'cytoskeleton') which has been thought to be involved in attachment to host cells, and which (in motile species) appears to have a role in GLIDING MOTILITY – the tip always pointing in the direction of motion.

The attachment organelle and cytoadherence proteins of *M. pneumoniae* have been detected by immunofluorescence microscopy [JB (2001) *183* 1621–1630].

The trilaminar cytoplasmic membrane contains sterols (in addition to e.g. phospholipids and proteins) – thus rendering the cells susceptible to POLYENE ANTIBIOTICS and to lysis by e.g. digitonin (which complexes sterols). Some species bear a capsule or slime layer – that in *M. mycoides* subsp *mycoides* being a galactan.

Replication of the genome may precede cytoplasmic division; hence, 'multinucleate' filaments may exist for a time before individual cells are delimited by constriction. Budding can also occur.

Most *Mycoplasma* spp are facultatively anaerobic, some apparently being obligately anaerobic on primary isolation. All species are chemoorganotrophic. 'Fermentative' species can use sugars such as glucose (metabolized to e.g. lactic acid via the EMP pathway), while 'non-fermentative' species can use e.g. arginine. All species need cholesterol or related sterols (e.g. cholestanol or stigmasterol). The organisms have a flavin-terminated electron transport chain which lacks both quinones and cytochromes. NADH oxidase occurs in the cytoplasm (cf. ACHOLEPLASMA). Growth occurs on complex media (e.g. HAYFLICK MEDIUM); fastidious mycoplasmas may be grown on diphasic SP-4 medium [recipe: Book ref. 22, p. 746]. Colonies (usually <1 mm diam.) are typically of the 'fried egg' type: an opaque, granular central region, embedded in the agar, surrounded by non-granular surface growth. Optimum growth temperature of mammalian strains: 36–37°C. Many species produce weak or clear haemolysis; haemolysis appears to be due to the secretion of H_2O_2 (a product which is believed to account for some aspects of pathogenicity). Mycoplasmas are commonly sensitive to chloramphenicol and to tetracyclines; most species can tolerate 1:2000/4000 thallous acetate. Broth cultures of *Mycoplasma* spp (supplemented with DMSO or glycerol) can be stored at −70°C; alternatively, broth cultures may be lyophilized. GC%: ca. 23–40. Type species: *M. mycoides*.

The genus currently contains over 60 species which are differentiated on the basis of certain tests: e.g., utilization of glucose and mannose, arginine hydrolysis, phosphatase production, the FILM AND SPOTS reaction, and haemadsorption.

M. glycophilum. An avian species [JGM (1984) *130* 597–603].

M. laidlawii. Re-classified as *Acholeplasma laidlawii*.

M. mycoides. Non-motile cells which often form repeatedly branching filaments. Under certain conditions a culture may contain cells called *rho*-forms; a *rho*-form contains an intracellular organelle (function unknown) which consists essentially of an axial fibre (ca. 40–120 nm diam.) extending the length of the cell and occupying a major part of the cell's volume. *M. mycoides* subsp *mycoides* causes contagious bovine pleuropneumonia.

M. pneumoniae (Eaton's agent). A slowly-growing species which causes a primary atypical pneumonia in man. On primary isolation, the colonies (after 5–10 days' incubation) are ca. 50–100 μm in diameter and are entirely granular, i.e., they are not typical 'fried egg' colonies; fried egg colonies generally develop on subculture. The organisms are generally highly sensitive to erythromycin.

M. pneumoniae respiratory tract infection is commonly investigated serologically by: (i) a complement-fixation test (CFT) that measures titres of anti-*Mycoplasma* antibodies, or (ii) a more rapid (and more sensitive) immunofluorescence assay for specific IgM antibodies. Following comparison of serology, culture and a PCR-based assay in the diagnosis of *M. pneumoniae* respiratory tract infection in children, it was considered appropriate to use PCR for rapid diagnosis and to confirm/reject any PCR-negative specimens on the basis of CFT examination of paired sera [JCM (1999) *37* 14–17].

T-strain mycoplasmas. See UREAPLASMA.

[Book ref. 22, pp. 742–770. *Mycoplasma* characterization: Book ref. 98. *Mycoplasma* evolutionary tree from 5S rRNA sequencing data: PNAS (1985) *82* 1160–1164.]

(See also MYCOPLASMAVIRUSES.)

mycoplasma virus type 1 phages *Syn.* PLECTROVIRUS.

mycoplasma virus type 2 phages *Syn.* PLASMAVIRIDAE.

Mycoplasmataceae A family of non-helical, sterol-requiring, cell wall-less bacteria of the order MYCOPLASMATALES. Two genera: MYCOPLASMA (urease-negative) and UREAPLASMA (urease-positive).

Mycoplasmatales An order of cell wall-less bacteria of the class MOLLICUTES; it comprises three families: MYCOPLASMATACEAE (non-helical cells which require sterols for growth), ACHOLEPLASMATACEAE (non-helical cells which do not require sterols), and SPIROPLASMATACEAE (cells often helical; sterols required for growth). [Book ref. 22, pp. 741–787.]

mycoplasmaviruses BACTERIOPHAGES which infect members of the MYCOPLASMATALES: see MV-L3 PHAGE GROUP, PLASMAVIRIDAE, PLECTROVIRUS, SPIROPLASMAVIRUSES. [Review: Intervirol. (1982) *18* 177–188.]

mycoplasmosis Any disease caused by a species of MYCOPLASMA (q.v.)

mycorrhiza A stable, usually mutualistic association between a fungus and the root (or rhizoid) of a plant. Mycorrhizas occur in the majority of plants, including vascular and some non-vascular species (e.g. liverworts). The fungi involved (e.g. basidiomycetes, ascomycetes, deuteromycetes) are always associated with the primary cortex of the root, and many appear never to occur as free-living saprotrophs. The formation of mycorrhizas leads to improved uptake of nutrients by the host plant; nutrients are apparently absorbed by hyphae (which may extend some distance from the root) and are transported back to the root to be released into the host tissue. Mycorrhiza formation and efficacy is greatest in nutrient-poor soils, and may be reduced or eliminated by application of soil fertilizers. Three major types of mycorrhiza are recognized.

Ectomycorrhizas ('ectotrophic mycorrhizas') occur mainly in temperate forest trees; the fungi involved include basidiomycetes (e.g. agarics, boletes), ascomycetes (e.g. *Tuber* spp) and zygomycetes (*Endogone*). A given tree may associate with more than one species of fungus. In an ectomycorrhiza the fungal hyphae occur on the root surface and may penetrate between the cortical cells of the root, but the cortical cells themselves are not penetrated. Typically, the host root becomes completely enclosed by a sheath of pseudoparenchymal fungal tissue (the *mantle*); hyphae from the mantle may penetrate the soil surrounding the root and also penetrate between the cortical cells of the root to enmesh individual cortical cells in a network of hyphae (the *Hartig net*). The root is morphologically distinct from an uninfected root: e.g., it lacks root hairs and a root cap; it is thicker than an uninfected root and may be a different colour; it may branch extensively and characteristically – e.g. pinnately (in *Fagus* spp) or dichotomously (in *Pinus* spp) – or not at all (e.g. in *Quercus* spp). In certain cases an ectomycorrhiza may develop in the form of nodules (= tubercles), each consisting of a rounded, dense mass of mycorrhizal roots.

Ectomycorrhizal fungi appear to have only limited (or no) ability to use complex carbohydrates (e.g. cellulose); they obtain simple sugars (e.g. glucose, fructose, sucrose) from the plant and store them (e.g. as mannitol, trehalose or glycogen) in the mantle. Benefits to the plant, in addition to improved uptake of nutrients (particularly phosphate), include increased protection from certain pathogenus – e.g., the fungus may produce antibiotic(s), or the mantle may function as a mechanical barrier to infection. Many ectomycorrhizal fungi produce PHYTOHORMONES, but the significance of this to the plant is unknown. In certain cases a mycorrhizal association appears to be essential for the normal development of the plant (e.g. in certain *Pinus* spp).

Endomycorrhizas ('endotrophic mycorrhizas') involve the development of the fungus *within* the cells of the root cortex; there is usually little or no change in root morphology, and an external fungal sheath is usually not formed. Typically, the fungal hyphae penetrate the cortical cells of the root and develop intracellularly; subsequently, the hyphae are digested by the root cell, leaving a knot of undigested hyphal wall material in the cell. As the root grows, the fungus invades new cells behind the root meristem; thus, a balance is set up between fungal invasion of the plant and plant digestion of the fungus. There are three main types of endomycorrhiza. (a) The *vesicular-arbuscular* (VA) type, found in a very wide range of plants, in which the (aseptate) fungal hyphae spread through the primary cortex of unsuberized roots and penetrate the cortical cells. A characteristic, much-divided haustorium (the *arbuscule*) is formed, and both intracellular and extracellular hyphae usually develop spherical, lipid-rich, intercalary or terminal swellings (*vesicles*). VA mycorrhizal fungi belong to the genera *Acaulospora*, *Gigaspora*, *Glomus* and *Sclerocystis* (all formerly *Endogone* spp); they have not yet been grown in pure culture. VA mycorrhizas improve uptake of nutrients (particularly phosphate) by the host plant. (b) The *ericoid* type, found in members of the Ericaceae, in which the fungi colonize the fine terminal roots of the host plant and form coils or loops within the host cells. The fungi involved all belong to, or are closely related to, the species *Pezizella ericae*. (c) The *orchid* type, found in embryos and roots in members of the Orchidaceae, in which the fungi penetrate the host cells and form intracellular hyphal coils. All known orchid mycorrhizal fungi are also normal soil saprotrophs or parasites of other plants; they are usually basidiomycetes (e.g. *Armillaria*, *Ceratobasidium*, *Marasmius*, *Thanatephorus*, *Tulasnella*). Mycorrhizal associations appear to be essential to the orchid – at least for germination and seedling growth; orchid seeds are very small and have little or no food reserve, so that (under natural conditions) nutrients must be supplied by an invading fungus for successful germination to occur. Saprotrophic orchids may depend on mycorrhizal fungi throughout their lives, associating with fungi (e.g. *Armillaria mellea*) which can degrade substrates such as cellulose and pectin to simple compounds which the orchid can assimilate. Green orchids generally associate with *Rhizoctonia (Thanatephorus)* spp, but may lose their mycorrhizal fungi when mature.

Ectendomycorrhizas ('ectendotrophic mycorrhizas') are intermediate in form between ecto- and endomycorrhizas; an organized fungal sheath is formed, and inter- and intracellular penetration of the root cortex occurs. This type of mycorrhiza is formed by only a limited number of plants, including certain members of the Ericaceae – e.g. *Monotropa* (a genus of achlorophyllous herbaceous plants) and *Arbutus* – and the seedlings of certain conifers. The fungi involved are apparently basidiomycetes; some fungi which form ectendomycorrhizas in plants of the Arbutae and Monotropaceae may form ectomycorrhizas in other plants.

[Host–fungus interactions: Book ref. 55, pp. 225–253. General: Book ref. 56. A check list of British mycorrhizas: New Phyt. (1987) *105* (Supplement, 1–102).]

mycosamine An aminosugar (3-amino-3,6-dideoxy-D-mannose) found e.g. in NYSTATIN and AMPHOTERICIN B.

mycoside C A peptidoglycolipid, distinct from WAX D, formed by some species of *Mycobacterium* (e.g. *M. avium*); in *M. lepraemurium* mycoside C occurs as a fibrillar or crystalline capsule. [Book ref. 54, pp. 28–30.]

mycosin *Syn.* CHITOSAN.

mycosis Any human or animal disease resulting from infection with a fungus (cf. MYCOTOXICOSIS and MYCETISM). See e.g. ADIASPIROMYCOSIS; ASPERGILLOSIS, BLACK PIEDRA; BLASTOMYCOSIS, CANDIDIASIS; CHROMOBLASTOMYCOSIS; COCCIDIOIDOMYCOSIS; HISTOPLASMOSIS; LOBOMYCOSIS; MADUROMYCOSIS; ONYCHOMYCOSIS; PARACOCCIDIOIDOMYCOSIS; PHAEOHYPHOMYCOSIS; PITYRIASIS NIGRA; PITYRIASIS VERSICOLOR; RHINOSPORIDIOSIS; RINGWORM; SPOROTRICHOSIS; WHITE PIEDRA; ZYGOMYCOSIS.

mycosis fungoides A chronic human cutaneous T-cell lymphoma. HTLV-I provirus has been detected in a very small proportion of cases, but HTLV-I is not thought to be an aetiological agent. (cf. ADULT T-CELL LEUKAEMIA.)

Mycosphaerella A genus of fungi (order DOTHIDEALES) which include saprotrophs and plant parasites (see e.g. BANANA LEAF SPOT). (See also CLADOSPORIUM.)

mycotête See TERMITE–MICROBE ASSOCIATIONS.

mycotic dermatitis *Syn.* LUMPY WOOL.

mycotoxicosis Any disease of man or animals resulting from the ingestion of MYCOTOXINS (cf. MYCETISM; MYCOSIS). See e.g. ALIMENTARY TOXIC ALEUKIA; DENDRODOCHIOTOXICOSIS; ERGOTISM; LUPINOSIS, PASPALUM STAGGERS; STACHYBOTRYOTOXICOSIS; YELLOW RICE.

mycotoxin A TOXIN produced by a fungus. The term is usually reserved for fungal metabolites that are toxic to man and/or animals and are produced by moulds growing on foodstuffs (cf. MYCETISM). See e.g. AFLATOXINS; ASTELTOXIN; AUROVERTINS; CHLOROPEPTIDE; CITREOVIRIDIN; CITRININ; CYCLOPIAZONIC ACID; ERGOCHROMES; ERGOT ALKALOIDS; OCHRATOXINS; PATULIN; PENITREMS; RUBRATOXINS; SLAFRAMINE; SPORIDESMINS; STERIGMATOCYSTIN; TRICHOTHECENES; XANTHOMEGNIN; ZEARALENONE. (cf. IPOMEAMARONE and 4-IPOMEANOL.)

mycotrophein See CONTACT BIOTROPHIC MYCOPARASITE.

mycotrophic (1) (*of plants*) Refers to the plant partner in a MYCORRHIZA. (2) (*of animals*) Refers to an animal (e.g. an invertebrate) that feeds on fungi.

Mycotypha See MUCORALES.

mycovirus Any VIRUS which can infect one or more fungi. Mycoviruses have been observed in a wide range of fungi; the majority are isometric (diam. ca. 25–50 nm) dsRNA viruses either with monopartite genomes or with segmented genomes, each segment being encapsidated in a separate particle. Many of the dsRNA virions have been shown to contain an RNA-DEPENDENT RNA POLYMERASE. Six groups of isometric dsRNA mycoviruses have been defined [Intervirol. (1984) *22* 17–23]. *Group A*: genome monopartite (4.8–6.1 kbp), virion diam. ca. 35–43 nm; members: *Saccharomyces cerevisiae* viruses ScV-L1 (= ScV-LA) and ScV-L2; probable members: *Ustilago maydis* viruses UmV-P1-H1, UmV-P4-H1 and UmV-P6-H1. *Group B*: genome monopartite (8.3 kbp), virion diam. ca. 48 nm; member: *Drechslera* (formerly *Helminthosporium*) *maydis* virus (HmV). *Group C*: genome bipartite (1.6 and 1.4 kbp), virion diam. ca. 30–35 nm; member: *Penicillium stoloniferum* virus S (PsV-S); probable member: *Diplocarpon rosae* virus (DrV). *Group D* (*Gaeumannomyces graminis* virus group 1): genome bipartite (1.8–1.9 and 1.7–1.8 kbp), virion diam. ca. 35 nm; members: *G. graminis* viruses GgV-019/6-A, GgV-OgA-B and GgV-F6-C. *Group E* (*Gaeumannomyces graminis* virus group 2): genome bipartite (2.2–2.3 and 2.0–2.1 kbp), virion diam. ca. 35 nm; members: *G. graminis* viruses GgV-F6-B, GgV-T1-A and GgV-OgA-A. *Group F*: genome tripartite (ca. 3.2, 3.0 and 2.9 kbp), virion diam. ca. 35–40 nm; members: *Penicillium chrysogenum* virus (PcV), *P. brevicompactum* virus (PbV) and '*P. cyaneofulvum*' (= *P. chrysogenum*) virus (Pc-fV). (Pc-fV actually has four dsRNA segments, the 4th probably being a variant, satellite or defective dsRNA.) Two families, Totiviridae and Partitiviridae, have been created to accommodate isometric dsRNA viruses with, respectively, monopartite and bipartite genomes [Intervirol. (1986) *25* 141–143].

Most dsRNA mycoviruses appear to have a single-layered capsid composed of a single polypeptide species. However, a dsRNA virus from *Lentinula edodes* appears to have a double-shelled capsid [Virol. (1982) *123* 93–101] and to resemble a PHYTOREOVIRUS. Other mycoviruses include e.g. '*Boletus* virus' (a possible member of the POTEXVIRUSES) and mushroom virus 3 (virions bacilliform, ca. 50 × 19 nm); DNA-containing lytic viruses have been observed in certain lower fungi (e.g. *Rhizidiomyces* [Virol. (1983) *130* 10–20, 21–28]).

Mycoviruses are apparently transmissible only by intracellular routes (e.g. during mating, via spores, or by hyphal anastomosis). In many cases the host fungus shows little evidence of infection (cf. MUSHROOM DISEASES); in certain cases virus infection may affect the virulence of a pathogenic fungus: e.g. an isometric dsRNA virus appears to be associated with virulence in strains of *Rhizoctonia solani* [JGV (1985) *66* 1221–1232] – cf. CHESTNUT BLIGHT and DUTCH ELM DISEASE. (See also KILLER FACTOR.)

mycovore A fungus-eating animal.

myelitis Inflammation of the spinal cord.

myeloblastosis-associated viruses See AVIAN LEUKOSIS VIRUSES.

myeloma A tumour consisting of a single clone of B LYMPHOCYTES; myelomas can occur spontaneously or they can be induced. A given myeloma secretes Ig ('paraprotein') molecules, all of which are structurally and electrophoretically homogeneous (monoclonal); some myeloma Igs react with environmental antigens. Myeloma cells are used e.g. for HYBRIDOMA formation. (See also BALB/C MICE.)

myeloperoxidase (MPO) A haemoprotein (MWt ca. 150000) found e.g. in MACROPHAGES and in the azurophilic granules of NEUTROPHILS; MPO catalyses the oxidation of halide ions to hypohalite. [Role of MPO in oxygen-dependent microbicidal activity of human neutrophils: Blood (1983) *61* 483–492.]

***myo*-inositol** See INOSITOL.

myonecrosis (clostridial) See GAS GANGRENE.

myoneme (*ciliate protozool.*) A bundle, ribbon or sheet of intracellular protein microfibrils, numbers of which occur in certain ciliates; myonemes are apparently contractile and may thus account for the observed contractility in those organisms which contain them. Myonemes occur below the infraciliature in e.g. *Stentor* and *Spirostomum*; in *Stentor* they are also known as *M bands* or *M fibres*. (See also SPASMONEME.) Myoneme contraction appears to be related to environmental and/or intracellular levels of calcium ions.

myosin A rod-shaped protein which occurs in most types of eukaryotic cell; the molecule consists of six polypeptides: two 'heavy chains' and four 'light chains', the light chains occurring at one end of the molecule and contributing to its bulbous 'head'. Myosin molecules can aggregate into bundles, each bundle consisting of a number of longitudinally adjacent molecules arranged so that head regions occur at each end. When contact occurs between a myosin head and an ACTIN microfilament, the head apparently undergoes a conformational change which causes the myosin molecule (or bundle) to move relative to the microfilament; moreover, when interposed between suitably aligned microfilaments, a myosin bundle can apparently cause one microfilament to move relative to the other. Continuing movement of a microfilament relative to a myosin head seems to involve a continual sequence of reversible changes in the conformation of the myosin head – the head repeatedly making and breaking contact with the microfilament, and repeatedly binding and hydrolysing ATP.

Myoviridae A family of BACTERIOPHAGES which are characterized by the possession of a long, complex, contractile tail consisting of a central tube surrounded by a coaxial contractile sheath; the phage head may be isometric or elongated. Type species: T2. Other members include e.g. the T-EVEN PHAGES, BACTERIOPHAGE P1, BACTERIOPHAGE P2, and many other enterobacterial phages; *Bacillus* phages PBS1 (q.v.), SP3, SP8 and SP50; *Clostridium* phage HM3; *Pseudomonas* phage φW-14 (q.v.); etc.

Myriangium See DOTHIDEALES.

***Myrica* root nodule** See ACTINORRHIZA.

Myriodesma See PHAEOPHYTA.

Myriostoma See LYCOPERDALES.

Myrmecia A genus of unicellular green algae (division CHLOROPHYTA) which may occur as photobionts in certain lichens (see e.g. BAEOMYCES and LOBARIA) or as free-living organisms. The cells are spherical, and each contains a single cup-shaped chloroplast with no pyrenoid; autospores and zoospores may be formed.

myrotheciotoxicosis *Syn.* DENDRODOCHIOTOXICOSIS.

Myrothecium See HYPHOMYCETES; see also DENDRODOCHIOTOXICOSIS and TRICHOTHECENES.

myxamoeba A non-flagellated amoeboid cell which occurs in the life cycles of the CELLULAR SLIME MOULDS and MYXOMYCETES.

Myxidium A genus of protozoa (class MYXOSPOREA) which are coelozoic parasites of e.g. sticklebacks (*M. gasterostei*), salamanders (*M. serotinum*) etc. The mature spore contains two polar capsules, one at each end of the spore.

myxin An antibiotic produced by *Lysobacter* sp.

myxo- A prefix derived from the Greek word for *mucus* or *slime*.

myxobacter Any member of the MYXOBACTERALES.

Myxobacterales An order of GLIDING BACTERIA which form *fruiting bodies* (see below) and which have GC values of ca. 68–72%; all species are Gram-negative. (cf. CYTOPHAGALES.) Myxobacteria occur e.g. in soil, on decaying vegetation, and on dung. Vegetative cells, generally 0.4–1.4 × 2–15 μm according to species and growth conditions, may be blunt-ended rods (*Chondromyces*, *Nannocystis*, *Polyangium*) or tapered rods (*Archangium*, *Cystobacter*, *Melittangium*, *Myxococcus*, *Stigmatella*); they bear extracellular slime. The cell wall resembles that of other Gram-negative bacteria (see CELL WALL), although in at least one species (*Myxococcus xanthus*) the peptidoglycan occurs in patches rather than in a continuous layer; the myxobacteria are generally much more sensitive than enteric bacteria to hydrophobic antibiotics such as actinomycin D and novobiocin – suggesting structural differences in the outer membrane. The cells generally contain non-carotenoid pigments and membrane-bound, photoinducible CAROTENOID glycosides (e.g. *myxobacton*: 1′,2′-dihydro-1′-glucosyl-4-oxotorulene, and *myxobactin*: 1′,2′-dihydro-1′-glucosyl-3,4-dehydrotorulene).

Studies on the motility apparatus of *Myxococcus xanthus* have been carried out with shock-frozen gliding cells [Microbiology (2001) *147* 939–947].

Myxobacteria are obligate aerobic chemoorganotrophs. Most species release enzymes which lyse bacteria, yeasts and moulds – organisms which form the primary food source; *Polyangium cellulosum* is atypical in being strongly cellulolytic. Myxobacteria generally grow on solid media containing live or dead microorganisms; the colony or aggregate of cells ('swarm' or 'pseudoplasmodium') is typically thin, flat and wrinkled. Reproduction occurs by transverse binary fission.

Fruiting in myxobacteria may be induced by starvation (or e.g. by the addition of cAMP); the vegetative cells aggregate, and about 80% of them lyse – possibly to provide nutrients for the remainder which become the resting cells (*myxospores*) from which the fruiting body is formed. [Development in *Myxococcus xanthus*: QRB (1984) *59* 119–138.] In some genera (e.g. *Chondromyces*, *Nannocystis*, *Polyangium*) the myxospores resemble vegetative cells, while in others (e.g. *Myxococcus*, *Stigmatella*) they are shorter, or spherical, and refractile; in at least some genera (e.g. *Myxococcus*) the myxospores are more resistant than vegetative cells to e.g. heat and desiccation. (In e.g. *Myxococcus xanthus* myxospore formation – without fruiting – can be induced in broth cultures e.g. by high concentrations of glycerol.) Myxospores can, under suitable conditions, germinate to form vegetative cells – thus completing the life cycle. The myxobacterial fruiting bodies vary in size (from microscopic to ca. 1 mm), shape and composition, according to species; they are often brightly coloured. In some myxobacteria (e.g. *Archangium*, *Myxococcus xanthus*) the fruiting body is merely a small, sessile (non-stalked), rounded or irregularly shaped mound of myxospores set in slime; in *M. stipitatus* it consists of a spherical mass of myxospores raised on a stalk of hardened slime. In many genera the myxospores are enclosed within one or more hard, sac-like structures termed *cysts* or *sporangia*. In *Nannocystis* individual sporangia are embedded in the substrate, while in *Cystobacter* and *Polyangium* the fruiting body consists of a single sessile sporangium or a group of sessile sporangia; in *Chondromyces*, *Melittangium* and *Stigmatella* one or more sporangia occur on simple or branched stalks.

Other myxobacterial genera include e.g. *Angiococcus*, *Corallococcus*, *Haploangium* and *Sorangium*.

myxobactin See MYXOBACTERALES.

myxobacton See MYXOBACTERALES.

Myxobionta *Syn.* MYXOMYCOTA.

Myxobolus A genus of protozoa (class MYXOSPOREA) which are histozoic parasites in fish. *M. pfeifferi* causes 'boil disease' in cyprinid fish, in which the sporonts develop in large (up to 7 cm) tumour-like cysts. *M. cyprini* is a common parasite of the common carp (*Cyprinus carpio*); it develops primarily in the muscle fibres of skeletal muscle, the mature spores subsequently being liberated into the bloodstream where they are retained in the capillaries [Parasitol. (1985) *90* 549–555].

Myxococcus See MYXOBACTERALES.

myxoflagellate A flagellated amoeboid cell formed by members of the MYXOMYCETES.

Myxogastria See EUMYCETOZOEA.

myxoma subgroup See LEPORIPOXVIRUS.

myxoma virus See LEPORIPOXVIRUS and MYXOMATOSIS.

myxomatosis An acute infectious disease which primarily affects the European rabbit *Oryctolagus cuniculus*; it is caused by the myxoma virus (see LEPORIPOXVIRUS). In *O. cuniculus* the disease, which is usually rapidly fatal, is characterized by the development of gelatinous tumours (myxomas) in the skin – particularly at mucocutaneous junctions (around the nose, mouth, ears and genitals); the tumours result from accumulation of mucinous material together with cellular proliferation. Transmission occurs (mechanically) via arthropod vectors (rabbit fleas, mosquitoes etc), and by direct contact etc. In its natural host, *Sylvilagus brasiliensis* (a forest rabbit native to Brazil and Uruguay), the myxoma virus usually causes only a localized lesion. Hares and jackrabbits (*Lepus* spp) are generally resistant.

The myxoma virus has been used for the BIOLOGICAL CONTROL of rabbit populations – first in Australia in the early 1950s, later in France and the UK; initially, mortality rates in *O. cuniculus* were very high (up to 99%) but have subsequently declined owing to attenuation of the virus and natural selection of resistant rabbits.

Myxomycetes (acellular slime moulds, or 'true' slime moulds) A class of SLIME MOULDS of the MYXOMYCOTA (cf. EUMYCETOZOEA). The major vegetative stage is a multinucleate, diploid, migratory PLASMODIUM which typically exhibits shuttle streaming (rhythmic reversible flow) of its protoplasm. Multispored, usually macroscopic fruiting bodies develop from the plasmodium. Sexual reproduction occurs in most (if not all) species, and the life cycle typically exhibits alternate haploid and diploid phases. Myxomycetes occur in various habitats, e.g., soil, humus, rotting wood, treebark, dung, decaying vegetation, etc. [Ecologically significant features of myxomycetes: TBMS (1984) *83* 1–19.]

Life cycle. A haploid spore germinates to release one or more uninucleate myxamoebae or flagellated amoeboid cells (which are typically biflagellated and heterokont); the two types of cell are interconvertible – e.g. myxamoebae may become flagellated in the presence of water. The cells feed phagocytically and osmotrophically, and divide mitotically (flagellated cells first withdrawing their flagella); under unfavourable conditions they may form MICROCYSTS. The amoeboid/flagellated cells function as gametes: cells of compatible mating type fuse, and the zygote develops into the main vegetative stage of the organism, the *plasmodium*. Three main types of plasmodium are distinguished. A *protoplasmodium* is microscopic, resembling a multinucleate amoeba, and lacks reticulations or 'veins'. Protoplasmodia occur e.g. in ECHINOSTELIUM and many *Licea* spp; each gives rise to a single fruiting body. An *aphanoplasmodium* (formed e.g. by *Stemonitis* spp) is thin, delicate, and inconspicuous, with a reticulum of 'veins' leading into a broad, fan-shaped, advancing front; aphanoplasmodia lack a slime sheath, and their protoplasm is relatively non-granular. A *phaneroplasmodium* (formed e.g. in PHYSARUM) resembles an aphanoplasmodium but is much thicker and more robust, and is often conspicuous; its protoplasm is very granular, and it is surrounded by a gelatinous slime sheath – traces of which are left behind on the substratum over which the plasmodium migrates. Shuttle streaming (a rapid, rhythmic, back-and-forth flow) of the protoplasm occurs in aphanoplasmodia and phaneroplasmodia; in protoplasmodia streaming is very slow and irregular. Plasmodia are commonly yellow or white, but they may be e.g. orange, brown, red, violet or black; the larger plasmodia may measure several centimetres to a metre or more across. As it migrates over a substratum, the plasmodium feeds by ingesting bacteria, microalgae, protozoa, yeasts, particles of organic detritus etc, as well as myxamoebae and flagellated cells of its own species ('cannibalism'); as it grows, its nuclei divide synchronously. Coalescence of two or more plasmodia may occur.

Eventually, the plasmodium gives rise to fruiting bodies (or, in some species, to a SCLEROTIUM). Prior to fruiting, the plasmodium generally flows out onto an exposed surface; in some species light is apparently necessary to initiate fruiting. Three major types of fruiting body are formed, in each of which the spores develop in masses within a persistent or evanescent peridium. A *sporangium* is a relatively small, sessile or stalked structure which is more or less uniform in size and shape for a given species; many sporangia usually develop from a single plasmodium. (See e.g. ARCYRIA and STEMONITIS.) An *aethalium* is typically sessile, relatively large, hemispherical or pulvinate, derived from an entire plasmodium or from a major part of one; aethalia may vary in size and shape even within a single species. (See e.g. LYCOGALA.) (*Pseudoaethalia*, consisting of numerous compacted sporangia, occur in some species.) A *plasmodiocarp* is sessile, irregularly shaped and often branched or reticulate; it is derived from the major veins of an aphanoplasmodium or phaneroplasmodium. (See e.g. HEMITRICHIA.) Depending on species, fruiting bodies may contain e.g. a CAPILLITIUM, COLUMELLA, PSEUDOCAPILLITIUM and/or PSEUDOCOLUMELLA. In some species certain parts of the fruiting body (e.g. peridium, capillitium) bear deposits of lime. Meiosis is believed to occur in the maturing spores; three of the four resulting nuclei apparently degenerate.

Several hundred species of myxomycetes are recognized [descriptions and illustrations: Book ref. 145]. Orders: Echinosteliales (e.g. ECHINOSTELIUM); Echinosteliopsidales (*Bursulla, Echinosteliopsis*); Liceales (e.g. *Cribraria, Enteridium*, LYCOGALA, *Licea, Tubifera*); Physarales (e.g. *Badhamia, Didymium*, FULIGO, PHYSARUM); Stemonitales (e.g. *Comatricha, Lamproderma*, STEMONITIS); Trichiales (e.g. ARCYRIA, HEMITRICHIA, *Perichaena, Trichia*).

Myxomycota (Myxobionta) In some mycological classification schemes [e.g. Book ref. 64]: a division which includes the SLIME MOULDS and organisms which resemble slime moulds in certain respects. Classes: ACRASIOMYCETES, Ceratiomyxomycetes (see CERATIOMYXA), DICTYOSTELIOMYCETES, LABYRINTHULOMYCETES, MYXOMYCETES, PLASMODIOPHOROMYCETES, PROTOSTELIOMYCETES. (cf. EUMYCOTA.)

myxophage Any bacteriophage whose host is a member of the MYXOBACTERALES. Currently, only DNA myxophages are known; these are designated Mx1, Mx4 etc.

Myxophyceae Obsolete class for the CYANOBACTERIA.

Myxosarcina A genus of unicellular CYANOBACTERIA) (section II) in which growth and binary fission result in cubical aggregates of cells; baeocytes initially lack a fibrous outer cell wall layer and are motile (by gliding). GC%: 43–44.

Myxosoma A genus of protozoa (class MYXOSPOREA) which are histozoic parasites in fish (see WHIRLING DISEASE).

myxospore See MYXOBACTERALES.

Myxosporea A class of protozoa (phylum MYXOZOA) which are parasitic (coelozoic or histozoic) in poikilothermic ('cold-blooded') vertebrates, particularly fish; many members are economically important fish pathogens. The myxosporean spore contains between one and six (typically two) polar capsules (each containing a coiled filament), and one or two sporoplasms; the spore wall consists of two valves (order Bivalvulida: e.g. *Ceratomyxa, Chloromyxum*, HENNEGUYA, MYXOBOLUS, MYXOSOMA, MYXIDIUM, *Sinuolinea, Sphaeromyxa, Sphaerospora, Thelohanellus*) or of three or more valves (order Multivalvulida: e.g. *Hexacapsula, Kudoa, Trilospora*).

Generalized life cycle. When ingested by a host animal, the spore explosively discharges its coiled filament(s), the discharged filament apparently serving to attach the organism to the gut epithelium. The amoeboid sporoplasm(s) emerges and passes through the gut wall, subsequently localizing in a particular tissue (in histozoic species: e.g. species of *Henneguya, Kudoa, Myxobolus, Myxosoma*) or body fluid (in coelozoic species: e.g. *Chloromyxum* spp, *Leptotheca* spp, *Myxidium* spp, *Zschokkella floridanae*). The sporoplasm grows and its nucleus divides many times, resulting in the formation of a multinucleate plasmodium – the main vegetative stage of the organism. Eventually, many of the nuclei in the plasmodium each become segregated into cells (*sporonts*) destined to become spores; depending on species, a sporont may develop into a single spore (monosporous sporont) or into two or more spores within a common envelope (the whole being termed a *pansporoblast*). A sequence of nuclear divisions then ensues, each division apparently giving rise to a 'somatic' nucleus and a 'generative' nucleus. Cells containing somatic nuclei are involved in the formation of the spore wall, polar filament(s) and polar capsule(s), while generative nuclei are incorporated in the sporoplasm(s) (commonly two nuclei per sporoplasm). The mature spores may be released into the bloodstream or may remain in situ until the host is eaten e.g. by another fish or by a bird; in the latter case, the spores may be passed unharmed with the bird's faeces, thus possibly being transported to fresh aquatic habitats. (cf. WHIRLING DISEASE.)

myxosporidean (Pertaining to) a protozoon of the MYXOZOA (formerly of the class Myxosporidea, subphylum Cnidospora).

Myxotheca See FORAMINIFERIDA.

Myxotrichum See GYMNOASCALES.

myxoviruses A (non-taxonomic) group of viruses: the ORTHOMYXOVIRIDAE and PARAMYXOVIRIDAE.

myxoxanthophyll See CAROTENOIDS.

Myxozoa A phylum of PROTOZOA [JP (1980) *27* 37–58] which are mainly intracellular parasites of invertebrates and poikilothermic vertebrates. The organisms form spores of multicellular origin; a spore contains one or more extrusile *polar filaments*, each coiled within a separate sac (the *polar capsule*), and one or more sporoplasms (amoeboid cells which are the infective forms of the organism). The spore wall consists of one, two or three (rarely more) *valves*. Two classes are recognized: ACTINOSPOREA and MYXOSPOREA; however, since two apparently distinct organisms – one from each class – have been found to be different stages in the life cycle of the same organism (see WHIRLING DISEASE), a reappraisal of myxozoan taxonomy may be necessary [discussion: JP (1985) *32* 589–591]. (See also MICROSPORA and ASCETOSPORA.)

MZ B cells MARGINAL ZONE B CELLS (q.v.).

1. Words in SMALL CAPITALS are cross-references to separate entries.
2. Keys to journal title abbreviations and Book ref. numbers are given at the end of the Dictionary.
3. The Greek alphabet is given in Appendix VI.
4. For further information see 'Notes for the User' at the front of the Dictionary.

N

N Asparagine (see AMINO ACIDS).

N-end rule The observed relationship between the N-terminal amino acid of a protein and the half-life of that protein *in vivo*. In *Escherichia coli*, for example, proteins whose N-terminal amino acid is either arginine or lysine are characterized by a short half-life; early degradation of such proteins may be avoided in strains mutant in the AAT GENE.

n. gen. See GENUS NOVUM.

N-Serve See NITRAPYRIN.

n. sp. See SPECIES NOVA.

N-type starter See LACTIC ACID STARTERS.

NA Numerical aperture: see RESOLVING POWER.

Na^+-ATPase See ION TRANSPORT and SODIUM MOTIVE FORCE.

Na^+-motive force See SODIUM MOTIVE FORCE.

Na^+ pump See ION TRANSPORT.

nabam (dithane D-14) Disodium ethyleneBISDITHIOCARBAMATE, an agricultural antifungal agent sometimes used in conjunction with zinc sulphate and lime – when ZINEB is formed.

nactins See MACROTETRALIDES.

NAD (coenzyme I; cozymase) Nicotinamide adenine dinucleotide (formerly known as diphosphopyridine nucleotide, DPN). NAD is an important coenzyme, functioning as a hydrogen carrier in a wide range of redox reactions; the H is carried on the nicotinamide residue (see figure). The oxidized form of the coenzyme is written NAD^+, the reduced form as NADH (or $NADH + H^+$); hence:

$$XH_2 + NAD^+ \longrightarrow X + NADH + H^+$$

The E'_0 of the NAD^+/NADH couple is −320 mV.

NAD phosphate (NADP: formerly triphosphopyridine nucleotide, TPN, or coenzyme II) functions in the same way as NAD; the E'_0 of the $NADP^+$/NADPH couple is −317 mV. Many oxidoreductases are specific for either NAD or NADP, although some can function with either. (When either coenzyme can be used in a given reaction – or e.g. when different coenzymes serve the same function in different organisms – the designation NAD(P) is generally used.) As a broad generalization, NADP is more commonly associated with biosynthetic reactions, NAD with catabolic and energy-yielding reactions. Reduced NADP necessary for biosynthetic reactions may be generated e.g. by the HEXOSE MONOPHOSPHATE PATHWAY, by the isocitrate dehydrogenase reaction of the TCA CYCLE, and by PHOTOSYNTHESIS in photosynthetic eukaryotes. (See also TRANSHYDROGENASE.) [NADPH production and consumption in yeasts: JGM (1983) *129* 953–964.]

In addition to its role as a hydrogen carrier, NAD^+ also functions as an ADP-ribosyl donor in ADP-RIBOSYLATION reactions.

Biosynthesis of NAD and NADP. NAD may be synthesized de novo from aspartate and dihydroxyacetone phosphate or from tryptophan (see NICOTINIC ACID; cf. V FACTOR). Nicotinamide released during the degradation of NAD^+ (by NAD^+ nucleosidase) may be re-cycled (i.e., incorporated into new NAD molecules) via the 'pyridine nucleotide salvage cycle'; such a cycle may also be operative in certain organisms which cannot synthesize nicotinic acid/nicotinamide de novo (e.g. *Bordetella pertussis* [AvL (1984) *50* 33–37]). NADP is synthesized by the ATP-dependent phosphorylation of NAD by NAD^+ kinase.

NAD^+ kinase See NAD.

NADH–UQ oxidoreductase Complex I of the mitochondrial ELECTRON TRANSPORT CHAIN.

NADP See NAD.

Nadsonia A genus of yeasts (family SACCHAROMYCETACEAE) in which the cells are usually lemon-shaped; vegetative reproduction occurs by bipolar bud-fission. Pseudomycelium is not formed. Ascus formation follows conjugation between a bud and its mother cell; ascospores contain a prominent lipid globule and are spherical, brownish, warty or spiny, 1 or 2 per ascus. Glucose is fermented by *N. elongata* and *N. fulvescens*, but not by *N. commutata*. Nitrate is not assimilated. Maximum growth temperature: ca. 24–26°C. Species have been isolated from soil and from the slime flux of trees. [Book ref. 100, pp. 279–284.]

Naegleria A genus of amoeboflagellates (order SCHIZOPYRENIDA). Some species (e.g. *N. gruberi*) are entirely free-living. *N. fowleri* (= *N. aerobia*, *N. invades*) is an opportunist pathogen (see MENINGOENCEPHALITIS); it is ca. 10 μm in diameter when rounded, and its single nucleus contains a large central nucleolus. (See also AMOEBOSTOME.) In *N. fowleri* interconversion can occur between the amoebic form and a transient, non-dividing biflagellate form [JB (1981) *147* 217–226]. Amoebae can encyst. [ARM (1982) *36* 101–123.]

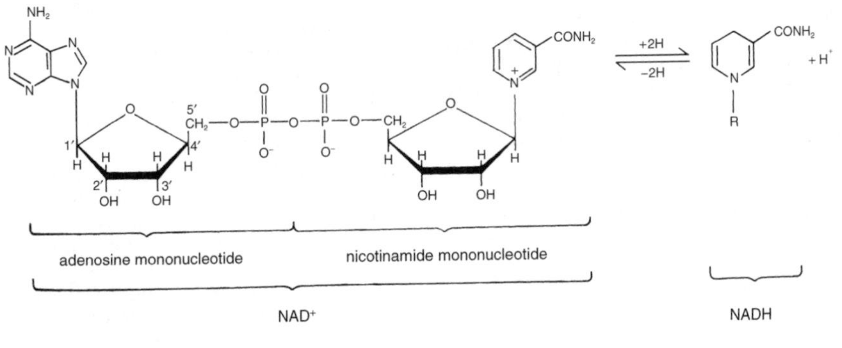

NAD (nicotinamide adenine dinucleotide). NADP = NAD 2′-phosphate.

nafcillin See PENICILLINS.

naftifine See ALLYLAMINES.

NAG vibrios Non-agglutinable vibrios – see NON-CHOLERA VIBRIOS.

nagana Any of a range of trypanosomiases which affect domestic animals (e.g. cattle) in various parts of Africa and in which the vector is a biting fly, often a species of *Glossina*; the pathogens include *Trypanosoma brucei*, *T. congolense*, *T. suis* and *T. vivax*. Symptoms typically include anaemia, cachexia, and damage to e.g. heart, kidneys and liver; affected cattle typically exhibit a generalized immunosuppression. Abortion and infertility are common in certain areas.

Nagler's reaction A test used to detect strains of *Clostridium perfringens* which produce a diffusible LECITHINASE (the α-toxin). (An antigenically similar phospholipase C, which also cleaves lecithin, is produced by *C. bifermentans*.) When grown on EGG-YOLK AGAR such strains give rise to colonies each of which is surrounded by a zone of opacity. In the Nagler reaction, one half of a plate of egg-yolk agar is spread with anti-α-toxin, and the plate is inoculated with the unknown strain – the inoculum forming a line which extends into both halves of the plate; following incubation, strains which produce α-toxin produce zones of opacity only in that half of the plate which does not contain antitoxin – although growth occurs in both halves of the plate.

Nairobi sheep disease A disease of sheep and goats caused by a virus of the genus NAIROVIRUS. It occurs mainly in East Africa. Symptoms include haemorrhagic gastroenteritis, splenomegaly, nephritis, and myocardial degeneration; mortality rates are high. The virus is transmitted by the tick *Rhipicephalus appendiculatus*.

Nairovirus A genus of viruses of the BUNYAVIRIDAE. Host range: various vertebrates; vectors: primarily ticks. MWts of L, M and S RNAs: ca. 4.1–4.9, 1.5–1.9, and 0.6–0.7 × 10^6, respectively. MWts of proteins L, G1, G2 and N: ca. 145–200, 72–84, 30–40 and 48–54 × 10^3, respectively. The genus comprises ca. 6 serogroups, including e.g. the Crimean–Congo haemorrhagic fever group, NAIROBI SHEEP DISEASE group, and the Qalyub group.

naive (*immunol.*) Unprimed (cf. PRIMED).

naked (1) (of viruses) Not enveloped. (2) (of nucleic acids, VIROIDS etc) Not associated with protein.

***nalA* gene** See GYRASE.

nalidixic acid A QUINOLONE ANTIBIOTIC. Nalidixic acid inhibits bacterial DNA replication (and, to a lesser extent, RNA and protein synthesis) and causes degradation of DNA in the cell. It acts as a specific inhibitor of GYRASE, interfering with the function of the A subunits (cf. NOVOBIOCIN) and preventing the breakage-and-reunion activity – and hence the supercoiling and relaxing activities – of the enzyme. The drug apparently causes the formation of a very stable gyrase–DNA complex. In vitro, treatment of the complex with e.g. SDS triggers the cleavage of DNA strands with consequent covalent binding of the 5′ ends to the A subunits; in vivo, the complex may block the movement of replication forks [ARB (1981) *50* 901–902]. A related quinolone, norfloxacin, has been shown to inhibit gyrase by binding not to the enzyme itself but to regions of ssDNA, suggesting that gyrase is actually inhibited by a DNA–quinolone complex [PNAS (1985) *82* 307–311].

[Gyrase-targeted antibiotics: TIM (1998) *6* 269–275.]

name bearer *Syn.* NOMENCLATURAL TYPE.

naming of microorganisms See NOMENCLATURE.

NANA *N*-AcetylNEURAMINIC ACID.

NANB hepatitis NON-A–NON-B HEPATITIS.

Nannizzia A genus of heterothallic fungi (order GYMNOASCALES) which occur e.g. in soil; anamorphs occur in the genus *Microsporum* (see DERMATOPHYTES).

Nannoceratopsis A fossil DINOFLAGELLATE (lower Jurassic).

Nannochloris A genus of green algae. *N. bacillaris* can tolerate a wide range of salinities [stepwise adaptation to salinity: CJB (1985) *63* 327–332].

Nannocystis See MYXOBACTERALES.

Nannomonas A subgenus of TRYPANOSOMA within the SALIVARIA; species occur, in parts of Africa, as parasites and pathogens in a wide range of wild and domestic animals including cattle, equines and suids (see also NAGANA). In the vertebrate host, bloodstream forms are trypomastigotes, <20 μm in length, in which the kinetoplast is subterminal (and typically close to the pellicle) and in which there is no free flagellum. In the vector (*Glossina* spp), trypomastigotes and epimastigotes – which also lack a free flagellum – may reach ca. 40 μm in length. Species include *T.* (*N.*) *congolense* and *T.* (*N.*) *simiae* (an important pathogen of the pig).

Nannophrya See GYMNOSTOMATIA.

nannoplankton See PLANKTON.

nanobacteria Minute bacteria, detected e.g. in humans and bovines, which have been classified in the α-2 subgroup of Proteobacteria; the cells are generally about 0.2–0.5 μm, but smaller cells may be seen under the electron microscope (accounting for the ability of these organisms to pass through membrane filters of pore size 0.1 μm). The cells can form a carbonate apatite deposit on their envelope at near-neutral pH, and it has been suggested that they may contribute e.g. to the development of kidney stones [PNAS (1998) *95* 8274–8279] (see also UROLITHIASIS).

Nanochlorum A genus of very small, unicellular, possibly primitive, eukaryotic green algae (division CHLOROPHYTA).

nanometre (nm) 10^9 m, or 10^{-3} μm.

nanoplankton See PLANKTON.

NANP See PLASMODIUM.

naphthazarins (naphthazarines) A class of red pigments (5,8-dihydroxynaphthoquinones substituted at position C-2 and sometimes at C-3) produced by FUSARIUM spp; they readily chelate multivalent metal ions, and are toxic to plants and insects. Naphthazarins include e.g. fusarubin, isomarticin, javanicin and marticin.

naphthoquinones See QUINONES.

naramycin B See CYCLOHEXIMIDE.

narasin An ionophore antibiotic, produced by *Streptomyces aureofaciens*, which is used as an anti-coccidial agent in poultry and as a FEED ADDITIVE for ruminants.

narbomycin See MACROLIDE ANTIBIOTICS.

***Nar*I** See RESTRICTION ENDONUCLEASE (table).

narrow groove *Syn.* minor groove (see DNA).

NARTC Nalidixic acid-resistant thermophilic campylobacters: CAMPYLOBACTER strains which are related to but distinct from *C. jejuni* and *C. coli*.

NASBA (nucleic acid sequence-based amplification) A method used for the isothermal amplification of RNA target sequences; NASBA, which is based on the replication strategy of retroviruses, involves several enzymes (including REVERSE TRANSCRIPTASE) and is typically run at ~41°C for e.g. ~90 min. The products include ssRNA amplicons (which are complementary to the target RNA) and double-stranded cDNA.

An outline of NASBA is given in the figure.

(*Transcription-mediated amplification* is a very similar process – see TMA.)

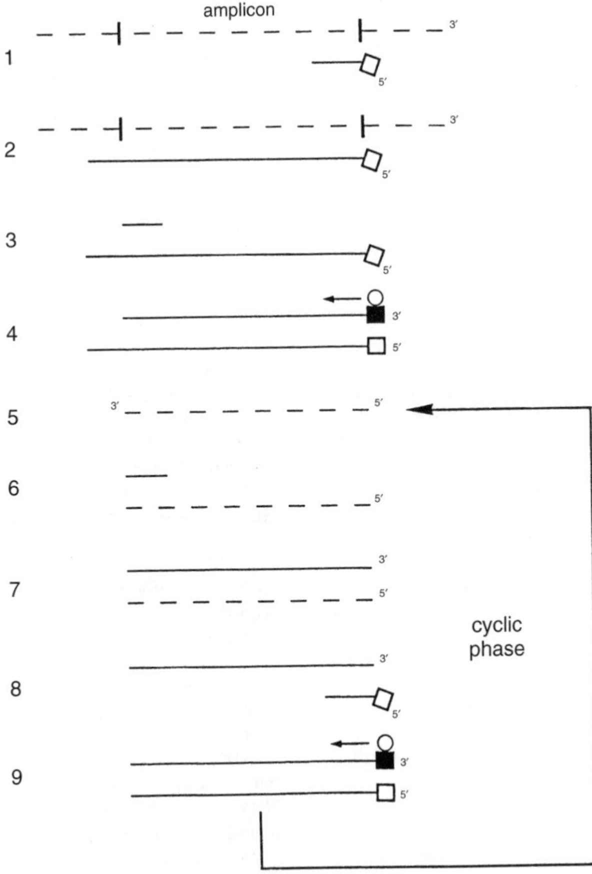

NASBA (nucleic acid sequence-based amplification): isothermal amplification of an RNA target; amplification is carried out at ~41°C.

The dashed lines are strands of RNA, the solid lines strands of DNA. For each strand the polarity is indicated by a 3′ and/or 5′ label.

With clinical specimens, amplification of targets is preceded by treatment of the sample to ensure that target sequences, if present, are made accessible. Such pre-treatment necessarily includes release of nucleic acid from organisms and heating to denature target RNA; although RNA is generally single-stranded, intra-strand base-pairing can occur in mRNA and occurs extensively in rRNA. Pre-treatment may also involve inactivation of inhibitory substances. The following stages are involved in amplification:

1. A strand of RNA showing the target sequence (amplicon) delimited by two, short vertical bars. A primer (primer 1) has bound to the 3′ end of the amplicon. The 5′ end of this primer is tagged with a short sequence (□) containing the *promoter* of an RNA polymerase.
2. The enzyme reverse transcriptase has extended the primer to form a strand of cDNA, thus giving rise to a hybrid RNA:DNA duplex intermediate.
3. The enzyme RNase H has degraded (removed) the RNA strand. Another primer, primer 2, has bound to the 3′ end of the cDNA.
4. Reverse transcriptase (which can also synthesize DNA on a DNA template) has extended primer 2 to form double-stranded cDNA; notice that the promoter sequence in primer 1 has also become double stranded as primer 2 has been fully extended on the template to form the complementary strand (■) of the promoter. Now double-stranded, the promoter is functional; an RNA polymerase (○) has bound to the promoter and will synthesize an RNA strand in the direction of the arrow, i.e. in a 5′-to-3′ direction.
5. The newly synthesized strand of RNA. Note the strand's polarity: it is *complementary* to the amplicon in the sample RNA (compare 5 with 1).
6. The RNA strand has bound primer 2.
7. Reverse transcriptase has synthesized cDNA on the RNA strand.
8. RNA has been removed by RNase H. Primer 1 has bound to the cDNA.
9. Reverse transcriptase has synthesized a complementary strand of cDNA; note that the 3′ end of the template cDNA has also been extended to form a functional (double-stranded) promoter for the RNA polymerase. RNA polymerase has bound to the promoter and will (repeatedly) synthesize RNA strands identical to the one shown at stage 5. These strands can participate in the cyclic phase, leading to high-level amplification of the target. (*Continued on page 512.*)

As in other nucleic-acid-amplification methods, products can be detected by gel electrophoresis and staining. Alternatively, the amplicons can be monitored by means of MOLECULAR BEACON PROBES [see e.g. NAR (1998) *26* 2150–2155].

In clinical microbiology, NASBA is useful e.g. for monitoring the development of transcriptional activity in latent DNA viruses. For example, by monitoring transcripts of the late genes of cytomegalovirus (CMV) it is possible to identify patients at risk for CMV-related disease [JCM (1998) *36* 1341–1346]; a commercial assay kit for this purpose has been marketed.

[NASBA in diagnostic microbiology, research and other non-commercial applications: RMM (1999) *10* 185–196.]

[Isothermal nucleic-acid-amplification methods (NASBA, TMA, SDA): Book ref. 221, pp 126–151.]

nasopharyngeal carcinoma (NPC) A carcinoma arising from the epithelial lining of the post-nasal space. Undifferentiated NPC is associated with EPSTEIN–BARR VIRUS (EBV) infection; the epithelial tumour cells contain multiple copies of EBV DNA, but chromosomal abnormalities have not been detected (cf. BURKITT'S LYMPHOMA). NPC is generally rare, but occurs with high incidence among certain ethnic groups, particularly Cantonese Chinese and Eskimos. Since EBV is ubiquitous, genetic and/or environmental factors are believed to play a role in pathogenesis; nitrosamines (present in salted and dried fish eaten by Cantonese people) and certain components (e.g. phorbol esters) of Chinese herbal medicines have been implicated as potentiating factors.

nasopharynx microflora See RESPIRATORY TRACT MICROFLORA.

nasse (*protozool.*) *Syn.* CYRTOS.

Nassellarida See RADIOLARIA.

Nassula A genus of freshwater protozoa (subclass HYPOSTOMATIA). Cells: ovoid to elongate, ca. 150–200 μm in length, completely covered with cilia. Food: e.g. cyanobacteria.

Nassulida See HYPOSTOMATIA.

natamycin *Syn.* PIMARICIN.

Natronobacterium A genus of halophilic, alkalophilic archaeans [FEMS Reviews (1986) *39* 9–15].

Natronococcus A genus of halophilic and alkalophilic archaeans [FEMS Reviews (1986) *39* 9–15].

natto A Japanese food made by fermenting soybeans. Soybeans are soaked, steam-cooked, wrapped in rice straw, and incubated at e.g. ca. 40°C for 1–2 days. '*Bacillus natto*' (a strain of *B. subtilis*), supplied by the straw or added as a pure-culture starter, develops and coats the beans with a viscous, sticky polymer; the product develops a strong persistent flavour. [Book ref. 8, pp. 520–521.]

natural antibodies Immunoglobulins, present in plasma, which act as antibodies towards antigens with which the body has apparently not been in immunological contact; they may derive e.g. from previous or ongoing subclinical infections or from immunological contact with CROSS-REACTING ANTIGENS.

natural antigen *Syn.* ANTIGEN (= IMMUNOGEN sense 1).

natural immunological tolerance *Syn.* SELF TOLERANCE.

natural killer cells See NK CELLS.

Naumanniella A genus of bacteria found e.g. in soil and well-waters [see e.g. Book ref. 45, pp. 1049–1059].

Navicula A genus of pennate DIATOMS in which the cells are characteristically spindle- or boat-shaped and show gliding motility. Species occur on mud or other substrates in freshwater, marine and terrestrial habitats.

NBT Nitroblue TETRAZOLIUM.

NCAM See IMMUNOGLOBULIN SUPERFAMILY.

NCCLS National Committee for Clinical Laboratory Standards (USA).

Ncd cells See PROTOPLAST FUSION.

NCDO National Collection of Dairy Organisms, Food Research Institute (Reading), Shinfield, Reading, Berkshire RG2 9AT, UK.

NCIB National Collection of Industrial Bacteria, Torry Research Station, P.O. Box 31, 135 Abbey Road, Aberdeen AB9 8DG, Scotland, UK.

NCMB National Collection of Marine Bacteria, Torry Research Station, P.O. Box 31, 135 Abbey Road, Aberdeen AB9 8DG, UK.

***Nco*I** See RESTRICTION ENDONUCLEASE (table).

NCPPB National Collection of Plant Pathogenic Bacteria, Plant Pathology Laboratory, Hatching Green, Harpenden, UK.

NCS NEOCARZINOSTATIN (q.v.).

NCTC National Collection of Type Cultures, Central Public Health Laboratory, Colindale Avenue, London NW9 5HT, UK.

NCV NON-CHOLERA VIBRIO.

NCVP Non-O1 *Vibrio cholerae* pilus (non-cholera-vibrio pilus): a filamentous cell-surface appendage found in non-O1 strains of *Vibrio cholerae*, including many strains of the O139 Bengal serogroup. (See also CHOLERA.)

NDC Nuclear dehydrogenating clostridia: see GASTROINTESTINAL TRACT FLORA.

2′-NDG 2′-Nor-2′-deoxyguanosine: see DHPG.

Ndumu virus See ALPHAVIRUS.

NDV NEWCASTLE DISEASE virus.

neamine (neomycin A) An AMINOGLYCOSIDE ANTIBIOTIC whose molecule contains an aminosugar linked to DEOXYSTREPTAMINE.

Nebraska calf diarrhoea See ROTAVIRUS.

necridium (sacrificial cell) In certain filamentous bacteria (e.g. certain cyanobacteria, *Beggiatoa*): a specialized intercalary cell which, by death and lysis, allows fragmentation of the trichome and liberation of hormogonia.

necrobacillosis Any of various diseases of man or animals caused by *Fusobacterium necrophorum*.

In humans, *F. necrophorum* can cause e.g. an acute systemic disease characterized by sore throat, submandibular lymphadenopathy, fever with rigors, and subsequently formation of lung abscesses with consequent pleuritic pain and haemoptysis.

In animals, *F. necrophorum* may invade the skin or mucous membranes, usually via damaged tissues, causing necrosis and sometimes abscess formation. In cattle and lambs the pathogen may infect via the navel or rumen and then localize e.g. in the liver. Infections localized to the mouth and larynx ('calf diphtheria') or to the mouth only ('necrotic stomatitis') may also occur. (See also FOOT-ROT.)

NASBA (*continued*)

The double-stranded cDNA templates in stage 9 are permanent products and are continually transcribed by the RNA polymerase. Operation of the cyclic phase produces many copies of the amplicon in the form of (i) complementary RNA (the major product), and (ii) cDNA.

In NASBA, amplification involves three enzymes: reverse transcriptase, RNA polymerase and RNase H; in a similar process, TMA, the role of RNase H is carried out by the reverse transcriptase, so that only two enzymes are used.

Reproduced from *Bacteria*, 5th edition, Figure 8.25, pages 226–227, Paul Singleton (1999) copyright John Wiley & Sons Ltd, UK (ISBN 0471-98880-4) with permission from the publisher.

necrosis Localized death and degeneration of tissues in a living organism – due e.g. to infection or injury. (See also CASEATION.)

necrotic stomatitis (*vet.*) See NECROBACILLOSIS.

necrotizing enteritis *Syn.* ENTERITIS NECROTICANS.

necrotizing enterocolitis (NEC) A severe disease of newborn infants; it is characterized by vomiting, abdominal distension, bloody stools, pneumatosis intestinalis, ischaemic necrosis of the intestinal wall and sloughing of the mucosa. The aetiology of NEC is unknown; microorganisms – particularly *Clostridium perfringens* – are believed to play a role in at least some cases. [Book ref. 107, pp. 15–26.]

necrotizing fasciitis A condition characterized by soft-tissue oedema and necrosis, with histology-based evidence of necrosis of the superficial fascia, fascial oedema, and an infiltrate of PMNs. A range of bacteria have been isolated from such tissues – including *Staphylococcus aureus*, enterobacteria, and group A streptococci; streptococci have been causally connected with the disease in about 10% of cases.

Identification of the causal agent in a given case facilitates the choice of appropriate chemotherapy; moreover, necrotizing fasciitis may be associated with a risk of familial transmission – a possible indication for chemoprophylaxis of close contacts.

Tissue samples/blood may be culture-negative owing to prior chemotherapy. A small PCR-based study, designed to detect streptococci, found that all 10 cases of confirmed streptococcal aetiology could be identified by testing for the *speB* gene (which encodes exotoxin B in almost all group A streptococci) [JCM (1998) *36* 1769–1771].

necrotizing toxin A TOXIN which can cause NECROSIS.

necrotroph (1) An organism which kills part or all of another organism before deriving nutrients from it; the term is usually applied to an association between a plant and a microorganism.

(2) An organism which derives nutrients from dead plant or animal tissues, whether or not it is responsible for the death of those tissues.

(cf. BIOTROPH; PERTHOTROPH.)

nectin Protein which forms the 'stalk' region of mitochondrial PROTON ATPASE.

Nectria A genus of fungi (order HYPOCREALES) which include plant-pathogenic species (see e.g. APPLE CANKER and BEECH BARK DISEASE; see also PISATIN); anamorphs occur in the genera ACREMONIUM, CYLINDROCARPON and FUSARIUM.

N. cinnabarina occurs mainly on the dead branches of woody plants; however, it can cause dieback (particularly in e.g. *Acer* spp), gaining access e.g. via wounds or via dead tissues. Unicellular conidia are formed on simple or sparsely branched conidiophores which develop on orange-pink sporodochia (ca. 1–5 mm diam.) – hence the popular name 'coral spot'. Dark red perithecia (the overwintering stage) subsequently develop on the same sporodochium; each ascus contains eight two-celled ascospores.

N. galligena differs from *N. cinnabarina* e.g. in that the conidia are multiseptate and the perithecia are formed singly, without the development of a stroma.

needle (*bacteriol.*) *Syn.* STRAIGHT WIRE.

needle complex Part of a type III PROTEIN SECRETION system: a multi-protein structure consisting of a base, which spans the cell envelope, and a needle-like extension from the surface of the bacterium; in overall structure, this complex resembles the hook–basal body portion of a bacterial FLAGELLUM. Studies on the type III protein secretion system of *Salmonella typhimurium* have found that the *prgI* gene product is the main component of the needle, and that the length of the needle is regulated by the *invJ* gene product [PNAS (2000) *97* 10225–10230].

[Structure and composition of the *Shigella flexneri* needle complex: Mol. Microbiol. (2001) *39* 652–663.]

Studies on *Yersinia enterocolitica* suggest that the needle is formed by polymerization of a 6 kDa protein, and that polymerization of this protein provides the force necessary for puncturing the cytoplasmic membrane of a eukaryotic cell [PNAS (2001) *98* 4669–4674].

Neethling virus See LUMPY SKIN DISEASE.

negative complementation See COMPLEMENTATION TEST.

negative contrast See ELECTRON MICROSCOPY (a).

negative control (of an operon) See OPERON.

negative interference (*genetics*) See INTERFERENCE (2).

negative Pasteur effect *Syn.* CUSTERS EFFECT.

negative phase (*immunol.*) The phase in which the in vivo titre of a given antibody is reduced following administration of the homologous antigen – due to combination of the antigen with circulating antibody.

negative-sense genome See VIRUS.

negative staining See STAINING and ELECTRON MICROSCOPY (a).

negative-strand virus A VIRUS (q.v.) with a negative-sense genome.

Negishi virus See FLAVIVIRIDAE.

Negri bodies Acidophilic, intracytoplasmic INCLUSION BODIES which develop in cells of the CNS in most cases of human or animal RABIES. They may occur only in the late stage of the disease, and are not formed in FIXED VIRUS infections.

negro coffee mosaic virus See POTEXVIRUSES.

neighbour exclusion principle See INTERCALATING AGENT.

Neisser–Wechsberg leucocidin See *staphylococcal* α-HAEMOLYSIN.

Neisseria A genus of aerobic, oxidase-positive, Gram type-negative bacteria (family NEISSERIACEAE) which occur as parasites or as primary or opportunist pathogens on the mucous membranes of mammals. Cells: typically cocci, 0.6–1.0 μm in diameter (which often occur singly, or in pairs with adjacent sides flattened or concave), but in one species (*N. elongata*) the cells are rod-shaped, ca. 0.5 μm in width, and often occur as diplobacilli; the cells tend to resist decolorization in the Gram stain. Cells may exhibit TWITCHING MOTILITY, and in some species the FIMBRIAE are associated with pathogenicity. (See also ANTIGENIC VARIATION and IGA1 PROTEASES.) *N. gonorrhoeae* and *N. meningitidis* are nutritionally fastidious (isolation and general culture media including e.g. THAYER–MARTIN AGAR, MUELLER–HINTON MEDIUM and CHOCOLATE AGAR); the other species of *Neisseria* are able to grow on unenriched nutrient agar. The growth of all species is improved by a high relative humidity (ca. 50%). Optimum growth temperature: ca. 35–37°C. All strains form CARBONIC ANHYDRASE (cf. MORAXELLA). Most strains form catalase – some strains of *N. elongata* are catalase-negative. (See also SUPEROXOL TEST.) *Neisseria* spp are highly susceptible to desiccation and to exposure to direct sunlight. GC%: ca. 46.5–53.5. Type species: *N. gonorrhoeae*.

N. canis. Some strains are haemolytic on rabbit blood; no acid from glucose or other sugars; nitrate is usually reduced; a yellow pigment is formed. Strains have been isolated e.g. from the throats of cats.

N. caviae. See FALSE NEISSERIAE.

N. cinerea. Non-haemolytic; no acid from glucose or other sugars.

N. cuniculi. See FALSE NEISSERIAE.

N. denitrificans. Non-haemolytic; acid is usually formed from glucose, fructose, mannose and sucrose, but not from maltose.

N. elongata. Cells rod-shaped; non-haemolytic; can grow in unenriched peptone media; some strains (subsp. *glycolytica*) cause weak acidification of glucose media and are catalase-positive, while others (subsp. *elongata*) do not acidify glucose media and are catalase-negative.

N. flavescens. Non-haemolytic; no acid is produced from glucose or other sugars; yellow pigment is formed on Loeffler's serum medium.

N. gonorrhoeae (the gonococcus). The causal agent of GONORRHOEA (and a common cause of ophthalmia neonatorum) in man, but apparently not pathogenic in other animals. Culture is carried out under 3–10% CO_2. Colonies (48-hour): typically round, smooth, convex, greyish-white, opaque, 0.6 mm to >1 mm in diameter, usually becoming mucoid on further incubation. Non-haemolytic. Acid (no gas) is usually formed from glucose but not from other sugars. (See also FIMBRIAE; PEPTIDOGLYCAN; TRANSFORMATION.)

Certain Opa^+ strains of *N. gonorrhoeae* (see OPA PROTEINS) are capable of TRANSCYTOSIS [see e.g. TIM (1998) *6* 489–495].

N. gonorrhoeae can be detected in clinical specimens by culture and e.g. by a PROBE-based test [PACE 2C test: JCM (1995) *33* 2587–2591] and by the ligase chain reaction [Abbott LCx assay (target: *opa* gene): JCM (1998) *36* 1630–1633].

[*N. gonorrhoeae* (various molecular TYPING methods): RMM (1998) *9* 1–8.]

N. lactamica. Some strains are haemolytic (horse blood); acid is usually formed from glucose and maltose, and (unlike other neisseriae) from lactose.

N. meningitidis (the meningococcus). A causal agent of MENINGITIS. Culture is carried out under 5–10% CO_2 (which enhances growth). Colonies (24-hour) on Mueller–Hinton agar are typically round, smooth, translucent and butyrous, 1 mm or more in diameter, becoming viscid on further incubation. Non-haemolytic. Acid (no gas) is usually formed from glucose and maltose but not from fructose, lactose, mannose or sucrose. Nitrate is not reduced.

Serogroups A, B, C, D, H–L, X, Y, Z, Z′ (= 29E) and W-135 have been distinguished on the basis of polysaccharide capsular antigens. Strains of serogroups A and C (sometimes B) are commonly associated with epidemics. Strains of serogroups H–L are apparently not pathogenic. Strains of serogroups Z and Z′ are reported to be killed by normal human serum, and to give rise to disease only in those with underlying health problems.

PCR-based typing of *N. meningitidis* (culture-independent typing, i.e. typing direct from the clinical specimen) may be achieved with specimens that have become culture-negative owing to pre-admission chemotherapy [JCM (1997) *35* 1809–1812].

For multilocus sequence typing (see MLST) of *N. meningitidis*, a database of allelic profiles may be found at the website:

http://mlst.zoo.ox.ac.uk

[MLST: TIM (1999) *12* 482–487.]

N. mucosa. Non-haemolytic; acid is usually formed from glucose, maltose, fructose and sucrose; colonies are typically mucoid; unlike most other neisseriae *N. mucosa* reduces nitrate.

N. ovis. See FALSE NEISSERIAE.

N. sicca. Haemolysis may be exhibited on horse, human, rabbit and/or sheep blood agar; acid is usually formed from glucose, fructose, maltose and sucrose; colonies are dry and wrinkled, and the growth agglutinates spontaneously in saline.

N. subflava. Non-haemolytic; acid is usually formed from glucose and maltose, and sometimes from fructose and sucrose. Some strains are saline-agglutinable.

Neisseriaceae A family of aerobic, chemoorganotrophic, non-motile, asporogenous Gram type-negative bacteria which are typically parasitic and/or pathogenic in warm-blooded hosts. Most or all strains are oxidase +ve.

The family Neisseriaceae formerly included the genera ACINETOBACTER, KINGELLA, MORAXELLA and NEISSERIA [Book ref. 22, pp 288–309]. The genera *Acinetobacter* and *Moraxella* have been transferred to the family MORAXELLACEAE.

nekton Collectively, the actively motile organisms in a body of water (e.g. lake, sea). (cf. PLANKTON; see also NEUSTON.)

nelfinavir See ANTIRETROVIRAL AGENTS.

Nelsonian illumination *Syn.* CRITICAL ILLUMINATION.

nematode-trapping fungi See NEMATOPHAGOUS FUNGI.

nematodesma One of a number of bundles of microtubules which reinforce the walls of the ciliate CYTOPHARYNX. (cf. TRICHITE.)

nematophagous fungi A heterogeneous group of fungi which derive nutrients from nematode worms. Many are predatory, trapping and killing their prey before invading its tissues, while others infect and apparently parasitize nematodes; most can also grow saprotrophically. Nematophagous fungi occur e.g. in soil and decaying organic matter.

(a) *Nematode-trapping fungi*. Most fungi in this category produce various types of specialized 'trap', generally in response to the presence of nematodes or of compounds such as certain peptides or amino acids (collectively termed 'nemin'). These traps may be simple adhesive knobs (e.g. in *Acaulopage pectospora* [CJB (1985) *63* 1386–1390], *Dactylaria candida* [CJB (1977) *55* 2956–2970] and *Monacrosporium drechsleri*), relatively undifferentiated adhesive lateral hyphal branches (e.g. in *Triposporina aphanopaga*) or adhesive hyphal branches which branch and anastomose to form two- or three-dimensional networks (e.g. in *Arthrobotrys oligospora*, *Dactylella cionopaga* [CJB (1984) *62* 674–679], and some *Monacrosporium* spp). [*A. oligospora*: AvL (1985) *51* 385–398 (EM study), 399–407 (microbodies in trap cells); FEMS Ecol. (1985) *31* 17–21 (trap formation in liquid culture).] Non-adhesive hyphal rings are formed by some species, and these rings may be constricting (e.g. in *Arthrobotrys dactyloides*, *Dactylaria brochopaga* [CJB (1977) *55* 2945–2955], *Dactylella* spp) or non-constricting (e.g. *Dactylaria candida* – which also forms adhesive knobs). When a nematode enters a ring of the constricting type, the ring constricts abruptly, trapping and killing the animal; the mechanism of constriction is unknown, but apparently involves the rapid swelling of cells of the ring in response to contact between the *inner* surface of the ring and the nematode. In the case of a non-constricting ring, a nematode of appropriate size may be trapped when it begins to move through the ring but is too wide to pass right through; in *D. candida* the rings are apparently not effective traps [CJB (1977) *55* 2956–2970].

Some fungi (e.g. *Stylopage*) can trap nematodes without the formation of specialized structures; when a nematode contacts a hypha, the hypha secretes at the point of contact an adhesive which traps the animal.

(b) *Fungi endoparasitic in nematodes*. Some species (e.g. *Meria coniospora*) produce adhesive conidia which adhere to a nematode and germinate, the hyphae then invading the animal; after the death of the host, conidiophores emerge from the body. *Harposporium* spp produce non-adhesive conidia, and a nematode becomes infected by ingestion of these conidia. Certain chytridiomycetes (e.g. *Catenaria*), oomycetes (e.g. *Nematophthora gynophila*) and zygomycetes (e.g. *Gonimochaete pyriforme* [EM study: CJB (1985) *63* 2326–2331]) can infect and parasitize nematodes; *N. gynophila* can apparently play a role in the control of the cereal cyst nematode, *Heterodera avenae* [Nematol. (1980) *26* 57–68].

Nematophthora See NEMATOPHAGOUS FUNGI.

Nematospora A genus of fungi (family METSCHNIKOWIACEAE) in which multilateral budding occurs; a well-developed pseudomycelium is usually formed, and true mycelium may be formed. Ascospores: fusiform with a whip-like appendage at one end, usually 8 per ascus. Species occur mainly as parasites and pathogens of plants (see e.g. STIGMATOMYCOSIS). [Book ref. 100, pp. 285–288.]

Nematosporaceae *Syn.* METSCHNIKOWIACEAE.

Nematostelium See PROTOSTELIOMYCETES.

nemin See NEMATOPHAGOUS FUNGI.

neoantigen An antigen formed in vivo by the combination of a protein with an exogenous substance. (See also CONTACT SENSITIVITY.)

Neocallimastix A genus of anaerobic fungi found in the RUMEN; species include *N. patriciarum* [TBMS (1986) *86* 178–181, 103–109] and *N. frontalis*. *N. frontalis* forms multiflagellate zoospores [ultrastructure: CJB (1983) *61* 295–307] and produces an extracellular cellulase which is more effective than any other known cellulases in solubilizing the extensively hydrogen-bonded cellulose in cotton fibres [FEMS (1986) *34* 37–40]. It has been proposed that *Neocallimastix* be placed in an amended class CHYTRIDIOMYCETES. [Mitosis in and phylogeny of the genus *Neocallimastix*: CJB (1985) *63* 1595–1604.]

neocarzinostatin (NCS) An antitumour agent, produced by *Streptomyces carzinostaticus*, which consists of an acidic 109-amino-acid protein (containing two disulphide bonds) linked non-covalently to a complex, highly labile, non-protein chromophore which is responsible for the activity of the drug. In vitro, NCS interacts with DNA (its chromophore behaving as an INTERCALATING AGENT), causing single-stranded breaks in which the 3′ and 5′ ends generally bear 3′-phosphate and 5′-aldehyde groups, respectively; strand breakage occurs mainly at thymidine and deoxyadenosine residues and requires the presence of a thiol and O_2 as cofactors. NCS-induced strand breakage in intact cells appears to occur by a reaction similar to that in vitro [Biochem. (1987) *26* 384–390]. NCS can also cause e.g. release of free bases (particularly thymine and adenine) from the DNA.

***Neodiprion sertifer* NPV** See BACULOVIRIDAE.

neogregarines See GREGARINASINA.

neomycin Either of two related AMINOGLYCOSIDE ANTIBIOTICS (neomycins B, C) produced by *Streptomyces fradiae*; the drug used clinically consists mainly of neomycin B with some neomycin C. (cf. NEAMINE.)

neonatal calf diarrhoea coronavirus See CORONAVIRIDAE.

neonatal herpes See HERPES SIMPLEX.

neoplasia Progressive, uncontrolled cell division which, if the progeny cells remain localized (at least initially), results in the formation of an abnormal growth (tumour, neoplasm). (cf. HYPERPLASIA.) In man or animals, a neoplasm may be BENIGN or MALIGNANT (sense 2). (See also CANCER; cf. LEUKAEMIA.)

neoplasm See NEOPLASIA.

Neorickettsia A genus of Gram-negative bacteria of the tribe EHRLICHIEAE. Cells: coccoid, often pleomorphic, non-motile, maximum dimension ca. 0.5 μm. Growth occurs intracytoplasmically in lymphoid cells of canine animals. Growth does not occur in cell-free media. Type species: *N. helminthoeca*.

N. helminthoeca, the sole species, is the causal agent of 'salmon poisoning' in dogs: a disease (usually fatal in untreated cases) which is transmitted by trematodes (flukes) and in which the pathogen occurs singly or in colonies in the cytoplasm of reticuloendothelial cells in lymphoid tissue; the pathogen has not been detected microscopically in circulating lymphocytes, but the blood of a diseased animal is infectious. The fluke-borne disease of dogs known as Elokomin fluke fever appears to be a different disease since immunity to this disease does not confer immunity to salmon poisoning. [Book ref. 22, pp. 710–711.]

neosaxitoxin See SAXITOXIN.

Neo-Sensitabs® Antibiotic-containing tablets which are used (in place of antibiotic-impregnated paper discs) in antibiotic-sensitivity tests; they are used e.g. in Denmark, Holland, Belgium, Norway, Finland and other countries. Neo-Sensitabs® (produced by A/S ROSCO, Taastrup, Denmark) are colour-coded for identification, and most are stable at room temperature for a minimum of 4 years.

neosolaniol See TRICHOTHECENES.

neostibosan See ANTIMONY.

neoteny The persistence, in a mature organism, of features which are characteristic of the immature organism.

neotype Any strain which has been described and published in accordance with the relevant nomenclatural code in order to serve as a NOMENCLATURAL TYPE in place of e.g. a nomenclatural type which is no longer extant.

Neovossia See USTILAGINALES and KARNAL BUNT.

neoxanthin See CAROTENOIDS.

nephelometry Any procedure in which the concentration of cells (or particles) in a suspension is estimated by passing a beam of (monochromatic) light through the suspension and measuring the intensity of the *scattered* light – commonly at 90° to the incident light (cf. TURBIDIMETRY). Light scattering is due to intracellular refraction, diffraction, and reflection from the cell surface; the amount of light scattered becomes greater with increases in cell concentration, cell size, and length of light path through the suspension, and with decreases in the wavelength of the incident light. The intensity of the scattered light varies with its angular displacement from the axis of the incident light – exhibiting several peaks of intensity; these peaks are due to interference phenomena. The relationship between the intensities of the incident and scattered light and the concentration of cells is given by:

$$\log_{10}(I_0/I_{scat}) \propto 1/(\text{cell concentration})$$

where I_0 is the intensity of incident light, and I_{scat} is the intensity of scattered light. The sensitivity of a nephelometer is lower when cell concentrations are high owing to the effects of secondary scattering, i.e., redirection of a deflected ray by other cell(s). (See also COUNTING METHODS.)

nephridial network See CONTRACTILE VACUOLE.

nephritic (1) Pertaining to, or affected with, nephritis. (2) Of or pertaining to the kidneys.

nephritis Inflammation of the kidney. (See PYELONEPHRITIS.)

nephritogenic Causing nephritis.

Nephroma A genus of foliose LICHENS (order PELTIGERALES); photobiont: *Coccomyxa* or *Nostoc*, according to species. Thallus: lobed, corticate on both surfaces, grey, green or brown; apothecia (rare e.g. in *N. expallidum*, common e.g. in *N. laevigatum*) are formed on the lower surface of the thallus near the margins; fertile lobes are usually held erect. Other structures which may be present include *Nostoc*-containing cephalodia (in *Coccomyxa*-containing species – e.g. *N. arcticum*), tubercles on the lower surface (*N. resupinatum*), soredia on the upper surface (*N. parile*) and tomentum on the lower surface (many species). *Nephroma* spp occur mainly on mossy trees and rocks e.g. in boreal regions.

Nephroselmis See MICROMONADOPHYCEAE.

nephrotoxic Toxic to the kidneys.

nepoviruses ('nematode-transmitted polyhedral viruses'; tobacco ringspot virus group) A group of ssRNA-containing PLANT VIRUSES which have a wide host range. Infected plants may be symptomless, or the leaves may initially develop ringspots, spots, mottling, curling etc; leaves produced subsequently may be symptomless. Vesicular inclusion bodies occur in the cytoplasm of infected cells, usually near the nucleus. Transmission occurs via seeds, via soil nematodes (e.g. *Longidorus* spp), and (experimentally) mechanically; replication does not occur in the nematode vector. Type member: tobacco ringspot virus (TobRV); other members include e.g. cherry leaf roll virus, chicory yellow mottle virus, cocoa necrosis virus, peach rosette mosaic virus, potato black ringspot virus, raspberry ringspot virus, tomato black ring virus, tomato ringspot virus. Possible members: cherry rasp leaf virus, satsuma dwarf virus, strawberry latent ringspot virus, tomato top necrosis virus.

Virion: icosahedral, ca. 28 nm diam.; genome: two linear, positive-sense ssRNA species, RNA1 (MWt ca. 2.8×10^6) and RNA2 (MWt ca. $1.3–2.4 \times 10^6$). Virus particles may contain either RNA1 (B particles) or RNA2 (M particles), M and B particles being distinguished by their sedimentation coefficients; T particles contain no RNA, while some (B-like) particles may contain two molecules of RNA2. Both M and B particles are necessary for infectivity. Particles of some nepovirus isolates may also contain a SATELLITE RNA.

RNA1 and RNA2 show little sequence homology; they are polyadenylated at the 3′ end and each is covalently linked to a small polypeptide (MWt ca. 3000–6000) which is necessary for infectivity. RNA1 encodes the genome-linked polypeptide and (possibly) the virus-specific replicase; RNA2 encodes the coat protein.

neritic Of or pertaining to coastal waters.

Nernst equation See e.g. ION TRANSPORT.

nested PCR (nPCR) A variant form of PCR carried out in two phases. In the initial phase, a standard form of PCR is run for e.g. ~25 cycles; in the second phase, an aliquot of product from the first phase is used as template for another phase of amplification using a (different) pair of primers that are complementary to subterminal (internal) sequences within the first-phase amplicon. Primers used in the first and second phases have been referred to as 'outer' and 'inner' primers, respectively.

The object of nPCR is to improve sensitivity and specificity in a PCR assay; such improvements may reflect e.g. the extra stage of amplification and the competitive advantage of a shorter amplicon during the second phase. nPCR has been particularly useful for amplifying samples of DNA which are in poor condition or which are of limited quantity.

nPCR assays may suffer from susceptibility to contamination during preparation of the second stage of amplification. However, the problem of contamination can be minimized by using so-called 'one-tube' protocols.

One-tube nPCR has been used e.g. for detecting *Trichomonas vaginalis* in vaginal discharge [JCM (1997) *35* 132–138]. Both sets of primers were included in a single reaction mixture (which used the *hot-start* approach). In the first phase of amplification (30 cycles) the annealing temperature was 62°C; at this temperature only the outer primers were able to bind. In the second phase of amplification (20 cycles) the annealing temperature was lowered to 45°C – permitting binding of the inner primers and leading to the formation of second-phase amplicons. Detection of the second-phase amplicons involved pre-labelling of inner primers. One type of primer was 5′-biotinylated while the other was 5′-labelled with DIGOXIGENIN – so that the double-stranded amplicon was labelled with both compounds. By means of the biotin tag, amplicons were captured on an AVIDIN-coated microtitre plate; an antidigoxigenin–alkaline phosphatase conjugate was used to label the amplicon (by digoxigenin–antidigoxigenin binding), and the label (enzyme) was detected by addition of a reagent (*p*-nitrophenylphosphate) and monitoring colorimetrically at 405 nm for the enzymic cleavage products.

Other examples of nPCR include detection of *Mycobacterium malmoense* [JCM (1999) *37* 1454–1458], and nested duplex PCR for the detection of *Bordetella pertussis* and *B. parapertussis* [JCM (1999) *37* 606–610].

net blotch A CEREAL DISEASE, caused by *Pyrenophora teres*, which affects mainly winter barley. Pale, yellowish blotches, later turning brown, appear on leaves in the autumn; the brown lesions, which have a net-like appearance, may coalesce to form longitudinal stripes. (See also FLUTRIAFOL, GUAZATINE, PROCHLORAZ, PROPICONAZOLE.)

net plasmodium See LABYRINTHULAS.

net slime moulds See LABYRINTHULAS.

netilmicin An AMINOGLYCOSIDE ANTIBIOTIC.

netropsin A basic oligopeptide ANTIBIOTIC (isolated from '*Streptomyces netropsis*') containing two *N*-methylpyrrole rings; the functionally similar *distamycin A* contains three such rings. Both netropsin and distamycin bind specifically and avidly to B-DNA, with a strong preference for AT-rich regions [DNA–drug specificity: PNAS (1985) *82* 1376–1380]; the drug appears to bind (non-intercalatively) in the minor groove (see DNA). Both drugs show antimicrobial and antitumour activity (apparently by inhibiting synthesis of RNA and DNA) and inhibit the development of DNA viruses, but they are too toxic for clinical use. Low concentrations of netropsin have been reported to inhibit sporulation without affecting growth in *Bacillus subtilis* [CJM (1980) *26* 420–426].

Neuberg's fermentations Neuberg's '1st form of fermentation' is the ALCOHOLIC FERMENTATION as normally carried out by yeasts such as *Saccharomyces*. Neuberg's '2nd form of fermentation' (= 'glycerol fermentation', 'sulphite fermentation') refers to the alcoholic fermentation perturbed by the presence of sodium bisulphite ($NaHSO_3$). Bisulphite forms an addition compound with acetaldehyde, preventing its reduction (by NADH) to ethanol; NADH (from the EMP pathway) instead reduces dihydroxyacetone phosphate to glycerol-3-phosphate – which is dephosphorylated to GLYCEROL. Neuberg's '3rd form of fermentation' refers to the alcoholic fermentation perturbed by imposed alkalinity; the end products are glycerol, ethanol and acetic acid.

Neufchatel cheese See CHEESE-MAKING.

Neufeld quellung phenomenon See QUELLUNG PHENOMENON.

neuraminic acid A C_9-compound (5-amino-3,5-dideoxy-D-glycero-D-galactononulosonic acid) derived from mannosamine and pyruvate; *N*-acetyl and *O*-acetyl derivatives (e.g. *N*-acetylneuraminic acid, NANA) are called *sialic acids*. Sialic acids occur widely in animal tissues and fluids – e.g. in glycolipids, mucopolysaccharides and mucoproteins in mucous secretions, serum, brain glycolipids (gangliosides), milk (*N*-acetylneuraminyllactose), etc. Terminal sialic acid residues in glycoprotein or glycolipid cell membrane components serve as receptor sites for the adsorption of certain viruses (see e.g. INFLUENZAVIRUS). (See also COLOMINIC ACID; NEURAMINIDASE; COMPLEMENT FIXATION (b).)

neuraminidase (acylneuraminyl hydrolase, EC 3.2.1.18; sialidase) An enzyme which hydrolyses α-ketosidic linkages between terminal *N*-acetylNEURAMINIC ACIDS and sugar residues in polysaccharides, glycoproteins, etc. Extracellular neuraminidases are produced by certain bacteria (e.g. *Streptococcus pneumoniae, Vibrio cholerae*) and occur in certain viruses – e.g. members of the Orthomyxoviridae (see INFLUENZAVIRUS) and Paramyxoviridae (genus PARAMYXOVIRUS). The viral neuraminidase (a RECEPTOR-DESTROYING ENZYME, RDE) lowers the viscosity of mucous secretions, exposing the cell surface receptor sites; it also assists the release of progeny virions by eluting them from their receptors. (See also HAEMAGGLUTINATION.)

Neuraminidases from different sources may have different specificities, but most cannot hydrolyse COLOMINIC ACID (q.v. for exceptions); they are generally inhibited by EDTA.

neuroexocytosis See TETANOSPASMIN.

Neurospora A genus of fungi (order SORDARIALES) which occur e.g. on burnt or decomposing vegetation; the organisms form longitudinally-ribbed ascospores in asci borne in perithecia. *N. crassa* and *N. sitophila* are both hermaphroditic and heterothallic; other species (e.g. *N. tetrasperma*) are homothallic or secondarily homothallic (see HOMOTHALLISM and HETEROTHALLISM). *N. sitophila* forms a TRICHOGYNE-bearing ascogonium but no antheridium, plasmogamy occurring by spermatization or conidiation; eight-spored asci are formed in dark, pyriform perithecia. Conidial stages occur in *N. crassa* and *N. sitophila*. (See also BREAD SPOILAGE, ONCOM, POKY MUTANT, RAGI, TEMPEH.)

[Control of growth and nuclear division cycle in *N. crassa*: MR (1981) *45* 99–122; *Neurospora* plasmids: Book ref. 179, pp. 245–258.]

neurotoxin A TOXIN which acts on nerve cells or tissues, or at neuromuscular junctions. See e.g. BOTULINUM TOXIN, SAXITOXIN, TETANOSPASMIN.

neurotransmitter See BOTULINUM TOXIN and TETANOSPASMIN.

neurotropism See TROPISM (sense 2).

neuston The community of organisms which live in the surface film of a body of water (e.g. lake, sea); some members of the community (the *epineuston*) are associated with the upper surface and project into the air, while others (the *hyponeuston*) are associated with the underside of the film and project into the water below. (cf. NEKTON, PLANKTON.)

neutral dye See DYE.

neutral petite See PETITE MUTANT.

neutral red A red, basic azine DYE used e.g. in VITAL STAINING and as a PH INDICATOR: pH 6.8–8.0 (red to yellow). (See also BIOFUEL CELL.)

neutralism (1) The coexistence of two different organisms in an association in which neither is affected by the presence of the other.

(2) SYMBIOSIS (sense 1) in which one symbiont is (permanently or temporarily) dependent on the association, and the other derives neither benefit nor harm from the association.

(cf. COMMENSALISM.)

neutralization test Any test in which antibody prevents or inhibits the expression of a biological function by a microorganism or toxin. Thus, e.g., a microorganism may be identified by its failure to exhibit certain features (e.g. growth, ability to infect cells or organisms) in the presence of known, specific antiserum; a toxin may be identified or quantified by its failure to produce toxic effects (in vitro or in vivo) following admixture with specific antitoxin. (See also DANYSZ' PHENOMENON.)

neutropenic enterocolitis (ileocaecal syndrome) (*med.*) A syndrome characterized by neutropenia (a diminished number of circulating neutrophils) and ulceration, haemorrhage and necrosis of the distal ileum or caecum; *Clostridium septicum* may play a major causative role in the symptoms.

neutrophil An amoeboid, phagocytic PMN (q.v.) responsive to chemotactic stimuli – often the major type of cell involved early in acute INFLAMMATION. Neutrophils contain many enzymes (e.g. LYSOZYME, MYELOPEROXIDASE) (see also BPI PROTEIN, LACTOFERRIN). Stimulation (e.g. by opsonized bacteria or concanavalin A) leads to a 'respiratory burst' (metabolic burst) of increased oxygen consumption (over several minutes) starting within seconds of stimulation; much of the oxygen may be used to form SUPEROXIDE radicals (see PHAGOCYTOSIS). (cf. MACROPHAGE.) (See also CALPROTECTIN.)

neutrophile An organism which has an optimum growth pH within the pH range 6–8. (cf. ACIDOPHILE and ALKALOPHILE.)

nevirapine See ANTIRETROVIRAL AGENTS and AIDS.

Nevskia A poorly characterized genus of aquatic bacteria [see e.g. Book ref. 45, pp. 520–523].

new combination In NOMENCLATURE: *syn.* COMB. NOV. (q.v.).

new variant CJD See CREUTZFELDT–JAKOB DISEASE.

New World arenaviruses See ARENAVIRIDAE.

Newcastle disease (avian pneumoencephalitis; Ranikhet disease) A highly infectious POULTRY DISEASE caused by a PARAMYXOVIRUS. Chickens, turkeys and pheasants may be affected, and mild infections have been reported in ducks, geese and pigeons. Transmission may occur by droplet inhalation or by ingestion of contaminated water or food (e.g. food contaminated with infected pigeon droppings [VR (1984) *115* 601–602]). Incubation period: ca. 4–12 days. Symptoms: variable, but commonly include e.g. loss of appetite, diarrhoea, laboured breathing, discharge from eyes, nostrils and mouth, often followed by neurological signs (drowsiness, weakness, paralysis). Mortality rates vary from 0 to ca. 100%. *Lab. diagnosis*: identification of the virus; an HAI test. *Control*: slaughter of infected flocks; vaccination using either live or inactivated vaccines.

The Newcastle disease virus has been reported to cause conjunctivitis in man.

Newcombe experiment An experiment by which Newcombe (in 1949) demonstrated that the apparent adaptation of a bacterial population to an altered environment involves the selection of pre-existing mutants.

Nutrient agar plates are each spread with the same quantity of inoculum from a single culture of phage-sensitive bacteria. The plates are incubated for a few hours, during which time each viable cell gives rise to a clone. The plates are then divided into two groups. In group A, the bacterial growth on each plate is redistributed (with a sterile glass spreader) so that the cells of each clone are dispersed over the agar surface. Group B plates are left undisturbed. All plates are then sprayed with a suspension of virulent phage and re-incubated. All the sensitive bacteria lyse; thus, colonies which subsequently develop on any plate must have arisen from cells which were, or had become, resistant to phage.

If phage-resistant bacteria developed adaptively, resistant cells would be formed only *after* exposure of phage; the redistribution of growth in group A plates would therefore not influence the subsequent development of resistant cells, and similar numbers of resistant cells – and hence colonies – should appear on all the plates. If, however, phage-resistant cells arose spontaneously *before* exposure to phage, each resistant (mutant) cell would form a discrete clone of resistant cells; subsequent redistribution of growth on the group A plates would disperse the resistant cells over the agar surface, so that each resistant cell on a group A plate would subsequently form a single colony. On the group B plates, however, each undisturbed clone of resistant cells would form only one colony. Hence, far more colonies would be formed on the group A plates than on the group B plates – as is found in practice. (cf. FLUCTUATION TEST.)

nexin See FLAGELLUM (b).

NF-κB See CYTOKINES.

ngaione See IPOMEAMARONE.

NGU NON-GONOCOCCAL URETHRITIS.

NHI protein NON-HAEM IRON PROTEIN.

N_i protein See ADENYLATE CYCLASE.

Nia A genus of marine fungi (GASTEROMYCETES, Melanogastrales). *N. vibrissa* occurs on submerged wood; it forms globose basidiocarps (2–4 mm diam.) and five-limbed basidiospores.

niacin NICOTINIC ACID.

niacin test A test used to determine whether or not a detectable amount of niacin is liberated into a solid medium when a given strain of *Mycobacterium* is grown on that medium. Water or saline (ca. 1 ml) is allowed to stand in contact with the medium for 15 min to extract any niacin present. To ca. 0.5 ml extract is added 0.5 ml aniline (4% in ethanol) followed by 0.5 ml cyanogen bromide (CNBr: 10% aqueous); an immediate yellow coloration indicates the presence of niacin (a positive test). (To avoid the formation of free HCN from CNBr, used materials should be treated with NaOH prior to disposal.) In an alternative method the extract is tested with commercial niacin-test strips instead of with aniline and CNBr. [Book ref. 53, p. 1708.]

niacinamide Nicotinamide.

NIAID National Institute of Allergy and Infectious Diseases (USA).

nic See CONJUGATION (b)(i).

NICED National Institute of Cholera and Enteric Diseases, Beliaghata, Calcutta 700010, India.

Nichols' treponeme (1) A virulent strain of *Treponema pallidum* subsp. *pallidum*. The organism has been reported to grow and divide on the surface of a tissue culture of cotton-tail rabbit epithelium cells (Sf1EP) under 1.5% oxygen [Inf. Immun. (1981) *32* 908–915]. (2) Strains of *Treponema refringens*.

nick A break in one strand of a dsDNA molecule due to the loss (or absence) of a phosphodiester linkage in an otherwise intact strand, as opposed to the loss (or absence) of one or more *nucleotides* from the strand, which is termed a *gap*. A nick can be repaired by a DNA LIGASE, a gap cannot.

nick-closing enzyme A type I TOPOISOMERASE.

nick translation technique A technique used e.g. for preparing a radioactively labelled DNA strand. A dsDNA molecule is nicked, using an appropriate endonuclease, and treated with DNA POLYMERASE I; this enzyme, using the nicks as starting points, can degrade the nicked strand using its $5' \rightarrow 3'$ exonucleolytic activity, replacing it with a new strand using its polymerase activity. The presence of plenty of radiolabelled dNTPs in the medium ensures that the new strand will be extensively labelled.

nickase An ENDONUCLEASE which makes a single-strand nick in dsDNA (e.g., gpA of BACTERIOPHAGE φX174).

nicking enzyme An enzyme that nicks one strand of dsDNA: see NICKASE and RESTRICTION ENDONUCLEASE.

Nicollia See BABESIA.

nicotinamide See NICOTINIC ACID.

nicotinamide adenine dinucleotide See NAD.

nicotinamide nucleotide transhydrogenase See TRANSHYDROGENASE.

nicotinic acid (niacin) Pyridine 3-carboxylic acid; the corresponding amide, *nicotinamide*, is a component of the important coenzymes NAD (q.v.) and NADP. Niacin may be biosynthesized (via quinolinic acid) from aspartate and dihydroxyacetone phosphate (e.g. in many bacteria, and in *Saccharomyces cerevisiae* growing under anaerobic conditions), or from tryptophan (e.g. in *Xanthomonas* spp and in most fungi – including *S. cerevisiae* growing under aerobic conditions). Organisms auxotrophic for niacin or nicotinamide include e.g. *Lactobacillus* spp (used in the BIOASSAY of niacin), *Morganella* and *Proteus* spp (but not *Providencia*), *Bordetella pertussis*, some yeasts and fungi (e.g. *Mucor rouxii* requires niacin under anaerobic conditions but not under aerobic conditions), and certain protozoa (e.g. *Tetrahymena pyriformis*, coccidia). (See also NIACIN TEST; V FACTOR; VITAMIN.)

niddamycin See MACROLIDE ANTIBIOTICS.

Nidula See NIDULARIALES.

Nidulariales An order of humicolous, lignicolous or coprophilous fungi (class GASTEROMYCETES) characterized by a type of basidiocarp in which the GLEBA occurs in several parts, each part contained within a closed chamber (*peridiole*) which is physically separate from the other chambers; in some species each peridiole is attached to the inside of the (typically cup-shaped or funnel-shaped) peridium by a complex, extensible structure. The appearance of peridioles at the bottom of the peridium (which is open at maturity) has earned these fungi the name 'bird's nest fungi'. Genera include *Crucibulum*, CYATHUS, *Mycocalia* and *Nidula*.

***nif* genes** Genes whose expression is necessary for NITROGEN FIXATION. In *Klebsiella pneumoniae* there are at least 17 *nif* genes clustered (close to the *his* operon) in 8 contiguous transcriptional units (*nifJ*, *nifHDK*, *nifY*, *nifENX*, *nifUSVM*, *nifF*, *nifLA*, and *nifBQ*) spanning ca. 23 kb on the chromosome. Genes *nifHDK* are structural genes for NITROGENASE: *nifD* and *nifK* specify the two subunits of the MoFe protein ('Kp1'), *nifH* specifies the subunit of the Fe protein ('Kp2'). The *nifJ* product is the pyruvate:flavodoxin oxidoreductase which transfers electrons from pyruvate to the flavodoxin (*nifF* product) that reduces the Fe protein of nitrogenase. The *nifB*, *nifE*, *nifN* and *nifV* genes all appear to be involved in the synthesis of the FeMo cofactor; *nifQ* appears to be involved in molybdate uptake; *nifM*, *nifS* and *nifU* may be involved in the processing of nitrogenase components; *nifL* and *nifA* are concerned with *nif* gene expression.

Expression of *nif* genes in *K. pneumoniae* is regulated at the transcriptional level and is repressed e.g. by alternative sources of nitrogen, by high oxygen levels, etc. The *nifA* gene specifies an activator of transcription which is necessary for the expression of most *nif* genes. The *nifL* gene specifies a negative regulator which represses transcription in the presence of e.g. amino acids, low levels of NH_4^+, and oxygen; it appears not to interact directly with DNA but may complex with and inactivate the NifA protein. Transcription of *nif* genes is regulated by the availability of combined nitrogen and is subject to control by the NTR GENES.

The *nif* genes of other bacteria are generally less well characterized: *Azotobacter* appears to have additional copies of *nifH*; in *Bradyrhizobium* strains *nifH* is separated from *nifDK*; *Rhodopseudomonas capsulata* has multiple copies of *nifHDK*, while its other *nif* genes are apparently more widely dispersed on the chromosome than are those of *K. pneumoniae*. In *Rhizobium* spp *nif* genes are located on the Sym plasmid (see entry ROOT NODULES).

In most nitrogen-fixing CYANOBACTERIA, *nif* genes are expressed only in HETEROCYSTS. In vegetative cells of e.g.

Anabaena, the *nifH* and *nifD* genes are contiguous, but *nifK* is separated from *nifD* by ca. 11 kb. It has been shown that, late in heterocyst differentiation, the DNA undergoes certain rearrangements, including the excision and circularization of the 11-kb segment between *nifD* and *nifK*; this results in the fusion of *nifK* with *nifHD* and allows coordination of transcription of the *nifHDK* genes.

[*nif* and associated genes in *Acetobacter diazotrophicus*: JB (2000) *182* 7088–7091.]

nifuratel See NITROFURANS.

nifuroxime See NITROFURANS.

nifurprazine See NITROFURANS.

nigeran (mycodextran) A linear α-D-glucan in which glucopyranose residues are linked by alternating (1 → 3)-α- and (1 → 4)-α-glycosidic linkages; it is soluble in hot water. Nigeran occurs in the hyphal walls of *Aspergillus* spp (but not in *A. nidulans*, which contains PSEUDONIGERAN) and *Penicillium* spp. [Nigeran as a phylogenetic marker: Mycol. (1978) *70* 1201–1211.]

nigericin A linear MACROTETRALIDE antibiotic which can act as a mobile carrier IONOPHORE for the transport of monovalent cations – K^+or Rb^+ (or, to a lesser extent, e.g. Na^+). In order to transport an ion the nigericin molecule loses a proton and binds the ion, and the (electrically neutral) nigericin–ion complex diffuses across the membrane; since nigericin can also cross the membrane in a protonated state, it can bring about the exchange of H^+ for K^+ (see ELECTRONEUTRAL TRANSPORT). Nigericin is used e.g. to alter a membrane pH gradient (see CHEMIOSMOSIS).

nigerose α-D-Glucopyranosyl-(1 → 3)-D-glucopyranose.

nigrosin A black composite indulin-based DYE used e.g. for negative STAINING.

***nikA* gene** See CONJUGATION (1b)(i).

nikkomycins A group of antibiotics which closely resemble POLYOXINS both in structure and in mode of action; nikkomycins have a wider spectrum of activity than polyoxins, being more active e.g. against the yeast forms of *Mucor rouxii* and *Candida albicans*.

Nikolsky sign Wrinkling of the skin (evident on gentle stroking) due to separation of the outer epidermis from the basal layer. (See also EPIDERMOLYTIC TOXIN.)

Nile blue A (Nile blue sulphate) A blue basic oxazine DYE used e.g. in VITAL STAINING and as a lipid stain. In aqueous solution some of the dye is spontaneously oxidized to the phenoxazone LYSOCHROME Nile red (soluble in aqueous Nile blue A); this allochromatic mixture stains fatty acids and phospholipids blue, and neutral lipids red. Nile blue A can be used as a fluorescent stain for poly-β-hydroxybutyrate [AEM (1982) *44* 238–241]; Nile red is useful as a vital stain for intracellular lipid droplets, being strongly fluorescent only in a hydrophobic environment [JBC (1985) *100* 965–973].

Nile red See NILE BLUE A.

nisin A low-MWt LANTIBIOTIC produced by strains of *Lactococcus*; genes for nisin synthesis appear to be plasmid-borne. [Nisin (review): AAM (1981) *27* 85–123.] Nisin is bactericidal against Gram-positive bacteria, and has some sporicidal activity; it is both heat-stable and acid-stable, but stability and solubility decrease with increasing pH, and nisin is inactivated by certain proteases (e.g. chymotrypsin). Nisin is not used clinically; it is employed as a food PRESERVATIVE, generally in conjunction with heat treatment, in e.g. milk and DAIRY PRODUCTS (in which it occurs naturally) and certain canned vegetables. Canned, low-acid foods in which nisin is used must still receive adequate heat treatment because *Clostridium botulinum* is relatively resistant to nisin. (cf. DIPLOCOCCIN and SUBTILIN.)

[Synergistic action of nisin and CO_2 against *Listeria monocytogenes*: AEM (2000) *66* 769–774. Bioassay for nisin in food: AEM (2003) *69* 4214–4218.]

Nitella See CHAROPHYTES.

nitrapyrin (N-Serve) 2-Chloro-6-(trichloromethyl)pyridine, a NITRIFICATION INHIBITOR.

nitratase *Syn.* nitrate reductase.

nitrate broth Nutrient broth (or peptone water) containing KNO_3 (0.1–0.2% w/v).

nitrate reductase See ASSIMILATORY NITRATE REDUCTION; DENITRIFICATION; NITRATE RESPIRATION.

nitrate reduction See ASSIMILATORY NITRATE REDUCTION and DISSIMILATORY NITRATE REDUCTION.

nitrate reduction test (nitrate test) A test used to indicate whether or not a given bacterial strain can reduce nitrate; the medium and test conditions used depend on the particular organism being examined. For many bacteria (including enterobacteria and pseudomonads, but excluding e.g. *Mycobacterium* spp) the organism is cultured for 2–5 days in NITRATE BROTH. The culture is then tested for the presence of nitrite by adding 0.5 ml 'reagent A' (0.8% w/v sulphanilic acid in 5 N acetic acid) followed by 0.5 ml 'reagent B' (0.6% w/v *N*,*N*-dimethyl-α-naphthylamine in 5 N acetic acid); if present, nitrite reacts with sulphanilic acid to form a diazonium salt which, with the naphthylamine, forms a soluble red azo dye. The absence of coloration could mean that either (a) nitrate has not been reduced, or (b) the nitrite has itself been reduced. To resolve this question a trace of zinc dust is added; if nitrate is present it is reduced (by the zinc dust) to nitrite (which gives a red coloration) – the absence of coloration indicating possibility (b).

For *Mycobacterium* spp, a small loopful of growth is dispersed in 0.02 M phosphate buffer (pH 7.0) containing 0.01 M $NaNO_3$, and the suspension is incubated (37°C/2hours). To the suspension is then added 1 drop of HCl (concentrated HCl:water mixed 1:1 v/v) followed by 2 drops of sulphanilamide solution (0.2% aqueous) and 2 drops of *N*-naphthylethylenediamine dihydrochloride (0.1% aqueous); pink or red coloration indicates reduction of nitrate to nitrite. If no colour develops, the addition of zinc dust causes a red coloration if the medium still contains nitrate – and fails to do so if the nitrate had been reduced beyond nitrite.

nitrate respiration ANAEROBIC RESPIRATION in which, typically, nitrate (as terminal electron acceptor) is reduced to nitrite in a reaction catalysed by an enzyme (*dissimilatory nitrate reductase*) located at the end of an ELECTRON TRANSPORT CHAIN; this reaction, which is equivalent to the first stage of DENITRIFICATION, generates proton motive force (pmf: see CHEMIOSMOSIS) which can be used e.g. for ATP synthesis. (cf. DISSIMILATORY NITRATE REDUCTION.) In some organisms the nitrite is excreted, but in others it can be reduced to ammonia. In nitrate respiration the energy yield is less than that obtainable with oxygen as electron acceptor.

In some bacteria (including e.g. *Escherichia coli*) nitrate can be used for the anaerobic oxidation of e.g. NADH, succinate and/or formate – pmf being generated in each case; the resulting nitrite inhibits aconitase and fumarase, thus inhibiting the TCA cycle [Appendix II(a)].

In e.g. *Escherichia coli* a soluble ('assimilatory') nitrite reductase can catalyse the anaerobic reduction of nitrite to ammonia in a reaction coupled to the oxidation of NADH. This reaction does not, in itself, yield energy; however, by regenerating NAD^+ from NADH (produced during the metabolism of e.g. glucose) it tends to prevent the (energetically wasteful) synthesis of ethanol – see

Appendix III(e) – thereby increasing the yield of acetic acid and, hence, conserving energy via acetyl phosphate. Since the oxidation of NADH by nitrite (i) depends on an exogenous electron acceptor, (ii) results in an overall energy yield greater than that obtainable by the mixed acid fermentation, and (iii) results in a net oxidation of the substrate, it conforms to the definition of RESPIRATION and may be described as 'nitrite respiration'. In *Desulfovibrio gigas* the coupling of nitrite reduction to the oxidation of hydrogen appears to generate pmf; accordingly, this process, too, may be referred to as nitrite respiration.

Some anaerobic bacteria (e.g. *Clostridium perfringens*) can carry out dissimilatory nitrate reduction [JGM (1975) *87* 120–128]. This process, in which nitrate is reduced to ammonia, apparently does not generate pmf; however, it is coupled to the regeneration of NAD^+ from NADH, and brings about an increase in the proportion of those intermediate metabolites which can participate in substrate-level phosphorylation. This process, which is sometimes referred to as 'fermentative nitrate reduction', may be regarded as another example of nitrate respiration.

So-called *dissimilatory reduction of nitrate to ammonia* (DRNA) is a respiratory process reported to take place e.g. in marine sediments, particularly where nitrate concentrations are low [see FEMS Ecol. (1996) *19* 27–38].

nitrate test See NITRATE REDUCTION TEST.

nitric oxide (NO) On suitable activation, various types of cell (including macrophages) synthesize an inducible form of the enzyme *nitric oxide synthase* (iNOS); synthesis of iNOS is positively regulated at the level of transcription by certain cytokines (e.g. TNF-α, IFN-γ). iNOS catalyses the NADPH-dependent oxidative deamination of L-arginine to L-citrulline and nitric oxide.

Nitric oxide gives rise to a range of reactive species by interaction with oxygen, water and/or other reactive intermediates; they include *peroxynitrite* ($ONOO^-$) (formed by the reaction between superoxide and nitric oxide), nitrogen dioxide and nitrogen trioxide. Peroxynitrite forms an adduct with carbon dioxide that is potentially toxic. In e.g. a target bacterium, nitric oxide and the various reactive species inhibit oxygen-binding haem-containing enzymes (such as terminal oxidases) and certain other iron-containing enzymes (e.g. the TCA cycle enzyme *aconitase*). Damage to an infecting bacterium may include failure of DNA replication and/or the loss of important metabolic processes, leading to death of the cell. The production of large amounts of nitric oxide by iNOS therefore serves a protective role. However, ongoing production of high levels of nitric oxide in a persistent, unresolved infection can suppress the host's immune system and bring about various forms of disease.

Inhibitors of iNOS inhibit the killing of *Mycobacterium tuberculosis* by IFN-γ-activated murine macrophages; moreover, studies in mice indicate that deletion of IFN-γ genes increases the susceptibility of the mice to *M. tuberculosis* – this being associated with a deficiency of iNOS.

Bacterial reaction to nitric oxide includes e.g. increased activity of enzymes involved in the antioxidant response to oxidative stress.

[Effects of nitric oxide on pathogens and host: RMM (1998) *9* 179–189. Niric oxide in sepsis and endotoxaemia: JAC (1998) *41* (supplement A) 31–39. Bacterial responses to nitric oxide and nitrosative stress: Mol. Microbiol. (2000) *36* 775–783.]

nitric oxide synthase See NITRIC OXIDE.

nitrification The oxidation of ammonia to nitrite, and nitrite to nitrate: a process which occurs e.g. in aerobic soils, in aquatic environments, and in the upper (aerobic) region of some sediments. (See also SEWAGE TREATMENT.) Nitrification is carried out by autotrophic bacteria of the family NITROBACTERACEAE and by many heterotrophic organisms, including some fungi. It has been generally supposed that only the autotrophic nitrifying bacteria carry out nitrification to an ecologically significant extent; however, while this appears to be the case e.g. in the sediments of a Scottish estuary [JGM (1984) *130* 2301–2308], the potential for heterotrophic nitrification was found to be greater than that for autotrophic nitrification in a Californian forest soil [AEM (1984) *48* 802–806].

Autotrophic nitrification occurs in two stages, each stage being carried out by one of the two categories of nitrifying bacteria. (a) *Ammonia to nitrite* ('nitrosofication'). Ammonia is initially oxidized to hydroxylamine (NH_2OH) in a MONOOXYGENASE-dependent, non-energy-yielding reaction; NH_2OH is then oxidized to nitrite by an enzyme complex, electrons being transferred to molecular oxygen with concomitant oxidative phosphorylation. (b) *Nitrite to nitrate*. The oxidation of nitrite to nitrate provides energy by oxidative phosphorylation: in at least some nitrite-oxidizing bacteria NO_2^- is oxidized in the reaction: $NO_2^- + H_2O \rightarrow NO_3^- + 2H^+ + 2e^-$; the electrons are believed to be transferred to cytochrome (cyt) a_1 which in turn transfers them via cyt aa_3 to molecular oxygen, 1ATP being formed for each pair of electrons transferred.

In the nitrite-oxidizing bacteria (and probably also in NH_4^+ oxidizers) pyridine nucleotide coenzymes are reduced by reversed electron transport. In at least some nitrite oxidizers, electrons appear to pass from cyt a_1, via cyts c and b, to a flavoprotein and thence to the pyridine nucleotides; this process is highly energy-dependent, so that nitrifying bacteria need large quantities of NO_2^- to sustain their relatively slow rates of growth.

Agriculturally, nitrification can be disadvantageous under certain conditions: NO_3^- is readily leached from the topsoil by rain – thereby becoming unavailable as a source of crop nitrogen. (By contrast, owing to its charge, NH_4^+ is more strongly adsorbed to soil particles and is relatively immobile in top-soils.) Nevertheless, when e.g. the crop's nitrogen demand is high and the rainfall (and hence risk of leaching) is likely to be low, rapid nitrification may actually be desirable since (a) optimal growth in many or most plants appears to occur only when both $NH_4^+$and NO_3^- are available, and (b) NH_4^+ may be assimilated (in preference to NO_3^-) by soil microorganisms so that, in the absence of nitrification, much of the available soil nitrogen may be lost to crops as a consequence of being immobilized in microbial biomass. However, when e.g. the risk of leaching is high, the use of NITRIFICATION INHIBITORS can be advantageous; these agents have a temporary effect (they are gradually lost by evaporation and/or decomposition), and they not only benefit soil fertility but also reduce the pollution of groundwater by nitrate. (See also NITROGEN CYCLE.)

nitrification inhibitors Agents which retard or block NITRIFICATION; nitrification inhibitors appear to be typically bacteriostatic under field conditions. The agriculturally important agents inhibit microbial oxidation of ammonia and may also inhibit the oxidation of nitrite – though nitrite oxidation is generally much less sensitive to these agents; agents which block only nitrite oxidation cannot be used since they could cause accumulation of (phytotoxic) nitrite ions. Commercial nitrification inhibitors – some of which also have useful antifungal properties – include e.g. DICYANDIAMIDE, ETRIDIAZOLE and NITRAPYRIN. (See also BARBAN and CARBON DISULPHIDE.) [Review: Book ref. 121, pp. 33–43.]

In laboratory studies, acetylene has been used to block ammonia oxidation, and chlorate to block nitrite oxidation, in autotrophic nitrifiers [AEM (1984) *48* 802–806].

nitrifying bacteria See NITROBACTERACEAE.

nitriloacetic acid See NTA.

nitrite (a) (as a food additive and preservative) See e.g. CURING (sense 1).

(b) (in the environment) See NITROGEN CYCLE.

nitrite reductase See ASSIMILATORY NITRATE REDUCTION; DENITRIFICATION; NITRATE RESPIRATION.

nitrite respiration See NITRATE RESPIRATION.

Nitrobacter A genus of Gram-negative bacteria of the NITROBACTERACEAE; strains occur in soils and in freshwater and marine habitats. Cells of the type species, *N. winogradskyi*, are rods ca. 0.6–0.8 × 1.0–2.0 μm; they are commonly non-motile but may have a single polar flagellum. Peripherally arranged cytomembranes form a cap at one pole of the cell. The cells reproduce by budding. Some strains are obligate chemolithoautotrophs which obtain energy by oxidizing nitrite to nitrate; other strains are facultatively chemoorganoheterotrophic. Storage compounds include PHB. Growth occurs optimally at ca. 25–30° C and at a pH of ca. 7.5–8.0. GC%: ca. 61. (See also CARBOXYSOMES.)

Nitrobacteraceae (nitrifying bacteria) A family of Gram-negative, asporogenous, obligately aerobic bacteria which occur e.g. in soil and in freshwater and marine habitats; the organisms obtain energy by the oxidation of ammonia or nitrite (see NITRIFICATION), and all except some strains of *Nitrobacter winogradskyi* are obligate chemolithoautotrophs. The family includes coccoid, rod-shaped, spiral and lobate bacteria. Cells may be non-motile or may have polar, subpolar or peritrichous flagella. The cells of many species contain characteristic cytomembranes. Most species divide by binary fission; budding occurs in *Nitrobacter winogradskyi*. Species which oxidize ammonia to nitrite are placed in the following genera: NITROSOCOCCUS, NITROSOLOBUS, NITROSOMONAS, NITROSOSPIRA and NITROSOVIBRIO; nitrite-oxidizing species are placed in the genera NITROBACTER, NITROCOCCUS and NITROSPINA.

nitroblue tetrazolium See TETRAZOLIUM.

nitrocefin A chromogenic cephalosporin used to test for β-LACTAMASES; when cleaved by a β-lactamase it changes from yellow to red. (See also PADAC.)

Nitrococcus A genus of Gram-negative marine bacteria of the NITROBACTERACEAE. Cells of the type species, *N. mobilis*, are coccoid, ca. 1.5–1.8 μm diameter, each having one or two subpolar flagella; the cells contain tubular cytomembranes arranged randomly in the cytoplasm. The cells reproduce by binary fission. GC%: ca. 61.

nitrofurans A family of synthetic ANTIBIOTICS which are derivatives of 2-substituted 5-nitrofurans; they are active against many Gram-positive and Gram-negative bacteria, certain fungi, and some protozoa. Some nitrofurans are used as chemotherapeutic agents, but many are known to be mutagenic/oncogenic. The *antibacterial* activity of a nitrofuran depends on the (stepwise) intracellular reduction of the nitro group, the intermediate(s) formed being able to cause strand breakage in DNA. (cf. NITROIMIDAZOLES.) In e.g. *Escherichia coli*, the reduction of a nitrofuran involves NADH-linked 'nitrofuran reductase 1' (NR1) and NADPH-linked 'nitrofuran reductase 2' (NR2); mutants lacking NR1 exhibit high-level resistance to nitrofurans. (Little is known about plasmid-encoded resistance to nitrofurans; if it exists, the incidence appears to be negligible.) [Resistance to nitrofurans: Book ref. 160, pp. 317–344.] Examples of nitrofurans:

Furaltadone is used e.g. against colibacillosis and salmonellosis in pigs and cattle. (See also PULLORUM DISEASE.)

Furazolidone (*furoxone*) has been used e.g. against bacillary dysentery and *Salmonella* infections in man, and is used e.g. in the treatment of human GIARDIASIS and certain diseases of animals (e.g. BLACKHEAD).

Furylfuramide (*AF-2*) is an oncogenic nitrofuran once used as a food preservative in Japan.

Nifuratel has been used e.g. against vaginal trichomoniasis.

Nifuroxime has been used, topically, as an antifungal agent, being active against e.g. *Candida albicans*; it has also been used (in conjunction with furazolidone) against vaginal trichomoniasis.

Nifurprazine is used as a topical, broad-spectrum antibacterial agent.

Nitrofurantoin (*furadantin*) is the most widely-used nitrofuran. It is used as a broad-spectrum antibacterial agent against infections of the (human) urinary tract; many strains of *Pseudomonas aeruginosa* are resistant, and resistant strains of e.g. *Proteus* also occur.

Nitrofurazone (*furacin*) is used, rarely, for the topical treatment of e.g. infected wounds.

nitrofurantoin See NITROFURANS.

nitrofurazone See NITROFURANS.

nitrogen cycle In nature: the cyclical interconversion of nitrogen and its compounds by living organisms and non-biological processes. The major microbial conversions are shown on page 522. (See also AMMONIA ASSIMILATION, AMMONIFICATION, ASSIMILATORY NITRATE REDUCTION, DENITRIFICATION, DISSIMILATORY NITRATE REDUCTION, DISSIMILATORY REDUCTION OF NITRATE TO AMMONIA, NITRATE RESPIRATION, NITRIFICATION, NITROGEN FIXATION.) (cf. CARBON CYCLE and SULPHUR CYCLE.)

nitrogen fixation (dinitrogen fixation) The reduction of gaseous nitrogen (dinitrogen, N_2) to ammonia. (See also NITROGEN CYCLE.)

Biological nitrogen fixation is carried out only by certain prokaryotes; these nitrogen-fixing organisms are called *diazotrophs*. The diazotrophs include various cyanobacteria (e.g. *Anabaena*, *Nostoc*, *Plectonema*, *Trichodesmium*); certain species of *Bacillus* and *Clostridium; Azospirillum; Klebsiella pneumoniae* (some strains); members of the Azotobacteriaceae, Methylococcaceae, Rhizobiaceae and Rhodospirillales; and certain species of the Archaea.

That *plants* are unable to fix nitrogen is interesting in that, during evolution, plants are believed to have incorporated prokaryotes as organelles (e.g. mitochondria). However, during that period of evolution, prokaryotes may not have yet developed nitrogen-fixing ability and, moreover, plants may not have been subjected to sufficient selection pressure [PTRSLB (1992) *338* 409–416].

The free-living diazotrophs occur e.g. in soil and water, while a number of diazotrophs form symbiotic associations with other organisms (which benefit from a supply of fixed nitrogen). AZOSPIRILLUM fixes nitrogen in the RHIZOSPHERE of various plants, including some important crop plants. Certain species of AZOTOBACTER fix nitrogen in association with the roots of a grass, *Paspalum notatum*, and the surface of the green seaweed *Codium*. *Clostridium* spp are found in waterlogged (anaerobic) soils, and may also fix nitrogen in the intestines of man and other animals.

Frankia spp form symbiotic associations with certain non-leguminous plants (see ACTINORRHIZA), while *Rhizobium/Bradyrhizobium* spp are found in leguminous ROOT NODULES and STEM NODULES.

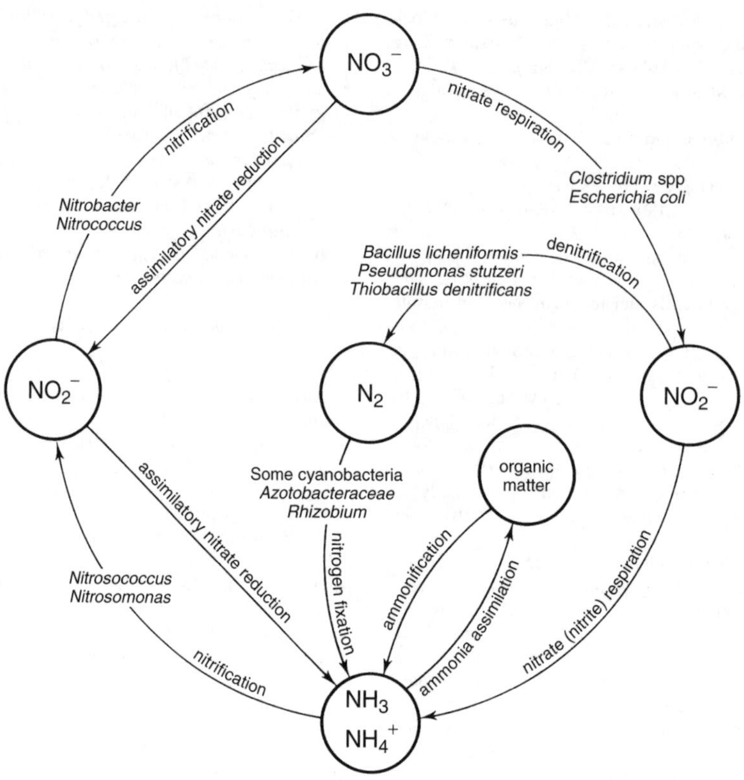

NITROGEN CYCLE – microbial aspects, with examples of organisms. (For cross-references see entry NITROGEN CYCLE.) Not shown is the activity of ANAMMOX BACTERIA (q.v.).

Reproduced from *Bacteria* 5th edition, Figure 10.2, page 258, Paul Singleton (1999) copyright John Wiley & Sons Ltd, UK (ISBN 0471-98880-4) with permission from the publisher.

Nitrogen-fixing cyanobacteria are an important source of nitrogen e.g. in rice paddies; for example, *Anabaena azollae* forms a symbiotic association with a small, floating fresh-water fern, *Azolla*, that is used as 'green manure' in rice-growing regions of the Far East. (See also NOSTOC.) Cyanobacteria can also form associations with certain LICHENS (see e.g. CEPHALODIUM).

Both free-living and symbiotic nitrogen-fixers occur in the open ocean, and such organisms may contribute significantly to the nitrogen budget of these waters. [Nitrogen-fixing microorganisms in tropical and subtropical oceans: TIM (2000) *8* 68–73.]

The benefits enjoyed by those plants that have nitrogen-fixing symbiotic partners have suggested that the growth of certain crop plants, such as cereals, may be improved by infecting them experimentally with nitrogen-fixing organisms. Experiments have been carried out to determine whether *wheat* can be infected with the nitrogen-fixing bacterium *Azorhizobium caulinodans* (which nodulates a tropical legume, *Sesbania rostrata*); *A. caulinodans* became established endophytically (in the xylem and root meristem), and the wheat showed significantly increased dry weight and nitrogen content compared with uninoculated control plants [Proc. RSLB (1997) *264* 341–346].

Nitrogen fixation is catalysed by the enzyme complex NITROGENASE. The reaction requires a strong reducing agent (source of electrons) and consumes ~12–16 molecules of ATP per molecule of nitrogen fixed; electrons from a source such as hydrogen or NADPH are transferred e.g. to a FERREDOXIN or FLAVODOXIN and thence to nitrogenase. (For in vitro work, the artificial reductant dithionite can be used.)

In some diazotrophs (e.g. *Bacillus polymyxa*, *Clostridium pasteurianum*, *Klebsiella pneumoniae*, some cyanobacteria) reducing power and ATP can both be obtained from pyruvate. Reducing power is provided by the oxidative decarboxylation of pyruvate to acetyl-CoA (catalysed by the enzyme pyruvate:flavodoxin oxidoreductase); the acetyl-CoA is converted to acetyl phosphate – from which phosphate can be transferred to ADP in a substrate-level phosphorylation.

The ammonia produced by nitrogen fixation may be assimilated e.g. via the GOGAT pathway (see AMMONIA ASSIMILATION).

Nitrogenase is highly sensitive to oxygen, and many diazotrophs (e.g. clostridia) fix nitrogen anaerobically or microaerobically. In some cyanobacteria the process is carried out within specialized cells (see HETEROCYST). Certain diazotrophs have specific mechanisms for protecting nitrogenase during aerobic nitrogen fixation; for example, rhizobium bacteroids in ROOT

NODULES are protected by LEGHAEMOGLOBIN. (See also TRICHODESMIUM.) Strains of *Azotobacter* and possibly *Gloeothece* [JGM (1984) *130* 495–503] exhibit oxygen-scavenging 'respiratory protection': they show high rates of respiration in the presence of high levels of oxygen, their branched respiratory chains being poorly coupled to ATP synthesis; respiration thus serves primarily to consume oxygen. In those organisms which have an 'uptake hydrogenase' (see NITROGENASE), the transfer of electrons from hydrogen – via an electron transport chain to oxygen (as terminal electron acceptor) – may also help to protect nitrogenase from oxygen. In unicellular, aerobic, nitrogen-fixing cyanobacteria of the genus *Cyanothece*, nitrogen fixation was found to occur only in the dark period during alternating 12-hour light/dark cycles, but was continuous during 24-hour illumination; nitrogen fixation during the dark period would avoid exposure to the oxygen evolved by photosynthesis, but continuous fixation during continual illumination requires additional explanation – and may involve the 'inclusion granules' that develop under nitrogen-fixing conditions [JB (1993) *175* 1284–1292].

[Cyanobacterial nitrogen fixation (oxygen relationships): MR (1992) *56* 340–373. Nitrogen fixation by non-heterocystous cyanobacteria: FEMS Reviews (1997) *19* 139–185.]

nitrogen mustards Highly reactive bifunctional ALKYLATING AGENTS, most of which correspond to the general formula $Cl(CH_2)_2.NR.(CH_2)_2Cl$. They are toxic (e.g. causing blisters on contact with the skin), mutagenic and carcinogenic, and promote IMMUNOSUPPRESSION. Nitrogen mustards (and the SULPHUR MUSTARDS) react with DNA e.g. at the N-7 position of guanine residues, forming mono-adducts as well as inter- and intra-strand cross-links.

nitrogen-regulated genes See NTR GENES.

nitrogenase (dinitrogenase) An enzyme complex (EC 1.18.2.1) which catalyses biological NITROGEN FIXATION: it combines with N_2 and reduces it to NH_3. Nitrogenase can also reduce other substrates – e.g., acetylene, azide, cyanide, nitrous oxide; the reduction of acetylene to ethylene (detectable by gas chromatography) is the basis of an assay for nitrogenase. The nitrogenase complex typically comprises two components: the MoFe (or FeMo) protein (also called e.g. component I, molybdoferredoxin, azofermo, or dinitrogenase), and the Fe protein (= e.g. component II, azoferredoxin, azofer, dinitrogenase reductase). The MoFe protein (MWt ca. 220000–250000) is an $\alpha_2\beta_2$ tetramer containing ca. 50 iron atoms, 2 molybdenum atoms, and ca. 30 acid-labile sulphide groups; the molybdenum occurs in an FeMo cofactor (FeMoco) containing at least 1Mo per 6Fe [ARB (1984) *53* 231–257]. (An alternative nitrogenase has been found in a strain of *Azotobacter chroococcum* in which the MoFe protein is replaced by a vanadium-containing protein [Nature (1986) *322* 388–390].) The Fe protein (MWt ca. 55000–60000) contains two identical protein subunits, 4 iron atoms and 4 acid-labile sulphide groups. The substrate (e.g. N_2) is bound by the MoFe protein and, in the presence of Mg-ATP, electrons are transferred from reduced ferredoxin or flavodoxin to the Fe protein which, in turn, reduces the MoFe protein with concomitant ATP hydrolysis. (At least 2ATP are hydrolysed for each electron transferred.)

In an apparently inevitable side-reaction of nitrogenase, electrons are transferred to protons, resulting in the ATP-dependent evolution of H_2; H_2 evolution thus competes with nitrogen fixation for energy and reducing power. Some organisms (e.g. some *Rhizobium* bacteroids [Book ref. 55, pp. 179–203], *Azotobacter* spp, and some cyanobacteria) possess an oxygen-dependent enzyme system termed 'uptake hydrogenase' which oxidizes the H_2 to yield ATP; this enzyme system also serves to remove H_2 (a competitive inhibitor of nitrogenase) from the active centre of nitrogenase.

Nitrogenase action is slow, and large amounts of the enzyme are necessary for effective nitrogen fixation; the enzyme complex may comprise up to 10% of the cellular protein in an actively diazotrophic cell.

Control of nitrogenase may occur at the transcriptional level (see NIF GENES) or at the level of the enzyme itself. Nitrogenase is inhibited e.g. by H_2, NH_3 (indirectly), and NO_3^- (probably via nitrite); addition of NH_3 to cultures of nitrogen-fixing photosynthetic bacteria or *Azospirillum* results in inactivation of the Fe protein by covalent modification. Nitrogenase is highly oxygen-labile (see NITROGEN FIXATION).

nitroimidazoles A group of synthetic antimicrobial agents which are maximally effective against anaerobes. Their activity appears to depend on at least partial reduction of the nitro group by metabolic pathways linked to low-redox-potential electron transport components (such as ferredoxin) in the susceptible organism; reduction results in the formation of short-lived cytotoxic intermediate(s) – see HYDROGENOSOME. (cf. NITROFURANS.) Nitroimidazoles include e.g. azomycin (2-nitroimidazole), carnidazole [1-(2-ethylcarbamothioic acid *O*-methyl ester)-2-methyl-5-nitroimidazole], DIMETRIDAZOLE, flunidazole [1-(2-hydroxyethyl)-2-(*p*-fluorophenyl)-5-nitroimidazole], METRONIDAZOLE, ornidazole [1-(3-chloro-2-hydroxypropyl)-2-methyl-5-nitroimidazole], panidazole [1-(2-(4-pyridyl)-ethyl)-2-methyl-5-nitroimidazole], secnidazole [1-(2-hydroxypropyl)-2-methyl-5-nitroimidazole], and tinidazole [1-(2-(ethylsulphonyl)-ethyl)-2-methyl-5-nitroimidazole]. [Comparative evaluation of the 2-methyl-5-nitroimidazoles against *Bacteroides fragilis*: AAC (1985) *28* 561–564.]

nitromersol See MERCURY.

nitrophenols (as disinfectants) See PHENOLS.

***N*-nitroso compounds** Compounds which contain the group –N–N=O. Many nitroso compounds are effective MUTAGENS: e.g., dialkylnitrosamines ($R_2N.N{=}O$), *N*-alkylnitrosoureas ($NH_2.CO.NR.N{=}O$), and *N*-alkyl-*N'*-nitro-*N*-nitrosoguanidines ($NO_2.NH.C{=}NH.NR.N{=}O$; see e.g. MNNG) act as ALKYLATING AGENTS (after metabolic activation in the case of dialkylnitrosamines). *Halonitrosoureas* – e.g. 1,3-bis-(2-chloroethyl)-1-nitrosourea (BCNU), bis-(2-fluoroethyl)-nitrosourea (BFNU), and 1-(2-chloroethyl)-3-cyclohexyl-1-nitrosourea (CCNU) – apparently transfer haloethyl, hydroxyethyl and aminoethyl groups to various positions in bases in DNA.

Nitrosococcus A genus of Gram-negative bacteria of the NITROBACTERACEAE; the organisms occur in soil and in freshwater and marine habitats. Cells: cocci, diameter 1.5 μm or greater, which divide by binary fission. *N. mobilis* and *N. oceanus* are motile with one or more flagella; they differ in the arrangement of their cytomembranes. (See also CARBOXYSOMES.) *N. nitrosus* is non-motile. GC% (*N. oceanus*): ca. 51.

nitrosofication See NITRIFICATION.

nitrosoguanidine See MNNG.

Nitrosolobus A genus of Gram-negative bacteria of the NITROBACTERACEAE; the organisms occur in soil. Cells: irregularly rounded ('lobular'), ca. 1.0–1.5 μm in diameter, with 1–20 flagella; the cells divide by binary fission. Cytomembranes compartmentalize the cell into several central regions and a larger number of peripheral regions. GC%: ca. 53–55.

Nitrosomonas A genus of Gram-negative bacteria of the NITROBACTERACEAE; strains occur e.g. in soils and in freshwater and marine habitats. Cells of the type species, *N. europaea*, are typically rod-shaped, ca. 0.8–1.0 × 1.0–2.0 μm, and either non-motile or motile with 1–2 subpolar flagella. The cells contain lamellar cytomembranes in a mainly peripheral arrangement. Marine strains (and some other strains) have an S LAYER. (See also CARBOXYSOMES.) GC%: ca. 47–51.

Nitrosospira A genus of Gram-negative bacteria of the NITROBACTERACEAE; the organisms occur in soil. Under the light microscope the cells appear as rods, ca. 0.3–0.4 × 0.8–1.0 μm, but they appear as tightly coiled spirals under the electron microscope; the cells may be non-motile, or motile with 1–6 flagella, and they divide by binary fission. Cytomembranes are absent. Type species: *N. briensis*. GC%: ca. 54.

nitrosoureas See *N*-NITROSO COMPOUNDS.

Nitrosovibrio A genus of Gram-negative bacteria of the NITROBACTERACEAE. The type species, *N. tenuis*, is represented by a single isolate from soil in the Hawaiian Islands; cells: straight to curved rods, ca. 0.3–0.4 × 1.1–3.0 μm (coccoid in stationary-phase cultures), with 1–4 lateral or subpolar flagella. Extensive cytomembranes are lacking. Cells divide by binary fission.

Nitrospina A genus of Gram-negative marine bacteria of the NITROBACTERACEAE. Cells of the type species, *N. gracilis*, are non-motile rods, ca. 0.3–0.4 × 2.6–6.5 μm, which divide by binary fission; the cells may contain numerous deposits of glycogen. GC%: ca. 58. (See also CARBOXYSOMES.)

nitrous acid (as a MUTAGEN) Nitrous acid (HNO_2) can cause the oxidative deamination of guanine (G), cytosine (C) and adenine (A) to form xanthine (X), uracil (U) and hypoxanthine (HX), respectively. When DNA treated with nitrous acid undergoes replication, X pairs with C which, on subsequent replication, pairs with G: i.e., no mutation results from G deamination. U pairs with A, and HX pairs with C: thus deamination of C and A leads to G·C-to-A·T and A·T-to-G·C transitions, respectively. (HNO_2 can also cause e.g. cross-linking between bases.)

In e.g. *Escherichia coli* the abnormal bases HX and U can be removed from DNA by specific hypoxanthine- and URACIL-DNA GLYCOSYLASES.

Nitschkea See SORDARIALES.

Nitzschia See DIATOMS.

nivalenol See TRICHOTHECENES.

NK cells Natural killer cells: lymphoid cells, derived from bone marrow stem cells, which are related to, but antigenically distinct from, T LYMPHOCYTES; NK cells do not express the T cell receptor (TCR), and do not express CD3, CD4 or CD8 antigens but they do express the CD16 antigen. NK cells are classified as *large granular lymphocytes* (LGLs).

NK cells have little phagocytic activity, but they can exert cytotoxic/cytolytic activity against various pathogens, against host cells infected with certain bacteria (e.g. *Listeria monocytogenes*) or viruses, and against certain tumours; moreover, NK cells can be stimulated to release CYTOKINES, including IFN-γ (see INTERFERONS).

NK cells may bind to target cells (e.g. virus-infected cells) via the CD16 receptor – i.e. CD16 binding to the Fc portion of antibodies that have bound to the target cell (see ADCC). Alternatively, binding to target cells may occur in an antibody-independent way – receptors on NK cells binding directly to ligands on the target cell.

The killing activity of NK cells appears to involve the release of perforins and granzymes which induce APOPTOSIS in target cells.

NK cells can attack virus-infected host cells which do *not* express viral antigen, linked to MHC class I molecules, at the cell surface; such cells would not normally be targeted by $CD8^+$ cytotoxic T cells – which are MHC class I-restricted. (NK cells are not MHC class I-restricted; in fact, class I molecules inhibit the killing activity of NK cells.) This role of NK cells appears to be particularly valuable in that many viruses are able to inhibit the display of MHC class I-linked viral antigens at the cell surface.

The killing activity of NK cells is stimulated by certain cytokines, including INTERLEUKIN 12.

NLF A non-lactose-fermenting bacterium.

nm (nanometre) 10^{-9} m, or 10^{-3} μm.

NMR NUCLEAR MAGNETIC RESONANCE.

NNN medium (Novy, Nicolle and MacNeal's medium) A DIPHASIC MEDIUM used e.g. for the culture of certain *Leishmania* spp and stercorarian trypanosomes. It is a blood agar SLOPE (containing defibrinated mammalian blood) to which has been added a small volume of e.g. Locke's solution containing antibiotics to inhibit bacterial growth. The inoculated slope is incubated (e.g. 25–28°C) for days or weeks; a loopful of fluid can be removed at intervals for examination.

In culture, pathogenic flagellates may form only certain morphological types; *Leishmania donovani* grows in the promastigote form, although *Trypanosoma cruzi* cultures contain e.g. epimastigotes and metacyclic forms.

NNRTI NON-NUCLEOSIDE REVERSE TRANSCRIPTASE INHIBITOR.

Noble agar A purified (washed) form of AGAR.

noble rot (*pourriture noble; Edelfäule*) A rot of grapes caused by *Botrytis cinerea*. On ripe or over-ripe white grapes the mould penetrates the grape skins and leads to loss of water; must from such grapes is thus more concentrated than normal and is used for making high-quality sweet wines (see WINE-MAKING). (The same mould on unripe or black grapes renders them useless for wine-making.)

nocardamine A colourless, trihydroxamate SIDEROPHORE produced e.g. by certain actinomycetes and by *Pseudomonas stutzeri* under iron-deficient conditions. [JGM (1980) *118* 125–129.]

Nocardia A genus of Gram-positive, aerobic, chemoorganotrophic, non-motile bacteria (order ACTINOMYCETALES, wall type IV) which occur e.g. in soil and as pathogens of man and other animals (see e.g. MADUROMYCOSIS, NOCARDIOSIS; see also NOCARDICINS); the degree of acid-fastness appears to be variable even within a given isolate. The organisms typically form a branched (substrate and aerial) mycelium (hyphae 1 μm or less in diam.) which, according to species, tends to fragment more or less readily into rods and coccoid forms, some of the latter being heat- and/or desiccation-resistant (see MICROCYST (1)). Cell walls contain MYCOLIC ACIDS. Growth occurs readily at 30°C and 37°C on enriched media; colonies are often wrinkled or ridged, and may be any of a variety of colours. Glucose is metabolized oxidatively. GC%: 64–72. Type species: *N. asteroides*.

The genus includes e.g. *N. amarae* (isolated e.g. from the foam on an activated-sludge sewage treatment plant) and also *N. asteroides*, *N. brasiliensis* and *N. farcinica* (Kyoto-1 group), *N. otitidis-caviarum*, and *N. vaccinii* (a species reported to produce galls in the blueberry plant). '*N. autotrophica*', '*N. mediterranea*' and '*N. rugosa*' each have type IV actinomycete walls but they do not contain mycolic acids and have been removed from the genus [Book ref. 73, pp. 89–91, 129].

(See also e.g. HYDROCARBONS and RUBBER SPOILAGE.)

nocardicins A class of monocyclic β-LACTAM ANTIBIOTICS (q.v. for structure) produced by a species of *Nocardia*.

nocardioform actinomycetes A category within the order ACTINOMYCETALES which includes *Caseobacter*, *Corynebacterium*,

Mycobacterium, *Nocardia* and *Rhodococcus* – organisms which form either non-mycelial growth or a fragmenting mycelium and which have a cell wall of chemotype IV. (cf. SPOROACTINOMYCETES.)

Nocardioides A genus of aerobic soil bacteria (order ACTINOMYCETALES, wall type I) which form branching substrate and aerial mycelia that fragment into coccoid and rod-shaped pieces. GC%: ca. 66–67. Type species: *N. albus*. [Book ref. 73, pp. 100–101.]

Nocardiopsis A genus of bacteria (order ACTINOMYCETALES, wall type III) which occur e.g. in soil. The organisms form branching, fragmenting substrate mycelium, and an aerial mycelium which gives rise to long chains of spores – each spore tending to form at an angle to its neighbours, thus giving the chain an undulating appearance. GC%: ca. 70–76. Type species: *N. dassonvillei* (formerly *Actinomadura dassonvillei*). [Book ref. 73, pp. 117–118.]

nocardiosis An acute or chronic disease of man and animals, caused by a species of *Nocardia* (commonly *N. asteriodes*). The disease may be systemic: suppurative lesions are formed mainly in the lungs, but the infection may disseminate to other organs (brain, kidney etc). Nodular, often suppurating, subcutaneous lesions may result from disseminated infection or from primary infection via a wound. (See also MADUROMYCOSIS.)

nocardomycolic acids MYCOLIC ACIDS of *Nocardia* spp.

Noctiluca A genus of marine, bioluminescent, heterotrophic (holozoic) DINOFLAGELLATES. The vegetative cells are large (e.g. 200–2000 μm diam.), roughly spherical (with no girdle), and have highly vacuolated cytoplasm; the flagella are very short, the transverse flagellum being reduced to a small projection. There is a deep sulcus from the region of which arises a retractable *tentacle*; the tentacle is apparently sticky and aids in the capture of prey. A cytostome is present near the base of the tentacle. Some strains of *Noctiluca* contain endosymbiotic unicellular green flagellates in vacuoles within the cell. (See also BIOLUMINESCENCE.)

Nodamura virus See NODAVIRIDAE.

Nodaviridae (Nodamura virus group) A family of ssRNA viruses which infect insects of the Diptera, Coleoptera and Lepidoptera. The virions are roughly spherical, naked, 29 nm diam., and each contains one major polypeptide species and two molecules of non-polyadenylated ssRNA (MWts ca. 1.15 and 0.46×10^6), both of which are necessary for infection. The type species, Nodamura virus, was isolated from mosquitoes (*Culex tritaeniorhynchus*) in Japan; it can replicate in various insects, can cause a lethal infection in suckling mice, and can replicate in various vertebrate and invertebrate cell cultures (including e.g. BHK cells). Other member: Black beetle virus, isolated from the New Zealand Black beetle, *Heteronychus arator*. Possible members include e.g. Arkansas bee virus.

Nodosaria See FORAMINIFERIDA.

Nodularia A genus of filamentous freshwater and marine (littoral) CYANOBACTERIA (section IV) in which the trichomes are made up of disc-shaped vegetative cells. Heterocysts may be intercalary or terminal. GC%: ca. 41. *N. spumigena* is a gas-vacuolate, primarily freshwater, bloom-forming species which may form toxin(s) responsible for severe vomiting and bloody diarrhoea, or chronic liver and kidney damage, in animals which drink the water.

nodules (*plant microbiol.*) See ACTINORRHIZAE; PHYLLOBACTERIUM; ROOT NODULES; STEM NODULES.

nodulins ROOT NODULE-specific proteins specified by host plant genes; they appear to be involved in symbiotic nitrogen fixation and nitrogen metabolism. *C-nodulins* ('common nodulins') occur in all (leguminous) root nodules; *S-nodulins* are specific to the nodules of certain plant species. [Book ref. 55, pp. 79–83; review of nodulins and nodulin genes in *Glycine max* (soybean): PMB (1986) 7 51–61.]

nogalamycin See ANTHRACYCLINE ANTIBIOTICS.

nom. cons. *Nomen conservandum*: a CONSERVED NAME.

nom. dub. NOMEN DUBIUM (q.v.).

nom. nud. NOMEN NUDUM (q.v.).

nom. rejic. NOMEN REJICIENDUM (q.v.).

nom. rev. NOMEN REVICTUM.

Nomarski interference-contrast microscopy See MICROSCOPY (d).

nomen conservandum (nom. cons.) A CONSERVED NAME.

nomen dubium (nom. dub.) A name which cannot, with certainty, be applied to a specific organism; the uncertainty may be due e.g. to an incomplete description of an organism in the original publication.

nomen nudum (nom. nud.) In NOMENCLATURE: a name which, when proposed, was not supported by an adequate description of the relevant taxon.

nomen rejiciendum (nom. rej.; nom. rejic.) Any name which has been rejected by a recognized taxonomic authority.

nomen revictum (nom. rev.) A revived name.

nomenclatural type (name bearer, nominifer, or 'type') Commonly, a viable specimen of a given organism to which a name has been attached and which is regarded as the permanent representative, or reference specimen, of a given taxon; otherwise, e.g., a preserved specimen or a photographic (or other) representation – sometimes simply a description – of an organism to which is permanently attached the name of a given taxon. (See also TYPE SPECIES and TYPE STRAIN.) [Bacterial types: Book ref. 22, p. 21.]

nomenclature (*microbiol.*) The naming of organisms: one of the functions of TAXONOMY. Nomenclature is formally governed by an array of rules and priorities promulgated by various ad hoc bodies – each concerned with a particular type of organism; thus, e.g. bacterial names are regulated by the International Committee on Systematic Bacteriology which promulgates the International Code of Nomenclature of Bacteria [Book ref. 71 and subsequent revisions]. (Nevertheless, such rules and regulations are not always universally adhered to – due e.g. to a lack of awareness or to disagreement with published opinions.)

With some exceptions (see later) each species of microorganism is referred to by a unique BINOMIAL. (The names of species and genera are usually printed in italics; in typewritten and handwritten manuscripts such names are usually underlined.) Where ambiguity cannot arise the generic name may be abbreviated; thus, e.g. *Staphylococcus aureus* may be written *Staph. aureus* or *S. aureus*. The name of a species is sometimes followed by an authority citation; this gives the name(s) of the author(s) who first proposed the name in a 'valid publication' and may also give the date of that publication – e.g. *Bacillus cereus* Frankland and Frankland 1887. (A *valid publication* is one which contains an adequate description of a newly proposed taxon and a correctly formulated Latin or Latinized name; it must be in printed form and must be available to the public or to the scientific community. A publication which lacks an adequate description but which fulfils the other requirements is called an *effective publication.*) Sometimes a double citation is used – e.g. *Escherichia coli* (Migula 1895) Castellani and Chalmers 1919; here, Migula is the author who described the organism under the name *Bacillus coli*, and Castellani and Chalmers are the authors who first described

NOMENCLATURE: some suffixes used in microbial nomenclature

	Algae	Bacteria	Fungi	Protozoa	Viruses
Division (phylum)	-phyta		-mycota	-a	
Subdivision (subphylum)	-phytina		-mycotina	-a	
Superclass				-a	
Class	-phyceae		-mycetes	-ea	
Subclass	-phycidae		-mycetidae	-ia	
Superorder				-idea	
Order	-ales	-ales	-ales	-ida	
Suborder	-ineae	-ineae	-ineae	-ina	
Superfamily				-oidea	
Family	-aceae	-aceae	-aceae	-idae	-viridae
Subfamily	-oideae	-oideae	-oideae	-inae	-virinae
Tribe	-eae	-eae	-eae	-ini	
Subtribe	-inae	-inae	-inae		
Genus					-virus

the organism as *Escherichia coli*. (Sometimes an organism is known by one or more trivial names, as well as by its scientific name; these may be descriptive – e.g. 'orange-peel fungus' – or may have historical or other interest.)

Taxa below the rank of species may be referred to by the binomial of the given species followed by a further designation(s): e.g., *Campylobacter sputorum* subsp *bubulus*, *Fusarium solani* f. sp. *pisi*, *Pseudomonas syringae* pv. *savastanoi*, *Puccinia graminis* var. *phlei-pratensis* Race 1. (See also TERNARY COMBINATION.)

A taxon of the rank genus or above is referred to by a unique uninomial (one-name designation); the rank of a particular named taxon is often indicated by a standardized suffix (see table above).

While an organism may bear only one valid taxonomic name, certain fungi may be referred to by a second binomial (see e.g. FORM GENUS); in deuteromycetes the name of the TELEOMORPH (where known) is the valid taxonomic name – that of the anamorph sometimes being written, in parenthesis, afterwards: e.g. *Ajellomyces dermatitidis* (*Blastomyces dermatitidis*).

nomenspecies (1) One or more strains designated by a given binomial, i.e., a strain or group of strains treated as a species but not necessarily recognized as such taxonomically. (2) A type species.

nominifer *Syn.* NOMENCLATURAL TYPE.

Nomuraea A genus of fungi of the HYPHOMYCETES. *N. rileyi* has been evaluated for the BIOLOGICAL CONTROL of caterpillars of noctuid moths; it can be grown on artificial media [inexpensive media and methods: J. Inv. Path. (1986) *48* 246–248], and its conidia can remain infective when sprayed onto crops. Conidia germinate on the surface of a caterpillar, and hyphae penetrate the insect, subsequently growing in the haemocoel; the caterpillar usually dies within 1–2 days and becomes covered with a dense mycelial mat from which conidia are eventually produced.

non-A–non-B hepatitis (NANB hepatitis) See HEPATITIS C.

nonactin See MACROTETRALIDES.

non-agglutinable vibrios *Syn.* NON-CHOLERA VIBRIOS.

non-agglutinating antibodies BLOCKING ANTIBODIES (sense 1).

non-brewed condiment See VINEGAR.

non-cellular *Syn.* ACELLULAR.

non-cholera vibrios (NCVs; non-agglutinable vibrios) Those strains of *Vibrio cholerae* which are not agglutinated by O1 (or O139) antiserum; strains of *V. cholerae* which cause CHOLERA (q.v.) are characteristically agglutinated by O1 (or O139) antiserum. NCVs are classified serologically into serogroups O2 to O138; they are generally regarded as less pathogenic than O1 (or O139) strains, although they may cause mild to severe, cholera-like disease.

(See also NCVP.)

non-circulative transmission (of viruses) (*plant pathol.*) (1) A mode of transmission in which viruses are temporarily associated with surface region(s) of an insect VECTOR; non-circulatively transmitted viruses can be acquired from an infected plant and transmitted to a healthy plant within seconds or minutes, i.e., there is no latent period (cf. CIRCULATIVE TRANSMISSION). A distinction is sometimes made between those cases in which virus is retained by the vector for minutes or hours (*non-persistent transmission*) and those cases in which virus is retained for up to several days (*semi-persistent transmission*).

In non-circulative transmission the insect's ability to transmit a previously acquired virus is lost during moulting. When an insect moults it replaces its external body covering, parts of the pharynx, and the anterior and posterior regions of the gut lining; hence, any virus carried superficially is shed, and the loss of infectivity during moulting is regarded as a diagnostic feature of non-circulative transmission.

In many cases, non-circulative transmission involves more than the simple mechanical contamination of the surface of an insect vector. For example, aphid-borne transmission of *purified* POTYVIRUSES occurs only if the aphid vector acquires a 'helper component' (HC) before or simultaneously with – but not subsequent to – acquiring the virus; HC is a virus-encoded protein (serologically distinct from either virus structural protein or 'pinwheel' protein) which occurs in potyvirus-infected leaves, and which can be obtained in the laboratory by cell-free translation of viral RNA. The *uptake* of virus by the aphid vector is apparently not influenced by the presence or absence of HC [MS (1984) *1* 191–193]. Interestingly, the HC of a given potyvirus can also promote the aphid-borne transmission of certain other potyviruses – and even some unrelated viruses (e.g. potato aucuba mosaic virus) – which are not aphid-transmissible in the absence of HC. (HC is not necessary for mechanical, i.e., manual, infection of plants by potyviruses.) In certain CAULIMOVIRUSES, aphid-borne transmission is – as in potyviruses – dependent on helper factors (termed 'transmission factors' in this virus group), and the aphid-borne transmission of certain caulimoviruses which are not normally aphid-transmissible can be promoted by transmission factor from other caulimoviruses [JGV (1985) *66* 921–923].

(2) The term 'non-circulative transmission' (or 'non-persistent transmission') is sometimes used to refer to the transmission of viruses by vectors other than insects; examples include transmission of SOIL-BORNE WHEAT MOSAIC VIRUS and TOBACCO NECROSIS VIRUS by species of *Polymyxa* and *Olpidium*, respectively, and the transmission of NEPOVIRUSES by nematodes of the genera *Longidorus* and *Xiphinema*.

non-coding strand (of dsDNA) See CODING STRAND.

non-cognate interaction (*immunol.*) Interaction between a B cell and a T cell in which each cell recognizes, and binds to, a separate (physically distinct) molecule (cf. COGNATE HELP); the T cell may influence the B cell via soluble factors (i.e. CYTOKINES).

non-complementing diploid See PROTOPLAST FUSION.

non-conjugative plasmid (1) As commonly used: a PLASMID which does not encode one or more of the functions needed for its own intercellular transmission by CONJUGATION (sense 1b). (cf. CONJUGATIVE PLASMID.)

(2) According to some authors [ARG (1979) *13* 99–125 (105–106)]: a plasmid which does not encode one or more of the functions (e.g. pilus formation) needed for specific cell–cell contact, and which may or may not be capable of being *mobilized*.

non-disjunction Atypical distribution of chromosomes or chromatids (or parts thereof) to the poles of the spindle during MEIOSIS or MITOSIS. (See e.g. PARASEXUAL PROCESSES.)

non-genetic reactivation (*virol.*) A phenomenon in which an inactivated virus (which cannot complete its replication cycle when infecting cells on its own) can produce viable progeny virions in the presence of a second, closely related virus as a result of COMPLEMENTATION between viral proteins. An example is the 'Berry–Dedrick phenomenon' in which rabbits infected with a mixture of heat-inactivated myxoma virus and infectious (non-damaged) rabbit fibroma virus may develop myxomatosis. (cf. REACTIVATION.)

non-gonococcal urethritis (NGU; non-specific urethritis, NSU) Sexually transmitted URETHRITIS caused by any organism other than *Neisseria gonorrhoeae* (cf. GONORRHOEA and URINARY TRACT INFECTION). The causal agent is commonly *Chlamydia trachomatis* which infects the epithelium of the urinary tract. Symptoms of chlamydial infection may be mild, or there may be severe urethritis with a profuse purulent discharge (usually in males); the infection may spread to the genital tract epithelium, causing e.g. epididymitis in males, salpingitis in females [BMB (1983) *39* 133–150]. Other causal agents of NGU include e.g. *Candida albicans*, *Trichomonas vaginalis*, *Mycoplasma hominis*, and (possibly) *Ureaplasma urealyticum*. *Chemotherapy*: tetracyclines.

non-haem iron protein A category of iron-containing proteins in which the iron is not present as part of a HAEM (sense 1) moiety; they were once believed to include all IRON–SULPHUR PROTEINS but some of the latter are now known to contain e.g. SIROHAEM.

non-histone proteins A range of different proteins, other than HISTONES, associated with CHROMATIN; many occur in relatively small amounts. Examples include those proteins involved in transcription (e.g. RNA polymerases) and so-called *HMG proteins* (high-mobility group proteins) – the latter being characterized by their DNA-binding domains (see HMG BOX).

The HMG1 protein is a small (25 kDa), highly charged protein which contains two similar HMG boxes (designated A and B). It occur in large numbers (e.g. 10^6 copies per nucleus) and binds to dsDNA in an apparently non-sequence-specific manner – although it binds strongly to specific DNA *structures* such as angular or kinked DNA or four-way junctions. HMG1 can also interact with certain other DNA-binding proteins and, in this way, can cause sharp bends to develop in dsDNA; this ability can enhance the activity of certain transcription factors, and HMG1 apparently affects the expression of particular genes.

HMG1 also has extra-nuclear activity. In vitro, it can be secreted by macrophages on stimulation by CYTOKINES or ENDOTOXIN, and it may have a role e.g. in inflammation. More recently it has been reported that HMG1 is released by certain damaged or necrotic cells, including endothelial cells, and that it can act as a chemoattractant, induce cell migration and bring about reorganization in the cytoskeleton in certain cells [JCB (2001) *152* 1197–1206].

Nonidet P-40 A non-ionic detergent of the polyoxyethylene *p-t*-octyl phenol series.

Nonion See FORAMINIFERIDA.

non-Mendelian inheritance *Syn.* CYTOPLASMIC INHERITANCE.

non-nucleoside reverse transcriptase inhibitors (NNRTIs) A category of ANTIRETROVIRAL AGENTS (q.v.) that includes delavirdine, efavirenz and nevirapine. The NNRTIs are structurally diverse compounds; they bind to reverse transcriptase and inhibit enzyme activity. The clinically used NNRTIs are reported to be effective against the reverse transcriptase of HIV-1 but not against that of HIV-2. (See also NU-1320.)

[NNRTIs and other antiretroviral agents in the treatment of AIDS: BMJ (2001) *322* 1410–1412.]

non-O1 *Vibrio cholerae* pilus See NCVP.

non-permissive conditions See CONDITIONAL LETHAL MUTANT.

non-persistent transmission See NON-CIRCULATIVE TRANSMISSION.

non-photochromogen A term used (very loosely) to refer e.g. to any strain of *Mycobacterium* which (i) forms little or no pigment, (ii) does not form pigment during early growth, and/or (iii) does not need to grow in the light in order to produce pigment (cf. SCOTOCHROMOGEN).

non-propagative viruses See CIRCULATIVE TRANSMISSION.

non-reciprocal recombination See RECOMBINATION.

nonsense codon See GENETIC CODE.

nonsense mutation A MUTATION in which a codon specifying an amino acid is altered to a nonsense (chain-terminating) codon: e.g. an amber codon (*amber mutation*) or an ochre codon (*ochre mutation*). (cf. MIS-SENSE MUTATION.)

nonsense suppressor See SUPPRESSOR MUTATION.

non-specific immunity The constitutive (or, in certain cases, acquired) resistance of the body to foreign materials in general – including those (living and non-living) materials with which it has had no previous contact. Non-specific immunity involves e.g. PHAGOCYTOSIS and the activities of INTERFERONS, LYSOZYME and PROPERDIN. (cf. SPECIFIC IMMUNITY.)

non-specific immunization The administration of certain immunostimulants (e.g. BCG) in order to elicit a heightened immune response to antigen(s) unrelated to the immunostimulant.

non-specific immunotherapy The therapeutic use of NON-SPECIFIC IMMUNIZATION.

non-specific urethritis *Syn.* NON-GONOCOCCAL URETHRITIS.

non-specific vaginitis *Syn.* BACTERIAL VAGINOSIS.

non-sterilizing immunity *Syn.* PREMUNITION.

non-steroidal anti-inflammatory drugs (NSAIDs) See PROSTAGLANDINS.

non-sulphur purple bacteria See RHODOSPIRILLACEAE.

non-treponemal tests *Syn.* STANDARD TESTS FOR SYPHILIS.

non-vibrio cholera Any CHOLERA-like disease caused by organisms other than *Vibrio cholerae* – e.g. enterotoxigenic strains of *Escherichia coli* (see ETEC).

nopaline An opine: N^2-(1,3-dicarboxypropyl)-L-arginine – see CROWN GALL.

nopalinic acid See CROWN GALL.

2′-nor-2′-deoxyguanosine See DHPG.

norfloxacin See QUINOLONE ANTIBIOTICS.

nori See LAVER.

Normandina A genus (*incertae sedis*) of LICHENS; photobiont: *Trebouxia*. The thallus is squamulose, the squamules being shell-shaped, 1–2 mm across, bluish-green, with raised, often sorediate margins. Fruiting structures are unknown; however, ascocarps of a (probably parasitic) fungus (*Sphaerulina chlorococca*) may be visible as dark dots on the surface of the thallus. The sole species, *N. pulchella*, occurs as an epiphyte on bryophytes.

North American blastomycosis *Syn.* BLASTOMYCOSIS.

Northern blot See BLOTTING.

Norwalk virus (Norwalk agent) A small (27 nm) non-enveloped virus categorized as a SMALL ROUND STRUCTURED VIRUS (q.v.); it was first described in Norwalk, Ohio, in 1972. Norwalk virus can cause outbreaks of diarrhoeal disease/gastroenteritis, and is transmitted by the faecal–oral route. (cf. SNOW MOUNTAIN AGENT; see also WATER SUPPLIES and FOOD POISONING (i).)

nose microflora See RESPIRATORY TRACT MICROFLORA.

Nosema A genus of protozoa (class MICROSPOREA) which are parasitic mainly in insects.

N. apis is parasitic in epithelial cells of the midgut in the honeybee (*Apis mellifera*); spores liberated by disruption of the epithelial cells are voided in the faeces. Infection occurs by ingestion; transovarial transmission apparently does not occur. (See also BEE DISEASES.)

N. bombycis causes *pébrine* – a commonly fatal disease of silkworms (larvae of the silk-moth, *Bombyx mori*). Infection may occur by ingestion of mulberry leaves contaminated with spore-containing faeces from an infected silkworm. The spores germinate in the gut; reproduction occurs first in gut epithelium–but later in various tissues. Infected silkworms have characteristic dark spots. Transmission can occur transovarially.

N. cuniculi: see ENCEPHALITOZOON.

Other species may parasitize mosquitoes, moths, fish, etc; *N. algerae* and *N. locustae* have been tested for use in the BIOLOGICAL CONTROL of mosquitoes and grasshoppers, respectively.

nosocomial disease A disease acquired in hospital. Infection rates of about 30% have been reported for patients in adult intensive-care units [review: Lancet (2003) *361* 2068–2077].

Nostoc A genus of filamentous CYANOBACTERIA (section IV) in which the vegetative cells are spherical, ovoid or cylindrical; young trichomes formed from hormogonia have terminal heterocysts, but in longer trichomes heterocysts may also occur in intercalary positions. Akinetes, if formed, often occur in chains and develop half-way between two heterocysts. In *N. muscorum*, at least, gas vacuoles occur only in hormogonia (see HORMOGONIUM). GC%: 39–46.

In nature, free-living *Nostoc* spp occur in various habitats (e.g. river-beds, wet rocks, damp soil), typically forming green to blackish gelatinous colonies up to several centimetres across; the colony form is rarely retained in culture.

Nostoc spp form symbiotic associations with a wide range of eukaryotes, the latter deriving benefit from the ability of the cyanobiont to fix nitrogen. In LICHENS, *Nostoc* may be the main photobiont (e.g. in *Collema*, some *Peltigera* spp) or may occur in discrete cephalodia (see CEPHALODIUM). In liverworts (e.g. *Blasia*, *Cavicularia*) and hornworts (*Anthoceros*), *Nostoc* occurs in cavities in the underside of the thallus. Among angiosperms, *Gunnera* (a genus of rhubarb-like marsh plants) contains '*Nostoc punctiforme*' in mucilage-filled glands in the plant stem at the bases of petioles; the cyanobiont penetrates between the cortical cells and enters meristematic cells, thus becoming an intracellular symbiont. [Development of *Nostoc* in *Gunnera* mucilage glands: Bot. Gaz. (1985) *146* 56–62.] Cycads are the only gymnosperms known to be capable – via symbionts – of nitrogen fixation; they develop coralloid root nodules which contain symbiotic nitrogen-fixing cyanobacteria (*Nostoc*) in mucilage-filled spaces in the root cortex [ultrastructural studies of *Nostoc–Zamia* association: New Phyt. (1985) *101* 707–716].

Nostocales See CYANOBACTERIA.

notatin A flavoprotein GLUCOSE OXIDASE produced intracellularly e.g. by '*Penicillium notatum*' (= *P. chrysogenum*). It can show antimicrobial activity due to the production of HYDROGEN PEROXIDE.

***Not*I** See RESTRICTION ENDONUCLEASE (table).

notifiable diseases Those diseases (of man, animals or plants) the occurrence (or suspected occurrence) of which must, by law, be notified to the appropriate authorities.

novel antigen Any weakly immunogenic or non-immunogenic molecule which can be made (more) immunogenic by CONJUGATION (sense 2) with an immunogenic molecule. (cf. HAPTEN; NEOANTIGEN.)

novobiocin (albamycin) An ANTIBIOTIC (produced e.g. by *Streptomyces spheroides* and *S. niveus*): 4,7-dihydroxy-3-amino-8-methylcoumarin linked via the 3-amino group to a substituted *p*-hydroxybenzoic acid and via the 7-hydroxy group to a substituted hexose (*noviose*). Novobiocin is more active against Gram-positive than Gram-negative bacteria; resistance develops readily. It primarily inhibits synthesis of DNA and, to a lesser extent, RNA. It selectively and competitively inhibits the binding of ATP to GYRASE B subunits; thus DNA supercoiling by gyrase is inhibited, but the (ATP-independent) DNA-relaxing gyrase activity is not. (See also CURING.) Secondary effects of novobiocin include inhibition of protein and cell wall synthesis. (See also COUMERMYCINS; cf. NALIDIXIC ACID.)

Novyella A subgenus of PLASMODIUM.

Nowakowskiella A genus of eucarpic, polycentric fungi (order CHYTRIDIALES) which are typically aquatic saprotrophs. The vegetative thallus includes a RHIZOMYCELIUM whose branching rhizoids penetrate the substratum. Operculate zoosporangia are formed, and each liberated zoospore germinates to form a rhizomycelium. Thick-walled resting spores are formed on pseudoparenchymatous structures, but whether or not these are sexual structures is unknown.

noxythiolin (*N*-methyl-*N*′-hydroxymethylthiourea) A broad-spectrum antimicrobial agent which has been used e.g. for the treatment of bacterial peritonitis; its activity appears to depend on the release of FORMALDEHYDE. Noxythiolin can kill blastospores of *Candida albicans* in vitro [JAB (1986) *60* 319–325].

nPCR See NESTED PCR.

NPV NUCLEAR POLYHEDROSIS VIRUS.

NR1 plasmid See R100 PLASMID.

***nrdA* gene** (*dnaF* gene) A gene which codes for ribonucleotide reductase.

***nrdB* gene** See SPLIT GENE (e).

***nrdD* gene** See SPLIT GENE (e).

NRRL Culture collection of the Northern Regional Research Laboratories, US Dept. Agriculture, 1815 North University St, Peoria, Illinois 61604, USA.

NRTIs NUCLEOSIDE REVERSE TRANSCRIPTASE INHIBITORS.

***Nru*I** See RESTRICTION ENDONUCLEASE (table).

N_S protein See ADENYLATE CYCLASE.

NSAIDs Non-steroidal anti-inflammatory drugs. (See also PROSTAGLANDINS.)

NSU See NON-GONOCOCCAL URETHRITIS.

nt Abbreviation for nucleotide(s).

NTA Nitriloacetic acid, $N(CH_2COOH)_3$ (= triglycollamic acid), a chelating agent which complexes a range of cations (e.g. Ca^{2+}, Zn^{2+}) – though less efficiently than does EDTA. It may be carcinogenic.

NTG Nitrosoguanidine: see MNNG.

NTM Non-tuberculous mycobacteria (see MOTT).

NTP NUCLEOSIDE-5′-triphosphate.

***ntr* genes** In enterobacteria, *rpoN* (= *ntrA*, *glnF*), *ntrB* (= *glnL*) and *ntrC* (= *glnG*): genes whose products control the expression of *nitrogen-regulated genes* – the latter including e.g. *glnA* (encoding glutamine synthetase, GS), NIF GENES, and genes concerned with transport and degradation of amino acids; transcription of nitrogen-regulated genes is typically repressed in the presence of high concentrations of ammonia and induced when ammonia becomes limiting. (See also AMMONIA ASSIMILATION.)

rpoN (*ntrA*) encodes an enhancer-dependent SIGMA FACTOR, σ^{54}, involved in the control of initiation of transcription of nitrogen-regulated genes. [σ^{54} (minireview): JB (2000) *182* 4129–4136.] NtrC is an *activator* protein that permits transcription from $E\sigma^{54}$ (see SIGMA FACTOR). NtrB appears to be involved in the activation of NtrC.

The *glnA* operon in *Salmonella typhimurium* contains at least two promoters. P2, nearest the structural genes, is the major nitrogen-regulated promoter; transcription from P2 requires the *rpoN* and *ntrC* gene products. ($E\sigma^{70}$ apparently cannot initiate transcription from P2.) P1 is a secondary promoter, upstream from P2, which is activated by the cAMP–CRP complex (see CATABOLITE REPRESSION); it is not dependent on the *rpoN* and *ntrC* gene products.

In *Klebsiella pneumoniae*, transcription of all the NIF GENES is dependent on the sigma factor σ^{54} (*rpoN* gene product). Transcription of the *first* operon, *nifLA*, requires the NtrC activator; after transcription of this operation, the NifA product functions as an activator for σ^{54} in all subsequent *nif* genes/operons (see figure). However, while NtrC and NifA are functionally related, NtrC – but not NifA – is activated by phosphorylation.

nu particle See PSEUDOCAEDIBACTER.

NU-1320 (*syn.* NU1320) A candidate anti-AIDS drug within the category NON-NUCLEOSIDE REVERSE TRANSCRIPTASE INHIBITORS. It is reported to be several million times more potent than AZT but to have a very low level of toxicity. The drug was discovered by Tronchet's team in Geneva and is being developed by the French company Mayoly Spindler.

nuclear area (nuclear body) (*bacteriol.*) See CHROMOSOME.

nuclear dehydrogenating clostridia See GASTROINTESTINAL TRACT FLORA.

nuclear factor-κB See CYTOKINES.

nuclear localization signal See CROWN GALL.

nuclear magnetic resonance (NMR) The absorption and consequent emission of electromagnetic radiation by the nuclei of certain types of atom in the presence of a strong magnetic field. This phenomenon is the basis of a non-destructive, non-invasive method used (e.g. in biochemistry, medicine and microbiology) for elucidating the structures, conformations and interactions of molecules in both in vitro and in vivo systems; NMR is used primarily for investigating small molecules (MWt <2000) in solution.

Principle. The types of nuclei particularly suitable for NMR studies occur in those atoms which have an *odd* mass number. In such atoms (e.g. ^1H, ^{13}C, ^{31}P) the nucleus has a magnetic moment, and in the presence of a strong external magnetic field the nuclear magnetic axis 'wobbles' about the field's axis with a characteristic *precession* frequency; exposure to a pulse of radiation at the precession frequency will increase the amplitude of the wobble (= nuclear magnetic resonance). When the resonance decays it generates electromagnetic radiation which

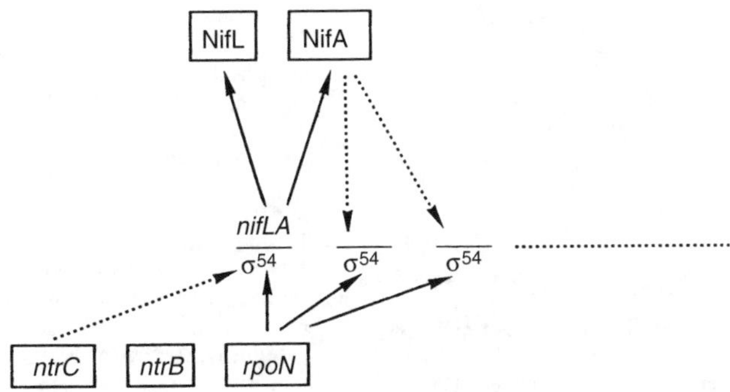

ntr GENES: control of the *nif* genes by the *ntr* genes in *Klebsiella pneumoniae* (diagrammatic) (see entry). Transcription of all the *nif* genes/operons requires the enhancer-dependent sigma factor σ^{54} (*rpoN* gene product); additionally, transcription *initially* requires an activated (phosphorylated) NtrC protein – which is present when combined nitrogen is limiting. NtrB is apparently involved in regulating the activity of NtrC.

The first *nif* genes to be transcribed are those in the *nifLA* operon. NifA, which is functionally related to NtrC, takes over the role of activator of E σ^{54} for all subsequent *nif* genes/operons. NifL can inhibit NifA in the presence of combined nitrogen; moreover, NifA is inactivated by oxygen. Nitrogen fixation is blocked completely by null mutations in *rpoN*, *ntrC* or *nifA*.

can be detected by its ability to induce a signal (a radiofrequency alternating voltage) in a coil tuned to the particular resonance frequency. Usually, a single excitatory pulse produces only a weak resonance signal, so that several hundred or several thousand intermittent pulses are commonly used, the resulting resonance signals being added together in a computer.

The resonance frequency of a given nucleus is influenced by the precise location of the atom within a molecule: if atoms of a given type (e.g. ^{1}H) are located in different molecular environments (e.g. CH_2, CH_3, OH) their nuclei will have slightly different resonance frequencies. When, due to a chemical reaction, an atom moves to a different molecular environment, the resonance frequency of its nucleus will change to that specified by the new location. A given type of nucleus within a particular molecular environment may be identified by its *chemical shift* (δ): the relative amount by which its resonance frequency is shifted from that of a given reference standard – which, for ^{1}H, is the agent tetramethylsilane (TMS), $(CH_3)_4Si$; chemical shift is expressed in the dimensionless unit 'parts per million' (ppm).

If, for a given type of nucleus (e.g. ^{1}H), the nuclear resonance frequency is determined (under standard conditions) for each type of molecular environment, it is possible e.g. to follow the progress of a chemical reaction by monitoring changes in signal strength at the various resonance frequencies; this is possible because the signal strength at a given resonance frequency is proportional to the number of nuclei occupying a given type of molecular environment.

An NMR signal is a composite radio-frequency trace which incorporates simultaneous signals at various resonance frequencies; it is called a *free induction decay* (FID) since signal intensity decreases to zero as all the nuclei relax into their former (pre-excited) state of thermal equilibrium. The various resonance signals are separated out from the FID by mathematical manipulation involving the *Fourier transform*: a function which relates a complex waveform to its component spectrum of frequencies.

The apparatus used for NMR includes an electromagnet or a superconducting magnet in order to obtain a strong magnetic field. Field strength directly influences the absolute resonance frequencies of nuclei; thus, e.g., resonance frequencies of ^{1}H nuclei are centred around ca. 42.6 megacycles (= megahertz, MHz) when the magnetic field is 1 tesla (= 10000 gauss), while at 4.7 tesla the ^{1}H nucleus resonates at ca. 200 MHz. Other resonance frequencies include e.g. (MHz per tesla): 10.7(^{13}C), 4.31(^{15}N) and 17.24 (^{31}P).

[Book ref. 151; NMR and its clinical applications: BMB (1984) *40* 113–201; NMR in biochemistry: Book ref. 152, pp. 1–67; NMR and drug design: Book ref. 153, pp. 267–281; microbiological applications of NMR: MS (1985) 2 203–211, 340–345.]

nuclear petite See PETITE MUTANT.

nuclear polyhedrosis viruses (NPVs) Viruses belonging to subgroup *A* of the BACULOVIRIDAE (q.v.). Virus replication and occlusion body formation occur exclusively in the host cell nucleus (cf. CYTOPLASMIC POLYHEDROSIS VIRUS GROUP). The NPV virion may contain a single nucleocapsid (*S*NPV) or many nucleocapsids (*M*NPV). Type species: *Autographa californica* NPV (Ac*M*NPV); other members include e.g. *Bombyx mori* (silkworm) NPV (Bm*S*NPV), *Orygia pseudotsugata* (Douglas Fir tussock moth) NPVs (Op*S*NPV and Op*M*NPV), and viruses from many other insects and from certain crustacea.

Nuclearia A genus of amoebae (class FILOSEA). *N. delicatula* feeds by ingesting filamentous cyanobacteria [feeding behaviour and structure: JP (1986) *33* 369–374].

nuclease Any enzyme which can cleave the sugar–phosphate backbone of a NUCLEIC ACID. (See also EXONUCLEASE and ENDONUCLEASE.)

nuclease S_1 *Syn.* ENDONUCLEASE S_1.

nucleation site See CELL CYCLE (b).

nucleic acid A polymer of NUCLEOTIDES in which the 3′ position of one nucleotide sugar is linked to the 5′ position of the next by a phosphodiester bridge. In a linear nucleic acid strand, one end typically has a free 5′-phosphate group, the other a free 3′-hydroxyl group. (See DNA and RNA.)

nucleic acid sequence-based amplification See NASBA.

nucleocapsid A structure containing nucleic acid and protein which forms part (or all) of a virion. Depending on virus, 'nucleocapsid' may refer e.g. to the CORE plus the CAPSID (see e.g. HERPESVIRIDAE, TOGAVIRIDAE) or to a helical nucleoprotein structure (see e.g. ARENAVIRIDAE, PARAMYXOVIRIDAE, TOBACCO MOSAIC VIRUS); it corresponds to the 'core' in the REOVIRIDAE.

nucleoid (1) (*bacteriol.*) The bacterial CHROMOSOME [review: MR (1994) *58* 211–232]. During cell division, the mechanism by which nucleoids are partitioned to daughter cells is unknown [partitioning in *Escherichia coli*: JB (1992) *174* 7883–7889; in *Bacillus*: JB (1998) *180* 547–555].

(2) (*virol.*) The ribonucleoprotein structure within the core, or the core itself, in the virion of a retrovirus (see RETROVIRIDAE).

(3) Mitochondrial or chloroplast DNA.

(4) The electron-dense central core in a PEROXISOME.

nucleolar organizer See NUCLEOLUS.

nucleolus Within most types of eukaryotic NUCLEUS: a distinct region, not delimited by a membrane, in which at least some species of rRNA are synthesized and assembled into ribonucleoprotein subunits of RIBOSOMES. In the nucleolus rRNA is transcribed from a *nucleolar organizer*, i.e., a group of tandemly repeated chromosomal genes which encode rRNA and which are transcribed by RNA polymerase I. (See also RRNA.) [Nucleolar structure and ribosome biogenesis: TIBS (1986) *11* 438–442.] In most cases nucleoli do not persist during MITOSIS (cf. ENDOSOME sense 3). Nucleoli do not occur e.g. in the ciliate MICRONUCLEUS.

nucleoside A compound consisting of a purine or pyrimidine base covalently linked to a pentose – usually ribose (in *ribonucleosides*) or 2-deoxyribose (in *deoxyribonucleosides*); the figure shows some of the bases commonly found in nucleosides.

Ribonucleosides containing the bases adenine, guanine, cytosine, uracil, thymine and hypoxanthine are called (respectively) adenosine, guanosine, cytidine, uridine, thymidine and inosine. The corresponding deoxyribonucleosides are called deoxyadenosine, deoxyguanosine etc. (Note that 'thymidine' usually refers to *deoxythymidine* as thymine is associated primarily with DNA.)

(cf. NUCLEOTIDE; see also Appendix V.)

nucleoside anitbiotic A NUCLEOSIDE-containing ANTIBIOTIC: see e.g. POLYOXINS, PSICOFURANINE, PUROMYCIN, SHOWDOMYCIN.

nucleoside reverse transcriptase inhibitors (NRTIs) A category of deoxyribonucleoside analogues which are used e.g. as ANTIRETROVIRAL AGENTS (q.v.); all NRTIs have a similar mode of action – see AZT for details. These agents are involved in chemotherapy against AIDS [BMJ (2001) *322* 1410–1412].

Examples of NRTIs: AZT (zidovudine); didanosine (2′,3′-dideoxyinosine); LAMIVUDINE; stavudine (2′,3′-dideoxythymidine); and zalcitabine (2′,3′-dideoxycytidine).

NRTIs can be mutually antagonistic: see e.g. AZT.

The anti-HIV activity of an NRTI may be raised by increasing the ratio of NRTIs to the cell's own dNTPs; this can be done e.g. by using hydroxyurea to inhibit the enzyme ribonucleotide

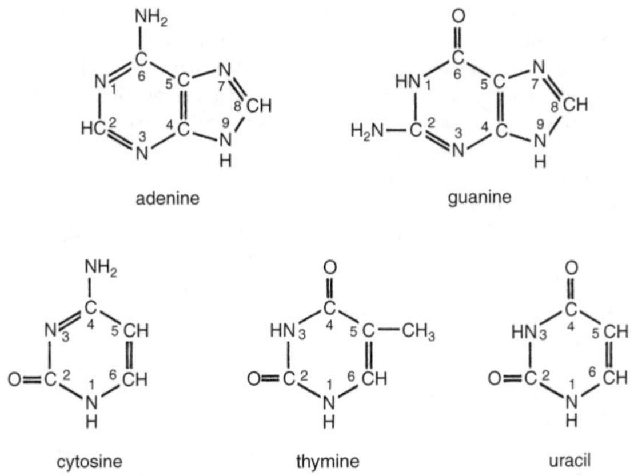

NUCLEOSIDE: some of the bases commonly found in nucleosides. Reproduced from *Bacteria* 5th edition, Figure 7.3, page 112, Paul Singleton (1999) copyright John Wiley & Sons Ltd, UK (ISBN 0471-98880-4) with permission from the publisher.

reductase (this causing a decrease in endogenous dNTPs). This approach was found to give clinical benefit when hydroxyurea was combined with didanosine in the Swiss HIV Cohort Study (1998).

At least some NRTIs have been used against DNA viruses, i.e. they also inhibit DNA-dependent DNA polymerases (see e.g. LAMIVUDINE).

nucleoskeleton In a eukaryotic cell: an intranuclear framework, analogous to the CYTOSKELETON, which is believed to include e.g. points of anchorage for complexes involved in the replication and transcription of DNA; the concept of a nucleoskeleton is supported e.g. by demonstrations of intranuclear actin and myosin, and by the demonstration that DNA polymerases are retained within nuclei, under isotonic conditions, following treatment of the encapsulated nuclei with restriction endonucleases [EMBO (1986) *5* 1403–1410]. There is also evidence that the cytoskeleton and nucleoskeleton are connected at the nuclear periphery. Transport between nucleus and cytoplasm may involve an ATP- and actomyosin-dependent control of the nuclear pore complex [JCB (1986) *102* 859–862].

[Organization beyond the gene: TIBS (1986) *11* 249–252.]

nucleosome See CHROMATIN.

nucleotide A NUCLEOSIDE in which the sugar carries one or more phosphate groups (see figure); nucleotides are the subunits of NUCLEIC ACIDS. (See also ATP.)

Pathways for the de novo biosynthesis of the main nucleotides are shown in Appendix V. Deoxyribonucleotides are synthesized by the reduction of ribonucleotides catalysed by the enzyme *ribonucleotide reductase*. In *Escherichia coli*, deoxyribonucleotides are synthesized mainly by a THIOREDOXIN-dependent reduction of ribonucleoside 5′-*di*phosphates, while in e.g. *Lactobacillus leichmannii* ribonucleoside 5′-*tri*phosphates are reduced in a VITAMIN B12-dependent reaction.

In addition to de novo synthesis, nucleotides can be synthesized using purine and pyrimidine bases derived from pre-formed nucleotides in a so-called 'salvage cycle'. (See also NAD and THYMIDINE KINASE.)

Antibiotics which affect nucleotide biosynthesis include AZASERINE; DON; HADACIDIN; MYCOPHENOLIC ACID and PSICOFURANINE.

nucleotide excision repair See EXCISION REPAIR.

nucleotropism See TROPISM (sense 2).

Nuclepore filter See FILTRATION.

nucleus Within a eukaryotic cell: a membrane-limited body which contains e.g. CHROMOSOMES and one or more nucleoli (see NUCLEOLUS). (See also NUCLEOSKELETON.) The number of chromosomes per nucleus depends e.g. on species. (See also PLOIDY.) The nuclear membrane consists of a double unit-type membrane which is perforated by a number of pores; the outermost membrane is continuous with the ENDOPLASMIC RETICULUM, and the space between the two membranes is called the *perinuclear space*.

A cell may contain more than one nucleus (see e.g. COENOCYTE and SYNCYTIUM), and ciliate protozoa typically have two *types* of nuclei (see MACRONUCLEUS and MICRONUCLEUS). The *fungal* nucleus is characteristically much smaller than the nuclei of most other eukaryotes, and in many cases fungal chromosomes cannot be seen by light microscopy [the fungal nucleus: Book ref. 168].

Nudaurelia β virus group A family of non-enveloped, icosahedral (ca. 35 nm diam.), ssRNA-containing viruses which infect lepidopteran insects; the genome is a single molecule of positive-sense ssRNA (MWt ca. 1.8–1.9 × 10^6).

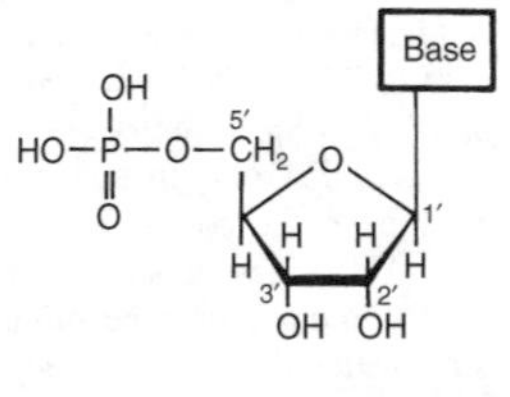

NUCLEOTIDE. The figure shows a *ribo*nucleotide; ribonucleotides are the subunits in RNA. In a *deoxy*ribonucleotide the 2′ position of ribose carries –H rather than –OH; deoxyribonucleotides are the subunits in DNA. The ribonucleotide in the figure is a ribonucleo*side mono*phosphate; a ribonucleoside triphosphate (such as ATP) carries a chain of three phosphate groups – α, β and (terminal) γ – at the 5′ position of the ribose.

In a *dideoxy*ribonucleotide (used e.g. in DNA SEQUENCING), the 2′ and 3′ positions in the ribose carry –H (instead of –OH).

Type species: *Nudaurelia* β virus (NβV), isolated from the pine emperor moth – *Nudaurelia cytherea capensis* (= *Imbrasia cytherea*) – whose larvae cause extensive defoliation of pine trees in South Africa. Other members of the family have been isolated from various lepidopteran hosts – e.g. *Darna trima* (a pest of coconut and oil palms in SE Asia), *Thosea asigna* (a pest of oil palms in Malaysia), *Trichoplusia ni* (cabbage looper). The viruses have not been propagated in cell culture, and must be cultivated by inoculation into suitable host larvae; virus replication occurs in the cytoplasm of gut cells. Infected larvae become flaccid. [Review in JGV (1985) *66* 647–659.]

null cells (*immunol.*) Lymphoid cells which lack characteristic B cell and T cell antigens; some are non-cytolytic, others are cytolytic – the latter including NK CELLS.

null mutation A MUTATION which results in complete loss of function of the sequence in which it occurs. (cf. LEAKY MUTATION.)

numerical aperture See RESOLVING POWER.

numerical taxonomy (taxometrics) A computer-assisted form of TAXONOMY in which the organisms in a given study are sorted into groups of mutually similar organisms on the basis of a large number (ideally >50) of observable properties (*characters*); typically, every character is regarded as having the same degree of importance, i.e. equal 'weighting', in the computation of similarities (cf. ADANSONIAN TAXONOMY; see also POLYTHETIC CLASSIFICATION). (The use of a computer is not strictly essential, but the calculations involved would otherwise take an inordinate length of time.) Numerical taxonomy is primarily a form of PHENETIC CLASSIFICATION in which the aim is to define TAXOSPECIES (see also PHENON); however, the availability of certain fundamental data (derived e.g. from DNA homology studies) has permitted the method to be modified for use in PHYLOGENETIC CLASSIFICATION.

In numerical taxonomy each strain – or OTU (q.v.) – among those to be classified is initially examined or tested in respect of each of the biochemical, morphological or other characters used as criteria in the study. The nature or value of a given test result is then expressed in a form such as + or – (or e.g. as 0 or 1); for example, 'growth at 45°C' could be written as '+', while '–' could be used for 'no growth at 45°C'. Each OTU in the study may thus be defined by a series of simple symbols; this is the form in which the data are fed into the computer.

In the computer the first step in classification is the comparison of each OTU with every other OTU – likenesses being expressed as percentage similarities; for example, two given OTUs which match in any 60 out of 80 characters would have a percentage similarity of 75%. (There are actually several different ways in which a given pair of OTUs can be compared: see entry S_{SM} and compare it with S_G, S_J and S_P; cf. entry D_P.) The result of these initial comparisons is a list of percentage similarities – each having been derived from the comparison of one given OTU with another given OTU; these data are usually printed out as an S MATRIX. The computer may be programmed to search the percentage similarities for pairs or groups of 'identical organisms', i.e., pairs or groups which exhibit a mutual similarity of 100%; such pairs or groups are called '100% clusters'. On the computer printout such clusters, if present, may appear thus:

100% 5–10–11–19–
43–47–
29–31–33–

in which each number (5, 10 etc) is the designation of one of the OTUs in the study; in this example three clusters are formed at the 100% level of similarity. Although strains within a given cluster are mutually 100% similar, no information is given here on the degree of similarity which exists between the OTUs of one cluster and those of another. The remaining OTUs are progressively added at successive, lower levels of similarity (e.g. 97%, 95%, 80%, 75%, 50%) to one or other of the initially formed clusters. Additionally, two or more clusters may unite at a given level of similarity; thus, e.g. in the above example the clusters 43–47 – and 29–31–33 – may form a single cluster at, say, 75% – meaning that one or more OTUs in a given cluster has a mutual similarity of at least 75% with one or more OTUs in the other cluster. The mode in which OTUs are added to a cluster, and in which clusters unite, may be varied according to the purpose of the taxonomist: see e.g. SINGLE LINKAGE and UPGMA. At the end of the study the result of clustering may be shown e.g. in a PHENOGRAM or in a SHADED MATRIX. The processed data can then be examined to determine whether or not a given phenon can be regarded as a TAXON.

Nummilites See FORAMINIFERIDA.

nummilitids See FORAMINIFERIDA.

nungham The freshwater red alga *Lemanea mamillosa* (division RHODOPHYTA) used as a food in India.

nu/nu mice A strain of congenitally athymic mice.

***nusA* gene** See TRANSCRIPTION and BACTERIOPHAGE LAMBDA.

nutrient agar NUTRIENT BROTH gelled with 1.5–2.0% AGAR: a general-purpose solid basal MEDIUM.

nutrient broth A liquid basal MEDIUM; the name may refer to either extract broth or infusion broth. *Extract broth* (the type most commonly used) may be prepared as an aqueous solution of beef extract and peptone (each ca. 0.5–1.0% w/v) and sodium chloride (ca. 0.5% w/v); it is sterilized by autoclaving, and the final pH is ca. 7.2–7.4. Powdered (dehydrated) extract broth is available commercially; it is reconstituted with distilled (or tap) water, dispensed into tubes, and autoclaved. *Infusion broth* is prepared by infusing overnight, at 4°C, 50 g of minced lean beef in 100 ml distilled water; peptone (ca. 1% w/v) and sodium chloride (ca. 0.5% w/v) are added, and the mixture is boiled, filtered, and sterilized by autoclaving. Final pH: ca. 7.3–7.5.

nutritionally variant streptococci (NVS) Strains of *Streptococcus*, present e.g. in the normal human throat and urogenital tract, which are characterized by a requirement for cysteine, or for pyridoxal hydrochloride or pyridoxamine hydrochloride, when grown on complex media; the organisms were originally detected as satellite colonies (see SATELLITE PHENOMENON) of viridans streptococci associated with colonies of *Staphylococcus*. Two species are distinguished: *S. adjacens* and *S. defectivus*, the latter species differing from *S. adjacens* e.g. in fermenting trehalose and forming β-galactosidase. The organisms may form a red pigment. NVS may be responsible for ~5% of those cases of endocarditis which involve oral streptococci.

Nuttallia See BABESIA.

nvCJD New variant CREUTZFELDT–JAKOB DISEASE.

NVDP See PLASMODIUM.

NVL No visible lesion.

NVS NUTRITIONALLY VARIANT STREPTOCOCCI.

Nycodenz See IODINATED DENSITY-GRADIENT MEDIA.

nystatin (fungicidin) An antifungal POLYENE ANTIBIOTIC, produced by *Streptomyces noursei* and named after the New York State Health Department (where it was discovered). It contains MYCOSAMINE and is usually classed as a tetraene, although it contains six double bonds (four conjugated double bonds separated from two others by two methylene groups). Nystatin is used primarily for the topical treatment of superficial mycoses (e.g. mucosal *Candida* infections); it is not absorbed when taken orally.

O

O antigens (Boivin antigens) The heat-stable, alcohol-resistant LIPOPOLYSACCHARIDE–protein somatic antigens of Gram-negative bacteria (particularly members of the Enterobacteriaceae); O antigens are important in the serological characterization of enterobacteria (see e.g. KAUFFMANN–WHITE CLASSIFICATION). Antigenic specificity is determined by the polysaccharide O-specific chains (see LIPOPOLYSACCHARIDE). O antigens may be modified e.g. by BACTERIOPHAGE CONVERSION or by mutation (e.g. in salmonellae a mutation may give rise to T1 side-chains, consisting of ribose and galactose residues, instead of normal O-specific chains – see also SMOOTH–ROUGH VARIATION). (See also VI ANTIGENS.)

O side chain See LIPOPOLYSACCHARIDE.

O-specific chain See LIPOPOLYSACCHARIDE.

O-type starter See LACTIC ACID STARTERS.

O1 strains (of *Vibrio cholerae*) See VIBRIO.

O/129 2,4-Diamino-6,7-diisopropylpteridine: a water-soluble agent bacteriostatic for e.g. many members of the Vibrionaceae (e.g. most *Vibrio* strains). Sensitivity to O/129 may be determined using O/129-impregnated discs on nutrient agar containing 0.5% w/v NaCl.

Some strains of *Vibrio cholerae* produce a trimethoprim-resistant dihydrofolate reductase (DHFR) and are resistant to both trimethoprim and O/129; the gene encoding the DHFR may be borne on a plasmid or on a transposon [Ann. Mic. (1985) *136B* 265–273].

O139 Bengal (*Vibrio cholerae*) See CHOLERA.

O157:H7 *E. coli* See EHEC.

oak butt rot See BUTT ROT.

oak-moss *Evernia prunastri*.

oak wilt A disease of oak trees (*Quercus* spp) caused by *Ceratocystis fagacearum*. Infection occurs via wounds (made e.g. by man or woodpeckers). The fungus forms pressure cushions (sclerotia) between the infected xylem and phloem, bursting open the overlying bark and allowing entry of insects, particularly sap-feeding beetles of the Nitidulidae, which act as vectors of the disease.

Oakley–Fulthorpe test See DOUBLE DIFFUSION.

oar weed Any species of *Laminaria* which has an oar-shaped thallus, e.g. *L. digitata* and *L. hyperborea*.

oat diseases See CEREAL DISEASES.

oat mosaic virus See POTYVIRUSES.

oat necrotic mottle virus See POTYVIRUSES.

oat sterile dwarf virus See FIJIVIRUS.

oat striate virus See RHABDOVIRIDAE.

Obesumbacterium A genus of non-motile, slow-growing bacteria of the ENTEROBACTERIACEAE (apparently related to *Hafnia alvei*), found as contaminants in fermenting beer wort. Cells: pleomorphic rods, 0.8–2.0 × 1.5–100 μm (short, fat rods predominate in wort). [Book ref. 22, pp. 506–509.]

objective (objective lens) See MICROSCOPY (a).

obligate Refers to an essential attribute of a given organism. For example, an obligate aerobe is an organism which grows *only* under aerobic conditions, and an obligate parasite is an organism which, in nature, can grow only as a parasite. (cf. FACULTATIVE.)

obligate proton reducers See ANAEROBIC DIGESTION.

Obodhiang virus See LYSSAVIRUS.

occlusion bodies See BACULOVIRIDAE.

Occlusosporida See STELLATOSPOREA.

Oceanospirillum A genus of marine (coastal), Gram-negative, asporogenous bacteria formerly included in the genus SPIRILLUM. Cells: rigid, helical, 0.3–1.4 × ca. 5–20 μm. POLAR MEMBRANES and intracellular PHB granules are present. COCCOID BODIES predominate in old cultures in all species except *O. japonicum*. Motile, with bipolar tufts of flagella (or, in *O. pusillum*, a single flagellum at each pole). Aerobic; metabolism is strictly respiratory. Nitrate respiration does not occur, but some species can reduce NO_3^-. Oxidase +ve; catalase and phosphatase variable; indole −ve; arylsulphatase −ve. Optimum growth temperature: 25–32°C. Seawater is required for growth. Carbon sources: amino acids or salts of organic acids, but not carbohydrates. GC%: 42–51. Type species: *O. linum*. Nine species are recognized [Book ref. 22, pp. 104–110].

ocellus See EYESPOT (sense 1).

Ochoterenaia See TRICHOSTOMATIDA.

ochratoxins A group of MYCOTOXINS produced by species of *Penicillium* and *Aspergillus*. Ochratoxin A is a dihydrocoumarin derivative linked to L-phenylalanine via an amide bond; it inhibits phosphoenolpyruvate carboxykinase. Ingestion of ochratoxin A in contaminated food causes nephropathy in e.g. cattle, pigs and man.

ochre codon See GENETIC CODE.

ochre mutation See NONSENSE MUTATION.

ochre suppressor See SUPPRESSOR MUTATION.

Ochrobium A genus of IRON BACTERIA. Cells: ellipsoidal to rod-shaped. Type species: *O. tectum*. [Population ecology of *Ochrobium*: JGM (1981) *125* 85–93.]

Ochrolechia A genus of crustose LICHENS (order PERTUSARIALES). Photobiont: a green alga. Thallus: often thick, warty, irregular in shape. Apothecia: lecanorine, with characteristically thick thalline margins and yellowish or pinkish discs. Species generally occur on rocks, sometimes also on bark; *O. frigida* occurs on the ground on mosses etc. *O. tartarea* has been harvested on a commercial scale for the production of CUDBEAR.

Ochromonas A genus of unicellular CHRYSOPHYTES. The cell is usually ovoid, contains (typically) two parietal chloroplasts, and has two flagella of unequal length. (See also FLAGELLAR MOTILITY (b).) The organisms are both photosynthetic and phagotrophic, apparently requiring organic nutrients (e.g. VITAMIN B_{12}) or prey for survival. Some forms have a delicate, cup-shaped LORICA composed of microfibrils of CHITIN; typically, the lorica adheres via a stalk to the substratum or to other cells. (Loricate forms of *O. malhamensis* are sometimes known as *Poterioochromonas stipitata*.)

Ockelbo disease See SINDBIS VIRUS.

OCT plasmid An IncP-2 *Pseudomonas* PLASMID (ca. 500 kb) which encodes the capacity to metabolize octane and decane. (cf. TOL PLASMID.) [Characterization of the OCT plasmid: JB (1986) *165* 650–653.]

octapeptins A group of ANTIBIOTICS closely related to the POLYMYXINS; they differ e.g. in having a shorter side-chain and a longer β-hydroxy fatty acid residue. A mixture of octapeptins A and B is up to ten times more effective than polymyxin B against certain Gram-positive bacteria.

Octomyxa See PLASMODIOPHOROMYCETES.

octopine N^2-(D-1-carboxyethyl)-L-arginine – see CROWN GALL. Other, related opines include histopine and lysopine.

octopinic acid See CROWN GALL.

***Octopus vulgaris* disease virus** See IRIDOVIRUS.

Octosporella See EIMERIORINA.

ocular *Syn.* eyepiece (see MICROSCOPY).

ocular herpes See HERPES SIMPLEX.

ocular micrometer See MICROMETER.

OD Optical density (see TURBIDIMETRY).

ODC Ornithine decarboxylase: see DECARBOXYLASE TESTS.

Odontostomatida An order of typically freshwater, 'polysaprobic' ciliate protozoa (class POLYHYMENOPHOREA) in which the cells are laterally compressed, have reduced somatic ciliature, and often have posterior spines or cirri. Genera include e.g. *Epalxella* (formerly *Epalxis*), *Pelodinium*, SAPRODINIUM.

oedema (edema; 'dropsy') The abnormal accumulation of fluid in body tissues. (cf. ASCITES.)

oedema disease (bowel oedema) A PIG DISEASE which typically attacks young animals subjected to stress (e.g. change of diet at weaning). It is apparently caused by toxigenic strains of *Escherichia coli*. Infection may result in sudden death, or there may be oedema of certain tissues (especially the stomach wall and mesentery) with signs of nervous disturbance (incoordination, loss of balance) usually followed by coma and death.

oedematolysin See THIOL-ACTIVATED CYTOLYSINS.

Oedogonium A genus of freshwater (commonly epiphytic), unbranched filamentous green algae (division CHLOROPHYTA). Each filament is composed of uninucleate cells and is usually attached to the substratum by a differentiated basal cell. The individual cells each contain a single parietal reticulate chloroplast with several pyrenoids. In some cells of the filament the cell wall has a characteristic series of ridges (the 'cap') at one end; these ridges result from an unusual mode of cell division: shortly after mitosis, the cell wall ruptures and the cell increases rapidly in size before the cross-wall is formed. Asexual reproduction occurs by the formation of multiflagellate, naked zoospores; sexual reproduction is oogamous. (See also BULBOCHAETE.)

Oerskovia A genus of asporogenous bacteria (order ACTINOMYCETALES, wall type VI) which occur e.g. in soil. In culture the organisms form a primary mycelium which fragments into small, monotrichously flagellated rods (ca. 0.4×1.1 μm) that develop into larger, peritrichously flagellated rods, the latter giving rise to a branched mycelium in which the hyphal diameter is ca. 0.5 μm. Growth occurs aerobically (organisms catalase-positive) or anaerobically (organisms catalase-negative) on rich media (e.g. trypticase–soy agar) with the formation of yellowish colonies; glucose can be metabolized either oxidatively or fermentatively. GC%: ca. 70–75. Type species: *O. turbata*. [Book ref. 73, pp. 52–53.]

O–F test OXIDATION–FERMENTATION TEST.

OFAGE Orthogonal-field-alternation gel electrophoresis: a form of PFGE (q.v.) used e.g. for separating chromosomes by size. OFAGE has been used e.g. to obtain a karyotype of the yeast *Saccharomyces cerevisiae* [PNAS (1985) *82* 3756–3760].

ofloxacin See QUINOLONE ANTIBIOTICS.

Ogawa variant See VIBRIO (*V. cholerae*).

ogi A Nigerian food: a type of sour porridge made by the natural fermentation of ground maize, sorghum or millet. A range of organisms is present, including e.g. *Lactobacillus plantarum* and various yeasts.

O'Hara's disease See TULARAEMIA.

Ohio corn stunt agent *Syn.* MAIZE CHLOROTIC DWARF VIRUS.

Oidiopsis See HYPHOMYCETES and ERYSIPHALES.

oidium (*mycol.*) A thin-walled THALLOSPORE; oidia may function as disseminative spores or as male gametes.

Oidium See HYPHOMYCETES and ERYSIPHALES.

Oikomonas A genus of colourless, unicellular CHRYSOPHYTES which resemble *Chromulina* spp; the organisms occur in organically polluted waters, activated sludge, etc.

oil biodeterioration See PETROLEUM.

oil-immersion condenser See RESOLVING POWER and MICROSCOPY (b).

oil-immersion objective See RESOLVING POWER.

oil shale A grey or black ancient sedimentary rock which consists primarily of KEROGEN with a small proportion of solvent-soluble organic matter.

O_J locus See FINOP SYSTEM.

Okazaki fragments See DNA REPLICATION.

OKT antigens The former name for surface antigens on human T LYMPHOCYTES. The designation 'OKT' was replaced by 'CD' – for example, OKT4 is now CD4.

old dog encephalitis See CANINE DISTEMPER.

old tuberculin See TUBERCULIN.

oleandomycin A MACROLIDE ANTIBIOTIC (from *Streptomyces antibioticus*) closely related to erythromycin. (cf. SIGMAMYCIN.)

olefin metabolism See HYDROCARBONS.

oligo(dT)-cellulose See MRNA (a).

oligodynamic effect The antibacterial (or antimicrobial) action exerted by low concentrations of certain metals in elemental form (see HEAVY METALS); the effect may be observed by placing small metal discs on a flood plate and measuring the zone of no growth surrounding each disc following appropriate incubation.

Oligohymenophorea A class of protozoa (phylum CILIOPHORA) in which, typically, the cytostome occurs at the base of a BUCCAL CAVITY, and the oral ciliature is obviously differentiated from the somatic ciliature; the PARORAL MEMBRANE is characteristically of the stichodyad type; and STOMATOGENESIS is buccokinetal or parakinetal. This class includes ciliates previously referred to as 'higher holotrichs' (e.g. the hymenostomes) together with e.g. the astomes. Subclasses: HYMENOSTOMATIA, PERITRICHIA.

oligomycin An antibiotic, produced by *Streptomyces* spp, which binds non-covalently to mitochondrial (F_0F_1)-type PROTON ATPASES, inhibiting the transfer of protons through the F_0 site and, hence, inhibiting both ATP hydrolysis and OXIDATIVE PHOSPHORYLATION. Oligomycin does not collapse pmf (cf. UNCOUPLING AGENTS).

oligonucleotide cataloguing See RRNA OLIGONUCLEOTIDE CATALOGUING.

oligonucleotide-directed mutagenesis See SITE-SPECIFIC MUTAGENESIS.

oligonucleotide fingerprinting Any FINGERPRINTING technique in which the fingerprint consists of a strain-specific pattern of oligonucleotides. For example, to differentiate strains of an RNA virus, the genome is digested with a sequence-specific enzyme (such as RNASE T1) and the resulting oligonucleotide fragments are subjected to gel ELECTROPHORESIS; the pattern (bands) of fragments in the gel constitute the characteristic fingerprint of the given strain of virus.

oligosaccharide A carbohydrate consisting of 2 to ca. 10 monosaccharides joined by glycosidic linkages.

oligosaprobic zone See SAPROBITY SYSTEM.

Oligotrichida An order of ciliate protozoa (class POLYHYMENOPHOREA) which are typically free-living, marine, pelagic organisms, but which include some freshwater and edaphic species. Cells: rounded to elongate with a thickened pellicle and (in many species) a perilemma; somatic ciliature is typically reduced, and in some species the oral ciliature is divided into two parts: one part within the buccal cavity, the other encircling the anterior end of the cell. Some species are loricate.

Members of the suborder Oligotrichina are typically small and rounded organisms whose somatic ciliature may consist of only a few short rows of 'bristles', but which may have groups of 'cirri' or 'bristles' forming an equatorially encircling band; genera include e.g. HALTERIA, *Strombidium*, *Strombilidium*, and *Tontonia*.

The other suborder, TINTINNINA (q.v.), comprises a distinctive group of organisms.

Oligotrichina See OLIGOTRICHIDA.

oligotroph Any organism which can grow in relatively nutrient-poor environments (cf. COPIOTROPH).

oligotrophic (1) Of or pertaining to an OLIGOTROPH. (2) Of lakes, rivers etc: containing low levels of nutrients, particularly of those nutrients which support the growth of aerobic photosynthetic organisms. (cf. EUTROPHIC.)

olivanic acids An ill-defined group of CARBAPENEM antibiotics derived from *Streptomyces olivaceus* and other *Streptomyces* spp.

olive-green mould *Chaetomium olivaceum*: a mould which frequently grows on mushroom compost after pasteurization (see MUSHROOM CULTIVATION) and which inhibits mycelial development in *Agaricus brunnescens* by competition for nutrients. [Biological control by a thermophilic *Bacillus* sp: AEM (1983) *45* 511–515.]

olive knot A disease of the olive, oleander and privet, in which small galls develop on young twigs and leaves. The causal agent is *Pseudomonas syringae* pv. *savastanoi*; the pathogen produces IAA and cytokinin which may be responsible for the formation of the galls. In at least some strains IAA production appears to be plasmid-mediated. Infection occurs via wounds. Transmission may occur via the olive fly (*Dacus oleae*).

olives (pickled) See PICKLING.

olivomycin See CHROMOMYCIN.

Olpidiopsis See LAGENIDIALES.

Olpidium A genus of unicellular, endobiotic, holocarpic fungi (order CHYTRIDIALES) which are parasitic on algae and other aquatic organisms and on some terrestrial plants; *Olpidium* spp are also involved in the transmission of the LETTUCE BIG VEIN AGENT and the TOBACCO NECROSIS VIRUS. In the asexual cycle of *O. viciae* a zoospore settles on a susceptible host (*Vicia*), encysts, and the cyst's contents enter a host cell; after intracellular growth and a number of nuclear divisions, the fungus eventually forms an inoperculate zoosporangium – the numerous zoospores subsequently escaping via one or more pores. In the sexual cycle, two morphologically indistinguishable zoospores fuse and then encyst on a susceptible host; the cyst's contents enter a host cell and develop into thick-walled resting sporangia. Karyogamy (and probably meiosis) occur before the germination of resting sporangia – each of which produces a number of zoospores.

OM OUTER MEMBRANE.

omasum See RUMEN.

O'Meara's method See VOGES–PROSKAUER TEST.

ω-oxidation See HYDROCARBONS.

ω (omega) protein See TOPOISOMERASE.

omicron An endosymbiotic, Gram-negative, rod-shaped bacterium which occurs in the cytoplasm of *Euplotes* spp; it does not confer a killer characteristic, and it appears to be essential for the viability of its protozoan host. [Book ref. 22, p. 798.]

omnivorous Refers to protozoa which feed on microscopic 'animals' (e.g. other protozoa) and 'plants' (e.g. algae, bacteria). (cf. CARNIVOROUS; HERBIVOROUS.)

Omp The designation of certain major OUTER MEMBRANE PROTEINS, each type of protein being distinguished as OmpA, OmpC etc. Some of these proteins are PORINS – for example OmpC and OmpF (see OSMOREGULATION). In *Escherichia coli*, OmpA (also called e.g. 3A, d and II*) is a product of the *ompA* gene (also called e.g. *con*, *tolG* and *tut*); it apparently spans the thickness of the outer membrane but does not seem to function as a porin. OmpA-deficient mutants have an unstable outer membrane and are deficient in conjugation and in sensitivity to colicin L and phages K3 and TuII. In *E. coli*, OmpT is a protease involved in type IV (autotransporter) protein secretion.

OMP Orotidine 5′-monophosphate [see Appendix V(b)].

OmpA See OMP.

Omphalotus See AGARICALES (Tricholomataceae).

OmpR See OSMOREGULATION.

OmpT protein See PROTEIN SECRETION (type IV systems).

Omsk haemorrhagic fever See VIRAL HAEMORRHAGIC FEVERS and FLAVIVIRIDAE.

onc See ONCOGENE.

Oncobasidium See TULASNELLALES and VASCULAR-STREAK DIE-BACK.

oncogen See ONCOGENIC.

oncogene A gene which can (potentially) induce neoplastic transformation in the cell in which it occurs or into which it is introduced. Oncogenes occur e.g. in certain viruses, particularly members of the RETROVIRIDAE. Viral oncogenes (v-*onc*), originally identified as the transforming determinants of acutely oncogenic retroviruses, were subsequently found to have closely related cellular homologues (designated c-*onc* or proto-*onc*) in the chromosomes of eukaryotes; c-*onc* sequences have been highly conserved through evolution, and homologous genes have been found in all vertebrates examined, as well as in lower eukaryotes such as *Drosophila* and *Saccharomyces*. In the normal cell, at least some oncogenes are known to have one or more major roles in cell growth/development. Retroviral v-*onc* sequences may have arisen from c-*onc* genes by recombination (see RETROVIRIDAE) and may differ from them e.g. in lacking introns; they are often expressed with adjacent viral sequences as fusion proteins.

Retroviral and cellular oncogenes are usually referred to by three-letter designations based on the retrovirus in which they were first identified: e.g. *myc* from myelocytomatosis virus, *fes* from feline sarcoma virus, etc; viral genes are prefixed by v- (e.g. v-*myc*), their cellular homologues by c- or proto- (e.g. c-*myc* or proto-*myc*).

A v-*onc* gene introduced into a cell by an infecting retrovirus may induce the neoplastic transformation of that cell in various ways. The products of many retroviral oncogenes (e.g. v-*abl*, v-*fes*, v-*fps*, v-*ros*, v-*src*, v-*yes*) have tyrosine-specific protein kinase activity, transferring γ-phosphate from ATP (or GTP) to tyrosine residues in proteins. (See ABL, FES and SRC; cf. ERB and MOS.) In normal (non-neoplastic) cells, tyrosine kinase activity is associated with the receptors for certain growth-promoting peptides: e.g. epidermal growth factor (EGF), platelet-derived growth factor (PDGF) and insulin. Oncogenic tyrosine kinase activity has been targeted by selective inhibitors of proteins encoded by e.g. *bcr-abl* and v-*abl*; some inhibitors can be effective against a range of oncogenic kinases [e.g. Pharmacology and Experimental Therapeutics (2000) *295* 139–145], and this has encouraged the screening of existing clinically useful drugs for activity against the emergent drug-resistant mutant kinases [PNAS (2005) *102* 11011–11016].

v-*myc* and v-*myb* products lack tyrosine kinase activity (see MYC and MYB). v-*sis* encodes a protein almost identical to the B chain of human PDGF.

Modification or activation of a c-*onc* sequence is apparently involved in certain cases of oncogenesis, and may even prove to be a common mechanism of oncogenesis in general. Oncogenesis could result from alterations in the properties of a c-*onc* gene product (e.g. as a result of mutation – see RAS), or from abnormal regulation of a c-*onc* gene; for example, a c-*onc* gene which is normally silent or expressed at very low levels in a particular cell may be activated by the insertion of a retroviral provirus adjacent to and upstream of the c-*onc* sequence, the sequence thus coming under the control of the viral transcriptional regulation signals (see e.g. AVIAN LEUKOSIS VIRUSES). The regulation of a c-*onc* gene may also be disturbed as a result of its translocation to a novel context by chromosomal rearrangements (see e.g. ABL and BURKITT'S LYMPHOMA).

Viral genes involved in oncogenesis induced by certain DNA viruses are sometimes referred to as oncogenes, although they differ in several respects from the oncogenes of retroviruses; they are viral genes which apparently have no cellular homologues and which encode proteins essential for virus replication. (See e.g. MASTADENOVIRUS and POLYOMAVIRUS; cf. EPSTEIN–BARR VIRUS.)

oncogenesis The development of a (benign or malignant) tumour (or other neoplastic condition: e.g. LEUKAEMIA).

A tumour may develop e.g. as a result of the effect on an ONCOGENE of a rearrangement of chromosomes or of mutation (see e.g. ABL, BURKETT'S LYMPHOMA, RAS).

oncogenic Strictly: refers to any agent which can induce the formation of a tumour; more generally, the term is often applied to any agent which can bring about the neoplastic transformation of cells – either in vivo (resulting in the formation of a tumour or e.g. LEUKAEMIA) or in TISSUE CULTURE. Such agents may be termed *oncogens* (cf. CARCINOGEN).

oncom (ontjom) An Indonesian food made by the fermentation of (usually) peanut press-cake from which at least some of the oil has been extracted. The presscake is broken into fragments, soaked, steamed and cooled, placed in shallow trays, inoculated with a starter (RAGI) of *Neurospora intermedia* or *Rhizopus oligosporus*, and incubated at ca. 30°C; the mould binds the fragments into a cake which becomes coated with pink-orange, powdery conidia (*Neurospora*) or black sporangia (*Rhizopus*). [Book ref. 8, pp. 513–515.] (cf. TEMPEH.)

oncornaviruses Former name for viruses of the ONCOVIRINAE.

Oncovirinae A subfamily of viruses of the RETROVIRIDAE; the subfamily includes all of the oncogenic retroviruses as well as many closely related but non-oncogenic retroviruses. On the basis of e.g. morphology and mode of development, the subfamily has been divided into the type B viruses (genus TYPE B ONCOVIRUS GROUP), type C viruses (genus TYPE C ONCOVIRUS GROUP), and type D viruses (proposed genus TYPE D RETROVIRUS GROUP) [Book ref. 23, p. 125]. (See also HTLV.)

one-carbon compounds See C_1 COMPOUNDS.

one-step growth experiment A procedure devised by Ellis and Delbrück [JGP (1939) *22* 365–384] for the quantitative study of lytic bacteriophage multiplication; it has since been modified for the study of other lytic viruses.

A suspension of susceptible cells is inoculated with enough viruses to infect every cell. After allowing time for virus adsorption, unadsorbed viruses are removed. The cell suspension is then incubated at a suitable temperature, and aliquots are periodically withdrawn and subjected to e.g. PLAQUE ASSAY. During the initial period (*latent period*) the number of plaques per aliquot typically remains fairly constant; in this period each virus-infected cell behaves as a single 'infectious centre' capable of giving rise to a single plaque – regardless of the number of virions it may contain. (The latent period lasts ca. 20–25 min in phage T2/*Escherichia coli* B systems, but may be of the order of hours in virus-infected animal cells.) Aliquots taken sequentially after the latent period give progressively increasing numbers of plaques; during this period the infected cells begin to lyse, releasing large numbers of infectious virions each capable of giving rise to a plaque. The number of plaques per aliquot reaches a plateau when all the cells have lysed. (See also BURST SIZE.) If the virions are labile the plateau is followed by a *decay period* characterized by a fall in plaque titres.

If, during the latent period, the cells in each aliquot are lysed artificially – e.g. by the addition of SDS (for animal cells) or chloroform (for Gram-negative bacteria) – a so-called *eclipse period* can be distinguished. In this period, which begins at the same time as the latent period, infectious virions cannot usually be detected owing to the 'uncoating' of the virus following infection; the eclipse period ends when the first progeny virions are assembled. (The latent period normally continues until the host cell lyses naturally, i.e., after the assembly of many progeny virions.)

one-tube nPCR See NESTED PCR.

one-way incompatibility INCOMPATIBILITY exhibited by two given plasmids only when one particular plasmid is introduced into a cell containing the other; if the second plasmid is introduced into a cell containing the first, incompatibility is not exhibited. [One-way incompatibility between IncH1 plasmids and the F plasmid: JGM (1985) *131* 1523–1530.]

onion rot *White rot* (which can also affect e.g. leeks) is caused by *Sclerotium cepivorum*; leaves become yellow and die, and the rotting roots and scale bases become covered with white, fluffy mycelium. Black sclerotia develop in or on necrotic tissue. *Neck rot*, caused by *Botrytis aclada* (= *B. allii*), is more common in stored onions. A *soft rot* can be caused by *Burkholderia cepacia* (formerly *Pseudomonas cepacia*). (See also SMUDGE).

onion yellow dwarf virus See POTYVIRUSES.

onium compounds *Syn.* QUATERNARY AMMONIUM COMPOUNDS.

ONPG test A test used to detect β-D-galactosidase activity in bacteria. A loopful of growth of the test organism (grown on e.g. TSI agar) is introduced into phosphate-buffered peptone water (pH 7.0) containing ONPG (*o*-nitrophenyl-β-D-galactopyranoside); the medium is then incubated at 37°C. β-GALACTOSIDASE hydrolyses ONPG to galactose and *o*-nitrophenol – the latter (being yellow) acting as a test indicator. Organisms which give a positive ONPG test do not *necessarily* metabolize lactose since they may not synthesize a lactose permease. (ONPG can enter cells without a specific permease.)

ontjom *Syn.* ONCOM.

ontogeny The range of stages which occur during the development of an individual organism to the mature or adult stage. The supposition that ontogeny recapitulates PHYLOGENY (the 'law of recapitulation' or the 'recapitulation theory') was made by the 19th-Century biologist Haeckel.

onychomycosis MYCOSIS of nails (cf. PARONYCHIA). [Review: Clinical Microbiology Reviews (1998) *11* 415–429.]

Onygena See GYMNOASCALES.

o'nyong-nyong fever (joint-breaker fever) A human disease caused by an ALPHAVIRUS and transmitted by anopheline mosquitoes; it occurs e.g. in Africa. Symptoms: fever, headache, myalgia, maculopapular rash, and severe arthralgia. Wild primates may act as a reservoir of the virus.

oocyst (1) (*mycol.*) *Syn.* OOGONIUM. (2) (*protozool.*) A structure in which sporozoites are formed. (3) An encysted fertilized female gamete.

Oocystis A genus of unicellular, non-motile, ellipsoidal or lemon-shaped, freshwater green algae (division CHLOROPHYTA); asexual reproduction occurs by autospore formation.

Oodinium See DINOFLAGELLATES and VELVET DISEASE.

oogamy Fertilization involving morphologically distinguishable male and female elements – typically a large non-motile female gamete and a small motile male gamete (see e.g. MONOBLEPHARIDALES; cf. OOMYCETES). (See also ANISOGAMY.)

oogonium A female structure in which one or more gametes (*oospheres*) develop; a fertilized oosphere is called an *oospore*.

ookinete A motile zygote.

Oomycetes A class of aquatic and terrestrial fungi (subdivision MASTIGOMYCOTINA) which typically give rise to zoospores having one anteriorly-directed tinsel flagellum and one posteriorly-directed whiplash flagellum (cf. PERONOSPORA). The organisms include saprotrophs and parasites – the latter including some important pathogens of cultivated crops. According to species, the thallus ranges from a simple, unicellular, holocarpic structure to a well-developed mycelium; in at least some species the thallus is diploid. The CELL WALL is generally believed to lack chitin and to contain CELLULOSE; however, there is evidence that CHITIN and cellulose both occur in the wall in e.g. certain members of the LEPTOMITALES and Saprolegniales. Some oomycetes exhibit DIPLANETISM, others MONOPLANETISM. Sexual reproduction is *oogamous*, involving the fusion of non-motile male and female elements within gametangia (cf. MONOBLEPHARIDALES). Orders: LAGENIDIALES, LEPTOMITALES, PERONOSPORALES, SAPROLEGNIALES.

Some authors include this class of fungi in the subdivision Diplomastigomycotina of the division Mastigomycota.

oophoritis Inflammation of the ovaries.

oosphere See OOGONIUM.

oospore See OOGONIUM.

Oosporidium A genus of non-fermentative imperfect yeasts (class HYPHOMYCETES). The cells are variously shaped, reproduce by multilateral bud-fission, and usually remain attached to each other to form long chains. Asexual endospores are formed, arthrospores are not. Pink or orange-yellow non-carotenoid pigments are synthesized. One species: *O. margaritiferum*, isolated from the slime flux of trees. [Book ref. 100, pp. 886–889.]

Opa proteins (of *Neisseria*) In pathogenic strains of *Neisseria*: OUTER MEMBRANE adhesins, encoded by *opa* genes, whose presence affects the opacity (hence Opa) and colour of the colonies (owing to increased bacterial aggregation); the Opa proteins mediate adhesion to (and uptake by) epithelial cells or opsonin-independent phagocytosis by PMNs. (Non-pathogenic strains of *Neisseria* contain genes with sequences homologous to those of the *opa* genes; it is possible that the products of these genes may mediate mucosal colonization by commensal strains.) (See also CD66.)

Some Opa$^+$ strains exhibit TRANSCYTOSIS.

Within a given cell, the chromosome contains several allelic (variant) forms of *opa* gene, and the organisms exhibit phase variation, i.e. a given cell may change (e.g.) from expressing a single type of *opa* gene to expressing more than one type (or no type) of *opa* gene. Phase variation is regulated by certain DNA re-arrangements in the 5′ leader sequence of the *opa* genes. The 5′ region of these genes contains multiple copies of a pentameric sequence (CTCTT), and the number of copies of this 'coding repeat' can vary; thus, while all the *opa* genes are transcribed constitutively, translation of a transcript will give rise to a functional Opa protein only if the number of (*penta*meric) coding repeats is such that the coding sequence of the gene is in the correct reading frame.

[Role of Opa proteins in interactions with host (eukaryotic) cells: TIM (1998) *6* 489–495.]

opacity-associated proteins (of *Neisseria*) See OPA PROTEINS.

opal codon See GENETIC CODE.

Opalina See OPALINATA.

Opalinata A subphylum of protozoa (phylum SARCOMASTIGOPHORA); the organisms are typically parasitic in amphibians. Individual cells are often oval or elongated, ca. 60–600 μm in length (according to species), and are generally flattened and leaf-like; each contains two or more (typically many) identical nuclei, and none has a cytostome. Each cell bears many short flagella (which are described as 'cilia' by some authors). Asexual reproduction typically occurs by longitudinal (INTERKINETAL) binary fission. In *Opalina* spp (rectal parasites of the frog) a phase of rapid asexual multiplication and encystment coincides with the breeding season of its amphibian host; it is believed that this phase is triggered by the host's hormonal activity. Cysts, the infective stages, are voided by the frog and ingested by tadpoles – in which excystment is followed by repeated longitudinal fission and the production of flagellated gametes. After syngamy, the encysted zygote is voided and ingested by another tadpole in which the zygote may undergo fission and gamete formation, or develop to form the mature, multinucleate trophic stage. (See also ENTAMOEBA (*E. paulista*).) Other genera include *Cepedea* and *Zelleriella*.

Opegrapha A genus of crustose LICHENS (order OPEGRAPHALES). Photobiont: *Trentepohlia*. Fruiting structures may be lirelliform (see LIRELLA) or rounded. Species occur on rocks, bark etc.

Opegraphales An order of fungi of the ASCOMYCOTINA; members include crustose to fruticose LICHENS (photobiont commonly trentepohlioid) and lichenicolous fungi. Ascocarp: APOTHECIOID, with a typically thick, carbonaceous excipulum. Asci: bitunicate, elongated to clavate. Genera: e.g. *Lecanactis*, OPEGRAPHA, ROCCELLA.

open complex See TRANSCRIPTION.

open culture *Syn.* CONTINUOUS CULTURE.

open reading frame (ORF) A sequence of codons (see GENETIC CODE), starting with an initiator codon and ending with a stop codon, which encodes a known or (as yet) unidentified polypeptide–or (see later) part of a polypeptide; the existence of an ORF is revealed by sequencing studies (see DNA SEQUENCING). A reading frame is said to be 'blocked' if a stop codon is located close to the initiator codon.

In some cases, two adjacent ORFs *jointly* encode a single polypeptide; translation of such a polypeptide requires some kind of mechanism (see e.g. READTHROUGH) that permits such coupling. In one case, the mechanism of translational bypassing of a 50-nucleotide gap between two ORFs appears to involve (i) a stem–loop structure between the first ORF and the coding gap, and (ii) a sequence of amino acid residues within the nascent peptide translated from the first ORF [EMBO (2000) *19* 2671–2680].

open-system culture *Syn.* CONTINUOUS CULTURE.

operational taxonomic unit See OTU.

operator (*mol. biol.*) Any specific nucleotide sequence to which a regulatory element can bind in order to control the expression (transcription) of a given gene or operon. An operator commonly consists of, or contains, a PALINDROMIC SEQUENCE, and it may lie e.g. between the PROMOTER and the coding sequence of a gene, or between the promoter and first structural gene in an operon – or (as in the TRP OPERON) it may occur entirely within the promoter region.

In operons under negative promoter control, the binding of repressor to operator sequence(s) inhibits initiation of transcription. The *gal* operon (encoding utilization of galactose, and biosynthesis of UDP-galactose) contains *two* operators (located on either side of the promoter region); it appears that, in the repressed state, these two sequences may be physically linked by an oligomer of the repressor protein (GalR) to form a *DNA loop* that (by sequestering the promoter region) inhibits initiation of transcription. (See also LAC OPERON.) Such DNA looping is also used for transcriptional control in the *araBAD* operon (see ARA OPERON). This operon contains four operator sequences, and the way in which the regulatory protein (AraC) binds to particular sequences (influenced by the presence/absence of arabinose) controls the activity of the *araBAD* operon. When arabinose is absent, AraC binds to the I_1 and O_2 operators, forming a loop that represses transcription. In the presence of arabinose, the arabinose–AraC complex binds to the I_1 and I_2 operators (which are close together and also referred to, jointly, as the initiator, *araI*); this promotes transcription of *araBAD*.

In *Pseudomonas aeruginosa*, synthesis of the siderophore PYOVERDIN is apparently controlled by a transcriptional regulator, the FUR PROTEIN, via an operator designated the 'Fur box' or 'iron box'; with adequate iron, Fur binds to the operator, blocking transcription of *pvdS* – a gene encoding a SIGMA FACTOR necessary for the synthesis of pyoverdin [see JB (1996) *178* 2299–2313].

Iron-regulated transcriptional control occurs at the operator of the *tox* gene encoding diphtheria toxin. Repression of *tox* by transitional metal ions apparently involves a helix-to-coil change in the repressor protein (DtxR); in this (iron-modified) state, dimers of DtxR inhibit transcription by binding on either side of the operator region [Nature (1998) *394* 502–506].

Opercularia A genus of sedentary ciliate protozoa (subclass PERITRICHIA) which occur in various freshwater habitats; some species (e.g. *O. coarctata*) are common in SEWAGE TREATMENT plants. *Opercularia* resembles CARCHESIUM, but the zooid lacks a flared rim, and the stalks are non-contractile.

operculate ascus An ASCUS whose wall has a large, annular, apical thickening that encloses a circular region of thinner wall tissue; the circular region of tissue ruptures prior to ascospore discharge.

operculum (*mycol.*) A hinged lid or an easily detachable portion of the wall in a sac-like spore-bearing structure (see e.g. OPERCULATE ASCUS).

operon In prokaryotes: two or more genes whose expression can be co-ordinated from (i) a common PROMOTER, the *contiguous* genes being transcribed as a single polycistronic mRNA (see e.g. LAC OPERON) *or* (ii) a common regulatory region – mRNAs being formed by DIVERGENT TRANSCRIPTION from different promoters.

The genes in an operon are often – though not necessarily – functionally related: e.g. they may encode enzymes of a particular metabolic pathway. The organization of genes into an operon may allow their expression to be co-ordinately 'turned on' (*induced*) or 'turned off' (*repressed*) according to the cell's needs. For example, enzymes of a catabolic pathway may be induced in the presence of the substrate of that pathway, and repressed again when the substrate has been used up; enzymes of an anabolic pathway must be induced when the end-product of the pathway becomes limiting and repressed when adequate levels of the product have been synthesized. Regulation of structural gene expression may occur at the level of transcription initiation (*promoter control*), transcription termination (*attenuator control*), or a combination of both; exceptionally, regulation occurs at the level of translation.

Promoter control. An operon under promoter control is commonly regulated by the product of a specific *regulator gene* which may or may not be contiguous with the structural genes. (Opinions differ as to whether or not the regulator gene should be regarded as part of the operon.) The regulator gene generally has its own promoter and may be expressed constitutively or may be subject e.g. to AUTOGENOUS REGULATION. The operon is said to be under *negative control* if the structural genes are expressed until 'turned off' by the regulator protein, and under *positive control* if they are not expressed until 'turned on' by the regulator protein. (The regulator protein may require a low-MWt *effector molecule* for activity.)

In a negatively controlled operon the regulator gene encodes a *repressor protein* which binds to a specific region of the DNA (the OPERATOR) upstream of the structural genes, thereby preventing the initiation of transcription. If the structural genes encode enzymes of a catabolic pathway, the presence of the substrate brings about the induction of gene expression. The molecule effecting induction is called an *inducer*; it is not necessarily the substrate of the pathway itself, but may be e.g. a derivative of it. (Certain analogues of the natural inducer may also act as inducers; an analogue which can act as an inducer but which cannot be metabolized by the cell is called a *gratuitous inducer*.) The inducer inactivates the repressor, causing it to release the operator and allow transcription to proceed. When the substrate is removed (e.g. when all of the substrate has been metabolized) the repressor again binds to the operator and transcription ceases; since mRNA is generally short-lived in bacteria (see MRNA), enzyme synthesis also rapidly ceases. (See e.g. LAC OPERON.)

Structural genes encoding enzymes involved in an anabolic pathway may also be subject to negative control. In this case, the regulator gene encodes an *aporepressor* which cannot itself bind to the operator of the operon, but which can do so when it interacts with the end-product of the pathway (or a derivative or analogue of it), the *corepressor*.

In a positively-controlled operon the regulator protein may be an *activator* (= *inducer protein*) which binds to a region of the operon (the *initiator*), thereby turning on (activating) transcription initiation. In the case of a catabolic operon the regulator gene may encode an *apo-activator* which must interact with the substrate (or a derivative of it), the *co-activator*, to form a functional activator. (See e.g. ARA OPERON.)

Attenuator control. Attenuation is a regulatory mechanism in which gene expression is prevented by the termination of transcription at a rho-independent terminator (see TRANSCRIPTION), the *attenuator*, in the LEADER region of the mRNA; thus, transcription is terminated before the RNA polymerase reaches the first structural gene of the operon. Whether or not transcription terminates at the attenuator depends on the secondary structure adopted by the leader region of the mRNA: in one conformation the stem-and-loop structure of the rho-independent terminator is formed and transcription is terminated, while in the alternative conformation the terminator structure is not formed and transcription continues into the structural genes.

Attenuation is involved e.g. in the regulation of various operons concerned with amino acid biosynthesis. In these operons the leader encodes a small peptide which contains a relatively high proportion of the amino acid whose biosynthesis is encoded by the operon. If the amino acid is present at levels adequate to meet the cell's requirements, the leader peptide is synthesized and the mRNA adopts the conformation which terminates transcription at the attenuator. However, when levels of the amino acid fall, the corresponding aa-tRNA becomes limiting and the ribosome translating the leader peptide stalls when it reaches the codon(s) for that amino acid; the presence of a stalled ribosome in this position causes the mRNA to adopt the alternative conformation (sometimes called the 'antiterminator'), and transcription continues through the attenuator and into the structural genes of the operon.

Operons subject to attenuator control include e.g. the HIS OPERON and the TRP OPERON. Attenuation is also involved in the regulation of e.g. genes involved in pyrimidine nucleotide biosynthesis [Appendix V(b)]; *pyrBI* (encoding subunits of aspartate carbamoyltransferase) and *pyrE* (encoding orotate phosphoribosyl transferase) are regulated in response to the level of cellular UTP [review: TIBS (1986) *11* 362–365]. (cf. TNA OPERON.)

Translational control. Translational control is exemplified by the IF3-L35-L20 operon (containing genes *infC-rpmI-rplT* respectively) in *Escherichia coli*; this operon (transcribed as a polycistronic mRNA) encodes two ribosomal proteins (L35, L20) and a protein factor involved in protein synthesis (IF3: translation initiation factor 3). Translation of L35 and L20 is repressed by L20, i.e. the concentration of L20 in the cell regulates gene expression – high levels of L20 repressing translation. Repression by L20 appears to involve the binding of L20 to mRNA at a site upstream of *rpmI*. The mechanism of repression by L20 is unknown – L20 could simply block mRNA–ribosome binding, or it could bind the ribosome to form an inactive complex. However, in vitro studies suggest the possibility that regulation may involve an RNA 'pseudoknot' formed by base-pairing between mRNA sites upstream of *rpmI*; it has been suggested that L20 may stabilize the pseudoknot (which includes the Shine–Dalgarno sequence and initiator codon of *rpmI*) so as to inhibit ribosome binding and, hence, prevent translation of both *rpmI* and *rplT* [EMBO (1996) *15* 4402–4413]. (See also RIBOSOME.)

operon fusion In genetic engineering: the fusion of two operons (which are transcribed in the same direction) e.g. by the deletion of a region of DNA between them; the structural genes of the second operon thus come under the control of the regulatory sequences (promoter, operator etc) of the first. (cf. GENE FUSION; see also MUDLAC SYSTEM.)

Ophiobolus See DOTHIDEALES. (For *O. graminis* see GAEUMANNOMYCES.)

ophiobolus patch disease See TAKE-ALL.

Ophiocytium A genus of unicellular algae (class XANTHOPHYCEAE) in which the vegetative cell is rod-shaped to filamentous and is either free-floating or attached to the substratum by a stalk; the cell is usually multinucleate and reproduces by zoospore or aplanospore formation.

Ophiostoma See OPHIOSTOMATALES.

Ophiostomatales An order of saprotrophic and plant-parasitic fungi of the subdivision ASCOMYCOTINA. Ascocarp: a black, long-necked, superficial or partly immersed perithecium; hamathecium: absent. Asci: unitunicate and thin-walled, rounded to ellipsoidal, evanescent. Genera: e.g. CERATOCYSTIS. (The genus *Ophiostoma* is distinguished by some authors.)

Ophryoglena See HYMENOSTOMATIDA.

ophthalmia Inflammation of any part of the eye, especially the conjunctiva. (See also CONJUNCTIVITIS.)

ophthalmia neonatorum Ophthalmia affecting newborn babies, usually contracted from the birth canal; it is commonly caused by *Neisseria gonorrhoeae* (see GONORRHOEA) or *Chlamydia trachomatis*. (The disease caused by *C. trachomatis* is also called *inclusion blennorrhoea* or *inclusion conjunctivitis*.) [Chlamydial ophthalmia neonatorum (aetiology, epidemiology, diagnosis, prophylaxis and treatment): Book ref. 193, pp. 297–304.]

opine See CROWN GALL.

opisthe (*ciliate protozool.*) The posterior of the two cells formed during HOMOTHETOGENIC cell division. (cf. PROTER.)

opisthomastigote See HERPETOMONAS and TRYPANOSOMATIDAE.

***oppBCDF* genes** See PEPTIDOGLYCAN.

opportunist pathogens Organisms which are normally free-living, or which occur as part of the normal body flora, but which may adopt a pathogenic role under certain conditions – e.g. when the normal antimicrobial defence mechanisms of the host have become impaired; thus, e.g. *Bacteroides* spp (common inhabitants of the human gut) may give rise to peritonitis following bowel surgery. (See also ENDOGENOUS INFECTION.)

opsin The apoprotein of e.g. BACTERIORHODOPSIN (bacterio-opsin), HALORHODOPSIN (halo-opsin) or SLOW-CYCLING RHODOPSIN (slow-opsin).

opsonic index The ratio of phagocytic activity of a patient's blood to that of a normal (non-immune) subject in respect of a specific opsonin. To determine the opsonic index: both (citrated) blood samples are incubated for 15 min with a standard number of the relevant antigenic particles (usually microbial cells); the average number of particles engulfed by the phagocytes is then determined by microscopy of smears from each sample – the ratio of the two averages being the opsonic index.

opsonin Any molecule which, when combined with a particulate antigen, increases the susceptibility of that antigen to PHAGOCYTOSIS. Opsonins include: (i) certain types of antibody (e.g. most types of IgG) – receptors for the FC PORTION of an antibody occurring e.g. on the macrophage surface; (ii) the C3b and C4b components of COMPLEMENT; (iii) certain ACUTE-PHASE PROTEINS.

opsonization The process in which a particulate antigen (e.g. a microbial cell) becomes more susceptible to PHAGOCYTOSIS as a result of combination with an OPSONIN. The term may refer to either a complement-dependent or complement-independent increase in susceptibility.

optical bleach *Syn.* FLUORESCENT BRIGHTENER.

optical brightener *Syn.* FLUORESCENT BRIGHTENER.

optical density See TURBIDIMETRY.

optical staining *Syn.* RHEINBERG ILLUMINATION.

optimal proportions (optimal ratio; equivalence) (*serol.*) In a PRECIPITIN TEST: the proportions in which antigen and antibody must be mixed in order to obtain the maximum amount of precipitate; for a given antigen–antibody system different optimal proportions are observed in the DEAN AND WEBB TITRATION and the RAMON TITRATION. (See also ANTIGEN EXCESS and LATTICE HYPOTHESIS.)

optochin Ethylhydrocupreine hydrochloride (EHC): a reagent which strongly inhibits the growth of *Streptococcus pneumoniae*; other streptococci which form colonies similar to those of *S. pneumoniae* are commonly resistant to ten or more times the concentration of optochin that inhibits *S. pneumoniae*. Optochin is used to detect the presence of *S. pneumoniae* in a mixed flora; the sample is streaked onto the surface of a solid medium, and an optochin-impregnated paper disc is placed on the medium prior to incubation. A significant decrease in the number of colonies in the vicinity of the disc may indicate the presence of *S. pneumoniae*.

oral hairy leukoplakia See HAIRY LEUKOPLAKIA.

oral herpes See HERPES SIMPLEX.

oral microbiology Microbiology of the mouth: see e.g. MOUTH MICROFLORA. (See also DENTAL CARIES, DENTAL PLAQUE and PERIODONTITIS.)

oral rehydration solution A solution of salts and glucose recommended by the WHO (World Health Organization) for use as an oral treatment in cases of severe diarrhoeal disease and dysentery. One litre contains: sodium chloride (3.5 g), sodium citrate (2.9 g), potassium chloride (1.5 g) and anhydrous glucose (20.0 g).

orange peel fungus See ALEURIA.

Orbivirus A genus of viruses of the REOVIRIDAE. The capsid contains at least seven polypeptide species, of which four are major structural components. Genome: 10 molecules of dsRNA. Type species: bluetongue virus. Orbiviruses infect a wide range of arthropods (mainly insects) and vertebrates, and some can cause disease in man or animals (see e.g. AFRICAN HORSE SICKNESS, BLUETONGUE; cf. COLORADO TICK FEVER); transmission is by vectors such as midges, mosquitoes, sandflies and ticks. The genus comprises many named viruses which are classified in the following serological subgroups (main vertebrate and invertebrate hosts in parentheses): African horse sickness (equines; *Culicoides*); bluetongue (ruminants; *Culicoides*); Changuinola (rodents; *Phlebotomus*); Colorado tick fever (man and rodents; ticks) – but see COLORADO TICK FEVER; Corriparta (birds, mosquitoes); epizootic haemorrhagic disease (deer; ?*Culicoides*); equine encephalosis (equines; unknown); Eubenangee (kangaroos, wallabies; mosquitoes); Kemerovo (various; ticks); Palyam (unknown; *Culicoides*, mosquitoes); Wallal (kangaroos, etc; *Culicoides*); Warrego (cattle, kangaroos, wallabies; *Culicoides*). [Review: Book ref. 83, pp. 287–357; biochemistry: Arch. Virol. (1984) *82* 1–18.]

orchid fleck virus See RHABDOVIRIDAE.

orchid mycorrhiza See MYCORRHIZA.

orchil (orseille) A purple dye obtained from species of the lichen ROCCELLA.

orchinol 2,4-Dimethoxy-6-hydroxy-9,10-dihydrophenanthrene: a wide-spectrum antifungal PHYTOALEXIN produced by the tubers (and, to a lesser degree, by the roots and stems) of certain orchids (e.g. *Orchis militaris*) in response to infection with parasitic or mycorrhizal fungi (e.g. *Rhizoctonia repens*) or with certain soil bacteria.

orchitis Inflammation of the testes.

orcinol 3,5-Dihydroxytoluene – a compound obtained from various lichen species.

order A taxonomic rank (see TAXONOMY and NOMENCLATURE).

orellanin poisoning MYCETISM, involving renal dysfunction, due to the ingestion of *Cortinarius orellanus*.

ores (microbial leaching) See LEACHING.

orf *Syn.* CONTAGIOUS PUSTULAR DERMATITIS (1).

ORF OPEN READING FRAME.

orf subgroup See PARAPOXVIRUS.

***orf18* gene** See CONJUGATIVE TRANSPOSITION.

organelle Any structure which forms part of a cell and which performs a specialized function: e.g. a FLAGELLUM, a MITOCHONDRION, or the PARORAL MEMBRANES of a ciliate.

organic acids (as antimicrobial agents) Certain organic acids and their salts or esters are used as PRESERVATIVES and e.g. as antifungal agents for the treatment of skin diseases (see e.g. BENZOIC ACID, PARABENZOATES, SALICYLIC ACID, SORBIC ACID and UNDECYLENIC ACID). Typically, their antimicrobial activity is maximum in the lower pH range; this is commonly taken as evidence that activity depends on the unionized molecule, although changes in pH may also affect e.g. transport systems and/or surface properties in the target cells. In membrane vesicles of *Escherichia coli*, parabens and sorbic acid were found to eliminate ΔpH – leaving $\Delta\psi$ much less disturbed [JGM (1985) *131* 73–76].

organism See MICROORGANISM.

organoleptic Affecting the organs of sense – particularly those of taste and smell.

organophosphorus compounds Phosphorus-containing organic compounds are widely used as insecticides and acaricides, and some are also effective agricultural ANTIFUNGAL AGENTS; the latter include e.g. ditalimfos (used against barley powdery mildew), HINOZAN, KITAZIN, and PYRAZOPHOS. [Accumulation, metabolism and effects of organophosphorus insecticides on microorganisms: AAM (1982) *28* 149–200.]

organotin See TIN.

organotroph An organism which obtains energy by the metabolism of an organic substrate (i.e., a *chemoorganotroph*) or which uses an organic substrate as an electron donor in phototrophic metabolism (i.e. a *photoorganotroph*). (cf. LITHOTROPH; HETEROTROPH.) Chemoorganotrophs may obtain energy from organic substrates e.g. by FERMENTATION (sense 1) and/or by RESPIRATION; substrates may include any of a wide range of compounds, e.g. (depending on species) various sugars, amino acids, fatty acids, cyclic compounds, etc. Photoorganotrophs include e.g. certain photosynthetic bacteria which use relatively simple substrates (e.g. acetate, caproate, lactate) as electron donors in PHOTOSYNTHESIS.

orgmet See BIOMIMETIC TECHNOLOGY.

ori A designation which usually indicates an origin of DNA REPLICATION (see e.g. *oriC*); *oriT* is an origin of transfer (see CONJUGATION sense 1b).

***ori*1** See F PLASMID.

***ori*2** See F PLASMID.

oriC An origin of chromosome replication. In *Escherichia coli* the *oriC* region is ca. 245 bp long and contains sequences for recognition by and interaction with initiation factors; these sequences are precisely separated by 'spacer' sequences [Book ref. 69, pp. 257–273; interaction of *oriC* with proteins: pp. 289–301]. (See also DNA REPLICATION.)

oriental sore See CUTANEOUS LEISHMANIASIS.

origin (*mol. biol.*) See DNA REPLICATION.

oriS See F PLASMID.

orisome See DNA REPLICATION.

oriT See CONJUGATION (1b) (i).

oriV The/an origin of vegetative replication in a PLASMID. (cf. *oriT* in CONJUGATION (1b) (i).)

Orleans process See VINEGAR.

ornidazole See NITROIMIDAZOLES.

L-ornithine See e.g. Appendix IV(a).

ornithine carbamoyltransferase See ARGININE DIHYDROLASE and Appendix IV(a).

ornithine decarboxylase test See DECARBOXYLASE TESTS.

Ornithocercus A genus of tropical marine DINOFLAGELLATES in which the cells have elaborate thecal extensions ('wings').

ornithocoprophilous Preferring a habitat rich in bird faeces.

ornithosis See PSITTACOSIS.

orotate Uracil 4-carboxylate: see Appendix V(b).

orotidine monophosphate See Appendix V(b).

Oroya fever See BARTONELLOSIS.

orphan virus A virus which is not known to cause any disease.

orseille *Syn.* ORCHIL.

Orthocerina See FORAMINIFERIDA.

orthochromatic dye A dye which does *not* give metachromatic effects (see METACHROMASY).

Orthodonella See HYPOSTOMATIA.

orthokinesis See KINESIS (sense 2).

orthomere See MACRONUCLEUS.

Orthomyxoviridae A family of pleomorphic, enveloped VIRUSES which have a segmented, negative-sense ssRNA genome (cf. PARAMYXOVIRIDAE). Virus replication in host cells is inhibited by actinomycin D and by α-amanitin. Orthomyxoviruses are mainly pathogens of the upper respiratory tract in man, animals and birds; they include influenza viruses A, B and C, distinguished on the basis of their ribonucleoprotein antigens. Influenza viruses A and B constitute one genus: INFLUENZAVIRUS; INFLUENZA VIRUS C probably constitutes a separate genus.

Two other viruses (DHORI VIRUS and THOGOTO VIRUS), previously considered to have affinities with the Bunyaviridae, have been shown to resemble orthomyxoviruses in structure; both viruses have a segmented genome of seven ssRNA molecules. [Virol. (1983) *127* 205–219.]

orthoplectic metachrony See METACHRONAL WAVES.

Orthopoxvirus (vaccinia subgroup) A genus of viruses of the CHORDOPOXVIRINAE. The orthopoxviruses are morphologically indistinguishable from each other and are very closely related serologically; they are distinguished chiefly on the basis of their biological properties. Orthopoxviruses grow on the CAM, producing 'pocks' characteristic for each virus; they can also replicate in certain cells in culture (e.g. chick embryo fibroblasts, Vero cells, HeLa cells, etc). All members produce a lipoprotein haemagglutinin in infected cells; this haemagglutinin apparently becomes incorporated in the host cell membrane, and may be present on the surface of (extracellular) virions. Recombination between different orthopoxviruses occurs readily: e.g. a recombinant has been derived from vaccinia and ectromelia viruses which exhibits specific pathogenicity markers of both parents [JGV (1985) *66* 621–626]. (See POXVIRIDAE for replication cycle etc.)

Type species: VACCINIA VIRUS; other members: buffalopox virus, camelpox virus, COWPOX VIRUS, ECTROMELIA VIRUS, MONKEYPOX virus, rabbitpox virus, VARIOLA VIRUS.

Orthoreovirus *Syn.* REOVIRUS.

***Oryctes rhinoceros* virus** See BACULOVIRIDAE.

***Orygia pseudotsugata* NPV** See NUCLEAR POLYHEDROSIS VIRUSES.

Oscillatoria A genus of filamentous CYANOBACTERIA (section III) in which the trichomes are straight, motile (see GLIDING MOTILITY), and composed of disc-shaped cells with little or no constriction between them; sheath formation is never pronounced. GAS VACUOLES may be present, and hormogonia are formed by fragmentation of whole trichomes. GC%: 40–50. Species include e.g. *O. agardhii*, *O. limnetica*, *O. princeps*, and *O. rubescens*; *O. redekei* strains may belong to the genus *Pseudanabaena*. (See also TRICHODESMIUM.) The organisms occur in a wide range of habitats, including aquatic (freshwater, brackish and marine) and damp terrestrial environments, often forming dense conspicuous mats. (See also GEOSMIN and PAINT SPOILAGE.) *O. limnetica* has a high sulphide tolerance and characteristically occurs, together with purple and green photosynthetic bacteria, in sulphide-rich anaerobic habitats, where it can carry out facultative anoxygenic phototrophy using H_2S as an electron donor for CO_2 assimilation (see CYANOBACTERIA).

Oscillochloris See CHLOROFLEXACEAE.

OSCP Oligomycin-sensitivity-conferring protein: see PROTON ATPASE.

Osler's nodes See ENDOCARDITIS.

osmic acid See OSMIUM TETROXIDE.

osmiophilic Having an affinity for OSMIUM TETROXIDE.

osmium tetroxide ('osmic acid'; OsO_4) A volatile compound used e.g. for post-FIXATION (1–2% w/v in phosphate or cacodylate buffer); the specimen is immersed in (or exposed to the vapour of) the solution. OsO_4 may cross-link and stabilize e.g. lipids by reacting with their double bonds. Longer exposure to OsO_4 may be used for positive staining in ELECTRON MICROSCOPY.

osmolyte *Syn.* OSMOTICUM.

osmophilic Refers to an organism which grows optimally in or on media of high osmotic pressure.

osmoregulation A form of ADAPTATION in which a cell increases or decreases its intracellular osmotic pressure in response to an increase or decrease, respectively, in extracellular osmotic pressure. Maintenance of internal osmotic pressure, within certain limits, is necessary for growth/viability; for example, osmotic withdrawal of water from a cell – without adaptive response – would tend to raise the concentrations of some solutes to levels that may be inhibitory or lethal.

In microorganisms osmoregulation involves the regulation of intracellular concentrations of certain types of solute and/or (in some eukaryotes) the action of CONTRACTILE VACUOLES.

Solutes used in osmoregulation include certain ions (e.g. K^+) and a number of low-MWt organic compounds such as glutamate and glycine betaine ($(CH_3)_3.N^+.CH_2COO^-$), a substance common in nature. (See also COMPATIBLE SOLUTE.)

Solutes (*osmotica*) may be taken up from the environment and/or synthesized in situ. Uptake or pumping out of ions through the cytoplasmic membrane involves e.g. the activity of particular types of ATPase; this form of osmoregulation is common in bacteria: see e.g. the *Escherichia coli* Kdp-based system in TWO-COMPONENT REGULATORY SYSTEMS. Another form of osmoregulation in *E. coli* is controlled by a different two-component system. In this system, increased osmolarity in the environment stimulates the kinase activity of a cell-envelope sensor, EnvZ, which transfers phosphate to the regulator protein, OmpR; activated OmpR (i) enhances transcription from *ompC* (the gene encoding OmpC PORIN) and (ii) inhibits transcription from the OmpF porin gene. Hence, increased osmolarity results in a modified outer membrane containing a higher proportion of OmpC (which forms a pore smaller than that of OmpF).

Following osmotic up-shift, uptake of glycine betaine is common in both Gram-positive and Gram-negative bacteria. In *E. coli* this compatible solute is taken up very efficiently by a binding-protein-dependent transport system (a so-called ABC importer) designated ProU [FEMS Reviews (1994) *14* 3–20].

The bacterium *Sinorhizobium meliloti* actually metabolizes most of the known osmoprotectants, including glycine betaine; in *S. meliloti*, osmotic up-shift triggers the accumulation of *endogenous* solutes (e.g. glutamate and *N*-acetylglutaminylglutamine amide) – accumulation being stimulated e.g. by exogenous sucrose [JB (1998) *180* 5044–5051].

[Regulation of compatible solute accumulation in bacteria: Mol. Microbiol. (1998) *29* 397–407.]

Osmotic down-shift triggers efflux systems. For example, K^+ ions and glutamate are released rapidly by *E. coli* (efflux apparently being energy-independent); much of the K^+ released is believed to pass through *mechanosensitive channels* in the CYTOPLASMIC MEMBRANE: stretch-activated channels which respond to increases in the cell's turgor pressure by increasing their pore size [JBC (1997) *272* 32150–32157]. In some bacteria (including *E. coli*) the membrane contains structures called *aquaporins* that may contribute to osmoregulation.

In the alga DUNALIELLA, glycerol appears to play a major role in osmoregulation. In this organism, the GLYCEROL cycle (glycerol → dihydroxyacetone (DHA) → DHA-phosphate (DHAP) → glycerol phosphate → glycerol) can be used to accumulate glycerol when external osmotic pressure is high – DHAP being derived from the Calvin cycle or from starch. In *Poterioochromonas malhamensis*, ISOFLORIDOSIDE is used as an osmoregulatory agent.

osmotic shock ('cold shock') In cells or subcellular organelles etc.: sudden change(s) in osmotic pressure sufficient to cause physical disruption. Osmotic shock has been used e.g. to release periplasmic enzymes from Gram-negative bacteria. The cells are pre-incubated in 20% sucrose containing EDTA and TRIS buffer, and then rapidly diluted in cold (e.g. 4°C), dilute $MgCl_2$ solution (e.g. 0.005 M); on centrifugation the liberated enzymes remain in the supernatant. During pre-incubation sucrose enters the periplasmic space, causing plasmolysis; on dilution, water enters the periplasmic space at a rate much greater than that of sucrose efflux via the porin channels – resulting in the rupture of the OUTER MEMBRANE.

osmotic shock-sensitive permease See BINDING PROTEIN-DEPENDENT TRANSPORT SYSTEM.

osmotica See OSMOTICUM.

osmoticum (*pl.* osmotica) A substance which contributes significantly to intracellular osmotic pressure. (See also OSMOREGULATION.)

osmotrophy A mode of nutrition in which soluble nutrients are absorbed through the cell surface – as occurs e.g. in bacteria and fungi. (cf. PHAGOTROPHY.)

osteomyelitis Inflammation of bone. *Acute osteomyelitis* is caused by bacterial infection via wounds or via the blood from another site of infection (e.g. skin, throat); it may occur as a complication of septicaemia. The commonest causal agent is *Staphylococcus aureus*. Acute osteomyelitis occurs most commonly in children and affects mainly the long bones of the arms and legs. Symptoms: e.g. fever and bone pain, with inflammation of the overlying skin; subsequent bone growth may be prevented. *Chronic osteomyelitis* may be due e.g. to tubercle bacilli, the vertebrae being most commonly involved (*Pott's disease*); spinal deformity may result. Chronic osteomyelitis may also occur as a complication of *Salmonella* infection (see e.g. FOOD POISONING (e)) – especially in patients with sickle-cell disease, bone defects, or prostheses; other causal agents include e.g. *Pseudomonas aeruginosa* (especially in drug addicts and open-heart surgery patients) and *Escherichia coli*.

osteopetrosis ('marble bone') A disease characterized by the formation of abnormally dense bone. In chickens, osteopetrosis ('big bone disease') may result from infection with an AVIAN LEUKOSIS VIRUS (ALV), and involves the abnormal thickening of bones – particularly the long bones of the legs – as a result of abnormal growth and differentiation of ALV-infected osteoblasts. Most exogenous ALVs appear to have the potential for inducing osteopetrosis, although strains differ in the time of onset of disease after infection, and in the severity of the disease induced. The time of onset and severity of disease seem to correlate with the amount of viral DNA in osteoblasts; affected cells each contain several to many copies of unintegrated viral DNA as well as integrated (proviral) DNA [Virol. (1985) *141* 130–143]. It has been reported that sequences near the 5′-LTR of the ALV genome are responsible for osteopetrosis induction [JV (1986) *59* 45–49], while 3′-*env*-LTR sequences influence the time of onset and severity of the disease [Virol. (1985) *145* 94–104].

osteosarcoma See SARCOMA.

ostiole (*mycol.*) A tubular or pore-like opening to the exterior in e.g. a perithecium, pseudoperithecium or pycnidium.

Ostracoblabe implexa A coenocytic, mycelial fungus (taxonomic position uncertain) which occurs on the shells of marine molluscs, obtaining nutrients from the organic matrix of the shell. In the European flat oyster (*Ostrea edulis*) the fungus causes *shell disease*: it penetrates between the shell and mantle of the oyster, causing severe irritation; the oyster responds by producing a highly proteinaceous shell which further encourages growth of the fungus. Shell disease occurs mainly in shallow waters where temperatures commonly exceed 18°C. [Book ref. 1, pp. 216–218.] (See also OYSTER DISEASES.)

Ostreobium A genus of siphonaceous green algae (division CHLOROPHYTA) which grow endolithically in shells and corals (and beneath the layer of zooxanthellae in corals) in tropical and temperate oceans; the cells contain siphonein and siphonoxanthin, and their photosynthetic pigments have absorption maxima in the far-red.

ostreogrycins See STREPTOGRAMINS.

ostropalean ascus An ASCUS whose wall contains a substantial, non-refractive apical thickening (apical cap); at maturity, the ascospores are discharged via a narrow channel in the cap. Such asci characteristically contain filiform ascospores.

Ostropales An order of saprotrophic, plant-parasitic and lichenized fungi of the ASCOMYCOTINA. Ascocarp: apothecioid or perithecioid. Asci: unitunicate, inoperculate, cylindrical. Ascospores: commonly filiform.

OT Old TUBERCULIN.

Oth-25/89/J virus A SMALL ROUND STRUCTURED VIRUS.

otitis media Inflammation of the middle ear. The condition is commonly suppurative: pyogenic bacteria from the throat infect the middle ear via the eustachian tube – often following an upper respiratory tract infection or tonsillitis. The eustachian tube becomes blocked, and pus formation causes bulging and sometimes rupture of the eardrum. Causal agents include e.g. *Haemophilus influenzae*, *Moraxella* (*Branhamella*) *catarrhalis*, *Streptococcus pneumoniae* and *S. pyogenes*.

otomycosis A MYCOSIS affecting the ears (see e.g. ASPERGILLOSIS).

Ottens' antigens Polysaccharide antigens which occur in many strains of *Streptococcus milleri* but which seldom occur in other streptococci.

OTU Operational taxonomic unit: a designation applied to each individual strain, species, genus etc used in classification carried out by NUMERICAL TAXONOMY; in bacteriology OTU commonly refers to a given strain.

ouabain (γ-strophanthin) A polycyclic, L-rhamnose-containing glycoside which inhibits Na^+/K^+-ATPase (see ION TRANSPORT).

Ouchterlony test A GEL DIFFUSION test of the DOUBLE DIFFUSION, double DIMENSION type. A pattern of holes ('wells') is cut into a sheet of agar (or similar material) contained (e.g.) in a Petri dish. The number and arrangement of wells depend on the number and nature of the samples to be tested, and on the purpose of the test. Each well is filled with an antigen, antibody (usually antiserum), or control preparation; the whole is covered (to prevent drying) and incubated (usually at room temperature) for hours or (more usually) days. The diagram shows the arrangement of wells, and some possible results, for the serological comparison of two samples of antiserum.

Oudemansiella See AGARICALES (Tricholomataceae).

Oudin test See SINGLE DIFFUSION.

***out* genes** See PROTEIN SECRETION (type II systems).

outer envelope See SPIROCHAETALES.

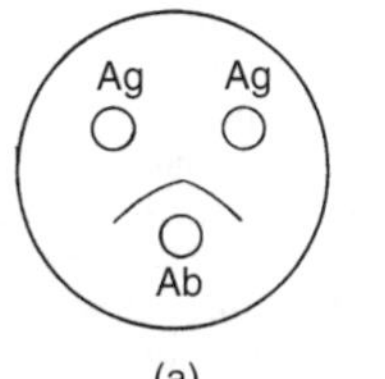

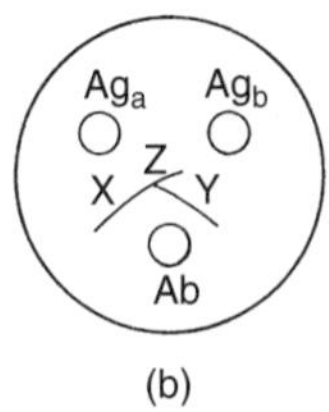

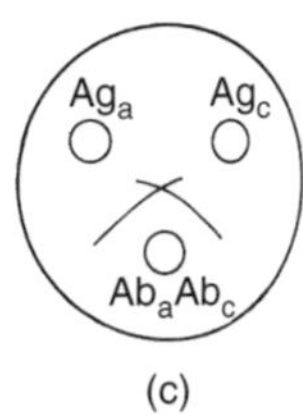

OUCHTERLONY TEST. (a) *Reaction of identity*. With wells located symmetrically, the lines of precipitate are symmetrical and do not overlap. Since the antigens (Ag) in the top two wells react identically with the antibody (Ab) they are taken to be serologically identical. (b) *Reaction of partial identity*. X is a line of precipitate formed where antibody and its homologous antigen (Ag_a) meet in optimal proportions. At Y, antibody and a *cross-reactive* antigen (Ag_b) meet in optimal proportions. In this instance, Ag_b has less affinity than Ag_a for the antibody, and antibody which has diffused beyond Y forms a precipitate with Ag_a at Z; Z, which is called a *spur*, points towards the cross-reactive antigen. (c) *Reaction of non-identity*. Two independent precipitating systems are present: each antibody reacts only with its homologous antigen.

outer membrane (OM) The outer layer of the CELL WALL in Gram-negative bacteria: a composite layer external to the PEPTIDOGLYCAN sacculus. The OM is an important permeability barrier (see later) which helps to control the cell's internal environment and give protection against certain antibiotics and other harmful agents. In pathogenic Gram-negative bacteria the hydrophilicity of the OM may help the cells to evade phagocytosis.

Composition of the OM. Most studies on the OM have been carried out on *Escherichia coli* and *Salmonella typhimurium*, and the following refers primarily to these and related enterobacteria. In electronmicrographs, the OM is seen as a trilaminar structure. Much of it consists of an asymmetrical lipid bilayer: a monolayer of phospholipid molecules forming the inner layer, and LIPOPOLYSACCHARIDE (LPS) molecules forming the outer (external) layer; the LPS molecules are so arranged that lipid A contributes to the hydrophobic interior of the membrane, and the (hydrophilic) O-specific chains are directed outwards.

Almost half the mass of the OM consists of proteins, and the various species of protein are distributed throughout the membrane – either embedded or partly embedded in the lipid bilayer. The quantitatively dominant 'major proteins' include the specialized channel-forming ('pore-forming') transmembrane proteins (see PORIN), the BRAUN LIPOPROTEIN, and (in *E. coli*) the OmpA protein (see OMP). Other proteins include those involved in the uptake of specific substance(s) (see e.g. BTUB PROTEIN, FEPA PROTEIN, LAMB PROTEIN, TONA PROTEIN, TSX PROTEIN), and enzymes such as proteases (see e.g. OmpT protein in type IV PROTEIN SECRETION) and phospholipases.

In members of the Enterobacteriaceae the outer surface of the OM carries a significant amount of ENTEROBACTERIAL COMMON ANTIGEN.

A stable and functional OM depends on ionic and other interactions between the various components. The core oligosaccharides of adjacent LPS molecules appear to be linked by divalent cations, particularly magnesium and calcium ions; in this context the phosphate group(s) of the heptose residue(s) may be important because their loss has been correlated with increased leakiness of the OM to hydrophobic molecules. Each LPS molecule is also hydrophobically bound (via lipid A) to the fatty acid chains of phospholipids. Some of the proteins appear to be linked to heptose residue(s) – or to the adjacent glucose residue – of the core oligosaccharide; thus, rough mutants which lack distal residues of the core oligosaccharide retain normal amounts of OM protein, while deep rough mutants which lack the heptose residue(s) have significantly lower amounts of OM protein. Porins (and e.g. OmpA protein) are bound to the underlying layer of peptidoglycan by ionic links; OmpA is readily released by treatment with SDS at temperatures below 56°C, but porins remain bound even after treatment with 2% SDS at 60°C – their removal requiring drastic treatment with SDS or e.g. with 0.5 M NaCl.

At a number of locations the OM is linked covalently to peptidoglycan via the Braun lipoprotein (see CELL WALL), the lipid portion of the Braun lipoprotein apparently being anchored in the hydrophobic interior of the OM.

At various localized regions there are apparent fusions between the OM and the cytoplasmic membrane (see ADHESION SITE); these sites are believed to be involved e.g. in the export of LPS and porin proteins from the cytoplasmic membrane to the OM. (Although widely accepted, a few authors question the existence of adhesion sites.)

Permeability of the OM. In wild-type (smooth) enteric bacteria the OM is permeable to small ions and small hydrophilic molecules; these entities can thus be taken up across the outer membrane in an energy-independent manner. The OM is generally much less permeable (or impermeable) to hydrophobic molecules – and to amphipathic molecules such as bile salts and SDS; such molecules do not pass easily through the porin channels, and their inability to penetrate the OM reflects the low permeability of the asymmetric lipid bilayer. In many cases, therefore, uptake across the outer membrane is energy-dependent (see e.g. TONB PROTEIN).

In deep rough mutants (e.g. the Rd and Re mutants of *Salmonella typhimurium*) the permeability of the OM to e.g. crystal violet, nafcillin, phenol and other hydrophobic molecules – and to anionic and cationic detergents – is much higher. This increased permeability may be due e.g. to the loss of some LPS molecules and their replacement by phospholipid molecules, resulting in the formation of phospholipid bilayers which are presumed to be more permeable to lipophilic substances. Alternatively, or perhaps additionally, rough mutants may be defective in LPS–LPS cross-linking; thus, the absence of phosphate group(s) in the core oligosaccharide may permit hydrophobic molecules to pass through gaps between the LPS molecules. [Review: CRM (1986) *13* 1–62.]

Experimental disruption of the OM. Some of the LPS molecules (up to ca. 50% of the total) can be removed from the OM by the action of metal ion chelators, such as EDTA (q.v.), in the presence of TRIS buffer – or by high concentrations of TRIS (e.g. 0.1 M, pH 7.2) without EDTA; by sequestering divalent cations, chelators increase the electrostatic repulsion between LPS molecules (and between LPS and acidic protein molecules), while TRIS may further destabilize the membrane by replacing LPS-bound cations. Cells thus treated become permeable to various hydrophobic molecules (e.g. actinomycin D, novobiocin). (See also OSMOTIC SHOCK.)

An EDTA–TRIS–LYSOZYME mixture can be used for the gentle lysis of Gram-negative bacteria, while the same procedure carried out in hypertonic sucrose solution can be used for SPHAEROPLAST formation.

POLYMYXINS, polylysines and polyornithines are examples of other agents which disrupt the OM and which increase its permeability. [Agents that increase the permeability of the OM: MR (1992) *56* 395–411.] (See also BPI PROTEIN.)

Low concentrations of magnesium ions (e.g. 0.1–5.0 mM) can help to stabilize the OM in deep rough mutants, but high concentrations of divalent cations (particularly calcium, but also magnesium) can destabilize/disrupt the wild-type OM; thus, e.g. 100 mM magnesium ions in the cold (0–4°C) causes the release of periplasmic β-lactamases in *Salmonella typhimurium.* High concentrations of cations may cause these effects by raising the melting point of the membrane lipids: at low temperatures the fluidity of the membrane may be decreased to such an extent that it readily ruptures; this effect is used to promote TRANSFORMATION and TRANSFECTION in enterobacteria.

outer membrane protein In Gram-negative bacteria: any of a range of proteins in the OUTER MEMBRANE – see e.g. OMP, FEPA PROTEIN, LAMB PROTEIN, TONA PROTEIN and PORIN. (cf. TONB PROTEIN.)

outer primers See NESTED PCR.

outer sheath See SPIROCHAETALES.

outer veil (*mycol.*) *Syn.* UNIVERSAL VEIL.

outgrowth (of bacterial endospores) See ENDOSPORE (sense 1(a)).

overgrowth syndrome See GASTROINTESTINAL TRACT FLORA.

overlapping genes Two (or more) GENES in which part or all of one gene is coextensive with part of another. The genes may be translated in different reading frames (see GENETIC CODE) or in the same reading frame with different start and/or stop points (see e.g. LEVIVIRIDAE and MICROVIRIDAE) or different splicing patterns. The phenomenon of overlapping genes maximizes the coding capacity of a genome, and can also provide a means for the regulation of expression of the genes.

overoxidation See ACETOBACTER.

overpressure retort See BATCH RETORT.

overproduction Synthesis of high concentrations of a given protein, usually a heterologous (i.e. 'foreign') protein, in a genetically modified organism. This procedure has been used e.g. for the synthesis of STREPTOKINASE and other therapeutic agents in *Escherichia coli.*

E. coli is commonly used for overproduction, but it cannot express certain genes – for example, eukaryotic genes whose products require specific post-translational modification in a eukaryotic cell. Moreover, genes containing introns would not be expressed (although this problem can be circumvented by using cDNA).

Efficient, high-level expression of heterologous genes in *E. coli* requires careful attention to specific aspects of the procedure; some of the relevant factors are considered below.

Transcription of the heterologous gene should be optimized by using a strong, tightly regulated promoter in the expression vector. The use of a tightly regulated promoter is necessary because gene expression should not occur until the population of *E. coli* has reached a high density; this avoids an early depression of growth rate and maximizes the yield of the required protein. Such tight regulation is clearly essential if the gene product is frankly toxic for *E. coli.* Moreover, the choice of transcription regulator should be appropriate; thus, e.g. a toxic inducer such as IPTG (isopropyl-β-D-thiogalactoside) would not be optimal for the overproduction of therapeutic proteins.

The efficiency of translation depends on factors such as the precise Shine–Dalgarno (SD) sequence and the number of nucleotides between the SD sequence and the start codon; these aspects of translational efficiency must be considered when constructing the expression vector. Additionally, some genes contain a DOWNSTREAM BOX whose precise sequence may be an important factor in efficient translation in *E. coli.* A further consideration is CODON BIAS.

One poorly understood problem is the appearance of 'inclusion bodies' – aggregates of unfolded or incorrectly folded proteins which form insoluble masses in the cytoplasm; this may result in little or no biologically active product. Approaches to this problem include (i) the use of a lower growth temperature, (ii) co-expression of so-called 'chaperone' proteins to promote correct folding, and (iii) use of gene fusion techniques; the latter method may involve fusion of the heterologous gene with another gene whose product promotes solubility of the fusion protein. (If, however, the inclusion body proteins can be correctly folded *in vitro*, the development of such bodies can be advantageous in that they may facilitate purification of the protein product and help to prevent degradation of the product by intracellular proteases.)

A further problem to consider is the risk of proteolysis by *E. coli* proteases. This risk may be avoided or minimized by targeting the protein product to the bacterial periplasm (in which there are fewer proteases). Another approach involves elimination of particular proteolytic site(s) in the product by re-coding the gene in a way which does not alter the product's biological activity. A further possibility is to fuse the heterologous gene with another gene whose product is secreted. Re-coding a gene may also help to increase the stability of certain proteins which are (inherently) unstable owing to the presence of particular amino acids (e.g. arginine, leucine, lysine, tryptophan) at the N-terminal. (In re-coding, it should be noted that the *penultimate* N-terminal amino acid in a bacterial protein is also important in that it may potentiate removal of the terminal *N*-formylmethionine (by methionine aminopeptidase) and thus itself become the terminal amino acid.)

A different approach to proteolysis involves the use of cells mutant in the *rpoH* gene (see HEAT-SHOCK PROTEINS); because RpoH promotes expression of the Lon protease these mutants are deficient in Lon, and such mutants have been found to give greatly increased yields of heterologous proteins.

[Achieving high-level expression of genes in *E. coli*: MR (1996) *60* 512–538.]

overwound DNA See DNA.

ovine diseases See SHEEP DISEASES.

ovine encephalomyelitis *Syn.* LOUPING-ILL.

ovine foot-rot See FOOT-ROT.

ovine progressive interstitial pneumonia See MAEDI.

Ovivora See EIMERIORINA.

Ovulariopsis See HYPHOMYCETES and ERYSIPHALES.

OXA-type β-lactamase See β-LACTAMASES.

oxacephems Compounds (e.g. MOXALACTAM) structurally resembling cephalosporins but having oxygen instead of sulphur at position 1 (see β-LACTAM ANTIBIOTICS, Figure (b)).

oxacillin See PENICILLINS.

oxadixyl See PHENYLAMIDE ANTIFUNGAL AGENTS.

Oxalobacter formigenes See UROLITHIASIS.

Oxalophagus A proposed genus of bacteria containing organisms which have been included in the genus *Clostridium* [IJSB (1994) *44* 812–826].

oxathiin antifungal agents See e.g. CARBOXIN.

oxazolidinones Synthetic antibiotics which inhibit PROTEIN SYNTHESIS in (mainly) Gram-positive bacteria; they include *linezolid* and *eperezolid*. Oxazolidinones bind to the 50S ribosomal subunit in the central region of domain V of 23S rRNA – a site similar to that targeted by chloramphenicol and lincomycin; this site is within the peptidyltransferase region of the ribosome, and it is thought that these drugs may affect function(s) associated with this location of the ribosome [JB (2000) *182* 5325–5331].

Linezolid (administered orally or intravenously) is a bacteriostatic agent which is effective (MIC 1–4 mg/l) against e.g. *S. pneumoniae*, staphylococci and enterococci; it can be useful against MRSA and other multidrug-resistant pathogens. [Review: TIM (1997) *5* 196–200.] Linezolid has been licenced for use in the USA.

[Linezolid (when to use?): JAC (2000) *46* 347–350.]

[Linezolid: review of use in the management of serious Gram-positive infections: Drugs (2001) *61* 525–551.] Clinical resistance to linezolid has been reported in vancomycin-resistant enterococci (VRE) [Lancet (2001) *357* 1179].

Oxford strain A strain of *Staphylococcus aureus* (NCTC strain 6571) which is sensitive to a wide range of antibiotics; it may be used as a control in antibiotic sensitivity tests.

oxic Refers to an environment in which oxygen is present.

oxiconazole See AZOLE ANTIFUNGAL AGENTS.

oxidase An OXIDOREDUCTASE which catalyses a reaction in which electrons removed from a substrate are donated *directly* to O_2.

oxidase reagent See OXIDASE TEST.

oxidase test A test used to detect the presence, in bacteria, of cytochrome *c* and its associated oxidase. In one form of the test, filter paper is moistened with a few drops of KOVÁCS' OXIDASE REAGENT, and a small amount of bacterial growth is smeared onto the moist filter paper with a glass spatula or a platinum loop. (Nichrome or iron-containing loops may give a false-positive reaction.) In a positive test (cyt *c* and cyt *c* oxidase present) a purple coloration (due to oxidation of the reagent) develops immediately or within 10 sec; electrons are transferred from the reagent to cyt *c* and thence, via cyt *c* oxidase, to molecular oxygen. In the so-called *cytochrome oxidase test* (= *indophenol oxidase test*) the reagent used is a mixture of equal volumes of 1% α-naphthol in 95% ethanol and 1% aqueous *p*-aminodimethylaniline oxalate; when the reagent is poured over bacterial growth, a positive result is indicated by a strong blue coloration (due to indophenol blue formation) within ca. 30 sec. (cf. BENZIDINE TEST.)

oxidation–fermentation test (Hugh and Leifson's test; O–F test) A test used to indicate whether a given carbohydrate (usually glucose) is attacked oxidatively or fermentatively (or not at all) by a given strain of bacterium. The peptone–agar medium includes NaCl, K_2HPO_4, the carbohydrate, and a pH indicator (usually BROMTHYMOL BLUE); the unused medium is green (if bromthymol blue is used) and has a pH of 7.1. In the test, each of two tubes of the (oxygen-free) medium is stab-inoculated with the test organism, and the medium in one of the tubes is immediately covered with a layer of sterile liquid paraffin ca. 1 cm in depth. Both tubes are then incubated and subsequently examined for evidence of carbohydrate utilization (acidification: yellowing of the medium). Acidification in only the uncovered medium indicates oxidative attack on the carbohydrate, while acidification in the covered tube indicates that the carbohydrate can be attacked fermentatively; no reaction in either tube usually indicates that the particular carbohydrate is not utilized by the test organism. [Modified medium for marine bacteria: AEM (1985) *49* 1541–1543.]

oxidation–reduction potential REDOX POTENTIAL.

oxidative metabolism See RESPIRATION.

oxidative pentose phosphate pathway *Syn.* HEXOSE MONOPHOSPHATE PATHWAY.

oxidative phosphorylation As commonly used: the phosphorylation of ADP to ATP, or the formation of PYROPHOSPHATE from inorganic phosphate, in a reaction which is driven by energy derived from proton motive force (see CHEMIOSMOSIS) generated by means of a *respiratory* chain; electron transport may be driven by aerobic RESPIRATION or by ANAEROBIC RESPIRATION. Oxidative phosphorylation may be catalysed by a PROTON ATPASE or a PROTON PPASE. (cf. ELECTRON TRANSPORT PHOSPHORYLATION and SUBSTRATE-LEVEL PHOSPHORYLATION.)

oxidative rancidity RANCIDITY due to AUTOXIDATION of a food's unsaturated fatty acids; such oxidation is encouraged e.g. by metal ions and by light. The occurrence of oxidative rancidity can be inhibited or delayed e.g. by refrigeration, by removing oxygen (e.g. by vacuum packing), or by the use of ANTIOXIDANTS.

oxidizing agents (as antimicrobial agents) Strong or moderately strong oxidizing agents can behave as effective antimicrobial agents when they are present in adequate concentrations under suitable conditions. However, their use as DISINFECTANTS is often limited by their high reactivity – which renders them liable to inactivation in the presence of excess organic matter. Inactivation by inert organic matter can be compensated for by using a concentration of the oxidizing agent sufficient to oxidize the inert matter and to leave a residual concentration which can act against viable microorganisms – see e.g. *chlorine demand* in WATER SUPPLIES. (For 'adaptation' of bacteria to oxidizing agents see HYDROGEN PEROXIDE.)

Antimicrobial oxidizing agents include e.g. CHLORINE, HYDROGEN PEROXIDE, HYPOCHLORITES, OZONE and POTASSIUM PERMANGANATE.

oxidoreductases ENZYMES (EC class 1) which catalyse oxidation–reduction reactions. (See also DEHYDROGENASE; OXIDASE; OXYGENASE; PEROXIDASE.)

oxine *Syn.* 8-HYDROXYQUINOLINE.

Oxobacter See CLOSTRIDIUM.

2-oxoglutarate dehydrogenase See TCA CYCLE.

oxolinic acid An analogue of NALIDIXIC ACID; it resembles nalidixic acid in its mode of action but is much more potent.

6-oxo-PGF$_{1\alpha}$ See PROSTACYCLIN.

oxycarboxin See CARBOXIN.

Oxyfume Sterilants A range of STERILANTS containing ETHYLENE OXIDE.

oxygenase An OXIDOREDUCTASE which catalyses the incorporation of one or both atoms of a molecule of oxygen into a molecule of substrate. A MONOOXYGENASE catalyses the incorporation of one atom of oxygen, the other being reduced to water; a dioxygenase (EC 1.13.11) catalyses the incorporation of both atoms of oxygen. Oxygenases are frequently involved e.g. in the first step of the bacterial degradation of HYDROCARBONS.

oxygenic photosynthesis See PHOTOSYNTHESIS.

oxyhydrogen reaction KNALLGAS REACTION.

Oxymonadida An order of protozoa (class ZOOMASTIGOPHOREA) which are parasitic in insects (e.g. termites). Cells have one to many axostyles (which, in some genera, are contractile) and one or more karyomastigonts, each containing four flagella arranged typically in two pairs; the kinetosomes of each flagellar pair are connected by a 'preaxostyle': a paracrystalline structure into which are inserted the anterior ends of the axostylar microtubules. Mitochondria and Golgi apparatus are absent. Genera: e.g. *Monocercomonoides*, *Oxymonas*, *Pyrsonympha*.

Oxymonas See OXYMONADIDA.

Oxyphotobacteria A proposed class of oxygenic, photosynthetic prokaryotes (division GRACILICUTES) comprising the CYANOBACTERIA and the PROCHLOROPHYTES.

***oxyR* locus** See HYDROGEN PEROXIDE.

Oxyrrhis See DINOFLAGELLATES.

oxysome *Syn.* FERNÁNDEZ–MORÁN PARTICLE.

oxysporum wilt See FUSARIUM WILT.

oxytetracycline See TETRACYCLINES.

Oxytricha See HYPOTRICHIDA.

oyster cap *Syn.* OYSTER FUNGUS.

oyster diseases For pathogens of oysters (e.g., *Ostrea* and *Crassostrea* spp) see e.g. LAGENIDIALES; MINCHINIA; OSTRACOBLABE IMPLEXA; PERKINSUS; VAHLKAMPFIA; VIBRIO (e.g. *V. anguillarum*). (See also INVERTEBRATE DISEASES.)

oyster fungus (oyster mushroom or oyster cap) Any species of *Pleurotus*, but especially *P. ostreatus*. A number of *Pleurotus* spp are edible and can easily be cultivated by inoculating logs (cf. SHII-TAKE); oyster fungi have also been grown on a number of waste materials such as sawdust, cereal straws, cotton waste, waste paper etc. [Mycol. (1983) *75* 351–360.]

ozaena (ozena) Rhinitis, caused by *Klebsiella pneumoniae* subsp *ozaenae*, characterized by a profuse, mucopurulent, fetid nasal discharge with crusting of the nasal mucosa.

ozena *Syn.* OZAENA.

ozone (as an antimicrobial agent) Ozone (O_3) is an OXIDIZING AGENT which reacts readily with most types of organic matter and which is active against a wide range of microorganisms; it is more efficient at lower temperatures, and requires a minimum relative humidity of ca. 60% for the disinfection of surfaces. Ozone has been used for the preservation of certain foods, but it is active against only surface contaminants, and it is not suitable for use with fats, meat etc which are likely to be oxidized. Ozone is sometimes used for the disinfection of potable WATER SUPPLIES; for this purpose it is superior to chlorine in that it is a stronger oxidizing agent, it is less affected by pH, and it does not impart a taste or odour. The treatment of wastewater (processed sewage) with ozone, rather than with chlorine, is advantageous e.g. in that it avoids the production of possibly toxic chlorinated organic compounds. The disinfection of water by ozone has been studied by monitoring the consumption of ozone-generated hydroxyl free radicals [Wat. Res. (1984) *18* 473–478]. (See also INFLUENZA.)

1. Words in SMALL CAPITALS are cross-references to separate entries.
2. Keys to journal title abbreviations and Book ref. numbers are given at the end of the Dictionary.
3. The Greek alphabet is given in Appendix VI.
4. For further information see 'Notes for the User' at the front of the Dictionary.

P

p (1) Prefix denoting the polypeptide product of a gene (see e.g. RETROVIRIDAE). (2) Designation for a precursor molecule (as e.g. in pTP in ADENOVIRIDAE). (3) Prefix denoting a PLASMID, usually a *recombinant* plasmid.

P Proline (see AMINO ACIDS).

P agar An agar medium containing peptone (1%), yeast extract (0.5%), NaCl (0.5%), glucose (0.1%) and agar (1.5%).

P antigen (parvovirus receptor) See ERYTHROVIRUS.

P fimbriae Chromosome-encoded FIMBRIAE found e.g. on many strains of *Escherichia coli* associated with human PYELONEPHRITIS or CYSTITIS. P fimbriae mediate mannose-resistant adhesion; they bind to α-D-galactopyranosyl-(1 → 4)-β-D-galactopyranoside receptors on ceramide-linked glycosphingolipids in human uroepithelium.

Adhesion of *E. coli* to urinary tract epithelium elicits the release of various CYTOKINES, including IL-6 and the chemokine IL-8. P fimbriae are reported to be more potent than other bacterial adhesins as an elicitor of IL-6; this cytokine can e.g. stimulate the synthesis of immunoglobulins and promote the release of ACUTE-PHASE PROTEINS.

The expression of P fimbriae is regulated by temperature, and is believed to involve the H-NS PROTEIN (q.v.).

P fimbriae are encoded by genes of the *pap* operon. These genes include *papA* (encoding the major fimbrial subunit), *papCD* (encoding, respectively, 'usher' and 'chaperone' proteins – see FIMBRIAE), and *papB* (encoding a regulatory protein). [P fimbriae: structure, genetics, assembly: FEMS Reviews (2000) *24* 21–44 (27–31).]

Interestingly, the PapB protein is involved in regulatory cross-talk between the *fim* and *pap* adhesin operons in *E. coli* – PapB being able to inhibit expression of type I fimbriae (see FIMBRIAE) [EMBO (2000) *19* 1450–1457].

P fimbriae are also called 'Pap pili' (Pap = pili associated with pyelonephritis).

987P fimbriae See F6 FIMBRIAE.

P ring See FLAGELLUM (a).

P segment See RECOMBINATIONAL REGULATION.

P-selectin See SELECTINS.

P site (of a ribosome) See PROTEIN SYNTHESIS.

P1 plasmid See BACTERIOPHAGE P1.

p15A See COLE1 PLASMID.

p21 family (Ras superfamily) See G PROTEINS.

P48 See PROTEIN SECRETION.

p55 (CD 120a) See TNF.

p75 (CD 120b) See TNF.

P-450 See CYTOCHROMES.

P582 A sirohaem-containing sulphite reductase which occurs in *Desulfotomaculum nigrificans.* (cf. DESULFOVIRIDIN.)

Pa See PASCAL.

PABA (PAB) *p*-AMINOBENZOIC ACID.

***pac* site** A packaging initiation site in the DNA of certain bacteriophages: see e.g. BACTERIOPHAGE P22 and BACTERIOPHAGE P1. (See also TRANSDUCTION.)

PA-CBT See IMMUNOSORBENT ELECTRON MICROSCOPY.

PACE 2C test See PROBE.

PA-CGT See IMMUNOSORBENT ELECTRON MICROSCOPY.

pachyman A water- and alkali-insoluble (1 → 3)-β-glucan comprising ca. 250–700 glucose residues with several (1 → 6)-β-linked branch points per molecule; it occurs in the sclerotia of *Poria cocos*.

Pachysolen A genus of yeasts (family SACCHAROMYCETACEAE). Cells are spheroidal to ellipsoidal and reproduce by multilateral budding; pseudomycelium is poorly developed or absent. Ascus formation is characteristic: an ascus develops at the tip of a tube which grows from a vegetative cell; the walls of the cell and tube become thickened and refractile except at the ascospore-bearing end – where the wall is thin and splits to release bowler-hat-shaped ascospores. Glucose is fermented; NO_3^- is assimilated. One species: *P. tannophilus*, isolated from concentrated tanning liquors. [Book ref. 100, pp. 289–291.]

pachytene stage See MEIOSIS.

Pachytichospora A genus of yeasts (family SACCHAROMYCETACEAE). [Book ref. 100, pp. 292–294.]

packet (of cells) A more or less regular arrangement of cells formed as a result of successive cell divisions along three mutually perpendicular axes, the progeny cells failing to separate. Packets are formed e.g. by *Sarcina* and *Sarcinosporon.* (cf. SARCINIFORM.)

pactamycin An antibiotic which specifically inhibits the initiation of translation during PROTEIN SYNTHESIS – apparently by inhibiting binding of the initiator tRNA to the initiation complex. Pactamycin has been used for genome mapping e.g. in picornaviruses which synthesize a POLYPROTEIN from a single initiation site (see PICORNAVIRIDAE). The antibiotic is added, together with radioactively-labelled amino acids, to virus-infected cells. Label is incorporated only into those polypeptides which were initiated before addition of pactamycin. A polyprotein molecule subsequently completed will contain an amount of label which depends on the stage of elongation reached when the drug and label were added; different polyprotein molecules will therefore contain different amounts of label – a minority bearing label near the N-terminus and a majority bearing label near the C-terminus (corresponding to the 5′ and 3′ ends of the mRNA, respectively). Thus, the amount of label present in the polypeptides derived from the polyprotein indicates the relative positions of their structural genes in the genome.

PADAC A violet-coloured pyridine-2-azo-*p*-dimethylanilide cephalosporin derivative which turns yellow on hydrolysis (e.g. by β-LACTAMASES); it is used in assays for β-lactamase levels. (See also PENICILLINASE.)

paddy-straw mushroom *Syn.* PADI-STRAW MUSHROOM.

PADGEM See SELECTINS.

padi-straw mushroom (paddy-straw, rice-straw or straw mushroom) Either of the tropical edible mushrooms *Volvariella volvacea* and *V. diplasia*, widely cultivated in S. E. Asia and the Far East. The mushrooms are traditionally grown on soaked (uncomposted) rice-straw, but the modern method – using e.g. composted rice-straw and cotton waste – resembles that for cultivating *Agaricus brunnescens* (see MUSHROOM CULTIVATION).

Paecilomyces A genus of fungi of the class HYPHOMYCETES; teleomorphs occur in e.g. BYSSOCHLAMYS. (See also COAL BIODEGRADATION; INDUSTRIAL ALCOHOL; LEATHER SPOILAGE; PAPER SPOILAGE; SINGLE-CELL PROTEIN.)

PAGE POLYACRYLAMIDE GEL ELECTROPHORESIS.

PAI See PATHOGENICITY ISLAND.

paint spoilage The susceptibility of liquid paints to microbial attack depends mainly on the nature of the solid phase and its solvent. Liquid emulsion paints may contain a variety of organic substances (e.g. as stabilizers, thickeners), inorganic constituents, and a high content of water, and may be susceptible to attack by bacteria (e.g. *Pseudomonas* spp) and fungi (e.g. *Fusarium* spp); biocides incorporated in such paints (to extend shelf-life) include e.g. organo-tins (such as tributyl tin oxide), certain QUATERNARY AMMONIUM COMPOUNDS, and mercurials such as phenylmercuric acetate. Oil-based liquid paints are less susceptible to spoilage, but the hardened films of both oil-based and emulsion paints are susceptible to attack – particularly under humid conditions and when surfaces are soiled; spoilage organisms may obtain nutrients from the film itself, from surface contamination, or from the substrate (wood etc) beneath the film. Spoilage organisms include species of the fungi *Aspergillus*, *Aureobasidium*, *Cladosporium* and *Penicillium*; cyanobacteria (e.g. *Nostoc*, *Oscillatoria*) and algae (e.g. *Trentepohlia*) may grow given adequate moisture and light. Biocides are incorporated in paints to help paint films resist microbial attack; these include e.g. organo-mercury compounds, SALICYLANILIDES, halogenated phenolics, and a variety of copper and zinc compounds.

Antifouling paints are used e.g. on ships' hulls to inhibit both spoilage and colonization by microorganisms and other organisms. In soluble-matrix paints the binder constituent is (slowly) soluble in seawater, low levels of biocide continually becoming available at the surface. In other paints, biocide diffuses slowly to the surface of an insoluble matrix. Biocides used include e.g. copper and zinc compounds, phenols and dithiocarbamates; some (e.g. tributyl tin oxide) inhibit algae as well as fungi.

paired organelles *Syn.* RHOPTRIES.

paired sera Two samples of serum from a given patient – the second sample taken e.g. days or weeks after the first. Paired sera are used e.g. to monitor changes in antibody titres.

Palade pathway In a eukaryotic cell: the path followed by many types of protein from their site of synthesis to their final cellular or extracellular location. Thus, e.g., a protein destined for secretion is synthesized on a ribosome bound to the cytoplasmic face of the rough ENDOPLASMIC RETICULUM (RER); it is translocated to the luminal face of the RER (see SIGNAL HYPOTHESIS) and is then enclosed within a vesicle (*cisterna*) which buds from the smooth endoplasmic reticulum and subsequently coalesces with a vesicle in the GOLGI APPARATUS. Finally, a secretory vesicle (containing the protein) buds from the Golgi apparatus and subsequently fuses with the cytoplasmic membrane to release the protein at the cell surface. Other types of protein – e.g. those destined for the cytoplasmic membrane itself, or for the lumen of a LYSOSOME – follow an analogous pathway; additionally, in cells infected with certain enveloped viruses the viral ENVELOPE proteins are transported to their appropriate membrane via this pathway. The production of secretory vesicles at the Golgi apparatus, and the subsequent fusion of these vesicles with the cytoplasmic membrane, can be inhibited e.g. by MONENSIN – which can therefore inhibit the replication of certain viruses.

Characteristically, proteins become progressively more glycosylated during their passage along the Palade pathway, and their intracellular location may even by determined by their degree of glycosylation. The glycosylation of newly synthesized (cell or viral) proteins may be inhibited e.g. by TUNICAMYCIN.

Proteins that are not transported via the Palade pathway include e.g. the small subunit of ribulose-1,5-bisphosphate carboxylase–oxygenase. This protein is synthesized (with an N-terminal signal peptide) on a cytoplasmic (i.e., not membrane-bound) ribosome and is taken up, post-translationally, by chloroplasts – being proteolytically processed during or after uptake.

***Pal*I** See RESTRICTION ENDONUCLEASE (table).

palindromic sequence A region of a nucleic acid which contains a pair of INVERTED REPEAT sequences. In a double-stranded molecule, such a region shows two-fold rotational (dyad) symmetry – or hyphenated dyad symmetry if the two IR sequences are separated by another sequence. A double-stranded palindromic sequence can adopt either of two possible conformations: a linear structure with interstrand hydrogen-bonding – e.g.

5′...TCCACATGTGGA...3′
3′...AGGTGTACACCT...5′

– or a cruciform structure in which the two strands each form HAIRPINS by intrastrand hydrogen-bonding. For steric reasons, non-paired bases occur at the proximal and distal ends of the hairpins in a cruciform, and the linear form is normally the more stable form. However, the formation of a cruciform tends to relieve the torsional strain of supercoiling in negatively supercoiled DNA, and purified negatively supercoiled DNA containing palindromic sequences tends to adopt a cruciform structure in vitro [JBC (1982) *257* 6292–6295]; nevertheless, cruciforms appear not to be formed in vivo [Book ref. 69, pp. 19–28].

Palindromic sequences occur e.g. in many operators, DNA replication origins, transcription terminators, etc, and are apparently important e.g. in protein–DNA recognition. (See also e.g. DEPENDOVIRUS.)

[Properties of palindromic sequences: JB (1985) *161* 1103–1111.]

palindromic unit *Syn.* REP SEQUENCE.

palisade A number of elongated cells (or other structures) arranged side by side in a row, adacent cells being in contact.

palisade plectenchyma See PLECTENCHYMA.

pallidins See LECTINS.

Palmaria A genus of algae of the RHODOPHYTA. *P. palmata* (formerly *Rhodymenia palmata*) is used as food ('dulse') in various countries.

Palmella A genus of freshwater green algae (division CHLOROPHYTA). (See also PALMELLOID PHASE.)

palmella phase *Syn.* PALMELLOID PHASE.

palmelloid phase (palmella phase) A vegetative phase formed under certain environmental conditions (e.g. on damp soil, agar media etc) by some unicellular green algae which are usually flagellated and motile (e.g. CHLAMYDOMONAS); the flagellated cells lose or withdraw their flagella and form a type of colony in which an indefinite number of cells is embedded in a gelatinous or mucilaginous matrix. The cells continue to divide within the matrix, and under suitable conditions (e.g. when the colony is flooded with water) they can develop flagella and escape from the colony.

A 'palmelloid' type of colony is the usual vegetative form in some *Chlamydomonas*-like algae (e.g. *Palmella*, TETRASPORA); in these algae motile cells are generally released from the colonies only as gametes or zoospores.

Paludrine See CHLORGUANIDE.

Palyam subgroup See ORBIVIRUS.

palynology The study of (fossil and living) pollen and spores.

pAMβ_1 See CONJUGATION (1b) (ii).

Panacide Dichlorophane (see BISPHENOLS).

Panaeolus See AGARICALES (Strophariaceae).

Panama disease See FUSARIUM WILT.

pandemic An EPIDEMIC affecting (simultaneously) large numbers of people in a major geographical area, or worldwide.

Pandorina A genus of freshwater flagellate green algae (division CHLOROPHYTA) which form motile, ball-like coenobia consisting of clusters of 4–32 cells with no central space. (cf. VOLVOX.)

panelling Refers to the condition of a can (see CANNING) in which, due to atmospheric pressure, part of the wall of the can has collapsed inwards. Panelling occurs e.g. in large high-vacuum cans, in cans cooled under excessive pressure, or in cans whose walls are made of thin tinplate.

pangola stunt virus See FIJIVIRUS.

panidazole See NITROIMIDAZOLES.

panleukopenia A significant reduction in the numbers of both lymphocytes and neutrophils – a condition brought about e.g. by the FELINE PANLEUKOPENIA VIRUS.

panmictic Of, or pertaining to, PANMIXIS.

panmixis (panmixia) Unrestricted interbreeding.

Pannaria A genus of LICHENS (order LECANORALES). Photobiont: *Nostoc*. Thallus: mainly squamulose, bluish-grey, with a dark hypothallus. Apothecia: lecanorine with reddish-brown discs. Species occur among mosses on bark, rocks etc.

panose α-Glucosyl-(1 $\rightarrow$ 6)-*D*-maltose.

Pansoil See ETRIDIAZOLE.

pansporoblast See MYXOSPOREA.

pantetheine See PANTOTHENIC ACID.

pantoic acid See PANTOTHENIC ACID.

Panton–Valentine leucocidin (PVL) A staphylococcal LEUCOCIDIN which specifically lyses human or rabbit PMNs and macrophages; no other type of cell appears to be affected. PVL consists of two synergistically acting components, designated 'fast' (F) and 'slow' (S) (indicating their migration rates on CMC columns). According to a recent model, the S component binds to GM_1 receptors on sensitive cells, causing activation of an endogenous (membrane-bound) phospholipase A_2 and stimulating uptake of Ca^{2+}; F then binds to products of phospholipase action and rapidly induces the formation of K^+-specific ion channels in the membrane. The resulting secondary effects lead to lysis. [Book ref. 44, pp. 732–737.] Insensitive cells may bind S but do not bind F – possibly because activation of phospholipase does not occur. The role of PVL in pathogenesis is unclear; in e.g. staphylococcal abscesses it may kill phagocytes and thus facilitate the survival and spread of the pathogen.

pantotheine See PANTOTHENIC ACID.

pantothenic acid A water-soluble VITAMIN which consists of *pantoic acid* (2,4-dihydroxy-3,3-dimethylbutyric acid) linked by an amide bond to β-alanine. Pantothenic acid is linked to 2-mercaptoethylamine (= cysteamine) to form *pantetheine* (= pantotheine); pantetheine 4′-phosphate is a component of COENZYME A (q.v. for formula) and is also the prosthetic group of acyl-carrier proteins (ACPs) involved e.g. in fatty acid biosynthesis.

Pantothenic acid is required as a growth factor e.g. by many yeasts (e.g. *Candida* spp, *Saccharomyces* spp), the filamentous fungus *Polyporus texanus*, certain bacteria (e.g. *Acetobacter* spp, *Lactobacillus* spp, *Morganella* spp), and certain protozoa (e.g. *Crithidia* spp, *Leishmania* spp, *Tetrahymena* spp). The requirement for pantothenic acid may be satisfied by its precursor(s) – e.g., β-alanine, pantoic acid.

pantropic (*virol*). Refers to a virus which can infect many types of cell. (See also TROPISM sense 2.)

PAP (1) 3′-Phosphoadenosine-5′-phosphate ('adenosine-3′,5′-diphosphate'): see ASSIMILATORY SULPHATE REDUCTION.

(2) Poly(A)polymerase: see MRNA.

***pap* operon** See FIMBRIAE.

Pap pili See P FIMBRIAE.

PAP technique (peroxidase–antiperoxidase technique) An IMMUNOPEROXIDASE METHOD which involves three antibody-mediated stages. Initially, separate antisera are raised (e.g. in rats) against (i) the specific antigen and (ii) a peroxidase; if these antisera are rat-derived, anti-rat Ig antiserum is raised e.g. in a rabbit. To locate specific antigen in a section, the section is first treated with rat antiserum homologous to the antigen ('primary' antiserum) and then washed free of uncombined antibody. The section is then treated with *excess* rabbit anti-rat Ig antiserum; under such antiserum-excess conditions, rabbit anti-rat Ig antibodies tend to bind to rat antibodies with only one of their combining sites. The section is washed and then treated with rat anti-peroxidase antiserum – these antibodies being bound by the free combining sites of the rabbit anti-rat Ig antibodies. (The rabbit anti-rat Ig antibodies are called 'bridging' or 'linking' antibodies.) Peroxidase is then added and combines with rat anti-peroxidase. Peroxidase (and hence antigen) is detected by incubating the section with reagents as in the ABC IMMUNOPEROXIDASE METHOD.

papain A heat-stable, basic thiol PROTEASE (EC 3.4.22.2) obtained from the latex and unripe fruit of the pawpaw (*Carica papaya*). (See also IMMUNOGLOBULINS.)

***Papaya* mosaic virus** See POTEXVIRUSES.

***Papaya* ringspot virus** See POTYVIRUSES.

PapB See P FIMBRIAE.

paper partition chromatography See CHROMATOGRAPHY.

paper spoilage Pulpwood (timber selected for pulping) may be attacked e.g. by cellulolytic fungi such as *Poria* and *Stereum* (see also TIMBER SPOILAGE). Wood *pulp* (produced from pulpwood by mechanical or chemical methods) is a watery suspension containing e.g. cellulose, hemicelluloses, sugars and proteins; it is susceptible to attack by a range of cellulolytic fungi (e.g. species of *Aspergillus*, *Cladosporium*, *Fusarium*, *Phoma* and *Stereum*) and fungi which cause darkening or staining (e.g. species of *Aureobasidium*, *Paecilomyces* and *Penicillium*). Chemically prepared pulps – e.g. those containing acidified $Ca(HSO_3)_2$ (*acid sulphite pulps*) or NaOH and Na_2SO_3 (*kraft pulps*) – are generally less susceptible to spoilage than are mechanically prepared pulps. Cellulolytic attack on either pulpwood or pulp shortens the cellulose fibres, resulting in inferior (weaker) paper. Slime formation in pulps is due e.g. to species of *Alcaligenes*, coliforms and other bacteria (*Serratia* forms red slime), and to fungi such as *Fusarium* and *Geotrichum*; slime may cause weaknesses, holes and/or discoloration in the paper. PRESERVATIVES used in pulps include e.g. dichlorophane, pentachlorophenol, and salicylanilides; slime formation has been controlled e.g. by chlorine and mercurials. Finished paper and board products have a low moisture content and are subject to fungal spoilage (particularly under high humidity) by e.g. *Chaetomium globosum*, *Stachybotrys atra* and other cellulolytic species; *Aspergillus* and *Penicillium* spp are common causes of discoloration. Paper and board preservatives include e.g. copper and zinc compounds, dichlorophane, and dithiocarbamates.

papilla A microcolony, within and visibly distinct from, a larger colony. For example, a colony of Lac^- mutant cells of *Escherichia coli* may contain a papilla that consists of a subpopulation of Lac^+ revertants; on MacConkey's agar, such a papilla would be pink/red in a colourless colony.

papilloma (*pl.* papillomas or papillomata) A small tumour of epithelial tissue (skin, mucous membranes, or glandular ducts), caused by a PAPILLOMAVIRUS; papillomas affecting the skin are commonly known as *warts* or *verrucas*. Papillomaviruses are largely host-specific, and different types of papillomavirus may cause morphologically distinct types of papilloma in a given host. In man, papillomas are caused by human papilloma viruses (HPVs): e.g. *plantar warts*, caused by HPV-1, occur on the soles of the feet and are flat, being pressed into the dermis by the weight of the body; *genital warts* (venereal warts, papillomata acuminata, condylomata acuminata), caused by HPV-6, occur in the anogenital region and are transmitted by sexual contact (cf. BUSCHKE–LÖWENSTEIN TUMOUR and MOLLUSCUM CONTAGIOSUM).

Papillomas are usually benign, but those caused by certain viruses can become malignant in the presence of certain 'cofactors' – e.g., certain animal papillomas can become malignant on exposure to sunlight. (cf. EPIDERMODYSPLASIA VERRUCIFORMIS.) In cattle, papillomas (caused by bovine papilloma virus 4) in the oesophagus and forestomachs can become malignant if the cattle eat bracken (*Pteridium aquilinum*). In humans, HPVs associated with genital warts are believed to play a role in the development of cancer of the genital tract (particularly of the cervix) [BMJ (1984) *288* 735–737; Lancet (1986) *i* 573–575].

Papillomavirus A genus of viruses of the PAPOVAVIRIDAE. Type species: cottontail rabbit (Shope) papilloma virus (CRPV) [genomic structure: PNAS (1985) *82* 1580–1584]; other members include human papilloma viruses (many types, designated HPV-1, HPV-2 etc), bovine papilloma viruses (BPV-1, BPV-2 etc), and papilloma viruses infecting deer, dogs, goats, horses, rats and sheep. Papilloma viruses can induce benign squamous epithelial tumours of the cutaneous and mucosal epithelium of their hosts (see PAPILLOMA). The virus can infect – but not replicate in – the basal cells of the epidermis or mucosa, establishing a persistent infection in which multiple copies of the viral DNA occur in the nucleus in the form of circular extrachromosomal molecules; however, production of virions can occur only in the non-dividing, differentiated squamous epithelial cells of the stratum spinosum and stratum corneum. (Since these cells cannot be grown in culture, papilloma viruses cannot be propagated in vitro, and molecular biological studies have depended largely on the use of cloned viral DNA.)

Although the papillomas induced by papilloma viruses are usually benign, certain human and animal papilloma viruses may be involved in the development of malignant tumours; in most of these systems infection with the papilloma virus is by itself insufficient for malignant transformation, and additional physical, chemical and/or biological factors are necessary for carcinogenesis (see PAPILLOMA and EPIDERMODYSPLASIA VERRUCIFORMIS). In experimental systems, several papilloma viruses can induce fibroblastic tumours in hamsters and can transform rodent cells in vitro (although such cells do not produce virions). The mechanism of transformation is unknown. In BPV-1, the presence of only a portion (69%) of the viral genome is necessary for transformation, and the expression of viral genes is apparently necessary; BPV-1 appears to carry multiple transforming genes [PNAS (1985) *82* 1030–1034]. There is evidence that viral DNA may integrate into the host chromosome in at least some cases of malignancy – e.g. integrated HPV-16 DNA has been detected in malignant tumours of the human genital tract [JGV (1985) *66* 1515–1522; see also *Gynecologic Oncology* (1992) *47* 263–266].

Papovaviridae (papovavirus group) A family of non-enveloped, small, icosahedral DNA viruses, most of which infect mammals (cf. BUDGERIGAR FLEDGLING DISEASE VIRUS); most or all members can induce tumours in certain host animals. Viral DNA replication and assembly occur in the host cell nucleus. The family contains two genera: PAPILLOMAVIRUS and POLYOMAVIRUS, distinguished on the basis of size, genome characteristics, a genus-specific antigen (detected in disrupted virions), and biological properties.

The papovavirus virion consists of an icosahedral capsid, diam. ca 45 nm (*Polyomavirus*) or 55 nm (*Papillomavirus*), which comprises 72 capsomers and which contains a supercoiled, covalently closed circular dsDNA genome; the DNA – MWt ca. 3×10^6 (ca. 5 kb) in *Polyomavirus*, ca. 5×10^6 (ca. 8 kb) in *Papillomavirus* – is associated with host-derived histones. (In the host cell, the viral DNA occurs in the nucleus – usually as a MINICHROMOSOME.) Papovavirus virions do not contain lipid or carbohydrates and are resistant to lipid solvents, low pH, heat treatment, etc.

papovaviruses Viruses of the PAPOVAVIRIDAE.

PAPS 3′-Phosphoadenosine-5′-phosphosulphate (also called adenosine-3′phosphate-5′-sulphatophosphate or 3′-phosphoadenylylsulphate). See ASSIMILATORY SULPHATE REDUCTION. (cf. APS.)

papulacandin B An antibiotic which is effective against yeasts but apparently not against other fungi, bacteria or protozoa; it inhibits the formation of CELL WALL β-D-glucan in whole cells and sphaeroplasts of e.g. *Saccharomyces cerevisiae* and *Candida albicans*. (cf. ACULEACIN A.)

papule On skin: a small, discrete, raised, solid lesion. (cf. MACULE, VESICLE.)

***par* locus** See PARTITION.

para-aminobenzoic acid See *p*-AMINOBENZOIC ACID.

parabactin See SIDEROPHORES.

Parabasalidea A superorder of protozoa (class ZOOMASTIGOPHOREA) which consists of the two orders HYPERMASTIGIDA and TRICHOMONADIDA.

parabens PARABENZOATES.

parabenzoates (parabens) Antifungal and antibacterial esters of *p*-hydroxybenzoic acid used e.g. as PRESERVATIVES in various foods, cosmetics and pharmaceuticals; they are more effective against fungi and Gram-positive bacteria than Gram-negative bacteria. Antimicrobial activity and insolubility in water both increase with increases in the chain length of the substituent group. Parabens are active at neutral as well as at acidic values of pH. (cf. BENZOIC ACID.)

paracentric (of a chromosome) To one side of the CENTROMERE; e.g., a paracentric inversion is a CHROMOSOME ABERRATION in which the inverted segment does not contain the centromere. (cf. PERICENTRIC.)

paracoagulation Clumping of cells (rather than clotting of plasma) which occurs e.g. when plasma (containing an anticoagulant) is mixed with a suspension of cells whose surface components include either staphylococcal clumping factor (see COAGULASE) or protamine sulphate.

Paracoccidioides A genus of fungi of the class HYPHOMYCETES. *P. brasiliensis*, the causal agent of PARACOCCIDIOIDOMYCOSIS, is a dimorphic fungus which occurs within tissues in a yeast-like form; the spherical, yeast-like cells (ca. 5–60 μm) can form several large buds (each bud attached to the parent cell by a narrow neck) or many small buds (forming the so-called ship's wheel form). When cultured at 37°C on blood-agar, *P. brasiliensis* grows in the yeast-like form; when cultured at room temperature or at 25°C on Sabouraud's dextrose agar (see SABOURAUD'S MEDIUM) it gives rise to a colourless, septate mycelium and to 'chlamydospores' and cells which resemble conidia. (See also CELL WALL (c) and PSEUDONIGERAN.)

paracoccidioidomycosis (South American blastomycosis) A chronic granulomatous human disease caused by *Paracoccidioides brasiliensis*; it occurs mainly in Central and South America. Infection may result in a benign or progressive lung disease, in a mucocutaneous disease (with formation of granulomatous ulcers chiefly on the mucous membranes of the mouth, nose and eyes, accompanied by regional lymphadenopathy), or in a systemic disease involving various internal organs. *Lab. diagnosis*: microscopy of infected tissues; serological tests (e.g. an EIA [SAB (1984) *22* 73–78]). (cf. BLASTOMYCOSIS.)

Paracoccus A genus (*incertae sedis*) of aerobic, oxidase-positive, catalase-positive, Gram-negative bacteria which occur e.g. in soil and in meat-curing brines. Cells: non-motile cocci or rods which typically contain granules of PHB. Metabolism: respiratory (oxidative), the terminal electron acceptor being O_2 or (in anaerobic respiration) nitrate, nitrite or nitrous oxide. One species (*P. halodenitrificans*) is chemoorganotrophic and halophilic, the other (*P. denitrificans*) is facultatively chemolithotrophic and methylotrophic (see METHYLOTROPHY) and is non-halophilic. Carbon sources include a wide range of organic compounds. Optimum growth temperature: 25–30°C. GC%: ca. 64–67. Type species: *P. denitrificans*.

P. denitrificans (formerly *Micrococcus denitrificans*). Cells (on e.g. nutrient agar): cocci, ca. 0.5–0.9 μm diam., or rods, ca. 0.9–1.2 μm in length. (See also PORINS.)

P. halodenitrificans. Cells (on e.g. nutrient agar containing 1.0 M (5.8%) NaCl): cocci, ca. 0.5–0.9 μm diam.; cells grown in lower concentrations of electrolyte (e.g. 0.6 M NaCl) are usually distorted, and no growth occurs below ca. 0.5 M NaCl. Isolated from meat-curing brines.

[Book ref. 22, pp. 399–402.]

paracolon An outdated term for those enterobacteria which do not metabolize lactose within 24/48 hours at 37°C.

paracytosis Invasive behaviour by certain pathogenic bacteria in which the pathogen crosses a layer of (eukaryotic) cells by passing *between* the cells; thus, e.g. *Haemophilus influenzae* can pass between the cells of lung epithelium [paracytosis by *H. influenzae* in tissue cultures of lung epithelial cells: Inf. Immun. (1995) *63* 4729–4737]. (See also TRANSCYTOSIS.)

paraffin metabolism See HYDROCARBONS.

paraflagellar rod *Syn.* PARAXIAL ROD.

paraformaldehyde ($HO(CH_2O)_nH$) A solid, white, polymeric form of FORMALDEHYDE which, when heated, yields gaseous formaldehyde.

parainfluenza viruses See PARAMYXOVIRUS.

parakinetal stomatogenesis See STOMATOGENESIS.

paralytic shellfish poisoning See SHELLFISH POISONING.

paramastigote A form assumed by the cells of species of *Herpetomonas* and *Leishmania* during certain stages in their life cycles: see TRYPANOSOMATIDAE.

Paramecium ('slipper animalcule') A genus of (typically) freshwater protozoa (order HYMENOSTOMATIDA). Cells: ovoid or elongate with uniform somatic ciliature; length varies with species: *P. aurelia* is ca. 120–200 μm, *P. bursaria* ca. 90–150 μm, *P. caudatum* ca. 180–300 μm, *P. multimicronucleatum* ca. 200–300 μm, and *P. trichum* ca. 50–120 μm. The buccal cavity is ventral and contains a single PARORAL MEMBRANE (= endoral membrane), two peniculi (see PENICULUS) and a QUADRULUS. STOMATOGENESIS is buccokinetal. Many species have one micronucleus and one macronucleus; *P. aurelia* has an additional micronucleus, and *P. multimicronucleatum* has three or four micronuclei. The presence of two CONTRACTILE VACUOLES is typical, and most species have large numbers of TRICHOCYSTS. Cysts are not formed. Paramecia are said to be negatively galvanotactic. Many strains of *Paramecium* contain endosymbionts: see CAEDIBACTER, HOLOSPORA, LYTICUM, PSEUDOCAEDIBACTER, TECTIBACTER; see also KILLER PARAMECIA, MU PARTICLE, ZOOCHLORELLAE. [Endonucleobiosis in ciliates: Int. Rev. Cytol. (1986) *102* 169–213.]

Asexual reproduction occurs by binary fission. AUTOGAMY and CONJUGATION occur in at least some species; mating types are divided into SYNGENS.

Paramecia consume bacteria and e.g. small flagellates. Certain species of *Paramecium* are a major food source for DIDINIUM.

paramere See MACRONUCLEUS.

Paramoeba A genus of marine amoebae (order AMOEBIDA). Species are generally free-living. However, *P. perniciosa* is apparently an obligate parasite of the blue crab (*Callinectes sapidus*) in which it can cause high mortality rates during moulting; another *Paramoeba* sp has been implicated as the cause of mass mortalities among sea urchins in Nova Scotia [J. Parasitol. (1985) *71* 559–565].

paramylon (*syn.* paramylum) A storage polysaccharide, a (1 → 3)-β-D-glucan, characteristic of the EUGLENOID FLAGELLATES. Paramylon occurs in refractile granules of various forms – discs, rings, rods etc – depending on species. Paramylon granules give no colour reaction with iodine. In *Euglena gracilis* paramylon is synthesized on a protein primer [Bioch. J. (1978) *174* 283–290].

paramylum *Syn.* PARAMYLON.

Paramyxa See PARAMYXEA.

Paramyxea A class of protozoa (phylum ASCETOSPORA) in which the spores are bicellular, consisting of a parietal cell and a sporoplasm. The spore wall has no orifice. One order: Paramyxida; genera include *Paramyxa*.

Paramyxoviridae A family of enveloped VIRUSes which have a non-segmented, negative-sense ssRNA genome. The family includes some important pathogens of humans, animals and birds – e.g. the causal agents of CANINE DISTEMPER, MEASLES, MUMPS, NEWCASTLE DISEASE, RINDERPEST, and various respiratory diseases. The viruses are transmitted mainly via the air (aerosols etc); vectors are unknown.

Virion: pleomorphic (usually roughly spherical), ca. 150–200 nm or more in diameter, containing 5–7 polypeptide species, one molecule of negative-sense linear ssRNA (MWt ca. $5–7 \times 10^6$) (some particles may contain positive-sense RNA strands), and 20–25% by weight of lipid. The envelope, which encloses a helical nucleocapsid, is derived from the host cell plasma membrane and contains virus glycoproteins (which form surface projections ca. 8 nm long) and a non-glycosylated virus protein (M). The virions contain various enzymes, including e.g. transcriptase, polyadenylate transferase, mRNA methyltransferase. The virions are inactivated by lipid solvents, non-ionic detergents, formaldehyde, oxidizing agents, etc.

Members of the Paramyxoviridae can be propagated in various types of cell culture. Infection occurs by fusion of the virus envelope with the plasma membrane of the host cell. Transcription and replication occur in the cytoplasm, the nucleocapsid functioning as template. Virions mature by budding through the host cell plasma membrane at sites containing the virus envelope proteins. Infected host cells commonly lyse, but temperate and persistent infections also occur, commonly resulting in cell fusion and syncytium formation, inclusions, haemadsorption, etc.

The family is divided into three genera – MORBILLIVIRUS, PARAMYXOVIRUS and PNEUMOVIRUS – on the basis of e.g. neuraminidase activity, virion structure, etc.

Paramyxovirus A genus of viruses of the PARAMYXOVIRIDAE in which the virions have both haemagglutinin and NEURAMINIDASE activities. The genus includes e.g. NEWCASTLE DISEASE virus (NDV) (type species), MUMPS virus, and parainfluenza viruses type 1 human, type 1 murine (= Sendai virus), type 2 (human), type 3 (human, bovine and ovine), type 4 (human), and type 5 (avian, canine and simian). (See also CALF PNEUMONIA, COMMON COLD, PNEUMONIA (g).)

The paramyxovirus virion consists of a helical nucleocapsid, composed of the genomic ssRNA and proteins NP, P and L, surrounded by an envelope containing a non-glycosylated protein (M) in the inner layer, and two glycoproteins which extend across the width of the envelope and beyond the outer surface to form spikes. Proteins P and/or L may have transcriptase activity. Protein M plays an important role in virus assembly. The larger of the envelope glycoproteins (designated HN) has cell-binding, haemagglutinating and neuraminidase activities, while the smaller (F) has haemolytic activity and promotes fusion between the virus envelope and the host plasma membrane. The F protein can also promote cell–cell fusion, a property exploited e.g. in the use of UV-inactivated Sendai virus in the production of hybrid cells (see e.g. HYBRIDOMA). In the Sendai and mumps viruses the F protein is synthesized as an inactive precursor, F_0, which is activated by proteolytic cleavage. [Sequence of the F protein gene in Sendai virus: JGV (1985) *66* 317–331.]

Host cell infection occurs by adsorption, via HN, to the cell surface, followed by F protein-mediated fusion between the virus envelope and the host plasma membrane; antibodies to HN and F proteins appear to be important in limiting or preventing infection in vivo. Within the host cell, virus RNA is transcribed into six (mostly monocistronic) mRNAs. In addition to the six structural proteins, two non-structural proteins (C and C′) are synthesized from a reading frame overlapping that of the P gene. HN and F proteins are synthesized on membrane-bound polysomes, glycosylated, and inserted into the host plasma membrane; during maturation, the virions bud through the region of the membrane containing these proteins.

paramyxoviruses (1) Viruses of the PARAMYXOVIRIDAE. (2) Viruses of the genus PARAMYXOVIRUS.

Parana virus See ARENAVIRIDAE.

Paranaplasma A genus of bacteria of the family Anaplasmataceae [Book ref. 21, pp. 908–909]; *P. caudatum* is now *Anaplasma caudatum* [Book ref. 22, p. 722].

parapertussis See WHOOPING COUGH.

paraphysis (1) A type of sterile, basally-attached, branched or unbranched filament (of unknown function), many of which occur among the hymenial asci in the ascocarps of certain ascomycetes; paraphyses may or may not project above the level of the asci. (cf. EPITHECIUM and HAMATHECIUM.) *Apical paraphyses* are sterile filaments which originate in the upper wall of a pseudothecial locule and grow downwards into the centrum, their tips remaining unattached. (cf. PSEUDOPARAPHYSIS.)

(2) *Syn.* BASIDIOLE (sense 3).

paraphysoid (1) In some (particularly lichenized) ascomycetes: a structure which resembles a PSEUDOPARAPHYSIS but which generally has few septa, and which may be formed before or after the asci have developed.

(2) In basidiomycetes: a sterile hymenial structure, e.g. a BASIDIOLE (sense 3).

Paraphysomonas A (probably heterogeneous) genus of freshwater and marine, non-pigmented CHRYSOPHYTES (q.v.) which bear various types (according to species) of silicified scales.

paraplectenchyma See PLECTENCHYMA.

Parapoxvirus (orf subgroup) A genus of viruses of the CHORDOPOXVIRINAE which infect ungulates and may infect man. Type species: orf virus (see CONTAGIOUS PUSTULAR DERMATITIS sense 1); other members: bovine pustular stomatitis virus, chamois contagious ecthyma virus, PSEUDOCOWPOX virus (= MILKERS' NODULE virus, paravaccinia virus).

Parapoxvirus virions are ovoid, ca. 250 × 160 nm (for orf virus) to 300 × 190 nm (for pseudocowpox virus); the outer surface of the virion exhibits a characteristic criss-cross pattern in negatively stained preparations. DNA MWt: ca. 85×10^6. Haemagglutinin is not formed. Infected cells contain only B-type inclusion bodies. Infectivity of virions can be destroyed e.g. by heating at 58–60°C for 30 min. Parapoxviruses do not produce pocks in CAM. They can bring about NON-GENETIC REACTIVATION of inactivated orthopoxviruses.

paraprotein Immunoglobulin produced by a neoplastic clone of B cells. (See also MYELOMA.)

parapyle See RADIOLARIA.

Paraquat *Syn.* METHYL VIOLOGEN.

pararosanilin A reddish TRIPHENYLMETHANE DYE.

pararotaviruses (group B rotaviruses) Viruses which are morphologically indistinguishable from ROTAVIRUSES but which lack the group-specific antigen common to all other known (avian and mammalian) rotaviruses; the 11 dsRNA segments also have a unique and characteristic electrophoretic migration pattern. Pararotaviruses have been isolated from human stool specimens, but their clinical significance – if any – is unknown.

parasexual processes (*mycol.*) Non-sexual processes which can result in the formation of recombinant nuclei. The process first reported in *Aspergillus nidulans* (and later found in many ascomycetes and in some deuteromycetes and basidiomycetes) has been called the standard parasexual process. This process starts with the formation of a heterokaryotic cell or mycelium (see HETEROKARYOSIS) and involves the fusion of two genetically dissimilar haploid nuclei (in the heterokaryon) to form a single diploid nucleus. Recombination can then occur through mitotic crossing over – i.e., the (infrequent) reciprocal exchange of chromatid segments between homologous chromosomes during *mitotic* propagation of the diploid nucleus – or (e.g. in *Saccharomyces cerevisiae*) predominantly by gene conversion [MR (1985) *49* 33–58 (43–47)]. Haploidization (the formation of haploid progeny nuclei) involves NON-DISJUNCTION in which there is a stepwise loss of one member of each pair of homologous chromosomes during successive mitoses.

For the imperfect fungi parasexuality brings the advantage of genetic recombination. The importance of parasexual processes in nature is uncertain.

[Review: Book ref. 168, pp. 191–214.]

***Para*Sight-F™ test** See MALARIA.

parasitaemia (parasitemia) The presence of parasites in the bloodstream.

parasite See PARASITISM. A parasite may be obligatorily or facultatively parasitic; it may be host- or strain-specific or able to parasitize a range of hosts. Parasites which remain external to the host's cells or tissues are called *ectoparasites*, while those that live intracellularly or within tissues are called *endoparasites*. If the host suffers significant harm, or is killed, through the activity of a microbial parasite, the parasite is referred to as a PATHOGEN.

Compared to its free-living (non-parasitic) relatives, a parasite often exhibits certain aspects of degeneracy – e.g. a reduction in, or loss of, structures, organelles, or metabolic mechanisms involving the ingestion and digestion of food, respiration, osmoregulation or excretion. Adaptation to the parasitic mode of life may be regarded as a form of specialization; some parasites are so well integrated into the economy of their hosts that they can be cultivated in vitro only with difficulty, or not at all.

parasitism SYMBIOSIS (sense 1) in which one organism (the PARASITE) benefits at the expense of the other (the *host*); typically, the parasite obtains its nutrients from the host, and the association is detrimental to the host. (cf. COMMENSALISM.)

parasitology The study of certain PARASITES: their life cycles, ecology, transmission, etc; those parasites studied in parasitology include some which are classified as microorganisms (e.g. amoebae, trypanosomes, malarial parasites) as well as various arthropods, helminths, etc. Pathogenic bacteria, fungi and viruses are traditionally not included within the scope of parasitology – even though they can live parasitically.

parasol ants See FUNGUS GARDENS.

parasol mushroom See LEPIOTA.

parasomal sac In many ciliates (e.g. *Paramecium*): a small flask-like invagination of the surface membrane near the ciliary bases, possibly involved in pinocytosis.

Parasponia See BRADYRHIZOBIUM.

parasporal crystal See DELTA-ENDOTOXIN and MILKY DISEASE.

paratenic Refers to an alternative host (of a parasite) which is incidental to the normal life cycle of the parasite: i.e. the paratenic host is neither essential for nor interferes with the completion of the parasite's life cycle.

parateny (*ciliate protozool.*) The condition in which part(s) of a cell have lines of kinetids arranged perpendicularly to the longitudinal axis of the cell; parateny is characteristic of some scuticociliates (e.g. *Loxocephalus*).

paratope (1) The COMBINING SITE of an antibody. (2) The site at which a cell-surface receptor combines with an exogenous molecule.

paratose (3,6-dideoxy-D-glucose) A sugar, first isolated from *Salmonella paratyphi*, which occurs in the O-specific chains of the LIPOPOLYSACCHARIDE in certain *Salmonella* serotypes and which contributes to the specificity of O antigen 2 in group A salmonellae (see KAUFFMANN–WHITE CLASSIFICATION).

paratrachoma See INCLUSION CONJUNCTIVITIS.

paratuberculosis *Syn.* JOHNE'S DISEASE.

paratyphoid fever (paratyphoid) (1) (*med.*) An acute infectious disease caused by *Salmonella paratyphi* A, B or C. It resembles TYPHOID FEVER in modes of transmission and infection, but symptoms are milder and are often accompanied by gastroenteritis.

(2) (*vet.*) Any disease (paratyphoid, salmonellosis) caused by a serotype of *Salmonella* and characterized by septicaemia (particularly in young animals) and/or enteritis. In cattle, paratyphoid is often caused by *S. dublin*, and sometimes e.g. by *S. typhimurium*; in the acute enteric form symptoms include fever and diarrhoea, while chronic enteritis may cause general unthriftiness. In pigs, the causal agents include *S. choleraesuis* and *S. typhimurium*; the acute enteric form, involving fever and diarrhoea, is often complicated by pulmonary involvement, while in the (usually fatal) septicaemic form there may be nervous involvement, red-purple skin discoloration (e.g. on the ears and abdomen), and subcutaneous petechial haemorrhages. In horses, *S. typhimurium* can cause e.g. an acute febrile condition with diarrhoea and dehydration, while in very young foals it can cause a fatal septicaemia. An asymptomatic carrier state can occur in cattle, pigs and (to a limited extent) horses.

Parauronema See SCUTICOCILIATIDA.

paravaccinia virus See PARAPOXVIRUS.

paraxial rod (paraflagellar rod) In the FLAGELLUM of many members of the KINETOPLASTIDA, DINOFLAGELLATES and EUGLENOID FLAGELLATES: a filament or rod which runs, within the flagellar membrane, parallel to the axoneme or (in the helical transverse flagellum of a dinoflagellate) parallel to the axis of the axoneme. [Paraxial rod in trypanosomatids: JP (1986) *33* 552–557.]

paracentric objectives In a given microscope: a set of objective lenses each of which gives a field of view whose centre is identical to that of the others. (cf. PARFOCAL OBJECTIVES.)

***parC* gene** See CELL CYCLE (b).

***parE* gene** See QUINOLONE ANTIBIOTICS.

parenteral Administered by any route *except* the alimentary canal.

parenthesome See DOLIPORE SEPTUM.

parfocal objectives In MICROSCOPY: a set of objective lenses which, in a given microscope, can be interchanged (by rotating the nosepiece) without the need to re-adjust the focusing controls. (cf. PARCENTRIC OBJECTIVES.)

parietal Lying near to, or attached to, the inner face of the wall of a structure; may refer e.g. to algal chloroplasts which lie close to the cell wall.

parietin An orange anthraquinone pigment found in lichens of the Teloschistaceae, in certain free-living fungi (e.g. *Aspergillus* and *Penicillium* spp), and in a few higher plants; it gives a purple colour with KOH. Parietin is present at higher concentrations in *Xanthoria parietina* thalli growing in bright light than in those growing in shade, and probably serves to protect the photobiont from high-intensity light.

pariobasidium In certain basidiomycetes: the distal, functional part of a mature METABASIDIUM whose proximal region consists of remnants of the probasidium [TBMS (1973) *61* 497–512].

Park nucleotide UDP-*N*-acetylmuramyl pentapeptide: an intermediate in the biosynthesis of PEPTIDOGLYCAN.

Parmelia A large genus of LICHENS (order LECANORALES); photobiont: a green alga. The thallus is foliose, usually with rhizines on the underside; according to species, the upper surface may be yellowish, grey, greyish-green, brown etc, and soredia, isidia, or pseudocyphellae may be formed. Apothecia are lecanorine. Species occur on rocks, bark, wood etc.

Parmesan cheese See CHEESE-MAKING.

PARNA See CUCUMOVIRUSES.

paromomycin Either of two related AMINOGLYCOSIDE ANTIBIOTICS, produced by *Streptomyces rimosus*, which may be regarded as derivatives of neomycins B and C.

paronychia Inflammation of the folds of tissue surrounding a fingernail or toenail. It may be caused e.g. by *Candida* spp (see CANDIDIASIS), by certain bacteria, or by HERPES SIMPLEX virus ('herpetic paronychia'). (cf. ONYCHOMYCOSIS; WHITLOW.)

paroral membrane In many ciliates: a membrane, composed of multiple cilia, which occurs on the right-hand side or border of the buccal cavity; such membranes are often undulatory and are sometimes called 'undulating membranes'. In ciliates of the OLIGOHYMENOPHOREA (e.g. hymenostomes) the paroral membrane is characteristically of the *stichodyad* type: a double row of kinetosomes arranged in a zigzag pattern, only the outermost kinetosomes commonly bearing cilia (see also HAPLOKINETY); in *Paramecium* the paroral membrane is also called the *endoral membrane*. In many or most ciliates of the POLYHYMENOPHOREA the paroral membrane is of the *stichomonad* (= *monostichomonad*) type in which the infraciliature consists of a single row of kinetosomes; other polyhymenophoreans have *diplostichomonad* membranes (the kinetosomes forming two parallel rows) or – in some hypotrichs – *polystichomonad* membranes (the kinetosomes forming more than two parallel rows). In some ciliates of the SCUTICOCILIATIDA the paroral membrane is clearly divided into two or three parts. (cf. AZM.)

parotitis Inflammation of the parotid gland. (cf. MUMPS.)

parrot fever *Syn.* PSITTACOSIS.

parsnip mosaic virus See POTYVIRUSES.

parthenogenesis The development of progeny from a female reproductive structure without the involvement of a male gamete. The term is sometimes used to include cases of AMIXIS and APOMIXIS.

partial veil (inner veil) (*mycol.*) In the immature fruiting bodies of certain agarics (e.g. *Agaricus* spp): a membranous tissue which extends from the margin of the pileus to the stipe; when intact, the partial veil seals off the gill cavity. As the fruiting body grows, the partial veil tears – leaving a ring of tissue (the *annulus*) around the stipe, and (often) a ragged ring of tissue (the *cortina*) attached to the periphery of the pileus. (cf. UNIVERSAL VEIL.)

partition (of plasmids) Segregation to daughter cells during cell division.

In high-COPY-NUMBER plasmids (e.g. COLE1 PLASMID) partition appears to depend on random segregation to daughter cells: there appear to be no special mechanisms for active partition.

In low-copy-number plasmids (e.g. the F plasmid, IncFII plasmids) active mechanisms are essential for accurate partition; such mechanisms prevent the segregation of plasmid-free daughter cells at cell division and help to maintain a stable copy number. These plasmids contain specific loci (designated e.g. *par*, *sop* or *sta*) associated with partition. The ParM protein of plasmid R1 polymerizes at a *parC*-ParR complex, the resulting filaments apparently forcing the plasmids to segregate [EMBO (2002) *21* 3119–3127]. Plasmids with similar/identical partition sites would, if present in the same cell, exhibit instability in partition and (hence) copy number, i.e. INCOMPATIBILITY.

Some plasmids which form multimers have efficient mechanisms for monomerization; thus, e.g. BACTERIOPHAGE P1 encodes a SITE-SPECIFIC RECOMBINATION system. Plasmid R46 encodes a recombinase at the *per* locus.

Partition in R1 appears to involve a mitosis-like process mediated by actin-like filaments. The *par* (partitioning) locus includes: *parC*, a centromere-like region, and the genes *parM* and *parR*. Partitioning seems to occur as follows. A pair of plasmids (formed at the cell's mid-cell replication machinery) become linked by the binding of ParR to the *parC* region of each plasmid; *parC*-ParR-*parC* is the *partitioning complex* that acts as a nucleation site on which ATP-ParM molecules polymerize to form actin-like filaments. Continued polymerization lengthens the filaments, forcing the plasmids to opposite poles [EMBO (2002) *21* 3119–3127].

partition chromatography See CHROMATOGRAPHY.

Partitiviridae See MYCOVIRUS.

partner gene See FUSION PROTEIN.

partridge wood Wood which has undergone a pocket rot caused e.g. by *Stereum frustulatum*.

parvaquone A naphthoquinone drug used e.g. in the treatment of EAST COAST FEVER.

parvo-like viruses See SMALL ROUND STRUCTURED VIRUSES.

parvobacteria A term sometimes used to refer to some or all of the small Gram-negative bacilli, e.g. species of *Brucella* and *Pasteurella*.

Parvoviridae A family of non-enveloped, isometric ssDNA viruses which infect invertebrates or vertebrates; virus replication occurs in the host cell nucleus. Virion: ~15–28 nm diameter, typically with icosahedral symmetry and usually containing three or four types of polypeptide; genome: one molecule of linear ssDNA. In some members of the family each virion contains negative-sense DNA; in other members, some of the virions in a population contain negative-sense DNA while others contain positive-sense DNA.

The virions are typically highly resistant to adverse conditions, surviving e.g. 56°C/>60 min, pH 3–9, lipid solvents, and many types of protease; they can generally be inactivated by formaldehyde, hydroxylamine, β-propiolactone, oxidizing agents or ultraviolet radiation.

Genera: DENSOVIRUS, DEPENDOVIRUS, ERYTHROVIRUS, PARVOVIRUS. [Cell infection: JV (2004) *78* 6709–6714.]

Parvovirus (parvovirus group) A genus of autonomous, helper-independent viruses (family PARVOVIRIDAE) which infect vertebrates, causing important diseases in domestic and other animals; parvoviruses responsible for human diseases are classified in the genus ERYTHROVIRUS. (So-called 'parvovirus-like agents' – associated with human food poisoning – are unrelated (ssRNA) viruses currently referred to as SMALL ROUND STRUCTURED VIRUSES.)

The type species is *r*-1 (= Kilham rat virus or 'rat virus'). Other members include e.g. Aleutian mink disease virus (see ALEUTIAN DISEASE OF MINK), bovine parvovirus, feline parvovirus (including the subspecies FELINE PANLEUKOPENIA VIRUS, CANINE PARVOVIRUS, mink enteritis virus), goose parvovirus (causal agent of a fatal hepatitis in goslings aged 8–30 days), H-1 virus, LuIII virus, porcine parvovirus (an important cause of infertility, abortion and stillbirth), and minute virus of mice (MVM). Several parvoviruses (e.g. H-1, LuIII) were originally isolated from cell cultures.

Most parvoviruses can agglutinate the erythrocytes (red blood cells) of at least one species of animal. [Differences in haemagglutinating activity among strains of canine parvovirus: JGV (1988) *69* 349–354.]

The autonomous parvoviruses are characterized by their need to replicate in dividing cells, and in at least some cases (see ERYTHROVIRUS) a virus may be highly tropic for cells in a particular phase of the cell cycle. (The frequent association of parvoviruses with tumours may be due to their preference for dividing cells – rather than to an oncogenic potential; in fact, parvoviruses are reported to have anti-neoplastic potential [Book ref. 97, pp 344–345].)

In some parvoviruses the virions contain only negative-strand DNA; in others, the virions in a given population may contain *either* negative-strand *or* positive-strand DNA.

In autonomous parvoviruses, replication of the DNA is believed to follow a 'rolling hairpin' model [modified rolling hairpin model: JV (1985) *54* 171–177], which is outlined below.

Replication of the single-stranded DNA genome is facilitated by the presence of a PALINDROMIC SEQUENCE at both ends of the strand; in B19 parvovirus (see ERYTHROVIRUS) the two terminal palindromes are identical. Replication begins at the 3′ end, where the palindromic sequence folds back on itself to form a T-shaped double-stranded structure; strand synthesis from the 3′ terminus gives rise to a double-stranded replicative form which extends to the 5′ terminus of the genome. (During strand synthesis, the T-shaped structure at the 5′ end is apparently 'opened up' by strand displacement.) The original 5′ end of the genome thus becomes paired with a newly synthesized strand that includes a (complementary) copy of the palindrome. The palindrome in each strand spontaneously forms a double-stranded T-shaped structure – so that the 3′ end is now positioned to use the newly synthesized strand as a template for ongoing synthesis. Strand synthesis along this (newly synthesized) template continues with synthesis along the original genome strand, resulting in a double-stranded genomic dimer. Nicking at specific sites leads to regeneration and recycling of intermediates and the formation of progeny genomes (see reference for further details).

parvoviruses (1) Viruses of the PARVOVIRIDAE. (2) Viruses of the genus PARVOVIRUS.

PAS (*p*-aminosalicylic acid) A chemotherapeutic agent which is bacteriostatic for some *Mycobacterium* spp, e.g., *M. tuberculosis*; it is used, together with e.g. ISONIAZID, in the treatment of TUBERCULOSIS. PAS is incorporated in FOLIC ACID (in place of PABA), giving rise to an analogue with impaired activity.

PAS reaction (periodic acid–Schiff reaction) A procedure for demonstrating the presence of certain carbohydrates – e.g. CELLULOSE, GLYCOGEN, STARCH. Adjacent hydroxyl groups in the carbohydrate are oxidized to aldehyde groups by periodic acid (HIO_4); the aldehyde groups react with SCHIFF'S REAGENT to give a purple coloration. (Tetraacetic acid or chromic acid may be used in place of periodic acid.) A positive PAS reaction is also given by sugars containing an unsubstituted amino group adjacent to a hydroxyl group (as in CHITOSAN). The PAS reaction has been used e.g. to detect certain pathogenic fungi (e.g. *Blastomyces dermatitidis*, *Histoplasma capsulatum*) in tissue sections.

pascal (Pa) The SI unit of pressure; 1 pascal is equal to 1 newton/metre^2. 1 mmHg = 1 torr = 133.3 Pa = 0.133 kPa. (1 atmosphere = 14.7 lb/inch^2 = 101.325 kPa.) (cf. BAR.)

paspalinines See PASPALUM STAGGERS.

paspalum staggers A MYCOTOXICOSIS which affects mainly cattle but also sheep and horses; it is caused by the ingestion of tremorgenic toxins (paspalinines) when grazing *Paspalum* grasses ergotized by *Claviceps paspali*. Symptoms: hypersensitivity to noise and movement, followed by muscle tremors and incoordination. Recovery is usually rapid following removal from affected pastures.

***Paspalum* striate mosaic virus** See GEMINIVIRUSES.

passage See SERIAL PASSAGE.

passive agglutination (*serol.*) Any form of AGGLUTINATION in which the combination of antibodies with antigens is made readily detectable by the prior adsorption of the antigen or antibody to an unrelated particulate entity – e.g. a cell or a colloidal particle. Thus, e.g., a soluble antibody or antigen may be adsorbed to bentonite or latex (see LATEX PARTICLE TEST) or to erythrocytes (see BOYDEN PROCEDURE); antibodies may be adsorbed e.g. to staphylococci (see SENSITIZATION (sense 5) and PROTEIN A). (If erythrocytes are used, the procedure is referred to as *passive haemagglutination*.)

passive agglutination test Any test, involving PASSIVE AGGLUTINATION, used for detecting or quantifying specific antibodies or antigens – e.g., antibodies homologous to soluble antigens, or particulate antigens homologous to particle-bound antibodies; in such tests the combination of particle-bound antigen (or antibody) with homologous antibody (or antigen) causes AGGLUTINATION of the particles. Passive agglutination tests are appreciably more sensitive than PRECIPITIN TESTS.

passive haemagglutination See PASSIVE AGGLUTINATION.

passive haemolysis See HAEMOLYSIS.

passive immunity (adoptive immunity) Temporary immunity brought about by the transfer of pre-formed antibody, or specifically sensitized lymphocytes, from an immune individual to a non-immune individual – the latter thus becoming immune without necessarily having had contact with the corresponding antigen(s). Passive immunity can be acquired e.g. by transfer of antibody via the placenta, in COLOSTRUM, or by passive IMMUNIZATION.

passive immunization See IMMUNIZATION.

Pasteur effect In organisms capable of both fermentative and respiratory modes of metabolism (e.g. *Saccharomyces cerevisiae*): the inhibition of glycolysis in the presence of O_2, manifested by the inhibition of ethanol formation and a decrease in the amount of glucose used. This effect reflects the higher energy yield obtained from glucose by RESPIRATION than by FERMENTATION. The Pasteur effect can apparently be explained, at least in part, by the allosteric modulation of phosphofructokinase activity in response to the ADENYLATE ENERGY CHARGE (see EMBDEN–MEYERHOF–PARNAS PATHWAY). (cf. CUSTERS EFFECT.)

The Pasteur effect is not observed e.g. in yeast cells growing in the presence of high concentrations of glucose: cf. CRABTREE EFFECT.

Pasteur pipette An open-ended glass tube (internal diam. ca. 5 mm) with one end drawn out to an internal diam. ca. 1 mm; the wide end is fitted with a rubber bulb. It is used e.g. for non-quantitative transfer of liquids, but may be roughly calibrated: e.g. if 50 drops are delivered per ml, each drop is 0.02 ml. For aseptic work Pasteur pipettes are plugged (see PLUGGING) and sterilized before use.

Pasteurella A genus of Gram-negative bacteria of the PASTEURELLACEAE. (The following account is based on data from Book ref. 22, pp. 552–558, but cf. final paragraphs of entry for proposals for a revised classification.) Cells: usually coccoid, rod-shaped or pleomorphic, ca. 0.3–1.0 × 1–2 μm, occurring singly or sometimes in pairs or short chains. Bipolar staining is common. On primary isolation virulent strains usually have a carbohydrate (e.g. hyaluronic acid) capsule which may be lost on subculture. Optimum growth temperature: 37°C. Only *P. aerogenes* and *P. haemolytica* can grow on MacConkey's agar. Most strains are non-haemolytic (cf. *P. haemolytica*). Glucose is usually fermented anaerogenically – gas being formed only by *P. aerogenes* and *Pasteurella 'gas'* (a urease +ve strain which may be a biotype of *P. pneumotropica*). MR −ve; VP −ve; oxidase +ve (some strains of *P. multocida* are −ve); gelatinase −ve; LDC −ve; ADH −ve; usually lactose −ve. GC%: 40–45. Type species: *P. multocida*.

P. aerogenes. Gas from glucose; indole −ve; urease +ve. Commensal in the intestines of pigs.

P. gallinarum. Indole −ve; urease −ve. Occurs e.g. in the upper respiratory tract of poultry and may be associated with chronic respiratory disease in poultry.

P. haemolytica. Haemolytic (e.g. on bovine blood agar); indole −ve; urease −ve. Can cause primary or secondary pneumonia in e.g. cattle and sheep, and is a major cause of 'shipping fever' (see BOVINE RESPIRATORY DISEASE); it can also cause e.g. mastitis in ewes and septicaemia in lambs. (See also RTX TOXINS.)

P. multocida (formerly *P. septica*). Indole +ve; urease −ve. Pathogenic in a range of animals, causing e.g. HAEMORRHAGIC SEPTICAEMIA (in cattle), FOWL CHOLERA, 'shipping fever', and secondary infections (see e.g. SWINE INFLUENZA). Can be an opportunist pathogen in man e.g. following animal bites (see e.g. CAT-BITE FEVER). [Genomic sequence of a common avian strain of *P. multocida*: PNAS (2001) *98* 3460–3465.]

P. pestis: see YERSINIA.

P. pneumotropica. Indole +ve; urease +ve. Occurs in the nasopharynx in rodents and other animals, and can cause e.g. respiratory tract infection in laboratory rodents.

P. pseudotuberculosis: see YERSINIA.

P. tularensis: see FRANCISELLA.

P. ureae. Indole −ve; urease +ve. Occurs occasionally in the upper respiratory tract in humans.

In a proposal for a revised classification [IJSB (1985) *35* 309–322] it was suggested that the genus should include the following species: *P. anatis* sp. nov.; *P. avium*; *P. canis* sp. nov. (includes e.g. the former *P. multocida* biotype 6); *P. dagmatis* sp. nov. (includes organisms previously called *P. 'gas'* or *P. pneumotropica* type Henriksen); *P. gallinarum*; *P. langaa* sp. nov.; *P. multocida* (includes three subspecies: *gallicida*, *multocida* and *septica*); *P. stomatis* sp. nov.; *P. volantium* sp.

nov. (V-factor-requiring avian and human strains). Organisms referred to as *P. haemolytica*, *P. pneumotropica* biotypes Jawetz and Heyl, *P. testudinis* and *P. ureae* were excluded from the genus, and were said to be more closely related to the *Actinobacillus* group; *P. aerogenes* and *P. multocida* biotype 1 were also excluded.

It was later proposed that *P. ureae* be transferred to the genus *Actinobacillus* (as *A. ureae*) [IJSB (1986) *36* 343–344].

Pasteurellaceae A family of Gram-negative, asporogenous, facultatively anaerobic bacteria. Cells: coccoid to rod-shaped, sometimes pleomorphic or filamentous; non-motile. Chemoorganotrophic (respiratory and fermentative); demethylmenaquinones (see QUINONES) are involved in aerobic and anaerobic electron transport. Nutritionally fastidious, requiring organic nitrogen and various growth factors. Chocolate or blood agars can be used for primary isolation of most strains. NO_3^- is reduced to NO_2^-. Acid (usually without gas) is usually formed from glucose but may be difficult to detect by conventional methods. Typically oxidase +ve; catalase +ve; alkaline phosphatase +ve. Species are parasitic in man, animals and birds; some are pathogenic. Type genus: PASTEURELLA; other genera: ACTINOBACILLUS and HAEMOPHILUS. [Book ref. 22, pp. 550–575.]

pasteurellosis (*vet.*) Any of a range of (primarily) septicaemic or pneumonic diseases caused by species of *Pasteurella*. (See e.g. shipping fever in BOVINE RESPIRATORY DISEASE; HAEMORRHAGIC SEPTICAEMIA (2).)

Pasteuria A genus of BUDDING BACTERIA which typically occur attached to particulate matter in aquatic environments; in these organisms the (proteinaceous) cell wall lacks peptidoglycan. *P. ramosa* is a parasite of water-fleas (*Daphnia* spp). [Book ref. 45, pp. 490–492.] A strain formerly designated *P. ramosa* was reclassified in a new genus, *Pirella* [IJSB (1984) *34* 492–495].

pasteurization A form of heat treatment which kills certain pathogens and/or spoilage organisms in e.g. milk and certain foods; temperatures below 100°C are used. (cf. APPERTIZATION.) For MILK, an early form of pasteurization involved holding the milk at a minimum of 63°C for at least 30 min (low temperature holding, LTH; low temperature long time, LTLT). In a more recent method milk is held at a minimum of 72°C for at least 15 sec (flash pasteurization; high temperature short time, HTST). Pasteurization is lethal for the causal agents of a number of milk-transmissible diseases (e.g. salmonellosis, tuberculosis) as well as for a proportion of the milk microflora. It also inactivates certain bacterial enzymes (e.g. lipases) which would otherwise cause milk spoilage. (Both the LTLT and the HTST methods may be inadequate for the inactivation of *Coxiella burnetii* (see Q FEVER).)

The term 'pasteurization' has also been used for the mild heat disinfection (80°C/5 min) of medical instruments.

pasture bloat See BLOAT.

patch stop See MICROSCOPY (b).

patch test Any form of SKIN TEST in which the antigen is applied to the *surface* of the skin – see e.g. VOLLMER PATCH TEST.

patching See CAPPING (sense 3).

patent period (*med., vet.*) The period during which parasitic organisms can be demonstrated in blood, tissues, faeces etc of the host. (cf. PREPATENT PERIOD.)

pathogen Any MICROORGANISM which, by direct interaction with (infection of) another organism (by convention, a multicellular organism), causes DISEASE in that organism. A pathogen is not, therefore, *any* microorganism which is causally connected with disease: for example, a microbe which produces a TOXIN that causes disease in the absence of the microbe itself would not be regarded as a pathogen. Thus, a toxinogenic strain of *Gonyaulax* may – indirectly – cause paralytic SHELLFISH POISONING in man, but it is not regarded as a human pathogen since it has never been known to cause disease by infecting man. Similarly, while *Claviceps purpurea* can be regarded as a pathogen of certain cereals, it cannot be called a human pathogen – even though the ergot alkaloids it produces can cause ERGOTISM in man. By analogy, *Clostridium botulinum* (often described as a pathogen in medical circles) is not acting as a pathogen when its toxin, in the absence of the bacterium, causes BOTULISM in an adult human, although it does act as a pathogen *sensu stricto* when it causes infant botulism, wound botulism, or certain (apparently atypical) *C. botulinum* infections in adults. (See also OPPORTUNIST PATHOGEN.)

By custom, disease-causing higher organisms, such as trematode worms, are referred to as parasites rather than pathogens. (See also PARASITOLOGY.)

pathogenesis The process or mechanism of disease development.

pathogenesis-related proteins (*plant pathol.*) A group of low-MWt, low-IEP proteins, resistant to proteases, formed de novo in plant tissues in response to infection with a pathogen; of unknown function, they are thought to have a role in resistance to disease. They were first found in tobacco cultivars exhibiting HYPERSENSITIVITY to tobacco mosaic virus infection, but occur in other plants infected with viruses, viroids, fungi or bacteria. (cf. PHYTOALEXINS.)

pathogenic Able to behave as a PATHOGEN.

pathogenicity island (PAI) In a bacterial pathogen: a cluster or 'cassette' of genes encoding a set of virulence factors; such sequences usually occur in the chromosome but their GC% is typically unlike that of the chromosome – suggesting acquisition by horizontal transfer. Many PAIs (e.g. *cag* in *Helicobacter pylori*) are flanked by direct repeats. (cf. AVIRULENCE GENE.) A PAI is often located at a site adjacent to a tRNA gene.

The LEE (locus of enterocyte effacement) PAI is found in all strains of EPEC and in many strains of EHEC. LEE is a 35.5 kb chromosomal sequence which is not flanked by direct repeats; in EPEC it is located 16 bp downstream of the *selC* gene (which encodes the tRNA for selenocysteine) [PNAS (1995) *92* 1664–1668]. The LEE PAI encodes components of a type III secretory system (see PROTEIN SECRETION) as well as various effector molecules. [Type III secretory systems and pathogenicity islands (review): JMM (2001) *50* 116–126.] Specific genes (and products) include:

eae – an adhesin (*intimin*).
espA – a cell–cell linking structure (?)
espB – an activator of epithelial cell enzymes (?)
escN – an ATPase for the secretory system (?)
tir – (*espE* in EHEC) Tir protein (see EPEC).

(The names of genes encoding components of the type III system in EPEC differ from those in the earlier literature; these genes are now designated *esc* – e.g. *escN* was formerly *sepB*.)

At least some genes in LEE (including *eae*, and those of secreted proteins, such as *espB*) are under transcriptional control from a regulatory region (*per*) in a plasmid (EPEC adherence factor plasmid; EAF plasmid) found in all strains of EPEC. This plasmid also encodes the so-called *bundle-forming pili* (BFP) which may be involved in the formation of microcolonies on the intestinal epithelium (see EPEC).

[Model of EPEC pathogenesis: TIM (1998) *6* 169–172.]

Other PAIs include *cag* in *Helicobacter pylori* [PNAS (1996) *93* 14648–14653; Nature (1997) *388* 539–547], and the island encoding so-called 'toxin co-regulated pili' (TCP) in *Vibrio cholerae*.

Salmonella spp contain at least five distinct PAIs. For example, SPI-1 includes the *inv-spa* genes that encode a type III secretory system used in the invasion of epithelial cells. SPI-3 encodes products needed for survival in macrophages; like LEE, this PAI is located downstream of *selC*. SPI-5 apparently contributes specifically to *entero*pathogenicity [Mol. Microbiol. (1998) *29* 883–891].

In *Staphylococcus aureus*, the gene for toxic shock toxin is carried by a *mobile* pathogenicity island [Mol. Microbiol. (1998) *29* 527–543].

pathognomonic Of symptoms: characteristic of, or specific to, a given disease.

PathoTec system See MICROMETHODS.

pathotype, pathovar See VARIETY.

patricins See STREPTOGRAMINS.

patronym (patronymic) In NOMENCLATURE: any name derived from that of a person.

Pattern coefficient See entry S_P.

Pattern difference See entry D_P.

patulin (clavacin, clavatin, claviformin) An antibacterial, antifungal, antimitotic (and possibly carcinogenic) MYCOTOXIN formed by species of *Penicillium*, *Aspergillus* and *Byssochlamys*. It is a heterocyclic compound which is soluble e.g. in water and alcohol and which reacts with sulphydryl groups. It occurs e.g. in animal feeds and has been implicated in fatalities in cattle. It is formed by *P. expansum* in apples and has been detected in commercial apple juice; it is destroyed by SO_2 and by alcoholic fermentation. [Mode of action on *Saccharomyces cerevisiae*: AEM (1983) *45* 110–115.]

Paul–Bunnell test A haemagglutination test used to detect and quantify a HETEROPHIL ANTIBODY (the 'Paul–Bunnell antibody') which usually appears in cases of INFECTIOUS MONONUCLEOSIS. (Heterophil antibodies may not be detectable in very young children (e.g. <4 years of age), so that the test may be unreliable in this group.) The antibody appears 1–3 weeks after the onset of symptoms and may persist for months; it agglutinates the red blood cells of the sheep, horse and OX. The test determines the titre of serum which agglutinates sheep RBCs.

In a modified form of the test, one sample of serum is pre-absorbed with OX RBCs (which bind the Paul–Bunnell antibody); another sample is absorbed with guinea-pig tissue (which binds e.g. Forssman antibodies but not the Paul–Bunnell antibody); as Forssman antibodies can agglutinate sheep RBCs, pre-absorption makes the test more specific for the Paul–Bunnell antibody.

Paulinella A genus of testate amoebae similar to EUGLYPHA spp. *P. chromatophora*, a small (20–30 μm long), ovoid, freshwater species, contains two CYANELLES (*Synechococcus*); on division, each daughter receives one cyanelle, which subsequently divides.

Paxillus See BOLETALES.

PBP PENICILLIN-BINDING PROTEIN.

PBP 2a (PBP 2′) See PENICILLIN-BINDING PROTEINS and MRSA.

***pbpA* gene** See PENICILLIN-BINDING PROTEINS.

***pbpB* gene** See PENICILLIN-BINDING PROTEINS.

pBR322 A small, multicopy, non-conjugative PLASMID (4362 bp) which carries genes for resistance to ampicillin and tetracycline; it resembles the COLE1 PLASMID in its mode of replication. pBR322 is widely used as a CLONING vector [review: Gene (1986) *50* 3–40]; insertion of DNA at restriction sites in resistance gene(s) is detected by loss of antibiotic resistance.

PBS (1) Phosphate-buffered saline. (2) Primer binding sites: see RETROVIRIDAE.

PCBs Polychlorinated biphenyls (see BIOREMEDIATION).

PCC Pasteur Culture Collection.

pCF-10 See SURFACE EXCLUSION.

PCNB See QUINTOZENE.

***pcnB* gene** See MRNA (a) and PLASMID (replication).

PCP See BIOREMEDIATION.

PCR (polymerase chain reaction) A method for copying ('amplifying') specific sequences of nucleotides (see AMPLICON senses 1 and 2) in DNA; a modified form of the process (REVERSE TRANSCRIPTASE PCR) is used for copying sequences in RNA. PCR depends on the ability of a (thermostable) DNA polymerase to extend a primer on a template containing the amplicon (see figure). (Patents covering PCR are owned by Hoffmann-La Roche.) Other methods for amplifying nucleic acids include LCR, NASBA and SDA; unlike PCR, the latter two methods are conducted isothermally.

Repeated rounds of replication (i.e. ongoing *cycling*) in PCR can yield millions of amplicons within hours; usually, the sequence being copied is less than 2000 base-pairs in length, but longer sequences have been copied.

In the basic form of PCR the reaction mixture includes: (i) the sample (dsDNA); (ii) the *primers*: oligonucleotides (~20–30 base-pairs in length) of two types – one type complementary to the 3′ end of the amplicon in one strand, the other type complementary to the 3′ end of the amplicon in the other strand; (iii) deoxyribonucleoside triphosphates of all four types; (iv) a thermostable DNA polymerase – e.g. the *Taq* DNA polymerase (see DNA POLYMERASE for examples of polymerases used in PCR).

Temperature cycling (see figure) can be controlled manually; however, it is usually achieved by carrying out the reaction in an instrument (a *thermal cycler* or *thermocycler*) in which timing is controlled automatically (see THERMOCYCLER). In most cases, each cycle of PCR involves a sequence of three levels of temperature: (i) e.g. 94/95°C for strand separation (denaturation of the sample duplex DNA); (ii) a temperature suitable for the binding of primers (*annealing*); and (iii) a temperature suitable for extension, i.e. ongoing synthesis of a DNA strand, from the primer, on the template strand. In some cases a two-temperature protocol has been used – e.g. a denaturing temperature of 94°C and combined annealing/extension at 68°C; this is possible e.g. with the GeneAmp® 9700 system (Perkin-Elmer) amplifying a 500-nucleotide amplicon of phage λ DNA.

While denaturation and extension temperatures are generally ~94/95°C and 72°C, respectively, the temperature used for annealing varies widely according to the particular assay. One factor which influences annealing temperature is the GC% of the primers – GC%-rich primers requiring a higher annealing temperature to avoid the formation of spurious products through 'mispriming' (binding of primers to inappropriate sequences in the template). (Examples of some primer-binding temperatures: GC% 33: 45°C [JCM (1999) *37* 772–774]; GC% 62: 63°C [NAR (1998) *26* 3614–3615].) One way of finding an optimal annealing temperature is TOUCHDOWN PCR.

For some purposes (e.g. ARBITRARILY PRIMED PCR) a primer is required to bind to a target sequence to which it is not fully complementary. In such cases, PCR can be run under so-called *low-stringency* conditions, i.e. particular values of temperature, pH and electrolyte concentration which optimize primer–target binding by helping to overcome, or minimize, the effects of mismatches between primer and target. *High-stringency* conditions are those which allow primer–target binding to occur only when the primer and target sequences are exactly complementary, or very nearly so. The greater the

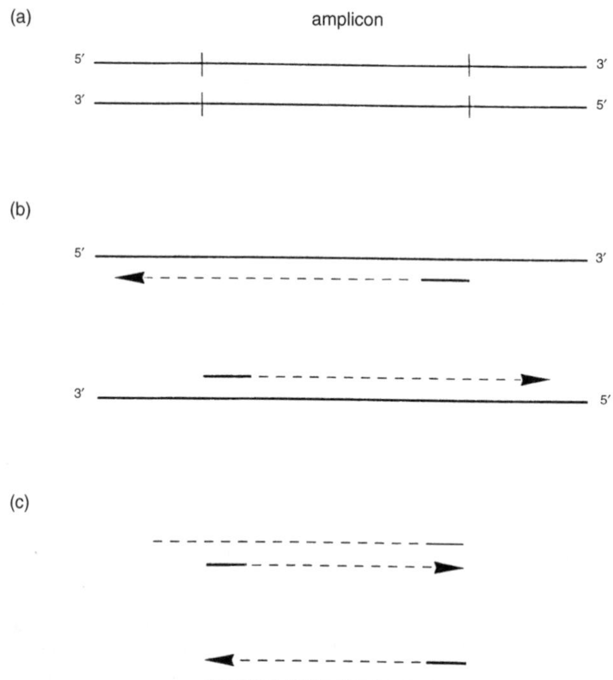

PCR (polymerase chain reaction): the basic method (principle, diagrammatic).

(a) Part of a double-stranded molecule of DNA: two antiparallel strands held together by base-pairing. The sequence to be amplified (the *amplicon*) is delimited by short vertical bars.

(b) Initial heating to e.g. 95°C has separated the two strands of DNA. Transient cooling to 45–65°C has permitted the *primers* – shown as short, solid lines – to bind (= *anneal*) to their respective (complementary) sites on the template strands; notice the locations of the bound primers in relation to the site of the amplicon. The temperature is then increased to e.g. 72°C, and DNA synthesis (dashed line) occurs from the 3′ end of each primer. Note that each of the newly synthesized strands in this first round of replication is longer than the amplicon.

(c) The second round of replication begins with initial heating (to ~95°C) so that the new daughter strands (synthesized in the first round of replication) are separated from their templates. Each new daughter strand (as well as the two parental strands) can then act as a template. In the diagram, a primer has bound to each new daughter strand and has been extended (dashed line). Note that the strand synthesized on a new daughter strand is the same length as the amplicon; such strands rapidly become numerically dominant in subsequent rounds of replication.

Reproduced from Figure 4.1, page 57, in *DNA Methods in Clinical Microbiology* (ISBN 07923-6307-8), Paul Singleton (2000), with kind permission of Kluwer Academic Publishers, Dordrecht, The Netherlands.

stringency (e.g. the higher the temperature) the greater will be the degree of matching needed between primer and target for successful binding to occur.

Difficulties in designing a primer – owing e.g. to insufficient data on the target sequence – may be approached by using a 'universal nucleotide' in place of one or more of the unknown nucleotides [universal nucleotides: Nature (1994) *369* 492–493].

Successful PCR requires close attention to detail and also exclusion of contamination by DNA from the environment and/or from the apparatus.

Contamination by extraneous DNA. Extraneous DNA may contain (i) sequences that are able to bind the primers being used and/or (ii) sequences able to prime irrelevant sites in the sample DNA; either way, unwanted sequences may be amplified – forming a troublesome mixture of products and reducing the efficiency of amplification of the required sequence.

There are two main approaches to the control of contamination. First, *amplicon containment*; this involves division of the working space into a number of dedicated areas – each area being used for only certain stage(s) of the procedure. In one common scheme there are five designated areas, two of which are reserved for (i) cycling and (ii) analysis of products.

The second approach, *amplicon inactivation*, can be achieved by either of two methods. In one method, *isopsoralen* (see PSORALENS) is added to the reaction mixture before cycling. After cycling, the reaction mixture is exposed to ultraviolet radiation – e.g. 365 nm/15 minutes/4°C; this results in covalently cross-linked double-stranded amplicons which – as they cannot be denatured into single-stranded templates – cannot jeopardize later assays in the same apparatus/laboratory. (Activation of isopsoralen at low temperature is more efficient than at room temperature [JCM (1999) *37* 261–262].)

In the second form of amplicon inactivation, all PCR assays are carried out routinely with deoxyuridine triphosphate (dUTP) instead of the usual deoxythymidine triphosphate (dTTP); as a result, dUMP is incorporated into all amplified products during cycling. (Such products can be detected and analysed in the usual way.) In each assay, the reaction mixture also includes the enzyme URACIL-DNA GLYCOSYLASE (= uracil-*N*-glycosylase, UNG); consequently, if amplicons from *previous* assay(s) contaminate a given assay, then those contaminating amplicons will be degraded by the dual effects of (i) enzymic removal of uracil by UNG, and then (ii) denaturation and strand breakage caused by the high initial temperature of the first cycle in the new assay. (This high temperature also inactivates UNG; this is necessary in order to avoid degradation of products in the new assay.)

Inhibitors and facilitators of PCR. Some (e.g. clinical) specimens may contain substance(s) which inhibit PCR – so that a false-negative result may be obtained even when a given specimen contains the particular amplicon of interest. Inhibitors are found in a wide range of clinical specimens, e.g. bone marrow aspirates, cerebrospinal fluid, faeces and urine. PCR-inhibitory components of blood cells include haemoglobin (in erythrocytes) and lactoferrin (in leukocytes). [Purification and characterization of PCR-inhibitory components in blood cells: JCM (2001) *39* 485–493.]

Inhibitors of PCR also include the blood anticoagulants SPS (sodium polyanetholesulphonate) [JCM (1998) *36* 2810–2816] and HEPARIN; haemin; and agar (see GELLAN GUM for an alternative to the agar component in transport media).

The problem of inhibitors may be approached in several ways. For example, specimens may be diluted in order to reduce the concentration of the inhibitory substance(s); however, this also lowers the number of copies of the target sequence (if any) in the specimen and may lower the sensitivity of the assay.

Inhibitors in faeces appear to include complex polysaccharides of dietary origin; these inhibitors can be removed (at least partially) by means of a modified form of a commercial process used for preparing purified DNA from tissues and viruses [JCM (1997) *35* 995–998]. The removal of some polysaccharide inhibitors of PCR may be achieved by means of agarose–DNA preparations [NAR (1998) *26* 3309–3310].

In another form of nucleic-acid amplification, TMA (q.v.), certain inhibitors can be inactivated by treating specimens with the detergent sodium dodecyl sulphate (SDS) [JCM (1997) *35* 307–310]. SDS has been used for a similar purpose in tests based on the ligase chain reaction (LCR) [JCM (1998) *36* 1324–1329].

A distinct approach to the problem of inhibitors is offered by the use of magnetic separation techniques (see e.g. DYNABEADS); following capture on Dynabeads, specific molecules may be washed free of inhibitory substance(s).

The presence of inhibitors may be detected in several ways. For example, an aliquot of the specimen may be 'spiked' with the given target DNA prior to assay; in general, this is not a good idea because it involves extra handling of the specific target sequence and, hence, increases the risk of contamination.

An alternative approach uses an internal control (IC) consisting of molecules of nucleic acid that (i) contain binding sites for the given primers, and (ii) are distinguishable from the actual target by the presence of a unique PROBE-binding region. On efficient amplification of a specimen containing specific target and IC (but lacking inhibitors), two types of product are likely to be obtained: (i) amplicons from the actual target, and (ii) copies of the IC (identifiable by a specific probe). Failure to obtain copies of both target and IC may indicate the presence of inhibitors; amplification of the IC in the absence of target amplification suggests lack of target sequences in the specimen. In one experiment, 20 copies of the IC were added to each test sample, and a positive IC signal was taken to indicate that amplification was satisfactory when targets were present at the limit of the test's sensitivity [JCM (1998) *36* 191–197].

Even when known inhibitors have been taken into account it is mandatory to include suitable controls to detect the effects of unknown/unexpected inhibitors (which may occur in only certain batches of a given type of specimen).

Amplification facilitators are agents which can enhance the efficiency of PCR in the presence of inhibitory substance(s); such agents may also e.g. improve the specificity of PCR and increase the fidelity of DNA synthesis. Amplification facilitators include e.g. bovine serum albumin (BSA) and betaine; some facilitators are effective against a range of inhibitors while others have more selective activity. [Effects of amplification facilitators on diagnostic PCR in the presence of blood, faeces and meat: JCM (2000) *38* 4463–4470.]

Detection of PCR products. In the 'standard' form of PCR the products consist of millions of amplicons whose *precise* length (exact number of base pairs) is determined by the target-binding sites of the primer pair (see figure). Such products, all the same size, can be readily detected/identified by gel electrophoresis as a single band at a precise location in the gel. Typically, electrophoresis is carried out in an agarose gel, and the band of products is detected by staining with ethidium bromide and examining the gel under ultraviolet radiation. To facilitate identification, another lane in the *same* gel is used for electrophoresis of a set of fragments of known size (e.g. 100, 200, 300... base pairs); this gives a scale ('ladder') with which the experimental band(s) can be compared.

Gel electrophoresis is also used for detecting the products in MULTIPLEX PCR (in which products of more than one size are formed) and in ARBITRARILY PRIMED PCR (in which there are products of very many sizes – the multiple bands of products forming a *fingerprint* of the sample DNA).

Band(s) of products within a gel strip may be further examined by Southern blotting followed by SOUTHERN HYBRIDIZATION.

The identity of amplicons from a given reaction can also be checked by subjecting them to a particular RESTRICTION ENDONUCLEASE and checking (by electrophoresis) to see whether fragments of *expected* sizes have been formed.

Amplicons can be rapidly detected by a plate-based hybridization assay. First, single-stranded amplicons are prepared by denaturing the products. These amplicons are then exposed to *detector probes* which are complementary to a given sequence in the amplicon; each PROBE is labelled with the enzyme alkaline phosphatase. The whole is then transferred to a microtitre plate whose wells have been coated with *capture probes* – probes complementary to a different sequence in the amplicon. Incubation at 37°C permits amplicons (if present) to bind to the (surface-bound) capture probes. After washing (to remove unbound detector probes), a substrate for alkaline phosphatase is added to detect any bound, enzyme-labelled amplicons; if present, the enzyme will cleave the substrate, and this can be detected by monitoring the optical density of the mixture. This method has been used for detecting *Bordetella pertussis* in clinical specimens (by detecting a sequence of nucleotides specific to the pathogen) [JCM (1997) *35* 117–120].

Products can also be detected rapidly by PCR DIG-ELISA. In this method, the PCR reaction mixture includes DIGOXIGENIN-tagged deoxyuridine triphosphate (dUTP) so that dUMP is incorporated into the products. Moreover, the primers are 5′-biotinylated so that, after cycling, products can be captured on the STREPTAVIDIN-coated surface of a well in a microtitre plate. Any products captured on the well's surface can then be detected by exposing the well to a *conjugate* consisting of antidigoxigenin (the antibody to digoxigenin) covalently bound to alkaline phosphatase; any bound, digoxigenin-tagged products will bind the conjugate (via the antibody) and, hence, become labelled with alkaline phosphatase. As in the previous method, the enzymic label is detected by adding an appropriate substrate and monitoring the optical density. This method has been used for screening salmonellae [JCM (1997) *35* 714–718]. Note that, unlike the previous method, this method does not confirm the identity of the products by specific hybridization(s).

Real-time detection of products is possible e.g. with TAQMAN PROBES (q.v.) and MOLECULAR BEACON PROBES.

The LightCycler™ (Roche) is a thermocycler in which detection (and/or quantification) of products can be achieved by at least two distinct methods. One method uses SYBR® Green I (Molecular Probes Inc., Eugene, OR, USA) – a dye which fluoresces strongly (530 nm) when it binds to the minor groove of dsDNA. In PCR, this dye can be used to monitor the cycle-by-cycle build-up of products by including it in the reaction mixture and measuring its fluorescence at the end of the *extension* stage in each cycle – that is, when all amplicons are double-stranded, the quantity of dsDNA is at a maximum in that cycle, and the maximum amount of dye has been bound. With ongoing cycling, there is a progressive increase in the amount of dsDNA at the end of extension and, hence, in fluorescence from the bound dye.

Detection of products by the LightCycler can also be achieved by a probe-based format. Two types of probe, included (in large numbers) in the reaction mixture, can bind to *adjacent* sites on an internal (non-terminal) sequence in the amplicon. When bound, the 3′ end of probe 1 (labelled with fluorescein) is juxtaposed to the 5′ end of probe 2 (labelled with the dye LightCycler Red 640). When probes are bound in this way, fluorescence from probe 1 provides excitation for the dye on probe 2 (by a process called *fluorescence resonance energy transfer*, FRET); when excited in this way, the dye on probe 2 emits radiation of wavelength 640 nm. During cycling, probes 1 and 2 bind to amplicons at the *annealing* stage, and the signal (at 640 nm) is read at this stage. Subsequent primer extension displaces the probes – so that (because the dyes are then no longer juxtaposed) fluorescence from probe 2 ceases. During ongoing cycling, probes bind to increasing numbers of amplicons at the annealing stage in each cycle and give rise to a progressively increasing signal at that stage.

Optionally, the LightCycler can work in duplex mode, i.e. detecting two targets (in the same sample) by using two pairs of primers and two (different) pairs of probes; one probe in one pair is labelled with LightCycler Red 640, and one probe in the other pair is labelled with LightCycler Red 705 – so that two independent signals are obtained from the one sample. [Duplex LightCycler assay for detecting (i) the *mecA* gene and (ii) a *Staphylococcus aureus*-specific marker: JCM (2000) *38* 2429–2433.]

Quantification in PCR. The number of target sequences in a sample *before* amplification can be estimated by several commercial systems. These include TaqMan® probes (PE Applied Biosystems) and the probe-based and SYBR® Green I formats of the LightCycler™ (Roche); in all cases, quantification in a given reaction requires identification of the *threshold cycle* – see TAQMAN PROBES for rationale. (See also MOLECULAR BEACON PROBES.)

Uses of PCR. PCR is an extremely flexible technique which has found use in fields as diverse as palaeobiology and forensic science.

In microbiology, PCR is used e.g. for detecting and identifying specific microorganisms in medical or environmental specimens; this use depends on the amplification and examination of sequences of nucleotides which are *specific* to the organisms of interest. Such an approach is particularly useful for those organisms which cannot be cultured – or which grow very slowly (e.g. *Mycobacterium tuberculosis*). For diagnostic purposes, primers must be specific to sequences found *only* in a given pathogen (for example, a sequence in a gene encoding a unique toxin). Examples of the use of PCR for detecting pathogens include: *Bacteroides forsythus* [JCM (1999) *37* 1621–1624], *Bordetella pertussis* [JCM (1999) *37* 606–610], chlamydiae [JCM (1999) *37* 575–580], *Escherichia coli* O157 [JCM (1998) *36* 1801–1804], *Fusarium* spp [JCM (1999) *37* 2434–2438], group A streptococci [JCM (1998) *36* 1769–1771], HIV-1, HIV-2, HTLV-I, HTLV-II [PNAS (1999) *96* 6394–6399], rhinoviruses [JCM (1999) *37* 524–530] and *Trichomonas vaginalis* [JCM (1997) *35* 132–138].

For PCR in TYPING see: AFLP FINGERPRINTING, ARBITRARILY PRIMED PCR, PCR-RIBOTYPING, REP-PCR, RFLP, SPOLIGOTYPING and SSCP.

PCR is also used e.g. for (i) detecting specific virulence factors – such as toxins (e.g. diphtheria toxin) [JCM (1997) *35* 1651–1655]; (ii) detecting genes that confer resistance to antibiotics (e.g. the *mecA* gene in staphylococci [JCM (1999) *37* 3029–3030]); (iii) surveying the diversity of species in environmental samples by means of BROAD-RANGE PRIMERS; (iv) DIFFERENTIAL DISPLAY.

(See also ARBITRARILY PRIMED PCR; ASYMMETRIC PCR; HOT-START PCR; MULTIPLEX PCR; NESTED PCR; REP-PCR; REVERSE TRANSCRIPTASE PCR.)

[PCR in clinical microbiology: Book ref. 221, pp 56–125.]

PCR DIG-ELISA See PCR.

PCR-ISH See IN SITU PCR.

PCR-RFLP See RFLP.

PCR-ribotyping An alternative to RIBOTYPING (q.v.) in which, initially, PCR is used to amplify a particular sequence in the *rrn* operon (usually the intergenic spacer region between the 16S and 23S genes); in many bacteria the chromosome contains multiple copies of this sequence and, moreover, the length of the sequence may vary even within the same chromosome. The amplicons are examined by gel electrophoresis, giving a fingerprint of the strain under test. Strains are classified into 'PCR ribotypes' on the basis of differences in their fingerprints. PCR-ribotyping has been used e.g. for typing strains of *Burkholderia* [JCM (1992) *30* 2084–2087] and *Clostridium difficile* [RMM (1997) *8* (supplement 1) S55–S56], and detecting species of *Rhizobium* [LAM (1999) *28* 137–141].

PCR-SSCP See SSCP.

PDGF Platelet-derived growth factor.

pea early-browning virus See TOBRAVIRUSES.

pea enation mosaic virus (PEMV) An ssRNA-containing PLANT VIRUS which has a narrow host range; transmission occurs (circulatively) via aphids (e.g. *Acyrthrosiphon pisi*). Virion: icosahedral, ca. 28 nm diam.; two types occur, differing in S_w^{20}. Genome: two molecules of linear positive-sense ssRNA (MWts ca. 1.7 and 1.3×10^6), sometimes with a third molecule of MWt ca. 0.3×10^6. PEMV replicates in the nucleus; vesicular cytopathological structures arise from the nuclear membrane.

pea leaf roll virus See LUTEOVIRUSES.

pea mosaic virus See POTYVIRUSES.

pea streak virus See CARLAVIRUSES.

peach leaf curl See TAPHRINA.

peach rosette mosaic virus See NEPOVIRUSES.

peanut (toxicity associated with fungal contamination) See AFLATOXINS.

peanut clamp virus See TOBRAVIRUSES.

peanut clump virus See SOIL-BORNE WHEAT MOSAIC VIRUS and TOBAMOVIRUSES.

peanut mottle virus See POTYVIRUSES.

peanut stunt virus See CUCUMOVIRUSES.

peanut yellow mottle virus See TYMOVIRUSES.

pear decline disease A YELLOWS disease of pear trees caused by an MLO; oxytetracycline hydrochloride is used for control.

peat scours See SCOURS.

pébrine See NOSEMA.

pectase See PECTIC ENZYMES.

pectates Salts of pectic acid – see PECTINS.

pectic acid See PECTINS.

pectic enzymes ('pectinases') Enzymes which degrade PECTINS; they are widely distributed among microorganisms. There are two broad categories of pectic enzyme: those which de-esterify pectins (pectinesterases) and those which degrade the galacturonan backbone chain (depolymerases).

Pectinesterase (pectin methylesterase, pectase, pectin pectylhydrolase, EC 3.1.1.11) de-esterifies pectins, producing methanol and pectic acid; pectinesterases occur in higher plants, numerous fungi (including some yeasts) and some bacteria.

Pectic depolymerases include both hydrolases, which hydrolyse glycosidic linkages in the galacturonan, and lyases, which cleave a glycosidic bond with the formation of a double bond between C-4 and C-5 in the residue at the newly formed non-reducing end (β-elimination). These enzymes vary in their specificities; most function mainly or only on non-esterified galacturonans (pectic acids). Depolymerases which act as endoenzymes include e.g. *endopolygalacturonase* (poly-(1 → 4)-α-D galacturonide glycanhydrolase, EC 3.2.1.15), produced by many plant-pathogenic and saprotrophic fungi and bacteria and by some yeasts; *endopectate lyase* (polygalacturonic acid *trans*-eliminase, endopolygalacturonate lyase, EC 4.2.2.2), produced by various bacteria – especially SOFT ROT (2) agents – and some plant-pathogenic fungi; and *endopectin lyase* (pectin lyase, polymethoxylgalacturonide lyase, EC 4.2.2.10 – the only depolymerase specific for highly esterified pectins), produced mainly by fungi. Exoenzymes include *exopolygalacturonase* (poly (1 → 4)-α-D-galacturonide galacturonohydrolase, EC 3.2.1.67), produced e.g. by plants, fungi and some bacteria; and *exopectate lyase* (EC 4.2.2.9 – an enzyme which liberates unsaturated dimers or, less commonly, trimers [e.g. JGM (1982) *128* 2661–2665] from the reducing end of a pectic acid), produced e.g. by certain bacteria.

Pectinolytic organisms can be important plant pathogens and causal agents of spoilage of fresh and preserved fruit and vegetables (see e.g. SOFT ROT (2); PICKLING; CANNING). Many plants have cell wall-associated proteins which act as inhibitors of endopolygalacturonases in vitro; these proteins are thought to serve a defensive role against invasion by pectinolytic pathogens, although it has been reported that they may not be effective in vivo [PP (1985) *34* 54–60]. (See also TANNINS.) Pectinolytic organisms also fulfil useful roles: see e.g. COFFEE; RETTING; RUMEN. Preparations of pectic enzymes, obtained mainly from *Aspergillus niger*, are used e.g. to facilitate pressing in the extraction of fruit juices, to clarify (reduce turbidity and viscosity of) fruit juices, etc. (See also ENZYMES.)

pectic polysaccharides (*syn.* pectic substances) Polysaccharides obtained from isolated plant cell walls by extraction with boiling water (with or without EDTA) or dilute acid (cf. HEMICELLULOSES); they generally include PECTINS, pectic acid and its salts (pectates), and certain neutral polysaccharides (e.g. arabinogalactans) which occur in association with pectins.

pectic substances *Syn.* PECTIC POLYSACCHARIDES.

pectin See PECTINS.

pectin methylesterase See PECTIC ENZYMES.

pectinases See PECTIC ENZYMES.

pectinate (1) (*adj.*) Arranged like the teeth of a comb. (2) (*noun*) A salt of 'PECTINIC ACID'.

Pectinatus A genus of Gram-negative bacteria (family BACTEROIDACEAE) which have been isolated from spoilt beers. Cells: round-ended, slightly curved or helical rods or filaments, 0.7–0.8 × 2.0–32.0 μm; lateral flagella arise in a line along the concave side of the cell. Acid is formed from the fermentation of e.g. arabinose, fructose, glucose, glycerol, lactate, maltose and mannitol; lactate fermentation yields acetic, propionic and succinic acids, and CO_2, as major products. GC%: ca. 40. Type species: *P. cerevisiiphilus*.

pectinella One of the circumferential bands of cilia which encircle some ciliates (e.g. *Didinium*). (cf. TROCHAL BAND.)

pectinesterases See PECTIC ENZYMES.

pectinic acid An outmoded term for colloidal galacturonans which have a methyl ester content intermediate between that of PECTINS and pectic acids.

pectinolytic (pectolytic) Able to degrade PECTINS – see PECTIC ENZYMES.

pectins The major component of the PECTIC POLYSACCHARIDES. Pectins consist essentially of (1 → 4)-α-D-linked galacturonic acid residues (= pectic acid); a variable number of the C-6 carboxyl groups are methyl-esterified, and, commonly, α-L-rhamnose residues occur at intervals within the chain (forming a rhamnogalacturonan). Other neutral sugars (e.g. L-arabinose, D-galactose) may occur in (usually short) side-chains. Pectins occur in the primary cell walls and intercellular regions of plants (many fruits have a high pectin content); they also occur in the cell walls of certain algae.

Pectins are powerful gelling agents and are widely used in the food industry (e.g. as a setting agent in jams) and in pharmaceuticals and cosmetics. Pectin-degrading enzymes (PECTIC ENZYMES) are produced by a wide range of microorganisms but not by higher animals or by man. (See also RUMEN.)

Pectobacterium See ERWINIA (Carotovora group).

pectolytic *Syn.* PECTINOLYTIC.

Pediastrum A genus of freshwater colonial green algae (division CHLOROPHYTA). Each colony is flat and disc-like (up to ca. 100 μm diam.) and is typically made up of a central cell surrounded by concentric rings of cells; the outermost cells each have two lobes projecting radially outwards, so that the colony resembles a miniature gear-wheel.

pedicel A small stalk. In the fruiting bodies of some myxobacteria: a branch of the common (main) stalk bearing a single sporangium.

Pedinomonas See MICROMONADOPHYCEAE.

pediocin PA-1 A 44-amino-acid, pore-forming BACTERIOCIN produced by species of *Pediococcus*; it is active against *Listeria* spp.

Pediococcus A genus of Gram-positive bacteria of the family Streptococcaceae. Cells: non-motile cocci (ca. 1 μm diam.), commonly in pairs and tetrads. Certain vitamins and amino acids are essential for growth. Growth occurs optimally under microaerobic conditions; glucose is fermented homofermentatively and anaerogenically to DL-lactic acid, L(+)-lactic acid generally predominating. (See also LACTIC ACID FERMENTATION.)

The nomenclature of the species of *Pediococcus* is confused [for discussion see Book ref. 143, pp. 179–211]. Some authors have recognized the following species as validly published: *P. acidilactici*, *P. damnosus* (formerly *P. cerevisiae*), *P. dextrinicus*, *P. halophilus*, *P. parvulus* and *P. pentosaceus*. (See also BEER SPOILAGE; IDLI; LACTIC ACID BACTERIA; MALOLACTIC FERMENTATION; MISO; PICKLING; SOY SAUCE.)

pedology Soil science.

Pedomicrobium A genus of budding PROSTHECATE BACTERIA found e.g. in soils; the life cycle is similar to that of HYPHOMICROBIUM but differs e.g. in that the mother cell may have many prosthecae and typically bears a coating of iron and/or manganese compounds.

peduncle disease See COLD WATER DISEASE.

PEF PULSED ELECTRIC FIELD.

pefloxacin See QUINOLONE ANTIBIOTICS.

PEG POLYETHYLENE GLYCOL.

pegylated Refers to the form of a given molecule, prepared in vitro, which consists of a construct containing a polyethylene glycol moiety.

peitschengeissel flagellum See FLAGELLUM (b).

Pekilo process See SINGLE-CELL PROTEIN.

Pekin duck hepatitis B virus *Syn.* DUCK HEPATITIS B VIRUS.

pelagic May refer to that region of an aquatic habitat (e.g. lake, sea) comprising the entire body of water but excluding the BENTHIC zone, or may refer to organisms living in that region. (See also NEKTON; NEUSTON; PLANKTON.)

Pelagophycus See PHAEOPHYTA.

***Pelargonium* blackleg** See BLACKLEG (2).

pellet (1) The packed sedimented material formed by CENTRIFUGATION. (2) A spherical colony formed by the growth of a mycelial fungus within a liquid medium.

pellicle (1) (*bacteriol., mycol.*) A continuous or fragmentary film, or a mat of organisms, formed at the surface of a liquid culture by certain bacteria or fungi.

A *bacterial* pellicle may comprise bacterial cells, extracellular products, or both. Strains of *Acetobacter xylinum* (see ACETOBACTER) form a tough pellicle of CELLULOSE; each cell produces a single cellulose ribbon composed of a number of microfibrils which apparently emanate from distinct sites along the outer membrane [Book ref. 38, pp. 526–528]. The cellulose ribbons from many cells aggregate to form a network which floats to the surface of the medium, carrying the cells with it – presumably a mechanism for maintaining the organisms in the upper, oxygenated regions of the medium.

Among *fungi*, pellicles are formed on liquid cultures e.g. by various yeasts (see e.g. 'flor' yeasts in WINE-MAKING) and by some mycelial fungi (e.g. *Penicillium* spp).

(2) (*mycol.*) Any superficial membranous structure which can easily be removed from the tissues beneath it.

(3) (*protozool.*) A composite membranous structure which forms the limiting envelope in many protozoa. In some flagellates (e.g. *Trypanosoma*) the pellicle consists essentially of the cytoplasmic membrane (= plasmalemma) supported from beneath by a system of microtubules; some authors do not regard such a simple type of cell envelope as a pellicle. (cf. EUGLENOID FLAGELLATES.) In coccidia (phylum APICOMPLEXA) the pellicle of the sporozoite and merozoite stages consists of three unit-type membranes, the outermost membrane (plasmalemma) enclosing the entire cell; the middle and inner membranes are closely apposed and together form the inner pellicular complex – a structure which underlies the plasmalemma (except at the MICROPORES and at regions delimited by the POLAR RINGS) and is separated from it by an electron-translucent layer ca. 15–20 nm thick. In the apicomplexan *Gregarina garnhami* the three-membrane-thick pellicle forms a series of longitudinal folds, the crests of which are associated with two sets of longitudinal filaments which have been postulated to have a role in gliding motility [JUR (1984) *88* 66–76].

The typical ciliate pellicle consists of three layers of unit-type membrane, the outermost of which (the plasmalemma) covers the entire organism, including the cilia. The two inner layers are sometimes called 'alveolar membranes'; the space between these membranes is partitioned, forming a number of vesicles or alveoli which lie beneath the plasmalemma. Adjacent alveoli may or may not be in communication; in some species (see e.g. COLEPS) the alveoli contain inorganic deposits and constitute a semi-rigid endoskeleton. The alveolar pellicle does not cover the entire organism, and in some ciliates it forms only a relatively small part of the cell envelope. Even when the alveolar pellicle forms the major part of the cell envelope it is absent in the region of the CYTOSTOME, and it is characteristically penetrated by the MUCOCYSTS and by the kinetosomes of the cilia – which penetrate below the level of the innermost membrane. The inner surface of the innermost membrane may be in contact with a dense fibrillar layer of peripheral cytoplasm (the *epiplasm*); the epiplasm may be the site of connection and/or anchorage of subpellicular structures and/or pellicular structures. Some authors consider the epiplasm to be part of the pellicle.

(See also CORTEX; KINETY; SILVER LINE SYSTEM.)

pellis (cutis; cuticle) (*mycol.*) Non-velar cortical layers of a basidiocarp.

pellucid Transparent or translucent.

Pelobacter A genus of Gram-negative, rod-shaped anaerobic bacteria. Metabolism: fermentative; sugars are not utilized. Species include *P. acidigallici*; *P. venetianus* [name validation: IJSB (1984) *34* 91–92]; *P. carbinolicus* and *P. propionicus* [name validation: IJSB (1984) *34* 355–357].

Pelobiontida An order of naked, typically multinucleate amoebae (class LOBOSEA) which, when in motion, are typically cylindrical or elongate-ovoid, and which lack mitochondria but contain endosymbiotic bacteria. Flagellate stages are unknown. The organisms are free-living in freshwater (typically microaerobic) habitats, feeding mostly on non-motile or slow-moving algae (e.g. diatoms). The order includes the genus *Pelomyxa*. (See also PSEUDOPODIUM and UROID.)

Pelochromatium roseo-viride See CHLOROCHROMATIUM AGGREGATUM.

Pelodictyon A genus of photosynthetic bacteria (family CHLOROBIACEAE). The cells are rod-like or coccoid and contain GAS VACUOLES. Cell division occurs predominantly by binary fission; cells of *P. clathratiforme* form chains which, owing to the occurrence of ternary fission in some cells in the chains, develop into colonial structures consisting of three-dimensional networks.

Pelodinium See ODONTOSTOMATIDA.

Pelomyxa See PELOBIONTIDA.

Pelosigma A proposed genus of filamentous bacteria which occur, in S-shaped bundles, on mud and in fresh and brackish waters. [Book ref. 22, p. 138.]

pelotons Coils or balls of fungal hyphae within root cortical cells in certain endomycorrhizas.

Peltier system (thermocyclers) See THERMOCYCLER.

Peltigera A genus of foliose LICHENS (order PELTIGERALES). Photobiont: *Coccomyxa* (e.g. in *P. aphthosa*) or *Nostoc* (e.g. in *P. canina*); *Coccomyxa*-containing species commonly have *Nostoc*-containing cephalodia (see CEPHALODIUM). Thallus: typically large and more or less deeply lobed, lobes ca. 1–10 cm or more in length; only the upper surface has a cortex, the lower surface often being tomentose and marked with 'veins'. Conspicuous rhizines are present on the lower surface in many species. Apothecia are formed on the upper surface – often borne on narrow, erect, stalk-like extensions of the thallus margin. Species occur e.g. on soil, rotting logs etc; many are widely distributed.

Peltigerales An order of fungi of the ASCOMYCOTINA; members include foliose LICHENS (photobiont green or cyanobacterial) which often have cephalodia. Ascocarp: APOTHECIOID, with filiform paraphyses not forming a distinct epithecium. Asci: bitunicate, cylindrical. Ascospores are often elongated and multiseptate. Genera: e.g. LOBARIA, NEPHROMA, PELTIGERA, PSEUDOCYPHELLARIA, SOLORINA, STICTA.

peltigeroside β-*D*-Galactofuranosyl-(1 $\rightarrow$ 3)-D-mannitol, found in *Peltigera* spp.

Pelvetia See PHAEOPHYTA.

pemphigus neonatorum See SCALDED SKIN SYNDROME.

penams Compounds which contain the penicillin nucleus (see β-LACTAM ANTIBIOTICS). (cf. PENEMS.)

penciclovir A guanine nucleoside analogue ANTIVIRAL AGENT; it is effective against alphaherpesviruses. Penciclovir is used e.g. as a topical agent (1% in creams) for the treatment of labial and genital herpes simplex infections. (See also FAMCICLOVIR.)

penems A class of antibiotics; a penem contains a penicillin-like nucleus in which a double bond occurs between C-2 and C-3 (see β-LACTAM ANTIBIOTICS, Figure (a)).

penicillic acid A secondary metabolite of certain *Penicillium* and *Aspergillus* spp; it is carcinogenic in rats when administered by subcutaneous injection.

penicillin Any of the PENICILLINS, or (specifically) penicillin G.

penicillin acylase *Syn.* PENICILLIN AMIDASE.

penicillin amidase (penicillin acylase; penicillin amidohydrolase) An enzyme (EC 3.5.1.11) produced by many Gram-positive and Gram-negative bacteria; it cleaves the side-chain of PENICILLINS to form 6-aminopenicillanic acid.

penicillin amidohydrolase *Syn.* PENICILLIN AMIDASE.

penicillin-binding proteins (PBPs) Proteins which occur in the cell envelope of Gram-positive and Gram-negative bacteria and which bind covalently to PENICILLINS and to related β-LACTAM ANTIBIOTICS; according to species, there may be up to eight distinct types of PBP in a given cell. Different PBPs vary in their ability to bind particular β-lactam antibiotics.

The PBPs of a given species are numbered in order of decreasing MWt (PBP-1, PBP-2 etc.); similar patterns of PBPs occur in taxonomically related species.

PBPs have enzymic roles in the biosynthesis of PEPTIDOGLYCAN; they are thus involved e.g. in cell elongation and septum formation. (See also CELL CYCLE and THREE-FOR-ONE MODEL.)

Some PBPs are essential for cell viability, and these constitute the killing sites of the β-lactam antibiotics; a β-lactam antibiotic acylates (thus inactivating) a catalytic site on a PBP.

Low-MWt (~40000–50000) PBPs are typically efficient *carboxypeptidases* (see PROTEASES) and/or endopeptidases. In vivo they appear to play a subordinate role, and are apparently non-essential for cell viability.

High-MWt (>60000) PBPs are less abundant in the cell but at least one type appears to be essential for cell viability; they are generally more sensitive than the low-MWt PBPs to penicillins. The enzymic roles of these PBPs include mediation in *transglycosylation* and *transpeptidation*, i.e. the assembly of glycan subunits and the formation of peptide cross-links, respectively.

In *Escherichia coli* the high-MWt PBPs are PBP-1A, PBP-1B, PBP-2 and PBP-3 (see below).

PBP-1A and PBP-1B (encoded by genes *mrcA* and *mrcB*, respectively) are involved in cell elongation; inactivation of both of these PBPs leads to rapid lysis of growing cells.

PBP-2 (encoded by gene *pbpA*) is essential for maintaining the rod shape of the cell during growth; a null mutation in *pbpA* gives rise to spherical cells that eventually lyse. PBP-2 may be capable of substituting for PBP-1B in mutant cells which lack a functional PBP-1B. (See also MECILLINAM.)

PBP-3 (encoded by gene *ftsI*; also called *pbpB*) is required for peptidoglycan synthesis in the SEPTUM. Inactivation of PBP-3 leads to formation of filaments and eventual lysis.

The low-MWt PBPs (PBP-4, PBP-5, PBP-6) are encoded by genes *dacB*, *dacA* and *dacC*, respectively.

PBPs 4 and 5 have carboxypeptidase activity, PBP-5 being responsible e.g. for cleavage of the terminal D-ala from the pentapeptide side chain.

[Location of PBPs in *E. coli*: JB (1986) *165* 269–275.]

In *Staphylococcus aureus*, PBP 2a (= PBP2′) is associated with high-level resistance to methicillin: see MRSA.

penicillin G *Syn.* BENZYLPENICILLIN.

penicillin N (*syn.* cephalosporin N; synnematin B) An antibiotic (6-(D-α-aminoadipyl)aminopenicillanic acid) produced by *Acremonium kiliense* (= *Cephalosporium acremonium*) (cf. PENICILLINS). Isopenicillin N has an L-α-aminoadipyl side-chain.

penicillin V *Syn.* PHENOXYMETHYLPENICILLIN.

penicillinase A β-LACTAMASE active mainly or solely against PENICILLINS, forming PENICILLOIC ACIDS.

Penicillinase production (e.g. by staphylococci) can be demonstrated e.g. with *N*-phenyl-1-naphthylamine-azo-*o*-carboxybenzene, a substance which is yellow and soluble in alkaline form, purple and insoluble in acid form. Agar plates bearing growth of the test organism are flooded with a solution of the alkaline (yellow) form, dried, and then flooded with a solution of penicillin; penicillinase-producing colonies develop a purple colour. Other methods are based on the ability of penicilloic acid to react with iodine and to decolorize a starch–iodine complex; for example, the bacteria are grown on an agar medium containing 1% w/v soluble starch, and the plate is flooded with a solution of benzylpenicillin and 0.1 N iodine in KI; penicillinase-producing colonies are each surrounded by a white halo against a blue-black background. (See also PADAC.)

[Staphylococcal penicillinase: Book ref. 44, pp. 768–775.]

penicillins A class of antibiotics, each characterized by a β-lactam ring fused to a thiazolidine ring (see β-LACTAM ANTIBIOTICS for structures and mode of action). The original penicillins (BENZYLPENICILLIN and PHENOXYMETHYLPENICILLIN) are not effective against β-LACTAMASE-producing Gram-positive bacteria, and they are generally rather ineffective against Gram-negative species owing to poor penetration of the outer membrane; they are produced by species of *Penicillium* (cf. PENICILLIN N). Cleavage of benzylpenicillin by PENICILLIN AMIDASE yields 6-aminopenicillanic acid (6-APA), a precursor from which can be prepared many semi-synthetic penicillins with improved antimicrobial properties. Semi-synthetic penicillins which are resistant to many β-lactamases (particularly those of staphylococci) include e.g. *methicillin* (6-(2,6-dimethoxybenzamido)penicillanic acid), *nafcillin* (6-(2-ethoxy-1-naphthamido)penicillanic acid), *cloxacillin*, *dicloxacillin*, *flucloxacillin* and *oxacillin*; typically, these penicillins are poorly effective against Gram-negative bacteria. Improved activity against Gram-negative species (coupled with resistance to some β-lactamases) is given by *ampicillin* (6-(α-aminobenzylamido)penicillanic acid) and its derivatives (e.g.

AMOXYCILLIN, APALCILLIN, *azlocillin*, *becampicillin*, *mezlocillin*, *pivampicillin*, and *talampicillin*), and by *carbenicillin* (6-(α-carboxybenzamido)penicillanic acid) and its derivatives (e.g. *carfecillin* and *ticarcillin*); typically, these penicillins are less effective than benzylpenicillin against Gram-positive species.

Penicillins for oral administration (which must be acid-stable) include ampicillin, phenoxymethylpenicillin and phenoxyethylpenicillin; benzylpenicillin, carbenicillin and methicillin are administered parenterally. Infrequently, penicillins can act as ALLERGENS, the various penicillins commonly exhibiting cross-allergenicity (cf. CEPHALOSPORINS). (See also MECILLINAM.)

[Clinical pharmacology and therapeutic uses of penicillins (review): Drugs (1993) *45* 866–894.]

penicilliosis A disease of man or animals caused by a species of *Penicillium*. For example, *P. marneffei* can cause a progressive, disseminated, fatal human disease characterized by proliferation of yeast-like fungal cells within histiocytes, followed by the development of foci of necrosis and eventually abscess formation [AJCP (1985) *84* 323–327].

Penicillium A genus of fungi of the class HYPHOMYCETES; teleomorphs occur e.g. in the genus TALAROMYCES. The organisms are widespread in nature and are characteristically saprotrophic; however, *P. marneffei* can cause PENICILLIOSIS (see also SUBEROSIS), and many species form toxins (see e.g. CHLOROPEPTIDE, CITREOVIRIDIN, CYCLOPIAZONIC ACID, ERGOCHROMES, FUMITREMORGIN, GLIOTOXIN, OCHRATOXINS, PATULIN, PENITREMS, RUBRATOXINS, VERRUCOLOGEN). Some species can cause spoilage of various types of material (see e.g. BREAD SPOILAGE, CHEESE SPOILAGE, PAPER SPOILAGE), some are used in manufacturing processes (see e.g. CHEESE MAKING), and some produce useful antibiotics (see e.g. GRISEOFULVIN, MYCOPHENOLIC ACID, PENICILLINS).

Penicillium spp form a well-developed septate mycelium; in most species the aerial mycelium is colourless, but mycelium within the substratum may exhibit distinctive colours – the underside ('reverse') of a colony often being some shade of yellow, orange, red or purple; diffusible pigments are formed by some species. The conidiophores each consist essentially of an erect or prostrate hypha (*stipe*), often ca. 100–250 μm in length, which may be unbranched or branched. The tip of an unbranched conidiophore may terminate in a cluster of phialides (see CONIDIUM) (as in e.g. *P. frequentans*) or in a cluster of metulae (see METULA) (as in e.g. *P. variabile*). In branched conidiophores, each branch (*ramus*) terminates in one or more metulae. (The term 'ramus' is generally reserved for any cell or hypha which occurs between a metula and its stipe; however, confusingly, the terms 'ramus' and 'metula' are sometimes used as though they were synonymous.) The terminal region of a conidiophore (including phialides, metulae and/or rami) is termed the *penicillus*. In most species the conidiophores occur as discrete structures, but in e.g. *P. claviforme* and *P. concentricum* they form synnemata, and in e.g. *P. expansum* they form coremia. The conidia are spherical to ellipsoidal, aseptate, up to ca. 5 μm; in masses, they may be e.g. blue-green, grey-green or yellow, according to species. Colonies may appear e.g. velvety or woolly, and in some species they typically bear surface drops of exudate. Sclerotia are formed e.g. by *P. raistrickii* and *P. thomii*.

Species include e.g. *P. brevicompactum*, *P. charlesii*, *P. chrysogenum*, *P. citreoviride*, *P. digitatum*, *P. frequentans*, *P. griseofulvum*, *P. islandicum*, *P. italicum*, *P. janthinellum*, *P. marneffei*, *P. oxalicum*, *P. roqueforti*, and *P. rubrum*.

[Identification manual and atlas of *Penicillium* spp: Book ref. 185.]

***Penicillium brevicompactum* virus** See MYCOVIRUS.

***Penicillium chrysogenum* virus** See MYCOVIRUS.

***Penicillium cyaneofulvum* virus** See MYCOVIRUS.

***Penicillium stoloniferum* virus S** See MYCOVIRUS.

penicilloic acids Compounds formed when the β-lactam (C–N) bond in penicillins (see β-LACTAM ANTIBIOTICS) is hydrolysed by a β-LACTAMASE.

penicillopepsin An acid aspartyl PROTEASE from *Penicillium janthinellum*; it has two catalytically important aspartyl residues in its active centre.

penicillus See PENICILLIUM.

Penicillus A genus of siphonaceous, calcified green seaweeds related to HALIMEDA; when mature, the thallus bears an apical tuft of branches and somewhat resembles a shaving brush.

peniculus A ribbon-like membranelle consisting of a number of rows of cilia; peniculi occur in the buccal cavity of *Paramecium* and related ciliates.

Peniophora See APHYLLOPHORALES (Corticiaceae) and HETEROBASIDION.

penitrems Steroid MYCOTOXINS produced by *Penicillium* spp – e.g. *P. verrucosum* (*P. cyclopium*); ingestion by e.g. sheep, cattle or horses of feed contaminated with penitrems can lead to weakness, tremors, convulsions and death. These toxins (produced by soil penicillia) may play a role in the pathogenesis of *ryegrass staggers*: a condition associated with grazing perennial ryegrass (*Lolium perenne*) pastures. [Book ref. 33, pp. 1161–1162, 1167–1168.]

Penium A genus of placoderm DESMIDS. The cells are cylindrical, round-ended, and have a median suture line but no sinus (cf. CLOSTERIUM).

pennate diatoms See DIATOMS.

Pennoxide A STERILANT containing ETHYLENE OXIDE (12%) and dichlorodifluoromethane (88%).

penny bun See BOLETALES.

pentachlorophenol See PHENOLS and BIOREMEDIATION.

pentamer See ICOSAHEDRAL SYMMETRY.

pentamidine isethionate An aromatic DIAMIDINE used e.g. in the treatment of African trypanosomiasis and *Pneumocystis carinii* pneumonia.

Pentatrichomonas See TRICHOMONAS.

Penthesta process See AIV PROCESS.

pentitols Five-carbon POLYOLS (e.g. ARABITOL, RIBITOL, xylitol). Pentitols can be metabolized e.g. by many yeasts and by certain bacteria; e.g., strains of *Lactobacillus casei* metabolize xylitol and ribitol as shown in Appendix III(d), the end-products including acetate, ethanol (cf. PENTOSES), and a mixture of D-and L-lactate (even though L-lactate is the main product of glucose fermentation in this species).

penton See ADENOVIRIDAE.

pentosans Polysaccharides consisting of PENTOSE residues: see e.g. XYLANS.

pentose phosphate cycle *Syn.* HEXOSE MONOPHOSPHATE PATHWAY. (cf. CALVIN CYCLE.)

pentose phosphate pathway *Syn.* HEXOSE MONOPHOSPHATE PATHWAY.

pentoses Five-carbon sugars. Aldopentoses (general formula $CH_2OH(CHOH)_3CHO$) include e.g. ARABINOSE, RIBOSE and XYLOSE. The corresponding ketopentoses (general formula $CH_2OH.CO.(CHOH)_2CH_2OH$) are called pentuloses (e.g. RIBULOSE, XYLULOSE). (cf. PENTITOLS and PENTOSANS.)

Pentoses can be used as substrates by many bacteria and fungi, being metabolized via different pathways in different species. Heterofermentative LACTIC ACID BACTERIA and 'homofermentative' streptococci, pediococci and streptobacteria ferment pentoses to equimolar quantities of lactic and acetic acids via xylulose 5-phosphate and the phosphoketolase reaction (Appendix III (d and b), [AvL (1983) *49* 209–224]); the pentoses are taken up either by specific permeases or by PEP-dependent phosphotransferase systems (see PTS). (Thermobacteria are unable to ferment pentoses.) In e.g. *Bacillus* spp, some enterobacteria and certain yeasts pentoses/pentuloses are converted [as in Appendix III(d)] to xylulose 5-phosphate, ribulose 5-phosphate or ribose 5-phosphate which can enter the HEXOSE MONOPHOSPHATE PATHWAY [Appendix I(b)].

(See also ARA OPERON and INDUSTRIAL ALCOHOL.)

pentostatin See VIDARABINE.

pentraxin proteins See ACUTE-PHASE PROTEINS.

pentulose See PENTOSES.

PEP Phosphoenolpyruvate.

PEP-dependent phosphotransferase system See PTS.

PEP potential The sum of the intracellular concentrations of phosphoenolpyruvate (PEP) and 2-phosphoglycerate [JB (1977) *130* 583–595]. In cells of e.g. streptococci, high intracellular levels of these compounds develop during starvation, ensuring that when sugars become available they can be taken up rapidly by the (PEP-dependent) PTS. (See also FLUORIDES.)

peplomer See ENVELOPE (sense 1).

peplos *Syn.* ENVELOPE (sense 1).

pepsin A PROTEASE (EC 3.4.23.1) which is present in the stomach in most vertebrates and which shows maximum activity at low pH (ca. 1–2); it is denatured at pH >6.

pepstatin A peptide inhibitor of aspartyl proteases such as PENICILLOPEPSIN.

peptidase *Syn.* PROTEASE.

peptide antibiotics See e.g. BACITRACIN, CECROPINS, DEFENSINS, EDEINES, GRAMICIDINS, POLYMYXINS, TYROCIDINS.

peptide nucleic acid (PNA) See PNA.

peptidoglycan (murein; mucopeptide; glycosaminopeptide) A heteropolymer which is found in the CELL WALL of most bacteria (cf. e.g. PASTEURIA and PLANCTOMYCES), and which is also present in the cortex of bacterial ENDOSPORES, but which is not found in members of the ARCHAEA (cf. PSEUDOMUREIN). Peptidoglycan (PG) is the component which is primarily responsible for the mechanical strength of the cell wall (protecting the cell e.g. against osmotic lysis) and for maintaining the shape of the cell.

Structure of PG. PG consists essentially of linear, heteropolysaccharide backbone chains that are cross-linked, via short peptides, to form a net-like molecule (the 'murein sacculus' – see (a) in the figure); in almost all cases the sacculus envelops the bacterial protoplast (incomplete, or 'patchy', peptidoglycan occurs e.g. in *Myxococcus xanthus*: see MYXOBACTERALES).

In *Escherichia coli*, the backbone chains consist of alternating (1 → 4)-β-linked residues of *N*-acetyl-D-glucosamine (GlcNAc) and its 3-*O*-D-lactyl ether derivative, *N*-acetylmuramic acid (MurNAc). A MurNAc residue may be linked, via its carboxyl group, to a short peptide (the *stem peptide*) – commonly the tetrapeptide L-alanine–D-glutamic acid–*meso*-diaminopimelic acid–D-alanine. Cross-links between the backbone chains are formed directly between the D-alanine of one stem peptide and the *meso*-diaminopimelic acid (*meso*-DAP) of another on an adjacent chain (see (b) in figure). The backbone chains run *perpendicular* to the long axis of the cell.

Each peptide bridge shown in the figure links a given backbone chain to *one* of the other backbone chains. However, as described in the legend, some peptide bridges link together more than two backbone chains, and such links are essential features in the THREE-FOR-ONE MODEL of peptidoglycan growth.

In many other Gram-negative bacteria the PG is more or less similar to that of *E. coli*, but many variations in PG structure occur among the Gram-positive bacteria; for example, backbone chains in *Mycobacterium* spp contain *N*-glycollylmuramic acid, and in *Staphylococcus aureus* 30–50% of the muramic acid residues are not acetylated. Even in strains of *Neisseria gonorrhoeae*, though, the PG is extensively *O*-acetylated, and it contains 1,6-anhydromuramyl residues [JGM (1985) *131* 3397–3400]. The stem peptides may contain e.g. L-lysine in place of *meso*-DAP (in *S. aureus*), and isoglutamine instead of D-glutamic acid (in *S. aureus* and mycobacteria). Cross-links between stem peptides may occur via a short 'interpeptide bridge' (e.g. a penta- or hexaglycine bridge in *S. aureus*), and may occur in different positions in different species. The degree of cross-linking also varies with species, being generally more extensive in Gram-positive species. [PG structure in staphylococci and mycobacteria: Book refs 44 and 54, respectively.]

Classification based on PG structure. The structure of PG in a given strain tends to be constant and independent of environmental conditions (although, exceptionally, the amino acid content of the interpeptide bridge in staphylococcal PG can be affected to some extent by the availability of amino acids in the growth medium). PG structure may therefore be a useful feature in taxonomy. Depending on the mode of cross-linking, two basic 'PG types' can be recognized. In *group A* PGs, cross-links occur between the distal amino group of the diamino acid in position 3 of one stem peptide and the D-alanine in position 4 of another stem peptide; the cross-link may be direct (as in *E. coli*) or may involve an interpeptide bridge (as in *S. aureus*). In *group B* PGs an interpeptide bridge occurs between the carboxyl group of D-glutamic acid in position 2 of one stem peptide and the carboxyl group of D-alanine in position 4 of another. Group B PGs are much less common than group A PGs; they are found in certain coryneforms (e.g. strains of *Agromyces*, *Arthrobacter*, *Brevibacterium*, *Curtobacterium* and *Microbacterium*), in certain anaerobes (e.g. *Acetobacterium woodii*) and in *Erysipelothrix rhusiopathiae*. PG is also classified e.g. on the basis of the nature of the peptide bridge and of the amino acid at position 3. (See also ACTINOMYCETALES.) [Analysis of PG composition and structure: Book ref. 138, pp 123–156.]

Peptidoglycan in relation to the cell envelope. In Gram-positive bacteria, PG is the major component of the cell wall and is covalently linked to e.g. TEICHOIC ACIDS. (See also e.g. WAX D.) The PG layer is significantly thicker than that found in most Gram-negative bacteria. Interestingly, the thickness of the PG layer in many cyanobacteria is similar to that typical of Gram-positive species, and in at least some strains of cyanobacteria the degree of cross-linking resembles that found in Gram-positive bacteria [JB (2000) *182* 1191–1199].

In Gram-negative bacteria PG occurs as a layer between the cytoplasmic and outer membranes (see CELL WALL), and is linked covalently to the outer membrane by the BRAUN LIPOPROTEIN. (See also ADHESION SITE.) The content of PG may depend on the phase of growth: e.g. in *E. coli* the PG is reported to be 0.7–0.8% or 1.4–1.9% of the dry weight of log-phase or stationary-phase cells, respectively [JB (1985) *163* 208–212]. It has been suggested that, in *E. coli*, the PG occurs as a gel which fills the entire space between the inner and outer membranes ('periplasmic gel') [JB (1984) *160* 143–152]; it was proposed that this gel may be more extensively cross-linked near the outer membrane, becoming progressively less cross-linked towards the cytoplasmic membrane, and that periplasmic

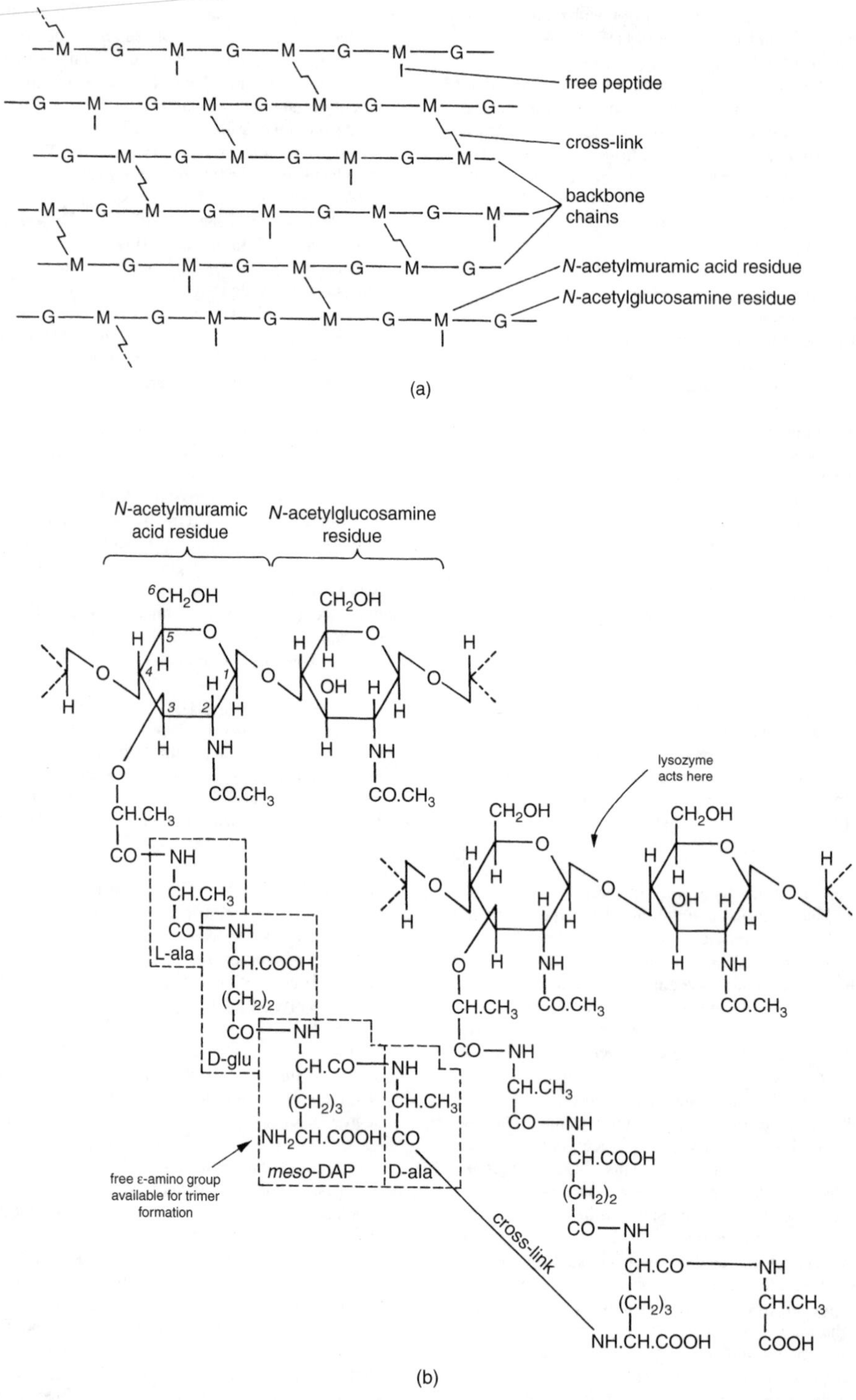

PEPTIDOGLYCAN. The structures shown are typical of those in *Escherichia coli*; similar types of peptidoglycan are found in many other Gram-negative bacteria. (*Continued on page 567*.)

proteins may diffuse freely in the gel. However, this would imply *multilayered* PG, and measurements of the turnover of PG in *E. coli* have suggested that the lateral wall of the cell may contain a *monolayer* of PG, with multilayered PG probably at the poles of the cell [JB (1993) *175* 7–11].

Biosynthesis of peptidoglycan in E. coli. The initial step in biosynthesis may be regarded as the formation of the UDP (uridine 5′-diphosphate) derivative of *N*-acetylglucosamine (here designated UDP-GlcNAc); this occurs in the cytoplasm.

Some of the UDP-GlcNAc is converted to UDP-*N*-acetylmuramic acid (UDP-MurNAc). This conversion begins with a reaction between UDP-GlcNAc and phosphoenolpyruvate catalysed by the enzyme phosphoenolpyruvate:UDP-GlcNAc enolpyruvyl transferase; the UDP-GlcNAc-3-enolpyruvyl ether intermediate is reduced to UDP-MurNAc in an NADPH-dependent reaction. (Conversion of UDP-GlcNAc to UDP-MurNAc is blocked by FOSFOMYCIN.)

Still within the cytoplasm, amino acids are added sequentially to UDP-MurNAc – by enzymes that require ATP and magnesium (or manganese) ions – to form UDP-*N*-acetylmuramylpentapeptide (the 'Park nucleotide'), which includes a terminal D-alanyl-D-alanine (added as a dipeptide). (See also CYCLOSERINE.) The Park nucleotide is then transferred, with release of UMP, to BACTOPRENOL monophosphate in the cytoplasmic membrane. UDP-GlcNAc is subsequently incorporated, with release of UDP, to complete the bactoprenol–disaccharide–pentapeptide subunit (a step which is inhibited by NISIN).

Disaccharide–pentapeptide subunits are transferred from the cytoplasmic membrane to the periplasmic region. Such transfer (blocked by VANCOMYCIN) involves release of the bactoprenol–pyrophosphate (bactoprenol–PP) – which thus remains in the cytoplasmic membrane. Bactoprenol–PP must be dephosphorylated (by a membrane-bound pyrophosphatase) to the monophosphate form before it can again act as an acceptor of the P–MurNAc–pentapeptide; dephosphorylation is inhibited by BACITRACIN.

In the periplasm, the disaccharide–pentapeptide subunits are inserted into the growing peptidoglycan sacculus. According to one scheme [Microbiology (1996) *142* 1911–1918], the disaccharides are polymerized (forming new backbone chains) by *transglycosylation* reactions that involve the enzymic activities of PENICILLIN-BINDING PROTEINS (see also MOENOMYCIN). (Although transglycosylation reactions seem to be catalysed primarily by penicillin-binding proteins, a penicillin-insensitive glycan polymerase, distinct from PBPs, has been reported in *E. coli* [FEBS (1984) *168* 155–160].) These new chains are inserted into pre-existing peptidoglycan by the formation of new peptide cross-links (*transpeptidation* – again, involving penicillin-binding proteins) in a manner described by the THREE-FOR-ONE MODEL (q.v.).

During transpeptidation reactions, the terminal D-alanine of a pentapeptide is cleaved, and the transpeptidase binds to the remaining D-alanine residue to form an acyl-enzyme complex; this conserves the energy of the peptide bond and allows the formation of a new peptide bond – in the absence of ATP – between the acyl group and an amino group of a peptide on another glycan chain. The liberated D-alanine re-enters the cytoplasm via a specific transport system (see also CYCLOSERINE). (See also β-LACTAM ANTIBIOTICS; LYSOSTAPHIN; LYSOZYME; MUROPEPTIDES.)

One report has linked peptidoglycan synthesis with the synthesis of membrane phospholipids [Microbiology (1996) *142* 2871–2877].

PG turnover during growth in E. coli. During growth in *E. coli* the PG is continually degraded (by periplasmic enzymes) and partly re-cycled. In one model, the PG is degraded to disaccharide–tripeptide subunits which complex with a periplasmic binding protein designated MppA; this complex is then taken up across the cytoplasmic membrane by a specific transport system (encoded by genes *oppBCDF*). Within the cytoplasm, the tripeptide (i.e. L-alanyl–D-glutamyl–*meso*-DAP) is cleaved from the subunit by an *amidase* and linked to UDP-MurNAc by a *ligase* [*mpl* gene product: JB (1996) *178* 5347–5352] – thus re-entering the pathway of peptidoglycan biosynthesis [JB (1998) *180* 1215–1223].

Biosynthesis of PG in Gram-positive bacteria. The general scenario is similar to that in Gram-negative species – differences including e.g. the use of other types of derivative in the glycan chains (such as *N*-glycollylmuramic acid in mycobacteria). When the PG contains an interpeptide bridge, this is added during the intra-membrane stage; in *Staphylococcus aureus*, for example, the pentaglycine bridge is incorporated from a specific glycyl-tRNA donor. The addition of new PG to the (thick) Gram-positive cell wall has been described by the INSIDE-TO-OUTSIDE MODEL.

Transport through the peptidoglycan barrier. The cell envelope must, on occasion, be permeable to certain large molecules (e.g. a DNA–protein complex during at least some types of conjugation); moreover, discontinuity in the PG is necessary for the assembly of various TRANSPORT SYSTEMS. It follows that the PG

PEPTIDOGLYCAN (*continued*) (a) The three-dimensional net-like peptidoglycan molecule: backbone chains of alternating residues of *N*-acetylglucosamine (G) and *N*-acetylmuramic acid (M) are held together by short peptides which link *N*-acetylmuramic acid residues. Each of the peptide bridges shown in the diagram links a given backbone chain to *one* other backbone chain. However, there are also peptide bridges which link together more than two backbone chains; the different types of peptide bridge in *E. coli* are briefly described in (b), below.

(b) Part of two adjacent backbone chains showing the mode of linkage between them. Each *N*-acetylmuramic acid residue bears a short *stem peptide* – in this example, the tetrapeptide L-alanine–D-glutamic acid–*meso*-diaminopimelic acid(*meso*-DAP)–D-alanine (shown in dotted boxes); in this kind of peptide bridge there is a direct, covalent link between the D-alanine of one stem peptide and the ε-amino group of *meso*-DAP in the other stem peptide. As this particular type of peptide bridge involves two stem peptides, each consisting of four amino acid residues, it has been referred to as a *tetra-tetra dimer* (or a tetra-tetra). Another type of peptide bridge, the *tetra-tri-dimer* (tetra-tri), is formed from one tetrapeptide and one tripeptide. A *trimer* peptide bridge is one which links together three of the glycan backbone chains; this occurs when a stem peptide on a *third* backbone chain forms a covalent linkage with the free ε-amino group (see diagram) of a dimer bridge. An appreciation of these different types of peptide bridge is essential for understanding the 3-for-1 model (THREE-FOR-ONE MODEL) of replication of the peptidoglycan sacculus.

The enzyme *lysozyme* hydrolyses the *N*-acetylmuramyl-(1→4) linkages in the backbone chain (see diagram); such activity weakens the cell envelope, and may lead to osmotic lysis.

barrier must be able to be breached without sacrificing the cell's structural integrity. The assembly of transport-specific or other structures which cross the cell envelope is therefore likely to require local modification of the PG, and the assembly of some of these structures has indeed been associated with peptidoglycan metabolism [JB (1996) *178* 5555–5562].

peptidyl transferase See PROTEIN SYNTHESIS.

Peptococcus A genus of Gram-positive, asporogenous, anaerobic bacteria which can ferment amino acids or peptones and which may or may not ferment carbohydrates. Cells: cocci (ca. 1 μm diam.) occurring in pairs, tetrads, clumps and short chains. The organisms have been isolated e.g. from lesions (abscesses etc) in man, often as part of a mixed microflora. (See also VAGINA MICROFLORA.) It has been proposed (mainly on the basis of GC values) that the species *P. asaccharolyticus*, *P. indolicus*, *P. magnus* and *P. prevotii* be transferred to the genus PEPTOSTREPTOCOCCUS [IJSB (1983) *33* 683–698]; the GC% of the type species, *Peptococcus niger*, is ca. 51. (For *P. saccharolyticus* see STAPHYLOCOCCUS (*S. saccharolyticus*).)

peptone A soluble product of protein hydrolysis; peptones are not coagulated by heat, are not precipitated by saturated $(NH_4)_2SO_4$, but are precipitated by phosphotungstic acid.

peptone water A MEDIUM containing 1–2% PEPTONE and 0.5% NaCl in water.

peptone-water sugars ('sugars') A range of liquid media each of which consists of PEPTONE WATER to which has been added a particular carbohydrate (final concentration often ca. 1% w/v) and a pH indicator (e.g. ANDRADE'S INDICATOR); such media are used to determine which (if any) of a series of carbohydrates are metabolized by a given strain of microorganism. The carbohydrates used include e.g. arabinose, galactose, glucose, lactose, maltose, mannitol, mannose, rhamnose, sucrose, and xylose. Bacterial growth in a given peptone-water sugar may give rise to acid, or acid and gas, as breakdown product(s) of the carbohydrate; acid is detected by the pH indicator, and gas production is detected by a DURHAM TUBE. 'Broth sugars' are used in the same way as peptone-water sugars; in broth sugars the carbohydrate is dissolved in 'broth' (e.g. peptone and meat extract) rather than in peptone water alone. For nutritionally fastidious organisms, peptone-water sugars or broth sugars are supplemented with e.g. serum ('serum sugars' or 'serum water sugars'), or with particular growth requirement(s), in order to permit growth to occur.

Some bacteria (e.g. certain *Bacillus* spp) produce an alkaline reaction in conventional sugars owing to the formation of excess alkaline breakdown products from the peptone; since this would mask acid production from carbohydrates, such organisms are tested in e.g. inorganic salts–carbohydrate media such as AMMONIUM SALT SUGARS.

peptone–yeast extract–glucose medium See PYG MEDIUM.

peptonization The formation of PEPTONE from protein.

Peptostreptococcus A genus of Gram-positive, asporogenous, anaerobic bacteria which can ferment amino acids or peptones and which may or may not ferment carbohydrates. Cells: cocci (ca. 1 μm diam.) which typically occur in chains. The organisms have been isolated e.g. from lesions in man, often as part of a mixed microflora. (See also GASTROINTESTINAL TRACT FLORA.) Species: *P. anaerobius* (the type species; GC%: ca. 33–34), *P. micros* (GC%: ca. 28), *P. parvulus* (GC%: ca. 44–45) and *P. productus* (GC%: ca. 44); it has been suggested that the saccharolytic species (*P. parvulus* and *P. productus*) be omitted from the genus [FEMS (1983) *17* 197–200]. It has been proposed that organisms previously referred to as *Gaffkya anaerobia* be included within the genus *Peptostreptococcus* as *P. tetradius* [IJSB (1983) *33* 683–698]. (For *P. elsdenii* see MEGASPHAERA.) (cf. PEPTOCOCCUS.)

***per* locus** See PARTITION.

Per phenotype Phage-encoded resistance phenotype (see LACTIC ACID STARTERS).

PER-1 ESBLs See β-LACTAMASES.

peracetic acid See HYDROGEN PEROXIDE (a).

peracids See HYDROGEN PEROXIDE (a).

Peranema A genus of colourless, heterotrophic (holozoic) EUGLENOID FLAGELLATES which ingest prey (unicellular algae and protozoa) by means of a specialized organelle (the *pharyngeal rod organ*) consisting of two parallel tapering rods (pharyngeal rods, trichites); essentially, ingestion of prey involves protrusion of the pharyngeal rods which adhere to the prey and then retract, the prey eventually being engulfed in a food vacuole at a cytostome situated subapically. (The rods can also be used to puncture or cut larger prey, allowing access to the cell contents.) *Peranema* has a characteristic smooth, gliding type of motility. One of the emergent flagella is extended forwards and held straight, movement being limited to a coiling or flickering of the tip, while a second is folded back along the cell and adheres to the pellicle. Genera closely related to *Peranema* include *Entosiphon* and HETERONEMA.

perborate See HYDROGEN PEROXIDE (a).

perchlorate degradation See BIOREMEDIATION.

percolating filter See SEWAGE TREATMENT.

percurrent proliferation In conidiogenesis (see CONIDIUM): the development, by the conidiogenous cell, of a new conidium-forming apex which grows through the previous apex; percurrent proliferation occurs e.g. in annellidic development (see CONIDIUM) and in the formation of new conidiogenous ampullae in *Gonatobotryum*.

perfect stage (*mycol.*) See TELEOMORPH.

perforins See APOPTOSIS.

perfringocins See BACTERIOCIN.

perfringolysin See THIOL-ACTIVATED CYTOLYSINS.

peribacteroid membrane See ROOT NODULES.

pericentric (of a chromosome) On both sides of, or including, the CENTROMERE; e.g., a pericentric inversion is a CHROMOSOME ABERRATION in which the inverted segment contains the centromere. (cf. PARACENTRIC.)

pericentriolar structure See CENTRIOLE.

Perichaena See MYXOMYCETES.

periclinal Parallel to a given surface. (cf. ANTICLINAL.)

Periconia See HYPHOMYCETES.

peridermioid Refers to a peridiate, typically tongue-shaped (flattened) aecium of the type formed e.g. by species of *Cronartium* and *Pucciniastrum*.

Peridermium See UREDINIOMYCETES.

peridiate Having a PERIDIUM.

peridinin See CAROTENOIDS.

Peridinium See DINOFLAGELLATES.

peridiole See NIDULARIALES and PISOLITHUS.

peridium The outer enveloping wall which encloses the spores, or the spore-bearing structures, in the fruiting bodies of various fungi (e.g. GASTEROMYCETES) and acellular slime moulds (MYXOMYCETES). (See also e.g. CLEISTOTHECIUM and PERITHECIUM.)

Périgord truffle See TRUFFLES.

perihepatitis Inflammation of the peritoneal coat of the liver; it can result from localized infection, trauma, or systemic inflammatory disease. [Review of chlamydial perihepatitis: BMB (1983) *39* 159–162.] (See also CURTIS–FITZ-HUGH SYNDROME.)

perilemma In many oligotrich ciliates: a membrane external to the pellicle.

perimycin (*syn.* fungimycin) A PEROSAMINE-containing heptaene POLYENE ANTIBIOTIC produced by a strain of *Streptomyces coelicolor*.

perinuclear space See NUCLEUS.

periodic acid–Schiff reaction See PAS REACTION.

periodontitis An acute or chronic inflammation of tooth-supporting tissues. The periodontium includes the *periodontal ligament*: a cup-like mass of fibrous connective tissue which binds the root of the tooth to the tooth socket in the jaw bone; the term 'periodontium' may also include the *cementum* (a bone-like substance which invests the root of the tooth), the *free gingiva* (see GINGIVITIS), and the *alveolar bone* (i.e. that part of the jaw bone which forms the tooth socket that binds the periodontal ligament.

Periodontitis is promoted by the accumulation of DENTAL PLAQUE in or near the *gingival sulcus*: the shallow groove (1–2 mm deep) at the tooth–gum junction; under these conditions the gum may shrink away from the tooth (the gingival sulcus enlarging to form a periodontal *pocket*) – thus exposing the cementum and the periodontal ligament to microbial attack.

In periodontal disease there is an increased flow of CREVICULAR FLUID.

Various bacterial species have been causally connected with periodontitis, but the aetiology of the disease is not yet fully understood; the human oral microflora has not yet been fully characterized, and it is possible that currently uncultured organisms may contribute to periodontitis. In one study, samples of healthy and periodontic material were examined with the object of investigating (from PCR-determined sequence data) whether any specific types of organism correlate with disease; one organism, whose data suggested a close relationship to asaccharolytic *Eubacterium* spp, was found only in deep pockets in periodontitis patients, possibly indicating an association with advanced periodontitis [JCM (1999) *37* 1469–1473].

Bacteria which have been aetiologically linked to periodontitis include species of BACTEROIDES (e.g. *B. gingivalis*) and FUSOBACTERIUM; *Actinobacillus actinomycetemcomitans* and *Capnocytophaga* sp have been implicated in 'juvenile periodontitis'. (See also CENTIPEDA.)

Multiplex PCR has been used for the simultaneous detection of *Bacteroides forsythus* and *Prevotella intermedia* in 152 clinical samples, including 38 specimens of plaque [JCM (1999) *37* 1621–1624].

peripheral membrane cylinder See ASCUS.

peripheral membrane protein See CYTOPLASMIC MEMBRANE.

periphyses (*sing.* periphysis) Short, sterile filaments which form a lining in the ostiolar canal in a perithecium or pseudoperithecium.

periphyton community *Syn.* AUFWUCHS.

periplasm *Syn.* PERIPLASMIC REGION.

periplasmic binding protein *Syn.* BINDING PROTEIN (sense 1).

periplasmic fibrils See SPIROCHAETALES.

periplasmic flagella See SPIROCHAETALES.

periplasmic gel See PEPTIDOGLYCAN.

periplasmic region (periplasmic space; periplasm) In Gram-positive bacteria and archaeans: the region between the CYTOPLASMIC MEMBRANE and the CELL WALL; in Gram-negative bacteria: the region between the cytoplasmic membrane and the OUTER MEMBRANE – or, according to some authors, between the cytoplasmic membrane and the layer of PEPTIDOGLYCAN.

Certain enzymes and other proteins are located partially or entirely in the periplasmic region and may be released from the cell e.g. by OSMOTIC SHOCK. (See also BINDING PROTEIN-DEPENDENT TRANSPORT SYSTEM and EXTRACYTOPLASMIC OXIDATION.)

periplasmic space *Syn.* PERIPLASMIC REGION.

periplast The cell envelope, or particular layer(s) of the cell envelope, in certain algae and protozoa; see e.g. PELLICLE (sense 3) and CHOANOFLAGELLIDA.

Peripylaria A category ('section') of LEISHMANIA species which multiply in the anterior part of the hindgut of the arthropod vector, and then migrate to the pharynx, cibarium and proboscis; it includes e.g. *L. braziliensis* and '*L. tarentolae*'. (cf. HYPOPYLARIA, SUPRAPYLARIA.)

periseptal annulus In Gram-negative bacteria: a continuous circumferential 'fusion' between the OUTER MEMBRANE and the cytoplasmic membrane, one annulus occurring on each side of the site of cell division – thus delimiting a small periplasmic compartment at the site of future septum formation. [PNAS (1983) *80* 1372–1376.] (cf. ADHESION SITE.)

peristome (1) The buccal cavity, or the extreme distal part of the buccal cavity, together with the somatic region adjacent to it; the term is used primarily for the appropriate parts in peritrichs and in certain spirotrichs, but it is sometimes used for the convoluted distal region of some chonotrichs. (cf. INFUNDIBULUM.) (2) *Syn.* BUCCAL CAVITY.

perithecioid Having the general characteristics of a PERITHECIUM.

perithecioid pseudothecium See ASCOSTROMA.

perithecium A hollow, spheroidal or flask-shaped ASCOCARP (ca. 100–300 μm) which opens to the exterior via an OSTIOLE; asci develop inside the perithecium, commonly in the form of a hymenial layer or as a basal tuft. According to species, perithecia may develop on the surface of a substratum or fungal STROMA, or embedded within a fungal stroma, and the wall (*peridium*) of a perithecium may be membranous or brittle, brightly coloured or dark. Some perithecia contain (in addition to asci) sterile structures such as paraphyses. (For ascospore discharge see ASCOSPORE.) Perithecia are formed e.g. by members of the SORDARIALES and SPHAERIALES.

(cf. ASCOSTROMA (pseudoperithecium).)

peritonitis Localized or generalized inflammation of the peritoneum (the membrane which lines the abdominal cavity and covers the viscera).

peritrich A member of the PERITRICHIA.

Peritrichia A subclass of freshwater, estuarine and marine ciliate protozoa (class OLIGOHYMENOPHOREA); one order: Peritrichida. In *mature* peritrichs, the somatic ciliature is limited or absent, but the oral ciliature is well developed and includes rows of cilia which – looking *into* the buccal cavity – spiral downwards towards the cytostome in an *anticlockwise* (counterclockwise) direction. According to species, individual cells may be shaped like an egg, bell or bottle etc; in some species the cell (*zooid*) is attached to the substratum either directly or by means of a contractile or non-contractile stalk. Some species are loricate, and some form colonies. STOMATOGENESIS is buccokinetal. Asexual reproduction in sedentary species may occur e.g. by budding with the formation of a motile, ciliated 'larva' (*telotroch*).

Predominantly sedentary peritrichs are placed in the suborder Sessilina (genera: e.g. *Apiosoma*, *Campanella*, CARCHESIUM, *Ellobiophrya*, *Epistylis*, *Lagenophrys*, OPERCULARIA, *Rhabdostyla*, *Scyphidia*, *Vaginicola*, VORTICELLA, *Zoothamnium*). The mobile peritrichs (all ecto- or endoparasitic, occasionally pathogenic, in freshwater or marine vertebrates and invertebrates) are placed in the suborder Mobilina (genera: e.g. *Leiotrocha*, *Trichodina*, *Vauchomia*).

Peritrichida See PERITRICHIA.

peritrichous Refers to flagella (or e.g. fimbriae) which are distributed more or less uniformly over the cell surface.

perkinetal (*ciliate protozool.*) Refers to the plane of cell division: perpendicular to the kineties. (cf. HOMOTHETOGENIC.)

Perkinsasida A class of parasitic, homoxenous protozoa (phylum APICOMPLEXA) in which the form of the conoid is atypical; reproduction is asexual. One species: *Perkinsus marinus* (see PERKINSUS).

Perkinsea A class of protozoa [JP (1980) *27* 37–58] equivalent to the PERKINSASIDA.

Perkinsus A genus of protozoa of the PERKINSASIDA [J. Parasitol. (1978) *64* 549] formerly regarded as fungi of the genus *Dermocystidium. P. marinum* is a pathogen of oysters.

permeaplast A cyanobacterial cell which has been treated with lysozyme and EDTA; permeaplasts have been used for the TRANSFORMATION of certain cyanobacteria.

permease (porter; transporter) A TRANSPORT SYSTEM or the (non-enzymic) component of a multicomponent transport system that effects transmembrane translocation of the substrate. (See also ION TRANSPORT.)

permissive cell (*virol.*) A cell which permits the replication of a given virus, with the consequent production of infectious progeny virions. (cf. ABORTIVE INFECTION.)

permissive conditions See CONDITIONAL LETHAL MUTANT.

Peronophythora See PERONOSPORALES.

Peronosclerospora See PERONOSPORALES.

Peronospora A genus of obligately plant parasitic fungi of the order PERONOSPORALES. Within the host plant, the (intercellular) mycelium gives rise to short club-shaped or branching filamentous haustoria. Conidia develop singly on curved, tapering and branching conidiophores which emerge from the stomata. On germination, the conidia form germ tubes – as do the (sexually-produced) oospores. *P. destructor* and *P. parasitica* cause DOWNY MILDEWS on onions and crucifers, respectively. *P. tabacina* causes the TOBACCO DISEASE *blue mould* – a disease which occurs worldwide; blue mould is encouraged by wet weather, and has been controlled e.g. by dithiocarbamates or metalaxyl.

Peronosporales An order of aquatic and terrestrial fungi (class OOMYCETES) which include both saprotrophs and parasites – some of the latter being major plant pathogens (see e.g. DOWNY MILDEWS). (See also PHENYLAMIDE ANTIFUNGAL AGENTS.) The thallus is characteristically a well-developed, coenocytic, branching mycelium which, in parasitic species, may give rise to haustoria. Asexually-derived zoospores are formed by most species; they are typically produced in rounded sporangia which are borne either on undifferentiated hyphae or on distinct sporangiophores, according to species. In *Bremia* and *Peronospora* the asexually-derived spores are conidia (see CONIDIUM) rather than sporangiospores. Sexual reproduction is oogamous (see OOMYCETES).

According to species, oogonia and antheridia are borne on the same or on different hyphae, and each may be uninucleate or multinucleate. Fertilization involves gametangial contact, the oosphere subsequently developing into a thick-walled oospore. Genera include ALBUGO, BASIDIOPHORA, BREMIA, *Bremiella*, *Peronophythora*, *Peronosclerospora*, PERONOSPORA, PHYTOPHTHORA, PLASMOPARA, *Pseudoperonospora*, PYTHIUM, *Sclerophthora*, SCLEROSPORA, and ZOOPHAGUS.

perosamine An aminosugar (4-amino-4,6-dideoxy-D-mannose) found e.g. in PERIMYCIN.

peroxidase An OXIDOREDUCTASE which catalyses a reaction in which the oxidation of a substrate is coupled with the reduction of hydrogen peroxide. See e.g. LIGNIN; cf. CATALASE.

peroxidase–antiperoxidase technique See PAP TECHNIQUE.

peroxin See PEX MUTATION.

peroxynitrite See NITRIC OXIDE.

peroxisome A type of MICROBODY found in many kinds of eukaryotic cell, including certain fungi, protozoa and algae. Peroxisomes contain a range of enzymes – including e.g. peroxidase, CATALASE, D-amino acid oxidase, dehydrogenase and acyl transferase. Each peroxisome is typically rounded, with a granular matrix and commonly an electron-dense core (= *nucleoid*); it is bounded by a single trilaminar membrane ~4.5–8 nm in thickness. Peroxisomes apparently develop by incorporating their component proteins (synthesized on free polysomes) and eventually dividing into daughter peroxisomes; this contrasts with the earlier view of biogenesis which supposed that peroxisomes budded from the endoplasmic reticulum.

In animal-type cells, peroxisomes carry out a range of functions which (depending on type of cell) include respiration, oxidation of alcohol and fatty acids, synthesis of plasmalogens, and purine metabolism; in plant/algal cells, functions include GLUCONEOGENESIS and PHOTORESPIRATION.

In methylotrophic yeasts (see METHYLOTROPHY) the peroxisomes, which contain methanol oxidase, occur in greater number and size when the yeasts are grown on methanol (rather than e.g. glucose). [Peroxisomes in methylotrophic yeasts: AMP (1983) *24* 1–82.] (See also HYDROCARBONS.)

[Recent developments in peroxisome biology: Endeavour (1996) (new series) *20* 68–73.]

[Peroxisome membrane proteins (role in biogenesis and function of the peroxisome): FEMS Reviews (2000) *24* 291–301.]

persistence (*virol.*) The phenomenon in which a virus remains within a cell, organism or population [Book ref. 90] for a prolonged period of time, during which it may or may not replicate to produce infectious progeny virions; if the virus persists in cryptic form, without production of infectious virions and without causing symptoms in the host, it is said to be *latent*. (Some authors reserve the term 'persistence' for those cases in which infectious virions are continually produced [Book ref. 148, p. 161].)

In a host organism, persistence (sensu lato) may be established directly or may follow an acute phase of disease. Various categories of persistent infection can be distinguished. (a) Infectious virions are not produced and infection is asymptomatic ('latent infection'); in some cases the virus may, under certain conditions, be 'reactivated' to produce infectious virions and cause sporadic recurrences of illness (see e.g. HERPES ZOSTER). (b) Infectious virions are not produced, but disease may nevertheless eventually occur (see e.g. SUBACUTE SCLEROSING PANENCEPHALITIS). (c) Infectious virions are continually produced but no symptoms are apparent – resulting in a CARRIER (sense 1). (See also LCM VIRUS.) (d) Infectious virions are continually produced, and ongoing chronic disease results.

Viral strategies for persistence in cells/hosts include e.g. integration of viral nucleic acid into host chromosomal DNA (as in the RETROVIRIDAE), or maintenance of viral nucleic acid in a stable extrachromosomal form (see e.g. EPSTEIN–BARR VIRUS); in some cases impaired immunological responses or immunological tolerance may be important factors. In cell cultures, persistence may result from the formation of viral mutants or of DEFECTIVE INTERFERING PARTICLES. (See also CARRIER CULTURE.)

persistent generalized lymphadenopathy See AIDS.

persistent virus (1) See CIRCULATIVE TRANSMISSION. (2) A virus which can establish a persistent infection (see PERSISTENCE).

pertactin A cell-surface adhesin found in species of *Bordetella*.

perthophyte *Syn.* PERTHOTROPH.

perthotroph (perthophyte) (1) A parasite which kills tissues within a living host, subsequently deriving nutrients from the dead tissues. (2) An organism which derives nutrients from dead tissues within a living host – whether or not it kills those tissues itself. (The term is usually applied to plant/microorganism interactions.) (cf. BIOTROPH; NECROTROPH.)

Pertusaria A genus of crustose LICHENS (order PERTUSARIALES); photobiont: a green alga. Apothecia frequently occur several together in warts on the thallus; they are initially perithecium-like with a pore-like opening, but in some species they subsequently open out to form flat or concave discs. In *P. pertusa*, the periphery of the thallus is surrounded by a white hypothallus; the apothecia do not open out as they mature. In *P. amara*, the thallus bears scattered white soralia (0.5–1.5 mm diam.) which have a bitter taste due to the presence of picrolichenic acid; apothecia are rare, but when present they are solitary and the mature discs are open and sorediate. Species occur e.g. on bark.

Pertusariales An order of fungi of the ASCOMYCOTINA; members include crustose LICHENS (photobiont commonly green), and saprotrophic and lichenicolus fungi. Ascocarp: APOTHECIOID, sometimes lirelliform. Asci: bitunicate. Genera: e.g. OCHROLECHIA, PERTUSARIA.

pertussigen *Syn.* PERTUSSIS TOXIN.

pertussis *Syn.* WHOOPING COUGH.

pertussis toxin (pertussigen; islet-activating protein; histamine-sensitizing factor; leucocytosis-promoting factor) An exotoxin, produced by *Bordetella pertussis*, which is one of the virulence factors associated with WHOOPING COUGH (see also CYCLOLYSIN, DERMONECROTIC TOXIN, TRACHEAL CYTOTOXIN).

The structure of pertussis toxin (PT) resembles that of e.g. CHOLERA TOXIN and SHIGA TOXIN: all of these toxins have an AB_5 arrangement of subunits in which five B subunits form a pentameric ring, and the A subunit occupies a central position to one side of the plane of the ring.

The B subunits of PT are not identical; some of them have binding sites for specific receptor molecules on eukaryotic target cells.

The A subunit is a 26 kDa protein with ADP-ribosyl transferase activity. Within the target cell it ADP-ribosylates a particular heterotrimeric G protein, G_i (= N_i), which normally inhibits the activity of ADENYLATE CYCLASE; ADP-ribosylation inhibits G_i, leading to an increase in the intracellular concentration of cyclic AMP (cAMP). (cf. CHOLERA TOXIN.)

B. parapertussis apparently contains a silent (i.e. non-expressed) gene for pertussis toxin.

Pesotum See CERATOCYSTIS and HYPHOMYCETES.

Pestalotia See MELANCONIALES.

peste des petits ruminants (*syn.* pseudorinderpest) A disease of sheep and goats which occurs in W. Africa. Symptoms include fever, nasal and ocular discharge, and necrotic stomatitis, followed by severe enteritis and pneumonia; mortality rates may be >90%.

The causal agent is a virus (the 'Kata virus') of the genus MORBILLIVIRUS.

pesticin I A BACTERIOCIN produced by strains of *Yersinia pestis*; it is active against *Y. pseudotuberculosis* and strains of *Escherichia coli*. Production of pesticin I correlates with the production of a fibrinolytic factor and a coagulase.

pestis minor A mild form of PLAGUE in which buboes are formed without systemic symptoms.

Pestivirus A genus of viruses (family TOGAVIRIDAE) which infect pigs (SWINE FEVER virus) and ruminants (BOVINE VIRUS DIARRHOEA/MUCOSAL DISEASE (BVD-MD) virus, and the closely related BORDER DISEASE virus). The viruses do not replicate in invertebrates and are not vector-borne. Pestiviruses can be transmitted transplacentally, and may cause abortion or malformation of the fetus.

petal breaking *Syn.* COLOUR BREAKING.

petechia See PURPURA.

Petersenia See LAGENIDIALES and ALGAL DISEASES.

petite mutant In certain yeasts (including e.g. *Saccharomyces cerevisiae*): a mutant strain characterized by the formation of abnormally small colonies; petite mutants are respiration-deficient owing to functionally defective mitochondria (see MITOCHONDRION), and they exhibit a low growth rate since they are restricted to the less efficient fermentative mode of metabolism.

In *nuclear petite* (= *segregational petite*) mutants the mutant characteristic is due to mutation(s) in the nuclear (chromosomal) DNA which affect mitochondrial function; such mutations arise with frequencies similar to those of other chromosomal mutations, and they can be induced by mutagenic agents such as ultraviolet radiation. Crosses between nuclear petites and wild-type strains yield petite and wild-type progeny in the ratio 1:1.

In *cytoplasmic petite* mutants the mutant characteristic is due to mutation(s) in the mitochondrial DNA. *Neutral petite* (= rho^0 *petite*) mutants are completely devoid of mtDNA; they are the main or sole types of petite mutant which arise in the presence of high concentrations of INTERCALATING AGENTS. In crosses between neutral petites and wild-type strains the rho^0 genotype is recessive. In *suppressive petite* (= rho^- *petite*) mutants the mtDNA has suffered deletion(s) such that, in general, much less than 50% of the wild-type mtDNA remains. In crosses between rho^- and wild-type strains, some of the progeny have only the petite genotype, the wild-type genotype having been lost – an effect called *suppression*; in highly suppressive petites, loss of wild-type mtDNA is apparently due to the preferential replication of petite mtDNA – possibly owing to higher concentrations of origin-like sequences in the petite mtDNA.

(cf. POKY MUTANT.)

Petractis See GYALECTALES.

Petri dish A (usually) round, shallow, flat-bottomed, glass or plastic dish (often ca. 10 cm diam.) with a vertical side, together with a similar, slightly larger structure which forms a loosely-fitting lid. Petri dishes are used in microbiology e.g. for the preparation of PLATES. (See also VENT.)

Petriellidium See PSEUDALLESCHERIA.

Petroderma See PHAEOPHYTA.

petroleum Crude oil, i.e., oil in the form obtained from oil deposits within the earth. (See also KEROGEN.) Petroleum is an extremely complex mixture of components, including aliphatic, alicyclic and aromatic hydrocarbons together with e.g. various heterocyclic compounds. It can be converted into various 'refined' products – used e.g. as fuels for internal combustion engines – which are still mostly complex mixtures (e.g. kerosene may contain 5000–10000 different compounds). Many of the components of petroleum and its products can be degraded by microorganisms (see e.g. HYDROCARBONS).

Contamination of petroleum-derived fuels and oils by certain microorganisms can be a serious economic problem, resulting in e.g. corrosion of fuel tanks, blockage of fuel pipes and filters, alteration of chemical and functional properties of lubrication and cutting oils, etc. Microbial growth requires the presence of at least some water (which may result e.g. from condensation), growth occurring in the water or at the oil–water or fuel–water interface. Organisms commonly involved include hydrocarbon-utilizing microbes such as *Hormoconis resinae* (formerly *Cladosporium resinae*) (particularly in aviation kerosene), strains of *Aspergillus* and *Pseudomonas*, etc. However, not all spoilage organisms are necessarily hydrocarbon utilizers: some may obtain nutrients from solutes in the aqueous phase, from impurities in the fuel, etc. (See also BIOBOR and EGME.)

[Petroleum microbiology: Book ref. 155.]

***Petunia* vein-clearing virus** See CAULIMOVIRUSES.

***pex* mutation** Any mutation which results in a deficiency in PEROXISOME biogenesis; *pex* genes encode various *peroxins*, i.e. protein constituents of the peroxisome. Most peroxins so far identified are components of the peroxisome membrane.

pexicyst An EXTRUSOME which occurs e.g. in *Didinium*; on discharge it adheres to the prey.

Peziza A genus of fungi (order PEZIZALES) which form light-brown, sessile, discoid or cup-shaped apothecia, often several centimetres across; the ascospores are typically biguttulate.

Pezizales An order of fungi (subdivision ASCOMYCOTINA) which occur e.g. on wood, soil and dung. Ascocarp: typically APOTHECIOID (but closed and hypogean in some species), with paraphyses. Asci: unitunicate, cylindrical to globose, with a non-thickened apex which typically opens, at maturity, by a slit or operculum (see OPERCULATE ASCUS). Genera: e.g. ALEURIA, *Ascobolus*, *Carbomyces*, *Gyromitra*, HELVELLA, *Monascus*, MORCHELLA, PEZIZA, PYRONEMA, *Sarcoscypha*, SCUTELLINIA, TUBER, VERPA.

Pezizella See HELOTIALES.

PF4 (chemokine) See CHEMOKINES.

Pf155 See PLASMODIUM.

PFA PHOSPHONOFORMIC ACID.

Pfeiffer phenomenon The rapid lysis of *Vibrio cholerae* by specific antibody and COMPLEMENT.

Pfeifferinella See EIMERIORINA.

Pfeiffer's bacillus *Haemophilus influenzae*.

Pfeiffer's disease INFECTIOUS MONONUCLEOSIS.

PFGE Pulsed-field gel electrophoresis: any of several forms of gel ELECTROPHORESIS used for separating large molecules of nucleic acid (up to $\sim 10^6$ base pairs in length). Essentially, electric fields are applied to the gel alternately from different angles so that the net migration of a given molecule depends on the speed with which it can change its direction of movement; this separates the nucleic acid molecules according to size. The range of sizes of molecule which can be separated is determined by the pulse *length*: short pulses are used for separating the (relatively) smaller molecules, while longer pulses are used for separating the larger molecules. (It may also be possible to choose a pulse length which includes plasmids in the assay; such a pulse length may exclude the largest chromosomal fragments.)

Following PFGE, bands of fragments may be detected in the gel e.g. by staining with ethidium bromide.

Various forms of PFGE differ e.g. in the number, type and location of electrodes; in some forms, electrodes can periodically switch polarity.

In contour-clamped homogeneous electric field electrophoresis (CHEF), electrophoresis is carried out in a hexagonal chamber in which electrodes are located on sides 2, 3, 5 and 6; this procedure is reported to be reproducible and highly discriminatory. (See also OFAGE.)

PFGE is used e.g. in TYPING. For example, this approach is used for separating large fragments of DNA produced when bacterial chromosomes are cleaved with so-called 'rare-cutting' restriction endonucleases (see DNA FINGERPRINTING). According to some authors, PFGE is the 'gold standard' among molecular typing methods. [Guidelines for typing bacterial strains with PFGE: JCM (1995) *33* 2233–2239.]

PFGE has been used e.g. for subtyping strains of the Penner HS11 serotype of *Campylobacter jejuni* [JMM (1998) *47* 353–357], isolates of *Shigella dysenteriae* type 1 [JMM (1999) *48* 781–784], and strains of *Serratia marcescens* [JCM (1997) *35* 325–327] and *Vibrio parahaemolyticus* [JCM (1999) *37* 2473–2478]. CHEF has been used e.g. for typing strains of *Neisseria gonorrhoea* [JMM (1993) *38* 366–370].

[Comparison of PFGE with five other molecular methods for typing methicillin-resistant strains of *Staphylococcus aureus*: JMM (1998) *47* 341–351.]

pFM739 See INCOMPATIBILITY.

pfmdr1 See AMINOQUINOLINES.

pfu PLAQUE-FORMING UNIT.

***Pfu* DNA polymerase™** See DNA POLYMERASE.

PG (1) PROSTAGLANDIN. (2) PEPTIDOGLYCAN.

PGD_2 See PROSTAGLANDINS.

PGE_2 See PROSTAGLANDINS.

$PGF_{2\alpha}$ See PROSTAGLANDINS.

Pgh1 See AMINOQUINOLINES.

PGI_2 See PROSTACYCLIN.

pGK*l* plasmids See KILLER PLASMIDS.

PGL (persistent generalized lymphadenopathy) See AIDS.

pH indicator A substance which changes colour over a particular pH range(s): see e.g. ANDRADE'S INDICATOR; BROMCRESOL GREEN; BROMCRESOL PURPLE; BROMPHENOL BLUE; BROMTHYMOL BLUE; CHLORPHENOL RED; CONGO RED; CRESOL RED; LITMUS; METHYL ORANGE; METHYL RED; NEUTRAL RED; PHENOL RED; PHENOLPHTHALEIN; THYMOL BLUE; THYMOLPHTHALEIN. (See also e.g. LITMUS MILK; MACCONKEY'S AGAR; METHYL RED TEST.)

pH-stat See CONTINUOUS CULTURE.

PHA (1) PHYTOHAEMAGGLUTININ. (2) Passive haemagglutination (see PASSIVE AGGLUTINATION). (3) Poly-β-hydroxyalkanoate: see POLY-β-HYDROXYBUTYRATE.

Phacidium See HELOTIALES.

Phacus A genus of EUGLENOID FLAGELLATES in which the cell has a rigid pellicle and is usually flattened with a posterior point, often appearing somewhat leaf-shaped.

Phaeodactylum A genus of pennate DIATOMS. *P. tricornutum* is polymorphic, occurring in three main morphotypes: oval, fusiform, and triradiate (cells with three corners – often extended into arms); the latter two forms may occur in chains, and the fusiform type is the most stable of the three. Daughter cells are (initially) of the same morphotype as the parent cell. The cell wall is almost entirely organic, the only siliceous structures being the gridle bands (up to 10); one valve is partially silicified in the oval morphotype.

Phaeodarea See RADIOLARIA.

phaeodium See RADIOLARIA.

Phaeographis See GRAPHIDALES.

phaeohyphomycosis Any of a range of human and animal diseases caused by any of several black moulds (e.g. *Dactylaria gallopava*, *Drechslera* spp, *Exophiala* spp, *Wangiella dermatitidis*, *Xylohypha bantiana* (formerly *Cladosporium bantianum*)) which form dark-walled septate mycelium in the tissues (cf. CHROMOBLASTOMYCOSIS). Such diseases generally affect only debilitated or immunosuppressed individuals; they usually involve the formation of chronic, progressive, subcutaneous abscesses or warty lesions in various parts of the body. The disease may be systemic, particularly when caused by *X. bantiana* (which tends to localize in the brain, but which may also cause pulmonary infections) or by *W. dermatitidis*.

Phaeophyceae The sole class within the PHAEOPHYTA.

Phaeophyta (brown algae) A division of predominantly marine ALGAE (freshwater members including species of e.g. *Bodanella* and *Heribaudiella*). The thallus ranges from an adherent disc or crust to hollow bulbous or whip-like forms, branched filaments and giant complex forms – the latter reaching 50 metres or more in length and containing sieve tubes within which photosynthate can be translocated; there are no unicellular species. The CELL WALL contains cellulose and e.g. FUCOIDIN and ALGINATES. Within the CHLOROPLASTS the THYLAKOIDS characteristically occur in groups of three. Brown algae contain CHLOROPHYLLS a, c_1 and c_2; certain CAROTENOIDS (e.g. fucoxanthin); carbohydrates such as MANNITOL and LAMINARIN; sterols (e.g. fucosterol and cholesterol); and sometimes large amounts of TANNINS ('fucosan') stored in 'fucosan vesicles' (= *physodes*) in the cytoplasm. Sexual processes include isogamy and oogamy, and many species exhibit a heteromorphic or isomorphic ALTERNATION OF GENERATIONS; the biflagellate, pyriform gametes each have one anteriorly-directed tinsel FLAGELLUM, and one posteriorly-directed whiplash flagellum. Some species form PHEROMONES. Life cycles are often complex.

Orders include [Book ref. 130]: Ascoseirales (e.g. *Ascoseira*); Chordariales (e.g. *Chordaria, Cladosiphon, Halothrix*); Cutleriales (e.g. *Cutleria*); Desmarestiales (e.g. *Desmarestia*); Dictyosiphonales (e.g. *Dictyosiphon, Striaria*); Dictyotales (e.g. *Dictyota, Zonaria*); Durvillaeales (*Durvillaea*); Ectocarpales (e.g. *Ectocarpus, Petroderma, Ralfsia*); Fucales (e.g. *Ascophyllum, Cystoseira*, FUCUS, *Halidrys, Himanthalia, Myriodesma, Pelvetia, Sargassum, Turbinaria*); Laminariales (e.g. *Alaria, Chorda, Costaria*, LAMINARIA, *Macrocystis, Pelagophycus, Thalassiophyllum*, UNDARIA); Scytosiphonales (e.g. *Colpomenia, Hydroclathrus, Scytosiphon*); Sphacelariales (e.g. *Cladostephus, Halopteris, Sphacelaria*); Sporochnales (e.g. *Sporochnus*); Tilopteridaceae (e.g. *Haplospora, Tilopteris*).

[Evolution of the Phaeophyta (particularly Fucales): Book ref. 167, pp. 11–46.]

phaeophytin A derivative of CHLOROPHYLL in which Mg^{2+} is replaced by two protons. (cf. BACTERIOPHAEOPHYTIN.)

phaeosporae See SACCARDOAN SYSTEM.

Phaffia A genus of imperfect yeasts (class HYPHOMYCETES). The cells are ellipsoidal, occurring singly, in pairs, or occasionally in short chains; they reproduce by budding. Spheroidal chlamydospores containing refractile granules are formed e.g. after several days in malt extract. Colonies on e.g. malt agar are orange-red or salmon-pink due to the synthesis of carotenoid pigments, predominantly astaxanthin (3,3′-dihydroxy-4,4′-diketo-β-carotene, a red pigment common among animals – particularly crustaceans, echinoderms and tunicates – but rare in plants). Glucose and certain other sugars are fermented. Neither inositol nor NO_3^- is assimilated. One species: *P. rhodozyma*, isolated from tree exudates. [Book ref. 100, pp. 890–892.]

phage Abbreviation for BACTERIOPHAGE. (For 'phage G4', 'phage λ' etc see entries BACTERIOPHAGE G4, BACTERIOPHAGE λ etc.)

phage conversion *Syn.* BACTERIOPHAGE CONVERSION.

phage typing A form of TYPING in which strains of bacteria are distinguished on the basis of differences in their susceptibilities to a range of BACTERIOPHAGES. Initially, a FLOOD PLATE is prepared from a culture of the strain under test; the surface of the plate is then dried (by leaving the plate in a 37°C incubator, with the lid partly off, for 10–20 minutes) and a grid is drawn on the base of the plate. The agar surface corresponding to each square of the grid is inoculated with one drop of a phage suspension (at ROUTINE TEST DILUTION) – each square being inoculated with a different phage; the plate is then incubated. Each of the phages used for typing is lytic for one strain, or a limited number of strains, of the species under test; a phage which lyses the particular strain being tested will form a clear macroscopic area amid the opaque layer of bacterial growth. The test strain can then be defined in terms of the phages to which it is sensitive. (The results of phage typing may be affected if the bacterium under test carries a PLASMID encoding a RESTRICTION ENDONUCLEASE which destroys phage nucleic acid.) Strains of *Salmonella typhi* which bear the VI ANTIGEN may be typed by a range of phages each of which is specific for one or more of the Vi strains of *S. typhi*; this procedure is often termed Vi phage typing.

phagemid A CLONING vector which includes (i) a plasmid's origin of replication; (ii) a multiple cloning site (MCS), i.e. a POLYLINKER; (iii) a marker gene (e.g. a gene encoding resistance to kanamycin); and (iv) the origin of replication of a filamentous phage (e.g. that of the ssDNA phage f1). The DNA sample (to be cloned) is inserted via the MCS, and the phagemid is then inserted into a bacterium e.g. by transformation.

A phagemid can be used e.g. for making double-stranded copies of the given fragment (i.e. used as a conventional cloning vector). It can also be used for making single-stranded copies if the phagemid-carrying bacteria are infected with a 'helper phage', e.g. phage M13, which promotes replication from the *phage* origin in the phagemid; the single-stranded copies are packaged into phage coat proteins (encoded by the helper phage) and exported to the medium. When freed from phage coat proteins, the single-stranded copies of the insert can be used e.g. for sequence analysis (see DNA SEQUENCING).

A phagemid may contain a promoter on each strand, either side of the MCS, so that, if required, the cloned fragment can be transcribed.

phagocyte Any cell which can carry out PHAGOCYTOSIS – particularly a neutrophil or macrophage.

phagocytosis The process in which particulate matter is ingested, and usually digested, by certain types of cell or microorganism (cf. PINOCYTOSIS). In mammals, phagocytosis by certain leucocytes is an important aspect of the immune defence system, while in some microorganisms phagocytosis is a mode of feeding.

(a) In mammals, phagocytosis is carried out e.g. by NEUTROPHILS and MACROPHAGES. Essentially, the particle (e.g. an opsonized bacterium) initially binds to the surface of the phagocyte; the phagocyte's cytoplasmic membrane then invaginates to form a pocket which contains the particle. (Different intracellular signals may be generated in the phagocyte according to whether engulfment was triggered from antibody- or complement-binding sites.) Subsequently, the invaginated membrane forms a closed, intracellular sac or vacuole, the *phagosome*. (A phagosome which contains a parasite is called a *parasitophorous vacuole*.) The pH within the phagosome decreases, and the phagosome coalesces with one or more LYSOSOMES to form a *phagolysosome* (= *lysophagosome*) within which the bacterium etc. may be killed and degraded.

The maturation of a phagosome into a phagolysosome can be monitored from changes in (i) its internal pH, and (ii) the associated proteins; for example, a phagolysosome is characterized by so-called lysosome-associated membrance proteins (LAMPs).

Antimicrobial activity within phagocytes is promoted by (i) the wide range of lysosomal enzymes, and (ii) the intracellular formation of certain toxic derivatives of oxygen – see e.g. SUPEROXIDE, HYDROGEN PEROXIDE, SINGLET OXYGEN and HYDROXYL RADICAL (species formed during the 'oxidative burst' (= 'respiratory burst') characteristic of some types of phagocyte); such derivatives are formed e.g. by neutrophils and macrophages. In NEUTROPHILS (q.v.), the oxygen derivatives are generated within 30–60 seconds of cell-surface stimulation. Ferric ions bound to lysosomal LACTOFERRIN may serve to promote the formation of hydroxyl radical from hydrogen peroxide and superoxide. During the oxidative burst, the consumption of oxygen by the phagocyte increases sharply. (See also CHEMILUMINESCENCE.) Reactive oxygen derivatives are also released to the extracellular environment.

Some bacteria can resist ingestion by phagocytes as a consequence of their capsular or cell-wall components. For example, the capsule of *Bacillus anthracis* is anti-phagocytic (cf. STERNE STRAIN). Again, the PROTEIN A cell-wall component of *Staphylococcus aureus* may serve to inhibit opsonization by binding the Fc portion of antibodies.

Some pathogens can kill phagocytes: see e.g. APOPTOSIS and LEUCOCIDINS.

Certain pathogens (e.g. *Coxiella*, *Leishmania* spp, *Listeria monocytogenes* and *Mycobacterium lepraemurium*) can survive or grow within phagosomes, or phagolysosomes, avoiding destruction by various mechanisms. For example, in the case of *Chlamydia* spp, *Legionella pneumophila* and *Mycobacterium tuberculosis* the phagosome does not acidify and does not fuse with the lysosome. Under certain circumstances, some pathogens (e.g. *Mycobacterium leprae*, *Rickettsia mooseri*, *Trypanosoma cruzi*) can escape from the phagosome and grow within the cytoplasm of the phagocyte. (See also REOVIRUS.)

Phagosomes containing live *Toxoplasma gondii* resist fusion with lysosomes, and *T. gondii* may also fail to trigger the phagocyte's oxidative mechanism and may inbibit development of the normal acidification of the phagosome. *M. leprae* apparently fails to stimulate the generation of superoxide anion by the phagocyte.

In some cases a pathogen is not taken up and killed because it has actually *invaded* the phagocyte. Invasion may involve uptake via adhesins and receptors which differ from those involved in routine phagocytosis. The *initial* contact between pathogen and phagocyte may be important in determining the pathogen's fate; for example, of the various modes of binding, some may favour survival by failing to trigger an oxidative burst within the phagocyte. In the case of *Legionella pneumophila*, the phagosome in a macrophage maintains near-neutral pH, and the pathogen is able to multiply, killing the phagocyte. In this organism the *icm* (intracellular multiplication) genes – also called *dot* (defect in organelle trafficking) genes – are believed to encode a secretory system (and associated effector molecules) involved e.g. in blocking phagosome–lysosome fusion; mutations in these genes have been found to affect the pathogen's ability to multiply within, and kill, macrophages [see e.g. TIM (1998) *6* 253–255].

It has been reported that the zinc metalloprotease secreted by *L. pneumophila* inhibits (i) the oxidative burst, and (ii) chemotaxis in phagocytes [JMM (2001) *50* 517–525].

In the case of *Mycobacterium tuberculosis*, invasion/survival may be influenced by (i) the cell-surface chemistry of the given strain; (ii) the site of infection (e.g. lung); and (iii) the timing of pathogen–phagocyte contact (e.g. at initial infection, or later). The opsonization of *M. tuberculosis* may occur via any of four pathways of COMPLEMENT FIXATION (q.v.); however, non-opsonic (complement- and antibody-independent) binding of the pathogen to the phagocyte may be important for *initial* infection of the normal lung. [*M. tuberculosis*–phagocyte interactions: TIM (1998) *6* 328–335.]

[Bacterial invasion of host cells: Book ref. 218, pp. 223–272.]

(b) Among microorganisms, particulate food is ingested e.g. by many sarcodines (see also PSEUDOPODIUM), ciliates and flagellates; in (at least) most of the ciliates, the essential features of phagocytosis appear to resemble those described in (a) above: particle(s) are enclosed within an invagination of the cytoplasmic membrane which gives rise to a closed intracellular vesicle (FOOD VACUOLE). In many ciliates there is a single, specialized site for phagocytosis (see CYTOSTOME), while in suctorians ingestion may occur at the tip of each of the tentacles.

phagolysosome See PHAGOCYTOSIS.

Phagomyxa A genus of endoparasitic, phagocytic, plasmodial organisms. *P. algarum* occurs in the cells of marine algae; it appears to resemble members of the PLASMODIOPHOROMYCETES but apparently does not form cysts.

phagosome See PHAGOCYTOSIS.

phagotrophy A mode of nutrition in which nutrients are ingested in particulate form, as in some protozoa. (cf. OSMOTROPHY.)

phagotype See VARIETY.

phagovar See VARIETY.

Phalacrocleptes A genus of protozoa (placed tentatively in the subclass SUCTORIA [Book ref. 135, p. 239]) which have short tentacles but which lack both cilia and infraciliature at all stages of the life cycle. The organisms are parasitic on polychaete annelids.

phallacidin See PHALLOTOXINS.

Phallales An order of humicolous and lignicolous fungi (class GASTEROMYCETES) which include both epigean and hypogean species. The immature basidiocarp ('egg') is rounded, ca. one to several centimetres in size, and consists essentially of a gleba enveloped in gelatinous material and enclosed within a membranous peridium. In some species the peridium is indehiscent, the basidiospores within being dispersed, when mature, on rupture of the peridium e.g. by burrowing animals. In other species (the 'stinkhorns') the 'egg' contains a mass of spongy tissue (the *receptacle*) which, during maturation of the basidiocarp, expands (rupturing the peridium) to form a typically hollow, columnar structure – also called the receptacle (or *pseudostem*) – which carries the hymenium-bearing gleba at its apex; autolysis of the glebal material gives rise to an often evil-smelling liquid which attracts certain insects, the latter dispersing the basidiospores. The stinkhorns include e.g. species of *Clathrus*, *Mutinus* and *Phallus*.

phallin B See PHALLOTOXINS.

phallisin See PHALLOTOXINS.

phalloidin See PHALLOTOXINS.

phalloin See PHALLOTOXINS.

phallotoxins A group of cyclic heptapeptide toxins present in *Amanita phalloides* and e.g. in *A. verna*. Phallotoxins include e.g. phallacidin, phallin B, phallisin, phalloidin and phalloin; they generally contain unusual amino acids: e.g. phalloidin, a bicyclic heptapeptide, contains D-threonine, 4(*cis*)-hydroxy-L-proline and γ,δ-dihydroxy-L-leucine, and is resistant to all proteases which have been tested. Phallotoxins produce clinical symptoms within a few hours of ingestion and are generally less toxic than AMATOXINS; symptoms of phallotoxin poisoning include weakness, severe vomiting and diarrhoea, and sometimes death. Degenerative changes occur in the cells of the liver. Phalloidin binds specifically and strongly to F-actin [binding site: EMBO (1985) *4* 2815–2818], effectively accelerating G- to F-actin polymerization (see ACTIN) and increasing the stability of actin filaments; consequently, cellular functions associated with the microfilament system (e.g. cytoplasmic streaming, cell motility) are irreversibly blocked. (See also MYCETISM.)

Phallus See PHALLALES.

Phanerochaete See APHYLLOPHORALES (Corticiaceae).

phaneroplasmodium See MYXOMYCETES.

pharyngeal basket *Syn.* CYRTOS.

pharyngeal rod organ See PERANEMA.

pharyngitis Inflammation of the pharynx. Acute pharyngitis is often caused by *Streptococcus pyogenes*; symptoms include e.g. sore throat, fever, pharyngeal exudate. Other causes include any of a range of viruses – e.g. adenovirus (see PHARYNGOCONJUNCTIVAL FEVER), COMMON COLD viruses, coxsackieviruses (see also HERPANGINA), EB virus (see INFECTIOUS MONONUCLEOSIS).

pharyngoconjunctival fever A (usually mild) disease, affecting mainly children and young adults, caused by certain adenoviruses (see MASTADENOVIRUS). Symptoms: fever, inflammation of conjunctivae and/or pharynx, and enlargement of cervical lymph nodes.

phase-contrast microscopy See MICROSCOPY (c).

phase degradation Phase variation in BORDETELLA.

phase-shift mutation *Syn.* FRAMESHIFT MUTATION.

phase variation Spontaneous switching from the synthesis of a given cell-surface component to the synthesis of a different form of that component – or the starting, or stopping, of synthesis of such a component; each form of a given component depends on the expression of a specific allele, so that the number of different forms of that component depends on the allelic complement of the organism (cf. ANTIGENIC VARIATION).

The term 'phase variation' was initially applied to the alternation of flagellar components in *Salmonella*, but is now applied more generally to a range of phenomena involving variation in cell surface components. Many of these phenomena are known to involve re-arrangements of DNA.

Some well-studied examples of phase variation include the prototypical flagellar phase variation in *Salmonella* (see below); the on/off switching of type I FIMBRIAE (q.v.) in *Escherichia coli*; and the expression of OPA PROTEINS in *Neisseria*. Variation in the Vi antigen of *Citrobacter freundii* may be a further example.

In most strains of *Salmonella* the filament of the FLAGELLUM contains either of two distinct types of FLAGELLIN encoded by genes H1 (= *fliC*) and H2 (= *fljB*); normally, only one of these genes is expressed at any given time, so that the flagellar filament contains either the H1 gene product or the H2 gene product. (Note that there are many different alleles of gene H1; the product of a *given* H1 allele – i.e. a particular *phase 1 antigen* (formerly called a 'specific-phase antigen') – is found only in certain serotypes of *Salmonella*. Gene H2 also occurs in a number of allelic forms. However, in many cases, a given product of gene H2 – i.e. a particular *phase 2 antigen* (formerly called a 'group-phase antigen') – occurs in a wide range of serotypes of *Salmonella*; for example, the phase 2 antigen '1' occurs in many, if not most, of the serotypes. Some salmonellae, including *S. typhi*, do not have phase 2 antigens. (See also H ANTIGEN, KAUFFMANN–WHITE CLASSIFICATION and CRAIGIE'S TUBE METHOD.)

One of the consequences of phase variation in *Salmonella* is the formation of flagella which may differ e.g. antigenically and/or in susceptibility to FLAGELLOTROPIC PHAGES.

The switch from H1 to H2 gene expression is reversible and occurs at a frequency of about 10^{-5} to 10^{-3} per cell per generation.

Switching involves the inversion of a specific, 995-bp sequence of nucleotides which includes the promoter of the H2 gene. When this sequence is in *one* orientation, the promoter can be used for transcription of the H2 gene – so that, in this case, the H2 gene product is incorporated into the flagellar filament. The H2 gene actually forms part of an operon that includes a gene (*rh1* or *rH1*) whose product represses the H1 gene; thus, with the sequence orientated this way, the H2 product is formed and the H1 product is repressed.

Inversion of the control sequence prevents transcription of H2 – and also prevents formation of the H1 repressor; hence,

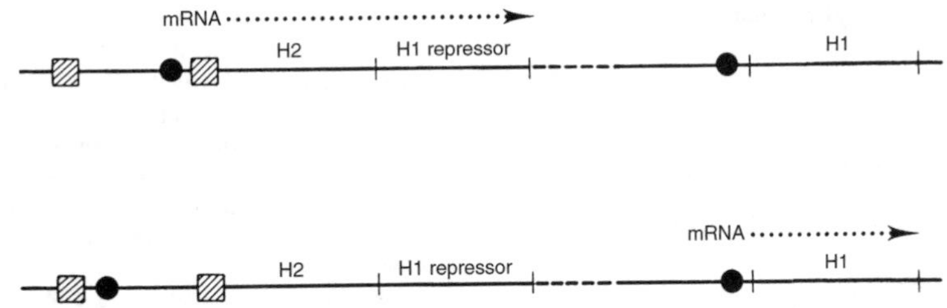

PHASE VARIATION: alternation of flagellar proteins in *Salmonella* (principle, diagrammatic; see entry).

In most strains of *Salmonella* the flagellar filament contains either of two distinct types of protein, encoded by genes H1 and H2. Normally, only one of these genes is expressed at any given time, so that the flagellar filament contains either the H1 gene product or the H2 gene product.

The promoter (●) of the H2 operon is flanked by two short inverted repeats (▨) which are recognized by a site-specific recombinase; that is, the sequence containing the promoter can be inverted by site-specific recombination.

Top. The position of the H2 promoter is such that the H2 operon can be transcribed. The H2 gene product is incorporated in the flagellar filament, and the H1 repressor protein blocks transcription of the H1 gene.

Below. Site-specific recombination has occurred between the two inverted repeats, so that the sequence containing the H2 promoter has been inverted. In this new orientation the H2 promoter is no longer functional. Because the H2 operon is not transcribed, the H1 repressor is not formed, and this permits transcription of the H1 gene – so that the flagellar filament now contains the H1 gene product.

Reproduced from *Bacteria*, 5th edition, Figure 8.3, page 168, Paul Singleton (1999) copyright John Wiley & Sons Ltd, UK (ISBN 0471-98880-4) with permission from the publisher.

with the sequence in *this* orientation the product of the H1 gene is incorporated into the flagellar filament.

Inversion of the control sequence – in either direction – involves SITE-SPECIFIC RECOMBINATION between two short INVERTED REPEATS, one of which is located at each end of the control sequence. The recombinase which mediates inversion is encoded by gene *hin* (= *vH2* or *vh2*) which, like the H2 promoter, occurs on the control sequence. The scheme for flagellar phase variation in *Salmonella* is shown diagrammatically in the figure.

(See also RECOMBINATIONAL REGULATION.)

phaseollidin An isoflavonoid PHYTOALEXIN produced by the French bean (*Phaseolus vulgaris*).

phaseollin An isoflavonoid PHYTOALEXIN (7-hydroxy-3′,4′-dimethylchromanocoumarin) produced by the French bean (*Phaseolus vulgaris*). It can be elicited e.g. by the polypeptide elicitor *monilicolin A* which can be obtained from culture filtrates of *Monilinia fructicola*.

phaseolotoxin A chlorosis-inducing TOXIN produced by *Pseudomonas syringae* pv. *phaseolicola*; it irreversibly inactivates ornithine carbamoyltransferase, an enzyme which catalyses the formation of citrulline from ornithine and carbamoyl phosphate. The toxin–enzyme interaction involves the so-called 'Kcat' mechanism; in this mechanism the target molecule (enzyme in this case) activates the toxin, and the activated toxin then brings about specific alkylation of the enzyme. *P. syringae* pv. *phaseolicola* can produce a form of ornithine carbamoyltransferase which is insensitive to the toxin.

Phaseolotoxin may also inhibit HYPERSENSITIVITY (sense 2) in the host plant.

phasins See e.g. POLY-β-HYDROXYBUTYRATE.

phasmid (1) A construct formed by recombination between e.g. a phage λ genome and a plasmid containing the λ *att* site and *cos* site. Integration of the plasmid into the phage genome is mediated (within a host bacterial cell) by the phage integrase site-specific recombination system (see BACTERIOPHAGE LAMBDA). The resulting genome–plasmid hybrid can be encapsidated during the assembly of phage virions. Phasmids are useful tools in genetic engineering since genes carried by the original plasmid can readily be introduced into other host cells by infection of those cells by phasmid-containing phage virions. (cf. COSMID.)

(2) (*biol.*) A caudal chemoreceptor in nematodes of the class Phasmidia.

PHB POLY-β-HYDROXYBUTYRATE.

Phellinus See APHYLLOPHORALES (Hymenochaetaceae).

phenanthridine An INTERCALATING AGENT which has trypanocidal and antitumour activity. Phenanthridine derivatives include e.g. ETHIDIUM BROMIDE.

phenethicillin *Syn.* PHENOXYETHYLPENICILLIN.

phenetic classification Any form of classification which is based on the observable (phenotypic and genotypic) characteristics of organisms – regardless of their ancestral lineage. (cf. PHYLOGENETIC CLASSIFICATION.) The relationship between two given organisms may be expressed as the proportion or percentage of similarities or differences on the basis of a certain minimum number of specified criteria (see e.g. NUMERICAL TAXONOMY).

phenocopy A cell or organism whose phenotype has become modified in a way which mimics a phenotypic modification that is due to genetic change. For example, F^+ donors of *Escherichia coli* (see F PLASMID) may lose their donor characteristics under certain conditions – e.g. when grown to maximum density in aerated broth; such transfer-deficient cells mimic F^- cells and are therefore called F^- phenocopies. (See also SURFACE EXCLUSION.) When cultured in a fresh medium, F^- phenocopies revert to the donor state.

Studies on F^- phenocopies reveal e.g. a low level of nicking at *oriT* in the late stationary phase – although the F pilus transfer apparatus appears to be maintained in the cell envelope even after cessation of transcription of the main transfer region [Microbiology (1998) *144* 2579–2587].

phenogram Any dendrogram which expresses phenetic relationships. (cf. CLADOGRAM.)

phenol (as a disinfectant) See PHENOLS.

phenol coefficient A number which expresses the antimicrobial activity of a given *phenolic* DISINFECTANT relative to that of phenol under standard conditions; it can be determined by e.g. the RIDEAL–WALKER TEST or the CHICK–MARTIN TEST.

phenol red A PH INDICATOR: pH 6.8–8.4 (yellow to red); pK_a 7.9.

phenolates (in iron uptake) See SIDEROPHORES.

phenolic disinfectants See PHENOLS.

phenolphthalein A PH INDICATOR: pH 8.3–10.0 (colourless to red); pK_a 9.6.

phenols (as ANTISEPTICS and/or DISINFECTANTS) Phenol and its derivatives ('phenolics' – cresols, xylenols etc) are microbistatic or microbicidal, depending on e.g. concentration and temperature; they may act e.g. by affecting bacterial membrane potentials or membrane permeability, or by coagulating the cytoplasm. The antimicrobial activity of phenolics increases with boiling point, but so too does their insolubility in water and (in general) their susceptibility to inactivation by organic matter; high-boiling-point phenolics tend to be less toxic to tissues. Phenolic disinfectants typically have a high DILUTION COEFFICIENT, and they can be inactivated by non-ionic surfactants; those which are poorly soluble in water may be solubilized by alkali or prepared as suspensions with soaps.

Phenol (carbolic acid, C_6H_5OH) is soluble in water and (under appropriate conditions) may be bactericidal, fungicidal and virucidal; its activity against bacterial endospores and acid-fast bacteria tends to be very slow. It is toxic to tissues. Phenol (now produced mainly by chemical synthesis) is used e.g. as a preservative in animal glues.

In general, alkyl- and halogen-substituted phenolics exhibit enhanced antimicrobial activity, tend to be less toxic to tissues, and are typically less water-soluble than phenol. In 4-alkylphenols, activity increases with alkyl chain length up to 6 carbon atoms. Among halogenated phenols, 4-halophenols are the most active; in phenolics containing both alkyl and halogen substituents, activity is maximum when these are in the 2 and 4 positions, respectively. Nitrophenols tend to be more toxic than other phenolics to tissues. Polyhydric phenols generally have weak antimicrobial activity. (See also BISPHENOLS.)

Lysol consists of 2-, 3- and 4-cresols (methylphenols) solubilized by soap. Aqueous Lysol (0.5%) kills many non-sporing pathogens in 15 min, but endospores may survive in 2% Lysol for several days. *Amyl-3-cresol* is used e.g. in antiseptic lozenges and mouthwashes. *Sudol* consists of xylenols (dimethylphenols) and ethylphenols; it is more active and less corrosive than Lysol and is active against e.g. staphylococci and tubercle bacilli. Aqueous Sudol (2%) can inactivate *Clostridium perfringens* endospores in about 4 hours, but inactivation of *Bacillus subtilis* endospores has been found to require 6 hours in a 66% solution. *Hexylresorcinol* (4-*n*-hexyl-1,3-dihydroxybenzene) has been used as a skin wound cleanser, a urinary antiseptic, and a constituent of throat lozenges. *Menthol* (3-methyl-6-isopropylphenol) and *carvacrol* (2-methyl-5-isopropylphenol) are both used in antiseptic lozenges and mouthwashes. *Thymol* (5-methyl-2-isopropylphenol) is used as a disinfectant, as an antiseptic (e.g. in mouthwashes), and as a preservative (e.g. in urine samples pending biochemical tests). *2-Phenylphenol* (2-hydroxydiphenyl) is primarily an antifungal agent but also has antibacterial activity; it has been used e.g. as a fungicide in the paper industry, as a preservative in waxed papers, and as a constituent of PINE DISINFECTANTS. *Sodium-o-phenylphenate* is used e.g. for the initial antifungal disinfection of stored fruits. *2-Benzyl-4-chlorophenol* is active against Gram-positive and Gram-negative bacteria, protozoa and fungi; it is used e.g. in antiseptic soaps. *Chlorocresol* (4-chloro-3-methylphenol) is used e.g. as a preservative for pharmaceutical products and e.g. for glues and paints. It is also used as a biocide in a low-temperature (98–100°C) STERILIZATION process. *Chloroxylenols* (e.g. 4-chloro-3,5-dimethylphenol) are used e.g. as topical antiseptics; they are active ingredients in *Dettol*. *Pentachlorophenol* is used e.g. as a fungicide in wood pulp and as a preservative for leather, paper and textiles. *Picric acid* (2,4,6-trinitrophenol), which has antibacterial and antifungal properties, is also used as a chemical FIXATIVE. *4-Nitrophenol* is used as a preservative in the leather industry. (See also DINITROPHENOLS.)

(See also BLACK FLUIDS; CREOSOTE; SALICYLIC ACID; WHITE FLUIDS.)

phenon Any group established on the basis of PHENETIC CLASSIFICATION; in some cases a phenon may be considered to be more or less equivalent to a TAXON.

phenosafranine A green redox dye – see REDOX POTENTIAL.

phenotype The observable characteristics of an organism, either in total or with respect to one or more particular named characteristics. The phenotype of an organism is the manifestation of gene *expression* in that organism. (cf. GENOTYPE.)

phenotypic adaptation See ADAPTATION.

phenotypic lag A delay in the phenotypic expression of a new allele – acquired e.g. by mutation or by gene transfer (e.g. transduction) – due e.g. to SEGREGATION LAG.

phenotypic mixing In a cell simultaneously infected by genetically different viruses: the formation of progeny virions which either incorporate the genome of one virus within the capsid of another (*transcapsidation* or *genomic masking*) or contain structural components derived from different 'parent' viruses. (See also PSEUDOTYPE.)

phenotypic suppression The suppression of a mutant phenotype by non-genetic (environmental) factors. For example, during mRNA synthesis, 5-fluorouracil (5-FU) can be incorporated instead of uracil and can be misread as cytosine during subsequent translation; thus, e.g., the nonsense codon UAG may occasionally be misread as CAG (specifying glutamine), thus simulating a nonsense SUPPRESSOR MUTATION. The accuracy of translation may also be affected by alteration of ribosomes e.g. in the presence of STREPTOMYCIN; the resulting mis-reading of certain codons may lead to the suppression of certain mutations.

phenoxetol See ALCOHOLS.

phenoxyethanol See ALCOHOLS.

β-phenoxyethyldimethyldodecylammonium bromide Domiphen bromide (see QUATERNARY AMMONIUM COMPOUNDS).

phenoxyethylpenicillin (phenethicillin) A semisynthetic penicillin; R = $C_6H_5.O.CH(CH_3).CO$ (see β-LACTAM ANTIBIOTICS). Its activity is similar to that of benzylpenicillin, but it can be administered orally.

phenoxymethylpenicillin (penicillin V) One of the first natural PENICILLINS to be produced (from *P. chrysogenum*); in phenoxymethylpenicillin R = $C_6H_5.OCH_2.CO$ (see β-LACTAM ANTIBIOTICS). Its activity is similar to that of BENZYLPENICILLIN, but it can be administered orally.

phenylalanine agar A medium (usually prepared as a slope) containing (g/100 ml water): yeast extract (0.3), L-phenylalanine (0.1), Na_2HPO_4 (0.1), NaCl (0.5), agar (1.2).

L-phenylalanine biosynthesis See AROMATIC AMINO ACID BIOSYNTHESIS.

phenylalanine deaminase test A test used to determine the ability of an organism to deaminate phenylalanine to phenylpyruvic acid. In one form of the test, the organism is grown overnight on PHENYLALANINE AGAR, and 0.2 ml of 10% aqueous $FeCl_3$ is added to the growth; phenylpyruvic acid, if present, gives a green coloration with the $FeCl_3$. Bacteria which usually give a positive test include e.g. *Proteus vulgaris* and *Moraxella phenylpyruvica*.

phenylamide antifungal agents A group of ANTIFUNGAL AGENTS which includes the acylalanines *benalaxyl*, *furalaxyl* and *metalaxyl*, the butyrolactones *milfuram* and *cyprofuram*, and the oxazolidinone ring-containing *oxadixyl*. The phenylamides show in vitro activity against members of the Peronosporales and certain other fungi [Mycol. (1985) *77* 424–432]. Metalaxyl (trade name: e.g. Ridomil) is widely used as an agricultural antifungal agent against a range of peronosporalean plant pathogens; it is a systemic fungicide which is most effective in young, actively growing foliage, and is often mixed with a contact fungicide such as copper, captan or mancozeb for maximum efficacy.

phenylethanol See ALCOHOLS.

phenylmercuric salts See MERCURY.

phenylmethanol See ALCOHOLS.

phenylpenicillin A semi-synthetic penicillin; R = $C_6H_5.CO–$ (see β-LACTAM ANTIBIOTICS). It is less active than benzylpenicillin.

2-phenylphenol See PHENOLS.

pheromone A type of molecule, released by an organism, which can cause one or more (non-harmful, non-nutritional) effects in another individual – typically an individual of the same species. (cf. ANTIBIOTIC; HORMONE). For example, a 'sex pheromone' can encourage sexual interaction between individuals e.g. by acting as a chemoattractant (see CHEMOTAXIS), by causing a TROPISM, and/or by promoting cell–cell adhesion (see below).

Bacterial pheromones. Pheromones (small linear peptides) are secreted e.g. by those cells of *Enterococcus faecalis* that are potential recipients for certain plasmids; a given pheromone can act on other (potential donor) cells of *E. faecalis* that are carrying a particular CONJUGATIVE PLASMID – causing them to form a cell-surface adhesin called 'aggregation substance' which promotes donor–recipient adhesion. (Because culture filtrates of recipients also cause donor–donor aggregation, these pheromones have been called 'clumping-inducing agents'.) On receipt of a plasmid (by CONJUGATION), a recipient stops secreting the corresponding pheromone (but may continue to secrete the pheromones corresponding to other plasmids). Each type of *E. faecalis* pheromone is designated according to the particular plasmid to which its function was originally linked; thus, e.g. pheromones cAD1 and cAM373 act on (donor) cells carrying the plasmids pAD1 and pAM373 respectively. [Review: CRM (1986) *13* 309–334.]

In *Bacillus subtilis*, pheromones are involved in the development of competence for transformation and in the initiation of endospore formation (see QUORUM SENSING).

Fungal pheromones. Many fungi form sex pheromones. For example, in *Allomyces* spp, female gametangia and gametes release the pheromone *sirenin* (a sesquiterpene diol); male gametes take up and inactivate sirenin, and move chemotactically towards its source.

In dioecious species of *Achlya*, the vegetative female thallus forms a steroid pheromone (*antheridiol*, 'hormone A') which stimulates the male thallus to form antheridial hyphae; the male thallus takes up, and inactivates, the pheromone and forms another steroid pheromone ('hormone B') which stimulates the female thallus to form oogonia.

In some bipolarly heterothallic zygomycetes (e.g. strains of *Blakeslea trispora* and *Mucor mucedo*), development of sexual structures in the 'plus' and 'minus' mating types (see HETEROTHALLISM) is governed by *trisporic acid* derivatives (formerly referred to as GAMONE) – terpene carboxylic acids synthesized from β-carotene. The plus strains synthesize and secrete the pheromone *prohormone* P^+ ('plus-progamone'), a methyldihydrotrisporate; minus strains synthesize and secrete a trisporol pheromone called *prohormone* P^- (formerly 'minus-progamone'). It appears that when each pheromone is taken up by strains of the opposite mating type it is converted to trisporic acid in situ. The trisporic acid promotes sexual development and increases prohormone synthesis.

Pheromones produced by *Saccharomyces cerevisiae* include the α-factor and the **a**-factor (see MATING TYPE). The α-factor is a tridecapeptide (13 amino acid residues) synthesized as a high-MWt precursor (which, interestingly, resembles the hormone precursors in higher eukaryotes); the precursor is sequentially cleaved by three proteinases: yscF, yscIV (formerly dipeptidyl-aminopeptidase A, or X-prolyldipeptidylaminopeptidase), and (apparently) yscα. The **a**-factor, an undecapeptide (11 residues), is apparently also synthesized in precursor form. [MS (1986) *3* 107–114.]

Algal pheromones. Pheromones are produced by various algae. In some cases a given pheromone can act on algae of different genera; for example, among brown algae, the pheromone *ectocarpene* (a cyclic hydrocarbon) can act on *Ectocarpus* and *Sphacelaria*, while *multifidene* is effective in *Chorda* and *Cutleria*. The pheromone *desmarestene* (structurally similar to ectocarpene) acts as an attractant to male gametes in the brown alga *Cladostephus* [Naturwissenschaften (1986) *73* 99–100].

ϕ Unwinding angle: see INTERCALATING AGENT.

ϕ6 phage group *Syn.* CYSTOVIRIDAE.

phialide (phialid) See CONIDIUM.

phialidic development See CONIDIUM.

Phialoascus See ENDOMYCETALES.

Phialophora A genus of fungi (class HYPHOMYCETES) which include plant-pathogenic species – e.g. *P. cinerescens* (causal agent of carnation wilt). (See also CHROMOBLASTOMYCOSIS.) [Species parasitic on bdelloid rotifers: Mycol. (1984) *76* 1107–1110.] The mycelium is septate, and the conidia (produced from phialides) are non-septate, colourless or pigmented. (For *P. dermatitidis* see WANGIELLA.)

phialopore See VOLVOX.

Phialospora See DOTHIDEALES.

Philadelphia chromosome An abnormal human chromosome 22 associated with chronic myelogenous leukaemia: see ABL.

Philaster See SCUTICOCILIATIDA.

Philomiragia bacterium A non-motile bacterium pathogenic in muskrats. It was formerly called *Yersinia philomiragia*, but has been excluded from *Yersinia* on the basis of DNA homology studies [Book ref. 22, p. 506].

pHisoHex See BISPHENOLS.

ϕX phage group *Syn.* MICROVIRIDAE.

Phlebia See APHYLLOPHORALES (Corticiaceae).

Phlebotomus A genus of blood-sucking sandflies which includes the vectors of certain diseases.

Phlebovirus (sandfly fever group) A genus of viruses of the BUNYAVIRIDAE. Host range: various vertebrates; vectors: primarily phlebotomine sandflies. The genus appears to include only one serological group with >30 viruses (including e.g. Punta Toro virus, RIFT VALLEY FEVER virus, Rio Grande virus, Sicilian sandfly fever virus). MWts of L, M and S RNAs: ca. 2.6–2.8, 1.8–2.2, and 0.7–0.8 × 10^6, respectively; MWts of proteins L, G1, G2 and N: ca. 145–200, 55–70, 50–60, and 20–30 × 10^3, respectively. The phlebovirus M RNA appears to resemble that of the bunyaviruses (see BUNYAVIRIDAE) in its genetic organization; however, the S RNA of Punta Toro virus appears to be an AMBISENSE RNA, encoding the NS_s protein on a viral-sense subgenomic mRNA and the N protein on a viral-complementary subgenomic mRNA.

Phleogena See AURICULARIALES.

phlobaphenes See TANNINS.

phloem necrosis (of coffee) A disease of the coffee plant (e.g. *Coffea liberica*, *C. arabica*) caused by *Phytomonas* sp. Symptoms: leaves become yellow and fall prematurely, and CALLOSE is deposited in the sieve tubes. Only plants older than ca. two years are affected. Vectors: unknown, but heteropterans are suspected.

phloeophagous 'Bark-eating' (see e.g. AMBROSIA FUNGI).

Phlogiotis See TREMELLALES.

phlogistic Able to cause inflammation.

phloretin See PHLORIDZIN.

phlorhizin *Syn.* PHLORIDZIN.

phloridzin (phlorhizin) A phenolic glycoside formed in certain tissues in apple trees (*Malus* spp) and in other members of the Rosaceae. The aglycone part of phloridzin, *phloretin* (β-(4-hydroxyphenyl)-propionophlorophenone), can be liberated by β-glycosidases and oxidized by polyphenol oxidase to yield polymers that are toxic to e.g. *Venturia inaequalis*.

PHLS Public Health Laboratory Service, 61 Colindale Avenue, London NW9, United Kingdom.

***pho* regulon** (phosphate regulon) In *Escherichia coli*: a REGULON comprising the *pho* genes: phosphate-controlled genes whose activity depends on the level of available Pi (inorganic orthophosphate). The cell's response to Pi starvation is complex and not fully understood, but the following model has been suggested for regulation of the *pho* genes.

Regulation is exerted partly through the products of genes *phoR* and *phoB*. PhoR and PhoB are associated in a TWO-COMPONENT REGULATORY SYSTEM in which PhoR is a sensor protein in the cytoplasmic membrane, and PhoB acts as a regulator protein. With low levels of phosphate, PhoR undergoes ATP-dependent autophosphorylation and transfers phosphate to PhoB. PhoB∼P behaves as a transcriptional activator by binding to the so-called *phosphate box*: a nucleotide sequence associated with the promoters of e.g. the *phoA*, *phoE* and *phoS* genes; PhoA is a periplasmic alkaline PHOSPHATASE (which releases Pi from organic phosphate esters), PhoE is a PORIN (for phosphate uptake), and PhoS is a Pi-binding protein. If phosphate levels rise, PhoB∼P is de-phosphorylated by the joint activity of PhoR and the product of gene *phoU*; this switches off the *pho* regulon.

Regulation of *pho* genes appears also to involve element(s) of the *pst* (phosphate-specific transport) operon because phosphatase (PhoA) is produced constitutively in *pst* mutants.

***phoA* gene** See PHO REGULON.

***phoB* gene** See PHO REGULON.

phobic response (shock movement; shock reaction; avoiding reaction) A transient behavioural response of a motile organism to a noxious stimulus, where the stimulus is an abrupt change in an external parameter (e.g., a light/dark boundary). The nature of the response depends on the organism, but typically the organism stops and then either reverses or reorientates and proceeds in a new, randomly selected direction. If conditions in the new direction are not significantly better, the phobic response may be repeated until the organism eventually moves into a more favourable region – when smooth swimming tends to supervene; thus, repeated phobic responses may result in a net progress in the more beneficial direction ('phobotaxis').

phobotaxis See PHOBIC RESPONSE.

***phoE* gene** See PHO REGULON and PORIN.

Pholiota See AGARICALES (Strophariaceae).

Phoma A genus of fungi (order SPHAEROPSIDALES) which include many plant pathogens (see e.g. BLACKLEG sense 2, and GANGRENE sense 2). Colourless, ovoid conidia develop singly on conidiophores borne in brown, spherical or spheroidal, thin-walled, papillate, uniloculate, ostiolate pycnidia immersed in the substratum.

Phomopsis See SPHAEROPSIDALES.

***phoR* gene** See PHO REGULON.

phorbol ester 12-*O*-tetradecanoylphorbol-13-acetate. (See also ZEBRA.)

-phore A suffix which signifies 'carrier of' or 'bearer of' – e.g. SPOROPHORE, IONOPHORE.

Phormidium A genus of cyanobacteria of the LPP GROUP. Species occur e.g. with other filamentous cyanobacteria (e.g. *Lyngbya*, *Oscillatoria*) in mats on sediments in marine lagoons, saltmarshes etc. (See also STROMATOLITE and EMULCYAN.)

phoront See TOMITE.

phorophyte The host plant of an epiphyte.

***phoS* gene** See PHO REGULON.

phosphatase (orthophosphoric monoester phosphohydrolase) Any enzyme which hydrolyses esters of phosphoric acid; phosphatases are classified (according to their pH optima) as *acid phosphatases* (EC 3.1.3.2) or *alkaline phosphatases* (EC 3.1.3.1). (See also ENZYMES.) They are widely distributed among living organisms, including many bacteria.

Staphylococcal phosphatases. An *acid phosphatase* (pH optimum 5.2) is produced by most or all coagulase-positive staphylococci and (generally in smaller amounts) by some coagulase-negative strains. The enzyme may be secreted into the medium or may be loosely or tightly cell-bound – the proportions of each depending on strain and medium composition. An *alkaline phosphatase* (pH optimum 10.8) is apparently cell-bound. Phosphatase production may be demonstrated by growing the test organism on a solid medium containing the sodium salt of phenolphthalein diphosphate. After incubation, the plate is exposed to gaseous ammonia; colonies of phosphatase-producing strains become deep pink owing to the presence of free PHENOLPHTHALEIN. (Coagulase-negative strains which initially appear phosphatase-negative may produce phosphatase on prolonged incubation.) Phosphatase may be assayed e.g. using the artificial substrate *p*-nitrophenyl phosphate; *p*-nitrophenol (yellow at alkaline pH) is released by acid or alkaline phosphatases and can be assayed spectrophotometrically (at 400 nm). (See also XP.)

Staphylococcal phosphatase production appears to have little or no significance in pathogenicity. [Staphylococcal phosphatases: Book ref. 44, pp. 775–780.]

In e.g. *Escherichia coli* a periplasmic alkaline phosphatase is produced in response to inorganic phosphate starvation: see PHO REGULON.

phosphatase test (for milk) A test used to detect the presence of alkaline PHOSPHATASE in *pasteurized* MILK; the enzyme, which is normally present in raw (i.e., untreated) milk, is inactivated by PASTEURIZATION – so that the test determines the efficiency of the pasteurization process. Milk containing alkaline-buffered disodium *p*-nitrophenyl phosphate is incubated at 37°C for 2 hours and then examined (e.g. by colorimetry) for the presence or intensity of yellow coloration due to *p*-nitrophenol. A false-positive result can be obtained if, subsequent to pasteurization, the milk has been contaminated with phosphatase-producing organisms.

Enzymic activity may reappear ('reactivation') in e.g. cream which has been treated by HTST pasteurization; reactivation can be promoted by the addition of $MgCl_2$.

phosphate box See PHO REGULON.

phosphate-controlled genes See PHO REGULON.

phosphate potential See CHEMIOSMOSIS.

phosphate regulon See PHO REGULON.

phosphatidases *Syn.* PHOSPHOLIPASES.

phosphatidic acid See LECITHIN.

phosphatidylcholine See LECITHIN.

phosphatidylethanolamine A product of esterification of ethanolamine ($NH_2(CH_2)_2OH$) with phosphatidic acid (see LECITHIN); it occurs e.g. in the CYTOPLASMIC MEMBRANE in many bacteria.

phosphoadenosine phosphosulphate See PAPS.

phosphoenolpyruvate carboxykinase See Appendix II(b).

phosphoenolpyruvate carboxylase See Appendix II(b).

phosphoenolpyruvate carboxytransphosphorylase See Appendix II(b).

phosphoenolpyruvate-dependent phosphotransferase system See PTS.

phosphoenolpyruvate synthase See Appendix II(b).

phosphofructokinase See EMBDEN–MEYERHOF–PARNAS PATHWAY.

6-phosphogluconate pathway May refer to either HETEROLACTIC FERMENTATION or the HEXOSE MONOPHOSPHATE PATHWAY.

phosphoketolase See HETEROLACTIC FERMENTATION.

phospholipases (phosphatidases) Enzymes that hydrolyse phospholipids, releasing fatty acids or other groups (cf. LIPASES). Phospholipases A_1 and A_2 release fatty acids from, respectively, positions 1 and 2 of a phospholipid; phospholipases C and D release the base – in phosphorylated and unphosphorylated form, respectively – from e.g. phosphatidylcholines (lecithins), sphingomyelins etc. Phospholipases may act specifically (or preferentially) on particular phospholipids – see e.g. LECITHINASE and SPHINGOMYELINASE.

phosphonoacetic acid (PAA) An ANTIVIRAL AGENT (used as disodium phosphonoacetate, = Fosfonet sodium) which is similar to, but more toxic than, PHOSPHONOFORMIC ACID.

phosphonoformic acid (PFA) An ANTIVIRAL AGENT (used as trisodium phosphonoformate, NaO.CO.PO.$(ONa)_2$, = Foscarnet sodium) which acts by selectively inhibiting the DNA polymerases of a number of viruses (and the reverse transcriptase of retroviruses) – apparently by acting as an analogue of pyrophosphate. It is effective e.g. in the topical treatment of herpesvirus infections.

phosphonomycin *Syn.* FOSFOMYCIN.

phosphorelay See ENDOSPORE (sense 1).

phosphorescence (1) The emission of light which occurs when certain substances absorb radiation; emission occurs as electrons excited to the triplet state (T_1) are returning to the (unexcited) ground state (S_0). The light emitted is of wavelength longer than that of the exciting radiation; emission typically lasts for 10^{-6} sec to 1 sec or longer. (cf. FLUORESCENCE.) (2) The term is sometimes applied, incorrectly, to BIOLUMINESCENCE.

phosphoribulokinase See CALVIN CYCLE.

phosphoroclastic split A reaction, analogous to hydrolysis, in which a molecule is cleaved with the addition of the components of phosphoric acid. The term is usually applied to the 'pyruvic phosphoroclasm' – the reaction pyruvate + $H_3PO_4 \rightarrow$ acetyl phosphate + formate (or CO_2 and H_2), a reaction which occurs e.g. in the MIXED ACID FERMENTATION and in the BUTYRIC ACID FERMENTATION, and which is also carried out by many SULPHATE-REDUCING BACTERIA (including species of *Desulfobacter*, *Desulfotomaculum* and *Desulfovibrio*). However, the initial product of pyruvate cleavage is actually acetyl-CoA, acetyl phosphate being formed from acetyl-CoA by the action of phosphotransacetylase; acetyl phosphate is subsequently dephosphorylated in the presence of ADP to yield acetate and ATP (enzyme: acetokinase).

phosphorylation potential See CHEMIOSMOSIS.

phosphotransferase transport system See PTS.

photic zone That zone of an aquatic habitat which is penetrable by sunlight and in which PHOTOSYNTHESIS can occur. In the upper part of the zone (*euphotic zone*) the amount of oxygen produced by photosynthesis exceeds that consumed by respiration (over a 24-hour period); the opposite situation occurs in the lower part of the zone (*dysphotic zone*), while between these zones (at the *compensation level*) photosynthetic oxygen production and respiratory oxygen consumption are balanced. (cf. APHOTIC ZONE.)

photoautotroph A phototrophic AUTOTROPH.

Photobacterium A genus of Gram-negative bacteria of the family VIBRIONACEAE. Cells: straight rods, 0.8–1.3 × 1.8–2.4 μm. Motile, with one to three unsheathed polar flagella. All species grow at 20°C but not at 40°C; *P. phosphoreum* and some strains of *P. angustum* can grow at 4°C. Growth requires Na^+ (opt. ca. 3% NaCl). Many strains are oxidase −ve. PHB accumulates in cells under certain conditions, but exogenous β-hydroxybutyrate is not utilized. Strains are sensitive to O/129. *P. phosphoreum* and *P. leiognathi* (formerly *P. mandapamensis*) exhibit blue-green BIOLUMINESCENCE and occur in seawater, on and in marine animals (see also FISH SPOILAGE), and in specialized luminous organs in certain marine fish. (See also SUPEROXIDE DISMUTASE.) GC%: 40–44. Type species: *P. phosphoreum*.

It has been suggested that the genus should include *P. leiognathi* and *P. phosphoreum*, and that *Vibrio fischeri* and *V. logei* should be transferred from the genus VIBRIO (q.v.) as *P. fischeri* and *P. logei*, respectively; it was also suggested that *P. angustum* is a biovar of *P. leiognathi* [SAAM (1985) *6* 171–182].

photobiont A photosynthetic symbiont. (See also PHYCOBIONT.)

photochromogen Any strain of MYCOBACTERIUM in which a period of exposure to light is necessary for the development of pigmentation. (cf. SCOTOCHROMOGEN; NON-PHOTOCHROMOGEN.)

Photocyta (photocytes) A (suggested) URKINGDOM (sense 2) comprising the halobacteria (e.g. *Halobacterium* spp) and the EUBACTERIA, the proposed relatedness of these organisms being based on three-dimensional RIBOSOME structure [PNAS (1985) *82* 3716–3720]; it has been proposed that the photocytes, together with the methanogens, gave rise to the Eocyta (see ARCHAEBACTERIA) and the eukaryotes [Nature (1986) *319* 626].

photocytes See PHOTOCYTA.

photodissociation Light-induced chemical dissociation (see e.g. CARBON MONOXIDE (b)).

photodynamic effect Light-induced damage or death which occurs when some microorganisms (particularly bacteria) are stained with certain fluorescent dyes (e.g. ACRIDINE ORANGE, EOSIN, ROSE BENGAL) and subsequently exposed to strong light under aerobic conditions; this effect appears to be due largely to the formation of SINGLET OXYGEN [JAB (1985) *58* 391–400]. ETHIDIUM BROMIDE can cause a light-induced nicking of DNA.

photoheterotroph A phototrophic HETEROTROPH.

photoinduction and photoinhibition (in fungi) In many fungi, light is needed for, or encourages, the development of fruiting bodies or spores, or the formation of particular compounds (*photoinduction*); in other fungi light inhibits growth and/or sporulation (*photoinhibition*). Commonly, light in the blue-violet region of the spectrum is effective in photoinduction/photoinhibition.

In some cases light may be needed only for the completion of a particular phase of development, e.g., the initiation of fruiting in *Pyronema omphalodes*, or the development of the pileus in *Lentinus lepideus* (cf. STAG'S HORN FUNGUS). Conidium formation is photoinducible in e.g. *Trichoderma viride*, and in *Choanephora cucurbitarum* it is stimulated by darkness *preceded* by a period of exposure to light. In some species of *Coprinus* fruiting bodies develop more rapidly in the light, i.e., light encourages fruiting but is not essential for it. In *Phycomyces blakesleeanus* carotenoid synthesis is stimulated by light, while some species of *Fusarium* and *Verticillium* which normally form negligible amounts of carotenoids form significant amounts of these compounds in the presence of light.

Photoinhibition of growth and/or sporulation occurs e.g. in at least some species of *Alternaria* and *Penicillium*. In e.g. *Alternaria cichorii*, *A. tomato* and *Botrytis cinerea*, conidium formation is controlled by a photoreceptor system referred to as *mycochrome*. On exposure to blue light mycochrome adopts the M_{NUV} form, in the presence of which conidia are not formed and conidiophores may revert to vegetative hyphae; on exposure to near-ultraviolet mycochrome adopts the M_B form, in the presence of which conidia can be formed. In *A. cichorii* the blue-light inhibition of conidium formation can be reversed by a period of darkness [CJB (1986) *64* 1016–1017].

photoinhibition See PHOTOINDUCTION AND PHOTOINHIBITION.

photokinesis A KINESIS (sense 1 or 2) in which the stimulus is light intensity.

photolithotroph See LITHOTROPH.

photolyase In some bacteria (including *Escherichia coli*): an enzyme which can repair thymine dimers in DNA damaged by ULTRAVIOLET RADIATION; the energy needed for repair is derived from (visible) light. In vitro studies on the repair of thymine dimers may be facilitated by using the reagent potassium permanganate ($KMnO_4$) [NAR (1998) *26* 3940–3943].

photomicrograph A photograph obtained by PHOTOMICROGRAPHY.

photomicrography Photography of microscopic objects through a microscope; it may involve the attachment of a camera to a microscope, or the use of a *photomicroscope*: a microscope which incorporates a camera. (cf. MICROPHOTOGRAPHY.)

photomicroscope See PHOTOMICROGRAPHY.

photomorphogenesis The formation of new cells or tissues under the stimulus of light.

photoorganotroph See ORGANOTROPH.

photophobic response A light-induced PHOBIC RESPONSE.

photophosphorylation See PHOTOSYNTHESIS and PURPLE MEMBRANE.

photoreactivation (photorestoration) A mechanism of DNA REPAIR in which pyrimidine dimers (formed as a result of exposure to ULTRAVIOLET RADIATION) are cleaved by exposure of the cells to light of wavelength ca. 300–600 nm; in *Escherichia coli* the reaction is mediated by the product of the *phr* gene (*DNA photolyase*) which, when bound to a pyrimidine dimer, absorbs light energy and cleaves the bonds between the bases ('monomerization'). [Action mechanism of *E. coli* DNA photolyase: JBC (1987) *262* 478–485, 486–491, 492–498.] The *phr* gene product may also play a role in UvrABC-dependent repair of UV damage in the dark [JB (1985) *161* 602–608]. Photoreactivation does not occur e.g. in *Bacillus subtilis*.

photoreceptor An organelle or region specialized for receiving light stimuli: see e.g. EYESPOT (sense 1), STENTORIN and STENTOR.

photoreduction In photosynthetic organisms (see PHOTOSYNTHESIS): any light-dependent reduction – e.g., the reduction of NAD^+, or of CO_2 via the Calvin cycle; electron donors used in photoreductions include e.g. water and H_2S.

In certain algae (e.g. *Chlorella, Chlamydomonas, Scenedesmus*) hydrogen can act as electron donor for the reduction of CO_2; HYDROGENASE is involved, and the process can occur only after the algae have been incubated anaerobically with hydrogen, followed by illumination with low levels of light.

photorespiration (C_2 carbon oxidation cycle) In photosynthetic organisms which fix CO_2 via the CALVIN CYCLE: a light-dependent process in which oxygen is consumed and CO_2 is evolved, but which is distinct from true (energy-yielding) RESPIRATION; photorespiration occurs under conditions of relatively high O_2 and low CO_2 concentrations and high light intensities, and results from the ability of RIBULOSE 1,5-BISPHOSPHATE CARBOXYLASE–OXYGENASE (RuBisCO) to act as an oxygenase under these conditions. RuBisCO catalyses the oxygenation and cleavage of ribulose 1,5-bisphosphate (RuBP) to 3-phosphoglycerate and 2-phosphoglycolate. (The dependence of photorespiration on light is thus explained by the need for photosynthesis to drive the CALVIN CYCLE and hence supply RuBP.) The 2-phosphoglycolate is dephosphorylated to glycolate which may be excreted – e.g. by certain microalgae and cyanobacteria (often resulting in relatively high levels of glycolate in aquatic environments) – and/or may be metabolized further: e.g. glycolate → glyoxylate → glycine → serine → hydroxypyruvate → glycerate → 3-phosphoglycerate (the 'glycolate pathway'). The step glycolate → glyoxylate is also an O_2-consuming step (catalysed by glycolate oxidase); H_2O_2 generated during the reaction is rapidly degraded by catalase. (The glycolate → glyoxylate conversion occurs in PEROXISOMES in photosynthetic eukaryotes.) The 3-phosphoglycerate generated by the glycolate pathway and by the original cleavage of RuBP by RuBisCO may then enter the Calvin cycle. An alternative pathway for glycolate metabolism in some organisms involves the oxidation of glycolate to glyoxylate which is then converted to glycerate via tartronic semialdehyde (TSA). In either pathway, CO_2 is lost: in the glycolate pathway during the glycine → serine reaction (during which two glycine molecules are converted to one of serine with loss of both CO_2 and NH_3), and in the TSA pathway during the formation of TSA from glyoxylate (2CHO.COOH → CHO.CHOH.COOH + CO_2). Photorespiration thus results in a net loss of photosynthetic productivity. Its physiological significance is still not understood. In green algae, photorespiration is suppressed by CO_2-concentrating mechanisms which become operative at low CO_2 levels. [Review of photorespiration: ARPphys (1984) *35* 415–442.] (cf. WARBURG EFFECT.)

A process very similar to photorespiration (although independent of light) occurs in some chemolithoautotrophic bacteria (e.g. *'Alcaligenes eutrophus', Thiobacillus neapolitanus*) which fix CO_2 via the Calvin cycle; by analogy with photorespiration, this process has been called 'chemorespiration'. [Book ref. 115, pp. 129–173.]

photorestoration *Syn.* PHOTOREACTIVATION.

photosynthate Any carbohydrate product of photosynthesis.

photosynthesis In certain (photosynthetic) organisms: the process in which radiant energy is absorbed by specialized CHLOROPHYLL-containing pigment system(s) and is converted to forms of energy (including chemical energy) which can be used for metabolic and other purposes (cf. PURPLE MEMBRANE); radiation of wavelengths in the visible spectrum is used by plants, ALGAE and photosynthetic bacteria (including CYANOBACTERIA and members of the RHODOSPIRILLALES), while some photosynthetic bacteria can also use infrared radiation. *Light reaction* refers, collectively, to those photochemical events involved in the conversion of radiant energy; *dark reaction* (= *light-independent reaction*) generally refers, collectively, to those reactions in which photosynthetically derived chemical energy is used for the synthesis of carbohydrate(s).

In all cases, the conversion of radiant energy occurs in a specialized ENERGY-TRANSDUCING MEMBRANE containing CHLOROPHYLLS or bacteriochlorophylls, accessory pigments, and components of ELECTRON TRANSPORT CHAIN(S). Within such a photosynthetic membrane there are well defined chlorophyll-containing REACTION CENTRES towards which radiant energy is channelled via LIGHT-HARVESTING COMPLEXES. On excitation of a reaction centre with radiant energy, the mid-point REDOX POTENTIAL of the reaction centre chlorophyll changes from a high positive value to a high negative value, and electrons are consequently ejected ('charge separation') from the chlorophyll (the 'primary electron donor'). The ejected electrons, by virtue of the high negative redox potential of their source, are associated with a certain amount of energy; these electrons can therefore pass down an electron transport chain in the direction of less negative (or more positive) potentials and, in doing so, can provide the energy for (a) the transmembrane pumping of protons (thereby contributing to a *proton motive force* – see CHEMIOSMOSIS) or (b) the reduction of pyridine nucleotides. The precise nature of photosynthetic energy conversion differs among the different groups of photosynthetic organisms.

(a) *Photosynthesis in bacteria of the Rhodospirillales* (anoxygenic photosynthesis) occurs only under anaerobic conditions, and only one type of reaction centre appears to occur in a given species.

In the 'purple' photosynthetic bacteria (RHODOSPIRILLINEAE), electrons which have been ejected from the bacteriochlorophyll follow a *cyclic* path: e.g. from the bacteriochlorophyll to BACTERIOPHAEOPHYTIN and thence – probably via QUINONES, *b*-and *c*-type cytochrome(s) – back to the reaction centre; this cyclic electron flow causes the transmembrane pumping of protons, i.e., it generates pmf. The pmf can be used, by means of membrane-bound PROTON ATPASES, for the synthesis of ATP (*photophosphorylation*). At least some of these organisms can obtain reducing power by REVERSE ELECTRON TRANSPORT; in this process, photosynthetically generated pmf is used to reduce NAD^+ at a membrane-bound NADH dehydrogenase – electrons being obtained from an external electron donor (such as hydrogen or sulphide). NADPH is obtained via a TRANSHYDROGENASE.

In the 'green' photosynthetic bacteria (CHLOROBIINEAE), the light-energized reaction centre reaches a much lower (more negative) potential (ca. −550 mV) than is the case in the purple bacteria, and – consequently – the ejected electrons are able to effect the reduction of NAD^+ (the $E_{m,7}$ of the $NAD^+/NADH$ redox couple being −320 mV); the reduction of NAD^+ involves a *non-cyclic* electron transport path which is believed to include iron–sulphur proteins. Electrons which are removed from the

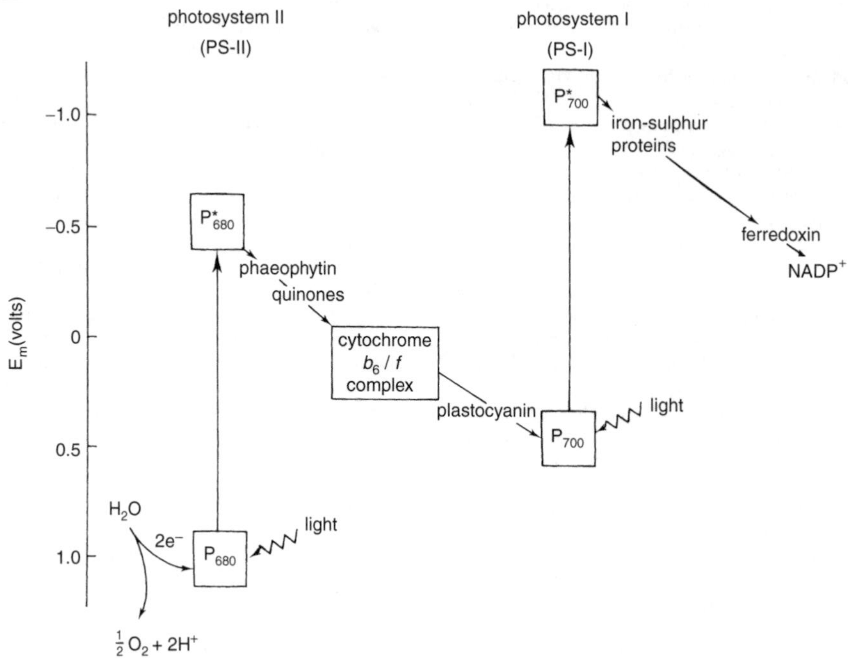

PHOTOSYNTHESIS. Simplified 'Z scheme' for non-cyclic, oxygenic photosynthetic electron transport (scale approximate). Each photosystem (PS-I and PS-II) consists of a reaction centre together with its associated accessory pigments. P_{680} and P_{700} represent the chlorophylls in the reaction centres of PS-II and PS-I, respectively; P^*_{680} and P^*_{700} represent the excited state in P_{680} and P_{700}, respectively. The cytochrome b_6/f complex includes cyts b_6 and *f*, quinone, and a Rieske iron–sulphur protein.

reaction centre during non-cyclic electron flow must be replaced if photosynthesis is to continue; these electrons are obtained (via a pathway containing cytochrome c_{555}) from certain substrates (e.g. sulphide). Water is never used as an electron donor, so that this type of photosynthesis is always anoxygenic.

The green photosynthetic bacteria appear to be also capable of *cyclic* elctron flow – electrons returning to the reaction centre via a menaquinone and cytochrome c_{555}.

In many photosynthetic bacteria the dark reaction involves the fixation of carbon dioxide via the CALVIN CYCLE; however, at least some of the green photosynthetic bacteria use the REDUCTIVE TRICARBOXYLIC ACID CYCLE for this purpose. Apart from photochemical reactions, most of the reactions which occur in photosynthetic bacteria are known to occur in non-photosynthetic species; thus, e.g. the Calvin cycle occurs in chemolithotrophic bacteria.

(b) *Photosynthesis in algae and cyanobacteria.* Photosynthesis in these organisms differs from that described in (a) in that: (i) it occurs *aerobically*; (ii) it involves a composite photosynthetic system (as shown in the figure); (iii) it involves non-cyclic electron flow in which the terminal electron donor, water, is oxidized to oxygen – such photosynthesis thus being *oxygenic*. [How does PSII split water? (review) TIBS (1996) *21* 44–49.] (Some species of cyanobacteria, e.g. *Oscillatoria limnetica*, can also carry out DCMU-insensitive, anaerobic, anoxygenic photosynthesis in which H_2S is used as external electron donor.) (See also PHOTOREDUCTION.) Oxygenic photosynthesis in algae and cyanobacteria (and *Prochloron*) closely resembles that in higher green plants.

The characteristic (oxygenic, non-cyclic) mode of photosynthesis in algae and cyanobacteria is generally depicted by the so-called *Z scheme* (see figure). In this scheme, the midpoint redox potential of chlorophyll in the reaction centre of photosystem II (PS-II) changes from ca. $+1.0$ volt to between ca. -550 and -600 mV when the reaction centre receives a photon of light; electrons which are ejected from PS-II (and which are replaced by the oxidation of water) flow 'downhill' (i.e., towards less negative or more positive potentials) via QUINONES (generally believed to include plastoquinones), the CYTOCHROME b_6/f complex, and PLASTOCYANIN, to the reaction centre of photosystem I (PS-I). This electron flow is associated with the development of a proton gradient across the thylakoid membrane. In chloroplasts the inner surface of the thylakoid membrane becomes positive with respect to the outer (stromal) surface; the associated pmf can be utilized by membrane-bound PROTON ATPASES for ATP synthesis (photophosphorylation) – the CF_1 part of the ATPase, and (hence) ATP synthesis, occurring at the outer surface of the thyalkoid membrane.

On excitation of PS-I by light, the midpoint redox potential of the reaction centre chlorophyll changes to a value which exceeds ca. -1.0 volt; electrons ejected from the reaction centre of PS-I can therefore flow downhill to the ferredoxin-NADP$^+$ reductase complex and can be used for the reduction of NADP$^+$ to NADPH. A revised eukaryotic Z scheme includes the reduction of NADP by PSII [TIBS (1996) *21* 121–122].

[Excited-state redox potentials in the Z scheme: TIBS (1985) *10* 382–383.]

In addition to non-cyclic electron flow, a pmf-generating cyclic flow of electrons can occur via PS-I and cytochromes b_6 and *f*. There is also evidence for a cyclic flow of electrons around PS-II, possibly via cyt b_{559}. [Photosynthetic electron transfer: Book ref. 85, pp. 95–148.]

In algae and cyanobacteria the dark reaction characteristically involves the CALVIN CYCLE.

Inhibitors of the photosynthetic electron transport chain include e.g. dibromomethyl-isopropyl *p*-benzoquinone (DBMIB) and DCMU, both of which inhibit electron flow between PS-II and PS-I, and hydroxylamine, which blocks electron flow from water to the reaction centre of PS-II.

Experimental electron donors to PS-II include e.g. benzidine and catechol, while electron acceptors for PS-II include e.g. ferricyanide; electron donors for PS-I include e.g. reduced phenylenediamines, while electron acceptors for PS-I include e.g. ferricyanide and METHYL VIOLOGEN.

[Molecular biology of the photosynthetic apparatus: Book ref. 163.]

See also CAROTENOID BAND SHIFT; ENHANCEMENT EFFECT; HILL REACTION; PHOTORESPIRATION; RED DROP; WARBURG EFFECT.)

photosynthetic bacteria A term which usually refers to bacteria of the RHODOSPIRILLALES (cf. CYANOBACTERIA).

photosystems I and II See PHOTOSYNTHESIS.

phototaxis A TAXIS in which an organism moves, directly or indirectly, from a region of low light intensity to one of higher light intensity (positive phototaxis) or vice versa (negative phototaxis). (A phototactic organism may be positively phototactic at low light intensities but negatively phototactic at high light intensities.)

In phototactic bacteria, changes in light intensity may affect the frequency of tumbling (in peritrichously flagellated organisms) and/or speed of swimming (cf. KINESIS sense 1). In purple photosynthetic bacteria, at least, phototaxis appears to be mediated by changes in proton motive force (cf. AEROTAXIS and CHEMOTAXIS) resulting from changes in the rate of energy conversion in the photosynthetic reaction centre. [Role of pmf in sensory transduction: ARM (1983) *37* 551–573.] *Halobacterium salinarium* responds differently to light of different wavelengths. The cells of this species are bipolarly flagellated; they swim in the direction of the long axis of the cell, with spontaneous reversals in direction at intervals of ca. 10–50 sec. A sudden decrease in yellow-green light or a sudden increase in blue/UV light elicits an extra reversal in direction, while an increase in yellow-green light or a decrease in blue/UV light leads to a suppression of the spontaneous reversals. (See also SLOW-CYCLING RHODOPSIN.) Thus cells tend to acumulate in regions illuminated by yellow-green or white light (areas optimum for photophosphorylation by the purple membrane) but avoid areas exposed to (possibly damaging) UV radiation [SEBS (1983) *36* 207–222].

See also STENTOR.

phototopotaxis A light-induced TOPOTAXIS.

phototroph An organism which can use light (and, in some species, infrared radiation) as a primary source of energy for metabolism and growth. (See also PHOTOSYNTHESIS and PURPLE MEMBRANE.) Some phototrophs are facultative CHEMOTROPHS.) (See also LITHOTROPH and ORGANOTROPH.)

phototropism See TROPISM (sense 1).

***phoU* gene** See PHO REGULON.

***phr* gene** See PHOTOREACTIVATION.

Phragmidium See UREDINIOMYCETES.

Phragmobasidiomycetidae A subclass of fungi (class HYMENOMYCETES) characterized by the formation of *phragmobasidia* (see BASIDIUM). Orders [Book ref. 64, p. 189]: AURICULARIALES, SEPTOBASIDIALES, TREMELLALES.

phragmobasidium See BASIDIUM.

phragmoplast In e.g. some green algae (e.g. *Chara*): a plate which develops in the cleavage plane between two newly forming cells during the telophase stage of MITOSIS; it consists of microtubules of the persistent mid-region of the mitotic spindle together with vesicles (derived from dictyosomes) which collect between the microtubules. (cf. PHYCOPLAST.)

phragmosporae See SACCARDOAN SYSTEM.

phthalimide antifungal agents A group of agricultural non-systemic ANTIFUNGAL AGENTS which includes CAPTAFOL, CAPTAN, FOLPET etc.

phthalylsulphathiazole See SULPHONAMIDES.

phthiocerols Complex waxes found in mycobacteria [JGM (1983) *129* 859–863].

phthisis *Syn.* TUBERCULOSIS.

phycobilin See PHYCOBILIPROTEINS.

phycobiliproteins (biliproteins) Water-soluble pigments which occur in CYANOBACTERIA, in algae of the RHODOPHYTA, and in CRYPTOPHYTES. In cryptophytes the pigments occur between the thylakoids and are antigenically unrelated to the cyanobacterial and rhodophytan pigments (which occur bound to thylakoids – see PHYCOBILISOMES). A phycobiliprotein consists of a protein 'monomer' comprising two distinct polypeptide chains (α and β), each of which is linked covalently (via a thioether bond) to an open-chain tetrapyrrole chromophore (*phycobilin* or *bilin*). The monomers tend to form trimers or hexamers which are the structural units of phycobilisomes. Phycobiliproteins function as light-harvesting pigments for photosystem II (see PHOTOSYNTHESIS); there are three main classes: *allophycocyanins* (λ_{max} 650–671 nm), *phycocyanins* (λ_{max} 617–620 nm), and *phycoerythrins* (including *phycoerythrocyanin*: λ_{max} 545–568 nm). Allophycocyanins and phycocyanins apparently occur in all cyanobacteria and red algae; phycoerythrins occur in most red algae and in many cyanobacteria, phycoerythrocyanin occurs in some cyanobacteria (e.g. *Anabaena variabilis*, *'Mastigocladus laminosus'*). Cryptophytes contain either phycocyanin or phycoerythrin, according to species. In cyanobacteria, the content of phycobiliproteins can be affected by many environmental factors, including temperature, CO_2 concentration, nitrogen starvation, wavelength and intensity of light [Book ref. 76, pp. 182–188]; in addition to its function as a photopigment, phycocyanin serves as a nitrogen reserve in cyanobacteria, being degraded by a specific protease induced during nitrogen starvation both in vegetative cells and in differentiating HETEROCYSTS. [Structure and function of cyanobacterial phycobiliproteins: Book ref. 75, pp. 23–42; the phycocyanin operon and light regulation of its expression in *Anabaena*: EMBO (1987) *6* 871–884.]

phycobilisomes Structures, each ca. 20–70 nm across, which are attached in regular arrays to the surface of thylakoid membranes in CYANOBACTERIA (cf. GLOEOBACTER) and red algae (RHODOPHYTA). Phycobilisomes are composed mainly of PHYCOBILIPROTEINS together with 'linker' polypeptides; they vary in composition and morphology (being typically hemidiscoid or hemispherical) according e.g. to species. The (mainly non-pigmented) linker polypeptides seem to have many functions – e.g. to stabilize the specific structure of phycobilisomes, to modify the phycobiliproteins so as to promote unidirectional energy flow within the phycobilisomes and to the reaction centres, and to link phycobilisomes to the photosynthetic membranes.

In general, each phycobilisome comprises two major structural domains: (i) a 'core' (consisting of two or three cylindrical elements), and (ii) five or six 'rods' (composed of stacked discs) that radiate from the core to form the hemidiscoid or hemispherical structure. The core is composed of allophycocyanin (APC) and is anchored to the thylakoid membrane by a linker polypeptide. Where the rods are adjacent to the core, they consists of discs of phycocyanin (PC), but discs of phycoerythrin (PE) and/or phycoerythrocyanin (PEC), when present, occur at the periphery.

Absorbed light energy is transferred from the periphery to the core, and thence to chlorophyll *a* – thus: PE/PEC → PC → APC → Chl *a*; this transfer is almost 100% efficient.

[Structure of a 'simple' phycobilisome (in *Synechococcus*): Ann. Mic. (1983) *134*B 159–180. Phycobilisome structure: JB (1993) *175* 575–576.]

phycobiont An algal symbiont. In lichenology, 'phycobiont' commonly refers to the main algal *or* cyanobacterial partner in a lichen, although in the latter case the term 'cyanobiont' is more appropriate. The term *photobiont* is a useful non-specific term for the principal photosynthetic partner in a LICHEN. (See also PHYCOZOAN.)

phycocyanins See PHYCOBILIPROTEINS.

phycoerythrins See PHYCOBILIPROTEINS.

phycoerythrocyanin See PHYCOBILIPROTEINS.

phycology The study of ALGAE.

Phycomyces See MUCORALES.

phycomycetes See LOWER FUNGI.

phycomycosis See ZYGOMYCOSIS and EQUINE PHYCOMYCOSIS.

Phycopeltis A genus of algae closely related to TRENTEPOHLIA. The thallus is typically a circular monostromatic disc (ca. 1–3 mm diam.) of compacted, branched filaments. Species occur mainly as epiphytes, growing on – not beneath – the leaf cuticle (cf. CEPHALEUROS), and also as photobionts in certain (mainly tropical) lichens.

phycoplast In many green algae: a plate which develops in the cleavage plane between two newly forming cells in species in which the mitotic spindle is non-persistent; it contains microtubules orientated perpendicular to the axis of the former spindle. (cf. PHRAGMOPLAST.)

phycosymbiodeme See CEPHALODIUM.

phycotoxin Any TOXIN produced by an alga (see e.g. BREVETOXINS, CIGUATERA, GONYAUTOXINS, PRYMNESIUM, SAXITOXIN, ULVA) or – traditionally – by a cyanobacterium ('blue-green alga': see e.g. ANABAENA, APHANIZOMENON, LYNGBYA, MICROCYSTIS, NODULARIA).

phycovirus A VIRUS which infects one or more ALGAE. (cf. CYANOPHAGE.) Several polyhedral dsDNA viruses have been observed in green algae: e.g. a tailed polyhedral virus, resembling (at least superficially) a tailed bacteriophage, has been found in *Uronema gigas* [Virol. (1980) *100* 156–165, 166–174], and various polyhedral dsDNA viruses have been observed in *Chlorella*-like algae, including the zoochlorellae of *Paramecium bursaria* and *Hydra viridis* [PNAS (1982) *79* 3867–3871].

phycozoan A term proposed to refer to a 'compound organism' in which cells (or chloroplasts) of an alga (the phycobiont) live within the tissues or cells of an animal (the zoobiont: an invertebrate) [Book ref. 129, pp. 5–17].

(See e.g. ZOOCHLORELLAE and ZOOXANTHELLAE; cf. LICHEN.)

phyletic classification *Syn.* PHYLOGENETIC CLASSIFICATION.

Phyllactinia See ERYSIPHALES.

Phyllobacterium A genus of Gram-negative bacteria of the RHIZOBIACEAE. Cells (in vitro) are straight rods which, in liquid media, form characteristic star-shaped clusters. Acid is formed from a wide range of carbohydrates. GC%: 59.6–61.3. Type species: *P. myrsinacearum*.

P. myrsinacearum and *P. rubiacearum* occur as pleomorphic cells (rod-shaped, ellipsoidal or branched) within leaf nodules which they induce in certain tropical plants of the Rubiaceae (e.g. madder), Myrsinaceae, Myrtaceae and Dioscoreaceae (e.g. yam). They have not been shown to be capable of nitrogen fixation.

[Book ref. 22, pp. 254–256.]

phyllocladia See STEREOCAULON.

phyllody (*plant pathol.*) A malformation in which normal floral components are replaced by leaf-like structures; it is a symptom of certain plant diseases: e.g. *green ear disease* of millet (*Pennisetum typhoides*) and other grasses caused by *Sclerospora graminicola*.

***Phyllopertha horticola* EPV** See ENTOMOPOXVIRINAE.

phylloplane The surface(s) of a leaf. (cf. PHYLLOSPHERE.) [Book ref. 158.]

Phylloporus See BOLETALES.

phylloquinone See QUINONES.

phyllosphere (1) *Syn.* PHYLLOPLANE. (2) The region immediately surrounding, and influenced by, a leaf. [Quantification of bacterial sugar consumption in the phyllosphere: PNAS (2001) *98* 3446–3453.] (See also RHIZOSPHERE.)

Phyllosticta See SPHAEROPSIDALES.

phylogenetic classification (phyletic classification) Any form of classification in which the aim is to group organisms according to their ancestral lineage (i.e. evolutionary relationships). (cf. PHENETIC CLASSIFICATION.)

phylogeny The range of developmental stages in the evolution of an organism. (cf. ONTOGENY.)

phylotype A taxon which is defined solely by nucleic acid sequence data (no representative species of the taxon having been cultured). [Example of use: JCM (1999) *37* 1469–1473.]

phylum A taxonomic rank (see TAXONOMY and NOMENCLATURE).

Physarales See MYXOMYCETES.

Physarum A large genus of slime moulds (class MYXOMYCETES) which usually form sporangia (some species form plasmodiocarps). *P. polycephalum* is widely used in laboratory studies. It generally forms a robust, conspicuous yellow phaneroplasmodium in which a network of major and subsidiary 'veins' leads to the advancing front edge. The sporangia are stalked, greyish, and typically multilobed.

Physcia A genus of small foliose LICHENS (order LECANORALES); photobiont: a green alga. The thallus generally consists of grey or greyish-brown, often pruinose, narrow branching lobes which are often radially orientated to form a rosette. Several species have cilia; some have soredia. Apothecia: lecanorine, with grey or blue-black, often pruinose discs. Ascospores: brown, one-septate. Species are generally nitrophilous, occurring e.g. on bark and/or on rocks.

physiological heterothallism See HETEROTHALLISM.

physiological race (*mycol.*) Organisms which are morphologically similar to, but which differ in physiological (and/or other) characteristics from, other members of the same FORMA SPECIALIS. A given race may be designated by a number, letter etc (see e.g. NOMENCLATURE).

physode See PHAEOPHYTA.

phytoalexins Low-MWt antimicrobial compounds which are synthesized by, and accumulate in, higher plants exposed to certain (pathogenic and non-pathogenic) microorganisms – or to heavy metals (e.g. copper, mercury), detergents or certain other chemicals, ultraviolet radiation or physical (e.g. freeze–thaw) damage. Phytoalexins are formed by many angiosperms (particularly by members of the Leguminosae and Solanaceae) and by some gymnosperms; they have not been detected in lower plants.

Under pre-stimulus conditions, phytoalexins may be either absent from plant tissues or present in extremely low concentrations.

The production of phytoalexins appears to be a local effect in the plant – their translocation has yet to be conclusively demonstrated; if phytoalexins are involved in a plant's natural resistance to disease, their localized accumulation may be an important factor in such resistance. In some cases the ability to produce phytoalexin(s) appears to enable a plant to resist the development of disease following infection by a pathogen; thus, e.g. resistance of the French bean to *Colletotrichum lindemuthianum* seems to correlate with the accumulation of a phytoalexin, phaseollin. In other cases the accumulation of phytoalexins appears not to be the sole factor which determines the plant's resistance to disease. Pathogens which are able successfully to invade a plant may (i) fail to induce the production of phytoalexins; (ii) be relatively insensitive to the phytoalexin(s) produced; or (iii) be able to convert a phytoalexin to an inactive form (see e.g. KIEVITONE and PISATIN). (See also HYPERSENSITIVITY sense 2.)

Elicitors of phytoalexins, i.e. specific chemical entities which promote the synthesis of phytoalexins, include the polypeptide *monilicolin A* (which elicits PHASEOLLIN) and various polysaccharides (such as β-1,3- and β-1,6-glucans and CHITOSAN) present e.g. in fungal cell walls or culture filtrates. Physical damage to plant tissues may cause release of the plant's own elicitors (*constitutive* or *endogenous* elicitors). [Phytoalexins and their elicitors: ARPphys. (1984) *35* 243–275.]

Phytoalexins are structurally/chemically diverse compounds; they include e.g. CAMALEXIN; CAPSIDIOL; DEOXYHEMIGOSSYPOL; FALCARINDIOL; GLYCEOLLIN; HIRCINOL; IPOMEAMARONE; KIEVITONE; LUBIMIN; MOMILACTONES (see also WL 28325); ORCHINOL; PHASEOLLIN; PISATIN; PTEROSTILBENE; RISHITIN; VIGNAFURAN; VINIFERINS; and WYERONE.

(cf. AVENACIN; CAFFEIC ACID; CHLOROGENIC ACID; PATHOGENESIS-RELATED PROTEINS.)

[Phytoalexins (structure, role etc.): ARPpath. (1999) *37* 285–306.]

phytobiont A plant symbiont – e.g. the plant partner in a MYCORRHIZA.

phytoflagellates See PHYTOMASTIGOPHOREA.

phytoflavin See FLAVODOXINS.

phytoglycogen A type of GLYCOGEN-like polysaccharide found e.g. in certain algae. (cf. CYANOPHYCEAN STARCH.)

phytohaemagglutinin (PHA) (1) A mitogenic LECTIN from the red kidney bean, *Phaseolus vulgaris*; it can agglutinate erythrocytes, and it binds to both B and T cells but is mitogenic mainly in T cells. PHA activity requires Ca^{2+} and Mg^{2+} and is inhibited e.g. by *N*-acetyl-D-galactosamine. (2) Any plant lectin which can agglutinate erythrocytes.

phytohormones (plant hormones) Compounds which are produced by plants and which regulate the plants' own metabolism, cell division, seed germination etc; they include ABSCISIC ACID, AUXINS, CYTOKININS, ETHYLENE and GIBBERELLINS.

phytokinins See CYTOKININS.

Phytomastigophorea (the phytoflagellates) A class of flagellated protozoa (subphylum MASTIGOPHORA) which either have chloroplasts or bear an obvious relationship to other members which have chloroplasts. Most phytoflagellates can also be regarded as algae and are alternatively classified in algal taxonomic schemes. Orders: Chloromonadida (the CHLOROMONADS); Chrysomonadida (the CHRYSOPHYTES); Cryptomonadida (the CRYPTOPHYTES); Dinoflagellida (the DINOFLAGELLATES); Euglenida (the EUGLENOID FLAGELLATES); Heterochlorida (unicellular, flagellate and/or amoeboid members of the XANTHOPHYCEAE); Prasinomonadida (cf. PRASINOPHYCEAE); Prymnesiida (= PRYMNESIOPHYCEAE); Silicoflagellida (the SILICOFLAGELLATES); Volvocida (including genera such as CHLAMYDOMONAS and VOLVOX).

Phytomonas (1) A genus of protozoa (family TRYPANOSOMATIDAE) parasitic in certain plants – particularly lactiferous plants (e.g. Euphorbiaceae) but also e.g. coffee plants (see e.g. PHLOEM NECROSIS) and palms; they are transmitted by sap-sucking bugs. The organisms occur in the promastigote form and are usually ca. 20 μm in length; some species have been grown in vitro. [Plant diseases caused by *Phytomonas*: ARPpath. (1984) 22 115–132; cultivation of *P. françai*, a species associated with a root rot disease of cassava (*Manihot esculenta*): JP (1986) *33* 511–513.]

(2) Obsolete name for *Xanthomonas*.

phytoncide (*plant pathol.*) A broad term for any substance which is produced (constitutively or inducibly) by a plant and which confers on that plant resistance to disease or infestation by killing or inhibiting the growth of a potential pathogen or parasite.

phytone A proteolytic digest (by PAPAIN) of plant material (e.g. soybean meal), used in certain media (see e.g. TRYPTICASE-SOY AGAR).

phytopathology Plant pathology.

Phytophthora A genus of fungi (order PERONOSPORALES) which include some important phytopathogenic species.

In many geographical regions, *P. infestans* (causal agent of LATE BLIGHT of potato) appears to overwinter as mycelium in infected tubers. In the spring, when the tubers produce shoots, the fungus grows into the developing shoots and subsequently gives rise to distinctive sporangiophores – bearing lemon-shaped sporangia – which emerge from the stomata of the leaves; the sporangia are dispersed e.g. by wind. A sporangium germinates to form either zoospores or a germ tube, depending on conditions. A zoospore subsequently encysts and later germinates, forming a germ tube; a germ tube may enter a host plant by means of an APPRESSORIUM or, directly, via a stoma. Within the new host a branching mycelium develops intercellularly, and elongated haustoria are formed; later, sporangiophores emerge from the stomata. Sexual reproduction occurs between strains of appropriate MATING TYPE. During gametangial development the oogonial initial penetrates and grows through the antheridium such that, when fully developed, the expanded, globose oogonium carries the antheridium as an encircling collar around its base; a fertilization tube develops between the two gametangia. The sexually-derived thick-walled oospore typically germinates to form a germ tube that terminates in a zoosporangium. Oospore formation is very rare in most geographical regions – though not in Mexico.

Other pathogenic species include e.g. *P. cactorum* (see CROWN ROT); *P. cinnamomi*, a worldwide pathogen of various trees and other plants (see e.g. JARRAH DIEBACK); *P. citricola* (see BLEEDING CANKER); *P. fragariae* (see RED CORE); *P. megasperma* var. *sojae* (causal agent of root and stem rot of soybean) [review: Book ref. 58, pp. 19–30]; *P. palmivora* (see BLACK POD DISEASE); and *P. parasitica* (see GUMMOSIS). *P. syringae* causes e.g. storage rot of apples and pears.

phytoplankton See PLANKTON.

phytoplasma See MLOS.

Phytoreovirus (plant reovirus subgroup 1) A genus of PLANT VIRUSES of the REOVIRIDAE. Genome: 12 linear dsRNA molecules. Replication occurs in cytoplasmic viroplasms; mRNA transcripts are capped. Transmission occurs propagatively via cicadellid leafhoppers; transovarial transmission occurs in the insect vector. The genus includes WOUND TUMOUR VIRUS (type species), RICE DWARF VIRUS, and RICE GALL DWARF VIRUS. [Review: Book ref. 83, pp. 505–563.]

phytotoxin (1) A TOXIN produced by a microorganism and active against a plant or against plant cells/tissues. (2) A toxin produced by a plant.

phytotron An apparatus consisting of a chamber within which plants etc can be grown under a variety of accurately controlled conditions of temperature, light intensity, humidity, etc.

pI See ISOELECTRIC POINT.

Pi Inorganic orthophosphate ($PO_4{}^{3-}$)

pi particle See PSEUDOCAEDIBACTER.

π protein (Pi protein) See R6K PLASMID.

pian *Syn.* YAWS.

pian bois (forest yaws) A form of LEISHMANIASIS in which the pathogen spreads, typically via the lymphatic route, producing a chain of ulcers in the peripheral lymphatic system; pian bois occurs in South America.

Pichia A large, apparently polyphyletic genus of yeasts (family SACCHAROMYCETACEAE). Cells are spheroidal, ellipsoidal or elongate; vegetative reproduction occurs by multilateral budding. Pseudomycelium and, to a limited extent, true mycelium may be formed. Arthrospores may be formed by some species. Species are homothallic or heterothallic. Asci are usually dehiscent. Ascospores: bowler-hat-shaped, Saturn-shaped, hemispheroidal, or spheroidal, smooth-surfaced, usually 1–4 per ascus. Sugars may or may not be fermented; $NO_3{}^-$ is not assimilated. More than 56 species are recognized; type species: *P. membranaefaciens*. Several species have anamorphs in the form-genus *Candida*: e.g. the anamorph of *P. burtonii* is *C. variabilis* (and *Trichosporon variabile*), that of *P. guilliermondii* is *C. guilliermondii*, that of *P. membranaefaciens* is *C. valida*, etc. *Pichia* species are found in a wide range of habitats: e.g. tunnels and frass of wood-boring beetles, tree exudates, pickling brines, tanning liquors, grain and flour (e.g. *P. farinosa*), wine (e.g. *P. carsonii* [formerly *P. vini*], *P. membranaefaciens*), naturally fermented apple juice (e.g. *P. delftensis*), cacti (e.g. *P. cactophila, P. opuntiae*), human skin, faeces, etc. [Book ref. 100, pp. 295–378.]

(See also ALCOHOL OXIDASE, ALCOHOLIC FERMENTATION, BEER SPOILAGE, CIDER, METHYLOTROPHY, WINE SPOILAGE.)

Pichinde virus See ARENAVIRIDAE.

pickling A traditional method of FOOD PRESERVATION in which the pH is lowered either by direct addition of acid (e.g. VINEGAR, LACTIC ACID) or by a LACTIC ACID FERMENTATION of the food; cabbage (see SAUERKRAUT), cucumbers and olives are often preserved by fermentation [Book ref. 5, pp. 227–258]. Fresh vegetable material has a large and varied microflora, of which lactic acid bacteria form only a very small component; fermentation conditions must therefore specifically encourage the growth of the lactic acid bacteria. The vegetables are graded and washed (Spanish-style olives may be treated with NaOH to remove the bitter phenolic glucoside oleuropein), and salt is added; the salt extracts the sugar-containing juices, encouraging the growth of the lactic acid bacteria (mainly *Lactobacillus brevis*, *L. plantarum* and *Pediococcus pentosaceus* in the case of olives and cucumbers – cf. SAUERKRAUT). [Lactic acid bacteria in vegetable fermentations: Food Mic. (1984) *1* 303–313.] Fermentative yeasts may also be present; these may ferment any sugars remaining after the lactic acid bacteria have been inhibited by acidity. The combination of salt, anaerobiosis and low pH serves to discourage the growth of undesirable organisms.

Microbial spoilage of pickled vegetables may be due e.g. to oxidative pectinolytic moulds and yeasts which cause softening of the vegetable tissue; these organisms form a surface film unless precautions are taken to exclude air. Oxidative yeasts may also metabolize lactic acid and hence reduce acidity, allowing the subsequent growth of other spoilage organisms. Spoilage bacteria (e.g. coliforms, clostridia) tend to grow when the pH is too high and/or the salt concentration too low; they can cause softening, off-flavours and odours, and/or the formation of gas pockets within the vegetable (called 'bloater damage' in cucumbers, 'fisheye spoilage' in olives).

picodnaviruses Rejected name for the PARVOVIRIDAE.

α-picolinic acid Pyridine 2-carboxylic acid, a toxin produced by *Pyricularia oryzae*. (See also BLAST DISEASE and PIRICULARIN.)

Picornaviridae A family of small, ssRNA-containing VIRUSes in which the icosahedral virion (ca. 22–30 nm diam.) is naked and ether-resistant. Most picornaviruses have a narrow host range, and some are important pathogens of man and animals; transmission generally occurs mechanically. The family includes four genera: APHTHOVIRUS, CARDIOVIRUS, ENTEROVIRUS and RHINOVIRUS; within the genera, species are distinguished by their susceptibility to specific neutralizing antibodies. Some picornaviruses are currently not classified into genera; these include equine rhinovirus types 1 and 2, and insect picornaviruses such as cricket paralysis virus, *Drosophila* C virus and *Gonometa* virus. Possible picornaviruses include 'unclassified small RNA viruses of invertebrates': e.g., bee acute paralysis virus, bee slow paralysis virus, bee virus X, black queen cell virus, *Drosophila* P and A viruses, FLACHERIE virus and SACBROOD virus. [Review of small RNA viruses of insects: JGV (1985) *66* 647–659.]

Virion structure. The capsid comprises 60 subunits, a subunit consisting of one molecule each of the 4 major capsid polypeptides (VP1–VP4); portions of VP1, VP2 and VP3 are exposed at the virion surface, while VP4 is probably internal and may be associated with the RNA. [Virion structure in polioviruses: Science (1985) *229* 1358–1365, and in rhinoviruses: Nature (1985) *317* 145–153.] The RNA genome is a single molecule of linear positive-sense ssRNA (MWt ca. 2.5×10^6) which is polyadenylated at its 3′ end and covalently linked at its 5′ end to a virus-encoded protein, VPg. In cardioviruses and aphthoviruses (but not in enteroviruses or rhinoviruses) the RNA contains a poly(C) tract near the 5′ end, the length of which varies with strain. [Poly(C) secondary structure in aphthoviruses: JGV (1985) *66* 1919–1929.]

Replication cycle. Virus replication occurs in the host cell cytoplasm. Infection begins when a virion attaches to receptors in the host cell plasma membrane; in e.g. poliovirus type 1 and mouse Elberfeld virus, entry into the host cell apparently occurs by endocytosis, and uncoating apparently occurs in endosomes and/or lysosomes [JGV (1985) *66* 483–492]. In polioviruses, translation of the viral RNA is initiated at one site near the 5′ end of the RNA, resulting in the formation of a polyprotein; the polyprotein is initially cleaved, during translation, into three precursor proteins (P1, P2 and P3), each of which undergoes further cleavage. P1 gives rise to 1A, 1B, 1C and 1D (capsid polypeptides); P2 gives rise to 2A, 2B (a host-range determinant), and 2C (involved in RNA synthesis); P3 undergoes intramolecular self-cleavage to form 3A, 3B (VPg), 3C (a protease which carries out most of the proteolytic cleavages), and 3D (an RNA polymerase which can elongate nascent RNA chains on an RNA template). [Nomenclature of picornavirus proteins: JV (1984) *50* 957–959.] In contrast to polioviruses, aphthoviruses have two distinct translation initiation sites [JGV (1985) *66* 2615–2626].

Replication of the RNA genome is associated with the smooth endoplasmic reticulum; (−)-strand synthesis may involve a primer formed by hairpin folding of the (+)-strand template, resulting in a product up to twice the length of the genome (at least in vitro) [JV (1986) *58* 790–796].

Virus assembly is preceded by cleavage of P1 to form an immature 5S protomer containing VP0, VP1 and VP3. The immature protomers become associated with an RNA genome to form a provirion. During maturation most or all of the VP0 molecules are cleaved to form VP2 and VP4. The mature virions are eventually released by host cells lysis. (Some picornaviruses – e.g. hepatitis A virus – can cause non-lytic infections.)

Most picornaviruses can be propagated in cell cultures. Infected cells undergo drastic changes in their macromolecular metabolism: the rate of RNA synthesis declines shortly after infection, and cellular protein synthesis is inhibited soon after. [Review of inhibition of protein synthesis: Book ref. 150, pp. 177–221.] Characteristic CPEs generally develop in the infected cells: e.g., accumulation of chromatin on the inside of the nuclear envelope; appearance of membranous vesicles in the cytoplasm; formation of crystalline arrays of virions in the cytoplasm; alterations in membrane permeability leading to shrivelling of the cell; etc.

picornaviruses Viruses of the PICORNAVIRIDAE.

picric acid (2,4,6-trinitrophenol) An antimicrobial agent (see PHENOLS), a fixative (see e.g. BOUIN'S FLUID), and a DYE.

piedra See BLACK PIEDRA and WHITE PIEDRA.

Piedraia A genus of fungi of the DOTHIDEALES (anamorph: *Trichosporon*). (See BLACK PIEDRA.)

piericidin An antibiotic, produced by *Streptoverticillium mobaraense*, which acts as a RESPIRATORY INHIBITOR in mitochondrial and bacterial ELECTRON TRANSPORT CHAINS (although *intact* cells may be impermeable and hence insensitive to piericidin); the (non-covalent) binding site(s) and mechanism of action of piericidin are apparently similar to those of ROTENONE.

pif A locus in the F PLASMID associated with the inability of certain phages (e.g. T7) to replicate in cells containing the F plasmid. (*pif* = phage-inhibitory function.) (See also FEMALE-SPECIFIC PHAGE.)

pig bel *Syn.* ENTERITIS NECROTICANS.

pig diseases (a) *Bacterial diseases*: see e.g. EPERYTHROZOONOSIS, FOOT-ROT, GLASSER'S DISEASE, GREASY PIG DISEASE, OEDEMA DISEASE, PARATYPHOID FEVER (sense 2), SWINE DYSENTERY, SWINE ERYSIPELAS, TULARAEMIA. (b) *Viral diseases*: see e.g. AFRICAN SWINE FEVER, AUJESZKY'S DISEASE, FOOT AND MOUTH DISEASE, INCLUSION BODY RHINITIS, PARVOVIRUS, SWINE FEVER, SWINE INFLUENZA, SWINE POX, SWINE VESICULAR DISEASE, TALFAN DISEASE, TESCHEN DISEASE, TRANSMISSIBLE GASTROENTERITIS, VESICULAR EXANTHEMA, VESICULAR STOMATITIS, VOMITING AND WASTING DISEASE.

pigeonpox virus See AVIPOXVIRUS.

***pil* genes** The designation of genes which encode type IV FIMBRIAE in various Gram-negative bacteria – e.g. *Neisseria gonorrhoeae* and *Pseudomonas aeruginosa*; for example, in *N. gonorrhoeae*, *pilC* encodes the terminal adhesin, *pilE* encodes the major subunit (see also ANTIGENIC VARIATION), and *pilQ* is a SECRETIN which is stabilized by an outer membrane lipoprotein encoded by *pilP*.

Pilaira A genus of fungi (order MUCORALES) which occur on the dung of herbivores. The long, positively phototropic sporangiophore bears a single non-projectile sporangium which, on maturity, becomes adhesive. (cf. PILOBOLUS.)

pileate (*mycol.*) Having a PILEUS.

pileus The structure on which the spore-bearing tissue is carried in the sexually-derived fruiting bodies of certain basidiomycetes and ascomycetes. Examples include the entire lamella-bearing structure ('cap') carried by the stipe in mushroom-shaped basidiocarps; the corky or leathery, non-stipitate fruiting bodies formed by certain wood-rotting fungi; and the ridged and pitted stipe-borne cylindrical structures of the morels (*Morchella* spp).

pili (*singular*: pilus) (1) (conjugative pili; sex pili) Elongated or filamentous, proteinaceous, plasmid-encoded structures which extend from the surface of those (Gram-negative) bacteria which contain an (expressed) CONJUGATIVE PLASMID. Some types of pili have been shown to play an essential role in CONJUGATION (sense 1b), and it is generally believed (though not proven) that all pili have essential roles in conjugation. (cf. FIMBRIAE; SPINA; FLAGELLUM.)

Commonly, only a few pili (sometimes only one) occur on a given donor cell (cf. FIMBRIAE). At least some types of pili appear to be retractable, but the mechanism of retraction is unknown.

In general, thin, flexible pili mediate so-called 'universal' conjugation systems, i.e. systems in which conjugation can occur equally well within a body of liquid (e.g. broth) or on a moist (but not submerged) solid surface (e.g. an agar plate or membrane filter). Rigid pili often mediate 'surface-obligatory' mating, i.e. conjugation which occurs at a negligible frequency in liquid media but which may occur at high frequency on a moist, non-submerged solid surface. Thick, flexible pili may mediate in either 'surface-preferred' mating (in which conjugal transfer occurs more readily on solid surfaces than within liquids) or universal mating. [Surface mating systems in *Escherichia coli*: JB (1980) *143* 1466–1470; surface mating systems in *Pseudomonas* spp: JGM (1983) *129* 2545–2556.]

F pili (encoded by the F PLASMID) are thin, flexible filaments that are one to several micrometres in length, ca. 8 nm in diameter; each F-pilus has an axial channel, ca. 2 nm diameter, with a hydrophilic lumen. It has been assumed that the base of a pilus is associated with an ADHESION SITE or similar structure.

F pili are composed mainly or solely of a single type of subunit, *pilin*; the 70-amino acid pilin molecule is a single polypeptide containing one residue of D-glucose and two phosphate residues. The way in which the pilin subunits are assembled to form the pilus is not known. In one model, rings of five pilin subunits are stacked along the axis of the pilus, each ring being rotated 29° relative to the previous ring. In a second model, the pilin molecules are arranged helically. [Structure of F-pilin and a model for the organization of pilin subunits in the F pilus: Mol. Microbiol. (1997) *23* 423–429.]

Bacteriophages which can bind to F pili include f1, fd and M13 (see INOVIRUS), and MS2 and Qβ (see LEVIVIRIDAE). (See also ANDROPHAGE.)

(Note that the term *pilin* is also used to refer to the subunits of other types of pili – and also to the subunits of fimbriae.)

F-like pili are serologically similar or identical to F pili, and they bind similar types of phage. They are encoded e.g. by the colicin plasmids ColV and ColI-K94.

I-like pili resemble F pili morphologically but they differ serologically and they bind different phages – e.g. If1 and If2 (see INOVIRUS). They are encoded e.g. by the ColIb-P9 plasmid.

Short, rigid, nail- or thorn-like pili are encoded e.g. by IncN enterobacterial plasmids and *Pseudomonas* IncP-7 plasmids.

The precise role of the pilus is as yet unknown – see CONJUGATION. The currently favoured model (at least for the F pilus and related pili) is one in which the pilus makes initial contact with the recipient and then retracts to draw donor and recipient into wall-to-wall contact. Evidence supportive of this model includes the demonstration that a mutant donor with a temperature-dependent block in DNA transfer can achieve wall-to-wall contact with a recipient independently of DNA transfer; DNA transfer can occur subsequently at the permissive temperature, thus excluding the need for an extended pilus as a conducting tube for DNA [JB (1985) *162* 584–590].

The role of the short, rigid pili encoded e.g. by IncN plasmids remains unknown.

(2) Collectively, pili (sense 1, above) and FIMBRIAE. Use of the term in this way should be discouraged owing to the confusion it causes (see FIMBRIAE for rationale).

Pilimelia A genus of keratinophilic bacteria (order ACTINOMYCETALES, wall type II) which occur e.g. in soil. The organisms resemble *Actinoplanes* spp but differ e.g. in that they form parallel chains of rod-shaped zoospores in spherical or cylindrical sporangia. Type species: *P. terevasa*. [Book ref. 73, pp. 65–66.]

pilin See FIMBRIAE and PILI.

Pillotina See SPIROCHAETALES.

Pilobolaceae See MUCORALES.

Pilobolus A genus of fungi (order MUCORALES) which occur on the dung of herbivores. In *Pilobolus* spp a single sporangium is carried on a subsporangial vesicle at the distal end of each long, positively phototropic sporangiophore; the (non-dehiscent) sporangium is propelled from the sporangiophore by a jet of liquid which escapes, explosively, from the subsporangial vesicle. (cf. PILAIRA.)

Pilophorus A genus of LICHENS (order LECANORALES); photobiont: a green alga. Erect pseudopodetia (see PSEUDOPODETIUM) – which may bear terminal, convex, black apothecia – arise from a granular primary thallus which may be evanescent; cephalodia are generally present. Species grow e.g. on rocks, particularly in mountainous regions.

pilot protein A protein which is believed to aid the transfer of DNA from a donor cell to a recipient during bacterial CONJUGATION (sense 1b) or from a bacteriophage virion to a host cell during certain types of phage infection.

pilus See PILI.

pimaricin (natamycin; tennecetin) A MYCOSAMINE-containing tetraene POLYENE ANTIBIOTIC produced by *Streptomyces natalensis*. It is used e.g. in the treatment of mycotic keratitis and as an antifungal food preservative e.g. in sausages, fruit juices, cheese etc; it can be inactivated by enzymes produced by certain moulds (e.g. *Aspergillus flavus*).

***pin* gene** See RECOMBINATIONAL REGULATION.

pine diseases Diseases of pine trees (*Pinus* spp) include BLISTER RUST and PITCH CANKER. *Pine wilt* may be caused by a species of the fungus *Ceratocystis* (vector: bark beetles) or by the nematode *Bursaphelenchus xylophilus* (vector: the pine sawyer). (See also TREE DISEASES.)

pine disinfectants (pine fluids) DISINFECTANTS which include certain volatile oils obtained synthetically or by steam-distillation of pine wood. The early pine fluids were made by solubilizing pine oil with soap; they were active mainly against Gram-negative bacteria, but were poorly active against Gram-positive species. Modern pine fluids (with RW coefficients of ca. 3–5) typically contain phenolics (e.g., 2-phenylphenol, 4-chloro-3,5-xylenol) together with e.g. terpineol (an antimicrobial constituent of pine oil) and a solubilizing agent, e.g., potassium ricinoleate; these fluids are active against both Gram-negative and Gram-positive bacteria. The activity of at least one pine fluid against *Pseudomonas aeruginosa* is greatly increased at temperatures above 30°C [Book ref. 13, pp. 85–90]. The recommended maximum dilution of a pine fluid is 20 times its RW coefficient.

pink-eye (pinkeye) (1) *Syn.* CONJUNCTIVITIS. (2) An acute contagious conjunctivitis caused by *Haemophilus aegyptius*. (3) *Syn.* INFECTIOUS KERATITIS (of cattle).

Pinnularia See DIATOMS.

pinocytosis The ingestion, by various types of eukaryotic cell (e.g. amoebae, macrophages), of minute droplets of fluid; in pinocytosis (as in PHAGOCYTOSIS) a small region of the CYTOPLASMIC MEMBRANE invaginates and is subsequently pinched off to form a closed, intracellular membranous sac (*pinocytotic vesicle, pinosome*) containing the ingested fluid. In one form of pinocytosis (*fluid-phase pinocytosis*) pinocytotic vesicles are formed continually at non-specific sites in the cytoplasmic membrane – samples of extracellular fluid being taken randomly; such vesicles often fuse with LYSOSOMES.

Receptor-mediated pinocytosis (= *absorptive pinocytosis*) is a selective process in which specific types of macromolecule (or e.g. virus particle) are taken up. The macromolecules bind to specific receptors located at the outward-facing surface of the cytoplasmic membrane, groups of such receptors occurring in discrete regions of the cytoplasmic membrane known as *coated pits*; a coated pit bears on its inner (i.e., cytoplasmic) surface a characteristic array of molecules of the fibrous protein *clathrin* (MWt ca. 215000). Clathrin occurs in propeller-shaped ('three-legged') trimeric units (*triskelions*) which assemble into pentagons and hexagons, forming a polygonal network which comprises the cytoplasmic face of the coated pit. (Often, bundles of ACTIN microfilaments – *stress fibres* – occur in the cytoplasm beneath a coated pit.) Following receptor-mediated binding of macromolecules to a coated pit, the coated pit region invaginates and eventually pinches off to form a pinocytotic vesicle – the outer (cytoplasm-facing) surface of the vesicle being covered by a lattice-like network of clathrin; the vesicle, enclosed within its clathrin 'cage', is called a *coated vesicle*. Before fusing with e.g. a lysosome, a coated vesicle must have its clathrin removed in an ATP-dependent process. [Enzymatic recycling of clathrin from coated vesicles: Cell (1986) *46* 5–9.]

pinosome See PINOCYTOSIS.

pinosylvin 3,5-Dihydroxy-*trans*-stilbene: an antifungal compound formed by many species of pine (*Pinus*); it appears to be at least partly responsible for the relative resistance of pine heartwood to fungal attack. (See also TIMBER PRESERVATION.)

pinta (mal del pinto; *syn.* spotted sickness) A chronic infectious human disease caused by *Treponema carateum*; it occurs in Central and South America. Symptoms: spots or patches of abnormally pigmented or depigmented skin; other tissues are rarely affected. *Chemotherapy*: e.g. penicillins.

p*Inv* plasmid See EIEC.

pionnotes In *Fusarium* spp: a flat spore mass which has a fat-like or greasy appearance.

pipemidic acid See QUINOLONE ANTIBIOTICS.

Piptocephalis A genus of fungi (order ZOOPAGALES) which are parasitic on other fungi, often other zygomycetes. The vegetative hyphae develop extracellularly on the host – from which nutrients are abstracted by means of haustoria. Asexually derived spores are formed in merosporangia carried on (typically) deciduous 'head cells' at the distal ends of dichotomously branched sporangiophores; each merosporangium contains one or a few spores. Most species are apparently homothallic.

Piptoporus A genus of lignicolous fungi of the APHYLLOPHORALES (family Polyporaceae) which form typically annual, non-stipitate basidiocarps having a whitish or pale brown, corky, dimitic, non-xanthochroic context; the bracket-type basidiocarp has a porous hymenophore on the underside. *P. betulinus* ('razor-strop fungus') is saprotrophic and parasitic on birch (*Betula*); the upper surface of the basidiocarp has a thin, pale brown, separable outer layer, and the lower porous surface is whitish. Basidiospores: colourless, cylindrical, ca. 5 × 2 μm. (See also BROWN ROT and TIMBER SPOILAGE; cf. POLYPORUS.)

***pir* gene** See R6K PLASMID.

Pirella See PASTEURIA.

piricularin (pyricularin) A toxin produced by *Pyricularia oryzae* (see BLAST DISEASE). Piricularin is toxic to the fungus itself, but *P. oryzae* produces a copper-containing 'piricularin-binding protein' which complexes the toxin and renders it non-toxic for the fungus without affecting its toxicity for the rice plant. CHLOROGENIC ACID can complex piricularin and inactivate it.

Piromonas A genus of fungi found in the RUMEN.

piroplasm (1) A member of the PIROPLASMASINA. (2) The intra-erythrocytic stage of *Theileria* spp.

Piroplasmasina A subclass of protozoa (class SPOROZOASIDA) in which the APICAL COMPLEX is less well developed than in other members of the Sporozoasida, and in which the conoid is absent; reproduction occurs asexually (by binary fission or schizogony) and sexual reproduction occurs in at least some species. Motility involves e.g. cell flexion. Species are parasitic in the erythrocytes (and other cells) of vertebrates but do not form pigment from host cell haemoglobin (cf. HAEMOSPORORINA). Piroplasms are heteroxenous and are transmitted by ticks. Genera: BABESIA, *Dactylosoma*, THEILERIA.

Piroplasmea A class of parasitic protozoa classified within the subphylum Sarcomastigophora [JP (1964) *11* 7–20] and subsequently, as the subclass PIROPLASMASINA, in the class SPOROZOASIDA.

Piroplasmia A subclass of protozoa [JP (1980) *27* 37–58] equivalent to the PIROPLASMASINA.

Piry virus A virus (genus VESICULOVIRUS) antigenically related to VSV.

pisatin An isoflavonoid PHYTOALEXIN (3-hydroxy-7-methoxy-4′,5′-methylenedioxychromanocoumaran) produced by the pea plant (*Pisum sativum*). Some isolates of *Nectria haematococca* which are pathogenic for *P. sativum* can tolerate pisatin and can 'detoxify' the phytoalexin by demethylating it.

Pisolithus A genus of saprotrophic or mycorrhizal fungi (order SCLERODERMATALES). *P. arhizus* (= *P. tinctorius*) forms large, club-shaped, partly hypogean, ochre to brownish basidiocarps (up to ca. 25 cm high) which contain closed pockets or chambers (*peridioles*) that eventually desiccate and crumble to release their content of basidiospores; this organism forms mycorrhizal associations with various trees and can apparently enhance the growth of e.g. pine trees under conditions which are normally suboptimal for their growth.

pit connection In many multicellular red algae: a structure located between two adjacent cells. A *primary* pit connection forms between two daughter cells during cell division: a central hole in the intercellular septum becomes blocked by a plug of electron-dense material, and a plug cap (formed from a flattened vesicle) appears on each side of the plug; the cytoplasmic membrane thus remains continuous from one cell to another. (cf. PLASMODESMA.) A *secondary* pit connection may develop between two cells which have become juxtaposed. (See also CHOREOCOLAX.)

pitch canker A disease of pine trees (*Pinus* spp) caused by *Fusarium moniliforme* var. *subglutinans*; symptoms include dieback of terminal and lateral branches, and the formation of CANKERS characterized by a copious flow of resin.

Pitelka convention A system for numbering the microtubular triplets in a kinetosome. Looking at a cross-section of a ciliary kinetosome from the direction of the *interior* of the cell, the triplet associated with the ribbon of POST-CILIARY MICROTUBULES is designated number 5; the numbering sequence is clockwise – as in the GRAIN CONVENTION.

Pithomyces See HYPHOMYCETES; see also SPORIDESMINS.

pitted keratolysis A human skin disease characterized by focal erosion of the stratum corneum, usually on the soles and heels of the feet; it appears to be caused by an organism resembling *Dermatophilus congolensis*.

Pittsburg pneumonia agent *Legionella micdadei*.

pityriasis nigra (tinea nigra) A benign superficial human dermatomycosis in which brownish or black macules are formed (usually on the palms of the hands). Causal agents: *Exophiala werneckii* (= *Cladosporium werneckii*) or *Stenella araguata* (= *Cladosporium castellanii*).

pityriasis versicolor (tinea versicolor) A mild, chronic, superficial human dermatomycosis caused by *Malassezia furfur* ('*Pityrosporum orbiculare*'); flat or slightly raised, scaly, brownish or fawn spots (which gradually enlarge and coalesce) develop mainly on the chest, back, arms and neck. Lesions may fluoresce greyish-yellow under Wood's lamp.

Pityrosporum *Syn.* MALASSEZIA.

pivampicillin See PENICILLINS.

Pixuna virus See ALPHAVIRUS.

P–K test See PRAUSNITZ–KÜSTNER TEST.

PKA See CYCLIC AMP.

PKC kinase See CHEMOKINES.

PKD PROLIFERATIVE KIDNEY DISEASE.

PKDL POST-KALA-AZAR DERMAL LEISHMANIASIS.

pKM101 See SOS SYSTEM and AMES TEST.

PKR kinase See INTERFERONS.

Placidiopsis See VERRUCARIALES.

placoderm desmid See DESMIDS.

placodioid (placoid) (*lichenol.*) Refers to a crustose thallus which is generally circular in outline with radial lobes at the periphery.

placoid *Syn.* PLACODIOID.

Plagiacantha See RADIOLARIA.

Plagonium See RADIOLARIA.

plague An acute infectious disease which primarily affects rodents but which can be transmitted to man; the causal agent is *Yersinia pestis* (see also VIRULON).

Before the availability of antibiotic therapy, epidemics and pandemics of plague caused enormous loss of life; for example, an outbreak in Europe during the 14th century (the 'Black Death') involved ca. 25 million deaths. The World Health Organization has recently recognized plague as a 're-emerging disease'.

An epidemic of human plague is commonly preceded by an epizootic among urban rats; *Rattus rattus* populations are rapidly decimated in such epizootics, but wild rodents in rural or wooded areas are less susceptible and provide a reservoir of infection.

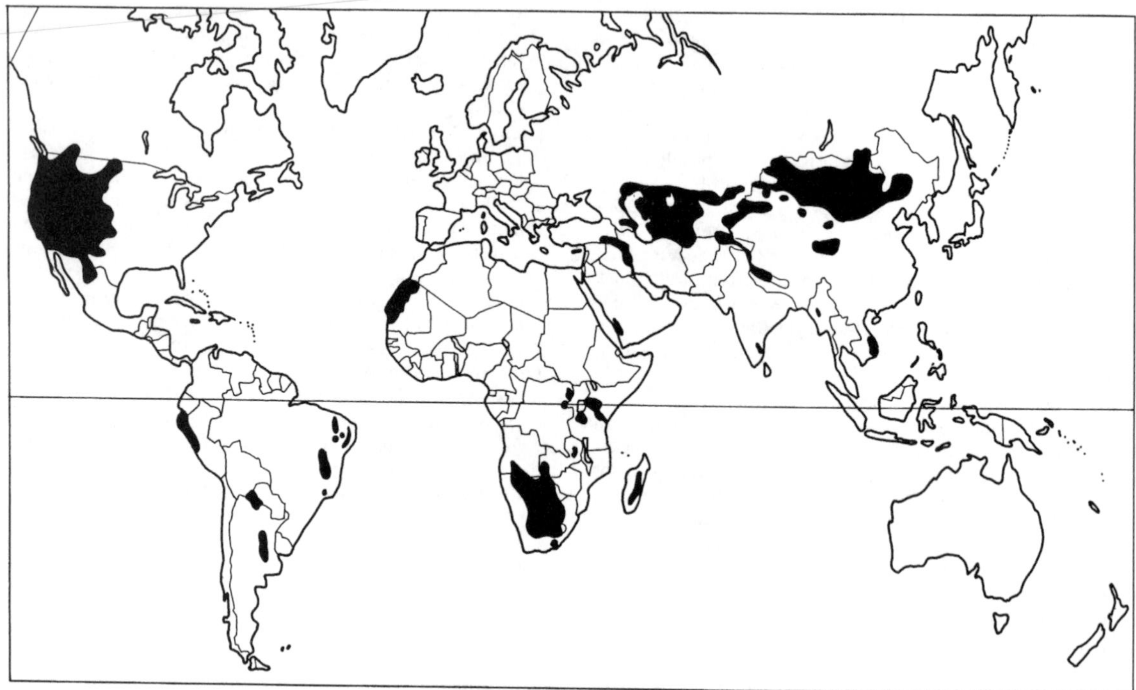

PLAGUE: areas where endemic plague is known to have persisted near the end of the 20th century. Areas in black are regions in which there are populations of wild rodents that carry the plague bacillus.

Reproduced from *Medical Microbiology* 14th edition, Figure 36.1, page 406, David Greenwood, Richard Slack & John Peutherer (eds) (1992) copyright Churchill Livingstone (Harcourt Publishers Ltd, Edinburgh) (ISBN 0-443-04256-X) by courtesy of the author, Dr J. D. Coghlan, and with permission from the publisher.

Transmission from rat to rat, and from rat to man, occurs via the bite of the rat flea (especially *Xenopsylla cheopsis*); the gut of an infected flea becomes blocked by plague bacilli which are regurgitated into the next bite. Person-to-person transmission may occur via the human flea, *Pulex irritans* (a less effective vector), or, in pneumonic plague, by droplet infection.

Plague bacilli have been reported to occur in the throats of symptomless human carriers.

In man, plague occurs in three clinical forms. In *bubonic plague*, the commonest form, the incubation period is 2–6 days; onset is sudden, with fever, prostration, haemorrhages and the development of *buboes*: swollen, inflamed, necrotic lymph nodes. Septicaemia may result in e.g. meningitis or secondary pneumonia. (cf. PESTIS MINOR.)

In *primary septicaemic plague* (incubation period: 2–6 days) there is a sudden onset with high fever, haemorrhages, vomiting and bloody diarrhoea, but no buboes.

Primary pneumonic plague results from infection by inhalation (incubation period: 2–3 days). There is high fever, prostration, severe pneumonia and frequently delirium and coma; the sputum contains large numbers of plague bacilli.

Mortality rates in untreated cases: e.g. bubonic plague 25–50%; pneumonic and septicaemic plague ~100%; early antibiotic therapy is associated with significantly lower mortality.

Lab. diagnosis. Demonstration of the pathogen in fluid from buboes, blood, CSF or sputum (by microscopy and/or culture) and/or serology for antibodies against the F1 ('fraction 1') antigen (see YERSINIA). PCR-based diagnosis (amplification of a sequence of the *Y. pestis caf1* gene) under field conditions in Madagascar was found to be less sensitive than culture or serology and was not recommended [JCM (2000) *38* 260–263].

[Rapid mAb/F1-based test for bubonic/pneumonic plague: Lancet (2003) *361* 211–216.]

Chemotherapy: e.g. tetracyclines, streptomycin or chloramphenicol.

Vaccines containing either inactivated virulent cells or live avirulent cells may offer short-term protection.

[Molecular and cell biology aspects of plague: PNAS (2000) *97* 8778–8783.]

Endemic plague persists in various regions of the world (see figure) and, significantly, transferable (plasmid-mediated) multidrug resistance has been reported in *Yersinia pestis* [NEJM (1997) *337* 677–680]. [Dynamics of plague in the rat population, and transmission of the disease to humans: Nature (2000) *407* 903–906. Bubonic plague (a metapopulation model of a zoonosis): Proc. RSLB (2000) *267* 2219–2230.]

planachromat See ACHROMAT.

planapochromat See APOCHROMAT.

Planctomyces A genus of planktonic freshwater bacteria. Cells: cocci, or coccoid to pear-shaped forms, ca. 0.3–2.0 μm, which develop non-prosthecate, bristle-like appendages when mature; the cell wall contains no peptidoglycan. Reproduction occurs by budding. Types species: *P. bekefii*. Other species include e.g. *P. guttaeformis* and *P. stranskae* [IJSB (1984) *34* 470–477];

'*P. gracilis*' has been excluded from the genus [IJSB (1984) *34* 465–469].

plankter See PLANKTON.

plankton Collectively, the microorganisms (chiefly microalgae and protozoa), small invertebrates (e.g. copepods and other small crustacea) and fish larvae etc which drift more or less passively in water currents within the pelagic zone of lakes, seas and other bodies of water; although many planktonic organisms are actively motile, even these organisms depend on water currents for significant amounts of displacement. The plant and plant-like components of plankton (including e.g. the photosynthetic flagellates) are referred to as *phytoplankton*, while the animal or animal-like components (including e.g. protozoa) are called *zooplankton*. An individual planktonic organism is called a *plankter*.

Plankton, particularly phytoplankton, is important as the basis of the aquatic food chain; it is consumed by a range of invertebrates and by certain fish and whales.

Microbial *marine* plankton includes e.g. DIATOMS, DINOFLAGELLATES, RADIOLARIA, members of the PRYMNESIOPHYCEAE and TINTINNINA, and certain foraminifera (e.g. GLOBIGERINA, GLOBIGERINOIDES). Microbial *freshwater* plankton includes e.g. CYANOBACTERIA, DESMIDS, DIATOMS, and members of the HELIOZOEA.

Various schemes have been proposed for the classification of planktonic organisms according to size; thus, e.g. *nannoplankton* (= *nanoplankton*) may include (depending on author) organisms which are less than 20 μm, less than 50 μm, 5–50 μm, 5–60 μm, or greater than 75 μm. Similarly, *mesoplankton* may include organisms of 0.5–1.0 mm, 0.2–2.0 mm, greater than 1.0 mm or 1–5 mm.

(cf. NEKTON, NEUSTON; see also BLOOM.)

Planktothrix A genus of gas-vacuolate CYANOBACTERIA. [Variation in gas vesicle genes: Microbiology (2000) *146* 2009–2018.]

Planobispora A genus of bacteria (order ACTINOMYCETALES, wall type III; group: maduromycetes) which occur in soil. The organisms form branching substrate and aerial mycelium, the latter bearing elongated sporangia, singly or in groups, each sporangium containing a serially-arranged pair of spores; liberated spores become motile (flagellated). Type species: *P. longispora*. [Book ref. 73, pp. 111–112.]

Planococcus A genus of Gram-positive, asporogenous, catalase-positive, chemoorganotrophic, aerobic bacteria which occur as marine saprotrophs. Cells: cocci (each having 1–3 flagella) which occur in pairs and tetrads; the cell wall contains no teichoic acids. A yellow-brown pigment is formed. Carbohydrates are not utilized. Growth occurs e.g. on nutrient agar containing 12% NaCl. GC%: ca. 39–51. Species: *P. citreus* (type species) and *P. halophilus*.

planofluorite See FLUORITE OBJECTIVE.

planogamete A motile GAMETE.

Planomonospora A genus of bacteria (order ACTINOMYCETALES, wall type III; group: maduromycetes) which occur in soil. The organisms form branching substrate and aerial mycelium, the latter bearing rows of parallel, elongated sporangia, each sporangium being attached at one end to the hypha; each sporangium contains one spore which becomes motile (flagellated) after release. Type species: *P. parontospora*. [Book ref. 73, 112–113.]

planospore A motile SPORE.

plant viruses VIRUSes which can infect and replicate in plants. (cf. VIROID; VIRUSOID.) Commonly, plant viruses show a low degree of host specificity, and some can also replicate in insects. Infection of a plant by a given virus may be systemic – spreading rapidly via the phloem (or occasionally xylem) or slowly from cell to cell via plasmodesmata – or may remain localized at the site of infection. (Spread is apparently a virus-specific function involving virus-encoded or virus-induced proteins [review: AVR (1984) *29* 313–364].) Infection may be symptomless (silent, 'latent'), or may involve the development of symptoms ranging from mild to severe or lethal; symptoms, when present, often depend on the species and variety of the host plant, on the particular virus strain, and on environmental conditions, and may be affected by the presence of another virus or of a

PLANT VIRUSES: basic characteristics

Genome	Envelope	Nucleocapsid morphology	Virus group (or unclassified virus)[a]
dsDNA	–	isometric	caulimoviruses
ssDNA	–	paired isometric particles	geminiviruses
dsRNA	–	isometric	plant reoviruses (Reoviridae), rice ragged stunt virus, some cryptic viruses
ssRNA	+	± rod-shaped	plant rhabdoviruses (Rhabdoviridae), tomato spotted wilt virus
ssRNA			
monopartite	–	isometric	luteoviruses, maize chlorotic dwarf virus, sobemoviruses, tobacco necrosis virus, tombusviruses, tymoviruses
monopartite	–	rod-shaped or filamentous	carlaviruses, closteroviruses, potexviruses, potyviruses, tobamoviruses
bipartite	–	isometric	comoviruses, dianthoviruses, nepoviruses, pea enation mosaic virus, velvet tobacco mottle virus
bipartite	–	rod-shaped or filamentous	furoviruses?, hordeiviruses (some strains), tobraviruses
tripartite	–	± isometric	bromoviruses, cucumoviruses, ilarviruses
tripartite	–	± rod-shaped	alfalfa mosaic virus, hordeiviruses (some strains)
multipartite	–	filamentous	rice stripe virus group

[a] See separate entries for details.

SATELLITE VIRUS or SATELLITE RNA. Virus-infected plants may exhibit one or more of the following symptoms: CHLOROSIS, COLOUR-BREAKING, curling or distortion of leaves, ENATIONS, GALL-formation, MOSAIC, MOTTLE, NECROSIS, RINGSPOT, ROSETTE-formation, STREAK, stunting of growth, TOP NECROSIS, VEIN BANDING, VEIN CLEARING, VIRESCENCE, yellowing (see YELLOWS), etc. (Plant structures which are fully matured at the time of infection may fail to develop symptoms.)

Transmission of many plant viruses can occur mechanically – e.g. by contact between plants or by contamination of a plant with sap from an infected plant (present e.g. on pruning implements, hands etc); in the laboratory mechanical transmission can often be achieved by rubbing the leaves of a plant with a virus preparation – often mixed with a mild abrasive. Some viruses can be transmitted via soil, via seeds or tubers, or by grafting (e.g. from an infected stock to an uninfected scion). Many plant viruses can be transmitted only by specific VECTORS such as sap-sucking or biting insects, mites, nematodes, fungi, or parasitic plants such as the dodder (*Cuscuta* spp). (See CIRCULATIVE TRANSMISSION and NON-CIRCULATIVE TRANSMISSION.)

Identification of a plant virus is seldom possible from symptoms exhibited by infected plants in the field, but may be achieved e.g. by light or electron microscopy of infected tissues (many viruses induce characteristic inclusion bodies in infected cells), by inoculation of the virus into a range of suitable DIFFERENTIAL HOSTS, and/or by serological tests; plant viruses are generally effective immunogens, and specific antibodies against a known virus can be raised in animals and used for the identification of an unknown virus (e.g. by ELISA, IMMUNOELECTRON MICROSCOPY, IMMUNOSORBENT ELECTRON MICROSCOPY, precipitin tests, etc). A plant virus may be assayed e.g. using a LOCAL LESION host.

Cultivation in vitro of many plant viruses can be achieved in plant protoplasts derived e.g. from cell suspension cultures or mesophyll tissue of leaves [AVR (1984) *29* 215–262] or from shoot cultures [e.g. of potatoes: JGV (1985) *66* 1341–1346].

Control of virus diseases of plants may involve any of various preventive measures: e.g. use of virus-free seed (obtained e.g. by meristem culture), control of vector populations, ROGUING, removal of volunteer plants or weeds in which the virus may overwinter, selection of resistant or tolerant plant cultivars, etc. Once a plant is infected with a virus there is no effective chemical treatment; however, certain viruses can be inactivated by subjecting the infected plant (or seeds in the case of seed-borne viruses) to air temperatures of 30–35°C (for days or weeks), to water immersion at ca. 40 or 50°C for shorter periods, etc ('thermotherapy'). [Heat treatment of lettuce seeds carrying lettuce mosaic virus: Ann. Appl. Biol. (1985) *107* 137–145.]

Nomenclature and classification. Individual plant viruses are generally given names based on the common, first-recognized, or most important host plant, and on the main symptom(s) produced by the virus in that plant under natural conditions. (Examples: apple chlorotic leafspot virus; barley yellow dwarf virus.) In the majority of plant viruses the virus genome consists of RNA which is often segmented and sometimes distributed between two or more separate virus components (see MULTICOMPONENT VIRUSES). On the basis of the nature of the genome, shape of the virus particle(s), presence or absence of an envelope, etc, plant viruses are classified into 'groups' (see table); the name of a group is commonly a sigla derived e.g. from the name of the type virus (e.g. *caulimo*viruses from *cauli*flower *mo*saic virus). A few plant viruses are related to certain animal viruses and are included in animal virus classification schemes: see e.g. REOVIRIDAE and RHABDOVIRIDAE. (cf. TOMATO SPOTTED WILT VIRUS.)

***Plantago* mottle virus** See TYMOVIRUSES.

***Plantago* severe mottle virus** See POTEXVIRUSES.

***Plantago* virus** See CAULIMOVIRUSES.

***Plantago* virus X** See POTEXVIRUSES.

plantar wart See PAPILLOMA.

plantaricin S See BACTERIOCINS.

plaque (1) A discrete, macroscopic or microscopic, usually circular region in a cell monolayer (see TISSUE CULTURE) or in a film or layer of bacterial growth (see LAWN PLATE) in which some or all of the cells have been lysed, or their growth has been strongly inhibited, as a result of the activity of a virus or other agent (e.g. BDELLOVIBRIO). In virology, it is generally assumed that each plaque develops following the replication of a single virion; thus, e.g., when the first-infected cell lyses, the progeny virions diffuse outwards, progressively infecting and lysing other cells until a visibly depopulated area (the plaque) is produced. Plaque characteristics – e.g. size, shape, nature of the margin – can be useful e.g. in identifying the infecting virus, or in determining whether or not a cell population is homogeneous in its susceptibility to the virus; thus, e.g., clear plaques (total cell lysis) on a lawn of susceptible bacteria are typically produced by virulent bacteriophages, while turbid plaques are characteristic of temperate phages – which lyse only some of the cells in the plaque area – and of non-lytic phages (see e.g. INOVIRIDAE).

(2) (*verb*) To inoculate a lawn plate or cell monolayer with a virus suspension in order to obtain plaques (sense 1).

(3) See DENTAL PLAQUE.

(4) (*med., vet.*) A patch or flat lesion e.g. on skin or mucous membranes (see e.g. DOURINE).

plaque assay A procedure for determining the PLAQUE TITRE of a given viral suspension.

Bacteriophages. Log dilutions of the phage suspension are initially prepared, and a measured volume of a given dilution is added to a known volume of molten (ca. 45–46°C) semi-solid agar containing an excess of sensitive bacteria; the whole is well mixed and poured onto a plate of solid nutrient agar, forming an upper layer 1–2 mm thick. An identical procedure is carried with each dilution; the plates are incubated at the growth temperature of the bacterium and subsequently examined for PLAQUES. Since each plaque is assumed to have been caused by a single phage, the plaque titre of the original phage suspension can be calculated from the volumes and dilutions used.

Other viruses. Plaque assay can be carried out for other viruses if suitably sensitive monolayer cultures (see TISSUE CULTURE) are available. In one method, each of a number of monolayer cultures is drained of growth medium and inoculated with a known volume of one of a range of log dilutions of the viral suspension; the cultures are rocked to enable viruses to adsorb throughout the cell sheet. The monolayers are then overlaid with buffered growth medium in semi-solid purified agar, incubated, and subsequently examined for plaques. (The agar overlay inhibits the spread of viruses from the plaque areas.)

plaque-forming unit (pfu) An entity – usually a virion, but also e.g. an 'infectious centre' (see ONE-STEP GROWTH EXPERIMENT) – which can give rise to a single PLAQUE under appropriate conditions. (See also EOP and PLAQUE TITRE.)

plaque mutant *Syn.* PLAQUE-TYPE MUTANT.

plaque-reduction assay See INTERFERONS.

plaque titre The number of PLAQUE-FORMING UNITS per unit volume.

plaque-type mutant (plaque mutant; plaque morphology mutant) A mutant virus which forms PLAQUES (sense 1) that differ from those formed by the wild-type virus in a population of sensitive cells.

plasma (1) (in vivo) The fluid (non-particulate) part of blood. (2) (in vitro) Fluid obtained (e.g. by centrifugation) from blood which has been pre-treated with an ANTICOAGULANT. (cf. SERUM.)

plasma cell (*immunol.*) An antibody-secreting cell that develops from a B LYMPHOCYTE on appropriate antigenic stimulation (see ANTIBODY FORMATION); plasma cells occur e.g. in the spleen and (particularly) in lymph nodes which are draining an infected site. In the earliest stage of the immune response the secreted antibodies are of the IgM class; however (for TD antigens), antibodies of other classes are secreted subsequently following *class switching* (see ANTIBODY FORMATION).

Typical plasma cells are ovoid, ca. 10 × 15 μm, with an off-centre nucleus which, when stained, may have a 'clock-face' or 'cartwheel' appearance due to the arrangement of chromatin. The (non-granular) cytoplasm contains a well-developed, ribosome-rich endoplasmic reticulum (see also METHYL GREEN–PYRONIN STAIN).

plasma clotting See FIBRIN; COAGULASE; ANTICOAGULANT.

plasma membrane *Syn.* CYTOPLASMIC MEMBRANE.

plasma volume extenders (for blood transfusion) See DEXTRANS.

plasmagel See e.g. SARCODINA.

plasmalemma *Syn.* CYTOPLASMIC MEMBRANE.

plasmalogen A glycerophospholipid in which the glycerol bears a 1-alkenyl ether group. Plasmalogens occur e.g. in the cytoplasmic membrane in anaerobic bacteria.

plasmasol See e.g. SARCODINA.

Plasmaviridae (mycoplasma virus type 2 phages; L2 acholeplasmavirus group) A family of enveloped, pleomorphic MYCOPLASMAVIRUSES containing dsDNA. One genus: *Plasmavirus*; type species: BACTERIOPHAGE MV-L2. Other members: bacteriophage 1307 and possibly v1, v2, v4, v5 and v7.

Plasmavirus See PLASMAVIRIDAE.

plasmid In many types of prokaryotic and eukaryotic cell: a linear or covalently closed circular (ccc) molecule of DNA (q.v.) – distinct from chromosomal DNA, mtDNA, ctDNA or kDNA – which can replicate autonomously (i.e., independently of other replicons) and which is commonly dispensable to the cell; the term 'plasmid' is sometimes used to refer also to the dsRNA elements which occur in certain yeasts: see KILLER FACTOR. (cf. EPISOME.) (The stable, extrachromosomal (i.e., non-integrated) prophages of certain bacteriophages (e.g. BACTERIOPHAGE P1) conform to the definition of plasmid and are commonly referred to as plasmids.) Some plasmids appear to have arisen from chromosomal or mitochondrial DNA: see e.g. THREE-MICRON DNA PLASMID.

The following refers primarily to *prokaryotic* plasmids.

Plasmids occur in both Gram-positive and Gram-negative bacteria; in size they range from ca. 1.5 kb (MWt ca. 10^6) to ca. 300 kb (MWt ca. 2×10^8). Most bacterial plasmids are ccc dsDNA molecules, and these are normally *supercoiled* (see DNA). However, some plasmids can occur as ccc ssDNA molecules [PNAS (1986) *83* 2541–2545], while linear plasmids occur e.g. in *Streptomyces* and in *Borrelia* [Mol. Microbiol. (1993) *10* 917–922]. Many bacterial plasmids carry one or more TRANSPOSABLE ELEMENTS. (cf. KILLER PLASMIDS; TWO-MICRON DNA PLASMIDS.)

Certain obligate pathogens in the alpha subclass of Proteobacteria normally *lack* plasmids, although plasmids occur in other members of the same subclass. It has been suggested that species of *Anaplasma*, *Bartonella*, *Brucella* and *Rickettsia* evolved without plasmids because their relatively constant (intracellular) environment does not select for the genetic diversity needed by species in more challenging environments [FEMS Reviews (1998) *22* 255–275]; however, some intracellular bacteria (e.g. *Chlamydia* spp) are known to contain plasmids [see e.g. Microbiology (1997) *143* 1847–1854].

Many plasmids encode product(s) and/or functions(s) which modify the phenotype of the host cell. Almost any feature may be encoded by plasmids, including e.g. resistance to particular ANTIBIOTIC(S) and/or HEAVY METAL(S) (see R PLASMID), and the synthesis of AGROCINS, BACTERIOCINS, GAS VACUOLES (e.g. in certain strains of *Halobacterium*), restriction endonucleases (see also PHAGE TYPING), SIDEROPHORES, toxins (see e.g. ANTHRAX TOXIN, ETEC) or virulence factors (see e.g. CROWN GALL and VW ANTIGENS). Some plasmid-encoded products alter the antigenic characteristics of a cell (see e.g. ANTIGEN GAIN), and some confer the ability to utilize particular substrate(s) (see e.g. LACTOSE PLASMID, OCT PLASMID, TOL PLASMID). One plasmid may counteract the effects of another; for example, the pSa plasmid can abolish the specific adherence properties of *Agrobacterium* strains which are due to the presence of the Ti plasmid [JGM (1983) *129* 3657–3660]. (See also CRYPTIC PLASMID.)

The presence of a plasmid within a cell may often be inferred from the cell's phenotype: certain characteristics are commonly plasmid-specified. Whether or not a plasmid is present may be determined e.g. by gel ELECTROPHORESIS of a CLEARED LYSATE, or by treating the cells with certain agents that eliminate plasmids (see CURING sense 2) and observing for the loss of particular phenotypic properties. [Plasmid technology: Book ref. 177.]

Some plasmids can promote their own intercellular transfer by CONJUGATION (sense 1b): see CONJUGATIVE PLASMID; some conjugative plasmids contain a TRANSFER OPERON which is DEREPRESSED for conjugal transfer. (See also EPIDEMIC SPREAD.) In addition to their own intercellular transfer, some conjugative plasmids can *mobilize* the host cell's chromosome (or certain types of non-conjugative plasmid) for intercellular transfer (see CONJUGATION sense 1b).

More than one type of plasmid can occur in a given cell; plasmids which have different modes of replication (see below) and of PARTITION are said to exhibit *compatibility*, i.e., they can co-exist stably in the same cell. Plasmids which have similar or identical replication/partition systems are incompatible, i.e., they cannot co-exist stably in the same cell; the phenomenon of incompatibility enables plasmids to be classified into a number of 'incompatibility groups' (Inc groups): see INCOMPATIBILITY.

Plasmid replication. A plasmid controls its own replication, using the cell's biosynthetic machinery to make any plasmid-encoded proteins etc. that are needed; plasmid DNA REPLICATION requires various host proteins – commonly e.g. the product of the DNAB GENE, the DNAC GENE, the *dnaE* gene (= *polC* gene: see DNA POLYMERASE) and the *dnaG* gene (see PRIMASE) in *Escherichia coli*. (See also DNAA GENE.)

Replication is not coupled to any specific stage or event in the CELL CYCLE, and in some plasmids it occurs throughout the cell cycle.

Replication and PARTITION appear to be independent processes, although both can determine INCOMPATIBILITY.

The rate-limiting step in replication is *initiation*, and the rate of initiation is controlled in ways that differ according to plasmid.

In some plasmids the control of initiation involves constitutive synthesis of small, *trans*-acting ANTISENSE RNA molecules. Thus, e.g. in plasmid ColE1 control involves the synthesis of two, free (i.e. non-duplexed) strands of plasmid-encoded RNA, one of which, RNA II, can bind at the plasmid's origin and act as a primer, thus permitting replication. The other strand, RNA I, can interfere by base-pairing with RNA II – such that the outcome of RNA I–RNA II interaction determines the occurrence, or otherwise, of replication (for further details see COLE1 PLASMID). (It appears that RNA I is normally polyadenylated, such polyadenylation facilitating its degradation; in *Escherichia coli*, mutations in the *pcnB* gene, which encodes a poly(A)polymerase, result in RNA I molecules that are stabilized through lack of polyadenylation – the effect of which is to inhibit replication of ColE1, i.e. to reduce its copy number.) Once initiated, replication in ColE1 proceeds as in the *E. coli* chromosome, although in this plasmid replication occurs *uni*directionally from the origin.

In the R1 PLASMID (q.v.), antisense RNA molecules bind to, and inhibit translation of, the mRNA of the RepA protein – which, in turn, is involved in regulating the frequency of plasmid replication.

Many of the small circular plasmids in Gram-positive bacteria (and some in Gram-negative bacteria) replicate by a ROLLING CIRCLE MECHANISM. Replication is initiated by a plasmid-encoded *Rep protein* which, by nicking a specific strand at the origin, generates a 3′ end for synthesis. One round of replication produces a complete dsDNA plasmid and a circular ssDNA copy, the latter then being replicated to the dsDNA state from an RNA primer. In some of these plasmids initiation is controlled by regulation of the synthesis/activity of the Rep protein (as the quantity of Rep protein affects the frequency of replication); in some cases small, plasmid-encoded RNA molecules bind to the *rep* mRNA and bring about either premature termination of transcription or inhibition of translation. Alternatively, the Rep protein may be inactivated *after only one use* by binding to a small plasmid-encoded oligonucleotide. [Replication control in rolling circle plasmids: TIM (1997) *5* 440–446.]

In many plasmids of Gram-negative bacteria, the Rep initiator protein binds to 'direct repeats' of nucleotides (*iterons*) in the region of the plasmid's origin of replication, thus contributing to a nucleoprotein complex that initiates replication. For example, in the F PLASMID, RepE (the E protein) carries out this function; replication proceeds by a *bi*directional Cairns-type mechanism (as in the *E. coli* chromosome). (See also R6K PLASMID.) For each plasmid, the iterons have a characteristic number, spacing and composition. In some plasmids, Rep can also repress transcription of its *own* gene by binding to inverted repeat sequences overlapping the promoter; in some cases (including the F plasmid) it has been shown that, while *monomers* of Rep bind to iterons (promoting replication), *dimers* bind to the promoters of their own genes (inhibiting replication). RepA monomers (as opposed to dimers) of the *Pseudomonas* plasmid pPS10 may be structurally suitable for RepA–iteron binding [EMBO (1998) *17* 4511–4526].

In general, plasmid replication and PARTITIONing seem likely to depend on some kind of association between the plasmid and the cytoplasmic membrane [Mol. Microbiol. (1997) *23* 1–10].

Under normal growth conditions, replication is initiated at a frequency which, for a given plasmid in a given type of cell under given conditions, leads to the establishment of a characteristic COPY NUMBER. During steady-state conditions each plasmid replicates, on average, once per cell division – although the inter-replication period may vary widely from one plasmid to another.

If protein synthesis is inhibited (e.g. by CHLORAMPHENICOL), some plasmids can continue to replicate, and can achieve copy numbers much higher than those achieved under normal growth conditions. Such plasmids (e.g. ColE1) are said to be under 'relaxed control'; a typical 'relaxed' plasmid is a small, non-conjugative MULTICOPY PLASMID whose copy number under *normal* conditions is ca. 10–30. (A 'relaxed' plasmid can exceed its normal copy number only under abnormal conditions. See also CLONING.) Plasmids which fail to replicate on inhibition of protein synthesis are said to be under 'stringent control'; a typical 'stringent' plasmid is a large, low-copy-number CONJUGATIVE PLASMID.

Whether a plasmid is 'relaxed' or 'stringent' depends on its system of replication control. Thus, e.g. in the 'relaxed' COLE1 PLASMID (q.v.), not only is ongoing protein synthesis not required for replication, but the inhibition of protein synthesis eliminates a protein (Rom/Rop) which is inhibitory to plasmid replication. Conversely, in 'stringent' plasmids replication involves e.g. the synthesis of proteins that are essential for the initiation of replication.

[Control of plasmid replication: Book ref. 161, pp. 189–214.]

Some plasmids have mechanisms for ensuring their persistence within a bacterial strain: see e.g. *ccd* mechanism in entry F PLASMID.

Nomenclature. Plasmids are designated in various ways. COLICIN PLASMIDS are designated according to the colicins they encode (e.g. ColE1-K30 encodes colicin E1, ColV,I-K94 encodes colicins V and Ia). Many antibiotic-resistance plasmids are designated by the letter 'R' followed by a number, e.g. R1, R46 etc; others include e.g. R6K, RK2, RP1 etc. Recombinant plasmids are often indicated by the prefix 'p', e.g. pML31. A number of plasmids are given trivial names (e.g., Δ), while the names of some indicate particular functions (see e.g. Ti plasmid under CROWN GALL, and TOL PLASMID).

(See also entries for individual plasmids: e.g. COLE1 PLASMID; DELTA; F PLASMID; PBR322; PMB1; PML31; PSC101; R1 PLASMID; R46 PLASMID; R100 PLASMID; R6K PLASMID; RP1 PLASMID.)

plasmid p*Inv* See EIEC.

plasmid pYV See FOOD POISONING (*Yersinia*).

plasmin See FIBRINOLYSIN.

plasminogen (profibrinolysin) An inactive precursor of FIBRINOLYSIN normally present in blood; it is a large (~90 kDa) glycoprotein.

Plasminogen has been found to bind disease-associated PRIONS, but not the normal protein (PrP^C); this property may be useful in diagnostic tests for TRANSMISSIBLE SPONGIFORM ENCEPHALOPATHIES [Nature (2000) *408* 479–483].

plasmodesma (*pl.* plasmodesmata) A fine channel in a plant cell wall or a fungal SEPTUM through which cytoplasm can pass; plasmodesmata allow continuity between the cytoplasm of adjacent cells. (See also DESMOTUBULE, PIT CONNECTION and PLANT VIRUSES.)

plasmodiocarp See MYXOMYCETES.

Plasmodiophora See PLASMODIOPHOROMYCETES and CLUBROOT.

Plasmodiophorea See RHIZOPODA.

Plasmodiophorina See GYMNOMYXA.

Plasmodiophoromycetes ('endoparasitic slime moulds') A class of eukaryotic organisms (division MYXOMYCOTA) which are obligate intracellular parasites in plants, algae and fungi. The vegetative stage of the organisms is a plasmodium (a naked, multinucleate protoplast); hyphae and fruiting bodies are apparently never formed. Although traditionally regarded as fungi, the true taxonomic position of these organisms remains uncertain; they are apparently unrelated to other members of the Myxomycota. rRNA data studies support the idea that they may be at the base of an evolutionary line leading to the

Chytridiomycetes, Ascomycotina and Basidiomycotina, but do not exclude the possibility of a protozoal ancestry [TBMS (1983) *80* 107–112]. (cf. RHIZOPODA and GYMNOMYXA.) Genera include *Ligniera, Octomyxa, Plasmodiophora, Polymyxa, Sorodiscus, Sorosphaera, Spongospora, Tetramyxa, Woronina.* (cf. ENDEMOSARCA; PHAGOMYXA.) Species are parasitic in e.g. higher plants (e.g. *Plasmodiophora brassicae* in crucifers – see CLUBROOT; *Spongospora subterranea* in potatoes – see POWDERY SCAB; *Ligniera betae* in beet; *L. junci* in various monocots and dicots); in algae (e.g. *Woronina* in *Vaucheria, Sorodiscus* in *Chara*); and in certain aquatic fungi (e.g. *Woronina* in *Achlya, Pythium* or *Saprolegnia*). Infected tissues often exhibit hypertrophy and sometimes hyperplasia.

In e.g. *Plasmodiophora brassicae* and *Polymyxa betae*, infection of a new host is initiated when a ('primary') zoospore attaches to the wall of a root hair, withdraws its flagella, and encysts. Within the encysted cell there develops a tubular structure, the *Rohr*, containing an osmiophilic spine-like structure, the *Stachel*. The *Rohr* evaginates rapidly, and the *Stachel* punctures the host cell wall; the protoplast of the encysted zoospore then enters the host cell and develops into a vegetative primary (= sporangiogenous) plasmodium. When mature, this plasmodium gives rise to uninucleate 'secondary' zoospores, each with two smooth flagella of different lengths. [Secondary zoospore development in *P. brassicae*: TBMS (1984) *82* 339–342.] The zoospores are released either by disintegration of the host cell or via an 'exit tube' arising from the zoosporangium. The role of these zoospores in the life cycle is unclear; fusion between two zoospores has been reported, but karyogamy – when it occurs – is apparently delayed. A zoospore (or fused pair of zoospores?) infects another host cell, in which it develops to form a (diploid?) secondary (= cystogenous) plasmodium. When mature, this plasmodium eventually cleaves to form a number of uninucleate, thick-walled, resistant cysts (resting spores); meiosis has been observed at this stage in several species. [Ultrastructure of meiosis in *Woronina pythii*: Mycol. (1984) *76* 1075–1088.] In some species the cysts are arranged in characteristic clusters (cystosori). [Fine structure of the cystosorus in *Spongospora subterranea*: CJB (1985) *63* 2278–2282.] On disintegration of the host cell the cysts are liberated into the soil where they may remain viable for long periods (e.g. years). On germination, each cyst gives rise to a biflagellate (heterokont) zoospore which can infect another host, thus initiating a new cycle.

Some members of the Plasmodiophoromycetes are important as vectors of certain plant viruses: see e.g. BEET NECROTIC YELLOW VEIN VIRUS, SOIL-BORNE WHEAT MOSAIC VIRUS, POTYVIRUSES, TOBAMOVIRUSES.

plasmodium A multinucleated, usually motile mass of protoplasm which is usually naked (i.e., bounded only by a plasma membrane) and which is generally variable in size and form; plasmodia are the main vegetative forms in e.g. members of the ACARPOMYXEA, MYXOMYCETES (q.v. for types), MYXOSPOREA and PLASMODIOPHOROMYCETES. (cf. PSEUDOPLASMODIUM.)

Plasmodium A genus of protozoa of the suborder HAEMOSPORORINA. The genus includes the common causal agents of MALARIA in man: *P. falciparum* [genome: Nature (2002) *419* 498–511], *P. malariae* (formerly *P. rodhaini*), *P. ovale* and *P. vivax*. *P. cynomolgi* and *P. knowlesi* are parasitic in non-human primates; they occasionally cause malaria in man, and have been widely used as models in research on human malaria. *P. berghei* and *P. yoelii* (also spelt *P. yoelli*) cause rodent malaria; they have been used for testing antimalarial drugs.

The life cycle is basically similar in all species. In *Plasmodium* spp, sexual reproduction occurs in a mosquito and leads to the formation of large numbers of fusiform or needle-shaped uninucleate cells (*sporozoites*), ca. 6–18 μm in length, that are infective for the vertebrate host. (Below ca. 15°C *P. falciparum* will not develop within the mosquito, and below ca. 18°C its development is retarded; *P. vivax* can develop in mosquitoes at somewhat lower temperatures – which probably accounts for its presence in temperate as well as tropical and subtropical regions.) Transmission of the parasite occurs when the (female) mosquito takes a blood meal by inserting her proboscis into the tissues of a vertebrate host: sporozoites, present in the insect's saliva, quickly reach the bloodstream and initiate the cycles of asexual reproduction.

In man, the sporozoites rapidly enter parenchymatous cells in the liver; there, they typically grow and eventually undergo SCHIZOGONY, rupturing the liver cells and releasing into the bloodstream large numbers of minute cells (*merozoites*), each ca. 1 μm in diameter. This stage is called *pre-erythrocytic* schizogony or *exoerythrocytic schizogony*; its duration ranges from ca. $5^1/_2$ days (in *P. falciparum*) to ca. 15 days (in *P. malariae*), and is ca. 8 and 9 days respectively, in *P. vivax* and *P. ovale*. Instead of growth and schizogony, some or all of the sporozoites of some species – e.g. *P. cynomolgi*, *P. ovale*, *P. vivax* – form specialized cells (*hypnozoites*) which can remain dormant within the liver for e.g. 8–10 months; on resuming development, the hypnozoites follow the normal course of growth and schizogony and give rise to the (periodic) relapses which are characteristic of malaria caused by these species.

Merozoites bind to, and enter, erythrocytes – except e.g. in *P. berghei* and *P. yoelii*, in which the merozoites parasitize reticulocytes (erythrocyte precursor cells). Within an erythrocyte a merozoite forms a central vacuole and develops into the characteristic 'ring' or 'signet-ring' stage; in stained preparations this commonly appears as a ring of cytoplasm containing a bulging nucleus. The growing cell (now called a *trophozoite*) obtains nutrients from e.g. the protein part of the host cell's haemoglobin; *P. falciparum* apparently obtains its iron from outside the parasitized erythrocyte (see TRANSFERRIN; cf. HAEMOZOIN). At this stage the erythrocyte and/or parasite may exhibit certain inclusions: see HAEMOZOIN, MAURER'S CLEFTS, SCHÜFFNER'S DOTS and ZIEMANN'S DOTS. A protein designated exp-1 was detected in a cytoplasmic compartment in *P. falciparum*-infected erythrocytes; it was suggested that parasite-encoded proteins pass through this compartment on their way to the erythrocyte's plasmalemma [EMBO (1987) *6* 485–491].

The fully grown parasite (now called a *schizont*), which no longer appears ring-shaped, subsequently undergoes schizogony to form a number of merozoites (each ca. 1 μm diam.) which are released from the ruptured erythrocyte; these merozoites can infect fresh erythrocytes. Typically, merozoites are released more or less synchronously from large numbers of parasitized erythrocytes, thus accounting for the periodic symptoms of malaria; the duration of the erythrocytic cycle (i.e., from infection to rupture) typically ranges from ca. 2 days (*P. falciparum*) to ca. 3 days (*P. malariae*), but is only 1 day in *P. knowlesi*.

After one or more erythrocytic cycles, some of the intraerythrocytic merozoites develop into gametocytes (rather than schizonts); within the bloodstream, gametocytes may remain viable for one month or more, but they do not develop further unless ingested by an anopheline mosquito. On ingestion by a mosquito the gametocytes develop into male and female gametes. In this process the (male) microgametocyte undergoes *exflagellation*: several nuclear divisions occur, up to eight thread-like appendages are formed, and the microgametocyte eventually breaks up to form uninucleate, uniflagellate male gametes (microgametes). A microgamete fuses with a (female) macrogamete to form a zygote which subsequently becomes motile (and is then called an *ookinete* or *vermicule*). The ookinete passes through the mosquito's gut wall, rounds off, and forms an oocyst in the outer layer of the gut wall (adjacent to the haemocoele). Sporozoites develop within the oocyst; they are subsequently released into the haemocoele, and many find their way to the insect's salivary glands, thus completing the cycle.

Plasmodial antigens and antimalarial vacines. Certain plasmodial antigens have been used in the preparation of antimalarial vaccines. One such antigen is the *circumsporozoite protein* (*CS protein*), a protein (MWt ca. 60000) which forms a thick surface coat on the (mature) sporozoite. [CS proteins: Cell (1985) *42* 401–403.] Monoclonal antibodies to CS proteins inhibit infectivity of sporozoites both in vitro and in vivo. (cf. CSP REACTION.) The CS proteins from different species of *Plasmodium* all contain a block of tandemly repeated peptides (though these vary in nature according to species); in e.g. *P. falciparum* there are multiple copies of the tetrapeptide asparagine–alanine–asparagine–proline (designated NANP), and these sequences are involved in the binding of anti-CS protein monoclonal antibodies. One experimental antimalarial vaccine consisted of a suspension of polypeptides each containing multiple repeats of NANP (and of another tetrapeptide, asparagine–valine–aspartic acid–proline (NVDP), also found in *P. falciparum*). Another experimental vaccine was prepared from an NANP trimer conjugated to tetanus toxoid.

The *ring-infected erythrocyte surface antigen* (RESA) is a parasite-encoded protein found in the plasmalemma in erthrocytes infected with the ring stage of *P. falciparum*; it is a protein of MWt ca. 155000, and is sometimes referred to as Pf155. Antibodies to a repeated amino acid sequence in RESA inhibit the invasion of erythrocytes by merozoites in vitro.

The possibility of a *transmission-blocking vaccine* arose from the observation that anti-gamete antibodies raised in an animal host can neutralize the gametes (of *Plasmodium*) in a mosquito which has taken a blood meal from that host; transmission of the parasite to a new host is therefore inhibited.

The problems of developing a satisfactory antimalarial vaccine are exacerbated by ANTIGENIC VARIATION in *Plasmodium*.

For more recent data on anti-malaria vaccines see the entry MALARIA.

Subgenera of Plasmodium. Subgenera are defined on the basis of (a) the vertebrate host(s), and (b) the morphology of the erythrocytic schizont and the gametocyte. Subgenera include:

Haemamoeba. Hosts: birds. Vector: the mosquito *Mansonia crassipes* (experimental vectors include *Aedes aegypti*, *Anopheles* spp, *Culex* spp). Schizonts are large, gametocytes spherical.

Laverania. Hosts: primates. Vector: anopheline mosquitoes. Schizonts are large, gametocytes elongated and crescentic. Species include *Plasmodium* (*Laverania*) *falciparum*.

Plasmodium. Hosts: primates. Vector: anopheline mosquitoes. Schizonts are large, gametocytes spherical. Species include *Plasmodium* (*Plasmodium*) *cynomolgi*, *P.* (*P.*) *knowlesi*, *P.* (*P.*) *malariae*, *P.* (*P.*) *ovale*, and *P.* (*P.*) *vivax*.

Vinckeria. Hosts: non-primate mammals. Vector: anopheline mosquitoes. Schizonts are small, gametocytes spherical.

Other (avian) subgenera: *Giovannolaia*, *Novyella*, *Huffia*.

plasmogamy Coalescence of the cytoplasm of two or more cells; in fertilization, plasmogamy is followed by KARYOGAMY.

Plasmopara A genus of fungi of the PERONOSPORALES; species include *P. viticola* and *P. halstedi*, the causal agents of DOWNY MILDEWS on grape and sunflower, respectively.

plasmotomy In certain multinucleate protozoa (e.g. *Actinosphaera*, *Pelomyxa*): an asexual reproductive process in which fission results in the formation of two or more multinucleate daughter cells, the nuclei of the parent cell being more or less evenly distributed between them; nuclear and cytoplasmic divisions appear to occur independently.

plastics (from bacteria) See BIOPOL.

plastid Any of various types of membrane-limited, typically DNA-containing organelle which occur within the cells of plants and algae but not in those of animals or prokaryotes. Examples: AMYLOPLAST, CHLOROPLAST, CHROMOPLAST, ELAIOPLAST, ETIOPLAST and LEUKOPLAST. (See also PLASTOME.)

plastocyanin A copper-containing BLUE PROTEIN which acts as an electron donor to photosystem I (see PHOTOSYNTHESIS). In certain algae (e.g. *Scenedesmus*) plastocyanin can apparently be replaced by a *c*-type cytochrome in the absence of copper. [Structure and function of plastocyanin: Book ref. 146, pp. 19–31.]

plastoglobuli Osmiophilic lipid globules in the stroma of a CHLOROPLAST.

plastome The genetic complement of a PLASTID.

plastoquinone See QUINONES and PHOTOSYNTHESIS.

plate (1) A solid MEDIUM in a PETRI DISH prepared (using an ASEPTIC TECHNIQUE) by pouring a sterile molten (AGAR- or GELATIN-based) medium into a Petri dish to a depth of 3–5 mm and allowing it to set. Freshly-poured plates to be used e.g. for STREAKING should be left for 10–20 min in a 37°C hot-air incubator with the lid partly off so that the surface moisture can evaporate; such 'drying' before inoculation prevents unwanted spreading of the inoculum in the surface film of moisture. (See also SPREAD PLATE.) (2) A *plate* CULTURE. (3) (*verb*) To inoculate a plate: see PLATING.

plate dilution test See DILUTION TEST.

platelet factor 4 (PF4) See CHEMOKINES.

platelets (thrombocytes) Flat, disc-like, membrane-bounded bodies (ca. 2.5 μm diam.) which occur in mammalian blood (ca. 250000/mm^3 in human blood); they are formed by fragmentation of megakaryocytes in red bone marrow. Platelets adhere to rough or damaged surfaces and assist in blood clotting (though clotting can occur in their absence). Platelets contain e.g. HISTAMINE and SEROTONIN.

plating (1) The act of distributing an INOCULUM on the surface of a PLATE – see e.g. SPREAD PLATE and STREAKING. (2) (*serol.*) The use of specifically coated plastic (or other) surfaces for the separation of specific cells (or other entities) from heterogeneous populations. The procedure is based on antigen–antibody specificity; thus, e.g., antibody, immobilized (by its Fc portion) on an inert support, will bind, and thus remove from a mixture, those cells which bear homologous cell-surface antigens.

Platyamoeba See AMOEBIDA.

Platymonas See TETRASELMIS.

***plcB* gene** See LISTERIOSIS.

plectenchyma (*mycol.*) A tissue composed of hyphae or cells; such tissues occur e.g. in LICHEN thalli, in fungal fruiting bodies, etc. There are two principal types of plectenchyma: *prosenchyma* (prosoplectenchyma), in which the hyphae remain distinguishable, are rather loosely woven, and are more or less parallel to each other; and *pseudoparenchyma* (paraplectenchyma), in which hyphae are usually not distinguishable as such, the tissue consisting of closely packed, isodiametric or oval cells (resembling the parenchyma of higher plants). 'Palisade plectenchyma' occurs in the cortex of many lichen thalli, and consists of short hyphae orientated more or less perpendicular to the thallus surface.

Plectonema A genus of cyanobacteria of the LPP GROUP. Although non-heterocystous, *P. boryanum* can carry out NITROGEN FIXATION under microaerobic (but not under aerobic) conditions.

Plectrovirus (mycoplasma virus type 1 phages; L1 acholeplasmavirus group) A genus of BACTERIOPHAGES of the INOVIRIDAE. Host: *Acholeplasma laidlawii*; plaques turbid (sometimes with a clear centre), 0.5–6.0 mm across. Other strains of *Acholeplasma* and *Mycoplasma* may be infected. The virions are short, naked, bullet-shaped rods (70–90 × 14–16 nm) comprising ss cccDNA (MWt 1.5×10^6) and four proteins. Proposed type member: MV-L51 (= L1 strain L51). Other member: MV-L1 (= L1 strain L1). Possible member: SV-C1 (see SPIROPLASMAVIRUSES). See also MYCOPLASMAVIRUSES.

pleiotropic Refers to e.g. a gene or a mutation which has multiple effects: e.g., expression of a pleiotropic gene affects more than one phenotypic characteristic.

Pleistophora See MICROSPOREA.

pleomorphic Refers to e.g. an organism which exhibits PLEOMORPHISM.

pleomorphism (1) In general: an inherent variability in size and shape e.g. among the cells in a pure culture of a given organism. Cells of a pleomorphic organism (e.g. *Corynebacterium* spp, *Propionibacterium* spp) may occur in a wide range of indeterminate forms which are often variations of a single basic shape: e.g. the cells may be basically rod-shaped but may be swollen at one or both ends, curved or straight, branched or unbranched, etc. (cf. POLYMORPHISM sense 1.)

(2) *Syn.* POLYMORPHISM sense 1. (3) (*mycol.*) Degenerative changes – e.g. loss of the ability to form conidia and/or to synthesize pigment – which occur in some fungi (e.g. certain dermatophytes) when propagated in laboratory cultures.

Pleospora See DOTHIDEALES.

Plesiomonas ('C27 organisms') A genus of Gram-negative bacteria of the VIBRIONACEAE. Cells: straight, round-ended rods, 0.8–1.0 × 1–3 μm. Usually motile with 2–5 lophotrichous unsheathed flagella; lateral flagella may be formed in young cultures on solid media. Optimum growth temperature: ca. 35–38°C. Growth does not require NaCl and cannot occur with 7.5% NaCl. Growth media: e.g. SS, XLD or Hektoen agars [media, identification etc: Book ref. 46, pp. 1285–1288]. Oxidase +ve; acid but not gas is formed from glucose and e.g. inositol; extracellular DNase, gelatinase and lipase are not formed; most strains are sensitive to O/129. Strains may possess the enterobacterial common antigen, and a few share an O antigen with *Shigella sonnei*. GC%: 51. Type species: *P. shigelloides* – found in association with fish and other aquatic animals and in mammals; strains can cause e.g. gastroenteritis in man. [Book ref. 22, pp. 548–550.]

According to 5S rRNA sequence analysis, *P. shigelloides* is more closely related to *Proteus mirabilis* than is *Proteus vulgaris* and should therefore be included in the genus PROTEUS as *Proteus shigelloides* [SAAM (1985) *6* 171–182]. (See also ENTEROBACTERIAL COMMON ANTIGEN.)

Pleurasiga See CHOANOFLAGELLIDA.

Pleurastrophyceae See TETRASELMIS and TREBOUXIA.

Pleurastrum A genus of filamentous green algae related to TREBOUXIA; the filaments are branched and lack holdfasts. *P. terrestre* may also grow as unicells under certain conditions.

Pleuroascus See EUROTIALES.

Pleurocapsa See PLEUROCAPSA GROUP.

Pleurocapsa group A large and diverse group of unicellular CYANOBACTERIA (section II) in which growth and binary fission lead to the formation of irregular, sometimes pseudofilamentous cell aggregates; baeocytes initially lack a fibrous outer cell wall layer and are motile (by gliding). In some members (subgroup I) growth of the baeocyte is symmetrical, resulting in a spherical vegetative cell, while in others (subgroup II) growth is asymmetrical and vegetative cells are elongated prior to the onset of binary fission. The group includes organisms previously included in many phycological genera (e.g. *Hyella*, *Pleurocapsa*) which were based on characters such as endolithic habit, arrangement of vegetative cells in aggregates, etc. GC%: 39–47.

pleurocapsalean (1) Refers to BAEOCYTE-forming CYANOBACTERIA (section II). (2) Refers to any member of the phycological order Pleurocapsales.

Pleurocapsales See CYANOBACTERIA.

Pleurococcus A genus of unicellular green algae (division CHLOROPHYTA) which are extremely common on wood, bark etc; the cells generally occur in characteristic packets. The organisms have been referred to by various names (*Pleurococcus viridis*, *P. vulgaris*, *Protococcus viridis*, etc), but apparently should be called *Desmococcus olivaceus* [Taxon (1985) *34* 671–672].

pleurodynia See BORNHOLM DISEASE.

Pleuronema See SCUTICOCILIATIDA.

Pleurosigma See DIATOMS.

Pleurotus A genus of fungi of the APHYLLOPHORALES (family Polyporaceae) in which the basidiocarp is commonly pileate and stipitate (stipe often eccentric or lateral); the hymenium is borne on decurrent lamellae (see LAMELLA). *P. ostreatus* is an edible species (see OYSTER FUNGUS) which grows on wood and which can be pathogenic for certain trees, particularly beech (*Fagus*).

plicate Folded.

***Plodia interpunctella* GV** See GRANULOSIS VIRUSES.

ploidy The number of (complete) sets of chromosomes (or AUTOSOMES) in a cell. (See also DIPLOID, HAPLOID, POLYPLOID, ANEUPLOID.)

plugging (of pipettes) The insertion of a small plug of COTTON WOOL into the wide end of a pipette or PASTEUR PIPETTE *before* sterilization; liquids being pipetted are thus protected from contamination from the rubber bulb, pi-pump etc.

plum pox (sharka disease) A disease affecting many species of *Prunus* (e.g. plum, apricot, peach, blackthorn); it is caused by a virus of the POTYVIRUSES and is transmitted by aphids and by infected rootstocks. Symptoms commonly include diffuse mottling or ring-spots on the leaves, premature fruit-drop, and (depending on fruit type) dark bands or rings, or grooving and pitting, on the fruits.

plumbagin 5-Hydroxy-2-methyl-1,4-naphthoquinone.

plunging jet See JET LOOP FERMENTER.

pluracidomycins CARBAPENEM antibiotics.

plurivorous Able to use a range of hosts or substrates.

Pluronic polyol F127 A copolymer of ethylene oxide and polypropylene oxide used e.g. as a solidifying agent for culture media. (See also MEDIUM.) A polyol-solidified medium can be liquefied by cooling, the gel–liquid transition temperature depending on the concentration of polyol used. The clear, paste-like medium is apparently unsuitable for streaking, but is reported to be suitable for use e.g. in the pour plate technique for heat-sensitive organisms, and in gradient gel systems. [JGM (1984) *130* 731–733.]

plus-progamone See PHEROMONE.

plus strand (of a gene) (*mol. biol.*) See CODING STRAND.

plus strand (*virol.*) See VIRUS.

Pluteaceae See AGARICALES.

PM2 phage group *Syn.* CORTICOVIRIDAE.

pMB1 A small, multicopy PLASMID which carries genes for colicin E1, the colicin E1 immunity protein, and the *Eco*RI restriction and modification enzymes. pMB1 belongs to the same incompatibility group as the COLE1 PLASMID, and its *ori* region shares extensive homology with that of the ColE1 plasmid.

pmf Proton motive force: see CHEMIOSMOSIS.

pMG25 See INCOMPATIBILITY.

PML PROGRESSIVE MULTIFOCAL LEUKOENCEPHALOPATHY.

pML31 An autonomous 9-kb plasmid obtained by digestion of the F PLASMID with *Eco*RI restriction endonuclease and the ligation of a part of the plasmid (containing an origin of replication) with an antibiotic-resistance determinant. (See also MINI-F PLASMID.)

PMN Polymorphonuclear leucocyte ('polymorph'; granulocyte): a type of LEUCOCYTE which has a lobed nucleus and granular cytoplasm (hence 'granulocyte'); diam. ca. 10–15 μm. PMNs include NEUTROPHILS, BASOPHILS and EOSINOPHILS – so named because of the affinity of their cytoplasmic granules for particular dyes; they are short-lived cells which do not replicate and which are continually replaced by cells derived from bone marrow. 'Polymorph' may refer to PMNs in general, but is sometimes used to refer solely to neutrophils.

PNA (peptide nucleic acid) An analogue of DNA, the backbone chains of which are modified polypeptides (rather than sugar–phosphate units). Two strands of PNA can hybridize to form a DNA-like helical duplex [Nature (1994) *368* 561–563]; the existence of such a molecule has prompted the suggestion that pre-biotic nucleic acids may not necessarily have had a sugar–phosphate backbone.

A strand of PNA can hybridize with a strand of DNA; this may be of use in DNA CHIP technology.

PNA probes have been used for identifying and enumerating specific microorganisms by an *in situ* hybridization technique [JAM (2001) *90* 180–189].

***pncA* gene** See PYRAZINAMIDE.

pneumatocyst (air bladder/vesicle; bladder; gas bladder) A gas-filled structure normally present in the thallus of certain algae (e.g. species of *Fucus*, *Pelagophycus* and *Sargassum*); a given thallus may contain many pneumatocysts. Pneumatocysts, which are generally observable as distinct swellings, contribute buoyancy to the thallus. (cf. GAS VESICLE.)

pneumococcal pneumonia See PNEUMONIA (a).

pneumococcus *Streptococcus pneumoniae*.

pneumococcus capsule swelling reaction *Syn.* QUELLUNG PHENOMENON.

Pneumocystis A genus of fungi (formerly regarded as protozoa) now classified as members of the subdivision Ascomycotina. *P. carinii* is an important causal agent of pneumonia in HIV-infected patients; however, it can also give rise to infection in e.g. children and elderly individuals in the absence of HIV infection [BCID (1995) *2* 461–470]. *P. carinii* is also a causal agent of disease in animals (e.g. pigs).

The cells of *P. carinii* are ~1–5 μm; asci (10–12 μm) each contain up to eight ascospores. No satisfactory culture technique has been developed.

Diagnosis of *Pneumocystis* pneumonia typically involves microscopy of induced sputum or of fluid from broncho-alveolar lavage. Asci can be seen by staining with either TOLUIDINE BLUE or Gomori's silver stain. Individual cells are detected with Giemsa or Wright's stain [AJCP (1984) *81* 511–514]. Immunofluorescence stains are also available for *P. carinii*.

[*Pneumocystis carinii* (epidemiology, pneumonia, laboratory diagnosis, therapy etc.): BCID (1995) *2* 409–576. Immune response to *P. carinii* infection: TIM (1998) *6* 71–75. *P. carinii* pneumonia in pigs (immunohistochemical study): VR (2000) *147* 544–549.]

pneumolysin See THIOL-ACTIVATED CYTOLYSINS.

pneumonia Inflammation of the lungs. It may involve the alveoli (cf. ALVEOLITIS) and/or the interstitial tissues (*interstitial pneumonia*); inflammation may affect most of the parenchyma in one or more lobes (*lobar pneumonia*) or may be diffuse. Pneumonia is commonly due to bacterial infection, often secondary to e.g. viral URTI (such as INFLUENZA) or other infection (e.g. MEASLES); other causal agents include e.g. viruses (see also PNEUMOCYSTIS). The pathogen may gain access to the lungs by inhalation or via the blood or lymph systems (see e.g. *bubonic* PLAGUE). Conditions which predispose to pneumonia include chronic lung disease, debilitating illness, immunosuppression, etc. Symptoms of pneumonia generally include fever, chills, rigors, malaise, dyspnoea and chest pain; cough may be initially dry, later productive of purulent, sometimes bloody sputum. *Lab. diagnosis*: examination of sputum, pleural fluid etc by microscopy, culture, and/or serological techniques (e.g. CIE). Sputum may be obtained by percutaneous aspiration from the trachea, rather than by expectoration, to reduce contamination by members of the normal respiratory tract flora.

(a) *Pneumococcal pneumonia* ('classical lobar pneumonia'), caused by certain strains of *Streptococcus pneumoniae* (the 'pneumococcus'), is the commonest form of bacterial pneumonia. Infection may be primary or secondary. The pathogen, present e.g. in the upper respiratory tract of carriers, gains access to the lungs by inhalation. Incubation period: 1–3 days. Affected lobes typically become consolidated, i.e., solidifed with inflammatory material (leucocytes, fibrin, pneumococci etc). Sputum is initially purulent, becoming 'rusty' or blood-stained as the consolidation breaks down. Complications include e.g. pericarditis, meningitis etc. *Chemotherapy*: e.g penicillins, erythromycin. Polyvalent vaccines containing polysaccharide antigens of various common pneumococcal serotypes are available for certain high-risk patients [NEJM (1984) *310* 651–653; AIM (1986) *104* 106–109, 110–112, 118–120].

(b) *Haemophilus influenzae* (usually type b) may cause primary pneumonia (mainly in children) or secondary pneumonia in patients with e.g. viral respiratory tract infection. The pneumonia may be diffuse or lobar, and abscesses may develop. *Chemotherapy*: e.g. ampicillin; tetracycline; chloramphenicol.

(c) *Staphylococcus aureus* can cause lobar or diffuse interstitial pneumonia, usually following e.g. influenza or staphylococcal septicaemia. Abscesses may develop in lungs and/or pleura. Sputum is yellow, purulent, and sometimes bloody. Proteases from some strains of *S. aureus* may activate influenza virus virions (by cleaving their haemagglutinins: see INFLUENZAVIRUS), thus promoting combined viral–bacterial pneumonia [Nature (1987) *325* 536–537]. *Chemotherapy*: e.g. β-lactamase-resistant β-lactam antibiotics.

(d) *Friedländer's pneumonia* (caused by *Klebsiella pneumoniae*) is a severe form which usually occurs as a complication of e.g. chronic lung disease. Typically, one or more lobes become densely infiltrated, and thin-walled abscesses occur. Sputum is often viscid, gelatinous and blood-stained. *Chemotherapy*: e.g. gentamicin; amikacin; cephalosporins.

(e) *Streptococcus pyogenes* may cause primary or secondary pneumonia which is usually diffuse and often leads to empyema. *Chemotherapy*: e.g. penicillins.

(f) Various other Gram-negative bacteria may cause primary or secondary pneumonia – particularly in the very young or elderly, debilitated, or immunosuppressed, often following antibiotic therapy. Nosocomial pneumonia may be due e.g. to species of *Enterobacter*, *Klebsiella*, *Proteus*, *Serratia*, *Escherichia coli* or *Pseudomonas aeruginosa*. Anaerobes (e.g. *Bacteroides* spp) may infect the lungs, often with abscess formation; infection originates from the mouth or – via the lymph or blood systems – from e.g. intra-abdominal abscesses. (See also LEGIONELLOSIS.)

(g) Certain viruses (e.g. adenoviruses, coxsackieviruses, parainfluenza viruses, respiratory syncytial virus) can cause pneumonia, particularly in small children. (See also (c), above.)

Cf. PNEUMONITIS and PRIMARY ATYPICAL PNEUMONIA. See also e.g. ASPERGILLOSIS; COCCIDIOIDOMYCOSIS; HISTOPLASMOSIS; PLAGUE; PNEUMOCYSTIS.

For pneumonia in *animals* see e.g. ATYPICAL INTERSTITIAL PNEUMONIA; BOVINE RESPIRATORY DISEASE; CALF PNEUMONIA; CONTAGIOUS BOVINE PLEUROPNEUMONIA: JAAGSIEKTE; MAEDI.

pneumonia of mice virus See PNEUMOVIRUS.

pneumonic pasteurellosis See BOVINE RESPIRATORY DISEASE.

pneumonitis (1) *Syn.* PNEUMONIA. (2) Inflammation of the lungs as part of a more generalized syndrome, or resulting from chemical or physical injury or allergy (as opposed to infection). (cf. ALVEOLITIS.)

pneumotropic Having an affinity for the lungs.

Pneumovirus A genus of viruses of the PARAMYXOVIRIDAE in which the virions have neither haemagglutinin nor neuraminidase activity. Type species: (human) respiratory syncytial virus (RSV), an important cause of respiratory disease in humans (particularly infants and young children) – see e.g. BRONCHIOLITIS, COMMON COLD, CROUP, PNEUMONIA (g). [Review of RSV: Arch. Virol. (1985) *84* 1–52. Experimental models for RSV infections: RMM (1996) *7* 115–122.]

Other members of the genus: bovine respiratory syncytial virus (see BOVINE RESPIRATORY DISEASE) and pneumonia of mice virus (common in laboratory mice, in which infection is latent or mild).

PNPase POLYNUCLEOTIDE PHOSPHORYLASE.

P/O ratio (P:O ratio) A measure of the efficiency of OXIDATIVE PHOSPHORYLATION: the number of molecules of ATP formed per atom of oxygen used. (cf. H^+/O RATIO.)

POC Particulate organic carbon. (cf. DOC.)

Pocheina See ACRASIOMYCETES.

pock (1) (*med.*) A cutaneous pustule, as formed e.g. in SMALLPOX.

(2) (*virol.*) A lesion produced on the CAM (see EMBRYONATED EGG) by certain viruses; the nature and number of pocks formed can be useful in the identification and assay of the virus (see e.g. COWPOX VIRUS, VACCINIA VIRUS and VARIOLA VIRUS).

pocket rot (of timber) A form of WHITE ROT in which decay is limited to small, discrete regions ('pockets') and LIGNIN is attacked preferentially. Pocket rot fungi include *Phellinus pini* (in conifers) and *Inonotus* sp (in oaks). [Mycol. (1983) *75* 552–556.]

pod rot (frosty pod; Quevedo disease) A CACAO DISEASE caused by *Moniliophthora roreri* and characterized by a whitish ('frosted') powdery covering of spores on the pods; the spores subsequently darken. [Phytopathol. Paper number 24 (March, 1981).]

Podaxales See GASTEROMYCETES.

podetium (*lichenol.*) A vertical, stem-like, lichenized structure of generative tissue (sometimes called a 'secondary thallus') which develops from a crustose, squamulose or foliose basal ('primary') thallus and which supports one or more hymenial discs ('apothecia'); the entire podetium – together with its hymenial disc(s) – should be regarded as a fruticose APOTHECIUM [Lichenol. (1982) *14* 105 – 113]. (cf. PSEUDOPODETIUM.) Podetia are formed e.g. by most species of CLADONIA; they are usually hollow and may be branched or unbranched or – in some *Cladonia* spp – may terminate in a cup- or funnel-shaped structure (the 'scyphus'). (According to the above definition, structures resembling podetia but lacking algae are not true podetia; such structures are formed e.g. by *Baeomyces* spp and *Cladonia caespiticia*.)

Podophrya See SUCTORIA.

podophyllotoxin A substance of plant origin (obtained from *Podophyllum peltatum*) which has antitumour activity. Podophyllotoxin, which contains a trimethoxy-substituted aromatic ring, binds to TUBULIN at a site close to (or identical with) that at which COLCHICINE binds – inhibiting the assembly of MICROTUBULES. Some *epipodophyllotoxins* (which have been used as antitumour agents) appear to affect DNA synthesis (rather than microtubule assembly) – apparently by inhibiting topoisomerase activity.

Podoscypha See APHYLLOPHORALES (Stereaceae).

Podosphaera See ERYSIPHALES.

Podospora A genus of fungi (order SORDARIALES) which occur e.g. on soil and dung; the organisms form dark ascospores, each bearing gelatinous appendages, in persistent asci formed in perithecia. In *P. anserina* the four binucleate ascospores per ascus each contain nuclei of compatible mating types (cf. secondary HOMOTHALLISM). (See also AMIXIS.)

Podoviridae A family of DNA-containing BACTERIOPHAGES which have isometric or elongated heads and short (ca. 20 nm) non-contractile tails. Members include BACTERIOPHAGE N4, P22, ϕ29, T3, T7 (type species), and TBILISI PHAGE (see separate entries). (See also MV-L3 PHAGE GROUP and SPIROPLASMAVIRUSES (SV-C3).)

Pogosta disease See SINDBIS VIRUS.

Poikilovirus See ALPHAHERPESVIRINAE.

point mutation (1) A type of MUTATION in which a single nucleotide is replaced by another: see TRANSITION MUTATION and TRANSVERSION MUTATION. A point mutation in a structural gene may result in a MIS-SENSE MUTATION, a NONSENSE MUTATION, or a SAME-SENSE MUTATION. Mutagens which can induce point mutations include e.g. 5-BROMOURACIL, HYDROXYLAMINE and NITROUS ACID.

(2) Any mutation involving a single nucleotide, including the gain or loss of a nucleotide (resulting in a FRAMESHIFT MUTATION) as well as transition and transversion mutations.

(cf. MULTISITE MUTATION.)

poising system See REDOX POTENTIAL.

Poitrasia See MUCORALES.

pokeweed mitogen (PWM) Any of five mitogenic LECTINS (designated Pa-1 to Pa-5) obtained from the roots, leaves and berries of the pokeweed, *Phytolacca americana*. All can stimulate T cells; Pa-1, in the presence of T cells and macrophages, can stimulate Ig secretion by B cells. PWM binds to oligomers of β-D-acetylglucosamine.

poky mutant A mutant strain of *Neurospora* which has a subnormal rate of growth owing to a deficiency in its respiratory chain. The mutant gene (= *mi*-1 or maternal inheritance-1 gene) appears to occur in the mitochondrial DNA, and is transmitted to sexually-derived progeny only from a female (ascogonial) parent; such MATERNAL INHERITANCE is presumed to be due to the absence of mitochondria in the (male) microconidium. (cf. PETITE MUTANT.)

***pol* genes** (1) See DNA POLYMERASE. (2) See RETROVIRIDAE.

***polA* gene** See DNA POLYMERASE.

polar body An intracellular granule at one or each pole of a cell – e.g. the deposits of PHB in *Beijerinckia*.

polar cap See MICROSPORA.

polar capsule See MYXOZOA.

polar fibre See MITOSIS.

polar filament See MICROSPORA and MYXOZOA.

polar membrane A multilaminar intracellular membrane, attached to the inside of the cytoplasmic membrane by bar-like links, present at the poles (sites of flagellar attachment) of the cells in species of helical and vibrioid bacteria (e.g. *Aquaspirillum*, *Campylobacter*, *Ectothiorhodospira*, *Oceanospirillum*, *Rhodospirillum*, *Spirillum*).

polar microtubule See MITOSIS.

polar mutation A mutation which, in addition to affecting the gene in which it occurs, reduces the expression of any gene(s) in the same (polycistronic) operon on the promoter-distal side of the mutation – a phenomenon known as *polarity*; genes between the promoter and the mutation are not affected. Thus, e.g., in the LAC OPERON a nonsense mutation in the *y* gene may eliminate permease synthesis, does not affect β-galactosidase synthesis, and reduces the amount of transacetylase synthesis. The mutation is said to be strongly polar when the expression of promoter-distal genes is strongly inhibited; weakly polar mutations cause only slight inhibition. The strength of polarity depends on the distance between the mutation and the end of the gene in which it occurs (i.e., between the mutation and the next translation initiation site), being greatest when this distance is greatest.

According to one model for polarity, cessation of translation (due e.g. to the presence of a nonsense mutation) may expose in the mRNA a ρ-dependent TRANSCRIPTION termination site which is normally protected from the ρ factor by the presence of ribosomes; interaction of ρ with this site results in premature termination of transcription and hence loss of expression of promoter-distal genes. Polar mutations can be suppressed by mutations in the *rho* gene, the mutant ρ factor apparently failing to interact with the premature termination signal; polarity suppressors thus restore expression of genes downstream of the mutant gene but not of the mutant gene itself. (cf. SUPPRESSOR MUTATION.) The polar effects of certain TRANSPOSABLE ELEMENTS (e.g. IS2) have been attributed to the presence of a ρ-dependent terminator within the TE sequence.

In some cases a polar mutation does not appear to affect transcription; in such cases ribosomes, having dissociated from the mRNA at a nonsense mutation, may be unable to re-initiate translation at the next initiation site – possibly because of the formation of a secondary structure in the untranslated mRNA.

polar plaque *Syn.* SPINDLE POLE BODY.

polar ring An osmiophilic annular structure, one or more of which occur at the anterior end in members of the APICOMPLEXA. Polar ring(s) may act as support(s) for the opening in the inner layer of the pellicle, and they may be involved in the function of the conoid.

polarilocular Refers to a lichen spore consisting of two cells which are separated by a thick septum that has a central canal. (See also TELOSCHISTALES.)

polaroplast See MICROSPORA.

***polB* gene** See DNA POLYMERASE.

***polC* gene** See DNA POLYMERASE.

poliomyelitis (polio; infantile paralysis) An acute infectious disease of humans, particularly children, caused by any of three serotypes of human poliovirus (see ENTEROVIRUS). (A few non-human primates, e.g. chimpanzees, are also susceptible.) Transmission occurs by the faecal–oral route (e.g. by person-to-person contact or via sewage-contaminated water). Incubation period: 3–35 (usually 7–14) days. Infection may be asymptomatic; clinical forms are usually mild and self-limiting, with fever, headache, nausea and vomiting, and pharyngitis (*abortive poliomyelitis*). In a minority of cases the virus may invade the CNS, causing inflammation and destruction of the lower motor neurones in the brain stem and spinal cord. In *non-paralytic poliomyelitis* symptoms include those of MENINGITIS (e.g. stiffness of the neck and back). In *paralytic poliomyelitis* there is weakness and eventually flaccid paralysis of muscles (commonly of the legs); the paralysis may or may not be permanent. Polio is not usually fatal, but may be if respiratory muscles become involved. *Lab. diagnosis*: serological tests; culture of the virus from pharyngeal secretions and faeces. (See also SALK VACCINE and SABIN VACCINE.)

polioviruses See ENTEROVIRUS.

Polish infectious dropsy (of carp) See CARP ERYTHRODERMATITIS.

pollen mould *Bettsia alvei* (q.v.).

pollution (environmental) See ENVIRONMENTAL POLLUTION.

poly(A) tail See MRNA.

polyacetylenes (polyynes) Linear compounds, containing both double and triple bonds, which are produced by certain higher fungi (particularly basidiomycetes) and which, in some cases, exhibit antifungal and antibacterial activity; the polyacetylene molecule usually has an oxygen-containing terminal group (e.g. a carboxyl or hydroxyl group). For example, mycorrhizal *Leucopaxillus cerealis* produces diatretyne nitrile (HOOC–CH=CH–C≡C–C≡C–C≡N) which is active e.g. against *Phytophthora cinnamomi*.

polyacrylamide gel electrophoresis ELECTROPHORESIS using a polyacrylamide gel – see e.g. SDS-PAGE.

polyadenylated mRNA See MRNA.

polyamines Polycationic compounds containing two or more amino groups; they appear to occur in all animal, plant and microbial cells and in certain viruses.

Putrescine [$NH_2(CH_2)_4NH_2$], spermidine [$NH_2(CH_2)_3NH(CH_2)_4NH_2$] and spermine [$NH_2(CH_2)_3NH(CH_2)_4NH(CH_2)_3NH_2$] appear to be almost universally present in prokaryotic and eukaryotic microorganisms. However, their cellular location(s) and biological function(s) are still not known, although numerous effects on e.g. nucleic acid structure and enzyme activity have been observed in vitro; polyamines may have a role e.g. in protein synthesis in bacteria, and are necessary for the replication of at least some bacteriophages, including the T-even phages (which contain putrescine and spermidine) and bacteriophage λ. (See also BACTERIOPHAGE φW-14.)

Polyamines are synthesized primarily from amino acids – e.g. putrescine may be formed by the decarboxylation of ornithine or by the combined action of arginine decarboxylase and agmatine ureohydrolase. (See also DECARBOXYLASE TESTS.) In bacteria the polyamine content is significantly affected by the growth conditions. A number of novel polyamines have been described from extreme thermophiles: e.g. '*Thermus thermophilus*' contains polyamines such as thermine

[= *sym*-norspermine: $NH_2(CH_2)_3NH(CH_2)_3NH(CH_2)_3NH_2$], *sym*-homospermidine [$NH_2(CH_2)_4NH(CH_2)_4NH_2$], thermospermine [$NH_2(CH_2)_3NH(CH_2)_3NH(CH_2)_4NH_2$] and also caldopentamine [$NH_2(CH_2)_3NH(CH_2)_3NH(CH_2)_3NH(CH_2)_3NH_2$]. These substances have been implicated in the maintenance of thermostability in cell components, enzymes, etc; however, *sym*-homospermidine has also been found in mesophilic bacteria [FEMS (1983) *20* 159–161].

[Reviews: MR (1985) *49* 81–99; ARB (1984) *53* 749–790.]

Polyangium See MYXOBACTERALES.

poly(A)polymerase See MRNA.

polybetahydroxybutyrate See POLY-β-HYDROXYBUTYRATE.

polycentric Refers to a thallus in which a number of regions, in different parts of the thallus, each have a reproductive function. (cf. MONOCENTRIC.)

polychlorinated biphenyls (PCBs) See BIOREMEDIATION.

polychrome methylene blue See METHYLENE BLUE.

polycistronic mRNA See MRNA.

polyclonal activator (*immunol.*) Any *non-specific* activator of lymphocytes, i.e. any agent which can bring about differentiation/proliferation in each of a number of different clones.

Polyclonal activators for B LYMPHOCYTES include lipopolysaccharides and various mitogens; such agents are THYMUS-INDEPENDENT ANTIGENS.

SUPERANTIGENS are polyclonal activators for T LYMPHOCYTES.

polyclonal antiserum Any antiserum prepared by injecting an animal with a given antigen; it contains a heterogeneous population of antibodies – each antibody being specific to one of the (usually many) determinants present on the antigen. (cf. MONOCLONAL ANTIBODIES.)

polycycline See TETRACYCLINES.

Polycystinea See RADIOLARIA.

Polydnaviridae A family of viruses [Intervirol. (1986) *25* 141–143], previously classified as subgroup *D* of the BACULOVIRIDAE, characterized by their polydisperse ccc dsDNA genomes. Virus replication occurs in host cell nuclei; virions are not occluded. The viruses infect certain species of parasitic hymenopteran insects. Two distinct groups (genera) are recognized. In one group (genus unnamed) the virion has a cylindrical nucleocapsid of variable length, and virions are surrounded – either individually or in groups – by a unit membrane envelope; these viruses infect braconid wasps. In the other group (genus *Polydnavirus*) the virions have fusiform nucleocapsids surrounded by two envelopes; these viruses infect ichneumonid wasps (e.g. *Hyposoter* spp). Viruses of both groups can apparently be transmitted vertically [Virol. (1986) *155* 120–131]. Type species: polydnavirus type 1 (*Hyposoter exiguae* virus).

polyene antibiotics ANTIBIOTICS which contain region(s) of conjugated double bonds. There are two structurally and functionally distinct types.

(a) A group of MACROLIDE ANTIBIOTICS (produced by streptomycetes) characterized by a large lactone ring containing a rigid, lipophilic region of (generally) unsubstituted, all-*trans* conjugated double bonds and a flexible, hydrophilic, hydroxylated region; the lactone may be substituted with aminosugar, carboxyl, aliphatic or aromatic groups. These polyenes are classified according to the number of conjugated double bonds they contain (e.g. tetraenes, heptaenes). They are poorly soluble in water but are soluble in organic solvents; they are unstable in solution in the presence of light owing to photo-oxidation of the double bonds.

The macrolide polyenes interact with sterols in the cytoplasmic membrane of sensitive organisms, causing leakage of ions and small molecules; the polyene molecule may form a pore in the membrane through which such leakage may occur. These antibiotics are microbistatic at low concentrations, microbicidal at higher concentrations.

Macrolide polyenes are effective against yeasts and other fungi and also against protozoa, but not against most bacteria (whose membranes lack sterols – cf. MYCOPLASMA).

The selective toxicity of the clinically useful macrolide polyenes is apparently due, at least in part, to their greater affinity for ergosterol (a principal sterol in fungal cytoplasmic membranes) compared with cholesterol (the main sterol in mammalian cell membranes) (cf. FILIPIN).

Macrolide polyenes include e.g. AMPHOTERICIN B; AUREOFUNGIN; CANDICIDIN B; ETRUSCOMYCIN; FILIPIN; HAMYCIN; NYSTATIN; PERIMYCIN; PIMARICIN; and TRICHOMYCIN.

(b) A group of structurally novel polyenes (produced by actinomycetes [FEMS (1984) *25* 121–124]) which are active against prokaryotes, inhibiting PROTEIN SYNTHESIS. In bacteria, *aurodox, azdimycin, dihydromocimycin, efrotomycin, factumycin, kirromycin* (= mocimycin) and *kirrothricin* all act by binding to EF-Tu and preventing its dissociation from the ribosome; the structurally related antibiotic *pulvomycin* binds EF-Tu-GTP and prevents the formation of the aa-tRNA-GTP-EF-Tu complex, thus preventing the binding of aa-tRNA to the ribosomal A site. In archaeans, elongation factors are generally insensitive to kirromycin, and the elongation factors of only some archaeans are sensitive to pulvomycin [JB (1986) *167* 265–271].

polyether antibiotics See MACROTETRALIDES.

polyethylene glycol (PEG; Carbowax) A polymer, $H(OCH_2.CH_2)_nOH$, available in a range of grades of mean MWt between ca. 200 and ca. 20000; above MWt ca. 1000 it is a solid, below it is a liquid. PEG is used e.g. in cell fusion (see e.g. HYBRIDOMA), in PROTOPLAST FUSION, in artificial TRANSFORMATION systems, and as a precipitating agent in clinical chemistry. PEG is also used for concentrating, by osmosis, serum and other aqueous solutions and suspensions; serum etc is placed in a U-shaped tube of semipermeable material (e.g. Visking tubing) and the tubing partly submerged in a concentrated aqueous solution of PEG.

poly(ethylene oxide) A high-MWt polymer $(-CH_2CH_2O-)_n$ used e.g. to increase the viscosity of a medium in order to slow down rapidly motile organisms – see COUNTING METHODS. (cf. FICOLL.)

polygenic mRNA See MRNA.

polyhedra (*virol.*) Large inclusion bodies formed in the cells of insects infected with certain viruses (see CYTOPLASMIC POLYHEDROSIS VIRUS GROUP and BACULOVIRIDAE); a polyhedron is composed of a matrix of crystalline, virus-specific protein (*polyhedrin*) in which mature virions are embedded ('occluded'). Polyhedra are stable structures which appear to give some protection to the virions they contain; they serve as vehicles for virus transmission: ingestion of polyhedra by a new host is followed by degradation of polyhedrin in the gut and the release of infectious virions. Polyhedra can be isolated from infected cells e.g. by differential centrifugation.

polyhedral bodies See CARBOXYSOMES.

polyhedrin See POLYHEDRA and BACULOVIRIDAE.

polyhedrosis An INSECT DISEASE caused by a NUCLEAR POLYHEDROSIS VIRUS or by a virus of the CYTOPLASMIC POLYHEDROSIS VIRUS GROUP.

poly-β-hydroxyalkanoate See POLY-β-HYDROXYBUTYRATE.

poly-β-hydroxybutyrate (PHB) A linear polymer of β-hydroxybutyrate (D(−)-3-hydroxybutyrate): $H-[O.CH(CH_3).CH_2.CO]_n-OH$. It occurs as refractile granules in the cells of various

PROKARYOTES, including cyanobacteria and members of the Chromatiaceae and Rhodospirillaceae. PHB synthase (and possibly also de-polymerizing enzymes) occur at the surface of the granules. Each mature granule is bounded by a layer apparently <4 nm in thickness – believed to be a monolayer consisting mainly of amphiphilic proteins (*phasins*) and phospholipids, all the molecules orientated with their hydrophilic regions facing the cytoplasm [JBM (1997) *37* 45–52]. PHB granules can be stained *in situ* (i.e. in the cell) by dyes such as NILE BLUE A.

In many bacteria PHB accumulates when nutrients other than carbon sources become limiting, being degraded when these conditions are reversed; PHB thus acts as a reservoir of carbon and/or energy. *Azotobacter beijerinckii* accumulates very large quantities (up to 80% of the cell mass) of PHB under conditions of oxygen limitation, PHB serving as a 'sink' for excess reducing power under these conditions.

Typically, PHB is synthesized via the pathway: acetyl-CoA → acetoacetyl-CoA → β-hydroxybutyryl-CoA → PHB (enzymes: β-ketothiolase, β-hydroxybutyryl-CoA dehydrogenase, PHB synthase). PHB degradation: PHB → β-hydroxybutyrate → acetoacetate → acetoacetyl-CoA → acetyl-CoA; the first enzyme in this pathway, PHB depolymerase, is reported to be granule-bound in some bacteria, soluble in others. Some bacteria, e.g. certain pseudomonads, can degrade exogenous PHB.

Other poly-β-hydroxyalkanoates (PHAs): e.g. *Bacillus megaterium* may contain small amounts of β-hydroxyheptanoate together with PHB, and *Pseudomonas oleovorans* forms poly-β-hydroxyoctanoate from *n*-octane.

[PHAs, including PHB and Biopol: FEMS Reviews (1992) *103* 91–376.]

(See also BIOPOL.)

Polyhymenophorea A class of protozoa (phylum CILIOPHORA) in which the oral ciliature is typically well developed and conspicuous (often extending out of the buccal cavity); the PARORAL MEMBRANE is of the stichomonad, diplostichomonad or polystichomonad type; the somatic ciliature is often reduced or replaced by cirri; kinetodesmata are rare; STOMATOGENESIS is apokinetal or parakinetal; and neither toxicysts nor trichocysts are formed. Some species are sedentary, and some are loricate. Most species are free-living. All members are placed in the single sublcass Spirotrichia which is divided into four orders: HETEROTRICHIDA, HYPOTRICHIDA, ODONTOSTOMATIDA, OLIGOTRICHIDA.

poly(I:C) See INTERFERONS.

polykaryocyte A multinucleate cell (e.g. a GIANT CELL).

polyketide A compound (often a secondary metabolite) which is, or may be regarded as, a condensation product of subunits such as acetate, propionate and/or malonate; polyketides, which may be classified according to the number of C_2 units they contain (e.g. a tetraketide is a C_8 compound), may undergo cyclization, glycosylation, oxidation etc.

Metabolites derived from polyketides include some antibiotics (e.g. erythromycin, tetracyclines) and mycotoxins (e.g. PATULIN – a tetraketide, AFLATOXINS, CYTOCHALASINS). Such compounds are synthesized from precursors such as acetyl-CoA and malonyl-CoA; for example, in aflatoxin biosynthesis, malonyl-CoA subunits condense with an acetyl-CoA 'starter', with concomitant decarboxylation, in a process analogous to fatty acid biosynthesis but minus dehydration and reduction [AAM (1983) *29* 53–92].

The molecule of erythromycin, a MACROLIDE ANTIBIOTIC obtained from actinomycetes (e.g. *Streptomyces erythreus* and *Saccharopolyspora erythraea*), has a cyclic polyketide nucleus which is synthesized by a multifunctional enzyme, polyketide synthase. The possibility of developing new drugs by genetic engineering of the polyketide synthetic pathway has been hindered by the difficulty of carrying out such work with actinomycetes; to address this problem, synthetic potential (including polyketide synthase activity) has been incorporated into a strain of *Escherichia coli* – which can now synthesize the nucleus of the erythromycin molecule [Science (2001) *291* 1790–1792; commentary: Science (2001) *291* 1683].

polylinker A region of DNA containing close/overlapping sites for *different* RESTRICTION ENDONUCLEASES. Synonym: multiple cloning site (MCS).

polymerase chain reaction See PCR.

polymetaphosphate See POLYPHOSPHATE.

polymorph (polymorphonuclear leucocyte) (1) *Syn.* PMN. (2) *Syn.* NEUTROPHIL.

polymorphic Refers to e.g. an organism which exhibits POLYMORPHISM.

polymorphism (1) The ability of an organism to occur in two or more morphologically distinct forms (MORPHOTYPES), according e.g. to conditions (as in PHAEODACTYLUM) or to the stage in the life cycle (e.g. TRYPANOSOMATIDAE). (cf. PLEOMORPHISM sense 1).

(2) *Syn.* PLEOMORPHISM (sense 1).

(3) (in nucleic acids) A variant region of nucleotides in a *related* nucleic acid molecule – e.g. a variant sequence found in a given allele in one or more strains of an organism but not in all strains of that organism. Some authors use the term even when the variation in sequence consists of only a point mutation.

Polymyxa See PLASMODIOPHOROMYCETES.

polymyxins A group of peptide ANTIBIOTICS produced by various species of *Bacillus* (e.g. *B. polymyxa*); they include e.g. antibiotics formerly classified as circulins and colistins. Polymyxins are active against many Gram-negative bacteria – exceptions including e.g. many strains of *Neisseria* and *Proteus*, and the *eltor* biotype of *Vibrio cholerae* (the *cholerae* biotype is susceptible); most Gram-positive bacteria and most fungi are resistant, but some strains of e.g. *Candida* are susceptible. Against bacteria, polymyxins are bactericidal to both growing and non-growing cells.

Polymyxins have MWts of ca. 1000–1200, and they contain a high proportion of α,γ-diaminobutyric acid (DAB). The polymyxin molecule is a cyclic heptapeptide with a peptide side-chain that terminates in a fatty acid residue; the fatty acid is 6-methyloctanoic acid in e.g. polymyxins A, B1, D1 and E1 (= colistin A), and is 6-methylheptanoic acid in e.g. polymyxins B2, D2 and E2. Variations also occur in the constituents of the cyclic and linear peptides. [Structure of polymyxins: Book ref. 14, pp. 189–194.]

In Gram-negative bacteria, polymyxins act by increasing the permeability of the CYTOPLASMIC MEMBRANE to small molecules and by damaging the OUTER MEMBRANE. Polymyxin B nonapeptide, a derivative of polymyxin B which lacks the fatty acid residue, lacks the bactericidal activity of polymyxin B but is able to damage the outer membrane [CJM (1986) *32* 66–69].

(See also OCTAPEPTINS.)

polynucleotide kinase An enzyme used e.g. for the 5′-end labelling of a PROBE (q.v.).

polynucleotide phosphorylase (PNPase) An enzyme, present e.g. in *Escherichia coli* and other prokaryotes, which in vitro can catalyse e.g. the phosphorolysis of poly- or oligoribonucleotides in the presence of phosphate, and the polymerization of ribonucleoside diphosphates.

polyol (polyhydric alcohol) Strictly: any compound containing more than one alcohol group. As commonly used, the term refers to a sugar alcohol, i.e., the compound obtained when the aldo or keto group of a sugar is reduced to a hydroxy group. For example, reduction of galactose gives galactitol (dulcitol), ribose gives ribitol, etc. (cf. PLURONIC POLYOL F127.) [Polyol metabolism in fungi: AMP (1984) *25* 149–193.] (See also e.g. PENTITOLS.)

polyoma virus (polyomavirus) (1) Mouse (murine) polyoma virus: see POLYOMAVIRUS. (2) Any virus of the genus *Polyomavirus* (q.v.).

Polyomavirus A genus of viruses of the PAPOVAVIRIDAE. Type species: mouse (murine) polyoma virus ('polyoma virus'); other members include e.g. BK virus and JC virus (human viruses), K virus (a mouse virus), SIMIAN VIRUS 40 (SV40), and stump-tailed macaque virus (STMV). (See also BUDGERIGAR FLEDGLING DISEASE VIRUS.) Polyoma viruses can be propagated readily in suitable cell cultures (see e.g. SIMIAN VIRUS 40; cf. PAPILLOMAVIRUS). The viruses generally appear to be common in their natural hosts, in which infection is usually asymptomatic (although JC virus – which seems to be common in humans – can cause PROGRESSIVE MULTIFOCAL LEUKOENCEPHALOPATHY in certain circumstances); however, many polyoma viruses can be oncogenic in other hosts: e.g. SV40, JC and BK can induce tumours in newborn rodents and can transform several types of cell in culture. JC virus can also induce brain tumours in adult monkeys. Cells transformed by SV40 or murine polyoma virus contain viral DNA integrated into the host chromosomal DNA; integration occurs at non-specific sites and may involve the entire viral genome or only a portion of it. However, integration appears not to be essential for transformation by BK virus, and its significance in SV40 and polyoma virus transformation is unknown. Expression of early viral genes appears to be necessary for transformation in all cases.

[Polyomaviruses and disease of the central nervous system: RMM (1998) *9* 79–85.]

In murine polyoma virus, at least three distinct genes appear to act in a coordinated manner to induce transformation in primary rat embryo fibroblasts; the product of each gene induces particular alterations in cell structure and growth control [Book ref. 113, pp. 109–116]. The three products are 'large T protein' or 'large T antigen' (MWt ca. 105000), 'middle T protein' or 'middle T antigen' (MWt ca. 56000), and 'small T (or t) protein' or 'small T antigen' (MWt ca. 22000). The large T protein is a DNA-binding protein which localizes in the host cell nucleus in association with the chromatin. The middle T protein localizes in the host cell plasma membrane; it is tightly associated with a tyrosine-specific protein kinase – identified as the c-*src* product $pp60^{c\text{-}src}$ (see SRC [EMBO (1984) *3* 585–591]. The small T protein is found in the soluble cytoplasmic fraction of the cell; its activities are unknown, but it may disrupt actin cables in the cell.

Viruses SV40, BK and JC appear to encode proteins corresponding to large T and small T proteins but do not encode a middle T protein. In SV40, the large T antigen ('T antigen') is a multifunctional phosphoprotein MWt ca. 92000) which can alone induce neoplastic transformation in cells from a variety of species; it also acts as a key regulatory protein in the lytic cycle of SV40 in permissive monkey cells. The protein binds to SV40 DNA at multiple sites near the replication origin, has ATPase activity, and binds to a host protein of MWt 53000. The SV40 small T protein may also have a role in transformation; it has been detected in the host cell cytoplasm and in the nucleus [Book ref. 113, pp. 369–375].

polyoxins A group of antifungal nucleoside–oligopeptide antibiotics produced by *Streptomyces cacaoi*. Polyoxins are competitive inhibitors of chitin synthase, being structural analogues of UDP-*N*-acetylglucosamine; other enzymes using this substrate seem not to be affected. Susceptible fungi take up polyoxins via peptide transport systems; resistant strains, which emerge readily, appear to have altered peptide permeases. Polyoxins have been used for the control of certain plant pathogens: e.g. *Alternaria kikuchiana* (black spot of pear); they have no clinical application. (See also GROWTH (fungal).)

polypectate *Syn.* pectate (see PECTINS).

Polyphagus See CHYTRIDIALES and MYCOPARASITE.

polyphasic taxonomy See TAXONOMY.

polyphenoloxidases See e.g. TANNINS.

polyphosphate (volutin) A linear phosphate polymer (polymetaphosphate: $[PO_3{}^{2-}]\text{–}O\text{–}[PO_3{}^{-}]_n\text{–}[PO_3{}^{2-}]$) which occurs in bacteria, fungi, algae, protozoa, and certain higher eukaryotes. In eukaryotic microorganisms (e.g. yeasts) polyphosphate appears to occur chiefly in the periplasmic region and/or in the cell vacuole; in most bacteria (including cyanobacteria) polyphosphate occurs mainly in the cytoplasm. (See METACHROMATIC GRANULES.) Polyphosphate is believed to function as a reserve of phosphate and, in at least some organisms, of energy; it may also play a role in the storage or sequestration of cations.

Polyphosphate is synthesized in some bacteria by the transfer of γ-phosphate from ATP to nascent polyphosphate, catalysed by an Mg^{2+}-dependent polyphosphate kinase (ATP-polyphosphate phosphotransferase). Other bacteria, and e.g. *Neurospora crassa*, synthesize polyphosphate by the transfer of phosphate from the 1-position of 1,3-bisphosphoglycerate. Some organisms (e.g. *Escherichia coli*) possess both enzyme systems. Degradation of polyphosphate can be achieved by several enzymes, including e.g. polyphosphate kinase, and an Mg^{2+}-dependent polyphosphate glucokinase – present in certain bacteria – which can catalyse the transfer of phosphate from polyphosphate to glucose to form glucose 6-phosphate. [Polyphosphate metabolism: AMP (1983) *24* 83–171.]

(See also PYROPHOSPHATE.)

polyphyletic Refers to taxa which have not arisen from a common evolutionary line but which are classified together because they share certain feature(s) – due e.g. to convergent evolution. (cf. MONOPHYLETIC.)

polyplanetism The phenomenon in which *secondary* zoospores (see DIPLANETISM) undergo repeated cycles of encystment and excystment, successive populations of secondary zoospores being formed.

polyploid Having a PLOIDY greater than two.

polypodial amoebae Amoebae which usually have several pseudopodia at a given time. (cf. MONOPODIAL AMOEBAE.)

Polyporaceae See APHYLLOPHORALES.

Polyporales *Syn.* APHYLLOPHORALES.

polypore Any species of the APHYLLOPHORALES (particularly bracket fungi of the Polyporaceae) in which the basidiocarp has a porous hymenophore.

Polyporus (1) A genus of typically lignicolous fungi of the APHYLLOPHORALES (family Polyporaceae) which form bracket-type to stipitate (mushroom-shaped) fruiting bodies in which the stipe may be central, eccentric, lateral, or virtually non-existent; the hymenophore is porous. *P. squamosus* is the causal agent of heart rot in a range of deciduous trees, particularly elm (*Ulmus*); the fruiting body (context: dimitic) has a light brown, scaly upper surface and a pale underside, and the basidiospores are colourless, cylindrical, ca. 12 × 5 μm. For *P. dryadeus* see INONOTUS. (See also BLACKFELLOWS' BREAD.)

(2) *Polyporus* is sometimes used as a form genus to include those polypores which form annual, fleshy to leathery fruiting

bodies; thus, e.g. *Piptoporus betulinus* is equivalent to the form species *Polyporus betulinus*. (cf. FOMES sense 2.)

polyprotein (*virol.*) A protein which is produced by the uninterrupted translation of two or more adjacent genes on a polycistronic mRNA; the products of individual genes are separated by post- or co-translational cleavage of the polyprotein. (See e.g. COWPEA MOSAIC VIRUS, PICORNAVIRIDAE and RETROVIRIDAE.) (Certain polypeptides of eukaryotes are also synthesized as polyproteins.)

polyribosome See PROTEIN SYNTHESIS.

polysaccharide (glycan) Any macromolecule that consists of monosaccharide (glycose) residues joined by glycosidic linkages. Polysaccharides containing only one kind of monosaccharide are called *homopolysaccharides* (= *homoglycans*), those containing two or more different kinds are called *heteropolysaccharides* (= *heteroglycans*). [Nomenclature: JBC (1982) *257* 3352–3354.]

Several polysaccharides obtained from microorganisms have important commercial uses: see e.g. AGAR, ALGINATE, CARRAGEENAN, CURDLAN, DEXTRAN, FUNORAN, FURCELLARAN, GELLAN GUM and XANTHAN GUM.

polysaprobic zone See SAPROBITY SYSTEM.

Polysiphonia A large genus of filamentous marine algae (division RHODOPHYTA). (See also CHOREOCOLAX.)

polysome See PROTEIN SYNTHESIS.

Polysphondylium A genus of cellular slime moulds (class DICTYOSTELIOMYCETES) in which the sorocarp stalk tube resembles that of DICTYOSTELIUM in containing empty cells, but differs in bearing whorls of branches along its length; globose sori are borne on the tips of the stalks and at the ends of the branches. The spores are purple in *P. violaceum*, white to hyaline in other species (e.g. *P. pallidum*). In *P. violaceum* the ACRASIN is not cAMP but an unusual dipeptide, 'glorin' (*N*-propionyl-γ-L-glutamyl-L-ornithine-δ-lactam ethyl ester [PNAS (1982) *79* 7376–7379]). In this species a single cell, the *founder* cell, serves as the focus for aggregation. Six species are recognized [descriptions and key: Book ref. 144, pp. 368–392].

Polyspira See HYPOSTOMATIA.

polystichomonad membrane See PARORAL MEMBRANE.

Polystigma See POLYSTIGMATALES.

Polystigmatales An order of primarily plant-pathogenic fungi of the subdivision ASCOMYCOTINA. Ascocarp: perithecioid, immersed. Asci: unitunicate, typically clavate. Ascospores: aseptate to multiseptate. Genera: e.g. *Magnaporthe* (see also PYRICULARIA), *Polystigma*, *Sphaerodothis*.

polystromatic Refers to a structure which is more than two cells thick. (cf. MONOSTROMATIC; DISTROMATIC.)

polysulphide See SULPHUR.

polytenization The repeated replication of a chromosome without separation of the daughter chromosomes; a polytene chromosome is thus a ribbon-like arrangement of multiple chromosomes lying parallel to one another.

polythetic classification Any form of classification based on a large number of criteria. (cf. MONOTHETIC CLASSIFICATION.)

Polythrix See RIVULARIACEAE.

polytropic *Syn.* DUALTROPIC.

polyurethanase Any enzyme which can degrade polyurethane. Such enzymes are produced by certain types of microorganism. Bacterial polyurethanase activity may be detected by a simple plate assay in which organisms are tested for their ability to degrade a colloidal form of polyurethane in the presence of rhodamine B, hydrolysis of the substrate being indicated by the development of orange fluorescence in the medium [LAM (2001) *32* 211–214].

polyvalent antiserum Any antiserum which contains antibodies to a number of different antigens.

polyvalent vaccine A single VACCINE containing the PROTECTIVE ANTIGENS and/or toxoids from each of several different strains of one species of pathogen. (cf. MIXED VACCINE.)

polyvinyl alcohol fixative A FIXATIVE: SCHAUDINN'S FLUID with five times the usual concentration of glacial acetic acid, supplemented with polyvinyl alcohol (5% w/v) and glycerol (1.5 ml per 100 ml fluid). Specimens can remain in PVA fixative, without damage, for long periods of time.

polyvinyl pyrrolidone–iodine See IODINE (a).

polyynes *Syn.* POLYACETYLENES.

pomona fever LEPTOSPIROSIS caused by *L. interrogans* serotype *pomona*.

Pontiac fever See LEGIONELLOSIS.

porcine circovirus (PCV) A small isometric virus (diam. 17 nm) containing ccc ssDNA (MWt 0.58×10^6). PCV was isolated from a porcine cell line, and is presumed to infect pigs since specific antibodies to the virus were present in randomly collected pig sera but not in the sera of other animals. PCV apparently represents a new family of ANIMAL VIRUSES. [Nature (1982) *295* 64–66.]

porcine haemagglutinating encephalitis virus See CORONAVIRUS and VOMITING AND WASTING DISEASE.

porcine icteroanaemia See EPERYTHROZOONOSIS.

porcine parvovirus See PARVOVIRUS.

porcine reproductive and respiratory syndrome See BLUE-EARED PIG DISEASE.

porcine transmissible gastroenteritis virus See CORONAVIRIDAE and TRANSMISSIBLE GASTROENTERITIS.

pored epicortex (*lichenol.*) A polysaccharide layer (ca. 0.6 μm thick) which overlies the upper cortex in certain foliose lichen thalli and which is perforated by numerous more or less regular perforations ca. 15–25 μm or more in diameter. The underlying cortex is continuous but is generally thin (e.g. 2–3 cell layers thick), the cells being relatively loosely packed; these cortical features, together with the pored epicortex, apparently allow sufficient gas exchange to obviate the need for pseudocyphellae or cyphellae [Lichenol. (1981) *13* 1–10].

porfiromycin The *N*-methyl derivative of MITOMYCIN C.

Poria (1) A genus of lignicolous fungi of the APHYLLOPHORALES (family Polyporaceae) which form resupinate fruiting bodies (ranging from millimetres to centimetres in thickness) in which the hymenophore is porous; the whitish context is monomitic and lacks clamp connections. In perennial fruiting bodies the tubules are stratified (cf. HETEROBASIDION). (See also BROWN ROT, COAL BIODEGRADATION, DRY ROT, PACHYMAN and PAPER SPOILAGE.)

(2) *Poria* is sometimes used as a form genus to include those polypores which form annual or perennial resupinate fruiting bodies.

porin (formerly: matrix protein) In the OUTER MEMBRANE of Gram-negative bacteria: a type of protein molecule which – usually in trimeric form (but sometimes dimeric or monomeric) – forms a water-filled transmembrane channel ('pore') that permits the passage of certain ions and small molecules; porins are also found e.g. in the mitochondrial outer membrane. (See also LAMB PROTEIN and TSX PROTEIN.) Note that the term 'porin' is also used to refer to the *complete* pore structure – whether it be composed of one, two or three porin proteins.

Bacterial porins. Porins have been demonstrated in many types of Gram-negative bacteria, but most studies have been carried out on *E. coli* and *Salmonella typhimurium*. Although porins are typically trimeric, dimeric porins have been found e.g. in *Paracoccus denitrificans* [JB (1985) *162* 430–433]; monomeric porins occur in the cyanobacteria [JB (2000) *182* 1191–1199].

The number of types of porin in a cell may vary according to conditions. For example, in *E. coli* the PhoE porin is induced when phosphate levels are low (see PHO REGULON). Moreover, extra porins may be acquired through LYSOGENY: the protein 2 (Lc protein) porin of *E. coli* K12 is present only when the cells are lysogenized by coliphage PA2.

The porins of *E. coli* typically include the products of genes *ompC*, *ompF* and (if phosphate is limiting) *phoE*; these products are designated OmpC (= Ib), OmpF (Ia) and PhoE, respectively. (See also OMP.) The OmpC and OmpF porins play a part in OSMOREGULATION (q.v.).

In addition to their role as channel components, porins may serve e.g. as receptor sites for phages; for example, in *Escherichia coli* the OmpF porin acts as a receptor for phages T2 and TuIa, while OmpC is a receptor for T4 and TuIb. During infection by Gram-negative bacteria, porins are reported to induce the release of CYTOKINES from monocytes and lymphocytes.

In *E. coli*, each transmembrane pore is formed by a group of three porin proteins. Structurally, the OmpC and OmpF pores are similar: a single channel from the inner (periplasmic) side of the pore branches into three channels which exit from the outer (exterior) face of the pore; the OmpF channel is somewhat wider than the OmpC channel. When extracted with SDS, these porins can be isolated as undenatured trimers.

The PhoE pore differs in that it contains three channels which remain separate across the width of the pore; a similar structure is found e.g. in the protein F porin of *Pseudomonas aeruginosa*.

Cyanobacterial porin proteins are larger than those of other Gram-negative bacteria (50–70 kDa versus 30–40 kDa), and they occur as monomers [JB (2000) *182* 1191–1199 (1193)].

Salmonella typhimurium typically contains the OmpC, OmpD and OmpF porins (previously referred to as 36K, 34K and 35K proteins, respectively).

In *Pseudomonas aeruginosa*, the porin channels have been reported to be larger [MR (1985) *49* 1–32] and relatively smaller [FEMS (1984) *21* 119–123] than those of *E. coli*.

The rate of diffusion of molecules through a porin channel depends strongly on the size, hydrophobicity and charge of the molecules. *E. coli* porin channels permit the passage of hydrophilic molecules of MWt up to ca. 650–850 (i.e. the 'exclusion limit' is ca. 650–850), but they are poorly permeable or impermeable to hydrophobic molecules (including many β-lactam antibiotics). By contrast, the pores in *Haemophilus influenzae* type b appear to be larger, and they offer little hindrance to the influx of a range of penicillins and cephalosporins [JB (1985) *162* 918–924].

In *E. coli* the main porin channels are not equally permeable: glucose (for example) diffuses through the OmpC channel at about half the rate at which it diffuses through the OmpF channel, and the differential is even greater with larger, hydrophobic or charged molecules. Mutants which lack OmpF channels exhibit increased resistance to e.g. chloramphenicol, tetracyclines and β-lactam antibiotics. [In *Enterobacter aerogenes*, resistance to e.g. cefepime and other cephalosporin antibiotics has been associated with altered porins: Microbiology (1998) *144* 3003–3009.]

The main porin channels in *E. coli* are preferentially permeable to cations (as opposed to anions), and they permit the entry of glucose at rates much higher than those of negatively charged glucose derivatives; a similar preference is shown by porin channels in *Pseudomonas aeruginosa*, but those in e.g. *Neisseria gonorrhoeae* (and those in the mitochondrial outer membrane) are preferentially permeable to anions. Negative charge may make little difference (or enhance) the passage of a molecule through the PhoE channel in *E. coli*. [Role of porins in outer membrane permeability (review): JB (1987) *169* 929–933.]

Porin channels in LIPOSOMES can be opened and closed by adjusting the value of the transmembrane potential; in this way they resemble the voltage-controlled mitochondrial porin channels (referred to as 'voltage-dependent anion channels' – VDACs). However, while the outer membrane can exhibit a transmembrane potential under certain conditions, the opening and closing of porin channels in intact cells has not been shown to be regulated in this way. (See also LIPOSOME SWELLING ASSAY.)

poroconidia See CONIDIUM.

porogenous development See CONIDIUM.

porospore See CONIDIUM.

porphin A cyclic TETRAPYRROLE in which the four pyrrole groups are linked by their α-carbon atoms via methene (–CH=) bridges; porphin is the parent compound of the PORPHYRINS.

porphobilinogen A substituted monopyrrole: an intermediate in the biosynthesis of PORPHYRINS.

Porphyra A genus of marine algae (division RHODOPHYTA). The nature of the thallus varies according to the stage in the life cycle; the thallus can be sheet-like (one or two cells thick) or filamentous – the filamentous form being called the '*Conchocelis* stage' since it was previously thought to be a different organism (designated *Conchocelis*). The organisms, which range from reddish-purple to olive green, attach to rocks etc by means of a holdfast. (See also LAVER and PORPHYRAN.)

porphyran A group of complex, mucilaginous, sulphated and methylated galactans (containing D- and L-galactose and 3,6-anhydro-L-galactose) produced by *Porphyra* spp.

Porphyridium A genus of unicellular, spherical algae (division RHODOPHYTA) which occur e.g. in freshwater habitats and on soil. Each cell contains a stellate chloroplast with a central pyrenoid; there is no cell wall, each cell being surrounded by a gelatinous polysaccharide capsule. When viewed in masses, the cells may be blood-red (e.g. *P. purpureum* – formerly *P. cruentum*) to blue-green (e.g. *P. aerugineum*), depending on the relative proportions of phycoerythrin and phycocyanin which they contain.

porphyrin test A test for determining the X FACTOR requirement of a *Haemophilus* strain by determining the ability of the strain to convert δ-aminolaevulinic acid (ALA) to porphyrins or porphobilinogen. A medium (0.5-ml aliquot) containing ALA (2 mM) and $MgSO_4$ (0.8 mM) in phosphate buffer (0.1 M, pH 6.9) is inoculated with a loopful of bacteria from an agar plate culture and incubated for ca. 4 hours at 37°C. A red fluorescence from the culture (or culture fluid) under WOOD'S LAMP (360 nm) indicates the formation of porphyrins (a *positive* test, i.e., the strain *does not* require the X factor); alternatively, 0.5 ml Kovács' indole reagent may be added, shaken vigorously, and the phases allowed to separate – a red coloration in the lower (aqueous) phase indicating the formation of porphobilinogen (a positive test). [Book ref. 22, pp. 561–562.]

porphyrins PORPHIN derivatives in which the pyrrole β-carbons are variously substituted: e.g., *aetioporphyrins* (= etioporphyrins) have 4 methyl and 4 ethyl substituents, *mesoporphyrins* have 4 methyl, 2 ethyl and 2 propionic acid substituents, and *protoporphyrins* have 4 methyl, 2 vinyl and

2 propionic acid substituents. Isomers of a given porphyrin are usually denoted by Roman numerals: e.g. aetioporphyrin I is 1,3,5,7-tetramethyl-2,4,6,8-tetraethylporphin, and aetioporphyrin III is 1,3,5,8-tetramethyl-2,4,6,7-tetraethylporphin. *Protoporphyrin IX* (= protoporphyrin III) is 1,3,5,8-tetramethyl-2,4-divinylporphin-6,7-dipropionic acid. (The numbering system used in the Dictionary – illustrated in the figure for CHLOROPHYLLS – is the widely used Fischer system; alternative systems have been proposed in which, e.g., all carbon atoms in the tetrapyrrole ring are numbered sequentially, 1–20, starting with the α-carbon of ring I next to the δ-methene position [PAC (1979) *51* 2251–2304].)

Porphyrins can readily chelate various metals, the metalloporphyrins being components of several important biological pigments: see e.g. CHLOROPHYLLS, CYTOCHROMES, HAEM. (See also CHLORIN; cf. PHYCOBILIPROTEINS.)

port See WINE-MAKING.

Port Salut cheese See CHEESE-MAKING.

porter See PERMEASE.

positive control (of an operon) See OPERON.

positive interference See INTERFERENCE (2).

positive-sense genome See VIRUS.

positive staining See ELECTRON MICROSCOPY (a).

positive-strand virus A VIRUS (q.v.) with a positive-sense genome.

post-antibiotic effect (PAE) The suppression of bacterial growth (typically for hours) which follows a short exposure to some types of antimicrobial agent – e.g. aminoglycosides, 4-quinolones, β-lactams, streptogramins and macrolides. PAE is a factor when considering the dose/dosage frequency of these antibiotics; it can be advantageous when coincident with low concentrations of an intermittently administered drug. PAE may involve different mechanisms in different organism–antibiotic combinations [AAC (1995) *39* 1314–1319].

The post-antibiotic sub-MIC effect may be more relevant, clinically, than the PAE [JAC (1999) *43* 71–77].

post-ciliary microtubules (*ciliate protozool.*) A ribbon of MICROTUBULES which arises from the right-hand posterior side of a kinetosome (close to triplet 9, GRAIN CONVENTION), extends upwards (towards the pellicle) and to the right, and then (sometimes) runs posteriorly between the kineties, the wider face of the ribbon being perpendicular to the body surface. (See also PITELKA CONVENTION.)

posterior station (*parasitol.*) The hindgut and/or rectum of an arthropod vector; infective forms of parasites in the posterior station can be transmitted to the vertebrate host via the faeces of the vector (see e.g. CHAGAS' DISEASE). (cf. CONTAMINATIVE INFECTION.)

post-fixation See FIXATION.

post-herpetic neuralgia See HERPES ZOSTER.

post-kala-azar dermal leishmaniasis A cutaneous leishmaniasis which may occur in a patient up to two years after that patient has been cured of VISCERAL LEISHMANIASIS. [PCR-based diagnosis of post-kala-azar dermal leishmaniasis: JCM (1998) *36* 1621–1624.]

post-meiotic segregation The segregation of alleles during the first nuclear division after MEIOSIS; it results from the separation of strands of uncorrected heteroduplex DNA (in a chromatid) formed previously during recombinational events in meiosis (see RECOMBINATION).

postoral suture See SYSTÈME SÉCANT.

post-replication repair (1) A DNA REPAIR process which operates after DNA replication: e.g. daughter-strand gap repair (see RECOMBINATION REPAIR) and some cases of MISMATCH REPAIR. (2) *Syn.* daughter-strand gap repair.

post-streptococcal glomerulonephritis (PSGN) GLOMERULONEPHRITIS which occurs as a late complication of *Streptococcus pyogenes* infection (usually IMPETIGO – cf. RHEUMATIC FEVER); PSGN is caused only by certain ('nephritogenic') strains of streptococci, and occurs mainly in young children. There may be a latent period of ca. 10–21 days between the original infection and the onset of PSGN.

post-transfusion hepatitis HEPATITIS transmitted by blood transfusion.

post-zygotic exclusion See PRE-ZYGOTIC EXCLUSION.

potassium cyanide broth See KCN BROTH.

potassium permanganate (as an antimicrobial agent) Potassium permanganate ($KMnO_4$) is an OXIDIZING AGENT which is active against a wide range of microorganisms at concentrations ranging from 0.001% to 1% w/v; it has been used for the treatment of certain superficial mycoses, but it can act as an irritant at high concentrations. As an algicide, potassium permanganate has been reported to be superior to copper sulphate. At concentrations >ca. 0.002% w/v, potassium permanganate imparts a pink coloration to water; this, together with the toxicity of its manganese component, makes potassium permanganate unsuitable for the disinfection of potable water.

potassium pump See ION TRANSPORT.

potassium transport See ION TRANSPORT.

potato aucuba mosaic virus See POTEXVIRUSES.

potato black ringspot virus See NEPOVIRUSES.

potato blight See EARLY BLIGHT and LATE BLIGHT.

potato diseases See e.g. BLACKLEG (2); EARLY BLIGHT; GANGRENE (2); LATE BLIGHT; POTATO SCAB; POTATO VIRUS Y; POWDERY SCAB; SILVER SCURF; SPRAING; WART DISEASE.

potato leaf roll virus See LUTEOVIRUSES.

potato mop top virus See SOIL-BORNE WHEAT MOSAIC VIRUS, SPRAING and TOBAMOVIRUSES.

potato paracrinkle virus See CARLAVIRUSES.

potato scab (common scab) A POTATO DISEASE caused by *Streptomyces scabies*. The disease occurs particularly in light, sandy, alkaline soils; corky scabs of various sizes typically develop on the surface of tubers, deeper lesions sometimes occurring. In heavy, wet soils POWDERY SCAB may develop.

potato slope A wedge-shaped piece of potato autoclaved (in a test-tube or bottle) and used as a MEDIUM.

potato spindle tuber viroid See VIROID.

potato virus A See POTYVIRUSES.

potato virus M See CARLAVIRUSES.

potato virus S See CARLAVIRUSES.

potato virus T A filamentous (about 640 × 12 nm), ssRNA-containing virus (possible member of the CLOSTEROVIRUSES) which has a rather limited host range; in potato plants, infection may be symptomless (latent) or may result in a mild leaf mottle. Transmission occurs via seeds in some solanaceous plants.

potato virus X See POTEXVIRUSES.

potato virus Y (PVY) The type member of the POTYVIRUSES. Virion: ca. 730 × 11 nm. PVY infects mainly plants of the Solanaceae, causing important diseases in several commercial crop plants (e.g. potato, capsicum, tobacco, tomato). In the potato, symptoms depend on virus strain, host cultivar, and environmental conditions. PVY Y^N strains may cause necrotic rings or spots, with mild mottling appearing later in the growing season; symptoms appearing in the second and subsequent years may include mild to severe mottling. PVY Y^O strains may cause necrosis, mottling or yellowing of leaflets, leaf-drop, and premature death (see also RUGOSE MOSAIC DISEASE). PVY Y^C strains may cause necrosis, mottling, crinkling, and necrotic streaks on leaves, petioles and stem; necrosis may occur in the tubers. Symptoms of PVY infection in capsicum, tobacco and tomato typically include mild mottling, but Y^N strains can cause a severe disease in tobacco ('tobacco veinal necrosis disease') which may result in total loss of the crop.

potato yellow dwarf virus See RHABDOVIRIDAE.

Poterioochromonas See OCHROMONAS.

potexviruses (potato virus X group) A group of PLANT VIRUSES in which the virions are flexuous filaments (ca. 470–580 × 13 nm) comprising one type of coat protein (MWt ca. 18000–23000) and one molecule of linear positive-sense ssRNA (MWt ca. 2.1×10^6). The RNA is capped at the 5′ end (sequence: $m^7G^{5'}$pppGpA...) but is not polyadenylated at the 3′ end; the adenine content of the RNA is ca. 30%. The host range for the group is wide, including monocots and dicots, but individual members infect only a narrow host range. Plants infected with potexviruses typically develop mosaic and ringspot symptoms. Virus particles form fibrous, sometimes banded, often large aggregates in the cytoplasm of infected cells; some members also induce nuclear inclusions. Transmission occurs mechanically.

Type member: potato virus X (PVX). Other members include e.g. cactus virus X, cassava common mosaic virus, clover yellow mosaic virus, *Cymbidium* mosaic virus, *Papaya* mosaic virus, *Plantago* severe mottle virus, *Plantago* virus X, white clover mosaic virus. Possible members include e.g. artichoke curly dwarf virus, bamboo mosaic virus, barley B-1 virus, *Boletus* virus, negro coffee mosaic virus, potato aucuba mosaic virus.

Potomac horse fever (equine monocytic ehrlichiosis) A HORSE DISEASE caused by *Ehrlichia risticii*, a species antigenically related to *E. canis* and *E. sennetsu* (and which can be transmitted, experimentally, to cattle [VR (2001) *148* 86–87]). Symptoms are variable, and may include e.g. fever, depression, anorexia, colic, mild to severe diarrhoea, oedema, etc; mortality rates may exceed 30%.

Potter–Elvehjem homogenizer A type of blender used e.g. for preparing subcellular fractions of eukaryotic cells.

Pott's disease Tuberculosis of the spine – see OSTEOMYELITIS.

potyviruses (potato virus Y group) A large group of PLANT VIRUSES in which the virions are flexuous filaments (ca. 680–900 × 11 nm) comprising one type of coat protein (MWt ca. 32000–36000) and one molecule of linear positive-sense ssRNA (MWt ca. 3.0–3.5 × 10^6). As a group, the viruses infect a wide range of host plants, including monocots and dicots, but individual members typically have a narrow host range; some members can cause important economic losses in crop plants. Symptoms depend on plant cultivar, virus strain, and environmental conditions, but typically include mosaic or mottle in the leaves, colour-breaking in flowers, mottling and/or distortion in the fruits; characteristic proteinaceous inclusion bodies (serologically unrelated to virus coat protein) are formed in the cytoplasm of infected plant cells, appearing as 'pinwheels' in transverse section or as 'bundles' in longitudinal section. A few members also form crystalline nuclear inclusions. Potyviruses are transmitted (non-persistently) by aphids (see NON-CIRCULATIVE TRANSMISSION sense 1) and can be transmitted mechanically; some viruses tentatively included in the group are transmitted by whitefly, mites or fungi.

Type member: POTATO VIRUS Y. Other members include e.g. bean common mosaic virus, bean yellow mosaic virus (of which pea mosaic virus is a strain), beet mosaic virus, blackeye cowpea mosaic virus, carnation vein mottle virus, carrot thin leaf virus, celery mosaic virus, cocksfoot streak virus, cowpea aphid-borne mosaic virus (of which Azuki bean mosaic virus is a strain), leek yellow stripe virus, lettuce mosaic virus, onion yellow dwarf virus, *Papaya* ringspot virus, parsnip mosaic virus, peanut mottle virus, PLUM POX virus, potato virus A, soybean mosaic virus, sugarcane mosaic virus (of which maize dwarf mosaic virus is a strain), tobacco etch virus, tulip breaking virus, turnip mosaic virus.

Many other viruses have been tentatively included as 'possible members' of the potyviruses. These include (a) certain aphid-borne viruses: e.g. carrot mosaic virus, celery yellow mosaic virus, groundnut eyespot virus, tobacco vein mottling virus, tomato (Peru) mosaic virus, wheat spindle streak virus, wheat streak virus; (b) a group of viruses which are transmitted by eriophyid mites: e.g., *Agropyron* mosaic virus, oat necrotic mottle virus, ryegrass mosaic virus, *Spartina* mottle virus, wheat streak mosaic virus; (c) the whitefly-transmitted virus sweet potato mild mottle virus; and (d) a group of viruses transmitted by *Polymyxa graminis*: barley yellow mosaic virus, oat mosaic virus, rice necrosis mosaic virus, wheat spindle streak mosaic virus, wheat yellow mosaic virus. Some or all of these 'possible members' may actually belong to or constitute separate groups.

pouch (food processing) See RETORT POUCH.

poultry diseases Poultry (chickens, ducks, turkeys, etc) can be affected by a wide range of diseases – some specific for one species of bird, others affecting a range of birds. (a) *Bacterial diseases*: see e.g. AIR SACCULITIS; ARIZONOSIS; BUMBLEFOOT; COLISEPTICAEMIA; DIPHTHEROID STOMATITIS; erysipelas (see SWINE ERYSIPELAS); FOWL CHOLERA; FOWL TYPHOID; HJÄRRE'S DISEASE; PSITTACOSIS; PULLORUM DISEASE. (b) *Fungal diseases*: see e.g. ASPERGILLOSIS; CANDIDIASIS; FAVUS. (c) *Protozoal diseases*: see e.g. BLACKHEAD; COCCIDIOSIS; HEXAMITIASIS. (d) *Viral diseases*: see e.g. AVIAN ACUTE LEUKAEMIA VIRUSES; AVIAN ENCEPHALOMYELITIS; AVIAN INFECTIOUS BRONCHITIS; AVIAN INFECTIOUS LARYNGOTRACHEITIS; AVIAN RETICULOENDOTHELIOSIS VIRUSES; AVIAN SARCOMA VIRUSES. DUCK VIRUS ENTERITIS; DUCK VIRUS HEPATITIS; FOWL PLAGUE; FOWL POX; HAEMORRHAGIC ENTERITIS OF TURKEYS; INCLUSION BODY HEPATITIS; INFECTIOUS BURSAL DISEASE VIRUS; LYMPHOID LEUKOSIS; MAREK'S DISEASE; NEWCASTLE DISEASE; OSTEOPETROSIS. (See also EGG-DROP SYNDROME, PARVOVIRUS and STUNTING SYNDROME.) (e) Diseases due to ingestion of *microbial toxins* include e.g. LIMBERNECK; TURKEY X DISEASE; WESTERN DUCK DISEASE.

poultry spoilage See MEAT SPOILAGE.

pour plate (1) A procedure (used e.g. as a COUNTING METHOD) in which a measured volume of liquid INOCULUM is placed in a PETRI DISH, a molten medium (e.g. an agar medium at 45–48°C) is added, and the whole swirled to disperse the inoculum throughout the medium; when set, the plate is incubated. (The inoculum may instead be mixed with the medium before the latter is poured into the Petri dish.) The medium may incorporate a reducing agent (e.g. sodium thioglycollate) so that even under aerobic incubation some anaerobes can grow *within* the medium. (2) A CULTURE prepared as in (1).

pourriture noble (French) *syn.* NOBLE ROT.

povidone–iodine See IODINE (a).

POW virus See POWASSAN ENCEPHALITIS.

Powassan encephalitis An acute, occasionally fatal, human ENCEPHALITIS caused by a flavivirus (see FLAVIVIRIDAE); it occurs e.g. in forested areas of Canada and Northern USA. The Powassan virus (POW virus) occurs in small mammals and is transmitted by ticks (usually *Ixodes* spp). [ARE (1981) *26* 84–85.]

powdery mildews Plant diseases caused by members of the ERYSIPHALES and characterized by the formation of at least partially superficial hyphal growth carrying a dense layer of conidia in which individual conidium-bearing structures are difficult to distinguish under low magnification (cf. DOWNY MILDEWS); dark-coloured cleistothecia may develop among the hyphae during the summer – being the overwintering stage in some species.

Powdery mildews of cereals (including barley and wheat) are caused by *Erysiphe graminis*. White to grey (later brownish) mycelium develops on small chlorotic spots on leaves, and may also occur on the glumes (particularly in wheat); lodging may occur. Control: antifungals such as CARBENDAZIM and PROCHLORAZ.

powdery scab A POTATO DISEASE caused by *Spongospora subterranea* (PLASMODIOPHOROMYCETES). Small wart-like swellings develop on the stolons, roots and tubers of infected plants; a dry brownish powder (consisting of masses of cystosori) forms within the tuber lesions. The cysts can persist in the soil for many years. The disease is favoured by wet soil conditions. The pathogen can also infect tomato roots.

powdery yeasts See FLOCCULATION (sense 2).

pOX1 plasmid See ANTHRAX TOXIN.

pOX2 plasmid See ANTHRAX TOXIN.

Poxviridae (the poxvirus group) A family of dsDNA-containing VIRUSES which infect mammals (including man), birds or insects; some members are important pathogens (e.g. the causal agents of SMALLPOX, MONKEYPOX, CONTAGIOUS PUSTULAR DERMATITIS (sense 1), etc). The family includes two subfamilies, CHORDOPOXVIRINAE (vertebrate poxviruses) and ENTOMOPOXVIRINAE (insect poxviruses), together with several unclassified viruses (e.g. MOLLUSCUM CONTAGIOSUM virus, TANAPOX VIRUS, YABA MONKEY TUMOUR POXVIRUS) [Book ref. 23, pp. 42–46].

The poxvirus virion is the largest and structurally the most complex of all animal viruses. Depending on virus, it may be ovoid or brick-shaped, ca. 200–450 × 170–260 nm. It contains a central *core* (incorporating the genome) which is commonly bilaterally concave; on either side of the core – coincident with each concavity – is an ellipsoidal structure called the *lateral body*. (cf. ENTOMOPOXVIRINAE.) The core and its two associated lateral bodies are enclosed within a lipoprotein *outer membrane* which bears a characteristic pattern of ridges or tubules on its surface. (In some poxviruses the extracellular virions have an extra lipoprotein layer, the *envelope*, acquired from the host cell; these are believed to be the infective forms of the virus.) The virion contains a variety of enzymes, including enzymes involved in e.g. DNA replication, transcription, and RNA processing. In general, poxvirus virions are stable at room temperature and are resistant to desiccation; most are ether-resistant but may be inactivated e.g. by chloroform and by heating (e.g. 60°C/20 min).

The genome is a single molecule of dsDNA (MWt in the range $85–250 \times 10^6$) in which – in at least some poxviruses – the two strands are covalently joined at each end of the molecule (i.e., the molecule is actually a single uninterrupted circular polynucleotide chain). The genome contains highly conserved sequences in the central region, and terminal hypervariable regions. [Genomes: orf virus, bovine papular stomatitis virus [JV (2004) *78* 168–177]; goatpox, sheeppox [JV (2002) *76* 6054–6061]; deerpox [JV (2005) *79* 966–977.]

Replication cycle. Poxvirus replication occurs in the cytoplasm of the host cell; studies on replication processes have been carried out mainly with orthopoxviruses (particularly VACCINIA VIRUS). Infection begins when the virion adsorbs to a host cell; vaccinia virus appears to bind to the host cell-surface epidermal growth factor (EGF) receptor [Nature (1985) *318* 663–665]. The virus apparently enters the cell by fusion between viral and cell membranes and/or by viropexis. An early consequence of infection is the inhibition of host cell DNA, RNA and protein synthesis [review: Book ref. 150, pp. 391–429]. The viral core is uncoated in the cytoplasm, and enzymes of the core begin transcription of the viral DNA; early mRNA is synthesized by an α-amanitin-resistant viral RNA polymerase, and viral enzymes then process the RNA to form the functional polyadenylated, capped and methylated mRNA. (Early mRNAs, at least, are apparently not spliced.) DNA replication then begins; this may involve nicking of the DNA to provide free 3′-ends which can act as primers. The onset of DNA replication allows initiation of the synthesis of late mRNA which codes for most of the structural proteins of the virion. Proteolytic processing of the structural proteins appears to be coupled to virus assembly. In the case of vaccinia virus, the mature virions each become enclosed in a double membrane derived from the Golgi apparatus and are transported to the cell periphery; it was suggested that the outer viral membrane may fuse with the cell membrane, resulting in the release of virions enclosed by a single layer of Golgi membrane [JGV (1985) *66* 643–646].

The discrete cytoplasmic sites of virus replication are termed 'virus factories', 'factory areas', or 'viroplasms'; they are visible in suitably stained preparations as *B-type inclusion bodies* (= GUARNIERI BODIES). Some poxviruses (e.g. COWPOX VIRUS) form a second type of inclusion body, the *A-type inclusion body*: acidophilic structures which may or may not contain virions.

poxvirus group See POXVIRIDAE.

pp Prefix denoting (1) phosphoprotein (see e.g. RETROVIRIDAE), or (2) diphospho- or pyrophospho- (as in ppGpp).

PPAR Peroxisome proliferator-activated receptor: see PROSTAGLANDINS.

PPD See TUBERCULIN.

PPFM Pink-pigmented facultatively methylotrophic bacterium.

ppGpp See (i) CELL CYCLE; (ii) STRINGENT CONTROL (sense 1); and (iii) SPOT GENE.

ppGpp has been classified as an ALARMONE.

PPi Inorganic PYROPHOSPHATE ($P_2O_7^{4-}$).

PPLOs Pleuropneumonia-like organisms: an early name for *Mycoplasma* spp.

PQQ See QUINOPROTEIN.

Pr Prefix denoting a polyprotein product of transcription (see e.g. RETROVIRIDAE).

PR proteins PATHOGENESIS-RELATED PROTEINS.

Pr65 *gag* precursor protein (of FeLV) See FELINE LEUKAEMIA VIRUS.

PRAS media Pre-reduced anaerobically-sterilized media.

Prasinocladus See TETRASELMIS.

Prasinomonadida See PHYTOMASTIGOPHOREA.

Prasinophyceae A class of unicellular, flagellated green algae which have scaly flagella arising from an anterior flagellar pit. The class, which is no longer recognized [Book ref. 123, pp. 65–66], included e.g. TETRASELMIS and many of the genera in the MICROMONADOPHYCEAE.

Prasiola A genus of marine and freshwater green algae (division CHLOROPHYTA). The thallus is foliose or filamentous. Asexual reproduction occurs by the formation of diploid aplanospores. In at least some species sexual reproduction is preceded by meiosis in the cells of the upper portion of the blade followed by several mitotic divisions; this results in the formation of polystromatic micro- and macrogametangial regions which give rise to biflagellate microgametes and non-motile macrogametes, respectively.

Prausnitz–Küstner antibodies *Syn.* REAGINIC ANTIBODIES.

Prausnitz–Küstner test (P–K test) A test used for demonstrating the presence of IgE antibodies homologous to a given allergen. Serum from a hypersensitive subject is injected intradermally into a normal subject; subsequent injection of the normal subject with the given allergen causes a WHEAL-AND-FLARE response to develop within minutes (a positive test). The P–K reaction can be inhibited by the prior injection of a fragment of ε-chain [Nature (1985) *315* 577–578].

PRD See CATABOLITE REPRESSION.

PRD1 phage group *Syn.* TECTIVIRIDAE.

preaxostyle See OXYMONADIDA.

pre-B cell See B LYMPHOCYTE.

prebiotic (1) A non-digestible part of a food (often an oligosaccharide) which, by selectively stimulating certain type(s) of colon bacteria, benefits the host.

(2) (*adj.*) The time before life appeared on Earth.

prebuccal area In some ciliates (e.g. *Paramecium*): a depression in the body surface leading to the BUCCAL CAVITY; its ciliature is essentially somatic (cf. VESTIBULUM).

precipitation (*serol.*) The formation of a precipitate when antibodies and *soluble* antigens react in suitable proportions (see e.g. OPTIMAL PROPORTIONS; cf. ANTIGEN EXCESS and AGGLUTINATION). (See also GEL DIFFUSION.)

precipitin An ANTIBODY which forms a precipitate with its homologous soluble antigen. (cf. AGGLUTININ.)

precipitin test Any serological test in which the interaction of antibodies with *soluble* antigens is detected by the formation of a precipitate.

precipitinogen (1) An antigen which elicits a PRECIPITIN. (2) Any agent (e.g. PROTEIN A) which can combine with antibody to form a precipitate.

precyst See CYST (a).

prednisone See IMMUNOSUPPRESSION.

pre-erythrocytic schizogony (in *Plasmodium*) See PLASMODIUM.

pre-filtration See FILTRATION.

pre-fixation See FIXATION.

pre-genome See HEPADNAVIRIDAE.

pregnancy test An IMMUNOASSAY in which urine is examined for human chorionic gonadotrophin (hCG), a hormone that appears during pregnancy. Urine is added to serum containing anti-hCG antibodies, which combine with hCG. Then, the absence of *uncombined* antibodies (a *positive* test) is shown by non-agglutination when latex-bound hCG is added.

pre-initiation complex See PROMOTER.

Preisz–Nocard bacillus See CORYNEBACTERIUM.

pre-lesion terminus In damaged DNA: the nucleotide immediately 5′ of the lesion site.

premunition (concomitant immunity; non-sterilizing immunity) (*immunol.*) Protective immunity, in respect of a given pathogen, due to the persistence of small numbers of that pathogen in the host; the host can resist superinfection (i.e., infection by the same or closely related pathogens) but cannot achieve STERILIZING IMMUNITY. In an apparently analogous phenomenon, cells of a given tumour type fail to develop when injected into an animal suffering from the same type of tumour.

preoral suture See SYSTÈME SÉCANT.

prepatent period In a parasitic infection: the time interval between the initial infection and the appearance of parasites, or their cysts, in blood, tissues or faeces.

prephenate See AROMATIC AMINO ACID BIOSYNTHESIS.

preprimosome See PRIMOSOME.

pre-protein See SIGNAL HYPOTHESIS.

pre-reduced medium See ANAEROBE.

preservation (1) (of materials, products) The use of physical and/or chemical means to kill or prevent the growth of those microorganisms which, by their growth and/or activities, may cause BIODETERIORATION of a given material or product. Physical methods (e.g. heat STERILIZATION, freezing – see e.g. FOOD PRESERVATION) either kill contaminating organisms or render the materials or products unsuitable for the (rapid) growth of likely contaminants. Chemical agents (PRESERVATIVES) may be microbistatic or microbicidal.

(2) (of microorganisms) Viable populations of microorganisms may be preserved (i.e. maintained for reference purposes etc) by e.g. FREEZING, FREEZE-DRYING or DESICCATION. Some bacterial cultures can be preserved by sealing with a layer of sterile mineral oil and storing at ca. 4°C. A culture may be preserved by periodic SUBCULTURE (see also STABILATE).

preservative Any chemical used for the PRESERVATION of materials or products; in some industries such chemicals are referred to as BIOCIDES. For examples of preservatives and their usage see e.g. FOOD PRESERVATION, PAINT SPOILAGE, PAPER SPOILAGE, SAUSAGE, TEXTILE SPOILAGE, TIMBER PRESERVATION; see also ALCOHOLS, BISPHENOLS, CHLORBUTANOL, EGME, FORMALDEHYDE, GERMALL 115, PHENOLS, QUATERNARY AMMONIUM COMPOUNDS, SALICYLANILIDES, THIOMERSAL. A preservative may be inactivated e.g. by colloids (such as magnesium trisilicate or bentonite – cf. BENZOIC ACID; CHLORBUTANOL) or by certain container materials (polyethylene can inactivate e.g. dichlorophenol and other phenolics, while polypropylene can inactivate e.g. dichlorophenol and sorbic acid).

prespore See ENDOSPORE (sense 1(a)).

pressure-cycle fermenter See AIRLIFT FERMENTER.

presumptive coliform count See COLIFORM TEST.

pretibial fever *Syn.* FORT BRAGG FEVER.

pretrichocyst See TRICHOCYST.

pretyrosine See AROMATIC AMINO ACID BIOSYNTHESIS.

Prevotella A genus of Gram-negative, asporogenous, anaerobic, rod-shaped bacteria found e.g. in the human oral cavity; some species form black or brown pigments when grown on blood-containing media. Most species ferment sucrose and lactose.

Using quantitative culture on a non-selective blood-agar medium, studies on the anaerobic oral microflora in the first year of life found that organisms in the *Prevotella melaninogenica* group were among the early colonizers (18% at 2 months, rising to 57% at 6 months and 77% at 12 months); members of the *P. intermedia* group were apparently late colonizers, whereas isolation of non-pigmented species of *Prevotella* increased from 'occasional' at 2 months to 75% at 12 months [RMM (1997) *8* (supplement 1) S19–S20].

Species of *Prevotella* have been associated with e.g. abscesses and PERIODONTITIS. [Simultaneous detection of *Prevotella intermedia* and *Bacteroides forsythus* by multiplex PCR in clinical samples: JCM (1999) *37* 1621–1624.]

pre-zygotic exclusion During processes such as CONJUGATION (sense 1b) and TRANSDUCTION: the absence of expression of a given donor gene in the recipient as a result of the failure of that gene to enter the recipient. If the gene entered the recipient but failed to recombine with the recipient's genome, the absence of expression of that gene is referred to as *post-zygotic exclusion.*

***prgI* gene** (*Salmonella typhimurium*) See NEEDLE COMPLEX.

Pribnow box See PROMOTER.

Pril A quaternary ammonium detergent used e.g. in certain media to prevent swarming by *Proteus* spp.

prill A solid in granular form.

primaquine See MALARIA (*chemotherapy*).

primary amoebic meningoencephalitis See MENINGOENCEPHALITIS.

primary atypical pneumonia (1) Any of a group of pneumonias characterized by symptoms distinct from those of 'classical' lobar PNEUMONIA and by a causal agent which cannot readily be isolated by routine laboratory methods; the group includes LEGIONELLOSIS; *Mycoplasma pneumoniae* pneumonia (see (2) below); PSITTACOSIS; Q FEVER.

(2) PNEUMONIA caused by *Mycoplasma pneumoniae*; it affects mainly children. Infection occurs by droplet inhalation. Incubation period: 2–3 weeks. Onset is insidious, with fever and malaise followed by symptoms of an upper respiratory tract infection; a persistent, non-productive cough typically develops, and ear infection is common. The disease is usually mild and self-limiting, with few chest signs; in only a minority of cases does frank pneumonia develop. *Chemotherapy*: e.g. erythromycin.

primary constriction *Syn.* CENTROMERE.

primary culture (1) A CULTURE prepared by inoculating a medium directly from a natural source of microorganisms (cf. SUBCULTURE). For example, a bacteriological medium may be inoculated directly from a faecal specimen. In TISSUE CULTURE a primary culture is one prepared from cells or tissues taken directly from an animal or plant. (2) The process of preparing a culture as in (1).

primary fixation *Syn.* pre-FIXATION.

primary fluorescence See FLUORESCENCE.

primary host (of rust fungi) See UREDINIOMYCETES.

primary hyphae See HYPHA.

primary metabolism Metabolism which is essential for and geared towards growth – i.e., energy metabolism, synthesis of cell components, etc. (cf. SECONDARY METABOLISM.)

primary mycelium (in actinomycetes) *Syn.* SUBSTRATE MYCELIUM.

primary production The production of new biomass by photosynthetic organisms as a result of their use of light (solar) energy. The phrase 'chemosynthetic primary production' has been used to refer to the production of biomass by those non-photosynthetic organisms which are believed to derive energy from the oxidation of e.g. sulphide and/or methane in the vicinity of HYDROTHERMAL VENTS in the ocean floor; these organisms (which include species of *Thiobacillus* and *Thiomicrospira*) appear to be at the start of a food chain which is unique in that it is not based on photosynthetic organisms.

primary septum See SEPTUM (b).

primary zoospore See e.g. DIPLANETISM and PLASMODIOPHOROMYCETES.

primase (DNA primase) An enzyme which, in vitro, can polymerize molecules of ribonucleoside triphosphate (rNTPs) and/or dNTPs in the 5′-to-3′ direction to form short oligonucleotides on an ssDNA template. In vivo, *Escherichia coli* primase (*dnaG* gene product) synthesizes short RNA primers for the formation of Okazaki fragments in DNA REPLICATION, and also synthesizes primers for the initiation of DNA replication of isometric ssDNA phages.

Primase either recognizes a specific region of DNA directly (as in BACTERIOPHAGE G4) or, more commonly, it requires the co-operation of additional proteins (see PRIMOSOME).

Certain phages (e.g. T4, T7) encode their own primases.

Primase action is resistant to rifampicin (cf. RNA POLYMERASE).

primate T-cell leukaemia virus See SIMIAN T-CELL LEUKAEMIA VIRUS.

prime plasmid A PLASMID which has undergone aberrant excision from a bacterial chromosome. In some types of prime plasmid ('type I'), part of the plasmid has been left behind in the chromosome, and the plasmid carries a sequence of bacterial DNA; a 'type II' prime plasmid consists of the entire plasmid carrying a sequence of bacterial DNA. The particular sequence of bacterial DNA carried by a prime plasmid depends on the site within the chromosome occupied by the plasmid before excision; thus, e.g. it is possible to isolate various F′ plasmids (derived from the F PLASMID) because this plasmid can integrate at various sites in the bacterial chromosome.

(See also specialized TRANSDUCTION.)

primed (*immunol.*) (1) (of persons, animals) Refers to those individuals who have had an initial immunological contact with a given antigen; such individuals may respond e.g. with ANTIBODY FORMATION. Subsequent exposure of a primed individual to the given antigen may bring about a heightened response (see ANAMNESTIC RESPONSE and HYPERSENSITIVITY (sense 1)). (See also IMMUNE RESPONSE.) (2) (of cells) Refers to those cells whose responsiveness to a given antigen has been heightened as a consequence of their previous contact with that antigen (or of previous contact between their precursor cells and that antigen). (cf. PRIMING.)

primer See (i) DNA REPLICATION; (ii) PCR; (iii) NASBA; (iv) SDA; and (v) FORWARD PRIMER.

primer-dimer In a PCR reaction mixture: an artefact resulting from the use of a pair of primers whose 3′ terminal sequences are mutually complementary; during extension, each primer uses the other as a template – forming products that consist essentially of two double-stranded primers in tandem.

priming (1) (*mol. biol.*) The initiation of synthesis of a DNA strand e.g. by the synthesis of an RNA primer (see DNA REPLICATION). (For 'specific' and 'general' priming see DNAB GENE.)

(2) (*immunol.*) The process in which an individual becomes PRIMED.

primite The anterior organism of a pair of gregarines in SYZYGY. (cf. SATELLITE.)

primociliatid gymnostomes Presumptively primitive ciliates (subclass GYMNOSTOMATIA, order Primociliatida) which are homokaryotic; species of STEPHANOPOGON are the only known examples.

primordium (*microbiol.*) *Syn.* INITIAL(s).

primosome A multiprotein complex involved in certain cases of primer synthesis (see DNA REPLICATION): e.g. in BACTERIOPHAGE ϕX174 DNA synthesis, and in Okazaki fragment synthesis during *Escherichia coli* chromosome replication and e.g. COLE1 PLASMID replication. Primosome formation on ssDNA coated with SINGLE-STRAND BINDING PROTEIN is initiated by protein n′(= factor Y) which recognizes and binds to specific short base sequences in the DNA. In the presence of ATP, a complex of DnaB (see DNAB GENE) and DnaC proteins is formed which, together with proteins i (= factor X), n, and n″ combine with n′ to form a prepriming intermediate (preprimosome) at or near the n′ recognition site. The preprimosome is recognized and bound by the DnaG protein (PRIMASE) to complete the primosome. The primosome migrates along the ssDNA template in a 5′-to-3′ (anti-elongation) direction (i.e., in the direction of replication fork movement in dsDNA replication), driven by ssDNA-dependent ATP hydrolysis by n′, and primase synthesizes short RNA primers at many different sites on the template DNA. The direction of primase action is opposite to that of primosome migration; primer synthesis is assumed to occur within the domain of the primosome. [JBC (1981) *256* 5273–5286.]

primulin A polycyclic fluorescent dye. (See also SCAR.)

prion (1) In man/animals: an aberrant form of a normal, chromosome-encoded glycoprotein; prions are now believed to be the aetiological agents of TRANSMISSIBLE SPONGIFORM ENCEPHALOPATHIES. [Mapping the parameters of prion-induced neuropathology: PNAS (2000) *97* 10573–10577.]

The normal protein, which is designated PrP^c, is encoded (in humans) by the *PRNP* gene (on chromosome 20). PrP^c is a component of the cytoplasmic membrane in certain types of cell – e.g. neurones, astrocytes, dendrocytes, lymphocytes and macrophages.

Synthesis of PrP^c involves intranuclear transcription of *PRNP* and translation on the endoplasmic reticulum; post-translational modification includes glycosylation (although some molecules remain unglycosylated) and attachment to a glycolipid phosphatidylinositol (GPI) 'anchor'. The PrP^c molecule is apparently anchored at the cell surface for a short period of time and then internalized and degraded in a lysosome. The function of PrP^c has not been established.

When a prion infects a cell, it appears that normal breakdown of (at least some) PrP^c does not occur, and that the secondary structure of these PrP^c molecules is changed to that of the prion. This is thought to occur either by interaction between individual PrP^c and prion molecules (heterodimer model), or by interaction between individual PrP^c molecules and organized aggregates of prion molecules (the 'seed' or 'nucleus' model). Accumulation of prions in the form of crystalloid structures eventually leads to the death of the cell. (Interestingly, specific parts of PrP^c (particularly the central region: residues 90–120) undergo *spontaneous* re-arrangement to a conformation resembling that of the infectious form when immobilized on a surface with recombinant antibodies [EMBO (2001) *20* 1547–1554].)

In the prion model of pathogenesis, prions may be present through *infection* – either via ingestion (as in KURU) or through their introduction by medical or surgical procedures. The *inherited* prion diseases, in which the *PRNP* gene is almost invariably heterozygous (wild-type and mutant alleles), are believed to result from the effect of mutant (prion) proteins on wild-type proteins. In the so-called 'sporadic' prion diseases, conversion of normal protein to prion protein is considered to result e.g. from random, spontaneous events.

The presence of normal (PrP^c) protein appears to be necessary for the development of prion diseases: knockout mice which are unable to form PrP^c do not develop disease on challenge with prion-containing tissue; this suggests that a prion cannot direct the production of further prions without a supply of the normal protein.

Prions do not differ, biochemically, from the normal protein, i.e. the amino acid sequence is identical. However, the conformation is different: about ~40% of normal protein is in the soluble, protease-sensitive α-helical state – with little or no β-sheet formation; in contrast, the prion form (isoform) is insoluble and protease-resistant, and is composed of ~50% β-sheet and ~20% α-helix forms. (See also PLASMINOGEN.)

Prions are resistant to e.g. heat, ionizing and ultraviolet radiation, formalin, β-propiolactone and bleach (as well as to proteases).

The prion form of PrP^c is often designated PrP^{res} (indicating resistance). Some authors use PrP^{sc} (which, strictly, indicates the SCRAPIE prion) as a generalized designation for any prion; others use superscripts which indicate the specific disease (e.g. 'GSS' for Gerstmann–Sträussler–Scheinker syndrome).

[Protein misfolding in prion disease: Cell (1997) *89* 499–510; human prion diseases: Book ref. 215, pp. 39–77.]

(2) Strains of the yeast *Saccharomyces cerevisiae* contain a protein *release factor* (see PROTEIN SYNTHESIS) designated eRF3 (= Sup35, product of gene *SUP35*) which, through variation in its effective concentration in the cytoplasm, can influence the fidelity with which translation is terminated on the ribosome; thus, a lowered concentration of the release factor promotes readthrough at certain stop codons. A non-genetic mechanism which can control the concentration of eRF3 may therefore permit heritable variation in the organism without changes in the genome [see e.g. Nature (2000) *407* 477–483].

Sup35 can exist in an aggregated, prion form (the [PSI^+] state) or in a soluble form (the [psi^-] state). In [PSI^+] cells, aggregation of Sup35 protein reduces the amount available for terminating translation on the ribosomes, and this allows readthrough at certain stop codons.

Propagation of [PSI^+] involves protein–protein interaction following the mixing of cytoplasm from different cells (e.g. during mating); in this process the soluble form of Sup35 is converted to the aggregated (prion) form. (Interestingly, if a [PSI^+] strain of *S. cerevisiae* is grown in the presence of 2 M glycerol (or certain other agents) the aggregates of Sup35 break down to the soluble form of Sup35, i.e. a [psi^-] strain of the yeast is formed.) [Elimination of [PSI^+] by guanidine hydrochloride: Mol. Microbiol. (2001) *40* 1357–1369.]

[eRF3 concentration in the termination of translation: Microbiology (2001) *147* 255–269 (262–263).]

In vitro studies involving fusion between the prion domain of Sup35 proteins from *S. cerevisiae* and *Candida albicans* may help to explain factor(s) which determine the ability of a prion to cross a species barrier [Nature (2001) *410* 223–227].

pristane An isoprenoid alkane: 2,6,10,14-tetramethylpentadecane.

pristinamycins See STREPTOGRAMINS.

pRNA See BACTERIOPHAGE ϕ29.

***PRNP* gene** See PRION.

probasidium (1) A developing BASIDIUM at the stage at which karyogamy occurs; in many basidiomycetes the terms probasidium and METABASIDIUM refer to the same structure at different developmental stages.

(2) That *part* of a developing basidium in which karyogamy occurs – e.g. the teliospore (or its contents) of the rust and smut fungi, or, in *Septobasidium*, the typically thick-walled structure from which arises, by outgrowth, the septate, basidiospore-bearing metabasidium.

probe A short strand of DNA or RNA (often 10–25 nucleotides in length) which is *complementary* to a given target sequence of nucleotides; because a probe can hybridize specifically with its target sequence, it can be used e.g. to detect, identify or locate the target. To make the probe detectable it is 'labelled' either before or after it hybridizes to the target sequence (see below). (In addition to DNA or RNA, probes may also be prepared from PNA [JAM (2001) *90* 180–189].)

Probe–target hybridization is affected e.g. by temperature, pH and concentration of electrolyte. Conditions can be chosen such that the probe will hybridize only to the *exact* target sequence, i.e. conditions under which mis-matches are not tolerated; such *high-stringency* conditions are used to promote specificity in probe–target binding. Under *low-stringency* conditions the effect of mis-matched base(s) can be minimized so that hybridization may occur even when the probe–target duplex contains mis-matched base(s).

The detection of *RNA* targets may be achieved with greater rapidity and specificity by means of 2′-*O*-methyl oligoribonucleotide probes (compared with 2′-deoxy oligoribonucleotide probes) [NAR (1998) *26* 2224–2229].

Labelling of probes. Probes may be labelled directly and covalently with radioactive isotopes (e.g. ^{32}P); such isotopic labels can be detected by autoradiography. Both DNA and RNA probes can be labelled at the 5′ end with e.g. [γ^{32}P]ATP (i.e. ATP with ^{32}P in the γ position); this reaction is mediated by polynucleotide kinase. Radioactive labels are robust, and are associated with good sensitivity (but limited resolution).

Non-radioactive (non-isotopic) labelling with e.g. enzymes or fluorescent molecules is common. Labelling may be either direct or indirect. Thus, for example, *existing* probes may be labelled with a reactive fluorophore which binds covalently to the probe; such labelling systems are available commercially (e.g. FluoReporter®, marketed by Molecular Probes Inc., Eugene, OR, USA). In the second form of direct fluorescent labelling, probes are *synthesized* in a reaction mixture containing fluorophore-conjugated nucleotides (which are thus incorporated into the probe). Such fluorescent nucleotides are available commercially (e.g. ChromaTide®; Molecular Probes Inc.). Fluorescent labels are detected by examining the preparation under ultraviolet light.

CHEMILUMINESCENCE is used in probe detection in certain commercial diagnostic tests (see e.g. *hybridization protection assay* in the entry TMA).

Note that, in direct labelling, the probe is already labelled prior to hybridization.

For *indirect* labelling (e.g. with an enzyme or fluorophore), a small molecule (e.g. a biotin or DIGOXIGENIN 'tag') is incorporated into the probe before use. *After* hybridization, the probe is detected by using a *conjugate* to link the tag with a fluorophore or enzyme – see e.g. DOT-BLOT. (It has been reported that short (~10-nucleotide) 5′-biotinylated probes give satisfactory results when labelled with a STREPTAVIDIN–alkaline phosphatase conjugate *prior to* the hybridization step [NAR (1999) *27* 703–705].)

Uses of probes. Probes can be used e.g. to detect, identify or locate particular pathogens in clinical specimens (by demonstrating the presence of pathogen-specific sequences); specimens are commonly subjected to pre-test procedures in which any target nucleic acids that may be present are exposed (in single-stranded form) to specific probes. One example is the PACE 2C test (Gen-Probe, San Diego, USA) which, in a single assay, can detect *Chlamydia trachomatis* and/or *Neisseria gonorrhoea* in urogenital specimens [see e.g. JCM (1995) *33* 2587–2591].

In situ hybridization (ISH) involves the use of strain- or species-specific probes e.g. for detecting pathogens in situ – i.e. at the actual site of infection within tissues or cells; bacterial, viral, fungal and protozoal pathogens may be detected. ISH is particularly useful for demonstrating nucleic acids which are focally distributed – e.g. virus-infected tissue. For example, ISH was found to be as sensitive as PCR for the detection of human papillomavirus in cervical scrapings [HJ (1995) *27* 54–59]. Much of the early work in ISH was conducted with radioactive probes; currently, fluorescent probes are common (fluorescence in situ hybridization: FISH). FISH is both rapid and flexible. For example, the target area can be interrogated simultaneously with several different types of probe (each labelled with a different fluorophore); thus, when different probes bind to closely adjacent targets they can be distinguished by their emission characteristics.

Probes are also used in a *line probe assay* for detecting certain point mutations which confer resistance to antibiotics in *Mycobacterium tuberculosis.* Essentially, PCR is used to amplify a sequence in the chromosome associated with common resistance-conferring mutations, and the amplicons are denatured to the single-stranded state; these products are then tested for hybridization against a range of probes which include non-mutant and mutant copies of the given chromosomal sequence. The presence/absence of hybridization between the amplicons and each of the probes provides an indication of the probable resistance or susceptibility of the test strain to given antibiotic(s). [Evaluation of a line probe assay: JCM (1997) *35* 1281–1283.]

In nucleic acid amplification processes, specialized probes can provide a real-time indication of progress in the reaction, and can also be used to assess the concentration of target molecules present in the sample prior to amplification (see TAQMAN PROBES). (See also MOLECULAR BEACON PROBES.)

[Probe-based methods in clinical microbiology: Book ref. 221, pp 44–55.]

probenecid A drug which e.g. delays the excretion of – and (thus) helps to maintain blood levels of – certain penicillins and cephalosporins.

Probenecid, which inhibits glucuronidation, may affect the metabolism of the antiretroviral drug AZT.

probiotic Any potentially beneficial preparation consisting typically of a culture of bacteria (or bacterial spores) of a type normally found in the healthy gut microflora. Oral administration of probiotics has been reported to restore normal gut microecology and e.g. to enhance intestinal IgA responses; perinatal and postnatal administration of the probiotic *Lactobacillus rhamnosus* (= *Lactobacillus* GG; ATCC 53103) has been associated with a significant reduction in the incidence of atopic disease (eczema) in at-risk infants [Lancet (2001) *357* 1076–1079; see also commentary: Lancet (2001) *357* 1057–1059].

The Shirota strain of *Lactobacillus casei* is reported to be useful therapeutically against an *Escherichia coli* urinary tract infection when administered intraurethrally [AAC (2001) *45* 1751–1760].

Probiotic activity has been reported for various strains of *Bacillus*. For example, the product Enterogermina (an aqueous suspension of spores) is used in Italy with the object of preventing or treating bacterial diarrhoea. [Molecular characterization of bacteria (*Bacillus clausii*) from samples of Enterogermina: AEM (2001) *67* 834–839.]

(See also YAKULT and VSL#3; cf PREBIOTIC.)

procapsid A capsid precursor formed during the assembly of a virus.

Procaryotae An obsolete kingdom which comprised the PROKARYOTES; proposed divisions of the Procaryotae: FIRMICUTES; GRACILICUTES; MENDOSICUTES; TENERICUTES.

procaryote *Syn.* PROKARYOTE.

processive enzyme An enzyme which binds to a (macromolecular) substrate and then moves along it, functioning as it goes. (See e.g. HELICASES; DNA POLYMERASES.)

Prochlorales See PROCHLOROPHYTES.

prochloraz (trade name: e.g. Sportak) An agricultural systemic ANTIFUNGAL AGENT which inhibits sterol biosynthesis, thereby disrupting the cytoplasmic membrane in susceptible fungi. It is used e.g. against rhynchosporium, net blotch and powdery mildew of cereals, and – mixed with e.g. mancozeb – for the control of yellow rust (*Puccinia striiformis*) on wheat. Mixed with CARBENDAZIM ('Sportak Alpha') it is useful e.g. against eyespot, sharp eyespot, septoria and other CEREAL DISEASES.

Prochlorococcus marinus See PROCHLOROPHYTES.

Prochloron See PROCHLOROPHYTES.

Prochlorophyta See PROCHLOROPHYTES.

prochlorophytes A category of unicellular prokaryotes (order Prochlorales, division Prochlorophyta) which contain eukaryotic-type chlorophylls *a* and *b* (but lack phycobiliproteins) and carry out oxygenic photosynthesis.

The first member identified was *Prochloron didemni* (formerly *Synechocystis didemni*). The cells are more or less spherical, 6–25 μm in diameter, and the photosynthetic pigments are attached to THYLAKOIDS. The organisms occur as ectosymbionts on marine didemnid ascidians (sea-squirts) in tropical and subtropical coastal waters.

Species identified later include *Prochlorothrix hollandica* and *Prochlorococcus marinus*.

Taxonomically (in terms of genomic sequence data), the prochlorophytes have been grouped with the cyanobacteria and chloroplasts.

[Photosynthetic machinery in prochlorophytes: FEMS Reviews (1994) *13* 393–414.]

Prochlorothrix hollandica See PROCHLOROPHYTES.

prodigiosin A red, water-insoluble, ethanol-soluble tripyrrole derivative. Prodigiosin and similar pigments are produced by some species and biotypes of SERRATIA and by other bacteria (e.g. *Streptomyces* spp).

prodromal Refers to e.g. symptom(s) which precede the main symptoms of an infectious disease (e.g. Koplik's spots in MEASLES).

prodrug The precursor form of a drug – see e.g. FAMCICLOVIR.

proenzyme *Syn.* ZYMOGEN.

professional phagocytes PHAGOCYTES (including NEUTROPHILS and MACROPHAGES) which can kill and digest microorganisms which they have ingested.

profibrinolysin See FIBRINOLYSIN.

profilin See ACTIN.

proflavine (proflavin) See ACRIDINES.

progametangium In zygomycetes: an enlarged ZYGOPHORE which differentiates into a distal gametangium and a proximal SUSPENSOR. (See also ZYGOSPORE.)

progamone See PHEROMONE.

progenote A (hypothetical) primitive organism; progenotes are presumed to be the phylogenetic progenitors of both prokaryotes and eukaryotes.

progressive multifocal leukoencephalopathy (PML) A human demyelinating disease that occurs primarily in those with an underlying immunological dysfunction (e.g. Hodgkin's disease, chronic lymphatic leukaemia, AIDS); multiple small areas of demyelination develop in the white matter of the brain, resulting in progressive paralysis, mental deterioration etc. – death usually occurring within 6–12 months of the first symptoms. Characteristic of PML are enlarged nuclei in the oligodendrocytes (glial cells associated with myelin formation).

It is widely belived that the causal agent of PML is a POLYOMAVIRUS: the JC virus (JC = the initials of a patient who died from PML); it is also believed that some form of immunosuppression is a prerequisite for the development of the disease. Other factor(s) may be involved; for example, one in vitro study found that the Tat protein, encoded by the AIDS virus (HIV-1), could activate the late promoter of JC virus in glial cells [PNAS (1990) *87* 3479–3483].

[Murine model for PML: Cell (1986) *46* 13–18. Polyomaviruses and disease of the central nervous system: RMM (1998) *9* 79–85.]

progressive pneumonia virus See LENTIVIRINAE.

proguanil See CHLORGUANIDE.

prohead A DNA-free phage head precursor formed during bacteriophage assembly. (See e.g. BACTERIOPHAGE T4; see also SCAFFOLDING PROTEIN.)

prohormone See PHEROMONE.

prokaryote (procaryote) A type of CELLULAR (sense 2 or 3) MICROORGANISM in which the CHROMOSOME(s) are not separated from the cytoplasm by a specialized membrane; the CYTOPLASMIC MEMBRANE is typically devoid of sterols; chloroplasts and mitochondria are absent (cf. INTRACYTOPLASMIC MEMBRANES); RIBOSOMES have a sedimentation coefficient of ca. 70S; a CELL WALL is typically present and may contain structural components such as PEPTIDOGLYCAN or PSEUDOMUREIN, or may consist e.g. of an S LAYER; storage compounds may include e.g. POLY-β-HYDROXYBUTYRATE; flagella (when present) are structurally relatively simple (see FLAGELLUM (a) and (c)).

CHEMOLITHOAUTOTROPHS, and the ability to carry out NITROGEN FIXATION, occur only among the prokaryotes.

The prokaryotes comprise two DOMAINS: the domain ARCHAEA (formerly referred to as the kingdom Archaebacteria) and the domain BACTERIA [JB (1994) *176* 1–6]. Members of the two domains differ in their 16S rRNA and e.g. in the composition, structure and/or mode of assembly of the cytoplasmic membrane, cell wall and flagella; differences also occur e.g. in the composition of the ribosomal proteins.

According to one hypothesis, EUKARYOTES arose from an energy-based symbiotic relationship between cells from the two prokaryotic domains [Nature (1998) *392* 37–41].

prolamellar body In an ETIOPLAST: a paracrystalline array of tubules bearing a small lamellar structure (*prothylakoid*) which contains certain thylakoid components (e.g. the CF_1 unit of PROTON ATPASE); on illumination, prolamellar bodies disperse, and prothylakoids eventually develop into THYLAKOIDS.

prolate Of a phage head: having a shape in which the distance from one pole to the other is greater than the diameter at the mid-point of the head; an example of a prolate head occurs in BACTERIOPHAGE ϕ29.

proliferative kidney disease (PKD) A FISH DISEASE which can be economically important in young salmonid fish. Symptoms include exophthalmia, anaemia, abdominal swelling and kidney hypertrophy; mortality rate: 10–95%. PKD is caused by an unclassified protozoon which apparently has affinities with members of the MYXOZOA [JP (1985) *32* 254–260].

Proliferobasidium See BRACHYBASIDIALES.

L-proline biosynthesis See Appendix IV(a).

proloculus See FORAMINIFERIDA.

promastigote A form assumed by the cells of many species of the TRYPANOSOMATIDAE (q.v.) during at least certain stages of their life cycles.

prometaphase See MITOSIS.

Promicromonospora A genus of asporogenous bacteria (order ACTINOMYCETALES, wall type VI) which occur e.g. in soil. The organisms form a yellow fragmenting mycelium. GC%: ca. 73. Type species: *P. citrea*. [Book ref. 73, pp. 53–54.]

promiscuous plasmids Those CONJUGATIVE PLASMIDS (sense 1) – e.g. IncP1 plasmids – which are capable of self-transmission between bacteria of a wide range of different species and genera.

promitochondrion See MITOCHONDRION.

promoter In a DNA strand: a nucleotide sequence which is recognized (directly or indirectly) and bound by a DNA-dependent RNA POLYMERASE (RPase) during the initiation of TRANSCRIPTION. Transcription is initiated at a position (the *start* *point* or *start site*) which is commonly within the promoter sequence. The first nucleotide to be transcribed is designated +1; nucleotides *downstream* of this position (i.e. those in the direction of RNA elongation: 5′-to-3′ with respect to the RNA) are numbered +2, +3, +4 etc., and nucleotides in the opposite (*upstream*) direction are numbered −1, −2, −3 etc.

Bacteria have several RPase holoenzymes (differing in their SIGMA FACTORS), each one recognizing a distinct class of promoter. In *Escherichia coli* most transcription is carried out by $E\sigma^{70}$; the enzyme binds to a region of DNA extending from ca. 50 bp upstream to ca. 20 bp downstream of the start point. Comparison of the sequences of many promoters recognized by $E\sigma^{70}$ has revealed that certain short sequences are more or less conserved. Thus, a CONSENSUS SEQUENCE, TATAAT (the *Pribnow box* or *−10 sequence*), is centred ca. 10 bp upstream from the start point, and another consensus sequence, TTGACA (the *−35 sequence*) is centred ca. 35 bp upstream of the start point. The start point itself is usually a purine, often an adenine residue in the sequence CAT.

Promoters recognized by $E\sigma^{70}$ in *E. coli* differ to varying extents from the theoretical 'consensus promoter', and may also differ from one another in functional efficiency (efficient promoters being described as 'strong', inefficient ones as 'weak'). Efficiency may be affected e.g. by changes in the consensus sequences or in the transcribed region downstream of the start point. Changes at one site may be compensated for by changes at another: thus, promoters with the same efficiencies may have different base sequencies; furthermore, a (synthetic) promoter containing the theoretical consensus sequences has been found to be relatively inefficient.

Bacterial promoters of other classes (recognized by different RPase holoenzymes) also typically have two short, distinctive conserved sequences upstream from the start point; however, the actual sequences vary from one class of promoter to another. For example, in *Bacillus subtilis* the main holoenzyme, $E\sigma^{43}$, recognizes the same type of promoter as does the $E\sigma^{70}$ of *E. coli*, but $E\sigma^{37}$ (see SIGMA FACTOR) recognizes a consensus sequence AGGATTTNA (N = any nucleotide) in the −35 region and GGAATTNTTT in the −10 region; the corresponding sequences for $E\sigma^{29}$ are TTNAAA and CATATT. (The phage T4 $E\sigma^{gp55}$ recognizes a promoter which apparently lacks a −35 sequence; in promoters recognized by the $E\sigma^{gpntrA}$ of e.g. *Klebsiella* a −35 sequence is apparently absent, but there is a consensus sequence in the −20 region.)

In eukaryotes, promoters are not recognized directly by the RNA polymerases; transcription initiation factors (TIFs) first bind to a promoter to form a *pre-initiation complex*, and only then does an RPase bind to form an *initiation complex*. Promoters for RPase II generally contain a short consensus sequence, TATA(A/T)A(A/T) (the *TATA box*, *Goldberg–Hogness box*, or *Hogness box*), ca. 25–30 bp upstream from the transcription start point; this appears to be the only well-conserved sequence in RPase II promoters [NAR (1986) *14* 10009–10026], although other sequences upstream of the TATA box may be important sites for the action of promoter-specific TIFs [review: Nature (1985) *316* 774–778]. (See also ENHANCER.)

The TATA box, found in many RPase II promoters, also occurs in a minority of type III promoters. The 'classical' RPase III promoters occur *within* the associated gene, i.e. intragenically, *downstream* of the start point; binding of the RPase at the promoter leads to initiation of transcription at a particular distance upstream of the binding site.

One factor, the TATA box-binding protein (TBP), is necessary for the formation of an initiation complex by each of the three types of RPase (I, II and III) – regardless of whether or not their promoters include a TATA box; TBP was first identified as a DNA-binding element in complexes assembled on type II, TATA-containing promoters, but in TATA-lacking promoters TBP appears to bind not to DNA but (directly or indirectly) to a DNA-binding protein (as well as to the polymerase) in the initiation complex. [TBP (minireview): Cell (1993) *72* 7–10.]

promoter control See OPERON.

promycelium See METABASIDIUM.

Pronase (proprietary name) A mixture of endopeptidases and exopeptidases (including carboxypeptidase(s) and aminopeptidases) which has non-specific protease activity. It is obtained from *Streptomyces griseus*.

Prontosil An early sulphonamide, first used clinically in the 1930s; its activity was due to its breakdown in vivo to sulphanilamide. It has been superceded by the more effective, less toxic modern SULPHONAMIDES.

pronucleus (1) The haploid nucleus of a gamete. (2) (*protozool.*) See AUTOGAMY and CONJUGATION.

proof-reading (*mol. biol.*) A system in which the accurary of a process is increased by the removal of 'errors' immediately after they have occurred: see e.g. DNA REPLICATION and PROTEIN SYNTHESIS.

propagative viruses See CIRCULATIVE TRANSMISSION.

propagule Any disseminative unit of an organism, e.g. a spore, a mycelial fragment.

propamidine An aromatic DIAMIDINE used as an antiseptic; it is active mainly against asporogenous Gram-positive bacteria and certain fungi.

propeller loop fermenter A LOOP FERMENTER in which a motor-driven propeller promotes circulation in the column by impelling the culture vertically up (or down) the DRAFT TUBE.

proper margin (excipulum proprium) (*lichenol.*) *Non*-lichenized (i.e., wholly fungal) tissue forming the excipular margin (rim) of a lichen APOTHECIUM. A proper margin is often the same colour as the hymenial disc. (cf. THALLINE MARGIN; see also LECIDEINE APOTHECIUM.)

properdin A γ-globulin (MWt ca. 220000) which occurs in normal serum (ca. 25 μg/ml). Properdin promotes the *alternative pathway* of COMPLEMENT FIXATION by stabilizing the C3 convertase (C3bBb) in the presence of specific activators of this pathway.

prophage See LYSOGENY.

prophage immunity See SUPERINFECTION IMMUNITY.

prophase See MITOSIS and MEIOSIS.

prophylaxis Measure(s) taken to prevent the occurrence of disease – e.g. DISINFECTION, IMMUNIZATION. (See also PROTECTANT.)

propiconazole (1-[2-(2,4-dichlorophenyl)-4-propyl-1,3-dioxolan-2,2-yl-methyl]-1H-1,2,4-triazole; trade names: e.g. Radar, Tilt) An agricultural AZOLE ANTIFUNGAL AGENT which has both contact and systemic action against a wide range of plant pathogenic fungi; it has both eradicant and protectant properties against e.g. cereal diseases such as eyespot, net blotch, powdery mildew, rhynchosporium, septoria, yellow and brown rusts, etc. It may be mixed with e.g. CARBENDAZIM to give a preparation with an even broader antifungal spectrum.

propidium A phenanthridine trypanocidal agent and INTERCALATING AGENT (apparent unwinding angle: 26°).

β-propiolactone (BPL) A water-miscible and non-inflammable cyclic ether (b.p. 155°C) used e.g. (in vapour form) as a surface STERILANT; it has low powers of penetration, needs a high relative humidity (ca. 70%) for maximum activity, and may be poorly effective against e.g. dried spores (cf. ETHYLENE

OXIDE). BPL acts as an ALKYLATING AGENT, substituting proteins etc with propionic acid residues. Above 25°C it is readily hydrolysed to β-hydroxypropionic acid (hydracrylic acid); under acid conditions it forms an open-chain ester-linked polymer. BPL is an irritant and may be carcinogenic.

Propionibacterium A genus of Gram-positive, asporogenous, chemoorganotrophic, anaerobic bacteria which occur e.g. in dairy products (see e.g. CHEESE-MAKING) and on the human skin, and which form propionic acid as a main product in the PROPIONIC ACID FERMENTATION of hexoses or lactate. Cells: non-motile, pleomorphic, branched or unbranched rods or coccoid forms which may exhibit a 'Chinese letter' arrangement in stained preparations; some form pigments. The organisms can be cultured e.g. on yeast extract-lactate-peptone media. GC%: ca. 57–67. Type species: *P. freudenreichii*.

P. acnes (formerly *Corynebacterium parvum*). The organisms ferment glucose (but not maltose or sucrose), do not hydrolyse aesculin, and are indole-positive; they digest casein and liquefy gelatin. [Extracellular proteolysis: JAB (1983) *54* 263–271.] Colonies older than ca. 4 days may become reddish. (See also ACNE and SKIN MICROFLORA.) [Pathogenicity: RMM (1994) *5* 163–173.] [Possible association between *P. acnes* and sciatica: Lancet (2001) *357* 2024–2025.]

Other species: *P. acidi-propionici*; *P. freudenreichii* (which incorporates *P. shermanii*) (see also PYROPHOSPHATE); *P. jensenii* (which incorporates *P. peterssonii* and *P. raffinosaceum*); *P. thoenii*. [Book ref. 46, pp. 1894–1902.]

propionic acid fermentation A FERMENTATION (sense 1), carried out e.g. by *Propionibacterium* spp, *Clostridium propionicum* and *Megasphaera elsdenii*, in which e.g. glucose and/or lactate yield propionic acid and acetic acid as the main end products. *Propionibacterium* spp ferment glucose or lactate via succinate [Appendix III(h)]; when lactate is the substrate the reduction of (endogenous) fumarate generates proton motive force (pmf), permiting ATP synthesis by electron transport phosphorylation. (cf. FUMARATE RESPIRATION.) *C. propionicum* and *M. elsdenii* ferment lactate via a different pathway (the *acrylate pathway*): lactate → lactyl-CoA → acrylyl-CoA → propionyl-CoA → propionic acid.

Propionigenium A genus of anaerobic, asporogenous bacteria. *P. modestum* converts succinate to propionate by reactions which include the decarboxylation of methylmalonyl-CoA; this decarboxylation is coupled to the outward pumping of Na^+ – generating a SODIUM MOTIVE FORCE which can be used for ATP synthesis by a membrane-bound ATPase [EMBO (1984) *3* 1665–1670].

Propionispira A genus of Gram-negative, obligately anaerobic, asporogenous, nitrogen-fixing bacteria which occur in trees affected with WETWOOD. The cells are curved rods or long spiral filaments; peritrichously flagellate. Various compounds are fermented: e.g. galacturonate, lactate, mannitol, and a number of sugars (e.g. glucose, lactose, sucrose); propionate, CO_2 and acetate are major metabolic products. GC%: ca. 37. Type species: *P. arboris*. [JGM (1982) *128* 2771–2779.]

proplastid See e.g. CHLOROPLAST.

propylene glycol See ALCOHOLS.

propylene oxide A water-miscible cyclic ether (b.p. 34°C) used (in vapour form) e.g. as a disinfectant for treating certain foods; on hydrolysis it yields propylene glycol. It is also used in the preparation of specimens for ELECTRON MICROSCOPY.

Prorocentrum A genus of 'primitive' DINOFLAGELLATES in which the cell is flattened, lacks a girdle and sulcus, and is divided longitudinally into two halves, each half containing a relatively thick, convex thecal plate. The two flagella are inserted apically.

Prorodon A genus of freshwater and marine carnivorous ciliates (subclass GYMNOSTOMATIA). Cells: ovoid, up to ca. 200 μm in length, with uniform somatic ciliature, an apical cytostome, and a cytopharynx with associated nematodesmata. Encystment is promoted e.g. by desiccation.

prosenchyma See PLECTENCHYMA.

prosome A stable ribonucleoprotein particle (ca. 19S) present in the nucleus and cytoplasm of various types of eukaryotic cell. Prosomes are believed to be involved in the post-transcriptional regulation of gene expression. [Prosomes – ubiquity and interspecies variation: JMB (1986) *187* 479–493.]

prosoplectenchyma See PLECTENCHYMA.

prostacyclin (PGI_2) A potent vasodilator, derived from PROSTAGLANDIN PGH_2, which degrades spontaneously (in minutes) to 6-oxo-$PGF_{1\alpha}$; PGI_2 promotes vascular permeability and induces pain. 6-Oxo-$PGF_{1\alpha}$ levels are raised in exudates from chronic granulomas.

prostaglandins (PGs) Compounds formed from ARACHIDONIC ACID via the activity of *cyclo-oxygenase*; this enzyme, which exists in two isoforms (COX-1 and COX-2), catalyses a two-step reaction in which the central region of the linear arachidonic acid molecule undergoes oxidative cyclization to form PGH_2, a precursor of $PGF_{2\alpha}$, PGD_2, PGE_2, PGI_2 and thromboxane.

Prostaglandins are included within the category *eicosanoids*.

Cyclo-oxygenases are inhibited e.g. by compounds designated non-steroidal anti-inflammatory drugs (NSAIDs). Many of these drugs can block the activity of *both* COX-1 and COX-2 by binding to the enzyme in the region of an arginine residue at position 120. (The common drug aspirin (acetylsalicylic acid) – made synthetically, but also found naturally in the willow (*Salix* spp) – inhibits cyclo-oxygenase by acetylating a serine residue in the enzyme: Ser-530 in the COX-1 isoform, Ser-516 in COX-2.) Some NSAIDs (e.g. the tri-cyclic *coxibs* celecoxib and rofecoxib) *specifically* inhibit COX-2, such inhibition apparently involving the region near a valine residue at position 523 (isoleucine occupies 523 in COX-1); such specific inhibitors of COX-2 may be useful e.g. against certain types of cancer.

Cyclo-oxygenase is apparently activated in vitro by NITRIC OXIDE [PNAS (1993) *90* 7240–7244].

COX-2 is inducible (e.g. by ENDOTOXIN, immune complexes, certain CYTOKINES – e.g. INTERLEUKIN-1 and TNF-α (although expression is inhibited by IL-4 and IL-10) – and phorbol esters). COX-1 is typically produced constitutively and seems to have a 'housekeeping' function. (It is generally thought that NSAIDs work by inhibiting COX-2 at sites of inflammation, and that their adverse gastrointestinal effects result from inhibition of COX-1.)

Particular PGs tend to be synthesized by particular types of cell. PGD_2 is formed e.g. by mast cells, PGE_2 and $PGF_{2\alpha}$ by macrophages and monocytes, and PGI_2 by endothelial cells.

Following synthesis, prostaglandins may exit the cell and then act on cell-surface receptors in an autocrine or paracrine fashion; for example, PGE_2 may promote an increase in the concentration of cAMP (cyclic AMP) by binding to the EP1 or EP4 receptor. Receptors for prostaglandins also occur in the cell nucleus – the so-called peroxisome proliferator-activated receptor (PPAR).

Prostaglandins have many physiological roles/effects – including vasodilation (e.g. PGE_2), vasoconstriction (e.g. $PGF_{2\alpha}$) and the febrile response. Prostaglandins, which occur e.g. in inflammatory exudates (see also INFLAMMATION), are associated with both inflammatory and anti-inflammatory activities. Prostaglandins have other roles e.g. in kidney function and certain aspects of reproduction.

(See also PROSTACYCLIN.)

prostheca In certain bacteria (PROSTHECATE BACTERIA): a narrow extension of the cell, bounded by the cell wall and cytoplasmic membrane; a cell may have one or more prosthecae – which may be long and narrow (e.g. in *Ancalomicrobium*) or short and conical (in *Prosthecomicrobium*), unbranched or branched. (A long, narrow prostheca is sometimes called a *hypha*.) In some genera prosthecae are involved in reproduction (see e.g. RHODOMICROBIUM); in *Caulobacter* the prostheca functions in ADHESION. (cf. STALK.)

prosthecate bacteria Those bacteria which form one or more prosthecae during at least some stage of their life cycle. Genera in which the prostheca appears to have an essential reproductive role include HYPHOMICROBIUM, RHODOMICROBIUM, and some species of RHODOPSEUDOMONAS. See also ANCALOMICROBIUM, ASTICCACAULIS, CAULOBACTER, HYPHOMONAS, PEDOMICROBIUM, PROSTHECOMICROBIUM and STELLA.

Prosthecochloris See CHLOROBIACEAE and CHLOROBIINEAE.

Prosthecomicrobium A genus of chemoorganotrophic, strictly aerobic PROSTHECATE BACTERIA found in natural waters. The cells are coccoid, ca. 1 μm, each bearing ca. 10–30 conically-shaped prosthecae. *P. pneumaticum* (type species) is non-motile and contains GAS VACUOLES; *P. enhydrum* is monotrichously flagellated and lacks gas vacuoles. *P. hirschii* is motile (monotrichously flagellated), lacks gas vacuoles, and forms short and/or long (*Ancalomicrobium*-like) prosthecae [IJSB (1984) *34* 304–308]. Cell division apparently occurs by budding [ibid.].

prosthetic group In an enzyme: any low-MWt, non-protein component which is firmly bound to the APOENZYME and which is necessary for the activity of the enzyme. (cf. COENZYME sense 1.)

protamines Low-MWt, strongly basic proteins found e.g. in association with DNA in the sperm of fish and birds. (See also PARACOAGULATION.)

Protargin See SILVER.

Protargol A silver proteinate (see SILVER) used e.g. as an antiseptic and as a reagent for staining protozoan flagella.

protease IV See SIGNAL HYPOTHESIS.

protease inhibitors (*med.*) A category of ANTIRETROVIRAL AGENTS that includes the drugs amprenavir, indinavir, nelfinavir, ritonavir and saquinavir. Protease inhibitors bind to, and inhibit, the viral protease. The target enzyme, encoded by the *pol* region of HIV, is necessary for production of mature virions as it functions in post-translational processing of precursor proteins. Protease inhibitors have been associated with side-effects such as lipodystrophy.

proteases (peptidases) ENZYMES which hydrolyse peptide bonds in proteins and peptides. *Endopeptidases* cleave bonds within the peptide chain with varying degrees of specificity for particular amino acyl residues (e.g. the pancreatic enzyme trypsin hydrolyses arginyl and lysyl residues). Endopeptidases include e.g. *serine proteases* (characterized by a catalytically active serine residue in the active centre) and *thiol proteases* (in which free –SH groups occur). Serine proteases include e.g. trypsin, chymotrypsin and SUBTILISINS; they are inactivated by organic phosphate esters (e.g. diisopropylfluorophosphate, DFP) which acylate the active serine residue. Thiol proteases include e.g. BROMELAIN and PAPAIN; they are inhibited by SH-reagents such as heavy metals. *Exopeptidases* remove amino acids sequentially from one end of a peptide chain and include N-terminal exopeptidases (*aminopeptidases*) and C-terminal exopeptidases (*carboxypeptidases*). Proteases are termed acid, alkaline or neutral if they are active at low, high or netural pH, respectively.

Microbial proteases have a number of commercial applications [Book ref. 31, pp. 49–114]. A major commercial use is the addition of microbial proteases to domestic detergents for the digestion of proteinaceous stains in fabrics. Since most commercial detergents contain tripolyphosphate as a sequestering agent, Ca^{2+}-requiring proteases (e.g. THERMOLYSIN) are not suitable for this purpose, and alkaline serine proteases (see SUBTILISINS) from *Bacillus* spp are usually used. Enzymes from *Rhizomucor pusillus*, *R. miehei* and *Endothia parasitica* have some use as rennin (rennet) substitutes in CHEESE-MAKING (see also CASEIN); the proteolytic-to-coagulating activity ratio of these enzymes is higher than that of rennin, and their use requires some modification of the cheese-making process in order to avoid the excessive formation of bitter peptides. Microbial proteases have some application in the manufacture of leather, although chemical methods are still cheaper; alkaline proteases from alkalophilic *Bacillus* spp may be used with lime for de-hairing hides, and proteases from e.g. *Aspergillus oryzae* or *Bacillus* spp may be used for bating hides (a process which makes the leather softer and more elastic). Other uses include e.g. treatment of flour to adjust its gluten composition (and hence baking qualities); as a digestive aid; and in the débridement of wounds (experimental) and dissolution of blood clots (see e.g. BRINASE). (See also PRONASE.)

Proteases produced by certain pathogenic microorganisms may act as virulence factors (see e.g. ELASTASE).

proteasome A hollow, intracellular proteolytic structure formed by the self-assembly (*autocompartmentalization*) of proteases and ATPases; proteolytic sites occur in the *interior* of the proteasome so that protein degradation can be carried out intracellularly without risk to the cell's integrity. A narrow entrance to the proteasome's interior restricts access to unfolded proteins.

In eukaryotes, proteasomes are used for the ATP-dependent degradation of various UBIQUITIN-bound proteins. For example, the ubiquitin–proteasome pathway is reported to be essential during changes in morphology (e.g. trypomastigote → amastigote) in *Trypanosoma cruzi* [Biochem. (2001) *40* 1053–1062].

Proteasomes also occur in prokaryotes – e.g. the archaeans *Methanosarcina thermophila* and *Thermoplasma acidophilum*, and the bacteria *Mycobacterium tuberculosis* and *Rhodococcus erythropolis*.

[Proteasomes in prokaryotes: TIM (1999) *7* 88–92.]

(See also DEGRADOSOME.)

protectant (prophylactic) (*plant pathol.*) Any chemical agent which prevents the occurrence of particular disease(s) among plants treated with that agent. (cf. ERADICANT.)

protective antigens Those antigens of a pathogen which can elicit an immune response that gives protective immunity against the pathogen.

protective immunity *Syn.* IMMUNITY (3).

Proteeae A tribe of bacteria (family ENTEROBACTERIACEAE) divided (e.g. on the basis of DNA relatedness) into three genera: MORGANELLA, PROTEUS and PROVIDENCIA. All members can oxidatively deaminate a range of amino acids. Most strains are motile (but may be non-motile at temperatures >30°C), urease +ve (cf. PROVIDENCIA), lactose –ve (although some strains may contain a Lac plasmid), indole +ve (cf. PROTEUS), VP –ve, MR +ve. Acid (± gas) is produced from glucose, and a reddish-brown pigment is produced on nutrient agar containing 5% tryptophan. Species are generally resistant to bacitracin, polymyxins and erythromycin. [Book ref. 46, pp. 1204–1224.]

protein 2 A PORIN of *Escherichia coli*.

protein A (SpA) A cell wall protein (MWt ca. 42000) found in most strains of *Staphylococcus aureus*; it is covalently bound to the wall peptidoglycan and can be released (intact) e.g. by LYSOSTAPHIN. SpA binds to the FC PORTION of IgG_1, IgG_2 and IgG_4; one SpA molecule can bind two IgG molecules. (cf. PROTEIN G.) SpA can also bind, apparently via the Fab region, to IgE, IgA, IgM, and some IgG_3. SpA can elicit Arthus- and anaphylaxis-type reactions, histamine release from granulocytes, and complement fixation by the classical pathway.

Applications. (a) SpA immobilized on e.g. sepharose can be used in affinity CHROMATOGRAPHY for IgG purification; IgGs are eluted at pH ca. 4. (IgM and IgA elute at a lower pH.) (b) SpA-containing staphylococci can bind to specific (IgG) antibodies (via their Fc portions) and will then agglutinate on exposure to the homologous antigen (CO-AGGLUTINATION); such antibody-coated staphylococci can be used e.g. in slide co-agglutination tests for detecting and/or identifying bacteria, toxins, etc. (c) SpA may be bound to red blood cells (e.g. with glutaraldehyde) and used e.g. to detect IgG on the surface of lymphocytes (by rosette formation). (d) SpA conjugated with an enzyme (e.g. β-lactamase, horseradish peroxidase) can be used e.g. for the assay of IgG in serum. Antigen is immobilized, incubated with the serum, and any homologous antibody binding to the antigen is detected with the SpA–enzyme conjugate; the amount of SpA bound, and hence the amount of homologous IgG in the serum, is determined by an assay of the enzymic activity (e.g. PADAC hydrolysis for β-lactamase conjugates) [PTRSLB (1983) *300* 399–410].

[Properties and applications of protein A: Book ref. 44, pp. 429–480.] (See also SORTASE.)

protein G A cell wall protein from group G streptococci; it resembles staphylococcal PROTEIN A in binding specifically to the Fc region of an IgG molecule, but it binds a broader range of IgG subclasses. [Structure of the IgG-binding regions of protein G: EMBO (1986) *5* 1567–1575.]

protein i See PRIMOSOME.

protein kinase An enzyme which phosphorylates particular amino acid residues in a protein. The functions of certain proteins are regulated by phosphorylation/dephosphorylation both in eukaryotic cells (see e.g. ONCOGENE) and in *Escherichia coli* (see e.g. TCA CYCLE (isocitrate dehydrogenase)). (See also CYCLIC AMP.)

protein kinase A See CYCLIC AMP.

protein n (also n′, n″) See PRIMOSOME.

protein phosphatase An enzyme which removes covalently bound phosphate from a protein; protein phosphatases have important roles e.g. in the regulation of the eukaryotic CELL CYCLE – in which they complement the activity of PROTEIN KINASES. Some protein phosphatases can cleave phosphate from more than one type of amino acid residue. (See also CYCLIN.)

protein priming See DNA REPLICATION (linear genomes).

protein secretion (and export) (in Gram-negative bacteria) At least five major modes of protein secretion occur in Gram-negative bacteria; they have been designated types I–V (but see note on types IV and V at the end of this entry).

Type I systems. Secretion by type I systems is energized by ATP hydrolysis and involves one-step translocation from cytoplasm to exterior without a free (periplasmic) intermediate; it is exemplified by secretion of α-haemolysin by the ABC EXPORTER of *Escherichia coli.*

Type II systems. Secretion by type II systems (the *general secretory pathway*, GSP) requires energy from both ATP and pmf – and, in some cases, from GTP. The first stage, translocation across the cytoplasmic membrane (CM), depends on the products of genes analogous to the *sec* genes of *E. coli* (hence, the alternative name for type II systems: the '*sec*-dependent pathway'); the products of certain *sec* genes form a secretory channel through the CM.

Proteins secreted by the GSP are synthesized with an N-terminal *signal sequence* (see SIGNAL HYPOTHESIS).

In one form of type II secretion, the initial phase (transport across the CM: *general export pathway*, GEP) begins in the cytoplasm when the secretory protein (i.e. the one being transported) binds to a *chaperone* protein, SecB; binding may occur during or after translation, but before folding (folding of *sec*-exported proteins occurs in the periplasm). SecB is targeted to a CM-linked ATPase, SecA. SecA is associated with a SecYEG protein complex (the *translocon*) [JB (1997) *179* 5699–5704; Nature (2002) *418* 662–665] which forms a secretory channel through the CM. SecB is released (to chaperone another protein), and translocation of the protein through the CM is associated with ATP hydrolysis at SecA. During, or directly after, translocation, the signal sequence is cleaved by a specific enzyme (a *signal peptidase*). *E. coli* has two distinct signal peptidases; *leader peptidase* (= *leader peptidase I*) is functional on a wide range of proteins, while *lipoprotein signal peptidase* (= *signal peptidase II*) acts only on certain modified proteins (e.g. BRAUN LIPOPROTEIN). (Signal peptidases may constitute a class of serine proteases mechanistically related to β-lactamases [TIBS (1992) *17* 474–478].) Excised signal peptides are degraded by a *signal peptide peptidase*; in *E. coli* at least one enzyme ('protease IV') carries out this function.

In a variant type II pathway, transport to the CM necessarily begins at an early stage in translation. One scheme for this pathway (in *E. coli*) is as follows. Initially, the nascent polypeptide (with its ribosome) binds, via the N-terminal sequence, to a *signal recognition particle* (SRP) that consists of (i) a 4.5S RNA molecule, and (ii) a 48 kDa multifunctional protein (with GTPase activity) designated P48 (or Ffh). It is not known whether protein–SRP binding inhibits further translation on the ribosome (as occurs in eukaryotic systems); it has been suggested that, in prokaryotes, translation arrest may not be necessary, and that the SRP pathway may simply accelerate association of nascent polypeptide chains with the translocon. A free cytosolic FtsY protein binds to the complex, and the FtsY–SRP–ribosome–polypeptide entity binds transiently, via FtsY, to the CM; hydrolysis of GTP energizes dissociation of the components, and the nascent polypeptide (with its ribosome) binds at a SecYEG (or similar) translocon. [Model for the SRP pathway in *E. coli*: EMBO (1998) *17* 2504–2512.] (Interestingly, in *Escherichia coli*, reduced levels of SRP lead to the induction of a heat-shock response (see HEAT-SHOCK PROTEINS); this may help to maintain viability by increasing the cell's capacity to degrade mislocalized CM proteins [JB (2001) *183* 2187–2197].)

For any given nascent polypeptide, the choice of pathway – i.e. SecB–SecA or SRP – is believed to depend on the hydrophobicity of the N-terminal sequence. Among *secreted* proteins, the SRP pathway may be used preferentially by those (e.g. β-lactamases) which have relatively hydrophobic N-terminal sequences. However, the SRP pathway may be used mainly for targeting and elaboration of CM proteins – whose N-terminal sequences are characteristically very hydrophobic; for these proteins, this pathway may be advantageous in that completion of translation at a membrane-associated site could avoid the risk of aggregation of hydrophobic domains which may occur in the cytoplasm. In vitro studies involving replacement or modification of the signal peptide sequence indicate that the targeting pathway of *Escherichia coli* pre-secretory and integral membrane proteins is specified by the hydrophobicity of the targeting signal [PNAS (2001) *98* 3471–3476].

The second stage of the GSP, translocation across the OUTER MEMBRANE, depends on a number of specific secretion factors. Although the GEP occurs in *E. coli*, this organism apparently does not *secrete* proteins via the GSP, i.e. it does not carry out the second stage of the GSP. However, other species of the Enterobacteriaceae do secrete proteins via this pathway. For example, *Klebsiella oxytoca* secretes a lipoprotein enzyme, pullulanase (see DEBRANCHING ENZYME), whose translocation across the outer membrane requires the products of 14 *pul* genes. (Interestingly, if all the *pul* genes from *K. oxytoca* are expressed in *E. coli*, this organism can secrete pullulanase.) Again, the GSP is used by *Erwinia chrysanthemi* for secreting the enzymes pectinase and cellulase, secretion needing expression of the *out* genes. (*E. coli* encoding a gene for pectinase can secrete this enzyme if supplied with *out* genes.) The secretory systems in *K. oxytoca* and *E. chrysanthemi* are similar though not identical: *K. oxytoca* can synthesize – but not secrete – a pectinase encoded by a gene from *E. chrysanthemi*. (See also SECRETON.)

Pseudomonas aeruginosa secretes an exotoxin, and the enzymes elastase and alkaline phosphatase, via a secretory system involving the *xcp* genes. Interestingly, XcpA is identical to the PilD protein (involved in the formation of fimbriae) and is also homologous to PulO, a protein involved in the secretion of pullulanase, and it has been suggested that, in at least some cases, the second phase of the GSP may involve a protein complex similar to the fimbrial assembly apparatus [TIG (1992) *8* 317–322; see also EMBO (2000) *19* 2221–2228].

Folded proteins may cross the CM via the pmf-dependent Tat system [see Microbiology (2003) *149* 547–556].

Type III systems. Secretion by type III systems occurs as a one-step process via a channel/pore which crosses the cell envelope from cytoplasm to cell surface; unlike type I systems, the type III pathway may involve 20 or more distinct proteins. (See also NEEDLE COMPLEX.) Type III systems occur in certain pathogens (including pathogens of plants [JB (1997) *179* 5655–5662]). (See also VIRULON.) [Type III secretory systems and pathogenicity islands (review): JMM (2001) *50* 116–126.]

Components of type III systems in various species exhibit homology. Moreover, some type III components exhibit homology with components of other secretory systems – e.g. components of the basal body of the (bacterial) FLAGELLUM.

Proteins secreted via type III systems include an N-terminal sequence which may be targeted to the pore or channel – but which it is not cleaved.

The energy requirements of these systems are unknown, but a pore-associated protein (YscN) in a type III system of *Yersinia* includes binding sites typical of ATPases.

In some type III systems, protein secretion seems to be activated by contact with a eukaryotic cell; at least some proteins are believed to be secreted *directly* into the eukaryotic cytoplasm because e.g. (i) proteins of the kind secreted by type III systems typically have little or no effect on eukaryotic cells in the absence of the secreting bacterium; (ii) some secreted proteins have been detected within eukaryotic cells; and (iii) some proteins (Yops) secreted by *Yersinia* have been associated with specific effects within eukaryotic cells (see VIRULON). In EHEC and EPEC strains of *Escherichia coli*, type III systems are reported to be regulated by QUORUM SENSING (q.v.).

Proteins secreted by type III systems include e.g. the IpaB protein of *Shigella* (which promotes APOPTOSIS in macrophages); the EspB and EspE proteins of enteropathogenic *E. coli* (EPEC); and the various Yops of *Yersinia* [the *Yersinia* Yop virulon: Mol. Microbiol. (1997) *23* 861–867].

[Type III secretion systems (review): TIM (1997) *5* 148–156.]

Type IV systems are mechanistically distinct and not fully understood; the following is an outline of a current model. In these systems the secreted protein is synthesized as part of a larger protein that consists of (i) an N-terminal signal peptide, (ii) an α domain, and (iii) a C-terminal β domain. Initially, the whole protein is translocated across the CM in a *sec*-dependent manner, with cleavage of the N-terminal signal peptide. To pass through the outer membrane, the β domain is believed to form a 'β-barrel' pore in the membrane through which the α ('passenger') domain is translocated.

At the bacterial surface, the α domain may (i) remain attached (exposed to the environment); (ii) undergo cleavage, but remain (non-covalently) attached; (iii) undergo cleavage and release to the environment. Cleavage may be autocatalytic, or it may involve another outer membrane protein such as the SopA protease of *Shigella* or the OmpT protease of *Escherichia coli*.

In a type IV system, the translated protein (i.e. N-terminal + α domain + β domain) has been termed an *autotransporter*. (This term has also been used by some authors to refer specifically to the β domain; other authors have used the term to refer specifically to the α domain.)

Proteins transported by type IV systems include the (cell-surface-retained) IcsA protein, associated with *Shigella* invasiveness in dysentery; the IGA1 PROTEASES of *Haemophilus influenzae*, *Neisseria gonorrhoeae* and *N. meningitidis*; and the AIDA-I system of *Escherichia coli*. The latter system has been used for AUTODISPLAY.

Type V systems. These include the mechanism by which T-DNA is transferred from *Agrobacterium tumefaciens* to the host cell in CROWN GALL, and the secretion of pertussis toxin by *Bordetella pertussis*.

Note on the designation of type IV and type V systems. The name 'type IV' secretion system has been widely used for autotransporters by a number of authors [see e.g. TIM (1998) *6* 370–378; Book ref. 218 (1999), p 42; Book ref. 223 (1999), p 94]. Some authors [see Inf. Immun. (2001) *69* 1231–1243] have suggested that (i) autotransporters be called type V systems, and that (ii) secretory mechanisms referred to above as type V systems be called type IV systems. In this dictionary the designation 'type IV' has been retained for autotransporters.

protein synthesis A protein consists essentially of one or more *polypeptides* – a polypeptide being a chain of amino acids linked by peptide bonds (–CO.NH–). Polypeptides are folded into a three-dimensional structure that is stabilized mainly by hydrogen and/or disulphide bonds formed between amino acids in different parts of the chain; the specific three-dimensional structure of a given protein is associated with its biological role.

The biosynthesis of a protein is a complex process in which the composition of the protein (i.e. the number, nature and sequence of its amino acids) is decoded from information contained in specific sequence(s) of nucleotides. In the majority of organisms (including all *cells*), proteins are encoded by DNA; in certain viruses, proteins are encoded by RNA. The information in nucleic acids is expressed via a GENETIC CODE (q.v.).

Unlike the polymerization of nucleotides (see e.g. DNA SYNTHESIS), polypeptide chains are not synthesized by the polymerization of amino acids on a template. In fact, protein synthesis occurs in several stages.

In all cases (in both prokaryotic and eukaryotic cells) the first stage is the synthesis of an RNA copy of the particular DNA sequence that encodes the given protein (see TRANSCRIPTION); as this RNA transcript carries the genetic 'message' from DNA it is called *messenger RNA* (or, more usually, mRNA) – see MRNA.

In some cases (commonly in eukaryotes, less commonly in prokaryotes) the initial transcript must be processed to remove non-coding part(s) (called *introns*) from the nucleotide sequence – see SPLIT GENE; the resulting (mature) mRNA is then used to guide assembly of the polypeptide chain – the stage of protein synthesis called *translation*. mRNAs which lack introns can be used directly for translation. (See also RNA EDITING.)

Along the length of the mRNA molecule, groups of three consecutive bases each encode a particular amino acid. Each of these three-base groups is called a *codon*; thus, e.g. the codon UCA (uracil-cytosine-adenine) encodes the amino acid serine (see table in GENETIC CODE). Hence, the sequence of codons in mRNA encodes the sequence of amino acids in a polypeptide.

For the synthesis of a polypeptide, each amino acid must first bind to a small 'adaptor' molecule of RNA which is specific for that amino acid and which also has a binding site (a three-base *anticodon*) for the appropriate codon. These RNA adaptor molecules are called *transfer RNA* (tRNA) – see TRNA. The binding of a given amino acid to its tRNA (i.e. the 'charging' of a tRNA molecule) occurs in a two-step reaction. First, the amino acid reacts with ATP in the presence of a specific aminoacyl-tRNA synthetase (forming an enzyme-bound aminoacyl-AMP complex); the aminoacyl group is then transferred to the 2′ or 3′ position of the terminal adenosine residue in the tRNA (with release of AMP and synthetase). (See also MUPIROCIN.)

The synthesis of a polypeptide on mRNA (i.e. the translation process) takes place on a RIBOSOME. In prokaryotes, correct positioning of mRNA on the ribosome is facilitated by a particular sequence of nucleotides in the mRNA (the *Shine–Dalgarno sequence*) located 'upstream' of the coding region; this sequence base-pairs with part of a 16S rRNA molecule in the 30S subunit of the ribosome.

Essentially, following mRNA–ribosome binding, the ribosome moves along the mRNA molecule in such a way that each codon, in turn, binds the relevant tRNA, and each successive amino acid is linked, covalently, to the previous one.

The following outlines a generally accepted scenario for protein synthesis in *Escherichia coli*; some details of the process are still unknown. The process appears to be essentially similar in most or all other bacteria.

Translation is initiated when mRNA binds to the ribosome. Each ribosome has two (adjacent) binding sites for charged tRNA molecules: the *A site* (= acceptor or entry site) and the *P site* (= peptidyl or donor site). At the start of translation, the first codon of mRNA (the *initiator codon*) coincides with the P site. The initiator codon is usually AUG, but is sometimes e.g. GUG, UUG or AUU.

The initiator codon is recognized by a distinct *initiator tRNA* ($tRNA_i$) which is charged with methionine (Met). In *E. coli* there are two tRNAs specific for methionine: $tRNA_m$ and $tRNA_f$; only $tRNA_f$ can function as $tRNA_i$ ($tRNA_m$ being used for incorporation of Met in subsequent location(s) in the polypeptide chain). In the charged $tRNA_f$ the α-amino group of the methionyl residue is formylated (via N^{10}-formyltetrahydrofolate), giving fMet-$tRNA_f$; hence, the first (N-terminal) amino acid to be incorporated in the polypeptide chain is N-formylmethionine. (The incorporation of N-formylmethionine as N-terminal amino acid is typical of prokaryotes – and of mitochondria and chloroplasts; in the eukaryotic cytoplasm, and in at least some archaeans, the methionine residue is not formylated, so that the N-terminal amino acid in these cases is methionine.)

Before discussing subsequent stages in protein synthesis, further consideration can be given to the initial binding of mRNA to the ribosome. In *E. coli* this preliminary step requires three protein *initiation factors*: IF-1, IF-2 and IF-3. The timing and mode of action of these factors is not yet fully understood, and there are several models for this phase of protein synthesis. According to one model: (i) mRNA binds to the (separate) 30S subunit of the ribosome; (ii) fMet-$tRNA_f$ binds to AUG at the P site on the 30S subunit before addition of the 50S subunit; (iii) IF-2 and GTP are needed for the binding of fMet-$tRNA_f$ to AUG; (iv) all three initiation factors will have been released by the time the 50S subunit has been added. During assembly of the complete 70S ribosome, the association of 30S and 50S subunits is linked to the hydrolysis of GTP; IF-2 acts as a GTPase. [Bacterial initiation factors in protein synthesis (review): Mol. Microbiol. (1998) *29* 409–417.]

Initiation can apparently also occur with a 70S monosome. This may happen e.g. when initiation occurs at an internal initiator codon in a polycistronic mRNA, the ribosome terminating synthesis of one polypeptide and initiating synthesis of the next without dissociating from the mRNA.

Once $tRNA_i$ is bound to AUG at the P site on a 70S ribosome, the *elongation* process proceeds by sequential addition of amino acids according to the codons that follow AUG. Elongation beings when an aminoacyl-tRNA binds to the vacant A site – which is adjacent to the P site and coincident with the *second* codon; this *decoding* step requires GTP and the *elongation factor T* (EF-T; 'transfer factor'). EF-T comprises two components: EF-Tu (= EF-1A; see also TUFA, TUFB GENES) and EF-Ts (= EF-1B). A complex consisting of these two components interacts with GTP, EF-Ts being displaced and the EF-Tu–GTP complex being involved in binding of the incoming tRNA to the A site; during this process GTP is hydrolysed, EF-Tu–GDP and phosphate being released. EF-Ts displaces GDP, re-forming the original complex – which can then react with GTP to repeat the cycle.

The binding of an incoming tRNA to the A site is followed by *transpeptidation*, i.e. the formation of a peptide bond between the carboxyl group of fMeT-$tRNA_f$ and the α-amino group of the second amino acid (at the A site); transpeptidation is catalysed by the enzyme *peptidyltransferase*. (Peptidyltransferase is part of the 50S ribosomal subunit.) At this stage a dipeptidyl-tRNA occupies the A site.

In the next stage, *translocation*, the ribosome moves along the mRNA by a distance equal to one codon (in the 5′-to-3′ direction), the tRNA at the P site being displaced in the process; the result of this movement is that the dipeptidyl-tRNA entity now occupies the P site – i.e. the A site is left vacant (and therefore ready to receive the third aminoacyl-tRNA). Translocation, which involves elongation factor EF-G, is accompanied by GTP hydrolysis (catalysed by EF-G); GTP hydrolysis may be necessary to effect detachment of EF-G from the ribosome [JMB (1986) *189* 653–662]. (Release of EF-G from the ribosome is essential for continuation of elongation as EF-G and EF-Tu apparently bind to similar or identical sites.)

The third aminoacyl-tRNA now enters the A site (coincident with the third codon of mRNA); transpeptidation follows, the carboxyl of the dipeptide being covalently joined to the α-amino group of the third amino acid.

The cycle of events outlined above – decoding, transpeptidation, translocation – is repeated for each incoming aminoacyl-tRNA during elongation.

The scheme described above is the *two-site model* for translocation. In the *three-site model* [MGG (1986) *204* 221–228, q.v.

for references], described for *E. coli* and e.g. '*Halobacterium halobium*', tRNAs remain bound to the ribosome both before and after translocation, i.e. deacylated tRNA is not released from the P site during translocation; instead, it is transferred to a third ribosomal site (the *E site*) concomitantly with translocation. Release of deacylated tRNA from the E site is triggered by the entry into the A site of the next (incoming) aminoacyl-tRNA.

Base-pairing between mRNA codons and tRNA anticodons is not sufficiently stable, in itself, to guarantee the degree of accuracy and efficiency needed in translation. It has been suggested that *proof-reading* may occur e.g. at the level of amino acid selection by aminoacyl-tRNA synthetases; when synthetase specificity is not high enough to prevent binding of the 'wrong' amino acid, the latter may be recognized and removed either at the aminoacyl-AMP stage (*pre-transfer proof-reading*) or at the aminoacyl-tRNA stage (*post-transfer proof-reading*) [see e.g. NAR (1986) *14* 7529–7539]. Proof-reading may also occur during elongation; aminoglycosides such as STREPTOMYCIN may exert their effect on translational accuracy by interfering with this proof-reading system. The ribosome itself may provide help, and this appears to be given by the rRNA [see e.g. Nature (1994) *370* 597–598].

Termination of translation (i.e. cessation of polypeptide synthesis) occurs when a specific termination codon (UAA, UAG or UGA in *E. coli* – see GENETIC CODE) appears at the A site. These codons are recognized by protein *release factors* (RFs); RF-1 recognizes UAA and UAG, while RF-2 recognizes UAA and UGA. The recognition of a termination codon by an RF is followed by hydrolysis of the ester bond between the tRNA (at the P site) and the polypeptide chain it carries – thus releasing the completed polypeptide; hydrolysis is catalysed by peptidyltransferase. A third factor, RF-3 (= 'S protein'), may stimulate the activities of RF-1 and RF-2.

Although the 'stop' signal is traditionally regarded as a *triplet* of bases (e.g. UAA), there is evidence that the base *following* a stop codon influences the efficiency of termination, and it may be that the actual stop signal is a four-base (or even longer) sequence [Mol. Microbiol. (1996) *21* 213–219]. In *E. coli* the efficiency of termination at a stop codon is also influenced by the 5′ (upstream) nucleotide context of the codon, the mechanism involving the C-terminal amino acid residues in the nascent polypeptide. [Translation termination and stop codon recognition (review): Microbiology (2001) *147* 255–269.]

As one ribosome translocates along the mRNA another can fill the vacated initiation site and start translation of another molecule of the polypeptide; thus, a given mRNA molecule may carry a number of ribosomes along its length – forming a *polyribosome* or *polysome*. (A single ribosome is sometimes called a *monosome*.)

At some time after the start of elongation, the formyl group of formylmethionine (see earlier) – or formylmethionine itself – is excised from the polypeptide. In *E. coli*, formylmethionine is cleaved from about 50% of proteins; whether or not cleavage occurs in a given protein appears to depend on the identity of the *penultimate* N-terminal amino acid residue – the longer the side-chain on this residue the smaller the probability of cleavage of formylmethionine. When such cleavage does occur it is effected by METHIONINE AMINOPEPTIDASE (q.v.). When the formyl group (only) is cleaved, the reaction is catalysed by methionine deformylase (= peptide deformylase, PDF). (PDF is inhibited by certain derivatives of hydroxamic acid, but these compounds are unlikely to be useful as broad-spectrum antibacterial agents because (i) they were bacteriostatic in the organisms tested, and (ii) resistance develops readily [AAC (2001) *45* 1058–1064].)

Signal sequences. Commonly, a protein which will form part of the cytoplasmic membrane, or pass through it, is synthesized with a special N-terminal sequence of amino acids (called a *signal sequence*) which facilitates passage into, or through, the membrane (see SIGNAL HYPOTHESIS).

Protein folding. As mentioned earlier, polypeptide chains must be folded correctly in order to form functional, biologically active proteins. Folding of periplasmic, membrane and secreted proteins, frequently involves the formation of disulphide bonds between specific amino acid residues in the polypeptide chain. (In *E. coli* the *cytoplasmic* proteins generally do not contain disulphide bonds as structural features; in the cytoplasm, disulphide bonds are usually reduced by enzymes such as the NADPH-dependent thioredoxin system [but see EMBO (1998) *17* 5543–5550].)

In *E. coli*, disulphide bonds are catalysed in the periplasmic region by the Dsb proteins (encoded by genes *dsbA* and *dsbB*). Mutations which affect *dsb* genes are *pleiotropic* – i.e. they have multiple effects; the reason for this is that disulphide bonds stabilize a range of different proteins (with widely differing functions) – for example, flagellar proteins, membrane proteins and various secreted proteins. In pathogens, mutations in the *dsb* genes may affect secreted protein virulence factors (and thus possibly reduce the virulence of the organism).

Proteins containing *proline* residues may need peptidyl–prolyl isomerases for normal folding; several such isomerases (e.g. FkpA [function: Mol. Microbiol. (2001) *39* 199–210], PpiA) are found in the periplasmic region in *E. coli*.

At least some protein-folding enzymes appear to be synthesized under the regulation of a TWO-COMPONENT REGULATORY SYSTEM [see e.g. GD (1997) *11* 1169–1182]. Transcription of the genes encoding certain protein-folding enzymes appears to involve sigma factor σ^E [GD (1997) *11* 1183–1193].

Protein folding is facilitated by so-called *molecular chaperones* (= 'chaperones'): molecules which bind to, and stabilize, nascent proteins – promoting correct folding/inhibiting incorrect folding; one or more chaperones may be involved in a given stage of folding or translocation. Chaperones may bind to proteins co-translationally [JBC (1997) *272* 32715–32718] or post-translationally. (See also HEAT-SHOCK PROTEINS.) Although chaperones are typically *proteins*, the membrane lipid phosphatidylethanolamine acts as a chaperone for the *E. coli* protein LacY [EMBO (1998) *17* 5255–5264].

[Disulphide bond formation (review): Mol. Microbiol. (1994) *14* 199–205. Protein folding and misfolding (some concepts and themes): EMBO (1998) *17* 5251–5254.]

Some proteins are folded in the cytoplasm and may be exported via the Tat system [Microbiology (2003) *149* 547–556].

[Protein folding (review): TIBS (2000) *25* 611–618.]

Peptide synthesis without ribosomes. Some very short polypeptides are synthesized by a multi-enzyme complex (rather than by ribosomes and RNA molecules). Such a process is used e.g. for the synthesis of certain antibiotics (including GRAMICIDINS and TYROCIDINS). [Non-ribosomal biosynthesis of peptide antibiotics (review): Eur. J. Bioch. (1990) *192* 1–15.]

Protein synthesis in eukaryotes. Some differences between prokaryotic and eukaryotic protein synthesis are listed below.

Aminoacyl-tRNA synthetases occur in high-MWt multienzyme complexes [review: Bioch. J. (1986) *239* 249–255].

There is apparently no equivalent to the Shine–Dalgarno sequence of prokaryotes.

In the cytoplasmic proteins of eukaryotes (and at least some archaeans) the N-terminal methionine is not formylated.

Initiation of synthesis of these proteins usually occurs at the AUG codon nearest the 5′ end of the mRNA, and generally requires a capped mRNA (see MRNA) for maximum efficiency. At least nine initiation factors appear to be required.

The eukaryotic eEF-1, equivalent to EF-G in *E. coli*, differs from the bacterial factor in being susceptible to ADP-ribosylation by DIPHTHERIA TOXIN. (In at least some members of the Archaea the corresponding factor is susceptible to diphtheria toxin [Book ref. 157, pp 379–410].)

The sole eukaryotic release factor, designated eRF, requires GTP for activity and can recognize any termination codon.

Antibiotics affecting protein synthesis: AMINOGLYCOSIDE ANTIBIOTICS, ANISOMYCIN, AURINTRICARBOXYLIC ACID, CHLORAMPHENICOL, CYCLOHEXIMIDE, FUSIDIC ACID, LINCOSAMIDES, MACROLIDE ANTIBIOTICS, OXAZOLIDINONES, PACTAMYCIN, POLYENE ANTIBIOTICS (b), PUROMYCIN, SPARSOMYCIN, STREPTOGRAMINS (including e.g. QUINUPRISTIN/DALFOPRISTIN), TETRACYCLINES, TIAMULIN, THIOSTREPTON, VIOMYCIN.

proteinase (1) *Syn.* protease (see PROTEASES). (2) *Syn.* endopeptidase (see PROTEASES).

proteinase K A non-specific PROTEASE obtained from *Tritirachium album*. (cf. PRONASE.)

Proteobacteria A taxonomic category (= PURPLE BACTERIA, sense 2) within the domain BACTERIA defined on the basis of 16S rRNA sequences. Some constituent genera are listed below.

Alpha subdivision. *Anaplasma, Agrobacterium, Azospirillum, Bartonella, Beijerinckia, Brucella, Ehrlichia, Hyphomicrobium, Methylobacterium, Rickettsia, Rochalimaea, Rhodopseudomonas, Rhodospirillum, Wolbachia.*

Beta subdivision. *Alcaligenes, Bordetella, Chromobacterium, Neisseria, Nitrosomonas, Rhodocyclus, Spirillum.*

Gamma subdivision. *Acinetobacter, Coxiella, Erwinia, Escherichia, Haemophilus, Legionella, Methylobacter, Methylocaldum, Methylococcus, Methylomicrobium, Methylomonas, Nitrosococcus, Pseudomonas, Proteus, Vibrio.*

Delta and epsilon subdivisions. *Bdellovibrio, Campylobacter, Desulfuromonas, Helicobacter, Myxococcus, Wolinella.*

proteomics (1) The study of a cell's *proteome* (collectively, all the proteins that can be expressed) as an integrated whole. (2) As in (1) with genetic and other aspects. [Microbiology and Molecular biology Reviews (2002) *66* 39–63.]

proteose A soluble product of protein hydrolysis; proteoses are not coagulated by heat but are precipitated in saturated $(NH_4)_2SO_4$.

proter (*ciliate protozool.*) The anterior of the two cells formed during HOMOTHETOGENIC binary fission. (cf. OPISTHE.)

Proteromonadida An order of parasitic protozoa (class ZOOMASTIGOPHOREA). The cells have one or two pairs of flagella, without paraxial rods, and a single mitochondrion which lacks a kinetoplast; cysts are formed. Genera: e.g. *Karatomorpha, Proteromonas.*

Proteromonas See PROTEROMONADIDA.

Proteus A genus of Gram-negative bacteria of the tribe PROTEEAE. Cells: ca. 0.4–0.8 × 1–3 μm. Most strains swarm at 37°C, typically forming characteristic concentric zones of growth on the moist surface of an agar or gelatin medium (see SWARMING and DIENES PHENOMENON). (Some strains may form a spreading, *uniform* film of growth.) Typical reactions: H_2S (in TSI) +ve (delayed in *P. myxofaciens*); gelatin is liquefied at 22°C; lipase +ve; urease +ve; mannose and sugar alcohols are not attacked. Nicotinic acid (but not pantothenic acid) is required for growth. GC%: 38–41. Type species: *P. vulgaris*.

P. inconstans. See PROVIDENCIA.

P. mirabilis. Indole −ve; maltose −ve, ODC +ve.

P. morganii. See MORGANELLA.

P. myxofaciens. Indole −ve; maltose +ve; ornithine decarboxylase −ve; abundant slime produced e.g. in trypticase-soy broth.

P. rettgeri. See PROVIDENCIA.

P. shigelloides. See PLESIOMONAS.

P. vulgaris. Indole +ve; maltose +ve; ornithine decarboxylase −ve.

P. mirabilis and *P. vulgaris* occur e.g. in soil, polluted water, intestines of (healthy) man and animals, etc. *P. mirabilis*, in particular, is an important opportunist pathogen, causing nosocomial URINARY TRACT INFECTION, pneumonia, septicaemia, etc. *P. myxofaciens* was isolated from gypsy moth larvae (*Porthetria dispar*).

[Book ref. 22, pp. 491–494.]

prothallus (*lichenol.*) (1) *Syn.* HYPOTHALLUS (sense 1). (2) A lichen thallus in the initial stages of development.

prothylakoid See PROLAMELLAR BODY.

proticins BACTERIOCINS from *Proteus* spp.

protist A member of the PROTISTA.

Protista (1) A taxon (kingdom) which includes the ALGAE, FUNGI and PROTOZOA (collectively, the eukaryotic protists), and the PROKARYOTES. (2) A KINGDOM comprising the *eukaryotic* protists – see (1) above. (cf. MONERA.)

protoaecidium See UREDINIOMYCETES stage I.

protoaecium See UREDINIOMYCETES stage I.

Protococcidiorida An order of protozoa of the subclass COCCIDIASINA.

Protococcus See PLEUROCOCCUS.

protocooperation In a mixed population of microorganisms: an interaction between two (or more) different microorganisms in which each organism benefits from the activities of the other (e.g. each organism may produce a substance which stimulates the growth of the other), but in which the interaction is not obligatory for either organism. (See e.g. YOGHURT; cf. MUTUALISM.)

protofilament See MICROTUBULES.

Protogonyaulax tamarensis *Syn. Gonyaulax tamarensis*.

protohaem See HAEM.

protokaryote Archaic term for PROKARYOTE.

protokerogen See KEROGEN.

protomerite In a cephaline gregarine: the anterior of the two main regions of the (septate) cell (cf. DEUTOMERITE). In the late trophozoite stage the anterior part of the protomerite may bear an EPIMERITE.

protometer In a cell: a hypothetical sensor which can respond to changes in proton motive force.

protomite See TOMITE.

Protomonas A genus of methylotrophic bacteria; the organisms have been re-classified in the genus *Methylobacterium* [IJSB (1985) *35* 209].

proton ATPase (proton-translocating ATPase; H^+-ATPase) Any ATPASE which can couple ATP hydrolysis to the energy-dependent translocation of protons across an ENERGY-TRANSDUCING MEMBRANE (see CHEMIOSMOSIS) and/or which can catalyse the pmf-dependent phosphorylation of ADP to ATP. (cf. PROTON PPASE.)

The most-studied H^+-ATPases are the (F_0F_1)-type H^+-ATPases (also referred to as (F_1F_0)-type H^+-ATPases) which occur e.g. in the bacterial cytoplasmic membrane, in the mitochondrial (inner) membrane, and in the thylakoid membranes of chloroplasts; these enzymes have a MWt of ca. 500000. Bacterial (F_0F_1)-type H^+-ATPases consist of two distinct parts: (a) the

F_0 moiety, a hydrophobic protein (typically containing three kinds of subunit – designated a, b and c in *Escherichia coli*) embedded in the membrane bilayer, and (b) the F_1 moiety, a hydrophilic protein (typically consisting of five kinds of subunit) which has ATPase activity, and which projects – like a knob on a short 'stalk' – from the cytoplasmic face of the membrane. (Some authors regard the *projecting* stalked knob as an artefact [JUR (1985) *93* 138–143].) The 'knob' of F_1 consists of two types of subunit (designated α and β), and the 'stalk' consists of three types of subunit (γ, δ and ε). Mitochondrial (F_0F_1)-type H^+-ATPases are essentially similar in structure but are somewhat more complex; the stalk region of the enzyme complex includes an OLIGOMYCIN-sensitivity-conferring protein (OSCP), which does not occur in bacterial or chloroplast ATPases, and a protein (called 'inhibitor protein', 'I', or 'IF_1') which inhibits the hydrolysis of newly-made ATP. The mitochondrial, bacterial and chloroplast ATPases all contain, in the F_0 moiety, a DCCD-binding protein which confers sensitivity to DCCD.

Nomenclature of H^+-ATPases. H^+-ATP-ases from different sources are designated as follows: CF_0F_1 (from chloroplasts); EF_0F_1 (from *Escherichia coli*); MF_0F_1 (from mitochondria); TF_0F_1 (from certain thermophilic bacteria, e.g. 'PS3'). Individual components may be designated e.g. CF_0, EF_1, MF_0 etc. The three types of subunit in TF_0 may be referred to as bands 4, 6 and 8 (from the SDS–gel electrophoresis pattern). Other types of H^+-ATPase include the (E_1E_2)-type enzymes which occur e.g. in the cytoplasmic membrane in certain fungi; these enzymes differ from (F_0F_1)-type ATPases e.g. in that their activity involves the formation of a phosphorylated intermediate. (See also ATPASE for a general note on the nomenclature of ATPases.)

ATP synthesis and hydrolysis. H^+-ATPase-catalysed ATP synthesis requires a minimum pmf of ca. 100–200 mV. During ATP synthesis protons flow through the enzyme system, passing down the proton concentration gradient; the actual route of proton flow appears to include the side-chains of particular amino acids – e.g. aspartic acid, glutamine, lysine and tyrosine. (Interestingly, it has been proposed that the proton-conducting pathway through another energy-transducing system, the purple membrane (see BACTERIORHODOPSIN), involves a similar range of amino acids.) Much or most of the energy derived from proton translocation may be used to effect the release of ATP from the tightly cohesive ATP–H^+-ATPase complex – rather than to carry out the phosphorylation step itself. In one model it is proposed that protons passing through the F_1 moiety cause conformational changes in the enzyme which decrease the affinity of ATP for the catalytic site. This type of model is believed to be compatible with variable H^+/P ratios – the number of protons needed to cause a given conformational change (with concomitant release of ATP) being postulated to vary with the prevailing pmf.

Under some conditions H^+-ATPases may catalyse either the synthesis or hydrolysis of ATP in order that G_{ATP} and pmf be more or less balanced (see CHEMIOSMOSIS). However, when pmf is below a certain level, H^+-ATPases are rapidly (and reversibly) inactivated; such inactivation prevents depletion of ATP e.g. under those conditions in which, for any reason, the membrane is temporarily de-energized. In chloroplasts and bacteria inactivation of H^+-ATPases occurs as a result of the tight binding between ADP and the catalytic site, while in mitochondria the enzyme is inactivated by the inhibitor protein. [Regulation of H^+-ATPase activity: TIBS (1986) *11* 32–35.]

Inhibitors of (F_0F_1)-type H^+-ATPases include e.g. AUROVERTINS, CITREOVIRIDIN, DCCD, EFRAPEPTIN, OLIGOMYCIN, QUERCETIN and tributyltin chloride; TF_1 (in contrast to the F_1 moieties of mesophilic bacteria) is resistant to e.g. aurovertins, efrapeptin and quercetin.

[H^+-ATPases, structure and function: ARB (1983) *52* 801–824; H^+-ATPase as an energy-converting enzyme: Book ref. 127, pp. 149–176.]

proton circuit See CHEMIOSMOSIS.

proton motive force See CHEMIOSMOSIS.

proton PPase (H^+-PPase) Proton pyrophosphatase: a membrane-bound enzyme or enzyme complex which couples the hydrolysis or synthesis of PYROPHOSPHATE to the energy-linked transmembrane translocation of protons; H^+-PPases occur widely e.g. in mammalian and yeast mitochondria, in chloroplasts, and in certain photosynthetic bacteria.

H^+-PPases function in a way analogous to that of PROTON ATPASES. Thus, e.g. PPi hydrolysis by an H^+-PPase can generate or augment a proton motive force (see CHEMIOSMOSIS) which can be used for energy-requiring processes such as ION TRANSPORT, reverse electron transport, and ATP synthesis (via an H^+-ATPase). Conversely, an H^+-PPase can use pmf, directly, for the synthesis of PPi (see also OXIDATIVE PHOSPHORYLATION); such synthesis has been reported to occur e.g. in mitochondria, in the chloroplasts of the alga *Acetabularia mediterranea* and of the pea plant, and in the bacterium *Rhodospirillum rubrum*.

(See also ELECTRON TRANSPORT CHAIN (figure).)

proton pump Any biochemical process which promotes the energy-requiring translocation of protons across an ENERGY-TRANSDUCING MEMBRANE: see e.g. ELECTRON TRANSPORT CHAIN and PROTON ATPASE.

proton reducers See ANAEROBIC DIGESTION.

proton-reducing acetogen See ACETOGEN.

proton translocators (protonophores) IONOPHORES which effectively increase the permeability, to protons, of the lipid bilayer in biological and/or artificial membranes; such agents act as mobile carriers of (specifically) protons, and can dissipate proton motive force (see CHEMIOSMOSIS) by effectively 'short circuiting' the membrane. (See also UNCOUPLING AGENTS.)

Proton translocators include a number of lipophilic weak organic acids which retain lipophilicity even in the deprotonated (dissociated) state (owing to delocalization of charge within the molecule); in the protonated state proton translocators deliver protons to one side of the membrane, while in the deprotonated state they accept protons from the other side. Examples include e.g. CCCP (carbonylcyanide-*m*-chlorophenylhydrazone), 2,4-dinitrophenol, and FCCP (carbonylcyanide-*p*-trifluoromethoxyphenylhydrazone).

protonmotive force Proton motive force (see CHEMIOSMOSIS).

protonmotive Q-cycle See Q-CYCLE.

protonophore *Syn.* PROTON TRANSLOCATOR.

proto-*onc* See ONCOGENE.

protoplasmic cylinder See SPIROCHAETALES.

protoplasmodium See MYXOMYCETES.

protoplast (1) The spherical or near-spherical, osmotically sensitive structure formed by the *complete* removal of the cell wall (e.g. by enzymic action) from a cell suspended in an isotonic (or hypertonic) medium; a protoplast thus consists of the cytoplasmic membrane and cell contents. Protoplasts can continue to metabolize – and may be able to revert to normal cells – under appropriate cultural conditions; however, protoplasts cannot divide. Bacterial protoplasts are prepared more easily from Gram-positive cells (e.g. by using LYSOZYME) than from Gram-negative cells; they are resistant to phage infection, but those from previously infected cells can support phage replication. (See also SPHAEROPLAST; AUTOPLAST; L-FORM; PROTOPLAST FUSION.)

(2) In an *intact* cell: the cytoplasmic membrane and all structures internal to it.

protoplast fusion A technique for in vitro genetic transfer in which a hybrid PROTOPLAST (and ultimately a hybrid cell) is formed by fusing protoplasts derived from different strains, species or genera. Protoplasts are initially prepared e.g. by treating cells with wall-degrading enzymes (e.g. LYSOZYME for some Gram-positive bacteria); populations of two types of protoplast are then mixed. Treatment of the mixture with e.g. POLYETHYLENE GLYCOL (PEG) increases the (low) spontaneous rate of fusion. (cf. ELECTROFUSION.) The protoplasts are then transferred to an appropriate (hypertonic) solid medium for cell wall regeneration. The resulting cells can be screened for those with genetic markers from both parents; e.g., if both parents were auxotrophic (with different growth requirements) the cells can be screened for prototrophs.

Hybrid bacterial protoplasts, containing a genome from each parent, are called *biparentals*. The two genomes may undergo recombination, or the diploids may be unstable, the two genomes segregating when the cell divides. In e.g. *Bacillus subtilis*, many of the progeny cells are unstable *complementary diploids* (i.e., genes on both genomes are expressed and can exhibit COMPLEMENTATION); however, some progeny cells are unstable *non-complementing diploids* (Ncd cells) in which one of the two parental genomes is expressed while the other is inactive until segregated on cell division. (cf. CONJUGATION sense 1b (ii).)

(See also TRANSFORMATION.)

protoplast membrane *Syn.* CYTOPLASMIC MEMBRANE.

protoporphyrins See PORPHYRINS.

Protosporangium See PROTOSTELIOMYCETES.

Protostelia See EUMYCETOZOEA.

protostelids See PROTOSTELIOMYCETES.

Protosteliia See EUMYCETOZOEA.

Protosteliomycetes (protostelids) A class of primitive slime moulds (division MYXOMYCOTA – cf. EUMYCETOZOEA) in which the vegetative phase consists of a small reticulate plasmodium or of simple amoebae which form filose – and sometimes also lobose – pseudopodia; in some species the amoebae bear one or more smooth flagella. 'Shuttle streaming' (i.e., rhythmic reversible flow of protoplasm) does not occur. An individual amoeba, or a segment of a plasmodium, gives rise to a fruiting body consisting of one to several spores borne on a slender tubular stalk, the whole being bounded by a sheath. Ballistospores are formed e.g. by *Protostelium nocturnum* [Mycol. (1984) *76* 443–447], *P. expulsum* [TBMS (1981) *76* 303–309], and *Schizoplasmodium cavostelioides* [sporocarp ultrastructure and development: Mycol. (1985) *77* 848–860]. [Ultrastructural study of trophozoite and cyst stages of *P. pyriformis*: JP (1986) *33* 405–411.] Protostelids are found in various habitats, including bark, rotting wood, etc. Other genera include e.g. *Cavostelium*, *Ceratiomyxella*, *Endostelium* [Mycol. (1984) *76* 884–891], *Nematostelium* and *Protosporangium*. (cf. CERATIOMYXA.)

Protostelium See PROTOSTELIOMYCETES.

Prototheca A genus of unicellular eukaryotic microorganisms which are generally regarded as achlorophyllous strains of CHLORELLA. The cells are hyaline and roughly spherical or ovoid (ca. 1.3–13.4 × 1.3–16.0 μm, depending on species, growth conditions etc); when mature, a cell divides to form 2–20 or more 'endospores' which are released by rupture of the mother cell. Species occur e.g. in soil and water, and some can behave as opportunist pathogens (see PROTOTHECOSIS).

protothecosis A disease, which affects man and animals, caused by a species of *Prototheca* (commonly *P. zopfii*); typically, it involves the formation of localized skin lesions – which may be nodular, ulcerating and papular, or (rarely) granulomatous. (See also MASTITIS.) A severe, often fatal, systemic form of the disease also occurs, particularly in animals.

prototroph A strain of microorganism whose nutritional requirements do not exceed those of the corresponding wild-type strain. (cf. AUXOTROPH.)

prototunicate ascus An ASCUS whose wall consists of a thin, delicate membrane; the ascospores are released when the ascus wall ruptures or deliquesces. Such asci are formed e.g. by ascogenous yeasts.

protoxin A precursor form of a TOXIN: see e.g. DELTA-ENDOTOXIN.

protozoa A diverse group of eukaryotic, typically unicellular microorganisms; they are classified as a subkingdom (Protozoa) within the kingdom Animalia or the kingdom PROTISTA. (Some of these organisms – e.g. the SLIME MOULDS and members of the PHYTOMASTIGOPHOREA – are also classified in botanical taxonomic schemes.) The earliest known protozoa lived in the late Precambrian (see e.g. FORAMINIFERIDA).

The protozoa exhibit a great variety of forms, structures and life styles, and are divided into seven phyla: APICOMPLEXA, ASCETOSPORA, CILIOPHORA, LABYRINTHOMORPHA, MICROSPORA, MYXOZOA and SARCOMASTIGOPHORA [JP (1980) *27* 37–58]. Many protozoa are free-living organisms which occur in freshwater, brackish or marine habitats. Some protozoa occur e.g. in soil, some are common in SEWAGE TREATMENT plants, and some inhabit the gut in vertebrates or invertebrates (see e.g. RUMEN and TERMITE–MICROBE ASSOCIATIONS). Parasitic species include e.g. members of the Apicomplexa, Ascetospora, Microspora and Myxozoa; some protozoa are pathogenic in vertebrates (including man) or invertebrates (see e.g. BABESIA, BALANTIDIUM, EIMERIORINA, ENTAMOEBA, GIARDIA, HISTOMONAS, NAEGLERIA, NOSEMA, PLASMODIUM, TOXOPLASMA, TRYPANOSOMA). PHYTOMONAS is parasitic in certain plants.

Protozoa range in size from ca. 1 μm to several millimetres; some have internal or external 'skeletons' of e.g. calcareous, siliceous and/or organic material. Many exhibit MOTILITY but some (see e.g. PERITRICHIA) are sedentary. Probably most protozoa are aerobic organisms, but some can grow microaerobically or anaerobically (see e.g. RUMEN and SAPROBITY SYSTEM). Feeding may occur saprotrophically and/or by pinocytosis – and/or it may occur holozoically (e.g. in many amoebae and ciliates), these organisms being classified as CARNIVOROUS, HERBIVOROUS or OMNIVOROUS. Some protozoa (see PHYTOMASTIGOPHOREA) can carry out PHOTOSYNTHESIS.

Asexual reproduction may involve binary fission, multiple fission and/or budding. Sexual processes occur in some species (see e.g. AUTOGAMY and CONJUGATION). According to species, protozoa may be haploid, diploid or polyploid; some (the foraminifera) exhibit an ALTERNATION OF GENERATIONS.

Populations of viable protozoa may be preserved by FREEZING (suitable e.g. for *Plasmodium*, *Tetrahymena*, *Trypanosoma*) or by repeated subculturing. CYST-forming species (e.g. *Didinium*, certain amoebae) may be encouraged to encyst (and thus survive for protracted periods) e.g. by slow desiccation or by starvation.

protrichocyst *Syn.* EJECTOSOME.

Providencia A genus of Gram-negative bacteria of the tribe PROTEEAE. Cells: ca. 0.6–0.8 × 1.5–2.5 μm. Swarming does not occur. Species do not produce H_2S or lipase, do not liquefy gelatin at 22°C, and can metabolize mannose and one or more sugar alcohols; colonies on DCA typically develop yellow-orange centres due to the precipitation of $Fe(OH)_3$ by alkaline metabolic products. Nicotinic and pantothenic acids are not required for growth. GC%: 39–42. Type species: *P. alcalifaciens*.

P. alcalifaciens (formerly included in *Proteus inconstans*). Urease −ve (weak or delayed +ve in a few strains); trehalose −ve; inositol. −ve; adonitol +ve.

P. rettgeri (including most of the strains formerly regarded as *Proteus rettgeri*). Urease +ve; trehalose −ve; inositol +ve; adonitol +ve.

P. stuartii (incorporating strains formerly included in *Proteus inconstans* and *Proteus rettgeri*). Urease −ve (weak or delayed +ve in a few strains); trehalose +ve; inositol +ve; adonitol −ve.

Providencia spp are rare in the normal human intestine, and their natural habitat is unknown. Some strains can be opportunist pathogens – *P. rettgeri* and *P. stuartii* causing e.g. nosocomial UTIs, *P. alcalifaciens* being more often associated with diarrhoeal illnesses.

[Book ref. 22, pp. 494–496.]

provirus Viral DNA which has become integrated into the chromosomal DNA of the host cell; in viruses of the RETROVIRIDAE the DNA is a transcript of the RNA genome, while in some DNA viruses the genome itself, or a portion of the genome, may integrate (see e.g. DEPENDOVIRUS, HEPATITIS B VIRUS and MASTADENOVIRUS).

prozone (*serol.*) (1) In a titration involving antibodies and *particulate* antigens: absence of visible AGGLUTINATION in tubes containing the higher concentration(s) of antibody. A prozone may be due e.g. to (i) relatively low titres of BLOCKING ANTIBODIES (sense 1) which are diluted out in tubes containing agglutinating titres of normal homologous antibodies; (ii) gross ANTIBODY EXCESS; or (iii) the presence in the antiserum of substances which inhibit agglutination non-specifically.

(2) In a titration involving antibodies and *soluble* antigens: the absence of visible precipitate in those tubes in which there is an ANTIGEN EXCESS.

prozygosporangium See ZYGOSPORE.

PrP 27–30 (PrP) See SCRAPIE.

PrP^c, PrP^{sc} See PRION.

PRRS Porcine reproductive and respiratory syndrome: see BLUE-EARED PIG DISEASE.

pruinose (of surfaces) Powdery in appearance.

***Prunus* necrotic ringspot virus** See ILARVIRUSES.

pruritis Itching.

Pruteen See SINGLE-CELL PROTEIN.

Prymnesiida See PHYTOMASTIGOPHOREA.

Prymnesiophyceae (Haptophyceae) A class of unicellular, uninucleate, commonly biflagellated algae (sometimes regarded as protozoa – see PHYTOMASTIGOPHOREA). Typically, the motile cell has two smooth flagella of more or less equal length, together with a HAPTONEMA; the golden-brown chloroplasts (usually two per cell) contain chlorophylls *a*, c_1 and c_2, and fucoxanthin. Storage compounds include CHRYSOLAMINARIN. Many members can ingest food phagocytically. Most prymnesiophytes bear a covering of delicate, typically oval scales embedded in a mucilaginous matrix external to the plasmalemma; these scales are produced in the Golgi apparatus (or in vesicles probably derived therefrom) before being passed to the cell exterior (cf. CHRYSOPHYTES). The scales may be entirely organic (e.g. in *Chrysochromulina* and *Prymnesium*) or may be calcified (in certain stages of the COCCOLITHOPHORIDS). Some of the scales may bear long or short spines, and – according to species – different types of scales may be present on the same cell and/or on different cells in different stages of the life cycle. The nature of the scales (determined by electron microscopy) is an important taxonomic characteristic.

Prymnesiophytes are mostly marine organisms, forming an important component of the nannoplankton. Some species (e.g. PRYMNESIUM) occur in brackish waters, others in fresh water.

Prymnesium A genus of unicellular algae (class PRYMNESIOPHYCEAE) in which the HAPTONEMA is short and immobile. Species occur in brackish water. *P. parvum* produces a potent, cation-activated, pH-dependent, haemolytic PHYCOTOXIN (comprising at least 6 components) which can cause mass mortalities in fish and other gill-bearing vertebrates (as well as in invertebrates) in e.g. brackish-water culture ponds.

PS3 A strain of thermophilic bacteria of uncertain taxonomic affiliation.

PS-5 A CARBAPENEM antibiotic.

PSA fimbriae ('PSA pili') See FIMBRIAE.

Psalliota See AGARICUS.

Psammetta See XENOPHYOPHOREA.

Psammina See XENOPHYOPHOREA.

Psammomitra See HYPOTRICHIDA.

psammophilic Preferring a sandy habitat.

Psathyrella See AGARICALES (Coprinaceae).

pSC101 A 9.4-kb, low-copy-number PLASMID which encodes tetracycline resistance and which has a single *Eco*RI cleavage site in a non-essential region, making it a useful CLONING vector. Replication of pSC101 occurs unidirectionally, does not require DNA polymerase I (cf. e.g. COLE1 PLASMID) or the host cell's DnaA protein (see DNAA GENE) but requires a pSC101-encoded initiator protein. [Nucleotide sequence of pSC101: NAR (1984) *12* 9427–9440.]

PSE-type β-lactamases See β-LACTAMASES.

Pseudallescheria (*syn. Petriellidium*) A genus of fungi of the order MICROASCALES. *P. boydii* ('*Allescheria boydii*') (anamorph: *Scedosporium apiospermum*) occurs e.g. in soil and in faecally contaminated cattle feed. It can cause mycotic abortion e.g. in cattle; in man it is a causal agent of MADUROMYCOSIS and may be implicated in infections of the respiratory tract, sinuses etc, particularly in immunosuppressed patients. [Clinical significance of *P. boydii*: Mayo Clin. Proc. (1985) *60* 531–537.]

Pseudanabaena A genus of filamentous CYANOBACTERIA (section III) in which the trichomes are composed of cylindrical cells, with constrictions between adjacent cells; trichomes are motile and non-sheathed, and the cells contain polar GAS VACUOLES. GC%: 44–52. The genus probably includes at least some strains of '*Oscillatoria redekei*'.

pseudoaethalium See MYXOMYCETES.

pseudoappendicitis See e.g. FOOD POISONING (j); see also PSEUDOTUBERCULOSIS.

Pseudocaedibacter A genus of Gram-negative bacteria which occur as endosymbionts in strains of *Paramecium*. Cells: non-motile rods, ca. 0.25–0.7 × 0.5–4.0 μm, which do not contain R bodies (cf. CAEDIBACTER) and which may or may not (according to species) confer a killer characteristic on their host paramecia. GC%: ca. 35–39. Type species: *P. conjugatus*.

P. conjugatus (see also MU PARTICLE) confers the MATE KILLER characteristic; in vitro cultivation has been reported.

P. falsus (nu and pi particles) does not confer a killer characteristic.

P. minutus (gamma particles) confers the killer characteristic. [Book ref. 22, pp. 807–808.]

pseudocapillitium In the aethalia or pseudoaethalia of certain MYXOMYCETES: a system of threads (generally irregular in width) and/or of membranous or perforated plate-like structures present among the spores; it has been suggested that these structures may be remnants of sporangial walls.

pseudocatalase A non-haemoprotein substance which catalyses the breakdown of H_2O_2 into water and O_2; pseudocatalases occur e.g. in various lactic acid bacteria, including strains of *Lactobacillus*, *Enterococcus faecalis*, and in some species of *Veillonella*. Unlike CATALASE, pseudocatalases are not inhibited by low concentrations of azide or cyanide. A manganese-containing protein pseudocatalase (a 'manganicatalase') has been isolated from a strain of *Lactobacillus plantarum* [JBC (1983) *258* 6015–6019].

Pseudocercosporella See HYPHOMYCETES; see also EYESPOT (sense 2).

pseudocilia See TETRASPORA.

pseudocoagulase In certain strains of *Staphylococcus*: any protease that mimics the effects of staphyloCOAGULASE by proteolytic activation of prothrombin, thus causing a false-positive reaction in a COAGULASE TEST. 'Pseudocoagulase' is believed to be a metalloprotease(s) and can be inhibited by adding EDTA to the plasma. [Book ref. 44, p. 530.]

pseudocolumella In the fruiting bodies of certain MYXOMYCETES: a COLUMELLA-like, spherical or rod-shaped, calcareous structure which occurs in the centre of the sporangium; it is not attached to the peridium, but may be attached to capillitial threads.

pseudocowpox A mild, usually benign CATTLE DISEASE characterized by the formation of papular, scab-forming lesions on the teats which may predispose to MASTITIS; the causal agent is a PARAPOXVIRUS. (See also MILKERS' NODULE; cf. COWPOX.)

pseudocyphella (*lichenol.*) A small pore (ca. 0.1–2.0 mm diam.) in the (upper and/or lower) cortex of the thallus in various lichens (e.g. species of *Cetraria*, *Parmelia*, *Pseudocyphellaria*). Pseudocyphellae have no definite margin (cf. CYPHELLA); they appear to be formed by the disintegration of the pseudoparenchymatous cortex, and are believed to facilitate gas exchange [Lichenol. (1981) *13* 1–10]. (cf. PORED EPICORTEX.)

Pseudocyphellaria A genus of foliose LICHENS (order PELTIGERALES) which closely resemble STICTA spp but differ in that the tomentose underside is perforated with PSEUDOCYPHELLAE. The upper surface may bear soredia which are yellow in *P. crocata*, bluish-grey in *P. intricata*. Species occur on soil, trees, rocks, etc.

pseudocyst (*protozool.*) See e.g. CHAGAS' DISEASE and TOXOPLASMOSIS.

pseudoepithecium In some types of APOTHECIUM: a layer, at the surface of the hymenium, which consists of the tips of paraphyses immersed in an amorphous matrix. (cf. EPITHECIUM.)

pseudo-Fiji disease See FIJI DISEASE.

pseudogene In a eukaryotic chromosome: a DNA segment which resembles a known functional gene but differs in carrying deleterious mutations and in having no apparent function. Many pseudogenes resemble cDNA copies of cellular mRNAs: they have a short A–T tract at the 3′ end (corresponding to the 3′-poly(A) of mRNA), lack introns, and are often flanked by direct repeats; such pseudogenes are believed to have arisen by reverse transcription of cellular mRNAs.

Pseudogenes have also been found in prokaryotes. For example, in *Mycobacterium leprae* about 27% of the genome is reported to consist of pseudogenes [Nature (2001) *409* 1007–1011].

pseudoglanders *Syn.* EPIZOOTIC LYMPHANGITIS.

pseudohaemoptysis (pseudohemoptysis) The apparent expectoration of blood due to the coloration of sputum by red-pigmented bacteria (e.g. *Serratia* sp).

Pseudohydnum See TREMELLALES.

pseudohyphae *Syn.* SPROUT MYCELIUM.

Pseudoklossia See EIMERIORINA.

pseudolumpy skin disease See LUMPY SKIN DISEASE.

pseudolysogeny A phenomenon in which an association between a BACTERIOPHAGE and its host mimics LYSOGENY in that host cell lysis is delayed or does not occur, but true lysogeny is not established. A bacterial population exhibiting pseudolysogeny can be freed of phage by prolonged incubation with neutralizing anti-phage antiserum or by serial subculture from isolated colonies. Different types of pseudolysogeny occur in different phage–bacterium systems. For example, when a population of susceptible bacteria (e.g. *Shigella dysenteriae*) is exposed to a virulent bacteriophage (e.g. T7) at a low multiplicity of infection, a few cells become infected. On lysis, these cells may release an enzyme which destroys the phage receptors on other cells in the population; phage virions therefore fail to adsorb to these cells – which, as a result, become (temporarily) resistant to infection (phenotypic resistance). These cells regain their susceptibility to infection after subculture in the absence of phage.

In other cases the phage genome may be carried, but not expressed or replicated, within a cell in a susceptible population of bacteria, and the genome is passed to only one of the daughter cells at each cell division; a culture containing a proportion of such 'carrier cells' is known as a CARRIER CULTURE. After several successive transits from cell to daughter cell, the phage genome enters the lytic cycle and progeny phages are released by cell lysis. These may infect other cells which may in turn become carrier cells. This type of pseudolysogeny may occur e.g. when most of the host cells are repressed for one or more functions needed for phage replication.

In *Azotobacter* strain O certain phages can apparently establish a pseudolysogenic state accompanied by the phenotypic conversion of the host cells (see BACTERIOPHAGE CONVERSION); the cells lose their polysaccharide capsule, become flagellated and motile, and acquire a yellow pigmented appearance [Virol. (1980) *102* 267–277].

pseudomembranous colitis A severe, acute form of COLITIS. There is a profuse, watery diarrhoea, cramps, fever, and the formation of pseudomembranous patches of inflammatory exudate on the colonic mucosa (sometimes also in the small intestine). It may follow chemotherapy (e.g. with some β-LACTAM ANTIBIOTICS, clindamycin) and is linked to CLOSTRIDIUM *difficile*; toxins A and B kill gut epithelial cells e.g. by disrupting tight junction proteins. Inadequate IgG response may promote disease/relapse. [Review: BCG (2003) *17* 475–493.]

pseudomethylotroph An organism which oxidizes C_1 compounds to CO_2 and which assimilates C_1 carbon only via the Calvin cycle.

Pseudomicrothorax See HYPOSTOMATIA.

Pseudomonadaceae A family of Gram-negative, aerobic, chemoorganotrophic or facultatively chemolithotrophic, typically motile (usually polarly flagellate), rod-shaped bacteria. GC%: 58–71. Genera: FRATEURIA, PSEUDOMONAS, XANTHOMONAS, ZOOGLOEA. [Book ref. 22, pp. 140–219.]

Pseudomonadales A poorly defined, obsolete order of bacteria.

Pseudomonas A genus of Gram-negative, aerobic, chemoorganotrophic or facultatively chemolithoautotrophic bacteria of the family PSEUDOMONADACEAE; species occur as free-living organisms in soil and aquatic habitats, and as pathogens in man, other animals, and plants – see e.g. BLUE PUS, BROWN BLOTCH, GINGER BLOTCH, OLIVE KNOT and TABTOXIN. [Mechanisms of plant pathogenesis: CJM (1985) *31* 403–410.] (See also BUTTER and MEAT SPOILAGE.) The cells are straight or slightly curved rods, 0.5–1.0 × 1.5–5.0 μm. In most species the majority of strains are motile with one or several unsheathed polar flagella per cell; however, sheathed flagella are formed e.g. by *P. andropogonis*, and some species produce both polar and lateral flagella – particularly when grown on solid media. [Biochemical and serological properties of flagella of some *Pseudomonas* spp: JGM (1985) *131* 873–883.] Polarly fimbriate strains may exhibit TWITCHING MOTILITY. Cell walls and cytoplasmic membranes resemble those of other Gram-negative bacteria. (See also PORIN, OUTER MEMBRANE and S LAYER.) Catalase +ve; many strains are oxidase +ve. GC%: 58–70. Type species: *P. aeruginosa*.

Nutrition, metabolism, growth. Typically, the organisms are nutritionally and metabolically highly versatile: most species will grow on inorganic salts with an organic carbon source, while some can grow chemolithoautotrophically; however, certain species (e.g. *P. diminuta*, *P. vesicularis*) require factors such as biotin, cyanocobalamin and pantothenate. Metabolism is respiratory (oxidative); many species can carry out NITRATE RESPIRATION. The TCA CYCLE operates in all species investigated. In e.g. *P. aeruginosa*, *P. fluorescens* and *P. putida*, glucose can be oxidized by EXTRACYTOPLASMIC OXIDATION to gluconate which may then be metabolized via the ENTNER–DOUDOROFF PATHWAY. Carbohydrates are metabolized anaerogenically. [Alternative pathways of carbohydrate metabolism in *Pseudomonas*: ARM (1984) *38* 359–387.] HYDROCARBONS can be metabolized by some strains. Some metabolic activities are plasmid-dependent: e.g., the ability to degrade camphor, salicylate, toluate and xylene is associated with plasmids CAM, SAL, TOL and XYL, respectively. Under nitrogen-limiting conditions some species accumulate POLY-β-HYDROXYBUTYRATE. Under iron-deficient conditions, siderophores (some fluorescent) may be secreted (see NOCARDAMINE, PYOVERDIN). (See also PYOCYANIN.) Some species produce bacteriocins (see e.g. PYOCIN). In many species growth occurs optimally at ca. 28–30°C and is generally inhibited at or below ca. pH 4.5.

Sensitivity to antimicrobial agents. In general, pseudomonads are resistant to a range of common antimicrobial agents; resistance may or may not be plasmid-mediated. *P. aeruginosa* may be sensitive to e.g. certain AMINOGLYCOSIDE ANTIBIOTICS (e.g. amikacin, gentamicin); it is resistant to many common DISINFECTANTS (cf. DETTOL CHELATE) but is usually lysed rapidly by EDTA.

Susceptibility to phages. Pseudomonas spp are susceptible to a variety of phages: see e.g. BACTERIOPHAGE ϕ6; BACTERIOPHAGE ϕW-14; INOVIRUS; TECTIVIRIDAE. The transducing phages F116 and G101 were used for mapping the chromosome of *P. aeruginosa*. Lysogeny is very common in *P. aeruginosa*, but is uncommon in other species.

Genetic aspects. Genetic studies have often been carried out with *P. aeruginosa* strain PAO. [Chromosomal maps of *P. aeruginosa* PAO and *P. putida* PPN compared: JGM (1985) *131* 885–896.] Gene transfer can occur by transduction and conjugation; most conjugative plasmids transfer better on solid media than in liquids [JGM (1983) *129* 2545–2556]. Transformation in *P. aeruginosa* has been achieved with $CaCl_2$-treated cells.

Taxonomy. A number of pseudomonads were classified by rRNA oligonucleotide cataloguing (q.v.) into rRNA groups I–V. Group I included fluorescent species (e.g. *P. aeruginosa*, *P. aureofaciens*, *P. chlororaphis*, *P. putida*, *P. syringae*) and non-fluorescent species (e.g. *P. alcaligenes*, *P. pseudoalcaligenes*, *P. stutzeri*), all of which (apart from some strains of *P. pseudoalcaligenes*) do not accumulate poly-β-hydroxybutyrate (PHB). Species of groups II and III do accumulate PHB; group II included pathogens (e.g. *P. cepacia*, *P. mallei*, *P. solanacearum*), while group III consisted of non-pathogens (e.g. *P. acidovorans*, *P. testosteroni*) and included some autotrophs which can grow chemolithotrophically (e.g. *P. facilis*, *P. flava*, *P. paleronii* and *P. saccharophila*). Group IV consisted of *P. diminuta* and *P. vesicularis*, while group V contained only *P. maltophila*; these three species require growth factors, and the last two are unable to assimilate nitrate. For group II see BURKHOLDERIA. For group V see *P. maltophila*, below.

Many species have not been assigned to rRNA groups, and their relationship to rRNA-grouped species is uncertain; such species include e.g. *P. agarici*, *P. amygdali*, *P. andropogonis*, *P. aurantiaca*, *P. avenae*, *P. cattleyae*, *P. fragi*, *P. fulva*, *P. lemoignei*, *P. marina* (see DELEYA), *P. nautica*, *P. spinosa* and *P. woodii*.

P. aeruginosa (formerly *P. pyocyanea*). Occurs in soil, river water etc, and is an important opportunist pathogen in man and other animals – being isolated from infected burns, urinary tract infections, etc. (See also BLUE PUS; EXOENZYME S; EXOTOXIN A.) It can occasionally be pathogenic in (stressed) plants. Motile strains usually have one polar flagellum per cell; rarely non-motile. Polarly fimbriate. Colonies on nutrient agar are often roundish, flat, dull and butyrous (R-type), but can be e.g. smooth and umbonate (S-type). Mucoid (M-type) colonies are formed by ALGINATE-producing strains isolated e.g. from CYSTIC FIBROSIS patients; only some mucoid strains can produce alginate on minimal agar. Most strains produce PYOCYANIN, and may also produce e.g. PYOVERDIN, while a few strains can form a brown pigment (*pyomelanin*) or a red pigment (*pyorubin*). (Pyomelanin production is enhanced in mineral salts media containing 1% L-tyrosine; 1% DL-glutamate enhances pyorubin production.) All strains are catalase +ve and oxidase +ve; many are urease +ve. Optimum growth temperature: 37°C; all strains can grow at 42°C. GC%: ca. 67.

The production of virulence factors by *P. aeruginosa* may be inhibited through inhibition of the organism's QUORUM SENSING mechanisms by the macrolide antibiotic azithromycin [AAC (2001) *45* 1930–1933].

P. alcaligenes. A non-fluorescent species which cannot utilize e.g. glucose, fructose, maltose or gluconate, but which can use e.g. L-arginine, malate, and propionate. Monotrichously flagellate. Growth can occur at 41°C; optimum temperature ca. 35°C. GC%: 64–68.

P. andropogonis (= *P. stizolobii*). A non-fluorescent, typically oxidase −ve species pathogenic e.g. for sorghum. The organisms form sheathed flagella at ca. 28°C but are non-flagellate when grown at 34°C.

P. carboxydohydrogena. See CARBOXYDOBACTERIA.

P. carboxydovorans. See CARBOXYDOBACTERIA.

P. cepacia. A former species (see BURKHOLDERIA) found e.g. in soil, water. Oxidase +ve. Lophotrichously flagellate. Non-fluorescent pigments often formed. Opt. growth: 30–35°C. Pathogenic in man (see CYSTIC FIBROSIS) and a causal agent of soft rot in onions.

P. cichorii. A fluorescent, PYOVERDIN-forming species pathogenic for chicory: Lophotrichously flagellate. Weakly oxidase +ve. Optimum growth temperature: ca. 30°C. GC%: 59.

'*P. cocovenenans*' (*incertae sedis*). See BONGKREKIC ACID.

P. facilis (formerly *Hydrogenomonas facilis*). A non-fluorescent, non-pigmented species which occurs e.g. in soil. Monotrichously flagellate. Oxidase +ve. Glucose, fructose and malonate can be utilized; extracellular PHB is hydrolysed. The organisms can grow chemolithoautotrophically using hydrogen and oxygen (see KNALLGAS REACTION). Optimum growth temperature: ca. 28°C. GC%: 62–64.

P. fluorescens. A fluorescent, PYOVERDIN-forming species which occurs e.g. in soil and water. Lophotrichously flagellate. Oxidase +ve. Gelatinase +ve. Growth can occur at 4°C but not at 41°C; optimum growth temperature: 25–30°C. GC%: ca. 59–61. (See also MUPIROCIN.)

P. fragi. A psychrophilic, peritrichously fimbriate species which can cause spoilage of refrigerated foods (e.g. BUTTER).

P. mallei. A non-motile, non-pigmented species which causes GLANDERS. Oxidase +ve, gelatinase +ve. D-xylose is utilized, D-ribose is not. (See BURKHOLDERIA.)

P. maltophila. A former species found in e.g. soil, water and milk. Lophotrichously flagellate. Most strains need cystine or methionine; PHB is not accumulated. Strongly lipophilic. No growth at 4°C or 41°C. Re-classified initially as *Xanthomonas maltophila* [IJSB (1983) *33* 409–413] and then as *Stenotrophomonas maltophila* (see CYSTIC FIBROSIS).

P. marina. See DELEYA.

P. nigrifaciens. See ALTEROMONAS.

P. phaseolicola. See *P. syringae*.

P. pseudoalcaligenes. A non-fluorescent species, some strains of which are pathogenic for watermelon. Monotrichously flagellate. Fructose can be utilized (cf. *P. alcaligenes*). GC%: 62–64.

P. pseudomallei. Occurs e.g. in soil and is the causal agent of MELIOIDOSIS. Motile, lophotrichously flagellate. Oxidase +ve. Gelatinase +ve. D-Ribose can be utilized, D-xylose cannot. Yellow or orange pigments may be formed. (See BURKHOLDERIA.)

P. putida. A fluorescent species which occurs e.g. in soil and water. Lophotrichously flagellate. Oxidase +ve. Optimum growth temperature: 25–30°C. Two biovars: A and B; strains of biovar B can utilize L-tryptophan and testosterone.

P. putrefaciens. See ALTEROMONAS.

P. pyocyanea. See *P. aeruginosa*.

P. saccharophila. The sole strain of *P. saccharophila* was isolated from mud. Monotrichously flagellate. Growth can occur chemolithoautotrophically with hydrogen as electron donor. PHB is accumulated. Oxidase +ve. Optimum growth temperature: ca. 30°C. GC%: ca. 69.

P. solanacearum. A plant-pathogenic species which causes wilts in e.g. members of the Solanaceae. Lophotrichously flagellate. Oxidase +ve. Some strains form brown pigments. GC%: ca. 66–68. (See BURKHOLDERIA.)

P. stizolobii. See *P. andropogonis*.

P. stutzeri. Occurs e.g. in soil and water. Non-fluorescent. PHB is not accumulated. Monotrichously flagellate, but lateral flagella are formed under certain conditions. Oxidase +ve. Starch is hydrolysed. GC%: 60–66. (See also BIOMIMETIC TECHNOLOGY.)

P. syringae. A fluorescent, plant-pathogenic species. [Genetic determinants of pathogenicity in pathovar *syringae*: PNAS (1985) *82* 406–410.] Lophotrichously flagellate. Oxidase −ve. Optimum growth temperature: ca. 25–30°C. GC%: 59–61. The species is divided into a large number of pathovars; these include (with their hosts) e.g. *P. syringae* pv. *coronafaciens* (e.g. oats); *P. syringae* pv. *glycinea* (soybean); *P. syringae* pv. *morsprunorum* (*Prunus* spp – see CANKER); *P. syringae* pv. *phaseolicola* (formerly *P. phaseolicola*) (e.g. beans); *P. syringae* pv. *pisi* (peas); *P. syringae* pv. *savastanoi* (olive trees – see OLIVE KNOT); *P. syringae* pv. *syringae* (lilac) (see also SYRINGOMYCIN); *P. syringae* pv. *tabaci* (tobacco – see TABTOXIN).

P. thermocarboxydovorans. See CARBOXYDOBACTERIA.

P. tolaasii. The causal agent of BROWN BLOTCH in mushrooms.

[Book ref. 22, pp. 140–199; ecology, isolation, culture: Book ref. 45, pp. 655–741; the biology of *Pseudomonas*: Book ref. 198.]

pseudomonic acid A *Syn.* MUPIROCIN.

pseudomurein A CELL WALL polymer present in some members of the ARCHAEA (METHANOBACTERIALES); it consists of (1 → 3)-β-linked backbone chains – containing *N*-acetyl-D-glucosamine (and/or *N*-acetyl-D-galactosamine, according to species) and *N*-acetyl-L-talosaminuronic acid – cross-linked by peptides which contain only L-amino acids (lys, glu, ala or thr). (cf. PEPTIDOGLYCAN.) Pseudomurein is resistant to LYSOZYME and to antibiotics such as penicillins and vancomycin. [Book ref. 157. pp. 416–423.]

pseudomycelium *Syn.* SPROUT MYCELIUM.

pseudonavicella See MONOCYSTIS.

pseudonigeran A water-insoluble, alkali-soluble (1 → 3)-α-D-glucan present in the CELL WALL in certain fungi: e.g. *Aspergillus nidulans*, *Paracoccidioides brasiliensis* (yeast form), *Schizophyllum commune*. Pseudonigeran can act as a reserve carbohydrate for cleistothecium formation in the teleomorph of *A. nidulans* (*Emericella nidulans*). (cf. NIGERAN.)

Pseudonocardia A genus of bacteria (order ACTINOMYCETALES, wall type IV) which occur e.g. in manure and soil. The organisms form non-fragmenting substrate mycelium, and unbranched aerial hyphae which give rise to chains of cylindrical spores. Mycolic acids are not formed. GC%: ca. 79. Type species: *P. thermophila*. [Book ref. 73, pp. 125–126; ecology, isolation, cultivation: Book ref. 46, pp. 2103–2117.]

pseudoparaphysis A type of sterile, typically branched and anastomosing, usually septate filament (of unknown function) which originates in the upper wall of a pseudothecial locule and grows downwards between the asci, subsequently fusing with the lower wall. (cf. PARAPHYSIS; see also HAMATHECIUM.)

pseudoparenchyma See PLECTENCHYMA.

pseudoperithecium See ASCOSTROMA.

Pseudoperonospora See PERONOSPORALES.

pseudopilins Certain components of the SECRETON which exhibit a region of homology with the pilins of type IV fimbriae; homology is limited to the 30 (hydrophobic) amino acids found at the N-terminal of the protein, i.e. those residues which, in the pilins, interact during fimbrial assembly.

Studies on those bacteria which encode both the secreton and type IV fimbriae suggested that a functional relationship may exist between the operation of the secreton and the mechanism of fimbriation. Subsequently, when genes encoding the pullulanase secreton of *Klebsiella oxytoca* were expressed in *Escherichia coli* K-12, one of the pseudopilins, PulG, was found to assemble into a pilus-like structure [EMBO (2000) *19* 2221–2228].

pseudoplasmodium (1) A multicellular, commonly motile structure formed by the aggregation of amoeboid cells (prior to fruiting) in CELLULAR SLIME MOULDS. (2) The ectoplasmic net of members of the LABYRINTHULOMYCETES. (3) In the MYXOBACTERALES: a number of individual cells embedded in a slime matrix.

pseudopodetium (*lichenol.*) A vertical, fruticose, branched or unbranched, usually solid secondary thallus which develops by growth from the primary thallus and is thus a purely thalline (vegetative, somatic) structure (cf. PODETIUM); ascocarp initials develop on the pseudopodetia. Pseudopodetia occur e.g. in species of PILOPHORUS and STEREOCAULON.

pseudopodium An extension of an amoeboid cell formed by extrusion of the cytoplasm (bounded by the plasmalemma). Pseudopodia are characteristically formed by protozoa of the SARCODINA; according to species, pseudopodia may function in locomotion and/or in feeding, and one or more may be formed at a time by a given cell. Various types of pseudopodia are formed by sarcodines. A *lobopodium* (lobose pseudopodium) is blunt-ended and generally broad, and often has a clear ectoplasmic area (the *hyaline cap*); lobopodia may be formed eruptively, i.e., by a burst of cytoplasmic movement, or more slowly by a steady cytoplasmic flow. A *filopodium* (filose or filiform pseudopodium) is slender, filamentous, sometimes tapered and sometimes branched, but lacks rigid axial elements (cf. AXOPODIUM) – although it may contain microtubules. Fine, filamentous pseudopodia which branch and anastomose to form a network are called *reticulopodia* or *rhizopodia*. In some sarcodines more or less fine 'subpseudopodia' may arise from a broader pseudopodial lobe (see e.g. ACANTHAMOEBA).

In e.g. *Amoeba proteus* (see AMOEBA) the lobose pseudopodia function in both feeding and locomotion. When a prey organism comes into the vicinity of a pseudopodium, the pseudopodium flows around it, initially forming a 'cup' (*food cup*) containing the prey; the plasmalemma eventually closes around the prey to form a FOOD VACUOLE. (cf. AXOPODIUM.) Locomotion in *A. proteus* involves the extension of pseudopodia which apparently adhere to the substratum; cytoplasm then continues to flow into one of the pseudopodia, which expands to accommodate it, and further pseudopodia are extended, etc. This type of locomotion is often regarded as typical 'amoeboid movement', although other forms of motility occur in other sarcodines. For example, in *Pelomyxa* the cell moves along by steady cytoplasmic flow without the formation of obvious pseudopodia. (See also HELIOZOEA and TAXOPODIDA.) Apart from functioning in active motility, pseudopodia in some amoebae (e.g. *Vannella* spp) can apparently function as flotation devices; such amoebae can, under certain conditions, produce distinct 'flotation forms' characterized by fine, usually hyaline pseudopodia which radiate from the central cell mass.

The mechanism of pseudopodial movement is not understood. It appears to involve microfilaments of ACTIN which lie just beneath – and which interact with – the plasmalemma.

pseudorabies *Syn.* AUJESZKY'S DISEASE.

pseudoraphe See DIATOMS.

pseudorinderpest *Syn.* PESTE DES PETITS RUMINANTS.

pseudorubella *Syn.* EXANTHEM SUBITUM.

pseudoseptum See SEPTUM (b).

pseudospores See USTILAGINALES.

pseudostem See PHALLALES.

pseudotemperate bacteriophage A BACTERIOPHAGE which can establish PSEUDOLYSOGENY.

pseudothecium See ASCOSTROMA.

Pseudotrebouxia A genus of sarcinoid unicellular green algae (division CHLOROPHYTA) which occur as photobionts in many LICHENS. The cells are more or less spherical (up to ca. 30 μm in diam.). *Pseudotrebouxia* spp closely resemble TREBOUXIA spp except in that asexual reproduction occurs not only by the formation of zoospores and aplanospores but also by 'vegetative cell division', i.e., basically, the division of one cell into two non-motile vegetative cells, etc – the resulting cells remaining in contact (at least initially) to form sarciniform groupings. [*Pseudotrebouxia* species in lichens, and discussion of vegetative cell division: Lichenol. (1981) *13* 65–86.]

pseudotuberculosis A self-limiting or fatal (septicaemic) disease of animals (especially rodents) due to *Yersinia pseudotuberculosis*; symptoms e.g.: mesenteric adenitis, chronic diarrhoea, internal caseous lesions. (Infected humans may have mesenteric adenitis or, if debilitated, septicaemia.) Wild animals form a reservoir of the pathogen.

pseudotype (*virol.*) A virus particle which, as a result of PHENOTYPIC MIXING, contains the genome of one virus and polypeptide(s) encoded by another virus; e.g., the genome of a replication-defective virus may be encapsidated by proteins encoded by a replication-competent helper virus.

pseudotyping (*genetic engineering*) A technique in which e.g. the envelope proteins of a retrovirus are replaced by the G glycoprotein of vesicular stomatitis virus; the pseudotyping of retroviruses enables them to infect a wider range of types of cell. [Example of use (in gene therapy): PNAS (1999) *96* 10379–10384.]

Some viruses used for gene delivery (in GENE THERAPY) have the disadvantage that they infect *too many* types of cell. Thus, if a gene is to be delivered to a specific type of cell, then viruses with a wide spectrum of target cells may have to be administered in high doses in order to allow for losses during interaction with other types of cell; such high doses of virus may expose the patient to the risk of adverse reactions. One such virus with a wide tropism is adenovirus type 5; in vitro, plasmid-mediated modification of the virion's attachment site (pseudotyping) has been able to de-target the virus from its (widely expressed) host receptor, allowing the development of an engineered virion which is more specifically targeted [JV (2001) *75* 2972–2981].

pseudovirion A 'virion' which contains only host-cell nucleic acid. (See e.g. TRANSDUCTION.)

pseudovitellus The mycetome of an aphid.

pseudo-wild phenotype See e.g. SUPPRESSOR MUTATION.

ψ Pseudouridine (see TRNA).

[*PSI*$^+$] See PRION (sense 2).

psicofuranine (9-β-D-psicofuranosyladenine) An antitumour, antimicrobial NUCLEOSIDE ANTIBIOTIC from *Streptomyces hygroscopicus*. It (and its close relative *decoyinine*) inhibits GMP synthesis by mimicking the allosteric inhibition by adenosine of XMP aminase [see Appendix V(a)].

psilocin See HALLUCINOGENIC MUSHROOMS.

Psilocybe See AGARICALES (Strophariaceae) and HALLUCINOGENIC MUSHROOMS.

psilocybin See HALLUCINOGENIC MUSHROOMS.

Psilolechia See LECIDEA.

psittacine beak and feather disease A disease of psittacines due to an ssDNA circovirus; feather, beak and immune cells are killed and the birds may die of other infections.

psittacosis (parrot fever; ornithosis*) An acute infectious disease which primarily affects birds – both psittacine (parrots, parakeets etc) and non-psittacine (e.g. pigeons, sparrows, canaries, domestic fowl) – but which is also transmissible to man; it is caused by *Chlamydia psittaci*. In birds, *C. psittaci* usually causes a mild or asymptomatic infection but can cause a fatal disease. The spleen and liver may be enlarged, with areas of focal necrosis; symptoms may include e.g. ocular and nasal discharge, weakness and diarrhoea. *C. psittaci* is present in the faeces, nasal secretions, feathers etc of infected birds. Man usually becomes infected by inhalation of dust from contaminated birds; person-to-person transmission is also known. Incubation period: 4–15 days. The disease may be mild and self-limiting, resembling influenza, but it can be fatal; symptoms may include e.g. fever, cough, headache, photophobia, myalgia, and lobular pneumonitis with an alveolar and interstitial exudate (cf. PRIMARY ATYPICAL PNEUMONIA). *Lab. diagnosis*: serological tests. *Chemotherapy*: e.g. tetracyclines. [Review: BMB (1983) *39* 163–167.]

*Although the terms 'psittacosis' and 'ornithosis' are used by some authors as synonyms for the disease caused by *C. psittaci* in both psittacine and non-psittacine birds and in man, others reserve the term 'ornithosis' for the disease in (i) psittacine and non-psittacine birds (i.e., excluding man); (ii) non-psittacine birds and humans infected from non-psittacine birds; or (iii) non-psittacine birds only.

Psora See LECIDEA.

psoralens (furocoumarins) Three-ringed heterocyclic compounds found e.g. in certain fungi and tropical fruits. A psoralen binds intercalatively (see INTERCALATING AGENT) and non-covalently to DNA or RNA in the dark. On irradiation of this complex with light (365 nm) the psoralen becomes covalently linked to a pyrimidine residue in one strand; it may then react photochemically with a second pyrimidine to form a cross-link. Owing to the small size of a psoralen molecule, two pyrimidines can be cross-linked only when they are very close (e.g. on complementary strands in a double helix or a hairpin loop). Psoralens are useful in studies on the secondary structures of DNA and RNA molecules [ARBB (1981) *10* 69–86].

psoroptic mange A disease of animals caused by mite bites (see e.g. SHEEP SCAB).

***pst* operon** See PHO REGULON.

***Pst*I** A RESTRICTION ENDONUCLEASE from *Providencia stuartii*; CTGCA/G.

PSTV Potato spindle tuber VIROID.

Psychrobacter A genus of Gram type-negative bacteria formerly proposed as a member of the family Neisseriaceae [IJSB (1986) *36* 388–391] but later classified within the family Moraxellaceae [IJSB (1991) *41* 310–319].

Strains of *P. immobilis* are oxidase +ve and catalase +ve coccobacilli found e.g. in fish and processed meat products; most are psychrotrophic (optimum growth temperature: 20–25°C). GC% 44–46.

psychrophile An organism which grows optimally at or below 15°C, which has an upper limit for growth of ca. 20°C, and which has a lower limit for growth of 0°C or below [Bact. Rev. (1975) *39* 144–167]. (This is currently the most widely accepted definition; however, some authors still use the term, very loosely, to include PSYCHROTROPHS.) Psychrophilic organisms include certain algae and fungi, a number of Gram-negative bacteria (e.g. some species of *Pseudomonas* and *Vibrio*), and a few Gram-positive bacteria (e.g. some *Clostridium* spp). [Ecology and physiology of psychrophilic bacteria: Book ref. 191, pp. 1–23.] (cf. MESOPHILE and THERMOPHILE.)

psychrotroph An organism which can grow at low temperatures (e.g. 0–5°C) but which has an optimum growth temperature >15°C and an upper limit for growth >20°C. (cf. PSYCHROPHILE.) Psychrotrophs include e.g. certain algae and fungi, and various Gram-negative and Gram-positive bacteria.

pT181 plasmids See INCOMPATIBILITY.

Pterosperma See MICROMONADOPHYCEAE.

pterostilbene A stilbene PHYTOALEXIN produced by the grapevine (*Vitis vinifera*).

pteroylglutamic acid *Syn.* FOLIC ACID.

PTFE Polytetrafluoroethylene (Teflon).

PTLV Primate T-cell leukaemia virus: see SIMIAN T-CELL LEUKAEMIA VIRUS.

PTS (phosphoenolpyruvate-dependent phosphotransferase system) A TRANSPORT SYSTEM, found in both Gram-negative and Gram-positive bacteria, in which the substrate (e.g. a hexose, disaccharide or hexitol) is phosphorylated in a reaction which is an essential part of the transport process. The source of energy (and of phosphate) is normally phosphoenolpyruvate (PEP); the formula of PEP is given in Appendix I(a). (Certain enzymes can generate PEP from ATP or GTP.)

In members of the Enterobacteriaceae, sugars whose uptake is PTS-dependent ('PTS sugars') include glucose, fructose, mannose, mannitol, sorbitol, glucosamine, *N*-acetylglucosamine and *N*-acetylmannosamine. (cf. LACTOSE.) (Lactose is a PTS sugar e.g. in *Staphylococcus aureus*.)

Because sugars are often metabolized via their phosphate derivatives – e.g. glucose 6-phosphate in Appendix I(a) – phosphorylation during uptake is a positive aspect of PTS transport. (Note, however, that a *non-metabolizable* substrate may be phosphorylated and transported by a PTS – e.g. the sugar analogue methyl α-glucoside can be transported by the glucose-PTS of *Escherichia coli* and of *Clostridium pasteurianum*.) (See also STREPTOZOTOCIN.)

A PTS consists of both cytoplasmic (soluble) and membrane-bound protein components; the early nomenclature for PTS components (enzymes I, II and III) has been somewhat modified. In the simplest case (mannitol uptake in *E. coli*: see diagram), PEP feeds energy and phosphate into the PTS by sequential phosphorylation of two soluble energy-coupling proteins: enzyme I (here designated I) and a histidine-containing protein, HPr (here designated H). (Genes encoding I and H occur in the same (*pts*) operon.) The phosphate and energy are transferred from H to a sugar-specific permease (the so-called II complex) in the cytoplasmic membrane; the permease includes a transmembrane component (IIC), involved in the binding and translocation of substrate, and two phosphorylation sites (IIA, IIB) – phosphorylation and *concomitant* transport (uptake) of mannitol apparently occurring only at the IIB site.

Uptake of glucose by PTS (see diagram) differs in that the IIA component is soluble – adding a further cytoplasmic step to the chain of phosphorylation.

Phosphorylated forms of components I and H have high free energies of hydrolysis (approximately that of PEP). The I and H components are not substrate-specific, i.e. they are required for the phosphorylation and transport of all PTS substrates; thus, mutation causing loss of one or both of these components will be pleiotropic, i.e. the mutant will be unable to transport any PTS-dependent substrate. In contrast, inactivation of a given II complex will affect the transport of only the relevant permease-specific substrate(s).

[PTS (review): Mol. Microbiol. (1994) *13* 755–764.]

In enterobacteria, some complex II components apparently contain oxidizable sulphhydryl groups, and it seems that only reduced forms are active [Biochem. (1985) *24* 47–51]; this may indicate an energy-dependent regulation of the PTS, and may account for the observed pmf-related regulation of PTS activity.

In some II complexes the components are grouped in other ways. For example, the fructose permease of enteric bacteria consists of a membrane-bound protein, containing IIC and IIB, and a separate protein containing the functions of both IIA and HPr. In *Rhodobacter capsulatus*, the fructose permease includes a single protein which carries out the functions of IIA, HPr and enzyme I.

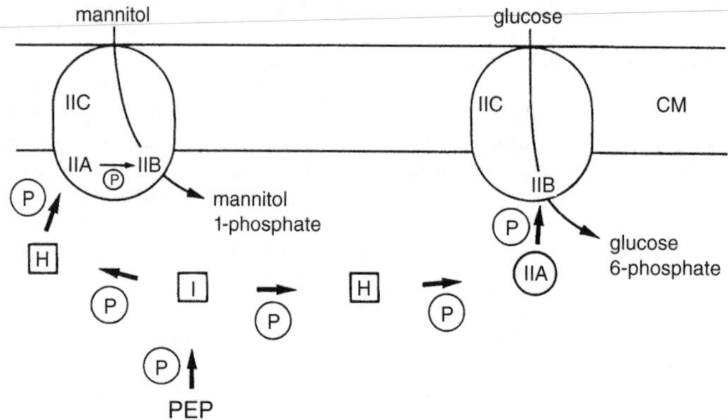

PTS (phosphoenolpyruvate-dependent phosphotransferase system). Transport of mannitol and glucose across the cytoplasmic membrane (CM) by the PTS in *Escherichia coli* (simplified, diagrammatic); transport is from the periplasm (above) to the cytoplasm (below). Note that mannitol and glucose are taken up by different membrane-associated systems (*permeases*).

The source of energy for transport is phosphoenolpyruvate (PEP). Energy is fed into the system by the sequential transfer of phosphate (P) from PEP to enzyme I (shown here as I) and then to the HPr protein (shown here as H). In mannitol transport, phosphate is transferred from H direct to a membrane-bound permease (enzyme II complex, here designated II). The permease for mannitol consists of three domains: IIA, IIB and IIC. IIC is a hydrophobic, transmembrane domain which binds the specific substrate and appears to contain, or contribute to, a channel through which the substrate is translocated. The IIA and IIB domains each contain a phosphorylation site, but phosphorylation of the substrate seems directly to involve only the IIB domain.

In the glucose permease, IIA is a separate (cytoplasmic, soluble) protein; however, the sequence of phosphorylation is the same as that in the mannitol permease, i.e. IIA → IIB → substrate.

In all cases, phosphorylation of the substrate is an integral part of the transport mechanism.

Reproduced from *Bacteria*, 5th edition, Figure 5.12, page 91, Paul Singleton (1999) copyright John Wiley & Sons Ltd, UK (ISBN 0471-98880-4) with permission from the publisher.

In the proposed nomenclature of PTS permeases [JB (1992) *174* 1433–1438], the mannitol permease of *E. coli* is designated $IICBA^{Mtl,Eco}$, and the glucose permease is $IICB^{Glc,Eco}$ + $IIA^{Glc,Eco}$ (see diagram).

In some cases a cell responds chemotactically to substrates taken up by a PTS (see CHEMOTAXIS). Moreover, a PTS can control some aspects of carbon *utilization*; thus, PTS phosphate carriers (see diagram) can regulate the activity of certain operons involved in the catabolism of carbon sources – sometimes via regulator sites designated 'PTS regulation domains' (PRD) [Mol. Microbiol. (1998) *28* 865–874] (see CATABOLITE REPRESSION).

PTS regulation domain (PRD) See CATABOLITE REPRESSION.

***pts* operon** See PTS.

Ptychodiscus See RED TIDE and BREVETOXINS.

Pu Purine (nucleotide).

PU Palindromic unit: see REP SEQUENCE.

pubescent Downy.

Puccinia A genus of typically macrocylic and heteroxenous rust fungi (class UREDINIOMYCETES) which characteristically form cupulate aecia, non-peridiate uredia, and pedicellate, two-celled teliospores. *Puccinia* species are parasitic on a wide range of plant hosts, including monocotyledonous and dicotyledonous species. See UREDINIOMYCETES for the life cycle of *P. graminis tritici*. (See also BLACK STEM RUST, BROWN RUST, CHRYSANTHEMUM WHITE RUST, CROWN RUST and YELLOW RUST.)

Pucciniastrum See UREDINIOMYCETES and PERIDERMIOID.

puerperal fever (childbed fever; puerperal sepsis) An acute, febrile condition, following childbirth, due to infection of the uterus and/or adjacent regions – usually by streptococci but sometimes by e.g. *Clostridium* spp; complications (e.g. septicaemia) may occur, and mortality rates are high in untreated cases.

puerperal mastitis See MASTITIS.

puerperal sepsis *Syn.* PUERPERAL FEVER.

puffball See LYCOPERDALES and GASTEROMYCETES (Tulostomatales).

puffing (*mycol.*) See ASCOSPORE.

***pul* genes** See PROTEIN SECRETION (type II systems).

pullorum disease (bacillary white diarrhoea, BWD) An acute infectious POULTRY DISEASE caused by *Salmonella pullorum*; it affects mainly chicks, but also occurs e.g. in turkey poults and pheasants. Symptoms: e.g. lethargy, anorexia, white diarrhoea, weakness and neurological symptoms; lungs or joints may be affected. Mortality rates may be high. Survivors become chronic carriers: the pathogen localizes in the ovaries and infects a proportion of the eggs; chicks hatching from these eggs develop the disease, and infection can then spread to other chicks – either by ingestion of food contaminated with faeces from infected (or carrier) birds, or by inhalation of fluff from infected chicks. *Lab. diagnosis*: identification of *S. pullorum* from e.g. liver, spleen or ovaries. Carriers are detected by an agglutination test (cf. FOWL TYPHOID). *Treatment*: e.g. furazolidone, furaltadone.

pullulan An extracellular, water-soluble, linear D-glucan synthesized by *Aureobasidium* (*Pullularia*) *pullulans*; it consists predominantly of MALTOTRIOSE units linked by (1 → 6)-α-glucosidic bonds.

pullulanase See DEBRANCHING ENZYMES.

Pullularia *Syn.* AUREOBASIDIUM.

pullulation Asexual reproduction by BUDDING.

pulmonary adenomatosis (*vet.*) *Syn.* JAAGSIEKTE.

pulpwood spoilage See PAPER SPOILAGE.

pulpy kidney An acute toxaemic disease of ruminants (especially lambs) caused by growth and toxigenesis by *Clostridium perfringens* type D in the intestines. Symptoms: diarrhoea, convulsions, paralysis, sudden death.

pulque A white, viscous, acidic, alcoholic Mexican beverage made by fermenting the juices of *Agave* spp. The alcohol is produced by *Saccharomyces cerevisiae* ('*S. carbajali*') and *Zymomonas mobilis*. Acidity and viscosity are contributed, respectively, primarily by *Lactobacillus plantarum* and dextran-producing strains of *Leuconostoc*. Pulque is rich in B vitamins. *Tequila* is made by the distillation of pulque. (See also WINE-MAKING; SPIRITS.)

pulse–chase technique A technique in which e.g. living cells are exposed to a pulse of labelled substrate for a fixed period of time – followed immediately by a high concentration of the unlabelled substrate (the 'chase' phase); the chase effectively stops further uptake/use of the labelled substrate, thus defining the duration of the pulse. The technique has been used e.g. to follow the intracellular movement of a secreted protein from the time of its synthesis to the time of secretion; for this experiment, the period of the pulse is increased in a stepwise manner.

pulsed electric field (PEF) A technique involving electric fields of typically >20 kV/cm applied in pulses of up to several μs (microseconds) for the inactivation of susceptible microorganisms in a liquid medium; exposure to PEF treatment appears to bring about irreversible structural damage to the cells. In one experiment, the use of 2500 pulses of a 30 kV/cm field at 50°C reduced the level of viable cells of *Mycobacterium paratuberculosis* in milk by ca. 5.9 $\log_{10}$ cfu/ml [AEM (2001) *67* 2833–2836].

This technique may have applications e.g. in the food industry.

pulsed-field gel electrophoresis See PFGE.

pulverulent Powdery.

pulvinate Cushion-like; forming a swelling.

pulvomycin See POLYENE ANTIBIOTICS (b).

punctate Having surface dot(s), pore(s) etc.

punctiform Dot-like.

Punta Toro virus See PHLEBOVIRUS.

PUO Pyrexia (fever) of unknown origin.

pure culture See CULTURE.

purified protein derivative See TUBERCULIN.

purine nucleotide biosynthesis See NUCLEOTIDE and Appendix V(a).

purity plate A plate inoculated from a given culture and subsequently incubated to determine whether the culture was pure or contaminated – the presence of two or more types of colony on the purity plate indicating a contaminated culture.

puromycin (6-dimethyl-3′-deoxy-3′-*p*-methoxyphenylalanylamino adenosine) A NUCLEOSIDE ANTIBIOTIC obtained from *Streptomyces alboniger* or synthesized chemically; it is active against prokaryotes and eukaryotes. It acts primarily as an inhibitor of PROTEIN SYNTHESIS, acting as an analogue of the aminoacyl-adenyl portion of an aminoacyl-tRNA. When a peptidyl-tRNA occupies a ribosomal P site, puromycin can enter the A site and a peptide bond can be formed between the amino group of puromycin and the C-terminus of the peptide. However, puromycin does not have a hydrolysable ester bond like that linking the aminoacyl group to a tRNA, and puromycin binds only weakly to the ribosome; thus, nascent polypeptide chains (in which the C-terminus is blocked by puromycin) are released.

purple bacteria (1) Bacteria of the RHODOSPIRILLINEAE. (2) A category (= PROTEOBACTERIA – q.v.) within the domain BACTERIA originally distinguished on the basis of 16S rRNA sequence analysis [SAAM (1985) *6* 143–151]. Organisms in this category are believed to have evolved from photosynthetic bacteria similar to members of the Rhodospirillineae; they are divided into the alpha, beta, gamma, delta and epsilon subdivisions.

purple laver See LAVER.

purple membrane In some strains of *Halobacterium salinarium*: functionally and structurally differentiated, purple-pigmented regions of the cytoplasmic membrane which develop under microaerobic or anaerobic conditions in the light; purple membrane may occupy up to 50% of the total membrane area – the remaining cytoplasmic membrane (which contains red carotenoid pigments) often being called the 'red membrane'. The purple membrane utilizes light energy to effect transmembrane translocation of protons, thus generating or augmenting a *proton motive force* (pmf) (see CHEMIOSMOSIS); cells can therefore grow as phototrophs under anaerobic conditions (see HALOBACTERIUM).

The protein component of the purple membrane consists almost entirely of BACTERIORHODOPSIN (q.v.) but includes another retinal-containing purple pigment, HALORHODOPSIN, and (in at least some cases) SLOW-CYCLING RHODOPSIN. (Strains of *H. salinarium* which cannot synthesize retinal may form an apoprotein-containing 'white membrane'.)

The lipids of the purple membrane are similar to those in the remainder of the cytoplasmic membrane, but they include a glycosulpholipid which is apparently unique to the purple membrane.

While photocycling in bacteriorhodopsin causes an outward efflux of protons, photocycling in halorhodopsin appears to cause an inward pumping of chloride ions (rather than an outward pumping of sodium ions as was previously thought).

The combined activities of light-energized bacteriorhodopsin and halorhodopsin may account for the various ion fluxes across the membrane and for the complex pattern of changes in transmembrane pH differential.

When whole cells are illuminated, the steady-state or 'resting' membrane potential (typically ca. 100–110 mV) is increased by ca. 10–40 mV, and the 'resting' transmembrane pH differential (ca. 2 units at pH 5, ca. 0 units at pH 8) changes by 0–1 unit; the increased membrane potential serves to energize e.g. photophosphorylation (phosphorylation of ADP), uptake of potassium ions and efflux of sodium ions.

purple non-sulphur bacteria See RHODOSPIRILLACEAE.

purple sulphur bacteria See CHROMATIACEAE.

purpura (*med.*) A condition characterized by reddish or purple patches due to haemorrhage into skin or mucous membrane and underlying tissues. A pin-point-sized haemorrhage is called a *petechia*, a larger one is known as an *ecchymosis*.

purse (*mycol.*) See CYATHUS.

purulent Containing, consisting of, or forming PUS.

pus A thick, usually yellowish, fluid product of inflammation formed at the site of an infection. Pus contains proteins, leucocytes, cell fragments, and e.g. living and/or dead bacteria. (cf. BLUE PUS.)

pustulan A (1 → 6)-β-D-glucan produced by *Lasallia* (*Umbilicaria*) *pustulata*.

pustule A small, raised, PUS-filled skin lesion. (cf. VESICLE.)

pusule In many DINOFLAGELLATES: an intracellular vesicular structure which opens to the exterior via a canal and a pore; it is believed to function in osmoregulation but is apparently non-contractile (cf. CONTRACTILE VACUOLE).

putidaredoxin See FERREDOXINS.

putrefaction The microbial degradation of proteinaceous materials with the formation of evil-smelling products such as amines (e.g. cadaverine and putrescine formed by the decarboxylation of lysine and ornithine, respectively) and HYDROGEN SULPHIDE (from sulphur-containing amino acids). Putrefaction is typically an anaerobic process and may be carried out e.g. by proteolytic clostridia; however, certain pseudomonads can bring about the aerobic putrefaction of e.g. meat.

putrescine See POLYAMINES, DECARBOXYLASE TESTS and PUTREFACTION.

Puumala virus See HANTAVIRUS.

pv. Pathovar: see VARIETY.

PVA See POLYVINYL ALCOHOL FIXATIVE.

***pvdS* gene** See PYOVERDIN.

PVP–iodine See IODINE (a).

***Pvu*I** See RESTRICTION ENDONUCLEASE (table).

PWM POKEWEED MITOGEN.

Py Pyrimidine (nucleotide).

pyaemia (pyemia) A particular form of SEPTICAEMIA in which *pyogenic* bacteria are disseminated via the bloodstream.

pycnidiospore A CONIDIUM formed in a PYCNIDIUM.

pycnidium A hollow, typically spherical, flattened, or flask-shaped fungal fruiting body (generally ca. 100–500 μm in size) within which conidia (= pycnidiospores) are produced, and which, at maturity, is – or becomes – open to the environment e.g. via one or more pores. Pycnidia, which may superficially resemble perithecia or cleistothecia, are formed by many members of the SPHAEROPSIDALES. According to species, a pycnidium may be thick-walled or thin-walled, unilocular or multilocular, black, brown or brightly coloured (e.g. orange or yellow), glabrous or setose; it may develop superficially or may be immersed – or partly immersed – within the substratum or within a fungal stroma. [Symphogenous development of pycnidia in *Chaetomella acutiseta*: CJB (1980) *58* 1129–1137.]

pycniospore A mononucleate, haploid SPERMATIUM formed in a pycnium (see UREDINIOMYCETES).

pycnium See UREDINIOMYCETES stage 0.

Pycnomonas A subgenus of TRYPANOSOMA within the SALIVARIA. *T.* (*P.*) *suis* occurs in wild and domestic pigs in parts of Africa (e.g. Zaire) and is transmitted by *Glossina* spp. The organisms are stout trypomastigotes, ca. 10–20 μm, in which there is a subterminal kinetoplast and a free flagellum.

pycnosis *Syn.* PYKNOSIS.

pycnospore An obsolete term for a pycnidiospore or a pycniospore.

Pycnothyriales An order of fungi (class COELOMYCETES) in which the vegetative mycelium and conidiomata may be superficial or immersed in the substratum; conidiomata may be e.g. multiloculate, and in some species each is elevated on a central columnar structure.

pyelitis See PYELONEPHRITIS.

pyelonephritis Inflammation of the kidney (nephritis) and the renal pelvis (pyelitis). (The renal pelvis is the region at the kidney–ureter junction.) Pyelitis and/or nephritis may be caused by infection with e.g. enterobacteria (particularly strains of *Escherichia coli*), staphylococci or streptococci; infection may occur either via the bloodstream or via the urethra, bladder and ureter. Symptoms usually include fever and pain in the loins, and there may be bacteraemia.

(See also CYSTITIS, GLOMERULONEPHRITIS, UPEC and URINARY TRACT INFECTION.)

pyemia See PYAEMIA.

PYG medium Peptone–yeast extract–glucose medium, used e.g. for culture of anaerobes. It contains peptone, yeast extract, cysteine-HCl, NaCl, resazurin, and glucose (added after autoclaving), and may be used as a broth (PYGB) or as an agar medium (PYGA). [Recipe: Book ref. 53, p. 1428.]

Py-GLC PYROLYSIS GAS–LIQUID CHROMATOGRAPHY.

pyknosis (pycnosis) (*histopathol.*) The shrinkage of a cell's nucleus, resulting in the formation of a smaller, densely staining body. (cf. KARYOLYSIS; KARYORRHEXIS.)

pyo- Prefix meaning *pus*.

pyocin (pyocine; aeruginocin) Any BACTERIOCIN produced by *Pseudomonas aeruginosa*; pyocins are bactericidal – log-phase target cells apparently being the most sensitive. R-type pyocins (MWt ca. 10^7) resemble phage tails and are relatively insensitive to proteases; S-type pyocins (MWt ca. 10^5) are amorphous and are sensitive to proteases.

Pyocin typing: see BACTERIOCIN TYPING.

pyocyanin (pyocyanine) A blue-green, water-soluble and non-fluorescent phenazine pigment synthesized by *Pseudomonas aeruginosa* from chorismate (an intermediate in the biosynthesis of aromatic amino acids: see Appendix IV(f)); it becomes red-purple when acidified. Pyocyanin has oxygen-dependent antimicrobial activity which appears to involve the (enzyme-independent) oxidation of NADH with production of SUPEROXIDE and H_2O_2 [JB (1980) *141* 156–163]; it also inhibits the beating of human respiratory cilia in vitro [JCI (1987) *79* 221–229].

(cf. CYANOMYCIN; see also IRON.)

pyoderma Any purulent skin disease – e.g. IMPETIGO.

pyofluorescein *Syn.* PYOVERDIN.

pyogenic Able to cause formation of pus.

pyomelanin See PSEUDOMONAS (*P. aeruginosa*).

pyorubin See PSEUDOMONAS (*P. aeruginosa*).

pyoverdin (pyoverdine; pyofluorescein; 'fluorescein') Any of a range of yellowish, fluorescent, water-soluble pigments produced by certain species of *Pseudomonas* under iron-deficient conditions; the pyoverdins act as SIDEROPHORES.

A model for the iron-regulated induction of pyoverdin in *P. aeruginosa* proposes that, under iron-deficient conditions, the (regulatory) Fur protein dissociates from a transcriptional control site upstream of the *pvdS* gene, allowing synthesis of PvdS; PvdS is a sigma factor which promotes transcription of genes involved in the synthesis of pyoverdin [JB (1996) *178* 2299–2313].

The pyoverdin of *P. fluorescens* contains a quinoline chromophore linked to a cyclic peptide; this compound represses synthesis of another siderophore, quinolobactin [AEM (2000) *66* 487–492].

[Functional analysis of PvdS: JB (2000) *182* 1481–1491.]

PYR A substrate (*N,N*-dimethylaminocinnamaldehyde) which is apparently hydrolysed by all strains of *Streptococcus pyogenes* but by no other streptococci. In an identification test for *S. pyogenes*, the organisms are grown aerobically on agar plates overnight at 35°C; when PYR is added to the surface growth, colonies of *S. pyogenes* become red, while those of other streptococci become yellow or do not change colour.

PYR is also hydrolysed by strains of ENTEROCOCCUS.

Pyr(6–4)Pyo See ULTRAVIOLET RADIATION.

pyracarbolid (2H-3,4-dihydro-6-methylpyran-5-carboxanilide) An agricultural systemic ANTIFUNGAL AGENT which is active e.g. against various rusts, smuts, and *Rhizoctonia* spp.

Pyramimonas See MICROMONADOPHYCEAE.

pyrazinamide 1,4-Diazine carboxamide: a drug active against *Mycobacterium tuberculosis* in vivo (including cells within phagosomes) and (at e.g. pH 5.6) in vitro. *M. bovis* is not susceptible.

Pyrazinamide is a prodrug which is converted, in vivo, to the active compound pyrazinoic acid; conversion involves the enzyme pyrazinamidase, which is encoded by gene *pncA* in *Mycobacterium tuberculosis*. Most pyrazinamide-resistant clinical isolates of *M. tuberculosis* have been found to have mutations in the *pncA* gene.

Expression of *M. smegmatis* pyrazinamidase in mutant (pyrazinamide-resistant) *M. tuberculosis* confers hypersensitivity to pyrazinamide and related amides [JB (2000) *182* 5479–5485].

pyrazinoic acid See PYRAZINAMIDE.

pyrazophos An antifungal ORGANOPHOSPHORUS COMPOUND which is active against various powdery mildews. It is readily absorbed by leaves and shoots, and is translocated within the plant; it is poorly absorbed by roots.

***pyrBI, pyrE* genes** See OPERON.

pyrenoid A dense proteinaceous body which occurs within a CHLOROPLAST; in some cases a pyrenoid may be 'stalked' – protruding from the chloroplast but always contained by the chloroplast envelope. The pyrenoid appears to be the site of synthesis of storage polysaccharide; the enzyme RuBisCO has been isolated as the main protein component of the pyrenoids in some species.

Pyrenomycetes A class of ascomycetes characterized by the formation of unitunicate asci in a perithecium or a cleistothecium. The class is not recognized in most modern taxonomic schemes.

Pyrenophora See DOTHIDEALES.

Pyrenula See PYRENULALES.

Pyrenulales An order of saprotrophic and lichenized fungi of the ASCOMYCOTINA. Ascocarp: ascolocular and perithecioid (see ASCOCARP). Asci: bitunicate. Ascospores: multiseptate. Genera: e.g. *Acrocordia, Pyrenula*.

Pyricularia A genus of fungi of the HYPHOMYCETES; teleomorph: *Magnaporthe*. *P. oryzae*, the causal agent of rice BLAST DISEASE, forms a septate mycelium and gives rise to typically pyriform, commonly biseptate conidia.

pyricularin *Syn.* PIRICULARIN.

pyridine haemochrome See CYTOCHROMES.

pyridine nucleotide coenzymes NAD (q.v.) and NADP.

pyridine nucleotide salvage cycle See NAD.

pyridoxal phosphate See PYRIDOXINE.

pyridoxamine See PYRIDOXINE.

pyridoxine (pyridoxin; vitamin B_6; pyridoxol) A water-soluble and photolabile VITAMIN: 2-methyl-3-hydroxy-4,5-di-(hydroxymethyl)pyrimidine; related compounds such as pyridoxal and pyridoxamine (bearing, respectively, –CHO and $-CH_2NH_2$ at the 4-position) may also be referred to as 'vitamin B_6'. The coenzyme form of the vitamin is *pyridoxal 5-phosphate* ('codecarboxylase'); *pyridoxamine 5-phosphate* is involved in some reactions. The coenzyme is important in amino acid metabolism – e.g., in reactions involving transamination, decarboxylation, racemization (interconversion of D-and L-amino acids), serine–glycine interconversion, etc.

Certain microorganisms (e.g. *Clostridium* spp, *Lactobacillus* spp, some yeasts, *Crithidia fasciculata*, *Tetrahymena* spp) require an exogenous supply of pyridoxine; the various derivatives are not necessarily equally effective in this respect.

pyridoxol *Syn.* PYRIDOXINE.

pyriform Pear-shaped.

pyrimethamine (2,4-diamino-5-(*p*-chlorophenyl)-6-ethylpyrimidine) A FOLIC ACID ANTAGONIST used – usually in combination with e.g. a sulphonamide – in the treatment of MALARIA and of toxoplasmosis and other coccidioses; it is active against the trophozoite, but resistance tends to develop readily.

pyrimidine dimer See ULTRAVIOLET RADIATION.

pyrimidine nucleotide biosynthesis See NUCLEOTIDE and Appendix V(b).

pyrimine See SERRATIA.

Pyrocystis See BIOLUMINESCENCE.

Pyrodictium A genus of chemolithoautotrophic prokaryotes (order THERMOPROTEALES) isolated from a submarine volcanic region. Growth occurs as a network of filaments associated with 'discs', each ca. 0.3–2.5 μm in diameter. Optimum growth temperature: 105°C. Growth occurs in up to 12% NaCl.

Pyrodinium See BIOLUMINESCENCE.

pyrogen (1) A fever-inducing agent. (2) LIPOPOLYSACCHARIDE.

pyrogenic exotoxin C See TOXIC SHOCK SYNDROME.

pyrogram See PYROLYSIS GAS–LIQUID CHROMATOGRAPHY.

pyrolysate See PYROLYSIS.

pyrolysis A technique, used e.g. in microbial taxonomy, in which a small sample of a microbial culture is thermally degraded, in a vacuum or in an inert atmosphere (e.g. nitrogen), to form a range of volatile low-MWt compounds characteristic of the organism(s); the products (*pyrolysate*) may be analysed by e.g. Py-GLC or Py-MS (see following entries). In *filament pyrolysers* the sample may be coated onto a platinum filament which is then heated by the transmission of an electric current. (See also CURIE POINT PYROLYSIS.) Laser pyrolysers have also been used.

pyrolysis gas–liquid chromatography (Py-GLC) An analytical technique, intended e.g. for the rapid characterization of microorganisms, in which microbial cells are subjected to PYROLYSIS and the pyrolysate components are separated by gas–liquid CHROMATOGRAPHY and recorded as a complex trace (*pyrogram*).

pyrolysis–mass spectrometry (Py-MS) An analytical technique, intended e.g. for the rapid characterization of microorganisms, in which microbial cells are subjected to PYROLYSIS and the pyrolysate is analysed by mass spectrometry.

Pyronema A genus of fungi (order PEZIZALES) which occur e.g. on burnt soils. The apothecium (commonly 1–2 mm diam.), which develops on a subiculum, may be white, pale orange or pinkish. Ascospores: uninucleate, colourless, eguttulate.

pyronin A red basic xanthene DYE.

pyrophosphate (inorganic pyrophosphate; PPi) Pyrophosphate ($P_2O_7^{4-}$) is involved in various metabolic reactions in both prokaryotic and eukaryotic organisms. In many organisms PPi is released from nucleoside triphosphates e.g. during the biosynthesis of nucleic acids and proteins; in some organisms it is synthesized by photophosphorylation (e.g. in *Rhodospirillum rubrum*) or OXIDATIVE PHOSPHORYLATION (e.g. in animal and plant mitochondria) involving a PROTON PPASE.

In e.g. *Propionibacterium freudenreichii* (formerly *P. shermanii*) and *Entamoeba histolytica*, PPi is used for the (reversible) phosphorylation of fructose 6-phosphate catalysed by PPi:phosphofructose dikinase – a reaction which appears to be important in the glycolytic pathway in these organisms. In DESULFOTOMACULUM (q.v.) PPi is used for the SUBSTRATE-LEVEL PHOSPHORYLATION of ADP; sulphate-reducing bacteria which lack the enzyme acetate:PPi phosphotransferase (e.g. *Desulfovibrio*) cannot use PPi in this way.

pyrophosphotransferase See STRINGENT CONTROL (1).

Pyrosequencing™ A rapid method for sequencing short DNA templates based on detection of the pyrophosphate released when a series of (known) nucleotides sequentially extend a primer [principle, details: Science (1998) *281* 363–365].

Pyrrhophyta *Syn.* PYRROPHYTA.

pyrrolo-(1,4)-benzodiazepine antibiotics See ANTHRAMYCIN.

pyrrolo-quinoline quinone See QUINOPROTEIN.

Pyrrophyta (Pyrrhophyta) A division of the algae which includes the DINOFLAGELLATES and sometimes also the CRYPTOPHYTES.

Pyrsonympha See OXYMONADIDA.

pyruvate An intermediate in a wide range of (aerobic and anaerobic) metabolic pathways; for examples of formation and fates of pyruvate see Appendices I, II, III and IV(b, e).

pyruvate carboxylase A BIOTIN ENZYME which is an important anaplerotic enzyme (see Appendix II (b)). [Structure: FEMS Reviews (1993) *104* 330–332.]

pyruvate dehydrogenase complex See e.g. Appendix II(a).

pyruvate synthase See TCA CYCLE and REDUCTIVE TRICARBOXYLIC ACID CYCLE.

pyruvic phosphoroclasm See PHOSPHOROCLASTIC SPLIT.

Pythium A genus of aquatic and terrestrial fungi (order PERONOSPORALES) which include saprotrophs, parasites, and pathogens of higher plants (see e.g. DAMPING OFF), certain algae (see ALGAL DISEASES), and mammals (see EQUINE PHYCOMYCOSIS). (See also MYCOPARASITE.) The thallus is a well-developed, branching mycelium, and sporangia are borne on sporangiophores which typically resemble somatic hyphae.

The generalized life cycle of a representative species, *P. debaryanum*, is as follows. A reniform zoospore (flagella arising from the concave surface) encysts and later forms a germ tube which penetrates the host plant – giving rise to both intercellular and intracellular mycelium; haustoria are not formed. Later, spherical or ovoid, non-deciduous sporangia develop, in terminal or intercalary positions, on undifferentiated sporangiophores. A sporangium germinates by extruding a thin tube which terminates in a thin-walled, bubble-like vesicle; the multinucleate contents of the sporangium flow into the vesicle where they differentiate into zoospores.

In sexual reproduction, a spherical oogonium and a smaller, club-shaped or elongated antheridium develop on the hyphae in a terminal or intercalary position. (Most species of *Pythium* appear to be homothallic, but at least one species, *P. sylvaticum*, is heterothallic.) Meiosis is generally thought to occur in the gametangia. A fertilization tube develops between antheridium and oogonium, and a male nucleus passes into the oogonium where fertilization occurs. The resulting thick-walled oospore germinates to form a germ tube which may terminate in a zoosporangium.

***Pythium* red rot** See ALGAL DISEASES.

Pythonella See EIMERIORINA.

pyuria The presence of pus in the urine. 'Sterile pyuria' (= 'abacterial pyuria') is a condition in which pus is repeatedly found in samples of urine from which no bacteria can be cultured by routine methods; common causes are probably chlamydiae or mycoplasmas.

pYV plasmid See VIRULON.

1. Words in SMALL CAPITALS are cross-references to separate entries.
2. Keys to journal title abbreviations and Book ref. numbers are given at the end of the Dictionary.
3. The Greek alphabet is given in Appendix VI.
4. For further information see 'Notes for the User' at the front of the Dictionary.

Q

Q (1) See TEMPERATURE COEFFICIENT. (2) QUEUOSINE. (3) Glutamine (see AMINO ACIDS).

Q-band (in ESR) See ELECTRON SPIN RESONANCE.

Q bases Bases formed by the modification of guanine in tRNA; they contain a pentenyl ring attached (via NH) to the methyl group of *N*-methylguanine. An example is queuine, the Q base of QUEUOSINE. (See also Y BASES.)

Q-cycle (protonmotive Q-cycle) A hypothetical pathway originally proposed (as part of the classical respiratory LOOP MODEL) to account e.g. for proton extrusion at Complex III of the mitochondrial ELECTRON TRANSPORT CHAIN; this pathway requires the presence of oxidized and reduced forms of ubiquinone (UQ and UQH_2, respectively) and the free radical, semiquinone (UQH). In one form of the Q-cycle, an electron which enters the matrix side of Complex III is taken up, together with one proton from the matrix side, by UQH – forming UQH_2. On passing to the cytoplasmic side, UQH_2 undergoes oxidation to UQH – one proton being extruded, and one electron passing back to the matrix side via *b*-type cytochromes; this electron, together with a proton from the matrix side, reduces UQ, thus regenerating UQH at the matrix side. At the cytoplasmic side, the UQH formed from UQH_2 undergoes oxidation to regenerate UQ, one proton being extruded, and one electron passing to cyt c_1.

[Thermodynamic and kinetic considerations of Q-cycle mechanisms: JBB (1985) *17* 51–64.]

Q enzyme A BRANCHING ENZYME which can convert amylose into an amylopectin-type polysaccharide; it cannot introduce branches into amylopectin. It occurs in certain algae and in higher plants.

Q fever In man, an acute disease caused by *Coxiella burnetii* (see COXIELLA). Infection may occur e.g. by the inhalation of contaminated dust or the ingestion of contaminated milk. After an incubation period of 2–3 weeks there is a sudden onset with headache, malaise, fever, muscular pain, and (often) respiratory symptoms (pneumonitis); there is no rash. Complications (e.g. endocarditis) may occur but the disease is rarely fatal. Diagnosis may include e.g. serological tests and/or attempts to culture *C. burnetii* from blood or sputum samples. Tetracyclines and chloramphenicol have been used therapeutically. Reservoirs of infection occur e.g. in sheep and cattle and in argosid and ixodid ticks. [Minireview: JMM (1996) *44* 77–78.]

[Goat-associated Q fever in Newfoundland: EID (2001) *7* 413–419.]

QAC See QUATERNARY AMMONIUM COMPOUNDS.

Qalyub group See NAIROVIRUS.

Qiagen® plasmid kit Any of several commercial kits (marketed by Qiagen GmbH, Hilden, Germany) used for isolating plasmids from bacteria (e.g. *Escherichia coli*); up to 20 μg of plasmid DNA can be isolated with the 'mini' kit, and up to 10 mg can be isolated with larger kits.

For *E. coli* (and other Gram-negative bacteria) the outer membrane is initially disrupted by a Tris–EDTA buffer. Then, controlled lysis of the cytoplasmic membrane with SDS (sodium dodecyl sulphate) permits release of plasmids and soluble proteins etc. (but not chromosomal DNA); contaminating RNA is digested with RNase A. SDS and proteins (but not plasmids) are precipitated in buffer, and a cleared lysate (containing the plasmids) is obtained by centrifugation. The spun lysate is passed through Qiagen anion-exchange resin; the resin is washed to remove contaminants (and to remove any DNA-binding proteins), and the plasmids are eluted with buffer of appropriate pH and ionic strength. Plasmid DNA is precipitated with isopropanol; after centrifugation, the supernatant is discarded and the DNA is washed with 70% ethanol, air-dried, and suspended in buffer.

qinghaosu (artemisinine, artemisinin, or arteannuin) An orally administered ANTIMALARIAL AGENT obtained from qinghao (*Artemisia annua*): a Chinese medicinal herb which has been used in China for at least 2000 years for the treatment of malaria [Chinese Medical Journal (1979) *92* 811–816, cited in BMB (1982) *38* 197]. Qinghaosu, a sesquiterpene lactone endoperoxide, is structurally distinct from all other known antimalarials. [Structure and chemical synthesis: PAC (1986) *58* 817–824.]

Qinghaosu is a rapidly acting blood schizonticide. It appears to be taken up selectively by infected erythrocytes [TRSTMH (1984) *78* 265–270], and is apparently effective against *Plasmodium falciparum* (including strains resistant to chloroquine) and *P. vivax*. In culture, the drug rapidly inhibits protein synthesis and, subsequently, nucleic acid synthesis in *P. falciparum*. It seems that iron, within the parasite, activates the drug and gives rise to potent free radicals [TRSTMH (1994) *88* (supplement 1) 31–32].

Parenteral derivatives of qinghaosu, *artemether* and *artesunate*, have been used successfully for treating severe malaria e.g. in Indochina [BCID (1995) *2* 309–330].

QPCR Quantitative PCR.

***qscR* gene** See QUORUM SENSING.

quadrulus A four-rowed ribbon of cilia which runs, in a spiral, down the buccal cavity in *Paramecium*.

quail disease Ulcerative enteritis in chickens, quail and pheasants; the causal agent is *Clostridium colinum* [IJSB (1985) *35* 155–159].

quail pea mosaic virus See COMOVIRUSES.

quailpox virus See AVIPOXVIRUS.

Quantiplex™ See BDNA ASSAY.

quarantine (1) The isolation of an individual (usually an animal), e.g. prior to entering a country, in order to determine whether or not that individual is suffering from a particular infectious disease; the period of isolation should be equal to or longer than the longest incubation period of the suspected disease. (From Italian *quarantina* = 40 days.) (2) The isolation of an individual suffering from an infectious disease in order to prevent the transmission of the disease to others.

quarg (quark) A soft, unripened German cheese made from e.g. skim milk in much the same way as is cottage cheese (see CHEESE-MAKING (e)); a B-type LACTIC ACID STARTER and rennin are used to coagulate the milk protein. [AvL (1983) *49* 83–84.]

quartan malaria See MALARIA.

quasi-equivalence See ICOSAHEDRAL SYMMETRY.

quasispecies Of a given RNA virus: a population or sample which, as a consequence of the *normal* rate of replication error, is genetically heterogeneous; the range of genomes present will depend e.g. on the effects of selection. [See JINF (1997) *34* 201–203.]

quaternary ammonium compounds (QACs; quats; *syn.* onium compounds) A group of cationic surfactants used as ANTISEPTICS, DISINFECTANTS and PRESERVATIVES; they appear to act by disrupting the cytoplasmic membrane and (at high concentrations) by denaturing proteins. A QAC may be regarded as an ammonium halide with four substituents: a long-chain alkyl group (C_8 to C_{18} for high antimicrobial activity) and short-chain alkyl and/or benzyl or heterocyclic groups. QACs are bacteriostatic at low concentrations, bactericidal at high concentrations. They are typically more active against Gram-positive than Gram-negative bacteria, are reported to be fungistatic, trypanocidal, and active against certain viruses, but have little or no activity against *Mycobacterium tuberculosis*, *Pseudomonas aeruginosa* and bacterial endospores. They have a DILUTION COEFFICIENT of 1, and are inhibited e.g. by low pH, organic matter (serum, faeces etc), anionic detergents, soaps, phospholipids, and certain cations (e.g. Ca^{2+}, Mg^{2+}) which may compete for sites on the (negatively charged) cell surface. Various substances, (e.g. agar, glass) strongly adsorb QACs.

QACs are used e.g. for pre-operative skin cleansing and surgical dressings, and for the disinfection of equipment used in the food and dairy industries. They include e.g. *benzalkonium chloride* (= *Zephiran*, a mixture of alkyldimethylbenzylammonium chlorides) – used e.g. as a preservative in ophthalmic solutions; *benzethonium chloride* (a substituted benzalkonium chloride); *Cetrimide* (= *Cetavlon*, a mixture of hexadecyl- (cetyl-), tetradecyl- and dodecyl-trimethylammonium bromides); *cetylpyridinium chloride* (1-hexadecylpyridinium chloride); *CTAB* (cetyltrimethylammonium bromide); *Domiphen bromide* (dodecyldimethyl-2-phenoxyethylammonium bromide).

quats QUATERNARY AMMONIUM COMPOUNDS.

Quayle cycle *Syn.* RMP PATHWAY.

Queensland tick typhus In man, a tick-borne disease caused by *Rickettsia australis*.

Quellkörper Within the closed ascocarps of certain fungi: a gelatinous mass of cells believed to be involved in the rupture of the (mature) ascocarp wall.

quellung phenomenon A phenomenon – originally described by Neufeld for *Streptococcus pneumoniae* – in which the bacterial CAPSULE becomes more easily observable (appears darker) in the presence of antiserum specific for the capsular material. Antibodies mark the outer limit of the capsule by combining with its outermost layer. Since in the absence of antibodies the capsule may be invisible by microscopy, antibodies were originally thought to cause the capsule to swell (*Quellung* = swelling); in fact, little actual swelling appears to occur.

quench-freezing The process of rapidly FREEZING a specimen by plunging it into a suitable CRYOGEN.

Querbalken Lysozyme-sensitive crossbands in the prosthecae of species of *Asticcacaulis* and *Caulobacter*.

quercetin A mutagenic flavonol glycoside pigment widely distributed in plants; it binds non-covalently to and inhibits (F_0F_1)-type PROTON ATPASES.

queuine See QUEUOSINE.

queuosine (Q) A hypermodified guanosine derivative (7-[4,5-*cis*-dihydroxy-1-cyclopentene-3-aminomethyl]-7-deazaguanosine) present in the wobble position (see WOBBLE HYPOTHESIS) of tRNAs for aspartic acid, asparagine, histidine and tyrosine in most prokaryotic and eukaryotic organisms investigated (but not in *Saccharomyces cerevisiae*). The Q BASE queuine pairs with either C or U. Modification of G to Q in the wobble position of a tRNA anticodon can influence codon selection by the tRNA in vivo [EMBO (1985) *4* 823–827]. In eukaryotes, the tRNA Q content is variable and correlates with certain developmental changes [e.g. in *Dictyostelium discoideum*: JGM (1984) *130* 135–144].

Quevedo disease See POD ROT.

quinacrine A weakly basic, yellow fluorescent ACRIDINE dye (chromophore: 2-methoxy-6-chloro-9-aminoacridine) which has been used in the chemotherapy of e.g. malaria and giardiasis, and which is used as a chromosome stain.

quinapyramine An anti-trypanosomal agent used e.g. for the treatment of dourine.

quinghaosu See QINGHAOSU.

quinic acid See CHLOROGENIC ACID.

quinidine See QUININE.

quinine An alkaloid obtained from the bark of the *Cinchona* tree; the sulphate and hydrochloride are ANTIMALARIAL AGENTS effective against the intraerythrocytic stage (blood schizont) of the parasite (*Plasmodium*).

The diastereoisomer of quinine, *quinidine*, is used in place of quinine e.g. in the USA; compared to quinine it is less protein-bound but more cardiotoxic.

Quinine appears to interfere with the parasite's detoxification of ferriprotoporphyrin IX (FP). FP, formed when the parasite digests haemoglobin, is able to lyse the parasite unless it is detoxified by polymerization to insoluble granules of pigment (see HAEMOZOIN); such polymerization seems to involve a 'haem polymerase' [Nature (1992) *355* 108–109], and the quinoline-type drugs (including e.g. quinine) possibly inhibit this enzymic activity. In a photoaffinity labelling experiment, quinine competitively inhibited the binding of a chloroquine analogue to specific proteins (perhaps 'haem polymerase'?) within parasitized erythrocytes – suggesting the presence of targets common to the quinoline-type antimalarial agents [JBC (1994) *269* 6955–6961].

8-quinolinol *Syn.* 8-HYDROXYQUINOLINE.

quinolobactin A SIDEROPHORE, from *Pseudomonas fluorescens*, whose synthesis is repressible by pyoverdin [AEM (2000) *66* 487–492].

quinolone antibiotics A group of synthetic ANTIBIOTICS whose targets are GYRASE [gyrase-targeted antibiotics: TIM (1997) *5* 102–109] and topoisomerase IV; each antibiotic contains a substituted 4-quinolone ring.

The original antibiotics in this group – cinoxacin, NALIDIXIC ACID, OXOLINIC ACID and pipemidic acid – are active mainly against Gram-negative bacteria (but not against *Pseudomonas aeruginosa*); being rapidly excreted in urine, they have been used for treating UTIs caused by members of the Enterobacteriaceae. However, these drugs are not well absorbed when given orally and/or are readily inactivated in the body; they have a limited antibacterial spectrum, and resistance develops readily.

Antibiotics subsequently added to the group were fluorinated, piperazinyl-substituted derivatives (so-called *fluoroquinolones*); they include ciprofloxacin, enoxacin, norfloxacin, ofloxacin and pefloxacin. Compared to the earlier quinolones, these drugs have a wider spectrum of activity (being active against e.g. *Pseudomonas aeruginosa* and certain Gram-positive cocci, including MRSA), are effective at much lower in vivo concentrations and are more stable in the body; moreover, resistance to these drugs develops less readily. [Comparison of activities of various quinolones: JAC (1985) *16* 475–484, 485–490. Laboratory and clinical evaluation of pefloxacin: JAC (1986) *17* (supplement B) 1–118.]

The fluoroquinolone *danofloxacin* has been used e.g. against *Mycoplasma bovis* in the treatment of CALF PNEUMONIA; little resistance to danofloxacin has developed in *M. bovis*, although significant levels of resistance have developed against other anti-*M. bovis* antibiotics, including oxytetracycline, spectinomycin and the macrolide tilmicosin [VR (2000) *146* 745–747].

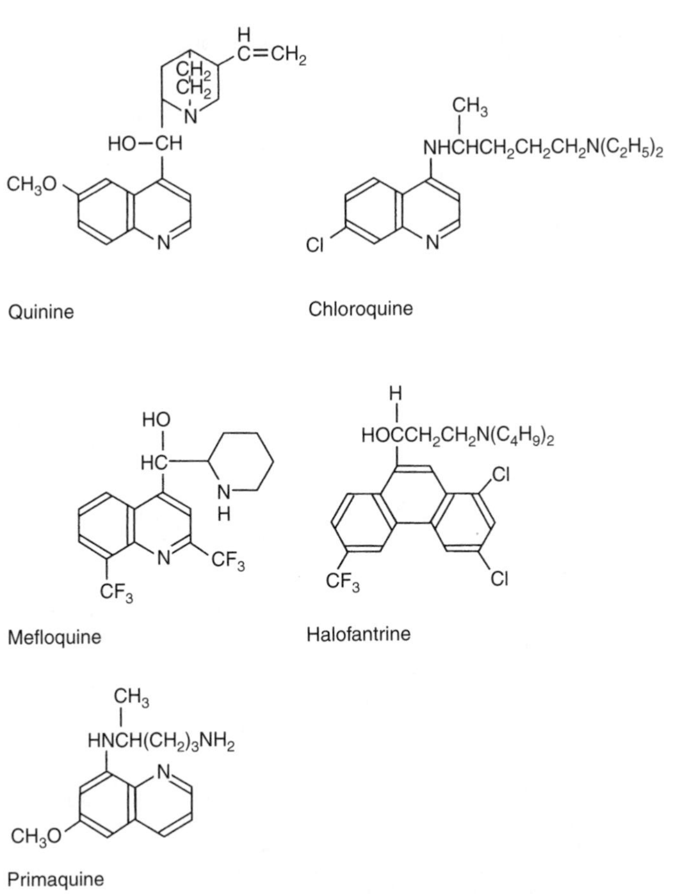

Quinine

Chloroquine

Mefloquine

Halofantrine

Primaquine

QUININE and some other antimalarial drugs (see MALARIA for further details of these and other antimalarials).

Reproduced from *Antimicrobial Drug Action*, Figure 7.3, page 123, R.A.D. Williams, P.A. Lambert & P. Singleton (1996) copyright Bios Scientific Publishers, UK (ISBN 1-872748-81-3) with permission from the publisher.

Some strains of *Acinetobacter baumannii* resistant to quinolone antibiotics (ciprofloxacin, nalidixic acid) have mutations in the *gyrA* (gyrase) and/or *parC* (topoisomerase IV) genes [RMM (1998) *9* 87–97]; a similar mechanism of resistance has also been found in other species. Mutations in gyrase and topoisomerase IV are commonly found in the so-called *quinolone resistance-determining regions*: a particular sequence in the *gyrA* gene and a homologous sequence in *parC*. However, in *Streptococcus pneumoniae*, mutations in *gyrA* and *parE* (the gene encoding the other subunit in topoisomerase IV) have been found in a strain showing increased resistance to certain of the newer fluoroquinolones (*moxifloxacin*, *sparfloxacin*, *grepafloxacin*) but with only slight (or no) increased resistance to older fluoroquinolones (ciprofloxacin, pefloxacin) – and with greater sensitivity to novobiocin [AAC (2001) *45* 952–955].

Other modes of resistance to quinoline antibiotics include reduced permeability of the cell envelope and efflux mechanisms. [Efflux-mediated resistance to fluoroquinolones in Gram-negative bacteria: AAC (2000) *44* 2233–2241.]

The activities of some of the newer fluoroquinolones (including gatifloxacin, grepafloxacin, levofloxacin, moxifloxacin and trovafloxacin) have been ascertained against ciprofloxacin-resistant *Streptococcus pneumoniae*; these newer drugs were found to have improved activities compared with that of ciprofloxacin [AAC (2001) *45* 1654–1659].

Various drugs (including gatifloxacin, gemifloxacin, grepafloxacin, levofloxacin and moxifloxacin) are discussed in the Report of the 7th International Symposium on New Quinolones (Edinburgh, June 10–12th, 2001) [JAC (2001) *47* (supplement S1)].

(See also CcdB in F PLASMID.)

quinomycins See QUINOXALINE ANTIBIOTICS.

quinone antifungal agents A group of agricultural ANTIFUNGAL AGENTS which includes e.g. BENQUINOX, CHLORANIL, DICHLONE and DITHIANON.

quinones Aromatic dioxo (diketo) compounds which are typically coloured (usually yellow, orange or red) and are e.g. constituents of many natural pigments. Quinones are classified as benzoquinones, naphthoquinones etc according to the nature of the aromatic ring system. Microorganisms contain a range of quinones, the nature of which can be a useful taxonomic feature [in bacteria: MR (1981) *45* 316–354].

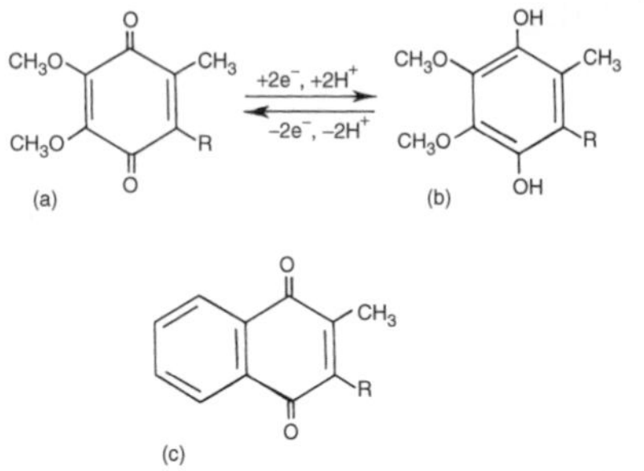

QUINONES. (a) A ubiquinone. (b) The reduced form of a ubiquinone: a hydroquinone. (c) A menaquinone (a vitamin K_2). $R = -[CH_2.CH{=}C(CH_3).CH_2]_nH$.

Benzoquinones include plastoquinones (2,3-dimethylbenzoquinones) and ubiquinones (coenzymes Q); the latter are 2,3-dimethoxy-5-methyl-1,4-benzoquinones with a variable-length isoprenoid side-chain at the 6-position. (Ubiquinones are given various designations based either on the number of isoprenoid units or on the number of carbon atoms in the side-chain: e.g., a ubiquinone with a side-chain containing 8 isoprenoid units may be designated ubiquinone-8, UQ-8, UQ_8, coenzyme Q_8, or ubiquinone-40, UQ-40, etc.) [Book ref. 89.]

Naphthoquinones include e.g. various pigments and toxins (see e.g. NAPHTHAZARINS and XANTHOMEGNIN), as well as the K vitamins: derivatives of 2-methyl-1,4-naphthoquinone (vitamin K_3, menadione) in which the 3-position may be substituted with a mono-unsaturated phytyl chain (in vitamin K_1, = phylloquinone) or with a variable-length polyisoprene chain (in vitamins K_2, = menaquinones). (Menaquinone terminology resembles that for ubiquinones: hence, e.g., menaquinone-6, MK-6, MK-30, vitamin $K_{2(30)}$ etc.) Some bacteria (e.g. members of the PASTEURELLACEAE) contain menaquinones which lack the 2-methyl group (2-demethyl-vitamins K_2). In mammals, vitamins K are important components of the blood coagulation system; mammals cannot synthesize these vitamins, depending on diet (particularly green plants) and on the bacterial flora of the gut (particularly *Escherichia coli* and *Bacteroides* spp) for an adequate supply.

Quinones can be reversibly reduced to hydroquinones; they function e.g. in aerobic and anaerobic ELECTRON TRANSPORT CHAINS, in PHOTOSYNTHESIS, etc. For example, *E. coli* synthesizes both ubiquinone-8 and menaquinone-8 in proportions which depend on growth conditions; the quinones function as carriers of reducing equivalents between dehydrogenases and terminal enzyme complexes (cytochrome oxidases, nitrate reductase, or fumarate reductase) – ubiquinone reduction involving the gain of two electrons followed by the addition of two protons. The menaquinone is specifically required for FUMARATE RESPIRATION, and is also required for the anaerobic synthesis of e.g. pyrimidines (linking dihydro-orotate dehydrogenase with fumarate reductase in the anaerobic synthesis of uracil). (See also QUINOPROTEIN.)

quinoprotein A class of ENZYME in which the prosthetic group is pyrrolo-quinoline quinone (PQQ): 2,7,9-tricarboxy-1*H*-pyrrolo(2,3*f*)-quinoline-4,5-dione. Quinoproteins include e.g. methanol dehydrogenase (in which PQQ is called *methoxatin*), and glucose dehydrogenase (EC 1.1.99.17). [PQQ and quinoproteins in microbial oxidations: FEMS Reviews (1986) *32* 165–178.] (See also EXTRACYTOPLASMIC OXIDATION and METHYLOTROPHY.)

quinovosamine 2-Amino-2,6-dideoxyglucose: a TRIFOLIIN A-binding component of LPS in strains of *Rhizobium leguminosarum* biotype *trifolii*.

quinoxaline antibiotics Bifunctional INTERCALATING AGENTS (ϕ ca. 48°), each having two quinoxaline 2-carboxylic acid chromophores linked by a cyclic octapeptide dilactone containing D- and L-amino acids; the peptide has a central cross-bridge (thioacetal in the *quinomycins*, disulphide in the *triostins*). The two parallel chromophores project from the (relatively rigid) peptide ring and intercalate at sites two base pairs apart; the peptide ring seems to fit into the minor groove of the DNA. The quinoxalines *echinomycin* (= quinomycin A) and *triostin A* differ only in the nature of the cross-bridge; both show some preference for GC-rich natural DNAs. TANDEM (des-*N*-tetramethyltriostin A: a synthetic analogue of triostin A) shows a marked preference for AT-rich sequences. Quinoxalines are antimicrobial and antitumour agents; they inhibit the chain elongation stage of DNA-directed RNA synthesis.

Quinqueloculina See FORAMINIFERIDA.

quinsy Peritonsillar ABSCESS (usually due to *Streptococcus pyogenes*) – a complication of TONSILLITIS.

quintozene (PCNB) Pentachloronitrobenzene, an agricultural antifungal agent used mainly for the treatment of soil; it is effective against a number of soil- and seed-borne diseases, e.g. DAMPING OFF, various rots of bulbs. Quintozene is insoluble in water and is used as a dust; it is very stable and has low volatility – hence it is very persistent in the soil. (cf. TECNAZENE, DICLORAN.)

quinupristin/dalfopristin (RP 59500; Synercid®) A bactericidal antibiotic (within the STREPTOGRAMIN group) which is active against e.g. most streptococci (including *S. pneumoniae*), MRSA, and *Enterococcus faecium* (*E. faecalis* is resistant); it has a long POST-ANTIBIOTIC EFFECT [Drugs (1996) *51* (supplement 1) 31–37]. [In vitro activity against *Staphylococcus aureus*: JAC (1997) *39* 53–58.]

Quinupristin/dalfopristin (administered intravenously) has been conditionally licenced in Europe and the USA; it has been useful e.g. for treating vancomycin-resistant infections. Side-effects include arthralgia and myalgia.

[Quinupristin/dalfopristin (when to use?): JAC (2000) *46* 347–350.]

qundai-cai See UNDARIA.

quorum sensing The phenomenon in which cells express particular characteristic(s) only when present as a population whose density is above a certain minimum (*quorum*). (Here, the 'density' of a population refers to the number of cells per unit volume.) Thus, in some cases, cells in a high-density population exhibit characteristics which are absent when the *same* cells are in a low-density population.

One example of quorum sensing is that of *Photobacterium fischeri* (sometimes referred to as *Vibrio fischeri*). *P. fischeri* can occur either as a free-living organism (in low-density populations), or as a symbiont in high-density populations within the light-emitting organs of certain fish; as a symbiont, *P. fischeri* produces a blue-green BIOLUMINESCENCE – whereas in the free-living state (low-density population) it produces little or no light.

(Bacterial bioluminescence is used by the fish to signal to one another.)

The mechanism of quorum sensing involves certain secreted molecules which, only in high-density populations of the secreting cell, reach a threshold concentration which is sufficient to trigger specific gene(s) within the cells. The secreted signalling molecule is referred to as an *autoinducer* because the cells themselves produce it.

Autoinducers are low-molecular-weight molecules. Different bacteria may produce different types of autoinducer that regulate different characteristics; however, in some cases different species may produce the same type of autoinducer for regulating different genes. A given species may produce a range of autoinducers to control the expression of various characteristics.

In many Gram-negative bacteria, the autoinducer is an *N*-acyl-L-homoserine lactone (AHL). [AHL-based quorum sensing: TIBS (1996) *21* 214–219.] In general, AHLs appear to act by diffusing into the cell and binding to an appropriate cytoplasmic protein involved in regulating the transcription of relevant gene(s). In the case of *P. fischeri*, an AHL triggers the *lux* operon (see BIOLUMINESCENCE). [Evolution of LuxI and LuxR as regulatory molecules in quorum sensing: Microbiology (2001) *147* 2379–2387.]

Strain CV026 of *Chromobacterium violaceum* is an inducer-negative mutant which produces VIOLACEIN when exposed to exogenous inducers, including all tested molecules of AHL and AHT (*N*-acylhomocysteine thiolactone) with *N*-acyl side-chains between C_4 and C_8; this mutant strain can therefore be used as a biosensor for detecting a range of inducer molecules [Microbiology (1997) *143* 3703–3711].

In *Pseudomonas aeruginosa* the genes encoding certain virulence factors are regulated by at least two quorum sensing systems – the *las* and *rhl* systems that involve different AHL autoinducers; these systems appear to act hierarchically, i.e. one system is always activated before the other [TIM (1997) *5* 132–134]. The product of gene *qscR* appears to be a negative regulator of quorum-sensing-controlled genes; it probably acts by repressing gene *lasI* (which encodes *N*-3-(oxododecanoyl) homoserine lactone) [PNAS (2001) *98* 2752–2757].

Quorum sensing has been reported to regulate expression of the type III PROTEIN SECRETION system in both EHEC and EPEC strains of *Escherichia coli* – this involving induction of genes in the LEE PATHOGENICITY ISLAND; regulation of the LEE operons involves an autoinducer encoded by gene *luxS*. To account for the unusually low infectious dose (low-density population) of *E. coli* O157:H7, it has been suggested that inducer may be derived from non-pathogenic strains of *E. coli* in the gut [PNAS (1999) *96* 15196–15201]. The existence of quorum sensing mechanisms in pathogens may permit therapy based on inhibition of autoinducers. Interestingly, the macrolide antibiotic AZITHROMYCIN has been reported to inhibit quorum sensing in *Pseudomonas aeruginosa* strain PAO1, the result (possibly advantageous e.g. in CYSTIC FIBROSIS patients [Lancet (1998) *351* 420]) being an inhibition of the pathogen's production of virulence factors [AAC (2001) *45* 1930–1933].

In *Burkholderia cepacia* there is evidence that swarming, and the maturation of biofilms, are aspects of the organism's physiology that are regulated by a quorum-sensing mechanism [Microbiology (2001) *147* 2517–2528].

In *Escherichia coli* O157:H7 quorum sensing has been reported to be a global regulatory mechanism involved in basic physiological functions as well as in the formation of virulence factors [JB (2001) *183* 5187–5197].

Interestingly, studies on *Pseudomonas aeruginosa* indicate that the stringent response (see STRINGENT CONTROL sense 1) can activate quorum sensing independently of cell density, a feature which may be important in nutrient-limited conditions during infection [JB (2001) *183* 5376–5384].

In Gram-positive bacteria, the signalling molecules used in quorum sensing are generally *peptides* (= *pheromones*). Thus, for example, when cells of *Bacillus subtilis* grow to high density, at least two types of pheromone accumulate in the extracellular environment: ComX and CSF. ComX promotes competence in TRANSFORMATION by activating a TWO-COMPONENT REGULATORY SYSTEM; it causes autophosphorylation of a membrane-bound histidine kinase (ComP) which, in turn, phosphorylates (activates) the transcription factor ComA – a regulator of the *comS* gene that is required for competence.

The pheromone CSF appears to be one of a number of factors involved in the initiation of sporulation. It is taken up by cells via an ATP-dependent oligopeptide permease transport system in the cytoplasmic membrane; within the cytoplasm CSF apparently inhibits the activity of certain phosphorylases that would otherwise de-phosphorylate Spo0F~P, thereby promoting the phosphorelay and sporulation.

In *Enterococcus faecalis*, conjugation involves secretion of pheromones by potential recipients.

[Peptide signalling (review): TIM (1998) *6* 288–294.]

In the (dimorphic) fungus *Candida albicans*, farnesol (a sesquiterpene alcohol) acts as a quorum-sensing molecule whose effect is reported to be inhibition of yeast-to-mycelium conversion [AEM (2001) *67* 2982–2992].

1. Words in SMALL CAPITALS are cross-references to separate entries.
2. Keys to journal title abbreviations and Book ref. numbers are given at the end of the Dictionary.
3. The Greek alphabet is given in Appendix VI.
4. For further information see 'Notes for the User' at the front of the Dictionary.

R

R (1) RÖNTGEN (q.v.). (2) (in cell cycle) See CELL CYCLE. (3) Arginine (see AMINO ACIDS).

R antigen (streptococcal) *Syn.* R PROTEIN.

R body A refractile, intracellular structure which occurs e.g. in certain bacterial endosymbionts of protozoa (see CAEDIBACTER) and which appears to confer killer characteristics on the protozoan host cell. An R body consists of a proteinaceous ribbon, ca. 0.2–0.5 μm wide and ca. 10–15 μm in length, which is rolled up to form a cylinder; under negative phase-contrast MICROSCOPY it appears as a bright ring or (when viewed in section from the side) as a pair of parallel rods. Its presence within a bacterial cell is apparently associated with the presence of plasmid(s) and/or phage(s). R bodies appear to unroll in the food vacuoles of sensitive paramecia; unrolling appears to occur from the inside or outside of the coil according to the species of the bacterium of origin.

R bodies have also been found in free-living strains of *Pseudomonas* which are toxic for sensitive paramecia [Arch. Micro. (1979) *121* 9–15] and in a free-living *Pseudomonas*-like bacterium [JGM (1986) *132* 2801–2805].

r-chromatin Chromatin containing rRNA genes.

R factor (1) See R PLASMID. (2) Release factor: see PROTEIN SYNTHESIS.

R-glucan Alkali-resistant (i.e. alkali-insoluble) glucans of fungal CELL WALLS.

R loop (*mol. biol.*) A single-stranded loop of DNA formed when a short ssRNA molecule pairs with a complementary region of one strand of a dsDNA molecule, displacing the corresponding region of the homologous strand (the R loop). An R loop can be observed by electron microscopy using the KLEINSCHMIDT MONOLAYER TECHNIQUE. If the ssRNA used is a mature mRNA derived from a SPLIT GENE, an intron in the DNA – having no homologous region in the mRNA – will appear as a loop of dsDNA extruded between two R loops (each R loop corresponding to one of the two adjacent exons); thus, R-looping can be used e.g. to detect introns in genes. (cf. D LOOP.)

r plasmid See R PLASMID.

R plasmid (formerly R factor, drug resistance factor etc) Any PLASMID which encodes resistance to one or more ANTIBIOTICS and/or to other antimicrobial agents (including e.g. heavy metal ions). A bacterium containing an R plasmid may express resistance e.g. by synthesizing a (plasmid-encoded) enzyme which destroys/modifies an antibiotic (see e.g. CHLORAMPHENICOL, β-LACTAMASES) or which modifies the target site (see e.g. MLS ANTIBIOTICS), or by synthesizing a drug-resistant form of a target enzyme (see e.g. SULPHONAMIDES).

A *conjugative* (= *transmissible*) R plasmid also contains genes which specify the intercellular transfer of the plasmid by CONJUGATION (sense 1b) (see also RTF); a *non-conjugative* (= *non-transmissible*) R plasmid does not specify its own conjugal transfer but it may be 'mobilized' by a conjugative plasmid. (Some authors refer to a non-conjugative R plasmid as an 'r plasmid'.) R plasmids belong to various Inc groups (see INCOMPATIBILITY).

R point (in cell cycle) See CELL CYCLE.

r-protein Ribosomal protein: see RIBOSOME.

R protein (R antigen) (in streptococci) A cell-surface protein found in streptococci of groups A, B, C and G. R protein is only moderately immunogenic and is apparently not associated with virulence. It occurs in two antigenically distinct forms: R28 (trypsin-resistant) and R3 (trypsin-sensitive). (See also M PROTEIN and T PROTEIN.)

R strain See SMOOTH–ROUGH VARIATION.

r-strand (in adenoviruses) See ADENOVIRIDAE.

R-TEM β-lactamase *Syn.* TEM-1 β-LACTAMASE.

R1 plasmid A low-COPY NUMBER, IncFII CONJUGATIVE PLASMID (MWt ca. 60000) which occurs in enterobacteria; it encodes resistance to e.g. ampicillin, chloramphenicol, sulphonamides and streptomycin. Replication is promoted, at the origin, by the (plasmid-encoded) RepA protein, and the frequency of plasmid replication is controlled by the regulation of RepA synthesis. The main negative control exerted on RepA synthesis is carried out by the *copA* product, an unstable, constitutively produced micRNA (see ANTISENSE RNA), of half-life ca. 1–2 minutes, which binds to the complementary sequence (*copT*) on the *repA* mRNA, thus inhibiting translation of the RepA protein; the inhibitory influence of the *copA/copT* system decreases with decrease in copy number – thereby providing a mechanism for stabilizing copy number. (CopA also has a role in INCOMPATIBILITY in IncFII plasmids.) It has been proposed that the level of *repA* expression is also affected by a region (designated *7k*) within the *repA* leader sequence; translation of (i.e., the presence of ribosomes in) this region, which overlaps the *copT* sequence, is believed to counteract the negative control of the *copA/copT* system [EMBO (1987) *6* 515–522].

R6 plasmid An enterobacterial IncFII CONJUGATIVE PLASMID (COPY NUMBER: 1–2); it encodes resistance to e.g. chloramphenicol, streptomycin, sulphonamides and mercury.

R6K plasmid A multicopy enterobacterial CONJUGATIVE PLASMID (ca. 38 kb) which carries genes specifying resistance to ampicillin and streptomycin. R6K apparently undergoes *sequential bidirectional replication* from any of three origins (designated α, β and γ), all of which are located within a 4-kb region of the plasmid. Replication initiated at an origin apparently proceeds first in one direction, stopping at a specific termination site (*ter*); re-initiation then occurs at the same origin and proceeds in the opposite direction to *ter*, thus completing replication of the plasmid.

Initiation of replication at any of the three origins requires a *trans*-acting R6K-encoded DNA-binding protein (the π or Pi protein) and a region within the γ-origin containing seven tandemly-arranged 22-bp direct repeats (ITERONS). Plasmids from which the γ-origin has been deleted cannot be replicated even when π is supplied in *trans*; the α- and β-origins each appear to have a minimum functional size of ca. 2 kb which includes the seven iterons in the region of the γ-origin. (Deletion of three or more of the iterons results in loss of plasmid replicability.) The π protein binds preferentially to the iterons and positively regulates initiation of replication from one of the origins (α, β or γ). At high concentrations, π can also play a role in the negative regulation of R6K replication, apparently by specifically repressing replication from the γ origin. (Experiments involving variation in the levels of π over a wide range suggest that the level of π in the cell is *not* responsible for controlling the copy number of R6K.)

The π protein can also bind to the promoter–operator region of its own gene (*pir*, located near the γ origin); this region contains an 8th iteron together with a pair of imperfect inverted repeats. The π-binding site overlaps the RNA polymerase-binding site at the *pir* promoter, and at high concentrations π can act as a repressor of its own synthesis.

[Review: Book ref. 161, pp. 125–140; binding of π to R6K DNA: JMB (1986) *187* 225–239.]

R10 medium RAPPAPORT–VASSILIADIS BROTH.

R18 plasmid See RP1 PLASMID.

R46 plasmid An IncN PLASMID (ca. 52 kb) which encodes resistance to ampicillin, streptomycin, sulphonamides and tetracycline. (See also PARTITION and SOS SYSTEM.)

R68 plasmid See RP1 PLASMID.

R100 plasmid (*syn.* NR1 plasmid; R222 plasmid) An Inc-FII enterobacterial CONJUGATIVE PLASMID (COPY NUMBER: 1–2) that encodes resistance to e.g. chloramphenicol, streptomycin, sulphonamides and mercury. (The tetracycline-resistance gene is carried on a TRANSPOSON – see entry for Tn*10*.) The R100.1 derivative is a DRD PLASMID.

R222 plasmid See R100 PLASMID.

R1822 plasmid See RP1 PLASMID.

Ra strain See SMOOTH – ROUGH VARIATION.

rabbit fibroma virus See LEPORIPOXVIRUS.

rabbit herpesvirus See GAMMAHERPESVIRINAE.

rabbitpox virus See ORTHOPOXVIRUS.

rabies (hydrophobia; lyssa) An acute, almost invariably fatal disease of man and animals – particularly carnivores, though most mammals can be infected. The rabies virus (see LYSSAVIRUS) is transmitted (in saliva) mainly by the bite of a rabid animal. The virus remains at the site of infection for a time before entering peripheral nerves and travelling along the nerves to the CNS.

Rabies in man. Incubation period: usually 2–13 weeks, but may be 6 days to a year or more, depending e.g. on the distance between the site of infection and the CNS. Early symptoms are non-specific: fever, headache, malaise etc. These are followed by neurological symptoms, including excitability and pharyngeal spasms (triggered e.g. by attempts to drink or even by the sight of water – hence 'hydrophobia'); periods of hyperactivity may alternate with periods of relative normality. Convulsions and paralysis precede death. (Occasionally, paralytic symptoms predominate, with little hyperactivity.) *Lab. diagnosis*: detection of the virus in saliva, CSF etc – e.g., by electron microscopy or by fluorescent antibody staining; when possible, confirmation of rabies in the animal which inflicted the bite – e.g., by postmortem detection of (pathognomonic) NEGRI BODIES and rabies virus. Rabies virus can be detected in brain samples by an rtPCR-based approach; an rtPCR–ELISA test has been evaluated on 60 isolates of rabies and rabies-related viruses [JVM (1997) *69* 63–72]. *Treatment*: none is effective once the disease is established. Post-exposure prophylaxis includes thorough washing of the wound followed by treatment with both rabies antiserum (part of which is instilled directly into the wound) and rabies vaccine; pre-exposure vaccination may be given to individuals at special risk (e.g. veterinarians, animal handlers). Inactivated vaccines are used for human treatment; these include *duck embryo vaccine* (DEV) (prepared from virus grown in embryonated duck eggs), and *human diploid cell vaccine* (HDCV) (prepared from virus grown in cultures of human diploid embryo lung fibroblasts). These vaccines are safer than those prepared from animal nerve tissue (e.g. SEMPLE VACCINE).

Rabies in animals. Dogs and other carnivores (wolves, foxes etc) may initially show excitement, violence etc ('furious rabies'), followed by depression and paralysis ('dumb rabies'). In cattle and horses, symptoms are variable. In vampire bats (important reservoirs of infection in Central and South America) only the dumb form occurs. Vaccines for use in animals may be live attenuated (see e.g. FLURY VIRUS) or inactivated. (cf. AUJESZKY'S DISEASE.)

***rac* prophage** See RECF PATHWAY.

race (1) (*mycol.*) PHYSIOLOGICAL RACE. (2) A non-specific designation which may refer e.g. to a STRAIN or a VARIETY.

racket mycelium *Syn.* RACQUET MYCELIUM.

racquet mycelium (racket mycelium) In DERMATOPHYTES: mycelium composed of hyphal cells which have terminal swellings and which thus resemble long-handled tennis racquets.

rad Radiation absorbed dose: a unit of absorbed radiation equal to 100 ergs of energy absorbed by 1 g of material (0.01 joules absorbed by 1 kg). (See also GRAY and IONIZING RADIATION; cf. RÖNTGEN.)

RadA protein In members of the Archaea: a protein apparently analogous to the bacterial RecA protein [GD (1998) *12* 1248–1253].

radappertization Irradiation of food with levels of IONIZING RADIATION sufficient to inactivate the spores of *Clostridium botulinum*. (cf. RADICIDATION; RADURIZATION.)

Radar See PROPICONAZOLE.

radiation (in disinfection and sterilization) See IONIZING RADIATION; MICROWAVE RADIATION; ULTRAVIOLET RADIATION.

radicidation Irradiation of food with levels of IONIZING RADIATION sufficient to inactivate certain non-sporing pathogens (e.g. *Salmonella* spp). (cf. RADURIZATION; RADAPPERTIZATION.)

radioallergosorbent test See RAST.

radioautography *Syn.* AUTORADIOGRAPHY.

radioimmunoassay (RIA) A highly sensitive IMMUNOASSAY by which antigens or antibodies are quantified using radioactive labelling.

Antibody assay. Essentially, the antiserum under test is allowed to react with excess homologous antigen (radiolabelled e.g. with ^{125}I), and the immune complex is separated from any uncombined antigen – e.g. by precipitation (as in the FARR TECHNIQUE); the amount of antigen in the immune complex is then determined from the level of radioactivity in the complex, and the amount of antibody can hence be calculated.

Antigen assay. This is based on the fact that the proportion of a fixed amount of radiolabelled antigen which combines with a fixed amount of homologous antibody depends on the amount of unlabelled antigen present: the more unlabelled antigen present, the smaller the proportion of labelled antigen – and the lower the level of radioactivity – in the immune complex. Thus it is possible to construct a standard curve for the amount of unlabelled antigen corresponding to given levels of radioactivity in the immune complex. In antigen assays, the antigen to be quantified is the unlabelled antigen, the amount of which can be derived from the standard curve. (cf. IMMUNORADIOMETRIC ASSAY.)

radioimmunosorbent test See RIST.

radiolaria A group of free-living, marine, mostly planktonic protozoa (superclass ACTINOPODA) in which the cells are typically more or less spherical (<100 μm to several millimetres in diameter) or thimble-shaped, and are generally solitary, although some species form aggregates or colonies up to 10 cm or more across. Most radiolaria have a siliceous 'skeleton' consisting of a delicate internal latticed shell and/or an array of radiating spines. The cell characteristically contains a *central capsule*: a spherical membranous structure, concentric with the cell, which separates the ectoplasm (extracapsular zone) from the endoplasm (intracapsular zone); the capsule is perforated by few to many pores. The endoplasm contains the nucleus or nuclei (according to species), mitochondria, food reserves, etc. The ectoplasm may comprise three more or less distinct regions. The *sarcomatrix*, next to the capsule, is rich in food vacuoles and appears to be the site of digestion. Next to the sarcomatrix is the *calymma*: a highly vacuolated region; the vacuoles increase the buoyancy of the cell, and the cell can apparently control its vertical position in the water by collapsing and regenerating these vacuoles. The *peripheral zone* of the ectoplasm gives rise to radially arranged axopodia and/or to an elaborate *rhizopodial network* consisting of branching and anastomosing protoplasmic filaments.

Radiolaria are classified (e.g. on the basis of skeletal structure and nature of the central capsule) into two classes: Phaeodarea and Polycystinea [JP (1980) *27* 37–58]. In the Phaeodarea the skeleton, when present, is composed of a silicified organic material and usually consists of hollow spines and/or latticed shells. The capsular membrane is very thick and has three pores, the largest of which is called the *astropyle*, the other two (the *parapyles*) occurring more or less opposite the astropyle. A brownish mass of waste material (*phaeodium*) collects in the ectoplasm near the astropyle. Six orders are recognized; genera include e.g. *Astracantha* and *Aulacantha*. In the Polycystinea most species have a skeleton composed of solid siliceous elements which either form one or more latticed shells, with or without radiating spines, or occur as isolated spicules. The capsular membrane usually consists of polygonal plates and is perforated by many pores; in the order Nassellarida (e.g. *Eucoronis*, *Plagiacantha*, *Plagonium*) these pores are clustered in one region, whereas in the order Spumellarida (e.g. *Coccodiscus*, *Collosphaera*, *Thalassicolla*) they are uniformly distributed. [Radiolarian fine structure and silica deposition: Book ref. 137, pp. 347–379.]

Radiolaria typically feed on protozoa, small crustacea etc which are trapped by the axopodia/rhizopodial network. Many radiolaria contain endosymbiotic algae – commonly dinoflagellates, but sometimes 'prasinophytes' – which occur within vacuoles in the ectoplasm. Under certain conditions (e.g. prolonged darkness) some of these endosymbiotic cells may be digested. [Radiolarian symbioses: Book ref. 129, pp. 69–89.]

Reproduction occurs by binary or multiple fission. Small flagellated cells ('flagellospores') may be produced by some species, but the nature of these is uncertain; they may be gametes, or they may be escaped cells of endosymbiotic dinoflagellates.

Fossil radiolarian tests have been found e.g. in ocean deposits (e.g. *radiolarian ooze*, a deep-sea ooze containing 30% or more of radiolarian tests) and in certain sedimentary rocks. Fossil radiolaria date from the Precambrian onwards; they are useful in the biostratigraphy of Mesozoic and Cenozoic deep-sea sediments. [Stratigraphy of planktonic radiolaria: Book ref. 136, pp. 573–712.]

radiolarian ooze See RADIOLARIA.

radiomimetic Refers to a compound whose physiological effects (particularly mutagenic effects) mimic those of IONIZING RADIATIONS.

Radiomycetaceae See MUCORALES.

Radiophrya See ASTOMATIDA.

radurization Irradiation of food with levels of IONIZING RADIATION sufficient to inactivate the more sensitive spoilage organisms. (cf. RADICIDATION; RADAPPERTIZATION.)

raffinose A non-reducing trisaccharide: α-D-galactopyranosyl-$(1 \rightarrow 6)$-α-D-glucopyranosyl-$(1 \leftrightarrow 2)$-β-D-fructofuranoside.

Raffinose is second to sucrose as the most abundant free sugar in plants. It can be metabolized by many yeasts and fungi and by certain bacteria; the ability to ferment raffinose is used to distinguish between top- and bottom-fermenting strains of BREWERS' YEAST.

ragi A type of STARTER culture made in Asian countries by growing mould(s) on cakes of rice or wheat flour. A ragi is used as an inoculum in the preparation of various foods, and may contain e.g. species of *Mucor*, *Rhizopus*, *Neurospora*, various yeasts etc. (cf. ONCOM; KOJI.)

Rahnella A genus of bacteria of the ENTEROBACTERIACEAE, found in fresh water and occasionally in human clinical specimens. Cells are motile when grown at 25°C, non-motile at 36°C. VP +ve; MR usually +ve. [Book ref. 22, p. 513.]

Ralfsia See PHAEOPHYTA.

Ramalina A genus of fruticose LICHENS (order LECANORALES). Thallus: strap-like (but not dorsiventral) and erect or pendulous, attached to the substratum by a discoid holdfast. Apothecia: terminal and/or marginal. Species occur in unpolluted regions e.g. on coastal rocks (*R. polymorpha*, *R. siliquosa*) or on tree-bark (*R. farinacea*, *R. fastigiata*, *R. fraxinea*).

Ramibacterium ramosum See CLOSTRIDIUM (*C. ramosum*).

Ramon titration A serological titration in which a constant volume of a given antigen solution is added to each of a number of serial dilutions of homologous antiserum; precipitation occurs most rapidly in that dilution in which antibody and antigen are present in OPTIMAL PROPORTIONS.

Ramsay Hunt syndrome HERPES ZOSTER affecting the facial nerve, causing pain in the ear and face.

Ramularia See HYPHOMYCETES.

ramus A branch. (See also PENICILLIUM.)

Ranavirus A genus of viruses (family IRIDOVIRIDAE) which infect amphibians. Type species: FROG VIRUS 3.

rancidity In certain lipid-containing foods: a form of spoilage characterized by the presence of organoleptically detectable amounts of the products of lipid autoxidation (see OXIDATIVE RANCIDITY) and/or of the products of the microbial hydrolysis of lipids (see e.g. BUTTER, CHEESE SPOILAGE).

random amplified polymorphic DNA See RAPD.

random coil See HELIX–COIL TRANSITION.

random walk A pattern of motility, exhibited e.g. by (peritrichously flagellated) enterobacteria, in which periods of smooth swimming are interrupted by episodes of tumbling (see FLAGELLAR MOTILITY for mechanism); tumbling causes a cell to adopt a new direction entirely at random, so that – in a uniform environment – cells move ('walk') randomly in three dimensions. Although tumbling results in a random change of direction, a cell can nevertheless bias the net direction of its motility in response to certain directional stimuli (see TAXIS) by increasing or decreasing the frequency of tumbling when travelling in (respectively) an unfavourable or favourable direction. During CHEMOTAXIS, for example, cells of e.g. *Escherichia coli* which encounter an unfavourable gradient of nutrients increase their frequency of tumbling – thus reducing the time they swim in an unfavourable direction; conversely, the tumbling response is repressed if a more favourable location is encountered, and a longer period of smooth swimming supervenes. Thus, there is a net movement towards a favourable stimulus and away from an unfavourable one – a type of tactic response known as a *biased random walk*.

Ranikhet disease *Syn.* NEWCASTLE DISEASE.

RANTES (chemokine) See CHEMOKINES.

RapA phosphatase See ENDOSPORE (sense 1).

RapB phosphatase See ENDOSPORE (sense 1).

RAPD Random amplified polymorphic DNA: a form of ARBITRARILY PRIMED PCR (q.v. for mechanism) which is used e.g. for TYPING. When typing strains by RAPD it is common to repeat the assay with each of a range of different primers because different primers produce different fragments (as they bind to different 'best-fit' sequences); this increases the chance of detecting differences between the strains.

RAPD has been used e.g. for typing strains of *Listeria monocytogenes* [AEM (1993) *59* 304–308], *Staphylococcus haemolyticus* [LAM (1994) *18* 86–89], *Campylobacter* spp [LAM (1996) *23* 167–170], *Vibrio vulnificus* [AEM (1999) *65* 1141–1144] and *Serratia marcescens* [LAM (2000) *30* 419–421].

raphe A slit or fissure, e.g., that which develops in the pycnidium of *Chaetomella acutiseta* (allowing release of conidia) [CJB (1980) *58* 1129–1137], or that present in the frustule in certain DIATOMS.

Raphidiophrys A genus of small uninucleate heliozoa (order CENTROHELIDA). The cell (ca. 20–60 μm diam.) has an outer gelatinous layer in which numerous siliceous spicules are embedded – particularly at the bases of the axopodia. *R. pallida* occurs among vegetation in fresh water.

Raphidophyceae See CHLOROMONADS.

rapid lysis mutant A mutant of BACTERIOPHAGE T4 (designated *r*) which does not show LYSIS INHIBITION in at least some host strains; such mutants form plaques which are larger and have sharper edges than those of wild-type (r^+) strains. The *r* phenotype may result from a mutation in genes *r*I, *r*II (actually two genes, *r*IIA and *r*IIB) or *r*III. [*r*II genes: Book ref. 99, pp. 327–333.]

rapid sand filter See WATER SUPPLIES.

Rappaport–Vassiliadis broth (R10 medium; RV medium) A medium (containing tryptone, $MgCl_2$ and malachite green) used for the ENRICHMENT of salmonellae; the medium is inhibitory to lactose- and sucrose-negative organisms which can form colonies similar to those of salmonellae. [Review: JAB (1983) *54* 69–76; evaluation of commercial RV media: J. Hyg. (1986) *96* 425–429.]

ras Designation for a family of ONCOGENES first discovered in HARVEY MURINE SARCOMA VIRUS (H-*ras* or Ha-*ras*) and KIRSTEN MURINE SARCOMA VIRUS (K-*ras* or Ki-*ras*); cellular *ras* genes have been highly conserved during evolution and occur e.g. in humans, rodents and *Saccharomyces cerevisiae* [Book ref. 113, pp. 419–423]. The *ras* genes code for highly related proteins (generic designation: p21) containing 189 amino acid residues; p21 proteins have GTP-binding, GTP-hydrolysing, and autophosphorylating activities. A *ras* oncogene can be activated e.g. by chemical carcinogens; transforming *ras* genes can differ from normal cellular *ras* genes by a single point mutation which results in a single amino acid substitution involving the 12th or 61st amino acids in p21. [Human *ras* genes and transformation: Book ref. 110, pp. 291–313.]

Ras superfamily See G PROTEINS.

raspberry cane blight See CANE BLIGHT.

raspberry diseases See e.g. CANE BLIGHT, ELSINOË and GREY MOULD.

raspberry leaf curl virus See LUTEOVIRUSES.

raspberry ringspot virus See NEPOVIRUSES.

raspberry vein chlorosis virus See RHABDOVIRIDAE.

RAST (radioallergosorbent test) An IMMUNOASSAY used for detecting specific IgE antibodies. The allergen, bound to the surface of a paper disc, is exposed to the test serum and washed; bound IgE is detected by the addition of radiolabelled anti-IgE antibodies.

rat-bite fever Either of two human diseases contracted from the bite of a rat or, less commonly, a cat, dog etc. One form (*Haverhill fever*) is caused by *Streptobacillus moniliformis*; symptoms include e.g. the development of a fluid-filled sore at the site of the bite, a skin rash, recurrent fever, lymphadenopathy, and more or less generalized arthritis. The second form (*sodoku*) occurs more commonly in Japan and other Eastern countries and is caused by *'Spirillum minus'* (see SPIRILLUM); symptoms resemble those of Haverhill fever, but there is usually no arthritis.

rat leukaemia viruses Ecotropic type C retroviruses (subfamily ONCOVIRINAE) which are apparently ubiquitous in rats; none is known to be leukaemogenic or to have any oncogenic potential. [Book ref. 114, pp. 101–104.]

rat virus See PARVOVIRUS.

rate–zonal centrifugation See CENTRIFUGATION.

Ratkowsky plot The square root of growth rate plotted against absolute temperature; in various bacteria, this plot shows a linear relationship over the full biokinetic range [JB (1983) *154* 1222–1226].

Rauscher virus (RV) A virus complex consisting of at least two components: a replication-competent MURINE LEUKAEMIA VIRUS (Ra-MuLV) and a replication-defective SFFV (cf. FRIEND VIRUS); within ca. 14 days of infection of newborn or adult mice, RV induces splenomegaly and anaemia (erythroblastosis), followed after several weeks by the onset of lymphocytic leukaemia.

raw water See WATER SUPPLIES.

ray body (*Strahlenkörper*) A microgamont or an isogamete of *Babesia* or *Theileria*.

ray fungus Archaic term for an ACTINOMYCETE.

razor-strop fungus *Piptoporus betulinus*.

RB Reticulate body (see CHLAMYDIA).

RBC Red blood cell (erythrocyte).

Rc strain See SMOOTH–ROUGH VARIATION.

RCF (relative centrifugal field) See CENTRIFUGATION.

RCM REINFORCED CLOSTRIDIAL MEDIUM.

Rd_1 strain See SMOOTH–ROUGH VARIATION.

Rd_2 strain See SMOOTH–ROUGH VARIATION.

RD-114 viruses See FELINE LEUKAEMIA VIRUS.

RDE RECEPTOR-DESTROYING ENZYME.

rDNA DNA coding for rRNA.

RE RESTRICTION ENDONUCLEASE.

Re strain See SMOOTH–ROUGH VARIATION

REA See DNA FINGERPRINTING.

reaction centre In photosynthetic membranes: a type of pigment-containing complex, distinct from LIGHT-HARVESTING COMPLEXes, to which radiant energy is channelled and at which charge separation occurs (see PHOTOSYNTHESIS).

In algae and cyanobacteria, the photosynthetic membranes contain two types of reaction centre (RC). The RC in photosystem I includes a specialized form of CHLOROPHYLL *a* designated P_{700} (indicating its absorption maximum of 700 nm); the RC chlorophyll represents only ca. 1/300th of the chlorophyll content of the membrane. The RC of photosystem II also appears to contain a chlorophyll *a* moiety (designated P_{680} – again, to indicate its absorption maximum) together with e.g. phaeophytin *a*.

The RCs of several anoxygenic photosynthetic bacteria have been characterized. Thus, e.g. that in *Rhodopseudomonas sphaeroides* includes bacteriochlorophyll *a* (P_{870}), bacteriophaeophytin *a*, ubiquinone and ferrous iron; the RC in *R. viridis* includes bacteriochlorophyll *b*, bacteriophaeophytin *b*, and (apparently) a *c*-type cytochrome, while that in *Rhodospirillum rubrum* includes bacteriochlorophyll *a* (P_{870}), bacteriophaeophytin, ubiquinone, ferrous iron and spirilloxanthin.

reactivation (*virol.*) (1) (genetic reactivation) A phenomenon in which a DEFECTIVE VIRUS can produce infectious progeny virions in the presence of another (defective or non-defective) virus as a result of recombination between or reassortment of the two genomes. If only one of the two 'parent viruses' is defective, reactivation is called *cross-reactivation* or *marker rescue*; if both parents are defective it is called *multiplicity reactivation*. Obviously, multiplicity reactivation can occur only if the genetic lesions are present at different sites on the two genomes. [Book ref. 148, pp. 115–116.] (cf. NON-GENETIC REACTIVATION.)

(2) The re-establishment of replication by a virus previously in a latent (non-replicating) form – see e.g. HERPES ZOSTER.

reactive arthritis ARTHRITIS which commonly develops in certain genetically predisposed (HLA B27$^+$) people after an acute infection with *Salmonella*, particularly *S. typhimurium* or *S. enteritidis*; in these people, certain microbial antigens – especially outer membrane components – seem to persist in the body and to provoke an inflammatory response in the joints. [Review: Lancet (1992) *339* 1096–1098.]

reactive lysis (*immunol.*) The lysis of a cell as a consequence of the attachment to its surface of complex C5b67 (see COMPLEMENT FIXATION) which was formed at another site; lysis follows the development of the membrane attack complex.

Reactive lysis is also called *bystander lysis*.

reactor (1) (*biotechnol.*) *Syn.* BIOREACTOR. (2) (*med., vet.*) One who gives a reaction to a given test procedure, e.g. a SKIN TEST.

reading frame (*mol. biol.*) See GENETIC CODE. (See also OPEN READING FRAME.)

Reading unit (RU) A unit used in respect of the antibiotic NISIN; 40 RU correspond to ca. 1 μg of pure nisin.

readthrough (1) A phenomenon in which, during PROTEIN SYNTHESIS, polypeptide chain elongation is continued past a stop codon, an amino acid being inserted at the stop codon. Readthrough may occur as a result of an intergenic nonsense SUPPRESSOR MUTATION, but can also occur in the presence of certain normal tRNAs; thus, e.g., the A1 protein of bacteriophage Qβ (see LEVIVIRIDAE) is produced by occasional readthrough of a UGA stop codon at the end of the coat protein gene, translation continuing until a UAG codon 195 codons further on is reached.

(2) The continuation of TRANSCRIPTION through a transcription terminator. (See also ANTITERMINATION.)

reagin (1) *Syn.* REAGINIC ANTIBODY. (2) See STANDARD TESTS FOR SYPHILIS.

reaginic antibodies CYTOPHILIC ANTIBODIES which are elicited by an ALLERGEN and which can bind to the surface of MAST CELLS (see TYPE I REACTION). The most effective reaginic antibodies are of the IgE type, although IgG antibodies can also function in this way. (cf. REAGIN and HOMOCYTOTROPIC ANTIBODIES.)

real-time PCR See PCR, TAQMAN PROBES, Book ref. 221.

reassortant virus In a virus with a SEGMENTED GENOME: a hybrid virion which contains genome segments from different parents; such a virion is produced in a cell coinfected with different (e.g. mutant) strains of a given virus. (See e.g. INFLUENZAVIRUS.)

REBASE® A source of information on RESTRICTION ENDONUCLEASES etc.: http://rebase.neb.com/rebase/rebase.html.

***recA* gene** See RECA PROTEIN.

RecA protein A multifunctional protein (MWt ca. 37800) encoded by the *recA* gene in *Escherichia coli* and other enterobacteria. (*RecA*-like genes or gene products have also been detected e.g. in *Agrobacterium tumefaciens* [MGG (1986) *204* 161–165], *Bacillus subtilis* [JBC (1985) *260* 3305–3313], *Methylophilus methylotrophus* [Gene (1986) *44* 47–53], *Vibrio cholerae* [MGG (1986) *203* 58–63], and *Ustilago maydis* (the *rec1* gene product) [Cell (1982) *29* 367–374].) In members of the Archaea, the RadA protein is homologous to RecA [GD (1998) *12* 1248–1253].

The RecA protein plays an essential role in general RECOMBINATION, and can promote various interactions between DNA molecules (including synapsis, strand exchange and branch migration) provided that at least one of the DNA molecules is at least partly single-stranded. For example, in vitro, RecA can promote base-pairing between a ccc ssDNA molecule and its complementary strand in a (linear) dsDNA molecule; the ssDNA 'invades' the dsDNA molecule and displaces the homologous strand (a phenomenon known e.g. as 'single-strand assimilation'). This reaction is initiated when RecA binds stoichiometrically and cooperatively to ssDNA in the presence of ATP to form a helical nucleoprotein filament [structure: JMB (1986) *191* 677–697] (*presynaptic stage*); polymerization is accelerated in the presence of SINGLE-STRAND BINDING PROTEIN (SSB protein) and is inhibited by ADP. The RecA–ssDNA complex then interacts with dsDNA to form a ternary dsDNA–ssDNA–RecA complex in which the dsDNA becomes extensively unwound. The ternary complex can be formed in the absence of homology; it is generally believed that a 'search' for homology between the ssDNA and dsDNA must then occur. Juxtaposition of complementary regions (*synapsis*) results in the formation of a nascent heteroduplex joint. Recent studies suggest that (contrary to an earlier view) the 'search' for homology is unlikely to involve a triplex DNA intermediate – but may involve unwinding of the target duplex coupled with local strand exchange; it also appears that the polarity of RecA–ssDNA does not affect its ability to interact with a circular target duplex [NAR (2001) *29* 1389–1398].

RecA-promoted strand exchange can continue through regions of mismatched bases, past thymine dimers, and past deletions or insertions in one of the strands.

The single-strand assimilation reaction *per se* is probably of limited significance in vivo (cf. TRANSFORMATION). However, in vitro, the RecA protein can also promote pairing between two dsDNA molecules provided that there is a gap in one of them, a reaction which may resemble more closely a homologous recombination event in vivo; the necessary regions of ssDNA may be provided e.g. by the RECBC PATHWAY.

[Role of RecA in recombination: CSHSQB (1984) *49* 507–580; mechanistic aspects of DNA strand exchange activity of RecA: TIBS (1987) *12* 141–145.]

The RecA protein is normally synthesized at a low level in the cell, being repressed to this level by the product of the *lexA* gene. However, under certain conditions – e.g. in the presence of damaged DNA – the RecA protein becomes 'activated', acquiring the ability to trigger proteolytic cleavage of the LexA protein (thus inducing the SOS SYSTEM q.v.) and of the repressor proteins of BACTERIOPHAGE LAMBDA and other LAMBDOID PHAGES (thus initiating prophage induction).

RecA* See SOS SYSTEM.

recapitulation theory See ONTOGENY.

***recB* gene** See RECBC PATHWAY.

RecBC pathway A pathway for general RECOMBINATION in *Escherichia coli* and related bacteria; it appears to be the major pathway for recombination following CONJUGATION. The pathway requires the products of the *recA* gene (see RECA PROTEIN) and of the *recB* and *recC* genes, the latter two being components of a multifunctional enzyme formerly known as the RecBC enzyme or exonuclease V. More recently, this enzyme (~330 kDa) has been re-designated the RecBCD enzyme.

The RecBCD enzyme carries out functions in one of two major pathways ('recombination machines') for homologous recombination in *Escherichia coli*. In this pathway the enzyme is responsible e.g. for repairing double-stranded breaks in DNA; repairs can be carried out only on *linear* substrates – including circular molecules with a double-stranded break.

The RecBCD enzyme has helicase and nuclease activities, and it also has a *synaptogenic* function (see later). One of the differences between the RecBCD pathway and the RecF pathway is that the former involves a single, multi-subunit enzyme while the latter involves the activities of a number of separate enzymes.

As in the RecF pathway, recombination can be *initiated* in the RecBCD pathway; in RecF, initiation is triggered e.g. by a single-stranded gap in the chromosome, while in the RecBCD pathway initiation is prompted e.g. by DNA damage involving a double-stranded break. Given a double-stranded break, the (ATP-dependent) RecBCD nuclease may degrade a 5′ terminal, leaving a free (single-stranded) 3′ terminal – to which molecules of the RecA protein bind, forming the *nucleoprotein filament*. (RecBCD mediates the *exclusion* of single-strand binding proteins from the 3′ terminal; this is necessary in order to permit binding of RecA monomers.)

A nucleoprotein filament is apparently the initial structure formed in all cases of homologous recombination (including recombination mediated by RecBCD and RecF systems).

The nucleoprotein filament is believed to 'search' another (target) duplex for a sequence of nuleotides homologous to the sequence in the nucleoprotein itself. This may involve local unwinding of the target duplex. The juxtaposition of homologous sequences (*synapsis*) leads to the formation of a hybrid duplex (*heteroduplex*). Events may then proceed as described for the Holliday model: see RECOMBINATION.

Recombination via the RecBCD pathway is stimulated by the presence of certain sequences of nucleotides within the chromosomal DNA. For example, the *E. coli* chromosome has about 1000 copies of the sequence:

5′–GCTGGTGG–3′

i.e. about one copy per 5 kb of DNA; these sequences, which are called chi (χ) sites, enhance homologous recombination in their vicinity.

The RecBCD enzyme has an affinity for chi sites and, as a result of interaction between RecBCD and a chi site, a chi site tends to be found at or near the 3′ terminus of DNA in the nucleoprotein filament.

Chi-dependent activity has been reported in other enterobacteria, and e.g. in some species of the family Vibrionaceae, but appears to be absent in various other bacteria; it may be that different sequences of nucleotides play a similar role in other types of bacteria.

More recently it has been found that certain functions of the RecBCD and RecF pathways are interchangeable; if, for example, a null mutation occurs in the *recD* (nuclease) subunit of RecBCD then this function can be compensated for by the RecJ nuclease of the RecF pathway. [Interchangeable parts of the *Escherichia coli* recombination machinery: Cell (2003) *112* 741–744.]

***recC* gene** See RECBC PATHWAY.

***recE* gene** See RECF PATHWAY.

RecE pathway See RECF PATHWAY.

receptacle (*mycol.*) See e.g. PHALLALES.

receptive hyphae *Syn.* FLEXUOUS HYPHAE.

receptor-destroying enzyme (RDE) An enzymic component of a virion which destroys the (virion-specific) surface receptors of the host cell. (See e.g. NEURAMINIDASE; cf. e.g. INFLUENZA VIRUS C.)

receptor-mediated pinocytosis See PINOCYTOSIS.

recessive allele See DOMINANCE.

***recF* gene** See RECF PATHWAY.

RecF pathway A *recA*-dependent, *recBC*-independent pathway for general RECOMBINATION in *Escherichia coli*.

Components of the RecF pathway are used in *recBC* mutants of *E. coli* which have reverted to recombination proficiency as a result of a mutation in the *sbcB* gene (= *xonA* gene, encoding exonuclease I – an enzyme that degrades ssDNA from the 3′ end). This pathway apparently involves (as well as *recA*) at least genes *recF*, *recJ*, *recN*, *recQ* and *ruv*, the first four of these genes are apparently from the RecF pathway.

Another pathway (= RecE pathway) involving components of the RecF pathway occurs in *recBC* mutants which have reverted to recombination proficiency by a mutation in *sbcA* (a gene in a prophage designated *rac*); this pathway involves *recE* – encoding exonuclease VIII, an enzyme that degrades one strand of dsDNA (5′→3′) – and *recA*, *recF* and *recJ*. The *sbcA* mutation appears to induce RecE.

Apart from the instances mentioned above, the RecF pathway is itself a major 'recombination machine' in *E. coli* which e.g. repairs single-stranded gaps in the chromosome. Unlike the recBC pathway, its activity involves a number of separate proteins – some of which can compensate for functional deficiencies in the recBC enzyme [interchangeable parts of the *E. coli* recombination machinery: Cell (2003) *112* 741–744].

At least some genes in the RecF pathway are inducible and are regulated by *lexA* (see SOS SYSTEM).

recipient Any cell or organism which receives genetic information e.g. by CONJUGATION, TRANSDUCTION or TRANSFORMATION.

reciprocal recombination A process of RECOMBINATION involving the mutual exchange of sequences by CROSSING OVER between the two participating nucleic acid molecules (or between two regions in the same molecule).

***recJ* gene** See RECF PATHWAY.

reckless DNA degradation The rapid and extensive degradation of DNA, carried out by the RecBC nuclease, which occurs in cells of *recA* mutants of *Escherichia coli* on exposure to UV radiation or other DNA-damaging treatment.

***recN* gene** See RECF PATHWAY.

recombinant (1) (*noun*) An organism whose genome is a product of RECOMBINATION. (2) (*adj.*) Refers to an organism as in (1) above, or to a nucleic acid which has resulted from recombination; the term *recombinant DNA* is commonly used specifically for DNA molecules which have been constructed in vitro using various GENETIC ENGINEERING techniques.

recombinant immunoblot assay (RIBA) See HEPATITIS C.

recombinase See SITE-SPECIFIC RECOMBINATION.

recombination (genetic recombination) A process in which one or more nucleic acid molecules are re-arranged to generate new combinations or sequences of genes, alleles, or other nucleotide sequences; it may involve e.g. the physical exchange of material between two molecules, the integration of two molecules to form a single molecule, the inversion of a segment within a molecule, etc. There are several categories of recombination, including *general recombination* (see below), SITE-SPECIFIC RECOMBINATION, and transpositional recombination (see TRANSPOSABLE ELEMENT). (cf. ILLEGITIMATE RECOMBINATION.)

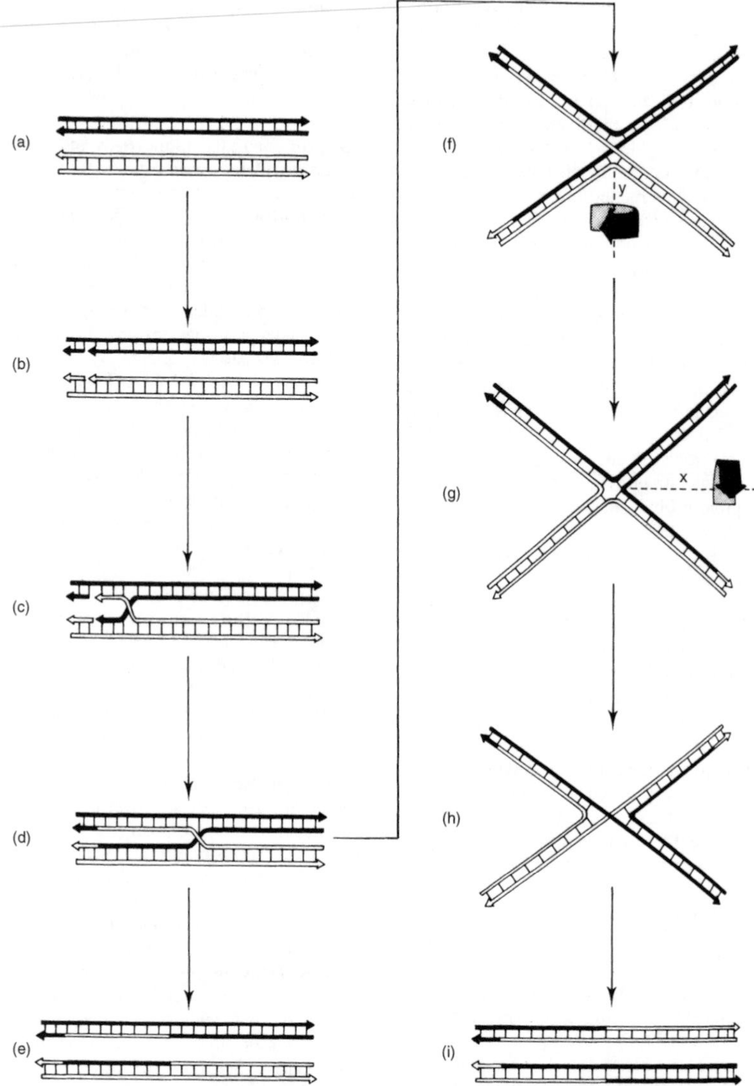

RECOMBINATION: Figure 1. *Holliday model* [Genet. Res. (1964) *5* 282–304]. (a) Two homologous DNA duplexes; the 3′ ends are identified by arrowheads. (b) A nick occurs at corresponding sites in one strand of each duplex. (c) One strand from each duplex begins to pair with the complementary strand of the other duplex to form a *heteroduplex joint*. (d) Ligation results in a branched 'half-chiasma' (a *Holliday structure* or *Holliday junction*); the position of the branch may move (*branch migration*) in either direction and may increase the length of the heteroduplex region (as shown). Resolution of the Holliday structure by nicking, exchange and re-ligation of the crossed strands results in the molecule shown at (e); in this case markers flanking the heteroduplex region are not exchanged (i.e., there is no CROSSING OVER). (f) Unstacking of the bases at the site of the joint allows the formation of a symmetrical 'chi structure' which can undergo the *isomerizations* shown at (g) and (h): rotation through 180° of the two lower arms about axis y gives the structure at (g), and a subsequent 180° rotation of the two right-hand arms about axis x gives the structure at (h). This results in the crossing of the previously un-nicked (outer) strands; resolution of the Holliday structure by nicking, exchange and ligation of these strands results in reciprocal exchange of flanking markers (crossing over) to give the recombinant duplexes shown at (i).

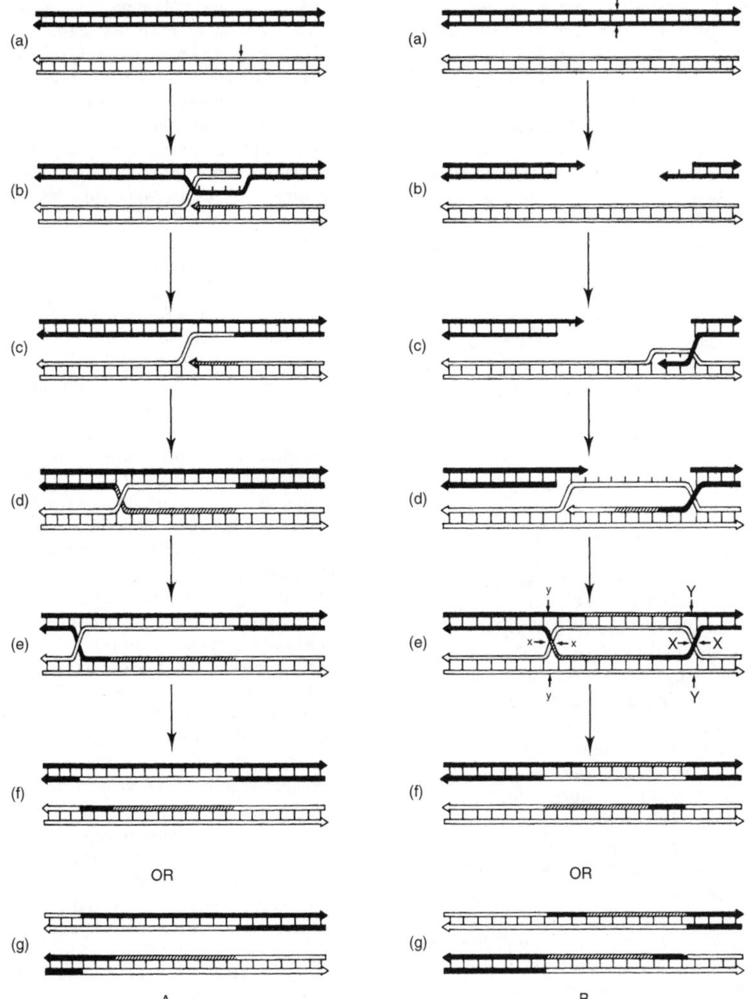

RECOMBINATION: Figure 2. *A. Meselson–Radding model (Aviemore model)* [PNAS (1975) *72* 358–361]. (a) Recombination is initiated by a nick in one strand of one of the two homologous duplexes (the 'donor' duplex). (All 3′ ends are identified by arrowheads.) (b) The 3′ end generated by the nick serves as a primer for DNA synthesis and is extended (shaded portion), displacing the 5′ end which then 'invades' a homologous region of the other ('recipient') duplex to generate a D LOOP. (c) The donor strand is ligated in place, and the D loop is degraded. (d) The heteroduplex region can be extended by further nuclease action on the recipient duplex and DNA polymerase action on the donor duplex. Ligation completes the Holliday structure. (e) Branch migration may occur (e.g. as shown). Resolution of the Holliday structure may occur by exchange between the inner strands, with no crossing over (f), or – following isomerization (see Fig. 1) – between the outer strands, resulting in crossing over (g).

B. Double-strand break – repair model [Cell (1983) *33* 25–35]. (a) Recombination is initiated by nicks in both strands of one of the two homologous duplexes (the 'recipient' duplex). (b) The nicks are enlarged to gaps by exonuclease action, and (c) a free 3′ end thus generated 'invades' a homologous region of the donor duplex, giving rise to a D loop. (d) Extension of the 3′ end of the invading strand by repair synthesis enlarges the D loop until a sequence homologous to the second recipient fragment is exposed. (e) Further repair synthesis and ligation complete two Holliday junctions; branch migration may occur (not shown). Cutting and ligation at x and X, as in (f), or at y and Y (not shown) resolve the Holliday structures without crossing over; cutting and ligation at X and y, as in (g), or at x and Y (not shown), result in crossing over.

General recombination (*homologous recombination*) occurs only between two sequences which have fairly extensive regions of homology; the sequences may be in different molecules or in different regions of the same molecule. Early models for the mechanism of general recombination were basically of two types. *Copy-choice models* supposed that recombination occurs during nucleic acid synthesis and involves switching of the polymerase from one template strand to another (homologous) template strand to form a new recombinant daughter strand. *Breakage-and-reunion models* supposed that recombination involves strand breakage, exchange of the broken strands, and rejoining of the exchanged strands to complete the recombinant molecule(s); when strand exchange is symmetrical the process is called *reciprocal recombination* (see CROSSING OVER). The copy-choice model may be applicable to recombination in (RNA) viruses of the RETROVIRIDAE, but recombination between DNA molecules appears generally to occur by a breakage-and-reunion-type mechanism – often in conjunction with limited DNA synthesis; some models are depicted in Figures 1 and 2 (on pages 646 and 647).

In addition to homology, another requirement for homologous recombination is the availability of single-stranded DNA in at least one of the molecules. Moreover, the RECA PROTEIN has an essential role in the process, being able to promote various interactions between DNA molecules. In the Holliday model (Figure 1), a free end of ssDNA (derived from a nicked duplex) is juxtaposed to homologous ssDNA in the un-nicked strand of another nicked duplex, both strands then forming a hybrid duplex (*heteroduplex*) that may contain one or more mismatched bases; similarly, the free end displaced from the second duplex may base-pair with ssDNA in the first duplex – again, forming a heteroduplex (Figure 1(c)). The structure at (c) is called a *heteroduplex joint*. If each free end is now ligated, the result is a branched structure (a 'half chiasma') called a *Holliday structure*, *Holliday junction* or *chi structure* (from its χ-like shape in electronmicrographs); this is a region where two duplexes are held together by two single strands crossing between the duplexes (Figure 1(d)). Each strand may continue to separate from its parent duplex and base-pair with ssDNA of the other duplex – so that the position of the branch migrates between the duplexes in the corresponding direction (*branch migration*): compare the position of the branch in Figure 1(c) and Figure 1(d). The extent of branch migration determines the final length of the heteroduplex. In *Escherichia coli* branch migration is driven by two ring-shaped helicases (one on either side of the Holliday junction) through which DNA passes; each helicase is a hexamer of the RuvB protein. Resolution of a Holliday junction (i.e. separation of duplexes) involves an endonuclease, RuvC, which cuts each strand, thus allowing ligation to occur in each duplex. In *E. coli*, RuvB and RuvC appear to interact co-operatively [EMBO (1998) *17* 1838–1845].

The precise role of RecA in homologous recombination is not understood. Apparently, this protein initially binds to a free ssDNA strand to form a helical nucleoprotein 'filament' containing six monomers of RecA per helical turn. The filament must then 'search' the target DNA duplex for a region homologous to the ssDNA; the searching process appears to involve unwinding of the target duplex, and local strand exchange, leading to juxtaposition of homologous strands (*synapsis*) and the formation of a heteroduplex [NAR (2001) *29* 1389–1398].

The region of heteroduplex DNA in a heteroduplex joint may be extensive. If the two strands in the joint carry different alleles (e.g. one wild-type, one mutant) the heteroduplex will have a region of mispaired bases at the mutation site. Such mismatched regions may be recognized by a MISMATCH REPAIR system in which nucleotides are removed from one of the strands and replaced, by repair synthesis, using the other strand as template; thus, the nucleotide sequence of one strand is converted to a sequence complementary to that of the template, a phenomenon known as *gene conversion* (an example of *non-reciprocal recombination*). (In a bacterial merozygote, the term 'gene conversion' may be restricted to cases in which a sequence in the recipient's DNA is converted to a sequence of the donor fragment.) Extensive mismatch repair within a long region of heteroduplex DNA can lead to the conversion of two or more adjacent alleles (*co-conversion*). (cf. INTERFERENCE (sense 2) and POST-MEIOTIC SEGREGATION.)

Although it was originally thought that general recombination could occur with equal probability between any regions of homology of a given length, it has been found that certain nucleotide sequences ('recombinators' or 'recombinogenic sequences') can specifically enhance general recombination in their vicinity: see e.g. *chi site* under RECBC PATHWAY.

In prokaryotes, general recombination may occur e.g. between donor and recipient DNA in a MEROZYGOTE, between homologous regions of different plasmids or of a plasmid and the chromosome, etc; in *Escherichia coli*, at least, several possible pathways of general recombination are known, all of which apparently require the product of the *recA* gene (see RECA PROTEIN and e.g RECBC PATHWAY).

In species of the domain ARCHAEA the RadA protein appears to be equivalent, in function, to the RecA protein.

In viruses, general recombination may occur e.g. between genomes of viruses that are co-infecting the same cell (cf. REASSORTANT VIRUS). In certain bacteriophages recombination is apparently an essential step in the viral replication cycle (see e.g. BACTERIOPHAGE T4).

In eukaryotes, general recombination may occur in meiosis – and occasionally in mitosis: see PARASEXUAL PROCESSES.

recombination frequency Following general RECOMBINATION (e.g. after meiosis): the number of recombinant progeny, in respect of the alleles of two different genes, as a proportion of the total progeny (expressed as a percentage). Recombination frequency (RF) gives an indication of the relative positions of markers (genes or mutations) in the genome. An RF of 50% indicates that the markers segregate independently between progeny, e.g., they occur on different chromosomes; lower frequencies indicate that the markers occur on the same chromosome – the lower the frequency, the smaller the distance between them. (See also INTERFERENCE (sense 2) and RECOMBINATOR.)

recombination nodule See SYNAPTONEMAL COMPLEX.

recombination repair In e.g. *Escherichia coli*: a *recA*-dependent DNA REPAIR process which can repair e.g. thymine dimers or double-strand breaks. During replication of a dsDNA molecule containing e.g. a thymine dimer, the DNA polymerase stops at the site of the dimer; DNA synthesis may be re-initiated by primer synthesis at a site beyond the lesion, leaving a gap in the daughter strand opposite the dimer in the damaged parent strand. This gap can be filled by RECOMBINATION between the gapped daughter strand and an undamaged homologous region of the strand complementary to the damaged parent strand (*daughter-strand gap repair*). The resulting gap in the homologous parent strand can then be filled by repair synthesis, its daughter strand acting as template. The dimer is thus not removed, but is effectively 'diluted out'. Genes apparently involved in daughter-strand gap repair include *recA*, *ruv*, *lexA* and *recF*; levels of the products of these genes, and hence the potential for recombination repair, are increased by induction of the SOS SYSTEM.

E. coli also has an inducible *double-strand break repair* system which requires the presence of at least one other dsDNA molecule homologous to the damaged molecule. This system requires e.g. *recA* and *recN* (SOS genes) and possibly involves a mechanism similar to that shown in RECOMBINATION (Fig. 2).

recombinational regulation A type of genetic regulation in which intramolecular recombination acts as a mechanism for switching gene expression on and off. In several cases a segment of DNA is inverted by SITE-SPECIFIC RECOMBINATION catalysed by a specific recombinase: see e.g. PHASE VARIATION (in *Salmonella*), BACTERIOPHAGE MU (G segment) and BACTERIOPHAGE P1 (C segment). A similar system – involving an invertible *P segment* (formerly known as the e14 element) and a recombinase encoded by *pin* – occurs in the chromosome of *Escherichia coli* K12 strains. The amino acid sequences of the Hin, Gin, Cin and Pin recombinases show high degrees of homology, and the enzymes are functionally interchangeable.

Other types of recombinational regulation are known: see e.g. FIMBRIAE; MATING TYPE; NIF GENES; VSG. (See also SHUFFLON.)

recombinator (recombinogenic element) A 'recombinational hotspot', i.e., a nucleotide sequence in a DNA molecule which enhances general RECOMBINATION in its vicinity. An example is the chi site in *Escherichia coli*: see RECBC PATHWAY.

recombinogenic element *Syn.* RECOMBINATOR.

***recQ* gene** See RECF PATHWAY.

rectus–flexibilis See SPIRA.

recurrent fever *Syn.* RELAPSING FEVER.

red algae Algae of the RHODOPHYTA.

red bread mould See BREAD SPOILAGE.

red clover mottle virus See COMOVIRUSES.

red clover necrotic mosaic virus See DIANTHOVIRUSES.

red clover vein mosaic virus See CARLAVIRUSES.

red core (of strawberry) A soil-borne STRAWBERRY DISEASE that is caused by *Phytophthora fragariae*. Plants become obviously stunted, and within the darkened, rotting roots the stele becomes red or purplish-brown. *P. fragariae* forms resistant oospores which can remain viable in soil for many years; in wet conditions the oospores can germinate to form zoospores, aiding the spread of disease.

red diaper syndrome The (apparently non-pathological) presence of red-pigmented *Serratia* spp in the faeces of neonates and infants.

red disease of pike virus See RHABDOVIRIDAE.

red drop In algae and green plants: an abrupt decrease in PHOTOSYNTHESIS which occurs when the wavelength of incident monochromatic light increases beyond ca. 680 nm; photosynthetic activity is virtually abolished at ca. 700 nm. (cf. ENHANCEMENT EFFECT.)

red leg A (usually fatal) disease of frogs and other freshwater aquatic animals, caused by *Aeromonas hydrophila*. Symptoms include haemorrhagic skin lesions and septicaemia.

red membrane See PURPLE MEMBRANE.

red mouth A FISH DISEASE common in trout farms when fish are stressed e.g. by overcrowding. The causal agent varies but is often *Aeromonas hydrophila*. Symptoms: reddening of the mouth, throat, and gill opercula. (cf. ENTERIC REDMOUTH.)

red nose *Syn.* INFECTIOUS BOVINE RHINOTRACHEITIS.

red pest Either of two diseases of eels involving haemorrhagic ulceration of the skin. Pathogens: *Listonella anguillarum* in marine and brackish waters, *Aeromonas hydrophila* in fresh water. (See also FISH DISEASES.)

red rice *Syn.* ANG-KAK.

red rot (1) A SUGARCANE DISEASE caused by *Colletotrichum falcatum*; fungal attack is generally concentrated within the stem and midribs of the leaves, with reddening of internal tissues and withering of the tips and margins of the leaves.

(2) See ALGAL DISEASES.

red rust (algal rust) A disease of tea (*Camellia sinensis*) and various other commercially important plants (e.g. avocado, cacao, citrus, coffee, mango) caused by a species of CEPHALEUROS. The alga grows beneath the cuticle of the leaves and stems, resulting in the formation of elevated, velvety, reddish or orange-brown lesions. Infected leaves may exhibit a 'wound response', with necrosis of tissue around the lesions, which may help to prevent the spread of the alga. Infections of leaves and fruit generally cause little harm to the plant, apart from the unsightly lesions and – in heavily infected plants – premature leaf-drop. However, infections of young stems may be more serious, and plants may even be killed if the stems are completely girdled. [Review: Book ref. 129, pp. 173–204.] The disease has been controlled with copper sprays. (See also BLISTER BLIGHT.)

red snow In mountainous areas of western North America: snow which appears reddish owing to the abundant growth of *Chlamydomonas nivalis*. *C. nivalis* grows – in spring and summer – mainly near the surface of the melting snow; it forms the basis of a fairly extensive food-chain involving protozoa, rotifers, nematodes and crustaceans.

red tide (red water) A phenomenon in which regions of the sea or of an estuary become discoloured (often red or reddish) owing to extensive BLOOMS of certain algae – particularly DINOFLAGELLATES. (Blooms of the cyanobacterium TRICHODESMIUM are sometimes called red tides.) For example, *Gonyaulax catenella* and *G. tamarensis* (= e.g. *G. excavata*, *Protogonyaulax tamarensis*, *Alexandrium tamarense*) are common causes of red tides on the coasts of North America, while *Ptychodiscus brevis* (= *Gymnodinium breve*) forms red tides in the Gulf of Mexico and along the Florida coast. Blooms of *G. tamarensis* are seasonal; evidence has been obtained for the existence of an endogenous annual 'clock' (circannual rhythm) in *G. tamarensis* from deep coastal waters [Nature (1987) *325* 616–617].

A red tide can cause massive mortalities in fish and other aquatic animals – due to oxygen depletion and/or to the production of various toxins (e.g. BREVETOXINS, GONYAUTOXINS, SAXITOXIN). These toxins can also cause poisoning in man, particularly when shellfish from affected waters are eaten (see *paralytic* SHELLFISH POISONING).

red vent disease *Syn.* ENTERIC REDMOUTH.

red wasting disease See ALGAL DISEASES.

red water (1) (of calves) An acute, febrile, often fatal disease of young calves, characterized by jaundice and haemoglobinuria; it is caused by *Leptospira interrogans* serotype *pomona* (cf. LEPTOSPIROSIS).

(2) Any of various diseases characterized by haemoglobinuria. (See also REDWATER FEVER.)

(3) *Syn.* RED TIDE.

red water fever See REDWATER FEVER.

redox couple See REDOX POTENTIAL.

redox dye See REDOX POTENTIAL.

redox potential (oxidation–reduction potential; E_h; E) A measure of the tendency of a given system to donate electrons (i.e., to act as a reducing agent) or to accept electrons (i.e., to act as an oxidizing agent); the E_h of a given system may be determined e.g. by measuring the electrical potential difference between that system and a standard hydrogen electrode whose potential is, by convention, arbitrarily taken to be zero volts. (E_h may be recorded in volts (V) or in millivolts (mV).) The redox state of certain microbial components can determine their activity: see e.g. TRANSPORT SYSTEMS.

A *redox couple* is a pair of mutually interconvertible substances (ions and/or atoms or molecules) which are present, in a given system, in proportions which can be altered by the addition or removal of electrons (and, in some cases, protons). An example is the Fe^{2+}/Fe^{3+} couple ('iron electrode'). A solution containing only ferric ions is fully oxidized; such a solution has no reducing power – but it has maximum oxidizing capacity and, by convention, it is said to have the highest (most positive) E_h for the Fe^{2+}/Fe^{3+} system; conversely, a solution containing only ferrous ions has maximum reducing power and the lowest possible (most negative) E_h for the system. Solutions containing both ferric and ferrous ions have intermediate values of E_h; the actual E_h in a given system depends e.g. on the ratio of *activities* (i.e., effective concentrations) of the two species of ion. If electrons are added to the Fe^{2+}/Fe^{3+} system, some or all of the Fe^{3+} ions are reduced, and the E_h falls to a lower value; similarly, removal of electrons raises the E_h.

Other redox couples include e.g. oxidized cytochrome/reduced cytochrome, $NAD(P)^+/NAD(P)H$, and $H^+/{}^1\!/_2\,H_2$.

Standard redox potential (half-reduction potential; mid-point potential; E_0; E^0; E_m). In any redox couple: when the ratio of the activities of the reduced and oxidized species is 1:1, the E_h (relative to the standard hydrogen electrode) is referred to as the standard redox potential, E_0, of that couple. The value of E_0 varies with pH; the E_0 at pH 7 is usually written E_0' or $E_{m.7}$. (For the hydrogen electrode, E_0 at pH 0 is 0 V, while E_0' is −420 mV.)

E_0 can be used to calculate the E_h of a given redox couple when the ratio oxidized:reduced species is not unity:

$$E_h = E_0 - \frac{RT}{nF}\log_e \frac{\text{activity of reduced species}}{\text{activity of oxidized species}}$$

where R is the gas constant (8.31 J/mol/degree abs.); T is the absolute temperature; F is the Faraday (96,490 C/mol); and n is the number of electrons transferred to/from each atom/ion/molecule – e.g. 1 in the Fe^{2+}/Fe^{3+} system.

E_0 can also be used to predict possible interactions between redox couples; thus, e.g. a given redox couple may oxidize a second of lower (less positive, or more negative) E_0. A redox system which can maintain a given E_h, within certain limits, in the presence of other redox couple(s) is referred to as a *poising system*; in a redox couple dominated by a poising system, the ratio oxidized:reduced species adjusts to conform to the imposed E_h.

Measurement of E_h. The most accurate determination of E_h involves the use of a hydrogen electrode (or other standard electrode). The hydrogen electrode consists of a suitably prepared platinum plate immersed in 1.228 N HCl (activity of $H^+ = 1$ gram ion/litre); when a stream of hydrogen gas (under standard conditions) impinges on the plate the following equilibrium is set up:

$$H_2 \longleftrightarrow 2H^+ + 2\text{ electrons}$$

and the potential difference between plate and acid is taken to be zero volts. The potential difference between the hydrogen electrode and a given redox couple (i.e. the E_h of the given couple) can be measured by a potentiometric (zero-current) method – i.e. a method which does not disturb the equilibrium in either redox couple.

Certain dyes (*redox dyes*, or *indicators*) can indicate E_h by changing from a coloured to a colourless state (and vice versa) within a given E_h range; thus, e.g. a dye may become completely colourless (at pH 7) when the E_h is $\leq$ ca. −40 mV (e.g. METHYLENE BLUE), ca. −70 mV (e.g. toluidine blue), ca. −110 mV (e.g. resazurin), and ca. −310 mV (e.g. phenosafranine). (See also TETRAZOLIUM.)

E_h of microbiological media. In general, an obligate ANAEROBE can grow only on or in media which are at or below a certain value of E_h. For this reason media are often *poised* at a certain E_h; poising is analogous to buffering in the context of pH – i.e., while the E_h produced depends on the ratio oxidized:reduced poising agent, the stability of E_h is influenced by the absolute amount of poising agent in the medium. Poising agents used in media include e.g. ASCORBIC ACID, cysteine hydrochloride and thioglycollate.

reduction division *Syn.* MEIOSIS.

reductive carboxylic acid cycle *Syn.* REDUCTIVE TRICARBOXYLIC ACID CYCLE.

reductive pentose phosphate cycle *Syn.* CALVIN CYCLE.

reductive tricarboxylic acid cycle A pathway for CO_2 fixation in certain prokaryotic AUTOTROPHS (mainly members of the Chlorobiaceae – cf. SULFOLOBUS). The pathway is effectively the reverse of the TCA CYCLE [see Appendix II], one molecule of acetyl-CoA being formed for every two of CO_2 fixed; the irreversible steps of the TCA cycle are bypassed by other enzymes: the 2-oxoglutarate dehydrogenase complex is replaced by 2-oxoglutarate:ferredoxin oxidoreductase (= 2-oxoglutarate synthase), and citrate synthase by ATP-citrate lyase (citrate + coenzyme A + ATP → oxaloacetate + acetyl-CoA + ADP + Pi). Succinate dehydrogenase is replaced by a (probably NADH-dependent) fumarate reductase. Acetyl-CoA can be converted to pyruvate (by ferredoxin-dependent pyruvate synthase) and thence e.g. to phosphoenolpyruvate – which may in turn be carboxylated to oxaloacetate or used for GLUCONEOGENESIS etc, depending on the requirements of the cell.

redundant (of genes etc) Present in more than one copy per chromosome or per genome. (cf. TERMINAL REDUNDANCY.)

redwater fever (murrain; cattle tick fever; Texas fever) An acute, sometimes fatal disease of cattle and other animals caused by species of BABESIA; it is transmitted by blood-sucking ixodid ticks (e.g. species of *Boophilus*, *Dermacentor*, *Rhipicephalus*). The incubation period is ca. 4–14 days. *Babesia* grows within and destroys the host's erythrocytes, causing e.g. fever and haemoglobinuria (hence 'redwater fever').

Reed–Muench method See END-POINT DILUTION ASSAY.

reference strain In bacteriology: any strain which is not a type strain or a neotype but whose name (or other designation) is used to define or identify products, antigens etc produced by that strain.

refractile Refers to structures which appear bright when viewed by bright-field MICROSCOPY.

refrigeration (of foods) See FOOD PRESERVATION (b).

regulator gene A GENE whose product is a regulator protein which regulates the transcription of one or more STRUCTURAL GENES (see e.g. OPERON).

regulon A system in which two or more (usually non-contiguous) structural genes and/or OPERONS, each with its own promoter, are subject to coordinated regulation by a common regulator molecule (e.g., a repressor or activator); the genes/operons share common or related regulatory sequences (e.g. operators or initiators) which can each be recognized by the regulator molecule.

(cf. OPERON.) Examples of regulons in *Escherichia coli* include the *araBAD/araFG/araE* system (see ARA OPERON); the *arg* regulon (see ARGININE BIOSYNTHESIS); CATABOLITE REPRESSION; HEAT-SHOCK PROTEINS; NTR GENES; PHO REGULON; SOS SYSTEM.

Reichstein process See ASCORBIC ACID.

reindeer moss See CLADONIA.

re-infection A second or subsequent infection of a given individual by a given type of exogenous pathogen. (cf. AUTOINFECTION.)

reinforced clostridial medium (RCM) A medium used (e.g. with a supplement of blood, egg yolk or milk) for the culture of *Clostridium* spp. RCM contains yeast extract, meat extract, peptone, starch, glucose, cysteine hydrochloride, sodium chloride and sodium acetate; it may be used in liquid form or gelled with 1.5% (w/v) agar.

Reiter antigens Group treponemal antigens, i.e., antigens found in a range of pathogenic and non-pathogenic treponemes, including e.g. the Reiter treponeme and *Treponema pallidum*. (See also FTA-ABS TEST.)

Reiter protein CFT See TREPONEMAL TESTS.

Reiter treponeme The reference strain of *Treponema phagedenis*; it contains significant amounts of antigen(s) common to a number of species of *Treponema*, including *T. pallidum*.

Reiter's syndrome A post-infection, often recurrent syndrome characterized by the triad of symptoms: acute polyarthritis, conjunctivitis, and urethritis; it commonly affects young males after infection of the genitourinary tract (particularly chlamydial urethritis) or intestinal tract. The majority of patients with this syndrome have tissue antigen HLA-B27. [*Yersinia enterocolitica* and Reiter's syndrome: Imm. Rev. (1985) *86* 27–45.]

***relA* gene** See STRINGENT CONTROL (sense 1).

relapsing fever (recurrent fever; borreliosis) Any one of several human infectious diseases caused by *Borrelia* spp and transmitted by ticks (*Ornithodoros* spp) or, contaminatively, by lice (*Pediculus* spp). Incubation period: ca. 3–10 days. Symptoms: fever, usually with severe headache, myalgia and arthralgia, and sometimes jaundice and rash. The fever lasts ca. 2–4 days and then subsides, but is then followed 3–10 days later by another bout of symptoms. This cycle may be repeated several times; the fever typically becomes progressively less severe, but fatalities occasionally occur. The relapses are due to ANTIGENIC VARIATION in the surface antigens of the pathogen; bacteria isolated from the bloodstream during each relapse carry different major surface antigens. In *B. hermsii*, the genes encoding the surface antigens are carried on a multicopy *linear* PLASMID; antigenic variation apparently involves a mechanism in which different structural genes are periodically juxtaposed (by a single site-specific recombination event between different plasmid molecules) with a single 'expression site' [Nature (1985) *318* 257–263].

Lab. diagnosis: detection of the pathogen in blood or urine samples taken during febrile periods. *Chemotherapy*: e.g. penicillins; tetracyclines.

relative centrifugal field See CENTRIFUGATION.

relative light units (RLUs) See TMA.

relative molecular mass (M_r) The ratio of the mass of a given molecule to one-twelfth of the mass of a neutral ^{12}C atom; M_r is unitless. The M_r of a molecule is commonly referred to as its 'molecular weight'. (See also ATOMIC MASS UNIT.)

relaxase See CONJUGATION (1b)(i).

relaxation complex See CONJUGATION (1b)(i).

relaxed control See PLASMID (replication).

relaxed DNA DNA which is not supercoiled (see DNA).

relaxed mutant See STRINGENT CONTROL (sense 1).

relaxing enzyme (1) A type I TOPOISOMERASE.

(2) Any enzyme capable of relaxing supercoiled DNA.

relaxosome See CONJUGATION (1b)(i).

release factors (in protein synthesis) See PROTEIN SYNTHESIS (termination).

Relenza™ See ZANAMIVIR.

relomycin See MACROLIDE ANTIBIOTICS.

Renibacterium A genus of aerobic, asporogenous, catalase-positive, oxidase-negative bacteria (order ACTINOMYCETALES, wall type VI); the type species, *R. salmoninarum* (formerly '*Corynebacterium salmoninus*') is the causal agent of KIDNEY DISEASE in salmonid fish. The organisms are rods or coccobacilli, 0.3–1.0 × 1.0–1.5 μm, which frequently occur in pairs. Growth occurs optimally at ca. 15–18°C on e.g. cysteine–serum agar (cysteine being required for growth). [Selective isolation of *Renibacterium*: FEMS (1983) *17* 111–114.] GC%: ca. 53–54.

reniform Kidney-shaped.

rennet substitutes See PROTEASES.

rennin See CASEIN.

rennin substitutes See PROTEASES.

Reoviridae A family of viruses in which the (non-enveloped) virion is typically icosahedral or amorphous (~60–80 nm in diam.) and contains a segmented genome of 10–12 different linear dsRNA molecules. The virion may contain up to 10–13 species of protein; it consists of a *core* (characteristically a double-layered structure incorporating the genome) commonly or always enclosed by an 'outer capsid layer', the latter being lost during viral penetration of the host cell. (The outer capsid layer is generally thought to be important for initial attachment to the target cell; however, cores of the bluetongue virus – from which the outer capsid layer has been removed – are still infective for the insect vector, even though poorly infective for mammalian cells [Virology (1996) *217* 582–593].) The viral core, released into the cytoplasm of the target cell, may include e.g. transcriptase, nucleotide phosphohydrolase, guanylyltransferase and transmethylase – enzymes responsible for synthesis and modification of the viral transcripts. In several reoviruses, the core's diameter has been reported to be ~700 Å [bluetongue virus: Nature (1998) *395* 470–478; reovirus: Nature (2000) *404* 960–967]. Reoviruses replicate in viroplasms in the cytoplasm, sometimes forming crystalline arrays.

The family has been divided into genera on the basis of capsid structure, number of genome segments and nature of host. Hosts include vertebrates (ORBIVIRUS, REOVIRUS, ROTAVIRUS), insects (CYTOPLASMIC POLYHEDROSIS VIRUS GROUP – cf. HOUSEFLY VIRUS) and plants (FIJIVIRUS, PHYTOREOVIRUS).

Reovirus ('respiratory enteric orphan virus'; *Orthoreovirus*) A genus of viruses of the family REOVIRIDAE. The viruses can infect various animals, including primates (e.g. man), birds, cattle and bats – and can be cultured in cells obtained from many types of vertebrate.

The reovirus virion comprises an inner *core* enclosed by a heat- and protease-labile 'outer shell' or 'outer capsid layer'. The outer capsid layer of the virion is lost during viral penetration of the host cell.

The core consists of a double-layered structure associated with (i) the genome: 10 (apparently tightly coiled) dsRNA segments, and (ii) enzymes responsible for synthesis and modification of viral transcripts; each segment of genomic RNA may be associated with a transcriptase complex. Of the eight species of protein comprising the complete virion, five occur in the core. The core is about 700 Å in diameter. [Structure of the reovirus core at 3.6 Å resolution: Nature (2000) *404* 960–967.]

The cell-surface receptor for reoviruses is reported to be the *junction adhesion molecule* (JAM) – an inegral tight-junction (zona occludens) protein which binds the viral attachment protein, $\sigma 1$, and which may also be involved in intracellular signalling [Cell (2001) *104* 441–451].

Following viral penetration, the core of the virion remains intact and becomes transcriptionally active [see also Book ref. 83]. Each transcript is capped but not polyadenylated; mature transcripts are exported to the cytoplasm. Initially, only four of the genome segments are expressed, the other six being derepressed by the product(s) of early genes. Early in the cycle, viral mRNA is translated by the host cap-dependent translational machinery; later, uncapped mRNA is produced, and translation becomes cap-independent. Mature virions accumulate in the cell, which eventually undergoes lysis.

reoviruses (1) Members of the REOVIRIDAE. (2) Members of the genus REOVIRUS.

rep-PCR Repetitive sequence-based PCR: a PCR-based method for TYPING those bacteria whose chromosome contains multiple copies of a 'repetitive sequence' of nucleotides – such as the REP SEQUENCE, ERIC SEQUENCE or SERE.

rep-PCR (principle, diagrammatic; see entry). In the diagram, three REP (R) sequences are shown in one strand of chromosomal DNA. A primer (P) has bound to each REP sequence, and elongation has proceeded from left to right; an arrowhead marks the end of a newly formed fragment. Note that elongation from a given primer cannot continue beyond the next primer. (The fragments do not join together because the reaction mixture does not contain the type of enzyme (ligase) which could make such a join.) After a number of cycles of PCR the different-sized molecules are separated by electrophoresis, yielding a fingerprint characteristic of the given strain. Differences between the fingerprints of different strains reflect differences in REP → REP distances in their respective chromosomes.

Reproduced from *Bacteria* 5th edition, Figure 16.4(d), Paul Singleton (1999) copyright John Wiley & Sons Ltd, UK (ISBN 0471-98880-4) with permission from the publisher.

When typing is based on the REP sequence (= REP-PCR; cf. rep-PCR), primers are designed to bind to a CONSENSUS SEQUENCE at the 5′ end of the REP region in a range of strains. Extension of primers on a given chromosomal template gives rise to a number of fragments which reflect a range of REP → REP distances (see figure); these inter-REP distances can vary between strains, so that amplicons from a given strain, on electrophoresis, will give rise to a fingerprint characteristic of that strain.

The analogous method based on ERIC sequences is referred to as ERIC-PCR, while that based on SERE is called SERE-PCR.

rep-PCR has been used e.g. for typing strains of *Haemophilus somnus* (ERIC-PCR) [JCM (1997) *35* 288–291], *Vibrio parahaemolyticus* (ERIC-PCR) [JCM (1999) *37* 2473–2478], viridans streptococci (REP-PCR, ERIC-PCR SERE-PCR) [JCM (1999) *37* 2772–2776] and *Listeria monocytogenes* (REP-PCR, ERIC-PCR) [JCM (1999) *37* 103–109].

Rep protein (1) See HELICASES. (2) A plasmid-encoded protein which functions as an activator of replication of that plasmid. (See e.g. INCOMPATIBILITY (pT181); R1 PLASMID; R6K PLASMID (π protein).)

REP sequence (repetitive extragenic palindromic sequence; palindromic unit) A highly conserved PALINDROMIC SEQUENCE with hypenated dyad symmetry, ca. 500–1000 copies of which are distributed in the chromosome in e.g. *Escherichia coli*, *Salmonella typhimurium* and other enterobacteria; REP sequences have also been found in e.g. *Listeria monocytogenes*. REP sequences occur between genes, or in the untranslated promoter-distal region (beyond the genes) in many operons, but they have not been detected within coding sequences. (See also ERIC SEQUENCE.)

REP sequences can potentially form cruciform structures in DNA, or stem-and-loop structures in the corresponding mRNA; thermodynamically, the latter is more likely.

REP sequences may play a role in nucleoid structure (earlier, evidence was obtained that REP sequences have protein-binding sites [FEBS (1986) *206* 323–328], and they are now known to be associated with HU PROTEIN).

REP sequences have been used in the typing of bacteria (by REP-PCR) [e.g. JCM (1999) *37* 103–109; JCM (1999) *37* 2772–2776].

Repetitive sequences – either clustered or dispersed through the genome – have long been recognized in higher eukaryotes: see e.g. ALU SEQUENCES.

***REP1*, *REP2* genes** See TWO-MICRON DNA PLASMID.

repair (of DNA) See DNA REPAIR.

repeating spore (*mycol.*) A spore from which arises mycelium of the same type as that from which the spore was formed.

repetitive extragenic palindromic sequence See REP SEQUENCE.

repetitive sequence-based PCR See REP-PCR.

replacement vector A CLONING vector which has a pair of cleavage sites flanking a dispensable sequence (the 'stuffer'); the stuffer sequence is removed and replaced by a sequence of exogenous DNA. (cf. INSERTION VECTOR.)

replica (in microscopy) See ELECTRON MICROSCOPY (a) and (b).

replica plating A technique for isolating mutants from a population of microorganisms grown under *non-selective* conditions. For the isolation of e.g. auxotrophic mutants from a population of prototrophic bacteria, a *master plate* is prepared by inoculating a plate of complete medium with an inoculum from the prototrophic population. Following incubation, a disc of sterile velvet (attached to one end of a cylindrical wooden block) is pressed lightly onto the surface of the master plate; the disc, to which has adhered growth from the master plate, is then used to inoculate one or more plates of minimal medium (the *replica plates*) – a careful record being made of the orientation of the disc relative to the master plate and to the replica plates during inoculation. The replica plates are then incubated; only prototrophic cells form colonies. The positions of colonies on the master and replica plates are compared, and presumptive auxotrophs can be identified by their absence from the replica plates.

replicase See RNA-DEPENDENT RNA POLYMERASE.

replication fork See DNA REPLICATION.

replicative form See RF.

replicative intermediate See RI.

replicative recombination See TRANSPOSABLE ELEMENT.

replicombinase A TRANSPOSASE which catalyses replicative recombination (see TRANSPOSABLE ELEMENT).

replicon Any DNA sequence or molecule which possesses a replication origin and which is therefore potentially capable of being replicated in a suitable cell (see DNA REPLICATION).

repliconation The process in which a circular dsDNA molecule, capable of replication, is formed within a cell from DNA that has been taken up e.g. by CONJUGATION (sense 1b).

replisome See DNA REPLICATION.

reporter gene See IVET (legend) and FUSION PROTEIN.

repressor protein See e.g. OPERON.

repressor titration A selective procedure used for detecting those cells which contain certain plasmid vectors. Use is made of (i) a specially constructed strain of *Escherichia coli* whose chromosome contains a kanamycin-resistance gene under the regulatory control of the *lac* operator–promoter system, and (ii) a *multicopy* plasmid vector that incorporates the *lac* operator sequence (see LAC OPERON). Cells which lack the plasmid vector will fail to grow on a kanamycin-containing medium because (in the absence of lactose or IPTG) transcription from the *lac* promoter is blocked by the LacI repressor protein. A cell which contains the multicopy plasmid will have many extra copies of the (plasmid-borne) *lac* operator sequence, and these extra copies will compete with the chromosomal *lac* operator for LacI; under these conditions, LacI fails to repress the kanamycin-resistance gene so that the cell will grow on kanamycin-containing media.

Because the vector does not carry an antibiotic-resistance gene (as is common in other selective methods) it can be smaller – and therefore transformable with greater efficiency. [Repressor titration: NAR (1998) *26* 2120–2124.]

reptation During gel ELECTROPHORESIS: the movement of a (linear) nucleic acid molecule in an end-on orientation, i.e. aligned parallel to the current.

RER Rough ENDOPLASMIC RETICULUM.

res In insertion sequence $\gamma\delta$: the region corresponding to the IRS of Tn*3* (see Tn*3*).

RESA See PLASMODIUM.

resazurin test (for milk) A test which resembles the METHYLENE BLUE TEST (q.v.) in principle and method but which employs *resazurin* instead of methylene blue as the redox indicator. Resazurin (a member of the quinone–imine group of dyes) is blue when fully oxidized; when the REDOX POTENTIAL is lowered sufficiently the compound becomes irreversibly reduced to the pink *resorufin*. On further reduction the colourless *hydroresorufin* is formed.

reservoir (1) (*epidemiol.*) Human and/or animal population(s) in which a given pathogen is maintained in a viable, infective state and which can serve as a source of infection for susceptibles (see EPIDEMIOLOGY); the pathogen may or may not cause overt symptoms of disease within the reservoir.

(2) (flagellar pocket) See TRYPANOSOMA.

(3) (in euglenids) See EUGLENOID FLAGELLATES.

residual body See LYSOSOME.

residual chlorine See WATER SUPPLIES.

resistance–nodulation–division superfamily See RND.

resistance transfer factor See RTF.

resolvase (in transposable elements) See Tn*3*.

resolvase-type recombinase See SITE-SPECIFIC RECOMBINATION.

resolving power (limit of resolution) In MICROSCOPY: the ability of a microscope to reveal fine detail in a specimen (cf. MAGNIFICATION). In light microscopy resolving power depends mainly on the characteristics of the objective lens, but correct illumination of the specimen is also necessary to exploit the resolving power of a given lens. The resolving power (d_{min}) of an objective is the minimum distance between two details that can be distinguished with the lens; with a lens of inferior resolving power the two details would appear as one. For a given objective:

$$d_{min} \propto \frac{\lambda}{\mu \sin\theta}$$

in which λ is the wavelength of light forming the image, μ is the refractive index of the medium between objective and specimen, and θ is *half* the apical angle of the widest cone of light which can enter the objective from the specimen plane. The quantity $\mu \sin\theta$ is the *numerical aperture* (NA) of the objective; the higher the NA the better the resolving power. (The NA of a CONDENSER is also given by $\mu \sin\theta$; here, θ is half the apical angle of the widest cone of light which can be focused in the specimen plane.) When the medium between objective and specimen is air ($\mu = 1.0$) the NA is less than 1.0. A greater NA is obtainable with an *immersion lens*: a lens designed to work with a liquid bridge between it and the specimen; an *oil-immersion objective* uses IMMERSION OIL (μ ca. 1.5), and a *water-immersion objective* uses distilled water (μ ca. 1.3). The use of oil or water enables the objective to receive a wider cone of light from the specimen plane; this increases resolving power by increasing the amount of diffracted light received from the specimen (see MICROSCOPY (a)). When using immersion oil, one drop is placed on the objective lens and another on the cover-glass; the objective is racked down until the drops coalesce, and the specimen is then brought into focus. This method avoids the risk of trapping air bubbles between lens and cover-glass – a risk greater with objectives that have a concave front lens; air bubbles cause flare and distortion. Usually, an *oil-immersion substage condenser* is used when using an oil-immersion objective lens; in this case the top lens of the condenser should be oiled to the underside of the slide in order to obtain maximum resolving power from the objective lens.

The resolving power of a good light microscope is about 200 nm; with an ultraviolet microscope it is about 130 nm, and with an electron microscope it can be less than 1 nm.

resorcinol 1,3-Dihydroxybenzene.

resorufin See RESAZURIN TEST.

respiration Energy-yielding metabolism in which the oxidation of an energy substrate involves an *exogenous* electron acceptor (i.e., an externally derived oxidizing agent); the participation of an exogenous electron acceptor results in the *net* oxidation of the energy substrate, and makes possible an energy yield significantly greater than that obtainable by the fermentation of the given substrate (cf. FERMENTATION sense 1). Respiration can occur aerobically or anaerobically (see ANAEROBIC RESPIRATION), and may involve an inorganic electron acceptor (e.g. oxygen or nitrate – see also NITRATE RESPIRATION) or an organic electron acceptor (see e.g. FUMARATE RESPIRATION).

In the respiratory (= 'oxidative') metabolism of an energy substrate, the exogenous electron acceptor commonly, but not always, undergoes reduction at the (most positive) end of an ELECTRON TRANSPORT CHAIN (ETC); electron flow from the substrate, via an ETC, to the electron acceptor typically generates proton motive force (see CHEMIOSMOSIS) which can be used for e.g. ATP synthesis (see OXIDATIVE PHOSPHORYLATION). Respiratory metabolism can occur without the involvement of an electron transport chain, ATP being synthesized entirely by substrate-level phosphorylation: see e.g. 'nitrite respiration' in *Escherichia coli* in entry NITRATE RESPIRATION.

Respiratory substrates include organic compounds (see ORGANOTROPH) and inorganic compounds or ions (see LITHOTROPH); according to species and conditions, the electron acceptor may be e.g. oxygen, nitrate, CO_2, sulphate or fumarate. In that the exogenous electron acceptor is involved in the final oxidative step in a respiratory pathway it is commonly called the *terminal electron acceptor*.

respiratory burst See NEUTROPHIL.

respiratory chain See ELECTRON TRANSPORT CHAIN.

respiratory control Control of the electron flow in a respiratory chain by proton motive force (see CHEMIOSMOSIS). According to its value, pmf may inhibit or promote the further extrusion of protons across the membrane – usually with a concomitant decrease or increase, respectively, in electron flow. In the so-called *State 3* (active state) the rate of respiration is high owing to decreased pmf; in the presence of UNCOUPLING AGENTS (which can abolish pmf entirely) respiration is maximal (*State* 3_{unc} or *3u*). In *State 4* (the controlled or resting state) respiration is minimal owing to the maximal thermodynamic back-pressure of the pmf; however, a low level of respiration may occur in order e.g. to balance the effects of proton leakage across the membrane.

In *Escherichia coli* the adsorption of certain bacteriophages has been shown to cause a transient release from respiratory control, phage adsorption apparently inducing an inflow of protons similar to that induced by protonophores [JB (1985) *161* 179–182].

respiratory inhibitors Substances which inhibit the flow of electrons along a respiratory chain: see e.g. AMYTAL, ANTIMYCIN A, AZIDE, BAL, CARBON MONOXIDE (b) CYANIDE, DCMU, HYDROXYQUINOLINE-N-OXIDES, PIERICIDIN and ROTENONE. (cf. UNCOUPLING AGENTS; OLIGOMYCIN.)

respiratory syncytial virus See PNEUMOVIRUS.

respiratory tract microflora The human nasal passages commonly harbour species of *Corynebacterium* and *Staphylococcus* (usually *S. epidermidis*, sometimes *S. aureus* also) and occasionally species of e.g. *Moraxella* and *Streptococcus*. In the nasopharynx the predominating organisms are often streptococci; various opportunist pathogens (e.g. *Haemophilus influenzae*, *Streptococcus pneumoniae*, strains of *Moraxella*) may also be present. (See also MOUTH MICROFLORA.)

The trachea, bronchi, bronchioli and lungs may be sterile or may have small numbers of contaminating organisms. The ability of extraneous air-borne particles (see AIR) to reach the trachea, or to penetrate further into the respiratory tract, is governed largely by the size and shape of such particles. For example, spores > ca. 10 μm (e.g. those of *Alternaria* and *Fusarium*) tend to be trapped in the nose, but may reach the bronchi during mouth-breathing; such spores may induce e.g. a TYPE I REACTION. Spores 5–10 μm (formed e.g. by many species of *Cladosporium* and *Mucor*) may be deposited in the bronchi or bronchioles where they may incite attacks of asthma in susceptible individuals; spores of this size may even penetrate into the alveoli if an individual is exposed to very large numbers of spores. Spores < 5 μm (e.g. those formed by species of *Aspergillus* and *Penicillium*, and by many actinomycetes) may reach the alveoli; the spores of actinomycetes are commonly implicated in certain types of EXTRINSIC ALLERGIC ALVEOLITIS.

(See also BODY MICROFLORA and e.g. BRONCHITIS; COMMON COLD; CYSTIC FIBROSIS; PNEUMONIA.)

response regulator See TWO-COMPONENT REGULATORY SYSTEM.

resting spore (winter spore) (*mycol.*) A thick-walled spore, particularly one formed by a sexual process, which germinates only after an extended period of dormancy – e.g. an overwintering teliospore. (cf. SUMMER SPORE.)

restricted transduction See TRANSDUCTION.

restriction endonuclease (REase, RE; restriction enzyme) An ENDONUCLEASE which binds to dsDNA, usually at a specific *recognition sequence* (= *recognition site*), and typically makes a single cut in each strand – *provided that* specific bases at that site are *un*methylated (for some REs) or methylated (for others); the cleaved duplex may exhibit STICKY ENDS or BLUNT-ENDED DNA, according to RE. Some REs have no unique recognition site, some have several. DNA may be cleaved within the site or outside it; some REs (e.g. *Bae*I) cut on both sides, excising a segment that includes the recognition site. (See also STAR ACTIVITY.)

REs with various specificities have been isolated from a wide range of prokaryotes. (In some cases, REs from different species recognize the same nucleotide sequence: see ISOSCHIZOMERS.)

An RE has a three-letter designation (based on the name of the host species) followed by a strain designation and/or a Roman numeral that indicates a particular RE from a given strain or species: see table for examples; there is currently a trend to avoid italics [RE nomenclature: NAR (2003) *31* 1805–1812]. Recognition sequences are written 5′-to-3′ (sequence for one strand only); an arrow (or stroke) shows a cleavage site (if within the recognition site).

What purpose do REs serve in their host cells? The main role seems to be protection against 'foreign' DNA – particularly phage DNA [MR (1993) *57* 434–450]. However, this protective mechanism is not always effective; for example, some phages encode specific inhibitors of REs. Moreover, antirestriction enzymes are encoded by certain plasmids [JB (1992) *174* 5079–5085], and resistance to restriction enzymes is a typical feature of conjugative transposons [e.g. ARM (1995) *49* 367–397].

Several distinct types of RE are recognized.

Type I REs can methylate and cut DNA. ATP-dependent cleavage occurs outside the recognition sequence at a non-specific site; the enzyme may bind to its recognition site while DNA is passed through another part of the enzyme until a cleavage site is reached. Example: *Eco*K1. [New type 1 REs from *Escherichia coli*: NAR (2005) *33* (13) e114.]

Type II REs generally recognize a specific sequence and cleave both strands, independently of ATP, at a particular location in or near the recognition site; each strand is left with a 5′-phosphate terminus and a 3′-hydroxyl terminus. These REs are useful in recombinant DNA technology owing to their ability to cut DNA precisely; >3500 are now known. Subtypes of type II REs include:

IIA. REs that recognize asymmetric sites.

IIB. REs that cut on both sides of the recognition site – e.g. *Bae*I, whose recognition (underlined)/cutting sites are:

/(10/15)<u>AC(4N)GTAYC</u>(12/7)/

in which 4N = any four nucleotides, Y = C or T; this strand is cut 10 nucleotides upstream, its complementary strand 15 nucleotides upstream – 12/7 indicating downstream cutting sites for both strands.

IIC. REs with cutting and methylation functions combined in a single polypeptide.

IIE. REs which cut at the recognition site but need interaction with another copy of that site in order to function.

IIF. REs that bind to, and cut, both copies of the recognition site.

IIM. REs which cut at (fixed) sites where there is a specific pattern of methylation – e.g. *Dpn*I. (cf. *Type IV*, below.)

IIP. REs with a fixed cutting site within, or very close to, the symmetrical (palindromic) recognition site – e.g. *Eco*RI.

Type III. REs that interact with two inversely orientated, non-palindromic sites; cleavage, at a distance from one of the sites, requires ATP-dependent translocation. Examples: *Eco*P1l, *Eco*P15l.

Type IV. REs with methylated recognition sites but which (unlike the IIM REs) apparently cut without specificity.

RESTRICTION ENDONUCLEASES (REs): some examples[d]

RE	Source (organism)	Recognition sequence[a]
*Aat*II	*Acetobacter aceti*	GACGT/C
*Alu*I	*Arthrobacter luteus*	AG/CT
*Apa*I	*Acetobacter pasteurianus*	GGGCC/C
*Bam*HI	*Bacillus amyloliquefaciens*	G/GATCC
*Bcl*I	*Bacillus caldolyticus*	T/GATCA
*Bgl*II	*Bacillus globigii*	A/GATCT
*Bss*HII	*Bacillus stearothermophilus*	G/CGCGC
*Dra*I	*Deinococcus radiophilus*	TTT/AAA
*Eco*RI	*Escherichia coli*	G/AATTC
*Eco*RV	*Escherichia coli*	GAT/ATC
*Hind*III	*Haemophilus influenzae*	A/AGCTT
*Hinf*I	*Haemophilus influenzae*	G/ANTC
*Hpa*I	*Haemophilus parainfluenzae*	GTT/AAC[b]
*Hpa*II	*Haemophilus parainfluenzae*	C/CGG
*Kpn*I	*Klebsiella pneumoniae*	GGTAC/C
*Mlu*I	*Micrococcus luteus*	A/CGCGT
*Msp*I	*Moraxella* sp	C/CGG
*Nar*I	*Nocardia argentiensis*	GG/CGCC
*Nco*I	*Nocardia corallina*	C/CATGG
*Not*I	*Nocardia otitidis*	GC/GGCCGC[c]
*Nru*I	*Nocardia rubra*	TCG/CGA
*Pal*I	*Providencia alcalifaciens*	GG/CC
*Pst*I	*Providencia stuartii*	CTGCA/G
*Pvu*I	*Proteus vulgaris*	CGAT/CG
*Sac*II	*Streptomyces achromogenes*	CCGC/GG
*Sal*I	*Streptomyces albus*	G/TCGAC
*Sau*3AI	*Staphylococcus aureus*	/GATC
*Sca*I	*Streptomyces caespitosus*	AGT/ACT
*Sma*I	*Serratia marcescens*	CCC/GCC[b]
*Sna*BI	*Sphaerotilus natans*	TAC/GTA
*Spe*I	*Sphaerotilus* sp	A/CTAGT
*Srf*I	*Streptomyces* sp	GCCC/GGGC[c]
*Taq*I	*Thermus aquaticus*	T/CGA
*Xba*I	*Xanthomonas campestris* var *badrii*	T/CTAGA
*Xho*I	*Xanthomonas campestris* var *holcicola*	C/TCGAG
*Bae*I	*Bacillus sphaericus*	/(10/15) AC(4N)GTAYC(12/7)/

[a] Recognition sequence in one strand (in the 5′-to-3′ direction); the cutting site is shown as '/'. A = adenine; C = cytosine; G = guanine; T = thymine.
[b] Produces 'blunt-ended' cuts.
[c] The eight-nucleotide sequence of a 'rare-cutting' enzyme.
[d] The proposed new nomenclature uses a non-italic format [NAR (2003) *31* 1805–1812]. Currently, the literature refers to REs in both italic and non-italic formats.
Modified from Table 2.1, page 18, in *DNA Methods in Clinical Microbiology*, Paul Singleton (2000) copyright Kluwer Academic Publishers, Dordrecht, The Netherlands (ISBN 0-7923-6307-8) with kind permission from the publisher.

Some *nicking enzymes* (which cut *one* strand of dsDNA) resemble REs [e.g. *Bst*NBI: NAR (2001) *29* 2492–2501]. In some cases the recognition site of an RE can be chemically modified such that the enzyme cuts one specific strand of dsDNA; this approach is used e.g. in SDA.

restriction endonuclease analysis See DNA FINGERPRINTING.

restriction fragment length polymorphism See RFLP.

restriction–modification system (R–M system) The system, present in many bacteria, in which DNA is modified by specific enzyme(s) in a manner characteristic of the particular bacterial strain, the modification serving to protect the DNA against enzymic degradation ('restriction') by the cell's own endonucleases. Modification generally involves a specific pattern of *methylation* of nucleotide residues in the DNA (see DNA METHYLATION); using *S*-adenosylmethionine as methyl donor, the methylation enzymes ('methylases') act on dsDNA in which one or both strands are unmodified. The sites at which modification occurs are also recognized by corresponding

RESTRICTION ENDONUCLEASES (REs); the REs cannot cleave DNA in which these sites have been modified in one or both strands. Modification does not interfere with normal base-pairing, so that semiconservative replication of fully modified dsDNA will result in two duplexes, each containing one (modified) parent strand and one unmodified daughter strand. The modified parent strand protects the duplex from restriction, and the daughter strand is modified before further DNA replication; thus, DNA susceptible to restriction does not normally occur in the cell. However, if 'foreign' DNA lacking the modification pattern of the strain enters the cell (e.g. by conjugation, transformation, phage infection etc) it will generally be degraded. (See also HOST-CONTROLLED MODIFICATION.)

Certain bacteriophages encode their own R–M systems: see e.g. T-EVEN PHAGES and BACTERIOPHAGE MU.

restriction point (in cell cycle) See CELL CYCLE.

restrictive conditions See CONDITIONAL LETHAL MUTANT.

resupinate (*mycol.*) Refers to a fruiting body which forms a crust on the surface of the substratum, the spore-bearing surface being outermost.

reticle See MICROMETER.

reticulate body See CHLAMYDIA.

reticule See MICROMETER.

reticuloendotheliosis viruses See AVIAN RETICULOENDOTHELIOSIS VIRUSES.

reticulopodium See PSEUDOPODIUM.

reticulum See RUMEN.

retinaculum apertum See SPIRA.

retinal The purple component of various photosensitive proteins, e.g. BACTERIORHODOPSIN, HALORHODOPSIN, SLOW-CYCLING RHODOPSIN. The free retinal molecule is $R-[CH=CH-C(CH_3)=CH]_2-CHO$ where R = 2,6,6-trimethyl-1-cyclohexenyl. (The corresponding alcohol, retinol, is vitamin A.) Linkage of the aldehyde group of retinal to a lysine residue in a halobacterial opsin creates a Schiff base. The synthesis of retinal in *Halobacterium salinarium* is inhibited e.g. when the organism is grown in the presence of nicotine.

retort (in canning) See BATCH RETORT.

retort pouch A flexible container used in the food processing industry as an alternative to the can. Typically, the pouch is made of a laminate consisting of aluminium foil sandwiched between an outer layer of polyester and an inner layer of e.g. polypropylene and nylon – the layers being bonded together by e.g. a polyester–isocyanate adhesive. (See also CANNING.)

Retortamonadida An order of parasitic protozoa (class ZOOMASTIGOPHOREA). The cells, which are not bilaterally symmetrical, have two to four flagella, one of which is directed posteriorly along the ventral surface; cells lack mitochondria, Golgi apparatus, an axostyle and an undulating membrane. Genera: e.g. CHILOMASTIX, *Retortamonas*.

Retortamonas See RETORTAMONADIDA.

retrogressive development See CONIDIUM.

retrohoming See INTRON HOMING.

retroid viruses Viruses which resemble members of the RETROVIRIDAE in that they appear to employ a replication mechanism which involves reverse transcription of an RNA intermediate; retroid viruses include CAULIFLOWER MOSAIC VIRUS [review: TIBS (1985) *10* 205–209] and members of the HEPADNAVIRIDAE. [Possible common evolutionary origin of hepatitis B virus and retroviruses: PNAS (1986) *83* 2531–2535.]

retron In some strains of *Escherichia coli* (and certain other Gram-negative bacteria): a chromosomal genetic element encoding a REVERSE TRANSCRIPTASE structurally similar to that encoded by retroviruses. Retron-containing strains can synthesize a unique, branched, hybrid RNA–DNA molecule called *multicopy single-stranded DNA* (msDNA); in this molecule, the 5′ end of single-stranded DNA is linked (via a phosphodiester bond) to the 2′ position of an *internal* (i.e. non-terminal) guanosine residue in the RNA. Synthesis of msDNA involves the reverse transcriptase. A cell may contain many copies (up to 500) of msDNA, but the function of this molecule is unknown.

Retrons seem to be 'foreign' or imported elements in *E. coli*.

Retrons are referred to by a system of nomenclature in which the initials of the host species are followed by the number of bases in the DNA part of the corresponding msDNA; for example, Mx162 is a retron in *Myxococcus xanthus* in which the DNA of the msDNA contains 162 bases.

Retrons also occur e.g. in *Klebsiella*, *Proteus*, *Rhizobium* and *Salmonella* [JB (1993) *175* 4250–4254].

retroposon *Syn.* RETROTRANSPOSON.

retroregulation (*mol. biol.*) A mode of post-transcriptional regulation in which expression of a gene is influenced by a *cis*-acting sequence *downstream* of the gene; if transcription proceeds into the *cis*-acting sequence, the resulting RNA is degraded and the gene is thus not expressed, whereas if transcription stops before the *cis*-acting sequence the RNA is not degraded and the gene is expressed. (See BACTERIOPHAGE λ (*int* gene expression).)

retrotranscription See REVERSE TRANSCRIPTASE.

retrotransfer The flow of genetic material from recipient to donor during bacterial CONJUGATION. In *Escherichia coli*, retrotransfer was reported to be inhibited by streptomycin, and it occurred to a lesser extent during conjugation mediated by plasmids encoding strong SURFACE EXCLUSION; it has been suggested that retrotransfer is newly initiated conjugation between transconjugants and donors [JB (1993) *175* 583–588] – an hypothesis supported by subsequent studies [JB (1996) *178* 1457–1464].

retrotransposon (retroposon) A TRANSPOSABLE ELEMENT in which transposition involves a retrovirus-like process of reverse transcription with the formation of an RNA intermediate (see e.g. TY ELEMENTS). It has been argued that such TEs may be degenerate retroviruses [Cell (1985) *40* 481–482, and *42* 507–517]. (See also A-TYPE PARTICLES.)

Retroviridae A family of enveloped ssRNA-containing animal viruses in which the genome is replicated via a dsDNA intermediate. Retroviruses have been isolated from a wide range of vertebrates – including mammals, birds and reptiles – and retrovirus-like particles have been observed e.g. in tapeworms and insects. Retrovirus infection in a given host may be asymptomatic or may result in any of various diseases, including e.g. pneumonia, anaemia, and malignant tumours and/or leukaemias. (See also AIDS.) Interestingly, retroviral RNA has been detected in the brain and CSF of individuals suffering from schizophrenia, suggesting a possible link [PNAS (2001) *98* 4634–4639]. Transmission may occur horizontally and/or vertically (see below).

The retrovirus virion is more or less spherical, ca. 80–120 nm diam., consisting of a nucleoprotein core (sometimes called the 'nucleoid') surrounded by an envelope derived largely from the host plasma membrane (or intracytoplasmic membranes). The core contains the viral REVERSE TRANSCRIPTASE and other viral proteins, the viral genome complexed with protein (also called the 'nucleoid') and low-MWt RNAs of host origin; the envelope contains a virus-encoded glycoprotein which determines the host

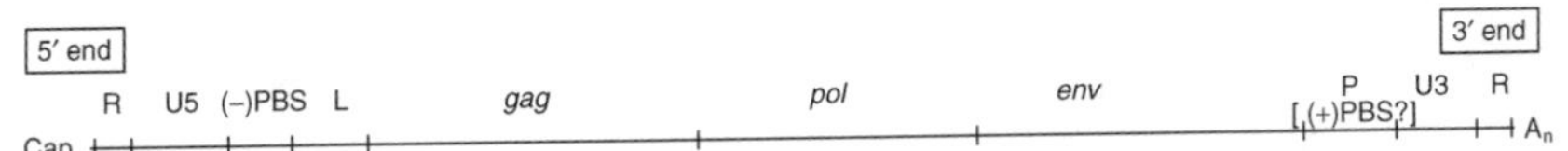

RETROVIRIDAE: Figure 1. The retrovirus genome, showing generalized features (*not to scale*). The genomes of some retroviruses have additional coding sequences not shown here.

R	=	Terminal redundancy.
U5	=	Sequence unique to the 5′ end (present at each end in proviral DNA).
(–) PBS	=	Primer binding site for the (–) DNA strand primer (a host tRNA).
L	=	Leader sequence.
gag	=	Group-specific antigen genes (coding region for nucleocapsid proteins).
pol	=	Coding region for e.g. reverse transcriptase.
env	=	Coding region for envelope glycoprotein.
P	=	Polypurine tract and probable primer binding site for (+) DNA strand synthesis.
U3	=	Sequence unique to the 3′ end (present at each end in proviral DNA).
Cap	=	5′ Cap nucleotide ($m^7G^{5'}ppp^{5'}$ ...), added after transcription and not copied during DNA synthesis.
A_n	=	3′ Polyadenylate tract (e.g. ca. 100–200 nucleotides long), added after transcription and not copied during DNA synthesis.

range of the virus and is the target for neutralizing antibodies. All of the genetic information of the virus is encoded by a single piece of positive-sense ssRNA (ca. 3.5–9.8 kb) which has the structural features characteristic of eukaryotic mRNA (see Figure 1); however, in the virion the genome occurs in a 60S–70S dimer in which two identical (or similar) 30S–35S viral RNA molecules ('monomers') are held together at their 5′ ends by hydrogen bonds (i.e., the virion is 'diploid'). The virions are sensitive to heat, lipid solvents and detergents, but are relatively resistant to UV- and X-irradiation.

Retroviruses are classified according to their virion structure, host range, pathological effects (both on the host and in cell cultures) etc. Three subfamilies are recognized: LENTIVIRINAE, ONCOVIRINAE and SPUMAVIRINAE.

Replication cycle. The mode of replication appears to be generally similar in the various types of retrovirus. Infection is initiated by interaction between the virus envelope glycoprotein and specific host cell surface receptors. In the host cytoplasm, a linear viral DNA is synthesized by the viral REVERSE TRANSCRIPTASE acting within the (presumably modified) infecting virus particle. Synthesis of the first (–) DNA strand is initiated at the (–) PBS at the 3′ boundary of the U5 region (see Figures 1 and 2), a host tRNA acting as a primer for initiation; most retroviruses use host $tRNA^{pro}$ as primer, but mouse mammary tumour virus, visna virus, and the AIDS virus use $tRNA^{lys}$. The viral RNA is degraded by viral RNase H (see RNASE H), and a second strand of DNA is synthesized using the (–) DNA strand as template. The resulting linear dsDNA molecule is slightly longer than the RNA template due to the duplication of regions U3 and U5 and the consequent generation of the long terminal repeats (LTRs: see Figure 2); the LTRs contain sequences necessary for the next stage in the cycle – integration of the viral DNA into the host cell DNA to form a *provirus* – and for the control of virus gene expression.

In e.g. HIV (q.v.) the dsDNA is circularized, and one or more copies are inserted into the host cell DNA. Integration appears to be an essential step for replication of the retroviral genome, and, in general, integration may occur at random sites in the host DNA. For at least some retroviruses, two or more base pairs are lost from each LTR, prior to or during integration, and a short sequence of cellular DNA is duplicated such that the provirus is flanked by short direct repeats. (cf. TRANSPOSABLE ELEMENT.)

Once established in the host chromosome, a provirus may be quiescent or – depending e.g. on its location, on the strength of its promoters, on the physiological state of the cell, and on the presence or absence of specific regulators – may be transcribed to form progeny RNA genomes and mRNAs. The LTRs contain most of the signals for the regulation of transcription, including a promoter, an enhancer (in some retroviruses), and signals for termination of transcription and for polyadenylation of the transcript. Transcription appears to be carried out by host RNA polymerase II and is highly sensitive to α-amanitin. Primary transcripts are approximately the length of a viral RNA monomer; they are polyadenylated, and then processed further (e.g. by splicing).

By convention, retroviral gene products are designated by their MWts $\times 10^{-3}$ prefixed by p, pp or gp for protein, phosphoprotein or glycoprotein, respectively: e.g. $pp12^{gag}$ = a phosphorylated product of *gag*, MWt ca. 12×10^3; polyprotein precursors are prefixed by Pr.

The 'genes' *gag*, *pol* and *env* (all necessary for virus replication) encode polyproteins. The *gag*-encoded polyprotein is cleaved to form the 4 or 5 components of the virus core. The *pol* sequence is apparently expressed with *gag* such that a *gag-pol* polyprotein precursor is formed ($Pr180^{gag-pol}$ in e.g. avian leukosis and murine leukaemia viruses); this precursor undergoes a series of cleavages, and the mature protein complex has reverse transcriptase, RNase H and DNA endonuclease activities. The *env*-encoded product is synthesized via a subgenomic spliced mRNA; the product is cleaved to remove a signal peptide, and is then glycosylated and cleaved to form the major protein component of the viral envelope. Envelope glycoproteins may continue to be synthesized until they saturate the plasma membrane receptors and thus prevent superinfection by any virus competing for those receptors.

Nucleoprotein cores assembled in the cytoplasm subsequently bud through the plasma or internal membranes of the host cell; the released virions apparently undergo further maturation, involving e.g. condensation of the nucleoid, cleavage of *gag* and *gag-pol* proteins, etc.

Transmission. Retroviruses may be transmitted by three distinct routes, not all of which can occur in all retroviruses. Horizontal transmission may occur e.g. by direct contact, via saliva, via insect vectors, etc, depending on virus. Vertical transmission may occur e.g. in avian retroviruses by infection of progeny via

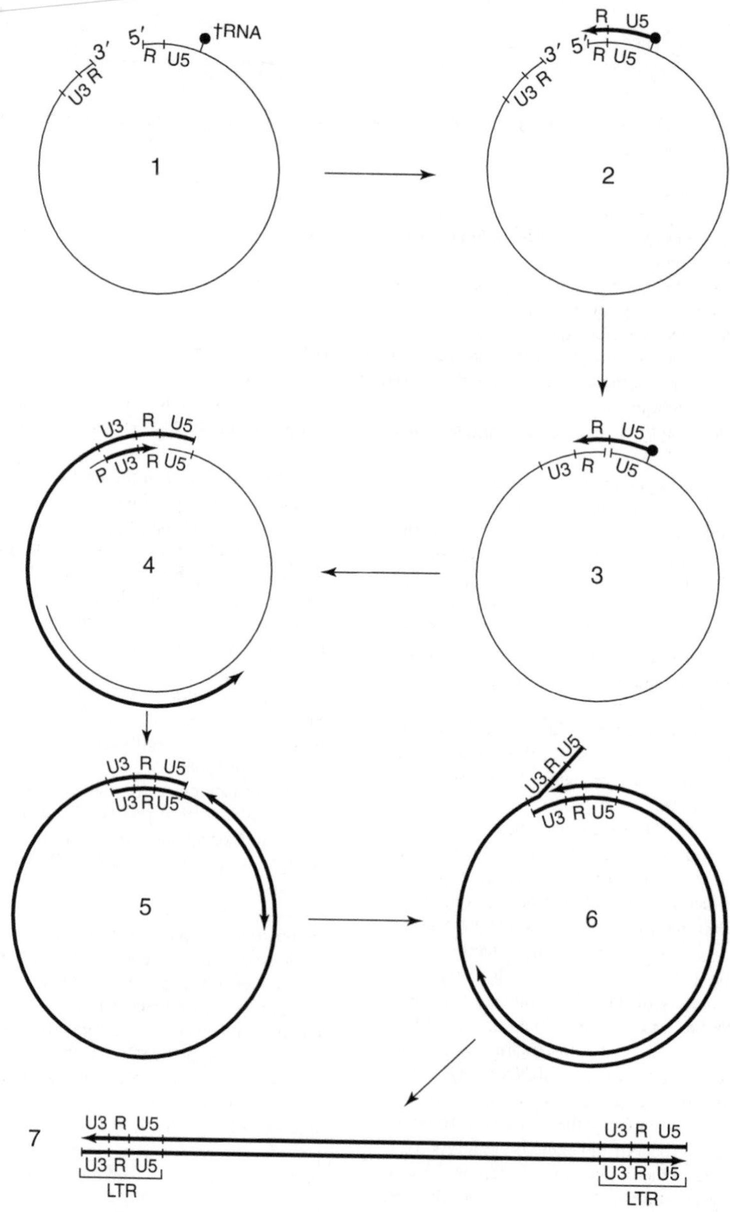

RETROVIRIDAE: Figure 2. Model for retroviral dsDNA synthesis from the ssRNA genome. (Thin lines = RNA; thick lines = DNA.) (*Continued on page 659*.)

the egg albumen, or in mammalian retroviruses by infection of the fetus via the placenta. Some authors regard this as a form of horizontal transmission, as distinct from the third route of transmission: vertical transmission by inheritance of the provirus through the germ line – i.e., the provirus is inherited as is any other chromosomal sequence. (cf. ENDOGENOUS RETROVIRUS and EXOGENOUS RETROVIRUS.)

Genetic interactions. Retroviruses appear to be genetically unstable and give rise to defective and recombinant variants at high frequencies. For example, recombination may occur between the genomes of closely related retroviruses infecting the same cell. An intermediate in such recombination may be a heterozygous virus particle in which the two RNA 'monomers' are contributed by different viruses; recombination may result from a 'copy-choice' mechanism involving the transfer of nascent DNA strands from one template to another during reverse transcription. Recombination can also occur e.g. between the nucleic acid of an infecting exogenous retrovirus and that of an endogenous virus, or between retroviral and host nucleic acids. The mechanism for such recombination is unknown; it could occur by a copy-choice mechanism involving RNA transcripts of the relevant sequences, or by a breakage-and-reunion mechanism involving e.g. proviral DNA and chromosomal DNA. Recombination between retroviral and chromosomal sequences may lead to the incorporation of host genes into the viral genome. Such a recombinant viral genome is usually replication-defective since the cellular sequences are usually acquired at the expense of viral coding sequences. These viruses can thus replicate only in the presence of a replication-competent helper virus (i.e., one with intact *gag*, *pol* and *env* genes) which can provide the missing replication functions; the defective genome may then be packaged into particles (PSEUDOTYPES) which can infect other cells, thereby resulting in the *transduction* of cellular DNA to other host cells.

Oncogenesis. The transducing ability of retroviruses was discovered when certain replication-competent viruses – which are normally capable of inducing neoplastic disease in their hosts only at very low frequencies and after long latent periods – were found to give rise to acutely oncogenic, usually replication-defective variants; many of these variants were found to contain sequences derived from cellular ONCOGENES, and appear to have arisen by recombination between sequences of the infecting retrovirus and host c-*onc* sequences. The viral oncogenes (v-*onc*) differ from their cellular counterparts e.g. in that they lack introns; they are subject to viral, rather than cellular, control signals, and appear to confer on the virus the ability to induce neoplastic disease efficiently and after a short latent period. (Not all v-onc^+ viruses are replication-defective: certain strains of ROUS SARCOMA VIRUS contain the oncogene v-*src* in addition to a full complement of viral genes.) All known v-onc^+ viruses can transform at least some type(s) of cell in culture; by contrast, very few v-onc^- viruses can transform cells in culture (cf. HTLV), although a few can cause CPE (e.g. vacuolization, syncytium formation, cell death) – usually only in certain types of cell. (See also e.g. AVIAN LEUKOSIS VIRUSES, FELINE LEUKAEMIA VIRUS, MURINE LEUKAEMIA VIRUSES; cf. FRIEND VIRUS and MCF VIRUSES.)

retroviruses Viruses of the RETROVIRIDAE.

retting The process by which cellulose fibres are released from the stems of flax (*Linum usitatissimum*), hemp (*Cannabis sativa*), or jute (*Corchorus* spp, Tiliaceae) by the action of pectinolytic fungi and bacteria; these organisms soften the stems, and the fibres are then separated by beating ('scutching'). The fibres are used in the textile industry e.g. for the manufacture of linen (from flax), hessian (from hemp or jute), twine (from hemp) etc. In aerobic retting processes ('dew retting') the organisms involved include e.g. *Cladosporium herbarum*, *Aureobasidium pullulans*, and species of *Cryptococcus*, *Rhizopus*, *Rhodotorula*, *Bacillus* and *Pseudomonas*. In anaerobic processes the stems are submerged in water-tanks, and the main retting agents are *Clostridium* spp, especially *C. felsineum*.

REV (REV-T, REV-A) See AVIAN RETICULOENDOTHELIOSIS VIRUSES.

reversal reaction See LEPROSY.

reverse CAMP test See CAMP TEST.

reverse electron transport Energy-*dependent* ('uphill') electron flow along an ELECTRON TRANSPORT CHAIN (or part of an ETC); the energy for reverse electron transport is derived from proton motive force (see CHEMIOSMOSIS), and, in e.g. animal mitochondria, the electrons may derive from the oxidation of succinate. Reverse electron transport is an essential process in e.g. photosynthetic bacteria of the Rhodospirillineae; in these organisms, pmf derived from PHOTOSYNTHESIS is used not only for the synthesis of ATP but also for the generation of reducing equivalents by uphill electron flow to NAD^+. In *Thiobacillus ferrooxidans*, reduction of pyridine nucleotides is reported to involve uphill electron flow through the bc_1/NADH-Q oxidoreductase complex acting in reverse [JB (2000) *182* 3602–3606].

reverse gyrase A TOPOISOMERASE which can introduce *positive* supercoiling into cccDNA, using the energy of ATP hydrolysis; the enzyme has been found e.g. in the archaean *Sulfolobus acidocaldarius* [Nature (1984) *309* 669–681] and in bacteria of the order Thermotogales [JB (1991) *173* 3921–3923] – all being *thermophilic* organisms. The function of reverse gyrase is unknown, but one suggestion is that it may have an important

RETROVIRIDAE: Figure 2 (*continued*)

1. The ssRNA retroviral genome.
2. (−)-Strand DNA synthesis by reverse transcriptase is initiated on the tRNA primer at the (−) PBS, and (−)-strand DNA is synthesized to the 5′ end of the template.
3. The 5′ R region of the RNA template is removed by RNase H. The DNA R region can then pair with the RNA 3′-R region, thus providing the template for continued DNA elongation.
4. As the (−) DNA strand elongates, the RNA template is degraded by RNase H; a short sequence of RNA at P [(+)PBS?], apparently being resistant to RNase H action, remains to function as a primer for the initiation of (+)-strand DNA synthesis [JV (1984) *52* 314–319].
5. (−)-Strand DNA synthesis reaches the end of the RNA template, thus providing a template for continued (+)-strand DNA synthesis.
6. (−)-Strand synthesis continues beyond the RNA genome length to duplicate the U3–R–U5 region.
7. Continued (+)-strand synthesis into the U3–R–U5 region results in the completed dsDNA molecule with a U3–R–U5 sequence (long terminal repeat, LTR) at each end.

role in life at high temperatures. Possibly, the enzyme pre-dates the evolutionary split between archaeans and bacteria.

reverse mutation *Syn.* BACK MUTATION.

reverse passive latex agglutination test See LATEX PARTICLE TEST.

reverse primer See FORWARD PRIMER.

reverse transcriptase RNA-dependent (i.e. RNA-directed) DNA polymerase: an enzyme which synthesizes DNA on an RNA template. Reverse transcriptases are encoded by viruses of the RETROVIRIDAE, by RETROID VIRUSES, and by certain retrovirus-like elements (e.g. A-TYPE PARTICLES and TY ELEMENTS); a reverse transcriptase has been isolated from the protozoan *Paramecium tetraurelia* [Eur. J. Biochem. (1987) *163* 569–575], and genes encoding a reverse transcriptase have been demonstrated in certain myxobacteria and in *Escherichia coli* [Nature (1989) *339* 254] (see also RETRON). (The DNA polymerase I of *Escherichia coli* has reverse transcriptase ability but, in this capacity, it has rather poor processivity [EMBO (1993) *12* 387–396].)

Like DNA-dependent DNA POLYMERASES, a reverse transcriptase requires a short RNA primer for the initiation of DNA synthesis, and catalyses polymerization in the 5′-to-3′ direction; divalent cations (Mg^{2+}or Mn^{2+}) are necessary for activity. The initial result of DNA synthesis on an RNA template (a process which is called reverse transcription or *retrotranscription*) is a DNA–RNA hybrid (see CDNA); at least some reverse transcriptases have RNase H activity (see RNASE H) and are able to degrade the RNA strand of the hybrid, allowing a complementary strand of DNA to be synthesized on the newly formed strand (giving rise to ds cDNA).

Reverse transcriptases lack 3′-to-5′ exonuclease activity, i.e. they have no proof-reading capacity (cf. DNA POLYMERASE), so that their synthesis of DNA is error-prone. These enzymes seem able to switch from one template to another during polymerization, possibly allowing recombination by a 'copy-choice' mechanism (see RETROVIRIDAE).

Reverse transcriptase can be inhibited e.g. by AZT, PHOSPHONOFORMATE and SURAMIN but is insensitive to many other inhibitors of DNA and RNA.

Reverse transcriptases are employed in several techniques (e.g. REVERSE TRANSCRIPTASE PCR (rtPCR), NASBA) which are used for the in vitro amplification of nucleic acids. Commonly used enzymes include avian myeloblastosis virus reverse transcriptase (AMV reverse transcriptase) and Moloney murine leukaemia virus reverse transcriptase (MMLV reverse transcriptase).

MMLV reverse transcriptase is a cloned enzyme of MWt ~71000 which has RNase H activity.

The AMV reverse transcriptase comprises two types of subunit: α (MWt 68000) and β (MWt 92000). The $\alpha\beta$ form has both DNA-dependent DNA polymerase and bidirectional RNase H activity (these two functions residing in the α subunit). [Example of use of AMV reverse transcriptase: JCM (1999) *37* 524–530].

reverse transcriptase PCR (rtPCR, rt-PCR, RT-PCR; formerly: RNA PCR) A procedure in which (i) a strand of DNA is synthesized on an RNA target molecule by a REVERSE TRANSCRIPTASE – forming a hybrid RNA/DNA duplex; (ii) the strand of RNA is degraded, either by RNase H or by the RNase H activity of reverse transcriptase; (iii) a complementary strand of DNA is synthesized on the first strand, forming double-stranded cDNA; (iv) the dsDNA copy of the original RNA amplicon is amplified by standard PCR methodology. The initial phase (reverse transcription) may be carried out as a separate stage in a 'two-tube' approach [e.g. JVM (1997) *69* 63–72]; alternatively, both phases (reverse transcription and PCR proper) may be carried out in a single tube [e.g. JCM (1999) *37* 524–530].

Reverse transcription, which requires a primer, is carried out isothermally at e.g. 50°C. Such relatively low temperatures can be problematic if the target RNA contains secondary structures that are destabilized only at a higher temperature: the presence of secondary structures can block the synthesis of cDNA by physically impeding the polymerase. The problem of secondary structures may be overcome by carrying out reverse transcription at a higher temperature. This approach requires a thermostable polymerase such as the r*Tth* DNA polymerase (Perkin-Elmer) – a recombinant form of the *Thermus thermophilus* enzyme which can operate as a reverse transcriptase at 60–70°C in the presence of Mn^{2+} and can also amplify DNA in the cycling phase of PCR in the presence of Mg^{2+}.

rtPCR can be used e.g. to detect specific types of mRNA molecule. It is particularly useful for studying RNA viruses; examples include: dengue virus [JCM (1999) *37* 2543–2547]; hepatitis G virus [JCM (1997) *35* 767–768]; hepatitis C virus and HIV-1 [Lancet (1999) *353* 359–363].

reverse transcription See REVERSE TRANSCRIPTASE.

reversion Restoration of the original phenotype of a mutant organism by means of a BACK MUTATION or a SUPPRESSOR MUTATION; the resulting organism is called a *revertant*.

revertant See REVERSION.

Reyes syndrome An acute disease of children and adolescents; it involves e.g. non-inflammatory encephalopathy with fatty infiltration and dysfunction of the liver, and is characterized clinically by persistent and profuse vomiting and neurological dysfunction – sometimes progressing to delirium, coma and death. The aetiology is unknown. The syndrome characteristically follows infection with certain viruses, particularly influenza A or B or varicella-zoster viruses. An association has also been recognized between the occurrence of Reyes syndrome and the administration of aspirin during the prodromal phase of the viral infection; in the USA, decreasing use of salicylates in children has apparently correlated with a decrease in the incidence of Reyes syndrome [Pediatrics (1986) *77* 598–602].

RF Replicative form, an intermediate formed during the replication of certain ssDNA or ssRNA viral genomes; an RF is a double-stranded structure consisting of the viral strand base-paired with its (newly synthesized) complementary strand. (cf. RI.) In e.g. SSDNA PHAGES the RF is a circular dsDNA molecule; the supercoiled ccc form is designated RFI (cf. DNA I), while the relaxed form – nicked in one strand – is designated RFII (cf. DNA II).

RF-1, RF-2, RF-3 See PROTEIN SYNTHESIS (termination).

***rfb* gene** See EHEC.

RFLP Restriction fragment length polymorphism: a phrase which refers to differences in fragment length which may be obtained when related sequences of DNA are exposed to the same RESTRICTION ENDONUCLEASE(S); such differences may arise e.g. if sequence(s) have lost or gained one or more (relevant) restriction sites as a result of mutation, or if particular fragment(s) are longer or shorter, respectively, as a result of insertion or deletion of nucleotides. The concept of RFLP is illustrated in the figure.

RFLP analysis is used e.g. for TYPING. For typing short sequences (e.g. a plasmid, viral DNA, or a short sequence in a bacterial chromosome), the fragments formed by restriction are examined by electrophoresis and staining (thus providing a *fingerprint*). This approach has been used e.g. for detecting variation in the genome of herpes simplex virus type 1 [RMM (1998) *9* 217–224].

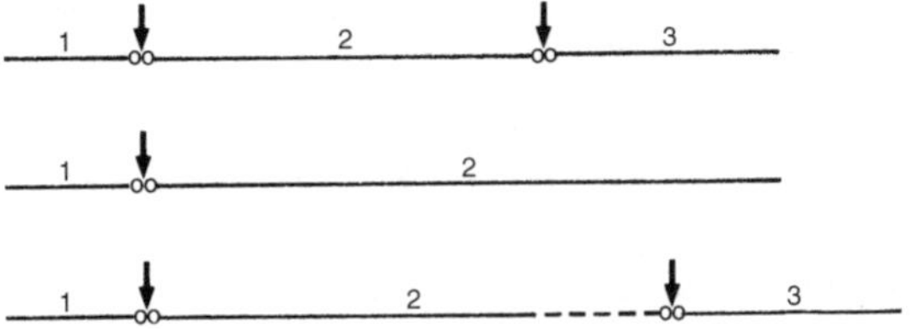

RFLP (principle, diagrammatic). Horizontal lines represent related DNA duplexes. The top duplex has two sites for a given restriction endonuclease; enzymic cleavage (arrow) at each of these two sites produces three restriction fragments. In the centre duplex, the second cleavage site has been lost through mutation; enzymic cleavage produces only two fragments, fragment 1 being the same as before. The lower duplex contains an extra short sequence of nucleotides (dashed line); enzymic cleavage of this duplex produces three fragments, but fragment 2 is longer than that in the top duplex. Electrophoresis of the fragments from each duplex will produce different fingerprints.

RFLP-based typing proceeds differently when large sequences (e.g. a whole bacterial chromosome) are involved. Initially, chromosomes of the test strain are cleaved by restriction enzyme(s) and the fragments separated by electrophoresis in polyacrylamide gel. Bands of fragments in the gel are 'blotted' onto a membrane, and the membrane is then exposed to labelled probes complementary to a specific target sequence in the chromosome; thus, probes will hybridize only to those bands of fragments which include the target sequence, and only these bands will be made visible by the probe's label. Suppose, for example, that the probe's target is a sequence in a single-copy gene. In this case probes will hybridize to only one band of fragments on the membrane; however, this may be sufficient to reveal differences between test strains as the target-containing fragment may vary in length among the different strains (giving rise to bands in different relative positions in the gel). In some cases the probe is complementary to a *multicopy* target sequence; any target which occurs in multiple copies in the genome has the advantage that it will probably give rise to a fingerprint consisting of more than one band and (for this reason) is likely to have superior discriminatory power as a typing procedure. One example of this is IS*6110*-based typing of *Mycobacterium tuberculosis*; most strains have multiple copies of IS*6110* (strain H37Rv has 16 copies [Nature (1998) *393* 537–544]), and, because IS*6110* is a mobile genetic element, it can insert at different sites in the chromosome and (thus) give rise to new strains (with new fingerprints).

Note that RIBOTYPING (q.v.) is a particular form of RFLP-based typing.

A possible problem in RFLP-based typing arises because chromosomal DNA isolated from a test strain will have undergone modification (i.e. methylation of bases in certain sequences); such modification will inhibit certain restriction endonucleases so that a careful choice of enzymes is necessary. This problem can be avoided by amplifying a chromosomal target sequence in vitro by PCR; the amplicons (which lack methylation owing to synthesis in vitro) can then be used for RFLP analysis. PCR-RFLP analysis has been used e.g. for typing isolates of *Staphylococcus aureus* from cows and sheep with mastitis [JCM (1999) *37* 570–574].

RGD motif See INTEGRINS.

***rhl* gene** (*rHl* gene) See PHASE VARIATION.

rhabdocyst A rod-like EXTRUSOME which occurs in some gymnostome ciliates.

Rhabdomonas See EUGLENOID FLAGELLATES.

rhabdomyosarcoma See SARCOMA.

rhabdos A type of (non-curved) CYTOPHARYNGEAL APPARATUS whose walls are strengthened externally by nematodesmata and are frequently lined by longitudinally arranged TRANSVERSE MICROTUBULES arising from peri-oral kinetosomes. Toxicysts are sometimes present. The rhabdos is believed to be more primitive than the CYRTOS.

rhabdosarcoma See SARCOMA.

Rhabdostyla See PERITRICHIA.

Rhabdoviridae A family of enveloped, ssRNA-containing VIRUSES which can infect vertebrates, invertebrates or plants; the family includes some important pathogens, e.g. the causal agents of RABIES and VESICULAR STOMATITIS. The rhabdovirus virion contains 4–5 proteins designated L (large), G (glycoprotein), N (nucleocapsid), NS (non-structural) and M (matrix); 15–25% lipid (composition depending on host cell); and ca 3% carbohydrate. Genome: one molecule of linear negative-sense ssRNA (MWt ca. $3.5–4.6 \times 10^6$). Maturation involves budding through host membranes. Rhabdoviruses are stable at pH 5–10 (unstable at pH 3), but are rapidly inactivated by heating to 56°C, by UV and X-irradiation, and by lipid solvents.

Animal rhabdoviruses are typically 'bullet-shaped' (i.e. cylindrical with one end flat, the other rounded), measuring e.g. ca. 170×70 nm (VSV). The virion consists of a helical nucleocapsid composed of an RNA-N protein complex, associated with proteins L and NS, and surrounded by protein M – the whole being enclosed within the lipoprotein (ether-sensitive) envelope. The G protein forms surface projections ca. 5–10 nm long; antibodies to the G protein neutralize the virus. Various enzymic activities are associated with the virion, including e.g. transcriptase, protein kinase, 5′ capping enzymes, etc. Virus replication occurs in the cytoplasm. Two genera: LYSSAVIRUS and VESICULOVIRUS. Other 'probable members' include e.g. BOVINE EPHEMERAL FEVER virus, EGTVED DISEASE virus, INFECTIOUS HAEMATOPOIETIC NECROSIS virus, red disease of pike virus, rhabdovirus of blue crab, rhabdovirus of *Entamoeba*, rhabdovirus of grass carp, SIGMA VIRUS of *Drosophila*, SPRING VIRAEMIA OF CARP virus.

Plant rhabdoviruses are usually bacilliform, sometimes bullet-shaped, typically ca. $200–350 \times 70–95$ nm; all contain proteins N and G, some have proteins L, M and NS, others have proteins 'M1' and 'M2'. Each member generally infects only one or a few plant species, and is transmitted (circulatively) by one or more particular vectors (aphids, leaf-hoppers, plant-hoppers, bugs or mites); some members (not those infecting gramineous hosts) can be transmitted mechanically under experimental conditions. Two subgroups of plant rhabdoviruses (A and B) have been distinguished on the basis of the site of virus assembly, virion protein composition, and transcriptase activity in vitro. Members of subgroup A – e.g. lettuce necrotic yellows virus (type member), broccoli necrotic yellows virus, wheat striate mosaic virus – share some properties with *Vesiculovirus*: virions mature and accumulate in the host cell cytoplasm, contain protein M, and have readily detectable transcriptase activity in vitro. Members of subgroup B – e.g. potato yellow dwarf virus (type member), eggplant mottled dwarf virus – bud at the inner membrane of the nuclear envelope and accumulate in the perinuclear space, contain proteins M1 and M2, and have low transcriptase activity in vitro. Other probable (but ungrouped) members of the plant rhabdoviruses include e.g. carrot latent virus, lucerne enation virus, raspberry vein chlorosis virus, and strawberry crinkle virus (all aphid-borne); barley yellow striate mosaic virus (= cereal striate virus), cereal chlorotic mottle virus, finger millet mosaic virus, maize mosaic virus, oat striate virus, rice transitory yellowing virus, Russian winter wheat mosaic virus, and wheat chlorotic streak virus (all hopper-borne); beet leaf curl virus (lace bug-borne); coffee ringspot virus (mite-borne). Listed by the ICTV [Book ref. 23, pp. 112–114] as 'possible members' are some non-enveloped viruses (e.g. *Citrus* leprosis virus and orchid fleck virus) and many virus-like particles; these appear to resemble rhabdovirus nucleocapsids (ca. 100–120 × 35 nm) and form characteristic 'spoked-wheel' structures in the plant cell nucleus.

rhabdovirus A virus of the RHABDOVIRIDAE.

rhamnogalacturonan See PECTINS.

rhamnolipid (2-*O*-α-L-rhamnopyranosyl-α-L-rhamnopyranosyl-β-hydroxydecanoyl-β-hydroxydecanoate) An extracellular glycolipid BIOSURFACTANT produced by certain strains of *Pseudomonas aeruginosa* during growth on HYDROCARBONS.

L-rhamnose 6-Deoxy-L-mannose; it occurs e.g. in many plant glycosides and PECTINS, many streptococcal C SUBSTANCES, and some LIPOPOLYSACCHARIDES (e.g. the O-specific chains of certain *Salmonella* serotypes).

Rhaphidophyceae See CHLOROMONADS.

rhapidosomes Rod-shaped or tubular structures, 20–30 nm in diameter, which occur in many species of bacteria (e.g. *Flexibacter*, *Pseudomonas* spp) and cyanobacteria (e.g. *Spirulina*). Rhapidosomes (in different organisms) have been variously regarded as bacteriocins, defective phage tails, particles of mesosomal membrane, and cell components involved in motility.

Rheinberg illumination (coloured field illumination; optical staining) Illumination analogous to that used in low-power *dark-field* MICROSCOPY (q.v.). Instead of an opaque stop, the condenser is fitted with a *Rheinberg disc*: a transparent disc with a central circular area of one colour and a surrounding annulus of one or more lighter colours; the central colour forms the background light while the specimen is lit by the peripheral rays of the lighter colour(s).

rheotaxis A TAXIS in which the stimulus is a current; a positively rheotactic organism swims against a prevailing current (i.e., it swims 'upstream').

rheotropism See TROPISM (sense 1).

rheumatic fever A condition which occurs as a late complication of SCARLET FEVER or a group A streptococcal infection of the throat ('strep throat', tonsillitis) or middle ear (OTITIS MEDIA); it may follow an asymptomatic streptococcal infection, but rarely or never occurs after streptococcal skin infection (e.g. IMPETIGO, ERYSIPELAS) – cf. POST-STREPTOCOCCAL GLOMERULONEPHRITIS. Rheumatic fever occurs most commonly in children aged between 5 and 15. An asymptomatic latent period of 1–3 weeks may occur between the original infection and the onset of rheumatic fever. Onset is usually rather insidious, with malaise, fever, and inflammation of joints; other symptoms may include e.g. subcutaneous nodules, rash, and/or involuntary twitching movements (Sydenham's chorea, 'St Vitus's dance'). Endocarditis may develop and may lead to chronic rheumatic valvular heart disease. An individual who has suffered an attack of rheumatic fever has greatly increased susceptibility to subsequent streptococcal infection. Rheumatic fever is apparently unrelated to the presence of living streptococci and is believed to be caused by an immunopathological mechanism; it may be due to an autoimmune reaction between anti-group A streptococcal antibody and a host tissue component.

rheumatoid arthritis A chronic, systemic disease involving inflammatory changes in the body's connective tissues, resulting in a progressive, deforming ARTHRITIS. Pathogenesis appears to involve an autoimmune reaction (in genetically predisposed individuals) which is triggered by an unknown environmental factor (possibly a virus); IgM autoantibodies ('rheumatoid factor') to IgG are formed, and IgG–IgM immune complexes are deposited in the synovium and blood vessels, causing inflammation. Whether or not viruses have a role in the pathogenesis of rheumatoid arthritis is controversial, but parvovirus-like particles have been detected in synovial tissue from the joints of a rheumatoid arthritis patient [Science (1984) *223* 1425–1428]. (See also ROSE–WAALER TEST.)

rheumatoid factor See RHEUMATOID ARTHRITIS and ROSE–WAALER TEST.

rhexolysis See CONIDIUM.

rhinitis Inflammation of the nasal mucous membranes. It occurs e.g. in the COMMON COLD and in allergic reactions such as hay fever.

Rhinocladiella See HYPHOMYCETES; see also CHROMOBLASTOMYCOSIS.

rhinophycomycosis Nasofacial or rhinocerebral ZYGOMYCOSIS.

rhinosporidiosis A chronic disease of man and animals, caused by *Rhinosporidium seeberi*, characterized by the formation of polyp-like growths chiefly on the mucosae of the nose and upper respiratory tract; breathing may be obstructed. Dissemination is rare.

Rhinosporidium A genus of fungi of uncertain taxonomic position; *R. seeberi*, causal agent of RHINOSPORIDIOSIS, apparently resembles 'Hyphochytridiomycetes' rather than Chytridiomycetes [JMM (1985) *20* x (abstr.)]. In infected tissues, *R. seeberi* forms spherical sporangia (up to ca. 300 μm, depending e.g. on stage of development) which, when mature, develop thick refractile walls and undergo nuclear and cytoplasmic division to form numerous uninucleate endospores (6–7 μm); endospores are released on rupture of the cell wall.

rhinotracheitis Inflammation of the mucous membranes of the nose and trachea. See e.g. FELINE RHINOTRACHEITIS and INFECTIOUS BOVINE RHINOTRACHEITIS.

Rhinovirus A genus of viruses (family PICORNAVIRIDAE) which are acid-labile (being rapidly inactivated at pH < 6) and which infect the mammalian upper respiratory tract (cf. ENTEROVIRUS). Human rhinoviruses are the major causal agents of the COMMON COLD; many serotypes are known. They can be propagated in various human cell cultures (e.g. WI-38) and have an optimum growth temperature of 33°C. Most strains are stable at or below room temperatures and can withstand freezing; some strains are stable at 50°C for 1 hour. Most heat-labile serotypes (but not type 30) can be stabilized to heat by the presence of $MgCl_2$. Rhinoviruses can be inactivated by e.g. citric acid (see COMMON COLD), tincture of iodine or phenol/alcohol mixtures, but generally not by e.g. alcohols, bile salts, hexachlorophene, organic solvents, or quaternary ammonium compounds. [Clinical and epidemiological aspects: Book ref. 148, pp. 795–816. Nucleotide sequences of the genomes of human rhinovirus types 2 and 14: NAR (1985) *13* 2111–2126 and PNAS (1985) *82* 732–736, respectively.]

Other members of the genus include the bovine rhinoviruses; 'equine rhinoviruses' may represent a separate genus.

Rhipicephalus A genus of ixodid ticks which are vectors of certain diseases: e.g. REDWATER FEVER.

Rhipidium See LEPTOMITALES.

Rhipocephalus A genus of siphonaceous, calcified green seaweeds (division CHLOROPHYTA).

Rhizamoeba See ACARPOMYXEA.

Rhizidiomyces A genus of unicellular fungi (class HYPHOCHYTRIOMYCETES) which occur in soil and as parasites of certain water moulds (members of the Saprolegniaceae) and algae (e.g. *Vaucheria*).

Rhizidiomycetaceae See HYPHOCHYTRIOMYCETES.

rhizine (rhizina) (*lichenol.*) A root-like structure consisting of a compact bundle of hyphae. Rhizines (rhizinae) arise mainly from the lower surface of the (usually foliose) thallus and serve to anchor the thallus to the substratum; in some species they may also facilitate the uptake of water and nutrients by the thallus [Bryol. (1981) *84* 1–15]. Rhizines may be unbranched or may be sparsely or extensively branched – branching being e.g. 'squarrose' (short branchlets emerging at ca. 90°) or dichotomous.

Rhizobiacea A family of Gram-negative, asporogenous, rod-shaped, motile, aerobic, chemoorganotrophic bacteria; all (except *Agrobacterium radiobacter*) can induce tumour-like growths in plants. Genera: AGROBACTERIUM, BRADYRHIZOBIUM, PHYLLOBACTERIUM, RHIZOBIUM.

Rhizobium A genus of Gram-negative bacteria of the RHIZOBIACEAE. Cells: rods, 0.5–0.9 × 1.2–3.0 μm; pleomorphic under adverse conditions. Motile: one (usually subpolar) flagellum (in *R. loti*) or several lateral or sometimes polar flagella (in *R. leguminosarum* and *R. meliloti*); in some strains 'complex flagella' occur: see FLAGELLUM. Refractile PHB granules are common, especially in older cells. Colonies: mucilaginous, usually 2–4 mm diameter in 3–5 days on yeast extract–manitol–mineral salts agar (cf. BRADYRHIZOBIUM). Acid (no gas) is formed from a range of carbohydrates (e.g. mannitol). Abundant extracellular slime (including a neutral (1 → 2)-β-glucan) is produced in carbohydrate-containing media. [Media and methods: Book ref. 45, pp. 824–834.] GC%: 59–64. Type species: *R. leguminosarum*.

Rhizobium spp are common in soil [ecology: Book ref. 45, pp. 818–822] and can incite ROOT NODULE formation in certain leguminous plants (mainly in temperate regions – cf. STEM NODULES); the free-living organisms appear to be incapable of nitrogen fixation. Strains which are both *infective* (able to induce nodule formation) and *effective* (able to fix nitrogen) tend to become ineffective (but may remain infective) e.g. after continual laboratory cultivation.

R. phaseoli, *R. trifolii* and *R. viceae* are currently regarded as biotypes of *R. leguminosarum*. *R. lupini* is currently not a recognized species; it may be a species or biotype of *Bradyrhizobium*. *R. japonicum*: see BRADYRHIZOBIUM.

[Book ref. 22, pp. 235–242.]

Rhizocarpon A genus of LICHENS (order LECANIDIALES); photobiont: a green alga. The thallus is crustose, typically forming a mosaic of yellow or grey areolae between which the black hypothallus is visible. Saxicolous. Apothecia: black, lecideine; ascospores: septate to muriform. [Ascus structure: Lichenol. (1980) *12* 157–172.]

Rhizoclonium A genus of filamentous, siphonocladous green algae (division CHLOROPHYTA) in which the filaments are unbranched or bear laterally inserted rhizoidal branches.

Rhizoctonia A genus of fungi (order AGONOMYCETALES) which include some important plant pathogens. *R. leguminicola*, the causal agent of blackpatch disease of clover (see SLAFRAMINE), can form sclerotia under appropriate conditions. *R. solani* can cause DAMPING OFF and EYESPOT (sense 2). (See also FOOT-ROT sense 2.) [Systematics and phylogeny of the *Rhizoctonia* complex: Bot. Gaz. (1978) *139* 454–466; genetics and pathology of *R. solani*: ARPpath. (1982) *20* 329–347.]

rhizogenic Capable of inducing root formation.

rhizoid A root-like structure which forms part of the thallus in certain algae and fungi; it may anchor the organism to the substratum and/or act as an absorptive organ or organelle. (cf. RHIZOMYCELIUM.)

rhizomania See BEET NECROTIC YELLOW VEIN VIRUS.

rhizomorph (*mycol.*) A macroscopic, typically dark-coloured thread of hard, compacted tissue formed by certain higher fungi, e.g. *Armillaria mellea*; the tissue generally appears to consist of individual cells rather than hyphae. Rhizomorphs are enduring structures which can remain dormant under adverse conditions and which can enable a fungus to spread into the surrounding environment – e.g. those of *A. mellea* permit the fungus to spread from an infected tree to an uninfected tree via the soil. (See also DRY ROT.)

Rhizomucor A genus of fungi of the MUCORALES. *R. pusillus* (formerly *Mucor pusillus*) can cause ZYGOMYCOSIS. (See also COMPOSTING.) *R. miehei* (formerly *Mucor miehei*) and *R. pusillus* form PROTEASES which are used as rennin substitutes.

rhizomycelium (*mycol.*) Branching, anucleate or sparsely nucleate, rhizoidal filaments of variable width which form part of the thallus e.g. of *Nowakowskiella* and of some species of the Laboulbeniales.

Rhizophidium See RHIZOPHYDIUM.

Rhizophydium (*Rhizophidium*) A genus of eucarpic fungi (order CHYTRIDIALES) which occur as saprotrophs or as parasites of aquatic or terrestrial plants; *R. graminis* parasitizes the root cells of mono- and dicotyledonous plants (apparently without causing them harm), while *R. couchii* has been reported to parasitize *Spirogyra*. In the asexual cycle of *R. couchii*, a zoospore encysts on the surface of a host and gives rise to a thallus whose rhizoids penetrate the host; the extracellular part of the thallus subsequently becomes converted into a zoosporangium. In the sexual cycle, zoospores develop into extracellular, rhizoid-bearing gametangia; after plasmogamy (and karyogamy?) the zygote becomes a thick-walled resting zoosporangium.

rhizoplane The root surface – cf. RHIZOSPHERE.

rhizoplasmodium See LABYRINTHULAS.

rhizopod An organism of the RHIZOPODA.

Rhizopoda A superclass of protozoa (subphylum SARCODINA) which are motile by means of lobopodia, filopodia or reticulopodia (see PSEUDOPODIUM) or by flow of protoplasm without the formation of discrete pseudopodia. Classes: ACARPOMYXEA; Acrasea (see ACRASIOMYCETES); EUMYCETOZOEA; FILOSEA; GRANULORETICULOSEA; LOBOSEA; Plasmodiophorea (= PLASMODIOPHOROMYCETES); XENOPHYOPHOREA.

rhizopodium See PSEUDOPODIUM.

Rhizopogon See GASTEROMYCETES (Hymenogastrales).

Rhizopus A genus of fungi (order MUCORALES) which occur on soil, fruit etc. The organisms form a branched, aseptate mycelium. *R. nigricans* (= *R. stolonifer*) can attach to the substratum by means of branching rhizoids; a *stolon* (a surface hypha analogous to the stolons in higher plants) grows for some distance and then gives rise to more rhizoids and to one or more vertical sporangiophores. The organisms form ZYGOSPORES (see also COMPATIBILITY sense 2). *Rhizopus* spp are used commercially e.g. in the preparation of ONCOM, RAGI and TEMPEH and in certain STEROID BIOCONVERSIONS. (See also BREAD SPOILAGE and ZYGOMYCOSIS.)

Rhizosolenia A genus of freshwater and marine centric DIATOMS. Some (marine) species contain a nitrogen-fixing cyanobacterial symbiont, *Richelia intracellularis*, which occurs as short, *Calothrix*-like filaments – each with a heterocyst at one end or, in the dividing diatom, at both ends. The association is not obligatory, and *Richelia* may also be found free-living or as an epiphyte on diatoms of the genera *Chaetoceros*, *Hemiaulus* and *Rhizosolenia*. (cf. RHOPALODIA.)

rhizosphere An environment regarded, variously, as e.g. (a) that region of the soil which is modified as a result of the uptake and deposition of substances by a growing root [AEE (1985) *12* 99–116]; (b) the root itself, together with that volume of soil which it influences [Book ref. 112, p. 237]; or (c) the root surface (*rhizoplane*) together with that region of the surrounding soil in which the microbial population is affected, qualitatively and/or quantitatively, by the presence of a root. The rhizosphere may extend a few millimetres, or centimetres, from the rhizoplane.

From a growing root, a number of substances pass out into the soil. These substances include various carbohydrates (see also MUCIGEL), vitamins, amino acids and sugars – many of which serve as nutrients and/or as sources of energy for microorganisms. Presumably as a result of this, the microflora of the rhizosphere differs appreciably from that of the surrounding soil. In the rhizosphere, bacterial numbers are often 10- to 50-fold higher than they are in the surrounding soil; the bacteria are mainly Gram-negative rods (e.g. pseudomonads), Gram-positive bacteria typically being less numerous than they are outside the rhizosphere. It has been reported that the presence of a vesicular-arbuscular MYCORRHIZA (involving *Glomus fasciculatum*) reduces the proportion of fluorescent pseudomonads and increases the proportion of facultatively anaerobic bacteria in the rhizosphere – but does not affect the absolute number of Gram-negative bacteria [SBB (1986) *18* 191–196].

Numerically, fungi may be present in similar or slightly greater abundance in the rhizosphere as compared with the surrounding soil. However, there may be qualitative (species) differences between rhizosphere and non-rhizosphere fungi, and in at least some cases there is evidence that the rhizospheres of certain plants encourage, or select, particular fungi.

The extent to which the plant benefits from the rhizosphere microflora remains unclear. It seems possible that the microflora may play some part in protecting the plant from the incursions of soil-borne pathogens; additionally, the plant may benefit from the mineralizing activities of these organisms, and it is generally believed that the presence of the rhizosphere microflora promotes the uptake of minerals (e.g. phosphate) by the root.

rhizothamnium A *Myrica*-type ACTINORRHIZA.

***rhl* genes** See QUORUM SENSING.

rho factor (ρ factor) See TRANSCRIPTION.

***rho*-form** See MYCOPLASMA (*M. mycoides*).

***rho* gene** In e.g. *Escherichia coli*: the gene encoding the ρ factor (see TRANSCRIPTION).

***rho*$^-$ petite** See PETITE MUTANT.

***rho*^{0} petite** See PETITE MUTANT.

rhodamine A generic term for a group of substituted xanthene FLUOROCHROME dyes; examples: rhodamine B (fluorescence red – see also AURAMINE-RHODAMINE STAIN and LISSAMINE RHODAMINE), rhodamine S (fluorescence yellow), rhodamine O (red), rhodamine 3G (orange). The rhodamines include cationic, anionic and neutral dyes. Uptake of *cationic* rhodamines (e.g. rhodamine 123) appears to depend on membrane potential in viable, functioning mitochondria [JCB (1981) *88* 526–535] and bacteria [FEMS (1984) *21* 153–157]. (See also VITAL STAINING.)

rhodanese (thiosulphate sulphurtransferase; EC 2.8.1.1) An enzyme that catalyses e.g. the formation of thiocyanate (SCN^-) and sulphite (SO_3^{2-}) from THIOSULPHATE and cyanide; it occurs e.g. in certain bacteria (including *Bacillus subtilis*, *Desulfotomaculum nigrificans*, *Thiobacillus* spp), in plants, and in most mammalian tissues. It has been suggested that one of the roles of rhodanese is to donate sulphur to a thiophilic anion (e.g. dihydrolipoate) which can complex iron and promote the formation of IRON-SULPHUR PROTEINS [TIBS (1986) *11* 369–372; discussion: TIBS (1987) *12* 56–57].

Rhodobacter A proposed genus of bacteria [IJSB (1984) *34* 340–343] which includes species transferred from RHODOPSEUDOMONAS (q.v.) and re-designated *Rhodobacter adriaticus*, *R. capsulatus* (type species), *R. sphaeroides* and *R. sulfidophilus*. The genus includes motile and non-motile organisms; the cells divide by binary fission.

Rhodochaete See RHODOPHYTA.

rhodocladonic acid A bright-red naphthoquinone pigment found in the hymenial discs of certain CLADONIA spp.

Rhodococcus A genus of aerobic, chemoorganotrophic, non-motile bacteria (order ACTINOMYCETALES, wall type IV) which occur e.g. in soil, in aquatic habitats, and as pathogens of man and other animals. In culture, the cells of all species are cocci, or coccobacilli, which – depending on species – give rise e.g. to pleomorphic rods or to a branching mycelium; the rods or mycelium subsequently fragment to form the next generation of coccoid forms. Acid-fastness appears to be variable. The cell wall contains MYCOLIC ACIDS. The organisms grow well on nutrient media at 30°C; some strains need thiamine. Colonies are often pigmented (e.g. yellow, orange, red). Glucose is metabolized oxidatively. GC%: ca. 63–73. Type species: *R. rhodochrous*.

The genus includes *R. bronchialis*; *R. coprophilus* (a mycelial species); *R. equi* (formerly *Corynebacterium equi*: a non-mycelial species which is the causal agent of e.g. suppurative bronchopneumonia in the horse); *R. erythropolis*; *R. fascians* (formerly *Corynebacterium fascians* [SAAM (1984) *5* 225–229]); *R. globerulus*; *R. luteus*; *R. maris*; *R. rhodnii*; *R. rhodochrous* (a species which forms branching rods, but not mycelium); *R. ruber* (a mycelial species); *R. rubropertinctus*; and *R. terrae*. [Book ref. 73, p. 93.] [A marine species, *R. marinonascens*: IJSB (1984) *34* 127–138.] (See also CORYNECINS.)

Rhodocyclus A proposed genus of non-motile, spiral-shaped photosynthetic bacteria (family RHODOSPIRILLACEAE) consisting of a single species, *R. purpureus* [IJSB (1978) *28* 283–288]. It was later proposed that certain species be transferred from the genera RHODOPSEUDOMONAS (q.v.) and RHODOSPIRILLUM (q.v.) and re-designated *Rhodocyclus gelatinosus* and *R. tenuis*, respectively [IJSB (1984) *34* 340–343].

Rhodomicrobium A genus of photosynthetic bacteria (class RHODOSPIRILLACEAE); the sole species, *R. vannielii*, occurs e.g. in freshwater muds and in estuarine and marine habitats. *R. vannielii* contains Bchl *a* and carotenoids (mainly β-carotene and spirilloxanthin) in lamellar intracytoplasmic membrane systems; cell suspensions are typically pink-brown. Growth occurs photoorganotrophically under anaerobic conditions, or chemoorganotrophically under microaerobic to aerobic conditions in the dark.

R. vannielii undergoes one of two types of vegetative cell cycle, depending on growth conditions. In one type, a non-motile, ovoid cell (ca. 1–3 μm) becomes prosthecate, and a daughter cell is formed by budding at the tip of the PROSTHECA. The daughter cell remains attached, forms a prostheca at the opposite pole, and buds in turn. This leads eventually to an array of cells which are joined together by branched and unbranched filaments (ca. 0.3 μm wide) derived from the prosthecae. (Branches arise from the filaments.) Each cell is separated from its neighbour by a plug formed within the filament joining them. Such arrays – apparently the commonest form under natural conditions – may be produced under conditions of high light intensity and low concentrations of CO_2. If the light dims and CO_2 levels rise (e.g. as the array becomes large), motile (peritrichous) *swarmer* cells form at the ends of filaments and are released by binary fission; under suitable conditions swarmers shed their flagella and become non-motile prosthecate cells, thus completing the cycle. If conditions of low light and high CO_2 persist, a second ('simplified') cell cycle occurs: arrays are not formed – budding cells giving rise only to swarmers; array formation can subsequently occur if CO_2 concentration decreases and light intensity increases. Under starvation conditions, cell arrays give rise to *exospores*: angular structures, resistant to heat and drying, which form at the tips of filaments and are released by binary fission; the exospores can germinate under aerobic conditions in the dark or anaerobically in the light, forming up to four vegetative cells. (Exospores are not formed in the 'simplified' cycle; late exponential phase cultures contain 'tiny' cells which are not resistant to stress [JGM (1980) *117* 47–55].)

[Reviews: ARM (1981) *35* 567–594; Book ref. 28, pp. 188–210.]

(See also HYPHOMICROBIUM.)

rhodomycin See ANTHRACYCLINE ANTIBIOTICS.

Rhodophyceae See RHODOPHYTA.

Rhodophyta (red algae) A division of ALGAE. Over 95% of the ca. 5000 species of red algae are marine organisms; those which occur in freshwater and/or soil habitats include species of *Audouinella*, *Bangia*, *Batrachospermum*, *Chroodactylon*, *Hildenbrandia*, *Lemanea* and *Porphyridium*. (Some genera, e.g. *Bangia*, *Bostrychia* and *Hildenbrandia*, contain both marine species and freshwater species.) A few red algae are parasitic on other algae (see e.g. CHOREOCOLAX and HOLMSELLA).

In most red algae the thallus is a branched filament or ribbon. However, some species are sheet-like (e.g. *Porphyra*) or crust-like (e.g. *Lithophyllum*), while a few (e.g. PORPHYRIDIUM) are unicellular. Some red algae are calcified – e.g. *Corallina* spp (thalli: erect, articulated) and *Lithophyllum*. Multicellular thalli (which are commonly 5–50 cm in length) are typically attached to rocks etc by a holdfast. The CELL WALL may contain cellulose or xylans. Plasmodesmata appear to be absent, but many multicellular red algae contain PIT CONNECTIONS. There are no flagellated species, and flagellated gametes are not produced. Marine species are usually some shade of red, while freshwater species are typically blue-green, yellow-green, brown or grey.

Red algae contain CHLOROPHYLL *a*, and some also contain chlorophyll *d*; the THYLAKOIDS occur singly (i.e., unassociated) and bear PHYCOBILISOMES containing phycoerythrins and/or phycocyanins. Various CAROTENOIDS (e.g. xanthophylls and β-carotene) are usually present. Storage products include FLORIDEAN STARCH and FLORIDOSIDE (see also ISOFLORIDOSIDE).

Sexual reproduction occurs in most species (but apparently not in *Porphyridium*) and oogamy is common; life cycles are often complex.

Certain red algae are used as sources of AGAR and CARRAGEENAN, while some are used as food (see e.g. LAVER, NUNGHAM, PALMARIA).

All species are placed in one class: Rhodophyceae. Orders [Book ref. 130]: Bangiales (e.g. *Bangia*, PORPHYRA); Ceramiales (e.g. *Bostrychia; Ceramium, Griffithsia, Polysiphonia*); Compsopogonales (e.g. *Compsopogon*); Cryptonemiales (e.g. CHOREOCOLAX, *Corallina*, *Gloiopeltis*, *Hildenbrandia*, HOLMSELLA, *Lithophyllum*); Gigartinales (e.g. *Chondrococcus*, *Chondrus*, *Eucheuma*, *Furcellaria*, *Gardneriella*, GIGARTINA, *Gracilaria*, *Iridaea*); Nemalionales (e.g. *Audouinella*, *Batrachospermum*, *Gelidium*, *Lemanea*); Palmariales (e.g. *Palmaria*); Porphyridiales (e.g. *Chroodactylon*, *Cyanidium*, PORPHYRIDIUM); Rhodochaetales (e.g. *Rhodochaete*); Rhodymeniales (e.g. *Coeloseira*, *Rhodymenia*).

[Biology of freshwater red algae: Book ref. 167, pp. 89–157.]

Rhodopila A proposed genus of bacteria [IJSB (1984) *34* 340–343] which includes a species transferred from RHODOPSEUDOMONAS (q.v.) and re-designated *Rhodopila globiformis*; the coccoid, flagellated cells of *R. globiformis* grow only at low pH.

Rhodopseudomonas A genus of photosynthetic bacteria (family RHODOSPIRILLACEAE). Cells: flagellated (or non-motile) rods or cocci, typically 1–5 μm; most species contain Bchl *a*, *R. sulfoviridis* and *R. viridis* contain Bchl *b*. *R. acidophila*, *R. palustris* (the type species), *R. sulfoviridis* and *R. viridis* divide by budding; other species divide by binary fission. *R. palustris* and *R. viridis* form prosthecae. The oval to rod-shaped species *R. adriatica*, *R. capsulata*, *R. sphaeroides* and *R. sulfidophila* all have vesicular-type photosynthetic membranes, and it has been proposed that they be transferred to a new genus: RHODOBACTER; it has also been proposed that the coccoid species *R. globiformis* (which has a vesicular photosynthetic membrane) and the curved, rod-shaped species *R. gelatinosa* be transferred to the genera RHODOPILA and RHODOCYCLUS, respectively [IJSB (1984) *34* 340–343].

[Gene transfer mechanisms: Ann. Mic. (1983) *134*B 195–204.] (See also CAPSDUCTION.)

Rhodospirillaceae (purple non-sulphur bacteria, or non-sulphur purple bacteria; Athiorhodaceae) A heterogeneous family (see RHODOSPIRILLALES) of photosynthetic bacteria (suborder RHODOSPIRILLINEAE); they occur typically in those anaerobic habitats in which a range of simple organic substrates has been made available by the metabolic activities of chemoorganotrophic bacteria – e.g. in the mud of ponds and rivers, and in sewage lagoons. The organisms occur in freshwater, brackish and marine habitats and in moist soils. Most species contain only Bchl *a* (two species of *Rhodopseudomonas* have only Bchl *b*) – see CHLOROPHYLLS; these pigments, and various carotenoids, occur in lamellar, tubular or vesicular intracytoplasmic membrane systems, according to species. The organisms include cocci, rods and spiral forms; most species divide by binary fission, but several divide by budding (see e.g. RHODOPSEUDOMONAS). Heat-resistant forms are produced by *Rhodomicrobium vannielii*. Most species (not e.g. *Rhodocyclus purpureus*) exhibit flagellar motility. Gas vacuoles are absent. The organisms are primarily photoorganotrophic heterotrophs under anaerobic conditions, but a few species, e.g. *Rhodopseudomonas sulfidophila*, can grow as photolithotrophic autotrophs (using sulphide as electron donor), and some other members of the family can use sulphide as electron donor only when it is present in low concentrations. It has been shown that at least some species can oxidize elemental sulphur [JGM (1985) *131* 791–798]; sulphide (or thiosulphate) can be oxidized to sulphate (without sulphur formation) by e.g. *Rhodopseudomonas sulfidophila*. Many species can oxidize sulphide with the production of extracellular deposits of sulphur. A number of species can use hydrogen as an electron donor for photolithotrophic growth. Typically, growth requires one or more factors such as biotin, *p*-aminobenzoic acid and thiamine; many species can carry out nitrogen fixation, but a good rate of growth generally requires a source of organic nitrogen. Under aerobic or microaerobic conditions most species are chemoorganotrophic.

Genera: RHODOMICROBIUM, RHODOPSEUDOMONAS, RHODOSPIRILLUM (the type genus); see also RHODOBACTER, RHODOCYCLUS, RHODOPILA.

Rhodospirillales An order of Gram-negative anoxygenic photosynthetic bacteria (cf. CYANOBACTERIA). All species can carry out anoxygenic PHOTOSYNTHESIS under anaerobic conditions, but some (e.g. *Chloroflexus aurantiacus*, *Thiocapsa roseopersicina*, most members of the Rhodospirillaceae) can also grow chemoorganotrophically under aerobic or microaerobic conditions. All species contain one or more bacterioCHLOROPHYLLS and CAROTENOIDS; maximum concentrations of the pigments occur only under anaerobic conditions in low levels of light. (cf. ERYTHROBACTER.) The order includes cocci, rods, filaments and spiral forms; motile species may have flagella or may exhibit GLIDING MOTILITY. Endospores are not formed. In most species cell division occurs by binary fission; *Pelodictyon clathratiforme* can divide by ternary fission, and BUDDING occurs in some species of *Rhodopseudomonas* and in *Rhodomicrobium vannielii*. Many species can carry out NITROGEN FIXATION. Suborders: CHLOROBIINEAE and RHODOSPIRILLINEAE.

The order appears to be taxonomically unsound; thus, e.g. not only is there a wide divergence between the two suborders, but some members of the family Rhodospirillaceae seem to be more closely related to certain non-photosynthetic bacteria than to other members of the family. (See also CHLOROBIINEAE.)

(See also HELIOBACTERIUM.)

Rhodospirillineae (purple bacteria) A suborder of photosynthetic bacteria (order RHODOSPIRILLALES); species contain bacteriochlorophyll *a* (usually) or *b* (in a few cases) in intracellular membrane systems that are continuous with the cytoplasmic membrane (cf. CHLOROBIINEAE). Most species are motile (flagellate). The suborder includes metabolically diverse species; CO_2 fixation occurs mainly via the CALVIN CYCLE. Families: CHROMATIACEAE and RHODOSPIRILLACEAE.

Rhodospirillum A genus of photosynthetic bacteria (family RHODOSPIRILLACEAE). Cells: spiral-shaped, polarly flagellated, 3–10 μm long and 0.5–1.5 μm wide; they contain vesicular, lamellar or stacked photosynthetic membranes. The cells divide by binary fission. Species: *R. fulvum*, *R. molischianum*, *R. photometricum*, *R. rubrum* (type species), and *R. salexigens*; it has been proposed that *R. tenue* be transferred to the genus RHODOCYCLUS [IJSB (1984) *34* 340–343].

***Rhodospirillum* haem protein** See RHP.

Rhodosporidium See SPORIDIALES and RHODOTORULA.

Rhodotorula A genus of imperfect yeasts (class HYPHOMYCETES). The cells are spheroidal, ovoid or elongate; vegetative reproduction occurs by multilateral budding. Pseudomycelium and true mycelium may be formed by some strains of some species. Ballistospores are not produced. Visible carotenoid pigments are synthesized by cultures on e.g. malt agar; the colonies are typically red, pink, orange or yellow, and may be mucoid (capsulated strains), pasty, or dry and wrinkled. Metabolism is strictly respiratory; inositol is not assimilated. NO_3^- is assimilated by all species except *R. minuta* and *R. rubra*. Other species include e.g. *R. aurantiaca*, *R. glutinis* (a complex of anamorphs of at least three teleomorphic species: *Rhodosporidium diobovatum*, *Rhodosporidium sphaerocarpum* and *Rhodosporidium toruloides*) and *R. graminis*. Strains have been isolated from e.g. plants and plant debris, beverages, pickling brines (see also SAUERKRAUT), seawater, fresh water, etc. [Book ref. 100, pp. 893–905.]

Rhodymenia See RHODOPHYTA (cf. PALMARIA).

Rhopalodia A genus of pennate DIATOMS. The freshwater species *R. gibba* and *R. gibberula* contain, in addition to the typical diatom chloroplast, intracellular inclusion bodies which closely resemble thin-walled unicellular cyanobacteria and which carry out light-dependent nitrogen fixation. (cf. RHIZOSOLENIA.)

Rhopalomyces See ZOOPAGALES.

rhoptries (*sing.* rhoptry) In members of the APICOMPLEXA: elongate (e.g. club-shaped) osmiophilic organelles found at the anterior end of the cell; two such organelles per cell occur e.g. in the merozoites of *Isospora* spp and many *Eimeria* spp, while more than two per cell occur e.g. in some *Eimeria* merozoites and in *Sarcocystis* and *Toxoplasma*. Rhoptries appear to secrete enzymes which assist in host cell penetration [JUR (1983) *83* 85–98].

RHP (*Rhodospirillum* haem protein; cytochromoid *c*) Original name for CYTOCHROMES *c′* and *cc′* which occur in at least some purple phototrophic bacteria and in *Pseudomonas denitrificans*.

Rhynchodida See HYPOSTOMATIA.

Rhynchoidomonas A genus of monoxenous protozoa (family TRYPANOSOMATIDAE) parasitic in the gut (particularly the Malpighian tubules) of flies. The organisms, 10–50 μm long, occur in the trypomastigote form but lack a free flagellum.

Rhynchomonas See BODONINA.

rhynchosporium A plant disease caused by *Rhynchosporium* sp: see e.g. LEAF BLOTCH.

Rhynchosporium See HYPHOMYCETES; see also LEAF BLOTCH.

Rhytidhysteron See ASCOSTROMA.

Rhytisma A genus of fungi of the RHYTISMATALES. *R. acerinum* causes *tar spot* disease in *Acer* spp (e.g. maple): conspicuous, flat, roundish, black stromata (ca. 1–2 cm diam.) develop on the upper surfaces of leaves from mid-summer onwards, the stromata bearing conidiophores of the imperfect (*Melasmia*) stage during the autumn; ascocarps (the over-wintering stage) develop within the stromata, each stroma splitting radially in the spring to expose the apothecial hymenia of asci containing filiform ascospores.

Rhytismatales An order of fungi (subdivision ASCOMYCOTINA) which include saprotrophic and plant-parasitic species. Ascocarp: APOTHECIOID, immersed in a stroma or in host tissue; hamathecium: filiform paraphyses. Asci: unitunicate; sometimes stipitate. Ascospores: septate or aseptate. Genera: e.g. *Coccomyces*, *Hypoderma*, RHYTISMA.

RI Replicative intermediate, an intermediate formed during the replication of certain ssRNA viral genomes; an RI is a partially double-stranded structure with many (growing) single-stranded 'tails' formed as a result of the concomitant synthesis of many new RNA strands on the same RNA template strand. (cf. RF.)

Ri plasmid See HAIRY ROOT.

RIA RADIOIMMUNOASSAY.

RIBA See HEPATITIS C.

ribavirin (1-β-D-ribofuranosyl-1,2,4-triazole-3-carboxamide; Virazole) An ANTIVIRAL AGENT which is active against a wide range of viruses and which has been used therapeutically against infections caused e.g. by the hepatitis C, Lassa fever and SARS (coronavirus) viruses.

Ribavirin is phosphorylated in cells; the monophosphate inhibits IMP dehydrogenase (and hence GMP and GTP synthesis: see Appendix V(a)), while the triphosphate directly inhibits e.g. influenzavirus RNA-dependent RNA polymerase. A suggestion that ribavirin may act as a cap analogue (see MRNA (b)) was not confirmed [RNA (2005) *11* 1238–1244].

Analogues and derivatives of ribavirin include e.g. ribavirin 2′, 3′, 5′-triacetate (RTA) and selenazofurin (= selenazole, 2-β-D-ribofuranosylselenazole-4-carboxamide), both of which are apparently more effective than ribavirin itself.

Ribi cell fractionator An apparatus similar in principle to the FRENCH PRESS but capable of higher pressures.

ribitol (adonitol) An optically inactive PENTITOL formed e.g. by the reduction of RIBOSE; it is a component of e.g. RIBOFLAVIN and some TEICHOIC ACIDS, and is a major product of photosynthesis in certain green algae (e.g. *Coccomyxa*, *Myrmecia*, *Trentepohlia*). (See also HONEYDEW and Appendix III(d).)

riboflavin (riboflavine; vitamin B_2; lactoflavin) A water-soluble, photolabile, heat-stable VITAMIN: 6,7-dimethy;-9-(1′-D-ribityl)-isoalloxazine. The important coenzyme forms are riboflavin 5′-phosphate (= *flavin mononucleotide*, FMN) and *flavin adenine dinucleotide* (FAD) – see figure; these are the prosthetic groups of *flavoproteins*: proteins which act as hydrogen-carriers in a wide range of redox reactions. The hydrogen atoms are carried at the N^1 and N^{10} positions of the isoalloxazine ring. Some reduced flavoproteins may be re-oxidized directly by molecular oxygen, HYDROGEN PEROXIDE being a product; others may be re-oxidized via an ELECTRON TRANSPORT CHAIN. Riboflavin may act as a photoreceptor in the phototropic responses of certain fungi; FMN is involved in bacterial BIOLUMINESCENCE.

Most microorganisms can apparently synthesize riboflavin, but it is required as a growth factor e.g. by *Lactobacillus* spp, *Crithidia fasciculata* and *Tetrahymena* spp. Under certain cultural conditions some organisms (e.g. *Ashbya gossypii*, *Eremothecium ashbyii*, certain yeasts, *Clostridium* spp) secrete large quantities of riboflavin into the medium; *A. gossypii* and *E. ashbyii* are important commercial sources of the vitamin.

ribonuclease See RNASE.

ribonucleic acid See RNA.

ribonucleoside See NUCLEOSIDE.

ribonucleotide See NUCLEOTIDE.

ribonucleotide reductase See NUCLEOTIDE. (See also IRON.)

ribophage A BACTERIOPHAGE with an RNA genome.

ribophorins Glycoproteins which, in eukaryotes, appear to play a role in binding RIBOSOMES to the rough ENDOPLASMIC RETICULUM.

RiboPrinter™ An automated system (manufactured by Qualicon, Wilmington, DE, USA) designed for rapidly characterizing a bacterium to strain level on the basis of its ribotype. The apparatus is intended primarily for use in the food industry, and its database contains ribotype patterns of hundreds of strains of food-borne pathogens (with which the ribotypes of new isolates can be compared). [Comparison of the RiboPrinter™ with traditional ribotyping: LAM (1999) *28* 327–333.]

[Use of automated riboprinter and PFGE for epidemiological studies of *Haemophilus influenzae* in Taiwan: JMM (2001) *50* 277–283.]

ribose An aldopentose (see PENTOSES) which plays many important roles in cell structure and metabolism. Phosphorylated derivatives of ribose and 2-deoxyribose are components of NUCLEOTIDES and NUCLEIC ACIDS [see also Appendix V(a) and (b)], and ribose phosphates are important intermediates in certain metabolic pathways: e.g. CALVIN CYCLE, HEXOSE MONOPHOSPHATE PATHWAY [Appendix I(b)], biosynthesis of phenylalanine, tyrosine and tryptophan [Appendix IV(f)] and of histidine [Appendix IV(g)], etc. [See also Appendix III(d).]

ribose phosphate pathway *Syn.* RMP PATHWAY.

ribosome A ribonucleoprotein intracellular organelle (about 25–30 nm in diameter) which mediates PROTEIN SYNTHESIS (q.v. for function). Some ribosomes occur in the cytoplasm; others (see e.g. ENDOPLASMIC RETICULUM) are membrane-bound. Ribosomes are usually present in large numbers in the cell.

Each ribosome consists of two subunits – one larger than the other; both subunits contain RNA and protein.

Different types of ribosome are characterized by different sedimentation coefficients (S) on ultracentrifugation (see also SVEDBERG UNIT). Bacterial ribosomes have a sedimentation coefficient of ca. 70S; each consists of one 30S subunit and one 50S subunit. Ribosomes of the eukaryotic cell cytoplasm have a sedimentation coefficient of ca. 80S; each consists of one 40S subunit and one 60S subunit. Other ribosomes reported: those in mammalian mitochondria (ca. 60S), in chloroplasts of plants and algae (ca. 70S), in mitochondria of plants (ca. 78S), and those in the mitochondria of yeasts and other lower eukaryotes (ca. 73S).

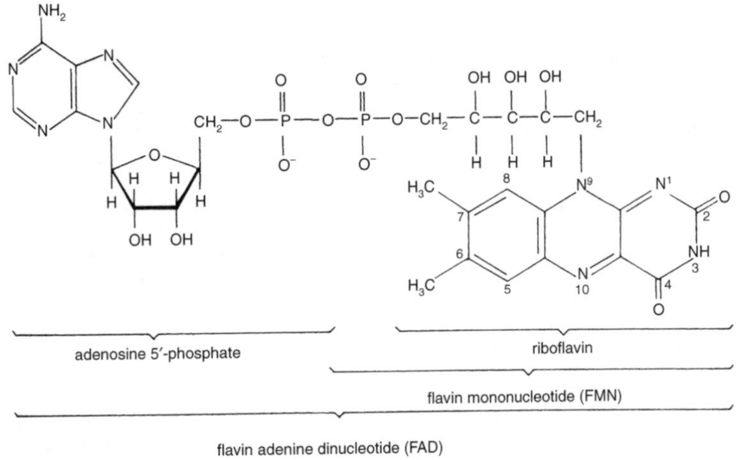

adenosine 5′-phosphate

riboflavin

flavin mononucleotide (FMN)

flavin adenine dinucleotide (FAD)

RIBOFLAVIN and its corresponding coenzymes.

Ribosomal RNA (rRNA) comprises ca. 65% (by mass) of the bacterial ribosome, and ca. 50–60% of the eukaryotic 80S ribosome.

The bacterial ribosome contains one molecule of 16S rRNA (in the 30S subunit), and one molecule each of 5S rRNA and 23S rRNA in the 50S subunit; the 16S and 23S rRNA molecules contain modified nucleotides. All three species of bacterial rRNA – together with tRNAs – are transcribed as a single molecule of RNA that is cut at specific sites; in this cutting process the ribozyme RNASE P is required for trimming the 5′ ends of tRNA molecules.

Crystallographic analysis of the 30S subunit from the bacterium *Thermus thermophilus* has shown that (as predicted) most of the interfacial region of the subunit, i.e. that part in contact with the 50S subunit, consists of RNA [Nature (1999) *400* 833–840]. (Results from the study also suggest that some of the proteins may be more directly involved in ribosomal function than has hitherto been assumed.)

The eukaryotic 80S ribosome contains one molecule of 18S rRNA (in the 40S subunit), and one molecule each of 28S, 5.8S and 5S rRNA in the 60S subunit; the 28S and 18S rRNAs contain modified nucleotides (the level of modification being greater than that in bacterial 23S and 16S rRNAs). Genes which encode the 5.8S, 18S and 28S rRNAs occur in the NUCLEOLUS, the gene encoding 5S rRNA being found outside the nucleolus. (For details of rRNA synthesis see entry RRNA.)

Ribosomal proteins (r-proteins). The r-proteins are closely associated with rRNA.

In *Escherichia coli* the small (30S) ribosomal subunit contains 21 distinct r-proteins which are designated S1–S21 ('S' for 'small') – according to their electrophoretic mobilities (S1 having the highest MWt). (These proteins were also designated ES1–ES21 – 'ES' for 'eubacterial small'.) Only one molecule of each type of protein is present in the subunit, and all except S1, S2 and S6 are basic proteins. (S1 is reported to be only loosely associated with the 30S subunit; there is apparently no S1 protein in the ribosomes of *Bacillus* spp.)

The large (50S) subunit of *E. coli* was originally reported to contain 34 r-proteins – which were designated L1–L34 (or EL1–EL34). Subsequently, the protein initially designated 'L8' was found to be a complex of L7, L10 and L12; 'L6' was identified as S20; and L7 and L12 were found to be almost identical – differing only in that L7 is acetylated at the amino terminus. The other L proteins retained their original L numbers. All the L proteins are present in single copy except L7/L12, which is present in four copies; this is the only acidic protein in the 50S subunit.

Eukaryotic 80S ribosomes are more complex; they contain >70 r-proteins.

Archaeal ribosomes resemble (at least superficially) those of bacteria; for example, they contain only three types of rRNA and have a sedimentation coefficient of ca. 70S. However, those of certain archaeans (e.g. *Halobacterium cutirubrum* and some methanogens) contain many acidic r-proteins which show little or no homology with those of bacteria; moreover, some of the r-proteins from *H. cutirubrum* appear to share some homology with eukaryotic r-proteins. Furthermore, archaeal ribosomes are insensitive to certain antibiotics (e.g. CHLORAMPHENICOL) which interfere with bacterial ribosome function, while some are sensitive to the 80S (eukaryotic) ribosome inhibitor ANISOMYCIN. [Archaeal ('archaebacterial') ribosomes: Book ref. 157, pp 345–377.]

Ribosome structure. The rRNA molecules can adopt complex secondary structures – extensive intramolecular base-pairing resulting in the formation of HAIRPINS and STEM-AND-LOOP STRUCTURES etc. [rRNA structure: ARB (1984) *53* 119–162.]

The integrity of a ribosome appears to involve hydrogen-bonding and both ionic and hydrophobic interactions, magnesium ions generally playing an important role in maintaining the structure.

The basic architecture of a ribosome has been strongly conserved during evolution, although ribosomes from e.g. bacteria, the eukaryotic cytoplasm and archaeans appear to show certain distinctive morphological features (cf. PHOTOCYTA.) [Evolving ribosome structure: ARB (1985) *54* 507–530. Three-dimensional model of the *E. coli* ribosome: Prog. Biophys. Mol. Biol. (1986) *48* 67–101.]

Taxonomic role of rRNA. Certain regions of rRNA have been very highly conserved during evolution, and sequence homology studies in rRNAs are widely used to indicate evolutionary relationships among organisms. [Evolutionary changes in 5S rRNA higher order structure: NAR (1987) *15* 161–179.] (See also RIBOTYPING.)

Biogenesis of ribosomes. Biogenesis appears to involve an assembly process in which, initially, some of the r-proteins bind to particular regions of rRNA; other r-proteins then bind co-operatively to this 'core' structure. [Protein–rRNA recognition and ribosome assembly: Book ref. 84, pp 331–352.]

In rapidly growing cells of *E. coli* the number of ribosomes per cell is essentially proportional to the growth rate; this necessitates control mechanisms for co-ordinated expression of the genes encoding r-proteins and rRNAs.

Genes encoding the r-proteins are grouped into a number of distinct OPERONS. For example, the *str* operon contains genes for (in order of transcription) S22, S7 and translation elongation factors EF-G and EF-Tu (see PROTEINS SYNTHESIS); the α operon contains genes for S13, S11, S4, RNA polymerase subunit α and L17; the β operon contains genes for L10, L7/12 and RNA polymerase subunits β and β'. Co-ordination of the synthesis of r-proteins involves post-transcriptional AUTOGENOUS REGULATION (*translational feedback regulation*). One model for the control of these operons postulates that one of the r-proteins encoded by a given operon acts as a regulatory molecule (translational repressor) for that operon by binding to its own (polycistronic) mRNA and blocking translation of some or all of the encoded proteins; thus, for example, in the β operon L10 represses the translation of L10 and L7/L12, but not of the co-transcribed RNA polymerase β and β' subunits. (See also RPO GENES.) The *E. coli* IF3–L35–L20 operon contains the genes *infC–rpmI–rplT* which encode (respectively) initiation factor IF3 (see PROTEIN SYNTHESIS) and the r-proteins L35 (*sic*) and L20; IF3 acts as a repressor of its own gene, and L20 apparently acts as a translational repressor (in a concentration-dependent way) by binding to a site (a *translational operator*) upstream of the *rpmI* sequence in the mRNA, thus blocking translation of both r-proteins. The mechanism of repression by L20 is unknown, but translational control of the operon apparently involves a pseudoknot which affects the control region of *rpmI* [EMBO (1996) *15* 4402–4413].

Some of the operon-regulatory proteins also have binding sites on either 16S or 23S rRNA. It is thought that, if the growth rate decreases, the decreased amount of 16S and 23S rRNA available in nascent ribosomes will leave these proteins free to inhibit translation of their respective transcripts. Note that, in this model, synthesis of ribosomal proteins is linked to the availability of rRNA, regulation of the rRNA genes being the key step in regulating the biogenesis of ribosomes; a *ribosome feedback regulation* model has been proposed in which rRNA synthesis is repressed (directly or indirectly) by the presence of 'free' (i.e. non-translating) ribosomes. [Review of regulation of ribosome biogenesis: Book ref. 188, pp 199–220.]

Synthesis of rRNA is linked to the cell's translational needs and is thus susceptible to up-regulation or down-regulation according to prevailing conditions. In *E. coli* the 16S, 23S and 5S rRNAs are co-transcribed (in that order) as a single transcript from the *rrn* operon; the *E. coli* chromosome contains seven copies of the *rrn* operon (designated A–E, G, H), each operon containing one copy each of the three kinds of rRNA – except the D operon, which contains two copies of the 5S gene. If (e.g. through mutation) any reduction occurs in the levels of 16S or 23S rRNA, a compensatory mechanism up-regulates the *rrn* operons. Interestingly, however, deletion of two or more copies of the 5S rRNA gene in *E. coli* brings about a sharp drop in growth rate (i.e. there is no compensatory effect) – such a reduction in growth rate being almost reversible by insertion of a plasmid-borne 5S rRNA gene [NAR (1999) *27* 637–642].

ribostamycin An AMINOGLYCOSIDE ANTIBIOTIC.

riboswitch In an mRNA molecule: a regulatory region consisting of an *aptamer*, which can bind a specific ligand, and an *expression platform*, which, when the ligand is bound, adopts a conformation that affects (typically inhibits) the mRNA's function. Riboswitches occur in prokaryotes [e.g. JB (2005) *187* 791–794; JB (2005) *187* 8127–8136] and at least in some eukaryotes. [RibEx (for locating riboswitches): NAR (2005) *33* (Web server issue) W690–W692.]

ribotype See RIBOTYPING.

ribotyping A method of TYPING in which: (i) chromosomes of the test strain are fragmented by a RESTRICTION ENDONUCLEASE; (ii) the fragments are subjected to gel electrophoresis; and (iii) the gel is examined with a labelled PROBE complementary to a region in the *rrn* (rRNA-encoding) operon. Only those bands of fragments which bind the probe are made visible in the gel by the probe's label (cf. DNA FINGERPRINTING).

Most bacteria contain multiple copies of the *rrn* operon (which includes the 16S, 23S and 5S rRNA genes); moreover, the space between any two of these three genes (the *intergenic spacer region*) can vary in length in different copies of the operon. Consequently, hybridization of probes to fragments containing the *rrn* target commonly yields a fingerprint consisting of a number of bands (often 3–6). (Species which do not contain multiple copies of *rrn* include *Mycobacterium tuberculosis* and *Tropheryma whipplei*.)

A strain defined by ribotyping is a *ribotype*; the ribotypes of a species reflect variation in *rrn* operons among strains. In *Vibrio cholerae*, ribotype variation may have arisen by recombination between *rrn* operons in the same chromosome [Microbiology (1998) *144* 1213–1221]; over time, such events may be reversed, affecting long-term monitoring by ribotyping. Even so, ribotyping can be useful for monitoring under short-term or outbreak conditions – e.g. distinguishing strains of *Legionella pneumophila* serogroup 1(identical by routine serotyping) [FEMS Imm. (1994) *9* 23–28], *Vibrio parahaemolyticus* [JCM (1999) *37* 2473–2478] and *Pseudomonas syringae* [AEM (2000) *66* 850–854].

(See also PCR-RIBOTYPING and RIBOPRINTER.)

ribovirus A VIRUS with an RNA genome.

ribozyme Any RNA molecule which can function as a catalyst (cf. ENZYME). Ribozymes occur in both eukaryotic and prokaryotic cells and e.g. in some viruses (including phages) and viroids; in nature, they are involved in processing RNA precursor molecules (see e.g. RNASE P and self-splicing introns in SPLIT GENE). (A shortened form of the self-splicing pre-rRNA intron of *Tetrahymena thermophila* can, in vitro, act as an RNA polymerase and (under different conditions) as an RNase [Science (1986) *231* 470–475].)

The roles of ribozymes have been adapted, in the laboratory, for the cleavage or modification of specific RNA or DNA target molecules.

[Characteristics and properties of ribozymes: FEMS Reviews (1999) *23* 257–275.]

ribulose A ketopentose (a pentulose). Phosphorylated ribulose derivatives are important intermediates in various metabolic pathways: e.g. CALVIN CYCLE, HETEROLACTIC FERMENTATION [Appendix III(b)], HEXOSE MONOPHOSPHATE PATHWAY [Appendix I(b)], PENTOSE metabolism [see e.g. Appendix III(d)], RMP PATHWAY.

D-ribulose 1,5-bisphosphate carboxylase–oxygenase (RuBisCO, Rubisco, or RUBISCO; D-ribulose 1,5-bisphosphate carboxylase, RuBPCase; D-ribulose 1,5-diphosphate carboxylase; carboxydismutase) A bifunctional enzyme (EC 4.1.1.39) which catalyses the carboxylation of ribulose 1,5-bisphosphate (RuBP) in the CALVIN CYCLE and, in the presence of high levels of O_2 and low levels of CO_2, the oxygenation and cleavage of RuBP to form phosphoglycolate and 3-phosphoglycerate (see PHOTORESPIRATION). RuBisCO from most sources consists of two types of subunit: 'large' (L) and 'small' (S). The L subunits appear to bear the active site; the function of the S subunits is unknown. RuBisCO containing 8 L and 8 S subunits (8L8S) has been isolated from a wide range of autotrophs, including algae, higher plants, cyanobacteria, *'Alcaligenes eutrophus', Paracoccus denitrificans, Pseudomonas facilis, Thiobacillus* spp, etc. (In plants, the S subunits are specified by nuclear genes, the L subunits by chloroplast genes.) *Rhodopseudomonas sphaeroides* and *R. capsulata* each contain two distinct forms of RuBisCO: an 8L8S form and a 6L form (containing six L subunits only). *Chlorobium thiosulfatophilum* contains a 6L enzyme, *Methylococcus capsulatus* contains a 6L6S enzyme, while RuBisCO from *Rhodospirillum rubrum* contains only two L subunits. [Book ref. 115, pp. 129–173.]

(See also CARBOXYSOME.)

ribulose monophosphate pathway *Syn.* RMP PATHWAY.

ribulose 5-phosphate kinase See CALVIN CYCLE.

rice black-streaked dwarf virus See FIJIVIRUS.

rice blast disease See BLAST DISEASE.

rice diseases See CEREAL DISEASES.

rice dwarf virus (RDV) A virus of the genus PHYTOREOVIRUS which infects rice (*Oryza sativa*) and certain other grasses. On rice, RDV causes e.g. yellowish spots or streaks on young leaves, stunting, and the formation of many small tillers which give the plant a rosette-like appearance. Vector: mainly *Nephotettix cincticeps*.

rice gall dwarf virus (RGDV) A virus of the genus PHYTOREOVIRUS [Intervirol. (1985) *23* 167–171]; RGDV is an important cause of disease in rice e.g. in Malaysia and Thailand. Infected plants are dark green, stunted, and develop small tumours on the lower surfaces of the leaves. Vector: *Nephotettix* spp. [RGDV structural proteins: JGV (1985) *66* 811–815.]

rice grassy stunt virus See RICE STRIPE VIRUS GROUP.

rice hoja-blanca virus See RICE STRIPE VIRUS GROUP.

rice necrosis mosaic virus See POTYVIRUSES.

rice ragged stunt virus (RRSV) An unclassified virus which may have affinities with the genus FIJIVIRUS but which appears to lack an outer capsid layer; genome: 10 dsRNA segments.

[Component proteins and structure of RRSV: JGV (1986) *67* 1711–1715.] RRSV is an important pathogen of rice in many parts of Asia, causing stunting, leaf distortion, excessive tillering, and enations. Transmission occurs propagatively via the brown rice planthopper *Nilaparvata lugens*. Viroplasms occur in the cells of infected plants and vectors.

rice-straw mushroom *Syn.* PADI-STRAW MUSHROOM.

rice stripe virus group A group of planthopper-transmitted PLANT VIRUSES which have characteristic filamentous virions (ca. 8 nm wide) which may adopt circular, spiral, pleomorphic, or branched configurations; the virions contain four or more pieces of ssRNA, and those of rice stripe virus (RSV) apparently contain an RNA-dependent RNA polymerase [JGV (1986) *67* 1247–1255]. RSV causes serious damage to Japonica-type rice varieties. Other members of the group include maize stripe virus, rice grassy stunt virus, and rice hoja-blanca virus. [Review of RSV: MS (1986) *3* 347–351.]

rice transitory yellowing virus See RHABDOVIRIDAE.

rice tungro virus A virus which resembles MAIZE CHLOROTIC DWARF VIRUS e.g. in morphology, S_w^{20}, and vector relationships, but which is serologically unrelated to MCDV.

rice vinegar See VINEGAR.

rice-water stools See CHOLERA.

rice yellow mottle virus See SOBEMOVIRUSES.

Richelia See RHIZOSOLENIA.

ricin See SHIGA TOXIN.

ricinoleic acid (12-hydroxy-9-octadecenoic acid) A fatty acid present (in glycerides) in castor oil and certain other vegetable oils, and in immature sclerotia and mycelium of *Claviceps purpurea*.

Rickettsia A genus of Gram-negative bacteria of the tribe RICKETTSIEAE; species are obligate intracellular parasites or pathogens in vertebrates (including man) and arthropods (ticks, mites, fleas etc). The cells are non-flagellated, non-fimbriate bacilli, 0.8–2 × 0.3–0.6 μm, which generally stain well with the GIMÉNEZ STAIN; the outer membrane is similar to that of other Gram-negative bacteria. Growth occurs only in living systems, e.g. in chick embryo cells, in the yolk sacs of hens' eggs, or in certain tissue cultures (such as HeLa, HEp-2); rickettsiae grow mainly in the cytoplasm, sometimes in the nucleus. Optimum growth temperature is 32–35°C. Metabolism appears to be oxidative (respiratory) with glutamate as the main energy-yielding substrate; glucose is not metabolized. Division occurs by transverse binary fission. The organisms are sensitive to disinfectants such as phenol and hypochlorite, and are killed within 30 min at temperatures above 56°C; tetracyclines and chloramphenicol have been used to treat diseases of rickettsial aetiology. GC% range for species examined: 29–33. Type species: *R. prowazekii*. Species (which can be distinguished by serological and other means) include e.g. *R. akari* (see RICKETTSIALPOX), *R. conorii* (see BOUTONNEUSE), *R. prowazekii* (see TYPHUS FEVERS and BREINL STRAIN), *R. rickettsii* (see ROCKY MOUNTAIN SPOTTED FEVER), *R. tsutsugamushi* (see SCRUB TYPHUS), *R. typhi* (= R. mooseri) (causal agent of murine typhus – see TYPHUS FEVERS). Other species: *R. australis*, *R. canada*, *R. montana*, *R. parkeri*, *R. rhipicephali*, *R. sibirica*. (For *R. quintana* see ROCHALIMAEA; for *R. sennetsu* see EHRLICHIA.) [Book ref. 22, pp. 688–698.]

Rickettsiaceae A family of bacteria (RICKETTSIALES) with tribes RICKETTSIEAE, EHRLICHIEAE, WOLBACHIEAE [Book ref. 22]. For recent taxonomic changes see ANAPLASMATACEAE; *in this edition of the dictionary all lower taxa in this family reflect earlier classifications.*

rickettsiae (1) Organisms of the RICKETTSIACEAE. (2) Organisms of the genus RICKETTSIA.

Rickettsiales An order of Gram-negative bacteria which includes the families ANAPLAMATACEAE and RICKETTSIACEAE.

rickettsialpox An acute, usually mild disease of man caused by *Rickettsia akari*. House mice seem to be the main reservoir of infection; the mite *Allodermanyssus sanguineus* transmits the pathogen to man. Incubation period: 10 days to 3 weeks. There may be a lesion at the site of the mite bite, and this is followed by sudden onset of fever with headache, chills, anorexia and photophobia; there may be a sparse maculopapular rash. The disease is usually self-limiting; no deaths have been reported.

Rickettsieae A tribe of bacteria of the RICKETTSIACEAE comprising the genera COXIELLA and RICKETTSIA. The genus *Rochalimaea* (including *R. quintana*, formerly *Rickettsia quintana*) has been unified with the genus BARTONELLA (q.v.).

Rickettsiella A genus of Gram-negative bacteria of the tribe WOLBACHIEAE; species grow intracellularly in, and are pathogenic in, arthropod hosts (including arachnids, crustaceans and insects). Growth does not occur in cell-free media. Cells: rod- or disc-shaped, maximum dimension typically <0.8 μm. The species: *R. chironomi* (parasitic e.g. in midges and spiders); *R. grylli* (e.g. in crickets – *Gryllus* spp – and the desert locust *Schistocerca gregaria*); *R. popilliae* (in various beetles – including the Japanese beetle, *Popillia japonica* – and the crane fly).

rickettsiosis Any disease caused by a species of RICKETTSIA.

Rickia See LABOULBENIALES.

Rideal–Walker test A SUSPENSION TEST used to determine the PHENOL COEFFICIENT of a (phenolic) disinfectant; the test determines that dilution of the test disinfectant which gives the same rate of kill of a test organism as does a standard dilution of phenol. Serial dilutions (in water) of the test disinfectant are equilibrated at 17–18°C. Each dilution (5 ml) is inoculated with a 24-hour broth culture (0.2 ml) of the test organism (e.g. *Salmonella typhi* NCTC 786) and incubated at 17–18°C. After 2.5, 5, 7.5 and 10 minutes, a standard loopful of the test suspension (from each dilution) is transferred to a separate volume of sterile broth which is incubated at 37°C for 2–3 days to detect surviving cells. A similar procedure is carried out with phenol. For one particular dilution of the test disinfectant, growth will occur in broths inoculated at 2.5 and 5 minutes but not in those inoculated at 7.5 and 10 minutes; a similar result will be obtained for one particular dilution of phenol (within the range 1/95 to 1/115 w/v). The phenol coefficient (*Rideal–Walker coefficient*) is given by the ratio of the dilution factors of these two particular dilutions:

$$\frac{\text{dilution factor of test disinfectant}}{\text{dilution factor of phenol}}$$

[Book ref. 13, pp. 1–15.] (cf. CHICK–MARTIN TEST.)

Ridomil See PHENYLAMIDE ANTIFUNGAL AGENTS.

Rieske protein Any of a range of [2Fe–2S] IRON–SULPHUR PROTEINS which occur e.g. in mitochondria and in various bacteria, and which characteristically have an E_m value within the range ca. +150 to +330 mV.

rifamide See RIFAMYCINS.

rifampicin See RIFAMYCINS.

rifampin Rifampicin: see RIFAMYCINS.

rifamycins A family of highly substituted macrocyclic ANTIBIOTICS (ANSAMYCINS) produced by '*Nocardia mediterranei*'. Mixtures of rifamycins A–E are generally produced in cultures of '*N. mediterranei*', but the proportion of rifamycin B can be increased by the addition of sodium diethylbarbiturate to the medium. In aqueous oxygenated solutions rifamycin B spontaneously gives rise to e.g. rifamycins O and S which show much higher antibacterial activity than rifamycin B itself. Rifamycin S is an important starting point for the production of semisynthetic rifamycins – which include e.g. rifamycin SV (= 'rifamycin', obtained by mild reduction of rifamycin S), rifamide, and rifampicin (= rifampin).

Rifamycins are generally highly active against Gram-positive bacteria (including mycobacteria, staphylococci and streptococci) and against some Gram-negative bacteria (e.g. *Brucella*, *Chlamydia*, *Haemophilus*, *Legionella* and *Neisseria* spp); other Gram-negative bacteria (e.g. enterobacteria) are less sensitive, spirochaetes and mycoplasmas are insensitive. Rifamycins specifically inhibit bacterial DNA-dependent RNA POLYMERASE, binding to the β subunit and inhibiting initiation of transcription. Resistance to rifamycins may result from e.g. a mutation which modifies the RNA polymerase β subunit such that it no longer binds to the drug. Rifamycins do not inhibit mammalian RNA polymerase or bacterial PRIMASE.

In some countries (including the USA), strains of *Mycobacterium tuberculosis* that are resistant to rifampicin are usually also resistant to at least certain other anti-tuberculosis drugs, so that resistance to rifampicin is a useful marker for multidrug-resistant strains of the pathogen [see e.g. JAMA (1994) *271* 665–671].

Resistance-conferring mutations occur at a number of sites in the *rpoB* gene (which encodes the β subunit of RNA polymerase), and specific mutations may occur with different frequencies in different geographical areas [e.g. Italy: JCM (1999) *37* 1197–1199].

Using DNA chip technology, high-density arrays have been used to identify *M. tuberculosis* and, simultaneously, to test for resistance to rifampicin [JCM (1999) *37* 49–55].

Rift Valley fever An acute, febrile disease of man and e.g. cattle, sheep, camels, rats, mice etc; it occurs in Africa. The causal agent, a virus of the genus PHLEBOVIRUS, is transmitted mainly by insects (chiefly mosquitoes); rodents may provide a reservoir of the virus. In lambs and calves, the incubation period is ca. 12 hours; onset is sudden, with high fever, hepatitis, incoordination, collapse and (usually) death within 36 hours. In adult sheep and cattle the main symptom is abortion, but there may be high fever and a mortality rate of ca. 20–30% in sheep, 10% in cattle. In man, the disease resembles influenza; onset is abrupt, with chills, fever, headache, myalgia and arthralgia. The disease is rarely fatal in man, but retinal haemorrhages may lead to (temporary or permanent) visual impairment.

Riftia See TROPHOSOME.

right-handed DNA See DNA.

right splice site See SPLIT GENE.

Rigidoporus See APHYLLOPHORALES (Polyporaceae).

Riley virus *Syn.* LACTATE DEHYDROGENASE VIRUS.

rimantadine See AMANTADINE.

rinderpest (bovine typhus; cattle plague) An acute, often fatal CATTLE DISEASE which occurs e.g. in parts of Africa and Asia; other animals, e.g. pigs, are also susceptible. The causal agent is a *Morbillivirus*, and infection apparently occurs by inhalation of aerosols. Incubation period (in cattle): ca. 1 week. Symptoms: fever and anorexia, followed by lesions in the mouth and a purulent nasal discharge; necrotic lesions appear on the lips and gums, and severe diarrhoea follows the development of intestinal lesions. Mortality rates may be very high. (cf. PESTE DES PETITS RUMINANTS.)

ring-infected erythrocyte surface antigen See PLASMODIUM.

ring slide A slide used e.g. in some serological tests; it is larger and thicker than a normal microscope SLIDE, and on one face it has a number of fixed, raised rings (1 to 3 cm in diameter) made of ceramic or other material.

ring spot See RINGSPOT.

ring stage (of *Plasmodium*) See PLASMODIUM.

ring test (*serol.*) A qualitative PRECIPITIN TEST, carried out in a narrow tube, in which a solution containing antigen is layered over one containing antibody. A positive reaction is indicated by a band of white precipitate which forms at the interface of the two solutions.

ring vaccination (*vet.*) Vaccination of animals in the annular region surrounding an isolated outbreak of disease; the object is to contain (i.e. prevent the spread of) the disease.

Ringer's solution A solution used e.g. as a general diluent; composition (g/100 ml distilled water): NaCl (0.9), KCl (0.042), $CaCl_2$ (0.048), $NaHCO_3$ (0.02).

ringspot (*plant pathol.*) A type of LOCAL LESION consisting of single or concentric rings of discoloration or necrosis, the regions between the concentric rings being green; the centre of the lesion may be chlorotic or necrotic.

ringworm (dermatophytosis; tinea) Any MYCOSIS of man and animals in which keratinized tissues (skin, hair, nails etc) are infected by a DERMATOPHYTE (q.v.). Infection occurs by direct contact with an infected individual or with fomites. Ringworm infections in man may be categorized according to the area of the body involved: tinea barbae = beard ringworm; tinea capitis (tinea tonsurans) = scalp ringworm, caused e.g. by *Microsporum audouinii*, *M. canis* or *Trichophyton schoenleinii* (cf. FAVUS); tinea corporis (tinea circinata) = body ringworm, caused e.g. by *Microsporum* or *Trichophyton* spp; tinea cruris = groin ringworm, caused e.g. by *Epidermophyton cruris*; tinea manus = hand ringworm; tinea pedis (ATHLETE'S FOOT) = foot ringworm; tinea unguium = nail ringworm, caused e.g. by *T. mentagrophytes* or *T. rubrum*. Lesions on skin are typically circular or ring-shaped (sometimes coalescing to become confluent), dry and scaling (but cf. KERION), and may or may not be painful or pruritic. Usually only the superficial layers of skin are affected. Adults are generally less susceptible to skin infection than are children owing to the fungistatic properties of fatty acids in the sebum. Infected nails become opaque and discoloured, distorted, and hard or flaky. Infected hairs usually break. *Lab. diagnosis*: e.g. microscopic and cultural examination of epidermal scales and hairs. Under WOOD'S LAMP, hairs infected with certain dermatophytes show characteristic fluorescence: e.g. hairs infected with *M. audouinii* or *M. canis* fluoresce bright green; those infected with *Trichophyton* spp may appear a (non-diagnostic) bluish-white colour. *Chemotherapy*: e.g. oral GRISEOFULVIN or POLYENE ANTIBIOTICS; topical ointments containing e.g. tolnaftate or salicylic, benzoic or propionic acids.

Rio Bravo virus See FLAVIVIRIDAE.

Rio Grande virus See PHLEBOVIRUS.

RISE-resistant tuberculosis TUBERCULOSIS which is resistant to treatment with a number of antibiotics, including – at least – *r*ifampin, *i*soniazid, *s*treptomycin and *e*thambutol. [RISE-resistant tuberculous meningitis in an AIDS patient: Lancet (1993) *341* 177–178.]

riser In a LOOP FERMENTER: the ascending column of liquid *or* that part of the fermenter which contains it. (cf. DOWNCOMER.)

rishitin A sesquiterpenoid PHYTOALEXIN produced by potato tubers.

RIST (radioimmunosorbent test) An IMMUNOASSAY used e.g. for quantifying IgE in serum. Essentially, immobilized anti-IgE antibodies are exposed to (i) a standard amount of radio-labelled IgE and (ii) the serum under test; any IgE in the serum competes with the radio-labelled IgE, and the quantity of serum IgE can be estimated from the amount of bound radio-labelled IgE.

ristocetin A glycopeptide antibiotic related to VANCOMYCIN (q.v.); it is not used as a clinical antimicrobial agent owing to its ability to cause aggregation of platelets.

RITARD model Reversible ileal tie adult rabbit diarrhoeal disease model: an in vivo model for studying the effects of enteric bacteria. With a ligated caecum, bacteria are injected into the ileum at a location proximal to a temporary ligature; after a few hours the ligature is removed and the animals observed for a few days for the development of diarrhoea.

ritonavir See ANTIRETROVIRAL AGENTS.

Ritter's disease See SCALDED SKIN SYNDROME.

river blindness See WOLBACHIA.

Rivulariaceae A phycological family of 'blue-green algae' (CYANOBACTERIA) in which genera (e.g. *Dichothrix*, *Fremyella*, GLOEOTRICHIA, *Polythrix*, *Rivularia*) are distinguished on the basis of field characteristics: e.g., colony form, false branching, etc. However, in culture these characteristics are often lost or variable; thus, all cyanobacteria with tapered trichomes and terminal heterocysts were placed in a single genus: CALOTHRIX [JGM (1979) *111* 1–61 (p. 36)].

RK2 plasmid See RP1 PLASMID.

RLUs See TMA.

RM organism *Yersinia ruckeri*.

R-M system RESTRICTION-MODIFICATION SYSTEM.

RMP pathway (allulose phosphate pathway; hexulose phosphate pathway; Quayle cycle; ribose phosphate pathway; ribulose monophosphate pathway; RuMP pathway) A cyclic metabolic pathway used by some types of methylotrophic bacteria for the assimilation of formaldehyde (see METHYLOTROPHY). The pathway may be considered in three phases (1–3 or a–c).

In phase 1 (formaldehyde fixation) the regenerated intermediate ribulose 5-phosphate (Ru5P) condenses with formaldehyde (enzyme: hexulose phosphate synthase, EC 4.1.2.-) to form a hexulose 6-phosphate (allulose 6-phosphate) which is converted, by an isomerase, to fructose 6-phosphate (F6P).

In phase 2 some of the F6P is cleaved into two 3-carbon compounds via one of two routes. In the 'glycolytic' or 'Embden–Meyerhof variant' route F6P is phosphorylated, and fructose 1,6-bisphosphate is cleaved (by fructose bisphosphate aldolase) to glyceraldehyde 3-phosphate (G3P) and dihydroxyacetone phosphate (DHAP) – the latter being assimilated into cell biomass. In the 'Entner–Doudoroff variant' route F6P is converted, via glucose 6-phosphate and 6-phosphogluconate, to 2-oxo-3-deoxy-6-phosphogluconate which is cleaved (by 2-oxo-3-deoxy-6-phosphogluconate aldolase) to G3P and pyruvate – the latter being assimilated into cell biomass.

Phase 3 ('rearrangement reactions') involves regeneration of the formaldehyde acceptor, Ru5P. As in phase 2, there are two possible routes. One route is similar to part of the HEXOSE MONOPHOSPHATE PATHWAY, a key enzyme being transaldolase. In this route G3P with F6P (enzyme: transketolase, EC 2.2.1.1) forms erythrose 4-phosphate (E4P) and xylulose 5-phosphate (Xu5P). E4P with F6P (transaldolase, EC 2.2.1.2) forms G3P and sedoheptulose 7-phosphate (S7P). S7P with G3P (transketolase) forms Xu5P and ribose 5-phosphate – each of which is then converted (by an epimerase and an isomerase, respectively) to Ru5P.

The alternative route in phase 3 resembles part of the CALVIN CYCLE. In this route F6P is phosphorylated, and fructose 1,6-bisphosphate is cleaved to G3P and DHAP. G3P with F6P forms E4P and Xu5P; Xu5P is converted (by an epimerase) to Ru5P, and E4P condenses with DHAP to form sedoheptulose 1,7-bisphosphate – which is then converted (by sedoheptulose bisphosphatase) to S7P. S7P with G3P forms ribose 5-phosphate and Xu5P, each of which is converted (by an isomerase and an epimerase, respectively) to Ru5P.

In some methylotrophic bacteria the RMP pathway can be used to generate energy from certain substrates. This involves a modified 'Entner–Doudoroff' route in phase 2: a dehydrogenase (EC 1.1.1.44) converts 6-phosphogluconate to Ru5P with concomitant elimination of CO_2 and reduction of NAD(P).

In the assimilatory mode, the consumption of energy by the RMP pathway varies according to the particular routes taken in phases 2 and 3. Thus, e.g. a combination of the 'Entner–Doudoroff' route in phase 2 and the sedoheptulose bisphosphate route in phase 3 consumes the largest amount of energy; no organism has been reported to use this combination. The (energetically) most economical routes of the RMP pathway consume less energy than does either the SERINE PATHWAY or the XMP PATHWAY.

Rms148, Rms149 plasmids See INCOMPATIBILITY.

RNA Ribonucleic acid: a NUCLEIC ACID consisting of ribo-NUCLEOTIDES, each of which contains one of the bases adenine, guanine, cytosine or uracil, or, in some RNAs, a modified form of one of these bases. (cf. DNA.) The 2′-OH group in the RNA backbone makes the molecule more reactive and less flexible than DNA. RNA molecules are usually single-stranded, but an RNA strand can form a duplex (resembling A-DNA in conformation) with a complementary strand of RNA or DNA; ssRNA molecules commonly adopt secondary structures (e.g. HAIRPINS, STEM-AND-LOOP STRUCTURES) by base-pairing between complementary regions of the same molecule.

RNA (double- or single-stranded) constitutes the genome in various ANIMAL VIRUSES, BACTERIOPHAGES and PLANT VIRUSES. In prokaryotic and eukaryotic cells the major RNAs are involved in all stages of PROTEIN SYNTHESIS (see MRNA, RRNA, TRNA), and many other types of RNA play regulatory, catalytic or other roles: see e.g. ANTISENSE RNA, RIBOZYME, SCRNA, SIGNAL RECOGNITION PARTICLE, SNRNA.

RNA I See COLE1 PLASMID.

RNA II See COLE1 PLASMID.

RNA III See AGR LOCUS.

RNA blotting See BLOTTING.

RNA degradosome See DEGRADOSOME.

RNA-dependent DNA polymerase *Syn.* REVERSE TRANSCRIPTASE.

RNA editing In animals, plants and certain microorganisms: an in vivo mechanism which can generate new functional transcripts from existing ones. One (mammalian) example is the editing of the mRNA for apolipoprotein B (apoB) by cytidine deaminase (an enzyme which converts C to U); the unedited mRNA yields a product of 4536 amino acids, while the edited mRNA (which contains an internal stop codon formed by cytidine deaminase) yields a product of 2152 amino acids.

Studies have suggested that RNA editing may be involved in the differentiation/maturation of antigen-stimulated B cells in germinal centres – a period of development that includes *affinity maturation* and *class switching* (see ANTIBODY FORMATION). Thus, a deficiency in *activation-induced cytidine deaminase* (AID) – an enzyme found specifically in germinal centre B cells – has been found to inhibit both affinity maturation and class switching [Cell (2000) *102* 553–563].

Deficiency in the AID enzyme has also been reported to be responsible for the autosomal recessive form of *hyper-IgM syndrome* (HIGM2), a rare human immunodeficiency with normal/raised IgM, no IgG (or other isotypes), lymph node hyperplasia and high vulnerability to bacterial infection [Cell (2000) *102* 565–575]. X-linked HIGM1, due e.g. to a mutated CD40L gene (see CD40), involves lack of B cell development.

In the KINETOPLAST, minicircles encode *guide RNAs* (gRNAs: ~ 55–70 bases) which direct the editing of RNAs. [Editing in *Trypanosoma brucei*: NAR (2001) *29* 703–709.]

RNA interference (RNAi) In plant and animal cells: an antiviral response triggered by dsRNA; dsRNA is cleaved by an endonuclease ('Dicer') to *small interfering RNA* (siRNA). siRNA complexes with proteins, forming the *RNA-induced silencing complex* which can degrade *homologous* mRNA – silencing the corresponding gene(s). [Possible therapeutic use: BMJ (2004) *328* 1245–1248; see GENE THERAPY.]

RNA PCR See REVERSE TRANSCRIPTASE PCR.

RNA-dependent RNA polymerase A type of RNA POLYMERASE, encoded by (probably all) RNA viruses (see VIRUS), which synthesizes RNA on an RNA template. Viral RNA-dependent RNA polymerases are responsible for replicating the genome and for synthesizing mRNA on a (−)-strand RNA template; the terms *replicase* and *transcriptase* are sometimes used specifically for enzymes with the former and latter functions, respectively, but may also be used as synonyms for any RNA-dependent RNA polymerase, regardless of function. (A clone of cDNA specific for an RNA-dependent RNA polymerase has been isolated from the tomato plant [Plant Cell (1998) *10* 2087–2101].)

RNA phages See LEVIVIRIDAE and BACTERIOPHAGE $\phi 6$.

RNA polymerase Any enzyme which can polymerize ribonucleoside 5′-triphosphates (with elimination of PPi) to form an RNA strand on a (DNA or RNA) template; however, when used without qualification, the term 'RNA polymerase' (RPase) usually refers specifically to a *DNA-dependent RNA polymerase* responsible for the TRANSCRIPTION of complete RNA molecules from a DNA template. (cf. RNA-DEPENDENT RNA POLYMERASE and PRIMASE.)

Escherichia coli contains one major type of (DNA-dependent) RPase which is responsible not only for RNA synthesis in the cell but also for some cases of priming of DNA REPLICATION. The RPase consists of a *core enzyme* comprising two α subunits, a β subunit and a β' subunit; the core enzyme can elongate an RNA strand, but for *initiation* of TRANSCRIPTION it must combine with another protein, a SIGMA FACTOR, to form the RPase *holoenzyme*. (See also RPO GENES.) RPases from other bacteria also have the $\alpha_2\beta\beta'\sigma$ holoenzyme structure; however, additional proteins may be more or less tightly associated with the enzyme in some species: e.g. a *δ factor* is associated with the RPase of *Bacillus subtilis* (vegetative and sporulating cells) and plays a role in suppressing non-specific binding to DNA and/or transcription initiation by the polymerase. Bacterial RPases are generally inhibited by e.g. RIFAMYCINS and STREPTOLYDIGIN.

Eukaryotes have three distinct nuclear RPases, each comprising many polypeptide subunits. RPase I (or A) occurs in the NUCLEOLUS and is responsible for synthesizing most rRNA species; RPases II (or B) and III (or C) both occur in the nucleoplasm, the former synthesizing mRNA and U1–U5 snRNAs, the latter tRNA, 5S rRNA, and U6 snRNA.

Eukaryotic RPases require certain protein factors (*transcription initiation factors*) in order to form an operational complex at the PROMOTERS. One such factor, needed by all the eukaryotic nuclear RPases (types I, II and III), is the TATA box-binding protein (TBP) – which acts in association with other initiation factors (see PROMOTER).

Eukaryotic nuclear RPases are not inhibited by rifamycins or streptolydigin (cf. α-AMANITIN).

RPases from mitochondria and chloroplasts are distinct from the nuclear enzymes, being e.g. much smaller.

The RPases of members of the ARCHAEA (formerly Archaebacteria) appear to be much more closely related to those of the eukaryotic nucleus than to the bacterial enzyme; they contain many components, and are relatively insensitive to rifamycins, streptolydigin and α-amanitin. [Archaebacterial (archaeal) RPases: Book ref. 157, pp. 499–524.]

Some bacteriophages encode their own RPases; those of e.g. T3 and T7 each consist of a single polypeptide chain. (See also BACTERIOPHAGE N4 and BACTERIOPHAGE SP6.) In some cases a phage modifies (and alters the specificity of) its host's RNA polymerase: see e.g. BACTERIOPHAGE T4 and BACTERIOPHAGE SPO1.

RNA splicing See SPLIT GENE.

RNA thermometer See HEAT-SHOCK PROTEINS.

RNA tumour viruses Viruses of the ONCOVIRINAE.

RNAase See RNASE.

RNAi See RNA INTERFERENCE.

RNAP RNA POLYMERASE.

RNase (RNAase; ribonuclease) An enzyme which cleaves RNA. An RNase may be an ENDONUCLEASE (= endoribonuclease) or an EXONUCLEASE (= exoribonuclease); enzymes in the latter category may act randomly (dissociating from the substrate after each catalytic event) or processively (see PROCESSIVE ENZYME).

A given RNase generates 3′-hydroxyl and 5′-phosphate *or* 3′-phosphate and 5′-hydroxyl termini. Most RNases rely on the conformation of the RNA substrate (rather than nucleotide sequence per se) for their specificity (cf. RNASE E).

RNases have important roles in RNA processing (see e.g. MRNA, RRNA, TRNA), in certain types of regulatory process (e.g. RETROREGULATION), and in degrading RNA molecules which have fulfilled their function.

For examples of RNases see following entries.

RNases can be inactivated by DIETHYLPYROCARBONATE.

RNase* See MRNA (a).

RNase I In *Escherichia coli*: an endoribonuclease (see RNASE) which can degrade most RNA molecules to oligonucleotides with 3′-phosphate and 5′-hydroxyl termini.

RNase II In *Escherichia coli*: an exoribonuclease (see RNASE) which removes nucleoside 5′-phosphate residues from the 3′ end of an RNA molecule which has little or no secondary structure.

RNase III In *Escherichia coli*: an endoribonuclease (see RNASE) which cleaves double-stranded regions in RNA molecules. RNase III is involved in rRNA precursor processing (see RRNA), in the RETROREGULATION of *int* gene expression in BACTERIOPHAGE LAMBDA, and in the release of monocistronic mRNAs from the polycistronic early-gene transcript in BACTERIOPHAGE T7; the target sites for RNase III action typically occur in stem-and-loop or hairpin structures in the RNA.

RNase BN See TRNA.

RNase D See TRNA.

RNase E In *Escherichia coli*: an endoribonuclease (see RNASE) which is involved e.g. in the formation of 5S rRNA from pre-rRNA (see RRNA). RNase E also cleaves RNA I (involved in the replication of the COLE1 PLASMID and related plasmids), cleavage occurring between the 5th and 6th nucleotides from the 5′ end of the molecule. The cleavage site occurs in a sequence of 10 nt which is almost identical in the two RNA substrates [JMB (1985) *185* 713–720].

(See also DEGRADOSOME.)

RNase F See INTERFERONS.

RNase H An RNASE which specifically cleaves an RNA strand base-paired to a complementary DNA strand, generating 3′-OH and 5′-phosphate termini. Such enzymes have been isolated from a wide range of prokaryotic and eukaryotic organisms and are encoded by certain viruses. For example, RNase H from *Escherichia coli* is an endoribonuclease which is apparently involved e.g. in certain types of plasmid replication (see COLE1 PLASMID). RNase H exoribonuclease activity is associated with the reverse transcriptase of viruses of the RETROVIRIDAE (q.v.); this enzyme cleaves oligonucleotides (ca. 6–12 nt long) from either end of the RNA strand.

RNase L See INTERFERONS.

RNase P An endoribonuclease (see RNASE) which, in most organisms, generates the 5′ termini of tRNAs from precursor tRNA molecules (see TRNA). The catalytic activity of RNase P is mediated by an RNA component (M1 RNA) 375 nt long in e.g. *Escherichia coli*. Under certain conditions in vitro the purified M1 RNA of the *E. coli* enzyme can itself catalyse the cleavage of precursor tRNA [reaction mechanism: Biochem. (1986) *25* 1509–1515]; however, both protein and RNA components are apparently necessary in vivo since a mutation in either of the genes *rnpA* and *rnpB* (genes for the protein and RNA components, respectively) leads to loss of RNase P function. In *Schizosaccharomyces pombe* RNase P apparently contains two RNA components, K1 and K2 (285 and 270 nt long, respectively) [EMBO (1986) *5* 1697–1703].

[Synthetic inhibitors of the processing of pre-tRNA by RNase P: Biochem. (2001) *40* 603–608.]

RNase T An exoribonuclease (see RNASE) of *Escherichia coli*; it specifically catalyses 'end turnover' of tRNA: RNase T removes the terminal AMP residue from the 3′ CCA end of an uncharged (non-aminoacylated) tRNA molecule – the residue being replaced by tRNA nucleotidyltransferase [PNAS (1985) *82* 6427–6430].

RNase T1 (RNase T_1) An endoribonuclease obtained commercially from *Aspergillus oryzae*. It cleaves RNA specifically at guanosine residues (Gp ↓ N), and is used e.g. in the analysis of RNA (see e.g. OLIGONUCLEOTIDE FINGERPRINTING and RRNA OLIGONUCLEOTIDE CATALOGUING).

RND Resistance–nodulation–division superfamily: a category of TRANSPORT systems which mediate e.g. efflux-dependent resistance to antibiotic(s) in certain Gram-negative bacteria; these systems typically include a MEMBRANE FUSION PROTEIN and apparently involve (pmf-dependent) drug–proton antiport.

***rnh* gene** In e.g. *Escherichia coli*: a gene encoding RNASE H.

***rnpA*, *rnpB* genes** See RNASE P.

rNTP RibONUCLEOSIDE-5′-triphosphate.

Robbins device An apparatus used in corrosion studies: a metal pipe whose wall contains a series of removable panels; after a sample has been cycled through the pipe, removal of the panel(s) permits examination of e.g. adherent microorganisms.

Robertson's cooked meat medium See COOKED MEAT MEDIUM.

Roccella A genus of fruticose LICHENS (order OPEGRAPHALES). Photobiont: *Trentepohlia*. The thallus consists of blue- or grey-brown to violet, terete or flattened, strap-like branches which form erect or pendulous tufts; bluish-white soralia may be present. Species occur mainly on rocks by the sea; they have been used commercially as sources of LITMUS, ORCHIL and ORCINOL.

Rochalimaea A (former) genus of Gram-negative bacteria which included *R. quintana* (formerly *Rickettsia quintana*) and *R. vinsonii*; on the basis of nucleic acid studies, members of this genus were transferred to the genus BARTONELLA (q.v.).

Rocio virus See FLAVIVIRIDAE.

rock tripe *Umbilicaria* spp, or the edible species *U. esculenta* eaten (as 'iwatake') in Japan.

rocket immunoelectrophoresis (electroimmunoassay) IMMUNOELECTROPHORESIS in which (typically) antibody, at or near its isoelectric point, is uniformly distributed in a gel layer, and homologous antigen (in a well in the layer) is subjected to an electric field. Antigen moves towards one of the poles, and forms (with the antibody) a 'rocket-shaped' line of precipitate – an extended paraboloid with the wide end towards the well; the length of the 'rocket' is proportional to the initial concentration of antigen in the well.

Rocky Mountain spotted fever A rickettsial disease of man which occurs in parts of North America, particularly in late spring and summer; the causal agent, *Rickettsia rickettsii*, is transmitted by ticks – chiefly species of *Dermacentor*. Infection, by the bite of an infected tick, is followed by an incubation period of 2–14 days; symptoms include headache, joint and muscular pains, and a sustained fever. A rash develops on the limbs and trunk (in that order); the lesions may become haemorrhagic and necrotic. Mortality in untreated cases may be 5–90%. Reservoirs of infection occur e.g. in the dog (which can develop clinical signs) and in wild rodents. *Lab. diagnosis*: e.g. a CFT and/or the WEIL–FELIX TEST.

rod (*bacteriol.*) *Syn.* BACILLUS sense 2.

rod organ (pharyngeal rod organ) See PERANEMA.

roentgen See RÖNTGEN.

roestelioid (cornute) Refers to an elongated aecium which projects through the epidermis of the host; at maturity the exposed portion of the peridium typically ruptures by a number of longitudinal slits, the torn strips of peridium often curling back in a rosette-like pattern. Such aecia are formed e.g. by species of *Gymnosporangium*.

rofecoxib See PROSTAGLANDINS.

roguing (*plant pathol.*) The removal of diseased plants from a crop in order to (e.g.) prevent the spread of the disease.

Rohament P A commercial PECTIC ENZYME preparation consisting mainly of endopolygalacturonase.

Rohr See PLASMODIOPHOROMYCETES.

roll-tube technique (Hungate technique; *syn.* Virginia Polytechnic Institute (VPI) technique) A technique used for the isolation, culture and enumeration of strict, O_2-sensitive ANAEROBES. Pre-reduced, agar-based media are sterilized in tubes, under anaerobic conditions, and the processes of inoculation, incubation and subculture are also carried out anaerobically; anaerobic conditions are maintained within the tube during e.g. inoculation by inserting a sterile cannula into the tube and passing in a gentle stream of O_2-free gas (e.g. CO_2) during any period when the tube is unstoppered. For inoculation, the medium is melted, cooled to ca. 45°C, and mixed with the inoculum; the tube is then held horizontally and rotated about its long axis so that the agar solidifies in a thin layer around the inner surface of the tube. The tube is then incubated until individual colonies develop on and within the medium. (If a heat-labile reducing agent is used in the medium it is incorporated immediately prior to inoculation.)

roller culture In TISSUE CULTURE, a method of incubating bottles and other vessels of circular cross-section: the culture vessels are placed on a horizontal array of parallel, closely spaced rotating rollers such that each vessel rotates about its long axis; during incubation a monolayer develops as a continuous band on the vessel's inner surface. Culture vessels may be rotated at ca. 0.1–3.0 r.p.m. on commercially available apparatus. Benefits of roller culture include a more efficient use of culture vessels, and the various advantages of a continually mixed medium.

rolling (of leukocytes) See INFLAMMATION.

rolling circle mechanism A mechanism for the replication of a cccDNA molecule (or a cccRNA molecule – see VIROID). In e.g. a dsDNA molecule, one strand of the parent molecule is nicked, and DNA synthesis proceeds by elongation of the 3′-OH end (with progressive displacement of the 5′ end), the unbroken circular strand acting as template (see DNA REPLICATION). The partly replicated intermediate is thus a double-stranded circular DNA with a single-stranded displaced tail (σ structure – cf. CAIRNS' MECHANISM). The displaced strand may itself serve as a template for Okazaki fragment synthesis. When one round of replication has been completed, elongation may continue past the original initiation site (yielding a CONCATEMER), or the daughter strand may be cleaved by endonuclease at this site (yielding a parent-length strand). (See e.g. BACTERIOPHAGE ϕX174.)

rolling hairpin model (DNA replication) See PARVOVIRUS.

***rom* gene** See COLE1 PLASMID.

Romaña's sign Conjunctivitis, with parotid lymphadenopathy, after contamination of eyes by *Trypanosoma cruzi*. (See also CHAGAS' DISEASE.)

Romanowsky stains A range of composite stains (e.g. Giemsa's stain, Leishman's stain, Wright's stain) each of which includes METHYLENE BLUE, azure A and/or azure B, EOSIN, and (sometimes) methylene violet; they are used e.g. to detect/identify blood parasites (e.g. *Plasmodium* spp, trypanosomes) in blood smears. The stains are commonly used in solution in methanol, or in methanol and glycerol, and are often diluted in buffer before use. Typical staining reactions: erythrocytes and SCHÜFFNER'S DOTS pink; protozoan nuclei and KINETOPLASTS red; protozoan cytoplasm bluish-purple.

Romney Marsh disease *Syn.* STRUCK.

rondomycin See TETRACYCLINES.

röntgen (R; roentgen) A unit of radiation exposure (i.e., dose): the dose which causes ionization that liberates 2.58×10^{-4} coulombs of charge, of each sign, per kilogram of air. (cf. RAD.)

Röntgen rays *Syn.* X-rays. (See also IONIZING RADIATION.)

root nodules GALL-like (hypertrophic, hyperplastic) structures, containing endosymbiotic nitrogen-fixing bacteria, formed on the roots of certain plants – particularly members of the Leguminosae (cf. ACTINORRHIZAE and NOSTOC). Legumes (e.g. peas, beans, clovers, alfalfa) form root nodules in association with strains of RHIZOBIUM and BRADYRHIZOBIUM, often with a degree of plant–bacterium specificity. The endosymbionts occur within the nodules as modified, membrane-enclosed, often pleomorphic cells (*bacteroids*). The plant provides the bacteroids with nutrients (including fixed nitrogen), while the bacteroids reduce atmospheric nitrogen to ammonia (see NITROGEN FIXATION) which is assimilated by the plant, enabling it to thrive in nitrogen-deficient soils. (The bacteroids themselves are apparently incapable of utilizing NH_3.) [Nitrogen assimilation in legumes: Book ref. 55, pp. 129–178.] Increased levels of exogenous fixed nitrogen inhibit root nodule formation (cf. STEM NODULES), and soil conditions may influence the effectiveness of nitrogen fixation by the nodules [Book ref. 45, pp. 820–822].

In legumes, there are two principal types of root nodule, indeterminate and determinate, although other types and infection strategies are also known. [Book ref. 55, pp. 95–128.]

Indeterminate (= *meristematic*) nodules – formed e.g. in clovers, peas, alfalfa (lucerne), and some beans – are elongate-cylindrical in shape. Nodulation begins when capsulated rhizobia become polarly attached to a root hair. Attachment may involve interaction between root surface LECTINS (e.g. TRIFOLIIN A) and bacterial surface polymers (e.g. polysaccharide, LPS) [Book ref. 55, pp. 3–31, cf. pp. 116–119]. Attachment may be followed by a tight curling of the root hair tip (due to differential cell wall synthesis in the root hair), resulting in the formation of a pocket which encloses the bacteria. The bacteria then penetrate the root hair via a tubular *infection thread* which, originating at the site of infection, grows to the base of the root hair and penetrates the underlying cortex. [Infection process: ARM (1983) *37* 399–424.] The wall of the infection thread seems to be of plant origin, resembling the primary plant cell wall in composition; mucopolysaccharide (apparently of bacterial origin) in the lumen may serve to protect the invading bacteria from the host's defence mechanisms [Book ref. 55, pp. 58–93]. The infection thread ramifies extensively within the cortex, and cortical cells ahead of it proliferate – apparently in response to a diffusible substance(s) (?phytohormones). (Rhizobia can produce IAA and cytokinins in vitro, but the significance of this in nodule formation is unclear.) The invading bacteria are released from the tips of the ramifying infection thread into the proliferating cortical cells, within which they undergo morphological and other changes to become pleomorphic bacteroids enveloped in a membrane of plant origin (the *peribacteroid membrane*). Once infected, host cells cease to divide and begin to swell; they remain vacuolated. Cells beyond the infection zone continue to divide until they too are invaded; thus, throughout the development of the nodule there is an advancing zone of meristematic cells. The bacteroid-containing cells contain LEGHAEMOGLOBIN and form a pink zone between the white zones of the meristem and of the ramifying infection thread. In indeterminate nodules the fixed nitrogen is exported from the nodule to the rest of the plant mainly as *asparagine*.

Determinate nodules occur in certain tropical legumes, soybeans, cowpeas etc. Infection threads enter the root cortex from an infected root hair but do not ramify extensively after entering the cortex; bacteria are spread mainly by division of infected cortical cells. The nodules are spherical, with a cortical cell layer surrounding a central, pink-pigmented zone containing both infected and non-infected cells; the meristematic stage is transient, restricted to the initial stage of nodule development. Infected cells are not vacuolated, and the bacteroids are (non-pleomorphic) rods and swollen rods which divide to form up to ca. 16 bacteroids within the peribacteroid membrane. Fixed nitrogen is exported from the nodule as *ureides* (mainly allantoin and allantoic acid).

An alternative infection strategy occurs in e.g. the peanut (*Arachis hypogea*): bacteria enter the root at the junction between the root hair and epidermal cells ('crack entry'), and an infection thread is not formed. Interestingly, crack entry has been exploited in the experimental infection of wheat with the nitrogen-fixing bacterium *Azorhizobium caulinodans*; *A. caulinodans* became established within the xylem and root meristem such that experimental plants showed significantly increased dry weight and nitrogen content in comparison to uninoculated control plants [Proc. RSLB (1997) *264* 341–346].

Nodulation involves coordinated expression of plant and bacterial genes, with derepression of certain plant genes (see e.g. NODULINS) and bacteroid genes (e.g. NIF GENES). (See also LEGHAEMOGLOBIN.) In *Rhizobium* spp many of the genes involved in the initiation of root hair infection, nodule formation, and nitrogen fixation are located on a large plasmid (*Sym plasmid*), although chromosomal genes are also involved (at least in nitrogen fixation). Sym plasmids may code for additional functions such as bacteriocin production, plasmid transfer, pigment production, etc. Sym plasmids appear not to occur in *Bradyrhizobium* spp. [Nodulation genes in *Rhizobium*: TIBS (1986) *11* 296–299.]

Nodulating strains of *Rhizobium* and *Bradyrhizobium* have been used commercially to inoculate legume seeds, prior to planting, in order to establish suitable strains in soils which lack them; this may be necessary e.g. when introducing plants (e.g. soybeans) into new geographical regions [Book ref. 45, pp. 822–824].

rootlet See BASAL BODY (b).

***rop* gene** See COLE1 PLASMID.

ropiness A form of FOOD SPOILAGE in which long, stringy threads of (usually) polysaccharide – e.g. DEXTRANS, LEVANS – are formed by spoilage bacteria: e.g. species of *Acetobacter* (in beer), *Bacillus* (e.g. *B. subtilis* in bread), *Leuconostoc* (see WINE SPOILAGE) and *Lactococcus* (but cf. TAETTE).

Roquefort cheese See CHEESE-MAKING.

roridins See TRICHOTHECENES and STACHYBOTRYOTOXICOSIS.

Rosalina See FORAMINIFERIDA.

rosanilin A bluish-red TRIPHENYLMETHANE DYE.

Rosculus See AMOEBIDA.

rose bengal A deep-pink fluorescent dye – sodium tetraiodotetrachloroFLUORESCEIN; it is sometimes used in media for the isolation of certain soil fungi owing to its antibacterial properties and its ability to inhibit some rapidly-growing fungi. (See also PHOTODYNAMIC EFFECT.)

rose black spot See BLACK SPOT (3).

Rose–Waaler test A serological test used to detect and quantify the rheumatoid factor (see RHEUMATOID ARTHRITIS). Essentially, the test determines the ability of the patient's serum to agglutinate sheep erythrocytes which have been sensitized (SENSITIZATION sense 3) with a subagglutinating amount of homologous IgG antibody; rheumatoid factor agglutinates such sensitized erythrocytes by combining with the Fc portions of the sensitizing IgG antibodies. Non-sensitized erythrocytes are used as a control.

roseola infantum *Syn.* EXANTHEM SUBITUM.

Roseolovirus A genus of viruses in which the type species is human (beta) herpesvirus 6 (see BETAHERPESVIRINAE) [ratification of genus name: Arch. Virol. (1993) *133* 491–498].

rosette (1) (*bacteriol.*) A radially arranged cluster of cells or filaments (see e.g. CAULOBACTER).

(2) (*protozool.*) An intracellular structure present near the cytostome in some members of the HYPOSTOMATIA.

(3) (*plant pathol.*) In a plant: an abnormal condition in which the leaves form a radial cluster on the stem due e.g. to stunting (involving a shortening of internodes) – a symptom of infection with certain PLANT VIRUSES.

(4) (*serol.*) See IMMUNOCYTE ADHERENCE TECHNIQUE.

(5) (*med.*) See MALARIA.

rosette technique IMMUNOCYTE ADHERENCE TECHNIQUE.

rosetting See MALARIA.

Ross River virus An ALPHAVIRUS which causes epidemics of polyarthritis (often with maculopapular rash) in humans; it occurs e.g. in Australasia and in the western Pacific region. Various other vertebrates – including domestic animals – can be infected, and wild rodents may be reservoirs of infection. The virus has been isolated from e.g. *Aedes vigilax* and *Culex annulirostris* mosquitoes.

Rotaliina See FORAMINIFERIDA.

Rotavirus (*Duovirus*, *Stellavirus*) A genus of viruses of the family REOVIRIDAE. Hosts include mammals and birds. Rotaviruses are important causes of acute, often severe, diarrhoeal disease, particularly in infants, young children and young animals – includng calves (e.g. Nebraska calf diarrhoea), foals, lambs, mice and piglets. (See also FOOD POISONING (i) and SUDDEN INFANT DEATH SYNDROME.) Rotaviruses replicate in the intestinal epithelial cells; the faeces of infected individuals may contain up to e.g. $10^9–10^{10}$ virions per gram.

A serological classification system for rotaviruses was proposed in 1985 [Ann. Vir. (1985) *136*E 5–12]. Vaccination has been used against rotavirus infection [see e.g. JID (1986) *153* 815–822, 823–831 and 832–839]; however, the tetravalent vaccine was withdrawn in 1999 (owing to side-effects), so that there is currently no vaccination available against rotavirus infections.

The capsid of the rotavirus virion is icosahedral in shape and is triple-layered, the outermost layer being lost during entry of the virion into a host cell. The capsid comprises various types of polypeptide – the innermost layer consisting of 120 copies of the VP2 protein, while the adjacent (middle) layer consists of 260 trimers of VP6.

The rotavirus genome consists of 11 dsRNA molecules.

Inside infected cells the double-layered viral particles become transcriptionally active, mRNAs passing out through pores in the capsid. VP6 is necessary for trancription, but its exact role is unknown. The crystal structure of VP6 from a group A rotavirus has been determined to 2 Å resolution; the results suggest that, during assembly of the capsid, the addition of the VP6 layer begins with the binding of VP6 trimers to the inner layer at the 20 icosahedral three-fold axes [EMBO (2001) *20* 1485–1497].

(See also PARAROTAVIRUSES.)

rotenone A mitochondrial RESPIRATORY INHIBITOR which inhibits electron flow between the Fe–S centres of Complex I and ubiquinone (see ELECTRON TRANSPORT CHAIN); it binds, non-covalently, to a site which is apparently close to that at which ubiquinone is reduced. Many bacterial systems are insensitive to rotenone.

Rothia A genus of aerobic, facultatively anaerobic, catalase-positive, asporogenous bacteria (order ACTINOMYCETALES, wall type VI) which occur e.g. in the human mouth. The organisms form a rudimentary mycelium which fragments into non-motile rods and coccoid forms, the latter being ca. 1–5 μm in diameter. Growth occurs on rich media, e.g. trypticase–soy agar, at 37°C; no growth occurs on Sabouraud's dextrose agar. Various sugars can be fermented; glucose metabolism yields mainly lactic acid. GC%: ca. 65–70. Type species: *R. dentocariosa*. [Strain differentiation in *R. dentocariosa*: IJSB (1984) *34* 102–106.]

Rotorod An instrument used e.g. for sampling the airborne microflora. (See also AIR.) Essentially, two vertically-orientated adhesive- or grease-coated rods, at opposite ends of a horizontal bar, are whirled at ca. 2500 rpm; the rods may be detached and examined e.g. by microscopy.

rotund body See THERMUS.

rough endoplasmic reticulum See ENDOPLASMIC RETICULUM.

rough strain See SMOOTH–ROUGH VARIATION.

Rous-associated viruses See AVIAN LEUKOSIS VIRUSES.

Rous sarcoma virus (RSV) An AVIAN SARCOMA VIRUS originally isolated from a naturally occurring sarcoma of a chicken; when inoculated into chickens or other fowl, RSV can cause sarcomas and a fatal acute erythroblastic leukaemia within 2–3 weeks. There are several strains of RSV which differ e.g. in oncogenicity, host range, etc. Most strains are replication-defective; all carry the ONCOGENE v-*src* (see SRC). Some strains can induce tumorigenesis on inoculation into mammals (e.g. mice, rabbits, monkeys). RSV can be cultivated in chick embryo cells or in embryonated eggs – defective strains requiring the presence of an appropriate helper virus (Rous-associated virus: see AVIAN LEUKOSIS VIRUSES).

routine test dilution (RTD) The highest dilution of a virus suspension (i.e. the lowest concentration of viruses) which can bring about confluent lysis in a monolayer of sensitive cells or (in the case of bacteriophage) in a lawn plate prepared with a sensitive strain of bacteria.

Roux flask (Roux bottle) A flat-sided, rectangular, glass or plastic vessel which opens at one end via a short wide tube; it is used, on its side, for TISSUE CULTURE.

roxithromycin A MACROLIDE ANTIBIOTIC.

Royce indicator sachet See ETHYLENE OXIDE.

Rozella See CHYTRIDIALES and MYCOPARASITE.

RP 59500 See QUINUPRISTIN/DALFOPRISTIN.

RP1 β-lactamase *Syn.* TEM-2 β-LACTAMASE.

RP1 plasmid (*syn.* R1822 plasmid) A CONJUGATIVE PLASMID (COPY NUMBER: 1–2) which occurs in both enterobacteria and *Pseudomonas* spp (IncP/IncP-1); it encodes resistance to carbenicillin, kanamycin and tetracycline. Plasmids R18, R68, RK2 and RP4 are very similar or identical to RP1.

RP4 plasmid See RP1 PLASMID.

RPase RNA POLYMERASE.

RPCF test See TREPONEMAL TESTS.

RPG effect The phenomenon (discovered by R. P. Gunsalus) in which METHANOGENESIS from CO_2 and H_2 (in crude extracts of certain methanogens) is stimulated by methyl-coenzyme M.

***rpl* genes** Genes encoding r-proteins of the large ribosomal subunit: see RIBOSOME.

***rpo* genes** Genes which encode DNA-dependent RNA POLYMERASE subunits; in e.g. *Escherichia coli*, *rpoA*, *rpoB*, *rpoC* and *rpoD* encode subunits α, β, β', and the major SIGMA FACTOR, respectively. [Autogenous control of *rpo* genes: Genet. Res. (1986) *48* 61–64.] (See also RIBOSOME (biogenesis).)

***rpoB* gene** In *Mycobacterium tuberculosis*: a gene encoding the β subunit of RNA polymerase. Mutations in *rpoB* are associated with resistance to the drug rifampicin; such mutations are detected by various genotypic methods which are used for testing isolates of *M. tuberculosis* for resistance to rifampicin – for further details see SSCP, PROBE and TUBERCULOSIS.

***rpoH* gene** (*htpR* gene) See HEAT-SHOCK PROTEINS.

***rpoN* gene** See NTR GENES.

***rpoS* gene** See SIGMA FACTOR.

RPR test Rapid plasma reagin test: a qualitative or quantitative STANDARD TEST FOR SYPHILIS in which the cardiolipin–lecithin–cholesterol antigen is modified by the incorporation of choline chloride; this modification permits serum to be tested without prior heating (cf. VDRL TEST). In the RPR (circle) card test the antigen suspension contains particles of charcoal; this permits flocculation to be assessed macroscopically. (cf. RST.)

***rps* genes** Genes encoding r-proteins of the small ribosomal subunit: see RIBOSOME.

***rrn* operons** See RRNA and RIBOTYPING.

rRNA Ribosomal RNA: the major component of a RIBOSOME (q.v. for structure etc). In e.g. *Escherichia coli* rRNA is encoded by 7 separate operons (designated *rrnA* to *rrnG*), each of which is transcribed (see TRANSCRIPTION) to give a single RNA precursor molecule (pre-rRNA, ca. 30S) which contains all three rRNA species. An *rrn* operon has two promoters (P1 and P2) separated by 109–119 bp, and has the overall structure: P1-P2-leader–16S rRNA gene–spacer–23S rRNA gene–5S rRNA gene–terminator; one or two tRNA genes lie in the 'transcribed spacer' between the 16S and 23S rRNA genes, and in some cases one or two tRNA genes also occur at the 3′ (distal) end of the operon.

The 30S pre-rRNA contains internal inverted repeats which can base-pair to form secondary structures. The 16S and 23S rRNA sequences each form the loop in a separate stem-and-loop structure, and are released from the pre-rRNA by the endoribonuclease RNASE III; RNase III makes a staggered break in each of the double-stranded stem regions to liberate individual 16S and 23S rRNA precursors (designated p16S and p23S) which are slightly longer than the mature molecules. The RNase III cleavage sites show no homology; the enzyme presumably recognizes secondary and/or tertiary structure in the RNA. Subsequent processing of the individual precursors to form the mature rRNA molecules is not well understood; that of the 5S rRNA involves the endonuclease RNASE E. (For processing of the tRNA(s) see TRNA.)

Because rRNA genes are transcribed but not translated, they might be expected to show a high incidence of premature transcription termination owing to the absence of the protective effect of ribosomes (see POLAR MUTATION). Apparently, specific ANTITERMINATION mechanisms operate in the *rrn* operons of *E. coli*: modification of the RNA polymerase as it transcribes certain sequences in the *rrn* leader region may occur, allowing the polymerase to continue transcription until it reaches a strong *rrn* terminator [minireview: JB (1986) *168* 1–5].

Eukaryotes synthesize an rRNA precursor (ca. 45S) which contains a leader sequence and sequences for the 28S, 18S, and 5.8S rRNAs separated by transcribed spacers; 45S transcripts are transcribed from many tandemly repeated sets of genes separated by non-transcribed spacer sequences. The 45S precursor is synthesized by RNA polymerase I in the NUCLEOLUS [minireview: Cell (1986) *47* 839–840]; the 5S rRNA genes are organized separately and are transcribed by RNA polymerase III. (In certain lower eukaryotes – e.g. in *Dictyostelium discoideum*, and in the macronucleus of *Tetrahymena* – rRNA genes occur on extrachromosomal DNA molecules present in many copies per nucleus.) (See also SPLIT GENE (b).)

rRNA oligonucleotide cataloguing A procedure once used in bacterial TAXONOMY. Essentially, rRNA (usually 16S rRNA) is cleaved by a sequence-specific endonuclease (e.g. RNase T1); the fragments are end-labelled, separated by electrophoresis, and then sequenced – giving a 'catalogue' of sequences characteristic of the species. The catalogues of various species can then be compared and used as a basis for classification. [16S rRNA oligonucleotide cataloguing: Book ref. 138, pp 75–107.]

rRNA oligonucleotide cataloguing was one of the earlier methods used in molecular taxonomy. Subsequently, more informative methods (e.g. sequencing of the entire rRNA molecule) were developed.

rRNA superfamily VI See HELICOBACTER.

RS layer *Syn.* S LAYER.

Rsd protein See ANTI-SIGMA FACTOR.

RSF1010 plasmid See INCOMPATIBILITY.

RSF1030 plasmid See COLE1 PLASMID.

RSSE RUSSIAN SPRING–SUMMER ENCEPHALITIS.

RST Reagin screen test: a qualitative or quantitative STANDARD TEST FOR SYPHILIS in which the antigen has been dyed blue in order to permit macroscopic assessment of flocculation (cf. RPR TEST).

RSV (1) Respiratory syncytial virus. (2) Rous sarcoma virus.

Rta See ZEBRA.

RTD ROUTINE TEST DILUTION.

RTEM β-lactamase *Syn.* TEM-1 β-LACTAMASE.

RTF (1) Resistance transfer factor: that part of a *conjugative* R PLASMID which contains the genes specifying CONJUGATION; thus, a conjugative R plasmid consists of the RTF together with the gene(s) encoding resistance (resistance determinants). The RTF is usually self-repressed. (See also SEX FACTOR.)

(2) *Syn.* R plasmid, or conjugative R plasmid.

rtPCR See REVERSE TRANSCRIPTASE PCR.

***rTth* DNA polymerase** See REVERSE TRANSCRIPTASE PCR.

RTX toxins A category of TOXINS, secreted by certain Gram-negative bacteria, which have in common various structural and functional characteristics; the name 'RTX' derives from the series of calcium-binding, glycine-rich nonapeptide repeat units (**r**epeats in **t**o**x**in) which occur near the C-terminal end in each toxin. At least two of the toxins – the HlyA (haemolysin) of *Escherichia coli* and the CYCLOLYSIN of *Bordetella pertussis* – are secreted by an ABC EXPORTER.

A feature shared by all RTX toxins is the ability to form pores in eukaryotic cell membranes – in some cases leading to osmotic lysis. Although high concentrations of toxin can kill effector cells of the immune system (leading e.g. to local inflammation and tissue damage), other modes of activity have been recorded. For example, quite low levels of HlyA can bring about a loss of killing power in PMNs, while cyclolysin can inhibit the oxidative burst in phagocytes; in the latter case, the pathogen may use this advantage to establish an intracellular carrier state. At lower, sub-lytic concentrations, RTX toxins may behave as MODULINS. Cyclolysin differs from the other RTX toxins in that it has ADENYLATE CYCLASE activity; in phagocytes, increased levels of intracellular cAMP can result in a reduction or loss of phagocytic activity. In general, the RTX toxins apparently behave primarily as virulence factors directed against the host's immune response.

Other RTX toxins include the LktA (leukotoxin) of *Pasteurella haemolytica*, the causal agent of an acute pneumonia in ruminants (cattle, sheep); this toxin is a particularly potent lytic agent for the white blood cells and platelets of ruminants. The cytolytic activity of this toxin is reported to be enhanced by lipopolysaccharide [MP (2001) *30* 347–357].

The pathogen *Actinobacillus pleuropneumoniae*, associated with pleuropneumonia in pigs, forms two haemolysins, designated ApxIA and ApxIIA, while a leukotoxin, AktA, is secreted by *Actinobacillus actinomycetemcomitans*, a pathogen linked to oral disease and abscesses in man.

[RTX toxins (review): RMM (1996) *7* 53–62.]

RU READING UNIT.

rubber spoilage Natural rubber is essentially a long-chain polymer of isoprene units; *vulcanized* rubber contains a high proportion of sulphur. Rubber may also contain fillers (e.g. clay, carbon black), antioxidants, antimicrobial preservatives etc. Both natural and vulcanized rubbers are susceptible to microbial spoilage – e.g. staining or discoloration due to surface colonization (e.g. by species of *Bacillus*, *Serratia*, *Aspergillus*, *Cladosporium*, *Penicillium*) or degradation of the polymer itself. In general, organisms cannot attack the long isoprene chains enzymically, and initial deterioration is usually chemical (e.g. oxidation); short-chain fragments can then be metabolized e.g. by species of *Nocardia*, *Streptomyces*, *Aspergillus*, *Fusarium* and *Penicillium*. Vulcanized rubber can be degraded e.g. by *Nocardia* sp [AEM (1985) *50* 965–970]; *Thiobacillus thiooxidans* can produce sulphuric acid from the sulphur, the acid causing extensive degradation of the polyisoprene. Rubber is particularly susceptible to spoilage when damp, in contact with soil, etc; microbial deterioration can cause problems e.g. with rubber insulation on buried cables, rubber joints in water and sewer pipes etc. Various antimicrobial preservatives (e.g. pentachlorophenol, cresylic acid, organoarsenicals) may be added to delay spoilage. Synthetic rubbers such as butyl, silicone and polychloroprene (neoprene) rubbers are more resistant than natural rubber to both chemical and microbial degradation. (See also LATEX.)

rubella (German measles) (1) An acute infectious human disease, which affects mainly children and young adults, caused by a virus (see RUBIVIRUS).

Acquired rubella. Infection occurs by droplet inhalation. Incubation period: 2–3 weeks. Early symptoms may include mild upper respiratory tract symptoms, mild fever, and enlargement of the lymph nodes of the head and neck. Following a stage of viraemia, a maculopapular rash appears first on the face, then on the rest of the body. Patients are infectious for ca. 1 week before and ca. 1 week after the onset of symptoms. The disease is usually mild and self-limiting; occasional complications include e.g. arthritis and encephalitis.

Congenital rubella. Maternal rubella acquired during pregnancy can lead to a persistent infection of the fetus and neonate, resulting in teratogenic effects which may range from mild (e.g. low birth weight) to fetal death; intermediate manifestations include permanent defects in the CNS (e.g. mental retardation), ears (nerve deafness), eyes (e.g. cataracts, retinal damage), heart etc (cf. TORCH DISEASES). Damage is most severe if infection is acquired during the first 3 months of pregnancy. A progressive rubella panencephalitis resembling SUBACUTE SCLEROSING PANENCEPHALITIS has been reported as a late manifestation of congenital rubella [NEJM (1975) *292* 990–993, 994–998].

Lab. diagnosis: detection of rubella-specific antibodies e.g. by a haemagglutination-inhibition test, radial haemolysis, ELISA etc. Rubella virus can be isolated from e.g. nasopharyngeal swabs. *Prophylaxis*: vaccination with a live, attenuated vaccine (see also MR and MMR).

(2) *Syn.* INFECTIOUS DROPSY (of carp).

rubelliform Refers to a RUBELLA-like disease or rash.

rubeola (1) (English, German) *Syn.* MEASLES. (2) (French, Spanish) *Syn.* RUBELLA.

rubidomycin See ANTHRACYCLINE ANTIBIOTICS.

RuBisCO RIBULOSE 1,5-BISPHOSPHATE CARBOXYLASE–OXYGENASE.

Rubivirus A genus of viruses (family TOGAVIRIDAE) which currently includes only the RUBELLA virus. Man is apparently the only natural host; some laboratory animals can be infected, but none develops a disease analogous to that in man. (There is no invertebrate host.) The rubivirus virion is relatively unstable at room temperatures, loses infectivity within ca. 30 min at 56°C, but is stable for ca. 1 week in protein-containing solutions at 4°C; it can be stored at or below −60°C, and can withstand freeze-drying. It can be inactivated by pH <6.8 or >8.1, by organic solvents, UV radiation, deoxycholate, ethylene oxide, etc. The virions promote Ca^{2+}-dependent haemagglutination of RBCs from various species (those from newly hatched chick, adult goose or pigeon, or human O-type RBCs being commonly used in HI tests). Rubella virus can grow in various types of cell culture; primary African green monkey kidney cells have been widely used for primary isolation. The virus is non-cytopathogenic in most vertebrate cells, and can be detected e.g. by immunofluorescence techniques and by its ability to prevent the replication of a superinfecting 'challenge virus' such as echovirus type 11 (see INTERFERENCE). [Clinical aspects: Book ref. 148, pp. 1005–1020.]

RuBP Ribulose 1,5-bisphosphate.

RuBPCase See RIBULOSE 1,5-BISPHOSPHATE CARBOXYLASE–OXYGENASE.

rubratoxins MYCOTOXINS produced e.g. by strains of *Penicillium rubrum*; toxin production is associated with active fungal growth. Consumption of rubratoxins by animals causes congestion of tissues (often with haemorrhage) – especially in the stomach and intestine; liver damage and brain lesions can also occur. Rubratoxins and AFLATOXINS act synergistically.

rubredoxins See IRON–SULPHUR PROTEINS.

Rudimicrosporea A class of protozoa (phylum MICROSPORA) which are hyperparasites of gregarines in annelid worms. The spores have a simple (rudimentary) extrusion apparatus consisting of a polar cap and thick polar tube terminating in a funnel (the infundibulum). There is no polaroplast. Genera: e.g. *Amphiacantha, Metchnikovella.*

rugose Wrinkled.

rugose mosaic disease (1) The disease of potatoes caused by POTATO VIRUS Y (Y^o strains) in the second and subsequent years of infection. (2) A severe disease of potatoes caused by mixed infection with potato virus X and POTATO VIRUS Y (Y^o strains).

rule of desmodexy (*ciliate protozool.*) The rule, based on observation, that each kinetodesma lies on the right-hand side of its corresponding kinetosome: see KINETODESMA.

rum See SPIRITS.

rumen One of four compartments which form the stomach in *ruminants* (cattle, deer, giraffes, goats, sheep etc); the other compartments are the *reticulum* (an outpocketing of the rumen usually included with the rumen proper when reference is made to the 'rumen'), the *omasum*, and the *abomasum* or 'true stomach'.

In the adult or weaned ruminant, food passes from the oesophagus to the rumen. The rumen (capacity ca. 5–10 litres in sheep, 100–150 litres in cattle) is essentially an anaerobic muscular sac which contains a large population of microorganisms (on average ca. 10^{10} cells/ml) – primarily bacteria and protozoa, but sometimes including a significant proportion of fungi; these microorganisms play an essential role in ruminant nutrition by breaking down the ruminant's diet of plant material (see ANAEROBIC DIGESTION) and providing the animal with assimilable nutrients and sources of energy. (The microflora of the rumen is indispensable to the ruminant since many of the components of plant material are not attacked by the enzymes of animal digestive systems. The rumen itself does not secrete enzymes.) To allow adequate time for microbial action, large particles of food may remain in the rumen for several days. Periodically a bolus of partly digested material (a *cud*) is regurgitated to the mouth, chewed, and swallowed; cuds return to the rumen where they undergo further digestion. The net effect of digestion in the rumen is the conversion of dietary materials to a mixture of fatty acids (mainly acetic, butyric and propionic acids), gases (primarily CO_2 and CH_4 – which are voided by eructation), and microbial biomass. Fatty acids are absorbed by the rumen and are used by the ruminant as primary sources of carbon and energy. Microbial biomass passes from the rumen to the omasum – where residual fatty acids, water and salts are absorbed – and then to the abomasum where the biomass undergoes true ('gastric') digestion; digested material passes into the small intestine where it is absorbed by the ruminant – amino acids derived from the digestion of *microbial* protein constituting the principal source of nitrogen for the ruminant. Certain vitamins synthesized by the rumen bacteria (e.g. VITAMIN B_{12}) are also absorbed as the rumen contents pass through the abomasum and intestines.

Ruminants produce large amounts of saliva; the saliva contains bicarbonate and phosphate which buffer the rumen contents – maintaining a pH of ca. 6.0–6.5 despite the ongoing formation of acidic products by microbial fermentations. The temperature of the rumen (in the cow) is ca. 39–40°C, slightly higher than the body temperature (ca. 38°C) due partly to the exothermic microbial metabolism. The E'_h of the rumen has been reported to be in the range −145 mV to −260 mV. [Influence of E_h on in vitro culture of rumen bacteria: JGM (1984) *130* 223–229.]

The types and numbers of microorganisms in the rumen of an adult or weaned ruminant depend primarily on species, on the animal's diet, and on the period of time since the last intake of food; thus, for example, the proportion of non-bacterial microorganisms in the rumen is much higher in Red deer and reindeer – which feed on e.g. heather and lichens – than it is in sheep consuming more easily digestible materials. Protozoa common in the rumen include species of *Dasytricha*, *Diplodinium*, *Entodinium*, *Epidinium* and *Isotricha*. Ciliates usually adhere to the ends of broken fibres, or to the rumen wall; some are cellulolytic and/or amylolytic, while others depend on soluble sugars. At least some protozoa ingest bacteria in the rumen. [Rumen ciliates: AP (1980) *18* 121–173; rumen holotrich ciliates: MR (1986) *50* 25–49.] Some rumen microorganisms which were once thought to be protozoa are now known to be fungi – e.g. the species *Neocallimastix frontalis* (see NEOCALLIMASTIX), *Piromonas communis* and *Sphaeromonas communis*. Various bacteria (mainly Gram-negative anaerobes) occur in the rumen (see later). [New developments in rumen microflora: TIM (1997) *5* 483–488.]

In a given ruminant, some species of microorganism are commonly present in large numbers and appear to play more or less well-defined roles in digestive processes within the rumen; although protozoa are normally present, experiments with defaunated animals indicate that they are not always essential. Disturbances of the overall balance of microorganisms in the rumen can occur e.g. in certain dietary regimes – leading to disorders such as ACIDOSIS and BLOAT.

Carbohydrate utilization in the rumen. The main dietary carbohydrates of a ruminant typically include plant polysaccharides such as CELLULOSE, HEMICELLULOSES and PECTINS which occur in leaves and stems; the free sugar and starch content of leaves and stems is relatively small. Cellulose is degraded by a range of bacteria, including species of BACTEROIDES (e.g. *B. succinogenes*) and RUMINOCOCCUS, and '*Micromonospora ruminantium*'; by protozoa (e.g. *Epidinium*); and by fungi (e.g. *Piromonas communis*). (See also CELLULASES.) CELLOBIOSE is attacked by *Bacteroides ruminicola* and species of BUTYRIVIBRIO (e.g. *B. fibrisolvens*) and SELENOMONAS (e.g. *S. ruminantium*). Pectins are degraded by some organisms (e.g. *Ruminococcus flavefaciens*) which do not use the breakdown products; other organisms (e.g. *Bacteroides ruminicola*, *Lachnospira multiparus*) degrade pectins and utilize the products. Other pectinolytic organisms include *Treponema saccharophilum* [AEM (1985) *50* 212–219]. XYLANS are attacked e.g. by *Bacteroides ruminicola* and *Butyrivibrio fibrisolvens* [rumen xylanolytic systems: FEMS Reviews (1993) *104* 68–70].

Animals which are fed on cereal concentrates have a higher-than-normal intake of STARCH, and this may be reflected in the nature of the rumen microflora. Organisms which can degrade starch include *Bacteroides amylophilus* and *B. ruminicola, Succinimonas amylolytica*, and *Succinivibrio dextrinosolvens; S. dextrinosolvens* seems to be a major fermenter of dextrins in high-starch diets. Starch can also be digested by some ciliates.

The presence in the diet of material with high levels of tannins can affect the composition of the rumen microflor – tannins depressing the growth rate of some organisms more than that of others (see TANNINS).

Products of the initial breakdown of polysaccharides include a range of mono- or disaccharides which are fermented to volatile fatty acids (VFA) – particularly acetic, butyric and propionic acids – and to other products such as lactic and succinic acids; these primary fermentations are carried out by many of the bacteria involved in the initial cleavage of polysaccharides – and e.g. by *Streptococcus bovis* and certain LACTIC ACID BACTERIA. VFA are absorbed by the rumen (and used by the ruminant) while e.g. lactic acid may be used as a substrate for a secondary fermentation (e.g. by *Veillonella parvula* and species of *Selenomonas* and MEGASPHAERA) with the formation of acetic and propionic acids. Other rumen bacteria (e.g. species of DESULFOBULBUS and DESULFOTOMACULUM) can oxidize lactic and propionic acids to acetic acid. (See also ANAEROBIC DIGESTION.)

Some of the CO_2 and H_2 derived from fermentations is used, by rumen METHANOGENS, for the production of methane (see METHANOGENESIS). In general, methane is produced in molar proportions greater than those of either propionate or butyrate. Theoretically, suppression of methanogenesis should increase yields of the other products – thus increasing feed conversion efficiency; this has been found to be the case in practice, and various agents (e.g. MONENSIN) have been used specifically for this purpose.

Lipid utilization in the rumen. Dietary lipids consist mainly of mono- and digalactoglycerides (in e.g. grasses) and triglycerides (in cereal concentrates); these lipids contain a high proportion of polyunsaturated long-chain fatty acids. In the rumen some of these lipids undergo lipolysis and hydrogenation to form free, more or less saturated, C_{16}–C_{18} fatty acids, while others are used for the synthesis of microbial lipids. (Owing to the hydrogenation of dietary lipids by certain rumen bacteria, the storage fats of ruminants are less affected by the nature of the dietary lipids than is the case in non-ruminants.) Lipolytic rumen bacteria include *Anaerovibrio lipolytica* and some strains of *Butyrivibrio*.

Nitrogen metabolism in the rumen. In the rumen, most dietary protein is initially degraded to peptides and amino acids. Many species of microorganism can carry out proteolysis in the rumen, some of the most actively proteolytic bacteria being closely associated with the rumen epithelium. In sheep, *Bacteroides ruminicola* has been found to be the major proteolytic bacterium in the rumen fluid [JGM (1985) *131* 821–832]. Although some of the amino acids are taken up by the rumen microorganisms, most are rapidly deaminated – the resulting ammonia constituting the principal source of nitrogen for the rumen microorganisms. (cf. FOG FEVER.) The metabolism of protein in this way is a relatively inefficient use of feedstuffs (by the ruminant) since, e.g., some of the ammonia is absorbed by the rumen, and some of this is excreted (as urea) in the urine (but see next paragraph). Attempts have been made to inhibit proteolysis in the rumen by modifying dietary protein (by heating or by treatment with e.g. formaldehyde); the modified protein is resistant to degradation in the rumen (near-neutral pH) but is readily digested in the (acidic) abomasum. (Some unmodified proteins, e.g. albumins and γ-globulins, are resistant to degradation in the rumen.) An alternative strategy is to inhibit the deamination of amino acids in the rumen; inhibition of in vitro deamination has been achieved with sodium arsenite (suggesting the involvement of the STICKLAND REACTION), and inhibition both in vitro and in vivo can be achieved e.g. by the use of diaryliodonium compounds. However, deamination of amino acids yields certain fatty acids (e.g. isobutyric acid from valine) which are needed by some of the rumen bacteria (e.g. *Ruminococcus* spp).

Other nitrogenous compounds (e.g. urea) are also used as sources of nitrogen by rumen microorganisms. Urea formed by the ruminant itself can be excreted (in urine) but much passes to the rumen (via the rumen wall, or in the saliva); here, the urea is cleaved by microbial ureases and the resulting ammonia is incorporated into microbial biomass. Thus, dietary nitrogen is conserved, and ruminants can survive on low-nitrogen diets. Feedstuffs low in nitrogen can be supplemented with e.g. urea.

In the *suckling* ruminant the rumen is poorly developed; milk passes from the oesophagus, via the oesophageal groove, directly to the omasum and abomasum (i.e., by-passing the rumen) and is digested by processes similar to those in other young mammals; the intestinal flora includes species of *Lactobacillus* and *Streptococcus*.

[Review of microbial ecology and activities in the rumen: CRM (1982) *9* 165–225 (part I) and 253–320 (part II).]

rumenitis Inflammation of the RUMEN. Primary rumenitis, involving erosion of the rumen wall, can occur as a result of ACIDOSIS; the weakened rumen wall may subsequently be invaded by pathogenic microorganisms which may give rise e.g. to necrotic foci in the rumen and/or in other tissues (such as the liver).

rumensin *Syn.* MONENSIN.

ruminal Of or pertaining to the RUMEN.

ruminal tympany *Syn.* BLOAT.

ruminate (1) (*verb*) To chew the cud. (See also RUMEN.) (2) (*adj.*) Much folded – as though having been chewed.

Ruminococcus A genus of GRAM TYPE-positive, asporogenous, anaerobic, typically cellulolytic bacteria which occur e.g. in the RUMEN; GC%: ca. 40–45. Cells: cocci (ca. 1 μm diam.) which occur in pairs or chains. Metabolism is typically heterofermentative; the main products of carbohydrate metabolism include acetic and formic acids (lactic acid may be formed in small amounts). CELLOBIOSE and (commonly) CELLULOSE are fermented; glucose and other simple sugars may or may not be fermented, according to strain (many strains lack a glucose transport system). Ammonia is essential as a source of nitrogen for most or all strains. Other growth requirements may include e.g. B vitamins. Type species: *R. flavefaciens*. Other species: *R. albus*.

RuMP pathway *Syn.* RMP PATHWAY.

rumposome In zoospores of certain members of the CHYTRIDIOMYCETES (e.g. *Monoblepharella*): a body of unknown function (though thought to be light-sensitive) which occurs beneath the cytoplasmic membrane at the posterior end of the cell.

Runella A genus of pink-pigmented bacteria (family SPIROSOMACEAE). The cells, 0.5–0.9 × 2.0–4.5 μm, are typically moderately or strongly curved – the ends of curved cells sometimes overlapping to form 'ring-shaped' structures. Acid is formed from few carbohydrates. GC%: 49–50. Type species: *R. slithyformis*.

Runyon groups In the genus *Mycobacterium*: groups of species distinguished on the basis of growth rate and pigmentation; slow-growing species are placed in groups I (photochromogens), II (scotochromogens), and III (others), while fast-growing species are placed in group IV.

Russian autumn encephalitis *Syn.* JAPANESE B ENCEPHALITIS.

Russian flu See INFLUENZAVIRUS.

Russian spring–summer encephalitis (RSSE) An acute human ENCEPHALITIS caused by a flavivirus (see FLAVIVIRIDAE); it occurs in epidemics, chiefly in and around the forests of Siberia. It is transmitted by ticks (e.g. *Ixodes* sp); human infection can also occur by ingestion of meat or milk from infected animals. The disease varies from mild to fatal; encephalitis may be followed by chronic progressive neuromuscular disorders or paralysis months or years after the original infection, apparently due to the persistence of the RSSE virus in the CNS. [ARE (1981) *26* 80–82.]

Russian winter wheat mosaic virus See RHABDOVIRIDAE.

Russula A genus of fungi (family RUSSULACEAE) which form mushroom-shaped, lamellate fruiting bodies that do not form 'latex' (cf. LACTARIUS). *R. emetica* is a poisonous species ('the sickener') which forms a fruiting body having a bright red, shiny cap; it is commonly found under or near pine trees. *R. mairei* ('the beechwood sickener') is another red-capped, poisonous species found near beech trees.

Russulaceae A family of fungi (order RUSSULALES) which form ballistospores on stipitate, agaricoid to gasteroid fruiting bodies. Genera include LACTARIUS and RUSSULA.

Russulales An order of fungi (subclass HOLOBASIDIOMYCETIDAE) in which a characteristic feature is the presence of SPHAEROCYSTS in the tissues of the fruiting body. The two families are Elasmomycetaceae (members form statismospores in gasteroid fruiting bodies) and RUSSULACEAE. (See also GASTEROMYCETES (Hymenogastrales) and SECOTIOID FUNGI.)

rust See RUSTS.

rusticyanin A BLUE PROTEIN (MWt ca. 16500) which occurs in *Thiobacillus ferrooxidans* (see also EXTRACYTOPLASMIC OXIDATION); E_m: ca. 680 mV.

rusts (1) Fungi of the class UREDINIOMYCETES. (2) Any of various plant diseases caused e.g. by members of the Urediniomycetes or by species of ALBUGO. (See also WHITE RUST and RED RUST.) Diseases caused by urediniomycetes are called 'rusts' because many of the causal agents form rust-coloured spores on affected plants; these diseases include e.g. BLISTER RUST, COFFEE RUST, various CEREAL DISEASES, CROWN RUST and FLAX RUST.

rutamycin An agent which binds non-covalently to and inhibits mitochondrial (F_0F_1)-type PROTON ATPASES.

ruthenium red A dye used for staining mucopolysaccharides and e.g. as a specific inhibitor of certain ATPases [JGM (1980) *120* 183–198].

***ruv* gene** See RECF PATHWAY.

RV medium RAPPAPORT–VASSILIADIS BROTH.

RW coefficient See RIDEAL–WALKER TEST.

rye diseases See CEREAL DISEASES.

ryegrass mosaic virus See POTYVIRUSES.

ryegrass staggers See PENITREMS.

1. Words in SMALL CAPITALS are cross-references to separate entries.
2. Keys to journal title abbreviations and Book ref. numbers are given at the end of the Dictionary.
3. The Greek alphabet is given in Appendix VI.
4. For further information see 'Notes for the User' at the front of the Dictionary.

S

s   Sedimentation coefficient: see SVEDBERG UNIT.

S   (1) SVEDBERG UNIT. (2) Serine (see AMINO ACIDS).

S fimbriae   FIMBRIAE which occur on many of the strains of *Escherichia coli* which cause meningitis in human neonates; they bind to sialyl galactoside receptors on human erythrocytes.

S-glucan   Alkali-soluble glucans of fungal CELL WALLS. (cf. R-GLUCAN.)

s. lat.   SENSU LATO (q.v.).

S layer   In a prokaryotic cell: a continuous layer – often the outermost layer of the cell – consisting of a repeating pattern of protein or glycoprotein subunits ('S proteins') arranged in squares, hexagons etc. S layers occur in bacteria (including strains of e.g. *Aeromonas*, *Campylobacter*, *Clostridium*, DEINOCOCCUS, *Lactobacillus*, *Nitrosomonas*, *Pseudomonas*, *Treponema*, some cyanobacteria) and archaeans (including strains of e.g. *Desulfurococcus*, *Halobacterium*, *Methanococcus*, *Sulfolobus*).

In bacteria, the S layer usually overlays the cell wall proper – e.g. in Gram-negative species it is external to the outer membrane. A *Bacillus* sp with an S layer on *both* sides of the cell wall has been reported [FEMS (1987) *40* 75–79]. Double S layers, containing the same or different subunits, occur e.g. in strains of *Aquaspirillum*, *Bacillus* and *Corynebacterium*. (See also GLYCOCALYX.)

In a strain of *Bacillus stearothermophilus* it has been found that variation in the S layer involves a chromosome–plasmid re-arrangement of DNA [JB (2001) *183* 1672–1679].

In some archaeans the S layer is the only cell wall component, being immediately external to the cytoplasmic membrane.

S proteins are typically rich in acidic residues. In Gram-negative bacteria the S proteins interact with LPS – in at least some cases by non-covalent links; changes in composition of LPS have been associated with the loss of an S layer. The mRNAs of S protein genes are typically long-lived (~10–20 minutes).

S proteins typically have an N-terminal signal peptide, indicating secretion by a type II (*sec*-dependent) system (see PROTEIN SECRETION). However, the way in which these proteins aggregate to form an S layer, and their mode of binding to LPS in Gram-negative bacteria, is not understood.

[Review: ARM (1983) *37* 311–339. Principles of S layer organization: JMB (1986) *187* 251–253. Secretion of S layers in Gram-negative bacteria: FEMS Reviews (2000) *24* 21–44 (35–38).]

S matrix   In NUMERICAL TAXONOMY, *similarity matrix*: the percentage similarity between each OTU and each of every other OTU (in a given study) expressed in tabular form – the ordinate and abscissa of the matrix each comprising a complete list of designations of the OTUs under study.

S organism   See METHANOBACILLUS OMELIANSKII.

S phase (of cell cycle)   See CELL CYCLE.

S protein (*immunol.*)   A protein (MWt ca. 80000) found in normal plasma. S protein binds to the fluid-phase (i.e. non-membrane-bound) complex C5b67 formed during COMPLEMENT FIXATION, destroying its membrane-binding and haemolytic potential. (See also REACTIVE LYSIS.)

S→R variation   See SMOOTH–ROUGH VARIATION.

S ring   See FLAGELLUM (a).

s. str.   SENSU STRICTO (q.v.).

S strain   See SMOOTH–ROUGH VARIATION.

***s* zonal centrifugation**   See CENTRIFUGATION.

S1 (S_1) endonuclease   *Syn.* ENDONUCLEASE S_1.

S1–S21 proteins   See RIBOSOME.

Sabin–Feldman dye test   TOXOPLASMA DYE TEST.

Sabin vaccine   An oral anti-poliomyelitis vaccine containing attenuated strains of three types of poliovirus (see ATTENUATION (1)); each strain may be administered separately to avoid mutual interference. The vaccine encourages an active viral infection of the gut and stimulates local secretory IgA formation. (cf. SALK VACCINE.)

sabkha   A hypersaline pool.

Sabouraud's medium   A medium used for the culture of various fungi pathogenic in man or animals. It contains e.g. 1–1.5% peptone (or phytone), 2–4% D-glucose ('Sabouraud's dextrose medium') or maltose ('Sabouraud's maltose medium'), and – for solid media – ca. 2% agar. The pH is adjusted to 5.4–5.6 (which inhibits the growth of many bacteria), and various antibiotics may be added to further suppress bacteria and unwanted fungi.

***Sac*II**   See RESTRICTION ENDONUCLEASE (table).

sacbrood   A BEE DISEASE which affects *Apis mellifera* and *A. cerana*; it is caused by an unclassified RNA virus (cf. PICORNAVIRIDAE). Infected larvae fail to pupate, apparently because they cannot shed their last larval skin; they become yellow and surrounded by fluid, and death occurs within a few days. The virus rapidly loses infectivity in the dried remains of the larvae; adult bees appear to act as carriers.

Saccammina   See FORAMINIFERIDA.

Saccamoeba   See AMOEBIDA.

Saccardoan system   A system of classification, introduced by Saccardo, in which the form genera of the DEUTEROMYCOTINA are grouped into sections on the basis of the characteristics of their spores (conidia). The sections are as follows. Amerosporae: spores rounded or slightly elongate, aseptate; didymosporae: spores ovoid or slightly elongate, uniseptate; phragmosporae: spores elongate with two or more transverse septa; dictyosporae: spores ovoid or elongate, having both transverse and longitudinal septa; scolecosporae: spores filamentous, septate or aseptate; helicosporae: spores helical, septate or aseptate; staurosporae: spores stellate.

Subsections may be formed e.g. on the basis of colour; thus, e.g. spores ovoid, uniseptate, colourless: hyalodidymae; spores rounded, aseptate, pigmented or dark: phaeosporae.

saccharase   *Syn.* INVERTASE.

saccharification   The degradation of polysaccharides (such as STARCH, CELLULOSE) to simple sugars (e.g. glucose).

Saccharomonospora   A genus of bacteria (order ACTINOMYCETALES, wall type IV) which occur e.g. in composts and soil. The organisms form aerial mycelium and a non-fragmenting substrate mycelium; spores are borne singly on unbranched sporophores which arise at short intervals on the aerial hyphae. Strains grown at 40–50°C usually form a white aerial mycelium which becomes grey-green as sporulation occurs; sporulation is typically accompanied by the formation of a soluble green pigment. Mycolic acids are not formed. GC%: ca. 74–75. Type species: *S. viridis*. [Book ref. 73, pp. 126–127.]

Saccharomyces   A genus of yeasts (family SACCHAROMYCETACEAE). Cells are spheroidal, ellipsoidal or cylindrical; vegetative reproduction occurs by multilateral budding. Pseudomycelium may be formed by some species, but true hyphae are absent.

The vegetative cells are normally predominantly diploid (or polyploid) – predominantly haploid species having been placed in the genera ZYGOSACCHAROMYCES and TORULASPORA. Asci, which are persistent, develop directly from the diploid vegetative cells. Ascospores are spheroidal, usually smooth, 1–4 per ascus. Conjugation occurs on, or shortly after, germination of the ascospores, i.e., it occurs between ascospores or between haploid somatic cells derived therefrom. Species can ferment one or more sugars, and can also carry out respiratory metabolism of a range of substrates. NO_3^- is not assimilated. Seven species have been recognized [Book ref. 100, pp. 379–395]; type species: *S. cerevisiae*.

S. cerevisiae (including many varieties and physiological races formerly accorded species status – e.g. *S. aceti*, *S. carlsbergensis*, *S. diastaticus*, *S. ellipsoideus*, *S. globosus*, *S. intermedius*, *S. inusitatus*, *S. sake*, *S. uvarum* etc). Cells (e.g. after 3 days at 25°C in malt extract) are ca. 3–10 × 4–20(–30) µm, occurring singly or in pairs, short chains or clusters. (See also CELL WALL.) Pseudohyphae may be formed e.g. on cornmeal agar. Many strains are heterothallic (see MATING TYPE). Sporulation is encouraged by media which are low in nitrogen and which contain acetate as the main or sole carbon source. All strains can ferment glucose, some can ferment galactose, sucrose and maltose, none can ferment lactose. (See also ALCOHOLIC FERMENTATION.) In the presence of O_2, strains can metabolize (oxidatively) substrates such as glycerol, ethanol and lactate. All strains are sensitive to cycloheximide (100 ppm). Strains have been isolated from e.g. alcoholic beverages, fruits and fruit juices, soil, human skin and sputum, etc.

Selected strains of *S. cerevisiae* are widely used in the manufacture of alcoholic beverages and fermented foods: see e.g. BREAD-MAKING (and BAKERS' YEAST), BREWING (and BREWERS' YEAST), CIDER, KEFIR, PULQUE, SAKE, WINE-MAKING, YEAST EXTRACT. (See also CHEESE-MAKING (c), IMMOBILIZATION (sense 1), STEROID BIOCONVERSIONS, WINE SPOILAGE.) *S. cerevisiae* is also widely used in genetic studies and for the propagation of genetically engineered genes. (See also e.g. KILLER FACTOR, KILLER PLASMIDS, TWO-MICRON DNA PLASMID, THREE-MICRON DNA PLASMID.) [Genetic map of *S. cerevisiae*: MR (1985) *49* 181–212.]

S. dairensis (= *S. castellii*). Cells (in malt extract): ca. 3.0–5.5 × 3.5–7.0 µm, occurring singly or in pairs. Glucose and galactose are fermented; sucrose, maltose and lactose are not. Most strains can grow in the presence of cycloheximide at 100 ppm but not at 1000 ppm. Strains have been isolated from buttermilk, soil and fermenting cucumber brines.

S. exiguus (anamorph: *Candida holmii*). Cells (in malt extract): ca. 2.5–5.0 × 3.5–6.5 µm, occurring singly or in pairs. Glucose, galactose and sucrose are fermented; maltose and lactose are not. Most strains can grow in the presence of cycloheximide at 100 ppm but not at 1000 ppm. Strains have been isolated from soil, fruit, fermenting cucumber brines etc. (See also BREAD-MAKING.)

S. kluyveri. Cells (in malt extract): ca. 3.5–7.0 × 4.0–11.0 µm, occurring singly or in pairs or clusters; pseudohyphae are usually formed e.g. on cornmeal agar. Glucose, galactose and sucrose are fermented, maltose and lactose generally are not, although some strains can ferment maltose weakly. All strains are sensitive to cycloheximide (100 ppm). The organisms closely resemble *S. cerevisiae* but can be distinguished by their ability to grow on ethylamine, cadaverine or L-lysine as sole nitrogen source. Strains have been isolated from *Drosophila* spp, tree exudates, soil, and fruit juice.

S. lactis. Now *Kluyveromyces marxianus* var. *lactis*: see KLUYVEROMYCES.

S. rouxii. Now *Zygosaccharomyces rouxii* (q.v.).

S. servazzii. Cells (in malt extract): ca. 3.0–4.5 × 4.0–5.5 µm, occurring singly or in pairs. Glucose and galactose are fermented; sucrose, maltose and lactose are not. Growth can occur in the presence of cycloheximide (100 ppm or 1000 ppm). Strains have been isolated from soil.

S. telluris (anamorph: *Candida pintolopesii*). Cells (in malt extract): 3.5–6.5 × 4.5–8.5(–17.0) µm, occurring singly or in pairs, short chains or small clusters. Asci contain one, occasionally two, ascospores which are slightly rough to distinctly spiny. Glucose is fermented; galactose, sucrose, maltose and lactose are not. All strains are sensitive to cycloheximide (100 ppm). Strains have been isolated from poultry, pigeons, rodents and rodent faeces, soil, etc.

S. transvaalensis. Now *Pachytichospora transvaalensis*.

S. unisporus (= *S. mongolicus*). Cells (in malt extract): ca. 2.5–4.5 × 3.0–6.0 µm, occurring singly or in pairs. Asci contain one, occasionally two, smooth ascospores. Glucose and galactose are fermented; sucrose, maltose and lactose are not. Growth occurs in the presence of cycloheximide (100 ppm and 1000 ppm). Strains have been isolated from e.g. cheese and KEFIR.

S. uvarum. See *S. cerevisiae*.

Saccharomycetaceae A family of fungi (order ENDOMYCETALES) which are predominantly unicellular, the cells being spherical, ovoid, cylindrical etc. (See YEASTS; see also CELL WALL.) The cells may occur singly, in pairs, or in short chains, and reproduce asexually by BUDDING, by BUD FISSION, or by FISSION. Many members can form pseudomycelium, at least under certain conditions, and some can form a true mycelium which may give rise to arthrospores or blastospores. The vegetative cells may be predominantly haploid or predominantly diploid. According to species, sexual reproduction may involve conjugation between two somatic cells, between two ascospores, between a bud and its mother cell, etc; species may be homothallic or heterothallic (see also MATING TYPE). An ascus develops meiotically either directly from the zygote or from a vegetative cell derived mitotically from the zygote. Asci typically contain 1–4 (occasionally 8) ascospores (those of *Kluyveromyces polysporus* each contain hundreds of spores); ascospores are e.g. spherical, ovoid, hemispherical, Saturn-shaped, bowler-hat-shaped, etc, but are rarely fusiform and never acicular (cf. METSCHNIKOWIACEAE). Asci may be persistent (i.e. spores are not liberated but germinate within the ascus) or evanescent (ascospores are liberated by partial or complete dissolution of the ascus wall). Most members of the family can carry out fermentative as well as respiratory metabolism of glucose and certain other sugars; a few members are non-fermentative. Genera are distinguished on the basis of mode of sexual reproduction, nature of the ascospores, ability to assimilate nitrate, ability to ferment particular sugars, ability to assimilate various carbon compounds, coenzyme Q content, etc. They include: AMBROSIOZYMA, ARTHROASCUS, CITEROMYCES, CLAVISPORA, CYNICLOMYCES, DEBARYOMYCES, DEKKERA, GUILLIERMONDELLA, HANSENIASPORA, HANSENULA, ISSATCHENKIA, KLUYVEROMYCES, LIPOMYCES, LODDEROMYCES, NADSONIA, PACHYSOLEN, *Pachytichospora*, PICHIA, SACCHAROMYCES, SACCHAROMYCODES, SACCHAROMYCOPSIS, SCHIZOSACCHAROMYCES, SCHWANNIOMYCES, SPOROPACHYDERMIA, STEPHANOASCUS, TORULASPORA, WICKERHAMIA, WICKERHAMIELLA, WINGEA, *Yarrowia*, ZYGOSACCHAROMYCES.

Members of the Saccharomycetaceae occur in a wide range of habitats, e.g. on fruits, in tree exudates, in the tunnels of wood-boring beetles, in soil, in faeces, etc. Apparently, no species is significantly pathogenic in either plants or animals, but some are important spoilage organisms (see e.g. BEER SPOILAGE and WINE SPOILAGE). Some species have important commercial uses – e.g. for fermenting foods and beverages (see e.g. BREAD-MAKING, BREWING, SOY SAUCE, WINE-MAKING), as foods (see e.g. SINGLE-CELL PROTEIN and YEAST EXTRACT), for the production of certain enzymes (see e.g. INVERTASE), for carrying out biochemical conversions (see e.g. STEROID BIOCONVERSIONS), etc.

Saccharomycodes A genus of yeasts (family SACCHAROMYCETACEAE). Cells are usually lemon-shaped or elongated (ca. 4–7 × 8–23 μm) and reproduce by bipolar budding on a broad base (bud-fission); pseudomycelium is absent or poorly developed. An ascus develops from a (diploid) somatic cell which has undergone meiosis. Asci contain 4 smooth, spherical spores occurring in pairs; on germination, the members of each pair may conjugate, each fused pair giving rise to a 'sprout mycelium' from which vegetative cells arise by budding. (In some strains germination may occur without conjugation.) Glucose, sucrose and raffinose are fermented; galactose, lactose and maltose are not. NO_3^- is not assimilated. One species: *S. ludwigii*, isolated e.g. from grape juice, wine, slime flux of trees. [Book ref. 100, pp. 396–398.]

Saccharomycopsis A genus of fungi (family SACCHAROMYCETACEAE) which form yeast cells, pseudomycelium, and abundant true mycelium with blastospores; arthrospores may be formed. Hyphal septa have plasmodesmata. According to species, ascus formation is preceded by conjugation between yeast cells, between hyphal cells, and/or between blastospores; ascospores: spherical, bowler-hat-shaped, or Saturn-shaped, smooth or warty, 1–4 per ascus. Some species can ferment e.g. glucose, others are non-fermentative; NO_3^- is not assimilated. Species include *S. capsularis* (type species; formerly e.g. *Endomycopsis capsularis*), *S. crataegensis*, *S. fibuligera* (formerly e.g. *Endomycopsis fibuligera*). *S. malanga*, *S. synnaedendra* and *S. vini* (formerly *Endomycopsis vini*). Species occur e.g. on fruit, foods, soil, and in tunnels of wood-boring beetles. [Book ref. 100, pp. 399–413.]

'*S. lipolytica*' (formerly e.g. *Endomycopsis lipolytica*, anamorph: *Candida lipolytica*) lacks plasmodesmata in its septa and was classified in a new genus: *Yarrowia* [AvL (1980) *46* 517–521].

(See also SINGLE-CELL PROTEIN.)

saccharopine See Appendix IV(e).

Saccharopolyspora A genus of bacteria (order ACTINOMYCETALES, wall type IV) isolated e.g. from sugar-cane bagasse. The organisms form a substrate mycelium which fragments into rod-shaped forms, and sparse white aerial hyphae which give rise to chains of rounded spores. Mycolic acids are not formed. GC%: ca. 77. Type species: *S. hirsuta*. [Book ref. 73, p. 128.]

saccharose *Syn.* SUCROSE.

saccoderm desmid See DESMIDS.

sacculus Any small, sac-like structure – e.g. the 'murein sacculus' in bacteria (see PEPTIDOGLYCAN).

sacrificial cell *Syn.* NECRIDIUM.

***sad* mutants** Mutants defective in ribosomal subunit assembly.

saddle-back fungus *Polyporus squamosus* (q.v.).

saddle fungi See HELVELLA.

SAF SCRAPIE-associated fibril.

safety cabinet ('sterile cabinet') A cabinet within which microbiological work is carried out so as to prevent contamination (by airborne microorganisms) of (a) cultures etc, and/or (b) personnel and the environment. (See also AEROSOL.) *Class I cabinet*. Air is sucked into the cabinet and passes out via a filter. Work is conducted through an open front panel or by means of arm-length rubber gloves fitted into holes in the panel. The operator and environment are given some protection, but cultures etc are not. *Class II cabinet*. Sterile (filtered) air enters the cabinet and passes downward onto the work surface; air (some of which may be re-circulated) passes to the exterior via an exit port (situated below the work surface) and is filtered before discharge to the environment. Work is conducted via an open front panel; the downward, laminar air-flow inside the cabinet, and the entry of air via the front panel, together form a moving 'curtain' of air which helps to protect the user, the cultures etc. and the environment. *Class III cabinet*. Sterile, filtered air enters the *gas-tight* cabinet, and air is filtered before discharge; a negative pressure is maintained inside during use. Work is conducted via arm-length rubber gloves fitted into the front panel. Access to the interior of the cabinet is via a separate two-door sterilizing/disinfecting chamber. [Book ref. 2, pp. 495–499.]

safranin O A red basic composite azine DYE.

SAG Sammlung von Algenkulturen (algal culture collection), Pflanzenphysiologisches, Universität Göttingen, Göttingen, Germany.

SAg SUPERANTIGEN.

sagenogen See LABYRINTHULAS and THRAUSTOCHYTRIDS.

sagenogenetosome See LABYRINTHULAS and THRAUSTOCHYTRIDS.

Sagiyama virus See ALPHAVIRUS.

saguaro cactus virus See TOMBUSVIRUSES.

SAIDS SIMIAN AIDS.

saimiriine herpesvirus See e.g. GAMMAHERPESVIRINAE ('*Herpesvirus saimiri*').

Saint Anthony's fire (1) *Syn.* ERGOTISM. (2) *Syn.* ERYSIPELAS.

Saint Louis encephalitis See ST LOUIS ENCEPHALITIS.

Saint Vitus's dance See RHEUMATIC FEVER.

sakacin A A 41-amino-acid, pore-forming class IIa BACTERIOCIN produced by *Lactobacillus sake*; synthesis is reported to be temperature-dependent and regulated by a three-component system [Microbiology (2000) *146* 2155–2160].

saké Japanese rice wine. Rice starch is degraded to fermentable sugars by *Aspergillus oryzae* in a steamed rice KOJI; this is inoculated with '*Saccharomyces sake*', a strain of *S. cerevisiae* with a high alcohol tolerance. A succession of organisms follows *A. oryzae*: nitrate-reducing bacteria, followed by lactic acid bacteria, followed by yeasts – '*S. sake*' finally becoming dominant [Book ref. 6, pp. 417–419]. The alcohol content reaches 20% v/v or higher.

Saksenaea See MUCORALES.

Saksenaeaceae See MUCORALES.

SAL plasmid An IncP-9 *Pseudomonas* PLASMID (ca. 85 kb) which encodes the capacity to metabolize salicylate. (cf. TOL PLASMID.)

***Sal*I** A RESTRICTION ENDONUCLEASE from *Streptomyces albus*; G/TCGAC.

salami Salami and salami-type sausages are manufactured by the fermentation of comminuted meat (usually salted and spiced) by (mainly) LACTIC ACID BACTERIA; traditional methods rely on the microflora of the meat, equipment, environment etc, but commercial starter cultures may be used. The product is extremely resistant to microbial spoilage owing to its low pH (usually ca. 4.8–5.2) and low a_w (usually ca. 0.7–0.8); however, it may be susceptible to spoilage by moulds and xerophilic yeasts. The lactic acid bacteria contribute to the preservation of salami by lowering the pH (by LACTIC ACID FERMENTATION of carbohydrates) and e.g. by the production of antibiotics (see FOOD PRESERVATION (e)); they also contribute to the flavour and texture of the product. Some salami-type sausages are mould-ripened, usually with *Penicillium* spp.

salicin 2-Hydroxymethylphenyl-β-D-glucopyranoside. It can be hydrolysed by β-glucosidase to D-glucose and saligenin (2-hydroxymethylphenol).

salicylanilides (as antimicrobial agents) Salicylanilide (2-hydroxy-*N*-phenylbenzamide) and its halogenated derivatives have antifungal and antibacterial activity and are used e.g. as antiseptics and industrial PRESERVATIVES. Salicylanilide itself (*Shirlan*) is used e.g. as an antifungal agent in textiles and wood pulps and for the antifungal treatment of walls and ceilings; it has also been used for the treatment of cutaneous mycoses.

salicylic acid (2-hydroxybenzoic acid) An antimicrobial, mildly keratinolytic ORGANIC ACID used for the topical treatment of e.g. ACNE and (with other agents) superficial mycoses. (cf. SALICYLANILIDES.)

saligenin See SALICIN.

saline agglutination *Syn.* AUTO-AGGLUTINATION.

salinomycin An ionophore antibiotic, produced by *Streptomyces albus*, which is used as an anti-coccidial agent in poultry and as a FEED ADDITIVE for e.g. ruminants and pigs.

Salivaria One of two groups of subgenera of mammalian trypanosomes; salivarian trypanosomes infective for the vertebrate host develop in the ANTERIOR STATION. The group comprises DUTTONELLA, NANNOMONAS, PYCNOMONAS and TRYPANOZOON. (cf. STERCORARIA.)

Salk vaccine An anti-poliomyelitis vaccine containing formalin-inactivated strains of three types of poliovirus; it is administered intramuscularly. (cf. SABIN VACCINE.)

salmon disease See ULCERATIVE DERMAL NECROSIS.

salmon poisoning (in dogs) See NEORICKETTSIA.

Salmonella A genus of Gram-negative bacteria of the ENTEROBACTERIACEAE (q.v. for general features). Cells: straight rods, ca. 0.5–1.0 × 1–5 μm; most strains are motile. Reactions for most salmonellae: indole −ve; MR +ve; VP −ve; citrate +ve; acid and gas from glucose at 37°C; lactose usually −ve; H_2S +ve (in TSI); urease −ve; lysine decarboxylase +ve; ornithine decarboxylase +ve (cf. serotypes listed below). Reactions may be altered e.g. by the presence of metabolic plasmids (e.g. a plasmid carrying genes for lactose utilization [FEMS (1983) *17* 127–130]). Salmonellae can grow e.g. on NUTRIENT AGAR, MACCONKEY'S AGAR, and DCA, but usually not in KCN media. Enrichment media: e.g. MÜLLER–KAUFFMANN BROTH, RAPPAPORT–VASSILIADIS BROTH, SELENITE BROTH, TETRATHIONATE BROTH. Selective media: e.g. BRILLIANT GREEN AGAR, HEKTOEN MEDIUM, SS AGAR. GC%: 50–52. Type species: *S. choleraesuis*.

Note on the pronunciation of 'Salmonella'. This genus was named after the American bacteriologist D. E. Salmon; the correct pronunciation of the genus name is therefore 'Salmon-ella', i.e. with the first 'l' silent. Vocalization of the first 'l' in *Salmonella* (as is frequently done e.g. by television newsreaders) is equivalent to vocalization of the 'p' in 'cupboard'.

Salmonellae occur mainly in the intestines of man and animals, where they can be important pathogens (see SALMONELLOSIS and FOOD POISONING (e)); at least some strains are toxigenic. In a number of salmonellae, virulence factors are encoded by genes on PATHOGENICITY ISLANDS, including e.g. type III secretion systems (see PROTEIN SECRETION) specified by the *inv-spa* genes; some pathogenicity islands may contain genes that specifically encode *entero*pathogenicity [e.g. Mol. Microbiol. (1998) *29* 883–891].

Some salmonellae are apparently host-specific ('host-adapted') (e.g. *S. typhi* in man, *S. abortusovis* in sheep), but most have a wider host range.

Salmonellae from faeces or sewage can survive for weeks in water and for months in soil under favourable conditions. Salmonellae are normally killed by routine chlorination of WATER SUPPLIES, by PASTEURIZATION, and by many common DISINFECTANTS.

Nomenclature of the salmonellae does not follow the rules of the international Code of Nomenclature of Bacteria in that binomials usually refer to *serotypes* rather than to species. Salmonellae were originally named according to the disease (or host) with which they were associated – e.g. *S. typhi* for typhoid, *S. abortusovis* for strains causing abortion in sheep. Subsequently, serotypes have been named according to the place where they were first isolated (e.g. *S. london*, *S. dublin*). Each serotype is also given an ANTIGENIC FORMULA and classified by the KAUFFMANN–WHITE CLASSIFICATION scheme. (See also SMOOTH–ROUGH VARIATION.)

The genus *Salmonella* is sometimes divided into five subgenera (I–V) on the basis of biochemical tests. Subgenus I (sometimes regarded as a species, '*S. kauffmannii*') contains most of the named serotypes isolated from man and warm-blooded animals; members of subgenera II ('*S. salamae*'), III ('*S. arizonae*'), IV ('*S. houtenae*') and V ('*S. bongor*') are typically found in the intestines of poikilotherms (e.g. reptiles). In an alternative scheme, members of subgenus III – which (typically) can carry out a delayed fermentation of lactose – are regarded as a separate genus, *Arizona*, with one species, *A. hinshawii*; *S. typhi*, *S. choleraesuis* and *S. enteritidis* are given species status, and all remaining named serotypes are regarded as serotypes of *S. enteritidis*. In yet another scheme, the genus comprises a single species, *S. choleraesuis*, with six subspecies corresponding to the earlier subgenera I, II, III (monophasic), III (diphasic), IV, and V – named, respectively, subspecies *choleraesuis*, *salamae*, *arizonae*, *diarizonae*, *houtenae* and *bongori*. Serotypes may be further subdivided by biochemical, antibiotic, bacteriocin and bacteriophage TYPING.

More recently, it has been suggested that all the serotypes of *Salmonella* be considered as members of a single species, *S. enterica*. (The name of the type species, *S. choleraesuis* was not chosen as this was considered likely to cause confusion.) Subspecies of *S. enterica* are listed in the table [see also JMM (1992) *37* 361–363].

In this scheme, individual serotypes within subspecies I are referred to by *name*; for example: *S. enterica* subsp *enterica* serotype Typhimurium; *S. enterica* subsp *enterica* serotype Agona; *S. enterica* subsp *enterica* serotype Dublin etc. Less cumbersome forms of designation, suitable for routine use,

Salmonella enterica: designation of subspecies[a,b]

Subspecies name	Subspecies number
enterica	I
salamae	II
arizonae	III(a)
diarizonae	III(b)
houtenae	IV
bongori	V
indica	VI

[a] See text.

[b] Subspecies I (*enterica*) contains the majority of salmonellae pathogenic in man and warm-blooded animals.

would be e.g. *Salmonella* serotype Typhimurium, *Salmonella* Typhimurium, or simply Typhimurium etc. [see JMM (1992) *37* 361–363].

For serotypes of subspecies II–VI it has been suggested that individual serotypes be designated by a combination of (i) subspecies number and (ii) antigenic formula.

Over 2000 named serotypes have been listed in the Kauffmann–White classification scheme; the following (antigenic formulae in brackets) are some of medical and veterinary importance.

S. abortusovis (4,12:c:1,6). Host: sheep; causes abortion.

S. agona (1,4,12:f,g,s:-). Associated e.g. with fishmeal from South America. (See also FOOD POISONING.)

S. choleraesuis (6,7:c:1,5). Gas from glucose; H_2S −ve (var. *kunzendorf* is H_2S +ve); citrate (delayed) +ve. Isolated from pigs (see PARATYPHOID FEVER sense 2) but can be pathogenic e.g. in man.

S. dublin (1,9,12,[Vi]:g,p:-). Aerogenic, but anaerogenic variants are common. Isolated mainly from cattle (see PARATYPHOID FEVER (2)).

S. gallinarum (probably = *S. pullorum* 1,9,12:-:-). Non-motile. No gas from glucose; H_2S – ve; citrate −ve. Host: fowl (see FOWL TYPHOID).

S. hadar (6,8:z_{10}:e,n,x). Associated with turkeys. (See also FOOD POISONING.)

S. paratyphi. Gas from glucose. Host: man (see PARATYPHOID FEVER (1)). *S. paratyphi-A* (1,1,12:a: [1,5]): H_2S (usually) −ve; citrate −ve. *S. paratyphi-B* (= *S. schottmuelleri*) (1,4, [5], 12:b:1,2): H_2S+ve; citrate +ve; produces a slime layer on media containing 0.5% glucose and 0.2 M sodium phosphate (pH 7). *S. paratyphi-C* (= *S. hirschfeldii*) (6,7, [Vi]: c:1,5): H_2S+ve; citrate +ve.

S. pullorum (probably = *S. gallinarum*) (1,9,12:-:-). Non-motile. Gas from glucose. Host: fowl (see PULLORUM DISEASE and FOWL TYPHOID).

S. typhi (formerly *S. typhosa*) (9,12, [Vi]:d:-). Wild strains may have H antigen z_{66} instead of d. No gas from glucose; H_2S weakly +ve; citrate −ve; ornithine decarboxylase −ve. Host: man (see TYPHOID FEVER).

S. typhimurium (1,4,[5],12:i:1,2). Gas from glucose. Host range includes man and other animals (see also FOOD POISONING (e) and ENTERIC FEVER).

***Salmonella*/microsome assay** *Syn.* AMES TEST.

salmonella–shigella agar See SS AGAR.

salmonellosis Any of certain diseases, in man or animals, in which the causal agent is a serotype of *Salmonella*. In man, salmonellosis generally refers to *Salmonella* FOOD POISONING (cf. PARATYPHOID FEVER (1) and TYPHOID FEVER). For salmonellosis in animals see e.g. PARATYPHOID FEVER (2). (See also FOWL TYPHOID and PULLORUM DISEASE.)

Salpingella See TINTINNINA.

salpingitis Inflammation of a tube – usually the Fallopian tube of the uterus. [Chlamydial salpingitis: BMB (1983) *39* 145–150.]

salt flotation See FLOTATION.

saltant An organism which exhibits SALTATION.

saltation (dissociation) (*mycol.*) A phenotypic change in a thallus or colony (or part thereof) due e.g. to a mutation or to the segregation of nuclei in a heterokaryon.

saltatorial cilia In certain ciliates (e.g. *Halteria*): long cilia which are sparsely distributed on the cell and which are used to produce a characteristic darting motility.

salting (of food) See FOOD PRESERVATION (c).

salvage cycle (salvage pathway) See e.g. NUCLEOTIDE and NAD.

salvage pathway (complement fixation) See COMPLEMENT FIXATION (d).

salvarsan See ARSENIC.

same-sense mutation A POINT MUTATION in a structural gene which results in a mutant codon specifying the same amino acid as did the original codon (owing to the degeneracy of the GENETIC CODE). (cf. SILENT MUTATION.)

San Joaquin Valley fever *Syn.* COCCIDIOIDOMYCOSIS.

San Miguel sea lion virus (SMSV) A virus (family CALICIVIRIDAE) isolated from sea lions (in which it may cause abortion) and apparently widespread among marine mammals, fish etc in the Pacific. SMSV is very closely related to vesicular exanthema of swine virus, and can cause a disease similar to vesicular exanthema in pigs; it is only distantly related to FELINE CALICIVIRUS. SMSV can be propagated e.g. in Vero cells.

sand filter See WATER SUPPLIES.

sandfly fever group See PHLEBOVIRUS.

sandwich technique A type of indirect IMMUNOFLUORESCENCE technique used e.g. to demonstrate the presence of specific antibodies at the surface of antibody-forming cells. Essentially, a suitably prepared section or smear is incubated with antigen homologous to the given antibody, and subsequently washed free of uncombined antigen; the preparation is then exposed to a fluorescent antibody of the same specificity as that being sought in the specimen, washed, and examined by fluorescence MICROSCOPY. The antigen is thus sandwiched between homologous antibody in the specimen and homologous, fluorescent antibody.

Sanger's chain-termination method See DNA SEQUENCING.

sanitizer Any preparation with both cleansing and antimicrobial properties (e.g. TEGO COMPOUNDS).

sap-stain Discoloration of freshly felled (moist) timber due to growth of fungi primarily or solely in the sapwood; wood may become e.g. green, purple or blackish, but more often appears bluish-grey (see BLUE-STAIN). Sap-staining fungi include e.g. *Aureobasidium pullulans*, *Cladosporium herbarum*, and species of *Diplodia* and *Graphium*. (See also TIMBER PRESERVATION.)

sap transmission Transmission of a plant virus by a mechanical VECTOR.

saponins A diverse group of powerful glycoside surfactants produced by a wide range of plants. Saponins have antifungal properties which depend on their ability to affect membrane permeability by complexing with sterols in the CYTOPLASMIC MEMBRANE (see e.g. TOMATINE); certain bacteria (e.g. MYCOPLASMA spp) can be lysed by saponins. (See also AVENACIN, BIOSURFACTANTS and DIGITONIN.)

saprobe *Syn.* SAPROTROPH.

saprobity system The system by which an organically polluted river or stream undergoing SELF PURIFICATION may be divided into a number of distinct zones, each zone being characterized e.g. by its level of pollution, content of dissolved oxygen, and types of microorganism (indicator-organisms) present. The *polysaprobic zone* is the zone most heavily polluted with organic matter (sewage etc), i.e., the zone nearest the source of pollution. In this zone oxygen consumption is high (dissolved O_2 is minimal) and there is a vigorous production of ammonia and H_2S. The organisms found in this zone may include e.g. the flagellates *Bodo* and *Euglena*, the ciliate *Glaucoma scintillans*, and various bacteria such as *Beggiatoa*, *Thiothrix* and *Oscillatoria*. In the *α-mesosaprobic zone* MINERALIZATION and oxygen consumption continue vigorously, but the content of dissolved O_2 is higher than that in the polysaprobic zone; there is some in situ utilization of ammonia, and there is no free H_2S. Organisms in this zone may include e.g. *Oscillatoria*, and the ciliates

Carchesium, *Stentor* and *Vorticella*. In the *β-mesosaprobic zone* oxygen consumption is low since mineralization is almost complete; the dissolved O_2 levels are considerably higher than those in the α-mesosaprobic zone, and levels of free ammonia are low. Organisms in this zone may include e.g. the cyanobacterium '*Phormidium*' and the ciliates *Halteria grandinella* and *Vorticella striata*. In the *oligosaprobic zone* (= *xenosaprobic zone*) mineralization is complete and dissolved O_2 is abundant. Aerobic, photosynthetic organisms may predominate in this zone.

In rivers, flow rates and dilution factors are important factors which determine the existence or extent of the various zones.

Saprodinium A genus of protozoa (order ODONTOSTOMATIDA) in which the cells have posteriorly directed terminal spines and isolated tufts of somatic cilia; species occur e.g. in organically polluted and stagnant waters.

Saprolegnia A genus of fungi (order SAPROLEGNIALES) which occur in aquatic environments and soil; the organisms are typically saprotrophic, but some species are parasitic on fish (see SAPROLEGNIASIS). *Saprolegnia* spp appear to be commonly hermaphroditic and homothallic. (See also DIPLANETISM and GEMMA.) [*Saprolegnia diclina-parasitica* complex: CJB (1983) *61* 603–625.]

Saprolegniales An order of fungi (class OOMYCETES) which occur typically as saprotrophs in water and soil – though some species are parasitic in e.g. fish, fish eggs, invertebrates, algae, and higher plants. According to species, the thallus ranges from a single cell to a well-developed mycelium; hyphae, when formed, lack constrictions (cf. LEPTOMITALES). Asexual reproduction commonly involves the formation of long, cylindrical zoosporangia at the ends of hyphae. In e.g. *Saprolegnia*, once a zoosporangium has released its zoospores another zoosporangium may develop inside it; such zoosporangia remain attached to the hyphae which bear them. In e.g. *Dictyuchus*, zoosporangia are typically deciduous at maturity. In addition to zoosporangia, some species form *gemmae* (see GEMMA). Genera include ACHLYA, APHANOMYCES, ATKINSIELLA, *Calyptralegnia*, *Dictyuchus*, *Ectrogella*, *Leptolegnia* (*L. baltica* and *L. chapmanii* have been associated with mortality in copepods and mosquito larvae, respectively), and SAPROLEGNIA. (See also BRANCHIOMYCES.)

saprolegniasis A freshwater FISH DISEASE caused by *Saprolegnia* spp; any of a wide range of fish, both wild and cultivated, may be affected. In salmonids the pathogen is usually *S. diclina* Type 1 (formerly *S. parasitica*). Conspicuous fluffy white patches or rings develop on the skin, increasing skin permeability and leading to osmoregulatory imbalance which is usually rapidly fatal. Predisposing factors: other infections, wounding. The fish's main defence seems to be the removal of fungal spores and cysts by epidermal mucus secretion. [Book ref. 1, pp. 271–297.]

sapropel The anaerobic, putrefying matter – rich in H_2S – which forms the bottom mud of some aquatic habitats. (See also ANAEROBIC DIGESTION.)

saprophyte (1) A 'plant-like' SAPROTROPH: e.g., a saprotrophic bacterium or fungus. (2) *Syn.* SAPROTROPH. (Hence *adj.* saprophytic – cf. SAPROTROPHIC.)

Saprospira A genus of non-sheathed, helical, filamentous, pigmented, chemoorganotrophic GLIDING BACTERIA (see CYTOPHAGALES) which occur in aquatic habitats.

saprotroph (saprobe; saprovore) An organism which obtains its nutrients from non-living organic matter (commonly dead and decaying plant or animal matter) by absorbing soluble organic compounds; a saprotroph may produce extracellular enzymes (e.g., CELLULASES, XYLANASES) which release soluble components from insoluble materials. Saprotrophs are important agents of decay and MINERALIZATION, and are thus essential links in the cycles of matter (see e.g. CARBON CYCLE). Some saprotrophs are facultative PARASITES. (cf. SAPROPHYTE; SAPROZOITE.)

saprotrophic (saprobic) Refers to a SAPROTROPH or to the mode of nutrition of a saprotroph.

saprovore *Syn.* SAPROTROPH.

saprozoite An 'animal-like' SAPROTROPH: e.g., a saprotrophic protozoon. (Hence *adj.* saprozoic – cf. SAPROTROPHIC.)

saquinavir See ANTIRETROVIRAL AGENTS.

saramycetin A polypeptide antifungal antibiotic produced by *Streptomyces saraceticus*; it has been used e.g. against aspergillosis, blastomycosis and histoplasmosis.

Sarcina A genus of Gram-positive, asporogenous, anaerobic bacteria which occur e.g. in the gut in man and other animals. Cells: non-motile cocci (ca. 2–3 μm diam.) which typically occur in cubical packets. The type species, *S. ventriculi*, commonly has a CELLULOSE microcapsule. Metabolism is fermentative; the main products of glucose fermentation are acetic acid and ethanol (in *S. ventriculi*) or butyric and acetic acids (in *S. maxima*). GC%: ca. 29–31.

sarcina sickness See BEER SPOILAGE.

sarciniform (sarcinaeform) Arranged in PACKETS.

sarcinoid Refers to an alga in which the cells are more or less SARCINIFORM. (See e.g. FRIEDMANNIA and PSEUDOTREBOUXIA.)

Sarcinosporon A genus of imperfect, non-fermentative fungi in which both budding yeast cells and septate hyphae are formed; blastospores develop from the mycelium, and septation in different planes in the blastospores and hyphal cells results in the formation of sarciniform clusters of cells. Asexual endospores are formed. One species: *S. inkin*, isolated from a human skin lesion. [Book ref. 100, pp. 906–908.]

sarcocyst See SARCOCYSTIS.

Sarcocystis A genus of coccidian protozoa (suborder EIMERIORINA) which are obligately heteroxenous parasites in a range of animals, including man (see SARCOSPORIDIOSIS). The vegetative cells resemble those of TOXOPLASMA. In the intermediate host, ingestion of oocysts or sporocysts is followed by schizogony in internal organs, and merozoites later localize in muscle tissues; a merozoite becomes rounded to form a *metrocyte* which, by repeated division, gives rise to an immature *sarcocyst*: a collection of slowly-dividing metrocytes bounded by a cyst wall. Over a period of ca. 2 months the contents of the sarcocyst change from metrocytes only to a mixture of metrocytes and merozoites, and finally (in the mature sarcocyst) to merozoites only. Ingestion of sarcocyst-containing flesh by the final host is followed by sexual reproduction (only) in intestinal tissues: gametogony is followed by ENDOSPORULATION, resulting in disporic, tetrazoic oocysts, or free sporocysts, which are shed in faeces; the oocysts lack a MICROPYLE, and the sporocysts lack a STIEDA BODY. (cf. FRENKELIA.) [Taxonomy of *Sarcocystis*: J. Parasitol. (1986) *72* 372–382.]

sarcocystosis *Syn.* SARCOSPORIDIOSIS.

Sarcodina A subphylum of protozoa (phylum SARCOMASTIGOPHORA) in which the vegetative cells form one or more pseudopodia (see PSEUDOPODIUM) or are motile by cytoplasmic flow without the formation of discrete pseudopodia. Flagella are formed only in temporary stages of certain species. In many sarcodines the cytoplasm is more or less clearly differentiated into an inner, typically granular, *endoplasm* (= *granuloplasm* or *plasmasol*) and a surrounding, typically hyaline layer (*ectoplasm*, *hyaloplasm*, *plasmagel*) immediately beneath the plasmalemma. Sarcodine morphology ranges from the simple, naked, amoeboid cell of e.g. *Amoeba* spp to relatively complex forms with external tests (e.g. foraminifera) or internal skeletal structures (e.g. heliozoa, radiolaria). CYSTS are commonly formed. Asexual reproduction occurs by fission; sexual processes occur in some species. Sarcodines are commonly solitary, free-living organisms in aquatic habitats, soil, etc; some are parasitic. There are two superclasses: ACTINOPODA and RHIZOPODA.

sarcoidosis A chronic, granulomatous human disease which resembles disseminated tuberculosis, with lesions (rarely necrotic) in e.g. liver, lungs and spleen; other sites, including skin, salivary glands and eyes may also be affected. The disease is apparently self-limiting in the majority of cases.

Cause: unknown; *Propionibacterium* may be a better candidate than *Mycobacterium* [JCM (2002) *40* 198–204].

sarcoma A malignant tumour derived from connective tissue – e.g. bone (osteosarcoma), cartilage (chondrosarcoma), muscle (rhabdomyosarcoma, = rhabdosarcoma), fat cells (liposarcoma), lymphoid tissue (lymphosarcoma), etc. (cf. FIBROSARCOMA and FIBROMA.) Sarcomas may be induced by infection with certain viruses: see e.g. KAPOSI'S SARCOMA, MURINE SARCOMA VIRUSES, ROUS SARCOMA VIRUS, SIMIAN SARCOMA VIRUS.

Sarcomastigophora A phylum of PROTOZOA [JP (1980) *27* 37–58] which have flagella or pseudopodia or both; cells typically have a single type of nucleus. Subphyla: MASTIGOPHORA, OPALINATA and SARCODINA.

sarcomatrix See RADIOLARIA.

sarconeme A MICRONEME in *Sarcocystis*.

Sarcoscypha See PEZIZALES.

sarcosine See SARKOSYL.

sarcosporidiosis Any disease of man and other animals – e.g. cattle, pigs, sheep – caused by species of SARCOCYSTIS. In cattle and sheep the experimentally-induced disease can involve lesions in a wide range of tissues and can lead to abortion (cf. DALMENY DISEASE). In man the disease typically involves parasitization of the muscles, but experimentally-induced induced disease, involving ingestion of meat infected with *S. hominis*, leads to abdominal pain, diarrhoea, and the excretion of sporocysts. [AP (1982) *20* 352–396.]

sarcotoxins Antibacterial proteins produced in the haemolymph of larvae of the flesh fly (*Sarcophaga peregrinia*) either on exposure to bacteria or following mechanical injury; sarcotoxin I appears to affect cytoplasmic membrane functions in *Escherichia coli*. [Sarcotoxin II: Biochem. (1987) *26* 226–230.] (cf. CECROPINS.)

Sargassum See PHAEOPHYTA.

Sarkosyl An anionic detergent: *N*-lauroylsarcosine. (Sarcosine = *N*-methylaminoacetic acid.)

SARS Severe acute respiratory syndrome: a sometimes-fatal coronavirus infection [Lancet (2003) *362* 263–270; EID (2004) *10* (whole issue) 167–387].

Sartorya A genus of fungi (order EUROTIALES) which include the teleomorph of *Aspergillus fumigatus*.

satellite The posterior gregarine in SYZYGY. (cf. PRIMITE.)

satellite colonies See SATELLITE PHENOMENON.

satellite lesions Lesions remote from the original site of infection.

satellite phenomenon (satellitism) The phenomenon in which certain growth-factor-requiring organisms can grow on a medium which lacks the required growth factor but which is supporting the growth of another ('feeder') organism which can supply the required growth factor – a manifestation of SYNTROPHISM. For example, to determine the V factor requirement of a *Haemophilus* strain, blood agar (in which V FACTOR is unavailable) is inoculated with the *Haemophilus* strain, and the inoculated area is crossed with a single streak of *Staphylococcus aureus*; V factor from the growing staphylococci diffuses into the medium and allows the growth of *satellite colonies* of a V-factor-requiring *Haemophilus* strain in the immediate vicinity of the line of staphylococcal growth.

satellite RNA A small, linear ssRNA molecule which may be encapsidated within a specific plant virus and which is totally dependent on that ('helper') virus for its replication; unlike a VIRUSOID, a satellite RNA is not necessary for the replication of the helper virus. (cf. VIROID and SATELLITE VIRUS.) The presence of a satellite RNA may significantly affect both plant (altering symptoms) and virus (reducing yield) – see CUCUMOVIRUSES.

satellite virus A virus which is inherently entirely dependent on the presence of a particular 'helper virus' for its replication in a host cell. (cf. SATELLITE RNA.) Satellite viruses may be associated with certain plant viruses (see e.g. TOBACCO NECROSIS VIRUS), certain animal viruses (see e.g. DELTA VIRUS and DEPENDOVIRUS), or certain bacteriophages (see e.g. BACTERIOPHAGE P4). [Satellites of plant viruses: ARPpath. (1982) *20* 49–70.]

satellitism *Syn.* SATELLITE PHENOMENON.

satratoxins See STACHYBOTRYOTOXICOSIS.

satsuma dwarf virus See NEPOVIRUSES.

***Sau*3AI** See RESTRICTION ENDONUCLEASE (table).

sauerkraut Cabbage pickled by a natural lactic acid fermentation (see PICKLING). The cabbage is wilted, washed, shredded, salted (to ca. 2.25% w/w NaCl), and packed into tanks which are then covered; the temperature should be ca. 18–21°C. Early in the fermentation, *Leuconostoc mesenteroides* grows and produces lactic and acetic acids (which rapidly lower the pH and discourage the growth of spoilage organisms) and CO_2 (which helps to generate anaerobic conditions). Fermentation is then continued by *Lactobacillus brevis* and *L. plantarum*, the latter being responsible for the final high levels of acidity. The product may be marketed raw or canned. (cf. KIMCHI.)

Spoilage by pectinolytic bacteria may occur if salt levels are low. Salt levels above ca. 3% may allow the growth of yeasts – e.g. *Rhodotorula* spp may cause a pink discoloration.

Saug–Kappe process A method of TIMBER PRESERVATION, similar to the BOUCHERIE PROCESS, in which preservative is sucked into the sapwood of newly felled timber by application of a vacuum to one end of each log. (cf. BETHELL PROCESS.)

sausage (microbiological aspects) Raw, non-fermented sausage is susceptible to spoilage by typical MEAT SPOILAGE organisms; preservatives used include e.g. sodium metabisulphite. For fermented sausage see SALAMI.

Savillea See CHOANOFLAGELLIDA.

Savlon See CHLORHEXIDINE.

saxicolous Growing on and/or in rock or stone etc (cf. ENDOLITHIC and EPILITHIC).

saxitoxin A potent heterocyclic (nitrogen-containing) neurotoxin produced e.g. by the dinoflagellates *Gonyaulax catenella* and *Gonyaulax tamarensis* (RED TIDE organisms) and apparently also by the cyanobacterium *Aphanizomenon flos-aquae*. In warm-blooded animals, saxitoxin blocks the transmission of nerve impulses, causing paralysis and eventually death by asphyxiation; poikilotherms are generally less sensitive to saxitoxin, while many filter-feeding bivalve molluscs are apparently insensitive, accumulating the toxin in their tissues (see *paralytic* SHELLFISH POISONING). (Saxitoxin was originally isolated from the Alaskan butter clam, *Saxidomus giganteus*, a frequent cause of PSP.) Several toxins structurally related to saxitoxin have been isolated from *Gonyaulax* and other dinoflagellates: e.g. *neosaxitoxin* (a hydroxylated derivative of saxitoxin) and the

GONYAUTOXINS. [Biosynthesis of saxitoxin and related toxins: PAC (1986) *58* 257–262.] (See also PHYCOTOXIN.)

***sbcA* gene** See RECF PATHWAY.

***sbcB* gene** See RECF PATHWAY.

SBE Subacute bacterial ENDOCARDITIS.

***Sca*I** See RESTRICTION ENDONUCLEASE (table).

scab (1) (*plant pathol.*) Any of a wide range of unrelated plant diseases: see e.g. APPLE SCAB, POTATO SCAB, WHEAT SCAB. (2) (*vet.*) See e.g. SHEEP SCAB.

scabby mouth *Syn.* CONTAGIOUS PUSTULAR DERMATITIS sense 1.

scabrous (scabrid) With a rough surface.

scaffolding protein (1) A protein – or complex of proteins – that forms a 'scaffolding' (i.e., a temporary structural framework) for the assembly of e.g. a bacteriophage head, but which is absent from the mature head. The head precursor (prohead or procapsid) is assembled on the scaffolding protein; the latter is then eliminated and either re-cycled (e.g. in phages P22 and ϕ29) or degraded (e.g. in phages λ and T4).

(2) (scaffold) See CHROMOSOME (b).

scalariform Ladder-shaped. *Scalariform conjugation*: see SPIROGYRA.

scald (1) (benign foot-rot) (of sheep) A mild form of FOOT-ROT caused by a weakly virulent strain of *Bacteroides nodosus*; the lesion is odourless, and severe lameness does not occur.

(2) General term for any of various plant diseases (see e.g. LEAF BLOTCH).

scalded skin syndrome (Lyell's syndrome) A rare exfoliative disease which occurs mainly in children but which can affect adults; in neonates it may be called *Ritter's disease* or *pemphigus neonatorum*. The disease is caused by strains of *Staphylococcus aureus* (primarily of phage group II) which produce an exfoliative toxin (see EPIDERMOLYTIC TOXIN). The characteristic feature is a generalized exfoliative dermatitis, but there are manifestations of the disease which occur in only some patients – e.g. an erythematous rash similar to that seen in scarlet fever. Outer layers of epidermis separate at the stratum granulosum from the underlying tissue, and sterile bullae form quickly. Healing generally occurs within a few weeks. In this disease, the toxin may be acting as a serine protease in the epidermis and/or as a SUPERANTIGEN – the exfoliative toxin ETA (see EPIDERMOLYTIC TOXIN) having known ability to stimulate particular subsets of T cells. Treatment can include β-lactam antibiotics or (if resistant) vancomycin. (See also TOXIC EPIDERMAL NECROLYSIS.)

[Scalded skin syndrome (review): RMM (1998) *9* 9–15.]

scales (of an agaric) See UNIVERSAL VEIL.

scanning electron microscopy See ELECTRON MICROSCOPY (b).

scar (*mycol.*) In certain fungi: a differentiated region of the cell wall marking the location at which cell separation occurred during BUDDING or FISSION. In e.g. *Saccharomyces cerevisiae* two types of scar can be distinguished: the *bud scar* on the parent cell, and the *birth scar* on the daughter cell. If cells of *S. cerevisiae* are stained with primulin and examined by fluorescence microscopy, the bud scar appears as a ring of bright fluorescence on the non-fluorescent (or weakly fluorescent) cell wall; the ability of the bud scar to take up primulin may reflect a local change in the microfibrillar structure of the cell wall. In the multipolar budding yeasts, one, several or many bud scars may occur on the wall of a given cell, and in *S. cerevisiae* the scar rings appear never to overlap. In *S. cerevisiae* the region of the bud scar seems to contain CHITIN in concentrations greater than those found elsewhere in the cell wall.

scarlatina *Syn.* SCARLET FEVER.

scarlet caterpillar fungus See CORDYCEPS.

scarlet fever (scarlatina) An acute infectious human disease which affects mainly children; it is caused by erythrogenic toxin-producing strains of *Streptococcus pyogenes*. Transmission may occur by direct contact, by droplet infection, or via contaminated milk etc; infection usually occurs via nose or mouth, occasionally via wounds or via the birth canal ('puerperal scarlet fever'). Incubation period: 2–5 days. Symptoms: e.g. sore throat, fever, swelling of the cervical lymph nodes, and a characteristic bright-red punctate rash on the body followed by desquamation. The tongue often develops a greyish coating through which the swollen red papillae can be seen ('strawberry tongue'). The disease ranges from mild to severe or occasionally fatal. Complications: e.g. RHEUMATIC FEVER. *Chemotherapy*: e.g. penicillins, erythromycin. (See also DICK TEST; SCHULZ–CHARLTON TEST.)

Non-streptococcal scarlatiniform rash (in which e.g. 'strawberry tongue' does not occur) is caused e.g. by staphylococci.

SCAT SHEEP CELL AGGLUTINATION TEST.

Scatchard plot A graph which can be used to analyse molecular interactions in a mixed population of molecules in which molecules of different types bind reversibly to one another. In serology, for example, a Scatchard plot can indicate (i) the association constant (= AFFINITY) of antigen–antibody binding (given by the slope of the curve); (ii) the nature of the antibody population: MONOCLONAL ANTIBODIES give a straight-line plot, while POLYCLONAL ANTISERUM gives a curved plot; and (iii) the number of binding sites per molecule.

Scedosporium See PSEUDALLESCHERIA and HYPHOMYCETES.

Scenedesmus A genus of non-motile, coenobial green algae (division CHLOROPHYTA) which occur – often abundantly – in a wide range of freshwater habitats. The coenobia each consist of 2, 4, 8, or occasionally 16 elongate or fusiform cells in a palisade arrangement (i.e., side-by-side). In some species the two outermost cells bear a spine or tuft of bristles at each corner of the free side of the cell; however, morphology may vary considerably according to environmental conditions. Each cell contains one nucleus and one chloroplast which usually has a single conspicuous pyrenoid. Asexual reproduction occurs by the formation of a complete daughter coenobium within each cell of a parent coenobium; the daughter coenobia are released – often explosively – by rupture of the parent cell wall. Biflagellate zoospores may also be formed, at least in some species. Sexual reproduction involves the formation of biflagellate isogametes.

Schaeffer–Fulton stain See ENDOSPORE STAINS.

Schardinger dextrins (cycloamyloses or cyclodextrins) Cyclic DEXTRINS formed from starch or glycogen by an extracellular enzyme, cyclodextrin glucosyltransferase, from *Bacillus macerans* (a similar enzyme is produced e.g. by *B. megaterium*); α-, β- and γ-cyclodextrins contain 6, 7 and 8 $(1 \rightarrow 4)$-α-linked D-glucose residues, respectively.

Schaudinn's fluid A FIXATIVE: mercuric chloride (100 ml saturated, aqueous) and 95% ethanol (50 ml) supplemented immediately before use with glacial acetic acid (1.5 ml).

scheduled diseases NOTIFIABLE DISEASES of animals.

Schellackia See EIMERIORINA.

Schick test A test which determines a person's susceptibility to DIPHTHERIA. Diphtheria toxin is injected intradermally into one arm; as a control, inactivated toxin is injected into the other arm. A *positive* test (person *susceptible*) is indicated, after several days, by an inflammatory response at the site of injection of the potent toxin. No reaction (*negative* test) indicates the presence of antitoxin and hence resistance to the disease.

Schiff's reagent Basic FUCHSIN decolorized with sulphurous acid; it reacts with aldehydes to give a red-purple colour. (See FEULGEN REACTION and PAS REACTION.)

Schizoblastosporion A genus of non-fermentative yeasts (class HYPHOMYCETES). The cells are ellipsoidal to cylindrical; they reproduce by bipolar budding on a broad base. Pseudomycelium may be formed. One species: *S. starkeyi-henricii*, isolated from soil. [Book ref. 100, pp. 909–910.]

Schizochytrium See THRAUSTOCHYTRIDS.

schizogenous (schizogenic; schizogenetic) Formed by splitting, cracking, or fission.

schizogony (merogony) Asexual reproduction in which the nucleus of a cell undergoes division several or many times, resulting in a multinucleate *schizont* which subsequently gives rise to a number of uninucleate cells (*merozoites*).

schizogregarines See GREGARINASINA.

schizolysis See CONIDIUM.

Schizomycetes Obsolete class for the BACTERIA.

schizont See SCHIZOGONY.

schizonticide Any agent which can kill a schizont. For example, most ANTIMALARIAL AGENTS can kill the intraerythrocytic (blood schizont) stage of *Plasmodium*, while a few (primaquine, proguanil) can kill tissue schizonts.

schizophrenia See RETROVIRIDAE.

Schizophyceae Obsolete class for the CYANOBACTERIA.

Schizophyllaceae See APHYLLOPHORALES.

Schizophyllum A genus of fungi of the APHYLLOPHORALES (family Schizophyllaceae) which form non-stipitate, fan-shaped or lobed, bracket-type fruiting bodies. In *S. commune* the upper surface of the fruiting body is grey-brown and finely hairy, and the underside bears a number of pale, lamella-like, longitudinally grooved ridges ('split gills') which radiate out from the point of attachment of the fruiting body; the hymenium lines the outer surfaces of each split gill. Basidiospores: elongated, colourless, ca. 6 × 3 μm. *S. commune* grows on various types of wood; it has also been isolated as a human pathogen: e.g. in a case of maxillary sinus infection [JCM (1986) *23* 1001–1005]. (See also CHOLESTEROL OXIDASE and COMPATIBILITY sense 2.)

Schizophyta Obsolete division for the PROKARYOTES.

Schizoplasmodium See PROTOSTELIOMYCETES.

Schizopyrenida An order of naked, roughly cylindrical, typically uninucleate amoebae (class LOBOSEA) which are motile by the more or less eruptive formation of (usually) a single hemispherical 'bulge'. Many members can form flagellate stages. Genera include NAEGLERIA, TETRAMITUS and VAHLKAMPFIA.

Schizosaccharomyces A genus of fungi (family SACCHAROMYCETACEAE) which form single globose to cylindrical cells that reproduce by FISSION, or true hyphae that tend to fragment into arthrospores. (See also CELL WALL.) The organisms are homothallic. Sexual reproduction typically involves the fusion of (haploid) somatic cells and the development of a 4-, 6- or 8-spored ascus direct from the zygote; asci are evanescent. All species ferment glucose but not galactose or lactose. Nitrate is not assimilated. Species (*S. japonicus, S. malidevorans, S. octosporus, S. pombe*) are distinguished on the basis of e.g. number of spores per ascus, ability to ferment particular sugars, etc [Book ref. 100, pp. 414–422]; they occur e.g. in fruit-juices, molasses, fermented beverages, etc. (See also WINE-MAKING.)

Schizothrix See LPP GROUP.

Schizotrypanum A subgenus of TRYPANOSOMA, within the STERCORARIA, which includes the causal agent of CHAGAS' DISEASE, *T. (S.) cruzi*; two biochemically distinct subspecies are recognized: *T. (S.) cruzi cruzi* (in man) and *T. (S.) cruzi marinkellei* (in bats). These organisms occur in the Americas in the approximate belt 40°N–40°S.

In man, *T. (S.) c. cruzi* occurs in the blood as non-dividing trypomastigotes, typically ca. 20 μm in length (range 12–30 μm), in which there is a large kinetoplast situated close to the pointed posterior end of the cell, and a free flagellum; intracellular amastigote and epimastigote forms occur in pseudocysts. In the vector, epimastigotes, sphaeromastigotes and metacyclic forms occur in the gut and rectum. [Cell biology of *T. cruzi*: Int. Rev. Cytol. (1984) *86* 197–283.] *T. (S.) c. cruzi* can be cultured in e.g. NNN MEDIUM. [A minimal medium for the cultivation of infective *T. cruzi* epimastigotes: JGM (1983) *129* 285–291; improved method for purification of metacyclics from the insect vector *Triatoma infestans*: JP (1986) *33* 132–134.]

schlepper (*immunol.*) Any substance which, by combining with a second substance, promotes the immunogenicity of the latter.

schlieren system An optical system for measuring a refractive index *gradient*.

Schmutzdecke The film or layer of microorganisms which develops e.g. in a slow sand filter (see WATER SUPPLIES); it usually contains algae, bacteria and protozoa.

Schüffner's dots In erythrocytes infected with *Plasmodium vivax* or *P. ovale*: numerous fine dots observed on staining with Romanowsky stains. (cf. ZIEMANN'S DOTS.)

Schulz–Charlton test A test sometimes used in the diagnosis of SCARLET FEVER. Antibody to the streptococcal erythrogenic toxin is injected into the skin of a suspected case; in a positive test (scarlet fever present) there is a localized blanching of the rash after 6–12 hours.

Schwagerina See FORAMINIFERIDA.

Schwanniomyces A genus of yeasts (family SACCHAROMYCETACEAE) in which the (haploid) vegetative cells are spheroidal or ovoid (occasionally elongate to cylindrical) and reproduce by multilateral budding; pseudomycelium is absent or rudimentary. Ascospores: globose, with a warty surface, an equatorial ridge, and a conspicuous lipid globule. Glucose and e.g. sucrose are fermented; NO_3^- is not assimilated. One species: *S. occidentalis*; two varieties: var. *occidentalis* (which can assimilate xylose and *n*-alkanes) and var. *persoonii* (which cannot assimilate xylose or *n*-alkanes). Strains have been isolated from soil. [Book ref. 100, pp. 423–426.]

Schwartzman reaction (Schwartzmann reaction) Common misspellings for SHWARTZMAN REACTION.

sciaphilic *Syn.* SKIAPHILIC.

SCID See SEVERE COMBINED IMMUNODEFICIENCY.

SCIDXl See SEVERE COMBINED IMMUNODEFICIENCY.

scintillon See BIOLUMINESCENCE.

scirpene See TRICHOTHECENES.

Sclerocystis See ENDOGONALES.

Scleroderma See SCLERODERMATALES.

Sclerodermatales An order of saprotrophic (humicolous or lignicolous) or mycorrhizal fungi (class GASTEROMYCETES) which typically form globose or reniform, non-stipitate basidiocarps ('earth balls'), ca. several centimetres across, in which the basidia are scattered throughout the gleba (i.e., there is no distinct hymenium) and in which, in most species, the gleba forms a dark, powdery mass at maturity; the release of basidiospores usually depends on the rupture of the characteristically tough peridium. Genera include *Astraeus*, PISOLITHUS, *Scleroderma* and SPHAEROBOLUS.

Sclerophthora See PERONOSPORALES.

Sclerospora A genus of fungi (order PERONOSPORALES) which include organisms parasitic on members of the Gramineae (see e.g. PHYLLODY). The lemon-shaped zoosporangia are borne on sporangiophores which have multibranched ends.

sclerotan A (1 → 3)-β-D-glucan found in *Sclerotinia* spp.

sclerotic bodies See CHROMOBLASTOMYCOSIS.

Sclerotinia A genus of fungi (order HELOTIALES) which include various plant-parasitic species – e.g. *S. fructigena* (see BROWN ROT sense 2), *S. homeocarpa* (see DOLLAR SPOT), and *S. trifoliorum* (see CLOVER ROT). *Sclerotinia* can sometimes cause serious loss in oilseed rape (*Brassica napus*); symptoms: whiteheading, white stem lesions, and black sclerotia within the stem cavity.

Sclerotinia spp form stipitate apothecia directly from sclerotia; the ascospores are typically ellipsoidal.

sclerotinia crown and stem rot See CLOVER ROT.

sclerotium (1) Of certain phaneroplasmodial MYXOMYCETES: a dormant, resistant structure formed by the conversion of the plasmodium into a hardened, irregularly shaped mass. In e.g. *Physarum polycephalum* the sclerotium is a thick-walled, yellowish mass which is divided internally into a number of (usually) multinucleate sections called *spherules*; the walls of the spherules apparently contain a rare polysaccharide, a polymer of galactosamine (found also in the spore walls in this species). Sclerotia may be formed in response to adverse environmental conditions (e.g. desiccation, starvation, low temperatures), and in some species the sclerotium is the overwintering form of the organism; under favourable conditions the sclerotium germinates to give rise to a vegetative plasmodium.

(2) (*mycol.*) A resting form produced by certain fungi; sclerotia are hard, resistant, plectenchymatous bodies which, under favourable conditions, may give rise to mycelium or to sexual or asexual fruiting bodies. The sclerotium of e.g. *Claviceps purpurea* (see ERGOT) is a black, elongated body ca. 1 cm in length. (See also BLACKFELLOW'S BREAD.)

Sclerotium A genus of fungi (order AGONOMYCETALES) which include some important plant pathogens (see e.g. ONION ROT and FOOT-ROT sense 2). *S. rolfsii* causes disease in various host plants, particularly in the tropics and subtropics. [Biology, ecology and control of *S. rolfsii*: ARPpath. (1985) *23* 97–127.] Some species have a teleomorph in the Ascomycotina or the Basidiomycotina.

scolecosporae See SACCARDOAN SYSTEM.

Scolytidae A family of wood- or bark-boring beetles, many of which are associated (specifically or non-specifically) with fungi: see e.g. AMBROSIA FUNGI and DUTCH ELM DISEASE.

scombroid poisoning See FISH SPOILAGE.

scopoletin 6-Methoxy-7-hydroxycoumarin. Scopoletin occurs in a number of higher plants, often as *scopolin* (scopoletin 7-glucoside). It accumulates in the tissues of certain microbially infected plants – e.g. in potato tubers affected with LATE BLIGHT, and in tobacco plants affected by GRANVILLE WILT. Scopoletin, which inhibits 'IAA-oxidase' (see AUXINS), may be a factor in the pathogenesis of certain diseases.

scopolin See SCOPOLETIN.

scopula In certain (particularly sessile) peritrichs: an aboral, frequently cup-shaped organelle consisting of a large number of kinetosomes, each typically carrying a short, immobile cilium; presumed functions include the formation of a stalk.

Scopulariopsis See HYPHOMYCETES.

Scorias See SOOTY MOULDS.

Scotobacteria A proposed class of the GRACILICUTES comprising the non-photosynthetic Gram-negative bacteria.

scotochromogen Any strain of *Mycobacterium* which forms carotenoid pigment when growth occurs either in the light or in the dark (cf. PHOTOCHROMOGEN). Scotochromogens include some slow-growing species (e.g. *M. scrofulaceum*) and some fast-growing species (e.g. *M. phlei*).

Scourfieldia See MICROMONADOPHYCEAE.

scours (*vet.*) DIARRHOEA – a common symptom in many types of infectious disease (see e.g. COLIBACILLOSIS, JOHNE'S DISEASE, PARATYPHOID FEVER (2)), but also a symptom e.g. of certain mycotoxicoses and of secondary copper deficiency syndromes (e.g. peat scours, scouring disease, teart) in ruminants. (See also CALF SCOURS; TRANSMISSIBLE GASTROENTERITIS; WINTER DYSENTERY.)

SCP SINGLE-CELL PROTEIN.

scrapie A TRANSMISSIBLE SPONGIFORM ENCEPHALOPATHY (q.v.) which, in nature, affects sheep and goats. After an incubation period of months or years the infected animal exhibits neurological symptoms – e.g. evidence of intense itching, incoordination and trembling – followed by emaciation and (after weeks or months) death. Histopathology in the central nervous system: e.g. vacuolation, gliosis and astrocytosis. No inflammatory response is elicited at any stage, interferons are not induced, and there is no specific immune response. Different breeds of sheep differ markedly in their susceptibility to scrapie.

The causal agent is a PRION designated PrP^{sc}. It can be transmitted experimentally e.g. to mice and hamsters – in which animals the disease follows a much shorter course and may involve different symptoms (for example, scrapie-infected mice do not scratch). PrP^{sc} is initially demonstrable in the spleen and lymphoid tissues; eventually it reaches maximum titres in the brain. Abnormal, amyloid-like fibrillar structures can be seen in synaptosomal and spleen preparations of infected mice; these *scrapie-associated fibrils* (SAFs) contain a protein subunit formerly designated 'PrP 27–30' (because its MWt was apparently 27000–30000). SAFs appear to be closely associated with infectivity.

screening test (1) Any test used to detect the presence of a given agent (e.g. a specific antibody or microorganism) in each of a large number of specimens; such tests are typically simple and inexpensive but are usually not completely specific for the antibody or microorganism. A certain proportion of false-positive results may be tolerated as a positive screening test may be followed up by other tests which are more specific. (2) Any test which enables a given (unknown) organism to be assigned to one of a possible range of taxa.

scRNA Small cytoplasmic RNA. scRNAs occur in the cytoplasm of eukaryotic cells in the form of ribonucleoprotein particles (scRNPs or *scyrps*), including PROSOMES. (cf. SNRNA.)

scrofula Cervical lymphadenitis caused by *Mycobacterium scrofulaceum* or by other members of the MAC.

scrub typhus (tsutsugamushi disease) An acute, systemic human disease caused by *Rickettsia tsutsugamushi* (= *R. orientalis*) and transmitted by the bite of the larval stage of a trombiculid mite; rodents are a reservoir of infection. The disease occurs in Asia, Australia, and the Pacific area. Incubation period: 1–3 weeks. A primary lesion (an ESCHAR) often develops at the location of the bite; subsequent symptoms resemble those of epidemic typhus (see TYPHUS FEVERS). *Lab. diagnosis*: e.g. the WEIL–FELIX TEST. *Chemotherapy*: tetracyclines, chloramphenicol.

SCSR See CLOVER ROT.

Scutellinia A genus of fungi (order PEZIZALES) which form apothecia bearing dark, pointed setae. *S. scutellata* (common on rotting wood) forms brick-red apothecia typically several millimetres across.

scutica In members of the SCUTICOCILIATIDA: a transient, hook-like structure, consisting of a number of (generally non-ciliated) kinetosomes, formed during STOMATOGENESIS.

Scuticociliatida An order of protozoa (subclass HYMENOSTOMATIA) in which somatic ciliature is uniform to sparse (sometimes involving distinctive caudal ciliature), and in which the buccal ciliature often includes a large, protruding and conspicuous PARORAL MEMBRANE; STOMATOGENESIS is buccokinetal and involves the appearance of a SCUTICA. Cyst formation is common. Genera include e.g. *Ancistrum*, *Boveria*, CYCLIDIUM, *Loxocephalus*, *Paraurונema*, *Philaster*, *Pleuronema*, and *Uronema*.

scutulum See FAVUS.

ScV *Saccharomyces cerevisiae* virus (see MYCOVIRUS).

scyllitol See INOSITOL.

Scyphidia See PERITRICHIA.

scyphus (*lichenol.*) See PODETIUM.

scyrps See SCRNA.

Scytonema A genus of filamentous CYANOBACTERIA (section IV) in which the mature trichomes are uniform in width (cf. CALOTHRIX) and are composed of disc-shaped, isodiametric or cylindrical vegetative cells; young trichomes from hormogonia each have a heterocyst at only one end. GC%: ca. 44.

In nature, trichomes are heavily ensheathed and exhibit frequent FALSE BRANCHING. A distinction is often made between *Tolypothrix* (false branches occur singly) and *Scytonema* (false branches occur in pairs); however, since false branching is variable in culture, these two genera are generally considered as one, *Scytonema* [Book ref. 45, p. 254]. (See also STROMATOLITE.)

Scytosiphon See PHAEOPHYTA.

SD sequence (SD region) Shine–Dalgarno sequence: see PROTEIN SYNTHESIS.

SDA Strand displacement amplification: a method for copying ('amplifying') a given sequence of nucleotides in DNA. Like NASBA (but unlike LCR or PCR), SDA is carried out isothermally; originally the operating temperature was ~40°C but, owing to suboptimal specificity, it was later raised to ≥50°C (= thermophilic SDA; tSDA).

SDA is considerably more complex than other nucleic-acid-amplification procedures. It has two distinctive features: (i) the use of a DNA polymerase which can displace an existing strand of DNA by using the complementary strand as a template, and (ii) the use of a RESTRICTION ENDONUCLEASE for repeatedly *nicking* a modified cutting site in a mechanism for regenerating templates.

Only certain types of DNA polymerase are able to carry out strand displacement – one of which is the exonuclease-deficient form of the KLENOW FRAGMENT. As illustrated in the figure, the strand-displacing activity of a polymerase can be manifested in two ways. In one mode, a so-called *bumper primer* binds upstream of the strand to be displaced; extension of the 3′ end of the bumper primer by a (strand-displacing) DNA polymerase displaces the strand downstream. (In SDA, displacement of the 5′ end of the strand begins while the 3′ end is still being synthesized.) The second mode of strand displacement occurs at a NICK site; here, the polymerase extends the 3′ end of the nick and, in so doing, displaces the downstream region of the same strand. (In effect, the 3′ end at the nick site functions as a primer.)

The restriction endonuclease used in SDA is *Hinc*II; it has the (generalized) recognition site:

5′.....GTPy/PuAC.....3′
3′.....CAPu/PyTG.....5′

(Py = a pyrimidine, Pu = a purine). The oblique (/) indicates the cleavage site in each strand; thus, *Hinc*II *normally* makes a 'blunt-ended' cut across both strands.

The specific *Hinc*II recognition site used in SDA is:

5′.....GTT/GAC.....3′
3′.....CAA/CTG.....5′

and the sequence 5′-GTTGAC-3′ is used as a terminal tag in both primers (S_1 and S_2).

During amplification of the target in SDA, all strand synthesis is carried out in a reaction mixture that contains a chemically modified form of deoxyadenosine triphosphate: α-thiophosphoryl dATP (dATPαS). Hence, when a complementary strand is synthesized on the *Hinc*II site of a primer, the sequence (3′...CAACTG...5′) contains two of the modified nucleotides (AA) adjacent to the cleavage site; *this* strand is resistant to cleavage by *Hinc*II – i.e. the double-stranded *Hinc*II site contains one cleavable sequence (in the primer) and one non-cleavable sequence (in the newly synthesized, complementary strand). Such a 'hemi-modified' or 'hemiphosphorothioate' *Hinc*II site can be *nicked* between 'T' and 'G' in the primer strand. When such a nick is made, the polymerase can extend the 3′ end and displace the strand downstream of the nick site. Importantly, note that the strand formed by extension from the nick site (5′-GTT →→→) will incorporate a modified nucleotide at the 'A' position of GAC; despite this, the *Hinc*II site in the newly synthesized strand 5′-GTTGAC.....3′ is 'nickable' – i.e. incorporation of a single modified nucleotide in GAC does not affect nickability (unlike the incorporation of two modified nucleotides in the sequence 3′...CAACTG...5′ in the complementary strand).

In practical terms the method is uncomplicated. Essentially, sample dsDNA is heat-denatured in a reaction mixture containing all components except the polymerase and restriction endonuclease; the mixture is cooled to an appropriate temperature, the enzymes are added, and incubation is continued for the required period.

SDA-based diagnostic procedures are being developed by the Becton Dickinson organization. The BDProbe Tec system has been tested for its ability to detect *Mycobacterium tuberculosis* in respiratory specimens [JCM (1999) *37* 137–140], and a more recent system, BDProbe Tec™ ET, has been evaluated for the detection of *Chlamydia trachomatis* and *Neisseria gonorrhoeae* [CC (1999) *45* 777–784].

[SDA, NASBA, TMA in clinical microbiology: Book ref. 221, pp 126–151.]

SDS Sodium dodecyl (= lauryl) sulphate.

SDS-PAGE ('Laemmli electrophoresis' [Nature (1970) *227* 680–685]) Sodium dodecyl sulphate polyacrylamide gel electrophoresis: a specialized form of ELECTROPHORESIS used e.g. for determining the molecular weight of a protein, or for separating mixtures of proteins.

Molecular weight determination by SDS-PAGE. At concentrations > ca. 10^{-4} to 10^{-3} M, sodium dodecyl sulphate (SDS; $CH_3(CH_2)_{11}OSO_3Na$) binds cooperatively to proteins and denatures them, forming long rod-like SDS-polypeptide complexes; virtually all proteins bind a similar amount of SDS per unit weight of protein. Each SDS molecule binds via its hydrophobic end, and the coating of SDS molecules effectively masks the charges on the polypeptide. Since the length of an SDS-polypeptide complex is roughly proportional to the molecular weight of the polypeptide, the electrophoretic mobility of such a complex will give an approximate indication of the molecular weight of the polypeptide. Thus, if the SDS complex of an unknown protein is subjected to PAGE together with the SDS complexes of reference proteins (of known molecular weights), the molecular weight of the unknown protein can be calculated by comparing the electrophoretic mobilities.

Some proteins (e.g. pepsin, many viral capsid proteins) form SDS complexes only if they are *heated* in the presence

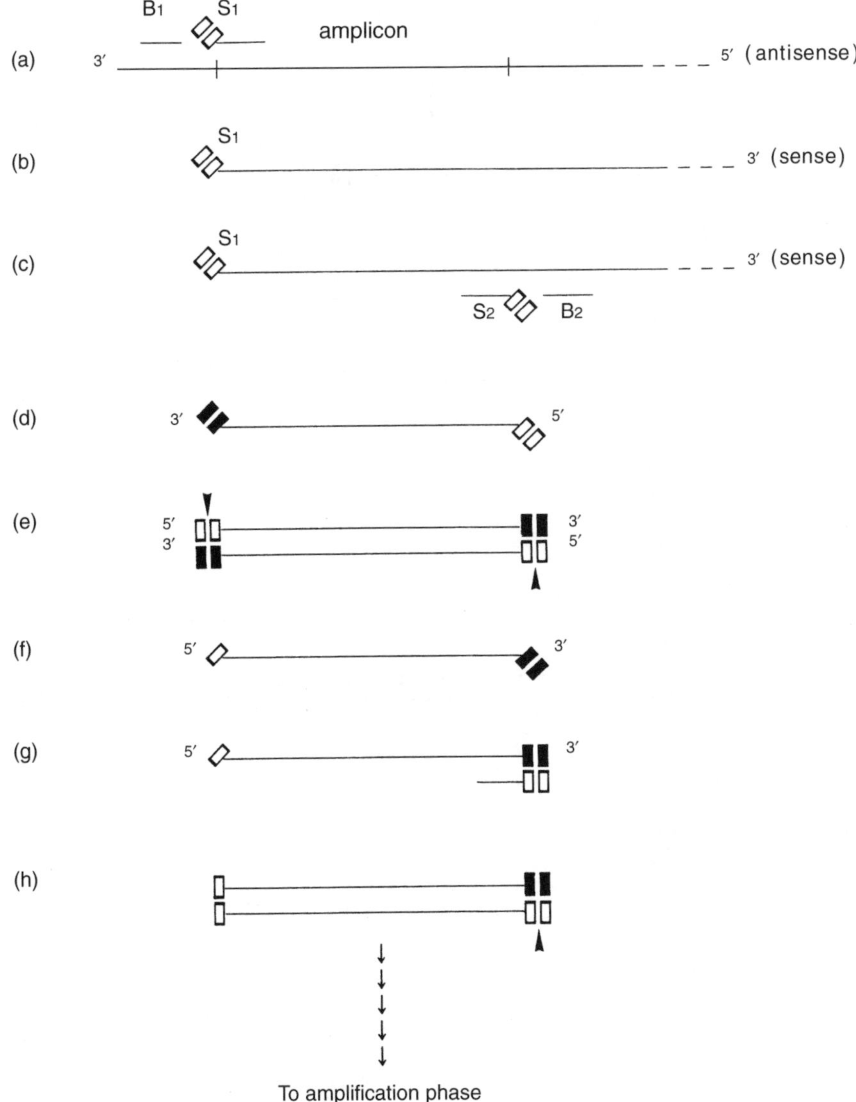

SDA (strand displacement amplification) – I: the target-generation phase (diagrammatic). The target DNA is denatured by heat to the single-stranded form prior to SDA.

(a) The antisense (3′-to-5′) strand of the target duplex showing the amplicon delimited by two short, vertical bars. (For clarity, and economy of space, only one strand of the amplicon is considered; corresponding events occur on the complementary strand.) A primer (S_1) has bound at the 3′ end of the amplicon; the 5′ end of this primer is tagged with (one strand of) the *Hinc*II recognition sequence: (5′-GTTGAC-3′) (▯▯). A bumper primer (B_1, see entry) has bound upstream of primer S_1. Extension of S_1 will produce a 'sense' strand on the antisense template, and this strand will be displaced when B_1 is extended; the displaced sense strand is shown at (b).

(c) Primer S_2 has bound at the 3′ end of the amplicon on the sense strand; like S_1, its 5′ end is tagged with the recognition sequence of *Hinc*II. The bumper primer B_2 has bound upstream of S_2. Extension of S_2 produces an antisense amplicon which is tagged at both ends by a *Hinc*II recognition sequence; this strand, which is displaced by the extension of B_2, is shown at (d). Notice that the 3′ *Hinc*II sequence in this strand, having been synthesized with dATPαS, is modified – as indicated by (▮▮); this sequence is not susceptible to *Hinc*II.

(d) Primer S_1 (not shown) binds at the 3′ end of this strand, the *Hinc*II sequence in S_1 hybridizing with the modified sequence in the strand. Extension of S_1 forms the double-stranded amplicon at (e). Each end of this amplicon consists of a hemi-modified recognition site for *Hinc*II; a hemi-modified site can be nicked in the *non*-modified strand (arrowhead). Nicking of the upper strand at (e) is followed by extension of the 3′ end of the nick – this displacing the strand downstream of the nick site; this displaced strand is shown at (f).

(g) Primer S_2 binds to the displaced strand. Extension of S_2 forms the product at (h).

(h) This double-stranded product feeds into the amplification phase (see part II of figure).

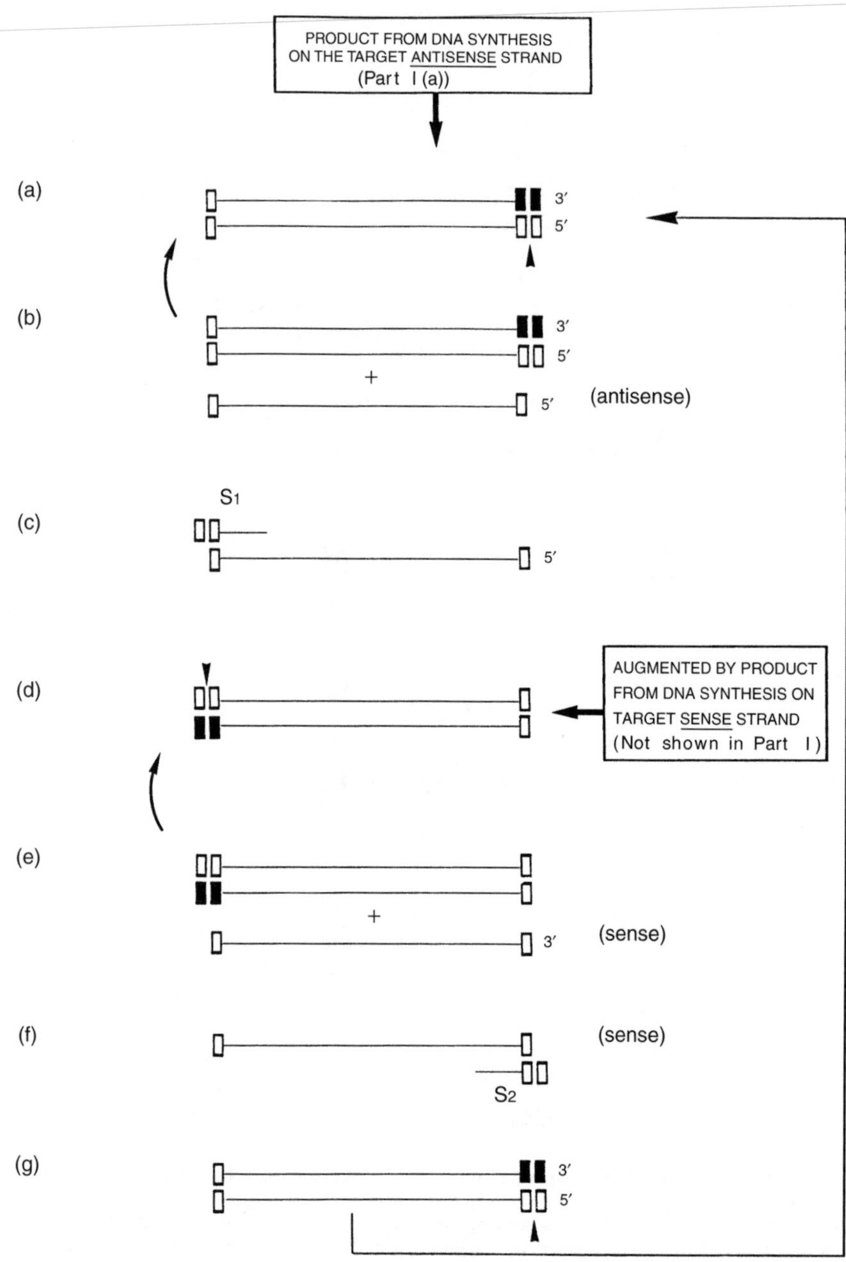

SDA (strand displacement amplification) – II: the amplification phase (diagrammatic).

(a) A double-stranded product from the target-generation phase; notice that this product derives from S_1-primed DNA synthesis on the *antisense* strand of the target sequence (see part I(a)). The antisense strand in this product contains a nickable *Hin*cII restriction site (arrowhead) derived from the tag in primer S_2 (see part I (g)). When this restriction site is nicked, DNA synthesis proceeds from the 3′ end at the nick site. During synthesis, the polymerase displaces the downstream section of the strand, producing an antisense copy of the target sequence which is flanked, on each side, by half a restriction site; moreover, the newly synthesized strand regenerates the original duplex – and the restriction site in this duplex is still nickable because a modified nucleotide occurs only in the 5′-GAC-3′ sequence (see text). Both of these products (the displaced strand and the regenerated duplex) are shown at (b).

Thus, the product shown at (a) can undergo repeated nicking, strand extension and regeneration in a cyclical fashion, displacing many copies of the antisense strand. (*Continued on page 695*.)

of SDS – or e.g. treated with 2-mercaptoethanol (or similar reagent) to cleave intraprotein disulphide bonds.

[Effect of SDS on proteins: BBA (1975) *415* 29–79 (47–50).]

sea ivory *Ramalina siliquosa.*

sea lettuce *Ulva lactuca*: see ULVA.

sea lion virus See SAN MIGUEL SEA LION VIRUS.

sea sawdust See TRICHODESMIUM.

seasoning (of timber) See TIMBER PRESERVATION.

seaweed Any macroscopic marine alga of the CHLOROPHYTA ('green seaweeds'), PHAEOPHYTA ('brown seaweeds') or RHODOPHYTA ('red seaweeds').

seaweed diseases See ALGAL DISEASES.

***sec*-dependent pathway** See PROTEIN SECRETION.

***sec* genes** See PROTEIN SECRETION.

SecA, SecB See PROTEIN SECRETION.

secalonic acids *Syn.* ERGOCHROMES.

secnidazole See NITROIMIDAZOLES.

second-site reversion See SUPPRESSOR MUTATION.

secondary fixation *Syn.* post-FIXATION.

secondary fluorescence See FLUORESCENCE.

secondary homothallism See HOMOTHALLISM.

secondary hyphae See HYPHA.

secondary metabolism Metabolism which is not essential for, and plays no part in, growth; it commonly occurs maximally under conditions of restricted growth or absence of growth (e.g. at the end of log-phase or trophophase growth in BATCH CULTURES). Secondary metabolites ('idiolites') include substances such as ANTIBIOTICS and MYCOTOXINS and are produced from substrates provided by PRIMARY METABOLISM – particularly shikimic acid (precursor of many aromatic compounds), amino acids (precursors of many alkaloids and antibiotics), and acetate (precursor of isoprenoids and many toxins – see also POLYKETIDE).

[Review: AMP (1984) *25* 1–60; secondary metabolism in fungi: Book ref. 117, pp. 336–367. Secondary metabolites: Book ref. 205.]

secondary response (*immunol.*) See ANTIBODY FORMATION.

secondary zoospore See e.g. DIPLANETISM and PLASMODIOPHOROMYCETES.

secotioid fungi Fungi whose sexually-derived fruiting bodies typically resemble AGARICOID fruiting bodies but differ from them in that the margin of the pileus may not separate from the stipe, the lamellae are characteristically convoluted, and the spores are not forcibly discharged; the secotioid fruiting body may be epigean or hypogean, the latter type closely resembling the fruiting body of a typical gasteromycete. Secotioid fungi are found in various taxa; they include e.g. certain members of the Russulales. *Secotium*, formerly classified in the Agaricales, is now included in the Gasteromycetes (Podaxales). [The secotioid syndrome: Mycol. (1984) *76* 1–8.]

Secotium See SECOTIOID FUNGI.

secretin Any one type of a family of proteins which oligomerize to form pores in the OUTER MEMBRANE; in at least some cases the multimeric pores are stabilized by a specific outer membrane lipoprotein.

One example of a secretin is the PilQ protein of *Neisseria gonorrhoeae*; this protein, which is involved in the formation of type IV FIMBRIAE, forms an outer membrane pore that is stabilized by the PilP protein. Another example is the *usher* protein, PapC, involved in the formation of P fimbriae (type I fimbriae); the corresponding structure is stabilized by the PapH protein. The YscC protein of the *Yersinia* VIRULON is a further example.

secretion (of a protein) The transmembrane translocation of a protein, towards the cell's exterior, with eventual release of the protein to the external environment. (cf. EXPORT.)

secretion of proteins (by Gram-negative bacteria) See PROTEIN SECRETION.

secretion sequence See ABC EXPORTER.

secreton In type II PROTEIN SECRETION systems: the complex of proteins which, together, enable proteins to cross the OUTER MEMBRANE, i.e. the 'terminal branch' of the type II system; thus, a protein may be translocated across the cytoplasmic membrane in a *sec*-dependent manner and then secreted to the cell's exterior via the secreton.

The secreton is functionally analogous to the protein transport system involved in the formation of certain fimbriae [EMBO (2000) *19* 2221–2228].

secretory ampulla *Syn.* AMPULLA sense 1 or 2.

secretory component (secretory piece) A glycoprotein (MWt ca. 60000) produced (as a membrane protein) in hepatocytes and in epithelial cells of various mucosal surfaces in the body; it is a component of sIgA (see IgA). Secretory component binds to

SDA (strand displacement amplification) – II (*continued*) (c) Each antisense strand from (b) can bind primer S_1. Extension of S_1, and of the (recessed) 3′ end of the antisense strand, forms the product at (d). Extension of the 3′ end of the antisense strand has formed a modified, non-nickable *Hin*cII restriction sequence (▮▮) because the three terminal nucleotides in this strand (3′-CAA-5′) include two modified nucleotides (dATPαS).

(d) This product has a nickable restriction site (arrowhead) derived from primer S_1. Like the product at (a), it can undergo a cyclical process of nicking, strand extension and regeneration, concurrently displacing copies of a single-stranded product; however, this single-stranded product is a *sense* strand of the target sequence. The regenerated product and sense strand are shown at (e).

Recall that part I illustrates the events arising from DNA synthesis on the *antisense* strand of the target duplex. If events are followed on the sense strand, the resulting product is identical to that shown here at (d).

(e) The single-stranded (sense) product shown here is able to bind primer S_2.

(f) The single-stranded product from (e) has bound primer S_2. DNA synthesis (i.e. extension from S_2, and also from the recessed end of the sense strand) then gives rise to the product seen at (g). (DNA synthesis at the 3′ end of the sense strand forms a non-nickable restriction site for the reason given under (c), above.)

(g) This product is equivalent to the one shown at (a).

Summarizing, two types of double-stranded product from the target-generation phase, derived from different strands of the target, undergo cyclical nicking, strand synthesis and regeneration, concurrently displacing numerous copies of antisense and sense strands of the target sequence. Antisense strands bind primer S_1 and form the (double-stranded) product which yields sense strands; sense strands bind primer S_2 and form the (double-stranded) product which yields antisense strands. Nickable *Hin*cII restriction sites are provided by the 5′-tags on S_1 and S_2 primers (which are present in excess in the reaction mixture).

Parts I and II reproduced from Figures 5.3 and 5.4, pages 134–137, in *DNA Methods in Clinical Microbiology* (ISBN 07923-6307-8), Paul Singleton (2000), with kind permission from Kluwer Academic Publishers, Dordrecht, The Netherlands.

the dimeric IgA–J chain complex at the basal surface of (e.g.) a mucosal epithelial cell; the $(IgA)_2$–J chain–secretory component complex is then taken into the cell by endocytosis and passes through the cell cytoplasm (in an endocytotic vacuole), sIgA being released at the mucosal surface by proteolytic cleavage of the (transmembrane) secretory component.

secretory IgA See entry IgA.

section *(histol.)* See MICROTOME.

sectoring (in a colony) The development of a sector or wedge-shaped region of growth which differs, in e.g. appearance, from the rest of the colony.

SecYEG See PROTEIN SECRETION.

sedentary *Syn.* SESSILE (sense 1).

sedimentation basin See WATER SUPPLIES.

sedimentation coefficient See SVEDBERG UNIT.

sedoheptulose A 7-carbon ketosugar originally isolated from *Sedum* sp (stonecrop: Crassulaceae) in which it occurs as the free monosaccharide. Phosphorylated derivatives of sedoheptulose are intermediates in the CALVIN CYCLE, the HEXOSE MONOPHOSPHATE PATHWAY [Appendix I(b)], and the RMP PATHWAY.

seed *(verb)* To inoculate.

seeligerolysin See THIOL-ACTIVATED CYTOLYSINS.

segmented genome In a virus: a genome which occurs as two or more pieces of nucleic acid. (See e.g. ARENAVIRIDAE, BACTERIOPHAGE ϕ6, BUNYAVIRIDAE, ORTHOMYXOVIRIDAE, REOVIRIDAE.)

segregation *(genetics)* The separation of homologous chromosomes or chromatids, or homologous or non-homologous (incompatible) plasmids, or of particular named genes or alleles, etc, into different daughter cells during cell division.

segregation lag A delay in the phenotypic expression of a newly acquired genotype owing to the time required for the segregation of the new allele(s) to a cell which has no corresponding wild-type allele(s). Thus, e.g., if a mutation occurs in one chromosome in a bacterial cell which contains two or more chromosomes, the mutant phenotype will be expressed only when the mutant chromosome is segregated (by cell division) into a cell which lacks a wild-type chromosome.

segregational petite See PETITE MUTANT.

Seitz filter See FILTRATION.

***selC* gene** See PATHOGENICITY ISLAND.

selectins A family of calcium-dependent CELL ADHESION MOLECULES with the properties of LECTINS. E-selectin (see CD62E) is expressed on endothelial cells during INFLAMMATION. L-selectin (LECAM-1; = CD62L) occurs on leukocytes, and is involved in lymphocyte homing and in the interaction between leukocytes and endothelium during inflammation. P-selectin (PADGEM; = CD62P) is found e.g. on (activated) platelets and endothelial cells during inflammation; in the unstimulated state, P-selectin occurs in α granules in platelets and within endothelial cells. P-selectin is the largest of the selectin molecules (140 kDa); the lymphocyte-associated form of L-selectin is the smallest ($\sim$75 kDa).

selection synchrony See SYNCHRONOUS CULTURE.

selective medium See MEDIUM.

selenate respiration ANAEROBIC RESPIRATION, found in certain (Gram-positive and Gram-negative) bacteria (e.g. *Thauera selenatis* [IJSB (1993) *43* 135–142]), in which the terminal electron acceptor is selenate. Some species of bacteria can use arsenate as terminal electron acceptor. [Arsenic- and selenium-respiring bacteria: FEMS Reviews (1999) *23* 615–627.]

selenazofurin See RIBAVIRIN.

selenazole See RIBAVIRIN.

Selenidium A genus of the GREGARINASINA.

selenite broth (selenite F broth) A MEDIUM used e.g. for the ENRICHMENT of strains of *Salmonella* in samples of faeces; selenite inhibits many of the common enteric bacteria. The medium consists of an aqueous solution of peptone (0.5%), lactose or mannitol (0.4%), $NaHSeO_3$ (0.4%), and either Na_2HPO_4 (1.0%) or a combination of Na_2HPO_4 and NaH_2PO_4 to give a final pH of ca. 7.0. The medium should not be autoclaved, but may be steamed for 30 min. Incubation of the inoculated medium should not exceed ca. 12–18 hours.

Selenococcidium See EIMERIORINA.

Selenomonas A genus of Gram-negative bacteria (family BACTEROIDACEAE) which occur e.g. in the RUMEN and in the human gingival crevice. Cells: curved (crescent- or kidney-shaped) rods, 0.9–1.1 × 3.0–6.0 μm, with a tuft of flagella arising from the concave side of the cell. Substrates include sugars and, sometimes, amino acids and lactate; glucose fermentation yields propionic and acetic acids as major products together with e.g. CO_2 and/or lactate. GC%: ca. 54–61. Type species: *S. sputigena* (which does not ferment cellobiose, dulcitol or salicin, and which occurs in the human mouth); other species: *S. ruminantium* (which ferments cellobiose, dulcitol and salicin, and which occurs in the rumen).

Selenophoma A genus of fungi of the COELOMYCETES.

self-cloning experiment A CLONING experiment in which the DNA to be cloned is derived from the same strain or species as that in which the recombinant DNA is to be replicated; a recombination-deficient strain may be used to prevent recombination between the cloned DNA and the organism's genome.

self fertilization *(protozool.)* *Syn.* AUTOGAMY.

self purification The natural process in which organic material (faeces etc) in rivers, streams etc undergoes MINERALIZATION, and the resulting simple substances (e.g. nitrates) are made available to photosynthetic and other organisms. Self purification can occur only if the polluting load is not excessive. (See also SAPROBITY SYSTEM.)

self-splicing introns See SPLIT GENE (b) and (c).

self tolerance (natural immunological tolerance) Unresponsiveness of the immune system to an individual's own (autologous) antigens.

self-transmissible plasmid See CONJUGATIVE PLASMID.

selfing *(ciliate protozool.)* *Syn.* CYTOGAMY.

Seliberia A genus of Gram-negative, iron-accumulating, BUDDING BACTERIA which occur e.g. in soil; the organisms typically occur as radial clusters of rod-shaped cells. [Book ref. 45, pp. 516–519.]

Selysina See EIMERIORINA.

SEM See ELECTRON MICROSCOPY (b).

semiapochromatic objective *Syn.* FLUORITE OBJECTIVE.

semiconservative replication Replication of a double-stranded nucleic acid such that each daughter duplex contains one parental strand and one new strand. (See DNA REPLICATION.)

semipermissive cells *(virol.)* A cell population in which only some of the cells are permissive for lytic infection by a given virus. (cf. PERMISSIVE CELLS; see also CARRIER CULTURE.)

semipersistent transmission See NON-CIRCULATIVE TRANSMISSION.

semi-rough mutant See SMOOTH–ROUGH VARIATION.

semi-solid agar See AGAR and MEDIUM.

Semliki Forest virus (SFV) An ALPHAVIRUS which was first isolated from mosquitoes in Uganda. Its natural hosts and vectors are unknown, but antibodies to the virus have been detected in humans and wild primates in various parts of Africa, Malaya and North Borneo. The virus is apparently non-pathogenic in man. When injected into mice, SFV can cause an acute, lethal encephalitis; virulent strains appear to damage neurones (or particular neurone subsets) directly, while other strains infect oligodendrocytes and cause demyelination – apparently by triggering an autoimmune response [JGV (1985) *66* 2365–2373; review: JGV (1985) *66* 395–408]. SFV can be transmitted transplacentally in mice, causing e.g. abortion or malformation of infected fetuses. (See TOGAVIRIDAE for replication cycle etc.)

Semple vaccine An anti-RABIES vaccine consisting of a phenol-inactivated preparation of rabbit-fixed rabies virus (i.e. rabies virus passaged in rabbit brain).

***sen* gene** See EIEC.

Sendai virus See PARAMYXOVIRUS.

senior synonym See SYNONYM.

sennetsu rickettsiosis See EHRLICHIA.

sens. lat. SENSU LATO (q.v.).

sense codon A codon which specifies an amino acid: see GENETIC CODE.

sense strand (of DNA) See CODING STRAND.

sensitin Any preparation (e.g. an extract of a given microorganism, as in brucellin, coccidioidin etc) which can act as an ALLERGEN; sensitins are used in SKIN TESTS.

sensitivity agar Any of various standardized, solid, agar-based media used for carrying out ANTIBIOTIC-sensitivity tests.

sensitization (*immunol.*) (1) (of a person or animal) The initial exposure of an individual to an allergen such that a manifestation of HYPERSENSITIVITY occurs on subsequent exposure(s) of the individual to that allergen. (2) (of persons, animals or cells) The process of *priming* (see PRIMED). (3) The combination of cell-surface antigens with their homologous antibodies (see e.g. HAEMOLYTIC SYSTEM). (4) The coating of erythrocytes with soluble antigens (see BOYDEN PROCEDURE). (5) Any procedure of the type in which certain strains of staphylococci are coated ('sensitized') with unrelated antibodies (see PROTEIN A) for use in a PASSIVE AGGLUTINATION TEST for the homologous antigen. (6) The combination of cells with cytophilic antibodies (see e.g. REAGINIC ANTIBODIES).

sensu lato (s.l.; sens. lat.) Of a term or name: (1) used in a broad or more inclusive sense, or (2) used with a meaning other than the original meaning.

sensu stricto (s.s.; s. str.) Of a term or name: (1) used with a precise or narrow meaning, or (2) used with the original meaning.

Seoul virus See HANTAVIRUS.

***sepB* gene** See PATHOGENICITY ISLAND.

Sephadex See DEXTRANS and GEL FILTRATION.

Sepik virus See FLAVIVIRIDAE.

sepsis (*med., vet.*) (1) The state in which pathogen(s) are present in particular tissue(s). (cf. ASEPSIS; SEPTICAEMIA.) (2) The symptoms associated with microbial infection of tissues. (See also SIRS.)

septa Plural of SEPTUM.

septal pore cap See DOLIPORE SEPTUM.

septate Having one or more septa.

septic (1) (*med., vet.*) Refers to SEPSIS (generally sense 2). (2) Refers to the presence or activities of microorganisms involved in putrefaction or decay (see e.g. SEPTIC TANK).

septic diphtheria See DIPHTHERIA.

septic shock *Syn.* ENDOTOXIC SHOCK.

septic tank A ventilated tank into which domestic sewage flows and within which the sewage undergoes primary purification (settlement of suspended solids) and some degree of ANAEROBIC DIGESTION; solids (sludge and scum) are periodically removed for disposal. The highly polluting supernatant overflows and must be treated, e.g. by aerobic biological oxidation or (when permissible, and when underground water supplies would not be contaminated) allowed to disperse in the soil below the surface ('subsurface irrigation'). A septic tank can serve e.g. 1–50 houses. (cf. CESSPOOL; see also SEWAGE TREATMENT.)

septicaemia (septicemia; blood poisoning) A particular form of BACTERAEMIA in which there are clinical symptoms (e.g. fever). The term may also refer to the presence of *Candida albicans* or other fungi in the bloodstream. (cf. VIRAEMIA.)

(See also SEPSIS and SIRS.)

septicemia *Syn.* SEPTICAEMIA.

septicolysin See THIOL-ACTIVATED CYTOLYSINS.

Septobasidiales An order of fungi (subclass PHRAGMOBASIDIOMYCETIDAE) which form non-gelatinous fruiting bodies; they include parasites of scale insects (insects of the superfamily Coccoidea). Genera: *Septobasidium* and *Uredinella*.

Septobasidium See SEPTOBASIDIALES.

septoria A plant disease caused by a species of *Septoria*: see e.g. GLUME BLOTCH and HALO SPOT.

Septoria A genus of fungi (order SPHAEROPSIDALES) which include many plant pathogens – e.g. *S. apiicola*, the causal agent of late blight (leaf spot) of celery. (See also e.g. GLUME BLOTCH and HALO SPOT.) Conidia are filiform and typically multiseptate and colourless; they are formed in dark, ostiolate pycnidia immersed in the substratum.

septum (cross-wall) (a) (*bacteriol.*) A partition which (i) divides a parent cell into daughter cells during BINARY FISSION, (ii) occurs between adjacent cells in a filament (see e.g. ACTINOMYCETALES), and (iii) separates a developing ENDOSPORE from the rest of the mother cell.

Septum formation is an essential phase in the CELL CYCLE (q.v. for details of septum formation in bacteria).

During sporulation in *Bacillus subtilis*, one of the early stages is the formation of an *asymmetric* septum, i.e. one which (unlike that formed during cell division) develops near one pole of the cell. Initially, an FtsZ ring (see CELL CYCLE) forms at the mid-cell position but then become helical and finally forms an FtsZ ring at both polar sites; septation occurs at only one site [Science (2002) *298* 1942–1946].

(b) (*mycol.*) A partition, one or more of which divides certain fungal structures (hyphae, spores etc.) into cells; septa are also formed in fission yeasts (e.g. *Schizosaccharomyces*) during cell division.

A *primary* septum is one formed in association with nuclear division, i.e. one serving to separate daughter cells. An *adventitious* septum develops independently of nuclear division.

Septate hyphae are characteristic of the higher fungi, but septa are also formed in certain lower fungi – although often only in ageing hyphae and/or in order to delimit reproductive structures (such as sporangia).

Septate (multicellular) spores are formed by various fungi (e.g. macroconidia in *Fusarium*).

In general, 'true' septa (*eusepta*) have essentially the same composition as the hyphal CELL WALL, but certain fungal structures (e.g. the mycelium of some of the more complex chytridiomycetes, and the conidia in certain hyphomycetes) have septum-like *pseudosepta* (= *distosepta*) whose composition is distinct from that of the cell wall.

Hyphal septa appear always to be formed by centripetal growth from annular rims which develop on the inner surface of the hyphal wall. In some cases growth of the septum continues inwards until a complete plate is formed. In other cases, growth stops before the septum is complete, so that a small pore (ca.

0.4–1.0 μm diam.) remains in the centre of the septum; such pores (characteristic of ascomycetes) may permit passage of cytoplasm and nuclei from one hyphal cell to the next, but in many fungi the pores may be plugged or occluded (see e.g. WORONIN BODIES).

Plasmodesmata (fine, cytoplasm-filled channels) occur in the septa of certain fungi.

(See also DOLIPORE SEPTUM.)

sequence capture PCR See PCR and DYNABEADS.

sequencing (of DNA and RNA) See DNA SEQUENCING.

Sequestrene *Syn.* VERSENE.

SER Smooth ENDOPLASMIC RETICULUM.

sera Plural of SERUM.

SERE *Salmonella enteritidis* repetitive element: a widely dispersed bacterial repetitive DNA element [JMM (1998) *47* 489–497]. (See also REP-PCR.)

SERE-PCR See REP-PCR.

Serény test A test used with the intention of determining the invasiveness of a bacterial pathogen. The conjunctival sac of a guinea pig is inoculated with the pathogen; invasiveness is indicated by ulceration of the cornea. An HEp-2 tissue culture test may be used as a more humane alternative [JCM (1981) *13* 596–597.]

serial dilutions A set of dilutions (of a given sample) prepared by initially diluting an aliquot of the sample, then diluting an aliquot of this dilution, and so on. In *doubling dilutions* the dilution factor progressively doubles: 1/2, 1/4, 1/8 etc. In *log dilutions* the dilution factors are related thus: 1/10, 1/100, 1/1000 etc.

serial passage (1) (*syn.* passage) Any procedure in which a pathogen (usually a virus) is transferred from one to another of a succession of animals, eggs or tissue cultures etc – growth (or replication) of the pathogen taking place before each transfer. Serial passage is sometimes used e.g. for the ATTENUATION (sense 1) of a pathogen. For example, the rabies virus may be attenuated by adapting it to chick embryo tissues – the virus being serially passaged through a series of hens' eggs. (See also AVIANIZED VACCINE and FLURY VIRUS.) For use in a vaccine, a successfully passaged pathogen should (a) be non-pathogenic for particular host(s), and (b) retain its specific immunogenicity in order to stimulate the formation of protective antibodies.

(2) In TISSUE CULTURE, 'passage' refers to the transfer of an inoculum of cells from an existing cell culture to fresh growth medium in another vessel (i.e. subculture); 'serial passage' refers to repeated subculture. (This contrasts with the serial passage of a virus in tissue cultures: when a virus is passaged the *supernatant* is transferred.)

series A taxonomic rank (sometimes used in mycology) between subclass and order.

L-serine biosynthesis See Appendix IV(c).

serine pathway A cyclic metabolic pathway used by some types of methylotrophic bacteria for the assimilation of 1-C substrates (see METHYLOTROPHY). (A strain of *Streptomyces* has been reported to use both the serine pathway and the RMP PATHWAY, while at least some moulds (as opposed to yeasts) apparently use the serine pathway [Book ref. 3, p. 180].) In those methylotrophic bacteria which use the serine pathway, 1-C substrates are assimilated partly as formaldehyde (HCHO) and partly as CO_2.

HCHO condenses with tetrahydrofolate (THF), thus forming 5,10-methylene-THF (MTHF); MTHF with glycine (enzyme: serine hydroxymethyltransferase, EC 2.1.2.1) yields serine and regenerates THF. Serine is deaminated (serine glyoxylate aminotransferase, EC 2.6.1.45) to hydroxypyruvate – this reaction being linked to the conversion of glyoxylate to glycine. The reaction sequence continues thus: hydroxypyruvate → glycerate → 2-phosphoglycerate, the latter either being converted to 3-phosphoglycerate (phosphoglycerate mutase, EC 2.7.5.3) and assimilated into biomass, or converted to phosphoenolpyruvate (PEP). CO_2 enters the cycle at this point: with PEP (PEP carboxylase, EC 4.1.1.31) it forms oxaloacetate. The cycle continues oxaloacetate → malate → malyl-CoA, the latter being split (by a lyase, EC 4.1.3.24) to glyoxylate (used to regenerate glycine – see above) and acetyl-CoA. Acetyl-CoA with oxaloacetate (citrate synthase) initiates a subsidiary cycle, first forming citrate and then isocitrate – which is split (isocitrate lyase, EC 4.1.3.1) to glyoxylate and succinate, the latter being converted, in several steps, to oxaloacetate which re-enters the subsidiary cycle [see Appendix II(b)]; as before, glyoxylate is used to regenerate glycine.

The pathway described above is referred to as the icl or icl^+ (isocitrate lyase) serine pathway. Some organisms which use the serine pathway (e.g. '*Pseudomonas* AM1') contain neither malate thiokinase (and hence cannot form malyl-CoA) nor isocitrate lyase (and hence cannot regenerate glyoxylate in the subsidiary cycle); the pathway in these organisms is called the icl^- serine pathway. The way in which acetyl-CoA is used to regenerate glyoxylate (and glycine) in the icl^- serine pathway is not yet understood; in the proposed 'homo-isocitrate pathway' acetyl-CoA condenses with 2-oxoglutarate to form homocitrate and then homoisocitrate which is cleaved to glyoxylate and glutarate – the latter being used to regenerate 2-oxoglutarate.

serine proteases See PROTEASES and type II systems in PROTEIN SECRETION.

seroconversion The development of antibodies in response to exposure to antigen.

seroconversion illness See AIDS.

serodiagnosis Diagnosis based on serological tests.

serofactor 1 See YERSINIA (*Y. pestis*).

serofast Refers to a patient whose serum (sampled at intervals of time) consistently gives positive results in a given serological test.

serological tests for syphilis *Syn.* STANDARD TESTS FOR SYPHILIS.

serological typing See TYPING.

serology The study of in vitro reactions involving one or more of the constituents of SERUM (e.g. a particular type of ANTIBODY, or a component of COMPLEMENT) or of PLASMA (see e.g. clumping factor in COAGULASE); serology is used e.g. in medical diagnostic procedures to detect and/or quantify particular antigens or antibodies. See e.g. HAEMAGGLUTINATION-INHIBITION TEST, IMMUNOASSAY, WIDAL TEST.

serotonin 5-Hydroxytryptamine: a base (derived from tryptophan) with physiological activity similar to that of HISTAMINE; it occurs e.g. in the platelets of many species, and is also found e.g. in the MAST CELLS and basophils of some species.

(See also ENTEROCHROMAFFIN CELLS.)

serotype (serological type; serovar) A serologically (antigenically) distinct VARIETY (usually) within a bacterial species.

serotyping A method of TYPING in which strains are distinguished on the basis of differences in their surface ANTIGENS (e.g. cell wall, flagellar and/or capsular antigens). Essentially, each strain to be typed is tested with a variety of antibodies (different antibodies being specific for antigens on different strains of the organism). If the cells of an unknown (i.e. untyped) strain bind a particular antibody (indicated in vitro by agglutination or precipitation of the cells), the unknown strain is placed in the same category (type) as the strain homologous to that antibody.

Serotyping may be carried out in small test tubes (see also DREYER'S TUBE) or on slides (see SLIDE AGGLUTINATION TEST). (See also LATEX PARTICLE TEST.)

Strains distinguished mainly on the basis of their antigens are called *serotypes*.

serovar *Syn.* SEROTYPE.

Serpens A genus (*incertae sedis*) of catalase-positive, oxidase-positive, microaerophilic, chemoorganotrophic, Gram-negative bacteria which occur e.g. in sediments in freshwater ponds. Cells: flexible rods or filaments, 0.3–0.4 × 8.0–12.0 μm, which have bipolar tufts of flagella (4–10 flagella in each tuft) and some lateral flagella; the cells can exhibit motility in liquids of low and high viscosity, and a 'serpentine-like' motility within agar gels. Metabolism is respiratory (oxidative), with oxygen as terminal electron acceptor. The principal source of carbon and energy is lactate; carbohydrates are not used. NH_4Cl or e.g. peptone (but not nitrite or nitrate) can serve as a source of nitrogen. Optimum temperature: 28–30°C. GC%: ca. 66. Type species: *S. flexibilis*. [Book ref. 22, pp. 373–375.]

Serpula A genus of fungi of the APHYLLOPHORALES (family Coniophoraceae). *S. lacrymans* (formerly *Merulius lacrymans*) causes DRY ROT; it typically forms a thick, resupinate, leathery, initially pale grey fruiting body which, when the basidiospores develop, becomes rusty red with a pale mycelial margin. The hymenium may be wrinkled or shallowly pitted, or it may occur on a layer of irregularly shaped, pendulous, tooth-like processes (the so-called 'stalactite' fruiting body); basidiospores: ellipsoidal, ca. 5 × 10 μm.

serrated wrack *Fucus serratus*.

Serratia A genus of Gram-negative bacteria of the ENTEROBACTERIACEAE (q.v.). Cells: 0.5–0.8 × 0.9–2.0 μm, usually motile. Some species and strains produce the non-diffusible pigment PRODIGIOSIN, forming colonies that are red or have red centres, margins or sectors; prodigiosin production is optimum on peptone–glycerol agar at 20–35°C, and occurs only under aerobic conditions. In the presence of Fe^{2+}, some strains of *S. marcescens* produce a water-soluble pink pigment *pyrimine*, L-2-(2-pyridyl-Δ′-pyrroline-5-carboxylic acid), which diffuses into the agar surrounding the colonies. Typical reactions: citrate +ve; MR −ve; VP +ve at 30°C, but may be −ve at 37°C (O'Meara's method may be −ve); extracellular enzymes include gelatinase, lipase and DNase. Glucose is fermented by the ENTNER-DOUDOROFF PATHWAY. Lactose +ve or −ve, according to species. GC%: 52–60. Type species: *S. marcescens*.

S. ficaria. Prodigiosin −ve. Cultures have a musty, potato-like odour. Occurs mainly in association with figs and fig-wasps [Curr. Micro. (1979) 2 277–282].

S. fonticola. Atypical (species *incertae sedis*).

S. liquefaciens (formerly *Enterobacter liquefaciens*). Prodigiosin −ve. A heterogeneous species which may be split into three: *S. liquefaciens* sensu stricto, *S. proteamaculans* and *S. grimesii*.

S. marcescens. Biogroups A1 and A2 produce prodigiosin. Some strains produce pyrimine. *S. marcescens* is reported to exhibit a flagellum-independent *spreading* motility in which surface tension may be involved [JB (1995) *177* 987–991].

S. odorifera. Prodigiosin −ve. Cultures have a musty, potato-like odour.

S. plymuthica. Most strains produce prodigiosin.

S. rubidaea (= *S. marinorubra*). Most strains produce prodigiosin.

Serratia spp occur in water and soil, on plants, in insects, and in man and animals (see also DFD MEAT). *S. marcescens* and *S. liquefaciens* may cause lethal septicaemia in insects [JGM (1983) *129* 453–464]. Species have also been associated with disease in reptiles, spoilage of hens' eggs, bovine mastitis etc. In man, *S. marcescens* (especially non-pigmented strains) are increasingly frequent causes of nosocomial infections (e.g. pneumonia, UTI). (See also PSEUDOHAEMOPTYSIS; RED DIAPER SYNDROME.)

[Book refs. 22, pp. 477–484, and 46, pp. 1187–1203.]

serum The fluid fraction of coagulated (clotted) blood; it differs from PLASMA e.g. in that it does not contain FIBRINOGEN (a factor in the clotting mechanism). Normal serum contains e.g. various nutrients, electrolytes, albumins, immunoglobulins, waste products.

serum amyloid A See ACUTE-PHASE PROTEINS.

serum hepatitis *Syn.* HEPATITIS B.

serum killing The killing of some strains of Gram-negative bacteria, by either immune or non-immune serum, due mainly or solely to activation of the alternative pathway of COMPLEMENT FIXATION; such *serum sensitivity* may be greater in rough mutants and less or non-existent in capsulated strains [JGM (1983) *129* 2181–2191]. (See also SURFACE EXCLUSION.)

serum sensitivity See SERUM KILLING.

serum sickness A systemic condition, involving a TYPE III REACTION, which may occur ca. 8 days after the initial administration of a large amount of antigen (given e.g. during passive immunization); in such cases antigen is still circulating when antibody first appears in the plasma, and soluble antigen–antibody complexes are formed due to ANTIGEN EXCESS. A generalized inflammatory reaction occurs, with fever, enlargement of lymph nodes, swelling of joints, a generalized urticarial rash, and sometimes renal dysfunction. (cf. GLOMERULONEPHRITIS.)

serum sugars See PEPTONE-WATER SUGARS.

serum water sugars See under PEPTONE-WATER SUGARS.

sessile (1) Of an organism (e.g. a protozoon): attached to the substratum (with or without a stalk), i.e., not free-swimming. (2) Of a fruiting body, spore etc: attached directly, i.e., *without* a stalk, to the substratum, sporophore etc; e.g., the sessile basidiospores of the smut fungi lack sterigmata, arising directly from the basidia.

Sessilina See PERITRICHIA.

sessilinid A member of the Sessilina.

seston All the fine particulate matter which drifts passively in lakes, seas and other bodies of water; it includes PLANKTON and TRIPTON.

***set2* gene** See EIEC.

seta (*pl.* setae) A bristle-like structure. Setae occur e.g. in some types of fungal fruiting body; in some species of *Colletotrichum*, some of the setae which occur in acervuli have truncated, near-colourless apices which give rise to conidia, while the darker, usually pointed setae are sterile [Mycol. (1984) *76* 359–362].

setose Bearing setae – see SETA.

severe combined immunodeficiency (SCID) A heterogeneous group of immune deficiency diseases in which the common feature is a block in the development of T lymphocytes; in some of these diseases there is also a deficiency in B lymphocytes and/or NK cells. A consequence of such immunodeficiency is a marked susceptibility to opportunist pathogens coupled with a poor prognosis. (cf. DIGEORGE SYNDROME.)

One of the SCID diseases is ADENOSINE DEAMINASE DEFICIENCY (q.v.), resulting from mutation in the *ADA* gene.

Another of these diseases is caused by mutation in the gene encoding the γc subunit that is common to the cell-surface receptors of various CYTOKINES – including the interleukins IL-2, IL-4, IL-7, IL-9 and IL-15 [see e.g. Cell (1993) *73* 147–157];

deficiency in the receptor for IL-7 inhibits development of T cells, while deficiency in the receptor for IL-15 inhibits development of NK cells. The gene encoding the γc subunit occurs on the X chromosome, and this (X-linked) disease is designated XSCID of SCIDX1.

SCIDX1 can be treated successfully by GENE THERAPY: $CD34^+$ cells were infected, ex vivo, with a retrovirus-derived vector carrying the gene for γc [Science (2000) *288* 669–672]. In one patient, however, this therapy apparently caused a leukaemia-like illness [Nature (2002) *420* 116–118].

During normal stimulation of cell-surface receptors containing the (wild-type) γc subunit, a kinase, designated JAK-3, binds to the intracellular domain of γc as part of the intracellular signalling process. A phenotype similar to that of SCIDX1 can therefore be caused by mutation in the *JAK-3* gene which inactivates the kinase; unlike the gene encoding γc, *JAK-3* is an autosomal gene, and mutations in *JAK-3* give rise to an autosomal recessive form of the disease in which (like the X-linked form) both T cells and NK cells are affected.

[Primary immunodeficiency diseases (an experimental model for molecular medicine): Lancet (2001) *357* 1863–1869 (SCID: 1864–1865).]

severin See ACTIN.

sewage fungus Slimy, macroscopic masses of mixed microbial growth which develop on rocks etc in many organically polluted (e.g. sewage-polluted) freshwater habitats. The organisms which contribute to sewage fungus are typically those which occur as part of the normal microflora of the habitat; they include bacteria (e.g. *Sphaerotilus natans*, *Zoogloea* spp), fungi (e.g. *Fusarium* spp, *Geotrichum candidum*), protozoa (e.g. *Carchesium polypinum*), and algae (e.g. *Stigeoclonium* spp). Usually, a particular sample of sewage fungus contains one, or a few, dominant species; the dominance of a given species may correlate e.g. with the availability of a particular carbon source.

sewage treatment Sewage includes domestic wastes (from drains, water closets etc.) – sometimes with varying amounts of agricultural and/or industrial effluent – and (often) the contents of rain-water drains; it contains substances in suspension, in solution and in colloidal form.

If discharged to environmental waters, sewage can be harmful in several ways. For example, it can act as a source of infection – encouraging the spread of water-borne diseases such as CHOLERA and TYPHOID FEVER. (See also SHELLFISH POISONING.) Another problem is its content of biodegradable organic matter. In metabolizing these nutrients, the large numbers of sewage organisms can rapidly deplete the oxygen at a locally polluted site, especially in slow-moving or static waters; the consequent development of a microaerobic or anaerobic environment means a loss of habitat for all oxygen-dependent organisms (e.g. fish) in the vicinity. Moreover, anaerobic conditions allow the growth of SULPHATE-REDUCING BACTERIA and other microorganisms whose metabolic products include sulphide and other malodorous substances.

Sewage treatment has two main objectives: (i) to eliminate (or reduce the numbers of) pathogens which cause water-borne diseases, and (ii) to diminish the oxygen-depleting capacity of sewage, i.e. to reduce its BOD.

Small quantities of sewage (e.g. from one or several houses in isolated rural areas) may be treated in a SEPTIC TANK.

Large-volume sewage (from urban areas) is treated by a two-stage process, described below.

Primary treatment may involve simply passing the raw (i.e. untreated) sewage through a screen of metal bars to separate, and dispose of, the grosser debris; the screened sewage may then be passed through a *comminutor* (a rotary shredding device which breaks up the smaller solids). The screened, comminuted sewage effluent then passes slowly through a sedimentation (settlement) tank in which some particulate matter settles out (and is removed as sludge). Sedimentation is sometimes assisted by the addition of alum as the effluent enters the sedimentation tank.

Secondary treatment is designed to reduce the BOD of the primary effluent to acceptable levels by microbial oxidation of the dissolved organic content. This is usually achieved by one of three types of (aerobic) process: the trickle filter, the activated sludge process and the (more recent) biological aerated filter (BAF); these processes are considered below.

The *trickle filter* (*biological filter*, *percolating filter*) consists of a bed of crushed rock, ca. 2 m deep, over which the primary sewage effluent is sprayed. The bed of rock may be enclosed within a circular wall, the sewage being sprayed through holes in the arms of a rotary sprinkler; with a rectangular bed of rock, sewage is sprayed from a distributor arm moving backwards and forwards. The rock surface bears a film of microorganisms – for example, the bacterium *Zoogloea ramigera* and species of ciliate protozoa (e.g. *Carchesium*, *Chilodonella*, *Opercularia* and *Vorticella*). (Also commonly present are e.g. rotifers, crustaceans, insects and arachnids.)

Percolating slowly through the crushed rock, the sewage makes close contact with surfaces that bear the biofilm. The sprayed sewage carries with it dissolved oxygen, so that some of the dissolved organic matter can be oxidized by organisms in the sewage and by those in the biofilm; moreover, some dissolved organic carbon is assimilated, as biomass, by these organisms. The system is not intended to act as a mechanical sieve – but rather to permit close contact between the biofilm and sewage under aerobic conditions; this reduces the level of dissolved organic matter. Moreover, large numbers of sewage bacteria are ingested by protozoans in the biofilm.

Effluent leaving the bed usually contains small particles of biofilm washed from the rock; these particles may be allowed to settle in a *humus tank* before the supernatant is discharged as the final effluent. The final effluent has a much lower BOD – so that, if discharged to a river etc., it will take less oxygen from the water.

The *activated sludge* process is another form of aerobic secondary treatment. Effluent from the primary treatment stage enters a vessel containing activated sludge – a mass of organisms consisting mainly of bacteria (e.g. species of *Acinetobacter* and *Alcaligenes*, *Sphaerotilus natans* and *Zoogloea ramigera*) and protozoa; the latter include ciliates (e.g. *Aspidisca*, *Carchesium*, *Opercularia*, *Trachelophyllum*, *Vorticella*), flagellates, and the testate amoebae *Cochliopodium* and *Euglypha* – amoebae often being found in large numbers and sometimes forming the major component of the biomass. Other organisms present include fungi, rotifers and nematodes. Effluent and sludge are vigorously agitated and aerated for e.g. 6–12 hours so that much of the soluble organic matter in the effluent is oxidized, or assimilated, by the biomass; the BOD is thus greatly reduced, and large numbers of sewage bacteria are ingested by protozoa. The treated effluent is then left in a sludge-settling tank. Good-quality final effluent depends on efficient *flocculation* (= aggregation) of the organisms, this facilitating clarification of the effluent by sedimentation. Flocculation is encouraged by cell-surface hydrophobicity which promotes (i) adherence of cells to flocs, and (ii) *penetration* of the flocs by cells via channels/pores within the flocs [see Microbiology (1998) *144* 519–528]. [Microbial communities in sewage treatment plants: FEMS Ecol. (1998) *25* 205–215.]

In this process, the sludge increases in mass due to microbial growth; following sedimentation, most is removed for disposal, some being retained for treating the next batch of sewage.

The *biological aerated filter* (BAF) is a more efficient approach to aerobic secondary treatment. It consists of a *submerged* bed of fine granular material coated with biofilm; the sewage passes downwards through the bed while air is pumped in at the base of the bed. Because the granules are small, the system can function as a mechanical filter (for fine particulate matter) as well as allowing mineralization of dissolved organic matter. The biofilm in a BAF contains up to five times more biomass than that in a trickle filter of equivalent size – so that, for a given treatment capacity, a BAF can be much smaller.

Appropriate flow conditions in the BAF allow adequate aeration and the establishment of nitrifying bacteria in the biofilm. The importance of this is that NITRIFICATION facilitates the elimination of nitrogen from sewage; thus, if nitrification can be achieved it can be followed by DENITRIFICATION. To encourage denitrification the BAF can be operated anaerobically; much of the nitrogen in sewage can therefore be eliminated by operating aerobic and anaerobic BAF reactors in series. [Combined nitrification–denitrification: FEMS Reviews (1994) *15* 109–117.]

Anaerobic treatment is useful for sewage containing a high proportion of solids (e.g. farm waste, sludge from the activated sludge process). In this process (ANAEROBIC DIGESTION), the stirred sludge is digested in a tank maintained at ca. 35°C. This reduces the bulk of sludge, giving a less offensive material which can be de-watered in sludge-drying beds; much of the carbon is eliminated as methane (which can supply most or all of the energy needs of the plant). (See also IMHOFF TANK.)

Anaerobic treatment can also be used e.g. for wastewater containing terephthalic acid (1,4-benzenedicarboxylic acid) and its isomers – compounds used in the manufacture of polyester products and plastic bottles; in one study, organisms associated with the anaerobic granular sludge system included a high proportion of unidentified bacteria of the δ-Proteobacteria and a number of archaeans related to *Methanosaeta* and *Methanospirillum* [Microbiology (2001) *147* 373–382].

Tertiary treatment is sometimes used to reduce still further the BOD of the effluent; this may be needed e.g. if discharge of final effluent to a river is associated with a low dilution factor. To reduce the amount of fine particulate matter, the effluent can be passed through a *microstrainer*: a hollow cylinder of fine-mesh stainless steel fabric, closed at one end, which rotates on a horizontal axis; effluent is pumped into the cylinder, and strained effluent passes out through the mesh – material retained on the inner surface of the cylinder being constantly removed by jets of water.

The *Immedium filter* is a sand filter in which the grain size increases progressively from top to bottom; effluent flows *upwards* through the filter. Other forms of filter, and grassland irrigation, are also used for tertiary treatment.

The main purpose of tertiary treatment is often regarded as the reduction of BOD by elimination of *carbon* from the effluent. However, removal of *nitrogen* (present mainly as ammonia and nitrate) and *phosphorus* is also desirable in order to discourage the development of algal BLOOMS in the receiving waters. (See also EUTROPHICATION.) The elimination of nitrogen was considered earlier (see BAF). Phosphorus has been removed by using a cycle of alternating anaerobic and aerobic treatments. Anaerobically, some organisms increase in biomass but release phosphorus; aerobically, the (increased) biomass assimilates phosphorus – and is subsequently separated and disposed of. The bacterium *Acinetobacter calcoaceticus* is reported to accumulate phosphate (up to ~10% dry weight), >50% as POLYPHOSPHATE. [Biological phosphorus removal: MS (1984) *1* 149–152. Identification of polyphosphate-accumulating bacteria for biological phosphorus removal in activated sludge plants: AEM (2000) *66* 1175–1182.]

Disinfection of final effluent has been carried out with e.g. CHLORINE or OZONE. In at least one plant in the USA, de-watered sludge (50% solids) has been disinfected with IONIZING RADIATION (dose: 1 Mrad from a ^{137}caesium *gamma* source).

[Bacterial activities in sewage treatment plants: AvL (1984) *50* 665–682. Environmental health engineering in the tropics: Book ref. 216. Low-cost sewerage for developing countries: Book ref. 217.]

sewer gas See ANAEROBIC DIGESTION.

sewer swab *Syn.* MOORE SWAB.

sex factor (1) A general term for a CONJUGATIVE PLASMID (sense 1). (2) Collectively, those genes in a conjugative plasmid which specify conjugation (see e.g. RTF). (3) *Syn.* F PLASMID.

sex pheromone See PHEROMONE.

sex pili See PILI.

sex-ratio organism See SRO.

sexduction The transfer, by CONJUGATION (sense 1b), of *chromosomal* genes from one bacterium to another. Sexduction may involve an HFR DONOR or a donor strain carrying a PRIME PLASMID; if the chromosome of the Hfr donor is mobilized by the F PLASMID, or if conjugal transfer is promoted by an F′ plasmid, the process is called *F-duction*.

sexually transmitted disease (STD) Any disease which can be transmitted by intimate contact with e.g. genitals, mouth or rectum. In addition to the 'classical' VENEREAL DISEASES, STDs include e.g. AIDS, HEPATITIS B, HERPES SIMPLEX, MOLLUSCUM CONTAGIOSUM, NON-GONOCOCCAL URETHRITIS, TRICHOMONIASIS, and genital warts (see PAPILLOMA), as well as parasitic infestation by lice ('crab') or mites (scabies, 'the itch').

Sézary syndrome (SS) A cutaneous T-cell lymphoma which resembles – but is distinct from – ADULT T-CELL LEUKAEMIA; HTLV-I provirus has been detected only rarely in SS, and does not appear to be a primary causal agent. The neoplastic T cells in SS may resemble those of ATL, but the nucleus is more commonly cerebriform than multilobed; ATL and SS cells are antigenically distinct and are functionally distinct in vitro [Book ref. 106, pp. 275–284].

SFFV See FRIEND VIRUS and RAUSCHER VIRUS.

***sfiA*, *sfiB* genes** See SOS SYSTEM.

SFV SEMLIKI FOREST VIRUS.

S_G Gower coefficient. In the comparison of two strains by NUMERICAL TAXONOMY: an expression of the degree of similarity in which the comparison involves both simple, discontinuous data (e.g. + and −, or 0 and 1) and continuous (quantitative) data. (cf. entry S_{SM}.)

SH-activated cytolysins THIOL-ACTIVATED CYTOLYSINS.

SH antigen *Syn.* HBsAg: see HEPATITIS B VIRUS.

shaded matrix In NUMERICAL TAXONOMY: a graphic representation of the interrelationships between the OTUs in a given study. It is a triangular tabulation, the ordinate and abscissa each consisting of a complete list of designations of the OTUs under study; the degree of mutual similarity between the OTUs in any given cluster is indicated by the intensity of shading: the highest degree of similarity corresponds to the darkest shading.

shadow-casting See ELECTRON MICROSCOPY (a).
shadow yeasts *Syn.* MIRROR YEASTS.
shadowing See ELECTRON MICROSCOPY (a).
shake culture (1) A procedure used e.g. for the isolation of anaerobic bacteria. The inoculum is dispersed (by gentle shaking) in a molten agar medium (at ca. 48°C) in a long glass test-tube or a VEILLON TUBE; the agar is allowed to set, and the tube is incubated. An individual colony in the lower (anaerobic) part of the agar can be removed for subculture. (2) A culture in a *liquid* medium which is continually shaken to ensure aerobiosis in the medium.
sharka disease *Syn.* PLUM POX.
sharp eyespot See EYESPOT (sense 2).
sheath (1) In SHEATHED BACTERIA: a secreted, tubular structure formed around a chain of cells or around a bundle of filaments; cells may or may not subsequently separate from the sheath.

(2) In certain bacteria: a layer of outer membrane covering the flagellum (see FLAGELLUM (a)).

(3) See 'outer sheath' in SPIROCHAETALES.
sheath blight (of rice) CEREAL DISEASES.
sheathed bacteria Species of e.g. HALISCOMENOBACTER, LEPTOTHRIX, SPHAEROTILUS, THIOPLOCA and THIOTHRIX which form sheathed trichomes or fascicles.
sheathed flagellum See FLAGELLUM (a).
sheep cell agglutination test (SCAT) (1) The PAUL–BUNNELL TEST or, in general, any test in which non-sensitized sheep erythrocytes are agglutinated. (2) The ROSE–WAALER TEST.
sheep diseases Sheep are susceptible to a wide range of diseases, some of which are specific to sheep. (a) *Bacterial diseases*: see e.g. ANTHRAX, BLACK DISEASE, BRAXY, BRUCELLOSIS, CAMPYLOBACTERIOSIS, ENZOOTIC ABORTION, FOOT ROT, JOHNE'S DISEASE, JOINT-ILL, LAMB DYSENTERY, LISTERIOSIS, LUMPY WOOL, PULPY KIDNEY, SCALD, STRUCK, SWELLED HEAD, TULARAEMIA; cf. WIMMERA GRASS POISONING. (b) *Fungal diseases*: see e.g. ASPERGILLOSIS. (c) *Mycotoxicoses*: see e.g. LUPINOSIS, PASPALUM STAGGERS, PENITREMS, SPORIDESMINS. (d) *Viral diseases*: see e.g. BLUETONGUE, BORDER DISEASE, CONTAGIOUS PUSTULAR DERMATITIS (sense 1), FOOT AND MOUTH DISEASE, JAAGSIEKTE, LOUPING-ILL, MAEDI, NAIROBI SHEEP DISEASE, PESTE DES PETITS RUMINANTS, RIFT VALLEY FEVER, SHEEP POX, VISNA; cf. SCRAPIE. (See also SHEEP SCAB.)
sheep pox (variola ovina) An acute, infectious disease specific to sheep; the causal agent, a *Capripoxvirus*, appears to infect via abrasions or e.g. by droplet infection. Incubation period: 2 days to 2 weeks. Adult sheep typically develop vesicular, scab-forming lesions on the skin; there is no systemic involvement and mortality rates are generally low, but in ewes the disease may predispose towards mastitis if the udder is affected. In lambs (and sometimes in adult sheep) the disease involves e.g. fever, nasal discharge, and the development of lesions on the skin and buccal mucosa and in the alimentary, respiratory and urogenital tracts; mortality rates can be high, and lambs may die even before the skin lesions develop. (cf. GOAT POX.)
sheep-pox subgroup See CAPRIPOXVIRUS.
sheep scab A (non-microbial) disease caused by bites of the mite *Psoroptes ovis*; symptoms: itching, skin lesions, and wool shedding.
shelf fungi *Syn.* BRACKET FUNGI.
shell (*virol.*) A term used by some authors to refer to a CAPSID.
shell disease (1) *Syn.* BURNED SPOT DISEASE. (2) See OSTRACOBLABE IMPLEXA.
shell vial assay A (qualitative or quantitative) method for detecting human cytomegalovirus (HCMV) in which the specimen is centrifuged onto a culture of fibroblasts and the preparation subsequently stained with fluorescent monoclonal antibodies specific for the 72 kDa phosphoprotein product (p72) of an immediate-early gene.
shellfish diseases See e.g. CRUSTACEAN DISEASES and OYSTER DISEASES.
shellfish poisoning A form of FOOD POISONING which results from the consumption of shellfish (molluscs, crustacea) contaminated with toxins or pathogens. Aquatic *bivalve molluscs* (clams, cockles, mussels, oysters etc) obtain their food by filtering microscopic organisms from the ambient water. Any human pathogens present in the water can thus be concentrated in the bodies of the shellfish and can cause disease when these are eaten by humans. Pathogens transmitted in this way include e.g. HEPATITIS A virus, enteroviruses, 'parvovirus-like agents' (see FOOD POISONING (i)), and salmonellae (including *S. typhi*); such pathogens typically derive from sewage, whereas *Vibrio parahaemolyticus* (see FOOD POISONING (h)) is a natural inhabitant of brackish and marine waters. *Gastropod molluscs* (e.g. whelks, winkles) and *crustacea* (e.g. lobsters, prawns) are not filter-feeders but may nevertheless become contaminated with pathogens which may be derived from the natural habitat of the animal (as e.g. in the case of *V. parahaemolyticus*) or may be introduced during the handling and processing of the shellfish (as e.g. in the case of staphylococci). (See also FISH SPOILAGE.)

Paralytic shellfish poisoning (PSP, 'mussel poisoning') is a severe, often fatal condition resulting from the consumption of shellfish (commonly mussels or clams) which have themselves consumed neurotoxin-forming dinoflagellates (e.g. *Gonyaulax* spp) and accumulated the toxin(s) – e.g. SAXITOXIN – in their tissues. (See also RED TIDE.) Symptoms of PSP (caused by saxitoxin) typically include diarrhoea, fatigue, and a tingling sensation beginning around the face and spreading to the arms, fingers, and toes; later, numbness may develop, followed by weakness, paralysis, and death (usually within 12 hours) from respiratory paralysis. (cf. CIGUATERA.)
sherry See WINE-MAKING.
ShET2 enterotoxin See EIEC.
Shewanella A genus of bacteria proposed to include *Alteromonas hanedai (S. hanedai), A. putrefaciens (S. putrefaciens)*, and strains of barophilic bacteria (see BAROPHILE) isolated from deep-sea waters (*S. benthica*) [SAAM (1985) *6* 171–182; name validation: IJSB (1986) *36* 354–356]. (cf. ALTEROMONAS.)
shield cell (*algol.*) See CHAROPHYTES.
shift up, shift down (Refers to) a change in (experimental) conditions which causes the growth rate of a microorganism to increase (shift up) or decrease (shift down).
Shiga–Kruse bacillus *Shigella dysenteriae* serotype 1.
shiga-like toxins See SHIGA TOXIN.
shiga toxin A protein toxin, produced by strains of *Shigella dysenteriae*, which apparently causes at least some of the symptoms of (bacillary) DYSENTERY; the toxin is cytotoxic in vivo, and is also toxic for some types of cultured cells (e.g. HeLa).

The shiga holotoxin consists of an A subunit associated with a pentameric ring of B subunits. The holotoxin binds, via B subunits, to glycolipid receptor sites on the target cell; these receptors – globotriosylceramide (Gb_3) – are found e.g. in the membrane of endothelial cells (cells that line blood vessels). Shiga toxin is reported to be active against e.g. endothelial cells, colonic/ileal epithelial cells, and B lymphocytes expressing the CD77 ANTIGEN.

On uptake of toxin by a cell, the A subunit undergoes proteolytic cleavage, the (catalytic) part of the subunit remaining attached temporarily by a disulphide bond – which is later reduced. The catalytic part of the A subunit has *N*-glycosidase activity: it cleaves a *specific* adenine residue from the 28S rRNA in eukaryotic ribosomes; this inhibits protein synthesis (by inhibiting the EF-1-mediated binding of tRNA), resulting in cell death. (This mechanism is identical to that of the plant toxin *ricin*.)

[Enteric bacterial toxins (mode of action and relevance to intestinal secretion): MR (1996) *60* 167–215.]

Shiga-like toxins (verotoxins; verocytotoxins; Vero cell cytotoxins) are produced by so-called *enterohaemorrhagic* strains of *Escherichia coli* (see EHEC). These (phage-encoded) toxins are classified as shiga-like toxins I and II (SLT-I, SLT-II); SLT-I is virtually identical to the shiga toxin of *Shigella dysenteriae* type 1. Note that these toxins are also called 'shiga toxins' or 'Shiga toxins'; thus, with this terminology, SLT-I has been abbreviated to Stx1, and SLT-II to Stx2. There are a number of variant forms of SLT-II, each designated by a lower-case letter. A new variant form of SLT-II (Stx2f) has been isolated from *E. coli* in the faeces of pigeons (*Columba livia*) [AEM (2000) *66* 1205–1208].

All members of the shiga toxin family appear to share a common membrane receptor site and apparently have the same mode of action.

The precise role of these toxins in pathogenesis is uncertain. However, it has been suggested that they may act as *vasculotoxins* whose primary target is the vascular endothelium. In one model of EHEC pathogenesis, macrophages, adherent at toxin-mediated vascular lesions, are stimulated to secrete certain pro-inflammatory cytokines (TNF-α, IL-1) which upregulate local expression of Gb_3 sites – resulting in further toxin binding and exacerbation of vascular damage [TIM (1998) *6* 228–233].

(See also STARFISH.)

Shiga's bacillus *Shigella dysenteriae* serotype 1.

Shigella A genus of Gram-negative bacteria of the ENTEROBACTERIACEAE (q.v.). Cells: straight rods, 0.5–1.0 × 1–3 μm, non-motile. Growth occurs e.g. on nutrient agar, MacConkey's agar and EMB agar, may or may not occur on more inhibitory media (e.g. DCA, SS agar), and does not occur on KCN media. Sugars are fermented to acid (usually without gas). MR +ve; VP −ve; citrate −ve; H_2S −ve (in TSI); phenylalanine deaminase −ve; lysine decarboxylase −ve. GC%: 49–53. Type species: *S. dysenteriae*.

Shigella spp are intestinal pathogens of man and other primates (see bacillary DYSENTERY); other animals are normally resistant. In animal studies, *Shigella* is reported to be taken up by the M cells of Peyer's patches [e.g. TIM (1998) *6* 359–365].

S. boydii (= subgroup C). 18 serotypes. Mannitol is fermented, lactose is not.

S. dysenteriae (= subgroup A). 12 serotypes. Serotype 1 is catalase −ve and produces SHIGA TOXIN. Mannitol and lactose are (usually) not fermented.

S. flexneri (= subgroup B). 6 serotypes. Most strains ferment mannitol but not sucrose or lactose; serotype 6 includes the Newcastle strain (mannitol not fermented, gas produced from glucose), the Manchester strain (acid and gas from glucose and mannitol), and the Boyd 88 strain (acid, no gas, from glucose and mannitol).

S. sonnei (= subgroup D). One serotype with two 'phases', I and II (resembling the SMOOTH–ROUGH VARIATION of other enterobacteria). Mannitol is fermented; fermentation of lactose and sucrose is detectable after 24 hours. Strains are identified by colicin TYPING.

[Book ref. 22, pp. 423–427; serology: Book ref. 68, pp. 113–142.]

[Recommendations on the classification of shigellae: IJSB (1984) *34* 87–88.]

shigellosis See DYSENTERY (a). (cf. SLEEPY FOAL DISEASE.)

shii-take The edible mushroom *Lentinula edodes* (formerly *Lentinus edodes* or *Cortinellus edodes*) traditionally cultivated in Japan. Hardwood logs are inoculated via drill-holes; fruiting, which occurs in spring and autumn, begins after ca. 8 months and may continue for several years.

shikimate pathway See Appendix IV (f).

Shine–Dalgarno sequence See PROTEIN SYNTHESIS.

shingles *Syn.* HERPES ZOSTER.

shipping fever See BOVINE RESPIRATORY DISEASE.

shipyard eye *Syn.* EPIDEMIC KERATOCONJUNCTIVITIS.

Shirlan See SALICYLANILIDES.

shmoo cell A large, asymmetrical, often pear-shaped yeast cell formed e.g. by the action of a sex PHEROMONE.

shock-sensitive transport system *Syn.* BINDING PROTEIN-DEPENDENT TRANSPORT SYSTEM.

shoestrings *Syn.* boot laces: see ARMILLARIA.

Shope fibroma virus See LEPORIPOXVIRUS.

Shope papilloma virus See PAPILLOMAVIRUS.

short-incubation hepatitis *Syn.* HEPATITIS A.

short patch repair See EXCISION REPAIR.

shotgun experiment A CLONING experiment in which the entire genome of an organism is cut into random fragments and cloned (e.g. to generate a genomic LIBRARY).

showdomycin A NUCLEOSIDE ANTIBIOTIC which structurally resembles uridine; it inhibits e.g. UMP kinase and other enzymes.

shoyu *Syn.* SOY SAUCE.

shufflon A term proposed for a cluster of DNA segments (within e.g. a plasmid) which can invert independently or in groups, resulting in complex rearrangements of the DNA. Such a 'clustered inversion region' has been observed in the IncIα plasmid R64; it appears to function as a biological switch, regulating the selection of one of 7 possible open reading frames [NAR (1987) *15* 1165–1172]. (cf. RECOMBINATIONAL REGULATION.)

Shukla's method See IMMUNOSORBENT ELECTRON MICROSCOPY.

shut-down cell *Syn.* GROWTH PRECURSOR CELL.

shuttle vector (bifunctional vector) A CLONING vector which can replicate in more than one type of organism; e.g., a shuttle vector which can replicate in both *Escherichia coli* and *Saccharomyces cerevisiae* can be constructed by linking sequences from an *E. coli* plasmid with sequences from the yeast 2μ plasmid.

SHV β-lactamase See β-LACTAMASES.

Shwartzman reaction The *local* Shwartzman reaction involves localized skin necrosis which occurs when an intravenous injection of e.g. ENDOTOXIN (sense 1) is given some hours after an intradermal injection of endotoxin.

The *generalized* Shwartzman reaction (GSR) is a poorly understood, complex syndrome exhibited when a second intravenous injection of endotoxin is given 24 hours after the first; it is characterized by intravascular coagulation and shock. Various mechanisms have been proposed to account for the intravascular coagulation; in one, COMPLEMENT-FIXATION leads to injury of immature PMNs which release factors that promote clotting. In other models macrophages are considered to be the primary cellular effectors. (cf. ENDOTOXIC SHOCK.)

sialic acids See NEURAMINIC ACID.

sialyl-Lewis x See INFLAMMATION.

***sib* sequence** See BACTERIOPHAGE λ.

sibiromycin See ANTHRAMYCIN.

Sicilian sandfly fever virus See PHLEBOVIRUS.

sickener See RUSSULA.

SID test (*vet.*) SINGLE INTRADERMAL TEST.

SIDD Supercoiling-induced duplex destabilization (= stress-induced duplex destabilization): destabilization (involving strand separation) of a particular region (a SIDD site) in a DNA duplex as a response to a given level of superhelicity. In a model for the activation of the *ilv*P_G promoter in *Escherichia coli*, destabilization of an upstream SIDD site is counteracted by the binding of INTEGRATION HOST FACTOR close to the site; this causes translocation of superhelical energy to the (downstream) −10 sequence of the promoter – thus facilitating the formation of an open complex and initiating transcription from the promoter [see e.g. Mol. Microbiol. (2001) *39* 1109–1115 (1112)].

sideramines (1) *Syn.* SIDEROPHORES. (2) *Syn. hydroxamate* SIDEROPHORES.

Siderocapsa A genus of IRON BACTERIA. Cells: cocci or coccoid forms which occur singly or in clusters within a common capsule. Type species: *S. treubii*.

siderochromes *Syn.* SIDEROPHORES.

Siderococcus A genus of bacteria found e.g. in soil [see e.g. Book ref. 45, pp. 1049–1059].

sideromycins Iron-chelating ANTIBIOTICS, formed by certain actinomycetes, which are structurally related to hydroxamate SIDEROPHORES; it appears that the iron-chelating part of the molecule permits transport into the target cell, and that the other part of the molecule, which is released by intracellular hydrolysis, is the (as yet uncharacterized) toxic moiety. Sideromycins include *albomycin* and the *ferrimycins* – analogues of ferrichrome and ferrioxamine, respectively.

siderophilins A family of ferric IRON-chelating glycoproteins which occur in vertebrates: see e.g. LACTOFERRIN and TRANSFERRIN. (cf. FERRITIN; see also SIDEROPHORES.) Certain pathogens (such as *Neisseria gonorrhoeae* and *N. meningitidis*) can obtain iron directly from siderophilins [acquisition of transferrin-bound iron by pathogens: Mol. Microbiol. (1994) *14* 843–850]; some pathogens (including the species of *Neisseria* referred to) have cell-surface receptors for human transferrin, and such receptors have been considered as potential targets for VACCINE production.

siderophores (*syn.* ironophores; siderochromes) Low-molecular-weight *ferric* iron-chelating compounds synthesized and exported by most microorganisms for the sequestration and uptake of IRON (q.v.). The two main structural classes of siderophores are catecholamides and hydroxamates; a given organism may produce siderophores of one or both classes. (Typically, iron is bound more tightly by catecholamides than by hydroxamates.) Some siderophores are PLASMID-encoded (see e.g. COLV PLASMID).

Catecholamides ('phenolates') are derivatives of catechol (*o*-dihydroxybenzene). One example is *enterobactin* (= *enterochelin*): a cyclic trimer of 2,3-dihydroxy-*N*-benzoyl-L-serine produced e.g. by various enterobacteria; in *Escherichia coli* it is synthesized by condensation of 2,3-dihydroxybenzoic acid (formed via chorismic acid) with L-serine – the synthesis involving products of genes *entA–entG*. In *E. coli* the cell-surface receptor for enterobactin is the FEPA PROTEIN. *Parabactin* is a catecholamide formed by *Paracoccus denitrificans* from N^1,N^8-bis(2,3-dihydroxybenzoyl)spermidine ('compound II'). *Vibrio anguillarum* forms a novel catechol-type siderophore, *anguibactin* [characterization of anguibactin: JB (1986) *167* 57–65].

Hydroxamates, derivatives of hydroxamic acid (R.CO.NHOH), include *aerobactin*, a siderophore formed e.g. by various enterobacteria; genes encoding the aerobactin system – the siderophore itself (*iucA–iucD*) and its (outer membrane) receptor protein (*iutA*) – are carried by many COLV PLASMIDS, although the system may also be chromosomally encoded. [Plasmid-specified aerobactin system: Book ref. 161, pp. 741–757; characterization of the *iucA* and *iucC* genes: JB (1986) *167* 350–355.] The products of the *fhuB*, *fhuC* and *fhuD* genes, located in the cytoplasmic membrane and/or periplasm, may form a binding system for the uptake of iron by aerobactin and other hydroxamate siderophores [JGM (1992) *138* 597–603].

In *Escherichia coli*, the outer membrane receptor for aerobactin, the IutA protein, can also bind CLOACIN DF13. In *E. coli*, the aerobactin and enterobactin systems can be induced independently [Inf. Immun. (1986) *51* 942–947]; both systems apparently require the TONB PROTEIN for transport into the periplasm [JGM (1992) *138* 597–603].

Other hydroxamate siderophores include e.g. *coprogen* (produced by many fungi); *ferrichrome* (formed e.g. by species of *Aspergillus*, *Neurospora* and *Ustilago*): a cyclic hexapeptide consisting of a glycine tripeptide and a tripeptide of δ-*N*-acetyl-L-δ-*N*-hydroxyornithine; *ferrioxamines* and DESFERRIOXAMINE (formed e.g. by certain bacteria); NOCARDAMINE; and the *terregens factor* (a growth requirement of e.g. *Arthrobacter* spp).

For extracellular growth, the pathogen *Mycobacterium tuberculosis* uses a peptide siderophore, EXOCHELIN, and a cell-wall-associated lipophilic compound: MYCOBACTIN; exochelin appears to obtain iron from SIDEROPHILINS and to pass it to mycobactin for internalization.

Certain siderophores, e.g. parabactin, have been found to be active, in vitro, against a murine leukaemia cell line and against herpes simplex type 1 virus [TIBS (1986) *11* 133–136].

(See also PYOVERDIN.)

SIDS SUDDEN INFANT DEATH SYNDROME.

sIg Any immunoglobulin at the surface of a B LYMPHOCYTE.

sIgA Secretory IgA (see entry IgA).

sigatoka disease *Syn.* BANANA LEAF SPOT.

***sigF* gene** See SIGMA FACTOR.

sigla Plural of SIGLUM.

siglum A designation consisting of letters (particularly initial letters) and/or other characters. Some sigla are acronyms.

σ (superhelix density) See DNA.

σ factor (sigma factor) A protein which can bind to a DNA-dependent RNA POLYMERASE (RPase) core enzyme and confer on it the ability to initiate TRANSCRIPTION from a *particular* type of PROMOTER.

Cells typically or always encode more than one type of sigma factor, and different types of sigma factor generally confer on the RNA polymerase the ability to transcribe (in vivo) from different promoters; by synthesizing particular type(s) of sigma factor, at appropriate times, the cell can thus control expression of particular gene(s).

In *Escherichia coli* the main sigma factor is σ^{70} (MWt ~70000) encoded by gene *rpoD*; an RPase holoenzyme containing σ^{70}($E\sigma^{70}$) can initiate transcription from most promoters in the *E. coli* genome – sometimes in association with positive regulatory factors (as e.g. in CATABOLITE REPRESSION). Other sigma factors in *E. coli* include e.g. σ^{32} (RpoH, formerly HtpR: see HEAT-SHOCK PROTEINS) and σ^{60} (the NtrA protein: see NTR GENES).

There are many examples of the way in which the timing of gene expression is controlled by the activity of particular sigma factor(s). In *E. coli*, assembly of the FLAGELLUM from a number of different proteins is a highly organized process in which components are added in strict sequence – the corresponding genes being expressed in a way that reflects this sequence.

Thus, class III genes are expressed only after genes of classes I and II, and the expression of class III genes depends on synthesis of a sigma factor encoded by the *fliA* gene; however, even when FliA has been synthesized, this sigma factor is temporarily inactivated by an *anti-σ factor* (FlgM, product of gene *flgM*) until the basal body and hook of the flagellum have been completed – completion of the hook apparently acting as a signal that allows release of FlgM and consequent transcription of the class III genes. [Genetic control of flagellar assembly: Cell (1995) *80* 525–527.]

The *E. coli* sigma factor σ^s (encoded by *rpoS*) was once associated solely with stationary-phase events but is now known to be involved in the regulation of various responses to stress under log-phase, as well as stationary-phase, conditions [Mol. Microbiol. (1996) *21* 887–893]. Under in vitro conditions, either $E\sigma^{70}$ or $E\sigma^s$ can be used for transcription from some promoters, but, in vivo, many promoters exhibit specificity for *one* of the two holoenzymes – $E\sigma^{70}$ *or* $E\sigma^s$; studies on sigma factor specificity in the *osmY* gene have indicated that specificity for σ^s in this gene is influenced strongly (perhaps even determined) by three global regulators – the cAMP–CRP complex (see CATABOLITE REPRESSION), the integration host factor (IHF) and the leucine-responsive regulator protein (Lrp) – which preferentially inhibit $E\sigma^{70}$-dependent expression from the *osmY* promoter [EMBO (2000) *19* 3028–3037].

Expression of the *rpoS* gene is reported to be regulated by the HU PROTEIN [Mol. Microbiol. (2001) *39* 1069–1079] and also by a small, novel RNA molecule which interacts with *rpoS* mRNA and relieves the inhibitory activity of a stem- and-loop structure [Mol. Microbiol. (2001) *39* 1382–1394].

In vegetative cells of *Bacillus subtilis* the major sigma factor is σ^{43} (formerly called σ^{55}; encoded by gene *rpoD*); this sigma factor shows some homology with the *E. coli* σ^{70} and recognizes the same class of promoters. There are also a number of minor sigma factors. A sigma factor may be present at low levels during normal growth but up-regulated under appropriate conditions. For example, σ^H (encoded by gene *spo0H*) is up-regulated during the initiation of sporulation in *Bacillus subtilis* as a consequence of repression of *abrB* by Spo0A~P (see figure (a) in ENDOSPORE). Later in the initiation process, the sporulation-specific sigma factors σ^E (formerly σ^{29}; encoded by *spoIIG*) and σ^F are synthesized to promote the next stage of sporulation.

The $E\sigma^{28}$ of *B. subtilis* can initiate transcription at *E. coli* heat-shock promoters.

In *B. subtilis*, many of the genes and operons so far investigated have two or more overlapping or adjacent promoters which are usually specific for RPase holoenzymes containing different sigma factors; this presumably allows control of gene expression under various conditions and/or during different stages of development.

In *Mycobacterium tuberculosis*, the discovery of a gene (*sigF*) [PNAS (1996) *93* 2790–2794] encoding a protein homologous to a sporulation-specific sigma factor in *Streptomyces coelicolor* has provided some evidence in support of the hypothesis that latency (persistence) in this pathogen is associated with the presence of a spore-like state. [Mechanisms of latency in *M. tuberculosis*: TIM (1998) *6* 107–112.]

Certain bacteriophages encode sigma factors that are necessary for the expression of some of their genes: see e.g. BACTERIOPHAGE SPO1 and BACTERIOPHAGE T4 (gp55).

In some cases, the expression of a sigma factor results from up-regulated translation of its mRNA (rather than increased transcription of the gene). In *E. coli*, for example, the increase in σ^{32} following heat shock is due mainly to increased translation of mRNA; translation is up-regulated because the higher temperature destabilizes a secondary structure in the mRNA which, under normal conditions, represses translation [GD (1999) *13* 633–636].

Enhancer-dependent sigma factors (σ^{54}; = σ^N). This unique type of sigma factor, once thought to occur only in higher organisms, is found in many species of bacteria (although not in all species [FEMS (2000) *186* 1–9]). When bound to a promoter, the σ^{54}-RNA polymerase holoenzyme must be *activated* by an energy-dependent process before DNA melting (formation of an 'open complex', and initiation of transcription) can occur at the promoter. Such activation involves a specialized *activator* protein which binds to DNA at an *enhancer* region remote from the promoter – but which can be brought close to the holoenzyme through *looping* of the intervening DNA; with activator and holoenzyme juxtaposed, the activator's ATPase activity (triggered e.g. by phosphorylation) provides the energy for localized melting (open complex formation) in the promoter region. (In some cases, DNA looping involves the INTEGRATION HOST FACTOR.)

One early example of an enhancer-dependent sigma factor is the *rpoN* (= *ntrA*) gene product involved in regulating nitrogen metabolism (see NTR GENES). A more recent example is a σ^{54}-regulated operon concerned with the processing of RNA [JBC (1998) *273* 25516–25526].

[σ^{54} (minireview): JB (2000) *182* 4129–4136.]

(See also ANTI-SIGMA FACTOR.)

sigma particles See LYTICUM.

σ structure See ROLLING CIRCLE MECHANISM.

sigma virus A probable member of the RHABDOVIRIDAE. Sigma virus infects fruit-flies (*Drosophila melanogaster*) and is transmitted by the female flies to their offspring. The virus is normally non-pathogenic; however, on brief exposure to CO_2 at certain concentrations, infected flies are paralysed and die within a few hours, whereas non-infected flies recover completely. [Sigma virus in cell culture: JGV (1984) *65* 91–99.] (See also DROSOPHILA X VIRUS.)

σ^E (AlgU) See ALGINATE.

sigmamycin A mixture of TETRACYCLINE and OLEANDOMYCIN.

signal amplification (bDNA) See BDNA ASSAY.

signal hypothesis The hypothesis that a protein which is destined to pass through (or into) a membrane is synthesized in a precursor form (*pre-protein*) with a specific N-terminal sequence of amino acid residues (= *signal sequence*, *signal peptide*, *leader peptide*) which is essential for the initiation of translocation and is typically excised (cleaved) during translocation. Translocation in this mode has been demonstrated for many (but not all) secreted and membrane proteins in prokaryotes and eukaryotes; in other cases, protein translocation involves distinct processes which lack the characteristics described above (see later).

A signal sequence often consists of several hydrophilic terminal amino acid residues followed by a sequence of >10 predominantly hydrophobic residues and a specific cleavage site; the hydrophobic residues have a tendency to form an α-helix conformation. It is generally assumed that a signal sequence adopts a conformation which spans the membrane and facilitates translocation of the remainder of the polypeptide.

[Signal sequences: Mol. Microbiol. (1994) *13* 765–773.]

In addition to a terminal signal sequence, the pre-protein may incorporate other sequence(s) which ensure that it reaches the correct final destination; for example, a membrane protein (as opposed to a secreted protein) may contain a *membrane anchor sequence* (= 'anchor' or 'stop transfer' sequence) that holds the mature protein on or within the correct membrane. Moreover, a protein may undergo post-translational modification to ensure that it is targeted to the correct membrane site (see e.g. BRAUN LIPOPROTEIN).

In eukaryotes, secretory proteins (i.e. those to be secreted) – and proteins destined e.g. for the plasma membrane, or for LYSOSOMES – are synthesized on ribosomes of the rough ENDOPLASMIC RETICULUM. Initially, synthesis begins on a free (cytosolic) ribosome, but the newly synthesized (N-terminal) signal sequence binds to a SIGNAL RECOGNITION PARTICLE (SRP) – and this stops polypeptide chain elongation. The ribosome–signal peptide–SRP complex then binds to a 'docking protein' at a site on the cytoplasmic surface of the rough endoplasmic reticulum; the SRP is released, and the signal sequence inserts into a transmembrane 'translocation complex' such that, on resumption of chain elongation, the nascent polypeptide is translocated through the ER co-translationally – apparently driven mechanically by chain elongation. The signal peptide is excised by an endopeptidase (*signal peptidase*) whose catalytic site occurs on the lumen-facing side of the ER. Secretion then follows the PALADE PATHWAY.

By contrast, most of the proteins destined to enter a MICROBODY (e.g. a PEROXISOME) are synthesized on free (cytosolic) ribosomes without a cleavable signal sequence [BBA (1986) *866* 179–203]. [Protein import into organelles (hierarchical targeting signals): Cell (1986) *46* 321–322.]

In prokaryotes the role of the signal sequence is essentially similar in that its function involves insertion into the transmembrane translocation system. (An earlier view – the 'membrane trigger' or 'trigger' hypothesis – supposed that the signal peptide confers on the pre-protein the ability to change conformation following contact with the membrane, thus initiating uptake by the membrane.)

In Gram-negative bacteria, signal sequence-dependent secretion is involved in type II and type IV forms of PROTEIN SECRETION (q.v.); in type II systems, the secretion of *some* proteins is mediated by a signal recognition particle (see PROTEIN SECRETION). During, or immediately after, translocation of the protein, the signal sequence is cleaved by an enzyme (*signal peptidase*); in *Escherichia coli*, one of the two signal peptidases (*leader peptidase*; = *leader peptidase I*) is active on a wide range of proteins, while the second enzyme (*lipoprotein signal peptidase*; = *signal peptidase II*) is active only on specific, modified proteins (such as the BRAUN LIPOPROTEIN). (In both of these signal peptidases the catalytic site is apparently on the periplasmic side of the cytoplasmic membrane.) Excised signal peptides are degraded by an enzyme (*signal peptide peptidase*) – e.g. *protease IV* in *E. coli*.

In some cases the signal sequence is not excised following transmembrane translocation of a protein. One such exception is the export of the TONB PROTEIN to the periplasm in *E. coli* [JBC (1988) *263* 11000–11007].

Secretion that is independent of a cleavable N-terminal signal sequence occurs in type I and type III forms of protein secretion in Gram-negative bacteria.

signal peptidase See SIGNAL HYPOTHESIS and PROTEIN SECRETION.

signal peptidase II See PROTEIN SECRETION.

signal peptide See SIGNAL HYPOTHESIS.

signal peptide peptidase See SIGNAL HYPOTHESIS.

signal recognition particle (SRP) In eukaryotes: a particle that is involved in translocation of proteins across the endoplasmic reticulum (see SIGNAL HYPOTHESIS); an SRP contains six polypeptides together with a 7S RNA which is essential for function [Nature (1982) *299* 691–698].

In prokaryotes: a particle involved in the translocation of certain proteins across the cytoplasmic membrane (see type II systems in PROTEIN SECRETION).

signal sequence See SIGNAL HYPOTHESIS.

signal transducers and activators of transcription (STATs) See CYTOKINES.

signal transduction pathway See TWO-COMPONENT REGULATORY SYSTEM.

signature-tagged mutagenesis A form of TRANSPOSON MUTAGENESIS which can be used to detect those genes, in a pathogen, which are necessary for the pathogen's growth within the host organism. The principle of signature-tagged mutagenesis is outlined in the figure. (See also IVET.)

signet-ring stage (of *Plasmodium*) See PLASMODIUM.

significant bacteriuria BACTERIURIA in which a sample contains potentially pathogenic bacteria in numbers high enough to indicate a URINARY TRACT INFECTION; the numbers of such bacteria must be significantly higher than those which might be expected to occur from contamination of the sample.

In the absence of infection, an MSU (taken correctly) should contain no more than $\sim 10^4$ (commonly $<10^3$) contaminating bacteria/ml, and these contaminants will typically consist of several species – giving rise to different types of colony on the culture plate.

By contrast, counts greater than $\sim 10^5$/ml – usually consisting of one species (but sometimes two species) – are likely to indicate an infection of the urinary tract; note that counts at this level may not be obtained after chemotherapy with an effective agent. (Note also that an inappropriately high count may be obtained if the specimen of urine has been stored, prior to culture, under conditions which allow multiplication of bacteria; ideally, the sample should be examined promptly.)

In some cases a significant bacteriuria can be demonstrated in an asymptomatic patient *prior to* development of pyelonephritis. Conversely, some samples from patients with a urinary tract infection may give low-level counts. Thus, interpretation of counts must reflect the overall clinical picture.

Some pathogens can persist *intra*cellularly in the facet cells of bladder epithelium (see UPEC), so that counts of bacteria in urine samples may not necessarily reflect the level of pathogenic organisms within the urinary tract.

[Automated detection of significant bacteriuria by flow cytometry: JCM (2000) *38* 2870–2872.]

silage A feed for animals (particularly ruminants) which is made by the anaerobic storage of large masses of finely chopped grasses, legumes or other green crops. Bacteria – mainly *Lactobacillus* spp – which are present on the vegetation and/or in the storage vessel (*silo*) metabolize the plant sugars (especially glucose, fructose and sucrose) primarily by a LACTIC ACID FERMENTATION, thereby causing a rapid fall in pH; the low pH (ca. 4.0) inhibits various putrefactive organisms (particularly *Clostridium* spp), thus helping to preserve the nutrient value of the crop, and inhibits the growth of *Listeria monocytogenes* (see also LISTERIOSIS). (cf. AIV PROCESS.) Ensilage may take ca. 1–4 months.

The crop to be ensiled is harvested when its nutrient value is high, and before ensilage it may or may not be 'wilted' (i.e., allowed to lose water); wilting to a dry-matter level (DM level) of ca. 300 g/kg is usually sufficient to inhibit the growth of putrefactive clostridia – which need a minimum WATER ACTIVITY

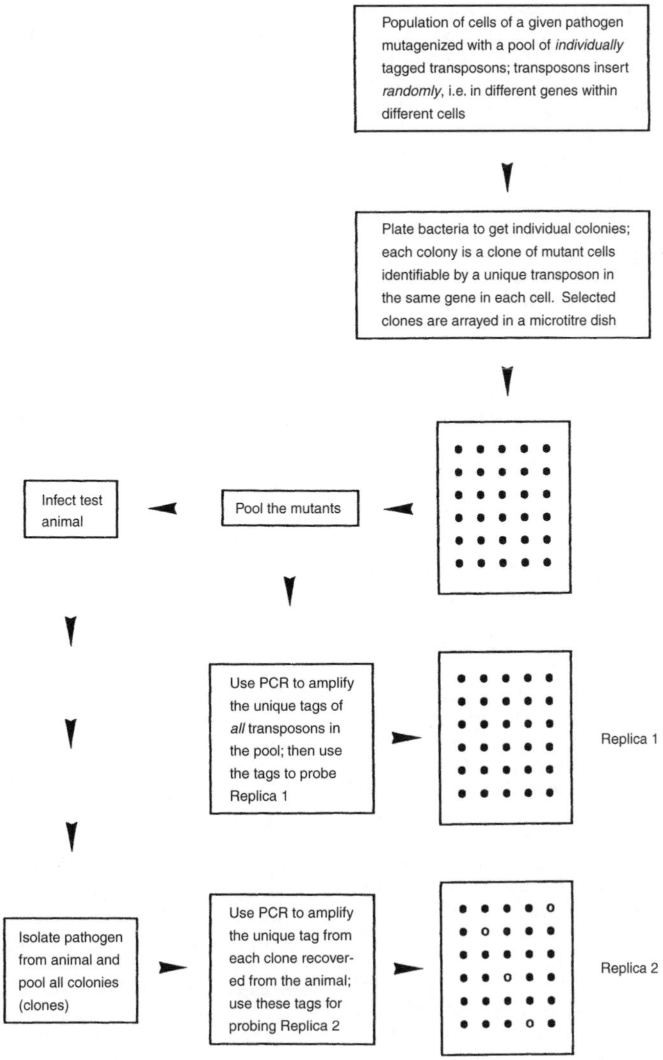

Population of cells of a given pathogen mutagenized with a pool of *individually* tagged transposons; transposons insert *randomly*, i.e. in different genes within different cells

Plate bacteria to get individual colonies; each colony is a clone of mutant cells identifiable by a unique transposon in the same gene in each cell. Selected clones are arrayed in a microtitre dish

Infect test animal

Pool the mutants

Use PCR to amplify the unique tags of *all* transposons in the pool; then use the tags to probe Replica 1

Replica 1

Isolate pathogen from animal and pool all colonies (clones)

Use PCR to amplify the unique tag from each clone recovered from the animal; use these tags for probing Replica 2

Replica 2

SIGNATURE-TAGGED MUTAGENESIS: a method for identifying virulence-associated genes in a pathogen (general principle).

A unique sequence ('tag') of about 40 base-pairs in length is inserted into each of a population of transposons, i.e. the tag in each transposon is different from that in all the other transposons. Each tag is flanked, on both sides, by a short primer-binding site – *these* sites being the same in all transposons. The transposons are used to mutagenize cells of the pathogen, and they insert, randomly, into different genes in different cells.

The mutagenized cells are plated to form individual colonies; a given colony consists of a clone of mutant cells, each containing the same uniquely tagged transposon in the same gene. A number of colonies are chosen, and an inoculum from each colony is arrayed, separately, in a microtitre dish. (The dish shown in the diagram has 30 wells, but larger dishes are normally used.) Two replica 'blots' of the array are made on membranes (for subsequent DNA hybridization studies); in these blots the cells are lysed and their chromosomal DNA is exposed and fixed to the membrane in single-stranded form.

Cells are taken from each of the wells and pooled. The pool is used in two ways. First, it provides an inoculum for infection of the test animal. Second, cells from this ('input') pool are lysed, and PCR is used – with labelled primers – to amplify the unique tags of all transposons in the pool. The amplified, labelled tags are then used as probes on one of the replica blots; in this blot (Replica 1), each (unique) tag should hybridize with the DNA from cells containing the corresponding transposon. This 'pre-screening' process (on Replica 1) checks for efficient amplification of each unique tag in the pool. (The need to pre-screen may be avoidable if a set of 'dedicated' tags is used [TIM (1998) *6* 51].)

The pathogen is recovered from the test animal by plating an appropriate specimen. The resulting colonies are pooled (forming the 'recovered' pool), and PCR is used (with labelled primers) to amplify the tag from each clone in this pool. The amplified, labelled tags are then used to probe the second replica blot (Replica 2). (*Continued on page 708.*)

higher than that required by lactic acid bacteria. The vegetation is chopped into pieces ca. 2.5 cm in length and may be treated with additives such as (a) a bacterial culture (usually certain *Lactobacillus* spp) to augment the natural microflora, (b) a carbohydrate (e.g. molasses) to augment the plant sugars, and/or (c) formic acid, which e.g. can quickly establish a low pH suitable for the growth of lactic acid bacteria. (At ca. pH 5, formic acid is much more inhibitory to clostridia and enterobacteria than it is to lactic acid bacteria; it thus tends to promote the correct, i.e., lactic acid, fermentation.) The chopped/treated vegetation is packed into the silo. When the silo is sealed, microbial metabolism quickly results in anaerobiosis.

Organisms dominant in fermenting silage include homo- and heterofermentative species of LACTOBACILLUS, particularly *L. brevis*, *L. buchneri*, *L. curvatus* and *L. plantarum*; other organisms may include e.g. *Enterococcus faecium*, *Lactococcus* lactis, and species of *Pediococcus* and *Leuconostoc*. (Enterobacteria and *Bacillus* spp can be dominant during the first day or so in vegetation of relatively low DM and high pH.) The heterofermentative organism *L. brevis* ferments glucose or fructose to lactic and acetic acids, but does not form ethanol since it apparently lacks aldehyde dehydrogenase [see Appendix III(b)]; in this species the excess reducing power can be used to reduce exogenous fructose to mannitol (cf. RESPIRATION). Interestingly, the mannitol and acetic acid formed by *L. brevis* can be used as substrate and electron acceptor, respectively, by the homofermentative species *L. plantarum* [Food Mic. (1986) *3* 73–81].

When the silage is eventually removed from the silo it can undergo aerobic deterioration if left uneaten; particularly prone to such deterioration is silage made from overwilted vegetation or made in a non-airtight silo – i.e., circumstances which tend to increase the aerobic silage microflora. Organisms particularly associated with aerobic deterioration include yeasts (e.g. *Hansenula*, *Pichia*, *Saccharomyces*), other fungi (e.g. *Aspergillus*, *Byssochlamys*, *Fusarium*, *Geotrichum* and *Mucor*), and *Bacillus* spp. The addition of certain compounds (e.g. propionic acid) to the vegetation at the beginning of ensilage can reduce subsequent aerobic deterioration in the finished silage. Sorbic acid can have a similar effect.

Various additives are used e.g. to increase the efficiency of the process and/or to enhance the performance of animals fed on the silage. One common additive includes strains of lactic acid bacteria that carry out a homolactic fermentation. Ideas for improved additives include (i) strains of lactic acid bacteria particularly suited to specific target crops; (ii) lactic acid bacteria that carry out a *hetero*lactic fermentation – the volatile fatty acids inhibiting growth of fungi when the silage is exposed to air at feeding time; (iii) genetically engineered lactic acid bacteria which can utilize polysaccharides in those silage crops with low levels of soluble carbohydrates [new trends in silage additives: FEMS Reviews (1996) *19* 53–68].

Choice of crops for ensilage. An ideal silage crop should have a DM level > ca. 200 g/kg and a good content of soluble, fermentable carbohydrates. (Among common grasses, rye-grasses (*Lolium* spp) tend to have the highest content of carbohydrate, and cocksfoot (*Dactylis glomerata*) the lowest.) A silage crop should also have a low buffering capacity (measured e.g. in mE of alkali needed to bring the pH from 4 to 6 in 1 kg DM); a low buffering capacity permits the in-silo pH to drop rapidly, while a high buffering capacity may allow the occurrence of a non-lactic acid fermentation and/or of clostridial spoilage. In e.g. maize (*Zea mays*) the buffering capacity can be as low as ca. 150–200 mE/kg, while in some grasses, and lucerne (*Medicago sativa*), it can be up to ca. 500–600 mE/kg; other factors which make maize a good silage crop include its high DM and its content of water-soluble sugars (which can reach e.g. ca. 300 g/kg DM). (A disadvantage of maize is its low content of crude protein; this can be overcome by adding urea – as a source of nitrogen – to the crop at the beginning of ensilage.)

Effluents. Effluent from a silo has a high BOD and can cause serious pollution if discharged into streams etc. Factors which affect effluent formation include the DM levels of the ensiled vegetation.

[Book ref. 195.]

(See also RUMEN.)

silent mutation A MUTATION which has no apparent effect on the phenotype of the organism in which it occurs (see also SAME-SENSE MUTATION and MIS-SENSE MUTATION). Note that a same-sense mutation may not be 'silent' if it introduces a detectable CODON BIAS: in *Escherichia coli*, a synonymous but infrequent (mutant) codon in the leader region of the *ompA* gene strongly affected *in vivo* translation, giving rise to a significant reduction in the synthesis of OmpA protein [NAR (1998) *26* 4778–4782].

Silent mutations may also affect the results of e.g. DNA-based identification tests [Book ref. 221, pp 16, 172 and 218].

silica deposition vesicle A vesicle, bounded by a unit-type membrane (the *silicalemma*), present in most – if not all – algae and protozoa which (endogenously) form siliceous structures: e.g. the scales of certain CHRYSOPHYTES, the frustules of DIATOMS, skeletal elements of RADIOLARIA. [Siliceous structures in biological systems: Book ref. 137.]

silica gel Hydrated silicon dioxide; silica gel can be prepared in an aqueous, jelly-like form for use e.g. as the basis of a solid MEDIUM. A silica-gel-based medium can be used for the culture of autotrophs (in the complete absence of organic material) and e.g. for testing the ability of heterotrophs to use particular substrates as the sole source of carbon. Powdered silica gel (10 g) is dissolved by heating in 100 ml 7% (wt/vol) aqueous KOH; the solution is dispensed in 20-ml aliquots and autoclaved. To each 20-ml aliquot is added 20 ml of a sterile double-strength liquid nutrient medium, followed by a (pre-determined) volume (ca. 4 ml) of a 20% solution of *o*-phosphoric acid sufficient to give a

SIGNATURE-TAGGED MUTAGENESIS (*continued*) Considering the original, mutagenized cells in the 'input' pool, the cells of interest are those which, through a (transposon-mediated) mutation, are *unable* to grow within the test animal, i.e. cells whose virulence has been lowered; such cells will be absent, or few in number, in the 'recovered' pool (compared with those cells which were able to grow normally in the test animal). Hence, the signature tags of these 'virulence-attenuated' cells will be present in the input pool but absent (or rare) in the recovered pool; consequently, such cells can be identified by hybridization on Replica 1 but an *absence* of (or weak) hybridization on Replica 2 (see figure). (If 'dedicated' tags had been used, virulence-attenuated cells would be identified by the absence of hybridization on a single blot.)

In a given 'virulence-attenuated' clone, the relevant gene (identifiable from the inserted transposon) can be isolated, cloned and sequenced for further study.

pH of 7.0. After mixing, the solution is immediately dispensed to Petri dishes. Solidification starts in ca. 1 min and is complete in ca. 15 min. The water of syneresis can be evaporated in an incubator. Incubation of inoculated plates should be carried out in a moist chamber to prevent excessive drying of the medium.

silicalemma See SILICA DEPOSITION VESICLE.

silicoflagellates A group of unicellular planktonic marine algae of uncertain taxonomic position (formerly included in the Chrysophyceae). (cf. PHYTOMASTIGOPHOREA.) The organisms possess a characteristic net-like, siliceous endoskeleton composed of tubular elements which are often arranged in a polygonal 'ring' with supporting struts and projecting spines. The cells each have a single emergent flagellum and many brownish chloroplasts. Present-day genera include *Dictyocha*. Fossil silicoflagellate skeletons are known from the Late Cretaceous onwards [Book ref. 136, pp. 811–846].

Silicoflagellida See PHYTOMASTIGOPHOREA.

silkworm diseases For diseases of the silkworm (*Bombyx mori*) see e.g. CYTOPLASMIC POLYHEDROSIS VIRUS GROUP, FLACHERIE, and NOSEMA (pébrine). (See also BEAUVERIA.)

silver (as an antimicrobial agent) Silver is a HEAVY METAL which, in elemental or compound form, is typically microbistatic in low concentrations; the functional antimicrobial moiety is the silver *ion* – which appears to act by binding to e.g. thiol, amine, phosphate and other groups in proteins, nucleic acids and/or other targets. In at least some cases binding is reversible. Antimicrobial agents consisting of metallic or other forms of silver appear to differ primarily in the rate at which they yield silver ions. Gram-positive bacteria are typically much less susceptible than Gram-negative species, and antimicrobial activity can be diminished by non-living organic matter. In bacteria, resistance to silver may be plasmid-borne.

Silver nitrate (lunar caustic, $AgNO_3$) has astringent properties and may be microbistatic or microbicidal according to concentration; aqueous solutions are used e.g. on dressings for burns (0.5% w/v) and for the treatment of ulcers on mucous membranes (10%). (See also Credé procedure in GONORRHOEA.) The *silver nitrate pencil* is a fused mass containing ca. 98% $AgNO_3$ – together with AgCl and KNO_3 – cast in a pencil-shaped mould; the moistened tip of the pencil is used e.g. for the cauterization of wounds and the destruction of warts. *Ammoniacal silver nitrate* has been used in dentistry for the disinfection of tooth cavities. *Colloidal silver* has been used in medicine (e.g. *Colsargen*: an isotonic suspension of colloidal silver (0.5%) used for treating mucous surfaces). (See also KATADYN SILVER.) *Silver proteinates* (e.g. *Argyn, Argyrol, Protargin, Protargol*) may be made by precipitating a soluble silver salt, or silver oxide, with protein and then solubilizing with excess protein; they have been used e.g. in various gynaecological preparations. (The antimicrobial activity of a silver proteinate depends on the amount of silver ions it liberates rather than on the proportion of silver in the preparation.) *Silver sulphadiazine* (1% in an ointment) is used e.g. for the treatment of burns; this compound (which can be bactericidal) releases silver ions, and – unlike $AgNO_3$ – causes blebs in the cell envelope of *Pseudomonas aeruginosa*.

silver dag A suspension of colloidal silver in a quick-drying organic solvent. (See also ELECTRON MICROSCOPY (b).)

silver leaf A disease of trees of the Rosaceae (particularly plum trees) caused by *Chondrostereum purpureum*. Infection occurs via wounds; the fungus grows in the wood, causing discoloration but little or no decay. Infected trees show a characteristic 'silvering' of the leaves; this is apparently due to the production of fungal toxin(s) which cause leaf tissues to separate, the resulting air spaces giving the silvery effect. The fungus itself does not occur in the leaves.

silver line system (argyrome) (*ciliate protozool.*) A regular pattern of lines observed in the superficial layer(s) of silver-stained ciliates; these lines of deposited silver were first described by Klein in the 1920s. While some of the lines appear to coincide with cell-surface detail, others appear to correspond to the locations of certain subpellicular structures. Thus, e.g. the main (most densely staining) lines in *Tetrahymena* correspond to the rows of kinetosomes which mark the locations of the kineties; these (longitudinal) lines have been termed *primary meridia*. Other, more weakly staining, lines observed between the primaries, have been termed *secondary meridia*. Cross-striations have also been observed; some workers believe that lines of deposited silver are formed at the junctions of the pellicular alveoli. (See also KINETY and PELLICLE.)

silver–methenamine stain See METHENAMINE–SILVER STAIN.

silver scurf A POTATO DISEASE caused by *Helminthosporium solani*; the skin of the tuber exhibits brown or silver patches which extend on storage.

silver stains See e.g. DIETERLE SILVER STAIN; FONTANA'S STAIN; METHENAMINE–SILVER STAIN. (See also SILVER LINE SYSTEM.)

silver top (*syn.* white top) A disease of turf grasses caused by *Fusarium poae*; the seedheads of infected plants wither before maturing, appearing at first silvery, later white. (cf. WHITEHEAD.)

Simbu group See BUNYAVIRUS.

simian AIDS (SAIDS) A disease which occurs in macaque monkeys; SAIDS closely resembles human AIDS e.g. in its clinical manifestations, including immune deficiency, progressive depletion of CD4 T cells. The causal agent is a lentivirus, the *simian immunodeficiency virus* (SIV); SIV differs from the human immunodeficiency virus e.g. in structure and antigenicity of envelope proteins. (Severe immunodeficiency is also caused by a betaretrovirus, the MASON-PFIZER MONKEY VIRUS, but the pathology differs.) Rapid pathogenesis in SAIDS may be accompanied by selection of gp120 variants with decreased ability to bind to CD4 T cells [JV (2002) *76* 7903–7909]. [Assay of SIV: JV (2004) *78* 5324–5337.]

simian foamy virus See SPUMAVIRINAE.

simian haemorrhagic fever virus (SHFV) A virus which can cause a fatal haemorrhagic disease in e.g. rhesus monkeys (*Macaca mulatta*); the natural reservoir of the virus is believed to be the African patas monkey (*Erythrocebus patas*).

SHFV is an enveloped ssRNA virus formerly linked with the family FLAVIVIRIDAE: it resembles flaviviruses e.g. in the 5′ cap structure on the viral RNA [Virol. (1986) *151* 146–150], but differs in that at least some of the RNA molecules are polyadenylated [Virol. (1985) *145* 350–355]. SHFV has been placed in the arterivirus group (see ARTERIVIRUS).

simian immunodeficiency virus See SAIDS.

simian malaria See MALARIA.

simian parvovirus See ERYTHROVIRUS.

simian sarcoma virus (SSV; woolly monkey sarcoma virus) A mammalian type C retrovirus (subfamily ONCOVIRINAE) isolated from a fibrosarcoma in a woolly monkey. It is a replication-defective, v-*onc*$^+$ virus which carries the *sis* oncogene (see SIS). SSV occurs in association with a replication-competent virus – simian sarcoma-associated virus, SSAV – which is serologically related to the GIBBON APE LEUKAEMIA VIRUS; the SSV/SSAV complex can cause fibrosarcomas or fibromas in various monkeys.

simian T-cell leukaemia virus (STLV; simian T-lymphotropic virus; primate T-cell leukaemia virus, PTLV) A generic term for exogenous retroviruses which infect the T cells of monkeys and other subhuman primates. STLV-I (formerly 'STLV') is closely related to HTLV-I (see HTLV); it infects Old World monkeys and apes, and can cause lymphomas in e.g. macaques. STLV-III has been isolated from rhesus monkeys (*Macaca mulatta*) and African green monkeys (*Cercopithecus aethiops*) and appears to be closely related to the human AIDS virus; a virus similar to STLV-III can apparently infect humans, and it has been suggested that the AIDS virus and STLV-III may have had a common origin [Science (1986) *232* 238–243]. (cf. SIMIAN AIDS.)

simian virus 40 (SV40; vacuolating agent) A virus of the genus POLYOMAVIRUS. SV40 was originally isolated from kidney cells of the rhesus monkey (*Macaca mulatta*) and is common (in latent form) in such cells; early polio and adenovirus vaccines prepared in rhesus monkey kidney cells were frequently contaminated with SV40. (See also MASTADENOVIRUS.) In kidney cells from the African green monkey (*Cercopithecus aethiops*) SV40 causes CPE, including characteristic vacuolation of the cytoplasm. SV40 is usually non-pathogenic in its natural host, but can induce fibrosarcomas and gliomas in newborn rodents (though not in immunocompetent adult animals): see POLYOMAVIRUS for details.

similarity matrix See S MATRIX.

Simkania negevensis See SPLIT GENE (bacterial).

Simmons' citrate agar KOSER'S CITRATE MEDIUM incorporating BROMTHYMOL BLUE (0.008%) and agar (1.5–2.0%).

Simonsiella A genus of GLIDING BACTERIA (see CYTOPHAGALES) which occur in the oral cavity in man and other vertebrates. The organisms are flat filaments (ca. 2–4 μm in length), each composed of elongated cells arranged side-by-side; the outer face of each terminal cell is rounded. Gliding occurs in a direction parallel to the long axis of the filament. Metabolism: chemoorganotrophic.

Simonsiellaceae See CYTOPHAGALES.

simple matching coefficient See entry S_{SM}.

Simplexvirus See ALPHAHERPESVIRINAE.

Sin Nombre virus See HANTAVIRUS.

SIN virus SINDBIS VIRUS.

Sindbis virus (SIN virus) An ALPHAVIRUS which occurs in Africa, Asia, Australia and Europe; it is apparently maintained primarily in birds and is transmitted by mosquitoes which feed on birds (e.g. *Culex univattatus* in Africa). Other vertebrates can be infected, and the virus is occasionally associated with human illness characterized by arthralgia and a maculopapular rash ('Sindbis fever'); human diseases caused by Sindbis or related viruses are known as Karelian fever in the former USSR, Ockelbo disease in Sweden, and Pogosta disease in Finland. (See TOGAVIRIDAE for replication cycle etc.)

sinefungin An ANTIBIOTIC (obtained from *Streptomyces griseolus*) which has antifungal, antiprotozoal and antiviral activity; it acts as an analogue of ornithine, inhibiting ornithine decarboxylase and hence putrescine formation.

single burst experiment A procedure for studying the lysis of a single cell by a virus (e.g. to determine the BURST SIZE). Essentially, a suspension of cells is infected with viruses and then diluted such that each of a large number of aliquots has a low probability of containing more than one virus-infected cell. Following incubation and lysis, the pfu counts in the aliquots can be statistically analysed e.g. to assess individual burst sizes. The single burst experiment can also be used e.g. to investigate recombination in bacteriophages, genetic analysis being made of the phage progeny from a single bacterial cell coinfected with two genetically distinct phages.

single-cell protein (SCP) SCP refers to the cells of microscopic organisms (bacteria, yeasts, moulds or microalgae) grown in large-scale cultures for use primarily as a source of protein in human or animal diets. Although not yet fully exploited commercially, SCP has many advantages over the more conventional sources of protein such as soybean meal and fishmeal. (i) It is rich in protein – ca. 40–85% crude protein, depending e.g. on source; typically, SCP is superior to soybean meal both in protein content and amino acid profile. (Protein content is usually given either as 'crude protein', i.e. Kjeldahl nitrogen ×6.25, which includes e.g. nitrogen in nucleic acids, or as 'true protein' determined e.g. by the Biuret reaction or by estimation of total amino acids.) (ii) The yield coefficient of SCP, Y_s (g cell mass produced/g carbon utilized), is much higher than that of either plants or animals. (iii) SCP can be produced rapidly (owing to the high growth rates of microorganisms) and, unlike agricultural crops, production can be continuous throughout the year – even in countries with little or no agricultural potential or expertise. (iv) A wide range of raw materials, including wastes from other industries, can be used as substrates. In some cases this has the additional advantage of reducing the BOD of certain trade effluents. However, SCP also has disadvantages as a food source. (i) Microbial proteins tend to be deficient in sulphur-containing amino acids, particularly L-methionine (an essential amino acid for man); this can be overcome by supplementing SCP products with L-methionine or its hydroxy analogue (see HMA). (ii) Some microbial components, e.g. cell walls, cannot be digested by non-ruminant animals or man. (iii) SCP is rich in nucleic acids (ca. 5–15%), mainly RNA. Digestion of purine nucleotides yields uric acid which can be broken down by most mammals; however, man has no uricase, and an intake of more than 2 g nucleic acid per day causes deposition of urates in joints and kidneys, leading to gout and kidney stones. Nucleic acids may be removed e.g. by enzymatic treatment. (iv) Some SCP can affect man adversely in ways not always indicated by animal feeding experiments. Such effects may be due e.g. to residues of toxic substances from the growth medium, or to cell products or components – e.g. fatty acids with odd numbers of carbon atoms (in yeasts grown on hydrocarbons), D-amino acids – usually absent in traditional diets. (v) Cost. Weight for weight, SCP is significantly more expensive than either soybean meal or fishmeal (at least on early cost analyses).

A wide range of microorganisms has been evaluated for SCP production, but few are currently used on a commercial scale. *Bacteria* have advantages such as rapid growth rates and the ability to use a wide range of substrates. They are rich in protein which has a better range of essential amino acids than that found in soybean protein; their content of methionine and cysteine is higher than that of yeasts, although still limiting. However, for human consumption, bacterial SCP needs extensive treatment to remove nucleic acids, LPS etc, and public acceptance is low owing to a tendency to associate bacteria with disease; additionally, the (small) cells are not easy to harvest. *Cyanobacteria* lack some of these problems (see e.g. SPIRULINA). *Yeasts*, being larger, are easier to harvest, and they can grow at low pH (reducing the risk of microbial contamination); they have a high content of lysine (an essential amino acid) and have good public acceptability. However, compared with bacteria, yeasts tend to have a lower content of protein and are poorer in methionine. (Yeasts grown primarily for food, rather than for the production of other foods – cf. BAKERS' YEAST – are called 'food yeasts': see e.g. TORULA YEAST.) *Mycelial fungi* are easy to harvest but they have relatively low growth rates and protein content. Protein from *microalgae* is inferior, nutritionally, to that of most other microbial sources, and algal SCP has poor acceptability in taste, colour, odour etc. The problem of indigestible algal cell walls may be overcome by cultivating protoplasts or wall-less algae (e.g. *Dunaliella* [AEM (1976) *31* 602–604] and *Cosmarium* [AEM (1976) *32* 436–437]).

Production of SCP. Bacteria, yeasts and moulds are generally cultured in well-aerated liquid media in a FERMENTER; typically, the process is continuous (see CONTINUOUS CULTURE) with limiting carbon (C:N less than ca. 10:1) to minimize production of storage compounds such as poly-β-hydroxybutyrate. Algae are grown in batch or semi-continuous culture in ponds with large surface areas. Yeasts and bacteria are harvested by centrifugation; in one process (Pruteen, see later) the bacteria are initially agglutinated so that a more concentrated slurry can be fed to the centrifuge – thus increasing the efficiency of centrifugation. Algae and moulds are harvested by filtration. The concentrated biomass is washed, dried, and used either directly (e.g. in animal feeds) or (for human consumption) its protein content may be extracted and used as a nutrient supplement or as a 'functional' ingredient (e.g. to alter the texture of traditional foods).

Substrates. The choice of substrate involves considerations such as cost, efficiency of utilization, continuity of supply, and risk of toxic residues. There is, with some exceptions, a good correlation between cell yield and substrate energy content for substrates with heats of combustion up to ca. 11 kcal/g of substrate carbon (approximately that of glycerol); above this value cell yields do not increase [Book ref. 11, pp. 33–34].

(a) *High-energy substrates.* Hydrocarbons, e.g. natural gas (80–95% methane) and *n*-alkanes (particularly C_{10}–C_{23}), can be utilized very efficiently by certain bacteria and yeasts, and the risk of microbial contamination is low owing to the inability of most potential contaminants to grow on these substrates. However, hydrocarbons are poorly soluble in (aqueous) growth media, and they may contaminate the final product – although methane, being gaseous, is easily eliminated; the *n*-alkanes can be used in low concentrations in media containing ammonium salts, phosphate, etc. A further (operational) problem is that hydrocarbon substrates require high rates of oxygenation and heat removal. Organisms which have been grown on hydrocarbons on an industrial scale include yeasts – e.g. *Yarrowia lipolytica* (= *Saccharomycopsis* (*Candida*) *lipolytica*) – and bacteria (e.g. *Pseudomonas* spp which utilize waxy residues of crude petroleum); they have been used as animal feeds. (See also HYDROCARBONS and METHANOTROPHY.) [Hydrocarbons as substrates in industrial fermentations: Book ref. 155, pp. 643–683.]

Methanol and ethanol are more soluble than hydrocarbons, are readily available in pure form, and are easily removed from the SCP – but they are more expensive. Of organisms grown on methanol, higher yields are given by those which use the RMP PATHWAY rather than the SERINE PATHWAY. One commercial product manufactured in the 1980s, was prepared by growing the obligate methylotroph *Methylophilus methylotrophus* on methanol in an AIRLIFT FERMENTER. The product ('Pruteen') (Imperial Chemical Industries, UK) was marketed (ca. 70000 metric tons annually) as an animal feed, with purified carbon dioxide as a by-product; the process was subsequently discontinued owing to a fall in the price of protein.

Owing to the high cost of ethanol, ethanol-derived SCP is likely to be prepared only for human consumption; it includes TORULA YEAST – which has been produced at the rate of over 7000 tons annually by one American company.

(b) *Waste products.* (i) *Molasses* (from sugar-refining) is used mainly for growing yeasts (see e.g. BAKERS' YEAST) but is also used in Taiwan and Japan for growing *Chlorella* heterotrophically. (ii) *Sulphite waste liquor* (from paper mills) can be used as a substrate (e.g. for torula yeast) after the hemicellulose has been hydrolysed, by sulphurous acid, to hexoses, pentoses, and acetic, galacturonic and formic acids; sulphite levels are reduced with lime, residual sulphur compounds are removed by steam stripping, and ammonia and various salts are added. In the Pekilo process (Finnish Pulp and Paper Institute), the mould *Paecilomyces varioti* has been grown in continuous culture and used as animal feed. (iii) *Whey* (from CHEESE-MAKING) contains ca. 4% lactose which can be utilized efficiently by only a few SCP organisms; the yeast *Kluyveromyces marxianus* (*K. fragilis*: 'fragilis yeast') is one such organism, ca. 5000 tons of which have been produced annually by one American company. Pilot-scale production of the mould *Penicillium verrucosum* (*P. cyclopium*) has been carried out in France. (iv) *Starch-rich wastes* (e.g. from potato and rice processing) are converted to SCP in the Symba process (Swedish Sugar Corporation) in which one yeast, *Saccharomycopsis fibuligera* (*Endomycopsis fibuligera*), produces α- and β-amylases which hydrolyse the starch to glucose and maltose, while a second yeast, *Candida utilis*, utilizes these sugars to form the bulk of the SCP. This process reduces the BOD of the waste by ca. 90% and thus reduces difficulties in disposal. Cassava starch has been used as a substrate for '*Cephalosporium eichhorniae*' [AEM (1982) *43* 403–411]. (v) *Sewage effluents* are used for the culture of e.g. *Chlorella*, *Scenedesmus* and photosynthetic bacteria (e.g. *Rhodopseudomonas capsulata*).

(c) *Renewable sources* (e.g. wood and other plant materials). Lignocellulosic materials need pre-treatment to separate the LIGNIN, HEMICELLULOSE and CELLULOSE components. Once free, cellulose can be hydrolysed to fermentable sugars by acid treatment or by CELLULASES (e.g. those from *Trichoderma* spp). Yeast SCP (*Candida* sp) has been grown on wood hydrolysates in the former USSR.

The future potential of SCP may depend e.g. on advances in technology (such as improved methods for harvesting and protein isolation), on the availability of cheap(er) substrates, and on strain improvement; thus, e.g., the ICI Pruteen strain had improved AMMONIA ASSIMILATION (with consequent improvement in yield) owing to the introduction of the *gdh* (glutamate dehydrogenase) gene from *Escherichia coli* [Nature (1980) *287* 396–401]. However, other (economic) factors, e.g. the price of 'conventional' protein, may be pivotal – although the link between SCP and pollution control may alter the balance of economic viability.

single diffusion (*serol.*) GEL DIFFUSION in which only one component of the system (e.g. antigen) diffuses through the gel – the other component being uniformly pre-distributed throughout the gel. (cf. DOUBLE DIFFUSION.) In the *Oudin test* (single diffusion, single DIMENSION), antibody is pre-distributed throughout a column of agar (or similar) gel, and a suspension of antigen is layered on top; a band of precipitate forms where the antigen, diffusing downward, meets homologous antibody in OPTIMAL PROPORTIONS. As more antigen reaches the precipitate an ANTIGEN EXCESS occurs at this location and the precipitate dissolves – the zone of equivalence (and the band of precipitate) thus moving down the gel column; the distance moved by the band of precipitate is proportional to the square root of the time taken for the movement. If several different antigen–antibody systems are present, several bands of precipitate will form (and move) independently.

In the *Mancini test* (= single radial diffusion test, or single radial immunodiffusion test) – a single diffusion, double dimension method – antibody is pre-distributed uniformly throughout a flat sheet of gel, and the antigen preparation is put into a small, cyclindrical hole ('well') cut into the gel sheet; antigen diffuses radially outwards and forms, with the antibody, a halo of precipitate; the area of the circle formed by the halo is proportional to the initial concentration of antigen.

single dimension (*serol.*) See DIMENSION.

single intradermal comparative tuberculin test *Syn.* COMPARATIVE SINGLE INTRADERMAL TUBERCULIN TEST.

single intradermal test (*vet.*) A TUBERCULIN TEST involving an injection of human or bovine tuberculin into an anal fold or into the neck. The test cannot distinguish between infections due to different species of *Mycobacterium* (cf. COMPARATIVE SINGLE INTRADERMAL TUBERCULIN TEST and STORMONT TEST).

single linkage In NUMERICAL TAXONOMY: a method in which two groups ('clusters') of OTUs unite (coalesce) at a similarity level determined by the pair of OTUs – one from each cluster – which exhibit the maximum mutual degree of similarity.

single radial diffusion test See SINGLE DIFFUSION.

single radial immunodiffusion test See SINGLE DIFFUSION.

single-site mutation *Syn.* POINT MUTATION.

single-step growth experiment *Syn.* ONE-STEP GROWTH EXPERIMENT.

single-strand binding protein (SSB protein; helix destabilizing protein; 'unwinding protein') A class of proteins which bind specifically and cooperatively to ssDNA; they protect the DNA from the action of many nucleases and e.g. can prevent transcription. They have no known enzymic activity, and do not unwind dsDNA. Their roles in the cell are believed to include the stabilization and protection of ssDNA formed e.g. during processes such as DNA replication and DNA repair – although they are now believed to be excluded e.g. during homologous recombination mediated by the RecBCD enzyme. SSB proteins include *Escherichia coli* binding protein I, phage T4 gp32, phage fd gp5 (see INOVIRUS), and proteins encoded by the F plasmid and other conjugative plasmids.

single strand conformation polymorphism See SSCP.

single-stranded DNA phage *Syn.* SSDNA PHAGE.

single-vent Petri dish See VENT.

singlet oxygen ($^1O_2$ or O_2^1) An energized and reactive, but uncharged, form of oxygen which can be toxic to cells. Singlet oxygen may be produced e.g. (i) during ULTRASONICATION; (ii) during the transfer of energy from photoexcited fluorescent dyes to oxygen (see also PHOTODYNAMIC EFFECT); (iii) during the formation of HYDROXYL RADICAL by the reaction between SUPEROXIDE and HYDROGEN PEROXIDE; (iv) during the reaction between hypohalite and H_2O_2 (see also MYELOPEROXIDASE). (Some of the oxygen produced during the spontaneous, i.e. non-enzymatic, dismutation of superoxide has been reported to be singlet oxygen.) Singlet oxygen is quenched e.g. by CAROTENOIDS.

***sinI* gene** See ENDOSPORE (figure (a) legend).

Sinorhizobium meliloti See OSMOREGULATION.

sintered glass filter See FILTRATION.

sinuate (1) Wavy; undulating. (2) (*syn.* emarginate) In an agaric: refers to a LAMELLA in which the lower edge is notched at its junction with the stipe.

Sinuolinea See MYXOSPOREA.

sinus Any groove, constriction or cavity in an organism or cell: see e.g. *placoderm* DESMIDS.

sinusitis Inflammation of one or more of the paranasal sinuses – often a complication of e.g. a COMMON COLD or a tooth infection. Symptoms include pain at the affected site, purulent nasal discharge, and fever. Common causes of acute sinusitis are *Haemophilus influenzae* or *Streptococcus pneumoniae* (or both); less commonly, *S. pyogenes*, *Staphylococcus aureus*, *Branhamella catarrhalis*, or anaerobes (e.g. *Bacteroides* spp) may be involved.

siomycin See THIOSTREPTON.

siphonaceous (siphonous) Refers to a tubular or vesicular, coenocytic, basically aseptate algal thallus in which the cytoplasm (containing the chloroplasts, nuclei etc) forms a peripheral layer around a large central vacuole. (Septa may separate reproductive structures from the rest of the thallus, and may be formed in response to wounding.) According to species, a siphonaceous thallus may be a simple or branched tube or vesicle (see e.g. BOTRYDIUM and VAUCHERIA); in some algae (e.g. CODIUM) the thallus is made up of a complex combination of such tubes and/or vesicles. (cf. SIPHONOCLADOUS.)

siphonein See CAROTENOIDS.

siphonocladous Refers to a filamentous algal thallus which is composed of multinucleate cells (as in e.g. CHAETOMORPHA, CLADOPHORA, RHIZOCLONIUM, SIPHONOCLADUS, VALONIA). (cf. SIPHONACEOUS.)

Siphonocladus A genus of marine, mainly tropical, siphonocladous green algae (division CHLOROPHYTA) in which the mature thallus is erect with lateral branches.

siphonous *Syn.* SIPHONACEOUS.

siphonoxanthin See CAROTENOIDS.

sirenin See PHEROMONE.

siRNA See RNA INTERFERENCE.

sirodesmins See EPIPOLYTHIAPIPERAZINEDIONES.

sirohaem A HAEM (sense 1) which is the prosthetic group in e.g. DESULFOVIRIDIN, in the sulphite and nitrite reductases of e.g. *Escherichia coli*, and in certain IRON–SULPHUR PROTEINS in green plants.

Sirolpidium See LAGENIDIALES.

SIRS Systemic inflammatory response syndrome: specific symptoms (e.g. temperature >38°C or <36°C; white blood cell count >12000/mm^3, <4000/mm^3 or >10% immature forms) used to define SEPSIS (sense 2) precisely. [Sepsis/SIRS: JAC (1998) *41* (suppl A) 1–112.]

SIRS virus See BLUE-EARED PIG DISEASE.

sis An ONCOGENE originally identified as the transforming determinant of SIMIAN SARCOMA VIRUS; the v-*sis* product has an amino acid sequence almost identical to that of human platelet-derived growth factor (PDGF), and may cause transformation by mimicking PDGF [EMBO (1986) *5* 1535–1541].

sisomycin 4′,5′-DehydroGENTAMICIN C_{1a}.

sister chromatid See CHROMATID.

site-directed mutagenesis *Syn.* SITE-SPECIFIC MUTAGENESIS.

site-specific mutagenesis The (in vitro) induction of MUTAGENESIS at a specific site in a given (target) DNA molecule.

In one method (D-loop mutagenesis), small fragments of ssDNA, each corresponding to the site to be mutated, are mixed (in vitro) with the full-sized supercoiled dsDNA target molecules in the presence of RecA protein (q.v.) and ATP; under these conditions the fragment invades the dsDNA, generating a single-stranded D-loop which is susceptible to the ssDNA-specific mutagen BISULPHITE. The bisulphite-treated DNA is then introduced into *ung*$^-$ cells (see URACIL-DNA GLYCOSYLASE) by TRANSFORMATION; during a subsequent round of DNA synthesis, the uracil residues (generated from cytosine by the bisulphite) pair with adenine, leading to G·C → A·T transitions.

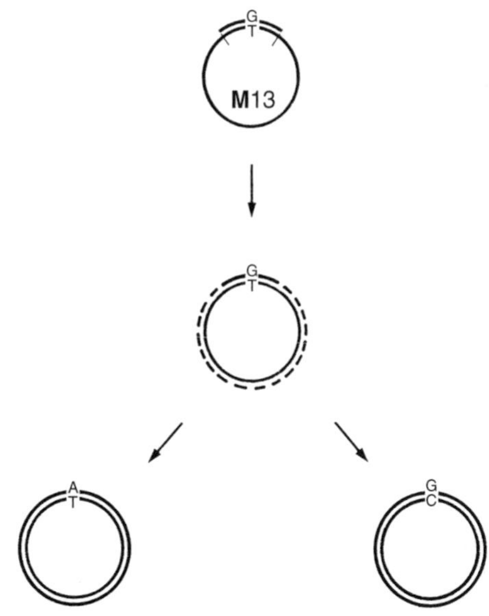

SITE-SPECIFIC MUTAGENESIS (oligonucleotide-directed mutagenesis) (diagrammatic).

Here, a phage M13 cloning vector is being used to create a point mutation at a known site in a given gene. The purpose of phage M13 is to make single-stranded copies of the relevant gene. Initially, the (double-stranded) gene is inserted into a circular, double-stranded form of the M13 genome – the so-called *replicative form* (an intermediate stage in phage replication); during the final stage of replication (in a bacterium), M13 produces ccc ssDNA genomes, each genome incorporating one single-stranded copy of the target gene.

The diagram (*top*) shows one single-stranded phage M13 genome. Within the M13 genome, a 20-nucleotide sequence of the target gene (its size exaggerated for clarity) is shown between two bars. Within this sequence of the target gene we wish to replace deoxythymidine (T) with deoxycytidine (C).

The first step is to synthesize an oligonucleotide that is *complementary* to the 20-nucleotide sequence of the target gene – except for one mismatch at the required site: deoxyguanosine (G) opposite deoxythymidine (T). Each of the synthesized oligonucleotides is then bound to a copy of the target gene (*top*) and used to prime in vitro DNA synthesis (*dashed line, centre*); subsequent ligation results in dsDNA molecules, each containing the single mismatch in the target gene. These dsDNA molecules are inserted into bacteria by transformation; within the bacteria, replication of the molecules will segregate mutant molecules (*bottom, right*) and non-mutant molecules (*bottom, left*).

Reproduced from *Bacteria* 5th edition, Figure 8.20, page 215, Paul Singleton (1999) copyright John Wiley & Sons Ltd, UK (ISBN 0471-98880-4) with permission from the publisher.

Another method involves the use of a chemically synthesized oligonucleotide containing the desired mutant base sequence; this method (*oligonucleotide-directed mutagenesis*) is outlined in the figure.

site-specific recombination (SSR) RECOMBINATION involving CROSSING OVER between short, *specific* dsDNA sequences, in the same or different molecules, with no synthesis or degradation of DNA – and without hydrolysis of phosphodiester bonds (i.e. the reaction is independent of ATP and other energy sources). The initial juxtaposition (*synapse*) of the sequences involves DNA-binding proteins; the two sequences are often identical, but in some cases (e.g. the *att*B and *att*P sites in phage λ integration) they are dissimilar. Cuts are made at staggered sites in the two strands of a given duplex, either concurrently or sequentially, and similar cuts are made in the juxtaposed duplex; strand exchange, between the duplexes, is followed by ligation.

Each particular type of SSR system involves a protein (a *recombinase*) which recognizes specific sequences and appears to mediate cutting, strand exchange and ligation; as well as a recombinase, accessory proteins (e.g. the HU PROTEIN) may be needed.

Recombinases are of two main types. One type is exemplified by Tn*3* resolvase (see TN3), the other by phage λ integrase [TIG (1992) *8* 432–439, and erratum in TIG (1993) *9* 45].

A resolvase-type recombinase (encoded e.g. by Tn*3*, Tn*21* and Tn*552*) forms an intermediate in which all four strands are cut concurrently, with 5′ phosphate bound to recombinase. (Phosphate–protein binding conserves energy for subsequent ligation.) The integrase-type recombinase (e.g. λ Int, and the *lox*/Cre system of BACTERIOPHAGE P1) cuts/joins strands pairwise; this enzyme forms an intermediate equivalent to a *Holliday junction* (see RECOMBINATION) in which only one pair of strands has been exchanged.

Examples of SSR include: integration/excision of BACTERIOPHAGE λ (q.v.); RECOMBINATIONAL REGULATION systems of *Salmonella* and phage Mu; *lox*/Cre in BACTERIOPHAGE P1; the resolvase of Tn*3* (q.v.) and other transposable elements; gene splicing; monomerization of plasmid dimers; some cases of plasmid–chromosome interaction [JB (1992) *174* 7495–7499]; and insertion of the gene cassette into an integron [JAC (1999) *43* 1–4].

Some authors regard transposition as a form of (non-conservative) site-specific recombination.

sitophilous Living or growing on food (especially cooked or prepared food).

Sivatoshella See EIMERIORINA.

sixth disease *Syn.* EXANTHEM SUBITUM.

S_J Jaccard coefficient (sometimes spelt Jacquard coefficient). In NUMERICAL TAXONOMY: a coefficient which is similar to S_{SM} (q.v.) but which is calculated on the basis of matching *positive* characteristics; if a given characteristic is negative in both OTUs it is discounted.

skeletal hyphae See HYPHA.

Skeletonema See DIATOMS.

skiaphilic (skiaphilous) Shade-loving.

skin microflora Human skin is a relatively hostile environment for most microorganisms due e.g. to its low a_w and the presence of antimicrobial fatty acids released from sebum by the resident flora. Resident bacteria normally include staphylococci (e.g. *S. epidermidis*) and *Propionibacterium acnes*; these organisms generally occur in scattered colonies on the skin and in the pilosebaceous follicles [JGM (1984) *130* 797–801; 803–807]. (See also ACNE.) Other organisms commonly present on skin include yeasts (e.g. *Malassezia* spp), aerobic coryneforms, and micrococci. Transient organisms depend e.g. on standards of personal hygiene, and may include coliforms, streptococci, pseudomonads etc.

A reduction in the numbers of microorganisms on the skin may be desirable for various reasons (e.g. to prevent wound infection, body odour etc) and may be achieved by washing or by treatment with ANTISEPTICS. Washing with SOAPS may remove large numbers of organisms, particularly Gram-negative bacteria, but may simply redistribute the resident microflora; organisms in pilosebaceous follicles are almost impossible to eliminate entirely. 'Body odour' is due to the metabolism of apocrine sweat gland secretions by the skin microflora with the formation of malodorous substances (e.g. amines); this can be countered by washing and by the use of deodorants containing aluminium salts (to inhibit sweating) and antiseptics to inhibit microbial action.

skin substantivity The extent to which a residue of an antiseptic remains on the skin after washing with a soap containing that antiseptic.

skin test Any test in which an antigenic preparation (derived e.g. from a pathogen) is injected into the skin, or applied to the surface of the skin (a *patch test*), and in which the ensuing reaction (if any) may be used e.g. to assist in a medical diagnosis, to indicate susceptibility to a given disease, or to detect an allergy. (See e.g. BRUCELLIN TEST; DICK TEST; FREI TEST; HISTOPLASMIN TEST; LEPROMIN TEST; SCHICK TEST; TUBERCULIN TEST; VOLLMER PATCH TEST.)

skin tuberculosis A chronic disease of cattle in which hard lesions develop in the skin, often on the legs. Acid-fast organisms are often found in the lesions, but the aetiology is uncertain. Affected animals may react to tuberculin (see COMPARATIVE SINGLE INTRADERMAL TUBERCULIN TEST).

Skrjabinella See EIMERIORINA.

Skulachev ions See CHEMIOSMOSIS.

s.l. SENSU LATO (q.v.).

slaframine A parasympathomimetic alkaloid MYCOTOXIN produced by *Rhizoctonia leguminicola*, a fungus which causes 'blackpatch disease' of red clover; consumption of the infected red clover by livestock results in a MYCOTOXICOSIS characterized by excessive salivation ('slobber syndrome') and feed refusal. Slaframine itself has no biological activity but it is converted in the liver to an active form which, chemically and physiologically, closely resembles acetylcholine. (The active form can also be generated photochemically in the presence of FMN.)

slant *Syn.* SLOPE.

slapped cheek syndrome *Syn.* ERYTHEMA INFECTIOSUM.

sleeping sickness (African trypanosomiasis) A human TRYPANOSOMIASIS which occurs in Africa and which is transmitted by species of GLOSSINA. The causal agent is *Trypanosoma* (*Trypanozoon*) *brucei gambiense* (in West and West Central Africa, and sometimes also in e.g. Sudan and Uganda), or *T.* (*T.*) *brucei rhodesiense* (mainly in East Africa, from Ethiopia to Botswana); the disease caused by *T.* (*T.*) *b. rhodesiense* tends to be more acute. [See book ref. 72, pp 97–128.]

Weeks or months after the initial infected bite there is an irregular, recurrent fever, enlargement of lymph nodes (see also WINTERBOTTOM'S SIGN), malaise, joint pains and loss of weight; later symptoms may include e.g. insomnia or drowsiness, epileptic attacks and coma. The disease is usually fatal if untreated.

Lab. diagnosis: e.g. microscopy of thick blood smears, lymph node aspirate or cerebrospinal fluid (CSF); MAECT; serological methods – e.g. an indirect fluorescent antibody test.

Chemotherapy. SURAMIN has been used in the early stages of the disease when there is no CNS involvement; in later stages (when the CNS is involved) agents such as *melarsoprol* (see ARSENIC) and *eflornithine* are used because they can pass the blood–brain barrier.

Eflornithine (DL-α-difluoromethylornithine; DFMO) is used against *Trypanosoma* (*T.*) *brucei gambiense* but is not effective against *T.* (*T.*) *brucei rhodesiense*; this agent inhibits the enzyme ornithine decarboxylase (ODC), thus inhibiting the synthesis of *trypanothione* (see ARSENIC). The ability of eflornithine to act therapeutically may depend on a slow turnover of trypanosomal ODC (relative to that of mammalian ODC); this appears to be the case with a cattle-infecting trypanosome [JBC (1990) *265* 11823–11826]. Eflornithine is useful in late-stage trypanosomiasis as it readily enters the CSF.

sleepy foal disease ('shigellosis') An acute, usually fatal, septicaemic disease of newborn foals, caused by *Actinobacillus equuli*. Infection may occur in utero or postnatally.

slide (microscope slide) A piece of flat, transparent, colourless glass, about 75 × 25 × 1 mm, on which a specimen is placed (or a SMEAR made) for microscopical examination; the specimen may be overlaid with a COVER-GLASS. (The slide *plus its specimen* is also referred to as a 'slide'.) A suitably prepared specimen (e.g. a fixed, dehydrated, cleared section of tissue) may be made permanent by mounting it (e.g. in CANADA BALSAM) and overlaying it with a cover-glass; when dry, the balsam cements cover-glass to slide. (cf. CAVITY SLIDE, RING SLIDE.)

slide agglutination test Any agglutination test in which antigen and antibody preparations are mixed and allowed to interact on the surface of a slide; one or more dilutions of the antigen and/or antibody preparations may be used in the test – a given dilution reacting with a standard amount of the other reactant. Such tests are generally used only when detectable agglutination occurs within minutes. (Agglutination may be detected e.g. by low-power microscopical examination.)

slim disease In Africa: the name for a condition characterized by a profound loss of weight and constant diarrhoea; the condition is commonly (or always) AIDS.

slime bacteria Bacteria of the MYXOBACTERALES.

slime layer See CAPSULE.

slime moulds A category of eukaryotic organisms which typically have some fungus-like attributes (e.g., the production of spores in or on fruiting bodies) and some animal-like attributes (e.g., a phagocytic, amoeboid vegetative phase); as commonly used, the term 'slime moulds' refers specifically to the so-called 'true' or acellular slime moulds (see MYXOMYCETES) and the CELLULAR SLIME MOULDS. Taxonomically, these organisms are variously classified – together with certain other groups of organisms – as fungi (division MYXOMYCOTA), as protozoa (mostly in classes of the RHIZOPODA – the labyrinthulas and thraustochytrids being placed in a separate phylum, Labyrinthomorpha), or as a distinct phylum (GYMNOMYXA) of the kingdom PROTISTA (sense 2).

slimicide An antimicrobial agent used to inhibit slime-forming organisms.

slipper animalcule The ciliate *Paramecium*.

slit sampler Any instrument, used for sampling the airborne microflora, in which AIR is drawn in through a narrow slit onto an adhesive collecting surface – see e.g. HIRST SPORE TRAP.

slobber syndrome (*vet.*) See SLAFRAMINE.

slope (slant) (1) A solid medium which has been allowed to set (in the case of e.g. agar or gelatin media) or which has been inspissated (in the case of e.g. DORSET'S EGG) in a diagonally oriented test-tube or bottle. (See also BUTT and POTATO SLOPE.)

(2) A CULTURE prepared by the inoculation and incubation of a slope (sense 1).

sloppy agar (semi-solid agar) See AGAR and MEDIUM.

slow-cycling rhodopsin (SCR) A RETINAL-containing protein, small amounts of which have been detected in some of the strains of *Halobacterium salinarium* which contain a PURPLE MEMBRANE. On illumination with yellow light, a 375-nm-absorbing form of SCR accumulates, while illumination with blue (or near-UV) light causes an accumulation of a 590-nm-absorbing form. It has been suggested that SCR may act as the photosensitive pigment during PHOTOTAXIS in halobacteria.

slow disease (slow virus infection) An infectious human or animal disease which is characterized by a long, usually asymptomatic, incubation period (months or years) and a prolonged, progressive course, usually or always ending in death. See e.g. PROGRESSIVE MULTIFOCAL LEUKOENCEPHALOPATHY, SUBACUTE SCLEROSING PANENCEPHALITIS and VISNA. [Pathogenesis of slow virus diseases: JID (1986) *153* 441–447.] (See also LATENT INFECTION and PERSISTENCE.)

'Unconventional slow virus infections' was the former name for the PRION-mediated TRANSMISSIBLE SPONGIFORM ENCEPHALOPATHIES.

slow-opsin See OPSIN.

slow sand filter See WATER SUPPLIES.

slow virus (1) A virus of the LENTIVIRINAE. (2) Any virus (or subviral agent) responsible for a SLOW DISEASE in man or animals.

SLT-I See SHIGA TOXIN.

SLT-II See SHIGA TOXIN.

sludge gas See ANAEROBIC DIGESTION.

slug (of slime moulds) See DICTYOSTELIUM.

SM medium A medium widely used for the culture of cellular slime moulds (e.g. DICTYOSTELIUM). SM medium generally contains (w/v) peptone (1%); D-glucose or lactose (1%); $Na_2HPO_4.12H_2O$ (0.1%); KH_2PO_4 (0.15%); agar (2%). The pH is adjusted to 6.5. Other ingredients – e.g. yeast extract (0.1%) and $MgSO_4$ (0.05%) – may be added. The medium is inoculated with suitable prey (e.g. *Escherichia coli*) and with spores or myxamoebae of the slime mould; incubation is commonly carried out at ca. 20–25°C. [Cultivation of dictyostelids: Book ref. 144, pp. 48–86.]

***Sma*I** A RESTRICTION ENDONUCLEASE from *Serratia marcescens*; CCC/GGG.

SMAC medium See EHEC.

small cytoplasmic RNA See SCRNA.

small iridescent insect viruses See IRIDOVIRUS.

small nuclear RNA See SNRNA.

small round structured viruses (SRSVs) A heterogenous group of small (~20–40 nm) viruses that include common causal agents of food-borne and water-borne diarrhoeal disease/gastroenteritis; the name SRSVs reflects their appearance under the electron microscope. Viruses with this description have been referred to as e.g. ASTROVIRUSES, human caliciviruses (HCVs) and parvo-like viruses; many have been named after the location of particular outbreaks of disease – e.g. NORWALK VIRUS.

Because the SRSVs have not been cultured, most of the information about them has come from epidemiological studies, studies with volunteers, and laboratory investigations – the latter relying heavily on electron microscopy. Specialized applications of electron microscopy that have been useful include immunoelectron microscopy (IEM) (in which viruses and antibodies react in the fluid phase), and solid-phase immunoelectron microscopy (SPIEM) (in which viruses are captured by grid-bound antibodies).

SRSVs have a genome of positive-sense ssRNA (~7.5 kb in several which have been sequenced). They have been classified in the CALICIVIRIDAE; however, comparison of the sequence and genome organization of Manchester virus (a calicivirus) with those of SRSVs suggests that SRSVs are taxonomically distinct from caliciviruses. Sequencing data also show that SRSVs are genetically diverse. Phylogenetic analysis indicates at least two genogroups, each consisting of a number of genotypes; for example, Norwalk virus, Southampton virus and Desert Shield virus comprise genogroup 1, while three clusters (one consisting of Hawii virus and Melksham virus) form genogroup 2.

[SRSVs (review): RMM (1997) *8* 149–155.]

small T antigen See POLYOMAVIRUS.

smallpox (variola major) An acute infectious human disease caused by the variola major virus (see VARIOLA VIRUS). Infection occurs by droplet inhalation and by direct contact with infected persons or fomites. (Contaminated clothing, scabs etc can remain infective for years.) Incubation period: 7–17 days. Typically, symptoms begin abruptly, with fever, headache, nausea, and muscle and joint pains; 2–4 days later the characteristic skin lesions appear. The lesions are most numerous on exposed parts of the body (face, hands etc); they begin as macules, becoming vesicular, then pustular, finally drying to form thick crusts which eventually drop off, leaving scars (pockmarks). The disease may be fatal; recovery is followed by long-term immunity. Other forms of the disease occur. *Haemorrhagic smallpox* is the most severe, with generalized haemorrhages and mortality rates of ca. 100%. *Variola sine eruptione* is a mild, feverish illness (without skin lesions) resembling the prodromal stage of classical smallpox; it occurs mainly in vaccinated individuals. *Alastrim* (variola minor, Kaffir pox, amaas) is a mild form resembling classical smallpox; it is caused by the variola minor virus. *Complications* of smallpox include e.g. secondary bacterial infection of skin lesions followed by septicaemia; lesions in the eyes may lead to blindness. *Lab. diagnosis*: identification and/or culture of the virus from vesicle fluid, smears from skin lesions etc (see also GUARNIERI BODIES); serological tests (e.g. precipitin test, CFT). *Vaccines* against smallpox contain live VACCINIA VIRUS which is injected into the skin to establish a localized viral infection in which a smallpox-like lesion forms at the vaccination site. Complications of vaccination include e.g. eczema vaccinatum (see ECZEMA), post-vaccinal ENCEPHALITIS, and generalized vaccinia. Immunity induced by vaccination wanes after several years, although an allergic sensitivity to viral proteins may persist.

In the past, smallpox occurred in epidemics and pandemics throughout the world, with enormous suffering and loss of life. A world-wide eradication programme (based on vaccination, isolation of cases etc), launched by the WHO in 1967, has led to the complete extinction of the disease in nature – the first disease to have been conquered on a global scale. The last recorded naturally acquired cases occurred in October 1975 (variola major) and October 1977 (variola minor). Eradication was possible since there is no asymptomatic human carrier state and apparently no animal reservoir (cf. MONKEYPOX).

Smc protein See CELL CYCLE (b).

smear (film) Any material (e.g. bacterial culture, wound exudate) prepared for microscopical examination as a thin film on a SLIDE; a smear is often dried, fixed (see e.g. FLAMING), and stained prior to microscopical examination.

SMEDI viruses See ENTEROVIRUS.

smf SODIUM MOTIVE FORCE.

smoking (of meats, fish) A method of FOOD PRESERVATION in which the foodstuff is exposed to smoke for hours or days; smoking has a drying effect, lowering the WATER ACTIVITY and raising the salt concentration of the food. *Wood* smoke contains various phenolic, cresylic and aldehyde components which have antibacterial and antifungal properties.

smooth endoplasmic reticulum See ENDOPLASMIC RETICULUM.

smooth–rough variation (S→R variation) In many types of (Gram-positive and Gram-negative) bacteria: a change in cell-surface composition which occurs spontaneously during in vitro or in vivo growth. Some types of S→R variation are reversible; in some cases reversion from the rough to the smooth condition occurs if the rough strain is grown in a living animal. In a number of cases S→R variation appears to involve selection of mutants.

S→R variation was first recorded in the enterobacteria. In these organisms, *smooth* (glossy) colonies may be formed on primary isolation, and *rough* (dull) colonies may develop on subculture; however, because cell-surface components differ widely in different bacteria, the colonies of strains designated 'smooth' or 'rough' do not always conform to this description.

In pathogenic bacteria an S→R transition is, by definition, associated with a lowering of virulence; thus, e.g. smooth (capsulated) strains of *Streptococcus pneumoniae* are virulent, whereas rough (non-capsulated) strains are found to be avirulent in experimental animals.

In general, an S→R transition involves some or all of the following changes:

(a) the formation of colonies of altered appearance;
(b) loss of specific smooth-strain antigens and (hence) loss of agglutinability in specific antiserum;
(c) an increase in cell-surface hydrophobicity;
(d) an increased susceptibility to hydrophobic antibacterial agents;
(e) an increased susceptibility to phagocytosis;
(f) an increased autoagglutinability in saline;
(g) altered suceptibility to certain phage(s) and bacteriocin(s);
(h) loss of, or reduction in, virulence.

Enterobacterial *smooth* (S) strains are those in which the cell wall LIPOPOLYSACCHARIDES are entire, i.e. they include the (complete) O-specific chains. In *semi-rough* (SR) mutants only one oligosaccharide subunit is present – such strains apparently lacking the ability to polymerize the (normally repeating) subunits of the O-specific chains. In *rough* (R) mutants the O-specific chains are lacking altogether; such strains are deficient in antigenic specificity (see O ANTIGEN) because their cell-surface constituents (core oligosaccharide + lipid A) may be similar or identical to those in organisms of related species or genera. In some mutants, not only is the O-specific chain missing, but part of the core oligosaccharide is also missing; such mutants are referred to as *deep roughs*.

In *Salmonella typhimurium*, rough mutants designated Ra, Rc, Rd_1, Rd_2 and Re strains lack, respectively, the O-specific chain, terminal glucose and galactose residues, the proximal glucose residue, the distal heptose residue, and the proximal heptose residue of the core oligosaccharide (see figure in entry LIPOPOLYSACCHARIDE). Because rough strains fail to agglutinate with anti-O antisera, smooth strains of enterobacteria are used in agglutination tests.

Smooth strain See SMOOTH–ROUGH VARIATION.

SMP Submitochondrial particle: see ETP.

smudge A disease of e.g. onions (mainly white varieties) and leeks caused by *Colletotrichum circinans*. Fungal stromata form dark patches on the bulb; in wet conditions the stromata may form a mass of conidiophores. (See also ONION ROT.)

smut fungi See USTILAGINALES.

smut spores See USTILAGINALES.

smuts (1) Fungi of the order USTILAGINALES, particularly those of the family Ustilaginaceae.

(2) Plant diseases caused by the smut fungi (Ustilaginales); those caused by members of the Tilletiaceae on cereals are called 'stinking smuts' or bunts (see e.g. COMMON BUNT and KARNAL BUNT). In *covered smuts* of cereals, the smut sori remain intact within the glumes, so that the spores are not normally released until the grain is threshed; in *loose smuts*, disruption of the host's tissues by the sori permit the spores to be easily dispersed by wind etc. Some smut fungi incite GALLS in the host plant.

Covered smuts on barley and oats are caused by *Ustilago hordei* (formerly *U. kolleri* on oats) and are characterized by blackened ears and (often) the absence of awns; control: antifungal seed treatment. Loose smuts of barley and wheat are caused by (different strains of) *U. nuda* and are characterized by the development of powdery masses of black spores in place of the grains – diseased ears emerging slightly early, and the stems which bear them often being taller than normal stems; the dispersed spores can infect healthy grain which, if used as seed grain, perpetuates the disease. Loose smut was once controlled by immersing seed grain in warm water (e.g. 49°C for 1.5 hours); it is now controlled by the use of certified (pathogen-free) seed grain or by the use of a systemic antifungal agent (e.g. CARBOXIN). Common smut of maize is caused by *U. maydis* and is characterized by the development of whitish, glistening galls on various parts of the plant (including the ears); the galls rupture to release a mass of black spores.

***Sna*BI** See RESTRICTION ENDONUCLEASE (table).

SNAP-25 See TETANOSPASMIN.

snap-freezing *Syn.* QUENCH-FREEZING.

snow algae See e.g. RED SNOW.

snow blight A disease of Scots pine and other conifers caused by *Phacidium infestans*. It occurs in regions of prolonged deep snow cover, the fungus apparently spreading among the needles of young trees under snow; in spring the affected needles turn brown and fall.

snow mould Any of several fungi which cause disease in various winter-sown crop plants, developing preferentially under a cover of snow. Common examples: *Fusarium nivale*, which can cause a severe disease in winter-sown cereals such as wheat and rye (sometimes causing complete crop loss), and *Typhula* spp which can cause disease ('snow rot', 'typhula blight') in snow-covered turf-grasses and cereals. (The term 'snow mould' is sometimes used specifically for *F. nivale*.)

Snow Mountain agent (SMA) A Norwalk-like virus (see NORWALK VIRUS) which can cause acute gastroenteritis in humans. [RIA for the detection of SMA: JMV (1986) *19* 11–18.]

snow rot See SNOW MOULD.

snowshoe hare virus See BUNYAVIRUS.

SNP Single-nucleotide polymorphism.

SNPV See NUCLEAR POLYHEDROSIS VIRUSES.

snRNA Small nuclear RNA. The eukaryotic nucleus contains numerous small RNA molecules which are typically <300 nt long. In e.g. metazoa 6 snRNAs are particularly abundant (e.g. $2–10 \times 10^5$ of each type per nucleus), and these are designated U1–U6 (owing to their high content of uridylic acid). These U-RNAs have been strongly conserved during evolution. U1–U5 RNAs are synthesized by RNA POLYMERASE II, and each has a 2,2,7-trimethylguanosine cap at the 5′ end (cf. MRNA (b)). U6 RNA is synthesized by RNA polymerase III [JBC (1987) *262* 75–81]; it is capped and has a poly(U) sequence at its 3′ end. In *Saccharomyces cerevisiae* there are apparently at least 24 distinct snRNAs which vary in size and abundance (at least four are >300 nt long), but are generally much less numerous than the metazoan U-RNAs.

snRNAs function in the nucleus in the form of 'small nuclear ribonucleoprotein particles' (snRNPs or *snurps*). Some proteins are common to several snurps, and some snurps contain more than one snRNA (e.g. U4 and U6 both occur – apparently base-paired together – in the same snurp). Snurps are believed to play important roles in mediating and regulating post-transcriptional RNA processing events: e.g. the splicing of introns from nuclear pre-mRNA (see SPLIT GENE (a)). [Nature (1985) *316* 105–106; functions for snRNAs in yeast: TIBS (1986) *11* 430–434; conservation of U-snRNPs in fungi, plants and vertebrates: EMBO (1987) *6* 469–476.] (cf. SCRNA.)

snRNP See SNRNA.

snurp See SNRNA.

Snyder–Theilen feline sarcoma virus See FES.

soaps (as antimicrobial agents) In general, soaps (sodium or potassium salts of long-chain fatty acids) have little or no intrinsic antimicrobial activity, but their surfactant properties can help to reduce the number of microorganisms on the skin. Some soaps are given positive antimicrobial properties by the incorporation of certain antiseptics (see e.g. BISPHENOLS); carbolic soap contains phenol, and some antiseptic soaps contain TCC (see CARBANILIDES). Critical concentrations of soap are needed for the activity of some disinfectants (e.g. Lysol – see PHENOLS). Some antimicrobial agents (e.g. QUATERNARY AMMONIUM COMPOUNDS) may be inactivated by soaps.

SOB system See SOS SYSTEM.

sobemoviruses (southern bean mosaic virus group) A group of ssRNA-containing PLANT VIRUSES, each of which has a relatively narrow host range; transmission occurs via beetles, via seeds (in some hosts), and mechanically. Type member: southern bean mosaic virus (SBMV); other member: turnip rosette virus. Possible members include e.g. blueberry shoestring virus, cocksfoot mottle virus, and rice yellow mottle virus.

Virion: icosahedral (ca. 30 nm diam.) stabilized by divalent cations. The structure of SBMV is similar to that of TOMBUSVIRUSES, but the coat protein (MWt ca. 30000) lacks a P domain. Genome: one molecule of linear positive-sense ssRNA (MWt ca. 1.4×10^6); the 3′ end is neither polyadenylated nor tRNA-like, and the 5′ end is associated (in SBMV and TRV) with a small protein which is necessary for infectivity. Virions occur in both nucleus and cytoplasm in infected plant cells, sometimes forming crystalline arrays in the cytoplasm.

SOD SUPEROXIDE DISMUTASE.

Sodalis A genus of Gram-negative bacteria of the family Enterobacteriaceae which occur as intra- and intercellular symbionts in various tissues of the tsetse fly (GLOSSINA) (but which can also be grown in vitro). Studies on the genome of *S. glossinidius* have indicated that, compared with the genome of *Escherichia coli*, many of the genes concerned with energy metabolism and with the assimilation of carbon compounds appear to be missing; it was suggested that this may indicate an adaptatin to the use of those energy sources which are available in blood [genome size and coding capacity of *S. glossinidius*: JB (2001) *183* 4517–4525].

(See also WIGGLESWORTHIA and WOLBACHIA.)

sodium azide See AZIDE.

sodium dodecyl sulphate See SDS-PAGE.

sodium motive force (smf) The energy associated with a transmembrane electrochemical gradient of sodium ions; smf is analogous to proton motive force (pmf) (see CHEMIOSMOSIS).

In some organisms smf is generated by Na^+/H^+ antiport and is used as a driving force for certain TRANSPORT SYSTEMS; for example, the Na^+/ melibiose symport system is used by *Escherichia coli* for uptake of the sugar melibiose – melibiose being transported together with sodium ions (i.e. transport at the expense of smf); in such cases smf is secondary to (i.e. dependent on) pmf. (See also ION TRANSPORT.)

The marine bacterium *Vibrio alginolyticus* can generate smf by a primary sodium pump: in this organism respiration can be directly coupled to the outward pumping of sodium ions at the NADH–quinone oxidoreductase segment of the respiratory chain; smf thus generated is insensitive to PROTON TRANSLOCATORS such as CCCP. In *V. alginolyticus* the primary sodium pump (which is inhibited by HOQNO) operates when the extracellular medium is alkaline, but under acidic conditions smf is generated by CCCP-sensitive (pmf-dependent) Na^+/H^+ antiport; $\Delta\psi(Na^+)$ is maximum at ca. pH 8.5, minimum at ca. pH 6.0. In this organism smf can be used e.g. for the uptake of a range of amino acids and for rotation of the FLAGELLUM. [Na^+-motive respiratory chain in marine bacteria: MS (1985) *2* 65–71, and in *Vibrio parahaemolyticus*: JB (1985) *162* 794–798.] Smf is also used to energize rotation of the polar flagellum in *Vibrio cholerae* [JB (1999) *181* 1927–1930]; moreover, the exogenous and endogenous levels of sodium ions appear to be regulatory factors in the expression of virulence factors in *V. cholerae* [PNAS (1999) *96* 3183–3187].

Propionigenium modestum apparently lacks a respiratory chain, but it can generate smf (see PROPIONIGENIUM); it seems that this organism can live independently of pmf – ATP being generated by a membrane-bound Na^+-ATPase [EMBO (1984) *3* 1665–1670].

In the higher range of growth temperatures, the cytoplasmic membrane tends to become more permeable to both protons and sodium ions; however, the increase in permeability is greater for protons than it is for sodium ions. At these higher temperatures, organisms whose energy metabolism is based on protons have to use more energy to generate a given level of pmf – simply in order to compensate for the higher rate of inward diffusion of protons. In some organisms, the composition of the cytoplasmic membrane is such that this problem is minimized. However, some thermophilic bacteria do not use protons in their energy metabolism; for example, *Clostridium fervidus* uses sodium ions as the sole energy-coupling ion (i.e. pmf is not used in this species). [Ion permeability at high temperatures: FEMS Reviews (1996) *18* 139–148.]

[Enzymes that translocate sodium ions: BBA (1997) *1318* 11–51.]

sodium perborate See HYDROGEN PEROXIDE (a).

sodium *o*-phenylphenate See PHENOLS.

sodium polyanetholsulphonate See SPS.

sodium pump See ION TRANSPORT.

sodoku See RAT-BITE FEVER.

soft chancre *Syn.* CHANCROID.

soft core ham A canned ham (see CANNING) which has undergone spoilage (e.g. souring) due to the growth of e.g. *Enterococcus faecalis* var. *liquefaciens*. (See also FLAT SOUR.)

soft rot (1) (of timber) A softening – initially superficial, becoming progressively deeper – in wood which is very moist, or which is kept wet and aerobic (e.g. the wooden slats in cooling towers); it is due to the growth of various cellulolytic ascomycetes (e.g. *Chaetomium globosum*) and deuteromycetes (e.g. *Humicola*, *Monodictys*) within the cell walls (S_2 layer) of the wood. Soft rot fungi can cause serious economic losses of timber, but they do not become established if there is competition from other wood-rotting fungi (e.g. basidiomycetes). [Book ref. 39, pp. 165–169; ultrastructure of soft rot fungi: Mycol.

(1985) *77* 447–463, 594–605.] (See also TIMBER SPOILAGE; TIMBER PRESERVATION.)

(2) (of fruit and vegetables) A type of rot, caused by pectinolytic organisms, in which fruit or vegetables degenerate into a wet slimy mass. In acidic fruits (e.g. tomatoes, apples) soft rot is commonly caused by pectinolytic fungi (e.g. species of *Botrytis*, *Penicillium*, *Rhizopus*), while in non-acid fruits and vegetables (e.g. potatoes, carrots) it is more commonly caused by bacteria (e.g. species of *Clostridium*, *Erwinia*, *Pseudomonas*). (See also BROWN ROT (sense 2) and PECTIC ENZYMES; cf. DRY ROT (2).)

soft swell See SWELL.

***sog* gene** See CONJUGATION (1b) (i).

soil-borne wheat mosaic virus (SBWMV) A rod-shaped ssRNA-containing virus which causes mosaic, stunting and yield reduction in winter wheat in parts of Europe, Japan and the USA. It is transmitted by the 'fungus' *Polymyxa graminis*. Isolates of SBWMV consist of at least two components, virion I (281–300 nm long) and virion II (138–160 or 92–110 nm long) – both of which are necessary for infectivity. SBWMV has been classified with the TOBAMOVIRUSES, but in view of its bipartite nature and mode of transmission it has been proposed as the type member of the *furoviruses* (fungus-borne rod-shaped viruses); other members of the group may include e.g. BEET NECROTIC YELLOW VEIN VIRUS, potato mop top virus, and peanut clump virus [JGV (1984) *65* 119–127].

***Solanum nodiflorum* mottle virus** (SNMV) See VELVET TOBACCO MOTTLE VIRUS and VIRUSOID.

solfatara A dormant or decadent volcanic vent which (gently) emits sulphurous gases.

solid-phase immunoelectron microscopy See SMALL ROUND STRUCTURED VIRUSES.

Solorina A genus of foliose LICHENS (order PELTIGERALES). Photobiont: *Coccomyxa*, usually also with *Nostoc*. In e.g. *S. spongiosa* the thallus consists of two distinct parts: rounded, bright-green, *Coccomyxa*-containing lobes, each with a large, central, pitcher-shaped apothecium, and blue-black, *Nostoc*-containing squamules. In *S. crocea*, *Nostoc* occurs within the thallus in a layer below the green algal layer. In *S. saccata*, *Nostoc* occurs in discrete internal cephalodia. *Solorina* spp occur mainly on calcareous soils, particularly at high altitudes.

solvent fermentation *Syn.* ACETONE–BUTANOL FERMENTATION.

somatic (1) Refers to an assimilatory, 'vegetative' (i.e., non-reproductive) structure or function.

(2) Refers to the 'body' or main part of a cell. Thus, e.g. a *somatic antigen* is a molecule which forms part of the body of a cell, usually at the cell surface, rather than one which occurs e.g. in a capsule or flagellum. (See e.g. O ANTIGENS.)

(3) Refers to 'common pili', i.e., FIMBRIAE, as opposed to PILI.

somatic cell hybridization See HYBRIDOMA.

somatic hypermutation See ANTIBODY FORMATION.

somatogamy The fusion of somatic (vegetative) cells or hyphae involving PLASMOGAMY but not KARYOGAMY.

Sonacide An acidic GLUTARALDEHYDE solution.

Sonderia See TRICHOSTOMATIDA.

sonicate (1) (*verb*) To carry out SONICATION. (2) (*noun*) The product of sonication.

sonication The use of audible sound waves, produced by a SONICATOR, e.g. for the disintegration of cells in a liquid medium (cf. ULTRASONICATION).

sonicator (sonifier) An instrument which provides sound energy for SONICATION or ULTRASONICATION. Sonicators commonly work on the principle of magnetostriction: the resonant deformation exhibited by certain types of metal in an alternating magnetic field; the resonating metal is mechanically linked to a metal *probe* (a cylindrical metal rod) which dips into the liquid and acts as the transmitter of sound energy. The sound waves used are often between 10 kHz and 25 kHz with amplitudes of ca. 10–50 μm. (1 kHz = 1 kc/s = 1000 cycles/second.)

sonifier *Syn.* SONICATOR.

sooty bark A disease of *Acer* spp (maple, sycamore) caused by *Cryptostroma corticale*; the bark of standing trees peels away to reveal dark-brown masses of conidia. (See also MAPLE BARK STRIPPERS' DISEASE.)

sooty moulds Fungi of the family Capnodiaceae and of certain other families of the order DOTHIDEALES. The organisms grow epiphytically, utilizing HONEYDEW, and form dark, spongy, hyphal mats on the surfaces of certain plants. Genera: e.g. *Capnodium*, *Scorias*. (cf. BLACK MILDEWS.)

***sop* locus** See PARTITION.

SopA protein See PROTEIN SECRETION (type IV systems).

SopB See FOOD POISONING (*Salmonella*).

sophorolipids Extracellular glycolipid BIOSURFACTANTS produced by *'Torulopsis'* sp during growth on hydrocarbons; they consist of sophorose covalently linked to the hydroxy group of a hydroxy fatty acid.

sophorose β-D-Glucopyranosyl-(1 $\rightarrow$ 2)-D-glucopyranose – an effective inducer or certain CELLULASES. It occurs e.g. in certain plant glycosides and as an impurity in commercial glucose.

soralium (*lichenol.*) A delimited mass of soredia (see SOREDIUM) present on the thallus in certain lichens. Soralia may occur in positions and forms which are characteristic for a given lichen species: they may be e.g. marginal (at the edges of the thallus), capitate (at the tips of lobes or branches), laminal (on the upper or lower surface of the thallus), etc.

Sorangium See MYXOBACTERALES.

sorbates See SORBIC ACID.

sorbic acid ($CH_3(CH{=}CH)_2COOH$) An antibacterial and antifungal ORGANIC ACID used (as the free acid or e.g. as the potassium salt) as a PRESERVATIVE (e.g., in soft drinks, wines, pickles, cheese, pharmaceuticals). It has been reported that both dissociated and undissociated forms of sorbic acid have antimicrobial activity, but that organisms differ greatly in their susceptibilities [JAB (1983) *54* 383–389]. Potassium sorbate (1–2% w/v) has been shown to reduce the germination rate of *Clostridium botulinum* type E endospores under acidic conditions [AEM (1982) *44* 1212–1221].

sorbitol (glucitol) The POLYOL corresponding to glucose; D-sorbitol may be obtained e.g. by reduction of D-glucose, D-fructose or L-sorbose (see also ASCORBIC ACID). Sorbitol occurs naturally e.g. in certain plants, including a few unicellular green algae and a single red alga (*Bostrychia scorpioides*) [Book ref. 37, pp. 165–167]. D-Sorbitol is used e.g. as a food sweetener for diabetics, and as a substrate in biochemical characterization tests for bacteria (it is attacked by many species).

sorbitol–MacConkey agar See EHEC.

sorbose fermentation A commercial FERMENTATION (sense 2) in which D-sorbitol (see SORBITOL) is oxidized to the 2-ketohexose L-sorbose by *Acetobacter* or *Gluconobacter* spp, especially *G. oxydans* (*'Acetobacter suboxydans'*). L-Sorbose is used mainly as an intermediate in the manufacture of ASCORBIC ACID.

Sordaria A genus of fungi (order SORDARIALES) which occur e.g. on dung and soil; the organisms form brown or black ascospores in persistent asci within dark perithecia – see ASCOSPORE for mode of spore discharge. Many species apparently do not form conidia. *S. fimicola* is homothallic.

Sordariales An order of fungi (subdivision ASCOMYCOTINA) which occur typically as saprotrophs on wood, soil or dung; the organisms typically form perithecia which are not immersed in a stroma and in which a hamathecium is absent. Asci: unitunicate, cylindrical to clavate, sometimes deliquescent. Ascospores: often dark, sometimes appendaged or bearing a gelatinous coat. Genera include e.g. *Cercophora*, CHAETOMIUM, *Melanospora*, NEUROSPORA, *Nitschkea*, PODOSPORA, SORDARIA, *Sphaerodes*, *Triangularia*.

sordelliolysin See THIOL-ACTIVATED CYTOLYSINS.

sorediate Bearing or consisting of soredia.

soredium (*lichenol.*) A vegetative propagule (25–100 μm diam.) consisting of a few algal cells enmeshed in a few fungal hyphae; there is no cortex. Soredia appear, macroscopically, as powdery or granular structures on the thallus surface – often occurring in clusters (see SORALIUM). They originate in the medulla and are pushed up through pores or cracks in the cortex.

Soret band See CYTOCHROMES.

sori See SORUS.

sorocarp A fruiting body, formed by certain SLIME MOULDS, consisting of a stalk bearing a mass of spores or chains of spores. (cf. SOROCYST.)

sorocyst A fruiting body, formed by certain SLIME MOULDS, which resembles a SOROCARP but lacks a stalk.

Sorodiscus See PLASMODIOPHOROMYCETES.

sorogen In cellular slime moulds (e.g. DICTYOSTELIUM): the total cell mass which is undergoing differentiation to form a sorocarp.

Sorosphaera See PLASMODIOPHOROMYCETES.

sortase In *Staphylococcus aureus*: an enzyme, associated with the cytoplasmic membrane, which catalyses a reaction in which certain proteins, secreted through the membrane, undergo site-specific cleavage and covalent linkage to the cell envelope peptidoglycan. Such proteins, which are thus tethered to the cell wall (cf. AUTODISPLAY in *Escherichia coli*), are characterized by a C-terminal *cell wall sorting signal* that includes the conserved sequence LPXTG (see AMINO ACIDS); sortase cleaves between threonine and glycine and links the threonine residue to peptidoglycan. PROTEIN A is an example of a cell-wall-anchored protein.

[Sortase-catalysed anchoring of surface proteins to the cell wall of *S. aureus*: Mol. Microbiol. (2001) *40* 1049–1057.]

sorus (*pl.* sori) A type of fruiting structure composed essentially of a mass of spores (e.g. in fungi of the Uredinales and Ustilaginales, and in certain SLIME MOULDS) or sporangia (e.g. in SYNCHYTRIUM).

SOS box See SOS SYSTEM.

SOS chromotest A colorimetric assay for GENOTOXINS using a genetically engineered strain of *Escherichia coli*. In this strain the SOS gene *sulA* (= *sfiA*) is fused with *lacZ* such that *lacZ* (structural gene for β-galactosidase) behaves as an SOS gene, being induced in response to DNA damage (see SOS SYSTEM). The strain also has mutations in *uvrA* (preventing excision repair) and in *rfa* (rendering it LPS-deficient and more permeable to certain chemicals); it also synthesizes alkaline phosphatase constitutively. Basically, the test involves incubating cultures of the 'tester' strain with various concentrations of a suspected pected genotoxin, and assaying for β-galactosidase activity using (e.g.) ONPG as substrate. The possibility that the chemical under test may inhibit protein synthesis can be excluded by assaying for alkaline phosphatase activity. [Mut. Res. (1985) *147* 65–78, 79–95.] (cf. AMES TEST.)

SOS mutagenesis See SOS SYSTEM.

SOS response In *Escherichia coli*: expression of genes of the SOS SYSTEM in response to conditions which induce that system.

SOS system A system in which the expression of certain genes ('SOS genes') is induced (turned on, or enhanced) in response to damaged DNA or to inhibition of DNA replication – caused e.g. by the effects of ULTRAVIOLET RADIATION or the action of an ALKYLATING AGENT or quinolone antibiotic (such as NALIDIXIC ACID). (The SOS system can also be induced e.g. by the CcdB toxin: an agent, encoded by the F plasmid, which affects gyrase at sites other than those targeted by the quinolone antibiotics [TIM (1998) *6* 269–275].)

The SOS response results in e.g. inhibition of cell division, an increased capacity for DNA repair, alleviation of restriction, an altered pattern of energy metabolism, and (in those strains which contain a COLICIN PLASMID) production of colicin; it also gives rise to an increased rate of mutation.

The SOS system has been studied mainly in *Escherichia coli*, and the following account refers primarily to the operation of the SOS system in that organism.

Control of the SOS response. The SOS system consists of about 30 unlinked genes which are normally repressed by the LexA protein (*lexA* gene product). LexA (perhaps as a dimer) inhibits transcription by binding to an operator sequence (an *SOS box*) in the promoter region of each gene or operon; the binding site for LexA is apparently the consensus sequence 5′-CTGTN_8ACAG-3′ (in which N_8 is an 8-nucleotide sequence). Different SOS boxes have different levels of affinity for LexA, and many of the SOS genes are expressed at low levels even in the repressed state. (At least one SOS gene, *uvrB*, has a second promoter which is not subject to control by LexA.) The LexA protein also represses its own synthesis; however, the SOS box of the *lexA* gene has a low affinity for LexA, so that there is always enough LexA to repress the other SOS genes.

Damaged DNA 'activates' the RECA PROTEIN in an unknown way, i.e. the precise nature of the SOS-inducing signal is not known – although it is generally believed to involve a region of ssDNA and/or product(s) of DNA degradation.

Activated RecA (designated RecA*) behaves as a CO-PROTEASE which triggers the autocatalytic cleavage of LexA – permitting expression of the SOS genes. In vitro studies on the cleavage of LexA have suggested that cleavage occurs more rapidly when a chi (χ) site (5′-GCTGGTGG-3′) occurs near a double-stranded break in DNA; it may be that the presence of χ promotes activation of RecA [Cell (1998) *95* 975–979].

Weak SOS-inducing signals (due e.g. to low-level damage to DNA) may lower the intracellular concentration of LexA to only a limited extent – so that only those SOS genes with low-affinity SOS boxes will be induced.

Inhibition of cell division. One aspect of the SOS response is a temporary inhibition of septum formation (and, hence, cell division); mutants in which the SOS system is constitutively de-repressed form aseptate filaments.

During induction of the SOS system, septum formation is prevented as a result of inhibition of polymerization of the FtsZ protein (product of the *ftsZ* gene: see 'Z ring' in CELL CYCLE (b)). Polymerization of FtsZ is blocked by SulA, product of the SOS gene *sulA* (= *sfiA*) [JB (1998) *180* 3946–3953]; cells may continue to grow as septum-less filaments. (Previously, some mutant alleles of *ftsZ* were referred to as *sulB* and *sfiB*.)

SulA is normally very unstable, so that when repression of *sulA* is re-imposed normal septation is rapidly restored. At least part of the instability of SulA is due to its proteolytic inactivation by the product of the *lon* gene; in *lon* mutants SulA is much more stable, and induction of the SOS response – even by mild ultraviolet radiation – may lead to lethal filamentation.

(Lon is a HEAT-SHOCK PROTEIN; note that the heat-shock response can be triggered by certain factors – such as ultraviolet radiation – which induce the SOS system.)

A temporary inhibition of cell division may be advantageous to the cell in that the cell is given time to carry out repairs to damaged DNA.

DNA repair. The UvrABC-dependent nucleotide excision repair system (including genes *uvrA* and *uvrB*) and the RECOMBINATION REPAIR system are enhanced in the SOS response; these repair systems have a high degree of accuracy ('error-free' repair).

Another repair system, which is apparently expressed only when the SOS system is induced, involves genes *umuC* and *umuD*; the operation of this so-called 'error-prone' repair system (also called *mutagenic repair*) results in an increased rate of mutation (which is referred to as *SOS mutagenesis*). In this process, RecA*, acting as a co-protease, mediates autocatalytic cleavage of UmuD to an active fragment, UmuD′ (see DNA POLYMERASE V). SOS mutagenesis is believed to involve DNA (repair) synthesis on a template strand that contains a 'lesion' – e.g. a pyrimidine dimer (two adjacent, covalently linked pyrimidines on the same strand); such *translesion* synthesis is likely to involve specific DNA polymerase(s) induced as part of the SOS response – including DNA POLYMERASE V, DNA POLYMERASE IV and DNA polymerase II (*polB* gene product) [TIBS (2000) *25* 74–79]. Because of a local loss of base-pairing specificity at the lesion, repair synthesis is likely to introduce incorrect bases and give rise to a mutant genome. (See also WEIGLE REACTIVATION.)

Once the damaged DNA has been repaired, RecA becomes inactive as a co-protease; as the *lexA* gene is induced during the SOS response, the high level of LexA being synthesized means that, when activated RecA disappears, repression of the SOS genes (and restoration of normal cell function) can occur rapidly.

Systems resembling the SOS system in *E. coli* have been observed in some other Gram-negative bacteria – e.g. species of *Bacteroides*, *Rhizobium* and *Salmonella* – as well as in *Bacillus subtilis* ('SOB system').

Many bacteria appear to be inherently resistant to mutagenesis induced by ultraviolet radiation, and seem not to possess a *umuDC* function; such bacteria include *Deinococcus radiodurans*, *Haemophilus influenzae*, *Proteus mirabilis*, *Streptococcus pneumoniae* and *Salmonella typhimurium*.

Functional homologues of *umuC* and *umuD* are carried by certain conjugative plasmids (e.g. ColI, R46 and its derivative pKM101, R205); the genes *mucA* and *mucB* in pKM101 are able to suppress the non-mutability of *E. coli umuC* and *umuD* mutants, and when introduced into cells which normally lack a *umuCD* system, pKM101 can decrease the susceptibility of those cells to killing while increasing their susceptibility to mutagenesis on exposure to *umuCD*-dependent mutagens. (See also AMES TEST.) [Plasmid genes affecting DNA repair and mutation: JCS (1987) *6* (supplement) 303–321.]

The SOS system may be viewed as an adaptive genetic response to specific types of unfavourable environment. Under these conditions it is advantageous for bacteria to be able to make essential repairs and to adapt rapidly by making use of any (fortuitous) new combinations of nucleotides that may enhance their survival. This is reflected in increased repair activity and also in SOS mutagenesis – the latter offering the possibility of beneficial as well as lethal mutations. Suppression of restriction may favour exploitation of imported DNA, while inhibition of cell division may help e.g. to conserve energy and avoid the formation of non-viable daughter cells.

Error-prone repair in in vitro mutagenesis. A variety of mutagenic agents depend on the *umuDC* system for their mutagenic effects; thus, *umuDC* mutants of *E. coli* are resistant to the mutagenic effects of e.g. ultraviolet radiation and methylmethane sulphonate (see ALKYLATING AGENTS) – but can still be mutated by agents such as MNNG and ethylmethane sulphonate (see ALKYLATING AGENTS) which cause lesions leading directly to base mispairing.

souma NAGANA caused (usually or exclusively) by *Trypanosoma vivax*.

sour cream Cultured sour cream is made by the fermentation of pasteurized and homogenized cream (18–20% milk fat) using the same organisms as for BUTTERMILK.

sourdough bread See BREAD-MAKING.

South American blastomycosis *Syn.* PARACOCIDIOIDOMYCOSIS.

Southampton virus See SMALL ROUND STRUCTURED VIRUSES.

Southern bacterial wilt *Syn.* GRANVILLE WILT.

southern bean mosaic virus See SOBEMOVIRUSES.

Southern blot (Southern blotting) (1) A BLOTTING technique in which *DNA* is transferred from a gel to a nitrocellulose matrix (see figure).

(2) Erroneously, a synonym of SOUTHERN HYBRIDIZATION.

Southern hybridization A procedure used e.g. to detect a specific sequence of nucleotides among fragments of DNA which have been separated by gel electrophoresis. Initially, using the SOUTHERN BLOT technique (q.v.), single-stranded forms of the fragments are bound to a matrix. In the next step (Southern hybridization) the matrix is exposed to a labelled PROBE complementary to the sequence of interest; HYBRIDIZATION (sense 1) between the probe and a specific region of the blotted DNA (detected by the probe's label) indicates the required sequence.

Southwestern blotting A technique for detecting DNA-binding proteins in a cell lysate etc. The sample is initially subjected to gel electrophoresis, and the *proteins* thus separated are electroblotted (see BLOTTING) onto a nitrocellulose filter. The affinity of any blotted protein for a specific DNA sequence is then determined by using labelled DNA sequences as probes and detecting the label of any probe which has bound to a given protein.

soy paste *Syn.* MISO.

soy sauce (shoyu) A condiment traditionally prepared by fermenting soybeans and (usually) wheat (cf. TAMARI SAUCE). A KOJI (made from whole or defatted soybeans and crushed roasted wheat) is transferred to a deep vessel and is suspended in brine (to 17–19% NaCl) to form 'moromi'; the moromi is allowed to ferment, usually with occasional agitation and aeration, for e.g. 6–8 months or more. During this stage, enzymes from the koji mould hydrolyse proteins and starch in the raw materials to provide low-MWt substrates which are initially fermented primarily by lactic acid bacteria (e.g. *Lactobacillus delbrueckii*, *Pediococcus halophilus*); the pH falls rapidly to ca. 5.0 or lower, after which fermentation is continued primarily by yeasts (e.g. *Zygosaccharomyces rouxii*). *'Torulopsis'* spp (*Candida etchellsii*, *C. versatilis*) may grow at later stages in the fermentation; these yeasts produce substances which contribute to the flavour and aroma of the final product. The soy sauce fermentation may be carried out by microorganisms naturally present in the raw materials, or pure cultures of selected strains may be used. After the fermentation, the raw soy sauce is decanted or pressed from the solid residue and is pasteurized, filtered and bottled. [Book ref. 5, pp. 39–86; microbes used in soy sauce manufacture: Food Mic. (1984) *1* 339–347.]

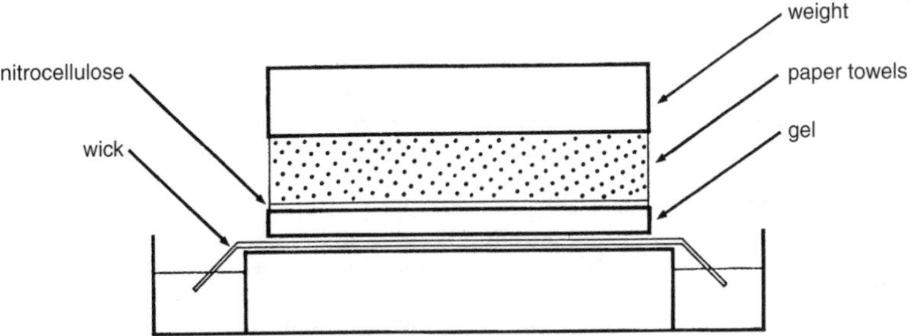

SOUTHERN BLOT (Southern blotting): the transfer of DNA fragments from a gel strip to a sheet of nitrocellulose by capillary action.

Within the gel, the fragments are distributed in discrete zones, according to size, having been previously separated by gel electrophoresis. The fragments are first denatured (made single-stranded) by exposing the gel to alkali; the gel is then exposed to neutral buffer and arranged as shown in the diagram. Driven by capillary action, the neutral, saline solution in the dish rises, via the wick, into and through the gel, through the (permeable) nitrocellulose, and into the stack of paper towels; this upward stream of liquid carries the DNA fragments from the gel to the sheet of nitrocellulose. Importantly, the relative positions of fragments on the sheet are the same as they were in the gel. The sheet is removed and baked at 70°C under vacuum to bind the DNA. The fragments can then be examined e.g. by exposing the sheet to a specific labelled probe; probe–target binding (= *Southern hybridization*), detected by the probe's label, identifies the required sequence.

One alternative to nitrocellulose is APT PAPER (q.v.). This not only avoids the need for baking, it also permits removal of probes – so that different probes can be subsequently used on the same sheet.

Northern blotting refers to the transfer of RNA from gel to matrix.

Western blotting refers to the transfer of protein from gel to matrix.

Electroblotting is a faster method which uses an electric field for effecting transfer.

Figure reproduced from *Bacteria* 5th edition, Figure 8.13, page 195, Paul Singleton (1999) copyright John Wiley & Sons Ltd, UK (ISBN 0471-98880-4) with permission from the publisher.

soybean casein digest agar *Syn.* TRYPTICASE–SOY AGAR.

soybean dwarf virus See LUTEOVIRUSES.

soybean mosaic virus See POTYVIRUSES.

S_p Pattern coefficient. In the comparison of two strains by NUMERICAL TAXONOMY: a coefficient which indicates the degree of similarity corrected for any difference(s) due solely to inter-strain differences in metabolic vigour. (cf. entry D_p.)

sp An unspecified SPECIES: e.g. *Bacillus* sp (a species of *Bacillus*); spp indicates two or more unspecified species or *may* indicate all the species of the given genus.

sp. n. See SPECIES NOVA.

sp. nov. See SPECIES NOVA.

SpA PROTEIN A.

Sparassidaceae See APHYLLOPHORALES.

Sparassis A genus of fungi (family Sparassidaceae). *S. crispa* (the 'brain fungus' or 'cauliflower fungus') is an edible species which typically grows at the base of conifer trees; the thallus (branching, flattened lobes) is pale buff, up to ca. 50 cm across, and the basidiospores are whitish or cream-coloured, ca. 6 × 4 μm.

sparfloxacin See QUINOLONE ANTIBIOTICS.

sparrowpox virus See AVIPOXVIRUS.

sparsomycin An ANTIBIOTIC which inhibits PROTEIN SYNTHESIS in both prokaryotic and eukaryotic cells; it binds to the larger ribosomal subunit and inhibits the peptidyl transferase reaction.

***Spartina* mottle virus** See POTYVIRUSES.

spasmoneme The MYONEME within the stalk of a contractile peritrich (such as VORTICELLA).

Spathularia See HELOTIALES.

Spathulospora See SPATHULOSPORALES.

Spathulosporales An order of fungi (subdivision ASCOMYCOTINA) which are parasitic on red algae. The thallus is a crustose or intracellular stroma. Ascocarp: perithecioid. Asci: unitunicate, deliquescent. Ascospores: appendaged, some spoon-shaped (spathulate). The sole genus, *Spathulospora*, is classified in the LABOULBENIALES by some authors.

spawn (*mycol.*) Mycelium – usually of *Agaricus brunnescens (A. bisporus)* – growing e.g. in a block ('brick') of horse manure, used as an inoculum in MUSHROOM CULTIVATION. It may be derived from a previous culture or may be grown by inoculating sterile compost with spores of a selected strain.

***spc* operon** In e.g. *Escherichia coli*: an OPERON containing genes for r-proteins L14, L24, L5, S14, S8, L6, L18, S5, L30, and L15, and gene *prlA* (= *secY*: involved in protein secretion), listed in order of transcription. (See RIBOSOME.)

***Spe*I** See RESTRICTION ENDONUCLEASE (table).

***speB* gene** (*Streptococcus*) See NECROTIZING FASCIITIS.

special form *Syn.* FORMA SPECIALIS.

specialized transduction See TRANSDUCTION.

speciation (1) The evolutionary process which leads to the formation of new species. (2) The taxonomic process by which species are delimited.

species (*microbiol.*) (singular and plural: species) One of the less-inclusive categories in TAXONOMY – the individuals of a given species displaying a high degree of mutual similarity; a universally applicable and acceptable definition of 'species' is currently not available.

Among microorganisms, SPECIATION (sense 2) was particularly unsatisfactory for prokaryotes prior to the mid-1960s. Since then, molecular methods have progressively provided a basis for what is thought to be a PHYLOGENETIC CLASSIFICATION of these organisms. Such methods include DNA HOMOLOGY studies and the comparison of organisms on the basis of nucleotide sequences in their 16S rRNA.

DNA homology studies suggest e.g. that bacterial strains whose DNA exhibits >70% relatedness are likely to belong to the same species. Moreover, 16S rRNA sequences in different species of a given genus have generally been found to differ by at least 1.5%. These two measurements therefore help to determine which strains belong to a given species and which strains belong to different species. However, certain species of *Bacillus* – which are clearly differentiated at the species level by DNA homology studies – were shown to have virtually identical sequences in their 16S rRNA molecules [IJSB (1992) *42* 166–170]; a possible explanation is that the genomes of these species have diverged recently (on an evolutionary scale) and that their 16S rRNA sequences have had insufficient time to reflect the change.

Strains within a given species may exhibit variation in character due e.g. to the effects of MUTATION (see also ECOTYPE; cf. ECAD) or to the acquisition (or loss) of a PLASMID. Strains can be differentiated by various TYPING techniques.

The term 'species' is also used, for convenience, to refer to the *organisms* belonging to one or more given species – as, for example, in: 'Gram-negative species were isolated from the sample'.

(See also BINOMIAL, TYPE SPECIES; TYPE STRAIN.)

species nova (n. sp.; sp. n.; sp. nov.) A designation used to indicate a newly proposed species at the time of its initial publication.

specific epithet See BINOMIAL.

specific growth rate (μ) Of a given organism: the number of grams of biomass formed per gram of biomass per hour; the unit of μ is hour^{-1}. In any given case, μ depends on factors which include the strain of organism, the temperature, and the concentration of nutrients. Under steady-state conditions in CONTINUOUS CULTURE μ can often be calculated from the Monod equation:

$$\mu = \mu_{max}\frac{s}{k_s + s}$$

in which s is the concentration of a potentially limiting nutrient, μ_{max} is the value of μ when the given nutrient is not limiting, and k_s is numerically equal to the concentration of the given nutrient when $\mu = 0.5\ \mu_{max}$. (See also DILUTION RATE.)

specific immunity Any form of IMMUNITY specific to a given antigen – including e.g. antibody-mediated protective immunity, hypersensitivity, immunological tolerance. (cf. NON-SPECIFIC IMMUNITY.)

specific pathogen free (SPF) Refers to the condition of an animal which was born (or removed from the uterus) and reared under conditions in which specific pathogen(s) have been rigorously excluded. (cf. GNOTOBIOTIC.)

specific-phase antigen See PHASE VARIATION.

spectinomycin An aminocyclitol usually classified as an AMINOGLYCOSIDE ANTIBIOTIC (even though it does not contain an aminosugar). Spectinomycin is produced by *Streptomyces spectabilis* and is (reversibly) bacteriostatic. Spectinomycin-*dependent* strains of *Escherichia coli* have been found to be double mutants – one mutation conferring resistance to the drug by modifying ribosomal protein S5, the other conferring drug dependence [MGG (1977) *151* 261–267]; spectinomycin does not support the growth of STREPTOMYCIN-dependent bacteria.

spermagonium (*syn.* spermogonium, spermatogonium) In certain fungi: a structure within which male reproductive cells (spermatia) are formed. In rust fungi, a spermagonium is called a *pycnium*.

spermatiophore A modified or undifferentiated hypha which bears a SPERMATIUM.

spermatium A non-motile male reproductive cell which can function in SPERMATIZATION. (cf. MICROCONIDIUM (sense 1) and PYCNIOSPORE.)

spermatization In certain higher fungi: the union of a SPERMATIUM with a female reproductive structure (e.g. a TRICHOGYNE or a FLEXUOUS HYPHA).

spermatogonium *Syn.* SPERMAGONIUM.

spermidine See POLYAMINES.

spermine See POLYAMINES.

spermogonium *Syn.* SPERMAGONIUM.

Spermophthora See METSCHNIKOWIACEAE.

Spermophthoraceae *Syn.* METSCHNIKOWIACEAE.

SPF SPECIFIC PATHOGEN FREE.

SPf66 vaccine See MALARIA.

Sphacelaria See PHAEOPHYTA.

Sphacelia A genus of fungi of the HYPHOMYCETES; teleomorphs occur e.g. in the genus CLAVICEPS.

Sphaceloma A genus of fungi of the COELOMYCETES. Teleomorph: *Elsinoë*.

Sphacelotheca See USTILAGINALES.

Sphaeractinomyxon See ACTINOSPOREA.

Sphaeriales An order of fungi (subdivision ASCOMYCOTINA) which typically form carbonaceous perithecia (often containing filiform paraphyses) immersed in a stroma. Asci: unitunicate, cylindrical or ellipsoidal. Ascospores: aseptate or septate, colourless or dark. Genera: e.g. *Ascotricha*, *Camarops*, DALDINIA, *Hypoxylon*, XYLARIA.

Sphaerobolus A genus of lignicolous and coprophilous fungi (order SCLERODERMATALES) formerly classified in the Nidulariales. *S. stellatus* forms globose basidiocarps (ca. 1.5–2.5 mm diam.) in which the peridium is composed of six layers. At maturity, the glebal mass is violently discharged following rupture of the peridium and the sudden eversion of certain inner layers; eversion occurs apparently as a result of increased osmotic pressure in one of the inner layers (due to the breakdown of glycogen to soluble sugars) and the consequent stress upon this (and an adjacent) layer on absorption of water. [Culturing for fruiting bodies: Mycol. (1984) *76* 944–946.]

sphaerocyst A spherical or ovoid cell, numbers of which occur in the TRAMA, and in other parts of the CONTEXT, e.g. in fungi of the genera *Lactarius* and *Russula*.

sphaerocyte An (abnormal) spherical or subspherical form of an erythrocyte formed, e.g., as a result of the activity of *Escherichia coli* α-haemolysin.

Sphaerocytophaga A genus of bacteria; the organisms are similar or identical to species of CAPNOCYTOPHAGA [Book ref. 45, pp. 356–379 (365)].

Sphaerodes See SORDARIALES.

Sphaerodothis See POLYSTIGMATALES.

sphaeromastigote A form assumed by the cells of species of *Trypanosoma* during certain stages of their life cycles: see TRYPANOSOMATIDAE.

Sphaeromonas A genus of fungi found in the RUMEN.

Sphaeromyxa See MYXOSPOREA.

Sphaerophoraceae See CALICIALES.

Sphaerophorus (1) (*mycol.*) See CALICIALES. (2) (*bacteriol.*) Obsolete genus of bacteria which are currently included in the genera BACTEROIDES and FUSOBACTERIUM.

Sphaerophrya See SUCTORIA.

sphaeroplast (spheroplast) A spherical or near-spherical, osmotically sensitive structure which resembles a PROTOPLAST but is formed by the disruption or *partial* removal of the cell wall of a cell suspended in an isotonic (or hypertonic) medium – i.e., cell wall fragments remain adhering to the cytoplasmic membrane. *Bacterial* sphaeroplasts are typically formed from Gram-negative cells, since it is generally difficult to effect the complete removal of the cell wall from such cells. Sphaeroplasts from Gram-negative bacteria may be prepared by suspending the cells in a solution containing EDTA, TRIS, LYSOZYME, and (hypertonic) sucrose, or by exposing the cells (in a hypertonic medium) to antibiotics (e.g. PENICILLINS) which inhibit cell wall synthesis. Sphaeroplasts may be formed from Gram-positive cells by limited action of wall-lysing enzymes. (cf. AUTOPLAST.) *Yeast* sphaeroplasts may be prepared by treating yeast cells with e.g. ZYMOLYASE.

Sphaeropsidales An order of fungi (class COELOMYCETES) in which the vegetative mycelium and the conidiomata may develop on the surface or within the substrate or host, and the conidiogenous cells line discrete pycnidia (see PYCNIDIUM) or locules in a stroma. Genera include e.g. ASCHERSONIA, ASCOCHYTA, *Chaetomella*, *Coniothyrium*, *Dendrophoma*, *Diplodia* (see also BLUE-STAIN and SAP-STAIN), PHOMA, *Phomopsis* (see also LUPINOSIS), *Phyllosticta*, SEPTORIA, *Sphaeropsis*, ZYTHIA.

Sphaeropsis See SPHAEROPSIDALES.

Sphaerospora See MYXOSPOREA.

Sphaerotheca See ERYSIPHALES.

Sphaerotilus A genus of Gram-negative, rod-shaped, obligately aerobic, asporogenous bacteria which occur in flowing, organically polluted fresh water (see SEWAGE FUNGUS) and in activated sludge (see SEWAGE TREATMENT). Individual cells ('swarmers') of the type (sole) species, *S. natans*, are ca. $1-2 \times 3-10$ μm, each with a subpolar or polar tuft of flagella; the cells commonly occur as sheathed trichomes, the SHEATH (sense 1) consisting of a polysaccharide–protein–lipid complex. (Sheaths are not formed in nutritionally rich media, and trichomes in activated sludge have thin, colourless sheaths. In some cases a sheath bears a slime layer containing an accumulation of $Fe(OH)_3$.) Sheathed trichomes are commonly attached (at one end) to the substratum. The organisms can utilize a wide range of carbon and nitrogen sources; VITAMIN B_{12} (replaceable by methionine) is a growth requirement. PHB may be accumulated. GC%: ca. 70. [Book ref. 45, pp. 425–440 (428–431).]

Sphaerulina See DOTHIDEALES.

spheroid (*virol.*) See ENTOMOPOXVIRINAE.

spheroidin See ENTOMOPOXVIRINAE.

spheroidosis (*pl.* spheroidoses) See ENTOMOPOXVIRINAE.

spheroplast *Syn.* SPHAEROPLAST.

spherule (1) See COCCIDIOIDES. (2) See SCLEROTIUM (sense 1). (3) See ENTOMOPOXVIRINAE.

spherulin An antigen, prepared from the spherule stage of *Coccidioides immitis*, said to be more sensitive than COCCIDIOIDIN in skin tests for COCCIDIOIDOMYCOSIS.

Sphingobacterium A genus proposed for sphingolipid-containing strains of *Flavobacterium* [see e.g. FEMS (1983) *20* 375–378].

sphingolipid A lipid derived from the aminoalcohol *sphingosine*. (Sphingosine is $CH_3.(CH_2)_{12}.CH{=}CH.CH(OH).CH(NH_2).CH_2OH$.) Sphingolipids include the phospholipid SPHINGOMYELIN and the *non*-phospholipids *cerebroside* (in which the C-1 hydroxyl of sphingosine is linked to a sugar residue) and GM gangliosides (in which the hydroxyl group is linked e.g. to several sugar residues and a residue of *N*-acetylneuraminic acid).

Sphingomonas chlorophenolica See BIOREMEDIATION.

sphingomyelin A SPHINGOLIPID (q.v.) which occurs e.g. in the plasma membrane in erythrocytes; in sphingomyelin, the amino group at C-2 of sphingosine forms an amide bond with a long-chain fatty acid (forming *ceramide*), and the terminal (C-1) hydroxyl is esterified with phosphorylcholine.

sphingomyelinase A PHOSPHOLIPASE (q.v.) which cleaves SPHINGOMYELIN. An example is the staphylococcal β-HAEMOLYSIN, a C-type phospholipase which cleaves sphingomyelin to phosphorylcholine and ceramide. (See also HOT–COLD LYSIS.) *Corynebacterium pseudotuberculosis* (formerly *C. ovis*) forms a sphingomyelinase (a D-type phospholipase) which cleaves choline from sphingomyelin.

sphingosine See SPHINGOLIPID.

SPI *Salmonella* PATHOGENICITY ISLAND.

spiculum A narrow, apical extension of a STERIGMA on which a spore is borne.

spin killing A type of killing of sensitive strains of *Paramecium* by certain types of killer paramecia (e.g. strains containing *Caedibacter varicaedens*); affected cells swim with rapidly reversing rotational movements.

spin label See ELECTRON SPIN RESONANCE.

spina (*pl.* spinae) In certain prokaryotes (e.g. some pseudomonads): a hollow, rigid, apparently non-prosthecate appendage, one or more of which project from the cell surface; it is a thin-walled structure formed from a cross-linked helical filament. [Conical spinae: CJM (1984) *30* 716–718. Possible function: JGM (1991) *137* 1081–1086.]

spinach latent virus See ILARVIRUSES.

spindle (1) See MITOSIS and MEIOSIS. (2) (*virol.*) See ENTOMOPOXVIRINAE.

spindle pole body (*syn.* centriolar plaque; polar plaque) In some types of eukaryotic cell (including e.g. *Saccharomyces cerevisiae*): a body which is embedded in the nuclear membrane and which can act e.g. as a MICROTUBULE-ORGANIZING CENTRE during MITOSIS.

spinning disease A FISH DISEASE which affects the Atlantic menhaden (*Brevoortia tyrannus*); the disease tends to occur in large-scale annual epizootics with massive mortalities. Terminal symptoms include loss of coordination and erratic swimming behaviour. The causal agent is a virus which closely resembles (and may be identical to) IPN VIRUS.

spinoculation Centrifugally-assisted enhancement of transduction efficiency (a procedure in which the virus vector and cells are centrifuged together).

spira Refers to helical aerial hyphae – one of three categories used in a system for the morphological classification of streptomycetes; the other two categories are rectus–flexibilis (straight to flexuous hyphae) and retinaculum apertum (hook-shaped, looped, or spiral hyphae with one or two turns). [Disadvantages of the system: Book ref. 46, p. 2067.]

spiral plate count method A method used for counting bacteria in samples of milk. Essentially, the inoculum is delivered from a syringe onto a rotating plate so that it is deposited along a spiral path; the volume of inoculum delivered to a particular region of the plate can be read off from the apparatus. A colony count is made after incubation of the plate.

spiramycins See MACROLIDE ANTIBIOTICS.

Spirillaceae A family of bacteria which formerly included the genera *Campylobacter* and *Spirillum* but which has been abandoned pending further taxonomic studies [Book ref. 22, p. 71].

Spirillospora A genus of bacteria (order ACTINOMYCETALES, wall type III; group: maduromycetes) which occur e.g. in soil and leaf litter. The organisms form thin (<1 μm diam.) branching substrate and aerial mycelium, the latter giving rise to spherical sporangia (up to ca. 24 μm diam.); in water, mature sporangia release motile (flagellated) rod-shaped spores. Type species: *S. albida*. [Book ref. 73, p. 113.]

spirilloxanthins See CAROTENOIDS.

spirillum (1) Any spiral-shaped, relatively rigid bacterial cell. (2) Any species or strain of *Spirillum*.

Spirillum A genus of Gram-negative, asporogenous bacteria. Cells: rigid, helical, ca. 1.4–1.7 × 14–60 μm. POLAR MEMBRANES and PHB granules are formed, COCCOID BODIES are not. Motile: bipolar tufts of flagella of long wavelength. Microaerophilic. Metabolism is respiratory; nitrate respiration does not occur. Oxidase +ve; catalase −ve; phosphatase +ve; indole −ve; arylsulphatase −ve. Growth is inhibited by NaCl levels >0.02%, and by phosphate >0.01 M. Carbohydrates are not utilized; carbon sources: salts of organic acids (e.g. succinate). GC%: 38. Type species: *S. volutans*, found in stagnant freshwater habitats. (cf. AQUASPIRILLUM; AZOSPIRILLUM; OCEANOSPIRILLUM; SPOROSPIRILLUM.)

Species *incertae sedis*: *'S. minus'* (= *'S. minor'*), causal agent of one form of RAT-BITE FEVER; cells: ca. 0.2 × 3–5 μm, usually spiral, actively motile by one or more flagella at each pole. *'S. pulli'*, apparent causal agent of DIPHTHEROID STOMATITIS in chickens; cells: rigid spirals, ca. 1.0 × 5–12 μm, motile with one flagellum at each pole.

[Book ref. 22, pp. 90–93. Culture, media etc: Book ref. 45, pp. 597–599.]

spirits Alcoholic beverages prepared by the distillation of fermented liquors or mashes. The characteristic flavours and qualities of the various spirits are determined by the nature of the fermented raw materials, conditions of distillation, effects of ageing, etc. The fermentation stage generally resembles that in BREWING or WINE-MAKING. Use is made of yeast (*Saccharomyces*) strains which effect maximum *attenuation* (i.e. they convert the maximum amount of carbohydrate to alcohol); these strains are characterized by high alcohol-tolerance. Distillers' yeasts produce a range of higher alcohols (FUSEL OIL) which contribute to the aroma and flavour of the product. Spirits are not subject to microbial spoilage owing to their high alcohol content.

Bourbon is an American whisky (whiskey) made by distilling a fermented mash containing maize and rye. *Brandy* (*cognac*) is prepared by distilling grape wine. *Gin* is made by distilling fermented rye or maize worts and redistilling the product in a still containing herbs and juniper berries ('botanicals'). *Rum* is a distillate of fermented molasses or sugarcane products. *Tequila*: see PULQUE. *Whisky* is distilled either from a wort derived from malted barley (*malt whisky*) or from a mash of unmalted grain (often barley or maize) mixed with a proportion of malted barley (*grain whisky*).

Spirochaeta A genus of Gram-negative bacteria of the family SPIROCHAETACEAE; the organisms, which include both obligately and facultatively anaerobic species, occur as free-living inhabitants of freshwater and marine environments, including muds, and are common in H_2S-containing habitats. The cells are 0.2–0.75 × 5–250 μm; the type species, *S. plicatilis*, contains many periplasmic flagella per cell, but the other species contain only two per cell. The organisms are motile in liquids and can also creep over solid surfaces and migrate through agar (≤1%) – the latter feature being useful in isolation procedures. In most species the anaerobic metabolism of carbohydrates yields ethanol, acetate, CO_2 and H_2; *S. zuelzerae* forms lactate and succinate. GC%: ca. 51–65.

Species differentiation is based on e.g. cell diameter, ability to grow aerobically, NaCl requirement and pigmentation. Six species are recognized [Book ref. 22, pp. 39–46]: *S. aurantia* and *S. halophila* (both facultative anaerobes; *S. halophila* is halophilic and forms a red pigment); *S. litoralis* and *S. stenostrepta* (both obligate anaerobes; *S. litoralis* is halophilic); *S. plicatilis* (cell diameter ca. 0.75 μm); and *S. zuelzerae* (obligate anaerobe, forms succinate). (All species except *S. plicatilis* have been grown in pure culture.)

Spirochaetaceae A family of anaerobic, facultatively anaerobic or microaerophilic bacteria of the order SPIROCHAETALES; the cells typically do not have hooked ends, their cell wall PEPTIDOGLYCAN contains the diamino acid L-ornithine, and their carbon and energy sources are carbohydrates and/or amino acids (cf. LEPTOSPIRACEAE). Genera: BORRELIA, CRISTISPIRA, SPIROCHAETA, TREPONEMA.

Spirochaetales An order of helical, flexible, asporogenous, Gram type-negative bacteria which include free-living, parasitic and pathogenic species. The cells are 0.1–3.0 μm in width and 5–250 μm in length, according to species, and they divide by transverse binary fission; motility in liquid media involves periplasmic flagella (see below and FLAGELLAR MOTILITY), and the organisms can also creep over solid surfaces. Depending on size, spirochaetes may be seen by bright-field MICROSCOPY (e.g. after staining with dyes or silver-deposition techniques), by dark-field microscopy, or by fluorescence microscopy following treatment with fluorochrome-labelled antibodies. Species include anaerobes (obligate and facultative) and microaerophiles (family SPIROCHAETACEAE) and aerobes (family LEPTOSPIRACEAE). The organisms are chemoorganotrophs which have fermentative and/or respiratory (oxidative) type(s) of metabolism.

Ultrastructure. The spirochaetal cell consists of a helical *protoplasmic cylinder* (= PROTOPLAST (sense 2) covered by a 'cell wall' of PEPTIDOGLYCAN), the whole being enclosed within an *outer sheath* (= outer envelope) similar to the OUTER MEMBRANE of other Gram type-negative bacteria; one or more *periplasmic flagella* (= axial fibrils, axial filaments, endoflagella, periplasmic fibrils) arise at *each* end of the protoplasmic cylinder and wind around the protoplasmic cylinder – i.e. they are located between the protoplasmic cylinder and the outer sheath. The periplasmic flagellum is similar in ultrastructure to the prokaryotic FLAGELLUM; the number of periplasmic flagella per cell is a stable characteristic of the various species and genera of spirochaetes.

[Structure/motility of spirochaetes: JB (1996) *178* 6539–6545.]

Habitats. Spirochaetes occur in host tissues (BORRELIA, CRISTISPIRA, LEPTOSPIRA, TREPONEMA) and in soil and aquatic habitats (LEPTOSPIRA, SPIROCHAETA). Spirochaetes occur also in the hindgut of termites (see TERMITE–MICROBE ASSOCIATIONS); these have not been isolated in pure culture, and the names proposed for these organisms (e.g. *Diplocalyx*, *Hollandina*, *Pillotina*) have not been validly published [Book ref. 22, pp. 67–70].

spirochaete (spirochete) A member of the SPIROCHAETALES.

Spirochona See HYPOSTOMATIA.

Spirocystis A genus of protozoa (suborder EIMERIORINA) parasitic in oligochaetes.

Spirogyra A genus of unbranched filamentous green algae (division CHLOROPHYTA) in which sexual reproduction occurs by a process of conjugation; flagellated cells are never formed. Species are common in standing fresh water; the filaments are usually free-floating, but are sometimes attached to stones etc.

Each filament is composed of a chain of cylindrical cells (arranged end-to-end); the filaments are coated with mucilage and may be capable of GLIDING MOTILITY. Each cell has a thin peripheral layer of cytoplasm surrounding a large central vacuole; a single nucleus is suspended in the centre of the vacuole by fine cytoplasmic threads. The cytoplasm contains one or more (according to species) ribbon-like, spirally arranged chloroplasts (containing many pyrenoids) extending throughout the length of the cell.

Asexual reproduction occurs by fragmentation of the filaments; zoospores are not formed.

Sexual reproduction occurs by conjugation: typically, two filaments lie side-by-side (held together by mucilage), and a conjugation tube is formed between two juxtaposed cells (one cell in each filament: *scalariform* conjugation). (Conjugation can also occur between adjacent cells in a single filament: *lateral* conjugation.) One protoplast (designated male) passes from one cell to the other via the conjugation tube, and syngamy occurs. (In some species a zygote is formed within the conjugation tube.) The zygote develops a thick, resistant wall and undergoes a period of dormancy. Meiosis occurs before germination, and 3 of the 4 resulting nuclei disintegrate; on germination, a single haploid filament develops.

Genera related to *Spirogyra* include e.g. MOUGEOTIA, ZYGNEMA and ZYGOGONIUM; see also DESMIDS.

Spiromyces See KICKXELLALES.

Spiroplasma A genus of facultatively anaerobic bacteria (family SPIROPLASMATACEAE) which occur as epiphytes or as intracellular or extracellular parasites or pathogens in a range of invertebrates and plants (see e.g. YELLOWS). (See also MLO and SRO.) Cells: pleomorphic, ranging from helical filaments (commonly 0.1–0.2 μm × 3.0–5.0 μm) which typically occur in the logarithmic phase of growth (and which may persist into the stationary phase), to branched non-helical filaments and to coccal and coccoid forms (ca. 0.3 μm diam.); in some infected tissues the cells of some strains may be indistinguishable from MLOs. Helical filaments contain a system of fibrils [JGM (1985) *131* 983–992] which may be involved in the characteristic flexing and other movements of which these forms are capable; the cells lack flagella. NADH oxidase occurs in the cytoplasm (cf. ACHOLEPLASMA). The organisms are chemoorganotrophs which require sterols for growth; with the exception of SROs, they can be cultured e.g. on serum-containing media. GC%: ca. 25–31. Type species: *S. citri*.

S. apis. The causal agent of a May-disease-like disorder of the honey bee (*Apis mellifera*) in France; isolated also e.g. from flowers.

S. citri. Isolated e.g. from citrus plants suffering from STUBBORN DISEASE, from other naturally-infected plants (e.g. broad beans; see also CORN STUNT DISEASE), and from leafhoppers; some strains are pathogenic for honey-bees.

S. floricola. Isolated e.g. from magnolia flowers; can cause disease in beetles of the genus *Melolontha*.

S. kunkelii. See CORN STUNT DISEASE.

S. mirum. Isolated e.g. from rabbit ticks; can cause disease in e.g. rodents and chick embryos under experimental conditions.

Spiroplasma spp are divided into serogroups I (e.g. *S. citri*, *S. melliferum*), II (SROS), III (*S. floricola*), IV (*S. apis*) and V (*S. mirum*).

(See also SPIROPLASMAVIRUSES.)

Mycoplasma-like organisms have been referred to by the trivial name 'phytoplasma' [see ARM (2000) *54* 221–255].

Spiroplasmataceae A family of (typically) helical and filamentous, cell-wall-free bacteria (order MYCOPLASMATALES) which are associated with various invertebrates and plants. Sole genus: SPIROPLASMA. [Book ref. 22, pp. 781–787.]

spiroplasmaviruses BACTERIOPHAGES which infect *Spiroplasma* spp. SV-C1 phages are filamentous (230–280 × 10–15 nm) and contain DNA; they may belong in the INOVIRIDAE. In SV-C2 phages there is an icosahedral-type head (52–58 × 48–51 nm) with a tail (75–83 × 6–8 nm) attached at one vertex. SV-C3 phages (= C3 strain C3 phages) probably belong to the PODOVIRIDAE; the icosahedral-type head (ca. 40 × 35–37 nm) contains linear dsDNA (MWt 1.4×10^7) and has a short tail (13–18 × 6–8 nm) attached at one vertex. SV-C3 progeny virions seem to be released in host-derived membrane vesicles. (See also MYCOPLASMAVIRUSES).

Spirosoma A genus of yellow-pigmented bacteria (family SPIROSOMACEAE). The cells, 0.5–1.0 × 1.5–6.0 μm, are commonly helical or curved rods – 'ring-shaped' forms sometimes being seen as a result of overlap of the ends of highly curved cells. Acid is formed from many carbohydrates. GC%: 51–53. Type species: *S. linguale*.

Spirosomaceae A family of Gram-negative, obligately aerobic, chemoorganotrophic, pink- or yellow-pigmented bacteria which occur in soil and aquatic habitats; the organisms are rigid, straight, curved or spiral, non-motile rods. Growth (optimum temperature range: 20–30°C) occurs e.g. on peptone–yeast extract–glucose agar. Genera: FLECTOBACILLUS, RUNELLA, SPIROSOMA. [Book ref. 22, pp. 125–132.]

Spirostomum A genus of ciliate protozoa (order HETEROTRICHIDA) which occur e.g. in microaerobic freshwater and estuarine habitats. Cells: elongate, cylindrical, up to ca. 4 mm in length – but able to contract to a fraction of their normal length; somatic ciliature is uniform, and the cytostome is lateral. A large posterior contractile vacuole discharges via a long canal which opens to the exterior near the anterior end. In some species (e.g. *S. ambiguum*) the macronucleus is moniliform, in others (e.g. *S. teres*) it is ovoid.

spirotrich A member of the Spirotrichia.

Spirotrichia See POLYHYMENOPHOREA.

Spirotrichonympha See HYPERMASTIGIDA.

Spirulina A genus of filamentous CYANOBACTERIA (section III) in which the trichomes are helical with little or no constriction between adjacent cells. GC%: 44–53. The trichomes show GLIDING MOTILITY, and the tips are capable of a twitching or jerking motion which may be brought about by a contractile (actin-like?) protein [Book ref. 76, pp. 420–421]. *Spirulina* typically occurs in warm, saline, alkaline lakes, where it forms dense tangled masses. For centuries it has been used as food by people in the Lake Chad region of Africa, and was also eaten by the Aztecs of ancient Mexico; in Africa, the masses of growth are collected, sun-dried, and made into thin cakes which may be eaten directly or after cooking.

Spitzenkörper (apical granule) A small, densely staining body which occurs, close to the cytoplasmic membrane, in the tip of an actively growing fungal hypha; in at least some fungi it consists of a cluster of vesicles associated with microfilaments. (See also GROWTH (fungal).)

spleen focus-forming virus (SFFV) See FRIEND VIRUS and RAUSCHER VIRUS.

splenic fever See ANTHRAX.

splice sites See SPLIT GENE.

spliceosome See SPLIT GENE (a).

splicing (1) (of RNA) See SPLIT GENE. (2) (of DNA) See e.g. CLONING.

split gene (interrupted gene) A structural gene (encoding e.g. a protein, rRNA or tRNA) that contains one, several or many specific sequences of nucleotides (*intervening sequences*; *introns*) which, although represented in the primary RNA transcript of the gene, are absent from the mature RNA molecule (mRNA, tRNA etc.) and, hence, do not encode any part of the gene product. Thus, *maturation* of the primary RNA transcript of a split gene must involve a process of *splicing* in which sequence(s) corresponding to intron(s) are deleted ('spliced out') and the remaining sequences (termed *exons*) are joined together; for example, in the case of mRNA (q.v.), the amalgamated exons form the complete coding sequence of the gene together with any non-coding leader and/or trailer sequences.

In some cases a given gene can be spliced in different ways to yield different versions of the encoded product, i.e. splicing can give rise to different combinations of exons – producing correspondingly different coding sequences. Such *alternative splicing* can be advantageous to a cell e.g. in that synthesis of multiple products from a single gene permits economy in the size of the genome.

Introns are often non-coding sequences that are degraded following excision – but see (b) and (c), below, and INTRON HOMING.

Split genes occur in eukaryotes and in certain prokaryotes (including bacteria and archaeans) and viruses. Many (or most) eukaryotic genes contain at least one intron – sometimes many; the gene encoding type I collagen appears to contain about 50 introns. By contrast, the (human) insulin gene contains only one intron, while the gene encoding interferon-β is intron-less.

In higher eukaryotes, many (perhaps most) of the nuclear structural genes are split genes, but in typical eukaryotic microorganisms (e.g. *Dictyostelium*, *Saccharomyces*) fewer of these genes are reported to contain introns, and the introns tend to be smaller than those in higher eukaryotes; by contrast, no intron-less protein-encoding genomic sequences were seen in the database of *Physarum polycephalum* – each gene containing, on average, 3.7 introns [NAR (2000) *28* 3411–3416].

Splicing of the RNA transcript of a split gene involves cleavage of a precise phosphodiester bond at both ends of each intron (at the exon–intron junctions) – the 5′ and 3′ *splice sites*; this is followed by the formation of a phosphodiester bond between the exon at the 5′ end of the intron (the *5′ exon*) and the exon at the 3′ end (the *3′ exon*). Details of the splicing mechanism differ in different types of split gene.

Split genes in eukaryotes are considered in (a) to (d), below; split genes in prokaryotes and viruses are considered in (e).

(a) *Nuclear mRNA introns.* In nuclear protein-encoding genes of e.g. *Saccharomyces cerevisiae*, the introns have been found to contain three CONSENSUS SEQUENCES that are necessary for splicing:

GTATGT

at the 5′ end of the intron,

PyAG

at the 3′ end, and

TACTAAC

(the 'TACTAAC box') near the 3′ end. These sequences are those in the DNA strand which has the same sequence as the RNA transcript; they may also be written in terms of the RNA itself, e.g. TACTAAC = UACUAAC.

The consensus sequences mentioned above may differ somewhat in other eukaryotes, but nearly all introns of this type have the 5′ terminal GT (the 5′, *left* or *donor* splice site) and the 3′-terminal AG (the 3′, *right* or *acceptor* splice site); this is the so-called *GT...AG (or GU... AG) rule*.

Excision of an intron from a pre-mRNA molecule (transcribed by RNA polymerase II) depends on *trans*-acting factors and initially involves cleavage at the 5′ splice site, thus releasing the 3′ end of the 5′ exon, and formation of an intron–3′-exon intermediate in the form of a *lariat* (a tailed, circular molecule); the lariat results from the formation of a 2′–5′ phosphodiester bond between the (free) 5′ end of the intron and a site within the intron near its 3′ end (the *branch point*). In yeast, this branch occurs at the last adenine residue in the TACTAAC box.

Subsequently, the 3′ splice site is cleaved, and the 5′ and 3′ exons are ligated. The intron, released as a lariat, is 'debranched', enzymically, to release a free linear intron (which is degraded in vivo). Debranching enzymes from *S. cerevisiae* and *Schizosaccharomyces pombe* exhibit some homology with the human RNA lariat debranching enzyme [NAR (2000) *28* 3666–3673].

Secondary or tertiary structure in the intron appears not to be important for the juxtaposition of splicing sites in nuclear introns (cf. other mechanisms below).

The splicing reaction depends on certain small molecules of RNA – snRNAs (see SNRNA) – which apparently act as *external guide sequences* (EGSs) to locate the relevant sites in the pre-mRNA. In higher eukaryotes, U1 snRNP binds to the 5′ splice site (the 5′ end of U1 snRNA probably base-pairing with the conserved sequence GUNAGU), U2 snRNP binds to the branch point, and U5 snRNP probably binds to the 3′ splice site.

Interactions between the snRNPs, and between snRNPs and the RNA substrate, presumably bring about juxtaposition of the 3′ and 5′ splice sites. Splicing of the pre-mRNAs in yeasts and in higher eukaryotes involves a large (40S–60S) ribonucleoprotein particle (the *spliceosome*) which, in e.g. human cells, contains U snRNPs; in yeasts, the spliceosome contains at least three snRNAs, and its assembly is dependent on the presence of a 5′ splice site and a TACTAAC box within the RNA [Cell (1986) *45* 869–877].

(b) *Group I introns* (= *class I introns*). Introns of this type occur in nuclear pre-rRNA of *Tetrahymena thermophila* and *Physarum polycephalum*, in several fungal mitochondrial pre-rRNAs and pre-mRNAs, and in certain chloroplast tRNA genes. They are characterized by a common splicing mechanism and by a number of conserved sequences; interaction between these sequences apparently folds the intron into a 'core structure' in which the 5′ and 3′ splice sites are brought into close proximity. Excision of this type of intron is an intrinsic property of the RNA itself, i.e. group I introns are *self-splicing* or *autocatalytic*; in some cases splicing can occur in vitro when the RNA is incubated with divalent cations and guanosine or a guanosine nucleotide (GMP, GDP or GTP – or a G at the 3′ end of an oligo- or polynucleotide) in the absence of protein. (cf. RIBOZYME.)

Splicing involves two distinct and independent cleavage–ligation steps – and transesterification reactions in which the total number of phosphodiester bonds remains constant (so that splicing does not require an additional source of energy).

Cleavage occurs first at the 5′ splice site concomitantly with the formation of a phosphodiester bond between the 3′-OH of a guanine nucleoside or nucleotide and the 5′ end of the intron. Subsequently, cleavage of the 3′ splice site is coupled with ligation of the 5′ and 3′ exons. In many cases the excised linear intron circularizes by a transesterification reaction in which the 3′-terminal residue (usually or always a guanosine residue) forms a covalent bond at a site near the 5′ end, the terminal oligonucleotide being released.

It has been proposed that, during the splicing process, a nucleotide sequence within the 5′ exon (adjacent to the 5′ splice site) base-pairs with a complementary sequence within the intron (the *internal guide sequence*, IGS) to form a secondary structure that defines the 5′ splice site and holds the 5′ exon in place for the second (exon-ligation) step [Nature (1986) *322* 86–89, (discussion 16–17)]; the actual nucleotide sequence of the IGS is different in different introns (i.e. it is not a conserved sequence).

Some group I introns are not capable of autocatalytic splicing in vitro in the absence of protein – protein(s) apparently being necessary for the correct folding of the molecule. Certain introns in fungal mitochondrial genes encode a small protein (a *maturase*) which is apparently synthesized by readthrough from the preceding exon; in at least some cases the maturase is necessary for the splicing of the intron encoding it – possibly being involved in folding the intron to align the splice sites correctly.

(c) *Group II introns* (*class II introns*). Introns of group II are distinct from those of group I in their conserved sequences and secondary structure (e.g. they have a conserved stem-and-loop or hairpin structure near the 3′ end). Group II introns occur e.g. in fungal mitochondrial protein-encoding genes and in *Euglena* chloroplast genes.

Certain yeast mitochondrial group II introns are self-splicing in vitro in the presence of magnesium ions and spermidine; however, the reaction does not require guanosine or any other nucleotide, and the excised intron initially occurs in the form of a *lariat* similar to that formed by nuclear pre-mRNA splicing [Cell (1986) *44* 213–223, 225–234; *46* 557–565].

Group II introns also differ from those in group I in having at their splice sites conserved sequences which are similar to those in nuclear pre-mRNA introns. Nevertheless, splicing of group II introns resembles that of group I introns e.g. in being autocatalytic and in not requiring *trans*-acting RNA molecules (an IGS may be involved); furthermore, certain group II introns apparently encode 'maturases'. Thus, group II introns appear to be intermediate in character between group I and nuclear pre-mRNA introns. [Splicing of nuclear pre-mRNA and groups I and II introns (minireview): Cell (1986) *44* 207–210.]

Interestingly, certain group II introns encode potential products which appear to exhibit homology (over part of their length) with several reverse transcriptases encoded by viruses and transposable elements [Nature (1985) *316* 641–643 (discussion 574–575)].

(d) *Nuclear tRNA introns*. In yeast, at least some nuclear tRNAs contain a single intron in the anticodon arm of the pre-tRNA (see TRNA); the intron contains a sequence that is complementary to the anticodon of the tRNA, and this may allow the anticodon arm to adopt a secondary structure which can be recognized by the splicing enzymes. Splicing requires at least two distinct enzymes. The first catalyses endonucleolytic cleavage with the generation of 5′-OH ends and 3′ ends bearing a 2′,3′-cyclic phosphate group; the second (an *RNA ligase*) catalyses exon ligation in an ATP-dependent reaction.

(e) *Introns in viruses and prokaryotes*. Introns have been found in the genomes of certain viruses – including bacteriophages. For example, in phage T4 there are three group I introns: a 598-bp intron in gene *nrdB* (encoding ribonucleotide reductase), a 1033-bp intron in *nrdD* (= *sunY*) (encoding anaerobic ribonucleotide reductase), and a 1016-bp intron in gene *td* (encoding thymidylate synthase); efficient splicing of *td* is reported to require a ribosomal function [GD (1998) *12* 1327–1337]. All three introns in the T4 genes can self-splice in vitro. [Phage T4 introns: ARG (1990) *24* 363–385.] Similar introns also occur in other T-even phages.

In adenovirus-infected cells, different patterns of RNA splicing can generate different proteins from the same RNA transcript. Thus, the very long polycistronic viral RNA transcripts produced in these cells can be spliced in different ways – generating functional mRNAs in which the same 5′ cap and leader sequence is spliced onto different coding sequences. (See also SUBGENOMIC MRNA.) Certain small adenovirus-encoded RNAs (the VA RNAs) which associate with cellular proteins – forming RNP particles – may be involved in splicing viral RNA; VA1 RNA is apparently complementary to splicing junctions in some adenovirus genes.

Among archaeans, some tRNA genes contain an intron which, when transcribed, gives rise to a corresponding sequence within the anticodon arm of the pre-tRNA molecule; the position of the intron is generally similar to that in yeast tRNA genes (i.e. on the 3′ side of the anticodon), but in the extreme thermophile *Thermoproteus tenax* a $tRNA^{Leu}$ contains an intron within the anticodon sequence itself, while $tRNA^{Ala}$ has an intron in a unique position on the 5′ side of the anticodon in the stem of the anticodon arm [EMBO (1987) *6* 523–528].

Bacterial introns appear to be mainly or solely of the autocatalytic type (group I or group II). The bacterial group I introns seem to occur primarily within tRNA genes; they have been found e.g. in the purple bacterium *Azoarcus* and the cyanobacterium *Synechococcus* [structure–function relationships of the intron ribozymes of *Azoarcus* and *Synechococcus*: NAR (2000) *28* 3269–3277], in *Anabaena* and other cyanobacteria, in *Agrobacterium tumefaciens* and in *Simkania negevensis* (a member of the Chlamydiales in which the intron interrupts a 23S rRNA gene).

Bacterial group II introns (like group I introns) are mobile genetic elements but are not primarily associated with tRNA genes. Group II inrons have been found e.g. in *Lactococcus* spp and in *Escherichia coli*, *Pseudomonas alcaligenes* and *Streptococcus pneumoniae*. [Bacterial group II introns (review): Mol. Microbiol. (2000) *38* 917–926.]

In some cases, splicing in bacteria may occur while the transcript is still bound to a ribosome [splicing and translation in bacteria (perspective): GD (1998) *12* 1243–1247]. Splicing in vivo has been demonstrated for a group II intron in the conjugative transposon Tn*5397* [JB (2001) *183* 1296–1299].

Bacterial introns (of both groups I and II) can propagate by INTRON HOMING. [Barriers to intron promiscuity in bacteria: JB (2000) *182* 5281–5289.]

split-product vaccine (split-virus vaccine) A VACCINE which contains *some* of the components of a virus (as opposed to whole virions).

split-virus vaccine See SPLIT-PRODUCT VACCINE.

splitter See TAXONOMY.

***spo* mutant** See ENDOSPORE (sense 1(a)).

Spo0A∼P See ENDOSPORE (sense 1).

Spo0E See ENDOSPORE (sense 1).

Spo0F See ENDOSPORE (sense 1).

SpoIIB See ENDOSPORE (figure (a) legend).

SpoIID See ENDOSPORE (figure (a) legend).

spoIIG See ENDOSPORE (figure (a) legend).

SpoIIM See ENDOSPORE (figure (a) legend).

SpoIIP See ENDOSPORE (figure (a) legend).

SpoIIIE See ENDOSPORE (figure (a) legend).

spoilage See BIODETERIORATION.

spoligotyping A rapid, PCR-based method for simultaneously detecting and typing strains of the *Mycobacterium tuberculosis* complex [original description: JCM (1997) *35* 907–914]. Essentially, PCR is used to amplify spacer sequences in the direct repeat (DR) locus of the chromosome (which apparently

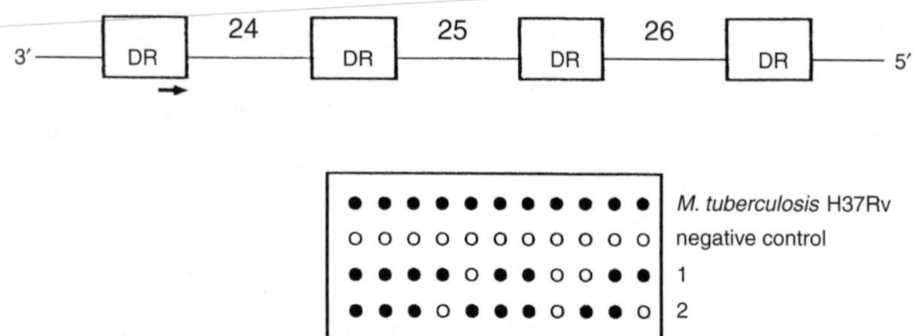

SPOLIGOTYPING (simplified protocol; principle, diagrammatic) (see entry).

The direct repeat (DR) locus is a distinct region in the chromosome of members of the *Mycobacterium tuberculosis* complex. It consists of a variable (strain-dependent) number of highly conserved 36-bp direct repeats (DRs) between which are spacer sequences of 34–41 base-pairs in length; the presence/absence and length of spacer sequences varies from strain to strain. (Such variation in the DR locus may result from homologous recombination between different sections of the locus and/or from integration of the insertion sequence IS*6110*.) In the reference strain *M. tuberculosis* H37Rv, the chromosome contains 48 DRs; each intervening spacer is identified by a number.

Top. Part of a strand of chromosomal DNA in the direct repeat (DR) locus of *Mycobacterium tuberculosis*. The (non-repetitive) sequences ('spacers') which separate the DRs vary in length from 34 to 41 nucleotides and are numbered consecutively.

For spoligotyping, all the spacers of a given strain are amplified by PCR. Here, a primer (small arrow) is shown binding to a (conserved) sequence in a DR; primers bind to the same sequence in all other DRs. During cycling, primers are extended across the subsequent spacers. The amplicons from a given test strain are then examined for their ability to hybridize to one row of immobilized oligonucleotides which represent selected spacer sequences of *M. tuberculosis* strain H37Rv. (In the actual method used by Kamerbeek et al (1997) there were 43 25-mer oligonucleotides in each horizontal row; these oligonucleotides represented spacers 1–19, 22–32 and 37–43 of *M. tuberculosis* strain H37Rv, and spacers 20–21 and 33–36 of *M. bovis* BCG.)

Below. Simplified protocol showing a membrane to which has been covalently bound 11 identical horizontal rows of oligonucleotides; in each row the oligonucleotides represent sequences from selected spacers of *M. tuberculosis* strain H37Rv. Each horizontal row was exposed to amplicons from one of the test strains; hybridization (dark spot) has occurred when a DR spacer in the test strain matched the sequence of a spacer in strain H37Rv. Note clustering of strains 6, 7 and 8 (the kind of result which might be obtained with a group of strains isolated from a single outbreak of disease).

Reproduced from Figure 7.7, page 195, in *DNA Methods in Clinical Microbiology* (ISBN 07923-6307-8), Paul Singleton (2000), with kind permission of Kluwer Academic Publishers, Dordrecht, The Netherlands.

occurs only in members of the *M. tuberculosis* complex); spacer amplicons from the test strain are then hybridized to a set of oligonucleotides which represent spacer sequences from a standard reference strain (*M. tuberculosis* strain H37Rv). (The name 'spoligotyping' is thus a contraction of 'spacer oligotyping'.) For a given strain, the pattern of matching and mis-matching of spacers (compared with those of the reference strain) is the *spoligotype*.

A simplified form of the method is outlined in the figure.

In addition to typing strains of the *M. tuberculosis* complex, spoligotyping is also useful for distinguishing between *M. tuberculosis* and *M. bovis* – a task which can be difficult by traditional methods; such differentiation is based on the absence, in *M. bovis*, of the five 3′ outermost spacers (which are present in *M. tuberculosis* strain H37Rv – and in a large number of other strains of this species which have been examined). This feature of spoligotyping is particularly useful as *M. bovis* commonly contains only a single copy of the insertion sequence IS*6110* – making it difficult to distinguish between strains of *M. bovis* by IS*6110* fingerprinting. Spoligotyping may help to promote investigations on the animal reservoirs of *M. bovis* which are potential sources of infection for both animals and humans.

Spoligotyping has also been used for checking the validity of IS*6110*-based typing of *M. tuberculosis* [JCM (1999) *37* 788–791] and for investigating laboratory cross-contamination [JCM (1999) *37* 916–919].

[Spoligotype database of *Mycobacterium tuberculosis* (biogeographic distribution of shared types, and epidemiological and phylogenetic perspectives): EID (2001) *7* 390–396.]

Spondweni virus See FLAVIVIRIDAE.

spongiform encephalopathies See SUBACUTE SPONGIFORM ENCEPHALOPATHIES.

Spongilla See ZOOCHLORELLAE.

spongiome See CONTRACTILE VACUOLE.

spongioplasm See CONTRACTILE VACUOLE.

Spongospora See PLASMODIOPHOROMYCETES and POWDERY SCAB.

spontaneous generation *Syn.* ABIOGENESIS.

spontaneous mutation (background mutation) Any MUTATION which occurs 'naturally', i.e., in the absence of any obvious external MUTAGEN; such mutations normally occur at low frequencies (cf. MUTATOR GENE). Mutations which are apparently spontaneous may arise e.g. as a result of the effects of environmental mutagens (e.g. natural gamma radiation, ULTRAVIOLET RADIATION, heat) or of endogenous mutagens produced during metabolism (e.g. peroxides, and possibly the endogenous alkylating agent *S*-adenosylmethionine). However, mutations are usually regarded as truly spontaneous only when they result from errors in DNA replication or repair, or from chemical reactions which tend to occur spontaneously in the DNA itself. For example, deamination of bases and depurination (or, less commonly, depyrimidination) of nucleotides are spontaneous reactions of DNA; thus, e.g., deamination of cytosine to uracil leads to mutagenesis since, during subsequent DNA synthesis, uracil pairs with adenine (resulting in a G·C-to-A·T transition). (cf. URACIL-DNA GLYCOSYLASE.) Certain regions of a given DNA molecule may be particularly prone to spontaneous mutagenesis: e.g., in *Escherichia coli* such 'hotspots' occur e.g. at sites occupied by the modified base 5-methylcytosine (see DNA METHYLATION); spontaneous deamination of this base generates thymine, resulting in a G·C-to-A·T transition. (The thymine residue, being a normal component of DNA, cannot be recognized by any DNA REPAIR system.) Other hot-spots may contain e.g. DIRECT REPEATS or INVERTED REPEATS, and these are apparently associated with insertions or deletions (and hence e.g. FRAMESHIFT MUTATIONS). In *E. coli*, error-prone DNA repair (see SOS SYSTEM) is apparently responsible for a low level of random (non-targeted) mutagenesis. Deletions, insertions, inversions etc may also be attributable to the activities of TRANSPOSABLE ELEMENTS.

[Analysis of spontaneous mutations in the *lacI* gene of *E. coli*: JMB (1986) *189* 273–284.]

spoonleaf A disease of redcurrants caused by the raspberry ringspot virus (a NEPOVIRUS).

spora That fraction of the particulate matter in AIR (q.v.) consisting of, or including, SPORES.

sporabola (*mycol.*) The trajectory of a forcibly-discharged basidiospore.

sporadin The gametocyte form of a gregarine.

sporangiole *Syn.* SPORANGIOLUM.

sporangiolum (sporangiole) (*mycol.*) A SPORANGIUM-like structure containing a single spore or a small number of spores; it characteristically lacks a columella. Some fungi form both sporangiola and sporangia (see e.g. Choanephoraceae in MUCORALES). The sporangiolum may be an evolutionary intermediate between the sporangium and the CONIDIUM.

sporangiomycin See THIOSTREPTON.

sporangiophore A modified or undifferentiated, simple or branched hypha which bears at least one SPORANGIUM or SPORANGIOLUM. In some fungi (e.g. *Pilaira*, *Pilobolus*) the sporangiophores are positively phototropic.

sporangiospore (*mycol.*) A thin-walled, motile or non-motile spore formed in a SPORANGIUM or in a SPORANGIOLUM; such spores are characteristic of the lower fungi. (cf. CONIDIUM.)

sporangium (1) (*mycol.*) In some lower fungi: a sac-like structure whose contents are converted into motile or non-motile spores (sporangiospores). (cf. MEROSPORANGIUM; SPORANGIOLUM; ZOOSPORANGIUM.) Sporangia are often globose or elongated, and may occur e.g. singly and terminally on a sporangiophore, in clusters on branched sporangiophores, or (e.g. in *Albugo*) in basipetally formed chains; in many fungi (e.g. *Mucor*) the sporangium contains a COLUMELLA. A sporangium may be deciduous at maturity (e.g. in *Dictyuchus*) or it may remain attached to the sporangiophore (e.g. in *Saprolegnia*); the spores may be released via a pore or by dissolution of the sporangial wall. In e.g. *Pilobolus* the entire sporangium is forcibly discharged. (See also *meiosporangium* and *mitosporangium* under ALLOMYCES, and *zygosporangium* under ZYGOSPORE.)

(2) (*bacteriol.*) The cell in which an ENDOSPORE is formed.

(3) (*bacteriol.*) That part of a cell which subsequently develops into an endospore.

(4) (*bacteriol.*) In e.g. *Actinoplanes* and members of the MYXOBACTERALES: a specialized structure containing one or more spores.

spore A differentiated form of an organism which may be (a) specialized for dissemination; (b) produced in response to, and characteristically resistant to, adverse environmental conditions; and/or (c) produced during or as a result of an asexual or sexual reproductive process. (Not all microorganisms can produce spores.) A spore may be unicellular (i.e., it may contain only one protoplast), bicellular, or multicellular; it may be thick-walled or thin-walled, pigmented or non-pigmented, motile or non-motile. Under suitable conditions, disseminative and resistant forms of spore typically give rise to a vegetative organism(s); a spore formed in a reproductive process may e.g. give rise to a vegetative organism or act as a gamete.

Bacterial spores. Bacterial ENDOSPORES (q.v.) are resistant (and may be disseminative) forms rather than reproductive forms, while the EXOSPORES formed e.g. by species of the ACTINOMYCETALES (q.v.) are characteristically reproductive and disseminative forms. (See also *myxospores* in MYXOBACTERALES.)

Fungal spores. See e.g. ASCOSPORE; AZYGOSPORE; BALLISTOSPORE; BASIDIOSPORE; CHLAMYDOSPORE; CONIDIUM (and SACCARDOAN SYSTEM); GEMMA; OIDIUM; SPORANGIOSPORE; STATISMOSPORE; THALLOSPORE; ZYGOSPORE. (See also UREDINIOMYCETES and USTILAGINALES.)

Some fungal spores exhibit DORMANCY. *Exogenous* dormancy is that due to the effects of environmental conditions; germination occurs only under conditions favourable for vegetative growth. *Endogenous* (= *constitutive*) dormancy is due to internal factors; these may include (a) the presence of a permeability barrier to nutrients, (b) the existence of a (reversible) metabolic block, and/or (c) the presence of an endogenous chemical inhibitor of germination. The self-inhibitor in e.g. uredospores of *Puccinia graminis* is methyl-*cis*-ferulate (which inhibits removal of the germination plug in the spore wall), while in the uredospores of some other rusts the inhibitor is methyl-*cis*-3,4-dimethoxycinnamate (which blocks initiation of the germ tube). In general, self-inhibitors may account for the low rate of germination of spores in masses; the leaching of inhibitor may account for the enhancement in germination when such spores are washed with water.

Dormancy can be broken by *activation* e.g. by certain chemical agents (*activators*) that include detergents and organic solvents; by heating the spores to ca. 50–60°C for 10–20 min (e.g. *Neurospora* spp); or by subjecting the spores to cold (e.g. teliospores of *Puccinia graminis*). Some spores can be induced to germinate by damaging the spore wall.

[Physiology and biochemistry of fungal sporulation: ARPpath. (1982) *20* 281–301.]

Protozoal spores. See e.g. ASCETOSPORA; MICROSPORA; MYXOZOA.

spore ball See USTILAGINALES.

spore coat See ENDOSPORE (sense 1 (a)).

spore print (*mycol.*) A deposit of spores formed when the cap of a mature agaric is left, gills-down, on a piece of plain white paper; it is useful e.g. for determining spore colour.

spore stains See e.g. ENDOSPORE STAINS.

spore strip A strip of filter paper, foil etc on which a suspension of endospores (e.g. those of *Bacillus stearothermophilus*) has been allowed to dry; it is used e.g. for monitoring the performance of an AUTOCLAVE. [Commercial spore strip performance: AEM (1982) *44* 12–18.]

sporic meiosis MEIOSIS in a life cycle characterized by an ALTERNATION OF GENERATIONS.

Sporichthya A genus of bacteria (order ACTINOMYCETALES, wall type I) which occur e.g. in soil. The organisms form an aerial mycelium consisting of sparingly-branched hyphae (0.5–1.2 μm in diameter, up to ca. 25 μm in length) which are anchored to the medium by holdfasts; no substrate mycelium is formed. Aerial hypae divide into coccoid and rod-shaped forms which, in the presence of water, give rise to motile (flagellated) propagules. Growth occurs at room temperatures and at 37°C. Type species: *S. polymorpha*. [Book ref. 73, pp. 101–102.]

sporicide Any chemical agent which inactivates spores (particularly bacterial ENDOSPORES) irreversibly.

sporidesmins Cyclic depsipeptide MYCOTOXINS produced by *Pithomyces chartarum*. Sporidesmins can cause a mycotoxicosis, 'facial eczema', in sheep or cattle grazing pastures contaminated with *P. chartarum*; facial eczema occurs in parts of Australasia, Africa and the USA. Symptoms, which are strongly dose-dependent, include e.g. anorexia, diarrhoea, dehydration, photosensitivity, jaundice, and inflamed oedematous swellings on the lips, face and vulva; death may occur. Thiabendazole has been used to control the fungus in pastures. Sporidesmin and sporidesmin E inhibit the growth of certain bacteria (e.g. *Bacillus subtilis*) and are extremely toxic (at <1 ng/ml) to a range of mammalian cell cultures (an early effect apparently being the disruption of microfilaments in the cells [JCS (1986) *85* 33–46]). [Book ref. 16, pp. 29–68.]

Sporidiales An order of fungi (class USTILAGINOMYCETES) which are typically saprotrophic, although the anamorph of at least one species, *Filobasidiella neoformans* (see FILOBASIDIELLA), is a human pathogen; the organisms characteristically occur in a yeast-like (unicellular) form, but most or all species can form a true mycelium or a pseudomycelium. Teliospores are formed by species of *Leucosporidium*, *Rhodosporidium* and SPORIDIOBOLUS but not by members of the FILOBASIDIACEAE.

Sporidiobolus A genus of fungi (order SPORIDIALES) which form budding, yeast-phase cells, pseudomycelium and true mycelium, and which give rise to ballistoconidia and to intercalary or terminal *teliospores*. The teliospores germinate to form haploid basidiospores (in heterothallic species) or diploid basidiospores (in homothallic species) on basidia which may be phragmobasidia or holobasidia. (In both homothallic and heterothallic species, karyogamy occurs in the teliospore; in heterothallic species meiosis occurs during germination of the teliospore.) Carotenoid pigments (e.g. pink, red) are formed. *S. salmonicolor* and *S. pararoseus* (both heterothallic) are the teleomorphs of two species of SPOROBOLOMYCES: *S. salmonicolor* and *S. shibatanus*, respectively. [Book ref. 100, pp. 532–540.]

sporidium (1) A BASIDIOSPORE formed by the rust or smut fungi. (2) In smut fungi: a spore formed by GERMINATION BY REPETITION.

sporistasis The state in which germination and outgrowth of a viable spore are prevented by chemical and/or other factors.

sporoactinomycetes A category within the order ACTINOMYCETALES which includes spore-forming organisms that are characterized by a morphology more complex than that of the NOCARDIOFORM ACTINOMYCETES.

Sporobolomyces A genus of fungi (see SPOROBOLOMYCETACEAE) which form spheroidal, ovoid or elongate budding yeast cells; some species can form pseudomycelium and true mycelium. The cell walls lack xylose, and growth on malt agar is pink, red or orange due to the presence of carotenoid pigments (cf. BULLERA). Ballistospores are formed on hyphal and yeast cells; they are typically bilaterally symmetrical (e.g. reniform or sickle-shaped) and develop obliquely at the tips of branched or unbranched sterigmata. Metabolism is strictly respiratory. NO_3^- is assimilated by *S. holsaticus*, *S. puniceus* (formerly *Candida punicea*), *S. roseus* and *S. salmonicolor*, but not by *S. albo-rubescens*, *S. gracilis* or *S. shibatanus*. Species may be anamorphs of SPORIDIOBOLUS spp; they have been isolated e.g. from plant material, from beer (*S. roseus*), etc. [Book ref. 100, pp. 911–920.]

Sporobolomycetaceae A family of anamorphic fungi (see BLASTOMYCETES) which are characterized by the formation of BALLISTOSPORES. The organisms, which have basidiomycetous affinities, form budding yeast cells; some can form pseudomycelium and/or true mycelium. The family includes BULLERA and SPOROBOLOMYCES. (See also MIRROR YEASTS.)

sporocarp *Syn.* FRUITING BODY.

Sporochnus See PHAEOPHYTA.

sporocladium See KICKXELLALES.

sporocyst (1) In most coccidia: a sac, formed within an oocyst, containing one or more (often 2, 4 or 8) sporozoites; there may be one or more sporocysts per oocyst. Sporocysts may be ovoid, spherical or elongated, and those of some species have a STIEDA BODY. (See also EIMERIORINA and SPORULATION.) (2) See MONOCYSTIS.

Sporocytophaga A genus of GLIDING BACTERIA of the CYTOPHAGALES. The cells are pigmented rods, up to ca. 8 μm long, from which resting cells (*microcysts*) develop; the microcyst is a spherical, desiccation-resistant structure which has a thick capsule. The organisms are chemoorganotrophic; growth occurs aerobically on cellulose (see CELLULASES), cellobiose, or glucose (mannose is used by some strains). Growth is inhibited by high concentrations of organic nitrogen or of sugars.

sporodochium A pad- or cushion-like fungal STROMA which bears a surface layer of conidiophores. Sporodochia are formed e.g. by species of *Fusarium* and *Nectria*.

sporogony (1) The division of a zygote (= *sporont*) into a number of haploid cells (*sporozoites*). (2) The (mitotic) division of a cell (*sporont*) into a number of spores or sporozoites. (3) A process which involves the formation and fusion of gametes and the division of the zygote to form spores or sporozoites.

Sporolactobacillus A genus of Gram-positive, chemoorganotrophic, microaerophilic, catalase-negative, ENDOSPORE-forming bacteria which occur e.g. in soil. Cells: flagellated rods, ca. 1×3–5 μm. Lactic acid is formed, anaerogenically, from e.g. glucose and other sugars and from sugar alcohols. Type species: *S. inulinus*.

Sporomusa A genus of Gram-negative, anaerobic, ENDOSPORE-forming bacteria isolated e.g. from river-mud, soil, waste water etc. Cells: mainly banana-shaped, motile (with up to 15 lateral flagella). The organisms can use e.g. *N*-methyl compounds (e.g. betaine, sarcosine), primary alcohols, hydroxy fatty acids, and 2,3-butanediol as substrates, acetate being the characteristic product of metabolism; H_2 and CO_2 are converted to acetate.

NO_3^- and SO_4^{2-} are not used as electron acceptors. Cells contain a cytochrome *b* and ubiquinone. GC%: ca. 41.3–47.4. Species: *S. ovata* and *S. sphaeroides*. [Arch. Micro. (1984) *139* 388–396.]

sporont See SPOROGONY.

Sporopachydermia A genus of yeasts (family SACCHAROMYCETACEAE) in which the cells are ovoid, ellipsoidal, or elongate; vegetative reproduction occurs by multilateral budding. Pseudomycelium and true mycelium are not formed. Asci contain 1–4 spheroidal or ellipsoidal ascospores which are smooth-surfaced and have very thick walls; the spores are embedded in a highly refractile material – both within the ascus and after their release. Non-fermentative; NO_3^- is not assimilated. Inositol can be used as sole carbon source. A characteristic offensive odour is produced by cultures on a range of media. The species are *S. cereana* (formerly *Cryptococcus cereanus*) and *S. lactativora* (type species; formerly *Cryptococcus lactativorus*); they have been isolated from (respectively) rotting cacti (*Cereus* spp) and e.g. Antarctic sea-water. [Book ref. 100, pp. 427–430.]

sporophore Any SPORE-bearing structure – e.g. a CONIDIOPHORE, a HYMENOPHORE, or a spore-bearing hyphal branch of *Actinoplanes*.

sporophyte See ALTERNATION OF GENERATIONS.

sporoplasm (1) (*protozool.*) In the ASCETOSPORA, MICROSPORA and MYXOZOA: an amoeboid, initially intrasporal cell which, when released from the spore, functions as the infective/invasive form of the parasite. (2) (*mycol.*) The cytoplasm of a developing ascus or sporangium.

sporopollenin A polymer which is highly resistant to chemical, physical and enzymic degradation; it can be degraded by strong oxidizing agents such as chromic acid. Sporopollenin or sporopollenin-like polymers occur e.g. in the exines of pollen grains, in the zygospore walls of e.g. *Mucor mucedo*, and in the cell walls of certain algae (e.g. *Coccomyxa* and other lichen photobionts [CJB (1985) *63* 2221–2230], *Chlorella* spp, *Scenedesmus* spp). Sporopollenin is believed to be formed by the oxidative polymerization of CAROTENOIDS.

Sporosarcina A genus of Gram-positive, chemoorganotrophic, obligately aerobic, ENDOSPORE-forming bacteria which occur e.g. in soil. The type species, *S. ureae*, can grow in media containing 5–10% urea; the cells are flagellated cocci (max. diam. ca. 2.5 μm) which occur e.g. in tetrads or packets. Each cell may give rise to a centrally located endospore. GC%: ca. 40–43. [*S. halophila*: SAAM (1983) *4* 496–506; motility, behaviour and aggregation of *S. ureae*: FEMS (1983) *17* 201–204.]

Sporospirillum A genus (*incertae sedis*) of motile, rigid, helical bacteria (1.8–4.8 × 40–100 μm) which have been observed in the intestinal contents of tadpoles; the cells contain one or two ENDOSPORE-like structures. [Book ref. 22, pp. 89–90.]

sporothallus See ALTERNATION OF GENERATIONS.

Sporothrix A genus of dimorphic fungi of the HYPHOMYCETES. Species are saprotrophs e.g. in soil, on timber, etc; *S. schenckii* is an opportunist pathogen (see SPOROTRICHOSIS). In *S. schenckii*, branched, septate hyphae are formed at room temperature; single-celled, pyriform conidia (2–3 × 3–6 μm) are borne (on sterigmata) along the hyphae, and on terminal conidiophores. The yeast form occurs in culture at 37°C and in host tissues; cells are round, oval or cigar-shaped (2–3 × 3–10 μm) and reproduce by budding.

sporotrichosis A MYCOSIS, affecting man and animals (particularly horses), caused by *Sporothrix schenckii*. Infection occurs via wounds or by inhalation. Typically, chronic subcutaneous nodules are formed – often in a line following a lymphatic vessel; the nodules eventually ulcerate, and lymphangitis may occur. Occasionally, internal organs may be infected. Potassium iodide has been used for treatment.

Sporotrichum A genus of fungi of the HYPHOMYCETES. *S. pulverulentum* is the anamorph of the WHITE ROT basidiomycete *Phanerochaete chrysosporium*.

Sporozoa A subphylum of parasitic protozoa comprising the classes Telosporea, Toxoplasmea and Haplosporea [JP (1964) *11* 7–20]; organisms in this taxon (excluding those of the Haplosporea) have since been classified, with the piroplasms, in the APICOMPLEXA.

Sporozoasida A class of protozoa of the phylum APICOMPLEXA (q.v. for ref.) in which the APICAL COMPLEX is generally well developed, and the conoid, when present, forms a complete truncate cone; typically, cyst formation and both sexual and asexual reproduction occur. Motility involves e.g. cell flexion; the microgametes of some species are flagellated. Some species are parasitic in vertebrates, some in invertebrates. Subclasses: COCCIDIASINA, GREGARINASINA and PIROPLASMASINA.

Sporozoea A class of protozoa [JP (1980) *27* 37–58] equivalent to the SPOROZOASIDA.

sporozoite See SPOROGONY.

sporulation (1) The process of development of a SPORE (see e.g. ENDOSPORE sense 1(a)).

(2) In coccidia: the formation, within the oocyst, of (haploid) sporozoites from the zygote (see EIMERIORINA); in most species the sporozoites are formed in SPOROCYSTS. In many species (e.g. most *Eimeria* spp) sporulation occurs *outside* the host, i.e. only unsporulated oocysts are found in fresh faeces; for identification purposes the oocysts in faeces can be concentrated (e.g. by FLOTATION) and sporulated in a 2–3% solution of potassium dichromate for 2–7 days at e.g. 20–30°C. (cf. ENDOSPORULATION.)

***spoT* gene** In *Escherichia coli*: a gene encoding an enzyme which can synthesize and degrade ppGpp.

spotted fevers A group of human rickettsioses which includes e.g. ROCKY MOUNTAIN SPOTTED FEVER, BOUTONNEUSE, QUEENSLAND TICK TYPHUS and RICKETTSIALPOX. In spotted fevers the rash develops first on the extremities, later on the trunk. (In TYPHUS FEVERS and SCRUB TYPHUS the rash appears first on the trunk, later on the extremities.)

spotted sickness *Syn.* PINTA.

spp Plural form of 'sp' (q.v.).

SPR See BIACORE.

spraing A POTATO DISEASE caused either by the tobacco rattle virus (see TOBRAVIRUSES) or the potato mop top virus (see TOBAMOVIRUSES). TRV causes brown, crescent-shaped corky regions to develop in the tubers, and small yellow rings to form on the leaves. PMTV may cause e.g. yellowish or brown V-shaped lesions on leaves, and crescent-shaped lesions within tubers.

spread plate (1) A method of INOCULATION: a liquid INOCULUM (e.g. 0.05–0.1 ml) is spread over the surface of a 'dried' PLATE with a sterile SPREADER or SWAB; the inoculated plate is then dried before incubation. The spread plate procedure is used e.g. as a COUNTING METHOD. (2) A plate inoculated as in (1).

spreader A simple L-shaped tool used in preparing a SPREAD PLATE; it can easily be made by fusing the tip of a Pasteur pipette and bending the fine-bore tube to a convenient angle.

spreading factor *Syn.* HYALURONATE LYASE.

spring viraemia of carp (German INFECTIOUS DROPSY) A FISH DISEASE important among cultured carp (*Cyprinus carpio*) in

Europe. The causal agent is a virus (a probable member of the RHABDOVIRIDAE). Symptoms include abdominal distension, internal haemorrhage, and oedema.

springer A can (see CANNING) in which one end is distended (convex) but can be pressed back into the normal position – the opposite end concomitantly becoming distended. (cf. SWELL.)

sprout mycelium (pseudohyphae; pseudomycelium) A mycelium-like structure consisting of chains of cells formed by sequential sprouting (budding).

SPS Sodium polyanetholsulphonate (or sodium polyanetholesulphonate): a blood anticoagulant used e.g. (at ~0.025%) in BLOOD CULTURE; its toxicity for e.g. *Neisseria gonorrhoeae* or *Peptostreptococcus anaerobius* may be nullified by gelatin.

SPS is reported to inhibit PCR [JCM (1998) *36* 2810–2816].

Spumavirinae (foamy viruses) A subfamily of viruses of the RETROVIRIDAE. Spumaviruses have been isolated from various mammalian species; viral infection is persistent but appears not to be associated with any disease. In tissue cultures the spumaviruses induce characteristic CPE, including syncytium formation and the development in infected cells of a highly vacuolated ('foamy') cytoplasm. Spumaviruses do not transform cells in vitro and are non-oncogenic in their animal hosts. The subfamily includes bovine syncytial virus, feline syncytial virus, human foamy viruses, and simian foamy virus [macaque → human transfer of SFV: Lancet (2002) *360* 387–388].

Spumellarida See RADIOLARIA.

spur (*serol.*) See OUCHTERLONY TEST.

Spurr's medium An embedding medium containing, *inter alia*, the epoxy resin vinylcyclohexene dioxide. (See also ELECTRON MICROSCOPY.)

sputter-coating A method for coating a specimen with metal prior to examination in the SEM – see ELECTRON MICROSCOPY (b). The apparatus consists essentially of two electrodes in a chamber; the cathode is made of the metal (e.g. gold, platinum) to be used for coating the specimen. The specimen, on a stub, is placed on the anode. The chamber is evacuated and then filled with argon at a pressure of ca. 0.1 torr (ca. 0.1 mm Hg). A potential difference of ca. 1 kV is applied across the electrodes; the argon ionizes, and (positive) argon ions strike the cathode, dislodging atoms of metal which then permeate the chamber. Metal atoms impinge on the specimen from all angles – penetrating fine crevices and forming a continuous metal film.

sputum (*med.*) Secretions ejected from the lungs, bronchi and trachea by coughing and spitting (Latin *spuere*: spit). Sputum consists of mucus and may contain pus and/or microorganisms; specimens obtained via the mouth normally contain saliva. Sputum is examined microbiologically in the diagnosis of lung infections (see e.g. PNEUMONIA and TUBERCULOSIS).

squalene synthase inhibitors See ZARAGOZIC ACIDS.

squamulose Bearing or consisting of small scales. Of a LICHEN: having a thallus consisting of small scales or lobes (squamules), each of which is attached to the substratum by one edge; the squamules may be imbricated.

square bacteria Flat, often square bacteria (side: 2–5 μm) found e.g. in hypersaline pools; the cell wall consists of regularly arranged hexagonal and tetragonal subunits [JB (1982) *150* 851–860]. (cf. BOX-LIKE BACTERIA.)

squash mosaic virus See COMOVIRUSES.

squirrel fibroma virus See LEPORIPOXVIRUS.

SR mutant See SMOOTH–ROUGH VARIATION.

SRBC Sheep red blood cell (erythrocyte).

src An ONCOGENE originally identified as the transforming determinant of ROUS SARCOMA VIRUS. The v-*src* product ($pp60^{v-src}$) has tyrosine-specific protein kinase activity. The c-*src* product also has tyrosine kinase activity, but c-*src* is expressed at very low levels in most normal cells; cells transformed by RSV usually contain relatively high levels of $pp60^{v-src}$. The c-*src* and v-*src* products are similar but not identical, differing in their C-terminal amino acid sequences; this difference in structure may be at least partly responsible for the transforming capacity of $pp60^{v-src}$ [Book ref. 113, pp. 19–25].

***Srf*I** See RESTRICTION ENDONUCLEASE (table).

SRID Single radial immunodiffusion: see SINGLE DIFFUSION.

sRNA 'Soluble RNA', an early term for tRNA.

SRO Sex-ratio organism: any of several strains of *Spiroplasma* (serogroup II) which can infect *Drosophila* spp, inducing a trait in which infected female flies produce only female progeny. SROs can be transmitted transovarially; they have not been cultivated in vitro.

SRP SIGNAL RECOGNITION PARTICLE.

***srrAB* genes** See AGR LOCUS.

SRSVs SMALL ROUND STRUCTURED VIRUSES.

s.s. SENSU STRICTO (q.v.).

ss (of DNA or RNA) Single-stranded.

SS agar (salmonella–shigella agar) An agar medium containing e.g. beef extract, lactose, thiosulphate, ferric citrate, bile salts, neutral red and brilliant green [recipe: Book ref. 2, p. 137]. SS agar is used as a differential and selective medium for salmonellae and shigellae. Colonies are pink for lactose +ve strains, colourless for lactose −ve strains; for H_2S-producing strains, each colony has a black centre.

SSA See SUPERANTIGEN.

SSB protein SINGLE-STRAND BINDING PROTEIN.

SSC See STANDARD SALINE CITRATE.

ss(c)DNA Single-stranded circular DNA.

SSCP (single strand conformation polymorphism) SSCP analysis is a method (used e.g. for TYPING) in which *related* samples of single-stranded DNA (e.g. homologous sequences from related strains) are compared by comparing their electrophoretic speeds in polyacrylamide gel; strands in which even one base differs may be distinguishable if this change in sequence affects intra-strand base-pairing (conformation) in such a way that electrophoretic speed is altered. (A *non*-denaturing gel is used for SSCP analysis in order to promote/preserve intra-strand base-pairing.) Typically, strands of ~100–300 bp in length are examined by SSCP analysis.

Samples of single-stranded DNA can be produced by PCR; such a protocol (PCR-SSCP) has been used e.g. for typing *Neisseria meningitidis* [JCM (1997) *35* 1809–1812].

PCR-SSCP has also been used e.g. for detecting point mutations that confer resistance to the antibiotic rifampicin in *Mycobacterium tuberculosis*. In this procedure, PCR amplifies a specific mutation-prone region in the *rpoB* gene (which encodes the β subunit of RNA polymerase – the target for rifampicin); amplicons from a given test strain are then compared, by SSCP analysis, with the corresponding sequences from wild-type (i.e. rifampicin-sensitive) and mutant strains. Resistance to other antibiotics can be assessed similarly by analysing relevant sequences of the genome; in one study, SSCP analysis was used for detecting resistance to rifampicin and isoniazid by examining four chromosomal loci [JCM (1997) *35* 719–723].

One problem with SSCP analysis is that some silent mutations produce conformations that simulate those of resistant strains, giving rise to false-positive results [JCM (1997) *35* 492–494].

(See also DGGE.)

ssDNA phage (single-stranded DNA phage) Any BACTERIOPHAGE which contains ss cccDNA: see INOVIRIDAE and MICROVIRIDAE.

ssDNA phages are widely studied as models for bacterial DNA REPLICATION. Replication of the phage genome involves three stages. *Stage I*: synthesis of a complementary (*c* or *minus*) strand on the viral (*v* or *plus*) ss cccDNA to form a ds cccDNA replicative form (RF). Stage I requires only host proteins and involves the synthesis of RNA primer(s) on SINGLE-STRAND BINDING PROTEIN-coated *v* DNA by one of three mechanisms (see INOVIRUS, BACTERIOPHAGE G4, BACTERIOPHAGE ϕX174); primers are elongated with DNA by host DNA POLYMERASE III holoenzyme, and the RNA is then excised and replaced with DNA by host DNA polymerase I. The ends of the *c* strand are then joined by DNA ligase. *Stage II*: semiconservative replication of RF to yield progeny RF. Usually the *v* strand is nicked at a specific site by a phage-coded enzyme (cf. BACTERIOPHAGE G4); elongation of the 3′-end thus formed is carried out by host DNA polymerase III holoenzyme (using *c* strand as template) in a ROLLING CIRCLE MECHANISM, the RF strands being unwound by host Rep protein (see HELICASES). The displaced *v* strand serves as template for *c* strand synthesis. *Stage III*: asymmetric synthesis in which only *v* ss cccDNA is produced from RF. This stage requires phage-coded proteins (see INOVIRUS and BACTERIOPHAGE ϕX174). [DNA replication in ssDNA phages: BBA (1985) *825* 111–139.]

SSF Simultaneous saccharification and fermentation: a batch-type process in which dextrins are enzymically hydrolysed to glucose (saccharification), the glucose being immediately taken up and fermented by organisms in the dextrin solution.

S_{SM} Simple matching coefficient. In the mutual comparison of two OTUs by NUMERICAL TAXONOMY: the number of characteristics which match – i.e. which are the same (both positive or both negative) – in the given pair of OTUs, expressed as a percentage of the total number of characteristics (criteria) used in the comparison. S_{SM} can be used only when the characteristics can be expressed in simple, discontinuous form, e.g. + or −, 0 or 1 (cf. entry S_G).

ssp Subspecies.

SsrA–SsrB See TWO-COMPONENT REGULATORY SYSTEM.

SSSS Staphylococcal SCALDED SKIN SYNDROME.

SST See STANDARD TESTS FOR SYPHILIS.

ST See ETEC.

St Anthony's fire (1) *Syn.* ERGOTISM. (2) *Syn.* ERYSIPELAS.

St Louis encephalitis (SLE) An acute, viral ENCEPHALITIS caused by a flavivirus (see FLAVIVIRIDAE); it occurs in epidemics in urban regions of central and southern USA. The virus is prevalent in wild birds and is transmitted by *Culex* spp (mosquitoes) which feed on birds and man. The disease may be benign with few sequelae, but can be fatal, particularly in the elderly.

St Vitus's dance See RHEUMATIC FEVER.

ST-I *Syn.* STa (see ETEC).

ST-II *Syn.* STb (see ETEC).

STa *Syn.* ST-I (see ETEC).

***sta* locus** See PARTITION.

STAA medium A selective medium used e.g. for isolating *Brochothrix thermosphacta* from meat and meat products; it consists of a basal growth medium containing streptomycin, thallous acetate and Actidione (cycloheximide).

stab culture See CULTURE.

stabilate A population of viable microorganisms preserved by FREEZE-DRYING or by FREEZING so that their characteristics remain unchanged. (Populations preserved by repeated subculture are subject to genetic change by selection of mutants.)

stabilized erythrocyte See HAEMAGGLUTINATION.

stable RNA In prokaryotes: a term sometimes used to refer to rRNA and tRNA – as opposed to the (typically) short-lived mRNA.

Stachel See PLASMODIOPHOROMYCETES.

stachybotryotoxicosis A MYCOTOXICOSIS (affecting e.g. horses, cattle, poultry, man) caused by the ingestion or inhalation of (or skin contact with) toxins of *Stachybotrys atra* (present e.g. in mouldy hay); the toxins involved are TRICHOTHECENES, including e.g. roridin E and satratoxins F, G and H. Symptoms include e.g. irritation and ulceration of the mucous membranes of the mouth, throat and nose, widespread haemorrhages, leucopenia; death may occur.

Stachybotrys A genus of fungi (class HYPHOMYCETES) which form septate mycelium; dark, spherical or elongated conidia are produced in chains from conidiophores which each consist of a straight hypha bearing a terminal cluster of phialides. (See also CELLULASES, PAPER SPOILAGE and STACHYBOTRYOTOXICOSIS.)

stachyose A non-reducing tetrasaccharide: α-D-galactopyranosyl-$(1 \rightarrow 6)$-α-D-galactopyranosyl-$(1 \rightarrow 6)$-α-D-glucopyranosyl-$(1 \leftrightarrow 2)$-β-D-fructofuranoside (6^G-α-galactobiosylsucrose); it occurs e.g. in soybeans, peanuts etc.

stage 0, I, II etc (1) (of sporulation) See ENDOSPORE sense 1(a). (2) (of rust fungi) See UREDINIOMYCETES.

stage micrometer See MICROMETER.

staggers (megrims) (*vet.*) A general term for certain conditions characterized by incoordination and a loss of balance in an animal. (See e.g. PASPALUM STAGGERS.)

staghorn calculi See STRUVITE.

stag's horn fungus Abnormal (branching, antler-like) fruiting bodies formed in the dark by *Lentinus lepideus* (see LENTINUS).

stain See DYE and STAINING.

staining In microbiology: any process which imparts colour, opacity (or electron density), contrast, or the ability to fluoresce to part(s) of a specimen – usually prior to examination by MICROSCOPY (or ELECTRON MICROSCOPY); staining is commonly used to facilitate detection or observation of specific organisms or intracellular features. Staining is generally carried out on dead, fixed cells or tissues (cf. VITAL STAINING). Extra contrast between different features in a given specimen may be achieved by COUNTERSTAINING; in *negative staining* the background is stained or made opaque (see e.g. CAPSULE STAIN; ELECTRON MICROSCOPY (a)).

In light microscopy most substances used for staining ('stains') are themselves coloured, and they impart their colour to the specimen in various ways (cf. DYE and LYSOCHROME). (See also METACHROMATIC DYE.) Some staining procedures depend on in situ chemical reaction to produce colour (see e.g. FEULGEN REACTION and PAS REACTION). Others involve the deposition of metallic silver (e.g. FONTANA'S STAIN) or of a dye-containing complex (e.g. LEIFSON'S FLAGELLA STAIN); such methods generally involve the use of a *mordant*, i.e. a substance which helps to bind a stain to the target surface.

The usefulness of a given stain may depend on its ability to stain some types of cell or cell component more effectively than others; thus e.g. certain stains are commonly used for staining protein (e.g. amido black 10B, COOMASSIE BRILLIANT BLUE), nucleic acids (e.g. ACRIDINES, CHROMOMYCIN, DAPI, METHYL GREEN–PYRONIN STAIN), lipid (e.g. NILE BLUE A, SUDAN BLACK B), and cellulose (e.g. CONGO RED). General purpose stains include e.g. EOSIN, FUCHSIN, METHYLENE BLUE; see also LACTOPHENOL COTTON BLUE. Whether or not a given dye can stain a particular cell or intracellular feature may depend not only on the nature and solubility of the dye but also on factors such as the permeability or integrity of the cell membrane (see e.g. TOXOPLASMA DYE TEST). Another important factor is pH; thus, e.g., a protein at pH below its ISOELECTRIC POINT can be stained by an acidic dye, and at pH above its IEP by a basic dye. The differential staining effect of some composite stains (e.g. methyl green–pyronin stain) depends on pH and on the admixture of component dyes in their correct proportions. (See also ACID FAST STAIN; ALBERT'S STAIN; ENDOSPORE STAINS; GIMÉNEZ STAIN; GRAM STAIN; KOSTER'S STAIN; METHENAMINE–SILVER STAIN; ROMANOWSKY STAINS.)

staling Adverse, growth-inhibitory changes in the environment in the immediate vicinity of a colony, or in a batch culture, due, apparently, to the accumulation of waste metabolic products of the organism.

stalk (1) (*bacteriol.*) *Syn.* PROSTHECA. (2) (*bacteriol.*) A non-prosthecate appendage such as that formed by species of GALLIONELLA. (3) (*bacteriol.*) The supportive part of the fruiting body formed by certain members of the MYXOBACTERALES. (4) (*mycol.*) *Syn.* STIPE. (5) (*protozool.*) Part of the vegetative cell in species of e.g. CARCHESIUM and VORTICELLA; some species form contractile stalks.

standard redox potential See REDOX POTENTIAL.

standard saline citrate (SSC) A saline solution (0.15 M NaCl) containing sodium citrate (0.015 M); pH 7.0.

standard serological tests *Syn.* STANDARD TESTS FOR SYPHILIS.

standard tests for syphilis (STS; serological tests for syphilis; standard serological tests, SST; non-treponemal tests) Tests in which a non-treponemal antigen is used to detect or quantify non-treponemal antibodies (*reagins*) which occur in the serum of patients suffering from SYPHILIS; the antigen used is a suspension of fine particles of CARDIOLIPIN–lecithin–cholesterol. STS include COMPLEMENT-FIXATION TESTS (e.g. Kolmer CFT, WASSERMANN REACTION) and flocculation tests (e.g. ART, RPR TEST, RST, USR TEST, VDRL TEST); flocculation tests are typically carried out on slides or on cards. Because reagins also occur in the serum of patients with certain other diseases (e.g. infectious mononucleosis, leprosy, malaria, measles), and during pregnancy and certain other conditions, STS may give BIOLOGICAL FALSE POSITIVE REACTIONS. (cf. TREPONEMAL TESTS.)

Stannophyllum See XENOPHYOPHOREA.

staphyla (*mycol.*) See FUNGUS GARDENS.

staphylocoagulase See COAGULASE.

staphylococcal food poisoning See FOOD POISONING (g).

staphylococcal scalded skin syndrome See SCALDED SKIN SYNDROME.

staphylococcal toxic shock syndrome See TOXIC SHOCK SYNDROME.

staphylococcin Any BACTERIOCIN produced by *Staphylococcus* sp; some staphylococcins are reported to be bacteriostatic. [Effect of staphylococcin 1580 on *Staphylococcus aureus*: AvL (1978) *44* 35–48; a staphylococcin-like substance bacteriostatic for Gram-positive and Gram-negative bacteria: JGM (1984) *130* 2291–2300.]

Staphylococcus A genus of Gram-positive, catalase-positive, chemoorganotrophic, characteristically halotolerant bacteria of the family MICROCOCCACEAE which are usually highly sensitive to LYSOSTAPHIN; most staphylococci are facultatively anaerobic (*S. saccharolyticus* is obligately anaerobic). The cell wall generally contains appreciable amounts of TEICHOIC ACIDS, and the ratio of glycine to lysine in PEPTIDOGLYCAN is characteristically >2 (cf. MICROCOCCUS).

Cells: spherical, ca. 1 μm diam., occuring in characteristic clusters (the name *Staphylococcus* derives from the Greek for 'bunch of grapes'). Non-motile. Orange/yellow CAROTENOID pigments occur in *S. aureus* and in strains of some other species. A CAPSULE may be present. (Capsulation can e.g. interfere with phage typing, prevent the clumping factor reaction (in the COAGULASE TEST), and affect the morphology of the colony.) Most species appear to contain only *a*- and *b*-type CYTOCHROMES (*c*-type cytochromes occur in *S. sciuri*) [Book ref. 44, p 399].

Staphylococci occur as commensals/parasites and pathogens in man and other animals. Diseases caused by species of *Staphylococcus* in man and animals include e.g. ARTHRITIS, BLACK POX, BOIL, BRONCHITIS, BUMBLEFOOT, CARBUNCLE, CYSTITIS, ENDOCARDITIS, FOOD POISONING, IMPETIGO, JOINT-ILL, KERATOCONJUNCTIVITIS, MENINGITIS, OSTEOMYELITIS, PNEUMONIA, SCALDED SKIN SYNDROME and TOXIC SHOCK SYNDROME. (See also EAR MICROFLORA, GENITOURINARY TRACT FLORA, RESPIRATORY TRACT MICROFLORA, SKIN MICROFLORA, VAGINA MICROFLORA.)

GC%: ca. 30–39. Type species: *S. aureus*.

Classification of the staphylococci. Staphylococci are divided into COAGULASE-positive and coagulase-negative strains. Strains of *S. aureus* and *S. intermedius* are coagulase-positive, as are some strains of the animal pathogen *S. hyicus* subsp *hyicus*; most species of *Staphylococcus* are coagulase-negative. In terms of pathogenic potential, *S. aureus* is considered to be the most important species; however, other species, including a number of coagulase-negative staphylococci (CNS), are also associated with disease, often as opportunist pathogens.

Species include:

S. albus. See *S. epidermidis*.

S. aureus. A major pathogen of man and other animals. The cells, ca. 1 μm in diam., are often pigmented (see STAPHYLOXANTHIN), particularly in freshly isolated strains. The cell wall usually contains *ribitol* teichoic acids, and most strains form PROTEIN A; the PEPTIDOGLYCAN contains little or no L-serine. The cell wall lipoteichoic acid is a virulence factor: it acts as an adhesin which binds to fibronectin (a glycoprotein component of e.g. epithelial cells). Protein adhesins of *S. aureus* bind to mammalian targets such as collagen and fibronectin [TIM (1998) *6* 484–488]. (See also SORTASE.)

The expression of virulence factors by *S. aureus* is controlled by various mechanisms which involve e.g. the AGR LOCUS. The expression of certain virulence factors, including exotoxin, appears to involve a TWO-COMPONENT REGULATORY SYSTEM that responds to levels of environmental oxygen [JB (2001) *183* 1113–1123].

In general, the expression of virulence factors in *S. aureus* depends on the growth phase and/or on conditions of growth. Regulation of exotoxins and wall-associated proteins (e.g. PROTEIN A) is complex [FEMS Reviews (2004) *28* 183–200].

The organism forms a thermostable DEOXYRIBONUCLEASE, and produces both free coagulase and clumping factor (see COAGULASE and COAGULASE TEST). [Evaluation of the STAPHYLOSLIDE and other agglutination tests: JCM (1985) *21* 726–729.] All strains coagulate e.g. human and rabbit plasma.

Glucose is usually fermented to DL-lactic acid, and lactose metabolism involves the tagatose 6-phosphate pathway (see Appendix III(a)). Mannitol is utilized aerobically and anaerobically, and maltose is utilized aerobically. VP +ve. (See also HYALURONATE LYASE and PHOSPHATASE.)

S. aureus is characteristically sensitive to NOVOBIOCIN (MIC <1.6 μg/ml). Most strains are resistant to benzylpenicillin (owing to the action of staphylococcal β-LACTAMASES), and some strains are resistant to methicillin (see MRSA). MRSA has been detected by multiplex PCR, target sequences being in *mecA* (encoding PBP 2a) and the *coa* gene (encoding coagulase, and used as a marker for *S. aureus*) [JMM (1997) *46* 773–778].

Antibiotics that have been used therapeutically against *S. aureus* include various β-lactams (e.g. flucloxacillin), clindamycin, erythromycin and vancomycin.

S. aureus is susceptible to effective concentrations of phenolic disinfectants, HYPOCHLORITES, CHLORHEXIDINE and povidone–iodine (see IODINE). The organism can withstand 60°C/30 min. The D VALUE of *S. aureus* at 70°C is often less than 1 minute (although much longer in some strains).

S. aureus has been divided into four biotypes: A–D (biotypes E and F are now classified as *S. intermedius*). The biotypes usually correlate with source: man (A), poultry and pigs (B), bovines and sheep (C), and hares (D). The biotypes of *S. aureus* are distinguishable e.g. by FIBRINOLYSIN production (A strains only); coagulation of bovine plasma (C strains only); formation of α-haemolysin (see HAEMOLYSIN, sense 2e) (A strains, some B and C strains); formation of β-haemolysin (C and D strains, most B strains, few A strains); mannitol fermentation within 5 days (A and B strains, some C strains).

PHAGE TYPING of *S. aureus* is carried out with five groups of phages. A strain of *S. aureus* which can be lysed by phage(s) of lytic group I is referred to as *S. aureus* phage type I; a strain lysed by phage(s) of group II is referred to as *S. aureus* phage type II etc. A given strain may be lysed by the phages of more than one group; for example, many of the (phage-typable) strains of MRSA are lysed by phages of groups I and III (while other strains of MRSA may be lysed by group III phages only). The phages used for typing *S. aureus* have been standardized: they belong to the 'international set'; for example, group I phages include those designated 29, 52, 52A, 79 and 80. The phages of group IV are used for typing isolates of *S. aureus* derived from animals. (Some strains of *Staphylococcus* are not typable by this system; such strains presumably lack the receptors for relevant phages.)

S. aureus can also be typed by MLST [database: JCM (2000) *38* 1008–1015]. A further method is PCR-RFLP with a species-specific target – e.g. a sequence in the gene *aroA* (which encodes the enzyme 5-enolpyruvylshikimate-3 phosphate, used in the synthesis of aromatic amino acids and folate) [JCM (1999) *37* 570–574]. Other typing methods include PFGE. [A comparison of PFGE with five other molecular methods for typing *S. aureus* (MRSA): JMM (1998) *47* 341–351.]

S. capitis. Coagulase-negative. The cell wall contains glycerol teichoic acids. Sensitive to novobiocin (MIC <1.6 μg/ml). Acid is not formed, aerobically, from e.g. maltose, trehalose or melezitose. Occurs e.g. on the normal human scalp, but has been associated with several cases of infective endocarditis and with urinary tract infections.

S. caprae. [Original species proposal: IJSB (1983) *33* 480–486.] Isolated from both animal and human specimens. Evidence for pathogenicity in man [see e.g. RMM (1995) *6* 94–100 (97)].

S. carnosus. [Original species proposal: IJSB (1982) *32* 153–156.]

S. caseolyticus. [Original species proposal: IJSB (1982) *32* 15–20.]

S. cohnii. Coagulase-negative. The cell wall contains glycerol teichoic acids. Resistant to novobiocin (MIC ≥1.6 μg/ml). Acid is not formed, aerobically, from sucrose or melezitose. Occurs on human skin.

S. epidermidis (formerly *S. albus*). Coagulase-negative. The cell wall contains glycerol teichoic acids. Novobiocin-sensitive (MIC <1.6 μm/ml). Acid is not formed, aerobically, from L-arabinose, D-mannitol, trehalose or D-xylose but is produced from sucrose. Anaerobic growth occurs in thioglycollate medium. *S. epidermidis* is found on human skin; it can be pathogenic. The cell wall lipoteichoic acid acts as an adhesin which binds to host-cell fibronectin. [Epidemiological typing of *S. epidermidis*: MR (1985) *49* 126–139.]

S. gallinarum. [Original species proposal: IJSB (1983) *33* 480–486.] Referred to ENTEROCOCCUS.

S. haemolyticus. Coagulase-negative. The cell wall contains glycerol teichoic acids. Novobiocin-sensitive (MIC <1.6 μm/ml). Bovine blood is commonly haemolysed. Occurs on the skin of humans and non-human primates.

S. hominis. Similar to *S. haemolyticus*, but haemolysis of bovine blood is weak or absent.

S. hyicus. *S. hyicus* subsp *chromogenes* contains only coagulase-negative strains; *S. hyicus* subsp *hyicus* contains coagulase-negative and coagulase-positive strains. The coagulase-positive strains contain glycerol teichoic acids, and their peptidoglycan contains little or no L-serine; like *S. aureus*, they form a thermostable deoxyribonuclease, but they form L-lactic acid from glucose, are VP –ve, and do not form clumping factor or utilize maltose or mannitol aerobically. Strains of *S. hyicus* subsp *hyicus* have been implicated in GREASY PIG DISEASE.

S. intermedius. Coagulase-positive with bovine or rabbit plasma (some strains also coagulate human plasma); clumping factor is formed by only some strains. The cell wall contains glycerol teichoic acids. A β-haemolysin and a thermostable deoxyribonuclease are formed. Mannitol is not utilized within 5 days. VP –ve. Strains occur on the skin and/or nares of e.g. dogs, foxes, horses and mink.

S. lentus. [Proposed species (re-classified from *S. sciuri* subsp *lentus*): SAAM (1983) *4* 382–387.]

S. lugdunensis. [Original species proposal: IJSB (1988) *38* 168–172.] Coagulase-negative. The organism resembles *S. aureus* in producing a thermostable deoxyribonuclease. Resistance to cadmium has been recorded in many isolates. Almost all strains isolated in Europe are reported to be sensitive to β-lactam antibiotics, but those isolated in the USA are often resistant (many produce β-lactamases). *S. lugdunensis* is one of the most pathogenic species of CNS, and has been identified as the cause of a number of cases of infective endocarditis and e.g. abscesses [see RMM (1995) *6* 94–100 (95–96)].

S. saccharolyticus. A species re-classified from *Peptococcus saccharolyticus* [name validation: IJSB (1984) *34* 91–92]. Anaerobic. Coagulase-negative. The organism has been reported to cause infective endocarditis [JCM (1990) *28* 2818–2819].

S. saprophyticus. Coagulase-negative. The cell wall contains ribitol teichoic acids. Resistant to novobiocin (MIC ≥1.6 μm/ml). Acid is formed, aerobically, from maltose, sucrose and trehalose. Bovine blood is not haemolysed. Occurs e.g. on the skin of humans and pigs, and can cause e.g. urinary tract infections (see e.g. CYSTITIS).

S. schleiferi. [Original species proposal: IJSB (1988) *38* 168–172.] Coagulase-negative. Two subspecies: *schleiferi* and *coagulans*. Some strains have been reported to clot plasma [JCM (1994) *32* 388–392]. Isolated e.g. from bacteraemic patients and from an infected dog's ear.

S. sciuri. Coagulase-negative. Occurs e.g. on certain squirrels.

S. simulans. Coagulase-negative. Occurs e.g. on human skin.

S. warneri. Coagulase-negative. Occur e.g. on the skin of humans and non-human primates. The organism has been associated e.g. with endocarditis [JCM (1992) *30* 261–264] and urinary tract infections.

S. xylosus. Coagulase-negative. Isolated e.g. from the skin of dogs, goats and sheep.

staphylokinase A protein, produced by many strains of coagulase-positive staphylococci, which activates profibrinolysin (from various species) to fibrinolysin, thus causing indirect FIBRINOLYSIS. Its mechanism of action appears to be similar to that of STREPTOKINASE. [Assay, properties etc: Book ref. 44, pp. 790–796.]

staphylomycins See STREPTOGRAMINS.

Staphyloslide A test used to detect clumping factor (see COAGULASE) in strains of *Staphylococcus aureus*; essentially, a given strain is examined (on a slide) for its ability to clump fibrinogen-coated sheep erythrocytes.

Staphylothermus A genus of heterotrophic, coccoid archaeans. The type species, *S. marinus*, grows optimally at ca. 92°C under anaerobic conditions; it was isolated from a HYDROTHERMAL VENT. [Book ref. 196, pp. 69–71.]

staphylothrombin See COAGULASE.

staphyloxanthin The main CAROTENOID pigment in *Staphylococcus aureus*. [Structure and synthesis: JB (1981) *147* 900–919.]

star activity (of a restriction endonuclease) A change, or reduction, in the specificity of certain RESTRICTION ENDONUCLEASES (e.g. *Bam*HI, *Eco*RI, *Sal*I) which occurs under non-optimal conditions.

starch A composite D-glucan consisting of amylose (ca. 20–25%, depending on source) and amylopectin. *Amylose* is a water-soluble linear $(1 \rightarrow 4)$-α-D-glucan which forms blue-black complexes with iodine. *Amylopectin* is a highly branched $(1 \rightarrow 4)$-α-D-glucan with $(1 \rightarrow 6)$-α-D-glucosidic branch points; it is water-insoluble and forms violet or red-brown complexes with iodine (cf. GLYCOGEN). Starch is abundant as the main storage carbohydrate in higher plants and in certain algae, e.g., members of the Chlorophyta ('chlorophycean starch'); an amylose-like glucan occurs in the cell walls of certain fungi, and yeasts of the genus *Cryptococcus* characteristically form an extracellular CAPSULE of an amylose-like polysaccharide. (See also CYANOPHYCEAN STARCH; FLORIDEAN STARCH; LICHENIN.) Starch may be stained e.g. with iodine or (non-specifically) by the PAS REACTION.

Starch is attacked by a wide range of microorganisms (see e.g. AMYLASES and DEBRANCHING ENZYMES). That from higher plants is an important raw material for a number of biotechnological processes – see e.g. INDUSTRIAL ALCOHOL and SINGLE-CELL PROTEIN.

STARFISH A water-soluble multivalent carbohydrate ligand (a starfish-shaped molecule) which can bind to, and neutralize, the shiga-like toxins (SLT-I and SLT-II) of *Escherichia coli* O157:H7 with subnanomolar activity; the structure of STARFISH is related to that of the (glycolipid) toxin-binding sites (Gb_3) on mammalian vascular endothelium. (See also SYNSORB P^k.) [STARFISH: Nature (2000) *403* 669–672.]

starlingpox virus See AVIPOXVIRUS.

Starria A genus of filamentous CYANOBACTERIA in which the trichomes are triradiate in cross-section; *S. zimbabweënsis* has been recorded in only one geographical location. [J. Phycol. (1977) *13* 288–296.]

start (in cell cycle) See CELL CYCLE.

start point (start site) See TRANSCRIPTION.

starter (starter culture) An INOCULUM, usually consisting of a pure or mixed culture of microorganisms, used to initiate a commercial FERMENTATION (sense 2) (see e.g. LACTIC ACID STARTERS). Starter cultures are often grown under aseptic conditions in a laboratory (but cf. e.g. KEFIR and *sourdough* BREAD-MAKING).

State 3, State 3u, State 4 (*bioenergetics*) See RESPIRATORY CONTROL.

stathmokinesis Inhibition of cell division (e.g. by COLCHICINE).

stationary phase (of growth) See BATCH CULTURE.

statismospore (*mycol.*) A spore which is not forcibly discharged. (cf. BALLISTOSPORE.) In basidiomycetes, a BASIDIOSPORE which is a statismospore is either formed in a *symmetrical* orientation on its sterigma, or it is formed directly from the basidium, i.e., with no sterigma. (cf. GASTEROID BASIDIOSPORE.)

statolith See e.g. MÜLLER'S VESICLE.

statospore (algol.) (1) Any resting cell. (2) Any resting cell ('spore' or 'cyst') which is formed *within* a vegetative cell: see e.g. CHRYSOPHYTES and XANTHOPHYCEAE.

STATs See CYTOKINES.

Staurastrum A large, probably artificial genus of placoderm DESMIDS. The cells are morphologically highly variable, but each semicell appears triangular when viewed from one end; sometimes spines or projections extend from the corners.

Staurodesmus A genus of placoderm DESMIDS.

Stauroneis See DIATOMS.

staurosporae See SACCARDOAN SYSTEM.

Staurothele See VERRUCARIALES.

stavudine See NUCLEOSIDE REVERSE TRANSCRIPTASE INHIBITORS.

STb *Syn.* ST-II (see ETEC).

STD SEXUALLY TRANSMITTED DISEASE.

steam (for disinfection and sterilization) See e.g. AUTOCLAVE; LOW-TEMPERATURE STEAM DISINFECTION; LOW-TEMPERATURE STEAM–FORMALDEHYDE STERILIZATION; STEAMER. (See also CANNING.)

steam trap In an AUTOCLAVE: a valve which permits the passage of air and/or water ('condensate') but which blocks the passage of saturated steam. One form of steam trap incorporates a metal bellows which contains a small amount of water. When steam at the required shut-off temperature flows around the bellows to the outlet line the vapour pressure in the bellows causes the bellows to expand and thus close the valve.

steamer A vessel in which objects or materials may be subjected to steam at *atmospheric* pressure (cf. AUTOCLAVE); it is used e.g. for TYNDALLIZATION.

STEC See EHEC.

Steers' replicator A multipoint plate inoculator.

Stefansky's bacillus *Mycobacterium lepraemurium.*

steganacin See MICROTUBULES.

Stella A genus of Gram-negative, asporogenous, non-motile, flat, prosthecate (typically 'six-pronged star-shaped') bacteria found in fresh water, sewage and soil. The organisms are chemoorganotrophic, growing on amino acids and organic acids. *S. vacuolata* contains GAS VACUOLES, *S. humosa* does not. [IJSB (1985) *35* 518–521.]

stellacyanin See BLUE PROTEINS.

stellate Star-shaped.

Stellatosporea A class of protozoa (phylum ASCETOSPORA) in which the spore contains one or more sporoplasms and 'haplosporosomes' (membrane-bounded, osmiophilic, ovoid or spherical bodies – possibly polysaccharide reserves). Two orders are recognized: Occlusosporida (>1 sporoplasm per spore, spore wall entire (imperforate); genera: e.g. *Marteilia*) and Balanosporida (1 sporoplasm per spore, spore wall with an orifice covered externally by an operculum or internally by a diaphragm; genera: e.g. HAPLOSPORIDIUM, MINCHINIA, *Urosporidium*).

Stellavirus *Syn.* ROTAVIRUS.

STEM See ELECTRON MICROSCOPY (c).

stem-and-loop structure (*mol. biol.*) A secondary structure in a nucleic acid molecule in which complementary sequences within a strand base-pair to form the 'stem', nucleotides between these sequences forming the single-stranded 'loop'; such structures can occur in regions of hyphenated dyad symmetry (see PALINDROMIC SEQUENCE).

stem blight (stem spot) (of lupins) See LUPINOSIS.

stem nodules Nitrogen-fixing nodules formed, as a result of association with *Rhizobium* strains [JGM (1983) *129* 3651–3655], on the stems of certain (mainly tropical or subtropical) leguminous plants (genera *Aeschynomene*, *Sesbania* and *Neptunia*); the plants may form nodules either on stems or on roots, or both. Nitrogen fixation in stem nodules is comparable in efficiency with that in soybean ROOT NODULES, and stem nodulation is not significantly inhibited by increased concentrations of applied nitrogen. [Book refs 55, pp. 255–268, and 112, pp. 161–169.] (See also *Gunnera* under NOSTOC.)

Stemonitales See MYXOMYCETES.

Stemonitis A genus of slime moulds (class MYXOMYCETES) which form aphanoplasmodia that give rise to clusters of long, narrow, often brownish sporangia borne on slender, glossy black stalks. Capillitial threads ramify through the spore mass, extending from a central black columella (continuous with the stalk) and anastomosing at the periphery of the spore mass to form a surface network. Species occur on rotting wood, humus, etc.

Stenella See HYPHOMYCETES; see also PITYRIASIS NIGRA.

steno- A prefix signifying 'narrow' or 'limited' – e.g. a *stenohaline* organism is one able to tolerate only a narrow range of salt concentrations. (cf. EURY-.)

Stenocybe See CALICIALES.

Stenophora See GREGARINASINA.

Stenotrophomonas maltophila See CYSTIC FIBROSIS.

stenoxenous Having a narrow host range.

Stentor A genus of freshwater ciliate protozoa (order HETEROTRICHIDA). Cells: funnel- or trumpet-shaped, up to ca. 2 mm in length, and contractile; the cytostome is at the broader end, and the narrow end of the cell is often attached to the substratum. A mucilaginous lorica is formed by some species. The macronucleus may be moniliform (e.g. in *S. coeruleus*), ovoid (e.g. in *S. igneus*) or elongate and of uniform width (e.g. in *S. roeseli*). A contractile vacuole at the broader end of the cell communicates with a longitudinal canal. *S. coeruleus* contains a blue pigment; other species may be e.g. yellow, red or brown. (See also ZOOCHLORELLAE.)

S. coeruleus exhibits PHOTOTAXIS (see also STENTORIN). [The photoreceptor in *S. coeruleus*: SEBS (1983) *36* 503–520.]

stentorin The presumed photoreceptor in *Stentor coeruleus* (see STENTOR); it consists of the chromophore *hypericin* bound to protein. Stentorin occurs in rows of subpellicular membranous vesicles or granules.

Stephanoascus A genus of fungi (family SACCHAROMYCETACEAE) which form budding yeast cells, pseudomycelium, and abundant branched, septate, true mycelium; blastospores are formed, arthrospores are not. Hyphal septa contain plasmodesmata. Ascospores: initially bowler-hat-shaped, becoming hemispherical with a thick convex wall. Non-fermentative; NO_3^- is not assimilated. One species: *S. ciferrii* (anamorph: *Candida ciferrii*), isolated e.g. from soil and from clinical specimens from man and animals. [Book ref. 100, pp. 431–433.]

Stephanodiscus See DIATOMS.

Stephanoeca See CHOANOFLAGELLIDA.

stephanokont Refers to a cell which bears a ring of flagella (of equal length) encircling the cell near one end. (See e.g. DERBESIA; cf. ISOKONT; HETEROKONT.)

Stephanopogon A genus of marine benthic ciliates (subclass GYMNOSTOMATIA). *Stephanopogon* is considered to be a primitive ciliate owing e.g. to its possession of an apical cytostome and (unlike any other known ciliate) the possession of only one type of nucleus.

Stephanopyxis See DIATOMS.

sterco- Indicates a connection with faeces, e.g., *stercorous*: resembling faeces.

Stercoraria One of two groups of subgenera of mammalian trypanosomes; stercorarian trypanosomes infective for the vertebrate host develop in the POSTERIOR STATION. The group comprises HERPETOSOMA, MEGATRYPANUM and SCHIZOTRYPANUM. (cf. SALIVARIA.)

Stereaceae See APHYLLOPHORALES.

Stereocaulon A genus of LICHENS (order LECANORALES); photobiont: a green alga. Erect pseudopodetia (see PSEUDOPODETIUM) – typically richly branched and fruticose, covered with scale-like, finger-like or coralloid outgrowths (*phyllocladia*) – arise from a granular, usually rapidly evanescent primary thallus. Dark-brown apothecia may occur terminally or laterally on the pseudopodetia, and dark-coloured cephalodia are commonly present. Species occur on soil and rocks, some being associated with habitats rich in heavy metals.

stereocilium *Syn.* CLAVATE CILIUM.

Stereomyxa See ACARPOMYXEA.

Stereum A genus of lignicolous fungi of the APHYLLOPHORALES (family Stereaceae) which form typically resupinate or effused-reflexed, leathery or woody fruiting bodies in which the context lacks clamp connections and is usually dimitic. (See also PAPER SPOILAGE, PARTRIDGE WOOD, WOODWASP FUNGI and XYLANASES.)

Steriflamme process A form of heat treatment used for canned foods (see CANNING). Cans are initially heated to a given temperature by steam at atmospheric pressure. The rapidly spinning cans are then exposed directly to gas flames so that their contents are quickly heated to ca. 120–130°C. The cans are held at the processing temperature for the required time and are then cooled by water sprays.

sterigma (*pl.* sterigmata) (*mycol.*) (1) In certain fungi: a small or substantial protrusion from which a spore or daughter cell is produced – see e.g. BASIDIUM and STERIGMATOMYCES. (See also SPICULUM.)

(2) See ASPERGILLUS.

sterigmatocystin A carcinogenic, hepatotoxic MYCOTOXIN which contains a bifuran moiety fused at the 4,5 positions to a substituted xanthone (cf. AFLATOXINS); it is produced e.g. by *Aspergillus nidulans* and *A. versicolor*.

Sterigmatomyces A genus of fungi (class HYPHOMYCETES) which form spherical to ovoid yeast cells and which may form a true mycelium. A vegetative cell produces one or more fine projections (sterigmata), at the end of each of which a daughter cell forms by budding. When mature, the daughter cell may in turn produce sterigmata, often resulting in the formation of short, branching chains of cells; cells may separate by the formation of a septum in the middle or at the distal end of the sterigma, according to species. Metabolism is strictly respiratory. Six species are recognized; type species: *S. halophilus*. [Book ref. 100, pp. 921–929; cell cycle in *S. halophilus*: JGM (1983) *129* 2129–2162.]

sterilant Any chemical agent which, under carefully controlled conditions, can sterilize objects, materials etc (see STERILIZATION; cf. DISINFECTANT); examples include e.g. ETHYLENE OXIDE, GLUTARALDEHYDE, β-PROPIOLACTONE (cf. FORMALDEHYDE). If conditions are not carefully controlled, a sterilant may fail to sterilize [JAB (1983) *54* 91–99]. (cf. BIOCIDE.)

sterile (1) Refers to the condition of an object, or an environment, which is free of all living cells, all viable spores (and other resistant and disseminative forms), and all viruses and sub-viral agents capable of replication. (An effective STERILIZATION process must therefore irreversibly inactivate all organisms.) For most practical purposes the main criterion of sterility is satisfied by the failure to detect viable cells, spores, viruses etc on or in a given object, material or environment; it should be appreciated that a viable organism can be detected only if it is provided with conditions suitable for its growth or replication or other form of expression.

(2) Refers to any structure or tissue which does not or cannot have a reproductive function.

sterile cabinet *Syn.* SAFETY CABINET.

sterile immunity *Syn.* STERILIZING IMMUNITY.

sterile pyuria See PYURIA.

sterile technique *Syn.* ASEPTIC TECHNIQUE.

sterility The state or condition of being STERILE. (cf. COMMERCIAL STERILITY.)

sterilization (1) Any process by which objects, materials, or environments may be rendered STERILE (sense 1). (Since sterility is an absolute condition, the expression 'partial sterilization' is meaningless – cf. DISINFECTION.) When contaminating organisms are known to consist entirely of vegetative or other vulnerable forms, sterility can usually be achieved by mild heat or chemical treatments; however, 'sterilization' commonly refers to those processes which guarantee sterility regardless of the nature of the contaminating microflora.

(a) *Methods of sterilization.*

(i) *Physical methods.* Heat is the most reliable (and generally the preferred) sterilizing agent; it can reach organisms that may be protected from the action of chemical sterilizing agents or from radiation. The lethal action of heat is due largely to the denaturation of microbial proteins and nucleic acids and to the lability of membranes; at sterilizing temperatures nucleic acids may be deaminated, depurinated or otherwise degraded.

Dry heat. Sterilization of clean glassware etc may be carried out in a HOT-AIR OVEN (q.v.); under such conditions proteins are denatured, cytoplasm desiccated, and various cell and virus components are oxidized. *Infrared radiation* (λ within the range 7.5×10^{-5} to 4×10^{-2} cm) has been used e.g. for the sterilization of small items of medical glassware. *Incineration* (combustion) is used e.g. for the disposal of used surgical dressings, and is the basis of FLAMING.

Moist heat (steam) is more rapidly lethal than dry heat at the same temperature. The relative ease with which moist heat denatures proteins may be due to facilitated rupture of the protein-stabilizing hydrogen bonds; these bonds are broken more easily when water molecules are available for hydrogen bonding. Moist heat also effects an efficient transfer of heat since steam which condenses on a cooler object (during the heating-up period) delivers up a large amount of latent heat. Moist heat sterilization is generally carried out in an AUTOCLAVE (q.v.).

Radiation. See IONIZING RADIATION; MICROWAVE RADIATION; ULTRAVIOLET RADIATION.

Filtration. See FILTRATION.

Ultrasonication. ULTRASONICATION may achieve sterility in some cases, but it is not a reliable method of sterilization.

(ii) *Chemical methods.* Among the few reliable STERILANTS are ETHYLENE OXIDE, GLUTARALDEHYDE and β-PROPIOLACTONE; such agents must be used in effective concentrations under carefully controlled conditions.

(iii) *Combined physical and chemical methods.* In the pharmaceutical industry, the vapour of e.g. chlorocresol (see PHENOLS) is used at 98–100°C in a procedure that has been called 'heating with a bactericide'. (See also e.g. LOW-TEMPERATURE STEAM–FORMALDEHYDE STERILIZATION.)

(b) *Kinetics of sterilization (some basic ideas).* In a single-species microbial population, the death rate produced by many types of lethal agent (including heat, radiation, various chemical disinfectants and sterilants) follows approximately first-order kinetics. However, close approximations to first-order kinetics are obtained only when sterilizing temperatures are relatively high, or when disinfectants or sterilants are used in relatively high concentrations. Under these conditions, the decrease in numbers of microbial contaminants may be summarized by the equation:

$$\log_e N_0 - \log_e N = kt$$

where N_0 is the initial number of viable organisms, N is the number of survivors at time t, and k is the *death rate constant.* For a given organism, different values of k can be obtained for different sterilizing temperatures or different concentrations of a disinfectant or sterilant. In a mixed microbial population the cells will have different degrees of resistance to heat, chemicals etc, and a sterilization process will not follow first-order kinetics; in such cases the curve obtained by plotting $\log_e N$ versus t (to find k) will be a composite curve.

(See also APPERTIZATION; DISINFECTION; D VALUE; F_0 VALUE; MPED; THERMAL DEATH POINT; THERMAL DEATH TIME; Z VALUE.)

(2) (*med., vet.*) Complete elimination of a parasite or pathogen from a host either by the immune response (see STERILIZING IMMUNITY) or by chemotherapy.

sterilizer (1) Any apparatus used for STERILIZATION, e.g., an AUTOCLAVE or a HOT-AIR OVEN. (2) A STEAMER. (3) Any of certain types of apparatus used in the CANNING industry for APPERTIZATION.

sterilizing immunity (sterile immunity) An immune response which results in the total elimination of a parasite or pathogen from the body. (cf. PREMUNITION.)

Sterne strain A toxin-producing strain of *Bacillus anthracis* which lacks the poly-D-glutamic acid capsule; it is used as a live vaccine against ANTHRAX. The lack of a capsule allows the elimination of the organism by phagocytosis.

steroid bioconversions Certain steroid hormones and related compounds are produced commercially by a process in which most of the steps are effected chemically, but some rely on the selective action of certain microorganisms; the microbial reactions have the advantages of being site-specific and stereo-specific *Examples.* Production of corticosteroid hormones (e.g. cortisol) involves the chemical conversion of stigmasterol (from various plants) or diosgenin (e.g. from yams) to progesterone; the 11α-hydroxylation step is achieved e.g. by *Rhizopus nigricans* or *Aspergillus* spp. (The remainder of the synthesis is achieved chemically.) In the synthesis of testosterone, the reduction of 4-androstene-3,17-dione to the 17β-hydroxy derivative can be achieved e.g. by strains of *Saccharomyces cerevisiae.* [Book ref. 31, pp. 369–465.] (See also CHOLESTEROL OXIDASE and BILE ACIDS.)

Ster-zac See BISPHENOLS.

stibamine See ANTIMONY.

stibanilic acid See ANTIMONY.

stibophen See ANTIMONY.

stichobasidial Refers to a *longitudinal* arrangement of nuclear spindles in a developing BASIDIUM. (cf. CHIASTOBASIDIAL.)

Stichococcus A genus of freshwater or terrestrial green algae (division CHLOROPHYTA) which are apparently closely related to KLEBSORMIDIUM. When undisturbed, the organisms grow as filaments, but these readily fragment into bacilliform, uninucleate cells which each contain a single large chloroplast. Asexual reproduction occurs by cell division and fragmentation of filaments.

stichodyad membrane See PARORAL MEMBRANE.

Sticholonche See TAXOPODIDA.

stichomonad membrane See PARORAL MEMBRANE.

stichonematic Refers to a eukaryotic FLAGELLUM which bears a single row of fine hairs along its length.

Stickland reaction In many proteolytic species of *Clostridium*: an energy-yielding reaction in which, typically, the oxidation of one amino acid (the reductant) is coupled to the reduction of another (the oxidant). Essentially, the reductant undergoes oxidative deamination to an α-ketoacid (= α-oxoacid), NAD^+ being reduced. The ketoacid is oxidatively decarboxylated to a fatty acid in a reaction requiring coenzyme A and inorganic phosphate, NAD^+ being reduced and ATP synthesized. NAD^+ is regenerated from NADH by the reduction or reductive deamination of the oxidant.

Some amino acids (e.g. alanine, valine) act preferentially as reductants, while others (e.g. glycine) act preferentially as oxidants; according to organism, leucine and the aromatic amino acids phenylalanine and tryptophan may act either as oxidants or reductants. Some species (e.g. *C. sporogenes*) can use ornithine by converting it to proline (an oxidant). In some cases leucine can be used as both oxidant and reductant, and some species can use e.g. BETAINE or sarcosine (*N*-methylaminoacetic acid) as oxidant, betaine yielding acetate and trimethylamine.

sticky ends (cohesive ends) (*mol. biol.*) Complementary single-stranded ends of a double-stranded nucleic acid molecule – generated e.g. by staggered cleavage of a circular dsDNA molecule by certain RESTRICTION ENDONUCLEASES; base-pairing between sticky ends of the same molecule or of two identical molecules generates (after ligase action) cccDNA or a CONCATEMER, respectively. (See e.g. BACTERIOPHAGE λ.)

Sticta A genus of foliose LICHENS (order PELTIGERALES). Photobiont: *Nostoc* (e.g. in *S. sylvatica*) or a green alga (e.g. in *S. canariensis*); the mycobiont of *S. canariensis* can alternatively associate with a cyanobacterium to form a distinct morphotype ('*S. dufourii*') – cf. CEPHALODIUM. Thallus: large, lobed, corticate on both surfaces; the upper surface may bear isidia (e.g. *S. sylvatica*) or soredia (e.g. *S. limbata*), the lower surface is tomentose and is perforated by characteristic cyphellae (see CYPHELLA). Species occur mainly on mossy trees and rocks.

Stieda body In some species of the APICOMPLEXA (e.g. some *Isospora* spp): a structure which may form part of, or is intimately associated with, the polar wall of the sporocyst; it is probably involved in excystation.

stiff agar See AGAR and MEDIUM.

Stigeoclonium A genus of polymorphic, filamentous (often heterotrichous) green algae related to CHAETOPHORA. The filaments are uniseriate, branched, and may terminate in multicellular hyaline hairs; both prostrate and erect filaments may form rhizoids. Each cell contains a single parietal chloroplast. Reproduction is predominantly asexual, quadriflagellate zoospores and aplanospores being produced. Sexual reproduction is isogamous. Species occur worldwide in standing or running fresh water, growing attached to rocks, plants, etc. (See also SEWAGE FUNGUS.)

stigma (*algol.*) *Syn.* EYESPOT (sense 1).

Stigmatella See MYXOBACTERALES.

stigmatomycosis An insect-transmitted disease of cotton (*Gossypium*) and other plants, caused by members of the METSCHNIKOWIACEAE (e.g. *Ashbya gossypii*, *Nematospora gossypii*).

Stigonematales See CYANOBACTERIA.

stilbaceous Having synnemata.

Stilbellales See HYPHOMYCETES.

Stilton cheese See CHEESE-MAKING.

stimulatory hypersensitivity *Syn.* TYPE V REACTION.

stinkhorns See PHALLALES.

stinking smut *Syn.* bunt: see COMMON BUNT and KARNAL BUNT.

stipe A stalk or stem. In e.g. an agaric-type fungal fruiting body the stipe is composed of hyphae which are characteristically orientated parallel to the axis of the stipe (cf. *receptacle* in PHALLALES). In PENICILLIUM spp 'stipe' refers to the (single) hypha which supports the penicillus.

stipitate Having a STIPE.

stirred tank reactor (STR) A FERMENTER in which the culture is agitated by a mechanical device (e.g. a turbine) but in which there is no circulatory loop (cf. LOOP FERMENTER). The traditional STR is a squat, cylindrical vessel, but the *multistage STR* is a tall ('column') fermenter in which the mechanical agitators are arranged in a vertical series on a common drive shaft. The STR is generally more flexible than other fermenters, i.e., it is suitable for a wider range of operating conditions.

STLV SIMIAN T-CELL LEUKAEMIA VIRUS.

***stn* gene** See FOOD POISONING (*Salmonella*).

stock Any uncharacterized population of trypanosomes derived from a natural host animal; a given stock may contain different strains of trypanosomes.

Stoffel fragment See DNA POLYMERASE.

stolon (*mycol.*) See e.g. RHIZOPUS.

stomach (human) See GASTROINTESTINAL TRACT FLORA.

stomate Having a 'mouth', stoma(ta), or CYTOSTOME.

stomatitis Inflammation of the tissues of the mouth. The causes of stomatitis include e.g. irritation of the tissues by ill-fitting dentures, vitamin deficiency, and infection by any of a range of microorganisms: see e.g. CANDIDIASIS, GINGIVITIS, HAND FOOT AND MOUTH DISEASE, HERPES SIMPLEX (oral herpes), PERIODONTITIS and VINCENT'S ANGINA. (See also LUDWIG'S ANGINA.)

Stomatochroon A genus of obligately endophytic algae related to TRENTEPOHLIA. The organisms typically inhabit the stomata of leaves, their prostrate filaments growing between the cells of the leaf mesophyll.

Stomatococcus A genus of facultatively anaerobic, Gram-positive cocci of the family MICROCOCCACEAE. *S. mucilaginosus*, the sole species, occurs in the human mouth; it has been associated with endocarditis and (in immunosuppressed individuals) with bacteraemia. Catalase weakly +ve. Oxidase −ve. Carbohydrate metabolism: fermentative. Colonies are small, dark and mucoid. The cells are resistant to LYSOSTAPHIN. Unlike *Staphylococcus*, the organism is sensitive to bacitracin.

stomatogenesis The formation of the oral structures and infrastructures during the overall process of binary fission in a stomate ciliate; morphogenetic details – particularly the origin of kinetosomes of the new oral region – differ among the various groups of ciliates. There are four principal modes of stomatogenesis: apokinetal, buccokinetal, parakinetal and telokinetal (listed in the order which approximates an evolutionary sequence from most advanced to most primitive). *Apokinetal* ('de novo') *stomatogenesis* involves the appearance of a group of oral kinetosomes which do not appear to have arisen either from those of the parental oral apparatus or from those of the somatic ciliature. In *buccokinetal stomatogenesis* at least some of the kinetosomes appear to be derived from the division of kinetosomes in the parental buccal area; in *parakinetal stomatogenesis* they may be derived from kinetosomes of somatic cilia posterior to the buccal area. In *telokinetal stomatogenesis* (which occurs in 'primitive' ciliates of the KINETOFRAGMINOPHOREA) oral kinetosomes derive from those which occur at – or which (presumably) had an evolutionary origin in – the anterior end of the somatic kineties.

stone-brood A BEE DISEASE caused by *Aspergillus flavus* or *A. fumigatus*. Infected larvae appear white and fluffy at first, becoming pale brownish or greenish yellow; most die after being sealed in their cells prior to pupation.

stoneworts See CHAROPHYTES.

stop See MICROSCOPY (b).

stop codon See GENETIC CODE.

stop transfer sequence See SIGNAL HYPOTHESIS.

stopping filter See MICROSCOPY (e).

Stormont test (*vet.*) A TUBERCULIN TEST in which an injection of PPD is given in the animal's neck, and a second injection is given at the same site seven days later; the test is designed to detect weakly reactive animals. In a positive test, skin thickening is detectable one day after the second injection. The test is negative for *M. avium* infections, but positive e.g. in animals with SKIN TUBERCULOSIS.

stormy clot (stormy fermentation) A type of clot obtained when certain bacteria (e.g. *Clostridium perfringens*) are cultured in a milk-containing liquid medium (e.g. LITMUS MILK): the clot has a characteristic turbulent appearance owing to the formation of copious amounts of gas within the medium.

STR STIRRED TANK REACTOR.

***str* operon** See RIBOSOME (biogenesis).

Strahlenkörper *Syn.* RAY BODY.

straight wire An instrument used in bacteriology e.g. to SUBCULTURE from an isolated colony, or to prepare a stab-CULTURE; the rod-shaped metal handle holds, at one end, a straight piece of platinum or nickel-steel wire commonly 5–8 cm long. Before *and* after *each* use the wire is sterilized by FLAMING and allowed to cool. In use, the *tip* of the sterile wire is brought into contact with e.g. the centre of a colony; the small quantity of bacterial growth which adheres to the wire can be used as an INOCULUM. (cf. LOOP; CITRATE TEST.)

strain An organism, or population of organisms, distinguishable from at least some of the other organisms or populations within a given named taxon, or within a given category of organisms; the members or cells of a given strain are generally regarded as being genetically similar or identical.

strain T virus See AVIAN RETICULOENDOTHELIOSIS VIRUSES.

stramineous Straw-coloured.

strand (of DNA and RNA) A single chain of nucleotide subunits. Double-stranded nucleic acids result from base-pairing between the complementary bases of two strands. For triple-stranded DNA see TRIPLEX DNA.

(See also CODING STRAND.)

strand displacement amplification See SDA.

strangles (distemper) A HORSE DISEASE characterized by inflammation of the upper respiratory tract, abscess formation in adjacent lymph nodes, fever, anorexia, and a purulent nasal discharge; the causal agent, *Streptococcus equi*, enters the body on contaminated feedstuffs or by droplet infection. The disease occurs world-wide, and typically affects young horses (<5 years). *Treatment*: antibiotic therapy. *Control*: vaccination.

stratigraphy See BIOSTRATIGRAPHY.

straw mushroom *Syn.* PADI-STRAW MUSHROOM.

strawberry crinkle virus See RHABDOVIRIDAE.

strawberry diseases See e.g. CROWN ROT; GREY MOULD; RED CORE; XANTHOMONAS (*X. fragariae*).

strawberry foot-rot A proliferative dermatitis of sheep, affecting chiefly the lower parts of the limbs, caused by *Dermatophilus congolensis*. (See also DERMATOPHILOSIS.)

strawberry latent ringspot virus See NEPOVIRUSES.

strawberry vein-banding virus See CAULIMOVIRUSES.

streak (*plant pathol.*) Roughly parallel lines of discoloration or necrosis which develop on the stems or leaves of plants infected e.g. with certain PLANT VIRUSES.

streaking INOCULATION of the surface of a solid MEDIUM in such a way that, during INCUBATION, individual colonies form on at least part of the surface. One method is as follows. A LOOP carrying an INOCULUM is drawn back and forth (i.e. *streaked*) on the surface of a 'dried' PLATE as at A (see figure). The loop is then flamed (see FLAMING), allowed to cool, and streaked as at B. Streakings C, D, and E are similarly made, the loop being flamed and cooled between each streaking and after the last. In this way the inoculum is progressively thinned out as it is distributed along the streak lines. On incubation, each well-isolated viable cell capable of growth on the medium gives rise to a single COLONY.

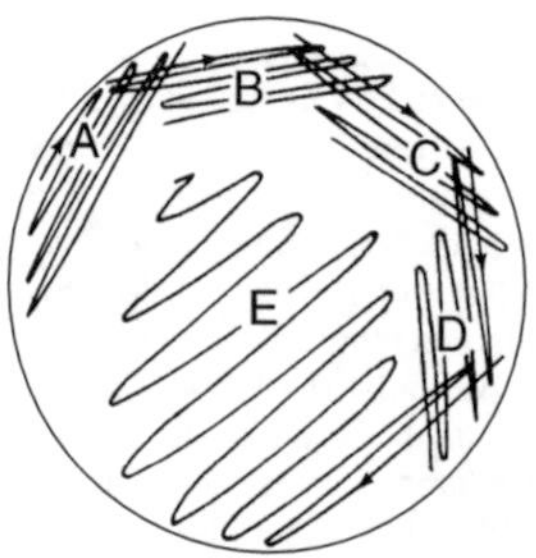

STREAKING. The pattern of inoculation used in one method of streaking. (See text for explanation.)

street virus A wild-type, virulent strain of rabies virus, i.e., one present in an animal or person with naturally acquired rabies. (cf. FIXED VIRUS.)

strep throat A sore throat caused by *Streptococcus pyogenes* infection; other symptoms may include fever, rash etc. (See also RHEUMATIC FEVER.)

streptavidin A protein, obtained from *Streptomyces avidinii*, which resembles AVIDIN in binding with high affinity to BIOTIN; it can be used instead of avidin e.g. in the ABC IMMUNOPEROXIDASE METHOD.

streptidine Diguanidylinositol, a component of certain AMINOGLYCOSIDE ANTIBIOTICS.

Streptoalloteichus A genus of bacteria (order ACTINOMYCETALES, wall type III). The organisms form substrate mycelium which bears spheroidal sporangia (each containing 1–4 rod-shaped, motile spores), and an aerial mycelium which gives rise to chains of spores. Type species: *S. hindustanus*. [Book ref. 73, pp. 118–119.]

Streptobacillus A genus (*incertae sedis*) of anaerobic, facultatively aerobic, chemoorganotrophic, Gram-negative bacteria which occur e.g. in the throat and nasopharynx in rats and which cause one form of RAT-BITE FEVER. Cells: non-motile rods, ca. $0.1–0.7 \times 1.0–5.0$ μm (occurring singly or in chains), or filaments, which may be highly pleomorphic – cell shape depending e.g. on cultural conditions. L-FORMS may arise spontaneously. Metabolism is primarily fermentative. Growth requires e.g. serum or blood. Acid (no gas) is formed from glucose and other carbohydrates. Nitrate is not reduced. Oxidase −ve. Catalase −ve. GC%: ca. 24–26. Type species: *S. moniliformis*. [Book ref. 22, pp. 598–600.]

Streptobacterium See LACTOBACILLUS.

streptobiosamine A disaccharide component of STREPTOMYCIN, consisting of L-streptose (5-deoxy-3-formyl-L-lyxose) and 2-deoxy-2-methylamino-L-glucose.

Streptococcaceae A family of Gram-positive (or Gram type positive), asporogenous, facultatively anaerobic, chemoorganotrophic, spherical or coccoid bacteria; genera: *Aerococcus*, *Gemella*, *Leuconostoc*, *Pediococcus* and *Streptococcus* [Book ref. 21, pp. 490–517].

streptococcal superantigen A See SUPERANTIGEN.

Streptococcus A genus of Gram-positive, asporogenous, chemoorganotrophic, facultatively anaerobic, catalase-negative and oxidase-negative cocci or coccoid bacteria, typically ca. 1 μm diam., which generally occur in pairs or chains; non-motile. CAPSULE-formation is common in some species. Growth usually occurs optimally in the range 30–37°C. Typically fermentative, with carbohydrates metabolized anaerogenically and homofermentatively – glucose usually yielding L-lactic acid as the main product; however, in the presence of oxygen, some streptococci (e.g. strains of *S. mutans*) can oxidize NADH by NADH oxidase and NADH peroxidase activities [AvL (1985) *51* 577–578]. On blood agar some streptococci give rise to clear HAEMOLYSIS ('β-haemolysis'); the VIRIDANS STREPTOCOCCI characteristically give rise to zones of partial haemolysis with greenish to brownish discoloration (so-called 'α-haemolysis').

The organisms occur as commensals/parasites and as pathogens in man and other animals.

Streptococci are killed by PASTEURIZATION and by many common disinfectants.

GC%: 34–46. Type species: *S. pyogenes*.

Classification of the streptococci. Streptococcus is a large genus (~40 species) which is divided into three main groups of species on the basis of 16S rRNA studies: the pyogenic group, the oral group and 'other species'.

The pyogenic group includes species associated with pyogenic infections in man and other animals – e.g. *S. agalactiae*, *S. dysgalactiae*, *S. equi* and *S. pyogenes*.

The oral group contains species (commonly isolated from the oral cavity) which are associated with pyogenic and other infections in various sites – including mouth, heart (infective endocarditis), joints, skin, muscle and central nervous system (e.g. brain abscesses). The species in this group include *S. anginosus*, *S. mutans*, *S. oralis* and *S. salivarius*. (See also VIRIDANS STREPTOCOCCI.) [Oral streptococci (taxonomy, and relation to infections): RMM (1994) *5* 151–162.]

The 'other' species include e.g. *S. acidominimus*, *S. bovis*, *S. equinus* and *S. suis*. Most of these species occur in domestic animals (although e.g. *S. bovis* is an infrequent cause of endocarditis in man).

A number of organisms, previously classified as streptococci, have been transferred to the genera ENTEROCOCCUS and LACTOCOCCUS. (See also ATOPOBIUM.)

In addition to the three main groups of species (mentioned above), many streptococci have been classified by LANCEFIELD'S STREPTOCOCCAL GROUPING TEST – which identifies certain group-specific cell-wall compounds; thus, e.g. species in the pyogenic group (see above) include those in Lancefield group A (*S. pyogenes*), B (*S. agalactiae*), C (strains of *S. dysgalactiae*) and G (*S. canis*). Note that the cell wall in some species does not contain components which react in the Lancefield grouping test, so that these species cannot be assigned to a Lancefield group. (For group R see *S. suis*, below.)

Other components of the streptococcal cell wall (e.g. the M PROTEIN and T PROTEIN) are also involved in streptococcal typing.

Species (and some former species) include:

S. acidominimus. Viridans streptococci, isolated e.g. from cattle, from raw milk and from the human mouth. No Lancefield group.

S. adjacens. See NUTRITIONALLY VARIANT STREPTOCOCCI.

S. agalactiae. Pyogenic group, isolated from man and cattle. Lancefield group B. α-, β- and non-haemolytic strains occur, some being pigmented. A causal agent of bovine MASTITIS (see also CAMP TEST) and e.g. pneumonia or meningitis in human neonates/infants. [Phage typing of *S. agalactiae*: Book ref. 143, pp 1–22.] β-Lactam antibiotics are often used as chemoprophylaxis against infection of the neonate, while e.g. erythromycin may be used as an alternative. However, while most strains of *S. agalactiae* are currently susceptible to β-lactams, a recent study found that >20% of 126 clinical isolates expressed resistance to macrolides. [Antibiotic susceptibility and mechanisms of erythromycin resistance in clinical isolates of *S. agalactiae* (a multicentre study): AAC (2001) *45* 2400–2402.]

S. anginosus. See '*S. milleri*' (below).

S. avium. Lancefield group D. Typically α-haemolytic organisms which occur in the intestine in man and fowl. Referred to ENTEROCOCCUS.

S. bovis. Lancefield group D (most strains). Strains occur e.g. in the human gut (also found in domestic animals); infrequent cause of endocarditis in man. (See also BLOAT; THIOPEPTIN.) [Taxonomic study: SAAM (1984) *5* 467–482.]

S. canis. [Species proposal: IJSB (1986) *36* 422–425.] Pyogenic group, isolated from cows with mastitis and from dogs with genital, skin or wound infections. Lancefield group G. β-Haemolytic.

S. constellatus. See '*S. milleri*' (below).

S. cremoris. Lancefield group N organisms (see LACTOCOCCUS).

S. cricetus. A species isolated from hamsters (formerly referred to as serotype a of *S. mutans*).

S. durans. Lancefield group D. α-, β- and non-haemolytic strains. Referred to ENTEROCOCCUS.

S. dysgalactiae. Pyogenic group, isolated from cattle, pigs, man. Includes strains of Lancefield groups C, G and L.

S. equi. Pyogenic group. Lancefield group C. Two subspecies: *equi* and *zooepidemicus*. Causal agents of e.g. STRANGLES in equines and glomerulonephritis and pharyngitis in humans. (See also MASTITIS.)

S. faecalis. Lancefield group D. Referred to ENTEROCOCCUS.

S. faecium. Lancefield group D. Referred to ENTEROCOCCUS.

S. ferus. Oral group. Formerly referred to as strains of *S. mutans*. Isolated from rats.

S. gallinarum. Lancefield group D. β-Haemolytic on horse-blood agar. Occurs e.g. in the intestines of fowl. Referred to ENTEROCOCCUS.

S. gordonii. Oral group. Organisms previously referred to as strains of *S. sanguis*.

S. hyointestinalis. Pyogenic group, isolated from pigs. No Lancefield group.

S. intermedius. See '*S. milleri*' (below).

S. lactis. Lancefield group N. Referred to LACTOCOCCUS.

Streptococcus MG. See STREPTOCOCCUS MG.

'*S. milleri*'. Streptococci associated with various types of endogenous pyogenic infection in humans; they occur e.g. in the oropharyngeal region and the gastrointestinal tract. The organisms include α-, β- and non-haemolytic strains, and strains of Lancefield groups A, C, F and G (as well as some strains which are non-typable by the Lancefield method). Strains of group C can be distinguished from the others e.g. in that they agglutinate human platelets.

These organisms often form acid from lactose, salicin, sucrose and trehalose (but not sorbitol) and hydrolyse arginine and aesculin. All are VP +ve.

Bacteria within the '*S. milleri*' group have been classified in three species: *S. anginosus*, *S. constellatus* and *S. intermedius* [RMM (1997) *8* 73–80]. [Iliac osteomyelitis and gluteal muscle abscess caused by *S. intermedius*: JMM (2001) *50* 480–482.]

S. mutans. Oral group. Organisms which occur e.g. in the human mouth and which are associated with DENTAL PLAQUE (see also MUTAN). Some of the strains originally included in this species have been transferred to other species on the basis of e.g. differences in GC%, type of murein and metabolic characteristics (*S. cricetus*, *S. downei*, *S. ferus*, *S. macacae*, *S. rattus*, *S. sobrinus*) [see e.g. IJSB (1985) *35* 482–488].

S. oralis. Oral group, viridans streptococci found e.g. in the human mouth. The species includes some strains previously designated e.g. *S. sanguis* II and *S. viridans* I, II and IV [IJSB (1985) *35* 482–488].

S. parauberis. Pyogenic group, isolated e.g. from cattle, and a causal agent of bovine MASTITIS.

S. plantarum. Referred to LACTOCOCCUS.

S. pluton. See *Melissococcus pluton* in EUROPEAN FOULBROOD.

S. pneumoniae (the 'pneumococcus'). Oral group, related to *S. oralis*. Strains of *S. pneumoniae* occur as part of the human RESPIRATORY TRACT MICROFLORA. The organism is associated with various diseases: e.g. BRONCHITIS, MENINGITIS, OTITIS MEDIA, PNEUMONIA and SINUSITIS. [Diagnosis of *S. pneumoniae* infections: RMM (1994) *5* 224–232.] Until quite recently most isolates were susceptible to penicillin but many strains are now resistant [Drugs (1996) *51* (supplement 1) 1–5]. Mutations in some strains are associated with enhanced resistance to certain of the newer fluoroquinolone antibiotics [AAC (2001) *45* 952–955].

[Activities of newer fluoroquinolones against ciprofloxacin-resistant *Streptococcus pneumoniae*: AAC (2001) *45* 1654–1659.]

Cells: ovoid, often in pairs (diplococci) with the longitudinal axes of the cells aligned (hence the common, but erroneous, designation '*Diplococcus pneumoniae*'). No Lancefield antigen. Characteristically there is a polysaccharide CAPSULE which may exist in any of a number of antigenically distinct forms (this variation being used as a basis for pneumococcal TYPING); the type III pneumococcal polysaccharide, for example, consists of repeating subunits of cellobiuronic acid linked by β-(1 $\rightarrow$ 3) glucosidic bonds. The existence of a large number of serotypes (~100) has tended to undermine the value of anti-*S. pneumoniae* CONJUGATE VACCINES – which are useful against only a small proportion of the serotypes. (Multivalent conjugate vaccines that cover more serotypes would be expensive.) As an alternative approach to anti-*S. pneumoniae* vaccines it has been suggested that certain cell-surface *proteins* of the pathogen (which show little antigenic variation) may offer better protection [TIM (1998) *6* 85–87].

[Complete genome sequence of a virulent isolate of *Streptococcus pneumoniae*: Science (2001) *293* 498–506.]

On blood-agar the typical colonial form is the DRAUGHTSMAN COLONY. *S. pneumoniae* is sensitive to OPTOCHIN. (See also BILE SOLUBILITY and QUELLUNG PHENOMENON.)

Some strains of *S. pneumoniae* form products such as HYALURONATE LYASE, NEURAMINIDASE and THIOL-ACTIVATED CYTOLYSIN.

S. porcinus. A species associated with e.g. abscesses in the cervical lymph nodes of pigs and with porcine pneumonia and septicaemia. The species includes strains of Lancefield groups E, P, U and V [SAAM (1984) *5* 402–413].

S. pyogenes (formerly *S. haemolyticus*). Pyogenic group. Lancefield group A. Characteristically β-haemolytic. In broth cultures, cells commonly form long chains, sometimes with a HYALURONIC ACID capsule. (See also PYR.) Extracellular products may include e.g. HYALURONATE LYASE, STREPTODORNASE, STREPTOKINASE, STREPTOLYSIN O, STREPTOLYSIN S. [Complete genome sequence of an M1 strain of *S. pyogenes*: PNAS (2001) *98* 4656–4663.]

S. pyogenes is associated with various diseases: see e.g. ERYSIPELAS, GLOMERULONEPHRITIS, IMPETIGO, NECROTIZING FASCIITIS, OTITIS MEDIA, PHARYNGITIS, PNEUMONIA (e), QUINSY, SCARLET FEVER, SINUSITIS, STREP THROAT, TONSILLITIS and TOXIC SHOCK SYNDROME. (See also SUPERANTIGEN.)

Some group A streptococci have the ability to enter (and survive within) human respiratory-tract cells; invasion is promoted by a protein encoded by the pathogen's *prtF1* gene. An intracellular location would effectively protect the pathogen from those antibiotics (including e.g. β-lactams) which remain largely extracellular; it is therefore of importance that invasive group A streptococci have been associated with resistance to erythromycin (a drug which can enter eukaryotic cells) [Lancet (2001) *358* 30–33].

S. raffinolactis. Lancefield group N. Referred to LACTOCOCCUS.

S. rattus. See *S. mutans* (above).

S. saccharolyticus. A non-haemolytic species isolated from e.g. cows and straw bedding [SAAM (1984) *5* 467–482].

S. salivarius. Oral group, isolated from e.g. the human mouth; viridans streptococci related to *S. thermophilus*. Some strains form extracellular LEVANS. [Identification of *S. salivarius* by PCR and DNA probe: LAM (2001) *32* 394–397.]

S. sanguis. Viridans streptococci which occur e.g. in the human mouth. Some strains form extracellular DEXTRANS. (See also CO-AGGREGATION.)

S. sobrinus. See *S. mutans* (above).

S. suis. Associated with disease in pigs (e.g. meningitis, endocarditis, bronchopneumonia, reproductive disorders); serotype 2 is the most common serotype found in infections. [Serotypes 3–28 associated with disease in pigs: VR (2001) *148* 207–208.] Serotype 2 (also referred to as the group R streptococcus) can cause meningitis, respiratory infections and septicaemia in people exposed to infected pigs/pork.

S. thermophilus. Oral group, related to *S. salivarius*; the organisms are isolated from dairy products. No Lancefield group. Characteristically weakly α-haemolytic. Growth occurs at 50°C, and the organisms can survive 65°C/30 min. (See also LACTIC ACID STARTERS, LEBEN, YOGHURT.)

S. zooepidemicus. Re-classified as a subspecies of *S. equi*.

***Streptococcus* MG** The designation used for certain non-β-haemolytic streptococci which have been included in a species referred to as '*Streptococcus milleri*' [JCM (1985) *22* 772–777]. At least some strains are agglutinated by the sera of patients who are suffering from mycoplasmal pneumonia. [The '*Streptococcus miller*' group: RMM (1997) *8* 73–80.]

streptodornase (streptococcal nuclease) An extracellular nuclease produced by *Streptococcus* spp. Such enzymes are frequently synthesized e.g. by strains of Lancefield groups A and

E – group A strains producing at least four serologically distinct nucleases designated A, B, C and D; all four types of nuclease cleave the backbone of DNA, while types B and D are also active against RNA. The activity of streptodornases is promoted by divalent cations, particularly Ca^{2+} and Mg^{2+}. In contrast to staphylococcal DEOXYRIBONUCLEASE (q.v.), the products of streptococcal nucleases are 5′-nucleotides.

streptogramins A group of closely related MLS ANTIBIOTICS that includes mikamycins, ostreogrycins, patricins, pristinamycins, staphylomycins, synergistins, vernamycins and VIRGINIAMYCINS (at least some of these names are synonyms). (cf. GRISEOVIRIDIN, VIRIDOGRISEIN.)

Each of these antibiotics is a complex of at least two different components. The 'A' (formerly 'M') component is a polyunsaturated macrolactone ring, while the 'B' (formerly 'S') component is a cyclic hexadepsipeptide which contains unusual amino acids (e.g. pipecolic acid).

Both the A and B molecules inhibit PROTEIN SYNTHESIS by binding to the 50S ribosomal subunit. In at least some cases, the A and B components are (jointly) bactericidal owing to synergistic action; individually, the components may be bacteriostatic. The A component inhibits transpeptidation, while the B component inhibits the binding of aminoacyl-tRNA complexes to the ribosome and/or binding of the growing peptide chain at the P site during translation of certain amino acids (e.g. proline), thus causing premature release of the polypeptide chain. In at least some cases, binding of the A component distorts the ribosome in a way that facilitates binding of the B component.

Resistance to A-type streptogramins may be due to *O*-acetylation of the antibiotic (in at least some cases by a plasmid-encoded enzyme).

Resistance to B-type streptogramins (and other MLS antibiotics) in e.g. staphylococci and streptococci can be due to N^6-methylation of adenines in 23S rRNA (preventing antibiotic–ribosome binding); this site may be methylated by inducible, plasmid-encoded enzymes or by constitutively expressed enzymes. (Strains designated $MLS_{B/c}$ contain constitutively expressed methylases.) In at least some cases (e.g. quinupristin), resistance is reported to occur only if methylases are expressed constitutively.

QUINUPRISTIN/DALFOPRISTIN is a recently introduced member of the group which is useful e.g. against vancomycin-resistant strains of certain Gram-positive pathogens.

streptokinase A protein (produced e.g. by certain streptococci of Lancefield groups A and E) which activates profibrinolysin (plasminogen) and thus has *indirect* fibrinolytic activity (see FIBRINOLYSIS). Streptokinase interacts with profibrinolysin to form a complex which releases active fibrinolysin. [Mechanism: JBC (1978) *253* 1090–1094.] Streptokinase is assumed to be an AGGRESSIN, assisting the invasiveness of pathogenic streptococci by lysing the fibrin barrier which may enclose streptococcal lesions [TIM (1997) *5* 466–467].

Streptokinase is used clinically e.g. in the treatment of thrombosis.

streptolydigin An ANTIBIOTIC which binds reversibly to the β-subunit of bacterial RNA POLYMERASE, inhibiting elongation of nascent RNA chains; at low concentrations the antibiotic slows the rate of polymerization without significantly affecting the accuracy of transcription. (Tirandamycin appears to be structurally and functionally similar.) (cf. RIFAMYCINS.)

streptolysin O A THIOL-ACTIVATED CYTOLYSIN (q.v.) produced by strains of streptococci of groups A, C and G. In haemolytic activity, monomers of streptolysin O bind to cholesterol-containing sites on red blood cells (RBCs) and then oligomerize into ring-shaped structures which form pores of ~30 nm (larger than the pores formed e.g. by the MAC in COMPLEMENT FIXATION). Haemolysis is inhibited by free cholesterol.

Streptolysin O is immunogenic (cf. STREPTOLYSIN S), and it cross-reacts with the *theta* toxin of *Clostridium perfringens*. In certain diseases (e.g. RHEUMATIC FEVER) the titre of antistreptolysin O antibodies is typically elevated (see ANTISTREPTOLYSIN O TEST).

[Mechanism of assembly of the streptolysin O pore: EMBO (1998) *17* 1598–1605.]

streptolysin S An oxygen-stable, non-immunogenic leucocidin and haemolysin produced e.g. by many group A streptococci (cf. STREPTOLYSIN O); it appears that some streptolysin S may remain cell-bound while the rest is released by the cell. The haemolytic activity of streptolysin S is inhibited by low levels of certain phosphatides, e.g. phosphatidylethanolamine.

Streptomyces A genus of Gram-positive, aerobic bacteria (order ACTINOMYCETALES, wall type I) which occur e.g. in soil and aquatic habitats – a few species being pathogenic in plants (e.g. potato, sugar beet) and man (see MADUROMYCOSIS). The organisms form both AERIAL MYCELIUM and a non-fragmenting, branched SUBSTRATE MYCELIUM, and in most species spores are formed only by the aerial mycelium which fragments to give rise to non-verticillate chains of spores [differentiation and spore formation: Book ref. 67, pp. 89–115]; the genus includes organisms which have been referred to as species of *Actinopycnidium*, *Actinosporangium* and *Chainia*, and the genus definition has been widened to include *Elytrosporangium* and *Microellobosporia* (organisms which form spores on both substrate and aerial mycelium) [Book ref. 73, pp. 94–105; JGM (1983) *129* 1743–1813]. (cf. STREPTOMYCETE.) The organisms are chemoorganotrophic (metabolism: oxidative) and are not nutritionally fastidious; usable carbon sources include e.g. chitin, glucose, lactate, starch. Colonies are typically coloured, and many species form a diffusible pigment. ANTIBIOTIC production is common: see e.g. CARBAPENEMS, CHLORAMPHENICOL, CLAVULANIC ACID, MACROLIDE ANTIBIOTICS and STREPTOMYCIN. [Book ref. 201.] (See also COMPLESTATIN.) GC%: 69–78. Type species: *S. albus*.

Studies on the cell envelope of *S. griseus* have revealed the presence of a hydrophilic channel that runs through the cell wall, and have further identified a streptomycin-binding site within the channel [Mol. Microbiol. (2001) *41* 665–673].

Many of the hundreds of named species of *Streptomyces* have been amalgamated to form a much smaller number of species [see e.g. JGM (1983) *129* 1743–1813]. [Ecology, isolation and cultivation of streptomycetes: Book ref. 46, pp. 2028–2090.]

streptomycete Any organism of the (related) genera *Intrasporangium*, *Kineosporia*, *Nocardioides*, *Sporichthya*, *Streptomyces* and *Streptoverticillium* [Book ref. 73, pp. 94–105].

streptomycin A broad-spectrum AMINOGLYCOSIDE ANTIBIOTIC produced by *Streptomyces griseus*; it consists of the base STREPTIDINE linked glycosidically to STREPTOBIOSAMINE. Streptomycin is active against e.g. *Mycobacterium tuberculosis*, some staphylococci, *Brucella* spp and many other Gram-negative bacteria (including some strains of *Pseudomonas aeruginosa*); bactericidal activity increases with concentration. The antibiotic is less active, or inactive, under anaerobic conditions. Streptomycin also inhibits vegetative growth and sporulation of certain fungi [JGM (1983) *129* 3401–3410].

When used as a chemotherapeutic agent, streptomycin is associated with certain side-effects, including dose-dependent damage to the 8th cranial nerve (resulting in impairment of hearing); hypersensitivity reactions may also occur.

Streptomycin affects many bacterial cell functions – e.g., it binds to protein S12 (also called P10) in the 30S subunit of bacterial 70S RIBOSOMES and inhibits PROTEIN SYNTHESIS by causing the formation of aberrant initiation complexes. Sublethal concentrations of streptomycin cause misreading of mRNA; misreading occurs mainly at the pyrimidine (rarely the purine) bases in codons, and may be due to conformational changes in the ribosome caused by streptomycin binding.

Resistance to streptomycin may be due to a mutation in the *strA* gene (= *rpsL* gene) resulting in a modified S12 protein which may fail to bind streptomycin or which may alter the effect of the bound antibiotic; such one-step high-level resistance to streptomycin occurs quite commonly. Resistance may also be due e.g. to a plasmid-specified enzyme *streptomycin phosphotransferase*. [Resistance to aminoglycoside antibiotics: TIM (1997) *5* 234–240.]

In certain ('streptomycin-dependent') *str* mutants the mutation results in non-functional ribosomes – such mutants being able to grow only in the presence of streptomycin or certain other aminoglycoside antibiotics (e.g. kanamycin, neomycin, paromomycin), or e.g. ethanol; these substances apparently distort the ribosome in such a way that translation can occur. *Conditional* streptomycin dependence is exhibited by certain auxotrophs which can grow in the absence of their required growth factor only when supplied with (sublethal) concentrations of e.g. streptomycin – 'conditional' referring to the fact that streptomycin dependence is exhibited only when the growth factor is absent.

streptonigrin An antitumour antibiotic which, in the presence of O_2 and a reducing agent (e.g. Fe^{2+}), causes DNA strand breakage; streptonigrin action is inhibited in the presence of free radical scavengers (including superoxide dismutase).

L-streptose 5-Deoxy-3-formyl-L-lyxose (cf. STREPTOBIOSAMINE).

Streptosporangium A genus of bacteria (order ACTINOMYCETALES, wall type III; group: maduromycetes) which occur e.g. in soil, leaf litter, and dung. The organisms form branching substrate and aerial mycelium, the latter bearing clusters of spherical sporangia (each 1–40 μm diam., according to species). GC%: ca. 69–71. Type species: *S. roseum*. [Book ref. 73, pp. 114–115.]

streptotrichosis *Syn.* DERMATOPHILOSIS.

streptovaricins See ANSAMYCINS.

Streptoverticillium A genus of aerobic bacteria (order ACTINOMYCETALES, wall type I) which occur e.g. in soil. The organisms form a non-fragmenting substrate mycelium, and an aerial mycelium characterized by the formation of spore chains in verticils. GC%: ca. 69–73. Type species: *S. baldaccii*.

streptovitacins See CYCLOHEXIMIDE.

streptozotocin An antibiotic – 2-deoxy-2-(3-methyl-3-nitrosoureido)-D-glucopyranoside – which can cause diabetes in experimental animals (e.g. rats) [FEBS (1980) *120* 1–3] and which is bactericidal in certain (growing) Gram-positive and Gram-negative bacteria. In bacteria, susceptibility to streptozotocin requires a functional PTS for *N*-acetylglucosamine which is responsible for the uptake of the streptozotocin; the antibiotic enters the cytoplasm as streptozotocin 6-phosphate, a substance which gives rise to the highly toxic and mutagenic compound diazomethane. Since non-growing cells are unaffected, and cells killed by streptozotocin do not lyse, this antibiotic can in some cases be a more useful agent than penicillin for the selection of e.g. rare AUXOTROPHS.

stress fibre See PINOCYTOSIS.

stretch-activated channel See MECHANOSENSITIVE CHANNEL.

Striaria See PHAEOPHYTA.

Striatella See DIATOMS.

strict (1) *Syn.* OBLIGATE. (2) A term used to emphasize the extent to which a given term is applicable. For example, a strict anaerobe is an organism which grows only in the complete absence of oxygen (and which may also require a pre-reduced medium), and strict anaerobiosis refers to the condition in which there is a complete absence of oxygen. Similarly, a strict autotroph can grow in the complete absence of organic carbon, i.e. its carbon requirements can be satisfied by CO_2 alone.

Strigomonas *Syn.* CRITHIDIA.

Strigula See CEPHALEUROS.

string of pearls effect See BACILLUS (*B. anthracis*).

stringency (in PCR) See PCR.

stringent control (1) (of rRNA and protein synthesis) In bacteria: a control mechanism in which cells lacking an essential amino acid respond with e.g. decreased synthesis of rRNA and proteins. The response is triggered by an uncharged tRNA at a ribosomal A site (see PROTEIN SYNTHESIS); this stimulates a ribosome-bound enzyme, pyrophosphotransferase (RelA; the *relA* gene product; stringent factor), to synthesize ppGpp (in which G is guanosine, p is phosphate) from GTP (or GDP) and ATP. ppGpp (= guanosine 5′-diphosphate 3′-diphosphate; 'guanosine tetraphosphate') is an example of an ALARMONE. (Cells with a null mutation in *relA* do not exhibit the stringent response and are said to be *relaxed*.)

ppGpp can apparently enhance transcription from certain operons that control the synthesis of particular amino acids (e.g. histidine). ppGpp may thus help to maintain a correct balance of amino acids in the cell; for example, given a level of histidine low enough to trigger the stringent response, increased synthesis of ppGpp may help to restore the balance by (i) slowing down protein synthesis, and (ii) stimulating the *his* operon.

A high-resolution study revealed that ppGpp can bind to RNA polymerase in different orientations. This suggested that base-pairing between ppGpp and cytosines in non-template DNA might be important in transcriptional control by ppGpp [Cell (2004) *117* 299–310].

(2) (of plasmid replication) See PLASMID.

stringent factor See STRINGENT CONTROL (sense 1).

stripe rust *Syn.* YELLOW RUST.

stroma (1) (*mycol.*) A compact mass of prosenchymatous or pseudoparenchymatous fungal tissue (or of intermingled fungal and non-fungal tissue) on or within which spore-bearing or spore-containing structures often develop. Stromata are formed by many fungi – e.g. by species of *Nectria* and *Xylaria*. (See also ASCOSTROMA, SCLEROTIUM, SPORODOCHIUM.)

(2) See CHLOROPLAST.

stromatolites Rock-like accretions, often columnar or domed, generally believed to have been formed by the incorporation of sediment and/or precipitated calcareous/siliceous material in MICROBIAL MATS within marine and non-marine habitats; the oldest of such 'organosedimentary' structures, ca. 3.5×10^9 years old, occur in Western Australia.

Stromatolites commonly exhibit a laminar (i.e. stratified) pattern of construction; some authors use the term 'thrombolite' for analogous, non-laminar structures.

Calcified stromatolites apparently do not yield fossil microorganisms; however, cyanobacterium-like microfossils have been detected in silicified stromatolites – such organisms including filaments (e.g. *Gunflintia minuta*) and coccoid organisms (e.g. *Huroniospora microreticulata*). [Book ref. 154.]

Studies have been carried out on stromatolites which are currently forming around the Bahamas. [Role of microbes in accretion, lamination and early lithification of modern marine stromatolites: Nature (2000) *406* 989–992.]

Strombidium See OLIGOTRICHIDA.

Strombilidium See OLIGOTRICHIDA.

Strongylidium See HYPOTRICHIDA.

γ-strophanthin *Syn.* OUABAIN.

Stropharia See AGARICALES (Strophariaceae).

Strophariaceae See AGARICALES.

struck (*syn.* Romney Marsh disease) A SHEEP DISEASE caused by *Clostridium perfringens* type C; it occurs in one- to two-year-old animals and involves intestinal congestion and enterotoxaemia.

structural gene A GENE whose product is e.g. an enzyme, structural protein, tRNA, rRNA etc – cf. REGULATOR GENE.

structured granules CYANOPHYCIN granules.

struvite Magnesium ammonium phosphate ($MgNH_4PO_4.6H_2O$), a major component of 'infectious stones' of the urinary tract, i.e. calculi typically associated with alkaline urine resulting from infection with urease-forming bacteria (such as *Proteus* spp). (See also UROLITHIASIS.) Struvite-based calculi may also include calcium phosphate, forming so-called 'triple phosphate'. When calculi are branched (*staghorn* calculi), their development has involved deposition of minerals within the collecting system of the kidneys.

STS STANDARD TESTS FOR SYPHILIS.

Stuart's transport medium A TRANSPORT MEDIUM used e.g. for a range of anaerobic bacteria, delicate organisms such as *Neisseria gonorrhoeae*, and *Trichomonas vaginalis*. Most formulations include e.g. sodium thioglycollate or thioglycollic acid (a reducing agent), methylene blue (a redox indicator), and agar (0.2–1.0% giving a liquid or semi-solid medium, respectively). (See also GELLAN GUM.) The medium is commonly prepared in bijou bottles which are filled nearly to the brim; the sterilized medium should be colourless.

[Recipe: Book ref. 53, p. 1411.]

stub See ELECTRON MICROSCOPY (b).

stubborn disease A disease of citrus trees caused by strains of *Spiroplasma citri*. The disease affects mainly young trees and is transmitted by insects (particularly leaf-hoppers); it occurs e.g. in the USA (California) and in the Middle East.

stuffer (*mol. biol.*) See REPLACEMENT VECTOR.

stump-tailed macaque virus See POLYOMAVIRUS.

stumpy form A short, broad TRYPOMASTIGOTE which has an UNDULATING MEMBRANE but lacks a *free* flagellum.

stunting syndrome (infectious stunting syndrome) A POULTRY DISEASE which affects chickens ca. 3–6 days old. Infected birds develop diarrhoea and a hunched appearance. The causal agent is unknown, but virus particles have been observed in the enterocytes in jejunal villi of infected birds [VR (1986) *118* 303–304].

Stx1, Stx2 See SHIGA TOXIN.

sty *Syn.* STYE.

stye (sty; hordeolum) An inflamed, suppurative lesion on the eyelid due to infection (usually by staphylococci) of the sebaceous gland at the base of an eyelash.

stylet-borne transmission *Syn.* non-persistent transmission (see NON-CIRCULATIVE TRANSMISSION).

Stylochona See HYPOSTOMATIA.

Stylonychia See HYPOTRICHIDA.

Stylopage See ZOOPAGALES and NEMATOPHAGOUS FUNGI.

Styloviridae A proposed family of BACTERIOPHAGES in which the virion consists of a polyhedral (isometric or elongated) head and a long, non-contractile tail. Type species: BACTERIOPHAGE λ (q.v.). Other members: e.g., ACTINOPHAGES ϕC, R1, R2 and VP5; *Bacillus* phages BPB1, ϕ105 (q.v.), SPβ and SPP1; *Caulobacter crescentus* phage ϕCbK; enterobacterial phages χ (a FLAGELLOTROPIC PHAGE), PA2, ϕ80, ϕD328, and T5; *Pseudomonas* phage F116 (q.v.). Possible member: (virulent) coliphage T1. (cf. LAMBDOID PHAGES.)

stylovirus A member of the STYLOVIRIDAE.

subacute bacterial endocarditis See ENDOCARDITIS.

subacute sclerosing panencephalitis (SSPE) An uncommon disease of children and adolescents associated with persistent measles virus (MV) infection (see MORBILLIVIRUS). Symptoms begin insidiously, with intellectual deterioration and memory loss followed by convulsions, spasticity, coma and death. MV nucleocapsids can be detected as inclusion bodies in CNS tissues, but budding viruses and infectious virions usually cannot be detected. Persistence apparently involves defective MV gene expression which may lead e.g. to a defect in MV maturation. It is frequently associated with the absence or reduced amounts of the virion M protein and/or with loss of haemadsorption – possibly resulting from M protein instability and failure to process the H protein, respectively [Virol. (1985) *143* 536–545]. [Altered transcription of a defective MV genome derived from a case of SSPE: EMBO (1987) *6* 681–688.] SSPE viruses have been propagated by co-culture of infected brain cells with e.g. Vero cells or human embryonic lung cells. [Neurovirulence of cultured SSPE virus: JGV (1985) *66* 373–377.]

subacute spongiform encephalopathies (SSEs) The former name for PRION-mediated diseases which are now referred to as TRANSMISSIBLE SPONGIFORM ENCEPHALOPATHIES.

subclinical infection (*med.*) INFECTION (sense 2) in which symptoms are not apparent – or the stage before symptoms become apparent.

subculture (1) A CULTURE prepared by inoculating a medium from an existing culture (cf. PRIMARY CULTURE). (2) The process of preparing a culture as in (1). (3) (*verb*) To prepare a culture as in (1).

suberin An aromatic polymer, somewhat similar to LIGNIN, to which are attached (by ester linkages) aliphatic components, e.g. ω-hydroxy acids, dicarboxylic acids, C_{20}–C_{30} alcohols. Suberin occurs, with waxes, in the cell walls of underground parts of higher plants, and is deposited e.g. at wound sites. (cf. CUTIN.) It provides protection against infection etc, but the cutinases of certain plant-pathogenic fungi may have some action on suberin. [Book ref. 58, pp. 79–100.]

suberization The deposition of SUBERIN by a plant e.g. at a site of injury caused by fungi, insects etc.

suberosis An EXTRINSIC ALLERGIC ALVEOLITIS associated with inhalation of cork dust contaminated with *Penicillium frequentans*.

subgenomic mRNA Of an RNA virus: an mRNA, produced in an infected (eukaryotic) cell, which is shorter than the genomic RNA. In eukaryotic cells, translation generally cannot be initiated from an internal site in an mRNA strand; the formation of a subgenomic mRNA allows an initiation site which is located internally in the viral genome to be located terminally, thus permitting independent translation of genes within the genome (cf. POLYPROTEIN). Subgenomic mRNAs may be monocistronic or polycistronic. (See e.g. CORONAVIRIDAE, RETROVIRIDAE, TOBACCO MOSAIC VIRUS, TOGAVIRIDAE; cf. e.g. FLAVIVIRIDAE.)

subhymenium Generative tissue immediately underlying a HYMENIUM. (See also APOTHECIUM.)

subiculum (*mycol.*) A pad or felt of hyphae on which, or partly within which, fruiting bodies develop.

submitochondrial particle See ETP.

subpellicular tubules In various protozoa: a system of MICROTUBULES beneath the PELLICLE. In e.g. coccidia the subpellicular tubules extend posteriorly from a POLAR RING (see also APICAL COMPLEX).

subpseudopodia See PSEUDOPODIUM.

subset (*immunol.*) A category or group within a population of cells (see e.g. T LYMPHOCYTE).

subsp Subspecies.

subspecies (subsp; ssp) (1) A taxonomic rank immediately below species. (2) Any given taxon with the rank of subspecies (sense 1); such a taxon is designated by the binomial of the species followed by the subspecific epithet: e.g. *Campylobacter sputorum* subsp *bubulus*. (3) Organisms belonging to one or more subspecies.

substage condenser See CONDENSER.

substage diaphragm See CONDENSER.

substantivity (of antiseptics) See SKIN SUBSTANTIVITY.

substrate-accelerated death The phenomenon in which a population of starving bacteria (suspended in a non-nutrient medium) exhibits a very high death rate when provided with more of the energy source (substrate) whose depletion had led to starvation. In at least some cases the phenomenon is associated with a sharp fall in the intracellular concentration of cyclic AMP, and is inhibited by magnesium ions.

substrate-level phosphorylation The phosphorylation of ADP to ATP, or the phosphorylation of other nucleoside diphosphates, by a process which occurs in the cytoplasm and in which the energy is derived, directly, from an energy-yielding chemical reaction; the latter reaction must be characterized by a yield of free energy greater than that required for phosphorylation. (cf. OXIDATIVE PHOSPHORYLATION.) Substrates involved in substrate-level phosphorylation include 'high-energy' phosphates (e.g. phosphoenolpyruvate) and acyl-coenzyme A thioesters (e.g. succinyl-CoA). Examples of substrate-level phosphorylation include the pyruvate kinase and phosphoglycerate kinase reactions of the EMBDEN–MEYERHOF–PARNAS PATHWAY, the succinyl-CoA synthetase reaction of the TCA CYCLE, and the acetokinase reaction in DESULFOTOMACULUM. (See also FERMENTATION sense 1.)

substrate mycelium (primary mycelium; vegetative mycelium) In many actinomycetes: mycelium which occurs on the surface of, and which may penetrate, the medium (cf. AERIAL MYCELIUM). Spores are formed on the substrate mycelium e.g. in species of *Micromonospora* (see also STREPTOMYCES). [Book ref. 73, pp. 166–169.]

subterranean clover mottle virus See VELVET TOBACCO MOTTLE VIRUS.

subterranean clover red leaf virus See LUTEOVIRUSES.

subterranean clover stunt virus See LUTEOVIRUSES.

subtilin An antibiotic, structurally related to NISIN, produced by *Bacillus subtilis*.

subtilisins Extracellular, alkaline serine proteases (EC 3.4.21.4) produced by *Bacillus* spp. Subtilisin Carlsberg, from *B. licheniformis*, is the most widely used enzyme in 'biological' washing powders (see PROTEASES). It has a broad specificity, hydrolysing most types of peptide bond and some ester bonds. It does not require Ca^{2+} for stability and hence is not inactivated by chelating agents. It is stable over a broad pH range (e.g. 6–10). Subtilisin Novo (= subtilisin BPN), produced e.g. by *B. amyloliquefaciens*, differs in specificity from, but shows extensive homology with, subtilisin Carlsberg and is more dependent on Ca^{2+} for stability. [Production and properties: Book ref. 31, pp. 57–69.]

subtilysin *Syn.* SURFACTIN.

subviral agents Acellular infectious agents which lack one or more of the essential features of a VIRUS; such agents include SATELLITE RNAS, VIROIDS, VIRUSOIDS, PRIONS and the (hypothetical) VIRINOS.

subviral particle A particle – formed by certain viruses as an intermediate in the replication cycle – which consists of the virion minus certain of its components (see e.g. REOVIRUS and BACTERIOPHAGE $\phi 6$).

succinate thiokinase See Appendix II(a).

Succinimonas A genus of Gram-negative bacteria (family BACTEROIDACEAE) which occur e.g. in the RUMEN. Cells: coccobacilli or short rods, 1.0–1.5 × 1.0–3.0 μm, which are monotrichously (polarly) flagellated. Glucose, maltose or e.g. dextrins or starch can be fermented; succinate and acetate are major products of glucose fermentation. Fermentation is anaerogenic, and butyrate is not formed. Type species: *S. amylolytica*.

Succinivibrio A genus of Gram-negative bacteria (family BACTEROIDACEAE) which occur e.g. in the RUMEN. Cells (when freshly isolated): curved or helical rods, 0.4–0.6 × 1.0–7.0 μm, with pointed ends; monotrichously (polarly) flagellated. The main products of glucose metabolism include succinate, acetate and formate; glucose fermentation occurs anaerogenically, and butyrate is not formed. The type species, *S. dextrinosolvens*, appears to be a major fermenter of dextrins in ruminants fed on high-starch diets.

succinylsulphathiazole See SULPHONAMIDES.

sucrase *Syn.* INVERTASE.

sucrose A non-reducing disaccharide: α-D-glucopyranosyl-β-D-fructofuranoside. It is produced by almost all photosynthetic eukaryotes (but not by e.g. some diatoms and some red and green algae). Sucrose can be metabolized by a range of microorganisms. Certain bacteria (e.g. *Leuconostoc* spp, *Streptococcus* spp) split sucrose into its constituent monosaccharides, one of which is metabolized while the other is polymerized to form DEXTRAN or levan (see FRUCTANS). (See also INVERTASE.) High concentrations of sucrose can inhibit microorganisms and are used e.g. in FOOD PRESERVATION.

Suctoria A subclass of freshwater and marine ciliates (class KINETOFRAGMINOPHOREA) in which the adult cells are characteristically sedentary (commonly on a non-contractile stalk), lack cilia (though they generally contain an infraciliature), and generally have few to many haptocyst-bearing tentacles. Prey organisms are ingested at the tips of the tentacles. Reproduction occurs by budding, typically with the formation of motile, ciliated larval forms which lack tentacles. (See also BROOD POUCH.) Some species are loricate. Genera include e.g. *Acineta*, *Dactylostoma*, *Dendrosoma*, *Podophrya*, *Sphaerophrya*, and *Tokophyra*. (See also PHALACROCLEPTES.)

suctorian A ciliate of the subclass SUCTORIA.

Sudan black B A blue-black diazo LYSOCHROME used (0.3% w/v in 70% ethanol) for staining intracellular lipid: see BURDON'S STAIN.

sudanophilic Having an affinity for lipid stains (e.g. SUDAN BLACK B).

sudden infant death syndrome (SIDS; *syn.* cot death, crib death) The sudden death of an infant (typically aged between ca. 3 weeks and 5 months) with no apparent cause. Various infections have been implicated as causes of SIDS, including e.g. rotavirus infection [JMV (1982) *10* 291–296] and BOTULISM, but often there is no evidence of microbial involvement.

Sudol See PHENOLS.

suffocation disease A disease of the rice plant which is apparently due to toxic levels of hydrogen sulphide produced by microorganisms in the flooded soils used for rice cultivation. Plants may be protected from sulphide by the activities of sulphide-utilizing microorganisms (e.g. *Beggiatoa*). (See also CEREAL DISEASES and SULPHATE-REDUCING BACTERIA.)

sufu A Chinese fermented soybean food resembling soft cream cheese. Soybeans are ground, boiled or steamed, strained, and a coagulant (e.g. $CaSO_4$, $MgSO_4$, acetic acid) is added to the liquid fraction to precipitate proteins. The precipitate is pressed to form a cubical cake-like curd (*tofu*), surface-inoculated with a fungus (e.g. *Actinomucor elegans*, *Mucor* spp), and incubated at e.g. 20°C for 3–7 days. The cubes are salted, rinsed, and immersed in brine containing alcohol, flavouring etc, and aged for e.g. 1–3 months. [Book ref. 8, pp. 516–520.]

sugar alcohol See POLYOL.

sugar beet yellows virus (SBYV) The type member of the CLOSTEROVIRUSES. SBYV can infect a wide range of plants; in sugar beet, symptoms include yellowing and necrosis spreading inwards from the leaf tips and margins.

sugarcane diseases Diseases of sugarcane (*Saccharum officinarum*) include e.g. FIJI DISEASE and RED ROT. [Details of various sugarcane diseases classified according to causal agent: Phytopathol. Paper number 29, February 1986.]

sugarcane Fiji disease virus See FIJIVIRUS.

sugarcane mosaic virus See POTYVIRUSES.

sugars See PEPTONE-WATER SUGARS.

suicide gene Any gene whose expression in a cell is lethal for that cell.

suicide substrate Any substrate whose uptake or metabolism by a cell is lethal for that cell.

suid (alpha) herpesvirus 1 See ALPHAHERPESVIRINAE.

suid herpesvirus 1 group See ALPHAHERPESVIRINAE.

suid herpesvirus 2 See BETAHERPESVIRINAE.

Suipoxvirus (swinepox subgroup) A genus of viruses of the CHORDOPOXVIRINAE; the genus includes the swinepox virus (see SWINE POX).

***sulA* gene** See SOS SYSTEM.

***sulB* gene** See SOS SYSTEM.

sulbactam A β-LACTAMASE INHIBITOR. (See also SULTAMICILLIN.)

sulconazole See AZOLE ANTIFUNGAL AGENTS.

sulcus A groove: see e.g. DINOFLAGELLATES.

sulfate- See sulphate-.

sulfazecin See MONOBACTAMS.

Sulfolobus A genus of thermophilic, aerobic and facultatively anaerobic sulphur-dependent prokaryotes (domain ARCHAEA, formerly Archaebacteria) which occur e.g. in certain hot springs (see also COAL BIODEGRADATION). Cells: cocci, coccoid or irregularly shaped forms; the CELL WALL is an S LAYER. Growth occurs between ca. 50 and 90°C, at pH ca. 1–5. Many strains are facultatively CHEMOLITHOAUTOTROPHIC (q.v.), but some are obligately heterotrophic. Energy is obtained by the oxidation of sulphur (or Fe^{2+}) and/or by SULPHUR RESPIRATION. CO_2 is assimilated via a reductive tricarboxylic acid pathway. Species include *S. acidocaldarius*, *S. ambivalens* and *S. solfataricus*. (See also REVERSE GYRASE and SULPHUR CYCLE.)

sulfur- See sulphur-.

sulpha drugs (1) *Syn.* SULPHONAMIDES. (2) The sulphonamides and the SULPHONES.

sulphacetamide See SULPHONAMIDES.

sulphadiazine 4-Amino-*N*-2-pyrimidinylbenzenesulphonamide: one of the more effective and less toxic of the antibacterial SULPHONAMIDES; it is rapidly absorbed from the gut, and is effective against e.g. streptococci, *Neisseria* spp, *Escherichia coli*, etc. Sodium sulphadiazine may be given intravenously e.g. in the treatment of meningococcal meningitis; silver sulphadiazine (see SILVER) is used topically in the treatment of burns.

sulphadimethoxine See SULPHONAMIDES.

sulphadimidine See SULPHONAMIDES.

sulphadoxine A SULPHONAMIDE which has been used (e.g. in combination with pyrimethamine) for the treatment of clinically uncomplicated, chloroquine-resistant MALARIA. Sulphadoxine and pyrimethamine act synergistically to block (i) the synthesis of dihydrofolate (DHF) and (ii) the reduction of DHF to tetrahydrofolate (THF). Resistance to sulphadoxine may arise through the effects of mutation on the target enzyme [Eur. J. Bioch. (1994) *224* 397–405].

sulphaguanidine See SULPHONAMIDES.

sulphamethazine See SULPHONAMIDES.

sulphamethizole See SULPHONAMIDES.

sulphamethoxazole See SULPHONAMIDES.

sulphamylon *Syn.* MARFANIL.

sulphanilamide See SULPHONAMIDES.

sulphate assimilation See ASSIMILATORY SULPHATE REDUCTION.

sulphate-reducing bacteria A phrase traditionally used to refer to a (non-taxonomic) category of strictly anaerobic bacteria which are characterized by their ability to carry out DISSIMILATORY SULPHATE REDUCTION. The organisms include cocci, rods and filaments, most stain Gram-negatively, and the cells of one genus (DESULFOTOMACULUM) form endospores. They occur e.g. in soil; in anaerobic muds and sediments in freshwater, brackish and marine habitats; in the RUMEN, and in the intestinal tract in man and other animals; and in sewage. Some species tolerate or need salt (NaCl); the spore-forming sulphate-reducing bacteria tolerate a maximum of ca. 2% NaCl. In vitro culture requires the use of pre-reduced media in which the REDOX POTENTIAL is more negative than ca. −100 mV.

Substrates used as electron donors include e.g. lactate, propionate, pyruvate and certain alcohols; according to species these may be oxidized to acetate and CO_2 or (completely) to CO_2. [Use of amino acids as energy substrates by marine strains of *Desulfovibrio*: FEMS Ecol. (1985) *31* 11–15.] Electron acceptors include sulphate, sulphite (see also DESULFOVIRIDIN), thiosulphate and tetrathionate; such oxidized compounds of sulphur may be derived from biological activity (see e.g. SULPHURETUM) or from the autoxidation of sulphide [FEMS Ecol. (1985) *31* 39–45]. At least some strains can use nitrate and/or fumarate as electron acceptors instead of oxidized sulphur compounds.

Some strains can grow mixotrophically by oxidizing H_2 (see e.g. DESULFOVIBRIO), some can gain energy from pyrophosphate (see DESULFOTOMACULUM), and many can derive energy from pyruvate via the PHOSPHOROCLASTIC SPLIT. In the absence of an oxidized sulphur compound some species exhibit fermentative metabolism, although carbohydrates are rarely used. (See also FERMENTATION sense 1.)

The activities of the sulphate-reducing bacteria are responsible for much of the H_2S produced in organically polluted waters. In rice paddies H_2S can be beneficial (e.g. by killing parasitic nematodes) but high concentrations of H_2S are phytotoxic and can damage crops (see also SUFFOCATION DISEASE); nitrate (which inhibits sulphate reduction) has been used to control this problem. Sulphide from these bacteria is often at least partly responsible for the corrosion of underground ferrous pipes (see also CATHODIC DEPOLARIZATION THEORY and TUBERCLE (sense 2)) and for the degeneration of concrete and stonework; metals can be affected by the sulphide itself, but damage to concrete and stone derives primarily from the action of sulphuric acid produced from the sulphide by e.g. thiobacilli.

[Microbial sulphate reduction in the early Archaean era: Nature (2001) *410* 77–81.]

(See also DESULFOBACTER, DESULFOBULBUS, DESULFOCOCCUS, DESULFOTOMACULUM, DESULFOMONAS, DESULFONEMA, DESULFOSARCINA, DESULFOVIBRIO.)

[Book ref. 102.]

sulphate reduction The microbial reduction of sulphate may involve ASSIMILATORY SULPHATE REDUCTION or DISSIMILATORY SULPHATE REDUCTION. (See also SULPHUR CYCLE.)

sulphate respiration *Syn.* DISSIMILATORY SULPHATE REDUCTION.

sulphathiazole See SULPHONAMIDES.

sulphide See HYDROGEN SULPHIDE.

sulphite fermentation See NEUBERG'S FERMENTATIONS.

sulphonamides A group of chemotherapeutic agents. The antimicrobial sulphonamides are mostly derivatives of *sulphanilamide* (*p*-aminobenzenesulphonamide: $NH_2.C_6H_4.SO_2.NH_2$) (cf. MARFANIL); they are microbistatic for a wide range of Gram-positive and Gram-negative bacteria, and for various protozoa (e.g. coccidia, *Plasmodium* spp). Sulphonamides are used – usually in combination with other chemotherapeutic agents (see e.g. FOLIC ACID ANTAGONISTS) – for treating certain urinary tract infections, MALARIA (q.v.), coccidioses, etc.

Sulphonamides act as structural analogues of *p*-aminobenzoic acid (PABA), competitively inhibiting the incorporation of PABA during the formation of dihydropteroic acid in FOLIC ACID synthesis. Those organisms which synthesize their own folic acid and which cannot use an exogenous supply of the vitamin are sensitive to sulphonamides (provided that the cells are permeable to the drugs), while organisms which require exogenous folic acid for growth are insensitive. A lag period of several generations occurs between exposure of sensitive cells to sulphonamide and growth inhibition; during this time the cells exhaust their pre-formed supply of endogenous folic acid. This delayed effect allows sulphonamides to be used in combination with antibiotics (e.g. penicillins) which are active only against growing organisms.

The inhibitory effect of sulphonamides may be neutralized by supplying the cells with those metabolites which normally require folic acid for their synthesis (e.g. purines, certain amino acids); such substances may be present e.g. in pus, so that sulphonamides may be ineffective in the treatment of certain suppurative infections. Bacteria readily develop resistance to sulphonamides: e.g., in *Streptococcus pneumoniae* modification of dihydropteroic acid synthetase by a single-step mutation may reduce the affinity of the enzyme for sulphonamides without significantly reducing its affinity for PABA. Plasmid-borne resistance also occurs, and can involve e.g. a plasmid-encoded sulphonamide-resistant dihydropteroic acid synthetase. (See also entries Tn*21* and Tn*2410*.)

The many sulphonamides differ e.g. in their clinical properties, toxicities, etc. Most are derivatives bearing substituents at the nitrogen of the sulphonamide group ($NH_2.C_6H_4.SO_2.NHR$). Substitution at the *p*-amino group results in the loss of antibacterial activity; however, such derivatives may be hydrolysed in vivo to an active derivative. For example, *p-N*-succinylsulphathiazole and phthalylsulphathiazole are inactive and are poorly absorbed from the gut, but they are hydrolysed in the lower intestine to release the active component sulphathiazole; these drugs have been used e.g. pre- and post-operatively in abdominal surgery.

Sulphonamides include e.g. *sulphacetamide* (*N*-[(4-aminophenyl)sulphonyl]-acetamide); SULPHADIAZINE; *sulphadimethoxine* (4-amino-*N*-(2,6-dimethoxy-4-pyrimidinyl)benzenesulphonamide); *sulphadimidine* (= *sulphamethazine*: 4-amino-*N*-(4,6-dimethyl-2-pyrimidinyl)benzenesulphonamide); *sulphaguanidine* (4-amino-*N*-(aminoiminomethyl)benzenesulphonamide); *sulphamethizole* (4-amino-*N*-(5-methyl-1,3,4-thiadiazol-2-yl) benzenesulphonamide); *sulphamethoxazole* (4-amino-*N*-(5-methyl-3-isoxazolyl)benzenesulphonamide); *sulphathiazole* (4-amino-*N*-2-thiazolylbenzenesulphonamide); etc.

(See also PAS; PRONTOSIL; SULPHONES.)

sulphonation See WATER SUPPLIES.

sulphones Antibacterial agents (general formula: R_2SO_2) active against a range of species; owing to toxicity they are not widely used in medicine. An example is DAPSONE.

sulphorhodamine 101 acid chloride See TEXAS RED.

sulphorhodamine B See LISSAMINE RHODAMINE.

sulphur (as an ANTIFUNGAL AGENT) Elemental sulphur is generally effective against certain plant diseases of fungal aetiology, e.g., apple scab, powdery mildews, black spot of roses; it may be applied as a fine dust, as a WETTABLE POWDER, or in colloidal form. The degree of fungitoxicity is related to the size of the sulphur particles, the smallest particles generally being the most toxic.

Sulphur may be applied as *polysulphide*: a product (formed by the reaction of certain metal sulphides with sulphur under alkaline conditions) which, when acidified e.g. by atmospheric CO_2, liberates elemental sulphur. An early preparation, *liver of sulphur* (potassium polysulphide), was made by the fusion of potassium hydroxide and sulphur. *Lime sulphur* is prepared by boiling sulphur with lime – the resulting solution containing calcium thiosulphate and calcium polysulphide; it tends to be phytotoxic and is incompatible with many chemical pesticides owing to its alkalinity. *Green sulphur* contains finely divided sulphur and certain iron compounds; it is formed as a by-product of a coal-gas purification process in which hydrogen sulphide reacts with ferric oxide.

Elemental sulphur can cause premature defoliation and dropping of fruit in certain varieties of fruit tree – a phenomenon called *sulphur shyness*.

sulphur bacteria Those bacteria which contribute to stage(s) of the SULPHUR CYCLE *other than* assimilatory sulphate reduction and the decomposition of *organic* sulphur compounds to sulphide; they include the SULPHATE-REDUCING BACTERIA, certain photosynthetic bacteria, and species/strains of *Beggiatoa*, *Desulfuromonas*, *Oscillatoria* and *Thiobacillus*. (cf. COLOURLESS SULPHUR BACTERIA.) [Biochemistry and ecology of the sulphur bacteria: PTRSLB (1982) *298* 429–602.]

(See also SULPHUR-DEPENDENT ARCHAEANS.)

sulphur cycle In nature: the cyclical interconversion of sulphur and its compounds. The figure shows some of the interconversions carried out by bacteria (see also SULPHUR-DEPENDENT ARCHAEANS).

In addition to biological interconversions, sulphide can be oxidized by *autoxidation* [see e.g. FEMS Ecol. (1985) *31* 39–45] to e.g. elemental sulphur, thiosulphate and sulphate.

Sulphur is required e.g. for the synthesis of certain amino acids, ferredoxins and cofactors such as coenzyme A. Many bacteria assimilate sulphur in the form of sulphate – which is usually available in adequate quantities in the environment; the sulphate is initially reduced to sulphide (see ASSIMILATORY SULPHATE REDUCTION).

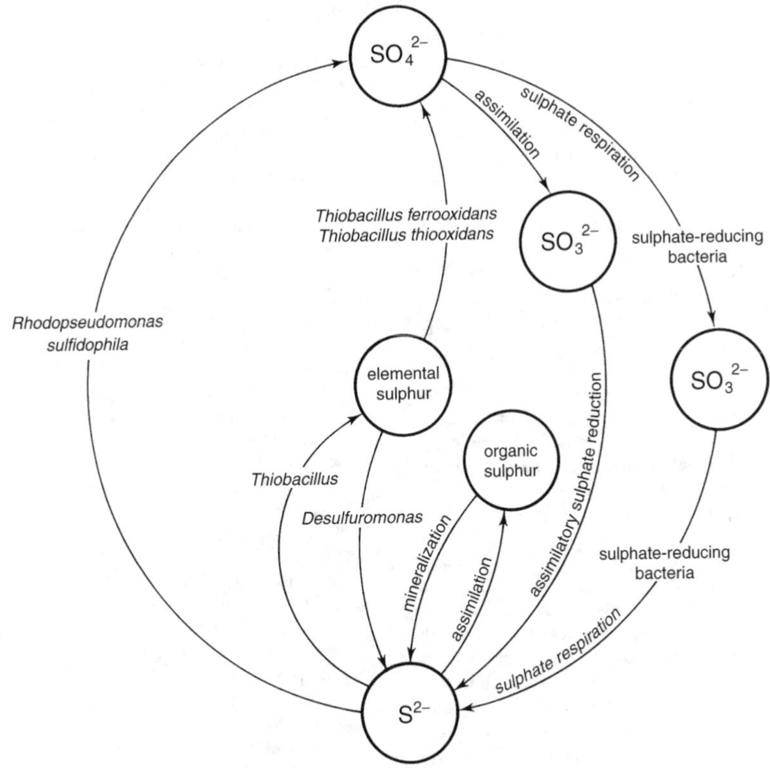

SULPHUR CYCLE: some interconversions carried out by bacteria. Many species can use sulphate as a source of sulphur (shown here as assimilatory sulphate reduction). The sulphate is reduced, intracellularly, to sulphide, and the sulphide is incorporated in different ways by different species; in e.g. *Escherichia coli*, sulphide is incorporated into *O*-acetylserine to form cysteine.

In a dissimilatory mode, sulphate respiration is carried out e.g. by the 'sulphate-reducing bacteria' – sulphate being used as terminal electron acceptor in anaerobic respiration. Elemental sulphur is used as terminal electron acceptor by *Desulfuromonas. Thiobacillus* spp carry out aerobic respiration in which e.g. sulphide or elemental sulphur is oxidized. *Rhodopseudomonas sulfidophila* is one of a number of species which use sulphide as an electron donor in anaerobic phototrophic metabolism.

Elemental sulphur and sulphur compounds can also be used in a dissimilatory mode (see e.g. DISSIMILATORY SULPHATE REDUCTION and SULPHUR RESPIRATION).

A range of compounds, not shown in the figure, are associated with certain stages of the sulphur cycle. For example, thiosulphate and polythionates are thought to be important metabolites in the bacterial oxidation of pyrite (metal sulphides) [AEM (1996) *62* 3424–3431]. [Metabolism of sulphonates and sulphate esters by Gram-negative bacteria: FEMS Reviews (2000) *24* 135–175.]

[Antiquity of the biological sulphur cycle (evidence from sulphur and carbon isotopes in 2700-million-year-old rocks of the Belingwe Belt, Zimbabwe): Proc. RSLB (2001) *268* 113–119.]

See: BEGGIATOA; CHLOROBIACEAE; CHROMATIACEAE; HYDROGEN SULPHIDE; HYDROTHERMAL VENT; MINERALIZATION; OSCILLATORIA; RHODANESE; RHODOSPIRILLACEAE; SULFOLOBUS; SULPHATE-REDUCING BACTERIA; SULPHUR BACTERIA; SULPHUR-OXIDIZING BACTERIA; SULPHUR-REDUCING BACTERIA; SULPHURETUM; THIOBACILLUS.

(See also CARBON CYCLE and NITROGEN CYCLE.)

sulphur-dependent archaeans Collectively, members of the order THERMOPROTEALES and of the genus SULFOLOBUS, many of which can obtain energy only by the metabolism of elemental sulphur (S^0). (See also SULPHUR RESPIRATION.) [Review: Book ref. 157, pp. 85–170.]

sulphur dioxide (as antimicrobial agent) Sulphur dioxide (SO_2) has antibacterial and antifungal properties which are believed to depend on its ability to reduce disulphide bonds in enzymes. It is used for FUMIGATION and as a food PRESERVATIVE in e.g. fruit juices and pulps, jams, beers and wines; it is also used as an antioxidant to prevent browning in e.g. fruit and vegetables. In WINE-MAKING, SO_2 inhibits the growth of contaminating yeasts and bacteria – the wine-making yeasts being relatively resistant; it also protects wine from excessive oxidation during storage, and combines with acetaldehyde (produced during fermentation) which has undesirable organoleptic properties.

Acid solutions of sulphites, bisulphites, and metabisulphites (e.g. sodium metabisulphite, $Na_2S_2O_5$) – all of which release SO_2 – are often used in place of gaseous SO_2. (See also FOOD PRESERVATION.)

sulphur granules See ACTINOMYCOSIS (1).

sulphur mustards Structural and functional analogues of the NITROGEN MUSTARDS – e.g. mustard gas: $Cl(CH_2)_2.S.(CH_2)_2Cl$.

sulphur-oxidizing bacteria Bacteria which can use elemental sulphur (S^0) as an electron donor in lithotrophic metabolism – e.g. certain photosynthetic species and some species of THIOBACILLUS.

sulphur polypore *Laetiporus sulphureus.*

sulphur-reducing bacteria Bacteria which can use elemental sulphur (S^0) as terminal electron acceptor in anaerobic respiratory metabolism – e.g. species of DESULFUROMONAS. (See also SULPHUR RESPIRATION.)

sulphur respiration Energy-yielding metabolism in which elemental sulphur (S^0) is used anaerobically as terminal electron acceptor. Such metabolism is found in certain bacteria (see DESULFUROMONAS) and in some archaeans (e.g. many strains of SULFOLOBUS and species of the THERMOPROTEALES); in at least some species of *Thermoproteus*, cystine or malate can be used instead of sulphur.

sulphur shyness See SULPHUR.

sulphur tuft *Hypholoma fasciculare.*

sulphuretum A temporary, or more or less stable, region of e.g. a pond or a marine sediment containing a number of metabolically interrelated organisms which, collectively, metabolize sulphur and its compounds through cycles of oxidation and reduction. For example, a pond polluted by leaf litter would initially support a population of aerobic bacteria which, by their mineralizing activities (see MINERALIZATION), enrich the water with inorganic salts, and – by using oxygen – create an anaerobic environment in the lower layers of the pond. Sulphide, formed by SULPHATE-REDUCING BACTERIA, may then be oxidized in the upper (aerobic or microaerobic) water by e.g. species of *Beggiatoa* and *Thiobacillus*, and in the lower (anaerobic) water by e.g. photosynthetic bacteria such as *Chromatium*; sulphate (or other oxidized compounds of sulphur) released by these organisms may then be re-used by the sulphate-reducers. Eventually, through e.g. loss of H_2S to the atmosphere, sulphate supplies become exhausted and the cycling of sulphur ceases.

sultamicillin (VD-1827) An antibiotic formed by linking a molecule of penicillanic acid sulphone (sulbactam) to one of ampicillin by a hydrolysable ester bridge.

Sulzberger–Chase phenomenon General, systemic, immunological unresponsiveness to challenge by certain types of antigen as a consequence of the prior oral administration of those antigens; the phenomenon appears to be due, at least in part, to the enhancement of suppressor T cell activity. Despite systemic unresponsiveness, antigenic challenge may lead to enhanced antibody secretion at mucosal sites.

summer mastitis See MASTITIS.

summer spore (*mycol.*) A type of spore (e.g. a CONIDIUM), typically produced in large numbers in the season of rapid growth, which can germinate soon after its formation. (cf. RESTING SPORE.)

summer truffles See TRUFFLES.

sun animalcules See HELIOZOEA.

***sunY* gene** See SPLIT GENE (e).

***sup* genes** See SUPPRESSOR MUTATION.

superantigen (SAg) A protein which can bind *simultaneously* to (i) the T cell receptor (TCR) of a T lymphocyte (T cell) that expresses a particular Vβ chain (see T CELL RECEPTOR), and (ii) an MHC class II molecule on an antigen-presenting cell (e.g. a macrophage or B cell); such binding, in which the superantigen acts as a cross-link between the Vβ chain and the class II molecule, causes (i) T cell proliferation, and (ii) release of cytokines from the T cell and from the antigen-presenting cell. Because many different strains of T cell express a given Vβ chain, a superantigen causes *polyclonal* activation of T cells (as though the corresponding range of antigens had been present); this is accompanied by a massive release of cytokines. During the joint stimulation of a T cell and a macrophage, the cytokines released include IL-1, IL-2, IL-4, IL-6, IL-8, TNF-α, TNF-β and IFN-γ.

Superantigens are able to activate both $CD4^+$ and $CD8^+$ cells, binding to the Vβ chain at a site outside the normal antigen-binding region. Activation and proliferation of T cells naturally gives rise to an initial increase in the number of T cells of the given Vβ subset; however, there is evidence to suggest that many of the T cells which bind superantigen either become unresponsive (= *T cell anergy*) or undergo apoptosis. Thus, while the T cell population of a given subset may initially rise, there may be a net loss of functional cells of that subset in the longer term.

Superantigens include the five major ENTEROTOXINS of *Staphylococcus aureus* (SEA, SEB, SEC, SED and SEE), the toxic shock syndrome toxin 1 (TSST-1; formerly enterotoxin F) of *S. aureus*, and the exotoxin A of *Streptococcus pyogenes* (= streptococcal superantigen A; SSA). Different superantigens bind preferentially to different Vβ chains; for example, TSST-1 binds to Vβ type 2, SED binds to Vβ 5 and 12, and SEA binds to Vβ types 1, 5, 6, 7 and 9.

Superantigens are apparently associated with some viruses, e.g. rabies virus.

The ability of a superantigen to induce massive release of cytokines is apparently involved in the pathogenesis of TOXIC SHOCK SYNDROME. It has been suggested that this mechanism may also be responsible for at least some of the symptoms of certain other diseases (such as KAWASAKI DISEASE).

'B cell superantigens' are analogous molecules; they bind to the V_H chain of the B cell receptor.

[Clinical significance of T cell superantigens: TIM (1998) *6* 61–65.]

superchlorination See WATER SUPPLIES.

supercoiled DNA See DNA.

supercoiling-induced duplex destabilization See SIDD.

superelongation disease (of cassava) See GIBBERELLINS.

superhelical DNA See DNA.

superhelix density (σ) See DNA.

superhelix winding number (τ) See DNA.

superinfection exclusion See SUPERINFECTION IMMUNITY.

superinfection immunity A phenomenon in which a virus-infected cell can resist infection by a second virus of the same or a similar type. The second virus may be prevented from adsorbing to or penetrating the cell (*superinfection exclusion*), or it may enter the cell but be unable to replicate. For example, if a bacterial cell contains a λ prophage (see LYSOGENY), any incoming λ genome will immediately be subject to repression by the repressor encoded by the incumbent prophage. When phage P22 establishes a lysogenic infection in *Salmonella typhimurium*, it causes glycosylation of the LPS of the host cell, resulting in modification of the P22 receptor sites and preventing further infection by P22. These are examples of *lysogenic* (or *prophage*) *immunity*; superinfection immunity may also be exhibited by lytic phages such as T4. (See also e.g. RETROVIRIDAE.)

Phages which belong to the same 'immunity group' (i.e., one cannot infect a cell already infected by any of the others in the group) are said to be *homoimmune*; phages of different immunity groups are said to be *heteroimmune*. (See e.g. LAMBDOID PHAGES.)

(The term superinfection immunity has also been applied to some cases of INCOMPATIBILITY in plasmids.)

superoxide (superoxide radical) A term used to refer either to O_2^- (superoxide anion, also written $\cdot O_2^-$ or $O_2^{\bar{}}$ etc) or to the more highly oxidizing species $HO_2\cdot$; these two species are interconvertible, O_2^- being the conjugate base of the weak acid $HO_2\cdot$ (pK_a 4.8). The conversion of O_2^- to $HO_2\cdot$ is encouraged e.g. by low pH, and is influenced by the dielectric constant of the microenvironment.

O_2^- is generated, under aerobic conditions, in both prokaryotic and eukaryotic cells e.g. during autoxidation of reduced cytochrome *c* (or ferredoxins) and during oxidations catalysed by e.g. aldehyde oxidase, dihydroorotate dehydrogenase, or xanthine oxidase (see also PYOCYANIN); it is also formed during the 'respiratory burst' in activated NEUTROPHILS. The generation of O_2^- can also occur extracellularly; thus, e.g., suspensions of *Escherichia coli* can take up and reduce paraquat, the reduced paraquat (which is released into the medium) undergoing autoxidation and giving rise to O_2^- [JBC (1979) *254* 10846–10852].

O_2^- does not act indiscriminately on cell constituents; however, it inactivates CATALASE and it reacts with the NADH–lactate dehydrogenase complex and e.g. with catechols and oxyhaemoglobin. The protonated form, $HO_2\cdot$, is significantly more reactive. In aerobic organisms, and in most aerotolerant anaerobes, the potentially harmful effects of endogenously formed O_2^- are prevented by the action of intracellular SUPEROXIDE DISMUTASE. [Superoxide as an endogenous toxicant: ARPT (1983) *23* 239–257.] Superoxide and other reduced oxygen species, e.g. HYDROXYL RADICAL, are produced e.g. during PHAGOCYTOSIS and are generally believed to function as antimicrobial agents. (See also SINGLET OXYGEN.)

[Oxygen toxicity and oxygen radicals (review): Bioch. J. (1984) *219* 1–14.]

superoxide anion See SUPEROXIDE.

superoxide dismutase (SOD; superoxide oxidoreductase; EC 1.15.1.1) Any of a range of metalloenzymes which catalyse the dismutation of SUPEROXIDE (e.g. $HO_2\cdot + O_2^- + H^+ \rightarrow H_2O_2 + O_2$); SODs occur widely in both prokaryotic and eukaryotic cells and are believed to have as their main or sole function the protection of aerobic or aerotolerant organisms from the toxic effects of superoxide. (SODs occur in some anaerobic organisms, e.g., *Bacteroides melaninogenicus*.)

The SODs of prokaryotes characteristically contain iron or manganese (in e.g. *Propionibacterium shermanii* Fe and Mn appear to be interchangeable [JBC (1982) *257* 13977–13980]) and are typically insensitive to cyanide, while those of eukaryotes characteristically contain copper and zinc and are typically sensitive to cyanide; however, a bacterial CuZnSOD (*bacteriocuprein*) occurs e.g. in *Caulobacter crescentus* [JBC (1982) *257* 10283–10293], in both *Pseudomonas diminuta* and '*P. maltophila*' [JB (1985) *162* 1255–1260], and (possibly as a result of gene transfer from its eukaryotic host) in *Photobacterium leiognathi* [PNAS (1985) *82* 149–152], while MnSODs occur e.g. in the cytosol of human liver cells and in the mitochondrial matrix.

SODs have been reported to be inducible (in both prokaryotes and eukaryotes) e.g. by increases in intracellular superoxide anion or by increases in the partial pressure of oxygen; however, in *Escherichia coli* – which synthesizes a constitutive, periplasmic (*sodB*-encoded) FeSOD and an inducible, cytoplasmic (*sodA*-encoded) MnSOD – the intracellular concentration of superoxide anion appears not to regulate the oxygen-dependent synthesis of the MnSOD [PNAS (1984) *81* 4970–4973].

In *Lactobacillus plantarum* (in which SODs are absent) the dismutation of O_2^- is carried out by manganese – a metal accumulated to an intracellular concentration of ca. 25 mM; the low efficiency of manganese as a 'dismutase' appears to be offset by its high intracellular concentration. (*L. plantarum* becomes oxygen-sensitive in manganese-deficient media.) Certain other organisms (e.g. *Mycoplasma pneumoniae* [BBRC (1980) *96* 98–105]) lack SODs but their functional equivalent(s) (if any) are unknown.

[Varieties, distribution, physiological functions of SODs: ARPT (1983) *23* 239–257 (pp. 244–249); SOD, an evolutionary puzzle: PNAS (1985) *82* 824–828; SOD mutants of *Escherichia coli*: EMBO (1986) *5* 623–630.]

superoxide radical *Syn.* SUPEROXIDE.

superoxol test A form of CATALASE TEST in which 20% or 30% H_2O_2 is used in place of the more usual 3%. It may be used e.g. in the presumptive identification of *Neisseria gonorrhoeae* [BJVD (1984) *60* 87–89].

super-repressor See e.g. LAC OPERON.

supertwisted DNA Supercoiled DNA.

supervital staining See VITAL STAINING.

support film See ELECTRON MICROSCOPY (a).

suppressive petite See PETITE MUTANT.

suppressive soils Soils which suppress the survival, growth, and/or pathogenic activity of particular plant-pathogenic fungi. In at least some cases suppression is due to the presence of an antagonistic microflora: e.g., mycophagous amoebae may be involved in the suppression of TAKE-ALL [SBB (1984) *16* 197–199].

suppressor mutation (suppressor) A MUTATION which alleviates the effects of a previous (*primary*) mutation at a different locus ('second-site reversion'); commonly, the wild-type phenotype is only partially restored, resulting in a *pseudo-wild* phenotype. (cf. BACK MUTATION.)

In *intragenic suppression*, both primary and suppressor mutations occur in the same gene. For example: (a) A (primary) MIS-SENSE MUTATION – which results in an inactive gene product – can be countered by a second (suppressor) mutation elsewhere in the same gene which allows the gene product to adopt a conformation capable of at least some activity. (b) A primary FRAMESHIFT MUTATION may be suppressed by a second nucleotide deletion/insertion which restores the correct reading frame; the nucleotide sequence between the two mutations will still be read out of phase.

In *intergenic suppression* the suppressor mutation occurs in a gene (designated *sup*) other than that containing the primary mutation. A *direct* intergenic suppressor (*informational suppressor*) usually affects a tRNA molecule and results in e.g. the insertion of an amino acid at the site of a nonsense mutation (*nonsense suppressor*) or a different amino acid at the site of a mis-sense mutation (*mis-sense suppressor*). The activity of the resulting gene product will depend on the suitability of the amino acid inserted. Various types of direct intergenic suppressor are known. For example: (a) A mutation in a tRNA gene may alter the anticodon of the tRNA such that it recognizes a codon containing a nonsense mutation or a mis-sense mutation. Thus, e.g., an *ochre suppressor* in a gene for $tRNA^{Gln}$ may alter the anticodon from NUG to NUA (N = nucleotide), allowing the tRNA to recognize the ochre codon UAA instead of a glutamine codon; thus glutamine is inserted at the site of an ochre mutation, and premature polypeptide chain termination is avoided. (Owing to 'wobble pairing' – see WOBBLE HYPOTHESIS – an ochre suppressor can also suppress an amber mutation, albeit more weakly; however, amber suppressors can suppress only amber mutations.) (b) A mutation in a tRNA gene may alter the nucleotide sequence of the tRNA at a site other than the anticodon; the resulting change in tRNA structure may either interfere with the accuracy of codon recognition (e.g., a nucleotide change in the D arm of $tRNA^{Trp}$ allows it to recognize UGA (opal) and UGU (Cys) as well as the correct codon UGG), or may cause the tRNA to be recognized by the wrong aminoacyl-tRNA synthetase and hence to be charged with the wrong amino acid. (c) Intergenic suppressors of *frameshift mutations* may involve e.g. the insertion of an extra nucleotide in the anticodon of a tRNA such that it reads a quadruplet instead of a triplet codon, restoring the correct reading frame in the case of a +1 frameshift. Alternatively, the structure of a tRNA may be altered so that it interferes with the reading of an adjacent codon by another tRNA when both tRNAs are occupying the A and P sites of the ribosome. In either case the sequence between the primary mutation and the site of action of the mutant tRNA will still be read out of phase.

In general, an intergenic suppressor can function only if it occurs in a gene present in more than one copy per cell, or in a gene encoding one of a family of tRNAs which recognize the same codon(s) (i.e. enough normal tRNA must be present for normal translation to occur). Intergenic suppressors operate with varying degrees of efficiency, depending not only on the nature of the suppressor (e.g. amber suppressors are generally much more efficient than ochre suppressors) but also e.g. on the context in which the primary mutation occurs. This may, in part, explain why e.g. a nonsense suppressor does not necessarily seriously impair normal polypeptide chain termination, since natural termination codons may occur in a distinct context which allows them to operate even in the presence of nonsense suppressors. (cf. READTHROUGH.) [Precision in protein synthesis: Book ref. 60, pp. 23–63.]

Indirect suppression involves the bypass of the effects of a primary mutation: e.g., the product of the gene containing the suppressor mutation may activate an alternative metabolic pathway which may bypass that blocked by the primary mutation.

(cf. PHENOTYPIC SUPPRESSION.)

suppressor-sensitive mutation As commonly used: an amber mutation, i.e., a mutation which can be suppressed by an intergenic amber SUPPRESSOR MUTATION.

suppuration The formation and/or discharge of PUS.

Suprapylaria A category ('section') of LEISHMANIA species which multiply in the midgut of the arthropod vector and then migrate to the pharynx, cibarium and proboscis; it includes e.g. *L. aethiopica*, *L. donovani*, *L. major*, *L. mexicana* and *L. tropica*. (cf. HYPOPYLARIA, PERIPYLARIA.)

supravital staining See VITAL STAINING.

suramin A drug, 8-(3-benzamino-4-methyl-benzamino)naphthalene-1,3,5-trisulphonic acid, used e.g. for the treatment and prophylaxis of TRYPANOSOMIASIS; it cannot pass the blood–brain barrier and is thus not effective in the later stages of the disease.

In the alternative pathway of COMPLEMENT FIXATION, suramin blocks the inactivation of C3b by Factors H and I; it apparently acts on the C3b molecule, making it resistant to cleavage by Factors H and I.

Suramin inhibits the reverse transcriptase of a number of animal retroviruses, including the AIDS virus [review: Antiviral Res. (1987) *7* 1–10]; however, its potential in the treatment of AIDS is probably limited [see e.g. AIM (1986) *105* 32–37].

surface electrophoresis See ELECTROPHORESIS.

surface exclusion (entry exclusion) The phenomenon in which the presence of a conjugative plasmid in a cell reduces the ability of that cell to receive an identical or related plasmid, by CONJUGATION (sense 1b), from another cell. (cf. SUPERINFECTION IMMUNITY.) In the *Escherichia coli*/F PLASMID conjugal system, surface exclusion involves the products of genes *traS* (a cytoplasmic membrane protein) and *traT* (an OUTER MEMBRANE protein); the *traT* product appears to inhibit stable contact between donor cells. The effect of surface exclusion may be lost if e.g. donor cells progress to the stationary phase of growth or are starved of amino acids; such cells, which can act as recipients, are called *F^- phenocopies*. (Cells which express the *traT* gene are, incidentally, less susceptible to PHAGOCYTOSIS and to SERUM KILLING; resistance to serum killing apparently derives from the ability of the TraT protein to inhibit the assembly or action of the *membrane attack complex* (see COMPLEMENT FIXATION) [JGM (1985) *131* 1511–1521].)

An analogous effect appears to occur in cells of *Enterococcus faecalis* carrying the conjugative plasmid pCF-10; when these donor cells are exposed to recipient PHEROMONES they form a cell-surface protein which inhibits their ability to act as recipients for a derivative of pCF-10 [PNAS (1985) *82* 8582–8586].

surface-obligatory conjugation system See PILI.

surface plasmon resonance See BIACORE.

surface-preferred conjugation system See PILI.

surfactin ('subtilysin') A highly effective lipopeptide BIOSURFACTANT produced by *Bacillus subtilis*; it consists of a cyclic heptapeptide covalently linked to a hydroxy fatty acid. Surfactin is haemolytic and can also lyse the protoplasts of certain bacteria; it is not immunogenic.

Surirella See DIATOMS.

surra A disease of e.g. camels and horses which occurs in Africa (including Madagascar and Mauritius), Asia (e.g. India) and South America (cf. MAL DE CADERAS); it is caused by *Trypanosoma evansi* and is transmitted non-cyclically by biting flies (e.g. *Stomoxys*, *Tabanus*) or, in South America, by blood-sucking vampire bats (*Desmodus*). (Various animals, e.g. buffaloes, cattle, deer, sheep, may act as reservoirs of infection.) Symptoms include weakness, recurrent fever, and wasting; the disease is often fatal. *Lab. diagnosis*: e.g. indirect fluorescent antibody and haemagglutination tests. *Treatment*. In horses, the disease has been treated with e.g. arsenicals (e.g. tryparsamide), while prophylaxis has been achieved e.g. with quinapyramine-SURAMIN complexes.

surrogate light chains See B LYMPHOCYTE.

***sus* mutant** A suppressor-sensitive (amber) mutant.

suspension test A test, used to determine the efficacy of a DISINFECTANT, in which the disinfectant is allowed to act for a specified time(s) on organisms suspended in a liquid medium and/or diluent. (cf. CARRIER TEST.) In a *qualitative* suspension test (e.g. the RIDEAL–WALKER TEST) the effect of the disinfectant is determined by subculturing the test suspension to a liquid growth medium to detect the presence of viable organisms; in *quantitative* suspension tests (e.g., the FIVE-FIVE-FIVE TEST) the effect of the disinfectant is determined by counting the viable organisms remaining in the test suspension.

suspensor (*mycol.*) A clavate or conical structure formed by a PROGAMETANGIUM. In some species (e.g. *Phycomyces blakesleeanus*) the suspensor gives rise to short filaments or spines.

suture (1) A line which marks the junction of two parts or structures. (2) (*ciliate protozool.*) See SYSTÈME SÉCANT.

suzukacillin See ALAMETHICIN.

SV40 SIMIAN VIRUS 40.

SVD SWINE VESICULAR DISEASE.

Svedberg unit (S) The unit in which the *sedimentation coefficient* of a particle (e.g. a macromolecule) is commonly quoted. The sedimentation coefficient (= sedimentation constant), *s*, is given by:

$$s = \frac{dx}{dt} \cdot \frac{1}{\omega^2 r}$$

in which dx/dt is the measured sedimentation rate (in the ultracentrifuge), ω is the angular velocity (radians sec^{-1}), and r is the distance (cm) between a point within the sample and the axis of rotation; when values are substituted in the above expression, *s* is obtained in terms of reciprocal seconds. The unit is the Svedberg unit (10^{-13} sec).

swab (*microbiol.*) A device used for sampling the microflora at a given site. Swabs suitable for laboratory or clinical use typically consist of a compact piece of COTTON WOOL securely attached to one end of a thin wooden stick or a piece of wire; a *sterile* swab may be used for transferring material e.g. from the throat to a bacteriological culture medium. (See also SPREAD PLATE.) (Calcium ALGINATE wool may be used instead of cotton wool; it dissolves in e.g. aqueous sodium citrate, or in 1% sodium hexametaphosphate in quarter-strength Ringer's solution – thus releasing all organisms trapped in its fibres.) (See also MOORE SWAB.)

swamp cancer *Syn.* EQUINE PHYCOMYCOSIS.

swamp fever (equine infectious anaemia) An infectious HORSE DISEASE caused by a retrovirus (generally considered to be a member of the LENTIVIRINAE). Swamp fever occurs e.g. in Europe, Canada and the USA. It is transmitted by biting flies and mosquitoes. Incubation period: 2–4 weeks. The initial stage of the disease is typically acute, with e.g. anaemia, depression, incoordination, intermittent fever, myocarditis and splenomegaly. Animals may recover temporarily (only to relapse after 2–3 weeks with similar but usually less severe signs), or they may become progressively weaker and die after ca. 10–14 days. In survivors, relapses – which apparently result from ANTIGENIC VARIATION in virion surface glycoprotein – may continue to occur, and the animals probably remain carriers for life; viraemia occurs even during remissions. The virus replicates in and destroys host macrophages, and many of the symptoms may result from the deposition of immune complexes. *Lab. diagnosis*: e.g. an agar gel immunodiffusion test for serum precipitins. *Control*: e.g. slaughter of infected animals. [Book ref. 33, pp. 713–717.]

swan animalcule The ciliate *Lacrymaria olor*.

swan-neck flask Apparatus used by Pasteur to refute the doctrine of ABIOGENESIS. Each spherical glass flask had a long narrow neck which was bent downwards then upwards, thus inhibiting the access of airborne microorganisms to the sterilized contents of the flask.

swarm (1) (*verb*) To exhibit SWARMING (sense 1 or 2). (2) (*bacteriol.*) A motile colony formed e.g. by gliding bacteria. (3) A PSEUDOPLASMODIUM (sense 3).

swarm cell (1) *Syn.* SWARMER. (2) See e.g. SWARMING.

swarm period See e.g. DIPLANETISM.

swarm spores Motile spores.

swarmer (swarm cell) An actively motile cell which has developed from a non-motile precursor – see e.g. CAULOBACTER, RHODOMICROBIUM.

swarming (1) A phenomenon, involving morphological and functional differentiation, exhibited when certain species of bacteria (for example, *Proteus mirabilis* and *P. vulgaris*) are cultured on suitable solid media (e.g. media containing a concentration of AGAR lower than that found in normal growth media). In the case of *Proteus* spp, the first-formed progeny cells at the point of inoculation are short, sparsely flagellated bacilli ca. 2–4 μm in length, and these cells form a colony in the usual way. After several hours of growth, some of the cells at the edge of the colony grow to lengths of 20–80 μm and develop numerous flagella; these cells are called *swarm cells*. Swarm cells migrate outwards to positions a short distance from the edge of the colony; migration then stops, and each swarm cell divides into several short bacilli of the type present in the original colony. The short bacilli grow and divide normally for several generations, forming a ring of growth which surrounds, and is concentric with, the original colony. Subsequently, another generation of swarm cells is produced and the cycle is repeated. (See also DIENES PHENOMENON.)

Swarming can be prevented e.g. by culturing on stiff AGAR, on certain inhibitory media (such as MacConkey's agar) or on CLED MEDIUM. (See also PRIL.)

Swarming also occurs in other (Gram-negative and Gram-positive) bacteria, including e.g. *Escherichia coli* and *Serratia marcescens*. Unlike *P. mirabilis*, which forms concentric rings of growth due to discrete periods of swarming, some bacteria swarm continuously, forming a thin layer of growth, while others give rise to individual microcolonies.

Interestingly, although swarming in *E. coli* involves components of the CHEMOTAXIS system, the substances/signals that provoke swarming differ from the known chemoeffectors that mediate chemotaxis [PNAS (1998) *95* 2568–2573].

[Swarming (review): Mol. Microbiol. (1994) *13* 389–394.]

(2) In general, the process in which motile organisms actively spread on the surface of a suitably moist solid medium.

sweet clover necrotic mosaic virus See DIANTHOVIRUSES.

sweet curdling See MILK SPOILAGE.

sweet potato mild mottle virus See POTYVIRUSES.

swell (blower; blown can) A can (see CANNING) whose ends are distended by internal pressure due to gas production by microbial or chemical activity within the can; in a *hard swell* the ends of the can cannot be moved by thumb pressure, while in a *soft swell* the ends are movable by thumb pressure but they cannot be pressed back into their normal positions. A can becomes a FLIPPER and then a SPRINGER before becoming a swell. Organisms which can cause swells include e.g. *Bacillus polymyxa*, *Clostridium thermosaccharolyticum*, and certain non-sporing species (e.g. *Staphylococcus* and *Streptococcus* spp). (cf. HYDROGEN SWELL; PANELLING.)

swelled head In sheep, MALIGNANT OEDEMA which is often confined to the head region and which may follow fighting among rams.

swimmers' itch See LYNGBYA.

swine diseases See PIG DISEASES.

swine dysentery Infectious mucohaemorrhagic colitis involving invasion of the colonic epithelium by *Treponema hyodysenteriae* (which may require the presence, in the gut, of other bacteria – e.g. *Fusobacterium necrophorum*, *Bacteroides vulgatus*); the disease occurs world-wide. Infection occurs orally. Onset is generally insidious, with reduction in appetite, followed e.g. by fever, (typically) grey to black diarrhoea, oedema and haemorrhage. Death may occur (within days) from dehydration and toxaemia or from shock. A carrier state occurs. (See also DYSENTERY.)

swine erysipelas (diamond skin disease) An acute or chronic PIG DISEASE caused by *Erysipelothrix rhusiopathiae*. Infection occurs via the mouth or wounds; it may lead to sudden death, or symptoms may include e.g. fever, prostration, abortion, and (pathognomonic) diamond-shaped skin lesions. Affected animals may die, may recover (but remain carriers), or may develop chronic disease with e.g. progressive unthriftiness, arthritis, and valvular lesions in the heart. *Treatment*: e.g. penicillins. *Prophylaxis*: e.g. vaccination.

E. rhusiopathiae can also cause 'erysipelas' in poultry (particularly turkeys); the disease is commonly fatal. Birds may be protected by vaccination.

swine fever (European swine fever; hog cholera) A highly infectious, mild or severe PIG DISEASE caused by a virus of the genus *Pestivirus* (see TOGAVIRIDAE). Infection occurs via the mouth or nose; the incubation period is usually ca. 5–10 days. In the acute disease there is fever, lethargy, conjunctivitis with discharge, and constipation followed by diarrhoea; reddish patches appear on the skin, and there are signs of CNS involvement (convulsions, circling, incoordination, ataxia). Haemorrhagic lesions of internal organs are found post mortem. Mortality rates may be high. In the chronic disease (low-virulence virus) there may be transient fever with few other clinical signs; 'button ulcers' (1–2 cm diam.) may be found (mainly) in the large intestine post mortem. The virus may cross the placenta and infect fetuses, causing abortion or congenital tremor. (cf. AFRICAN SWINE FEVER.)

swine infertility and reproductive syndrome virus See BLUE-EARED PIG DISEASE.

swine influenza A PIG DISEASE, affecting the respiratory tract, apparently caused jointly by an influenza A virus (see INFLUENZAVIRUS) and *Haemophilus parasuis* [Book ref. 33, pp. 790–791]. Symptoms: coughing, dyspnoea and prostration. Recovery is usually rapid, but may be complicated by secondary infection with e.g. *Pasteurella multocida*. Virus commonly persists in carrier pigs and can be carried e.g. by the lungworm *Metastrongylus apri*.

swine pox A (usually benign) PIG DISEASE caused by a *Suipoxvirus*; typical vesicular pox lesions develop mainly on the abdomen. Mortality rates may be high in suckling piglets. The virus is commonly transmitted by the pig louse (*Hematopinus suis*).

swine vesicular disease (SVD) An infectious PIG DISEASE clinically indistinguishable from FOOT AND MOUTH DISEASE (although cattle and sheep are not affected); it occurs e.g. in Europe and Japan. SVD is caused by a porcine ENTEROVIRUS. Infection occurs mainly via the tonsils, but also via the gut, wounds, abrasions etc. Incubation period: 2–7 days; the mortality rate is usually low. Transmission is mainly by pig–pig contact; a carrier state is unknown. The virus can survive for months e.g. in moist pig faeces; it is stable between pH 2.5 and 12, but can be inactivated by moist heat (60°C/30 min).

swineherds' disease LEPTOSPIROSIS caused by *L. interrogans* serotype *pomona*.

swinepox subgroup See SUIPOXVIRUS.

Swiss cheeses See CHEESE-MAKING.

swivelase A type I TOPOISOMERASE.

SXT A combination of sulphamethoxazole and trimethoprim; SXT-impregnated discs are used in an identification test for streptococci – β-haemolytic streptococci of Lancefield groups A and B usually being resistant to SXT.

SYBR® Green I See PCR.

Syc Specific Yop chaperone: see VIRULON.

Sydenham's chorea See RHEUMATIC FEVER.

Sym plasmid See ROOT NODULES.

Symba process See SINGLE-CELL PROTEIN.

Symbiodinium See ZOOXANTHELLAE.

symbiogenesis The (presumed) evolutionary development of chloroplasts and mitochondria from endosymbiotic microorganisms.

symbiont One of the participants in a symbiotic relationship (SYMBIOSIS sense 1).

symbiosis (1) A term used originally for any stable condition in which two different organisms (*symbionts*) live in more or less close physical association – regardless of the nature of the relationship. Subsequently, the term came to be used, in a more restricted sense, to refer only to those instances in which both organisms derive benefit from the association; however, there is currently a strong tendency to revert to the original meaning – which is considered here as sense 1.

For a given symbiont, the symbiotic mode of life may be optional or obligatory. The nature of the relationship between symbionts ranges from mutually beneficial (MUTUALISM) to antagonistic (PARASITISM). (See also COMMENSALISM and NEUTRALISM.)

The continued association of symbionts, from generation to generation, may be achieved in various ways. Thus, e.g., in many alga–protozoon relationships the symbionts divide synchronously, or near-synchronously, and in many LICHENS dispersal involves the formation of propagules (e.g. soredia) containing both symbionts.

For examples of (primarily mutualistic) symbioses see FUNGUS GARDENS, MYCETOCYTE, MYCORRHIZA, ROOT NODULES, RUMEN, TERMITE–MICROBE ASSOCIATIONS, WOODWASP FUNGI, ZOOCHLORELLAE and ZOOXANTHELLAE. (See also CYTOBIOSIS, ECTOSYMBIONT, ENDONUCLEOBIOSIS and ENDOSYMBIONT.)

(2) A stable condition in which two different organisms form a close association to their mutual benefit, i.e., MUTUALISM. This meaning is tending to be superseded by that given in sense 1, above.

***sym*-homospermidine** See POLYAMINES.

symmetrogenic Refers to the type of cell division typical of flagellates: longitudinal fission, the two daughter cells being mirror images of one another. (cf. HOMOTHETOGENIC.)

***sym*-norspermine** See POLYAMINES.

sympatric Existing in the same environment or geographical region (cf. ALLOPATRIC).

symphogenous development (*mycol.*) The development of certain fungal structures (e.g. pycnidia) by the convergence, interweaving, growth and differentiation of hyphal branches from a number of different hyphae. (cf. MERISTOGENOUS DEVELOPMENT.)

symplasma (*pl.* symplasmata) In certain bacteria (e.g. *Enterobacter agglomerans/Erwinia herbicola*): a rounded or elongated aggregate of cells which may be surrounded by a translucent sheath.

symplectic metachrony See METACHRONAL WAVES.

sympodial (*adj.*) (1) Refers to the form or mode of development of a branching structure whose long axis consists of a number of sections – each being a branch of the preceding section. (cf. MONOPODIAL.)

(2) Refers to entities which are formed or united on a common structure or base.

Sympodiomyces A genus of marine (possibly ascomycetous) yeasts in which an elongated conidiophore may develop from each vegetative yeast cell; one species: *S. parvus*. [Book ref. 100, pp. 930–932.]

symport (co-transport) The coupled (linked) transmembrane transport of two solutes (e.g. ions) in the same direction. Symport may be electrogenic (see e.g. END-PRODUCT EFFLUX) or non-electrogenic – as e.g. when an anion is transported in symport with a cation. (cf. ANTIPORT, UNIPORT; see also ION TRANSPORT.)

synanamorph (*mycol.*) One of two or more ANAMORPHS of a given TELEOMORPH.

synapsis The pairing of homologous chromosomes during MEIOSIS, or the juxtaposition of homologous regions of DNA prior to general RECOMBINATION e.g. in prokaryotes, or the juxtaposition of appropriate sequences prior to SITE-SPECIFIC RECOMBINATION.

synaptobrevin See TETANOSPASMIN.

synaptonemal complex A structure, visible by electron microscopy, which lies between homologous chromosomes in the early stages of MEIOSIS; it consists of three parallel electron-dense strands (a 'central element' sandwiched between two 'lateral elements') – the whole being aligned with the long axis of the homologous pair of chromosomes, and interposed between them. One or more rounded or elongated structures (*recombination nodules*) may be embedded within, or may bridge, the synaptonemal complex; these nodules may be involved e.g. in the initiation of recombination since their number and distribution appear to coincide with those of the chiasmata.

syncaryon *Syn.* SYNKARYON.

Syncephalastraceae See MUCORALES.

Syncephalastrum See MUCORALES.

synchronous culture A form of CULTURE (sense 2) in which all the cells pass through the same stage of the CELL CYCLE at the same time. Synchronous culture is used e.g. to study particular biochemical events at specific stage(s) of the cell cycle.

Synchrony may be achieved in various ways. For example, synchronous cultures of certain algae can be obtained by the use of controlled cycles of light and dark. Synchrony in bacterial cultures may be achieved by subjecting the cells, for a given period of time, to an inhibitor of protein synthesis (e.g. chloramphenicol). Yeast cultures have been made synchronous by the use of EDTA or the IONOPHORE A23187 [JGM (1979) *114* 391–400]. Synchronous cultures of *Tetrahymena pyriformis* have been obtained by heat shock: the brief exposure of a non-synchronous culture to an elevated temperature (e.g. 40°C). All of these methods, however, tend to perturb normal metabolic activity so that they may be unsuitable for use in studies on cell-cycle-linked physiology. A method which avoids this problem can be used for endospore-forming bacteria; in this method a population of endospores is induced to germinate at a given time so that the resulting vegetative cells – at least initially – behave synchronously.

A more generally applicable method of obtaining synchronous cultures is termed *selection synchrony*; in this method, cells at a given stage of the cell cycle are separated out from a non-synchronous culture by purely physical means. In one selection method (used e.g. to obtain synchronous bacterial cultures) advantage is taken of the fact that the mass:volume ratio of the cell varies during the cell cycle; a non-synchronous, log-phase culture is subjected to density gradient CENTRIFUGATION so that the fraction of maximum density (consisting of cells about to divide, and new daughter cells) can be separated from the least-dense fraction – which consists of a population of cells at a particular stage of the cell cycle. [Selection synchrony method for bacterial cultures: Book ref. 2, pp. 166–168.] A selection method has also been used e.g. with cultures of *Dictyostelium discoideum* [JGM (1982) *128* 2449–2452].

Synchronization of *mammalian* cells (e.g. in tissue cultures) may be achieved e.g. by the *double thymidine blockade*. A non-synchronous culture is exposed to a high concentration of thymidine. Cells passing through the *S* phase of the CELL CYCLE remain blocked in that phase, while all other cells continue their development to the end of G_1. (Development beyond G_1 does not occur during the blockade.) The cells are then exposed to a thymidine-free medium for a period of time equal to that of the *S* phase; during this time all the cells continue their development, and, at the end of the period, no cells are passing through phase *S*. The cells are then re-exposed to excess thymidine until *all* the cells have reached the end of G_1 – i.e., the culture has become synchronous.

In a simpler, 'selective' method for synchronization of mammalian cells, a tissue culture is gently agitated; cells undergoing mitosis are more easily detached from the glass, and can be decanted with the growth medium, giving a suspension of approximately synchronous *M* phase cells.

In the absence of a synchronizing influence, synchronous cultures normally return to the asynchronous condition.

Synchytrium A genus of unicellular, endobiotic, holocarpic fungi (order CHYTRIDIALES) which are parasitic on various plants (see e.g. WART DISEASE). The asexual cycle of *S. endobioticum* resembles that of OLPIDIUM species except that e.g. each zoospore may give rise to a group of sporangia (a *sorus*). Sexually-formed thick-walled resting sporangia are also produced.

syncilium A group or tuft of cilia which resembles other forms of COMPOUND CILIATURE but which differs in the infraciliary organization; syncilia occur in the ENTODINIOMORPHIDA.

syncytium (1) A multinucleate cell or structure formed by the cytoplasmic fusion, without nuclear fusion, of a number of individual mononucleate protoplasts or cells (see e.g. GIANT CELLS). (2) *Syn.* COENOCYTE.

syndrome (*med., vet.*) A group of symptoms and/or signs which, taken together, characterize or are indicative of a distinct disease or abnormality.

Synechococcus A genus of unicellular CYANOBACTERIA (section I) in which the cells are often more or less rod-shaped, occur singly, contain thylakoids, and lack a sheath; division occurs by equal binary fission in one plane. This genus [see JGM (1979) *111* 1–61] includes organisms with a wide range of types of morphology, physiology, GC% (39–71), etc, and proposals have been made to subdivide the genus into four genera [Ann. Mic. (1983) *134*B 21–36]: *Cyanobacterium* (GC%: 39–41, cells 1.7–2.3 μm wide); *Cyanobium* (GC%: 66–71, cells ca. 0.8–1.4 μm wide); *Cyanothece* (GC%: ca. 42, cells ca. 4–6 μm wide); *Synechococcus* (GC%: 47–56, cells 1–2(–3.5) μm wide). These organisms occur in a wide range of habitats: e.g., as plankton in the open ocean [Nature (1979) *277* 293–294], in hypersaline environments such as salt evaporation ponds (e.g. *Cyanothece* (*Aphanothece*) *halophytica*), and in hot springs – e.g. *S. lividus* can grow at temperatures up to ca. 70°C.

Certain strains of *Synechococcus* (*sensu lato*) exhibit a swimming (not gliding) motility – even though the organisms lack flagella; swimming speeds are 5–25 μm/sec, and the type of swimming is influenced by cell morphology: coccoid cells carry out looping and spiral movements, while rods move in relatively straight paths [Science (1985) *230* 74–76]. Flagellum-less, calcium-dependent motility (of unknown mechanism) has been reported in strains of *Synechococcus* [JB (1997) *179* 2524–2528].

Circadian regulation of the gene *psbAI* (which encodes a component of photosystem II) has been studied in *Synechococcus* [TIM (1998) *6* 407–410].

Synechocystis A genus of unicellular CYANOBACTERIA (section I) in which the cells are coccoid and divide by equal binary fission in two or three planes, giving rise to clusters of cells; a sheath is not formed. GC%: 35–48. This definition tentatively includes the phycological genera *Aphanocapsa*, EUCAPSIS, MERISMOPEDIA and MICROCYSTIS [JGM (1979) *111* 1–61]. (See also PROCHLOROPHYTES.)

Synercid® See QUINUPRISTIN/DALFOPRISTIN.

syneresis The spontaneous expulsion of liquid from a gel.

synergism (in antibiotic action) The phenomenon in which a mixture of two antibiotics exerts an antimicrobial effect (on a given organism) which is (1) greater than that exerted by the more active antibiotic acting alone, or (2) greater than the additive effect of the two antibiotics acting independently. (cf. ANTAGONISM).

synergistins See STREPTOGRAMINS.

syngamy The fusion of gametes.

syngen (*ciliate protozool.*) A group of individuals, within a given species, in which two or more 'mating types' can usually be distinguished – CONJUGATION normally taking place only between members of different mating types within a given syngen. *Paramecium aurelia* has at least 12 syngens, and *P. bursaria* has at least 6; syngens occur also e.g. in *Euplotes* and *Tetrahymena*. Each syngen may be referred to by a number. It has been suggested that each syngen should be recognized as a distinct species.

syngeneic Derived from a genetically identical individual. (cf. ALLOGENEIC; XENOGENEIC.)

Synhymeniida See HYPOSTOMATIA.

synkaryon (syncaryon) (1) The diploid *nucleus* of a zygote. (2) A zygote.

synnema In some fungi: an erect tuft of spore-bearing hyphae in which there is characteristically an appreciable degree of hyphal fusion. (cf. COREMIUM and SPORODOCHIUM; see also CONIDIUM.) Analogous structures are formed by certain bacteria (e.g. ACTINOSYNNEMA).

synnematin B *Syn.* PENICILLIN N.

synnematospores Spores produced on a SYNNEMA.

synonym (taxonomy) (1) Objective synonym: any of two or more names which refer to the nomenclatural type of one particular taxon. (2) Subjective synonym: any of two or more names which do not refer to a common nomenclatural type but which, in the opinion of some taxonomist(s), refer to organisms which should be included in the same species.

A synonym is designated 'senior' or 'junior' according to whether it was published earlier or later, respectively, than a particular co-synonym.

Synsorb P^k A synthetic carbohydrate whose molecules, which mimic the binding sites of shiga-like toxins, are linked to insoluble particles of silica. Synsorb P^k is a potential therapeutic agent being evaluated for its ability to prevent toxin-mediated damage e.g. in HAEMOLYTIC URAEMIC SYNDROME; it is used with the object of binding (sequestering) toxin in the gut – thereby preventing uptake and (hence) denying the toxin access to its binding sites on vascular endothelium. (See also STARFISH.)

syntaxin See TETANOSPASMIN.

synthase A term applied to various enzymes *excluding* any belonging to EC class 6 (these being termed *synthetases* or LIGASES).

synthetase *Syn.* LIGASE. (cf. SYNTHASE.)

syntrophism The phenomenon in which the growth of one organism depends on (or is improved by) the provision of one or more growth factors or substrates by another organism growing in the vicinity. (See e.g. SATELLITE PHENOMENON; cf. CROSS-FEEDING.) *Syntrophism tests* can be used e.g. in the elucidation of biosynthetic pathways. AUXOTROPHS which have blocks at different points in the same pathway may exhibit a syntrophic relationship: a substrate of a particular blocked reaction tends to accumulate and diffuse into the surrounding medium, supporting the growth of any auxotrophs blocked at *earlier* stages in the common pathway.

Syntrophobacter A genus of Gram-negative, anaerobic, non-motile bacteria which oxidize propionate to acetate, CO_2 and H_2; growth occurs only in the presence of H_2-scavenging systems – e.g. in co-culture with *Desulfovibrio* in the presence of sulphate. [Genus proposed in AEM (1980) *40* 626–632.] (cf. SYNTROPHOMONAS.)

Syntrophomonas A genus of Gram-negative, anaerobic bacteria which can use only protons as electron acceptors in the oxidation of short-chain fatty acids (e.g. butyrate) to acetate (or to a mixture of acetate and propionate); the organisms can grow only in the presence of H_2-scavenging systems – e.g. in co-culture with methanogens (see ANAEROBIC DIGESTION). Cells: round-ended, curved or helical rods with 2–8 flagella arising from the concave side of the cell. (cf. SYNTROPHOBACTER.)

Syntrophus A proposed genus of Gram-negative anaerobic bacteria. *S. buswellii*, the type species, was isolated from anaerobic digester sludge. Cells: round-ended rods with monotrichous polar flagella in the early stages of growth. Benzoate is metabolized with the formation of acetate, CO_2 and H_2 (or formate). Growth is inhibited by H_2 and requires the presence of appropriate hydrogenotrophic bacteria. [IJSB (1984) *34* 216–217.]

syntype (cotype) Each of several specimens – none having been designated a HOLOTYPE – on which an author has based a published description of a new taxon.

Synura A genus of colonial CHRYSOPHYTES. Each elongate-pyriform cell has two flagella of similar length. The cells are compactly arranged – flagellate (anterior) ends outwards – in a globular, motile colony (100–400 μm diam.). Intricate siliceous spicules occur at the anterior end of each cell. Species occur in freshwater lakes and reservoirs with hard water, sometimes forming blooms; *S. petersenii* has been implicated as a cause of obnoxious 'fishy' taste and odour in lake waters [CJB (1985) *63* 1482–1493].

syphilis A chronic human disease caused by *Treponema pallidum*. Infection generally occurs by direct contact with lesions of primary or secondary syphilis (see *T. pallidum* under TREPONEMA; see also VENEREAL DISEASE); *T. pallidum* can apparently penetrate undamaged mucous membranes, and is rapidly disseminated through the body via the blood and lymphatic systems. *Primary syphilis*. The first clinical manifestation is the *chancre*: a relatively painless, indurated lesion which develops at the infection site ca. 10–90 days after infection; in women the chancre may develop on the cervix and may thus be undetected. The chancre heals, without treatment, after ca. 10–40 days. *Secondary syphilis* begins one to several months after the disappearance of the chancre, and involves e.g. a rash of cutaneous lesions (which may be papular, macular, wart-like etc), thin white sores on the mucous membranes ('mucous patches'), loss of hair, lymphadenopathy, and mild general symptoms (slight fever, malaise etc). GLOMERULONEPHRITIS may occur. *Latent syphilis* occurs between the (spontaneous) disappearance of secondary symptoms and the development of late syphilis. The latent infection may persist for 1–40 years or more; there is no clinical evidence of infection (although symptoms of secondary syphilis may recur), but serological tests may be positive. *Late syphilis (tertiary syphilis)* may involve any tissue or organ – e.g., skin, bones, eyes (often leading to blindness), CNS (leading e.g. to insanity and death), and cardiovascular system (which may result in heart failure, aortic rupture following aortitis, etc). The characteristic lesion is the *gumma*: a granuloma with caseating necrosis, formed as a result of a localized allergic reaction.

Congenital syphilis. Syphilis is rarely acquired by a fetus from an infected mother before the 4th month of gestation; fetal infection may actually occur before this, but disease does not develop – possibly because the early fetus is not immunologically capable of mounting the inflammatory response which leads to congenital syphilis. The likelihood of fetal infection decreases with increasing duration of the mother's disease, being greatest during primary or secondary syphilis in the mother. Consequences of fetal infection include miscarriage, stillbirth, or a severely affected infant. The neonate may show no symptoms at birth, but ca. 3 weeks later may develop skin lesions (mainly on the face, anogenital region, palms and soles), 'snuffles', hepatosplenomegaly etc; death may occur in the first year. Sequellae of congenital syphilis include notched incisors (Hutchinson's teeth), bossed ('mulberry') molars, saddle-shaped nose, mental deficiency, nerve deafness etc. Congenital syphilis can be prevented by treating the mother before the 4th month of pregnancy.

Lab. diagnosis: material from lesions etc may be examined by dark-field microscopy. For serological tests see STANDARD TESTS FOR SYPHILIS and TREPONEMAL TESTS. *Chemotherapy*: penicillins; tetracyclines or erythromycin. (See also JARISCH–HERXHEIMER REACTION.)

When diagnosis is uncertain (e.g. borderline serology), a fine-needle aspirate of inguinal lymph node may be examined for treponemal DNA by PCR; this method confirmed syphilis in three patients with suspected secondary or latent disease. PCR amplified regions within *tpn47* (a gene encoding the major treponemal membrane lipoprotein), fragments being analysed by (i) gel ELECTROPHORESIS and (ii) Southern hybridization (see SOUTHERN BLOT) [Lancet (2002) *360* 388–389].

syringomycin A peptide-containing phytotoxin, produced by *Pseudomonas syringae* pv. *syringae*, which causes a rapid, detergent-like lysis of plant and fungal cytoplasmic membranes. (cf. SYRINGOTOXIN.) [Regulation of syringomycin production: JAB (1985) *58* 167–174.]

syringotoxin A phytotoxin produced by citrus-infecting strains of *Pseudomonas syringae* [PPP (1981) *18* 41–50]. (cf. SYRINGOMYCIN.)

systematics (1) *Syn.* TAXONOMY. (2) The range of studies (theoretical and practical) involved in the classification of organisms.

système sécant (*ciliate protozool.*) A 'line', marked by the absence of kinetosomes, which delimits adjacent groups or fields of kineties. Examples include e.g. the *preoral suture* (which typically runs along the ventral surface between the oral region and the anterior pole) and the *postoral suture* (which typically runs, ventrally, between the oral region and the posterior pole).

systemic (1) (*med., vet.*) Involving the whole body. (2) (*plant pathol.*) Refers to a chemical agent (e.g. an ANTIFUNGAL AGENT) which penetrates and is translocated within the tissues of a plant.

systole See CONTRACTILE VACUOLE.

syzygy In certain protozoa; a stage of sexual reproduction in which gametocytes become physically joined without losing their identity as separate cells. A pair of cells 'in syzygy' usually encysts and gives rise to male and female gametes.

1. Words in SMALL CAPITALS are cross-references to separate entries.
2. Keys to journal title abbreviations and Book ref. numbers are given at the end of the Dictionary.
3. The Greek alphabet is given in Appendix VI.
4. For further information see 'Notes for the User' at the front of the Dictionary.

T

T (1) Thymine (or the corresponding nucleoside or nucleotide) in a nucleic acid. (2) (*virol.*) See ICOSAHEDRAL SYMMETRY. (3) Threonine (see AMINO ACIDS).

T antigen (1) Tumour antigen: i.e., an antigen specific to a tumour cell. (See e.g. POLYOMAVIRUS.) (2) (in streptococci) *Syn.* T PROTEIN.

T cell (*immunol.*) *Syn.* T LYMPHOCYTE.

T cell anergy See SUPERANTIGEN.

T cell receptor (TCR) An antigen-specific receptor which occurs on every T LYMPHOCYTE; the same receptor occurs on all T cells of the same clone, while different receptors are found on T cells of different clones. The role of the TCR is to interact with its specific antigen. (See also IMMUNOGLOBULIN SUPERFAMILY.)

A TCR consists of two distinct, transmembrane peptide chains, linked by a disulphide bond, which project from the surface of the cell. In some TCRs the two chains are referred to as α and β; in other TCRs the two chains have sequences which differ from those of the α and β chains, and in these TCRs the chains are called γ and δ. Thus, a given TCR may consist of either $\alpha\beta$ chains or $\gamma\delta$ chains.

In a given TCR, the extracellular region of each chain includes a distal variable (V) region and a proximal constant (C) region; thus, for example, in an $\alpha\beta$-type TCR the α chain includes $V\alpha$ and $C\alpha$ regions, and the β chain contains $V\beta$ and $C\beta$ regions. The two variable (V) regions constitute the antigen-binding part of the TCR.

In humans, the $V\beta$ region of the TCR is encoded by a gene which is formed by random selection and assembly of small gene segments during maturation of the T cell; consequently, this gene occurs in about 25 different variant forms, each identified by a number (e.g. $V\beta1$, $V\beta2$ etc.). Note that those T cells which exhibit any given variable region (e.g. $V\beta1$) will include a very large number of different T cell clones. (See also SUPERANTIGEN.)

The TCR is part of a small group of transmembrane cell-surface structures called the *T cell receptor complex*. This complex comprises (i) the TCR, (ii) CD3 (q.v.) and (iii) a molecule consisting of two identical disulphide-linked ζ (zeta) peptides or a heterodimer of one ζ peptide and one η (eta) peptide (the $\zeta\zeta$ homodimer occurs in most TCRs).

T-dependent antigen THYMUS-DEPENDENT ANTIGEN.

T-DNA See CROWN GALL.

T-even phages A category of virulent (lytic) enterobacterial dsDNA-containing BACTERIOPHAGES which includes T2, T4 and T6. (See also MYOVIRIDAE.) These phages resemble one another morphologically, they cross-react serologically, and their genomes can undergo recombination with each other when present in the same host cell. They differ e.g. in their host cell receptor sites: the T4 receptor on *Escherichia coli* B strains is the α-glucosyl-(1 $\rightarrow$ 3)-glucose terminus of the rough LPS core, while on K strains the LPS and OUTER MEMBRANE protein OmpC serve as separate, independent receptors; the T2 receptor is the OmpF protein, that of T6 is the Tsx protein.

The virions of T-even phages are structurally extremely complex, containing many components. Each consists of an elongated head attached to a long, complex, contractile tail. The head (ca. 110×85 nm) apparently contains between 9 and 19 polypeptide species and has icosahedral symmetry elongated along a 5-fold axis by the insertion of an extra row of capsomers. It is attached – via a complex neck region which includes a 'collar' bearing six 'whiskers' – to a tail (ca. 100×22 nm) consisting of two coaxial tubes (an inner 'core' and an outer contractile sheath). The distal end of the tail terminates in a hexagonal base-plate consisting of a central hub or 'core' surrounded by six wedges. In T4, the hub contains at least 8 proteins – including a thymidylate synthase (gp*td*), a glutamyl carboxypeptidase (gp28), an ATP-requiring folyl polyglutamyl synthetase (gp29), and a lysozyme (gp5) which is apparently able to induce LYSIS FROM WITHOUT as well as lysis from within [JV (1985) *54* 460–466]; the hub also contains dihydropteroyl (dihydrofolyl) hexaglutamate. The wedges each contain at least six proteins, including a dihydrofolate reductase (gp*frd*). Each corner of the base-plate bears a long jointed tail fibre (ca. 150×3–4 nm) and a short tail fibre (ca. 35×3–4 nm) – the latter containing zinc.

The phage genome consists of a linear, circularly permuted, terminally redundant dsDNA molecule ca. 170 kb long (including the terminal redundancy of ca. 2.3%). The DNA contains hydroxymethylcytosine (HMC) instead of cytosine, and may also be modified by glucosylation of HMC residues and (in T2 and T4, but not in T6) N^6-methylation of a small proportion of the adenine residues.

See BACTERIOPHAGE T4 for infection cycle.

T helper cell (T_h; T_H) See T LYMPHOCYTE.

T-independent antigen THYMUS-INDEPENDENT ANTIGEN.

T lymphocyte (T cell) A type of LYMPHOCYTE involved in e.g. ANTIBODY FORMATION, CELL-MEDIATED CYTOTOXICITY, INFLAMMATION and DELAYED HYPERSENSITIVITY. There are various types (subsets) of T cell, each subset being distinguishable by function and/or cell-surface antigens. Like B LYMPHOCYTES, T cells develop from haemopoietic stem cells (in the bone marrow) via the lymphoid pathway of development. The (morphologically distinguishable) large granular lymphocytes known as NK CELLS also develop from T cell precursors; NK cells are thus related to, but distinct from, T cells. (Other leukocytes, including macrophages and PMNs, develop from haemopoietic stem cells via the myeloid pathway.)

An antigen-specific T CELL RECEPTOR enables each T cell to respond to a unique antigen (T cells in the same *clone* responding to the same antigen). Unlike B cells (which bind soluble, i.e. free or isolated, antigen), T cells react to specific antigen only when it appears in suitable form on the surface of another cell; for example, during the process of ANTIBODY FORMATION, a T cell of a particular subset interacts with antigen presented in association with an MHC class II molecule on the surface of a B cell. *Cytotoxic* T cells (see later) interact with ligands on their target cells; cytotoxic cells have the important function of killing host cells that are infected with e.g. viruses or intracellular bacteria. In general, when a T cell interacts with its specific antigen the T cell is induced to proliferate (i.e. multiply), thus forming a clone (or a larger clone) of cells reactive to that specific antigen; such proliferation of a T cell of given antigenic specificity is called *clonal expansion*.

Although T cells respond to *antigens* only when the latter are cell-associated, they can respond to other isolated (soluble) molecules such as CYTOKINES.

Unlike B cells, most or all T cells undergo their process of maturation in the thymus gland. (The mouse is reported to have a subset of mature, thymus-independent T cells.) Individuals with a non-functional thymus gland have a defective immune system.

Most *mature* (i.e. thymus-processed) lymphocytes of the $\alpha\beta$ type (see T CELL RECEPTOR) express a cell-surface *co-receptor* (or *accessory molecule*) that is closely (but non-covalently) associated with the TCR. The co-receptor is either a CD4 molecule (q.v.) or a CD8 molecule. An *immature* cell has the potential to develop either as a $CD4^+$ or $CD8^+$ cell, i.e. to express either the CD4 or the CD8 molecule, respectively; whether a given cell develops as $CD4^+$ or $CD8^+$ is reported to depend on the duration of antigen receptor signalling during the maturation process [Nature (2000) *404* 506–510]. CD4 and CD8 subsets of T cells are considered separately below.

$CD4^+$ T cells are also referred to as T helper cells, helper T cells, T_H cells or T_h cells.

When a $CD4^+$ T cell *initially* binds antigen it differentiates into one of two main types of helper cell – Th1 and Th2, the cells of each type being characterized by their secretion of a particular range of CYTOKINES (q.v.). (*Prior* to binding antigen the T cell is described as a *naive* cell; such cells are often referred to as Th0.) (See also CD28.) [The expanding universe of T-cell subsets: IT (1996) *17* 138–146.]

Th1 and Th2 cells also interact in characteristic ways with other cells of the immune system. For example, a Th1 cell may interact with a macrophage that has engulfed a bacterium. The macrophage acts as an ANTIGEN-PRESENTING CELL, presenting bacterial antigen, linked to an MHC class II molecule, to the (antigen-specific) T cell; the T cell binds antigen via its TCR, and binds the class II molecule with its co-receptor (CD4). Following such binding, the T cell releases IFN-γ (see INTERFERONS) – which promotes activation of the macrophage; when activated, the macrophage behaves more aggressively towards the engulfed bacterium (e.g. by producing NITRIC OXIDE and toxic free radicals) and it also secretes e.g. INTERLEUKIN 12.

Th2 cells characteristically interact with B cells during ANTIBODY FORMATION (q.v) and also give rise to specific cytokines (e.g. IL-4, IL-5, IL-10) that promote development of the Th2 subset of T cells and stimulate defences against parasitic infections.

$CD8^+$ T cells are generally referred to as *cytotoxic* (or *cytolytic*) T cells (sometimes written as T_{cyt}). (See also CELL-MEDIATED CYTOTOXICITY.) These T cells kill host cells that are infected with viruses or certain other pathogens; in such cases it appears that the T cell makes direct physical contact with the target cell in an antigen-specific manner. Killing may involve interaction between the target cell's Fas receptor (see FAS) and the T cell's Fas L ligand (leading to APOPTOSIS). Alternatively, the T cell may release (i) proteins (*perforins*) that damage the membrane of the target cell, and (ii) serine proteases (*granzymes*) that inflict damage on the target cell and/or promote apoptosis.

Cells bearing the CD8 co-receptor are restricted by MHC class I antigens.

T cells with a $\gamma\delta$ TCR (see T CELL RECEPTOR) include some which are $CD4^+$ and some which are $CD8^+$; however, it appears that many or most have neither a CD4 nor a CD8 antigen. T cells of the $\gamma\delta$ type are found among the so-called *intra-epithelial lymphocytes* (IELs) associated with mucosal epithelium.

The proportion of $\gamma\delta$ T cells in the overall T cell population is much lower than that of the $\alpha\beta$ T cells; $\gamma\delta$-type cells are reported to occur in all mammalian species examined so far.

At least some of the $\gamma\delta$-type T cells, and a small subset of $\alpha\beta$-type T cells which lack both CD4 and CD8 antigens (*double negatives*), can recognize non-peptide antigens; for example mycobacterial cell components, including mycolic acids, can be recognized by the $\alpha\beta$-type cells.

T-lysin See FURUNCULOSIS (sense 1).

T number See ICOSAHEDRAL SYMMETRY.

T odd phages Bacteriophages T1 and T5 (STYLOVIRIDAE) and T3 and T7 (PODOVIRIDAE).

T protein (T antigen) (in streptococci) A cell-surface protein found in certain group A streptococci. T proteins are not associated with virulence, but are highly antigenic and may be used in typing of group A streptococci. Particular T-antigenic types are usually associated with particular M-antigenic types (see M PROTEIN sense 1). (cf. R PROTEIN.)

T-strain mycoplasmas Earlier name for members of the genus UREAPLASMA.

T-strand See CONJUGATION (1b)(i).

T1 side-chains See O ANTIGENS.

T-2 toxin A TRICHOTHECENE mycotoxin.

T2H test (TCH test) A test used in the identification of strains of *Mycobacterium*. Essentially, the test determines the ability of a given strain to grow (*positive* result) on a medium (e.g. Löwenstein–Jensen medium) containing thiophene-2-carboxylic acid hydrazide (T2H, TCH) – often at 10 μg/ml, a concentration inhibitory to *M. bovis* but not to most strains of *M. tuberculosis*; those strains of *M. tuberculosis* which are inhibited by 10 μg/ml T2H may be able to grow on media containing 1–5 μg/ml (concentrations still inhibitory to *M. bovis*) [Book ref. 53, p. 1710].

T7 phage group A genus of phages of the PODOVIRIDAE. Type species: BACTERIOPHAGE T7; other members: e.g., bacteriophages T3, W31, ϕI and ϕII.

T_{90} The time required for 90% mortality to occur in a population of microorganisms exposed to a given hostile regime.

TAB An INACTIVATED VACCINE containing the cells of *Salmonella typhi* and *S. paratyphi* A and B.

Tabellaria See DIATOMS.

TABT TAB vaccine incorporating tetanus toxoid.

tabtoxin A β-lactam-containing dipeptide toxin produced by e.g. *Pseudomonas syringae* pv. *tabaci*. In the tobacco plant, tabtoxin is enzymically hydrolysed to tabtoxinine-β-lactam which inhibits glutamine synthetase; the consequent accumulation of ammonia appears to be an important factor in the pathogenesis of WILDFIRE DISEASE. [APRpath. (1984) *22* 218–221; susceptibility of bacterial glutamine sythetase to tabtoxinine-β-lactam: JGM (1985) *131* 1061–1067; contribution of tabtoxin to the pathogenicity of *Pseudomonas syringae* pv. *tabaci*: PPP (1984) *25* 55–69.]

Tabtoxin-producing strains of *P. syringae* can be detected rapidly by PCR-based methodology [LAM (2001) *32* 166–170].

Tac antigen A T lymphocyte surface antigen associated with the receptor for INTERLEUKIN-2.

***tac* promotor** A hybrid PROMOTER constructed in vitro by fusion between elements of the *lac* and *trp* promoters of *Escherichia coli* (see LAC OPERON and TRP OPERON). The *tacI* promoter contains the Pribnow box from the *lac* promoter and the −35 sequence from the *trp* promoter; *tacII* contains the *trp* −35 sequence but its Pribnow box is a hybrid between those of the *trp* and *lac* promoters. The *tac* promoters are stronger than either 'parent' promoter; they can be repressed by the *lac* repressor and induced by IPTG.

Tacaribe virus See ARENAVIRIDAE.

tâche noir See BOUTONNEUSE.

tachyzoite *Syn.* ENDOZOITE.

TACTAAC box See SPLIT GENE (a).

tactic response *Syn.* TAXIS.

tactophily A tendency to adhere to, and to grow on, solid surfaces.

taette A Scandinavian food made by fermenting milk with yeasts and a rope-forming strain of *Lactococcus lactis*. (cf. ROPINESS.) (See also DAIRY PRODUCTS.)

***tag* gene** See ADAPTIVE RESPONSE.

tagatose 6-phosphate pathway See LACTOSE and Appendix III(a).

tailing (*mol. biol.*) The addition of nucleotides to the 3′ end of a DNA strand using dNTPs and the enzyme *terminal deoxynucleotidyltransferase* (= terminal transferase); if only *one* type of dNTP is used (for example, dATP), the resulting tail is homogeneous (e.g. –A–A–A–) and the process is called *homopolymer tailing*.

Taka-amylase *Syn.* TAKADIASTASE.

Takadiastase (Taka-amylase) A commercial enzyme preparation, containing mainly α-AMYLASES, obtained from *Aspergillus oryzae* grown on the surface of a solid substrate (usually moist wheat-bran or rice).

Takatsi technique A technique for preparing doubling dilutions (see SERIAL DILUTIONS) using an instrument consisting of a small spiral of wire attached to the end of a metal rod (handle). When immersed in a liquid and withdrawn, the wire carries a fixed volume of liquid. The wire is then immersed in diluent (equal in volume to that carried by the wire) and the handle is rotated between forefinger and thumb; this mixes the liquids, giving a 1-in-2 dilution. The wire is then withdrawn and immersed in a fresh volume of diluent, and so on. Another version of the instrument consists of a small slotted metal block at the end of a metal handle – liquid being carried in the slots of the block.

take-all A CEREAL DISEASE, caused by *Gaeumannomyces graminis* (formerly *Ophiobolus graminis*), which can affect wheat, barley, rye and oats; it can also affect turf grasses ('ophiobolus patch disease'). Symptoms on wheat include stunting of seedlings, the formation of fewer (or no) tillers, and a dry, black rot in the roots and in the lower stem (where perithecia may develop); the heads are characteristically bleached and lack grain ('whiteheads'). Infection apparently occurs by means of hyphae which grow out from the roots of previously infected plants. The disease is currently not susceptible to chemical control (cf. SUPPRESSIVE SOILS).

talampicillin See PENICILLINS.

Talaromyces A genus of fungi of the order EUROTIALES; anamorphs occur in the genus PENICILLIUM. (For some authors, *Talaromyces* includes only those species which form asci in chains – species forming asci singly, from croziers, being placed in the genus *Hamigera*.)

Talfan disease (benign enzootic paresis) A mild form of TESCHEN DISEASE affecting young pigs. It occurs in e.g. Australia, Denmark, North America and the UK. Symptoms include fever and impaired control of the hind legs. Affected pigs may recover, or muscular weakness may persist; fatalities are unusual.

tamari sauce (1) *Syn.* SOY SAUCE. (2) SOY SAUCE made entirely from soybeans.

Tamiami virus See ARENAVIRIDAE.

Tamm–Horsfall protein (THP; urinary mucoprotein; uromucoid) A mucoprotein component of the mucus on the surface of human uroepithelial cells; it binds e.g. *Escherichia coli* type 1 (mannose-sensitive) FIMBRIAE – such binding possibly being an important host defence mechanism [Lancet (1980) *i* 887]. (See also URINARY TRACT INFECTION.) [Possible relevance of THP in chronic pyelonephritis: YJBM (1985) *58* 91–100.] THP is a leukocyte adhesion molecule [BBRC (1994) *200* 275–282].

Tanapox virus A virus of the POXVIRIDAE which can infect humans, causing an acute, febrile illness with localized, pock-like skin lesions; the illness has occurred in epidemics among people in the region of the Tana River, Kenya. Tanapox virus can also infect monkeys. It does not grow in the CAM but can be cultivated in monkey kidney cell cultures. In serological tests the Tanapox virus shows partial cross-reactivity with YABA MONKEY TUMOUR POXVIRUS.

TANDEM See QUINOXALINE ANTIBIOTICS.

tane koji See KOJI.

tanned cells See BOYDEN PROCEDURE.

tanned ox haemagglutination See FIMBRIAE.

tannic acid (1) Penta-(*m*-digalloyl)-glucose, a hydrolysable tannin. (2) Any hydrolysable tannin (see TANNINS).

tannins A heterogeneous group of complex polyhydroxybenzoic acid derivatives which occur e.g. in plants and in brown algae (see PHAEOPHYTA). *Hydrolysable tannins* consist of a sugar (often glucose) partially or wholly esterified with e.g. gallic acid (3,4,5-trihydroxybenzoic acid), digallic acid (a DEPSIDE), or ellagic acid; hydrolysis yields the component sugar and acids. *Condensed (non-hydrolysable) tannins* are believed to be oligomers of flavanoid compounds; they tend to polymerize (e.g. under acid conditions) to form insoluble reddish compounds (*phlobaphenes*).

Tannins can precipitate and inactivate proteins by forming cross-links (a property exploited e.g. in leather tanning and in the BOYDEN PROCEDURE). In plants, tannins occur mainly in bark and woody tissues but also in certain leaves (e.g. tea-leaves); they may occur in vacuoles, in cell walls, or in specialized cells. They are also formed in damaged plant tissues as a result of the oxidation (in the presence of air) of phenolic compounds in the cell vacuoles by polyphenoloxidases normally sequestered in lysosomes. Tannins are believed to play a role in the resistance of plants to disease: e.g., by inhibiting enzymes (such as PECTIC ENZYMES) produced by invading pathogens, and by binding to and reducing the availability of proteins in the damaged plant tissue.

A comparison of two species of rumen bacteria, *Streptococcus bovis* and *S. gallolyticus* (= *S. caprinus*), found that the latter species can adapt to higher levels of tannins e.g. by upregulating the enzyme gallate decarboxylase and forming an extracellular polysaccharide matrix [Microbiology (2001) *147* 1025–1033].

***tap* gene** See CHEMOTAXIS (*Escherichia coli*).

Taphrina A genus of dimorphic fungi (order TAPHRINALES) which include species parasitic e.g. on various trees – including alder (*Alnus*); birch (*Betula*); cherry, peach and plum (*Prunus*); and oak (*Quercus*). *Taphrina* spp form an intercellular or subcuticular septate mycelium; some species form haustoria. Asci (see ASCUS) may develop as a subcuticular, erumpent layer, each ascus (containing ovoid or spheroidal ascospores) developing from a cell produced by hyphal fragmentation. In culture, *Taphrina* spp may grow only in a budding, yeast-like form.

T. deformans can cause e.g. WITCHES' BROOM and peach (and almond) *leaf curl*. In leaf curl, the young leaves become distorted and yellowish with reddish patches, and a white or silvery film develops on their upper surfaces; the leaves later become brown and drop prematurely. Young shoots, flowers and fruit may also become distorted.

Taphrinales An order of plant-parasitic fungi of the subdivision ASCOMYCOTINA. Ascocarps are not formed; the asci develop, directly or indirectly, from the hyphae, or from individual cells formed by the fragmentation of hyphae. Genera: e.g. TAPHRINA.

***Taq* DNA polymerase** See DNA POLYMERASE.

***Taq*I** See RESTRICTION ENDONUCLEASE (table).

TaqMan® probes Oligonucleotide probes (marketed by PE Applied Biosystems) which are used for (i) monitoring the progress of amplification in PCR, and (ii) estimating the initial (pre-amplification) concentration of target sequences in the sample. The structure and activity of these probes are shown diagrammatically in the figure.

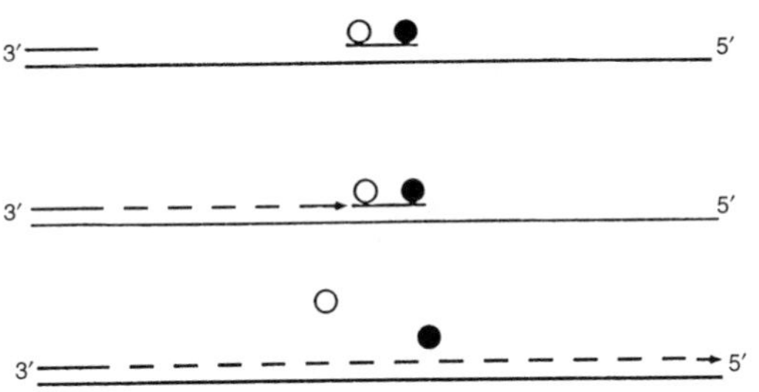

TaqMan® probes: structure and activity (diagrammatic).

Top. One strand of an amplicon (3′-to-5′). A primer has bound to the 3′ end, and a TaqMan probe has bound to an internal target site. The TaqMan probe consists of a target-specific oligonucleotide to which are covalently linked (i) a fluorophore ('reporter' dye) (○), and (ii) a quencher of fluorescence (●); the close proximity of the quencher to the reporter dye inhibits fluorescence.

Centre. Extension from the primer (dashed line) has reached the probe. The polymerase now exerts 5′-to-3′ exonuclease activity: it displaces and degrades the probe – separating reporter and quencher so that the reporter can now produce a fluorescent signal; note that the release of one molecule of reporter is linked to the synthesis of one product strand. (Not all polymerases have 5′-to-3′ exonuclease activity; the use of a polymerase with this function is thus essential when working with TaqMan probes.)

Below. Ongoing extension completes the strand.

Reproduced from Figure 4.4, page 64, in *DNA Methods in Clinical Microbiology* (ISBN 07923-6307-8), Paul Singleton (2000), with kind permission from Kluwer Academic Publishers, Dordrecht, The Netherlands.

TaqMan® probes are added (in large numbers) to the reaction mixture prior to amplification. During cycling, ongoing synthesis of amplicons is paralleled by a rise in the number of fluorescent (unquenched) reporter molecules from degraded probes. Amplification can therefore be monitored at any given time by measuring the level of fluorescence in the reaction mixture.

Quantification (estimation of the initial number of target sequences in the sample). From the beginning of cycling, the level of fluorescence is monitored on a cycle-by-cycle basis. The first cycle in which reporter fluorescence exceeds background fluorescence is the *threshold cycle*. Identification of the threshold cycle in a given reaction allows calculation of the number of target sequences in the sample prior to amplification. Such calculation depends on the inverse linear relationship between (i) threshold cycle and (ii) $\log_{10}$ of the initial number of target sequences – a relationship seen as a straight-line graph when threshold cycle (1, 2, 3...25, 26, 27...etc.) is plotted against $\log_{10}$ of the initial number of target sequences. This graph, initially constructed from experimental data (with known numbers of target molecules), permits estimates to be made from samples in which the initial target numbers are unknown.

Threshold cycles in PCR can also be determined with MOLECULAR BEACON PROBES (q.v.); fluorescence levels from these probes are recorded, cycle-by-cycle, at the primer annealing stage (when the amplicon-bound probes are fluorescent).

TaqMan® probes have been used e.g. for the quantitative estimation of *Borrelia burgdorferi* in specimens of tissue [JCM (1999) *37* 1958–1963].

***tar* gene** See CHEMOTAXIS (a).

tar spot See RHYTISMA.

tartar emetic See ANTIMONY.

tartronic semialdehyde pathway See TCA CYCLE [and Appendix II(b)] and PHOTORESPIRATION.

Tat See PROTEIN SECRETION (end of type II systems).

TATA box See PROMOTER.

TATA box-binding protein See PROMOTER.

Tatlockia See LEGIONELLACEAE.

Tatumella A genus of bacteria of the ENTEROBACTERIACEAE, isolated from human clinical (usually respiratory tract) specimens. Non-motile at 36°C; many strains are motile at 25°C [JCM (1981) *14* 79–88].

τ (superhelix winding number) See DNA.

***tau* protein** See MICROTUBULE-ASSOCIATED PROTEINS.

Tauber trap A simple instrument used for sampling airborne particles. (See also AIR.) It consists of a cylindrical vessel with an aerodynamically shaped lid containing a central hole.

Taumelkrankheit In fish: a loss of muscular coordination leading to a spiral swimming motion; it may be a (uncommon) symptom of ICHTHYOPHONOSIS or of thiamine deficiency.

taurine An aminosulphonic acid ($NH_2CH_2CH_2SO_3H$) derived from cysteine; it occurs e.g. conjugated with BILE ACIDS.

taurocholic acids See BILE ACIDS.

taurolin A wide-spectrum FORMALDEHYDE-releasing antimicrobial agent which has been used e.g. for the treatment of peritonitis. [Review: JAB (1980) *48* 89–96.]

taxa Plural of TAXON.

taxis A locomotive response to an external stimulus, exhibited by certain (motile) cells or organisms, in which the direction – or net change in direction – of locomotion is related to the direction or orientation of the stimulus. (cf. KINESIS; TROPISM.) In general, a tactic response is triggered only when the level of the stimulus is above a minimum (threshold) value; however, extreme levels of stimulation may promote tactic responses different from those produced by moderate stimulation (see e.g. PHOTOTAXIS). (See also ADAPTATION.)

In bacteria, some taxes depend on changes in the cell's proton motive force (see e.g. AEROTAXIS and PHOTOTAXIS; cf. CHEMOTAXIS). [Pmf-dependent taxis: ARM (1983) *37* 551–573.]

See also e.g. ELASTICOTAXIS; GALVANOTAXIS; GEOTAXIS; MAGNETOTACTIC BACTERIA; RHEOTAXIS; THIGMOTAXIS.

taxol A complex substance of plant origin (obtained from *Taxus brevifolia*) which has antitumour activity, and which can e.g. inhibit the growth of HeLa cells at a concentration of 0.25 μM. Taxol apparently enables the assembly of MICROTUBULES to occur with very low concentrations of tubulin (ca. 0.01 mg/ml in vitro) – i.e., it decreases the value of the critical concentration of tubulin – thereby promoting microtubule assembly; it also stabilizes MTs and (hence) prevents a cell from disassembling its MTs preparatory to the formation of a mitotic spindle. Taxol permits in vitro assembly of MTs in the absence of GTP.

taxometrics *Syn.* NUMERICAL TAXONOMY.

taxon (*pl.* taxa) In a given system of biological classification: any category, consisting of one or more kinds of organism, regarded as having an identity distinct from that of any other category in that system; a taxon may be a group of strains, species, genera, etc.

taxonomy The science of biological classification, i.e. the grouping of organisms according to their natural affinities. (cf. SYSTEMATICS.) A classified arrangement of organisms permits a logical and informative system of naming (see NOMENCLATURE) and can also be useful e.g. for identifying newly isolated organisms.

Ideally, biological classification should reflect the natural (evolutionary) relationships between organisms (i.e. PHYLOGENETIC CLASSIFICATION; cf. PHENETIC CLASSIFICATION). Such a phyletic approach has been used traditionally in taxonomic schemes for higher animals and plants (in which evolutionary relationships can be deduced from structural and other features). However, among microorganisms, the prokaryotes have (until recently) presented particular difficulties for taxonomists, and earlier forms of classification tended to suffer from artificiality and subjectivity; attempts to introduce a more rational basis for the taxonomy of these organisms lead to increased interest in e.g NUMERICAL TAXONOMY. (An objective approach was also sought in areas such as enzymological analysis [e.g. CRM (1982) *9* 227–252], protein structure (MICROCOMPLEMENT FIXATION), RIBOSOME structure, and the *sequence* of loci in the genome [IJSB (1985) *35* 217–220].)

The modern approach to prokaryotic taxonomy began with molecular studies in which organisms are compared and classified according to the sequences of nucleotides in their nucleic acids (see e.g. RRNA OLIGONUCLEOTIDE CATALOGUING and DNA HOMOLOGY). Thus, the concept of a SPECIES (q.v.) may be considered in terms of the similarity (%) between chromosomal DNA from different strains; moreover, different species in a given genus commonly exhibit a certain minimum level of difference in their 16S rRNA sequences. However, reliance on 16S rRNA sequences as a guide to species identity has been questioned [IJSB (1992) *42* 166–170]. (16S rRNA is widely used for the phyletic classification of prokaryotes for two main reasons. First, it occurs in all prokaryotes. Second, it contains highly conserved (i.e. stable) sequences as well as more variable sequences; the highly conserved sequences are considered to be appropriate for classification at the higher end of the taxonomic spectrum – e.g. kingdom, domain – while variable sequences are deemed to be useful for lower-rank taxa.) 16S rRNA oligonucleotide cataloguing initially revealed two kingdoms – Archaebacteria and Eubacteria – each later elevated to domain status (Archaea and Bacteria, respectively).

There is also a 'consensus' form of taxonomy – *polyphasic taxonomy* – in which the aim is to base classification on the *maximum* amount of information available, including both phenotypic and molecular data [polyphasic taxonomy: MR (1996) *60* 407–438].

A *taxonomic hierarchy* is a classificatory system in which a single, all-inclusive category (e.g. 'bacteria' or 'fungi') is progressively divided into smaller and less-inclusive categories, all the organisms in a given category conforming to the same criteria (i.e. all having, or lacking, certain feature(s)). A taxonomic hierarchy may have the following principal ranks (listed in descending order of inclusiveness): *division* (in botanical schemes) or *phylum* (in zoological schemes); *class; order; family; genus*; SPECIES; one or more species constitute a genus, one or more genera constitute a family, and so on. There are also intermediate ranks: see e.g. the table in the entry NOMENCLATURE. (See also FORM GENUS and SACCARDOAN SYSTEM.) A species can be subdivided into STRAINS which are distinguishable by TYPING procedures. (See also FORMA SPECIALIS; SEROTYPE; VARIETY.)

Taxonomists themselves are sometimes classified as *lumpers* or *splitters*. In any given system of classification, a 'lumper' would tend e.g. to 'lump together' a number of more or less similar species to form a single genus; a 'splitter' – emphasizing differences between the species – would tend to establish two or more genera from those same species. (The scope for lumping and splitting would seem to be less in the era of modern molecular taxonomy.)

(See also ALGAE; ARCHAEA; BACTERIA; CYANOBACTERIA; FUNGI; PROTOZOA; VIRUS.)

Taxopodida An order of protozoa (class HELIOZOEA) which are bilaterally symmetrical; the axopodia are arranged in parallel longitudinal rows, and the organisms apparently swim by a 'rowing' action of the axopodia. Siliceous spines are present. Genera include *Sticholonche*.

taxospecies Any given group of strains which exhibit a high degree of mutual phenetic similarity.

tazobactam A substituted penicillanic acid sulphone which acts as a β-LACTAMASE INHIBITOR; it is used e.g. with piperacillin.

TB TUBERCULOSIS.

Tb phage TBILISI PHAGE.

TBE virus Tick-borne encephalitis virus: see FLAVIVIRIDAE.

Tbilisi phage (bacteriophage Tb; Tb phage) A group 1 *Brucella* phage (family PODOVIRIDAE) which can replicate efficiently only in smooth (S), smooth-intermediate (SI) and intermediate (I) strains of *B. abortus*; some replication can occur in S, SI and I strains of *B. neotomae* but the EOP is low. High concentrations of Tb phage can lyse S, SI and I strains of *B. suis*.

TBP See PROMOTER.

TbpA, TbpB See VACCINE.

TBSV Tomato bushy stunt virus – see TOMBUSVIRUSES.

TBTO (tributyltin oxide; bis(tri-*n*-butyltin) oxide) A compound of TIN used e.g. for TIMBER PRESERVATION and as an anti-algal constituent in ships' anti-fouling paints. (The green alga *Ankistrodesmus falcatus* concentrates TBTO with an apparent bioconcentration factor of 3×10^4, converting it to the dibutyl derivative and to small amounts of the monobutyl derivative and inorganic tin [CJFAS (1984) *41* 537–540].)

TCA cycle (tricarboxylic acid cycle; citric acid cycle; Krebs cycle) A cyclic sequence of reactions [see Appendix II(a)] which plays a central role in the metabolism of many microorganisms, both eukaryotic and prokaryotic, as well as of higher organisms. In eukaryotes the enzymes of the TCA cycle are located in the MITOCHONDRION; in both prokaryotes and eukaryotes most of the enzymes are soluble (in the cytoplasm or matrix, respectively), but the succinate dehydrogenase is membrane-bound in both cases. The TCA cycle can have both catabolic and anabolic functions, the relative importance of which often depends largely on the physiological state of the cell.

When the TCA cycle is functioning *catabolically*, acetyl-CoA – derived from carbohydrate metabolism (via pyruvate: see Appendix I) or e.g. from fatty acid degradation – enters the cycle by condensation with oxaloacetate (OAA) to form citrate; with one complete turn of the cycle, 2 molecules of CO_2 are eliminated, 3 molecules of $NAD(P)^+$ and one of FAD are reduced, one molecule of ATP (or GTP) is synthesized (by SUBSTRATE-LEVEL PHOSPHORYLATION), and one molecule of OAA is regenerated. Thus, the acetyl group is effectively completely oxidized. The NADH and $FADH_2$ may be reoxidized via a respiratory chain (see ELECTRON TRANSPORT CHAIN) to yield energy. The enzyme isocitrate dehydrogenase is specific for NADP in most bacteria and e.g. in yeasts, and this reaction – together with the HEXOSE MONOPHOSPHATE

PATHWAY – is probably an important source of NADPH for biosynthesis in many organisms [e.g. yeasts: JGM (1983) *129* 953–964].

When functioning *anabolically*, the TCA cycle provides precursors for various biosynthetic pathways [see Appendix II(b)]. Since intermediates withdrawn from the cycle cannot be used to regenerate OAA, OAA must be generated in some other way if the cycle is to continue to operate; various reactions for achieving this occur in microorganisms: e.g., OAA may be formed directly by the carboxylation of pyruvate or of phosphoenolpyruvate (PEP) [see Appendix II(b)]. Such replenishing reactions are known as *anaplerotic sequences*. (See also GLUCONEOGENESIS.)

Many bacteria can use TCA cycle intermediates (di- and tricarboxylic acids) as sole sources of carbon and energy. (However, possession of the TCA cycle does not *necessarily* enable an organism to grow on such compounds; the presence of a TRANSPORT SYSTEM for the substrate is also necessary – see also CIT PLASMID.) Citrate, for example, can be converted to OAA via reactions of the TCA cycle; however, acetate must then be formed (from another molecule of citrate) if the cycle is to continue. This may be achieved by the decarboxylation of malate (by the 'malic enzyme') to pyruvate which, in turn, is converted to acetyl-CoA by the pyruvate dehydrogenase complex [Appendix II(b)]. (See also CITRIC ACID.)

If acetate (or an acetate-generating substrate such as a fatty acid) is used as the sole source of carbon and energy under aerobic conditions, acetate can be converted to acetyl-CoA by acetyl-CoA synthetase (with concomitant formation of AMP and PPi from ATP) and can thus be metabolized via the TCA cycle. However, if intermediates are withdrawn from the cycle for biosynthesis, OAA cannot be generated by carboxylation of pyruvate or PEP since in aerobic organisms pyruvate cannot be synthesized directly from acetate. Under these conditions dicarboxylic acids are synthesized by the *glyoxylate cycle* (*glyoxylate shunt*) involving two reactions, catalysed by isocitrate lyase (ICL) and malate synthase, which bypass the decarboxylation reactions of the TCA cycle [see Appendix II(b)]. Thus, in effect, one molecule of a 4-C dicarboxylic acid is generated from two of acetate for each turn of the glyoxylate cycle; pyruvate – required e.g. for GLUCONEOGENESIS – can be formed by the decarboxylation of OAA generated in this way. The key enzyme in controlling whether isocitrate is metabolized via the TCA cycle or via the glyoxylate cycle is *isocitrate dehydrogenase* (ICDH); ICDH has a much higher affinity for isocitrate than does ICL. In *Escherichia coli*, growth on acetate induces the synthesis of a bifunctional enzyme, *ICDH kinase–phosphatase*, which catalyses the phosphorylation and dephosphorylation of ICDH; phosphorylation inactivates ICDH, allowing isocitrate to reach concentrations sufficient to permit its flow into the glyoxylate cycle. The phosphatase function of ICDH kinase–phosphatase is stimulated, and the kinase function inhibited, by various metabolites – particularly pyruvate [JGM (1986) *132* 797–806]. (See also PROTEIN KINASE and GLYOXYLATE CYCLE.)

Some bacteria can grow on glyoxylate (formed e.g. by purine degradation), or on glyoxylate precursors such as glycolate, by using the glyoxylate cycle. In this case acetyl-CoA is formed via 3-phosphoglycerate produced by the *tartronic semialdehyde pathway* (= *glycerate pathway*) [see Appendix II(b)]; acetyl-CoA may then condense with another molecule of glyoxylate to form malate and hence OAA.

TCA cycle enzymes are subject to various control mechanisms affecting their biosynthesis and/or activity. For example, in *E. coli* they are repressed by anaerobiosis. However, when conditions are such that some of the enzymes are necessary for biosynthesis, the levels of those enzymes are increased despite anaerobiosis; under such conditions 2-oxoglutarate dehydrogenase (= α-ketoglutarate dehydrogenase) remains repressed and the pathway ceases to be cyclic, functioning in biosynthesis as two linear pathways: an oxidative pathway from citrate to 2-oxoglutarate, and a reductive pathway from OAA to succinyl-CoA. 2-Oxoglutarate dehydrogenase is also repressed by anaerobiosis in many other facultatively anaerobic organisms. The TCA cycle is inherently incomplete in some bacteria: e.g. in cyanobacteria, in *Gluconobacter*, and in some members of the Methylococcaceae; in at least some cyanobacteria succinyl-CoA is formed via the glyoxylate cycle rather than via the reductive pathway from OAA. (See also GLUTAMIC ACID.) A complete cycle is apparently involved in the oxidation of acetate by sulphate in the obligately anaerobic sulphate-reducing bacterium *Desulfobacter postgatei* (see DESULFOBACTER), although some of the enzymes involved are unusual in certain respects: e.g., malate and possibly succinate dehydrogenases appear to use quinones instead of pyridine nucleotides as coenzymes, and 2-oxoglutarate dehydrogenase is specific for a ferredoxin instead of NAD^+ as electron acceptor. In *D. postgatei*, pyruvate (and hence OAA) can be synthesized from acetyl-CoA by direct reductive carboxylation catalysed by a ferredoxin-dependent pyruvate synthase.

[Bacterial TCA cycle enzymes (particularly citrate synthase and succinate thiokinase): AMP (1981) *22* 185–244.]

(See also REDUCTIVE TRICARBOXYLIC ACID CYCLE.)

TCBS agar Thiosulphate–citrate–bile salts agar: a medium containing peptone, sucrose, thiosulphate, citrate, cholate, oxgall, ferric citrate, NaCl, thymol blue, and bromthymol blue [recipe: Book ref. 46, p. 1276]. It is used for the isolation of *Vibrio* spp, especially *V. cholerae*; sucrose-fermenting strains form yellow colonies. The medium is strongly inhibitory for most faecal enterobacteria, pseudomonads, and Gram-positive bacteria.

TCC See CARBANILIDES.

TCCR (cytokine receptor) See CYTOKINES.

TCD_{50} An alternative expression for $TCID_{50}$ (q.v.).

***tcdA, tcdB, tcdE* genes** See LYSIS PROTEIN (sense 3).

TCGF See INTERLEUKIN-2.

TCH test *Syn.* T2H TEST.

$TCID_{50}$ (TCD_{50}) Tissue culture infective dose (50%): that dose of a viral suspension which, when used to inoculate each of a number of TISSUE CULTURES, causes observable effects in 50% of those cultures. (See also END-POINT DILUTION ASSAY.)

TCNB See TECNAZENE.

TCP Toxin co-regulated pili: see CHOLERA, BACTERIOPHAGE CTXΦ and PATHOGENICITY ISLAND.

TCR See T CELL RECEPTOR.

t_D See THERMAL DEATH TIME.

TD antigen THYMUS-DEPENDENT ANTIGEN.

***td* gene** (in bacteriophage T4) See SPLIT GENE.

TDH See FOOD POISONING (*Vibrio*).

tDNA DNA which encodes tRNA.

T-DNA See CROWN GALL.

tea diseases Diseases of the tea plant, *Camellia sinensis* (= *Thea sinensis*), include e.g. BLISTER BLIGHT and RED RUST.

teart See SCOURS.

teat (*microbiol.*) A rubber bulb used e.g. for operating a pipette.

tecnazene (TCNB) An agricultural antifungal agent (2,3,5,6-tetrachloronitrobenzene) used e.g. (as a dust) for the control of dry rot (*Fusarium coeruleum*) of stored potatoes; it also inhibits the sprouting of stored potatoes. TCNB is also used (as a smoke) in glasshouses for the control of *Botrytis* infections. (cf. QUINTOZENE.)

Tectibacter A genus of Gram-negative bacteria which occur as endosymbionts in the cytoplasm of strains of *Paramecium aurelia*. Cells: peritrichously flagellated straight rods, ca. 0.4–0.7 × 1.0–2.0 μm. No strains appear to be toxic to protozoa. Type species: *T. vulgaris* (earlier name: delta particles).

[Book ref. 22, p. 811.]

Tectiviridae (PRD1 phage group) A family of lipid-containing, icosahedral BACTERIOPHAGES (diam. (53-)65(-80) nm) which contain linear dsDNA (MWt ca. 10^7). One genus: *Tectivirus*; type species: phage PRD1. Host range for the family includes Gram-negative and Gram-positive bacteria. Phage progeny are released by cell lysis. Plaques: clear or turbid, minute to large. The virion contains a dense core, probably solely of DNA, surrounded by a lipid bilayer – the whole being enclosed by the icosahedral protein shell; single spikes (ca. 20 nm long) are sometimes seen on the vertices. Some virions have a 'tail' (length variable) which appears when DNA is ejected or virions are damaged and which may be part of a mechanism for DNA ejection. Virions are sensitive to organic solvents and detergents.

Phages PRD1, PR3, PR4, PR772 (probably = PR4), PR5 and L17 are closely related and are specific for Gram-negative bacteria (e.g. *Escherichia coli*, *Salmonella typhimurium*, *Pseudomonas aeruginosa*) which harbour IncN, IncP1 or IncW plasmids. The site of phage adsorption is controversial: phage may adsorb to pili (which may then retract) or directly to the cell surface.

Other tectiviruses attack *Bacillus* spp. Phages AP50 (host: some strains of *B. anthracis*) and Bam35 (hosts: *B. cereus*, *B. megaterium*, *B. thuringiensis*) are very similar and are both highly sensitive to heat. Phage ϕNS11 is acidophilic and thermophilic (as is its host, *B. acidocaldarius*). ϕNS11 is stable at pH 2–5 but is rapidly inactivated at pH above 6; it contains the unusual ω-cyclohexyl fatty acids characteristic of *B. acidocaldarius* cell membranes.

Tectivirus See TECTIVIRIDAE.

tectum (*protozool.*) See COCHLIOPODIUM.

teeth (diseases of) See DENTAL CARIES and PERIODONTITIS.

Teflon filter See FILTRATION.

Tego compounds A range of antimicrobial ampholytic surfactants used e.g. as antiseptics and disinfectants; an example is Tego 103S (dodecyl-di(aminoethyl)-glycine hydrochloride). [Book ref. 65, pp. 335–345.]

tegulicolous Growing on tiles.

tegument See HERPESVIRIDAE.

teichoic acids (TAs) Polymers which typically consist mainly of glycerol phosphate or ribitol phosphate (or rarely mannitol phosphate) substituted more or less extensively with amino acids and/or sugars; they occur in the cell envelope in most Gram-positive bacteria. (TAs are very rare in Gram-negative bacteria, but have been found in strains of *Butyrivibrio* and of *Escherichia coli* – see below.)

Glycerol teichoic acids (GTAs) occur in the CYTOPLASMIC MEMBRANE, in the periplasmic region, and/or in the CELL WALL; they are commonly composed of glycerol residues linked 1 → 3 by phosphodiester bridges and substituted with e.g. D-alanine, *N*-acetylglucosamine (GlcNAc), glucose, KOJIBIOSE, etc. (See also C SUBSTANCE.) In some GTAs a sugar forms an integral part of the polymer backbone. (The K2 capsular antigen of *E. coli* contains alternating galactofuranosylglycerol phosphate and galactopyranosylglycerol phosphate residues [Biochem. (1982) *21* 1279–1284].)

Ribitol teichoic acids (RTAs) occur mainly in the cell wall, and consist of ribitol residues linked 1 → 5 by phosphodiester bridges; sugar residues may occur within the chain. The ribitol residues are usually substituted with D-alanine and sometimes also with e.g. GlcNAc, glucose, etc.

In both types of TA the nature and extent of substitution, and e.g. the ratio of GTA to RTA, may vary greatly between strains and even in the same strain under different growth conditions. (See also TEICHURONIC ACIDS.) TAs may account for ca. 20–60% of the dry weight of the cell wall in Gram-positive bacteria. Wall TA tends to be concentrated towards the cell surface, and is linked covalently to PEPTIDOGLYCAN via a 'linkage unit' attached to the 6-position of a muramic acid residue; in the RTAs of e.g. *Staphylococcus aureus* strains the linkage unit consists of tri-(glycerol phosphate)-*N*-acetylglucosamine 1-phosphate, while in the GTAs of strains of *Bacillus subtilis* and *B. licheniformis* it is *N*-acetylmannosaminyl-(1 → 4)-*N*-acetylglucosamine [JB (1984) *158* 990–996]. In the cytoplasmic membrane, TAs are linked covalently to glycolipid, forming *lipoteichoic acid* (LTA); membrane LTAs are found in most Gram-positive bacteria, and have been isolated from *Butyrivibrio fibrisolvens* [JGM (1976) *94* 126–130] (cf. BUTYRIVIBRIO). In e.g. group A streptococci the LTA appears to be anchored to the cytoplasmic membrane by the glycolipid and to extend through the cell wall into the medium. (TAs may also be released into the medium.)

[Teichoic acids in the genus *Nocardiopsis*: FEMS Reviews (2001) *25* 269–283.]

TAs bind cations and may function as a cation-exchange system, regulating the flow of ions through the cell envelope and controlling the availability of ions (particularly Mg^{2+}) necessary for the function of membrane-bound enzymes (e.g. autolysins). (cf. BILE SOLUBILITY.) Streptococcal LTA may function as an ADHESIN. TAs at the cell surface can also act as phage receptors.

Antibodies to staphylococcal RTAs can be found in serum from patients with staphylococcal infections, and the detection of such antibodies can be useful e.g. in the diagnosis of staphylococcal endocarditis. [Evaluation of commercial antibody assay ENDO-STAPH: Arch. Int. Med. (1984) *144* 250–252, 261–264.]

Biosynthesis of wall TAs begins with the synthesis of the linkage unit. In the case of the wall RTAs of *S. aureus*, GlcNAc phosphate is transferred (in a TUNICAMYCIN-sensitive step) from its UDP derivative to a cytoplasmic membrane-bound polyisoprenol phosphate carrier (see] BACTOPRENOL); three glycerol phosphate residues are then added sequentially from three CDP-glycerol molecules (synthesized from glycerol 1-phosphate and CTP). Ribitol phosphate units are then transferred from CDP-ribitol molecules (synthesized from ribitol 5-phosphate and CTP) and polymerized on the linkage unit. After the addition of the various substituents, the entire molecule is transferred to peptidoglycan. The transmembrane translocation of wall polymers apparently requires a pmf [JB (1984) *159* 925–933]. [Review of TA biosynthesis: Book ref. 125, pp. 279–307.]

teichomycin A2 *Syn.* TEICOPLANIN.

teichuronic acids Cell wall polymers which contain glycuronic acid(s) and other sugars, but not phosphate; they are formed – either in place of or in addition to TEICHOIC ACIDS – by many Gram-positive bacteria growing in phosphate-deficient media. Teichuronic acids may contain e.g. D-glucuronic acid and *N*-acetylgalactosamine (in strains of *Bacillus licheniformis*); glucose, rhamnose and D-glucuronic acid (*B. megaterium*); or *N*-acetylmannosaminuronic acid and glucose (*Micrococcus luteus*). As with teichoic acids, teichuronic acids are covalently linked by phosphodiester bridges to muramic acid residues in peptidoglycan – via a 'linkage unit' in *M. luteus* but apparently directly in *B. licheniformis*. [Biosynthesis: Book ref. 125, pp. 279–307.]

teicoplanin (teichomycin A2) A glycopeptide ANTIBIOTIC produced by *Actinoplanes teichomyceticus*. It resembles VANCOMYCIN both structurally and functionally, but is more active against e.g. enterococci [AAC (1986) *29* 52–57].

teleology The doctrine which states that developments or phenomena tend to be determined by the purpose or function which they serve.

teleomorph (*mycol.*) The sexual (= perfect) stage or state of a given fungus. (cf. ANAMORPH and HOLOMORPH.)

teleutosorus See UREDINIOMYCETES stage III.

teleutospore See UREDINIOMYCETES stage III and USTILAGINALES.

Teliomycetes (Hemibasidiomycetes) An obsolete class of fungi which included the rusts and smuts – organisms now included in the UREDINIOMYCETES and USTILAGINOMYCETES respectively.

teliospore A type of thick-walled spore characteristic of fungi of the UREDINIOMYCETES and USTILAGINALES, and of some fungi of the SPORIDIALES; on germination, a teliospore usually gives rise to a basidium.

telium See UREDINIOMYCETES stage III.

***Tellina tenuis* virus** See IPN VIRUS.

tellurite media Media which are inhibitory to many species of bacteria owing to their content of potassium tellurite (K_2TeO_3); in sensitive cells of *Escherichia coli*, tellurite is apparently taken up by a phosphate transport system, and resistance to tellurite can result from mutation in the *phoB–phoA* region of the PHO REGULON [AAC (1986) *30* 127–131]. (Tellurite resistance in e.g. *Citrobacter*, *Klebsiella* and *Pseudomonas* spp may be plasmid-mediated.) Cystine–tellurite–blood agar [recipe: Book ref. 2, pp. 132–133], which contains 0.0375% (w/v) K_2TeO_3, is used e.g. for the primary isolation of *Corynebacterium diphtheriae*, a species inherently resistant to tellurite. Certain bacteria can reduce K_2TeO_3 to (black) metallic tellurium. In a test for strains of the *Mycobacterium avium* complex, 2 drops of sterile aqueous K_2TeO_3 (0.2%) are added to a 7-day culture in Middlebrook 7H-9 broth (5 ml); a black precipitate within ca. 4 days is a positive reaction. (See also MONSUR MEDIUM.)

telokinetal stomatogenesis See STOMATOGENESIS.

telomere A terminal section of a eukaryotic chromosome – comprising perhaps a few hundred base pairs – which has a specialized structure [ARB (1984) *53* 175–194] and which is involved in chromosomal replication and stability.

telophase See MITOSIS and MEIOSIS.

Teloschistales An order of fungi of the ASCOMYCOTINA; members include LICHENS and lichenicolous fungi. Ascocarp: APOTHECIOID, with paraphyses. Asci: bitunicate (lecanoralean). Ascospores: ellipsoidal, aseptate or POLARILOCULAR. Genera: e.g. CALOPLACA, TELOSCHISTES, XANTHORIA (these three genera being in the family Teloschistaceae).

Teloschistes A genus of fruticose LICHENS of the order TELOSCHISTALES. The thallus contains PARIETIN and is usually yellow or orange (but may be grey in some species); it is more or less branched, the branches being flattened (e.g. in *T. chrysophthalmus*) or terete (e.g. in *T. flavicans*). Species occur on rocks, trees etc.

Telosporea A class of parasitic protozoa (comprising the subclasses Coccidia and Gregarinia) classified in the subphylum SPOROZOA; these organisms, together with those of the TOXOPLASMEA and the PIROPLASMEA, have been incorporated in the class SPOROZOASIDA.

telotroch The motile (ciliated) immature stage of a sedentary peritrich. (See also TROCHAL BAND.)

TEM See ELECTRON MICROSCOPY (a).

TEM-1 β-lactamases See β-LACTAMASES.

TEM-2 β-lactamases See β-LACTAMASES.

tempeh (tempe kedelee) A food, originally from Indonesia, consisting of soybeans fermented and bound into a cake by mycelium of *Rhizopus* spp (e.g. *R. oligosporus*); a strain of *Klebsiella pneumoniae* is also present and contributes vitamin B_{12} to the product [Book ref. 6, pp. 414–416]. Soybeans are soaked, dehulled, boiled, surface-dried, and mixed with an inoculum of *Rhizopus* sp; the beans are then packed tightly into a container and incubated until they are coated and bound by an abundant white mycelium (e.g. 32°C for ca. 20 hours). *Bongkrek* (*tempe bongkrek*) is a similar product made from coconut presscake or grits remaining after the extraction of coconut oil; fermentation may be carried out by *R. oligosporus* or by *Neurospora intermedia*. (See also BONGKREKIC ACID; cf. ONCOM.)

[Tempe (recent data): Book ref. 228.]

temperate bacteriophage A BACTERIOPHAGE which can lysogenize its host cell (see LYSOGENY). (cf. VIRULENT BACTERIOPHAGE.)

temperature coefficient (θ; Q) Of a given DISINFECTANT acting on a single-species population of microorganisms: a coefficient which indicates the factor by which the death rate increases when the temperature rises by 1°C – or, more commonly, 10°C, in which case the temperature coefficient is designated θ^{10} or Q_{10}. The temperature coefficient for a given temperature range (e.g. 10–20°C) is not necessarily the same as that for other ranges (e.g. 0–10°C; 20–30°C).

temperature-sensitive mutant (*ts* mutant) A mutant (e.g. a CONDITIONAL LETHAL MUTANT) in which the mutated gene is functional at certain temperatures (*permissive* temperatures) but not at others (*non-permissive* or *restrictive* temperatures); *ts* mutations are commonly MIS-SENSE MUTATIONS which alter an essential protein such that it functions at the permissive temperature but not at the restrictive (usually higher) temperature.

template strand (of DNA) See CODING STRAND.

temporal temperature gradient gel electrophoresis (TTGE) See DGGE.

TEN TOXIC EPIDERMAL NECROLYSIS.

−10 sequence See PROMOTER.

Tenericutes A proposed division of the PROCARYOTAE comprising one class (MOLLICUTES).

tennecetin *Syn.* PIMARICIN.

tentoxin A cyclic tetrapeptide TOXIN produced by *Alternaria* spp; it binds non-covalently to, and inhibits, (F_0F_1)-type PROTON ATPASES, inducing chlorosis in various plants.

tequila See PULQUE.

teratogen Any agent which causes developmental abnormalities in an embryo or fetus. (See also TORCH DISEASES; SYPHILIS.)

teratoma A neoplasm composed of more than one type of tissue.

terbinafine A synthetic allylamine antifungal agent which is used clinically e.g. against dermatophyte infections.

The activity of terbinafine against *Aspergillus* spp has been compared, in vitro, with the activities of itraconazole and amphotericin B; compared with these two agents, terbinafine had greater activity against *A. flavus*, *A. terreus* and *A. niger* but it was less effective against *A. fumigatus* [AAC (2001) *45* 1882–1885].

terbutryne An ALGICIDE and herbicide which is active against a range of algae (e.g. species of *Cladophora*, *Enteromorpha*, *Rhizoclonium*, *Spirogyra*) and against certain aquatic vascular plants. It is useful for the control of algae and water-weeds in static and slow-moving waters (lakes, canals, farm ponds, etc).

terconazole See AZOLE ANTIFUNGAL AGENTS.

terete Circular in transverse section.

terminal deoxynucleotidyl transferase See TAILING.

terminal electron acceptor See e.g. RESPIRATION.

terminal protein (TP) See BACTERIOPHAGE ϕ29.

terminal redundancy (*mol. biol.*) The presence of identical base sequences at the two ends of a linear nucleic acid molecule. In e.g. a terminally redundant dsDNA molecule, removal of nucleotides from e.g. the 3′ end of each strand results in the formation of STICKY ENDS.

terminal transferase See TAILING.

termination codon See GENETIC CODE.

terminator See TRANSCRIPTION.

Termitaria A genus of fungi of the COELOMYCETES; *T. snyderi* is parasitic on termites.

termite–microbe associations Termites are social insects of the Isoptera, found mainly in tropical regions; there are ca. 2000 species, many of which can feed on wood and can cause severe economic damage in buildings etc. Many species depend on symbiotic associations with microorganisms for the digestion of cellulose and other components in their diet.

In 'lower' termites (Hodotermitidae, Kalotermitidae, Mastotermitidae, Rhinotermitidae, Serritermitidae) the hindgut is densely packed with microorganisms, including cellulolytic flagellate protozoa (e.g. TRICHONYMPHA) and various facultatively or strictly anaerobic bacteria (e.g. strains of *Bacillus*, *Bacteroides*, enterobacteria, spirochaetes (see SPIROCHAETALES), *Staphylococcus* and *Streptococcus*). The protozoa phagocytose particles of cellulosic material ingested by the termite and ferment the cellulose to acetate, CO_2 and H_2. Acetate is absorbed by the termite and used as the major oxidizable energy source for termite metabolism. The H_2 and CO_2 are used by methanogens and bacteria (possibly including *Acetobacterium*-like acetogens: cf. ANAEROBIC DIGESTION). Other bacteria may generate acetate and other volatile fatty acids from mono-, di- and oligosaccharides present in material ingested by the termite or released from the protozoa. Thus, the protozoa are the main agents of cellulose degradation, while prokaryotes serve to remove H_2 (which would inhibit further cellulose degradation), maintain anaerobiosis, and possibly provide essential nutrients for the protozoa which feed on them. Some of the bacteria are also important in the nitrogen economy of the termite. The termites' diet is often low in nitrogen, and the gut bacteria may play an important role by NITROGEN FIXATION and/or by recycling the nitrogen of uric acid produced in the termite tissues; uric acid is passed into the hindgut where it is used as an energy source by e.g. *Bacteroides*, *Citrobacter*, and *Streptococcus* spp, being fermented to NH_3, CO_2 and acetate. (cf. RUMEN.) The gut microflora is also apparently involved in the degradation of other plant components (e.g. hemicelluloses, lignin), but the mechanisms are less well understood. [Review: Book ref. 77, pp. 173–203.]

In 'higher' termites (Termitidae) the hindgut contains a large population of bacteria [AEM (1985) *49* 1226–1236] but no cellulolytic protozoa. Members of the subfamily Macrotermitinae 'cultivate' a fungus (*Termitomyces* sp) in *fungus combs* analogous to the FUNGUS GARDENS of the parasol ants; the termites construct a comb from small fragments of plant tissue and termite faeces, and the surface of this comb becomes covered with a sparse fungal mycelium which develops numerous white, spherical nodules (*mycotêtes* or *gongylidia*) 0.5–2.0 mm in diameter. The termites feed on these nodules and on the comb itself; fungal cellobiohydrolases (exoglucanases – see CELLULASES) consumed with the nodules are used in the digestion of cellulose components in the diet [Book ref. 77, pp. 156–161]. When new termite colonies are established, the inoculum of *Termitomyces* necessary for establishing a new fungus comb may be either carried (as conidia) in the guts of the winged insects (e.g. in *Macrotermes bellicosus*, *Microtermes* spp) or collected (as basidiospores) by foraging workers (e.g. in *Macrotermes subhyalinus*, *Ancistrotermes*, *Odontotermes*).

(See also INSECT–MICROBE ASSOCIATIONS.)

Termitomyces A genus of white-rot fungi (family Amanitaceae) found only in certain termite nests (see TERMITE–MICROBE ASSOCIATIONS).

ternary combination In NOMENCLATURE: a designation which consists of a binomial together with a subspecific epithet – see e.g. SUBSPECIES. (cf. TRINOMIAL.)

ternary fission An asexual reproductive process in which more than two cells are formed from one cell (cf. BINARY FISSION); ternary fission occurs e.g. in PELODICTYON.

terramycin See TETRACYCLINES.

Terrazole See ETRIDIAZOLE.

terregens factor See SIDEROPHORES.

terricolous Growing on soil, peat, sand etc.

tertian malaria See MALARIA.

Teschen disease A PIG DISEASE caused by a porcine ENTEROVIRUS; it occurs e.g. in Central Europe and is named after the Teschen (Cieszyn) district of (the former) Czechoslovakia. Symptoms include fever followed by signs of encephalomyelitis: incoordination, tremors, convulsions and paralysis; the virus can be isolated from the faeces of infected pigs. The mortality rate may be 70% or higher, death occurring in 3–4 days. The disease may be controlled by slaughter or by vaccination using a live, attenuated vaccine. (cf. TALFAN DISEASE.)

tesla (T) A unit of magnetic flux density; 1 tesla = 10000 gauss.

TESPA See AAS.

test (*protozool.*) (1) A close-fitting protective 'shell' which surrounds the cell e.g. in certain (testate) amoebae; the test may be predominantly or entirely organic (as in e.g. ARCELLA) or may incorporate inorganic materials such as sand grains (e.g. DIFFLUGIA), silica (e.g. EUGLYPHA), etc. (See also FORAMINIFERIDA.) (2) *Syn.* LORICA.

test medium See MEDIUM.

Testacealobosia A subclass of testate amoebae (see ARCELLINIDA and TRICHOSIDA). (cf. GYMNAMOEBIA.)

testate Possessing a TEST.

testosterone synthesis See STEROID BIOCONVERSIONS.

Tet protein See TETRACYCLINES.

***tetA* gene** See TETRACYCLINES.

tetaine *Syn.* BACILYSIN.

tetanolysin A THIOL-ACTIVATED CYTOLYSIN produced by *Clostridium tetani*; it is distinct from TETANOSPASMIN.

tetanospasmin (tetanus toxin) A plasmid-encoded neurotoxin, produced by *Clostridium tetani*, which causes the symptoms of TETANUS.

Tetanospasmin is a single polypeptide of ~150 kDa. Activation involves proteolytic cleavage to form a di-chain structure consisting of an H chain (~100 kDa) and an L chain (~50 kDa) held together by a disulphide bond.

The activated toxin binds, via the H chain, to specific sites on the cell membrane of a neurone at the nerve–muscle junction; receptor sites on the membrane may include (i) polysialogangliosides and (ii) a specific protein.

Surface-bound toxin is internalized by uptake within a vesicle. Toxin-containing vesicles are then translocated back along the axon to interneurones in the spinal cord. According to one model, release of the active component (the L chain) occurs after acidification of the vesicle has brought about a conformational change in the H chain; the H chain may then insert into the vesicle's membrane and bring about the translocation of the L chain from the vesicle to the cytoplasm of the neurone. Reduction of the disulphide bond then releases the active toxin.

Normally, in the absence of toxin, contraction of a given muscle stimulates the interneurones to release an inhibitory neurotransmitter (?glycine) which prevents contraction of the corresponding antagonist muscle. Neurotransmitter, present in vesicles within the interneurone, is released into the (extracellular) synaptic cleft by a process in which the vesicles first 'dock' at specific sites on the (inner) surface of the cell's membrane and then fuse with the membrane (= *exocytosis* or *neuroexocytosis*); this process is triggered by appropriate electrical stimulation and the resulting intracellular rise in levels of Ca^{2+}.

During neuroexocytosis, docking and fusion of vesicles involves specific interaction between molecules in the vesicle membrane and others at the inner surface of the neurone's membrane. These molecules include VAMP/synaptobrevin (VAMP = vesicle-associated membrane protein), and (on the neurone's membrane) SNAP-25 and syntaxin.

Tetanospasmin acts as a zinc-endopeptidase which recognizes and cleaves a specific molecule required for vesicle-docking in neuroexocytosis, i.e. it prevents the docking of vesicles and (thus) inhibits the release of neurotransmitter to the synaptic cleft. Absence of the *inhibitory* neurotransmitter permits simultaneous contraction of both protagonist and antagonist muscles in a given pair, thus giving rise to spastic (rigid) paralysis. The target for tetanospasmin is apparently VAMP/synaptobrevin.

Tetanospasmin is also able to act peripherally, at the nerve–muscle junction, giving rise to a local, flaccid, paralysis.

[Mechanism of action of tetanus neurotoxin: Mol. Microbiol. (1994) *13* 1–8.]

tetanus (lockjaw) An acute disease, affecting man and animals, caused by toxinogenic strains of *Clostridium tetani*. The disease develops when *C. tetani* contaminates wounds or other lesions, especially those which have become anaerobic due e.g. to necrosis or to infection with other bacteria; common sources of *C. tetani* are faeces and soil. The incubation period may be days or weeks; short incubation periods correlate with high mortalities, and vice versa. Symptoms – caused by TETANOSPASMIN – may begin with stiffness or spasms of the jaw muscles, typically followed by a generalized tetanus: movements or external stimuli (e.g. noise) trigger violent, painful, convulsive spasms of the muscles in the trunk and limbs, and there is a characteristic contorted facial expression (*risus sardonicus*). Death may result from asphyxia or exhaustion. Less common forms of the disease include *local tetanus* (rigidity and increased muscle tone in the vicinity of the infected wound) and *cephalic tetanus* (in which certain cranial nerves becomes paralysed, usually following infection of a head lesion). *Treatment* may involve e.g. administration of (preferably human) antitoxin, anticonvulsant drugs, and/or neuromuscular blockade. Antitoxin may also be used for short-term prophylaxis; longer-term (but still temporary) protection is achieved by vaccination with tetanus TOXOID. (cf. DPT and TABT.)

tetanus toxin TETANOSPASMIN. (cf. TETANOLYSIN.)

tethering (of leukocytes) See INFLAMMATION.

tetracyclines A family of broad-spectrum ANTIBIOTICS whose structure is based on the naphthacene skeleton (see figure); they are produced by species of *Streptomyces* (e.g. *S. aureofaciens*, *S. rimosus*) and/or made semi-synthetically.

The tetracyclines are used primarily as antibacterial (bacteriostatic) agents in the treatment of human and animal diseases caused e.g. by species of *Brucella*, *Chlamydia*, *Mycoplasma*, *Rickettsia* and *Vibrio*. (Tetracyclines have also been used against certain plant diseases – e.g. COCONUT LETHAL YELLOWING and FIREBLIGHT.)

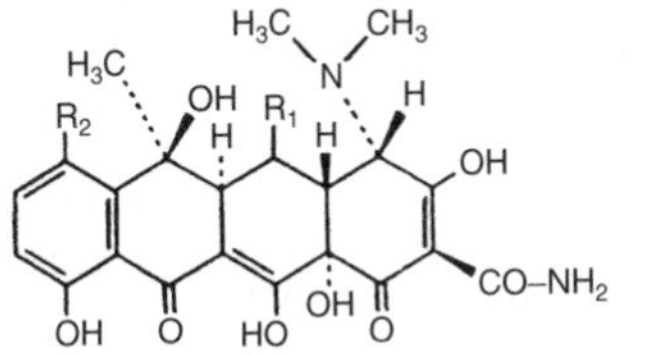

TETRACYCLINES: tetracycline (R_1, R_2 = H), chlortetracycline (R_1 = H, R_2 = Cl), and oxytetracycline (R_1 = OH, R_2 = H).

In most or all susceptible bacteria, tetracyclines are taken up in an energy-dependent manner. The drugs bind to ribosomes, inhibiting PROTEIN SYNTHESIS by inhibiting the binding of aminoacyl-tRNA complexes to the A site on the 30S ribosomal subunit. It appears that only one drug molecule per ribosome is sufficient for inhibition; in *Escherichia coli* tetracyclines bind preferentially to protein S7 near the A site.

Bacterial resistance to tetracyclines. Resistance to tetracyclines can arise in several distinct ways, and resistance to one of the drugs commonly means resistance to most or all of them.

Resistance may involve bacterial modification (i.e. detoxification) of the drugs, but this appears to occur only rarely.

In another mechanism, the Tet(M) protein appears to interact with the machinery of protein synthesis, reducing the influence of tetracyclines on aminoacyl-tRNA–ribosome binding [JB (1993) *175* 7209–7215]. (The Tet(M) protein is encoded by gene *tet*(M); *tet*(M) occurs in all conjugative transposons (see CONJUGATIVE TRANSPOSITION) from pathogenic species.)

Resistance can be due to an EFFLUX MECHANISM. For example, in some strains of *Escherichia coli* an inducible protein in the cytoplasmic membrane, Tet, appears to pump tetracycline outwards into the periplasm [JB (1995) *177* 998–1007]; the gene encoding this pump (*tetA*) is found e.g. in transposon Tn*10*. (See also TN1721.) Resistance determinants may also be carried on R PLASMIDS.

[Tetracycline resistance: FEMS Reviews (1996) *19* 1–24.]

A tetracycline referred to as 'GAR-936' (the 9-*t*-butylglycylamido derivative of minocycline) has been designed with the object of evading bacterial defences, including efflux [in vitro and in vivo antibacterial activities of GAR-936: AAC (1999) *43* 738–744].

Clinically useful tetracyclines. These include e.g. *chlortetracycline* (aureomycin): 7-chlorotetracycline; *demeclocycline* (declomycin, demethylchlortetracycline): 6-demethyl-7-chlorotetracycline; *doxycycline* (vibramycin): 5-hydroxy-6-deoxytetracycline; *methacycline* (rondomycin): 6-demethyl-6-deoxy-5-hydroxy-6-methylenetetracycline; *minocycline*: 6-demethyl-6-deoxy-7-dimethylaminotetracycline; *oxytetracycline* (terramycin): 5-hydroxytetracycline; and *tetracycline* (achromycin, polycycline, tetramycin): see figure. These tetracyclines differ mainly in lipophilicity, the more lipophilic members being more readily absorbed via the gastrointestinal tract.

tetrad A group of four – e.g. four cells (particularly a regular arrangement of four cocci), or the four chromatids of a bivalent in MEIOSIS.

tetra-*O*-di(biphytanyl) diglycerol See ARCHAEA.

Tetragoniomyces A genus of fungi which have been placed in the order TREMELLALES [CJB (1981) *59* 1034–1040]. *T. uliginosus* forms structures referred to as basidia, each of which gives rise, *internally*, to four cells that may be regarded as spores [BBMS (1983) *17* 82–94 (92)] – although these cells are not described as basidiospores in the original genus description.

tetrahydrofolic acid See FOLIC ACID.

tetrahydromethanopterin See METHANOGENESIS.

tetrahydrosarcinapterin See METHANOGENESIS.

Tetrahymena A genus of ciliate protozoa (order HYMENOSTOMATIDA); species occur in various freshwater habitats, and some are endoparasitic in metazoa. Of the many species, *T. pyriformis* has been studied extensively; this organism can be grown in a sterile, defined medium (yielding an axenic culture) and its cell division can be readily synchronized (see SYNCHRONOUS CULTURE).

T. pyriformis is commonly pear-shaped, ca. 50 μm in length (range ca. 30–80 μm) with a uniform somatic ciliature of some 15 to 26 longitudinal kineties. One micronucleus and one macronucleus are commonly present, but amicronucleate strains are known. A contractile vacuole lies near the blunt posterior end, and the cytoproct is also posterior. Trichocysts are not formed. Fine structural details differ among the various SYNGENS. The cytostome is lateral, near the tapered anterior end; the buccal cavity contains one undulating membrane (= PARORAL MEMBRANE) on the right, and a group of three membranelles (AZM) on the left – an arrangement referred to, in other organisms, as 'tetrahymenal'. *Microstome* strains have a small buccal cavity; *macrostomes* have a large buccal cavity. The growth requirements of *T. pyriformis* include FOLIC ACID, LIPOIC ACID, NICOTINIC ACID, PANTOTHENIC ACID, PYRIDOXINE, RIBOFLAVIN and THIAMINE. Cultures can be kept viable at −196°C by the use of a cryoprotectant (e.g. DMSO). (See also BIOASSAY.)

tetramethyl-*p*-phenylenediamine See TMPD.

tetramethylsilane (TMS) See NUCLEAR MAGNETIC RESONANCE.

Tetramitus A genus of amoeboflagellates of the SCHIZOPYRENIDA; the flagellate stage has four flagella and a cytostome-like opening.

tetramycin See TETRACYCLINES.

Tetramyxa See PLASMODIOPHOROMYCETES.

tetraphenylborate See CHEMIOSMOSIS.

tetrapolar heterothallism See HETEROTHALLISM.

tetrapyrrole See e.g. PORPHIN, PHYCOBILIPROTEINS, VITAMIN B_{12}, and METHANOGENESIS (factor 430).

Tetraselmis A genus of unicellular, quadriflagellate, marine and freshwater green algae (division CHLOROPHYTA); the genus currently includes the former genera *Platymonas* and *Prasinocladus*. The cell has a THECA (sense 1), and the flagella – which arise from a deep pit at the anterior end of the cell – bear minute scales. The interzonal mitotic spindle is not persistent. (cf. MICROMONADOPHYCEAE.) Sexual reproduction is unknown. (See also CONVOLUTA.) [Batch and continuous culture of the marine species *T. suecica*: Wat. Res. (1985) *19* 185–190.]

Tetraselmis (formerly a member of the PRASINOPHYCEAE) is a genus in the order Tetraselmidales of the class Pleurastrophyceae [Book ref. 123, pp. 29–72]. (cf. TREBOUXIA.)

Tetraspora A genus of freshwater green algae (division CHLOROPHYTA) which are apparently closely related to CHLAMYDOMONAS spp. The vegetative stage is a colony of non-motile cells embedded in a gelatinous matrix (a 'tetrasporal' or 'palmelloid' colony: cf. PALMELLOID PHASE). In some species the cells bear characteristic long flagellum-like (but motionless) *pseudocilia*; these are similar in structure to the eukaryotic FLAGELLUM but lack one or both of the central microtubules of the axoneme. Biflagellate isogametes are released from the colonies.

tetrasporal colony See TETRASPORA.

tetraspore A type of spore which develops in groups of four.

tetrasporic Of a coccidian oocyst: containing four sporocysts.

tetra-tetra dimers (in peptidoglycan) See PEPTIDOGLYCAN (figure legend).

tetrathionate broth A medium containing peptone, bile salts, sodium thiosulphate, and $CaCO_3$; immediately before use, iodine (in KI solution) is added (to convert thiosulphate, $S_2O_3^{2-}$, to tetrathionate, $S_4O_6^{2-}$), after which the medium should not be heated. [Recipe: Book ref. 47, p. 666.] It is used e.g. for the enrichment of salmonellae, although *Proteus* spp also grow well in it.

tetrazoic Of a coccidian oocyst: containing four sporozoites per sporocyst.

tetrazolium Tetrazolium salts – e.g. nitroblue tetrazolium (NBT), 2,3,5-triphenyltetrazolium chloride (TTC, = tetrazolium red) – can be reduced to coloured formazan compounds, each salt being reduced at a specific REDOX POTENTIAL. Tetrazolium salts have a wide range of applications as redox indicator dyes. For example, they may be used to demonstrate reducing microenvironments generated as a result of microbial activity: e.g., regions of concentrated bacterial growth on the surfaces of meat or other foods may be demonstrated by the development of a colour in these regions following treatment with a tetrazolium salt solution. (Reduction of tetrazolium salts may be prevented by low pH and hence may not occur in the presence of fermentative organisms.) NBT reduction in the cytoplasm of NEUTROPHILS (resulting in the formation of bluish-black granules) is indicative of a normal 'metabolic burst' associated with PHAGOCYTOSIS, and is used in a test for normal neutrophil function [test: Book ref. 53, pp. 1255–1256]. TTC is used in commercial kits for the diagnosis of bacteriuria, TTC reduction being taken as indicative of the presence of significant numbers of actively metabolizing bacteria.

tetronomycin See MACROTETRALIDES.

tetrose A 4-carbon sugar (e.g. ERYTHROSE).

Texas fever *Syn.* REDWATER FEVER.

Texas red Sulphorhodamine 101 acid chloride: a FLUOROCHROME (excitation at ca. 568 nm) used e.g. in FLOW CYTOMETRY.

Tex-KL bacterium *Legionella dumoffii.*

textile spoilage Textiles made from vegetable (cellulosic) fibres – e.g. cotton, linen, hessian (cf. RETTING) – are susceptible to attack by a range of cellulolytic fungi (e.g. *Aspergillus* spp, *Chaetomium globosum*, *Cladosporium herbarum*), especially under damp conditions, resulting in loss of strength and flexibility of the fabric, odour production, discoloration, etc. *Zalerion* spp can cause rotting of twines and ropes exposed to seawater. Preservatives for textiles (usually for outdoor use) include e.g. copper (or zinc) naphthenate, copper–oxine, cuprammonium hydroxide, dichlorophane, SALICYLANILIDES. (See also WOOL SPOILAGE.)

Textularia See FORAMINIFERIDA.

Textulariina See FORAMINIFERIDA.

$TF_0F_1H^+$-ATPase See PROTON ATPASE.

TFT TRIFLUOROTHYMIDINE.

TGE TRANSMISSIBLE GASTROENTERITIS.

T_h (T_H) (*immunol.*) Helper T cell (see T LYMPHOCYTE).

Th1 T cells See T LYMPHOCYTE.

Th2 T cells See T LYMPHOCYTE.

Thalassicolla See RADIOLARIA.

Thalassiophyllum See PHAEOPHYTA.

Thalassiosira See DIATOMS.

thalidomide (as therapy for mycobacterial and HIV infections) See INTERFERONS.

thallic conidiogenesis See CONIDIUM.

thalline (1) Of or pertaining to a THALLUS. (2) *Syn.* SOMATIC (sense 1).

thalline margin (excipulum thallinum) (*lichenol.*) In a LECANORINE APOTHECIUM: *lichenized* tissue (i.e., tissue containing both algal and fungal cells) external to (surrounding, and sometimes obscuring) the PROPER MARGIN; the thalline margin is generally the same colour as the thallus, but may differ in colour from the hymenial disc.

thallium salts (inhibitory action) Low concentrations of thallium salts (e.g. 0.01% thallous acetate) inhibit the growth of many species of bacteria. Some species of *Mycoplasma* are not inhibited, and thallous acetate is used in a selective medium for the primary isolation of these organisms. (Thallium salts inhibit the growth of *M. genitalium*, some strains of *M. hominis*, and the related organism, *Ureaplasma urealyticum*.)

Thallobacteria A proposed class of bacteria (division FIRMICUTES) comprising branching, thread-like organisms (ACTINOMYCETES and related organisms) [Book ref. 22, p. 33]. (cf. e.g. THERMOACTINOMYCES.)

Thallophyta A division of the plant kingdom containing those organisms whose vegetative form is a thallus, i.e., the fungi, algae etc.

thallose Consisting of or resembling a thallus.

thallospore (*mycol.*) Any SPORE formed by fragmentation of part or all of a fungal thallus – e.g. an ARTHROSPORE, CHLAMYDOSPORE or OIDIUM.

thallus A vegetative form of an organism which is not differentiated into stem and leaves, as in fungi, algae and lichens.

THAM See TRIS.

Thamnidiaceae See MUCORALES.

Thamnidium See MUCORALES.

Thamnocephalis See MUCORALES.

Thamnolia A genus of fruticose LICHENS; photobiont: a green alga. The thallus is hollow and may be erect or prostrate; fruiting structures are unknown.

Thanatephorus See TULASNELLALES.

Thauera selenatis See SELENATE RESPIRATION.

Thayer–Martin agar (VCN agar) An enriched growth medium containing vancomycin, colistin and nystatin; it is used for the selective culture of *Neisseria* spp from swabs taken e.g. from the genitourinary tract. For culture of rectal swabs, trimethoprim may be added to suppress swarming by *Proteus* spp. [Recipe: Book ref. 2, p. 138.]

theca (1) In certain algae of the Chlorophyta: a type of cell 'wall' which lacks the microfibrillar structure typical of a true algal CELL WALL, being composed of very small, more or less isodiametric units or scales; such thecae appear to have evolved as modifications of scaly coverings, and occur in certain flagellates (e.g. TETRASELMIS). [Book ref. 123, p. 31.]

(2) See DINOFLAGELLATES.

Thecamoeba See AMOEBIDA.

thecate Possessing a THECA.

thecium An ambiguous term which can refer e.g. to the HYMENIUM of an ascomycete or to the entire ascocarp.

Theileria A genus of protozoa (subclass PIROPLASMASINA) parasitic in animals and transmitted by ticks (see EAST COAST FEVER for life cycle); *Theileria* spp infect both red and white blood cells of the vertebrate host (cf. BABESIA). The cells resemble those of *Babesia* but are typically smaller. [Characterization of some East African *Theileria* species by isoenzyme analysis: IJP (1985) *15* 271–276.]

Theiler's murine encephalomyelitis virus (TMEV; or murine poliovirus) A murine ENTEROVIRUS. TMEV strains are common in mice; infection is usually asymptomatic, but occasional individuals develop flaccid paralysis of the hind limbs ('mouse poliomyelitis'). Young mice injected intracerebrally with TMEV may develop a chronic inflammatory demyelinating disease; in nu/nu mice demyelination apparently results from lytic infection by TMEV of oligodendrocytes, and is independent of functional T cells [Lab. Inv. (1986) *54* 515–522] (cf. MULTIPLE SCLEROSIS).

Thelidium See VERRUCARIALES.

Thelohanellus See MYXOSPOREA.

Thelohania A genus of protozoa (class MICROSPOREA) which are parasitic e.g. in mosquitoes (e.g. *T. legeri* in anophelines, *T. opacita* in culicines), in shrimps, etc.

Thelotrema See GRAPHIDALES.

2-thenoyltrifluoroacetone See TTFA.

theophylline 1,3-Dimethyl-2,6-dihydroxypurine – a specific phosphodiesterase inhibitor.

therapeutic index For a given drug, the ratio maximum tolerated dose/minimum effective dose; medically useful drugs should ideally have a high therapeutic index.

thermal cycler See THERMOCYCLER.

thermal death point The lowest temperature at which a single-species population of microorganisms, in neutral aqueous suspension, can be heat-killed in ten minutes; thermal death point depends e.g. on the size of the population.

thermal death time (t_D) The time required, at a given temperature, for the heat-killing of a single-species population of microorganisms – or of a given proportion of that population – in aqueous suspension; t_D depends e.g. on the size of the population and on the pH of the suspension.

thermal inactivation point (TIP) (*virol.*) The temperature at which a virus preparation must be held for 10 min in order to inactivate all the viruses. For many plant viruses the TIP is 50–60°C; for e.g. tobacco mosaic virus it is >90°C.

thermal melting profile (of DNA) A plot of UV absorption against temperature for a given sample of dsDNA. The absorption of UV radiation (at 260 nm) by two single strands of DNA is about 40% greater than that by the same strands united in a double helix. dsDNA is progressively converted to ssDNA as the temperature increases (from a value which depends e.g. on the ionic strength and pH of the solution and on the solvent used); this conversion results in a corresponding increase in the level of UV absorption. The thermal melting profile is essentially linear between minimum absorption (all dsDNA) and maximum absorption (all ssDNA); the temperature corresponding to the midpoint between these two extremes is the T_m (= Tm). The T_m (typically ca. 85–95°C) depends on the proportion of GC pairs in the DNA – GC pairs, having three hydrogen bonds, being more stable than AT pairs which have only two. (See also GC%.) In the presence of a reagent (e.g. formamide) which destabilizes hydrogen bonds, the T_m is greatly reduced, and this allows strand separation to occur at much lower temperatures (e.g. 40°C); under these conditions much of the damage to the DNA which may occur at high temperatures can be avoided.

thermine See POLYAMINES.

Thermoactinomyces A genus of (typically) thermophilic bacteria which occur e.g. in composts, soil and sediments. The organisms form substrate and aerial mycelium and, for this reason, were classified (on a morphological basis) in the order ACTINOMYCETALES; however, the organisms form endospores and have a relatively low GC% (ca. 53), and these features, together with e.g. rRNA oligonucleotide cataloguing data, indicate that

Thermoactinomyces spp have an affinity with genera such as *Bacillus*. Endospores are formed in both aerial and substrate mycelium; they occur singly, though a given hypha may contain many individual endospores. *T. vulgaris* (*syn. T. candidus*) has been associated with FARMERS' LUNG, while *T. sacchari* has been implicated in BAGASSOSIS. Type species: *T. vulgaris*.

Thermoascus A genus of fungi of the order EUROTIALES; anamorphs occur e.g. in the genus PAECILOMYCES. [Ascocarp morphology and development: TBMS (1981) *76* 457–478.]

Thermobacterium See LACTOBACILLUS.

Thermobacteroides A genus of Gram-negative, obligately anaerobic, chemoorganotrophic, rod-shaped (non-motile or peritrichously flagellated) bacteria which have an optimum growth temperature within the range 55–70°C; growth does not occur below 35°C. [IJSB (1985) *35* 425–428.]

thermocline *Syn.* METALIMNION.

Thermococcus A genus of anaerobic, marine and freshwater archaeans (order THERMOPROTEALES) which metabolize peptides and sugars, growth being stimulated by the reduction of sulphur; growth occurs optimally >75°C. Cells: cocci, ca. 1 μm diam. *T. celer* is polarly flagellate. GC%: ca. 40–60%.

[Strains of *Thermococcus* from sites in the Pacific Ocean: FEMS Ecol. (2001) *36* 51–60.]

thermocycler (thermal cycler) An instrument used in certain nucleic-acid-amplification techniques (LCR and PCR) for automatically making the cyclical changes in temperature. Such instruments can simultaneously process a number of reaction mixtures and can usually be programmed for any of a range of time/temperature protocols.

The way in which reaction mixtures are heated and cooled is based on one of two methods. In the Peltier system, each reaction mixture is processed in a thin-walled tube which fits snugly in a hole in a metal block; temperature cycling involves heating and cooling the block according to the particular protocol. Cycling times are affected by the rate at which the block can be heated and cooled; in one commercial instrument, heating can occur at a rate of up to 3°C/second while the maximum rate of cooling is ~2°C/second. In these (open-tube) methods the problem of evaporation (loss of sample) is tackled in two ways. One approach involves the use of an oil overlay. Another solution is to use a *heated lid* cycler in which a temperature of up to 120°C is maintained over the sample tubes.

Alternatively, the sample may be sealed into a capillary glass tube which is heated and cooled by means of a forced air current; in the LightCycler™ (Roche), the air temperature around each tube can be changed at a rate of 20°C/second, permitting rapid cycling.

Some instruments (including the LightCycler) incorporate a system for detecting and/or quantifying the products of amplification.

Thermodiscus A genus of prokaryotes of the order THERMOPROTEALES. Cells: disc-shaped, ca. 0.3–3.0 μm diameter, ca. 0.2 μm thick.

thermoduric (thermotolerant) Refers to an organism which can survive temperature–time combinations that are normally lethal for many or most vegetative microorganisms. In dairy microbiology the term refers to those organisms which survive PASTEURIZATION (e.g., many strains of *Microbacterium* can survive 70–80°C/15 minutes). (See also THERMOPHILE.)

Thermofilum A genus of chemolithoheterotrophic prokaryotes (order THERMOPROTEALES) which occur in Icelandic solfataras; the organisms metabolize e.g. peptides. Cells: rods or filaments, ca. 0.2 × 5–100 μm. *T. pendens* is non-flagellated. GC% ca. 57.

thermolysin A heat-stable, neutral metallo-PROTEASE (EC 3.4.24.4) produced by a strain of *Bacillus stearothermophilus* ('*B. thermoproteolyticus*'). It is the most heat-stable protease available commercially, retaining 50% activity after 1 hour at 80°C. Thermolysin contains zinc and four Ca^{2+} ions/molecule. It has a broad specificity.

Thermomicrobium A genus (*incertae sedis*) of aerobic, catalase-positive, chemoorganotrophic, thermophilic, Gram-negative bacteria which occur e.g. in hot springs. Cells: non-motile, pleomorphic, pink-pigmented rods, ca. 1.3–1.8 × 3.0–6.0 μm. Optimum growth temperature: 70–75°C; optimum pH: 8.2–8.5. GC%: 64. Type species: *T. roseum*. [Book ref. 22, pp. 338–339.]

Thermomonospora A genus of (typically) thermophilic bacteria (order ACTINOMYCETALES, wall type III) which occur e.g. in soil and composts. The organisms form branching substrate and aerial mycelium; of the five recognized species, *T. alba*, *T. curvata*, *T. fusca* and *T. mesophila* form white aerial mycelium, while *T. chromogena* forms yellowish-brown aerial mycelium. Spores are borne singly at the ends of short branched or unbranched sporophores on the aerial hyphae (or, in at least *T. fusca*, on both aerial and substrate mycelium). The organisms are metabolically versatile – e.g. *T. alba* and *T. curvata* can degrade agar, cellulose, and keratin, while *T. alba* can also degrade pectin. Type species: *T. curvata*. [Book ref. 73, pp. 119–121; taxonomy and identification: JGM (1984) *130* 5–25.] (cf. ACTINOBIFIDA.)

Thermomyces See HYPHOMYCETES; see also COMPOSTING.

thermonuclease See DEOXYRIBONUCLEASE.

thermophile An organism whose optimum growth temperature is >45°C; this definition is widely used [see e.g. Book ref. 45, p. 236, and Book ref. 196, p. 40] and is compatible with the definition of MESOPHILE. (See also PSYCHROPHILE.) Many published ad hoc definitions of 'thermophile' are mutually incompatible, often because a given definition may be applicable in only a limited context; moreover, such definitions are frequently incompatible with the generally accepted upper limit of 'mesophile'.

Thermophiles occur e.g. in hot springs and other habitats (see e.g. COMPOSTING, LEACHING, HYDROTHERMAL VENT); they include e.g. species of METHANOCOCCUS, PYRODICTIUM, SULFOLOBUS, THERMOBACTEROIDES, THERMOFILUM, THERMOMICROBIUM, THERMOPLASMA, THERMOPROTEUS, THERMOTHRIX and THERMUS.

A super-thermophile archaean, designated 'strain 121', has been cultured from samples taken from a HYDROTHERMAL VENT; it grows at 85–121°C [Science (2003) *301* 934].

(See also CALDOACTIVE.)

Thermoplasma A genus of aerobic, heterotrophic prokaryotes (order THERMOPLASMALES, formerly class MOLLICUTES) which occur e.g. in hot springs and self-heating coal refuse piles. *T. acidophilum* grows on yeast extract at 50–64°C at a pH of 0.8–3.0. Cells: pleomorphic cocci, ca. 0.1–0.3 μm diameter, to filamentous. GC%: ca. 46. [Book ref. 157, pp. 88–91.]

Thermoplasmales An order of aerobic, thermophilic prokaryotes of the domain ARCHAEA (formerly Archaebacteria) characterized by the absence of a cell wall. One genus: THERMOPLASMA.

Thermoproteales An order of thermophilic, anaerobic sulphur-dependent prokaryotes (domain ARCHAEA, formerly Archaebacteria) which occur e.g. in hot springs; growth is typically chemolithoautotrophic and/or chemolithoheterotrophic. Cells: cocci, rods or irregularly shaped forms, the cell wall typically being a detergent-resistant S LAYER. Genera: DESULFUROCOCCUS, PYRODICTIUM, THERMOCOCCUS, THERMODISCUS, THERMOFILUM and THERMOPROTEUS.

Thermoproteus A genus of thermophilic prokaryotes (order THERMOPROTEALES) found e.g. in Icelandic solfataras. Cells: rods or filaments, ca. 0.5 × 1–80 μm. Chemolithoautotrophic and/or chemolithoheterotrophic, according to species; heterotrophs use e.g. ethanol, formate, glucose, methanol, sucrose. Energy is obtained by SULPHUR RESPIRATION. Species include *T. neutrophilus* (obligate autotroph) and *T. tenax* (facultative autotroph). GC%: ca. 55–56.

thermospermine See POLYAMINES.

thermotaxis A TAXIS in which the stimulus is a temperature gradient.

thermotherapy (*plant virol.*) See PLANT VIRUSES.

Thermothrix A genus of thermophilic bacteria. *T. thiopara* is a facultatively anaerobic, rod-shaped or filamentous organism which occurs in hot springs (45–75°C); it is facultatively chemolithoautotrophic, electron donors including e.g. sulphide and elemental sulphur. [Ecology and metabolism: AAM (1986) *31* 233–270.]

Thermotoga See BACTERIA (taxonomy).

Thermotogales See BACTERIA (taxonomy).

thermotolerant *Syn.* THERMODURIC.

Thermus A genus (*incertae sedis*) of aerobic, chemoorganotrophic, thermophilic, Gram-negative bacteria which occur e.g. in hot springs. Cells: non-motile rods, 0.5–0.8 × 5.0–10.0 μm, or filaments, which typically contain yellow, orange or red carotenoid pigments and novel POLYAMINES. The cell ultrastructure resembles that of other Gram-negative bacteria, but the 'outer wall' (external to the peptidoglycan layer) seems to be more substantial than the typical outer membrane of Gram-negative cells; the peptidoglycan lacks DAP, contains ornithine, and has a high proportion of glycine and glucosamine. Most strains give rise to 'rotund bodies': membrane-enclosed spherical aggregates of cells (average: 14 cells per aggregate), ca. 10–20 μm in diameter; the limiting membrane of a rotund body is formed by fusion of the outer walls of adjacent cells at the periphery of the body, the inward-facing surface of these cells lacking an outer wall. Metabolism is respiratory (oxidative), with O_2 as terminal electron acceptor. Carbon sources include e.g. sugars and organic acids; nitrogen sources: e.g. NH_4^+, amino acids. Optimum growth temperature: ca. 66–75°C; optimum pH: ca. 7.0. Oxidase +ve. Catalase +ve GC%: ca. 61–71. Type species: *T. aquaticus*. [Book ref. 22, pp. 333–337.] [Proposed species – *T. ruber* (revised name): IJSB (1984) *34* 498–499.]

theront In the polymorphic life cycles of certain protozoa: a cell which 'searches for' or 'hunts' a new host organism or a source of food; see e.g. ICHTHYOPHTHIRIUS.

θ See TEMPERATURE COEFFICIENT.

θ (theta) replication *Syn.* CAIRNS' MECHANISM.

THF TetrahydroFOLIC ACID.

thiabendazole 2-(4′-Thiazolyl)-benzimidazole. Originally used as an anti-helminthic agent, thiabendazole is now regarded as an efficient fungicide and is used e.g. to protect against seed-borne smuts and bunts of cereals, and to prevent rots of stored fruit and vegetables. Thiabendazole has been widely used as the principal agent for the protection of stored citrus fruit against *Penicillium digitatum* and *P. italicum*; however, thiabendazole-resistant strains of these and other citrus rot fungi have emerged [Book ref. 121, pp. 149–162].

(See also BENZIMIDAZOLES; GANGRENE (sense 2); SPORIDESMINS.)

thiacetazone See THIOSEMICARBAZONES.

thiamine (thiamin; vitamin B_1; aneurin) A water-soluble VITAMIN: 3-(2-methyl-4-amino-5-pyrimidinylmethyl)-5-(β-hydroxyethyl)-4-methylthiazole (see figure); the molecule carries a net positive charge and is usually supplied commercially as thiamine chloride hydrochloride ('thiamine hydrochloride'). The coenzyme form is *thiamine pyrophosphate* (TPP, cocarboxylase). TPP functions in decarboxylation reactions (e.g. pyruvate → acetaldehyde); in oxidative decarboxylations – e.g. (in conjunction with LIPOIC ACID) of pyruvate and 2-oxoglutarate [see Appendix II(a)]; in α-ketol formation – e.g. acetoin formation [Appendices III(c) and III(f)]; in transketolase and phosphoketolase reactions [Appendices I(b) and III(b), respectively]; etc.

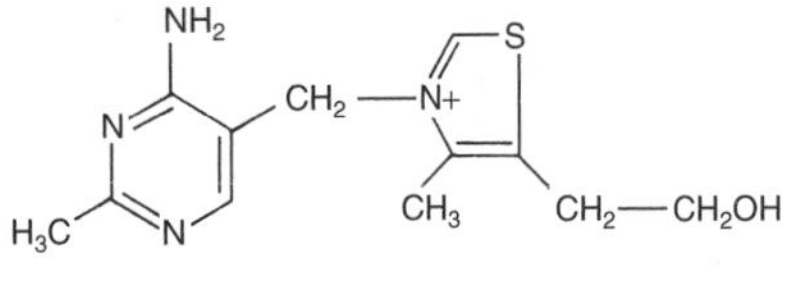

NH_2 — CH_2 — N^+ — S — N — H_3C — N — CH_3 — CH_2—CH_2OH

THIAMINE

Microorganisms which require exogenous thiamine (or one or both of the pyrimidine and thiazole precursors) include e.g. *Euglena* spp, certain fungi (e.g. *Phycomyces blakesleeanus*, *Phytophthora* spp, *Trichophyton* spp, many yeasts), some bacteria (e.g. *Lactobacillus* spp, *Leuconostoc* spp, *Staphylococcus aureus*), and various protozoa (e.g. *Acanthamoeba castellanii*, *Crithidia fasciculata*, *Tetrahymena* spp).

Thielaviopsis See HYPHOMYCETES; see also FOOTROT (sense 2).

thienamycin An antibiotic produced by *Streptomyces cattleya* and by *S. penemifaciens*; it is a CARBAPENEM (see β-LACTAM ANTIBIOTICS for structure) in which R′ = CH_3.CHOH and R″ = $NH_2.(CH_2)_2$.S. Thienamycin binds to the PENICILLIN-BINDING PROTEINS of *Escherichia coli* (preferentially to PBP-2); it is highly active against Gram-positive and Gram-negative bacteria (including species of *Pseudomonas*) and is resistant to many β-LACTAMASES, but is chemically unstable. Many naturally occurring and semi-synthetic derivatives of thienamycin have been isolated; the *N*-formimidoyl derivative (*imipenem*; MK-0787) is less unstable than thienamycin, is resistant to many β-lactamases, has a similar level of activity against most bacteria, and is more active against *Pseudomonas* spp. [Imipenem (review): AIM (1985) *103* 552–560.]

Thiéry staining A procedure for detecting polysaccharides by electron microscopy.

thigmotaxis A TAXIS in which the stimulus is 'touch' (physical contact); an organism may move towards the contact (positive thigmotaxis) or away from it (negative thigmotaxis).

thimerosal *Syn.* THIOMERSAL.

thin-layer chromatography See CHROMATOGRAPHY.

Thiobacillus A genus of Gram-negative, obligately or facultatively chemolithoautotrophic (or mixotrophic) bacteria which occur e.g. in soil, marine muds, mine drainage and hot springs. The cells are typically polarly flagellated rods, ca. 0.5 × 1.0–3.0 μm. Most species obtain energy by oxidizing sulphur and/or reduced sulphur compounds (sulphide, sulphite and thiosulphate); in at least some cases *c*-type cytochromes are involved. GC%: ca. 50–68. Type species: *T. thioparus*.

Thiobacillus A2. See *T. versutus* (below).

T. delicatus. Non-motile. Facultative anaerobe. Capable of nitrate respiration. Electron donors: e.g. sulphide, thiosulphate. Optimum temperature: 30–35°C. Optimum pH: 5.5–6.0. [IJSB (1984) *34* 139–144.]

T. denitrificans. Electron donors: e.g. sulphide and thiosulphate. Under aerobic conditions oxygen is the electron acceptor; anaerobically, nitrate, nitrite or nitrous oxide can function as electron acceptor, and is reduced to nitrogen. Optimum pH: 6–8.

T. ferrooxidans. Aerobic; growth also occurs anaerobically with ferric ions as electron acceptor for the oxidation of reduced sulphur compounds. Electron donors include sulphur, thiosulphate, sulphide and ferrous ions. Capable of NITROGEN FIXATION. Optimum growth temperature: ca. 20°C. Growth occurs within the approximate pH range 1.4–6.0. (See also LEACHING.) [Molecular genetics of *T. ferrooxidans*: MR (1994) *58* 39–55.]

T. thiooxidans. Electron donors include e.g. thiosulphate and (particularly) elemental sulphur. Optimum pH: ca. 2–4; growth occurs within the approximate range 0.5–6.0. Can cause RUBBER SPOILAGE. (See also LEACHING.)

T. thioparus. Motile rods. Electron donors include sulphide and thiosulphate; some strains can use thiocyanate (CNS^-). A sulphur-containing pellicle may be formed on thiosulphate-containing liquid media. Nitrate is reduced to nitrite during anaerobic growth. Optimum growth temperature: ca. 28°C. Optimum pH: ca. 6–8; growth ceases below ca. pH 4.5.

T. versutus (formerly *Thiobacillus* A2). A species capable of chemolithoautotrophic or mixotrophic growth. Oxidation of organic substrates (but not inorganic sulphur compounds) can be coupled to the reduction of nitrate.

Other species include *T. intermedius* and *T. perometabolis* [IJSB (1984) *34* 139–144], *T. acidophilus*, *T. neapolitanus*, *T. novellus* and *T. organoparus*.

[Media and culture: Book ref. 45, pp. 1023–1036.]

Thiocapsa A genus of photosynthetic bacteria (family CHROMATIACEAE). The cells are non-motile cocci; in addition to various carotenoids, *T. roseopersicina* (1.2–3.0 μm) contains Bchl *a*, and *T. pfennigii* (ca. 1.5 μm) contains Bchl *b* – cell suspensions typically being reddish and orange-brown, respectively. In *T. pfennigii* the pigments occur in tubular intracytoplasmic membrane systems. Gas vacuoles are absent. *T. roseopersicina* is moderately tolerant of organic pollution, and may give rise to reddish blooms in sewage lagoons; it is capable of chemoorganotrophic growth under microaerobic conditions in the dark.

thioctic acid *Syn.* LIPOIC ACID.

Thiocystis See CHROMATIACEAE.

Thiodictyon See CHROMATIACEAE.

thioglycollate broth See e.g. BREWER'S THIOGLYCOLLATE MEDIUM.

thiol-activated cytolysins (SH-activated cytolysins) A class of cytolysins, formed by certain bacteria, which are active only in the reduced state, and which can be (reversibly) activated by thiols; the cytolytic activity of these toxins is confined to those cells (including e.g. erythrocytes) whose cytoplasmic membrane contains cholesterol. Thiol-activated cytolysins are formed by species of the (Gram-positive) bacteria *Bacillus*, *Clostridium*, *Listeria* and *Streptococcus*. (A thiol-activated cytolysin was reported to be formed by the (Gram-negative) species *Klebsiella pneumoniae* [CJM (1985) *31* 297–300].)

The thiol-activated cytolysins are proteins of ~47–60 kDa; they are optimally active at similar values of pH and temperature, they cross-react serologically, and their haemolytic activity is irreversibly lost in the presence of cholesterol or stereochemically related sterols.

The thiol-activated cytolysins are pore-forming agents. They bind reversibly, in a temperature-independent manner, to cholesterol-containing membranes, and oligomerize to form (typically) ring-shaped formations (pores in the membrane) of up to ~40 nm in diameter. Oligomerization is temperature-dependent, activity being reduced at low temperatures.

Studies with animal models have indicated that at least some of the toxins (e.g. pneumolysin) can contribute to virulence.

The thiol-activated cytolysins include:
alveolysin (*Bacillus alvei*)
bifermentolysin (*Clostridium bifermentans*)
botulinolysin (*C. botulinum*)
cereolysin (*B. cereus*)
chauveolysin (δ-toxin of *C. chauvoei*)
histolyticolysin (ε-toxin of *C. histolyticum*)
ivanolysin (*Listeria ivanovii*)
laterosporolysin (*B. laterosporus*)
listeriolysin (*L. monocytogenes*)
oedematolysin (δ-toxin of *C. novyi*)
perfringolysin (θ-toxin of *C. perfringens*)
pneumolysin (*Streptococcus pneumoniae*)
seeligerolysin (*L. seeligeri*)
septicolysin (δ-toxin of *C. septicum*)
sordelliolysin (*C. sordelli*)
STREPTOLYSIN O (q.v.)
tetanolysin (*C. tetani*)
thuringiolysin (*B. thuringiensis*)

[Thiol-activated cytolysins: RMM (1996) *7* 221–229.]

thiol protease See PROTEASES.

thiomersal (merthiolate; thimerosal) Sodium ethylmercurithiosalicylate, $C_2H_5.Hg.S.C_6H_4.COONa$. An antibacterial and antifungal agent used as a preservative e.g. in biological laboratory reagents; it is effective in low concentrations (e.g. 0.01–0.02%).

Thiomicrospira A genus of bacteria which occur e.g. in estuarine muds and in the vicinity of HYDROTHERMAL VENTS. Cells: curved or spiral rods; many strains are motile by means of a single polar flagellum. The organisms resemble THIOBACILLUS in general physiology. *T. denitrificans* and *T. pelophila* are both obligate chemolithotrophs. [Culture: Book ref. 45, pp. 1023–1036; *T. crunogena*, an obligately chemolithoautotrophic species from a hydrothermal vent: IJSB (1985) *35* 422–424.]

thionin (Lauth's violet) A blue–purple dye structurally related to METHYLENE BLUE. (See also BRUCELLA.)

Thiopedia See CHROMATIACEAE and COENOBIUM.

thiopeptin A peptide antibiotic related to THIOSTREPTON; it is active against e.g. *Streptococcus bovis*. Thiopeptin inhibits experimentally-induced lactic ACIDOSIS in ruminants.

thiophanate-methyl (1,2-bis(3-methoxycarbonyl-2-thioureido) benzene) An agricultural antifungal agent (see BENZIMIDAZOLES) which is effective e.g. against *Rhizoctonia solani* infections in seedlings, *Pyricularia oryzae* in rice, eyespot and *Rhynchosporium* infection in cereals, and black spot of roses; it also has some effect in controlling clubroot in crucifers.

thiophene-2-carboxylic acid hydrazide test *Syn.* T2H TEST.

α-thiophosphoryl dATP See SDA.

Thioploca A genus of GLIDING BACTERIA (see CYTOPHAGALES) which occur e.g. in sulphide-containing muds (freshwater to marine). The organisms occur as sheathed fascicles, each fascicle (macroscopic in some species) consisting of a longitudinally arranged bundle of filaments; individual filaments (ca. 1–10 μm wide; wider in marine species) can glide in the sheath and emerge from it. In at least some species metabolism involves oxidation of sulphide. Sulphur is deposited intracellularly. Type species: *T. schmidlei*. Other species: e.g. *T. ingrica* from lake sediments [IJSB (1984) *34* 344–345], *T. araucae*, *T. chileae* (marine benthic species) [IJSB (1984) *34* 414–418].

Strains from a HYDROTHERMAL VENT have vacuoles that seem not to be involved in the accumulation of nitrate for use in anoxic conditions [AEM (2004) *70* 7487–7496].

thioredoxin A heat-stable 109-amino-acid protein which occurs ubiquitously in cells; in *Escherichia coli* it is encoded by the *trxA* gene. Thioredoxin functions as a hydrogen donor e.g. for the reduction of ribonucleotides by ribonucleotide reductase (see nucleotide), in assimilatory sulphate reduction, in the reduction of thiol groups in proteins, etc; the active site of the oxidized form of thioredoxin contains a disulphide bridge between two cysteine residues, and this disulphide bridge can be reduced by the enzyme thioredoxin reductase, using NADPH as hydrogen donor. Thioredoxin also functions as an essential component of BACTERIOPHAGE T7-specific DNA polymerase, and is required for filamentous phage assembly (see INOVIRUS). Thioredoxin can be released from *E. coli* cells by osmotic shock, and may be located at the ADHESION SITES in the cell envelope.

Thiorhodaceae See CHROMATIACEAE.

Thiosarcina See CHROMATIACEAE.

thiosemicarbazones Derivatives of thiosemicarbazone ($R_2C{=}N.NH.CS.NH_2$) – e.g. *thiacetazone*: 4-acetamidobenzaldehyde thiosemicarbazone – have been utilized as chemotherapeutic agents against tuberculosis. The earlier drugs have been largely superseded by less toxic drugs, but 2-acetylpyridine thiosemicarbazones were synthesized for potential use in antiparasite (e.g. antimalarial), antibacterial and antineoplastic therapy; they apparently function by inhibiting ribonucleoside diphosphate reductase. Some of these thiosemicarbazones also have antiviral activity, inactivating the ribonucleoside diphosphate reductase of e.g. herpes simplex virus type 1 in a time-dependent manner [JGV (1986) *67* 1625–1632]. (See also ISATIN-β-THIOSEMICARBAZONE.)

Thiospira A genus of Gram-negative, helical, polarly flagellated bacteria which occur in freshwater and marine H_2S-containing habitats; the cells typically contain globules of sulphur.

Thiospirillum A genus of photosynthetic bacteria (family CHROMATIACEAE). The cells are spiral, ca. 10–100 μm long by up to 4 μm wide, and motile by lophotrichous flagella; in addition to various carotenoids, they contain Bchl *a* – cell suspensions being typically orange-brown. All strains are strict anaerobes and require sulphide and vitamin B_{12}.

thiostrepton A sulphur-containing peptide antibiotic, produced by *Streptomyces azureus*, which is active against Gram-positive bacteria and certain members of the Archaea (methanogens); it binds strongly to the 50S ribosomal subunit and inhibits protein synthesis. Antibiotics which appear to be structurally and functionally related include siomycin, sporangiomycin and THIOPEPTIN.

thiosulphate A salt of thiosulphuric acid; the thiosulphate ion is $S_2O_3^{2-}$. Thiosulphate can be used as a substrate for lithotrophic metabolism e.g. by *Thiobacillus* spp, and as an electron acceptor e.g. by SULPHATE-REDUCING BACTERIA; it is used as a source of sulphur in e.g. KLIGLER'S IRON AGAR. (See also RHODANESE.)

thiosulphate–citrate–bile salts agar See TCBS AGAR.

Thiothrix A genus of GLIDING BACTERIA (see CYTOPHAGALES); species occur e.g. attached to surfaces in sulphide-containing freshwater and marine habitats, particularly in flowing water, where dissolved oxygen levels are ca. 10% of the saturation value. The organisms are sheathed, non-gliding filaments which form gliding gonidia from the unattached end of the filament; filaments adhere to the substratum or form rosettes. Metabolism appears to be mixotrophic: in those strains examined, small amounts of simple carbon compounds are required, and energy seems to be derived from the oxidation of sulphide or thiosulphate; sulphur is deposited intracellularly. One strain of *Thiothrix* has been found to contain carboxysome-like structures, suggesting that it may be autotrophic. [Review: ARM (1983) *37* 354–359.]

thiourea derivatives (as antimicrobial agents) Antimicrobial derivatives of thiourea, $(NH_2)_2CS$, include e.g. *diphenylthiourea*, $(C_6H_5NH)_2CS$, which has been used as an antimycobacterial agent, and the THIOSEMICARBAZONES.

Thiovulum A genus of bacteria (Proteobacteria) found e.g. in marine habitats characterized by the simultaneous presence of sulphide and oxygen; the organisms obtain energy by oxidizing sulphide/sulphur, and typically form a whitish layer at the interface of sulphidogenic sediments and oxygenated water. Cells of the sole species, *T. majus*, are spherical, with a diameter reported to range from 5 to 25 μm; the (flagellate) organisms can reach speeds of ~600 μm/second [Microbiology (1994) *140* 3109–3116].

thiram (TMTD) TetramethylTHIURAM DISULPHIDE: $(CH_3)_2.N.CS.S.S.CS.N.(CH_3)_2$; an ANTIFUNGAL AGENT which is also active against (mainly) Gram-positive bacteria. (cf. DMDC.) Thiram has been used as an antiseptic in the topical treatment of dermatophyte infections, and as an agricultural antifungal agent (e.g. as a seed dressing, and for the control of apple scab, *Botrytis* fruit rots, etc). Thiram (being insoluble in water) is used in aqueous suspensions.

−35 sequence See PROMOTER.

thistle mottle virus See CAULIMOVIRUSES.

thiuram disulphides These compounds ($R_2.N.CS.S.S.CS.N.R_2$) are formed by the mild oxidation of DITHIOCARBAMATES. See e.g. METIRAM and THIRAM.

Thogoto virus An unclassified virus (see ORTHOMYXOVIRIDAE) which has been isolated from ticks and mammals; it has been implicated in a severe human disease (optic neuritis, meningoencephalitis) in Africa.

Thoma chamber See COUNTING CHAMBER.

thraustochytrids A category of eukaryotic organisms which are generally classified together with the LABYRINTHULAS (q.v.). (The thraustochytrids were formerly included within the Oomycetes – either as a family, Thraustochytriaceae, of the Saprolegniales, or as a distinct order, Thraustochytriales.) The organisms are widely distributed in marine and esturaine waters, living – apparently mainly as saprotrophs – on algae, plants, organic detritus, etc. The vegetative stage consists of a rounded or oval, somewhat amoeboid cell from which arise ectoplasmic filaments that branch and anastomose to form a rhizoid-like ectoplasmic net. The net is produced from sagenogenetosomes (sagenogens, 'bothrosomes') which are essentially similar to those of labyrinthulas; it appears to function in adhesion and nutrition (which is osmotrophic) and to some extent in motility, but the cell does not move into the net (cf. LABYRINTHULAS). The cell body serves as a sporangium which, in most species, gives rise to laterally biflagellate zoospores resembling those of LABYRINTHULAS (although apparently lacking an eyespot). Genera include *Aplanochytrium* (which produces aplanospores instead of zoospores), *Japonochytrium*, *Labyrinthuloides*, *Schizochytrium* and *Thraustochytrium*.

Thraustochytrium See THRAUSTOCHYTRIDS.

three-component regulatory system See BACTERIOCINS.

three-for-one model (3-for-1 model) A model which describes the mode in which newly synthesized PEPTIDOGLYCAN (q.v.) is incorporated in the sacculus of a growing cell [Microbiology (1996) *142* 1911–1918].

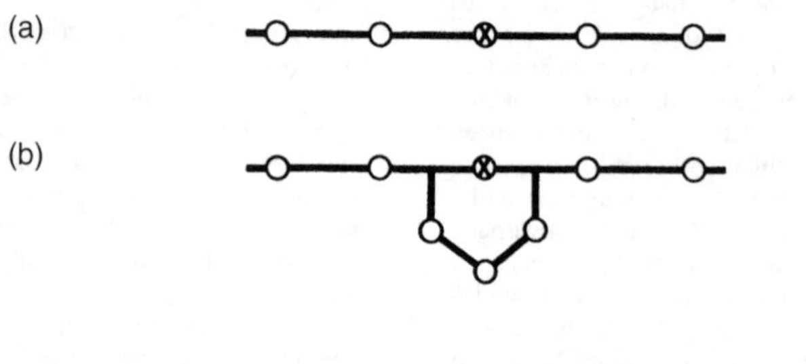

THREE-FOR-ONE MODEL: incorporation of newly synthesized peptidoglycan in the sacculus of a growing cell of *Escherichia coli* (simplified outline, diagrammatic). In PEPTIDOGLYCAN (q.v.), the glycan backbone chains are *perpendicular* to the long axis of the cell; the diagram shows an *end-on* view of backbone chains (open circles) joined by peptide bridges (straight lines).

(a) A peptidoglycan monolayer; the monolayer is a stress-bearing structure and is under considerable tension. The chain in the centre (⊗) – a 'docking' chain – is to be replaced by three chains (hence 3-for-1).

(b) A triplet of three linked chains is located below (i.e. on the cytoplasmic side of) the peptidoglycan monolayer, with the flanking chains of the triplet covalently bound either side of the docking chain; in this particular plane, a stem peptide from each flanking chain has bound to the ε-amino group of a dimeric peptide bridge to form a *trimeric* bridge (see legend to figure in entry PEPTIDOGLYCAN).

(c) The docking chain has been removed by enzymic action, and – because of the tension in the monolayer – the incoming triplet of chains has taken up its position in the sacculus; note that this contributes to cell *elongation*. The model assumes that the central chain in the incoming triplet will, when incorporated in the sacculus, function as a new docking chain; for this reason, the dimeric bridge to each flanking chain is assumed to have a free ε-amino group located in the stem peptide of the flanking chain.

For cell length to *double* during the cell cycle, the model assumes that (i) docking and non-docking chains alternate in the sacculus, and (ii) only those docking chains which are present at the *start* of the new cell cycle will be replaced by an incoming triplet; that is, 'new' docking chains inserted *during* the current cell cycle will not be replaced until the next cell cycle. For this purpose, the growth mechanism must be able to distinguish 'old' from 'new' docking chains. This is possible because newly synthesized peptidoglycan is linked to the sacculus solely by tetra-tetra peptide bridges; it is therefore postulated that 'new' chains (identified by their tetra-tetra links) are not recognized as docking chains, but that, at the start of each new cell cycle, all tetra-tetra links will have been changed enzymically (by an LD-carboxypeptidase) to tetra-tri links, and that this allows recognition of the functional docking chains.

Reproduced from *Bacteria*, 5th edition, Figure 3.1, page 42, Paul Singleton (1999) copyright John Wiley & Sons Ltd, UK (ISBN 0471-98880-4) with permission from the publisher.

An outline of the model is shown in the diagram. It is postulated that co-ordination of the synthetic and lytic aspects of the process involves enzymes that include PENICILLIN-BINDING PROTEINS and 'lytic transglycosylases' (enzymes that degrade a glycan strand processively).

three-micron DNA plasmid (3μ plasmid; also 3μm plasmid) A cccDNA plasmid, of unknown biological role, found e.g. in strains of *Saccharomyces cerevisiae* and other yeasts; 3μ plasmids appear to be excised portions of chromosomal DNA – each corresponding to a unit of the host cell's repetitive sequence of rRNA genes. [ARM (1983) *37* 266.] (See also TWO-MICRON DNA PLASMID.)

three-site model (of translation) See PROTEIN SYNTHESIS.

threitol A 4-carbon POLYOL. D-Threitol occurs in certain fungi, e.g. *Armillaria mellea.*

L-threonine biosynthesis See Appendix IV(d).

threshold cycle (in quantitative PCR) See TAQMAN PROBES.

-thrix A suffix signifying a *hair* or *thread.*

thrombin See FIBRIN.

thrombocytes *Syn.* PLATELETS.

Thrombocytozoons ranarum An intracellular parasite of certain frogs (e.g. *Rana septentrionalis*, the mink frog) which occurs exclusively in the thrombocytes of its host; *T. ranarum* is apparently a Gram-type-positive prokaryote. [J. Parasitol. (1984) *70* 454–456.]

thrombolite See STROMATOLITES.

thrush See CANDIDIASIS.

thujaplicins Isopropyltropolones, fungitoxic compounds formed by the Western Red Cedar (*Thuja plicata*); they appear to be responsible for the high resistance to fungal attack of timber from *T. plicata*. (See also TIMBER PRESERVATION.)

thuringiensin (β-exotoxin) An exotoxin produced by *Bacillus thuringiensis*; it is apparently a specific inhibitor of DNA-dependent RNA polymerase, acting as an ATP analogue. Thuringiensin is toxic to a wide range of insects and to some vertebrates; strains of *B. thuringiensis* used commercially for biological control (see DELTA-ENDOTOXIN) are therefore usually not thuringiensin-producers.

thuringiolysin 0 See THIOL-ACTIVATED CYTOLYSINS.

thylakoids Flattened, membranous vesicles which occur in the cytoplasm of nearly all CYANOBACTERIA (cf. GLOEOBACTER) and in the CHLOROPLASTS of ALGAE and higher plants; thylakoid membranes contain the CHLOROPHYLLS and electron carriers etc involved in PHOTOSYNTHESIS.

In cyanobacteria, thylakoids typically occur near, and parallel to, the cell envelope, but seem to be structurally distinct from the cytoplasmic membrane; unlike the thylakoids in higher plants and some algae, those in cyanobacteria are not stacked (cf. PROCHLOROPHYTES). (See also PHYCOBILIPROTEINS.) In at least some cyanobacteria, thylakoid membranes are the sites of respiratory (as well as photosynthetic) electron transport; during photosynthetic or respiratory activity protons are pumped *into* the thylakoid vesicles – generating a proton motive force (see CHEMIOSMOSIS) across the thylakoid membrane [Book ref. 75, pp. 199–218].

Algal chloroplasts may contain individual (unassociated) thylakoids (in the Rhodophyta); longitudinally-adjacent pairs of thylakoids (in the cryptophytes); bands of three thylakoids (in the chrysophytes, diatoms, dinoflagellates, euglenoid flagellates, and members of the Phaeophyta, Prymnesiophyceae and Xanthophyceae); or bands of 2–6 thylakoids (Chlorophyta). A stack of ca. four or more thylakoids is termed a *granum*.

thylaxoviruses Viruses of the ONCOVIRINAE.

thymic aplasia (congenital) *Syn.* DIGEORGE SYNDROME.

thymidine See NUCLEOSIDE and Appendix V(b). (See also *double thymidine blockade* in SYNCHRONOUS CULTURE.)

thymidine kinase (TK; ATP:thymidine 5′-phosphotransferase; EC 2.7.1.21) An enzyme which catalyses the phosphorylation of thymidine to thymidine 5′-monophosphate; it is an important enzyme in the pyrimidine salvage pathway. TKs occur in various bacteria [JGM (1985) *131* 3091–3098] and in eukaryotic cells (in the cytoplasm and mitochondria), and are encoded by many DNA viruses [review: MS (1985) 2 369–375]. (See also e.g. ACYCLOVIR and HYBRIDOMA.)

thymine dimer See ULTRAVIOLET RADIATION.

thymocyte An immature T cell undergoing development in the thymus gland.

thymol See PHENOLS.

thymol blue A PH INDICATOR. Acid range: pH 1.2–2.8 (red to yellow); pK_{a1} 1.5. Alkaline range: pH 8.0–9.6 (yellow to blue); pK_{a2} 8.9.

thymolphthalein A PH INDICATOR: pH 9.3–10.5 (colourless to blue); pK_a 9.9.

thymus-dependent antigen (TD antigen; T-dependent antigen) Any antigen (e.g. most soluble proteins) which can trigger ANTIBODY FORMATION in antigen-specific B LYMPHOCYTES only with the help of T LYMPHOCYTES. (TD antigens do not elicit antibody in congenitally athymic individuals.)

thymus-independent antigen (TI antigen; T-independent antigen) Any antigen which can trigger ANTIBODY FORMATION in antigen-specific B LYMPHOCYTES without help from T LYMPHOCYTES. TI antigens include e.g. lipopolysaccharides, polymerized flagellin, dextrans and levans. (See also POLYCLONAL ACTIVATOR.)

TI antigen THYMUS-INDEPENDENT ANTIGEN.

Ti plasmid See CROWN GALL.

tiamulin A semisynthetic antibiotic which inhibits PROTEIN SYNTHESIS in bacteria but which has little or no effect against archaeans. It apparently interacts with the 50S subunit of the ribosome, inhibiting the peptidyltransferase reaction.

ticarcillin See PENICILLINS.

tick-borne encephalitis virus See FLAVIVIRIDAE.

Tilletia A genus of plant-parasitic fungi (order USTILAGINALES) which typically attack the ovaries of the host plant, though the leaves may also be parasitized; *Tilletia* spp are the causal agents of e.g. COMMON BUNT and dwarf bunt.

In a typical life cycle, a teliospore germinates to give rise to an elongated, aseptate or uniseptate structure bearing an apical tuft of eight or more thread-like basidiospores (sometimes called 'primary sporidia' or 'primary conidia'); in e.g. *T. caries* (which exhibits HETEROTHALLISM) the basidiospores on a given structure are of two different mating types. While still attached, the basidiospores conjugate in pairs, giving rise to H-shaped forms within which plasmogamy occurs; subsequently, each (detached) H-shaped binucleate form, or mycelium derived from it, gives rise to binucleate ballistoconidia (sometimes called 'secondary sporidia' or 'secondary conidia'). The conidia germinate to form dikaryotic hyphae which initiate infection of a fresh host and subsequently give rise to teliospores.

Tilletiaceae See USTILAGINALES.

tilmicosin A MACROLIDE ANTIBIOTIC which has been used e.g. against *Mycoplasma bovis* in the treatment of CALF PNEUMONIA; in vitro tests on field isolates of *M. bovis* indicate that significant levels of resistance have developed against tilmicosin [VR (2000) *146* 745–747].

Tilopteris See PHAEOPHYTA.

tilorone (2,7-bis-[2-(diethylamino)ethoxyl]-9-fluorenone) An antiviral, antitumour agent which seems to (a) induce INTERFERON synthesis, and (b) interfere directly with virus expression by functioning as an INTERCALATING AGENT (ϕ ca. 13°) with a preference for AT-rich regions of dsDNA.

Tilsit cheese See CHEESE-MAKING.

Tilt See PROPICONAZOLE.

timber decay See TIMBER SPOILAGE and TREE DISEASES.

timber preservation The *seasoning* (i.e. drying) of timber renders it less susceptible to fungal attack (see TIMBER SPOILAGE), and may facilitate penetration by chemical preservatives; once dried, timber can be preserved by e.g. preventing the access of air and moisture by the use of sealants or paint films (see also PAINT SPOILAGE). In some cases (e.g. elm wood) decay can be inhibited by waterlogging which limits the supply of oxygen to potential spoilage fungi.

Decay of timber may also be prevented or delayed by the use of PRESERVATIVES – e.g. CREOSOTE, sodium fluoride, pentachlorophenol, borates (alone or with QACs), copper naphthenate, and tributyl tin oxide (alone or with QACs). Most of these substances are inherently fungitoxic, but tributyl tin oxide (TBTO) inhibits fungal attack by binding strongly to cellulose, rendering it insusceptible to fungal CELLULASES; wood treated with TBTO may still be susceptible to lignin-utilizing WHITE ROT fungi unless TBTO is used in conjunction with e.g. a phenolic preservative to protect the lignin. The control of SOFT ROT can be achieved only by the use of preservatives which penetrate into the cell walls of the wood. Preservatives may be applied either as surface coatings or by the use of specialized procedures (see e.g. BETHELL PROCESS; BOUCHERIE PROCESS; SAUG–KAPPE PROCESS). Aqueous preservatives may be retained in treated timbers e.g. by the formation of insoluble precipitates; the insolubility of some preservatives prevents leaching under damp or wet conditions. (Certain types of wood contain natural fungitoxic substances – e.g. PINOSYLVIN in pine, TANNINS in oak, THUJAPLICINS in Western Red Cedar.) (See also HETEROBASIDION.)

Decay of chemically treated timber may occur e.g. through splitting – which permits fungal penetration through a treated zone. Rhizomorphs may enable a fungus to cross treated (or non-nutrient) zones (see e.g. DRY ROT), while some fungi can detoxify copper preservatives by converting them to copper oxalate.

timber spoilage Microbial spoilage of timber is caused primarily by fungi – particularly basidiomycetes – and involves decay (see e.g. BROWN ROT, DRY ROT, POCKET ROT, SOFT ROT, WET ROT, and WHITE ROT) and/or staining (see TIMBER STAINING). Commonly, only wood having a moisture content greater than ca. 20% is susceptible to fungal attack (cf. DRY ROT); the moisture content is calculated as the weight of water in a sample of timber divided by the dry weight of that sample. In fallen or freshly felled timber the sapwood is particularly vulnerable to SAP-STAIN and early decay by any of a range of fungi; thus e.g. fallen birch branches rapidly succumb to a brown cubical rot (caused by e.g. *Piptoporus betulinus*) or to a soft white fibrous rot (caused e.g. by *Heterobasidion annosum* – see HETEROBASIDION). Timber may also support a surface growth of non-cellulolytic, non-lignolytic ascomycetes and deuteromycetes (cf. SOFT ROT (1)); such growth generally causes little or no damage to the timber. Fungi of importance in the decay of domestic and industrial timbers include e.g. *Serpula lacrymans* (see DRY ROT) and *Coniophora puteana* (see WET ROT); *Lentinus lepideus* is moderately resistant to creosote and can cause a brown rot in e.g. railway sleepers. (See also PAPER SPOILAGE; TIMBER PRESERVATION; TREE DISEASES.)

timber staining (by fungi) Various wood-infecting fungi cause discoloration in felled timber and in standing trees; some infections are superficial, but in others the fungus grows into the timber or is carried in by wood-boring beetles. Superficial staining can be caused by a range of fungi (e.g. species of *Alternaria*, *Fusarium*, *Penicillium*) and typically does not affect the strength or value of the timber; deep staining may weaken the timber (see e.g. BROWN OAK). Staining may be due to the production of a fungal pigment or to an optical effect (see BLUE-STAIN). (See also GREEN OAK; TIMBER PRESERVATION.)

Timentin A composite antibiotic consisting of ticarcillin (see PENICILLINS) and CLAVULANIC ACID. [Clinical evaluation: Am. J. Med. (1985) *79* Supplement 5B 1–196; laboratory and clinical perspective: JAC (1986) *17* Supplement C 1–244.] (cf. AUGMENTIN.)

timothy-grass bacillus *Mycobacterium phlei*.

tin (as an antimicrobial agent) Tin is a HEAVY METAL which, in elemental or compound form, can exert antimicrobial activity. The presence of metallic tin has been reported to inhibit growth and toxin production by *Clostridium botulinum* in food contained in tin-plate cans [cited in Book ref. 35, p. 282]. (If the inside of a can is significantly de-tinned – e.g. by CURING salts, or by certain acidic fruits – the underlying steel may rapidly corrode and give rise to a HYDROGEN SWELL.)

Organotins (i.e. tin-containing organic compounds) are typically more active than inorganic compounds of tin – trialkyltins (e.g. TBTO) being among the most effective. (See also FENTIN.) Trisubstituted organotins exhibit microbistatic activity at low concentrations – apparently by inhibiting ATP synthesis; at high concentrations they can be microbicidal, possibly by disrupting the cytoplasmic membrane.

tinangaja disease A disease of coconut palms which occurs on the island of Guam; it is apparently caused by the COCONUT CADANG-CADANG VIROID.

tincture of iodine See IODINE (a).

tinder fungus *Fomes fomentarius*.

tinea See RINGWORM. (See also e.g. FAVUS; PITYRIASIS NIGRA; PITYRIASIS VERSICOLOR.)

tinidazole See NITROIMIDAZOLES.

Tinopal AN A FLUORESCENT BRIGHTENER used e.g. as a selective antibacterial agent; in e.g. *Escherichia coli* it apparently inhibits RNA polymerase [JGM (1984) *130* 1999–2005]. [Use for distinguishing phytopathogenic from saprotrophic *Pseudomonas* spp: JAB (1985) *58* 283–292.]

tinsel flagellum See FLAGELLUM (b).

tintinnid See TINTINNINA.

Tintinnidium See TINTINNINA.

Tintinnina ('tintinnids') A suborder of ciliate protozoa (order OLIGOTRICHIDA); most species are marine pelagic organisms, but some occur in freshwater habitats. Cells: cylindrical, conical or vase-shaped, typically one hundred to several hundred micrometres in length; all are loricate and highly contractile. Somatic ciliature is reduced, but oral ciliature is conspicuous and includes a closed ring of cilia around the apical oral region. A perilemma is commonly or always present, and in some cases the lorica is covered by small inorganic particles. Genera include e.g. *Codonella*, *Salpingella*, *Tintinnidium*, *Tintinnopsis*, and *Tintinnus*. [Grazing, respiration, excretion and growth rates of tintinnids: Limn. Ocean. (1985) *30* 1268–1282.]

Tintinnopsis See TINTINNINA.

Tintinnus See TINTINNINA.

tioconazole An AZOLE ANTIFUNGAL AGENT. [Review of antimicrobial activity and therapeutic use in superficial mycoses: Drugs (1986) *31* 29–51.]

TIP (1) THERMAL INACTIVATION POINT. (2) Tumour-inducing principle: the Ti plasmid responsible for CROWN GALL formation.

***Tipula* iridescent virus** See IRIDOVIRUS.

***tir* gene** See PATHOGENICITY ISLAND.

tirandamycin An antibiotic which appears to be structurally and functionally similar to STREPTOLYDIGIN.

Tissierella A proposed genus of bacteria which includes organisms currently classified as *Bacteroides praeacutus* (proposed new name: *T. praeacuta*) [IJSB (1986) *36* 461–463].

tissue culture (1) The in vitro culture (growth) or maintenance of isolated mammalian (or other) tissues or organs, or of populations of individual cells obtained by disruption of such tissues. Those procedures which involve populations of individual cells are also referred to as *cell culture*; the following account refers specifically to cell culture.

Cell cultures (TISSUE CULTURE sense 2) are widely used e.g. for the culture of viruses, and for studies in cancer research, immunology, toxicology etc; they are also used to culture certain bacteria (e.g. species of *Chlamydia* and *Rickettsia*).

Cell cultures are prepared and handled with an ASPETIC TECHNIQUE. Culture vessels include LEIGHTON TUBES, medical flats, roller bottles (see ROLLER CULTURE), Roux flasks, test-tubes etc; vessels may be made of (neutral-pH) glass (e.g. borosilicate or flint glass) or of certain plastics (e.g. polystyrene).

Media. A *growth medium* permits cellular growth and division; it consists essentially of a BALANCED SALT SOLUTION supplemented with e.g. glucose and/or peptone, various amino acids and vitamins, a pH indicator (e.g. PHENOL RED), a buffer system (often bicarbonate or HEPES), antibiotic(s), and (usually) 5–10% calf serum. (Serum-less media are also used.) Antibiotics are included to prevent the development of microbial contaminants which may be present; however, the continual suppression of contaminants during SERIAL PASSAGE is undesirable, and periodic subcultures should be made in antibiotic-free media in order to detect contaminants. Common bacterial contaminants include species of *Acholeplasma* and *Mycoplasma* (PCR-based detection e.g. with Mycoplasma *Plus*™ PCR Primer Set from Stratagene). Common growth media include various modifications of EAGLE'S MEDIUM – e.g. basal medium Eagle (BME), and minimal essential medium (MEM) – and medium 199. A *maintenance medium* is used to preserve the viability and properties of a population of cells while restricting to a minimum their growth and division; growth media containing lower levels of serum (e.g. 1%) are often used as maintenance media.

In *monolayer* culture a population of cells is incubated in a shallow layer of growth medium until the cells form a more or less confluent layer, one cell thick, on the inner surface of the culture vessel; macroscopically, the monolayer appears as a translucent film. The initial cell suspension may be derived from fresh tissue (see PRIMARY CULTURE) or from an ESTABLISHED CELL LINE.

Preparation of a monolayer culture from fresh tissue. Selected tissue is cut into small pieces and washed several times with pre-warmed (30–37°C) phosphate-buffered saline (PBS). The tissue is then subjected to trypsinization, i.e., digested in PBS containing trypsin (see PROTEASES) (0.05–0.1% w/v) – the whole being kept at 30–37°C and stirred mechanically for 20–30 min. (Trypsin weakens the intercellular bonds; excessive trypsinization can damage cells.) The turbid supernatant (containing damaged cells) is discarded, and the tissue fragments are again digested in fresh PBS–trypsin. The supernatant, containing single cells and small aggregates, is centrifuged (ca. 150–200 *g*/5 min) and the pellet of cells is resuspended in fresh growth medium. The concentration of viable cells is determined (see COUNTING CHAMBER and VITAL STAINING) and is adjusted (by diluting with growth medium) to ca. 10^5 cells/ml. A small volume of the suspension is then incubated (37°C for mammalian cells) for several days, or a week or more, until a monolayer develops on the inner surface of the culture vessel; for flat-surfaced culture vessels (e.g. medical flats, Roux flasks) ca. 15–20 ml of cell suspension is required for an area of 100 cm^2. (Test-tubes or bottles of circular cross-section are used for ROLLER CULTURE.) When cultured, animal cells generally become either spindle-shaped ('fibroblast-like') or polygonal ('epithelioid'). For SERIAL PASSAGE, the cells are first *stripped*, i.e. detached from the surface on which they have been grown: the monolayer is subjected for 1–2 min to a stripping agent – e.g. PBS containing VERSENE (ca. 0.02–0.05%) and ca. 0.1% trypsin; the stripping agent is decanted, and the detached, disaggregated cells are washed and resuspended in fresh growth medium, and can be used to prepare fresh cultures. Cells, in ampoules, can be preserved for long periods e.g. by FREEZING.

Transformation. After a certain number of serial passages, cells may die out or may develop into an ESTABLISHED CELL LINE. In the latter case the cells undergo certain alterations (usually termed *transformation*) such that they exhibit some or all of the properties characteristic of tumour cells: an apparent capacity for unlimited in vitro growth and division; the ability to grow in soft agar; the development of new surface antigens; the ability to form tumours when injected into animals; an increase in the rate of nutrient uptake; a loss of *contact inhibition* (i.e. inhibition of movement and cell division due to contact with neighbouring cells). (Loss of contact inhibition in a monolayer culture permits the development of multilayer colonies from the clones of transformed cells.) These changes may also be brought about by certain viruses, e.g. SV40 and Rous sarcoma virus.

The characteristics of cultured cells may be monitored e.g. by serological means or by examining the KARYOTYPE of the (COLCHICINE-treated) cells.

(2) Any population of cells involved in tissue culture (sense 1).

(See also HYBRIDOMA and SYNCHRONOUS CULTURE.)

tissue cyst (of coccidia) See e.g. TOXOPLASMOSIS.

tissue-destructive factor See LEGIONELLOSIS.

tissue factor (TF) See DISSEMINATED INTRAVASCULAR COAGULATION.

tissue factor pathway inhibitor (TFPI) See DISSEMINATED INTRAVASCULAR COAGULATION.

titer See TITRE.

titration (*mol. biol.*) Concentration-dependent interaction between a regulatory molecule and other molecule(s) or specific nucleotide sequence(s) by means of which the level of activity of the regulatory molecule is controlled. (See e.g. F PLASMID (*replication*).)

titre (titer) (*serol., virol.*) A numerical expression for the 'concentration' of specific antibodies, antigens, virions etc in a given sample. If the titre of antibodies or antigens is determined by an END-POINT TITRATION, the titre is equal, numerically, to the dilution factor of the highest dilution of sample giving a positive serological reaction. For example, if the end-point reaction mixture consisted of a 1/200 dilution of serum (antibody) and an equal volume of antigen, the titre of antibody can be given as 200 ('initial serum dilution') or 400 ('final serum dilution'); these titres may, alternatively, be quoted as 1/200 and 1/400, respectively.

Virus titres may be determined e.g. by the PLAQUE ASSAY method (see also END-POINT DILUTION ASSAY).

TK THYMIDINE KINASE.

TLC Thin-layer CHROMATOGRAPHY.

Tm (T_m) See THERMAL MELTING PROFILE.

TMA Trimethylamine (see e.g. FISH SPOILAGE).

TMA (transcription-mediated amplification) A method used for the isothermal amplification of RNA target sequences; the principle of TMA is the same as that of NASBA (q.v.) – one practical difference being that, in the commercial applications of TMA, use is made of the RNase H activity of reverse transcriptase, so that only two enzymes are used.

Commercial TMA- and NASBA-based tests differ also in the way in which amplified target sequences are detected. In TMA-based tests, the targets are detected by a so-called *hybridization protection assay* (HPA). HPA involves the use of DNA probes labelled with acridinium ester. After allowing time for probe–target hybridization, an added reagent hydrolyses the acridinium ester label on all *unbound* probes – the label on bound probes being protected as a consequence of its location within the probe–target duplex; subsequent addition of other reagents elicits CHEMILUMINESCENCE from the *bound* probes – emitted light being measured by a luminometer in *relative light units* (RLUs).

Commercial applications of TMA include several diagnostic tests – one of which, the amplified *Mycobacterium tuberculosis* direct test (AMTDT; Gen-Probe, San Diego, USA), was approved by the U.S. Food & Drug Administration (FDA) for detection of *Mycobacterium tuberculosis* in (smear-positive) respiratory specimens (e.g. sputa). A new-format 'enhanced' AMTDT, which uses a larger volume of sample, was approved by the FDA in 1998.

A TMA-based test for *Chlamydia trachomatis* in urogenital specimens was approved by the FDA in 1996; the principle of this test (AMP CT) is identical to that of the AMTDT.

[Evaluation of AMP CT for the detection of *Chlamydia trachomatis* in a high-risk population of women: JCM (1997) *35* 676–678. Evaluation of AMTDT for non-respiratory specimens: JCM (1997) *35* 307–310. Comparison of original and 'enhanced' versions of AMTDT for respiratory and non-respiratory specimens: JCM (1998) *36* 684–689. Evaluation of 'enhanced' AMTDT for rapid diagnosis of pulmonary tuberculosis in a high-risk prison population: JCM (1999) *37* 1419–1425. False-positive results with AMTDT in patients with *Mycobacterium avium* and *M. kansasii* infections: JCM (1999) *37* 175–178.]

[Isothermal nucleic-acid-amplification methods (TMA, NASBA, SDA): Book ref. 221, pp 126–151.]

TMAO Trimethylamine-*N*-oxide (see e.g. FISH SPOILAGE).

TMEV THEILER'S MURINE ENCEPHALOMYELITIS VIRUS.

TMPD Tetramethyl-*p*-phenylenediamine: a reagent which can e.g. reduce cytochrome oxidase (see ELECTRON TRANSPORT CHAIN); oxidized TMPD can be reduced by ascorbic acid. (See also KOVACS' OXIDASE REAGENT.)

TMS Tetramethylsilane (see NUCLEAR MAGNETIC RESONANCE).

TMTD See THIRAM.

TMV TOBACCO MOSAIC VIRUS.

Tn*1* See Tn*3*.

Tn*2* See Tn*3*.

Tn*3* A 4957-bp class II TRANSPOSON which has identical 38-bp terminal inverted repeats and carries a gene (*bla*) for a TEM-type β-LACTAMASE; Tn*3* occurs e.g. in plasmid R1*drd-19*. Tn*3* insertion tends to occur preferentially at target sites in AT-rich regions of DNA, and results in a 5-bp duplication of target DNA. Two Tn*3* genes (*tnpA* and *tnpR*) are necessary for transposition, which occurs by a two-stage mechanism via a COINTEGRATE intermediate. The *tnpA* gene specifies a transposase responsible (at least in part) for the formation of a cointegrate. The *tnpR* gene specifies a resolvase which can catalyse resolution of the cointegrate by SITE-SPECIFIC RECOMBINATION involving a region – termed IRS (internal resolution site) – between the *tnpA* and *tnpR* genes in each Tn*3* copy. Tn*3* resolvase acts specifically on a replicon containing two IRS regions in the same orientation, and cannot act on IRS regions in different replicons; the resolution reaction is therefore irreversible. [Model for Tn*3* resolvase action: Cell (1985) *40* 147–158.] (Tn*3* cointegrate intermediates can also be resolved by general recombination involving the host *recA* system.) The *tnpR* resolvase also regulates the frequency of Tn*3* transposition by acting as a repressor and preventing expression of both *tnpA* and *tnpR* genes.

Many bacterial TRANSPOSABLE ELEMENTS are closely related to Tn*3*: all have related short (35–40 bp) inverted repeats, all induce a 5-bp duplication in target DNA on insertion, all are transposed by a two-stage mechanism involving the formation and resolution of a cointegrate, and most show TRANSPOSITION IMMUNITY. Tn*3*-like elements include e.g. Tn*1* and Tn*2* (both ca. 5 kb, encoding TEM-like β-lactamase), Tn*4* (encoding resistance to ampicillin, streptomycin and sulphonamides), Tn*21* (q.v.), Tn*501* (q.v.), Tn*551* (5.3 kb, encoding resistance to erythromycin, derived from a *Staphylococcus aureus* plasmid), Tn*1721* (q.v.), Tn*2603* (22 kb, encoding an OXA-1 β-lactamase and resistance to streptomycin, sulphonamides and mercury – cf. Tn*2410*), and γδ (= IS*1000*, a 5.8-kb INSERTION SEQUENCE which shares extensive DNA homology with Tn*3*). Tn*3* and the Tn*3*-like elements are sometimes collectively termed Tn*A*.

Two subgroups of 'Tn*A*' elements are recognized e.g. on the basis of whether the *tnpA* and *tnpR* genes are transcribed in opposite directions (as in Tn*3* and γδ) or in the same direction (as in Tn*4*, Tn*21*, Tn*501*, Tn*1721*, Tn*2603*).

[Review of Tn*3* and related transposons: Book ref. 20, pp. 223–260.]

Tn*4* See Tn*3*.

Tn*5* A 5818-bp composite TRANSPOSON containing three antibiotic-resistance genes, one of which encodes aminoglycoside 3′-phosphotransferase (conferring resistance to kanamycin and other aminoglycoside antibiotics). The three, contiguous antibiotic-resistance genes are flanked by a pair of terminal inverted repeats: two IS*50* insertion sequences (IS*50*L and IS*50*R) that differ slightly from one another. Each IS*50* sequence is flanked by a pair of 19-bp sequences referred to as the outer end (OE) and inner end (IE); the two OE sequences thus form the two ends of the transposon.

IS*50*R encodes two proteins: P1 (a *cis*-acting transposase, Tnp) and P2 (an inhibitor of the transposase, Inh) – P2 being encoded by the same reading frame as P1 but lacking the N-terminal 55 amino acids. (IS*50*L encodes two corresponding proteins, P3 and P4; however, IS*50*L contains an ochre codon so that P3 and P4 are non-functional proteins.)

Tnp can catalyse transposition of (i) the entire transposon (Tn*5*) or (ii) IS*50*; transposition of Tn*5* involves the two OE sites, whereas transposition of IS*50* involves one OE site and one IE site.

Tnp is characterized by a low level of activity which is further down-regulated by Inh; the relative levels of Tnp and Inh influence the frequency of transposition. In vivo, host factors tend to favour transposition from the OE sites (in preference to the IE sites). Other regulatory factors in vivo include *dam* methylation (see DAM GENE) which negatively affects the function of the P1 promoter (and also results e.g. in coupling transposition to DNA replication).

[Tn*5* (review): ARM (1993) *47* 945–963.]

Early studies suggested that transposition of Tn*5* may involve a cut-and-paste (rather than a replicative) mechanism (see TRANSPOSABLE ELEMENT) [CSHSQB (1984) *49* 215–226], but subsequent work suggested the involvement of cointegrates (i.e. a replicative mechanism) [JMB (1986) *191* 75–84]. However, more recent evidence favours a cut-and-paste model [JBC (1998) *273* 7367–7374].

The latter study [JBC (1998) *273* 7367–7374] demonstrated that transposition could be achieved in an in vitro system consisting of: (i) transposase, (ii) DNA flanked by OE sequences, and (iii) target DNA. Such in vitro transposition is optimized by a 'hyperactive' form of transposase transcribed from IS*50*R containing three mutations – one of which blocks transcription of the inhibitor (Inh) while another enhances the binding of transposase to OE sequences.

Tn*7* A TRANSPOSON which can insert into the chromosome of *Escherichia coli* with a high degree of target specificity; the target site is designated *att*Tn*7*. Insertion of Tn*7* involves several Tn*7*-encoded proteins: TnsA, TnsB, TnsC and TnsD; TnsA and TnsB jointly provide transposase activity. Initially, TnsD binds to *att*Tn*7* in a sequence-specific way, and such binding distorts the 5′ end of the binding site; it has been suggested that this distortion of DNA acts as a signal for the recruitment of TnsC to the site [EMBO (2001) *20* 924–932].

A second, distinct pathway of transposition of Tn*7* involves the transposon-encoded TnsA, TnsB, TnsC and TnsE; TnsE, which binds DNA in a structure-dependent way, promotes insertion of Tn*7* into (i) certain plasmids, and (ii) the chromosome of *E. coli* at sites proximal to double-stranded breaks in DNA and also at locations where DNA replication terminates [GD (2001) *15* 737–747].

Tn*9* A 2638-bp composite TRANSPOSON which includes terminal, directly repeated IS*1* elements and a gene (*cat*) for resistance to CHLORAMPHENICOL (*cat* encodes chloramphenicol acetyltransferase – an enzyme which also inactivates fusidic acid). Tn*9* apparently transposes by a replicative process involving cointegrate formation, *recA* function being necessary for cointegrate resolution [JMB (1986) *191* 75–84]. (cf. Tn*3* and Tn*5*.) Tn*9* occurs e.g. in the R plasmid pSM14 (= R14), a plasmid which also contains Tn*10*. Tn*9*-derived chloramphenicol resistance from Gram-negative bacteria can be cloned in, but not phenotypically expressed in, *Bacillus subtilis* [PNAS (1982) *79* 5886–5890]. (See also IS*1*.)

Tn*10* A 9.3-kb composite TRANSPOSON which carries a gene for tetracycline resistance; Tn*10* occurs e.g. in the transmissible plasmid R100 (= R222). The mechanism of Tn*10* transposition is distinct from that of Tn*3* (q.v.); it apparently involves a non-replicative 'cut-and-paste' mechanism in which excision of the transposon from the donor molecule (by double-stranded breaks at each end) is followed by insertion of the transposon into the new target site. The remainder of the donor molecule is probably lost ('donor-suicide' model). [Genetic evidence for non-replicative Tn*10* transposition: Cell (1986) *45* 801–815.]

The two ends of Tn*10* are IS elements (IS*10*, 1.4 kb) in opposite orientation; the two IS*10* elements are similar but not identical, and are designated IS*10*-Right (IS*10*-R) and IS*10*-Left (IS*10*-L). IS*10*-R encodes at least one protein (transposase) which acts at the ends of Tn*10* and is necessary for its transposition; the transposase is apparently preferentially *cis*-acting, possibly migrating along the DNA molecule rather than being freely diffusible in the cell cytoplasm. IS*10*-R can promote normal levels of transposition (e.g. ca. 10^{-7} transpositions/TE/bacterial generation) even when IS*10*-L is inactive; however, IS*10*-L is functionally defective and can provide only ca. 1–10% of the transpositional activity of IS*10*-R. Insertion of Tn*10* into a new target site – presumably recognized by the IS*10*-R transposase – results in the duplication of a 9-bp target sequence. The preferred ('hot-spot') target site for Tn*10* contains a symmetrical 6-bp sequence within the 9-bp sequence (NGCTNAGCN) duplicated during insertion; insertion can also occur, less efficiently, at many other sites, with concomitant duplication of a different 9-bp sequence. (See also DAM GENE and MULTICOPY INHIBITION.)

Tn*21* A 19.3-kb Tn*3*-like transposon (see Tn*3*) present e.g. in plasmid R100 from *Shigella flexneri*; it encodes resistance to sulphonamides, streptomycin and mercuric ions. Tn*21* transposition is regulated by a *modulator protein* encoded by a gene (*tnpM*) located upstream of the Tn*21* IRS. The *tnpA*, *tnpR* and *tnpM* genes are all transcribed in the same direction (cf. Tn*3*). The Tn*21* modulator apparently enhances Tn*21* transposition and suppresses cointegrate resolution by Tn*21* resolvase. A sequence homologous to the Tn*21* *tnpM* gene has been observed in an analogous site in Tn*501*. [Cell (1985) *42* 629–638.]

Tn*501* An 8.2-kb Tn*3*-like transposon (see Tn*3*) from *Pseudomonas* spp (plasmid pUS1); it encodes resistance to mercury. The frequency of transposition and efficiency of cointegrate resolution are both substantially increased in the presence of low levels of mercury, suggesting that the *tnpA* and *tnpR* genes can be transcribed from the same promoter as the inducible gene for mercury resistance. Tn*501* may encode a modulator protein similar to that of Tn*21* (q.v.). Tn*501* provides 'hot-spots' for Tn*3* insertion.

Tn*551* See Tn*3*.

Tn*554* A site-specific, repressor-controlled TRANSPOSON from *Staphylococcus aureus*; it carries (inducible) genes for resistance to erythromycin and spectinomycin.

Tn*916* See CONJUGATIVE TRANSPOSITION.

Tn*925* See CONJUGATIVE TRANSPOSITION.

Tn*951* A defective transposon of the Tn*3* group (see Tn*3*) which encodes genes for lactose metabolism (*lacZ* and *lacY*); its ends are perfect inverted repeats (40 bp), but it apparently lacks genes for its transposition – transposition requiring Tn*3*-encoded transposase and resolvase. (cf. IS*101*.) Tn951 contains an IS*1* sequence [MGG (1980) *178* 367–374].

Tn*1000* *Syn.* γδ (see Tn*3*).

Tn*1207.1* See MACROLIDE ANTIBIOTICS.

Tn*1545* See CONJUGATIVE TRANSPOSITION.

Tn*1681* A TRANSPOSON which carries a gene for the heat-stable enterotoxin STa of enterotoxigenic *Escherichia coli* (see ETEC); Tn*1681* contains two copies of IS*1* in inverted orientation.

Tn*1721* An 11.4-kb Tn*3*-like transposon (see Tn*3*) encoding resistance to tetracycline. In the presence of tetracycline the resistance gene is duplicated many times, resulting in increased levels of tetracycline resistance.

Tn*2410* An 18.5-kb (probably Tn*3*-like) TRANSPOSON derived from plasmid R1767 from *Salmonella typhimurium*; it carries genes encoding an OXA-2 β-LACTAMASE and resistance to sulphonamides and mercury [JGM (1983) *129* 2951–2957].

Tn*2603* See Tn*3*.

Tn*3701* See CONJUGATIVE TRANSPOSITION.

Tn*5253* See CONJUGATIVE TRANSPOSITION.

Tn*5397* A conjugative transposon (see CONJUGATIVE TRANSPOSITION) which is found e.g. in *Clostridium difficile* (and which is transferable to e.g. *Bacillus subtilis*); it encodes resistance to tetracycline. Although related to Tn*916*, Tn*5397* encodes a dual-function protein, TndX, that mediates both insertion and excision (in contrast to the products of the *int* and *xis* genes of Tn*916*). Tn*5397* contains a group II intron which has been shown to undergo splicing in vivo [JB (2001) *183* 1296–1299].

Tn*A* See Tn*3*.

***tna* operon** An OPERON which includes the structural gene (*tnaA*) for TRYPTOPHANASE; *tnaA* is separated from the *tna* promoter by a long LEADER (*tnaL*) which contains several rho-dependent terminators [JB (1986) *166* 217–223] (see TRANSCRIPTION) and a sequence (*tnaC*) encoding a 24-amino-acid leader peptide. The operon is subject to CATABOLITE REPRESSION and is inducible by tryptophan; translation of *tnaC* apparently plays an essential role in operon expression [JB (1986) *167* 383–386]. In the absence of tryptophan, rho-dependent transcription termination occurs in the *tnaL* region, preventing expression of *tnaA*; however, in the presence of tryptophan, termination is prevented (see ANTITERMINATION), allowing *tnaA* gene expression. (cf. OPERON (attenuator control).)

TNase Thermonuclease (see DEOXYRIBONUCLEASE).

TNF Tumour necrosis factor: a pro-inflammatory cytokine (see CYTOKINES) synthesized by many types of cell on appropriate stimulation. The two forms of TNF, TNF-α and TNF-β, are both encoded by three-intron, single-copy genes.

TNF-α (formerly 'cachectin') is produced e.g. by activated macrophages, by the Th1 subset of T lymphocytes, and by endothelial cells and leukocytes stimulated by lipopolysaccharides (see also CD14).

TNF-α appears to have a wide range of activities which include (i) induction of APOPTOSIS in target cells, (ii) cytolysis of certain types of tumour, (iii) regulation of proliferation and differentiation in lymphocytes, (iv) upregulation of CELL ADHESION MOLECULES (E-selectins, ICAM-1) on endothelial cells during INFLAMMATION, (v) promotion of ADCC activity in neutrophils, (vi) stimulation of monocytes and macrophages to synthesize e.g. TNF-α, IL-1 and IL-6, and (vii) induction of certain CHEMOKINES (e.g. IL-8).

Injection of TNF into animals gives rise to a syndrome indistinguishable from ENDOTOXIC SHOCK.

In vitro studies suggest that TNF-α may upregulate endothelial toxin receptors during EHEC infection and (thus) promote the vascular damage associated with HAEMOLYTIC URAEMIC SYNDROME

[TIM (1998) *6* 228–233]. (See also DISSEMINATED INTRAVASCULAR COAGULATION and other cytokine-associated conditions listed under CYTOKINES.)

In some individuals, susceptibility to a severe form of MALARIA has been found to correlate with a variant form of the TNF-α promoter [Nature (1994) *371* 508–510], and an increased susceptibility to endotoxic shock appears to correlate with enhanced synthesis of TNF [JAMA (1999) *282* 561–568]. The apparent link between levels of TNF and pathogenesis has suggested that benefit may be obtained, in certain cases, from the use of ligands that sequester TNF and prevent its binding to specific cell-surface receptors; such anti-TNF ligands include *etanercept* and the monoclonal antibody *infliximab*.

TNF-α exists in both soluble (secreted) and membrane-associated forms, both forms being active.

TNF-β (also called lymphotoxin α) is produced by activated lymphocytes, and is characteristic of the Th1 subset of T cells. TNF-β resembles TNF-α in its activities – many of which coincide with the activities of INTERLEUKIN-1; unlike IL-1, however, TNF can trigger apoptosis in a target cell.

The cell-surface receptors of tumour necrosis factor are designated p55 (= CD120a) and p75 (= CD120b) and they occur on a wide range of cells. Each type of receptor can apparently serve as a binding site for both forms of TNF; however, the two receptors may have distinct roles in TNF activity, and p75 may have a subordinate role. Following endotoxaemia, the peripheral circulation contains raised levels of soluble (cell-free) TNF receptors; this may be a protective mechanism which down-regulates the effects of TNF on host cells.

The binding of TNF to its receptor may give rise to various effects ranging from stimulation to apoptosis – according to the particular signalling pathway which is activated within the target cell. Currently, these pathways (and the factors which influence the choice of pathway) are incompletely understood, but apoptosis seems to involve part of the cytoplasmic side of the TNF receptor – a region called the 'death domain'. Following TNF–receptor binding, the death domain appears to associate with certain cytoplasmic factors – including TRADD (TNF receptor-associated death domain protein) and a serine threonine kinase called RIP (receptor interacting protein); RIP may activate caspase(s) such as IL-1β-converting enzyme (ICE) (leading to APOPTOSIS) or it may phosphorylate (and inactivate) the inhibitor IκB, thus activating the nuclear transcription factor NF-κB.

***tnpA* gene** See Tn*3*.

Tn*phoA* See TRANSPOSON MUTAGENESIS.

***tnpM* gene** See Tn*21*.

***tnpR* gene** See Tn*3*.

toadstool Any umbrella-shaped basidiocarp, but particularly one which is inedible or poisonous.

tobacco diseases For diseases of the tobacco plant see e.g. GRANVILLE WILT, PERONOSPORA (blue mould), POTATO VIRUS Y (e.g. veinal necrosis), TOBACCO MOSAIC VIRUS, TOBACCO NECROSIS VIRUS and WILDFIRE DISEASE.

tobacco etch virus See POTYVIRUSES.

tobacco leafcurl virus See GEMINIVIRUSES.

tobacco mosaic virus (TMV) The type member of the TOBAMOVIRUSES. The TMV virion is a tubular filament (ca. 300 × 20 nm, central hole approx. 2 nm diam., S_w^{20} ca. 194); it comprises coat protein subunits (MWt ca. 17500) arranged in a single right-handed helix with the ssRNA intercalated between the turns of the helix (ca. three nucleotides per protein subunit). TMV infects tobacco (*Nicotiana tabacum*) and other plants. The common, 'Vulgare' or field strain (U1 strain) usually causes a systemic disease of tobacco in which leaves are distorted, blistered, and marked with a mosaic of light and dark green patches; intracellular crystalline arrays of virus particles are commonly visible by light microscopy. Other TMV strains may affect other plants (e.g. strain Cc affects legumes, TMV-L affects tomatoes). (See also PATHOGENESIS-RELATED PROTEINS.) TMV is transmitted mechanically; it may remain infective for a year or more in soil or dried leaf tissue. The virions may be inactivated e.g. at pH <3 or >8 or by formaldehyde, iodine, etc; TIP: ca. 95°C. Preparations of TMV may be obtained from plant tissues e.g. by $(NH_4)_2SO_4$ precipitation followed by differential centrifugation.

The TMV ssRNA genome is ca. 6400 nucleotides long [sequence: PNAS (1982) *79* 5818–5822] and is capped at the 5′ end (sequence: $m^7G^{5'}ppp^{5'}Gp$) but is not polyadenylated. The genomic RNA can serve as mRNA for a protein of MWt ca. 130000 (130 K) and another, produced by readthrough, of MWt ca. 180000 (180 K); however, it cannot function as messenger for the synthesis of e.g. coat protein. Other genes are expressed during infection by the formation of monocistronic, 3′-coterminal SUBGENOMIC MRNAS, including one (LMC) encoding the 17.5 K coat protein and another (I_2) encoding a 30 K protein. (Other putative subgenomic mRNAs may be artefacts generated during electrophoresis [Book ref. 80, pp. 69–72].) The 30 K protein has been detected in infected protoplasts [Virol. (1984) *132* 71–78]; it may be involved in the cell-to-cell transport of the virus in an infected plant. The functions of the two large proteins are unknown.

Several dsRNA molecules – including dsRNAs corresponding to the genomic, I_2 and LMC RNAs – have been detected in plant tissues infected with TMV; these are presumably intermediates in genome replication and/or mRNA synthesis – processes which appear to occur by different mechanisms [Book ref. 80, pp. 84–87].

TMV assembly apparently occurs in the plant cell cytoplasm, although it has been suggested that some TMV assembly may occur in chloroplasts since transcripts of ctDNA have been detected in purified TMV virions [Book ref. 80, pp. 73–79]. Initiation of TMV assembly occurs by interaction between ring-shaped aggregates ('discs') of coat protein (each disc consisting of two layers of 17 subunits) and a unique internal nucleation site in the RNA: a hairpin region ca. 900 nucleotides from the 3′ end in the common strain of TMV. (Any RNA – including e.g. subgenomic RNAs – containing this site may be packaged into virions.) The discs apparently assume a helical form on interaction with the RNA, and assembly (elongation) then proceeds in both directions (but much more rapidly in the 3′-to-5′ direction) from the nucleation site. [Review of structure and assembly: JGV (1984) *65* 253–279.]

tobacco necrosis virus (TNV) A PLANT VIRUS which can infect a range of angiosperms; transmission occurs via *Olpidium* spp and can occur mechanically under experimental conditions. Symptoms of TNV infection range from severe necrosis (e.g. in Augusta disease of tulips) to local necrotic leaf lesions (e.g. in tobacco seedlings). Virion: icosahedral, ca. 28 nm diam., containing one polypeptide species (MWt ca. 22600); genome: a single molecule of linear positive-sense ssRNA (MWt ca. $1.3–1.6 \times 10^6$) with the 5′-terminal sequence: ppApGpU. . . .

TNV-infected plant cells may also contain a SATELLITE VIRUS: an icosahedral, ssRNA-containing virus [structure: JMB (1982) *159* 93–108] which is completely dependent on the presence of

TNV for its replication. The satellite can be transmitted by the same mechanisms as TNV, and may influence the nature of the symptoms in the infected plant.

TNV (A strain) is the type member of a taxonomic group (the tobacco necrosis virus group); a possible member of this group is cucumber necrosis virus.

tobacco necrotic dwarf virus See LUTEOVIRUSES.

tobacco rattle virus See TOBRAVIRUSES.

tobacco ringspot virus See NEPOVIRUSES.

tobacco streak virus See ILARVIRUSES.

tobacco vein distorting virus See LUTEOVIRUSES.

tobacco vein mottling virus See POTYVIRUSES.

tobacco veinal necrosis disease See POTATO VIRUS Y.

tobacco yellow dwarf virus See GEMINIVIRUSES.

tobacco yellow net virus See LUTEOVIRUSES.

tobamoviruses A group of PLANT VIRUSES in which the virion is a rigid filament consisting of one molecule of linear, positive-sense ssRNA associated with a single type of coat polypeptide. Type member: TOBACCO MOSAIC VIRUS; other members: e.g. cucumber green mottle mosaic virus, cucumber 4 virus, tomato mosaic virus. A group of morphologically similar viruses which are transmitted by members of the Plasmodiophoromycetes (*Polymyxa*, *Spongospora*) – including e.g. beet necrotic yellow vein virus, peanut clump virus, potato mop top virus, soil-borne wheat mosaic virus – were tentatively included in the tobamoviruses, but see SOIL-BORNE WHEAT MOSAIC VIRUS.

tobramycin 3′-DeoxyKANAMYCIN B.

tobraviruses (tobacco rattle virus group) A group of bipartite ssRNA-containing PLANT VIRUSES which have a wide host range (including monocots and dicots) and which are transmitted primarily by nematodes (*Paratrichodorus* and *Trichodorus* spp) but also via seeds and (sometimes) mechanically; the viruses persist in, but do not replicate in, the nematode vector. Type member: tobacco rattle virus (TRV) (PRN isolate); other member: pea early-browning virus. Possible member: peanut clamp virus. Symptoms of tobravirus infection typically include necrosis; TRV can cause e.g. SPRAING in potatoes.

Virion: tubular, consisting of a helix (pitch 2.5 nm) of protein-coated linear positive-sense ssRNA; two types of particle occur: one (L particle) is ca. 180–215 nm long and contains RNA1 (MWt ca. 2.4×10^6), the other (S particle) is 46–114 nm long (depending on isolate) and contains RNA2 (MWt ca. $0.6–1.4 \times 10^6$). RNA1 can be replicated in plant cells in the absence of RNA2, and L particles (but not S particles) can alone cause lesions in the host plant; however, both L and S particles are necessary for the formation of progency virions (RNA2 encodes the coat protein).

Todd–Hewitt broth A medium used for the culture of certain *Streptococcus* spp. It contains an infusion of fat-free beef heart, peptone, glucose, sodium chloride, and a buffer system (sodium bicarbonate and disodium hydrogen phosphate); pH: 7.8.

Todd unit (TU) A unit used to express the antibody titre in an ANTISTREPTOLYSIN O TEST; it has been defined as the minimum amount of serum which can neutralize 2.5 minimum haemolytic doses of a standard preparation of streptolysin O.

tofu An intermediate in SUFU production. Tofu may itself be used as a food.

Togaviridae A family of enveloped ssRNA-containing VIRUSes, most of which can infect a wide range of vertebrates; most members also infect arthropods which act as VECTORS (cf. ARBOVIRUSES). The family includes a number of important pathogens of man and animals, some of which can be transmitted transplacentally and can cause abortion or abnormalities in the infected fetus. Arthropod vectors (which remain infected for life) are apparently unharmed. The Togaviridae contains four genera (members of a genus being serologically related to each other but unrelated to those of other genera): ALPHAVIRUS, ARTERIVIRUS, PESTIVIRUS and RUBIVIRUS; possible members of the family include LACTATE DEHYDROGENASE VIRUS and carrot mottle virus. [Taxonomy: Intervirol. (1985) *24* 125–139.] (cf. FLAVIVIRIDAE.)

The togavirus virion is spherical (ca. 50–70 nm diam.) and consists of a nucleocapsid (ca. 28–35 nm diam.) – which is icosahedral in at least some species – surrounded by a lipoprotein envelope. The nucleocapsid is composed of a single type of core protein (C protein) associated with the RNA genome. The envelope bears surface projections (spikes) composed of two major glycoproteins, E1 and E2; in Semliki Forest virus a third glycoprotein, E3, is associated with the spikes, but in other alphaviruses E3 is released into the culture fluid in infected cell cultures. The nature of the glycosylation of these proteins is variable, depending at least in part on the nature of the host cell. The envelope glycoproteins are responsible for haemagglutinating activity and for virus infectivity. [Structure of alphaviruses: see e.g. Cell (2001) *105* 5–8.]

The virus genome consists of a single molecule of linear positive-sense ssRNA (MWt ca. 4×10^6); in those viruses which have been investigated, the RNA is capped and polyadenylated. In alphaviruses, at least, the gene sequence is 5′-nsP1–nsP2–nsP3–nsP4–C–E3–E2–E1-3′, where nsP1–nsP4 are non-structural proteins (apparently RNA-dependent RNA polymerase components). [Nucleotide sequence of Sindbis virus: Virol. (1984) *133* 92–110.]

Togavirus replication occurs in the host cell cytoplasm; the process has been studied chiefly in the alphaviruses Semliki Forest virus and Sindbis virus. In these viruses, the non-structural proteins are synthesized directly from the (42S) genome-length (+)-strand RNA, while structural proteins are translated from capped and polyadenylated 26S subgenomic RNA which is initiated internally on a full-length (−)-strand template. (Rubella virus employs a similar strategy, producing 40S genome-length and 24S subgenomic RNAs [JV (1984) *49* 403–408].) Translation is initiated at a single site, and the proteins are generally cleaved from the nascent polyprotein during translation. The synthesis of each type of viral RNA (full-length (−)-strand, full-length (+)-strand, and subgenomic) is apparently regulated independently; 26S RNA is produced in excess of 42S RNA, and genomic RNA is rapidly encapsidated, so that a large excess of structural over non-structural proteins is achieved. Virus maturation involves budding through the plasma membrane or through intracytoplasmic membranes. [Overview of the replication cycle: Book ref. 148, pp. 1021–1032.]

Many alphaviruses induce rapid and dramatic CPE in vertebrate cells (cf. RUBIVIRUS); host cell RNA and protein synthesis is rapidly arrested, and the cells eventually lyse. Some alphaviruses can establish persistent infections in vertebrate cells. [Effects of alphavirus infection on vertebrate cells: Book ref. 150, pp. 465–499.] Invertebrate cells are generally infected persistently with no apparent adverse effects.

togaviruses Viruses of the TOGAVIRIDAE.

Tokophrya See SUCTORIA.

TOL plasmid An IncP-9 *Pseudomonas* PLASMID (ca. 117 kb) which encodes the capacity to metabolize toluene and xylene. (cf. CAM PLASMID; OCT PLASMID; SAL PLASMID.) [Degradative plasmids of *Pseudomonas*: Book ref. 198, pp. 295–323.]

TOL encodes the enzymes of a two-stage catabolic pathway for the mineralization of toluene and xylene. Initially, these substrates are oxidized to their respective carboxylic acids. Such oxidation involves enzymes encoded by the *upper* operon of the TOL plasmid; the Pu promoter of the upper operon is regulated by an enhancer-dependent sigma factor (σ^{54}; see SIGMA FACTOR) in association with the activator protein XylR. The *meta* operon of TOL encodes enzymes involved in aromatic ring fission and subsequent formation of pyruvate and acetaldehyde (which enter the TCA cycle). [Regulation of the *upper* and *meta* operons of the TOL plasmid: EMBO (2001) *20* 1–11 (4–6).]

In strains of *Pseudomonas* carrying the Tol plasmid, the availability of iron may be an important factor in the oxidative metabolism of toluene [AEM (2001) *67* 3406–3412].

TolA protein See COLICINS.

TolB protein See COLICINS.

toleragen See IMMUNOLOGICAL TOLERANCE.

tolerance (*immunol.*) See IMMUNOLOGICAL TOLERANCE.

***tolG* gene** See OMP.

tolnaftate (2-naphthyl-*N*-methyl-*N*-(*m*-tolyl) thiocarbamate) An ANTIFUNGAL AGENT which is effective in the topical treatment of dermatophyte infections but which has little or no activity against most other fungi or bacteria.

***tolQ* gene** See TONB PROTEIN.

TolQ protein See COLICINS.

***tolR* gene** See TONB PROTEIN.

TolR protein See COLICINS.

toluidine blue A blue metachromatic basic thiazine DYE used e.g. in ALBERT'S STAIN and for VITAL STAINING.

Tolypocladium See HYPHOMYCETES; see also CYCLOSPORIN A.

tolypomycins See ANSAMYCINS.

Tolyposporium See USTILAGINALES.

Tolypothrix See SCYTONEMA.

***tolZ* gene** See FTSH.

tomatine A glycoalkaloid SAPONIN present in high concentrations in green tomatoes. It is toxic to many fungi; in mutant strains of *Fusarium solani* which are resistant to tomatine the cytoplasmic membrane has been found to have lower-than-normal levels of sterols.

tomato apical stunt viroid See VIROID.

tomato aspermy virus See CUCUMOVIRUSES.

tomato black ring virus See NEPOVIRUSES.

tomato bunchy top viroid See VIROID.

tomato bushy stunt virus See TOMBUSVIRUSES.

tomato golden mosaic virus See GEMINIVIRUSES.

tomato mosaic virus See TOBAMOVIRUSES.

tomato (Peru) mosaic virus See POTYVIRUSES.

tomato planta macho viroid See VIROID.

tomato ringspot virus See NEPOVIRUSES.

tomato spotted wilt virus (TSWV) A PLANT VIRUS which appears to be unrelated to other known viruses, although it has some properties resembling those of bunyaviruses. Virion: spherical, enveloped, ca. 82 nm diam., containing 4 major and up to 3 minor proteins. Genome: ssRNA (one negative-sense and two AMBISENSE pieces).

TSWV can infect a wide range of host plants. Infected cells contain granular cytoplasmic inclusion bodies. Immature virions are each surrounded by two lipoprotein membranes, but the mature virions each have a single envelope and occur in groups within membranous vesicles. Maturation resembles that of the BUNYAVIRIDAE. Transmission occurs (circulatively) via thrips (Thysanoptera); only the larval stage of the insect can acquire the virus from an infected plant, but once acquired the virus is retained by the insect for life. TSWV can also be transmitted mechanically under experimental conditions.

tomato top necrosis virus See NEPOVIRUSES.

tomato yellow dwarf virus See GEMINIVIRUSES.

tomato yellow leafcurl virus See GEMINIVIRUSES.

tomato yellow mosaic virus See GEMINIVIRUSES.

tomato yellow net virus See LUTEOVIRUSES.

tomato yellow top virus See LUTEOVIRUSES.

tomaymycin See ANTHRAMYCIN.

tombusviruses (tomato bushy stunt virus group) A group of ssRNA-containing PLANT VIRUSES which can infect a wide range of angiosperms. Transmission occurs mechanically and possibly via the soil. Type member: tomato bushy stunt virus (TBSV); other members: e.g. artichoke mottled crinkle virus, carnation Italian ringspot virus, *Cymbidium* ringspot virus, and eggplant mottled crinkle virus. (Other 'possible members' – e.g. carnation mottle virus, saguaro cactus virus, and turnip crinkle virus – appear on the basis of subgenomic dsRNA analysis to have a different genome organization and may constitute a separate group [Book ref. 80, pp. 80–83].) [Serological relationships among tombusviruses: JGV (1986) *67* 75–82.]

Virion: icosahedral, 30 nm diam., containing one molecule of linear positive-sense ssRNA (MWt ca. 1.5×10^6). The capsid is composed primarily of 180 molecules of a major coat protein. Each coat protein molecule (386 amino acid residues) is folded into three distinct domains (P, S and R): the S domains form the tightly bonded icosahedral shell, the P domains form surface projections, and the (N-terminal) R domains apparently project inwards and may bind the genomic RNA; the P and S domains are linked via a flexible hinge region. [Virion structure: Book ref. 81, pp. 43–50.] Within infected plant cells, the virus particles occur in the cytoplasm and nucleus (often associated with the nucleolus), sometimes forming crystalline arrays; compact membranous inclusion bodies ('multivesicular bodies') occur in the cytoplasm.

tomentose Downy; woolly.

tomentum (*lichenol.*) A downy or felted mat of hyphae which occurs (usually) on the lower surface of the thallus in certain foliose lichens.

tomite A typically small, motile, non-feeding stage in the life cycles of certain protozoa; it is generally formed, within a CYST, by the division of a cell called a *tomont*. In e.g. ICHTHYOPHTHIRIUS the tomite acts as a THERONT. In members of the Apostomatida the tomite itself eventually encysts at a site on a suitable host (becoming a *phoront*) and subsequently develops into a cell capable of feeding (a *trophont*) – which later develops into a tomont; in some apostomes a stage known as a *protomite* occurs between the tomont and tomite stages.

tomont See TOMITE.

TonA protein (FhuA protein) In *Escherichia coli*: an OUTER MEMBRANE protein encoded by the *tonA* gene; it acts as a receptor for e.g. colicin M and bacteriophages T1 and ϕ80, and is involved in the uptake of albomycin and ferrichrome.

TonB protein In e.g. *Escherichia coli*, a periplasmic protein that *shuttles* between the cytoplasmic membrane and outer membrane [Mol. Microbiol. (2003) *49* 869–882], supplying energy (from pmf) to the outer membrane for the uptake of e.g. ferric iron–chelate complexes (see SIDEROPHORES); it is also involved e.g. in the uptake of VITAMIN B_{12} by BINDING PROTEIN-DEPENDENT TRANSPORT.

The uptake of vitamin B_{12} across the outer membrane in *E. coli* is pmf-dependent [JB (1993) *175* 3146–3150]; it appears that maximum activity of this (and other) TonB-dependent systems requires the products of *exbB* and *exbD* (proteins in the cytoplasmic membrane) which may stabilize TonB. ExbB and

ExbD seem to be replaceable, to some extent, by the products of *tolQ* and *tolR*.

TonB is involved in the susceptibility of the cell to bacteriophage ϕ80. It is also required for the uptake of group B colicins: colicins B, D, Ia, Ib, M, V, 5 and 10 [colicin import into *Escherichia coli* cells: JB (1998) *180* 4993–5002]. (See also FEPA PROTEIN and TSX PROTEIN.)

tonoplast The membrane which forms the boundary of an intracellular vacuole.

tonsillitis Inflammation of the tonsils, commonly caused by *Streptococcus pyogenes*. (See also OTITIS MEDIA, QUINSY, RHEUMATIC FEVER.)

Tontonia See OLIGOTRICHIDA.

tooth diseases See DENTAL CARIES and PERIODONTITIS.

top-fermenting yeast See BREWING.

top necrosis (*plant pathol.*) Necrosis (death) of the terminal bud or of the entire top of a plant.

topical (*med., vet.*) Local, i.e., restricted to a particular region of the body; not SYSTEMIC.

topo I, topo II, etc. *Syn.* TOPOISOMERASE type I, topoisomerase type II, etc.

topoisomer Topoisomers are molecules which differ from one another only in their topological properties. For example, a given dsDNA molecule may occur in a number of topologically distinct forms: linear, relaxed circular, or supercoiled circular (ds cccDNA molecules with different linking numbers, but otherwise identical, are topoisomers); interconversion of topoisomers necessitates strand breakage and/or joining. (See also DNA.)

topoisomerase (DNA topoisomerase) Any enzyme which can convert one topological isomer of cccDNA to another; topoisomerases can e.g. alter the linking number (*Lk*) of a cccDNA molecule (see DNA) and (at least in vitro) can form and resolve knots and CATENANES – i.e. the reactions catalysed by a topoisomerase can be intermolecular or intramolecular. All such topological conversions involve transient breakage of either one or both strands of DNA.

Type I (= *type 1*) topoisomerases (also called untwisting, relaxing or nick-closing enzymes, or swivelases) break only one strand of DNA; in a ds cccDNA molecule, the unbroken strand is passed through the break before re-sealing, thus changing the value of *Lk* by one for each such event.

Type II (= *type 2*) topoisomerases break both strands of DNA in a ds cccDNA molecule; a (double-stranded) segment of DNA from elsewhere in the molecule is then passed through the break (without strand rotation) before re-sealing – each such event changing the value of *Lk* by two.

Both types of enzyme are found in both prokaryotic and eukaryotic cells.

Certain topoisomerases are the targets of QUINOLONE ANTIBIOTICS (q.v.).

An example of a type I topoisomerase is the ω (omega) protein of *Escherichia coli* (*E. coli* topoisomerase I, or 'topo I'); in vitro, this enzyme can e.g. partially relax negatively (but not positively) supercoiled DNA, can introduce topological knots into ss cccDNA, and can convert complementary ss cccDNA circles into completely base-paired ds cccDNA. The primary in vivo role of topo I appears to be the relaxation of negative supercoiling; topo I is necessary e.g. for chromosomal DNA REPLICATION. Topo III (encoded by gene *topB*) is another type I topoisomerase in *E. coli*; like topo I, this enzyme can relax negative supercoiling (and e.g. form and resolve knots) in vitro, but its in vivo role has not been established.

Type II topoisomerases include GYRASE, REVERSE GYRASE, bacteriophage T4 topoisomerase (gp39, 52, 60), and topoisomerase II′ (topo II′, an enzyme from *E. coli* which contains gyrase subunit A and a part of gyrase subunit B). The T4 enzyme and topo II′ have been reported to relax positively or negatively supercoiled DNA. In vitro, type II topoisomerases can form and resolve catenanes and can also form and resolve knots. In a model for the mechanism of type II topoisomerases, binding of the enzyme initially introduces a sharp bend in the DNA, the enzyme adopting a specific orientation with respect to the bend and subsequently promoting unidirectional passage of the strands through the break [PNAS (2001) *98* 3045–3049].

Topoisomerase IV is a type II enzyme encoded by genes *parC* and *parE* in *E. coli*. This enzyme can e.g. relax more fully those negatively supercoiled molecules of DNA which have been partially relaxed by topo I; topo IV appears to be responsible – solely – for decatenation and for the resolution of knots in *E. coli* [GD (2001) *15* 748–761].

topological winding number See DNA.

topotaxis A behavioural response in which a motile organism moves (directly or indirectly) towards or away from a directional stimulus. (cf. PHOBIC RESPONSE.)

Torbal jar See MCINTOSH AND FILDES' ANAEROBIC JAR.

TORCH diseases A group of diseases – comprising TOXOPLASMOSIS, RUBELLA, CYTOMEGALIC INCLUSION DISEASE, and HERPES SIMPLEX – each of which can cause abortion, stillbirth, severe neonatal disease, and sometimes fetal tissue damage with consequent developmental abnormalities.

toroidal DNA See DNA TOROID.

Toroviridae A family of enveloped RNA viruses proposed to include the BERNE VIRUS and serologically related viruses: e.g. BREDA VIRUS and some human gastroenteritis viruses.

torr See PASCAL.

Torula A genus of fungi of the HYPHOMYCETES. *T. herbarum* is common e.g. on dead grasses – forming dark, velvety colonies, and dark conidia in chains.

torula yeast A food yeast, *Candida utilis* (= *Torulopsis utilis*), grown e.g. on ethanol or sulphite liquor waste (see SINGLE-CELL PROTEIN); the yeast is pasteurized and dried before use. It is a rich source of protein and vitamins and is used medicinally and as a food additive.

torularhodin See CAROTENOIDS.

Torulaspora A genus of yeasts (family SACCHAROMYCETACEAE) in which the cells are globose or ellipsoidal; vegetative reproduction occurs by multilateral budding. Pseudomycelium is not formed. Vegetative cells are predominantly haploid (cf. SACCHAROMYCES). Ascus formation is preceded by conjugation – usually between a cell and its bud, but sometimes between independent cells; asci are persistent. Ascospores: globose or ellipsoidal, smooth- or rough-surfaced, 1–4 per ascus. Glucose and certain other sugars are fermented vigorously; NO_3^- is not assimilated. Species: *T. delbrueckii* (anamorph: *Candida colliculosa*), *T. globosa* (formerly e.g. *Saccharomyces kloeckerianus*), and *T. pretoriensis* (formerly e.g. *Saccharomyces pretoriensis*); strains have been isolated from fruit juice, wines, fermenting cucumber brines, soil, faeces, etc. [Book ref. 100, pp. 434–439.]

Torulopsis A genus of yeasts (class HYPHOMYCETES), now generally included within the genus CANDIDA [Book ref. 100, p. 841]. Hyphae and pseudohyphae are generally not formed.

torulopsosis A MYCOSIS, now regarded as a form of CANDIDIASIS, caused by *Candida glabrata* (= *Torulopsis glabrata*). The disease (in man) is usually systemic, involving lungs, kidney, heart, CNS etc.

torulosis *Syn.* CRYPTOCOCCOSIS.

total cell count See COUNTING METHODS.

Totiviridae See MYCOVIRUS.

touchdown PCR Repetition of a given PCR assay with a stepwise decrease in annealing temperature in each run; the object is to find the lowest annealing temperature (for the given assay) which will permit normal primer binding but avoid the problem of mispriming (i.e. binding of primers to inappropriate sequences). (The procedure thus seeks a level of stringency which promotes maximum specificity of primer binding.) [Examples of use: NAR (1991) *19* 4008; JCM (1999) *37* 1274–1279.]

tower fermenter See FERMENTER.

toxaemia (toxemia) The condition in which TOXINS are present in the blood. (cf. PYAEMIA; SEPTICAEMIA.)

toxemia *Syn.* TOXAEMIA.

toxic Poisonous; harmful. (cf. TOXIN.)

toxic epidermal necrolysis (TEN) A syndrome characterized by erythema followed by separation of the outer epidermis from the basal layer of the skin; it may be caused by ET-producing staphylococci (staphylococcal TEN – see SCALDED SKIN SYNDROME) or by a hypersensitivity reaction to certain drugs (allergic TEN).

toxic food poisoning See FOOD POISONING.

toxic shock syndrome (TSS) A severe, often fatal, illness characterized by fever, vomiting, diarrhoea and hypotension; late symptom: a sunburn-like rash with peeling of the skin, especially on palms and soles. In menstruating women, TSS is associated with the use of vaginal tampons. In children, TSS may follow burns or scalds etc.; encephalitic symptoms may occur in very young children in whom cytokines may cross the blood–brain barrier.

TSS is caused by certain of the toxins that are classified as SUPERANTIGENS: the B and C enterotoxins of *Staphylococcus aureus*, the toxic shock syndrome toxin 1 (TSST-1) of *S. aureus* (known earlier as enterotoxin F, pyrogenic exotoxin C and toxic shock toxin), and the superantigen A of *Streptococcus pyogenes*. The gene that encodes TSST-1 occurs on a *mobile* PATHOGENICITY ISLAND – possibly accounting for the spread of toxigenicity in *S. aureus* [Mol. Microbiol. (1998) *29* 527–543].

Bacteraemia is uncommon in staphylococcal TSS, common in streptococcal TSS. In the acute phase of staphylococcal TSS there is a higher proportion of e.g. Vβ2 T cells, and raised levels of e.g. IL-2, IL-4, TNF-α, TNF-β and IFN-γ.

[Multiplex PCR for detection of the genes of *S. aureus* enterotoxins B and C, and TSST-1: JMM (1998) *47* 335–340. Multiplex PCR for detection of genes for *S. aureus* enterotoxins, exfoliative toxins, toxic shock syndrome toxin 1 and methicillin resistance: JCM (2000) *38* 1032–1035.]

toxic shock syndrome toxin 1 (TSST-1) See TOXIC SHOCK SYNDROME AND SUPERANTIGEN.

toxicoinfection See *infant* BOTULISM.

toxicosis Any human or animal disease caused by poisoning (see e.g. MYCOTOXICOSIS).

toxicyst A type of tubular EXTRUSOME which occurs in many carnivorous ciliates (e.g. *Actinobolina*, *Didinium*, *Dileptus*); on activation, the tubular structure everts, penetrates the prey, and apparently introduces toxin(s) and proteolytic enzymes.

toxigenic (toxinogenic) Refers to an organism which can produce one or more TOXINS.

toxin In a microbiological context: any of various microbial products or components which, at low concentrations, can act in a specific way on cells or tissues in a higher (multicellular) organism and cause local and/or systemic damage, or death – such an agent (alone, or with other toxin(s) and/or virulence factor(s)) being responsible, directly or indirectly, for at least some aspect of pathogenesis. In many cases there is insufficient information on the role of a given agent in pathogenesis to decide whether or not it conforms to the above definition.

In some cases the known activity of a toxin is sufficient to account for all the observed features of pathogenesis in a given disease. For example, the zinc-endopeptidase activity of tetanus toxin, which blocks neuroexocytosis (see TETANOSPASMIN), adequately explains why inhibitory neurotransmitter is not released into the synaptic cleft.

In other cases, the activity of a toxin depends on one or more additional virulence factors. For example, toxigenic strains of *Bacillus anthracis* can cause ANTHRAX only if the cells have an antiphagocytic capsule. Again, the virulence of ETEC depends on plasmid-encoded fimbriae with which the bacteria adhere to the epithelium of the small intestine.

While tetanus toxin is involved *directly* in the mechanism of pathogenesis, the pathogenic role of some toxins is due partly or solely to their ability to behave as MODULINS – e.g. the B and C enterotoxins and TSST-1 of *Staphylococcus aureus* (see TOXIC SHOCK SYNDROME and SUPERANTIGEN). Many types of bacterial toxin can induce cytokines, the exotoxins being particularly potent inducers.

Not all toxic microbial substances are called 'toxins'. For example, generally *not* regarded as toxins are those products (e.g. lactic acid, hydrogen sulphide) which are significantly toxic to animals or plants only when present in relatively high concentrations (cf. ACIDOSIS sense 1 and SUFFOCATION DISEASE) – or which have an essentially physical role in causing or enhancing disease (e.g. alginate in CYSTIC FIBROSIS).

Also traditionally excluded from the category 'toxin' are those microbial products which, at low concentrations, are toxic to other *microorganisms* (e.g. ANTIBIOTICS, BACTERIOCINS).

The definition of a toxin may be based on criteria which differ in different disciplines. Thus, medical and veterinary workers may accept as toxins: (i) certain microbial enzymes which act as AGGRESSINS; (ii) substances toxic to cells in vitro but whose role in causing or exacerbating disease under natural conditions is unknown; (iii) substances produced at a site remote from the 'target' organism (see e.g. BOTULISM and *paralytic* SHELLFISH POISONING). Plant pathologists may not accept that *enzymes* can be toxins, and may not accept as toxins substances in categories (ii) and (iii) [discussion: Book ref. 58. pp 139–140].

A given toxin may be categorized according to the nature of the organism producing it (e.g. mycotoxin, phycotoxin), the nature of the organism affected by it (e.g. ichthyotoxin), the type of cell or tissue affected (e.g. ENTEROTOXIN, HAEMOLYSIN sense 2, hepatotoxin, LEUCOCIDIN, NEUROTOXIN) etc. (See also ENDOTOXIN and EXOTOXIN.) However, such categories are sometimes imprecise and may be misleading: e.g. an 'ichthyotoxin' may be toxic to organisms other than fish, and terms such as 'hepatotoxin' and 'enterotoxin' may be applied to toxins which can affect other types of cell and tissue.

For algal toxins see: PHYCOTOXIN.

For fungal toxins see: MYCOTOXIN.

For protozoal toxins see *amoebic* DYSENTERY.

For bacterial toxins see: ANTHRAX TOXIN, BOTULINUM TOXIN, CHOLERA TOXIN, DELTA-ENDOTOXIN, DIPHTHERIA TOXIN, EPIDERMOLYTIC TOXIN, EXOTOXIN A, PERTUSSIS TOXIN, PHYTOTOXIN, RTX TOXINS, SHIGA TOXIN, TETANOSPASMIN. (See also HAEMOLYSIN (sense 2), HYALURONATE LYASE, LEUCOCIDIN, THIOL-ACTIVATED CYTOLYSINS.)

[Bacterial protein toxins: Book ref. 206.]

toxin co-regulated pili (TCP) See CHOLERA, BACTERIOPHAGE CTXΦ and PATHOGENICITY ISLAND.

toxinogenic *Syn.* TOXIGENIC.

toxoid A TOXIN which has been modified (e.g. by treatment with formalin) so as to destroy its toxicity without affecting its specific immunogenicity. Thus, antibody formed against a toxoid can inactivate the corresponding toxin; toxoids are therefore useful in VACCINES.

toxoneme A MICRONEME in *Toxoplasma*.

Toxoplasma A genus of parasitic, facultatively heteroxenous protozoa (suborder EIMERIORINA); the organisms have been classified by some authors [AP (1982) *20* 403–406] in the genus ISOSPORA. *T. gondii* (= *I. gondii*) is an intracellular parasite and the causal agent of TOXOPLASMOSIS (q.v. for life cycle; see also PHAGOCYTOSIS); the sporozoites of *T. gondii* are uninucleate, banana-shaped, motile cells ca. 5–8 × 1–2 μm, but smaller, stumpy or ovoid cells occur within cysts. In the INTERMEDIATE HOST (birds, non-felines, man) only asexual reproduction occurs; in the FINAL HOST (members of the cat family) reproduction occurs asexually in extra-intestinal tissues as well as sexually in the intestinal tissues. *T. gondii* forms disporic, tetrazoic oocysts (10–15 μm) which undergo SPORULATION outside the (feline) host. Once sporulated, oocysts remain infective in soil for e.g. 1–2 years; they can be rendered uninfective e.g. by heating to 90°C for 30 sec [AP (1982) *20* 310].

***Toxoplasma* dye test** (Sabin–Feldman dye test) A serological test used to detect and quantify antibodies to *Toxoplasma gondii*. The patient's serum is heated (56°C/30 min) and a range of dilutions is prepared. To each dilution is added (i) live laboratory-cultured cells of *T. gondii*, and (ii) normal serum containing a known amount of 'accessory factor' – i.e., COMPLEMENT [JID (1980) *141* 366–369]; each test dilution is then incubated. In the presence of antibodies (*positive test*) the cells are damaged and are unable to take up an indicator dye (alkaline methylene blue); this is determined by microscopy. In the absence of antibodies the (undamaged) cells take up the dye.

Toxoplasmea A class of parasitic protozoa (subphylum SPOROZOA); organisms of this class (e.g. *Sarcocystis*, *Toxoplasma*), together with those of the TELOSPOREA and PIROPLASMEA, have been incorporated in the class SPOROZOASIDA.

toxoplasmosis An acute or chronic disease of man and other animals (e.g. cat, goat, pig, sheep) caused by *Toxoplasma gondii* (see TOXOPLASMA). In man, infection occurs by ingestion of sporulated oocysts (present e.g. in infected cat faeces) or of 'tissue cysts' in raw or insufficiently cooked infected meats. Infection may be asymptomatic, there may be mild lymphadenopathy, or the disease may be generalized with e.g. hepatitis, pneumonia, myalgia, meningoencephalitis etc; latent infection may persist for years. In the acute phase of the disease the parasite multiplies endodyogenously in cells of various tissues (including e.g. macrophages), forming *pseudocysts*; a pseudocyst (= *group stage*) is the remains of a host cell packed with rapidly dividing cells (*endozoites*, *tachyzoites*) of the parasite. (See also PHAGOCYTOSIS.) If the host survives, the disease enters the chronic phase in which the parasite localizes in certain tissues (e.g. brain, eye, skeletal muscles, heart), forming *tissue cysts*; a tissue cyst (also called a *pseudocyst* or *meront*) is a structure (ca. 50–100 μm) containing cells of the parasite (*cystozoites*, *bradyzoites*) which are undergoing slow endodyogenous division. Toxoplasmosis may be transmitted transplacentally; the congenital disease often involves lesions in the brain and/or eyes and may lead to blindness, mental retardation, or death (cf. TORCH DISEASES). *Lab. diagnosis*: e.g. (i) inoculation of material from lesions into mice and examination of smears or sections post mortem; (ii) serological tests – e.g. a CFT and the TOXOPLASMA DYE TEST; (iii) an indirect immunofluorescence test (or an ELISA-based test) for antibodies. A PCR-based assay for *T. gondii* (in HIV-positive patients) was reported to have a specificity of 100% but poor sensitivity (only 25%); it was nevertheless suggested that this assay may find use in differentiating between cerebral toxoplasmosis and clinically similar disease [JCM (1997) *35* 2639–2641]. *Chemotherapy*: e.g. PYRIMETHAMINE with sulphonamides.

In the cat infection with *T. gondii* may occur on ingestion of mature oocysts (from other cats) or tissue cysts (in avian or mammalian prey); *T. gondii* develops in both extra-intestinal and intestinal tissues. In the intestine the parasite multiplies by endodyogeny, endopolygeny and schizogony and forms macrogametes and biflagellate microgametes. Macrogametes are fertilized in situ, and the zygote develops into an oocyst which is shed, unsporulated, in the faeces. Adult cats are usually symptomless; kittens may respond with diarrhoea and death or with anorexia and wasting. Congenital toxoplasmosis occurs e.g. in goats and sheep, and is one cause of abortion in sheep. [AP (1982) *20* 296–332.]

Toxothrix A genus of colourless GLIDING BACTERIA which occur e.g. in cold, iron-containing springs; the trichomes give rise to bundles of thin, iron-oxide-containing inanimate filaments. [Habitat and culture: Book ref. 45, pp. 409–411.]

ToxR, ToxS, ToxT See BACTERIOPHAGE CTXΦ.

TPA 12-*O*-tetradecanoylphorbol-13-acetate. (See also ZEBRA.)

TPB$^-$ Tetraphenylborate: see CHEMIOSMOSIS.

TPHA test *Treponema pallidum* haemagglutination test. A TREPONEMAL TEST, involving passive haemagglutination, in which an *absorbed* serum (cf. FTA-ABS TEST) is tested for its ability to agglutinate tanned erythrocytes sensitized with antigens from the Nichol's strain of *Treponema pallidum*. In *primary* SYPHILIS the TPHA test is less sensitive than the FTA-ABS test, but these tests have similar sensitivities in other stages of the disease.

TPI test (*Treponema pallidum* immobilization test) A TREPONEMAL TEST previously used extensively as a confirmatory test in the diagnosis of SYPHILIS; the antibodies detected by the test are specific to *Treponema pallidum* and some other *Treponema* spp. Essentially, the test depends on the immobilization of living (motile) cells of *T. pallidum* in the presence of specific antibodies and COMPLEMENT. To the serum under test is added a standardized suspension of *T. pallidum* and a volume of fresh guinea-pig serum as a source of complement; the whole is incubated anaerobically for 18 hours at 37°C and examined microscopically. In a positive reaction a specified proportion of treponemes is immobilized.

TPMP$^+$ Triphenylmethylphosphonium: see CHEMIOSMOSIS.

TPN Triphosphopyridine nucleotide (NADP): see NAD.

TPP THIAMINE pyrophosphate.

***tra* gene** See TRANSFER OPERON and F PLASMID.

***tra* operon** See TRANSFER OPERON.

***traA* gene** See F PLASMID.

tracheal antimicrobial peptide See DEFENSINS.

tracheal cytotoxin In WHOOPING COUGH: a toxin (apparently a fragment of PEPTIDOGLYCAN from the pathogen) which kills epithelial cells to which the pathogen is adherent.

Trachelomonas A genus of EUGLENOID FLAGELLATES in which the cells closely resemble those of EUGLENA except that each is enclosed within a yellowish to reddish-brown, round or ovoid LORICA composed mainly of ferric hydroxide and manganese oxides; the lorica is ornamented with spines in some species. The emergent flagellum protrudes through the lorica, and the organisms are free-swimming.

Trachelophyllum A genus of carnivorous ciliates (subclass GYMNOSTOMATIA) which occur e.g. in some SEWAGE TREATMENT plants. Cells: elongated, with uniform body ciliature and a cytostome at the end of the narrower anterior region.

Tracheloraphis See KARYORELICTID GYMNOSTOMES.

Trachelostyla See HYPOTRICHIDA.

Trachipleistophora See MICROSPORIDIOSIS.

trachoma A potentially blinding disease of the eye which, in nature, affects only man; the causal agent is *Chlamydia trachomatis* (see CHLAMYDIA). Infection occurs contaminatively. Initially there is a follicular CONJUNCTIVITIS affecting particularly the margin of each upper eyelid (tarsal conjunctivae) – but also affecting the inner surfaces of the eyelids (palpebral conjunctive) and the (bulbar) conjunctiva on the anterior surface of the eyeball. Discharging follicles, containing accumulations of macrophages, polymorphs and lymphocytes, develop within the conjunctival tissues, the latter subsequently becoming scarred. The contraction of scarred tissues leads to inturned eyelids. The consequent abrasion of the cornea by the eyelashes causes ulceration, scarring and impairment/loss of vision. A thin fibrovascular membrane (*pannus*) develops on the surface of the cornea. Lymphoid follicles may develop at the cornea–sclera junction; on healing, these leave characteristic depressions (*Herbert's pits*). The mechanism of pathogenesis is not understood; it has been suggested that it may involve cell-mediated HYPERSENSITIVITY. Chemotherapeutic agents commonly used include e.g. TETRACYCLINES. Laboratory diagnosis typically involves the examination of specimens by e.g. IMMUNOFLUORESCENCE or enzyme immunoassay techniques.

[Book ref. 193, pp. 135–170.]

Trachyspora See UREDINIOMYCETES.

trachytectum *Syn.* EXOSPORIUM (sense 2).

tractellum A (eukaryotic) FLAGELLUM which 'pulls' the cell forwards.

TRADD See TNF.

traffic ATPase (1) *Syn.* ABC TRANSPORTER. (2) *Syn.* BINDING PROTEIN-DEPENDENT TRANSPORT SYSTEM.

Trager duck spleen necrosis virus See AVIAN RETICULOENDOTHELIOSIS VIRUSES.

***traI* gene** See F PLASMID.

trailer (*mol. biol.*) See MRNA.

***traJ* gene** See F PLASMID.

***traM* gene** See F PLASMID.

trama The sterile (i.e., non-generative) inner tissue of a LAMELLA or DISSEPIMENT or of the 'teeth' of members of the Hydnaceae. (See also BILATERAL TRAMA.)

Trametes A genus of lignicolous fungi of the APHYLLOPHORALES (family Polyporaceae) which form basidiocarps in which the hymenophore is porous. *T. pini* is parasitic on certain trees. For *T. versicolor* see CORIOLUS. (See also XYLANASES.)

***trans*-acting** See CIS-DOMINANCE.

***trans* complementation** See CIS–TRANS TEST.

transaldolase See Appendix I(b) and RMP PATHWAY.

transaminases (of *Escherichia coli*) See AMMONIA ASSIMILATION.

transcapsidation See PHENOTYPIC MIXING.

transcipient A cell which has received DNA from another cell.

transconjugant See CONJUGATION (1b) and CONJUGATIVE TRANSPOSITION.

transcriptase An RNA POLYMERASE involved in TRANSCRIPTION. (cf. RNA-DEPENDENT RNA POLYMERASE.)

transcription The synthesis of an RNA strand in a process in which ribonucleoside 5′-triphosphates (rNTPs) base-pair sequentially with nucleotides in a template strand and are polymerized in the 5′-to-3′ direction (with elimination of PPi) by an RNA POLYMERASE; the template strand is DNA in cells and DNA viruses, RNA in e.g. RNA viruses (see RNA-DEPENDENT RNA POLYMERASE). (cf. REVERSE TRANSCRIPTASE.)

(Until recently it has been observed that only one of the two strands of DNA in a gene acts as template in transcription. Interestingly, it has been reported that a certain gene in the fruitfly (*Drosophila*) contains protein-encoding information in *both* strands – RNAs synthesized on both template strands being subsequently joined to form a single molecule of mRNA [Nature (2001) *409* 1000].)

In e.g. *Escherichia coli* (and other bacteria) *initiation* of transcription begins when the RNA POLYMERASE (RPase) holoenzyme binds to a PROMOTER. The polymerase core enzyme itself has a high affinity for dsDNA and can bind at (apparently) any site to form a stable, 'closed' enzyme–DNA complex in which the DNA strands are not unwound. Interaction of a SIGMA FACTOR with the core enzyme confers on it specificity for a particular class of PROMOTER while greatly reducing its affinity for other DNA sequences. The holoenzyme binds very tightly to its corresponding promoter; initially a 'closed' complex is formed, but this is subsequently converted to an 'open' complex in which a short region of the DNA bound by the enzyme becomes unwound. The first rNTP can then pair with a base on one of the DNA strands (the *start point* or *start site*). A few nucleotides are then incorporated, each being added to the 3′-OH of the preceding nucleotide; the σ factor then dissociates. (The first nucleotide retains its 5′-triphosphate group.) [Polymerase–promoter interactions: JB (1998) *180* 3019–3025.]

Elongation is carried out by the RPase core enzyme (possibly in association with the product of the *nusA* gene, a protein which can bind to the core enzyme but not to the holoenzyme); the polymerase moves along the dsDNA, locally unwinding the strands to expose the ssDNA template, 'supervising' the correct base-pairing between incoming rNTPs and the template, and linking the nucleotides in the 5′-to-3′ direction (antiparallel to the template) with elimination of pyrophosphate. In this way, a transient RNA–DNA hybrid duplex is formed in the region of the enzyme–DNA complex; as the enzyme proceeds, the RNA peels away from the template and the DNA duplex is restored. [Role of RNA polymerase in elongation: JB (1998) *180* 3265–3275.]

In some cases, transcription of genes or operons occurs divergently (in opposite directions) from closely spaced promoters. This causes an increase in the level of negative supercoiling in the region between polymerases (negative supercoiling being generated behind each advancing polymerase). Such localized accumulation of negative superhelicity may affect the expression of genes/operons [Mol. Microbiol. (2001) *39* 1109–1115].

Elongation continues until a specific *termination* signal (*terminator, t*) is reached. Terminators in bacteria vary in efficiency and in mechanism of action; secondary structure in the transcript itself appears to be important in effecting termination. In a *rho-independent* ('simple') terminator the RNA transcript of the termination region contains a GC-rich PALINDROMIC SEQUENCE (which can form a stem-and-loop structure) continuous with a run of consecutive uridine (rU) residues at the 3′ end. The stem-and-loop structure causes the RPase to pause at the oligo-rU region, and the sequence of rU·dA base pairs (which is relatively unstable) may facilitate the release of the transcript and/or dissociation of the transcription complex. In a *rho-dependent* ('complex') terminator, the transcript may contain a stem-and-loop structure (which is not particularly GC-rich), but there is generally no oligo-rU sequence or any other apparent consensus sequence. In this case termination requires the participation of a protein, the *rho* (ρ) *factor*; the ρ factor is apparently active in hexameric form, has RNA-dependent NTPase activity (which is necessary for termination), and apparently interacts directly with the RNA transcript. The precise mode of action of ρ is unknown; it has been proposed that there is a ρ recognition site in the RNA, and that termination occurs at a relatively fixed distance downstream of this recognition site [Book ref. 188, pp. 155–178]. The NusA protein also appears to play some role in termination. [Review of transcription termination: Book ref. 60, pp. 123–161.]

The initial product of transcription (the *primary transcript*) may undergo extensive processing and/or modification to give the mature RNA product: see e.g. entries for mRNA, rRNA and tRNA.

Initiation and termination of transcription are important control points for gene expression: see e.g. ANTITERMINATION, CATABOLITE REPRESSION, OPERON. (See also POLAR MUTATION.)

transcription antitermination In certain Gram-positive bacteria (e.g. *Bacillus subtilis*): a control mechanism in which genes whose products are involved in the synthesis of amino acids (and in linking amino acids to tRNAs) are switched on when limiting amounts of those amino acids give rise to uncharged tRNAs. It appears that an uncharged tRNA interacts with the leader region of the gene's mRNA, causing reversal of a transcription terminator structure (formed when the cell has adequate amounts of the amino acid) to an antiterminator structure, thus enabling synthesis of the amino acid. [tRNA-directed transcription antitermination: Mol. Microbiol. (1994) *13* 381–387.]

transcription initiation factors See RNA POLYMERASE.

transcription-mediated amplification See TMA.

transcriptional coupling Coupling between closely spaced promoters which results from *divergent* transcription from the promoters and the consequent transcription-dependent increase in negative superhelicity in the region between the active RNA polymerases; such local accumulation of negative superhelicity may activate or repress a promoter, or it may have little or no effect, depending on the intrinsic properties of the given promoter. [Possible role of transcriptional coupling in the *ilv* regulon of *Escherichia coli*: Mol. Microbiol. (2001) *39* 1109–1115.]

transcytosis In certain Opa^{+} strains of *Neisseria*: translocation through a layer of (mammalian) epithelial cells without disruption of the eukaryotic cell–cell junctions; transcytosis involves an initial phase in which the bacteria bind to specific receptors on the host cells and are then engulfed by the host cells [TIM (1998) *6* 489–495]. (See also PARACYTOSIS.)

transductant See TRANSDUCTION.

transduction The virus-mediated transfer of host DNA (chromosomal or plasmid) from one host cell (the *donor*) to another (the *recipient*). Transduction was first observed in bacteriophage/bacterium systems, but has since also been found to be mediated by certain viruses infecting eukaryotic cells (see RETROVIRIDAE). The account below concerns only phage/bacterium systems.

Essentially, when a phage replicates in a (donor) cell, a few progeny virions encapsidate pieces of host DNA instead of – or in addition to – phage DNA; these virions can adsorb to a new host cell and introduce their DNA in the usual way. There are two basic types of phage-mediated transduction.

(a) *Generalized transduction*: any of a wide range of donor genes may be transduced, and the transducing phage particles contain *only* donor DNA. In some systems any of the host genes has a more or less equal chance of being transduced, but in other systems some genes are transferred at higher frequencies than others. For example, in the BACTERIOPHAGE P22/*Salmonella* system, certain regions of the host chromosome resemble the P22 *pac* site, and packaging of chromosomal DNA may be initiated at and proceed from these sites [MGG (1982) *187* 516–518]; thus, the probability of a given gene being transduced depends on its distance from a *pac*-like site. (Certain 'high-frequency transduction' (HT) mutants of P22 have been shown to be defective in *pac* recognition [JMB (1982) *154* 551–563].)

The fate of the transduced DNA in the recipient cell (now called a *transductant*) depends on various factors. If the DNA is a complete replicon (e.g. a plasmid) it may be stably inherited by the transductant. If the DNA is a fragment of a chromosome or plasmid, it may undergo one of three possible fates. (i) It may be completely degraded by the recipient cell's RESTRICTION ENDONUCLEASE system. (ii) It may undergo RECOMBINATION with a homologous region of the recipient's chromosome (or plasmid), so that at least some of the genes it carries can be stably inherited (*complete transduction*). (iii) It may persist in the cell in a stable but non-replicating form (*abortive transduction*). (The transduced DNA in an abortive transductant may exist as a circular DNA–protein complex [Virol. (1980) *106* 30–40].) Any donor genes present (and linked to promoters) may be expressed, and (if they are dominant alleles) the transductant may express the donor phenotype in respect of these genes. However, when an abortive transductant divides, only one of the daughter cells will receive the donor fragment; the other may nevertheless receive sufficient donor-gene products to permit expression of the donor phenotype for one or a few generations. Such abortive transduction is normally manifest by the formation of minute (often microscopic) colonies on medium selective for transductants; only one cell in the colony actually contains donor DNA.

The transduction of one particular donor gene is a rare event (e.g. 10^{-7}–10^{-5}, depending e.g. on phage). If two or more genes are transduced together (*cotransduction*) they are assumed to occur on the same fragment of DNA and are thus closely linked in the donor; generalized transduction has thus been used in the detailed mapping of donor chromosomes, distances between markers being estimated by determining their contransduction frequencies.

(b) *Specialized* (*restricted*) *transduction* is mediated only by temperate phages which integrate into the host chromosome (see LYSOGENY); only bacterial genes immediately adjacent to the prophage can be transduced, and the transduced DNA is covalently linked to some or all of the phage genes. For example, when a population of *Escherichia coli* cells lysogenized by BACTERIOPHAGE λ is induced, a small proportion of the progeny virions may carry either *gal* or *bio* host genes – often at the expense of certain phage genes at the opposite end of the prophage (virion component genes in *gal*-containing particles, control genes in *bio*-containing particles); such virions (specialized transducing particles, STPs) arise as a result of rare aberrant excision events in which recombination occurs between sites other than the hybrid *att* sites (*attR* and *attL*). (A lysate from an induced lysogen will normally contain a heterogeneous population of STPs resulting from different aberrant excision events in different cells.) An STP can infect a recipient cell and introduce its DNA into the cell; however, if it lacks phage genes essential for replication (i.e., if it is *defective*) it will require the presence of a (non-defective) helper phage in order to produce progeny STPs. (cf. RETROVIRIDAE.)

A lysate obtained by induction of a lysogen contains only a small (and mixed) population of STPs (e.g. one STP per 10^6

wild-type particles); consequently, when the lysate is used to infect a culture of recipient cells only a few of the cells will be infected by an STP. Such lysates are therefore termed *low-frequency transducing* (LFT) lysates. If e.g. a gal^- recipient culture is infected with an LFT lysate derived from a gal^+ donor population, a few recipients will become $gal^-/\lambda gal^+$ (merodiploid heterogenotes). In some of these cells homologous recombination may occur between the gal^- and gal^+ regions to result in a stable, non-lysogenic gal^+ transductant ('replacement transduction'). Alternatively, the DNA of the STP may integrate, intact, into the host chromosome by homologous recombination, producing a lysogenic heterogenote ('addition transduction'). (Owing to the manner of its formation, an STP will normally contain a hybrid *attR* or *attL* site instead of the usual *attP*, and so integrates at *attB* only with low efficiency.) However, since non-transducing (wild-type) phage particles are present in great excess in an LFT lysate, at high multiplicities of infection a recipient is likely to be simultaneously infected with both wild-type and transducing phages. In this case both phage genomes can integrate into the recipient's chromosome to form a *double lysogen*; this probably occurs by normal site-specific integration of the wild-type genome, followed by homologous recombination between it and the defective phage genome. If the double lysogen is subsequently induced, both phages will replicate (the wild-type phage supplying the missing functions of the defective STP); the resulting lysate may thus contain approximately equal numbers of λ and λgal^+ particles. Furthermore, the population of STPs will be homogeneous, unlike that in an LFT lysate. Such a lysate can be used to infect another population of cells, and the gal^+ genes will be transduced with high frequency; the lysate is thus termed a *high-frequency transducing* (HFT) lysate.

Rarely, the genome of a wild-type λ integrates at sites other than *attB* in the host chromosome, and STPs from such lysogens will carry genes other than *gal* or *bio*.

(See also MINI-MU.)

transductional shortening The phenomenon observed in the TRANSDUCTION of a large plasmid: the transduced plasmid is smaller than the original donor plasmid, probably because the phage particles encapsidate deletion mutants of the plasmid (which arise spontaneously at low frequencies).

transfection Originally, the introduction of viral nucleic acid into an intracellular environment and the subsequent formation of normal virions.

Currently, the term is used for any *in vitro* procedure in which nucleic acid is introduced into cells in order to modify those cells genetically or to carry out intracellular experimentation. Transfection is also used to refer to the introduction of proteins or peptides into cells.

transfer DNA synthesis *Syn.* DCDS: see CONJUGATION (1b)(i).

transfer operon (*tra* operon) In a CONJUGATIVE PLASMID: an OPERON containing those genes which specify functions necessary for CONJUGATION (sense 1b). (cf. SEX FACTOR sense 2.) In some conjugative plasmids (see e.g. F PLASMID) the genes of the transfer operon encode e.g. PILI, certain proteins involved in DNA mobilization, and proteins involved in SURFACE EXCLUSION. Some conjugative plasmids (e.g. the F plasmid) are DEREPRESSED for conjugal transfer, but many or most are repressed. (See also FINOP SYSTEM.)

transfer RNA See TRNA.

transferases ENZYMES (EC class 2) which catalyse the transfer of a group from one molecule to another. Subclasses are recognized and numbered according to the nature of the group transferred (e.g., subclass 1: one-carbon groups such as methyl, formyl, hydroxymethyl).

transferosome See CONJUGATION (1b)(i).

transferrin A SIDEROPHILIN which occurs in vertebrate plasma; its main function is to transport iron into cells. The iron–transferrin complex binds to a cell-surface receptor and is internalized via a coated vesicle (see PINOCYTOSIS); the iron is subsequently released, within the cell, and may be stored as an iron–FERRITIN complex. Although iron is tightly bound by transferrin it can be removed e.g. by certain bacteria or their products: see IRON. Interestingly, the intraerythrocytic stage of *Plasmodium falciparum* appears to obtain iron from iron–transferrin complexes by synthesizing transferrin receptors that become localized in the cell surface of the infected erythrocyte [Nature (1986) *324* 388–391].

transformant A cell or organism which has undergone genetic TRANSFORMATION.

transformation (1) (*genetics*) A process in which exogenous DNA is taken up by a (*recipient*) cell, sphaeroplast or protoplast, in which it may be incorporated into the chromosome (or e.g. into a plasmid) by homologous RECOMBINATION or converted into an autonomous replicon. The DNA (*transforming* or *donor* DNA) may be e.g. a fragment of chromosomal DNA from a related strain, a plasmid, or a viral genome (see TRANSFECTION). Transformation can occur under natural conditions in some bacteria (e.g. *Bacillus*, *Haemophilus*, *Neisseria* and *Streptococcus* spp), but in many bacteria (including e.g. *Escherichia coli*) and in certain eukaryotic microorganisms it can occur only in cells 'permeabilized' to DNA by artificial methods.

It has been suggested that transformation (among other mechanisms) may contribute to the spread of antibiotic resistance in certain pathogenic bacteria [Science (1994) *264* 375–382].

Natural bacterial transformation systems. Cells which are able to take up DNA are said to display *competence*. Competence is apparently constitutive in some species – e.g. *Neisseria gonorrhoeae* (in which it is reported to occur only in fimbriated cells [JGM (1984) *130* 3165–3173]). However, in other species competence is a transient phenomenon that is dependent on certain factors – such as nutritional status and/or the phase of growth (in batch cultures). In *Haemophilus influenzae*, for example, competence is induced by growth-inhibiting conditions; in this species it is promoted by high levels of intracellular cAMP (cyclic AMP). In both *Bacillus subtilis* and *Streptococcus pneumoniae* competence is affected e.g. by the population density of the bacteria – an example of QUORUM SENSING. For example, when *B. subtilis* grows to a high density of cells, a secreted pheromone (designated ComX) accumulates in the extracellular environment; at an appropriate concentration, ComX activates a TWO-COMPONENT REGULATORY SYSTEM which, in turn, leads to the activation of certain genes (including *comS*) whose function is needed for the establishment of competence. In *S. pneumoniae*, competence has also been associated with the transmembrane transport of calcium [JB (1994) *176* 1992–1996]. [Competence in transformation: TIG (1996) *12* 150–155.]

To be effective in transformation, a fragment of chromosomal DNA must be double-stranded and larger than a certain minimum size – which depends on recipient. The DNA binds to the surface of a competent cell; initially binding is reversible and dependent on the concentration of the donor DNA, but subsequently a limited number of DNA molecules become irreversibly bound to specific cell-surface receptor sites. In *S. pneumoniae* and *B. subtilis* binding is non-specific with respect to the DNA, DNA from other species being bound as readily as DNA from the same species. By contrast, in *Haemophilus* and *Neisseria* spp DNA binding is sequence-specific; *H. influenzae* takes up only dsDNA containing a 'DNA uptake site', a sequence (5′-A–A–G–T–G–C–G–G–T–C–A-3′ [Gene (1980) *11* 311–318]) which occurs frequently in the genomes of *H. influenzae* and *H. parainfluenzae* but much less so in DNA from other organisms.

In *B. subtilis* and *S. pneumoniae* the bound DNA is cleaved by an envelope-associated endonuclease. Subsequently, one strand of a fragment is internalized while its complement is degraded. It has been suggested that binding and uptake are achieved by a membrane-bound nuclease-containing protein complex, the nuclease processively degrading one strand of the DNA, thereby (possibly) driving the other through a pore in the membrane formed by other components of the complex. Within the recipient cell, the single-stranded fragment rapidly associates with cellular proteins to form a presynaptic *eclipse complex*; this probably protects the DNA from degradation. Synapsis between the donor (chromosomal) DNA and a homologous region of the recipient's chromosome rapidly ensues, and a single-stranded fragment of recipient DNA is effectively excised and replaced by the donor ssDNA to form a region of *hybrid DNA*; this could occur e.g. by invasion of the recipient duplex by the donor strand followed by 'strand assimilation' (cf. RECA PROTEIN). In *Haemophilus influenzae* and *H. parainfluenzae* a different type of uptake mechanism apparently occurs [PNAS (1982) *79* 6350–6374], and *ds*DNA enters the recipient cell; however, recombination apparently still involves the pairing of a single donor strand with a complementary region of the resident DNA.

Resolution of the heteroduplex region may occur by separation of the strands during a subsequent round of semiconservative DNA synthesis, yielding some progeny resembling the donor, others the recipient, with respect to markers in this region. Alternatively, one of the strands may be corrected by a MISMATCH REPAIR system (*gene conversion*: cf. RECOMBINATION) to yield a homoduplex and hence a single type of progeny. In *S. pneumoniae* (but not in *B. subtilis*) some single-site mutations in the donor DNA are integrated into the recipient's chromosome more efficiently than others; it seems that this organism encodes a mismatch repair system (*hex* genes) which has different affinities for different mismatched base pairs, eliminating low-efficiency (LE) markers more efficiently than high-efficiency (HE) markers. (In hex^- mutants all single-site mutations are integrated with the same high efficiency.) [*Hex* system in *S. pneumoniae*: MR (1986) *50* 133–165.]

Plasmid DNA (ccc, ds) may also be taken up by competent cells; however, in natural systems transformation with plasmid monomers generally occurs only at very low frequencies, if at all, whereas plasmid oligomers (trimers or larger) can transform at high frequencies. Apparently the ccc dsDNA binds to the surface of a competent cell and undergoes cleavage and some degradation, and single strands enter the cell with the same polarity; if the plasmid is an oligomer, enough complementary ssDNA can eventually enter the cell to allow annealing and circularization and hence, with repair synthesis and ligation, regeneration of a plasmid monomer.

Artificial transformation systems. Non-competent cells of e.g. *Bacillus* or *Streptococcus*, or cells of those bacteria (e.g. *E. coli*) which cannot undergo transformation under natural conditions, may be induced to take up transforming DNA by various procedures involving laboratory-induced competence. For example, treatment of *E. coli* and other Gram-negative bacteria with high (millimolar) concentrations of calcium chloride, often in association with heat shock, induces artificial competence; calcium ions may promote binding of DNA to the cell and increase the permeability of the cell envelope. In one scheme, calcium chloride solution (approx. 50 mM, 0.2 ml) containing 10^8–10^9 washed, mid-log-phase cells of *E. coli* is chilled on ice, and a DNA suspension (10 μl) is added to give a final concentration of DNA of approx. 0.2 μg/ml; after further chilling at 0°C (15–30 minutes), the suspension is heat-shocked (42°C/1.5–2 minutes) and allowed to recover – e.g. returned to ice and then incubated in Luria–Bertani broth (1 ml) at 37°C for 1 hour.

The acquisition of competence by *E. coli* in ice-cold solutions of calcium is associated with the presence of a high concentration of poly-β-hydroxybutyrate/calcium polyphosphate complexes in the cytoplasmic membrane. It has been suggested that these complexes may form transmembrane channels which facilitate DNA transport, and that divalent cations may act as links between DNA and phosphate at the mouth of such channels [JB (1995) *177* 486–490].

Small, circular plasmids tend to transform more readily than do larger ones. In wild-type *E. coli*, linear dsDNA transforms poorly (if at all) as it is degraded by the RecBC enzyme; linear dsDNA can transform some *recBC* mutants which lack a functional enzyme.

High-efficiency transformation of *E. coli* (and various other bacteria) with plasmids may be achieved by ELECTROPORATION.

Protoplasts from Gram-positive bacteria can be induced to take up DNA by treatment with POLYETHYLENE GLYCOL (PEG); viable cells can then be regenerated from the transformed protoplasts (and from certain types of sphaeroplast). (See also PROTOPLAST FUSION.)

PERMEAPLASTS have been used for the transformation of certain cyanobacteria. [DNA uptake by '*Anacystis nidulans*': MGG (1986) *204* 243–248.]

In artificially competent cells, sphaeroplasts and protoplasts, DNA appears to be taken up intact.

[Methods for bacterial transformation by plasmid DNA: Book ref. 177, pp. 61–95.]

Certain fungi can be transformed artificially. In e.g. *Saccharomyces cerevisiae*, one method involves mixing sphaeroplasts (prepared e.g. with Zymolyase) with the DNA in the presence of Ca^{2+}, and then adding PEG; in another method, whole cells are treated with alkali metal ions (usually lithium ions) and PEG, a method which is simpler but less efficient. [Transformation in fungi: Book ref. 179, pp. 161–195 (yeast) and pp. 259–278 (*Aspergillus*); Book ref. 178, pp. 468–472 (*Cephalosporium acremonium*).]

Certain unicellular green algae can also be transformed: e.g. *Chlamydomonas reinhardtii* has been transformed with yeast DNA [Nature (1982) *296* 70–72].

(2) (of cells in tissue culture) See TISSUE CULTURE.

(3) (of lymphocytes) See BLAST TRANSFORMATION.

transforming DNA ('transforming principle') See TRANSFORMATION (1).

transfusion-transmitted infection Any infectious disease which can be transmitted via a transfusion of blood or blood-related products. Transfusion-transmissible agents include certain bacteria (e.g. *Treponema pallidum*), protozoa (e.g. *Plasmodium* spp) and viruses (e.g. hepatitis viruses B and C, human immunodeficiency virus (HIV) and, more recently, TT VIRUS). (See also CREUTZFELDT–JAKOB DISEASE.)

[Transfusion-associated hepatitis G virus infection: NEJM (1997) *336* 747–754. Transfusion-transmitted malaria in the USA: NEJM (2001) *344* 1973–1978.]

Other risks from transfused blood (apart from infection) include the possible presence of pro-inflammatory cytokines released from white blood cells during prolonged storage [e.g. Transfusion (1996) *36* 960–965].

Avoidance of transfusion-transmitted infection is promoted e.g. by careful screening of donors and by appropriate tests on donated blood for specific infectious agents. Even so, infections still occur through transfusion. (*Autologous* blood transfusion involves re-infusion of the patient's own blood taken prior to anticipated requirement; one risk is that contamination may occur during donation, storage or use.)

For viral diseases, a major risk is that blood may be donated at a time when the donor has been recently infected with a given agent, is able to transmit that agent via blood, but has not yet become seropositive for the agent (i.e. the agent is still undetectable serologically). The period between initial infectivity and development of seropositivity (known as a diagnostic *window*) varies from one virus to another but is generally of the order of weeks/months. (In this context it is noteworthy that, normally, donor blood can be stored for a maximum of 35 days in the UK and 42 days in the USA.) To reduce the risk from window-phase blood, PCR-based tests have been used to detect the nucleic acid of hepatitis B and C viruses and the HIV-1 virus [Lancet (1999) *353* 359–363]; nevertheless, transmission of hepatitis C virus via a blood donation negative in nucleic-acid-amplification tests has been reported [Lancet (2000) *355* 41–42].

Other risk factors associated with transfused blood include the possibility that tests for a given agent may be insufficiently sensitive or may not detect genetically variant forms of an infectious agent. Additionally, human (laboratory) error, though rare, is always a possibility.

[Genome detection versus serology in blood screening for microbial agents: BCH (2000) *13* (4) chapter 8.]

transgenesis Incorporation of *alien* DNA, heritably, into living organisms by in vitro methods. [Book ref. 204.]

transhydrogenase Nicotinamide nucleotide transhydrogenase (EC 1.6.1.1): an enzyme which catalyses the (reversible) reduction of $NADP^+$ by NADH. *AB-transhydrogenases* are membrane-bound enzymes which occur e.g. in some bacteria and in mammalian mitochondria; *BB-transhydrogenases* are soluble flavoprotein enzymes found only in certain bacteria. Transhydrogenation of $NADP^+$ by AB-transhydrogenases appears to be regulated by pmf (see CHEMIOSMOSIS); transhydrogenation by BB-transhydrogenases is promoted e.g. by 2′-AMP and Ca^{2+}.

transition mutation A type of POINT MUTATION in which one purine nucleotide is replaced by another, or one pyrimidine nucleotide is replaced by another. (cf. TRANSVERSION MUTATION.)

transition state (sporulation) See ENDOSPORE (sense 1).

transketolase See e.g. CALVIN CYCLE and Appendix I(b).

translation See PROTEIN SYNTHESIS.

translational attenuation A control mechanism in which the expression of certain genes is regulated by inhibition of their translation under given conditions. For example, in certain Gram-positive bacteria the *cat* gene (encoding chloramphenicol acetyltransferase) is induced by CHLORAMPHENICOL but normally remains unexpressed in the absence of this antibiotic. In the absence of chloramphenicol, the *cat* gene is transcribed but not translated; this is because the ribosome-binding site of the coding sequence is inaccessible owing to an inhibitory secondary structure in the mRNA formed by base-pairing between certain ribonucleotides. The presence of chloramphenicol inhibits development of the secondary structure, thus relieving repression of translation. (This mechanism operates with levels of chloramphenicol below those which inhibit protein synthesis.)

translational enhancer See DOWNSTREAM BOX.

translational feedback regulation See e.g. RIBOSOME (biogenesis).

translational frame-shifting A control mechanism in which the expression of certain genes is regulated during the *translocation* stage of PROTEIN SYNTHESIS. In this process, a specific sequence of nucleotides in the transcript acts as a signal that causes the ribosome to 'slip' along the mRNA; such a movement usually involves a '+1 slip' (i.e. a 1-nucleotide shift in the direction of translation) or a '−1 slip' (a 1-nucleotide shift in the opposite direction). Such a shift necessarily affects all subsequent codons.

[Review: Mol. Microbiol. (1994) *11* 3–8.]

translational operator See e.g. RIBOSOME.

translesion synthesis See SOS SYSTEM.

translocase *Syn.* translocon: see PROTEIN SECRETION (type II systems).

translocation (1) See PROTEIN SYNTHESIS. (2) Passage of viable bacteria across the epithelial barrier from the gut lumen [review: BCG (2003) *17* 397–425].

translocon See PROTEIN SECRETION.

transmissible disease (1) Any disease which can be transmitted from one individual to another, or to a fetus, by any means, including experimental infection; 'transmissible disease' is therefore broader (more inclusive) than INFECTIOUS DISEASE.

(2) Loosely, an infectious disease.

transmissible gastroenteritis (in pigs) An acute, typically non-febrile PIG DISEASE, characterized by diarrhoea and dehydration, which is often fatal in very young pigs; the causal agent is a coronavirus. The mode of transmission is uncertain. Incubation period: ca. 1–2 days. The virus replicates in distal regions of the villi, causing malabsorption and osmotically induced, typically yellowish-green diarrhoea.

transmissible hypovirulence See HYPOVIRULENCE.

transmissible mink encephalopathy A disease of mink which is probably a form of SCRAPIE; it appears to have arisen as a result of feeding farmed mink with scrapie-infected sheep meat. The disease can be transmitted experimentally to various other animals (but not to mice); in monkeys, infection results in a disease resembling experimental CREUTZFELDT–JAKOB DISEASE.

transmissible spongiform encephalopathies (TSEs) Fatal diseases of the nervous system characterized by spongiform degeneration of the brain without inflammation or specific immune response; in each disease the causal agent is a PRION (q.v.).

Unique features of TSEs include: (i) a novel mechanism of pathogenesis, (ii) manifestation in sporadic, inherited and transmitted forms, and (iii) (in contrast to e.g. Alzheimer's disease) manifestation in a range of histopathological patterns.

Human TSEs include CREUTZFELDT–JAKOB DISEASE, GERSTMANN–STRÄUSSLER–SCHEINKER SYNDROME (GSS syndrome) and KURU (all previously referred to as 'transmissible viral dementias'), and *fatal familial insomnia* (FFI; [NEJM (1992) *326* 444–449]). FFI and GSS syndrome are both inherited diseases.

TSEs acquired by infection include kuru and (apparently) nvCJD (as well as the iatrogenic forms transmitted via medical/surgical treatment). Where infection involves ingestion of prion-contaminated food, it is believed that conversion of PrP to the prion form occurs initially within tissues of the gastrointestinal tract, and that such conversion continues to occur progressively in the lymphatic and/or peripheral nervous systems until it reaches the central nervous system. [Book ref. 215.]

The deposition of prions, in itself, may not be sufficient to cause neuropathy [Nature (1996) *379* 339–343], suggesting that the development of disease may involve some kind of interaction between prions and the host's tissues.

Prions can be distinguished from the normal form of the protein by their ability to bind to plasminogen (a precursor of FIBRINOLYSIN), and this may be used as a basis for the development of a diagnostic test [Nature (2000) *408* 479–483].

In animals, the TSEs include BOVINE SPONGIFORM ENCEPHALOPATHY (BSE; 'mad cow disease'), chronic wasting disease of elk and mule-deer, SCRAPIE, and TRANSMISSIBLE MINK ENCEPHALOPATHY.

Transmission of TSEs across species barriers has been demonstrated experimentally in a number of cases (e.g. sheep → mice; bovines → monkeys).

transmissible viral dementias See TRANSMISSIBLE SPONGIFORM ENCEPHALOPATHIES.

transmission-blocking vaccine See e.g. PLASMODIUM.

transmission factor (*plant virol.*) See NON-CIRCULATIVE TRANSMISSION.

transovarial transmission Transmission of a parasite from a host (or vector) to its offspring via the egg. (See e.g. BLACKHEAD and PULLORUM DISEASE; cf. VERTICAL TRANSMISSION.)

transpeptidation (in protein synthesis) See PROTEIN SYNTHESIS.

transport medium Any medium used specifically for the transportation and/or temporary storage of material (e.g. swabs) from which the isolation of particular organism(s) is subsequently to be attempted. The purpose of such a medium is to maintain the viability and/or infectivity of the organism(s) during the delay between collection and culture of the specimen. An effective transport medium is necessary e.g. for anaerobes (which may be killed by short or prolonged exposure to oxygen), for delicate organisms (e.g. *Neisseria gonorrhoeae*), and for certain viruses. The commonly used transport media are non-nutrient media which include STUART'S TRANSPORT MEDIUM and its modifications, e.g. Cary–Blair transport medium and Amies transport medium – the latter incorporating charcoal. (Charcoal-containing media obviate the need for collecting specimens on charcoal-impregnated swabs – the charcoal, in each case, serving to adsorb inhibitory substances in the medium and/or specimen.) Some transport media have been devised for particular organisms: see e.g. MONSUR TRANSPORT MEDIUM. For many types of virus, HANKS' BSS supplemented with protein (e.g. 0.5–2.0% bovine albumin) may be an effective transport medium; in such a medium some viruses can be rapidly frozen and stored at ca. −70°C.

transport systems Systems which permit the uptake or efflux of molecules or ions across membranes (e.g. ENERGY-TRANSDUCING MEMBRANES or the OUTER MEMBRANE in Gram-negative bacteria) – which are otherwise impermeable; the term 'transport' is used in a way which excludes transmembrane translocations concerned *exclusively* with energy conversion.

Some transport systems are specific for a particular substrate (or a few substrates) while others may be used for a range of substrates. Transport systems involved in the acquisition of essential component(s) are of course vital for the survival/growth of an organism. Such systems include e.g. the various IRON-uptake mechanisms in a range of organisms. Again, the inability of parasitic protozoa to synthesize their own purine nucleotides *de novo* means that these organisms must import pre-formed purines from their hosts; such uptake involves the operation of nucleoside transport systems which have been identified in various parasitic protozoa (including *Leishmania donovani*, *Plasmodium falciparum*, *Toxoplasma gondii* and *Trypanosoma brucei*) [see e.g. TIP (2001) *17* 142–145].

Typically, transport requires energy (= *active transport*). Energy-independent transport may involve simple diffusion – for example, uptake of small molecules or certain types of ion through a PORIN – or FACILITATED DIFFUSION (see also *glycerol facilitator* in MIP CHANNEL). In some cases transport and energy conversion are linked: see e.g. END-PRODUCT EFFLUX.

Energy-dependent transport may permit a cell or organelle to accumulate a substrate against a concentration gradient; this is necessary e.g. when a required molecule or ion occurs in low concentrations in the external environment. Such transport includes the following categories:

(a) *Group translocation.* In this form of transport the transmembrane translocation of substrate is obligatorily linked with chemical modification of that substrate. Transport systems of this type are commonly used e.g. for the uptake of sugars by bacteria: see PTS.

(b) *ATP hydrolysis-dependent transport.* The hydrolysis of ATP provides energy for various types of transport. For example, in *Escherichia coli*, ATP hydrolysis is needed for osmoregulatory uptake of potassium ions at a particular membrane ATPASE (potassium pump). (See also ION TRANSPORT (b) and (d); TWO-COMPONENT REGULATORY SYSTEM.)

Uptake of IRON by the *E. coli* SIDEROPHORE enterochelin involves initial binding of the iron–enterochelin complex by the FEPA PROTEIN. FepA is energized via the TONB PROTEIN (q.v.), and the iron–siderophore complex, released from FepA, passes through a system of proteins in the cytoplasmic membrane – transfer of the complex to the cytoplasm depending on ATP hydrolysis at a specific ATPase (FepC protein) in the cytoplasmic membrane.

ATP hydrolysis is also an essential feature of a large family of transport systems known as ABC TRANSPORTERS (q.v.) which include both importers and exporters.

Certain types of bacterial PROTEIN SECRETION involve energy derived from ATP hydrolysis.

(c) *Pmf-dependent transport.* Proton motive force (see CHEMIOSMOSIS) can be used, directly, for the transport of certain charged and uncharged species. Examples include the uptake of calcium ions by mitochondria (see ION TRANSPORT), and the proton–lactose symport system for lactose uptake in *Escherichia coli*. Such transport is inhibited by PROTON TRANSLOCATORS.

The uptake of vitamin B_{12} across the outer membrane in *E. coli* is pmf-dependent [see JB (1993) *175* 3146–3150].

As well as acting as a source of energy for transport, pmf can also *regulate* other transport systems. For example, by controlling the redox state of certain redox-sensitive membrane carrier proteins, pmf (as well as the redox potential of the environment) can control the transport of e.g. glucose by the PTS [AvL (1984) *50* 545–555].

(d) *Smf-dependent transport.* See SODIUM MOTIVE FORCE and ION TRANSPORT.

Transport in the control of metabolism. The transport of a given substrate may be the rate-limiting step in the metabolism of that substrate; hence, the transport process itself may constitute an important control mechanism. According to the particular system involved, transport may be dependent on e.g. the level of pmf, the availability of a given ion species (for ion–substrate symport) or (as in the PTS) the availability of 'high-energy' compounds. In some cases, transport is reported to require the operation of the electron transport chain, i.e. to depend on electron flow rather than on pmf [AvL (1984) *50* 545–555].

Effect of transport on yield coefficient. Differences in the YIELD COEFFICIENT of a given organism grown on different substrates may be attributed, at least partly, to differences in the energy costs of substrate transport. For example, in *E. coli* the transport of lactose by proton–lactose symport has been calculated to cost ca. 0.5 ATP per molecule transported, while maltose transport via a binding protein-dependent mechanism has been calculated to cost ca. 1.0–1.2 ATP per molecule transported [JB (1985) *163* 1237–1242].

Transport in other cell functions. The above account refers primarily to transport involved in metabolism, secretion and osmoregulation. Transport is also required for e.g. PTS-dependent CHEMOTAXIS and for certain aspects of pathogenicity (see type III systems in PROTEIN SECRETION); it also appears to be involved in GLIDING MOTILITY in certain cyanobacteria [see e.g. JB (2000) *182* 1191–1199 (1195–1196)].

transporter See PERMEASE.

transposable element (TE; 'jumping gene') A discrete DNA segment, within a larger DNA replicon, which can (actually or effectively) translocate to another (*target*) site in the same replicon, to a target site in another replicon in the same cell, or (see CONJUGATIVE TRANSPOSITION) to a target site in another cell; such translocation (termed *transposition*) does not require extensive DNA sequence homology between the TE and its target site. The following account refers to 'classical' transposable elements, i.e. TEs other than conjugative transposons.

TEs are normal components of e.g. chromosomes, plasmids and phage genomes. They are found in prokaryotes (see INSERTION SEQUENCE and TRANSPOSON) and in eukaryotes – including yeasts (see e.g. TY ELEMENT), protozoa, *Drosophila* and higher plants. (See also INTRON HOMING.)

By convention, the presence of a TE in a given replicon is indicated by a double colon; for example, the presence of transposon Tn*3* within a phage λ genome is designated λ::Tn*3*.

Different TEs employ different mechanisms for transposition. In bacteria, some TEs transpose by a *rec*-independent 'replicative' process, while others employ a non-replicative ('simple') mechanism – terms illustrated in the figure. (See also entries for Tn*10* and Tn*5*.)

Some TEs can apparently employ different transposition mechanisms under different conditions: see e.g. BACTERIOPHAGE MU.

In eukaryotes, certain TEs appear to transpose via an RNA intermediate: see TY ELEMENT.

Transposition is normally a rare event, the actual frequency depending e.g. on the TE and on the physiological state of the cell (see e.g. entry for Tn*501*).

Functions necessary for transposition are usually encoded by the TE itself, although certain host cell functions may also be required – for example, in *Escherichia coli*, GYRASE has been implicated in the transposition of e.g. Tn*5* [Cell (1982) *30* 9–18]. Most TEs have a reading frame for at least one protein (e.g. a TRANSPOSASE), and each TE includes terminal INVERTED REPEAT sequences which apparently permit recognition of the TE by its transposase.

Target-site specificity is an intrinsic property of a TE. Some TEs (see entry for Tn*7*) are highly site-specific, while others appear to insert almost randomly; for example, Tn*3* and IS*1* show a preference for targets in (easily denaturable) AT-rich regions (cf. BACTERIOPHAGE MU), while IS*5* inserts only at sites containing the sequence C(T/A)A(G/A). Nevertheless, even when insertion seems random, some kind of selective process is likely to be involved [ARB (1997) *66* 437–474]. Studies on the transposition of IS*903* into a large (55 kb) plasmid indicate that, while insertion can occur at many different sites, there are certain *preferred* regions into which IS*903* will insert more than once on different occasions [JB (1998) *180* 3039–3048].

Certain targets, termed 'hot-spots', are especially prone to insertion by particular TEs; for example, hot-spots for Tn*3* insertion have been found to contain regions of homology with the Tn*3* termini – although hot-spots for Tn*10* insertion (see entry Tn*10*) bear no obvious relationship to the Tn*10* termini.

Insertion of a TE at a target site results in the formation of direct repeats flanking the inserted TE; this is illustrated in the figure – which shows a scheme for the insertion of a TE by both 'simple' and 'replicative' mechanisms. The length of the target duplication is usually specific for the TE involved – for example, 5 bp for Tn*3* and 9 bp for Tn*10* (cf. IS*1*).

The transposition of some TEs (e.g. bacteriophage Mu, members of the Tn*3*/γδ family) is regulated by a *trans*-acting repressor (cf. Tn*5*); when a TE of this type enters a 'naive' host (i.e. a cell lacking a TE of the same type), transposition functions are expressed at a high level until repression of the TE has been established. (See also TRANSPOSITION IMMUNITY and MULTICOPY INHIBITION.) The transposition of e.g. Tn*10* from a chromosomal location in *Escherichia coli* tends to be linked to the cell cycle; this is because Dam methylation inhibits transcription of the Tn*10* transposase gene, and such inhibition is transiently relieved during chromosomal replication (immediately after the replication fork has passed the transposase gene but prior to Dam methylation of the gene).

TEs may be more or less 'host-specific'; for example, IS*1* appears to be limited mainly to strains of *Escherichia coli* K12.

[TEs in prokaryotes (review): ARB (1985) *54* 863–896.]

The consequences of insertion of a TE at a new site depend on the TE and on the nature of the target site etc. If the insertion site is in a structural gene, an insertion mutation will result. If insertion occurs in an OPERON, strong polar effects may occur because TEs carry transcriptional and/or translational stop signals which prevent expression of all genes downstream of the insertion site. Insertion of a TE may also activate an operon – for example, the TE may carry a functional promoter, or may insert into, and inactivate, a sequence involved in the negative control of the operon. Precise excision of a TE from a site of insertion (an event that occurs at low frequency) leads to reversion of the corresponding mutation; however, imprecise excision (e.g. with deletion of adjacent host DNA) can also occur.

See also TRANSPOSON MUTAGENESIS.

TEs can also mediate a wide range of DNA rearrangements, including deletions and inversions of adjacent host DNA, formation and resolution of COINTEGRATES, gene fusion and amplification etc. These processes may be mediated by the mechanisms responsible for transposition per se; however, in some cases they may occur by transposition of a TE to another site in the same (or a different) replicon followed by *recA*-dependent recombination between the homologous sites thus created.

(See also entries for individual TEs under 'IS' and 'Tn'.)

transposase An enzyme which is responsible for at least part of the process of transposition of a TRANSPOSABLE ELEMENT; a transposase (or a component of it) is usually encoded by the TE itself and is specific for that TE or for TEs with closely related terminal inverted repeats. (See e.g. Tn*3*; cf. IS*101* and Tn*951*.)

transposition See TRANSPOSABLE ELEMENT.

transposition (conjugative) See CONJUGATIVE TRANSPOSITION.

transposition immunity A phenomenon in which a replicon containing a copy of Tn*3* (or a Tn*3*-related transposon) is immune to the insertion of a second copy of the same transposon; the frequency of transposition of the transposon to other replicons in the same cell is unaffected. In the case of Tn*3* the immunity

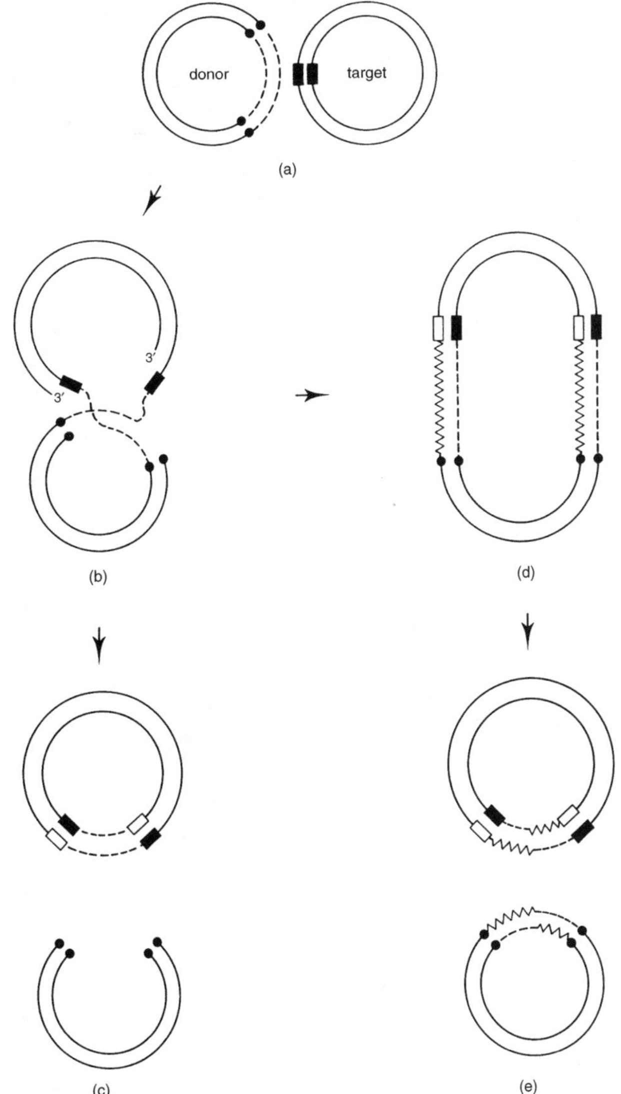

TRANSPOSABLE ELEMENT: a (diagrammatic) scheme for a transposon undergoing 'simple' and 'replicative' transposition.

(a) Two circular double-stranded DNA molecules. On the left, the donor molecule includes a transposon (dashed lines) – either side of which is an old 'target' site (●); the donor's target site was duplicated when the transposon was originally inserted into the donor molecule (see later). On the right, the molecule which will receive the transposon has a single target site (■) where the transposon will be inserted.

(b) An enzyme (a *transposase* – not shown) mediates at least the initial stages of transposition. A staggered break has been made at the target site. In the donor molecule, a nick has been made in each strand of the transposon (at opposite ends), and the free ends have been ligated to the target molecule, as shown. In 'simple' transposition (which occurs e.g. in transposon Tn*10*) the next (and final) stage is shown at (c).

(c) The result of 'simple' transposition. DNA synthesis has occurred from each free 3′ end in the target molecule, using the (single-stranded) target site as template, to form the complementary strand of each target site (□). The remaining strand-ends of the transposon have been nicked and ligated as shown. Note that the target molecule's target site has been duplicated – compare with the donor molecule at (a). The rest of the donor molecule may be non-viable ('donor suicide').

(d) 'Replicative' transposition (which occurs e.g. in Tn*3*) involves stages (b), (d) and (e). At (d), new DNA synthesis (from each 3′ end in the target molecule) has continued beyond the target site, using each strand of the transposon as template; that is, the transposon has been replicated. The end of each new strand has been ligated to a free strand-end in the donor molecule. The structure shown at (d) is called a *cointegrate*; the zigzag line represents newly synthesized DNA. Progress from stage (d) to stage (e) involves the activity of an enzyme called a *resolvase* (which is encoded by the transposon); this enzyme 'resolves' the cointegrate by promoting site-specific recombination between sites in the two copies of the transposon – forming the molecules shown at (e). (*Continued on page 794*.)

is highly specific: a copy of Tn*3* in a replicon confers immunity to insertion of a second copy of Tn*3* but not to insertion of a closely related transposon such as Tn*501*. Apparently, only the terminal 38 bp of Tn*3* need be present in a replicon to confer immunity. (cf. MULTICOPY INHIBITION.)

transposon (Tn) A TRANSPOSABLE ELEMENT (q.v.) which encodes not only those functions necessary for transposition but also functions that are unrelated to transposition – for example, resistance to antibiotics (see e.g. entries for Tn*3*, Tn*5*, Tn*9*, Tn*10*, Tn*1721*) or heavy metals (e.g. Tn*501*) or both (e.g. Tn*2410*), toxin production (e.g. ETEC heat-stable toxin: see Tn*1681*), and lactose metabolism (e.g. Tn*951*). (See also *conjugative transposons* in the entry CONJUGATIVE TRANSPOSITION.)

Some transposons (referred to as *class I*, *composite* or *compound* transposons) consist of a sequence containing structural genes flanked on either side by an INSERTION SEQUENCE – the two insertion sequences forming identical or non-identical, direct or inverted repeats; one or both IS elements mediate transposition of the entire transposon (see e.g. Tn*10*) and may also be able to transpose independently of each other and of the rest of the transposon.

Other transposons (*class II*, *simple* or *complex* transposons) consist of a sequence containing one or more structural genes flanked on either side by a short inverted repeat (see e.g. Tn*3*).

The use of transposons in laboratory studies.
Transposons are valuable research tools that have many uses in the research laboratory. Depending on requirements, transposons can be inserted into the genomes of *living* cells (see e.g. TRANSPOSON MUTAGENESIS and SIGNATURE-TAGGED MUTAGENESIS) or inserted into isolated DNA molecules in an in vitro system.

The use of transposons has been facilitated e.g. by the development of a genetically modified form of a natural transposon, Tn*5* (see TN5); this system can be used in the following approaches.

Transposons can be inserted, at random sites, into a population of plasmids (or other molecules) simply by incubating together the target DNA (i.e. population of plasmids), copies of the transposon and the relevant transposase; incubation is carried out at 37°C for a few hours in the presence of Mg^{2+}. After incubation, the plasmids (most or all now containing a transposon inserted at a random location) are inserted into bacteria (e.g. *Escherichia coli*) by TRANSFORMATION or ELECTROPORATION; most or all of the bacteria will receive a transposon. If the transposon includes an antibiotic-resistance gene, plating the (transformed) bacteria on a medium containing the given antibiotic will select for those cells which contain a transposon. If, after incubating the plate, several colonies are chosen, each colony will contain cloned copies of the plasmid containing the transposon at a given, unique location.

How may such clones be used? If the plasmid is a large molecule, then each of a number of clones can be used for sequencing different parts of the molecule – using primers complementary to end-sequences in the transposon for bi-directional sequencing (see DNA SEQUENCING); as plasmids in different clones contain transposons in different (random) locations, sequencing can start from different sites around the molecule.

In vitro insertion of transposons, described above, may also be used for modifying a plasmid. Thus, the randomly inserting transposons may, in one or a few plasmids, insert into (and inactivate) a particular gene, causing a *required* 'knock-out'; such a mutant may be subsequently selected for in the population of transformed bacteria.

Transposons are also useful for carrying a specific gene into target cells; such a gene is inserted into the transposon.

In vivo insertion of transposons may be carried out with the same (Tn*5*-based) system: transposons, complexed with transposase in an Mg^{2+}-free environment, are electroporated into living cells and insert (randomly) into the genome. This procedure may be used e.g. to insert a specific gene (carried by the transposon) and/or to create an insertional mutation in the genome.

transposon mutagenesis TRANSPOSON-mediated MUTAGENESIS. Various approaches involving transposon mutagenesis have been used e.g. for detecting/identifying potential virulence genes and/or gene products in pathogenic bacteria. One example is the detection of secreted and/or cell-surface-associated proteins – i.e. proteins which, because of their possible direct interaction with the host organism, are of interest as potential virulence-associated factors. To detect secreted/cell surface proteins, a population of the given pathogen is mutagenized with transposons that contain a promoter-less gene for alkaline phosphatase (*phoA*); these transposons (Tn*phoA*) insert randomly into the genome in cells of the bacterial population – inserting into different genes in different cells; the cells are then plated so that each cell forms an individual colony. Colonies are tested with a substrate for alkaline phosphatase which, if cleaved by the enzyme, generates colour. This substrate can be cleaved only if alkaline phosphatase has been secreted by the cells or if it is associated with the cell surface; hence, any colony which gives a positive (colour) reaction with the enzyme's substrate indicates a secreted or cell-surface phosphatase. Because *phoA* (in the transposons) is promoter-less and lacks a signal sequence (see SIGNAL HYPOTHESIS), production of an extracellular alkaline phosphatase by cells in a given colony indicates that, in these cells, Tn*phoA* has inserted 'in frame' into the gene of a secreted or cell-surface protein, and that the signal sequence of the gene has enabled secretion of the phosphatase.

The gene containing Tn*phoA* is likely to have been inactivated by insertion of the transposon. If the gene is a virulence gene, the virulence of cells in the colony may be demonstrably reduced when the cells are tested in experimental animals; if so, the relevant (transposon-mutated) gene can be isolated, sequenced and characterized.

A system for transposon-mediated mutagenesis in *Haemophilus influenzae* takes advantage of the specific uptake-signal sequence which facilitates import of DNA by TRANSFORMATION [AEM (1998) *64* 4697–4702].

Transposon mutagenesis, using *luxAB* genes (see BIOLUMINESCENCE) as a reporter system, has been used to identify genes induced in the COLD-SHOCK RESPONSE in *Sinorhizobium meliloti* [AEM (2000) *66* 401–405].

(See also SIGNATURE-TAGGED MUTAGENESIS.)

trans-stadial transmission In a tick VECTOR (or, less commonly, an insect vector): the retention of infectious agents (e.g. viruses or bacteria) during the transition from larva to nymph, from nymph to adult, or throughout the entire period of transition from larva to adult. (See e.g. COWDRIA.)

TRANSPOSABLE ELEMENT (*continued*) (e) Donor and recipient molecules each contain a copy of the transposon; note that, in this model, each molecule contains parts of the original transposon (dashed lines) as well as newly synthesized DNA (zigzag lines).

Adapted from Fig. 11.29, p. 336, in *Molecular Biology of the Gene* 4th edn by Watson et al.

transverse binary fission See BINARY FISSION.

transverse microtubules (*ciliate protozool.*) A ribbon of microtubules which arises at the left-hand anterior side of a kinetosome (near triplets 3–4(–5), GRAIN CONVENTION), extends upwards towards the pellicle, and then passes leftwards at right angles to the kineties, the wider face of the ribbon being parallel to the body surface. (See also RHABDOS.)

transversion mutation A type of POINT MUTATION in which a purine nucleotide is replaced by a pyrimidine nucleotide, or vice versa. (cf. TRANSITION MUTATION.)

***traO* locus** See FINOP SYSTEM.

Trapelia See LECIDEA.

***traS* gene** See F PLASMID and SURFACE EXCLUSION.

***traT* gene** See F PLASMID and SURFACE EXCLUSION.

travellers' diarrhoea DIARRHOEA in persons outside their normal environment. Enterotoxigenic strains of *Escherichia coli* (see ETEC) are common causal agents, but almost any FOOD POISONING organism may be responsible.

***traY* gene** See CONJUGATION (sense 1b).

***traZ* gene** See CONJUGATION (sense 1b).

treadmilling See e.g. MICROTUBULES.

Trebouxia A genus of unicellular, coccoid, non-motile green algae (division CHLOROPHYTA); species are common photobionts in LICHENS, but they also occur (sparsely) in the free-living state e.g. on wood, bark, etc. The cells are variable in shape and size (up to ca. 35 μm diam.), free-living forms being larger than those in lichen thalli; each cell contains a single large central chloroplast with at least one central pyrenoid. 'Vegetative cell division' does not occur (cf. PSEUDOTREBOUXIA). Asexual reproduction involves the formation of biflagellate wall-less zoopores – at least in cells growing in liquid media; in lichen thalli zoospore formation is usually (but not always [Lichenol. (1980) *12* 173–187]) suppressed, asexual reproduction occurring by autospore formation. [*Trebouxia* species from lichens: Lichenol. (1981) *13* 65–86.]

Trebouxia has been placed – together with e.g. FRIEDMANNIA, MICROTHAMNION, PLEURASTRUM and PSEUDOTREBOUXIA – in an order, Pleurastrales, in the class Pleurastrophyceae (division Chlorophyta) [Book ref. 123, pp. 29–72]. (cf. TETRASELMIS.)

tree diseases Decay of standing trees is generally due to fungal attack on the (dead) heartwood, the (living) sapwood being much less frequently attacked. Rotting of the heartwood generally does not affect the vital processes of the tree, but the tree eventually becomes structurally weakened and may be easily toppled in high winds. The fungi which attack heartwood (causing 'heart rots') normally gain access to it via wounds – e.g., damaged roots or broken branches. Important tree-rotting fungi include e.g. species of *Armillaria*, *Ganoderma*, *Heterobasidion*, *Piptoporus*, *Polyporus* and *Trametes*. (cf. TIMBER SPOILAGE; see also BROWN OAK; BROWN ROT; BUTT ROT and WHITE ROT.)

Other tree diseases include e.g. APPLE CANKER, BEECH BARK DISEASE, BLEEDING CANKER, BLISTER RUST, CHESTNUT BLIGHT, DUTCH ELM DISEASE, FIREBLIGHT, JARRAH DIEBACK, OAK WILT, PITCH CANKER, SOOTY BARK, SNOW BLIGHT, WATERMARK DISEASE and WETWOOD.

tree lungwort *Lobaria pulmonaria.*

tree squirrel hepatitis B virus See HEPADNAVIRIDAE.

trehalose (α,α-trehalose) A non-reducing disaccharide: α-D-glucopyranosyl-α-D-glucopyranoside. It occurs e.g. in the haemolymph of insects, in certain algae (particularly red algae), in a few plants (commercial source: *Selaginella*), in actinomycetes and certain other bacteria (see e.g. CORD FACTOR), and as the major storage disaccharide in many fungi and yeasts and some lichens. Trehalose is hydrolysed by an α-glucosidase, *trehalase*, present in all trehalose-producing organisms. [Regulation of trehalose mobilization in fungi: MR (1984) *48* 42–59.]

Tremella See TREMELLALES.

Tremellales An order of saprotrophic, lignicolous fungi (subclass PHRAGMOBASIDIOMYCETIDAE) which typically form gymnocarpous fruiting bodies in which the (usually) globose or clavate basidia each consist of a *longitudinally* septate metabasidium from which arise four elongated sterigmata. According to species, the basidiocarp may be e.g. crust-like, sheet-like and convoluted, or erect and stalked, it may be gelatinous (see JELLY FUNGI), waxy or leathery, and it may be e.g. orange, pink, brown or black. Genera include e.g. *Aporpium*, *Exidia* (*E. glandulosa* = 'witches' butter'), *Phlogiotis*, *Pseudohydnum*, *Tremella*. (See also TETRAGONIOMYCES.)

tremorgenic Capable of inducing tremors. (See e.g. PENITREMS.)

trench fever (Wolhynian fever) A louse-borne disease of man caused by *Bartonella quintana* (formerly called *Rochalimaea quintana*). The incubation period may be a few days or longer than a month. Symptoms include fever, severe muscular pain (particularly in the back and legs) and sometimes a maculopapular rash that occurs on the chest and abdomen; splenomegaly is common. Mortality rates are low, but complications (e.g. cardiac dysfunction) may occur, and relapses may occur in untreated patients. XENODIAGNOSIS has been used in the past, but diagnosis can now be achieved by culture and/or by complement-fixation tests and other serological approaches. The disease usually responds to chemotherapy with e.g. tetracyclines or chloramphenicol.

trench mouth *Syn.* VINCENT'S ANGINA.

Trentepohlia A genus of non-aquatic filamentous algae (division CHLOROPHYTA) which shares many features with e.g. CEPHALEUROS, PHYCOPELTIS and STOMATOCHROON [Book ref. 123, pp. 233–250]: the thallus is commonly heterotrichous (except in *Phycopeltis*), i.e., it is composed of more or less well-developed erect and prostrate systems of filaments; an orange-red pigment ('haematochrome') is stored in the cytoplasm; pyrenoids are absent; and zoospores and gametes are produced in specialized cells from which they are released via a single papillate exit pore. The flagella of the motile cells are characteristically bilaterally 'keeled' or 'winged'. In *Trentepohlia* spp, the erect system of orange-red filaments is well-developed, the prostrate system weakly developed. Species grow on rocks, wood or bark, or as photobionts in certain lichens (see e.g. OPEGRAPHA and ROCCELLA). (See also ERYTHRITOL; PAINT SPOILAGE; RIBITOL.)

Trepomonas See DIPLOMONADIDA.

Treponema A genus of Gram-negative bacteria (family SPIROCHAETACEAE) which occur as parasites or pathogens in the mouth, intestinal tract, and genital regions in man and other animals; some species are obligate anaerobes, but others (including *T. carateum* and *T. pallidum*) are now generally considered to be microaerophiles. The cells are helical, 0.1–0.4 × 5–20 μm, and motile; they stain poorly by the Gram stain but satisfactorily with silver deposition methods (see e.g. FONTANA'S STAIN). *T. pallidum*, *T. carateum* and *T. paraluiscuniculi* have not been grown in cell-free media, and are propagated e.g. intratesticularly in rabbits (see also NICHOLS' TREPONEME). The other (obligately anaerobic) species have been grown in complex media and have a fermentative metabolism; isolation procedures are based e.g. on insensitivity to rifampicin, or on the ability of the cells to migrate through a membrane filter (pore size 0.2 μm) into, and through, a suitable agar medium. GC%: 25–54. Type species: *T. pallidum*. [Book ref. 22, pp. 49–57.]

T. bryantii. Occurs in the bovine RUMEN; cells ca. 0.3 μm in diameter. Cultivable in media containing e.g. isobutyrate, pyridoxal, folic acid, biotin, thiamine and niacinamide. GC%: ca. 36.

T. carateum. The causal agent of PINTA. The cells are similar to those of *T. pallidum* (q.v.).

T. denticola. Occurs in the mouth (particularly the tooth – gum margin) in man and primates; cell diameter <0.2 μm. Cultivable in e.g. peptone–yeast extract–serum medium. GC%: 37–38.

T. hyodysenteriae. A pathogen of swine (see SWINE DYSENTERY). Cell diameter up to ca. 0.4 μm; 8 or 9 periplasmic flagella arise at each end of the cell. Cultivable in e.g. trypticase–soy broth containing 10% fetal calf serum and incubated under 10% CO_2; produces a clear ('β') haemolysis. (cf. *T. innocens*.)

T. innocens. Occurs in the intestinal tract in dogs and swine but is apparently non-pathogenic. Cells are similar in size and structure to those of *T. hyodysenteriae*. Cultivable; weakly haemolytic. [Comparison of lipids in *T. hyodysenteriae* and *T. innocens*: IJSB (1984) *34* 160–165.]

T. minutum. A non-pathogenic species which occurs on the genitals. Cells are 0.2 μm or less in diameter; 2 or 3 periplasmic flagella arise at each end of the cell. Cultivable on e.g. peptone–yeast extract–serum agar. GC%: 37.

T. pallidum. Cells helical, <0.2 μm (commonly 0.13–0.15 μm) in diameter, with pointed ends; 3 or 4 periplasmic flagella arise at each end of the cell. Motile. Apparently microaerophilic. GC%: 52–54. Three subspecies: *T. pallidum* subsp *pallidum*, subsp *pertenue* (formerly *T. pertenue*), and subsp *endemicum*. *T. pallidum* subsp *pallidum* is the causal agent of sexually- and congenitally-transmitted SYPHILIS which occurs worldwide. *T. pallidum* subsp *pertenue* is the causal agent of YAWS. *T. pallidum* subsp *endemicum* is the causal agent of human non-venereal endemic syphilis – which occurs in Africa, the Middle East, SE Asia and the former Yugoslavia, and which can be transmitted e.g. via fomites. All three subspecies (and *T. carateum*) give similar reactions in serological tests for syphilis.

T. paraluiscuniculi. The causal agent of a syphilis-like disease in rabbits. Cells are similar to those of *T. pallidum*.

T. pertenue. See *T. pallidum*.

T. phagedenis. A non-pathogenic species. Cells helical, 0.2–0.25 μm in diameter, with blunt ends; 3 to 8 periplasmic flagella arise at each end of the cell. Can be grown in e.g. peptone–yeast extract–serum media. Two biovars (*reiter* and *kazan*) are distinguishable antigenically and also by minor biochemical differences. (See also REITER TREPONEME.) GC%: 38–39.

T. refringens. A non-pathogenic species, occurring e.g. in the normal microflora of the human genitals; isolated from condyloma acuminatum and syphilitic lesions. Cells 0.2–0.25 μm in diameter, with 2–4 periplasmic flagella arising at each end. Cultivable in e.g. peptone–yeast extract–serum media. GC%: 39–43.

T. saccharophilum. See RUMEN.

T. scoliodontum. Occurs in the human mouth. Cells ≤0.2 μm in diameter. Cultivable in serum-enriched media.

T. socranskii. An oral species, three subspecies of which have been isolated from human periodontia [IJSB (1984) *34* 457–462].

T. succinifaciens. Occurs in the intestine in swine. Cells up to 0.3 μm in diameter. GC%: 36.

T. vincentii (formerly *Borrelia vincentii*). Occurs in the human mouth and is associated with VINCENT'S ANGINA. Cells 0.2–0.25 μm in diameter, with 4–6 periplasmic flagella arising at each end. Cultivable in peptone–yeast extract–serum media.

[Pathogenesis and immunology of treponemal infection: Book ref. 78.]

***Treponema pallidum* immobilization test** See TPI TEST.

treponemal tests Tests in which treponemal antigens or intact cells are used to detect specific anti-treponemal antibodies which occur in the serum of patients suffering from SYPHILIS (or PINTA or YAWS). These tests are more specific than STANDARD TESTS FOR SYPHILIS and are used as confirmatory tests (e.g., when results of STS are equivocal); however, as they remain positive for long periods after effective therapy, treponemal tests are less suitable than STS for assessing the progress of therapy. Treponemal tests include e.g. the FTA-ABS TEST, TPHA TEST, Reiter protein complement fixation test (RPCF test), and the TPI TEST.

treponematosis (*pl.* treponematoses) Any disease of man or animals in which the causal agent is a species of *Treponema* (see e.g. SYPHILIS, PINTA, YAWS).

tretic development See CONIDIUM.

TRF (*immunol.*) T cell replacing factor: a term originally used for the factor(s) which are present in the supernatant of cultures of activated T cells and which are necessary for the in vitro development of antibody-secreting B cells.

***trg* gene** See CHEMOTAXIS (a).

Triactinomyxon See ACTINOSPOREA and WHIRLING DISEASE.

triadimefon (trade name: e.g. Bayleton) An agricultural AZOLE ANTIFUNGAL AGENT which is particularly active against powdery mildews and rusts in e.g. cereals; it has both eradicant and protectant properties, and can apparently be translocated both acropetally and basipetally in the plant.

triadimenol An agricultural AZOLE ANTIFUNGAL AGENT used mainly as a seed treatment – sometimes mixed with e.g. fuberidazole – for the control of various cereal diseases (e.g. barley powdery mildew, rhynchosporium).

Triangularia See SORDARIALES.

triangulation number See ICOSAHEDRAL SYMMETRY.

triarylmethane dyes See TRIPHENYLMETHANE DYES.

triatomine bugs A category of obligate blood-feeding insects of the Hemiptera (true bugs); they include the vectors of CHAGAS' DISEASE. A major aspect of the control of Chagas' disease is the use of insecticidal sprays on bug-infected premises. Hence, characterization of triatomine populations and improvement in the accuracy of vector identification are likely to optimize control; in this context certain molecular methods (such as RAPD and SSCP) have contributed valuable information [molecular tools and triatomine systematics (a public health perspective): TIP (2001) *17* 344–347].

triazole antifungal agents See AZOLE ANTIFUNGAL AGENTS.

tribavirin *Syn.* RIBAVIRIN.

tribe A taxonomic rank (see TAXONOMY and NOMENCLATURE).

Tribonema A genus of freshwater filamentous algae (class XANTHOPHYCEAE) in which the filaments consist of elongated cells joined end to end; each cell contains several parietal chloroplasts.

tributyltin chloride See PROTON ATPASE.

tributyltin oxide See TBTO.

TRIC agents Trachoma and inclusion conjunctivitis agents, i.e. strains of CHLAMYDIA which cause these diseases.

tricarboxylic acid cycle See TCA CYCLE and REDUCTIVE TRICARBOXYLIC ACID CYCLE.

Triceratium See DIATOMS.

Trichamoeba A genus of freshwater amoebae of the AMOEBIDA; cells are monopodial and commonly have a bulbous and/or filamentous UROID.

Trichia See MYXOMYCETES.

Trichiales See MYXOMYCETES.

trichite (1) Former name for NEMATODESMA. (2) Former name for a particular type of TRICHOCYST. (3) See PERANEMA.

3, 4, 4′-trichlorocarbanilide See CARBANILIDES.

Trichochloris See MICROMONADOPHYCEAE.

trichocyst A term formerly used to refer to many different types of EXTRUSOME but currently applicable to a particular type: a subpellicular, capsule-like extrusile organelle of the type which occurs e.g. in *Paramecium*; on appropriate stimulation (chemical, mechanical or electrical) the trichocyst rapidly extrudes a tipped thread (filament, shaft), ca. 15–25 μm in length, which has osmiophilic cross-striations of periodicity ca. 60 nm. Trichocysts may act as anchoring devices and/or as a form of defence, but their precise role is unknown. Electron microscopic studies of *Frontonia* and *Paramecium* indicate that trichocysts develop from minute membrane-bound vesicles within the cytoplasm, and that the developing structures (*pretrichocysts*) migrate to their final, subpellicular positions only shortly before reaching maturity. In *Paramecium aurelia* the mature trichocysts are ovoid, ca. 5 × 2 μm, orientated approximately perpendicular to the pellicle; the pointed, crystalline tip of the filament appears to be the only osmiophilic structure within the undischarged trichocyst. Trichocysts may be few to numerous, according to species. (See also FIBROCYST.)

Trichoderma A genus of fungi (class HYPHOMYCETES) which form septate mycelium and give rise to colourless or pigmented conidia from phialides on branched conidiophores; teleomorphs occur in the genus *Hypocrea*. *Trichoderma* spp grow rapidly and can use a wide range of substrates (see e.g. CELLULASES, XYLANASES); they are primarily saprotrophs, but some are MYCOPARASITES, and many are antagonistic to other fungi e.g. by forming antibiotics and/or by competing for nutrients. [Biology and potential for biological control: ARPpath. (1985) *23* 23–54.]

Species include e.g. *T. hamatum*, *T. harzianum*, *T. koningii* and *T. viride*. *T. viride* strain QM 6a has been called *T. reesei*, but its correct name is apparently *T. longibrachiatum* [discussion and references: MS (1986) *3* 285–286].

Mutant strains of *Trichoderma*, resistant to heavy metal stress, have been isolated and may be suitable for development as biocontrol agents against plant-pathogenic fungi e.g. in the presence of heavy-metal-containing pesticides [LAM (2001) *33* 112–116].

(See also GLIOTOXIN and TRICHOTHECENES.)

trichodermin See TRICHOTHECENES.

Trichodesmium A genus of filamentous planktonic marine CYANOBACTERIA which are closely related to – and regarded by some as congeneric with – *Oscillatoria*. The filaments contain phycoerythrin and may be red, purple or brown; they typically occur in well-defined sickle-shaped, spherical or fusiform colonies which are buoyed up by peripherally arranged, unusually strong GAS VACUOLES. Cells in the centre of a colony are less strongly pigmented; the immediate environment of these cells may be characterized by low levels of oxygen (at least in calm seas), and it has been suggested that this may permit the cells to carry out NITROGEN FIXATION, despite the absence of heterocysts. However, nitrogen fixation in culture apparently does not depend on colony formation, and nitrogenase activity appears to be light-dependent (and insignificant in the dark); the role of light may be either (i) to provide ATP and reductant (via photosynthesis) for nitrogen fixation, or (ii) to decrease intracellular levels of oxygen by promoting PHOTORESPIRATION [oxygen relationships in nitrogen fixation: MR (1992) *56* 350–351].

Trichodesmium forms conspicuous, dense, floating blooms ('sea sawdust') covering vast areas of open ocean in tropical and subtropical regions (cf. RED TIDE).

Species that are sometimes distinguished on the basis of the arrangement of filaments in colonies include *T. contortum*, *T. erythraeum*, *T. hildenbrandtii* and *T. thiebautii*.

Trichodina See PERITRICHIA.

Trichoglossum See HELOTIALES.

trichogyne In certain eukaryotic organisms: a (typically) hair-like structure which arises from a female reproductive organ (e.g. gametangium) and which receives male nuclei or male reproductive cells.

Tricholoma See AGARICALES (Tricholomataceae).

Tricholomataceae See AGARICALES.

Trichomaris invadens See BLACK MAT SYNDROME.

trichome In many CYANOBACTERIA and in e.g. members of the CYTOPHAGALES, CARYOPHANON etc: a thread-like, usually uniseriate filament composed of closely juxtaposed cells which may or may not be surrounded by a common SHEATH. The cells in a trichome are separated by septa, the positions of which may or may not be marked by constrictions in the trichome; new septa arise by centripetal growth of the peptidoglycan layer from an annulus on the inside of the cell wall. In many cyanobacteria, at least, the contents of adjacent cells in a trichome communicate via small pores called microplasmodesmata (see also HETEROCYST).

Trichomonadida An order of protozoa of the ZOOMASTIGOPHOREA. Cells typically have 4–6 flagella, one of which is recurrent and may be associated with an undulating membrane; cysts are rarely formed. Reproduction occurs by binary fission. The organisms occur typically in the intestinal, respiratory and/or reproductive tracts in animals. Genera: e.g. DIENTAMOEBA, HISTOMONAS, *Monocercomonas*, TRICHOMONAS.

Trichomonas A genus of flagellate protozoa (order TRICHOMONADIDA) which are parasitic in man and other animals; some species can be pathogenic (see TRICHOMONIASIS). Cells: ovoid or pear-shaped, each having a single anteriorly located nucleus, a cytostome, endoskeletal structure(s), and several flagella which arise from the blunt anterior end of the cell. Cysts are not formed.

T. foetus ('*Tritrichomonas foetus*'). Cells: pear-shaped, ca. 20 × 10 μm, with well-developed COSTA and AXOSTYLE. Typically, three of the four flagella are directed anteriorly, while the fourth runs back along the margin of an undulating membrane and extends beyond the posterior limit of the body. The organism has a jerky form of motility. *T. foetus* is pathogenic in cattle (see TRICHOMONIASIS).

T. gallinae. Similar to *T. vaginalis* (see below) but smaller; it is pathogenic in birds.

T. hominis ('*Pentatrichomonas hominis*'). Cells: pear-shaped, ca. 8–20 μm in length, with 3–5 anteriorly directed flagella and one flagellum that forms the margin of an undulating membrane; the undulating membrane runs the entire length of the cell, and its associated flagellum extends beyond the posterior limit of the cell. The organism has a jerky form of motility. *T. hominis* is a common, non-pathogenic inhabitant of the human colon.

T. tenax. Occurs e.g. in the human mouth; it is normally non-pathogenic (but see TRICHOMONIASIS).

T. vaginalis. Cells: pear-shaped, ca. 15 × 10 μm, with costa and axostyle; there are four anteriorly directed flagella, and a short fifth flagellum that forms the margin of an undulating membrane extending approximately half the length of the cell. *T. vaginalis* apparently occurs only in the human genitourinary tract and may or may not cause disease (see TRICHOMONIASIS).

trichomoniasis Any disease in which the causal agent is a species of TRICHOMONAS.

In humans, the commonest form of trichomoniasis is that which affects the genitourinary tract and which is caused by *T. vaginalis*. Infection in men is often asymptomatic (but may result e.g. in epididymitis, prostatitis or NON-GONOCOCCAL URETHRITIS), but infection in women commonly results in VAGINITIS, with vaginal itching and burning, and sometimes a malodorous yellowish-green discharge. The infection can be transmitted sexually. [Cyto- and histopathological changes in the cervical epithelium in *T. vaginalis* infection: Obstet. Gynec. (1984) *64* 179–184.]

Diagnosis. Wet mounts (of e.g. vaginal discharge) are reported to detect only ~60% of culture-positive specimens, while culture (a sensitive technique) may take up to 7 days. [Simplified diagnostic technique permitting both immediate examination and culture of the specimen: JCM (1992) *30* 2265–2268. PCR-based diagnosis: JCM (1997) *35* 132–138.]

Chemotherapy: e.g. METRONIDAZOLE, TRICHOMYCIN or furoxone (see NITROFURANS).

A *pulmonary* form of human trichomoniasis also occurs and is usually caused by aspirated *T. tenax* [JMM (1985) *20* 1–10].

In cattle, *T. foetus* causes a sexually transmissible bovine trichomoniasis which can lead to early abortion; infection is usually self-limiting, the aborted fetus carrying with it all the parasites. Following abortion, the cow is usually immune to re-infection.

In young birds – particularly pigeons, but sometimes chickens and turkeys – *T. gallinae* causes an inflammatory and ulcerative condition of the upper digestive tract; 2-amino-5-nitrothiazole has been found to be an effective treatment.

Trichomycetes A class of fungi (subdivision ZYGOMYCOTINA) which typically live in the alimentary tract in aquatic or terrestrial arthropods. The thallus is commonly simple, branched or unbranched, with a basal holdfast – the thallus being septate in the orders Asellariales and Harpellales, aseptate in the Amoebidiales and Eccrinales. The cell wall appears to lack chitin. Asexual reproduction may involve the formation of sporangiospores (which may be amoeboid in the Amoebidiales) or arthrospores. (See also TRICHOSPORE.) Sexual reproduction is unknown in members of the Amoebidiales (e.g. *Amoebidium*) and Asellariales, and is known in only one genus (*Enteropogon*) in the Eccrinales. [Book ref. 64, p. 387.]

[Book ref. 172.]

trichomycin A MYCOSAMINE-containing heptaene POLYENE ANTIBIOTIC produced by *Streptoverticillium hachijoense*; it has been used in the treatment of candidiasis and *Trichomonas vaginalis* infections.

Trichonympha A genus of cellulolytic protozoa (order HYPERMASTIGIDA) which occur in the gut of wood-eating termites. (See also CELLULASES and TERMITE–MICROBE ASSOCIATIONS.) The bell-shaped *T. campanula* (long axis ca. 100–300 μm) contains a single central nucleus and bears numerous flagella at the narrow (anterior) end of the cell; wood fragments are ingested at the naked posterior end.

Trichophyton See DERMATOPHYTES.

***Trichoplusia ni* GV** See GRANULOSIS VIRUSES.

Trichosida An order of marine testate amoebae (class LOBOSEA) in which the test is fibrous with, in at least one stage of the life cycle, calcareous spicules and multiple apertures through which short, conical pseudopodia extend. The order includes the genus *Trichosphaerium*.

Trichosphaerium See TRICHOSIDA.

trichospore A dehiscent one-spored sporangiolum formed by fungi of the Harpellales.

Trichosporon A genus of fungi (class HYPHOMYCETES) which may form variously-shaped budding cells, scanty to well-developed pseudomycelium, and (usually abundant) septate true mycelium which breaks up into arthrospores; blastospores may be formed, and some species (e.g. *T. capitatum*, *T. beigelii* [= *T. cutaneum*], *T. fermentans*) may form spherical or ellipsoidal asexual endospores. Most species are non-fermentative; only *T. terrestre* and *T. pullulans* can assimilate NO_3^-. On the basis of e.g. CELL WALL structure, some species appear to have ascomycetous affinities (e.g. *T. capitatum*, *T. eriense*, *T. fermentans*, *T. sericeum*, *T. terrestre*, *T. variabile* [see PICHIA]), others to have basidiomycetous affinities (e.g. *T. aquatile*, *T. beigelii*, *T. pullulans*). Species have been isolated from a range of habitats, including e.g. water, plants, wood pulp, soil, human skin (see also WHITE PIEDRA), faeces, etc. [Book ref. 100, pp. 933–962; DNA relatedness among *Trichosporon* spp: AvL (1984) *50* 17–32.]

(See also ACICULOCONIDIUM.)

Trichostomatida An order of protozoa (subclass VESTIBULIFERIA) in which the vestibular ciliature differs little from the somatic ciliature. In one suborder (Trichostomatina) the somatic ciliature covers the cell more or less completely; genera: e.g. BALANTIDIUM, DASYTRICHA, *Isotricha*, *Sonderia*. In the other suborder (Blepharocorythina) somatic ciliature is greatly reduced; these organisms (e.g. *Ochoterenaia*) occur in herbivorous mammals, particularly horses.

Trichostomatina See TRICHOSTOMATIDA.

trichothecenes A group of sesquiterpenoid MYCOTOXINS produced by species of e.g. *Fusarium*, *Myrothecium*, *Trichothecium*. The parent compound is *scirpene* (12,13-epoxy-Δ^9-trichothecene), produced e.g. by *Trichothecium roseum*. Other trichothecenes include e.g. calonectrin (produced e.g. by *Fusarium nivale*), crotocin (e.g. *T. roseum*, '*Cephalosporium crotocinigenum*'), deoxynivalenol (*Fusarium* spp), diacetoxyscirpenol (= anguidine) (*Fusarium* spp), fusarenon-X (*Fusarium* spp), HT-2 toxin (e.g. *F. solani*, *F. tricinctum*), neosolaniol (*Fusarium* spp), nivalenol (e.g. *F. nivale*), T-2 toxin (*Fusarium* spp, e.g. *F. tricinctum*), trichodermin (*Trichoderma viride*), and trichothecin (*Trichothecium roseum*). The roridins and verrucarins (produced mainly by *Myrothecium roridum* and/or *M. verrucaria*) are trichothecenes which contain a (UV-absorbing) conjugated diene bridge.

Trichothecenes are toxic to a wide range of eukaryotic cells and are potent inhibitors of eukaryotic protein synthesis. In animals and man the toxins can, on contact with the skin, cause a severe local irritation with inflammation and desquamation. On ingestion (e.g. in contaminated mouldy cereals) they can cause a severe mycotoxicosis, involving haemorrhagic gastroenteritis with haemorrhages in other organs ('haemorrhagic syndrome' in animals; see also ALIMENTARY TOXIC ALEUKIA and STACHYBOTRYOTOXICOSIS). Trichothecenes are stable within the ranges of pH and temperature (including cooking) normally associated with foods.

[Book ref. 15.]

trichothecin See TRICHOTHECENES.

Trichothecium See HYPHOMYCETES; see also TRICHOTHECENES.

trickle filter See SEWAGE TREATMENT.

trickling generator See VINEGAR.

triclocarban See CARBANILIDES.

triclosan Irgasan DP 33; 5-chloro-2-(2,4-dichlorophenoxy phenol) A phenolic ANTISEPTIC (see PHENOLS) used in medicated soaps, hand cleansers and a wide range of other products; it is active against bacteria (particularly staphylococci) and certain fungi. Triclosan has been reported to act primarily by inhibiting fatty acid biosynthesis – specifically by inhibiting the enzyme enoyl-acyl carrier protein reductase (ENR) [Nature (1999)

398 383–384]. However, it appears that the relevant enzyme in *Streptococcus pneumoniae* is insensitive to triclosan, even though *S. pneumoniae* itself is sensitive; moreover, triclosan can be partially mineralized by a *Sphingomonas*-like organism [FEMS Ecol. (2001) *36* 105–112].

Tricornaviridae A proposed family of PLANT VIRUSES which have tripartite, ssRNA genomes [Intervirol. (1981) *15* 198–203].

tridemorph (*N*-tridecyl-2,6-dimethylmorpholine) A MORPHOLINE ANTIFUNGAL AGENT used – either alone or mixed with e.g. carbendazim, maneb or propiconazole – against powdery mildew, rhynchosporium, eyespot and yellow rust in cereals; tridemorph apparently inhibits haustorium formation.

trifluorothymidine (F_3T; trifluridine; 5-trifluoromethyl-2′-deoxyuridine; TFT; Viroptic) An ANTIVIRAL AGENT which is effective e.g. in the topical treatment of herpes simplex keratitis. It is too toxic for systemic use.

trifluralin See CHAGAS' DISEASE.

trifluridine *Syn.* TRIFLUOROTHYMIDINE.

trifoliin A A LECTIN found at the root surface of white clover (*Trifolium repens*); it binds e.g. 2-deoxy-D-glucosyl residues, and can agglutinate e.g. *Rhizobium leguminosarum* biotype *trifolii*. (See also ROOT NODULES and QUINOVOSAMINE.)

triforine (*N*,*N*′-bis(1-formamido-2,2,2-trichloroethyl)-piperazine) An agricultural systemic ANTIFUNGAL AGENT used e.g. as a seed dressing for the control of powdery mildew of barley, and – mixed with e.g. CARBENDAZIM and/or DITHIOCARBAMATES – as a broad-spectrum antifungal preparation against various other CEREAL DISEASES; it is also effective against powdery mildews of apples, cucurbits etc and against apple scab. Triforine inhibits haustorium formation in *Erysiphe graminis* f. sp. *hordei* and conidium germination in various ascomycetes and deuteromycetes.

trigger hypothesis See SIGNAL HYPOTHESIS.

trigger mechanism (bacterial uptake) See DYSENTERY.

triglycollamic acid *Syn.* NTA.

Trigonopsis A genus of yeasts (class HYPHOMYCETES) in which the cells are triangular or ellipsoidal, sometimes tetrahedral or rhombohedral, occurring singly or in pairs; cell shape is affected by growth conditions (e.g. nitrogen and carbon sources, growth temperature, presence of detergents). Vegetative reproduction occurs by budding – multilaterally in ellipsoidal cells, and from the angles in angular cells. Pseudomycelium is not formed. Metabolism is strictly respiratory; NO_3^- is not assimilated. One species: *T. variabilis*, isolated from beer and from grape must. [Book ref. 100, pp. 963–965.]

Triloculina See FORAMINIFERIDA.

Trilospora See MYXOSPOREA.

trimethoprim (5-(3,4,5-trimethoxybenzyl)-2,4-diaminopyrimidine) A FOLIC ACID ANTAGONIST (q.v.) which has much higher affinity for bacterial dihydrofolate reductase than for the mammalian enzyme; it has a broad spectrum of (bacteriostatic) activity, being active against e.g. at least some enterobacteria, *Haemophilus* spp, staphylococci and streptococci, but not e.g. against mycoplasmas, *Pseudomonas aeruginosa* or *Treponema pallidum*. It is used – generally together with a sulphonamide – mainly for treating infections of the upper respiratory tract and urinary tract.

trimethylamine-*N*-oxide See FISH SPOILAGE.

trimethylene glycol See ALCOHOLS.

trimitic See HYPHA.

Trimitus See DIPLOMONADIDA.

trinactin See MACROTETRALIDES.

trinomial A three-name designation such as *Puccinia graminis tritici*. (cf. TERNARY COMBINATION.)

triostins See QUINOXALINE ANTIBIOTICS.

trioxsalen 4,5′,8-TrimethylPSORALEN.

triphenylmethane dyes A family of DYES each of which may be regarded as an oxidation product of the corresponding colourless triarylmethane *leucobase*: $C(Ar)_3H$; each dye contains a quinonoid ring (the chromophore). Triphenylmethane dyes include: *crystal violet* (*tris*(4-dimethylaminophenyl)methyl chloride), *malachite green* (4,4′-*bis*(dimethylamino)triphenylmethyl chloride), *pararosanilin* (*tris*(4-aminophenyl)methyl chloride), and *rosanilin* (4,4′,4″-triamino-3-methyltriphenylmethyl chloride). Some triphenylmethane dyes are acidic, some basic (see also FUCHSIN). Some common stains (e.g. METHYL VIOLET) are mixtures of triphenylmethane dyes.

Many triphenylmethane dyes are used for STAINING. Some are used in media owing to their ability, at very low concentrations, to inhibit many Gram-positive bacteria; much higher concentrations are generally required to inhibit Gram-negative species, while acid-fast species are typically even more resistant. (See also BRILLIANT GREEN and BRUCELLA.) Some of the dyes are also fungistatic – e.g. for species of *Candida*, *Torula* and *Trichophyton*. (See also AURINTRICARBOXYLIC ACID.)

triphenylmethylphosphonium See CHEMIOSMOSIS.

triphenyltetrazolium chloride See TETRAZOLIUM.

triphosphopyridine nucleotide NADP (see NAD).

triple phosphate See STRUVITE.

triple-stranded DNA See TRIPLEX DNA.

triple-sugar–iron agar See TSI AGAR.

triple vaccine A VACCINE containing diphtheria and tetanus TOXOIDS and killed cells of *Bordetella pertussis* (pertussis vaccine). (See also ANTIGENIC COMPETITION.)

triple-vent Petri dish See VENT.

triplex DNA (triple-stranded DNA) A triple-stranded structure in which pyrimidine or purine bases in a strand of nucleic acid, located within the major groove of a DNA duplex, bind to purine bases in the duplex; bases in the strand interact with purines in the duplex via *Hoogsteen hydrogen bonding*.

A DNA triplex may be formed by interaction between a duplex and a *triplex-forming oligonucleotide* (TFO). Alternatively, a DNA triplex may be formed by interaction between certain regions of a duplex and a single-stranded region from another part of the same (supercoiled) molecule; thus, e.g. in supercoiled plasmids, a homopurine–homopyrimidine tract may adopt a conformation (termed H-DNA) which depends primarily on the development of a triplex structure.

A triplex structure may also be formed by PNA; it has been found that two molecules of PNA can displace one strand of a DNA duplex – forming a triplex with the other strand.

[Triplex DNA structures: ARB (1995) *64* 65–95.]

The specificity with which TFOs bind to DNA duplexes has suggested the use of these oligonucleotides as agents in GENE THERAPY. Thus, e.g. a TFO may inhibit transcription of a given gene by binding to a polypurine–polypyrimidine sequence in the promoter region. A triplex approach has been used successfully to downregulate expression of the *bcl-2* proto-oncogene in HeLa cells, suggesting that this technique may have value for oligonucelotide-based gene therapy [NAR (2001) *29* 622–628].

Triposporina See HYPHOMYCETES and NEMATOPHAGOUS FUNGI.

tripton The non-living components of SESTON – e.g. organic detritis, mineral particles.

TRIS (THAM; Tris; tris; Tromethamine) Tris (hydroxymethyl) aminomethane; with e.g. HCl, TRIS forms the buffering system $(CH_2OH)_3CNH_2/(CH_2OH)_3CNH_3^+$ which is effective between ca. pH 7 and pH 9 with a pK_a (at 25°C) of 8. TRIS-buffered pH is significantly temperature-dependent.

triskelion (of clathrin) See PINOCYTOSIS.

trisodium phosphonoformate See PHOSPHONOFORMIC ACID.

trisomic See CHROMOSOME ABERRATION.

trisporic acid See PHEROMONE.

tristeza disease See CITRUS TRISTEZA VIRUS.

Tritirachium See HYPHOMYCETES; see also PROTEINASE K.

Triton X-100 A non-ionic detergent of the polyoxyethylene *p-t*-octyl phenol series.

Tritrichomonas See TRICHOMONAS.

Trk system See ION TRANSPORT.

tRNA Transfer RNA: a type of RNA which, during PROTEIN SYNTHESIS, acts as an adaptor molecule, matching amino acids to their condons on mRNA. (tRNA is also involved in the priming of reverse transcription in viruses of the RETROVIRIDAE and in certain control functions.)

During protein synthesis, a given tRNA must be able to bind covalently to a specific amino acid, recognize the appropriate codon via an anticodon triplet (see GENETIC CODE), and fit into the ribosomal A and P sites – i.e. all tRNAs must be similar in size and shape. A given amino acid may be recognized by two or more different tRNAs (*isoaccepting tRNAs*).

A tRNA molecule is single-stranded, is ~74–95 nt long (ca. 4S), and contains a high proportion of 'unusual' (modified and hypermodified) nucleosides – e.g. dihydrouridine (D), pseudouridine (ψ), QUEUOSINE (Q), 4-thiouridine (S^4U), ribothymidine (T) – at specific positions in the chain. (See e.g. WOBBLE HYPOTHESIS.)

The 3′-OH end of a tRNA molecule always ends with the sequence pCpCpA (p = phosphate), the terminal residue being the site at which the amino acyl residue is bound. Base-pairing between short complementary regions in the chain results in the formation of a characteristic secondary structure, the *cloverleaf* structure; this structure generally comprises the *acceptor arm* (= CCA arm, the 'stalk' of the cloverleaf) and three stem-and-loop structures: (i) the *D arm* (which often contains dihydrouridine), (ii) the *anticodon arm* or *A arm* (which includes the anticodon), and (iii) the *TψC arm* (the 'common arm' or 'T arm' which often includes the sequence ribothymidine–pseudouridine–cytidine) – together with a so-called *extra arm* (= variable arm) between the anticodon and TψC arms.

Variation in size among different tRNA molecules occurs chiefly in the loop of the D arm and (particularly) in the extra arm – the latter generally containing either 4–5 nt (in *class I* tRNAs) or 13–21 nt (in *class II* tRNAs).

The positions of nucleotides in a tRNA molecule are given standardized numbers, 1–76, in the 5′ → 3′ direction; the additional nucleotides present in some tRNAs are numbered in relation to the preceding standard position: for example, nucleotides 20:1 and 20:2 occur between nucleotides 20 and 21. [Numbering system: Book ref. 186, pp. 518–519.] In the cloverleaf, nucleotides 1–7 pair with 72–66 to form the acceptor arm, the 3′ end extending beyond the 5′ end by (usually) 4 nt. Nucleotides 10–25 form the D arm (stem: 3 or 4 bp), 27–43 form the anticodon arm (stem: 5 bp; anticodon: nucleotides 34–36), and 49–65 form the TψC arm (stem: 5 bp); nucleotides 47:1 to 47:n form the extra arm.

The cloverleaf structure is itself folded into a specific, compact, apparently L-shaped tertiary form which is essential for biological activity.

Variations in tRNA structure include e.g. the absence of dihydrouridine and ribothymidine in tRNAs from archaeans [tRNAs in archaebacteria (archaeans): Book ref. 157, pp. 311–343], and the absence of a D arm in $tRNA^{Ser}$ from mammalian mitochondria [Book ref. 84, pp. 259–278]. *Initiator* tRNAs are structurally distinct from those tRNAs which are involved in polypeptide chain *elongation* – for example, initiator tRNAs characteristically have three sequential G·C pairs in the stem of the anticodon arm (adjacent to the anticodon loop).

tRNA synthesis. TRANSCRIPTION of tRNA genes results in a large precursor molecule which is subsequently processed by cleavage and by modification of appropriate bases. (The primary transcript may contain sequences for several tRNA molecules.) In *Escherichia coli* (and most other organisms) the 5′ end of a tRNA molecule is generated by the action of an endoribonuclease, RNASE P, on the precursor. Processing of the 3′ end appears to involve exonucleolytic cleavage; in some tRNAs (apparently most or all in *E. coli*) the mature 3′ end is generated by cleavage to the potential 3′-terminal CCA (possibly by the exoribonuclease RNase D), but in others (e.g. some tRNAs encoded by bacteriophage T4) there is no CCA – exoribonuclease action (possibly by RNase BN [PNAS (1983) *80* 3301–3304]) being followed by the addition of CCA catalysed by *tRNA nucleotidyltransferase* (product of the *cca* gene in *E. coli*). (See also RNASE T.)

The organization of the tRNA genes differs in different organisms; for example, in *Bacillus subtilis* (but not in *E. coli*) tRNA genes are highly clustered. [tRNA genes in *B. subtilis*: MR (1985) *49* 71–80.] (See also RRNA.) Some tRNA genes in archaeans [e.g. the $tRNA^{Trp}$ gene of *Halobacterium volcanii*: JBC (1985) *260* 3132–3134] and in eukaryotes contain *introns* (see SPLIT GENE). (See also SUPPRESSOR MUTATION and PATHOGENICITY ISLAND.)

tRNA-directed transcription antitermination See TRANSCRIPTION ANTITERMINATION.

trochal band (locomotor fringe) A ring of cilia surrounding a TELOTROCH, or encircling the basal disc of an adult mobiline (see PERITRICHIA). (cf. PECTINELLA.)

troleandomycin See MACROLIDE ANTIBIOTICS.

Tromethamine See TRIS.

Tropheryma See WHIPPLE'S DISEASE.

trophont *Syn.* TROPHOZOITE.

trophophase In BATCH CULTURE, that phase in which primary (growth-directed) metabolism is dominant (cf. IDIOPHASE); it corresponds to the latter part of the lag phase and the early and middle periods of the log phase.

trophosome An organ which occurs in the giant tube worm, *Riftia pachyptila* (see HYDROTHERMAL VENT), and which may account for ca. 60% of the mass of the worm; it contains masses of (endosymbiotic) prokaryotic organisms of various types, including sulphide-oxidizing chemolithoautotrophs. Trophosome tissue contains two key enzymes of the CALVIN CYCLE (RuBisCo and phosphoribulokinase) as well as e.g. adenosine phosphosulphate reductase, ATP sulphurylase and RHODANESE.

trophozoite (trophont) A cell capable of (or specialized for) feeding.

tropical canine pancytopenia See EHRLICHIA.

tropism (1) A response to a directional stimulus exhibited by bending or growth of an organism – or of part of an organism – in an orientation dictated by that of the stimulus; growth towards the stimulus is called positive tropism, away from the stimulus negative tropism. (cf. TAXIS.) The stimulus may be e.g. a chemical gradient (*chemotropism*), gravity (*geotropism*), light (*phototropism*), water or air currents (*rheotropism*), etc. Examples: fruiting bodies of bracket fungi exhibit geotropism, developing horizontally in response to gravity; sporangiophores of *Pilobolus* spp are positively phototropic; chemotropism is exhibited by certain fungi in response to PHEROMONES.

(2) The tendency of a parasite or pathogen to localize in a particular type of cell, tissue, or intracellular location. Thus, e.g., a virus may occur preferentially in lymphoid cells (*lymphotropism*), the skin (*dermotropism* or *epitheliotropism*), liver (*hepatotropism*), nerves (*neurotropism*) – and in the nucleus (*nucleotropism*) or cytoplasm (*cytotropism*) of the host cell. (cf. PANTROPIC.) A cell tropism may be due e.g. to the presence of specific cell-surface receptors on the target cell.

tropocollagen See COLLAGEN.

***trp* operon** (tryptophan operon) An OPERON (mapping at ca. 27.5 min on the chromosome of *Escherichia coli*) which contains five contiguous structural genes encoding enzymes involved in the biosynthesis of tryptophan [Appendix IV(f)]. The operon is subject to both negative promoter control and attenuator control, and has the structure: promoter(–operator)–leader(–attenuator)–*trpE*–*trpD*–*trpC*–*trpB*–*trpA*–*t*–*t′*. The *trpE*. . .*trpA* genes are structural genes encoding enzymes for the conversion of chorismate to tryptophan; *t* is a rho-independent transcription terminator, *t′* a rho-dependent terminator (see TRANSCRIPTION).

An unlinked gene, *trpR*, encodes an aporepressor which, when activated by tryptophan (the corepressor), binds to the operator (*trpO*) and reduces (but does not abolish) transcription of the *trp* genes. The OPERATOR contains a palindromic sequence and occurs entirely within the PROMOTER, occupying (roughly) positions −23 to −3. The activated *trpR* repressor also represses the *trpR* gene itself (i.e., *trpR* is autoregulated) and the *aroH* gene (which encodes a 3-deoxyarabinoheptulosonate 7-phosphate (DAHP) synthase: see Appendix IV(f)). (In *E. coli* there are three DAHP synthases; one (the *aroH* gene product) is subject to feedback regulation by tryptophan, one by phenylalanine, and one by tyrosine.) The three operators recognized by the *trpR* repressor are closely related, having a consensus sequence which presumably reflects points of interaction with the repressor. (See also DAM GENE.)

Further reduction in *trp* gene transcription is achieved by attenuation (see OPERON); in the presence of tryptophan the attenuator in the leader region of the mRNA adopts the stem-and-loop structure characteristic of a rho-independent terminator, and transcription stops before *trpE*. However, in the absence of tryptophan, a ribosome beginning translation of the leader peptide stalls when it reaches two contiguous tryptophan codons, preventing the formation of the terminator stem-and-loop structure and allowing transcription to proceed into the *trp* structural genes. (Termination at the attenuator also occurs in the absence of ribosome binding at the leader sequence.)

(cf. TNA OPERON.)

true slime moulds See MYXOMYCETES.

truffles The typically hypogean, edible fruiting bodies of certain fungi – e.g. *Tuber aestivum* (summer truffle), *T. melanosporum* (Périgord truffle) and *T. uncinatum* (Burgundy truffle). 'False truffles' include the fruiting bodies of species of *Elaphomyces* (hart's truffles) and those of the gasteromycetes *Hymenogaster*, *Melanogaster* and *Rhizopogon*.

***trxA* gene** See THIOREDOXIN.

trypan blue A blue acid diazo DYE used e.g. for VITAL STAINING.

trypan red A red acid diazo DYE formerly used for treating trypanosomiasis.

trypanocide Any agent which can kill trypanosomes – e.g. BERENIL, ETHIDIUM BROMIDE, PROPIDIUM, quinapyramine and SURAMIN. Pentamidine has been used for decades in the treatment of SLEEPING SICKNESS (q.v.); it is a substituted dibenzamidine which appears to inhibit the trypanosomal enzyme *S*-adenosyl-L-methionine decarboxylase. (See also *melarsoprol* in ARSENIC (a) and *eflornithine* in SLEEPING SICKNESS; see also ANTIMONY.)

Trypanosoma A genus of parasitic flagellate protozoa (family TRYPANOSOMATIDAE); certain species cause human and animal TRYPANOSOMIASIS – primarily in parts of Africa and in Central and South America.

Most species have a life cycle which involves alternate phases of growth in a vertebrate and an invertebrate host; typically, trypomastigotes occur in the vertebrate while other forms develop in the invertebrate host (see illustrations in the entries TRYPANOSOMATIDAE and TRYPANOZOON). Cell sizes range from small intracellular amastigotes, ca. 3–6 μm in diameter, to the largest trypanosomes which measure over 1 mm in length; however, the cells are commonly ca. 10–100 μm in length.

In the vertebrate host, the trypomastigote is bounded by a typical unit-type membrane beneath which lies a layer of (apparently) cytoskeletal MICROTUBULES; often there is an external, electron-dense coat with which is associated the variant surface glycoprotein (see VSG) involved in ANTIGENIC VARIATION. The nucleus is rounded, and there is an elongated mitochondrion which runs most of the length of the body; within the mitochondrion is the KINETOPLAST which is situated close to the flagellar basal body. The FLAGELLUM arises from an invagination (*flagellar pocket*, or *reservoir*). In most species (cf. *Nannomonas*) there is a 'free flagellum' (i.e. that part of the flagellum which does not form the outer limit of the UNDULATING MEMBRANE) projecting anteriorly – that is, in the direction of locomotion; in the undulating membrane, the flagellar and cell body membranes are attached by desmosome-like structures. (Within the invertebrate host the flagellum can serve as a means of attachment; at the region of attachment the interior of the flagellum develops an electron-dense structure, the whole attachment complex being termed a *hemidesmosome*.)

The organisms are chemoorganotrophs. Soluble nutrients may be absorbed through the cell surface, and pinocytosis can occur via the cytoplasmic membrane in the region of the reservoir. In (at least) some stercorarian trypanosomes (e.g. *T. brucei*) the energy-yielding metabolism varies according to the phase of the life cycle – changes in this context being associated with changes in the structure of the mitochondrion. Some species can be cultured in blood-based media (e.g. NNN MEDIUM); see also SCHIZOTRYPANUM.

Within the vertebrate host, reproduction occurs asexually by longitudinal binary fission in the trypomastigote, and e.g. by multiple fission in certain other stages; binary fission involves duplication of the basal body and kinetoplast, nuclear division, and finally separation of the two daughter cells. A sexual process may occur. (See also METACYCLIC FORMS.) [Trypanosomatid genomes (special issue): Science (2005) *309* 337–520.]

Trypanosoma spp parasitize a variety of animals, including birds, reptiles, amphibians and mammals; the invertebrate hosts include insects of the orders Diptera and Hemiptera, and are commonly involved (as VECTORS) in the cyclical transmission of *Trypanosoma* spp. Species parasitic in mammals are divided into the SALIVARIA and STERCORARIA.

For anti-trypanosomal agents see TRYPANOCIDE.

Trypanosoma spp include:

T. brucei: see TRYPANOZOON.

T. congolense: see NANNOMONAS.

T. cruzi: see SCHIZOTRYPANUM.

T. equinum: see TRYPANOZOON.

T. equiperdum: see TRYPANOZOON.
T. evansi: see TRYPANOZOON.
T. gambiense: see TRYPANOZOON.
T. lewisi: see HERPETOSOMA.
T. melophagium: see MEGATRYPANUM.
T. microti: see HERPETOSOMA.
T. nabiasi: see HERPETOSOMA.
T. rangeli: see HERPETOSOMA.
T. rhodesiense: see TRYPANOZOON.
T. platydactyli: see LEISHMANIA.
T. simiae: see NANNOMONAS.
T. suis: see PYCNOMONAS.
T. theileri: see MEGATRYPANUM.
T. theodori: see MEGATRYPANUM.
T. uniforme: see DUTTONELLA.
T. vivax: see DUTTONELLA.
T. zapi: see HERPETOSOMA.

trypanosomal form Obsolete *syn.* TRYPOMASTIGOTE.

trypanosomatid Any member of the family TRYPANOSOMATIDAE.

Trypanosomatidae A family of parasitic protozoa (suborder TRYPANOSOMATINA); constituent genera: BLASTOCRITHIDIA, CRITHIDIA, ENDOTRYPANUM, HERPETOMONAS, LEISHMANIA, LEPTOMONAS, PHYTOMONAS, RHYNCHOIDOMONAS and TRYPANOSOMA. (Monoxenous genera have been referred to as the 'Lower Trypanosomatidae'.) The diagram shows the range of types of cell which occur in the life cycles of species of the Trypanosomatidae.

Trypanosomatina A suborder of (typically) uniflagellate parasitic protozoa (order KINETOPLASTIDA) which comprises one family: TRYPANOSOMATIDAE.

trypanosome Any member of the genus *Trypanosoma*. (cf. TRYPANOSOMATID.)

trypanosome adhesion test A serological test used for detecting the presence of antibodies homologous to particular species of trypanosomes. The test depends on the adhesion of erythrocytes or platelets (of certain species) to trypanosomes in the presence of homologous antibodies and complement.

trypanosomiasis Any human or animal disease in which the causal agent is a species of TRYPANOSOMA; such diseases include CHAGAS' DISEASE, DOURINE, SURRA, and the 'African trypanosomiases' SLEEPING SICKNESS and NAGANA. [Human trypanosomiasis: BMB (1985) *41* 103–194; TDB (1986) *83* R1–R60.]

trypanothione See ARSENIC (a).

Trypanozoon A subgenus of TRYPANOSOMA within the SALIVARIA; it includes cyclically transmitted species (vector: *Glossina*) and mechanically transmitted species – the latter, though not truly salivarian, exhibiting certain affinities with the former.

The cyclically transmitted members traditionally include *Trypanosoma* (*Trypanozoon*) *brucei* (= *T. (T.) brucei*), *T.* (*T.*) *gambiense*, and *T.* (*T.*) *rhodesiense*. Some authors regard these three species as subspecies of a single species, *T.* (*T.*) *brucei*; *Trypanosoma* (*Trypanozoon*) *brucei brucei* (= *T. (T.) b. brucei*), though not infective for man, is a causal agent of NAGANA, while *T.* (*T.*) *b. gambiense* and *T.* (*T.*)

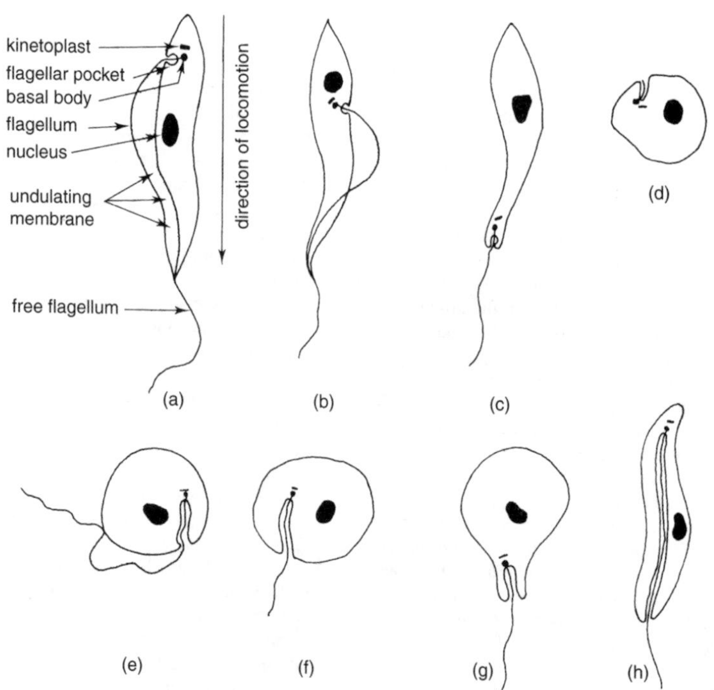

TRYPANOSOMATIDAE. Some forms which occur in the life cycles of species of the Trypanosomatidae. (a) Trypomastigote. The undulating membrane extends for most of the length of the body with the flagellum forming its outer margin. (b) Epimastigote. (c) Promastigote. (d) Amastigote; amastigotes of the type shown (i.e., with a reduced, non-free flagellum) occur in *Leishmania* spp. (e) Sphaeromastigote. (f) Paramastigote. (g) Choanomastigote. (h) Opisthomastigote.

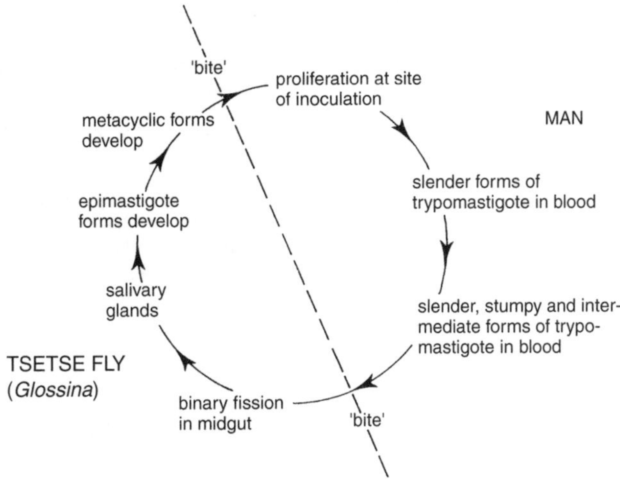

TRYPANOZOON. The life cycle of *Trypanosoma* (*Trypanozoon*) brucei gambiense.

b. rhodesiense are both causal agents of SLEEPING SICKNESS. In the vertebrate host these organisms are 12–42 μm in length; they may occur as *slender* trypomastigotes (commonly ca. 25 μm in length), as short, *stumpy* trypomastigotes (which typically lack a *free* flagellum), and/or as some intermediate form. (See figure in the entry TRYPANOSOMATIDAE.)

Mechanically transmitted species include *T.* (*T.*) *evansi*, the causal agent of SURRA, and dyskinetoplastic trypanosomes formerly called '*T. equinum*' but now regarded as strains of *T. evansi*; bloodstream forms are trypomastigotes, 15–36 μm in length. *T. equiperdum*, morphologically indistinguishable from *T. evansi*, causes DOURINE.

tryparsamide See ARSENIC.

trypomastigote A form assumed by the cells of species of *Endotrypanum*, *Rhynchoidomonas* and *Trypanosoma* during at least certain stages of their life cycles: see TRYPANOSOMATIDAE.

trypsin See PROTEASES.

trypsinization (in tissue culture) See TISSUE CULTURE.

trypticase *Syn.* TRYPTONE.

trypticase–soy agar (tryptone soya agar; tryptic soy agar; soybean casein digest agar; etc) An agar-based medium containing a pancreatic digest of casein (trypticase, tryptone), a papain digest of soybean meal (phytone), and NaCl (0.5% w/v). It is used e.g. as a general-purpose medium, as a BLOOD AGAR base, etc.

trypticase–soy broth (TSB) A general-purpose broth medium similar in composition to TRYPTICASE–SOY AGAR (without the agar), sometimes with the addition of D-glucose.

tryptone A tryptophan-rich PEPTONE: e.g. a pancreatic digest of casein.

tryptone soya agar *Syn.* TRYPTICASE–SOY AGAR.

tryptone water A MEDIUM containing 1–2% TRYPTONE and 0.5% NaCl in water.

L-tryptophan biosynthesis See AROMATIC AMINO ACID BIOSYNTHESIS, Appendix IV(f), and TRP OPERON.

tryptophanase L-Tryptophan indole-lyase (deaminating); EC 4.1.99.1. See e.g. Appendix IV(f), INDOLE TEST, and TNA OPERON.

***ts* mutant** TEMPERATURE-SENSITIVE MUTANT.

TSA TRYPTICASE–SOY AGAR.

TSB TRYPTICASE–SOY BROTH.

tSDA See SDA.

TSEs TRANSMISSIBLE SPONGIFORM ENCEPHALOPATHIES.

tsetse fly See GLOSSINA and SODALIS.

TSI agar Triple-sugar–iron agar: a medium used e.g. in a SCREENING TEST (sense 2) for members of the Enterobacteriaceae. The medium is prepared as a slope with a deep BUTT; it contains peptone (2%), glucose (0.1%), lactose (1%), sucrose (1%), ferrous ammonium sulphate (0.02%), NaCl (0.5%), sodium thiosulphate (0.02–0.03%), PHENOL RED (0.0025%), and agar (ca. 1.5%). The strain under test is inoculated onto the surface of the slope and is also stab-inoculated (with a straight wire) deep into the butt. Following incubation the slope is examined for (a) acid/alkaline reaction at the (aerobic) surface of the slope; (b) acid/alkaline reaction in the (anaerobic or microaerobic) butt; (c) H_2, CO_2 production – detectable as gas pockets, or splitting of the butt; (d) H_2S production – detectable as blackening of the medium (due to ferrous sulphide formation). Thus, e.g. *Escherichia coli* (which typically attacks both glucose and lactose) commonly produces an acid slope and butt, and forms small amounts of H_2 and CO_2, though (usually) no H_2S. Non-lactose-fermenting strains of enterobacteria (e.g. most strains of *Salmonella*) form an acid butt and an alkaline slope – though some organisms (e.g. most strains of *Salmonella*) typically form large amounts of H_2S which may obscure these reactions by blackening most or all of the medium. *S. typhi* typically forms an acid butt, an alkaline slope, and very small amounts of H_2S. (cf. KLIGLER'S IRON AGAR.)

***tsr* gene** See CHEMOTAXIS (a).

TSS TOXIC SHOCK SYNDROME.

TSST-1 See TOXIC SHOCK SYNDROME and SUPERANTIGEN.

tsutsugamushi disease *Syn.* SCRUB TYPHUS.

Tsx protein In *Escherichia coli*: and OUTER MEMBRANE protein encoded by the *tsx* gene; it acts as a receptor for bacteriophage T6 and for COLICINS K, 5 and 10, and may be a channel-forming protein involved in the uptake of e.g. glycine, phenylalanine, serine and possibly nucleosides (cf. PORIN). Synthesis of the Tsx protein is subject to CATABOLITE REPRESSION.

TT virus A non-enveloped ssDNA virus, 30–50 nm diam., which has been associated with post-transfusion hepatitis of unknown aetiology. The sequence of the viral genome appears to be unlike that of other viruses; it has been suggested that the TT virus be classified in a new family (Circinoviridae) [PNAS (1999) *96* 3177–3182]. A number of distinct genotypes of TT virus have been reported [JMV (1999) *57* 252–258].

TTC See TETRAZOLIUM.

TTFA 2-Thenoyltrifluoroacetone (4,4,4-trifluoro-1-[2-thienyl]-1,3-butanedione): an inhibitor of reconstituted mitochondrial electron transport systems (see ELECTRON TRANSPORT CHAIN).

TTGE See DGGE.

TU (1) Tuberculin unit: 1 TU = 0.02 μg of PPD (see TUBERCULIN). (2) See TODD UNIT.

tube dilution test See DILUTION TEST.

Tuber A genus of fungi (order PEZIZALES) which form closed, hypogean ascocarps containing clavate to globose asci. (See also ASCOSPORE, ASCUS, MYCORRHIZA and TRUFFLES.)

Tuberales An order of fungi (recognized by some authors) which include certain hypogean ascomycetes, e.g. *Tuber* spp.

tubercle (1) (*med.*) See TUBERCULOSIS.

(2) A deposit formed on the inner surfaces of iron or steel water-pipes; tubercles may contain e.g. oxides and hydroxides of iron, as well as sulphate- and nitrate-reducing bacteria [AEM (1982) *44* 761–764] and e.g. IRON BACTERIA such as *Crenothrix*, *Gallionella* and *Leptothrix* spp.

(3) (*lichenol.*) In the thallus of certain lichens (e.g. *Nephroma resupinatum*): a wart-like protuberance in which there may be a discontinuity in the cortex; it may be analogous in function to e.g. a CYPHELLA.

tubercle bacillus (1) *Mycobacterium tuberculosis*. (2) Any *Mycobacterium* sp which causes TUBERCULOSIS.

tubercular (1) Resembling, due (directly or indirectly) to, or pertaining to TUBERCULOSIS. (2) Consisting of or resembling a TUBERCLE (any sense). (cf. TUBERCULOUS.)

Tuberculariales See HYPHOMYCETES.

tuberculin Any of various preparations, containing tuberculoprotein, obtained by filtration of a culture of tubercle bacilli. *Old tuberculin* (OT) is a filtrate of a steamed, evaporated, glycerol-broth culture. *Purified protein derivative* (PPD) is prepared from the filtrate of a steamed culture grown in a synthetic medium; the filtrate is treated with e.g. trichloroacetic acid to precipitate tuberculoprotein which is then washed and prepared in standard aqueous dilutions. ('Tuberculin' may refer specifically to PPD.) *Mammalian* tuberculin (used in the COMPARATIVE SINGLE INTRADERMAL TUBERCULIN TEST) is PPD prepared from bovine tubercle bacilli; *avian* tuberculin is PPD prepared from the avian tubercle bacillus.

tuberculin test A SKIN TEST in which, typically, PPD (see TUBERCULIN) is injected intradermally; a positive test, indicated by induration and erythema (i.e. a hard, red bump at the site of injection), is given in ~24–48 hours and reflects DELAYED HYPERSENSITIVITY to PPD and/or to a cross-reactive entity. A positive test may indicate prior immunological contact with one or more of the following: (i) tubercle bacilli, (ii) BCG (through vaccination), (iii) certain other mycobacteria. Hence, a positive test is not proof of active infection by tubercle bacilli: sensitivity to PPD is induced not only by *Mycobacterium tuberculosis* and BCG but also by various non-tuberculous mycobacteria – for example, positive tests have been recorded in some cases of cervical adenitis caused by members of the *M. avium* complex (see MAC). Nevertheless, reactivity to 'environmental' mycobacteria may be weaker than that induced by tubercle bacilli, and the tuberculin test is still regarded as a useful diagnostic tool.

False-negative tests may be given by patients with overwhelming (e.g. miliary) tuberculosis or those with certain malignant conditions.

A negative test in an immunocompetent, healthy individual indicates a 'susceptible' who may benefit from vaccination with BCG.

Among the various forms of tuberculin test are the *Mantoux test*, in which PPD (commonly 5 TU) is injected intradermally; the HEAF TEST; the *Moro test*, in which tuberculin is smeared over a small area of skin; and the VOLLMER PATCH TEST. (cf. COMPARATIVE SINGLE INTRADERMAL TUBERCULIN TEST.)

(See also ESAT-6.)

tuberculin-type reaction A DELAYED HYPERSENSITIVITY reaction similar to that seen in a positive TUBERCULIN TEST but involving a different antigen – see e.g. HISTOPLASMIN TEST.

tuberculin unit See TU.

tuberculoprotein Protein derived from tubercle bacilli – see TUBERCULIN.

tuberculosis (phthisis; TB) A chronic infectious disease of man and animals caused by certain species of MYCOBACTERIUM – commonly *Mycobacterium tuberculosis*, sometimes *M. bovis*; *M. africanum* causes tuberculosis primarily in Africa. Tuberculosis is responsible for several million human deaths each year, and the incidence is rising; the re-emergence of TB in the USA was associated with (i) changes in social structure, (ii) the spread of AIDS (creating many immunocompromised, highly susceptible, individuals), and (iii) failure of public health programmes [review: Science (1992) *257* 1055–1064]. The problem of TB has been exacerbated by the appearance of multiply drug-resistant strains of the main causal agents; for example, one strain of *M. bovis*, resistant to 11 drugs, was reported to cause high mortality rates in HIV-positive patients [Lancet (1997) *350* 1738–1742]. [Multiple drug resistance (review): BCID (1997) *4* 77–96.]

Tuberculosis in the human population.

New infection arises commonly through inhalation of aerosols containing viable cells of the pathogen. In most immunocompetent people, the inhaled pathogen may remain viable but is constrained to quiescence by the body's immune system; such *latent* infection carries a life-time risk of active disease, this risk being increased e.g. by the use of immunosuppressive drugs or by infection with human immunodeficiency virus (HIV). The mechanism of latency is unknown. It is not known, for example, whether the pathogen survives in a non-acid-fast form or in a spore-like state, and the site(s) of (latent) infection within the body are also unknown; apparently, individuals with latent infection generally do not transmit the disease. [Latency in tuberculosis: TIM (1998) *6* 107–112.]

Epidemiological investigation (contact tracing) may be facilitated by the use of assays for T lymphocytes (T cells) that react to the tuberculosis-associated antigen ESAT-6 [Lancet (2001) *357* 2017–2021]. (See also ELISPOT ASSAY.)

In a proportion of cases infection leads to active disease; this may reflect a genetically based predisposition [BCID (1997) *4* 207–229]. Within the lungs, some bacilli may be taken up by phagocytosis and killed by activated MACROPHAGES. However, some bacilli *invade* macrophages and survive within the phagosome. In such cases the phagosome fails to acidify and it also fails to fuse with a lysosome; moreover, the phagosomal membrane apparently becomes more permeable – allowing the pathogen access to the host cell's nutrients and, hence, permitting growth and reproduction [PNAS (1999) *96* 15190–15195].

The ability of the pathogen to invade – and to survive and grow within – a macrophage may depend on factors such as (i) the cell-surface chemistry of the particular strain of pathogen; (ii) the site of infection (e.g. lung); and (iii) the timing of pathogen–macrophage contact (at initial infection, or subsequently). Opsonization of the pathogen via the COMPLEMENT system may involve one or more pathways of fixation [TIM (1998) *6* 47–49, 49–50]. However, for the *initial* infection of lung tissue it has been suggested that *non*-opsonic binding of the pathogen to a macrophage may be important; thus, *M. tuberculosis* may invade macrophages when certain polysaccharides on the pathogen's surface bind to the *lectin*-binding site of macrophage receptor CR3 [*M. tuberculosis*–macrophage interactions: TIM (1998) *6* 328–335].

[Pathogenesis of *Mycobacterium tuberculosis* (review): Cell (2001) *104* 477–485.]

In *pulmonary* tuberculosis, lesion(s) develop in the lung tissue (although tubercle bacilli may be disseminated, via lymph and blood, to various parts of the body). In the lungs, each focus of infection develops to form a granulomatous lesion called a *tubercle*, such development reflecting the activity of the body's CELL-MEDIATED IMMUNITY; one important factor in this process is the involvement of antigen-specific T LYMPHOCYTES. A mature tubercle consists of a necrotic centre, containing dead macrophages and dead/viable tubercle bacilli, surrounded by a layer of epithelioid macrophages and an outer mantle of monocytes/macrophages and lymphocytes; GIANT CELLS are commonly present. The necrotic centre of a tubercle may subsequently calcify and remain quiescent, although it may still contain viable tubercle bacilli.

Necrotic degeneration within tubercles, seen in human (and guinea-pig) lung tissue, is uncommon in mice. In mice, the lesions are penetrated by lymphocytes – which may thus continue CYTOKINE-mediated stimulation of the macrophages. By contrast, in human and guinea-pig lesions the lymphocytes remain peripherally distributed – so that macrophages in central parts of a lesion may remain unstimulated (and more likely to be killed by internalized tubercle bacilli); it has been suggested that this aspect of development may account for the large central region of necrosis observed in human and guinea-pig tubercles [TIM (1998) *6* 94–97].

In pulmonary TB ('consumption') there is a persistent cough, initially unproductive but later productive of mucopurulent sputum (containing tubercle bacilli); in later stages there is typically dyspnoea and haemoptysis. In general, symptoms of TB include weakness, fever, night sweats and weight loss. (When infection spreads beyond the lungs, symptoms of disease depend on the tissues affected and may include e.g. lymphadenitis, pyuria/haematuria, MENINGITIS, OSTEOMYELITIS, pericarditis, peritonitis etc.; the pleura and skin are other sites which may be affected.) In some cases tubercles may rupture into a blood vessel, leading to massive dissemination of bacilli and the formation of many synchronously progressive tubercles in sites throughout the body (*miliary tuberculosis*).

Traditionally, tuberculosis has been classified into *primary* disease (which follows the initial infection) and *post-primary* disease (which may follow months, years or decades after the primary disease has resolved) – post-primary disease resulting from *re-activation* of the latent pathogen. These two categories of disease are sometimes distinguishable by features such as level of lymphatic involvement, cavity formation in lung tissue, and site affected (usually the upper lobes in post-primary disease); however, the distinction between primary and post-primary disease is not always clear-cut.

Infection, re-infection and re-activation are beginning to be studied with molecular tools that throw some light on the various manifestations of tuberculosis; these studies have indicated, for example, that more cases than previously thought are due to recent transmission (rather than re-activation) and that re-infection (following earlier active disease) is not a rare

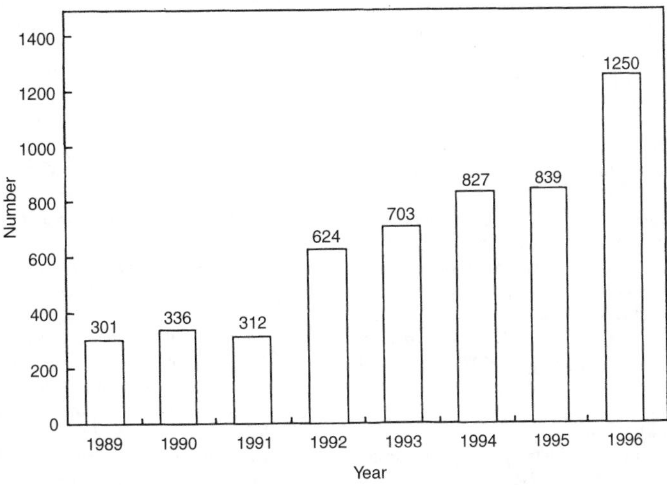

TUBERCULOSIS – the number of cases of tuberculosis diagnosed between 1989 and 1996 in Hlabisa hospital, a district hospital in a rural location in South Africa. It was estimated that, in 1995, at least 44% of the cases were linked to infection with the human immunodeficiency virus (HIV).

Reproduced from *Baillière's Clinical Infectious Diseases* (1997) *4* 97–105 (Figure 1, page 101) by courtesy of the author, Dr David Wilkinson, and with permission from Harcourt Publishers Ltd.

occurrence, especially among HIV-positive individuals [BCID (1997) *4* 173–183].

Tuberculosis in the immunocompromised. In AIDS, the fall in immunocompetence (particularly the lowered efficiency of cell-mediated immunity) is associated with susceptibility to various diseases, of which tuberculosis is one; moreover, in AIDS patients (particularly adults), clinical presentation of tuberculosis may be atypical and may include a higher proportion of cases of disseminated or extrapulmonary disease. It appears that infection with *M. tuberculosis* can promote HIV-mediated disease [J. Imm. (1996) *157* 1271–1278], perhaps by stimulation of viral replication (see HIV).

For AIDS patients with a positive TUBERCULIN TEST, the drug ISONIAZID may be useful as a prophylactic measure against tuberculosis. [TB in the AIDS era (the role for preventive therapy): BCID (1997) *4* 63–76.]

Lab. diagnosis of tuberculosis may include e.g. identification of the pathogen by acid-fast staining and culture from sputum, CSF, urine sediments, biopsy specimens etc. (See also BACTEC.) Sputum may be treated with a MUCOLYTIC AGENT prior to examination.

Molecular methods of diagnosis. Smear-positive respiratory specimens may be examined – for confirmation of the diagnosis – by methods that involve in vitro amplification of specific target sequences in the pathogen's nucleic acids. These methods include the PCR-based Amplicor™ system (Roche) which amplifies a DNA target in the pathogen's 16S rRNA gene, and AMTDT (Gen-Probe) (see TMA) which amplifies an rRNA target; both of these methods have been approved by the US Food and Drug Administration (FDA) for smear-*positive* respiratory specimens. (A further method, based on the ligase chain reaction (LCR), is being evaluated [JCM (1999) *37* 229–232].) [Comparison of Amplicor with AMTDT for direct detection of *M. tuberculosis* in respiratory specimens: JCM (1997) *35* 193–196.]

The molecular methods of diagnosis are useful, although some problems have arisen. For example, false-positive results in AMTDT were obtained when lung infections involved *M. avium* and/or *M. kansasii* – suggesting the need for an alteration in the interpretation of luminometer readings in the test [JCM (1999) *37* 175–178]. Moreover, the original form of AMTDT suffered from low sensitivity; this problem was addressed in a later version by increasing the volume of sample used in the test. [Comparison of the new and original AMTDT: JCM (1998) *36* 684–689. Clinical evaluation of the new-format AMTDT: JCM (1999) *37* 1419–1425.]

AMTDT has been evaluated for the direct detection of *M. tuberculosis* in *non*-respiratory specimens [JCM (1997) *35* 307–310]. [Evaluation of new and original forms of AMTDT for the detection of *M. tuberculosis* in respiratory and non-respiratory specimens: JCM (1998) *36* 684–689.]

The simultaneous detection and TYPING of *M. tuberculosis* in clinical specimens may be achieved by SPOLIGOTYPING [JCM (1997) *35* 907–914].

Skin test. See TUBERCULIN TEST.

Antibiotic resistance testing of isolates. Given the slow rate of growth of *M. tuberculosis* (and the inapplicability of routine tests) special methods have been devised. Isolates can be tested in liquid media using the BACTEC system. Alternatively, there are several molecular ('genotypic') methods; all of these are used to detect specific, resistance-conferring mutation(s) by initially amplifying the relevant sequence of genomic DNA and then testing the amplicons for the presence/absence of the given mutation(s).

Resistance to isoniazid, or streptomycin, can arise through mutation in more than one gene. However, resistance to rifampicin (= rifampin) is commonly due to one (or more) of several mutations within a single gene, *rpoB*, which encodes the β subunit of RNA polymerase; technically, therefore, resistance to rifampicin is easier to detect (by molecular methods), and most studies have concentrated on detecting mutations in *rpoB*. (In some regions, including the USA [JAMA (1994) *271* 665–671], resistance to rifampicin commonly correlates with resistance to at least some other antituberculosis drugs, so that rifampicin resistance may be used as a marker for multidrug resistance.)

In *dideoxy fingerprinting*, amplicons of the target (*rpoB*) sequence are examined by a process resembling dideoxy sequencing (see DNA SEQUENCING) – mutation(s) in the target sequence being detected by variations in the nature of the chain-terminated products (compared with the products from wild-type amplicons) [JCM (1995) *33* 1617–1623]. Another genotypic method is *line probe assay* (see PROBE), and a further method is SSCP analysis (see SSCP). (See also ANTIBIOTIC.)

Chemotherapy. A *combination* of drugs is used (i) for maximum effect and (ii) in order to discourage the emergence of drug-resistant strains. Major antituberculosis drugs include ETHAMBUTOL, ISONIAZID, PYRAZINAMIDE, rifampicin (see RIFAMYCINS) and STREPTOMYCIN. Given appropriate treatment, a patient whose disease is not caused by drug-resistant strains has an excellent prospect of cure.

In the mid-1990s the World Health Organization (WHO) initiated a comprehensive antituberculosis strategy known as *directly observed therapy, short-course* (DOTS). One aspect of DOTS involves direct supervision of patients during ingestion of medication – thus ensuring compliance (lack of which is a common cause of failed therapy). DOTS also promotes active searching for new cases among suspects, and requires short-course chemotherapy for all cases of smear-positive pulmonary TB.

The administration of antituberculosis drugs two or three times a week is reported to be as effective as daily treatment; this approach, combined with the DOTS strategy, is seen as a sensible therapeutic option in the treatment of tuberculosis [BCID (1997) *4* 97–105, 107–109].

Vaccines. The BCG vaccine has been used for many years, although its efficacy against pulmonary TB in adults is variable, and it has also failed to prevent primary TB in some children.

More recently, DNA VACCINES (q.v.) have been used prophylactically and therapeutically against tuberculosis in mice, and this approach, combined with chemotherapy, may be applicable in human tuberculosis.

Other mycobacteria causing pulmonary, extrapulmonary or disseminated disease. Such species include members of the *M. avium* complex (see MAC), *M. kansasii* and *M. malmoense* (both of the latter being able to cause disease in damaged lungs and/or in immunocompromised patients). There is also a (growing) number of infrequently isolated mycobacteria among which are species whose pathogenic potential is still undetermined [RMM (1997) *8* 125–135].

Tuberculosis in animals.

Many warm- and cold-blooded animals are susceptible to infection by one or more species of *Mycobacterium*. (See also SKIN TUBERCULOSIS.) *M. bovis* is the most important species among domestic animals, causing tuberculosis in cattle and pigs – less commonly in sheep, horses, dogs, cats etc.

Certain wild animals (e.g. the badger, *Meles meles*, in Great Britain [PTRSLB (1985) *310* 327–381]) may provide a reservoir for bovine tuberculosis; however, the role of the badger in this context is still controversial, and a cull of badgers in selected areas is now being conducted in the United Kingdom in an attempt to resolve this question. As of August, 2001, the question remains unresolved, and the recent outbreak of foot and mouth disease is one of the factors likely to extend the time to resolution [VR (2001) *149* 162–163].

Where tuberculosis has been detected in cattle, a policy of extensive slaughter of infected animals has been successful in reducing the incidence of this disease in developed countries. (See also COMPARATIVE SINGLE INTRADERMAL TUBERCULIN TEST.)

tuberculous (1) *Syn.* TUBERCULAR. (2) Infected with tubercle bacilli.

Tubifera See MYXOMYCETES.

tubular loop fermenter See AIRLIFT FERMENTER.

tubulin The acidic protein of which MICROTUBULES (q.v.) are composed; closely related (but not identical) tubulins have been isolated from a wide range of eukaryotic cells. The α and β subunits of the (dimeric) protein each have a MWt of ca. 50000. Tubulin has binding sites for guanine nucleotides and Ca^{2+} (and e.g. for COLCHICINE). Ca^{2+} can bind at one high-affinity site and at several low-affinity sites; binding to the high-affinity site is inhibited by Mg^{2+}.

Under in vitro (and experimental in vivo) conditions, tubulin can assemble into microtubules and into structures such as rings, ribbons and flat sheets; sheet formation is promoted e.g. by high concentrations of glycerol.

[Molecular biology and genetics of tubulin: ARB (1985) *54* 331–365.]

***tufA, tufB* genes** In *Escherichia coli*: unlinked genes which encode the EF-Tu component of elongation factor T (see PROTEIN SYNTHESIS); the genes are nearly identical, the products differing in only the C-terminal amino acid (glycine in EF-TuA, serine in EF-TuB).

tularaemia (tularemia) A disease of rodents, rabbits and other animals, also transmissible to man (O'Hara's disease), caused by *Francisella tularensis*; it occurs in Europe and N. America. The pathogen varies in virulence apparently according to the composition of its capsule (and hence its ability to resist phagocytosis). In man, infection occurs via wounds or mucous membranes (e.g. by direct contact with an infected animal or carcass) or via the bite of any of various arthropod vectors (e.g., biting flies, ticks). Incubation period: 2–7 days. Onset is sudden, with chills and fever, headache, nausea and vomiting, and prostration. In the *ulceroglandular* form an ulcerative lesion develops at the site of infection, and there is regional lymphadenopathy. The *glandular* form closely resembles bubonic PLAGUE. Less commonly, there may be septicaemia or (following infection by ingestion) severe gastrointestinal symptoms – either of which may be fatal. Untreated infections may persist for months. Complications may result from the dissemination of *F. tularensis* to the heart, lungs or meninges, or from secondary infection of lesions by other bacteria. *Lab. diagnosis*: e.g. microscopic examination, culture, and/or inoculation into guinea pigs of material from lesions, lymph node aspirates etc. *Chemotherapy*: streptomycin, or e.g. tetracyclines or chloramphenicol.

F. tularensis can cause a fatal (usually tick-borne) septicaemia in sheep, pigs, and – to a lesser degree – calves. In horses, tularaemia involves e.g. fever and oedema of the limbs; in foals, dyspnoea and incoordination may also occur.

Tulare apple mosaic virus See ILARVIRUSES.

Tulasnella See TULASNELLALES.

Tulasnellaceae See TULASNELLALES.

Tulasnellales An order of fungi (subclass HOLOBASIDIOMYCETIDAE) that form thin, resupinate, often inconspicuous basidiocarps in which the basidia each have four relatively large sterigmata that are swollen and sometimes deciduous (Tulasnellaceae) or neither swollen nor deciduous (Ceratobasidiaceae). The organisms include saprotrophs and plant parasites, and some species form mycorrhizal associations with orchids (see MYCORRHIZA). Genera include *Ceratobasidium*, *Oncobasidium* (see also VASCULAR-STREAK DIEBACK), *Thanatephorus* and *Tulasnella*.

tulip Augusta disease See TOBACCO NECROSIS VIRUS.

tulip breaking virus See POTYVIRUSES and COLOUR-BREAKING.

Tulostomatales See GASTEROMYCETES.

tumbling See FLAGELLAR MOTILITY (a) and RANDOM WALK.

tumor See TUMOUR.

tumorigenic Causing or giving rise to tumours.

tumour (tumor) (1) (*med.*) See NEOPLASIA and CANCER. (2) (*plant pathol.*) See GALLS.

tumour necrosis factor See TNF.

tunica Any of various coat-like or enveloping structures – e.g. a LORICA.

tunicamycin An ANTIBIOTIC, obtained from *Streptomyces lysosuperificus*, which is structurally complex – consisting of uracil, *N*-acetylglucosamine (GlcNAc), and a complex amino sugar (tunicamine) linked at its amino group to an unsaturated fatty acid of variable length. Tunicamycin inhibits the synthesis of GlcNAc-containing macromolecules by preventing the transfer of GlcNAc 1-phosphate from its nucleotide derivative to BACTOPRENOL phosphate or DOLICHOL phosphate; it thus inhibits e.g. the synthesis of TEICHOIC ACIDS in bacteria, and the glycosylation step in the synthesis of GlcNAc-containing glycoproteins in eukaryotes (see PALADE PATHWAY) – and hence maturation of certain enveloped viruses.

tuning-fork basidium See DACRYMYCETALES.

turanose See MELEZITOSE.

turbidimetry (turbidometry) Any procedure in which the concentration of cells (or particles) in a suspension is estimated by passing a beam of light through the suspension and comparing the intensity of the *transmitted* light with that obtained with a different suspension or a cell-free control; light transmitted by a suspension is of lower intensity than that of the control due mainly to light scattering – although pigmented cells may absorb (as well as scatter) a significant amount of light of particular wavelength(s). A measure of the decrease in transmitted light is the *optical density* (OD) – given by $\log_{10}(I_0/I)$ where I_0 is the intensity of the incident light, and I the intensity of the transmitted light; for a given type of cell the relationship between OD and cell concentration is approximately linear over a certain range of values. Turbidimeters (e.g. the Klett–Summerson colorimeter) consist essentially of a colourless, transparent sample holder (*cuvette*), a means of passing a narrow beam of monochromatic light through the cuvette, and a photoelectric cell to measure the intensity of the transmitted light. (See also COUNTING METHODS.)

turbidometry *Syn.* TURBIDIMETRY.

turbidostat See CONTINUOUS CULTURE.

Turbinaria See PHAEOPHYTA.

turimycin See MACROLIDE ANTIBIOTICS.

turkey bluecomb disease virus See CORONAVIRIDAE.

turkey diseases See POULTRY DISEASES.

turkey herpesvirus See GAMMAHERPESVIRINAE and MAREK'S DISEASE.

turkey X disease An outbreak of AFLATOXICOSIS which occurred among turkey poults in England during the 1960s; the source of the aflatoxins was found to be contaminated groundnut feed.

Symptoms: depression, anorexia, neurological symptoms, weakness and death; necrotic liver lesions were found postmortem. (See also POULTRY DISEASES.)

turkeypox virus See AVIPOXIVIRUS.

turnip crinkle virus See TOMBUSVIRUSES.

turnip mosaic virus See POTYVIRUSES.

turnip rosette virus See SOBEMOVIRUSES.

turnip yellow mosaic virus See TYMOVIRUSES.

turnip yellows virus See LUTEOVIRUSES.

***tut* gene** See OMP.

Tw Twist: see DNA.

TWAR strains Strains of bacteria isolated from patients with suspected trachoma and acute respiratory disease, and initially thought to be strains of the species *Chlamydia psittaci*; such strains were subsequently classified in a new species, *Chlamydia pneumoniae* (see CHLAMYDIA).

Tween Any of a range of non-ionic surfactants which are polyoxyalkylene derivatives of fatty acid esters of sorbitan: e.g., polyoxyethylene-sorbitan monolaurate (Tween 20), monopalmitate (Tween 40), monostearate (Tween 60), and mono-oleate (Tween 80).

Tween hydrolysis A TWEEN may be hydrolysed (with the liberation of free fatty acids) by certain LIPASES. A test of the ability of an organism to hydrolyse a Tween is used in the identification of certain bacteria. Essentially, the strain under test is grown in a liquid medium (pH 7) containing (usually) Tween 80 and NEUTRAL RED; the hydrolysis of Tween (a *positive* result) is indicated by a red coloration (due to the release of free fatty acids). For *Mycobacterium* spp, the test is read after e.g. 5 days for rapidly-growing strains, and after e.g. 10 days for slow-growing strains.

twist (of supercoiled DNA) See DNA.

twitching motility A jerky form of translocation exhibited by polarly (but not peritrichously) fimbriate bacteria in thin films of surface liquid – e.g. on freshly prepared plates incubated under high humidity. Formerly regarded as a passive form of locomotion, twitching motility (in *Neisseria gonorrhoeae*) is reported to involve retraction of type IV FIMBRIAE, at approx. 1 μm/sec, and to require protein synthesis and the PilT protein [Nature (2000) *407* 98–102]. (cf. MOTILITY.) Twitching motility occurs in both flagellate and non-flagellate bacteria of various genera; for example, it has been seen in *Acinetobacter calcoaceticus*, *Moraxella* spp and certain species of *Pseudomonas*, including *P. aeruginosa*.

two-component regulatory system In bacteria: a system that regulates gene expression in response to environmental signals (e.g. changes in extracellular osmolarity). The signal acts via a *signal transduction pathway* in the cell. The first component of the pathway is a HISTIDINE KINASE (located in the cell envelope); an appropriate signal causes the kinase to undergo ATP-dependent autophosphorylation and to transfer phosphate to an aspartate-containing site in the second component, a 'regulator protein' (= 'response regulator'). When the latter protein is phosphorylated, it regulates certain gene(s), e.g. by controlling their transcription, thus eliciting a response to the signal. [Histidine kinases in signal transduction: TIG (1994) *10* 133–138.] (See also *phosphorelay* in the entry ENDOSPORE.)

A signal may regulate the kinase *directly*, the kinase then being the *sensor* – see e.g. EnvZ in OSMOREGULATION. A further example of a kinase sensor occurs in the *E. coli* Kdp (high-affinity) transport system for K^+ ions; on osmotic up-shift (in the medium), the sensor KdpD apparently responds to decreased turgor pressure and transfers phosphate to KdpE, a cytosolic regulator protein. KdpE, in turn, promotes transcription of the Kdp operon; the product of gene *kdpB* is an ATPase at which ATP hydrolysis provides energy for the uptake of K^+ ions.

Alternatively, the sensor may be distinct from the kinase: signals act initially on the sensor which then regulates the kinase; thus, in CHEMOTAXIS in *Escherichia coli*, signals from the MCP sensor regulate the CheA kinase.

Two-component systems regulate a wide range of activities in bacteria. For example, in *Clostridium perfringens* a two-component system is involved in regulating transcription from toxin-encoding genes in GAS GANGRENE; the *virR/virS* operon encodes VirR (regulator) and VirS (histidine kinase), but the nature of the regulatory signal(s) is unknown [RMM (1997) *8* (suppl 1) S25–S27].

In *Salmonella typhimurium*, the SsrA–SsrB system regulates transcription of genes in the SPI-2 pathogenicity island; the products of these genes are needed for the pathogen's intracellular growth [Mol. Microbiol. (1998) *30* 175–188].

In *Bacillus subtilis*, growth to high-density populations promotes competence in TRANSFORMATION; this results from the accumulation of an extracellular peptide pheromone, ComX, which activates a two-component regulatory system. ComX causes autophosphorylation of the membrane-bound histidine kinase ComP; ComP, in turn, phosphorylates (activates) a transcription factor, ComA, that regulates the *comS* gene (whose expression is needed for competence) [TIM (1998) *6* 288–294].

In *Staphylococcus aureus*, a two-component regulatory system designated ArlS–ArlR is involved in adhesion, autolysis and extracellular proteolytic activity [JB (2000) *182* 3955–3964].

Another two-component system in *S. aureus* is reported to be involved in regulating the expression of certain virulence factors, including exotoxin, in response to environmental levels of oxygen [JB (2001) *183* 1113–1123]. (See also AGR LOCUS.)

Certain two-component systems appear to be *essential* for viability; these systems are involved e.g. in regulating gene expression in cell cycle control. Examples of essential two-component systems: the *yycF–yycG* system in *Bacillus subtilis* [Microbiology (2000) *146* 1573–1583] and the *mtrA–mtrB* system in *Mycobacterium tuberculosis* [JB (2000) *182* 3832–3838].

Studies on the *pho* regulon in *Escherichia coli* have indicated that autoamplification of the (PhoR/PhoB) two-component regulatory system leads to a faster response in cells previously exposed to phosphate limitation (i.e. the cells exhibit 'learning' behaviour) [JB (2001) *183* 4914–4917].

(See also *three-component* regulatory system in entry BACTERIOCIN.)

two-dimensional electrophoresis Any procedure which involves two stages (= dimensions) of ELECTROPHORESIS, the stages differing e.g. in the physical and/or chemical properties on which the electrophoretic separation of components is based. An example of such a procedure is the analysis of serum proteins by an initial stage of ISOELECTRIC FOCUSING and a second stage of IMMUNOELECTROPHORESIS at right-angles to the axis of focusing; *serological* procedures of this type are called *two-dimensional immunoelectrophoresis* or *crossed immunoelectrophoresis*.

two-micron DNA plasmid (2μ plasmid; 2μ circle; 2 μm circle) A 6318-bp ccc dsDNA PLASMID present in most strains of *Saccharomyces cerevisiae*; ca. 60–100 copies of the plasmid occur in the nucleus of a diploid cell. The plasmid in vivo occurs complexed with histones in a CHROMATIN-like structure. It appears to encode functions concerned only with its own partition and recombination, and has no known biological role in the yeast cell. The plasmid consists of two unique regions

(2774 bp and 2346 bp) separated by two 599-bp sequences which constitute a pair of INVERTED REPEATS. Intramolecular or intermolecular SITE-SPECIFIC RECOMBINATION between the inverted repeats results in, respectively, the inversion of one unique region with respect to the other (thus, the plasmid exists in two interconvertible monomeric forms) or the formation of plasmid multimers. Site-specific recombination is catalysed by a *trans*-acting recombinase encoded by the plasmid *FLP* gene (= *A* gene). Replication of the 2μ plasmid apparently occurs only during S phase, each plasmid normally replicating only once per cell cycle. The plasmid contains two genes, *REP1* (= *B*) and *REP2* (= *C*), which encode proteins apparently concerned with partition of the plasmid during nuclear division [MGG (1986) *203* 154–162]; the *REP1* protein is apparently associated with the NUCLEOSKELETON [JBC (1987) *262* 883–891]. [Minireview: Mol. Microbiol. (1987) *1* 1–4.]

(See also THREE-MICRON DNA PLASMID.)

two-site model (of translation) See PROTEIN SYNTHESIS.

Ty element A TRANSPOSABLE ELEMENT present in the genome of the yeast *Saccharomyces cerevisiae*. Ty elements show considerable sequence heterogeneity, but all known elements consist of an internal region ca. 5600 bp long, flanked by direct repeat sequences (δ sequences) each ca. 330 bp long. The haploid yeast genome contains ca. 30–35 Ty elements as well as 100 or more 'solo' δ sequences. Ty transposition involves non-homologous target sites and results in a 5-bp duplication of target DNA; transposition occurs via an RNA intermediate, apparently involving a reverse transcriptase encoded by the *tyb* gene of the Ty element [Cell (1985) *42* 507–517] and a mechanism similar to that in retroviruses [TIBS (1986) *11* 273]. (cf. RETROTRANSPOSON.)

***tyb* gene** See TY ELEMENT.

tylose (tylosis; *pl.* tyloses) (*plant pathol.*) A balloon-like outgrowth from a xylem parenchyma cell which expands into, and blocks, the lumen of a xylem vessel or a tracheid; tyloses are typically impregnated with TANNINS, and their formation occurs during abscission and in response to wounding or infection. Tyloses may play a role in resistance to disease by restricting the translocation of the pathogen; however, they may be responsible for the symptoms of certain diseases such as vascular wilts due e.g. to species of *Fusarium* or *Ceratocystis* (see DUTCH ELM DISEASE).

tylosin A MACROLIDE ANTIBIOTIC (from *Streptomyces fradiae*) used e.g. in veterinary medicine.

tylosis (1) *Syn.* TYLOSE. (2) Tylose formation.

tymoviruses (turnip yellow mosaic virus group) A category of ssRNA-containing PLANT VIRUSES which generally have a rather narrow host range (e.g. turnip yellow mosaic virus (TYMV), the type member, can infect various crucifers, including turnip, swede, cauliflower, Chinese cabbage); transmission occurs mechanically and via beetles. Other tymoviruses include e.g. belladonna mottle virus, cacao yellow mosaic virus, eggplant mosaic virus, peanut yellow mottle virus, *Plantago* mottle virus.

Virion: icosahedral, ca. 29 nm diam., containing a single species of coat protein (MWt ca. 20000). Genome: one molecule of linear, positive-sense ssRNA (MWt ca. 2×10^6) with a capped 5′ end (sequence: $m^7G^{5'}ppp^{5'}Gp\ldots$), and a 3′ end which has a (valine-accepting) tRNA-like structure. (Subgenomic mRNAs may also be encapsidated.) Virus RNA replication appears to occur in small, flask-shaped vesicles, each bounded by a double membrane, which develop at the periphery of chloroplasts in infected cells. Coat protein may be synthesized and assembled into hexamers and pentamers in the cytoplasm, virus assembly occuring at the necks of the vesicles. Clumping of the chloroplasts commonly occurs in infected cells.

TYMV Turnip yellow mosaic virus (see TYMOVIRUSES).

tyndallization ('fractional sterilization') A process sometimes used *with the aim of* sterilizing certain heat-labile materials (e.g. culture media which cannot be autoclaved). The material is heated to 80–100°C for periods of up to ca. 1 hour on each of three successive days, being incubated at room temperature, or 37°C, for the intervening periods. The first period of heating is intended to kill all vegetative cells initially present in the material. Incubation during the intervening periods is intended to permit germination of any spores which may be present; vegetative cells arising from such spores are killed during subsequent periods of heating. Tyndallization is based on the assumption that materials so treated are able to support the germination and outgrowth of any spores they may contain. (cf. STERILIZATION.)

type (*microbiol.*) (1) *Syn.* NOMENCLATURAL TYPE. (2) The name (per se) of the nomenclatural type of a given named taxon. (3) An infrasubspecific population (see VARIETY).

type 1 fimbriae See FIMBRIAE.

type I fimbriae See FIMBRIAE.

type 1 killer strain (of *Saccharomyces cerevisiae*) See KILLER FACTOR.

type I protein secretion See PROTEIN SECRETION.

type I reaction (anaphylaxis) A form of IMMEDIATE HYPERSENSITIVITY which involves the binding of ALLERGEN to homologous MAST CELL-bound REAGINIC ANTIBODIES and the consequent degranulation of the mast cells with release of HISTAMINE and other physiologically active agents; the reaction is complement-independent. Type I reactions, which develop within minutes of contact with the allergen, occur e.g. in hay fever and in ANAPHYLACTIC SHOCK.

type 1 septum See ACTINOMYCETALES.

type II protein secretion See PROTEIN SECRETION.

type II reaction (cytotoxic hypersensitivity) A form of IMMEDIATE HYPERSENSITIVITY which involves the binding of antibodies to antigenic sites on the surface of cells; such binding may result in (i) COMPLEMENT-dependent IMMUNE CYTOLYSIS or (ii) OPSONIZATION and phagocytosis of the target cells.

A type II reaction is involved e.g. in one form of haemolytic anaemia which may be associated with the continued use of certain drugs (e.g. chlorpromazine); in such cases the drug forms an antigenic complex at the surface of erythrocytes and thus renders these cells liable to complement-dependent cytolysis, or to phagocytosis, when homologous antibodies are formed. As a further example, haemolysis would occur if e.g. blood of group A or B were (incorrectly) transfused into a type O recipient; this is because type O individuals (normally) contain antibodies to antigens on type A and type B erythrocytes.

(See also ADCC.)

type 2 septum See ACTINOMYCETALES.

type III protein secretion See PROTEIN SECRETION.

type III reaction (complex-mediated hypersensitivity) A form of IMMEDIATE HYPERSENSITIVITY involving the development of complexes between soluble antigens and antibodies; the occurrence of COMPLEMENT FIXATION may cause local or systemic damage to tissues.

The effect of complex formation in vivo depends e.g. on the size and precipitability of the complexes. For example, if there are high levels of circulating complement-fixing antibodies, an intradermal injection of the corresponding antigen results in rapidly precipitating complexes which bring about a local inflammatory response at the site of the injection (see ARTHUS REACTION). (See also FARMERS' LUNG.)

With low levels of circulating complement-fixing antibodies, the presence of antigen can result in ANTIGEN EXCESS so that small, soluble (non-precipitating) complexes are formed; these complexes may circulate in the body and give rise to systemic symptoms (see e.g. SERUM SICKNESS). (See also GLOMERULONEPHRITIS.)

type IV fimbriae See FIMBRIAE.

type IV protein secretion See PROTEIN SECRETION.

type IV reaction See DELAYED HYPERSENSITIVITY.

type V protein secretion See PROTEIN SECRETION.

type V reaction (stimulatory hypersensitivity) A form of IMMEDIATE HYPERSENSITIVITY in which humoral antibody reacts with a cell-surface component (e.g. a hormone receptor) with consequent stimulation of the cell; thus, e.g., in certain thyrotoxic patients a thyroid-stimulating antibody (an autoantibody) mimics thyroid-stimulating hormone (TSH) in stimulating thyroid cells.

type B oncovirus group (B-type particles) A genus of retroviruses of the ONCOVIRINAE. B-type particles are assembled throughout the host cell cytoplasm (cf. TYPE C ONCOVIRUS GROUP); the mature B-type particle has prominent surface spikes and an electron-dense core located eccentrically within the envelope. Prototype of the type B oncovirus group: MOUSE MAMMARY TUMOUR VIRUS. (cf. A-TYPE PARTICLES.)

type C oncovirus group (C-type particles) A genus of retroviruses of the ONCOVIRINAE. C-type particles are assembled just beneath the plasma membrane of the host cell. Initially, an electron-dense crescent-shaped structure is formed; subsequently, this structure becomes a spherical core (with an electron-lucent centre) which buds through the plasma membrane to form an extracellular enveloped particle. The extracellular particle has a centrally located core which is electron-lucent in the immature particle, electron dense in the mature particle. Some C-type particles have discernible surface spikes, some do not.

Type C oncoviruses may be subdivided into *mammalian type C oncoviruses*: e.g. bovine leukosis virus (but see BOVINE VIRAL LEUKOSIS), FELINE LEUKAEMIA VIRUS, GIBBON APE LEUKAEMIA VIRUS, MURINE LEUKAEMIA VIRUSES, MURINE SARCOMA VIRUSES, RAT LEUKAEMIA VIRUSES, SIMIAN SARCOMA VIRUS; *avian type C oncoviruses*: e.g. AVIAN ACUTE LEUKAEMIA VIRUSES, AVIAN LEUKOSIS VIRUSES, AVIAN RETICULOENDOTHELIOSIS VIRUSES, AVIAN SARCOMA VIRUSES; *reptilian type C oncoviruses*: e.g. viper type C oncovirus, etc.

type culture (1) A culture of a TYPE STRAIN. (2) Any culture in a 'type culture collection' – whether or not a type strain.

type D retrovirus group (D-type particles) A proposed genus of the ONCOVIRINAE. 'Immature' intracellular D-type particles are ring-shaped (ca. 60–95 nm diam.) and occur most abundantly near the plasma membrane; the 'mature' extracellular particle (ca. 100–120 nm diam.) contains an electron-dense, eccentrically located nucleoid and bears surface spikes (shorter than those of the TYPE B ONCOVIRUS GROUP). Prototype of the type D retrovirus group: MASON–PFIZER MONKEY VIRUS. (See also SIMIAN AIDS.)

type species One of the species, or the sole species, included in the description of a newly proposed, validly published genus; the NOMENCLATURAL TYPE of a given genus. The type species must remain permanently within the genus of which it is the nomenclatural type; any other species removed from the genus must be given a different genus name. A type species is not necessarily typical (in characteristics) of those species which constitute the genus.

type strain One of the strains, or the sole strain, included in the description of a newly proposed, validly published species; the (viable) NOMENCLATURAL TYPE of a given species. A type strain must remain permanently within the species whose name it bears; it is not necessarily typical (in characteristics) of those strains which constitute the species. Type strains are often designated by the culture number attached to the strain in an established culture collection; e.g. the type strain of *Escherichia coli* is ATCC 11775.

typhoid fever (typhoid) An acute infectious human disease caused by *Salmonella typhi*. Transmission occurs via the faecal–oral route, e.g. via contaminated food or water. Asymptomatic carriers are an important source of infection; *S. typhi* usually localizes in the gall bladder and may be excreted in the stools for years. Incubation period: usually 7–15 days. Symptoms include malaise, myalgia, headache, and fever that increases stepwise over several days; there may also be abdominal discomfort with constipation or diarrhoea, a transient macular rash ('rose spots'), delirium or stupor, and enlargement of the spleen and mesenteric lymph nodes. Inflammation of the lymphoid tissues in the ileum (Peyer's patches) is characteristic, and may lead to necrosis and bowel perforation or haemorrhage which may be fatal. *Lab. diagnosis*: *S. typhi* may be isolated by BLOOD CULTURE during the first 7–10 days, after which it may be cultured from urine or stools. The WIDAL TEST is rarely positive before 10 days after onset. *Chemotherapy*: e.g. chloramphenicol, ampicillin, amoxycillin. Elimination of *S. typhi* from carriers may necessitate large doses of ampicillin together with surgical removal of the gall bladder. Inactivated typhoid vaccines are available.

typhoid Mary A typhoid CARRIER, Mary Mallon, who was responsible for transmitting the disease to ca. 30 persons. The name is sometimes used to refer to a potential, suspected, or actual carrier in an outbreak of typhoid or other disease.

Typhula See APHYLLOPHORALES (Clavariaceae) and SNOW MOULD.

typhula blight See SNOW MOULD.

typhus fevers (a) *Epidemic* (*classical* or *louse-borne*) *typhus*. In man: an acute systemic disease caused by *Rickettsia prowazekii*; infection occurs e.g. when louse faeces, containing *R. prowazekii*, contaminate a louse bite or other lesion. (Infection by inhalation of dried, contaminated louse faeces has also been reported.) The incubation period is 1–2 weeks. Typical symptoms include shivering, headache, muscular pains, malaise, and a sustained fever lasting 1–2 weeks – during which a macular rash develops on the trunk and limbs (rarely on the face); the lesions may become haemorrhagic. Delirium or stupor may occur. Mortality rates in untreated cases may be 5–70%. Diagnosis may involve a CFT and/or the WEIL–FELIX TEST. Tetracyclines and chloramphenicol have been used for treatment. Control measures include de-lousing and the use of prophylactic vaccines. (b) *Brill–Zinsser disease* (*Brill's disease*; *recrudescent typhus*). A mild form of classical typhus which occurs some time after a previous attack; during the supervening period the pathogen remains latent in the body. [Brill–Zinsser disease (historical perspective): Lancet (2001) *357* 1198–1200.] (c) *Murine typhus* (*endemic typhus*; *flea-borne typhus*). In man, this disease is similar to, but less severe than, epidemic typhus. It is primarily a disease of the rat. The causal agent, *R. typhi* (= *R. mooseri*), is spread among rats by the rat flea, *Xenopsylla cheopis*; man becomes infected contaminatively. Diagnosis and treatment are as for epidemic typhus. (See also BOUTONNEUSE, SCRUB TYPHUS.)

typing (of microorganisms) The process of (i) distinguishing between closely related strains of a microorganism on the basis of specified criteria (i.e. a form of classification), and (ii) using the same criteria to characterize an unknown strain of the organism by matching it with a strain already classified in the typing process (i.e. a form of identification). An unknown strain which has been matched in this way is said to have been 'typed'.

In one approach to typing, all known strains of a given species are classified into categories according to specified criteria. If the classification includes a truly comprehensive range of strains, such a scheme should permit most newly isolated strains to be typed.

Another, distinct, form of typing is the *ad hoc* scheme used for *non*-inclusive populations of strains – for example, strains of a given pathogen isolated from patients in a geographically localized region. In typing systems of this kind, strains are differentiated by comparing each strain with other strains in the same limited population.

It should be noted that no system of typing can prove that an unknown strain is identical to a particular known strain (although *non*-identity can be demonstrated); this is because a given typing system reveals similarity, or otherwise, in respect of only one (or a few) characteristics, and even if the known and unknown strains match in *these* characteristic(s) they may well differ in others. For a given species, different typing systems categorize on the basis of different criteria and, for this reason, tend to give different results; thus, for example, a population of unknown strains may be divided into a small number of groups by one typing system but into many groups by another (the latter system being said to be more *discriminatory*, i.e. able to reveal a greater number of fine differences between the strains).

To be widely applicable, a typing system should satisfy certain basic requirements. First, the criteria used should enable all strains of the given organism to the typed. (However, in practice, it is not uncommon to find that some strains are 'untypable' in a given system.) Second, the typing system should clearly distinguish each strain from the other strains. Third, results should be reproducible.

Applications of typing. In a medical/veterinary context, typing has various uses – for example:

(a) *Detection of laboratory cross-contamination*. In a clinical laboratory, cross-contamination among pre-test specimens can lead e.g. to false-positive results and incorrect diagnoses. In one published episode, typing revealed that 60 specimens in a reference laboratory had been contaminated with *Mycobacterium tuberculosis* from one true-positive specimen [JCM (1999) *37* 916–919].

(b) *Distinguishing between re-activation and new infection in tuberculosis*. DNA-based typing methods have been able to indicate the probable origins of strains isolated from patients with tuberculosis, and have indicated e.g. that recent transmission (as opposed to re-activation of a previous infection) is more common than previously supposed [BCID (1997) *4* 173–183].

(c) *Epidemiology* (*community-based and global*). In epidemiology, typing permits the investigator to identify sources of infection and to follow the routes of transmission; such investigation is possible only if individual strains have been characterized (typed) and are distinguishable. The underlying rationale for the epidemiological use of typing is that all isolates of a given pathogen from a particular *outbreak* of disease, or from within the same *chain of infection*, will be progeny of the same ancestral cell and will therefore be recognizable by close genetic similarity (clonal relationship) compared with other, randomly acquired isolates of the same pathogen. Thus, if strains of a pathogen have been stably typed, it is often possible to link a particular case or outbreak of disease with a known source of infection by comparing the causal strain with strains of the pathogen known to occur in specific locations. [Evaluation and use of epidemiological typing systems: CMI (1996) *2* 2–11.] Tracking of virulent and antibiotic-resistant strains of pathogenic bacteria on a global scale, via the Internet, is possible by multilocus sequence typing (see MLST).

Typing *direct* from a clinical specimen (i.e. without prior culture) has been used for strains of *Neisseria meningitidis*; this can be advantageous when a specimen is culture-negative owing to pre-admission chemotherapy [JCM (1997) *35* 1809–1812].

In a veterinary context, typing has been used e.g. to investigate outbreaks of *Pseudomonas aeruginosa* mastitis in dairy herds [AEM (1999) *65* 2723–2729].

Typing methods. There are two main categories: conventional (phenotypic) methods, and nucleic-acid-based (genotypic) methods. Conventional methods include e.g. BACTERIOCIN TYPING, GRIFFITH'S TYPING, MULTILOCUS ENZYME ELECTROPHORESIS, PHAGE TYPING and SEROTYPING. Any of a range of characteristics may be suitable for typing purposes; for example, strains of *Pseudomonas aeruginosa* may be typable on the basis of variability in the siderophore *pyoverdin* [Microbiology (1997) *143* 35–43].

Genotypic methods include AFLP FINGERPRINTING, ARBITRARILY PRIMED PCR, DNA FINGERPRINTING, DNA SEQUENCING, MLST, PFGE, RAPD, REP-PCR, RFLP, RIBOTYPING, SSCP and SPOLIGOTYPING.

[Nucleic-acid-based typing: Book ref. 221, pp 168–202.]

tyrocidins Cyclic decapeptide ANTIBIOTICS, produced by *Bacillus brevis*, which are bactericidal mainly for certain Gram-positive bacteria. The tyrocidin molecule contains one or more free amino groups, and includes at least one D-amino acid. Tyrocidins appear to disrupt the CYTOPLASMIC MEMBRANE, causing leakage of essential metabolites etc. (cf. GRAMICIDINS.)

Tyromyces See APHYLLOPHORALES (Polyporaceae).

tyrosinase See MELANIN and COPPER.

L-tyrosine biosynthesis See AROMATIC AMINO ACID BIOSYNTHESIS.

tyrosine kinase Commonly refers to a tyrosine-specific PROTEIN KINASE – see e.g. CYTOKINES.

tyrothricin A mixture of GRAMICIDINS (ca. 20%) and TYROCIDINS, produced by *Bacillus brevis*.

tyvelose (3,6-dideoxy-D-mannose) A sugar, first isolated from *Salmonella typhi*, which occurs in the O-specific chains of the LIPOPOLYSACCHARIDE in certain *Salmonella* serotypes and which contributes to the specificity of O antigen 9 in group D salmonellae (see KAUFFMANN–WHITE CLASSIFICATION).

Tyzzeria A genus of homoxenous coccidian protozoa (suborder EIMERIORINA) in which the oocysts contain 8 naked sporozoites (i.e. sporocysts are not formed). *T. perniciosa* causes an intestinal disease in ducks.

Tyzzer's disease A fatal disease which can affect e.g. rodents, cats, young foals and rhesus monkeys. The causal agent was reported to be an intracellular parasitic bacterium not cultivable in artificial media [JGM (1984) *130* 1757–1763]; formerly referred to as *'Bacillus piliformis'*, it is now designated *Clostridium piliforme* [IJSB (1993) *43* 314–318]. In foals, onset is sudden, and death may occur within a few hours or days; the disease involves a necrotizing hepatitis and colitis with high fever and, in some cases, jaundice and diarrhoea. At post mortem the (greatly enlarged) liver exhibits miliary necrotic foci.

U

U Uracil (or the corresponding nucleoside or nucleotide) in a nucleic acid.

U-RNA See SNRNA.

U1–U6 snRNAs See SNRNA.

UACUAAC box See SPLIT GENE (a).

UAS Upstream activation site: a sequence in the DNA of *Saccharomyces cerevisiae* which is involved in the regulation of transcription; it apparently resembles an ENHANCER except that it cannot function when located downstream of a promoter.

ubiquinones See QUINONES.

ubiquitin A polypeptide (containing 76 amino acids) present – either free or covalently bound to protein – in the cells of higher eukaryotes and in various microorganisms. Energy-dependent binding of ubiquitin to a protein is apparently an essential signal for the selective proteolysis of that protein in the cell.

The ubiquitin–PROTEASOME pathway has been reported to be essential for proteolysis during morphological changes in *Trypanosoma cruzi* [Biochem. (2001) *40* 1053–1062].

UDN See ULCERATIVE DERMAL NECROSIS.

Udotea A genus of siphonaceous, usually calcified, fan-shaped green seaweeds related to HALIMEDA.

UDP Uridine 5′-diphosphate [see Appendix V(b)].

UHT milk See APPERTIZATION and CASEIN.

ulcer A localized break in an epithelial surface (e.g. the skin or mucous membrane), often accompanied by inflammation, formed as a result of the sloughing of necrotic tissue. Ulcers occur in certain infectious diseases or may result from physical damage (e.g. bed-sores), defective circulation, malignant tumours etc.

ulcer disease (in salmonids) A FISH DISEASE involving skin ulceration and septicaemia. Pathogen: '*Haemophilus piscium*' – now thought to be an atypical strain of *Aeromonas salmonicida*. (cf. FURUNCULOSIS.)

ulceracin 378 A BACTERIOCIN produced by *Corynebacterium diphtheriae* var. *ulcerans*. [Purification and characterization: JGM (1985) *131* 707–713.]

ulcerative dermal necrosis (UDN; salmon disease) An infectious FISH DISEASE of unknown aetiology affecting mature salmonid fish re-entering fresh water from the sea. Skin lesions range from small grey patches to deep haemorrhagic ulcers. Secondary SAPROLEGNIASIS is common.

ulcerative enteritis (of fowl) See QUAIL DISEASE.

ulcerative lymphangitis (ulcerative cellulitis) A chronic infectious disease, mainly of horses but also e.g. of cattle, caused by *Corynebacterium pseudotuberculosis*. Infection occurs via wounds; flies may be mechanical vectors [VR (1983) *113* 496–497]. Progressive inflammation of the subcutaneous lymphatic vessels is typically accompanied by ulceration and/or induration; internal organs (e.g. lungs, kidneys) may be affected.

Ulothrix A genus of filamentous freshwater and marine green algae (division CHLOROPHYTA). The filaments are generally unbranched and attached to a substratum by a basal cell holdfast. Individual cells are uninucleate, and each contains a single chloroplast. Asexual reproduction occurs by fragmentation of filaments or by release of quadriflagellate zoospores. Sexual reproduction is isogamous; gametes are biflagellate.

ultimate carcinogen See CARCINOGEN.

UlTma™ DNA polymerase See DNA POLYMERASE.

ultracentrifugation See CENTRIFUGATION.

ultracytostome *Syn.* MICROPORE.

ultradian rhythms In certain parameters of a eukaryotic cell or organism: innate rhythmical changes which occur with a periodicity of less than 24 hours. (cf. CIRCADIAN RHYTHMS.)

ultrafiltration See FILTRATION.

ultramicrotome See MICROTOME.

ultrasonication The use of sound waves of frequency ca. 16 kHz or higher (ultrasound) produced by a SONICATOR. Ultrasonication is used, in a *liquid* medium, e.g. for sphaeroplast lysis, for dispersing the fine particulate matter in centrifuge pellets, and for CELL DISRUPTION. Sphaeroplast lysis requires only a short exposure to ultrasound, resulting in minimal damage to liberated cell components; breakage of whole cells requires longer exposure, with concomitant risk of damage to cell components. The efficacy of ultrasonication increases with increasing frequency, and decreases with increasing viscosity. Ultrasonication causes *cavitation*: the formation of minute bubbles of gas or vapour in those regions of the liquid corresponding (at a given instant) to rarefactions in the sound waves; the shearing forces produced by fast-moving bubbles together with the abrupt pressure changes which occur in these microenvironments probably account for the main disintegrative effect. Ultrasonicated material is subject to oxidation – probably due to the formation of hydrogen peroxide and/or SINGLET OXYGEN during cavitation in an aerated medium; oxidation can be lessened by enveloping the sample with an inert gas.

ultrasound See ULTRASONICATION.

ultraviolet microscopy See MICROSCOPY (f).

ultraviolet radiation (UVR) (a) *Effects on DNA*. UVR of wavelength 240–300 nm is strongly absorbed by the bases in DNA and can induce the formation of a variety of photoproducts which may lead to mutation or cell death. A major product is the cyclobutyl pyrimidine (particularly thymine) dimer which arises by the formation of covalent bonds between the 5,6-positions of two adjacent pyrimidine residues (a cyclobutane ring being formed between the residues). The linked bases can no longer form Watson–Crick base pairs, and localized distortion of the DNA duplex occurs. Pyrimidine dimers prevent replication of the DNA and can be lethal unless acted upon by a DNA REPAIR mechanism (see e.g. EXCISION REPAIR, PHOTOREACTIVATION and RECOMBINATION REPAIR). (See also MAXICELL.)

In e.g. *Escherichia coli*, *mutagenesis* induced by UVR is apparently dependent on the induction of the SOS SYSTEM; many types of mutation occur, including base substitutions, frameshifts and deletions [JMB (1985) *182* 45–68], and these are believed to result from *umuDC*-dependent error-prone repair of UV-induced lesions. The cyclobutyl pyrimidine dimers, although potentially lethal and capable of inducing the SOS response, are apparently not directly mutagenic. The main pre-mutagenic lesion is believed to be another UVR-induced product, the '(6–4) photoproduct' (Pyr(6–4)Pyo), formed between adjacent pyrimidines (usually 5′-T – C-3′ or 5′-C – C-3′) by covalent bonding between the 6-position of the 5′ pyrimidine and the 4-position of the 3′ pyrimidine [Cell (1983) *33* 13–17; JMB (1985) *184* 577–585; Mut. Res. (1986) *165* 1–7]; the (6–4) photoproduct is not susceptible to enzymic photoreactivation but can be removed e.g. by excision repair. Apurinic sites may also be important pre-mutagenic lesions.

(b) *In disinfection.* UVR 'germicidal' lamps (emitting UVR at ca. 254 nm) are used in certain circumstances e.g. for the DISINFECTION of air and of exposed surfaces in enclosed areas. However, UVR (a non-IONIZING RADIATION) has poor powers of penetration: it is readily absorbed by solids, and can penetrate liquids only to a limited extent; furthermore, its effects on microoganisms may, depending e.g. on dose, be reversed by various DNA repair processes (see (a) above). (See also BROWN–BOVERI UV-C LAMP; cf. WOOD'S LAMP.)

Ulva A genus of green foliose seaweeds (division CHLOROPHYTA). The sheet-like irregularly shaped thallus is generally anchored to rocks by a perennial, rhizoid-like holdfast; it consists of two layers of uninucleate cells, each cell containing a single chloroplast with one or more pyrenoids. The cells of the holdfast are multinucleate. The life cycle involves an isomorphic alternation of generations; sporophytes produce quadriflagellate zoospores, gametophytes produce biflagellate gametes.

U. lactuca (the 'sea lettuce', 'green LAVER') is reputed to be edible; it is common in coastal and estuarine waters and may be particularly abundant near freshwater outfalls and in sewage-polluted waters (cf. ENTEROMORPHA). *U. lactuca* can have a detrimental effect on invertebrates which share its environment – due both to its production of toxic exudates and to its consumption of O_2 at night [JEMBE (1985) *86* 73–83].

Ulvophyceae (Ulvaphyceae) See CHLOROPHYTA.

umber codon See GENETIC CODE.

Umbilicaria A genus of LICHENS (order LECANORALES); photobiont: a green alga. The thallus is foliose, discoid, entire or lobed, attached to the substratum via a single, more or less central holdfast (umbilicus) on the underside. Apothecia: lecideine, smooth or gyrose according to species. Species occur mainly on acidic rocks in arctic–alpine regions. (cf. LASALLIA; see also IWATAKE and ROCK TRIPE.)

umbilicate (1) Having a localized central depression. (2) (of a lichen thallus) Attached to the substratum by an umbilicus (see UMBILICARIA).

umbilicin β-D-Galactofuranosyl-(1 $\rightarrow$ 3)-D-arabitol, a compound found in *Umbilicaria* spp.

umbilicus See UMBILICARIA.

umbo A discrete swelling at the centre of (e.g.) the upper part of a pileus, or of an otherwise flat bacterial colony.

umbonate Having an UMBO.

umecyanin See BLUE PROTEINS.

UMP Uridine 5′-monophosphate (= uridylic acid) [see Appendix V(b)].

***umuC, umuD* genes** See SOS SYSTEM.

UmuD′ See SOS SYSTEM.

Una virus See ALPHAVIRUS.

unbalanced growth See GROWTH.

***unc* mutant** A mutant bacterium defective in the coupling between pmf (see CHEMIOSMOSIS) and PROTON ATPASE activity.

Uncinula See ERYSIPHALES.

uncoating When a virus infects a host cell: the process in which the viral nucleic acid is released from the capsid or is exposed by modification of the capsid (depending on virus) prior to its expression and replication.

unconventional slow virus infections See SLOW DISEASE.

uncoupling agents Those chemical agents (e.g. PROTON TRANSLOCATORS) which can collapse (or prevent the formation of) a transmembrane electrochemical gradient (see CHEMIOSMOSIS) and which can thus uncouple the link between pmf-generating electron transport and the pmf-utilizing PROTON ATPASE. (See also RESPIRATORY CONTROL.) Even when a gradient has been collapsed, the electron transport chain which generated the gradient may continue to operate, although no gradient develops in the continuing presence of the uncoupling agent. Uncoupling agents also tend to promote intracellular ATP hydrolysis ('uncoupler-stimulated ATPase activity'). (cf. OLIGOMYCIN.)

Undaria A genus of seaweeds of the PHAEOPHYTA. *U. pinnatifida* is widely used as food in Japan (where it is known as *wakame*) and in China (*qundai-cai*); in Japan it is cultivated by sowing spores on ropes which are submerged in the sea [Book ref. 130, pp. 690–692].

undecaprenol See BACTOPRENOL.

undecenoic acid *Syn.* UNDECYLENIC ACID.

undecylenic acid (undecenoic acid; $CH_2{=}CH(CH_2)_8COOH$) An antifungal ORGANIC ACID used (as the free acid or e.g. as the zinc salt) e.g. for the treatment of superficial dermatophyte infections.

underwound DNA See DNA.

undulant fever See BRUCELLOSIS.

undulating membrane (1) (*ciliate protozool.*) See PARORAL MEMBRANE. (2) In some flagellates (e.g. species of TRICHOMONAS and TRYPANOSOMA): a membranous structure which spans the region between the FLAGELLUM or the proximal part of the flagellum (according to species) and the cell body – see e.g. figure under TRYPANOSOMATIDAE. In e.g. *Trypanosoma* the undulating membrane is a double membrane.

***ung* gene** See URACIL-DNA GLYCOSYLASE.

UNG method (decontamination) See PCR.

ungulate (*adj.*) Shaped like a hoof.

unialgal Of e.g. an algal culture: containing only one species of alga.

unicell A single autonomous cell.

unicellular Consisting of a single cell.

unifactorial heterothallism See HETEROTHALLISM.

unilocular Consisting of or having a single LOCULE: see e.g. FORAMINIFERIDA.

uniport The transmembrane transport of one type of solute (e.g. a given ion) by a carrier. (cf. ANTIPORT, SYMPORT; see also e.g. VALINOMYCIN and ION TRANSPORT.)

uniseriate Arranged in or forming a single row; also used to refer to e.g. a filament or chain composed of a single row of cells or spores.

unit membrane A conceptual structure, proposed in the 1930s by Danielli and Davson, for biological membranes. Essentially, a membrane was supposed to consist of a bilayer of lipid molecules (orientated with their fatty acid side-chains perpendicular to the plane of the membrane, and their hydrophilic portions forming the surfaces of the bilayer) with a layer of protein coating each surface of the bilayer. The model has been superseded by the fluid mosaic model (see CYTOPLASMIC MEMBRANE), but *unit-type membrane* refers to any membrane based on a lipid bilayer.

unitunicate ascus An ASCUS whose wall consists of a functionally single layer, i.e., if the wall is bi- or multilayered the layers do not move relative to one another during the discharge of ascospores. Ascospores are discharged e.g. via an apical pore or an operculum. (cf. BITUNICATE ASCUS.)

univalent (*noun*) An unpaired chromosome at meiosis.

universal bottle A cylindrical, shoulderless, screw-cap glass jar of capacity ca. 25 ml. (cf. MCCARTNEY BOTTLE.)

universal conjugation system See PILI.

universal genetic code See GENETIC CODE.

universal nucleotide See PCR.

universal primers See BROAD-RANGE PRIMERS.

universal veil (*mycol.*) In certain agarics (e.g. *Amanita* spp): a membranous tissue which entirely covers the immature fruiting body; as the fruiting body develops, the universal veil tears – leaving a VOLVA and (sometimes) fragments of the membrane (*scales*) adhering to the upper surface of the pileus. (cf. PARTIAL VEIL.) (For some authors 'universal veil' refers also to the PERIDIUM of gasteromycetes.)

unlinked gene See LINKAGE.

Unna–Pappenheim stain *Syn.* METHYL GREEN–PYRONIN STAIN.

untwisting enzyme A type I TOPOISOMERASE.

unwinding angle See INTERCALATING AGENT.

unwinding protein See DNA UNWINDING PROTEIN.

up mutation (up-promoter mutation) A MUTATION in a PROMOTER which results in an increased level of transcription from that promoter. (cf. DOWN MUTATION.)

UPEC Uropathogenic *Escherichia coli*: strains of *E. coli* associated with a high proportion of human urinary tract infections. (See also CYSTITIS and PYELONEPHRITIS.)

Strains of UPEC typically encode a range of virulence factors which include a cytotoxic agent, a haemolysin and a number of ADHESINS such as type I FIMBRIAE, DR ADHESINS and P FIMBRIAE. In at least some strains of UPEC, virulence factors (including the haemolysin) are encoded by a 70 kb chromosomal PATHOGENICITY ISLAND, designated PAI-1, which is reported to be inserted downstream of the *selC* gene.

Effective adhesion of UPEC to uroepithelium is an essential first step in pathogenesis as there is a need to combat the flushing action of the urine flow. Type I fimbriae (the FimH adhesin) can bind to certain UROPLAKINS (UPIa, UPIb); moreover, the type I fimbria can act as an INVASIN, promoting uptake of the bacterial pathogen by bladder epithelial cells. (This role appears to be specific to type I fimbriae; adhesion mediated by the PapG adhesin of P fimbriae did not promote uptake [EMBO (2000) *19* 2803–2812].)

In the mouse model, strains of UPEC are able to persist within bladder tissue; if this reflects the human situation it could at least partly explain recurrent infections of the urinary tract.

[Interplay between uropathogenic *E. coli* and the innate host defences: PNAS (2000) *97* 8829–8835.]

[Disseminated multidrug-resistant clonal group A: Lancet (2002) *359* 2249–2251.]

UPGMA In NUMERICAL TAXONOMY, unweighted pair group method with averages: a method in which two groups ('clusters') unite (coalesce) at a similarity level determined by the arithmetical average of the similarities which exist between intercluster OTU pairs.

uphill electron transport See REVERSE ELECTRON TRANSPORT.

upstream (*mol. biol.*) See DOWNSTREAM.

upstream activation site See UAS.

uptake hydrogenase See NITROGENASE.

UQ Ubiquinone: see QUINONES.

uracil-DNA glycosylase (uracil-*N*-glycosylase) An enzyme (the product of the *ung* gene in *Escherichia coli*) which removes uracil from a dUMP residue in ssDNA or dsDNA (but not from dUMP itself); removal of uracil from dsDNA is followed by BASE EXCISION REPAIR of the AP SITE. The enzyme is believed to function primarily in protecting against mutation resulting from the deamination of cytosine to uracil in DNA (see SPONTANEOUS MUTATION).

In PCR, this enzyme is used in one approach for preventing the contamination of apparatus and environment by 'carryover' amplicons from previous amplifications.

uracil-*N*-glycosylase *Syn.* URACIL-DNA GLYCOSYLASE.

uranyl acetate See ELECTRON MICROSCOPY.

urea (as an antimicrobial agent) Urea (carbamide: $(NH_2)_2CO$) has mild antibacterial properties against certain bacteria (but see UREASES), as do certain of its derivatives (e.g. urethanes: $RNH.CO.OC_2H_5$). (See also CARBAMIC ACID DERIVATIVES; CARBANILIDES; *N*-NITROSO COMPOUNDS (nitrosoureas); THIOUREA DERIVATIVES.)

Ureaplasma A genus of urease-positive, microaerophilic bacteria (family MYCOPLASMATACEAE) which occur e.g. in the mouth and in the respiratory and urogenital tracts in man and other animals. Cells: non-motile, coccoid (ca. 0.1–0.85 μm diam.) in young cultures, or pleomorphic. *Ureaplasma* spp (formerly called T-strain mycoplasmas) resemble *Mycoplasma* spp in their general properties, and they can be cultured in serum- and yeast extract-containing media; however, colonies are small (ca. 15–60 μm diam.) and may lack the zone of surface growth (and hence the 'fried egg' appearance) typical of *Mycoplasma* spp. Infection with ureaplasmas has been associated with various diseases – e.g. chorioamnionitis, urethritis – although the causal role of ureaplasmas in at least some of these diseases remains uncertain. GC%: ca. 27–30. Type species: *U. urealyticum* (which includes strains isolated from man); other species: *U. diversum* (which includes bovine strains). [Book ref. 22, pp. 770–775. Ovine and caprine ureaplasmas: JGM (1983) *129* 3197–3202.]

urease An enzyme (EC 3.5.1.5) which hydrolyses UREA to ammonia and carbon dioxide; ureases occur e.g. in certain bacteria (e.g. *Helicobacter pylori*, *Proteus* spp, some other enterobacteria, and *Ureaplasma*), in some plants (e.g. soybean) and in certain invertebrates. [Bacterial ureases (structure, regulation of expression, and role in pathogenesis): Mol. Microbiol. (1993) *9* 907–913.]

Urease activity in enterobacteria can be detected by growing the organisms on Christensen's urea agar: a phosphate-buffered medium containing glucose, peptone, urea and PHENOL RED; the medium is inoculated heavily, incubated at 37°C, and examined after 2–4 hours, 18 hours, and daily for 4 days. In a positive reaction (urea hydrolysed) the resulting alkalinity causes the pH indicator to turn red.

Uredinales See UREDINIOMYCETES.

Uredinella See SEPTOBASIDIALES.

Urediniomycetes (rust fungi; rusts) A class of plant-parasitic fungi (subdivision BASIDIOMYCOTINA), all of which are placed in the order Uredinales; the rusts form sexual organs but do not form macroscopic basidiocarps (cf. USTILAGINOMYCETES). Some rusts cause economically important plant diseases: see RUSTS (sense 2). Rust fungi often exhibit a high degree of host specificity, and a given species of rust may consist of many PHYSIOLOGICAL RACES. Some rusts are HOMOXENOUS, others are HETEROXENOUS. Many rusts exhibit HETEROTHALLISM. Many species of rust have been grown in axenic culture.

Morphology and life cycles of the rust fungi. The vegetative (somatic) form is a hyaline, septate mycelium which generally occurs intercellularly in the host plant; haustoria are commonly formed.

The life cycle involves a series of distinct spore stages which occur in a definite sequence; some species exhibit up to five stages, while other exhibit four or fewer stages. The rust life cycle may be exemplified by that of *Puccinia graminis tritici*, a heteroxenous, heterothallic rust which exhibits all five spore stages and which causes black stem rust of wheat; wheat is the *primary host* (i.e., that host of a heteroxenous rust in which arise the (BASIDIUM-forming) *teliospores* – see later), while the secondary host of *P. graminis tritici* is an appropriate species of the barberry (*Berberis*). The complete life cycle of *P. graminis tritici* (not always exhibited in nature) is as follows.

Stage 0. A basidiospore germinates on a barberry leaf and gives rise, subsequently, to a mononucleate, haploid, intercellular, HAUSTORIUM-forming mycelium. After several days, erumpent, flask-shaped structures (*pycnia*) develop as small, raised, yellowish lesions on the upper surface of the leaf. A *pycnium* (= *spermagonium*, *spermatogonium* or *spermogonium*) is a wholly fungal structure lined with a layer of elongated cells or hyphae from the ends of which are produced minute, mononucleate, haploid cells – *pycniospores* (= *pycnospores* or *spermatia*) – that are exuded in drops of sweet, sticky nectar from the ostiole (opening) of the pycnium. Extending from the ostiole is a tuft of stiff, sterile, orange filaments, and a number of thin-walled *flexuous hyphae* (= *receptive hyphae*). The pycniospores function as male reproductive cells, while the flexuous hyphae function as female structures; however, in *P. graminis tritici* (and other heterothallic rusts) fertilization does not occur between pycniospores and flexuous hyphae from the same pycnium because both are of the same MATING TYPE. Fertilization occurs when pycniospores from another pycnium (of different mating type) fuse with the flexuous hyphae; the nucleus of the pycniospore then travels down the flexuous hypha and subsequently undergoes repeated division.

Stage I. Within the barberry leaf the mycelium forms a number of differentiated regions near the lower epidermis; these spheroidal regions are aecial initials (= *protoaecia* or *protoaecidia*). A *protoaecium* (= *protoaecidium*) does not develop further unless it receives nuclei of appropriate mating type – either via the flexuous hyphae or by fusion of hyphae of compatible strains which occupy adjacent regions in the leaf. On DIKARYOTIZATION each protoaecium develops into an *aecium* (= *aecidium*) (*pl.* aecia and aecidia, respectively). An aecium consists of a basal layer of sporogenous cells which gives rise to a mass of parallel, basipetally-formed chains of binucleate spores (*aeciospores*, *aecidiospores*) arranged perpendicularly to the leaf surface; the entire mass of spores is contained within a thin wall (*peridium*) of fungal origin. The mature aecium later breaks through the leaf's lower epidermis, and the exposed peridium ruptures to form a cup-like structure (*aecial cup* or *cluster cup*) from which the orange aeciospores are discharged.

Stage II. Aeciospores cannot re-infect the barberry; they can infect suitable strains of wheat, giving rise to a binucleate, intercellular mycelium. Later, fungal structures termed *uredia* (= *uredinia* or *uredosori*) develop subepidermally in stem and leaf tissues, and eventually become erumpent. A *uredium* (= *uredinium* or *uredosorus*) bears a layer of single-celled, binucleate, rust-coloured spores – *uredospores* (= *urediniospores* or *urediospores*) – each spore being borne on a thin stalk (*pedicel*); in this species the uredium is not bounded by a peridium. After the spore mass has ruptured the host's epidermis the uredospores are dispersed e.g. by wind. Heavy infection of stems can cause LODGING. Uredospores cannot infect the barberry; they can infect wheat (and may re-infect the original host plant), giving rise to a binucleate mycelium. Several crops of uredospores may be formed during the summer; uredospores are thus the main dispersal agents of this rust. (See also AMPHISPORE; for dormancy in uredospores see SPORE.)

Stage III. In late summer/early autumn the fungus (still within wheat) ceases to form uredia and gives rise to *telia* (= *teleutosori*). A *telium* (= *teleutosorus*) resembles a uredium in that it is an erumpent mass of stalk-borne spores unbounded by a peridium; it differs, however, in that each *teliospore* (= *teleutospore*) is a very dark, two-celled structure (each cell binucleate) with a wall much thicker than that of a uredospore. By the time the spore has reached maturity karyogamy has occurred in both cells. Teliospores are RESTING SPORES. The telial stage is regarded as the sexual stage of rust fungi – which are classified mainly on the basis of the characteristics of this stage.

Stage IV. In spring, teliospores germinate to form basidia (see BASIDIUM) which give rise to haploid, unicellular, thin-walled BASIDIOSPORES (= *sporidia*). Of the four basidiospores formed on each basidium, two are of one mating type, and the other two are of a different mating type. Basidiospores are dispersed e.g. by wind.

In some parts of the world (e.g. the eastern USA) *P. graminis tritici* overwinters on barberry. In e.g. Great Britain, however, outbreaks of black stem rust are commonly due to the arrival of wind-borne uredospores from southern Europe. In warm regions the pathogen can overwinter as uredospores and can re-infect wheat in the following year.

Characteristics of other rust fungi. Rusts may be classified according to the range of types of spore they produce: see e.g. DEMICYCLIC RUSTS, MACROCYCLIC RUSTS, MICROCYCLIC RUSTS. According to species, the pycnium may be any of twelve different morphological types; the aecium may be e.g. AECIDIOID, CAEOMATOID, PERIDERMIOID or ROESTELIOID; and the uredium may be non-peridiate (e.g. *Puccinia*) or peridiate (e.g. *Melampsora*). (The uredium of *Pucciniastrum agrimoniae* has a persistent peridium, the uredospores escaping via a pore.) Teliospores may be sessile or pedicellate, and may be one-celled (e.g. *Uromyces*), two-celled (e.g. *Puccinia*) or multicelled (e.g. *Xenodochus*).

The genera of rust fungi include e.g. *Arthuria*, *Coleosporium*, CRONARTIUM, *Cystomyces*, *Dasyspora*, *Endophyllum*, GYMNOSPORANGIUM, *Hemileia* (see also COFFEE RUST), *Hyalospora*, *Intrapes*, MELAMPSORA, *Peridermium*, *Phragmidium*, PUCCINIA, *Pucciniastrum*, *Trachyspora*, *Uredinopsis*, *Uromyces* and *Xenodochus*.

urediniospores See UREDINIOMYCETES stage II.

uredinium See UREDINIOMYCETES stage II.

Uredinopsis See UREDINIOMYCETES.

urediospores See UREDINIOMYCETES stage II.

uredium See UREDINIOMYCETES stage II.

uredosorus See UREDINIOMYCETES stage II.

uredospores See UREDINIOMYCETES stage II.

urethanes See UREA.

urethritis Inflammation of the urethra. Symptoms: frequent and painful micturition, sometimes with a purulent discharge. Urethritis may be caused by any of a range of organisms – see e.g. GONORRHOEA and NON-GONOCOCCAL URETHRITIS. (See also URINARY TRACT INFECTION.)

uridine See NUCLEOSIDE and Appendix V(b).

uridylic acid See UMP.

urinary mucoprotein *Syn.* TAMM–HORSFALL PROTEIN.

urinary tract infection (UTI) Infection of the human urinary tract (which is generally more frequent in women than in men) may involve any of a range of organisms. UTI is common as a nosocomial infection following catheterization; in these cases, causal organisms are usually endogenous faecal bacteria – frequently *Escherichia coli*, species of *Klebsiella* or *Proteus*, or enterococci. (See also CATHETER-ASSOCIATED INFECTION.)

Infection by strains of UPEC is believed to involve initial adhesion of the pathogen to the bladder UROPLAKINS [PNAS (2000) *97* 8829–8835].

The examination of a urine sample typically includes (i) microscopy of a WET MOUNT (to assess the likelihood of urinary tract infection from the numbers of leukocytes present), and (ii) a semi-quantitative form of culture, to determine whether the urine contains potential pathogen(s) in numbers high enough to be consistent with a pathogenic role (see SIGNIFICANT BACTERIURIA). The sample usually consists of a *mid-stream* urine (MSU) – this method of sampling tending to minimize contamination from e.g. the urethra.

In a murine model of an *E. coli* urinary tract infection, the PROBIOTIC *Lactobacillus casei* strain Shirota, administered intraurethrally, was reported to exhibit potent therapeutic activity [AAC (2001) *45* 1751–1760].

For 'stones' (calculi) in the urinary tract see UROLITHIASIS.

(See also CYSTITIS, PYELONEPHRITIS, URETHRITIS and GENITOURINARY TRACT FLORA.)

urinogenital tract flora See GENITOURINARY TRACT FLORA.

urkaryote A hypothetical evolutionary forerunner of eukaryotic cells.

urkingdom (1) A taxonomic category comprising the ARCHAEBACTERIA – from which the EUBACTERIA and eukaryotes were supposed to have evolved independently [Book ref. 157, pp. 561–564].

(2) A taxonomic category used for any of several ancient 'primary' groups of organisms [see e.g. PNAS (1985) *82* 3716–3720]. (See also PHOTOCYTA.)

urocanate See HISTIDINE DEGRADATION.

Urocystis See USTILAGINALES.

urogenital tract flora See GENITOURINARY TRACT FLORA.

Uroglena A genus of freshwater colonial CHRYSOPHYTES. The colony is motile and consists of cells arranged around the periphery of a ball of clear mucilage; the posterior end of each cell is drawn out into a long 'tail' which reaches the centre of the colony.

uroid (*protozool.*) The posterior end (relative to the direction of locomotion) of an amoeba. According to species, a uroid may or may not be clearly differentiated; it may have a knobby appearance (e.g. in some *Mayorella* spp), and/or trailing filaments may extend from it as a result of adhesion of small areas of its surface to the substratum during locomotion. In *Trichamoeba* the uroid is typically bulbous and usually bears short, fine, fairly rigid filaments (a *villous-bulb* uroid). In e.g. *Pelomyxa* food (e.g. non-motile algae) is ingested at the uroid.

urolithiasis The presence of mineral-based bodies ('stones'; calculi) within the urinary tract. 'Infectious stones' are commonly associated with certain infections (involving e.g. *Proteus* spp) that lead to the formation of alkaline urine: see STRUVITE. It has been suggested that calculi in urine of near-neutral pH may be associated with the presence of nanobacteria [PNAS (1998) *95* 8274–8279].

Urinary calculi may also be associated with reduced numbers of oxalate-degrading organisms (e.g. *Oxalobacter formigenes*) in the intestinal tract; these organisms seem to be important for regulating the absorption of oxalate. For example, loss of activity of *O. formigenes* has been associated with hyperoxaluria and possible urolithiasis. The population of *O. formigenes* in human faecal specimens has been assessed with a PCR-based method [JCM (1999) *37* 1503–1509].

uromucoid *Syn.* TAMM–HORSFALL PROTEIN.

Uromyces See UREDINIOMYCETES.

Uronema (1) A genus of filamentous freshwater and marine green algae related to CHAETOPHORA; the filaments are uniseriate, erect and unbranched. Quadriflagellate zoospores are formed. (2) A genus of protozoa of the SCUTICOCILIATIDA.

uronic acids Compounds with the general formula $CHO(CHOH)_n COOH$, formed e.g. by the oxidation of the primary alcohol group of an aldose; the uronic acid corresponding to glucose is glucuronic acid.

Uronychia See HYPOTRICHIDA.

uropathogenic *E. coli* See UPEC.

uroplakin Any of four types of membrane protein (designated UPIa, UPIb, UPII and UPIII) that form hexagonal complexes on the luminal surface of the mammalian bladder; such complexes occur on the lumen-facing surface of the multinucleate *facet* cells that line the bladder. These complexes form plaques that cover most of the luminal surface. Thus, the lumen of the bladder is delimited by a uroplakin-studded membrane which is referred to as the *asymmetric unit membrane* (AUM); the AUM is believed to have the dual function of acting as a permeability barrier and stabilizing the facet cells during their stretching in an active bladder.

UPIa and UPIb can bind the type I FIMBRIAE of *Escherichia coli*, apparently via the FimH adhesin; such binding, which is inhibited by D-mannose, is believed to be important in the initial colonization of the bladder by UPEC [PNAS (2000) *97* 8829–8835].

Urosoma See HYPOTRICHIDA.

Urospora (1) A genus of unbranched, filamentous, siphonocladous green algae (division CHLOROPHYTA). (2) See GREGARINASINA.

Urosporidium See STELLATOSPOREA.

Urostyla See HYPOTRICHIDA.

Urotricha See GYMNOSTOMATIA.

URTI Upper respiratory tract infection.

urticaria (hives) A condition, due e.g. to an allergy, in which swellings (wheals) appear (usually) in the skin; the wheals are typically redder than the surrounding skin and they characteristically itch or burn.

use-dilution For a given DISINFECTANT: that dilution which can be regarded as effective under specified conditions of usage.

use-dilution test Any test which, for a given DISINFECTANT, determines the USE-DILUTION – e.g. the KELSEY–SYKES TEST or the AOAC USE-DILUTION TEST.

usher protein See FIMBRIAE.

Usnea A genus of fruticose LICHENS (order LECANORALES); photobiont: a green alga. The thallus is greenish-grey, filamentous, extensively branched, usually pendulous, with a tough, whitish axial strand (cf. ALECTORIA). Isidia, soredia or papillae may be formed, according to species. *Usnea* spp occur chiefly on trees in regions of high humidity (particularly mountain woodlands). (See also USNIC ACIDS.)

usnic acids Yellow pigments (acetylphloroglucinol derivatives) which occur in the cortex in many lichens (e.g. *Cetraria islandica*, *Usnea* spp). Usnic acids have antibiotic activity against many Gram-positive bacteria (including *Mycobacterium tuberculosis*) and some fungi; they also have antitumour activity. Usnic acids have been used e.g. in the treatment of certain fungal diseases of plants, and in ointments for the topical treatment of burns or skin infections; they are relatively non-toxic to man, although they can cause an allergic skin reaction in sensitive individuals.

USR test Unheated serum reagin test: a qualitative STANDARD TEST FOR SYPHILIS in which samples of serum are tested without prior heating (cf. VDRL TEST).

Ussing chamber An in vitro arrangement of intestinal epithelium, or relevant cultured cells, used to examine changes in (energy-dependent) transmembrane ion flux promoted by enteric bacterial toxins. [Description and references: MR (1996) *60* 167–215 (p. 170).]

ustilagic acids Insoluble extracellular glycolipids produced by *Ustilago* spp e.g. when cultured in aerated liquid media; they are β-D-cellobiosides of ustilic acids (di- and trihydroxyhexadecanoic acids). Ustilagic acids inhibit e.g. Gram-positive and Gram-negative bacteria, and may be toxic to plants infected with *Ustilago* spp; they have been used in the perfume industry for preparing macrocyclic musks.

Ustilaginaceae See USTILAGINALES.

Ustilaginales (smut fungi; smuts) An order of characteristically plant-parasitic fungi of the class USTILAGINOMYCETES. Most species are parasitic only on flowering plants, infection usually being confined to the sexual organs, flowers, leaves and/or stem of the host plant (cf. ENTORRHIZA, MELANOTAENIUM); some smuts cause economically important plant diseases: see SMUTS (sense 2). Some species of smut fungi have been grown in vitro.

Morphology and life cycles of the smut fungi. The vegetative (somatic) form is a hyaline, septate, usually annual mycelium which generally occurs intercellularly in the plant host (intracellular hyphae are formed e.g. by *Ustilago maydis*), and which, according to species, may exhibit CLAMP CONNECTIONS and/or haustoria. Ballistoconidia (asexual spores) are formed by some smut fungi.

Basidiospores (= 'sporidia') are formed by most species of smut fungi (cf. e.g. *Ustilago nuda*); the life cycle from the basidiospore stage onwards appears to vary according to species. In some species the germination of a basidiospore can give rise to a hypha which is able to infect a host; some authors believe that such a hypha is monokaryotic and haploid, and that it subsequently undergoes DIKARYOTIZATION (by hyphal fusion) within the host, while others believe that the basidiospores which give rise to infective hyphae are themselves binucleate or diploid. In many other smuts, however, the mycelium derived from a basidiospore cannot infect a plant host; in such species dikaryotization must precede infection, and it may occur e.g. by the fusion of a basidiospore with a hypha, or by the fusion of buds formed by different basidiospores. Many smuts exhibit HETEROTHALLISM, and in such species the hyphae and/or basidiospores must be of appropriate MATING TYPES for fusion to occur.

Within an infected plant, the dikaryotic mycelium gives rise to TELIOSPORES (also called e.g. brand spores, chlamydospores, pseudospores, smut spores, teleutospores, ustilospores and ustospores); individual hyphal cells round up, each developing a thick wall and becoming a (typically dark-coloured) teliospore. Teliospores develop in masses (sori), a sorus being composed of numbers of individual teliospores (e.g. in *Tilletia* and *Ustilago*) or of pairs or clusters of teliospores (a cluster being called a *spore ball*). A spore ball is a definite structure composed of a characteristic arrangement of teliospores, or of teliospores and sterile cells; such arrangements are often genus-specific. Smut sori may occur on leaves, stems etc according to the species of smut fungus and host plant.

Teliospores may germinate immediately or behave as RESTING SPORES; those of some species can remain viable for many years. When the mature (uninucleate, diploid) teliospore germinates it generally gives rise to a basidium bearing basidiospores laterally and/or apically; the basidiospores are STATISMOSPORES.

The order contains three families [Book ref. 64, pp. 397–398]: Graphiolaceae, Tilletiaceae and Ustilaginaceae. Tilletiaceae: species characteristically form basidiospores apically on aseptate or uniseptate basidia; genera include e.g. ENTORRHIZA, ENTYLOMA, *Melanotaenium*, *Neovossia*, TILLETIA and *Urocystis*. Ustilaginaceae: species characteristically form transversely septate basidia bearing lateral basidiospores; genera include e.g. *Sphacelotheca*, *Tolyposporium* and USTILAGO.

[Ustilaginales of the British Isles: Mycol. Pap. (1984) Nr *154*.]

Ustilaginomycetes A class of typically plant-parasitic or saprotrophic fungi (subdivision BASIDIOMYCOTINA) which do not form sexual organs or give rise to macroscopic basidiocarps (cf. UREDINIOMYCETES); most species form phragmobasidia (cf. FILOBASIDIACEAE and SPORIDIOBOLUS). Two orders [Book ref. 64, pp. 398–399]: SPORIDIALES and USTILAGINALES.

Ustilago A genus of plant-parasitic fungi (order USTILAGINALES) which include some important pathogens of cereals (see e.g. SMUTS sense 2).

In a typical life cycle, the teliospore germinates to give rise to a septate promycelium from which bud (laterally and apically) a number of haploid basidiospores ('sporidia'); subsequent DIKARYOTIZATION may occur in various ways (see USTILAGINALES). In species which do not form basidiospores (e.g. *U. nuda*) the teliospores germinate to form hyphae between which dikaryotization occurs prior to infection of a fresh host.

[Ultrastructure of meiosis, postmeiotic mitosis, basidiospore development, and septation in *U. maydis*: Mycol. (1984) *76* 468–502.]

***Ustilago maydis* virus** See MYCOVIRUS.

ustilic acids See USTILAGIC ACIDS.

ustilospores See USTILAGINALES.

ustospores See USTILAGINALES.

uterotropic Having an affinity for the uterus.

UTI URINARY TRACT INFECTION.

utricle Any of various sac-like structures: e.g., a former term for an ASCUS. (See also CODIUM.)

Uukuniemi virus See UUKUVIRUS.

Uukuvirus A genus of viruses of the BUNYAVIRIDAE. Host range: various vertebrates (including birds and rodents); vectors: ticks. MWts of L, M and S RNAs: ca. 2.0–2.5, 1.0–1.3, and $0.4–0.6 \times 10^6$, respectively. MWts of proteins L, G1, G2 and N: ca. 180–200, 70–75, 65–70, and $20–25 \times 10^3$, respectively. The genus comprises a single serological group which includes the Uukuniemi virus (type species).

UV radiation See ULTRAVIOLET RADIATION.

uveitis Inflammation affecting part(s) of the *uvea* – the pigmented, vascular layer within the eyeball. There are various causes which include e.g. the varicella-zoster virus. (See also HTLV.)

***uvr* genes** Genes involved in the repair of DNA damaged by ULTRAVIOLET RADIATION (see EXCISION REPAIR); mutation in a *uvr* gene results in extreme sensitivity to UV radiation.

UvrABC-mediated repair See EXCISION REPAIR.

UWO University of Western Ontario Culture Collection, Department of Plant Sciences, London, Ontario N6A 587, Canada.

V

V Valine (see AMINO ACIDS).

V-DNA An in vitro form of ds cccDNA produced by annealing two complementary ss cccDNA molecules without strand breakage; V-DNA is thus characterized by a linking number of zero (see DNA). Any regions of right-handed double-helical conformation must be compensated for by negative supercoils and/or by regions of left-handed double-helical conformation. There is evidence that V-DNA of natural base sequence can adopt a combination of right-handed (B-DNA) and left-handed (Z-DNA) conformations [Nature (1982) *300* 545–546].

V factor A growth factor required by certain *Haemophilus* spp; the requirement is satisfied by NAD^+ or $NADP^+$ (or, minimally, by nicotinamide mononucleotide). 'V factor' occurs e.g. in yeast extract and in blood, although it is not available in whole blood since it occurs within the RBCs (which also contain NADase); heating of blood, as in the preparation of CHOCOLATE AGAR, releases the $NAD(P)^+$ and destroys the NADase. (V factor is destroyed e.g. by 121°C/30 min.) (See also X FACTOR and SATELLITE PHENOMENON.)

V forms See V–W TRANSITION.

v-onc See ONCOGENE.

V region (*immunol.*) See VARIABLE REGION.

v strand (V strand) (*virol.*) An ssDNA viral genome, or a strand homologous to it. (cf. C STRAND.)

V3 loop (of HIV) See AIDS.

VA mycorrhiza Vesicular-arbuscular MYCORRHIZA.

VA RNAs See SPLIT GENE (e).

vaccination (1) IMMUNIZATION (sense 1) by the parenteral administration of a VACCINE. (2) IMMUNIZATION (sense 1) by the oral or parenteral administration of a vaccine (cf. INOCULATION (2)). (3) *Syn.* IMMUNIZATION (1).

vaccine Any preparation administered with the object of stimulating the recipient's protective immunity to specific pathogen(s) and/or toxin(s). Until quite recently, all vaccines consisted of preparations of pathogen(s) or specific antigen(s); although many such vaccines are currently in use, there are now also vaccines which consist of the *DNA* which encodes relevant antigen(s): DNA VACCINES (q.v.).

Some vaccines (e.g. SABIN VACCINE) are given orally [oral vaccines: Microbiology (1994) *140* 215–224], while others (e.g. BCG) are administered parenterally.

A vaccine should elicit antibodies and/or a cell-mediated response to PROTECTIVE ANTIGENS. Antibodies must be formed in those parts of the body where they can efficiently counteract the specific pathogen or toxin (see, for example, SABIN VACCINE); moreover, antibodies must be present when required, and in sufficient quantity. Optimal protection against bacterial *enteric* pathogens is obtained when antigens act via the immune system of the gut mucosa; thus, for example, development of effective vaccines against enterotoxigenic *E. coli* (ETEC) depends on *oral* administration [RMM (1996) *7* 165–177].

(cf. *active* and *passive* IMMUNIZATION; see also VACCINATION.)

Vaccines may be categorized as follows:

(a) INACTIVATED VACCINES (see e.g. SEMPLE VACCINE and JAPANESE B ENCEPHALITIS). Inactivated vaccines against cholera (killed cells of *Vibrio cholerae* administered parenterally) give poor protection; currently they are rarely recommended.

(b) ATTENUATED VACCINES (see e.g. BCG and YELLOW FEVER). One problem with attenuated vaccines is their potential to cause active disease in immunocompromised patients. Moreover, such vaccines should not normally be given to pregnant women. [*Salmonella* vaccines: FEMS Reviews (2002) *26* 339–353.]

(c) TOXOIDS – used e.g. in the TRIPLE VACCINE.

(d) Synthetic vaccines and CONJUGATE VACCINES. The synthetic vaccines include synthetic peptides which correspond to antigenic determinants of toxins or pathogens (see INFLUENZA, FOOT AND MOUTH DISEASE, PLASMODIUM). Conjugate vaccines have been used against *Streptococcus pneumoniae*, but they are useful for only a limited range of serotypes; vaccines based on cell-surface proteins may offer better protection against this pathogen [TIM (1998) *6* 85–87]. Conjugate vaccines have been successful against *Haemophilus influenzae* type b [RMM (1996) *7* 231–241].

(e) Solutions/suspensions of particular antigens(s) or antigenic extracts from specific pathogen(s) – e.g. acellular vaccines (ACVs) against pertussis (whooping cough); these ACVs show reduced side-effects compared with the whole-cell vaccines (WCVs), but they may be less effective against *Bordetella parapertussis* than against *B. pertussis* [RMM (1996) *7* 13–21].

A pathogen's absolute requirement for IRON has prompted research on new vaccines which elicit antibodies against the pathogen's receptors for SIDEROPHILINS. For example, transferrin-receptor proteins (TbpA, TbpB) in *Neisseria meningitidis* are potential antigens for a new anti-meningitis vaccine, although problems exist with antigenic heterogeneity among strains of this pathogen [RMM (1998) *9* 29–37].

(f) Genetically engineered vaccines (e.g. VACCINIA VIRUS). A vaccine produced by recombinant DNA technology is used against hepatitis B.

(g) DNA vaccines: see DNA VACCINES.

Vaccines containing preservatives (for example, thiomersal, i.e. sodium ethylmercurithiosalicylate), or antibiotics, may cause allergic reactions in some patients.

(See also ARD; AUTOGENOUS VACCINE; CAPRINIZED VACCINE; KILLED VACCINE; LAPINIZED VACCINE; LIVE VACCINE; MIXED VACCINE; POLYVALENT VACCINE.)

[Vaccines, coming of age after 200 years: FEMS Reviews (2000) *24* 9–20.]

vaccinia subgroup See ORTHOPOXVIRUS.

vaccinia virus A virus of the genus ORTHOPOXVIRUS. Vaccinia virus has been used for many years as a 'live' vaccine against SMALLPOX (q.v.). It is usually assumed to have derived from the COWPOX VIRUS, but its evolutionary origin is uncertain. It could have resulted from genetic recombination between different orthopoxvirus strains, or from genetic changes occurring during serial passage of the cowpox (or variola) virus; alternatively, it could be a surviving representative of a virus now extinct in nature [Book ref. 148, p. 663]. Vaccinia virus has a broad host range and can replicate in many types of cell. On the CAM it typically produces large, opaque, white or haemorrhagic pocks; it is highly lethal for mice and for chick embryos. A-type inclusion bodies are not formed in infected cells. Genome MWt: ca. 120–124 × 10^6 (ca. 186000 bp). (See POXVIRIDAE for structure and replication cycle.)

Vaccinia virus is a useful vector for inserting foreign DNA into animal cells. It can infect a wide range of cells in culture, and even cell lines which do not allow completion of the viral replication cycle (and which are therefore not lysed by the virus) may be infected and may allow expression of viral early genes. Thus, e.g., a recombinant vaccinia virus containing a cDNA encoding Sindbis virus structural proteins has been used to infect cultured mosquito cells; infected cells were not lysed, and Sindbis virus proteins were synthesized and processed in their cytoplasm [JGV (1985) *66* 2761–2765]. Recombinant vaccinia viruses which can express antigens from unrelated pathogens (e.g. hepatitis B virus surface antigen [Nature (1983) *302* 490–495], or the circumsporozoite protein of a malarial parasite [Parasitol. (1986) *92* S109–S117]) have been suggested as 'live' vaccines against diseases caused by those pathogens [commentary: Nature (1986) *319* 549–550]. (See also AIDS.)

Vacuolaria See CHLOROMONADS.

vacuolating agent SIMIAN VIRUS 40.

vacuole Any of various membrane-delimited compartments within a cell: see e.g. CONTRACTILE VACUOLE, FOOD VACUOLE, GAS VACUOLE. In yeasts such as *Saccharomyces cerevisiae*, vacuoles function e.g. as repositories for the storage of e.g. POLYPHOSPHATE and certain amino acids, and apparently also function as LYSOSOMES, sequestering e.g. proteases and other hydrolytic enzymes. In mycelial fungi, vacuoles occupy a large proportion of the interior of older parts of the mycelium, possibly serving e.g. to concentrate the bulk of the cytoplasm at the growing hyphal tip (see GROWTH (b)).

vacuum autoclave See AUTOCLAVE.

vacuumizing See CANNING.

vagina microflora In adult women (post-puberty, pre-menopause) the vagina harbours a varied microflora in which species of *Lactobacillus* (particularly *L. acidophilus*) typically predominate; these bacteria metabolize glycogen and/or its breakdown products (present in the vaginal secretions), thus maintaining a low pH (< ca. 4.5) in the vagina – which helps to protect against CANDIDIASIS, TRICHOMONIASIS, and other vaginal infections (see also BACTERIAL VAGINOSIS). Other organisms which may be present in the vagina in a high proportion of healthy women include e.g. species of *Acinetobacter*, *Bacteroides*, *Bifidobacterium*, *Corynebacterium*, members of the Enterobacteriaceae, *Moraxella*, *Peptococcus*, *Peptostreptococcus*, *Staphylococcus*, *Streptococcus*, *Ureaplasma*, etc. [Changes in the vagina microflora during the menstrual cycle: AJRIM (1985) *9* 1–5.] Before puberty and after the menopause, the vagina is less acidic and it harbours a microflora which may include e.g. corynebacteria, enterobacteria, streptococci, etc. (See also GENITOURINARY TRACT FLORA.)

Vaginicola See PERITRICHIA.

vaginitis Inflammation of the vagina; it may involve any of a range of organisms: see e.g. CANDIDIASIS, TRICHOMONIASIS. (cf. BACTERIAL VAGINOSIS.) In *vulvovaginitis* both the vulva and vagina are affected. (See also ZEARALENONE.)

vaginosis See BACTERIAL VAGINOSIS.

Vahlkampfia A genus of amoebae (order SCHIZOPYRENIDA) in which flagellate stages are unknown. Cysts are commonly formed. Species are mainly free-living in freshwater, soil and marine habitats; *V. patuxent* is parasitic in the alimentary tract in oysters. (See also BEE DISEASES.)

VAHS (virus-associated haemophagocytic syndrome) See HAEMOPHAGOCYTIC SYNDROME.

valaciclovir An ANTIVIRAL AGENT used in the treatment of alphaherpesvirus infections; valaciclovir is the L-valyl ester of ACYCLOVIR.

valency (*immunol.*) The number of COMBINING SITES per monomeric or polymeric ANTIBODY; the term is sometimes also used to refer to the number of DETERMINANTS per antigen. (See also POLYVALENT ANTISERUM; POLYVALENT VACCINE.)

valid publication See NOMENCLATURE.

L-valine biosynthesis See Appendix IV(b).

valinomycin A DEPSIPEPTIDE ANTIBIOTIC, produced by *Streptomyces* sp, which can act as a mobile carrier IONOPHORE for the UNIPORT of Rb^+, K^+, Cs^+, or $NH_4{}^+$ across mitochondrial, thylakoid, bacterial or artificial membranes; valinomycin has a ca. 10^4-fold lower selectivity for Na^+ than for K^+. Valinomycin is a cyclic molecule containing 12 residues: (–D-valine–L-lactic acid–L-valine–D-α-hydroxyisovaleric acid–)$_3$; a single ion can be carried at the centre of the molecule – any ion being carried having, of necessity, previously lost its water of hydration. Valinomycin can be used e.g. to alter or abolish a membrane potential (see CHEMIOSMOSIS).

valley fever *Syn.* COCCIDIOIDOMYCOSIS.

Valonia A genus of marine, mainly tropical, siphonocladous green algae (division CHLOROPHYTA) in which the cells are inflated to form macroscopic vesicles.

VAMP See TETANOSPASMIN.

Vampirovibrio A genus of Gram-negative bacteria which are predatory on viable cells of *Chlorella*; a predatory bacterium attaches to the prey cell and grows without penetration (cf. BDELLOVIBRIO). The organisms are vibrioid, 0.3–0.6 μm in diameter, and each cell has a single, polar, non-sheathed flagellum. GC%: ca. 50. Type species: *V. chlorellavorus*.

Vampyrella See FILOSEA.

***van* genes** (*vanA*, *vanB*) See ENTEROCOCCUS.

vancomycin A complex glycopeptide ANTIBIOTIC produced by certain actinomycetes. Vancomycin inhibits PEPTIDOGLYCAN biosynthesis by binding to the D-alanyl-D-alanine of the pentapeptide in the bactoprenol-bound disaccharide–pentapeptide – thus preventing transfer of the disaccharide–pentapeptide from the membrane to the periplasmic site of incorporation.

Vancomycin is active against many Gram-positive bacteria; it is used clinically e.g. in the treatment of pseudomembranous colitis and infections involving MRSA. (Most Gram-negative bacteria are resistant as they are impermeable to the drug.)

Structurally related antibiotics (also obtained from actinomycetes) with apparently similar modes of action include actaplanin, actinoidin, AVOPARCIN, RISTOCETIN and TEICOPLANIN.

[Enterococci resistant to vancomycin: Drugs (1996) *51* (supplement 1) 6–12. Screening for vancomycin-resistant enterococci with multiplex PCR: JCM (1999) *37* 2090–2092. Reduced susceptibility of MRSA to vancomycin: JAC (1997) *40* 135–136; Lancet (1999) *353* 1587–1588.]

vanillin 3-Methoxy-4-hydroxybenzaldehyde.

Vannella See AMOEBIDA and PSEUDOPODIUM.

vapam *Syn.* METHAM SODIUM.

var. VARIETY (q.v.).

variable region (V region) (*immunol.*) Any of the four regions of a given type of (monomeric) Ig molecule (see IMMUNOGLOBULINS) in which compositional variation is most marked among antibodies of different specificity; one V region occurs (as a DOMAIN) at the N-terminal end of each heavy chain and light chain. (See also COMBINING SITE; HYPERVARIABLE REGION.)

variant A strain which differs from other related strains in a (specified or unspecified) way. (cf. VARIETY.)

variant antigen type See VSG.

variant surface glycoprotein See VSG.

varicella *Syn.* CHICKENPOX.

varicella-zoster virus See ALPHAHERPESVIRINAE.

variegation (*plant pathol.*) Patchy – or otherwise irregular – colour variation in leaves, petals etc; variegation may be genetically based or may result from e.g. virus infection. (cf. COLOUR-BREAKING; MOSAIC; MOTTLE.)

variety (var.) In mycology: a rank between subspecies and subvariety. In bacteriology: formerly a rank equivalent to subspecies; currently an infrasubspecific rank which has no official standing in nomenclature. A variety may be distinguished from other varieties within a given subspecies by e.g. metabolic and/or physiological properties; it is then called a *biovar* (= *biotype*). Other criteria used include e.g. morphological feature(s) (*morphovar*; *morphotype*); pathogenicity for specific host(s) (*pathovar*, abbreviated pv.; *pathotype*); susceptibility to lysis by specific bacteriophage(s) (*phagovar*; *phagotype*); serological (antigenic) characteristics (*serovar*; *serotype*).

N.B. All of the above terms are often used, loosely, in a non-taxonomic sense.

variola major See SMALLPOX.

variola minor See SMALLPOX.

variola ovina *Syn.* SHEEP POX.

variola sine eruptione See SMALLPOX.

variola virus A virus of the genus ORTHOPOXVIRUS; it has a narrow host range (apparently only man), and forms small, opaque, white pocks on the CAM. Two main types of variola virus are distinguished: variola major virus (causal agent of classical SMALLPOX), which can form pocks on CAM at 38.5°C, and variola minor virus (causal agent of alastrim), which does not form pocks on CAM at temperatures >38°C. Variola virus is not highly lethal in mice or in chick embryos; variola major virus is more lethal than variola minor virus for chick embryos. A-type inclusion bodies are not formed in infected cells. Genome MWt: ca. 120×10^6.

Varney jar A type of ANAEROBIC JAR within which O_2 is removed by the combustion of phosphorus.

vascular cell adhesion molecule-1 See CD106.

vascular-streak dieback A CACAO DISEASE which occurs e.g. in Papua-New Guinea; the aerial parts of the plant are affected, and the causal agent is believed to be *Oncobasidium theobromae*.

vascular wilt (wilt) (*plant pathol.*) Any of various plant diseases in which infection of the vascular system by a fungal or bacterial pathogen results in wilting (flaccidity) of the plant. Many fungi which cause wilts – e.g. *Fusarium* spp (see FUSARIUM WILT), *Verticillium* spp (e.g. *V. albo-atrum*, *V. dahliae*) – are soil-borne pathogens which infect via the roots and become established in the vascular system of the plant; these fungi can also be seed-borne. Other wilt pathogens – e.g. ERWINIA spp, *Pseudomonas solanacearum*, and the causal agents of DUTCH ELM DISEASE and OAK WILT – gain entry to the vascular system via wounds in the aerial parts of the plant. Factors which may be involved in pathogenesis include e.g. the formation of high-MWt gummy polysaccharides which block the vessels, blockage of vessels by the pathogen itself, TYLOSE formation, and the collapse of infected vessels – all of which cause water losses to exceed water uptake; in at least some cases TOXINS produced by the pathogen may also be involved (see e.g. FUSARIC ACID, LYCOMARASMIN).

vasculotoxin See SHIGA TOXIN.

Vaspar A mixture of Vaseline and paraffin wax, used e.g. to make an airtight seal in vessels used for anaerobic culture.

VAT Variant antigen type: see VSG.

Vaucheria A genus of siphonaceous algae (class XANTHOPHYCEAE). The vegetative organism consists of long, irregularly branched filaments (diam. ca. 50–200 μm) within which a thin peripheral layer of cytoplasm (containing numerous chloroplasts and nuclei) forms a 'tube' surrounding a large central vacuole; the vacuole extends the length of the filament, stopping short of the tip. The organisms may be terrestrial, being anchored to e.g. moist soil by rhizoid-like branches, or aquatic, forming bright-green mats e.g. in stagnant/slow-moving fresh waters.

Vaucheria spp differ from other xanthophytes in many respects: e.g., they form a unique type of zoospore, their sexual reproduction is oogamous, and the spermatozoid is the only cell to resemble the 'typical' xanthophycean heterokont cell. In zoospore formation, the swollen tip of a filament is closed off by a septum to form a zoosporangium within which the (multinucleate) protoplast develops one pair of flagella for each nucleus it contains. The resulting multinucleate, multiflagellate zoospore is released and remains (sluggishly) motile for a short time, after which it withdraws its flagella and germinates to form a new thallus. In terrestrial organisms, the entire contents of the sporangium may develop into a single aplanospore. *Sexual* reproduction involves the formation of antheridia and oogonia on lateral branches, each being separated from the rest of the thallus by a septum. After release of the spermatozoids and fertilization, the zygote develops a thick wall and remains dormant for some time; meiosis is believed to occur on germination.

Vauchomia See PERITRICHIA.

VBNC See VNC.

VCAM-1 See CD106.

VCN agar *Syn.* THAYER–MARTIN AGAR.

VD VENEREAL DISEASE.

VD-1827 SULTAMICILLIN.

VDAC Voltage-dependent anion channel: a PORIN channel in the mitochondrial outer membrane.

VDRL test (Venereal Disease Research Laboratory test) A qualitative or quantitative STANDARD TEST FOR SYPHILIS. A fixed volume of the test antigen is added to a fixed volume of the patient's INACTIVATED SERUM (or a known dilution of it) on a slide; mixing is conducted under standard conditions for a fixed period of time. Flocculation (aggregation of the antigen particles into clumps) occurs in the presence of reagins, and is detected under the microscope (×100); large clumps are reported as 'reactive', small clumps as 'weakly reactive', and a regular or finely granular suspension of antigen as 'non-reactive'. Strongly positive sera may exhibit a PROZONE.

vector (1) (*med., vet., plant pathol.*) Any living organism – conventionally, an invertebrate, or a microorganism (see e.g. PLASMODIOPHOROMYCETES) – which effects the transmission of a parasite (e.g. a pathogenic bacterium, fungus, protozoon or virus) from one individual (man, animal or plant) to another. (cf. CARRIER; VEHICLE.) In many cases a parasite can be transmitted by only one or a few species of vector; in other cases there may be little or no parasite–vector specificity. Common vectors of human and animal parasites include flies, lice, mosquitoes and ticks, while vectors of plant parasites include insects such as aphids, leafhoppers, and whiteflies, as well as mites, fungi (see e.g. OLPIDIUM), nematodes, etc. (See also TRIATOMINE BUGS.)

In a *biological vector* the parasite replicates and/or undergoes one or more stages of development in the vector (see e.g. CYCLICAL TRANSMISSION and *propagative* CIRCULATIVE TRANSMISSION). In a *mechanical vector* the parasite is transported more or less passively by the vector (cf. NON-CIRCULATIVE TRANSMISSION).

(See also TRANSOVARIAL TRANSMISSION and TRANS-STADIAL TRANSMISSION.)

(2) (*mol. biol.*) See e.g. CLONING, EXPRESSION VECTOR, PHAGEMID, SHUTTLE VECTOR.

vectorial electron transfer See EXTRACYTOPLASMIC OXIDATION.

vectorial group translocation Directional translocation of given molecular or ionic species across a membrane as a consequence of the presence of fixed pathways across the membrane; in this concept, translocation of the species does not involve conformational changes in the membrane proteins – the latter merely providing, passively, suitable transmembrane routes. This concept is exemplified in the LOOP MODEL of proton translocation by the respiratory chain (see ELECTRON TRANSPORT CHAIN and CHEMIOSMOSIS).

VEE VENEZUELAN EQUINE ENCEPHALOMYELITIS.

vegetable spoilage See e.g. CANNING; GANGRENE (2); PECTIC ENZYMES; PICKLING; SOFT ROT (2).

vegetative (1) (assimilative; somatic; trophic) Refers to a cell or organism, or to a stage in an organism's life cycle, in which nutrition and growth (as opposed to e.g. sexual reproduction or dormancy) predominate.

(2) (*virol.*) Refers to the 'growth phase' of a virus (i.e., the phase in which progeny virions are formed), or to a virus in the 'growth phase'.

vegetative mycelium (in actinomycetes) *Syn.* SUBSTRATE MYCELIUM.

vehicle (1) (*epidemiol.*) An inanimate medium in or on which a (usually pathogenic) microorganism may be transmitted: e.g., unpasteurized milk. (cf. VECTOR.)

(2) (*mol. biol.*) *Syn.* cloning vector (see CLONING).

(3) (*mol. biol.*) The organism in which a recombinant DNA molecule is replicated during CLONING experiments.

veil (*mycol.*) See PARTIAL VEIL and UNIVERSAL VEIL.

Veillon tube A glass tube (e.g. 25 cm long, internal diam. 1 cm) which can be sealed at both ends with rubber bungs; it is used for SHAKE CULTURE (sense 1).

Veillonella A genus of Gram-negative bacteria (family VEILLONELLACEAE) which occur e.g. in the mouth and in the intestinal and respiratory tracts of man and other animals. Cells: cocci, 0.3–0.5 μm diam., often in pairs. Some species (*V. criceti*, *V. dispar*, *V. ratti*) form a pseudocatalase. Colonies fluoresce red under ultraviolet light (360 nm). The organisms ferment e.g. lactate (forming acetate, propionate, CO_2 and H_2), fumarate and pyruvate; in general, carbohydrates are attacked weakly or not at all. GC%: (Bd) ca. 40–44. Type species: *V. parvula*. Other species: *V. atypica*, *V. caviae*, *V. rodentium*. (*V. alcalescens* has been included within *V. parvula*.)

Veillonellaceae A family of Gram-negative, oxidase-negative, chemoorganotrophic, anaerobic bacteria which occur e.g. in the intestinal tract in man, ruminants and other animals. Cells: non-motile cocci, ca. 0.3–2.5 μm diam., which usually occur in pairs (adjacent sides of cells may be flattened); cells may resist decolorization in the Gram stain. No species forms catalase, but some strains form a PSEUDOCATALASE. Genera: ACIDAMINOCOCCUS, MEGASPHAERA, VEILLONELLA. [Book ref. 22, pp. 680–685.]

vein-banding (*plant pathol.*) A narrow band of tissue of altered coloration on each side of the main vein(s) of a leaf – a symptom of certain plant diseases.

vein-clearing (*plant pathol.*) Increased transparency of the veins in a leaf – a symptom of certain plant diseases.

velar (*mycol.*) Of or pertaining to a VELUM.

velleral See LACTARIUS.

velum (*mycol.*) *Syn.* veil (see e.g. UNIVERSAL VEIL).

velvet disease A FISH DISEASE caused by *Oodinium* species; affected fish appear 'dusty' and gills may be heavily infested. *O. ocellatum* attacks marine fish, *O. limneticum* attacks freshwater fish. The disease may be fatal in aquaria.

velvet tobacco mottle virus (VTMoV) A PLANT VIRUS which infects plants of the Solanaceae; it is transmitted by insects (e.g. coccinellid beetles) and can be transmitted mechanically. Virion: icosahedral, ca. 30 nm in diam., containing one major polypeptide species. Genome: a linear ssRNA, RNA1 (ca. 4.5 kb), and a circular ssRNA, RNA2 (also known as a VIRUSOID q.v.); both RNA1 and the virusoid are necessary for infectivity. Closely related viruses include lucerne transient streak virus (LTSV), *Solanum nodiflorum* mottle virus (SNMV), and subterranean clover mottle virus (SCMoV).

venereal disease (VD) (1) Any disease transmissible by sexual intercourse. The 'classical' venereal diseases are CHANCROID, GONORRHOEA, GRANULOMA INGUINALE, LYMPHOGRANULOMA VENEREUM and SYPHILIS. (cf. SEXUALLY TRANSMITTED DISEASE.) (2) *Syn.* SEXUALLY TRANSMITTED DISEASE.

venereal wart See PAPILLOMA.

Venezuelan equine encephalomyelitis (VEE; Venezuelan equine encephalitis; Venezuelan encephalitis) An acute disease of man and horses caused by an ALPHAVIRUS; it occurs in epidemics and epizootics in northern South America, Central America, the West Indies, and southern USA. Small mammals appear to be a reservoir of infection; transmission occurs via mosquitoes (e.g. *Aedes* and *Psorophora* spp). In horses, the disease may be rapid and fulminating with high mortality rates. In man, the disease is usually benign; symptoms include fever, chills, headache, nausea, and pains in muscles, bones and joints. (Symptoms of encephalitis are uncommon in man.)

vent In a PETRI DISH: a small projection on the periphery of the inner surface of the lid; a Petri dish may have one or three vents ('single-vent' or 'triple-vent'). Vent(s) slightly raise the lid of the (closed) Petri dish, thus facilitating equilibration between the air/gas(es) inside and outside the Petri dish. (A non-vented plastic Petri dish can be vented by pressing a red-hot wire on the edge of the wall to make two or three vertical 2-mm slits at different locations.)

Venturia A genus of fungi (order DOTHIDEALES) which include plant parasites and pathogens.

V. inaequalis causes APPLE SCAB (q.v.). Within the host the mycelium gives rise to a subcuticular stroma from which arise numerous, short, erumpent conidiophores – each bearing a single, terminal, flask-shaped conidium. Conidia continue to be formed until early autumn when the ascocarps – individual pseudoperithecia – start to develop in stromata within the dead and dying leaves. During ascocarp formation, fertilization (which involves GAMETANGIAL CONTACT) occurs via a trichogyne which projects through the stromal initial; since *V. inaequalis* exhibits HETEROTHALLISM, male nuclei must be derived from an antheridium of appropriate mating type. Each pseudoperithecium contains a layer of asci and pseudoparaphyses, and it bears several setae around the rim of its ostiolar pore; the yellowish ascospores each have a septum which divides the spore into two cells of unequal sizes.

vermicule (1) The ookinete of PLASMODIUM. (2) A uninucleate, motile cell of *Babesia* or *Theileria* formed in the gut of the tick vector.

vernamycins See STREPTOGRAMINS.

Vero An ESTABLISHED CELL LINE derived from the kidney of an African green monkey (*Cercopithecus aethiops*); the cells are heteroploid and fibroblast-like.

Vero cell cytotoxin See SHIGA TOXIN.

verocytotoxins See SHIGA TOXIN and EHEC.

veronal *Syn.* BARBITAL.

verotoxin See SHIGA TOXIN.

Verpa A genus of fungi (order PEZIZALES) in which the ascocarp, an APOTHECIUM, consists of a stipitate, bell-shaped pileus – the tip of the stipe being attached to the inside of the pileus at its centre; the hymenium develops on the outer surface of the pileus.

verruca *Syn.* wart – see PAPILLOMA.

Verrucaria A genus of crustose LICHENS (order VERRUCARIALES). The thallus may be white, grey, brown or black, and may be epilithic or ENDOLITHIC (e.g. *V. hochstetteri* in calcareous rocks). The pseudothecia may be superficial or immersed in the thallus. Some species are aquatic (e.g. *V. aquatilis* occurs on rocks in streams), while e.g. *V. maura* forms an extensive, thin, black thallus covering rocks on the seashore in the intertidal and supratidal zones.

Verrucariales An order of (mainly lichenized) fungi of the ASCOMYCOTINA. Most members are crustose saxicolous LICHENS; some are foliose or squamulose lichens, and a few are lichenicolous or saprotrophic fungi. Bitunicate asci are formed in pseudothecia (often termed 'perithecia'). Genera include e.g. CATAPYRENIUM, DERMATOCARPON, *Endocarpon*, *Placidiopsis*, *Staurothele*, *Thelidium*, VERRUCARIA.

verrucarins See TRICHOTHECENES.

verrucologen A tremorgenic MYCOTOXIN produced e.g. by *Penicillium piscarium* and *P. estinogenum*.

verrucose dermatitis *Syn.* CHROMOBLASTOMYCOSIS.

verrucous (verrucose) Of or pertaining to warts (see PAPILLOMA); wart-like; bearing wart-like elevations.

verruga peruana See BARTONELLOSIS.

Versene A sodium salt of EDTA.

vertical illumination See MICROSCOPY (g).

vertical resistance (*plant pathol.*) In a given cultivar: the existence of significantly higher levels of resistance to some races of a given pathogen than to others. A cultivar which is resistant to an existing race of a pathogen may become susceptible when a new race of the pathogen emerges, i.e., resistance to the pathogen tends to be temporary. (cf. HORIZONTAL RESISTANCE.)

vertical transmission The transmission of a disease or parasite from a parent to its offspring via the egg (TRANSOVARIAL TRANSMISSION), via the placenta (transplacental transmission), or by genetic inheritance (see RETROVIRIDAE). (cf. HORIZONTAL TRANSMISSION.)

verticil Chains of spores arranged in a whorl, or a number of structures arranged in a whorl.

verticillate Whorled.

Verticillium A genus of fungi (class HYPHOMYCETES) which include some important plant-pathogenic species – see e.g. COTTON WILT and VASCULAR WILT; an entomopathogenic species, *V. lecanii* ('*Cephalosporium lecanii*'), has been used successfully against various insect pests under both glasshouse and field conditions [AEE (1985) *12* 151–156]. (See also BUBBLE DISEASES.) The mycelium is septate; colourless or pale, non-septate conidia (sometimes in slimy masses) are formed (from phialides) at the tips of whorled branches on the conidiophores.

verticillium wilt See VASCULAR WILT.

vesicle (*med., vet.*) In skin or mucous membranes: a raised lesion containing a clear watery fluid. (cf. PUSTULE.)

vesicle-associated membrane protein See TETANOSPASMIN.

vesicular-arbuscular mycorrhiza See MYCORRHIZA.

vesicular exanthema (of swine) An acute, febrile PIG DISEASE clinically indistinguishable from FOOT AND MOUTH DISEASE; outbreaks have occurred mainly in the USA. The causal agent is a calicivirus (see CALICIVIRIDAE).

vesicular stomatitis (*vet.*) An acute, infectious disease which affects e.g. cattle, horses and pigs. The causal agent is a VESICULOVIRUS; infection appears to occur via skin lesions or insect bites. Clinically, vesicular stomatitis can resemble FOOT AND MOUTH DISEASE but vesicles may or may not occur in the mouth and are very uncommon on the feet; in milking cows extensive lesions may occur on the teats and the milk yield is appreciably decreased.

In man, infection results in influenza-like symptoms.

Vesiculovirus A genus of viruses of the RHABDOVIRIDAE; type species: vesicular stomatitis virus (VSV), Indiana serotype. VSV can infect a very wide range of animals (including man) and birds; the principal natural hosts appear to be horses, cattle and pigs – see VESICULAR STOMATITIS. The virus can replicate rapidly in many types of cultured cells. The RNA genome (MWt ca. $3.6–4.2 \times 10^6$) contains 5 genes in tandem and a leader sequence of 47 nucleotides: 3′-leader–N–NS–M–G–L-5′. The RNA–N protein nucleocapsid functions as the template for transcription; it is associated with the L protein (which appears to be the transcriptase) and the NS protein (which may function in transcriptional control) [JGV (1985) *66* 1011–1023]. [The transcription complex of VSV (minireview): Cell (1987) *48* 363–364.] Transcription occurs in a linear and sequential manner from the 3′ end, generating a transcript of the leader sequence and 5 monocistronic mRNAs. The mRNAs are all capped and polyadenylated. The leader transcript is neither capped nor polyadenylated; it appears to be responsible for the inhibition ('shut-off') of host cell DNA, RNA and protein synthesis characteristic of VSV infection [review: MS (1985) *2* 152–156]. (However, host protein synthesis is apparently also necessary for inhibition of host RNA synthesis – possibly indicating that the active inhibitor is a nucleoprotein complex containing the leader RNA transcript [Virol. (1985) *140* 91–101].) VSV nucleocapsids are synthesized in the cytoplasm; budding occurs mainly through the plasma membrane.

Vestibuliferia A subclass of ciliates (class KINETOFRAGMINOPHOREA) in which the cytostome is at the base of a VESTIBULUM located apically, subapically or laterally. Cilia may cover the entire body or may be restricted to particular regions; oral ciliature is characteristically somatic in origin. The cytopharyngeal apparatus resembles the RHABDOS type. Orders: COLPODIDA, ENTODINIOMORPHIDA, TRICHOSTOMATIDA.

vestibulum (vestibule) (1) In ciliates of the VESTIBULIFERIA: a depression in the body surface leading to the CYTOSTOME; the ciliature is characteristically somatic in origin but may include distinctive membranelles. (2) *Syn.* PREBUCCAL AREA.

Vexillifera See AMOEBIDA.

V_H region See IMMUNOGLOBULINS.

***vh2* gene** (*vH2* gene) See PHASE VARIATION.

VHF VIRAL HAEMORRHAGIC FEVER.

Vi antigens Polysaccharide cell-surface (microcapsular) antigens which occur in some enterobacteria – e.g. certain strains of *Escherichia coli*, *Citrobacter freundii* (see also PHASE VARIATION), and *Klebsiella*, and three (of ca. 2000) serotypes of *Salmonella* (certain strains of *S. typhi*, *S. paratyphi-C*, and *S. dublin*); in *S. typhi* the Vi antigen is a $(1 \rightarrow 4)$-α-linked polymer of *N*-acetylgalactosaminuronic acid. Vi antigens may play a role in the virulence of a pathogen for a particular host(s).

Vi antigens are detected in the serological characterization of certain enterobacteria (see e.g. KAUFFMANN–WHITE CLASSIFICATION). Cells with Vi antigens are agglutinated by homologous Vi-antisera. However, the (cell-surface) Vi antigens may, to a variable extent, mask the O ANTIGENS and inhibit agglutination by O-antisera; the Vi antigens are generally destroyed by heat-treating the cells (100°C/10–30 min) which may then be agglutinated by homologous O-antiserum.

Vi phage typing See PHAGE TYPING.

viable In microbiology: refers to an organism capable of reproducing (or, in the case of a virus, replicating) under appropriate conditions.

viable but non-cultivable See VNC.

viable cell count See COUNTING METHODS.

vial (ampoule) (1) Any of a range of types of tubular glass vessel, closed at one end, in which (e.g.) a cell suspension can be freeze-dried (see FREEZE-DRYING) or frozen (see FREEZING) for storage. Before use, a vial is stoppered with a cotton wool plug and sterilized in a hot-air oven. (2) A vial as in (1) together with its freeze-dried or frozen contents.

vibramycin See TETRACYCLINES.

vibrio (1) A bacterium of the genus *Vibrio*. (2) Any curved, rod-shaped bacterial cell, regardless of taxon.

Vibrio A genus of Gram-negative bacteria of the family VIBRIONACEAE; the genus has been classified in the gamma group of Proteobacteria. The cells are straight or curved rods, 0.5–0.8 × 1.4–2.6 μm; INVOLUTION FORMS are common under appropriate conditions. In liquid media most species are motile with one sheathed polar flagellum; the sheath is continuous with – but compositionally somewhat different from – the OUTER MEMBRANE of the cell. On solid media, some species (e.g. *V. alginolyticus*, *V. campbellii*, *V. harveyi*, *V. parahaemolyticus*, some strains of *V. fluvialis*) form additional, unsheathed, lateral flagella (L-flagella); lateral and polar flagella are structurally and antigenically distinct and differ in wavelength (ca. 0.9 μm and ca. 1.4–1.8 μm, respectively). On solid complex media, *V. alginolyticus*, *V. parahaemolyticus* and *V. proteolyticus* form elongated cells capable of SWARMING (apparently involving the L-flagella). [Regulation of transcription of L-flagella genes in *V. parahaemolyticus*: JB (1986) *167* 210–218.]

At least low concentrations of NaCl (Na^+) are essential for growth in all species of *Vibrio* except *V. cholerae* and some strains of *V. metschnikovii* – although NaCl stimulates growth even in these species; *V. costicola* requires a minimum concentration of 600–700 mM NaCl.

All species can grow at 20°C, most at 30–35°C, some at 40°C; *V. logei* and *V. marinus* grow only at or below 20°C and can grow at 4°C.

Many species can tolerate moderately high pH (*V. cholerae* and *V. metschnikovii* can grow at pH 10). Enrichment media for clinical use include APW and various tellurite-containing media (see e.g. MONSUR MEDIUM). Selective media include e.g. TCBS AGAR. [Medium for differentiating *V. alginolyticus* and *V. parahaemolyticus*: AEM (1983) *45* 310–312.]

V. fischeri and *V. logei* form a cell-associated yellow-orange pigment after 3–4 days on complex solid media. *V. gazogenes* forms PRODIGIOSIN, and *V. nigripulchritudo* forms a characteristic insoluble blue-black pigment, crystals of which accumulate in colonies on basal media.

Most species are oxidase +ve. Glucose is fermented by the MIXED ACID FERMENTATION with the formation of acid but (usually) without gas. Various sugars can be fermented, but lactose is usually not fermented (exceptions: strains of *V. metschnikovii* and *V. vulnificus*). Most species do not require organic growth factors. PHB is usually not accumulated. Most strains are sensitive to O/129 (q.v.). Some species exhibit BIOLUMINESCENCE (e.g. *V. splendidus* biotype I, *V. fischeri*, *V. logei*, some strains of *V. harveyi* and *V. cholerae*); a strain of *V. logei* emits an unusual yellow light at temperatures <18°C but emits blue-green light at temperatures >23°C. Some strains have been reported to be capable of nitrogen fixation [e.g. *V. diazotrophicus*: IJSB (1982) *32* 350–357].

Vibrio spp are primarily aquatic, occurring in freshwater, estuarine and marine habitats and in association with aquatic organisms – including e.g. planktonic copepods (*V. cholerae*, *V. parahaemolyticus*). Some strains are pathogenic in man (*V. cholerae*, *V. parahaemolyticus*) and in shellfish (e.g. *V. tubiashii*, a pathogen of bivalve molluscs [IJSB (1984) *34* 1–4], *V. alginolyticus*, *V. logei*). The species formerly known as *V. anguillarum* (see LISTONELLA) is a pathogen of eels. The bioluminescent *V. fischeri* (see also PHOTOBACTERIUM) occurs e.g. in specialized luminous organs in certain teleost fish and squid. Most vibrios are CHITINolytic and may be important in the primary degradation of chitin in aquatic ecosystems. Vibrios also occur in salted foods and brines (see e.g. MEAT SPOILAGE and CURING sense 1).

The genus includes species formerly classified in the genera *Beneckea*, LUCIBACTERIUM and PHOTOBACTERIUM (q.v. in respect of *V. fischeri* and *V. logei*). (See also LISTONELLA and WOLINELLA.)

GC%: 38–51. Type species: *V. cholerae*.

V. alginolyticus is commonly isolated from temperate and tropical coastal and estuarine waters and seafoods; it has also been isolated from (human) wounds, from the stools of gastroenteritis patients, and from diseased and dead shellfish. L-flagella and swarming occur on complex solid media. Growth occurs at 30°C and 40°C but not at 4°C. Oxidase +ve. VP +ve. Sucrose +ve. D-Gluconate +ve. Arginine dihydrolase −ve. Respiration can be coupled to the development of SODIUM MOTIVE FORCE, smf being used e.g. for the uptake of amino acids.

V. anguillarum. See LISTONELLA.

V. cholerae. CHOLERA-causing strains produce two main virulence factors: (i) CHOLERA TOXIN (encoded by BACTERIOPHAGE CTXφ) and (ii) so-called 'toxin co-regulated pili' – appendages which mediate adhesion to the intestine (and which are encoded by a chromosomal PATHOGENICITY ISLAND). (Strains of *V. cholerae* which do not produce cholera toxin may produce the Ace and Zot toxins; such strains may give rise to a less severe form of disease.)

The cells do not form L-flagella; rotation of the polar flagellum is energized by SODIUM MOTIVE FORCE [JB (1999) *181* 1927–1930]. Growth can occur in the absence of NaCl, and at 30°C and 40°C but not at 4°C. Oxidase +ve. Sucrose +ve. D-Gluconate +ve. Arginine dihydrolase −ve. Nitrate is reduced.

The genome consists of two (dissimilar) chromosomes [Nature (2000) *406* 477–483].

Colonies on nutrient agar are round, entire, glossy and translucent, 1–2 mm in 24 hours. A thin PELLICLE is formed in peptone water.

V. cholerae may be killed e.g. by heating to 60°C for 15 minutes (although it is preferable to boil unsafe water.)

Serotypes are distinguished on the basis of O antigens (H antigens appear to be common to all strains and are therefore not suitable for typing.)

Prior to 1992, all cholera-causing strains of *V. cholerae* were of the O1 serogroup (also referred to as serovar 1, O1, O:1, O-1 etc.). The O1 strains include two biotypes: (i) El Tor (also called Eltor, or *eltor*) and (ii) *cholerae* (= classical biotype); the El Tor biotype is more resistant than the classical biotype to adverse conditions. The original El Tor isolates were strongly haemolytic for sheep blood, but strains subsequently isolated were less strongly haemolytic, or even non-haemolytic. Typical El Tor strains are VP +ve, agglutinate chicken erythrocytes, and are resistant to group IV phages and to polymyxin B; typical *cholerae* strains have the opposite characteristics.

O1 strains can be subdivided into variants designated Ogawa, Inaba and Hikojima [Book ref. 46, p 1280].

Multiple-drug-resistant O1 strains have been reported [Lancet (1997) *349* 924].

In 1992 non-O1 choleragenic strains were isolated in India; these strains are designated O139 Bengal [RMM (1996) *7* 43–51]. Vaccines against O1 strains do not protect against O139 strains; in one approach to the development of anti-O139 vaccines, genes for certain of the pathogen's virulence factors have been deleted [see e.g. JID (1994) *170* 278–283]. [A rapid screening test for O139 strains based on monoclonal antibodies: JCM (1998) *36* 3595–3600.]

Strains other than those of serogroups O1 and O139 may cause gastrointestinal/diarrhoeal disease – sometimes mild, but in some cases of cholera-like severity.

V. cholerae is believed to be a member of the autochthonous flora of estuarine and other brackish environments – e.g. in association with planktonic copepods [AEM (1983) *45* 275–283]. So-called 'viable but non-cultivable' (see VNC) strains of *V. cholerae* may retain virulence [Science (1996) *274* 2025–2031].

V. fischeri. This organism is now commonly classified within the genus PHOTOBACTERIUM as *P. fischeri*. It produces blue-green bioluminescence, and is associated e.g. with the luminous organs of certain fish. The development of bioluminescence in high-density populations (e.g. in the luminous organs of fish) is a manifestation of so-called QUORUM SENSING – the autoinducer being a member of the *N*-acyl-L-homoserine lactones [TIBS (1996) *21* 214–219].

V. metschnikovii (formerly *V. cholerae* biotype *proteus*) differs from *V. cholerae* e.g. in that it is oxidase −ve, lactose +ve (some strains), and is unable to reduce nitrate. It occurs in marine habitats and may cause gastroenteritis in man.

V. parahaemolyticus occurs (worldwide) in coastal and estuarine waters during the warmer months, apparently spending the winter in marine sediments; it also occurs in fresh and brackish waters rich in organic material.

L-flagella and swarming are seen on complex media. The organism grows with 1–8% NaCl (opt. 2–4%) but not with 0% or 10% NaCl. Growth occurs at 30°C and 40°C but not at 4°C. Oxidase +ve. VP −ve. Sucrose −ve. D-Gluconate +ve. Arginine dihydrolase −ve. Nitrate reduction +ve. Colonies on TCBS agar: large, and blue to green.

The organism may be killed e.g. by heating to 60°C/15 minutes, by acidity, and by desiccation.

V. parahaemolyticus can cause gastroenteritis: see FOOD POISONING (h). It is also associated with wound infection in swimmers and fish handlers, and with diseased shellfish.

[Molecular typing: JCM (1999) *37* 2473–2478; genome sequence: Lancet (2003) *361* 743–750.]

V. succinogenes. See WOLINELLA.

V. vulnificus (= '*Vibrio* sp biovar 6330'). A marine species [distribution: AEM (1983) *45* 985–998]. *V. vulnificus* can be pathogenic in man, causing fatal septicaemia, particularly in debilitated or immunosuppressed individuals; infection may occur via wounds or by ingestion of contaminated seafood. For the detection of *V. vulnificus* in environmental samples, an optimal approach is reported to be the combined use of a selective medium and a species-specific PROBE [JAM (2001) *91* 322–327].

L-flagella are not formed. Growth can occur at 30°C and 40°C but not at 4°C, and not in peptone water containing 8% NaCl. Oxidase +ve. VP −ve. Sucrose −ve. Lactose (commonly) +ve. D-Gluconate +ve. Arginine dihydrolase −ve. Nitrate reduction +ve.

[RAPD analysis of *V. vulnificus*: AEM (1999) *65* 1141–1144.]

Other species include: *V. aestuarianus*, *V. mimicus*, *V. natriegens*, *V. nereis*, *V. orientalis* and *V. pelagius*.

vibriocins BACTERIOCINS produced by *Vibrio* spp. Vibriocin production (by cells in the mid- or late-logarithmic phase) can be induced e.g. by anaerobiosis, UV radiation, or mitomycin C. A vibriocin is a tubular structure which resembles the tail section of certain phages; it has a contractile sheath. Vibriocins are active against e.g. strains of *Vibrio*, *Escherichia coli* and other enterobacteria, *Pseudomonas aeruginosa* – and certain mammalian cells (e.g. HeLa cells). In bacteria, vibriocins cause (bactericidal) changes in the permeability of the cytoplasmic membrane and/or inhibition of DNA synthesis; the lethal activity of vibriocins can be inhibited by the action of trypsin added within ca. 10 min of adsorption. [Review: AVR (1984) *29* 298–312.]

vibrioid Vibrio-like; curved with a twist, i.e., not in one plane, as in an incomplete turn of a helix.

Vibrionaceae A family of Gram-negative, asporogenous, facultatively anaerobic bacteria. Cells: straight or curved rods, coccobacilli, short filaments, etc. Most strains are motile with one or more polar flagella; lateral flagella may be formed on solid media. Metabolism is chemoorganotrophic (respiratory or fermentative); most strains can grow on media containing only salts, NH_4^+ and glucose, although a few require organic growth factors. Most strains are oxidase +ve and can reduce NO_3^- to NO_2^-. Species occur mainly in marine and freshwater habitats and in association with aquatic animals; some are pathogenic in man and/or animals (e.g. fish). GC%: 38–63. Genera: AEROMONAS, ALLOMONAS; PHOTOBACTERIUM; PLESIOMONAS; VIBRIO. [Book ref. 22, pp. 516–550.]

[Phylogeny and reorganization of the Vibrionaceae: SAAM (1985) *6* 171–182.]

vibriosis Any disease of man or animals (e.g. fish) caused by a VIBRIO sp; the term usually refers to diseases other than cholera.

vibriostatic Refers to any agent bacteriostatic for *Vibrio* spp.

vibunazole See AZOLE ANTIFUNGAL AGENTS.

Victoria blight (of oats) A seed- and soil-borne CEREAL DISEASE which affects the Victoria and related cultivars of oats (*Avena* spp); causal agent: see VICTORIN. Symptoms: necrosis and reddening of seedling leaves, and later rotting of stem bases.

victorin A TOXIN, synthesized by *Drechslera* (= *Helminthosporium*) *victoriae*, which is responsible for the symptoms of VICTORIA BLIGHT of oats; victorin causes leakage of electrolyte through the plasmalemma.

vidarabine (adenine arabinoside; 9-β-D-arabinofuranosyladenine; ara-A; Vira-A) An ARABINOSYL NUCLEOSIDE used e.g. as an ANTIVIRAL AGENT for the treatment of certain herpesvirus infections (e.g. topically for herpes simplex keratitis, systemically for varicella-zoster virus infections in immunocompromised patients). In vivo, vidarabine is readily converted to the less active arabinosylhypoxanthine by adenosine deaminase (an enzyme widely distributed in host cells); its antiviral activity can be potentiated by the adenosine deaminase inhibitor *pentostatin*. (Vidarabine was originally synthesized chemically, but was subsequently found to be produced e.g. by *Streptomyces antibioticus*.)

vignafuran A PHYTOALEXIN, 2-(2′-methoxy-4′-hydroxyphenyl)-6-methoxybenzofuran, produced by the cowpea (*Vigna unguiculata*).

vilia (viili, filia, filli) A Finnish food made by fermenting milk with capsulated ('rope-forming') strains of *Lactococcus lactis* which give the product flavour (LACTIC ACID and DIACETYL) and a stringy texture. A surface growth of *Geotrichum candidum* may be present; this organism lowers the acidity by utilizing some of the lactic acid. (See also DAIRY PRODUCTS.)

villin See ACTIN.

vimentin See INTERMEDIATE FILAMENTS.

vinblastine See VINCA ALKALOIDS.

***Vinca* alkaloids** (*Cantharanthus* alkaloids) A group of iridoid indole alkaloids obtained from the periwinkle *Cantharanthus roseus* (formerly *Vinca rosea*). Some are used in anticancer therapy; they include e.g. vinblastine (= vincaleukoblastine), vincristine and vindesine – each of which can combine with TUBULIN, inhibiting the assembly of MICROTUBULES (and, hence, inhibiting MITOSIS). The *Vinca* alkaloids can precipitate tubulin, in paracrystalline form, intracellularly.

vincaleukoblastine See VINCA ALKALOIDS.

Vincent's angina (trench mouth; Vincent's disease) An ulcerative disease of the mouth and pharynx, believed to be caused by *Treponema vincentii* together with (uncharacterized) anaerobic fusiform bacteria.

Vinckeria A subgenus of PLASMODIUM.

vinclozolin See DICARBOXIMIDES.

vincristine See VINCA ALKALOIDS.

vinculin See ACTIN.

vindesine See VINCA ALKALOIDS.

vinegar A condiment made by the ACETIFICATION (by *Acetobacter* spp) of various fermented liquors – e.g. beer, wine, cider, or distilled alcohol (for malt, wine, cider and spirit vinegars, respectively). *Slow processes* of manufacture involve the bulk acetification of naturally fermented liquor on exposure to air. The *Orleans process*, still used in some regions, is a continuous slow process in which vinegar (serving as an inoculum) is mixed with the fermented liquor which is allowed to acetify in partially filled kegs; acetifying bacteria form a film on the liquid surface. As vinegar is withdrawn it is replaced with fresh liquor. *Quick processes* are of two types. (a) In *trickling generators* the liquor trickles through a tank packed with e.g. twigs, corn-cobs, or wood-shavings which are coated with a film of *Acetobacter* spp. The commonest, the Frings trickling generator, is packed with beechwood shavings; the liquor is pumped from a reservoir at the base to the top of the tank and allowed to trickle back to the reservoir through the coated shavings. Air is forced in at the bottom and up through the shavings. Cooling is necessary to maintain a temperature of ca. 29–35°C. Trickling generators can become blocked by growth of the slime-producing *Acetobacter xylinum*, especially if acidity levels are low. (b) *Submerged culture generators* are becoming more widely used. The commonest is the Frings Acetator: a fermenter with a high-speed rotor that distributes fine air-bubbles which rise through the agitated liquor. Temperature control, aeration etc can be fully automated. Vinegar produced by any of these processes may be aged prior to filtration, bottling and pasteurization. ('Vinegar' made chemically, e.g. by the carbonylation of methanol at high temperature and pressure, is referred to in Europe as 'non-brewed condiment'; it is cheaper than microbially-produced vinegar.)

In Japan, one type of rice vinegar (*komesu*) is made from polished rice, while another type (*kurosu*) is made from unpolished rice – *Acetobacter pasteurianus* being important in both production processes. [Characterization of acetic acid bacteria in the traditional production of rice vinegar in Japan: AEM (2001) *67* 986–990.]

viniferins The (α- and ε-) viniferins are stilbene oligomer PHYTOALEXINS produced by the grapevine (*Vitis vinifera*).

Vioform See 8-HYDROXYQUINOLINE.

violacein A purple, water- and chloroform-insoluble pigment formed from tryptophan by species of *Chromobacterium* and *Janthinobacterium*; it consists of two substituted indole nuclei linked via a pyrrolone residue. An ethanolic solution of violacein has an absorption maximum at 579 nm and a minimum at 430 nm; it becomes green when acidified with H_2SO_4.

The synthesis of violacein may be used as an indicator system in the study of QUORUM SENSING (q.v.).

violaxanthin See CAROTENOIDS.

viomycin A peptide ANTIBIOTIC which inhibits PROTEIN SYNTHESIS in bacteria; it apparently binds to the ribosome at a site identical to the binding site of the 2-deoxystreptamine-containing AMINOGLYCOSIDE ANTIBIOTICS, inhibiting the initiation of translation and the translocation step of elongation. Viomycin has little or no activity against at least some archaeans (methanogens).

viper type C oncovirus See TYPE C ONCOVIRUS GROUP.

***vir* loci** See CROWN GALL.

Vira-A See VIDARABINE.

viraemia (viremia) The condition in which virions are present in the bloodstream.

viral haemorrhagic fevers A group of clinically similar diseases in which characteristic features are increased capillary permeability, leukopenia and thrombocytopenia; causal agents include arenaviruses (e.g. Argentinian and Bolivian haemorrhagic fevers and LASSA FEVER), flaviviruses (e.g. Omsk haemorrhagic fever), Ebola virus (see FILOVIRIDAE), Marburg virus (see MARBURG FEVER), and bunyaviruses (Crimean–Congo haemorrhagic fever – a tick-borne disease [ARE (1981) *26* 77–80]). (See also HANTAVIRUS; KOREAN HAEMORRHAGIC FEVER.)

The incubation period may be 7–14 days. Onset is insidious, with chills and fever, myalgia, headache, conjunctivitis, anorexia and vomiting; exanthem and oedema of the face, neck and upper thorax may occur. After a few days, hypotension and haemorrhages develop, and death may follow uraemic coma or hypovolaemic shock.

(See also DENGUE and SIMIAN HAEMORRHAGIC FEVER VIRUS.)

viral haemorrhagic septicaemia (of trout) See EGTVED DISEASE.

viral hepatitis See e.g. HEPATITIS A, HEPATITIS B, HEPATITIS C.

viral infectivity See END-POINT DILUTION ASSAY.

viral pneumonia of calves See CALF PNEUMONIA.

Virazole *Syn.* RIBAVIRIN.

Virchow cells See LEPROSY.

viremia *Syn.* VIRAEMIA.

virescence (*plant pathol.*) An abnormal greening (e.g. of petals): a symptom of certain plant diseases.

VirG protein See DYSENTERY.

virginiamycins Antibiotics (see STREPTOGRAMINS), produced by a strain of *Streptomyces virginiae*, which are active against primarily Gram-positive bacteria; they have been used e.g. as FEED ADDITIVES for pigs, poultry and ruminants.

viricidal *Syn.* VIRUCIDAL.

viridans streptococci Streptococci which are typically α-haemolytic (see STREPTOCOCCUS), which do not contain Lancefield grouping antigens, and which are characteristically found in the upper respiratory tract in man and other animals; they include e.g. *S. acidominimus*, some strains of *S. mutans* (*sensu lato*), *S. oralis*, *S. pneumoniae*, *S. salivarius* and *S. sanguis*. [Classification of some viridans streptococci as *S. oralis*: IJSB (1985) *35* 482–488.]

viridogrisein An antibiotic structurally related to streptogramin B (see STREPTOGRAMINS).

virino A hypothetical infectious particle consisting of a host-encoded protein complexed with a small nucleic acid which does not encode any protein but which can serve as a template for its own replication (catalysed by host enzymes). (cf. VIROID.) The nucleic acid could play a regulatory role by inducing the synthesis of host-encoded structural protein(s). It was once suggested that a virino could be the causal agent of SCRAPIE and other (PRION-mediated) diseases.

virion A single complete VIRUS particle.

virogenes Viral genes integrated in a (eukaryotic) host cell chromosome. (cf. PROVIRUS.)

virogenic Virus-producing.

viroid A plant-pathogenic infectious agent comprising small, naked (i.e., protein-free), circular ssRNA (240–380 nucleotides); the RNA appears not to code for any protein and depends entirely on the host cell for its replication. (cf. VIRUSOID; SATELLITE RNA.) Viroids currently known include: avocado sunblotch viroid (ASBV); burdock stunt viroid (BSV); chrysanthemum chlorotic mottle viroid (CCMV); chrysanthemum stunt viroid (CSV); *Citrus* exocortis viroid (CEV); COCONUT CADANG-CADANG VIROID (CCCV); cucumber pale fruit viroid (CPFV) – closely homologous with, and probably a strain of, hop stunt viroid (HSV); potato spindle tuber viroid (PSTV); tomato apical stunt viroid (TASV); tomato bunchy top viroid (TBTV); and tomato 'planta macho' viroid (TPMV). (A viroid-like RNA apparently related to HSV has been isolated from grapevine leaves [JGV (1985) *66* 333–338].) Many of these viroids cause commercially important losses among crop plants. Symptoms of viroid infection range from mild (e.g. discolouration and malformation of leaves) to severe and even lethal (e.g. in the case of CCCV). Transmission typically occurs mechanically, particularly by vegetative propagation (e.g., grafting has been an important mode of spread of CEV); PSTV transmission via seed and pollen has been recorded in certain host plants, and PSTV may also be transmitted at low frequencies by the aphids *Macrosiphum euphorbiae* and *Myzus persicae*.

All viroids seem to be essentially similar in structure. They are covalently closed circular ssRNA molecules that form highly base-paired rigid rods in which short double-stranded segments are separated by small unpaired internal loops; a small loop also occurs at each end of the rod. This secondary structure is highly conserved among the viroids and is apparently necessary for replication and possibly also for pathogenicity.

Since viroid RNA appears not to encode proteins, it must rely on host enzymes for its replication. Replication occurs in the host cell nucleus, where the viroid RNA is apparently associated with the nucleolus. Evidence from studies using in vitro systems and α-amanitin suggests that DNA-dependent RNA polymerase II may be involved, although a role for the RNA-dependent RNA polymerase present in many plants has not been excluded. Initially, infectious (+) viroid RNA is transcribed – apparently by a ROLLING CIRCLE MECHANISM – into an oligomeric complementary (−) RNA several times the unit length of the viroid RNA. The (−) RNA oligomers may serve as templates for the synthesis of (+) RNA oligomers which are then presumably cleaved into unit-length linear strands; processing and ligation of the linear molecules is then necessary to yield 'mature', infectious viroid RNA.

It has been demonstrated that viroids show some homology with group I introns (see SPLIT GENE), leading to the suggestion that they may have evolved from group I introns and/or that cleavage of viroid oligomers to form monomers may require structural features similar to those of group I introns [see e.g. PMB (1986) *7* 129–142]. It has been proposed that viroids may exert their pathogenic effects by interfering with pre-rRNA processing in host cells [see e.g. BBA (1986) *868* 190–197].

[General reviews: Book ref. 81, pp. 281–334; Intervirol. (1984) *22* 1–16.]

viropexis The uptake by a cell of virus particle(s) (bound to cell surface receptors) by a non-specific phagocytic process. (See e.g. REOVIRUS.)

viroplasm (*syn.* factory area; viroplasmic matrix; viroplast; virus factory; X body) A type of INCLUSION BODY which is – or is presumed to be – the site of virus replication and/or assembly in a virus-infected cell. A viroplasm generally consists of or contains accumulations of virions and/or virus components.

viroplasmic matrix *Syn.* VIROPLASM.

viroplast *Syn.* VIROPLASM.

Viroptic *Syn.* TRIFLUOROTHYMIDINE.

virosome A LIPOSOME which incorporates viral (usually ENVELOPE) proteins.

virucidal (viricidal) Able to inactivate viruses. (See also ANTIVIRAL AGENTS.)

virulence (of pathogenic microorganisms) The capacity of a PATHOGEN to cause disease – defined broadly in terms of the severity of symptoms in the host; thus, a highly virulent strain may cause severe symptoms in a susceptible individual, while a less virulent strain would produce relatively less severe symptoms in the same individual. The severity of the disease which actually develops may be regarded as the result of the joint expression of the degree of INFECTIVITY (sense 1) and virulence of the pathogen and the state of immunity (either natural or acquired) of the host in relation to that pathogen.

virulent bacteriophage A BACTERIOPHAGE which causes lysis of its host cell. (cf. TEMPERATE BACTERIOPHAGE.)

virulon In *Yersinia* spp: a system, encoded by a set of plasmid-borne genes, which consists essentially of (i) a number of virulence proteins (Yops), and (ii) a type III transport system (see PROTEIN SECRETION) for delivering Yops into the cytoplasm of a eukaryotic cell [Mol. Microbiol. (1997) *23* 861–867]. Analogous systems (i.e. effector molecules and a type III secretion system) occur in certain other bacteria (e.g. species of *Salmonella* and *Shigella*).

The 70-kb *Yersinia* virulence plasmid (pYV) which encodes the virulon is found in *Y. enterocolitica* and *Y. pseudotuberculosis* as well as in *Y. pestis*.

The type III secretion system is constructed from a number of different Ysc proteins. Thus, at least four of these proteins (YscDRUV) are known to span the cytoplasmic membrane, while YscC is a SECRETIN. YscN is believed to function as an ATPase. (See figure.)

Some of the Yop proteins are effector molecules that are secreted into the eukaryotic cell. Others (e.g. YopBD) are 'translocators' which mediate translocation of effectors through the cytoplasmic membrane of the host cell. In some cases, an effector or translocator has a specific CHAPERONE – a given chaperone being designated SycD, SycE etc. (Syc = specific Yop chaperone); translocation of effector molecules would be inefficient, or absent, in the absence of specific chaperones. The chaperones of the effector molecules may also be involved in determining the order in which Yops are delivered to the eukaryotic cell [JB (2000) *182* 4811–4821].

Among effector Yops, YopH (a phosphatase) can inhibit phagocytosis by de-phosphorylating certain macrophage proteins. YopE and YopT can disrupt eukaryotic actin-based structures. YopJ can down-regulate certain kinases in macrophages, thus inhibiting release of the pro-inflammatory cytokine TNF-α [Mol. Microbiol. (1998) *27* 953–965].

[The Yop virulon in the context of plague: PNAS (2000) *97* 8778–8783.]

viruria The presence of viruses in urine.

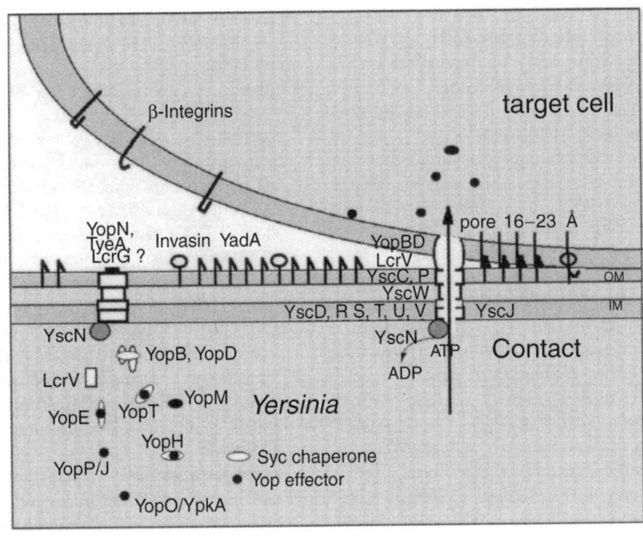

VIRULON: the Yop virulon of *Yersinia*. When cells of *Yersinia* are placed in a rich environment at 37°C, the Ysc secretion channel is installed. Proteins YscD, -R, -S, -T, -U and -V are localized in the inner membrane (IM), whereas YscC and YscP are exposed at the bacterial surface. Lipoprotein YscW stabilizes YscC. YscN belongs to the family of ATPases. A stock of Yop proteins is synthesized, and some of them are capped with their specific Syc chaperone. As long as there is no contact with a eukaryotic cell, a stop-valve, possibly made of YopN, TyeA and LcrG, blocks the Ysc secretion channel. On contact with a eukaryotic target cell, the bacterium attaches tightly by interaction between its YadA and Inv adhesins and β-integrins, and the secretion channel opens. The Yops are then transported through the Ysc channel, and the Yop effectors are translocated across the plasma membrane, guided by the translocators YopB, YopD and LcrV.

Reproduced, with permission, from *Proceedings of the National Academy of Sciences of the USA* (2000) *97* 8778–8783 (copyright 2000 National Academy of Sciences, USA) by courtesy of the author, Professor Guy R. Cornelis, Christian de Duve Institute of Cellular Pathology and Faculté de Médecine, Université catholique de Louvain, Brussels.

virus A non-cellular MICROORGANISM (q.v.) which consists minimally of protein and nucleic acid (DNA or RNA) and which can replicate only within particular (animal, microbial or plant) cells; the viral nucleic acid (genome) encodes viral structural proteins and (usually) one or more types of molecule necessary e.g. for viral gene expression, genome replication, etc. (cf. SUBVIRAL AGENTS.) A virus has no intrinsic metabolism, relying entirely on its host cell for energy, for precursor molecules (nucleotides, amino acids etc), and for all essential enzymes not encoded by the virus; host cell macromolecule synthesis is often inhibited by the infecting virus. Infection of an organism by a virus may or may not result in disease, depending e.g. on virus, host, and/or conditions. (See also ANIMAL VIRUSES, BACTERIOPHAGE, CYANOPHAGE, MYCOVIRUS, PHYCOVIRUS, PLANT VIRUSES.)

Depending on virus, an individual virus particle (*virion*) may be e.g. isometric, rod-shaped, filamentous, or pleomorphic, or it may consist of an isometric 'head' with a rod-shaped or filamentous 'tail'. Virions vary in structural complexity from relatively simple, with only one or a few types of major structural protein, to highly complex, with many components – including glycoproteins and various enzymes (see e.g. POXVIRIDAE and T-EVEN PHAGES). Nevertheless, in the majority of viruses the virion conforms essentially to one of two basic types of construction: the genome occurs either (a) within a shell-like, isometric CAPSID (see e.g. ADENOVIRIDAE, MICROVIRIDAE, TOMBUSVIRUSES), or (b) complexed with protein in a structure (often rod-shaped or filamentous) with helical symmetry (see e.g. TOBACCO MOSAIC VIRUS). (cf. NUCLEOCAPSID.) Virions of either type may have an external lipoprotein ENVELOPE.

In size, virions range from ca. 20–25 nm diam. or less (PARVOVIRIDAE, PICORNAVIRIDAE) to ca. 200–450 nm maximum dimension (POXVIRIDAE); filamentous viruses may reach lengths of ca. 2000 nm (CLOSTEROVIRUSES). Thus, some virions are larger than some bacteria: cf. CHLAMYDIA and MYCOPLASMA.

The virus *replication cycle* begins with the *attachment* (*adsorption*) of the virion to a specific receptor on the host cell surface, followed by *penetration* of the cell by the virion or by part of the virion (including at least its genome). (See e.g. BACTERIOPHAGE ϕ6; BACTERIOPHAGE T4; ENVELOPE; VIROPEXIS.) After partial or complete UNCOATING of the genome, viral *gene expression* begins. Viral genes are often expressed in a particular temporal sequence: 'early genes', the first to be expressed, commonly encode enzymes necessary for genome replication and/or for subsequent ('middle' and/or 'late') gene expression; 'late genes' generally encode virion structural proteins, and their expression is often coupled to or preceded by replication of the genome. (See e.g. ADENOVIRIDAE, BACTERIOPHAGE SPO1 and BACTERIOPHAGE T4.) *Genome replication* generally requires at least some virus-encoded enzymes: e.g. RNA viruses generally encode an RNA-DEPENDENT RNA POLYMERASE. *Virion assembly* may occur e.g. in the host cell cytoplasm or nucleus, depending on virus, and culminates in *maturation* which may involve e.g. elimination of SCAFFOLDING PROTEIN, acquisition of an ENVELOPE, conformational changes or cleavages of certain virion proteins (see e.g. RETROVIRIDAE), etc. *Release* of the progeny virions from the host cell may occur e.g. by host cell lysis, by exocytosis, or by budding of the virions through the plasma membrane. Hundreds or thousands of progeny virions may be formed per cell. For details of replication cycles see separate entries for virus families/groups (see ANIMAL VIRUSES, PLANT VIRUSES, BACTERIOPHAGE). (See also ONE-STEP GROWTH EXPERIMENT.)

The strategy for viral gene expression and replication depends largely on the nature of the viral genome – which may be DNA (ss or ds, linear or circular) or RNA (ss or ds linear, or ss circular), and may consist of one or more pieces of nucleic acid (i.e., it may be monopartite, bipartite, etc). In viruses with single-stranded genomes, the genome is said to be *positive-sense* or *negative-sense* according to whether it is *homologous with* or *complementary to*, respectively, the viral mRNA (which is arbitrarily defined as positive-sense, = plus- or(+)-strand). Thus, e.g., a virus with a *positive-sense ssDNA genome* (e.g. SSDNA PHAGES, GEMINIVIRUSES) must synthesize a complementary negative-sense DNA strand before mRNA can be transcribed. A *negative-sense ssDNA* genome (see e.g. PARVOVIRIDAE) can – at least theoretically – be transcribed directly into mRNA. *Positive-sense ssRNA* genomes can generally act directly as mRNA; however, since in eukaryotes translation cannot usually be initiated at internal initiation sites, internal genes in an ssRNA genome must be expressed e.g. by the synthesis of SUBGENOMIC MRNAS on a complementary (−) strand and/or by the synthesis of a POLYPROTEIN on a polycistronic mRNA. Viruses with *negative-sense ssRNA* (e.g. ORTHOMYXOVIRIDAE, RHABDOVIRIDAE) generally carry their own RNA-DEPENDENT RNA POLYMERASE in the virion, since a (+)-strand must be synthesized before any gene expression can occur; negative-strand RNA viruses may e.g. synthesize subgenomic mRNAs and/or have segmented genomes. (See also AMBISENSE RNA.) Viruses of the RETROVIRIDAE carry a REVERSE TRANSCRIPTASE which synthesizes a DNA strand complementary to the RNA genome; mRNA is transcribed from the (ds) DNA provirus. Viruses with *dsDNA* genomes may synthesize mRNA on only one strand, the (−)-strand (see e.g. HEPADNAVIRIDAE), or on both strands (e.g. in the ADENOVIRIDAE and BACTERIOPHAGE LAMBDA); in the latter case neither strand can be designated '+' or '−'. Most known *dsRNA* viruses have SEGMENTED GENOMES (cf. MYCOVIRUS), one strand (the (+)-strand) of each segment functioning as mRNA. (cf. BALTIMORE CLASSIFICATION; see also INFECTIOUS NUCLEIC ACID.)

Classification of viruses. Current classification schemes are based on features such as the composition, morphology and anatomy of the virion, the nature of the genome, replication strategy, etc. The ANIMAL VIRUSES and some BACTERIOPHAGES have been classified into families (names ending in -viridae), occasionally subfamilies (names ending in -virinae), genera (names ending in -virus), and species (corresponding to individual viruses) [Book ref. 23]; however, the species concept in virology has not been universally accepted [debate: Intervirol. (1985) *24* 61–98]. (cf. PLANT VIRUSES.) Strains or types of a given virus are recognized e.g. on the basis of virulence, antigenic determinants, etc.

Assay of viruses. See COUNTING METHODS (b).

Inactivation of viruses. Viruses may be inactivated e.g. by extremes of pH (cf. ENTEROVIRUS), by relatively mild heat treatment (e.g. 50–80°C for some viruses), and by UV and ionizing radiation; some viruses are sensitive to desiccation and to various ANTISEPTICS and DISINFECTANTS. Enveloped viruses (and some non-enveloped viruses) are generally inactivated by organic solvents (e.g. ether, chloroform) and by detergents. (See also ANTIVIRAL AGENTS and STERILIZATION.)

virus-associated haemophagocytic syndrome (VAHS) See HAEMOPHAGOCYTIC SYNDROME.

virus factory *Syn.* VIROPLASM.

virusoid A type of small (ca. 300–400-nucleotide) ccc ssRNA present (as 'RNA2') within the virions of VELVET TOBACCO MOTTLE VIRUS (VTMoV) and related viruses. In VTMoV and *Solanum nodiflorum* mottle virus the virusoid is an integral part of the genome, both RNA1 and the virusoid being necessary for infectivity. (However, the RNA1 of lucerne transient streak virus (LTSV) can replicate in the absence of its 'virusoid'; the LTSV 'virusoid' may thus be a SATELLITE RNA [JGV (1983) *64* 1167–1173].) A virusoid resembles a VIROID, at least superficially, in its secondary structure (which is apparently more flexible than that of a viroid), in its apparent lack of coding capacity, and possibly also in its mode of replication (concatemeric forms have been detected in infected plant cells); however, virusoids differ from viroids in being encapsidated and in being incapable of infecting plant cells in the absence of RNA1.

visceral leishmaniasis LEISHMANIASIS of man and other animals in which the pathogen, typically *Leishmania donovani*, infects the internal organs; *L. donovani* is transmitted by the bites of sandflies. The disease occurs primarily in Africa, Asia and South America; in India the disease in man is called *kala-azar*. In man, the incubation period may be 3–6 months (range: weeks–years); typically, there is a recurrent low-grade fever, splenomegaly, hepatomegaly, leucopenia and cachexia – with high mortality rates in untreated cases. Diagnosis may involve e.g. an indirect fluorescent antibody test; the Montenegro test is usually negative. Chemotherapeutic agents used for treatment include pentavalent ANTIMONY compounds; however, resistance to pentavalent antimony is now found e.g. in the Bihar region of India – where almost half of the world's cases occur, and where therapeutic use is being made of e.g. AMPHOTERICIN B.

[Clinical and experimental advances in the treatment of visceral leishmaniasis: AAC (2001) *45* 2185–2197.]

(See also POST-KALA-AZAR DERMAL LEISHMANIASIS.)

viscotaxis A TAXIS in which the stimulus is a change in viscosity; a positively viscotactic organism swims towards regions of higher viscosity.

visna A chronic, progressive, fatal meningoencephalitis of sheep; the causal agent is a virus of the LENTIVIRINAE (cf. MAEDI). Incubation period: months or years. Symptoms – which include e.g. abnormal gait, tremor of facial muscles, increasing emaciation and paralysis – are associated with inflammation and subsequent demyelination of tissues in the central nervous system. [Antigenic variation in the visna virus: JMB (1982) *158* 415–434.]

vital staining The STAINING of *living* cells or their components by dyes whose presence is tolerated by the cells for varying periods of time; 'vital staining' may also refer to the uptake of minute insoluble particles (e.g. carbon, carmine) by certain types of cell. Vital dyes include Bismarck brown, brilliant cresyl blue, Janus green, METHYLENE BLUE, NEUTRAL RED and NILE BLUE A; most vital dyes are basic dyes – EOSIN is an exception. (See also FLUORESCENT BRIGHTENER.) Some vital dyes (e.g. Janus green) are more toxic than others, neutral red being one of the least toxic; all vital dyes are used in low concentrations (e.g. less than 0.01% w/v). Vital staining has been used e.g. for studying differentiation in the pseudoplasmodium of cellular slime moulds, and is involved e.g. in the TOXOPLASMA DYE TEST; Janus green is used e.g. for staining mitochondria and for following redox changes in mitochondria (see also RHODAMINE). In animals, *intra vitam* (= *intravital*) staining involves the injection or ingestion of a dye which is taken up by certain types of cell in the (living) animal; thus, e.g. on injection into a mammal, TRYPAN BLUE is taken up by macrophages. Trypan blue stains some types of cell only when they are dead or damaged; hence, this dye is used to determine the proportion of dead/damaged cells in a cell suspension used to prepare a TISSUE CULTURE. The staining of living cells after their removal from the host has been referred to as *supervital* or *supravital* staining when the cells remain viable during the staining process.

vitamin In microbiology: a generic term for any of a range of unrelated organic compounds, many of which are necessary – in small quantities – for the normal metabolism and growth of certain microorganisms; they typically function as COENZYMES (sense 2) or as components of coenzymes. Most microorganisms can synthesize the vitamins they require. Those which cannot synthesize a particular vitamin must obtain it – or sometimes one or more of its precursors – from the growth medium; such organisms can be used in the BIOASSAY of that vitamin. Some microorganisms are important natural and/or commercial sources of certain vitamins. See also ASCORBIC ACID, BIOTIN, BIOPTERIN, FOLIC ACID, LIPOIC ACID, NICOTINIC ACID, PANTOTHENIC ACID, PYRIDOXINE, RIBOFLAVIN, THIAMINE, VITAMIN B_{12}.

vitamin A See RETINAL.

vitamin B_1 *Syn.* THIAMINE.

vitamin B_2 *Syn.* RIBOFLAVIN.

vitamin B_6 See PYRIDOXINE.

vitamin B_{12} A photolabile, water-soluble VITAMIN. The term 'vitamin B_{12}' may refer specifically to cyanocobalamin or to any of its various analogues (see below). These compounds are *corrinoids*, i.e. they contain a tetrapyrrole ring system (the corrin ring) which is highly substituted and which contains a central cobalt atom linked to the four pyrrole nitrogen atoms to form *cobinamide*; 3-phosphoribose linked to a side-chain on one of the pyrrole rings forms *cobamide*. In the *cobalamins*, 5,6-dimethylbenzimidazole forms a bridge between the cobalt atom and the α-1-position of the ribose residue; the sixth coordination position of the cobalt may be occupied by e.g. cyanide (in *cyanocobalamin*), hydroxyl (in *hydroxycobalamin*) etc. (cf. HAEM).

Cobamide coenzymes (= coenzymes B_{12}) typically consist of a cobalamin in which a 5′-deoxyadenosyl residue has been linked, through its 5′ position, to the cobalt atom, forming 5′-deoxyadenosylcobalamin. These coenzymes are involved in various intramolecular re-arrangements: e.g. glutamate/β-methylaspartate and methylmalonyl-CoA/succinyl-CoA interconversions, dehydration of 1,2-diols (e.g. glycerol$\rightarrow$ β-hydroxypropionaldehyde) etc. [Mechanism of coenzyme B_{12}-dependent re-arrangements: Science (1985) *227* 869–875.] (See also NUCLEOTIDE.)

Methylcobalamin, in which a methyl group occurs in place of the 5′-deoxyadenosyl group, functions as a coenzyme e.g. in the synthesis of methionine from homocysteine [see Appendix IV(d)], apparently acting as a methyl carrier.

In certain methanogens, a corrinoid-containing enzyme (methanol:corrinoid methyltransferase) is involved in the biosynthesis of methane from methanol (see METHANOGENESIS).

Vitamins of the B_{12} group are apparently synthesized primarily or solely by bacteria, and are obtained commercially from e.g. *Streptomyces* spp as a by-product of antibiotic production. In man, B_{12} synthesized by bacteria in the small intestine may – in at least some cases – be an important source of the vitamin [Nature (1980) *283* 781–782], but B_{12} synthesized by bacteria in the large intestine cannot be absorbed and is voided with the faeces. In ruminants, the microflora of the RUMEN is an important source of this vitamin for the animal; a 'wasting disease' of ruminants grazing cobalt-deficient pastures is due to the inability of the rumen microflora to synthesize adequate amounts of B_{12}.

B_{12} vitamins are required as growth factors e.g. by many algae, by certain protozoa, and by *Lactobacillus* spp; *Euglena gracilis* has been used for the BIOASSAY of B_{12}. (See also BLOOM.)

In *Escherichia coli*, B_{12} is taken up by (i) pmf-dependent transport across the OUTER MEMBRANE (see also BTUB PROTEIN and TONB PROTEIN) and (ii) a BINDING PROTEIN-DEPENDENT TRANSPORT SYSTEM in which the binding protein BtuF [PNAS (2002) *99* 16642–16647] transfers B_{12} to BtuCD (see ABC TRANSPORTER); negatively charged (glutamate) residues on BtuF may interact with positive (arginine) residues on BtuCD.

vitamin C *Syn.* ASCORBIC ACID.

vitamin H *Syn.* BIOTIN.

vitamins K See QUINONES.

Vitreoscilla A genus of GLIDING BACTERIA (see CYTOPHAGALES) which occur in e.g. soil, fresh water, and decaying vegetation. The organisms are chemoorganotrophs which resemble BEGGIATOA spp but differ e.g. in the width of the trichome (ca. 1–4 μm) and in that sulphur is not deposited intracellularly.

vitrification See ELECTRON MICROSCOPY (d).

Vittaforma See MICROSPORIDIOSIS.

vivotoxin (*plant pathol.*) A substance in a diseased plant – produced either by the pathogen or by the host – which causes damage to the plant tissues and is thus responsible for at least some of the symptoms of the disease.

V_L region See IMMUNOGLOBULINS.

VLP Virus-like particle.

VNC (VBNC) Viable but non-cultivable: refers to normally-cultivable bacteria which have become non-cultivable owing to stress (e.g. starvation) – but which still show features of living cells (e.g. metabolic activity and maintenance of structure). VNC pathogens (e.g. *Vibrio cholerae*) may retain virulence. Some view the VNC state as a kind of 'dormancy' adopted by non-spore-forming bacteria as a normal response to stress [FEMS Ecol. (1998) *25* 1–9]. Others believe that failure to culture may, in at least some cases, reflect inappropriate media/conditions; for example, starved cells transferred to a nutrient-rich aerobic medium may be damaged/killed by *endogenous* toxic radicals (e.g. SUPEROXIDE) which normal (adapted) cells can detoxify [Microbiology (1998) *144* 1–3].

VNC strains of enterotoxigenic *Escherichia coli* (ETEC) can produce toxin [AEM (1996) *62* 4621–4626].

VNC enterococci are reported to have changes in peptidoglycan, lipoteichoic acid and penicillin-binding proteins [AEM (2000) *66* 1953–1959] and to maintain resistance to vancomycin when division is resumed [AAC (2003) *47* 1154–1156].

vodka See SPIRITS.

Vogelbusch IZ fermenter See JET LOOP FERMENTER.

Voges–Proskauer test (VP test) An IMVIC TEST used to determine the ability of an organism to form acetoin and/or DIACETYL (formed e.g. during the BUTANEDIOL FERMENTATION) in a phosphate-buffered glucose–peptone medium. (For *Bacillus* spp the medium should not contain phosphate as this may prevent the formation of acetoin.) The test medium is inoculated and then incubated at 37°C for 48 h or at 30°C for 5 days or more. In *Barritt's method* 5% alpha-naphthol in ethanol (0.6 ml) and 40% KOH (0.2 ml) are added to ca. 1 ml of the culture; the (stoppered) tube is shaken vigorously and left in a sloping position. A positive reaction (acetoin/diacetyl present) is indicated by the development of a red coloration within 5–60 minutes. In *O'Meara's method* (which is somewhat less sensitive) 2 drops of creatine (1% in 0.1 N HCl) are used in place of the alpha-naphthol, and 1 ml of KOH is used; a positive reaction is indicated by a pink coloration.

In e.g. the *Westerfeld modification*, both creatine *and* alkaline alpha-naphthol are used – this combination of reagents apparently detecting aliphatic compounds (of minimum chain length 4 carbons) which have vicinal carbonyl groups in the 2- and 3-positions; acetoin gives a positive reaction because it is apparently oxidized by the alpha-naphthol to diacetyl [AEM (1982) *44* 40–43].

Vollmer patch test A form of TUBERCULIN TEST in which a patch of tuberculin-impregnated gauze is fastened to the skin with adhesive tape.

volunteer plant A self-propagated plant, i.e., a (normally cultivated) plant which is growing as a result of natural propagation.

volutin *Syn.* POLYPHOSPHATE.

volva The cup-shaped lower part of the UNIVERSAL VEIL or PERIDIUM surrounding, respectively, the base of the stipe in certain mature agarics (e.g. *Amanita* spp) or the base of the receptacle in certain mature gasteromycetes (e.g. *Mutinus* spp).

Volvariella See AGARICALES (Pluteaceae) and PADI-STRAW MUSHROOM.

Volvocida See PHYTOMASTIGOPHOREA.

Volvox A genus of motile, coenobial, freshwater green algae (division CHLOROPHYTA). (cf. PHYTOMASTIGOPHOREA.) The coenobium consists of a spherical or ovoid gelatinous matrix (up to ca. 1.0 mm diam. in certain species) in which ca. 500–20000 cells (depending on species) are arranged in a single peripheral layer. The individual cells are biflagellate and resemble CHLAMYDOMONAS; in the coenobium they are orientated with their flagella projecting outwards, and are often interconnected by fine cytoplasmic threads. The coenobium has a characteristic rolling motility.

Asexual reproduction. Certain cells within a vegetative coenobium enlarge to become *gonidia*; each gonidium divides many times to form a hollow sphere of progeny cells oriented such that their flagellated ends face towards the centre of the sphere. Subsequently the sphere of cells everts via a pore (the *phialopore*) such that the flagellated ends of the cells face outwards, thus forming a daughter coenobium. Daughter coenobia may not be released until the parental coenobium disintegrates.

Sexual reproduction is oogamous; male and female gametes may be formed in the same coenobium or in different coenobia, according to species. In dioecious species such as *V. carteri* male coenobia may be formed spontaneously, but female coenobia may be produced only in the presence of male coenobia which produce a highly potent glycoprotein sexual inducer. (This inducer can be produced in response to heat shock in *V. carteri* [Science (1986) *231* 51–54].) In some species (e.g. *V. carteri*) the male gametes (produced in male coenobia) are released as 'sperm bundles'. A sperm bundle is attracted towards, and adheres to, a female coenobium; the bundle then breaks down, an adjacent region of the female coenobium is also lysed, and individual sperm cells penetrate the female coenobium and fuse with female gametes. The resulting zygotes subsequently develop thick walls, are released by disintegration of the parent coenobium, and undergo a period of dormancy. On germination (during which meiosis occurs) a small 'juvenile' coenobium develops in a manner similar to that in which a vegetative coenobium develops from a gonidium. Subsequent daughter coenobia, produced asexually, have increased numbers of cells, so that eventually full-sized coenobia are formed.

(See also e.g. ASTREPHOMENE and PANDORINA.)

vomiting and wasting disease A PIG DISEASE, which affects suckling piglets, caused by porcine haemagglutinating encephalitis virus (a coronavirus). Symptoms: vomiting, constipation, and subsequent unthriftiness; an encephalitic form also occurs. Mortality rates may be high.

vomitoxin Deoxynivalenol: see TRICHOTHECENES.

von Economo's encephalitis (epidemic encephalitis) A rare form of ENCEPHALITIS believed to be of viral aetiology.

von Willebrand factor (vWF) A large, multimeric glycoprotein which normally occurs (i) in plasma, in which e.g. it stabilizes blood clotting factor VIII, (ii) in the α-granules of platelets, and (iii) in cytoplasmic granules (called *Weibel–Palade bodies*) in endothelial cells. vWF is named after a Finnish doctor, Erik von Willebrand.

When vascular endothelium is damaged, vWF binds to ligands (collagen?) in subendothelial connective tissue; platelets can then bind to the vWF via their membrane glycoprotein GP1b. Platelets thus adhere to subendothelial tissue, and to one another, forming a plug.

vWF also binds to coagulase-negative staphylococci [JMM (2000) *49* 217–225].

[vWF (biochemistry and genetics): ARB (1998) *67* 395–424.]

Vorticella A genus of sedentary ciliate protozoa (subclass PERITRICHIA) which occur in fresh, brackish and marine waters; some species (e.g. *V. convallaria*) are common in SEWAGE TREATMENT plants. (See also SAPROBITY SYSTEM.) The zooid is commonly bell-shaped or goblet-shaped, ca. 50–150 μm in length, the buccal cavity being at the wider (distal) end, and the narrow end being attached, by a stalk, to the substratum; both the zooid and stalk are contractile, the stalk (usually several hundred micrometres in length) being tubular and containing a SPASMONEME. Unlike CARCHESIUM, *Vorticella* is a solitary organism, only one zooid being carried on each (unbranched) stalk. The macronucleus is curved or U-shaped, and the micronucleus is spherical. One or more contractile vacuoles are usually present. Bacteria appear to be the main component of the diet for most species.

VP test See VOGES–PROSKAUER TEST.

VPI technique See ROLL-TUBE TECHNIQUE.

VRE Vancomycin-resistant enterococci (see ENTEROCOCCUS).

VSG Variant surface glycoprotein: in certain species of *Trypanosoma*, the material which forms the electron-dense cell-surface layer (the 'surface coat'); by periodically switching from the synthesis of one VSG to another of different antigenic specificity, pathogens of a given species can evade the host's immunological defence mechanisms.

A given trypanosome contains an estimated 1000 different VSG-specifying genes (*VSG* genes) of which only *one* can be expressed at any given time. (Earlier reports of the simultaneous expression of two *VSG* genes are now believed to be erroneous.)

To be expressed, a given *VSG* must be located within a specialized *expression site* situated in a TELOMERE; the genome appears to contain an estimated 20–40 copies of such expression sites, only one of which is functional at any given time. Some of the expression sites are used *specifically* by the bloodstream stage of the organism (i.e. the stage found within the vertebrate host); other expression sites are used *specifially* by METACYCLIC FORMS of the organisms (which occur within the tsetse fly).

Different trypanosomes expressing different VSGs are said to be different *variant antigenic types* (VATs).

Switching from one VSG to another appears to occur by two distinct mechanisms. The (apparently) less common mechanism is one in which transcription is switched from one expression site (containing the resident *VSG*) to another expression site (containing the 'new' *VSG*), the first expression site being inactivated. In the alternative switching mechanism, the resident *VSG*, within the active expression site, is replaced by a new *VSG* – the mechanism apparently involving homologous RECOMBINATION and gene conversion; such *duplicative transposition* seems to be the common form of switching mechanism.

[Antigenic variation in trypanosomes (*VSG* switching mechanisms): TIP (2001) *17* 338–343.]

VSL#3 A PROBIOTIC (Yovis; Sigma-Tau, Pomezia, Italy) consisting of strains of *Lactobacillus*, *Bifidobacterium* and *Streptococcus*. Results of a double-blind, placebo-controlled trial suggest that oral administration of VSL#3 is effective in preventing flare-up in *chronic pouchitis*, a complication following surgery for ulcerative colitis. [Oral bacteriotherapy with VSL#3 in chronic pouchitis: Gastroenterology (2000) *119* 305–309.]

VSV See VESICULOVIRUS.

VT1, VT2 See EHEC.

VTEC See EHEC.

vulvovaginitis See VAGINITIS.

VW antigens Surface antigens present in virulent strains of *Yersinia pestis*, *Y. pseudotuberculosis*, and some virulent strains of *Y. enterocolitica*. The V (protein) and W (lipoprotein) components together confer on the cells the ability to resist phagocytic digestion. The formation of VW antigens is plasmid-mediated (Vwa plasmid) and is associated with certain other features – e.g. colony morphology [FEMS (1983) *17* 121–126], autoagglutinability at 37°C [JB (1984) *158* 1033–1036], production of specific OMPs, requirement for Ca^{2+} for growth at 37°C.

V–W transition In certain strains of *Citrobacter*: a form of PHASE VARIATION which involves a reversible loss or re-establishment of the ability to synthesize VI ANTIGEN; cells which synthesize the antigen are called *V forms*, those which do not are called *W forms*.

Vwa plasmid See VW ANTIGENS and YERSINIA (*Y. pestis*).

vWF See VON WILLEBRAND FACTOR.

VZV Varicella-zoster virus (see ALPHAHERPESVIRINAE).

1. Words in SMALL CAPITALS are cross-references to separate entries.
2. Keys to journal title abbreviations and Book ref. numbers are given at the end of the Dictionary.
3. The Greek alphabet is given in Appendix VI.
4. For further information see 'Notes for the User' at the front of the Dictionary.

W Tryptophan (see AMINO ACIDS).

W forms See V–W TRANSITION.

W-reactivation WEIGLE REACTIVATION.

Wagatsuma agar A medium containing human or horse RBCs added to a basal medium of yeast extract, peptone, NaCl (7%), D-mannitol, crystal violet, and agar [recipe: Book ref. 46, p. 1284]: see KANAGAWA PHENOMENON.

wakame See UNDARIA.

Waldhof fermenter A type of FERMENTER in which aeration and agitation are provided by a hollow-bladed wheel-type impeller.

walking (*mol. biol.*) See CHROMOSOME WALKING.

wall-forming body A granule, present in a coccidian macrogamete, which subsequently contributes to the structure of the oocyst wall.

Wallal subgroup See ORBIVIRUS.

Wallemia A genus of osmophilic fungi which form reddish-brown mycelial growth e.g. on jams and salted fish; chains of spores develop acropetally. *W. sebi* may be related to the FILOBASIDIACEAE [AvL (1986) *52* 183–187].

Wangiella A genus of fungi of the class HYPHOMYCETES. *W. dermatitidis* (sometimes called '*Hormiscium dermatitidis*', '*Hormodendrum dermatitidis*' or '*Phialophora dermatitidis*') can grow saprotrophically in e.g. soil, but it can also act as a dermotropic and/or neurotropic pathogen in man (see PHAEOHYPHOMYCOSIS); dark, budding, yeast-like cells are formed in culture, but in tissues the organism occurs predominantly in the form of dark hyphae. Non-septate, spherical to ovoid, smooth conidia are produced from flask-shaped to cylindrical, smooth, brown phialides which lack collarettes (see CONIDIUM). [Human isolates of *W. dermatitidis*: Mycol. (1984) *76* 232–249.]

Warburg–Dickens pathway *Syn.* HEXOSE MONOPHOSPHATE PATHWAY.

Warburg effect The inhibition of CO_2 fixation and O_2 evolution during oxygenic PHOTOSYNTHESIS by the presence of high levels of O_2; the effect is due at least in part to the competitive inhibition of the carboxylase function of RIBULOSE 1,5-BISPHOSPHATE CARBOXYLASE–OXYGENASE by O_2 and an increased rate of PHOTORESPIRATION.

warhead delivery A mechanism whereby an antimicrobial compound which does not itself readily cross the cell envelope of a potentially susceptible cell can be made to do so by attachment to a carrier component (e.g. a peptide) which can be recognized and transported by a specific TRANSPORT SYSTEM in the target cell; once inside the cell, the active component is released from the carrier component by cellular enzymes. See e.g. ALAFOSFALIN and BACILYSIN. [Review: Book ref. 153, pp. 219–266.]

Warrego subgroup See ORBIVIRUS.

wart (*med., vet.*) A skin PAPILLOMA (q.v.).

wart disease (black wart disease) (of potato) A soil-borne POTATO DISEASE, caused by *Synchytrium endobioticum* (see SYNCHYTRIUM), characterized by cauliflower-like outgrowths (GALLS) from the tubers and lower stem; entire tubers may be converted into warty masses, the warts subsequently blackening. The warts may continue development during storage of the harvested tubers.

wash-out (in continuous culture) See DILUTION RATE.

washing powders (biological) See SUBTILISINS.

Wassermann antibody See WASSERMANN REACTION.

Wassermann reaction (WR) A STANDARD TEST FOR SYPHILIS carried out as a COMPLEMENT-FIXATION TEST in which the antibody (reagin, 'Wassermann antibody') undergoes a complement-fixing reaction with the cardiolipin–lecithin–cholesterol antigen. (The term 'Wassermann reaction' may be used loosely to refer to any STS or to *any* serological test used for the diagnosis of syphilis.)

wastewater treatment *Syn.* SEWAGE TREATMENT.

wasting disease (1) (of ruminants) See VITAMIN B_{12}. (2) (of eel-grass) See LABYRINTHULAS. (3) (of elk) See TRANSMISSIBLE SPONGIFORM ENCEPHALOPATHIES.

watchglass organelle *Syn.* LIEBERKÜHN'S ORGANELLE.

water activity (a_w) The amount of 'free' or 'available' water in a given substrate – e.g. a foodstuff or a medium; it may be defined as 1/100th the relative humidity (RH%) of the air in equilibrium with the substrate. Thus, e.g. an RH of 95% corresponds to an a_w of 0.95. Temperature variation between 0°C and 37°C has little effect on a_w; however, freezing lowers the a_w of water: for pure water the values are ca. 0.95 at −5°C, 0.90 at −10°C, 0.84 at −15°C, and 0.82 at −20°C. The presence of solutes also lowers the a_w.

For many species of bacteria the optimum a_w for growth is above ca. 0.99; for e.g. staphylococci it is ca. 0.93, and for some halophiles it is ca. 0.75. Many bacteria (e.g. species of *Escherichia, Clostridium, Pseudomonas*) may fail to grow at a_w below 0.95; staphylococci have a minimum of ca. 0.87. For many yeasts the minimum a_w for growth is ca. 0.88–0.92, but the xerophilic yeast *Zygosaccharomyces rouxii* has a minimum of ca. 0.65. A number of filamentous fungi have a_w minima between ca. 0.80 (e.g. *Aspergillus, Penicillium* spp) and ca. 0.90 (e.g. *Mucor* spp). Most algae have a minimum a_w of about 0.95, although for *Dunaliella salina* it is ca. 0.75. (See also FOOD PRESERVATION (c).)

water bath A covered glass or metal tank containing water at a thermostatically-controlled temperature; in microbiology, water baths are used for INCUBATION – the material to be incubated being placed in a vessel suspended in the water. A motor-driven propeller ensures rapid mixing of the water and the maintenance of a given temperature in all parts of the tank; far greater temperature stability can be achieved in a water bath (e.g. ±0.1°C) than in a hot-air incubator. A refrigerated water bath is used for maintaining temperatures below the ambient temperature; it includes a cooling coil (which operates continually) and maintains a steady temperature by means of the thermostat and heater of the water bath.

water bloom *Syn.* BLOOM.

water-immersion objective See RESOLVING POWER.

water moulds Traditionally, the aquatic, mycelial members of the orders Saprolegniales and Leptomitales, though the term is sometimes used to include all aquatic mycelial fungi.

water net See HYDRODICTYON.

water supplies Supplies of fresh (non-saline) water are obtained mainly from surface waters (rivers, streams, lakes) and underground layers of water-bearing rock (*aquifers*); water from aquifers is called *groundwater*. Typically, surface waters are more polluted. (Some supplies are obtained by desalination of seawater.)

To be made potable, i.e. drinkable, water from any natural source must be suitably treated – primarily to make it microbiologically safe. The type of treatment needed will depend on (i) the initial quality of the water (the lower the quality the more extensive the treatment), and (ii) the quantity of water involved – different approaches being used for small-scale (low-volume) supplies and large-scale (urban) supplies (see below).

[Microbiological safety of drinking water (emerging pathogens, cyanobacterial toxins, microbial ecology): ARM (2000) *54* 81–127.]

[Developments in microbiological risk assessment for drinking water (with particular reference to *Cryptosporidium parvum*, rotavirus and bovine spongiform encephalopathy): JAM (2001) *91* 191–205.]

Small-scale water supplies. Supplies in rural areas (e.g. for isolated houses) are often obtained from groundwater (wells) or from streams. Treatment may involve passing the water through an efficient fibre filter and then through a 'chlorinator' – i.e. a chamber containing tablets of calcium hypochlorite; the tablets gradually dissolve, releasing chlorine. Alternatively, the filtered water may be disinfected by ultraviolet radiation.

Small supplies (on expeditions etc.) can be disinfected by boiling or by using *halazone* tablets (*p*-carboxy-*N,N*-dichlorobenzenesulphonamide).

Novel methods have been used to improve the microbiological quality of drinking water in rural areas in some developing countries. For example, in Bangladesh, multi-layered local fabric has been used as a filter to retain a high proportion of plankton-associated *Vibrio cholerae*, and this may reduce the incidence/severity of cholera [AEM (1996) *62* 2508–2512]. In Kenya, water has been disinfected by exposure to solar radiation [AEM (1996) *62* 399–402; JMM (1999) *48* 785–787].

Large-scale (urban) water supplies. Water for use as a public supply must be adequately treated in order to (i) eliminate those pathogens capable of causing the major water-borne diseases (such as CHOLERA), and (ii) eliminate, or reduce to safe levels, any harmful substance (e.g. nitrate) which may be present. A further aim is to achieve a final product which is acceptable in terms of clarity, taste and odour. Note that treated water entering an urban distribution system is not necessarily sterile; moreover, treatment may fail to remove taints caused by certain microbial products: see e.g. GEOSMIN, METHYLISOBORNEOL and SYNURA.

The suitability of *raw water* (i.e. water to be made potable by treatment) is assessed according to various standards (including bacteriological, chemical) which are recommended by the World Health Organization and other bodies.

Some regions (e.g. Scotland, British Columbia) depend on surface waters; others (e.g. Austria, Denmark, Portugal) use mainly groundwater.

The type of treatment varies according to the origin (groundwater/surface water) and quality of the raw water.

(i) *Treatment of groundwater.* Groundwater is obtained via bore-holes and springs. Generally it is of good quality; however, because it derives from an underground source it commonly needs aeration, rapid sand filtration and disinfection (see later). Extra treatment may be necessary if it contains e.g. excess nitrates (see later).

(ii) *Treatment of surface water.* Initial storage of raw water for a number of days can be useful e.g. for allowing sedimentation of some particulate (suspended) matter. (In some cases, raw water is seeded with coagulant and left in a *sedimentation basin* until the coagulated matter has formed a sediment.) The water may then be microstrained through rotating stainless-steel mesh drums of pore size ~30 μm.

If the water contains little dissolved oxygen it may be aerated by a cascade or fountain process.

Most of the remaining fine particulate matter can be removed by adding a coagulant such as aluminium sulphate – which causes the particles to aggregate, forming *flocs*. Flocs can be removed (*clarification*) by passing the water upwards through a *floc blanket clarifier* – a tank within which the flocs form a 'sludge blanket' below a layer of clarified water; water is continually passed upwards through the blanket, clear water overflowing at the top of the tank and passing to the next stage of treatment.

Fine particles can be removed by allowing the water to flow downwards through a *rapid sand filter*: a bed of sand (grain size ~1 mm) which functions essentially as a mechanical sieve; when the filter begins to clog (lowering the filtration rate), air is blown upwards through the sand to dislodge particulates – this being followed by water (*back-washing*) to flush away the solids.

A *slow sand filter* is a bed of finer sand, the upper layer supporting a film of microorganisms – including bacteria and protozoa; it acts not only as a mechanical sieve but also (by virtue of the BIOFILM) as a means of mineralizing some of the dissolved organic matter and of removing some nitrogen and phosphorus as biomass. Moreover, a slow sand filter can remove certain taste- and odour-causing substances, and may also reduce the levels of any cyanobacterial toxins that may be present; dangerous levels of such toxins can be formed by BLOOMS of e.g. *Anabaena*, *Aphanizomenon*, *Microcystis*, *Nodularia* and/or *Oscillatoria* which may develop in reservoirs (and other bodies of water) if conditions are suitable. These toxins may cause gastroenteritis, liver disease and other conditions in man and various animals [see e.g. RMM (1994) *5* 256–264]. [Ecological and molecular investigations of cyanotoxin production (review): FEMS Ecol. (2001) *35* 1–9.]

The biofilm in a slow sand filter operates under *aerobic* conditions – essential e.g. for the activity of the (obligately aerobic) nitrifying bacteria (see NITROBACTERACEAE) which form part of the biofilm. Hence, water entering the slow sand filter must be adequately aerated. Unlike rapid sand filters, a slow sand filter is not back-washed; instead, the uppermost layer is periodically skimmed off (and replaced with new sand to maintain the depth of the filter).

Water leaving a slow sand filter is disinfected before entering the distribution system; disinfection usually involves chlorination and, because the activity of chlorine is affected by pH, the pH of the water may need to be adjusted (chemically).

Disinfection. CHLORINE, the common disinfecting agent, is very effective (at suitable concentrations) against vegetative pathogenic bacteria of faecal origin – e.g. *Escherichia coli* and species of *Campylobacter*, *Clostridium*, *Salmonella*, *Shigella* and *Vibrio*; chlorination is therefore an excellent measure against cholera, typhoid and other water-borne diseases.

Because chlorine reacts with impurities in the water, some chlorine is initially lost in satisfying this *chlorine demand*. Addition of further chlorine to the water leaves so-called 'residual chlorine' or *free residuals*; free residuals include elemental chlorine, hypochlorous acid (HClO) and ClO^- ions. The recommended level of free residuals is usually about 0.5–2 parts per million (p.p.m.) (= 0.5–2 mg/litre). Disinfection by chlorine is maximally effective at pH <7 because acidic conditions tend to depress the dissociation of hypochlorous acid.

If water contains ammonia, chlorine combines with it to form *chloramines* (*combined residuals*). Chloramines decompose slowly, releasing chlorine; as disinfectants, they are less effective – but more persistent – than chlorine. Ongoing addition of chlorine raises the concentration of chloramines to a maximum level – which depends on the initial concentration of ammonia. Further chlorine begins to oxidize the chloramines;

levels of combined residuals therefore begin to fall. On continued addition of chlorine, the combined residuals reach their lowest level (the *breakpoint*), and any further chlorine added will form free residuals in the water. Hence, to achieve free residuals, the amount of chlorine added must be beyond the breakpoint (*breakpoint chlorination*).

Superchlorination uses higher concentrations of chlorine in order to eliminate undesirable odours and/or tastes; excess chlorine is later removed by adding sulphur dioxide (*sulphonation*) until the desired level of free residuals is achieved.

(OZONE, compared with chlorine, is a stronger disinfectant, but it lacks residual activity. If ozone is used it is followed by low-level chlorination in order to provide residual disinfection.)

Testing for free residuals. With tablets of DPD (*N*,*N*-diethyl-*p*-phenylenediamine), chlorine levels can be determined by the level of coloration which develops; the intensity of coloration (and, thus, the level of chlorine) is indicated by comparison with a colour chart or 'comparator'.

Tests for the efficiency of disinfection. Certain bacteria, particularly faecal coliforms and 'faecal streptococci' (the latter including *Enterococcus faecalis* and related organisms) are used as indicators of faecal pollution; their presence in treated drinking water points to a failure of the treatment process or to contamination subsequent to treatment. Coliforms and 'faecal streptococci' are used as indicator organisms because they are common intestinal bacteria; their presence in water points to the potential presence of water-borne faecal pathogens. Usually, pathogens are not tested for directly as they may occur only intermittently; moreover, the indicator bacteria greatly outnumber pathogens (e.g. $\sim 10^8$ *E. coli* cells per gram faeces), and it is possible to detect very low levels of faecal pollution.

Indicator bacteria are detected/counted by the MULTIPLE-TUBE METHOD and by membrane filtration. In membrane FILTRATION, a known volume of water is filtered through a membrane filter (pore size ~ 0.2 μm); the membrane is then incubated, face upwards, on a pad saturated with appropriate growth medium. Indicator bacteria, if present, form colonies on the membrane; this allows enumeration from (i) the number of colonies, and (ii) the volume filtered. Confirmation of *E. coli* (a thermotolerant 'faecal' coliform) is obtained by two further tests carried out at 44°C/24 hours: the INDOLE TEST and the EIJKMAN TEST. (The medium used for the Eijkman test includes an agent that inhibits endospore-formers; this is particularly important when testing *chlorinated* water because spore-formers are more resistant than *E. coli* to chlorine.)

Note that fully treated water may still contain pathogens. For example, certain viruses (e.g. the NORWALK VIRUS [AEM (1985) *50* 261–264]) and thick-walled oocysts of CRYPTOSPORIDIUM can survive routine levels of chlorination. (The cysts of *Entamoeba histolytica* may survive chlorination but are likely to be retained by an efficient sand filter.) [Cysts of GIARDIA in raw and chlorinated water in Canada: AEM (1996) *62* 47–54.] Studies on the resistance of *Mycobacterium avium* to disinfectants (including chlorine and ozone) have suggested that resistance to these agents may be a factor which promotes persistence of *M. avium* in drinking water [AEM (2000) *66* 1702–1705].

Problems with nitrates, pesticides etc. Nitrates are common in surface waters and, increasingly, in groundwater; a major source is agricultural fertilizer. The upper limit recommended by the EC is 50 mg of nitrate ion per litre; the USEPA value is 44.29 mg/litre (= 10 mg/l as N). Water containing high levels of nitrate may be blended with low-nitrate waters. Alternatively, water may be stored for long periods to permit DENITRIFICATION. Nitrate can also be removed by ion-exchange.

Pesticides etc. may be removed by adsorption to activated charcoal.

Problems in the distribution system. Microbial growth/slime in pipelines may lower water quality and pumping efficiency. This problem has been linked to the presence of organic carbon, but a more stringent removal of phosphorus might offer a cost-effective method for control [Nature (1996) *381* 654–655]. [Drinking water biofilms: FEMS Ecol. (1997) *22* 265–279.]

Waterhouse–Friderichsen syndrome See MENINGITIS.

watermark disease A disease of willows (*Salix* spp) caused by *Erwinia salicis*; symptoms: leaves wither and become reddish, a watery liquid is discharged from infected shoots, and the wood is stained brown or black and may be rendered unfit for use as timber.

watery stipe (La France disease) A MUSHROOM DISEASE of viral causation. A consistent symptom is cessation of growth leading to a serious reduction in yield. The stipe is often long, slightly bent, and watery, and the cap may be very small and may fail to open. [Review: Book ref. 180, pp. 239–324.]

A disease known as *hard gill* has also been attributed to viral infection.

[Properties of two viruses from *Agaricus brunnescens*: JGV (1979) *42* 231–240.]

Watson–Crick base pair See BASE PAIR.

Watson–Crick double helix See DNA.

wax caps See AGARICALES (Hygrophoraceae).

wax D A peptidoglycolipid component of the cell wall in *Mycobacterium* spp. Wax D is a branched arabinogalactan which is esterified with MYCOLIC ACIDS and linked, via an acid-labile phosphodiester bridge, to an *N*-glycolylmuramic acid residue in the PEPTIDOGLYCAN. The glycolipid is soluble in chloroform or ether, but not in acetone, and can be hydrolysed by KOH; it appears to be released by autolysis in old cultures. Wax D can replace whole mycobacteria in Freund's complete adjuvant. (cf. MYCOSIDE C.)

Wb phage WEYBRIDGE PHAGE.

WBC White blood cell.

WCVs See VACCINE.

weal-and-flare See WHEAL-AND-FLARE.

wedeloside An analogue of ATRACTYLOSIDE which occurs in the Australian weed, *Wedelia asperrima*; it inhibits the mitochondrial ADP/ATP carrier system [FEBS (1985) *189* 245–249].

WEE WESTERN EQUINE ENCEPHALOMYELITIS.

Weibel–Palade bodies See VON WILLEBRAND FACTOR.

Weigle mutagenesis See WEIGLE REACTIVATION.

Weigle reactivation (W-reactivation) A phenomenon in which certain bacteriophages (e.g. λ) containing UV-damaged DNA show much higher survival rates in host cells which have been irradiated with UV before infection than in non-irradiated host cells. Furthermore, the mutation frequency in the UV-damaged phage is higher when the phage is grown in pre-irradiated host cells (*Weigle mutagenesis*). These observations are at least partly explicable on the basis that irradiation of the host cells induces the SOS SYSTEM which can bring about high levels of repair, including *umuDC*-dependent error-prone repair, of the UV-damaged phage DNA (the damaged phage DNA is apparently not an effective signal for SOS induction in non-irradiated cells). However, W-reactivation depends on UV irradiation of the host cells even when the cells are mutants in which the SOS genes are fully derepressed, implicating the involvement of another factor – possibly the 'active' form of the RecA protein [MGG (1985) *201* 329–333].

Weil–Felix test A serological test which is used for the presumptive diagnosis of certain rickettsial diseases; the test is based on the fact that antibodies against certain rickettsial antigens can agglutinate cells of *Proteus vulgaris* strains OX19, OX2 and OXK. In the test, a suspension of a given strain of *P. vulgaris* is mixed with the patient's serum, and the tube is subsequently examined for agglutination. OX19 and OX2 generally give positive results in cases of epidemic and murine typhus and spotted fevers, but negative results in cases of scrub typhus, Q fever, rickettsialpox, and (usually) Brill–Zinsser disease. OXK generally gives positive results *only* in cases of scrub typhus.

Weil's disease See LEPTOSPIROSIS.

Wenyonella See EIMERIORINA.

Wescodyne See IODINE (a).

Wesselsbron virus See FLAVIVIRIDAE.

West Nile virus See FLAVIVIRIDAE.

Westerfeld modification See VOGES–PROSKAUER TEST.

Western blot See BLOTTING.

western duck disease A type of BOTULISM which affects waterfowl that feed on decaying vegetation in shallow, alkaline waters harbouring *C. botulinum* type C_α. The disease occurs in the western USA, often in extensive epidemics.

western equine encephalomyelitis (WEE; western equine encephalitis; western encephalitis) An acute ENCEPHALITIS (or encephalomyelitis) of man and horses, caused by an ALPHAVIRUS. The WEE virus is widely distributed in North America in wild birds and can be transmitted from bird to bird by the mosquitoes *Culex tarsalis* and *Culiseta melanura*; however, WEE occurs chiefly in western regions, reflecting the distribution of *C. tarsalis* (common west of the Mississippi river) which bites mammals as well as birds. (*C. melanura* occurs in remote swamps and feeds primarily on birds.) In horses, WEE may have a high mortality rate. In man, WEE is usually benign but may be fatal; it can cause serious sequelae (e.g. mental impairment, paralysis) in young infants.

wet bubble See BUBBLE DISEASES.

wet mount (wet preparation) In microscopy: a temporary preparation in which the specimen is present in a thin layer of liquid sandwiched between a SLIDE and COVER-GLASS.

wet rot (of timber) A BROWN ROT of softwood and hardwood timbers which have a high moisture content (e.g. 40–60% – see TIMBER SPOILAGE); the common cause is *Coniophora puteana*. Infected timbers usually exhibit a dark, branching, surface mycelium. Wet rot may be controlled by effective ventilation and drying of the timbers. (See also TIMBER PRESERVATION.)

wettable powder (wp) A form in which water-insoluble antifungal agents (e.g. CAPTAN) etc may be prepared for agricultural use. The active ingredient, usually mixed with a filler (e.g. fine clay, diatomaceous earth) is finely ground, and a surface-active 'wetter' is added – usually a soluble salt of an organic acid. The filler permits the active ingredient to be ground into fine particles and prevents caking during storage in an aqueous medium; the wetter allows rapid dispersion of the powder in water.

wetwood A disease of living trees, common e.g. in elms and cottonwoods, associated with the presence of dense populations of hydrolytic, methanogenic and nitrogen-fixing prokaryotes in moist, alkaline, anaerobic dead xylem tissue. Within a given tree, the disease appears to spread as a result of the degradation of xylem vessel-to-ray pit membranes by pectinolytic bacteria; anaerobic cellulolytic bacteria have also been detected [AEM (1985) *50* 807–811]. Holes bored into the trunk of an infected tree yield methane and a fetid liquid. Organisms isolated from wetwood include species of *Bacteriodes*, *Clostridium*, *Erwinia*, *Lactobacillus* and *Propionispira*, and *Methanobrevibacter arboriphilus*. [Microbial populations in wetwood of European white fir (*Abies alba*): FEMS Ecol. (1986) *38* 141–150.] (See also TREE DISEASES.)

Weybridge phage (Wb phage) A group 3 *Brucella* phage which can replicate in smooth strains of *B. abortus*, *B. neotomae* and *B. suis*.

WGA WHEAT GERM AGGLUTININ.

Whataroa virus See ALPHAVIRUS.

WHBV See WOODCHUCK HEPATITIS VIRUS.

wheal-and-flare A localized oedematous swelling (wheal) and reddening (flare) on the body surface: a manifestation of certain forms of immediate hypersensitivity which is brought about by the release of histamine from degranulated mast cells (see e.g. PRAUSNITZ–KÜSTNER TEST).

wheat (nitrogen fixation) See NITROGEN FIXATION.

wheat chlorotic streak virus See RHABDOVIRIDAE.

wheat diseases See CEREAL DISEASES.

wheat dwarf virus See GEMINIVIRUSES.

wheat germ agglutinin (WGA) A non-mitogenic LECTIN from wheat, *Triticum vulgare*; it binds to oligosaccharides which contain *N*-acetylglucosamine.

wheat scab (Fusarium scab; Fusarium blight) A disease of wheat caused by *Fusarium graminearum* (see FUSARIUM). Consumption of grain harvested from infected plants ('scabby wheat') is a cause of TRICHOTHECENE poisoning.

wheat soil-borne mosaic virus *Syn.* SOIL-BORNE WHEAT MOSAIC VIRUS.

wheat spindle streak mosaic virus See POTYVIRUSES.

wheat spindle streak virus See POTYVIRUSES.

wheat streak mosaic virus See POTYVIRUSES.

wheat streak virus See POTYVIRUSES.

wheat striate mosaic virus See RHABDOVIRIDAE.

wheat yellow leaf virus See CLOSTEROVIRUSES.

wheat yellow mosaic virus See POTYVIRUSES.

whey The liquid by-product of CHEESE-MAKING; it contains ca. 4–5% lactose, 0.8% protein, and 0.2–0.6% lactic acid. It has been intensively investigated as a source of substrates for various commercial microbial products – e.g. INDUSTRIAL ALCOHOL and SINGLE-CELL PROTEIN.

whey proteins In MILK: non-CASEIN proteins which are not precipitated under acidic conditions.

whiplash flagellum See FLAGELLUM (b).

Whipple's disease Intestinal lipodystrophy: in man, a malabsorption syndrome characterized e.g. by diarrhoea, steatorrhoea, lymphadenopathy, CNS involvement, and infiltration of the intestinal mucosa by macrophages containing PAS-positive material. It is linked to a Gram-positive bacterium, *Tropheryma whipplei* (formerly *T. whippelii*) [culture in fibroblasts: NEJM (2000) *342* 620–625; genome: Lancet (2003) *361* 637–644].

Diagnosis of Whipple's disease has been based primarily on histology [VA (1996) *429* 335–343] with PCR-based confirmation using e.g. duodenal tissue or cerebrospinal fluid (CSF) [diagnosis and treatment: Drugs (1998) *55* 699–704]. PCR-positive tests for *T. whipplei* in patients with no clinical signs of Whipple's disease suggest that diagnosis of this disease should not be based solely on the results of PCR-based tests [Lancet (1999) *353* 2214].

Treatment (e.g.): parenteral penicillin + streptomycin for 2 weeks, or (long-term) trimethoprim–sulphamethoxazole.

[Whipple's disease (review): Lancet (2003) *361* 239–246.]

whirling disease A FISH DISEASE which affects salmonid fish; it is caused by a protozoon (see below). The pathogen causes little harm to the European brown trout (*Salmo trutta*), apparently its original host, but in the rainbow trout (*S. gairdneri*) it causes a severe infection, migrating to and developing in cartilaginous tissue and causing cartilage destruction and skeletal deformities; affected fish commonly swim with a whirling, 'tail-chasing' motion.

The causal agent of whirling disease is *Myxosoma cerebralis* (class Myxosporea). However, the spores released from an infected fish cannot infect another fish, but can infect certain tubificid oligochaete worms; an infected worm in turn produces a distinct stage of the pathogen – a stage previously regarded as belonging to the genus *Triactinomyxon*, class Actinosporea (see MYXOZOA). The '*Triactinomyxon*' stage cannot reinfect other worms but can infect salmonid fish, thus completing the life cycle. [Science (1984) *225* 1449–1452.]

whisky (whiskey) See SPIRITS.

white blister disease See ALBUGO.

white clover mosaic virus See POTEXVIRUSES.

white fluids Commerical DISINFECTANTS consisting of oil-in-water emulsions of certain phenolic coal tar fractions; emulsifying agents used include e.g. animal glue, casein. White fluids are more stable than BLACK FLUIDS in the presence of electrolyte.

white membrane See PURPLE MEMBRANE.

white piedra A chronic human MYCOSIS in which hair of the scalp, beard and/or pubic region is infected by *Trichosporon beigelii* (= *T. cutaneum*). Soft, gelatinous, white or fawn nodules, composed of a dense mycelial network, adhere to infected hairs; the hair shaft may be penetrated and may break in the vicinity of the nodule, but skin is not infected. (cf. BLACK PIEDRA.)

white pine blister rust *Syn.* BLISTER RUST.

white precipitate See MERCURY.

white rot (1) (of timber) A form of fungal TIMBER SPOILAGE in which the CELLULOSE, HEMICELLULOSES and LIGNIN are decomposed, leaving the wood soft, whitish and fibrous. The hyphae of at least some white-rot fungi (e.g. *Coriolus versicolor*, *Phanerochaete chrysosporium*) penetrate the wood cell lumen and lie in contact with the inner surface of the cell wall; degradation of the wall then occurs in the vicinity of the hypha, often resulting in the formation of a trough with a central ridge upon which the hypha rests. The enzymes (see CELLULASES) appear to be retained in a mucilaginous sheath surrounding the hyphae. [Book ref. 39, pp. 59–61; microstructural changes in wood due to white rot basidiomycetes: CJB (1986) *64* 905–911.] (cf. BROWN ROT; POCKET ROT.)

(2) (of onions) See ONION ROT.

white rust (1) See ALBUGO. (2) See CHRYSANTHEMUM WHITE RUST.

white scours Scouring in young calves caused by strains of *Escherichia coli*. (cf. CALF SCOURS.)

white smoker See HYDROTHERMAL VENT.

white spot May refer to ICHTHYOPHTHIRIASIS or to a similar disease of marine fish caused by the ciliate *Cryptocaryon irritans*.

whitehead (1) (*plant pathol.*) The dead and bleached upper part of a plant; in cereals: a white, narrow ear which is empty or contains only shrivelled grains. Whiteheads are formed as a result of certain diseases. (cf. SILVER-TOP.)

(2) (*med.*) See e.g. ACNE.

whitlow (felon) (1) *Syn.* PARONYCHIA. (2) An abscess affecting the pulp of a fingertip.

WHO plate A rectangular block of plastic (e.g. 18 × 15 × 1.5 cm), one face of which bears a number of rows of indentations (wells) – each well being approximately semicircular in (vertical) cross-section and having a capacity of e.g. 1.5 ml. WHO plates are used e.g. in various quantitative serological procedures which involve the preparation of serial dilutions of serum etc.

whole mount Any specimen mounted whole on a SLIDE for microscopical examination.

whooping cough (pertussis) An acute infectious disease which affects mainly children; it may be caused by *Bordetella pertussis* or *B. parapertussis* – the latter pathogen commonly (though not invariably) causing a less severe form of the disease ('parapertussis'). (*B. bronchiseptica* causes disease in dogs and other animals, and may also cause respiratory disease in humans, particularly the immunocompromised.)

Infection occurs by droplet inhalation. The incubation period is ~7–10 days. Early symptoms may resemble those of a common cold. The characteristic paroxysmal stage occurs ~7–14 days later; this involves paroxysms of coughing, each ending in a prolonged, audible inspiratory effort (the 'whoop') as air is forcibly drawn through the narrowed glottis. The whoop is often followed by vomiting. (In infants, the whoop may be replaced by cyanosis.) Complications include e.g. secondary infections (otitis media, pneumonia etc.) and CNS involvement (e.g. ENCEPHALITIS). The disease may be fatal, especially in infants. Recovery is usually followed by long-term immunity.

Pathogenesis. The pathogen does not invade respiratory epithelium. It adheres to the mucosa; adhesion to ciliated epithelium is apparently mediated primarily by (i) the pathogen's (non-fimbrial) *filamentous haemagglutinin* (FHA), and (ii) certain B subunits of PERTUSSIS TOXIN. Pathogenesis apparently involves the combined effects of several toxins: PERTUSSIS TOXIN, CYCLOLYSIN, TRACHEAL CYTOTOXIN (cf. DERMONECROTIC TOXIN).

Lab. diagnosis. In preference to the earlier COUGH PLATE method, a pernasal swab is (i) plated on either BORDET–GENGOU AGAR or CHARCOAL BLOOD AGAR, and (ii) used to prepare a smear for detecting the pathogen by an IMMUNOFLUORESCENCE test. (Improved isolation of *B. parapertussis* has been reported on Moredun medium [JCM (1996) *34* 638–640].) For serology, emulsified growth from several colonies is examined with specific antisera in a SLIDE AGGLUTINATION TEST.

A nested duplex PCR is able simultaneously to detect both *B. pertussis* and *B. parapertussis* [JCM (1999) *37* 606–610].

Chemotherapy. Antibiotics do not alter the clinical course once the paroxysmal stage has begun; erythromycin may reduce disease severity if give before this stage (and may reduce infectivity).

Vaccines. Whole-cell VACCINES (WCVs) do not confer complete immunity to pertussis, and have been associated with side-effects. Acellular pertussis vaccines (ACVs) are reported to have fewer side-effects but may give less protection against disease caused by *B. parapertussis* [RMM (1996) *7* 13–21].

WHV WOODCHUCK HEPATITIS VIRUS.

WI-38 (Wistar Institute 38) A diploid cell line derived from female embryonic lung tissue.

Wickerhamia A genus of yeasts (family SACCHAROMYCETACEAE). Cells: oval to elongate or lemon-shaped, occurring singly, in pairs, or in very short chains; vegetative reproduction occurs by bipolar budding. Ascus formation is not preceded by conjugation. Ascospores: sporting-cap-shaped, usually 1–2 per ascus. Glucose and other sugars are fermented; NO_3^- is not assimilated. One species: *W. fluorescens* (formerly *Kloeckera fluorescens*); only one strain has been isolated (from dung of a wild squirrel in Japan). [Book ref. 100, pp. 440–442.]

Wickerhamiella A genus of yeasts (family SACCHAROMYCETACEAE). Cells: spheroidal, ovoid or ellipsoidal; vegetative reproduction occurs by multilateral budding. Pseudomycelium is not formed. Ascus formation is preceded by conjugation. Ascospores: oblong with obtuse ends, minutely rugose (appearing smooth by light microscopy). Non-fermentative; NO_3^- is assimilated. One species: *W. domercqii* (anamorph: *Candida domercqii*), isolated e.g. from a wine vat and from effluent from a cane-sugar factory. [Book ref. 100, pp. 443–445.]

Widal test An AGGLUTINATION TEST used for detecting serum antibodies to salmonellae which cause human ENTERIC FEVERS (including TYPHOID FEVER). Serial dilutions of serum are tested for the presence of agglutinins to the O and H antigens of these salmonellae – suspensions of (killed) bacteria of each species or serotype being used as the test antigens. A rising titre in paired sera is indicative of infection. (Anti-H, but not anti-O, agglutinins may persist for years after infection.)

wide groove *Syn.* major groove (see DNA).

WIGA bacterium *Legionella bozemanii.*

Wigglesworthia A genus of endosymbiotic, Gram-negative bacteria of the family Enterobacteriaceae which occur within BACTERIOCYTES in the anterior midgut of the tsetse fly (GLOSSINA).

(See also SODALIS and WOLBACHIA.)

wild type Refers to a given strain of organism – or to a genotypic or phenotypic trait in a given organism – which is characteristic of the majority of naturally occurring members of its kind, i.e., a *non-mutant* strain or trait.

wildfire disease (of tobacco) A disease of the tobacco plant that involves severe chlorosis of the leaves; symptoms are caused by a toxin, TABTOXIN, secreted by the pathogen, *Pseudomonas syringae* pv. *tabaci*. (See also TOBACCO DISEASES.)

Wilson and Blair's agar (bismuth sulphite agar) A medium containing peptone, beef extract, glucose, disodium phosphate, ferrous sulphate, brilliant green, bismuth sulphite (as indicator) and agar [recipe: Book ref. 53, p. 1393]. The medium is used for the primary isolation of *Salmonella* strains (e.g. from faeces); typically, coliforms and some strains of *Shigella* are inhibited. Isolated colonies of *Salmonella* are black, but heavy or confluent growth may not exhibit blackening.

wilt (1) (*plant pathol.*) A disease (or symptom) characterized by a loss of turgidity in a plant: see e.g. VASCULAR WILT. (2) (*entomopathol.*) Any viral INSECT DISEASE.

Wimmera grass poisoning Poisoning of sheep and cattle following consumption of e.g. Wimmera ryegrass (*Lolium rigidum*) bearing galls (which resemble ergot sclerotia) induced by larvae of the nematode *Anguina lolii*; it appears that galls responsible for the poisoning contain a yellow-pigmented *Corynebacterium* sp [Australian Vet. J. (1976) *52* 242, cited in Book ref. 33, pp. 1168–1169]. Symptoms: ataxia, convulsions, sometimes with haemorrhages and abortion; the toxicosis is an important cause of death among livestock in western Australia.

window-phase blood See TRANSFUSION-TRANSMITTED INFECTION.

wine-making Wine is made by the ALCOHOLIC FERMENTATION of grapes or grape-juice by special strains of yeast – usually strains of *Saccharomyces cerevisiae* var. *ellipsoideus* ('*S. ellipsoideus*'). Grapes contain ca. 10–25% fermentable sugars (mainly glucose and fructose) together with organic acids (mainly tartaric and malic) and small amounts of nitrogenous compounds (e.g. amino acids), tannins, pigments, vitamins, minerals etc. The grapes are crushed and SULPHUR DIOXIDE is added to inhibit 'wild' yeasts and bacteria (wine yeasts are relatively resistant to SO_2). For white wines either black or white grapes may be used, the juice being extracted from the pulp by pressing before fermentation. For red wines black grapes are used, and pressing does not occur until after fermentation; during fermentation anthocyanin pigments are leached from the grape skins by the alcohol. Rosé wines may be made from black grapes (the skins being removed before all the pigment has been extracted), from less strongly pigmented grapes, or by blending red and white wines.

Before the grape-juice/pulp (*must*) is inoculated with the yeast, its composition may be adjusted e.g. by the addition of sugar, water, and/or traces of ammonium salts (to stimulate fermentation). Some musts may be excessively acidic. Levels of L-malic acid may be reduced either by the bacterial MALOLACTIC FERMENTATION or by the fermentation of malic acid to ethanol and CO_2 by species of *Schizosaccharomyces*; either fermentation may occur during or after the alcoholic fermentation. (Acidity may also be reduced by adding $CaCO_3$.) Alcoholic fermentation of the must is generally initiated by the addition of a starter culture of wine yeast (pre-grown e.g. in sterile grape-juice). The fermentation temperature is important; it affects the flavour of the finished product – lower temperatures favouring the formation of aromatic compounds (and hence bouquet) – and also the time necessary for fermentation to be completed (several days or longer). For white wines the starting temperature is ca. 7–16°C, rising to 20–23°C; red wines are fermented at e.g. 21–27°C. (Since fermentation generates heat, cooling is necessary.)

Fermentation is complete when all the fermentable sugars have been used or when the level of alcohol is high enough to inhibit further yeast action. Particulate matter is allowed to settle and the wine is drawn off as early as possible after fermentation to avoid the effects of yeast autolysis; autolysis can cause off-flavours, and the amino acids, vitamins etc released can encourage the growth of spoilage bacteria. (However, if a malolactic fermentation is required, some yeast autolysis may be desirable to encourage growth of the fastidious malolactic bacteria.) The wine is then clarified e.g. by centrifugation and/or 'fining' (e.g. by the addition of the adsorbent bentonite) and/or filtration; lower quality wines may be pasteurized. Finally, the wine is allowed to age (for several months or more) either in the bottle or in oak casks prior to bottling.

Table wines have an alcohol content of ca. 9–14% by volume and may be *dry* (i.e. with little or no remaining sugar) or *sweet*. Sweet wines may be made from musts with a high sugar content (see e.g. NOBLE ROT), fermentation being stopped (e.g. by chilling) before all the sugar has been used; alternatively, fermentation may be stopped by fortification (see below). *Sparkling wines* (e.g. champagne) contain CO_2 under pressure (ca. 2–5 atmospheres); cheaper wines may be carbonated directly, but quality wines undergo a secondary fermentation of added sugar either in the bottle or in large closed tanks prior to bottling under pressure.

Fortified wines (e.g. port, sherry) are made by adding wine spirit (brandy) to the primary fermentation at a stage determined by the level of sugar required in the finished product. The alcohol content of a fortified wine is usually ca. 16–20%. *Flor sherries* (e.g. Spanish fino sherries) are made by a more complex process. The primary fermentation resembles that for other wines and is carried out to completion; the required final alcohol content is ca. 15%. A secondary 'flor fermentation' (FERMENTATION sense 2) is then carried out in oak barrels and requires the presence of air. A thick film of special 'flor' yeasts (including various species and strains of *Saccharomyces*) forms on the surface of the wine; these yeasts slowly oxidize a small proportion of the ethanol to acetaldehyde which, together with acetal (a product of its reaction with ethanol), is an important flavour component. Completion of the traditional surface flor fermentation takes one to several years; however, more rapid methods have been developed in which the flor yeast is submerged in the wine in pressure tanks. Finally the sherry is fortified to the required alcohol content before bottling.

[Selection and modification of microorganisms used in winemaking: Food Mic. (1984) *1* 315–332.]

(See also PULQUE, SAKE; cf. BREWING, CIDER, SPIRITS.)

wine spoilage Microbial spoilage of wines may be due to various yeasts (e.g. species of *Saccharomyces*, *Pichia*, *Brettanomyces*) which may cause turbidity and off-flavours; off-flavours may also result from autolysis of the wine yeast itself (see WINE-MAKING) and from the reduction during fermentation of elemental sulphur (sprayed onto the grapes as a fungicide) to H_2S. Spoilage bacteria include e.g. *Acetobacter* spp (which may cause ACETIFICATION) and lactobacilli (which may cause off-flavours due e.g. to the production of DIACETYL from citric acid); *Leuconostoc mesenteroides* may cause ROPINESS if enough sugar is present for dextran formation. The MALOLACTIC FERMENTATION may be regarded as spoilage in wines already low in acid.

Wingea A genus of yeasts (family SACCHAROMYCETACEAE) in which the cells are spheroidal to ellipsoidal; vegetative reproduction occurs by multilateral budding. Pseudomycelium is not formed. Prior to ascus formation, cells with 'copulatory' protuberances are formed; asci (which are persistent) develop from these cells either directly or following fusion between two cells via their protuberances. Ascospores: lentiform, light brown, smooth, 1–4 per ascus. Glucose, sucrose, maltose and raffinose are fermented; NO_3^- is not assimilated. One species: *W. robertsii*, isolated e.g. from soil. [Book ref. 100, pp. 446–448.]

winter coccidiosis See COCCIDIOSIS (b).

winter dysentery (*syn.* black scours) An acute, sometimes fatal disease of (primarily) adult (and particularly housed) cattle; symptoms: fever and dark, watery diarrhoea, or sometimes DYSENTERY. Aetiology is uncertain: strains of *Campylobacter* may be the primary cause or may play a secondary role to a viral agent. Infection occurs via faecally-contaminated feed or water. A carrier state occurs.

winter mushroom *Syn.* ENOKITAKE.

winter spore (*mycol.*) *Syn.* RESTING SPORE.

winter vomiting disease See FOOD POISONING (i).

Winterbottom's sign Following infection from a bite on the *head* by a tsetse fly carrying the causal agent of SLEEPING SICKNESS: enlargement of the post-auricular or occipital lymph nodes.

wire stem See DAMPING OFF.

Wisconsin mastitis test See CALIFORNIA MASTITIS TEST.

Wiskott–Aldrich syndrome A primary immunodeficiency which affects young children and which is characterized e.g. by susceptibility to recurrent bacterial infections (meningitis, otitis media etc.). Mutation in the *WASP* gene has been found in cases of Wiskott–Aldrich syndrome, although mutant forms of this gene are also associated with e.g. X-linked thrombocytopenia.

[Primary immunodeficiency diseases: Lancet (2001) *357* 1863–1869.]

WISH An ESTABLISHED CELL LINE derived from human amnion; the cells are heteroploid and epithelioid.

witches' broom An abnormal form of plant growth, most common in woody plants, in which there is a profuse outgrowth of lateral buds to give a 'witches' broom' appearance; the shoots may be thickened and may bear abnormal leaves. Witches' brooms may result from infection by various microorganisms – e.g. *Taphrina* spp (e.g. *T. deformans* in the hairy birch, *Betula pubescens*, and *T. turgida* in the silver birch, *B. pendula*), *Crinipellis perniciosa* in cacao plants, and various rust fungi. Many of these organisms produce auxins in vitro, but the mechanism of witches' broom formation in vivo remains unclear.

witches' butter See TREMELLALES.

WL 28325 A fungicide (2,2-dichloro-3,3-dimethyl-cyclopropane carboxylic acid) which causes an increase in the amount of MOMILACTONE produced by rice plants infected with *Pyricularia oryzae*.

wobble hypothesis A hypothesis originally proposed by Crick [JMB (1966) *19* 548–555] to account for the observed pattern of degeneracy in the 3rd base of a codon (see GENETIC CODE), i.e., it explains how certain tRNAs can recognize more than one codon when the codons differ only in the 3rd position (the base at the 3′ end). According to this hypothesis, normal base-pairing can occur between bases in positions 1 and 2 of the codon and their complementary bases (3rd and 2nd, respectively) in the anticodon; however the base in the *1st* position of the anticodon (the 'wobble position') can undergo non-Watson–Crick base-pairing with the corresponding (3rd) base in the codon: G in the wobble position can pair with either C or U in the 3rd position of the codon, and similarly U can pair with either A or G. These recognition patterns can be affected by modification of the bases in the anticodon. For example, A seems never to occur in the wobble position in an anticodon, being always modified (by deamination) to inosine (I) in this position; I can pair with A, C or U. Similarly, unmodified U rarely occurs in the wobble position; 2-thiouridine in this position can pair with A but does not allow 'wobble pairing' with G. (See also QUEUOSINE.)

The wobble hypothesis is applicable to most, but not all, tRNAs; the decoding potential of a tRNA may be influenced e.g. by the conformation of its anticodon arm. (See also SUPPRESSOR MUTATION.)

Wolbachia A genus of Gram-negative bacteria of the tribe WOLBACHIEAE; the organisms grow intra- or extracellularly in various invertebrate hosts but are apparently not overtly pathogenic. The cells are coccoid or rod-shaped, morphologically similar to those of e.g. *Rickettsia*.

W. melophagi grows in the gut lumen of e.g. sheep keds (*Melophagus ovinus*) and in certain flies.

W. persica grows e.g. in ticks of the genera *Argas*, *Dermacentor* and *Rhipicephalus*.

W. pipientis grows in certain mosquitoes and in the almond moth.

Strains of *Wolbachia* also occur in reproductive tissues of the tsetse fly (GLOSSINA). (See also SODALIS and WIGGLESWORTHIA.)

Certain strains of *Wolbachia* have been found as obligate endosymbionts in the filarial nematode *Onchocerca volvulus*; *O. volvulus* causes human onchocerciasis – one manifestation of which is the disease known as *river blindness*. Onchocerciasis is transmitted to man via a blood-sucking insect (black fly). The adult forms of *O. volvulus* are characteristically found in subcutaneous tissues, where they give rise to nodular lesions, while the (juvenile) microfilariae (which are approx. 300 × 8 μm) may be isolated from the skin or bloodstream (those in the bloodstream potentially localizing in the eyes and causing river blindness).

Treatment of onchocerciasis with ivermectin, while effective against juvenile forms, does not affect the adult stage. However, treatment with a tetracycline (doxycycline) has been found to deplete the *Wolbachia* endosymbiont, causing long-term sterility in the nematode and inhibition of the worm's development and viability; a potential advantage of this treatment is interruption of the transmission of *O. volvulus*.

[Endosymbiotic bacteria in worms as targets for a novel chemotherapy in filariasis: Lancet (2000) *355* 1242–1243. Depletion of *Wolbachia* endobacteria in *Onchocerca volvulus* by doxycycline: Lancet (2001) *357* 1415–1416. Comment: TIP (2001) *17* 358. Onchocerciasis: BMJ (2003) *326* 207–210.]

Wolbachieae A tribe of bacteria of the family RICKETTSIACEAE; species are parasites or pathogens in invertebrate hosts. The genera: RICKETTSIELLA; WOLBACHIA. [Book ref. 22, pp. 711–717.]

Wolhynian fever *Syn.* TRENCH FEVER.

Wolinella A genus of oxidase-positive, Gram-negative bacteria (family BACTEROIDACEAE) which occur e.g. in the RUMEN and in the human gingival crevice, and which have been isolated from dental root canal infections. Cells: straight, curved or helical rods, 0.5–1.0 × 2.0–6.0 μm; monotrichously (polarly) flagellated. Energy is obtained by the anaerobic oxidation of formate and/or H_2 using e.g. fumarate as electron acceptor (see FUMARATE RESPIRATION); fumarate reductase, which appears to receive electrons from a menaquinone via a *b*-type cytochrome, catalyses the reduction of fumarate to succinate. L-Aspartate, L-malate or nitrate may be used instead of fumarate – although in at least some cases succinate must be supplied if nitrate is used as electron acceptor. Carbohydrates are not utilized. The organisms grow e.g. in complex or synthetic media supplemented with haemin, sodium formate and sodium fumarate. GC%: ca. 42–48. Type species: *W. succinogenes* (formerly *Vibrio succinogenes*). Other species: *W. curva* [a species of human origin: IJSB (1984) *34* 275–282], and *W. recta* (predominantly straight rods).

Woloszynskia See DINOFLAGELLATES.

wood blewit See BLEWIT.

wood-boring beetles See SCOLYTIDAE.

wood mushroom See AGARICUS.

wood pulp spoilage See PAPER SPOILAGE.

wood spoilage See TIMBER SPOILAGE.

wood sugar XYLOSE.

Wood–Werkman reaction (1) The carboxylation of pyruvate to yield oxaloacetate [see Appendix II(b)]. (2) Any carboxylation reaction which yields oxaloacetate, e.g., carboxylation of phosphoenolpyruvate [see Appendix II(b)].

woodchuck hepatitis virus (WHV; woodchuck hepatitis B virus, WHBV) A virus of the HEPADNAVIRIDAE which infects woodchucks (*Marmota monax*) and which closely resembles the human HEPATITIS B VIRUS (HBV); like HBV, WHV can cause chronic liver disease and hepatocellular carcinoma (HCC) in its host, HCC developing after ca. 2–4 years in up to 90% of chronically infected woodchucks. The chromosomes of HCC tumour cells contain integrated, extensively rearranged WHV DNA sequences (cf. HEPATITIS B VIRUS).

wooden tongue See ACTINOBACILLOSIS.

Woodruffia See COLPODIDA.

Wood's lamp (Wood's light) A lamp in which ultraviolet radiation (emitted from a mercury vapour source) is transmitted through a nickel oxide-containing soda-glass filter (Wood's filter) that allows passage of radiation of wavelength ca. 365 nm only. The lamp is used e.g. in the diagnosis of certain infectious diseases (see e.g. ERYTHRASMA and RINGWORM) and for detecting porphyrins (see PORPHYRIN TEST) or other fluorescent materials.

woodwasp fungi Fungi (usually species of *Amylostereum*, *Daedalea* or *Stereum*) which line the tunnels of the wood-boring larvae of woodwasps (Siricidae and Xiphydriidae). Adult wasps harbour fungal oidia in a pair of pouches (mycetangia, mycangia) at the base of the ovipositor; during egg-laying, fungal oidia are deposited with the eggs. The fungi degrade many of the components of wood (including cellulose and lignin). The wasp larvae feed on wood and fungal mycelium. Siricid larvae, at least, require fungal enzymes (cellulases, xylanases etc) for the digestion of wood components, and they acquire these enzymes by ingestion of the fungal mycelium [Book ref. 77, pp. 161–163]. (See also AMBROSIA FUNGI and INSECT–MICROBE ASSOCIATION.)

wool spoilage Raw wool is relatively resistant to microbial attack owing to its content of antimicrobial fatty acids. However, woollen textiles – especially when kept under humid conditions – are susceptible to attack by various bacteria (including actinomycetes) and fungi (e.g. *Microsporum* and *Trichophyton* spp) which may cause loss of strength of the wool fibres, discolouration (e.g. due to pigment production), odour production, etc. Preservatives for woollen goods may include e.g. phenolics (e.g. pentachlorophenyl laurate, *o*-phenylphenol), salicylanilide, sodium fluorosilicate. (See also TEXTILE SPOILAGE.)

woolly monkey sarcoma virus *Syn.* SIMIAN SARCOMA VIRUS.

woolsorters' disease Pulmonary ANTHRAX.

working distance In MICROSCOPY: the distance between the specimen (or upper surface of the cover-glass, when present) and the nearest face of the objective lens when the specimen is sharply in focus.

Woronin bodies In certain mycelial fungi: spheroidal, refractile and electron-dense, membrane-limited bodies (ca. 0.2 μm) located in the cytoplasm – commonly around septal pores; they may act as plugs, sealing the pores e.g. following hyphal damage. [Review: FEMS Reviews (1987) *46* 1–11.]

Woronin granules *Syn.* WORONIN BODIES.

Woronina See PLASMODIOPHOROMYCETES.

wort See BREWING.

wound botulism See BOTULISM (a).

wound tumour virus (WTV; clover wound tumour virus) A virus of the genus PHYTOREOVIRUS. WTV was originally isolated from the leafhopper *Agalliopsis novella*, but has never again been isolated from a natural source; the natural plant hosts (if any) are therefore unknown. Experimentally, WTV can infect a wide range of dicotyledonous plants; infection spreads systemically, and tumours are produced irregularly – mainly on the veins of leaves and stems and on the roots. (In e.g. *Melilotus alba* (sweet clover) and *M. officinalis* the number of tumours can be greatly increased by artificial wounding.) The virus also replicates in the insect vector. Prolonged culture of WTV in plant cells tends to result in a reduction in or loss of WTV's ability to infect insects, and vice versa; this is due to the deletion of one or more dsRNA segments or parts of segments. Such mutants are termed *subvectorial* and *exvectorial* according to whether the loss of ability to infect insect vectors is partial or complete, respectively. [Molecular biology: AVR (1984) *29* 57–93.]

wp WETTABLE POWDER.

Wr Writhe: see DNA.

WR WASSERMANN REACTION.

WR238605 See MALARIA.

Wright's stain See ROMANOWSKY STAINS.

writhe (of supercoiled DNA) See DNA.

wyerol See WYERONE.

wyerone A furanoacetylene PHYTOALEXIN produced by the broad bean (*Vicia faba*); other furanoacetylene phytoalexins produced by *Vicia faba* include wyerone acid, wyerone epoxide, wyerol, dihydrowyerone, dihydrowyerone acid, and dihydrowyerol. Wyerone and its derivatives can be metabolized to less fungitoxic compounds e.g. by *Botrytis fabae* (causal agent of CHOCOLATE SPOT). (cf. MEDICARPIN.)

wyosine See Y BASES.

X

X adhesin An ADHESIN, distinct from both mannose-sensitive adhesins and P fimbriae, which occurs on a proportion of pyelonephritic strains of *Escherichia coli*; some of these 'X-specific' strains of *E. coli* bind specifically to neuraminyl α-$(2 \rightarrow 3)$-galactosides [BBRC (1983) *111* 456–461].

X-bacteria Gram-negative endosymbiotic bacteria which occur in the cytoplasm of certain strains of *Amoeba proteus*; they do not confer a killer characteristic. Amoebae may be dependent on their endosymbionts. X-bacteria are sensitive to e.g. chloramphenicol, and are killed at temperatures above ca. 26°C. [Book ref. 22, p. 799.]

X-band (in ESR) See ELECTRON SPIN RESONANCE.

X body (*virol.*) *Syn.* VIROPLASM.

X factor A growth factor required for *aerobic* growth by certain *Haemophilus* spp; the requirement is usually satisfied by haemin or by protoporphyrin IX (essential for the synthesis of e.g. CYTOCHROMES and CATALASE). Some strains (e.g. of *H. aegyptius*) lack ferrochelatase and thus cannot insert iron into protoporphyrin IX; these strains therefore require haemin. 'X factor' is available in media containing lysed RBCs (e.g. chocolate agar). (See also PORPHYRIN TEST.)

X-gal See XGAL.

X-linked lymphoproliferative syndrome See XLP SYNDROME.

X-Press An apparatus for CELL DISRUPTION similar in principle to the HUGHES PRESS.

X-ray spectrochemical analysis A method used e.g. to determine the in vivo location of certain types of atom within a specimen. When the electron beam irradiates a specimen, atoms of a given element give rise to a characteristic X-ray emission spectrum; the specimen, as a whole, also responds with a non-specific background continuum radiation of X-rays (*Bremsstrahlung*). The emitted X-rays are analysed (in terms of either energy or wavelength) and their source(s) within the specimen can be mapped with an accuracy which depends e.g. on the diameter and energy of the electron beam and on the thickness of the specimen.

X-rays See IONIZING RADIATION.

xanthan gum An extracellular heteropolysaccharide produced by strains of *Xanthomonas campestris*. It consists of a linear backbone chain of $(1 \rightarrow 4)$-linked β-D-glucosyl residues in which alternate residues are substituted at the O-3 position with a trisaccharide side-chain (D-mannose – D-glucuronic acid – D-mannose); typically, at least some of the proximal mannosyl residues are acetylated, and ca. 50% of the terminal mannosyl residues bear a pyruvic acid acetal group bridging the O-4 and O-6 positions. (Structural details differ in gums from different strains of *X. campestris*.)

Xanthan gum has many commercial applications: e.g. in the food industry it is used as a gelling agent, gel stabilizer, thickener, and crystallization inhibitor in various foods; it is generally resistant to microbial degradation and is stable at temperatures up to ca. 120°C. In nature, xanthan gum may mediate adhesion of *X. campestris* to its plant host. (See also CAPSULE.)

Xanthidium A genus of placoderm DESMIDS; the cells resemble those of COSMARIUM except in that they bear conspicuous spines.

Xanthobacter A genus (*incertae sedis*) of aerobic, catalase-positive, Gram type-negative bacteria which occur e.g. in soil and water. Cells: straight or curved, pleomorphic, non-motile or peritrichously flagellated rods (0.4–1.0 × 0.8–6.0 μm), cocci, or filaments – cell shape depending primarily on the carbon source used for growth. The cells give a positive or variable reaction in the Gram stain, but have Gram-negative type cell walls which contain LPS, *meso*-DAP-linked peptidoglycan, and a water-insoluble yellow carotenoid pigment (zeaxanthin dirhamnoside). An α-polyglutamine capsule is formed during chemoorganotrophic growth; the α-polyglutamine can apparently be used by the cells e.g. under conditions of nitrogen limitation. Large amounts of polysaccharide slime are also produced during growth on e.g. lactate or carbohydrates. Metabolism is respiratory (oxidative), with oxygen as terminal electron acceptor. NITROGEN FIXATION occurs (usually only under microaerobic conditions). All strains can grow as chemolithoautotrophs with mineral salts and a mixture of H_2, O_2 and CO_2, or as chemoorganotrophs using e.g. methanol (see METHYLOTROPHY), ethanol, *n*-propanol, or some organic acids; some strains can use certain carbohydrates (e.g. fructose and/or sucrose). Carbohydrates and gluconate are metabolized via the ENTNER–DOUDOROFF PATHWAY and the HEXOSE MONOPHOSPHATE PATHWAY; CO_2 is assimilated via the CALVIN CYCLE. Optimum growth temperature: 25–30°C; optimum pH: 5.8–9.0. GC%: 66–68. Type species: *X. autotrophicus* (formerly known as *Corynebacterium autotrophicum*); other species: *X. flavus* (formerly known as *Mycobacterium flavum*).

[Book ref. 22, pp. 325–333.]

xanthochroic (*mycol.*) Refers to a hymenomycete basidiocarp in which the context is yellowish brown but becomes dark brown or olive-brown with KOH (10% aqueous).

xanthomegnin A naphthoquinone pigment and hepatotoxic MYCOTOXIN produced e.g. by species of *Aspergillus*, *Penicillium* and *Trichophyton*. It gives rise to an electron transport shunt from NADH dehydrogenase to cytochrome *c* in the mitochondrial ELECTRON TRANSPORT CHAIN.

xanthomonadins Water-insoluble pigments which occur in *Xanthomonas* spp; they were once thought to be carotenoids but are now known to be brominated aryl polyenes.

Xanthomonas A genus of Gram-negative, obligately aerobic, chemoorganotrophic, plant-pathogenic bacteria within the family PSEUDOMONADACEAE. The cells are straight, monotrichously (polarly) flagellated rods, usually 0.4–0.7 × 0.7–1.8 μm; most strains contain water-insoluble, yellow (non-carotenoid) pigment(s) (see e.g. XANTHOMONADINS) and some strains form extracellular ('capsular') polysaccharides (see e.g. XANTHAN GUM). Growth requirements typically include factors such as glutamic acid and methionine. In pathovars of *X. campestris* (the only strains investigated) glucose is metabolized mainly via the ENTNER–DOUDOROFF PATHWAY and to a lesser extent via the HEXOSE MONOPHOSPHATE PATHWAY; the TCA CYCLE and the glyoxylate cycle are reported to be present. Degradative enzymes produced by *Xanthomonas* spp may include e.g. carboxymethylCELLULASES, mannanases, PECTIC ENZYMES (e.g. endopectate lyase, pectin methylesterase) and XYLANASES. No species can carry out nitrate respiration (cf. *Pseudomonas*). Growth in some strains may be slow, visible colonies being formed only after a week or more; young colonies are typically round, smooth, entire and butyrous. Optimum growth temperature: usually 25–30°C. Growth is inhibited by 0.1% triphenylTETRAZOLIUM chloride. Catalase +ve; oxidase −ve or weakly +ve. GC%: 63–71. Type species: *X. campestris*.

X. albilineans. Hosts: certain species of the Gramineae. The causal agent of leaf scald of sugar cane. Growth is non-mucoid on nutrient agar containing 5% glucose. NaCl tolerance is low (0.5%).

X. ampelina. The causal agent of bacterial blight and canker of the grape vine (*Vitis vinifera*). Growth is non-mucoid on nutrient agar containing 5% glucose. NaCl tolerance: 1%. Urease +ve. Pigments are apparently not xanthomonadins; a brown, water-soluble pigment is formed in cultures on yeast extract–galactose–calcium carbonate agar.

X. axonopodis. The causal agent of gummosis in species of *Axonopus* (Gramineae). Growth is non-mucoid on nutrient agar containing 5% glucose. Starch is hydrolysed. H_2S is formed from peptone. NaCl tolerance: 1%.

X. campestris. The repository for a large number of nomenspecies, each distinguished on the basis of pathogenicity for particular host(s), and each now classified as a pathovar within the species. Typically, *X. campestris* pathovars form mucoid growth on nutrient agar containing 5% glucose, cause proteolysis in milk, form H_2S from peptone, and tolerate 2–5% NaCl. The pathovars (with examples of hosts) include e.g. *X. campestris* pv. *badrii* (members of the Compositae and Leguminosae, e.g. *Pisum sativum*); pv. *campestris* (*Brassica* spp); pv. *cassavae* (*Manihot* spp); pv. *cerealis* (Gramineae, e.g. *Hordeum* spp); pv. *fici* (*Ficus carica*); pv. *glycines* (Leguminosae, e.g. *Glycine* spp); pv. *mangiferaeindicae* (mango: see BLACK SPOT sense 3); pv. *musacearum* (*Musa* spp); pv. *oryzae* (Gramineae, e.g. *Oryza* spp); pv. *phaseoli* (Leguminosae, e.g. *Phaseolus vulgaris*).

X. fragariae. The causal agent of a leaf spot disease of strawberry. Growth is mucoid on nutrient agar containing 5% glucose. Milk protein is not hydrolysed, and H_2S is not formed from peptone. NaCl tolerance: 0.5–1%.

X. maltophila See PSEUDOMONAS (*P. maltophila*).

[Book ref. 22, pp. 199–210. Host range of the genus *Xanthomonas*: Bot. Rev. (1984) *50* 308–356.]

Xanthophyceae (yellow-green algae) A class of ALGAE characterized by the formation – during at least some stage of the life cycle – of motile biflagellated (heterokont) cells which have one long tinsel flagellum directed anteriorly, and one much shorter whiplash flagellum directed posteriorly; the flagella arise at or near the anterior end of the cell. The chloroplasts are yellow-green and contain chlorophyll *a* (*Vaucheria geminata* apparently also contains chlorophyll *c*); each is surrounded by two membranes of chloroplast endoplasmic reticulum, the outer membrane typically being continuous with the nuclear envelope. The EYESPOT, when present, occurs within a chloroplast. The cell wall in e.g. *Botrydium* and *Vaucheria* contains cellulose.

The vegetative organisms may be unicellular and either solitary (e.g. OPHIOCYTIUM) or colonial (e.g. GLOEOCHLORIS, MISCHOCOCCUS), multicellular and filamentous (e.g. TRIBONEMA), or siphonaceous and either vesicular (e.g. BOTRYDIUM) or tubular (e.g. VAUCHERIA). Certain unicellular xanthophytes (e.g. *Chloramoeba*, *Heterochloris*) form vegetative cells which are flagellated and/or amoeboid, and these organisms are sometimes classified as protozoa of the PHYTOMASTIGOPHOREA.

Asexual reproduction occurs by fragmentation (in filamentous species) or by the formation of zoospores and/or aplanospores; some members can form thick-walled resistant resting cells (akinetes), and some can form cysts (statospores) intracellularly in certain cells.

Sexual reproduction has been observed in *Botrydium*, *Tribonema* and *Vaucheria*, and is oogamous in *Vaucheria*, isogamous or anisogamous in other genera.

Xanthophytes occur mainly in freshwater habitats; a few are marine, while e.g. *Botrydium* and some *Vaucheria* spp can grow on damp soil.

xanthophylls See CAROTENOIDS.

xanthophytes Algae of the XANTHOPHYCEAE.

Xanthoria A genus of foliose LICHENS (order TELOSCHISTALES). Photobiont: *Trebouxia/Pseudotrebouxia*. [Evidence for free-living photobiont and mycobiont of *X. parietina*: New Phyt. (1984) *97* 455–462.] The thallus contains parietin and is yellow or orange (sometimes greenish-yellow when growing in shade – see PARIETIN); it consists of small to large lobes which in some species (e.g. *X. parietina*) are radially arranged to form a rosette. Apothecia: lecanorine, orange, abundant (e.g. in *X. parietina*) or scarce (e.g. in *X. fallax*); ascospores: polarilocular. Species are nitrophilous (ornithocoprophilous), growing primarily on rocks (e.g. *X. elegans*), on bark (e.g. *X. fallax*), or on any of a wide range of substrata (e.g. *X. parietina*). *X. parietina* can tolerate moderate levels of air pollution.

xanthosine 5′-monophosphate See e.g. Appendix V(a).

xanthosomes Granules present in the cytoplasm of certain foraminifera – believed to be waste products of digestion.

***Xba*I** A RESTRICTION ENDONUCLEASE from *Xanthomonas campestris* pv. *badrii*; T/CTAGA.

***xcp* genes** See PROTEIN SECRETION (type II systems).

xenoantibody Any antibody, raised in one species, whose homologous antigen derives from a different species.

xenobiotic Any chemical which is present in a natural environment but which does not normally occur in nature: e.g. a pesticide, industrial pollutant, drug, etc. (See also BIOREMEDIATION and CO-METABOLISM.)

Xenococcus A genus of unicellular CYANOBACTERIA (section II) in which the vegetative cells are rounded and variable in size; baeocytes are non-motile and have a fibrous outer cell wall layer at the time of their release. GC%: ca. 44.

xenodiagnosis A method for diagnosing a vector-transmitted disease (e.g. CHAGAS' DISEASE). A laboratory-bred, pathogen-free strain of the vector is allowed to suck blood from the patient; after e.g. several weeks the vector (or its faeces) is examined for the presence of the pathogen.

Xenodochus See UREDINIOMYCETES.

xenogeneic Derived from a different species. (cf. ALLOGENEIC and SYNGENEIC.)

xenoma A tumour-like lesion which develops in animal tissues infected with certain parasites; it consists of hypertrophied host tissue within which the parasite lives. (See e.g. GLUGEA.) [Morphogenesis of *Glugea*-induced xenomas: JP (1985) *32* 269–275.]

Xenophyophorea A class of marine organisms (superclass RHIZOPODA) which form a multinucleate plasmodium enclosed within a system of transparent, organic, branching tubes; barite crystals occur in the cytoplasm. Genera include *Psammetta*, *Psammina*, *Stannophyllum*.

Xenorhabdus A genus of motile bacteria of the ENTEROBACTERIACEAE [Book ref. 22, pp. 510–511]; species have been isolated from parasitic nematodes and their (insect larva) hosts. Growth is poor (or absent) at 36°C, optimum at ca. 25°C; results are negative in most of the biochemical tests used for enterobacteria. *X. luminescens* is weakly bioluminescent, *X. nematophilus* is non-luminescent. [DNA relatedness and phenotypic study of the genus: IJSB (1984) *34* 378–388.]

xenosaprobic zone See SAPROBITY SYSTEM.

xenosome (1) A bacterium-like endosymbiont in certain marine scuticociliates (e.g. *Parauronema*). [Characterization of xenosome DNA in *P. acutum*: JGM (1983) *129* 1317–1325.]

(2) In certain testate amoebae (e.g. *Difflugia*), 'xenosomes' are the sand particles in the (agglutinated) test that are acquired from the environment, the spaces between the 'xenosomes' being filled with apparently endogenously formed 'idiosomes' [Book ref. 137, pp. 272–275].

xenotropic Refers to an ENDOGENOUS RETROVIRUS which, although transmitted genetically in its host of origin, cannot replicate in that host, but can infect and replicate in cells of another species. An example is a xenotropic C-type murine retrovirus which cannot infect other mouse cells but which can infect and replicate in human, rabbit or duck cells. (cf. ECOTROPIC; AMPHOTROPIC.)

xero- A prefix meaning *dry*.

xerophilic Refers to organisms which grow best at low a_w (see WATER ACTIVITY) and which may not grow at all at high a_w.

Xgal (XGal) 5-Bromo-4-chloro-3-indolyl-β-D-galactoside. Hydrolysis of Xgal by β-GALACTOSIDASE yields a blue-green substance which adheres to bacterial cells; thus, cells which produce β-galactosidase form blue-green colonies on Xgal-containing plates. (cf. XP.)

***Xho*I** See RESTRICTION ENDONUCLEASE (table).

xid mice (*immunol.*) Genetically deficient mice which do not form Lyb 5^+ B LYMPHOCYTES.

***xis* gene** See e.g. BACTERIOPHAGE λ.

XLD agar XYLOSE–LYSINE–DEOXYCHOLATE AGAR.

XLP syndrome (Duncan disease) An inherited (X-linked recessive) lymphoproliferative syndrome in which infection with EPSTEIN–BARR VIRUS (EBV) leads to a fatal INFECTIOUS MONONUCLEOSIS; a genetically determined failure of specific T cell and other cellular immune responses results in a massive, uncontrolled proliferation of EBV-infected B cells.

XMP (1) Xanthosine 5′-monophosphate [see Appendix V(a)]. (2) Xylulose monophosphate (xylulose 5-phosphate).

XMP pathway (dihydroxyacetone pathway; XuMP pathway; xylulose monophosphate pathway) A cyclic metabolic pathway for the assimilation of formaldehyde by yeasts growing on methanol (see METHYLOTROPHY). The regenerated intermediate xylulose 5-phosphate (Xu5P) transfers a glycolaldehyde group to formaldehyde to yield glyceraldehyde 3-phosphate (G3P) and dihydroxyacetone (DHA); the reaction is catalysed by DHA synthase [properties of DHA synthase: JGM (1983) *129* 935–944]. DHA is phosphorylated, and DHAP condenses with G3P to form fructose 1,6-bisphosphate which is converted to fructose 6-phosphate (F6P); F6P and G3P undergo 'rearrangement reactions' (involving e.g. transketolase, transaldolase, epimerase – cf. RMP PATHWAY) to regenerate Xu5P. Some of the G3P is assimilated into cell biomass. The XMP pathway is energetically more efficient than the (icl) serine pathway but less efficient than the more economical routes of the RMP pathway.

***xonA* gene** (*sbcB* gene) See RECF PATHWAY.

XP 5-Bromo-4-chloro-3-indolylphosphate-*p*-toluidine: a reagent used e.g. for detecting alkaline PHOSPHATASE activity. (cf. XGAL.)

XSCID See SEVERE COMBINED IMMUNODEFICIENCY.

***xseA* gene** See EXONUCLEASE VII.

***xthA* gene** See EXONUCLEASE III.

XuMP pathway XMP PATHWAY.

xylan See XYLANS.

xylanases Enzymes which degrade XYLANS. There are several distinct types – differing in specificities and end-products. *β-Xylosidases*, produced by many fungi (e.g. *Aspergillus*, *Penicillium*, *Rhizopus*, *Trichoderma*) and bacteria (e.g. *Agrobacterium*, *Erwinia*, *Xanthomonas*, rumen bacteria), release xylose from short xylo-oligosaccharides from the non-reducing end. *Exo-xylanases* (formed e.g. by *Bacillus pumilus*, *Aspergillus niger*) can act on xylans to yield xylose and xylo-oligosaccharides. *Endoxylanases* (from e.g. *Stereum*, *Trametes*, *Streptomyces*) cleave bonds within xylan chains to form xylose and/or xylo-oligosaccharides; some can also cleave L-arabinose–xylose branch points. [Book ref. 26, pp. 111–129.]

xylans Polysaccharides consisting largely or solely of xylosyl residues (see XYLOSE). In angiosperms, xylans form the bulk of the HEMICELLULOSE component of the cell wall; they consist essentially of $(1 \rightarrow 4)$-β-linked D-xylopyranosyl residues with side-chains of other sugars – commonly single residues of (4-*O*-methyl)-α-D-glucopyranosyluronic acid (in dicots and gymnosperms) or one or more α-L-arabinofuranosyl residues (e.g. in grasses).

In hardwoods, 4-*O*-methylglucuronoxylans are partially acetylated.

In many members of the Chlorophyta (e.g. *Caulerpa*, *Halimeda*, *Udotea*) a linear $(1 \rightarrow 3)$-β-xylan replaces cellulose as the main structural component of the CELL WALL. A major polysaccharide in the red alga *Palmaria palmata* is a linear xylan containing a random sequence of $(1 \rightarrow 3)$- and $(1 \rightarrow 4)$-β-D-xylosidic linkages.

(See also XYLANASES.)

[Molecular biology of xylan degradation: FEMS Reviews (1993) *104* 65–82.]

Xylaria A genus of fungi (order SPHAERIALES) which occur e.g. on tree stumps and fallen timber. The organisms form an erect, branched or unbranched, prosenchymatous stroma (commonly ca. 1–8 cm high) which is initially covered, or partly covered, with conidiophores; perithecia subsequently develop in the superficial layer of the same stroma with their ostiolar pores at the surface.

X. hypoxylon (the 'candle-snuff fungus') forms branched, antler-like, somewhat flattened stromata which are black and velvety except at the distal end – which is initially covered with chalk-white conidia. Ascospores: black, bean-shaped, ca. 12×6 μm.

X. polymorpha ('dead man's fingers') forms clavate, typically unbranched, black stromata, each on a short, cylindrical stipe; the tissue inside the stromata is white.

xylenols (as disinfectants) See PHENOLS.

xylitol An optically inactive PENTITOL formed e.g. by the reduction of XYLOSE. [See also Appendix III(d).]

xylobiose β-D-Xylopyranosyl-$(1 \rightarrow 4)$-D-xylose: a common product of enzymic degradation of XYLANS.

Xylohypha A genus of fungi of the HYPHOMYCETES. *X. bantiana* (formerly *Cladosporium bantianum*) is an important causal agent of cerebral PHAEOHYPHOMYCOSIS; the organism forms conidiophores that are indistinguishable from vegetative hyphae, and gives rise to aseptate (rarely two-celled), smooth-walled conidia in long, sparsely branched chains. [Reclassification of *C. bantianum* as *X. bantiana*: JCM (1986) *23* 1148–1151.]

xylomycetophagous See AMBROSIA FUNGI.

xylophagous Wood-eating.

xylophilous Having an affinity for wood.

xylose (wood sugar) An aldopentose (see PENTOSES) which occurs e.g. in XYLANS. Xylose can be metabolized by some bacteria [see e.g. Appendix III(d)] but by few yeasts; xylose is heat-labile, and solutions for use in media (e.g. XLD agar) should be sterilized by filtration. (cf. XYLULOSE; see also INDUSTRIAL ALCOHOL.)

xylose isomerase See GLUCOSE ISOMERASE.

xylose–lysine–deoxycholate agar (XLD agar) An agar medium containing e.g. xylose, L-lysine, lactose, sucrose, deoxycholate, thiosulphate, NaCl, ferric ammonium citrate, phenol red; pH 7.4. [Recipe: Book ref. 47, p. 671.] On XLD agar, *Shigella* ferments no sugars (i.e. forms no acid) and give rise to red colonies (phenol red: pH 6.8 yellow → pH 8.4 red). *Salmonella* ferments xylose (forms acid) – but also decarboxylates lysine, giving a *net* alkaline reaction (red colonies); unlike *Shigella*, *Salmonella* usually gives rise to *black-centred* red colonies because it produces hydrogen sulphide that reacts with the ferric salt. *Escherichia* typically ferments all three sugars; it gives rise to *yellow* colonies, but tends to be inhibited by deoxycholate.

β-xylosidase See XYLANASES.

Xylosphaera Former name of XYLARIA.

xylulose A pentulose (see PENTOSES). Xylulose 5-phosphate is an important intermediate in various metabolic pathways: see e.g. CALVIN CYCLE, HEXOSE MONOPHOSPHATE PATHWAY [Appendix I(b)], HETEROLACTIC FERMENTATION [Appendix III(b)], pentose metabolism [see e.g. Appendix III(d)], XMP PATHWAY.

xylulose monophosphate pathway XMP PATHWAY.

1. Words in SMALL CAPITALS are cross-references to separate entries.
2. Keys to journal title abbreviations and Book ref. numbers are given at the end of the Dictionary.
3. The Greek alphabet is given in Appendix VI.
4. For further information see 'Notes for the User' at the front of the Dictionary.

Y

Y Tyrosine (see AMINO ACIDS).

Y YIELD COEFFICIENT.

Y bases Bases formed by the modification of guanine in tRNA; an example is *wyosine*, in which an additional ring is fused with the purine ring of guanine, the additional ring itself bearing a substituted hydrocarbon side-chain. Y bases occur e.g. in $tRNA^{Phe}$ in bacteria and yeasts. (cf. Q BASES.)

Yaba monkey tumour poxvirus (Yaba virus) A virus of the POXVIRIDAE which can cause benign tumours in monkeys. Monkey handlers may become infected, infection typically resulting in the formation of a small nodular lesion which eventually regresses. Yaba virus replicates in the CAM and in certain types of cell culture. (cf. TANAPOX VIRUS.)

YadA protein See FOOD POISONING (*Yersinia*).

yakult A 'health food' marketed in Japan and in other countries; skim-milk containing glucose and an extract of *Chlorella* is inoculated with *Lactobacillus casei* and allowed to ferment at 37° C for several days. Sweeteners and flavourings may be added. (See also PROBIOTIC.)

Yamaguichi-73 sarcoma virus See AVIAN SARCOMA VIRUSES.

yaourt *Syn.* YOGHURT.

Yarrowia See SACCHAROMYCOPSIS.

Y_{ATP} See YIELD COEFFICIENT.

yaws (framboesia; frambesia; pian) A chronic infectious human disease caused by *Treponema pallidum* subsp *pertenue*; it occurs e.g. in tropical regions of America, Africa and the Far East, chiefly in areas where standards of hygiene are poor. Transmission occurs mainly by direct contact with infected persons or fomites; infection occurs via wounds and abrasions. An ulcerative lesion develops at the site of infection, and after a few weeks open sores appear all over the body. After several years tissue destruction becomes extensive, involving skin, bones and joints. *Chemotherapy*: e.g. penicillins. (cf. PIAN BOIS.)

YDC agar *Syn.* YGC AGAR.

yeast A name often used specifically for *Saccharomyces cerevisiae* (see SACCHAROMYCES) – cf. YEASTS. (See also BAKERS' YEAST and BREWERS' YEAST.)

yeast extract Yeast extracts are water-soluble preparations (liquid, paste, powder or granules) obtained from e.g. brewers' or BAKERS' YEAST (*Saccharomyces cerevisiae*), *Kluyveromyces marxianus* var. *lactis*, *K. marxianus* var. *marxianus*, or *Candida utilis* (TORULA YEAST). Yeast extracts are rich in amino acids and peptides, B vitamins, trace elements, etc; they are used as condiments and food additives and as nutritional supplements in industrial fermentations, microbial growth media etc. (See also SINGLE-CELL PROTEIN.)

Yeast autolysates are made by subjecting yeast cells to ca. 40–60°C. Endogenous enzymes degrade many of the yeast cell components to soluble amino acids, peptides etc; some of the cell wall polymers and some glycogen remain insoluble. Autolysis can be accelerated by plasmolysing agents such as NaCl or sucrose, or by organic solvents such as ethanol (which also serve to inhibit bacterial putrefaction).

Yeast hydrolysates are made by treating a suspension of yeast cells with strong acid at ca. 100°C followed by neutralization and concentration; while yields are higher than those for autolysates in terms of hydrolysis of cell components, the product tends to be inferior in terms of amino acid content (tryptophan is destroyed, levels of tyrosine, methionine and cysteine are greatly reduced), vitamin content, salt content (hydrolysates contain up to 50% NaCl), and flavour.

yeast phase See DIMORPHIC FUNGI.

yeasts A *non*-taxonomic category of FUNGI defined in terms of morphological and physiological criteria. The 'typical' yeast (e.g. *Saccharomyces cerevisiae*) is a unicellular saprotroph which can metabolize carbohydrates by FERMENTATION (sense 1) and in which asexual reproduction occurs by BUDDING. However, some yeasts (e.g. *Schizosaccharomyces* spp) divide by fission, and some can form a pseudomycelium and/or a true mycelium; some (e.g. *Hansenula canadensis*, *Lipomyces* spp, *Sporobolomyces* spp) are non-fermentative, and some (e.g. *Candida albicans*) can be pathogenic. (cf. DIMORPHIC FUNGI; see also SACCHAROMYCETACEAE and SPOROBOLOMYCETACEAE.)

yellow-brown algae CHRYSOPHYTES.

yellow fever (yellow jack) An acute infectious disease of man and other primates, endemic in parts of Africa and S. America; it is caused by a flavivirus (see FLAVIVIRIDAE). In the *urban cycle* ('urban yellow fever') the virus is transmitted among humans by *Aedes aegypti*. The *jungle* or *sylvatic cycle* ('jungle yellow fever') involves mainly wild primates, the principal vectors being *Aedes simpsoni* or *A. africanus* (in Africa) or *Haemagogus* spp (in S. America). Man is incidental in the natural sylvan cycle. In man, the incubation period is 2–6 days. Onset is sudden, with fever, headache, myalgia and prostration; the fever abates but then recurs. Others symptoms typically include vomiting, jaundice, albuminuria, and haemorrhages (e.g. 'black vomit'); death (due usually to liver or kidney failure) may occur 6–10 days after onset. Long-term immunity follows recovery. *Lab. diagnosis*: identification of the virus from blood samples; detection of rising antibody titres; histological examination of the liver (see COUNCILMAN BODIES). *Prevention*: vaccination with attenuated virus (17D strain); control of mosquito vectors (where feasible).

[History of yellow fever research by the US Army: Lancet (2001) *357* 1772.]

yellow-green algae See XANTHOPHYCEAE.

yellow rice (1) Rice grains yellowed as a result of the growth and pigment formation of *Penicillium islandicum*. Consumption of such rice can lead to a MYCOTOXICOSIS ('yellow rice disease') due to the hepatotoxic, carcinogenic, anthraquinone pigment luteoskyrin and other toxins (e.g. CHLOROPEPTIDE).

(2) Rice yellowed by any of several pigment-forming *Penicillium* spp – e.g., *P. islandicum*, *P. citreoviride* (see CITREOVIRIDIN), *P. citrinum* (see CITRININ).

yellow rust (stripe rust) A CEREAL DISEASE caused by *Puccinia striiformis*. On wheat the disease is characterized by lines of lemon-yellow uredial pustules which occur between the veins on the upper (adaxial) surface of the leaf. On barley the symptoms are similar; in severe infections the pustules can also occur on the ears.

yellows (*plant pathol.*) Any of a wide variety of plant diseases in which a major symptom is a uniform or non-uniform (e.g. striped or mottled) yellowing of leaves and/or other plant components. (cf. CHLOROSIS.) Yellows may be caused by fungi (e.g. one form of CELERY YELLOWS), viruses (diseases in many plants – see e.g. SUGAR BEET YELLOWS VIRUS), bacteria (e.g. COCONUT LETHAL YELLOWING), or protozoa (e.g. HARTROT, PHLOEM NECROSIS).

(See also CORN STUNT DISEASE.) In yellows caused by wall-less prokaryotes (SPIROPLASMA spp, MLOS) the pathogen grows (intracellularly) within the sieve tubes; *Spiroplasma* gives rise e.g. to stunting and yellowing, while MLOs, in addition, may cause VIRESCENCE and PHYLLODY or WITCHES' BROOM formation. Some yellows previously attributed to viruses (e.g. aster yellows, and the 'viral' form of celery yellows) are now known to be caused by MLOs.

Yersinia A genus of Gram-negative bacteria of the family ENTEROBACTERIACEAE (q.v.). *Yersinia* spp are parasites and pathogens in man (see e.g. FOOD POISONING (j) and PLAGUE) and in other animals, including fowl and fish; some species are also found in food and water etc.

Cells: bacilli or coccobacilli, 0.5–0.8 × 1–3 μm; the cells (particularly those of *Y. pestis*) tend to show bipolar staining.

Phenotypic characteristics are largely dependent on temperature. For example, most species are motile below 30°C but non-motile at 37°C (*Y. pestis* is always non-motile). The VP test is always negative at 37°C but may be positive in some species at 25°C. The optimum growth temperature is ~28°C. The organisms can grow on unenriched nutrient agar, but generally grow better on e.g. blood agar. Colonies (after 24–30 hours at 28–37°C) may be ≤1 mm diam. (*Y. pestis*) or 2–3 mm (most other species); after 48 hours colonies may become umbonate.

MR +ve. Acid (little or no gas) from glucose. Typically lactose −ve, but ONPG TEST +ve. (Rarely, strains may harbour a Lac plasmid.) Urease −ve (e.g. *Y. pestis*) or +ve (e.g. *Y. enterocolitica, Y. pseudotuberculosis*). Oxidase −ve. Catalase +ve.

The major human pathogens (*Y. enterocolitica, Y. pesits* and *Y. pseudotuberculosis*) each encode a VIRULON (q.v.) which e.g. helps the organisms to evade phagocytosis. Invasion of intestinal tissue by *Y. enterolitica* is described in FOOD POISONING (j).

GC%: ~46–47. Type species: *Y. pestis*.

Y. aldovae. (Formerly '*Y. enterocolitica*-like group X2'.) [IJSB (1984) *34* 166–172.]

Y. enterocolitica. Urease +ve. Rhamnose −ve, maltose +ve, sucrose +ve. VP test may be +ve at 25°C. A causal agent of adenitis and diarrhoea (mainly in children) (see FOOD POISONING (j)), and of e.g. arthritis and erythema nodosum in adults). VW ANTIGENS may be present.

Y. enterocolitica is psychrotrophic and can be selected by cold enrichment (4°C); it occurs e.g. in meat (see MEAT SPOILAGE and DFD MEAT) and also in milk (but is usually destroyed by pasteurization [AEM (1982) *44* 517–519]).

Y. frederiksenii. Urease +ve. Rhamnose +ve, sucrose +ve, raffinose −ve. VP (25°C)+ve. May cause diarrhoea [GE (1984) *86* 1237 (abstr.)].

Y. intermedia. Urease +ve. Rhamnose +ve, sucrose +ve, raffinose +ve. VP (25°C) +ve. May be an opportunist pathogen.

Y. kristensenii. Urease +ve. Rhamnose −ve, sucrose −ve, raffinose −ve. VP −ve.

Y. pestis (formerly *Pasteurella pestis*). Urease −ve. Rhamnose −ve, sucrose −ve, glucose (acid, no gas) +ve. ONPG +ve. Non-motile.

The causal agent of PLAGUE (q.v.). (See also VIRULON.)

On primary isolation from an animal host, or on incubation in protein-rich media at 37°C, virulent strains have been associated with certain antigens: (i) a heat-labile (100°C/10 min) water-soluble capsular antigen (the Fraction 1 antigen, serofactor 1, F1 antigen) which comprises glycoprotein (F1A) and protein (F1B) components; (ii) heat-stable somatic VW ANTIGENS (encoded by plasmid Vwa); the Vwa plasmid is readily lost on subculture in nutrient media at 37°C, and such loss has been associated with a loss of virulence.

In some strains of *Y. pestis* COAGULASE activity (evident only in rabbit or guinea-pig plasma) and fibrinolytic activity are correlated with the production of a bacteriocin (PESTICIN I).

Y. philomiragia. See PHILOMIRAGIA BACTERIUM.

Y. pseudotuberculosis (formerly *Pasteurella pseudotuberculosis*). Urease +ve. Rhamnose +ve, sucrose −ve, salicin +ve. VP −ve. ONPG +ve. VP −ve. Virulent strains possess the F1 and VW antigens (see *Y. pestis*). *Y. pseudotuberculosis* is the causal agent of PSEUDOTUBERCULOSIS. It has been proposed that, on the basis on DNA homology, *Y. pseudotuberculosis* and *Y. pestis* be reclassified, respectively, as *Y. pseudotuberculosis* subsp *pseudotuberculosis* and *Y. pseudotuberculosis* subsp *pestis* [Curr. Micro. (1980) *4* 225–230], but this proposal was rejected [IJSB (1985) *35* 540].

Y. ruckeri. Urease −ve. Rhamnose −ve, sucrose −ve, raffinose −ve. VP variable. Pathogenic in fish, causing ENTERIC REDMOUTH. This organism may be sufficiently distinct from other yersiniae to warrant a separate genus.

yersiniosis Any disease of man or animals caused by a species of YERSINIA.

yes See AVIAN SARCOMA VIRUSES and ONCOGENE.

Y_G See YIELD COEFFICIENT.

YGC agar (YDC agar) An agar medium containing yeast extract, D-glucose, and precipitated chalk; it is used e.g. as a general-purpose medium for plant-pathogenic bacteria (e.g. *Erwinia* spp). [Recipe: Book ref. 46, p. 1267.]

yield coefficient (growth yield coefficient; Y) The amount of biomass (grams, dry weight) formed per mass (grams) of a given substrate; this 'observed' or 'overall' relationship may be designated e.g. Y, Y_0, Y_s or $Y_{x/s}$, and the theoretical or maximum yield coefficient may be written e.g. $Y_{x/s(MAX)}$ or Y_{MAX}. If the amount of substrate is expressed in moles (so that the growth-supporting potential of different substrates can be compared) the coefficient may be designated e.g. $Y_{x/m}$ and is referred to as the *molar growth yield* or the *molar yield coefficient*. $Y_{x/ATP}(= Y_{ATP})$ refers to the amount (grams) of biomass (or of a given metabolic product) formed per mole of ATP synthesized by the biomass. $Y_{x/O_2}(= Y_{O_2})$ refers to the amount of biomass (or product) formed per mole of oxygen consumed.

The observed yield coefficient (Y) differs from the 'true' yield coefficient (Y_G; *growth-specific yield coefficient)* in that Y_G is calculated from the amount of the given substrate which is actually incorporated into biomass – i.e., a quantity which excludes the amount of that substrate used for MAINTENANCE ENERGY. Y and Y_G are related:

$$\frac{1}{Y} = \frac{m}{\mu} + \frac{1}{Y_G}$$

where μ is the SPECIFIC GROWTH RATE and m is the *maintenance coefficient*; m can be calculated by plotting $1/Y$ (ordinate) versus $1/\mu$ (abscissa), the slope of the straight-line graph being numerically equal to m. (The intercept on the $1/Y$ axis gives the value of $1/Y_G$.) The maintenance coefficient gives a measure of the proportion of the total energy requirements needed to satisfy maintenance energy during periods of slow growth, i.e., high values of m indicate that maintenance energy is a relatively high proportion of the total energy requirements, and conversely.

Although widely used, parameters such as Y_{ATP} should be interpreted with care. Thus, for example, while in some species growth seems to be stoichiometrically related to the amount of ATP synthesized, this relationship is dependent on certain conditions: e.g. growth must be 'energy-limited', i.e., limited by the supply of the energy source, and none of the energy source must be incorporated into biomass. Moreover, some organisms can carry out 'energy-spilling' reactions and FUTILE CYCLES which necessarily affect parameters such as Y_{ATP}. [Y_{ATP} and maintenance energy as biologically interpretable phenomena: ARM (1984) *38* 459–486.] The growth yield can also be significantly influenced by the amount of energy needed to transport and phosphorylate a given exogenous substrate [JB (1985) *163* 1237–1242].

YM broth Yeast extract–malt extract broth, a medium for the culture of yeasts; it contains yeast extract, malt extract, peptone and glucose. YM agar is YM broth solidified with 2% agar. [Recipes: Book ref. 100, p. 51.]

YMA Yeast extract–mannitol agar, used e.g. for the culture of *Rhizobium* spp. [Recipe: Book ref. 45, p. 826.]

yoghurt (yogurt, yaourt) A food made by fermenting milk with a mixed culture of *Lactobacillus bulgaricus* and *Streptococcus thermophilus*; the bacteria produce lactic acid and other flavour components (traces of acetaldehyde, DIACETYL and acetic acid – see Appendix III(c)). Neither species can be used successfully without the other. Initially, *L. bulgaricus* breaks down milk proteins to amino acids and small peptides which stimulate the growth of *S. thermophilus*; formic acid produced by *S. thermophilus* stimulates growth of *L. bulgaricus* which produces most of the lactic acid and acetaldehyde. The mixed culture is sufficiently stable for the production of yoghurt by a continuous process [AvL (1983) *49* 84–85.] (See also PROTOCOOPERATION.)

MILK used for yoghurt-making is high in milk solids, usually low in fat, and may be treated with a stabilizer (e.g. gelatin); it is pasteurized, homogenized, cooled, and inoculated with roughly equal numbers of the two species of bacteria. Incubation is carried out at ca. 35–45°C for a few hours, during which the pH falls to about 4.2–4.6 and the milk proteins coagulate. The product is then cooled rapidly to ca. 4°C and stored at this temperature. Additional flavourings or whole fruit, pasteurized separately, may be added before fermentation.

(See also DAIRY PRODUCTS.)

Yop virulon See VIRULON.

Yops See VIRULON.

yscIV See PHEROMONE.

yscF See PHEROMONE.

YscN protein See VIRULON.

Yst See FOOD POISONING (*Yersinia*).

YZ endonuclease See MATING TYPE.

1. Words in SMALL CAPITALS are cross-references to separate entries.
2. Keys to journal title abbreviations and Book ref. numbers are given at the end of the Dictionary.
3. The Greek alphabet is given in Appendix VI.
4. For further information see 'Notes for the User' at the front of the Dictionary.

Z

Z The factor 2.3(RT/F): see e.g. CHEMIOSMOSIS.

Z-DNA A left-handed helical form of dsDNA (cf. DNA) first observed in oligomers of alternating deoxycytidine (dC) and deoxyguanosine (dG) in the presence of high salt concentrations or ethanol. In these oligomers the strands are connected by Watson–Crick BASE PAIRing, the sugar–phosphate backbones form an irregular zig-zag (hence Z-DNA), there are ca. 12 bp/turn, and the bases are relatively peripheral with the N7 and C8 positions of guanine exposed; there is a single very deep helical groove corresponding to the minor groove of B-DNA (see DNA). Stretches of alternating dC and dG in a plasmid can undergo transition from right- to left-handed helical form under physiological levels of ionic strength and superhelical density; the transition is driven by the torsional strain of negative supercoiling and is facilitated by C5-methylation of cytidine [Nature (1982) *299* 312–316].

Whether or not Z-DNA occurs in vivo remains controversial. Unlike B-DNA, Z-DNA is strongly immunogenic, and monoclonal antibodies have been used for its detection in cells; however, the specificity of these antibodies for Z-DNA is uncertain. The isolation of proteins which preferentially bind Z-DNA rather than B-DNA provides circumstantial evidence for the existence of Z-DNA in vivo.

(See also V-DNA.)

Z ring See CELL CYCLE (b).

Z scheme See PHOTOSYNTHESIS.

z value The increase in temperature (°C) required for a 10-fold decrease in the D VALUE.

Zadoks' code (Zadoks, Chang and Konzak decimal code) (*plant pathol.*) A code, designed primarily for computer use, in which the various growth stages of cereals are represented by numbers between 0 and 100: 0–9, germination; 10–19, seedling growth; 20–29, tillering; 30–39, stem elongation; 40–49, 'booting' (swelling of the leaf sheath enclosing the developing ear); 50–59, ear emergence; 60–69, flowering; 70–79, 'milky' stage of grain development; 80–89, 'doughy' stage of grain development; 90–99, grain ripening. [Review (with illustrations): Ann. Appl. Biol. (1987) *110* 441–454.] (cf. FEEKES' SCALE.)

zalcitabine See NUCLEOSIDE REVERSE TRANSCRIPTASE INHIBITORS.

Zalerion See HYPHOMYCETES; see also TEXTILE SPOILAGE.

zanamivir (Relenza™, Glaxo Wellcome) An anti-INFLUENZA drug: a sialic acid analogue which can inhibit the viral neuraminidase in influenza virus types A and B; the drug is inhaled. [Efficacy and safety of zanamivir: NEJM (1997) *337* 874–880; Lancet (1998) *352* 1877–1881.]

ZAP Zoster-associated pain: see HERPES ZOSTER.

zaragozic acids Metabolites of certain fungi which are potent inhibitors of the enzyme squalene synthase; they have potential uses as antifungal agents and/or as therapeutic agents for lowering the levels of plasma cholesterol. [Zaragozic acids: ARM (1995) *49* 607–639.]

ZDV Zidovudine (see AZT).

α-zearalenol See ZEARALENONE.

zearalenone (F-2 toxin) A MYCOTOXIN produced by *Gibberella zeae* (*Fusarium graminearum*) growing e.g. on damp cereal feedstuffs. It has oestrogenic activity and can cause hyperoestrogenism (manifest by vulvovaginitis and infertility) in sows; cattle and poultry can be affected, though to a lesser extent. Zearalenone can also act as a regulatory hormone in the sexual cycle of *Gibberella*. Chemical reduction of zearalenone yields *α-zearalenol*, which is 4.8 times more oestrogenic than zearalenone and has anabolic properties; it is used to promote rapid weight gain in cattle and as an oestrogen substitute in postmenopausal women. [RIA of zearalenone and zearalenol in human serum: AEM (1983) *45* 16–23.]

zeatin A naturally occurring CYTOKININ: 6-(4-hydroxy-3-methylbut-2-enyl)-aminopurine. Its derivatives (e.g. zeatin riboside) also function as cytokinins.

zeaxanthin See CAROTENOIDS.

zeaxanthin rhamnoside See XANTHOBACTER.

ZEBRA (Zta) Z EBV replication activator: the protein product of (immediate-early) viral gene *bzlf-1* which promotes replication of the EPSTEIN–BARR VIRUS in latently infected B lymphocytes. ZEBRA can be induced by treating latently infected B cells with e.g. anti-immunoglobulin, corticosteriods or 'phorbol ester' (12-*O*-tetradecanoylphorbol-13-acetate, TPA).

The lytic cycle in latently infected B cells is also promoted by another transcription factor, Rta, encoded by the immediate-early viral gene *brlf-1*; it has been reported that the promoter of *brlf-1* is activated by acetylation of histones [NAR (2000) *28* 3918–3925].

Zelleriella See OPALINATA.

Zenker's fluid (modified) A FIXATIVE: mercuric chloride (5 g) and potassium dichromate (2.5 g) in distilled water (100 ml) supplemented immediately before use with glacial acetic acid (5 ml).

Zephiran See QUATERNARY AMMONIUM COMPOUNDS.

Zeta Plus filter See FILTRATION.

ζ-potential (zeta potential; electrokinetic potential) Of e.g. a bacterial cell, or a charged colloid particle: the electrical potential at the surface of shear – i.e., the surface of the cell or particle, *including* adherent counterions and water molecules – which tends to move relative to the surrounding medium when an external electrical field is applied; it is a determinant of the electrophoretic migration rate of a cell etc.

In media of neutral or alkaline pH most bacteria carry a surface charge of negative polarity – due to the ionization of surface groups. In many cases the ISOELECTRIC POINT of a bacterium is ca. 3.0; at or near this pH such bacteria tend to agglutinate spontaneously.

[Electrical properties and topochemistry of bacteria: Adv. Coll. Int. Sci. (1982) *15* 171–221.]

(See also ADHESION.)

Zetapor membrane filter See FILTRATION.

zeugite An early term for a cell or structure in which KARYOGAMY ends DIKARYOPHASE.

zicai See LAVER.

zidovudine *Syn.* AZT (q.v.).

Ziehl–Neelsen's stain An ACID-FAST STAIN. A heat-fixed smear is flooded with concentrated CARBOLFUCHSIN, heated and kept steaming (not boiling) for 5 min, allowed to cool, and rinsed in running water; the slide is then passed through several changes of acid-alcohol (e.g. 3% v/v conc. HCl in 95% ethanol), washed in water, and counterstained with e.g. 0.5% aqueous malachite green. After a final washing in water the smear is dried and examined by microscopy. Acid-fast organisms stain red, others green.

Ziemann's dots In erythrocytes infected with *Plasmodium malariae*: fine dots sometimes observed on heavy staining with Romanowsky stains. (cf. SCHÜFFNER'S DOTS.)

Zika virus See FLAVIVIRIDAE.

zinc (a) (microbial requirement) Zinc is a HEAVY METAL which is needed, in trace amounts, for the activity of a number of microbial enzymes – e.g. alcohol dehydrogenase; de-acetylase (encoded by gene *lpxC* in *Escherichia coli*, and involved in lipid A synthesis); some β-lactamases and superoxide dismutases; the UvrA protein involved in DNA repair; the zinc-endopeptidase TETANOSPASMIN; and CARBONIC ANHYDRASE. (See also FTSH.)

(b) (as an antimicrobial agent) In effective concentrations, zinc, and certain of its compounds, are useful antimicrobial agents. For example, zinc undecylenate (see UNDECYLENIC ACID) and zinc oxide have been used for treating certain superficial mycoses (e.g. athlete's foot), and zinc oxide is used as a mould inhibitor in paints. Zinc naphthenate can replace copper naphthenate as a wood preservative but is apparently less effective. Zinc dimethyldithiocarbamate (see DMDC) has been used as a preservative in rubber (it also serves as a vulcanization catalyst) and as an agricultural antifungal agent (*ziram* – see also ZINEB).

The precise mechanism of antimicrobial action may vary with organism; thus, e.g. zinc appears initially to cause membrane damage in *Saccharomyces cerevisiae* but may act by inhibiting intracytoplasmic proteins in another yeast, *Sporobolomyces roseus* [JGM (1983) *129* 3421–3425].

(c) (zinc chelation) The mammalian protein CALPROTECTIN is a zinc-chelating agent whose bacteriostatic action presumably reflects its ability to sequester zinc ions [see RMM (1997) *8* 217–224].

zinc-endopeptidase See TETANOSPASMIN.

zinc sulphate flotation See FLOTATION.

zinc undecylenate See UNDECYLENIC ACID.

zinconazole See AZOLE ANTIFUNGAL AGENTS.

zineb (dithane Z-78) Zinc ethyleneBISDITHIOCARBAMATE; this important agricultural antifungal agent is used to control a wide range of plant pathogens, e.g. *Botrytis* spp, *Fulvia fulva* (*Cladosporium fulvum*), *Peronospora* spp, *Phytophthora infestans*; it may be prepared as a WETTABLE POWDER – or may be prepared in the field by mixing NABAM, zinc sulphate and lime.

ZipA protein See CELL CYCLE (b).

zippering See FOOD POISONING (*Yersinia*).

ziram See DMDC.

zoite In coccidia: any of a range of stages in the life cycle – e.g. cystozoite, endozoite, sporozoite etc.

zonal centrifugation See CENTRIFUGATION.

Zonaria See PHAEOPHYTA.

zone centrifugation See CENTRIFUGATION.

zone lines In decaying wood: dark lines which are the edges of sheets of pigmented fungal tissues in the wood [see e.g. Book ref. 39, pp. 109–128 (122–124)]. In wood rotted by e.g. *Phellinus weirii* the coloration is due to a melanin-like pigment which apparently inhibits the growth of microorganisms antagonistic to *P. weirii* [Mycol. (1983) *75* 562–566].

zoobiont An animal symbiont. (cf. PHYCOZOAN.)

zoochlorellae Green endosymbiotic algae present in various PHYCOZOAN associations. For example, *Chlorella* occurs e.g. in certain protozoa (e.g. foraminifera, *Mayorella viridis*, *Paramecium bursaria*, *Stentor polymorphus*), the freshwater sponge *Spongilla* sp, the coelenterate *Hydra viridis*, the freshwater flatworm *Dalyellia viridis* (see also CONVOLUTA), and the freshwater clam *Anodonta*. The phycobiont generally supplies the zoobiont with products of photosynthesis (e.g. MALTOSE in *Hydra* and *Paramecium*, glucose in *Spongilla*), although the zoobiont may continue to feed. (See also ELYSIA; cf. ZOOXANTHELLAE.)

zooflagellates Flagellates of the ZOOMASTIGOPHOREA.

zoogloea (zooglea) A mass or film of cells embedded in a slimy matrix; zoogloeae are formed e.g. by *Zoogloea ramigera*.

Zoogloea A genus of Gram-negative, aerobic, chemoorganotrophic bacteria (family PSEUDOMONADACEAE) which occur e.g. in organically polluted freshwater habitats, and in aerobically treated sewage (see also SEWAGE TREATMENT). Cells: straight or slightly curved, monotrichously (polarly) flagellated, non-pigmented rods, 1.0–1.3 × 2.1–3.6 μm, typically embedded, in masses, within an extracellular polysaccharide matrix – forming flocs or surface films (branching or amorphous zoogloeae); polysaccharide (and floc) formation tends to be enhanced when the carbon:nitrogen ratio is high, and may be inhibited by high levels of carbon and nitrogen – conditions which favour increased biomass production [AEM (1982) *44* 1231–1237]. Metabolism is respiratory (oxidative); the organisms can carry out nitrate respiration. Major carbon sources can include e.g. fumarate, lactate, pyruvate, dicarboxylic amino acids (e.g. aspartate, glutamate), and certain alcohols. PHB is accumulated. Optimum growth temperature: 28–37°C; optimum pH: 7.0–7.5. Growth is inhibited by 3% NaCl. Oxidase +ve. Catalase +ve (weak). Most strains are urease +ve. GC%: ca. 65. Type species: *Z. ramigera*. [Book ref. 22, pp. 214–219.]

zooid (*microbiol.*) (1) A motile spore. (2) Of a stalked ciliate (e.g. *Vorticella*): the cell body, as opposed to the stalk.

zoom microscope A compound microscope in which magnification can be varied continuously over a range of values.

Zoomastigophorea A class of flagellate protozoa (subphylum MASTIGOPHORA) which lack chloroplasts and which are non-photosynthetic. Orders: CHOANOFLAGELLIDA, DIPLOMONADIDA, HYPERMASTIGIDA, KINETOPLASTIDA, OXYMONADIDA, PROTEROMONADIDA, RETORTAMONADIDA, TRICHOMONADIDA.

zoonosis Any infectious disease which can be contracted by man and in which the pathogen is normally maintained in a reservoir consisting of animal (i.e. non-human) population(s). (cf. ANTHROPONOSIS.)

zoonotic Pertaining to ZOONOSIS.

Zoopagales An order of fungi (class ZYGOMYCETES) which occur e.g. in soil and water, and which are parasitic on e.g. certain amoebae, nematodes, and fungi; the organisms form zygospores and give rise to asexual spores which are not forcibly discharged (cf. ENTOMOPHTHORALES). Genera: e.g. *Bdellospora*, *Cochlonema*, PIPTOCEPHALIS, *Rhopalomyces*, and *Stylopage* (see also NEMATOPHAGOUS FUNGI).

Zoophagus A genus of fungi of the PERONOSPORALES; *Z. insidians* is an aquatic species which captures and feeds on rotifers.

zoophilic Refers to a parasite or pathogen which preferentially infects animals (cf. ANTHROPOPHILIC).

Zoophthora See ENTOMOPHTHORALES.

zooplankton See PLANKTON.

zoosporangium A SPORANGIUM in which motile spores are formed.

zoospore A motile (flagellated) SPORE.

Zoothamnium See PERITRICHIA.

Zooxanthella See ZOOXANTHELLAE.

zooxanthellae Endosymbiotic DINOFLAGELLATES found in various marine invertebrates, including sea anemones (e.g. *Anthopleura*), giant clams (*Tridacna*), jellyfish (*Cassiopeia*), reef-forming tropical corals [Book ref. 129, pp. 19–35], and members of the FORAMINIFERIDA and RADIOLARIA. (The loss of zooxanthellae from corals is a useful indicator of stress due e.g. to pollution

[AMB (1985) *22* 1–63].) All such endosymbiotic dinoflagellates have been regarded as belonging to a single species, variously called *Symbiodinium microadriaticum*, *Gymnodinium microadriaticum*, or *Zooxanthella microadriatica*, but according to one report [see Science (1985) *229* 656–658] there are probably several distinct species. (See also PHYCOZOAN.) [Evidence for heterotrophy by zooxanthellae in the sea anemone *Aiptasia pulchella*, and its detrimental effects on the host under certain conditions: Biol. Bull. (1986) *170* 267–278.]

zoster *Syn.* HERPES ZOSTER.

zoster-associated pain (ZAP) See HERPES ZOSTER.

Zot toxin (*Vibrio cholerae*) See BACTERIOPHAGE CTXΦ.

Zovirax *Syn.* ACYCLOVIR.

Zschokkella See MYXOSPOREA.

Zta *Syn.* ZEBRA.

Zwischenferment Glucose 6-phosphate dehydrogenase (see HEXOSE MONOPHOSPHATE PATHWAY).

zwoegerziekte virus (zwogerziekte virus) See LENTIVIRINAE.

Zygnema A genus of freshwater, unbranched filamentous green algae related to SPIROGYRA; each cell contains two stellate chloroplasts. Akinetes may be formed.

Zygoascus A genus of filamentous, heterothallic fungi that include the teleomorphic stages of *Candida hellenica*, *C. inositophila* and *C. steatolytica*; type species: *Z. hellenicus*. [AvL (1986) *52* 25–37.]

Zygoceros See DIATOMS.

Zygogonium A genus of filamentous green algae related to SPIROGYRA. The filaments contain a purple pigment and grow e.g. on damp peaty soils; some species are thermophilic, forming dense purple mats near hot springs.

Zygomycetes A class of fungi of the ZYGOMYCOTINA; most zygomycetes are terrestrial saprotrophs, but some are parasites or pathogens of animals (including insects), plants, and other fungi. (See also ZYGOMYCOSIS.) The typical thallus is a well-developed, branched, coenocytic (aseptate) mycelium, but some species form a septate mycelium, and some (see MUCORALES) are DIMORPHIC FUNGI (sense 1); the CELL WALL contains CHITIN and/or CHITOSAN. Characteristic asexual reproductive structures include the SPORANGIUM and the SPORANGIOLUM (cf. CONIDIUM). Sexual reproduction typically involves ZYGOSPORE formation by the fusion of two morphologically similar gametangia; some species exhibit HOMOTHALLISM, some HETEROTHALLISM. (See also PHEROMONE.)

Orders [Book ref. 64, pp. 411–412]: Dimargaritales (members form merosporangia each containing two spores arranged end-to-end); ENDOGONALES; ENTOMOPHTHORALES; KICKXELLALES; MUCORALES; ZOOPAGALES.

zygomycosis Any of a range of human and animal diseases caused by fungi of the ZYGOMYCETES; the term currently includes diseases formerly called phycomycosis (caused by 'phycomycetes'), mucormycosis (caused by fungi of the MUCORALES), and entomophthoromycosis (caused by fungi of the ENTOMOPHTHORALES). In all zygomycoses the invasive form of the pathogen is an aseptate or sparsely septate, broad (e.g. 6–25 μm diam.), hyaline mycelium; the hyphae may be branched and may assume bizarre forms.

Zygomycosis in man. Subcutaneous zygomycosis, in which a sharply delineated, pain-less nodule develops and grows to form a tumour-like mass, is caused usually by *Basidiobolus haptosporus* (= *B. meristosporus*). *Nasofacial* zygomycosis is caused by *Conidiobolus coronatus* and involves the development of tumefactions in the nasal mucosa and adjacent tissues. *Rhinocerebral* zygomycosis is caused by *Rhizopus oryzae* and is associated specifically with acidosis due to acute uncontrolled diabetes; infection occurs mainly via the nasal turbinates and paranasal sinuses, spreading rapidly to the eyes and brain, and the condition is rapidly fatal if untreated. *Systemic* zygomycosis may involve an initial pulmonary infection which, if untreated, may spread to other internal organs; causal agents include *Absidia corymbifera*, *Conidiobolus incongruus*, *Cunninghamella bertholletiae* (= *C. elegans*), *Rhizomucor pusillus*, *Rhizopus microsporus*, *R. oryzae*, and *Saksenaea vasiformis* [identification of *S. vasiformis*: SAB (1985) *23* 137–140].

Zygomycosis in animals. Causal agents include some human pathogens as well as species apparently non-pathogenic in man (e.g. *Mortierella* spp). Infection may lead e.g. to placentitis followed by abortion (in cattle), or to the formation of granulomatous, tuberculosis-like lesions in lymph nodes, intestines etc (e.g. in pigs and cattle). (See also EQUINE PHYCOMYCOSIS.)

Zygomycotina A subdivision of fungi (division EUMYCOTA) which form non-flagellate asexually derived spores and which typically reproduce sexually by gametangial copulation with the formation of ZYGOSPORES (q.v.). Classes: TRICHOMYCETES and ZYGOMYCETES.

zygophore A short hyphal branch which develops into a PROGAMETANGIUM.

Zygorhynchus See MUCORALES.

Zygosaccharomyces A genus of yeasts (family SACCHAROMYCETACEAE) in which the cells are globose, ellipsoidal or cylindrical; vegetative reproduction occurs by multilateral budding. Pseudomycelium may be formed. The vegetative cells are predominantly haploid (cf. SACCHAROMYCES). Ascus formation is preceded by conjugation between individual cells (occasionally between a cell and its bud); asci are persistent. Ascospores: globose to ellipsoidal, 1–4 per ascus. Sugars are fermented vigorously; NO_3^- is not assimilated. Eight species are recognized: *Z. bailii* (formerly e.g. *Saccharomyces elegans*), *Z. bisporus*, *Z. cidri*, *Z. fermentati*, *Z. florentinus*, *Z. microellipsoides*, *Z. mrakii*, and *Z. rouxii* (anamorph: *Candida mogii*; numerous synonyms, including e.g. *Saccharomyces rouxii*). [Book ref. 100 pp. 393–395 and 449–465.]

(See also e.g. MISO and SOY SAUCE.)

zygosporangium See ZYGOSPORE.

zygospore A thick-walled, sexually-derived resting spore characteristic of fungi of the ZYGOMYCOTINA. (Zygospores are apparently not formed by e.g. members of the Saksenaeaceae or of the Amoebidiales.) The fusion of two gametangia (see PROGAMETANGIUM) gives rise to a *prozygosporangium* which develops into a *zygosporangium* having a thick, multilayered, often highly ornamented wall; in at least some zygomycetes the zygospore is known to develop as a separate structure *within* the zygosporangium [CJB (1978) *56* 1061–1073]. The term 'zygospore' has been traditionally used to refer to the zygosporangium together with its enclosed zygospore.

zygote A single diploid cell formed by the fusion of two gametes.

zygotene stage See MEIOSIS.

zygotic induction The induction of a prophage when a chromosome containing that prophage is transferred from a lysogenic conjugal donor (see LYSOGENY and bacterial CONJUGATION) to a recipient not lysogenized by the same (or a closely related) bacteriophage; the induction occurs as a result of the absence of a phage repressor protein in the recipient cell.

zygotic meiosis MEIOSIS, in a zygote, preceding the formation of haploid vegetative cells in a life cycle in which HAPLOPHASE predominates. (cf. GAMETIC MEIOSIS.)

zymase Old term for the enzyme fraction isolated from disrupted yeast cells which is capable of catalysing ALCOHOLIC FERMENTATION.

zymodeme A subpopulation within a given taxon (e.g. genus) distinguished on the basis of one or more isoenzymes.

zymogen (proenzyme) An inactive enzyme precursor that is usually converted to the active form of the enzyme by proteolytic cleavage. (See e.g. FIBRINOLYSIN.)

zymogenous Refers to those (predominantly transient or alien) microorganisms in a given environment (e.g. soil) which exhibit an upsurge in growth (and hence in numbers or biomass) on those occasions when the levels of nutrients increase, or when a particular substrate becomes available; in the absence of suitable levels of nutrients, relatively small numbers of such organisms may be capable of existence, in that environment, in a dormant or starvation-resistant stage. (cf. AUTOCHTHONOUS sense 1.)

zymogram (1) A medium (e.g. a starch gel strip) which has been used for the ELECTROPHORESIS of a cell homogenate and which has been subsequently stained to detect or quantify a given enzyme. (2) A table showing the results of tests which determine the ability of one or more organisms to ferment each of a range of carbohydrates.

Zymolyase A commercial enzyme preparation obtained from culture filtrates of *Arthrobacter luteus*. It contains endo-(1→3)-β-glucanase activity ('Z-glucanase') – which is active against most (1→3)-β-glucans – and also some protease activity ('Z-protease') [JB (1984) *159* 1018–1026]. Zymolyase is used e.g. for investigating fungal (particularly yeast) CELL WALL structure, for preparing yeast sphaeroplasts, etc.

Zymomonas A genus (*incertae sedis*) of oxidase-negative, catalase-positive, chemoorganotrophic, Gram-negative bacteria which occur e.g. as spoilage organisms in alcoholic beverages (see e.g. CIDER spoilage) and which are used in specific fermentations (see e.g. PULQUE). [*Zymomonas* ethanol fermentations: MS (1984) *1* 133–136.] The genus includes obligately anaerobic and facultatively aerobic strains. Cells: round-ended rods, ca. 1.0–1.4 × 2.0–6.0 μm, non-motile or with 1–4 polar flagella. Metabolism: primarily fermentative. Glucose or fructose is fermented, via the ENTNER–DOUDOROFF PATHWAY, to ethanol, CO_2 and lactic acid with smaller amounts of acetaldehyde, acetoin and glycerol. Ethanol tolerance: at least ca. 5%. Under aerobic conditions, some strains metabolize glucose (in glucose–yeast extract medium) to ethanol, the latter then being oxidized to acetate; during such growth glucose provides only about 50% of the carbon, the remainder being supplied by the yeast extract. All strains need biotin and pantothenate. Nitrate is not reduced. Optimum growth temperature: 25–30°C. Many or all strains can grow at or below pH 4.0. Colonies (2 days, 30°C) on glucose–yeast extract agar: smooth, white to cream, 1–2 mm diam. GC%: ca. 47.5–49.5. Type species: *Z. mobilis*. *Z. mobilis* subsp *mobilis* (which includes e.g. strains previously named *Z. anaerobia* var. *anaerobia*) can grow in glucose–yeast extract broth at 36°C. *Z. mobilis* subsp *pomacii* (which includes e.g. strains previously named *Z. anaerobia* subsp *pomaceae*) cannot grow in glucose–yeast extract broth at 36°C.

[Book ref. 22, pp. 576–580.]

zymosan An insoluble polysaccharide, found in the cell wall in certain yeasts, which promotes COMPLEMENT FIXATION via the alternative pathway.

zymotype A BIOTYPE characterized on the basis of a ZYMOGRAM.

Zythia A genus of fungi (order SPHAEROPSIDALES) which include *Gnomonia* (*Zythia*) *fragariae*, a pathogen of the strawberry plant. Elongated conidia are formed in light-coloured pycnidia.

1. Words in SMALL CAPITALS are cross-references to separate entries.
2. Keys to journal title abbreviations and Book ref. numbers are given at the end of the Dictionary.
3. The Greek alphabet is given in Appendix VI.
4. For further information see 'Notes for the User' at the front of the Dictionary.

Appendices

Abbreviations and Conventions used in the Appendices

ADP	adenosine 5′-diphosphate
AMP	adenosine 5′-monophosphate
ATP	adenosine 5′-triphosphate (see entry ATP)
CoASH	coenzyme A (see entry COENZYME A)
e^-	electron
FAD	flavin adenine dinucleotide (see entry RIBOFLAVIN)
GDP	guanosine 5′-diphosphate
GTP	guanosine 5′-triphosphate
NAD	nicotinamide adenine dinucleotide; NAD^+ = oxidized form, $NADH + H^+$ = reduced form (see entry NAD)
NADP	nicotinamide adenine dinucleotide phosphate; $NADP^+$ = oxidized form, $NADPH + H^+$ = reduced form
Ⓟ	$-PO_3^{2-}$
PEP	phosphoenolpyruvate
Pi	inorganic phosphate (PO_4^{3-})
PPi	inorganic pyrophosphate ($P_2O_7^{4-}$)
PTS	phosphoenolpyruvate-dependent phosphotransferase system (see entry PTS)
TPP	thiamine pyrophosphate

A *bisphosphate* is a compound containing two phosphate groups which are not linked to one another (as in e.g. fructose 1,6-bisphosphate); in a *diphosphate* the two phosphate groups are linked to each other (as in e.g. adenosine 5′-diphosphate).

A *synthetase* is an enzyme belonging to EC class 6 (see entry ENZYME); a *synthase* may belong to any other EC class.

For simplicity and visual clarity we have followed the usual practice of giving the *formula* for an organic acid in the unionized form but *naming* it as the salt: e.g., CH_3COOH = acetate.

EMBDEN–MEYERHOF–PARNAS PATHWAY

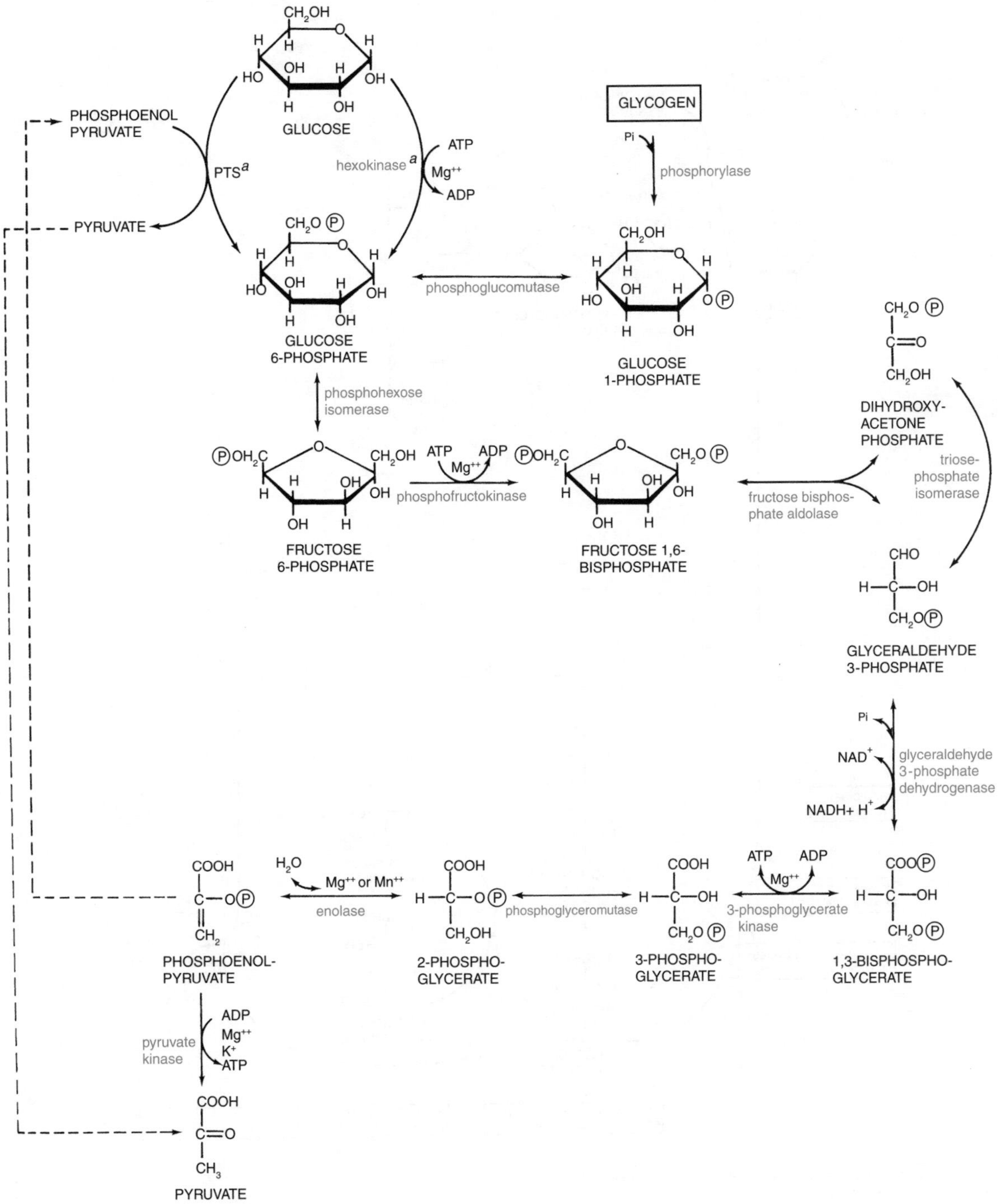

[a]Depending on organism (see entry EMBDEN–MEYERHOF–PARNAS PATHWAY).

PTS = phosphoenolpyruvate-dependent phosphotransferase system (see entry PTS).

HEXOSE MONOPHOSPHATE PATHWAY[a]

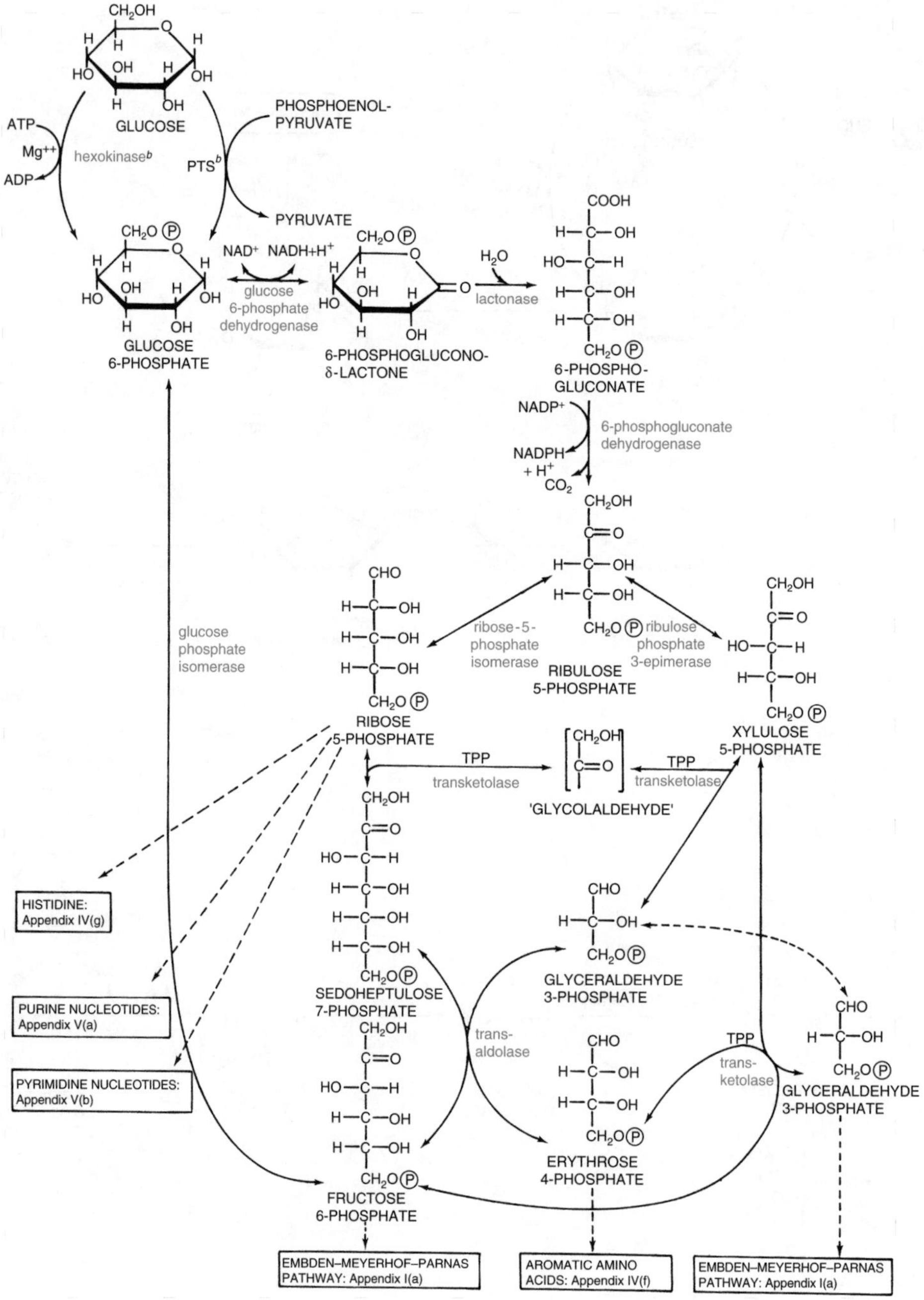

[a]See entry HEXOSE MONOPHOSPHATE PATHWAY.
[b]Depending on organism.

PTS = phosphoenolpyruvate-dependent phosphotransferase system (see entry PTS). TPP = thiamine pyrophosphate (see THIAMINE).

ENTNER–DOUDOROFF PATHWAY[a]

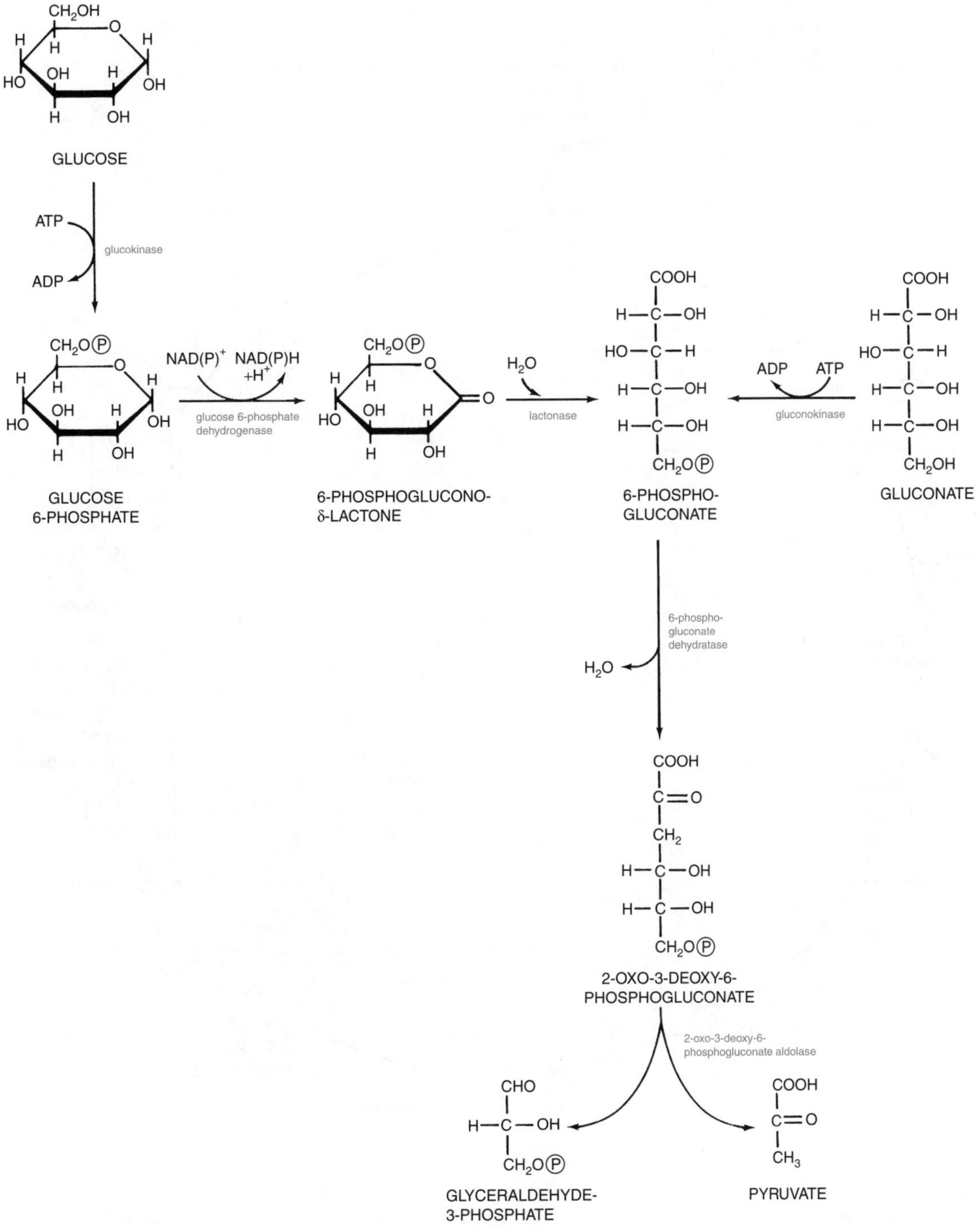

GLUCOSE

ATP → ADP (glucokinase)

GLUCOSE 6-PHOSPHATE

$NAD(P)^+$ → $NAD(P)H + H^+$ (glucose 6-phosphate dehydrogenase)

6-PHOSPHOGLUCONO-δ-LACTONE

H_2O (lactonase)

6-PHOSPHO-GLUCONATE

GLUCONATE

ATP → ADP (gluconokinase)

6-phospho-gluconate dehydratase → H_2O

2-OXO-3-DEOXY-6-PHOSPHOGLUCONATE

2-oxo-3-deoxy-6-phosphogluconate aldolase

GLYCERALDEHYDE-3-PHOSPHATE

PYRUVATE

[a]See entry ENTNER–DOUDOROFF PATHWAY.

TRICARBOXYLIC ACID CYCLE[a]

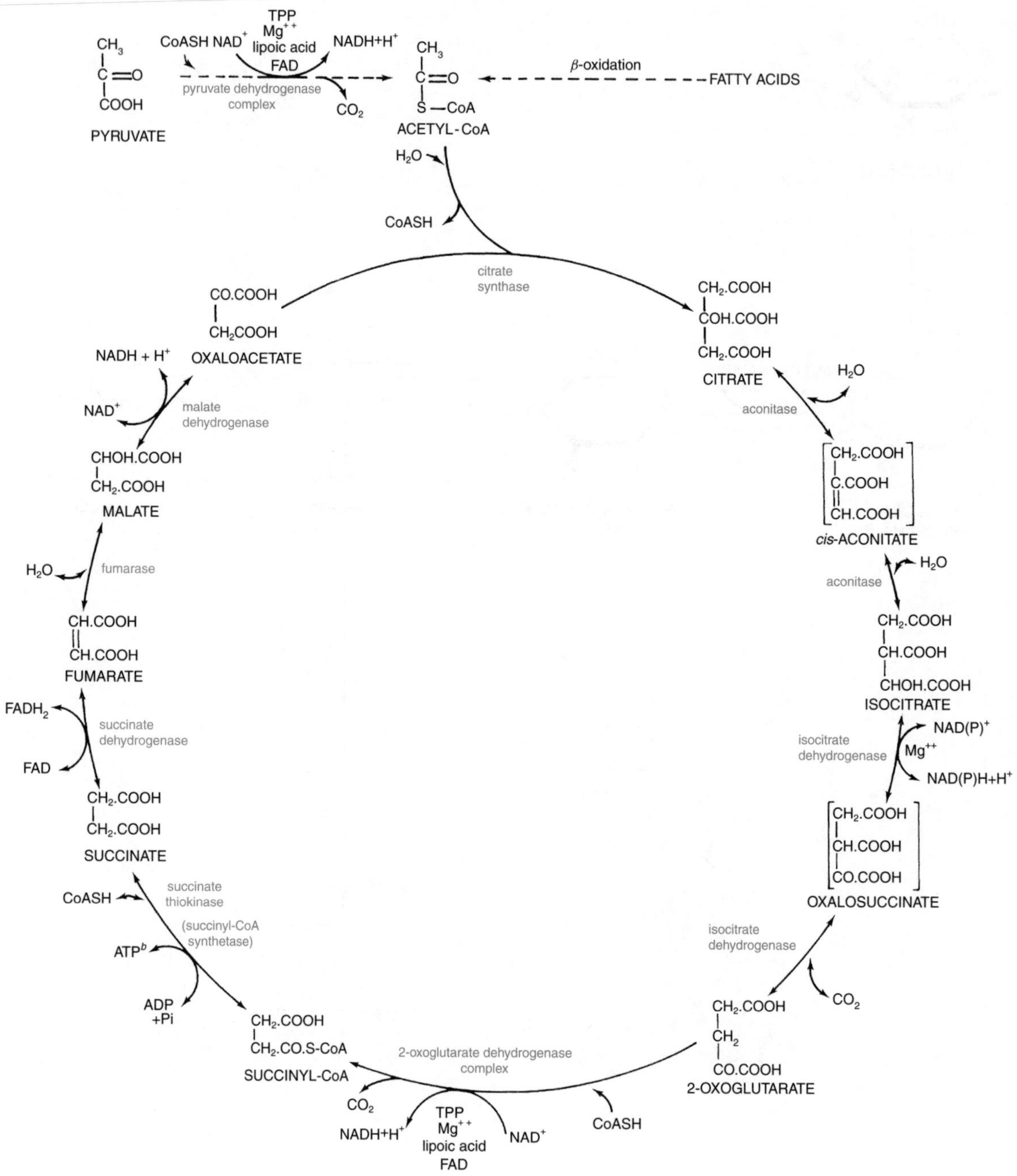

[a]See entry TCA CYCLE.
[b]ATP ⟷ ADP + Pi in e.g. *Escherichia coli*; GTP ⟷ GDP + Pi in e.g. mammalian systems; either reaction in some bacteria.
Square brackets indicate an enzyme-bound intermediate.

TRICARBOXYLIC ACID CYCLE – *showing some anaplerotic sequences and catabolic and anabolic interactions*[a]

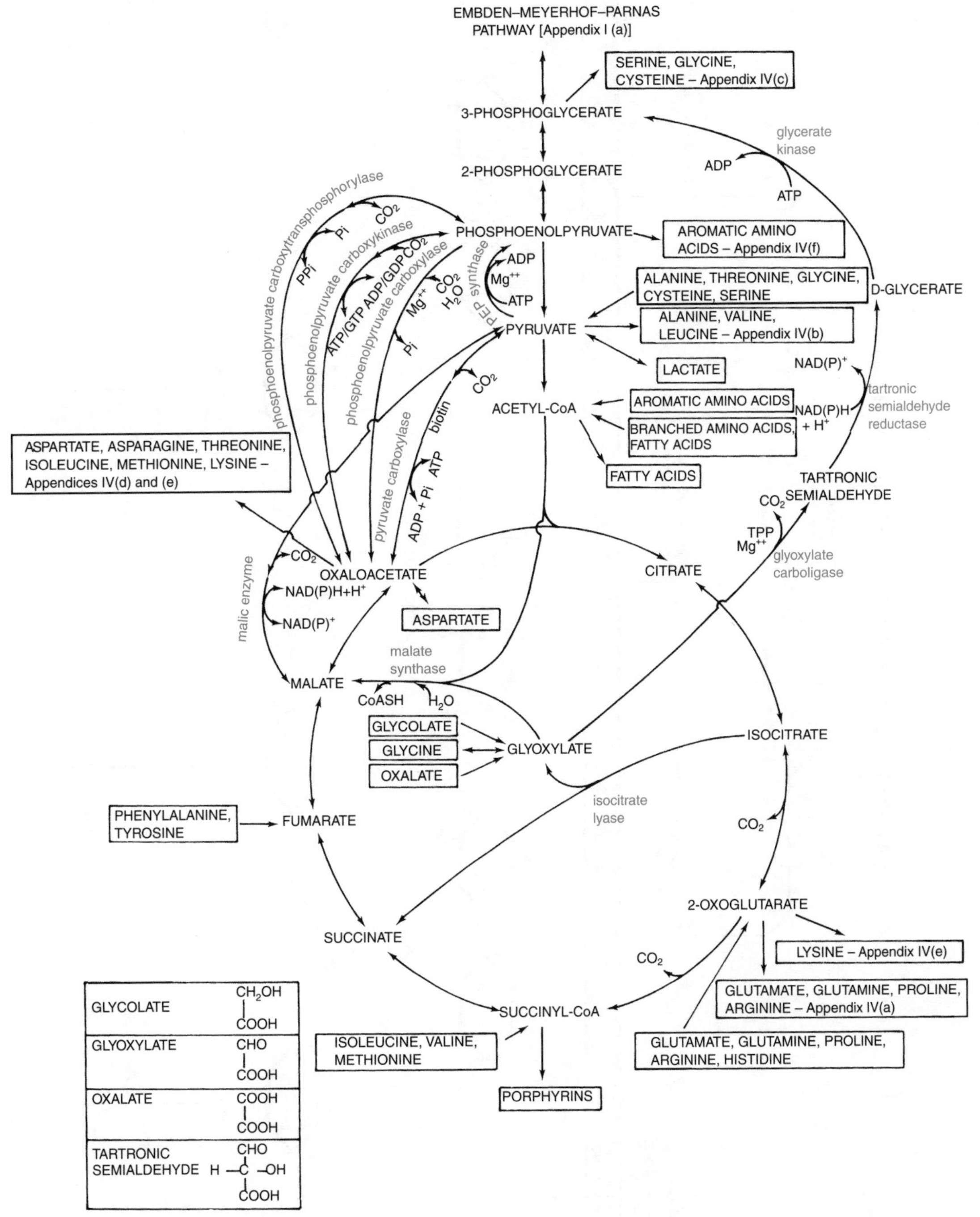

[a]See entry TCA CYCLE.

LACTOSE METABOLISM IN SOME BACTERIA[a]

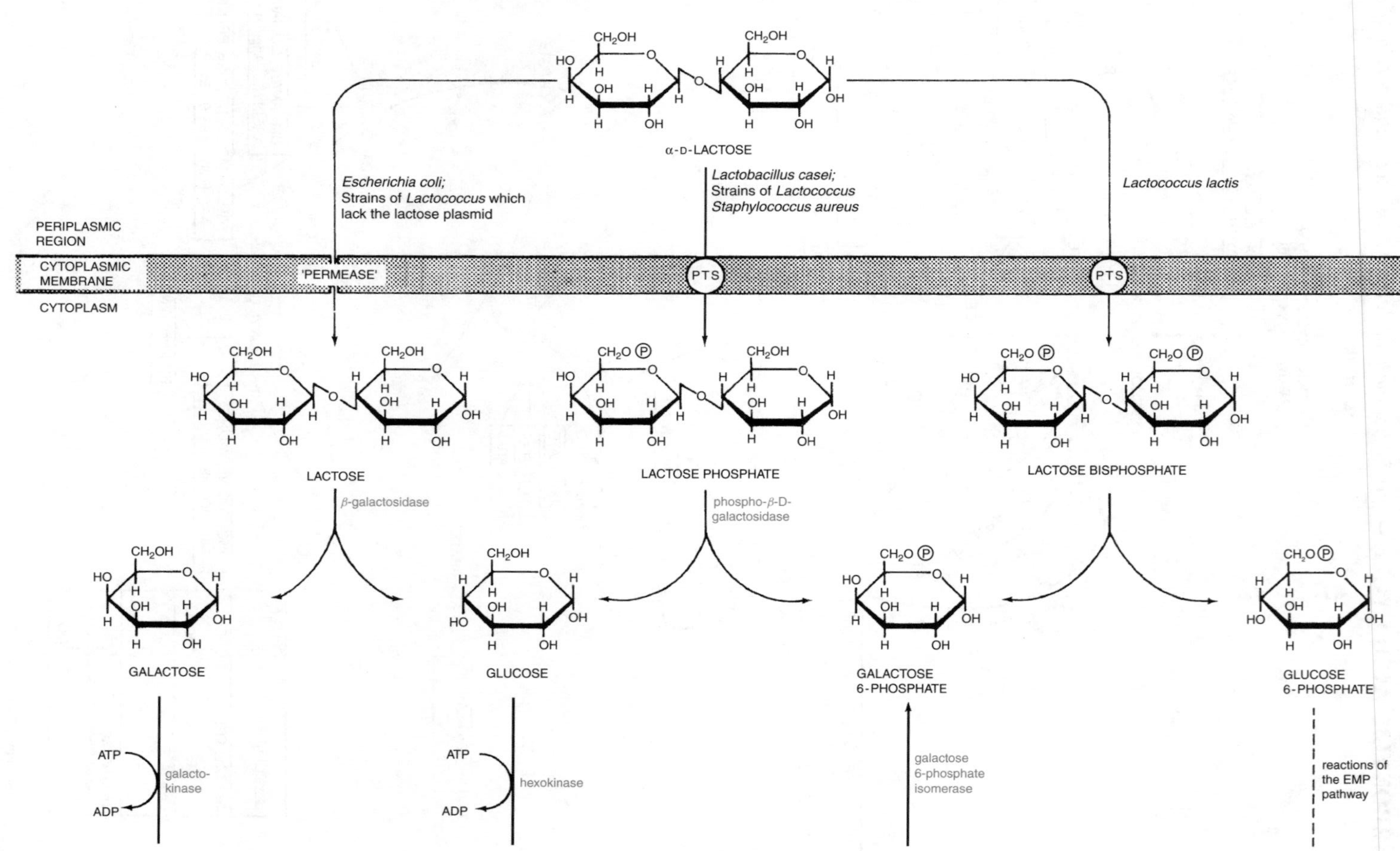

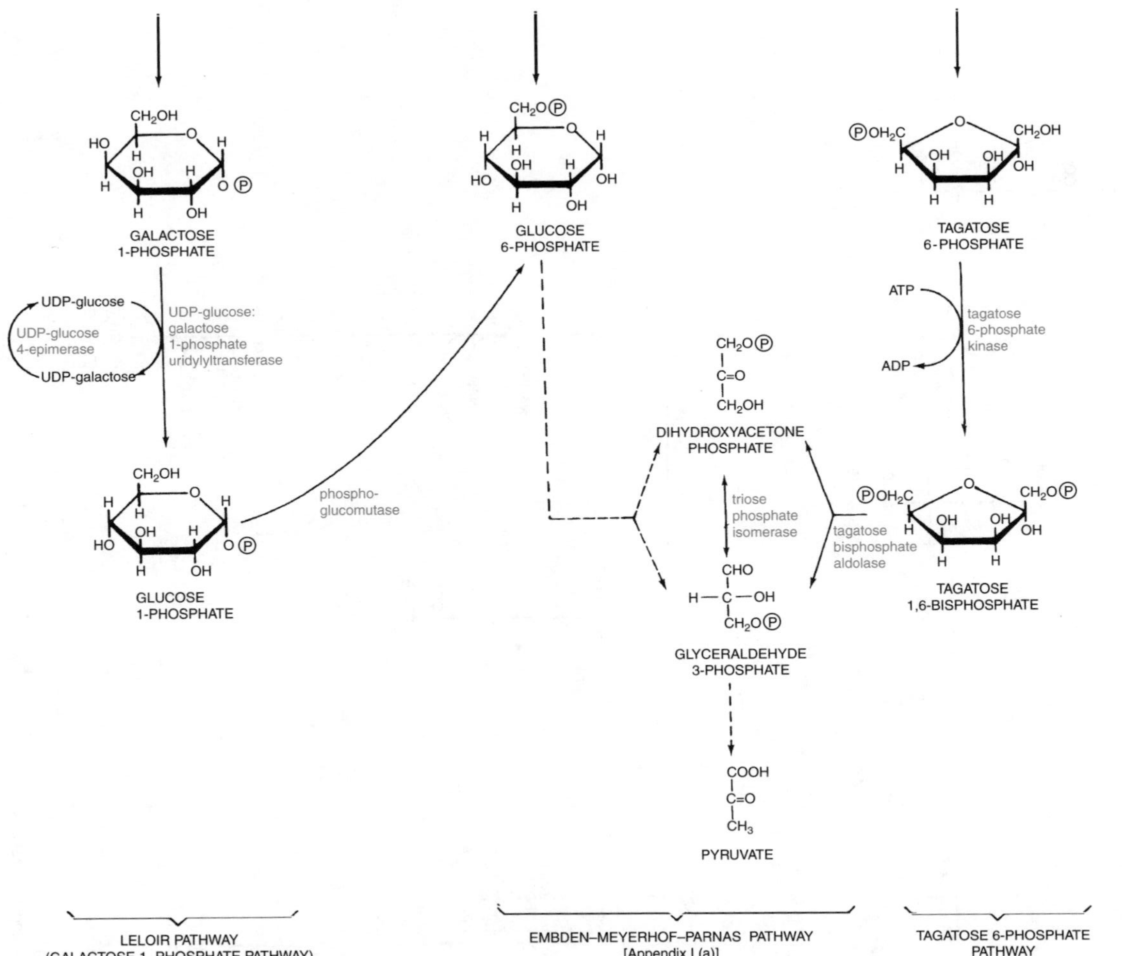

[a] See entry LACTOSE.

PTS = phosphoenolpyruvate-dependent phosphotransferase system (see entry PTS).

HETEROLACTIC FERMENTATIONS[a]

'Classical' pathway

GLUCOSE → (ATP, Mg^{++} → ADP; hexokinase) → GLUCOSE 6-PHOSPHATE → ($NAD(P)^+$ → $NAD(P)H+H^+$; glucose 6-phosphate dehydrogenase) → 6-PHOSPHOGLUCONO-δ-LACTONE → (H_2O; lactonase) → 6-PHOSPHO-GLUCONATE → ($NAD(P)^+$ → $NAD(P)H + H^+$, CO_2; 6-phospho-gluconate dehydrogenase) → RIBULOSE 5-PHOSPHATE → (ribulose phosphate 3-epimerase) → XYLULOSE 5-PHOSPHATE

PENTOSES - - - → Appendix III (d) → XYLULOSE 5-PHOSPHATE

XYLULOSE 5-PHOSPHATE → (Pi, TPP, Mg^{++}; phospho-ketolase) → GLYCERALDEHYDE 3-PHOSPHATE + ACETYL-PHOSPHATE

GLYCERALDEHYDE 3-PHOSPHATE → (NAD^+ → $NADH^+ + H^+$; reactions as in EMBDEN-MEYERHOF-PARNAS PATHWAY: Appendix I (a)) → PYRUVATE → (NADH $+H^+$ → NAD^+; lactate dehydrogenase) → LACTATE

ACETYL-PHOSPHATE → (ADP → ATP; acetokinase) → ACETATE

ACETYL-PHOSPHATE → (CoASH → Pi; phosphotransacetylase) → ACETYL-CoA → (NADH $+H^+$ → NAD^+, CoASH; aldehyde dehydrogenase) → ACETALDEHYDE → (NADH $+H^+$ → NAD^+; alcohol dehydrogenase) → ETHANOL

[a]See entry HETEROLACTIC FERMENTATION.

Bifidobacterium pathway

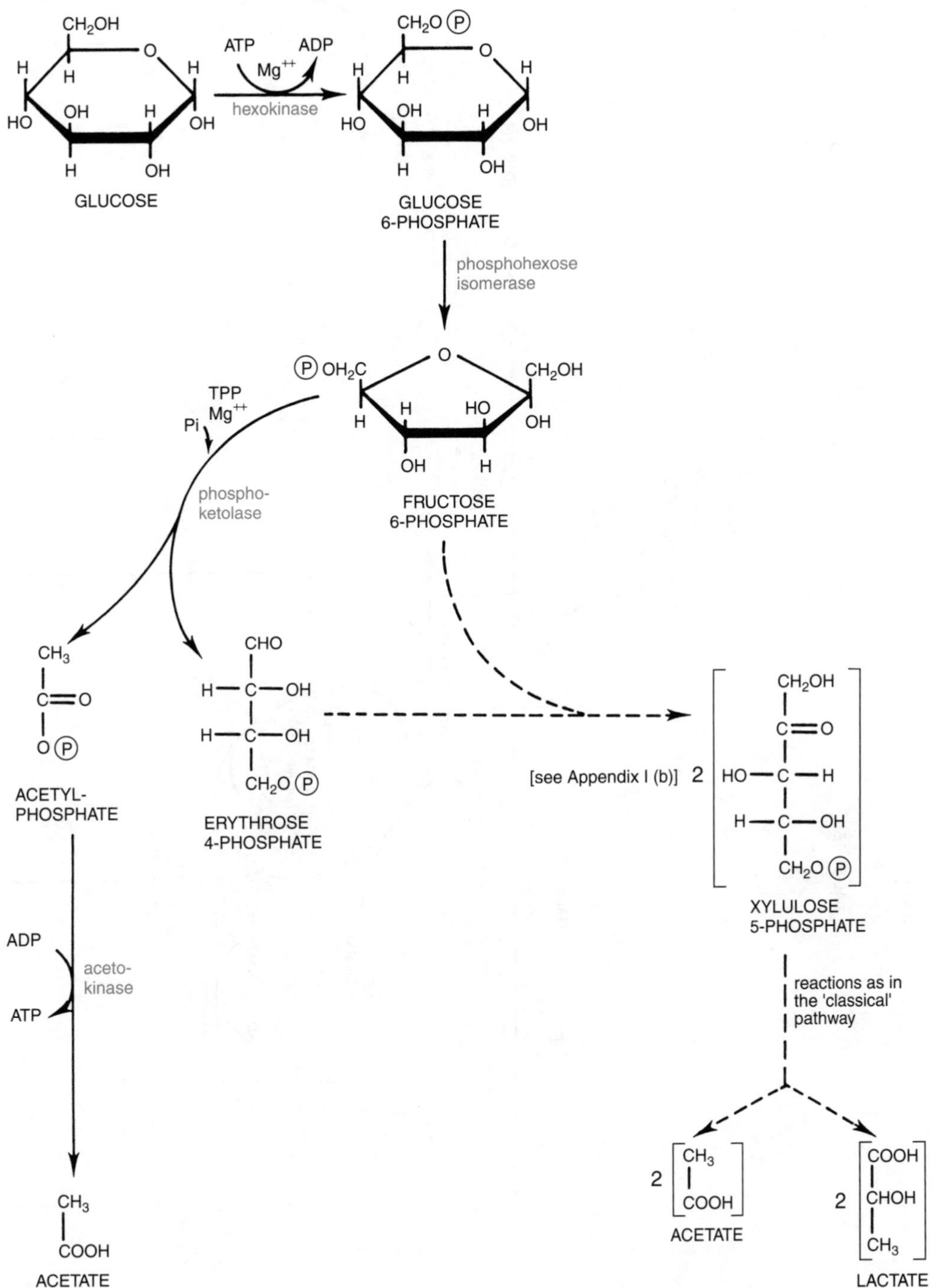

SOME ANAEROBIC FATES OF PYRUVATE IN LACTIC ACID BACTERIA[a]

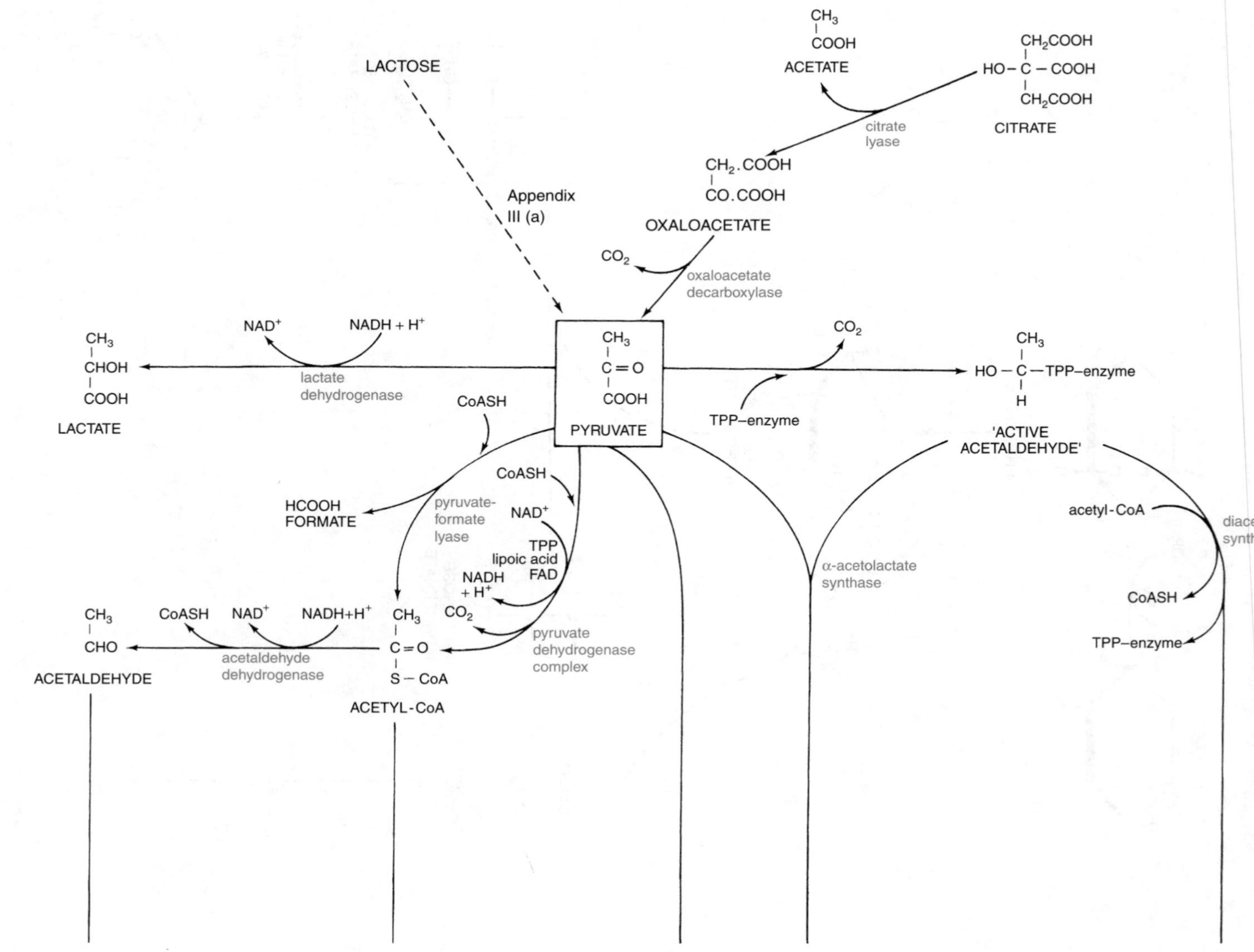

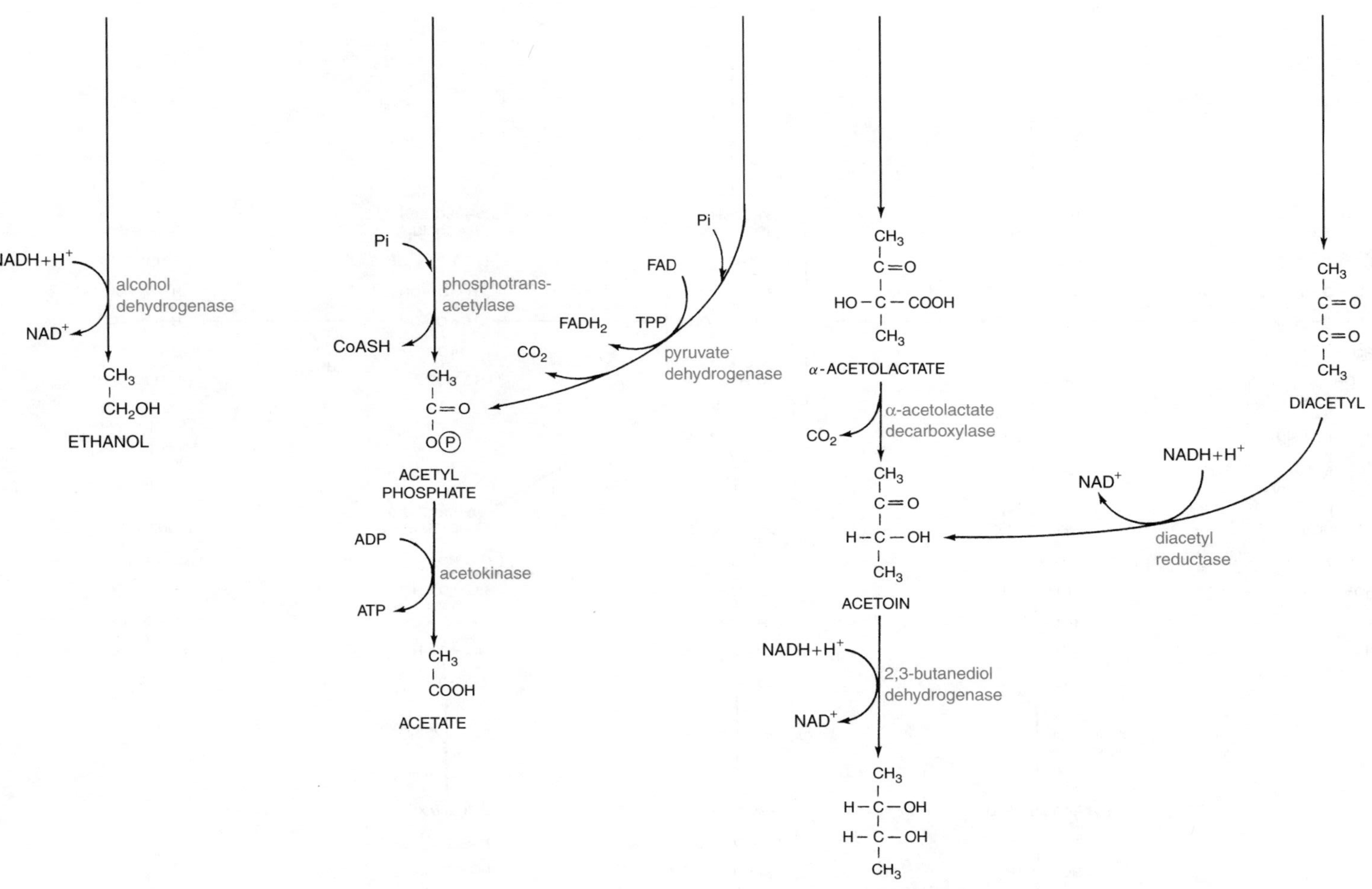

[a]Not all reactions shown occur in all lactic acid bacteria.

The relative proportions of the various end-products depend on organism and conditions. See e.g. entries for DIACETYL, HOMOLACTIC FERMENTATION, and YOGHURT.

PENTOSE AND PENTITOL DISSIMILATION IN LACTIC ACID BACTERIA[a]

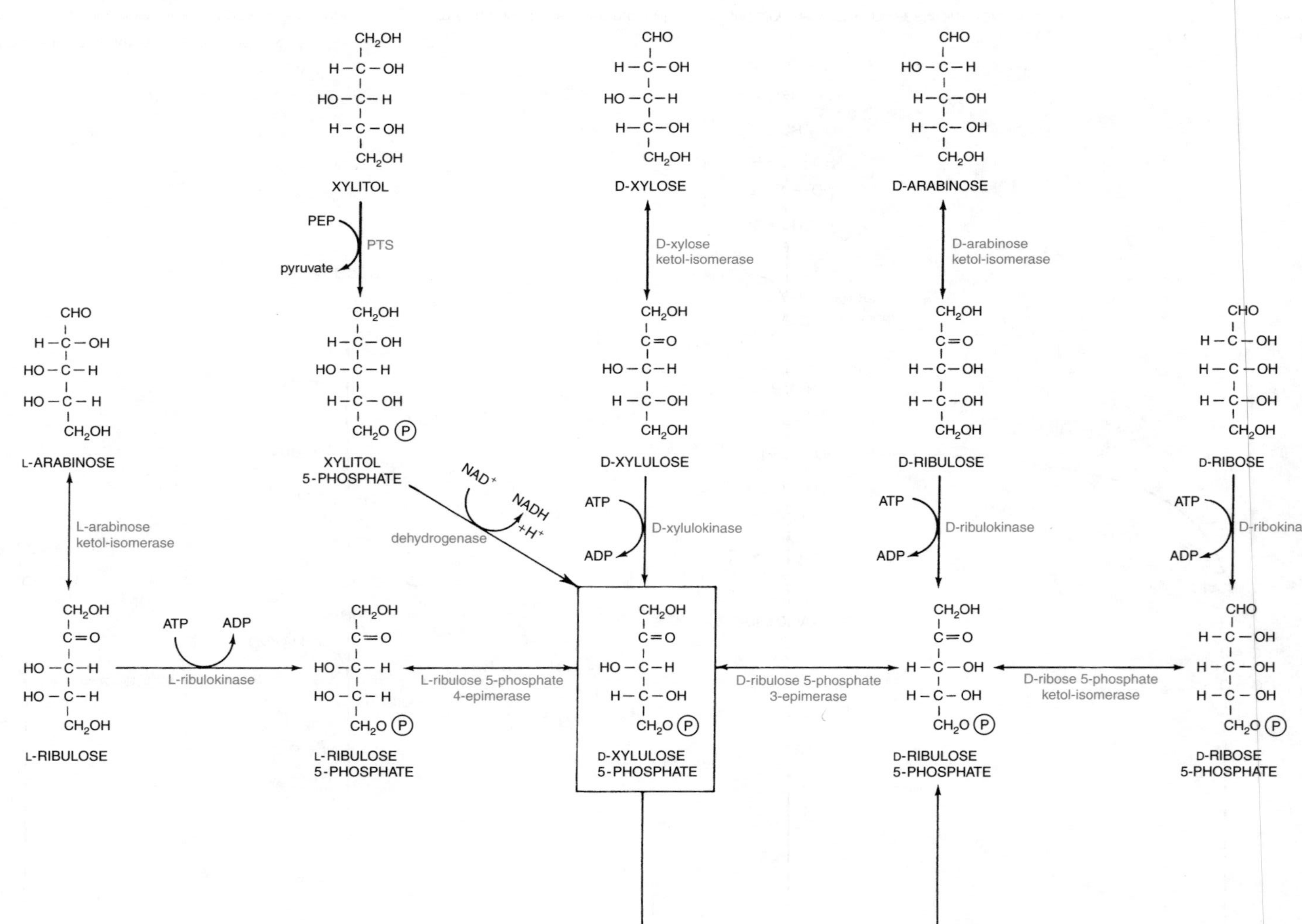

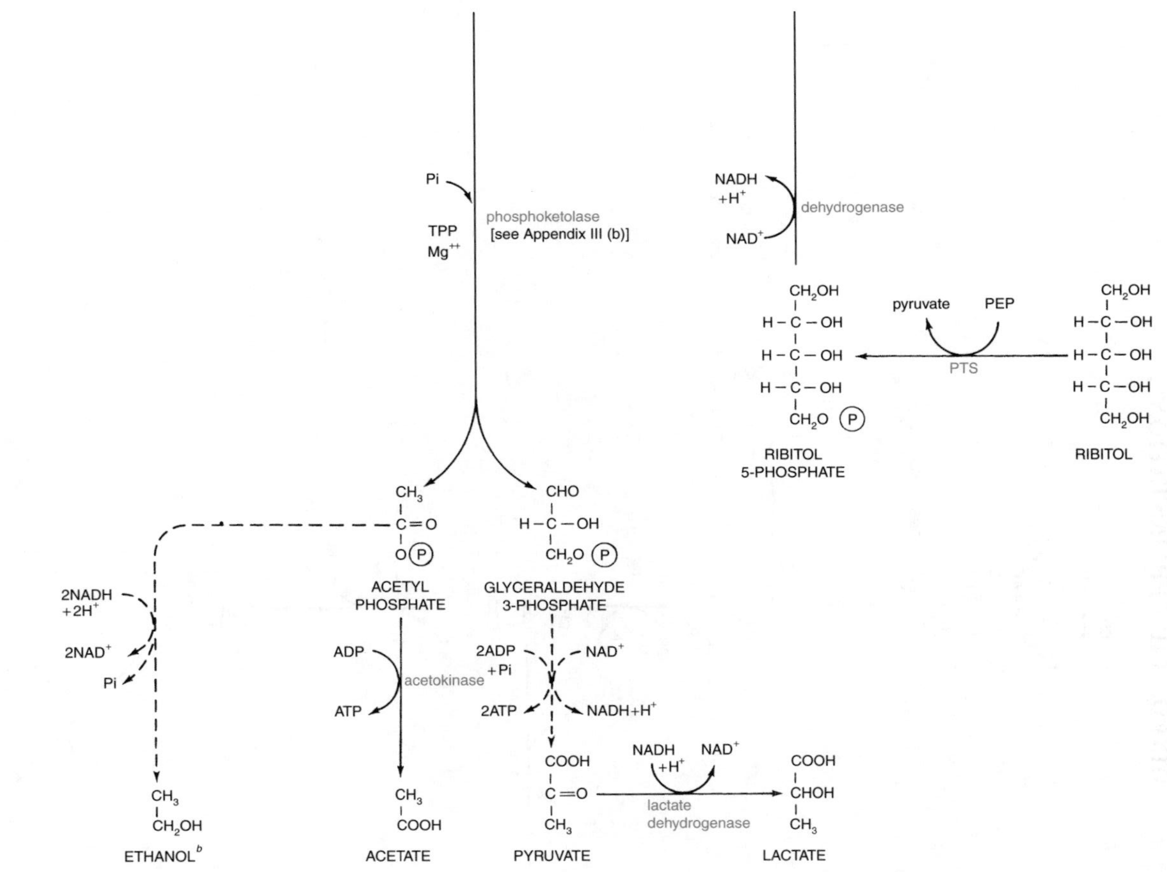

[a]Not all lactic acid bacteria carry out all of these conversions, and many of the pentose/pentulose/pentitol interconversions occur in other microorganisms: see entries PENTOSES and PENTITOLS.

[b]Ethanol is produced only when the substrate is a pentitol, the NADH formed by the pentitol phosphate dehydrogenase reaction being reoxidized by this means.

PTS = phosphoenolpyruvate-dependent phosphotransferase system (see entry PTS).

MIXED ACID FERMENTATION[a]

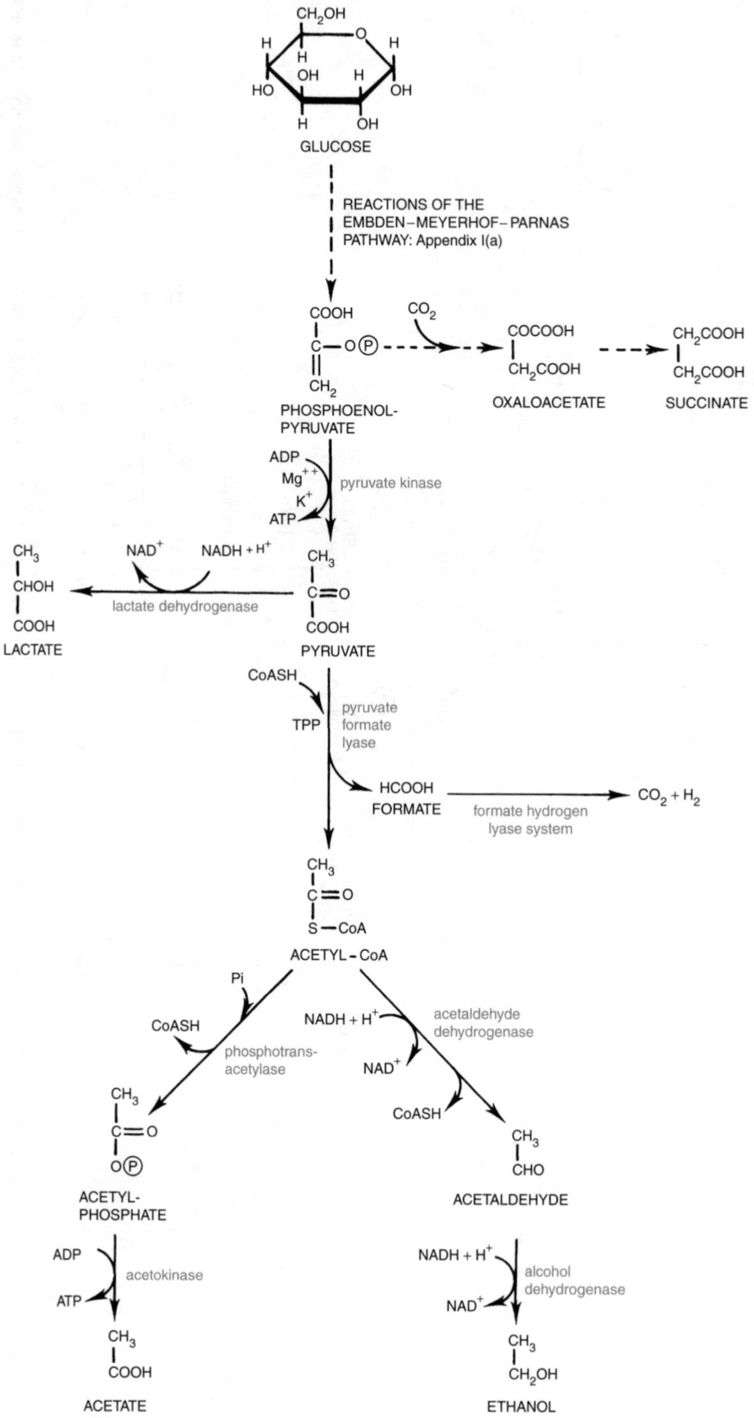

[a]See entry MIXED ACID FERMENTATION.

BUTANEDIOL FERMENTATION[a]

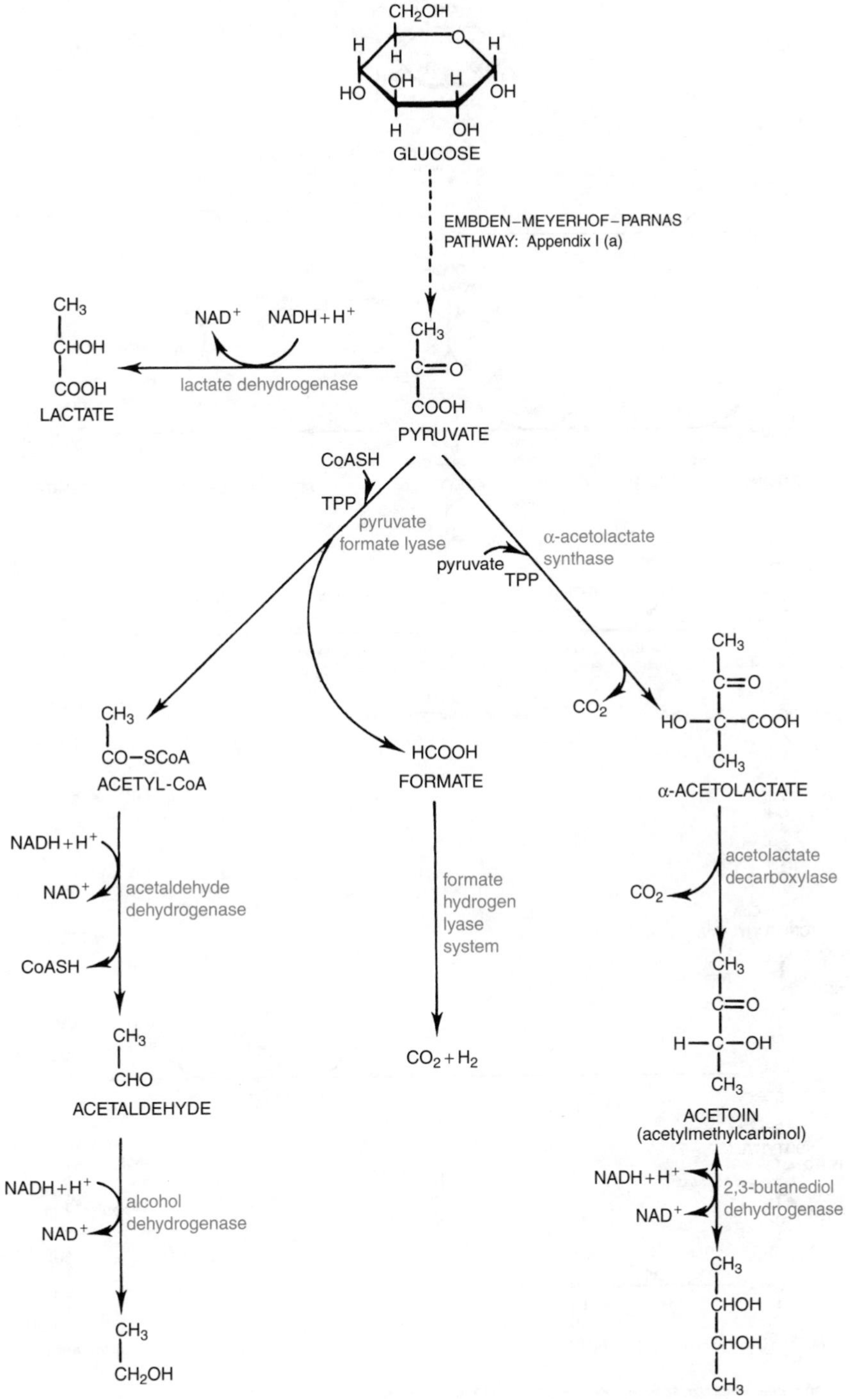

[a]See entry BUTANEDIOL FERMENTATION.

BUTYRIC ACID FERMENTATION AND ACETONE–BUTANOL FERMENTATION[a]

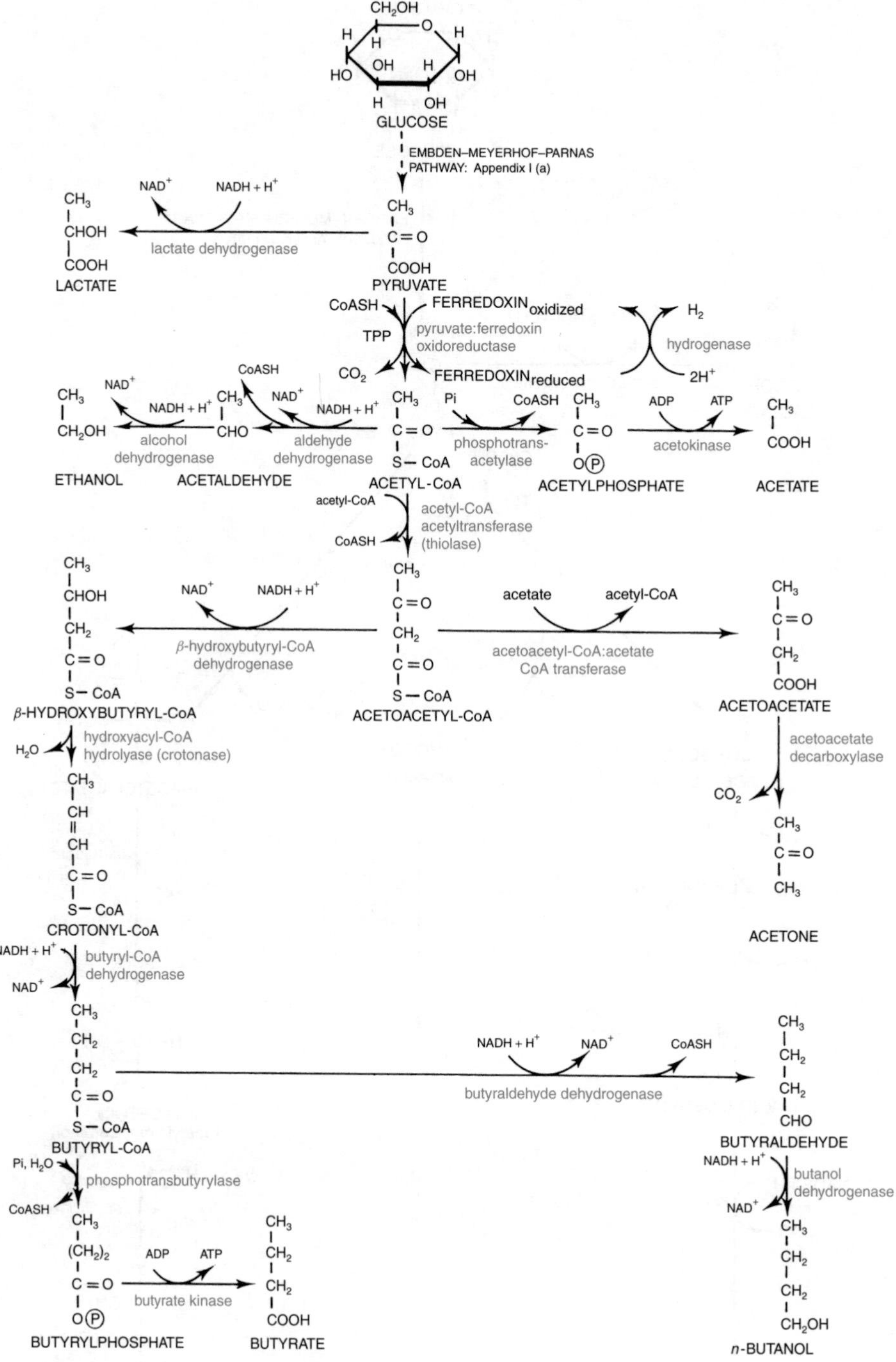

[a]See entries BUTYRIC ACID FERMENTATION and ACETONE–BUTANOL FERMENTATION.

PROPIONIC ACID FERMENTATION (in *Propionibacterium*)[a]

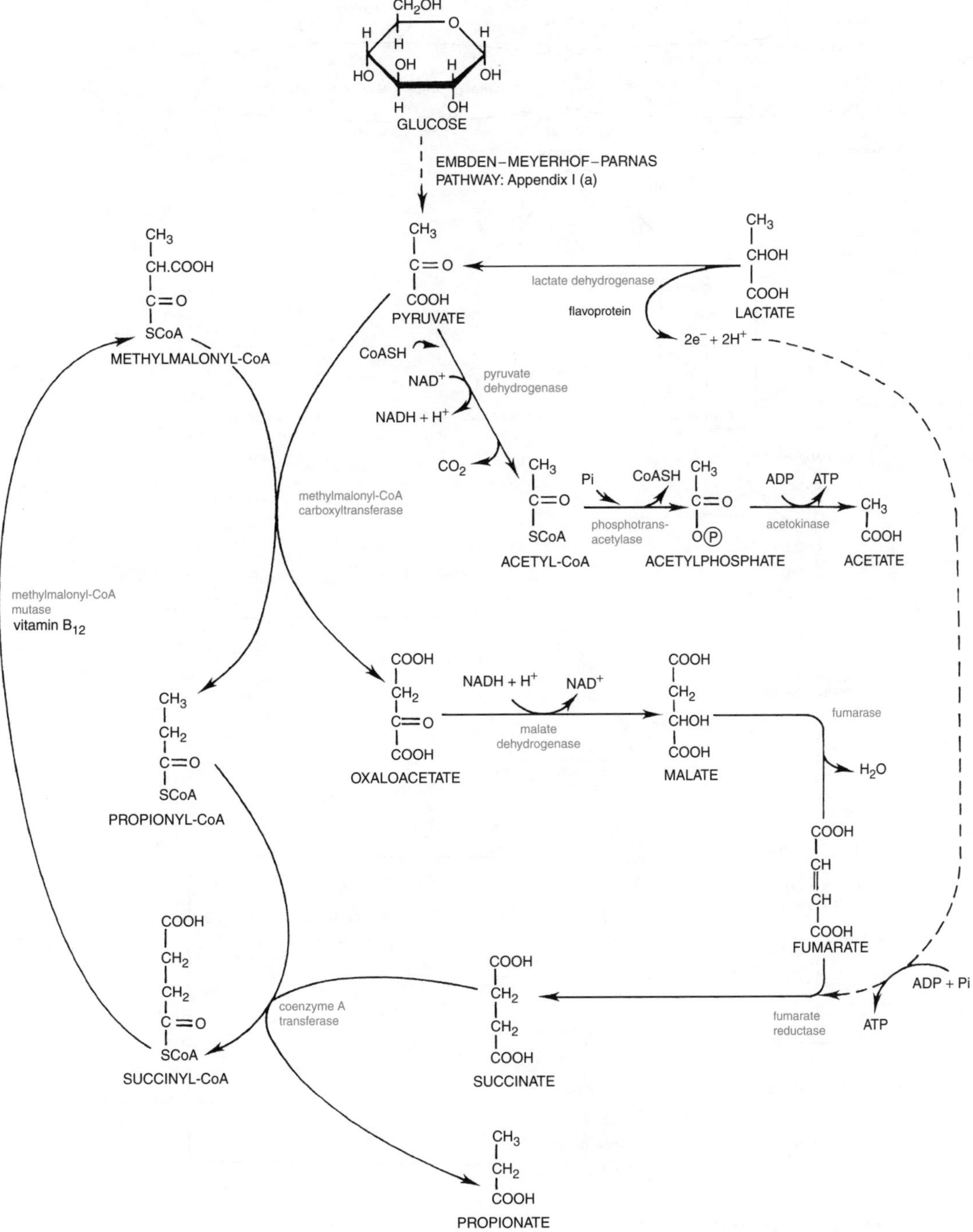

[a]See entry PROPIONIC ACID FERMENTATION.

BIOSYNTHESIS OF ARGININE[a], GLUTAMATE[b], GLUTAMINE[c], PROLINE

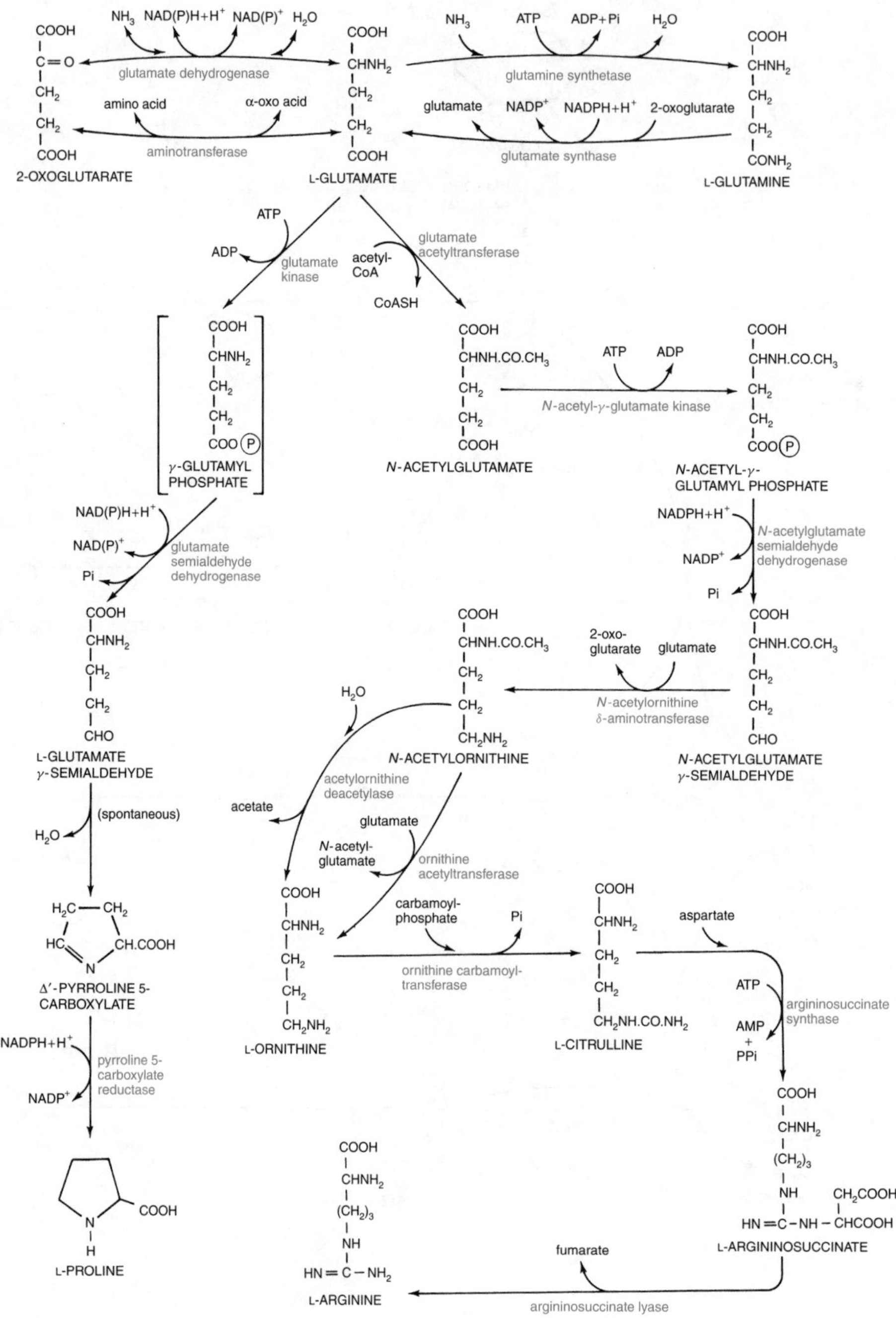

[a]See entry ARGININE BIOSYNTHESIS for reference.
[b]See entry GLUTAMIC ACID.
[c]See entry GLUTAMINE.

BIOSYNTHESIS OF ALANINE, LEUCINE, VALINE

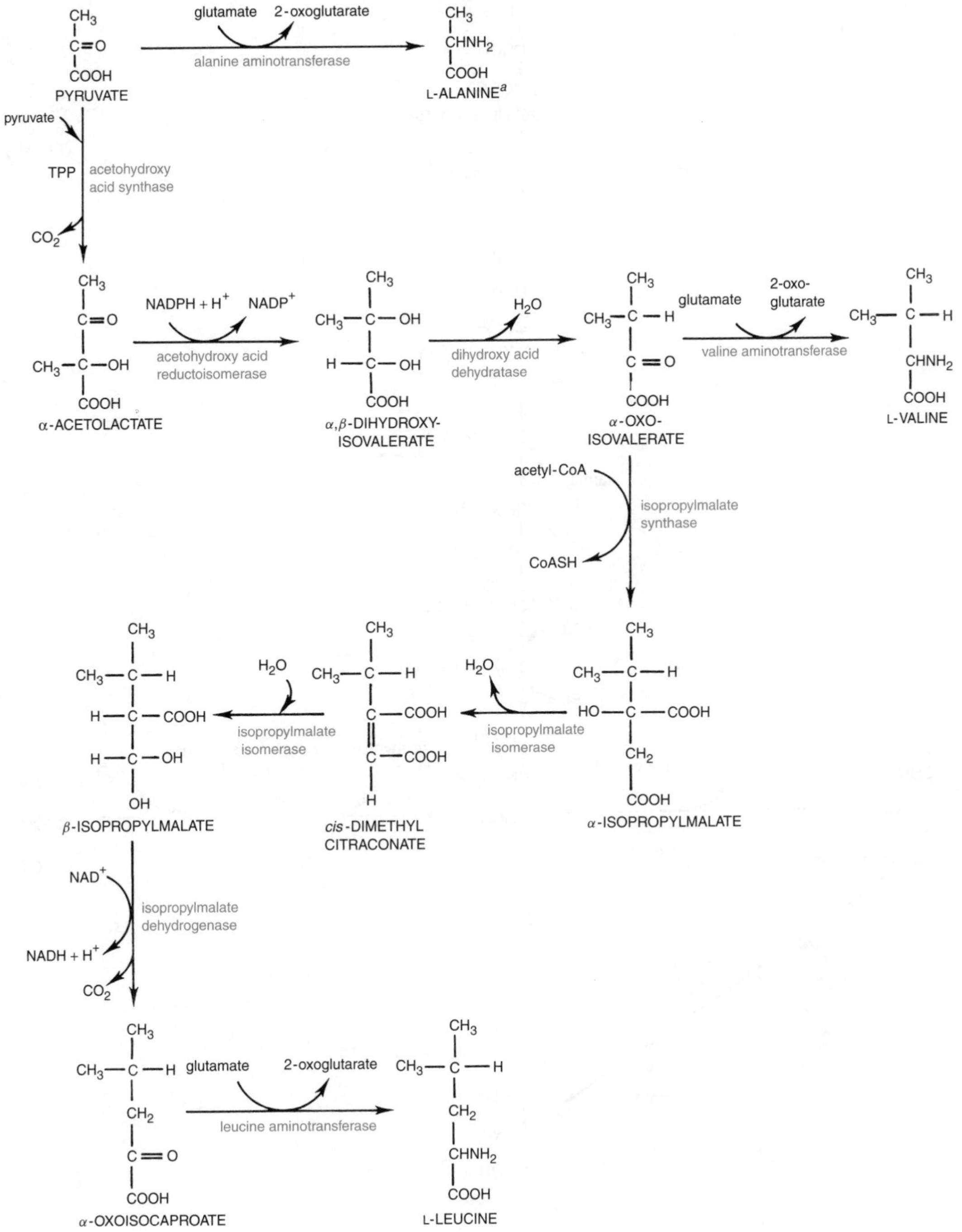

PYRUVATE → (glutamate → 2-oxoglutarate; alanine aminotransferase) → L-ALANINE[a]

PYRUVATE + pyruvate → (TPP; acetohydroxy acid synthase; − CO_2) → α-ACETOLACTATE

α-ACETOLACTATE → ($NADPH + H^+$ → $NADP^+$; acetohydroxy acid reductoisomerase) → α,β-DIHYDROXY-ISOVALERATE

α,β-DIHYDROXY-ISOVALERATE → (− H_2O; dihydroxy acid dehydratase) → α-OXO-ISOVALERATE

α-OXO-ISOVALERATE → (glutamate → 2-oxo-glutarate; valine aminotransferase) → L-VALINE

α-OXO-ISOVALERATE → (acetyl-CoA → CoASH; isopropylmalate synthase) → α-ISOPROPYLMALATE

α-ISOPROPYLMALATE → (− H_2O; isopropylmalate isomerase) → *cis*-DIMETHYL CITRACONATE

cis-DIMETHYL CITRACONATE → (+ H_2O; isopropylmalate isomerase) → β-ISOPROPYLMALATE

β-ISOPROPYLMALATE → (NAD^+ → $NADH + H^+$; isopropylmalate dehydrogenase; − CO_2) → α-OXOISOCAPROATE

α-OXOISOCAPROATE → (glutamate → 2-oxoglutarate; leucine aminotransferase) → L-LEUCINE

[a]cf. entry AMMONIA ASSIMILATION.

BIOSYNTHESIS OF CYSTEINE, GLYCINE, SERINE

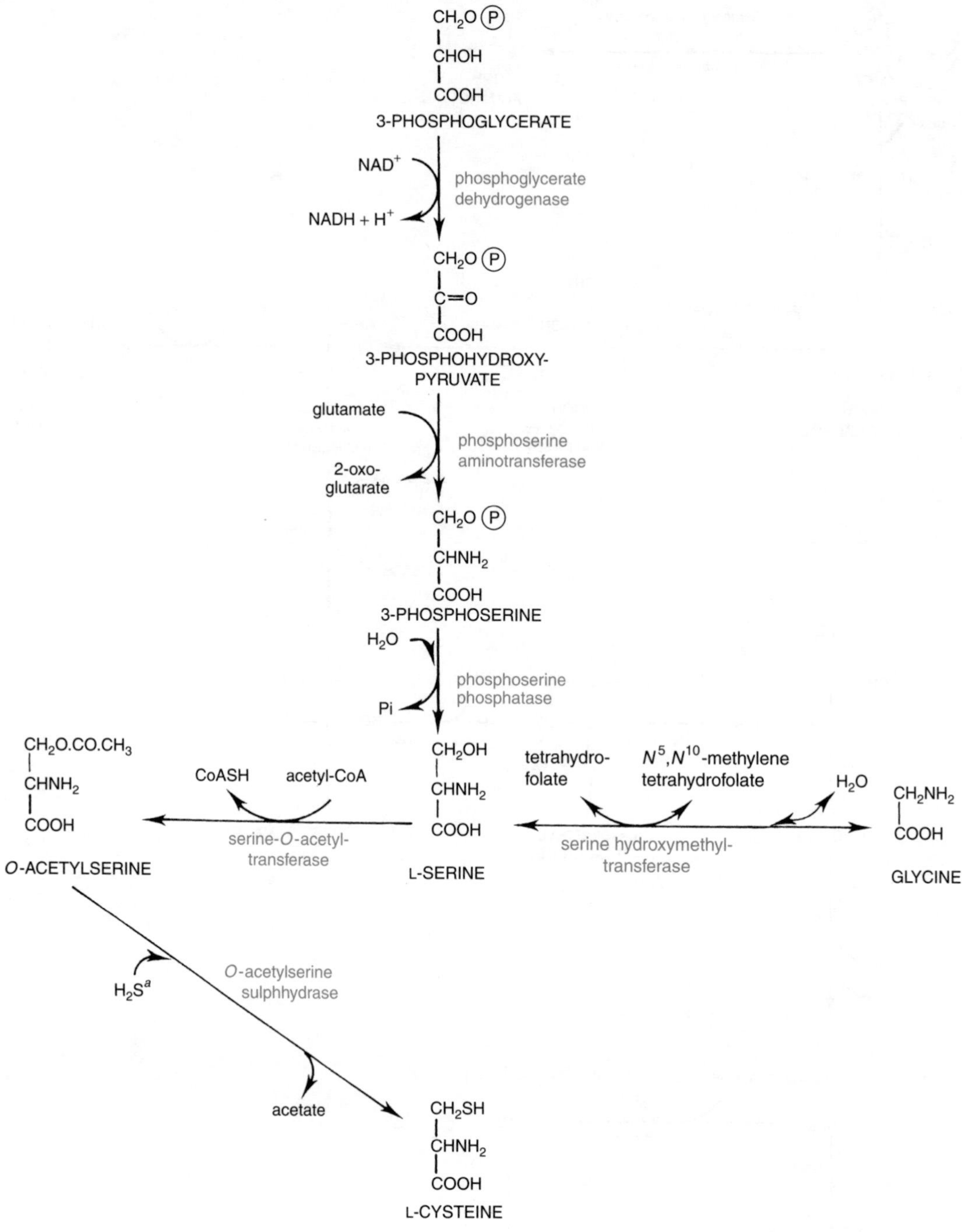

[a]Formed by ASSIMILATORY SULPHATE REDUCTION (q.v.).

BIOSYNTHESIS OF ASPARAGINE, ASPARTATE[a], ISOLEUCINE[b], METHIONINE, THREONINE

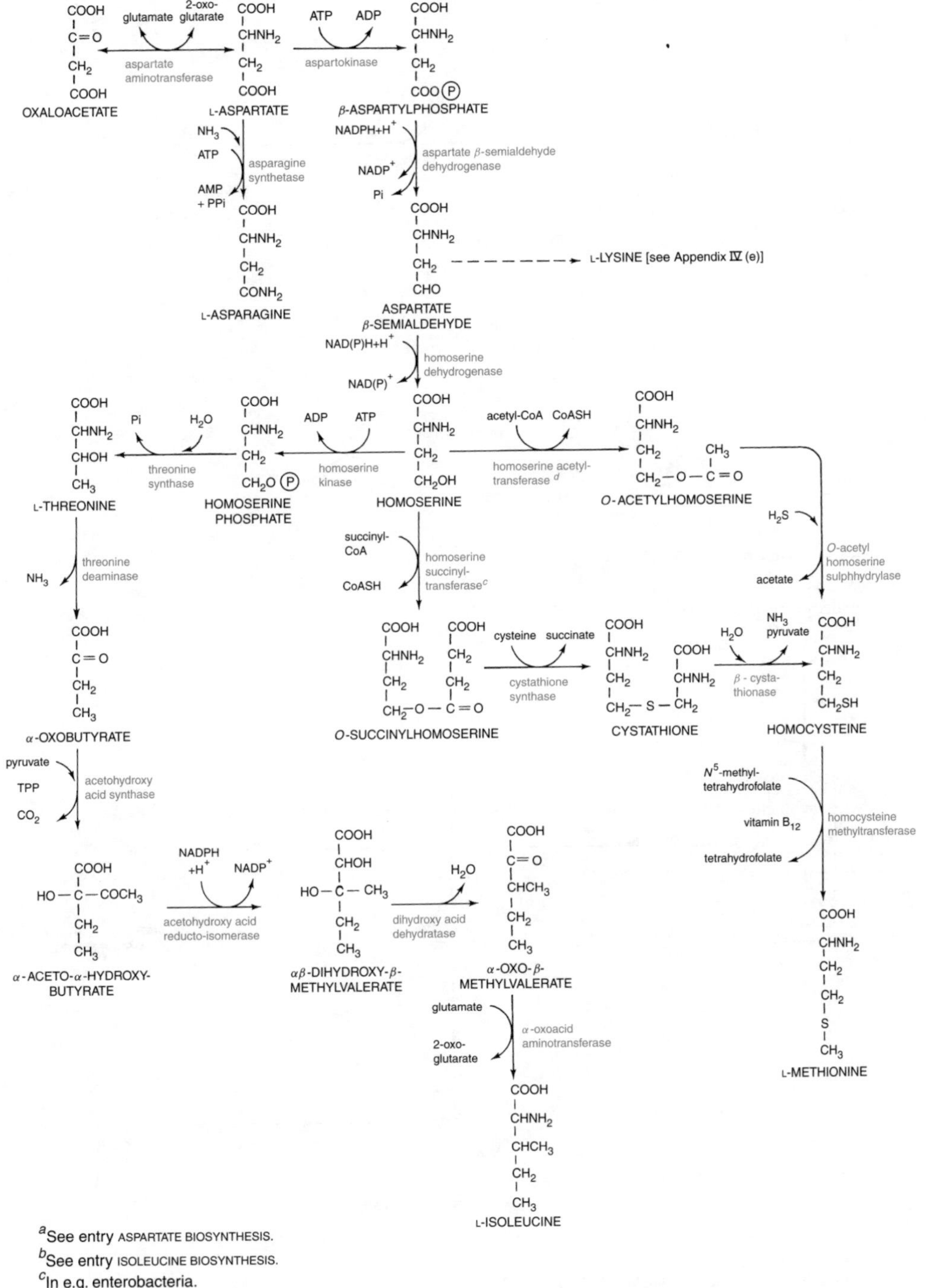

[a]See entry ASPARTATE BIOSYNTHESIS.
[b]See entry ISOLEUCINE BIOSYNTHESIS.
[c]In e.g. enterobacteria.
[d]In certain bacteria and fungi.

BIOSYNTHESIS OF LYSINE

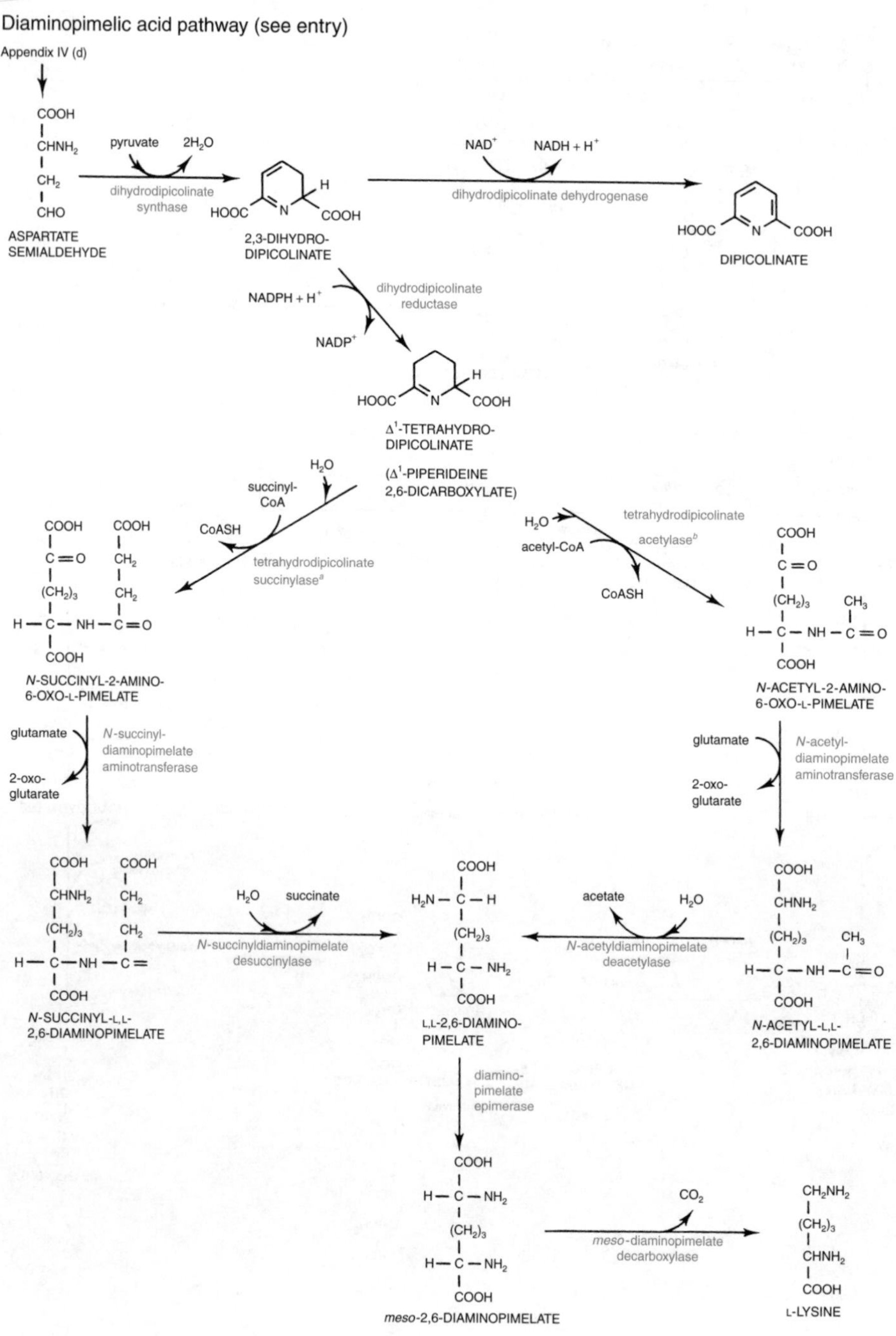

[a]In e.g. *Escherichia coli*.
[b]In e.g. *Bacillus megaterium* [JGM (1982) *128* 1073-1081].

Aminoadipic acid pathway (see entry)

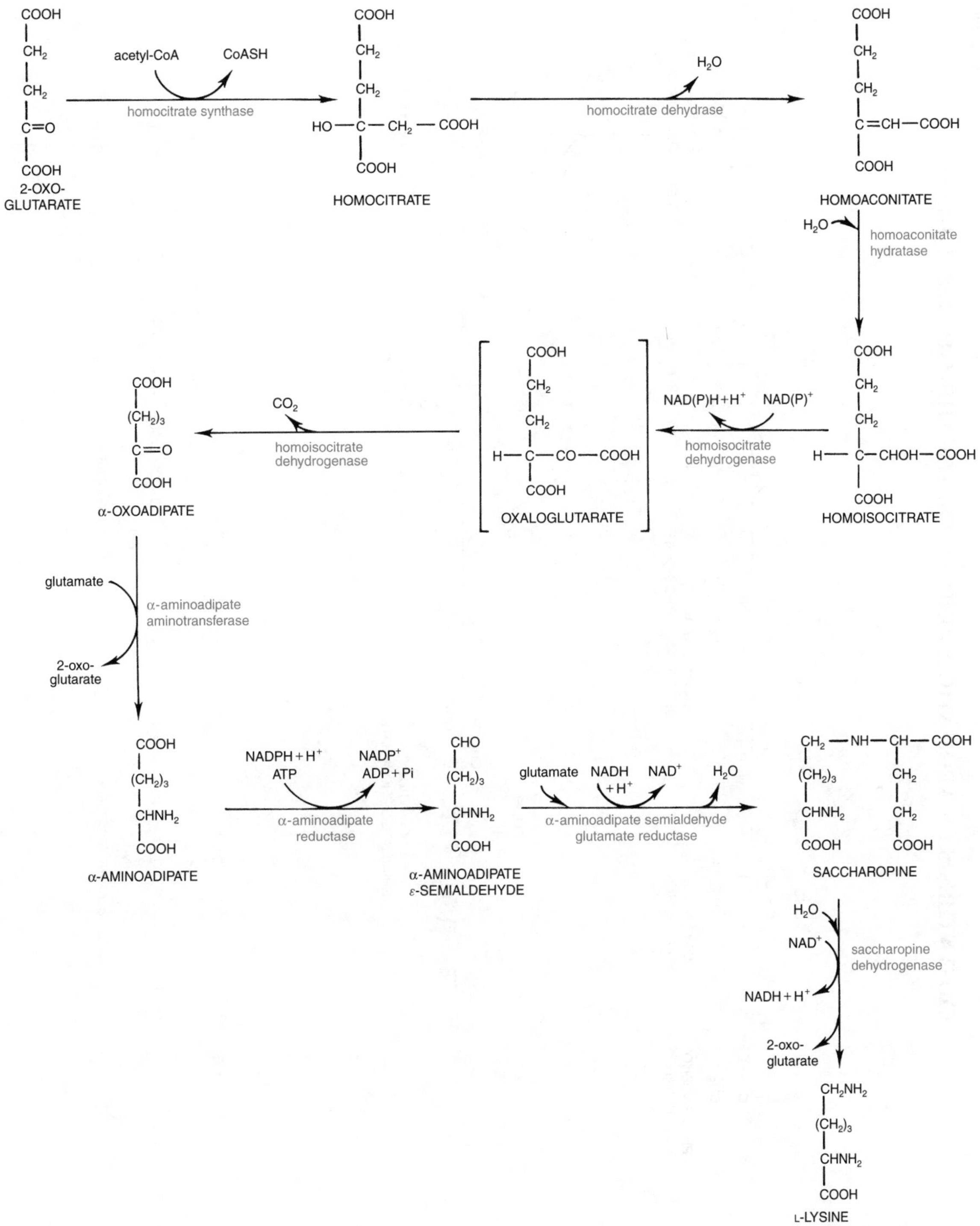

2-OXO-GLUTARATE → (acetyl-CoA → CoASH; homocitrate synthase) → HOMOCITRATE → (H_2O; homocitrate dehydrase) → HOMOACONITATE → (H_2O; homoaconitate hydratase) → HOMOISOCITRATE → ($NAD(P)^+$ → $NAD(P)H+H^+$; homoisocitrate dehydrogenase) → [OXALOGLUTARATE] → (CO_2; homoisocitrate dehydrogenase) → α-OXOADIPATE → (glutamate → 2-oxo-glutarate; α-aminoadipate aminotransferase) → α-AMINOADIPATE → ($NADPH+H^+$, ATP → $NADP^+$, ADP+Pi; α-aminoadipate reductase) → α-AMINOADIPATE ε-SEMIALDEHYDE → (glutamate, NADH $+H^+$ → NAD^+, H_2O; α-aminoadipate semialdehyde glutamate reductase) → SACCHAROPINE → (H_2O, NAD^+ → $NADH+H^+$, 2-oxo-glutarate; saccharopine dehydrogenase) → L-LYSINE

2-OXO-GLUTARATE: COOH–CH_2–CH_2–C=O–COOH

HOMOCITRATE: COOH–CH_2–CH_2–C(HO)(COOH)–CH_2–COOH

HOMOACONITATE: COOH–CH_2–CH_2–C(COOH)=CH–COOH

HOMOISOCITRATE: COOH–CH_2–CH_2–C(H)(COOH)–CHOH–COOH

OXALOGLUTARATE: COOH–CH_2–CH_2–C(H)(COOH)–CO–COOH

α-OXOADIPATE: COOH–$(CH_2)_3$–C=O–COOH

α-AMINOADIPATE: COOH–$(CH_2)_3$–$CHNH_2$–COOH

α-AMINOADIPATE ε-SEMIALDEHYDE: CHO–$(CH_2)_3$–$CHNH_2$–COOH

SACCHAROPINE: CH_2–NH–CH–COOH; CH_2–$(CH_2)_3$–$CHNH_2$–COOH; CH–CH_2–CH_2–COOH

L-LYSINE: CH_2NH_2–$(CH_2)_3$–$CHNH_2$–COOH

BIOSYNTHESIS OF AROMATIC AMINO ACIDS: THE SHIKIMATE PATHWAY[a]

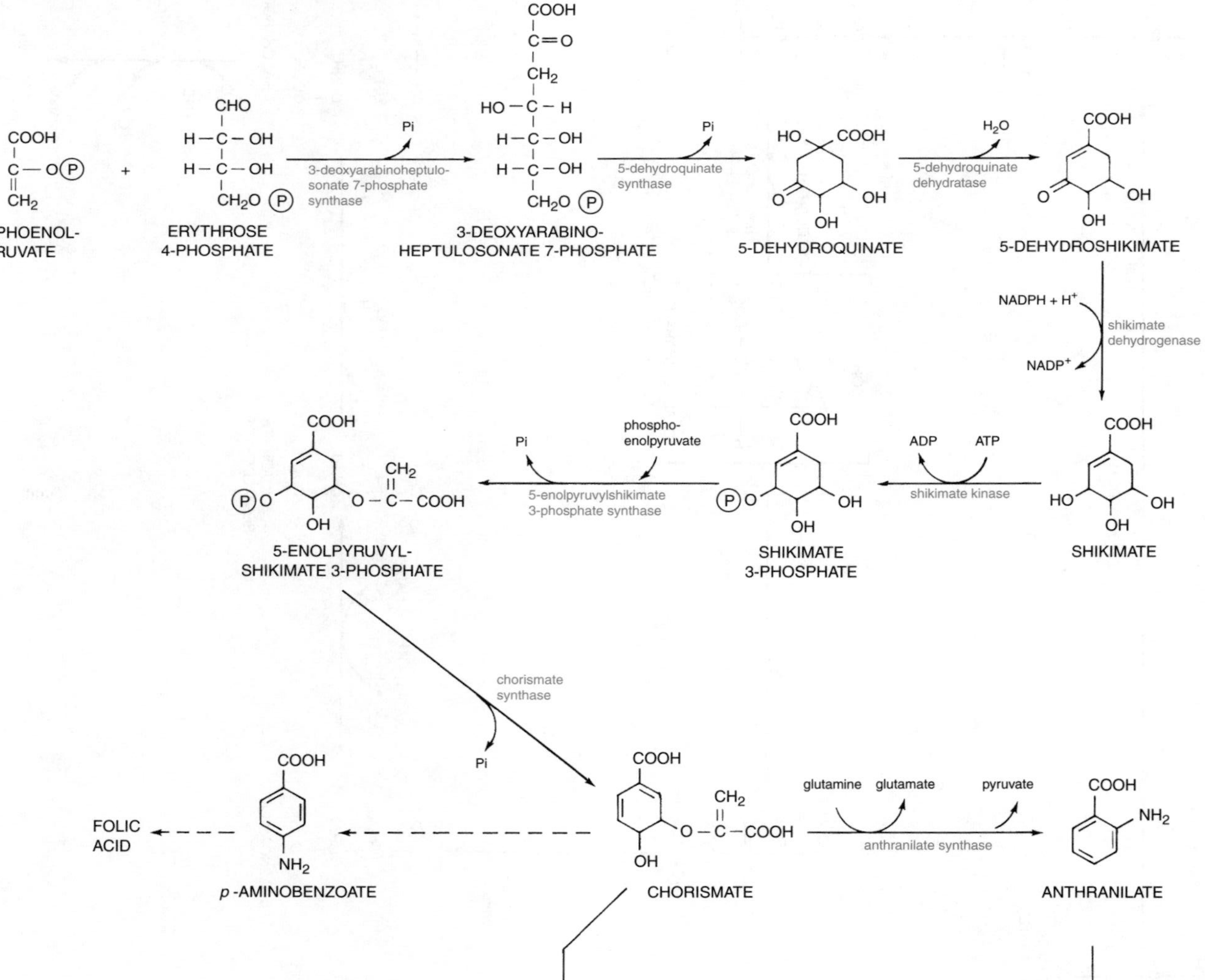

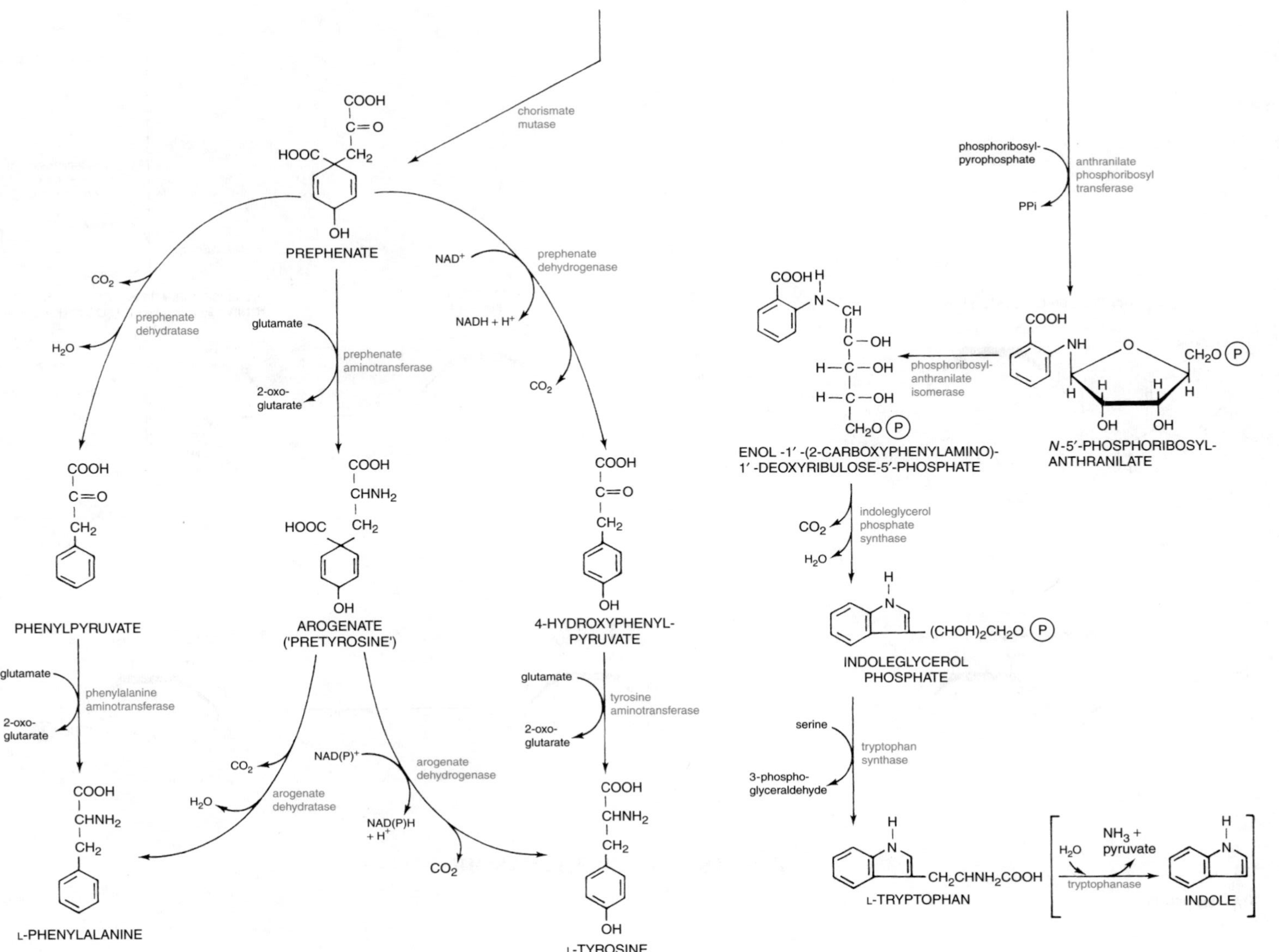

[a]See entry AROMATIC AMINO ACID BIOSYNTHESIS.

BIOSYNTHESIS OF HISTIDINE

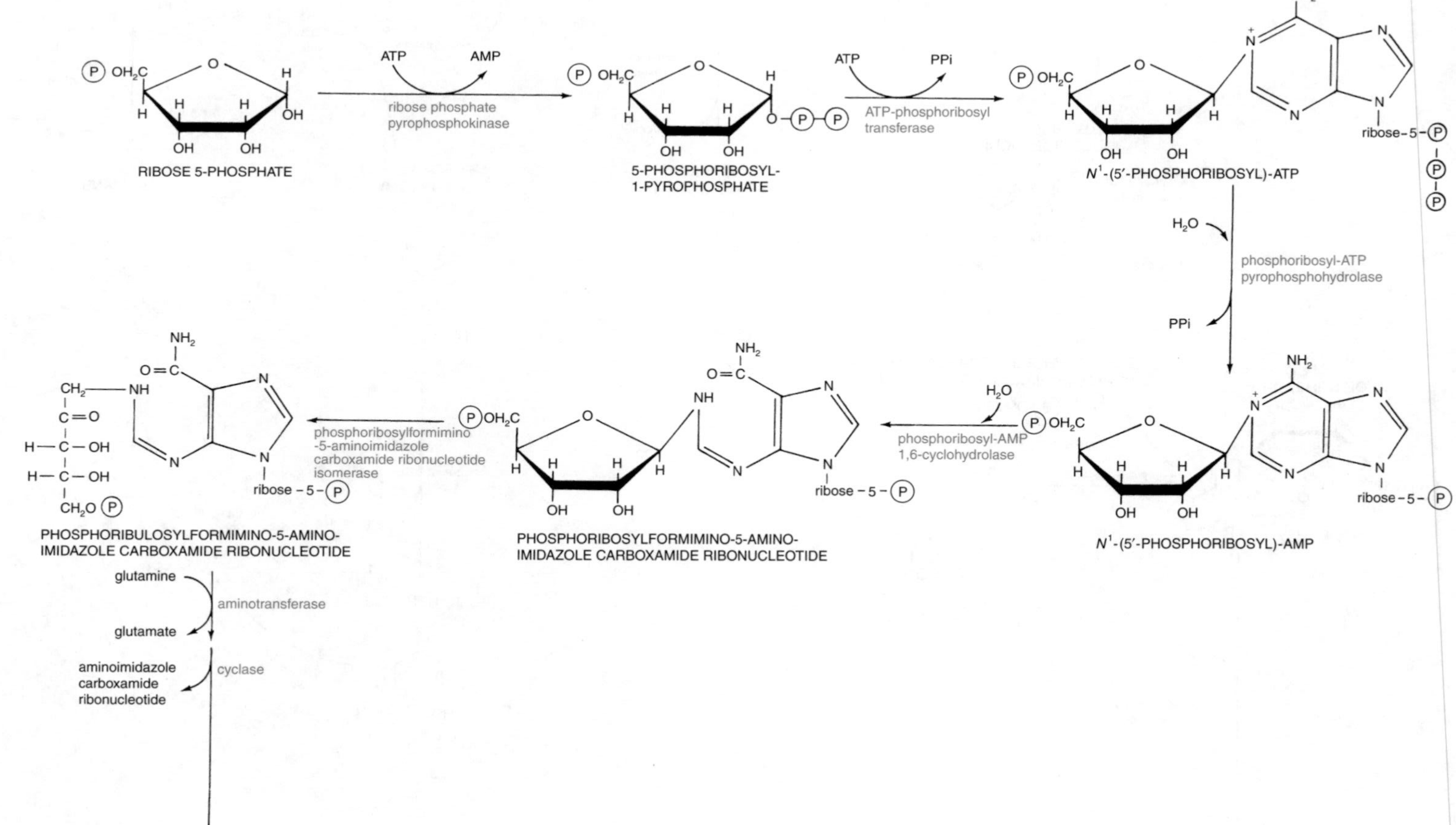

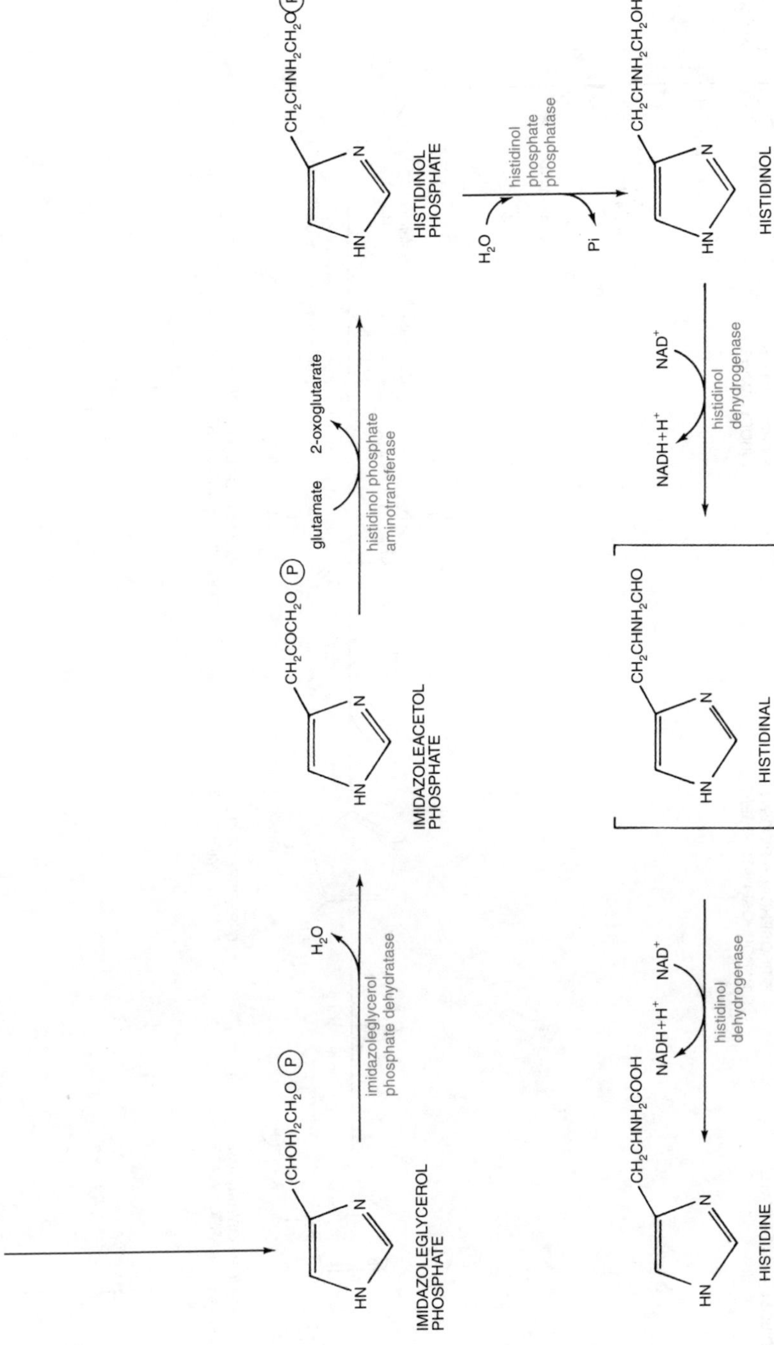

IMIDAZOLEGLYCEROL PHOSPHATE
(CHOH)2CH2O (P)
imidazoleglycerol phosphate dehydratase
H2O
IMIDAZOLEACETOL PHOSPHATE
CH2COCH2O (P)
glutamate
2-oxoglutarate
histidinol phosphate aminotransferase
HISTIDINOL PHOSPHATE
CH2CHNH2CH2O(P)
H2O
Pi
histidinol phosphate phosphatase
HISTIDINOL
CH2CHNH2CH2OH
NAD+
NADH+H+
histidinol dehydrogenase
HISTIDINAL
CH2CHNH2CHO
NAD+
NADH+H+
histidinol dehydrogenase
HISTIDINE
CH2CHNH2COOH
HN
N

BIOSYNTHESIS OF PURINE NUCLEOTIDES[a]

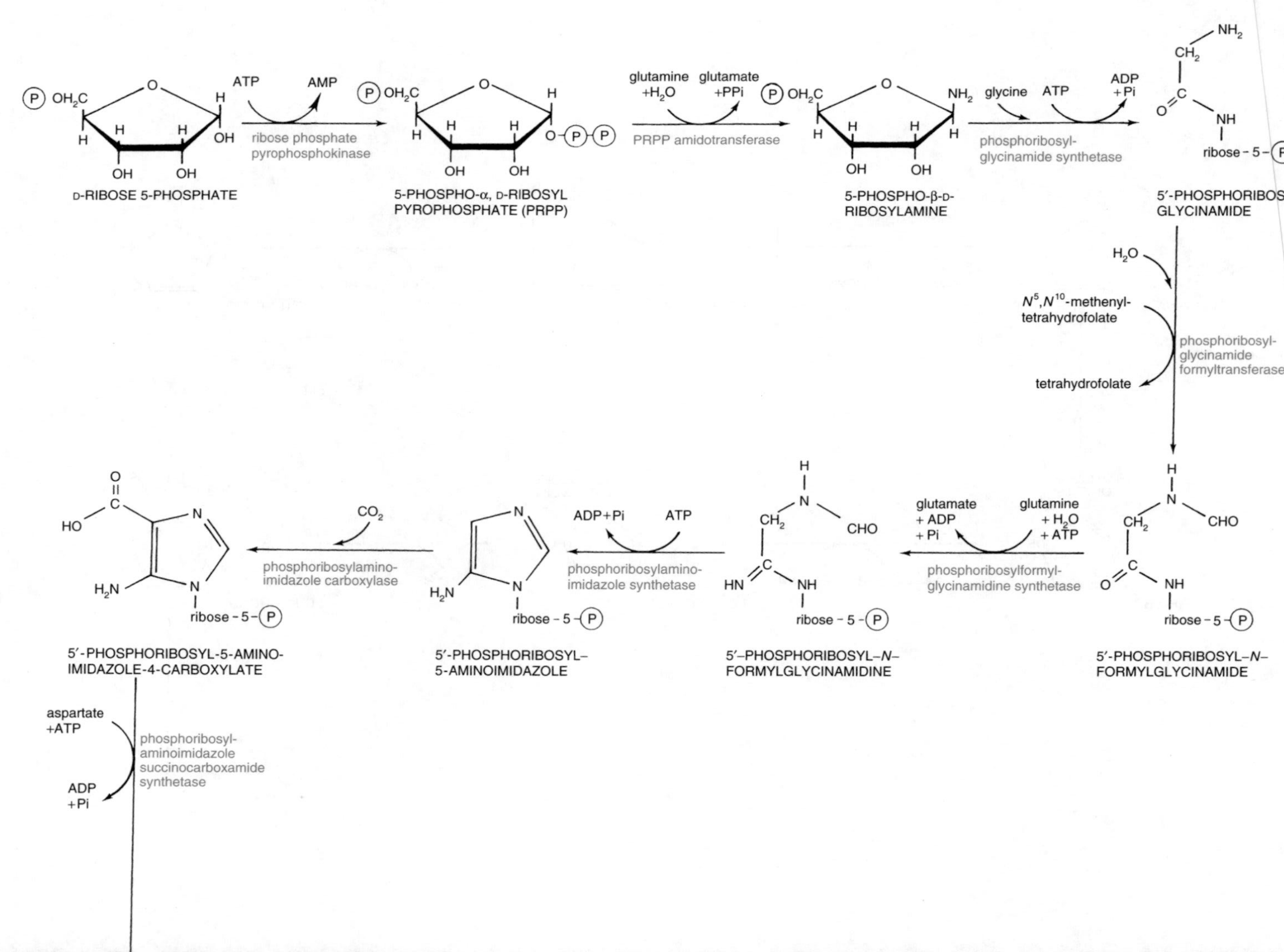

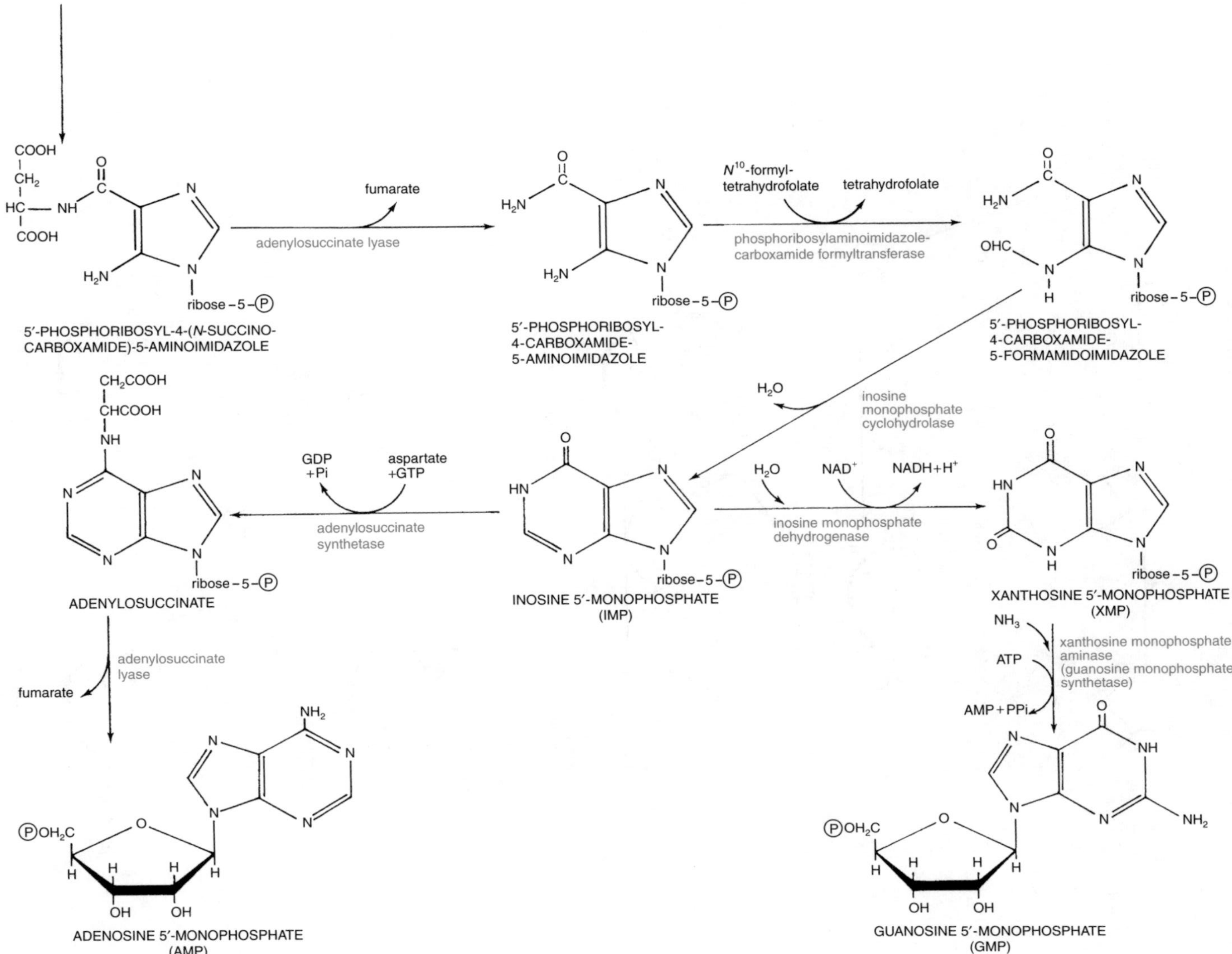
5′-PHOSPHORIBOSYL-4-(N-SUCCINO-CARBOXAMIDE)-5-AMINOIMIDAZOLE
fumarate
adenylosuccinate lyase
5′-PHOSPHORIBOSYL-4-CARBOXAMIDE-5-AMINOIMIDAZOLE
N10-formyl-tetrahydrofolate
tetrahydrofolate
phosphoribosylaminoimidazole-carboxamide formyltransferase
5′-PHOSPHORIBOSYL-4-CARBOXAMIDE-5-FORMAMIDOIMIDAZOLE
H2O
inosine monophosphate cyclohydrolase
ADENYLOSUCCINATE
GDP +Pi
aspartate +GTP
adenylosuccinate synthetase
INOSINE 5′-MONOPHOSPHATE (IMP)
H2O
NAD+
NADH+H+
inosine monophosphate dehydrogenase
XANTHOSINE 5′-MONOPHOSPHATE (XMP)
NH3
ATP
AMP+PPi
xanthosine monophosphate aminase (guanosine monophosphate synthetase)
fumarate
adenylosuccinate lyase
ADENOSINE 5′-MONOPHOSPHATE (AMP)
GUANOSINE 5′-MONOPHOSPHATE (GMP)

[a]See entry NUCLEOTIDE.

BIOSYNTHESIS OF PYRIMIDINE NUCLEOTIDES[a]

ASPARTATE + CARBAMOYL PHOSPHATE → (aspartate carbamoyltransferase; Pi) → N-CARBAMOYL ASPARTATE

N-CARBAMOYL ASPARTATE → (dihydroorotase; H_2O) → DIHYDROOROTATE

DIHYDROOROTATE → (dihydroorotate dehydrogenase (orotate reductase); NAD^+ → NADH + H^+) → OROTATE

OROTATE → (orotate phosphoribosyl transferase; phosphoribosyl-pyrophosphate → PPi) → OROTIDINE 5′-MONOPHOSPHATE (OMP)

OROTIDINE 5′-MONOPHOSPHATE (OMP) → (orotidine monophosphate decarboxylase; CO_2) → URIDINE 5′-MONOPHOSPHATE (UMP)

URIDINE 5′-MONOPHOSPHATE (UMP) → (nucleoside monophosphate kinase; ATP → ADP) → URIDINE 5′-DIPHOSPHATE (UDP)

URIDINE 5′-DIPHOSPHATE (UDP) → (nucleoside diphosphate kinase; ATP → ADP) → URIDINE 5′-TRIPHOSPHATE (UTP)

URIDINE 5′-TRIPHOSPHATE (UTP) → (cytidine triphosphate synthetase; NH_3, ATP → ADP + Pi) → CYTIDINE 5′-TRIPHOSPHATE (CTP)

URIDINE 5′-MONOPHOSPHATE (UMP) ⇢ DEOXYURIDINE 5′-MONOPHOSPHATE (dUMP)

DEOXYURIDINE 5′-MONOPHOSPHATE (dUMP) → (thymidylate synthase; N^5,N^{10}-methylene-tetrahydrofolate → dihydrofolate) → DEOXYTHYMIDINE 5′-MONOPHOSPHATE (dTMP)

[a]See entry NUCLEOTIDE.

THE GREEK ALPHABET

Α	α	alpha	Ν	ν	nu
Β	β	beta	Ξ	ξ	xi
Γ	γ	gamma	Ο	ο	omicron
Δ	δ	delta	Π	π	pi
Ε	ε	epsilon	Ρ	ρ	rho
Ζ	ζ	zeta	Σ	σ	sigma
Η	η	eta	Τ	τ	tau
Θ	θ	theta	Υ	υ	upsilon
Ι	ι	iota	Φ	ϕ	phi
Κ	κ	kappa	Χ	χ	chi
Λ	λ	lambda	Ψ	ψ	psi
Μ	μ	mu	Ω	ω	omega

Key to Journal Title Abbreviations

AAC	*Antimicrobial Agents and Chemotherapy*
AAM	*Advances in Applied Microbiology*
ABC	*Agricultural and Biological Chemistry*
Acta Tropica	*Acta Tropica*
ADC	*Archives of Disease in Childhood*
Adv. Coll. Int. Sci.	*Advances in Colloid and Interface Science*
Adv. Gen.	*Advances in Genetics*
Adv. Imm.	*Advances in Immunology*
Adv. Vet. Med.	*Advances in Veterinary Medicine* (Supplements of *Journal of Veterinary Medicine*)
AEE	*Agriculture, Ecosystems and Environment*
AEM	*Applied and Environmental Microbiology* (formerly *Applied Microbiology*)
AIM	*Annals of Internal Medicine*
AJCP	*American Journal of Clinical Pathology*
AJEBMS	*Australian Journal of Experimental Biology and Medical Science*
AJP	*American Journal of Pathology*
AJRIM	*American Journal of Reproductive Immunology and Microbiology*
AJVR	*American Journal of Veterinary Research*
AM	*Applied Microbiology* (cf. AEM)
AMB	*Advances in Marine Biology*
Am. J. Med.	*American Journal of Medicine*
Am. J. Path.	*American Journal of Pathology*
AMP	*Advances in Microbial Physiology*
Ann. Appl. Biol.	*Annals of Applied Biology*
Ann. Mic.	*Annales de Microbiologie (Institut Pasteur)* to 1984, renamed *Annales de l'Institut Pasteur/Microbiologie* in 1985
Ann. Vir.	*Annales de Virologie (Institut Pasteur)* to 1984, renamed *Annales de l'Institut Pasteur/Virologie* in 1985
Antiviral Res.	*Antiviral Research*
AP	*Advances in Parasitology*
ARB	*Annual Review of Biochemistry*
ARBB	*Annual Review of Biophysics and Bioengineering*
ARCB	*Annual Review of Cell Biology*
Arch. Derm.	*Archives of Dermatology*
Arch. Gen. Psych.	*Archives of General Psychiatry*
Arch. Int. Med.	*Archives of Internal Medicine*
Arch. Micro.	*Archives of Microbiology*
Arch. Virol.	*Archives of Virology*
ARE	*Annual Review of Entomology*
ARG	*Annual Review of Genetics*
ARM	*Annual Review of Microbiology*
ARMed.	*Annual Review of Medicine*
ARP	*Annual Review of Physiology*
ARPpath.	*Annual Review of Phytopathology*
ARPphys.	*Annual Review of Plant Physiology*
ARPT	*Annual Review of Pharmacology and Toxicology*
ASM News	*American Society for Microbiology News*
Australian Vet. J.	*Australian Veterinary Journal*
AvL	*Antonie van Leeuwenhoek*
AVR	*Advances in Virus Research*
BAB	*Biotechnology and Bioengineering*
Bact. Rev.	*Bacteriological Reviews* (cf. MR)
BBA	*Biochimica et Biophysica Acta*
BBMS	*Bulletin of the British Mycological Society*
BBRC	*Biochemical and Biophysical Research Communications*

Key to Journal Title Abbreviations

BC	*Biodiversity and Conservation*
BCEM	[formerly *Baillière's*, now *Best Practice & Research*] *Clinical Endocrinology & Metabolism*
BCG	[formerly *Baillière's*, now *Best Practice & Research*] *Clinical Gastroenterology*
BCH	[formerly *Baillière's*, now *Best Practice & Research*] *Clinical Haematology*
BCID	*Baillière's Clinical Infectious Diseases*
BCP	[formerly *Baillière's*, now *Best Practice & Research*] *Clinical Paediatrics*
Biochem.	*Biochemistry*
Bioch. J.	*Biochemical Journal*
Biol. Bull.	*Biological Bulletin*
Biol. Rev.	*Biological Reviews*
BIP	*Bulletin de l'Institut Pasteur*
BJCP	*British Journal of Clinical Pharmacology*
BJVD	*British Journal of Venereal Diseases*
Blood	*Blood*
BMB	*British Medical Bulletin*
BMJ	*British Medical Journal*
Bot. Gaz.	*Botanical Gazette*
Bot. J. Lin. Soc.	*Botanical Journal of the Linnean Society*
Bot. Rev.	*Botanical Review*
Br. J. Ophth.	*British Journal of Ophthalmology*
Br. J. Surg.	*British Journal of Surgery*
Bryol.	*Bryologist*
BWHO	*Bulletin of the World Health Organization*
CC	*Clinical Chemistry*
Cell	*Cell*
Cell Mot.	*Cell Motility*, renamed *Cell Motility and the Cytoskeleton* in 1986
Chest	*Chest*
CIA	*Clinics in Immunology and Allergy*
CJB	*Canadian Journal of Botany*
CJFAS	*Canadian Journal of Fisheries and Aquatic Sciences* (formerly *Journal of the Fisheries Research Board of Canada*)
CJM	*Canadian Journal of Microbiology*
CMI	*Clinical Microbiology and Infection*
COGD	*Current Opinion in Genetics and Development*
CRB	*CRC Critical Reviews in Biochemistry*
CRM	*CRC Critical Reviews in Microbiology*
CSHSQB	*Cold Spring Harbor Symposia on Quantitative Biology*
CTCR	*Current Topics in Cell Regulation*
CTMI	*Current Topics in Microbiology and Immunology*
CTMT	*Current Topics in Membranes and Transport*
Curr. Micro.	*Current Microbiology*
Drugs	*Drugs*
Econ. Bot.	*Economic Botany*
EID	*Emerging Infectious Diseases*
EMBO	*EMBO (European Molecular Biology Organization) Journal*
Epidem. Inf.	*Epidemiology and Infection* (formerly *Journal of Hygiene*)
Eur. J. Bioch.	*European Journal of Biochemistry*
Experientia	*Experientia*
FEBS	*FEBS Letters*
Fed. Proc.	*Federation Proceedings*
FEMS	*FEMS Letters*
FEMS Ecol.	*FEMS Microbiology Ecology*
FEMS Imm.	*FEMS Immunology and Medical Microbiology*
FEMS Reviews	*FEMS Microbiology Reviews*
Food Mic.	*Food Microbiology*
GD	*Genes and Development*
GE	*Gastroenterology*

Gene	*Gene*
Genetics	*Genetics*
Genet. Res.	*Genetical Research*
Gut	*Gut*
HJ	*Histochemical Journal*
IJP	*International Journal for Parasitology*
IJSB	*International Journal of Systematic Bacteriology*
Imm. Rev.	*Immunological Reviews*
Inf. Immun.	*Infection and Immunity*
Intervirol.	*Intervirology*
Int. Rev. Cytol.	*International Review of Cytology*
Israel J. Bot.	*Israel Journal of Botany*
Israel J. Med. Sci.	*Israel Journal of Medical Sciences*
IT	*Immunology Today*
JAB	*Journal of Applied Bacteriology*
JAC	*Journal of Antimicrobial Chemotherapy*
JAM	*Journal of Applied Microbiology*
JAMA	*Journal of the American Medical Association*
J. Antibiot.	*Journal of Antibiotics*
JB	*Journal of Bacteriology*
JBB	*Journal of Bioenergetics and Biomembranes*
JBC	*Journal of Biological Chemistry*
JBM	*Journal of Basic Microbiology* (Berlin)
JCB	*Journal of Cell Biology*
JCBiochem.	*Journal of Cellular Biochemistry*
JCI	*Journal of Clinical Investigation*
JCM	*Journal of Clinical Microbiology*
JCP	*Journal of Clinical Pathology*
JCS	*Journal of Cell Science*
JEMBE	*Journal of Experimental Marine Biology and Ecology*
J. Exp. Med.	*Journal of Experimental Medicine*
J. Fish Dis.	*Journal of Fish Diseases*
J. Food Protect.	*Journal of Food Protection*
JFRBC	*Journal of the Fisheries Research Board of Canada* (cf. CJFAS)
JGM	*Journal of General Microbiology*
JGP	*Journal of General Physiology*
JGV	*Journal of General Virology*
J. Hyg.	*Journal of Hygiene* (cf. Epidem. Inf.)
JID	*Journal of Infectious Diseases*
JIM	*Journal of Immunological Methods*
J. Imm.	*Journal of Immunology*
JINF	*Journal of Infection*
J. Inst. Wat. Eng. Scient.	*Journal of the Institute of Water Engineers and Scientists*
J. Inv. Path.	*Journal of Invertebrate Pathology*
JLB	*Journal of Leukocyte Biology*
JM	*Journal of Microscopy*
JMB	*Journal of Molecular Biology*
JMM	*Journal of Medical Microbiology*
JMV	*Journal of Medical Virology*
J. Nat. Prod.	*Journal of Natural Products* (formerly *Lloydia*)
JP	*Journal of Protozoology*
J. Parasitol.	*Journal of Parasitology*
JPed.	*Journal of Pediatrics*
J. Phycol.	*Journal of Phycology*
JSB	*Journal of Strucutural Biology*
J. Sed. Pet.	*Journal of Sedimentary Petrology*
J. Theor. Biol.	*Journal of Theoretical Biology*

Key to Journal Title Abbreviations

JUR	*Journal of Ultrastructure Research*
JV	*Journal of Virology*
JVM	*Journal of Virological Methods*
Lab. Inv.	*Laboratory Investigation*
LAM	*Letters in Applied Microbiology*
Lancet	*Lancet*
Lichenol.	*Lichenologist*
Limn. Ocean.	*Limnology and Oceanography*
Mayo Clin. Proc.	*Mayo Clinic Proceedings*
MCB	*Molecular and Cellular Biology*
ME	*Microbial Ecology*
MGG	*Molecular and General Genetics*
Microbiology	*Microbiology* (formerly *Journal of General Microbiology*)
MJA	*Medical Journal of Australia*
Mol. Immunol.	*Molecular Immunology*
Mol. Microbiol.	*Molecular Microbiology*
MP	*Microbial Pathogenesis*
MR	*Microbiological Reviews* (formerly *Bacteriological Reveiws*)
MS	*Microbiological Sciences*
Mut. Res.	*Mutation Research*
Mycol.	*Mycologia*
Mycol. Pap.	*Mycological Paper* (a numbered series published by the Commonwealth Mycological Institute, Kew, UK)
NAR	*Nucleic Acids Research*
Nature	*Nature*
Nature Biotech.	*Nature Biotechnology*
Nature Medicine	*Nature Medicine*
Nature Genetics	*Nature Genetics*
Naturwissenschaften	*Naturwissenschaften*
NEJM	*New England Journal of Medicine*
Nematol.	*Nematologica*
New Phyt.	*New Phytologist*
NJPP	*Netherlands Journal of Plant Pathology*
Obstet. Gynec.	*Obstetrics and Gynecology*
PAC	*Pure and Applied Chemistry* (IUPAC)
Parasitol.	*Parasitology*
Pediatrics	*Pediatrics*
Ped. Inf. Dis.	*Pediatric Infectious Disease*
PHLS Digest	*PHLS Microbiology Digest* (a publication of the Public Health Laboratory Service, UK)
Photochem. Photobiol.	*Photochemistry and Photobiology*
Physiol. Plant.	*Physiologia Plantarum*
Phytopath.	*Phytopathology*
Phytopath. Paper	*Phytopathological Paper* (a numbered series published by the Commonwealth Mycological Institute, Kew, UK)
PJ	*Pharmaceutical Journal*
Plasmid	*Plasmid*
PM	*Postgraduate Medicine*
PMB	*Plant Molecular Biology*
PNARMB	*Progress in Nucleic Acid Research and Molecular Biology*
PNAS	*Proceedings of the National Academy of Sciences of the USA*
PP	*Plant Pathology*
PPP	*Physiological Plant Pathology*
Proc. RSE	*Proceedings of the Royal Society of Edinburgh*
Proc. RSLB	*Proceedings of the Royal Society of London*, Series B
Prog. Allergy	*Progress in Allergy*
Prog. Biophys. Mol. Biol.	*Progress in Biophysics and Molecular Biology*
Prog. Med. Virol.	*Progress in Medical Virology*
Prog. Mol. Subcell. Biol.	*Progress in Molecular and Subcellular Biology*

PT	*Parasitology Today* (renamed *Trends in Parasitology*)
PTRSLB	*Philosophical Transactions of the Royal Society of London*, Series B
QRB	*Quarterly Review of Biology*
RID	*Reviews of Infectious Diseases*
SAAM	*Systematic and Applied Microbiology*
SAB	*Sabouraudia: Journal of Medical and Veterinary Mycology* (renamed *Journal of Medical and Veterinary Mycology* in 1986)
SBB	*Soil Biology and Biochemistry*
Sci. Am.	*Scientific American*
Science	*Science*
Sci. Prog.	*Science Progress*
SEBS	*Society for Experimental Biology Symposia*
Surgery	*Surgery*
Taxon	*Taxon*
TBMS	*Transactions of the British Mycological Society*
TDB	*Tropical Diseases Bulletin*
TIBS	*Trends in Biochemical Sciences*
TIBtech.	*Trends in Biotechnology*
TIG	*Trends in Genetics*
TIM	*Trends in Microbiology*
TIP	*Trends in Parasitology* (formerly *Parasitology Today*)
TRSTMH	*Transactions of the Royal Society of Tropical Medicine and Hygiene*
VA	*Virchows Archiv*
Virol.	*Virology*
VR	*Veterinary Record*
Wat. Res.	*Water Research*
WRB	*Water Resources Bulletin*
Yeast	*Yeast*
YJBM	*Yale Journal of Biology and Medicine*
Zbl. Bakt. Hyg. A.	*Zentralblatt für Bakteriologie Mikrobiologie und Hygiene: Series A: Medical Microbiology, Infectious Diseases, Virology, Parasitology*

Key to Book References

1. *Microbial Diseases of Fish* (Society for General Microbiology Special Publication No. 9), Roberts, R. J. (ed.); Academic Press, 1982.
2. *Manual of Methods for General Bacteriology*, Gerhardt, P. et al (eds); American Society for Microbiology, 1981.
3. *Advances in Biotechnological Processes*, Volume I. Mizrahi, A. & van Wezel, A. L. (eds); Alan R. Liss, 1983.
4. *Electron Microscopy and Cytochemistry of Plant Cells*, Hall, J. L. (ed.); Elsevier, 1978.
5. *Fermented Foods* (Economic Microbiology Volume 7), Rose, A. H. (ed.); Academic Press, 1982.
6. *Microbial Interactions and Communities*, Volume 1. Bull, A. T. & Slater, J. H. (eds); Academic Press, 1982.
7. *Problems in the Identification of Parasites and their Vectors* (Symposia of the British Society for Parasitology, Volume 17). Taylor, A. E. R. & Muller, R. (eds); Blackwell Scientific Publications, 1979.
8. *Prescott and Dunn's Industrial Microbiology*, 4th edition. Reed, G. (ed.); Macmillan (UK) and AVI Publishing Co. (USA), 1982.
9. *Diseases of Fishes*, Book 2 (bacterial diseases), Snieszko, S. F. & Axelrod, H. R. (eds); T. F. H. Publications (New Jersey, USA), 1971.
10. *Diseases of Fishes*, Book 6: *Fungal Diseases of Fishes*, Neish, G. A. & Hughes, G. C. (series eds Snieszko, S. F. & Axelrod, H. R.); T. F. H. Publications (New Jersey, USA), 1980.
11. *Microbial Technology: Current State, Future Prospects* (29th Symposium of the Society for General Microbiology), Bull, A. T., Ellwood, D.C. & Ratledge, C. (eds); Cambridge University Press, 1979.
12. *Principles and Practice of Disinfection, Preservation and Sterilization*, Russell, A. D., Hugo, W. B. & Ayliffe, G. A. J. (eds); Blackwell Scientific Publications, 1982.
13. *Disinfectants: Their Use and Evaluation of Effectiveness* (Society for Applied Bacteriology Technical Series No. 16), Collins, C. H., Allwood, M. C., Bloomfield, S. F. & Fox, A. (eds); Academic Press, 1981.
14. *The Molecular Basis of Antibiotic Action*, 2nd edition. Gale, E. F., Cundliffe, E., Reynolds, P. E., Richmond, M. H. & Waring, M. J.; John Wiley, 1981.
15. *Trichothecenes: Chemical, Biological and Toxicological Aspects*, Ueno, Y. P. (ed.); Elsevier, 1983.
16. *Mycotoxins*, Purchase, I. F. H. (ed.); Elsevier, 1974.
17. *Communicable and Infectious Diseases*, 9th edition. Wehrle, P. F. & Top, F. H. (eds); C. V. Mosby, 1981.
18. *The Biology of the Coccidia*, Long, P. L. (ed.); Edward Arnold, 1982.
19. *Cowan and Steele's Manual for the Identification of Medical Bacteria*, 2nd edition, revised by Cowan, S. T.; Cambridge University Press, 1974.
20. *Mobile Genetic Elements*, Shapiro, J. A. (ed.); Academic Press, 1983.
21. *Bergey's Manual of Determinative Bacteriology*, 8th edition. Buchanan, R. E. & Gibbons, N. E. (eds); Williams & Wilkins, 1974.
22. *Bergey's Manual of Systematic Bacteriology*, Volume 1. Krieg, N. R. (ed.); Williams & Wilkins, 1984.
23. *Classification and Nomenclature of Viruses* (4th Report of the International Committee on Taxonomy of Viruses, 1982, reprinted from *Intervirology* (1982) *17* Nos. 1–3), Matthews, R. E. F. (ed.); Karger, 1982.
24. *Legionella Infections*, Bartlett, C. L. R., Macrae, A. D. & Macfarlane, J. T.; Edward Arnold, 1986.
25. *Coccidioidomycosis: A Text*, Stevens, D. A.; Plenum, 1980.
26. *Trends in the Biology of Fermentations for Fuels and Chemicals*, Hollaender, A. et al (eds); Plenum Press, 1981.
27. *Aujeszky's Disease* (Current Topics in Veterinary Medicine and Animal Science, Volume 17), Wittmann, G. & Hall, S. A. (eds); Martinus Nijhoff, 1982.
28. *Microbes in their Natural Environments* (34th Symposium of the Society for General Microbiology), Slater, J. H., Whittenbury, R. & Wimpenny, J. W. T. (eds); Cambridge University Press, 1983.
29. *Psychrotrophic Microorganisms in Spoilage and Pathogenicity*, Roberts, T. A., Hobbs, G., Christian, J. H. B. & Skovgaard, N. (eds); Academic Press, 1981.
30. *Meat Microbiology*, Brown, M. H. (ed.); Applied Science Publishers, 1982.
31. *Microbial Enzymes and Bioconversions* (Economic Microbiology, Volume 5), Rose, A. H. (ed.); Academic Press, 1980.
32. *The Photosynthetic Bacteria*, Clayton, R. K. & Sistrom, W. R. (eds); Plenum Press, 1978.
33. *Veterinary Medicine*, 6th edition. Blood, D. C., Radostits, O. M. & Henderson, J. A.; Baillière Tindall, 1983.
34. *Developmental Biology of Prokaryotes*, Parish, J. H. (ed.); Blackwell Scientific Publications, 1979.
35. *Canned Foods: Thermal Processing and Microbiology*, 7th edition. Hersom, A. C. & Hulland, E. D.; Churchill Livingstone, 1980.
36. *Isolation and Identification Methods for Food Poisoning Organisms* (Society for Applied Bacteriology Technical Series No. 17), Corry, J. F. L., Skinner, F. A. & Roberts, D. (eds); Academic Press, 1982.
37. *Plant Carbohydrates I – Intracellular Carbohydrates* (Encyclopedia of Plant Physiology: New Series Volume 13A), Loeuws, F. A. & Tanner, W. (eds); Springer-Verlag, 1982.
38. *Plant Carbohydrates II – Extracellular Carbohydrates* (Encyclopedia of Plant Physiology: New Series Volume 13B), Tanner, W. & Loeuws, F. A. (eds); Springer-Verlag, 1981.
39. *Decomposer Basidiomycetes: their Biology and Ecology*, Frankland, J. C., Hedger, J. N. & Swift, M. J. (eds); Cambridge University Press, 1982.
40. *Genetics as a Tool in Microbiology* (31st Symposium of the Society for General Microbiology), Glover, S. W. & Hopwood, D. A. (eds); Cambridge University Press, 1981.
41. *The Direct Epifluorescent Filter Technique for the Rapid Enumeration of Microorganisms*, Pettipher, G. L.; Research Studies Press/John Wiley, 1983.

Key to Book References

42. *Fundamental Immunology*, Paul, W. E. (ed.); Raven Press, 1984.
43. *Staphylococci and Staphylococcal Infections*, Volume 1: *Clinical and Epidemiological Aspects*, Easmon, C. S. F. & Adlam, C. (eds); Academic Press, 1983.
44. *Staphylococci and Staphylococcal Infections*, Volume 2: *The Organism in Vivo and in Vitro*, Easmon, C. S. F. & Adlam, C. (eds); Academic Press, 1983.
45. *The Prokaryotes: A Handbook on Habitats, Isolation and Identification of Bacteria*, Volume I. Starr, M. P. et al (eds); Springer-Verlag, 1981.
46. *The Prokaryotes: A Handbook on Habitats, Isolation and Identification of Bacteria*, Volume II. Starr, M. P. et al (eds); Springer-Verlag, 1981.
47. *Manual of Clinical Microbiology*, Blair, J. E., Lennette, E. H. & Truant, J. P. (eds); American Society for Microbiology/Williams & Wilkins, 1970.
48. *Interferons: from Molecular Biology to Clinical Application* (35th Symposium of the Society for General Microbiology), Burke, D. C. & Morris, A. G. (eds); Cambridge University Press, 1983.
49. *Vibrios in the Environment*, Colwell, R. R. (ed.); John Wiley, 1984.
50. *Food Microbiology* (Economic Microbiology, Volume 8), Rose, A. H. (ed.); Academic Press, 1983.
51. *Mould Allergy*, Al-Doory, Y. & Domson, J. F. (eds); Lea & Febiger, 1984.
52. *Gradwohl's Clinical Laboratory Methods and Diagnosis*, 8th edition, Volume 1. Sonnenwirth, A. C. & Jarett, L. (eds); C. V. Mosby, 1980.
53. *Gradwohl's Clinical Laboratory Methods and Diagnosis*, 8th edition, Volume 2. Sonnenwirth, A. C. & Jarett, L. (eds); C. V. Mosby, 1980.
54. *The Biology of the Mycobacteria*, Volume 1: *Physiology, Identification and Classification*, Ratledge, C. & Stanford, J. (eds); Academic Press, 1982.
55. *Genes Involved in Microbe–Plant Interactions*, Verma, D. P. S. & Hohn, T. (eds); Springer-Verlag, 1984.
56. *Mycorrhizal Symbiosis*, Harley, H. L. & Smith, S. E.; Academic Press, 1983.
57. *Aerobiology*, Edmonds, R. L. (ed.); Dowden, Hutchinson & Ross, 1979.
58. *Biochemical Plant Pathology*, Callow, J. A. (ed.); John Wiley, 1983.
59. *Microbiological Classification and Identification*, Goodfellow, M. & Board, R. G. (eds); Academic Press, 1980.
60. *Gene Function in Prokaryotes*, Beckwith, J., Davies, J. & Gallant, J. A. (eds); Cold Spring Harbor Laboratory, 1983.
61. *A Guide to Identifying and Classifying Yeasts*, Barnett, J. A., Payne, R. W. & Yarrow, D.; Cambridge University Press, 1979.
62. *Progress in Industrial Microbiology*, Volume 18: *Microbial Polysaccharides*, Bushell, M. E. (ed.); Elsevier, 1983.
63. *The Aerobic Endospore-forming Bacteria: Classification and Identification*, Berkeley, R. C. W. & Goodfellow, M. (eds.); Academic Press, 1981.
64. *Ainsworth and Bisby's Dictionary of the Fungi*, 7th edition. Hawksworth, D. et al; Commonwealth Mycological Institute (Kew, Surrey, UK), 1983.
65. *Disinfection, Sterilization and Preservation*, 3rd edition. Block, S. S. (ed.); Lea and Febiger, 1983.
66. *Honeybee Pathology*, Bailey, L.; Academic Press, 1981.
67. *Microbial Development*, Losick, R. & Shapiro, L. (eds); Cold Spring Harbor Laboratory, 1984.
68. *Methods in Microbiology*, Volume 14. Bergan, T. (ed.); Academic Press, 1984.
69. *Mechanisms of DNA Replication and Recombination* (Proceedings of a UCLA Symposium, Keystone, Colorado, 1983), Cozzarelli, N. R. (ed.); Alan R. Liss, 1983.
70. *Ecological Aspects of Used-water Treatment*, Volume 2: *Biological Activities and Treatment Processes*, Curds, C. R. & Hawkes, H. A. (eds); Academic Press, 1983.
71. *International Code of Nomenclature of Bacteria* (*Bacteriological Code, 1976 Revision*), Lapage, S. P. et al (eds); American Society for Microbiology, 1976.
72. *The Biology of Trypanosoma and Leishmania*, Molyneux, D. H. & Ashford, R. W.; Taylor & Francis, 1983.
73. *The Biology of the Actinomycetes*, Goodfellow, M., Mordarski, M. & Williams, S. T. (eds); Academic Press, 1983.
74. *Interferon*, Volume 1: *General and Applied Aspects*, Billiau, A. (ed.), and Volume 2: *Interferon and the Immune System*, Vilček, J. & de Maeyer, E. (eds); Elsevier, 1984.
75. *Photosynthetic Prokaryotes: Cell Differentiation and Function*, Papageorgiou, G. C. & Packer, L. (eds); Elsevier Biomedical, 1983.
76. *The Biology of Cyanobacteria*, Carr, N. G. & Whitton, B. A. (eds); Blackwell Scientific Publications, 1982.
77. *Invertebrate–Microbial Interactions*, Anderson, J. M., Rayner, A. D. M. & Walton, D. W. H. (eds); Cambridge University Press, 1984.
78. *Pathogenesis and Immunology of Treponemal Infection*, Schell, R. F. & Musher, D. M. (eds); Marcel Dekker, 1983.
79. *Lambda II*, Hendrix, R. W. et al (eds); Cold Spring Harbor Laboratory, 1983.
80. *Plant Infectious Agents: Viruses, Viroids, Virusoids and Satellites* (Current Communications in Molecular Biology), Robertson, H. D. et al (eds); Cold Spring Harbor Laboratory, 1983.
81. *The Microbe 1984. Part I: Viruses* (36th Symposium of the Society for General Microbiology), Mahy, B. W. J. & Pattison, J. R. (eds); Cambridge University Press, 1984.
82. *The Microbe 1984. Part II: Prokaryotes and Eukaryotes* (36th Symposium of the Society for General Microbiology), Kelly, D. P. & Carr, N. G. (eds); Cambridge University Press, 1984.
83. *The Reoviridae*, Joklik, W. K. (ed.); Plenum Press, 1983.
84. *Gene Expression: the Translational Step and its Control* (Proceedings of the Alfred Benzon Symposium 19), Clark, B. F. C. & Petersen, H. U. (eds); Munksgaard (Copenhagen), 1984.

85. *Bioenergetics* (New Comprehensive Biochemistry, Volume 9), Ernster, L. (ed.); Elsevier, 1984.
86. *Segmented Negative Strand Viruses: Arenaviruses, Bunyaviruses and Orthomyxoviruses*, Compans, R. W. & Bishop, D. H. L. (eds); Academic Press, 1984.
87. *Non-segmented Negative Strand Viruses: Paramyxoviruses and Rhabdoviruses*, Bishop, D. H. L. & Compans, R. W. (eds); Academic Press, 1984.
88. *Viruses and Demyelinating Diseases*, Mims, C. A., Cuzner, M. L. & Kelly, R. E. (eds); Academic Press, 1983.
89. *Coenzyme Q: Biochemistry, Bioenergetics and Clinical Applications of Ubiquinone*, Lenaz, G. (ed.); John Wiley, 1985.
90. *Virus Persistence* (33rd Symposium of the Society for General Microbiology), Mahy, B. W. J., Minson, A. C. & Darby, G. K. (eds); Cambridge University Press, 1982.
91. *Genetics of Influenza Viruses*, Palese, P. & Kingsbury, D. W. (eds); Springer-Verlag, 1983.
92. *Handbook of Endotoxin*, Volume 1: *Chemistry of Endotoxin*, Rietschel, E. T. (ed.); Elsevier, 1984.
93. *Handbook of Endotoxin*, Volume 2: *Pathophysiology of Endotoxin*, Hinshaw, L. B. (ed.); Elsevier, 1985.
94. *Handbook of Endotoxin*, Volume 3: *Cellular Biology of Endotoxin*, Berry, L. J. (ed.); Elsevier, 1985.
95. *Handbook of Endotoxin*, Volume 4: *Clinical Aspects of Endotoxin Shock*, Proctor, R. A. (ed.); Elsevier, 1985.
96. *Clinical and Biochemical Luminescence*, Kricka, L. J. & Carter, T. J. N. (eds); Marcel Dekker, 1982.
97. *The Parvoviruses*, Berns, K. I. (ed.); Plenum Press, 1984.
98. *Methods in Mycoplasmology*, Volume I: *Mycoplasma Characterization*, Razin, S. & Tully, J. G. (eds); Academic Press, 1983.
99. *Bacteriophage T4*, Mathews, C. K. et al (eds); American Society for Microbiology, 1983.
100. *The Yeasts: a Taxonomic Study*, 3rd edition. Kreger-van Rij, N. J. W. (ed); Elsevier, 1984.
101. *Bacterial Outer Membranes*, Inouye, M. (ed.); John Wiley, 1979.
102. *The Sulphate-reducing Bacteria*, 2nd edition. Postgate, J. R.; Cambridge University Press, 1984.
103. *Candidiasis*, Bodey, G. P. & Fainstein, V. (eds); Raven Press, 1985.
104. *The Molecular Virology and Epidemiology of Influenza*, Stuart-Harris, C. H. & Potter, C. W. (eds); Academic Press, 1984.
105. *Leukaemia and Lymphoma Research*, Volume 1: *Mechanisms of Viral Leukaemogenesis*, Goldman, J. M. & Jarrett, O. (eds); Churchill Livingstone (for the Leukaemia Research Fund, UK), 1984.
106. *Human T-cell Leukemia/Lymphoma Virus. The Family of Human T-Lymphotropic Retroviruses: their Role in Malignancies and Association with AIDS*, Gallo, R. C., Essex, M. E. & Gross, L. (eds); Cold Spring Harbor Laboratory, 1984.
107. *Microbes and Infections of the Gut*, Goodwin, C. S. (ed.); Blackwell Scientific Publications, 1984.
108. *Microbial Adhesion and Aggregation*, Marshall, K. C. (ed.); Springer-Verlag, 1984.
109. *Adhesion of Microorganisms to Surfaces*, Ellwood, D. C., Melling, J. & Rutter, P. (eds.); Academic Press, 1979.
110. *Viruses and Cancer* (37th Symposium of the Society for General Microbiology), Rigby, P. W. J. & Wilkie, N. M. (eds); Cambridge University Press, 1985.
111. *Microbial Adhesion to Surfaces*, Berkeley, R. C. W. et al (eds); Ellis Horwood (Chichester, UK), 1980.
112. *Current Perspectives in Microbial Ecology* (Proceedings of the Third International Symposium on Microbial Ecology, Michigan State University, 1983), Klug, M. J. & Reddy, C. A. (eds); American Society for Microbiology, 1984.
113. *Cancer Cells. 2. Oncogenes and Viral Genes*, Vande Woude, G. F. et al (eds); Cold Spring Harbor Laboratory, 1984.
114. *RNA Tumor Viruses*, 2nd edition. Weiss, R. et al (eds); Cold Spring Harbor Laboratory, 1982.
115. *Aspects of Microbial Metabolism and Ecology* (Society for General Microbiology Special Publication No. 11), Codd, G. A. (ed.); Academic Press, 1984.
116. *The Adenoviruses*, Ginsberg, H. S. (ed.); Plenum Press, 1984.
117. *Fungal Nutrition and Physiology*, Garraway, M. O. & Evans, R. C.; John Wiley, 1984.
118. *Microbiology – 1984*, Leive, L. & Schlessinger, D. (eds); American Society for Microbiology, 1984.
119. *Pneumocystis carinii Pneumonia: Pathogenesis, Diagnosis and Therapy*, Young, L. S. (ed.); Marcel Dekker, 1984.
120. *Manual of Clinical Microbiology*, 4th edition. Lennette, E. H. et al (eds); American Society for Microbiology, 1985.
121. *Antimicrobials and Agriculture*, Woodbine, M. (ed.); Butterworths, 1984.
122. *Molecular Cytology of Escherichia coli*, Nanninga, N. (ed.); Academic Press, 1985.
123. *Systematics of the Green Algae* (The Systematics Association Special Volume No. 27), Irvine, D. E. G. & John, D. M. (eds); Academic Press, 1984.
124. *The Enzymes of Biological Membranes*, Volume 1: *Membrane Structure and Dynamics*, 2nd edition. Martonosi, A. N. (ed.) Plenum Press, 1985.
125. *The Enzymes of Biological Membranes*, Volume 2: *Biosynthesis and Metabolism*, 2nd edition. Martonosi, A. N. (ed.); Plenum Press, 1985.
126. *The Enzymes of Biological Membranes*, Volume 3: *Membrane Transport*, 2nd edition. Martonosi, A. N. (ed.); Plenum Press, 1985.
127. *The Enzymes of Biological Membranes*, Volume 4: *Bioenergetics of Electron and Proton Transport*, 2nd edition. Martonosi A. N. (ed.); Plenum Press, 1985.
128. *The Biology of Desmids* (Botanical Monographs Volume 16), Brook, A. J. Blackwell Scientific Publications, 1981.
129. *Algal Symbiosis: a Continuum of Interaction Strategies*, Goff, L. J. (ed.); Cambridge University Press, 1983.
130. *The Biology of Seaweeds* (Botanical Monographs Volume 17), Lobban, C. S. & Wynne, M. J. (eds); Blackwell Scientific Publications, 1981.
131. *The Applied Mycology of Fusarium*, Moss. M. O. & Smith, J. E. (eds); Cambridge University Press, 1984.
132. *Microbial Gas Metabolism* (Society for General Microbiology Special Publication No. 14), Poole, R. K. & Dow, C. S. (eds); Academic Press, 1985.
133. *An Illustrated Key to Freshwater and Soil Amoebae*, Page, F. C.; Freshwater Biological Association (UK) publication 34, 1976.

Key to Book References

134. *Approaches to Antiviral Agents*, Harnden, M. R. (ed.); Macmillan, 1985.
135. *The Ciliated Protozoa*, Corliss, J. O.; Pergamon Press, 1979.
136. *Plankton Stratigraphy*, Bolli, H. M., Saunders, J. B. & Perch-Nielsen, K. (eds); Cambridge University Press, 1985.
137. *Silicon and Siliceous Structures in Biological Systems*, Simpson, T. L. & Volcani, B. E. (eds); Springer-Verlag, 1981.
138. *Methods in Microbiology*, Volume 18. Gottschalk, G. (ed.); Academic Press, 1985.
139. *The Herpesviruses*, Volume 1. Roizman, B. (ed.); Plenum Press, 1982.
140. *The Herpesviruses*, Volume 2. Roizman, B. (ed.); Plenum Press, 1983.
141. *The Herpesviruses*, Volume 3. Roizman, B. (ed.); Plenum Press, 1985
142. *The Herpesviruses*, Volume 4: *Immunobiology and Prophylaxis of Human Herpesvirus Infections*, Roizman, B. & Lopez, C. (eds); Plenum Press, 1985.
143. *Methods in Microbiology*, Volume 16. Bergan, T. (ed.); Academic Press, 1984.
144. *The Dictyostelids*, Raper, K. B. & Rahn, A. W.; Princeton University Press, 1984.
145. *The Myxomycetes*, Martin, G. W. & Alexopoulos, C. J.; University of Iowa Press, 1969.
146. *Metalloproteins. Part 1: Metal Proteins with Redox Roles*, Harrison, P. M. (ed.); Macmillan, 1985.
147. *The Mycetozoans*, Olive, L. S.; Academic Press, 1975.
148. *Virology*, Fields, B. N. et al (eds); Raven Press, 1985.
149. *Papillomaviruses* (Ciba Foundation Symposium 120), Evered, D. & Clark, S. (eds); John Wiley, 1986.
150. *Comprehensive Virology*, Volume 19. Fraenkel-Conrat, H. & Wagner, R. R. (eds); Plenum Press, 1984.
151. *Magnetic Resonance in Medicine and Biology*, Foster, M. A. (ed.); Pergamon Press, 1984.
152. *Modern Physical Methods in Biochemistry*, Neuberger, A. & van Deenen, L. (eds); Elsevier, 1985.
153. *The Scientific Basis of Antimicrobial Chemotherapy* (38th Symposium of the Society for General Microbiology), Greenwood, D. & O'Grady, F. (eds); Cambridge University Press, 1985.
154. *Microbial Mats: Stromatolites*, Cohen, Y. et al (eds); Alan R. Liss, 1984.
155. *Petroleum Microbiology*, Atlas, R. M. (ed.); Macmillan, 1984.
156. *Chloroplast Biogenesis*, Ellis, R. J. (ed.); Cambridge University Press, 1984.
157. *The Bacteria*, Volume VIII: *Archaebacteria*, Woese, C. R. & Wolfe, R. S. (eds); Academic Press, 1985.
158. *Microbial Ecology of the Phylloplane*, Blakeman, J. P. (ed.); Academic Press, 1981.
159. *Subviral Pathogens of Plants and Animals: Viroids and Prions*, Maramorosch, K. & McKelvey, J. J. (eds); Academic Press, 1985.
160. *Antimicrobial Drug Resistance*, Bryan, L. E. (ed.); Academic Press, 1984.
161. *Plasmids in Bacteria*, Helinski, D. R. et al (eds); Plenum Press, 1985.
162. *The Molecular Biology of Polioviruses*, Koch, F. & Koch, G.; Springer-Verlag, 1985.
163. *Molecular Biology of the Photosynthetic Apparatus*, Steinback, K. E. et al (eds); Cold Spring Harbor Laboratory, 1985.
164. *Sporulation and Germination*, Levinson, H. S. et al (eds); American Society for Microbiology, 1981.
165. *Microtubules*, Dustin, P.; Springer-Verlag, 1984.
166. *Molecular Biology of the Cytoskeleton*, Borisy, G. G., Cleveland, D. W. & Murphy, D. B. (eds); Cold Spring Harbor Laboratory, 1984.
167. *Progress in Phycological Research*, Volume 3. Round, F. E. & Chapman, D. J. (eds); Biopress (Bristol, UK), 1984.
168. *The Fungal Nucleus*, Gull, K. & Oliver, S. G. (eds); Cambridge University Press, 1981.
169. *Microbial Growth on C_1 Compounds*, Crawford, R. L. & Hanson, R. S. (eds); American Society for Microbiology, 1984.
170. *The Molecular Biology of the Bacilli*, Volume 1: *Bacillus subtilis*, Dubnau D. A. (ed.); Academic Press, 1982.
171. *The Molecular Biology of the Bacilli*, Volume II. Dubnau, D. A. (ed.); Academic Press, 1985.
172. *The Trichomycetes: Fungal Associates of Arthropods*, Lichtwardt, R. W.; Springer, 1986.
173. *The Microbial Cell Cycle*, Nurse, P. & Streiblová, E. (eds); CRC Press Inc., Boca Raton, 1984.
174. *Zoosporic Plant Pathogens*, Buczacki, S. T. (ed.); Academic Press, 1983.
175. *The Fungal Spore: Morphogenetic Controls*, Turian, G. & Hohl, H. R. (eds); Academic Press, 1981.
176. *The Epstein–Barr Virus: Recent Advances*, Epstein, M. A. & Achong, B. G. (eds); Heinemann Medical Books, 1986.
177. *Methods in Microbiology*, Volume 17: *Plasmid Technology*, Bennett, P. M. & Grinsted, J. (eds); Academic Press, 1984.
178. *Microbiology–1985*, Leive, L. (ed.); American Society for Microbiology, 1985.
179. *Gene Manipulations in Fungi*, Bennett, J. W. & Lasure, L. L. (eds); Academic Press, 1985.
180. *Viruses and Plasmids in Fungi*, Lemke, P. A. (ed.); Marcel Dekker, 1979.
181. *Medical Microbiology*, Volume 1. Easmon, C. S. F. & Jeljaszewicz, J. (eds); Academic Press, 1982.
182. *Cereal Diseases*, Jones, D. G. & Clifford, B. C.; John Wiley, 1983.
183. *The Biology of Botrytis*, Coley-Smith, J. R., Verhoeff, K. & Jarvis, W. R. (eds); Academic Press, 1980.
184. *Identification of Enterobacteriaceae*, 4th edition. Ewing, W. H.; Elsevier, 1986.
185. *Manual and Atlas of the Penicillia*, Ramirez, C.; Elsevier Biomedical Press, 1982.
186. *Transfer RNA: Structure, Properties and Recognition*, Schimmel, P. R., Söll, D. & Abelson, J. N. (eds); Cold Spring Harbor Laboratory, 1979.
187. *The Powdery Mildews*, Spencer, D. M. (ed.); Academic Press, 1978.
188. *Regulation of Gene Expression 25 Years On* (39th Symposium of the Society for General Microbiology), Booth, I. R. & Higgins, C. F. (eds); Cambridge University Press, 1986.
189. *Eukaryotic Transcription: the Role of Cis- and Trans-acting Elements in Initiation* (Current Communications in Molecular Biology), Gluzman, Y. (ed.); Cold Spring Harbor Laboratory, 1985.
190. *The Genus Coelomomyces*, Couch, J. N. & Bland, C. E. (eds); Academic Press, 1985.

191. *Microbes in Extreme Environments*, Herbert, R. A. & Codd, G. A. (eds); Academic Press, 1986.
192. *Alkalophilic Microorganisms*, Horikoshi, K. & Akiba, T.; Japan Scientific Societies Press (Tokyo)/Springer-Verlag (Berlin), 1982.
193. *Chlamydial Infections*, Oriel, D. et al (eds); Cambridge University Press, 1986.
194. *Evolution of Prokaryotes*, Schleifer, K. H. & Stackenbrandt, E. (eds); Academic Press, 1985.
195. *The Biochemistry of Silage*, McDonald, P.; John Wiley, 1981.
196. *Thermophiles*, Brock, T. D. (ed.); John Wiley, 1986.
197. *Chrysophytes: Aspects and Problems*, Kristiansen, J. & Anderson, R. A. (eds); Cambridge University Press, 1986.
198. *The Bacteria*, Volume X: *The Biology of Pseudomonas*, Sokatch, J. R. (ed.); Academic Press, 1986.
199. *Molecular Cloning and Gene Regulation in Bacilli*, Ganesan, A. T., Chang, S. & Hoch, J. A. (eds); Academic Press, 1982.
200. *Molecular Cloning: a Laboratory Manual*, Maniatis, T., Fritsch, E. F. & Sambrook, J.; Cold Spring Harbor Laboratory, 1982.
201. *The Bacteria*, Volume IX: *Antibiotic-producing Streptomyces*, Queener, S. W. & Day, L. E. (eds); Academic Press, 1986.
202. *Ecology of Microbial Communities* (41st Symposium of the Society for General Microbiology), Fletcher, M., Gray, T. R. G. & Jones, J. G. (eds); Cambridge University Press, 1987.
203. *Structure and Function of Biofilms*, Characklis, W. G. & Wilderer, P. A. (eds); Wiley, 1989.
204. *Transgenesis*, Murray, J. A. H. (ed.); Wiley, 1992.
205. *Secondary Metabolites: their Function and Evolution*, CIBA Foundation Symposium 171; Wiley, 1992.
206. *Sourcebook of Bacterial Protein Toxins*, Alouf, J. E. & Freer, J. H. (eds); Academic Press, 1991.
207. *Gene Regulation: Biology of Antisense RNA and DNA*, Erickson, R. P. & Izant, J. G. (eds); Raven Press (New York), 1992.
208. *Antisense RNA and DNA*, Murray, J. (ed.); Wiley, 1992.
209. *Prospects for Antisense Nucleic Acid Therapy of Cancer and AIDS*, Wickstrom, E. (ed.); Wiley, 1991.
210. *PCR Protocols* (Current methods and applications), White, B. A.; Humana Press, 1993 (ISBN 0-89603-244-2).
211. *PCR* (Essential data), Newton, C. R. (ed.); John Wiley, 1995.
212. *PCR* (Essential techniques), Burke, J. (ed.); John Wiley, 1996.
213. *Antimicrobial Drug Action*, Williams, R. A. D., Lambert, P. A. & Singleton, P.; βios Scientific Publishers, Oxford, 1996 (ISBN 1-872748-81-3).
214. *Cystic Fibrosis*, Shale, D. J. (ed.); BMJ Publishing Group, 1996 (ISBN 0-7279-0826-X).
215. *Progress in Pathology*, Volume 4. Kirkham, N. & Lemoine, N. R. (eds); Churchill-Livingstone, 1998 (ISBN 0-443-06032-0).
216. *Environmental Health Engineering in the Tropics*, 2nd edn, Cairncross, S & Feachem, R; John Wiley, 1993 (ISBN 0-471-93885-8).
217. *Low-cost Sewerage*, Mara, D. (ed.); John Wiley, 1996 (ISBN 0-471-96691-6).
218. *Cellular Microbiology*, Henderson, B., Wilson, M., McNab, R. & Lax, A. J.; John Wiley, 1999 (ISBN 0-471-98681-X).
219. *Practical Medical Microbiology*, 14th edn, Collee, J. G., Fraser, A.G., Marmion, B.P. & Simmons, A.; Churchill-Livingstone, 1996 (ISBN 0-443-047219).
220. *Therapy with Botulinum Toxin*, Jancovic, J. & Hallett, M. (eds); Marcel Dekker (New York), 1994.
221. *DNA Methods in Clinical Microbiology*, Singleton, P.; Kluwer Academic Publishers (Dordrecht), 2000 (ISBN 07923-6307-8).
222. *DNA Methylation: Molecular Biology and Biological Significance*, Jost, J. P. & Saluz, H. P. (eds); Birkhäuser, 1993 (ISBN 3-7643-2778-2).
223. *Bacteria in Biology, Biotechnology and Medicine*, 5th edn, Singleton. P.; John Wiley, 1999 (ISBN 0-471-98880-4).
224. *External Eye Diseases: a Colour Atlas*, Watts, M. T. & Nelson, M. E. (eds); Churchill-Livingstone, 1992 (ISBN 0-443-04446-5).
225. *Plasmids for Therapy and Vaccination*, Schleef, M. (ed.); Wiley-VCH, 2000/2001 (ISBN 3527-30269-7).
226. *The Molecular Biology of Cytokines*, Meager, T.; John Wiley, 1998 (ISBN 0-471-98272-5).
227. *Immunology*, 3rd edn, Benjamini, E., Sunshine, G. & Leskowitz, S.; Wiley-Liss, 1996 (ISBN 0-471-59791-0).
228. *Tempe*, Baumann, U. & Bisping, B. (eds); Wiley-VCH, 2001 (ISBN 3527-30090-2).